Kuby

IMMUNOLOGY

Icons Used in This Book

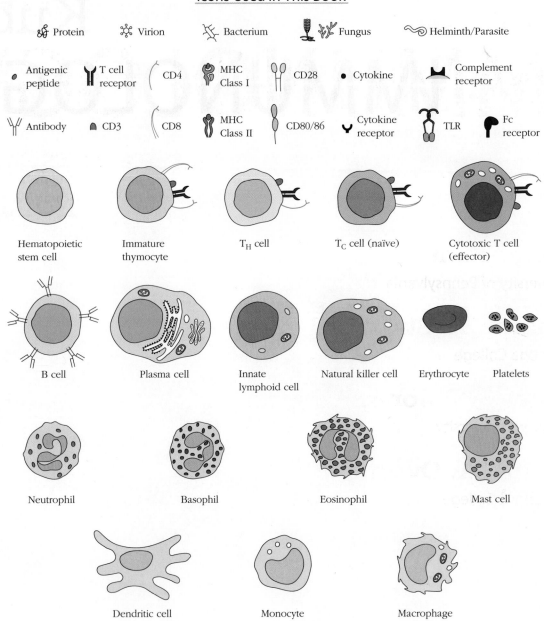

Protein	Virion	Bacterium	Fungus	Helminth/Parasite

Antigenic peptide   T cell receptor   CD4   MHC Class I   CD28   Cytokine   Complement receptor

Antibody   CD3   CD8   MHC Class II   CD80/86   Cytokine receptor   TLR   Fc receptor

Hematopoietic stem cell   Immature thymocyte T_H cell T_C cell (naïve)   Cytotoxic T cell (effector)

B cell   Plasma cell   Innate lymphoid cell   Natural killer cell   Erythrocyte   Platelets

Neutrophil   Basophil   Eosinophil   Mast cell

Dendritic cell   Monocyte   Macrophage

Kuby
IMMUNOLOGY

Eighth Edition

Jenni Punt
University of Pennsylvania

Sharon A. Stranford
Pomona College

Patricia P. Jones
Stanford University

Judith A. Owen
Haverford College

w.h.freeman
Macmillan Learning
New York

Vice President, STEM: Daryl Fox
Executive Editor: Lauren Schultz
Executive Marketing Manager: Will Moore
Marketing Assistant: Savannah DiMarco
Development Editor: Erica Champion
Development Editor: Erica Pantages Frost
Media Editor: Jennifer Compton
Assistant Editor: Kevin Davidson
Senior Content Project Manager: Liz Geller
Senior Media Project Manager: Jodi Isman
Permissions Manager: Jennifer MacMillan
Photo Researcher: Richard Fox
Director of Design, Content Management: Diana Blume
Designer: Blake Logan
Illustrations: Imagineering
Illustration Coordinator: Janice Donnola
Senior Workflow Project Supervisor: Susan Wein
Production Supervisor: Lawrence Guerra
Composition: Lumina Datamatics, Inc.
Printing and Binding: LSC Communications, Inc.

North American Edition
Cover Image: Courtesy of Audra Devoto and Xian-McKeon Laboratory

Library of Congress Control Number: 2018939693

North American Edition
ISBN-13: 978-1-4641-8978-4
ISBN-10: 1-4641-8978-1

North American Edition
W. H. Freeman and Company
One New York Plaza
Suite 4500
New York, NY 10004-1562
www.macmillanlearning.com

To all the students, fellows, and colleagues who have made our careers in immunology a source of joy and excitement, and to our families and mentors who made these careers possible. We hope that future generations of immunology students will find this subject as fascinating and rewarding as we have. And in memory of Shannon Moloney, who had too little time to finish her own life goals but who will be remembered for how she helped us to meet our goals in this project.

About the Authors

All four authors are active scholars and teachers who have been/are recipients of research grants from the NIH and the NSF. They have all served in various capacities as grant proposal reviewers for the NSF, NIH, HHMI, and other funding bodies and, as well, have evaluated manuscripts submitted for publication in immunological journals. In addition, they are all active members of the American Association of Immunologists (AAI) and have served that national organization in a variety of ways.

Jenni Punt

Jenni Punt received her A.B. from Bryn Mawr College, magna cum laude, with high honors in biology from Haverford College. She was a combined degree student at the University of Pennsylvania, graduating summa cum laude from the School of Veterinary Medicine (V.M.D.) with a Ph.D. in immunology. She pursued her interest in T-cell development as a Damon Runyon-Walter Winchell Physician-Scientist fellow with Dr. Alfred Singer at the National Institutes of Health and was appointed to the faculty of Haverford College in 1996. After 18 wonderful years there, working on T-cell and hematopoietic stem cell development, she accepted a position as associate dean for student research at Columbia University's College of Physicians and Surgeons. There she was the founding director of an M.D./M.Sc. dual degree program and co-ran a laboratory on hematopoiesis with her husband, Dr. Stephen Emerson. After being tempted back to the School of Veterinary Medicine at the University of Pennsylvania, she is now developing new educational programs as director of One Health Research Education. She has received multiple teaching awards over the course of her career and continues to find that students are her most inspirational colleagues.

Sharon Stranford

Sharon Stranford received her Ph.D. in microbiology and immunology from Hahnemann University (now Drexel), where she studied multiple sclerosis. She then spent 3 years exploring transplant immunology as a postdoctoral fellow at Oxford University, followed by 3 years at the University of California, San Francisco, conducting human HIV/AIDS research. In 2001 she was hired as a Clare Boothe Luce Assistant Professor at Mount Holyoke College, a small liberal arts college for women in Massachusetts, where she served in the Department of Biological Sciences and the Program in Biochemistry for 12 years. Sharon is now a professor of biology at Pomona College in Claremont, California, where she investigates immunologic markers that influence susceptibility to immune deficiency. She also studies the science of teaching and learning; in particular, initiatives within STEM that foster a sense of inclusion and that welcome first-generation college students, like herself. Her teaching repertoire, past and present, includes cell biology, immunology, advanced laboratories in immunology, and seminars in infectious disease, as well as a team-taught course blending ethics and biology, entitled "Controversies in Public Health."

Pat Jones

Pat Jones graduated from Oberlin College in Ohio with highest honors in biology and obtained her Ph.D. in biology with distinction from Johns Hopkins University. She was a postdoctoral fellow of the Arthritis Foundation for 2 years in the Department of Biochemistry and Biophysics at the University of California, San Francisco, Medical School, followed by 2 years as an NSF postdoctoral fellow in the Departments of Genetics and Medicine/Immunology at Stanford University School of Medicine. In 1978 she was appointed assistant professor of biology at Stanford and is now a full professor and currently holds the Dr. Nancy Chang Professorship in Humanities and Sciences. Pat has received several undergraduate teaching awards, was the founding director of the Ph.D. Program in Immunology, served as vice provost for faculty development and diversity, and in July 2011, she assumed the position of Director of Stanford Immunology, a position that coordinates immunology training activities across the university.

Judy Owen holds B.A. and M.A. (Hons) degrees in biochemistry from Cambridge University. She pursued her Ph.D. at the University of Pennsylvania with the late Dr. Norman Klinman and her postdoctoral fellowship with Dr. Peter Doherty in viral immunology. In 1981, she was appointed to the faculty of Haverford College, one of the first undergraduate colleges to offer a course in immunology. Judy teaches numerous laboratory and lecture courses in biochemistry and immunology; her teaching awards include the Excellence in Mentoring Award from the American Association of Immunologists. She is currently a participant in Haverford's First Year Writing Program and has been involved in curriculum development across the college. Judy served as director of the Marian E. Koshland Integrated Natural Sciences Center from 2013 to 2017 and currently holds the Elizabeth Ufford Green Professorship in Natural Sciences.

Together, Jenni Punt and Judy Owen developed and ran the first AAI introductory immunology course, which is

Brief Contents

Feature Boxes in *Kuby Immunology*, Eighth Edition

CLASSIC EXPERIMENT

ADVANCES

Contents

Preface

Like all of the previous authors of this book,

we are dedicated to the concept that immunology is best taught and learned in an experimentally based manner, and we have retained that emphasis with this edition. It is our goal that students should complete an immunology course not only with a firm grasp of content, but also with a clear sense of how key discoveries were made, what interesting questions remain, and how they might best be answered. We believe that this approach ensures that students master fundamental immunological concepts, internalize a vision of immunology as an active and ongoing process, and develop the ability to contribute to new knowledge, themselves. Guided by this vision, this new edition has been extensively updated to reflect the recent advances in all aspects of our discipline.

Rod Searcey

Pat Jones

New Co-Author, Pat Jones

The new edition of *Kuby Immunology* welcomes a new member to our author team, Patricia P. Jones, who had been a contributing author to the seventh edition. Dr. Jones is professor of biology at Stanford University and holds the Dr. Nancy Chang Professorship in Humanities and Sciences. Having earned her undergraduate degree in biology from Oberlin College and her Ph.D. in biology, with a focus on immunology, from Johns Hopkins University, Dr. Jones did postdoctoral training at both UCSF and Stanford University School of Medicine before joining the faculty at Stanford. She and her research group have made fundamental contributions to our understanding of the genetics, structure, and expression of MHC class II proteins and of mechanisms regulating adaptive and innate immune responses. Dr. Jones has served in various leadership positions at Stanford, including chairing the Department of Biology and the Faculty Senate, and serving as vice provost for faculty development and diversity. She was the founding director of the Ph.D. Program in Immunology and currently holds the position of Director of Stanford Immunology, which oversees all immunology training–related activities at Stanford. Dr. Jones has taught students at all levels, including teaching for many years the basic molecular and cellular immunology course for undergraduate and graduate students at Stanford. Her dedication to teaching and her enthusiasm for immunology shine through in her work.

Understanding the Big Picture

Two of the most challenging aspects of teaching immunology are the many important details (cell types, proteins, interactions, and terminology) and the interconnected or circular nature of the response. We find that students often fail to recognize how these pieces work together in an immune response that is dynamic. Our primary goal in the eighth edition is to bring this big picture to the forefront by providing a map or scaffold that both faculty and students can refer to in order to draw regular connections between concepts and individual players in the immune response.

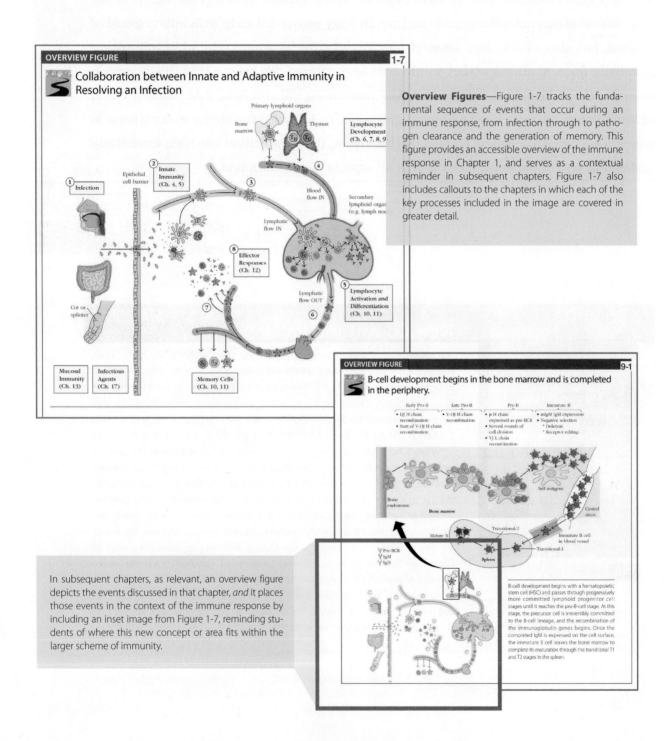

Overview Figures—Figure 1-7 tracks the fundamental sequence of events that occur during an immune response, from infection through to pathogen clearance and the generation of memory. This figure provides an accessible overview of the immune response in Chapter 1, and serves as a contextual reminder in subsequent chapters. Figure 1-7 also includes callouts to the chapters in which each of the key processes included in the image are covered in greater detail.

In subsequent chapters, as relevant, an overview figure depicts the events discussed in that chapter, *and* it places those events in the context of the immune response by including an inset image from Figure 1-7, reminding students of where this new concept or area fits within the larger scheme of immunity.

Concepts and Context

Cells, Organs, and Microenvironments of the Immune System

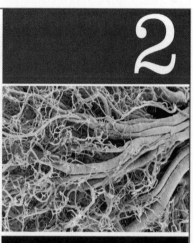

2

Scanning electron micrograph of blood vessels in a lymph node.
[Susumu Nishinaga/Science Source.]

Learning Objectives
After reading this chapter, you should be able to:

1. Describe the types of blood cells that make up the immune system and outline the main events that occur during hematopoiesis, the process that gives rise to immune cells.

2. Identify the primary, secondary, and tertiary immune organs in vertebrates and describe their function.

3. Recognize and describe the microenvironments where immune cells mature and the immune response develops.

4. Identify several experimental approaches used to understand how blood cells and immune responses develop.

Key Terms

Hematopoiesis	Secondary lymphoid organs	T-cell zone
Hematopoietic stem cell (HSC)	Lymph nodes	B-cell follicle
Myeloid lineage cells	Spleen	Germinal centers
Lymphoid lineage cells	Barrier tissues (MALT and skin)	Fibroblastic reticular cell conduit (FRCC) system
Primary lymphoid organs	Lymphatic system	Follicular dendritic cells (FDCs)
Bone marrow	Tertiary lymphoid tissue	
Thymus		

Key Concepts:

- HSCs reside primarily in the bone marrow, where stromal cells regulate their quiescence, proliferation, and trafficking. Long-term HSCs reside in the perivascular niche, in association with cells that line the blood vessels.

- In the bone marrow, HSCs differentiate into progenitors, which can become myeloid or lymphoid cell lineages. B lymphocytes complete their maturation in the bone marrow, but progenitors that can differentiate into T lymphocytes exit and complete their maturation in the thymus.

A Conceptual Approach to Signaling

Chapter 3 (Recognition and Response) now combines a description of the antigen receptors of innate and adaptive immunity with a brief introduction to cytokines, chemokines, and their respective receptors, formerly found in Chapter 4. Using a conceptual approach, Chapter 3 now foregrounds the major concepts required for understanding the processes of signal recognition and signal transduction throughout the immune system. We highlight the diverse roles of receptor diversity, multivalency, coreceptors, lipid rafts, and multiple signaling pathways in the regulation of immune responsiveness.

New Chapter—Barrier Immunity and the Microbiome

Research on the interaction between the microbiome and the immune response has flourished in recent years. Not only do our immune cells shape the diverse communities of microbes that live on our epithelial surfaces, but these communities have a powerful influence on the development and activity of a healthy immune system. The eighth edition of *Kuby Immunology* now includes a new chapter, **Barrier Immunity: The Immunology of Mucosa and Skin (Chapter 13)**, that reviews our new understanding of the interaction between microbes and immunity at epithelial surfaces, including mucosal tissues and skin.

Advances in Immunology—Other Notable Updates

Immunology is a rapidly growing field, with new discoveries, advances in techniques, and previously unappreciated connections coming to light every day. In addition to a new chapter on barrier immunity, the eighth edition of *Kuby Immunology* has been thoroughly updated throughout, and now includes the following material and concepts.

- Natural killer (NK) cells are now recognized to be a subset of a larger group of innate lymphoid cells (ILCs) with characteristics similar to T_H cell subsets, but that originate in the myeloid lineage. ILCs are introduced in Chapter 2 and their roles in the innate and adaptive immune responses are discussed in Chapters 4 and 10, respectively.

- Exciting new immunotherapeutic approaches for treating a variety of conditions are described in Chapters 12, 15, 18, and 20.

- The role of the microbiome and its interactions with the immune system in health and disease is discussed in Chapters 1, 11, 13, 15, and 16.

- Insights gained from advanced imaging technology continue to be updated. For example, Chapter 6 describes immunofluorescence techniques that reveal changes in chromosomal organization accompanying V(D)J recombination.

New boxes have been added on the following topics:

- **Classic Experiment Box 4-1:** Discovery of Invertebrate Toll and Vertebrate Toll-Like Receptors
- **Advances Box 5-2:** The role of complement in the development of the nervous system and vision
- **Evolution Box 6-3:** The evolution of V(D)J recombination and *RAG* genes
- **Clinical Focus Box 7-3:** MHC expression and Tasmanian devil facial tumor disease
- **Clinical Focus Box 10-2:** Checkpoint inhibitors and cancer therapy
- **Advances Box 10-4:** Jumping genes, T_{REG} cells and the evolution of immune tolerance during pregnancy
- **Advances Box 11-1:** Tracking the movements of B cells between the dark and light zones of the germinal center
- **Clinical Focus Box 12-1:** Therapeutic antibodies for the treatment of diseases
- **Advances Box 13-1:** Cells involved in barrier immunity
- **Clinical Focus Box 13-2:** Communication between the gut and the brain
- **Advances Box 13-3:** Germ-free animal model systems
- **Clinical Focus Box 17-1:** Zika virus and vaccine development
- **Advances Box 18-2:** Broadly neutralizing antibodies to HIV
- **Clinical Focus Box 19-2:** CAR-T cells as a potential cancer cure

LaunchPad for *Kuby Immunology*

The eighth edition of *Kuby Immunology* is fully supported in LaunchPad. We designed LaunchPad as a resource to help students achieve better results. Our goal was to increase their confidence by providing a place where they could read, study, practice, complete homework, and succeed. In addition, LaunchPad always provides instructors and students with superior service and support, based on Macmillan's legendary high-quality content. LaunchPad includes a suite of supplements that build on the text by engaging students inside and outside the classroom.

In-Class Activities—In many classrooms, student engagement is key to addressing misconceptions and reinforcing important concepts. The *Kuby Immunology* authors have provided instructions and materials for a variety of activities they use in their own classrooms to engage students. These tried-and-true activities range in length and complexity and can serve as a springboard for active learning in the classroom.

Case Studies—Interpreting experimental data is essential in understanding immunology. These case studies explore immune function, disease, and treatment through the application of primary research and data. Students are led through a series of experiments and challenged to interpret the data and draw conclusions. By integrating experimental techniques from immunology, molecular biology, and biochemistry, these case studies teach students to think critically and synthesize their knowledge of immunology and other branches of science.

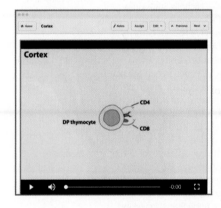

Animations—Many of the most difficult topics in immunology are multistep events that are best visualized through animations. We have created a suite of 2D animations for the eighth edition that walk students through these difficult topics, showing each step of the process. Each animation is accompanied by assessments.

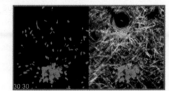

Videos—Dynamic imaging techniques allow immunologists to observe the immune system at work in vivo. These striking videos show a T cell crawling along a network of stromal cells, the change in behavior when a naïve B cell is activated, and the chemotactic response of neutrophils to a site of damage.

Lämmermann T., et al., "Neutrophil swarms require LTB4 and integrins at sites of cell death in vivo." *Nature* 2013, June 20; 498:371–75, Video 2.

Learning Curve—LearningCurve adaptive quizzing offers individualized question sets and feedback for each student based on his or her correct and incorrect responses. All the questions are tied back to the e-Book to encourage students to use the resources at hand.

e-Book—The *Kuby Immunology*, Eighth Edition, e-Book is available through Vital Source and LaunchPad. This fully enhanced e-Book includes embedded animations and videos, as well as web links to additional resources. e-Book access can be purchased through the Macmillan Student Store and represents a significant cost savings versus a printed copy of the book.

Advanced Online Material—Feature boxes within the text describe clinical connections, classic experiments, technological advances, and evolutionary aspects of the immunology topics discussed. Boxes and other content that have been retired from the print text are available for instructor download at the catalog site.

Test Bank—The *Kuby Immunology* test bank has been expanded to include more higher-order questions in both multiple choice and short answer formats. Over 700 dynamic questions in PDF and editable Word formats are rated by level of difficulty and Bloom's taxonomy level, and tagged to specific sections of the text.

Optimized Art—Fully optimized JPEG files of every figure, photo, and table in the text are available, featuring enhanced color, higher resolution, and enlarged fonts. Images are also offered in PowerPoint format for each chapter.

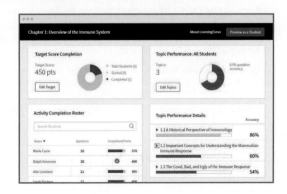

Acknowledgments

We owe special thanks to individuals who offered insightful ideas, contributed detailed reviews that led to major improvements, and provided the support that made writing this text possible. These notable contributors include Dr. Stephen Emerson, Dr. Scott Owen, Dr. Alexander Stephan, and the many students—undergraduates and graduates—who provided invaluable perspectives on our chapters. We hope that the final product reflects the high quality of the input from these experts and colleagues and from all those listed below who provided critical analysis and guidance.

We are also grateful to the previous authors of *Kuby Immunology*, whose valiant efforts we now appreciate even more deeply. Their commitment to clarity, to providing the most current material in a fast-moving discipline, and to maintaining the experimental focus of the discussions set the standard that is the basis for the best of this text.

We also acknowledge that this book represents the work not only of its authors and editors, but also of all those whose scientific experiments, papers, and conversations provided us with ideas, inspiration, and information. We thank you and stress that all errors and inconsistencies of interpretation are ours alone.

We thank the following reviewers for their comments and suggestions about the manuscript during preparation of this eighth edition. Their expertise and insights have contributed greatly to the book.

Jorge N. Artaza, University of California, Los Angeles
Roberta Attanasio, Georgia State University
Avery August, Cornell University
Kenneth Balazovich, University of Michigan
Amorette Barber, Longwood University
Scott Barnum, University of Alabama at Birmingham
Carolyn A. Bergman, Georgian Court University
Ashok P. Bidwai, West Virginia University
Jay H. Bream, Johns Hopkins University
Walter J. Bruyninckx, Hanover College
Eric Buckles, Dillard University
Peter Burrows, University of Alabama at Birmingham
Stephen K. Chapes, Kansas State University
Janice Conway-Klaassen, University of Minnesota
Jason Cyster, University of California, San Francisco
Kelley Davis, Nova Southeastern University
Brian DeHaven, La Salle University
Joyce Doan, Bethel University
Erastus C. Dudley, Huntingdon College
Uthayashanker Ezekiel, Saint Louis University
Karen Golemboski, Bellarmine University
Sandra O. Gollnick, University at Buffalo, SUNY
Elizabeth A. Good, University of Illinois, Urbana-Champaign
Susan M. R. Gurney, Drexel University
Kirsten Hokeness, Bryant University
Judith Humphries, Lawrence University
Seth Jones, University of Kentucky
George Keller, Samford University
Kevin S. Kinney, DePauw University

Edward C. Kisailus, Canisius College
David Kittlesen, University of Virginia
Ashwini Kucknoor, Lamar University
Narendra Kumar, Texas A&M Health Science Center
Courtney Lappas, Lebanon Valley College
Melanie J. Lee-Brown, Guilford College
Vicky M. Lentz, SUNY Oneonta
Marlee B. Marsh, Columbia College
James McNew, Rice University
Daniel Meer, Cardinal Stritch University
JoAnn Meerschaert, St. Cloud State University
Pamela Monaco, Molloy College
Rajkumar Nathaniel, Nicholls State University
Samantha Terris Parks, Georgia State University
Rekha Patel, University of South Carolina
Sarah M. Richart, Azusa Pacific University
James E. Riggs, Rider University
Vanessa Rivera-Amill, Ponce Health Sciences University-School of Medicine
Ryan A. Shanks, University of North Georgia
Laurie P. Shornick, Saint Louis University
Paul K. Small, Eureka College
Jennifer Ripley Stueckle, West Virginia University
Gabor Szalai, West Virginia School of Osteopathic Medicine
Clara Toth, St. Thomas Aquinas College
Vishwanath Venketaraman, Western University of Health Sciences
Barbara Criscuolo Waldman, University of South Carolina
Denise G. Wingett, Boise State University
Laurence Wong, Burman University

Finally, we thank our experienced and talented colleagues at W. H. Freeman and Company. Particular thanks to the production team members Liz Geller, Susan Wein, Janice Donnola, Diana Blume, Jennifer MacMillan, Richard Fox, and Mark Mykytiuk and his team at Imagineering. Thanks are also due to the editorial, media, and marketing teams of Lauren Schultz, Kevin Davidson, Jennifer Compton, and Will Moore.

However, we are particularly grateful for the insights, diplomacy, energy, and quality of judgment of our heroic developmental editors, Erica Champion and Erica Frost. "Our" Ericas have guided us from the beginning with probing vision and keen eyes for narrative and clarity. They have exhibited endless patience for our complex schedules, perspectives, and needs as authors and professors. In short, these two extraordinarily talented team members have made this edition, and its ambitious aspirations, possible.

Overview of the Immune System

1

Learning Objectives
After reading this chapter, you should be able to:

1. Trace the study of immunology from a desire to vaccinate against infectious disease to far-reaching applications in basic research, medicine, and other fields of study.

2. Examine and question prior assumptions related to immunology and categorize features unique to the immune system.

3. Practice and apply some immunology-specific vocabulary, while distinguishing cells, structures, and concepts important to the field of immunology.

4. Recognize the need for balance and regulation of immune processes and evaluate the consequences of dysregulation.

5. Begin to integrate concepts from immunity into real-world issues and medical applications.

The immune system evolved to protect multicellular organisms from pathogens. Highly adaptable, it defends the body against invaders as diverse as the tiny (~30 nm), intracellular virus that causes polio and as large as the giant parasitic kidney worm *Dioctophyme renale*, which can grow to over 100 cm in length and 10 mm in width. This diversity of potential pathogens requires a range of recognition and destruction mechanisms to match the multitude of invaders. To meet this need, vertebrates have evolved a complicated and dynamic network of cells, molecules, and pathways. Although elements of these networks can be found throughout the plant and animal kingdoms, the focus of this text will be on the highly evolved mammalian immune system.

A human macrophage (red) ingesting *Mycobacterium tuberculosis* (green), the bacterium that causes tuberculosis. *[Science Photo Library/ Science Source.]*

The fully functional immune system involves so many organs, molecules, cells, and pathways in such an interconnected and sometimes circular process that it is often difficult to know where to start! Recent advances in cell imaging, genetics, bioinformatics, as well as in cell and molecular biology, have helped us to understand many of the individual players in great molecular detail. However, a focus on the details (and there are many) can make it difficult to see the bigger picture, and it is often the bigger picture that motivates us to study immunology. Indeed, the field of immunology can be credited with the vaccine that eradicated smallpox, the ability to transplant organs between humans, and the drugs used today to treat asthma.

Our goal in this chapter is to present the background and concepts in immunology that will serve as a foundation for the cellular and molecular detail presented in subsequent chapters. Overview figures and immunology-specific

Key Terms

Immunity	Active immunity	Clonal selection	Innate immunity
Immunoglobulin	Cell-mediated immunity	Pathogens	Adaptive immunity
Antibodies	T lymphocytes (T cells)	B-cell receptors	Inflammatory response
Humoral immunity	B lymphocytes (B cells)	T-cell receptors	Primary response
Passive immunity	Antigen	Tolerance	Secondary response

concepts presented in this chapter will re-appear in later chapters where more detailed pathways are described. Our hope is that by presenting a conceptual scaffold here, the big picture can remain in focus in subsequent chapters, where details of the intricate coordination of the vertebrate immune system are presented.

The study of immunology has produced fascinating stories (some of which you will find in this text), where host and microbe engage in battles waged over both minutes and millennia. But the immune system is also much more than an isolated component of the body, merely responsible for search-and-destroy missions. In fact, it interleaves with many of the other body systems, including the endocrine, nervous, and metabolic systems, with more connections undoubtedly to be discovered in time. It has become increasingly clear that elements of immunity play key roles in regulating homeostasis throughout the body, establishing a healthy balance. Information gleaned from the study of the immune system, as well as its connections with other systems, will likely have resounding repercussions across many basic science and biomedical fields, not to mention in the future of clinical medicine.

We start with a historical perspective, charting the beginnings of the study of immunology, largely driven by the human desire to survive major outbreaks of infectious disease. This is followed by presentation of a few key concepts that are important hallmarks of the mammalian immune response, many of which may not have been encountered elsewhere in genetics and cell biology. In fact, one objective for this chapter is to address common misunderstandings or conceptual roadblocks that can serve as an impediment to understanding this unique field. We hope this introduction will whet the appetite and prepare the reader for a more thorough discussion of the specific components of immunity presented in the following chapters. We conclude by introducing a few challenging clinical situations, such as instances in which the immune system fails to act or becomes the aggressor, turning its awesome powers against the host. More in-depth coverage of these and other medical aspects of immunology can be found in the final chapters of this text.

A Historical Perspective of Immunology

The discipline of immunology grew out of the observation that individuals who had recovered from certain infectious diseases were thereafter protected from the disease. The Latin term *immunis,* meaning "exempt," is the source of the English word **immunity,** a state of protection from infectious disease. Perhaps the earliest written reference to the phenomenon of immunity can be traced back to Thucydides,

the great historian of the Peloponnesian War. In describing a plague in Athens, he wrote in 430 BC that only those who had recovered from the plague could nurse the sick because they would not contract the disease a second time. Thucydides and colleagues must therefore have realized that the human body was able to "learn" from exposure, acquiring some form of protection from future illness of the same kind. (The idea of disease caused by unseen infectious agents, or germ theory, did not arise until much later!) Although early societies recognized the phenomenon of acquired immunity, almost 2000 years passed before the concept was disseminated into the current common medical practice of vaccination.

Early Vaccination Studies Led the Way to Immunology

The first recorded attempts to deliberately induce immunity were performed by the Chinese and Turks in the fifteenth century. They were attempting to prevent smallpox, a disease that is fatal in about 30% of cases and that leaves survivors disfigured for life (**Figure 1-1**). Reports suggest that the dried crusts derived from smallpox pustules were either

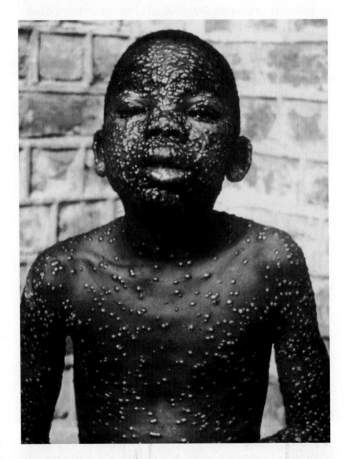

FIGURE 1-1 African child with rash typical of smallpox on face, chest, and arms. Smallpox, caused by the virus *Variola major,* has a 30% mortality rate. Survivors are often left with disfiguring scars. *[Centers for Disease Control.]*

inhaled or inserted into small cuts in the skin (a technique called *variolation*) in order to prevent this dreaded disease. In 1718, Lady Mary Wortley Montagu, the wife of the British ambassador in Constantinople, observed the positive effects of variolation on the native Turkish population and had the technique performed on her own children.

The English physician Edward Jenner later made a giant advance in the deliberate development of immunity, again targeting smallpox. In 1798, intrigued by the fact that milk-maids who had contracted the mild disease cowpox were subsequently immune to the much more severe smallpox, Jenner reasoned that introducing fluid from a cowpox pustule into people (i.e., inoculating them) might protect them from smallpox. To test this idea, he inoculated an 8-year-old boy with fluid from a cowpox pustule and later intentionally infected the child with smallpox. As predicted, the child did not develop smallpox. Although this represented a major breakthrough, as one might imagine, these sorts of human studies could not be conducted under current standards of medical ethics.

Jenner's technique of inoculating with cowpox to protect against smallpox spread quickly through Europe. However, it was nearly 100 years before this technique was applied to other diseases. As so often happens in science serendipity combined with astute observation led to the next major advance in immunology: the induction of immunity to cholera. Louis Pasteur had succeeded in growing the bacterium that causes fowl cholera in culture, and confirmed this by injecting it into chickens that then developed fatal cholera. After returning from a summer vacation, he and colleagues resumed their experiments, injecting some chickens with an old bacterial culture. The chickens became ill, but to Pasteur's surprise, they recovered. Interested, Pasteur then grew a fresh culture of the bacterium with the intention of trying this experiment again. But as the story is told, his supply of chickens was limited, and therefore he tested this fresh bacterial culture on a mixture of chickens; some previously exposed to the "old" bacteria and some new, unexposed birds. Unexpectedly, the chickens with past exposure to the older bacterial culture were completely protected from the disease and only the previously unexposed chickens died. Pasteur hypothesized and later showed that aging had weakened the virulence of the bacterial pathogen and that such a weakened or **attenuated** strain could be administered to provide immunity against the disease. He called this attenuated strain a **vaccine** (from the Latin *vacca*, meaning "cow"), in honor of Jenner's work with cowpox inoculation.

Pasteur extended his discovery to other diseases, demonstrating that it was possible to attenuate a pathogen and administer the attenuated strain as a vaccine. In a now classic experiment performed in the small village of Pouilly-le-Fort in 1881, Pasteur first vaccinated one group of sheep with anthrax bacteria (*Bacillus anthracis*) that were attenuated by heat treatment. He then challenged the vaccinated sheep, along with some unvaccinated sheep, with a virulent

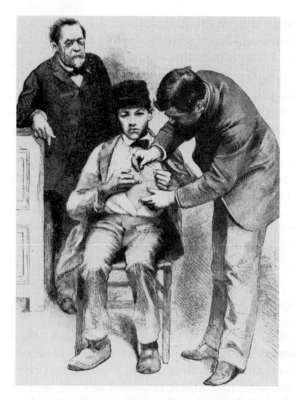

FIGURE 1-2 Wood engraving of Louis Pasteur watching Joseph Meister receive the rabies vaccine. *[From Harper's Weekly, 1885, Vol. 29:836; courtesy of the National Library of Medicine.]*

culture of the anthrax bacillus. All the vaccinated sheep lived and all the unvaccinated animals died.

In 1885, Pasteur administered his first vaccine to a human, a young boy who had been bitten repeatedly by a rabid dog (**Figure 1-2**). The boy, Joseph Meister, was inoculated with a series of attenuated rabies virus preparations. The rabies vaccine is one of very few that can be successful when administered shortly after exposure, as long as the virus has not yet reached the central nervous system and begun to induce neurologic symptoms. Joseph lived, and later became a caretaker at the Pasteur Institute, which was opened in 1887 to treat the many rabies victims that began to flood in when word of Pasteur's success spread; it remains to this day an institute dedicated to the prevention and treatment of infectious disease.

Key Concepts:

- Long before we understood much about the immune system, key principles of this system were already being studied and applied to solve public health issues associated with infectious disease.

- The principle behind vaccination is that exposure to safe forms of an infectious agent can result in future acquired protection, or immunity, to the real and more dangerous infectious agent.

Vaccination Is an Ongoing, Worldwide Enterprise

The emergence of the study of immunology and the discovery of vaccines are tightly linked. The goal of vaccination is to expose the individual to a pathogen (or a fragment of pathogen) in a safe way, allowing the immune cells to respond, developing and honing a strategy to fight this pathogen or others that are similar. When it works, this experiential learning process can produce extremely specific and long-lived memory cells, capable of protecting the host from the pathogen for many decades. However, the development of effective vaccines for some pathogens is still a major challenge, as discussed in Chapter 17. Yet, despite many biological and social hurdles, vaccination has yielded some of the most profound success stories in terms of improving mortality rates worldwide, especially in very young children.

In 1977, the last known case of naturally acquired smallpox was seen in Somalia. This dreaded disease was eradicated by universal application of a vaccine similar to that used by Jenner in the 1790s. One consequence of eradication is that universal vaccination becomes unnecessary. This is a tremendous benefit, as most vaccines carry at least a slight risk to those vaccinated. In many cases every individual does not need to be immune in order to protect most of the population. As a critical mass of people acquires protective immunity, either through vaccination or recovery from infection, they can serve as a buffer for the rest. This principle, called **herd immunity**, works by decreasing the number of individuals who can harbor and spread an infectious agent, significantly reducing the chances that susceptible individuals will become infected.

This presents an important altruistic consideration: although many of us could survive infectious diseases for which we receive a vaccine (such as the flu), this is not true for everyone. Some individuals cannot receive the vaccine (e.g., the very young or immune compromised), and vaccination is never 100% effective. In other words, the susceptible, nonimmune individuals among us can benefit from the pervasive immunity of their neighbors. For good reason, the balance of personal choice and public good is an area of heated debate (see **Clinical Focus Box 1-1**).

However, there is a darker side to eradication and the end of universal vaccination. Over time, the number of people with no immunity to the disease will begin to rise, ending herd immunity. Vaccination for smallpox largely ended by the early to mid-1970s, leaving well over half of the current world population susceptible to the disease. This means that smallpox, or a weaponized version, is now considered a potential bioterrorism threat. In response, new and safer vaccines against smallpox are still being developed today, most of which go toward vaccinating U.S. military personnel thought to be at greatest risk of possible exposure.

In the United States and other industrialized nations, vaccines have eliminated a host of childhood diseases that were the cause of death for many young children just 50 years ago. Measles, mumps, chickenpox, whooping cough (pertussis), tetanus, diphtheria, and polio, once thought of as an inevitable part of childhood, are now extremely rare or nonexistent in the United States because of current vaccination practices (**Table 1-1**). One can hardly estimate the savings to society resulting from the prevention of these diseases. Aside from suffering and mortality, the cost to treat these

TABLE 1-1	Cases of selected infectious disease in the United States before and after the introduction of effective vaccines		
	ANNUAL CASES/YR	CASES IN 2016	
Disease	Prevaccine	Postvaccine	Reduction (%)
Smallpox	48,164	0	100
Diphtheria	175,885	0	100
Measles	503,282	79^	99.98
Mumps	152,209	145*	98.90
Pertussis ("whooping cough")	147,271	964*	99.35
Paralytic polio	16,316	0	100
Rubella (German measles)	47,745	0*	100
Tetanus ("lockjaw")	1,314 (deaths)	1* (case)	99.92
Invasive *Haemophilus influenzae*	20,000	356*	98.22

Data from CDC Statistics of Notifiable Diseases (as of January, 2017).

The number of annual cases per year in 2016 increased^ or decreased* since 2010.

Vaccine Controversy: Weighing Evidence against Myth and Personal Freedom against Public Good

Despite the record of worldwide success of vaccines in improving public health, some opponents claim that vaccines do more harm than good, pressing for elimination or curtailment of childhood vaccination programs. There is no dispute that vaccines represent unique safety issues, since they are administered to people who are healthy. Furthermore, there is general agreement that vaccines must be rigorously tested and regulated, and that the public must have access to clear and complete information. Although the claims of vaccine critics must be evaluated, many can be addressed by careful and objective examination of records.

One example is the claim that vaccines given to infants and very young children contribute to the rising incidence of autism. This began with the suggestion that thimerosal, a mercury-based additive used to inhibit bacterial growth in some vaccine preparations since the 1930s, was causing autism in children. In 1999 the U.S. Public Health Service (USPHS) and the American Academy of Pediatrics (AAP) recommended that vaccine manufacturers begin to gradually phase out thimerosal with the goal of keeping children at or below Environmental Protection Agency (EPA)–recommended maximums in mercury exposure. With the release of this recommendation, parent-led public advocacy groups began a media-fueled campaign to build a case demonstrating a link between vaccines and an epidemic of autism, leading to declines in vaccination rates. However, cases of autism in children have continued to rise since thimerosal was removed from all childhood vaccines in 2001, dispelling this claim.

A 1998 study appearing in the *Lancet*, a reputable British medical journal, further fueled antivaccine organizations. The article, published by Andrew Wakefield, claimed the measles-mumps-rubella (MMR) vaccine caused pervasive developmental disorders in children,

including disorders on the autism spectrum. Almost two decades of subsequent research has been unable to substantiate these claims, and 10 of the original 13 authors on the paper later withdrew their support for the conclusions of the study. In 2010, the *Lancet* retracted the original article when it was shown that the data in the study had been falsified to reach desired conclusions. Nonetheless, in the years between the original publication of the *Lancet* article and its retraction, this case is credited with decreasing rates of MMR vaccination from a high of 92% to a low of almost 60% in certain areas of the United Kingdom. The resulting expansion in the population of susceptible individuals led to rising rates of measles and mumps infection and is credited with thousands of extended hospitalizations and several deaths of infected children.

Why has there been such a strong urge to cling to the belief that vaccines are linked autism in children despite much scientific evidence to the contrary? One possibility lies in the timing of the two events. Based on current AAP recommendations, most children receive 14 different vaccines and a total of up to 26 shots by the age of 2. In 1983, children received less than half this number. Couple this with the onset of the first signs of autism and other developmental disorders in children, which can appear quite suddenly and peak around 2 years of age. Furthermore, basic scientific literacy among the general public has decreased, while the number of ways to gather medical information (accurate or not) has increased. As concerned parents search for answers, one can begin to see how even scientifically unsupported links could begin to take hold.

Importantly, vaccination is not just a personal health choice; it's a public health issue. All states require childhood vaccinations before matriculation into the public school system, although medical exemptions are allowed for children who are immunocompromised, or who

have known allergic reactions to vaccines. Approximately 20 states also allow a range of personal, philosophical, or moral exemptions, which vary widely in their specifications and required documentation. In June 2015, California joined two other states (Mississippi and West Virginia) by enacting a controversial law (SB277) aimed at removing the religious exemptions clause, which allows families to opt out of vaccinating their children based on their religious beliefs. Research has shown that states with the most lenient exemptions have the lowest vaccination rates and that there is a significant correlation between ease of opting out and the rates of vaccine-preventable illness in that state.

This brings us to an important ethical question: how to draw the line between what is an allowable exemption and what is not? In a classic example of "tragedy of the commons", how do we weigh public good against personal freedom? Families who choose to opt out of vaccination for social or religious reasons tend to cluster with other such families. This clustering of unprotected individuals can escalate the spread of disease and lead to erosion of herd immunity, placing the entire community at risk. How do we weigh the rights of community members who are not eligible for vaccination, such as the very young, seriously ill, or immune compromised, against personal freedom?

The history of science and medicine is not without stories of bias and harm, vaccines included. However, while answers to these questions may be hard to find, opting for an exemption from rational scientific debate should not be one of them.

REFERENCES

Larson, H. J., L. Z. Cooper, J. Eskola, S. L. Katz, and S. Ratzan. 2011. Addressing the vaccine confidence gap. *Lancet* **378:**526.

Gostin, L. O. 2015. Law, ethics, and public health in the vaccination debates: politics of the measles outbreak. *JAMA* **313:**1099.

illnesses and their aftereffects or sequelae (such as paralysis, deafness, blindness, and developmental delays) is immense and dwarfs the costs of immunization.

Although these diseases have been largely eradicated in the United States, worldwide vaccination efforts continue. In 2000 the Global Alliance for Vaccines and Immunization (Gavi) was born. The goal of this international public-private partnership is to increase immunization coverage for children in poor countries and to accelerate access to new vaccines. In their first 15 years Gavi claims to have reached 500 million additional children, avoiding an estimated 7 million deaths. In addition to raising billions of dollars by the end of 2015, it may also be their unique approach that helps yield the greatest long-term success. The organization allows eligible developing countries to set their own agenda and monitor progress, while also requiring a financial commitment. This is sustained by both monetary and non-financial support through such entities as the World Bank, World Health Organization, donor countries, and the Bill & Melinda Gates Foundation. GAVI's goal is to create equal access to both established and new vaccines so that someday all nations will be able to pay the price for these vaccines in dollars rather than lives.

Despite the many successes of vaccine programs, such as the eradication of smallpox, many vaccine challenges still remain. Perhaps the greatest current challenge is the design of effective vaccines for major killers such as malaria and human immunodeficiency virus (HIV). As the tools of molecular and cellular biology, genomics, and proteomics improve, so will our understanding of the immune system, leaving us better positioned to make progress toward preventing these and other emerging infectious diseases. A further issue is the fact that millions of children in developing countries die from diseases that are fully preventable by available, safe vaccines. High manufacturing costs, instability of the products, and cumbersome delivery problems keep these vaccines from reaching those who might benefit the most. This problem could be alleviated in many cases by development of future-generation vaccines that are inexpensive, heat stable, and administered without a needle. Finally, misinformation and myth surrounding vaccine efficacy and side effects continue to hamper many potentially life saving vaccination programs (see Clinical Focus Box 1-1).

Key Concepts:
- Worldwide vaccination programs have effectively eradicated or protected us from many previously deadly infectious diseases, especially in young children.

- If many individuals in a group are protected from an infectious agent, either naturally or through vaccination, it is less likely to spread and unvaccinated individuals in the group are inadvertently protected as well.

Immunology Is about More than Just Vaccines and Infectious Disease

For some diseases, immunization programs may be the best or even the only effective defense. At the top of this list are infectious diseases that can cause serious illness or even death in unvaccinated individuals. Those transmitted by microbes that spread rapidly between hosts are especially good candidates for vaccination. However, vaccination, a costly process, is not the only way to prevent or treat infectious disease. Many infections are prevented, first and foremost, by other means. For instance, access to clean water, good hygiene practices, and nutrient-rich diets go a long way toward inhibiting transmission of infectious agents. In addition, some infectious diseases are self-limiting, easily treatable, and nonlethal for most individuals; these diseases are unlikely targets for expensive vaccination programs. They include the common cold, caused by rhinovirus infection, and cold sores that result from herpes simplex virus infection. Finally, some infectious agents are just not amenable to vaccination. This could be due to a range of factors, such as the number of different molecular variants of the organism, the complexity of the regimen required to generate protective immunity, or an inability to establish the needed immunologic memory responses (more on this later).

One major breakthrough in the treatment of infectious disease came when the first antibiotics were introduced in the 1920s. Antibiotics are chemical agents designed to destroy certain types of bacteria. They are ineffective against other types of infectious agents, as well as some bacterial species. At present there are more than 100 different antibiotics on the market, although most fall into just six or seven categories based on their mode of action. One particularly worrying trend is the steady rise in antibiotic resistance among bacterial strains traditionally amenable to these drugs, making the design of next-generation antibiotics and new classes of drugs increasingly important.

Although antiviral drugs are also available, most are not effective against many of the more common viruses, including influenza virus. This makes preventive vaccination the only real recourse against many debilitating infectious agents, even those that rarely cause mortality in healthy adults. For instance, because of the high mutation rate of the influenza virus, each year a new flu vaccine must be prepared based on a prediction of the prominent genotypes likely to be encountered in the next season. Some years this vaccine is more effective than others. If and when a more lethal and unexpected pandemic strain arises, there will be a race between its spread and the manufacture and administration of a new vaccine. With the current ease of worldwide travel, emergence of a pandemic strain of influenza today could dwarf the devastation wrought by the 1918 flu pandemic, which left up to 50 million dead.

However, the eradication of infectious disease is not the only worthy goal of immunology research. As we will see later, exposure to infectious agents is part of our evolutionary history. Wiping out all microbes from the bodies of their hosts could potentially cause more harm than good, both for the hosts and for the environment. Thanks to many technical advances allowing scientific discoveries to move efficiently from the bench to the bedside, clinicians can now manipulate the immune response in ways never before possible. For example, treatments to boost, inhibit, or redirect the specific efforts of immune cells are being applied to treat autoimmune disease, cancer, transplant rejection, and allergy, as well as other chronic disorders. These efforts are already extending and saving lives. Likewise, a clearer understanding of immunity has highlighted the interconnected nature of body systems, providing unique insights into areas such as cell biology, human genetics, and metabolism. For example, while a cure for acquired immune deficiency syndrome (AIDS) and a vaccine to prevent HIV infection are still the primary targets for many scientists who study this disease, a great deal of basic science knowledge came from the study of just this one virus and its interaction with the human immune system.

Key Concept:

- Beyond vaccination, it has become increasingly clear that elements of the immune system impact or regulate many other body systems and that these elements can be manipulated for the treatment of a range of human diseases.

Immunity Involves Both Humoral and Cellular Components

Pasteur showed that vaccination worked, but he did not understand how. Some scientists believed that immune protection in vaccinated individuals was mediated by cells, while others postulated that a soluble agent delivered protection. The experimental work of Emil von Behring and Shibasaburo Kitasato in 1890 gave the first insights into the mechanism of immunity, earning von Behring the Nobel Prize in Physiology or Medicine in 1901 (**Table 1-2**). Von Behring and Kitasato demonstrated that serum—the liquid,

TABLE 1-2	Nobel Prizes for immunologic research		
Year	**Recipient**	**Country**	**Research**
1901	Emil von Behring	Germany	Serum antitoxins
1905	Robert Koch	Germany	Cellular immunity to tuberculosis
1908	Elie Metchnikoff Paul Ehrlich	Russia Germany	Role of phagocytosis (Metchnikoff) and antitoxins (Ehrlich) in immunity
1913	Charles Richet	France	Anaphylaxis
1919	Jules Bordet	Belgium	Complement-mediated bacteriolysis
1930	Karl Landsteiner	United States	Discovery of human blood groups
1951	Max Theiler	South Africa	Development of yellow fever vaccine
1957	Daniel Bovet	Switzerland	Antihistamines
1960	F. Macfarlane Burnet Peter Medawar	Australia Great Britain	Discovery of acquired immunological tolerance
1972	Rodney R. Porter Gerald M. Edelman	Great Britain United States	Chemical structure of antibodies
1977	Rosalyn R. Yalow	United States	Development of radioimmunoassay
1980	George Snell Jean Dausset Baruj Benacerraf	United States France United States	Major histocompatibility complex
1984	Niels K. Jerne César Milstein Georges J. F. Köhler	Denmark Great Britain Germany	Immune-regulatory theories (Jerne) and technological advances in the development of monoclonal antibodies (Milstein and Köhler)
1987	Susumu Tonegawa	Japan	Gene rearrangement in antibody production
1990	E. Donnall Thomas Joseph Murray	United States United States	Transplantation immunology
1996	Peter C. Doherty Rolf M. Zinkernagel	Australia Switzerland	Role of major histocompatibility complex in antigen recognition by T cells

(continued)

TABLE 1-2   Nobel Prizes for immunologic research *(continued)*

Year	Recipient	Country	Research
2002	Sydney Brenner H. Robert Horvitz John E. Sulston	South Africa United States Great Britain	Genetic regulation of organ development and cell death (apoptosis)
2008	Harald zur Hausen Françoise Barré-Sinoussi Luc Montagnier	Germany France France	Role of HPV in causing cervical cancer (zur Hausen) and the discovery of HIV (Barré-Sinoussi and Montagnier)
2011	Jules Hoffmann Bruce A. Beutler Ralph M. Steinman	France United States United States	Discovery of activating principles of innate immunity (Hoffmann and Beutler) and role of dendritic cells in adaptive immunity (Steinman)
2015	William C. Campbell Satoshi Ōmura Youyou Tu	United States Japan China	Discoveries concerning novel therapies against parasitic diseases caused by roundworms (Campbell and Ōmura) and malaria (Tu)
2016	Yoshinori Ohsumi	Japan	Elucidation of the mechanisms underlying autophagy, involved in degradation of intracellular proteins during homeostasis and infection

noncellular component recovered from coagulated blood—from animals previously immunized with diphtheria could transfer the immune state to unimmunized animals.

In 1883, even before the discovery that a serum component could transfer immunity, Elie Metchnikoff, another Nobel Prize winner, demonstrated that cells also contribute to the immune state of an animal. He observed that certain white blood cells, which he termed **phagocytes**, ingested (phagocytosed) microorganisms and other foreign material (**Figure 1-3**, left). Noting that these phagocytic cells were more active in animals that had been immunized, Metchnikoff hypothesized that cells, rather than serum components, were the major effectors of immunity. The active phagocytic cells identified by Metchnikoff were likely blood monocytes and neutrophils (see Chapter 2), which can now be imaged using very sophisticated microscopic techniques (**Figure 1-3**, right).

Humoral Immunity

The debate over cells versus soluble mediators of immunity raged for decades. In search of the protective agent of immunity, various researchers in the early 1900s helped characterize the active immune component in blood serum. This soluble component could neutralize or precipitate toxins and could

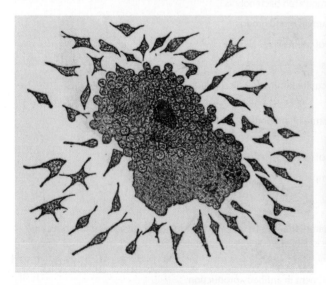

FIGURE 1-3 *Left:* **Drawing by Elie Metchnikoff of phagocytic cells surrounding a foreign particle.** *Right:* **Modern image of a phagocyte engulfing bacteria that cause tuberculosis.** Metchnikoff first described and named the process of *phagocytosis*, or ingestion of foreign matter by white blood cells. Today, phagocytic cells can be imaged in great detail using advanced microscopy techniques. *[Left: © The British Library Board. Lectures on the Comparative Pathology of Inflammation delivered at the Pasteur Institute in 1891. Translated by F. A. Starling and E. H. Starling with plates by Mechnikov, Il'ya Il'ich, 1893, p. 64, Fig. 32. Right: Science Photo Library/Science Source.]*

agglutinate (clump) bacteria. In each case, the component was named for the activity it exhibited: antitoxin, precipitin, and agglutinin, respectively. Initially, different serum components were thought to be responsible for each activity, but during the 1930s, mainly through the efforts of Elvin Kabat, a fraction of serum first called gamma globulin (now **immunoglobulin**) was shown to be responsible for all these activities. The soluble active molecules in the immunoglobulin fraction of serum are now commonly referred to as **antibodies**. Because these antibodies were contained in body fluids (known at that time as the body *humors*), the immunologic events they participated in was called **humoral immunity**.

The observations made by von Behring and Kitasato were quickly applied to clinical practice. **Antiserum**, the antibody-containing serum fraction from a pathogen-exposed individual, derived in this case from horses, was given to patients suffering from diphtheria and tetanus. A dramatic vignette of this application is described in **Clinical Focus Box 1-2**. Today there are still therapies that rely on transfer of immunoglobulins to protect susceptible individuals. For example, emergency use of immune serum containing antibodies against snake or scorpion venom, for treating the victims of certain poisonous bites or stings. This form of immune protection that is transferred between individuals is called **passive immunity** because the individual receiving it did not make his or her own immune response against the pathogen. Newborn infants benefit from passive immunity provided by the presence of maternal antibodies in their circulation. Passive immunity may also be used as a preventive (prophylaxis) to boost the immune potential of those with compromised immunity or who anticipate future exposure to a particular microbe.

While passive immunity can supply a quick solution, it is short-lived and limited, as the cells that produce these antibodies are not being transferred. On the other hand, natural infection, or the administration of a vaccine, is said to engender **active immunity** in the host: the production of one's own immunity. The induction of active immunity can supply the individual with renewable, long-lived protection from the specific infectious organism. As we discuss further below, this long-lived protection comes from memory cells, which provide protection for years or even decades after the initial exposure.

Cell-Mediated Immunity

As described above, a controversy developed between those who held to the concept of humoral immunity and those who agreed with Metchnikoff's concept of immunity imparted by specific cells, or **cell-mediated immunity**. The relative contributions of the two were widely debated at the time. It is now obvious that both are correct—the full immune response requires the action of both cells (cell-mediated) and soluble antibody components (humoral), the latter derived from white blood cells. Early studies of immune cells were hindered by the lack of genetically defined animal models and modern tissue culture techniques, whereas early

studies with serum took advantage of the ready availability of blood and established biochemical techniques to purify proteins. Information about cellular immunity therefore lagged behind the characterization of humoral immunity.

In a key experiment in the 1940s, Merrill Chase, working at the Rockefeller Institute, succeeded in conferring immunity against tuberculosis by transferring white blood cells between guinea pigs. Until that point, attempts to develop an effective vaccine or antibody therapy against tuberculosis had met with failure. Thus, Chase's demonstration helped to rekindle interest in cellular immunity. With the emergence of improved cell culture and transfer techniques in the 1950s, the **lymphocyte**, a type of white blood cell, was identified as the cell type responsible for both cellular and humoral immunity. Soon thereafter, experiments with chickens pioneered by Bruce Glick at Ohio State University indicated the existence of two types of lymphocytes: **T lymphocytes** (**T cells**), derived from the *t*hymus, and **B lymphocytes** (**B cells**), derived from the *b*ursa of Fabricius in birds (an outgrowth of the cloaca). In a convenient twist of nomenclature that makes B- and T-cell origins easier to remember, the mammalian equivalent of the *b*ursa of Fabricius is *b*one marrow, the home of developing B cells in mammals. *We now know that cellular immunity is imparted by T cells and that the antibodies produced by B cells confer humoral immunity.* The real controversy about the roles of humoral versus cellular immunity was resolved when the two systems were shown to be intertwined and it became clear that both are necessary for a complete immune response against most pathogens.

> **Key Concepts:**
>
> - Humoral immunity involves combating pathogens via antibodies, which are produced by B cells and can be found in bodily fluids. Antibodies can be transferred between individuals to provide passive immune protection.
>
> - Cell-mediated immunity involves the work of pathogen-specific T lymphocytes, which can act directly to eradicate the infectious agent as well as aiding other cells in their work.

How Are Foreign Substances Recognized by the Immune System?

One of the great enigmas confronting early immunologists concerned how the specificity of the immune response was determined for a particular pathogen or foreign material. Around 1900, Jules Bordet at the Pasteur Institute expanded the concept of immunity beyond infectious diseases, demonstrating that nonpathogenic substances, such as red blood cells from other species, could also elicit an immune response. Serum from an animal that had been inoculated with noninfectious but otherwise foreign (nonself) material would nevertheless react with the injected material in a specific manner.

Passive Antibodies and the Iditarod

In 1890, immunologists Emil Behring and Shibasaburo Kitasato, working together in Berlin, reported an extraordinary experiment. After immunizing rabbits with an attenuated form of tetanus and then collecting blood serum (*immune serum*) from these animals, they injected a small amount of the immune serum (a cell-free fluid) into the abdominal cavity of six mice. Twenty-four hours later, they infected the treated mice and untreated controls with live, virulent tetanus bacteria. All of the control mice died within 48 hours of infection, whereas the treated mice not only survived but showed no effects of infection. This landmark experiment demonstrated two important points. First, it showed that substances that could protect an animal against pathogens appeared in serum following immunization. Second, this work demonstrated that immunity could be passively acquired, or transferred from one animal to another by taking serum from an immune animal and injecting it into a nonimmune one. These and subsequent experiments did not go unnoticed. Both men eventually received titles (Behring became von Behring and Kitasato became Baron Kitasato). A few years later, in 1901, von Behring was awarded the first Nobel Prize in Physiology or Medicine (see Table 1-2).

These early observations, and others, paved the way for the introduction of passive immunization into clinical practice. During the 1930s and 1940s, passive immunotherapy, the endowment of resistance to pathogens by transfer of antibodies from an immunized donor to an unimmunized recipient, was used to prevent or modify the course of measles and hepatitis A. Subsequently, clinical experience and advances in the technology of immunoglobulin preparation have made this approach a standard medical practice. Passive immunization based on the transfer of antibodies is widely used in the treatment of immunodeficiency and some autoimmune diseases. It is also used to protect individuals against anticipated exposure to infectious and toxic agents against which they have no immunity. Finally, passive immunization can be lifesaving during episodes of certain types of acute infection, such as following exposure to rabies virus.

Immunoglobulin for passive immunization is prepared from the pooled plasma of thousands of donors. In effect, recipients of these antibody preparations are receiving a sample of the antibodies produced by many people to a broad diversity of pathogens—a gram of intravenous immune globulin (IVIG) contains about 10^{18} molecules of antibody that recognize more than 10^{7} different antigens. However, a product derived from the blood of such a large number of donors carries a risk of harboring pathogenic agents, particularly viruses. This risk is minimized by modern-day production techniques. The manufacture of IVIG involves treatment with solvents, such as ethanol, and the use of detergents that are highly effective in inactivating viruses such as HIV and hepatitis. In addition to treatment against infectious disease, or in acute situations, IVIG is also used today to treat some chronic diseases, including several forms of immune deficiency. In all cases, the transfer of passive immunity supplies only temporary protection.

One of the most famous instances of passive antibody therapy occurred in 1925, when an outbreak of diphtheria was diagnosed in what was then the remote outpost of Nome, Alaska. Lifesaving diphtheria-specific antibodies were available in Anchorage, but no roads were open and the weather was too dangerous for flight. History tells us that 20 mushers set up a dogsled relay to cover the almost 700 miles between Nenana, the end of the railroad run, and remote Nome. In this relay, two Norwegians and their dogs covered particularly critical territory and withstood blizzard conditions: Leonhard Seppala (**Figure 1**, left), who covered the most treacherous territory, and Gunnar Kaasen, who drove the final two legs in whiteout conditions, behind his lead dog Balto. Kaasen and Balto arrived in time to save many of the children in the town. To commemorate this heroic event, later that same year a statue of Balto was placed in Central Park, New York City, where it still stands today. This journey is memorialized every year in the running of the Iditarod Trail Sled Dog Race. A map showing the current route of this more than 1000-mile trek is shown in Figure 1, right.

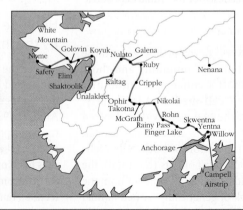

FIGURE 1 *Left:* Leonhard Seppala, the Norwegian who led a team of sled dogs in the 1925 diphtheria antibody run from Nenana to Nome, Alaska. *Right:* Map of the current route of the Iditarod Trail Sled Dog Race, which commemorates this historic delivery of lifesaving antibody. *[Corbis]*

The work of Karl Landsteiner and those who followed him showed that injecting an animal with almost any nonself organic chemical could induce production of antibodies that would bind specifically to the chemical. These studies demonstrated that antibodies have an almost unlimited range of reactivity, including being able to respond to compounds that had only recently been synthesized in the laboratory and were otherwise not found in nature! In addition, it was shown that molecules differing in the smallest detail, such as by a single amino acid, could be distinguished by their reactivity with different antibodies. To explain this high degree of specificity the *selective theory* was proposed.

The earliest conception of the selective theory dates to Paul Ehrlich in 1900. In an attempt to explain the origin of serum antibody, Ehrlich proposed that cells in the blood expressed a variety of receptors, which he called *side-chain receptors*, that could bind to infectious agents and inactivate them. Borrowing a concept used by Emil Fischer in 1894 to explain the interaction between an enzyme and its substrate, Ehrlich proposed that binding of the receptor to an infectious agent was like the fit between a lock and key. Ehrlich suggested that interaction between an infectious agent and a cell-bound receptor would induce the cell to produce and release more receptors with the same specificity or conformation (**Figure 1-4**). He thus coined the term **antigen**, any substance that elicits a specific response by B or T lymphocytes. In Ehrlich's mind, the cells were pluripotent, expressing a number of different receptors, each of which could be individually "selected" by the antigen. According to Ehrlich's theory, the specificity of the receptor was determined in the host before its exposure to the foreign antigen, and therefore the antigen *selected* the appropriate receptor. Ultimately, most aspects of Ehrlich's theory would be proven correct, with the following minor refinement: instead of one cell making many receptors, *each cell makes many copies of just one membrane-bound receptor (one specificity)*. An army of cells, each with a different antigen specificity, is therefore required. *The selected B cell can be triggered to proliferate and to secrete many copies of these receptors in soluble form (now called* antibodies) *once it has been selected by antigen binding.*

Through the insights of F. Macfarlane Burnet, Niels Jerne, and David Talmadge, this hypothesis was restructured into a model that came to be known as the **clonal selection** theory. This model has been further refined over the years and is now accepted as an underlying paradigm of modern immunology. According to this theory, an individual B or T lymphocyte expresses many copies of a membrane receptor that is specific for a single, distinct antigen. This unique receptor specificity is determined in the lymphocyte before it is exposed to the antigen. Binding of antigen to its specific receptor activates the cell, causing it to proliferate into a clone of daughter cells that have the same receptor specificity as the parent cell.

Overview Figure 1-5 presents a very basic scheme of clonal selection in the humoral (B-cell) and cellular (T-cell) branches of immunity. We now know that B cells produce

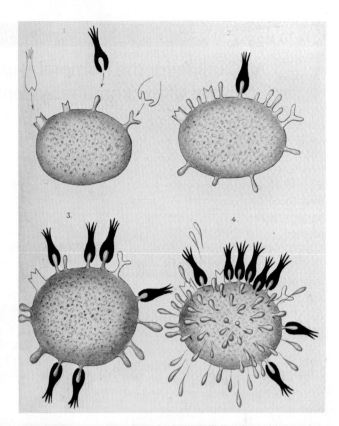

FIGURE 1-4 Representation of Paul Ehrlich's side-chain theory to explain antibody formation. In Ehrlich's initial theory, the cell is pluripotent in that it expresses a number of different receptors or side chains, all with different specificities. If an antigen encounters this cell and has a good fit with one of its side chains, synthesis of that receptor is triggered and the receptor will be released. *[The Royal Society.]*

antibodies, a soluble version of their receptor protein, that bind to foreign proteins, flagging them for destruction. T cells, which come in several different forms, also use their surface-bound T-cell receptors to sense antigen. These cells can perform a range of different functions once selected by antigen encounter, including the secretion of soluble compounds to aid other white blood cells (such as B lymphocytes) and the destruction of infected host cells.

Key Concepts:

- Antigen-specific immunity relies on surface molecules, called B- and T-cell receptors, unique to each individual lymphocyte. These receptors bind to a specific pathogenic structure called an *antigen*.

- Clonal selection is the process by which individual T and B lymphocytes are engaged by antigen and cloned to create a population of antigen-reactive cells with identical antigen specificity.

An Outline for the Humoral and Cell-Mediated (Cellular) Branches of the Immune System

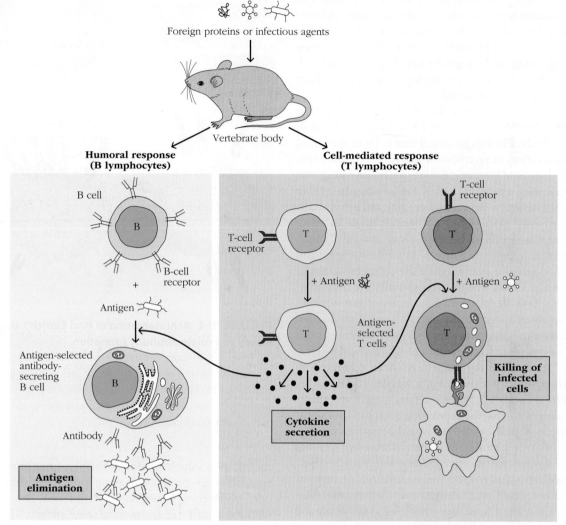

The humoral response involves interaction of B cells with foreign proteins, called antigens, and their differentiation into antibody-secreting cells. The secreted antibody binds to foreign proteins or infectious agents, helping to clear them from the body.

The cell-mediated response involves various subpopulations of T lymphocytes, which can perform many functions, including the secretion of soluble messengers that help direct other cells of the immune system and direct killing of infected cells.

Important Concepts for Understanding the Mammalian Immune Response

Today, more than ever, we are beginning to understand at the molecular and cellular levels how a vaccine or infection leads to the development of immunity. As highlighted by the historical studies described above, this involves a complex system of cells and soluble compounds that have evolved to protect us against an enormous range of invaders of all shapes, sizes, and chemical structures. In this section, we cover the range of organisms that challenge the immune system and

several of the important new concepts that are unique hallmarks of how the immune system carries out this task.

Pathogens Come in Many Forms and Must First Breach Natural Barriers

Organisms causing disease are termed **pathogens**, and the process by which they induce illness in the host is called **pathogenesis**. The human pathogens can be grouped into four major categories based on shared characteristics: viruses, fungi, parasites, and bacteria (**Table 1-3**). As we will see in the next

TABLE 1-3 Major categories of human pathogens

Viruses *Rotavirus*

Poliovirus	Poliomyelitis (polio)
Variola virus	Smallpox
Human immunodeficiency virus	AIDS
Measles virus	Measles
Influenza virus	Influenza
Rhinovirus	Common cold
Ebola virus	Hemorrhagic fever
Zika Virus	Zika fever/virus disease

Bacteria *Mycobacterium tuberculosis*

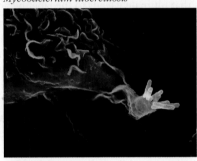

Mycobacterium tuberculosis	Tuberculosis
Bordetella pertussis	Whooping cough (pertussis)
Vibrio cholerae	Cholera
Borrelia burgdorferi	Lyme disease
Neisseria gonorrhea	Gonorrhea
Haemophilus influenzae	Bacterial meningitis & pneumonia

Fungi *Candida albicans*

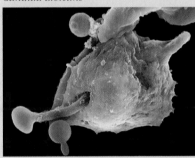

Candida albicans	Candidiasis (thrush)
Tinea corporis	Ringworm
Cryptococcus neoformans	Cryptococcal meningitis
Aspergillus fumigatus	Aspergillosis
Blastomyces dermatitidis	Blastomycosis

Parasites *Filaria*

Plasmodium species	Malaria
Leishmania major	Leishmaniasis
Entamoeba histolytica	Amoebic colitis
Schistosoma mansoni	Schistosomiasis
Wuchereria bancrofti	Lymphatic filariasis

PHOTOGRAPHS: (false color)

Virus: Transmission electron micrograph of multiple rotavirus virions, a major cause of infant diarrhea. Rotavirus accounts for approximately 1 million infant deaths per year in developing countries and hospitalization of about 50,000 infants per year in the United States. *[Dr. Linda M. Stannard, University of Cape Town/Science Source]*

Bacterium: *Mycobacterium tuberculosis* (orange), the bacterium that causes tuberculosis, being ingested by a human macrophage. *[Max Planck Institute for Infection Biology/Dr. Volker Brinkmann]*

Fungus: *Candida albicans,* a yeast inhabiting the human mouth, throat, intestines, and genitourinary tract; *C. albicans* commonly causes an oral rash (thrush) or vaginitis in immunosuppressed individuals or in those taking antibiotics that kill normal bacterial flora. *[SPL/Science Source]*

Parasite: The larval form of a parasitic filarial worm, being attacked by macrophages (yellow). Approximately 120 million persons worldwide have some form of filariasis. *[Oliver Meckes/Nicole Ottawa/Eye of Science/Science Source]*

section, some of the shared characteristics that are common to groups of pathogens, but not to the host, can be exploited by the immune system for recognition and destruction.

The microenvironment in which the immune response begins to emerge can also influence the outcome; the same pathogen may be treated differently depending on the context in which it is encountered. Some areas of the body, such as the central nervous system or the eye, are virtually "off limits" for the immune system because the immune response could do more damage than the pathogen. In other cases, the environment may come with inherent directional cues for immune cells. For instance, some foreign compounds that enter via the digestive tract, including the commensal microbes that help us digest food, are tolerated by the immune system. However, if these same foreigners enter the bloodstream they are typically treated much more aggressively. *Each encounter with pathogen thus engages a distinct set of strategies that depends on the nature of the invader and on the microenvironment in which engagement occurs.*

It is worth noting that immune pathways do not become engaged until foreign organisms first breach the physical barriers of the body. Obvious barriers include the skin and the mucous membranes. The acidity of the stomach contents, of the vagina, and of perspiration poses a further barrier to many organisms, which are unable to grow under low-pH conditions. Finally, soluble antimicrobial proteins secreted by the epithelial cells at the surfaces of the body help to hold would-be pathogens at bay. All these barriers are discussed in detail in Chapters 4 and 13. The importance of these barriers becomes obvious when they are surmounted. Animal bites can communicate rabies or tetanus, whereas insect puncture wounds can transmit the causative agents of such diseases as malaria (mosquitoes), plague (fleas), and Lyme disease (ticks). A dramatic example is seen in burn victims, who lose the protective skin at the burn site and must be treated aggressively with drugs to prevent the rampant bacterial and fungal infections that often follow.

> **Key Concepts:**
> - Pathogens fall into four major categories (viruses, bacteria, fungi, and parasites) and exist in many forms within each broad category.
> - The initial response by the immune system is determined by both the nature of the pathogen and the environment in which this encounter occurs.

The Immune Response Quickly Becomes Tailored to Suit the Assault

With the above in mind, an effective defense is one that is specifically designed to address the nature of the invading pathogen offense. The cells and molecules that become activated in a given immune response have evolved to meet the specific challenges posed by each pathogen, which include the structure of the pathogen and its location within or external to host cells. This means that different chemical structures and microenvironmental cues need to be detected and appropriately evaluated, initiating the most effective response strategy. *The process of pathogen recognition involves an interaction between the foreign organism and a recognition molecule (or molecules) expressed by host cells.* Although these recognition molecules are frequently membrane-bound receptors, soluble receptors or secreted recognition molecules can also be engaged. Ligands for these recognition molecules can include whole pathogens, antigenic fragments of pathogens, or products secreted by these foreign organisms. The outcome of this ligand binding is an intracellular or extracellular cascade of events that ultimately leads to the labeling and destruction of the pathogen—simply referred to as the *immune response*. The culmination of this response is engagement of a complex system of cells that can recognize and kill or engulf a pathogen (cellular immunity), as well as soluble proteins that help to orchestrate labeling and destruction of foreign invaders (humoral immunity).

The nature of the immune response will vary depending on the number and type of recognition molecules engaged. For instance, all viruses are tiny, obligate, intracellular pathogens that spend the majority of their life cycle residing inside host cells. An effective defense strategy must therefore involve identification of infected host cells along with recognition of the surface of the pathogen. This means that some immune cells must be capable of detecting changes that occur in a host cell after it becomes infected. This is achieved by a range of cytotoxic cells but especially by **cytotoxic T lymphocytes** (also known as **CTLs**, or T$_C$ **cells**), a part of the cellular arm of immunity. In this case, recognition molecules positioned *inside* cells are key to the initial response. These intracellular receptors bind to viral proteins present in the cytosol and initiate an early warning system, alerting the cell to the presence of an invader.

Sacrifice of virally infected cells often becomes the only way to truly eradicate this type of pathogen. In general, this sacrifice is for the good of the whole organism, although in some instances it can cause disruptions to normal function. For example, HIV infects a type of T cell called a **T helper cell (T$_H$ cell)**. These cells are called helpers because they guide the behavior of other immune cells, including B cells, and are therefore pivotal for selecting the pathway taken by the immune response. Once too many of these cells are destroyed or otherwise rendered nonfunctional, many of the directional cues needed for a healthy immune response are missing and fighting all types of infections becomes problematic. As we discuss later in this chapter, the resulting immunodeficiency allows opportunistic infections to take hold and potentially kill the patient.

Similar but distinct immune mechanisms are deployed to mediate the discovery of extracellular pathogens, such as fungi, most bacteria, and some parasites. These rely primarily on cell surface or soluble recognition molecules that probe the extracellular spaces of the body. In this case, B cells and the antibodies they produce as a part of humoral immunity play major roles. For instance, antibodies can squeeze into spaces in the body where B cells themselves may not be able to reach, helping to identify pathogens hiding in these out-of-reach places. Large parasites present yet another problem; they are too big for phagocytic cells to envelop. In cases such as these, cells that can deposit toxic substances or that can secrete products that induce expulsion (e.g., sneezing, coughing, vomiting) become a better strategy.

As we study the complexities of the mammalian immune response, it is worth remembering that a single solution does not exist for all pathogens. At the same time, these various immune pathways carry out their jobs with considerable overlap in structure and in function.

Key Concepts:

- During the initial stages of infection, the receptors that first recognize the foreign agent help the immune response categorize the offender and tailor the subsequent immune response.

- Unique pathways begin to emerge that are specific for different types of pathogens, such as cytotoxic T cells that kill virally infected host cells, T helper cells that assist other immune cells, and antibodies secreted by B cells to fight extracellular infection.

Pathogen Recognition Molecules Can Be Encoded as Genes or Generated by DNA Rearrangement

As one might imagine, most pathogens express at least a few chemical structures that are not typically found in mammals. **Pathogen-*a*ssociated *m*olecular *p*atterns** (or **PAMPs**) are common foreign structures that characterize whole groups of pathogens. It is these unique antigenic structures that the immune system frequently recognizes first. Animals, both invertebrates and vertebrates, have evolved to express several types of cell surface and soluble proteins that quickly recognize many of these PAMPs: a form of pathogen profiling. For example, encapsulated bacteria possess a polysaccharide coat with a unique chemical structure that is not found on other bacterial or human cells. White blood cells naturally express a variety of receptors, collectively referred to as **pattern recognition receptors** (**PRRs**), that specifically recognize these sugar residues, as well as other common foreign structures. When PRRs detect these chemical structures, a cascade of events labels the target pathogen for destruction. PRRs are proteins encoded in the genomic DNA and are always expressed by many different immune cells. These conserved proteins are found in one form or another in many different types of organism, from plants to fruit flies to humans, and represent a first line of defense for the quick detection of many of the typical chemical identifiers carried by the most common invaders. Collectively, these receptors and the cellular processes they help to enact

constitute a primitive and highly conserved response system known as *innate immunity* (discussed in more detail below).

A significant and powerful corollary to this is that it allows early categorizing or profiling of the sort of pathogen of concern. This is key to the subsequent immune response routes that will be followed, and therefore the fine tailoring of the immune response as it develops. For example, viruses frequently expose unique chemical structures only during their replication inside host cells. Many of these can be detected via intracellular receptors that bind exposed chemical moieties while still inside the host cell. This can trigger an immediate antiviral response in the infected cell that blocks further virus replication. At the same time, this initiates the secretion of chemical warning signals sent to nearby cells to help them guard against infection (a neighborhood watch system!). This early categorizing happens via a subtle tracking system that allows the immune response to make note of which recognition molecules were involved in the initial detection event and relay that information to subsequent responding immune cells, allowing the follow-up response to begin to focus attention on the likely type of assault underway.

Host-pathogen interactions are an ongoing arms race; pathogens evolve to express unique structures that avoid host detection, and the host recognition system co-evolves to match these new challenges. However, because pathogens generally have much shorter life cycles than their vertebrate hosts, and some use error-prone DNA polymerases to replicate their genomes, pathogens can evolve rapidly to evade host-encoded recognition systems. If this were our only defense, the host immune response would quickly become obsolete thanks to these real-time pathogen avoidance strategies. How can the immune system prepare for this? How can our DNA encode a recognition system for things that change in random ways over time? Better yet, how do we build a system to recognize new chemical structures that may arise in the future?

Thankfully, the vertebrate immune system has evolved a clever, albeit resource-intensive, response to this dilemma: to favor randomness in the design of some recognition molecules. This strategy, called **generation of diversity**, is employed only by developing B and T lymphocytes. *The result is a group of B and T cells in which each cell expresses many copies of one* unique *recognition molecule—collectively, a cell population with the theoretical potential to respond to any antigen that may come along* (**Figure 1-6**). This feat is accomplished by rearranging and editing the genomic DNA that encodes the antigen receptors expressed by each B or T lymphocyte. Not unlike the error-prone DNA replication method employed by pathogens, this system allows chance to play a role in generating a menu of responding recognition molecules. *Thus, B and T cells make surface receptors unique to each individual, which are then not passed on to offspring.* This is in direct contrast to the DNA that encodes PRRs, which are inherited and passed on to the next generation.

As one might imagine, however, this cutting and splicing of chromosomes is not without risk. Many B and T cells do not survive this DNA surgery or the quality control processes that

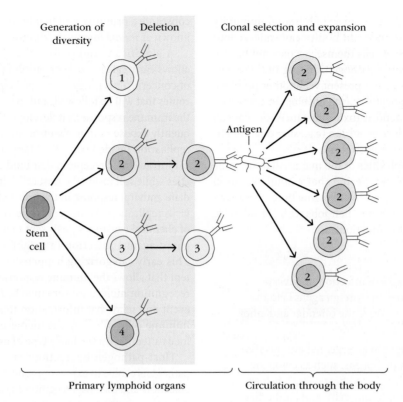

FIGURE 1-6 **Generation of diversity and clonal selection in T and B lymphocytes.** Maturation of T and B cells, which occurs in primary lymphoid organs (bone marrow for B cells and thymus for T cells) in the absence of antigen, produces cells with a committed antigenic specificity, each of which expresses many copies of surface receptor that binds to one particular antigen. Different clones of B cells (numbered 1, 2, 3, and 4) are illustrated in this figure. Cells that do not die or become deleted during this maturation and weeding-out process move into the circulation of the body and are available to interact with antigen. There, clonal selection occurs when one of these cells encounters its cognate or specific antigen. Clonal proliferation of an antigen-activated cell (number 2, or pink in this example) leads to many cells that can engage with and destroy the antigen, plus memory cells that can be called on during a subsequent exposure. The B cells secrete antibody, a soluble form of the receptor, reactive with the activating antigen. Similar processes take place in the T-lymphocyte population, resulting in clones of memory T cells and effector T cells; the latter include activated T$_H$ cells, which secrete cytokines that aid in the further development of adaptive immunity, and cytotoxic T lymphocytes (CTLs), which can kill infected host cells.

follow, all of which take place in primary lymphoid organs: the thymus for T cells and bone marrow for B cells. Surviving cells move into the circulation of the body, where they are available if their specific, or *cognate*, antigen is encountered. When antigens bind to the surface receptors on these cells, they trigger clonal selection (see Figure 1-6). The ensuing proliferation of the selected clone of cells creates an army of cells, all with the same receptor and responsible for binding more of the same antigen, with the ultimate goal of destroying the pathogen in question. In B lymphocytes, these recognition molecules are **B-cell receptors** when they are surface structures and *antibodies* in their secreted form. In T lymphocytes, where no soluble form exists, they are **T-cell receptors**. In 1976 Susumu Tonegawa, then at the Basel Institute for Immunology in Switzerland, discovered the molecular mechanism behind the DNA recombination events that generate B-cell receptors and antibodies (Chapter 6 covers this in detail). This was a true turning point in immunologic understanding; for this discovery he received widespread recognition, including the 1987 Nobel Prize in Physiology or Medicine (see Table 1-2).

Key Concepts:

- Initial immune responses rely on recognition molecules that are conserved and recognize common pathogenic structures. These are inherited.

- As the immune response progresses, antigen-specific recognition molecules that were generated randomly in each individual T and B cell via DNA rearrangement drive the bulk of the response. These are not inherited.

Tolerance Ensures That the Immune System Avoids Destroying the Host

One consequence of generating random recognition receptors is that some could recognize and target the host. In order for the immune system's diversity strategy to work effectively, it must somehow avoid accidentally recognizing and destroying host tissues. This principle, which relies on self/nonself discrimination, is called **tolerance**, another hallmark of the immune response. Sir Frank Macfarlane Burnet was the first to propose that exposure to nonself antigens during certain stages of life could result in an immune system that ignored these antigens later. Sir Peter Medawar later proved the validity of this theory by exposing mouse embryos to foreign antigens and showing that these mice developed the ability to tolerate these antigens later in life. Together, Burnet and Medawar were awarded the Nobel Prize in Physiology or Medicine in 1960 for their foundational work characterizing immune tolerance (see Table 1-2).

To establish tolerance, the antigen receptors present on developing B and T cells must first pass a test of nonresponsiveness against host structures. This process, which begins shortly after these randomly generated receptors are produced, is achieved by the destruction or inhibition of any cells that have inadvertently generated receptors with the ability to harm the host. *Successful maintenance of tolerance ensures that the host always knows the difference between self and nonself (usually referred to as "foreign").*

One recent re-envisioning of how tolerance is operationally maintained is the **danger** or **damage model.** This theory, proposed by Polly Matzinger at the National Institutes of Health, suggests that the immune system constantly evaluates each new encounter more for its potential to be dangerous versus safe to the host than for whether it is self versus nonself. For instance, cell death can have many causes, including natural homeostatic processes, mechanical damage, or infection. The former is a normal part of the everyday biological events in the body ("good death") and only requires a cleanup response to remove debris. This should not and normally does not activate an immune response. The latter two ("bad death"), however, come with warning signs that include the release of intracellular-only contents, expression of cellular stress proteins, and sometimes also pathogen-specific products. The host damage or danger-associated compounds released in these situations, collectively referred to as *alarmins*, can engage specific host recognition molecules (e.g., the same PRRs that recognize PAMPs) that deliver a signal to immune cells to get involved during these unnatural causes of cellular death. In other words, seeing "other" in some instances (without danger signals) may *not* lead to an immune response, while seeing "self" in the wrong context (*with* danger signals) can lead to a break in tolerance. In fact, there is significant support for this theory, including the coincidence between exposure to some infectious agents and the development of autoimmunity (immune reactivity against host structures).

As one might imagine, failures in the establishment or maintenance of tolerance can have devastating clinical outcomes. One unintended consequence of robust self-tolerance is that the immune system frequently ignores cancerous cells that arise in the body, as long as these cells continue to express self structures that the immune system has been trained to ignore. Dysfunctional tolerance is at the root of most autoimmune diseases, discussed further at the end of this chapter and in greater detail in Chapter 16.

Key Concepts:

- The phenomenon of self-tolerance, which prohibits immune responses to host tissue, is maintained through the elimination or inhibition of cells or receptors that could respond to self-structures.

- The danger or damage model of self-tolerance postulates that the immune response is not activated when host cell death occurs safely, but only when this death is accompanied by damage- or danger-associated signals produced by host cells.

The Immune Response Is Composed of Two Interconnected Arms: Innate Immunity and Adaptive Immunity

Although reference is made to "the immune system," it is important to appreciate that there are really two interconnected systems of response: innate and adaptive. These two systems collaborate to protect the body against foreign invaders. **Innate immunity** includes built-in molecular and cellular mechanisms that are evolutionarily primitive and aimed at preventing infection or quickly eliminating common invaders. This includes physical and chemical barriers to infection, as well as the DNA-encoded receptors recognizing common chemical structures of many pathogens (see PRRs, above; and Chapter 4). These are inherited from our parents and constitute a quick-and-dirty response; rapid recognition and subsequent phagocytosis or destruction of the pathogen is the outcome. Innate immunity also includes a series of preexisting serum proteins, collectively referred to as **complement**, that bind common pathogen-associated structures and initiate a cascade of labeling and destruction events (Chapter 5). This highly effective first line of defense prevents most pathogens from taking hold, or eliminates infectious agents within hours of encounter. The recognition elements of the innate immune system are fast, some occurring within seconds of a barrier breach, but they are not very specific and are therefore unable to distinguish between small differences in foreign antigens.

A second form of immunity, known as **adaptive immunity**, is much more attuned to subtle molecular differences. This part of the system, which relies on B and T lymphocytes, takes longer to come on board but is much more antigen specific. Typically, there is an adaptive immune response against a pathogen within 5 or 6 days after the barrier breach and initial exposure, followed by a gradual resolution of the infection. Adaptive immunity is slower partly because fewer cells possess the perfect receptor for the job: the antigen-specific receptors on T and B cells that are generated via DNA rearrangement, mentioned earlier. It is also slower because parts of the adaptive response rely on prior encounter and "categorizing" of antigens undertaken by innate processes. After antigen encounter, T and B lymphocytes undergo selection and

TABLE 1-4	Comparison of innate and adaptive immunity	
	Innate	**Adaptive**
Response time	Minutes to hours	Days
Specificity	Limited and fixed	Highly diverse; adapts to improve during the course of immune response
Response to repeat infection	Same each time	More rapid and effective with each subsequent exposure
Major components	Barriers (e.g., skin); phagocytes; pattern recognition molecules	T and B lymphocytes; antigen-specific receptors; antibodies

proliferation, according to the clonal selection theory of antigen specificity described earlier (see Figure 1-5). Although slow to act, once these B and T cells have been selected, replicated, and have honed their attack strategy, they become formidable opponents that can typically resolve the infection.

The adaptive arm of the immune response evolves in real time in response to infection and adapts (thus the name) to better recognize, eliminate, and remember the invading pathogen. Adaptive responses involve a complex and interconnected system of cells and chemical signals that come together to finish the job initiated during the innate immune response. *The goal of all vaccines against infectious disease is to elicit the development of specific and long-lived adaptive responses, so that the vaccinated individual will be protected in the future when the real pathogen comes along.* This arm of immunity is orchestrated mainly via B and T lymphocytes following engagement of their randomly generated antigen recognition receptors. How these receptors are generated is a fascinating story, covered in detail in Chapter 6 of this text. An explanation of how these cells develop to maturity (Chapters 8 and 9), become activated during an immune response (Chapters 10 and 11), and then work in the body to protect us from infection (Chapters 12–14) or sometimes fail us (Chapters 15–19) takes up the vast majority of this text.

The number of pages dedicated to discussing adaptive responses should not give the impression that this arm of the immune response is more important, or can work independently from innate immunity. In fact, the full development of the adaptive response is dependent on earlier innate pathways. The intricacies of their interconnections remain an area of intense study. The 2011 Nobel Prize in Physiology or Medicine was awarded to three scientists who helped clarify these two arms of the response: Bruce Beutler and Jules Hoffmann for discoveries related to the activation events important for innate immunity, and Ralph Steinman for his discovery of the role of dendritic cells in activating adaptive immune responses (see Table 1-2). Because innate pathways make first contact with pathogens, the cells and molecules involved in this arm of the response use information gathered from their early encounter with pathogen to help direct the process of adaptive immune development. *Adaptive immunity thus provides a second and more comprehensive line of defense, informed by the struggles undertaken by the innate system.*

It is worth noting that some infections are, in fact, eliminated by innate immune mechanisms alone, especially those that remain localized and involve very low numbers of fairly benign foreign invaders. (Think of all those insect bites or splinters in life that introduce bacteria under the skin!) **Table 1-4** compares the major characteristics that distinguish innate and adaptive immunity. Although for ease of discussion the immune system is typically divided into these two arms of the response, there is considerable overlap of the cells and mechanisms involved in each of these arms of immunity.

Key Concepts:

- The vertebrate immune response can be divided into two interconnected arms of immunity: innate and adaptive. Innate responses are rapid but less pathogen-specific, using inherited recognition molecules and phagocytic cells. Adaptive responses are slower (taking days to develop) but highly specialized to the pathogen, and rely on randomly generated recognition receptors made by B and T cells.

- Innate and adaptive immunity operate cooperatively; activation of the innate immune response produces signals that are required to stimulate and direct the behavior of subsequent adaptive immune pathways.

Immune Cells and Molecules Can Be Found in Many Places

For an immune response to be effective, the required cells and molecules need to be wherever the pathogen is. This means that unlike many of the body's other systems, which can be concentrated in one or a few specialized organs (e.g., the digestive and reproductive systems), the immune system is highly dispersed. Specialized depots of immune activity are positioned at strategic locations in the body, and immune cells can be found to reside as sentinels in most other tissues. White blood cells or their products are constantly circulating through the body visiting these depots in search of pathogen.

White blood cells, which mediate both innate and adaptive immune responses, come in many different types, and one or more of their members can be found in most of the spaces in the body. Some spaces get more than others, like the gut versus the nervous system, and this is frequently commensurate with the potential threat in terms of the sheer number of intimate daily exposures to potential invaders. Tissue-resident immune cells, sometimes referred to as *sentinel cells*, typically remain inconspicuous and relatively inactive unless a threat arises. Their job is to serve as a local alarm system and as first responders, kicking off the cascade of innate immune events to get the ball rolling. That cascade may begin at the site of infection, but in order for adaptive immunity to be initiated the rare lymphocytes with receptors specific for a particular pathogen need to be found. This means that the perfect lymphocytes for the job need to somehow end up in the right place at the right time.

To solve this issue of place and time, the immune system has evolved specialized organs such as lymph nodes (Chapter 2), where the transition from innate to adaptive immunity occurs. Through one route, the fluid bathing our tissues is funneled to and filtered through these sieve-like structures before it is returned to the blood. Through another route, antigen-specific lymphocytes enter these lymphoid organs, scanning for foreign antigens. This fluid and cell recirculation pattern allows relatively quick convergence of antigen and antigen-specific lymphocytes at the same location and in a microenvironment designed for the task. The result of this encounter is clonal selection and the start of an adaptive response.

Having a system that is spread throughout the body creates challenges regarding coordination and communication. In order for the cells involved in innate and adaptive immunity to work together, these two systems must be able to communicate with one another and coordinate a plan of attack. This communication is achieved both by direct cell-to-cell communication and by messenger proteins that are typically secreted and known by the general name **cytokines** (Chapter 3). Whether soluble or membrane-bound, these messengers bind to receptors on responding cells, inducing intracellular signaling cascades that can result in activation, proliferation, and differentiation of target cells. This is usually, but not always, mediated by changes in gene transcription that induce new functions in the target cell population. The target cells may now have the ability to make new factors or ligands of their own, or to migrate to new locations based on a fresh set of adhesion molecules.

A subset of these soluble signals are called **chemokines** because they have chemotactic activity, meaning they can recruit specific cells to the site—like a trail of molecular breadcrumbs. In this way, *cytokines, chemokines, and other soluble factors produced by immune cells recruit cells and draw fluid to the site of infection, providing help for pathogen eradication.* We've probably all felt this convergence in the form of swelling, heat, and tenderness at a site of infection. These events are part of a larger process collectively referred to as an **inflammatory response**, which is covered throughout this text in the context of a normal immune response, and in detail in Chapters 4 and 15. Frequently, more than one type of cytokine or chemokine is involved in these communication sessions between cells, and the unique set of receptors activated by this combination of signals helps to fine-tune the message and the resulting cellular response.

Overview Figure 1-7 highlights the major events of an immune response. In this example, bacteria are shown breaching a mucosal or skin barrier, where they are recognized and engulfed by a local phagocytic cell (step 1). As part of the innate immune response, the local phagocytic cell releases cytokines and chemokines that attract other white blood cells to the site of infection, initiating inflammation (step 2). A phagocytic cell that has engulfed pathogen or the infectious agent itself then migrates to a local lymph node or other secondary lymphoid structure through lymphatic vessels (step 3). Lymphocytes (B and T cells) that have developed in primary lymphoid organs like the bone marrow and thymus make their way to these secondary lymphoid structures (step 4), where they can now meet up with the pathogen. Those lymphocytes with receptors that are specific for the pathogen are selected, proliferate, and begin the adaptive phase of the immune response, as shown in an example lymph node (step 5). This results in many antigen-specific T and B cells (called effector cells), the latter releasing antibodies that are specific for the pathogen. Many of these cells will exit the secondary lymphoid structure and join with the blood circulating through the body (step 6). At sites in the body experiencing the effects of innate responses, or inflammation, these effector cells and molecules will exit blood vessels and enter the inflamed tissue (step 7), migrating towards the pathogen and first responder phagocytic cells. Antibodies and T cells can now attach to and or attack the intruder, directing its destruction (step 8). At the conclusion, the adaptive response leaves behind memory T and B cells that recall the strategy used to eradicate the pathogen and can employ this strategy again during subsequent encounters. It is worth noting that memory is a unique capacity that arises from adaptive responses; there is no memory component of innate immunity (see below).

> **Key Concept:**
> - Components of the immune system can be found throughout the body, as sentinel cells in most tissues, in the form of specialized lymphoid organs, and through the specific recruitment of immune cells and fluid to sites of infection.
>
> - Overview Figure 1-7 outlines the basic scheme of an immune response and serves as a preview of concepts essential to the stages of the immune response, discussed in detail in later chapters.

Collaboration between Innate and Adaptive Immunity in Resolving an Infection

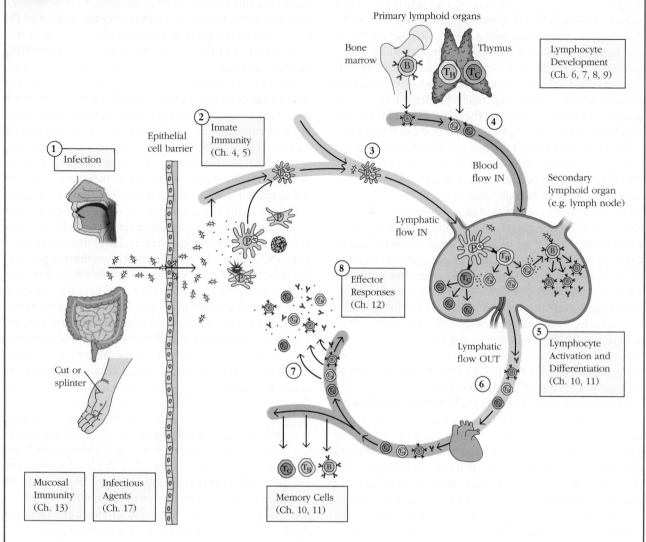

This very basic scheme shows the sequence of events that occurs during an immune response, highlighting interactions between innate and adaptive immunity. (1) Pathogens (e.g., bacteria) may enter the body through mucosal surfaces (e.g., lungs or intestines) or a breach in the skin. After breaching epithelial cell barriers (2), the pathogen is detected by resident phagocytic cells (yellow) and the innate stage of the immune response begins. The responding phagocytic cells undergo changes that allow them to fight the infection locally via release of antimicrobial compounds, chemokines, and cytokines (black dots) that also cause fluid influx that helps recruit other immune cells to the site (inflammation). (3) Free pathogen and some phagocytic cells that have engulfed the pathogen flow or migrate through lymphatic vessels toward secondary lymphoid structures (e.g., lymph nodes), (4) where they intersect with lymphocytes entering from the blood. Adaptive immunity is initiated in secondary lymphoid structures, where T helper cells (blue), T cytotoxic cells (red), and B cells (green) with the appropriate receptor specificity bind pathogen and are clonally selected, resulting in many rounds of proliferation and differentiation. (6) These specialized T and B cells, along with their products (e.g., antibodies generated by B cells), migrate out of the lymph node and eventually join the bloodstream, being pumped by the heart through the body. (7) As they identify areas of infection (signified by the inflammation from earlier innate responses) they exit the blood vessels and (8) migrate toward the infection, where they can help label and destroy any remaining pathogen (the effector phase). Residual long-term memory T and B cells take up residence in various locations in the body (not shown), from which they will be available if this pathogen is encountered again and can initiate a more rapid and antigen-specific secondary response. The relevant chapters for each stage of these responses are noted. (Abbreviations: T_C = T cytotoxic cell; T_H = T helper cell; B = B cell; P = phagocyte.)

Adaptive Immune Responses Typically Generate Memory

One particularly significant and unique attribute of the adaptive arm of the immune response is **immunologic memory**. This is the ability of the immune system to respond much more swiftly and with greater efficiency during a second exposure to the same pathogen. Unlike almost any other biological system, the vertebrate immune response has evolved not only the ability to learn from (adapt to) its encounters with foreign antigen in real time but also the ability to store this information for future use. During a first encounter with foreign antigen, adaptive immunity undergoes what is termed a **primary response**, during which the key lymphocytes that will be used to eradicate the pathogen are clonally selected, honed, and enlisted to resolve the infection. As mentioned above, these cells incorporate messages received from the innate players into their tailored response to the specific pathogen.

All subsequent encounters with the same antigen or pathogen are typically referred to as the **secondary response** (**Figure 1-8**). During a secondary response, **memory cells**, kin of the final and most efficient B and T lymphocytes trained during the primary response, are re-enlisted to fight again. These cells begin almost immediately and pick up right where they left off, continuing to learn and improve their eradication strategy during each subsequent encounter with the same antigen. Depending on the antigen in question, memory cells can remain for decades after the conclusion of the primary response. Memory lymphocytes provide the means for subsequent responses that are so rapid, antigen-specific, and effective that when the same pathogen infects the body a second or subsequent time, dispatch of the offending organism often occurs without symptoms. It is the remarkable property of memory that prevents us from catching many diseases a second time. Immunologic memory harbored by residual B and T lymphocytes is the foundation for vaccination, which uses crippled or killed pathogens as a safe way to "educate" the immune system to prepare it for later attacks by life-threatening pathogens. Memory cells then save the strategy used, not the pathogen (or vaccine), for later reference during repeat encounters with the same infectious agent.

Sometimes, as is the case for some vaccines, one round of antigen encounter and adaptation is not enough to impart protective immunity from the pathogen in question. In many of these cases, immunity can develop after a second or even a third round of exposure to an antigen. It is these sorts of pathogens that necessitate the use of vaccine booster shots. Booster shots are nothing more than a second or third episode of exposure to the antigen, each driving a new round of adaptive events (secondary response) and refinements in the responding lymphocyte population. The aim is to hone these responses to a sufficient level to afford protection against the real pathogen at some future date.

> **Key Concept:**
>
> • The first exposure to a pathogen results in a primary immune response, which culminates in the creation of memory cells, or B and T cells that remain after pathogen eradication and that can be activated during a subsequent exposure to that same pathogen (a secondary response).

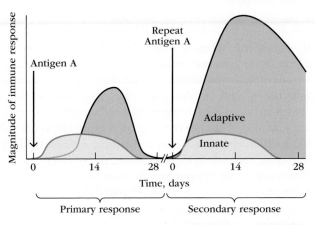

FIGURE 1-8 Differences in the primary and secondary adaptive immune response to injected antigen reflect the phenomenon of immunologic memory. When an animal is injected with an antigen, it produces a primary antibody response (dark blue) of low magnitude and short duration, peaking at about 10 to 20 days. At some later point, a second exposure to the same antigen results in a secondary response that is greater in magnitude, peaks in less time (1–4 days), and is more antigen specific than the primary response. Innate immune responses (light blue), which have no memory element and occur each time an antigen is encountered, are unchanged regardless of how frequently this antigen has been encountered in the past.

The Good, Bad, and Ugly of the Immune System

The picture we've presented so far depicts the immune response as a multicomponent interactive system that always protects the host from invasion by all sorts of pathogens. However, failures of this system do occur. They can be dramatic and often garner a great deal of attention, despite the fact that they are generally rare. Certain clinical situations also pose unique challenges to the immune system, including tissue transplants between individuals (probably not part of any evolutionary plan!) and the development of cancer. In this section we briefly describe some examples of common failures and challenges to the development of healthy immune responses. Each of these clinical manifestations is covered in much greater detail in the concluding chapters of this text (Chapters 15–19).

Inappropriate or Dysfunctional Immune Responses Can Result in a Range of Disorders

Most instances of immune dysfunction or failure fall into one of the following three broad categories:

- **Hypersensitivity (allergy):** Overly zealous attacks on common benign but foreign antigens
- **Autoimmune disease:** Erroneous targeting of self-proteins or tissues by immune cells
- **Immune deficiency:** Insufficiency of the immune response to protect against infectious agents
- **Immune imbalance:** Dysregulation in the immune system that leads to aberrant activity of immune cells, especially enhanced inflammation and/or and reduced immune inhibition

A brief overview of these situations and some examples of each are presented below. At its most basic level, immune dysfunction occurs as a result of improper regulation that allows the immune system either to attack something it shouldn't or fail to attack something it should. Hypersensitivities, including allergy, and autoimmune disease are cases of the former, where the immune system attacks an improper target. As a result, the symptoms can manifest as *pathological inflammation*—an influx of immune cells and molecules that results in detrimental symptoms, including chronic inflammation and rampant tissue destruction. In contrast, immune deficiencies, caused by a failure to properly deploy the immune response, usually result in weakened or dysregulated immune responses that can allow pathogens to get the upper hand. Immune imbalance, a less well-characterized phenomenon, can result from changes in the environment that disrupt immune homeostasis. Manifestations of this typically present as either allergic or autoimmune conditions, both examples of overly active immune response states.

Hypersensitivity Reactions

Allergies and asthma are examples of hypersensitivity reactions. These result from inappropriate and overly active immune responses to common innocuous environmental antigens, such as pollen, food, or animal dander. The possibility that certain substances induce increased sensitivity (hypersensitivity) rather than protection was recognized in about 1902 by Charles Richet, who attempted to immunize dogs against the toxins of a type of jellyfish. He and his colleague Paul Portier observed that dogs exposed to sublethal doses of the toxin reacted almost instantly, and fatally, to a later challenge with even minute amounts of the same toxin. Richet concluded that a successful vaccination typically results in *phylaxis* (protection), whereas **anaphylaxis** (antiprotection)—an extreme, rapid, and often lethal overreaction of the immune response to something it has encountered before—can result in certain cases in which exposure to antigen is repeated. Richet received the Nobel Prize in 1913 for his discovery of the anaphylactic response

FIGURE 1-9 Patient suffering from hay fever as a result of an allergic reaction. Such hypersensitivity reactions result from sensitization caused by previous exposure to an antigen in some individuals. In the allergic individual, histamines are released as a part of the hypersensitivity response and cause sneezing, runny nose, watery eyes, and such during each subsequent exposure to the antigen (in this context called an *allergen*). [Chris Rout/Alamy]

(see Table 1-2). The term is used today to describe a severe, life-threatening, allergic response.

Fortunately, most hypersensitivity or allergic reactions in humans are not fatal. There are several different types of hypersensitivity reactions; some are caused by antibodies and others are the result of T-cell activity (see Chapter 15). However, most allergic or anaphylactic responses involve a type of antibody called *immunoglobulin E* (IgE). Binding of IgE to its specific antigen (allergen) induces the release of substances that cause irritation and inflammation, or the accumulation of cells and fluid at the site. When an allergic individual is exposed to an allergen symptoms may include sneezing (**Figure 1-9**), wheezing and difficulty breathing (asthma); dermatitis or skin eruptions (hives); and, in more severe cases, strangulation due to constricted airways following extreme inflammation. A significant fraction of our health resources is expended to care for those suffering from allergies and asthma. One particularly interesting rationale to explain the unexpected rise in allergic disease, called the hygiene hypothesis and linked to immune imbalance, is discussed in **Clinical Focus Box 1-3.**

Autoimmune Disease

Sometimes the immune system malfunctions and a breakdown in self-tolerance occurs. This could be caused by a sudden inability to distinguish between self and nonself or by a misinterpretation of a self-component as dangerous, causing an immune attack on host tissues. This condition, called **autoimmunity**, can result in a number of chronic debilitating diseases. The symptoms of autoimmunity differ, depending on which tissues or organs are under attack. For example, multiple sclerosis is due to an autoimmune attack on a protein in nerve sheaths in the brain and central

BOX 1-3

The Hygiene Hypothesis

As of 2012, an estimated 334 million people worldwide had asthma, and approximately 14% of the world's children suffered from symptoms (see Chapter 15). In the United States the most common reason for a trip to a hospital emergency room (ER) is an asthma attack, accounting for up to one-third of all visits. Asthma is seen more frequently in the young and disproportionally affects minorities. Amongst African Americans, 15% of adults and over 18% of children in the United States report having suffered from asthma.

In the past 25 years, the prevalence of asthma in industrialized nations has doubled, and other types of allergic disease have increased as well. What accounts for this climb in asthma and allergy in the last few decades? One idea, called the *hygiene hypothesis,* suggests that a decrease in human exposure to previously common environmental microbes has had adverse effects on the human immune system. The hypothesis suggests that several categories of disorders caused by excessive immune activation, have become more prevalent in industrialized nations thanks to diminished exposure to particular classes of microbes following the widespread use of antibiotics and overall hygienic practices. This idea was first proposed by D. P. Strachan in an article published in 1989 suggesting a link between hay fever and household hygiene. More recently, this hypothesis has been expanded to include the view by some that it may be a contributing factor in many allergic diseases, several autoimmune disorders, and, more recently, inflammatory bowel disease.

What is the evidence supporting the hygiene hypothesis? The primary clinical support comes from studies that have shown a positive correlation between growing up under environmental conditions that favor microbe-rich (sometimes called "dirty") environments and a decreased incidence of allergy, especially asthma. To date, childhood exposure to cowsheds and farm animals, having

several older siblings, attending day care early in life, or growing up in a developing nation have all been correlated with a decreased likelihood of developing allergies. While viral exposures during childhood do not seem to favor protection, exposure to certain classes of bacteria and parasitic organisms may. Of late, the primary focus of attention has been on specific classes of parasitic worms (called *helminths*), spawning New Age allergy therapies involving intentional exposure. This gives whole new meaning to the phrase "Go eat worms"!

What are the proposed immunologic mechanisms that might underlie this link between a lack of early-life microbial exposure and allergic disease? Current dogma supporting this hypothesis posits that millions of years of coevolution of microbes and humans have favored a system in which early exposure to a broad range of common environmental bugs helps tune the immune system for the ideal balance between aggression and inhibition. These microbes have played a longstanding role in our evolutionary history, both as pathogens and as harmless microbes that make up our historical flora. Referred to as "old friends," these organisms may engage with the pattern recognition receptors (PRRs) present on cells of our innate immune system, driving them to warn cells involved in adaptive responses to tone it down. This hypothesis posits that without early and regular exposure of our immune cells to antigens derived from these old friends the development of "normal" immune regulatory or homeostatic responses is thrown into disarray, setting us up for an immune system poised to overreact in the future.

Animal models of disease lend some support to this hypothesis and have helped immunologists probe this line of thinking. For instance, certain animals raised in partially or totally pathogen-free environments are more prone

Hero Images/Getty Images

to type 1, or insulin-dependent, diabetes, an autoimmune disease caused by immune attack of pancreatic cells (see Chapter 16). The lower the infectious burden of exposure in these mice, the greater the incidence of diabetes. Animals specifically bred to carry enhanced genetic susceptibility favoring spontaneous development of diabetes (called NOD mice, for *non-obese diabetic*) and treated with a variety of infectious agents can be protected from diabetes. Meanwhile, NOD mice maintained in pathogen-free housing almost uniformly develop diabetes. Much like this experimental model, susceptibility to asthma and most other allergies is known to run in families, suggesting that genes and environment both play a role. While the jury may still be out concerning the verdict behind the hygiene hypothesis, animal and human studies clearly point to strong roles for both genes and environment in susceptibility to allergy. As data in support of this hypothesis continue to grow, the old saying concerning a dirty child—that "It's good for their immune system"—may actually hold true!

REFERENCES

Strachan, D. P. 1989. Hay fever, hygiene, and household size. *BMJ* **299**(6710):1259-60.

Liu, A. H., and J. R. Murphy. 2003. Hygiene hypothesis: fact or fiction? *Journal of Allergy and Clinical Immunology* **111**:471.

Sironi, M., and M. Clerici. 2010. The hygiene hypothesis: an evolutionary perspective. *Microbes and Infection* **12**:421.

nervous system that results in neuromuscular dysfunction. Crohn's disease is an attack on intestinal tissues that leads to destruction of gut epithelia and poor absorption of food. One of the most common autoimmune disorders, rheumatoid arthritis, results from an immune attack on joints of the hands, feet, arms, and legs.

Both genetic and environmental factors are likely involved in the development of most autoimmune diseases. However, the exact combination of genes and environmental exposures that favors the development of each particular autoimmune disease is difficult to pin down; immunologic research in this area is very active. Recent discoveries and the search for improved treatments are all covered in greater detail in Chapter 16.

Immune Deficiency

In most cases, when a component of innate or adaptive immunity is absent or defective, the host suffers from some form of **immunodeficiency**. Some of these deficiencies produce major clinical effects, including death, while others are more minor or even difficult to detect. Immune deficiency can arise due to inherited genetic factors (called **primary immunodeficiencies**) or as a result of disruption/damage by chemical, physical, or biological agents (termed **secondary immunodeficiencies**). Both of these forms of immune deficiency are discussed in greater detail in Chapter 18.

The severity of the disease resulting from immune deficiency depends on the number and type of affected immune response components. A common type of primary immunodeficiency in North America is a selective immunodeficiency in which only one type of antibody, called *immunoglobulin A*, is lacking; the symptoms may be an increase in certain types of infections, or the deficiency may even go unnoticed. In contrast, a rarer but much more extreme deficiency, called **severe combined immunodeficiency (SCID)**, affects both B and T cells and basically wipes out adaptive immunity. When untreated, SCID frequently results in death from infection at an early age. The most effective treatment for SCID is bone marrow transplantation, which can be long-lived and life-saving.

Secondary or acquired immunodeficiency can be caused by a number of factors including severe malnutrition, chronic diseases such as diabetes, and infection. By far, the most common cause of acquired immune deficiency worldwide is severe malnutrition, namely protein-calorie and micronutrient insufficiency. Estimates are that 30% to 50% of the world population suffers from some form of malnutrition, all of which can impact the potency of the immune response. Pneumonia, diarrhea, and malaria are among the most common infectious causes of death in populations suffering from malnutrition. These diseases, while caused by infectious agents, are much more likely to result in death when combined with malnutrition and the resulting immune suppression. Targeting this highly preventable condition might go further than any other global initiative to fight morbidity and mortality from infectious disease, especially in very young children.

While malnutrition tops the list in terms of number of affected individuals, the most well-known cause of secondary immunodeficiency is acquired immune deficiency syndrome (AIDS) resulting from chronic human immunodeficiency virus (HIV) infection. As discussed further in Chapter 18, humans do not effectively recognize and eradicate this virus, which takes up residence in T$_H$ cells. Over the course of the infection, so many T$_H$ cells are destroyed or otherwise rendered dysfunctional that a gradual collapse of the immune system ensues, resulting in a diagnosis of AIDS. The administration of anti-HIV drugs has vastly increased the life expectancy of those infected with HIV, although access is unequal; countries most impacted by AIDS, such as those in eastern and southern Africa, have the most limited access to these life saving medications.

It is important to note that many pervasive pathogens in our environment cause no problem for healthy individuals thanks to the immunity that develops following initial exposure. However, individuals with primary or secondary deficiencies in immune function become highly susceptible to disease caused by these ubiquitous microbes. For example, the fungus *Candida albicans*, present nearly everywhere and a nonissue for most individuals, can cause an irritating rash and a spreading infection on the mucosal surface of the mouth and vagina in patients suffering from immune deficiency. The resulting rash, called *thrush*, can sometimes be the first sign of immune dysfunction (**Figure 1-10**). If left unchecked, *C. albicans* can spread, causing systemic candidiasis, a life-threatening condition. Such infections by ubiquitous microorganisms that cause no harm in an immune-competent host, but that are often observed in

FIGURE 1-10 An immune-deficient patient suffering from oral thrush due to opportunistic infection with *Candida albicans*. *[Courtesy Dr. James Heilman (Wikipedia, CC BY SA)]*

cases of underlying immune deficiency, are termed **opportunistic infections**. Several rarely seen opportunistic infections identified in patients early in the AIDS epidemic were the first signs that these patients had seriously compromised immune systems, and helped scientists to identify the underlying cause.

Immune Imbalance

The immune response is so often described in "warfare" terms that it is hard to appreciate the gentler side to this system. The healthy immune system involves a constant balancing act between immune pathways leading to aggression and those requiring inhibition. While we rarely fail to consider erroneous attacks (such as autoimmunity) or failures to engage (such as immune deficiency) as dysfunctional, we sometimes forget to consider the significance of the inhibitory side of the immune response. Imperfections in the inhibitory arm of the immune response, present as a check to balance all the immune attacks we regularly initiate, can be equally profound. Healthy immune responses must therefore be viewed as a delicate balance, spending much of the time with one foot on the brake and one on the gas.

Many, maybe most, noncommunicable (noncontagious) diseases have now been linked to uncontrolled inflammation, like a stuck gas pedal (**Figure 1-11**). These include the usual suspects, such as the more common allergic and autoimmune disorders. More surprising is that some of the major life-threatening chronic medical conditions, including cardiovascular disease, insulin resistance, and obesity, have also been linked to inflammation. Recent additions to this list include neurologic and behavioral disturbances such as autism, depression, and bipolar disorder. If these observations hold true, what is tipping the balance toward uncontrolled inflammation over immune regulation or homeostasis? Likely candidates include the microbiome, diet, and stress, all of which have been shown to impact the immune, digestive, endocrine, and nervous systems. There is now clear evidence, both in mice and in humans, of a multidirectional interaction between diet, the microbiome, and immune function. In particular, it appears that the absence of certain gut **commensal organisms**, those microbes that live in and on us that cause no harm, and modern dietary changes may be linked to a paucity of "brakes" in the immune balance equation, leaving the inflammatory gas pedal stuck on!

Key Concepts:

- Dysfunctions of the immune system can include underperformance (immune deficiency) as well as overactivity or uncontrolled inflammation (allergy and autoimmune disease).

- Mounting evidence suggests that recent environmental and behavioral changes have tipped the immune balance toward uncontrolled inflammation and are contributing to many modern-day chronic conditions (e.g., diabetes, heart disease, autism).

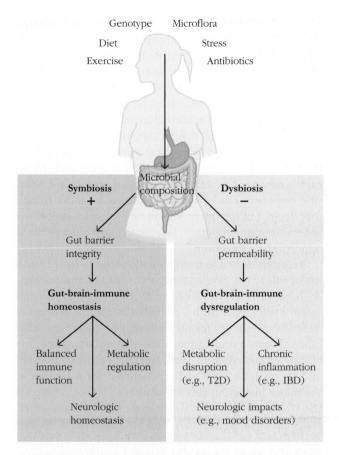

FIGURE 1-11 The proposed role of the microbiome in regulating immune, metabolic, and neurologic function. Diet, exercise, genotype, and environmental factors such as stress and the body microflora have a significant influence on the composition of the gut microbiome. In turn, this community of microbes helps to maintain gut integrity and "tune" the extensive gut immune system to create systemic homeostasis. Changes in diet and other lifestyle factors can lead to disruption of this community, or dysbiosis, resulting in immune imbalances that feed forward into a state of immune overstimulation (chronic inflammation, autoimmunity, and allergic disease). This state results in increased gut permeability and proposed disruptions to other body systems (metabolic and neurologic) and is believed to contribute to conditions such as type 2 diabetes, inflammatory bowel disease, and mood disorders, as well as others.

The Immune Response Renders Tissue Transplantation Challenging

Normally, when the immune system encounters foreign cells, it responds strongly to rid the host of the presumed invader. However, in the case of transplantation, these cells or tissues from a donor may be the only possible treatment for life-threatening disease. For example, it is estimated that more than 70,000 persons in the United States alone would benefit from a kidney transplant. The fact that the immune system will attack and reject any transplanted organ that is nonself, or not a genetic match, raises a formidable barrier to this potentially lifesaving treatment, presenting a unique challenge to clinicians who treat these patients. While the rejection of a transplant by a recipient's immune system may be seen as a "failure," in fact it is just a consequence of the immune system functioning properly. Normal tolerance processes governing self/nonself discrimination and immune engagement caused by danger signals (partially the result of the trauma caused by surgical transplantation) lead to the rapid influx of immune cells and coordinated attacks on the new resident cells. Some of these transplant rejection responses can be suppressed using immune-inhibitory drugs, but treatment with these drugs also suppresses general immune function, leaving the host susceptible to opportunistic infections.

Research related to transplantation studies has played a major role in the development of the field of immunology. A Nobel Prize was awarded in 1930 to Karl Landsteiner (mentioned earlier for his contributions to the concept of immune specificity) for the discovery of the human ABO blood groups, a finding that allowed blood transfusions to be carried out safely. In 1980, George Snell, Jean Dausset, and Baruj Benacerraf were recognized for discovery of the major histocompatibility complex (MHC). These are the tissue antigens that differ most between nongenetically identical individuals, and are thus one of the primary targets of immune rejection of transplanted tissues. Finally, in 1990 E. Donnall Thomas and Joseph Murray were awarded the Nobel Prize for treatment advances that paved the way for more clinically successful tissue transplants (see Table 1-2). The development of procedures that would allow a foreign organ or cells to be accepted without suppressing immunity to all antigens still remains a major goal, and a challenge, for immunologists today (see Chapter 16).

> **Key Concept:**
> • The rejection of a tissue transplant is an example of the immune system functioning properly, by identifying the graft as foreign.

Cancer Presents a Unique Challenge to the Immune Response

Just as graft rejection is the expected response of a healthy immune system to the addition of foreign (if benign) tissues, a tendency to ignore cancer cells might also be viewed as a normal response to what belongs and is accepted as self. Cancer, or **malignancy**, occurs in host cells when they begin to divide out of control. Since these cells are self in origin, self-tolerance mechanisms can inhibit the development of an immune response, making the detection and eradication of cancerous cells a continual challenge. That said, it is clear that many tumor cells do express unique or developmentally inappropriate proteins, making them potential targets for immune cell recognition and elimination, as well as targets for therapeutic intervention. However, as with many microbial pathogens, the increased genetic instability of these rapidly dividing cells gives them an advantage in terms of evading immune detection and elimination machinery.

We now know that the immune system actively participates in the detection and control of cancer in the body (see Chapter 19). The number of malignant disorders that arise in individuals with compromised immunity, such as those taking immune-suppressing medications, highlights the degree to which the immune system normally controls the development of cancer. Both innate and adaptive elements have been shown to be involved in this process, although adaptive immunity likely plays a more significant role. However, associations between inflammation and the development of cancer, as well as the degree to which cancerous cells evolve to become more aggressive and evasive under pressure from the immune system, have demonstrated that the immune response to cancer can have both healing and disease-inducing characteristics. As the mechanics of these elements are resolved in greater detail, there is hope that therapies can be designed to boost or maximize the antitumor effects of immune cells while dampening their tumor-enhancing activities.

Our understanding of the immune system has clearly come a very long way in a fairly short time. Yet much still remains to be learned about the mammalian immune response and the ways in which this system interacts with other body systems. With this enhanced knowledge, the hope is that we will be better poised to design ways to modulate these immune pathways through intervention. This would allow us to develop more effective prevention and treatment strategies for cancer and other diseases that plague society today, not to mention preparing us to respond quickly to the new diseases or infectious agents that will undoubtedly arise in the future.

> **Key Concept:**
> • The healthy immune system tolerates or ignores cells it identifies as self, which alas often includes those that become cancerous.

Conclusion

The mammalian immune response consists of a complicated and interconnected network of molecules, cells, and organs capable of protecting us from an equally complicated and increasingly diverse set of microbial invaders.

As a basic field of study immunology is relatively young, although societies have applied foundational immunologic principles to fight infectious agents for more than a millennium. While we are well on our way to understanding the inner workings of the immune system, it has only recently become clear that this system walks a daily tightrope of challenges to the immune balance of aggression versus regulation. Likewise, in contrast to common perceptions and earlier assumptions, we have come to appreciate the immune system as a highly evolved network that is sensitive to our environment as well as other body systems. With this new knowledge comes the prospect of innovative medical treatments and a wealth of new questions, many of which might not have been recognized as part of the purview of the immune response just a decade ago.

REFERENCES

Burnet, F. M. 1959. *The Clonal Selection Theory of Acquired Immunity.* Cambridge University Press, Cambridge, England.

Descour, L. 1922. *Pasteur and His Work* (translated by A. F. and B. H. Wedd). T. Fisher Unwin, London, England.

Kimbrell, D. A., and B. Beutler. 2001. The evolution and genetics of innate immunity. *Nature Reviews Genetics* **2**:256.

Landsteiner, K. 1947. *The Specificity of Serological Reactions.* Harvard University Press, Cambridge, MA.

Matzinger, P. 2012. The evolution of the danger theory: interview by Lauren Constable, Commissioning Editor. *Expert Review of Clinical Immunology* **8**:311.

Medawar, P. B. 1958. *The Immunology of Transplantation: The Harvey Lectures, 1956–1957.* Academic Press, New York.

Metchnikoff, E. 1905. *Immunity in the Infectious Diseases.* Macmillan, New York.

Paul, W., ed. 2012. *Fundamental Immunology,* 7th ed. Lippincott Williams & Wilkins, Philadelphia, PA.

Prescott, S. 2013. Early-life environmental determinants of allergic diseases and the wider pandemic of inflammatory noncommunicable diseases. *Journal of Allergy and Clinical Immunology* **131**:23.

Silverstein, A. M. 1979. History of immunology: cellular versus humoral immunity: determinants and consequences of an epic 19th century battle. *Cellular Immunology* **48**:208.

Useful Websites

www.aai.org The website of the American Association of Immunologists contains a good deal of information of interest to immunologists.

www.ncbi.nlm.nih.gov/PubMed PubMed, the National Library of Medicine database of more than 9 million publications, is the world's most comprehensive bibliographic database for biological and biomedical literature. It is also a highly user-friendly site.

www.aaaai.org The American Academy of Allergy, Asthma, & Immunology site includes an extensive library of information about allergic diseases.

www.who.int/en The World Health Organization directs and coordinates health-related initiatives and collects worldwide health statistics data on behalf of the United Nations system.

www.cdc.gov Part of the United States Department of Health and Human Services, the Centers for Disease Control and Prevention coordinates health efforts in the United States and provides statistics on U.S. health and disease.

www.nobelprize.org/nobel_prizes/medicine/laureates The official website of the Nobel Prize in Physiology or Medicine.

www.historyofvaccines.org A website run by the College of Physicians of Philadelphia with facts, articles, and timelines related to vaccine developments.

www.niaid.nih.gov The National Institute of Allergy and Infectious Diseases is a branch of the U.S. National Institutes of Health that specifically deals with research, funding, and statistics related to basic immunology, allergy, and infectious disease threats.

www.gavi.org The Global Alliance for Vaccines and Immunization (GAVI, or the Vaccine Alliance) is an international initiative aimed at bringing together both public and private sectors involved in vaccine access and delivery. It began in 2000 with the goal of making certain there is equal access to life saving vaccines in all nations, especially for children living in poor countries.

STUDY QUESTIONS

1. Why was Jenner's vaccine superior to previous methods for conferring resistance to smallpox?

2. Did the treatment for rabies used by Pasteur confer active or passive immunity to the rabies virus? Is there any way to test this?

3. Infants immediately after birth are often at risk for infection with group B *Streptococcus*. A vaccine is proposed for administration to women of childbearing years. How can immunizing the mothers help the babies?

4. Indicate to which branch(es) of the immune system the following statements apply, using *H* for the humoral branch and *CM* for the cell-mediated branch. Some statements may apply to both branches (*B*).

   a. Involves B cells
   b. Involves T cells
   c. Responds to extracellular bacterial infection
   d. Involves secreted antibody
   e. Kills virus-infected self cells

5. Adaptive immunity exhibits several characteristic attributes, which are mediated by lymphocytes. List four attributes of adaptive immunity and briefly explain how they arise.

6. Name three features of a secondary immune response that distinguish it from a primary immune response.

7. Give examples of mild and severe consequences of immune dysfunction. What is the most common cause of immunodeficiency throughout the world today?

8. For each of the following statements, indicate whether the statement is true or false. If you think the statement is false, explain why.

   a. Booster shots are required because repeated exposure to an antigen builds a stronger immune response.
   b. The gene for the T-cell receptor must be cut and spliced together before it can be expressed.
   c. Our bodies face the greatest onslaught from foreign invaders through our skin.
   d. Increased production of antibody in the immune system is driven by the presence of antigen.
   e. Innate immunity is deployed only during the primary response, and adaptive immunity begins during a secondary response.
   f. Autoimmunity and immunodeficiency are two different terms for the same set of general disorders.
   g. If you receive intravenous immunoglobulin to treat a snakebite, you will be protected from the venom of this type of snake in the future, but not from the venom of other types of snakes.
   h. Innate and adaptive immunity work collaboratively to mount an immune response against pathogens.
   i. The genomic sequences in our circulating T cells for encoding a T-cell receptor are the same as those our parents carry in their T cells.
   j. Both the innate and adaptive arms of the immune response will be capable of responding more efficiently during a secondary response.
   k. Memory cells save portions of the pathogen they encounter for later use during a secondary response.

9. What was the significance of the accidental re-inoculation of some chickens that Pasteur had previously exposed to the bacteria that causes cholera? Why do you think these chickens did not die after the first exposure to this bacterium?

10. Briefly describe the four major categories of pathogen. Which are likely to be the most homogeneous in form and which the most diverse? Why?

11. Describe how the principle of herd immunity works to protect unvaccinated individuals. What characteristics of the pathogen or of the host do you think would most impact the degree to which this principle begins to take hold?

12. Ehrlich's original idea of the selective theory for lymphocyte specificity postulated that a lymphocyte expresses many different antigen-specific receptors, with a foreign antigen or pathogen "selecting" one specific receptor. We now know that the outcome of clonal selection for B cells is the secretion of many copies of the same B-cell receptor in the form of a soluble antibody (humoral immunity). In what specific way was Ehrlich's original theory later refined? What are the challenges to aligning Ehrlich's original model with the above observation of humoral immunity? Does our current model of clonal selection fit this observation any better?

13. Compare and contrast innate and adaptive immunity by matching the following characteristics with the correct arm of immunity, using *I* for innate and *A* for adaptive:

   a. Is the first to engage on initial encounter with antigen
   b. Is the most pathogen specific
   c. Employs T and B lymphocytes
   d. Adapts during the response
   e. Responds identically during a first and second exposure to the same antigen
   f. Responds more effectively during a subsequent exposure
   g. Includes a memory component
   h. Is the target of vaccination
   i. Can involve the use of PAMP receptors
   j. Involves antigen-specific receptors binding to pathogens
   k. Can be mediated by antibodies

14. What is meant by the term *tolerance*? How do we become tolerant to the structures in our own bodies?

15. What is an antigen? An antibody? What is their relationship to one another?

16. How are PRRs different from B- or T-cell receptors? Which is most likely to be involved in innate immunity and which in adaptive immunity?

17. In general terms, what role do cytokines play in the development of immunity? How does this compare with chemokines?

18. **a.** The following statement is a common refrain in most genetics texts: "Every cell in your body contains the same DNA sequence and the same set of genes." Is there anything about this statement that specifically contradicts your understanding of the immune system?

    b. Likewise, all genetics texts will tell you that the two copies of each of your genes were inherited from your biological parents. Is this statement in conflict with any of your understanding of any specific cells involved in the immune response? Why or why not?

19. If you were to use war as a metaphor to think about the immune response and the development of memory, do you think that immunologic memory is more like carrying around a photograph of the enemy for quick future identification, or like making replicas of the most effective weapons from the previous battle to have on hand if needed, or both?

20. Do you inherit immunologic memory? Why or why not? What cell types are responsible for imparting memory?

21. The innate arm of immunity is responsible for the initial sorting of dangerous pathogens into categories based on common microbial features and microenvironmental cues. During the innate immune response, which of the following pathogen types would you expect to be treated more similarly: helminths and viruses *or* extracellular bacteria and fungi? Why?

22. Do you expect clonal selection to occur at the site of an infection or elsewhere? Explain your answer. Are there any sites in the body where you expect little or no immune response, even if a dangerous pathogen is present? What do these sites have in common?

23. What type of symptoms might you expect if the immune system failed to apply the brakes after eradicating a pathogen?

24. Antibiotics can be used to eradicate sometimes life-threatening bacterial infections. However, their overuse or liberal application, especially in infants and young children, has been linked to disease later in life. Specifically, what types of immunologic disorders would you expect to see in individuals with extensive exposure to antibiotics as children?

25. There are two different but not necessarily exclusive theories of what triggers an immune reaction: the self/nonself theory and the danger or damage theory. How do these two theories differ in terms of how they do or do not explain our response to the commensal microbes that reside in our guts?

CLINICAL FOCUS QUESTIONS

1. Despite decades of safe and effective vaccines to treat some of the most fatal infectious diseases in children, vaccine use varies greatly from one country to another. What barriers (physical, societal, cultural, logistic, morale, etc.) stand in the way of more widespread use of these established vaccines in developing countries? Are these same barriers influencing regional differences in vaccine application in developed nations, where pockets within affluent communities are sometimes more likely to experience vaccination gaps? Compare and contrast these two situations.

2. In 2015 Zika virus, transmitted via infected mosquitoes, was identified as the likely cause of microencephaly in some of the children born to mothers who became infected while pregnant. This frightening correlation has raised many important and urgent questions. How long has this virus been around? Is this a new phenomenon and/or is the current strain of Zika a new, more virulent genetic variant of earlier strains? When during pregnancy are women and their unborn children most vulnerable? Do women and their partners need to protect themselves from infection before conception, and if so how long before? Do we develop immunity to Zika after resolving an infection, and should women who have developed natural memory responses worry if they become pregnant? Using the recent Zika virus outbreak as an example, briefly explain how passive immunotherapy could or could not be used to protect those most at risk of disease from this virus. What do you think limits this procedure in terms of its more widespread use to fight this particular infectious disease?

Cells, Organs, and Microenvironments of the Immune System

Learning Objectives
After reading this chapter, you should be able to:

1. Describe the types of blood cells that make up the immune system and outline the main events that occur during hematopoiesis, the process that gives rise to immune cells.

2. Identify the primary, secondary, and tertiary immune organs in vertebrates and describe their function.

3. Recognize and describe the microenvironments where immune cells mature and the immune response develops.

4. Identify several experimental approaches used to understand how blood cells and immune responses develop.

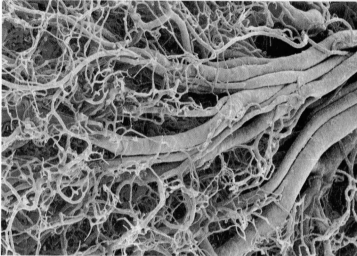

Scanning electron micrograph of blood vessels in a lymph node.
[Susumu Nishinaga/Science Source.]

A successful immune response to a pathogen depends on finely choreographed interactions among diverse cell types (see Figure 1-7): innate immune cells that mount the first line of defense against pathogen, antigen-presenting cells that communicate the infection to lymphoid cells, which coordinate the adaptive response and generate the memory cells that prevent future infections. The coordination required for a full immune response is made possible by the specialized anatomy and microanatomy of the immune system, which is dispersed throughout the body and organizes cells in time and space. **Primary lymphoid organs**—including the bone marrow and the thymus—are sites where immune cells develop from immature precursors. **Secondary lymphoid organs**—including the spleen, lymph nodes, and specialized sites in the gut and other mucosal tissues—are sites where the mature antigen-specific lymphocytes first encounter antigen and begin their differentiation into effector and memory cells. Two circulatory systems—blood and lymphatic vessels—connect these organs, uniting them into a functional whole.

Remarkably, all mature blood cells, including red blood cells, granulocytes, macrophages, dendritic cells, and lymphocytes, arise from a single cell type, the **hematopoietic stem cell (HSC)** (**Figure 2-1**). We begin this chapter with a description of **hematopoiesis**, the process by which HSCs differentiate into mature blood cells. We describe the features and function of the various cell types that arise from HSCs and then discuss the anatomy and microanatomy of the major primary lymphoid organs where hematopoiesis takes place. We feature the lymph nodes and the spleen in our

Key Terms

Hematopoiesis

Hematopoietic stem cell (HSC)

Myeloid lineage cells

Lymphoid lineage cells

Primary lymphoid organs

  Bone marrow

  Thymus

Secondary lymphoid organs

  Lymph nodes

  Spleen

  Barrier tissues (MALT and skin)

Lymphatic system

Tertiary lymphoid tissue

T-cell zone

B-cell follicle

Germinal centers

Fibroblastic reticular cell conduit (FRCC) system

Follicular dendritic cells (FDCs)

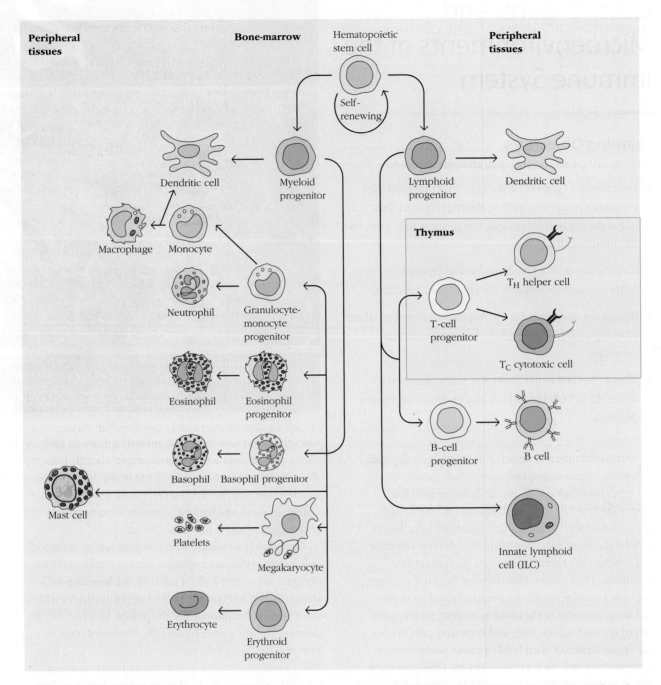

FIGURE 2-1 Hematopoiesis. Self-renewing hematopoietic stem cells give rise to lymphoid and myeloid progenitors. Most immune cells mature in the bone marrow and then travel to peripheral organs via the blood. Some, including mast cells and macrophages, undergo further maturation outside the bone marrow. T cells develop to maturity in the thymus.

description of secondary lymphoid organs. The secondary lymphoid tissue in the distinctive mucosal immune system is described in Chapter 13.

Four focused discussions are also included in this chapter. In two Classic Experiment Boxes, we describe the discovery of a second thymus and the history behind the identification of hematopoietic stem cells. In a Clinical Focus Box, we discuss the clinical use and promise of hematopoietic stem cells, and finally, in an Evolution Box, we describe some intriguing variations in the anatomy of the immune system among our vertebrate relatives.

Hematopoiesis and Cells of the Immune System

Stem cells are defined by two capacities: (1) the ability to regenerate or "self-renew" and (2) the ability to differentiate into diverse cell types. Embryonic stem cells have the capacity to generate almost every specialized cell type in an organism (in other words, they are *pluripotent*). Adult stem cells, in contrast, have the capacity to give rise to the diverse cell types that specify a particular tissue (they are *multipotent*). Multiple adult organs harbor stem cells that can give rise to cells specific for that tissue (*tissue-specific stem cells*). The HSC was the first tissue-specific stem cell identified and is the source of all of our red blood cells (erythroid cells) and white blood cells (leukocytes).

Hematopoietic Stem Cells Differentiate into All Red and White Blood Cells

HSCs originate in fetal tissues and reside primarily in the bone marrow of adult vertebrates. A small number can be found in the adult spleen and liver. Regardless of where they reside, HSCs are a rare subset—less than one HSC is present per 5×10^4 cells in the bone marrow. Their numbers are strictly controlled by a balance of cell division, death, and differentiation. Their development is tightly regulated by signals they receive in the microenvironments of primary lymphoid organs.

Under conditions when the immune system is not being challenged by a pathogen (steady state or **homeostatic** conditions), most HSCs are quiescent; only a small number divide, generating daughter cells. Some daughter cells retain the stem-cell characteristics of the mother cell—that is, they remain self-renewing and are able to give rise to all blood cell types. Other daughter cells differentiate into *progenitor cells* that have limited self-renewal capacity and become progressively more committed to a particular blood cell lineage. As an organism ages, the number of HSCs decreases, demonstrating that there are limits to an HSC's self-renewal potential.

When there is an increased demand for hematopoiesis, for example, during an infection or after chemotherapy, HSCs display an enormous proliferative capacity. This can be demonstrated in mice whose hematopoietic systems have been completely destroyed by a lethal dose of x-rays (950 rads). Such irradiated mice die within 10 days unless they are infused with normal bone marrow cells from a genetically identical mouse. Although a normal mouse has 3×10^8 bone marrow cells, infusion of fewer than 10^4 bone marrow cells from a donor is sufficient to completely restore the hematopoietic system. Our ability to identify and purify this tiny subpopulation has improved considerably, and in theory we can rescue the immune systems of irradiated animals with just a few purified stem cells, which give rise to progenitors that proliferate rapidly and repopulate the blood system.

Because of their rarity, investigators initially found it very difficult to identify and isolate HSCs. **Classic Experiment Box 2-1** describes experimental approaches that led to the first successful isolation of HSCs. Briefly, these efforts featured clever process-of-elimination strategies. Investigators reasoned that undifferentiated HSCs would not express surface markers specific for mature cells from the multiple blood lineages ("Lin" markers). They used several approaches to eliminate cells in the bone marrow that did express these markers (Lin$^+$ cells) and then examined the remaining (Lin$^-$) population for its potential to continually give rise to all blood cells over the long term. Other investigators took advantage of two technological developments that revolutionized immunological research—monoclonal antibodies and flow cytometry (see Chapter 20)—and identified surface proteins, including CD34, Sca-1, and c-Kit, that were expressed by the rare HSC population and allowed them to be isolated directly.

We now recognize several different types of Lin$^-$ Sca-1$^+$c-Kit$^+$ (LSK) HSCs, which vary in their capacity for self-renewal and their ability to give rise to all blood cell populations (pluripotency). *Long-term HSCs (LT-HSCs)* are the most quiescent and retain pluripotency throughout the life of an organism. These give rise to *short-term HSCs (ST-HSCs)*, which are also predominantly quiescent but divide more frequently and have limited self-renewal capacity. In addition to being a useful marker for identifying HSCs, c-Kit is a receptor for the cytokine SCF, which promotes the development of **multipotent progenitors (MPPs)**; these cells have a much more limited ability to self-renew, but proliferate rapidly and can give rise to both lymphoid and myeloid cell lineages.

Key Concepts:

- All red and white blood cells develop from pluripotent HSCs during a highly regulated process called *hematopoiesis*. In the adult vertebrate, hematopoiesis occurs primarily in the bone marrow, a primary lymphoid organ that supports both the self-renewal of stem cells and their differentiation into multiple blood cell types.

- The HSC is a rare cell type that is self-renewing and multipotent. HSCs have the capacity to differentiate and replace blood cells rapidly. First isolated by negative selection techniques that enriched for undifferentiated stem cells, they are now isolated by high-powered sorting techniques.

- HSCs include multiple subpopulations that vary in their quiescence and capacity to self-renew. Long-term HSCs are the most quiescent and long-lived. They give rise to short-term HSCs, which can develop into more proliferative MPPs, which give rise to lymphoid and myeloid cell types.

BOX 2-1

Isolating Hematopoietic Stem Cells

By the 1960s researchers knew that HSCs existed and were a rare population in the bone marrow. However, they did not have the technology or knowledge required to isolate HSCs for clinical study and applications. How do you find something that is very rare, whose only distinctive feature is its function—its ability to give rise to all blood cells? Investigators adopted clever strategies to find the elusive HSC and owed a great deal to rapidly evolving technologies, including the advent of monoclonal antibodies and flow cytometry (see Chapter 20).

Investigators recognized that HSCs were unlikely to express proteins specific for mature blood cells. Using monoclonal antibodies raised against multiple mature cells, they trapped and removed mature cells from bone marrow cell suspensions. They started with a process called *panning* (**Figure 1**), in which the heterogeneous pool of bone marrow cells was incubated with antibodies bound to plastic. Mature cells stuck to the antibodies and cells that did not express these surface markers were gently dislodged and collected. Investigators showed that cells that did not stick were enriched for stem cells by several thousand-fold with this approach. One of the first images of human stem cells isolated by panning is shown in Figure 1. This negative selection strategy remains very useful today, and stem cells enriched by removing mature blood cells are referred to as "Lin⁻" cells, reflecting their lack of lineage-specific surface markers.

Once investigators were able to identify surface proteins specifically expressed by HSCs, such as CD34, they could use techniques to positively select cells from heterogeneous bone marrow cell populations. The flow cytometer offered the most powerful way to pull out a rare population from a diverse group of cells. This machine, invented by the Herzenberg laboratory and its interdisciplinary team of inventors, has revolutionized immunology and clinical medicine. In a nutshell, it is a machine that can identify, separate, and recover individual cells on the basis of their unique protein and/or gene expression patterns. These patterns are revealed by fluorescent reagents, including antibodies. Irv Weissman and colleagues took advantage of each of these advances and, using a combination of positive and negative selection, developed an efficient approach to isolate HSCs (**Figure 2**).

At present, investigators agree that HSCs are enriched among cells that bear no mature (lineage-specific) markers, but express both the surface proteins Sca-1 and c-Kit. These are referred to as Lin⁻Sca-1⁺c-Kit⁺ or LSK cells. Even this subgroup, which represents less than 1% of bone marrow cells, is phenotypically and functionally heterogeneous and investigators routinely evaluate 10 or more additional protein markers to sort through the multiple types of cells that have stem cell capacities. This breakthrough is only one of many that emerge from a combination of technological and experimental creativity, a synergy that continues to drive experimental advances.

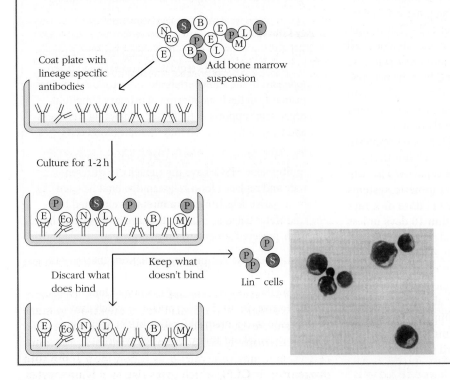

FIGURE 1 Panning for stem cells. Early approaches to isolate hematopoietic stem cells (HSCs) took advantage of antibodies that were raised against mature blood cells and a process called *panning*. Briefly, investigators layered a suspension of bone marrow cells onto plastic plates coated with antibodies that would bind multiple mature ("lineage-positive" [Lin⁺]) blood cells. Cells that did not stick were therefore enriched for HSCs (the "lineage-negative" [Lin⁻] cells desired). One of the first images of HSCs isolated in this way is shown. Abbreviations: S = stem cell; P = progenitor cell; M = monocyte; B = basophil; N = neutrophil; Eo = eosinophil; L = lymphocyte; E = erythrocyte. [*Republished with permission of The American Society for Clinical Investigation, from Emerson, S.G., et al. from "Purification and demonstration of fetal hematopoietic progenitors and demonstration of recombinant multipotential colony-stimulating activity,"* J. Clin. Invest., *Sept. 1985* **76**: *1286–1290, Figure 3. Permission conveyed through Copyright Clearance Center, Inc.*]

(continued)

CLASSIC EXPERIMENT *(continued)*   **BOX 2-1**

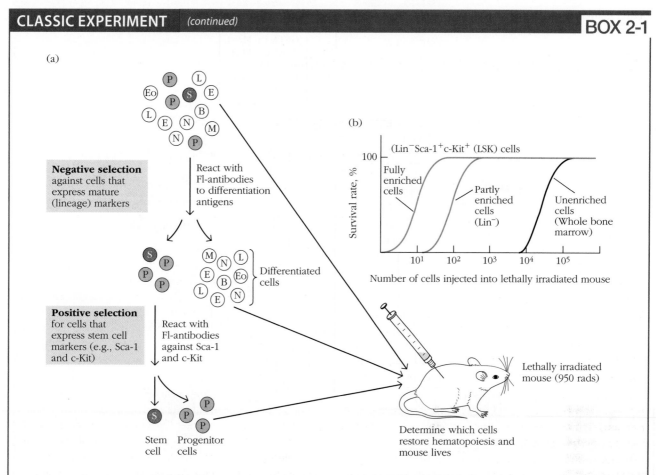

FIGURE 2 **Current approaches for enrichment of pluripotent stem cells from bone marrow.** Shown is a schematic of a commonly used, current approach to enrich stem cells from bone marrow, originated by Irv Weissman and colleagues. (a) Enrichment is accomplished first by *negative selection*: using antibodies to remove cells we don't want. In this case, the undesired cells are the more mature hematopoietic cells (indicated by the white circled letters), which bind to fluorescently labeled antibodies (Fl-antibodies). The next step is *positive selection*: using antibodies to isolate the cells we do want (the stem cells and progenitor cells, indicated by the blue and gray circled letters). In this case the Fl-antibodies are specific for Sca-1 and c-Kit. Abbreviations: S = stem cell; P = progenitor cell; M = monocyte; B = basophil; N = neutrophil; Eo = eosinophil; L = lymphocyte; E = erythrocyte. (b) Enrichment of stem cell preparations is measured by their ability to restore hematopoiesis in lethally irradiated (immunodeficient) mice. Only animals receiving pluripotent stem cells survive. Progressive enrichment for stem cells (from whole bone marrow, to Lin⁻ cells, to Lin⁻Sca-1⁺c-Kit⁺ [LSK] cells) is revealed by the decrease in the number of cells needed to restore hematopoiesis. An enrichment of about 1000-fold is possible by this procedure.

REFERENCES

Emerson, S. G., et al. 1985. Purification of fetal hematopoietic progenitors and demonstration of recombinant multipotential colony-stimulating activity. *Journal of Clinical Investigation* **76**:1286.

Spangrude, G. J., S. Heimfeld, and I. L. Weissman. 1988. Purification and characterization of mouse hematopoietic stem cells. *Science* **241**:58.

Shizuru, J. A., R. S. Negrin, and I. L. Weissman. 2005. Hematopoietic stem and progenitor cells: clinical and preclinical regeneration of the hematolymphoid system. *Annual Review of Medicine* **56**:509.

HSCs Differentiate into Myeloid and Lymphoid Blood Cell Lineages

An HSC that is induced to differentiate ultimately loses its ability to self-renew as it progresses from being an LT-HSC to an ST-HSC and then an MPP (**Figure 2-2**). At this stage, a cell makes one of two lineage commitment choices. It can become a **myeloid progenitor cell** (sometimes referred to as a **common myeloid progenitor** or **CMP**), which gives rise to red blood cells, platelets, and myeloid cells (granulocytes, monocytes, macrophages, and some dendritic cell populations). Myeloid cells are members of the innate immune system, and are the first cells to respond to infection or other insults. Alternatively, it can become a **lymphoid progenitor cell** (sometimes referred to as a **common lymphoid progenitor** or **CLP**), which gives rise to B lymphocytes,

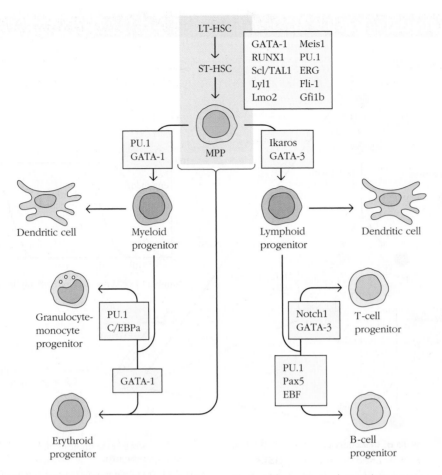

FIGURE 2-2 Regulation of hematopoiesis by transcription factors. A large variety of transcription factors regulate hematopoietic stem cell (HSC) activity (quiescence, self-renewal, multipotency) as well as differentiation to the several lineages that emerge from HSCs. Several key factors are shown here. Note that erythrocytes and megakaryocytes may arise not only from myeloid progenitors, but also from the earliest hematopoietic stem cell populations. Hematopoietic regulation is an active area of investigation; this is one possible schematic based on current information.

T lymphocytes, innate lymphoid cells (ILCs), as well as specific dendritic cell populations. B and T lymphocytes are members of the adaptive immune response and generate a refined antigen-specific immune response that also gives rise to immune memory. ILCs have features of both innate and adaptive cells.

Recent data suggest that precursors of red blood cells and platelets can arise directly from the earliest LT- and ST-HSC subpopulations (see Figure 2-2). Indeed, the details behind lineage choices are still being worked out by investigators, who continue to identify intermediate cell populations within these broad progenitor categories.

As HSC descendants progress along their chosen lineages, they also progressively lose the capacity to contribute to other cellular lineages. For example, MPPs that are induced to express the receptor Flt-3 lose the ability to become erythrocytes and platelets and are termed *lymphoid-primed, multipotent progenitors (LMPPs)* (**Figure 2-3**). As LMPPs become further committed to the lymphoid lineage, levels of the stem-cell antigens c-Kit and Sca-1 fall, and the cells begin to express RAG1/2 and TdT, enzymes involved in the generation of lymphocyte receptors. Expression of RAG1/2 defines the cell as an *early lymphoid progenitor (ELP)*. Some ELPs migrate out of the bone marrow to seed the thymus as T-cell progenitors. The rest of the ELPs remain in the bone marrow as B-cell progenitors. Their levels of the interleukin-7 receptor (IL-7R) increase, and the ELP now develops into a CLP, a progenitor that is now c-Kit^{low}Sca-1^{low}IL-7R$^+$ and has lost myeloid potential. However, it still has the potential to mature into any of the lymphocyte lineages: T cell, B cell, or ILC.

Genetic Regulation of Lineage Commitment during Hematopoiesis

Each step a hematopoietic stem cell takes toward commitment to a particular blood cell lineage is accompanied by genetic changes. HSCs maintain a relatively large number of genes in a "primed" state, meaning that they are accessible to transcriptional machinery. Environmental signals that induce HSC differentiation upregulate distinct sets of transcription factors that drive the cell down one of a number of possible developmental pathways. As cells progress down a lineage pathway, primed chromatin regions containing genes that are not needed for the selected developmental

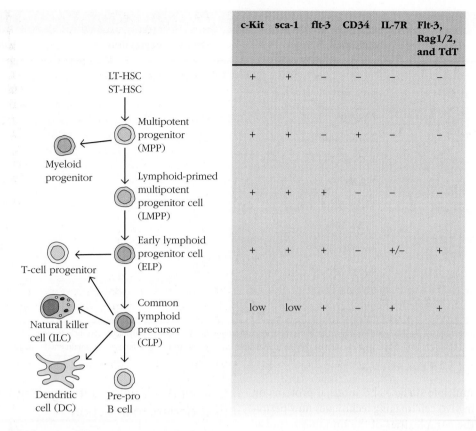

c-Kit	sca-1	flt-3	CD34	IL-7R	Flt-3, Rag1/2, and TdT
+	+	−	−	−	−
+	+	−	+	−	−
+	+	+	−	−	−
+	+	+	−	+/−	+
low	low	+	−	+	+

FIGURE 2-3 An example of lineage commitment during hematopoiesis: the development of B cells from HSCs. The maturation of HSCs into lymphoid progenitors, and the progressive loss of the ability to differentiate into other blood-cell lineages, is exemplified in this figure, which specifically traces the development of B lymphocytes from multipotent progenitors (MPPs). As cells mature from MPP to lymphoid-primed multipotent progenitors (LMPPs) to common lymphoid progenitors (CLPs) they progressively lose the ability to differentiate into other leukocytes. Pre-pro B cells are committed to becoming B lymphocytes. These changes are also accompanied by changes in expression of cell surface markers as well as by the acquisition of RAG and TdT activity.

pathway are shut down. Many transcription factors that regulate hematopoiesis and lineage choices have been identified. Some have distinct functions, but many are involved at several developmental stages and engage in complex regulatory networks. Some transcription factors associated with hematopoiesis are illustrated in Figure 2-2. However, our understanding of their roles continues to evolve.

A suite of factors appear to regulate HSC quiescence, proliferation, and differentiation (see Figure 2-2). Recent sequencing techniques have identified a "top ten" that include GATA-2, RUNX1, Scl/Tal-1, Lyl1, Lmo2, Meis1, PU.1, ERG, Fli-1, and Gfi1b, although others are bound to play a role. Other transcriptional regulators regulate myeloid versus lymphoid cell lineage choices. For instance, *Ikaros* is required for lymphoid but not myeloid development; animals survive in its absence but cannot mount a full immune response (i.e., they are severely *immunocompromised*). Low levels of PU.1 also favor lymphoid differentiation, whereas high levels of PU.1 direct cells to a myeloid fate. Activity of Notch1, one of four Notch family members, induces lymphoid progenitors to develop into T rather than B lymphocytes (see Chapter 8). GATA-1 directs myeloid progenitors toward red blood cell (erythroid) development rather than granulocyte/monocyte lineages. PU.1 also regulates the choice between erythroid and other myeloid cell lineages.

Distinguishing Blood Cells

Historically, investigators classified cells on the basis of their appearance under a microscope, often with the help of dyes. Their observations were especially helpful in distinguishing myeloid from lymphoid lineages, granulocytes from macrophages, and neutrophils from basophils and eosinophils. The pH-sensitive stains hematoxylin and eosin (H&E) are still commonly used in combination to distinguish cell types in blood smears and tissues. The basic dye hematoxylin binds basophilic nucleic acids, staining them blue, and the acidic dye eosin (named for Eos, the goddess of dawn) binds eosinophilic proteins in granules and cytoplasm, staining them pink.

Microscopists drew astute inferences about cell function by detailed examination of stained and unstained cells. Fluorescence microscopy enhanced our ability to identify more molecular details, and in the 1980s, inspired the development of the flow cytometer. This invention revolutionized the study of immunology by allowing us to rapidly measure

TABLE 2-1   Features of cells in human blood

Cell type	Cells/mm³	Total leukocytes (%)	Life span*
Myeloid cells			
Red blood cell	5.0×10^6		120 days
Platelet	2.5×10^5		5–10 days
Neutrophil	$3.7–5.1 \times 10^3$	50–70	6 hours to 2 days
Monocyte	$1–4.4 \times 10^2$	2–12	Days to months
Eosinophil	$1–2.2 \times 10^2$	1–3	5–12 days
Basophil	$<1.3 \times 10^2$	<1	Hours to days
Mast cell	$<1.3 \times 10^2$	<1	Hours to days
Lymphocytes	$1.5–3.0 \times 10^3$	20–40	Days to years
T lymphocytes	$0.54–1.79 \times 10^3$	7–24	
B lymphocytes	$0.07–0.53 \times 10^3$	1–10	
Total leukocytes	7.3×10^3		

*Life spans of cell types in humans are expressed in ranges. Life spans vary, cell populations are heterogeneous (lymphocytes include memory and naïve cells, monocytes circulating in blood could be brand new, or could be coming from tissues, etc.), and measurements depend on experimental conditions.

the presence of multiple surface and internal proteins on individual cells. In vivo cell imaging techniques now permit us to penetrate the complexities of the immune response in time and space. Together with our ever-increasing ability to edit animal and cell genomes, these technologies have revealed an unanticipated diversity of hematopoietic cell types, functions, and interactions. While our understanding of the cell subtypes is impressive, it is by no means complete. **Table 2-1** lists the major myeloid and lymphoid cell types, as well as their life spans and representation in our blood.

Key Concepts:

- HSCs that are induced to differentiate make one of two broad lineage choices. They can give rise to CMPs that develop into myeloid cell types or they can give rise to CLPs that develop into lymphoid cell types. As progenitors differentiate, they progressively lose their ability to self-renew as well as their ability to give rise to other cell lineages.

- Hematopoiesis and lineage choices are regulated by a network of transcription factors including GATA-2, Ikaros, PU.1, and Notch. Environmental signals influence the set of transcription factors expressed by HSCs and thereby determine HSC fate, allowing an organism to develop immune cell subsets according to demand.

- Hematopoietic cells can be distinguished visually using hematoxylin and eosin stains or fluorescent markers. Flow cytometry takes advantage of monoclonal antibodies to distinguish individual cells on the basis of the many surface and internal proteins they express.

Cells of the Myeloid Lineage Are the First Responders to Infection

Myeloid lineage cells include all red blood cells, granulocytes, monocytes, and macrophages. The white blood cells within this lineage are innate immune cells that respond rapidly to the invasion of a pathogen and communicate the presence of an insult to cells of the lymphoid lineage (below). As we will see in Chapter 15, they also contribute to inflammatory diseases (asthma and allergy).

Granulocytes

Granulocytes are often the first responders during an immune response and fall into four main categories: neutrophils, eosinophils, basophils, and mast cells. All granulocytes have multilobed nuclei that make them visually distinctive and easily distinguishable from lymphocytes, whose nuclei are round. Granulocyte subtypes differ by the staining characteristics of their cytoplasmic granules, membrane-bound vesicles that release their contents in response to pathogens (**Figure 2-4**). These granules contain a variety of proteins with distinct functions: some damage pathogens directly; some regulate trafficking and activity of other white blood cells, including lymphocytes; and some contribute to the remodeling of tissues at the site of infection. See **Table 2-2** for a partial list of granule proteins and their functions.

Neutrophils constitute the majority (50% to 70%) of circulating leukocytes (see Figure 2-4a) in adult humans and are much more numerous than eosinophils (1%–3%), basophils (<1%), or mast cells (<1%). After differentiation in the bone marrow, neutrophils are released into the peripheral blood and circulate for 7 to 10 hours before migrating into the tissues, where they have a life span of only a few

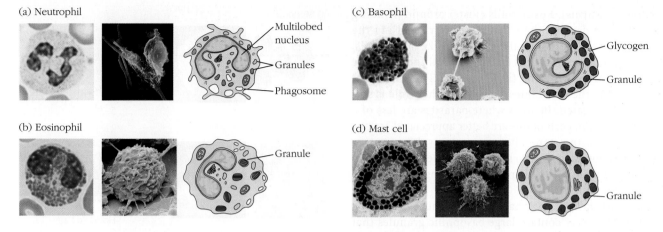

(a) Neutrophil
- Multilobed nucleus
- Granules
- Phagosome

(b) Eosinophil
- Granule

(c) Basophil
- Glycogen
- Granule

(d) Mast cell
- Granule

FIGURE 2-4 Examples of granulocytes. (a) Neutrophils, (b) eosinophils, (c) basophils, and (d) mast cells, shown via (*left*) hematoxylin and eosin (H&E) stains of blood smears, (*middle*) scanning electron microscopy (SEM), or (*right*) as a drawing depicting the typical morphology of the indicated granulocyte. Note differences in the shape of the nucleus and in the number, color, and shape of the cytoplasmic granules. *[(a) Neutrophil photos* *from Science Source/Getty Images (left) and Max Planck Institute for Infection Biology/Dr. Volker Brinkmann. (b) Eosinophil photos from Ed Reschke/ Getty Images (left) and Steve Gschmeissner/Science Source (middle). (c) Basophil photos from Michael Ross/Science Source (left) and Steve Gschmeissner/Science Source (middle). (d) Mast cell photos from Biophoto Associates/Science Source (left) and Eye of Science/Science Photo Library (middle).]*

days. In response to many types of infection, innate immune cells generate inflammatory molecules (e.g., chemokines) that promote the development of neutrophils in the bone marrow. This transient increase in the number of circulating neutrophils is called **leukocytosis** and is used medically as an indication of infection.

Neutrophils swarm in large numbers to the site of infection in response to inflammatory molecules (**Video 2-4v**). Once in the infected tissue, they phagocytose (engulf) bacteria and secrete a range of proteins that have antimicrobial effects and tissue-remodeling potential. Neutrophils are the main cellular components of pus, where they accumulate at the end of their short lives. Once considered a simple and "disposable" effector cell, neutrophils are now thought to play a regulatory role in shaping the adaptive immune response.

Eosinophils contain granules that stain a brilliant pink in standard H&E staining protocols. They are thought to be important in coordinating our defense against multicellular parasitic organisms, including helminths

TABLE 2-2	Examples of proteins contained in neutrophil, eosinophil, and basophil granules		
Cell type	**Molecule in granule**	**Examples**	**Function**
Neutrophil	Proteases	Elastase, collagenase	Tissue remodeling
	Antimicrobial proteins	Defensins, lysozyme	Direct harm to pathogens
	Protease inhibitors	α_1-antitrypsin	Regulation of proteases
	Histamine		Vasodilation, inflammation
Eosinophil	Cationic proteins	EPO	Induces formation of ROS
		MBP	Vasodilation, basophil degranulation
	Ribonucleases	ECP, EDN	Antiviral activity
	Cytokines	IL-4, IL-10, IL-13, TNF-α	Modulation of adaptive immune responses
	Chemokines	RANTES, MIP-1α	Attract leukocytes
Basophil/mast cell	Cytokines	IL-4, IL-13	Modulation of adaptive immune
	Lipid mediators	Leukotrienes	Regulation of inflammation
	Histamine		Vasodilation, smooth muscle activation

(parasitic worms). Eosinophils cluster around invading worms, and damage their membranes by releasing the contents of their eosinophilic granules. Like neutrophils, eosinophils are motile cells (see Figure 2-4b) that migrate from the blood into the tissue spaces. They are most abundant in the small intestines, where their role is still being investigated. In areas where parasites are less of a health problem, eosinophils are better appreciated as contributors to asthma and allergy symptoms. Like neutrophils, eosinophils may also secrete cytokines that regulate B and T lymphocytes, thereby influencing the adaptive immune response.

Basophils are nonphagocytic granulocytes (see Figure 2-4c) that contain large basophilic granules that stain blue in standard H&E staining protocols. Basophils are relatively rare in the circulation, but are potent responders. Like eosinophils, basophils are thought to play a role in our response to parasites, particularly helminths (parasitic worms). When they bind circulating antibody/antigen complexes basophils release the contents of their granules. Histamine, one of the best known compounds in basophilic granules, increases blood vessel permeability and smooth muscle activity, and allows immune cells access to a site of infection. Basophils also release cytokines that can recruit other immune cells, including eosinophils and lymphocytes. In areas where parasitic worm infection is less prevalent, histamines are best appreciated as a cause of allergy symptoms.

Mast cells (see Figure 2-4d) also play a role in combating parasitic worms and contribute to allergies. They are released from the bone marrow into the blood as undifferentiated cells. They mature only after they leave the blood for a wide variety of tissues, including the skin, connective tissues of various organs, and mucosal epithelial tissue of the respiratory, genitourinary, and digestive tracts. Like circulating basophils, these cells have large numbers of cytoplasmic granules that contain histamine and other pharmacologically active substances.

Basophils and mast cells share many features, and basophils were once considered the blood-borne version of mast cells. However, recent data suggest that basophils and mast cells have distinct origins and functions.

Myeloid Antigen-Presenting Cells

Myeloid progenitors also give rise to three groups of phagocytic cells—monocytes, macrophages, and dendritic cells—the cells of each of these groups have **professional antigen-presenting cell (pAPC)** function (**Figure 2-5**).

Professional APCs form important cellular bridges between the innate and adaptive immune systems. They become *activated* after making contact with a pathogen at the site of infection. They communicate this encounter to

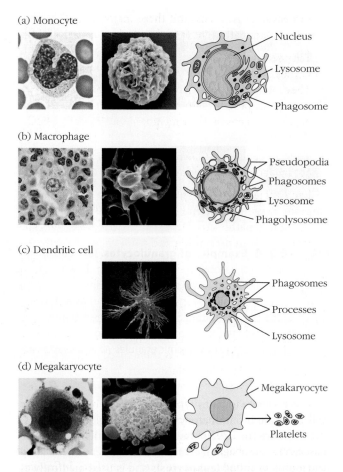

FIGURE 2-5 Examples of monocytes, macrophages, dendritic cells, and megakaryocytes. (a) Monocytes, (b) macrophages, (c) dendritic cells, and (d) megakaryocytes, shown via (*left*) H&E stains of blood smears, (*middle*) SEM, or (*right*) as a drawing depicting the typical morphology of the indicated cell. Note that macrophages are five- to tenfold larger than monocytes and contain more organelles, especially lysosomes. [(a) Monocyte photos from Michael Ross/Science Source (left) and Eye of Science/Science Source (middle). (b) Macrophage photos from Dr. Thomas Caceci, Virginia-Maryland Regional College of Veterinary Medicine, Blacksburg, Virginia (left) and SPL/Science Source (middle). (c) Dendritic cell photo from David Scharf/Science Source. (d) Megakaryocyte photos from Science Source/Getty Images (left) and Dr. Amar/Science Source (middle).]

T lymphocytes in the lymph nodes by displaying peptides from the pathogen to lymphocytes, a process called *antigen presentation* (discussed in Chapter 7). All cells have the capacity to present peptides from internal proteins using MHC class I molecules; however, pAPCs also have the ability to present peptides from external sources using MHC class II molecules (also discussed in Chapter 7). Only MHC class II molecules can be recognized by helper T cells, which initiate the adaptive immune response. (See "Cells of the Lymphoid Lineage" below and Chapter 7 for more information about MHC molecules.)

Professional APCs exhibit three major activities when they encounter pathogens (and thereby become activated):

1. They secrete proteins that attract and activate other immune cells.
2. They internalize pathogens via phagocytosis, digest pathogenic proteins into peptides, and then present these peptide antigens on their membrane surfaces via MHC class II molecules.
3. They upregulate *costimulatory molecules* required for optimal activation of helper T cells.

Each variety of pAPC plays a distinct role during the immune response, depending on its locale and its ability to respond to pathogens. Dendritic cells, for example, play a primary role in presenting antigen to—and activating—naïve T lymphocytes (lymphocytes that have not yet been activated by binding antigen). Macrophages are superb phagocytes and are especially efficient at removing both pathogen and damaged host cells from a site of infection. Monocytes regulate inflammatory responses at sites of tissue damage and infection. Investigators have now identified more varieties of APCs than ever anticipated. The functions of these subpopulations are under investigation; some will be described in more detail in coming chapters.

Monocytes constitute from 2% to 12% of white blood cells. They are a heterogeneous group of cells that migrate into tissues and differentiate into a diverse array of tissue-resident phagocytic cells (see Figure 2-5a). Two broad categories of monocytes have been identified. *Inflammatory monocytes* enter tissues quickly in response to infection. *Patrolling monocytes* crawl slowly along blood vessels,

monitoring their repair. They also provide a reservoir for tissue-resident monocytes in the absence of infection, and may quell rather than initiate immune responses.

Monocytes that migrate into tissues in response to infection can differentiate into **macrophages** (Figure 2-5b). These inflammatory macrophages are expert phagocytes and typically participate in the innate immune response. They undergo a number of key changes when stimulated by tissue damage or pathogens and have a dual role in the immune response: (1) they contribute directly to the clearance of pathogens from a tissue, and (2) they act as pAPCs for T lymphocytes.

Interestingly, recent work indicates that most tissue-resident macrophages actually arise early in life from embryonic cells rather than from circulating, activated monocytes. These resident macrophages, which include Kupffer cells in the liver, microglia in the brain, and alveolar macrophages in the lungs, have the ability to self-renew and form a committed part of the tissue microenvironment. They co-exist with circulating macrophages and share their function as pAPCs. However, they also assume tissue-specific functions. **Table 2-3** includes a more complete list of tissue-resident macrophages and functions.

Many macrophages express receptors for certain classes of antibody. If a pathogen (e.g., a bacterium) is coated with the appropriate antibody, the complex of antigen and antibody binds to antibody receptors on the macrophage membrane and enhances phagocytosis. In one study, the rate of phagocytosis of an antigen was 4000-fold higher in the presence of specific antibody to the antigen than in its absence. Thus, an antibody is an example of an **opsonin**, a molecule that binds an antigen and enhances its recognition

TABLE 2-3	Tissue-specific macrophages	
Tissue	**Name**	**Tissue-specific function (in addition to activity as pAPCs)**
Brain	Microglia	Neural circuit development (synaptic pruning)
Lung	Alveolar macrophage	Remove pollutants and microbes, clear surfactants
Liver	Kupffer cell	Scavenge red blood cells, clear particles
Kidney	Resident kidney macrophage	Regulate inflammatory responses to antigen filtered from blood
Skin	Langerhans cell	Skin immunity and tolerance
Spleen	Red pulp macrophage	Scavenge red blood cells, recycle iron
Peritoneal cavity	Peritoneal cavity macrophage	Maintain IgA production by B-1 B cells
Intestine	Lamina propria macrophage	Gut immunity and tolerance
	Intestinal muscularis macrophage	Regulate peristalsis
Bone marrow	Bone marrow macrophage	Maintain niche for blood cell development, clear neutrophils
Lymph node	Subcapsular sinus macrophage	Trap antigen particles
Heart	Cardiac macrophage	Clear dying heart cells

Data from Lavin, Y., A. Mortha, A. Rahman, and M. Merad. 2015. Regulation of macrophage development and function in peripheral tissues. *Nature Reviews Immunology* **15**:731; and Mass, E., et al. 2016. Specification of tissue-resident macrophages during organogenesis. *Science* **353**:aaf4238.

and ingestion by phagocytes. The modification of antigens with opsonins is called **opsonization**, a term from the Greek that literally means "to supply food" or "make tasty." Opsonization serves multiple purposes that will be discussed in subsequent chapters.

Ralph Steinman was awarded the Nobel Prize in Physiology or Medicine in 2011 for his discovery of the **dendritic cell (DC)** in the mid-1970s. Dendritic cells (Figure 2-5c) are critical for the initiation of the immune response and acquired their name because they extend and retract long membranous extensions that resemble the dendrites of nerve cells. These processes increase the surface area available for browsing lymphocytes. Dendritic cells are a more diverse population of cells than once was thought, and seem to arise from both the myeloid and lymphoid lineages of hematopoietic cells. The functional distinctions among dendritic cell populations are still being clarified, and each subtype is likely critically important in tailoring immune responses to distinct pathogens and targeting responding cells to distinct tissues.

Dendritic cells perform the distinct functions of antigen capture in one location and antigen presentation in another. Outside lymph nodes, immature forms of these cells monitor the body for signs of invasion by pathogens and capture intruding or foreign antigens. They process these antigens and migrate to lymph nodes, where they present the antigen to naïve T cells, initiating the adaptive immune response.

When acting as sentinels in the periphery, immature dendritic cells take in antigen in three ways. They engulf it by phagocytosis, internalize it by receptor-mediated endocytosis, or imbibe it by pinocytosis. Indeed, immature dendritic cells pinocytose fluid volumes of 1000 to 1500 mm³ per hour, a volume that rivals that of the cell itself. After antigen contact, they mature from an antigen-capturing phenotype to one that is specialized for presentation of antigen to T cells. In making this transition, some attributes are lost and others are gained. Dendritic cells that have captured antigen lose the capacity for phagocytosis and large-scale pinocytosis. They improve their ability to present antigen and express costimulatory molecules essential for the activation of naïve T cells. After maturation, dendritic cells enter the blood or lymphatic circulation, and migrate to regions containing lymphoid organs, where they present antigen to circulating T cells.

It is important to note that **follicular dendritic cells (FDCs)** do not arise from hematopoietic stem cells and are functionally distinct from dendritic cells. FDCs were named not only for their dendrite-like processes, but for their exclusive location in follicles, organized structures in secondary lymphoid tissue that are rich in B cells. Unlike dendritic cells, FDCs are not pAPCs and do not activate naïve T cells. Instead, they regulate the activation of B cells, as discussed in Chapters 11 and 14.

Erythroid Cells

Cells of the erythroid lineage—**erythrocytes**, or red blood cells—also arise from myeloid progenitors. Erythrocytes contain high concentrations of hemoglobin, and circulate through blood vessels and capillaries delivering oxygen to surrounding cells and tissues. Damaged red blood cells also release signals that induce innate immune activity. In mammals, erythrocytes are anuclear; their nucleated precursors, erythroblasts, extrude their nuclei in the bone marrow. However, the erythrocytes of nonmammalian vertebrates (birds, fish, amphibians, and reptiles) retain their nuclei. Erythrocyte size and shape vary considerably across the animal kingdom—the largest red blood cells can be found among some amphibians, and the smallest among some deer species.

Although the main function of erythrocytes is gas exchange, they may also play a more direct role in immunity. They express surface receptors for antibody and bind antibody complexes that can then be cleared by the many macrophages that scavenge erythrocytes. They also generate compounds, like nitric oxide (NO), that do direct damage to microbes.

Megakaryocytes

Megakaryocytes are large myeloid cells that reside in the bone marrow and give rise to thousands of **platelets**, very small cells (or cell fragments) that circulate in the blood and participate in the formation of blood clots (Figure 2-5d). Clots not only prevent blood loss, but when they take place at epithelial barriers, they also provide a barrier against the invasion of pathogens. Although platelets have some of the properties of independent cells, they do not have their own nuclei.

Key Concepts:

- Granulocytes, including neutrophils, eosinophils, basophils, and mast cells, respond to multiple extracellular pathogens, including bacteria and parasitic worms. When activated, they release the contents of granules, which directly and indirectly impair pathogen activity. These innate immune cells also release cytokines that influence the adaptive immune response and are potent contributors to allergic responses.

- Monocytes, macrophages, and dendritic cells are myeloid cells that, when activated by antigen, are professional antigen-presenting cells (pAPCs) that activate T lymphocytes. Macrophages can be found in all tissues and have two major origins. Some differentiate from circulating monocytes and continue to circulate among tissues. Others, known as tissue-resident macrophages, originate from embryonic cells and do not circulate. They adopt a variety of tissue-specific functions in addition to their role as pAPCs. Dendritic cells are the most potent antigen-presenting cells for naïve T cells.

- Erythrocytes (red blood cells) are anuclear and function primarily in carrying oxygen to cells and tissues. They may also play a direct role in immunity by regulating the clearance of immune complexes and generating antimicrobial compounds.

- Megakaryocytes give rise to platelets, which help generate clots when vessels are damaged.

Cells of the Lymphoid Lineage Regulate the Adaptive Immune Response

Lymphoid lineage cells, or lymphocytes (Figure 2-6), are the principal cell players in the adaptive immune response and the source of immune memory. They represent 20% to 40% of circulating white blood cells and 99% of cells in the lymph. Lymphocytes are broadly subdivided into three major populations on the basis of functional and phenotypic differences: B lymphocytes (B cells), T lymphocytes (T cells), and innate lymphoid cells (ILCs), which include the well-understood natural killer (NK) cells. In humans, approximately a trillion (10^{12}) lymphocytes circulate continuously through the blood and lymph and migrate into the tissue spaces and lymphoid organs. Large numbers of lymphocytes reside in the tissues that line our intestines, airways, and reproductive tracts, too. We briefly review the general characteristics and functions of each lymphocyte group and its subsets below.

Small, round, and dominated by their nucleus, lymphocytes are relatively nondescript cells. T and B lymphocytes, in fact, appear identical under a microscope. We therefore rely heavily on the profile of surface proteins they express to differentiate lymphocyte subpopulations.

Surface proteins expressed by cells of the immune system (as well as some other cells) are often referred to by the **cluster of differentiation (CD)** nomenclature. This nomenclature was established in 1982 by an international group of investigators who recognized that many of the new antibodies produced by laboratories all over the world (largely in response to the advent of monoclonal antibody technology) were binding to the same proteins, and hence some proteins were given multiple names by different labs. The group therefore defined clusters of antibodies that appeared to be binding to the same protein and assigned a name—a cluster of differentiation or CD—to each protein. Although originally designed to categorize the multiple antibodies, the CD nomenclature is now firmly associated with specific surface proteins found on cells of many types. **Table 2-4** lists some common CD molecules found on human and mouse lymphocytes. Note that the shift from use of a "common" name to the more standard "CD" name has taken place slowly. For example, investigators often still refer to the pan-T cell marker as "Thy-1" rather than CD90, and the costimulatory molecules as "B7-1" and "B7-2," rather than CD80 and CD86. Appendix I lists over 300 CD markers expressed by immune cells.

B and T cells express many different CD proteins on their surface, depending on their stage of development and state of activation. In addition, each B or T cell also expresses an antigen-specific receptor (the B-cell receptor or the T-cell receptor, respectively) on its surface. Although B and T cell populations express a remarkable diversity of antigen receptors (more than a billion), all antigen-specific receptors on an individual cell's surface are identical in structure and, therefore, are identical in their specificity for antigen. When a particular T or B cell divides, all of its progeny will also

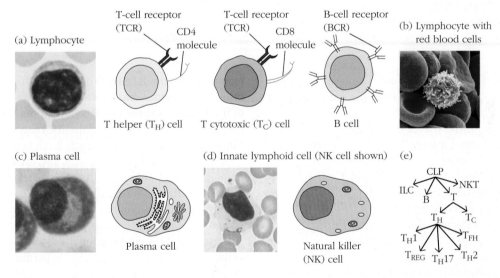

FIGURE 2-6 Examples of lymphocytes. (a) H&E stain of a blood smear showing a typical lymphocyte, with drawings depicting naïve T_H, T_C, and B cells. Note that naïve B cells and T cells look identical by microscopy. (b) SEM of a lymphocyte with red blood cells. (c) H&E stain of a blood smear showing a plasma cell, with a drawing depicting the plasma cell's typical morphology. The cytoplasm of the plasma cell is enlarged because it is occupied by an extensive endoplasmic reticulum network—an indication of the cell's dedication to antibody production. (d) H&E stain of a blood smear showing an innate lymphoid cell, the natural killer (NK) cell. NK cells have more cytoplasm than a naïve lymphocyte; this one is full of granules that are used to kill target cells. (e) A branch diagram that depicts the basic relationship among the lymphocyte subsets described in the text. (Abbreviations: CLP = common lymphoid progenitor; ILC = innate lymphoid cell; T_H1, T_H2, T_H17 = helper T type 1, 2, and 17 cells, respectively; T_{REG} = regulatory T cell.) *[(a) Lymphocyte H&E photo from Science Source/Getty Images. (b) Lymphocyte SEM from Steve Gschmeissner/Science Source. (c) Plasma cell photo © Benjamin Koziner/Phototake. (d) NK photo from Ira Ames, Ph.D., Department of Cell & Developmental Biology, SUNY Upstate Medical University.]*

TABLE 2-4 **Common CD markers used to distinguish functional lymphocyte subpopulations**

CD designation	Function	B cell	T$_H$ cell	T$_C$ cell	NK cell*
CD2	Adhesion molecule; signal transduction	−	+	+	+
CD3	Signal transduction element of T-cell receptor	−	+	+	−
CD4	Adhesion molecule that binds to MHC class II molecules; signal transduction	−	+ (usually)	− (usually)	−
CD5	Unknown	+ (subset)	+	+	+
CD8	Adhesion molecule that binds to MHC class I molecules; signal transduction	−	− (usually)	+ (usually)	Variable
CD16 (FcγRIII)	Low-affinity receptor for Fc region of IgG	−	−	−	+
CD19	Signal transduction; CD21 coreceptor	+	−	−	−
CD20	Signal transduction; regulates Ca^{2+} transport across the membrane	+	−	−	−
CD21 (CR2)	Receptor for complement (C3d) and Epstein–Barr virus	+	−	−	−
CD28	Receptor for costimulatory B7 molecule on antigen-presenting cells	−	+	+	−
CD32 (FcγRII)	Receptor for Fc region of IgG	+	−	−	−
CD35 (CR1)	Receptor for complement (C3b)	+	−	−	−
CD40	Signal transduction	+	−	−	−
CD45	Signal transduction	+	+	+	+
CD56	Adhesion molecule	−	−	−	+
CD161 (NK1.1)	Lectin-like receptor	−	−	−	+

Synonyms are shown in parentheses.

*NK cells are now considered a cytotoxic member of the innate lymphoid cell (ILC) family. ILCs include three groups of cells that differ by the cytokines they produce. Some classify NK cells within the ILC1 group; others have defined them as a distinct cytotoxic lineage of ILCs.

express this specific antigen receptor. The resulting population of lymphocytes, all arising from the same founding lymphocyte, is a clone (see Figure 1-6).

At any given moment, tens of thousands, perhaps a hundred thousand, distinct mature T- and B-cell clones circulate in a human or mouse, each distinguished by its unique antigen receptor. Newly formed B cells and T cells are considered **naïve**. Contact with antigen induces naïve lymphocytes to proliferate and differentiate into both effector cells and memory cells. **Effector cells** carry out specific functions to combat the pathogen, while **memory cells** persist in the host, and when rechallenged with the same antigen, respond faster and more efficiently. As you have learned in Chapter 1, the first encounter with antigen generates a **primary response**, and the re-encounter a **secondary response** (see Figure 1-8).

B Lymphocytes

The **B lymphocyte (B cell)** derived its letter designation from its site of maturation, in the *b*ursa of Fabricius in birds; the name turned out to be apt, as *b*one marrow is its major site of maturation in humans, mice, and many other

mammals. Mature B cells are definitively distinguished from other lymphocytes and all other cells by their expression of the **B-cell receptor (BCR)**, a membrane-bound immunoglobulin (antibody) molecule that binds to antigen (see **Figure 2-7a** and Chapter 3). Each B cell expresses a surface antibody with a unique specificity, and each of the approximately $1.5–3 \times 10^5$ molecules of surface antibody on a B cell has identical binding sites for antigen. B lymphocytes also improve their ability to bind antigen through a process known as *somatic hypermutation* and can generate antibodies of several different functional classes through a process known as *class switching*. Somatic hypermutation and class switching are covered in detail in Chapter 11.

Activated B lymphocytes are the only nonmyeloid cell that can act as a pAPC. They internalize antigen very efficiently via their antigen-specific receptor, and process and present antigenic peptides at the cell surface. Activated B cells also express costimulatory molecules required to activate T cells. By presenting antigen directly to T cells, B cells also receive T-cell help, in the form of cytokines that induce their differentiation into antibody-producing cells (plasma cells) and memory cells.

(a) Soluble antigen binding to a B cell   (b) APC presentation to a T cell

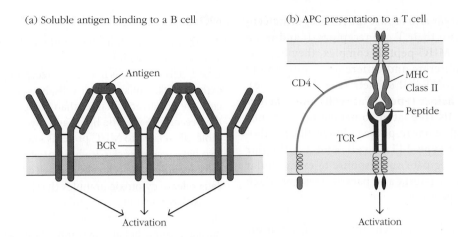

FIGURE 2-7 **Structure of the B-cell and T-cell antigen receptors.** (a) B-cell receptors recognize soluble antigens. (b) T-cell receptors (TCRs) recognize membrane-bound MHC–peptide complexes. This figure shows a TCR from a helper T cell that recognizes MHC class II (with the assistance of the CD4 molecule). A TCR from a cytotoxic T cell (not shown) would recognize MHC class I with the assistance of a CD8 molecule. (Abbreviations: APC = antigen-presenting cell; BCR = B-cell receptor; MHC = major histocompatibility complex.)

Ultimately, activated B cells differentiate into effector cells known as **plasma cells** (see Figure 2-6c). Plasma cells lose expression of surface immunoglobulin and become highly specialized for secretion of antibody. A single cell is capable of secreting from a few hundred to more than a thousand molecules of antibody per second. Plasma cells do not divide and, although some travel to the bone marrow and live for years, others die within 1 or 2 weeks.

T Lymphocytes

T lymphocytes (T cells) derive their letter designation from their site of maturation in the *t*hymus. Like the B cell, the T cell expresses a unique antigen-binding receptor called the **T-cell receptor (TCR;** see Figure 2-7b and Chapter 3). However, unlike membrane-bound antibodies on B cells, which can recognize soluble or particulate antigen, T-cell receptors recognize only processed pieces of antigen (typically peptides) bound to cell membrane proteins called **major histocompatibility complex (MHC) molecules.** MHC molecules are genetically diverse glycoproteins found on cell membranes. They were identified as the cause of rejection of transplanted tissue, and their structure and function are covered in detail in Chapter 7. The ability of MHC molecules to form complexes with antigen allows cells to decorate their surfaces with internal (foreign and self) proteins, exposing them to browsing T cells. MHC comes in two versions: **MHC class I molecules,** which are expressed by nearly all nucleated cells of vertebrate species, and **MHC class II molecules,** which are expressed primarily by pAPCs.

T lymphocytes are divided into two major cell types—**T helper (T_H) cells** and **T cytotoxic (T_C) cells**—that can be distinguished from one another by the presence of either **CD4** or **CD8** membrane glycoproteins on their surfaces. *T cells displaying CD4 generally function as helper (T_H) cells and recognize antigen in complex with MHC class II, whereas those displaying CD8 generally function as cytotoxic (T_C) cells and recognize antigen in complex with MHC class I (see* **Figure 2-8** *and Chapter 12).* The ratio of CD4$^+$ to CD8$^+$ T cells is approximately 2:1 in healthy mouse and human peripheral blood. A change in this ratio is often an indication of immunodeficiency disease (e.g., HIV infection), autoimmune disease, aging, and inflammation.

Naïve CD8$^+$ T_C cells browse the surfaces of antigen-presenting cells with their T-cell receptors. If and when they bind to an MHC-peptide complex, they become activated, proliferate, and differentiate into a type of effector cell called a **cytotoxic T lymphocyte (CTL).** The CTL has a vital function in monitoring the cells of the body and eliminating cells that display non-self-antigen complexed with MHC class I, such as virus-infected cells, tumor cells, and cells of a foreign tissue graft. To proliferate and differentiate optimally, naïve CD8$^+$ T cells also need help from mature CD4$^+$ T cells.

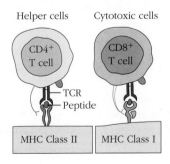

FIGURE 2-8 **T-cell recognition of antigen.** T cells displaying CD4 generally function as helper (T_H) cells and recognize peptide antigen associated with MHC class II, whereas those displaying CD8 generally function as cytotoxic (T_C) cells and recognize peptide antigen associated with MHC class I.

Naïve CD4$^+$ T$_H$ cells also browse the surfaces of antigen-presenting cells with their T-cell receptors. If and when they recognize an MHC-peptide complex, they become activated and proliferate and differentiate into one of a variety of effector T-cell subsets (see Figure 2-6e). In broad terms, **T helper type 1 (T$_H$1) cells and T helper type 17 (T$_H$17) cells** (the latter so named because they secrete IL-17) regulate our response to intracellular pathogens, and **T helper type 2 (T$_H$2) cells and T follicular helper (T$_{FH}$) cells** regulate our response to extracellular pathogens, such as bacteria and parasitic worms. **Each CD4$^+$T$_H$-cell subtype produces a different set of cytokines** that enable or "help" the activation of B cells, Tc cells, macrophages, and various other cells that participate in the immune response.

Which helper subtype dominates a response depends largely on what type of pathogen (intracellular versus extracellular, viral, bacterial, fungal, helminth) has infected an animal. The network of cytokines that regulate and are produced by these effector cells is described in detail in Chapter 10.

Another type of CD4$^+$ T cell, the **regulatory T cell (T$_{REG}$)**, has the unique capacity to inhibit immune responses. These cells, called natural T$_{REG}$ cells, arise during maturation in the thymus from cells that bind self proteins with high affinity (autoreactive cells). They also can be induced at the site of an immune response in an antigen-dependent manner (iT$_{REG}$ cells). Regulatory T cells are identified by the presence of CD4 and CD25 on their surfaces, as well as by the expression of the internal transcription factor FoxP3. T$_{REG}$ cells quell autoreactive responses and play a role in limiting our normal T-cell responses to pathogens.

Both CD4$^+$ and CD8$^+$ T-cell subpopulations may be even more diverse than currently described, and additional functional subtypes could be identified in the future.

NKT Cells

Another type of cell in the lymphoid lineage, the **NKT cell**, shares features with both adaptive and innate immune cells. Like T cells, NKT cells have T-cell receptors (TCRs), and some express CD4. Unlike most T cells, however, the TCRs of NKT cells are not diverse. Rather than recognize protein peptides, they recognize specific lipids and glycolipids presented by a molecule related to MHC proteins known as CD1. NKT cells also have receptors classically associated with innate immune cells, including the NK cells discussed below. Activated NKT cells release cytotoxic granules that kill target cells, but also release large quantities of cytokines that can both enhance and suppress the immune response. They appear to be involved in human asthma, but also may inhibit the development of autoimmunity and cancer. Understanding the exact role of NKT cells in immunity is one research priority.

Innate Lymphoid Cells (ILCs)

Investigators now recognize a group of cells that are derived from common lymphoid progenitors, but do not express antigen-specific receptors: **innate lymphoid cells (ILCs)**. Currently they are subdivided into three groups (ILC1, ILC2, and ILC3), distinguished by the cytokines they secrete, which mirror those produced by distinct helper T-cell subsets (**Table 2-5**). Many provide a first line of defense against pathogens in the skin and at mucosal tissues (Chapter 13). ILC helper subsets are the focus of active investigation, and the nomenclature describing ILC subtypes is still in flux as investigators work to determine their origins and their relationships to each other and to other blood cells.

Cytotoxic natural killer (NK) cells are the founding members of the innate lymphoid cell category and the best studied. Many investigators classify NK cells within the ILC1 group; some define them as a distinct cytotoxic lineage of ILCs. Regardless of their classification, NK cells are

TABLE 2-5	Cytokines secreted by ILC and T-cell subsets		
ILC	T-cell subset	Signature cytokines secreted by both	Master transcriptional regulators of both
Group 1 ILCs			
NK cell	CTL	IFN-γ, perforin, granzyme	T-bet
ILC1	T$_H$1	IFN-γ, TNF	
Group 2 ILCs			
ILC2	T$_H$2	IL-4, IL-5, IL-13, and amphiregulin	GATA-3
Group 3 ILCs			
LTi cell	T$_H$17, T$_H$22	IL-17A, IL-22, LTα, LTβ,	RORγt
ILC3		IL-22, IFN-γ	

Data from Walker, J. A., J. L. Barlow, and A. N. J. McKenzie. 2013. Innate lymphoid cells—how did we miss them? *Nature Reviews Immunology* **13**:75; and Gasteiger, G., and A. Y. Rudensky. 2014. Interactions between innate and adaptive lymphocytes. *Nature Reviews Immunology* **14**:631.

identified by their expression of the NK1.1 surface protein and constitute 5% to 10% of lymphocytes in human peripheral blood.

Efficient cell killers, they use two different strategies to attack a variety of abnormal cells. The first strategy is to attack cells that lack MHC class I molecules. Infection by certain viruses or mutations occurring in tumor cells often cause those cells to downregulate MHC class I. NK cells express a variety of receptors for self-MHC class I that, when engaged, inhibit their ability to kill. However, when NK cells encounter cells that have lost their MHC class I, these inhibiting receptors are no longer engaged and NK cells can release their cytotoxic granules, killing the target cell.

Second, NK cells express receptors (called *Fc receptors* or *FcRs*) for some antibodies. By linking these receptors to antibodies, NK cells can arm themselves with antibodies specific for pathogenic proteins, particularly viral proteins present on the surfaces of infected cells. Once such antibodies bring the NK cell in contact with target cells, the NK cell releases its granules and induces cell death, a process known as *antibody-dependent cell cytotoxicity* (ADCC). The mechanisms of NK-cell cytotoxicity are described further in Chapter 12.

Our understanding of ILCs is still in its infancy, and ongoing investigations are continually generating new insights into their origin and function.

Key Concepts:

- Lymphocytes include B cells, T cells, and innate lymphoid cells (ILCs) and come in many varieties that can be distinguished by patterns of expression of surface CD proteins.

- B and T cells clonally express unique antigen receptors on their surfaces: the B-cell receptor (BCR) and the T-cell receptor (TCR), respectively. Before encountering antigen they are referred to as naïve lymphocytes. After encountering antigen they differentiate into effector and memory lymphocytes.

- The BCR is a membrane version of an antibody. On activation by antigen binding, specificity for the antigen may improve.

- Activated B cells can operate as pAPCs, presenting antigen to T cells; the T cells then directly provide the B cells with the help they need to differentiate.

- B cells ultimately differentiate into antibody-producing cells called plasma cells.

- T lymphocytes express unique antigen receptors (TCRs), and include CD4$^+$ helper T cells (CD4$^+$ T$_H$), which recognize peptide bound to MHC class II, and CD8$^+$ cytotoxic T cells (CD8$^+$ T$_C$), which recognize peptide bound to MHC class I.

- Helper T cells come in a variety of subtypes that help tailor our response to distinct pathogens.

- A small population of T cells, called NKT cells, express less diverse TCRs and share features with innate immune cells.

- Innate lymphoid cells (ILCs) include cytotoxic natural killer (NK) cells and several helper cell subsets (ILC1, ILC2, ILC3); these cells do not synthesize antigen-specific receptors but regulate the immune system via the production of cytokines that resemble those generated by helper T cells.

- NK cells have the ability to kill some infected cells and tumor cells.

Primary Lymphoid Organs: Where Immune Cells Develop

HSCs reside in specialized microenvironments, or niches. **Stem cell niches** are sequestered regions lined by supportive cells that regulate stem cell survival, proliferation, differentiation, and trafficking. Microenvironments that nurture HSCs change over the course of embryonic development. By mid to late gestation, HSCs take up residence in the bone marrow, which remains the primary site of hematopoiesis throughout adult life. The bone marrow supports the maturation of all erythroid and other myeloid cells and, in humans and mice, the maturation of B lymphocytes (as described in Chapter 9).

HSCs are also found in blood and may naturally recirculate between the bone marrow and other tissues. This observation has simplified the process used to transplant blood cell progenitors from donors into patients who are deficient (e.g., patients who have undergone chemotherapy). Whereas once it was always necessary to aspirate bone marrow from the donor—a painful process that requires anesthesia—it is now sometimes possible to use enriched hematopoietic precursors from donor blood, which is more easily obtained (see **Clinical Focus Box 2-2**).

Unlike B lymphocytes, T lymphocytes do not complete their maturation in the bone marrow. Instead, T lymphocyte precursors leave the bone marrow and travel to unique microenvironments in the other primary lymphoid organ, the thymus. The structure and function of the thymus will be discussed briefly below and in more detail in Chapter 8.

The Site of Hematopoiesis Changes during Embryonic Development

The bone marrow niche develops late during embryonic development. However, the fetus still needs to generate red and white blood cells required for survival after birth. Where does this happen? During embryogenesis, the site of

Stem Cells—Clinical Uses and Potential

Stem cell transplantation holds great promise for the regeneration of diseased, damaged, or defective tissue. HSCs are already used to restore hematopoietic function, and their use in the clinic is described below. However, rapid advances in stem cell research have raised the possibility that other stem cell types may soon be routinely employed for replacement of a variety of cells and tissues. Two properties of stem cells underlie their utility and promise. They have the capacity to give rise to lineages of differentiated cells, and they are self-renewing—each division of a stem cell creates at least one stem cell. If stem cells are classified according to their descent and developmental potential, three levels of stem cells can be recognized: pluripotent, multipotent, and unipotent.

Pluripotent stem cells can give rise to an entire organism. A fertilized egg, the zygote, is an example of such a cell. In humans, the initial divisions of the zygote and its descendants produce cells that are also pluripotent. In fact, identical twins develop when pluripotent cells separate and develop into genetically identical fetuses. *Multipotent stem cells* arise from embryonic stem cells and can give rise to a more limited range of cell types. Further differentiation of multipotent stem cells leads to the formation of *unipotent stem cells*, which can generate only the same cell type as themselves. (Note that "pluripotent" is often used to describe the HSC. Within the context of blood cell lineages this is arguably true; however, it is probably strictly accurate to call the HSC a multipotent stem cell.)

Pluripotent cells, called embryonic stem (ES) cells, can be isolated from early embryos, and for many years it has been possible to grow mouse ES cells as cell lines in the laboratory. Strikingly, these cells can be induced to generate many different types of cells, including muscle

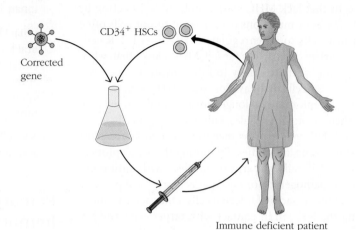

FIGURE 1 **The general strategy used to correct a defective gene by autologous HSC transplantation.** CD34+ HSCs from a patient are removed and infected with a viral vector (e.g., a retrovirus) that carries a corrected version of the gene. The cells with the corrected gene are re-introduced to the patient, who has been treated with chemotherapy or radiation to eliminate the defective blood cells.

cells, nerve cells, liver cells, pancreatic cells, intestinal epithelial cells, and hematopoietic cells.

Advances have made it possible to grow lines of human pluripotent stem cells and, most recently, to induce differentiated human cells to become pluripotent stem cells. These are developments of considerable importance to the understanding of human development, and they also have great therapeutic potential. In vitro studies of factors that determine or influence the development of human pluripotent stem cells along specific developmental paths are providing considerable insight into how cells differentiate into specialized cell types. This research is driven in part by the great potential for using pluripotent stem cells to generate cells and tissues that could replace diseased or damaged tissue. Success in this endeavor would be a major advance because transplantation medicine now depends entirely on donated organs and tissues, yet the need far exceeds the number of donations, and the need is increasing. Success in deriving cells, tissues, and organs

from pluripotent stem cells could provide skin replacement for burn patients, heart muscle cells for those with chronic heart disease, pancreatic islet cells for patients with diabetes, and neurons for the treatment of Parkinson's disease or Alzheimer's disease.

The transplantation of HSCs is an important therapy for patients whose hematopoietic systems must be replaced. It has multiple applications, including:

- Providing a functional immune system to individuals with a genetically determined immunodeficiency, such as severe combined immunodeficiency (SCID).
- Replacing a defective hematopoietic system with a functional one to cure patients with life-threatening nonmalignant genetic disorders in hematopoiesis, such as sickle-cell anemia or thalassemia.
- Restoring the hematopoietic system of cancer patients after treatment with doses of chemotherapeutic agents and radiation. This approach is particularly applicable to

(continued)

leukemias, including acute myeloid leukemia, which can be cured only by destroying the patient's own hematopoietic system—the source of the leukemia cells. Indeed, any patient receiving therapy based on irradiation or chemotherapeutic regimens that destroy the immune system will benefit from stem cell transplantation.

HSCs have extraordinary powers of regeneration. Experiments in mice indicate that as few as one HSC can completely restore the erythroid population and the immune system. In humans, for instance, as little as 10% of a donor's total volume of bone marrow can provide enough HSCs to completely restore the recipient's hematopoietic system. Once injected into a vein, HSCs enter the circulation and some find their way to the bone marrow, where they begin the process of engraftment. In addition, HSCs can be preserved by freezing. This means that hematopoietic cells can be "banked." After collection, the cells are treated with a cryopreservative, frozen, and then stored for later use. When needed, the frozen preparation is thawed and infused into the patient, where it reconstitutes the hematopoietic system. This cell-freezing technology even makes it possible for individuals to store their own hematopoietic cells for transplantation to themselves at a later time. Currently, this procedure is used to allow cancer patients to donate cells before undergoing chemotherapy and radiation treatments, and then reconstitute their hematopoietic system later, using their own cells.

Transplantation of stem cell populations may be **autologous** (the recipient is also the donor), **syngeneic** (the donor is genetically identical; i.e., an identical twin of the recipient), or **allogeneic** (the donor and recipient are not genetically identical). In any transplantation procedure, genetic differences between donor and recipient can lead to immune-based rejection reactions. Aside from host rejection of transplanted tissue (host versus graft), lymphocytes conveyed to the recipient via the graft can attack the recipient's tissues, thereby causing **graft-versus-host disease (GVHD; see Chapter 16)**, a life-threatening affliction. In order to suppress rejection reactions, powerful immunosuppressive drugs must be used. Unfortunately, these drugs have serious side effects, and immunosuppression increases the patient's risk of infection and susceptibility to tumors. Consequently, HSC transplantation has the fewest complications when there is genetic identity between donor and recipient.

At one time, bone marrow transplantation was the only way to restore the hematopoietic system. However, both peripheral blood and umbilical cord blood are now also common sources of HSCs. These alternative sources of HSCs are attractive because the donor does not have to undergo anesthesia or the highly invasive procedure used to extract bone marrow. Although peripheral blood may replace marrow as a major source of HSCs for many applications, bone marrow transplantation still has some advantages (e.g., marrow may include important stem cell subsets that are not as prevalent in blood). To obtain HSC-enriched

preparations from peripheral blood, agents are used to induce increased numbers of circulating HSCs, and then the HSC-containing fraction is separated from the plasma and red blood cells in a process called *leukapheresis*. If necessary, further purification can be done to remove T cells and to enrich the CD34+ population.

Umbilical cord blood contains an unusually high frequency of HSCs and is obtained from placental tissue that is normally discarded. Consequently, umbilical cord blood has become an attractive source of cells for HSC transplantation. For reasons that are still incompletely understood, however, cord blood stem cell transplants do not engraft as reliably as peripheral blood or bone marrow stem cell transplants.

Beyond its current applications in cancer treatment, autologous stem cell transplantation can also be useful for gene therapy, the introduction of a normal gene to correct a disorder caused by a defective gene. One of the most highly publicized gene therapy efforts—the introduction of the adenosine deaminase (ADA) gene to correct a form of severe combined immunodeficiency (SCID; see Chapter 18)—was performed successfully on HSCs (**Figure 1**). Unfortunately, in a number of patients, the retrovirus used to introduce the corrected ADA gene integrated into parts of the genome that resulted in leukemia. Investigators continue to work to improve the safety and efficiency of gene delivery and have met with more recent successes. New gene editing techniques, including those that involve the CRISPR/Cas9 system (see Chapter 20), may be part of the future for HSC transplantation.

blood cell generation shifts several times before moving into its final home (**Figure 2-9**).

Hematopoiesis begins when precursor cells in the **yolk sac** differentiate into primitive, nucleated erythroid cells that carry the oxygen the embryo needs for early development (7 days after fertilization in the mouse and 3 weeks after fertilization in the human). *Fetal HSCs* capable of generating all

blood cell types can be detected close to the developing kidney, specifically in the a*orta-gonad-mesonephros (AGM) region*, when the fetal heart starts beating. *Mature HSCs* capable of completely repopulating the hematopoietic system of irradiated animals can be isolated from multiple tissues, including the AGM, the *yolk sac*, *placenta*, and *fetal liver*. The placental HSC pool proliferates rapidly and ultimately contains more

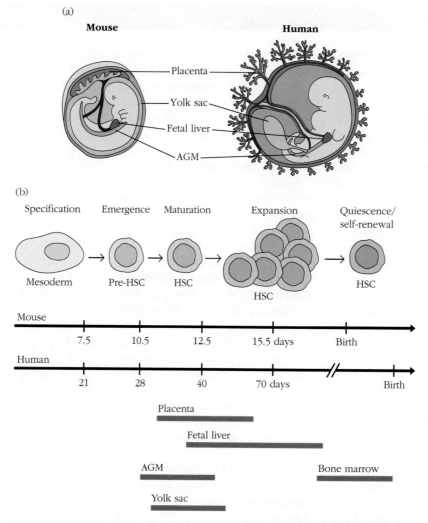

FIGURE 2-9 Sites of hematopoiesis during fetal development. (a) Blood cell precursors are initially found in the yolk sac (yellow) and then spread to the placenta (salmon), fetal liver (pink), and the aorta-gonad-mesonephros (AGM) region (green), before finding their adult home in the bone marrow. The mouse embryo is shown at 11 days of gestation; the human embryo at the equivalent 5 weeks of gestation. (b) The sequence of changes in sites of hematopoiesis during mouse and human embryogenesis. The bars below the time lines indicate when and where functional HSCs are formed.

HSCs than either the AGM or the yolk sac. However, the number of HSCs in the placenta drops as the HSC pool in the fetal liver expands. As an embryo completes its development, the fetal liver is the predominant site of HSC generation.

Within the fetal liver, HSCs form progenitor cells. At the earliest time points, hematopoiesis in the fetal liver is dominated by erythroid progenitors that give rise to the true, enucleated mature erythrocytes that ensure a steady oxygen supply to the growing embryo. Myeloid and lymphoid progenitors emerge gradually. HSCs first seed the bone marrow at late stages in fetal development, and the bone marrow finally takes over as the main site of hematopoiesis, where it will remain throughout postnatal life. Prior to puberty in humans, most of the bones of the skeleton are hematopoietically active, but by the age of 18 years only the vertebrae, ribs, sternum, skull, pelvis, and parts of the humerus and femur retain hematopoietic potential.

The Bone Marrow Is the Main Site of Hematopoiesis in the Adult

The **bone marrow** is the paradigmatic adult stem cell niche (**Figure 2-10**). It is responsible for maintaining the pool of HSCs throughout the life of an adult vertebrate and regulating their differentiation into all blood cell types.

Although the outside surface of a bone is hard, the inside or marrow, also known as the *medullary cavity*, is sponge-like and packed full of cells. A cross-section of a femur (see Figure 2-10b) reveals hematopoietic cells at every stage of differentiation. The medullary cavity can be divided into the **endosteal niche**, lining the bone, and the **perivascular niche**, lining the blood vessels that run through the center of the bone. These niches contain stromal cells that provide structure and guidance for hematopoiesis.

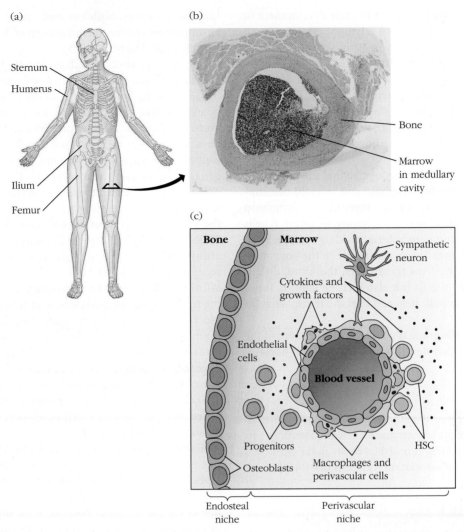

FIGURE 2-10 The bone marrow microenvironment. (a) Multiple bones support hematopoiesis in the adult, including the hip (ilium), femur, sternum, and humerus. (b) This figure shows a cross-section of a bone (femur) with a medullary (marrow) cavity. (c) A schematic of the cross section depicts the perivascular and endosteal niches in more detail. Various cells associated with the central blood vessels generate a niche that supports HSC self-renewal and differentiation. Perivascular cells secrete cytokines and growth factors, and express surface molecules that regulate HSC quiescence and differentiation. Osteoblasts line the bone and provide a niche for developing B cells (not shown). Differentiated cells exit the marrow via blood vessels, which are lined by endothelial cells. *[Photo courtesy of Indiana University School of Medicine, Medical Sciences Program.]*

The stromal cells in the marrow that regulate HSC quiescence, proliferation, trafficking, and differentiation include the following:

1. *Endothelial cells* that line the blood vessels
2. *Perivascular cells* that are diverse in function and interact with endothelial cells
3. *Sympathetic nerves* that transmit signals to other niche cells
4. *Macrophages*, which influence the activity of other niche cells
5. *Osteoblasts*, which generate bone and regulate the differentiation of lymphoid cells

Quiescent, long-lived HSCs are found in the perivascular niche, nurtured by perivascular and endothelial cells (see Figure 2-10c). Some HSCs remain quiescent, while others divide and differentiate into progenitors that develop into myeloid or lymphoid lineages. Specific niches that support myeloid development have not yet been identified. However, the main sites of lymphocyte differentiation are well understood.

B lymphocytes complete most of their development in the bone marrow. Progenitors of B lymphocytes are found in the endosteal niche in association with osteoblasts (Figure 2-10c). More mature B cells are found in the central sinuses of the bone marrow and exit the bone marrow to complete the final stages of their maturation in the spleen. Progenitors of T lymphocytes arise from bone marrow HSCs but exit at a very immature stage and complete their development in the thymus, the primary lymphoid organ for T-cell maturation.

Finally, it is important to recognize that the bone marrow is not only a site for lymphoid and myeloid development but is where fully mature myeloid and lymphoid cells can return. Many mature antibody-secreting B cells (plasma cells) become long-term residents in the bone marrow. Some mature T cells also reside in the bone marrow. Whole bone marrow transplants, therefore, do not simply include stem cells but also include mature, functional cells that can both help and hurt the transplantation effort.

With age, fat cells gradually replace 50% or more of the bone marrow compartment, and the efficiency of hematopoiesis decreases.

Key Concepts:

- HSCs reside primarily in the bone marrow, where stromal cells regulate their quiescence, proliferation, and trafficking. Long-term HSCs reside in the perivascular niche, in association with cells that line the blood vessels.

- In the bone marrow, HSCs differentiate into progenitors, which can become myeloid or lymphoid cell lineages. B lymphocytes complete their maturation in the bone marrow, but progenitors that can differentiate into T lymphocytes exit and complete their maturation in the thymus.

The Thymus Is the Primary Lymphoid Organ Where T Cells Mature

T-cell development is not complete until the cells undergo selection in the **thymus** (**Figure 2-11**). The importance of the thymus in T-cell development was not recognized until the early 1960s, when J. F. A. P. Miller, an Australian biologist, worked against the power of popular assumptions to advance his idea that the thymus was something other than a graveyard for cells. It was an underappreciated organ, very large in prepubescent animals, that was thought by some to be detrimental to an organism, and by others to be an evolutionary dead-end. The cells that populated it—small, thin-rimmed, featureless cells—looked dull and

inactive. However, Miller proved that the thymus was the all-important site for the maturation of T lymphocytes (see **Classic Experiment Box 2-3**).

T-cell progenitors, which still retain the ability to give rise to multiple hematopoietic cell types, travel via the blood from the bone marrow to the thymus. The thymus is a specialized environment where immature T cells, known as **thymocytes**, mature into functional T cells by passing through well-defined developmental stages in well-defined microenvironments. Thymocytes ultimately generate unique antigen receptors (T-cell receptors, or TCRs) and are selected to mature on the basis of their TCR reactivity to self-peptide/MHC complexes expressed on the surface of thymic epithelial cells. Thymocytes whose TCRs bind self-MHC/peptide complexes with too-high affinity are induced to die (**negative selection**), and thymocytes that bind self-MHC/peptides with intermediate affinity undergo **positive selection** and mature, migrating to the thymic medulla before entering the circulation. Most thymocytes do not navigate the journey through the thymus successfully; in fact, more than 95% of thymocytes die in transit. The majority of cells die because they have too-low affinity for the self-peptide/MHC combinations that they encounter on the surface of thymic epithelial cells and fail to undergo positive selection, a process called *death by neglect*. T-cell development is discussed in greater detail in Chapter 8.

T-cell development takes place in several distinct thymic microenvironments populated by epithelial cell subtypes (Figure 2-11d). T-cell precursors enter the thymus in blood vessels at the *corticomedullary junction* between the *thymic cortex*, the outer portion of the organ, and the *thymic medulla*, the inner portion of the organ. Thymocytes first travel to the *subcapsular cortex*, just beneath the capsule of the thymus, where they proliferate. They then travel to the cortex, where they first express mature TCRs and interact with cortical thymic epithelial cells (cTECs). Thymocytes that are positively selected in the cortex continue to mature and travel to the medulla, where they interact with medullary thymic epithelial cells (mTECs). Negative selection can happen in any of the microenvironments, although thymocytes are tested for reactivity to tissue-specific antigens in the medulla.

Thymocytes are also distinguished by their expression of two CD antigens, CD4 and CD8 (Figure 2-11d). The most immature thymocytes express neither and are referred to as *double negative* (DN). After entering the cortex, thymocytes upregulate both CD4 and CD8 antigens, becoming *double positive* (DP). As they mature, they lose one or the other CD antigen, becoming *single positive* (SP). CD4+ T cells are helper cells and CD8+ T cells are cytotoxic (killer) cells. Mature SP cells exit the thymus as they entered: via the blood vessels of the corticomedullary

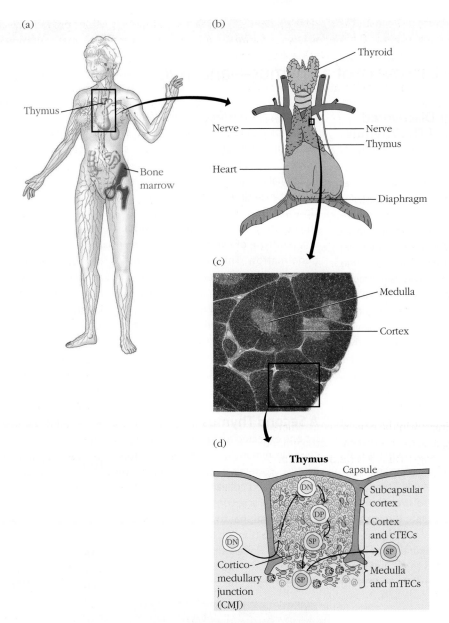

FIGURE 2-11 Structure of the thymus. The thymus is found just above the heart (a and b) and is largest prior to puberty, when it begins to shrink. (c) A stained thymus tissue section, showing the cortex and medulla. (d) A drawing of the microenvironments: the cortex, which is densely populated with double-positive (DP) immature thymocytes (blue cells) and the medulla, which is sparsely populated with single-positive (SP), mature thymocytes. These major regions are separated by the corticomedullary junction (CMJ), where cells enter from and exit to the bloodstream. The area between the cortex and the thymic capsule, the subcapsular cortex, is a site of much proliferation of the youngest, double-negative (DN) thymocytes. The route taken by a typical thymocyte during its development from the DN to DP to SP stages is shown. Thymocytes are positively selected in the cortex. Autoreactive thymocytes are negatively selected in the medulla; some may also be negatively selected in the cortex. (Abbreviations: cTECs = cortical thymic epithelial cells; mTECs = medullary thymic epithelial cells.) *[Photo (c) from Jose Luis Calvo/Shutterstock.]*

junction. Maturation is finalized in the periphery, where these new T cells (*recent thymic emigrants*) explore antigens presented in secondary lymphoid tissue, including the spleen and lymph nodes. T-cell development and positive and negative selection are discussed in more detail in Chapter 8.

BOX 2-3

The Discovery of a Thymus—and Two

J. F. A. P. Miller Discovered The Function of The Thymus

In 1961, Miller, who had been investigating the thymus's role in leukemia, published a set of observations in the *Lancet* that challenged notions that, at best, this organ served as a cemetery for lymphocytes and, at worst, was detrimental to health (**Figure 1**). He noted that when this organ was removed in very young mice (in a process known as *thymectomy*), the subjects became susceptible to a variety of infections, failed to reject skin grafts, and died prematurely. On close examination of their circulating blood cells, they also appeared to be missing a type of cell that another investigator, James Gowans, had associated with cellular and humoral immune responses. Miller concluded that the thymus produced functional immune cells.

Several influential investigators could not reproduce the data and questioned Miller's conclusions. Some speculated that the mouse strain he used was peculiar, others that his mice were exposed to too many pathogens and their troubles were secondary to infection. Miller responded to each of these criticisms experimentally, assessing the impact of thymectomy in different mouse strains and in germ-free facilities. His results were unequivocal, and his contention that this organ generated functional lymphocytes was vindicated. Elegant experiments by Miller, James Gowans, and others subsequently showed that the thymus produced a different type of lymphocyte than the bone marrow. This cell did not produce antibodies directly, but, instead, was required for optimal antibody production. It was called a *T cell* after the thymus, its organ of origin. Immature T cells are known as *thymocytes*. Miller is one of the few scientists credited with the discovery of the function of an entire organ.

A Second Thymus

No one expected a new anatomical discovery in immunology in the twenty-first century. However, in 2006 Hans-Reimer Rodewald and colleagues reported the existence of a second thymus in mice. The conventional thymus is a bilobed organ that sits in the thorax right above the heart. Rodewald and colleagues discovered thymic tissue that sits in the neck, near the cervical vertebrae, of mice. This cervical thymic tissue is smaller in mass than the conventional thymus, consists of a single lobe or clusters of single lobes, and is populated by relatively more mature thymocytes. However, it contributes to T-cell development very effectively and clearly contributes to the mature T-cell repertoire. Rodewald's findings raise the possibility that some of our older observations and assumptions about thymic function need to be re-examined. In particular, studies based on thymectomy that indicated T cells could develop outside the thymus may need to be reassessed. The cells found may have come from this more obscure but functional thymic tissue. The evolutionary implications of this thymus are also interesting—thymi are found in the neck in several species, including the koala and kangaroo.

REFERENCES

Miller, J. F. 1961. Immunological function of the thymus. *Lancet* **278:**748.

Miller, J. F. A. P. 2002. The discovery of thymus function and of thymus-derived lymphocytes. *Immunological Reviews* **185:**7.

Rodewald, H.-R. 2006. A second chance for the thymus. *Nature* **441:**942.

(a)

FIGURE 1 **(a) J. F. A. P. Miller in 1961 and (b) the first page of the *Lancet* article (1961) describing his discovery of the function of the thymus.**
[(a) © The Walter and Eliza Hall Institute of Medical Research. (b) Republished with permission of Elsevier, from J.F.A.P. Miller, "Immunological function of the thymus," from The Lancet, 1961 Sept.; 2(7205): 748-749. Permission conveyed through Copyright Clearance Center, Inc.]

(continued)

(b)

748   SEPTEMBER 30, 1961   PRELIMINARY COMMUNICATIONS   THE LANCET

Preliminary Communications

IMMUNOLOGICAL FUNCTION OF THE THYMUS

IT has been suggested that the thymus does not participate in immune reactions. This is because antibody formation has not been demonstrated in the normal thymus,[1] and because, even after intense antigenic stimulation plasma cells (the morphological expression of active antibody formation) and germinal centres have not been described in that organ.[2] Furthermore, thymectomy in the adult animal has had little or no significant effect on antibody production.[3]

On the other hand, there are certain clinical and experimental observations in man and other animals which suggest that the thymus may somehow be concerned in the control of immune responses. Thus, in acute infections, when presumably the need for antibody production is great, the thymus undergoes rapid involution; in patients with acquired agammaglobulinæmia the simultaneous occurrence of benign thymomas has been described,[4] and in fœtal or newborn animals, at a time when responsiveness to antigenic stimulation is deficient, the thymus is a very prominent organ.

The apparent contradiction between these two sets of observations may be partly explained by recent work,[5][6] which suggests that the thymus does not respond to circulating antigens because these cannot reach it owing to the existence of a barrier between the normal gland and the blood-stream. If the barrier is broken, for instance by local trauma, the histological reactions of antibody formation take place in the thymus.

In this laboratory, we have been interested in the role of the thymus in leukæmogenesis.[7] During this work it has become increasingly evident that the thymus at an early stage in life plays a very important part in the development of immunological response.

METHODS AND RESULTS

In the preliminary experiments mice of the C3H and Ak strains and of a cross between T_6 and Ak were used. The thymus was removed 1–16 hours after birth. Alternate littermates were used as sham-thymectomised controls—i.e., they underwent the full operative procedure, including excision of part of the sternum, but their thymuses were left intact. Mice in another group had thymectomy at 5 days of age. Wounds were closed with a continuous black silk suture and the baby mice were returned immediately to their mothers. No anti-

1. Fagreus, A. *J. Immunol.* 1948, **58**, 1.
2. Fagreus, A. *Acta med. scand.* 1948, suppl. 204, p. 3.
3. MacLean, L. D., Zak, S. J., Varco, R. L., Good, R. A. *Transplant. Bull.* 1957, **4**, 21.
4. Good, R. A., Varco, R. L. *Lancet*, 1955, i, 245.
5. Stoner, R. D., Hale, W. M. *J. Immunol.* 1955, **75**, 203.
6. Marshall, A. H. E., White, R. G. *Lancet*, 1961, i, 1030.
7. Miller, J. F. A. P. *Nature, Lond.* 1961, **191**, 248.

biotics were administered at any time either to the operated mice or to their mothers.

Mortality during and immediately after the operation ranged between 5 and 15% (excluding deaths due either to neglectful mothers or to cannibalism). Mortality in the thymectomised group was, however, higher between the 1st and 3rd month of life and was attributable mostly to common laboratory infections. This suggested that neonatally thymectomised mice were more susceptible to such infections than even sham-thymectomised littermate controls. When thymectomised and control groups were isolated from other experimental mice and kept under nearly pathogen-free conditions, the mortality in the thymectomised group was significantly reduced.

Absolute and differential white-cell counts were performed on tail blood at various intervals after thymectomy. The significant results of these estimations are summarised in fig. 1. In sham-thymectomised animals the lymphocyte/

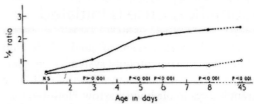

Fig. 1—Average lymphocyte:polymorph ratio of mice thymectomised in the neonatal period compared with sham-thymectomised controls. Statistical differences indicated.

○——————○ thymectomised mice.
●——————● sham-thymectomised mice.

polymorph ratio rose progressively in the first 8 days of life to reach the normal adult ratio of 2.5 ± 0.08. In the animals whose thymus was removed on the 1st day of life the ratio did not increase significantly and was only 1.0 ± 0.10 at 6 weeks of age.

Histological examination of lymph-nodes and spleens of thymectomised animals at 6 weeks of age revealed a conspicuous deficiency of germinal centres and only few plasma cells (figs. 2 and 3).

At 6 weeks of age, groups of thymectomised, sham-thymectomised, and entirely normal mice were subjected to skin grafting, Ak mice receiving C3H grafts and vice versa, and $(AkXT_6)F_1$ mice receiving C3H grafts. The median survival time of skin grafts in intact mice, sham-thymectomised mice, and mice thymectomised at 5 days of age ranged from 10 to 12 days. In more than 70% of mice whose thymus was removed on the 1st day of life the grafts were established and grew luxuriant crops of hair. Most of these grafts were tolerated for periods ranging from 6 weeks to 2 months and

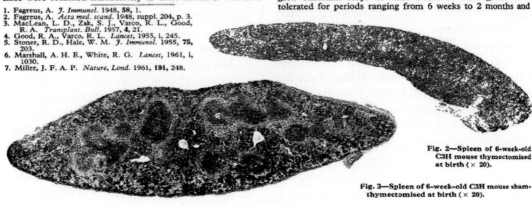

Fig. 2—Spleen of 6-week-old C3H mouse thymectomised at birth (× 20).

Fig. 3—Spleen of 6-week-old C3H mouse sham-thymectomised at birth (× 20).

Secondary Lymphoid Organs: Where the Immune Response Is Initiated

As described above, lymphocytes and myeloid cells develop to maturity in the primary lymphoid system: T lymphocytes in the thymus, and B cells, monocytes, dendritic cells, and granulocytes in the bone marrow. However, they encounter antigen and initiate an immune response in the microenvironments of secondary lymphoid organs and tissues.

Secondary Lymphoid Organs Are Distributed throughout the Body and Share Some Anatomical Features

Lymph nodes and the spleen are the most highly organized of the secondary lymphoid organs and are compartmentalized from the rest of the body by a fibrous capsule. Less organized secondary lymphoid tissues are associated with the linings of multiple organ systems, including the skin and the reproductive, respiratory, and gastrointestinal tracts, all of which protect us against external pathogens. These are collectively referred to as **barrier tissues**.

Although secondary lymphoid tissues vary in location and degree of organization, they share key features. All secondary lymphoid structures contain anatomically distinct regions of T-cell and B-cell activity. They also generate lymphoid follicles, highly organized microenvironments responsible for the development and selection of B cells that produce high-affinity antibodies.

Blood and Lymphatics Connect Lymphoid Organs and Infected Tissue

Immune cells are highly mobile and use two different systems to traffic through tissues: the blood and lymphatic systems (**Figure 2-12**). Blood vessels have access to virtually every organ and tissue and are lined by endothelial cells that are very responsive to inflammatory signals. Both red and white blood cells transit through the blood—flowing away from the heart via active pumping networks (arteries) and back to the heart via passive valve-based systems (veins)—within minutes. Arteries have thick muscular walls and depend on the beating of the heart to propel cells through vessels. Veins have thinner walls and rely on a combination of internal valves and the activity of muscles to return cells to the heart.

Endothelial cells cooperate with innate immune cells to recruit circulating white blood cells to infected tissue. These cells leave the blood by squeezing between endothelial cells and follow chemokine gradients to the site of infection (see Chapter 14).

Only white blood cells have access to the **lymphatic system**, a network of vessels filled with a protein-rich fluid (**lymph**) derived from the fluid component of blood (**plasma**). These vessels serve, or *drain*, many tissues and provide a route for activated immune cells and antigen to travel from sites of infection to secondary lymphoid organs, where they encounter and activate lymphocytes. Most secondary lymphoid tissues are, in fact, situated along the vessels of the lymphatic system. The spleen is an exception and appears to be served primarily by blood vessels.

Lymphatic vessels also return fluid that seeps from blood capillaries back to the circulatory system (see Figure 2-12c). Depending on the size and activity of an adult, seepage can generate 2.9 liters or more during a 24-hour period. This **interstitial fluid** permeates all tissues and bathes all cells. If this fluid were not returned to the circulation, tissues would swell, resulting in edema (specifically called *lymphedema*) that could become life-threatening. Some individuals are genetically predisposed to lymphedema and others experience it as a result of damage to lymphatic vessels by surgery or trauma.

The walls of the primary lymphatic vessels are thinner than those of blood vessels and more porous. They consist of a single layer of loosely apposed endothelial cells and allow fluids and cells to enter the lymphatic network relatively easily. Within these vessels, the fluid, now called *lymph*, flows into a series of progressively larger collecting vessels called **lymphatic vessels**.

All cells and fluid circulating in the lymph are ultimately returned to the blood system. The largest lymphatic vessel in our bodies, the **thoracic duct**, empties into the left subclavian vein. It collects lymph from all the body except the right arm and right side of the head. Lymph from these areas is collected into the *right lymphatic duct*, which drains into the right subclavian vein (Figure 2-12a). By returning fluid lost from the blood, the lymphatic system ensures steady-state levels of fluid within the circulatory system.

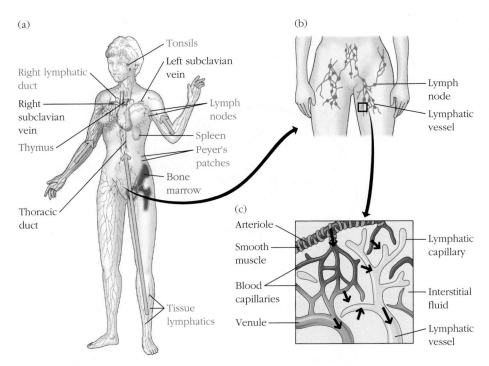

FIGURE 2-12 The human lymphatic system. The primary organs (bone marrow and thymus) are shown in red; secondary organs and tissues, in blue. These structurally and functionally diverse lymphoid organs and tissues are interconnected by the blood vessels (not shown) and lymphatic vessels (purple). Most of the body's lymphatics eventually drain into the thoracic duct, which empties into the left subclavian vein. The vessels draining the right arm and right side of the head (shaded blue) converge to form the right lymphatic duct, which empties into the right subclavian vein. Part (b) shows the lymphatic vessels in more detail, and (c) shows the relationship between blood and lymphatic capillaries in tissue. The lymphatic capillaries pick up interstitial fluid, particulate and soluble proteins, and immune cells from the tissue surrounding the blood capillaries (see arrows).

Like veins, lymphatic vessels rely on a series of one-way valves and the activity of surrounding muscles to establish a slow, low-pressure flow of lymph and cells. Therefore, activity enhances not just venous return, but lymph circulation.

All immune cells that traffic through lymph, blood, and tissues are guided by small molecules known as **chemokines** (see Chapter 3 and Appendix II). Chemokines are chemoattractants secreted by many different cell types including epithelial cells, stromal cells, antigen-presenting cells, lymphocytes, and granulocytes. Chemokine gradients are sensed by immune cells, which express an equally diverse set of chemokine receptors and migrate toward the source of chemokine production.

Key Concepts:

- Both blood and lymphatic vessels carry cells through and between tissues. Red and white blood cells travel from the heart in arteries and return via veins. Some white blood cells and fluid leave the blood to enter tissues. They can be picked up by the lymphatic system, which flows in one direction and ultimately connects back to the bloodstream via the thoracic duct.

- The lymphatic system also transports immune cells and foreign antigens from sites of infection to secondary lymphoid tissues and organs, where the adaptive immune response is activated.

The Lymph Node Is a Highly Specialized Secondary Lymphoid Organ

Lymph nodes (**Figure 2-13**) are the most specialized secondary lymphoid organs. Unlike the spleen, which also regulates red blood cell flow and fate, lymph nodes are fully committed to regulating an immune response. They are encapsulated, bean-shaped structures that include networks of stromal cells (i.e., support tissue) packed with lymphocytes, macrophages, and dendritic cells. Connected to both blood vessels and lymphatic vessels, lymph nodes are the first organized lymphoid structure to encounter antigens that enter the tissue spaces. The lymph node provides ideal microenvironments for encounters between antigen and lymphocytes and productive, organized cellular and humoral immune responses.

Structurally, a lymph node can be divided into three roughly concentric regions: the cortex, the paracortex, and the medulla, each of which supports a distinct microenvironment (see Figure 2-13a). The outermost layer, the **cortex**, contains lymphocytes (mostly B cells), macrophages, and follicular dendritic cells arranged in **follicles**. Beneath the cortex is the **paracortex**, which is populated largely by T lymphocytes but also contains dendritic cells that have migrated into the lymph node from the surrounding tissues (Figure 2-13b and c). The **medulla** is the innermost layer and the site where lymphocytes exit (*egress*) the lymph node through

(a)

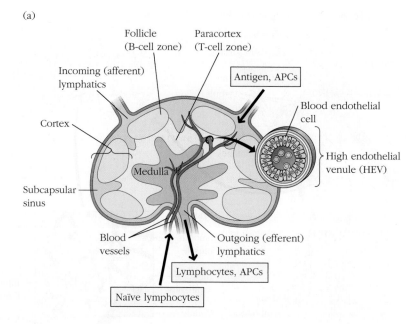

(b)

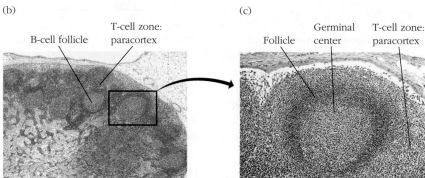

(c)

FIGURE 2-13 Structure of a lymph node. The microenvironments of the lymph node support distinct cell activities. (a) A drawing of the major features of a lymph node shows the major vessels that serve the organ: incoming (afferent) and outgoing (efferent) lymphatic vessels, and the arteries and veins. It also depicts the three major tissue layers: the cortex, the paracortex, and the innermost region, the medulla. Macrophages and dendritic cells, which trap antigen, are present in the cortex and paracortex. T cells are concentrated in the paracortex; B cells are primarily in the cortex, within follicles and germinal centers. The medulla is populated largely by antibody-producing plasma cells and is the site where cells exit via the efferent lymphatics. Naïve lymphocytes circulating in the blood enter the lymph node via high endothelial venules (HEVs), in a process called extravasation (see Chapter 14). Antigen and some leukocytes, including antigen-presenting cells (APCs), enter via afferent lymphatic vessels. All cells exit via efferent lymphatic vessels. (b) This stained lymph node section shows the cortex with a number of ovoid follicles, surrounding the T cell–rich paracortex. (c) Another stained lymph node section at higher magnification, showing the T-cell zone and a B-cell follicle that includes a germinal center (also referred to as a *secondary follicle*). [*Photo (b) from Jose Luis Calvo/Shutterstock. Photo (c) from Image Source/Alamy.*]

the outgoing (*efferent*) lymphatics. It is more sparsely populated with lymphoid lineage cells, which include plasma cells that are actively secreting antibody molecules.

Antigen travels from infected tissue to the cortex of the lymph node via the incoming (*afferent*) lymphatic vessels, which pierce the capsule of a lymph node at numerous sites and empty lymph into the *subcapsular sinus* (see Figure 2-13a). It enters either in particulate form or is processed and presented as peptides on the surface of *migrating* antigen-presenting cells. Particulate antigen can be trapped by *resident* antigen-presenting cells in the subcapsular sinus

or cortex, where it is passed to other antigen-presenting cells, including B lymphocytes in the follicles. Alternatively, particulate antigen can be processed and presented as peptide-MHC complexes on cell surfaces of resident dendritic cells that are already in the T cell–rich paracortex.

T Cells in the Lymph Node

It takes every naïve T lymphocyte about 16 to 24 hours to browse the MHC-peptide combinations presented by the antigen-presenting cells (APCs) in a single lymph node.

(a) Follicular reticular cell conduit system

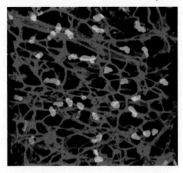
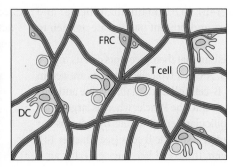

(b) Follicular dendritic cell

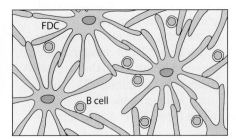

FIGURE 2-14 Stromal cell networks in secondary lymphoid tissue. T and B lymphocytes travel along distinct structures in secondary lymphoid microenvironments. (a) The paracortex is crisscrossed by processes and conduits formed by fibroblastic reticular cells (FRCs), which guide the migration of antigen-presenting cells and T cells, facilitating their interactions. *Left*: An immunofluorescence microscopy image with the FRCs shown in red and the T cells in green. *Right*: A drawing of the network and cell participants. (Abbreviations: DC = dendritic cell; HEV = high endothelial venule.) (b) The B-cell follicle contains a network of follicular dendritic cells (FDCs), which are shown (*left*) as an SEM image as well as (*right*) a drawing. FDCs guide the movements and interactions of B cells. *[Part (a) photo courtesy of Stephanie Favre and Sanjiv A. Luther, University of Lausanne, Switzerland. Part (b) photo courtesy of Mohey Eldin M. El Shikh.]*

Naïve lymphocytes typically enter the cortex of the lymph node via **high endothelial venules** (**HEVs**) of the blood stream. These specialized veins are lined with unusually tall endothelial cells that give them a thickened appearance (Figure 2-13a; and see Figure 14-2). The lymphocytes then squeeze between endothelial cells of the HEV, into the functional tissue of the lymph node.

Once naïve T cells enter the lymph node, they browse MHC–peptide antigen complexes on the surfaces of APCs in the paracortex, the lymph node's **T-cell zone**. The APCs position themselves on a network of fibers that arise from stromal cells called **fibroblastic reticular cells** (**FRCs**) (**Figure 2-14a**). This **fibroblastic reticular cell conduit** (**FRCC**) **system** guides T-cell movements via associated adhesion molecules and chemokines. Antigen-presenting cells wrap themselves around the conduits, giving circulating T cells ample opportunity to browse their surfaces as they are guided down the network. The presence of this specialized network elegantly enhances the probability that T cells will meet their specific MHC-peptide combination (see also Chapter 14 opening figure, Figure 14-6, and associated videos 14-Ov and 14-6v).

Although naïve T cells enter via the blood, they exit via the *efferent lymphatics* in the medulla of the lymph node (Figure 2-13a), if they do not find their MHC-peptide match. T cells expressing TCRs that bind an MHC-peptide complex stop migrating and take up residence in the node for several days. Here they proliferate and, depending on cues from the antigen-presenting cell itself, differentiate into effector cells with a variety of distinct functions. CD8+ T cells gain the ability to kill target cells. CD4+ T cells differentiate into several different kinds of effector cells, including those that further activate macrophages, CD8+ T cells, and B cells.

B Cells in the Lymph Node

The lymph node is also the site where B cells are activated and differentiate into high-affinity, antibody-secreting plasma cells. Optimal B-cell activation requires both antigen engagement by the B-cell receptor (BCR) and direct contact with an activated CD4+ T_H cell. Both events are facilitated by the anatomy of the lymph node. Like T cells, B cells circulate through the blood and lymph and visit the lymph nodes on a daily basis, entering via the HEVs. They respond to specific signals and chemokines that draw them not to the paracortex but to the lymph node follicles. Although they may initially take advantage of the FRCC system for guidance, they ultimately shift to tracts generated by follicular dendritic cells (FDCs) (Figure 2-14b). FDCs maintain follicular and germinal center structure and present particulate antigen to differentiating B cells.

B cells differ from T cells in that their receptors can recognize free, unprocessed antigen. A B cell typically meets its antigen in a lymph node follicle, or **B-cell follicle**. Small, soluble antigens can make their way directly into the follicle, whereas larger antigens are relayed to the follicular dendritic cells by subcapsular macrophages and non-antigen-specific B cells (see Chapter 14). If its BCR binds to antigen, the B cell becomes partially activated and engulfs the antigen and processes it, readying it for presentation as a peptide-MHC complex to CD4$^+$ T$_H$ cells.

B cells that have successfully engaged and processed antigen change their migration patterns and move to the T cell–rich paracortex, where they may encounter a previously activated CD4$^+$ T$_H$ cell. If this helper T cell recognizes the MHC-antigen complex presented by the B cell, the pair will maintain contact for a number of hours, during which the B cell receives signals from the T cell that induce B-cell proliferation and differentiation (see Figure 14-4 and accompanying videos).

Some activated B cells differentiate directly into antibody-producing cells (plasma cells), but others re-enter the follicle to establish a germinal center. A follicle that develops a germinal center is referred to as a **secondary follicle**; a follicle without a germinal center is referred to as a **primary follicle**. In **germinal centers**, B cells proliferate and undergo clonal selection (see Figure 1-6) to produce a colony of B cells with the highest affinity for a particular antigen. Some of these cells travel to the medulla of the lymph node and release antibodies into the bloodstream; others exit through the efferent lymphatics and take up long-term residence in the bone marrow, where they will continue to release antibodies into circulation.

Germinal centers are established within 4 to 7 days of the initial infection, but remain active for 3 weeks or more (Chapter 11). Lymph nodes swell visibly and sometimes painfully during those first few days after infection as immune cells migrate into the node and T and B cells proliferate.

The Generation of Memory T and B Cells in the Lymph Node

Interactions between T cells and APCs, and between activated T$_H$ cells and activated B cells, result in the generation of memory T and B cells. Memory T and B cells either take up residence in secondary lymphoid tissues or exit the lymph node and circulate to and among other tissues, including those that first encountered the pathogen. Memory T cells that reside in secondary lymphoid organs are referred to as *central memory cells* and are distinct in phenotype and functional potential from *effector memory* T cells that circulate among tissues. A third population, *tissue-resident memory* cells, settle in peripheral tissues for the long term and appear

to be the first cells to respond when an individual is re-infected with a pathogen. Memory cell phenotype, locale, and activation requirements are very active areas of investigation and will be discussed in more detail in Chapters 10 and 11.

Key Concepts:

- The lymph nodes organize the immune response to antigens that enter through lymphatic vessels. T cells and B cells are compartmentalized in different microenvironments in secondary lymphoid tissue. T cells are found in the paracortex of the lymph nodes, while B cells are organized in follicles in the cortex.

- Naïve T lymphocytes browse the surfaces of antigen-presenting cells in the T-cell zone or paracortex of the lymph node, guided by the FRC network. If they bind to an MHC-peptide combination, they are activated and undergo clonal expansion and differentiation into effector cells in secondary lymphoid organs. If they do not, they exit via the efferent lymphatics and continue browsing in other lymph nodes.

- T lymphocytes develop into mature killer CD8$^+$ and helper CD4$^+$ cells in secondary lymphoid organs. Some CD4$^+$ T cells help B cells to differentiate into antibody-secreting plasma cells; others activate macrophages and CD8$^+$ cytotoxic T cells.

- Naïve B cells that enter lymph nodes travel to B-cell follicles, where they meet their antigen on the FDC network. Those that bind antigen are activated, divide, and seek T-cell help. Those that do not exit the lymph node via the efferent lymphatics to browse another secondary lymphoid tissue.

- Activated B cells undergo further maturation into high-affinity, antibody-producing cells in specialized microenvironments called *germinal centers*, substructures that develop within B-cell follicles.

- B and T cells develop into long-lived memory cells in secondary lymphoid organs. Some memory cells remain in lymphoid tissue, some circulate, and some take up residence in other tissues, ready to respond quickly to a returning pathogen.

The Spleen Organizes the Immune Response against Blood-Borne Pathogens

The **spleen**, situated high in the left side of the abdominal cavity, is a large, ovoid secondary lymphoid organ that plays a major role in mounting immune responses to antigens in the bloodstream (**Figure 2-15**). Whereas lymph nodes are specialized for encounters between lymphocytes and antigen drained from local tissues, the spleen specializes in trapping

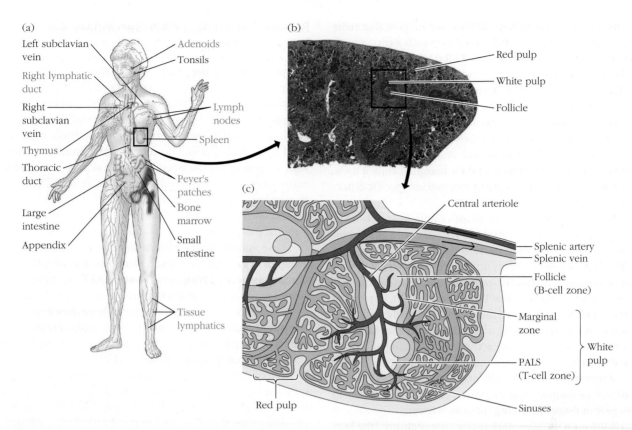

FIGURE 2-15 Structure of the spleen. (a) The spleen, which is about 5 inches long in human adults, is the largest secondary lymphoid organ. It is specialized for trapping blood-borne antigens. (b) A stained tissue section of the human spleen, showing the red pulp, white pulp, and follicles. These microenvironments are diagrammed schematically in (c). The splenic artery pierces the capsule and divides into progressively smaller arterioles, ending in vascular sinusoids that drain back into the splenic vein. The erythrocyte-filled red pulp surrounds the sinusoids. The white pulp forms a sleeve—the periarteriolar lymphoid sheath (PALS)—around the arterioles; this sheath is populated by T cells. Closely associated with the PALS are the B cell–rich lymphoid follicles that can develop into secondary follicles containing germinal centers. The marginal zone, a site of specialized macrophages and B cells, surrounds the PALS and separates it from the red pulp. *[Photo (b) courtesy Dr. Keith Wheeler/Science Source.]*

and responding to blood-borne antigens; thus, it is particularly important in the response to systemic infections. Unlike lymph nodes, the spleen is not supplied by lymphatic vessels. Instead, blood-borne antigens and lymphocytes are carried into the spleen through the **splenic artery** and out via the **splenic vein**. Experiments with radioactively labeled lymphocytes show that more recirculating lymphocytes pass daily through the spleen than through all the lymph nodes combined.

The spleen is surrounded by a capsule that extends into the interior, dividing the spleen into lobes, all of which function similarly. Two main microenvironmental compartments can be distinguished in each splenic lobe: the **red pulp** and **white pulp**, which are separated by a specialized region called the **marginal zone** (see Figure 2-15c). The splenic red pulp consists of a network of sinusoids populated by red blood cells, macrophages, and some lymphocytes. It is the site where old and defective red blood cells are destroyed

and removed; many of the macrophages within the red pulp contain engulfed red blood cells or iron-containing pigments from degraded hemoglobin. It is also the site where pathogens first gain access to the lymphoid-rich regions of the spleen, known as the white pulp. The splenic white pulp surrounds the branches of the splenic artery, and consists of B-cell follicles and the **periarteriolar lymphoid sheath (PALS)**, which is populated by T lymphocytes. As in lymph nodes, germinal centers are generated within these follicles during an immune response. The spleen also maintains a fibroblastic reticular network that provides tracts for T-cell and B-cell migration.

The marginal zone (MZ) is a specialized cellular border between the blood and the white pulp. A relatively recent development in the evolutionary history of the immune system, it is populated by specialized dendritic cells, macrophages, and unique B cells, referred to as *marginal zone B cells* (*MZ B cells*). These cells are the first line of defense against blood-borne pathogens, trapping antigens that enter via the splenic artery. Marginal zone B cells express both innate immune receptors (e.g., TLRs) and unique B-cell receptors that recognize conserved molecular patterns on pathogens. Once they bind antigen, MZ B cells differentiate rapidly and secrete high levels of antibodies. Although some MZ B cells require T-cell help for activation, others can be stimulated in a T cell–independent manner. Interestingly, mouse and human marginal zone anatomy differ, as does the phenotype and behavior of their marginal zone B cells. The basis for and significance of this species-specific difference are under investigation.

The events that initiate the adaptive immune response in the spleen are analogous to those that occur in the lymph node. Briefly, circulating naïve B cells encounter antigen in the follicles, and circulating naïve CD8$^+$ and CD4$^+$ T cells meet antigen as MHC-peptide complexes on the surface of dendritic cells in the T-cell zone (PALS). Once activated, CD4$^+$ T$_H$ cells then provide help to B cells, including some marginal zone B cells, and CD8$^+$ T cells that have also encountered antigen. Some activated B cells, together with some T$_H$ cells, migrate back into follicles and generate germinal centers. As in the lymph node, germinal center B cells can become memory cells or plasma cells, which circulate to a variety of tissues including the bone marrow.

Children who have undergone **splenectomy** (the surgical removal of a spleen) are vulnerable to *overwhelming post-splenectomy infection* (OPSI) characterized by systemic bacterial infections (sepsis) caused primarily by *Streptococcus pneumoniae*, *Neisseria meningitidis*, and *Haemophilus influenzae*. Although fewer adverse effects are experienced by adults, splenectomy can still lead to an increased vulnerability to blood-borne bacterial infections, underscoring the role the spleen plays in our immune response to pathogens that enter the circulation. Because the spleen also serves other functions in iron metabolism, platelet storage, and hematopoiesis, these are also compromised if it is removed.

Key Concepts:

- The spleen organizes the first immune response to blood-borne pathogens.

- The spleen is compartmentalized into the white pulp, which contains B and T lymphocytes, and the red pulp, which contains circulating red blood cells.

- The periarteriolar lymphoid sheath of the white pulp includes B-cell follicles and T-cell zones.

- The splenic marginal zone, a specialized region of macrophages and B cells, forms a boundary between the red and the white pulp and plays an important role in trapping and responding to blood-borne antigens.

Barrier Organs Also Have Secondary Lymphoid Tissue

Lymph nodes and the spleen are not the only organs with secondary lymphoid microenvironments. T-cell zones and lymphoid follicles are also found in barrier tissues, which include the skin and mucosal membranes of the digestive, respiratory, and urogenital tracts. Each of these organs is lined by epithelial cells. Our mucosal membranes are lined with a single epithelial layer, while our skin is protected by many layers of epithelial cells. Together, skin and mucosal membranes represent a surface area of over 400 m^2 (nearly the size of a basketball court) and are the major sites of entry for most pathogens.

These vulnerable membrane surfaces are defended by a group of organized lymphoid tissues known collectively as **mucosa-associated lymphoid tissue (MALT)**. Lymphoid tissues associated with different mucosal areas are sometimes given more specific names: for instance, **bronchus-associated lymphoid tissue (BALT), nasal-associated lymphoid tissue (NALT), gut-associated lymphoid tissue (GALT), and skin-associated lymphoid tissue (SALT)**.

Each of these tissues plays an important role in our innate immune defenses and recruits many different cell types to the effort. The epithelial cell layers provide more than just physical protection; they also respond actively to pathogens by secreting cytokines, chemokines, and even antimicrobial compounds. Many different types of immune cells reside in the deeper layers of barrier tissues and generate B-cell follicles. B cells that develop in these follicles tend to secrete IgA, which has the ability to cross epithelial barriers and interact with microbes in the lumen of our mucosal tracts.

Innate and adaptive immune cells in barrier organs not only organize our first response to invading pathogens, but they also play a critical role in maintaining tolerance to the diverse and abundant *commensal* microbes that contribute positively to our health. The distinct immune functions and cell residents of each barrier tissue are described in more detail in Chapter 13, but a preview of the organization of secondary lymphoid tissue in the intestine (GALT) is depicted in **Figure 2-16**.

Key Concepts:

- Barrier immune organs, which include the skin and mucosal tissues, contain secondary lymphoid tissue and mount an important first defense against pathogens that penetrate our epithelial layers. Epithelial cells play an active role and initiate the response of innate and adaptive immune cells, which can organize into B-cell follicles.

- Barrier immune systems also help us maintain tolerance to commensal microbes that coexist at our surfaces.

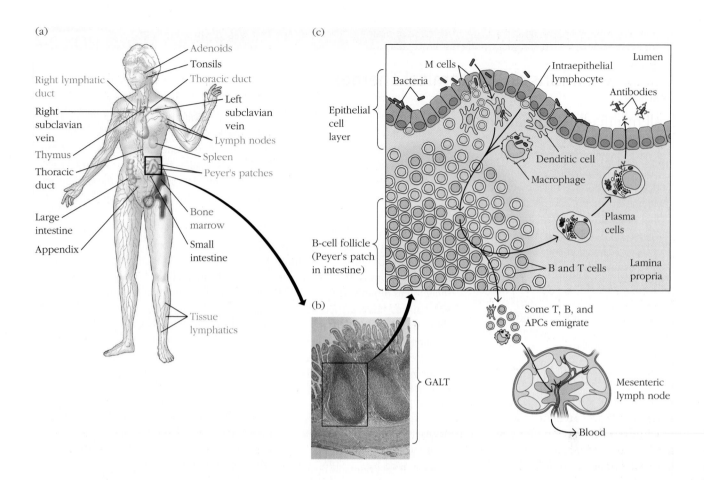

FIGURE 2-16 Example of secondary lymphoid tissue in barrier organs: gut-associated lymphoid tissue (GALT). (a) The Peyer's patch is a representative of the extensive GALT system that is found in the intestine. (b) A stained tissue cross-section of Peyer's patch lymphoid nodules in the intestinal submucosa is schematically diagrammed in (c). The single layer of epithelial cells includes specialized cells, called M cells, that convey antigens from the intestinal lumen to the inner layers (lamina propria) of the intestinal wall. Here they trigger the formation of B-cell follicles, which generate antibody-producing plasma cells. The antibodies pass back into the intestinal lumen and bind to pathogens, protecting the intestinal wall from inflammation and invasion. Other cells, including macrophages, dendritic cells, and intraepithelial lymphocytes, sample antigens from the lumen and, with the help of regulatory T cells, work to distinguish between beneficial commensal bacteria and more dangerous pathogens. Antigen-presenting cells and lymphocytes can travel to local lymph nodes, where they trigger a more systemic immune response to antigens. [Part (b) photo from Ed Reschke/Getty Images.]

Tertiary Lymphoid Tissues Also Organize and Maintain an Immune Response

A site of active infection and immune activity is often referred to as a **tertiary lymphoid tissue**. Lymphocytes activated by antigen in secondary lymphoid tissue return to these areas (e.g., lung, liver, brain, skin) as effector cells and can also reside there as tissue-resident memory cells. Tertiary lymphoid tissues can generate new microenvironments that organize lymphocyte responses. The brain, for instance, establishes reticular systems that guide lymphocytes responding to chronic infection with the protozoan that causes toxoplasmosis. Organized aggregations of lymphoid cells are especially prominent at sites of chronic infection and highlight the intimate relationship between immune and nonimmune cells, as well as the plasticity of tissue anatomy. This plasticity is also illustrated by the evolutionary relationships among immune systems and organs (see **Evolution Box 2-4**).

Key Concept:
- Tissues that are sites of infection are referred to as *tertiary tissues*. These sites can also develop organized lymphoid microenvironments, including B-cell follicles.

Variations on Anatomical Themes

All multicellular organisms

defend themselves against pathogens, and all have innate systems of immunity (see Chapter 4). The adaptive immune system appeared about 500 million years ago, with the emergence of vertebrate animals. Only vertebrates generate antigen-specific receptors. Interestingly, however, the location, organization, and function of lymphoid tissues vary widely across the vertebrate subphylum.

Vertebrates range from jawless fishes (Agnatha, the earliest lineages, which are represented by the lamprey eel and hagfish), to cartilaginous fish (e.g., sharks and rays, also called elasmobranchs), which represent the earliest lineages of jawed vertebrates (Gnathostomata), to bony fish, amphibians, reptiles, birds, and mammals. If you view these groups as part of an evolutionary progression, you see that, in general, immune tissues and organs evolved by earlier orders have been retained as newer organs of immunity, such as lymph nodes, have appeared (**Figure 1**). All vertebrates, for example, have gut-associated lymphoid tissue (GALT), but only jawed vertebrates

have well-developed thymi and spleens. All vertebrates have two different (B and T) lymphoid cell populations, suggesting that they were present in our common ancestor.

T cells were the first cell population to express a diverse repertoire of antigen receptors, and their appearance is directly and inextricably linked to the appearance of the primary immune organ, the thymus. This dependence is reflected in organisms today: all jawed vertebrates have a thymus, and the thymus is absolutely required for the development of T cells. Recent studies indicate that even jawless vertebrates, including the lamprey, harbor distinct thymic-like (thymoid) tissue in their gill regions (**Figure 2**).

In contrast, B-cell development is not bound to one particular organ. In many adult mammals, including mice and humans, B cells primarily develop in the bone marrow and mature further in the spleen. In cattle and sheep, B-cell maturation shifts from the fetal spleen to a patch of tissue embedded in the wall of the intestine called the ileal Peyer's patch.

The rabbit also uses gut-associated tissues, especially the appendix, as primary lymphoid tissue for important steps in the proliferation and diversification of B cells. Recent data suggest that the gut can act as a primary lymphoid organ for generating mature B lymphocytes in most vertebrates, even those that depend largely on the bone marrow.

The sites of B-cell development also vary among nonmammalian vertebrates. In jawless vertebrates, B lymphocyte–like cells develop in distinct regions (called *typhlosoles*) in the kidney and intestine. In sharks, B-cell development shifts from liver to kidney to spleen, and in amphibians and reptiles it shifts from liver and spleen to bone marrow. B-cell development in bony fish takes place primarily within the kidney. In birds, B cells complete their development in a unique lymphoid organ associated with the gut, the bursa of Fabricius (**Figure 3**).

Secondary lymphoid tissues appear to have increased in complexity throughout vertebrate evolution. The spleen is the most ancient secondary lymphoid organ, and lymph nodes are the most recent

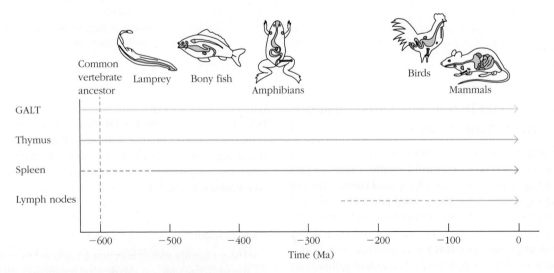

FIGURE 1 **Evolutionary distribution of lymphoid tissues.** The appearance and presence of primary and secondary lymphoid tissues in vertebrates over evolutionary time. (Abbreviations: GALT = **gut-associated lymphoid tissue**; Ma = million years ago.)

(continued)

EVOLUTION *(continued)* **BOX 2-4**

(a)　　　　　　　　(b)　　　　　　　　(c)

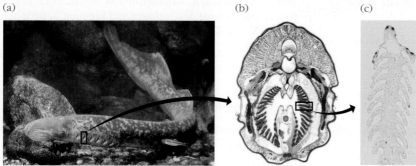

FIGURE 2 **Thymic tissue in the lamprey eel.** (a) Jawless vertebrates, including the lamprey eel, were once thought not to have a thymus. Recent work suggests, however, that they generate two types of lymphocytes analogous to B and T cells and have thymic tissue at the tip of their gills (b). The thymic tissue shown in (c) is stained for expression of CDA1, a gene specific for lymphocytes found in lampreys. [*Left: © Breck P. Kent/Animals Animals-Earth Scenes; all rights reserved. Middle: Dr. Keith Wheeler/Science Source. Right: Courtesy Thomas Boehm.*]

(a)　　　　　　　　　　　(b)

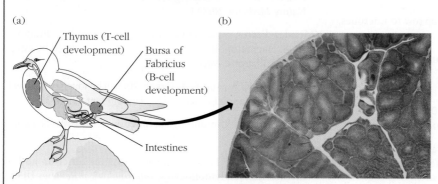

Thymus (T-cell development)

Bursa of Fabricius (B-cell development)

Intestines

FIGURE 3 **The avian bursa.** (a) Like most vertebrates, birds have a thymus, spleen, and lymph nodes. Although hematopoiesis occurs in their bone marrow, B-cell development takes place in a specialized organ, known as the *bursa*. The bursa is an outpouching of the intestine located close to the cloaca, the common end of the intestinal and genital tracts in birds. A stained tissue section of the bursa and cloaca is shown in (b). [*Photo courtesy Dr. Thomas Caceci, Associate Professor of Biomedical Sciences at the Virginia Tech/Carilion School of Medicine, Roanoke, VA.*]

immune adaptation. Lymph nodes likely arose from tissue associated with lymphatic vessels and appeared first in reptiles and birds. Mammals have the most highly organized lymph nodes, which centralize participants in the adaptive immune response. Secondary lymphoid structures are also found throughout epithelial tissues in vertebrates and vary widely in locale (e.g., although rodents do not have tonsils, they do have well-developed lymphoid tissue at the base of the nose). The general structural and functional features of secondary lymphoid tissue are shared by most vertebrates, with at least one interesting exception: pig lymph nodes exhibit a striking peculiarity—they are "inverted" anatomically, so that the medulla of the organ, where lymphocytes exit the organ, is on the outside, and the cortex, where lymphocytes meet their antigen and proliferate, is on the inside. Lymphocytes exit via blood vessels, rather than via efferent lymphatics. The adaptive advantages, if any, of these odd structural variations are unknown, but the example reminds us of the remarkable plasticity of structure-function relationships within biological structures and the creative opportunism of evolutionary processes.

REFERENCES

Boehm, T., I. Hess, and J. B. Swann. 2012. Evolution of lymphoid tissues. *Trends in Immunology* **33**:315.

Du Pasquier, A. L., and M. F. Flajnik. 2004. Evolution of the immune system. In: *Fundamental Immunology*, 5th ed., W. E. Paul, ed. Lippincott-Raven, Philadelphia.

Conclusion

All blood cells arise from hematopoietic stem cells, which reside primarily in the adult bone marrow. Immune cells differentiate in primary lymphoid organs, which include the bone marrow and, in the case of T lymphocytes, the thymus. Immune cells differentiate in the bone marrow and thymus (primary lymphoid organs), and then travel through the blood and lymphatics to lymph nodes and the spleen (secondary lymphoid organs), where they browse for antigen.

Lymphoid cells circulate to lymph nodes and spleen, secondary lymphoid organs where the adaptive immune response is initiated. Innate immune cells, including APCs and neutrophils, provide the first defense against pathogens that penetrate epithelial barriers. Antigen-presenting cells and antigen travel from the site of infection to the lymph nodes, where they meet and activate browsing T and B lymphocytes. Activated T and B cells differentiate into short-lived effector cells that help clear the infection and long-lived memory cells that protect us against repeat infections.

REFERENCES

Artis, D., and H. Spits. 2015. The biology of innate lymphoid cells. *Nature* **517**:293.

Bajenoff, M., et al. 2007. Highways, byways and breadcrumbs: directing lymphocyte traffic in the lymph node. *Trends in Immunology* **28**:346.

Bajoghil, B., et al. 2011. A thymus candidate in lampreys. *Nature* **47**:90.

Banchereau J., et al. 2000. Immunobiology of dendritic cells. *Annual Review of Immunology* **18**:767.

Boehm, T., I. Hess, and J. B. Swann. 2012. Evolution of lymphoid tissues. *Trends in Immunology* **33**:314.

Borregaard, N. 2011. Neutrophils: from marrow to microbes. *Immunity* **33**:657.

Boulais, P. E., and P. S. Frenette. 2015. Making sense of hematopoietic stem cell niches. *Blood* **125**:2621.

Catron, D. M., A. A. Itano, K. A. Pape, D. L. Mueller, and M. K. Jenkins. 2004. Visualizing the first 50 hr of the primary immune response to a soluble antigen. *Immunity* **21**:341.

Cerutti, A., M. Cols, and I. Puga. 2013. Marginal zone B cells: virtues of innate-like antibody-producing lymphocytes. *Nature Reviews Immunology* **13**:118.

Eberl, G., M. Colonna, J. P. Di Santo, and A. N. J. McKenzie. 2015. Innate lymphoid cells: a new paradigm in immunology. *Science* **348**:aaa6566.

Eberl, G., J. P. Di Santo, and E. Vivier. 2015. The brave new world of innate lymphoid cells. *Nature Immunology* **16**:1.

Ema, H., et al. 2005. Quantification of self-renewal capacity in single hematopoietic stem cells from normal and Lnk-deficient mice. *Developmental Cell* **8**:907.

Falcone, F. H., D. Zillikens, and B. F. Gibbs. 2006. The 21st century renaissance of the basophil? Current insights into its role in allergic responses and innate immunity. *Experimental Dermatology* **15**:855.

Fletcher, A. L., S. E. Acton, and K. Knoblich. 2015. Lymph node fibroblastic reticular cells in health and disease. *Nature Reviews Immunology* **15**:350.

Gasteiger, G., and A. Y. Rudensky. 2014. Interactions between innate and adaptive lymphocytes. *Nature Reviews Immunology* **14**:631.

Ginhoux, F., and S. Jung. 2014. Monocytes and macrophages: developmental pathways and tissue homeostasis. *Nature Reviews Immunology* **14**:392.

Godfrey, D. I., H. R. MacDonald, M. Kronenberg, M. J. Smyth, and L. Van Kaer. 2004. NKT cells: what's in a name? *Nature Reviews Immunology* **4**:231.

Halin, C., J. R. Mora, C. Sumen, and U. H. von Andrian. 2005. In vivo imaging of lymphocyte trafficking. *Annual Review of Cell and Developmental Biology* **21**:581.

Kiel, M., and S. J. Morrison. 2006. Maintaining hematopoietic stem cells in the vascular niche. *Immunity* **25**:852.

Lavin, Y., A. Mortha, A. Rahman, and M. Merad. 2015. Regulation of macrophage development and function in peripheral tissues. *Nature Reviews Immunology* **15**:731.

Liu, Y. J. 2001. Dendritic cell subsets and lineages, and their functions in innate and adaptive immunity. *Cell* **106**:259.

Mass, E. I., et al. 2016. Specification of tissue-resident macrophages during organogenesis. *Science* **353**:aaf4238.

Mebius, R. E., and G. Kraal. 2005. Structure and function of the spleen. *Nature Reviews Immunology* **5**:606.

Mendelson, A., and P. S. Frenette. 2014. Hematopoietic stem cell maintenance during homeostasis and regeneration. *Nature Medicine* **20**:833.

Mendez-Ferrer, S., D. Lucas, M. Battista, and P. S. Frenette. 2008. Haematopoietic stem cell release is regulated by circadian oscillations. *Nature* **452**:442.

Mikkola, H. K. A., and S. A. Orkin. 2006. The journey of developing hematopoietic stem cells. *Development* **133**:3733.

Miller, J. F. 1961. Immunological function of the thymus. *Lancet* **278**:748.

Miller, J. F. 2002. The discovery of thymus function and of thymus-derived lymphocytes. *Immunological Reviews* **185**:7.

Moon, J. J., et al. 2007. Naïve CD4+ T cell frequency varies for different epitopes and predicts repertoire diversity and response magnitude. *Immunity* **27**:203.

Pabst, R. 2007. Plasticity and heterogeneity of lymphoid organs: what are the criteria to call a lymphoid organ primary, secondary or tertiary? *Immunology Letters* **112**:1.

Piccirillo, C. A., and E. M. Shevach. 2004. Naturally occurring CD4+CD25+ immunoregulatory T cells: central players in the arena of peripheral tolerance. *Seminars in Immunology* **16**:81.

Picker, L. J., and M. H. Siegelman. 1999. Lymphoid tissues and organs. In *Fundamental Immunology*, 4th ed. W. E. Paul, ed. Philadelphia: Lippincott-Raven, p. 145.

Rodewald, H.-R. 2006. A second chance for the thymus. *Nature* **441**:942.

Rothenberg, M., and S. P. Hogan. 2006. The eosinophil. *Annual Review of Immunology* **24**:147.

Scadden, D. T. 2006. The stem-cell niche as an entity of action. *Nature* **441**:1075.

Shizuru, J. A., R. S. Negrin, and I. L. Weissman. 2005. Hematopoietic stem and progenitor cells: clinical and preclinical regeneration of the hematolymphoid system. *Annual Review of Medicine* **56**:509.

Walker, J. A., J. L. Barlow, and A. N. J. McKenzie. 2013. Innate lymphoid cells—how did we miss them? *Nature Reviews Immunology* **13**:75.

Weissman, I. L. 2000. Translating stem and progenitor cell biology to the clinic: barriers and opportunities. *Science* **287**:1442.

Useful Websites

www.bio-alive.com/animations/anatomy.htm A collection of publicly available animations relevant to biology. Scroll through the list to find videos on blood and immune cells, immune responses, and links to interactive sites that reinforce your understanding of immune anatomy.

www.ncbi.nlm.nih.gov/pubmedhealth/PMH0072579/ An accessible site from the U.S. National Library of Medicine that covers the fundamentals of immune system anatomy.

www.niaid.nih.gov/ An accessible site from the National Institute of Allergy and Infectious Diseases about immune system function and structure.

www.hematologyatlas.com/principalpage.htm An interactive atlas of both normal and pathological human blood cells.

https://stemcells.nih.gov Links to information on stem cells, including basic and clinical information, and registries of available embryonic stem cells. The direct link to information on blood stem cells is here:

https://stemcells.nih.gov/info/Regenerative_Medicine/2006Chapter2.htm

www.khanacademy.org/science/health-and-medicine/hematologic-system-diseases-2/leukemia/v/hematopoiesis The Khan Academy's mini-lecture on hematopoiesis.

www.immunity.com/cgi/content/full/21/3/341/DC1 A pair of simulations that trace the activities of T cells, B cells, and dendritic cells in a lymph node. These movies are discussed in more detail in Chapter 14.

www.hhmi.org/research/investigators/cyster.html Investigator Jason Cyster's public website that includes many videos of B-cell activity in immune tissue.

STUDY QUESTIONS

1. Each of the following statements is false. Please correct.

   a. Mature T cells are found only in the lymph nodes and spleen.

   b. The pluripotent stem cell is one of the most abundant cell types in the bone marrow.

   c. There are no stem cells in blood.

   d. Activation of macrophages increases their expression of MHC class I molecules, allowing them to present antigen to $CD4^+$ T_H cells more effectively.

   e. B cells develop in the thymus.

   f. Lymphoid follicles are present only in the spleen and lymph nodes.

   g. The FRC guides B cells to follicles.

   h. Infection has no influence on the rate of hematopoiesis.

   i. Follicular dendritic cells can process and present antigen to T lymphocytes.

   j. Dendritic cells arise only from the myeloid lineage.

   k. All lymphoid cells have antigen-specific receptors on their membrane.

   l. All vertebrates generate B lymphocytes in bone marrow.

   m. Only mammals have a thymus.

   n. Jawless vertebrates do not have lymphocytes.

2. Identify which of the following cells are myeloid and which are lymphoid.

   a. Dendritic cells

   b. Neutrophils

   c. NK cells

   d. Basophils

   e. Macrophages

   f. T_C cells

   g. B cells

   h. ILCs

3. List two primary and two secondary lymphoid organs and summarize their functions in the immune response.

4. What two characteristics distinguish HSCs from mature blood cells?

5. How does the thymus help us avoid autoimmune responses?

6. At what age does the thymus reach its maximal size?

   a. During the first year of life

   b. Teenage years (puberty)

   c. Between 40 and 50 years of age

   d. After 70 years of age

7. Preparations enriched in HSCs are useful for research and clinical practice. What is the role of immunodeficient mice (i.e., mice that are missing one or more immune cell type) in demonstrating the success of HSC enrichment?

8. Explain the difference between a monocyte and a macrophage.

9. What effect would removal of the bursa of Fabricius (bursectomy) have on chickens?

10. Indicate whether each of the following statements about the lymph node and spleen is true or false. If you think a statement is false, please explain.

   a. The lymph node is the first place that immune cells encounter blood-borne antigens.

   b. The lymph node paracortex is rich in T cells, and the splenic periarteriolar lymphoid sheath (PALS) is rich in B cells.

   c. Only the lymph node contains germinal centers.

   d. Fibroblastic reticular cell conduits enhance the probability of T cell–APC interactions.

   e. Afferent lymphatic vessels draining the tissue spaces enter the spleen.

   f. Lymph node, but not spleen, function is affected by a knockout of the *Ikaros* gene.

11. For each description below (1–15), select the appropriate cell type (a–o). Each cell type may be used once, more than once, or not at all.

Descriptions

1. Major cell type presenting antigen to naïve T cells

2. Phagocytic cell of the central nervous system

3. Granulocytic cells important in the body's defense against parasitic organisms

4. Gives rise to red blood cells

5. Generally first cells to arrive at site of inflammation

6. Supports maintenance of HSCs

7. Gives rise to thymocytes

8. Circulating blood cells that differentiate into macrophages in the tissues

9. An antigen-presenting cell that arises from the same precursor as a T cell but not the same as a macrophage

10. Cells that are important in sampling antigens of the intestinal lumen

11. Granulocytic cells that release various pharmacologically active substances

12. White blood cells that play an important role in the development of allergies

13. Cells that can use antibodies to recognize their targets

14. Cells that express antigen-specific receptors

15. Cells that share a common progenitor with T and B cells, but do not have antigen-specific receptors

Cell Types

a. Common myeloid progenitor cells
b. Monocytes
c. Eosinophils
d. Dendritic cells
e. Innate lymphoid cells
f. Mast cells
g. Neutrophils
h. M cells
i. Osteoblasts
j. Lymphocytes
k. NKT cells
l. Microglial cells
m. Myeloid dendritic cells
n. HSCs
o. Lymphoid dendritic cells

RESEARCH FOCUS QUESTION   Notch is a surface protein that regulates cell fate. When bound by its ligand, it releases and activates its intracellular region, which regulates new gene transcription. Investigators found that the phenotype of developing cells in the bone marrow differed dramatically when they overexpressed the active, intracellular portion of Notch. In particular, the frequency of BCR^+ cells plummeted, and the frequency of TCR^+ cells increased markedly. Interestingly, other investigators found that when Notch was knocked out, the phenotype of cells in the thymus changed: the frequency of BCR^+ cells increased and the frequency of TCR^+ cells decreased dramatically.

Propose a molecular model to explain these observations, and an experimental approach to begin testing your model.

CLINICAL FOCUS QUESTION   T and B cells that differentiate from HSCs recognize as self the bodies in which they differentiate. Suppose a woman donates HSCs to a genetically unrelated man whose hematopoietic system was totally destroyed by a combination of radiation and chemotherapy. Further, suppose that, although most of the donor HSCs differentiate into hematopoietic cells, some differentiate into cells of the pancreas, liver, and heart. Recall from **Clinical Focus Box 2-2** that transplanted lymphocytes can attack the recipient's tissues, causing a graft-versus-host (GVH) reaction. Decide which of the following outcomes is likely and justify your choice.

a. The T cells that arise from the donor HSCs do not attack the pancreatic, heart, and liver cells that arose from donor cells but mount a GVH response against all of the other host cells.

b. The T cells that arise from the donor HSCs mount a GVH response against all of the host cells.

c. The T cells that arise from the donor HSCs attack the pancreatic, heart, and liver cells that arose from donor cells but fail to mount a GVH response against all of the other host cells.

d. The T cells that arise from the donor HSCs do not attack the pancreatic, heart, and liver cells that arose from donor cells and fail to mount a GVH response against all of the other host cells.

Recognition and Response

Learning Objectives

After reading this chapter, you should be able to:

1. Explain how receptor clustering and localized secretion enhance the signaling of small molecules between immune cells.

2. Offer one example each of an adaptive immune receptor, an innate immune receptor, and a cytokine receptor in which one receptor protein chain may be used in combination with more than one partner chain to alter the nature of ligand specificity.

3. State where you would expect to find receptors of the adaptive and innate immune systems within the context of the cell and correlate the ligands and receptors of the two types of immune cells with their signaling outcomes.

4. Draw, and then compare and contrast, the structural features of T- and B-cell receptor complexes, indicating the presence of immunoglobulin domains, coreceptors, and signal transduction mediators.

5. Explain the common features of the cell signaling pathways used by innate, adaptive, and cytokine receptors.

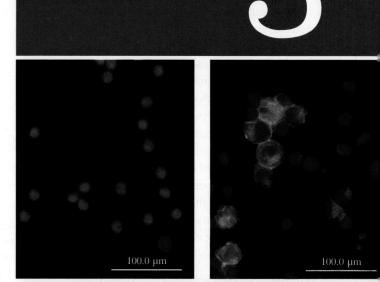

100.0 μm   100.0 μm

On stimulation (right), the interleukin-2 receptor alpha (IL-2Rα) chain (yellow) is up-regulated on T cells (blue), increasing affinity of the IL-2 receptor for IL-2. *[Courtesy R&D Systems, Inc., Minneapolis, MN.]*

All cells in a multicellular organism must receive and respond to signals derived from other cells to know when to grow, divide, differentiate, connect, disperse, or die, but cells of the immune system face particular challenges: (1) they must also interact with and eliminate a vast spectrum of potentially dangerous infectious organisms and toxins and (2) because of the widely separated distribution of immune cells among all the other organs of the body, they must be able to communicate with one another over long distances. In every type of immune stimulation, the immune system cell must receive a molecular signal via a membrane-bound or intracellular receptor, and translate the receipt of that signal into a meaningful cellular response such as cell division or differentiation.

In this chapter, we will briefly review common features of cellular receptors and highlight how those shared receptor properties have been adapted by the immune system for its particular use. Next, we will focus on the structure and function of the antigen receptors of the adaptive immune response, B- and T-cell receptors (BCRs and TCRs), and the molecules with which they interact. We will briefly discuss innate immune receptors and describe some of the properties that are shared by innate and adaptive immune receptors. The receptor section of this chapter will close with a discussion of cytokine receptors and their ligands and introduce a specialized family of cytokines, the chemokines, whose role it is to facilitate the movement of immune cells to the regions where they are needed.

The next section will address the concept of signal transduction. Subsequent to a binding interaction between a

Key Terms

Pattern recognition receptors (PRRs)

B-cell receptors (BCRs)

T-cell receptors (TCRs)

K_d, the dissociation constant

Multivalency

Avidity

Immunoglobulin superfamily

Variable (V) regions

Constant (C) regions

Complementarity-determining regions (CDRs)

Isotypes

Classes

Pathogen-associated molecular patterns (PAMPs)

Cytokines

Chemokines

Chemoattractants

Signal transduction

Lipid rafts

Src-family kinases

Adapter proteins

receptor and its cognate (matching) ligand, the responding cell must communicate knowledge of this receptor-ligand interaction to members of molecular signaling pathways that can evoke the appropriate cellular response. This process is referred to as **signal transduction**. The end results of signal transduction pathways are changes in the behavior of the responding cell. These changes, which collectively represent the outcome of (or the response to) the receptor's encounter with its ligand, might include some combination of cell division, differentiation, movement, altered metabolic status, alteration in the expression of surface or cytoplasmic molecules, or secretion of new compounds such as chemokines or cytokines.

The final section of the chapter will summarize some of the biological outcomes of immune system recognition, setting the stage for the chapters that follow.

General Properties of Immune Receptor-Ligand Interactions

Cells of the immune system must constantly recognize and respond to a myriad of molecular signals from the external environment. Immune system cells use several different types of receptors to recognize ligands. **Pattern recognition receptors (PRRs)** are found on innate immune cells (e.g., macrophages, dendritic cells, neutrophils, and others), as well as on adaptive immune cells (T and B lymphocytes) and have also been defined on cell types that are not part of the formal immune system, such as neuronal cells, epithelial cells, and others. In contrast, the antigen-specific **B-cell receptors (BCRs)** and **T-cell receptors (TCRs)** are found exclusively on the B and T lymphocytes of the adaptive immune system. All of these cells and many others can express cytokines, chemokines, and their receptors in order to communicate with one another as they respond to antigen encounters. Since B and T cells express both adaptive and innate immune receptor molecules, as well as many cytokine and chemokine receptors, they must therefore constantly interpret multiple simultaneous or sequential signals. It is the *integration* of these many receptor-ligand interactions at the cellular level that will ultimately determine the biological outcome of any immune stimulus.

The interactions of ligands with immune system receptor molecules can be extremely complex. Most students are familiar with the concept of enzyme-substrate binding interactions. Such interactions are usually monovalent (one enzyme molecule binds one molecule of substrate) and of moderate affinity, and only a small number of enzymes will have the capacity to interact with any one substrate. In contrast, interactions between antigens and their receptor molecules may be of extremely high affinity and the diversity of receptors capable of recognizing any given antigen is simply breathtaking. In addition, the affinity of cytokine receptors for their cognate cytokine as well as the level of expression

and the placement of receptors on the cell membrane may vary according to the activation status of the cell on which it is expressed (see the chapter opening photograph).

Receptor-Ligand Binding Occurs via Multiple Noncovalent Bonds

A receptor molecule attaches to its ligand by the same types of noncovalent chemical linkages that enzymes use to bind to their substrates. These include hydrogen and ionic bonds, and hydrophobic and van der Waals interactions. The key to a meaningful receptor-ligand interaction is that the sum total of the bonding interactions holds the two interacting surfaces together with sufficient *binding energy*, and for sufficient *time*, to allow a molecular signal to be received by the cell signifying that the receptor has bound its matching ligand. Because these noncovalent interactions are individually weak, many such interactions are required to form a biologically relevant receptor-ligand connection. Furthermore, since each of these noncovalent interactions operates only over a very short distance—generally about 1 angstrom ($1 \text{ Å} = 10^{-10}$ meters)—a high-affinity receptor-ligand interaction depends on a very close "fit," or degree of complementarity, between the receptor and the ligand (**Figure 3-1**).

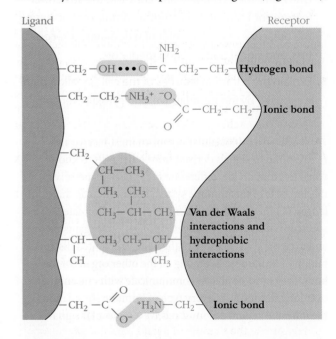

FIGURE 3-1   Receptor-ligand binding obeys the rules of chemistry. Receptors bind to ligands using the full range of noncovalent bonding interactions, including ionic and hydrogen bonds and van der Waals and hydrophobic interactions. For signaling to occur, the bonds must be of sufficient strength to hold the ligand and receptor in close proximity long enough for downstream events to be initiated. In B- and T-cell signaling, activating interactions also require receptor clustering. In an aqueous environment, noncovalent interactions are weak and depend on close complementarity of the shapes of receptor and ligand.

How Do We Describe the Strength of Receptor-Ligand Interactions?

We can describe the strength of the binding interaction between a receptor binding site, S, and a ligand, L, with the following expression:

$$K_d = \frac{[S][L]}{[SL]} \qquad \text{(Eq. 3-1)}$$

where K_d, the **dissociation constant** of the reaction, defines the relationship between the concentration of the receptor-ligand pair, [SL] and the product of the concentration of free receptor sites, [S] and the concentration of free ligand, [L]. The units of the dissociation constant are in molarity (M). *The lower* the K_d, the *higher* the affinity of the interaction. Note that when 50% of the binding sites are occupied, [SL] = [S] and the K_d = the concentration of free ligand.

For purposes of comparison, it is useful to consider that the K_d values of many enzyme-substrate interactions lie in the range of 10^{-3} to 10^{-5} M, which are analogous to the K_d values of rather low-affinity antigen-antibody interactions at the beginning of an immune response. However, since the antibody genes that are expressed on initial antigen stimulation are mutated and selected over the course of an immune response, antigen-antibody interactions late in an immune response may achieve a K_d as low as 10^{-12} M. This is an extraordinarily strong interaction; even if the antigen concentration is as low as 10^{-12} M, fully half of the antigen molecules will be bound to the receptor. Recent work with innate Toll-like receptors places the K_d of interaction of these receptors with their respective ligands in the range of 10^{-7} to 10^{-8} M.

The affinity of receptor-ligand interactions can be measured by equilibrium dialysis or surface plasmon resonance (SPR). Both of these methods are described in Chapter 20.

Interactions between Receptors and Ligands Can Be Multivalent

Many biological receptors, including B-cell receptors, have more than one ligand-binding site per molecule and are therefore characterized as **multivalent**. When both receptors and ligands are multivalent—as occurs when a bivalent B-cell receptor binds to two, identical ligands on a bacterial surface—the overall binding interaction is markedly stronger than that between identical, but univalent receptors and ligands. (Note, however, that the strength of binding through two identical receptor sites on one molecule to two identical ligands on the same cell may be somewhat less than twice the strength of binding through a single receptor site. This is because the bivalent binding may somewhat strain the geometry of the receptor or ligand and therefore slightly interfere with the "fit" of the individual interactions.)

Much of the benefit of multivalency results from the fact that noncovalent binding interactions are inherently reversible; the ligand spends some of its time binding to the receptor, and some of its time in an unbound, or "off" state. When more than one binding site is involved, it is less likely that all of the receptor sites will simultaneously be in the "off" state, and therefore that the receptor will release the ligand. Compare the univalent interaction in **Figure 3-2a** with the bivalent interaction in Figure 3-2b. The term **avidity** is used to describe the overall strength of the collective binding interactions that occur during multivalent binding.

(a)

Univalent
interaction

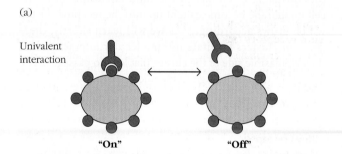

"On"   "Off"

(b)

Bivalent
interaction

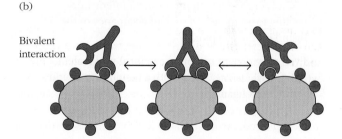

FIGURE 3-2 Univalent and bivalent (or multivalent) binding. (a) A univalent receptor approaches a multivalent antigen. The receptor exists in equilibrium with its ligand, represented here as a red circle. Part of the time it is bound (binding is in the "on" state), and part of the time it is unbound (binding is in the "off" state). The ratio of the time spent in the "on" versus the "off" state determines the affinity of the receptor-ligand interaction and is related to the strength of the sum of the noncovalent binding interactions between the receptor and the ligand. (b) Bivalent or multivalent binding helps to ensure that when one site momentarily releases the ligand, the interaction between the two molecules (or indeed, two cells) is not lost, as it is with monovalent binding.

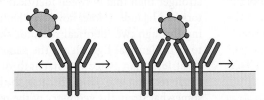

(a) Individual receptors bind a multivalent ligand and nucleate receptor cluster formation

(b) Multivalent ligand mediate cluster formation

FIGURE 3-3 Cell surface receptors cluster on binding multivalent antigens. (a) Receptors on a cell surface encounter ligands in a multivalent array, for example, on the surface of a bacterial cell. (b) The receptors migrate in the plane of the membrane to form clusters of occupied receptors.

Because of the fluid nature of the cell membrane, most membrane-bound antigen receptors function in a multivalent fashion, even if individual receptor molecules are monovalent in nature. **Figure 3-3** illustrates how a multivalent antigen, interacting with a membrane-bound receptor in a reversible way, incrementally stabilizes increasingly large receptor clusters on the cell membrane. This is because, as one or two stable connections are made, other receptors diffusing randomly in the fluid environment of the cell membrane become caught up into the receptor cluster, which grows accordingly. As we will learn shortly, these receptor clusters facilitate intramolecular interactions on the cytoplasmic side of the cluster that lead to passage of cell activation signals through to the nucleus.

Key Concepts:

- Multivalent binding interactions have increased avidity compared with monovalent interactions, leading to an increased time of occupancy of the ligand on the receptor.

- Even monovalent receptors can form multivalent clusters when interacting with a membrane-bound multivalent ligand.

Combinatorial Expression of Protein Chains Can Increase Ligand-Binding Diversity

Immune receptors are required to bind to an incredibly diverse range of antigens and signaling molecules, and the immune system has evolved an impressive number of strategies to do so within the constraints of a finite number of receptor coding sequences in the organism's DNA. One of those strategies involves re-using a single protein chain in combination with multiple partners to create a variety of different binding sites.

Some cytokine receptors are made up of two protein chains and, in most cases, both of these chains contribute

to the ligand-binding site. For example, three different class 1 cytokine receptor alpha (α) chains bind the same beta (β) chain to form receptors for interleukins 3 and 5 (IL-3 and IL-5) and granulocyte-macrophage colony-stimulating factor (GM-CSF), respectively (see **Figure 3-4**). Since the three-dimensional shape of the ligand-binding site depends on the manner in which the two chains interact with one another, this means that the β chain can bind to multiple cytokines, depending on the α chain with which it is paired. The IL-17 cytokine receptor family uses a similar strategy to bind multiple ligands, using a limited number of gene products.

An analogous strategy is adopted by the B- and T-cell receptors (BCRs and TCRs) of the adaptive immune

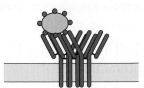

FIGURE 3-4 Combining one receptor chain with different partners allows increased receptor diversity and affinity while minimizing the need for new genetic information. A schematic diagram of the low-affinity and high-affinity receptors for the class 1 cytokines, interleukins 3 and 5 (IL-3 and IL-5), and granulocyte-macrophage colony-stimulating factor (GM-CSF) is shown. The cytokine receptor α subunits exhibit low-affinity binding and cannot transduce an activation signal from the cytokine to the interior of the cell. Noncovalent association of each subunit with a common β subunit yields a high-affinity dimeric receptor that can transduce a signal across the membrane on cytokine binding.

response as well as the Toll-like receptors (TLRs) of the innate immune response when binding antigens. Members of all three of these receptor families are made up of two protein chains per receptor molecule. In the case of the BCR, the two chains are called the heavy (H) chain and the light (L) chain. Both chains contact the antigen, and therefore contribute to antigen specificity. A similar diversity of antigen-combining sites can be generated by pairs of TCR chains or of TLR monomers.

Key Concept:

• By using different combinations of protein chains, the immune system can increase the variety of different receptor binding sites.

Adaptive Immune Receptor Genes Undergo Rearrangement in Individual Lymphocytes

So far, our discussions have not addressed the striking differences in the extents of diversity among innate versus adaptive immune receptor systems. We will describe these two types of receptor systems in detail in Chapters 4 and 6, respectively. However, we should note here that whereas the number of innate immune receptors of all types totals around 100 or less, the numbers of each of the two classes of adaptive immune receptors, BCRs and TCRs, are measured in units of *billions*.

The vast diversity of antigen-binding sites expressed by adaptive immune receptors is enabled by the unique ways in which T- and B-cell receptors are encoded in the genome. The DNA sequences that specify the antigen-binding sites in BCR and TCR protein chains are coded in short fragments in the germline DNA. These fragments are then stitched together in random combinations in each different B or T cell. A useful metaphor for thinking about the creation of the gene that encodes the entire antigen-binding site is that of creating a whole meal by selecting one item from each section of a menu and linking them together.

Adaptive immune receptor diversity is therefore generated by the random combinations of protein chains (heavy and light chains), *and* the random combinations of gene segments encoding each chain. These DNA segments are stitched together by *DNA recombination* in different combinations in each cell to create unique coding sequences for each heavy or light chain. Additional diversity at the junctions is generated during the DNA recombination process. A very similar process generates mature TCR and BCR genes and this mechanism is fully described in Chapter 6. For now, the point to grasp is that the extraordinary degree of the receptor diversity in the adaptive immune system is generated in part by recombination events *at the DNA level* (note—this is *not* RNA splicing!), the details of which are different in every adaptive immune cell!

Key Concepts:

• The number of different receptors in the adaptive immune system is strikingly large and measured in billions, as compared with the innate immune receptor repertoire that numbers about 100.

• The diversity in the adaptive receptor repertoire is achieved by the unique strategy of recombination between DNA sequences encoding small segments of receptor chains that are recombined in different ways in individual cells.

Levels of Receptor and Ligand Expression Can Vary during an Immune Response

One of the most striking features of the molecular logic of immune responses is that the level of cell surface expression of many immune receptors is coupled to the activation state of the cell. One very clear example of this phenomenon is the receptor for the cytokine interleukin 2 (IL-2), which was one of the earliest cytokines to be discovered and which provides a critical signal to lymphocytes to initiate proliferation and differentiation.

Most resting (i.e., non-antigen-activated) lymphocytes express a heterodimeric (two-chain), intermediate-affinity form of the IL-2 receptor, IL-2Rβγ. The affinity of the IL-2Rβγ form of the receptor is too low to enable it to bind IL-2 at physiological cytokine concentrations. However, once a lymphocyte has been activated by binding antigen through its BCR or TCR, the signal from the antigen-binding receptor causes an increase in the cell surface expression of a third chain of the IL-2 receptor, IL-2Rα, which then combines with the other two receptor chains on the cell surface (**Figure 3-5** and chapter opening photo).

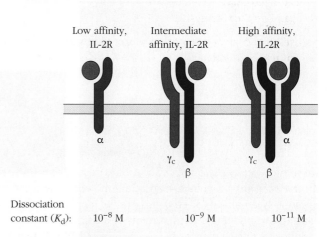

	Low affinity, IL-2R	Intermediate affinity, IL-2R	High affinity, IL-2R
Dissociation constant (K_d):	10^{-8} M	10^{-9} M	10^{-11} M

FIGURE 3-5 Comparison of the three forms of the IL-2 receptor. Signal transduction is mediated by the β and γ_c chains, but all three chains are required for high-affinity binding of IL-2. Dissociation constants reflect increased affinity of the three-chain IL-2R for IL-2.

Addition of this third chain converts the intermediate-affinity form of the IL-2 receptor to a high-affinity form, capable of responding to the cytokine levels found in the lymphoid organs. The functional corollary of this is that *only those lymphocytes that have already been activated by antigen binding have cytokine receptors of sufficiently high affinity to respond to physiological IL-2 concentrations.* In this way, the immune system both conserves energy and prevents the accidental initiation of an immune response to an irrelevant antigen.

Expression of some antigen receptors is also altered on cell activation. We will learn that B cells express two types of immunoglobulin antigen receptors on their cell membranes, IgM and IgD, that differ in the amino acid sequences of their *non–antigen-binding* regions. On cell activation, the expression of IgD drops significantly, whereas that of IgM remains constant. Some innate immune receptors also increase their expression patterns on cell activation.

Key Concept:

• Receptor expression patterns may change when a cell is activated, making it more or less responsive to particular signals.

Local Concentrations of Ligands May Be Extremely High during Cell-Cell Interactions

When considering the strength of the interactions between receptors and their ligands, it is important to consider the anatomical environment in which these interactions occur. This consideration becomes particularly important in the context of the adaptive immune system, where cells alter their locations and their binding partners multiple times during the induction and expression of an immune response. In particular, during the activation of a helper T (T_H) cell, the T_H cell and an antigen-presenting dendritic cell may remain in a complex with one another for 12 hours or more (see Figure 14-17). During this time, the two cells exchange cytokine signals.

Binding of the T cell to the dendritic cell induces redistribution of the microtubule-organizing center of the dendritic cell. That, in turn, causes redistribution of the cytokine-containing secretory organelles (the Golgi body and secretory vesicles) within the dendritic cell cytoplasm, such that the cytokine is released directly into the interface between the two cells before diffusing into the surrounding tissue fluid (**Figure 3-6**). In the T-cell/dendritic cell interaction, the activation is mutual, and over the course of the 12 hours of contact, the T cell will also redistribute its secretory apparatus and release cytokines directly into this tight intercellular interface that further activate the dendritic cell. An activated T_H cell might then dissociate from the dendritic cell and engage in a similar interaction with a B cell or a cytotoxic T cell, again redistributing its secretory apparatus so as to release cytokines directly into the new intercellular junction.

These cell-cell associations are common in the immune response, where the local concentration of cytokines at the cellular interface can be extremely high, much higher than in the tissue fluids generally. This concentration of activating cytokines at the junction between cells is an important mechanism for ensuring their efficient and accurate delivery.

Key Concept:

• Cell-cell interactions allow the directional release of ligands, creating locally high concentrations and increasing signal strength.

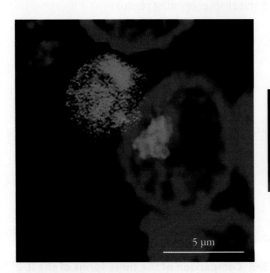
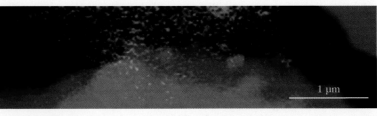

FIGURE 3-6 Polarized secretion of IL-12 (pink) by dendritic cells (blue) in the direction of a bound T cell (green). The higher magnification micrograph shows the secretion of packets of IL-12 through the dendritic cell membrane. *[Republished with permission of Rockefeller University Press from Pulecio et al., "Cdc42-mediated MTOC polarization in dendritic cells controls targeted delivery of cytokines at the immune synapse," Journal of Experimental Medicine **207**(2010): 2719-2732; Figure 3.]*

Many Immune Receptors Include Immunoglobulin Domains

There are few macromolecular structures in biology for which the relationship between structure and function is as clearly apparent as in the immunoglobulin domain. First described as a repeating unit in secreted antibody (immunoglobulin) molecules, this domain has since been characterized in a legion of molecules involved in recognition and adhesion functions (**Figure 3-7**).

Within each immunoglobulin (Ig) domain, several parallel β strands are arranged to form a pair of β sheets. Along the amino acid sequence of each β strand, hydrophobic and hydrophilic amino acids alternate so that the hydrophobic amino acids on one sheet are oriented toward those on the opposite sheet and the hydrophilic residues interact with the environment. The two β sheets thus form an extremely stable "hydrophobic sandwich" (see **Figure 3-8**) in which each domain is stabilized by hydrophobic interactions between the sheets and the protein as a whole is notably soluble.

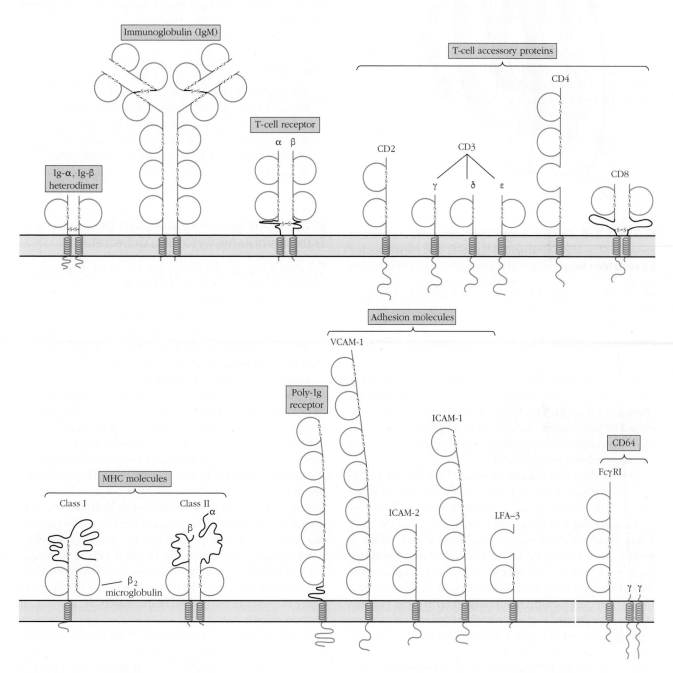

FIGURE 3-7 Some examples of proteins bearing immunoglobulin domains. Each immunoglobulin domain is depicted by a blue loop. Note the presence of the characteristic spacing of the disulfide bonds—about 67 amino acids separate the two cysteine residues—shown at the base of each amino acid domain loop.

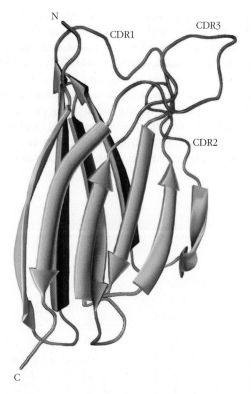

FIGURE 3-8 The immunoglobulin domain is made up of amino acid residues arranged in β sheets that are connected by variable loops. The two β pleated sheets are shown in two shades of blue. They are held together by hydrophobic interactions and by a conserved disulfide bond (not shown). The three loops of each variable domain, shown in red, vary considerably in length and amino acid sequence and make up the antigen-binding site. They are therefore referred to as *complementarity-determining regions* (*CDRs*).

Most immunoglobulin domains contain approximately 110 amino acids, and each β sheet contains three to six strands. The pairing of β sheets within each domain is stabilized by intrachain disulfide bonds. Neighboring domains are connected to one another by a stretch of relatively unstructured polypeptide chain.

So how does this domain structure facilitate the recognition functions of the many proteins that incorporate it? At the ends of each of the β sheets, more loosely folded polypeptide regions (loops) link one β strand to the next. These loosely folded regions can accommodate a variety of protein sequence lengths and structures without disrupting the overall backbone of the molecule. For example, in the BCR, those parts of the molecule that make contact with the antigen, the **complementarity-determining regions** (**CDRs**), are located in these loosely folded regions and are highlighted in red in Figure 3-8. It is clear that the CDRs can adopt a multitude of conformations without disrupting the essential β-sandwich framework structure of the molecule.

In other proteins, such as those illustrated in Figure 3-7, the same structural framework can also be used to support loops that contain amino acids important in cell adhesion,

or that operate as coreceptors. These properties explain why the immunoglobulin domain is found in so many proteins with recognition or adhesion functions. Together, these structurally related proteins comprise the **immunoglobulin superfamily**, a term that is used to denote proteins derived from a common primordial gene encoding the basic immunoglobulin domain structure.

Note that different types of recognition proteins may contain different numbers of immunoglobulin domains. For example, antibody heavy chains contain four or five domains, whereas the light chains contain only two. In each case, the antigen-binding function is located within the amino (N)-terminal domain of the protein.

Key Concept:

- The immunoglobulin superfamily of proteins includes BCRs, TCRs, adhesion molecules, and other receptors that function in the immune system. The immunoglobulin fold is composed of a pair of β sheets formed from β strands, which are connected by loops that define the protein-binding specificity.

Immune Antigen Receptors Can Be Transmembrane, Cytosolic, or Secreted

For most biologists, the word "receptor" conjures up images of a transmembrane protein, dutifully awaiting the diffusion of a soluble ligand into its vicinity so that it can respond to the ligand's signal. And indeed, the antigen receptors of the adaptive immune system, the BCR and TCR, are transmembrane proteins, as are many of the innate and cytokine receptors. However, this is not always the case.

First, it should be noted that the B-cell receptor exists in both membrane-bound and secreted forms, which will be described in more detail in the next section. The soluble form of the BCR is called an **antibody**, and is synthesized only after antigen stimulation of the relevant B cell. Both soluble antibodies and membrane-bound BCRs belong to the **immunoglobulin** family of proteins, and consist of two identical heavy (H) chains and two identical light (L) chains. In the membrane-bound form, hydrophobic residues at the carboxyl (C) terminus of the heavy chain anchor the receptor into the plasma membrane (**Figure 3-9a**). After antigen stimulation, the daughter cells of the original B cell begin to secrete soluble antibodies in which the hydrophobic residues at the C-terminal end of the antibody heavy chain are exchanged for more hydrophilic amino acid residues (Figure 3-9b).

Unlike their B-cell counterparts, TCRs are always found in membrane-bound form and are never secreted in soluble form. Moreover, as we will learn, the antigens that bind to TCRs are not normally soluble, but are usually located on the surface of antigen-presenting cells in the form of a complex with immune system molecules.

(a) Membrane-bound form (BCR)   (b) Secreted form (antibody)

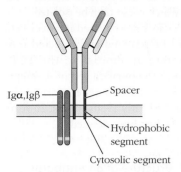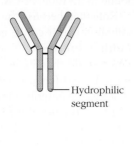

Igα,Igβ — Spacer

Hydrophobic segment

Cytosolic segment

Hydrophilic segment

FIGURE 3-9 The BCR exists in both membrane-bound (a) and soluble (b) forms. When a B cell is stimulated, it secretes a soluble form of its receptor that differs in sequence from the membrane-bound BCR at the C terminus but has the same antigen binding site as the membrane receptor. The hydrophobic region that anchors the membrane-bound form onto the surface of the B cell is replaced, by differential mRNA splicing, with a soluble (hydrophilic) amino acid sequence in the secreted antibody.

The locations of the antigen receptors of the innate immune system are significantly more variable than those of their adaptive immune system cousins. Some innate receptors, such as *mannose-binding lectin* (MBL), are entirely soluble, circulating in the tissue fluids. Others, such as members of the TLR family of innate receptors, may either be attached to the plasma membrane, or found in association with the cytosol-facing membrane of intracellular endosomes and lysosomes. Yet others are found within the cytosol, not associated with any intracellular membrane structures. These locations match the functions of the corresponding innate receptors: plasma membrane–bound TLRs bind antigens such as the lipopolysaccharides of gram-negative bacteria, whereas those located on the endosomal membranes are specific for molecules found only within the cytosol, such as CpG-rich bacterial DNA or viral RNA (see Chapter 4).

Key Concept:

• Immune receptors may be located on the plasma membrane, on intracellular membranes, in the cytosol, or even floating in the tissue fluids.

Immune Antigen Receptor Systems

The primary function of the immune system is to recognize and respond to pathogenic threats. The receptors of adaptive immunity, the TCR and BCR, are found on the surface of eukaryotic cells (lymphocytes) that can be grown in suspension and easily activated in vitro. The ease with which these receptors could be studied in the laboratory meant that many of the general concepts of eukaryotic signal transduction, as well as of receptor biochemistry, were first described for the lymphocyte receptor systems. Because of the central place occupied by the study of BCRs and TCRs in immunology, as well as the common structural features that they share with many immune receptor molecules, we have described them first, and in more detail than the receptors we discuss later.

Characterization of innate immune receptors is still ongoing. Although they are less diverse in number than the receptors of adaptive immunity, they are much more diverse in their cellular location, as described in the previous section. Later in this chapter, we will briefly explore their antigenic specificities and their scope, whetting the reader's appetite for more detailed discussion in Chapter 4.

The B-Cell Receptor Has the Same Antigen Specificity as Its Secreted Antibodies

The BCR is unique among the receptors of innate and adaptive immunity in that the same B cell is capable of making both membrane-bound and soluble forms of an immunoglobulin receptor that share the same antigen-binding site and are encoded by the same gene.

In Chapter 1, we described how experiments aimed at determining whether immunity resided in the soluble or cellular components of the blood resulted in the discovery of antibody molecules, proteins that conferred immunity to viruses, bacteria, and toxins, such as those secreted by diphtheria bacilli. However, it was not until the 1960s and 1970s that elegant experiments showed that these soluble antibody molecules were secreted by B lymphocytes bearing membrane-bound receptors that shared the same antigen-binding site as the secreted antibody (see Figure 3-9). As described above, the biochemical difference between the membrane-bound receptor and the secreted form of the antibody lies at the carboxyl terminus of the heavy chains. Secreted antibodies have a hydrophilic amino acid sequence of various lengths at the carboxyl terminus. In membrane-bound immunoglobulin receptors, this hydrophilic region is replaced by three sequentially arranged regions (see Figure 3-9a):

• An extracellular, hydrophilic "spacer" sequence of approximately 26 amino acids
• A hydrophobic transmembrane segment of about 25 amino acids
• A very short cytoplasmic tail

How is it possible to generate an antibody in two different forms that nonetheless share the same variable region and hence their antigen specificity? In the case of membrane versus secreted immunoglobulin, this puzzle is solved by differential mRNA splicing. The same B cell makes two different kinds of mRNA encoding the Ig heavy chain. One mRNA species expresses the coding information for the hydrophilic

C terminus of the secreted antibody and the other mRNA form expresses that for the hydrophobic C terminus of the membrane-bound receptor. When the B cell is resting, it makes the mRNA encoding only the membrane-bound form. On activation by antigen, signals from the antigen receptor instruct the RNA splicing machinery to make both types of heavy chain mRNA.

Experimentally, the ability to purify large quantities of soluble antibodies from the serum of vertebrates facilitated the isolation and characterization of the membrane-bound BCR. Those experiments were among the first to employ the common technique of immunoprecipitation, which takes advantage of the ability of antibody molecules to bind specifically to a target protein. Because antibodies are bivalent, one antibody can bind to more than one target molecule, which in turn can be bound to more than one antibody, forming a high molecular weight lattice that can be isolated as a precipitate (see Chapter 20 for a detailed description of immunoprecipitation).

In the experiments to purify the BCR, investigators first injected animals of one species (e.g., goat) with purified, soluble antibodies from a second species (e.g., mouse), generating goat antibodies to mouse immunoglobulins. These goat anti-mouse antibodies recognized all parts of the immunoglobulin (antibody) molecule. The goat anti-mouse antibodies were purified and then added to solubilized membrane preparations of mouse B lymphocytes that contained the BCR. Because antibodies share all the extracellular structures

of the BCR, these goat anti-mouse antibodies therefore bound to the mouse BCR, which could then be isolated.

Note that, because the TCR is not released in a secreted, soluble form, it could not be purified and injected into another species to generate antibodies to the TCR, which could then be used in the purification of its membrane-bound form. Hence, the basic biochemistry of the TCR was not described until the early 1980s, almost 20 years after the structure of the BCR had been defined.

The Three-Dimensional Structure of an Antibody Molecule

The Nobel Prize–winning experiments that first revealed the four-chain structure of antibodies are described in **Classic Experiment Box 3-1**.

All antibodies and B-cell receptors share a common structure of four polypeptide chains (**Figure 3-10a**), consisting of two identical light (L) chains and two identical heavy (H) chains. Each light chain is connected to its partner heavy chain by a disulfide bond between corresponding cysteine residues. The two heavy chains are also connected to one another via disulfide bonds located outside of the antigen-binding regions. The antigen-binding sites are made up of components of both the heavy and light chains, and the four-chain antibody molecule thus has two antigen-binding sites. Each light chain is made up of two immunoglobulin domains, whereas each heavy chain contains four or five Ig domains.

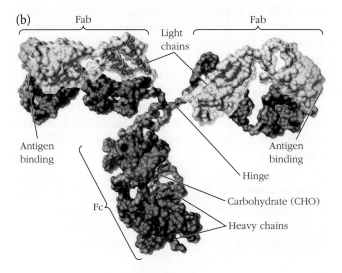

FIGURE 3-10 The structure of antibodies. (a) Each heavy chain (dark blue) and light chain (light blue) in an immunoglobulin molecule contains an amino-terminal variable (V) region that differs from one antibody to the next. The remainder of each chain in the molecule—the constant (C) regions—exhibits limited variation that defines two light-chain types and the five heavy-chain classes. Some heavy chains (γ, δ, and α) also contain a proline-rich hinge region that is flexible. The amino-terminal portions, corresponding to the V regions, bind to antigen; effector functions are mediated by the carboxyl-terminal domains. The

μ and ε heavy chains, which lack a hinge region, contain an additional domain in the middle of the molecule. CHO denotes a carbohydrate group linked to the heavy chain. (b) Clearly visible in this representation are the individual immunoglobulin domains, along with the open hinge structure in the center of the molecule. As in Figure 3-8, the surface-exposed location of the CDRs in the heavy and light chain is highlighted in red. (Abbreviations: Fab = antigen-binding portion of the antibody; contains paired V_L/V_H and C_L/C_H1 domains. Fc = non–antigen-binding region of the antibody, with paired C_H2/C_H2 and C_H3/C_H3 domains.)

Amino acid sequencing of antibody heavy and light chains revealed that the amino-terminal domain of each chain is extremely variable, while the sequence of the carboxyl-terminal sections can be classified into one of only a few major sequence types. The amino-terminal domains are thus referred to as the **variable**, or **V regions**, and the less variable, carboxyl-terminal regions are termed the **constant**, or **C regions**. Subscripts are used to identify the light-chain (V_L and C_L) and heavy-chain (V_H and C_H) regions. Since the heavy chains are significantly longer than the light chains, the variable region of the heavy chain occupies only one-quarter to one-fifth of the entire sequence, while the V_L segment occupies one-half of the light chain.

Visualized in three dimensions, the antibody molecule forms a Y shape with its two identical antigen-binding sites at the tips of the Y (Figure 3-10b). Each antigen-binding site is made up of amino acids derived from the variable domains of both the heavy chain and the light chain (V_H and V_L). The C_H1 and C_L domains serve to extend the antigen-binding arms of the antibody, maximizing the ability of the antibody to bind to more than one site on a multivalent antigen. An interchain disulfide bond between these two domains stabilizes the noncovalent interactions between the two chains. Figure 3-10b also shows how each of the chains is made up of a series of immunoglobulin domains, and highlights the location of the glycosylated region of the molecule. Glycosylation is important in maintaining the solubility of the secreted antibody and in preserving the overall structure and flexibility of the molecule.

Amino acid sequence analysis of antibody light chains further revealed that, within the variable region of the light chain, there were three regions of hypervariability. Similar regions of hypervariability were found in the heavy chain variable region sequences (**Figure 3-11a**). These hypervariable

(a)

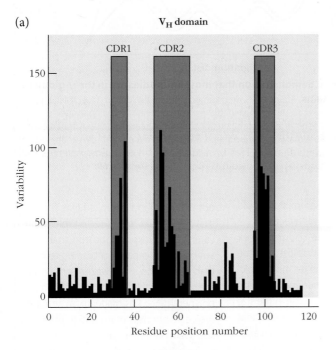

(b)

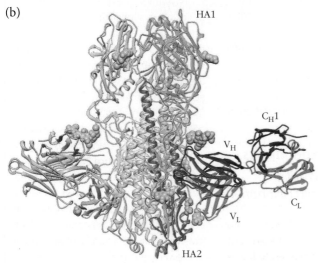

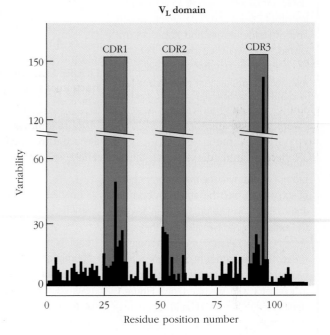

FIGURE 3-11 The presence of hypervariable regions in the amino acid sequences of antibody V_L and V_H domain complementarity-determining regions (CDRs). (a) Three hypervariable regions are present in both the heavy- and light-chain V domains. Variability (on the *y* axis) is defined as the number of different amino acids at each position divided by the frequency of the most common amino acid at that position. (b) X-ray analysis of antibody binding to the influenza hemagglutinin antigen shows that the hypervariable regions of the antibody represent the points of contact of the antibody with the antigen. *[PDB ID 4HLZ.]*

The Elucidation of Antibody Structure

First Set of Experiments

Arne Tiselius, Kai O. Pedersen, Michael Heidelberger, and Elvin Kabat

Since the late nineteenth century, it has been known that antibodies reside in the blood serum—that is, in that component of the blood which remains once cells and clotting proteins have been removed. However, the chemical nature of those antibodies remained a mystery until the experiments of Tiselius and Pedersen of Sweden and of Heidelberger and Kabat, in the United States, were published in 1939. They made use of the fact that, when antibodies react with a multivalent protein antigen, they form a multimolecular cross-linked complex that falls out of solution. This process is known as *immunoprecipitation* (see Chapter 20 for modern uses of this technique). They immunized rabbits with the protein ovalbumin (the major component of egg whites), bled the rabbits to obtain an anti-ovalbumin antiserum, and then divided the antiserum into two aliquots. They subjected the first aliquot to electrophoresis, measuring the amount of protein that moved different distances from the origin in an electric field. The blue plot in **Figure 1** depicts the four major protein subpopulations resolved by their technique. The first, and largest, is the albumin peak, the most abundant protein in serum, with responsibility for transporting lipids through the blood. They named the other peaks *globulins*. The two smaller peaks they denoted the α- and β-globulin peaks; the third globulin peak, γ-globulin, clearly represented a set of proteins in high concentration in the serum.

However, the most notable part of the experiment occurred when the investigators mixed the second serum aliquot with ovalbumin, the antigen. The antibodies in the serum bound to ovalbumin in a multivalent complex, which fell out of solution (i.e., it *precipitated*). The precipitate was then removed by centrifugation. Now that they had succeeded in removing the

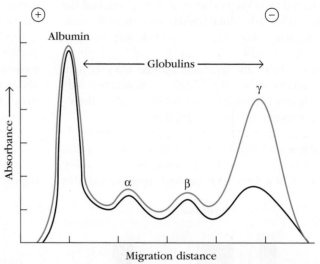

FIGURE 1 **Experimental demonstration that most antibodies are in the γ-globulin fraction of serum proteins.** After rabbits were immunized with ovalbumin (OVA), their antisera were pooled and electrophoresed, which separated the serum proteins according to their electric charge and mass. The blue line shows the electrophoretic pattern of untreated antiserum. The black line shows the pattern of antiserum that was first incubated with OVA to remove anti-OVA antibody and then subjected to electrophoresis. *[Data from Tiselius, A., and E. A. Kabat. 1939. An electrophoretic study of immune sera and purified antibody preparations.* Journal of Experimental Medicine **69:**119.]

antibodies from the antiserum, they could ask which protein peak was affected?

The black plot in Figure 1 illustrates their results. Very little protein was lost from the albumin peak, or from the α- and β-globulin peaks. However, immunoprecipitation resulted in a dramatic decrease in the size of the γ-globulin peak, demonstrating that the majority of their anti-ovalbumin antibodies could be classified as γ-globulins.

We now know that most antibodies of the IgG class are indeed found in the γ-globulin class. However, antibodies of other classes are found in the α- and β-globulin peaks, which may account for the slight decrease in protein concentration found after immunoprecipitation of these other protein peaks.

Second Set of Experiments

Rodney Porter, Gerald Edelman, and Alfred Nisonoff

Knowing the class of serum protein into which antibodies fall was a start, but immunochemists next needed to figure

out what antibodies looked like. The fact that they could form precipitable multivalent complexes suggested that each antibody was capable of binding to more than one site on a multivalent antigen. But the scientists still did not know how many polypeptide chains made up an antibody molecule nor how many antigen-binding sites were present in each molecule.

Two lines of experimentation conducted in a similar time frame on both sides of the Atlantic combined to provide the answers to these two questions. Ultracentrifugation experiments had placed the molecular mass of IgG antibody molecules at approximately 150,000 daltons (Da). Digestion of IgG with the enzyme papain produced three fragments, two of which were identical and a third that was clearly different (**Figure 2**). The third fragment, of approximately 50,000 Da, spontaneously formed crystals and was therefore named *Fragment crystallizable*, or Fc. By demonstrating that they could competitively inhibit the binding of antibodies to their antigen, the other two fragments

(continued)

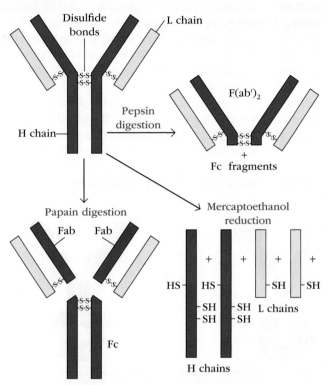

FIGURE 2 **Prototype structure of IgG, showing chain structure and interchain disulfide bonds.** The fragments produced by enzymatic digestion with pepsin or papain or by cleavage of the disulfide bonds with mercaptoethanol are indicated. Light (L) chains are in light blue, and heavy (H) chains are in dark blue.

were shown to retain the antigen-binding capacity of the original antibody. These fragments were therefore named Fab, or *Fragments antigen binding*. This experiment indicated that a single antibody molecule contained two antigen-binding sites and a third part of the molecule that did not participate in the binding reaction but could readily form crystals, so was likely to be very similar among different antibodies.

Use of another proteolytic enzyme, pepsin, resulted in the formation of a single fragment of 100,000 Da, which contained two antigen-binding sites that were still held together in a bivalent molecule. Because the molecule acted as though it contained two Fab fragments, but clearly had an additional component that facilitated the combination of the two fragments into one molecule, it was named F(ab')$_2$. Pepsin digestion does not result in a recoverable Fc fragment, which is apparently digested by the enzyme. However, F(ab')$_2$ derivatives of antibodies are often used in experiments in which scientists wish to avoid artifacts

resulting from binding of antibodies to Fc receptors on cell surfaces.

In the second set of experiments, investigators reduced the whole IgG molecule, using β-mercaptoethanol, in order to break disulfide bonds, and alkylated the reduced product so that the disulfide bonds could not spontaneously reform. They then used a technique called *gel filtration* to separate and measure the size of the protein fragments generated by this reduction and alkylation. (Nowadays, we would use sodium dodecyl sulfate [SDS]–polyacrylamide gels to do this experiment.) In this way, it was shown that each IgG molecule contained two heavy chains with molecular weights (MW) of 50,000 and two light chains of MW 22,000.

Now the challenge was to combine the results of these experiments to create a consistent model of the antibody molecule. To do this, the scientists had to determine which of the chains was implicated in antigen binding, and which chains contributed to the crystallizable fragments. Immunologists

often use immunological means to answer their questions, and this was no exception. They elected to use Fab and Fc fragments purified from rabbit IgG antibodies to immunize two separate goats. From these goats, they generated anti-Fab and anti-Fc antibodies, which they reacted, in separate experiments, with the heavy and light chains from the reduction and alkylation experiments. The answer was immediately clear.

Anti-Fab antibodies bound to both heavy and light chains, and therefore the antigen-binding site of the original rabbit IgG was made up of both heavy- and light-chain components. However, anti-Fc antibodies bound only to the heavy chains, not to the light chains of the IgG molecule, demonstrating that the Fc part of the molecule was made up of heavy chains only. Finally, careful protein chemistry demonstrated that the amino termini of the two chains resided in the Fab portion of the molecule. In this way, the familiar four-chain structure, with the binding sites at the amino termini of the heavy and light chain pairs, was deduced from some classically elegant experiments.

In 1972, Rodney Porter and Gerald Edelman were awarded the Nobel Prize in Physiology or Medicine for their work in discovering the structure of immunoglobulins.

REFERENCES

Edelman, G. M., et al. 1969. The covalent structure of an entire γG immunoglobulin molecule. *Proceedings of the National Academy of Sciences of the United States of America* **63:**78.

Fleishman, J. B., R. H. Pain, and R. R. Porter. 1962. Reduction of gamma-globulins. *Archives of Biochemistry and Biophysics* **Suppl 1:**174.

Heidelberger, M., and K. O. Pedersen. 1937. The molecular weight of antibodies. *Journal of Experimental Medicine* **65:**393.

Nisonoff, A., F. C. Wissler, and L. N. Lipman. 1960. Properties of the major component of a peptic digest of rabbit antibody. *Science* **132:**1770.

Porter, R. R. 1972. Lecture for the Nobel Prize for Physiology or Medicine 1972: structural studies of immunoglobulins. *Scandinavian Journal of Immunology* **34:**381.

Tiselius, A., and E. A. Kabat. 1939. An electrophoretic study of immune sera and purified antibody preparations. *Journal of Experimental Medicine* **69:**119.

regions correspond to the loosely folded polypeptide loops at the end of the variable region Ig domains (see Figure 3-9), and by x-ray crystallographic analysis these loops have been demonstrated to make direct contact with the bound antigen (Figure 3-11b). They have therefore been renamed the **complementarity-determining regions,** or **CDRs.** Of these CDRs, the CDR3 sequences of the heavy and light chains are more variable than CDRs 1 and 2, with CDR3 of the heavy chain being the most variable in sequence of all the Ig CDRs. The genetic basis of this finding lies at the heart of one of the most fascinating stories in immunology (Chapter 6).

To what extent does the amino acid sequence variability that characterizes the antigen-binding site extend to the non–antigen-binding portions of the heavy chain and light chains? Sequence analysis has taught us that there are only two major classes of light-chain constant region sequences, which are called **kappa** (κ) and **lambda** (λ). More extensive genetic analysis then further demonstrated that there are four subtypes of λ light chains, although the vast majority of λ light chains belong to subtype λ1. (We will revisit antibody classes in Chapter 6.)

Similarly, in contrast again to the extreme diversity of the variable region sequences, the remainder of the antibody heavy chain can be classified into one of only five major constant sequence types, or **isotypes** (**Figure 3-12**). Each isotype is designated by a particular Greek letter: mu (μ), delta (δ), gamma (γ), alpha (α), and epsilon (ε). Two antibody

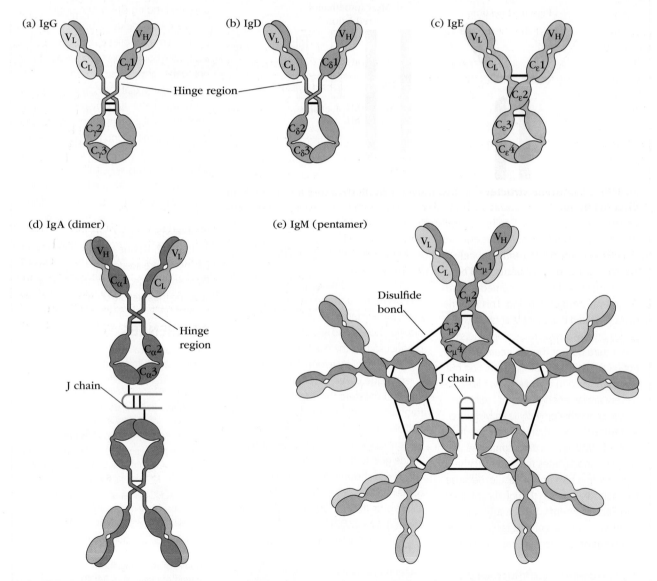

FIGURE 3-12 General structures of the five major classes of antibodies. Light chains are shown in lighter shades, and disulfide bonds are indicated by thick black lines. Note that the IgG, IgA, and IgD heavy chains contain four domains and a hinge region, whereas the IgM and IgE heavy chains contain five domains but no hinge region. The polymeric forms of IgM and IgA contain a polypeptide, called the *J chain*, that is linked by two disulfide bonds to the Fc region in two different monomers. Serum IgM is always a pentamer; most serum IgA exists as a monomer, although dimers, trimers, and even tetramers are sometimes present.

molecules that differ in their isotype expression are referred to as belonging to different antibody **classes**. For example, antibodies bearing μ isotype heavy chains belong to the IgM class; δ heavy chains define an antibody as belonging to the IgD class; γ, IgG class; α, IgA class; and ε, IgE class. Like the λ light chains, γ heavy chains are divided into four different subclasses (**Figure 3-13**), with the majority of γ chains belonging to the γ1 subclass. Each of these antibody constant regions is capable of binding to particular cell surface receptor(s), or to other immune molecules and mediating the effector functions of that antibody class.

Thus each antibody molecule is able to bind to a massively diverse array of antigens at its N terminus and to mediate a restricted number of different effector functions, such as phagocytosis (Chapter 4) or complement activation (Chapter 5), via the C-terminal part of the protein. The specific effector functions mediated by each antibody isotype are discussed in Chapter 12.

B cells express different classes of membrane immunoglobulins at particular developmental stages and under different stimulatory conditions. Immature B cells express only membrane IgM. Mature, unstimulated B cells express both membrane IgD and IgM. Interestingly, following antigen stimulation, IgD is lost from the cell surface. Differential expression of the membrane-bound and soluble forms of IgM and IgD is mediated via alternative RNA splicing.

However, expression of each of the other classes of antibody (IgG, IgA, and IgE) requires an additional, and irreversible, DNA recombination step. The regulation of the expression of particular heavy chain classes depends on cytokines released by T cells and antigen-presenting cells in the vicinity of the activated B cell and will be discussed further in Chapter 11.

BCR Coreceptors

When the BCR complex was isolated from the B lymphocyte membrane, using anti-immunoglobulin antibodies from a different species, additional molecules were co-immuno-precipitated with the heavy and light chains of the BCR. (*Co-immunoprecipitation* indicates that these molecules were noncovalently associated with the receptor on the B-cell membrane.) Further investigations showed that the BCR was noncovalently associated on the membrane with three transmembrane molecules: CD19, CD21, and CD81 (the latter also called TAPA-1) (**Figure 3-14**). Functional analysis of these molecules defined the CD21 molecule as participating in the antigen-binding activity of the BCR complex. CD21 is therefore referred to as a **coreceptor**.

Cooperative binding of antigen by the BCR and the CD21 coreceptor occurs when antigens are first identified as foreign by components of the innate immune system, prior to their interaction with B cells. The innate immune system can tag a pathogen for clearance by covalently binding a C3d protein fragment to the pathogen. C3d is a component of the complement system (described in detail in Chapter 5). The B-cell coreceptor, CD21, then specifically binds to C3d. Thus the same antigen is simultaneously bound *directly* via the BCR and *indirectly* through CD21 binding to C3d, thereby increasing the avidity of antigen binding to the cell. CD19 and CD81 do not specifically bind to antigens; rather, they participate in the passage of an antigen signal across the membrane of the B cell.

BCR Signal Transduction Mediators

Recall that the cytoplasmic tail of the BCR heavy chain is extremely short—for IgM, only three amino acids; how can such a short cytoplasmic tail pass a signal into the cytoplasm? This question was answered when each membrane-bound BCR molecule was also revealed to be noncovalently associated with a heterodimer, Igα,Igβ (CD79α,β; see Figure 3-14), that is responsible for transducing the antigen signal into the interior of the cell.

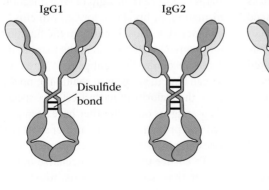
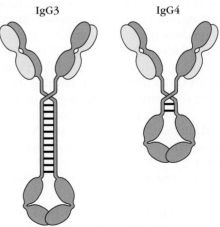
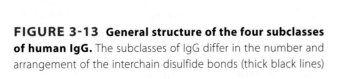

FIGURE 3-13 General structure of the four subclasses of human IgG. The subclasses of IgG differ in the number and arrangement of the interchain disulfide bonds (thick black lines) linking the heavy chains. A notable feature of human IgG3 is the high number of interchain disulfide bonds.

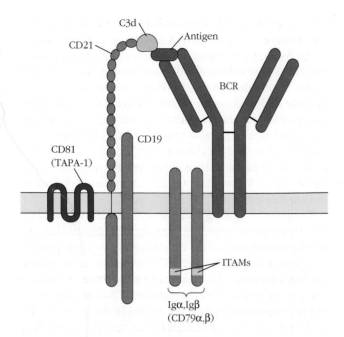

FIGURE 3-14 B-cell coreceptors require receptor-associated molecules and coreceptors for signal transduction. The CD21 coreceptor, which is associated with CD19, binds to the complement molecule C3d, which is covalently attached to the antigen. The interaction between CD21 on the B cell, and antigen-associated C3d, enhances the avidity of the B cell–antigen binding. BCRs require Igα, Igβ receptor-associated molecules for signal transduction. Phosphorylation of tyrosine residues in the ITAMs (yellow bands) of BCR signal transduction mediators allows the binding of downstream molecules and facilitates signal transduction from the receptors. Igα and Igβ bear intracytoplasmic ITAMs and, along with CD81 (TAPA-1), participate in downstream signaling events.

Igα and Igβ are transmembrane proteins with short, N-terminal regions and long intracytoplasmic tails containing regions called **immunoreceptor tyrosine-based activation motifs**, or **ITAMs**. ITAMs are short sequences of amino acids that include two tyrosine residues placed approximately 10 residues apart. These tyrosine residues become phosphorylated when the associated BCR molecule is activated by binding to its ligand. The phosphorylated tyrosine residues (pYs) can then serve as docking locations for the binding of downstream signaling molecules. Two different types of protein domains or motifs specifically bind to pY: SH2 domains and PTB domains. Proteins bearing these two domains will often participate in signal transduction relays.

We can therefore describe the B-cell receptor complex as being structurally and functionally divided into two components: a recognition component (the BCR and CD21), and a signal transduction component (Igα, Igβ). Note that antigen *signaling* of a B cell does not require the antigen (the receptor's ligand) to pass through the membrane. The receptor alters its conformation on ligand binding, and this conformational change enables the passage of the signal across the membrane

to the downstream molecular machinery. However, after the initial receptor-mediated signal has been transduced to the interior of the cells, antigen binding by the BCR then often also results in receptor-mediated endocytosis. Following introduction into the endosomal system, the antigen is broken down into peptides which are presented in association with immune system molecules for recognition by cooperating T cells (see Chapter 7).

Key Concepts:

- The antigen receptor on B cells (BCR) consists of two heavy chains and two light chains. Heavy chains contain one variable domain and four or five constant domains. Light chains contain one variable domain and one constant domain. Complementarity-determining regions (CDRs) in the variable domains make contact with antigen. Following antigen stimulation, B cells secrete immunoglobulins, also called *antibodies*, that bear the same antigen-binding site as the original BCR.

- The classes of antibody are defined by the sequence of their heavy chain constant regions. Antibodies of different classes perform different functions during an immune response.

- The BCR complex includes coreceptors such as CD21, which enhances the cell's ability to bind antigen that is complexed with complement components. The BCR complex also includes Igα and Igβ, ITAM-bearing proteins that transduce signals from antigen binding to the interior of the cell; and CD19 and CD81.

T-Cell Antigen Receptors Recognize Antigen in the Context of MHC Proteins

As discussed in Chapter 1, the function of B cells and antibodies is to rid the body of soluble or free toxins, viruses, and bacteria. In contrast, the function of T cells is to monitor the status of host cells for signs of viral infection, malignant transformation, or uptake of foreign proteins by pinocytosis and phagocytosis. This disparity in the functions of B and T cells is mirrored by differences in the manner in which B cells and T cells recognize their antigens. Whereas B cells and antibodies can bind to antigens in solution or on the surface of pathogens, the majority of naïve T cells are specialized to recognize their antigens only after they have been processed by the host cell and presented, in the form of short peptides, on the surface of a **professional antigen-presenting cell (pAPC)**.

To ensure that T cells specialize in binding cell-bound, rather than soluble antigens, most TCRs bind antigens only when they are in complex with plasma membrane–bound **major histocompatibility complex (MHC) proteins** (see Figure 2-7). Recall from Chapter 2 that APCs and other nucleated cells express MHC molecules, complexed with internal foreign and self-peptides, on their surface. This allows the T cell to browse those cells in search of its complementary MHC–peptide antigen.

The T cells that recognize MHC-presented peptides belong to the majority subclass of T cells that bear a heterodimeric receptor made up of one α chain and one β chain. Other T cells bear a heterodimeric γδ receptor and recognize specialized classes of antigens that may or may not be presented in a complex with MHC proteins.

When the TCR makes contact with its MHC–peptide antigen complex on the surface of a cell, the two cell membranes are brought into close apposition with one another, enhancing the binding of other molecules on the two cell surfaces. The biological consequences of this binding interaction are described in Chapter 10. Here, we briefly describe T-cell receptor structure. In **Classic Experiment Box 3-2,** we provide a description of the experiments that resulted in the isolation and characterization of the αβ TCR.

αβ and γδ TCRs

There are two types of TCRs, both of which are heterodimers. The majority of recirculating T cells bear heterodimeric receptors made up of an α chain and a β chain and are therefore called αβ TCRs. The αβ TCRs bind to complex antigens made up of an antigenic peptide fragment presented in a molecular groove on the surface of a MHC class I or class II molecule. A second subset of T cells expresses a different pair of protein chains of similar overall structure that generate the heterodimeric γδ T-cell receptor. Many T cells bearing γδ receptors have distinctive localization patterns, and many such cells home to tissues in the mucosa and skin. Some γδ T cells are capable of activation via nontraditional antigens that may, or may not, be localized on MHC platforms (see Chapter 7). Unless specified, the subsequent discussion of TCRs focuses on αβ TCRs.

TCR Structure

Like antibody light chains, all the TCR chains have two immunoglobulin domains (**Figure 3-15a**): one variable (V) domain

at the N terminus that serves as the antigen-binding site, and one constant (C) domain that lifts the antigen-binding site away from the plasma membrane. The TCR chains are held together by a disulfide bond that links two cysteine residues between the C domain and the plasma membrane. C-terminal to this disulfide is a transmembrane region of 21 or 22 amino acids, which anchors each chain in the plasma membrane. Both the α and the β chains have short, intracellular regions of five and nine residues, respectively.

As for BCRs, TCR V domains typically exhibit marked sequence variation, with the amino acid sequences of the remainder of each chain being conserved. Again analogous to immunoglobulins, each of the TCR V domains has three CDRs that make contact with the antigenic complex (Figure 3-15b and c). X-ray crystallographic analyses of TCR-antigen complexes show that the CDR1 and CDR2 regions of the αβ TCR make contact primarily with the MHC proteins, whereas the antigenic peptide appears to interact mainly with the CDR3 regions.

The extent of antigen-binding site diversity in αβ TCRs rivals that of immunoglobulins, whereas the number of different antigen-binding regions expressed by γδ TCRs may be more limited. Again, like antibodies, these diverse antigen-binding regions are generated by recombination of DNA fragments encoding short segments of the receptor V regions in different combinations so as to create a vast array of different binding site protein sequences (Chapter 6).

An important point of contrast between αβ and γδ TCRs lies in the relative length of the CDR3 of the TCR δ chain, which is significantly more variable and, on average, considerably longer than the CDR3 in TCR α, β, and γ chains. In this respect, the TCR δ chain most closely resembles the notably hypervariable CDR3 of the Ig heavy chain.

Like the BCR, the TCR is also glycosylated and recent experiments have indicated that the level of glycosylation may alter according to the degree of T-cell activation.

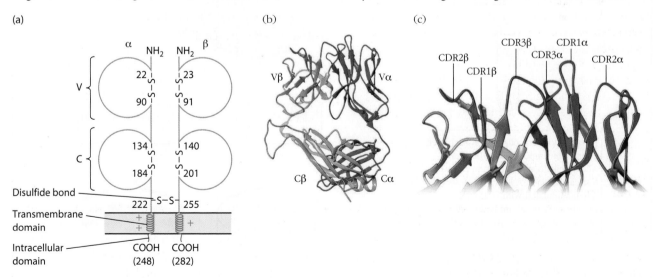

FIGURE 3-15 The three-dimensional structure of the αβ TCR. Parts (a) and (b) show the domain structure of an αβ TCR, illustrating the manner in which the domains of the two chains interact. Part (c) shows the relative location of the CDRs of the α and β chains. *[PDB ID 4GKZ.]*

The Discovery of the αβ T-Cell Receptor

Once scientists had established that the BCR was simply a membrane-bound form of the secreted antibody, the elucidation of BCR structure became a significantly more tractable problem. However, investigators engaged in characterizing the T-cell receptor (TCR) did not enjoy the same advantage, as the TCR is not secreted in soluble form. Understanding of TCR biochemistry therefore lagged behind that of the BCR until the 1980s, when an important scientific breakthrough—the ability to make monoclonal antibodies from artificially constructed B-cell tumors, or hybridomas—made the analysis of the TCR more technically feasible.

A *hybridoma* is a fusion product of two cells. B-cell hybridomas are generated by fusing antibody-producing, short-lived B lymphocytes with long-lived myeloma tumor cells (tumors of antibody-producing

---

① Generation of a T cell hybridoma with known antigen specificity

Immunize mouse with ovalbumin (OVA)

Wait several days
Remove lymph nodes
Culture lymph node T cells with OVA

② Add polyethylene glycol to induce fusion of antigen-specific T cells with long-lived T-cell line

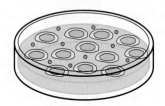

③ Dilute out fused cells so that each well contains a single T-cell hybridoma. Allow cells to divide to form clones and test each clone for its ability to secrete IL-2 when stimulated with ovalbumin peptides. Grow up individual clones of T-cell hybridomas that recognize OVA.

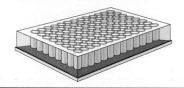

④ Production of antibodies that bind to TCR on the T-cell hybridoma

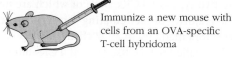

Immunize a new mouse with cells from an OVA-specific T-cell hybridoma

Wait several days
Isolate spleen

⑤ Fuse B cells from spleen with long-term B-cell line

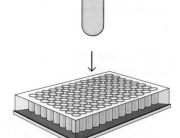

⑥ Selection and expansion of those long-term B-cell clones that secrete monoclonal antibodies that bind to the T-cell hybridoma

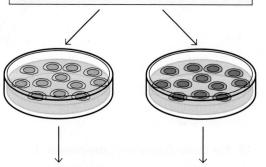

Collect monoclonal antibodies that bind to T-cell lines

(continued)

plasma cells) in order to generate long-lived daughter cells secreting large amounts of monoclonal antibodies. The term **monoclonal** refers to the fact that all of the cells in a given hybridoma culture are derived from a single clone of cells, and therefore carry the same DNA and produce the same antibody; details of the technology are described in Chapter 20. Although this technique was first developed for the generation of long-lived B cells, scientists working in the laboratory of John Kappler and Philippa Marrack also applied it to T lymphocytes.

The researchers began by immunizing a mouse with the protein ovalbumin (OVA), allowing OVA-specific T cells to divide and differentiate for a few days, and then harvesting the lymph nodes from the immunized animal. To enrich their starting population with as many OVA-specific T cells as possible, they cultured the harvested lymph node cells in vitro with OVA for several hours (**Figure 1**, step 1). After some time in culture, they fused these activated, OVA-specific T cells with cells derived from a T-cell tumor (Figure 1, step 2), thus

generating a number of long-lived **T-cell hybridoma** cultures that recognized OVA peptides bound to MHC proteins of the original mouse, which carried an MHC allele called *H-2^d*. They then diluted out the fused cells in each culture, generating several T-cell hybridoma lines in which all the cells in an individual hybridoma line derived from the product of a single fusion event (Figure 1, step 3). This is referred to as *cloning by limiting dilution*. In this way, they isolated a T-cell hybridoma that expressed a TCR capable of recognizing a peptide from OVA, in the context of MHC class II proteins from mice of the *H-2^d* strain.

These T cells could now be used as antigens and injected into a mouse (Figure 1, step 4). The spleen of this mouse was removed a few days later, and the mouse B cells were fused with B myeloma tumor cells (Figure 1, step 5). The investigators cloned the B-cell hybridomas and identified a B-cell hybridoma line that produced monoclonal antibodies that bound specifically to the T-cell hybridoma (Figure 1, step 6). Most important, these antibodies interfered with the T cell's ability to recognize its cognate

antigen (Figure 1, step 7). The fact that this monoclonal antibody inhibited TCR-antigen binding suggested that the antibody was binding directly to the receptor, and competing with the antigen for TCR binding. They then used these antibodies to immunoprecipitate the TCR from detergent-solubilized membrane preparations and purify the TCR protein (Figure 1, step 8).

Concurrent with these experiments, the laboratories of Stephen Hedrick and Mark Davis at the National Institutes of Health (NIH), and the laboratory of Tak Mak in Toronto, had been making headway searching for the genes encoding the T-cell receptor. These experiments, as well as subsequent work by Susumu Tonegawa, which completed the identification of the TCR genes, are described in detail in Chapter 6.

REFERENCES

Haskins, K., et al. 1983. The major histocompatibility complex–restricted antigen receptor on T cells. I. Isolation with a monoclonal antibody. *Journal of Experimental Medicine* **157:**1149.

Haskins, K., et al. 1984. The major histocompatibility complex–restricted antigen receptor on T cells. *Annual Review of Immunology* **2:**51.

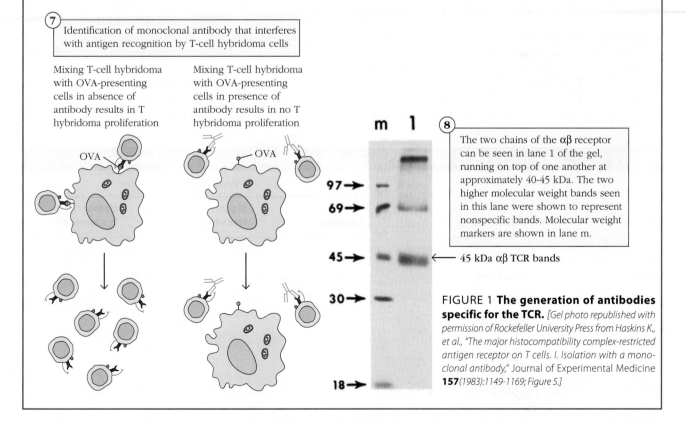

FIGURE 1 The generation of antibodies specific for the TCR. *[Gel photo republished with permission of Rockefeller University Press from Haskins K., et al., "The major histocompatibility complex-restricted antigen receptor on T cells. I. Isolation with a monoclonal antibody," Journal of Experimental Medicine* **157***(1983):1149-1169; Figure 5.]*

TCR Coreceptors

Like the BCR, the αβ TCR is also noncovalently associated with a number of accessory molecules on the cell surface (**Table 3-1**). Given the biochemical complexity of the antigen recognized by T cells, it makes sense that the molecules on the T cell charged with binding to the antigen-presenting cell should be similarly complex. However, only two of these accessory molecules, **CD4** and **CD8 (Figure 3-16),** have direct involvement in antigen recognition.

CD4 is a monomeric membrane glycoprotein that contains four extracellular immunoglobulin-like domains (D_1–D_4), a hydrophobic transmembrane region, and a long cytoplasmic tail. CD8 takes the form of a disulfide-linked αβ heterodimeric or αα homodimeric glycoprotein. Each chain consists of a single, extracellular, immunoglobulin-like domain, a stalk region, a transmembrane region, and a cytoplasmic tail. The extracellular domains of CD4 and CD8 bind to conserved regions of MHC class II and MHC class I molecules, respectively. The co-engagement of a single MHC molecule by both the TCR and its CD4 or CD8 coreceptor serves to enhance the avidity of TCR binding to its cellular target.

However, antigen binding through the αβ TCR, even when combined with binding by CD4 or CD8, is still insufficient to activate a naïve T cell that has had no prior contact with antigen. For full activation to occur, the **CD28** coreceptor must also engage its ligand, **CD80** or **CD86**, on the antigen-presenting cell (**Figure 3-17**). The status of CD28 as a coreceptor depends on its ability to engage a ligand on the antigen-presenting cell simultaneously with the TCR's engagement with the MHC-peptide complex. However, note that CD28 does not itself interact with the MHC–peptide antigen complex.

The cellular mechanisms that load peptides onto MHC molecules do not discriminate between self and foreign proteins. Therefore, as the T cell browses through the secondary lymphoid tissue, it will encounter a myriad of self-peptides presented on MHC molecules, and it is critically important that it not be inappropriately activated to induce an autoimmune response. The requirement that CD28 must engage its

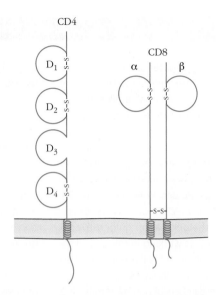

FIGURE 3-16 Structure of the CD4 and CD8 coreceptors. The monomeric CD4 molecule contains four Ig-fold domains; each chain in the CD8 molecule contains one. CD8 takes the form of an αβ heterodimer or an αα homodimer.

ligand on an antigen-presenting cell forms part of the safeguard against such inadvertent activation. How does this work?

The CD28 ligands, CD80 and CD86, are present only on professional antigen-presenting cells such as dendritic cells, macrophages, and activated B cells. What is more, the levels of their expression are significantly enhanced following antigen uptake resulting from innate immune receptor recognition by dendritic cells. Therefore, T cells will only recognize CD80 or CD86 to a significant extent on professional antigen-presenting cells that have been activated by, and are presenting biologically meaningful levels of foreign, MHC-complexed peptides.

The signaling events mediated through CD28, which include the stimulation of IL-2 synthesis by the T cell, are discussed fully in Chapter 10. Since inadvertent T-cell

TABLE 3-1	Selected T-cell accessory molecules participating in T-cell signal transduction			
		FUNCTION		
Name	Ligand	Adhesion	Signal transduction	Member of Ig superfamily
CD4	Class II MHC	+	+	+
CD8	Class I MHC	+	+	+
CD2 (LFA-2)	CD58 (LFA-3)	+	+	+
CD28	CD80, CD86	?	+	+
CTLA-4	CD80, CD86	?	+	−
CD45R	CD22	+	+	+
CD5	CD72	?	+	−

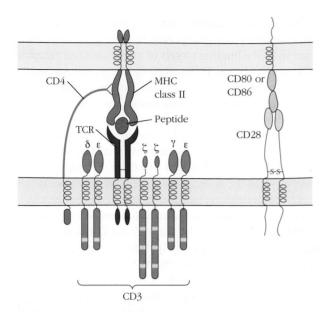

FIGURE 3-17 The T-cell receptor and coreceptor complex.
(a) T-cell receptors use CD3, a complex consisting of three dimers (a δε pair, a γε pair, and a ζζ or ζη pair) to transduce the signal to the interior of the cell. CD4 and CD8 coreceptors bind to MHC class II and MHC class I molecules, respectively. This figure shows CD4 binding to MHC class II. CD4 binding to MHC class II secures the connection between the T cell and the antigen-presenting cell, and also initiates a signal through the CD4 coreceptor. The CD28 coreceptor provides a critical component of the activation signal on binding to CD80 or CD86.

signaling can have such drastic physiological consequences, many checks and balances on T-cell activity have evolved that require the integration of multiple signaling pathways in order to achieve full T-cell function. Chapter 10 will describe how some of the same signaling molecules that are in play during T-cell activation are also used when the T-cell response is over and must be down-regulated.

TCR Signal Transduction Mediators

Again like its cousin the BCR, the TCR is paired with a molecular signal transduction complex. In T cells, this is composed of the CD3 complex, which is charged with the task of conveying signals received by the receptor into the interior of the cell (see Figure 3-17). The CD3 complex is made up of three dimers: a delta epsilon (δε) pair, a gamma epsilon (γε) pair, and a third pair that is made up of either a zeta zeta (ζζ) homodimer or a zeta eta (ζη) heterodimer.

Like Igα and Igβ, the cytoplasmic tails of the CD3 molecules are studded with ITAM motifs. Following activation-induced tyrosine phosphorylation, these ITAMs serve as docking sites for adapter proteins that transduce the antigenic signal into the cell's interior. Each of the γ, δ, and ε chains has a single ITAM unit and the ζ and η chains have three apiece. As we will see below, variability in the extent of phosphorylation of these tyrosine residues can result in variability in the strength of the signal that is produced.

Key Concepts:

- Most T cells bear a heterodimeric αβ T-cell receptor, which recognizes a complex ligand constituted of a peptide bound to an MHC protein. Some T cells instead express heterodimeric γδ receptors, which recognize nontraditional antigens. Each TCR chain (α, β, γ, or δ) has a variable domain and a constant domain. Hypervariable regions in the variable domains (CDRs) make contact with the MHC-peptide complex.

- The TCR coreceptors CD4 and CD8 interact with invariable regions of MHC class II and class I molecules, respectively. The TCR coreceptor, CD28, binds to its ligands, CD80 and/or CD86, on antigen-presenting cells. This binding is necessary for the activation of naïve T cells.

- The CD3 complex is noncovalently associated with the TCR and is responsible for transducing the antigen-binding signal to the interior of the cell. Like the Igα, Igβ complex on B cells, members of the CD3 complex bear ITAMs, which become phosphorylated on tyrosine residues when activated.

Receptors of Innate Immunity Bind to Conserved Molecules on Pathogens

One of the principal points of contrast between innate and adaptive immune responses is the speed with which they occur (see Table 1-4). The rapid innate immune response enables it to accomplish two things: first, in many antigen encounters, the innate immune system eliminates the invading pathogen by phagocytosis or cytotoxicity before the cells of the adaptive immune system are even aware of the incursion. Second, depending on the nature of the receptors that are engaged by the innate immune cells, the cells of innate immunity secrete different arrays of cytokines, molecular messengers that advise the adaptive immune system on what type of adaptive immune response, if any, might be required.

Chapter 1 described how the adaptive immune system tailors its response to fit the nature of the invading pathogen; B cells generate antibodies that neutralize toxins and viruses and opsonize bacteria for phagocytosis, while cytotoxic T cells kill virally infected host cells and activated helper T cells secrete cytokines that instruct the immune system on which cells or other defenses are needed to eliminate a particular invader.

As described in more detail in Chapter 4, cells of the innate immune system play a critical role in tailoring the nature of the adaptive immune response to best meet the challenges of a particular pathogen. For example, depending on which PRRs are engaged by the pathogen, dendritic cells, macrophages, and innate lymphoid cells will secrete different cytokines that advise the adaptive immune system on the optimal strategy.

Furthermore, if a dendritic cell secretes a cytokine that instructs a helper T cell, there will be two layers of cytokines involved in the signaling relay; one secreted by the dendritic cell that tells the helper T cell what cytokines to secrete, and one by the T cell that tells B and cytotoxic effector cells which of them is required and precisely what their function must be. The details of these relays of molecular "instructions" that are so central to immune cell differentiation will be revealed in later chapters. For now, it is important to grasp the concept that *innate immunity provides the mechanistic blueprint for adaptive immunity to follow that will best deal with each pathogen.*

In contrast to innate immune responses, adaptive immune responses are not optimized until 7–10 days following antigen encounter, and further refinements continue until deep into the response (see Chapter 11). In large measure, this is because of differences in the expression of innate versus adaptive immune system receptors. Specifically:

- Unlike the receptors of the innate immune system, BCRs and TCRs are expressed clonally (one cell, one receptor). The clonal nature of adaptive immune receptors coupled with the enormous diversity of the adaptive immune receptor repertoires means that any given adaptive immune receptor will be expressed only rarely. In order to generate sufficient cells of any particular specificity, B and T cells must first divide multiple times prior to differentiating and expressing their effector functions. In contrast, cells of the innate immune system do not need time to divide, and differentiate rapidly to a fully activated state.
- A single innate immune cell may express multiple types of innate receptors, and many innate cells can express the same innate immune receptor. This means that every antigen that is capable of binding to a particular innate receptor can be bound immediately by many cells.
- Many different cell types in addition to the classically recognized innate immune populations express innate immune receptors, including keratinocytes, epithelial cells, and even B and T lymphocytes.

Each innate immune receptor recognizes molecules that are shared among whole classes of microbes, whereas each BCR or TCR binds unique antigenic shapes specific to individual antigens. The molecules recognized by innate receptors are generally essential to microbial survival and therefore cannot be changed to evade an immune response. Their recognition by vertebrate host immune responses probably reflects evolutionary selection for these specificities, thus enabling the host to protect itself against broad classes of pathogens.

Interestingly, B and T lymphocytes express innate immune receptors in a nonclonal fashion simultaneously with the clonally expressed BCR or TCR. Lymphocytes therefore must integrate signals from both sets of receptors before embarking on a response.

Another important distinction between innate and adaptive immune system recognition has emerged from detailed analysis of the binding targets of innate immune receptors. Whereas BCRs and TCRs all recognize antigens expressed in solution or on the surfaces of cells, some innate receptors are expressed on internal cell membranes, such as those of the endosomal system, or even in the soluble portion of the cytosol, and are capable of recognizing antigens such as viral RNA that might never be expressed in the extracellular environment. Chapter 4 describes the innate immune receptors and the molecules they recognize in more detail.

The molecules recognized by innate immune cells are referred to as **pathogen-associated molecular patterns (PAMPs)**. Biochemically, many PAMPs are composed of recurrent molecular motifs (patterns) of pathogen molecules such as the cell wall carbohydrates of fungi or bacteria, the repeating protein motifs of flagellin, and nonmethylated CpG-rich sequences in DNA. (These patterns can be expressed by microbes whether or not they are pathogenic, hence they are sometimes referred to as *microbe-associated molecular patterns* [MAMPs].) Some innate immune receptors are also capable of recognizing antigens associated with dead or dying cells, referred to as *damage-associated molecular patterns* (DAMPs).

The receptors that recognize PAMPs, MAMPs, and DAMPs are collectively referred to as *pattern recognition receptors (PRRs)*. Several families of cellular PRRs contribute to the activation of innate immune responses, including the C-type lectin receptors, the Toll-like receptors, the RIG-like receptors, and the NOD-like receptors (described in Chapter 4). Each PRR has a distinct repertoire of specificities for conserved PAMPs; the PRR families and some of their known ligands are listed in **Table 3-2**.

Key Concepts:

- Receptors of innate immunity are expressed in a nonclonal fashion while those of adaptive immunity are clonally expressed. Whereas innate immune receptors are expressed on a diversity of cells, both immune and nonimmune in their primary functions, BCRs and TCRs are borne only by B and T cells, respectively.

- Receipt of signals from innate immune receptors instructs the innate immune cell to destroy the invader as well as to secrete cytokines that inform the subsequent adaptive immune response.

- Pattern recognition receptors (PRRs), receptors functioning in innate immunity, recognize pathogen-associated molecular patterns (PAMPs) that are common to whole classes of microbes, both those that are pathogenic and nonpathogenic. Some innate immune receptors recognize molecules released by damaged, dying, or dead host cells—damage-associated molecular patterns (DAMPs).

TABLE 3-2	Pattern recognition receptor families				
	Full name	**Cellular location(s)**	**Ligands**	**Cellular functions**	**Icon**
TLR	Toll-like receptor	Plasma membrane, endosomes, lysosomes	Microbial carbohydrates, lipoproteins, fungal mannans, bacterial flagellin, viral RNA, self-components of damaged tissues, etc.	Production of antimicrobials, antivirals, and cytokines; inflammation	TLR4/4
CLR	C-type lectin receptor	Plasma membrane	Carbohydrate components of fungi, mycobacteria, viruses, parasites, and some allergens	Phagocytosis, production of antimicrobials and cytokines; inflammation	
RLR	Retinoic acid-inducible gene-I (RIG-I)-like receptor	Cytosol	Viral RNA	Production of interferons and cytokines	MDA5, CARD, CARD
NLR	Nucleotide oligomerization domain (NOD)-like receptor	Cytosol	Fragments of intracellular or extracellular bacteria cell wall peptidoglycans	Production of antimicrobials and cytokines; inflammation	NLRP3, PYD NBD NAD LRR
ALR	Absent-in-melanoma (AIM)-like receptor	Cytosol and nucleus	Viral and bacterial DNAs	Production of interferons and cytokines	AIM2

Cytokines and Their Receptors

The hundreds of millions of cells that comprise the vertebrate immune system are distributed throughout the body of the host (see Chapter 2). In such a widely dispersed organ system, the various components must be able to communicate efficiently with one another, so that immune responses can be regulated and the right cells can home to the appropriate locations where they can take the necessary measures to destroy invading pathogens.

Proteins that communicate among cells of the immune system are referred to as **cytokines**. The interaction of a cytokine with its receptor on a target cell can induce a wide variety of responses. Cytokines can cause changes in the expression of adhesion molecules and **chemokine receptors** on the target membrane, thus allowing the cell to move from one location to another. Cytokines can also signal an immune cell to increase or decrease the activity of particular enzymes or to change its transcriptional program, thereby activating it to proliferate and differentiate, or to modulate its effector functions. Finally, cytokines can instruct a cell when to survive and when to die.

The sensitivity of a target cell to a particular cytokine is determined by the presence of specific cytokine receptors. In general, cytokines and their fully assembled receptors exhibit high affinity for one another, with dissociation constants ranging from 10^{-8} to 10^{-12} M. Because their receptor affinities can be so high and because cytokines are often secreted in close proximity to their receptors, the secretion of very few cytokine molecules can mediate powerful biological effects.

In this section, we briefly introduce the six major families of cytokines and their receptors by describing their cells of origin, their target cells, their receptors, and some fundamental information about their major effects (see Table 3-3). Additional information about many of the cytokines and their physiological effects will be found in subsequent chapters in the context of specific types of immune responses, as well as in Appendix II.

Cytokines Are Described by Their Functions and the Distances at Which They Act

In an early attempt to classify cytokines, immunologists began numbering them in the order of their discovery, and naming them **interleukins**. This name reflects the fact that interleukins communicate between (Latin, *inter*) white blood cells (*leukocytes*). Unfortunately, many cytokines that were named prior to this attempt at rationalizing nomenclature have resisted reclassification. Students will find a comprehensive list of known cytokines in Appendix II. **Chemokines** are cytokines that attract cells with the appropriate chemokine receptors to regions where the chemokine concentration is highest. The classification and nomenclature of chemokines is more logical than that of interleukins, and is based on their biochemical structures. A list of the known chemokines and their receptors can be found in Appendix III.

TABLE 3-3	The six major cytokine families		
Family name	**Representative members of family**	**Comments**	**Receptor icon**
Interleukin-1 family	IL-1α, IL-1β, IL-1Ra, IL-18, IL-33	IL-1 was the first noninterferon cytokine to be identified. Members of this family include important inflammatory mediators.	IL-1 — Interleukin-1-family receptors
Class 1 (hematopoietin) cytokine family	IL-2, IL-3, IL-4, IL-5, IL-6, IL-7, IL-12, IL-13, IL-15, IL-21, IL-23, GM-CSF, G-CSF, growth hormone, prolactin, erythropoietin/hematopoietin	Members of this large family of small cytokine molecules exhibit striking sequence and functional diversity.	IL-2 — Class 1 receptors — α, γ, β
Class 2 (interferon) cytokine family	IFN-α, IFN-β, IFN-γ, IL-10, IL-19, IL-20, IL-22, IL-24	While the IFNs have important roles in antiviral responses, all are important modulators of immune responses.	IFN-β — Interferon-type receptors (class 2)
Tumor necrosis factor family	TNF-α, TNF-β, CD40L, Fas (CD95), BAFF, APRIL, LT-β	Members of this family may be either soluble or membrane-bound; they are involved in immune system development, effector functions, and homeostasis.	TNF — TNF receptors
Interleukin-17 family	IL-17 (IL-17A), IL-17B, IL-17C, IL-17D, IL-17F	This is the most recently discovered family; members function to promote neutrophil accumulation and activation, and are proinflammatory.	IL-17A — IL-17 receptors
Chemokines (see Appendix III)	IL-8, CCL19, CCL21, RANTES, CCL2 (MCP-1), CCL3 (MIP-1α)	All serve chemoattractant function.	CCL19 — Chemokine receptors — G protein

Cytokines can be described according to the distance between the secreting cell and the recipient cell. Cytokines that must pass through the bloodstream before reaching their target are referred to as **endocrine**. Those that act on cells near the secreting cell, such that the cytokine merely has to diffuse a few angstroms to a nearby cell, are referred to as **paracrine**. Sometimes, a cell needs to receive a signal through its own membrane receptors from a cytokine that it, itself, has secreted. This type of signaling is referred to as **autocrine**. Many cytokines act over a short distance in an autocrine or paracrine fashion.

Key Concepts:

- Cytokines are proteins that communicate between cells of the immune system.

- Cytokines can be described according to their functions and the distances over which they act.

Cytokines Exhibit the Attributes of Pleiotropy, Redundancy, Synergism, Antagonism, and Cascade Induction

A cytokine that induces different biological effects depending on the nature of the target cells is said to have **pleiotropic** activity, whereas two or more cytokines that mediate similar functions are said to be **redundant**. Cytokine **synergy** occurs when the combined effect of two cytokines on cellular activity is greater than the additive effects of the individual cytokines. In some cases, the effects of one cytokine inhibit or **antagonize** the effects of another. **Cascade induction** occurs when the action of one cytokine on one target cell induces that cell to produce one or more additional cytokines (**Figure 3-18**).

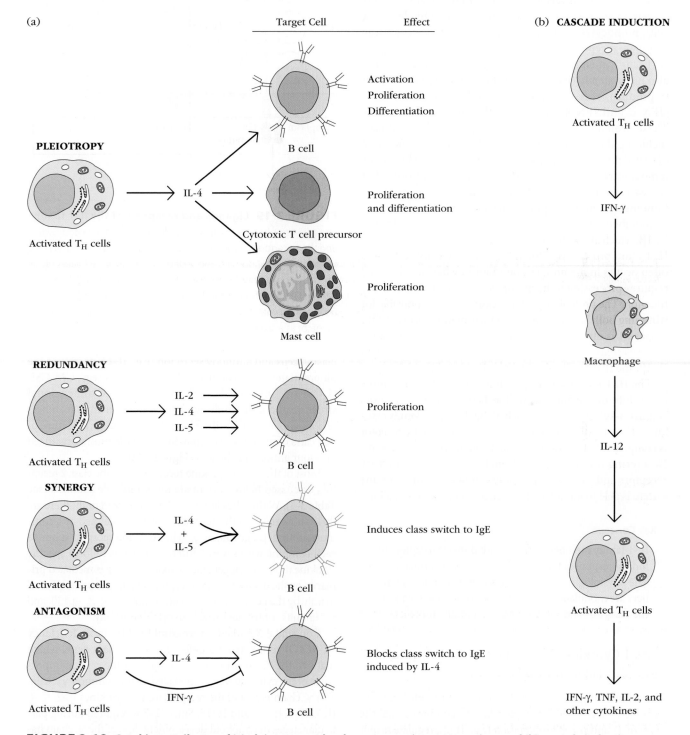

FIGURE 3-18 Cytokine attributes of (a) pleiotropy, redundancy, synergism, antagonism, and (b) cascade induction.

Cytokines of the IL-1 Family Promote Proinflammatory Signals

Cytokines of the **interleukin 1 (IL-1) family** are typically secreted very early in the immune response by dendritic cells, monocytes, and macrophages following recognition by innate receptors of viral, parasitic, or bacterial antigens. IL-1 family members are generally proinflammatory. IL-1 also has systemic (whole body) effects and signals the liver to produce other cytokines, such as the type I interferons (IFN-α and IFN-β), IL-6, and the chemokine CXCL8. These proteins further induce multiple protective effects, including the destruction of viral RNA and the generation of a systemic fever response (which helps to eliminate many temperature-sensitive bacterial strains). In addition, IL-1 serves as an intermediary between the innate and adaptive immune systems by helping to activate both T and B cells.

Members of the IL-1 cytokine family include IL-1α and IL-1β, which are both synthesized as 31-kDa precursors (pro-IL-1α and pro-IL-1β). Pro-IL-1α is biologically active and often occurs in membrane-bound form, whereas pro-IL-1β requires processing to its mature soluble form before it can function. The proteolytic enzyme caspase-1 is responsible for trimming both IL-1 precursors to their mature forms. Active caspase-1 is often found as part of a protein complex called an *inflammasome* (see Chapter 4). In addition to IL-1α and IL-1β, the IL-1 family also includes IL-18 and IL-33.

The IL-1 receptor family includes monomeric receptors and inhibitory ligands, as well as heterodimeric receptors (**Figure 3-19**). However, for each of the three IL-1 family cytokines, IL-1, IL-18, and IL-33, only one heterodimeric receptor is competent to transduce a signal from the cognate cytokine. By altering the ratio of functional receptors to inhibitory receptors and/or soluble inhibitory molecules, the immune system is able to modulate the strength of cytokine signaling.

Class 1 Cytokines Share a Common Structural Motif But Have Varied Functions

The cellular origins and target cells of **class 1 cytokines** are extremely diverse. For example, class 1 cytokines signal the onset of T- and B-cell proliferation (e.g., IL-2), regulate helper T-cell functions (e.g., IL-4), call for B-cell differentiation to

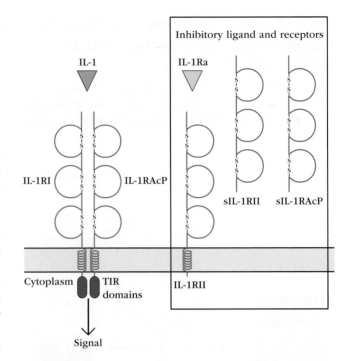

FIGURE 3-19 Ligands and receptors of the IL-1 family. The two agonist ligands, IL-1α and IL-1β, are represented by IL-1 and the antagonist ligand by IL-1Ra. The IL-1 receptor, IL-1RI, has a long cytoplasmic domain and, along with IL-1RAcP, activates signal transduction pathways. IL-1Ra acts as an IL-1 inhibitor. IL-1RII is a nonactivating IL-1 receptor. IL-1RAcP can also inhibit IL-1 signals by cooperating with IL-1RII in binding IL-1 either on the plasma membrane or as a soluble molecule (sIL-1RAcP).

plasma cells and antibody secretion (e.g., IL-6), or initiate the differentiation of particular leukocyte lineages (e.g., GM-CSF, G-CSF). Significant homology in the three-dimensional structure of class 1 family cytokines defines them as members of a single protein family. The defining structural feature of this class of cytokines is a four-helix bundle motif, organized into four antiparallel helices (**Figure 3-20**).

Most of the class 1 cytokine receptors (and the class 2 receptors discussed below) are made up of multiple subunits (see **Table 3-4**). Within the class 1 cytokine receptor family, there are three subfamilies, with each subfamily being defined by a common subunit: β_c, gp130, or γ_c. The common receptor subunits each combine with several different α chains to form specific cytokine receptors. In general, cytokine binding is mediated primarily by the α subunit, and receptor signaling is accomplished mainly by the accompanying chain, although the β_c, gp130, and γ_c subunits contribute variably to cytokine affinity.

Congenital **X-linked severe combined immunodeficiency (X-SCID)** results from a defect in the γ_c-chain gene, which maps to the X chromosome, highlighting the critical importance of the cytokine signals mediated by the common γ_c chain. The γ_c chain is part of the receptor for the cytokines IL-2, IL-4, IL-7, IL-9, IL-13, and IL-15. Since IL-7 is required during particular stages in lymphoid development, loss of the γ_c chain has a devastating effect on the generation of adaptive immunity.

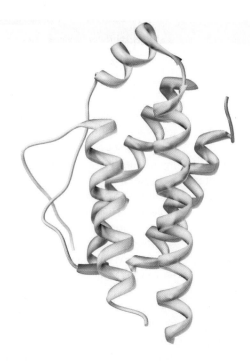

FIGURE 3-20 The four-helix bundle is the defining structural feature of the class 1 family of cytokines. Ribbon representation of the crystallographic structure of human IL-2, the defining member of the class 1 family—showing the four α helices of the class 1 cytokines pointing in alternating directions. *[PDB ID 1M47.]*

TABLE 3-4	Common subunits of class 1 family cytokine receptor subfamilies
Common cytokine receptor subunit	**Cytokines recognized by receptors bearing that common subunit**
γc	IL-2, IL-4, IL-7, IL-9, IL-15, IL-12
βc	IL-3, IL-5, GM-CSF
gp130	IL-6, IL-11, LIF, OSM, CNTF, IL-27

Key Concepts:

• The class 1 cytokine family is the largest family of cytokines, and members mediate diverse effects, including proliferation, differentiation, and antibody secretion. The class 1 cytokine family members share a common, four-helix bundle structure.

• Receptors for cytokines from the class 1 cytokine family are made up of at least two chains. In all cases, cytokine specificity is mediated by binding to the α chain and signaling is mediated by one of three alternative chains: γc, βc, or gp130.

Class 2 Cytokines Are Grouped into Three Families of Interferons

There are three subfamilies of interferons. **The type I interferon family** is composed of the *interferon-α* (IFN-α) members, a set of about 20 related proteins, and *interferon-β* (IFN-β). IFN-α and -β are dimeric, helical proteins of 18 to 20 kDa. They are secreted by activated macrophages and dendritic cells and also by virally infected cells following PRR-mediated recognition of viral components. Type I interferons bind to plasma membrane interferon receptors on many different cell types, inducing the production of ribonucleases that destroy viral RNA and inhibiting cellular protein synthesis. In this way, interferons prevent virally infected cells from replicating or from making new viral particles, thus limiting the spread of the viral infection. Type I interferons are used in the treatment of a variety of human diseases, most notably hepatitis infections (see **Clinical Focus Box 3-3**).

Type II interferon, more usually known as *interferon-γ* (IFN-γ), is produced by activated T cells and cytotoxic natural killer (NK) cells and is also released as a dimer. IFN-γ induces the activation of macrophages, with subsequent destruction of intracellular pathogens, and stimulates the differentiation of cytotoxic T cells. IFN-γ is also the key cytokine made by helper T cells of the T_H1 subset, which generally support cell-mediated immunity. IFN-γ is used clinically to bias the adaptive immune system toward a cytotoxic response in diseases such as leprosy and toxoplasmosis, in which antibody responses are less effective than those that destroy infected cells. If made in excess or inappropriately, it can also stimulate delayed-type hypersensitivity (DTH), which can result in tissue damage (see Chapter 15). The type II interferon family of cytokines also includes **IL-10**, which is secreted by monocytes and by T, B, and dendritic cells. IL-10 acts generally to regulate immune responses. It shares structural similarities with IFN-γ, and these similarities enable it to bind to the same class of receptors.

A third class of interferons, the **type III interferon**, *or* **interferon-λ** (IFN-λ) family, was discovered in 2003. This class of interferons is secreted by dendritic cells of a special type called *plasmacytoid* dendritic cells. Like type I interferons, the type III interferons upregulate the expression of genes controlling viral replication and host cell proliferation. All three types of interferon increase the expression of MHC proteins on the surface of cells, thus enhancing their antigen presentation capabilities.

Members of the interferon receptor family are heterodimers that share similarly located, conserved cysteine residues with members of the class 1 cytokine receptor family. Recent work has shown that the interferon receptor family consists of 12 receptor chains that, in their various assortments, bind no fewer than 27 different class 2 cytokines.

Members of the class 1 and class 2 cytokine receptor families share a signal transduction modality that will be described below (see Figure 3-25).

Cytokine-Based Therapies

The availability of purified cloned cytokines, monoclonal antibodies that bind and neutralize cytokines, and soluble cytokine receptors that inhibit cytokine binding to target cells offers the prospect of specific clinical therapies that modulate the immune response by modifying cytokine signaling. A number of strategies that interfere with cytokine signaling are now being employed in the clinic.

Some address disease-caused, or disease-related, cytokine deficiencies by supplementing the host's own cytokines. For example, cytokines from the interferon (IFN) family have been used clinically in this way.

High concentrations of IFN-α (also known by its trade names Roferon-A and Intron A) have been used for a number of years to treat hepatitis C and hepatitis B. Hepatitis C therapy commonly involves the use of IFN-α in combination with an antiviral drug such as ribavirin. The clearance time of IFN-α is lengthened by using it in a form complexed with polyethylene glycol (PEG), called *pegylated* interferon.

IFN-β has emerged as the first drug capable of producing clinical improvement in multiple sclerosis (MS). Young adults are the primary target of this autoimmune neurological disease, in which nerves in the central nervous system (CNS) undergo demyelination. This results in progressive neurological dysfunction, leading to significant and, in many cases, severe disability. The disease is often characterized by periods of nonprogression and remission alternating with periods of relapse. Treatment with IFN-β provides longer periods of remission and reduces the severity of relapses. Magnetic resonance imaging (MRI) studies of CNS damage in treated and untreated patients revealed that MS-induced damage was less severe in a group of IFN-β–treated patients than in untreated patients.

A particularly successful clinical application of IFN-γ is in the treatment of the hereditary immunodeficiency, chronic granulomatous disease (CGD). In these patients, there is a failure to generate microbicidal oxidants (H_2O_2, superoxide, and others). CGD features a serious impairment of the ability of phagocytic cells to kill ingested microbes, and patients with CGD suffer recurring infections with a number of bacteria (*Staphylococcus aureus, Klebsiella, Pseudomonas*, and others) and fungi such as *Aspergillus* and *Candida*. The administration of IFN-γ significantly reverses this defect. Before interferon therapy, the standard treatment for CGD included attempts to avoid infection, aggressive administration of antibiotics, and surgical drainage of abscesses. Administration of IFN-γ to patients with CGD significantly reduces the incidence of infections, the infections that are contracted are less severe, and the average number of days spent by patients in the hospital is reduced.

The use of interferons in clinical practice is likely to expand as more is learned about their effects in combination with other therapeutic agents. Although interferons, in common with other cytokines, are powerful modifiers of biological responses, the side effects accompanying their use are fortunately relatively mild. Typical side effects include flu-like symptoms, such as headache, fever, chills, and fatigue.

In addition to treatment programs in which administration of cytokines relieves clinical syndromes, cytokine-related reagents are now in frequent use to treat pathologies that result from an overabundance of cytokines. These reagents fall into two major categories: monoclonal antibodies that block the binding of cytokines to their receptors and soluble receptors that prevent cytokine binding to cell-bound, active receptors.

For example, soluble TNF-α receptor (etanercept [Enbrel]) and monoclonal antibodies against TNF-α (infliximab [Remicade] and adalimumab [Humira]) have been used to treat rheumatoid arthritis and ankylosing spondylitis in more than a million patients. These anti–TNF-α drugs reduce proinflammatory cytokine cascades; help to alleviate pain, stiffness, and joint swelling; and promote healing and tissue repair.

As powerful as these reagents may be, interfering with the normal course of the immune response is not without its own intrinsic hazards. Reduced cytokine activity brings with it an increased risk of infection and malignancy, and the frequency of lymphoma is slightly higher in patients who are long-term users of the first generation of TNF-α–blocking drugs.

In addition, the technical problems encountered in adapting cytokines for safe, routine medical use are far from trivial. As described above, during an immune response, interacting cells may produce extremely high local concentrations of cytokines in the vicinity of target cells, but achieving such high concentrations over a clinically significant time period, when cytokines must be administered systemically, is difficult. Furthermore, many cytokines have a very short half-life so frequent administration may be required. Finally, cytokines are extremely potent biological response modifiers, and they can cause unpredictable and undesirable side effects.

The use of cytokines and anticytokine therapies in clinical medicine holds great promise, and efforts to develop safe and effective cytokine-related strategies continue, particularly in those areas of medicine that have so far been resistant to more conventional approaches, such as inflammation, cancer, organ transplantation, and chronic allergic disease.

TNF Family Cytokines May Be Soluble or Membrane-Bound

Members of the *tumor necrosis factor* (**TNF**) **family** of cytokines have been shown to regulate the development, effector function, and homeostasis of cells of the skeletal and neuronal systems, as well as the immune system. Although some TNF family members are soluble proteins, others are transmembrane proteins, with short, intracytoplasmic N-terminal regions, and longer, extracellular C-terminal regions. The extracellular region typically contains a canonical TNF-homology domain responsible for interaction with the cytokine receptors. In some cases, the same cytokine exists in both soluble and membrane-bound forms.

TNF-α and lymphotoxin-α (LT-α, also known as TNF-β) are both secreted as soluble proteins. TNF-α (frequently referred to simply as TNF) is a proinflammatory cytokine, produced in response to infection, inflammation, and environmental stressors by activated macrophages, and also by lymphocytes, fibroblasts, and keratinocytes (skin cells). LT-α is produced by activated lymphocytes and can deliver a variety of signals. On binding to neutrophils, endothelial cells, and osteoclasts (bone cells), LT-α delivers activation signals; in other cells, binding of LT-α can lead to increased expression of MHC glycoproteins and of adhesion molecules.

We will also encounter five immunologically significant, membrane-bound members of the TNF cytokine family throughout this book. **Lymphotoxin-β** is important in lymphocyte differentiation. BAFF and APRIL deliver important signals in the context of B-cell development and homeostasis. CD40 ligand (CD40L) is a cytokine expressed on the surface of T cells that binds its receptor, CD40 on B cells, delivering a differentiation signal to the B cells. Fas ligand (FasL), or CD95L, induces cell death (apoptosis) on binding

to its cognate receptor, Fas, or CD95. Thus members of the membrane-bound TNF cytokine family regulate lymphocyte functions from development to death and everything in between, highlighting their importance in both basic and clinical immunology. It is therefore not surprising that TNF cytokine pathways are the targets of some of the most effective drugs (e.g., adalimumab [Humira] and etanercept [Enbrel]) that have been developed to counter diseases of the immune system.

Whether membrane-bound or in soluble form, cytokines of the TNF family assemble into trimers, which may be homo- or heterotrimeric. **Figure 3-21** shows how interaction of one trimeric TNF molecule with its receptors induces receptor trimerization and activation.

Although most TNF receptors are type 1 membrane proteins (their N termini are outside the cell), a few family members are cleaved from the membrane to form soluble receptor variants. Alternatively, some lack a membrane-anchoring domain at all, or are linked to the membrane only by covalently bound, glycolipid anchors. These soluble forms of TNF family receptors are known as "decoy receptors," as they are capable of intercepting the signal from the ligand before it can reach a cell, effectively blocking the signal. This is a theme that we have encountered before in our consideration of the IL-1 receptor family.

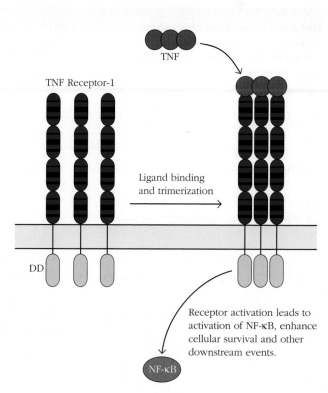

FIGURE 3-21 Binding of TNF to TNFR-1 induces trimerization and activation of downstream events. The precise nature of the downstream events that are stimulated depends on the constellation of adapter and signal-transducing proteins that are available in the particular cell type.

The IL-17 Family of Cytokines and Receptors Is the Most Recently Identified

IL-17A, the first member of the **IL-17 family** to be detected, is released by activated T cells and its receptors are found on neutrophils, keratinocytes, and other nonlymphoid cells. IL-17A binding to a cell instructs it to secrete cytokines that support a proinflammatory state. The T cells secreting these cytokines belong to a unique lineage, the T_H17 cell subset, which appears to occupy a locus at the interface of innate and adaptive immunity (see Chapter 4 and Chapter 10).

Table 3-5 shows the currently known IL-17A homologs, most of which promote the release of proinflammatory and neutrophil-mobilizing cytokines. However, IL-17E (IL-25) provides an exception to this general rule, instead promoting the differentiation of the anti-inflammatory T_H2 cell subclass, while suppressing further T_H17 cell responses.

In general, members of the IL-17 family exist as homodimers, although heterodimers of IL-17A and IL-17F have been described. The IL-17 receptor family is composed of five protein chains—IL-17RA, IL-17RB, IL-17RC, IL-17RD, and IL-17RE—which are variously arranged into homo- and heterodimeric and trimeric units to form the complete receptor molecules (**Figure 3-22**). Members of the IL-17 receptor family are all transmembrane proteins.

TABLE 3-5	Expression and known functions of members of the extended IL-17 cytokine family			
Family member	Other common names	Receptor(s)	Cell types expressing cytokines of the IL-17 family	Main function(s)
IL-17A	IL-17 and CTL-8	IL-17RA and IL-17RC	T_H17 cells, CD8$^+$ T cells, γδ T cells, NKT cells, LTi-like cells, neutrophils, Paneth cells	Induces expression of proinflammatory cytokines, neutrophil recruitment, and antimicrobial peptide induction; promotion of T-cell priming and antibody production
IL-17B	NA	IL-17RB	Chondrocytes, neurons	Induces expression of proinflammatory cytokines and neutrophil recruitment
IL-17C	NA	IL-17RE	T_H17 cells, DCs, macrophages, keratinocytes	Induces expression of proinflammatory cytokines and neutrophil recruitment
IL-17D	NA	Unknown	T_H17 cells, B cells	Proinflammatory cytokine production
IL-17E	IL-25	IL-17RA and IL-17RB	T_H17 cells, CD8$^+$ T cells, mast cells: eosinophils, epithelial cells, endothelial cells	Induces T_H2 and T_H9 responses; suppresses T_H1 and T_H17 responses; eosinophil recruitment
IL-17F	NA	IL-17RA and IL-17RC	T_H17 cells, CD8$^+$ T cells, γδ T cells, NK cells, NKT cells, LTi-like cells, epithelial cells	Neutrophil recruitment and immunity to extracellular pathogens; induces expression of proinflammatory cytokines

Data from Gaffen, S. L. 2009. Structure and signalling in the IL-17 receptor family. *Nature Reviews Immunology* **9**:556; and Iwakura, Y., H. Ishigame, S. Saijo, and S. Nakae 2011. Functional specialization of interleukin-17 family members. *Immunity* **34**:149.

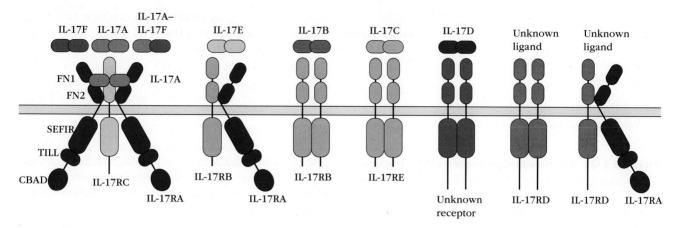

FIGURE 3-22 The IL-17 family of cytokines and their associated receptors. The cytokines that form the IL-17 family share a highly conserved, tightly folded dimeric structure, with four conserved cysteines. The five proteins that make up the IL-17 receptor family are IL-17RA, IL-17RB, IL-17RC, IL-17RD, and IL-17RE, which are arranged into homo- and heterodimers and trimers. Note the presence of fibronectin (FN), *SEF/IL-17R* (SEFIR), *TIR-like loop* domain (TILL), and *C/EBP* β activation domains (CBAD) that are important in mediating downstream signaling events.

Chemokines Induce the Directed Movement of Leukocytes

Chemokines are a structurally related family of small cytokines that bind to cell-surface receptors and induce the movement of leukocytes up a chemokine concentration gradient and toward the chemokine source. This soluble factor–directed cell movement is known as **chemotaxis**, and molecules that elicit such movement are referred to as **chemoattractants**. Some chemokines also display innate affinity for the carbohydrates named *glycosaminoglycans*, located on endothelial cell membranes. This property enables them to bind to the inner surfaces of blood vessels and set up a chemoattractant gradient along blood vessel walls, directing leukocyte movement to the site of an infection.

Leukocytes change their pattern of expression of chemokine receptors during the course of an immune response. Antigen stimulation induces the expression of chemokine receptors on activated leukocytes that direct them to the secondary immune organs, in which they undergo differentiation to mature effector cells. Leukocytes also alter the expression of chemokine receptors during the course of an immune response in order to facilitate movement within the secondary organs. For example, Chapter 11 describes how B cells move from lymph node follicles to the interfollicular zone under the influence of chemokines during B-cell activation. Once differentiation is complete, the leukocytes move out into the affected tissues to fight the infection, responding to different chemokine gradients with each movement.

Chemokines are relatively low in molecular weight (7.5–12.5 kDa) and structurally homologous. The tertiary structure of chemokines is constrained by a set of highly conserved disulfide bonds; the positions of the cysteine residues determine the classification of the chemokines into four different structural categories (**Figure 3-23**). Within any one category, chemokines may share 30% to 99% sequence identity. The grouping of chemokines into the subclasses shown in Figure 3-23 has functional, as well as structural, significance. For example, seven of the human CXC chemokines share the same receptor (CXCR2), attract neutrophils, are angiogenic (promote the formation of new blood vessels), and have greater than 40% sequence identity. See Appendix III for a more comprehensive tabulation of chemokines and the cells that express receptors for them and can respond to them.

Chemokine receptors bear structural homology to the receptors for the hormones adrenaline and glucagon. This class of receptors threads through the membrane seven times and transduces the ligand signal via interactions with a polymeric GTP/GDP-binding "G protein." Receptors of this type are known either as *G* **protein–coupled receptors (GPCRs)** or seven-pass transmembrane receptors. Chemokine GPCRs are classified according to the type of chemokine they bind. For example, the CC receptors (CCRs) recognize CC chemokines, the CXCRs recognize CXC chemokines, and so on. Chemokine receptors bind to their respective ligands quite tightly ($K_d \simeq 10^{-9}$ M). Interestingly, many chemokine receptors have been demonstrated to bind more than one chemokine from a particular family, and several chemokines are able to bind to more than one receptor. For example, the receptor CXCR2 recognizes seven different chemokines, and the CC chemokine CCL5 can bind to both CCR3 and CCR5.

> **Key Concept:**
> - Chemokines act on G protein–coupled receptors to promote chemoattraction, the movement of immune system cells into, within, and out of lymphoid organs.

Class	Structural signature	Names	Number (n) in class
CXC	............CX__C............C........C.........	CXCL#	15
CC	............C___C............C........C.........	CCL#	25
XC	...................C.....................C	XCL#	2
CX₃C	............CXXXC............C........C.........	CX3CL1	1

FIGURE 3-23 Disulfide bridges in chemokine structures. Chemokines are small proteins that share two or four conserved cysteine residues at particular points in their sequence that form intrachain disulfide bonds. The number of cysteines as well as the positions of the disulfide bonds determine the subclass of these cytokines, as shown. The overscores indicate the cysteines between which disulfide bonds are made. The naming of chemokines in part reflects the cysteine-determined class (see Appendix III).

A Conceptual Framework for Understanding Cell Signaling

If ligand binding to a receptor is to lead to a change in cellular function, the binding energy of the ligand-receptor interaction must be translated, or **transduced**, into a biochemical change in the cell's behavior, location, or metabolism. Changes induced by ligand binding may include variations in the activity of transcription factors in that cell, which in turn cause changes in the expression of intracellular, membrane-bound, or secreted proteins; initiation of cellular programs leading to differentiation and division; alterations in the activity of proteasomes that induce destruction of particular proteins; fluctuations in the secretory or phagocytic activity of the target cell, and/or changes in the cell's metabolic activity that ready it for division, differentiation, or even for death.

The process by which ligand binding to a cellular receptor is translated into a modification in cellular activity is referred to as **signal transduction**. Because the activity of the immune system is entirely dependent on ligand-induced alterations in immune cells, it should come as no surprise that many of the advances in our understanding of the biochemistry of signal transduction were made using the immune system as a model.

In the last decade or so, we have come to understand that common strategies, manifesting themselves as sequences of biochemical events, are shared across many signaling transduction pathways. For example, ligand binding will usually induce either an alteration in the conformation or in the polymerization status of the receptor molecule. As we discussed above in our description of B- and T-cell receptor binding, the primary receptor may be joined by coreceptors in strengthening the bonds between ligand and receptor, or between receptor- and ligand-bearing cells. We have also come to appreciate the vital role of tyrosine phosphorylation of the intracytoplasmic regions of either receptors or receptor-associated molecules in the passage of the signals from the exterior to the interior of the cell, as well as the function of adapter proteins that transmit the signal by bringing other proteins together without themselves being covalently altered.

Developing a grasp of these shared themes helps to provide a scaffold on which to hang the details of the specific signal transduction pathways that will be described in subsequent chapters. In this section, we therefore provide a conceptual framework for students to use when learning about the many signal transduction pathways they will encounter in their study of immunology and, more broadly, of cell biology. Where appropriate, examples will be drawn from particular immune receptor–ligand interactions (**Overview Figure 3-24**).

Note that the **upstream** components of a signaling pathway are those closest to the receptor; the **downstream** components are those closest to the effector molecules that determine the outcome of the pathway—for example, the transcription factors or enzymes whose activities are modified on receipt of the signal.

Ligand Binding Can Induce Dimerization or Multimerization of Receptors

The region of a receptor that makes molecular contact with a ligand is referred to as the *receptor binding site*. The binding site of most receptors represents only a small fraction of the size of the receptor molecule, and other parts of the receptor molecule are responsible for transmitting the signal induced by ligand binding across the cell membrane to the interior of the cell. To this end, binding of a ligand to a receptor binding site will often induce a conformational change in distal parts of the receptor molecule that alters its ability to bind to other receptor molecules. For example, in the case of some cytokine receptors, such as those that bind interferon molecules, interferon-induced receptor dimerization is the first step in the signal transduction cascade. For B- and T-cell receptors, binding of receptor to antigen facilitates the formation of clusters of receptor molecules on the cell surface.

Key Concept:

- Ligand binding can cause a conformational change in a receptor molecule that induces the receptor to bind to other receptors.

Concepts in Lymphocyte Signaling

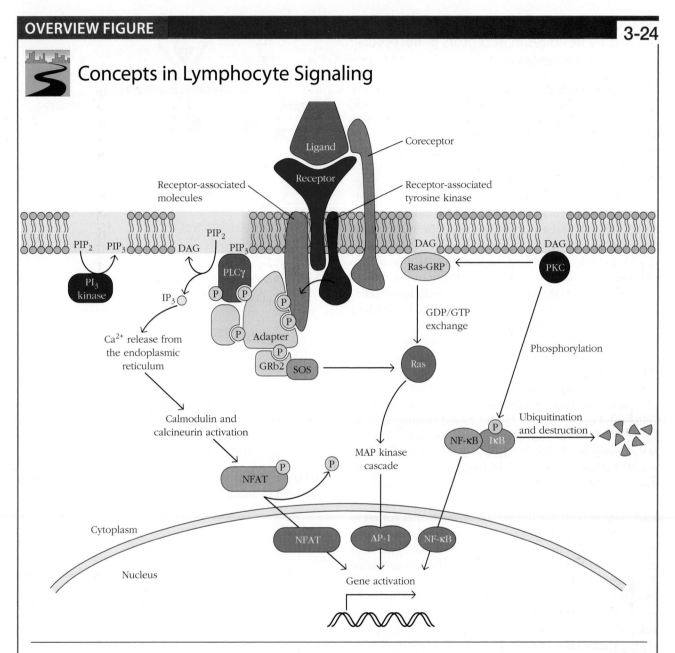

Ligand binding to receptors on a cell induces a variety of downstream effects, many of which culminate in transcription factor activation. Here we illustrate a few of the pathways that are addressed in this section. Binding of receptor to ligand induces clustering of receptors and signaling molecules into regions of the membrane referred to as *lipid rafts* (red). Receptor binding of ligand may be accompanied by binding of associated coreceptors to their own ligands, and causes the activation of receptor-associated tyrosine kinases, which phosphorylate receptor-associated proteins. Binding of downstream adapter molecules to the phosphate groups on adapter proteins creates a scaffold at the membrane that then enables activation of a variety of enzymes including phospholipase Cγ (PLCγ), PI$_3$ kinase, and additional tyrosine kinases. PLCγ cleaves phosphatidylinositol bisphosphate (PIP$_2$) to yield inositol trisphosphate (IP$_3$), which interacts with receptors on endoplasmic reticulum vesicles to cause the release of calcium ions. These in turn bind to calmodulin, which binds and activates the phosphatase calcineurin. Calcineurin dephosphorylates the transcription factor NFAT, allowing it to enter the nucleus. Diacylglycerol (DAG), remaining in the membrane after PLCγ cleavage of PIP$_2$, binds and activates protein kinase C (PKC), which phosphorylates and activates enzymes leading to the destruction of the inhibitor of the transcription factor NF-κB. With the release of the inhibitor, NF-κB enters the nucleus and activates a series of genes important to the immune system. Binding of the two GEF proteins, Ras-GRP and SOS, to the signaling complex allows for the activation of Ras, which in turn initiates the phosphorylation cascade of the MAP kinase pathways. This leads to the entry of a third set of transcription factors into the nucleus, and activation of the transcription factor AP-1.

Ligand Binding Can Induce Phosphorylation of Tyrosine Residues in Receptors or Receptor-Associated Molecules

The cytoplasmic regions of some ligand receptors, such as c-Kit, which recognizes the stem cell factor critical to early lymphocyte development, have an intrinsic tyrosine kinase activity that is activated on ligand binding. Such receptors are known as *receptor tyrosine kinases*, or RTKs. In RTKs, ligand binding induces receptor dimerization, which brings the tyrosine residues of one receptor within range of the kinase activity of the partner molecule. This is followed by the reciprocal phosphorylation of the cytoplasmic regions of each of the receptor molecules by its dimerization partner.

Other cytokine receptors, such as those that recognize interferons, associate noncovalently with tyrosine kinase enzymes via the cytoplasmic regions of the receptor molecules. These tyrosine kinases, like the RTKs, are activated on ligand-receptor binding and phosphorylation; they then phosphorylate transcription factors that dimerize and enter the nucleus, allowing for transmission of the cytokine signal (**Figure 3-25**).

Receptors, such as the BCR and TCR, have intracytoplasmic regions that are too small to allow for direct association with kinase enzymes. Instead, ligand binding by TCRs or BCRs induces the receptors to associate noncovalently (*oligomerize*) on the membrane surface and then move into specialized regions of the lymphocyte membrane known as **lipid rafts (Figure 3-26)**. These rafts are highly ordered, detergent-insoluble, cholesterol- and sphingolipid-rich membrane regions, populated by many molecules critical to receptor signaling. *Moving receptors and coreceptors into the lipid rafts renders them susceptible to the action of enzymes associated with those rafts.* For example, the raft-associated tyrosine kinase **Lyn** initiates the B-cell signaling cascade by phosphorylating the tyrosine residues on the ITAMs of the receptor-associated molecules Igα and Igβ (see Figure 3-14); **Lck** performs an analogous function in T cells by phosphorylating the ITAMs on CD3. These phosphorylation reactions then facilitate binding of downstream, signal-transducing molecules to the phosphorylated tyrosine residues in the receptor-associated molecules.

Key Concepts:

- Tyrosine kinase activation is a frequent early step in signal transduction. Phosphorylation may occur on the cytoplasmic domains of receptor proteins or on receptor-associated signaling proteins.

- Clustering of receptors at the cell membrane is an integral step in many signaling pathways. BCRs and TCRs, their coreceptors, and their accessory proteins cluster in lipid rafts, areas of the membrane that have a high concentration of cytosolic enzymes.

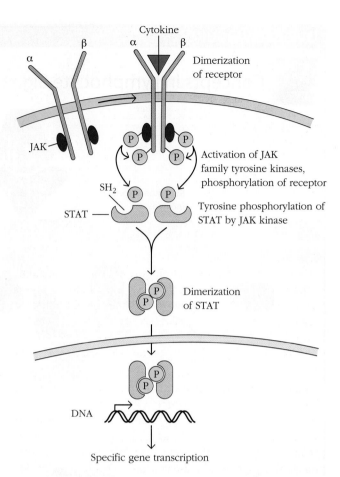

FIGURE 3-25 General model of signal transduction mediated by most class 1 and class 2 cytokine receptors. Binding of a cytokine induces dimerization of the receptor subunits, which leads to the activation of receptor subunit–associated JAK tyrosine kinases by reciprocal phosphorylation. Subsequently, the activated JAKs phosphorylate various tyrosine residues, resulting in the creation of docking sites for STATs on the receptor and the activation of the one or more STAT transcription factors. The phosphorylated STATs dimerize and translocate to the nucleus, where they activate transcription of specific genes. (Abbreviations: JAK = Janus kinase; STAT = signal transducer and activator of transcription. JAKs and STATs each exist in multiple isomeric forms, and each cytokine receptor signals through a pair of Janus kinases that may be either homo- or heterodimeric. A table illustrating which cytokine receptors use which JAK/STAT combinations is provided in Appendix II.)

Src-Family Kinases Play Important Early Roles in the Activation of Many Immune Cells

A particular family of tyrosine kinases, the **Src-family kinases,** which includes the enzymes Lck and Lyn, plays an important early role in the activation of many immune cells. Since inadvertent activation of these enzymes can lead to uncontrolled proliferation—a precursor to tumor formation—it is not surprising that their activity is tightly regulated by phosphorylation in not one, but two different and interconnected ways.

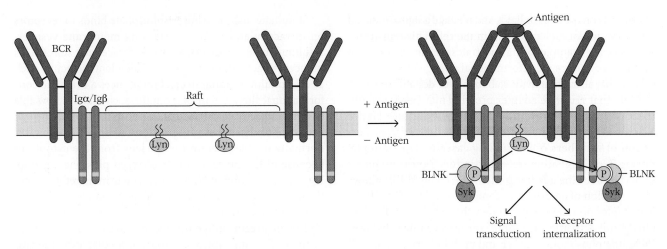

FIGURE 3-26 The role of lipid raft regions within membranes. In resting B cells, the B-cell receptor (BCR) is excluded from the lipid rafts, which are regions of the membrane high in cholesterol and rich in glycosphingolipids. The rafts are populated by tyrosine kinase signaling molecules, such as Lyn. Antigen binding induces the BCRs to oligomerize (form clusters), and increases their affinity for the lipid rafts. Movement of the BCRs into a lipid raft brings them into contact with the tyrosine kinase Lyn, which phosphorylates the receptor-associated proteins Igα,Igβ, thus initiating the activation cascade. A similar movement of TCRs into lipid rafts occurs on T-cell activation. Only two receptors are shown here for clarity.

Inactive Src-family tyrosine kinase enzymes exist in a closed conformation, in which a phosphorylated inhibitory tyrosine is tightly bound to an internal **SH2 domain** (*Src homology 2* domain) (**Figure 3-27**). (SH2 domains in proteins bind to phosphorylated tyrosine residues.) In lymphocytes, the tyrosine kinase enzyme Csk is responsible for maintaining phosphorylation of the inhibitory tyrosine, Y508. On cell activation, a tyrosine phosphatase removes the phosphate and the Src-family kinase opens up into a partially active conformation. Lipid raft–associated phosphatase enzymes help to keep it open and unphosphorylated. Full activity is achieved when the Src kinase phosphorylates itself on a second activating tyrosine residue, Y397.

Paradoxically, therefore, the first step in lymphocyte activation is not a tyrosine *phosphorylation*, but rather a tyrosine *dephosphorylation* reaction, from which many phosphorylation reactions then follow.

Tyrosine phosphorylation typically results in one or both of two outcomes. It can induce a conformational change in the phosphorylated protein, turning its enzymatic activity on or off. Alternatively, tyrosine phosphorylation can permit other proteins to bind to the phosphorylated protein via their SH2 or PTB domains as described earlier.

> **Key Concept:**
> • The cytosolic domains of receptors may be phosphorylated by Src-family tyrosine kinases, which have dual mechanisms of regulation involving both a dephosphorylation and a subsequent, different autophosphorylation reaction.

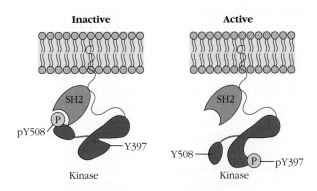

FIGURE 3-27 Activation of Src-family kinases. Src-family kinases are maintained in an inactive closed configuration by the binding of a phosphorylated inhibitory tyrosine residue (pY508 in this example) with an SH2 domain in the same protein. Dephosphorylation of this tyrosine opens up the molecule, allowing substrate access to the enzymatic site. Opening up the kinase also allows phosphorylation of a different internal tyrosine (pY397), which further stimulates the Src kinase activity.

Intracellular Adapter Proteins Gather Members of Signaling Pathways

The cytoplasm, far from being a random protein soup, is in fact an intricately organized environment in which three-dimensional arrays of proteins form and disperse as directed by cellular signaling events. Many of these reversible interactions between proteins are mediated by **adapter proteins**.

Adapter proteins have no intrinsic enzymatic or receptor function, nor do they act as transcription factors. They are characterized by having multiple surface domains, each of which possesses a precise binding specificity for a particular molecular structure, such as the phosphotyrosine residues mentioned above, or the proline-rich sequences recognized by SH3 domains. Their functions are to bind to specific motifs or domains on proteins or lipids and mediate a signal-induced redistribution of molecules within the cell. This redistribution may, for example, bring substrates within the range of enzymes or induce a conformational change in the bound protein that alters its activity, stabilizes it, or destabilizes it.

Note that multiple adapter proteins may participate in the formation of a **protein scaffold** (see Overview Figure 3-24) that provides a structural framework for the interaction of members of a signaling cascade. For example, during T-cell signaling, receptor-induced tyrosine phosphorylation of the adapter proteins LAT and SLP76 allows the formation of an extensive complex of at least five different adapter proteins. This complex then facilitates the clustering and activation of important intracellular enzymes including phospholipase Cγ1 and protein kinase C, as well as the proteins of the MAP kinase cascade. Activation of all of these enzymes ultimately leads to alterations in the transcriptional program of the cell.

Key Concept:

- Adapter proteins bind multiple proteins, bringing receptors into physical proximity with downstream effectors.

Common Sequences of Downstream Effector Relays Pass the Signal to the Nucleus

Many of the downstream effector molecules in immune receptor signaling pathways will be familiar as components of general cellular signaling pathways. In Overview Figure 3-24 we show how three signal transduction pathways in lymphocytes pass a molecular signal from the antigen receptor to the nucleus, resulting in antigen-mediated alterations in the cellular transcriptional program.

Those enzymes that cleave membrane phospholipids into the glycerol and phosphorylated head group components are collectively termed phospholipases, and enzymes that cleave the specific membrane phospholipid phosphatidyl inositol bisphosphate are referred to as phospholipases C. There are multiple members of the phospholipase C family, but those important in lymphocyte activation belong to the phospholipase Cγ classification, with phospholipase Cγ1 being active in T cells and phospholipase Cγ2 assuming the same role in B cells.

Activation of phospholipase Cγ as described above will facilitate the breakdown of phosphatidyl inositol bisphosphate into the soluble inositol trisphosphate and the membrane-bound lipid, diacylglycerol. Diacylglycerol binds and activates the serine/threonine kinase, protein kinase C, causing it to phosphorylate IκB, the inhibitory component of the transcription factor NF-κB. Once phosphorylated, IκB releases the active NF-κB transcription factor, which is now able to move through to the nucleus with resultant activation of transcription.

Simultaneously, inositol trisphosphate binds to receptors on intracellular, calcium-containing membrane vesicles, resulting in the release of Ca^{2+} from intracellular stores. This induces the activation of calcium-dependent molecules such as calmodulin. Binding to calcium induces a dramatic conformational alteration in calmodulin, such that it is now able to bind and activate calcineurin, a phosphatase. Calcineurin dephosphorylates the transcription factor NFAT, which is normally unable to enter the nucleus from the cytoplasm because of the presence of its charged phosphate group. Dephosphorylation of NFAT by calcineurin therefore allows the nuclear migration of NFAT and activation of NFAT-controlled transcription. Hence, activation of phospholipase Cγ1 by upstream, antigen-induced signals leads to changes in the cell's transcriptional program mediated by the two transcription factors, NF-κB and NFAT.

Receptor-mediated activation of the Ras/MAP kinase pathway is initiated in similar ways. Phosphorylation of receptor-associated molecules in B and T cells induces the binding of adapter molecules that, in turn, bind and activate the nucleotide exchange factor protein SOS. SOS binds to Ras, a small G protein in the lymphocyte membrane, inducing it to exchange GDP in its nucleotide-binding site for GTP. This exchange results in an activating conformational change in the Ras protein. The Ras-GTP complex can then bind and activate the cytoplasmic kinase Raf, which phosphorylates and activates MEK. MEK, in turn phosphorylates the kinase ERK. On phosphorylation, ERK is able to enter the nucleus, where it is responsible for phosphorylating and activating several other transcription factors including Fos, which is one of the components of an important transcription factor, AP-1. Activation of AP-1, NFAT, and NF-κB alter the transcriptional program of the cell.

However, if signaling of cells through a variety of receptors can give rise to increased activity of similar downstream signaling pathways, we must ask how different cells achieve distinct biological functions through intervention of similar enzyme cascades. For example, every time a cell increases the activity of phospholipase Cγ, is the biological end point of this up-regulation the same?

The answer to this question is most definitely no. First of all, many of these signal transduction pathway enzymes exist as multiple different isoforms that are differentially expressed in various cell types; they are subject to distinct means of regulation and act on different target populations.

Second, even when cells use shared pathways to mediate similar essential functions such as the induction of division or cell death, the transcription factors that specify such differentiated functions as antibody or cytokine secretion vary among discrete cell types. Furthermore, in addition to signaling for the production of mRNAs that will be translated into proteins, other alterations in transcriptional programming that occur on signaling may give rise to short, noncoding RNA sequences that vary from cell to cell, and that may have powerful regulatory effects on the differentiation of the activated cell.

Third, although when working in a laboratory environment one gets used to thinking of transcription as being either "on" or "off," in the in vivo situation cells are capable of fine levels of modulation of gene activity. For example, T and B cells can both be activated to greater or lesser extents following ligand binding. Both the CD3 complex in T cells and the Igα,Igβ pair in B cells have multiple phosphorylatable tyrosine residues, and these can be less or more fully phosphorylated depending on the strength of the initial ligand-receptor binding event, leading to stronger or weaker signal transduction; this is known as "scalable signaling." Scalable signaling allows for activation of different targets depending on the strength of the received signal.

Finally, each cell will be subject to interacting signal transduction cascades that begin with multiple ligand-receptor interactions. Recall that B cells, for example, express both a BCR and several Toll-like, innate immune receptors. A B cell whose receptor is specific for a gram-negative bacterium will thus have to integrate signals received through both its BCR and its TLR4 receptors, in addition to those received through coreceptors and receptors for modulating signals such as cytokines.

As you read about the activation of cells of the innate and adaptive immune systems in future chapters, you will learn where each of the strategies described above is used, separately and in combination, to induce immune responses in particular cell types. Overview Figure 3-24 illustrates how activation of the pathways discussed in this section, which are all signaled through the B- and T-cell receptors, combine to change the transcriptional program of the cell.

Key Concept:
- The components of a signaling pathway transduce a molecular signal from receptor-ligand binding at the cell surface to changes in enzyme and transcription factor activity. The specificity of the signal is generated by the specific combination of receptors, kinases, and downstream effector molecules activated by ligand binding.

Not All Ligand-Receptor Signals Result in Transcriptional Alterations

Although the majority of signaling events in an immune cell will result in alterations in the cell's transcriptome, it is important to remember that some immune signals, particularly those that interact with cells of the innate immune system, give rise to immediate cellular responses that do not require new transcription. For example, in the Paneth cells of the gut, sensing of bacteria in the gut lumen triggers the release of preformed antimicrobial peptides from intracellular vesicles. Another example: signaling through neutrophil surface receptors can induce immediate activation of the enzymes involved in the production of superoxides.

Interestingly, in this latter case, activation of a member of the phospholipase Cγ family by the neutrophil receptor is still implicated, demonstrating that a similar pathway can be used in both transcription-mediated and non–transcription-mediated signal transduction.

Other events that may be induced by immune signals include an increase in the destruction of particular proteins. For example, initiation of the apoptotic cascade is induced by proteolytic cleavage of procaspases to active caspases, whose activity leads ultimately to cellular destruction.

Finally, sometimes, immune signaling leads to modifications in the stability of mRNA that affects the levels of particular proteins in the cell. For example, the mRNA that encodes IL-2 is quite unstable in resting T cells. On T-cell activation, the IL-2 message is stabilized by the binding of proteins to the 3′ end of the mRNA, leading to an increase in translation and hence in the levels of secreted IL-2.

Key Concept:
- Signal transduction pathways can culminate in cellular functions such as the release of effector molecules from preformed vesicles, the destruction or modification of particular proteins, the alteration of mRNA stability, or the initiation of apoptosis.

Immune Responses: The Outcomes of Immune System Recognition

The biological outcomes of immune system recognition are described in detail in many of the chapters that follow. In this section we offer a framework that students can use to classify these outcomes in biologically meaningful ways, using examples from specific cell types.

Changes in Protein Expression Facilitate Migration of Leukocytes into Infected Tissues

The cells of the immune system are distributed throughout the body, and one of the greatest challenges to mounting an immune response is to bring the relevant cells into the required locations in a timely manner. The immune system uses a number of strategies to accomplish this. For example, cells in infected tissues, or in nearby lymph nodes, secrete chemokines that alert dendritic cells bearing the relevant chemokine receptors to leave the blood and enter the tissues. Simultaneously, epithelial cells lining blood capillaries in injured areas alter the expression of adhesion molecules on their cell surfaces to slow down blood flow and facilitate leukocyte exit into the tissues. Both of these events are signaled by recognition of PAMPs by innate immune receptors.

Once the dendritic cell has entered the tissues and is activated by antigen binding to one of its innate immune receptors, its immediate task is to migrate to the nearest lymph node, where it can activate a T cell. This migration is facilitated by the up-regulation of the chemokine receptor CCR7 on the dendritic cell surface. CCR7 receptors bind to chemokines that are secreted in the T-cell regions of the lymph node. Once the dendritic cell is expressing CCR7, it becomes susceptible to the chemoattractant properties of the lymph node chemokines and migrates into the T-cell regions of that lymph node.

So we see that signaling alters the expression of adhesion molecules in blood capillaries, and chemokine receptor expression on dendritic cells; this happens sequentially, such that the cells leave the bloodstream, enter the tissues, and then take their antigen on to the relevant lymph node. Within the lymph node, activation of T cells and B cells induces further chemokine receptor molecule alterations that result in T and B cells moving in and out of the relevant areas within the lymph node in order to maximize cooperation between these cells in the immune response.

> **Key Concept:**
> - Alterations in the expression of adhesion molecules and chemokine receptors facilitate the movement of immune cells to the sites of immune activity.

Activated Macrophages and Neutrophils May Clear Pathogens without Invoking Adaptive Immunity

When macrophages and neutrophils recognize their ligands via innate immune receptors, the cells are activated and their ability to eliminate pathogens is enhanced. Stimulation of both types of cells increases both their phagocytic capacity and their ability to destroy invading pathogens by digestion in lysosomal vesicles. In addition, neutrophils and macrophages share the capacity to mount a response known as an **oxidative burst**, in which cellular metabolism is shifted to produce large amounts of noxious chemicals including hypochlorous acid, peroxide and superoxide radicals, and reactive nitrogen species. These chemical species are located in the phagolysosomes of macrophages and neutrophils and are highly toxic to phagocytosed microbes.

Both types of cells also secrete cytokines such as IL-1, IL-6, and TNF-α on activation that act together to induce local vasodilation (widening of blood vessels to increase blood flow), and increased movement of cells and fluids out of the blood vessels and into the tissues. TNF-α also promotes clotting in nearby capillaries, thus minimizing the tendency of infections to spread from the immediate area of immunological insult. Local mast cells, which also accrue within the inflammatory environment, release other mediators including prostaglandins and histamines that add to the vasodilation and capillary leakage and induce fever.

The combination of vasodilation, capillary leakage, cytokine secretion, and movement of cells into the damaged tissue gives rise to the four cardinal signs of inflammation—*calor* (heat), *rubor* (redness), *tumor* (swelling), and *dolor* (pain). Inflammation is a sign of an ongoing local innate immune response. However, sometimes the localized tissue destruction that can accompany a vigorous immune response itself becomes a problem and the host then becomes a victim of chronic inflammation (see Chapter 15).

> **Key Concept:**
> - Macrophages and neutrophils of the innate immune system facilitate destruction of invading organisms by upregulating phagolysosome activity and cytokine secretion.

Antigen Activation Optimizes Antigen Presentation by Dendritic Cells

As described in Chapter 10, dendritic cells are the only types of cells capable of presenting antigen in such a way as to activate naïve T cells. Dendritic cells take up antigen via innate immune receptors (PRRs) and process it internally to create short peptide segments that are loaded onto the platforms of MHC class I and MHC class II proteins. These complex antigens, made up of MHC proteins plus the peptides they carry, are then recognized by T cells. On activation by ligand binding to PRRs, dendritic cells also express other costimulatory molecules that bind to the coreceptors on T cells, facilitating T-cell function.

Proteasomes are cylindrically shaped organelles that contain proteases located on the inner surface of the cylinder. These proteases normally break down cellular proteins that have outlived their usefulness. However, when a dendritic cell is activated, the proteases in its proteasomes are switched out in favor of proteases that create peptides suitable for loading onto the antigen-presenting platforms of MHC class I and MHC class II molecules. Thus signaling of dendritic cells enhances their antigen-presentation capacity.

> **Key Concept:**
> - Changes in the protease specificity of their proteasomes optimize the capacity of dendritic cells to mediate antigen presentation.

Cytokine Secretion by Dendritic Cells and T Cells Can Direct the Subsequent Immune Response

Antigen signaling of dendritic cells via innate receptors induces them to secrete cytokines, with different cytokine mixtures being secreted depending on the identities of the PRRs that are stimulated as well as the cytokine environment in which the stimulation occurs. The nature of the cytokines secreted by antigen-presenting dendritic cells in turn affects the type of T-cell response that is stimulated when the dendritic cell presents antigen to a T cell bearing receptors for its antigen.

For example, we will learn in Chapter 10 that a T cell interacting with an antigen-presenting dendritic cell that is simultaneously secreting IL-12 will be induced to differentiate into a type of T helper cell called a T_H1 cell. This T_H1 cell will then preferentially activate macrophages and cytotoxic T cells, as well as induce the secretion of particular classes of antibodies by B cells.

Key Concept:

- The nature of the innate immune receptor engaged in a dendritic cell determines what cytokines the dendritic cell secretes. These in turn determine the type of T-cell response that it initiates.

Antigen Stimulation by T and B Cells Promotes Their Longer-Term Survival

Naïve T and B cells have relatively short half-lives in the circulation. However, one of the most important outcomes of activation by antigen is the delivery of anti-apoptotic signals that lengthen cellular life spans so that the cells are able to mediate their respective functions of cytokine and antibody secretion. A subset of these activated cells go on to become memory T and B cells, and these cells may survive for the lifetime of the organism. How is this increased life span achieved?

One of the enzymes that is activated following antigen binding is phosphatidylinositol-3-kinase (PI3 kinase), whose activity is modulated after it binds to an adapter protein complex generated by antigen signaling. PI3 kinase adds a phosphate group to an inner membrane phospholipid that is then able to bind and participate in the activation of the protein kinase Akt. Akt plays a number of roles in cell activation, but one of its most important is the phosphorylation and resultant inactivation of molecules that promote apoptosis, thereby leading to an increased life span of antigen-activated lymphocytes.

Key Concept:

- Antigen activation increases the life span of T and B lymphocytes by inhibiting apoptosis.

Antigen Binding by T Cells Induces Their Division and Differentiation

Activation of a CD4-bearing T cell by antigen binding induces a series of signal transduction events that culminate in T-cell division and differentiation. Both these processes are mediated via pathways that alter the transcriptional program of the cell, as described above.

In the case of helper T cells, the end result of differentiation is an enhanced capacity to secrete an array of cytokines. The precise set of cytokines that is secreted by the differentiated T cell is determined by the antigen and (as indicated above) by the cytokines secreted by the antigen-presenting cell. Different subclasses of helper T cells secrete various combinations of cytokines that in turn facilitate different aspects of the immune response. For example, different cytokines promote the secretion of different classes of antibodies, cytotoxic T-cell activity, macrophage activation, and so on.

Activation of CD8-bearing cytotoxic T-cell precursors also leads to cell division, cell differentiation, and cytokine secretion. However, these cells also synthesize granules containing molecules that induce apoptosis. When the mature cytotoxic T cell binds to its target cell, the granule contents are released into the junction between the cytotoxic T cell and its target, with consequent target cell death.

Key Concept:

- Different subclasses of T cells secrete different cytokines that direct varying aspects of immune effector responses.

Antigen Binding by B Cells Induces Their Division and Differentiation

Activation of B cells leads, as for T cells, to cell division and cell differentiation. B cells use signaling pathways similar to those of T cells but, as described above, some of the enzymes of the signal transduction pathways are isozymes that are specific to B lymphocytes, rather than T lymphocytes. As the B cell differentiates, the level of antibody synthesis increases and the cell begins to synthesize the soluble, as well as the membrane-bound, form of the antibody molecule. Over the first several days after antigen contact, antigen-specific T cells direct their B-cell partners to make genetic modifications in the antibody genes that result in the generation of more efficient antibody molecules. Specifically, these changes result in the synthesis of antibodies of different classes (see Figure 3-12) and, in addition, the antigen-binding sites of these antibodies accrue mutations that are selected for enhanced antibody-antigen binding. Defects in the signaling proteins that participate in this pathway can have serious consequences, as demonstrated in **Clinical Focus Box 3-4**. Over the course of the immune response, long-lived memory B cells and memory plasma cells emerge, and take up residence in the lymph nodes and the bone marrow. These processes are further described in Chapter 11.

Key Concept:

- B-cell differentiation results in differentiation to antibody-secreting cells and memory B cells and the production of antibodies of different classes and enhanced binding affinity.

BOX 3-4

Defects in the B-Cell Signaling Protein Btk Lead to X-Linked Agammaglobulinemia

The characterization of the proteins necessary for B- and T-cell signaling opened up new avenues of exploration for clinicians working with patients suffering from immunodeficiency disorders. Clinicians and immunogeneticists now work closely together to diagnose and treat patients with immunodeficiencies, to the benefit of both the clinical and the basic sciences.

Characterization of the genes responsible for disorders of the immune system is complicated by the fact that antibody deficiencies may result from defective genes encoding either T- or B-cell proteins (since T cells provide helper factors necessary for B-cell antibody production), or even from mutations in genes encoding proteins in stromal cells important for healthy B-cell development in the bone marrow. However, no matter what the cause, all antibody deficiencies manifest clinically in increased susceptibility to bacterial infections, particularly those of the lung, intestines, and (in younger children) the ear.

In 1952, a pediatrician, Ogden Bruton, reported in the journal *Pediatrics* the case of an 8-year-old boy who suffered from multiple episodes of pneumonia. When the serum of the boy was subjected to electrophoresis, it was shown to be completely lacking in serum globulins, and his disease was therefore named *agammaglobulinemia*. This was the first immunodeficiency disease for which a laboratory finding explained the clinical symptoms, and the treatment that Bruton applied—administering subcutaneous injections of gamma globulin—is still used today. As similar cases were subsequently reported, it was noted that most of the pediatric cases of agammaglobulinemia occurred in boys, whereas when the disease was reported in adults, both men and women appeared to be similarly affected. Careful mapping of the disease susceptibility to the X chromosome resulted in the pediatric form of the disease being named XLA, for *X-linked agammaglobulinemia*.

With the characterization of the BCR signal transduction pathway components in the 1980s and 1990s came the opportunity to define which proteins are damaged or lacking in particular immunodeficiency syndromes. In 1993, 41 years after the initial description of the disease, two groups independently reported that many cases of XLA resulted from mutations in a cytoplasmic tyrosine kinase called Bruton's tyrosine kinase, or Btk; at this point, we now know that fully 85% of patients affected with XLA have mutations in the *Btk* gene.

Btk is a member of the Tec family of cytoplasmic tyrosine kinases, which is predominantly expressed in hematopoietic cells. Tec family kinases share a C-terminal kinase domain, preceded by SH2 and SH3 domains, a proline-rich domain, and an amino-terminal pleckstrin homology domain, which binds to *phosphatidylinositol trisphosphate* (PIP$_3$), a phospholipid generated by PI$_3$ kinase activity. Btk is expressed in both B cells and platelets and is activated following signaling through the BCR, the pre-BCR (which is expressed in developing B cells), the IL-5 and IL-6 receptors, and also the CXCR4 chemokine receptor. Its involvement in pre-BCR signaling explains why children with XLA suffer from defective B-cell development.

Following activation of B cells, Btk moves to the inner side of the plasma membrane, where it interacts with the adapter protein, BLNK, and PIP$_3$. There, it is phosphorylated by the *Src*-family kinase Lyn, and partially activated. Activation is completed when it autophosphorylates itself at a second phosphorylation site. Active Btk phosphorylates and activates phospholipase Cγ2, leading, as described earlier, to calcium flux and activation of the NF-κB and NFAT pathways. Btk therefore occupies a central position in B-cell activation, and it is no surprise that mutations in its gene result in such devastating consequences.

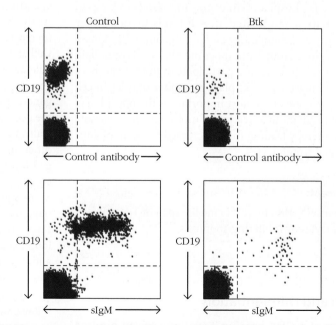

FIGURE 1 **Fluorescence-activated cell-sorting (FACS) profiles of a normal individual and a patient with XLA.** [*Data from Conley, M. E., et al. 2009. Primary B cell immunodeficiencies: comparisons and contrasts. Annual Review of Immunology* **27**:*199.*]

(continued)

Over 600 different mutations have been identified in the *btk* gene, with the vast majority of these resulting from single–base pair substitutions, or the insertion or deletion of fewer than five base pairs. As for other X-linked mutations that are lethal without medical intervention, XLA disease is maintained in the population by the generation of new mutations.

Patients with XLA are usually healthy in the neonatal (immediately after birth) period, when they still benefit from maternal antibodies. However, recurrent bacterial infections begin between the ages of 3 and 18 months, and currently the mean age at diagnosis in North America is 3 years. XLA is a so-called "leaky" defect; almost all children with mutations in *btk* have some serum immunoglobulin, and a few B cells in the peripheral circulation. The prognosis

for patients who are treated with regular doses of gamma globulin has improved dramatically over the last 25 years.

The B cells in patients with XLA have a distinctive phenotype that can be used for diagnostic purposes. CD19 expression is low and variable in patients with XLA, whereas membrane IgM expression, normally variable in mature B cells, is relatively high and consistent in patients with XLA. This phenotype can be observed in **Figure 1**, which shows the flow cytometric profiles of a normal control individual (left two graphs) and a patient with a defective *btk* gene (right two graphs). In the top two plots, we note that the patient with XLA has very few CD19⁺ B cells compared with the control individual, and that the levels of CD19 on the surface of those B cells that do exist are lower than those of

the control cells. In the lower two plots, we note that, although there are fewer CD19⁺ B cells overall in the *btk*-compromised patient, all of those CD19⁺ B cells have relatively high levels of surface IgM, whereas the levels of membrane IgM are much more variable in the healthy control cells.

REFERENCES

Bruton, O. C. 1952. Agammaglobulinemia. *Pediatrics* **9**:722.

Conley, M. E., et al. 2009. Primary B cell immunodeficiencies: comparisons and contrasts. *Annual Review of Immunology* **27**:199.

Tsukuda, S., et al. 1993. Deficient expression of a B cell cytoplasmic tyrosine kinase in human X-linked agammaglobulinemia. *Cell* **72**:279.

Vetrie, D., et al. 1993. The gene involved in X-linked agammaglobulinemia is a member of the *src* family of protein tyrosine kinases. *Nature* **361**:226.

Conclusion

As we noted at the beginning of this chapter, one of the biggest challenges to mounting an immune response is the coordination that must occur among cells at disparate locations, often separated from one another by physical barriers such as endothelial cell layers. This communication is mediated by small molecules, such as chemokines, that attract cells to the locations where they are needed and by cytokines that induce the correct cells to differentiate in such a way as to generate an immune response that appropriately targets the pathogenic insult. Alterations in the expression of cell surface adhesion molecules, induced by innate immune receptor recognition, enable cells to leave the circulation and enter areas of injury to mediate an immune response.

Interaction of T cells with antigen-presenting cells is mediated via antigen receptors and coreceptors that must be simultaneously occupied for activation to occur. This necessity for dual recognition of both antigen (by the antigen receptor) and coreceptor ligand (by the coreceptor) reduces the potential for a T cell to recognize a self-antigen on a nonactivated antigen-presenting cell and thereby inducing an autoimmune response.

We have learned that immune effector cells come in different varieties that include the neutrophil and macrophage (as well as the innate lymphoid cells) of the innate immune system, and the helper and cytotoxic T cells and B cells of adaptive immunity. Activation of the innate immune cells happens rapidly, whereas activation of cells of the adaptive immune system requires cell division and differentiation, and its effects take longer to manifest themselves. We have also learned that, as disparate as the cell surface receptors and the downstream effector functions of different cells may be, the immune system has evolved to recycle strategies of signal transduction, such that similar types of signaling molecule are used to transduce a variety of different stimuli into changes in cellular function. Whole sections of signaling pathways and signaling molecules recur in many different cell types, but in each cell they are connected to distinct upstream receptors and/or downstream effector molecules.

Finally, we have learned that cells that are engaged in the act of mounting an immune response must integrate signals from a variety of different receptors and signal transduction pathways so as to mount the immune response appropriate to the elimination of a particular pathogen. How an organism succeeds in this daunting task is the subject of the chapters that follow.

REFERENCES

Arend, W. P., G. Palmer, and C. Gabay. 2008. IL-1, IL-18, and IL-33 families of cytokines. *Immunological Reviews* **223**:20.

Boulanger, M. J., and K. C. Garcia. 2004. Shared cytokine signaling receptors: structural insights from the gp130 system. *Advances in Protein Chemistry* **68**:107.

Choudhuri, K., and M. L. Dustin. 2010. Signaling microdomains in T cells. *FEBS Letters* **584**:4823.

Crispin, J. C., and G. C. Tsokos. 2009. Transcriptional regulation of IL-2 in health and autoimmunity. *Autoimmunity Reviews* **8**:190.

Eisenbarth, S. C., and R. A. Flavell. 2009. Innate instruction of adaptive immunity revisited: the inflammasome. *EMBO Molecular Medicine* **1**:92.

Finlay, D., and D. Cantrell. 2011. The coordination of T-cell function by serine/threonine kinases. *Cold Spring Harbor Perspectives in Biology* **3**:a002261.

Gaffen, S. L. 2009. Structure and signalling in the IL-17 receptor family. *Nature Reviews Immunology* **9**:556.

Gaffen, S. L., R. Jain, A. V. Garg, and D. J. Cua. 2014. The IL-23–IL-17 immune axis: from mechanisms to therapeutic testing. *Nature Reviews Immunology* **14**:585.

Gee, K., C. Guzzo, N. F. Che Mat, W. Ma, and A. Kumar. 2009. The IL-12 family of cytokines in infection, inflammation and autoimmune disorders. *Inflammation & Allergy Drug Targets* **8**:40.

Gonzalez-Navajas, J. M., J. Lee, M. David, and E. Raz. 2012. Immunomodulatory functions of type 1 interferons. *Nature Reviews Immunology* **12**:125.

Kalliolias, G. D., and L. B. Ivashkiv. 2016. TNF biology, pathogenic mechanisms and emerging therapeutic strategies. *Nature Reviews Rheumatology* **12**:49.

Koretzky, G. A., and P. S. Myung. 2001. Positive and negative regulation of T-cell activation by adaptor proteins. *Nature Reviews Immunology* **1**:95.

Kufareva, I., C. L. Salanga, and T.M. Handel. 2015. Chemokine and chemokine receptor structure and interactions: implications for therapeutic strategies. *Immunology and Cell Biology* **93**:372.

Kurosaki, T. 2011. Regulation of BCR signaling. *Molecular Immunology* **48**:1287.

Love, P. E., and S. M. Hayes. 2010. ITAM-mediated signaling by the T-cell antigen receptor. *Cold Spring Harbor Perspectives in Biology* **2**:a002485.

Macian, F. 2005. NFAT proteins: Key regulators of T-cell development and function. *Nature Reviews Immunology* **5**:472.

Palacios, E. H., and A. Weiss. 2004. Function of the Src-family kinases, Lck and Fyn, in T-cell development and activation. *Oncogene* **23**:7990.

Raman, D., T. Sobolik-Delmaire, and A. Richmond. 2011. Chemokines in health and disease. *Experimental Cell Research* **317**:5755.

Rochman, Y., R. Spolski, and W. J. Leonard. 2009. New insights into the regulation of T cells by γ_c family cytokines. *Nature Reviews Immunology* **9**:480.

Rot, A., and U. H. von Andrian. 2004. Chemokines in innate and adaptive host defense: basic chemokinese grammar for immune cells. *Annual Review of Immunology* **22**:891.

Skalnikova, H., J. Motlik, S. J. Gadher, and H. Kovarova. 2011. Mapping of the secretome of primary isolates of mammalian cells, stem cells and derived cell lines. *Proteomics* **11**:691.

Stull, D., and S. Gillis. 1981. Constitutive production of interleukin 2 activity by a T cell hybridoma. *Journal of Immunology* **126**:1680.

Useful Websites

https://portal.genego.com/ This site offers a searchable database of metabolic and regulatory pathways.

www.nature.com/subjects/cell-signalling/research This is a collection of original research articles, reviews, and commentaries pertaining to cell signaling.

www.signalinggateway.org/molecule/ This gateway is powered by the University of California at San Diego and is supported by Genentech and *Nature*. An excellent resource, complete with featured articles.

www.cellsignallingbiology.org/csb/002/csb002.htm This is a comprehensive site, developed and maintained by Professor Sir Michael Berridge out of Cambridge University.

https://www.qiagen.com/us/shop/genes-and-pathways/pathway-central/ Qiagen; a useful commercial website.

www.biosignaling.biomedcentral.com Part of Springer Science+Business Media.

www.youtube.com Many excellent videos and animations of signaling websites are available on YouTube, too numerous to mention here. Just type your pathways into a web browser and go. But do be cognizant of the derivation of your video. Not all videos are accurate, so check your facts with the published literature. Similarly, YouTube has some excellent animations of chemotaxis.

www.abcam.com/pathways/chemokine-signaling-Interactive-pathway A useful commercial website that offers access to current chemokine signaling pathways.

www.hhmi.org/biointeractive/immunology/tcell .html Howard Hughes Medical Institute (HHMI) movie: *Cloning an Army of T Cells for Immune Defense.*

Many companies that sell recombinant cytokines or cytokine-related products provide useful information on their websites, or in print copy. The following are a few that are particularly helpful:

www.miltenyibiotec.com/cytokines

www.prospecbio.com/cytokines

www.peprotech.com

www.rndsystems.com

www.netpath.org A curated set of pathways, with information on interacting proteins. Many interleukin pathways are included.

STUDY QUESTIONS

1. The NFAT family is a ubiquitous family of transcription factors.
   a. Under resting conditions, where is NFAT localized in a cell?
   b. Under activated conditions, where is NFAT localized in a cell?
   c. How is it released from its resting condition and permitted to relocalize?
   d. Immunosuppressant drugs such as cyclosporin act via inhibition of the calcineurin phosphatase. If NFAT is ubiquitous, how do you think these drugs might act with so few side effects on other signaling processes within the body?

2. In the early days of experiments designed to detect the T-cell receptor, several different research groups found that antibodies directed against immunoglobulin proteins appeared to bind to the T-cell receptor. Given what you know about the structure of immunoglobulins and the T-cell receptor, why is this not completely surprising?

3. True or false? Explain your answers.

   Interactions between receptors and ligands at the cell surface:
   a. are mediated by covalent interactions.
   b. can result in the creation of new covalent interactions within the cell.

4. Describe how the following experimental manipulations were used to determine antibody structure.
   a. Reduction and alkylation of the antibody molecule
   b. Enzymatic digestion of the antibody molecule
   c. Antibody detection of immunoglobulin fragments

5. What is an ITAM, and what proteins modify the ITAMs in Igα and Igβ?

6. Define an adapter protein. Describe how an interaction between proteins bearing SH2 and phosphorylated tyrosine (pY) groups helps to transduce a signal from the T-cell receptor to downstream signal transduction pathway components.

7. IgM has 10 antigen-binding sites per molecule, whereas IgG only has two. Would you expect IgM to be able to bind five times as many antigenic sites on a multivalent antigen as IgG? Why/why not?

8. You and another student are studying a cytokine receptor on a B cell that has a K_d of 10^{-6} M. You know that the cytokine receptor sites on the cell surface must be at least 50% occupied for the B cell to receive a cytokine signal from a helper T cell. Your lab partner measures the cytokine concentration in the blood of the experimental animal and detects a concentration of 10^{-7} M. She tells you that the effect you have been measuring could not possibly result from the cytokine you're studying. You disagree. Why?

9. Activation of Src-family kinases is the first step in several different types of signaling pathways. It therefore makes biological sense that the activity of this family of tyrosine kinases is regulated extremely tightly. Describe how phosphorylation of Src-family kinases can deliver both activating and inhibitory signals to Src kinases.

10. You have generated a T-cell clone in which the Src-family tyrosine kinase Lck is inactive. You stimulate that clone with its cognate antigenic peptide, presented on the appropriate MHC platform, and test for interleukin-2 secretion as a measure of T-cell activation. Do you expect to see IL-2 secretion or not? Explain.

11. Name one protein shown to be defective in many cases of X-linked agammaglobulinemia, and describe how a reduction in the activity of this protein could lead to immunodeficiency.

12. The B- and T-cell receptor proteins have remarkably short intracytoplasmic regions of just a few amino acids. How can you reconcile this structural feature with the need to signal the presence of bound antigen to the interior of the cell?

13. Describe one way in which the structure of antibodies is superbly adapted to their function.

14. Your adviser has handed you (a graduate student) a T-cell clone that appears to be constitutively (i.e., always) activated, although at a low level, even in the absence of antigenic stimulation, and he has asked you to figure out why. Your benchmate suggests you start by checking out the sequence of its *lck* gene, or the status of the Csk activity in the cell. You agree that those are good ideas. What is your reasoning?

15. Define the terms pleiotropy, synergy, redundancy, antagonism, and cascade induction as they apply to cytokine action.

16. How might receipt of a cytokine signal result in the alteration of the location of a lymphocyte?

17. Describe one mechanism by which type I interferons "interfere" with the production of new viral particles.

18. The cytokine IL-2 is capable of activating all T cells to proliferate and differentiate.

    a. How does the immune system ensure that only T cells that have been stimulated by antigen are susceptible to IL-2 signaling?

The following diagram represents the results of a flow cytometry experiment in which mouse spleen cells were stained with antibodies directed against different components of the IL-2 receptor (IL-2R). The more antibody that binds to the cells, the further they move along the relevant axis. The number of cells stained with fluorescein-conjugated anti–IL-2Rβγ antibodies are shown along the *x* axis of the flow cytometry plot, and cells that stain with phycoerythrin-labeled antibodies to the α subunit of the IL-2 receptor move along the *y* axis. We have drawn for your reference a circle that represents cells that stain with neither antibody.

    b. On this plot, draw, as circles, and label where you would expect to find the populations representing unstimulated T cells and T cells after antigen activation, after treatment with the two fluorescent labels described above.

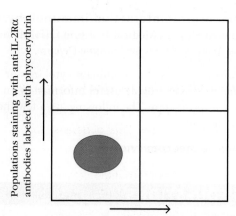

Populations staining with anti-IL-2Rβ and anti-IL-2Rγ antibodies labeled with fluorescein

19. Lymphocytes derived from a patient with a severe immunodeficiency disease known to affect only a single protein chain prove unable to respond to the cytokines IL-2, IL-4, IL-7, and IL-15, in addition to several others. Given what you know about the specificity of cytokine receptors, explain how a defect in a single protein chain can prevent the binding of so many different cytokines to their cellular receptors.

Innate Immunity

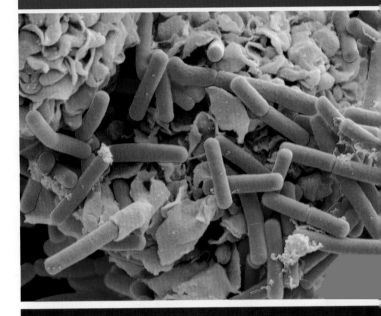

A macrophage (yellow) binds and phagocytoses *E. coli* bacteria (red). [*Science Source, Colorization by: Mary Martin.*]

Learning Objectives

After reading this chapter, you should be able to:

1. Identify and describe the components and characteristics of the two lines of defense that comprise the innate immune system.

2. Categorize the pattern recognition receptors in terms of the types of pathogen components that they bind, the basic mechanisms by which they stimulate responses, and the types of protective responses that result.

3. Describe effector mechanisms used by the innate immune system, the cells and molecules involved in each mechanism, and the type of pathogen destroyed by each mechanism.

4. Explain why the innate immune system is so highly regulated, with many types of both positive and negative regulation of responses.

5. Connect elements of the innate and adaptive immune systems and describe how innate responses help to ensure that an effective adaptive immune response is generated for a specific pathogen.

Vertebrates are protected by both **innate immunity** and **adaptive immunity**. In contrast to adaptive immune responses, which take days to arise following exposure to antigens, innate immunity consists of the defenses against infection that are ready for immediate action or are quickly induced when a host is attacked by a pathogen (viruses, bacteria, fungi, or parasites; see Table 1-3). The innate immune system includes anatomical barriers against infection—both physical and chemical—as well as cellular responses (**Overview Figure 4-1**). The main **physical barriers**— the body's first line of defense—are the epithelial layers of the skin and of the mucosal and glandular tissue surfaces connected to the body's openings; these epithelial barriers prevent infection by blocking pathogens from entering the body. **Chemical barriers** at these surfaces include specialized soluble substances that possess anti-microbial activity as well as acid pH.

If an infectious agent overcomes the initial epithelial physical and chemical barriers, **cellular innate immune responses** are rapidly activated, typically beginning within minutes of invasion. These responses, which constitute

Key Terms

Physical barriers

Chemical barriers

Cellular innate immune responses

Phagocytosis

Inflammation

Antimicrobial proteins and peptides

Defensins

Cathelicidin

Pattern recognition receptors (PRRs)

Pathogen-associated molecular patterns (PAMPs)

Damage-associated molecular patterns (DAMPs)

Toll-like receptors (TLRs)

C-type lectin receptors (CLRs)

NOD-like receptors (NLRs)

Autophagy

Inflammasomes

Pyroptosis

AIM2-like receptors (ALRs)

RIG-I-like receptors (RLRs)

cGAS

STING

Type I interferons

Opsonins

Reactive oxygen species (ROS) and reactive nitrogen species (RNS)

Regulated cell death

Innate lymphoid cells (ILCs)

Sepsis

Adjuvants

Innate Immunity

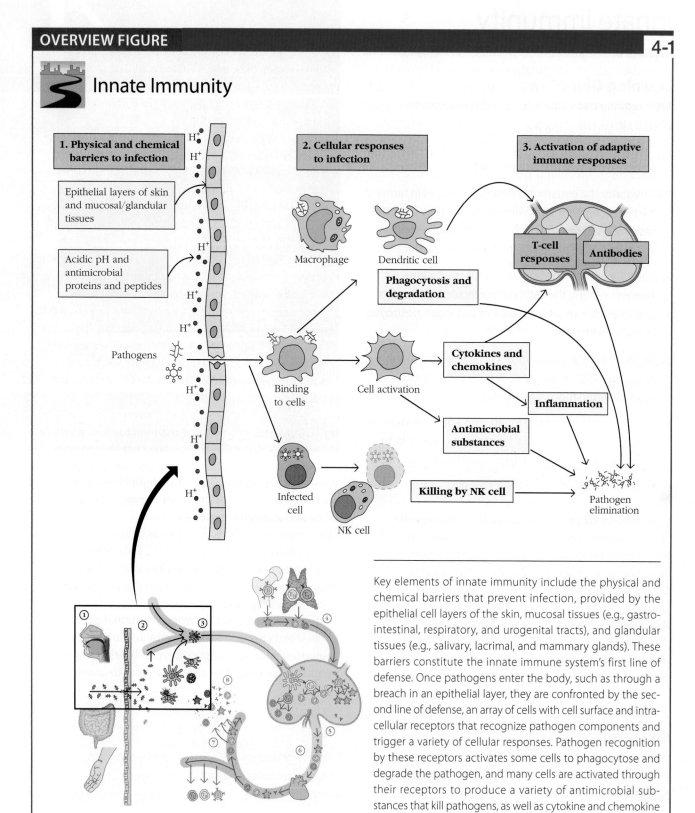

Key elements of innate immunity include the physical and chemical barriers that prevent infection, provided by the epithelial cell layers of the skin, mucosal tissues (e.g., gastrointestinal, respiratory, and urogenital tracts), and glandular tissues (e.g., salivary, lacrimal, and mammary glands). These barriers constitute the innate immune system's first line of defense. Once pathogens enter the body, such as through a breach in an epithelial layer, they are confronted by the second line of defense, an array of cells with cell surface and intracellular receptors that recognize pathogen components and trigger a variety of cellular responses. Pathogen recognition by these receptors activates some cells to phagocytose and degrade the pathogen, and many cells are activated through their receptors to produce a variety of antimicrobial substances that kill pathogens, as well as cytokine and chemokine proteins that recruit cells, molecules, and fluid to the site of infection, leading to swelling and other symptoms collectively known as *inflammation*. Innate lymphoid cells (ILCs) are activated by soluble mediators from nearby cells to also produce cytokines and chemokines. One type of ILC, the natural killer (NK) cell, recognizes and kills some virus-infected cells. Cytokines and chemokines can cause systemic effects that help to eliminate an infection, and also contribute—along with dendritic cells that carry and present pathogens to lymphocytes—to the activation of adaptive immune responses, the third line of defense in vertebrates. See Figure 1-7 for how these processes are integrated within the entire immune system.

the innate immune system's second line of defense, are triggered by cell surface or intracellular receptors that recognize conserved molecular components of pathogens. Some white blood cell types are activated to rapidly engulf and destroy extracellular microbes through the process of **phagocytosis.** Other receptors induce the production of proteins and other substances that have a variety of beneficial effects, including direct antimicrobial activity, as well as the recruitment of fluid, cells, and molecules to sites of infection. This influx causes swelling and other physiological changes that collectively are called **inflammation.** Such local innate and inflammatory responses usually are beneficial in that they eliminate pathogens and damaged or dead cells, promote healing, and help to activate adaptive immune responses.

As members of the innate lymphoid cell (ILC) lineage, natural killer (NK) cells recruited to the site can recognize and kill virus-infected, altered, or stressed cells. However, in some situations these innate and inflammatory responses can be harmful, leading to local or systemic consequences that can cause tissue damage and occasionally death. To prevent these potentially harmful responses, regulatory mechanisms have evolved that usually limit such adverse effects.

Despite the multiple layers of the innate immune system, some pathogens may evade the **innate immunity effector mechanisms**, the various chemical and cellular mechanisms by which the innate immune system eliminates pathogens. On call in vertebrates is the *adaptive immune system*, which counters infection with tailor-made responses specific for the attacking pathogen. These powerful responses, to be described in detail later in this text, consist of B cell–derived antibodies and effector T cells that specifically recognize and neutralize or eliminate the invaders but take longer to develop.

In many ways, innate and adaptive immunity are complementary systems (**Table 4-1**). Innate immunity is the most ancient form of defense, found in all multicellular plants and animals, while adaptive immunity is a much more recent evolutionary invention, having arisen in vertebrates. In these animals, adaptive immunity complements a well-developed system of innate immune mechanisms that share important features with those of our invertebrate ancestors. A growing body of research has revealed that as innate and adaptive immunity have co-evolved in vertebrates, a high degree of interaction and interdependence has arisen between the two systems. Recognition by the innate immune system not only kicks off the adaptive immune response but also helps to ensure that the type of adaptive response generated will be effective for the invading pathogen.

This chapter describes the components of the innate immune system—physical and chemical barriers, a battery of protective cellular responses carried out by numerous cell types, and inflammatory responses—and illustrates how they act together to defend against infection. We conclude with an overview of innate immunity in plants and invertebrates.

TABLE 4-1	Innate and adaptive immunity	
Attribute	**Innate immunity**	**Adaptive immunity**
Response time	Minutes/hours	Days
Specificity	Specific for molecules and molecular patterns associated with pathogens and molecules produced by dead/damaged cells	Highly specific; discriminates between even minor differences in molecular structure of microbial or nonmicrobial molecules
Diversity	A limited number of conserved, germ line–encoded receptors	Highly diverse; a very large number of receptors arising from genetic recombination of receptor genes in each individual
Memory responses	Some (observed in invertebrate innate responses and mouse/human NK cells)	Persistent memory, with faster response of greater magnitude on subsequent exposure
Self/nonself discrimination	Very good; no microbe-specific self/nonself patterns in host	Very good; occasional failures of discrimination result in autoimmune disease
Soluble components of blood	Many antimicrobial peptides, proteins, and other mediators, including cytokines	Antibodies and cytokines
Major cell types	Phagocytes (monocytes, macrophages, neutrophils, dendritic cells), natural killer (NK) cells, other leukocytes, epithelial and endothelial cells	T cells, B cells, antigen-presenting cells

Anatomical Barriers to Infection

The most obvious components of innate immunity are the external barriers to microbial invasion: the epithelial layers that insulate the body's interior from the pathogens of the exterior world. These epithelial barriers include the skin and the tissue surfaces connected to the body's openings: the mucous epithelial layers that line the respiratory, gastrointestinal, and urogenital tracts and the ducts of secretory glands such as the salivary, lacrimal, and mammary glands (which produce saliva, tears, and milk, respectively) (**Figure 4-2**). Skin and other epithelia provide a kind of living "plastic

Organ or tissue	Innate mechanisms protecting skin/epithelium
Skin	Antimicrobial peptides, fatty acids in sebum
Mouth and upper alimentary canal	Enzymes, antimicrobial peptides, and sweeping of surface by directional flow of fluid toward stomach
Stomach	Low pH, digestive enzymes, bile salts, antimicrobial peptides, fluid flow toward intestine
Small intestine	Digestive enzymes, antimicrobial peptides, fluid flow to large intestine
Large intestine	Normal intestinal flora compete with invading microbes, fluid/feces expelled from rectum
Airway and lungs	Cilia sweep mucus outward, coughing, sneezing expel mucus, macrophages in alveoli of lungs
Urogenital tract	Flushing by urine and mucus, low pH, antimicrobial peptides, and proteins
Salivary, lacrimal, and mammary glands	Flushing by secretions and mucus, antimicrobial peptides and proteins

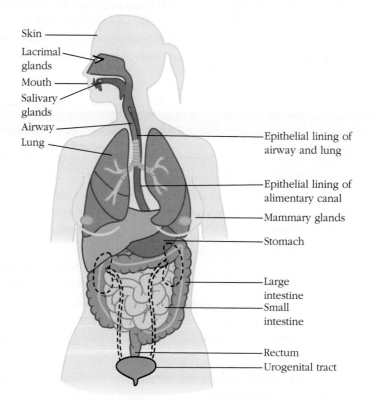

FIGURE 4-2 Skin and other epithelial barriers to infection. In addition to serving as physical barriers, the skin and the mucosal and glandular epithelial layers are defended against microbial colonization by a variety of mechanisms: mechanical (cilia, fluid flow, smooth muscle contraction), chemical (pH, enzymes, antimicrobial peptides), and cellular (resident macrophages and dendritic cells).

wrap" that encases and protects the inner domains of the body from infection. But these anatomical barriers are more than just passive wrappers. They contribute to physical and mechanical processes that help the body shed pathogens and also generate active chemical and biochemical defenses by synthesizing and deploying molecules, including peptides and proteins that have or induce antimicrobial activity.

Epithelial Barriers Prevent Pathogen Entry into the Body's Interior

The skin, the outermost physical barrier, consists of two distinct layers: a thin outer layer, the **epidermis**, and a thicker layer, the **dermis**. The epidermis contains several tiers of tightly packed epithelial cells; its outer layer consists mostly of dead cells filled with a waterproofing protein called *keratin*. The dermis is composed of connective tissue and contains blood vessels, hair follicles, sebaceous glands, sweat glands, and scattered myeloid leukocytes such as dendritic cells, macrophages, and mast cells. The epithelial surfaces of the respiratory, gastrointestinal, and urogenital tracts and the ducts of the salivary, lacrimal, and mammary glands are lined by strong barrier layers of epithelial cells stitched together by tight junctions that prevent pathogens from squeezing between them to enter the body.

A number of nonspecific physical and chemical defense mechanisms also contribute to preventing the entry of pathogens through the epithelia in these secretory tissues. For example, the secretions of these tissues (mucus, urine, saliva, tears, and milk) wash away potential invaders and also contain antibacterial and antiviral substances. Mucus, the viscous fluid secreted by specialized cells of the mucosal epithelial layers, entraps foreign microorganisms; mucins, glycoproteins found in mucus, can prevent pathogen adherence to epithelial cells. In the lower respiratory tract, **cilia**, hairlike protrusions of the cell membrane, cover the epithelial cells. The synchronous movement of cilia propels mucus-entrapped microorganisms from these tracts. Coughing is a mechanical response that helps us get rid of excess mucus, with trapped microorganisms, that occurs in many respiratory infections. The flow of urine sweeps many bacteria from the urinary tract.

With every meal, we ingest huge numbers of microorganisms, but they must run a gauntlet of defenses in the gastrointestinal tract that begins with the antimicrobial compounds in saliva and in the epithelia of the mouth and includes the hostile mix of digestive enzymes and acid found in the stomach. If infection in the gastrointestinal tract does occur, vomiting and diarrhea help remove pathogens from the stomach and intestine. The mucus and acidic pH of vaginal secretions are important in providing protection against bacterial and fungal pathogens. In addition to these chemical barriers, some mucosal epithelial layers, such as in the intestine and reproductive tract, have beneficial commensal microorganisms (normal microbiota) that limit infection by pathogens.

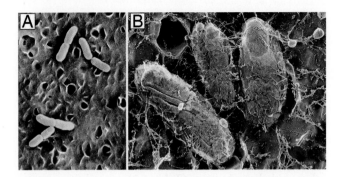

FIGURE 4-3 Electron micrograph of *Escherichia coli* bacteria adhering to the surface of epithelial cells of the urinary tract. *E. coli* is an intestinal bacterial species that causes urinary tract infections affecting the bladder and kidneys. [(a) Courtesy Kazuhiko Fujita, Juntendo University School of Medicine, Tokyo. (b) Matthew A. Mulvey, et al., Bad bugs and beleaguered bladders: Interplay between uropathogenic Escherichia coli and innate host defenses. PNAS USA August 1, 2000 vol. 97 no. 16 8829-8835. Copyright 2000 National Academy of Sciences, U.S.A.]

This may be accomplished by controlling the local microenvironment; an example is the maintenance of acidic pH in the vagina by the release of lactic acid by commensal lactobacteria.

Some organisms have evolved ways to evade these defenses of the epithelial barriers. For example, influenza virus has a surface molecule that enables it to attach firmly to cells in mucous membranes of the respiratory tract, preventing the virus from being swept out by the ciliated epithelial cells. *Neisseria gonorrhoeae*, the bacteria that causes gonorrhea, binds to epithelial cells in the mucous membrane of the urogenital tract. Adherence of these and other bacteria to mucous membranes is generally mediated by hairlike protrusions on the bacteria, called *fimbriae* or *pili*, that have evolved the ability to bind to certain glycoproteins or glycolipids expressed only by epithelial cells of the mucous membrane of particular tissues (**Figure 4-3**).

> **Key Concept:**
> • The epithelial layers that insulate the interior of the body from outside pathogens—the skin and epithelial layers of the mucosal tracts and secretory glands—constitute an anatomical physical barrier that is highly effective in preventing pathogens from entering the rest of the body.

Antimicrobial Proteins and Peptides Kill Would-Be Invaders

To provide strong defense at these barrier layers, epithelial cells secrete a broad spectrum of **antimicrobial proteins and peptides** that provide protection against pathogens. The capacity of skin and other epithelia to produce a wide variety of antimicrobial agents on an ongoing basis is important for controlling the microbial populations on these surfaces, as breaks in these physical barriers from wounds provide routes of infection that would be readily exploited by pathogenic microbes if not defended by biochemical means.

Antimicrobial Proteins

Among the antimicrobial proteins produced by the skin and other epithelia in humans (**Table 4-2**), several are enzymes and binding proteins that kill or inhibit growth of bacterial and fungal cells. **Lysozyme** is an enzyme found in saliva, tears, and fluids of the respiratory tract that cleaves the peptidoglycan components of bacterial cell walls. *Lactoferrin* and *calprotectin* are two proteins that bind and sequester metal ions needed by bacteria and fungi, limiting their growth.

Human skin produces a number of antimicrobial proteins, including *psoriasin*, a small protein of the S-100 family with potent antibacterial activity against *Escherichia coli*, an enteric (intestinal) bacterial species. This finding

TABLE 4-2	Some human antimicrobial proteins and peptides at epithelial surfaces	
Protein/peptide	**Location***	**Antimicrobial activities**
Proteins		
Lysozyme	Mucosal/glandular secretions (e.g., tears, saliva, respiratory tract)	Cleaves glycosidic bonds of peptidoglycans in cell walls of bacteria, leading to lysis
Lactoferrin	Mucosal/glandular secretions (e.g., milk, intestinal mucus, nasal/respiratory and urogenital tracts)	Binds and sequesters iron, limiting growth of bacteria and fungi; disrupts microbial membranes; limits infectivity of some viruses
Secretory leukocyte protease inhibitor	Skin, mucosal/glandular secretions (e.g., intestines, respiratory and urogenital tracts, milk)	Blocks epithelial infection by bacteria, fungi, viruses; antimicrobial
S100 proteins: - Psoriasin - Calprotectin	Skin, mucosal/glandular secretions (e.g., tears, saliva/tongue, intestine, nasal/respiratory and urogenital tracts)	- Disrupts membranes, killing cells - Binds and sequesters divalent cations (e.g., manganese and zinc), limiting growth of bacteria and fungi
Surfactant proteins SP-A, SP-D	Secretions of respiratory tract, other mucosal epithelia	Block bacterial surface components; promote phagocytosis
RegIII proteins	Intestinal epithelia	Bind cell wall carbohydrates and prevent bacterial binding to epithelial cells; produce membrane pores that kill cells
Peptides		
Defensins (α and β)	Skin, mucosal epithelia (e.g., mouth, intestine, nasal/respiratory tract, urogenital tract)	Disrupt membranes of bacteria, fungi, protozoan parasites, and viruses; additional toxic effects intracellularly; kill cells and disable viruses
Cathelicidin (LL-37)[†]	Mucosal epithelia (e.g., respiratory tract, urogenital tract)	Disrupts membranes of bacteria; additional toxic effects intracellularly; kills cells
Histatins	Saliva	Lectins that bind to fungal cell walls and enter the cytoplasm, where they have several harmful effects
Dermcidin	Skin (from sweat glands)	Antibacterial and antifungal; produces channels in membranes that disrupt ion gradients

*Examples listed in this table are all produced by cells in the epithelia of skin and mucosal and glandular tissues; examples of prominent epithelial sites are listed. Most proteins and peptides are produced constitutively at these sites, but their production can also be increased by microbial or inflammatory stimuli. Many are also produced constitutively in neutrophils and stored in granules. In addition, synthesis and secretion of many of these molecules may be induced by microbial components during innate immune responses by various myeloid leukocyte populations (monocytes, macrophages, dendritic cells, and mast cells).

[†]While some mammals have multiple cathelicidins, humans have only one.

answered a long-standing question: Why is human skin resistant to colonization by *E. coli* despite exposure to it from fecal matter resulting from lack of cleanliness or poor sanitation? As shown in **Figure 4-4**, incubation of *E. coli* on human skin for as little as 30 minutes kills the

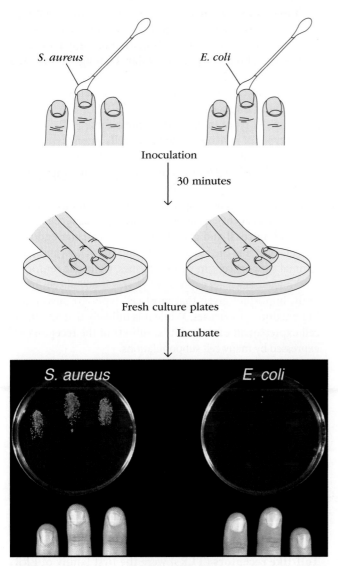

FIGURE 4-4 Psoriasin prevents colonization of the skin by *Escherichia coli* (*E. coli*). Skin secretes psoriasin, an antimicrobial protein that kills *E. coli*. Fingertips of a healthy human were inoculated with *Staphylococcus aureus* (*S. aureus*) and *E. coli*. After 30 minutes, the fingertips were pressed on a nutrient agar plate and the number of colonies of *S. aureus* and *E. coli* determined. Almost all of the inoculated *E. coli* were killed; most of the *S. aureus* survived. *[Reprinted with permission from Nature Publishing Group, from Gläser, R., et al, "Antimicrobial psoriasin (S100A7) protects human skin from Escherichia coli infection", Nature Immunology, 2004 November, 6: 57-64, supplementary figure 1, a & b, Permission conveyed through Copyright Clearance Center, Inc.]*

bacteria, but it does not kill *Staphylococcus aureus* (a major cause of food poisoning and skin infections), demonstrating the specificity of psoriasin for *E. coli*. In contrast, the related protein calprotectin kills *S. aureus* but not *E. coli*.

Additional antimicrobial proteins are made by mucosal epithelial tissues. Antimicrobial proteins produced by intestinal epithelia include members of a family of lectins (carbohydrate-binding proteins), the *RegIII proteins*, which bind carbohydrates on bacterial cell walls, preventing them from contacting the intestinal epithelial cells. RegIII proteins also are directly bactericidal; they generate membrane pores that kill the cells.

The epithelium of the respiratory tract secretes a variety of lubricating lipids and proteins called **surfactants**. Two surfactant proteins, SP-A and SP-D, which are present in the lungs as well as in the secretions of some other mucosal epithelia, are members of a class of microbe-binding proteins called **collectins**. SP-A and SP-D bind differentially to sets of carbohydrate, lipid, and protein components of microbial surfaces and help to prevent infection by blocking and modifying surface components and promoting pathogen clearance. For example, they differentially bind two alternating states of the lung pathogen *Klebsiella pneumoniae* that differ in whether or not they are coated with a thick polysaccharide capsule: SP-A binds the complex polysaccharides coating many of the capsulated forms, while SP-D only binds the exposed cell wall lipopolysaccharide of the nonencapsulated form.

Antimicrobial Peptides

Antimicrobial peptides differ from antimicrobial proteins in that they are generally less than 100 amino acids long. These peptides are an ancient form of innate immunity present in vertebrates, invertebrates, plants, and even some fungi. The discovery that vertebrate skin produces antimicrobial proteins came from studies in frogs, where glands in the skin were shown to secrete peptides called *magainins* that have potent antimicrobial activity against bacteria, yeast, and protozoans. Antimicrobial peptides generally are cysteine-rich, cationic, and amphipathic (containing both hydrophilic and hydrophobic regions). Because of their positive charge and amphipathic nature, they interact with acidic phospholipids in lipid bilayers, forming pores and disrupting the membranes of bacteria, fungi, parasites, and viruses. The peptides then can enter the microbes, where they have other toxic effects, such as inhibiting the synthesis of DNA, RNA, or proteins, and activating antimicrobial enzymes, resulting in cell death.

The main types of antimicrobial peptides found in humans are α- and β-**defensins**, cathelicidin, and histatins. Human defensins kill a wide variety of bacteria, including *E. coli*, *S. aureus*, *Streptococcus pneumoniae*, *Pseudomonas aeruginosa*, and *Haemophilus influenzae*.

Antimicrobial peptides also attack the lipoprotein envelope of enveloped viruses such as influenza virus and some herpesviruses. Defensins and **cathelicidin** LL-37 (the only cathelicidin expressed in humans) are secreted constitutively (i.e., continuously, without activation) by epithelial cells in many tissues, and are stored as well in neutrophil granules, where they contribute to killing phagocytosed microbes. Recent studies have shown that human α-defensin antimicrobial peptides secreted into the gut by intestinal epithelial Paneth cells, located in the deep valleys (crypts) between the villi, are important for maintaining beneficial bacterial flora that are necessary for normal intestinal immune system functions. As we will see later, production of these antimicrobial peptides also can be induced in many epithelial and other cell types by the binding of microbial components to cellular receptors.

Histatins, found in human saliva, are potent antifungal peptides. They bind to surface components on fungal cell membranes and enter the cytoplasm, where they interfere with mitochondrial ATP production and have several other harmful effects. Another antimicrobial peptide, *dermcidin*, is secreted by sweat glands onto the skin, where it has antibacterial and antifungal activities.

Despite the strong physical and chemical barriers of our protective epithelial layers, they may be disrupted by wounds, abrasions, and insect bites that may allow pathogens to pass through the epithelial barrier. Pathogens may also infect epithelial cells, allowing them to pass through this normally solid layer. Pathogens surviving their transit into the tissues below the epithelial layers are then targeted by the innate immune system's second line of defense, an array of cells expressing membrane receptors that recognize microbial components and activate a variety of cellular defense mechanisms against the invaders. The next several sections describe the receptors, the cellular responses that they activate, and their roles in combatting infections.

> **Key Concept:**
> - Epithelial layers provide a chemical barrier to infection, producing a variety of protective substances, including acidic pH, enzymes, binding proteins, and antimicrobial proteins and peptides.

Cellular Innate Response Receptors and Signaling

Several families of cellular **pattern recognition receptors** (**PRRs**) have essential roles in detecting the presence of a pathogen and activating innate immune responses that combat the infection. As introduced in Chapter 3, PRRs bind **pathogen-associated molecular patterns** (**PAMPs**) that trigger cellular responses. Some of these PRRs are expressed on the plasma membrane, where they bind and are activated by extracellular pathogens. Others are found *inside* our cells, either in endosomes/lysosomes where they bind PAMPs released by endocytosed pathogens, or in the cytosol, where they respond to PAMPs such as cytoplasmic bacteria and nucleic acids from replicating viruses. This range of PRR locations ensures that cells can recognize the PAMPs of virtually any pathogen, both extracellular and intracellular. **Damage-associated molecular patterns** (**DAMPs**) released by cell and tissue damage also can be recognized by both cell surface and intracellular PRRs.

Many cell types in the body express these PRRs, including all types of myeloid white blood cells (monocytes, macrophages, neutrophils, eosinophils, mast cells, basophils, dendritic cells) and subsets of three types of lymphocytes (B cells, T cells, and NK cells). PRRs are also expressed by some other cell types, especially those commonly exposed to infectious agents; examples include epithelial cells of the skin and mucosal and glandular tissues, vascular endothelial cells that line the blood vessels, and fibroblasts and other stromal support cells in various tissues. Cytosolic sensors of viral nucleic acids are expressed by most if not all cells in the body, important given that most cell types are susceptible to infection with viruses. While it is unlikely that any single cell expresses all of these PRRs, subsets of the receptors are expressed by many cell subpopulations.

This section will focus on PRR families, the PAMPs that they bind, and the signaling pathways that then are activated that induce the protective responses. The cellular innate immune effector mechanisms that result, including production of a variety of beneficial peptides and proteins, phagocytosis, and regulated cell death, will be described in subsequent sections.

Toll-Like Receptors Initiate Responses to Many Types of Molecules from Extracellular Pathogens

Toll-like receptors (**TLRs**) were the first family of PRRs to be discovered and are still the best characterized in terms of their structure, how they bind PAMPs and activate cells, and the extensive and varied set of innate immune responses that they induce. The story of their discovery (see **Classic Experiment Box 4-1**) shows how results from research in diverse organisms can contribute to revealing fundamental knowledge about human immune responses.

TLRs and Their Ligands

Intensive work over the last two decades has identified 13 TLRs that function as PRRs in humans and mice.

Discovery of Invertebrate Toll and Vertebrate Toll-like Receptors

In the 1980s, researchers in Germany discovered that *Drosophila* fruit fly embryos could not establish a proper dorsal-ventral (back to front) axis if the gene encoding the Toll membrane protein is mutated. (The name "Toll" comes from German slang meaning "weird," referring to the mutant flies' bizarrely scrambled anatomy.) For their subsequent characterization of the *toll* and related homeobox genes and the roles of these genes, which regulate embryonic development, Christiane Nüsslein-Volhard, Eric Wieschaus, and Edward B. Lewis were awarded the Nobel Prize in Physiology or Medicine in 1995. But what does this have to do with innate immune system receptors? Many mutations of the *toll* gene were generated, and in 1996 Jules Hoffmann and Bruno Lemaitre discovered that mutations in *toll* made flies highly susceptible to lethal infection with *Aspergillus fumigatus*, a fungus to which wild-type flies were immune (**Figure 1**). This striking observation led to other studies showing that Toll and related proteins are involved in the activation of innate immune responses in invertebrates. For his pivotal contributions to the study of innate immunity in *Drosophila*, Jules Hoffmann was a cowinner of the 2011 Nobel Prize in Physiology or Medicine.

Characterization of the Toll protein surprisingly revealed that its cytoplasmic signaling domain was homologous to that of the vertebrate receptor for the cytokine IL-1 (IL-1R). Through a search for human proteins with cytoplasmic

FIGURE 1 **Impaired innate immunity in fruit flies with a mutation in the Toll pathway.** Severe infection with the fungus *Aspergillus fumigatus* (yellow) results from a mutation in the signaling pathway downstream of the Toll pathway in *Drosophila* that normally activates production of the antimicrobial peptide drosomycin. *[Republished with permission of Elsevier, from, B. Lemaitre et al., "The dorsoventral regulatory gene cassette spätzle/Toll/cactus controls the potent antifungal response in Drosophila adults," Cell. 1996, Sept; 86 (6): 973-983, Figure 5. Courtesy of J. A. Hoffman, University of Strasbourg. Permission conveyed through Copyright Clearance Center, Inc.]*

domains homologous to those of Toll and IL-1R (now referred to as the Toll/IL-1R [TIR] domain; see Figure 4-5a), in 1997 Charles Janeway and Ruslan Medzhitov discovered a human gene for a protein similar to Toll that activated the expression of innate immunity genes in human cells. Appropriately, this and other vertebrate Toll relatives discovered soon thereafter were named *Toll-like receptors (TLRs)*.

Through studies with mutant mice, in 1998 Bruce Beutler obtained the important proof that TLRs contribute to normal immune functions in mammals. Mice homozygous for a mutant form of a gene called *lps* were resistant to the harmful responses induced by lipopolysaccharide (LPS; also known as *endotoxin*), a major component of the cell walls of *gram-negative bacteria* (see Figure 4-7). In humans, a buildup of endotoxin from severe bacterial infection can induce a too-strong innate immune response, causing septic shock, a life-threatening condition in which vital organs such as the brain, heart, kidneys, and liver may fail. Each year, about 20,000 people die in the United States of septic shock caused by gram-negative bacterial infections, so it was striking that some mutant strains of mice were resistant to fatal doses of LPS. Beutler found that the defective mouse *lps* gene encoded a mutant form of one TLR, TLR4, which differed from the normal form by a single amino acid so that it no longer was activated by LPS. This work provided an unequivocal demonstration that TLR4 is the cellular innate pattern recognition receptor that recognizes LPS and earned Beutler a share of the 2011 Nobel Prize.

Thus, this landmark series of experiments showed in rapid succession that invertebrates respond to pathogens, that they use receptors also found in vertebrates, and that one of these receptors is responsible for LPS-induced innate immune responses.

TLRs are membrane proteins that share a common structural element in their extracellular region called **leucine-rich repeats (LRRs)**; multiple LRRs make up the horseshoe-shaped extracellular ligand-binding domain of the TLR polypeptide chain (**Figure 4-5a**). When TLRs bind their PAMP or DAMP ligands via their extracellular LRR domains, they are induced to dimerize, either as a homodimer (e.g., TLR3/3) or as a heterodimer (e.g., TLR2/1) (Figure 4-5b).

As shown in **Figure 4-6**, TLRs exist both on the plasma membrane and in the membranes of endosomes and lysosomes; their cellular location is tailored to enable them

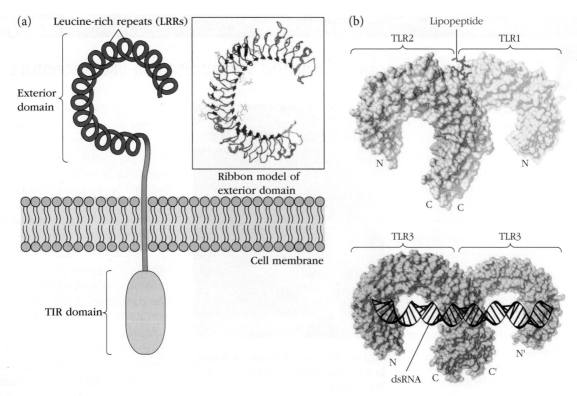

FIGURE 4-5 Toll-like receptor (TLR) structure and binding of PAMP ligands. (a) Structure of a TLR polypeptide chain. Each TLR polypeptide chain is made up of a ligand-binding exterior domain that contains many leucine-rich repeats (LRRs, repeating segments of 24–29 amino acids containing the sequence LxxLxLxx, where L is leucine and x is any amino acid), a membrane-spanning domain (blue), and an interior Toll/IL-1R (TIR) domain (yellow), which interacts with the TIR domains of other members of the TLR signal transduction pathway. In the presence of ligands, two such polypeptide chains pair to form TLR dimers (except for TLR8, which exists in a dimer form in the absence of ligand). (b) Crystal structures of TLR dimers (extracellular LRR domains only) with bound PAMP ligands. *Top*: TLR2/1 dimer with a bound lipopeptide molecule. *Bottom*: TLR3/3 dimer with bound double-stranded RNA (dsRNA) molecule. *[Part (b) data from Jin, M. S., and J.-O. Lee. 2008. Structures of the Toll-like receptor family and its ligand complexes.* Immunity ***29***:182. PDB IDs 2Z7X (top) and 3CIY (bottom).]*

to respond optimally to the particular microbial ligands they recognize. TLRs can bind via their LRR domains to a wide variety of conserved PAMPs from bacteria, including cell wall lipopolysaccharides (LPS) from gram-negative bacteria and peptidoglycans from gram-positive bacteria (**Figure 4-7**), as well as flagellin and bacterial nucleic acids. TLRs also recognize PAMPs from viruses (RNA, DNA, and protein), fungi (cell wall polysaccharides), and parasites (proteins and other components), as well as DAMPs from damaged cells and tissues.

Each TLR has a distinct repertoire of specificities (see **Table 4-3**). These PAMPs are usually from extracellular pathogens; plasma membrane TLRs recognize components on the outside of pathogens (e.g., LPS, peptidoglycan, and flagellin), while endosomal TLRs recognize components released during endosomal/lysosomal degradation (e.g., bacterial and viral nucleic acids). Similarly, extracellular DAMPs can be recognized by plasma membrane TLRs or endosomal TLRs after degradation. TLR4 is unique in that it can be found both on the plasma membrane and, after endocytosis, in endosomes, where it can bind PAMPs both on the outside and inside of pathogens.

TLR Signaling Pathways

Given the wide variety of potential pathogens that the innate immune system needs to recognize and combat, how does binding of a specific pathogen evoke an appropriate response for that pathogen? Signaling through TLRs uses many of the principles and some of the signaling molecules described in Chapter 3, along with some unique to pathways activated by TLRs (and by other PRRs, described below).

Studies of the signaling pathways downstream of all of the TLRs have revealed that they include some shared components and activate expression of many of the same genes. An important example of a shared component is the transcription factor **NF-κB**, which is of key importance for activating the expression of many innate and inflammatory genes. There are also signaling pathways and components

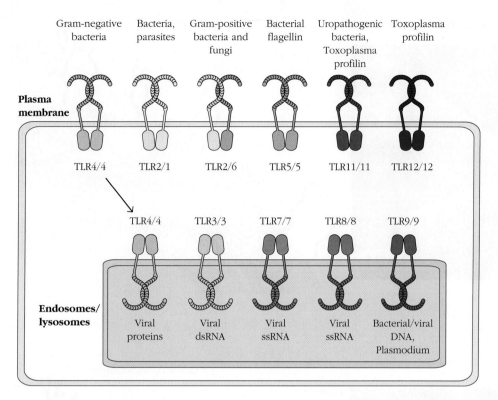

FIGURE 4-6 Cellular location of TLRs. TLRs that interact with extracellular ligands reside in the plasma membrane; TLRs that bind ligands released by endocytosed microbes localize to endosomes and/or lysosomes. On ligand binding, the TLR4/4 dimer moves from the plasma membrane to the endosomal/lysosomal compartment, where it activates different signaling components.

activated only by some TLRs that induce the expression of subsets of proteins, some of which are particularly effective in combating the type of pathogen recognized by the particular TLR(s).

An important example is expression of the potent antiviral **type I interferons**, IFN-α and IFN-β, induced by pathways downstream of the TLRs that bind viral components. As we will see, activation of the **interferon regulatory factors** (**IRFs**) is essential for inducing transcription of the genes encoding IFN-α and IFN-β. Combinations of transcription factors contribute to inducing the expression of many of these genes; examples include combinations of NF-κB, IRFs, and/or transcription factors downstream of MAP kinase (MAPK) pathways, such as AP-1, that can be activated by signaling intermediates downstream of certain TLRs.

TLR signaling is initiated following ligand-induced TLR dimerization. The particular signal transduction pathway(s) activated by a TLR dimer following PAMP or DAMP binding to the LRR extracellular domain is/are largely determined by the protein adaptor that binds to the TLR's cytoplasmic **TIR** (*Toll/IL-1 r*eceptor) domain (see Figure 4-5a). As described in Classic Experiment Box 4-1, the TLR signaling domain was given this name because it is common to both TLRs and the receptor for the cytokine IL-1. The two key adaptors that are recruited to TLR dimers are **MyD88** (*my*eloid *d*ifferentiation factor *88*) and **TRIF** (*TIR* domain–containing adaptor-inducing IFN-β factor). MyD88 is the adaptor used by most TLRs—all of the plasma membrane TLRs and most of those in endosomes. TRIF uniquely associates with TLR3 and also with TLR4 when it localizes to endosomes.

MyD88-dependent signaling pathways

The signaling pathways activated by plasma membrane TLR2/1 after binding a PAMP, such as lipoprotein, are typical of other plasma membrane TLRs, all of which use the MyD88 adaptor (**Figure 4-8**). After associating with the now-dimerized TIR domains, MyD88 initiates a signaling pathway that activates the NF-κB and MAPK pathways (see Figure 3-24. As shown in Figure 4-8, MyD88 recruits **IRAK1** (*IL-1 r*eceptor–*a*ssociated *k*inase *1*) and IRAK4. IRAK1 phosphorylates itself and TRAF6 (*t*umor necrosis factor *r*eceptor–*a*ssociated *f*actor *6*), activating it. TRAF6 creates a scaffold that serves as an organizing center for subsequent signaling components. The adapter proteins **TAB1** and **TAB2** (*TAK1-b*inding proteins *1* and *2*) bring associated

(a) Gram negative bacteria
   E. coli   Cell wall organization

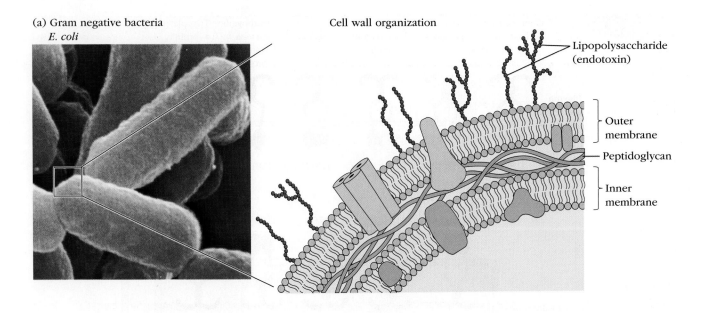

Lipopolysaccharide
(endotoxin)

Outer
membrane

Peptidoglycan

Inner
membrane

(b) Gram positive bacteria
   S. aureus   Cell wall organization

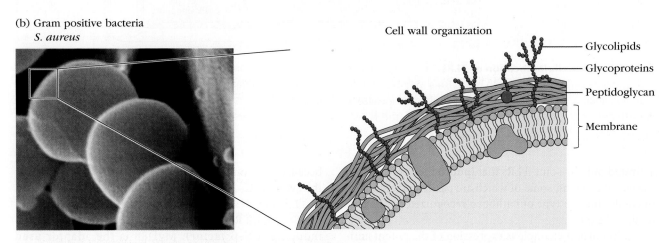

Glycolipids

Glycoproteins

Peptidoglycan

Membrane

FIGURE 4-7 Cell wall components of gram-negative and gram-positive bacteria. Cell wall structures differ for gram-negative (a) and gram-positive (b) bacteria. Because of their thick peptidoglycan layer, gram-positive bacteria retain the precipitate formed by the crystal violet and iodine reagents of the Gram stain, whereas the stain is easily washed out of the less-dense cell walls of gram-negative bacteria. Thus the Gram stain identifies two distinct sets of bacterial genera that also differ significantly in other properties. *[(a) Dr. Kari Lounatmaa / Science Source. (b) Eye of Science / Science Source.]*

TAK1 (*transforming growth factor-β–activated kinase 1*) into proximity with IRAK1, which phosphorylates and activates TAK1.

The **IKK** (*inhibitor of κB kinase*) complex, consisting of **NEMO** (*NF-κB essential modifier*), **IKKα**, and **IKKβ**, is then recruited, enabling TAK1 to phosphorylate and activate IKKβ. This leads to the final steps resulting in the activation of NF-κB. Inactive NF-κB is retained in the cytoplasm by its IκB (*inhibitor of NF-κB*) subunit. Activated IKK phosphorylates IκB, leading to its degradation and the release of NF-κB, allowing it to enter the nucleus and activate gene expression.

TAK1 does double duty in this TLR signaling cascade. After separating from the IKK complex, it activates MAPK signaling pathways that result in the activation of transcription factors including Fos and Jun, which make up the AP-1 dimer. Both NF-κB and AP-1 are essential for activating key antimicrobial proteins and peptides as well as proinflammatory cytokines and chemokines that are of key importance in the innate immune response.

Similar processes initiated by MyD88 activate NF-κB and MAPK pathways downstream of the endosomal TLRs, TLR7

TABLE 4-3 Toll-like receptors and their microbial ligands

TLR*	Ligand(s)	Microbes
TLR1	Triacyl lipopeptides	Mycobacteria and gram-negative bacteria
TLR2	Peptidoglycans GPI-linked proteins Lipomannan, lipoproteins Zymosan Phosphatidylserine	Gram-positive bacteria Trypanosomes Mycobacteria and other bacteria Yeasts and other fungi Schistosomes
TLR3	Double-stranded RNA (dsRNA)	Viruses
TLR4	LPS F protein G glycoprotein Mannans	Gram-negative bacteria Respiratory syncytial virus (RSV) Vesicular stomatitis virus (VSV) Fungi
TLR5	Flagellin	Bacteria
TLR6	Diacyl lipopolypeptides Zymosan	Mycobacteria and gram-positive bacteria Yeasts and other fungi
TLR7	Single-stranded RNA (ssRNA)	Viruses
TLR8	Single-stranded RNA (ssRNA)	Viruses
TLR9	CpG unmethylated dinucleotides Dinucleotides Herpesvirus components Hemozoin	Bacterial DNA Some herpesviruses Malaria parasite heme by-product
TLR10	Unknown	Unknown
TLR11	Unknown Profilin	Uropathogenic bacteria *Toxoplasma gondii*
TLR12	Profilin	*Toxoplasma gondii*
TLR13	rRNA Unknown	Gram-positive bacteria Vesicular stomatitis virus

*All function as homodimers except TLR1, TLR2, and TLR6, which form TLR2/1 and TLR2/6 heterodimers. Ligands indicated for TLR2 bind to both; ligands indicated for TLR1 bind to TLR2/1 dimers, and ligands indicated for TLR6 bind to TLR2/6 dimers. TLR10 is present in humans but not mice; TLR11, TLR12, and TLR13 are found in mice but not humans.

(**Figure 4-9**) as well as TLR8 and TLR9, all of which bind microbial nucleic acids. In addition, signaling downstream of these TLRs also activates pathways that induce production of IFN-α and IFN-β, potent antiviral agents. When triggered by these TLRs, the MyD88-associated IRAK1 directly phosphorylates IRF7. This allows IRF7 dimerization, activation, nuclear translocation, and induction of IFN gene expression.

TRIF-dependent signaling pathways

As mentioned above, after the endosomal TLR3 dimer binds viral double-stranded RNA, it associates with the TRIF adaptor instead of MyD88 (Figure 4-9). TRIF binds and activates TRAF3, which generates a scaffold that recruits a kinase complex containing adaptors NEMO and TANK (*T*RAF family member–*a*ssociated *N*F-κB activator) and

protein kinases IKKε and TBK1 (*TANK-b*inding *k*inase 1). TBK1 phosphorylates and activates IRF3 and IRF7, each of which dimerizes and enters the nucleus, inducing the transcription of the IFN-α and IFN-β genes. As a somewhat later response, TRIF also can activate TRAF6, initiating signaling events that lead to some NF-κB activation and inflammatory cytokine production.

The main message from the above description of TLR signaling pathways is that various TLRs may differentially activate distinct transcription factors (in particular, IRFs vs. NF-κB and AP-1), leading to variation in which genes are turned on. By and large, the proteins that are produced will best protect us against the invading pathogens. In particular, the TLRs that bind bacterial PAMPs stimulate production of antimicrobial proteins and peptides, enzymes, and

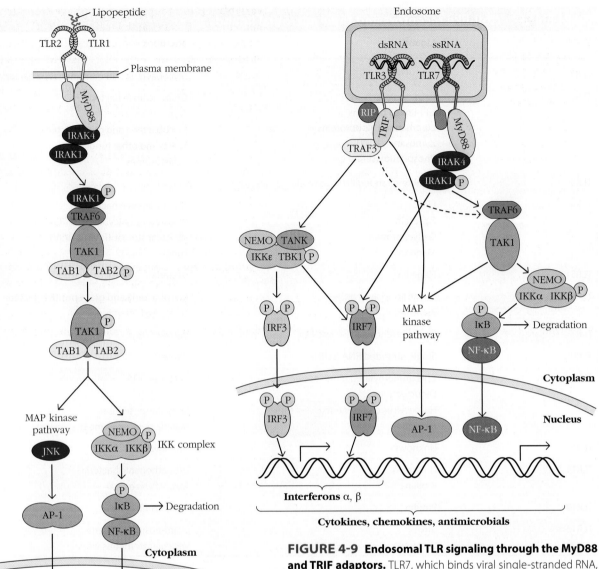

FIGURE 4-8 Plasma membrane TLR signaling through the MyD88 adaptor. Signaling pathways downstream of TLR2/1, which binds a bacterial triacyl lipopeptide PAMP, are shown, representative of signaling by other plasma membrane TLRs. After PAMP-induced TLR dimerization, the TIR-binding adaptor MyD88 initiates signaling by recruiting the IRAK1 and IRAK4 kinases. Additional proteins are recruited, including TRAF6 and the TAK1 kinase complex, leading to TAK1 phosphorylation and activation of MAP kinase pathways, which activate transcription factors such as AP-1, and the IKK complex, leading to the activation of NF-κB. *Note:* Not all signaling components are shown.

FIGURE 4-9 Endosomal TLR signaling through the MyD88 and TRIF adaptors. TLR7, which binds viral single-stranded RNA, activates signaling through MyD88. The pathway for activating NF-κB is similar to that shown for plasma membrane TLRs (see Figure 4-8; not all steps are shown). In addition, IRAK1 directly phosphorylates IRF7, which dimerizes and enters the nucleus, where it activates transcription of IFN-α and IFN-β genes. TLR3, which binds viral double-stranded RNA, signals through the adaptor TRIF. TRIF recruits and activates TRAF3, which activates an IKK complex that phosphorylates and activates both IRF3 and IRF7. Later in the response, TRIF also activates TRAF6, leading to NF-κB activation (dashed line).

proinflammatory cytokines including IL-1β and tumor necrosis factor (TNF), and chemokines important for antibacterial responses. In contrast, all of the intracellular TLRs that bind viral PAMPs following internalization and endosomal release of viral nucleic acids induce the synthesis and secretion of type I interferons, which inhibit the replication of the virus in infected cells.

C-Type Lectin Receptors Bind Carbohydrates on the Surfaces of Extracellular Pathogens

The second family of cell surface PRRs that activate innate and inflammatory responses is the **C-type lectin receptor (CLR)** family. CLRs are membrane receptors expressed variably on monocytes, macrophages, dendritic cells, neutrophils, B cells, and T-cell subsets. CLRs generally recognize carbohydrate components of fungi, mycobacteria, viruses, parasites, and some allergens (peanut and dust mite proteins). Humans have at least 15 CLRs that function as PRRs, most of which recognize one or more specific sugar moieties such as mannose (e.g., the mannose receptor and DC-SIGN), fucose (e.g., dectin-2 and DC-SIGN), and glucans (e.g., dectin-1).

CLRs trigger signaling pathways that activate transcription factors, which in turn induce effector gene expression. As illustrated for dectin-1, which binds its glucan PAMPs as a dimer (**Figure 4-10**), most CLRs initiate signaling through protein kinase–mediated phosphorylation of tyrosine residues in their cytoplasmic domains or associated signaling chains. Dectin-1 has one tyrosine, in what is a half-ITAM (*immunoreceptor tyrosine-based activation motif*), related to ITAMs in B-cell receptor (BCR) or T-cell receptor (TCR) signaling chains that become phosphorylated by tyrosine kinases following ligand binding. Activated tyrosine kinases trigger signaling cascades that activate phospholipase Cδ

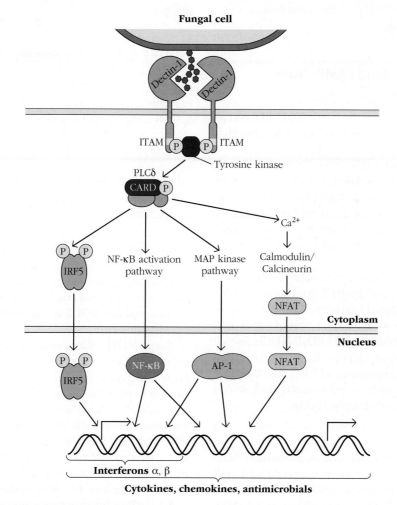

FIGURE 4-10 CLR signaling pathways. Signaling pathways downstream of the CLR dectin-1 are shown. Dectin-1 binds fungal glucans as a dimer. The tyrosine in the half-ITAM (*immunoreceptor tyrosine-based activation motif*), located in the cytoplasmic domain of each dectin-1, is phosphorylated, initiating signaling pathways that activate transcription factors NFAT, NF-κB, IRF5, and AP-1. *Note:* Not all signaling components are shown.

(PLCδ), which activates several CARD-containing complexes. These in turn lead via an increase in intracellular Ca²⁺ to NFAT activation, NF-κB activation, and MAPK pathways resulting in formation of the AP-1 transcription factor.

These transcription factors cooperate in inducing expression of genes for proinflammatory cytokines such as IL-1β and TNF, as well as IL-23, which promotes T-cell production of IL-17, an inflammatory cytokine that is important for antifungal responses. Dectin-1 also activates IRF5, leading to production of IFN-β, shown to enhance antifungal innate responses. As will be discussed below, early signaling events downstream of dectin-1 and several other CLRs also activate both phagocytosis of the bound fungal or mycobacterial cell and production of reactive oxygen species that kill the phagocytosed pathogen (see below).

> **Key Concept:**
>
> - C-type lectin receptors (CLRs) bind fungal and bacterial cell wall components, largely sugars and polysaccharides. Binding of these PAMPs triggers a variety of distinct signaling pathways that activate transcription factors that induce the expression of inflammatory cytokines.

NOD-Like Receptors Bind PAMPs from Cytosolic Pathogens

NLR is an acronym that stands for both **NOD-*like* receptor** and **nucleotide oligomerization domain/leucine-rich repeat–containing receptor**. The NLRs are a large family of cytosolic proteins activated by intracellular PAMPs and substances that alert cells to damage or danger (DAMPs and other harmful substances). They play major roles in activating beneficial innate immune and inflammatory responses, but, as we will see, some NLRs also trigger inflammation that causes extensive tissue damage and disease.

The human genome contains approximately 23 NLR genes, and the mouse genome up to 34. NLR proteins are divided into three major groups, based largely on their domain structure, as shown in **Figure 4-11**: NLRCs (some of which have *c*aspase *r*ecruitment *d*omains, or CARDs), NLRBs (which have *b*aculovirus *i*nhibitory *r*epeat [BIR] domains), and NLRPs (which have *py*rin *d*omains, or PYDs). The functions of many NLRs have not yet been well characterized; several NLRs and their functions are described below.

NOD1 and NOD2

NOD1 and NOD2 are well-characterized cytosolic NLRs that bind breakdown products of bacterial cell wall peptidoglycans. These PAMPs, diaminopimelic acid and muramyl dipeptides, which bind to NOD1 and NOD2, respectively, are generated during the synthesis or degradation of peptidoglycans of either cytosolic or endocytosed

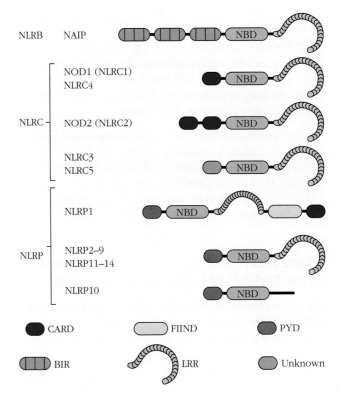

FIGURE 4-11 NLR protein domains. NLRs are characterized by distinct protein domains. Most NLRs have a leucine-rich repeat (LRR) domain, similar to those in TLRs, that functions in ligand binding in at least some NLRs, and a nucleotide-binding domain (NBD). The three main classes are distinguished by their N-terminal domain (the initial [*left*] domains in the figure): NLRC receptors have *c*aspase *r*ecruitment *d*omains (CARDs), NLRP receptors have *py*rin *d*omains (PYDs), and NLRB receptors have BIR (*b*aculovirus *i*nhibitor *r*epeat) domains. These domains function in protein-protein interactions, largely through homotypic domain interactions.

bacteria—peptides from the latter must enter the cytosol to activate NODs. Studies in mice have shown that NOD1 also provides protection against the intracellular protozoan parasite *Trypanosoma cruzi*, which causes Chagas disease in humans, and that NOD2 activates responses to some viruses, including influenza.

These NOD PRRs associate with the membrane of endosomes, where they efficiently bind bacterial components transported through endosomal membranes. PAMP binding to the LRR regions of NODs initiates signaling by activating NOD binding to RIP2 (*r*eceptor-*i*nteracting *p*rotein *k*inase *2*) through interactions between the CARDs of NOD and RIP2 (**Figure 4-12**). RIP2 then binds the TAK1/TAB complex, leading to its activation of MAPK pathways and of the IKK complex, which initiates the NF-κB activation pathway, as described earlier. The activated AP-1 and NF-κB induce transcription of inflammatory cytokines and antimicrobial and other mediators. In addition, in some cells RIP2 activates the TRAF3 complex, leading

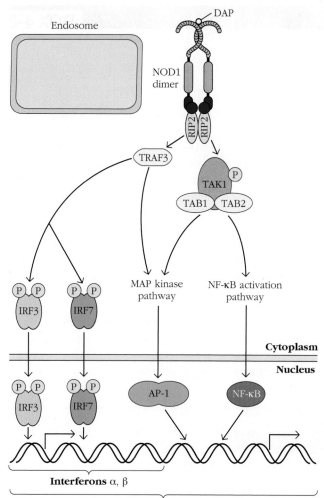

FIGURE 4-12 Signaling from the NOD1 NLR. NOD1 associates with endosomes, where it can bind PAMPs such as diaminopimelic acid (DAP, a fragment of cell peptidoglycans) from cytosolic or endocytosed bacteria. After dimerization NOD1 dimers recruit RIP2 (receptor-interacting protein kinase 2), which then binds the TAK1/TAB complex, activating MAPK pathways and also the NEMO/IKK complex, which initiates the NF-κB activation pathway. In dendritic cells NOD1 binding of RIP2 also activates TRAF3, leading to the phosphorylation and activation of IRF3 and IRF7 and IFN-β production.

to phosphorylation of IRF3 and IRF7 and to production of Type I interferons.

In addition to inducing expression of genes encoding antimicrobial proteins and peptides, NOD1 and NOD2 contribute to the elimination of cytosolic bacteria by initiating **autophagy**, in which membrane from the endoplasmic reticulum surrounds the bacteria, forming an autophagosome, which then fuses with lysosomes, killing the bacteria.

NLR Inflammasomes

Some NLRs contain domains that allow them to assemble with other proteins into large complexes. An example is the PYD of NLRPs (see Figure 4-11). Following binding of PAMPs or cellular proteins, these complexes activate associated proteases called *caspases*, which convert the inactive large precursor forms (procytokines) of the important cytokines IL-1β and IL-18 into the smaller, mature forms that are secreted by activated cells. Caspase activation can also induce the death of the activated macrophage through **pyroptosis**, allowing the release of the mature IL-1β and IL-18. Because of the very potent inflammatory effects of secreted IL-1β (and also to some extent IL-18), to be discussed below, these large complexes of NLRs with caspases and other proteins are referred to as **inflammasomes**. The discovery and properties of inflammasomes, which continue to surprise investigators as more is learned, are described in **Advances Box 4-2**.

> **Key Concepts:**
> - NOD-like receptors (NLRs) are a large family of cytosolic PRRs activated by intracellular PAMPs, DAMPs, and other harmful substances. The NLRs bind intracellular microbial components such as cell wall fragments and initiate signaling pathways that activate the NF-κB, MAPK, and IRF pathways.
> - Some NLRs assemble into inflammasomes, large protein complexes that cleave and activate the large precursors of the proinflammatory cytokines IL-1β and IL-18.

ALRs Bind Cytosolic DNA

The *AIM2-like receptors* (**ALRs**) are cytosolic receptors that bind DNA molecules from bacteria and viruses. Like members of the NLRP family, ALRs contain an amino (N)-terminal PYD. However, rather than having LRR domains, ALRs contain one or two copies of the oligonucleotide-binding **HIN** (*hematopoietic expression, interferon inducibility, nuclear localization*) **domain** at the carboxyl (C) terminus, which acts as the DNA-binding unit of the receptor. Thus ALRs are sometimes referred to as the *PYHIN family*.

In both NLRs and ALRs, the PYD serves as the effector domain transmitting downstream signals to the cellular machinery. The prototype ALR, AIM2, binds long double-stranded DNA (dsDNA) from cytosolic bacteria and viruses via its HIN domains. Binding of multiple ALRs to the same strand of dsDNA allows their PYDs to associate, driving assembly into long filaments that, together with the protein ASC and procaspase-1, form inflammasomes and generate mature IL-1β and IL-18 (see Advances Box 4-2). A second ALR, IFI16, also binds viral DNA in either the cytosol or the nucleus; its signaling pathways are still being characterized but may include inflammasome activation of inflammatory cytokines as well as other pathways activating IRFs and interferon production.

Inflammasomes

The cytokine interleukin (IL)-1 has been recognized as one of the most potent inducers of inflammation. (Note that there are actually two IL-1 cytokines, IL-1α and IL-1β, encoded by different genes. IL-1β is the major cytokine produced during innate and inflammatory responses, and most of what follows focuses on how active IL-1β is produced.) While it was known that the IL-1β gene is transcriptionally activated following exposure to PAMPs and DAMPs (pathogen-, damage-, and danger-associated molecules), additional steps clearly were needed to generate the mature IL-1β protein from its large pro-IL-1β precursor inside the cell.

An enzyme, initially called IL-1–converting enzyme (ICE), now known as caspase-1, was shown to carry out the cleavage of pro-IL-1β, but the enzyme itself existed in most cells as a large inactive precursor. The breakthrough in understanding how mature IL-1β is produced came in 2002 when Jürg Tschopp and others published biochemical studies showing that activation of cells by bacterial lipopolysaccharide (LPS) induced the formation of a large multiprotein aggregate containing an NLR and mature caspase-1 that cleaved pro-IL-1β into mature IL-1β, allowing its release from cells. Because of the importance of IL-1β in promoting inflammation, Tschopp and his colleagues coined the term *inflammasome*

for the large protein complex that activates caspase-1 to generate IL-1β. Three NLRs (NLRP1, NLRP3, and NLRC4) have been shown to form inflammasomes that activate caspase-1 to cleave the large precursors of both IL-1β and IL-18, generating the mature proinflammatory cytokines.

The NLRP3 inflammasome has been of great interest, as mutations in the NLRP3 gene are associated with several auto-inflammatory diseases in which the mutant NLRPs stimulate ongoing excessive caspase-1 activity, IL-1β production, and often debilitating inflammation. This inflammasome, which is expressed by monocytes, macrophages, neutrophils, dendritic cells, and some lymphocytes

(a)

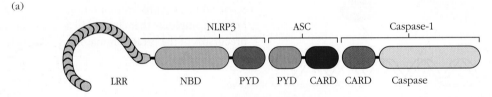

(b)

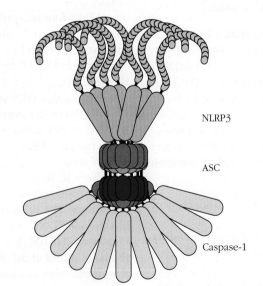

Sterile activators

Self-derived
ATP
Cholesterol crystals
Urate crystals
Glucose
Amyloid β
Hyaluronan

Environment-derived
Alum
Asbestos
Silica
Alloy particles
UV radiation
Skin irritants

NLRP3

ASC

Caspase-1

Pathogen activators

Bacteria-derived
Pore-forming toxins
(*S. aureus, Clostridia*)
Flagellin
Peptidoglycan fragment
RNA
DNA

Virus-derived
RNA
Influenza M2 protein

Fungus-derived
β-glucans
Mannan
Zymosan
Hyphae

Protozoa-derived
Hemozoin

FIGURE 1 The NLRP3 inflammasome and its activators. Assembly of the NLRP3 inflammasome is due to aggregation and homotypic domain interactions between the three component proteins. (a) Domains of NLRP3, ASC, and caspase-1. (b) Structure of an assembled NLRP3 inflammasome, with its activators. Activators are divided into two categories: sterile activators include nonmicrobial self- and environment-derived molecules; pathogen-associated activators include PAMPs derived from bacteria, viruses, fungi, and protozoa. (Abbreviations: ASC = apoptosis-associated speck-like protein containing a caspase recruitment domain; for others, see legend to Figure 4-11 and text.) *[Republished with permission of Annual Reviews, from Netea M.G., et al., Inflammasome-independent regulation of IL-1-family cytokines, from Annual Review of Immunology, 2015 March; 33:49–77, Figure 2. Permission conveyed through Copyright Clearance Center, Inc.]*

(continued)

and epithelial cells, is a large complex containing multiple copies each of NLRP3, the adapter protein ASC (which binds to NLRP3 via homotypic PYD-PYD interactions), and caspase-1 (which binds to ASC via homotypic CARD-CARD interactions) (**Figure 1**).

NLRP3 inflammasomes can be activated in cells by a variety of components from bacteria (including pore-forming toxins that allow ion flux through the plasma membrane), fungi, and some viruses. In addition to these pathogen components, NLRP3 can also be activated by nonmicrobial ("sterile") substances, including several DAMPs released by damaged tissues and cells, such as hyaluronan, β-amyloid (associated with Alzheimer's plaques), and extracellular ATP and glucose. Recent research has also implicated NLRP3 in mediating serious inflammatory conditions caused by an unusual class of harmful substances: crystals. Crystals of monosodium urate in individuals with hyperuricemia cause gout, an inflammatory joint condition, and inhalation of environmental silica or asbestos crystals causes the serious, often fatal, inflammatory lung conditions silicosis and asbestosis. When these crystals are phagocytosed, they damage lysosomal membranes, releasing lysosomal components into the cytosol. Similar effects may be responsible for the loosening (aseptic osteolysis) of artificial joints, caused by tiny metal alloy particles from the prosthesis that also activate NLRP3-mediated inflammation.

How these disparate PAMPs, DAMPs, crystals, and metal particles all activate the NLRP3 inflammasome is under intense investigation. To date there is no conclusive evidence for the binding of any of these substances directly to NLRP3; thus it may not act as a PRR for PAMPs and DAMPs per se but instead may sense changes in the intracellular milieu resulting from exposure to these materials. The current model is that two signals are required to activate NLRP3 inflammasomes and the generation of mature IL-1β and IL-18. Binding of PAMPs or DAMPs to TLRs, CLRs, or NOD NLRs primes the inflammatory response by inducing transcription of the genes for pro-IL-1β and pro-IL-18 and also for NLRP3 itself by the signaling pathways described earlier (signal 1 in **Figure 2**).

However, a second signal is required to activate inflammasome formation and function. Clues about the nature of signal 2 have come from a common set of intracellular signals triggered by many or all of these activators, such as potassium ion efflux, reactive oxygen species (ROS), and/or leakage of lysosomal contents, all of which appear to induce NLRP3 inflammasome assembly and caspase-1 activation. But how do these varied agents lead to NLRP3 inflammasome activation? Recent studies have shown that they may all act through ROS, which appears to induce the interaction of a protein kinase called NEK7 with NLRP3, triggering inflammasome formation and activation (Figure 2). Cells from mice defective in NEK7 do not generate mature IL-1β in response to the activators listed in Figure 1.

Other inflammasomes have been identified that are also large multiprotein aggregates containing NLRs or related proteins and caspases that generate mature IL-1 and IL-18. Some are activated directly by binding bacterial components.

The recent discovery of inflammasomes and their functions has provided a partial answer to one long-standing question: how is mature IL-1β produced in response to innate and inflammatory stimuli? One remaining puzzle is how the mature IL-1β and IL-18 cytokines are released from the cell after they are generated from their large precursors, as they are in the cytosol and not in secretory vesicles. Pyroptosis, the induced killing of cells such as macrophages by caspase-1–activated

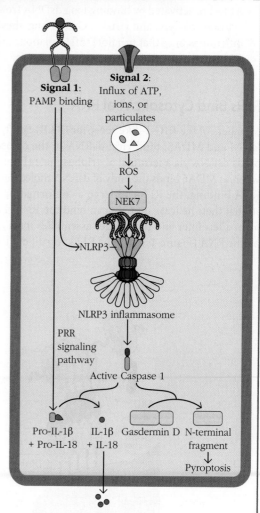

FIGURE 2 **Activation of inflammasomes.** *Left:* Binding of a PAMP to a PRR (e.g., a TLR) activates NF-κB and MAPK signaling pathways that induce the transcription and synthesis of large cytokine precursors pro-IL-1β and pro-IL-18, and of the NLR NLRP3. This constitutes signal 1 for the generation of mature IL-1β and IL-18. *Right:* Activation of the NLRP3 inflammasome. Various sterile and pathogen-derived activators (see Figure 1) initiate changes that constitute signal 2, eventually leading to the binding of protein kinase NEK7 to NLRP3, triggering its assembly with ASC and procaspase-1 into the NLRP3 inflammasome (schematized here). Procaspase-1 is cleaved, generating active caspase-1, which cleaves pro-IL-1β and pro-IL-18 into the mature cytokines. Caspase-1 also generates a fragment of gasdermin that induces cells to undergo pyroptosis.

gasdermin D (see Figure 2), is one mechanism for release of the cytokines, but whether there are any others is still uncertain.

REFERENCE

B. K. Davis et al. 2011. *Annual Review of Immunology* 29:707.

RLRs Bind Cytosolic Viral RNA

Members of the **RIG-I–like receptor** (**RLR**) family of PRRs, RIG-I and MDA5, bind viral dsRNA in the cytosol. RIG-I binds dsRNA via a terminal 5′-triphosphate (**Figure 4-13a**), whereas MDA5 binds the body of dsRNA molecules. On viral RNA binding, the RLRs undergo a conformational change so that their helicase domain can bind the RNA. Four RLRs form a tetramer and then larger assemblies form a filament on the RNA (Figure 4-13b). The RLRs recruit multiple copies

of their adapter molecule, the mitochondrial membrane-associated MAVS (*mitochondrial antiviral signaling*) protein, via association of their shared CARDs. MAVS proteins aggregate and recruit additional proteins such as TRAFs, leading to the activation of NEMO/IKKα/IKKβ and TBK1/IKKε. As in other PRR signaling pathways, these two intermediate protein kinase complexes activate NF-κB and IRF3 and IRF7, respectively, inducing expression of the potent antiviral proteins IFN-α and IFN-β, as well as cytokines.

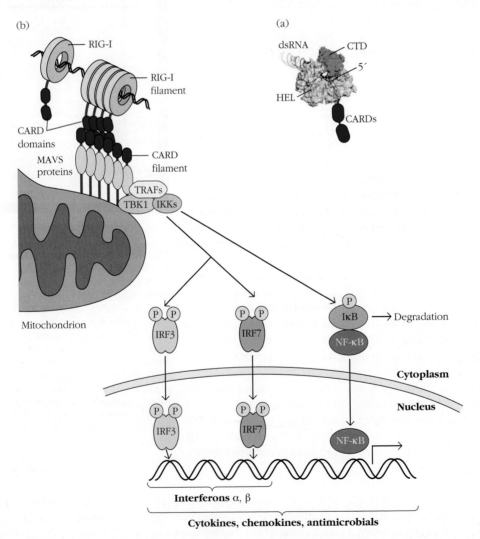

FIGURE 4-13 Signaling from the RIG-I RLR. (a) The recognition of viral RNA molecules by RIG-I. The C-terminal domain (CTD) interacts with the 5′ ends of short, duplex RNA helices bearing 5′ nucleoside triphosphates, inducing a conformational change allowing the helicase (HEL) domain to bind, as shown. (b) Multiple RIG-I RLRs associate with the RNA, forming a filament. Their CARDs bind to CARDs of MAVS proteins associated with the mitochondrial membrane, recruiting and activating TRAFs, TBK1, and IKKs, leading to the activation of NF-κB and IRF3 and IRF7.

cGAS and STING Are Activated by Cytosolic DNA and Dinucleotides

A recently discovered cytosolic PRR, **cGAS** (*cyclic GMP-AMP synthase*), recognizes cytosolic DNA, usually of viral or bacterial origin. cGAS is a nucleotidyltransferase; after it binds dsDNA it is activated to synthesize cGAMP (2′,5′-cyclic GMP-AMP dinucleotide) from GTP and ATP (**Figure 4-14**). cGAMP then serves as a second messenger, binding the endoplasmic reticulum (ER) membrane–associated protein **STING** (*st*imulator of *in*terferon *g*enes). STING had previously been identified because of its ability to bind the cyclic dinucleotide products c-di-GMP and c-di-AMP released from intracellular bacteria; hence STING is also a cytosolic PRR. Binding of cyclic dinucleotides alters the conformation of STING dimers, which relocate to Golgi complex membranes and recruit and activate TBK1. As in other PRR signaling pathways, TBK1 phosphorylates and activates IRF3 and NF-κB, leading to the synthesis of type I IFNs and cytokines.

Key Concept:

- Cytosolic DNA from viruses or bacteria activates the DNA sensor cGAS (cyclic GMP-AMP synthase), a nucleotidyltransferase, which synthesizes cGAMP (2′,5′-cyclic GMP-AMP dinucleotide). This DNA-derived dinucleotide, or other dinucleotides released by intracellular bacteria, activates the ER membrane–associated protein STING (*st*imulator of *in*terferon *genes*). STING then triggers signaling pathways activating IRF3 and NF-κB, leading to the synthesis of type I IFNs and cytokines.

Induced Innate Immunity Effector Mechanisms

Thus far in this chapter we have discussed the anatomical barriers that provide the first line of defense against infection and have introduced the six families of pattern recognition receptors

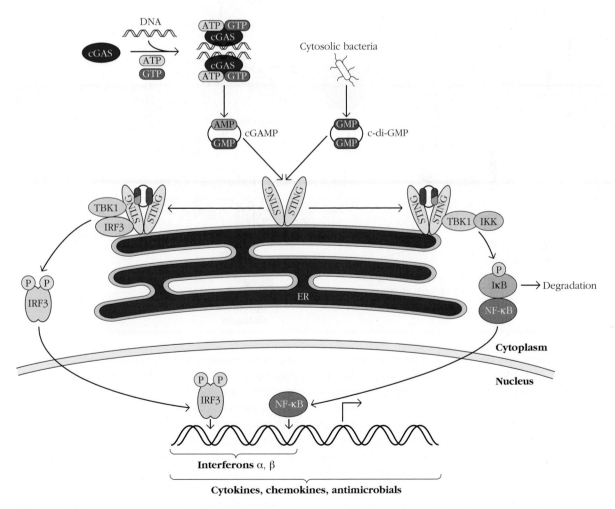

FIGURE 4-14 The cGAS/STING signaling pathway. After cGAS binds dsDNA, usually of viral or bacterial origin, it is activated to synthesize the dinucleotide cGAMP from GTP and ATP. cGAMP then binds to the ER-associated protein STING (*st*imulator of *inter*feron *genes*). STING also can bind the cyclic dinucleotide products c-di-GMP (and c-di-AMP, not shown) released from intracellular bacteria. Binding of cyclic dinucleotides alters the conformation of STING dimers, which recruit and activate TBK1. TBK1 phosphorylates and activates IRF3 and IKK, which activates NF-κB, leading to the synthesis of type I IFNs and cytokines.

and the signaling pathways they activate that induce the cellular innate immune responses that constitute the second line of defense. We will now present the induced effector mechanisms by which innate immune responses protect us. Some are molecules that are directly antimicrobial, while others are cellular responses that eliminate pathogens or infected cells. In general, these effector mechanisms are effective against the invading pathogen that induced them, such as antiviral interferons for viruses, phagocytosis for extracellular bacteria, or cell death for infected cells, but of course they are not specific for individual pathogens, as are adaptive immune responses. **Overview Figure 4-15** summarizes these innate immunity effector mechanisms.

OVERVIEW FIGURE 4-15

Effectors of Innate Immune Responses to Infection

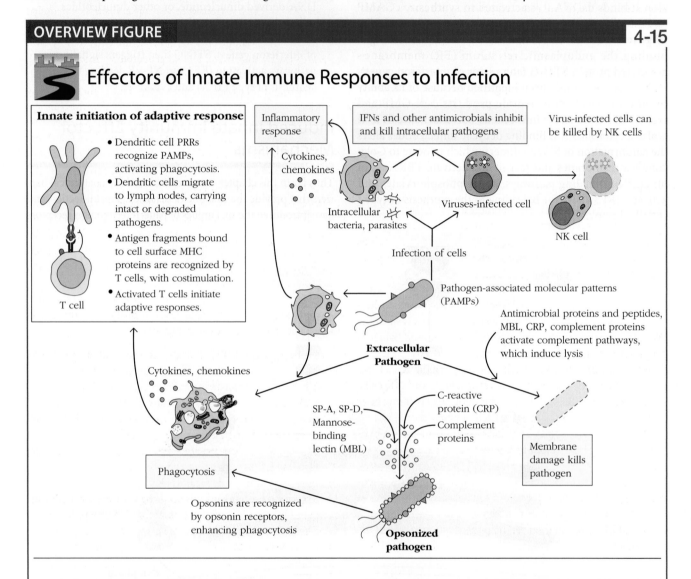

Innate initiation of adaptive response

- Dendritic cell PRRs recognize PAMPs, activating phagocytosis.
- Dendritic cells migrate to lymph nodes, carrying intact or degraded pathogens.
- Antigen fragments bound to cell surface MHC proteins are recognized by T cells, with costimulation.
- Activated T cells initiate adaptive responses.

T cell

Inflammatory response

Cytokines, chemokines

IFNs and other antimicrobials inhibit and kill intracellular pathogens

Virus-infected cells can be killed by NK cells

Intracellular bacteria, parasites

Viruses-infected cell

NK cell

Infection of cells

Pathogen-associated molecular patterns (PAMPs)

Antimicrobial proteins and peptides, MBL, CRP, complement proteins activate complement pathways, which induce lysis

Cytokines, chemokines

Extracellular Pathogen

SP-A, SP-D, Mannose-binding lectin (MBL)

C-reactive protein (CRP)

Complement proteins

Membrane damage kills pathogen

Phagocytosis

Opsonins are recognized by opsonin receptors, enhancing phagocytosis

Opsonized pathogen

Microbial invasion brings many effectors of innate immunity into play. Entry of microbial invaders through lesions in epithelial barriers exposes the invaders to attack by various effector molecules and cells and induces more complex processes of inflammation and the activation of adaptive responses. Expression of a variety of antimicrobial and proinflammatory molecules is induced by extracellular pathogens recognized by plasma membrane or endosomal PRRs and by intracellular pathogens in infected cells recognized by cytosolic PRRs. Extracellular pathogens with surface components recognized directly by certain PRRs or by soluble opsonins (e.g., C-reactive protein [CRP], mannose-binding lectin [MBL], complement components, or surfactant proteins A or D [SP-A and SP-D]) are cleared by phagocytosis. PRRs activate production of proinflammatory proteins including cytokines and chemokines; they increase vascular permeability, allowing influx of fluid and cells (neutrophils and macrophages) to cross the walls of blood vessels into the site of infection and further activate innate immune cells. Pathogen recognition by PRRs also initiates adaptive immune responses (see the *inset*). Dendritic cells bind microbes via receptors and are activated to mature; they also internalize and degrade microbes. These dendritic cells migrate through lymphatic vessels to nearby lymph nodes, where they present antigen-derived peptides on their MHC proteins to T cells. Antigen-activated T cells then initiate adaptive immune responses against the pathogen. Cytokines produced during innate immune responses support and direct the adaptive immune responses to infection.

Expression of Innate Immunity Proteins Is Induced by PRR Signaling

The PRR-activated signaling pathways described above activate transcription factors that turn on genes encoding an arsenal of proteins that help us to mount protective responses. Some of the induced proteins are antimicrobial and directly combat pathogens, while others serve key roles in activating and enhancing innate and adaptive immune responses. There is tremendous variation in what proteins are made in response to different pathogens, reflecting their PAMPs as well as the responding cell types and their arrays of PRRs. Some of the most common proteins and peptides that are secreted by cells following PAMP activation of PRRs and that contribute to innate and inflammatory responses are listed in **Table 4-4**.

Antimicrobial Peptides

Defensins and cathelicidins were mentioned earlier as being important in barrier protection, such as on the skin and the mucosal epithelial layers connected to the body's openings (see Table 4-2). Some cells and tissues constitutively express these peptides. For example, human intestinal epithelial Paneth cells constitutively express α-defensins and some β-defensins. In addition, some defensins and the cathelicidin LL-37 are constitutively synthesized and packaged in the granules of neutrophils, ready to kill phagocytosed bacteria, fungi, viruses, and protozoan parasites.

However, in some other cell types, such as mucosal and glandular epithelial cells, skin keratinocytes, and NK cells, the expression of these antimicrobial peptides is induced or enhanced by signaling through PRRs, in particular TLRs and NLRs. Macrophages do not produce these antimicrobial peptides following PRR activation; however, there is an indirect pathway by which microbes induce LL-37 in macrophages. Binding of microbial ligands to macrophage TLRs induces increased expression of receptors for vitamin D; binding of vitamin D to these receptors activates the macrophages to produce LL-37, which then can help the macrophages kill the pathogens.

Type I Interferons

Another major class of antimicrobial proteins transcriptionally induced directly by PRRs is the type I interferons, of which the major representatives are IFN-α and IFN-β. As summarized in Table 4-4, type I interferons are produced in two situations. When infected with a virus, many cell types are induced to make IFN-α and/or IFN-β following binding of cytosolic viral PAMPs, usually nucleic acids, to intracellular PRRs such as ALRs, RLRs, and cGAS. These PRRs activate the IRF transcription factors that induce expression of IFN genes. In addition, many uninfected cells express cell surface TLRs that recognize extracellular viral PAMPs, and/or they internalize virus without necessarily being infected, allowing endosomal TLRs to recognize viral components.

Signaling from these TLRs activates the IRFs and IFN-α and IFN-β production.

One type of dendritic cell, called the plasmacytoid dendritic cell (pDC) because of its shape, is a particularly effective producer of type I IFNs (and also type III IFNs such as IFN-λ, which have similar functions). The pDCs endocytose virus that has bound to various cell surface proteins (including CLRs such as DC-SIGN, which, e.g., binds HIV). TLR7 and TLR9 in the endosomes then are activated by viral PAMPs (ssRNA and viral DNA, respectively), leading to IRF activation and type I IFN production.

IFN-α and IFN-β exert their antiviral and other effects by binding to a specific receptor called IFNAR (*IFN-a*lpha *r*eceptor) that is expressed by most cell types. Like many cytokines, IFNs are dimers. Binding of the IFN dimer to IFNAR induces receptor dimerization and activation of the JAK/STAT signaling pathway, used by many cytokines to activate specific responses, as was introduced in Chapter 3 (see Figure 3-25). The IFNAR dimer activates the Janus kinases JAK1 and TYK2, which recruit and phosphorylate inactive STAT transcription factors (**Figure 4-16**). Phosphorylated STAT1 and STAT2 dimerize and change conformation, revealing a nuclear localization signal that allows the dimer to enter the nucleus, where it initiates transcription of specific genes.

Genes turned on by IFN are known as *interferon-stimulated genes* (ISGs). Four ISGs important for inhibiting viral replication are shown in Figure 4-16:

- Protein kinase R (PKR) binds and is activated by dsRNA; it then blocks viral (and cellular) protein synthesis by inhibiting the translation initiation factor eIF2α.
- 2′,5′-Oligoadenylate A synthetase (OAS) is a nucleotidyltransferase structurally related to cGAS. After its expression is induced by IFN, OAS binds cytosolic dsRNA, which activates it to generate 2′,5′-oligoadenylate from ATP. Oligoadenylate then binds RNase L and induces it to degrade viral RNA.
- Mx group proteins inhibit both the transcription of viral genes into mRNAs and the assembly of virus particles.
- The IFIT (*IFN-i*nduced proteins with *t*etratricopeptide repeats) proteins bind dsRNA, blocking viral RNA translation. Several IFITs also bind and inactivate the eIF3 translation initiation factor.

Reflecting the potent antiviral activities of type I interferons, they are used to treat some viral infections, such as hepatitis B and C. In addition to their key roles in controlling viral infections, type I interferons have several other beneficial immune-related activities. They increase expression of MHC class I proteins, making the cells better targets for T cell–mediated killing; activate NK cells; and regulate the activities of macrophages and T cells. Treatment with IFN-β has been shown to have beneficial effects in some forms of multiple sclerosis, a T cell–mediated autoimmune disease with inflammatory involvement, probably by inhibiting production of proinflammatory cytokines, including IL-1 and others produced by T cells (see Chapter 16).

TABLE 4-4	Secreted peptides and proteins induced by signaling through pattern recognition receptors			
	Peptides/proteins	**Produced by**	**Act on**	**Immune/inflammatory effects**
Antimicrobials	Defensins and cathelicidin*	Epithelia (e.g., oro/nasal, respiratory, intestinal, reproductive tracts; skin keratinocytes, kidney); neutrophils, NK cells	Pathogens Monocytes, immature dendritic cells, T cells Mast cells	Inhibit, kill Chemoattractant; activate cytokine production Activate degranulation
	Interferons α and β	Virus-infected cells, macrophages, dendritic cells, NK cells	Virus-infected cells NK cells Macrophages, T cells	Inhibit virus replication Activate Regulate activity
Cytokines	IL-1	Monocytes, macrophages, dendritic cells, keratinocytes, epithelial cells, vascular endothelial cells	Lymphocytes Bone marrow Vascular endothelium Liver Hypothalamus	Enhances activity Promotes neutrophil production Activates; increases vascular permeability Induces acute-phase response Fever
	IL-6	Monocytes, macrophages, dendritic cells, NK cells, epithelial cells, vascular endothelial cells	Lymphocytes Bone marrow Vascular endothelium Liver Hypothalamus	Regulates activity Promotes hematopoiesis → neutrophils Activates; increases vascular permeability Induces acute-phase response Fever
	TNF-α	Monocytes, macrophages, dendritic cells, mast cells, NK cells, epithelial cells	Macrophages Vascular endothelium Liver Hypothalamus Tumors	Activates Activates, increases vascular permeability, fluid loss, local blood clotting Induces acute-phase response Fever Cytotoxic for many tumor cells
	GM-CSF	Macrophages, vascular endothelial cells	Bone marrow	Stimulates hematopoiesis → myeloid cells
	IL-12, IL-18	Monocytes, macrophages, dendritic cells	Naïve CD4 T cells Naïve CD8 T cells, NK cells	Induce T$_H$1 phenotype, IFN-γ production Activate
	IL-10	Macrophages, dendritic cells, and mast cells; NK, T, and B cells	Macrophages, dendritic cells	Antagonizes inflammatory response, including production of IL-12 and T$_H$1 cells
Chemokines	Example: IL-8 (CXCL8)[†]	Macrophages, dendritic cells, vascular endothelial cells	Neutrophils, basophils, immature dendritic cells, T cells	Chemoattract cells to infection site

*Defensins and cathelicidin LL-37 vary among tissues in expression and whether constitutive or inducible.

[†]Other chemokines that are induced by PRR activation of cells in certain tissues, including various epithelial layers, may also specifically recruit certain lymphoid and myeloid cells to that site. See text, Chapter 14, and Appendix III.

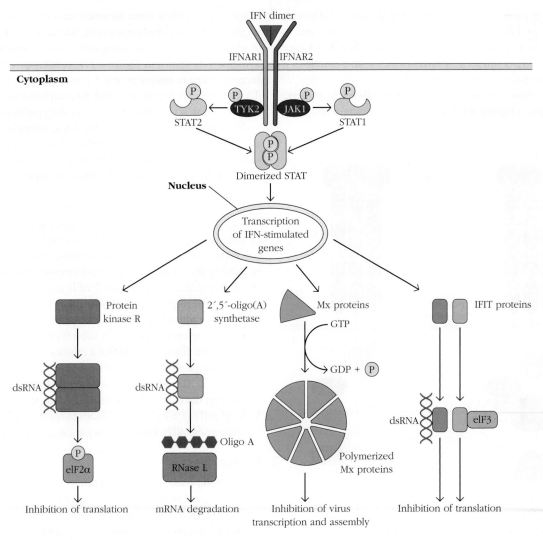

FIGURE 4-16 Induction of antiviral activities by type I interferons. Interferons α and β bind to and dimerize IFNAR (the *IFN-a*lpha *r*eceptor), which then recruits and activates the JAK1 and TYK2 protein kinases. They bind and phosphorylate STAT1 and STAT2, which dimerize, enter the nucleus, and stimulate expression of proteins that activate antiviral effects. Four are shown in this figure. Protein kinase R (PKR) binds viral dsRNA and inhibits the activity of the elF2α translation initiation factor. 2′,5′-Oligoadenylate synthetase synthesizes 2′,5′-oligoadenylate, which activates a ribonuclease, RNase L, that degrades viral and cellular mRNAs. Mx proteins self-assemble into ringlike structures that inhibit viral replication and the formation of new viral particles. IFIT proteins inhibit translation of viral proteins by binding to viral RNA and to elF3, a translation initiation factor.

Cytokines

Among the proteins transcriptionally induced by PRR activation are several key cytokines, which—while not directly antimicrobial—activate and regulate a wide variety of cells and tissues involved in innate, inflammatory, and adaptive responses. Cytokines function as the protein hormones of the immune system, produced in response to stimuli and acting on a variety of cellular targets. Several key examples of cytokines induced by PRR activation during innate immune responses are listed in Table 4-4, along with their effects on target cells and tissues.

Three of the most important cytokines are IL-1, TNF-α, and IL-6, the major proinflammatory cytokines. They act locally on blood vessels to increase vascular permeability and also on other cells, including lymphocytes, to recruit and activate them at sites of infection. They also have systemic effects (see below), including inducing fever and feeding back on bone marrow hematopoiesis to enhance production of neutrophils and other myeloid cells that will contribute to pathogen clearance.

IL-1 exists in two forms, IL-1α and IL-1β; the latter is more common and is what we refer to as IL-1. Its gene is activated by transcription factors downstream of many PRRs; as was discussed in Advances Box 4-2, the initial large pro-IL-1 precursor must be processed into the smaller form by caspases, usually part of activated inflammasomes. The IL-1 receptor (see Figure 3-19) has two chains that have immunoglobulin (Ig)-like domains and a cytoplasmic TIR domain (remember that TIR stands for *T*LR and *IL-1R*). Because of its TIR domain, after binding IL-1 the IL-1R recruits MyD88 to the receiver and activates the same signaling pathways, transcription factors, and genes for antimicrobial and proinflammatory proteins as do plasma membrane TLRs (see Figure 4-8). IL-1 induces its own synthesis, an example of a proinflammatory positive feedback loop.

TNF-α (often called just TNF) activates cells via the trimeric TNF receptor (**Figure 4-17**). Although it can induce apoptosis

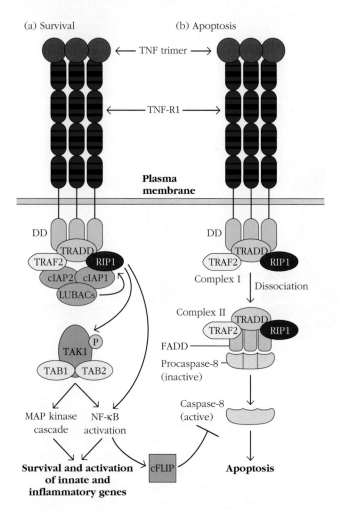

(a) Survival   (b) Apoptosis

FIGURE 4-17 **Signaling through TNF receptors.** Signaling through TNF receptors (TNFRs) can lead to different outcomes, depending on the cells and their environment. (a) Binding of TNF, which is a trimer, to the receptor activates the receptor's cytoplasmic DD (death domain) region and recruits the TRADD (TNF receptor–associated death domain) adaptor. TRADD then binds the kinase RIP1, TRAF2 (TNF receptor–associated factor 2), and several other proteins. The complex recruits and activates TAK1 as well as the IKK complex of the NF-κB pathway and MAPK pathways. NF-κB and AP-1 can activate the same genes as in TLR and IL-1 signaling pathways. Among other prosurvival effects, NF-κB activates the transcription of the cFLIP protein, which inhibits TNF-induced apoptosis. (b) In certain cells, the RIP1/TRADD/TRAF2 complex dissociates from the receptor and migrates to the cytoplasm, where it binds the adapter protein FADD, which in turn binds procaspase-8. Its clustering activates formation of the active caspase-8 protease, which cleaves other enzymes and initiates apoptosis.

of some cells, such as some tumors, it activates macrophages to be more active in innate immunity, becoming more efficient at phagocytosis and generating antimicrobial molecules and cytokines. As shown in Figure 4-17a, binding of TNF-α to its receptor triggers unique receptor-proximal steps that activate TAK1 and downstream signaling pathways and transcription factors common to TLR and IL-1R signaling pathways.

In contrast, IL-6, a class 1 cytokine, stimulates different responses. As is true for other class 1 cytokines and interferons, IL-6 binding to its receptor activates a JAK/STAT pathway and several other pathways, including MAPK pathways. IL-6 contributes to local and systemic inflammatory responses, as will be discussed below.

Activation of monocytes, macrophages, and dendritic cells by the binding of some PAMPs to certain TLRs also induces production of IL-12 and IL-18, cytokines that play key roles in driving the subsequent adaptive response by influencing the differentiation of activated T cells toward proinflammatory adaptive responses. IL-10 is another important cytokine specifically induced by some TLRs in macrophages, dendritic cells, other myeloid cells, and in subsets of T, B, and NK cells. IL-10 is anti-inflammatory, in that it inhibits macrophage activation and the production of proinflammatory cytokines by other myeloid cells. IL-10 levels increase over time and contribute to controlling the extent of inflammation-caused tissue damage. Roles of pathogen-induced cytokines in regulating T-cell adaptive responses will be discussed later in the chapter.

Chemokines

These small protein **chemoattractants** (agents that induce cells to move toward higher concentrations of the agent) recruit cells into, within, and out of tissues (see Chapter 14 and Appendix III). Some chemokines are responsible for constitutive (homeostatic) migration of white blood cells throughout the body. Other chemokines, produced in response to PRR activation, have key roles in the early stages of immune and inflammatory responses in that they attract cells that contribute both to clearing the infection or damage and to amplifying the response.

The first chemokine to be cloned, IL-8 (also called CXCL8), is produced in response to activation—by PAMPs, DAMPs, or some cytokines—of a variety of cells at sites of infection or tissue damage, including macrophages, dendritic cells, epithelial cells, and vascular endothelial cells. One of IL-8's key roles occurs in the initial stages of infection or tissue damage; it serves as a chemoattractant for neutrophils, recruiting them from the blood to sites of infection. Other chemokines are specifically induced by PRR activation of epithelial cells in certain mucosal tissues and serve to recruit cells specifically to those sites, where they generate immune responses appropriate for clearing the invading pathogen.

Enzymes: iNOS and COX2

Two enzymes produced in response to PRR-activated signaling pathways—**inducible nitric oxide synthase (iNOS)** and **cyclooxygenase-2 (COX2)**—have key roles in the generation of antimicrobial and proinflammatory mediators. The iNOS enzyme catalyzes an important step in the formation of nitric oxide, which kills phagocytosed microbes as will be discussed below. COX2, whose synthesis is induced by PRR activation in monocytes, macrophages, neutrophils, and mast cells, is key to converting the lipid intermediate arachidonic acid to prostaglandins, potent proinflammatory mediators.

> **Key Concept:**
>
> - The signaling pathways downstream of PRRs activate expression of a variety of genes, including those for antimicrobial peptides, type I interferons (IFN-α and IFN-β, which induce potent antiviral responses), cytokines (including proinflammatory IL-1β, TNF-α, and IL-6), chemokines, and enzymes that help to generate antimicrobial mediators (including reactive oxygen species and reactive nitrogen species) and inflammatory responses.

Phagocytosis Is an Important Mechanism for Eliminating Pathogens

Phagocytic cells make up an important line of defense against pathogens that have penetrated the epithelial cell barriers. Monocytes in the blood and macrophages, neutrophils, and dendritic cells in tissues are the main cell types that carry out *phagocytosis*, the cellular uptake (eating) and destruction of particulate materials greater than 0.5 microns (μm) in size, such as bacteria. This major role of phagocytic cells at the site of invading organisms is evolutionarily ancient, present in invertebrates as well as vertebrates. Elie Metchnikoff initially described the process of phagocytosis in the 1880s, using cells from starfish (echinoderm invertebrates), which are similar to vertebrate white blood cells (see Figure 1-3a). From these observations he concluded that phagocytosis has a major role in immunity. He was correct in this conclusion; we now know that defects in phagocytosis lead to severe immunodeficiency.

As described in Chapter 2, most tissues contain resident populations of macrophages that function as sentinels for the innate immune system. These macrophages are derived from embryonic precursors that seeded the tissues during early development. During an infection, monocytes (which circulate in the blood and can phagocytose blood-borne pathogens) are recruited into the infection site, where they differentiate into mature macrophages for fighting the infection. Through various cell surface receptors, macrophages recognize microbes, such as bacteria. This recognition activates signaling pathways that induce the polymerization of actin microfilaments, extending the phagocyte's plasma membrane to engulf and internalize the microbes into phagosomes (endosomes resulting from phagocytosis; see **Figure 4-18**). Lysosomes then fuse with

the phagosomes, delivering agents that kill and degrade the microbes. These lethal agents include hydrolytic enzymes that are activated by the increasing acidity of lysosomes as protons are pumped inside.

Neutrophils are a second major type of phagocyte, usually recruited early from the blood to sites of infection. Their granules contain pre-packaged antimicrobial proteins and peptides, which fuse with phagosomes. Finally, dendritic cells also can bind and phagocytose microbes.

(a)

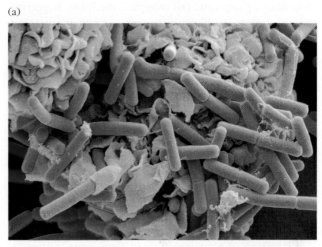

(b)

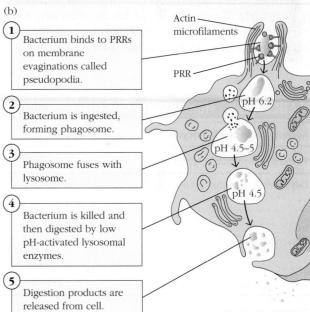

1. Bacterium binds to PRRs on membrane evaginations called pseudopodia.

2. Bacterium is ingested, forming phagosome.

3. Phagosome fuses with lysosome.

4. Bacterium is killed and then digested by low pH-activated lysosomal enzymes.

5. Digestion products are released from cell.

Actin microfilaments

PRR

pH 6.2

pH 4.5–5

pH 4.5

FIGURE 4-18 Phagocytosis. (a) Scanning electron micrograph of macrophage phagocytosis of bacteria. (b) Steps in the phagocytosis of a bacterium. PRRs on the macrophage directly recognize PAMPs on a bacterial cell, or opsonin receptors recognize opsonins attached to the bacterial cell. The receptors then trigger actin filament–mediated internalization and formation of a phagosome. The phagosome fuses with a lysosome; the acidic pH activates the lysosomal hydrolytic enzymes, which along with other mechanisms contribute to killing the bacterium (see text). Digestion products may be released from the macrophage. *[Science Source, Colorization by: Mary Martin.]*

As will be described later in this chapter and more extensively in Chapters 7 and 10, uptake and degradation of microbes by dendritic cells play key roles in the initiation of adaptive immune responses. In addition to triggering phagocytosis, various receptors on phagocytes recognize microbes and activate the production of a variety of molecules that contribute in other ways to eliminating infection.

Phagocytic Receptors

How does a phagocytic cell recognize microbes, triggering their phagocytosis? Phagocytes express on their surfaces a variety of receptors, many of which are PRRs that directly recognize PAMPs on the surfaces of microbes, such as cell wall components of bacteria and fungi. Some but not all PRRs induce phagocytosis; TLRs are a major class of PRRs that do not induce phagocytosis. PRRs that bind microbes and trigger phagocytosis of the bound microbes are listed at the top of **Table 4-5**, along with the PAMPs they recognize. As we will see later, there are other PRRs that, after PAMP binding, do not activate phagocytosis but trigger other types of responses. Most PAMPs that induce phagocytosis are cell wall components, including complex carbohydrates such as mannans and β-glucans, lipopolysaccharides (LPS), other lipid-containing molecules, peptidoglycans, and surface proteins.

Activation of phagocytosis can also occur indirectly, by phagocyte recognition of soluble proteins that have bound to microbial surfaces, thus enhancing phagocytosis; this process is called **opsonization** (from the Greek word for "to make tasty") (see Overview Figure 4-15). Many of these soluble phagocytosis-enhancing proteins (called **opsonins**) also bind to conserved, repeating components on the surfaces of microbes such as carbohydrate structures, lipopolysaccharides, and viral proteins; hence they are sometimes referred to as *soluble pattern-recognition proteins*. Once bound to microbe surfaces, opsonins are recognized by membrane opsonin receptors on phagocytes, activating phagocytosis (see Table 4-5, bottom).

A variety of soluble proteins function as opsonins; many also play other roles in innate immunity. **Mannose-binding lectin (MBL)**, a collectin with opsonizing activity, is found in the blood (where it can also activate the complement pathway) and respiratory fluids (**Figure 4-19a**). The complement component C1q also functions as an opsonin, binding bacterial cell wall components such as lipopolysaccharides and some viral proteins (see Figure 4-19b); C1q is recognized by the CR1 opsonin receptor, triggering phagocytosis. Another interesting example is the two surfactant collectin proteins, SP-A and SP-D, mentioned earlier in the context of their protective roles in mucosal secretions in the lungs and elsewhere. They are found in the blood, where they function as opsonins. After binding to microbes they are recognized by the CD91 opsonin receptor (see Table 4-5) and promote phagocytosis by alveolar and other macrophage populations. This function of SP-A and SP-D contributes to clearance of the fungal respiratory pathogen *Pneumocystis jirovecii*, a major cause of pneumonia in individuals with acquired

TABLE 4-5	Human receptors that trigger phagocytosis	
Receptor type on phagocytes	**Examples**	**Ligands**
Pattern recognition receptors		**Microbial ligands (found on microbes)**
C-type lectin receptors (CLRs)	Mannose receptor	Mannans (bacteria, fungi, parasites)
	Dectin-1	β-Glucans (fungi, some bacteria)
	DC-SIGN	Mannans (bacteria, fungi, parasites)
Scavenger receptors	SR-A	Lipopolysaccharide (LPS), lipoteichoic acid (LTA) (bacteria)
	SR-B	LTA, lipopeptides, diacylglycerides (bacteria), β-glucans (fungi)
Opsonin receptors		**Microbe-binding opsonins (soluble; bind to microbes)**
Collagen-domain receptor	CD91/calreticulin	Collectins SP-A, SP-D, MBL; L-ficolin; C1q
Complement receptors	CR1, CR3, CR4, CRIg, C1qRp	Complement components and fragments*
Immunoglobulin Fc receptors	FcαRs	Specific IgA antibodies bound to antigen†
	FcγRs	Specific IgG antibodies bound to antigen†; C-reactive protein

*See Table 5-5 for specific complement components or fragments that are bound by individual receptors.

†Opsonization of antibody-bound antigens is an adaptive immune response clearance mechanism.

(a) Mannose-binding lectin

(b) C1 bound to LPS

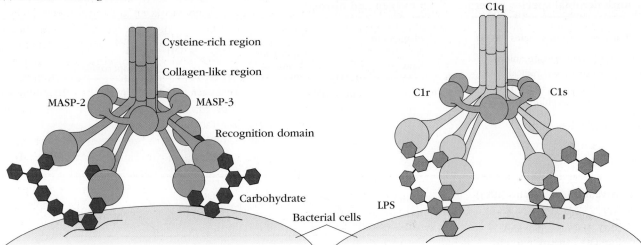

FIGURE 4-19 Structures of opsonins. (a) Mannose-binding lectin (MBL), a collectin, is a complex of multiple polypeptide chains, each containing an N-terminal cysteine-rich region followed by a collagen-like region, an α-helical neck region (not visible), and a recognition domain. Bound to the MBL are *MBL-associated serine proteases* (MASPs), which become active after the recognition domains bind to specific carbohydrate residues on pathogen surfaces. The MASPs then can activate the complement pathway. (b) C1, the first component of complement, has a multimeric structure similar to that of MBL. The C1q portion binds LPS on bacterial cell walls. MBL and C1q bound to microbes are then recognized by opsonin receptors, triggering phagocytosis.

immunodeficiency syndrome (AIDS). MBL (and other collectins), ficolins, and C1q share structural features, including similar polymeric structures with collagen-like shafts, but have recognition regions with different binding specificities (see Figure 4-19). As a result of their structural similarities, all are bound by the CD91 opsonin receptor (see Table 4-5) and activate pathogen phagocytosis.

Another opsonin, **C-reactive protein** (**CRP**), recognizes phosphocholine and carbohydrates on bacteria, fungi, and parasites (and some dead body cells) and is then bound by Fc receptors (FcRs), receptors that also bind the constant (Fc) region of antibodies and that are found on most phagocytes. Fc receptors are important for the opsonizing activity of several classes of antibodies, an important mechanism of adaptive immunity. As mentioned above, among the most effective opsonins are several components of the complement system, which is described in detail in Chapter 5. Present in both invertebrates and vertebrates, complement straddles both the innate and adaptive immune systems, indicating that it is ancient and important. As we will see in Chapter 5, phagocytosis is one of many important antimicrobial effects resulting from complement activation.

The importance of MBL as both an opsonin and a complement activator has been indicated by the effects of MBL deficiencies, which affect about 25% of the population. Individuals with MBL deficiencies are predisposed to severe respiratory tract infections, especially pneumococcal pneumonia. Interestingly, MBL deficiencies may be protective against tuberculosis, probably reflecting MBL's opsonizing role in enhancing the phagocytosis of *Mycobacterium tuberculosis*, the route by which it infects macrophages, potentially leading to tuberculosis.

Processes That Kill Phagocytosed Microbes

The binding of microbes—bacteria, fungi, protozoan parasites, and viruses—to phagocytes via pattern recognition receptors, or via opsonins and opsonin receptors, activates signaling pathways that initiate phagocytosis. The phagosomes then fuse with lysosomes and, in neutrophils, with preformed primary and secondary granules (see Figure 2-4a). The resulting phagolysosomes contain an arsenal of antimicrobial agents that then kill and degrade the internalized microbes. These agents include antimicrobial proteins and peptides (including defensins and cathelicidins), low pH (due to the activity of a vacuolar ATPase proton pump), hydrolytic enzymes including lysozyme and proteases activated by the increasing acidity of the phagolysosomes, and specialized molecules that mediate oxidative attack.

Oxidative attack on the phagocytosed microbes, which occurs in neutrophils, macrophages, and dendritic cells, employs highly toxic **reactive oxygen species** (**ROS**) and **reactive nitrogen species** (**RNS**), which damage microbial membranes and intracellular components (**Figure 4-20**). The reactive oxygen species are generated by the phagocytes' unique **NADPH oxidase enzyme complex** (also called **phagosome NADPH oxidase**), which is activated when microbes bind to the phagocytic receptors. The oxygen consumed by phagocytes to support ROS production by NADPH oxidase is provided by a metabolic process known as the **respiratory burst**, during which oxygen uptake by the cell increases severalfold. NADPH oxidase converts oxygen to superoxide ion ($\cdot O_2^-$); other ROS generated by the action of additional enzymes are hydrogen peroxide (H_2O_2), and hypochlorous acid (HClO), the active component of household bleach.

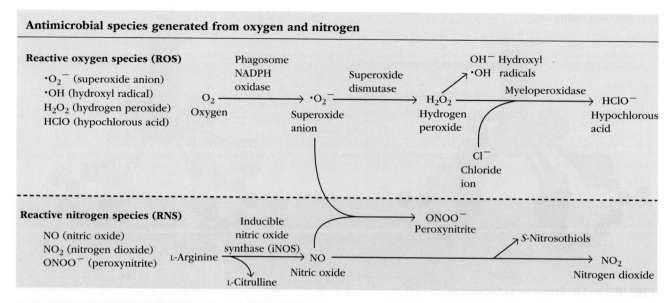

FIGURE 4-20 Generation of antimicrobial reactive oxygen and nitrogen species. In the cytoplasm of neutrophils, macrophages, and dendritic cells, several enzymes, including phagosome NADPH oxidase, transform molecular oxygen into highly reactive oxygen species (ROS) that have antimicrobial activity. One of the products of this pathway, superoxide anion, can interact with a reactive nitrogen species (RNS; in this case, nitric oxide [NO]) generated by inducible nitric oxide synthase (iNOS); the result is peroxynitrite (ONOO⁻), another RNS. NO can also undergo oxidation to generate the RNS nitrogen dioxide (NO₂).

The generation of RNS requires the transcriptional activation of the gene for the enzyme **inducible nitric oxide synthase (iNOS, or NOS2)**—called that to distinguish it from related nitric oxide synthases in other tissues. Expression of iNOS is activated by microbial PAMPs binding to various PRRs. iNOS oxidizes L-arginine to yield L-citrulline and nitric oxide (NO), a potent antimicrobial agent (see Figure 4-20). In combination with superoxide ion (•O₂⁻) generated by NADPH oxidase, NO produces an additional reactive nitrogen species, peroxynitrite (ONOO⁻), and toxic S-nitrosothiols. Collectively, ROS and RNS are highly toxic to phagocytosed microbes because of their alteration of microbial molecules through oxidation, hydroxylation, chlorination, nitration, and S-nitrosylation, along with the formation of sulfonic acids and destruction of iron-sulfur clusters in proteins. One example of how these oxidative species may be toxic to pathogens is the oxidation by ROS and RNS of cysteine sulfhydryl groups that are present in the active sites of many enzymes, inactivating the enzymes. ROS and RNS also can be released from activated neutrophils and macrophages and kill extracellular pathogens.

Evidence from genetic defects in humans and mice highlights the critical roles of these reactive chemical species in microbial elimination by phagocytic cells. The importance to antimicrobial defense of phagosomal NADPH oxidase and its products, ROS and RNS, is illustrated by **chronic granulomatous disease (CGD)**. Patients afflicted with this disease have dramatically increased susceptibility to some fungal and bacterial infections, caused by defects in subunits of NADPH oxidase that destroy its ability to generate oxidizing species. In addition, studies with mice in which the genes encoding iNOS were "knocked out" have shown that nitric oxide and substances derived from it account for much of the antimicrobial activity of macrophages against bacteria, fungi, and parasites. These mice lost much of their usual ability to control infections caused by such intracellular pathogens as *M. tuberculosis* and *Leishmania major*, the intracellular protozoan parasite that causes leishmaniasis.

Elimination of Intracellular Pathogens by Autophagy

Some bacterial pathogens, such as *Listeria*, escape immune effector mechanisms such as phagocytosis by replicating in the cytosol. However, they can't escape an intracellular form of phagocytosis called *autophagy*, in which membrane derived from the endoplasmic reticulum envelopes the bacteria, forming an autophagosome. This vesicle then fuses with lysosomes, leading to the destruction of the pathogens as discussed above. As mentioned earlier, autophagy is an innate effector function activated by the NLRs NOD1 and NOD2.

Cell Turnover and the Clearance of Dead Cells

Our discussion of phagocytosis thus far has focused on its essential roles in killing pathogens. As the body's main scavenger cells, macrophages also use their phagocytic receptors to take up and clear cellular debris, cells that have died from damage or toxic stimuli (necrotic cell death) or from apoptosis (programmed cell death), and aging red blood cells.

Considerable progress has been made in recent years in understanding the specific markers and receptors that trigger macrophage phagocytosis of dead, dying, and aging cells. Collectively, the components of dead/dying cells and damaged tissues that are recognized by PRRs, leading to their clearance, are sometimes referred to as *damage-associated molecular patterns (DAMPs)*. As the presence of these components may also be an indicator of conditions harmful to the body or may contribute to harmful consequences (such as autoimmune diseases), the "D" in DAMP also can refer to a "Danger" signal. Phagocytosis is the major mode of clearance of cells that have undergone apoptosis as part of developmental remodeling of tissues, normal cell turnover, or killing of pathogen-infected or tumor cells by innate or adaptive immune responses.

Apoptotic cells attract phagocytes by releasing the lipid mediator lysophosphatidic acid, which functions as a chemoattractant. These dying cells facilitate their own phagocytosis by expressing on their surfaces an array of molecules not expressed on healthy cells, including phospholipids (such as phosphatidylserine and lysophosphatidylcholine), proteins (annexin I), and altered carbohydrates. These DAMPs are recognized directly by phagocytic receptors such as the phosphatidylserine receptor and scavenger receptor SR-A1. Other DAMPs are recognized by soluble pattern recognition molecules that function as opsonins, including the collectins MBL, SP-A, and SP-D mentioned earlier; various complement components; and the pentraxins C-reactive protein and serum amyloid protein. These opsonins are then recognized by opsonin receptors, activating phagocytosis and degradation of the apoptotic cells.

An important additional activity of macrophages in the spleen and those in the liver (known as Kupffer cells) is to recognize, phagocytose, and degrade aging and damaged red blood cells. As these cells age, novel molecules that are recognized by phagocytes accumulate in their plasma membrane. Phosphatidylserine flips from the inner to the outer leaflet of the lipid bilayer and is recognized by phosphatidylserine receptors on phagocytes. Modifications of erythrocyte membrane proteins have also been detected that may promote phagocytosis.

Obviously it is important that normal cells not be phagocytosed, and accumulating evidence indicates that whether or not a cell is phagocytosed is controlled by sets of "eat me" signals—the altered membrane components (DAMPs) described above—and "don't eat me" signals expressed by normal cells. Young, healthy erythrocytes avoid being phagocytosed by not expressing "eat me" signals, such as phosphatidylserine, and also by expressing a "don't eat me" signal, the protein CD47. CD47, expressed on many cell types throughout the body, is recognized by the *SIRPα* (signal *regulatory protein α*) receptor on macrophages, which transmits signals that inhibit phagocytosis. Recent studies have shown that tumors use elevated CD47 expression to evade tumor surveillance and phagocytic elimination by the immune system. Increased expression of CD47 on all or most human cancers is correlated with tumor progression, probably because the CD47 activates SIRPα–mediated inhibition of the phagocytosis of tumor cells by macrophages. This understanding of the role of CD47 in preventing phagocytosis is being used to develop novel therapies for certain cancers, such as using antibodies to block CD47 on tumor cells, which should then allow them to be phagocytosed and eliminated.

Key Concepts:

- Phagocytosis—engulfment and internalization of particulate materials such as microbes—is mediated by receptors on phagocytes that either directly recognize PAMPs on the surface of microbes or recognize soluble proteins (opsonins) that bind to the microbes. PAMP binding triggers microbe uptake into phagosomes, which fuse with lysosomes or prepackaged granules, leading to their killing through the actions of lysosomal enzymes, antimicrobial proteins and peptides, and toxic effects of low pH and reactive oxygen species (ROS) and reactive nitrogen species (RNS).

- Intracellular bacteria may be killed by the process of autophagy, in which bacteria are surrounded with membrane to form an autophagosome that then fuses with lysosomes.

- Dead and dying cells express damage-associated molecular patterns (DAMPs), surface molecules that signal "eat me" to phagocytes. Receptors on phagocytes recognize these and clear the dead or dying cells.

- Tumor cells may escape phagocytosis by expressing the CD47 protein, which provides a "don't eat me" signal, inhibiting phagocytosis by macrophages.

Regulated Cell Death Contributes to Pathogen Elimination

In addition to the induced expression of antimicrobial and proinflammatory proteins and peptides, PAMPs also induce unusual responses, resulting in cell death, that can be beneficial in the control of infections. Cell death induced by receptor-activated signaling pathways is called **regulated cell death**. One form, apoptosis, is programmed cell death; its induction by TNF binding to the TNF receptor and by NK cells and cytotoxic T cells is an essential mechanism of cell-mediated immunity against altered self-cells (see Chapter 12). Several additional forms of regulated cell death exist; two of them—NETosis and pyroptosis—are forms of innate immune response.

NETs and NETosis

Several years ago neutrophils activated through various PRRs were shown to expel filaments of *chromatin* (the compacted DNA with bound proteins, including histones, that make up chromosomes) and other cellular debris that entrap and kill pathogens, with the neutrophil dying in the process. These filaments, which can extend to 10 to 15 times the size of the cell they originate from, are called **neutrophil extracellular traps (NETs)**, and the accompanying cell death is called **NETosis** (**Figure 4-21a and b**). NET formation requires activation of NADPH oxidase and the generation of ROS such as superoxides, which initiate damage of intracellular components, as mentioned earlier (and see Figure 4-21c). Enzymes, including neutrophil elastase and myeloperoxidase, enter the nucleus, modify histones, and trigger chromosome decondensation. Cell membranes disintegrate, and cytoplasmic and nuclear contents are expelled to form NETs. The NETs, containing DNA and other chromatin components, trap bacterial, fungal, and parasite cells, preventing their spread. In addition, associated with the fibrils of NETs are a variety of antimicrobial neutrophil proteins, including lysozyme, proteases, antimicrobial peptides, ion chelators, complement components, and histones (whose cationic properties give them antimicrobial activity). NETs also contribute to innate immunity indirectly, as DNA and other released chromatin components can serve as DAMPs, further activating local innate and inflammatory responses. Recent studies have shown that other granulocytes—eosinophils, mast cells, and basophils—can also form extracellular traps.

Pyroptosis

A second type of regulated cell death functioning in innate immunity, pyroptosis (typically of macrophages), is induced by inflammasome activation, described earlier in Advances

Box 4-2. Pyroptosis contributes to pathogen elimination in several ways. First, the death of infected macrophages prevents further spread of intracellular bacteria, such as *Salmonella* and *Listeria*, that replicate in these cells. Second, as mentioned earlier, pyroptosis appears to be an important mechanism for the release of mature IL-1β and IL-18, generated by inflammasome-associated caspase-1– or caspase-11–mediated cleavage from large precursors. These are important cytokines for promoting beneficial inflammatory responses.

> **Key Concepts:**
> - Cell death induced by PAMP-activated PRR signaling pathways is called *regulated cell death*.
> - Activated neutrophils release filaments made up of chromatin with associated antimicrobial proteins and peptides, called neutrophil extracellular traps (NETs), that catch and kill bacteria and fungal cells.
> - The regulated cell death of activated myeloid cells during NET formation (NETosis) or induced by activated inflammasomes (pyroptosis) can be beneficial, as it eliminates infected cells and allows release of mature IL-1β and IL-18 and of DAMPs that may enhance local inflammatory responses.

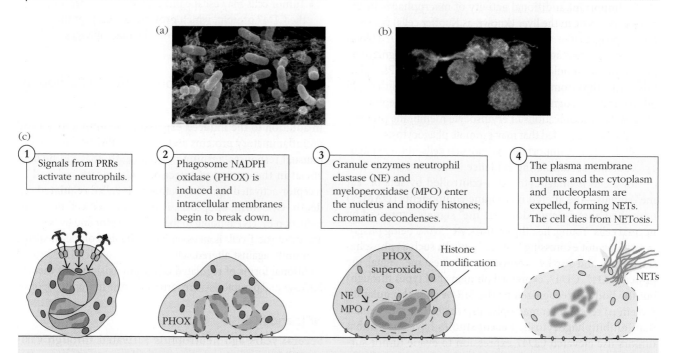

(a)

(b)

(c)

1 | Signals from PRRs activate neutrophils.

2 | Phagosome NADPH oxidase (PHOX) is induced and intracellular membranes begin to break down.

3 | Granule enzymes neutrophil elastase (NE) and myeloperoxidase (MPO) enter the nucleus and modify histones; chromatin decondenses.

4 | The plasma membrane ruptures and the cytoplasm and nucleoplasm are expelled, forming NETs. The cell dies from NETosis.

PHOX

PHOX superoxide

Histone modification

NE MPO

NETs

FIGURE 4-21 Neutrophil extracellular traps (NETs) and NETosis. (a) Fibers of NETs with trapped *Salmonella* bacteria. (b) A NET produced by an activated human neutrophil (dying due to NETosis), and four nonactivated neutrophils. Red stain: histones in decondensed chromatin of the NET. Green stain: Neutrophil elastase (visible in granules of nonactivated neutrophils and faintly in the NET). Blue stain: DNA. (c) Formation of NETs and NETosis. *[Republished with permission of The Rockefeller University Press, from Brinkmann, V. & Zychlinsky, A., from "Neutrophil extracellular traps: is immunity the second function of chromatin?" J. Cell Biol. 2012, 198, 773–783, Figure 2. Permission conveyed through Copyright Clearance Center, Inc.]*

Local Inflammation Is Triggered by Innate Immune Responses

When the outer barriers of the innate immune system—skin and other epithelial layers—are damaged, the resulting innate responses to infection or tissue injury can induce a complex cascade of events known as the **inflammatory response**. Inflammation may be acute (short-term effects contributing to combating infection, followed by healing)—for example, in response to local tissue damage—or it may be chronic (long term, not resolved), contributing to conditions such as arthritis, inflammatory bowel disease, cardiovascular disease, and type 2 diabetes.

The hallmarks of a localized inflammatory response were first described by the Roman physician Celsus in the first century A.D. as *rubor et tumor cum calore et dolore* (redness and swelling with heat and pain). An additional mark of inflammation added in the second century A.D. by the physician Galen is loss of function (*functio laesa*). Today we know that these symptoms reflect an increase in vascular diameter (vasodilation), resulting in a rise of blood volume in the area. Higher blood volume heats the tissue and causes it to redden. Vascular permeability also increases, leading to leakage of fluid from the blood vessels, resulting in an accumulation of fluid (**edema**) that swells the tissue. Within a few hours, leukocytes also enter the tissue from the local blood vessels. These hallmark features of inflammatory responses result from the activation of innate immune responses in the vicinity of the infection or wound.

When there is local infection, tissue damage, or exposure to some harmful substances (such as asbestos or silica crystals in the lungs), sentinel cells residing in the epithelial layer—macrophages, mast cells, and dendritic cells—are activated by PAMPs, DAMPs, crystals, and so on to start phagocytosing the offending invaders (**Figure 4-22**). The cells are also activated by the PRR signaling pathways discussed earlier in this chapter to release innate immunity mediators, including cytokines and chemokines that trigger a series of processes that collectively constitute the inflammatory response.

The recruitment of various leukocyte populations from the blood into the site of infection or damage is a critical early component of inflammatory responses. PRR signaling

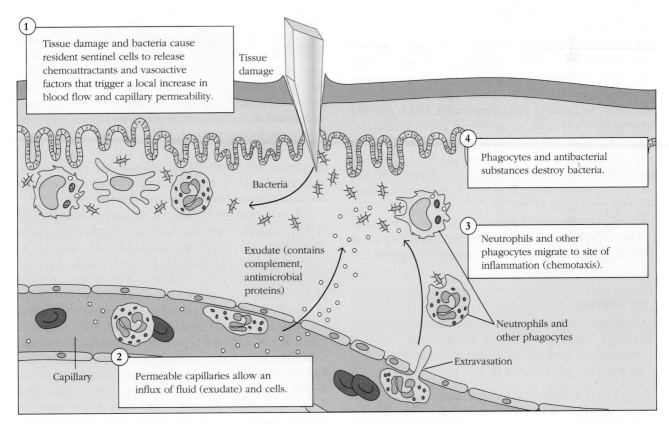

1. Tissue damage and bacteria cause resident sentinel cells to release chemoattractants and vasoactive factors that trigger a local increase in blood flow and capillary permeability.

Tissue damage

Bacteria

Exudate (contains complement, antimicrobial proteins)

Capillary

2. Permeable capillaries allow an influx of fluid (exudate) and cells.

4. Phagocytes and antibacterial substances destroy bacteria.

3. Neutrophils and other phagocytes migrate to site of inflammation (chemotaxis).

Neutrophils and other phagocytes

Extravasation

FIGURE 4-22 Initiation of a local inflammatory response. Bacterial entry through wounds activates initial innate immune mechanisms, including activation of phagocytosis by resident cells, such as macrophages and dendritic cells. Recognition of bacteria by cellular pattern receptors initiates production of cytokines, chemokines and other mediators, triggering changes in vascular endothelial cells that lead to an influx from the blood of fluid (containing antimicrobial substances) and phagocytes (first neutrophils and then monocytes) to the site of infection. These and subsequent events cause the redness, swelling, heat, and pain that are characteristic of local inflammatory responses.

activates resident macrophages, dendritic cells, and mast cells to release the initial components of cellular innate immune responses, including the proinflammatory cytokines TNF-α, IL-1β, and IL-6; chemokines; prostaglandins (following the induced expression of the COX2 enzyme); and histamine and other mediators released by mast cells. These factors act on the vascular endothelial cells of local blood vessels, increasing vascular permeability and the expression of **cell adhesion molecules (CAMs)** and chemokines such as IL-8. The affected epithelium is said to be inflamed or activated. Fluid enters the tissue, delivering antimicrobial molecules such as complement components and causing swelling. Cells flowing through the local capillaries are induced by chemokines and cell adhesion molecule interactions to adhere to vascular endothelial cells in the inflamed region and pass through the walls of capillaries and into the tissue spaces, a process called **extravasation** that will be described in Chapter 14. Neutrophils are the first to be recruited to a site of infection, where they enhance local innate responses, followed by monocytes that differentiate into macrophages; the macrophages participate in clearance of pathogens and cellular debris and help initiate wound healing.

In addition to these key events at the site of infection or damage, the key cytokines made early in response to innate and inflammatory stimuli—TNF-α, IL-1β, and IL-6—also have systemic effects, which will be described in more detail in Chapter 15. They induce fever (a protective response, as elevated body temperature inhibits replication of some pathogens) by inducing COX2 expression, which activates prostaglandin synthesis, as mentioned above. Prostaglandin E_2 (PGE$_2$) acts on the hypothalamus (the brain center controlling body temperature), causing fever. The three proinflammatory cytokines (TNF-α, IL-1β, and IL-6) also act on the liver, inducing the **acute-phase response**, which involves production by the liver of proteins that contribute to the elimination of pathogens (including opsonins and proteins that activate the complement system, such as MBL) and the resolution of the inflammatory response.

Key Concepts:

- Inflammatory responses are initiated by innate immune responses to local infection or tissue damage—in particular, by the proinflammatory cytokines IL-1β, TNF, and IL-6.

- Key early components of inflammatory responses are increased vascular permeability, allowing soluble innate mediators to reach the infected or damaged site, and the recruitment through the action of chemokines of neutrophils and monocytes from the blood into the site.

Innate Lymphoid Cells

In addition to the mechanisms of innate immunity described above, which largely are the responsibility of nonlymphoid cell types, especially myeloid and epithelial cells, a family of lymphocytes also is activated by infection, damage, or stress to enhance and regulate innate and inflammatory responses. Collectively, these cells are called **innate lymphoid cells (ILCs)**. Recall from Chapter 2 that ILCs lack the highly diverse antigen-specific receptors of B and T lymphocytes; instead they respond to other signals including infection, damage, or stress, to enhance and regulate innate and inflammatory responses.

Several examples of ILCs will be described below.

Natural Killer Cells Are ILCs with Cytotoxic Activity

Natural killer (NK) cells were discovered in the early 1970s as a population of cells that could kill some tumor cells without previous exposure. They were the first ILCs to be discovered.

Unlike B and T lymphocytes, whose receptors have tremendous diversity for foreign antigens, NK cells express a limited set of receptors that enable the cells to be activated by indicators of infection, cancer, or damage expressed by other cells. Again unlike B and T cells, which require days of activation, proliferation, and differentiation to generate their protective immune responses, NK cells are preprogrammed to respond immediately to appropriate stimuli, releasing from preformed secretory granules mediator proteins (perforin and granzymes) that kill altered cells by inducing apoptosis. This mechanism of cell-mediated cytotoxicity is also carried out by cytotoxic T cells, as a part of the adaptive response that occurs days later (see Chapter 12).

NK cell–mediated cytotoxic activity is enhanced by IFN-α produced early during virus infections, an example of positive feedback regulation in innate immunity. In addition to releasing cytotoxic mediators, some activated NK cells also secrete cytokines—most commonly the proinflammatory cytokine TNF-α, and IFN-γ (also known as type II interferon), a potent macrophage activator that also helps to activate and shape the adaptive response. Thus, along with the type I IFNs produced by virus-infected cells or induced during innate responses by viral PAMPs, NK cells are an important part of the early innate response to viral infections (as well as to malignancy and other indicators of danger). These early innate responses control the infection for the days to week it takes for the adaptive response (antibodies and cytotoxic T cells) to be generated.

How do NK cells sense that our cells have become infected, malignant, or potentially harmful in other ways?

As will be described more fully in Chapter 12, NK cells express a variety of novel receptors (collectively called *NK receptors*). Members of one group serve as *activating receptors* (of which more than 20 have been described in humans and mice) that have specificity for various cell surface ligands that serve as indicators of infection, cancer, or stress. While one of these activating ligands in mice has been shown to be a protein component of a virus (mouse cytomegalovirus), most NK activating receptors apparently have specificity not for a pathogen-associated component but for proteins specifically up-regulated on infected, malignant, or stressed cells, which serve as danger signals perceived by NK cells.

To limit the potential killing of normal cells in our body, NK cells also express *inhibitory receptors* that recognize membrane proteins (usually conventional MHC proteins) on normal healthy cells and inhibit NK cell–mediated cytotoxic killing of those cells. Many virus-infected or tumor cells lose expression of their MHC proteins and thus do not send these inhibitory signals. Receiving an excess of activating signals relative to the level of inhibitory signals tells an NK cell that a target cell is abnormal, and the NK cell is activated to kill the target cell. Thus NK cells are part of our innate sensing mechanisms that provide immediate protection, in this case recognizing and eliminating our body's own cells that have become harmful.

Recent studies have shown that some NK cells also may express some TLRs and other PRRs (including RIG-I and STING). These studies have suggested that TLR stimulation may play a role in activating NK-cell production of cytokines (such as IFN-γ) and cytotoxicity. In addition to signals from PRRs, additional stimulation may be required, often delivered by cytokines (including type I IFNs and IL-12) produced by dendritic cells, macrophages, neutrophils, and mast cells in response to signals from their own PRRs.

> **Key Concepts:**
> - Innate lymphoid cells (ILCs), which lack the antigen-specific receptors of B and T lymphocytes, play important roles in innate immune and inflammatory responses.
> - Natural killer (NK) cells, the first ILC discovered, have the unique function of killing cells that have become altered due to infection or stress. NK cells induce apoptosis of target cells if their activating receptors, which recognize markers of infection or stress on cells, send stronger signals than their inhibitory receptors, which recognize markers of normal cells, such as MHC proteins.

ILC Populations Produce Distinct Cytokines and Have Different Roles

Recent studies of lymphocytes that lack markers of B, T, and NK cells have revealed several additional distinct populations of ILC. All are derived from common lymphoid progenitors in the bone marrow (see Figure 2-1). While NK cells continue to be generated throughout life, as is true for B and T cells, the other ILCs found in adults are derived by the division of precursors that seeded peripheral tissues during embryonic development. At least six populations of ILCs have been identified, divided into three groups based largely on the types of proteins they produce and their functions (**Table 4-6**).

While many NK cells are found in lymphoid tissues and recirculate in the blood, the other ILC populations are found mainly in epithelial barrier tissues, including mucosal tissues such as the intestine and lungs, and glandular tissues such as salivary glands, where they have important roles in innate immune responses. ILC group 1 includes both NK cells and ILC1 cells, as they share production of IFN-γ and TNF-α; however, significant cytotoxic activity is restricted to

TABLE 4-6	Innate-like lymphoid cells (ILCs)		
ILC group	**ILC populations**	**Mediators**	**Functions**
1	NK cells	IFN-γ, TNF, perforin, granzymes	Immunity to viruses and intracellular pathogens, tumor surveillance, including cytotoxicity
	ILC1 cells	IFN-γ, TNF	Immunity to extracellular pathogens: viruses, bacteria, parasites
2	ILC2 cells	IL-4, IL-5, IL-9, IL-13; amphiregulin	Immunity to helminths, wound healing
3	LTi cells	LT-α, LT-β, IL-17A, IL-22	Lymphoid tissue development, intestinal homeostasis, immunity to extracellular bacteria
	ILC17 cells	IL-17, IFN-γ	Immunity to extracellular bacteria
	ILC22 cells	IL-22	Immunity to extracellular bacteria, homeostasis of epithelia

NK cells. Reflecting the roles of IFN-γ and TNF-α in macrophage activation and innate immune responses, intestinal ILCs have been shown to protect mice against infection by the intracellular protozoan parasite *Toxoplasma gondii* and *Clostridium difficile* bacteria.

The cytokines produced by ILC2 cells, IL-4, -5, -9, and -13, are important for innate protection against parasitic worms (helminths). IL-5 activates eosinophils to release mediators toxic for the parasites, while IL-13 promotes the smooth muscle contraction, mucus production, recruitment of activated macrophages, and other activities that stimulate expulsion of the worms. ILC2s also secrete amphiregulin, which mediates tissue repair during the resolution of infections.

As indicated in Table 4-6, ILC3 populations vary somewhat in the mediators they produce and the functions they perform. The LTi (*lymphoid tissue inducer*) population was identified several decades ago as being essential for the development of secondary lymphoid organs, in particular lymph nodes and Peyer's patches. This is an important role of LTi-produced lymphotoxin (LT)-α and -β cytokines (members of the TNF family). IL-22 made by the LTi and ILC22 populations induces intestinal epithelial cells to produce antimicrobial peptides and other molecules that help them to resist infection by bacteria such as *Salmonella typhimurium* as well as by rotavirus, a major cause of diarrhea. The IL-17 produced by some of the ILC3 populations promotes local inflammatory responses that are important in protection against yeast and other fungi.

Interestingly, the cytokines made by these ILC populations are similar to those of particular helper T-cell subsets, which originally were introduced in Chapter 2 and will be discussed in more detail in Chapter 10. Like ILC1 cells, T$_H$1 cells produce IFN-γ and TNF; like ILC2 cells, T$_H$2 cells produce IL-4, IL-5, and IL-13; like some ILC3 cells, T$_H$17 cells produce IL-17 and IL-22. This suggests that it is advantageous for individual lymphocytes, both T$_H$ and ILC, to produce specific sets of cytokines in both innate and adaptive immune responses in response to specific pathogens.

How are these ILC populations activated to produce their responses? Only NK cells and ILC3s in humans express TLRs, so in general other ILCs do not respond directly to pathogens, and most ILCs do not express activating NK receptors. In general, ILC production of the mediators listed in Table 4-6 appears to be induced by factors made by local cells such as epithelial cells, macrophages, and dendritic cells in response to direct activation by PAMP binding to their PRRs. These locally produced factors include cytokines such as IL-12, which activates ILC1s; IL-25, IL-33, and other cytokines and prostaglandin D$_2$, which activate ILC2s; and IL-1α (made by epithelial cells) and other cytokines, prostaglandin E$_2$, and leukotriene D$_4$, which activate ILC3s. Elucidating the mechanisms by which distinct ILC responses are activated should facilitate treatments for afflictions including infection-induced colitis, inflammatory bowel disease, allergies, asthma, and autoimmunity.

Key Concepts:

- ILCs are assigned to one of three groups (ILC1, ILC2, or ILC3), based on the cytokines they produce. Group 1 ILCs, which include NK cells, produce cytokines and other mediators contributing to cell-mediated immunity. Group 2 ILCs produce cytokines supporting immunity to helminth parasites and wound healing. Group 3 ILCs produce cytokines supporting lymphoid tissue development, epithelial integrity and homeostasis, and immunity to extracellular bacteria and fungi.

- Except for some NK and ILC3 cells, most ILCs do not have PRRs and hence are not activated directly by pathogens. They are instead activated by cytokines and other mediators produced by local epithelial cells, macrophages, and dendritic cells after PRR stimulation by PAMPs.

Regulation and Evasion of Innate and Inflammatory Responses

The importance of some of the individual molecules involved in the generation of innate and inflammatory responses is dramatically demonstrated by the impact on human health of genetic defects and polymorphisms (genetic variants) that alter the expression or function of these molecules (see **Clinical Focus Box 4-3**). As illustrated by these conditions, and by the many known roles (cited throughout this chapter) of innate and inflammatory mechanisms in protecting us against pathogens, these responses are essential to keeping us healthy. Some disorders show that innate and inflammatory responses can also be harmful, in that overproduction of various normally beneficial mediators and uncontrolled local or systemic responses can cause illness and even death. Therefore it is important that the occurrence and extent of innate and inflammatory responses be carefully regulated to optimize the beneficial responses and minimize the harmful responses.

Innate and Inflammatory Responses Can Be Harmful

To be optimally effective in keeping us healthy, innate and inflammatory responses should use their destructive mechanisms to eliminate pathogens and other harmful substances quickly and efficiently, without causing tissue damage or inhibiting the normal functioning of the body's systems. However, this does not always occur—a variety of conditions result from excessive or chronic innate and inflammatory responses.

The most dangerous of these conditions is **sepsis**, a systemic response to infection that includes fever, elevated heartbeat and breathing rate, low blood pressure, and compromised organ function due to circulatory defects. Several hundred thousand cases of sepsis occur annually in the United States, with mortality rates ranging from 20% to 50%, but sepsis can lead to **septic shock**—circulatory and respiratory collapse that has a 90% mortality rate. Sepsis results from **septicemia**, infections of the blood, in particular those involving gram-negative bacteria such as *Salmonella* and *E. coli*, although other pathogens can also cause sepsis.

The major cause of sepsis from gram-negative bacteria is the cell wall component LPS (also known as *endotoxin*), which as we learned earlier is a ligand of TLR4. As we have seen, LPS is a highly potent inducer of innate immune mediators, including the proinflammatory cytokines TNF-α, IL-1β, and IL-6; chemokines; and antimicrobial components. Systemic infections activate PRRs on blood cells including monocytes and neutrophils, vascular endothelial cells, and resident macrophages and other cells in the spleen, liver, and other tissues, to release these soluble mediators. They, in turn, systemically activate vascular endothelial cells, inducing them to produce cytokines, chemokines, adhesion molecules, and clotting factors that amplify the inflammatory response. Enzymes and reactive oxidative species released by activated neutrophils and other cells damage the vasculature. This damage, together with TNF-α–induced vasodilation and increased vascular permeability, results in fluid loss into the tissues that lowers blood pressure. TNF also stimulates release of clotting factors by vascular endothelial cells; locally this helps to limit the spread of infections, but systemically it results in blood clotting in capillaries. These effects on the blood vessels are particularly damaging to the kidneys and lungs, which are highly vascularized. High circulating TNF-α and IL-1 levels also adversely affect the heart. Thus the systemic inflammatory response triggered by septicemia can lead to circulatory and respiratory failure, resulting in septic shock and death.

As high levels of circulating TNF-α and IL-1β are highly correlated with morbidity, considerable effort is being invested in developing treatments that block the adverse effects of these normally beneficial molecules. Neutralizing these cytokines in early sepsis may be helpful, but by 24 hours following onset of sepsis other factors, including IL-6 and chemokines, become more important. Much still needs to be learned about sepsis and septic shock to enable the development of effective treatments.

While not as immediately dangerous as septic shock, chronic inflammatory responses resulting from ongoing activation of innate immune responses can have adverse consequences for our health. For example, a toxin from *Helicobacter pylori* bacteria damages the stomach by disrupting the junctions between gastric epithelial cells and also induces chronic inflammation that has been implicated in peptic ulcers and stomach cancer. Cytokines produced by intestinal ILCs in response to infection can cause colitis. Also, increasing evidence suggests that the noninfectious DAMPs cholesterol (as insoluble aggregates or crystals) and β-amyloid contribute, respectively, to atherosclerosis (hardening of the arteries) and Alzheimer's disease. Other examples of harmful sterile (noninfectious) inflammatory responses discussed earlier—including gout, asbestosis, silicosis, and aseptic osteolysis—are induced, respectively, by crystals of monosodium urate, asbestos, and silica, and by metal alloy particles from artificial joint prostheses. These varied substances are all potent inflammatory stimuli because of their shared ability to activate the NLRP3 inflammasome, resulting in the release of the proinflammatory cytokines IL-1β and IL-18. Additional examples of chronic inflammatory conditions will be presented in Chapter 15.

> **Key Concept:**
> • Infection of the blood can cause sepsis and the systemic expression of proinflammatory cytokines. Left unchecked, systemic inflammation leads to septic shock, a highly fatal condition.

Innate and Inflammatory Responses Are Regulated Both Positively and Negatively

Innate immune responses play essential roles in eliminating infections, but they also can be harmful when not adequately controlled. It is therefore not surprising that many regulatory processes have evolved that either enhance or inhibit innate and inflammatory responses. These mechanisms control the induction, type, and duration of these responses, in most cases resulting in the elimination of an infection without damaging tissues or causing illness.

Positive Regulatory Mechanisms

Innate and inflammatory responses are increased by a variety of mechanisms to enhance their protective functions. Signaling pathways downstream of multiple PRRs can work together to generate heightened responses. For example, in response to yeast, signaling pathways downstream of TLR2 and the CLR dectin-1 synergize to enhance protective cytokine production. Dengue virus RNA is recognized by TLR3, RIG-I, and MDA5, and signals from these three pathways synergize for heightened cytokine and IFN production. An important example, described earlier in this chapter, is the amplification of production of IL-1β and TNF-α, two of the initial cytokines induced by PAMP or DAMP binding to PRRs. As mentioned earlier, they activate pathways similar to those downstream of TLRs, and hence induce more of themselves, an example of positive feedback regulation.

Mutations in Components of Innate and Inflammatory Responses Associated with Disease

In combination with our rapidly expanding understanding of the mechanisms by which innate and inflammatory responses contribute to disease susceptibility and resistance, in recent years advances in human genetics have helped to identify a number of genetic defects that confer greater susceptibility to infectious and inflammatory diseases. The adverse effects of mutations in genes encoding essential components of innate and inflammatory processes highlight the critical roles of these proteins in keeping us healthy.

Since 2003, when the first mutations in innate immune components that predispose individuals to recurrent bacterial infections were discovered, a number of mutations interfering with the generation of protective innate immune responses have been identified. Two examples were mentioned earlier in this chapter: defects in NADPH oxidase, which cause chronic granulomatous disease, and MBL deficiencies, which predispose to respiratory infections. Also causing defects in innate immunity are mutations in two proteins—MyD88 and IRAK4—required for the MyD88-dependent signaling pathway downstream of all TLRs except TLR3 (see Figures 4-8 and 4-9). Children with these defects suffer from severe

invasive *Streptococcus pneumoniae* infections, some fatal, and are also susceptible to *Staphylococcus aureus* and *Pseudomonas aeruginosa* (**Figure 1a**). The MyD88 mutations completely prevent cytokine and chemokine induction by ligands for TLRs 2/1, 2/6, 5, 7, and 8. Not surprisingly, the effects of these mutations are less significant for TLR3 (which activates TRIF signaling pathways instead of MyD88) and TLR4 (which activates both MyD88 and TRIF signaling pathways). That the MyD88 mutations do not leave these children more susceptible to a wider variety of pathogens probably reflects the induction of protective immunity by other PRRs as well as by the adaptive immune system. In fact, the children become less susceptible to infections as they get older (see Figure 1b), consistent with the buildup of adaptive immunological memory to these pathogens.

Other genetic defects with clinical consequences have been identified in the pathways by which type I interferons (IFN-α, IFN-β) are induced by viral nucleic acid PAMPs and then block virus replication in infected cells. As highlighted in **Figure 2**, mutations that completely or partially block these pathways (red symbols) have been found in TLR3 and other components of the pathway that induce IFN-α

and IFN-β (see Figure 4-9). Mutations have also been found in TYK2 and STAT1, key components that activate the antiviral effects of the interferons in infected cells (see Figure 4-16). Interestingly, these mutations were all discovered in children presenting with herpes simplex virus (HSV) encephalopathy, a severe HSV infection of the central nervous system. Cells in the CNS express TLR3, and it may be that the mutations in these pathways severely disable innate responses that are critical to protection against CNS infection by this virus. Children with TYK2 and STAT1 mutations are also very susceptible to other infections, especially with mycobacteria, probably because macrophages must be activated by IFN-γ (which also uses TYK2 and STAT1 in its signaling pathway) to be able to kill these intracellular bacteria.

The final set of genetic defects associated with disease states to be discussed here involves the effects of genetic variants in NLRs (including inflammasomes) in promoting inflammatory diseases. Genome-wide genetic association studies have indicated that a number of allelic variants of TLRs and NLRs are associated with inflammatory disorders. Several variants are associated with inflammatory bowel disease, which includes ulcerative colitis and Crohn's disease. The most

(a)

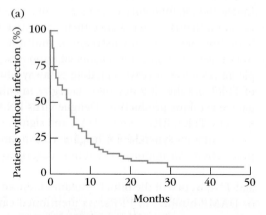

(b)

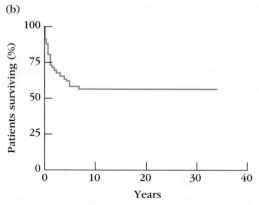

FIGURE 1 Severe bacterial infection and mortality among 60 children with MyD88 or IRAK4 deficiencies. (a) Decline in the percentage of children with these deficiencies who are asymptomatic reveals the incidence of the first severe bacterial infection during the first

50 months of life. (b) Survival curve of children with deficiencies shows reduced mortality after 5 years of age. *[Data from Picard, C., et al. 2010. Clinical features and outcome of patients with IRAK-4 and MyD88 deficiency. Medicine **89**:403.]*

(continued)

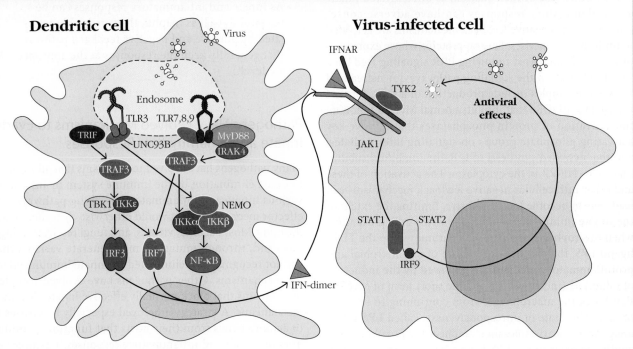

FIGURE 2 Genetic defects reducing the production and antiviral effects of IFN-α and IFN-β. This schematic shows some protein components involved in type I IFN production by dendritic cells and the type I IFN response in virus-infected cells. Proteins in which genetic mutations have been identified that result in defective functions and are associated with greater susceptibility to viral diseases are shown in red. Viruses are taken up by dendritic cells via specific receptors, and viral nucleic acids are detected by the various TLRs expressed in endosomes. Transport to endosomes of TLRs 3, 7, 8, and 9 is dependent on the ER protein UNC93B. Cytoplasmic signaling components activate transcription factors, including IRFs and NF-κB, leading to synthesis and secretion of IFN-α and IFN-β. In humans, TLR3, UNC93B, TRAF3, NEMO, and IRF7 deficiencies are associated with impaired IFN production in response to certain viruses. The binding of IFN-α and IFN-β to their receptor, IFNAR, induces the phosphorylation of JAK1 and TYK2, activating the signal transduction proteins STAT1, STAT2, and IRF9. This complex translocates as a heterotrimer to the nucleus, where it acts as a transcriptional activator, binding to specific DNA response elements in the promoter region of IFN-inducible genes. TYK2 and STAT1 deficiencies are associated with impaired IFN responses.

impressive genetic association was of Crohn's disease with mutations in NOD2 in or near its ligand-binding LRR region. As NOD2 is activated by bacterial cell wall fragments, investigators have hypothesized that intestinal epithelial cells with the mutant NOD2 PRRs are unable to activate adequate protective responses to gut bacteria and/or to maintain appropriate balance between normal commensals and pathogenic bacteria, and that these defective innate immune functions contribute to Crohn's disease pathogenesis. Consistent with this hypothesis, recent studies have found that intestinal Paneth cells from NOD2-defective individuals secrete reduced amounts of α-defensins, which, as mentioned earlier, are essential for maintaining normal commensal gut flora.

Genetic variants of the NLRP3 inflammasome have also been shown to be associated with Crohn's disease and other inflammatory disorders. In fact, mutations in NLRP3 (originally called *cryopyrin*) have been shown to be responsible for a set of autoinflammatory diseases (i.e., noninfectious inflammatory diseases affecting the body's tissues) collectively known as CAPS (for *cryopyrin-associated periodic fever syndromes*); one example is NOMID (*neonatal onset multisystem inflammatory disorder*). These devastating syndromes include many signs of systemic inflammation, including fever, rashes, arthritis, pain, and inflammation affecting the nervous system, with adverse effects on vision and hearing.

More than 70 inherited and novel mutations in NLRP3 associated with CAPS have been identified. Most are in the NRLP3 NBD element, although some are in the LRR domain (see Figure 4-11). What many have in common is their deregulating effect on NLRP3 activation of caspase-1, which may become constitutively active. Cells from patients with NOMID have recently been shown to secrete higher levels of IL-1 and IL-18, both spontaneously and when induced by PAMPs and DAMPs, promoting chronic inflammation. Another consequence of constitutive NLRP3 activation is the death of the activated macrophages by pyroptosis, releasing DAMPs that lead to more inflammation. Fortunately, new therapeutic approaches that inhibit IL-1 activity seem to alleviate these symptoms in some patients.

REFERENCES

Casanova, J.-L., L. Abel, and L. Quintana-Murci. 2011. Human TLRs and IL-1Rs in host defense: natural insights from evolutionary, epidemiological, and clinical genetics. *Annual Review of Immunology* **29**:447.

Bustamante, J., et al. 2008. Novel primary immunodeficiencies revealed by the investigation of paediatric infectious diseases. *Current Opinion in Immunology* **20**:39.

Negative Regulatory Mechanisms

On the other side of the equation, as uncontrolled innate and inflammatory responses can have adverse consequences, many negative feedback mechanisms are activated to limit these responses. Several proteins whose expression or activity is increased following PRR signaling feed back to inhibit steps in the signaling pathways downstream of the PRR. Examples include production of a short form of the adaptor MyD88 that inhibits normal MyD88 function; the activation of protein phosphatases that remove key activating phosphate groups on signaling intermediates; and the increased synthesis of IκB, the inhibitory subunit that keeps NF-κB in the cytoplasm. The activation of these and other intracellular negative feedback mechanisms can lead cells to become less responsive, limiting the extent of the innate immune response. In a well-studied example, when macrophages are exposed continuously to the TLR4 ligand LPS, their initial production of antimicrobial and proinflammatory mediators is followed by the induction of inhibitors (including IκB and the short form of MyD88) that block the macrophages from continuing to respond to LPS. This state of unresponsiveness, called **LPS tolerance** (or endotoxin tolerance), reduces the possibility that continued exposure to LPS from a bacterial infection will cause septic shock.

Other feedback pathways inhibit the inflammatory effects of TNF-α and IL-1β. Each of these cytokines induces production of a soluble version of its receptor or a receptor-like protein that binds the circulating cytokine molecules, preventing the cytokines from acting on other cells. In addition, the anti-inflammatory cytokine IL-10 is produced late in the macrophage response to PAMPs; it inhibits the production and effects of inflammatory cytokines and promotes wound healing.

In a recently described example, the induced production of IFN-β protects mice from the lethal hyperinflammatory effects of IL-1β during *Streptococcus pyogenes* infection. Following endocytosis by dendritic cells, *S. pyogenes* releases ribosomal RNA (rRNA), which binds to endosomal TLR13 and activates IFN-β production. The IFN-β binds to IFNAR on dendritic cells, macrophages, and neutrophils that have been activated by *S. pyogenes* PAMPs and reduces transcription of the IL-1 gene. This allows sufficient IL-1 to be produced to provide beneficial effects while avoiding excessive levels that would cause harmful hyperinflammatory responses.

However, these negative regulatory interactions sometimes may be disadvantageous. One example may explain how influenza virus infection causes increased susceptibility to bacterial infections that cause pneumonia. IRF3 activated by RLR signaling pathways triggered by influenza RNA binding reduces transcription of some cytokines normally induced by TLR signaling that promote protective antibacterial T-cell responses.

Key Concept:

- As innate and inflammatory responses can be harmful as well as helpful, they are highly regulated by positive and negative feedback pathways that generally keep the responses at the appropriate level.

Pathogens Have Evolved Mechanisms to Evade Innate and Inflammatory Responses

Many pathogens have evolved mechanisms that allow them to evade elimination by the immune system by inhibiting various innate and inflammatory signaling pathways and effector mechanisms that would otherwise clear them from the body. Most bacteria, viruses, and fungi replicate at high rates and, through mutation, may generate variants that are not recognized or eliminated by innate immune effector mechanisms. Other pathogens have evolved complex mechanisms that block normally effective innate clearance mechanisms. A strategy employed especially by viruses is to acquire genes from their hosts that function as inhibitors of innate and inflammatory responses. Examples of the wide range of mechanisms by which pathogens avoid detection by PRRs, activation of innate and inflammatory responses, or elimination by those responses are described in **Table 4-7**, and additional evasion mechanisms are presented in Chapter 17.

Key Concept:

- To escape elimination by innate immune responses, pathogens have evolved a wide array of strategies to block antimicrobial responses.

Interactions between the Innate and Adaptive Immune Systems

The many layers of innate immunity are important to our health, as illustrated by the illnesses seen in individuals with certain mutations in genes for one or another component of the innate immune system (see Clinical Focus Box 4-3 and other examples mentioned earlier in this chapter). However, innate immunity is not sufficient to protect us fully from infectious diseases, in part because many pathogens have features that allow them to evade innate immune responses, as discussed above. Hence the antigen-specific responses generated by our powerful adaptive immune system are usually needed to resolve infections successfully. While our B and T lymphocytes are the key producers of adaptive response effector mechanisms—antibodies and cell-mediated immunity (see Chapter 12)—it is becoming

TABLE 4-7	Pathogen evasion of innate and inflammatory responses
Type of evasion	**Examples**
Avoid detection by PRRs	Flagellin of proteobacteria has a mutation that prevents it from being recognized by TLR5
	Helicobacter, Coxiella, and *Legionella* bacteria have altered LPS that is not recognized by TLR4
	HTLV-1 virus p30 protein inhibits transcription and expression of TLR4
	Several viruses (Ebola, influenza, vaccinia) encode proteins that bind cytosolic viral dsRNA and prevent it from binding and activating RLR
Block PRR signaling pathways, preventing activation of responses	Vaccinia virus protein A46R and several bacterial proteins have TIR domains that block MyD88 and TRIF from binding to TLRs
	Several viruses block TBK1/IKK activation of IRF3 and IRF7, required for IFN production
	West Nile virus NS1 protein inhibits NF-κB and IRF transport into the nucleus
	Yersinia bacteria produce Yop proteins that inhibit inflammasome activity; the YopP protein inhibits transcription of the IL-1 gene
Prevent killing or replication inhibition	*Listeria* bacteria rupture the phagosome membrane and escape to the cytosol
	Mycobacterium tuberculosis blocks phagosome fusion with lysosomes and inhibits phagosome acidification
	M. tuberculosis and *Staphylococcus aureus* produce proteins that protect them from ROS and RNS
	Vaccinia virus encodes a protein that binds to type I IFNs and prevents them from binding to the IFN receptor
	Ebola virus blocks the antiviral effects of IFN by preventing the nuclear translocation of phosphorylated STAT1
	Hepatitis C virus protein NS3-4A and vaccinia virus protein E3L bind protein kinase R and block IFN-mediated inhibition of protein synthesis
	Herpesvirus and poxvirus encode versions of the anti-inflammatory cytokine IL-10 that reduces the local inflammatory response and T-cell activation

increasingly clear that our innate immune system plays important roles in helping to initiate and regulate adaptive immune responses so that they will be optimally effective. In addition, the adaptive immune system has co-opted several mechanisms by which the innate immune system eliminates pathogens, modifying them to enable antibodies to clear pathogens.

The Innate Immune System Activates Adaptive Immune Responses

When pathogens invade our body, usually by penetrating our epithelial barriers, the innate immune system not only reacts quickly to begin to clear the invaders but also plays key roles in activating adaptive immune responses. We have already seen that innate immune cells at the infection site (epithelial cells and resident macrophages, dendritic cells, and mast cells, and newly recruited neutrophils and monocytes) sense the invading pathogens through their PRRs and generate antimicrobial and proinflammatory responses that will slow down the infection. At the

same time, they also initiate steps to bring the pathogens to the attention of B and T lymphocytes and help activate responses that, days later, will generate the strong antigen-specific antibody and cell-mediated responses that will resolve the infection.

The first step in the generation of adaptive immune responses to pathogens is the delivery of the pathogen to lymphoid tissues, where T and B cells can recognize it and respond. As is discussed in more detail subsequent chapters, dendritic cells (usually immature) that serve as sentinels in epithelial tissues bind microbes through various pattern recognition receptors. The dendritic cells carry the bound microbes—either still attached to the cell surface or in phagosomes—via the lymphatic vessels to nearby secondary lymphoid tissues, such as the draining lymph nodes and Peyer's patches. There the dendritic cells can transfer or present the microbes or microbial components to other cells. In many cases the dendritic cell internalizes and degrades phagocytosed microbes, and microbe-derived peptides come to the cell surface bound to MHC class II proteins. Pathogens that replicate in the cytoplasm (viruses and some bacteria and protozoan parasites) are processed in the cytosol; their peptides come to the surface bound to MHC class I proteins. (Antigen processing and presentation on MHC proteins is described in detail in Chapter 7.) The binding of microbial PAMPs to the dendritic cell's PRRs activates the dendritic cell to mature so that it becomes a better antigen-presenting cell; for example, the mature dendritic cell expresses higher levels of MHC class II. In addition, the mature dendritic cell has turned on expression of costimulatory membrane proteins, such as CD80 or CD86 (see Chapter 3), that are recognized by receptors on T$_H$ cells and contribute to their activation, as will be described in more detail in Chapter 10.

As a result of these processes of microbe binding, processing, and maturation, mature dendritic cells are the most effective antigen-presenting cells, particularly for the activation of naïve (not previously activated) T cells. In contrast, recent evidence indicates that immature dendritic cells that have not been activated by PRR recognition of PAMPs may induce a state of unresponsiveness, in which the T cells *can't* respond, as will be discussed in future chapters.

> **Key Concept:**
>
> • Dendritic cells are a key cellular bridge between innate and adaptive immunity. Microbes bound by dendritic cells via their PRRs are brought from the site of infection to lymph nodes. Activation of a dendritic cell by PAMPs stimulates the cell to mature, so that it acquires the abilities to initiate the activation of naïve T cells that eventually will become mature cytotoxic and helper T cells.

Recognition of Pathogens by Dendritic Cells Influences Helper T-Cell Differentiation

Activation of dendritic cells by binding of certain PAMPs to PRRs has additional important consequences for adaptive immune responses. Depending on the particular type of dendritic cell and its location, the nature of the pathogen, and what PRRs and downstream signaling pathways are activated, dendritic cells are stimulated to secrete specific cytokines that regulate T-cell differentiation. These processes are essential for the immune system to generate the most appropriate kind of response to combat each invading pathogen.

The best example of this critical role of dendritic cells is their influence over the differentiation of naïve T cells into various helper T-cell subsets. As illustrated in **Figure 4-23**, various pathogen components interact with different PRRs, inducing signaling pathways that activate production of cytokines that induce naïve T cells to differentiate into one of several T-cell subsets: T$_H$1, T$_H$2, T$_H$17, and regulatory (T$_{REG}$) cells are the subsets shown in the figure. These T-cell

populations vary in the cytokines that they produce and hence in their immune functions. Most significantly, each T subset's functions are tailored to helping eliminate the pathogen that activated the dendritic cell.

As shown in Figure 4-23, extracellular bacteria and endosomal nucleic acids from internalized bacteria and viruses activate dendritic cells to secrete IL-12, which induces T-cell differentiation into the T$_H$1 subset. T$_H$1 cells secrete IFN-γ, which among other activities activates macrophages and NK cells to eliminate these pathogens and infected cells. In contrast, dendritic cells activated by PAMPs from helminths (parasitic worms) and some bacteria and fungi that bind to plasma membrane and cytosolic PRRs are blocked from producing IL-12 and instead produce IL-10, which helps (along with IL-4 and IL-13 from other nearby cells, such as basophils and ILC2) to induce naïve T cells to become T$_H$2 cells. Cytokines produced by T$_H$2 cells activate other leukocytes to release mediators that help to clear these pathogens. In a third example shown in Figure 4-23, fungal PAMPs bind and activate the CLR dectin-1 to produce cytokines that induce T-cell differentiation to T$_H$17 cells. IL-17 secreted by these T cells activates production of mediators that recruit inflammatory cells to the site and help to kill and clear fungal infections.

Finally, activation of TLRs such as TLR2/6 in the presence of vitamin A (such as in the intestine, where it comes from food) induces dendritic cells to convert the vitamin A into retinoic acid and to produce cytokines including IL-10 and TGF-β. Together these factors induce the formation of regulatory T cells, which inhibit other immune responses. This is one mechanism by which we develop tolerance to food substances and the beneficial microbiota of our intestine.

It is important to recognize that the schematic shown in Figure 4-23 is oversimplified, as it does not show some T$_H$-cell subsets and many complexities by which pathogen recognition by the innate immune system regulates T-cell differentiation. Also not shown in this figure are the contributions of various ILC populations to the cytokine milieu in tissues that influence T-cell differentiation (see above and Table 4-6). The molecular mechanisms by which naïve T$_H$ cells are influenced to differentiate into various mature T$_H$ subsets will be described in more detail in Chapter 10.

> **Key Concept:**
>
> • Various pathogens bind distinct PRRs on dendritic cells and activate signaling pathways that control what cytokines the dendritic cells will secrete. These cytokines induce the differentiation of naïve T$_H$ cells into mature T$_H$ subsets producing different cytokines and having distinct functions in immune responses. This role of dendritic cells helps to ensure that the type of adaptive response generated will be effective against the invading pathogen.

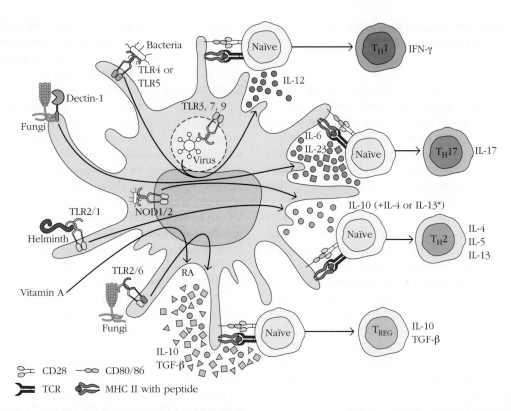

FIGURE 4-23 Pathogens induce differential signaling through dendritic cell PRRs, influencing helper T-cell functions. Naïve T cells receive multiple signals from dendritic cells that induce their differentiation into mature helper T cells. Dendritic cells are activated by pathogen binding to PRRs to degrade internalized microbes into peptides that are presented by the dendritic cell's MHC class II proteins, recognized by the T cell's TCR. Note that PRR signaling activates increased expression of MHC class II and costimulatory CD80 and CD86 proteins that provide additional stimulatory signals to the T cell. The distinct signaling pathways activated by PAMP binding to different PRRs differentially induce production of different cytokines and other mediators, such as retinoic acid (RA, derived from vitamin A, e.g., from food in the intestine). The particular cytokine(s) to which a naïve T cell is exposed induces that cell to turn on genes for certain cytokines, determining that cell's functions in pathogen elimination (see text).

Some Antigens Containing PAMPs Can Activate B Cells Independent of Helper T Cells

Typically, naïve B cells require multiple signals—antigen binding to the BCR, plus signals from helper T cells (cell-cell contact and T cell–derived cytokines)—to become activated and mature into antibody-secreting plasma cells. However, B cells express TLRs, and the binding of PAMPs to these TLRs activates signaling pathways that can add to or substitute for the signals normally required for B-cell activation.

One example has been well studied in mouse B cells. In combination with signals from the B cells' BCR after antigen binding, TLR4 binding of LPS (at low concentrations) can activate sufficient signals to induce the B cells to proliferate and differentiate into antibody-secreting plasma cells without T$_H$-cell help (via cell-cell contact and cytokines). At high concentrations of LPS, TLR4-activated signals are sufficient to activate all B cells (polyclonal activation), regardless of their antigen-binding specificity; hence

LPS has for many years been called a T-independent antigen (see Chapter 11). Human B cells do not express TLR4 and hence do not respond to LPS; however, they do express TLR9 and can be activated by microbial CpG DNA.

The ability of TLR signals to replace T$_H$ signals is often beneficial; cobinding of bacteria to both BCRs and TLRs on a B cell may activate that cell more quickly than if it had to wait for signals from a T$_H$ cell. Some T cells also express TLRs, which similarly function as costimulatory receptors to enhance protective responses.

> **Key Concept:**
>
> • Some PAMP-containing antigens can activate B cells independent of helper T cells. Signals from TLRs can substitute for costimulatory signals from T cells in activating B cells to differentiate into antibody-secreting plasma cells. Thus, recognition of pathogens helps activate these cells to generate adaptive immune responses.

Adjuvants Activate Innate Immune Responses That Increase the Effectiveness of Immunizations

Given these activating and potentiating effects of PRR ligands on adaptive immune responses, can they be used to enhance the efficacy of vaccines in promoting protective immunity against various pathogens? In fact, many of the materials—known as **adjuvants**—that have been shown over the years by trial and error to enhance immune responses in both laboratory animals and humans contain ligands for TLRs or other PRRs (see Chapter 17). For example, complete Freund's adjuvant, perhaps the most potent adjuvant for immunizations in experimental animals, is a combination of mineral oil and killed mycobacteria. The mineral oil emulsion produces a slowly dispersing depot of antigen, a property of many effective adjuvants, while fragments of the bacteria's cell wall peptidoglycans serve as activating PAMPs. Alum (a precipitate of aluminum hydroxide and aluminum phosphate) is used as an adjuvant in some human vaccines; it has recently been shown to activate the NLRP3 inflammasome, thereby enhancing IL-1β and IL-18 secretion and promoting inflammatory processes that enhance adaptive immune responses. Immunizations with alum usually lead the activated T cells to become T$_H$2 cells, which enhance antibody responses.

While many vaccines consist of killed or inactivated viruses or bacteria and hence contain their own PAMPs, which function as built-in adjuvants, some new vaccines consist of protein antigens that themselves are not very stimulatory to the immune system. Tumor antigens also tend to induce weak responses. Hence, considerable effort is being invested in developing new adjuvants based on current knowledge about PRRs. One approach for generating an effective vaccine for a pathogen protein is to fuse the protein to a TLR ligand, using genetic engineering. For example, fusions of pathogen proteins to the TLR5 ligand flagellin are currently being tested. In another example, LPS is a highly potent adjuvant but generates too much inflammation to use; less harmful versions of LPS are being developed as potential adjuvants.

Key Concept:
- Modern vaccines make use of the connection between innate and adaptive immune responses by including in the vaccine adjuvants, substances that have been shown to activate the innate immune response and enhance the adaptive immune response.

Some Pathogen Clearance Mechanisms Are Common to Both Innate and Adaptive Immune Responses

Adaptive immune responses—antibody responses in particular—have adopted and modified several effector functions by which the innate immune system eliminates

antigen, so that they are also triggered by antibody binding to antigens. While some will be discussed in more detail in Chapters 5 and 12, several examples are mentioned briefly here as illustrations of important interactions between innate and adaptive immune responses.

As discussed earlier in this chapter, several soluble proteins that recognize microbial surface components—including SP-A, SP-D, and MBL—function as opsonins; when they are bound to microbial surfaces they are recognized by receptors on phagocytes, leading to enhanced phagocytosis. Some classes of antibodies also serve as opsonins; after binding to microbial surfaces these antibodies can be recognized by receptors for immunoglobulin Fc regions that are expressed on macrophages and other leukocytes, triggering phagocytosis (see Chapter 12).

The complement pathway can be activated by both innate and antibody-mediated mechanisms. Components on microbe surfaces can be directly recognized by soluble pattern recognition proteins, including MBL and the complement component C1, leading to activation of the complement cascade (see Figure 4-19). Similarly, when certain classes of antibodies bind microbe surfaces, they can be recognized by the C1 component of complement, also triggering the complement cascade. The activation of the complement system by both innate and adaptive mechanisms will be discussed fully in Chapter 5. Once the complement pathway is activated by any of these microbe-binding proteins, it generates a common set of protective activities. Various complement components and fragments promote opsonization, lysis of membrane-bound microbes, and the generation of fragments that have proinflammatory and chemoattractant activities. Thus, the adaptive immune system makes good use of mechanisms that initially evolved to contribute to innate immunity, co-opting them for the elimination of pathogens.

Key Concept:
- The adaptive immune system has co-opted several pathogen clearance mechanisms, such as opsonization and complement activation, so that they contribute to antibody-mediated pathogen elimination.

Ubiquity of Innate Immunity

Determined searches among plant and invertebrate animal phyla for the signature proteins of the highly efficient vertebrate adaptive immune system—antibodies, T-cell receptors, and MHC proteins—have failed to find any homologs. Yet without them multicellular organisms have managed to survive for hundreds of millions of years. The interior spaces of organisms as diverse as

TABLE 4-8   Immunity in multicellular organisms

Taxonomic group	Innate immunity (nonspecific)	Adaptive immunity (specific)	Invasion-induced protective enzymes and enzyme cascades	Phagocytosis	Anti-microbial peptides	Pattern recognition receptors	Lympho-cytes	Variable lympho-cyte receptors	Anti-bodies
Higher plants	+	−	+	−	+	+	−	−	−
Invertebrate animals									
Porifera (sponges)	+	−	?	+	+	+	−	−	−
Annelids (earthworms)	+	−	?	+	+	+	−	−	−
Arthropods (insects, crustaceans)	+	−	+	+	+	+	−	−	−
Vertebrate animals									
Jawless fish (hagfish, lamprey)	+	+	+	+	+	+	+	+	−
Elasmobranchs (cartilaginous fish; e.g., sharks, rays)	+	+	+	+	+	+	+	−	+
Bony fish (e.g., salmon, tuna)	+	+	+	+	+	+	+	−	+
Amphibians	+	+	+	+	+	+	+	−	+
Reptiles	+	+	+	+	+	+	+	−	+
Birds	+	+	+	+	+	+	+	−	+
Mammals	+	+	+	+	+	+	+	−	+

Data from Flajnik, M. F., and L. Du Pasquier. 2008. Evolution of the immune system. In: *Fundamental Immunology*, 6th ed., W. E. Paul, ed. Lippincott, Philadelphia; and Wong, J. H., L. Xia, Ng, T. B. 2007. A review of defensins of diverse origins. *Current Protein and Peptide Science* 8:446.

the tomato, fruit fly, and sea squirt (an early chordate, without a backbone) do not contain unchecked microbial populations. Careful studies of these and many other representatives of nonvertebrate phyla have found arrays of well-developed processes that carry out innate immune responses. The accumulating evidence leads to the conclusion that multiple immune mechanisms protect all multicellular organisms from microbial infection and exploitation (**Table 4-8**).

Some Innate Immune System Components Occur across the Plant and Animal Kingdoms

In contrast to the adaptive immune system, components of the innate immune system are evolutionarily ancient, as evidenced by their presence in a virtually all multicellular organisms studied. For example, as mentioned early in this chapter, virtually all plant and animal species, and even some fungi, have antimicrobial peptides similar to defensins. Most multicellular organisms have pattern recognition receptors containing leucine-rich repeats (LRRs), although many organisms also have other families of PRRs. While the innate immune responses activated by these receptors in plants and invertebrates show both similarities and differences compared to those of vertebrates, innate immune response mechanisms are essential for the health and survival of these varied organisms.

Despite plants' tough outer protective barrier layers, such as bark and cuticle, and the cell walls surrounding each cell, plants can be infected by a wide variety of bacteria, fungi, and viruses, all of which must be combatted by the plant innate immune system. Plants do not have phagocytes or other circulating cells that can be recruited to sites of infection to mount protective responses. Instead, they rely on local innate immune responses for protection against infection. As described in **Evolution Box 4-4**, some resemble innate responses of animals, while others are quite distinct.

BOX 4-4

Plant Innate Immune Responses

In the plasma membrane under the cell wall, plant cells express pattern recognition receptors with LRR domains reminiscent of animal TLRs. These PRRs recognize what plant biologists refer to as *microbe-associated molecular patterns* (*MAMPs*), including bacterial flagellin, a highly conserved bacterial translation elongation factor, and various bacterial and fungal cell wall components (**Figure 1**). As is true for animal TLRs, some plant PRRs respond to danger-associated molecular patterns (DAMPs), which usually are created by pathogen enzymes that attack and fragment cell wall components. Some bacterial and fungal pathogens directly inject into plant cells toxin effector proteins that inhibit signaling through the plasma membrane PRRs. These toxins are recognized by a distinct class of LRR receptors in the cytoplasm called R proteins, which, like animal NLR proteins, have both LRR and nucleotide-binding domain (NBD) elements. After ligand binding, plant PRRs activate signaling pathways and transcription factors distinct from those of vertebrate cells (plants do not have NF-κB or IRF homologs), triggering innate responses.

The primary protective innate immune response mechanisms of plants to infection are the generation of reactive oxygen and nitrogen species, elevation of internal pH, and induction of a variety of antimicrobial peptides (including defensins) and antimicrobial enzymes that can digest the walls of invading fungi (chitinases) or bacteria (b-1,3-glucanase). Plants may also be activated to produce organic molecules, such as phytoalexins, that have antibiotic

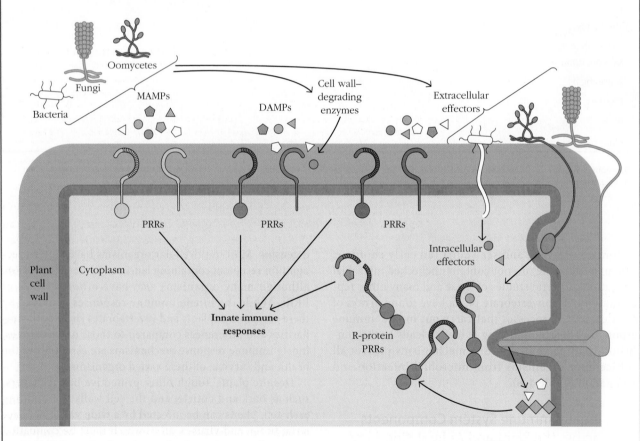

FIGURE 1 Activation of plant innate immune responses. Microbe-associated molecular patterns (MAMPs), danger-associated molecular patterns (DAMPs) generated by microbial enzymes (such as by degrading the cell wall), and microbial effectors (such as toxins) are recognized by plasma membrane LRR-containing pattern recognition receptors (PRRs). Microbial effectors that enter the cytoplasm are recognized by a class of PRRs called *resistance* (*R*) *proteins*. Recognition of MAMPs, DAMPs, and effectors by PRRs induces protective innate immune responses. While there are homologies between plant LRRs and those of animal TLRs, the cytoplasmic domains are very different. For example, some plant PRRs have cytoplasmic domains with tyrosine kinase activity and activate different signaling pathways.

(continued)

activity. In some cases, the responses of plants to pathogens even go beyond these protective substances to include structural responses. For example, to limit infection of leaves, PAMP binding to PRRs induces the epidermal guard cells that form the openings (stomata) involved in leaf gas exchange to close, preventing further invasion (**Figure 2**). Other protective mechanisms include the isolation of cells in the infected area by strengthening the walls of surrounding, noninfected cells and the induced death (necrosis) of cells in the vicinity of the infection to prevent the infection from spreading to the rest of the plant. Mutations that disrupt any of these processes usually result in loss of the plant's resistance to a variety of pathogens.

REFERENCE

Boller, T., and G. Felix. 2009. A renaissance of elicitors: perception of microbe-associated molecular patterns and danger signals by pattern-recognition receptors. *Annual Review of Plant Biology* **60**:379.

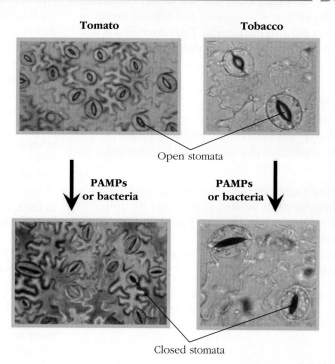

FIGURE 2 **Induced closure of leaf stomata following exposure to bacterial PAMPs.** Under normal light conditions, the openings (stomata) formed by pairs of guard cells in the leaf epidermis are open (visible as the football-shaped openings in the upper pair of leaf photographs), allowing normal gas exchange. However, as shown in the lower panels, exposure of the tobacco leaf to the bacterial plant pathogen *Pseudomonas syringae* and exposure of the tomato leaf to bacterial LPS induce the closure of the stomata (no openings visible). *[Republished with permission of Springer Science+Business Media, from, Melotto, M., et al., "Role of stomata in plant innate immunity and foliar bacterial diseases," Annual Review Phytopathology. 2008 Sept;46:101-122, Fig.3a. Permission conveyed through Copyright Clearance Center, Inc.]*

Key Concepts:

- Innate immunity appeared early during the evolution of multicellular organisms.

- Some innate immunity effector functions, such as antimicrobial peptides, are found in invertebrate and vertebrate animals, plants, and even some fungi.

- Innate immune defenses in plants include anatomical barriers as well as PRRs that activate antimicrobial innate immune responses, some of which are similar to those found in animals.

Invertebrate and Vertebrate Innate Immune Responses Show Both Similarities and Differences

Additional innate immune mechanisms evolved in animals, and we vertebrates share a number of innate immunity features with invertebrates (Table 4-8). PRRs (including

relatives of *Drosophila* Toll and vertebrate TLRs) that have specificities for microbial carbohydrate and peptidoglycan PAMPs are found in organisms as primitive as sponges. Together with soluble opsonin proteins (including some related to complement components), some of these early invertebrate PRRs function in promoting phagocytosis. Innate signaling has been well studied in *Drosophila*, and signaling proteins in flies have been identified that are homologous to several of those downstream of vertebrate TLRs (including MyD88 and IRAK homologs).

One significant difference is that fly Toll does not bind to PAMPs directly. Instead, pathogen binding to soluble pattern recognition proteins activates an enzymatic cascade, the end product of which activates fly Toll. The signaling pathways downstream of Toll are similar to those activated by plasma membrane TLRs of vertebrates. Thus, through the Toll pathway bacterial and fungal infections lead to the degradation of an IκB homolog and activation of NF-κB family members Dif and Dorsal, which induce the production of drosomycin, an insect defensin, and other antimicrobial peptides.

In addition to these and other pathways activated by PRRs, *Drosophila* and other arthropods employ other innate immune strategies not found in vertebrates, including the activation of phenoloxidase cascades that result in melanization—the deposition of a melanin clot around invading organisms that prevents their spread. Thus invertebrates and vertebrates have common as well as distinct innate immune response mechanisms.

Key Concept:

- Invertebrates and vertebrates share some innate immune response mechanisms, including phagocytosis and production of anti-microbial proteins and peptides, such as defensins.

- Invertebrate immune responses are best understood in *Drosophila melanogaster*, where the Toll protein responds to pathogens by activating pathways similar to those downstream of vertebrate TLRs, leading to the production of antimicrobial peptides.

- Invertebrates also use innate immune strategies not found in vertebrates, such as melanization.

Conclusion

It is appropriate that this text's introduction to the nature and mechanisms of immune responses begins with innate immunity, as the cells, tissues, and molecules of the innate immune system are charged with the responsibilities of providing early protection against infection. The first line of defense is provided by the epithelial layers that prevent the vast majority of pathogens in our environment from entering the body. The tightly stitched-together epithelial cells of the skin and the passageways connected to the body's openings in the mucosal and glandular tissues prevent easy entry to the body. They are also coated by a variety of chemical substances—from acid pH to antimicrobial peptides and proteins (including enzymes) that control pathogen populations at those sites. Consistent with its sheer size and critical role as a barrier, many call the skin the most important immunological organ in the body!

However, despite this normally effective first line of defense, infections can get established inside the body, be it via a skin wound, respiratory epithelial infection with influenza virus, or intestinal infection. Then it is up to the second line of defense, the cells of the innate immune system—especially the myeloid leukocytes macrophages, monocytes, neutrophils, and dendritic cells—to recognize the infection via their PRRs and mount an effective response that is appropriate for the particular pathogen. Given the great diversity of pathogens—viruses, bacteria, fungi, and parasites—and their ability to evolve through mutation and acquisition of host genes to evade innate responses, it is not surprising that the PRRs, their signaling pathways, and the responses that they stimulate are many and complex. Through animal evolution over millions of years, we have inherited genes encoding a limited but effective arsenal of PRRs, including those for TLRs, which go back to very early animal evolution. These receptors are located on the outside of the cell, in intracellular membrane compartments, and in the cytosol, poised to recognize multiple PAMPs on and inside individual pathogens and stimulate responses.

These local innate and inflammatory responses may be effective in clearing pathogens within hours or a few days. Thus many skin wounds heal on their own in a couple of days, and we undoubtedly are continually inhaling many respiratory viruses without getting sick. From phagocytosis and NETosis to complement activation to production of the many antimicrobial proteins and peptides to the activation of NK cells and other ILCs, the innate responses may be sufficient to eliminate or at least control an infection. In the first couple of days after viral infection, type I IFNs and NK cells limit the replication and spread of the virus. But when innate and inflammatory responses are not sufficient—perhaps because pathogens have evolved to evade innate responses—we mammals have our powerful adaptive immune responses to come to the rescue—our third and last line of defense.

As will be described in upcoming chapters, the randomly generated, highly diverse antigen-specific receptors of B cells and T cells are able to respond to virtually anything foreign that enters the body. But here, too, the innate immune system plays a critical role: helping to ensure that the adaptive antibody and cell-mediated immune responses are appropriate for the particular type of pathogen. This results from the selective activation through certain PRRs of dendritic and other innate immune cells to generate cytokines that influence the differentiation of T cells into those that will activate the type of adaptive response that will work against the pathogen. Without our innate immune system, our adaptive immune system would not be nearly as powerful! As you go on to learn about adaptive immunity, remember the many ways in which our innate immune system contributes to antibody and cell-mediated responses.

REFERENCES

Areschoug, T., and S. Gordon. 2009. Scavenger receptors: role in innate immunity and microbial pathogenesis. *Cellular Microbiology* **11**:1160.

Beutler, B., and E. T. Rietschel. 2003. Innate immune sensing and its roots: the story of endotoxin. *Nature Reviews Immunology* **3**:169.

Bowie, A. G., and L. Unterholzner. 2008. Viral evasion and subversion of pattern-recognition receptor signalling. *Nature Reviews Immunology* **8**:911.

Brubaker, S. W., K. S. Bonham, I. Zanoni, and J. C. Kagan. 2015. Innate immune pattern recognition: a cell biological perspective. *Annual Review of Immunology* **33**:257.

Bulet, P., R. Stöcklin, and L. Menin. 2004. Anti-microbial peptides: from invertebrates to vertebrates. *Immunology Review* **198**:169.

Cao, X. 2016. Self-regulation and cross-regulation of pattern-recognition receptor signalling in health and disease. *Nature Reviews Immunology* **16**:35.

Coffman, R. L., A. Sher, and R. A. Seder. 2010. Vaccine adjuvants: putting innate immunity to work. *Immunity* **33**:492.

Cording, S., J. Medvedovic, T. Aychek, and G. Eberl. 2016. Innate lymphoid cells in defense, immunopathology and immunotherapy. *Nature Immunology* **17**:755.

Dambuza, I. M., and G. D. Brown. 2015. C-type lectins in immunity: recent developments. *Current Opinion Immunology* **32**:21.

Davis, B. K., H. Wen, and J. P. Ting. 2011. The inflammasome NLRs in immunity, inflammation, and associated diseases. *Annual Review of Immunology* **29**:707.

DeFranco, A. L., R. M. Locksley, and M. Robertson. 2007. *Immunity: The Immune Response in Infectious and Inflammatory Disease* (*Primers in Biology* series). Sinauer Associates, Sunderland, MA.

Froy, O. 2005. Regulation of mammalian defensin expression by Toll-like receptor–dependent and independent signalling pathways. *Cellular Microbiology* **7**:1387.

Gordon, S. 2016. Phagocytosis: an immunobiologic process. *Immunity* **44**:463.

Grimsley, C., and K. S. Ravichandran. 2003. Cues for apoptotic cell engulfment: eat-me, don't-eat-me, and come-get-me signals. *Trends in Cellular Biology* **13**:648.

Hornung, V., R. Hartmann, A. Ablasser, and K.-P. Hopfner. 2014. OAS and cGAS: unifying concepts in sensing and responding to cytosolic nucleic acids. *Nature Reviews Immunology* **14**:521.

Iwasaki, A., and R. Medzhitov. 2015. Control of adaptive immunity by the innate immune system. *Nature Immunology* **16**:343.

Jimenez-Dalmaroni, M. J., M. E. Gerswhin, and I. E. Adamopoulos. 2016. The critical role of Toll-like receptors—from microbial recognition to autoimmunity: a comprehensive review. *Autoimmunity Reviews* **15**:1.

Kaufman, H. E., and A. Dorhoi. 2016. Molecular determinants in phagocyte-bacteria interactions. *Immunity* **44**:476.

Lemaitre, B. 2004. The road to Toll. *Nature Reviews Immunology* **4**:521.

Litvack, M. L., and N. Palaniyar. 2010. Soluble innate immune pattern-recognition proteins for clearing dying cells and cellular components: implications on exacerbating or resolving inflammation. *Innate Immunity* **16**:191.

Man, S. M., and T.-D., Kanneganti. 2016. Converging roles of caspases in inflammasome activation, cell death, and innate immunity. *Nature Reviews Immunology* **16**:7.

Medzhitov, R., P. Preston-Hurlburt, and C. A. Janeway, Jr. 1997. A human homologue of the Toll protein signals activation of adaptive immunity. *Nature* **388**:394.

Nathan, C., and A. Ding. 2010. SnapShot: reactive oxygen intermediates (ROI). *Cell* **140**:952.

Poltorak, A., et al. 1998. Defective LPS signaling in C3Hej and C57BL/10ScCr mice: mutations in *Tlr4* gene. *Science* **282**:2085.

Pulendran, B. 2015. The varieties of immunological experiences: of pathogens, stress, and dendritic cells. *Annual Review of Immunology* **33**:563.

Salzman, N. H., et al. 2010. Enteric defensins are essential regulators of intestinal microbial ecology. *Nature Immunology* **11**:76.

Saxena, M. and G. Yeretssian. 2014. NOD-like receptors: master regulators of inflammation and cancer. *Frontiers in Immunology* **5**:1.

Schroder, K., and J. Tschopp. 2010. The inflammasome. *Cell* **140**:821.

Sun, J. C., et al. 2011. NK cells and immune "memory." *Journal of Immunology* **186**:1891.

Tecle, T., S. Tripathy, and K. L. Hartshorn. 2010. Defensins and cathelicidins in lung immunity. *Innate Immunity* **16**:151.

Tieu, D. D., R. C. Kern, and R. P. Schleimer. 2009. Alterations in epithelial barrier function and host defense responses in chronic rhinosinusitis. *Journal of Allergy and Clinical Immunology* **124**:37.

Trejo-de la O, A., P. Hernández-Sancén, and C. Maldonado-Bernal. 2014. Relevance of single-nucleotide polymorphisms in human *TLR* genes to infectious and inflammatory diseases and cancer. *Genes and Immunity* **15**:199.

van de Vosse, E., J. T. van Dissel, and T. H. Ottenhoff. 2009. Genetic deficiencies of innate immune signalling in human infectious disease. *Lancet Infectious Diseases* **9**:688.

Willingham, S. B., et al. 2012. The CD47-signal regulatory protein alpha (SIRP1a) interaction is a therapeutic target for human solid tumors. *Proceedings of the National Academy of Sciences USA* **109**:6662.

Yoneyama, M., K. Onomoto, M. Jogi, T. Akaboshi, and T. Fujita. 2015. Viral RNA detection by RIG-I–like receptors. *Current Opinion in Immunology* **32**:48.

Yin, Q., T. M. Fu, J. Li, and H. Wu. 2015. Structural biology of innate immunity. *Annual Review of Immunology* **33**:393.

Useful Websites

www.biolegend.com/basic_immunology Presents summaries of key components of innate immunity, including the cells of the innate immune system (including innate lymphoid cells), roles of dendritic cells, pattern recognition receptors and their activation pathways, and products for research.

portal.systemsimmunology.org Systems Approach to Immunology (systemsimmunology.org) is a large collaborative research program formed to study the mechanisms by which the immune system responds to infectious disease by inciting innate inflammatory reactions and instructing adaptive immune responses.

www.bmcimmunol.biomedcentral.com/articles/
10.1186/1471-2172-9-7 Website of the Innate Immune
Database, an NIH-funded, multi-institutional project that
assembled microarray data on expression levels of more than
200 genes in macrophages stimulated with a panel of TLR ligands. The database is intended to support systems biology
studies of innate responses to pathogens.

www.immgen.org The Immunological Genome Project is
a new cooperative effort for deep transcriptional profiling of
all immune cell types.

www.ncbi.nlm.nih.gov/PubMed PubMed, the National
Library of Medicine database of more than 15 million publications, is the world's most comprehensive bibliographic
database for biological and biomedical literature. It is highly
user friendly, searchable by general or specific topics, authors,
reviews, and so on. It is the best resource to use for finding the
latest research articles on innate immunity or other topics in
the biomedical sciences.

wikipedia.org/wiki/Innate_immune_system The Wikipedia website presents a detailed summary of the innate
immune system in animals and plants; it contains numerous
figures and photographs illustrating various aspects of innate
immunity, plus links to many references.

**www.primaryimmune.org/about-primary-immuno-
deficiencies/specific-disease-types/innate-immune-
defects/** Page on the website of the Immune Deficiency
Foundation; presents information on deficiencies of the innate immune system.

STUDY QUESTIONS

1. Use the following list to complete the statements that follow.
Some terms may be used more than once or not at all.

Antibodies	NLRs
Arginine	NO
CARD	O_2
Caspase-1	OAS
C-reactive protein (CRP)	2′,5′-Oligoadenylate
cGAS	A synthetase
Complement	PAMPs
Costimulatory molecules	Phagocytosis
Cytokines	Proinflammatory
Defensins	cytokines
Dendritic cells	Protein kinase R
Ficolins	PRRs
IFIT proteins	Psoriasin
IL-1	Pyroptosis
Inflammasomes	RLRs
iNOS	ROS
Interferons-α, β	RNS
IRFs	Surfactant proteins
Lysozyme	(SP-A, SP-D)
Mannose-binding lectin	STING
(MBL)	TRIF
Mx proteins	T-cell receptors
MyD88	TLR2
NADPH	TLR3
Phagosome NADPH	TLR4
oxidase	TLR7
NF-κB	TLR9
NK cells	TNF-α

a. Examples of proteins and peptides with direct antimicrobial activity that are present on epithelial surfaces are
_____, _____, and _____.

b. Soluble pattern recognition proteins that function as
opsonins, which enhance _____, include _____, _____,
_____, and _____; of these, the ones that also activate
complement are _____ and _____.

c. The enzyme _____ uses _____ to generate microbe-
killing _____; one of these, plus the antimicrobial gas
_____ generated by the _____ enzyme from the amino
acid _____, are used to generate _____, which are also
antimicrobial.

d. As components of innate immune responses, both _____
(secreted proteins) and _____ (a type of lymphocyte)
defend against viral infection.

e. _____, receptors of innate immunity that detect _____,
are encoded by germline genes, whereas the signature
receptors of adaptive immunity, _____ and _____, are
encoded by genes that require gene rearrangements
during lymphocyte development to be expressed.

f. Among cell surface TLRs, _____ detects gram-positive
bacterial infections while _____ detects gram-negative
infections.

g. Some cells use the intracellular TLRs _____ and _____
to detect RNA virus infections and _____ to detect infections by bacteria and some DNA viruses.

h. _____ is unique among PRRs in that it functions both on
the plasma membrane and in endosomes and binds both
the _____ and _____ adapter proteins.

i. _____ include cytosolic receptors that detect intracellular bacterial cell wall components.

j. _____ is a cytosolic PRR that recognizes viral and bacterial DNAs.

k. _____ is a cytosolic protein that recognizes virus-
induced or bacteria-derived cyclic dinucleotides and initiates pathways activating IRFs and NF-κB.

l. Key transcription factors for inducing expression of proteins involved in innate immune responses are _____
and _____.

m. The production of the key proinflammatory cytokine
_____ is complex, as it requires transcriptional activation
by signaling pathways downstream of _____ followed by
cleavage of its large precursor protein by _____, which
is activated by _____, members of the _____ family of
innate receptors.

n. Four proteins induced by type I IFNs that mediate their
antiviral activity are _____, _____, _____, and _____.

o. After maturation induced by binding of _____ to their _____, cells known as _____ become efficient activators of naïve helper and cytotoxic T cells.

p. _____ produced by _____ in response to pathogen components that bind to their PRRs control the differentiation of naïve T cell into a specific T-cell subset that will contribute to the elimination of the pathogen.

q. _____ and _____ are innate immune system components common to both plants and animals.

2. What are the two lines of defense that comprise the innate immune system? For each give three examples of protective mechanisms.

3. Give three examples of receptors that induce phagocytosis of bacteria and how they trigger phagocytosis.

4. What were the two experimental observations that first linked TLRs to innate immunity in vertebrates?

5. What are the hallmark characteristics of a localized inflammatory response? How are they induced by the early innate immune response at the site of infection, and how do these characteristics contribute to an effective innate immune response?

6. What is regulated cell death? Give two examples, and explain how these forms of cell death may be beneficial.

7. Describe the roles of innate lymphoid cells (ILCs). How are they activated?

8. In vertebrates, innate immunity collaborates with adaptive immunity to protect the host. Discuss this collaboration, naming key points of interaction between the two systems. Include at least one example in which the adaptive immune response contributes to enhanced innate immunity.

9. As adaptive immunity evolved in vertebrates, the more ancient system of innate immunity was retained. Can you think of any disadvantages to having a dual system of immunity? Would you argue that either system is more essential?

CLINICAL FOCUS QUESTION

What infections are unusually prevalent in individuals with genetic defects in TLRs or the MyD88-dependent TLR signaling pathway? In individuals with defects in pathways activating the production or antiviral activities of IFN-α and IFN-β? Why is it thought that these individuals aren't susceptible to a wider range of diseases, and what evidence supports this hypothesis?

ANALYZE THE DATA

As a veterinary immunologist, you are an expert on the roles of innate and adaptive immune responses in animal infectious diseases, so when local residents in your town started finding large numbers of sick wild mice in their backyards, you were asked to investigate. After some months' investigation, you find that the mice are infected with a new mouse virus, which you call raccoon virus (RV), as mice become infected by contact with virus-containing urine from infected raccoons living in the same area.

While a virologist colleague is characterizing the virus, you initiate studies of the immune response to RV, using inbred mice from your colony. You find that RV infects the bladder epithelium, leading rapidly to an inflammatory infiltrate of neutrophils and monocytes. The virus can spread to the kidneys, and in your colony about one-third of the infected mice die of bladder and kidney damage within a week; the others gradually return to health.

You investigate what aspects of the immune response provide protection in mice that recover, using mice already available in your lab in which genes encoding some TLRs and some proteins in the TLR signaling pathways have been knocked out (indicated by "-/-", which means the mice are homozygous for that knock-out mutation). You infect groups of several dozen wild-type and knockout mice with the virus and monitor their survival for 21 days. The percentage of mice in each group that survive to 21 days is tabulated below.

Mouse	Percentage of mice surviving at 21 days
Wild-type	67
TLR3$^{-/-}$	62
TLR4$^{-/-}$	63
TLR7$^{-/-}$	5
TLR9$^{-/-}$	66
MyD88$^{-/-}$	3
TRIF$^{-/-}$	65

a. From the data in the table, what can you conclude about the role of the innate response in protection against innate immune response in protection against the virus? What do these data suggest is the viral component inducing the protective response? Justify your answer.

b. Further studies show that the inflammatory infiltrate is required for the protection of mice against lethal infections with the virus. Explain the likely process by which the innate immune response to RV is induced and provides protection against this virus.

c. Your virologist colleague shows that the virus replicates in the cytosol of infected bladder epithelial cells, with a double-stranded RNA (dsRNA) intermediate in the replication cycle. Would you expect the infected epithelial cells to make their own protective innate response? Explain how such a response might be generated. The fact that the inflammatory infiltrate is required indicates that the epithelial cells are not generating their own protective response. Why do you think this might be?

The Complement System

Learning Objectives

After reading this chapter, you should be able to:

1. Compare and contrast the three major pathways of complement activation with respect to the factors that initiate the pathways, the convertase enzymes that act in those pathways, and the regulatory factors that ensure that activation occurs only on microbial surfaces.

2. Teach a colleague how complement proteins mediate phagocytosis of apoptotic cells and clearance of immune complexes in the contraction phase of an immune response.

3. Describe three different ways in which microbes have evolved to evade the antimicrobial actions of the complement system.

4. Show why patients deficient in the early components of complement suffer an increased risk of autoimmune diseases such as systemic lupus erythematosus.

5. Explain the classification of the components of the complement system into evolutionarily related families.

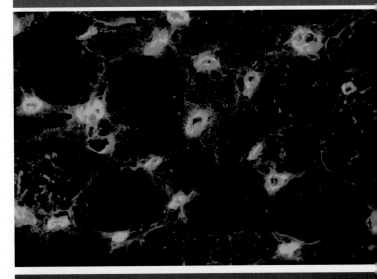

Complement fragment C3d is revealed by the green stain, indicating active rejection of a heart allograft. *[Republished with permission of American Society of Nephrology, from Collins, B. et al., 1999, Complement activation in acute humoral renal allograft rejection: diagnostic significance of C4d deposits in peritubular capillaries.* Journal of the American Society of Nephrology. *Oct 10 (10): 2208-14. Figure 2. Permission conveyed through Copyright Clearance Center, Inc.]*

The term **complement** refers to a set of 50-plus serum proteins that cooperates with both the innate and the adaptive immune systems to eliminate pathogens, dying cells, and immune complexes from the body. Complement proteins were first discovered by virtue of their ability to complex with antibodies and destroy the membranes of antibody-bound cells. They were therefore initially classified as part of the adaptive immune response. However, we have since come to realize that the capacity of complement proteins to punch holes in cell membranes represents only a small fraction of complement's role in immunity. The importance of complement proteins to the effective operation of both the innate and the adaptive branches of the immune system and the diversity of their roles within it are emphasized by the number of complement evasion strategies that have evolved in microbial pathogens. Indeed, the roles of complement proteins are so diverse that, more than a century after complement's initial discovery, we are still learning about novel aspects of their functions.

Research on complement began in the 1890s when Jules Bordet showed that sheep antiserum to the

Key Terms

Zymogens

Opsonins

Anaphylatoxins

Membrane attack complex (MAC)

C3 convertases

C5 convertases

Classical pathway of complement activation

Lectin pathway of complement activation

Alternative pathway of complement activation

Immune complex

Lectin

Properdin or factor P

bacterium *Vibrio cholerae* caused bacterial lysis (membrane destruction), and that heating the antiserum destroyed its bacteriolytic activity. Surprisingly, the ability to lyse the bacteria was restored to the heated serum by adding fresh serum that contained no antibacterial antibodies. Bordet reasoned that bacteriolysis required two different substances: the heat-stable, cholera-specific antibodies that bound to the bacterial surface, and a second, heat-labile (sensitive) component responsible for the lytic activity.

In an effort to purify this second, nonspecific component, Bordet developed antibodies specific for red blood cells and used these, in combination with purified fractions from serum, to identify those serum proteins that cooperated with antibodies to induce hemolysis (lysis of the red blood cells). The famous immunologist Paul Ehrlich, working independently in Berlin, conducted similar experiments and coined the term *complement*, defining it as "the activity of blood serum that *completes* the action of antibody."

In the ensuing years, researchers have discovered that the activities attributed to complement are mediated by more than 50 proteins and glycoproteins. Most complement components are synthesized in the liver by hepatocytes, although some are also produced by blood monocytes, tissue macrophages, fibroblasts, and epithelial cells of the gastrointestinal and genitourinary tracts. Complement components constitute approximately 15% of the globulin protein fraction in plasma. In addition, since several of the regulatory components of the system exist on cell membranes, the term *complement* now embraces proteins and glycoproteins distributed between the blood plasma and cell membranes. Complement components can be classified into seven functional categories (**Overview Figure 5-1**):

1. *Initiator complement components.* These proteins initiate their respective complement reactions by binding to particular soluble or membrane-bound molecules. Once activated by their ligand, they undergo conformational alterations resulting in changes in their biological activity.

2. *Enzymatic mediators.* Several complement components are proteolytic enzymes that cleave and activate the next member of a complement reaction sequence. Proteins that are inactive until cleaved by proteases are called **zymogens**. Some complement proteases become active by binding to other macromolecules and undergoing a conformational change; others are zymogens themselves, inactive until cleaved by another "upstream" protease. The two enzyme complexes that cleave the complement components C3 and C5 are called the *C3* and *C5 convertases*, respectively, and occupy places of central importance in complement biology. The sequence of proteins

in a complement pathway from the initiator protein to the biological effector is referred to as a "complement cascade."

3. *Phagocytosis-enhancing components, or opsonins.* On activation of the complement cascade, several complement proteins are cleaved into two fragments, each of which then takes on a particular role. For C3 and C4, the larger fragments, C3b and C4b, serve as **opsonins**, binding covalently to microbial cells and serving as ligands for phagocytic cells with receptors for C3b or C4b.

4. *Inflammatory mediators.* Some small complement fragments act as inflammatory mediators. These fragments bind to receptors on the endothelial cells lining small blood vessels and induce an increase in capillary diameter, thus enhancing blood flow to the affected area. They also attract other cells to the site of tissue damage. Since these effects can be harmful (even lethal) in excess, these fragments are called **anaphylatoxins**, derived from the Greek phrase meaning "against protection." C3a and C5a are examples of anaphylatoxins.

5. *Membrane attack proteins.* Proteins of the **membrane attack complex (MAC)** insert into the cell membranes of invading microorganisms and punch holes that result in lysis of the pathogen. The MAC has been extensively imaged by electron microscopy. The complex itself forms a ring-shaped multimer of complement proteins with a central hole through which cytoplasmic contents can escape. MACs can also form on infected host cells, although the complement system must first overcome the regulatory mechanisms designed to protect host cells from complement attack.

6. *Complement receptor proteins.* Receptor molecules on cell surfaces bind complement proteins and signal specific cell functions. For example, some complement receptors such as CR1 bind to complement components such as C3b that have opsonized pathogens, triggering phagocytosis of the C3b-bound pathogen. Binding of the anaphylatoxin complement component C5a to C5a receptors (C5aRs) on neutrophils stimulates neutrophil degranulation and inflammation.

7. *Regulatory complement components.* Host cells are protected from unintended complement-mediated damage by the presence of regulatory proteins. These regulatory proteins include factor I, which degrades C3b, and CD59 (protectin), which inhibits the formation of the MAC on host cells.

This chapter describes the components of the complement system, their activation via three major pathways, the effector functions of the molecules of the complement

Functional Categories of Complement Proteins

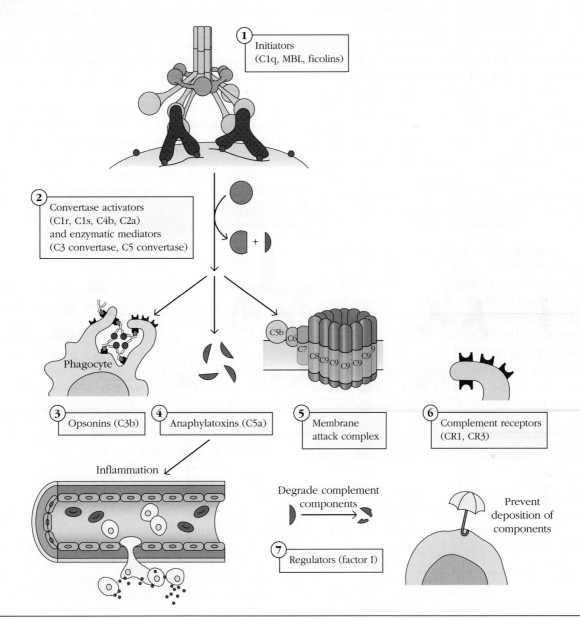

(1) Initiators
(C1q, MBL, ficolins)

(2) Convertase activators
(C1r, C1s, C4b, C2a)
and enzymatic mediators
(C3 convertase, C5 convertase)

Phagocyte

C5b C6 C7 C8 C9 C9 C9 C9 C9

(3) Opsonins (C3b)

(4) Anaphylatoxins (C5a)

(5) Membrane attack complex

(6) Complement receptors (CR1, CR3)

Inflammation

Degrade complement components

Prevent deposition of components

(7) Regulators (factor I)

(1) The complement pathways are initiated by proteins that bind to pathogens, either directly or via an antibody or other pathogen-specific protein. After a conformational change, (2) enzymatic mediators activate other enzymes that generate the central proteins of the complement cascade, the C3 and C5 convertases, which cleave C3 and C5, releasing active components that mediate all functions of complement, including (3) opsonization, (4) inflammation, and (5) generation of the membrane attack complex (MAC). Effector complement proteins can label an antibody-antigen complex for phagocytosis (opsonins), initiate inflammation (anaphylatoxins), or bind to a pathogen and nucleate the formation of the MAC. Often, these effectors act through (6) complement receptors on phagocytic cells, granulocytes, or erythrocytes. (7) Regulatory proteins limit the effects of complement by promoting their degradation or preventing their binding to host cells. (Note that a limited number of examples of each type of protein is shown in this figure. See text for a more complete description of each class of effector or regulatory complement protein.)

cascade, and their interactions with other cellular and molecular components of innate and adaptive immunity. In addition, it addresses the mechanisms that regulate the activity of these complement components, the evasive strategies evolved by pathogens to avoid destruction by complement, and the evolution of the various complement proteins. This chapter's Classic Experiment tells the tragic story of the scientist who discovered the alternative pathway of complement. In the Advances Box, we describe some interactions between complement and the nervous system. Finally, a Clinical Focus segment addresses various therapies that target elements of the complement cascades.

The Major Pathways of Complement Activation

Complement components represent some of the most evolutionarily ancient participants in the vertebrate immune response. As viruses, parasites, and bacteria have infected vertebrate hosts and learned to evade aspects of the complement system's function, new mechanisms of host immunity have evolved in an endless dance of microbial attack and host response.

There are three major pathways by which the complement cascade can be initiated: the classical pathway, the lectin pathway, and the alternative pathway, shown in **Figure 5-2**. Although the initiating event of each of the three

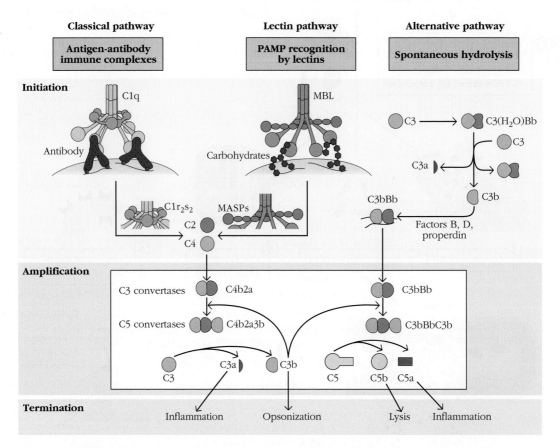

FIGURE 5-2 Generation of C3 and C5 convertases by the three major pathways of complement activation. The classical pathway is initiated when C1q binds to antigen-antibody complexes. The antigen is shown here in dark red and the initiating antibody in green. The C1r enzymatic component of C1 (shown in blue) is then activated and cleaves C1s, which in turn cleaves C4 to C4a and C4b. C4b attaches to the membrane and binds C2, which is then cleaved by C1s to form C2a and C2b. (C2b is then acted on further to become an inflammatory mediator.) C2a remains attached to C4b, forming the classical pathway C3 convertase (C4b2a). In the lectin pathway, mannose-binding lectin (MBL, green) binds specifically to conserved carbohydrate arrays on pathogens, activating the MBL-associated serine proteases (MASPs, blue). The MASPs cleave C2 and C4, generating the C3 convertase as in the classical pathway. In the alternative pathway, C3 undergoes spontaneous hydrolysis to C3(H$_2$O), which binds serum factor B. On binding to C3(H$_2$O), B is cleaved by serum factor D, and the resultant C3(H$_2$O)Bb complex forms a fluid-phase C3 convertase. Some C3b, released after C3 cleavage by this complex, binds to microbial surfaces. There, it binds factor B, which is cleaved by factor D, forming the cell-bound alternative pathway C3 convertase, C3bBb. This complex is stabilized by properdin. The C5 convertases are formed by the addition of a C3b fragment to each of the C3 convertases.

pathways of complement activation is different, they all converge in the generation of an enzyme complex that cleaves the C3 molecule. Enzymes that cleave C3 into two fragments, C3a and C3b, are referred to as **C3 convertases**. The classical and lectin pathways use the dimer *C4b2a* for their C3 convertase activity, while the alternative pathway uses *C3bBb* (see Figure 5-2). However, the final result of both C3 convertase activities is the same: a dramatic increase in the concentration of C3b, a critically important, multifunctional complement protein.

There is a second set of convertase enzymes generated in the early stages of complement activation. The **C5 convertases** are formed by the addition of a C3b component to each of the two C3 convertases. C5 convertases cleave C5 into C5a, an inflammatory mediator, or anaphylatoxin, and C5b, which is the initiating factor of the membrane attack complex.

We will now describe the three pathways of complement activation in more detail. The proteins involved in each of these pathways are listed in **Tables 5-1**, **5-2**, and **5-3**.

The Classical Pathway Is Initiated by Antibody Binding to Antigens

The **classical pathway of complement activation** is considered part of the adaptive immune response since it begins with the formation of antigen-antibody complexes. These complexes may be soluble, or they may be formed when an antibody binds to **antigenic determinants**, or **epitopes**, situated on viral, fungal, parasitic, or bacterial cell membranes. Soluble antibody-antigen complexes are often referred to as **immune complexes**. Only complexes formed by antigens with antibodies of the IgM class or certain subclasses of IgG antibodies are capable of activating the classical complement

TABLE 5-1	Initiating and amplifying proteins of the classical and lectin-mediated complement pathways		
Molecule	Biologically active fragments	Biological function	Active in which pathway
IgM, IgG		Binding to pathogen surface and initiating complement cascade	Classical pathway
Mannose-binding lectin (MBL), or ficolins		Binding to carbohydrates on microbial surface and initiating complement cascade	Lectin pathway
C1	C1q	Initiation of the classical pathway by binding Ig Binding to apoptotic blebs and initiating phagocytosis of apoptotic cells	Classical pathway
	(C1r)$_2$	Serine protease, cleaving C1r and C1s	
	(C1s)$_2$	Serine protease, cleaving C4 and C2	
MASP-1		MBL-associated serine protease 1. MASP-2 appears to be functionally the more relevant MASP protein	Lectin pathway
MASP-2		Serine protease. In complex with MBL/ficolin, cleaves C4 and C2	Lectin pathway
C2	C2a*	Serine protease. With C4b, is a C3 convertase	Classical and lectin pathways
	C2b*	Inactive in complement pathway. Cleavage of C2b by plasmin releases C2 kinin, a peptide that stimulates vasodilation	
C4	C4b	Binds microbial cell membrane via thioester bond. With C2a, is a C3 convertase	Classical and lectin pathways
	C4c, C4d	Proteolytic cleavage products generated by factor I	
C3	C3a	Anaphylatoxin. Mediates inflammatory signals via C3aR	Classical and lectin pathways
	C3b	Potent in opsonization, tagging immune complexes, pathogens, and apoptotic cells for phagocytosis With C4b and C2a, forms the C5 convertase	
	iC3b and C3f	Proteolytic fragments of C3b, generated by factor I iC3b binds receptors CR3, CR4, and CRIg; CR2 binds weakly	
	C3d and C3dg	Proteolytic fragments of iC3b generated by factor I and trypsin like proteases including plasmin, thrombin etc. C3d and C3dg bind to CR2, and when they are bound to both antigen and to CR2, this facilitates antigen binding to B cells	
	C3c	Proteolytic fragment of iC3b generated by factor I and trypsin-like proteases. C3c binds CRIg on fixed tissue macrophages	

*C2a in this text refers to the larger, active fragment of C2. Some writers have tried to alter the nomenclature in order to make C2 conform to the convention that the larger, active fragment of cleaved components of complement is designated with a "b," while the smaller fragment is denoted with an "a." However, this effort does not appear to be making headway. Note that the smaller fragment of C2, which we name C2b, is inactive in the complement pathway.

TABLE 5-2	Initiating and amplifying proteins of the alternative complement pathways	
Molecule	**Biologically active fragments**	**Biological function**
C3	C3a	Anaphylatoxin. Mediates inflammatory signals via C3aR
	C3b	Potent in opsonization, tagging immune complexes, pathogens, and apoptotic cells for phagocytosis With Bb, forms the C3 convertase With Bb and one more molecule of C3b (C3bBb3b), acts as a C5 convertase
	C3(H_2O)	C3 molecule in which the internal thioester bond has undergone hydrolysis With Bb, acts as a fluid-phase C3 convertase
	iC3b and C3f	Proteolytic fragments of C3b, generated by factor I iC3b binds receptors CR3, CR4, and CRIg; CR2 binds weakly
	C3d and C3dg	Proteolytic fragments of iC3b generated by factor I and trypsin-like proteases including plasmin, thrombin etc. C3d and C3dg both bind to CR2. When each is bound to both antigen and to CR2, this enhances the strength of antigen binding to B cells
	C3c	Proteolytic fragment of iC3b generated by factor I and trypsin-like proteases. C3c binds CRIg on fixed tissue macrophages
Factor B		Binds C3(H_2O) and is then cleaved by factor D into two fragments: Ba and Bb
	Ba	Smaller fragment of factor D–mediated cleavage of factor B May inhibit proliferation of activated B cells
	Bb	Larger fragment of factor D–mediated cleavage of factor B With C3(H_2O), acts as fluid-phase C3 convertase With C3b, acts as cell-bound C3 convertase With two molecules of C3b, acts as C5 convertase
Factor D		Proteolytic enzyme that cleaves factor B into Ba and Bb only when it is bound to either C3(H_2O) or to C3b
Properdin		Stabilizes the C3bBb complex on microbial cell surface

TABLE 5-3	The proteins of the complement membrane attack complex (MAC)	
Molecule	**Biologically active fragments**	**Biological function**
C5	C5a	Anaphylatoxin; binding to C5aR induces inflammation
	C5b	Component of membrane attack complex (MAC). Binds cell membrane and facilitates binding of other components of the MAC
C6		Component of MAC. Stabilizes C5b. In the absence of C6, C5b is rapidly degraded
C7		Component of MAC. Binds C5bC6 and induces conformational change allowing C7 to insert into interior of membrane
C8		Component of MAC. Binds C5bC6C7 and creates a small pore in membrane
C9		Component of MAC. Ten to 19 molecules of C9 bind C5bC6C7C8 and create large pore in membrane

pathway (see Chapter 12). The initial activation involves interaction of these antibody-antigen complexes with the complement components C1, C2, and C4, normally found in the plasma as inactive precursors or zymogens.

The formation of an antigen-antibody complex induces conformational changes in the non-antigen-binding (Fc) portion of the antibody molecule. This conformational change exposes a binding site on the antibody

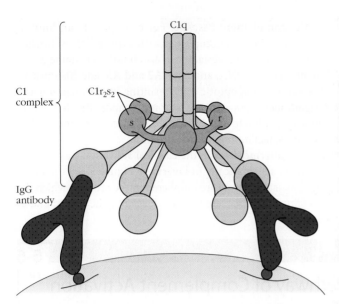

FIGURE 5-3 Structure of the C1 macromolecular complex. C1q interacts with two molecules each of C1r and C1s to create the C1 complex. The C1q molecule consists of 18 polypeptide chains in six collagen-like triple helices.

for the C1 component of complement. In serum, C1 exists as a macromolecular complex consisting of one molecule of C1q and two molecules each of the serine proteases C1r and C1s, held together in a Ca^{2+}-stabilized complex ($C1qr_2s_2$) (**Figure 5-3**). The C1q molecule itself is composed of 18 polypeptide chains that associate to form six collagen-like triple helical arms, the tips of which bind the C_H2 domain (see Figure 3-10a) of the antigen-bound antibody molecule.

Each C1 macromolecular complex must bind to at least two antibody constant regions for a stable C1q-antibody interaction to occur. In the serum, IgM exists as a pentamer of the basic four-chain immunoglobulin structure. In circulating, non-antigen-bound IgM, the antigen-binding arms project out in a star shape from the central, slightly raised core. **Figure 5-4a** shows the pentameric unit with the C1q-binding residues indicated in white. Figure 5-4b shows a projection of the same molecule from the side, illustrating the slight bulge in the center of the molecule. (Note that the dark blue monomer has been removed in this figure so as to allow better visualization of the rest of the molecule.) In this relatively planar conformation, the C1q-binding sites are not readily accessible to the C1q ligand. However, when pentameric IgM is bound to a multivalent antigen, it undergoes a substantial conformational change, assuming a "staple" configuration (Figure 5-4c), and revealing at least three binding sites for C1q. Thus, an IgM molecule engaged in an antibody-antigen complex can bind C1q, whereas circulating, non-antigen-bound IgM cannot.

In contrast to pentameric IgM, monomeric IgG contains only one C1q-binding site per molecule. Although this C1q-binding site is exposed, the affinity of this exposed

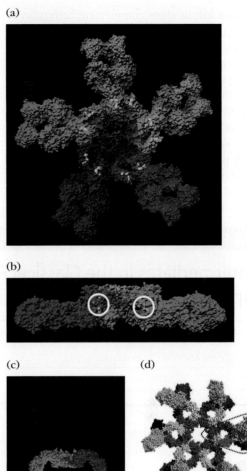

FIGURE 5-4 Models of pentameric IgM and hexameric IgG derived from x-ray crystallographic data. (a) The star-shaped form of pentameric IgM, showing the locations of the binding sites for C1q (in white), which are not readily accessible in this planar conformation. (b) Seen from the side, the shape of pentameric IgM in the absence of antigen is a quasiplanar, mushroom-shaped form. Note the slightly raised bulge in the center, which prevents the starfish-shaped molecule from adopting a completely flat configuration. In parts (b) and (c), the dark blue monomer from part (a) has been digitally removed so as to allow a clearer image of the molecule. (c) When IgM binds to more than one epitope of a multivalent antigen (red dots) it dramatically changes its conformation to a "staple" form, and exposes binding sites for C1q. (d) The shape of a hexameric complex of IgG is shown. The dotted line encloses a single IgG monomer and the C1q-binding lysine residue, located in the C_H2 region of the IgG molecule, is highlighted in red. *[Parts (a), (b), and (c) from Czajkowsky, D. M., and Z. Shao, 2009, The human IgM pentamer is a mushroom-shaped molecule with a flexural bias. Proceedings of the National Academy of Sciences USA **106**:14960. Figs 4a, 4b, and 5. Copyright (2009) National Academy of Sciences, USA. Part (d) republished with permission of the American Association for the Advancement of Science, from Diebolder, C., Beurskens, F, N. de Jong, R., et al., 2014, Complement is activated by IgG hexamers assembled at the cell surface. Science Mar 14, **343**(6176):1260–1263. Figure 1. Permission conveyed through Copyright Clearance Center, Inc.]*

binding site is too low to allow complement activation in the absence of antibody polymerization. Recently, structural analyses demonstrated that when IgG antibodies bind to their antigen, residues in the Fc portion of the antigen-complexed antibody participate in Fc-Fc binding to adjacent IgG molecules, leading to the formation of IgG hexamers that bind C1q with high affinity and activate complement (Figure 5-4d). Although smaller IgG complexes are capable of some degree of complement activation, their lower affinity for C1q will correspondingly reduce the extent of complement activity.

C1q is also capable of directly binding to a number of ligands independently of IgM or IgG, leading to activation of the complement cascade. For example, it can bind to C-reactive protein complexed with exposed phosphocholine residues on bacteria. It can also bind tissue damage elements such as DNA, annexins A2 and A5, and histones on the surface of apoptotic cells, resulting in opsonization and engulfment. As we will see later in Advances Box 5-2, it also appears to bind to as yet undefined molecules on incomplete synapses, facilitating their destruction.

The intermediates in the classical activation pathway are depicted schematically in **Overview Figure 5-5**. Proteins of the classical pathway are numbered in the order in which the proteins were discovered, which does not quite correspond to

---

OVERVIEW FIGURE   5-5

Intermediates in the Classical Pathway of Complement Activation up to the Formation of the C5 Convertase

① C1q binds antigen-bound antibody, and induces a conformational change in one C1r molecule, activating it. This C1r then activates the second C1r and the two C1s molecules.

② C1s cleaves C4 and C2. C4 is cleaved first and C4b binds to the membrane close to C1. C4b binds C2 and exposes it to the action of C1s. C1s cleaves C2, creating the C3 convertase, C4b2a.

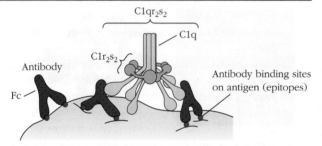

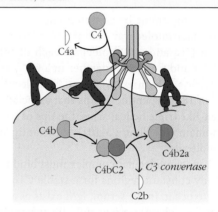

③ C3 convertase hydrolyzes many C3 molecules. Some combine with C3 convertase to form C5 convertase.

④ The C3b component of C5 convertase binds C5, permitting C4b2a to cleave C5.

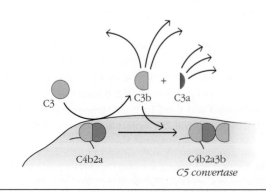

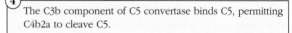
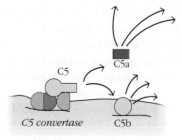

Antigenic determinants are shown in dark red, initiating components (antibodies and C1q) are shown in green, active enzymes are shown in blue, and anaphylatoxins in bright red.

the order in which the proteins act in the pathway (a disconnect that has troubled generations of immunology students). Note that binding of one component to the next always induces either a conformational change or an enzymatic cleavage that enables the next reaction in the sequence.

Binding of C1q to the C_H2 domains of the Fc regions of the antigen-complexed antibody molecule induces a conformational change in one of the C1r molecules. This conformational change in the C1r molecule converts it to an active serine protease enzyme that then cleaves and activates its partner C1r molecule. The two C1r proteases then cleave and activate the two C1s molecules (see Overview Figure 5-5, part 1).

Activated C1s has two substrates, C4 and C2. C4 is activated when C1s hydrolyzes a small fragment (C4a) from the amino terminus of one of its chains (see Overview Figure 5-5, part 2). The C4b fragment attaches covalently to the target membrane surface in the vicinity of C1, and then binds C2. C4b binding to the membrane occurs when an unstable, internal thioester on C4b is exposed on C4 cleavage and reacts with hydroxyl or amino groups of proteins or carbohydrates on the cell membrane. This reaction must occur quickly, before the unstable thioester is further hydrolyzed and can no longer make a covalent bond with the cell surface (**Figure 5-6**). Indeed, approximately 90% of C4b is hydrolyzed before it can bind the cell surface. C4b is also capable of forming covalent bonds with the constant regions of antibody molecules involved in antigen-antibody complexes.

On binding C4b at the membrane surface or on an immune complex, C2 becomes susceptible to cleavage by the neighboring C1s enzyme. A smaller C2b fragment diffuses away, leaving behind an enzymatically active C4b2a complex (Overview Figure 5-5, part 2). In this complex, C2a is the enzymatically active fragment, but it is active only when bound by C4b. This C4b2a complex, as we learned earlier, is the C3 convertase that converts C3 into its enzymatically active form.

The membrane-bound or immune complex–bound C3 convertase enzyme, C4b2a, now hydrolyzes C3, generating two unequal fragments: the small anaphylatoxin C3a, and the pivotally important fragment C3b. A single C3 convertase molecule can generate over 200 molecules of C3b, resulting in tremendous *amplification* at this step of the classical pathway.

The generation of C3b is an essential precursor to many of the subsequent reactions of the complement system. Deficiencies of complement components that act prior to C3 cleavage leave the host extremely vulnerable to both infectious and autoimmune diseases, whereas deficiencies of components later in the pathway are generally of lesser consequence. In particular, patients with deficiencies in C3 itself are unusually susceptible to infections with both gram-positive and gram-negative bacteria. This is because C3b acts in three important and different ways to protect the host:

- In a manner very similar to that of C4b, C3b binds covalently to microbial surfaces, providing a molecular "tag" that allows phagocytic cells with C3b receptors to engulf the tagged microbes. This process is called **opsonization**.

- C3b, like C4b, can attach to the Fc portions of antibodies participating in soluble antigen-antibody complexes. These C3b-tagged immune complexes are bound by C3b receptors on phagocytes or red blood cells, and are either phagocytosed, or conveyed to the liver where they are destroyed.

- Some molecules of C3b bind the membrane-localized C4b2a enzyme to form the trimolecular, membrane-bound, C5 convertase complex C4b2a3b (see Overview Figure 5-5, parts 3 and 4). As we will see later, the C3b component of this complex binds C5, and the complex then cleaves C5 into the two fragments: C5b and C5a. *C4b2a3b is therefore the C5 convertase of the classical pathway.*

The trio of tasks accomplished by the C3b molecule places it right at the center of complement attack pathways.

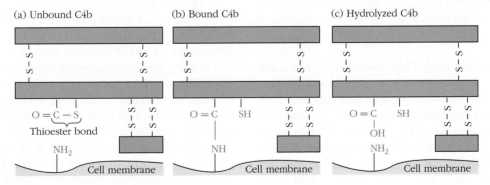

FIGURE 5-6 Binding of C4b to the microbial membrane surface occurs through a thioester bond via an exposed amino or hydroxyl group. (a) Both C3b and C4b contain highly reactive thioester bonds, which are subject to nucleophilic attack by hydroxyl or amino groups on cell membrane proteins and carbohydrates. (b) Breakage of the thioester bond leads to the formation of covalent bonds between the membrane macromolecules and the complement components. (c) If this covalent bond formation does not occur quickly after generation of the C3b and C4b fragments, the thioester bond will be hydrolyzed as shown.

C3b is thus a central component in all three complement activation pathways.

> **Key Concepts:**
>
> - Complement activation occurs by three pathways—classical, lectin, and alternative—which converge in a common sequence of events leading to membrane lysis.
>
> - The classical pathway is initiated by antibodies of the IgM or IgG class binding to a multivalent antigen. Next, C1 binds to an antibody, activating C1-associated serine proteases that cleave the second and fourth complement components, releasing C2a and C2b and C4a and C4b fragments. C2a and C4b combine to form an active serine protease, called C3 convertase, that cleaves C3 into C3a and C3b. The C2a4b complex then combines with one molecule of C3b, forming an active serine protease enzyme called C5 convertase that cleaves C5 into C5a and C5b.

The Lectin Pathway Is Initiated When Soluble Proteins Recognize Microbial Antigens

The **lectin pathway of complement activation**, like the classical pathway, proceeds through the activation of a C3 convertase composed of C4b and C2a. However, instead of relying on antibodies to recognize the microbial threat and to initiate the complement activation process, this pathway uses **lectins**—proteins that recognize particular carbohydrate components—as its specific receptor molecules (**Figure 5-7**). Because it does not rely on antibodies from the adaptive immune system, the lectin pathway is considered to be an arm of innate, rather than adaptive, immunity.

***Mannose-binding lectin* (MBL)**, the first lectin demonstrated to be capable of initiating complement activation, binds close-knit arrays of mannose (sugar) residues that are found on the surfaces of microbes such as *Salmonella*, *Listeria*, and *Neisseria* bacteria; *Cryptococcus neoformans* and *Candida albicans* fungi; and even on the membranes of some viruses such as HIV-1 and respiratory syncytial virus (RSV).

Further characterization of MBL demonstrated that it also recognizes *N*-acetylglucosamine, D-glucose, and L-fucose polymers on microbial surfaces. All these sugars, including mannose, present their associated hydroxyl groups in defined three-dimensional arrays and thus MBL is acting as a classic pattern recognition receptor (see Chapter 3). Consistent with MBL's place at the beginning of an important immune cascade, individuals with low levels of MBL suffer from repeated, serious bacterial infections.

MBL is constitutively expressed by the liver and belongs to the subclass of lectins known as *collectins*. More recently, other lectin receptors have been recognized as initiators of the lectin pathway of complement activation. These include

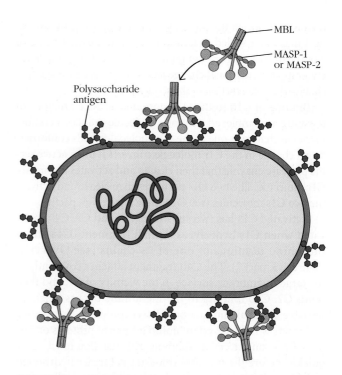

FIGURE 5-7 Initiation of the lectin pathway relies on lectin receptor recognition of microbial cell surface carbohydrates. Lectin receptors, such as MBL, bind microbial cell surface carbohydrates. They bind the MASP family serine proteases, which cleave C2 and C4 to mediate formation of a lectin-pathway C3 convertase.

collectin-10 and collectin-11 as well as several members of the ficolin family: ficolin-1, ficolin-2, and ficolin-3. These proteins share a common "stalk" consisting of a collagen-like triple helix, coupled to a carbohydrate recognition structure. The nature of the recognition structure defines the lectin as belonging to the collectin or the ficolin family. We will use MBL as our example of the initiating receptor for the lectin pathway in this section.

In the blood, MBL is associated with *MBL-associated serine proteases*, or MASP proteins. Three MASP proteins—MASP-1, MASP-2, and MASP-3—have been identified, but most studies of MASP function point to the MASP-2 protein as being the most important player in the next step of the MBL pathway. Interestingly, experiments performed with animals deficient in the genes encoding MASP-1 and MASP-3 indicate that these proteases may play important roles in the alternative pathway, discussed shortly.

MASP-2 is structurally related to the serine protease C1s, and when MBL binds to a microbial surface, associated MASP-2 molecules cleave both C2 and C4, giving rise to the C4b2a C3 convertase that we encountered in our discussion of the classical pathway (see Figure 5-2). From this point on, the lectin pathway uses all the same downstream components as the classical pathway. In comparing the lectin and the classical pathways, we note that the soluble lectin receptor replaces the antibody as the antigen-recognizing component, and MASP proteins take the place of C1r and C1s in cleaving C2 and C4, activating the C3 convertase. Once the C3 convertase is formed, the reactions of the lectin pathway are the same as for the classical pathway. The C5 convertase of the lectin pathway, like that of the classical pathway, is also C4b2a3b.

> **Key Concepts:**
> - The lectin pathway is initiated by binding of lectins such as mannose-binding lectin or members of the ficolin family to microbial surface carbohydrates.
> - The lectin provides the recognition function, binding to carbohydrate residues on a microbial surface. This binding activates an associated serine protease (MASP-2) molecule that cleaves C4 and C2 to create the C3 convertase, C2a4b.
> - As for the classical pathway, the end result of the initiating sequence of the lectin pathway is the generation of enzymes that cleave C3 into C3a and C3b, and C5 into C5a and C5b.

The Alternative Pathway Is Initiated in Three Distinct Ways

Initiation of the **alternative pathway of complement activation,** like the lectin pathway, is independent of antibody-antigen interactions. Therefore, this pathway is also considered to be part of the innate immune system. However, unlike the lectin pathway, the alternative pathway uses a different set of C3 and C5 convertases (see Figure 5-2). As we will see, the alternative pathway C3 convertase, C3bBb, is made up of one molecule of the C3b protein fragment and one molecular fragment unique to the alternative pathway, Bb. A second C3b is then added to make the alternative pathway C5 convertase, C3bBbC3b.

Recent investigations have revealed that the alternative pathway can be initiated in three distinct ways. The first mode of initiation to be discovered, the "tickover" pathway, uses the four serum components C3, factor B, factor D, and properdin (or factor P) (see Table 5-2). Two additional modes of activation for the alternative pathway have also been identified: one is initiated by properdin, and the other by proteases such as thrombin and kallikrein. The story of the discovery of properdin is addressed in the **Classic Experiment Box 5-1.**

The Alternative Tickover Pathway

The term *tickover* refers to the fact that C3 is constantly being made and spontaneously inactivated and is thus considered to be "ticking over." The **alternative tickover pathway** is initiated when C3, which is at high concentrations in serum, undergoes spontaneous hydrolysis at its

internal thioester bond (see Figure 5-6), yielding the molecule $C3(H_2O)$. The conformation of $C3(H_2O)$ is different from that of the C3 parent protein. $C3(H_2O)$ accounts for approximately 0.5% of plasma C3. In the presence of serum Mg^{2+}, $C3(H_2O)$ binds another serum protein, **factor B** (**Figure 5-8a**). When bound to $C3(H_2O)$, factor B becomes susceptible to cleavage by a serum protease, **factor D**. Factor D cleaves factor B, releasing a smaller Ba subunit that diffuses away, leaving a catalytically active Bb subunit bound to $C3(H_2O)$. The $C3(H_2O)Bb$ complex is referred to as the "fluid-phase C3 convertase" because it remains in the blood plasma and is not bound to any cells.

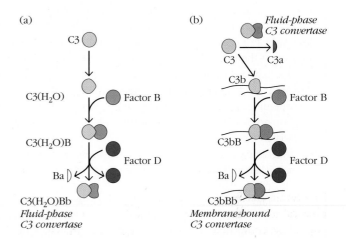

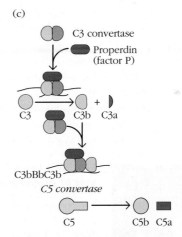

FIGURE 5-8 Initiation of the alternative tickover pathway of complement. (a) Spontaneous hydrolysis of soluble C3 to $C3(H_2O)$ allows the altered conformation of $C3(H_2O)$ to bind factor B, rendering it susceptible to cleavage by factor D. The resulting complex $C3(H_2O)Bb$ forms a fluid-phase convertase capable of cleaving C3 to C3a and C3b. (b) Some of the C3b molecules formed by the fluid-phase convertase bind to cell membranes. C3b, like $C3(H_2O)$, binds factor B in such a way as to make B susceptible to factor D–mediated cleavage. (c) The membrane-bound C3bBb is stabilized by properdin (factor P), which binds the C3bBb complex on the membrane. Addition of a second C3b molecule to the C3bBb complex forms the C5 convertase, which is also stabilized by properdin.

BOX 5-1

The Discovery of Properdin (Factor P)

The study of the history of science shows us that scientists, like any other professionals, are often tempted to think about problems only in ways that are already well trodden and familiar. Science, like art and fashion, has its fads and its "in crowds," and sometimes those whose work moves in directions too different from that of the mainstream in their field have difficulty gaining credibility until the thinking of the rest of their colleagues catches up with their own. Such was the case for Louis Pillemer, the discoverer of properdin; by the time others in the burgeoning field of immunochemistry appreciated the power of his discovery, it was quite simply too late.

Louis Pillemer (**Figure 1**) was born in 1908 in Johannesburg, South Africa. In 1909 the family emigrated to the United States and settled in Kentucky. Pillemer completed his bachelor's degree at Duke University and started medical school at the same institution. But in the middle of his third year, the emotional problems that would plague him for the rest of his life surfaced for the first time, and he left. At that time, deep in the Depression, individuals in Kentucky who could pass an examination in the rudiments of medical care were encouraged to care for patients not otherwise served by a physician.

Pillemer dutifully passed this examination and began to travel through Kentucky on horseback, visiting the sick and tendering whatever treatments were then available. In 1935, he quit this wandering life and entered graduate school at what was then Western Reserve University (now Case Western Reserve University).

There, Pillemer earned his Ph.D. and, except for some time at Harvard and a tour of duty at the Army Medical School in Washington D.C., he remained at Case Western Reserve for the rest of his life, developing a reputation as an excellent biochemist. Among his more noteworthy accomplishments were the first purifications of both tetanus and diphtheria toxins, which were used, along with killed pertussis organisms, in the development of the standard DPT vaccine.

After these successes, Pillemer turned his attention to the biochemistry of the complement system, which he had initially encountered during his graduate work. The antibody-mediated classical pathway had already been established. However, Pillemer was intrigued by some more recent experiments that showed that mixing human serum with zymosan, an insoluble carbohydrate extract from yeast cell walls, resulted in the selective loss of the vital third component of

FIGURE 1 **Louis Pillemer, the discoverer of properdin.** [*Republished with the permission of The American Association of Immunologists, Inc., from the Presidential Address to the AAI, 1980, by Lepow, I. H.,* Journal of Immunology *125: 471, 1980. Fig. 1.*]

complement, C3 (**Figure 2**). He was curious about the mechanism of this loss of C3, and initially interpreted the result to suggest that the C3 was being selectively adsorbed onto the zymosan surface. He reasoned that, if this were true, adsorption to zymosan might be used as a method to purify C3 from plasma. However, Pillemer's initial idea was proven incorrect. Next, he began to investigate whether the loss of C3 resulted from C3 cleavage that was occurring at the zymosan surface.

Pillemer succeeded in demonstrating that C3 was indeed cleaved at the zymosan surface, and went on to show that C3 cleavage happened only when his experiments were run at pH 7.0 and at 37°C. This suggested to him that perhaps an enzyme in the serum was binding to the zymosan and causing the inactivation of the C3 component. Consistent with this hypothesis, when he mixed the serum and zymosan at 17°C, no cleavage occurred. However, if he allowed the serum and zymosan to mix at 17°C and then warmed up the mixture to 37°C, the C3 was cleaved as effectively as if it had been incubated at 37°C all along.

Next, he incubated the serum and zymosan together at 17°C, spun down and removed the zymosan from the mixture, and then added fresh zymosan to the remaining serum containing the C3. He then raised the temperature to 37°C. Nothing happened. The C3 was untouched. Whatever enzymatic activity in the serum was responsible for the breakdown of C3 had been adsorbed by the serum-exposed zymosan and removed when the zymosan was removed.

Pillemer concluded that a factor present in serum and adsorbed onto the zymosan was necessary for the cleavage of C3. With his students and collaborators, he purified this component and named it properdin, from the Latin *perdere* meaning "to destroy." His flow sheet for these

(continued)

CLASSIC EXPERIMENT *(continued)*   **BOX 5-1**

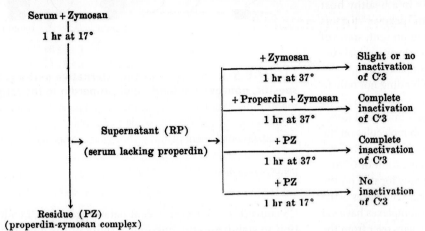

FIGURE 2 **The flow sheet of Pillemer's experiments, which showed that a substance in serum that is adsorbed onto zymosan, the yeast cell wall extract, is capable of catalyzing the cleavage of C3.** *[Republished with permission of the American Association for the Advancement of Science, from Pillemer, L., et al. The properdin system and immunity. I. Demonstration and isolation of a new serum protein, properdin, and its role in immune phenomena. Science, 1954 Aug 20;**120**(3112):279-85. Permission conveyed through Copyright Clearance Center, Inc.]*

initial experiments, as published in part of his landmark *Science* paper of 1954, is shown in Figure 2. In addition to properdin, Pillemer also identified a heat-labile factor in the serum that was required for C3 cleavage to occur.

Pillemer and colleagues went on to characterize properdin as a protein that represented less than 0.03% of serum proteins and whose activity was absolutely dependent on the presence of magnesium ions. In a brilliant series of experiments, Pillemer demonstrated the importance of his newly discovered factor in complement-related antibacterial and antiviral reactions, as well as its role in the disease known as paroxysmal nocturnal hemoglobinuria.

In light of today's knowledge, we can interpret exactly what was happening in his experiments. Properdin bound to the zymosan and stabilized the C3 convertase of the alternative pathway, resulting in C3 cleavage. Indeed, experiments performed as recently as 2007 show that properdin binds to zymosan in a manner similar to its binding to *Neisseria* membranes.

Pillemer's discovery of properdin coupled with his purification of the tetanus and diphtheria toxins should have sealed his reputation as a world-class biochemist. Indeed, his findings were deemed of sufficient general interest at the time that they were publicized in the lay press as well as in the scientific media; articles and editorials about them appeared in the *New York Times, Time* magazine, and *Collier's*. Pillemer did not oversell his case; scientists writing about this period are at pains to note that Pillemer did not make or broadcast any claims for his molecule beyond what appeared in the reviewed scientific literature.

However, in 1957 and 1958, scientist Robert Nelson offered an alternative explanation for Pillemer's findings. He pointed out that what Pillemer had described as a new protein could simply be a mixture of natural antibodies specific for zymosan. If that were the case, then all Pillemer had succeeded in doing was describing the classical pathway a second time. Sensitive biochemical experiments indeed demonstrated the presence of low levels of anti-zymosan antibodies in properdin preparations, and the immunological community began to doubt the relevance of properdin to the complement activation that Pillemer described.

Pillemer was devastated. Never completely stable emotionally, Pillemer's "behavior became erratic, he occasionally abused alcohol and he appeared to be experimenting with drugs," according to his former graduate student, Irwin H. Lepow, who went on to become a distinguished immunologist in his own right. On August 31, 1957, right in the midst of the controversy over properdin, Pillemer died of acute barbiturate intoxication. His death was ruled a suicide, although nobody can know whether he was merely seeking short-term relief from stress and his death was therefore accidental, or whether he truly meant to bring an end to his life.

Subsequent experiments demonstrated that antibodies to zymosan could be removed from partially purified properdin without loss of the ability of the preparation to catalyze the cleavage of C3, thus confirming Pillemer's finding. Furthermore, the heat-labile factor identified by Pillemer's early experiments was identified as a previously unknown molecule, which was subsequently named factor B. By the late 1960s other laboratories entered the arena, and slowly Pillemer's discovery was confirmed and extended into what we now know as the alternative pathway of complement activation. It is one of the great tragedies of immunology that Pillemer did not live to enjoy the validation of his elegant work.

REFERENCES

Lepow, I. H. 1980. Presidential address to the American Association of Immunologists in Anaheim, California, April 16, 1980: Louis Pillemer, properdin, and scientific controversy. *Journal of Immunology* **125**:471.

Pillemer, L., et al. 1954. The properdin system and immunity. I. Demonstration and isolation of a new serum protein, properdin, and its role in immune phenomena. *Science* **120**:279.

In the plasma, the fluid-phase convertase cleaves many molecules of C3 into C3a and C3b (Figure 5-8b). The C3(H₂O)Bb complex is not very stable in a healthy host and it is rapidly degraded, hence the term "tickover" for this pathway. However, if there is an infection present, some of the newly formed C3b molecules bind nearby microbial surfaces via their thioester linkages (see Figure 5-6).

It turns out that factor B is capable of binding not only to C3(H₂O) to form the fluid-phase convertase, but also to the C3b fragment as well. In the presence of a microbial infection, factor B binds the newly attached C3b molecules on the microbial cell surface (see Figure 5-8b), and becomes susceptible to cleavage by factor D, with the generation of C3bBb complexes. These C3bBb complexes are now located on the microbial membrane surface. Like the C4b2a complexes of the classical pathway, the cell-bound C3bBb complexes have C3 convertase activity, and this complex now takes over from the fluid-phase C3(H₂O)Bb as the predominant C3 convertase.

To be clear, there are two C3 convertases in the alternative tickover pathway: a fluid-phase C3(H2O)Bb, which initiates the pathway, and a membrane-bound C3bBb C3 convertase that amplifies it and results in microbial destruction.

The cell-bound alternative pathway C3 convertase is unstable until it is bound by **properdin** (otherwise known as factor P), a serum protein (Figure 5-8c). Once stabilized by properdin, these cell-associated, C3bBb C3 convertase complexes rapidly generate large numbers of C3b molecules on the microbial surface. Many of these then bind factor B, which is cleaved in turn by factor D, thus facilitating the cleavage of yet more molecules of C3 and amplifying the rate of C3b generation. This amplification pathway is rapid; once the alternative pathway has been initiated, more than 2 × 10⁶ molecules of C3b can be deposited on a microbial surface in less than 5 minutes.

All C3b molecules bound to the microbial surface are capable of recruiting factor B and amplifying the concentration of the C3 convertase, irrespective of whether the C3b was generated originally via the classical, the lectin, or the alternative pathway. Recent experiments indicate that, regardless of the initiating pathway, up to 90% of the deposited C3b molecules are generated via activation of the alternative pathway.

Just as the C5 convertase of the classical and lectin pathways was formed by the addition of C3b to the C4b2a C3 convertase complex, so the C5 convertase of the alternative pathway is formed by the addition of C3b to the alternative pathway C3bBb C3 convertase complex. The alternative C5 convertase complex therefore has the composition C3bBbC3b, and like the alternative C3 convertase, it is also stabilized by binding to properdin. Like the classical and lectin pathway C5 convertase, C3bBbC3b cleaves C5, which goes on to form the MAC (see Table 5-3 and Figure 5-8c).

The Alternative Properdin-Activated Pathway

In the previous section, we introduced properdin as a regulatory factor that stabilizes the C3bBb, membrane-bound

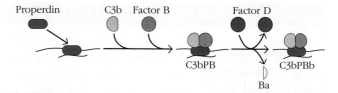

FIGURE 5-9 Initiation of the alternative pathway by specific, noncovalent binding of properdin to the target membrane. Properdin (factor P) binds to components of microbial membranes, and stabilizes the binding of C3bBb complexes of the alternative complement pathway. The difference between this and the tickover pathway is that properdin binds first and initiates complement deposition on the membrane.

C3 convertase. However, recent data suggest that, in addition to stabilizing the ongoing activity of the alternative pathway, *properdin may also serve to initiate it.*

In vitro experiments demonstrated that if properdin molecules were attached to an artificial surface and allowed to interact with purified complement components in the presence of Mg²⁺, the immobilized properdin bound C3b and factor B (**Figure 5-9**). This bound factor B proved to be susceptible to cleavage by factor D, and the resultant C3bPBb complex acted as an effective C3 convertase, leading to the amplification process discussed above. Thus, it seemed that properdin could initiate activation of the alternative pathway on an artificial substrate.

However, proving that a set of reactions *can* occur in vitro does not necessarily mean that it actually *does* occur in vivo. Investigators next investigated the ability of properdin to bind specifically to certain microbes, including *Chlamydia pneumoniae*, as well as to apoptotic and necrotic cell surfaces. Once bound, properdin was indeed able to initiate the alternative pathway, as indicated above.

Note that this pathway relies on the pre-existence of low levels of C3b, which must be generated by mechanisms such as the tickover pathway. However, the specific binding of properdin to *Chlamydia* membranes shows how the properdin pathway can provide greater selectivity than that available from the nonspecific C3b binding of the tickover pathway.

The Alternative Protease-Activated Pathway

The biochemical pathway that leads to complement activation is similar in concept to the blood coagulation pathway. Both use protease cleavage and conformational alterations of key proteins to modify enzyme activities, as well as amplification of various steps of the pathways by feed-forward loops. Recently, some elegant work has revealed the existence of functional interactions as well as theoretical parallels between these two proteolytic cascades.

Several decades ago, it was shown that protein factors involved in blood clotting, such as thrombin, could cleave the complement components C3 and C5 in vitro, with the release of the active anaphylatoxins C3a and C5a. Since these cleavage reactions required relatively high thrombin concentrations, they were at first thought not to be physiologically meaningful. More recently, however, it has been demonstrated in a mouse disease model that initiation of the coagulation cascade may result in the cleavage of physiologically relevant amounts of C3 and C5 to produce C3a and C5a.

Specifically, in an immune-complex model of acute lung inflammation, thrombin cleaved C5, with the release of active C5a, which interacted with its receptors to induce the release of inflammatory mediators. This thrombin-mediated C5 cleavage was also demonstrated in C3 knockout mice, in which C5 convertases could not possibly have been generated in the usual way. Given the proinflammatory role of this anaphylatoxin, C5a generation would result in further amplification of the inflammatory state. Additional experiments have since revealed that other coagulation pathway enzymes, such as plasmin, are capable of generating both C3a and C5a. However, it should be noticed that, although clinically relevant concentrations of C5a are formed following cleavage of C5 by blood-clotting enzymes, functionally meaningful C5b concentrations are *not* generated by this route.

Interestingly, when blood platelets are activated during a clotting reaction, they release high concentrations of ATP and Ca^{2+} along with serine/threonine kinases. These enzymes act to phosphorylate extracellular proteins, including C3b. Phosphorylated C3b is less susceptible to proteolytic degradation than its unphosphorylated form, and thus, by this route, activation of the clotting cascade enhances all of the complement pathways.

Key Concepts:

- The alternative tickover pathway is initiated when $C3(H_2O)$ binds to factor B, which then becomes susceptible to cleavage by factor D into Ba and Bb. The Bb fragment continues to bind to the hydrolyzed $C3(H_2O)$ and together they form a fluid-phase C3 convertase. Some of the C3b generated by this convertase adheres to microbial surfaces; there it binds factor B, which again, in the presence of factor D, is cleaved, resulting in the formation of the membrane-bound C3 convertase, C3bBb. This complex is stabilized by properdin.

- The alternative pathway may also be activated by the initial binding of properdin to a bacterial surface.

- Generation of the anaphylatoxin C5a can also be effected by thrombin cleavage of C5, linking the coagulation and complement cascades.

- The end result of the initiating sequence of the alternative pathway is the generation of enzymes that cleave C3 into C3a and C3b, and C5 into C5a and C5b.

The Three Complement Pathways Converge at the Formation of C5 Convertase and Generation of the MAC

All three initiation pathways culminate in the formation of C5 convertase. For the classical and lectin pathways, C5 convertase has the composition C4b2a3b; for the alternative pathway, C5 convertase has the formulation C3bBbC3b. However, *the end result of all types of C5 convertase activity is the same: cleavage of the C5 molecule into two fragments, C5a and C5b.*

The large C5b fragment is generated on the surface of the target cell or immune complex and provides a binding site for the subsequent components of the membrane attack complex (MAC). However, the C5b component is extremely labile and is not covalently bound to the membrane, as are C3b and C4b. Therefore, it is rapidly inactivated unless it is stabilized by the binding of C6.

All of the complement reactions we have described so far take place on the surfaces of microbial cells or on immune complexes in the fluid phase of blood, lymph, or tissues. But the MAC actually penetrates the cell membrane.

When C5b binds to the serum protein C6, the resulting complex interacts reversibly with the cell membrane via both ionic and hydrophobic bonds. Binding of C7 to C5bC6 induces a conformational change in C7 that exposes hydrophobic regions on its surface capable of inserting into the interior of the microbial membrane (**Figure 5-10a**). The insertion of C7 into the cell membrane is the triggering event for the formation of the membrane attack complex, which will ultimately cause cell death. If, however, C6 and C7 binding occurs on an antigen-antibody (immune) complex or other noncellular surface, then the hydrophobic binding sites will be unable to anchor the complex and it is released. Sometimes these complexes bind to C8 or even to C9 before being released. Released membrane attack complexes can potentially insert into the membrane of nearby cells and mediate "innocent bystander" lysis. However, under physiologic conditions, such lysis is usually minimized by regulatory proteins; the "orphan" complexes are bound by the regulatory protein S (also called vitronectin) and then destroyed.

C8 is made up of two peptide chains: C8β and C8αγ. Binding of C8β to the C5b67 complex induces a conformational change in the C8αγ dimer such that the hydrophobic domain of C8αγ can insert into the interior of the phospholipid membrane. The C5b678 complex is capable of creating a small membrane pore, 10 Å in diameter. The final step in the formation of the MAC is the binding and polymerization of C9 to the C5b678 complex. As many as 10 to 19 molecules of C9 can be bound and polymerized by a single C5b678 complex. During polymerization, the C9 molecules undergo a conformational transition, so that they, too, can insert into the membrane. The completed MAC, which has a tubular form and functional pore diameter of 70 to 100 Å,

(a)

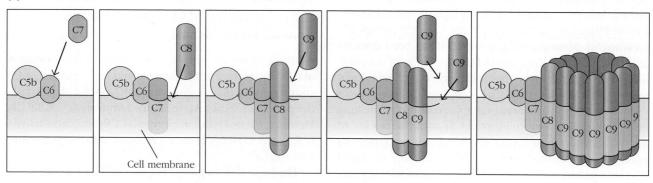

Cell membrane

(b)

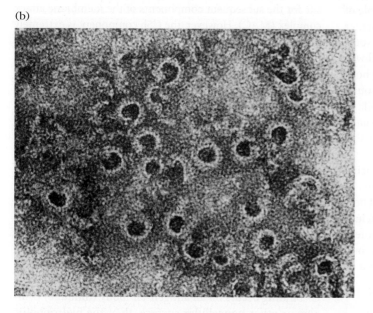

(c)

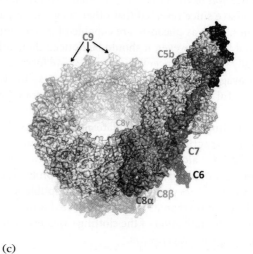

FIGURE 5-10 Formation of the membrane attack complex (MAC). (a) Formation of the MAC, showing the addition of C6, C7, C8, and C9 components to the C5b component. (b) Photomicrograph of poly-C9 complex formed by in vitro polymerization of C9 and complement-induced lesions on the membrane of a red blood cell. These lesions result from formation of membrane attack complexes. (c) Relative locations of the members of the membrane attack complex: C5b, C6, C7, C8, and C9. *[Part (b) republished with permission of Elsevier, from Humphrey, J. and Dourmashkin, R., 1969, The lesions in cell membranes caused by complement. Reprinted from* Advances in Immunology, *Volume 11, Copyright 1969, 75–115. Part (c) Drs. Robert Liddington and Alex Aleshin, SBP Medical Discovery Institute, La Jolla, CA.]*

consists of a C5b678 complex surrounded by a poly-C9 complex (Figures 5-10b and 5-10c). Loss of plasma membrane integrity leads irreversibly to cell death.

Key Concept:

- Activation of the terminal components of the complement cascade C5b, C6, C7, C8, and C9 results in the deposition of a membrane attack complex (MAC) onto the microbial cell membrane. This complex introduces large pores in the membrane, preventing it from maintaining osmotic integrity and resulting in the death of the cell.

The Diverse Functions of Complement

Table 5-4 lists the main categories of complement function, each of which is discussed in this section. In addition to its long-understood role in antibody-induced lysis of microbes, complement also plays important roles in innate immunity. The pivotal importance of C3b-mediated responses such as opsonization has been clearly demonstrated in C3 knockout animals, which display increased susceptibility to both viral and bacterial infections. In addition, recent experiments have explored the roles of various complement components at the interface of innate and adaptive immunity,

TABLE 5-4   The three main classes of complement activity in the service of host defense

Activity	Responsible complement component
Innate defense against infection	
Lysis of bacterial and cell membranes	Membrane attack complex (C5b-C9)
Opsonization	Covalently bound C3b, C4b
Induction of inflammation and chemotaxis by anaphylatoxins	C3a and C5a (anaphylatoxins) and their receptors on leukocytes
Interface between innate and adaptive immunity	
Augmentation of antibody responses	C3b and C4b and their proteolyzed fragments bound to immune complexes and antigen; C3 receptors on immune cells
Enhancement of immunologic memory	C3b and C4b and their fragments bound to antigen and immune complexes; receptors for complement components on follicular dendritic cells
Enhancement of antigen presentation	MBL, C1q, C3b, C4b, and C5a
Potential effects on T cells	C3, C3a, C3b, C5a
Complement in the contraction phase of the immune response	
Clearance of immune complexes from tissues	C1, C2, C4; covalently bound fragments of C3 and C4
Clearance of apoptotic cells	C1q; covalently bound fragments of C3 and C4. Loss of CD46 triggers immune clearance
Induction of regulatory T cells	CD46

This table has been considerably modified from the superb formulation of Walport, M. 2001. Complement: first of two parts. *New England Journal of Medicine* **344:**1058 [Table 1.].

and identified multiple mechanisms by which the release of active complement fragments acts to regulate the adaptive immune system. Complement also plays an important role in the contraction phase of the adaptive immune response, and recent work has even suggested that it is important in the elimination of excess synapses during the development of the nervous system (see **Advances Box 5-2**). These various functions of complement are described here.

Complement Receptors Connect Complement-Tagged Pathogens to Effector Cells

Many of the biological activities of complement depend on the binding of complement fragments to host cell surface receptors for complement components. The levels of a number of the complement receptors are subject to regulation by aspects of the innate and adaptive immune systems. For example, activation of phagocytic cells by various agents, including the anaphylatoxins of the complement system, has been shown to increase the number of complement receptors at the cell surface by as much as 10-fold. In addition, some complement receptors play an important role in regulating complement activity by mediating proteolysis of biologically active complement components.

Therefore, before we venture into a discussion of the biological functions of complement, we should first become familiar with the receptors for complement components, as well as their activities. The complement receptors and their primary ligands are listed in **Table 5-5**. Where receptors have more than one name, we have offered both on first introduction, and subsequently refer to that receptor by the more common name.

CR1 (CD35) is expressed on both leukocytes and erythrocytes and binds with high affinity to C3b and smaller C3b breakdown products, C4b, C1q, and MBL (Table 5-5). Erythrocytes bind immune complexes via their CR1 receptors and transport the complexes to the liver, where they are picked up by phagocytes and cleared from the body. CR1 receptors on phagocytes bind complement-opsonized microbial cells. This induces receptor-mediated phagocytosis and the secretion of proinflammatory molecules such as IL-1 and prostaglandins. CR1 on B cells mediates uptake of C3b-bound antigen, leading to its degradation in the B-cell lysosomal system and subsequent presentation of antigenic peptides to T cells. In addition, complement receptors on B cells have been shown to be important in the transport of antigens and antigenic fragments into the follicles of the lymph nodes and spleen, where they are transferred to macrophages or follicular

BOX 5-2

Complement and the Visual System

Generations of immunology students have viewed complement as a complex biological cascade to be quickly memorized and as quickly forgotten, of compelling interest only to immunologists with a decidedly biochemical bent. However, in the last decade or so, we have learned that the actions of various complement components may be pivotal in the development and maintenance of elements of the central nervous system. These initial results have stimulated additional questions about complement's role in the pathogenesis of nervous system diseases as devastating as glaucoma, macular degeneration, myasthenia gravis, Alzheimer's disease, autism, and schizophrenia. This area of research is extremely active at present. In this Advances Box, we will specifically address the role of complement in the development and pathogenesis of the visual system.

One of the first clues that complement may be important outside the study of the immune system came over two decades ago, when it was recognized that cells in the central nervous system (CNS), now known to be the microglia, are capable of synthesizing complement components. However, it wasn't until 2007 that investigators in the laboratory of Ben Barres at Stanford University showed that complement proteins play a critical role in organizing the nascent nervous system.

They were studying the path taken by neuronal signals from the eyes of developing mice. Retinal ganglion cells, or RGCs, are cells in the retina that receive information from the photoreceptors. The RGCs, in turn, send out long protrusions called *axons* that connect with neurons in the dorsal lateral geniculate nucleus (dLGN) of the brain. When these connections are

first established, a single cell in the dLGN makes contact with many RGC axons; at about 8 days after birth, before the neonatal animals open their eyes, each dLGN neuron may receive inputs from up to 10 RGCs. However, little distinction is made in these initial contacts between the RGCs and a particular dLGN cell as to whether the RGCs are sending signals from the right or the left eye (**Figure 1, left**).

In contrast, in the mature nervous system, RGC axons make contact *only* with dLGN cells on the *opposite* side of the brain. That is, RGCs receiving input from the left eye contact dLGN cells on the right side of the brain, and RGCs from the right eye synapse with dLGN cells on the left side of the brain. Figure 1 describes RGC position in relation to a particular dLGN cell as either "ipsilateral" (on the same side of the body) or "contralateral" (on opposite sides of the body). In Figure 1, the RGC axons are visualized as emanating from the bottom of the figure, and are making contact with dLGN cells at the top.

Note that, in the one-week-old animal, a single dLGN cell may receive incoming signals from both ipsilateral and the contralateral RGCs. In a mature animal, the

correct connections to a particular dLGN cell come from RGCs of the contralateral eye and the connections with RGCs from the ipsilateral eye have been lost. How does that happen?

During the second and third weeks after birth, the neurons of the visual system undergo a process known as "synaptic pruning." The RGCs that are coming from each eye and attempting to make synapses with neurons of the dLGN are identified and the inappropriate, ipsilateral synapses are eliminated. By approximately 30 days after birth, each dLGN cell is innervated by only one or two retinal nerves from the contralateral eye (Figure 1, right).

The researchers in Barres' group investigated which molecules were responsible for the loss of the "wrong," ipsilateral neurons during this pruning process. Their key observation was that particular brain cells, called *astrocytes*, make their first appearance over the time period during which pruning occurs. They showed that molecular signals released from these astrocytes induced the synthesis of C1q by RGCs and, furthermore, that C1q proteins appeared to colocalize with synaptic marker proteins (proteins found at the sites of neuronal connections) in the dLGN. After labeling pre- and postsynaptic nerve endings along with C1q with fluorescent markers, they performed a microscopic analysis of their work, using a sophisticated technique called *array tomography*. These experiments showed that C1q was found in association with pre- and postsynaptic markers, but not with complete synapses (where pre- and postsynaptic nerve endings are located in close apposition to one another) (**Figure 2**). This suggested to them that C1q might be involved in the pruning process.

One week after birth   Three weeks after birth

Lateral Geniculate Nucleus

● Ipsilateral RGC axon terminals
● Contralateral RGC axon terminals

FIGURE 1 During the first week after birth, axons from retinal ganglion cells (RGCs) grow toward the dorsal lateral geniculate nucleus in the thalamus, guided by molecular gradients. The diffuse array of synapses is then pruned over the course of the next 2 weeks, so that only inputs from the contralateral eye remain. Following engulfment of the inappropriate synapses, the neurons themselves are subsequently lost. *[From © Benjumeda et al., Flowers and weeds: cell-type specific pruning in the developing visual thalamus. BMC Biology 2014, 12:3, 1741-7007. Licensee BioMed Central Ltd. 2014. Figure 1.]*

(continued)

Next, using C1q knockout animals, the investigators demonstrated that, in the absence of C1q, synaptic pruning was significantly inhibited; in these knockout mice, the retinal projections from the two eyes of each mouse showed uncharacteristic overlap into adulthood. Interestingly, C3 knockout mice displayed a similar phenotype.

More recent work has identified the cytokine TGF-β as the molecular signal released by the astrocytes and received by receptors on RGCs that leads to RGC production of C1q. (It should be noted, however, that most of the C1q within the central nervous system is synthesized and released from microglial cells, and that it cannot be ruled out that some microglia-derived C1q may also be involved in synaptic pruning of RGCs.)

Although these experiments have taught us a great deal about the role of complement in synaptic pruning in the visual system, there is still much that we do not yet know. First and foremost, complement-mediated elimination is responsible for the loss of only a fraction of synapses in the developing brain. In other instances of neural circuit refinement elsewhere in the CNS, astrocytes have been shown to

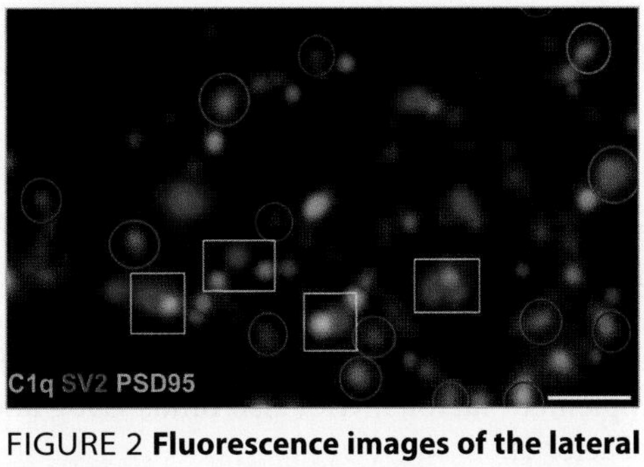

FIGURE 2 **Fluorescence images of the lateral geniculate nucleus, analyzed by array tomography.** C1q is labeled green; presynaptic neurons are labeled red, using antibodies to the SV2 protein; and postsynaptic neurons are labeled blue, using antibodies to the PSD-95 protein. Note that although C1q is often found in association with either pre- or postsynaptic neurons, when complete synapses are formed (red and blue labels are found in close apposition), C1q is absent. [Republished with permission of Elsevier, from Stevens, B. et al., 2007, The classical complement cascade mediates CNS synapse elimination. Cell 131(6):1164-1178, Figure 3b. Permission conveyed through Copyright Clearance Center, Inc.]

directly participate in complement-independent processes that culminate in synaptic engulfment. Other processes will undoubtedly be revealed with additional research. Hence, the frequency with which complement participates in synaptic pruning is still to be determined.

In addition, within the retinal synaptic pruning system, we still do not know what molecules stimulate the astrocytes to begin secreting TGF-β. Furthermore, the target of C1q binding has yet to be identified. And although it is clear that the generation of an appropriate, active contact protects a synapse from further pruning, we do not have a clear picture of how that protection is mediated.

With the appreciation of complement's importance in the elimination of nonfunctional synapses in healthy, developing animals, attention has been drawn to the possibility that complement may play a role in inappropriate synaptic loss in disease states.

The first observations that implicated complement-mediated pruning mechanisms in disease were made during investigations of the eye disease glaucoma. Up-regulation of C1q secretion has been shown to be an early event in mouse models of glaucoma. The timing of the increase in C1q concentration in the retina relative to disease onset suggests that patients with glaucoma may be experiencing inappropriate re-activation of complement-mediated synaptic destruction. This hypothesis is supported by the fact that genomic studies have since shown that individuals with a mutation in the gene encoding the C1q protein (C1qa) are protected from glaucoma.

Inappropriate complement activation has also been implicated in macular degeneration, an eye disease that is the third leading cause of blindness, and which preferentially affects aging individuals. Within the eye, retinal pigment epithelial (RPE) cells are responsible for the phagocytosis of worn-out components, in particular used photoreceptor

disks. Because of the constant turnover of these photoreceptor disks, the RPE cells are among the most active phagocytic cells in the body, and as they age they begin to lose their efficiency. This is particularly true of the RPE cells that are closest to the fovea (the center of the retina), since this is the area where light entering the eye is focused by the lens. As cellular debris accumulates under the RPE layer, it begins to activate the complement system, which in turn further damages the RPE cells, leading to localized loss of function. Imaging, biochemical, and genetic studies have all confirmed the involvement of the complement system in establishing and maintaining the disease processes of macular degeneration.

Even outside the visual system, investigators have shown that the expression of complement components is upregulated in a variety of disease conditions affecting the nervous system. Indeed, investigators are currently pursuing the possibility that complement-mediated synapse elimination may be reactivated in the aging brain, potentially contributing to degenerative pathologies associated with aging, such as Alzheimer's disease.

In conclusion, a clear understanding of the mechanisms and components of the complement system is becoming increasingly relevant to our understanding of nervous system physiology and pathology.

While this book was in the final stages of production, we were deeply saddened to learn of the untimely death of Professor Ben Barres of Stanford University.

REFERENCES

Bialas, A. R., and B. Stevens. 2013. TGF-β signaling regulates neuronal C1q expression and developmental synaptic refinement. Nature Neuroscience 16:1773.

Shi, Q., et al. 2015. Complement C3–deficient mice fail to display age-related hippocampal decline. Journal of Neuroscience 35:13029.

Stephan, A. H., B. A. Barres, and B. Stevens. 2012. The complement system: an unexpected role in synaptic pruning during development and disease. Annual Review of Neuroscience 35:369.

Stevens, B., et al. 2007. The classical complement cascade mediates CNS synapse elimination. Cell 131:1164.

TABLE 5-5   Receptors that bind complement components and their breakdown products

Receptor	Other name(s)	Ligand	Cellular expression pattern	Function
CR1	CD35	C3b, C4b, C1q, iC3b	Erythrocytes, neutrophils, monocytes, macrophages, eosinophils, FDCs, B cells, some T cells	Clearance of immune complexes, enhancement of phagocytosis, regulation of C3 breakdown
CR2	CD21, Epstein-Barr virus receptor	C3d, C3dg C3d, iC3b	B cells, FDCs	Enhancement of B-cell activation, B-cell coreceptor, and retention of C3d-tagged immune complexes
CR3	CD11b/CD18, Mac-1	iC3b and factor H	Monocytes, macrophages, neutrophils, NK cells, eosinophils, FDCs, T cells	Binding to adhesion molecules on leukocytes, facilitates extravasation; iC3b binding enhances opsonization of immune complexes
CR4	CD11c/CD18	iC3b	Monocytes, macrophages, neutrophils, dendritic cells, NK cells, T cells	iC3b-mediated phagocytosis
CRIg	VSIG4	C3b, iC3b, C3c	Fixed tissue macrophages	iC3b-mediated phagocytosis and inhibition of alternative pathway
C1qRp	CD93	C1q, MBL	Monocytes, neutrophils, endothelial cells, platelets, T cells	Induces T-cell activation; enhances phagocytosis
SIGN-R1	CD209	C1q	Marginal zone of spleen, lymph node macrophages	Enhances opsonization of bacteria by MZ macrophages
C3aR	(None)	C3a	Mast cells, basophils, granulocytes	Induces degranulation
C5aR	CD88	C5a	Mast cells, basophils, granulocytes, monocytes, macrophages, platelets, endothelial cells, T cells	Induces degranulation; chemoattraction; acts with IL-1β and/or TNF-α to induce acute-phase response; induces respiratory burst in neutrophils
C5L2	(None)	C5a	Mast cells, basophils, immature dendritic cells	Uncertain, but most probably down-regulates proinflammatory effects of C5a

Data from Zipfel, P. F., and C. Skerka. 2009. Complement regulators and inhibitory proteins. *Nature Reviews Immunology* **10**:729; Kemper, C., and J. P. Atkinson. 2007. T-cell regulation: with complements from innate immunity. *Nature Reviews Immunology* **7**:9; Eggleton, P., A. J. Tenner, and K. B. M. Reid. 2000. C1q receptors. *Clinical and Experimental Immunology* **120**:406; Ohno, M., et al. 2000. A putative chemoattractant receptor, C5L2, is expressed in granulocyte and immature dendritic cells, but not in mature dendritic cells. *Molecular Immunology* **37**:407; and Kindt, T. J., B. A. Osborne, and R. A. Goldsby. 2006. *Kuby Immunology*, 6th ed. W. H. Freeman, New York [Table 7-4].

dendritic cells for presentation to other B cells. Thus, CR1 is an actor in the adaptive, as well as the innate, immune response. As we will see later, CR1 also serves as a cofactor in the protection of host cells against the ravages of complement attack.

The C3b fragment is subject to breakdown by endogenous proteases, whether it is in solution or bound to the cell surface. The breakdown products of C3b—iC3b, C3d, and C3dg—are each bound specifically by the complement receptor **CD21** (**CR2**), which is expressed on B cells in noncovalent association with the B-cell receptor. Since C3b can form covalent bonds with antigens, and these bonds are not affected by breakdown of C3b to iC3b, C3d, and C3dg, the close association of CD21 with the B-cell receptor enables the B cell to simultaneously bind antigen via both the B-cell receptor and CD21 (**Figure 5-11**). This has the effect of reducing the antigen concentration necessary for B-cell activation by approximately 100-fold. Deficiencies in CD21 have been identified in patients

suffering from autoimmune diseases such as systemic lupus erythematosus.

CR3 (a complex of CD11b and CD18) and **CR4** (a complex of CD11c and CD18) are important in the phagocytosis of complement-coated antigens. CR3 and CR4 both bind iC3b, the cleavage product of C3b that results from breakdown by complement factor I. CR3 also binds C3dg and, interestingly, Factor H.

CRIg, expressed on macrophages resident in the tissues (i.e., fixed tissue macrophages), binds C3b. Its importance in clearing C3b-opsonized antigens by facilitating their removal from circulation in the liver is emphasized by the finding that CRIg-deficient mice are unable to efficiently clear C3-opsonized particles. Animals with this deficiency are therefore subject to higher mortality rates during infections.

C3aR, **C5aR**, and **C5L2** are all members of the G protein–coupled receptor (GPCR) family of receptors. C3aR and C5aR mediate inflammatory functions after binding the small

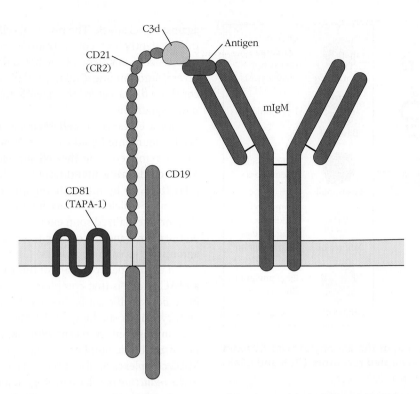

FIGURE 5-11 Coligation of antigen to B cells via the B-cell receptor and the B-cell coreceptor. The B-cell coreceptor is a complex of three cell membrane molecules: CD21 (CR2), CD81 (TAPA-1), and CD19. Antigen that has been covalently bound to fragments of the C3 complement component is bound by both the immunoglobulin BCR and the CD21 complement receptor, thus significantly increasing the avidity of the cell receptors for the antigen and allowing lower concentrations of antigen to trigger B-cell activation. The CD19 component is important in B-cell signaling by antigen.

anaphylatoxins C3a and C5a, respectively (**Figure 5-12**). The C5L2 receptor also binds C5a, is structurally similar to C5aR, and is expressed on some of the same cells. However, C5L2 is not functionally coupled to the G protein signaling pathway used by C5aR; instead, signaling through C5L2 appears to negatively modulate C5a signaling through the C5aR. As one might predict, C5L2 knockout animals express enhanced inflammatory responses on C5a release.

C1qRp is expressed on monocytes, neutrophils, endothelial cells, platelets, and T cells and has been shown to bind C1q and MBL and enhance phagocytosis of those proteins as well as of any molecules (or cells) to which they are attached.

More recently, the transmembrane lectin **SIGN-R1**, able to bind C1q, has been shown to be expressed on the surface of macrophages located in the marginal zone of the spleen. SIGN-R1 exists on the macrophage cell surface in aggregated form and is also able to bind carbohydrates present on the coat of the gram-positive bacterium *Streptococcus pneumoniae*. SIGN-R1 binding to streptococci activates the C1q-binding capacity of nearby SIGN-R1 molecules, resulting in complement-mediated opsonization of the bacteria. The opsonized bacteria are then released from these macrophages and bound by nearby phagocytes, B cells, or dendritic cells. This unusual mechanism explains a problem long noted with patients who have undergone splenectomy: an increased susceptibility to infection with *S. pneumoniae*.

> **Key Concept:**
>
> • Complement receptor proteins mediate the functions of complement components by acting as bridges between the complement components and the cells to which they bind.

Complement Enhances Host Defense against Infection

Complement proteins actively engage in host defense against infection by forming the MAC, by opsonizing potentially pathogenic microbes, and by inducing an inflammatory response that helps to guide leukocytes to the site of infection.

MAC-Induced Cell Death

The first function of complement to be described was its role in inducing cell death following insertion of the MAC into target cell membranes. Early experiments on MAC formation used erythrocytes as the target cells, and in this cellular

Anaphylatoxins and inflammatory response

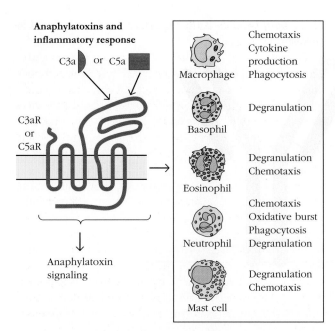

FIGURE 5-12 Binding of the anaphylatoxins C3a and C5a to the G protein–coupled receptors C3aR and C5aR. C3aR and C5aR are G protein–coupled receptors. Binding of the anaphylatoxins to these receptors stimulates the release of proinflammatory mediators from macrophages, neutrophils, basophils, eosinophils, and mast cells.

system large pores involving 17 to 19 molecules of C9 were reported. Formation of these holes in the cell membrane (see Figures 5-10b and 5-10c) facilitates the free movement of small molecules and ions through the membrane. The penetrated red blood cell membranes are unable to maintain osmotic integrity, and the cells lyse after massive influx of water from the extracellular fluid.

However, subsequent studies using nucleated eukaryotic cells indicated that smaller pores can be generated by only a few molecules of C9, and that death under these circumstances occurs via a type of apoptosis (programmed cell death), following calcium influx into the cytoplasm. Nuclear fragmentation, a hallmark of apoptotic death, was observed during MAC-induced lysis of nucleated cells, which further supported the notion that at least some MAC-targeted cells succumb to apoptosis. Additional experiments indicated that the apoptosis induced by the MAC does not share all the biochemical characteristics normally associated with programmed cell death, and so this MAC-induced apoptosis has been termed *apoptotic necrosis*.

Killing eukaryotic cells with complement is actually quite difficult, as the plasma membranes of such cells have a number of factors that inactivate the complement proteins, and thus protect the host cells from collateral damage during a complement-mediated attack on infectious microorganisms. However, when high concentrations of complement components are present, MACs can overwhelm the host defenses

against MAC attack. The resulting cell fragments, if present in sufficiently high concentrations, can induce autoimmunity. Indeed, complement-mediated damage is a problem in several autoimmune diseases, and the complement system is considered a target for therapeutic intervention in autoimmune syndromes.

Can a eukaryotic cell recover from a MAC attack? Well-documented studies have demonstrated that MACs can be removed from the cell surface, either by shedding MAC-containing membrane vesicles into the extracellular fluid, or by internalizing and degrading the MAC-containing vesicles in intracellular lysosomes. If the MAC is shed or internalized soon enough after its initial expression on the membrane, the cell can repair any membrane damage and restore its osmotic stability.

An unfortunate corollary of this capacity to recover from a MAC attack is that complement-mediated lysis directed by tumor-specific antibodies may be rendered ineffective by endocytosis or shedding of the MAC. Even more dramatically in this context, recent work has suggested that assembly of a sublytic number (i.e., not enough to cause lysis) of MAC complexes on the surface of a eukaryotic cell can lead to the *induction* of cell cycle progression and the *inhibition* of apoptosis! This seems an unfortunate application of the old maxim that "what doesn't kill you makes you stronger," and illustrates the dangers of attempting to use immunotherapy in the treatment of cancer without a thorough appreciation of all the underlying mechanisms.

Different types of microorganisms are susceptible to complement-induced lysis to varying degrees. Antibody and complement play an important role in host defense against viruses and can be crucial both in containing viral spread during acute infection and in protecting against reinfection. Most enveloped viruses, including herpesviruses, orthomyxoviruses such as those causing measles and mumps, paramyxoviruses including influenza, and retroviruses, are susceptible to complement-mediated lysis.

Although many bacteria are susceptible to MAC attack, some are not. In particular, gram-positive bacteria efficiently repel complement assault, as the complement proteins cannot penetrate the bacterial cell wall to gain access to the membrane beyond. When bound to gram-negative bacterial membranes, the MAC has been found to localize at regions of apposition of the inner and outer cell membranes, where it breaches them both simultaneously. *Neisseria meningitidis* is a gram-negative bacterium that is susceptible to MAC-induced lysis, and patients who are deficient in any of the complement components of the MAC are particularly vulnerable to potentially fatal meningitis caused by these bacteria.

Recent work has also demonstrated that sublytic MAC assembly is able to facilitate inflammasome activation and IL-1β production (see Chapter 4). This occurs via the MAC-induced increase in intracellular Ca^{2+} ion concentration, which leads, in turn, to an increase in the concentration of Ca^{2+} ions in the mitochondrial matrix, thus facilitating the assembly of the inflammasome.

Promotion of Opsonization

As we have learned, the term *opsonization* refers to the capacity of antibodies and complement components (as well as other proteins) to coat dangerous antigens that can then be recognized by Fc receptors (for antibodies) or complement receptors (for complement components) on phagocytic cells. Binding of complement-coated antigen by phagocytic cells results in complement receptor–mediated phagocytosis and antigen destruction (**Figure 5-13**). In addition, complement receptors on erythrocytes also serve to bind immune complexes, which are then transported to the liver for phagocytosis by macrophages. Although less visually compelling than MAC formation, *opsonization may*

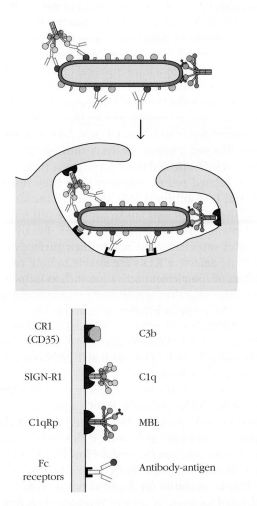

CR1 (CD35)	C3b
SIGN-R1	C1q
C1qRp	MBL
Fc receptors	Antibody-antigen

FIGURE 5-13 Opsonization of microbial cells by complement components and antibodies. Phagocytosis is mediated by many different complement receptors on the surface of macrophages and neutrophils, including CR1, SIGN-R1, and C1qRp. Phagocytes, using their Fc receptors, also bind to antigens opsonized by antibody binding.

be the most physiologically important function assumed by complement components.

Opsonization with antibody and complement also provides critical protection against viral infection. Antibody and complement can create a thick protein coat around a virus that neutralizes viral infectivity by preventing the virus from binding to receptors on the host cell. They then promote phagocytosis by activated macrophages via the Fc and complement receptors, followed by intracellular destruction of the digested particle.

Recent work has also suggested an additional mechanism by which complement can protect cells against intracellular infection by some viruses, such as adenoviruses. Viruses that enter the cell coated with complement, in the presence or absence of antibody, escape from the endosomal system and enter the cytosol. There, the shell of C3 and/or C4 molecules surrounding each viral particle activates the proteasomal system of degradation, leading to virus destruction. In addition, complement signals the cell that its defenses have been breached and activates the synthesis of protective proteins via the up-regulation of the transcription factors NF-κB, AP-1, and IRF3/5/7.

Promotion of Inflammation

So far, we have focused on the roles of the larger products of complement factor fragmentation: C3b and C4b in opsonization and C5b in the formation of the MAC. However, the smaller fragments of C3 and C5 cleavage—C3a and C5a—also mediate critically important events in immune responses, acting as anaphylatoxins or inducers of inflammation.

C3a and C5a are structurally similar, small proteins (about 9 kDa in size) that promote inflammation and also serve as chemoattractants for certain classes of leukocytes. C3a and C5a bind GPCRs (C3aR for C3a, C5aR for C5a) on granulocytes, monocytes, macrophages, mast cells, endothelial cells, and some dendritic cells (see Figure 5-12). Binding of these anaphylatoxins to their receptors triggers a signaling cascade that leads to the secretion of soluble, proinflammatory mediators such as IL-6 and TNF-α. These cytokines induce localized increases in vascular permeability that facilitate leukocyte migration into the site of infection, and a concomitant increase in smooth muscle motility that helps to propel the released fluid to the site of damage. In addition, anaphylatoxin-receptor binding promotes phagocytosis of offending pathogens by the anaphylatoxin-signaled localized degranulation of granulocytes (neutrophils, basophils, and eosinophils), with the resultant release of a second round of inflammatory mediators such as histamines and prostaglandins. Inflammatory mediators expedite the movement of lymphocytes into neighboring lymph nodes, where they are activated by the pathogen. This localized inflammatory response is further supported by systemic effects, such as fever, that decrease microbial viability.

Complement Acts at the Interface between Innate and Adaptive Immunities

There are multiple mechanisms by which components of the complement system modulate adaptive immunity. Many of these have been described very recently, and the study of how the binding of complement components and regulatory proteins may affect antigen-presenting cells, T cells, and B cells is still in its early phases.

Complement and Antigen-Presenting Cells

Dendritic cells (DCs), follicular dendritic cells (FDCs), and macrophages each express many of the known complement receptors. When bound to antigens, MBL, C1q, C3b, and C4b are each capable of engaging their respective receptors on antigen-presenting cells during the process of antigen recognition, and signaling through their respective receptors acts to *enhance antigen uptake.*

In addition, signaling of antigen-presenting cells through C5aR, the anaphylatoxin C5a receptor, has been shown to *modulate their migration* and *affect their production of interleukins*, particularly that of the cytokine IL-12. Production of IL-12 by an antigen-presenting cell normally skews the T-cell response toward the T_H1 phenotype (see Chapter 10). Since both induction and suppression of IL-12 production have been detected after activation of the anaphylatoxin receptor, depending on the route of antigen delivery, the nature of the antigen, and the maturation status of the antigen-presenting cell, we must infer that many signaling pathways are being integrated to arrive at the eventual biological response to antigen and complement components.

Complement and B Cell–Mediated Humoral Immunity

Early experiments showed that depleting mice of C3 impairs their T cell–dependent antigen-specific B-cell responses, thus implying that complement may participate in the initiation of the B-cell response. It now appears that this seminal observation was the first description of the complement receptor CD21 acting as a coreceptor in antigen recognition, which was described earlier (see the section "Complement Receptors Connect Complement-Tagged Pathogens to Effector Cells").

Complement and T Cell–Mediated Immunity

A role for complement in the quality control of T cells newly released from the thymus has recently been demonstrated. A small fraction of constitutively secreted IgM antibodies in the serum bind to recurrent patterns of carbohydrates. These antibodies are secreted by a particular subclass of B cells that lies at the interface of the innate and adaptive immune systems, and are referred to as "natural antibodies" (see Chapter 11). As T cells mature in the thymus (see Chapter 8), the number of sialic acid residues in their cell surface carbohydrates is increased. Under normal circumstances, the cell surface of recent thymic emigrants, or RTEs, is coated with these negatively charged sialic acid residues, which protect the RTEs against binding by natural IgM antibodies. The sialic acid residues also facilitate the binding of soluble inhibitors of complement activity, such as factor H (discussed shortly). This change in cell surface sialic acid density is regulated by a transcriptional repressor, NKAP.

RTEs with deficient NKAP and resulting low sialic acid levels leave the thymus at the usual rate, but fail to survive to complete maturation in the periphery. These T cells were shown to be bound by serum IgM and lysed by complement. Further analysis showed that the low levels of NKAP affected their survival via three different but interrelated routes. Not only were they susceptible to natural IgM antibody binding but, in the absence of NKAP, these RTEs also fail to express a cell membrane complement-inhibitory protein, CD55. Finally, in the absence of sufficient sialic acid in their surface carbohydrates, the defective RTEs are unable to bind to soluble inhibitors of complement activation such as factor H. This means that, once the complement cascade is engaged by natural IgM antibody binding, it runs quickly to completion in the absence of either membrane-bound or soluble inhibitory proteins. Thus complement is used to provide a quality control check to ensure that T cells about to leave the thymus are fully mature.

In addition, mice lacking the gene encoding the C3 protein have reduced $CD4^+$ and $CD8^+$ T-cell responses, and recent findings have provided clues to an unusual mechanism by which this may occur. T_H1 $CD4^+$ T cells continually produce low levels of intracellular C3a and C3b, and this C3a production has been shown to be essential for T-cell survival. When a T cell is activated through its antigen receptor, it secretes both C3a and C3b fragments. The small anaphylatoxin, C3a, binds to receptors on the T-cell membrane and induces that T cell to secrete proinflammatory cytokines that support a T_H1 response (see Chapter 10). Furthermore, the growth and survival of T_H1 T cells have also been shown to be dependent on the interaction of C3b fragments with the cell surface molecule CD46. Interestingly, at the close of a T_H1 immune response, there is also evidence that sustained production of these complement components by T cells is important in the contraction phase of an immune response (see the next section), as complement fragments have been shown to induce the formation of IL-10 in the T_H1 T-cell population as it begins to close down and approach apoptosis.

C5 has also been implicated in T-cell responses, as mice treated with inhibitors of C5aR signaling produced fewer antigen-specific CD8$^+$ T cells, following influenza infection, than did wild-type mice. This suggests that C5a may act as a costimulator during CD8$^+$ T-cell activation, possibly by increasing IL-12 production by antigen-presenting cells. These experiments demonstrate that signaling through complement components may have positive effects on T cell–mediated adaptive immune responses.

Thus, in addition to playing a surveillance role during T-cell development and aiding in the elimination of those T cells that fail to complete their maturation, complement has also been shown to affect mature T-cell activation.

Key Concepts:

- Binding of complement components to antigen-presenting cells enhances their phagocytic ability and modulates cytokine secretion.

- Complement components enhance the B cell–mediated immune response by increasing the avidity with which a B cell binds to a complement-bound antigen.

- Immature T cells are protected from natural antibody and complement-mediated lysis by the provision of additional sialic acid residues on their cell surface glycoproteins. Defective T cells do not have this protective layer, and so complement participates in quality control mechanisms during T-cell development.

- Binding of C3a, C5a, and C3b to their respective receptors on mature T cells facilitates their growth, differentiation, and survival.

Complement Aids in the Contraction Phase of the Immune Response

At the close of an adaptive immune response, most of the lymphocytes that were generated in the initial proliferative phase undergo apoptosis (programmed cell death), leaving only a few antigen-specific cells behind to provide immunological memory (see Chapters 10 and 11). We refer to this stage as the *contraction phase* of an immune response. At this stage, soluble antigen-antibody complexes may still be present in the bloodstream and

immune organs. If autoimmune disease is to be avoided, these excess lymphocytes and immune complexes must be disposed of safely, without the induction of further inflammation, and complement components play a major role in these processes.

Disposal of Apoptotic Cells and Bodies

Apoptotic cells express the phospholipid **phosphatidylserine** on the exterior surface of their plasma membranes. In healthy cells, this phospholipid is normally restricted to the cytoplasmic side of the membrane, and the change in its location as the cell enters apoptosis serves to signal the immune system that the cell is dying. Exposed phosphatidylserine is then bound by the serum protein, annexin A5, which in turn is recognized by C1q. Nuclear fragmentation, DNA cleavage, and the expression of DNA on the cell surface are also hallmarks of apoptosis, and recent work has demonstrated that C1q binds specifically to DNA as well as to glycoproteins and phospholipids exposed on the surfaces of dying cells and apoptotic fragments (**Figure 5-14**). Once apoptosis begins, the dying cell

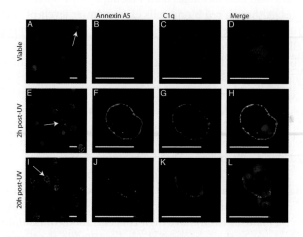

FIGURE 5-14 C1q colocalizes with annexin A5 on the surface of apoptotic cells. This figure demonstrates that both C1q and annexin A5 are deposited on the surface of human HeLa cells treated with lethal doses of ultraviolet light. The three pictures on the left show the locations of cells, stained purple to highlight nuclear material; some of these cells can also be seen in the merged images on the far right. The cells were also stained green for annexin A5 (which binds to exposed phosphatidylserine residues) and red for C1q before receiving lethal doses of ultraviolet light. The second and third rows show similar staining 2 hours and 24 hours following treatment, respectively. At 2 hours, C1q and annexin A5 binding are clearly visible. At 20 hours, nuclear fragmentation and nuclear blebbing, hallmarks of apoptotic cell death, can be seen. Scale bars represent 20 μm. *[Republished with permission of The American Association of Immunologists, Inc. from Darnault, C. et al., 2008, C1q binds phosphatidylserine and likely acts as a multiligand-bridging molecule in apoptotic cell recognition.* Journal of Immunology *180:2329. Figure 8.]*

is broken down into membrane-bound vesicles termed *apoptotic bodies*, which also express phosphatidylserine and/or DNA on their exterior membrane surfaces.

C1q binding promotes phagocytosis of both apoptotic cells and apoptotic bodies through recognition by the C1q receptor (C1qR) on phagocytic cells. It can also lead to initiation of the classical complement cascade, which leads to deposition of C3b on the dying cells and fragments. Phagocytosis is then also mediated through the recognition of deposited C3b by CR1 receptors on macrophages.

In the absence of C1q, apoptotic membrane blebs are released from the dying cells as apoptotic bodies, which can then act as antigens and initiate autoimmune responses. As a consequence, mice deficient in C1q show increased mortality and higher titers of auto-antibodies than control mice and also display an increased frequency of glomerulonephritis, an autoimmune kidney disease. Analysis of the kidneys of C1q knockout mice reveals deposition of immune complexes as well as significant numbers of apoptotic bodies.

Disposal of Immune Complexes

As mentioned earlier, the coating of soluble immune complexes with C3b facilitates their binding by CR1 on erythrocytes. Although red blood cells express lower levels of CR1 (100–1000 molecules per cell, depending on the age of the cell and the genetic constitution of the host) than do granulocytes (5×10^4 per cell), there are about 1000 erythrocytes for every white blood cell, and therefore erythrocytes account for about 90% of the CR1 in blood. Erythrocytes also therefore play an important role in clearing C3b-coated immune complexes by conveying them to the liver and spleen, where the immune complexes are stripped from the red blood cells and phagocytosed (**Figure 5-15**).

The significance of the role of complement in binding to immune-complexes is illustrated by the finding that patients with the autoimmune disease *systemic lupus erythematosus* (SLE) have high concentrations of immune complexes in their serum that are deposited in the tissues, rather than being cleared. Complement is then activated by these tissue-deposited immune complexes, and pathological inflammation is induced in the affected tissues (see Chapter 15).

Since complement activation is implicated in the pathogenesis of SLE, it may therefore seem paradoxical that the incidence of SLE is highly correlated to *C4 deficiency*. Indeed, 90% of individuals who completely lack C4 develop SLE. The resolution to this paradox lies in the fact that deficiencies in the early components of complement lead to a reduction in the levels of C3b that are deposited on the immune complexes. This reduction, in turn, inhibits their clearance by C3b-mediated opsonization and allows for the activation of the later inflammatory and cytolytic phases of complement activation.

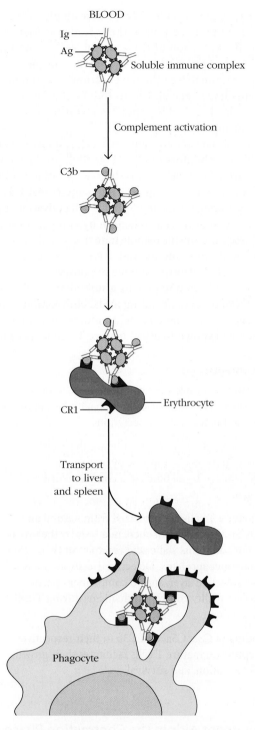

FIGURE 5-15 Clearance of circulating immune complexes by binding to erythrocyte complement receptors and subsequent stripping from these receptors by macrophage complement receptors in the liver and spleen. Because erythrocytes have fewer CR1 receptors than macrophages, the latter can strip the complexes from the erythrocytes as they pass through the liver or spleen. Deficiency in this process can lead to renal damage due to accumulation of immune complexes.

The Regulation of Complement Activity

All biological systems with the potential to damage the host are subject to rigorous regulatory mechanisms, and the complement system is no exception to this rule. Especially in light of the potent positive feedback mechanisms and the absence of antigen specificity of the alternative pathway, mechanisms must exist to ensure that the destructive potential of complement proteins is confined to the appropriate pathogen surfaces and that collateral damage to healthy host tissues is minimized.

Here, we discuss the different mechanisms by which the host protects itself against inadvertent complement activation. Protection of vertebrate host cells against complement-mediated damage is achieved by both general, passive regulatory mechanisms and specific, active regulatory mechanisms.

Complement Activity Is Passively Regulated by Short Protein Half-Lives and Host Cell Surface Composition

The *relative instability* of many complement components is the first means by which the host protects itself against extended periods of inadvertent complement activation. For example, the C3 convertase of the alternative pathway, C3bBbC3b, has a half-life of only 5 minutes, unless it is stabilized by reaction with properdin. A second passive regulatory mechanism depends on the difference in the cell surface carbohydrate composition of host versus microbial cells. For example, fluid-phase proteases that destroy C3b bind much more effectively to host cells that bear high levels of sialic acid, than to microbes that have significantly lower levels of

this sugar. (We encountered this mechanism in the section "Complement and T Cell–Mediated Immunity," in which we described complement's role in ensuring the destruction of improperly developed T cells). Hence, any C3b molecule that happens to alight on a host cell is likely to be degraded before it can cause significant damage.

In addition to these more passive environmental brakes on inappropriate complement activation, a series of active regulatory proteins work to inhibit, degrade, or reduce the activity of complement proteins and their fragments on host cells. The stages at which complement activity is subject to regulation are illustrated in **Figure 5-16**, and the regulatory proteins are listed in **Table 5-6**.

The C1 Inhibitor, C1INH, Promotes Dissociation of C1 Components

C1INH, the C1 inhibitor, is a plasma protein that binds in the active site of serine proteases, effectively poisoning them. C1INH belongs to the class of proteins called *serine protease inhibitors* (serpins), and it acts by forming a complex with the protease $C1r_2s_2$, causing it to dissociate from C1q and preventing further activation of C4 or C2 (see Figure 5-16a). C1INH inhibits both the classical pathway serine protease and that of the lectin pathway, MASP-2. It is the only regulatory protein capable of inhibiting the initiation of both the classical and lectin complement pathways, and its presence in plasma serves to limit the time period during which they can remain active.

Decay-Accelerating Factor Promotes Decay of C3 Convertases

Since the reaction catalyzed by the C3 convertase enzymes is the major amplification step in complement activation, the generation and lifetimes of the two C3 convertases, C4b2a and C3bBb, are subject to particularly rigorous control.

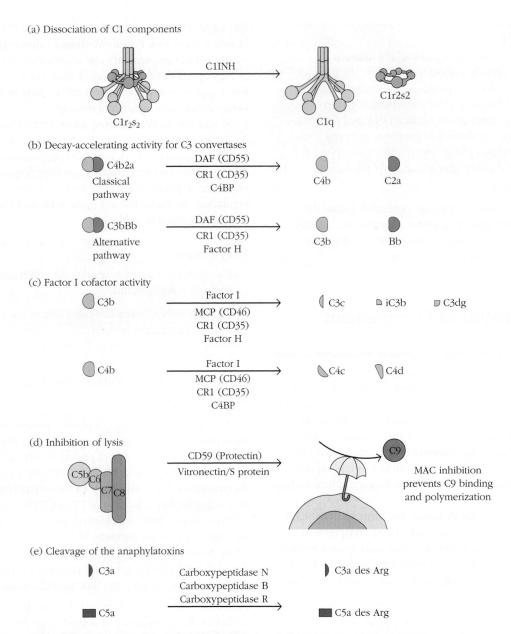

FIGURE 5-16 Regulation of complement activity. The various stages at which complement activity is subject to regulation are shown (see text for details).

The membrane-bound *decay-accelerating factor*, or DAF (CD55), accelerates the decay of the C4b2a C3 convertase on the surface of host cells. In order to complete its job, DAF requires the cofactors CR1 and C4BP (C4-binding protein) (see Figure 5-16b). These decay-accelerating proteins cooperate to *accelerate the breakdown of the C4b2a* complex into its separate components. C2a, inactive in the absence of C4b, diffuses away, and the residual membrane-bound C4b is degraded by another regulatory protein, factor I (see Figure 5-16c).

In the alternative pathway, DAF and CR1 function in a similar fashion. However, in place of C4BP they are joined by factor H in separating the C3b component of the alternative pathway C3 convertase from its partner,

Bb (see Figure 5-16b). Again, inactive Bb diffuses away, and residual C3b is degraded (see Figure 5-16c).

Whereas DAF and CR1 are membrane-bound components and their expression is therefore restricted to host cells, factor H and C4BP are soluble cofactors of regulatory complement components. Host-specific function of factor H is ensured by its binding to negatively-charged cell surface carbohydrates such as sialic acid and heparin, which are essential components of eukaryotic, but not prokaryotic, cell surfaces. Similarly, C4BP is preferentially bound by host cell membrane proteoglycans such as heparan sulfate. In this way, host cells are protected from the deposition of complement components; in contrast, microbial invaders that lack DAF and CR1

TABLE 5-6   Proteins involved in the regulation of complement activity

Protein	Soluble or membrane bound	Pathway(s) affected	Function
C1 inhibitor (C1INH)	Soluble	Classical and lectin	Induces dissociation and inhibition of $C1r_2s_2$ from C1q; serine protease inhibitor
Decay-accelerating factor (DAF; CD55)	Membrane bound	Classical, alternative, and lectin	Accelerates dissociation of C4b2a and C3bBb C3 convertases
CR1 (CD35)	Membrane bound	Classical, alternative, and lectin	Blocks formation of, or accelerates dissociation of, the C3 convertases C4b2a and C3bBb by binding C4b or C3b Cofactor for factor I in C3b and C4b degradation on host cell surface
C4BP	Soluble	Classical and lectin	Blocks formation of, or accelerates dissociation of, C4b2a C3 convertase Cofactor for factor I in C4b degradation
Factor H	Soluble	Alternative All pathways	Blocks formation of, or accelerates dissociation of, C3bBb C3 convertase Cofactor for factor I in C3b degradation
Factor I	Soluble	Classical, alternative, and lectin	Serine protease: cleaves C4b and C3b using cofactors shown in Figure 5-16
Membrane cofactor of proteolysis, MCP (CD46)	Membrane bound	Classical, alternative, and lectin	Cofactor for factor I in degradation of C3b and C4b
S protein (vitronectin)	Soluble	All pathways	Binds soluble C5b67 and prevents insertion into host cell membrane
CD59 (protectin)	Membrane bound	All pathways	Binds C5b678 on host cells, blocking binding of C9 and the formation of the MAC
Carboxypeptidases N, B, and R	Soluble	Anaphylatoxins produced by all pathways	Cleave and inactivate the anaphylatoxins C3a and C5a

expression and fail to bind factor H or C4BP are completely vulnerable to complement-mediated attack. However, as we will see, sometimes microbes hijack these mechanisms that are designed to ensure specificity of protection for host cells, and use them instead to protect themselves.

Key Concept:

- Any C3 convertase complexes of either the classical and lectin pathways (C4b2a) or the alternative pathway (C3bBb) that alight on host cells are degraded by the host cell membrane protein DAF, acting in concert with cofactors that are either expressed on, or bind specifically to, host cell membranes.

Factor I Degrades C3b and C4b

Factor H, C4BP, and CR1 also figure as co-factors in a second mechanism of complement regulation: that catalyzed by factor I. Factor I is a soluble, constitutively (always) active serine protease that can cleave membrane-associated C3b and C4b into inactive fragments (see Figure 5-16c).

However, if factor I is indeed soluble, constitutively active, and designed to destroy C3b and C4b, one might wonder how the complement cascades ever succeed in destroying invading microbes? The answer, once again, is that factor I requires the presence of these same, host cell–specific cofactors in order to function. Hence, cleavage of membrane-bound C3b on host cells is conducted by factor I in collaboration with the membrane-bound host cell proteins MCP and CR1, and the soluble cofactor factor H. Similarly, cleavage of membrane-bound C4b is achieved again by factor I, this time in collaboration with membrane-bound MCP and CR1 and soluble cofactor C4BP. Since these membrane-bound, or membrane-binding, cofactors are not found on microbial cells, C3b and C4b are thus destroyed if they alight on host cells, but are allowed to remain on microbial cells and exert their specific functions.

Recently, six proteins related to factor H have been identified with varying levels of complement-regulatory activity. Their activities and regulation are currently under careful investigation. Interestingly, genetic variations of factor H and its related proteins have been associated with

BOX 5-3

The Complement System as a Therapeutic Target

The involvement of complement-derived anaphylatoxins in inflammation makes complement an interesting therapeutic target in the treatment of inflammatory diseases, such as arthritis. In addition, autoimmune diseases that potentially result in complement-mediated damage to host cells, such as multiple sclerosis and age-related macular degeneration, are also potential candidates for complement-focused therapeutic interventions.

The last three decades have seen the crystallization and molecular characterization of several complement components, a necessary precursor to the development of designer drugs targeted to specific proteins. At the time of writing, three therapies directed at interfering with complement components are approved for clinical use, with several more in either clinical or preclinical trials. The therapies already in the clinic are either antibodies or parts of antibodies that specifically target and regulate the complement proteins C5 (two different drugs) or factor D.

Until now, the most successful complement-related treatment is specifically directed toward a disease that stems from a dysregulation of the complement cascade. Paroxysmal nocturnal hemoglobinuria (PNH) manifests as increased fragility of erythrocytes, leading to chronic hemolytic anemia, pancytopenia (loss of blood cells of all types), and venous thrombosis (formation of blood clots). The name PNH derives

from the presence of hemoglobin in the urine, most commonly observed in the first urine passed after a night's sleep. The cause of PNH is a general defect in the synthesis of cell surface proteins, which affects the expression of two regulators of complement: DAF (CD55) and CD59 (protectin).

DAF and CD59 are cell surface proteins that function as inhibitors of complement-mediated cell lysis, acting at different stages of the process. DAF induces dissociation and inactivation of the C3 convertases of the classical, lectin, and alternative pathways (see Figure 5-16). CD59 acts later in the pathway by binding to the C5b678 complex and inhibiting C9 binding, thereby preventing formation of the membrane pores that destroy the cell under attack. Deficiency in these proteins leads to increased sensitivity of host cells to complement-mediated lysis and is associated with a high risk of thrombosis.

CD59 and DAF are attached to the cell membrane via glycosylphosphatidylinositol (GPI) anchors, rather than by stretches of hydrophobic amino acids, as is the case for many membrane proteins. Most patients with PNH lack an enzyme subunit, called *phosphatidylinositol glycan class A* (PIG-A), that attaches the GPI anchors to the appropriate proteins. The gene encoding PIG-A is found on the X chromosome, and this part of the X chromosome is silenced in females, meaning that each cell has only one copy of the gene. A defect in this copy therefore

means that patients lack the expression of both DAF and CD59 needed to protect erythrocytes against lysis. The term *paroxysmal* refers to the fact that episodes of erythrocyte lysis are often triggered by stress or infection, which result in increased deposition of C3b on host cells. Episodes of hemoglobinuria result in dark red urine, and since urine is most concentrated first thing in the morning, the term *nocturnal* was applied to indicate that the release of hemoglobin into the urine was occurring overnight.

The defect identified in PNH occurs early in the enzymatic pathway leading to formation of a GPI anchor and resides in the *PIGA* gene. Transfection of cells from PNH patients with an intact *PIGA* gene restores the resistance of the cells to host complement lysis. Further genetic analysis revealed that the defect is not encoded in the germline genome (and therefore is not transmissible to offspring), but rather is the result of mutations that occurred within the hematopoietic stem cells themselves, such that any one individual may have both normal and affected cells. Those patients who are most adversely affected display a preferential expansion of the affected cell populations.

PNH is a chronic disease with a mean survival time post diagnosis of between 10 and 15 years. The most common causes of mortality in PNH are blood clots that affect veins in the liver, and progressive bone marrow failure.

(continued)

chronic inflammatory diseases such as age-related macular degeneration.

Variation in MCP expression has recently been implicated as a factor in the control of apoptosis followed by phagocytosis of dying T cells. When a T cell commits to apoptosis, it expresses DNA on its cell membrane that binds circulating C1q, as described earlier. It then begins to shed MCP from the cell surface. Only after MCP is lost can progression of the classical pathway occur, resulting in opsonization by C3b and eventual phagocytosis of the apoptotic T cells.

Key Concepts:

• Factor I is a soluble, constitutively active serine protease that cleaves C3b and C4b into inactive fragments only when it is associated on host cell membranes with the necessary cofactors.

• The cofactor, MCP, is lost as lymphocytes enter apoptosis, thereby allowing the deposition of C3b on the dying cell surface and subsequent phagocytosis.

A breakthrough in the treatment of PNH was reported in 2004; a humanized monoclonal antibody that targets complement component C5 was used to inhibit the terminal steps of the complement cascade and formation of the membrane attack complex. This antibody—eculizumab (Soliris)—was infused into patients, who were then monitored for the loss of red blood cells. A dramatic improvement was seen in patients during a 12-week period of treatment with eculizumab (**Figure 1**). Treatment of PNH patients with eculizumab relieves hemoglobinuria, reverses the kidney damage resulting from high levels of protein in the urine, and significantly reduces the frequency of thromboses. In 2007, eculizumab was approved for the treatment of PNH in the general population.

Since the control of infections with meningococcal bacteria *(Neisseria meningitidis)* relies on an intact and functioning membrane attack complex (MAC), patients being treated with eculizumab are routinely vaccinated against meningococcus.

The pathology of PNH underscores the potential danger to the host that is inherent in the activation of the complement system. Intricate systems of regulation are necessary to protect host cells from the activated complement complexes generated to lyse intruders, and alterations in the expression or effectiveness of these regulators have the potential to result in a pathological outcome.

The centrality of C3 within the complement cascades may suggest that it would be an excellent target for therapeutic intervention in complement-mediated disease. However, its very position at the crossroads of the three pathways means that interference with C3 activity places patients at increased risk for infectious diseases normally curtailed by the activities of the innate as well as the adaptive immune systems, and so systemic drugs specifically targeting C3 may prove to have too high a risk-benefit ratio.

However, scientists have recently succeeded in developing C3-targeted drugs that specifically address complement-mediated diseases that attack particular organs. One drug, compstatin, was developed as a result of experiments designed to discover compounds that bound specifically to C3. Chemical refinements based on structural and computational studies followed, and a compstatin derivative, POT-4, is now in phase 2 clinical trials for the treatment of age-related macular degeneration, a progressive and debilitating eye disease that leads inexorably and quickly to total blindness. Because POT-4 can be delivered directly into the vitreous humor of the eye, systemic side effects on the patient's immune system are minimized. A second C3-directed drug, AMY-101, is now in preclinical trials and has been granted orphan status by both the European Medicines Agency and the U.S. Food and Drug Administration for treatment of C3 glomerulonephropathy (C3G), a rare disease that affects kidney glomeruli and is caused by dysregulation of the alternative pathway of complement activation.

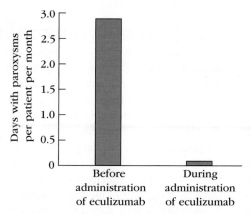

FIGURE 1 Treatment of PNH patients with eculizumab relieves hemoglobinuria. Shown are the number of days with paroxysms (sudden onsets) per patient per month in the month before treatment (left column) and for a 12-week period of treatment with eculizumab (right column). *[Data from P. Hillmen, et al. 2004. Eculizumab in patients with paroxysmal nocturnal hemoglobinuria.* New England Journal of Medicine *350:552.]*

CD59 (Protectin) Inhibits the MAC Attack

In the case of a particularly robust antibody response, or of an inflammatory response accompanied by extensive complement activation, inappropriate assembly of MACs on healthy host cells can potentially occur, and mechanisms have evolved to prevent the resulting inadvertent host cell destruction. A host cell surface protein, **CD59 (protectin)**, binds any C5b678 complexes that may be deposited on host cells and prevents their insertion into the host cell membrane (see Figure 5-16d). CD59 also blocks further C9 addition to developing MACs. In addition, the soluble complement *S protein*, otherwise known as *vitronectin*, binds any fluid-phase C5b67 complexes released from microbial cells, preventing their insertion into host cell membranes.

Both CD59 and DAF are membrane-associated molecules that are attached to the lipid bilayer via glycosylphosphatidylinositol anchors. Rare mutations in the gene *PIGA* prevent afflicted individuals from attaching these anchors, and so these patients suffer from dysregulation of complement activity. The development of drugs that specifically target complement components and help to address the symptoms of individuals suffering from complement-related diseases is described in **Clinical Focus Box 5-3**.

Carboxypeptidases Can Inactivate the Anaphylatoxins C3a and C5a

Anaphylatoxin activity is regulated by cleavage of the C-terminal arginine residues from both C3a and C5a by serum carboxypeptidases, resulting in rapid inactivation of their anaphylatoxin activity (see Figure 5-16e). Carboxypeptidases are a general class of enzymes that remove amino acids from the carboxyl termini of proteins; the specific enzymes that mediate the control of anaphylatoxin activity are carboxypeptidases N, B, and R. These enzymes remove arginine residues from the carboxyl termini of C3a and C5a to form the so-called *des-Arg* ("without arginine"), inactive forms of the molecules. In addition, as mentioned above, binding of C5a by C5L2 also serves to modulate the inflammatory activity of C5a.

Complement Deficiencies

Genetic deficiencies have been described for each of the complement components. Homozygous deficiencies in any of the early components of the classical pathway (C1q, C1r, C1s, C4, and C2) result in similar symptoms, notably a marked increase in immune-complex diseases such as SLE, glomerulonephritis, and vasculitis. The effects of these deficiencies highlight the importance of C3b's role in the clearance of immune complexes. In addition, as described earlier, C1q has been shown to bind apoptotic cells and cell fragments. In the absence of C1q binding, apoptotic cells can act as auto-antigens and lead to the development of autoimmune diseases such as SLE. Individuals with deficiencies in the early complement components may also suffer from recurrent infections with both gram-negative and gram-positive, pyogenic (pus-forming) bacteria such as streptococci and staphylococci. These latter organisms are normally resistant to the lytic effects of the MAC, but the early complement components are important in controlling such infections by mediating a localized inflammatory response and opsonizing the bacteria.

A deficiency in MBL, the first component of the lectin pathway, has been shown to be relatively common, and results in serious pyrogenic (fever-inducing) infections in babies and children. Children with MBL deficiency suffer from recurrent respiratory tract infections. MBL deficiency is also found with a frequency two to three times higher in patients with SLE than in normal subjects, and certain mutant forms of MBL are found to be prevalent in chronic carriers of hepatitis B. Deficiencies in factor D and properdin—early components of the alternative pathway—appear to be associated with *Neisseria* infections but not with immune-complex disease.

People with C3 deficiencies display serious clinical manifestations, reflecting the central role of C3 in opsonization and in the formation of the MAC. The first person identified with a C3 deficiency was a child suffering from frequent, severe bacterial infections leading to meningitis, bronchitis, and pneumonia, and who was initially diagnosed with agammaglobulinemia. After tests revealed normal immunoglobulin levels, a deficiency in C3 was discovered. This case highlighted the critical role of the complement system in converting a humoral antibody response into an effective host defense mechanism. The majority of people with C3 deficiency have recurrent bacterial infections and may also present with immune-complex diseases.

Levels of C4 vary considerably in the population. The genes encoding C4 are located in the major histocompatibility locus (see Chapter 7), and the number of C4 genes vary among individuals from two to six. Low gene copy numbers are associated with lower levels of C4 in plasma and with a correspondingly higher incidence of SLE, for the reasons described earlier. Patients with complete deficiencies of one or more of the components of the classical pathway, such as C4, contract more frequent infections with bacteria such as *S. pneumoniae*, *Haemophilus influenzae*, and *N. meningitidis*. However, even patients with low copy numbers of the C4 gene appear to be relatively well protected against such infections. Interestingly, C4 exists in two isoforms: C4A and C4B. C4B is more effective in binding to the surfaces of the three bacterial species mentioned above.

Individuals with deficiencies in components of the terminal complement cascade are more likely than members of the general population to suffer from meningitis, indicating that cytolysis by complement components C5 through C9 is of particular relevance to the control of *N. meningitidis*. This has resulted in the release of public health guidelines that highlight the need for vaccinations against *N. meningitidis* for individuals deficient in the terminal complement components.

Deficiencies of complement regulatory proteins have also been reported. As described previously, C1INH, the C1 inhibitor, regulates activation of the classical pathway by preventing excessive C4 and C2 activation by C1. However, as a serine protease inhibitor, it also controls two serine proteases in the blood clotting system. Patients with C1INH deficiency suffer from a complex disorder that includes excessive production of vasoactive mediators (molecules that control blood vessel diameter and integrity), which in turn leads to tissue swelling and extracellular fluid accumulation. The resultant clinical condition is referred to as *hereditary angioedema*. It presents clinically as localized tissue edema that often follows trauma, but sometimes occurs with no known cause. The edema can be in subcutaneous tissues; within the bowel, where it causes abdominal pain; or in the upper respiratory tract, where it can result in fatal obstruction of the airway. C1INH deficiency is an autosomal dominant condition with a frequency of 1 in 1000 in the human population.

Studies in humans and experimental animals with homozygous deficiencies in complement components have provided important information regarding the roles of individual complement components in immunity. These initial observations have been significantly enhanced by studies using knockout mice, genetically engineered to lack expression of specific complement components. Investigations of in vivo complement activity in these animals have allowed dissection of the complex system of complement proteins and the assignment of precise biologic roles to each.

Key Concepts:

- Patients suffering from complement deficiencies often present with immune-complex disorders and suffer disproportionally from infections by encapsulated bacteria such as *Neisseria meningitidis*.

- Animal models exist for most complement deficiencies, and knockout animals lacking particular complement components have been essential to the dissection of the roles of individual components in immune responses.

Microbial Complement Evasion Strategies

The importance of complement in host defense is clearly illustrated by the number and variety of strategies that have evolved in microbes, enabling them to evade complement attack (**Table 5-7**). Gram-positive bacteria have developed thick cell walls and capsules that enable them to shrug off the insertion of MACs, while other bacterial species escape into intracellular vacuoles to escape immune detection. However, these two general strategies are energy intensive for the microbe, and many microbes have adopted more specific complement evasion tactics to escape destruction. In this section, we address complement microbial evasion at a conceptual level, categorizing the various approaches that microbes have evolved to elude this effector arm of the immune response. It should be emphasized, however, that different microbes will use varying selections from this menu of strategies.

Many classes of viruses interfere with the classical complement pathway before it can be initiated, by synthesizing proteins and glycoproteins that specifically bind the Fc regions of antibodies, thus preventing complement binding. Some viruses also produce proteins that enhance the clearance of antibody-antigen complexes from the surfaces of virus-infected cells and/or manufacture proteins that induce rapid internalization of viral protein–antibody complexes.

Since highly specific protein-protein interactions between complement components are central to the functioning of the cascade, it is logical that microbes have evolved mechanisms that interfere with some of these binding reactions. The mechanisms adopted by *S. aureus* to protect itself from attack by complement have been particularly well studied, but the strategy of inhibiting the interactions between complement proteins is not restricted to bacteria. The production of molecules that inhibit the interactions between complement components has also been detected in certain human parasites. For example, a protein generated by some species of both *Schistosoma* and *Trypanosoma*, the complement C2 receptor trispanning protein, disrupts the interaction between C2a and C4b and thus prevents the generation of the classical pathway C3 convertase.

Some microbes produce proteases that destroy complement components. This strategy is used primarily by bacteria, and numerous bacterial proteases exist that are capable of

TABLE 5-7   Some microbial complement evasion strategies

Complement evasion strategy	Example
Interference with antibody-complement interaction	Antibody depletion by staphylococcal protein A Removal of IgG by staphylokinase
Binding and inactivation of complement proteins	*S. aureus* protein SCIN binds to and inactivates the C3bBb C3 convertase Parasite protein C2 receptor trispanning protein disrupts the binding between C2 and C4
Protease-mediated destruction of complement component	Elastase and alkaline phosphatase from *Pseudomonas* degrade C1q and C3/C3b ScpA and ScpB from *Streptococcus* degrade C5a
Microbial mimicry of complement-regulatory components	*Streptococcus pyogenes* M proteins bind C4BP and factor H to the cell surface, accelerating the decay of C3 convertases bound to the bacterial surface *Variola* and *Vaccinia* viruses express proteins that act as cofactors for factor I in degrading C3b and C4b

digesting a variety of complement components. For example, elastase and alkaline protease proteins from *Pseudomonas aeruginosa* target the degradation of C1q and C3/C3b, and two proteases derived from streptococcal bacteria, ScpA and ScpB, specifically attack the anaphylatoxin C5a. Streptococcal pyrogenic exotoxin B was found to degrade the complement regulator properdin, with resultant destabilization of the alternative pathway C3 convertase on the bacterial surface.

Fungi can also inactivate complement proteins. The opportunistic human pathogen *Aspergillus fumigatus* secretes an alkaline protease, Alp1, that is capable of cleaving C3, C4, C5, and C1q, as well as IgG. Since this pathogen tends to attack patients who are already immunocompromised, its capacity to further damage the immune system is particularly troubling in the clinic.

Several microbial species have exploited the presence of cell surface or soluble regulators of complement activity for their own purposes, either by mimicking the effects of these regulators, or by acquiring them directly. Indeed, recent work suggests that sequestration of host-derived regulators of complement activity may be one of the most widely utilized microbial mechanisms for complement evasion.

Many microbes have developed the ability to bind one or another of the fluid-phase inhibitors of complement, C4BP or factor H. For example, *Streptococcus pyogenes*, an important human pathogen, expresses a family of proteins called "M proteins" capable of binding to C4BP and factor H. Expression of these regulatory proteins on the bacterial surface results in the inhibition of the later steps of complement fixation.

More surprisingly, some microbes appear to acquire membrane-bound regulators from host cells. *Helicobacter pylori*, obtained from patients suffering from gastric ulcers, was found to stain positively for CD59. CD59 is normally attached to the host membrane by a glycosylphosphatidylinositol anchor, and so this anchor appears to be transferable in some way from the host to the bacterial cell membrane.

Many viruses have evolved the capacity to produce proteins that closely mimic the structure and function of complement regulatory proteins. For example, the *Variola* (smallpox) and *Vaccinia* (cowpox) viruses express complement-inhibitory proteins that bind C3b and C4b and serve as cofactors for factor I, thus preventing complement activation at the viral membrane. In addition to encoding complement-regulatory proteins within the viral genome, some viruses actively induce their synthesis within the host cells that harbor the viruses. Other viruses camouflage themselves during budding from the host cell by hiding within regions of the host membrane on which regulatory complement components are expressed. And finally, some viruses mimic eukaryotic membranes by incorporating high levels of sialic acid into the viral membrane, thus facilitating the binding of the cofactors for factor I that normally bind only to host cell membranes. This therefore results in the inhibition of complement activation on the viral surface.

Key Concepts:

- A broad variety of complement evasion strategies have evolved in viruses, bacteria, fungi, and parasites. The number of different complement evasion strategies used by microbes underscores the importance of complement as a host defense mechanism and many microbes use more than one evasion strategy.

- Some pathogens interfere with the first step of immunoglobulin-mediated complement activation by binding to the Fc regions of antibodies. Some viruses enhance the rate of clearance of antigen-antibody complexes from the surface of virally infected cells.

- Some microbes produce proteases that specifically destroy complement proteins. Some produce proteins that bind complement components, masking their sites of interaction. Some microbes mimic or directly acquire complement-regulatory proteins in order to evade complement-mediated destruction. And some induce the synthesis of regulatory proteins by the cells that harbor them.

The Evolutionary Origins of the Complement System

Although the complement system was initially characterized by its ability to convert antibody binding into pathogen lysis, studies of complement evolution have identified the components of the lytic MAC (C5b through C9) as relatively late additions to the animal genome. Activation of complement cascades in invertebrates and early vertebrates therefore most likely culminated in opsonization and phagocytosis by primitive hemocytes. It appears probable that, among the non-MAC complement components, the proteins of the alternative pathway were the first to appear, followed by those of the lectin recognition systems. A fully fledged MAC emerged only around the same time as the appearance of the adaptive immune system (**Figure 5-17** and **Table 5-8**).

Genomic analysis has classified complement components into five gene families, each of which possesses unique domain structures that have enabled investigators to trace their phylogenetic origins.

The first of these gene families encodes C1q, mannose-binding lectin (MBL), and ficolins. Genes for prototypical C1q molecules have been identified in species as primitive as lampreys (a jawless fishlike vertebrate), sea urchins, and ascidians (sea squirts). Analysis of C1q gene clusters in different species has demonstrated that C1q genes appeared prior to the generation of immunoglobulin

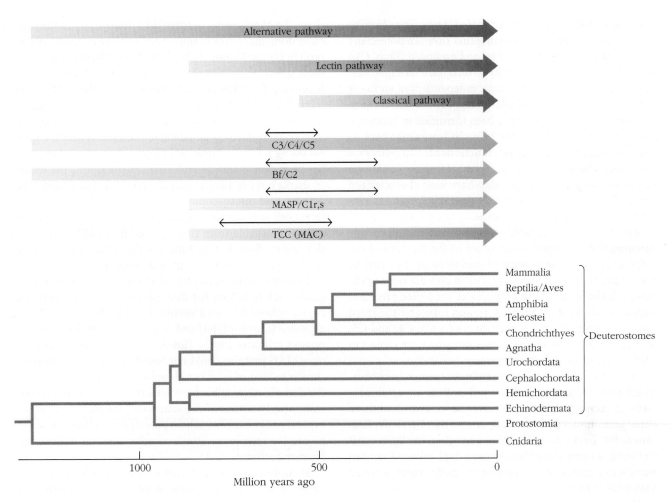

FIGURE 5-17 Evolution of complement components. The three gray arrows show the timing of appearance of each of the three pathways. The colored arrows reflect the time of first appearance of the first members of each of the families of complement proteins. The timing of the gene duplications that gave rise to the classical complement pathway is indicated by double-headed arrows. *[Reprinted with permission from Springer-Verlag, from Nonaka, M., and A. Kimura, 2006, Genomic view of the evolution of the complement system. Immunogenetics **58**:701. Figure 4. Permission conveyed through Copyright Clearance Center, Inc.]*

TABLE 5-8	Complement system pathways in the major groups of deuterostome animals				
Animal group	**Alternative pathway**	**Classical pathway**	**Lectin pathway**	**Membrane attack complex**	**Antibodies present?**
Mammals	+	+	+	+	+
Birds	+	+	+	+	+
Reptiles	+	+	+	+	+
Amphibians	+	+	+	+	+
Teleost fish	+	+	+	+	+
Cartilaginous fish	+	+	+	+	+
Agnathan fish	+	−	+	?	−
Tunicates	+	−	+	?	−
Echinoderms	+	−	?	?	−

Data from Sunyer, J. O., et al. 2003. Evolution of complement as an effector system in innate and adaptive immunity. *Immunologic Research* **27**:549 [Figure 2].

genes, and that at least some of these C1q proteins bind to specific carbohydrates, suggesting that they are potentially able to discriminate between host and pathogen. Thus, C1q may have expressed the capacity to recognize foreign molecules at a very early point in time, independent of antibody binding and in a manner similar to that of MBL and ficolins. Mannose-binding lectins have been identified in lampreys, thus placing the origin of the MBLs at least as far back as the early vertebrates. Indeed, functional lectin pathways have been characterized in ascidians. However, although genes encoding MBL-like proteins have been characterized in ascidian genomes, the nature of their relationship to vertebrate MBL proteins is still unclear.

The next three gene families to be considered all encode proteins that are somehow involved in the cleavage of the C3, C4, and C5 components of complement. The first of these families is composed of factor B and the serine protease C2; the second family comprises the serine proteases MASP-1, MASP-2, MASP-3, C1r, and C1s; and the third family contains the C3 family members C3, C4, and C5, which are not themselves proteases, but which are subject to protease cleavage and activation.

Studies of a number of invertebrate species have indicated that the genome of most, if not all, invertebrates contains only a single copy representative of each of these three gene families. In contrast, one or more *gene duplication events* have occurred within each gene family in most vertebrates, suggesting that the gene duplications found in vertebrates probably occurred in the early stages of jawed vertebrate evolution. Genes for each of the factor B, C3, and MASP families have been identified from all invertebrate deuterostomes that have so far been analyzed, with the single exception of the MASP family gene, which is missing from the sea urchin genome.

Interestingly, whereas members of the C3 and factor B families have been identified in some very early protostomes—for example, the horseshoe crab and the carpet-shell clam—analyses of whole genome sequences of *Drosophila melanogaster* or the nematode *Caenorhabditis elegans* have failed to locate any proto-complement protein sequences. This contrasts starkly with the presence of all three family genes in sea anemones and corals, and suggests that the ability to encode proteins of the complement system has been specifically *lost* from the genome of the common, model-system organisms *D. melanogaster* and *C. elegans*. The absence of genes encoding any of the later, cytolytic components of the complement cascade in protostomes and cnidarians (e.g., jellyfish, corals) supports the hypothesis that the proto-complement systems of cnidarians and protostomes function by opsonization.

The proteins encoded by the fifth gene family, C6, C7, C8α, C8βγ, and C9, together make up the terminal complement complex. They share a unique domain structure that enables molecular phylogenetic analysis. In particular, the mammalian components C6, C7, C8α, C8βγ, and C9 all share the MAC/perforin (MACPF) domain, in addition to other domains common among two or more members of this protein group. Mammals, birds, and amphibians share all of these genes (with the exception of the bird C9 gene, which is missing from the chicken genome sequence). Although the full set of MAC proteins has not yet been documented in cartilaginous fishes such as the sharks (the earliest known organisms to possess an adaptive immune system), a gene encoding a C8α subunit orthologous to mammalian C8α has been cloned and characterized. Furthermore, the serum of sharks has been known for decades to express hemolytic activity, and microscopic analysis of the pores formed on target cell surfaces reveals a transmembrane pore structure indistinguishable from that induced by an MAC attack. It thus seems likely that the functionality of the MAC emerged not long after the adaptive immune system.

Genomic analysis has traced the origin of this family of genes back to before the divergence of the urochordates, cephalochordates, and vertebrates (see Figure 5-17), as ascidians (urochordates) and amphioxus (cephalochordates) possess primitive copies. However, the ascidian and amphioxus MAC proteins could not have functioned in a manner similar to that of later vertebrate species because they lack the domain responsible for interacting with the C5 protein. Intriguingly, toxins of the venomous sea anemone, which express a very high hemolytic potential, are closest in structure to the membrane attack complex proteins of the complement pathway.

In summary, complement evolution is based on the diversification and successive duplications of members of five gene families that evolved first in response to the need for microbial recognition and opsonization and then, in response to the appearance of the adaptive immune system, by additional gene duplication and functional adaptation to form the complement system as we know it today.

Key Concepts:
- The genes encoding complement components belong to five families.

- Genes of the alternative pathway components appeared first in evolution, and those encoding the terminal complement components appeared last. Thus, prior to the emergence of adaptive immunity, complement served its protective functions by mediating phagocytosis.

Conclusion

The complement system comprises a set of serum proteins, many of which circulate in inactive forms that must first be cleaved or undergo conformational changes prior to activation. Complement proteins include initiator molecules, enzymatic mediators, membrane-binding components or opsonins, inflammatory mediators, membrane attack proteins, complement receptor proteins, and regulatory components.

Although genes encoding complement proteins appeared prior to those encoding the receptors of the adaptive immune system, complement proteins function in both innate and adaptive immunity. Perhaps the most important complement-mediated effector functions are those of the complement proteins C1q and C3b, which act as opsonins in phagocytosis. These proteins coat the surface of microbes and are recognized by complement-specific phagocytic receptors on macrophages. Other complement proteins serve as anaphylatoxins, bringing about the enhanced blood flow and capillary permeability characteristic of sites of inflammatory responses. Yet other complement proteins make up the membrane attack complex that punches holes in the membranes of microbial cells and sometimes microbially infected host cells, leading to their lysis and death.

Because of the destructive potential of the complement system, a rigorous system of regulatory proteins has co-evolved with the effector proteins. These regulatory proteins act to restrict complement-mediated injury to microbial targets and to minimize collateral damage to host cells.

The importance of the complement system to host defense is nowhere more clearly illustrated than in the plethora of strategies evolved by microbes to evade complement-mediated activity. Various microbes mimic or co-opt host regulatory proteins, specifically destroy complement proteins, or interfere with the interactions of complement proteins with one another or with antibody molecules, and as new microbes emerge, novel evasion strategies continue to evolve.

It is fitting that this chapter describing complement sits at the boundary between this book's descriptions of the innate and adaptive immune systems, since complement proteins act like the two-faced god, Janus, working with the ancient, innate immune system and the more recently evolved adaptive immune system with equal fluency and power.

REFERENCES

Aleshin, A. E., et al. 2012. Structure of complement C6 suggests a mechanism for initiation and unidirectional, sequential assembly of membrane attack complex (MAC). *Journal of Biological Chemistry* **287**:10210.

Alexander, J. J., A. J. Anderson, S. R. Barnum, B. Stevens, and A. J. Tenner. 2008. The complement cascade: yin-yang in neuroinflammation: neuro-protection and -degeneration. *Journal of Neurochemistry* **107**:1169.

Bajic, G., S. E. Degn, S. Thiel, and G. R. Andersen. 2015. Complement activation, regulation, and molecular basis for complement-related diseases. *EMBO Journal* **34**:2735.

Behnsen, J., et al. 2010. Secreted *Aspergillus fumigatus* protease Alp1 degrades human complement proteins C3, C4, and C5. *Infection and Immunity* **78**:3585.

Bialas, A. R., and B. Stevens. 2012. Glia: regulating synaptogenesis from multiple directions. *Current Biology* **22**:R833.

Bialas, A. R., and B. Stevens. 2013. TGF-β signaling regulates neuronal C1q expression and developmental synaptic refinement. *Nature Neuroscience* **16**:1773.

Blom, A. M., T. Hallström, and K. Riesbeck. 2009. Complement evasion strategies of pathogens: acquisition of inhibitors and beyond. *Molecular Immunology* **46**:2808.

Carroll, M. C., and D. E. Isenman. 2012. Regulation of humoral immunity by complement. *Immunity* **37**:199.

Chung, W. S., et al. 2013. Astrocytes mediate synapse elimination through MEGF10 and MERTK pathways. *Nature* **504**:394.

Czajkowsky, D. M., and Z. Shao. 2009. The human IgM pentamer is a mushroom-shaped molecule with a flexural bias.

Proceedings of the National Academy of Sciences USA **106**:14960.

Diebolder, C. A., et al. 2014. Complement is activated by IgG hexamers assembled at the cell surface. *Science* **343**:1260.

Dunkelberger, J. R., and W. C. Song. 2010. Complement and its role in innate and adaptive immune responses. *Cell Research* **20**:34.

Elward, K., et al. 2005. CD46 plays a key role in tailoring innate immune recognition of apoptotic and necrotic cells. *Journal of Biological Chemistry* **280**:36342.

Fearon, D. T., and M. C. Carroll. 2000. Regulation of B lymphocyte responses to foreign and self-antigens by the CD19/CD21 complex. *Annual Review of Immunology* **18**:393.

Ghannam, A., J. L. Fauquert, C. Thomas, C. Kemper, and C. Drouet. 2014. Human complement C3 deficiency: Th1 induction requires T cell–derived complement C3a and CD46 activation. *Molecular Immunology* **58**:98.

Hong, S., L. Dissing-Olesen, and B. Stevens. 2016. New insights on the role of microglia in synaptic pruning in health and disease. *Current Opinion in Neurobiology* **36**:128.

Kemper, C., J. P. Atkinson, and D. E. Hourcade. 2010. Properdin: emerging roles of a pattern-recognition molecule. *Annual Review of Immunology* **28**:131.

Kulkarni, P. A., and V. Afshar-Kharghan. 2008. Anticomplement therapy. *Biologics* **2**:671.

Lintner, K. E., et al. 2016. Early components of the complement classical activation pathway in human systemic autoimmune diseases. *Frontiers in Immunology* **7**:36.

Markiewski, M. M., B. Nilsson, K. N. Ekdahl, T. E. Mollnes, and J. D. Lambris. 2007. Complement and coagulation: strangers or partners in crime? *Trends in Immunology* **28**:184.

Rosen, A. M., and B. Stevens. 2010. The role of the classical complement cascade in synapse loss during development and glaucoma. *Advances in Experimental Medicine and Biology* **703**:75.

Schafer, D. P., et al. 2012. Microglia sculpt postnatal neural circuits in an activity and complement-dependent manner. *Neuron* **74**:691.

Sekar, A., et al. 2016. Schizophrenia risk from complex variation of complement component 4. *Nature* **530**:177.

Shi, Q., et al. 2015. Complement C3–deficient mice fail to display age-related hippocampal decline. *Journal of Neuroscience* **35**:13029.

Stephan, A. H., B. A. Barres, and B. Stevens. 2012. The complement system: an unexpected role in synaptic pruning during development and disease. *Annual Review of Neuroscience* **35**:369.

Stephan, A. H., et al. 2013. A dramatic increase of C1q protein in the CNS during normal aging. *Journal of Neuroscience* **33**:13460.

Stevens, B., et al. 2007. The classical complement cascade mediates CNS synapse elimination. *Cell* **131**:1164.

Tegla, C. A., et al. 2011. Membrane attack by complement: the assembly and biology of terminal complement complexes. *Immunologic Research* **51**:45.

Vlaicu, S. I., et al. 2013. Role of C5b-9 complement complex and response gene to complement-32 (RGC-32) in cancer. *Immunologic Research* **56**:109.

Zipfel, P. F., and C. Skerka. 2009. Complement regulators and inhibitory proteins. *Nature Reviews Immunology* **9**:729.

Useful Websites

www.complement-genetics.uni-mainz.de The Complement Genetics Homepage from the University of Mainz gives chromosomal locations and information on genetic deficiencies of complement proteins.

www.youtube.com/watch?v=XSjN_rq2tIE An overview of the complement system.

www.youtube.com/watch?v=mCNaHKnYhZU A carefully constructed lecture on all three pathways.

www.handwrittentutorials.com/videos.php?id=23 This is the first in a series of "handwritten tutorials" (like Khan academy but not) on complement. It offers a general introduction.

www.handwrittentutorials.com/videos.php?id=22 A handwritten tutorial about the alternative pathway.

www.handwrittentutorials.com/videos.php?id=24 A handwritten tutorial about the classical and MBL-initiated pathways.

www.youtube.com/watch?v=9ezkuJ08jMU A handwritten tutorial on the generation of the MAC.

www.youtube.com/watch?v=rBNzS5x_ok0 A brief animation (accompanied by a rhythm section) of the formation of an MAC. Note: it does not show the two monomeric units of C8.

STUDY QUESTIONS

1. Indicate whether each of the following statements is true or false. If you think a statement is false, explain why.

   a. A single molecule of bound IgM can activate the C1q component of the classical complement pathway.
   b. The enzymes that cleave C3 and C4 are referred to as *convertases*.
   c. C3a and C3b are fragments of C3 that are generated by proteolytic cleavage mediated by two different enzyme complexes.
   d. Nucleated cells tend to be more resistant to complement-mediated lysis than red blood cells.
   e. Enveloped viruses cannot be lysed by complement because their outer envelopes are resistant to pore formation by the membrane attack complex (MAC).
   f. MBL has a function in the lectin pathway analogous to that of IgM in the classical pathway, and MASP-1 and MASP-2 take on functions analogous to C1 components.

2. Explain why serum IgM cannot activate complement prior to antigen binding.

3. Genetic deficiencies have been described in patients for all of the complement components except factor B. Particularly severe consequences result from a deficiency in C3. Describe the consequences of an absence of C3 for each of the following:
   a. Initiation of the classical and alternative pathways
   b. Clearance of immune complexes
   c. Phagocytosis of infectious bacteria

4. Describe three ways in which complement acts to protect the host during an infection.

5. Complement activation can occur via the classical, alternative, or lectin pathway.
   a. How do the three pathways differ in the substances that can initiate activation?
   b. Which parts of the overall activation sequence differ among the three pathways, and which parts are similar?

c. How does the host ensure that inadvertent activation of the alternative pathway on its own healthy cells does not lead to autoimmune destruction?

6. Briefly explain the mechanism of action of the following complement regulatory proteins. Indicate which pathway(s) each protein regulates.
   a. C1 inhibitor (C1INH)
   b. C4b-binding protein (C4bBP)
   c. Decay-accelerating factor (DAF)
   d. Membrane cofactor of proteolysis (MCP; CD46)
   e. CD59 (protectin)
   f. Carboxypeptidase N

7. The importance of the complement system as a set of host defense machinery is underscored by the number of ways microbes have evolved to fight it. Please describe three mechanisms used by microbes to evade complement actions, giving examples of bacterial, viral, or fungal species that use that mechanism.

8. Explain why complement deficiencies in the early components of complement give rise to immune complex–mediated disorders such as systemic lupus erythematosus.

9. Are the following statements about the evolution of the complement system true or false? Explain your reasoning.
   a. The complement system evolved first as a component of the adaptive immune system.
   b. Genomic analysis has classified complement genes and the proteins they encode into five gene families on the basis of how many immunoglobulin domains they possess.
   c. C1q genes appeared prior to immunoglobulin genes and some C1q proteins specifically bind to particular carbohydrates.
   d. Gene duplication events have occurred in invertebrates in each of the five complement component gene families.

10. You have prepared knockout mice with mutations in the genes that encode various complement components. Each knockout strain cannot express one of the complement components listed across the top of the following table. Predict the effect of each mutation on the steps in complement activation and on the complement effector functions indicated in the table, using the following symbols: NE = no effect; D = process/function decreased but not abolished; A = process/function abolished.

	Complement component knocked out						
	C1q	**C4**	**C3**	**C5**	**C9**	**Factor B**	**MASP-2**
Formation of classical pathway C3 convertase							
Formation of alternative pathway C3 convertase							
Formation of classical pathway C5 convertase							
Formation of lectin pathway C3 convertase							
C3b-mediated opsonization							
Neutrophil chemotaxis and inflammation							
Cell lysis							

CLINICAL FOCUS QUESTION

As shown in the following figure, two flow cytometric histograms of red blood cells were obtained from a patient, stained with antibodies toward CD59 (protectin) (part a) and CD55 (decay-accelerating factor, or DAF) (part b). On the right of each histogram is a large population of cells expressing relatively high levels of CD59 (protectin) or CD55 antigens (population I). Populations II express midrange levels of the two antigens, and populations III express levels of antigen below statistical detectability. (Detecting laser voltages on flow cytometers are normally set such that negative staining populations show levels of fluorescence below 10^1 on the lower logarithmic scale. Note also the creep of cells up the 10^0 axis on each plot.)

a. Can you offer an explanation as to why this patient may have red blood cells expressing low levels of these two particular antigens? From what disease do you think this patient is suffering?

b. Why do you think a single patient can generate red blood cells expressing three different levels of CD59 and CD55?

c. Now look at parts c and d of the same figure, which display the expression of two other markers on red blood cells from the same patient. In these flow cytometric

(a)

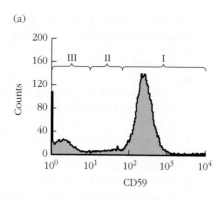

(b)

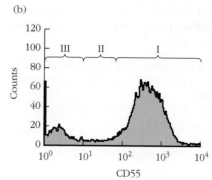

(c)

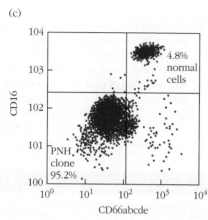

(d)

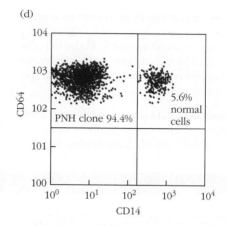

profiles, the investigators have shown you quadrant markers to indicate whether the cells they are investigating are considered to be positive or negative for the proteins labeled on the axes. For example, the cell population high and to the right in part c is considered to express both CD66abce and CD16, whereas the cell population low and to the left is negative for both of these markers.

Using your answer for part (a) as a starting point, what can you deduce about the biochemistry of the membrane proteins CD16, CD66abce, CD46, and CD14?

ANALYZE THE DATA

1. In the figure below, a flow cytometric histogram describes the numbers of human Jurkat T cells expressing low and high levels of CD46 after treatment with an apoptosis-inducing

reagent. On the figure, label the cell population that you think is undergoing apoptosis and explain your reasoning.

2. On the flow cytometric dot plot shown below, indicate the following and explain your labeling:

a. Where you would expect to find a healthy T-cell population?

b. Where you would expect to find a T-cell population undergoing apoptosis?

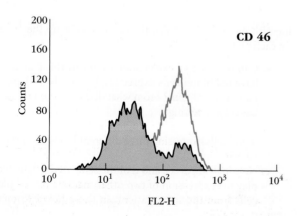

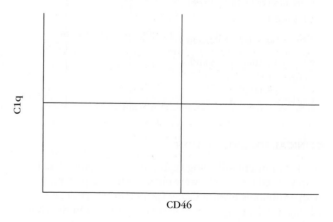

The Organization and Expression of Lymphocyte Receptor Genes

6

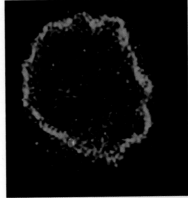
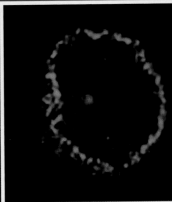

In T cells (*left*) the immunoglobulin heavy-chain locus (red) lies in the repressive environment around the nuclear lamina (green), whereas in developing B cells (*right*), the same genetic locus is placed in a more central and active location within the nuclear matrix. [*Republished with permission of the American Association for the Advancement of Science, from Kosak et al." Subnuclear compartmentalization of immunoglobulin loci during lymphocyte development," Science. 2002 Apr 5;296(5565):158–62, Figure 2. Permission conveyed through Copyright Clearance Center, Inc.*]

Learning Objectives
After reading this chapter, you should be able to:

1. Sketch the arrangement of the genetic elements (the V, D, and J regions) of the B- and T-cell receptors in the germ-line DNA and show how that arrangement is altered in mature B and T cells, respectively.

2. Differentiate between the roles of the RAG1/2 enzyme complex, terminal deoxyribonucleotidyl transferase, and DNA repair complex enzymes in V(D)J recombination.

3. Explain the mechanisms by which V region gene rearrangements are controlled and the role played by chromatin re-organization in the timing of V region gene rearrangement.

4. Show how the expression of membrane-bound and secreted IgM and IgD antibodies is controlled at the level of RNA splicing.

To protect its host, the immune system must recognize a vast array of rapidly evolving microorganisms. It must therefore be capable of generating a diverse and flexible repertoire of receptor molecules capable of recognizing microbial pathogens. At the same time, it must minimize the expression of receptors that bind self-antigens (autoreactive receptors). As described in Chapter 3, each B or T lymphocyte expresses a unique antigen-specific receptor. When these receptors bind to their corresponding antigens under the appropriate conditions (described in Chapters 10 and 11), T and B lymphocytes proliferate and differentiate into effector cells that eliminate the microbial threat (Chapter 12).

In Chapter 3, we described the biochemistry of the T- and B-lymphocyte receptors as well as that of the secreted antibodies elicited from B cells after antigenic stimulation. In this chapter, we address the question of how an organism can make use of a finite amount of genetic information to encode receptors capable of recognizing a constantly evolving and numerically overwhelming universe of potential pathogens. Since the mechanisms that give rise to B- and T-cell receptor diversity are almost identical, this chapter will focus first on the generation of diversity among immunoglobulin genes and then highlight those structures and

Key Terms

B-cell receptor (BCR)
Variable (V) region
Constant (C) region
Immunoglobulin heavy or light chain
Joining (J) gene segment
Diversity (D) gene segment

Kappa (κ) light chains
Lambda (λ) light chains
Recombination signal sequences (RSSs)
Nonamer
Heptamer

RAG1 (recombination activating gene 1)
RAG2 (recombination activating gene 2)
Terminal deoxyribonucleotidyl transferase (TdT)

Coding joints
Signal joints
Palindromic (P) nucleotides
Nontemplated (N) nucleotides
T-cell receptor (TCR)

mechanisms that differ between B- and T-cell receptor generation.

The production of specific lymphocyte receptors employs genetic mechanisms that are unique to the immune system. In 1987, Susumu Tonegawa won the Nobel Prize for Physiology or Medicine "for his discovery of the genetic principle for generation of antibody diversity," a discovery that challenged the fundamental precept that one gene encodes one polypeptide chain. His finding that DNA in developing B and T lymphocytes is cleaved and recombined to assemble intact antibody and T-cell receptor genes, thereby creating a diverse set of antigen receptor genes, was revolutionary. The experiments that led to this conclusion are described later in this chapter.

In addition to describing the organization of T- and B-cell receptor gene segments in germ-line cells, we also address the mechanisms employed to rearrange and express them in mature T and B lymphocytes. However, note that, in this chapter, we address only the mechanisms that shape the receptor repertoires of mouse and human B and T cells *before* antigen exposure, the so-called naïve receptor repertoires. Further receptor diversity is generated in B cells *after* they have been activated by antigen binding. In Chapter 11, we will describe these additional modifications to B-cell receptor genes, which include the generation of antibodies of different heavy-chain classes by the process of class switch recombination (CSR), as well as the enhancement of the affinity of specific antibodies by somatic hypermutation followed by antigen-driven selection. The various roles of antibodies bearing different constant region segments will be described in Chapter 12.

The Puzzle of Immunoglobulin Gene Structure

The immune system relies on a vast array of **B-cell receptors (BCRs)** that possess the ability to bind specifically to a correspondingly large number of potential pathogens. The first indication of the immense size of the antibody repertoire was provided by immunologists using synthetic organic molecules to stimulate antibody production. Changing the position of an amino or a nitro group on a phenyl ring was sufficient to alter the capacity of an antibody to bind. Investigators reasoned that if the immune system can discriminate in this exquisitely specific manner between small molecules that it had presumably never encountered during evolutionary selection, then the number of potential antibodies must be very large indeed. A series of experiments conducted in the late 1970s and early 1980s estimated the number of different BCRs generated in a normal mouse immune system to be at least 10^7; we now know that estimate was many orders of magnitude too small.

Investigators trying to understand immunoglobulin (Ig) genetics were also faced with an additional puzzle: protein sequencing of mouse and human antibody heavy and light chains revealed that approximately the first 110 amino-terminal amino acids of antibody heavy and light chains are extremely variable among different antibody molecules. This region was therefore defined as the **variable (V) region** of the heavy and light chains. In contrast, the remainder of the light and heavy chains could be classified into one of only a few sequences and was therefore named the **constant (C) region** of heavy and light chains (**Figure 6-1a**). So immunologists of this era were now faced with two seemingly intractable problems: First: how can a finite amount of genetic information encode a vast number of different antigen-binding sites? And second: how can an **immunoglobulin heavy or light chain** be identical to many other heavy or light chains in one part of its sequence, but be extraordinarily variable in another? Pertinent to the latter point, geneticists at the time pointed out that, if each antibody chain were separately encoded in the genome, the process of evolutionary drift would result in the accumulation of silent mutations in the constant regions, even in the absence of any particular evolutionary pressures toward diversity.

Further sequencing studies uncovered yet more complexity. Not only were different variable regions found to be associated with the same constant regions, as described above (Figure 6-1b), but additional studies identified antibodies in which the same heavy-chain variable (V_H) region was associated with more than one heavy-chain constant (C_H) region (Figure 6-1c). Together, these findings strongly implied that for each antibody chain, expression of the variable region and expression of the constant regions are independently controlled.

Investigators Proposed Two Early Theoretical Models of Antibody Genetics

The earliest theories of how antibody light- and heavy-chain genes were arranged in the genome were called **germ-line theories**. Germ-line theories suggest, as their name implies, that the genetic information for each antibody is separately encoded within the germ-line (inherited) genome. However, a quick calculation demonstrates that if there are 10^7 or more antibodies, each of which needs approximately 2000 nucleotides to specify both its heavy and light chains, encoding each antibody individually would require 2×10^{10} nucleotides. Since the size of the mouse genome is 2.8×10^9 nucleotides, this is clearly an impossibility. In addition, as more accurate estimates of the size of the antibody repertoire became available, it quickly became clear that 10^7 was a *minimal* estimate of the number of antibodies that could be made by an individual mouse. Immunologists had to face the fact that there simply was not enough DNA to encode all the relevant B-cell receptors, using the classical one gene–one polypeptide chain rule, and that new ideas had to be found.

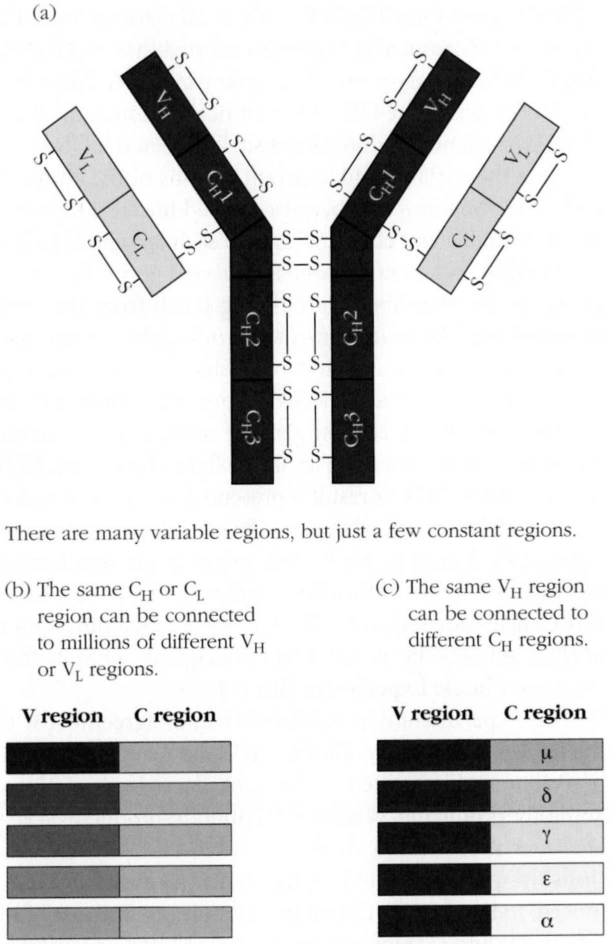

(a)

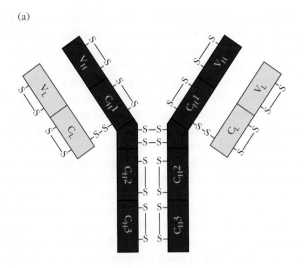

There are many variable regions, but just a few constant regions.

(b) The same C_H or C_L region can be connected to millions of different V_H or V_L regions.

(c) The same V_H region can be connected to different C_H regions.

V region	C region

V region	C region
	μ
	δ
	γ
	ε
	α

FIGURE 6-1 Early sequencing studies indicated that the variable and constant regions of both the immunoglobulin heavy and light chains are encoded separately in the germ-line genome. (a) An IgG antibody molecule consists of two heavy chains and two light chains. Each chain consists of an amino-terminal variable (V) region and a carboxy terminal constant (C) region. (b) For any given constant region there are millions of variable regions that can be associated with it. (c) For any given variable region there are only a few constant regions that can be associated with it. C_H = heavy-chain constant region; C_L = light-chain constant region; V_H = heavy-chain variable region; V_L = light-chain variable region.

(a) Germ-line DNA

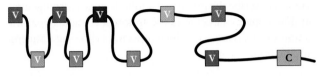

(b) B-cell DNA

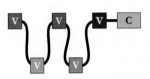

FIGURE 6-2 Dreyer and Bennett hypothesis. (a) Dreyer and Bennett suggested that one constant region (C) gene existed in the germ-line genome along with many different variable (V) region genes. (b) In antibody-producing cells (B cells), they hypothesized that a mechanism existed to bring together one of the variable region genes and the constant region gene. Different B cells would make different variable region genes contiguous with the constant region gene in order to generate a diverse repertoire of B-cell receptors and secreted antibodies.

In 1965, William Dreyer and J. Claude Bennett proposed that antibody heavy and light chains are each encoded in two separate segments in the germ-line genome (**Figure 6-2a**). To specify a complete heavy or light chain polypeptide, one of each of the V region– and C region–encoding segments must then be brought together in B-cell DNA to form one complete polypeptide-encoding gene (Figure 6-2b). The idea that DNA in somatic cells might engage in recombination was revolutionary—previously, only cells in meiosis had been shown to undergo recombination. However, Dreyer and Bennett's theory fitted the data so elegantly that germ-line theorists began to modify their ideas to embrace the idea that two genes might be necessary to encode one antibody polypeptide chain.

In the early 1970s, others suggested the innovative **somatic hypermutation theory** to account for the size of the mature antibody repertoire. The term "somatic" refers to all cells of the body with the exception of the germ-line sperm and egg cells, and therefore the somatic hypermutation theory suggested that a mutational process occurred *only in B cells* to alter the antibodies in an individual animal and thus increase the total number of different antibodies available to it. Since these mutations are not affecting germ-line genes, such mutations would not be passed on to offspring. According to the somatic mutation hypothesis, therefore, a limited number of germ-line antibody genes is acted on by unknown mutational mechanisms to generate a diverse receptor repertoire, only in mature B lymphocytes. This theory had the advantage of explaining how a large repertoire of antibodies could be generated from a relatively small number of genes, but the disadvantage that such a process had never before been observed.

Heated debates continued between the proponents of modified germ-line versus somatic mutation theories throughout the early 1970s, until a seminal set of experiments revealed that both sides were correct. We now know that multiple germ-line gene segments each encode a part of the antibody variable regions, and that these segments are rearranged differently in the formation of each naïve B cell to produce an extremely diverse primary receptor repertoire. In humans and mice, these rearranged genes are then further acted on *after antigen encounter* by somatic hypermutation and antigenic selection, resulting in an expanded and exquisitely honed population of antigen-specific B cells. We will describe the process of antibody gene rearrangement in this chapter, but delay a description of somatic hypermutation until Chapter 11, since it occurs following antigenic stimulation.

Key Concepts:

- The number of different antibody molecules that can be made by a single individual mouse or human is vast: approximately 10^{13-14}. Classical germ-line theories were insufficient to account for the immense size of the antibody variable gene repertoire.

- Dreyer and Bennett proposed that two genes, instead of one gene, might be required to encode antibody heavy and light chains, in order to explain the difference in sequence diversity between the variable and constant region genes.

- Immunogeneticists further suggested that somatic hypermutation may enhance the size of the antibody repertoire after antigenic stimulation.

Breakthrough Experiments Revealed That Multiple Gene Segments Encode the Immunoglobulin Light Chain

From the mid-1970s until the mid-1980s, a small group of brilliant immunologists completed a series of experiments that fundamentally altered the way in which scientists think about genetics. The first breakthrough occurred when Nobumichi Hozumi and Susumu Tonegawa showed that, as Dreyer and Bennett had predicted, multiple gene segments encode the antibody protein chains.

Tonegawa and his colleagues showed that the variable and constant regions of the antibody light-chain gene were encoded in segments located in two distinct places in the germ line. These segments were then brought together by DNA recombination only in mature, antibody-producing cells.

The variable (V) and constant (C) region gene segment families are located kilobases apart in the germ-line DNA. The work of Tonegawa and colleagues demonstrated that these two DNA segments, one containing coding information for the kappa light-chain variable (V_κ) region and the other containing coding information for the C_κ region gene segment families, are stitched together, only in B lymphocytes, to create the complete κ light-chain gene. Each B cell contains only one functional combination of κ gene segments, and different V_κ gene segments are used in each B cell.

How were these ground-breaking discoveries made? **Figure 6-3** outlines the experiment. They made use of the ability of the *BamH1* restriction endonuclease to cut DNA at a precise nucleotide sequence. (Each restriction

endonuclease cuts DNA at a characteristic nucleotide sequence.) Hozumi and Tonegawa showed that in the DNA from non–antibody-producing, embryonic liver cells (used in place of germ-line DNA for technical reasons), there is a *Bam*HI restriction endonuclease site between the DNA that encodes the variable and constant regions of the antibody molecule. We know this because labeled nucleotide probes for the variable and constant regions each recognized a different DNA fragment in *Bam*HI-digested germ-line DNA. However, in an antibody-producing B cell from the same mouse strain, the gene segments encoding the variable and constant regions appeared to be combined into a single fragment (later confirmed by DNA sequencing), thus demonstrating that DNA rearrangement must have occurred during the formation of an antibody light-chain gene. Note that in Figure 6-3 their result is presented as a Southern blot, in which DNA fragments cut by the enzyme are run out on a gel, blotted onto nitrocellulose paper to prevent further nucleotide fragment diffusion, and probed with a labeled short nucleotide sequence. For a more detailed description of their actual experiment, and the original methods they used, see **Classic Experiment Box 6-1**.

This experiment demonstrated that, in agreement with the Dreyer-Bennett hypothesis, the V and C regions of antibody genes were located in different places in the DNA of antibody-producing, versus non–antibody-producing, cells. However, it yielded no information about the physical relationship of the V and C segments in the animal's DNA; indeed, the initial experiment did not rule out the possibility that the V and C fragments could be encoded on different chromosomes in the embryonic cells. Subsequent sequencing experiments showed that the segments encoding the V and C segments of the κ light chains are on the same chromosome and that, in non-B cells, the V and C segments are separated by a long noncoding DNA sequence.

The impact of this result on the biological community was profound. *For the first time, mammalian DNA was shown to be cut and recombined during the process of cell differentiation.* Furthermore, this finding paved the way for the next surprise.

Scientists in Tonegawa's group sequenced the germ-line DNA segments that hybridized with their probes and analyzed the sequences for open reading frames that encode protein sequences. They found that the (unrearranged) embryonic liver V region segment had a short, protein-coding leader sequence at its 5′ terminus. This is a common feature of membrane proteins; amino-terminal leader sequences (also known as *signal peptides*) guide the nascent polypeptide chain to the endoplasmic reticulum membrane during translation (**Figure 6-4**). A 93-base pair (bp) sequence of noncoding DNA separated the leader sequence from a long stretch of DNA that encoded the first 97 amino acids of the V region. But the light-chain V region domain is approximately 110 amino acids long. Where was the coding information for the remaining 13 amino acids of the light-chain variable region?

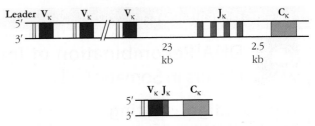

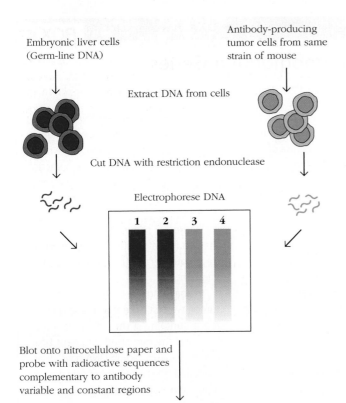

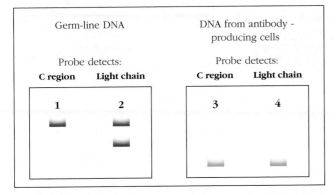

FIGURE 6-3 The κ light-chain gene is formed by DNA recombination between variable and constant region gene segments. DNA from embryonic liver cells was used as a source of germ-line type DNA, and DNA from a B-cell tumor cell line was used as an example of DNA from antibody-producing cells. Each DNA sample was digested with the *Bam*HI restriction endonuclease. The DNA fragments were separated by gel electrophoresis, and the location of fragments containing V and C regions were detected with specific probes for the entire κ light-chain (V and C region) and for the C region alone. In embryonic liver, the DNA sequences encoding the V and C regions, respectively, were located on different restriction endonuclease fragments. In contrast, these two sequences were colocated on a single restriction fragment in the myeloma DNA.

Sequencing of the light-chain constant region fragment from embryonic DNA provided the answer to this question. Upstream (toward the 5′ end) from the constant region coding sequence, and separated from it by a noncoding DNA

FIGURE 6-4 The antibody κ light-chain locus is composed of three families of DNA segments. In germ-line DNA a large family of V_κ segments is found upstream of a smaller group of J_κ segments. A single C_κ segment is located 2.5 kb downstream from the 3′-most J_κ segment. During B-cell development, DNA recombination events join one V_κ segment and one J_κ segment in each B cell.

segment of 2.5 kilobases (kb), were the 39 bp encoding the remaining 13 amino acids of the V region. This additional light-chain coding segment was named the **joining (J) gene segment** (shown in red in Figure 6-4), as it was the site of joining of the V-gene segment with the fragment containing the constant region during rearrangement.

Further sequencing of mouse and human light-chain variable and constant region genes confirmed this second, astonishing finding from Tonegawa's group. Not only had they proved that the variable and constant regions of antibody light chains are encoded by separate DNA segments, as Dreyer and Bennett had predicted; they had also shown, completely unexpectedly, that *the light-chain variable regions themselves are encoded by two separate gene segments, the V and J segments*, and that they are made contiguous only after the gene rearrangements that occur during the development of B cells. Analogous experiments performed later demonstrated that the lambda light-chain gene segments were similarly, although not identically, arranged (see below).

If the kappa light-chain variable region is encoded in two segments that are recombined in each antibody-producing cell to create a new combination of sequences, what about the heavy chain? By cloning and sequencing immunoglobulin (Ig) heavy-chain genes, Lee Hood's group showed that the heavy-chain variable region was encoded in not two, but *three separate gene segments* in the germ line, and that these segments were recombined during B-cell development to create the contiguous coding information for the heavy-chain variable region (**Figure 6-5**).

Specifically, the heavy-chain variable region was found to be encoded by a germ-line **heavy-chain variable (V_H) region** gene fragment that encodes amino acid residues 1–101 of the antibody heavy chain and a second fragment that included a **heavy-chain joining (J_H) region** gene segment that determined the sequence of amino acid residues 107–123. The DNA sequence necessary to encode residues 102–106 (approximately) of the heavy chain was discovered last. It was located 5′ of the J region in mouse embryonic DNA (see Figure 6-5). One of the features of the D region is

BOX 6-1

DNA Recombination of Immunoglobulin Genes Occurs in Somatic Cells

The paradigm-shifting experiment of Hozumi and Tonegawa was designed to determine whether the DNA that encodes immunoglobulin light-chain constant and variable regions existed in separate segments in non–antibody-producing cells, and are brought together in antibody-producing cells. Students are accustomed to thinking about the concept of alternative RNA splicing; however, this experiment addressed a radically different (and somewhat unnerving) concept. *Can a piece of DNA change its place on a chromosome in a somatic cell?*

To serve as their source of "germ-line DNA," Hozumi and Tonegawa used DNA from an organ in which immunoglobulin genes are typically not expressed, in this case, embryonic liver. (Sperm and egg DNA would have been much more difficult to obtain.) For "B-cell DNA," they used DNA from an antibody-producing tumor cell line, MOPC 321, which secretes fully functional κ light chains. They separately cut both sets of DNA with the same restriction enzyme and used radioactive probes to determine the sizes of the DNA fragments on which the variable and constant regions of the light-chain gene were found in the two sets of cells.

Below, we describe how Hozumi and Tonegawa conducted their experiment, indicating how the experimental protocol would be modified today using the reagents and techniques available to a modern molecular biologist.

Hozumi and Tonegawa:

a. Purified genomic DNA from embryonic liver cells and from the MOPC 321 tumor cell line.

b. Cut the two sets of genomic DNA with a restriction endonuclease (*Bam*HI) *that they had purified themselves* and separated the DNA fragments by electrophoresis. (Nowadays, we would use a polyacrylamide gel and electrophorese submicrogram quantities of samples over the course of a few hours; they used a *foot-long* agarose gel that needed 2 L of agarose. They loaded *5 mg* of DNA and ran the gel for *3 days*.)

c. Made ^{125}I-labeled mRNA probes specific for the two regions of mouse κ light chain. One of Tonegawa's probes was a full-length radiolabeled piece of mRNA encoding the entire κ-chain sequence. The other probe was designed to detect only the 3' half of the sequence, which would hybridize to the constant region— not to the variable region—of the κ-chain gene. (Nowadays, making enzymatically labeled, stable DNA probes for any particular sequence is a safe and relatively straightforward task, thanks to the advent of PCR. Hozumi and Tonegawa's task was significantly more challenging. PCR had not yet been invented, and long mRNA segments are notoriously unstable.)

d. Probed the nuclease-generated DNA fragments to determine the size of the fragments carrying the variable and constant region sequences. We would normally use a Southern blot procedure, running the fragments out on a gel, blotting the gel with nitrocellulose paper, and then probing the paper with enzyme-labeled fragments complementary to the sequences of interest. We would then develop the blots with substrates whose digestion yields fluorescent or luminescent products. Hozumi and Tonegawa's approach was much more time-consuming. They cut the gel into about 30 slices, melted the agarose, and separately eluted the DNA from each slice. To each DNA sample, they added radiolabeled RNA, allowed it to anneal, and removed the unannealed RNA using RNase. The radioactivity remaining in each fraction was then plotted against the size of the DNA in the slice (**Figure 1**).

Figure 6-3 shows the type of results that would be obtained from a modern-day Southern blot of Tonegawa's fragments.

In analyzing the blots from the B-cell tumor, we first note that the large fragment containing the constant region DNA and the mid-sized fragment bearing the variable region DNA have disappeared. Both the variable and constant region gene segments are now located on smaller fragments. This implies that a new pair of restriction endonuclease sites now forms the boundaries of each of the constant and variable regions. *Right away, we can tell that the DNA sequence environment around the light-chain genes changes as the B cell differentiates.*

Next, we note that the sizes of the DNA fragments on which the constant and variable region gene segments are located are apparently the same. This implies, although it does not yet prove, that the movement of the constant and/ or variable region gene segments has brought them into close proximity with one another, such that the constant and variable region gene segments colocate on the same fragment. (An alternative explanation is that they have each altered their locations, but in different ways, and the similarity in the size of the fragments is coincidental.) DNA sequencing experiments supported the first interpretation: *as the B cell differentiates, the variable and constant region gene segments are moved from distant regions of the chromosome into close apposition with one another.*

It is all too easy, with modern molecular biology technologies, to forget what a *tour de force* this experiment actually

(continued)

represented. Few of the reagents or pieces of apparatus we commonly encounter in the molecular biology laboratory today were available to Hozumi and Tonegawa—they had to make their own. It was an extraordinary piece of work and for it, Susumu Tonegawa was awarded the Nobel Prize in Physiology or Medicine in 1987.

What happened to the DNA on the alternative allele that did not encode the tumor cell secreted light chain? Subsequent analysis of the DNA from this tumor showed that the DNA from both chromosomes had undergone rearrangement. Hozumi and Tonegawa were fortunate that the variable regions used by both rearrangements were close by one another and so the fragment patterns overlapped, otherwise the analysis of this experiment might have taken years longer to decode the mysteries of antibody gene recombination.

REFERENCE

Hozumi, N., and S. Tonegawa. 1976. Evidence for somatic rearrangement of immunoglobulin genes coding for variable and constant regions. *Proceedings of the National Academy of Sciences USA.* **73**:3628.

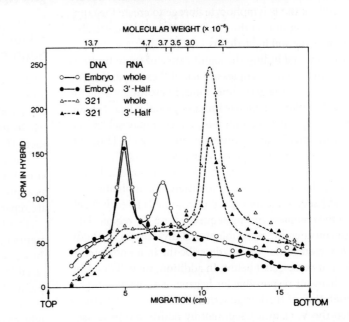

FIGURE 1 Original data from Hozumi and Tonegawa's classic experiment proving immunoglobulin gene recombination occurs in B cells. DNA from the sources shown was run out on an agarose gel for 3 days at 4°C. The gel was sliced, the DNA eluted, and then each sample was hybridized with radiolabeled probes for either the constant region of the κ light chain, or the whole chain. The plot shows the amount of radioactivity in each fraction as a function of the sample number, which reflects the molecular weight of the DNA fragment. See text for details. [*From Hozumi, N., and S. Tonegawa. 1976. Evidence for somatic rearrangement of immunoglobulin genes coding for variable and constant regions.* Proceedings of the National Academy of Sciences USA. **73**:3628.]

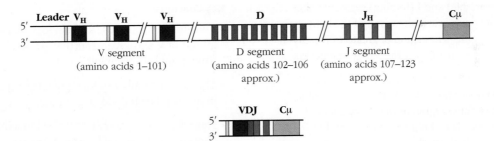

FIGURE 6-5 Variable region of antibody heavy chains is encoded in three segments—V, D, and J. During B-cell development, DNA recombination events bring together three DNA segments—a V$_H$, a D, and a J$_H$ segment—to form the variable region of the Ig heavy chain.

that it is quite variable in its length. The importance of the contribution made by this sequence, the D gene segment, to the diversity of antibody specificities is denoted by its name: the **diversity (D) region**. (Because there is no D region in the light chain, immunologists usually drop the subscript denoting the heavy chain.)

Significantly, DNA in the D gene segment encodes most of the amino acid residues shown to be part of the third complementarity-determining region of the antibody heavy chain, **CDR3** (see Figure 3-11a). This region provides amino acid residues that contact antigen for the binding of most antigens. In the light chain, the CDR3 portion of the antibody sequence is located at the V-J junction.

Thus, the variable region of the heavy chain of the antibody molecule is encoded by three discrete gene segments and the variable region of the light chain by two segments.

These segments in the germ-line genome are brought together by a process of DNA recombination that occurs only in cells of the B-lymphocyte lineage to create the complete and contiguous coding information for the variable regions of heavy and light chains.

Key to understanding the significance of these findings is the fact that multiple copies of each of these separate gene segments exist in the germ line. Thus, in creating a gene for the variable region of a mouse immunoglobulin heavy chain, a lymphoid-specific recombinase enzyme may select one of 100 different V_H gene segments; one of 12–14 D region gene segments, and one of four J_H gene segments. These segments are selected at random by the recombinase as B cells develop in the bone marrow, and different combinations of V, D, and J segments are ligated (i.e., joined) in each B cell to form the complete coding sequence for a V_H region. This leads to considerable recombinatorial diversity in antibody variable region genes. We will describe the immunoglobulin gene loci in more detail below. In addition, we will describe the mechanisms responsible for Ig gene rearrangements and the generation of further genetic diversity at the junctions between the V, D, and J segment by additional processes unique to the immune system.

Key Concepts:

- Tonegawa's key experiment showed that two gene segments encoding V and C light-chain regions, which are separate in the germ line, are brought together by recombination at the level of DNA in B cells, to form an intact light-chain gene. Light-chain V region diversity is achieved in part by combinatorial association of multiple gene segments termed V (variable) and J (joining) segments, that together encode the variable region of the light chain of the Ig molecule. The region of greatest diversity in the light chain is encoded at the V-J gene junction.

- Recombination between three gene segments—the V_H, D (diversity), and J_H segments—is required to encode the variable region of the heavy chain of the Ig molecule. The D region and V-D and D-J junctions encode the regions of greatest sequence diversity in the antibody heavy chain, corresponding to CDR3 regions of the antibody heavy chains.

Multigene Organization of Immunoglobulin Genes

Recall that Ig proteins consist of two identical heavy chains and two identical light chains (see Chapter 3). The light chains can be either **kappa (κ) light chains** or **lambda (λ) light chains**. The heavy-chain, κ, and λ gene families are each encoded on separate chromosomes.

The relative locations (or *loci*; singular, *locus*) of the various gene segments are illustrated in **Figure 6-6a**.

κ Light-Chain Genes Include V, J, and C Segments

The mouse immunoglobulin κ light-chain locus ($Ig_κ$) spans approximately 3.2 megabases (Mb) and, depending on the mouse strain, includes 120–140 $V_κ$ genes of which about 94–96 are functionally capable of encoding a $V_κ$ protein segment (Figure 6-6b). Note that the precise numbers of gene segments vary according to the strain in mice and the individual in humans, and so these numbers must be regarded as approximations. Individual $V_κ$ segments in both species are separated by noncoding gaps of 5 to 100 kb.

The transcriptional orientation of particular $V_κ$ segments may be in the same or in the opposite direction as the constant region segment. The relative orientations of the variable and constant region segments do not affect the frequency of use of the segments, but do alter some details of the recombinational mechanism that creates the complete light-chain gene, as will be discussed later.

Downstream of the mouse $V_κ$ region cluster are four functional $J_κ$ segments and one pseudo-$J_κ$ segment gene that contains a stop codon and therefore cannot be used. A similar arrangement is found in the human $V_κ$ locus, although the number of functional $V_κ$ gene segments in humans, 41, is a little lower than has been found in mice (**Table 6-1**). A single $C_κ$ segment is found downstream of the J region, and all κ light-chain constant regions are encoded by this segment.

Key Concepts:

- The human and mouse κ-chain loci are arranged in groups of V and J segments that are located upstream from a single $C_κ$ segment.

- The precise number of $V_κ$ segments varies by strain in mice and by individual in humans.

- Some $V_κ$ sequences are transcribed in a direction opposite to that of the $C_κ$ gene segment.

λ Light-Chain Genes Include Paired J and C Segments

The mouse immunoglobulin λ light-chain locus ($Ig_λ$) spans a much smaller region of approximately 240 kb (note the scale in Figure 6-6b). λ light chains are found in only 5% of mouse antibodies because of a deletion event in the mouse genome that has eliminated most of the λ variable gene segments. It was therefore not surprising to discover that there are only three fully functional $V_λ$ gene segments in laboratory mice and just a few more in wild mice. Surprisingly,

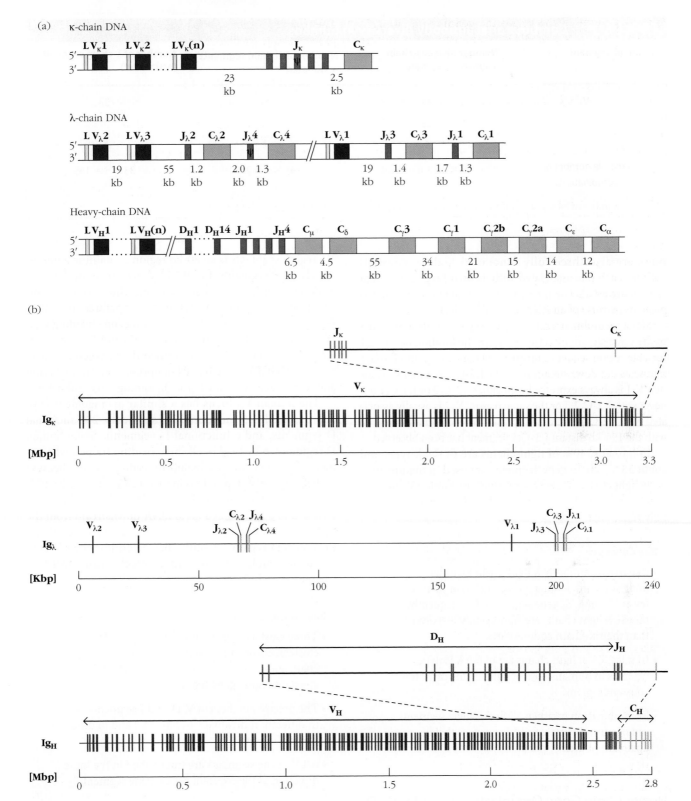

FIGURE 6-6 Organization of immunoglobulin germ-line gene segments in the mouse. (a) The κ and λ light chains are encoded by V, J, and C gene segments. The heavy chain is encoded by V, D, J, and C gene segments. (b) The κ, λ and H-chain gene segments in the mouse genome are illustrated in such a way as to indicate both the large numbers of V gene segments and the overall size of each of the V gene loci. Note the scale bars that describe the overall size of each locus. [*Data from Jhunjhunwala, S., M. C. van Zelm,*

*M. M. Peak, and C. Murre. 2009. Chromatin architecture and the generation of antigen receptor diversity. Cell **138***:435; Martinez-Jean, C., G. Folch, and M. -P. LeFranc. 2001. Nomenclature and overview of the mouse (Mus musculus and Mus sp.) immunoglobulin kappa (IGK) genes. Experimental and Clinical Immunogenetics **18***:255; and Lefranc, M.-P. and G. Lefranc. Immunoglobulin lambda (IGL) genes of human and mouse, in: Molecular Biology of B cells (Honjo, T., F. W. Alt, and M. S. Neuberger, eds). Academic Press, Elsevier Science, pp. 37–59 (2004).]*

TABLE 6-1   Combinatorial antibody diversity in humans

Nature of segment	Number of heavy-chain segments (estimated)	Number of κ-chain segments (estimated)	Number of λ-chain segments (estimated)
V	45	41	33
D	23		
J	6	5	5
Possible number of combinations	45 × 23 × 6 = 6210	41 × 5 = 205	33 × 5 = 165
Possible number of heavy- and light-chain combinations in the human = 6210 × (205 + 165) = 2.3 × 10⁶			

there are also three fully functional λ-chain constant regions, each one associated with its own J region segment (see Figure 6-6a). The $J_\lambda 4$-$C_\lambda 4$ segments are not expressed as proteins because of an RNA splice site defect.

Since recombination of Ig gene segments always occurs in the downstream direction (V to J), the location of the $V_\lambda 1$ variable region sequence upstream of the $J_\lambda 3$-$C_\lambda 3$ and $J_\lambda 1$-$C_\lambda 1$ segments but downstream from the $J_\lambda 2$-$C_\lambda 2$ segments means that $V_\lambda 1$ is always expressed with either $J_\lambda 3$-$C_\lambda 3$ or $J_\lambda 1$-$C_\lambda 1$ but never with $J_\lambda 2$-$C_\lambda 2$. For the same reason, $V_\lambda 2$ is usually associated with $J_\lambda 2$-$C_\lambda 2$, although occasional recombination of $V_\lambda 2$ with the 190-kb distant $J_\lambda 1$-$C_\lambda 1$ segment has been observed.

In humans, 40% of light chains are of the λ type, and about 33 V_λ-chain gene segments are used in mature antibody light chains. Downstream from the human V_λ locus is a series of seven J_λ-C_λ pairs, of which four or five pairs are fully functional.

Key Concepts:

• The mouse antibody λ light-chain locus has undergone a deletional event, such that there are fewer V_λ than V_κ gene segments. Consequently, mouse λ light chains are significantly less diverse than their κ-chain counterparts.

• In mice 5% of light chains are of the λ isotype, whereas in humans 40% of light chains carry the λ constant region.

• Recombination between V_λ and J_λ gene segments can occur only between V_λ sequences that lie upstream of their J_λ partners.

Heavy-Chain Gene Organization Includes V_H, D, J_H, and C_H Segments

Multiple V_H gene segments are located across a region of approximately 2.4 Mb in mice (see Figure 6-6b). Mice express approximately 100 different V_H segments, all of which are arranged in the same transcriptional orientation. At a distance 100 kb downstream from the cluster of mouse V_H region segments is an 80-kb region containing approximately 14 D segments. Just 0.7 kb downstream of the most 3′ D segment is the J_H region cluster, which contains four functional J_H regions. A further gap separates the last J_H segment from the first constant region exon encoding $C_\mu 1$. Within the 100-kb gap between the V_H and D_H regions is a critical regulatory element termed the *intergenic control region* 1 (IGCR1). The IGCR1 suppresses V_H transcription and rearrangement at the D_H-to-J_H joining stage (see below).

The human V_H locus has a similar arrangement with approximately 45 functional V_H segments, 23 functional D segments, and 6 functional J_H segments. Some human D regions can be read in all three reading frames, whereas mouse D regions are read mainly in reading frame 1, because of the presence of stop codons in their reading frames 2 and 3.

The eight constant regions of antibody heavy chains are encoded in a span of 200 kb of DNA downstream from the J_H locus in the order μ, δ, γ3, γ1, γ2b, γ2a, ε, α. Recall that the constant regions of antibodies determine their heavy-chain class and, ultimately, their effector functions (see Chapters 3 and 12).

Key Concepts:

• Three clusters of gene segments, the V_H, D, and J_H clusters, encode the variable region of the heavy chain and are located upstream of a set of eight constant region gene segments.

• The precise numbers of V, D, and J segments vary in different individuals (in humans) and in different strains (in mice).

• All V_H gene segments are transcribed in the same direction as the constant region gene segment.

The Antibody Genes Found in Mature B Cells Are the Product of DNA Recombination

By now, it should be clear that the puzzle of the enormous diversity of the antibody repertoire is at least partially solved by the ability of developing B cells to recombine the V, D, and J gene segments that encode the heavy and light-chain variable regions in many different combinations in developing B cells. Specifically, each B cell uses one V_κ or one V_λ coupled with one J_κ or one J_λ, respectively, to create a single light-chain variable region gene. It also recombines one V_H, one D, and one J_H segment to form a heavy-chain variable region gene. As we will see below, the large number of different combinations of V and J segments for light chains and of V, D, and J segments for heavy chains provides a major source of the diversity in variable regions that determines antigen-binding specificity. Clues about some of the additional mechanisms of generating diversity in antigen-binding regions of antibodies came directly from studies of the mechanism of V(D)J recombination.

Key Concept:

- Recombination among the various V region gene segments generates a diverse repertoire of antibody combining sites.

The Mechanism of V(D)J Recombination

In V(D)J recombination, the DNA encoding a complete antibody V region is assembled from V, D, and J (heavy chain) or from V and J (light chain) segments that are initially separated by many kilobases of DNA. Each developing B cell generates a novel pair of heavy- and light-chain variable region coding sequences by recombination of its genomic DNA. This recombinational event is catalyzed by a set of enzymes (**Table 6-2**), many of which are also involved in *nonhomologous end-joining* (NHEJ) DNA repair functions that occur in all cells. Because V(D)J recombination entails the cutting of DNA at both strands, and because inappropriate recombination is potentially catastrophic for the cell, mechanisms have evolved that restrict antigen receptor gene recombination events to the appropriate sites on the Ig genes and to ensure that they occur only during defined periods of B and T cell development.

TABLE 6-2	Proteins involved in V(D)J recombination	
Protein	**Function in V(D)J recombination**	**Immunological consequences of protein deficiency**
Lymphoid-Specific Proteins		
RAG1/2	Antigen receptor gene recombinase complex. DNA cleavage is mediated by RAG1. Epigenetic targeting is directed by RAG2.	Severe combined immuno-deficiency (SCID)
Terminal deoxyribo-nucleotidyl transfer-ase (TdT)	Adds nontemplated (N) nucleotides to V-D and D-J joints of Ig heavy chain and all joints of TCR chains in a template-independent manner.	Reduced N-nucleotide addition is seen at coding joints
Non-Lymphoid-Specific Proteins		
High mobility group B proteins 1 and 2 (HMGB1/2)	Stabilize binding of RAG1/2 to recombination signal sequences (RSSs). Stabilize introduction of bend into 23 RSS DNA by RAG1/2.	No information available
Non–Lymphoid-Specific Proteins of the Nonhomologous End-Joining (NHEJ) DNA Repair Pathway		
Ku70/80	Complex is recruited to DNA double-strand (DS) breaks. Stabilizes and aligns DNA ends prior to repair. Essential for both signal and coding joint repair in Ig and TCR genes. Recruits DNA-PKcs protein.	SCID occurs in the absence of either or both Ku proteins. Knockout mice are also small in size and sterile
DNA-PKcs	A protein kinase that forms a complex with Ku70/80. It phosphorylates and actives Artemis. It recruits the ligation machinery.	SCID occurs. Knockout mice otherwise develop normally
Artemis	Once Artemis has been phosphorylated by DNA-PKcs, it opens the hairpin on the coding end joint.	Artemis-deficient B and T cells have blocked formation of coding joints and accumulation of hairpin-sealed coding ends. Mice lacking Artemis have severely impaired B- and T-cell development
DNA ligation complex: DNA ligase IV, XRCC4, and XLF (Cernunnos)	XRCC4 maintains stability of DNA ligase IV and stimulates its catalytic activity. XRCC4 may also help to align DNA ends. DNA ligase IV is required for ligation of cut DNA ends, at both the coding and the signal joints.	Lack of DNA ligase IV or of XRCC4 causes a complete block in lymphoid development (SCID). Mice lacking XLF show enhanced sensitivity to radiation-induced DNA damage, but do not develop significant immunodeficiency

(continued)

TABLE 6-2	Proteins involved in V(D)J recombination *(continued)*	
Protein	**Function in V(D)J recombination**	**Immunological consequences of protein deficiency**
Unusual DNA polymerases, such as DNA Pol μ and DNA Pol λ	These polymerases add nucleotides at Ig heavy chain and TCR antigen receptor loci. In the absence of TdT activity, some N-nucleotide addition is observed, which is thought to be the result mainly of DNA Pol μ action. Whereas DNA Pol λ requires a template strand, DNA Pol μ, like TdT, can act in a template-independent manner. Pol μ participates primarily in heavy-chain and Pol λ in light-chain rearrangements.	Defects in the development of hematopoietic cells
Ataxia telangiectasia mutated (ATM) protein	A kinase that binds double-stranded breaks in DNA and blocks entry into the cell cycle until the breaks can be repaired. The ATM protein is recruited to the double-strand breaks by the MRN (Mre11/Rad50/Nbs1) complex.	Lymphopenia and predisposition to thymic lymphomas characterized by translocations involving TCR genes

This degree of accuracy is accomplished in part by the fact that the recombination enzymes recognize specific DNA sequence motifs called **recombination signal sequences (RSSs)**. These sequences also ensure that one of each type of segment (V and J for the light chain, or V, D, and J for the heavy chain) is included in the recombined heavy- and light-chain genes. During cleavage and ligation of the segments, the DNA is edited in various ways, adding further variability to the recombined gene. As you will see later in this chapter, similar, although not identical, mechanisms operate to generate complete T-cell receptor genes in developing thymocytes. We will also discuss the roles of epigenetic modifications and chromatin structure in restricting recombination to the relevant regions of antigen receptor genes.

V(D)J Recombination in Lymphocytes Is a Highly Regulated Sequential Process

Different immunoglobulin variable region gene segments are recombined at specific stages in lymphoid development (see Chapter 9, Figure 9-4). As B cells develop in the bone marrow, the first step in the creation of a mature immunoglobulin receptor is recombination that brings together a D and a J_H gene segment. This step occurs at a very early stage of development, termed the *pre-pro-B cell to the early pro-B-cell stage*, while the cell is still in the bone marrow and just beginning its journey of differentiation into a mature B cell (see Chapter 9). Recombination between the V_H and D-J_H segments follows during the pro-B-cell stage.

If V(D)J recombination is successful and a heavy-chain variable region is generated, the resulting heavy-chain protein is placed onto the cell surface in combination with a nonvariable pair of proteins, VpreB and λ5 (called a *surrogate* light chain), to form a pre-B-cell receptor (**Figure 6-7a**). Signaling from the pre-B-cell receptor halts heavy-chain recombination, initiates several rounds of proliferation, and then calls for the beginning of light-chain recombination. Light-chain recombination occurs at the small pre-B-cell stage of B-cell development. Light-chain recombination in the mouse initiates at the κ locus, and if this is not successful,

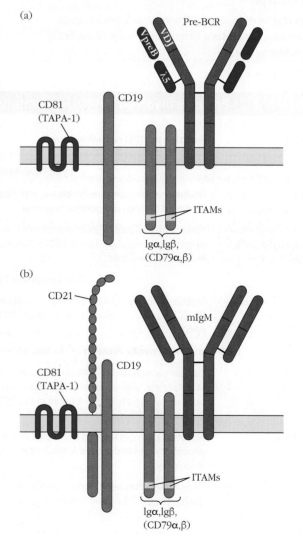

FIGURE 6-7   Pre-BCR and BCR complexes. (a) The pre-BCR: the μ heavy chain is expressed on the cell surface prior to light-chain rearrangement, in concert with the surrogate light chain, made up of VpreB and λ5, and the signaling components CD19 and Igα and Igβ (Igα, Igβ). The pre-BCR complex signals via Igα, Igβ that heavy-chain rearrangement is completed and initiates several rounds of proliferation, after which light-chain rearrangement begins. (b) The mature BCR complex is shown here for comparison.

continues at the λ locus. In each case, recombination occurs on one allele at a time. In humans, light-chain recombination may start at either the κ or the λ locus. Expression of an intact membrane IgM B-cell receptor (Figure 6-7b) shuts off further light-chain gene rearrangement.

Recombination Is Directed by Recombination Signal Sequences

In the late 1970s, investigators sequencing light-chain genes first described two blocks of conserved sequences—a **nonamer** (a set of 9 bp) and a **heptamer** (a set of 7 bp)—that are highly conserved and occur in the noncoding regions upstream of each J segment. The heptamer appeared to end *exactly* at the J region coding sequence. Further sequencing showed that the same motif was repeated in an inverted

manner on the downstream side of the V region coding sequences, again with the heptamer sequence ending flush with the V region gene segment (**Figure 6-8a**).

Between the nonamer and heptamer sequences, the researchers described a spacer sequence of either 12 or 23 bp in length. Although the nucleotide sequence of the RSS spacer is not well conserved, the significance of the spacer *lengths* was immediately clear; 12 nucleotides is about the length of one turn of the double helix, and 23 nucleotides is about two turns. In this way, the spacer sequence ensures that the ends of the nonamer and heptamer closest to the spacers would be on the same side of the double helix and therefore accessible to binding by the same enzyme. The investigators correctly concluded that they had discovered the DNA signal sequence that directs recombination between the V and J gene segments. They therefore termed this "heptamer-spacer-nonamer" motif the recombination signal sequence or RSS.

Nucleotide sequencing of the RSS demonstrated that it consists of three elements:

- An absolutely conserved, 7-bp (heptamer) consensus sequence: 5′-CACAGTG-3′
- A spacer of either 12 or 23 bp in which the sequence is not well conserved
- A second conserved, 9-bp (nonamer) consensus sequence: 5′-ACAAAAACC-3′

In the heavy-chain gene segments, a similar pattern was noted. The spacer regions separating the heptamer and nonamer pairs were 23 bp in length following the V segments and preceding the J segments, and 12 bp in length before and after the D segments. The relative locations of the 12- and 23-bp spacers (Figure 6-8b) suggested that the VDJ

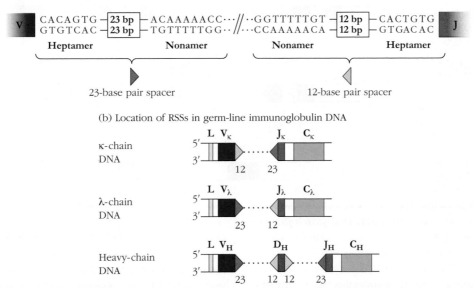

FIGURE 6-8 Two conserved sequences in light-chain and heavy-chain DNA function as recombination signal sequences (RSSs). (a) Both signal sequences consist of a conserved heptamer and conserved AT-rich nonamer separated by nonconserved spacers of 12 or 23 bp. (b) The two types of RSS have characteristic locations within λ-chain, κ-chain, and heavy-chain germ-line DNA. During DNA rearrangement of the Ig heavy and light chains, gene segments adjacent to the 12-bp RSS can join only with segments adjacent to the 23-bp RSS.

recombinase enzyme is designed to pair one RSS bearing a 12-bp spacer with a second RSS that includes a 23-bp spacer, something we now know to be the case. This is referred to as the **12/23 rule**.

Figures 6-9a and b illustrate the manner in which the RSSs act to bring together the appropriate gene segments during the generation of complete light-chain and heavy-chain variable region genes.

Key Concepts:

- Recombination signal sequences (RSSs) contiguous with each of the variable region gene segments serve to guide the recombination machinery to the correct locations in the genome.

- RSSs consist of conserved heptameric and nonameric sequences separated by either 12 or 23 bp. According to the 12/23 rule, recombination occurs between one region with a 12-bp spacer and a second region with a 23-bp spacer.

Gene Segments Are Joined by a Diverse Group of Proteins

In the early 1990s two proteins, encoded by **RAG1** (**recombination activating gene 1**) and **RAG2** (**recombination activating gene 2**), were shown to be required for recombining antibody variable region gene segments. The *RAG1* and *RAG2* genes are just 8 kb apart and are transcribed in opposite directions. *RAG* gene expression occurs only in cells of the immune system, is developmentally regulated in both T and B cells, and coincides with those periods during lymphoid development when receptor genes are being assembled (see Chapters 8 and 9). The RAG1/2 protein complex is required for RSS recognition and targeted cleavage of the DNA at the junction between the RSS and the respective variable region–coding segments.

The functional RAG1/2 occurs as a tetramer, with each protein being represented twice in the active protein complex. Recent x-ray crystallographic analysis has shed light on the relative locations of the two RAG1 and two RAG2 monomers within the tetrameric complex (**Figure 6-10a and b**).

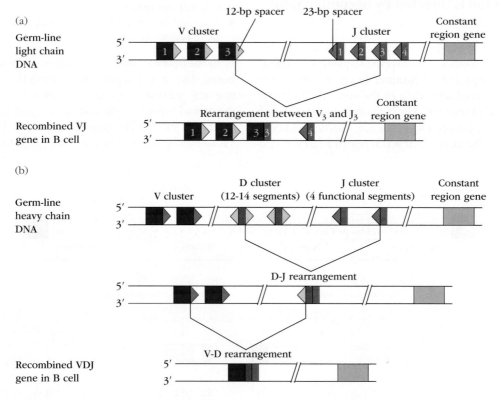

FIGURE 6-9 Recombination between gene segments is required to generate complete variable region light- and heavy-chain genes. (a) Recombination between a V region (in this case, V₃) and a J region (in this case, J₃) generates a single V-J light-chain gene in each B cell. The recombinase enzymes recognize the RSS downstream of the V region (orange triangle) and upstream of the J region (brown triangle). In every case, an RSS with a 12-bp (one-turn) spacer is paired with an RSS with a 23-bp (two-turn) spacer. This ensures that there is no inadvertent V-V or J-J joining in Ig genes. (b) Recombination of V (blue), D (purple), and J (red) segments creates a complete heavy-chain variable region gene. Again, the recombinase enzyme recognizes the RSS sequences downstream of the V region, up- and downstream of the D region, and upstream of the J region, pairing 23-bp spacers with 12-bp spacers.

(a)

(b)

(c)

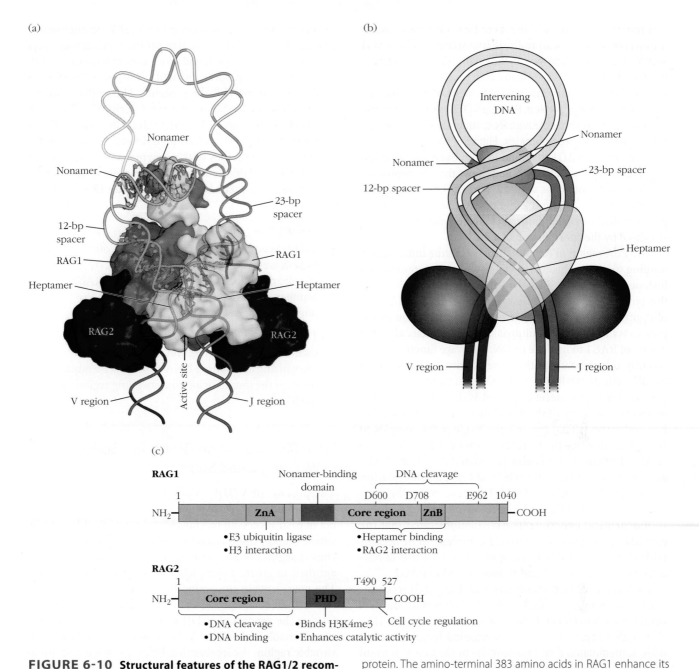

FIGURE 6-10 Structural features of the RAG1/2 recombinase proteins. (a) The tetrameric RAG1/2 proteins are shown in three dimensions drawn from x-ray crystallographic analysis, in complex with the RSSs, positioning the 12- and 23-bp spacer sequences to enable cleavage at the boundary of the coding sequence and the heptamer of the RSS. (b) A hypothetical model of how two coding regions to be joined may be arranged spatially, stabilized by the RAG1 and RAG2 recombinase complex. (c) The RAG1 protein is 1040 amino acids in length. The region of RAG1 that binds to the RSS nonamer lies to the amino-terminal side of the catalytic site, which contains three acidic amino acids, D600, D708, and E962, necessary for DNA cleavage. The heptamer binding site and several residues that interact with RAG2 also lie within the core region. The ZnA region is important for homodimerization with the RAG1 partner in the functional tetrameric protein. The amino-terminal 383 amino acids in RAG1 enhance its catalytic activity. The ubiquitin ligase activity ubiquitinylates histone H3, and may aid in releasing initial tight binding of RAG1 and allowing catalysis to proceed. RAG2 interacts with RAG1 and, like RAG1, also exists as a dimer in the RAG1/2 tetramer. The amino-terminal core region of RAG2 enhances the binding of RAG1 to DNA and is required for DNA cleavage to occur. A plant homeodomain (PHD) finger (so named because it was originally discovered in plants) binds specifically to a trimethylated lysine residue at position 4 in histone H3 (H3K4me3) and is critical for guiding RAG1/2 to regions of active chromatin. At the carboxyl-terminal end of RAG2 is a threonine residue, T490, which is phosphorylated during the S, G_2, and M phases of the cell cycle, inducing RAG2 proteolysis. Therefore RAG2 is active only in nondividing cells.

Figure 6-10c shows the location of functionally important sequences in the primary structures of RAG1 and RAG2. The active site of the recombinase complex is located in the RAG1 subunit and contains two aspartic acid residues and one glutamic acid residue (D600, D708, and E962, respectively). This active site "DDE motif" is found in many enzymes that cleave DNA, such as endonucleases, transposases, and recombinases. Figure 6-10c also identifies the domains on RAG1 that bind to both the nonameric and heptameric regions of the RSS and shows that the heptamer-binding region overlaps with the part of RAG1 that interacts with RAG2. ZnA and ZnB on RAG1 are regions of the proteins that form zinc fingers, elongated protein domains stabilized by the coordination of a Zn^{2+} ion.

The ZnA region of the protein assists in the initial RAG1 binding to active chromatin via its ability to interact with histone H3. However, once the RAG1/2 complex is in place, this binding then inhibits cleavage of DNA by RAG1/2. The ubiquitin ligase activity located in the same region of the protein is thought to ubiquitinylate the H3 protein, allowing release of RAG1 to mediate its DNA cleavage function. More recently, this same ubiquitin ligase activity has been shown to auto-activate the RAG1 recombinase activity.

The core region necessary for RAG2 activity lies at the amino-terminal end of the RAG2 molecule, while the plant homeodomain (PHD) region helps to guide the complex to active DNA bearing the H3K4me3 histone mark. Threonine residue T490 on RAG2 is phosphorylated during the S, G2, and M phases of the cell cycle, and this phosphorylation triggers the destruction of RAG2. This ensures that the complex does not cut DNA while the cell is undergoing division.

Biochemical experiments have demonstrated that the essential activities of the RAG1/2 complex can be accomplished by a so-called "core complex," which consists of residues 384–1008 of RAG1 and residues 1–383 of RAG2.

Only three of the proteins implicated in V(D)J recombination are unique to lymphocytes: RAG1, RAG2, and **terminal *deoxynucleotidyl* transferase (TdT)**. Like RAG1/2, TdT is also expressed only in developing lymphocytes. It adds *n*ontemplated ("N") nucleotides to the free 3′ termini of coding ends of heavy-chain V, D, and J segments following their cleavage by RAG1/2 recombinases. (These nucleotides are designated as "nontemplated" because they are not present in the germline, but rather are added to the DNA of a somatic cell.) TdT activity therefore contributes to the generation of additional receptor gene diversity in the CDR3 region of the antibody heavy chain.

Other proteins participating in the recombination process are not lymphoid specific. The high mobility group B proteins 1 and 2 (HMGB1 and HMGB2) act interchangeably to enhance RAG1/2 binding to the RSS and may also facilitate DNA bending at the recombination site. Whereas binding of the RSSs by RAG1/2 requires only RSS and HMGB proteins, other cellular factors, most of which are part of the nonhomologous end-joining (NHEJ) pathway of DNA

repair, are necessary to accomplish V(D)J recombination. The involvement of particular proteins at various steps in this process was deduced from observations of V(D)J recombination in natural and artificially generated systems lacking one or more of the proteins. The proteins known to participate in V(D)J joining are described in Table 6-2. **Clinical Focus Box 6-2** further describes some of the immunodeficiencies suffered by individuals with mutated or insufficient activities of the enzymes involved in V(D)J recombination. Additional descriptions of these immunodeficiency syndromes can be found in Chapter 18.

Key Concepts:
- V(D)J recombination occurs at the level of the DNA and is catalyzed by the lymphocyte-specific recombinase enzymes RAG1 and RAG2 acting in concert with TdT and enzymes of the nonhomologous end-joining pathway.
- The RAG1/2 complex contains two molecules each of RAG1 and RAG2 and is responsible for recognizing and cutting DNA at the precise junction between the immunoglobulin-encoding regions and the RSS.

V(D)J Recombination Occurs in a Series of Well-Regulated Steps

The process of V(D)J recombination occurs in several well-defined stages (**Overview Figure 6-11**). The end product of each successful rearrangement is an intact Ig gene, in which V and J (light chain) segments or V, D, and J (heavy chain) segments are made contiguous (flush) with one another, to create a complete heavy- or light-chain gene. The new joints in the antibody V region gene, created by this recombination process, are referred to as **coding joints**. During the process of V(D)J recombination of the heavy-chain variable region, or of V-J recombination of the λ-chain variable region, the intervening DNA is deleted and lost as an excision circle, or episome (Figure 6-11a). In the case of the κ light-chain gene, about 50% of the V_κ gene segments in the germ line are found in the opposite transcriptional orientation to the J_κ gene segments. In these cases, the intervening DNA is inverted and the excised sequences are retained on the chromosome upstream of the recombined gene (see Figure 6-11b). Regardless of whether the joints between the two RSS heptamers are lost as excision circles or retained in upstream DNA, they are referred to as **signal joints**.

The first phase of the recombination process, DNA recognition and cleavage, is catalyzed by the RAG1/2 proteins acting in concert with an HMGB1/2 protein. The second phase, end processing and joining, requires, in addition to RAG1/2, a more complex set of enzymatic activities: Artemis, other NHEJ proteins, and TdT (for heavy-chain

BOX 6-2

Some Immunodeficiencies Result from Impaired Receptor Gene Recombination

Since the RAG1/2 enzymes are responsible for both BCR and TCR gene rearrangements, mutations in the genes encoding RAG1 or RAG2 have catastrophic consequences for the immune system, resulting in patients with *severe combined immunodeficiency*, or SCID. The RAG1/2-encoding genes are located on human chromosome 11 (i.e., an autosome, not a sex chromosome), so such mutations are inherited in an autosomally recessive manner.

Babies born with a nonfunctional *RAG1* or *RAG2* gene have essentially no circulating B or T cells, although they do express normal levels of natural killer (NK) cells and their myeloid and erythroid cells are normal in number and function. Because infants receive antibodies passively from the maternal circulation, the first manifestation of this disease is the complete absence of T-cell function, and hence such infants suffer from severe, recurrent infections with fungi and viruses that would normally be combated by T cells in a healthy neonate. SCID used to be inevitably fatal within the first few months of life, unless babies were delivered directly into a sterile environment and remained there. The life span of a patient with SCID could be prolonged by preventing contact with all potentially harmful microorganisms. Air had to be filtered, all food sterilized, and they could have no direct contact with other people. Such isolation was feasible only as a temporary measure, pending treatment.

Nowadays, babies diagnosed sufficiently early with RAG1/2 deficiency can be successfully treated by bone marrow transplantation. If a suitable donor can be found, and transplantation is performed early in the patient's first year of life, the chances of the patient surviving to live a normal life are 97% or greater.

Patients who carry mutations resulting in a partially active or impaired *RAG1* or *RAG2* gene are diagnosed with Omenn syndrome. Patients with Omenn syndrome have no circulating B cells and abnormal lymph node architecture with deficiencies in the B-cell zones of the lymph nodes. Although some T cells are present, they are oligoclonal (derived from a very few precursors and hence have very few different receptors), and these T cells tend to be inappropriately activated. They do not respond normally to mitogens or to antigens in vitro. This condition, like RAG1/2 deficiency, is invariably fatal unless corrected by bone marrow transplantation.

If deficiency in *RAG1/2* genes causes such a catastrophic loss of immune function, it makes sense that loss of any of the other genes encoding proteins implicated in V(D)J recombination would also result in a SCID phenotype. Some of these proteins are described in Table 6-2. In 1998, it was determined that deficiency of the Artemis DNA repair enzyme also results in the loss of T- and B-cell function, in the face of normal NK cell activity. (Artemis, as described in the text, opens the hairpin formed after RAG1/2 creates the double-stranded break in receptor gene DNA). Loss of Artemis results in a SCID syndrome referred to as *Athabascan SCID*. Since Artemis is necessary for normal DNA repair, as well as V(D)J recombination, patients with Athabascan SCID also suffer from increased radiation sensitivity of skin fibroblasts and bone marrow cells. Similarly, human patients suffering from a deficiency in DNA ligase IV present with an immunodeficiency that affects T, B, and NK cells. As is the case with deficiencies in Artemis activity, DNA ligase IV is also implicated in DNA repair functions outside the immune system, and so such patients also suffer from generalized chromosomal instability, radiosensitivity, and developmental and growth retardation.

A SCID defect in mice, first noted and characterized by Melvin Bosma and colleagues, results from a nonsense mutation in the gene encoding DNA-PKcs. Although a similar recessive mutation has been found in Arabian horses, few, if any, human cases of SCID have been reported carrying this mutation.

REFERENCES

Mombaerts, P., et al. 1992. RAG-1-deficient mice have no mature B and T lymphocytes. *Cell.* **68**:869.

Abe, T., et al. 1994. Evidence for defects in V(D)J rearrangements in patients with severe combined immunodeficiency. *Journal of Immunology.* **152**:5504.

Villa, A., et al. 1998. Partial V(D)J recombination leads to Omenn syndrome. *Cell.* **93**:885.

Li, L., et al. 2002. A founder mutation in Artemis, an SNM1-like protein, causes SCID in Athabascan-speaking Native Americans. *Journal of Immunology.* **168**:6323.

Bosma, G. C., R. P. Custer, and M. J. Bosma. 1983. A severe combined immunodeficiency mutation in the mouse. *Nature.* **301**:527.

recombination only). The individual steps involved in the process of recombination between V_κ and J_κ segments are shown sequentially in **Figure 6-12**.

Step 1 *Recognition of the recombination signal sequence (RSS) by the RAG1/RAG2 enzyme complex*. The RAG1/2 recombinase tetramer forms a complex with the RSS next to one of the two gene segments to be joined. Binding is usually, but not always, initiated at the RSS containing the 12-bp spacer. Binding of the RAG1/2 complex is enhanced by the HMGB1/2 proteins, which may also serve to induce and stabilize bending of the DNA, facilitating its cleavage. The second RSS is then bound by the RAG1/2 complex and the two gene segments to be joined are brought into close contact (synapsis). Current models based on recent crystallographic

Recombination of Immunoglobulin Variable Region Genes

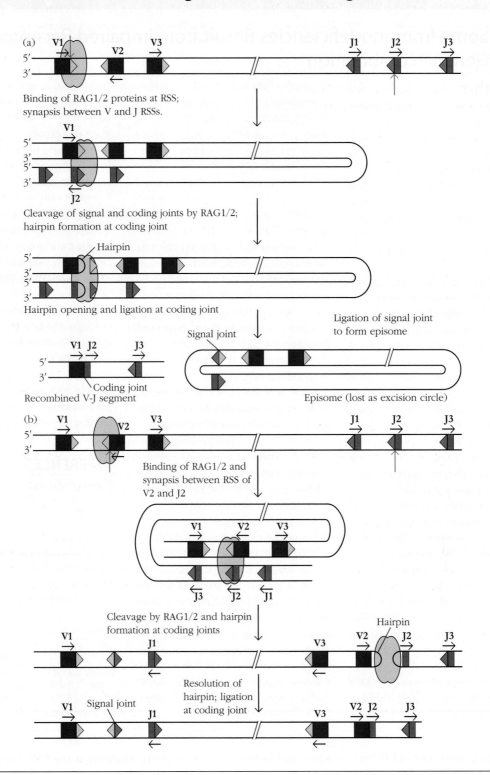

The tetrameric RAG1/2 complex (represented by the aqua ovals) binds the RSSs (usually binding first at the 12-bp RSS) and catalyzes recombination. Other enzymes belonging to the nonhomologous end-joining (NHEJ) pathway ligate the signal and coding joints. Additional mechanisms that generate diversity at the V-D and D-J joints of the Ig heavy-chain gene are not shown in this figure. See text for details. (a) Recombination between V_κ and J_κ segments arranged in the same transcriptional orientation in the germ-line genome results in excision of a circular DNA episome. (b) Recombination between V_κ and J_κ segments arranged in the opposite transcriptional orientation in the germ-line genome results in inversion of the intervening DNA.

studies suggest that binding of one type of spacer induces a conformational change in the RAG1/2 DNA-binding site that specifically accommodates the opposite type of spacer, thus enforcing the 12/23 rule.

Step 2   One-strand cleavage at the junction of the coding and signal sequences. The RAG1 protein then creates single-strand nicks, 5′ of the heptameric signal sequence on the coding strand of each V segment (i.e., at the junction between the V segment and the heptamer) and at the heptamer–J region junction. (Figure 6-12 shows this process for the V segment only.)

Step 3   Formation of V and J region hairpins and blunt signal ends. The free 3′-hydroxyl group at the end of the coding strand of the V segment now attacks the phosphate group on the opposite, noncoding V strand, forming a new covalent phosphodiester bond *across* the double helix and yielding a DNA hairpin structure on the V segment side of the break. This is called the *coding end.* Simultaneously, a blunt DNA end is formed at the edge of the heptameric signal sequence as a result of making a clean cut through both strands of DNA, with no overhang. This is the *signal end.* The same process occurs simultaneously on the J side of the incipient joint. At this stage, the RAG1/2 proteins and HMGB1/2 proteins are still associated with the coding and signal ends of both the V and J segments in a postcleavage complex. The serine/threonine kinase protein, *a*taxia *t*elangiectasia *m*utated (*ATM*), is thought to play an important role in stabilizing this complex and minimizing aberrant recombination events at this point in the process.

Step 4   Ligation of the signal ends. The NHEJ protein, DNA ligase IV, then ligates the free blunt ends to form the signal joint.

Step 5   Hairpin cleavage. Next, the hairpins at the ends of the V and J regions are opened by the endonuclease, Artemis, in one of three ways. The identical bond that was formed by the reaction described in step 3 may be reopened to create a blunt end at the coding joint. Alternatively, the hairpin may be opened asymmetrically either on the "top" or on the "bottom" strand, to yield a 5′ or a 3′ overhang, respectively. Artemis is a member of the NHEJ pathway and requires activation by the NHEJ kinase, DNA-PKcs, which binds to the DNA hairpin ends via its DNA-binding protein subunits Ku70/80. The most common overhang created by Artemis-mediated cleavage at immunoglobulin gene junctions is a 3′ overhang that leaves two unpaired residues. In addition to hairpin opening, the Artemis–DNA-PKcs complex also possesses both single- and double-stranded DNA endonuclease activity that is capable of removing several DNA bases or base pairs on each side of the nascent joint. This activity is rarely observed at the signal joint, but occurs often at the coding joint. The number of nucleotides that can be lost on each side of the joint ranges from 0 to 14.

Step 6   Overhang extension can lead to addition of palindromic nucleotides. In Ig light-chain rearrangements, nucleotide overhangs resulting from the steps described

previously can act as substrates for NHEJ DNA repair enzymes, leading to double-stranded **palindromic (P) nucleotides** at the coding joint. For example, the top row of bases in the V region shown in Figure 6-12, step 6, reading in the 5′ to 3′ direction, reads TCGA. Reading backward on the bottom strand from the point of ligation also yields TCGA. The palindromic nature of the bases at this joint is a direct function of an asymmetric hairpin-opening reaction. P-nucleotide addition can also occur at both the V-D and D-J joints of the heavy-chain gene segments but, as described below, other processes can intervene to add further diversity at the V_H-D and D-J_H junctions.

Step 7   Ligation of light-chain V and J segments. DNA ligase IV repairs the signal joints, as well as the coding joints. DNA ligase IV is usually found in complex with XRCC4, which helps to activate it. However, whereas at the signal joints, ligation almost always occurs without the addition or deletion of any nucleotides, the situation can be more complex at the coding joint. The enzymes of the NHEJ pathway include polymerases as well as Artemis and DNA ligase. As mentioned earlier, the endonuclease activity of Artemis will sometimes nibble at the coding ends after hairpin opening (see Table 6-2). In addition, the DNA polymerases associated with the NHEJ pathway, in particular DNA polymerase (Pol) λ and DNA Pol μ, are less faithful than the conventional DNA polymerase even when acting in a template-dependent manner. Even more dramatically, DNA Pol μ, like TdT, is capable of polymerizing DNA in a non–template-dependent manner and is therefore capable of adding random nucleotides at the coding joint.

Thus, NHEJ repair mechanisms can generate significant nucleotide diversity at the light-chain coding joint, even in the absence of TdT, which acts mainly at the heavy-chain joints.

Comparative sequence analysis of germ-line and mature B-cell Ig genes demonstrated that particularly extensive addition of nontemplated nucleotides could be identified in heavy-chain sequences. These additional nucleotide sequences occurred at both the V_H-D and D-J_H joints. In addition, careful comparative sequencing of germ-line versus somatic B-cell Ig heavy-chain sequences revealed that nucleotides were also often lost at these junctions. Two distinct types of enzyme-catalyzed activities are responsible for these findings in V_H sequences.

Step 8   Exonuclease trimming. Exonuclease activity trims back the edges of the V region DNA joints. Since the RAG proteins themselves can trim DNA near a 3′ flap, it is possible that the RAG proteins may cut off some of the lost nucleotides. Alternatively, as described in step 5, the Artemis–DNA-PKcs complex could be the enzyme responsible for the V(D)J-associated endonuclease function. Extensive exonuclease trimming is more common at the two heavy-chain V gene joints (V-D and D-J) than at the light-chain V-J joint. In cases where trimming is extensive, it can lead to the loss of the entire D region as well as the elimination of

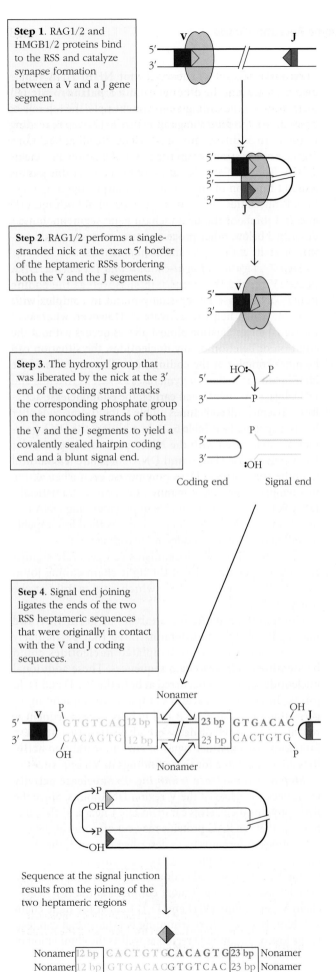

Step 1. RAG1/2 and HMGB1/2 proteins bind to the RSS and catalyze synapse formation between a V and a J gene segment.

Step 2. RAG1/2 performs a single-stranded nick at the exact 5′ border of the heptameric RSSs bordering both the V and the J segments.

Step 3. The hydroxyl group that was liberated by the nick at the 3′ end of the coding strand attacks the corresponding phosphate group on the noncoding strands of both the V and the J segments to yield a covalently sealed hairpin coding end and a blunt signal end.

Coding end   Signal end

Step 4. Signal end joining ligates the ends of the two RSS heptameric sequences that were originally in contact with the V and J coding sequences.

Nonamer

12 bp   23 bp

Nonamer

Sequence at the signal junction results from the joining of the two heptameric regions

| Nonamer | 12 bp | CACTGTG**CACAGTG** | 23 bp | Nonamer |
| Nonamer | 12 bp | GTGACAC**GTGTCAC** | 23 bp | Nonamer |

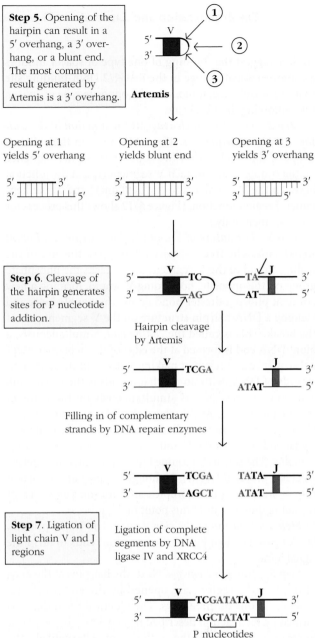

Step 5. Opening of the hairpin can result in a 5′ overhang, a 3′ overhang, or a blunt end. The most common result generated by Artemis is a 3′ overhang.

Artemis

Opening at 1 yields 5′ overhang

Opening at 2 yields blunt end

Opening at 3 yields 3′ overhang

Step 6. Cleavage of the hairpin generates sites for P nucleotide addition.

Hairpin cleavage by Artemis

Filling in of complementary strands by DNA repair enzymes

Step 7. Ligation of light chain V and J regions

Ligation of complete segments by DNA ligase IV and XRCC4

P nucleotides

FIGURE 6-12 Mechanism of V(D)J recombination, illustrated for V_κ-to-J_κ joining. The RAG1/2 tetramer (aqua ovals) and HMGB1/2 proteins bind to the RSSs and catalyze synapse formation. The coding (5′ → 3′) strand of DNA is drawn as a thick line, and the noncoding (3′ → 5′) strand as a thin line. For steps 2 to 5 we show only the events associated with the V_κ region gene segment, although the single-strand cleavage, hairpin formation, and templated nucleotide addition occur simultaneously at the borders of the V_κ and J_κ segments. In this example, the V and J regions are encoded in the same direction on the chromosome, and so the DNA encoding the RSSs and the intervening DNA is released into the nucleus as a circular episome and will be lost on cell division. The DNA that was on the coding strand of the V region prior to rearrangement is emboldened. The signal joint is between the residues that were in contiguity with the V and J regions, respectively. Only the heptamer sequence is written out, to preserve clarity. Nucleotides encoded in the germ-line genome are shown in black; P nucleotides are in blue; and nontemplated nucleotides added by TdT at heavy-chain VD and DJ joints are shown in red. Steps 8, 9, and 10, shown on the facing page, occur only in heavy-chain loci. See text for details.

(continued)

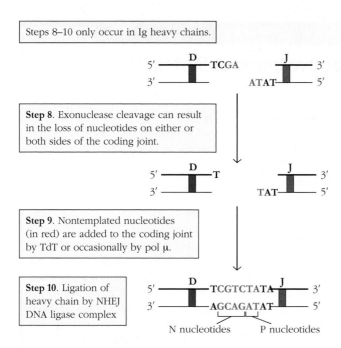

Steps 8–10 only occur in Ig heavy chains.

Step 8. Exonuclease cleavage can result in the loss of nucleotides on either or both sides of the coding joint.

Step 9. Nontemplated nucleotides (in red) are added to the coding joint by TdT or occasionally by pol μ.

Step 10. Ligation of heavy chain by NHEJ DNA ligase complex

N nucleotides   P nucleotides

FIGURE 6-12 *(continued)*

any P nucleotides formed as a result of asymmetric hairpin cleavage.

Step 9 N-nucleotide addition. (Most probably occurs simultaneously with step 8.) **Nontemplated (N) nucleotides** are added by TdT to the coding joints of heavy-chain genes after hairpin cleavage. This enzyme can add up to 20 nucleotides to each side of the joint. The two ends are held together throughout this process by the RAG1/2 enzyme complex. TdT-mediated N-nucleotide addition at the coding joints of the heavy-chain genes is more commonly observed than at light-chain joints, because TdT is expressed at the earliest phases of V(D)J recombination when the heavy chain, but not the light-chain, genes are being rearranged. TdT activity is then usually turned off before light-chain rearrangements begin in mice, although residual TdT activity is found during light-chain rearrangement in humans. In addition, as described above, some nontemplated nucleotide addition at light-chain joints may be mediated by the NHEJ DNA Pol μ.

Step 10 Ligation and repair of the heavy-chain gene. This final step is identical to the ligation and repair for the light-chain genes and is mediated by DNA ligase IV acting in concert with XRCC4.

When considering the creation of an immunoglobulin variable region gene, we must always take into account the fact that nucleotide addition and/or exonuclease trimming at the V(D)J joints does not necessarily occur in sets of three nucleotides, and so can lead to out-of-phase joining. Recombined V segment sequences in which trimming has caused the loss of the correct reading frame for the transcription process cannot encode antibody molecules, and such rearrangements are said to be **unproductive**.

If recombination at one heavy-chain locus is unproductive, rearrangement at the other allele is immediately initiated. Unproductive rearrangement at both alleles leads to apoptosis of the developing cell as it fails to receive necessary survival signals from the pre-BCR (see Chapter 9). Once light-chain rearrangements begin, sequential rearrangement of light-chain alleles occurs if prior rearrangements are unsuccessful.

The cleavage and rearrangement of DNA segments within somatic cells of a mammalian genome is an unusual occurrence and led scientists to question how such a mechanism might have evolved. Compelling evidence now supports the evolutionary origin of the genes encoding the RAG1/2 complex as a transposon unit that hopped into a primitive antigen receptor gene; this is discussed further in **Evolution Box 6-3**.

Key Concepts:

- The RAG1/2 and TdT proteins are lymphoid-specific. However, proteins of the NHEJ pathway stabilize the recombination complex (Ku70/80) and participate at the opening of the DNA hairpin (Artemis). DNA ligase IV closes both the signal and the coding joints during recombination. DNA polymerases μ and λ may also introduce additional diversity at the coding joints.

- At the sites of recombination, DNA hairpin cleavage can be asymmetric, resulting in the generation of P nucleotides, and in the heavy-chain genes, nontemplated nucleotide addition by TdT can result in N nucleotides. Exonuclease nibbling can reduce the number of N and P nucleotides in the final receptor gene products.

Five Mechanisms Generate Antibody Diversity in Naïve B Cells

The above description allows us to understand how such an immensely diversified antibody repertoire can be generated from a finite amount of genetic material. To summarize, the diversity of the naïve BCR repertoire is shaped by the following mechanisms (**Table 6-3**, first two columns):

1. *Multiple gene segments* exist at heavy (V, D, and J) and light-chain (V and J) loci. These can be combined with one another to provide extensive combinatorial diversity.

2. *Heavy-chain/light-chain combinatorial diversity:* The same heavy chain can combine with different light chains, and vice versa. The combination of different heavy- and light-chain pairs to form a complete antibody molecule provides further opportunities for increasing the number of available antibody combining sites.

BOX 6-3

A Central Mechanism of the Adaptive Immune System Has a Surprising Evolutionary Origin

The hallmark feature of the adaptive immune system is the existence of highly diverse antigen receptors generated by V(D)J gene rearrangements effected by the RAG1/2 recombinase. Thus, understanding how the RAG1/2 recombinase evolved is key to understanding the formation of the adaptive immune system during vertebrate evolution. A series of seminal observations made in the early 1990s gave rise to the notion that the process of V(D)J recombination may have its beginnings in the insertion of a transposase gene into an ancestral antigen receptor gene. Transposition and V(D)J recombination share a conceptual framework in that they are both systems in which DNA segments are "book-ended" by short signal sequences and subsequently moved around the genome by mechanisms that cleave and rejoin DNA. As more experiments have been conducted, this idea has been progressively fleshed out, and it now seems extremely likely that the RAG system has evolved from an ancient transposon.

Structurally, the RSSs bear all the hallmarks of a recognition unit for transposase activity. All transposons bear *cis*-acting sequences (sequences on the same strand of DNA that will be attacked) that must be recognized by a transposase enzyme. These sequences are usually short sequences (terminal inverted repeats, or TIRs) at either end of the transposon. The presence of the RSSs on either side of each of the variable gene segments in both T and B cells suggests their origin as recognition sequences for transposable elements. Sequence analysis of vertebrate RSSs demonstrated considerable sequence homology with the TIR sequences of the *Transib* family of transposons. This homology is most acute in the conserved heptameric sequence. Furthermore, the RAG1 enzyme itself has been shown to have considerable sequence homology to the *Transib* transposase enzyme, lending further support to the evolutionary origin of V(D)J

recombination as a form of transposition. However, notably, no homolog for RAG2 was found in the *Transib* transposase system.

Lending further support to the notion that V(D)J recombination has its origins in transposition, V(D)J recombination and transposon excision and repair have been shown to be very similar, mechanistically. During the movement of a transposon, the two transposon ends are held together in a nucleoprotein complex, perfectly analogous to the synaptic complex generated during V(D)J recombination. At the catalytic site of RAG1, V(D)J recombination uses three acidic residues: DDE (two aspartic acid residues and a glutamic acid residue). This same triad of acidic amino acids is characteristic of the DNA cleavage sites of the DDE family of transposases. The acidic triad coordinates metal ions (most probably magnesium ions in vivo) in both the DDE transposases and in RAG1. Investigators demonstrated that both the DDE transposases and RAG1 catalyze single-strand nicking followed by hairpin formation mediated by transesterification.

If RAG1 indeed evolved from a *Transib* transposase, the next obvious question

was whether the *Transib* transposase could mediate Ig gene segment recombination. To investigate this, scientists set up a test system in which mouse 3T3 fibroblasts were first transfected with a RAG1/2 recombination substrate (**Figure 1**). Note that, in this substrate, the transcriptional direction of the V_κ gene segment is opposite to that of the J_κ substrate; recall that this requires that the intervening sequence be inverted during the process of V-J recombination (Figure 6-11b). Inversion of this sequence in the test recombination substrate used in this experiment would then allow expression of the xanthine-guanine phosphoribosyltransferase (GPT) gene, which confers resistance of the cell to the antibiotic mycophenolic acid (MPA). Recombinational events could therefore easily be screened for by simply looking for the cells that survived in culture with MPA.

Initial control experiments showed that transfection with genes encoding RAG1 and RAG2 gave rise to the expected high frequency of recombination events that conformed to the 12/23 rule. Transfection with *RAG1* genes alone showed a lower number of recombination events, but,

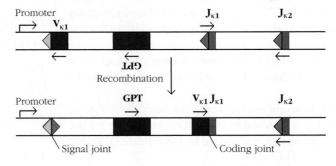

FIGURE 1 Elements of the recombination substrate used by Carmona and colleagues. The recombination substrate that was used to test the viability of the *Hztransib* transposon as a recombinase included one V_κ element flanked by a 12-bp RSS and two J_κ elements, each flanked by 23-bp RSSs. Transcription occurs from left to right. In terms of transcription, the V_κ element and the J_κ 1 sequence element are oriented in opposite directions, such that recombination between the V_κ segment and J_κ 1 leads to inversion of the intervening sequence. The intervening sequence contains an inverted gene encoding the enzyme xanthine-guanine phosphoribosyltransferase (GPT), which cleaves and inactivates the antibiotic mycophenolic acid (MPA). Successful recombination therefore enables the cell that contains the recombined elements to survive in medium containing MPA and enables rapid assessment of recombination frequency.

(continued)

surprisingly, transfection with *RAG1* genes alone continued to allow recombination of V_κ and J_κ segments even when both segments were flanked with the same RSS. It therefore appears that one of the roles of the RAG2 protein in the vertebrate immune system is to enforce the 12/23 rule in Ig gene recombination.

The next set of experiments tested whether the *Transib* transposase from *Helicoverpa zea*, *Hztransib*, could bring about recombination guided by the 12-bp RSS and 23-bp RSS signal sequences. Transfection of 3T3 cells with the recombination substrate and the *Hztransib* transposase enzyme gene alone yielded no recombination activity. Remarkably, however, when the *Hztransib* gene was simultaneously transfected along with the mouse *RAG2* gene, recombination occurred at a

level fully one-third of that catalyzed by the mouse RAG1/2 complex. Sequence analysis confirmed that this recombination yielded perfectly normal signal and coding joints.

These and other findings have led to a model of RAG evolution in which a *Transib* transposase gene served as the evolutionary precursor of *RAG1*, and the *RAG2* gene was separately acquired by a *Transib* element to form the vertebrate *RAG1-RAG2* transposon (**Figure 2**). The evolutionary origins of *RAG2* are currently unknown, although it may have existed as a host factor in the genome of the organism in which the transposon first arose.

Once the primitive *RAG* transposon was assembled, it is not difficult to envisage that it could insert into exons of primitive cell surface receptor genes. This insertion could be followed variously by

gene duplication events and genetic separation of the recombinase activity from between the terminal inverted RSS repeats to elsewhere in the genome. This would result in rapid evolution of receptor genes and lead to the complex receptor gene structure we know today.

REFERENCES

Carmona, L. M., S. D. Fugmann, and D. G. Schatz. 2016. Collaboration of RAG2 with RAG1-like proteins during the evolution of V(D)J recombination. *Genes and Development.* **30**:909.

Thompson, C. B. 1995. New insights into V(D)J recombination and its role in the evolution of the immune system. *Immunity.* **3**:531.

Koonin, E. V., and M. Krupovic. 2015. Evolution of adaptive immunity from transposable elements combined with innate immune systems. *Nature Reviews Genetics.* **16**:184.

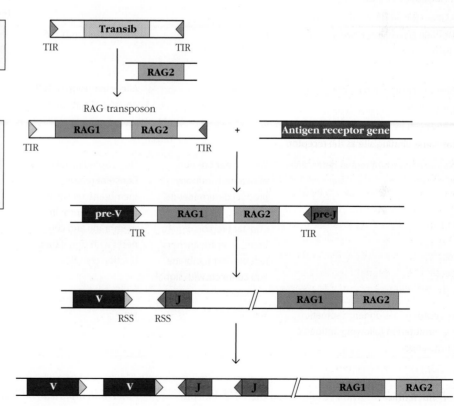

Step 1. Assembly of the RAG1/2 complex from a transposon and an unknown RAG2 ancestral gene.

Step 2. Insertion of the transposase into an ancestral receptor gene, followed by excision of the recombinase-coding genes and several rounds of duplication of the antigen receptor gene segments.

FIGURE 2 **A model for the evolution of the RAG1/2 recombinase.** A recent model for RAG1/2 evolution, developed by the Schatz laboratory, suggests that a transposable element from the *Transib* family acquired an ancestral *RAG2* gene. The origin of this *RAG2* gene is currently unknown. Combination of the *RAG1* and *RAG2* genes in a single, convergently transcribed transposon allowed the whole transposon to "hop" into the middle of a primordial antigen receptor gene. A subsequent event separated the enzyme-coding genes from the now split receptor gene, allowing for gene duplication of the receptor genes and recombination catalyzed by RAG1/2. The TIRs of the transposon became the RSSs.

TABLE 6-3	Comparison of the mechanisms for the generation and expression of diversity among B-cell and T-cell receptor molecules		
Mechanism	**Used in B cells**	**Used in T cells**	**Comments**
Multiple germ-line V(D)J genes	Yes	Yes	The mouse V_λ locus has undergone a severe contraction, and, therefore, only 5% of mouse light chains are of the λ type. The TCR γ-chain locus also has few V genes J region diversity is notably higher in TCR α-chain genes than in other TCR or Ig genes
Light-chain segment use	κ and λ variable regions encoded by V and J segments	α and γ variable regions encoded by V and J segments	
Heavy-chain segment use	V_H regions encoded by V, D, and J segments	β and δ variable regions encoded by V, D, and J segments	
Absolute dependence on RAG1/2 expression	Yes	Yes	
Junctional diversity: P-nucleotide and N-nucleotide addition	Yes	Yes	Many fewer N nucleotides found in Ig light chains because of developmental regulation of TdT
Multiple D regions per recombined chain	No	Present only in TCR δ	The presence of two D segments allows an additional site for N-nucleotide addition
Allelic exclusion of receptor gene expression	Absolute	Allelic exclusion of TCR α genes is not absolute	
On activation, secretes product with the same binding site as the receptor	Yes	No	
Nature of constant region determines function	Yes; constant region of secreted antibody product determines its function. Constant region of membrane receptor anchors receptor in membrane and connects with signal transduction complex	No secreted product. Constant region of membrane receptor anchors receptor in membrane and connects with signal transduction complex	
Receptor genes undergo somatic hypermutation following antigenic stimulation	Yes	No	

3. *P-nucleotide addition* results when the DNA hairpin at the coding joint of heavy and light chains is cleaved asymmetrically. Filling in the single-stranded DNA piece resulting from this asymmetric cleavage generates a short palindromic sequence.

4. *Exonuclease trimming* sometimes occurs at the V-D-J and V-J junctions, causing loss of nucleotides.

5. *Nontemplated (N)-nucleotide addition* by TdT in heavy-chain V-D and D-J junctions and from DNA polymerase μ in both heavy and light chains.

Mechanisms 3, 4, and 5 give rise to striking sequence diversity at the junctions between gene segments, and result in the formation of the highly variable CDR3 regions of the antibody heavy and light chains.

Together, these five mechanisms are responsible for the creation of the repertoire of BCRs that is available to organisms before any contact with pathogens or other antigens has occurred, the so-called **naïve BCR repertoire**.

Note that we have described the process of the generation of the primary Ig variable region repertoire as it occurs in humans and rodents. Although the same principles apply to most vertebrate species, different species have evolved their own variations. For example, the process of gene conversion is used in chickens and rabbits, and some species, such as sheep and cows, use somatic hypermutation in the generation of the primary as well as the antigen-experienced repertoire.

> **Key Concept:**
>
> - The mechanisms that generate a diverse naïve BCR repertoire include the occurrence of many alternative copies of V, D, and J Ig gene segments, combinatorial association of these gene segments within each receptor chain, combinatorial association of heavy and light chains with one another, and modifications of DNA sequences at the joints between the variable region gene segments.

The Regulation of V(D)J Gene Recombination Involves Chromatin Alteration

In attempting to understand the complex process of V(D)J recombination and its regulation, investigators must address the question of how Ig gene recombination happens only at particular stages of B-cell development and how two RSSs, located many kilobases or even megabases apart in the linear DNA sequence, are brought into sufficiently close apposition for accurate recombination to succeed.

Because of its capacity to induce genome instability by introducing double-strand breaks in DNA, the expression of the RAG1/2 complex must be tightly regulated. The enzyme complex is expressed only in lymphoid cells at specific periods in lymphoid development (see Chapters 8 and 9). Furthermore, as described earlier, it is inactivated prior to the cell's entry into S phase, when double-stranded breaks in the DNA might interfere with regulated chromatin distribution into daughter cells. Kinases coupled to the cell cycle phosphorylate RAG2 prior to the G_1-S cell cycle transition, targeting RAG2 for ubiquitin-dependent protein degradation prior to entry into S phase (see Figure 6-10c).

However, once RAG1/2 is expressed, how does the cell ensure that its activity is appropriately restricted to the correct sites on the chromatin? Although the native RAG recombinase complex is quite difficult to isolate, a core, catalytically active RAG1/2 heterodimer can be purified quite readily, and has been used to determine the binding specificity and orientation of the recombinase.

On isolated DNA fragments, the core RAG recombinase binds specifically to recombination signal sequences (RSSs), although it also binds to other sites on the genome that lack extensive sequence homology to the RSS. The nonamer-binding domains of RAG1 interact with the A-rich tract of the RSS nonamer. Additional regions in the RAG1 core also interact with the RSS heptamer and with the end of the V, D, or J coding sequence, as well as mediating the catalytic reaction. In general, the isolated core recombinase operating on purified DNA fragments tolerates more considerable sequence variation in the RSS spacer, the heptamer, and even the nonamer than is observed in vivo.

In vivo, the catalytic activity of RAG1/2 occurs in an extraordinarily complex chromosomal environment, and analysis of V(D)J recombination regulatory mechanisms in the native chromosomal context has required the refinement of techniques capable of analyzing the interactions between proteins and nuclear DNA *folded within its native chromatin structure*. Three techniques in particular: chromatin immunoprecipitation (ChIP); multicolor, three-dimensional fluorescence in situ hybridization (3-D FISH); and methods that identify DNA sequences that interact with one another within the context of active chromatin (e.g., Hi-C), are described in Chapter 20. These approaches were all used to generate the information described in this section.

Changes in Histone Marks

RAG1/2 binding is affected by particular epigenetic modifications on the histones associated with target sequences. Recall that eukaryotic DNA is wound around histone octamers to form nucleosomes. The core DNA that is directly associated with each nucleosome is 147 nucleotides in length, and nucleosomes are separated from one another by linker DNA sequences of up to 80 nucleotides in length that interact with histone H1. This "beads on a string" nucleosomal DNA is then coiled into structures of increasing complexity. Histone modifications, such as methylation or acetylation, can affect the degree to which the DNA in the associated chromatin is accessible to enzymatic activities, such as recombination or transcription, by altering the extent of nucleosome packing. The nature of histone modifications or epigenetic marks associated with a set of genes is referred to as its "histone code." Alterations in the histone code of chromatin associated with immunoglobulin DNA during B-cell development signal the onset of receptiveness of the Ig locus to transcription and recombination.

Analysis of the biochemical basis for RAG recombinase binding to chromatin shows that the RAG2 plant homeodomain region (see Figure 6-10c) interacts with histone H3 that has been trimethylated at the lysine in position 4 of the histone's amino acid sequence (H3K4me3). This H3K4me3 modification is typically found at transcriptional start sites in active chromatin. Disruption of this RAG2-histone interaction inhibits V(D)J recombination. Biochemical experiments have shown that RAG2 binding to

H3K4me3 increases the affinity of the RAG complex for its DNA substrates, possibly by inducing an activating conformational change in RAG1.

Furthermore, it has long been known that one of the earliest steps in Ig gene recombination is the transcription of noncoding RNA from promoters in DNA regions near the Ig gene segments. This germ-line transcription, irrespective of the nature of the RNA product, confirms that the DNA is now accessible for enzymatic manipulation. RNA polymerase II, the enzyme that transcribes the immunoglobulin genes and initiates the germ-line transcription process alluded to above, often travels with the histone methyltransferases, suggesting that the histone modifications that signal active chromatin are mechanistically linked to the germ-line transcription event and that together, they signal the readiness of the germ-line immunoglobulin DNA for recombination.

In immunoglobulin genes, both the trimethylated lysine histone modification and acetylation of histone residues, which also signals open chromatin, are concentrated in the J-gene segment regions of both heavy and light chain–encoding DNA, with a few trimethylated histones found in association with J-proximal D gene segments. Thus, the nature of the histone code directs the recombination apparatus first to the J regions of immunoglobulin heavy and light chains.

Changes in Higher Order Chromatin Structure

Because the V, D, and J gene segments are so spread out along the chromosome, higher order chromatin structure must also play a role in the regulation of V(D)J recombination. Chromatin visualization techniques have shown that chromatin folds extensively into loops of various lengths that cluster into the form of rosettes (**Figure 6-13a**). Clustering is regulated by the binding of proteins to specific sites on the DNA. Notable among these site-specific DNA-binding proteins is the factor CTCF, which binds specifically to regions with the DNA sequence CCCTC. Recent experiments have demonstrated that the three-dimensional structure of these loops is altered in real time in a surprisingly orderly fashion and affects variable region recombination.

Detailed analysis of the three-dimensional structure of the Ig heavy-chain gene locus has indicated that it initially appears to be arranged in space into three rosette-containing chromatin regions. One of these regions contains the distal V_H genes (those farthest from the D region); a second contains the proximal V_H genes (those nearest to the D region); and the third contains the gene segments of the D, J_H, and C_H regions. Since recombination events are topologically limited to include only genes within a rosette, the rosette loop that contains the D, J_H, and C_H regions defines the scope of RAG activity in the earliest B lymphoid precursors, pre-pro-B cells. Once D_H-J_H recombination has occurred, the loop structure is altered to allow V_H-D_H recombination at the pro-B-cell stage. Figure 6-13a illustrates the change in the structure of the Ig loci as development proceeds.

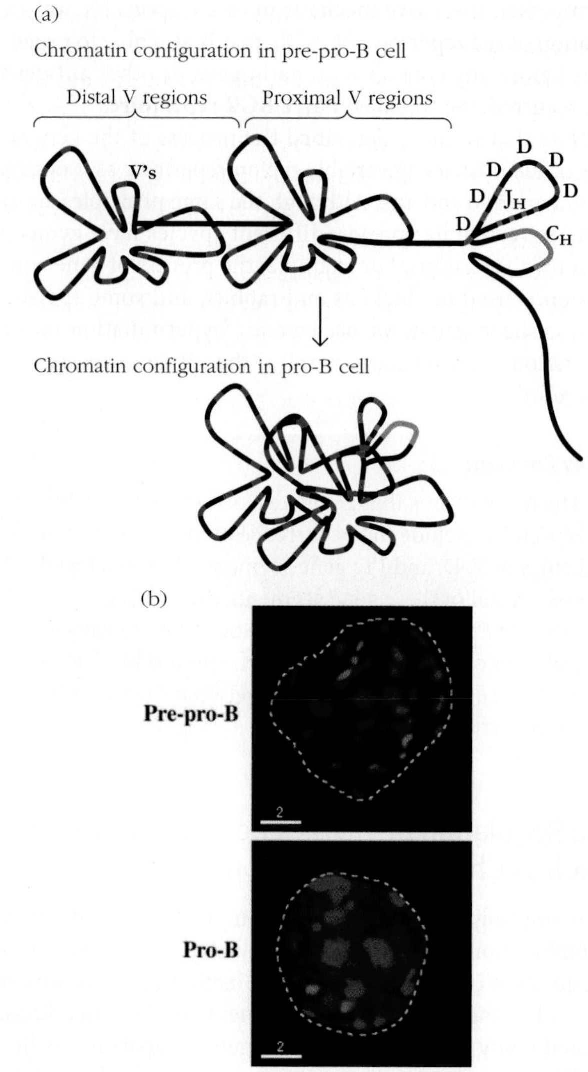

FIGURE 6-13 Three-dimensional organization of chromosomal regions containing V, D, and J segments changes during B-cell development. (a) In the earliest stage of B-cell development, the pre-pro stage, the region of the chromosome encoding the heavy chain of the Ig protein is folded into three clearly demarcated rosettes. One of these includes loops of DNA encoding the distal V_H regions (those farthest from the C_H complex); the second the more proximal array of V_H regions; and the third the D, J_H, and C_H regions. Since recombination occurs only within, but not between rosettes, this functionally restricts recombination to occurring at the D-J_H, but not the V_H-D, junctions during the earliest developmental stage. At the pro-B-cell stage, the rosette structure is altered and V_H-D recombination is permitted. (b) Chromosomes were labeled with baculovirus artificial chromosomal probes that include the entire Ig region. In pre-pro-B cells, the Ig genes could be seen clustered into two or three regions, whereas in pro-B cells, in which H-chain rearrangement is occurring, chromosomal contraction and re-localization bring the H-chain chromosome into a single cluster. *[Part (b) republished with permission from Elsevier, Jhunjhunwala et al. "The 3D structure of the immunoglobulin heavy-chain locus: Implications for long-range genomic interactions," Cell 133(2):265–79, 18 April 2008, Figure 6b. Permission conveyed through Copyright Clearance Center, Inc.]*

The alteration in chromatin topology can be visualized microscopically as a locus contraction event (Figure 6-13b) and has been shown to depend on the binding of transcription factors, including Pax-5, to the chromatin. It is thought that Pax-5, a key transcription factor inducing the formation of B lymphocytes (see Chapter 9), interacts with proteins that control the formation of the base of the loops. Once V(D)J recombination has occurred successfully on one allele, the inactive chromosome is decontracted.

Although it is tempting to speculate that selective placement of histone marks and regulated contraction of the chromosomal regions bearing Ig genes can together explain the exquisitely controlled ordering of Ig gene rearrangements, it is now clear that other factors are involved. Specifically, observations of the Ig_κ locus demonstrated that it is contracted in both pro-B cells and pre-B cells and has the potential for similar levels of long-range interactions at both of these cell stages. However, V_κ rearrangements do not occur at the pro-B-cell stage of development, but rather are delayed until after Ig_H rearrangements are complete, at the pre-B-cell stage. If the V_κ locus is contracted as early as the pro-B-cell stage, what is preventing V_κ rearrangement from occurring then?

The Intranuclear Localization of Antigen Receptor Chromatin

A further aspect of the regulation of RAG activity concerns the manner in which the *intra-nuclear localization* of the antigen receptor chromatin is altered in order to make available the relevant genes to the recombinase. Within the nucleus, inactive chromatin is found in regions associated with the nuclear lamina, which lies immediately inside the nuclear membrane. Indeed, some parts of the chromatin can be shown to be tethered to the nuclear lamina. Such inactive chromatin is unable to participate in either transcription or recombination.

In contrast, chromatin located in the general nucleoplasm tends to be more active. Considerable data now suggest that antigen receptor loci move away from the nuclear envelope prior to recombination and that those alleles that are excluded from productive rearrangement re-associate with the envelope once recombination terminates. The movement away from the nuclear lamina occurs subsequent to increased histone acetylation at Ig loci.

Figure 6-14 illustrates the sequence of movement of chromosomes within the nucleus during B-cell development. The Ig_H locus in hematopoietic progenitor cells and early pre-pro-B cells is associated with the inner nuclear lamina. Thus, colocalization with the nuclear lamina as well as the physical nature of the chromatin loops ensures that the only transcriptional and associated recombinational events that can occur are restricted to the heavy-chain D and J regions. As the B cell enters the pro-B-cell stage, the V gene locus moves away from the nuclear lamina and the entire locus contracts under the influence of Pax-5, facilitating rearrangements with distal as well as proximal V_H gene segments

Correspondingly, movement of the Ig_κ locus into the central nucleoplasmic region has also been demonstrated at the pre-B-cell stage, when light-chain rearrangement occurs,

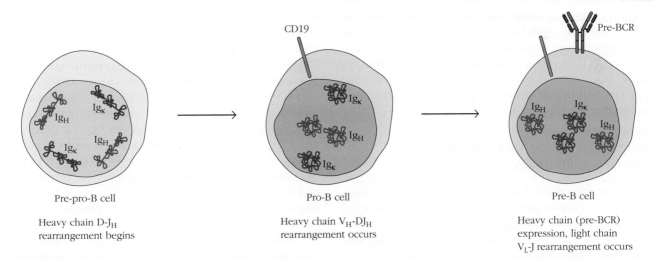

Pre-pro-B cell

Heavy chain D-J_H rearrangement begins

Pro-B cell

Heavy chain V_H-DJ_H rearrangement occurs

Pre-B cell

Heavy chain (pre-BCR) expression, light chain V_L-J rearrangement occurs

FIGURE 6-14 Nuclear positioning of *Ig_H* and *Ig_κ* loci alters during B-cell development. In pre-pro-B cells, both light- and heavy-chain immunoglobulin loci are located close to the nuclear lamina, in regions of heterochromatin. The Ig chromosomal regions are extended and do not permit recombination. In pro-B cells, in which Ig heavy-chain recombination has been initiated, the heavy-chain chromosomes can be found in the interior of the nucleus and the Ig chromosomal regions are contracted, so as to bring regions of recombination into closer proximity. The light-chain chromosome is also contracted, but it remains at the periphery by the nuclear envelope. In pre-B cells, light-chain recombination occurs. In these cells, again, both light- and heavy-chain chromosomes are contracted, but this time, the heavy-chain chromosomes are located at the periphery of the nucleus and the light chains are brought into a more central location. See text for details and chapter-opening photo for fluorescence microscopy data.

indicating that the positioning of the immunoglobulin loci in the nuclear environment, as well as the extent of locus contraction at Ig genes, together determine the capacity for recombination.

Following a productive recombination at one Ig receptor allele, the potential for recombination at the corresponding allele is shut down, a process known as *allelic exclusion* (discussed further below). At this point, it remains unclear whether allelic exclusion is correlated with movement of the excluded allele back toward the nuclear lamina. However, in the case of the Ig_H locus, suppression of the inactive allele has been associated with relocation to heterochromatic regions of the nucleus under the influence of the action of the *ataxia telangiectasia mutated* (ATM) protein.

Key Concepts:

- RAG recombinase activity depends on the availability of receptor RSSs within accessible chromatin.

- RSS accessibility depends on their location within the nucleus and within higher level chromatin loops as well as the nature of epigenetic modifications of nearby chromatin.

B-Cell Receptor Expression

As we have seen, the expression of a complete Ig receptor on the surface of a B cell is the end result of a complex and tightly regulated series of events. First, as described above, the cell must ensure that the various gene recombination events culminate in productive rearrangements at both the heavy- and light-chain loci. Second, only one heavy-chain allele and one light-chain allele must be expressed in each B cell. Finally, the receptor must be tested to ensure it does not bind self-antigens, in order to protect the host against the generation of an autoimmune response. This latter process will be described in greater detail in Chapter 9.

Each B Cell Synthesizes Only One Heavy Chain and One Light Chain

If all four of the alleles that encode both heavy-chain immunoglobulin receptor genes and their κ light-chain partners were to successfully undergo rearrangement, the host B cell could express up to four different antigen-binding sites, with each heavy chain binding to each κ light chain to create a new receptor type. Extending that idea, if successful rearrangements were to occur at both κ and both λ loci, even more antigen-binding sites could be expressed on each B-cell surface. The opportunity to increase the number of available receptors per B cell may initially sound advantageous to the organism. However, in practice, the presence of more than one receptor per B cell would make it very

difficult for the organism to prevent the expression of receptors that recognize self-antigens. Organisms have therefore evolved a mechanism by which B cells ensure that only one heavy-chain allele and one light-chain allele are transcribed and translated. This mechanism is referred to as **allelic exclusion**.

The rearrangement of Ig genes occurs in an ordered way, and begins, as described above, with recombination at one of the two homologous chromosomes carrying the heavy-chain loci. The production of a complete heavy chain and its expression on the B-cell surface in concert with a surrogate light chain as the pre-BCR (see Figure 6-7a) signals the end of heavy-chain gene rearrangement. At this point in B-cell development, the nascent B cell undergoes several rounds of cell division prior to initiating light-chain rearrangement. This ensures maximal use of those heavy-chain gene rearrangements that can successfully encode immunoglobulin heavy chains. It is therefore not unusual to find multiple B cells bearing the same heavy-chain variable region, but different light chains. A similar strategy is used by T cells, as we will see later.

Only one antibody heavy chain is allowed to complete the rearrangement process. If the first attempt at a heavy-chain rearrangement is unproductive (i.e., results in the formation of out-of-frame joints), rearrangement will initiate at the second heavy-chain allele. If this second attempt is also unsuccessful, the B cell will die (see Chapter 9). The choice of which allele is rearranged first appears to be random. The details of the mechanism by which the B cell prevents rearrangement of the second heavy chain if the first has successfully rearranged is still under investigation. However, we do know that, once successful rearrangement at one V_H locus has been achieved, the V_H gene locus on the nonrearranged or unsuccessfully rearranged chromosome is recruited to heterochromatic regions of the nucleus, where it remains inactive.

Once a complete heavy chain is expressed, light-chain rearrangement begins. In mice, light-chain rearrangement always begins on one of the κ alleles, and continues until a productive light-chain rearrangement is completed. In humans, light-chain recombination may begin at either a κ or a λ light-chain locus. Because successive rearrangements (using upstream $V_κ$ regions and downstream $J_κ$ regions) can occur on the same κ-chain chromosome (see below) and there are ultimately four alleles (two κ and two λ) to choose from, once a B cell has made a complete heavy chain, it will usually be able to rearrange a light chain and progress to maturity. Successful completion of light-chain rearrangement results in the expression of the BCR on the cell surface, and this signals the end of further Ig receptor gene rearrangement events (**Figure 6-15**).

The process of generating a fully functional set of B cells is energetically expensive for the organism because such a high fraction of B cells fail to undergo productive heavy-chain rearrangements. On average, two out of three attempts at the first heavy-chain chromosomal locus will result in an

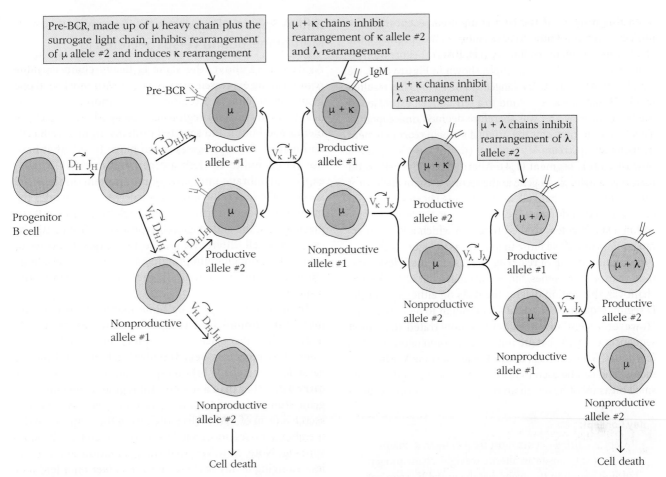

FIGURE 6-15 Generation of a functional immunoglobulin receptor requires productive rearrangement of heavy- and light-chain gene segments. Heavy-chain gene segments rearrange first, and once a productive heavy-chain gene rearrangement occurs, the pre-BCR containing the μ protein product prevents rearrangement of the other heavy-chain allele and initiates light-chain rearrangement. In the mouse, rearrangement of κ light-chain genes precedes rearrangement of the λ genes, as shown here. In humans, either κ or λ rearrangement can proceed once a productive heavy-chain rearrangement has occurred. Formation of IgM inhibits further light-chain gene rearrangement. If a nonproductive rearrangement occurs for one allele, then the cell attempts rearrangement of the other allele.

unproductively rearranged heavy chain because two-thirds of the rearranged sequences are out of frame, and the same is true for the rearrangement process at the second heavy-chain locus. Therefore the probability of successfully generating a functional heavy chain is only 1/3 (at the first allele) + (2/3 [probability the first rearrangement was not successful] × 1/3 [probability that the second rearrangement is successful]) = 0.55 or 55%.

Key Concepts:

- B cells express Ig heavy- and light-chain genes from only one allele (allelic exclusion) and one light-chain gene family.

- Heavy-chain rearrangement precedes light-chain rearrangement, and in mice (but not humans) κ-chain rearrangement precedes that of λ chains.

- The generation of a functional BCR is energetically expensive because of the considerable wastage involved, as a result of a significant frequency of nonproductive gene rearrangements.

Receptor Editing of Potentially Autoreactive Receptors Occurs in Light Chains

A given B cell can still make some changes in its receptor, even after it is expressed on the cell surface. If the receptor is found to be autoreactive, the cell can swap it out for a different one, in a process termed **receptor editing**. Since the receptor's binding site contains elements of both the heavy chain and the light chain, changing just one of these is often sufficient to alter the specificity, and light-chain receptor editing has been shown to occur quite frequently.

In this process, if the first completed receptor in the immature B cell is found to have autoimmune specificity, the DNA rearrangement machinery (i.e., RAG1/2 expression) is switched back on. In the example, shown in **Figure 6-16**, we stipulate that the initial rearrangement of V3 to J3 has resulted in a BCR that was able to bind to a self-antigen expressed in the bone marrow and was therefore deemed unacceptable. This B cell can now edit the original κ-chain rearrangement by using an upstream V_κ segment (in this case $V_\kappa2$) and a downstream J_κ segment ($J_\kappa4$). Alternatively, the cell could rearrange a light-chain gene at the second κ-chain allele, or it can rearrange a λ-chain allele. All of these choices have been demonstrated in different B cells. However, the term *receptor editing* usually refers to the process in which a B cell rearranges the same allele more than once, as in Figure 6-16.

Receptor editing of rearranged V_H segments is less common than with V_k genes. Since rearrangement of the heavy-chain variable region deletes the RSSs on either side of the D region sequences, it was at first thought to be impossible. However, recent advances have demonstrated that short sequences of DNA that are imperfectly homologous to RSSs (cryptic RSSs) exist (e.g., at the 3′ ends of some V-gene segments) and can be used in heavy-chain editing, and so some small measure of heavy-chain receptor editing does occur.

Key Concept:

- Receptor editing occurs most frequently in κ chains and is used to create an alternative light-chain antigen binding region in the event that the initial V_κ segment resulted in an autoreactive antibody. Receptor editing uses a secondary rearrangement that employs V_κ and J_κ segments upstream and downstream, respectively, from the original rearrangement.

mRNA Splicing Regulates the Expression of Membrane-Bound versus Secreted Ig

As described above, the same Ig heavy-chain variable region can be expressed in association with more than one heavy-chain constant region. Furthermore, each of the heavy-chain constant regions can be synthesized as both membrane-bound and secreted proteins. Immature B cells and mature B lymphocytes (both naïve B cells that have not yet encountered their antigen and memory B cells) express membrane immunoglobulin protein on their cell surface. These B cells only make mRNA that encodes the membrane-bound form of immunoglobulin. In contrast, plasma cells, the end-stage cell type arising from antigen-induced B-cell differentiation, secrete large quantities of soluble immunoglobulin protein and therefore make large amounts of mRNA encoding the secreted form of the protein.

Immature B cells in the bone marrow express only membrane-bound IgM receptors. However, as B cells mature, they also place IgD receptors on the cell surface. These IgD receptors bear the identical antigen-binding variable region as the IgM receptors on that same cell, but carry a δ, rather than a μ, constant region. Following antigenic stimulation of a mature B cell, the cell may express a secreted form of IgM. Cells switch from the synthesis of the membrane-bound to secreted forms of IgM and IgD by RNA splicing. Note, however, that the generation of antibodies belonging to any heavy-chain class *other than* IgM and IgD occurs by an additional DNA recombination process, referred to as **class switch recombination (CSR)**. Unlike RNA splicing, CSR results in the loss of the DNA encoding the μ and δ constant region gene segments. The process of CSR is described in Chapter 11.

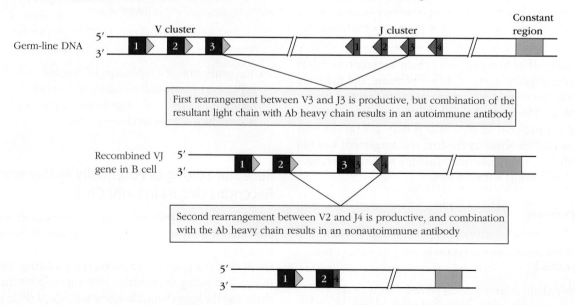

FIGURE 6-16 κ light-chain receptor editing. In the event of a productive light-chain rearrangement that leads to the formation of an autoreactive antibody, additional rounds of rearrangement may occur between upstream V_κ and downstream J_κ gene segments. Here, the primary rearrangement between V3 and J3 is edited out with a secondary rearrangement between V2 and J4.

The Ig heavy-chain RNA transcript in a mature naïve B cell encodes the rearranged V region and both the C_μ and the C_δ constant regions. Both constant regions are encoded in a series of exons separated by introns. The unspliced transcript is approximately 15 kb in length.

Analysis of this primary Ig transcript revealed that the $C\mu4$ exon (the 3′-most exon of the μ heavy chain) ends in a nucleotide sequence called S (secreted), which encodes the hydrophilic carboxyl-terminal end of the C_H4 domain of secreted IgM. The S portion of the mRNA sequence also includes a polyadenylation site, polyadenylation site 1. Two additional exons, M1 and M2 (M for membrane), are located 1.8 kb downstream of the 3′ end of the $C\mu4$ exon. M1 encodes the membrane-spanning region of the heavy chain and M2 encodes the three cytoplasmic amino acids at the carboxyl terminus of IgM, followed by a second site, polyadenylation site 2 (**Figure 6-17**).

Production of mRNA encoding the membrane-bound form of the μ chain occurs when cleavage of the primary transcript and addition of the poly-A tail occurs at polyadenylation site 2 (see Figure 6-17a). RNA splicing then removes the S sequence at the 3′ end of the $C\mu4$ exon and the RNA from the two introns, and joins the remainder of the $C\mu4$ exon to the M1 and M2 exons. This is the only splicing pattern that occurs in *immature* B cells, which express only membrane-bound IgM.

As the B cell matures, it begins to express membrane IgD in addition to membrane IgM. The exons encoding the membrane-bound and secreted forms of IgD are arranged similarly to those of IgM, with polyadenylation sites 3 and 4 at the 3′ termini of the sequences encoding the secreted and membrane-bound forms of IgD, respectively. If polyadenylation occurs at site 4, the C_μ RNA is spliced out and the J segment of the variable region is joined to the C_δ RNA which is processed to create the mature mRNA encoding the heavy chain of membrane bound IgD (see Figure 6-17a). Since mature B cells bear both mIgM and mIgD, it is clear that both these splicing patterns can occur simultaneously. IgD transcription is associated with the expression of the protein ZFP318.

Following antigenic stimulation, the B cell begins to generate the secreted form of IgM (Figure 6-17b). Polyadenylation shifts to site 1, and the mRNA sequences downstream of site 1 are degraded. Similarly, the use of splice site 3 generates secreted IgD (not shown).

At the protein level, the carboxyl terminus of the membrane-bound form of the Ig μ chain consists of a sequence of about 40 amino acids, which is composed of a hydrophilic segment that extends outside the cell, a hydrophobic transmembrane segment, and a very short hydrophilic segment at the carboxyl end that extends into the cytoplasm (Figure 6-17c). In the secreted form, these 40 amino acids are replaced by a hydrophilic sequence of about 20 amino acids in the carboxyl-terminal domain.

What mechanisms determine which splice sites are used? This question is not yet fully resolved. However, work with artificial constructs containing different combinations of the various splice sites suggests that there is an intrinsic "strength" to each of the splice sites that determines how frequently it is used, and that this strength reflects the binding affinity between the DNA sequence at the splice site and the spliceosome machinery. Analysis of the splice site sequences suggests that splice site 2 is considerably "stronger" than splice site 1, which helps to explain why immature B cells predominantly make the membrane-bound form of the μ chain. B-cell stimulation appears to cause an increase in the concentration of the splice site recognition proteins, as well as the early termination of transcription prior to the M1 and M2 sequences, thereby facilitating the use of splice site 1 and generation of the secreted form of the protein. A similar mechanism may control the differential splicing of the membrane-bound and secreted forms of the δ chain.

Current experiments are addressing questions of regulation of Ig mRNA splicing by methodologies such as placing the various splice sites on separate pieces of DNA in order to dissect out any intrachromosomal competition mechanisms between potential splice sites, and by performing site-specific mutagenesis experiments on the splice and poly-A recognition sites. A search for proteins differentially regulated at the mature B-cell and plasma-cell stages of development may also yield potential regulators of mRNA splicing.

Key Concepts:

- Mature B cells simultaneously express membrane-bound IgM and membrane-bound IgD receptors that bear the same binding site. The variable and constant regions of these two heavy-chain classes are synthesized on one long pre-mRNA, and differential RNA splicing is used to generate membrane forms of μ and δ heavy chains.

- Following antigenic stimulation, cells differentiating into antibody-secreting plasma cells generate secreted forms of IgM (or IgD). The shift from synthesizing membrane-bound to secreted IgM is controlled at the level of RNA splicing and is one of many changes that occur during B-cell differentiation into plasma cells.

- The expression of the constant regions of heavy-chain classes other than μ or δ requires a DNA recombination event, in which upstream constant region genes are deleted. This event is termed *class switch recombination* (CSR).

(a)

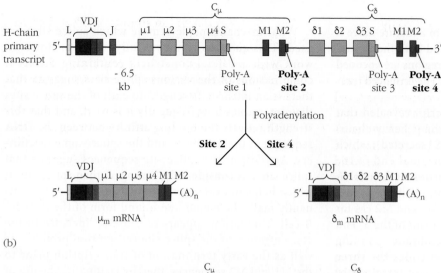

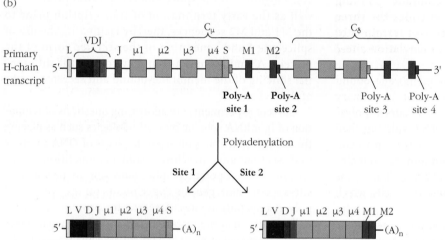

(b)

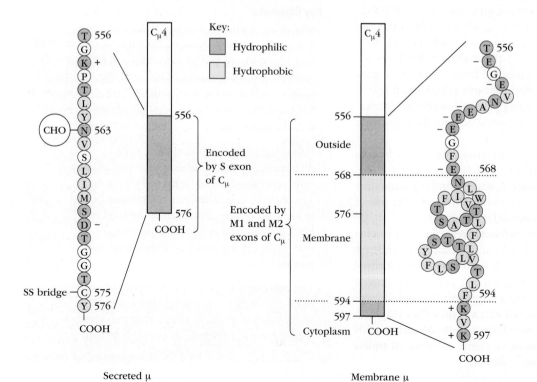

(c)

FIGURE 6-17 Differential expression of the secreted and membrane-bound forms of immunoglobulin μ and δ chains is regulated by alternative RNA processing. (a) Structure of H-chain primary transcript showing C_μ and C_δ exons and polyadenylation sites. Polyadenylation of the primary transcript at either site 2 or site 4, followed by splicing of the intervening DNA, gives rise to membrane Ig of the IgM or IgD class, respectively. (b) Polyadenylation of the primary transcript at either site 1 or site 2, and subsequent splicing, generates mRNAs encoding either secreted or membrane μ chains respectively, as shown. (c) Amino acid sequences of the carboxyl-terminal ends of secreted and membrane μ heavy chains. Residues are described by the single-letter amino acid code. Hydrophilic (pink) and hydrophobic (yellow) residues and regions are shown, charged amino acids are indicated with a + or −, and the site of carbohydrate attachment at an N-linked glycosylation site on the secreted form is labeled "CHO."

T-Cell Receptor Genes and Their Expression

The initial characterization of the BCR molecule was facilitated by the fact that the secreted Ig product of the B cell shares the antigen-binding region with the membrane receptor. The secreted Ig proteins could be used as antigens to generate antibodies in other species that recognized the B-cell surface receptor. These anti-receptor antibodies could then be employed as reagents to isolate and characterize the BCR. In addition, a significant number of naturally occurring and artificially generated antibody-secreting tumors were available to the investigators studying BCR structure.

Investigators attempting to purify and analyze the **T-cell receptor (TCR)** enjoyed no such advantages, and several decades elapsed between the characterization of the BCR and that of its T-cell counterpart. In Chapter 3, we described the structure of the TCR protein and its coreceptors. Below, we describe the story of the discovery of the TCR genes.

Understanding the Protein Structure of the TCR Was Critical to the Process of Discovering the Genes

In Chapter 3 we described how investigators characterized the TCR as an αβ heterodimeric protein (Figure 3-15). Each chain of the receptor consists of one variable and one constant region domain, and both chains contain regions arranged in the characteristic double β-sheet Ig fold. The presence of these variable and constant regions provided clues as to the potential arrangement of the genes that encode the TCR proteins, and enabled researchers to design the experiments that ultimately led to characterization of the TCR genes.

> **Key Concept:**
> • Limited knowledge of the TCR structure enabled investigators to develop strategies for TCR gene discovery.

The β-Chain Gene Was Discovered Simultaneously in Two Different Laboratories

Two different research groups published papers in the same issue of *Nature* in 1984 that described the discovery of the TCR genes in mice and in humans respectively. Stephen Hedrick, Mark Davis, and their collaborators used a uniquely creative approach, which we describe below, to isolate the β-chain receptor genes from several mouse T-cell hybridoma cell lines. The other group, led by Tak Mak, used more classical methods of molecular genetic analysis but achieved the same measure of success in isolating the β-chain gene of the human TCR.

The Davis-Hedrick group strategy was based on four assumptions about TCR genes:

- TCR genes will be expressed in T cells but not B cells;
- Since the genes encode a membrane-bound receptor, the transcribed mRNA will be found associated with membrane-bound polyribosomes;
- Like Ig genes, TCR genes will encode a variable region and a constant region;
- Like Ig genes, the genes that encode the TCR will undergo rearrangement in T cells.

Prior experiments had revealed that only about 2% of the genes expressed by lymphocytes differed between T and B cells. Furthermore, only a small proportion (about 3%) of T-cell mRNA had been shown to be associated with membrane-bound polyribosomes. Davis and Hedrick reasoned that if they could generate ^{32}P-labeled *DNA* copies (cDNA) of the membrane-bound polyribosomal fraction of antigen-specific T-cell mRNA, and remove from that population those sequences that were also expressed in B cells, they would be left with labeled, T cell–derived cDNA that would be greatly enriched in sequences encoding the TCR (**Figure 6-18**).

They therefore performed the following steps:

- Extracted membrane-bound polyribosomal mRNA (RNA sequences that encoded membrane-bound proteins) from a T-cell hybridoma line.
- Reverse-transcribed the polyribosomal mRNA in the presence of ^{32}P-labeled nucleotides, to generate radioactive cDNA copies of the mRNA.
- Mixed the ^{32}P-labeled T-cell cDNA with B-cell mRNA and removed all the cDNA that hybridized with B-cell mRNA, leaving behind only the T-cell mRNA that encoded proteins not expressed in B cells.
- Recovered the T cell–specific cDNAs and used them to identify hybridizing DNA clones from a T cell–specific library. This step confirmed the T-cell origin of the cDNA sequences and allowed the investigators to obtain cloned cDNA to use as reproducible probes in future experiments.
- Radioactively labeled the cloned DNA sequences that hybridized to the T cell–specific cDNAs. Used these cloned probes, individually, to probe Southern blots of DNA from liver cells and B-cell lymphomas (which would not be expected to rearrange TCR genes), as well as DNA from different helper T-cell hybridoma cell lines (which would be expected to show different patterns of rearrangements in different T-cell clones).

Probing with the T cell–specific TM90 cDNA probe (see bottom of Figure 6-18) resulted in a pattern of bands that did not vary depending on the origin of the cellular DNA. This indicated that, although the TM90 probe was derived from mRNA sequences that are specifically expressed in T cells, it is recognizing a DNA sequence that does *not*

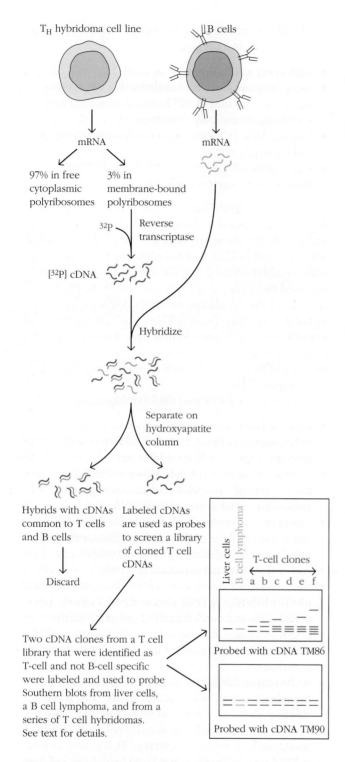

FIGURE 6-18 Production and identification of a cDNA clone encoding the T-cell receptor β gene. The flowchart outlines the procedure used by Hedrick, Davis, and colleagues to obtain ^{32}P-labeled cDNA clones corresponding to T cell–specific mRNAs. DNA subtractive hybridization was used to isolate T cell–specific cDNA fragments that were cloned and labeled with [32]P. The labeled cDNA clones were used to probe clones from a cDNA T-cell library, confirming the T-cell origin of the cDNA and providing cloned DNA with which they subsequently probed Southern blots of DNA isolated from embryonic liver (TCR genes should be in the germ-line configuration), from a B-cell lymphoma (TCR genes should also be in the germ-line configuration), and from a panel of T-cell clones (TCR genes should be differently arranged in each clone). One clone, TM86, was shown to encode a gene that rearranged differently in each T-cell clone analyzed. Comparison of its sequence with that of the protein sequence of the αβ TCR isolated by Kappler and Marrack revealed TM86 to encode the β chain of the TCR. TM90 cDNA identified the gene for another T-cell membrane molecule that does not undergo rearrangement. (For more information, see the following references: Hedrick, S. M., D. I. Cohen, E. A. Nielsen, and M. M. Davis. 1984. Isolation of cDNA clones encoding T cell–specific membrane-associated proteins. *Nature* **308**:149; and Haskins K., J. Kappler, and P. Marrack, 1984. The major histocompatibility complex-restricted antigen receptor on T cells. *Annual Review of Immunology* **2**:51).

of these patterns were different from the bands detected by this probe in either the liver or the B-cell DNA. This banding pattern is consistent with what would be expected of a DNA sequence that encoded a T-cell receptor gene, since, in each T-cell hybridoma, V(D)J gene rearrangement would alter the TCR DNA in a different way, leading to varying band patterns in a Southern blot. This proved Davis and Hedrick's stated assumption that the genes encoding the TCR recombine in T cells, but not in any other cell type, and demonstrated that they had cloned a TCR gene. In this way the mouse TCR gene encoding the receptor β chain was isolated.

Comparison of the DNA sequence of Mak's human clone with the amino acid sequence from a human hybridoma confirmed that Mak and colleagues, like Davis, Hedrick, and their coworkers, had isolated the gene encoding the TCR β gene.

> **Key Concept:**
>
> • Two groups identified the first TCR β-chain genes in mouse and humans, respectively. One of these groups used an approach that relied on the fact that rearrangement of these genes is unique to T cells.

A Search for the α-Chain Gene Led to the γ-Chain Gene Instead

Focus now switched to the search for the TCR α chain. But here, the science of immunology took one of its strange and wonderful twists.

rearrange in different ways in different cells, and therefore is probably *not* complementary to a TCR gene. (It was later learned that TM90 hybridized to the *Thy-1* gene, expressed in mouse T cells, testis, and brain).

In contrast, the TM86 cDNA probe showed different patterns of hybridization when tested against DNA derived from different T-cell hybridomas. This indicated that the sequence recognized by the TM86 probe is found in varying sequence contexts within the DNA of different T cells. Furthermore, all

Tonegawa's lab, using the subtractive hybridization approach pioneered by Hedrick and Davis, succeeded in cloning a gene that appeared at first to have all the hallmarks of the TCR α-chain gene. It was expressed in T cells, but not in B cells; it was rearranged in T-cell clones; and, on sequencing, it revealed regions corresponding to a signal peptide, two Ig family domains, a transmembrane region, and a short cytoplasmic peptide. Furthermore, the predicted molecular weight of the encoded chain appeared to be very close to that of the α-chain protein. Tonegawa's lab initially suggested that: "Thus, while direct evidence is yet to be produced, it is very likely the pHDS4/203 codes for the α subunit of the T cell receptor."

However, biochemical analysis of the TCR α- and β-chain proteins by another laboratory had clearly demonstrated that both the α and β chains of the TCR heterodimer were glycosylated. A search for potential sequences corresponding to sites of carbohydrate attachment on the putative α-chain sequence came up short. It therefore seemed that the gene Tonegawa's lab had isolated did *not* encode the α chain, as they had thought, but something very similar. Further work demonstrated that they had discovered a gene encoding a hitherto unknown receptor chain that contained both variable and constant regions and recombined only in T cells. They named this gene (and the receptor chain it encoded) γ.

Davis and colleagues next isolated a genomic sequence expressed in T and not B cells that encoded a protein with four potential N-linked glycosylation sites, and a molecular weight consistent with that of the TCR α chain.

With the genes for the α and β chains of the TCR fully characterized, those attempting to understand TCR genetics were then left with a series of intriguing questions. Was the γ chain expressed by the same T cells that expressed the αβ receptor? Did the γ chain that Tonegawa had discovered also exist as a heterodimer? If so, what was its partner, and who would find it first?

After 3 years of searching, Chien, Davis, and colleagues described a fourth TCR gene that they named δ. Intriguingly, the V_δ-chain gene was found to lie downstream of the previously described V_α genes and just 5′ to the J_α and C_α coding regions. This unusual location dictated that if a T cell rearranged the α chain, the intervening V_δ genes would be lost (**Figure 6-19**). Rearrangement of the δ locus was found to occur very early in thymic differentiation, and expression of δ-chain RNA appeared to occur in the same T cells in which γ-chain RNA was expressed. Study of the T cells expressing these γδ receptors demonstrated that γδ-bearing T cells represented an entirely new T-cell subset, with an antigen repertoire distinct from that of αβ T cells and a different anatomical distribution within the lymphoid system in the host.

Key Concept:

• The three remaining TCR chains, α, γ, and δ, were identified, and revealed that two classes of TCR, made up of αβ and γδ dimers, are expressed in distinct populations of T cells.

TCR Genes Are Arranged in V, D, and J Clusters of Gene Segments

Figure 6-19a shows the arrangement of the TCR genes in the mouse genome. The four TCR loci are organized in the germ line in a manner remarkably similar to the multigene organization of the Ig loci, shown in Figure 6-6; and as in the case of Ig genes, functional TCR genes are produced by rearrangements of V and J segments in the α- and γ-chain families and between V, D, and J segments in the β- and δ-chain genes.

As shown in Figure 6-19b, the mouse TCR β locus occupies approximately 450 kb of genomic DNA that includes approximately 35 V_β segments, of which about 22 are functional. These are located roughly 250 kb upstream from two clusters of D_β, J_β, and C_β gene segments. A further two V_β segments exist, one of which is positioned 115 kb upstream of the majority of the V_β segments whereas the other was found on the reverse strand, downstream of $C_\beta 2$. Each of the D_β and J_β clusters contains a single D_β and six J_β gene segments.

The mouse TCR α locus is composed of approximately 128 V_α gene segments, of which about 115 or so are functional. These are located in a gene region of approximately 1.5 Mb. A gap of about 200 kb separates the V_α cluster from a particularly extensive J_α cluster of approximately 40 functional J_α gene segments (and additional open reading frames or pseudo-J_α gene segments). Figure 6-19 also illustrates the fact that interspersed within the V_α gene segments are the gene segments that comprise the TCR δ locus, which contains approximately 15 V_δ segments, two D_δ segments, two J_δ segments, and one C_δ region. Interestingly, nine of the V_α segments have also been shown to be used with the D_δ, J_δ, and C_δ constant region gene segments, thus enhancing the number of options for generating a diverse TCR repertoire. Unlike the other three loci, the TCR γ locus is relatively small (approximately 175 kb; Figure 6-19b) and contains few V_γ (seven) segments and four J_γ-C_γ clusters.

Like the immunoglobulin loci, then, each TCR variable region consists of one chain that is encoded by V, D, and J segments (β and δ receptor chains) and one that is encoded only by V and J segments (α and γ receptor chains).

Key Concepts:

• Like BCR genes, TCR genes are arranged in clusters of V, D, and J segments.

• β and δ receptor genes include multiple V, D, and J segments; α and γ receptor genes, like Ig light-chain genes, include only V and J segments, of which there are again multiple different segments.

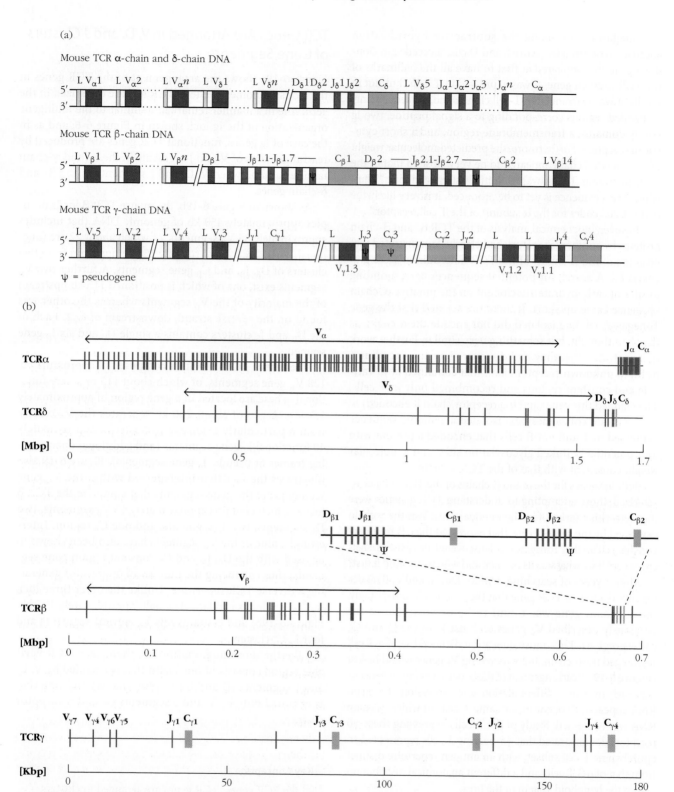

FIGURE 6-19 Germ-line organization of the mouse TCR α-, β-, γ-, and δ-chain gene segments. Each C gene segment is composed of a series of exons and introns, which are not shown. The organization of TCR gene segments in humans is similar. (a) Diagrammatic representation of the V, D, and J segments of the four families of TCR gene segments, showing their relative orientations with respect to one another on the three mouse chromosomes. (b) The V, D, and J segments of the four families of TCR gene families are illustrated so as to indicate both the approximate numbers of gene segments and the overall size of each of the V gene loci. Note the scale bars that describe the overall size of each locus. [Data from Jhunjhunwala, S., et al. 2008. Chromatin architecture and the generation of antigen receptor diversity. Cell. **138**:435.]

Recombination of TCR Gene Segments Proceeds at a Different Rate and Occurs at Different Stages of Development in αβ versus γδ T Cells

The development of T cells from common lymphoid progenitors occurs primarily in the thymus, and the two major T-cell lineages, αβ and γδ T cells, both develop there from T progenitor cells. This process will be described in detail in Chapter 8. However, in order to grasp the principles of TCR gene segment recombination it is useful to briefly introduce the stages of T-cell development here.

When the common lymphoid progenitor cell leaves the bone marrow and enters the thymus, it is referred to as a "double-negative" (DN) thymocyte. This is because it bears neither of the two T-cell differentiation antigens, CD4 or CD8, on its cell surface. The DN stage is further subdivided into four, sequential steps, DN1–4, on the basis of the expression of a number of cell surface markers (see Chapter 8). At the end of DN4, the thymocyte acquires first CD8, and then rapidly thereafter CD4, to become a double-positive (DP) thymocyte.

γδ T cells are the first population to appear in the thymus during the development of a mouse and can be found in the fetal mouse thymus as early as 14 days of gestation, declining in frequency after birth so that, in an adult animal, γδ T cells represent only approximately 0.5% of mature thymocytes. Many γδ TCRs recognize unusual antigens such as lipids and glycolipids. They do not need to recognize peptides presented as a molecular complex with proteins of the immune system, specifically the proteins encoded by the major histocompatibility complex, which is a defining feature of αβ TCR recognition (see Chapter 7). T cells bearing γδ receptors complete their differentiation and leave the thymus at the DN stage. Many of them find their eventual homes in mucosal and skin tissues.

Cells destined to become αβ T cells proceed to the DP stage where they are tested for their ability to recognize antigen displayed by immune system cells (positive selection) without inducing autoimmunity (negative selection). Ninety percent or more of the developing thymocytes will die in the thymus because of their inability to recombine functional, nonautoreactive receptor molecules. The process of receptor rearrangement and evaluation in T cells, as in B cells, is therefore energetically extremely expensive.

Unlike with B-cell light-chain genes, receptor editing does not appear to play a part in the thymic selection process. Although many T cells appear to rearrange their TCR α genes more than once during development, the frequency of additional rearrangements is not increased in the presence of an autoantigen, indicating that editing is not used as a strategy to rescue T cells from the generation of autoimmune receptors.

Key Concept:

- T cells bearing γδ receptors complete their differentiation at the DN stage, whereas cells destined to become αβ T cells proceed to the DP stage and undergo positive and negative selection.

The Process of TCR Gene Segment Rearrangement Is Very Similar to Immunoglobulin Gene Recombination

In DN thymocytes the chromatin regions on which the TCR α/δ and β variable region gene segments reside become accessible to the RAG1/2 recombinase, and locus contraction is initiated. At this point in development, variable region gene segment recombination begins simultaneously at all three loci. The mechanisms and rules for these recombination events appear to be similar to those established for B cells. Animals deficient in RAG1/2 or the other enzymes affecting Ig gene recombination, such as Artemis, are similarly deficient in the generation of both their BCRs and their TCRs (see Clinical Focus Box 6-2). Heptamer-nonamer RSSs are found at the 3' termini of the V sequences, the 5' and 3' termini of the D sequences, and the 5' termini of the J sequences, just as for the Ig variable region gene segments.

However, there is one important difference between the control of receptor gene segment recombination in B and T cells. Recall that the 12/23 rule for Ig gene segment movement specifies that recombination can occur only between gene segments flanked in one case with an RSS containing a 12 (one turn)-bp spacer and in the other with an RSS containing a 23 (two turn)-bp spacer (see Figure 6-8). By this rule, recombination between V_H and J_H immunoglobulin gene segments is disallowed, and hence all properly rearranged heavy-chain V regions must include V, D, and J segments.

In contrast, the D regions of the TCR β- and δ-chain genes are flanked on the upstream side by a 12-bp spacer and on the downstream by a 23-bp spacer (**Figure 6-20**). Downstream of $V_β$ is a 23 bp spacer and a 12 bp spacer is found upstream of the $J_β$ region. The same pattern recurs at the δ locus. Thus, according to the 12/23 rule, V and J segments from the β and δ TCR gene families could theoretically recombine directly, without an intervening D segment. Furthermore, the rule also allows for D-D segment recombination. Do these unusual recombination events occur in vivo?

In the case of the β-chain gene, the answer is that they do not. Evidence obtained from sequencing cDNA fragments containing precisely known V, D, and J sequences has demonstrated that functioning TCR β-chain genes always contain one each of the V, D, and J segments. The mechanisms that forbid the occurrence of $D_β$-$D_β$ recombination, despite the fact that such joints would obey the

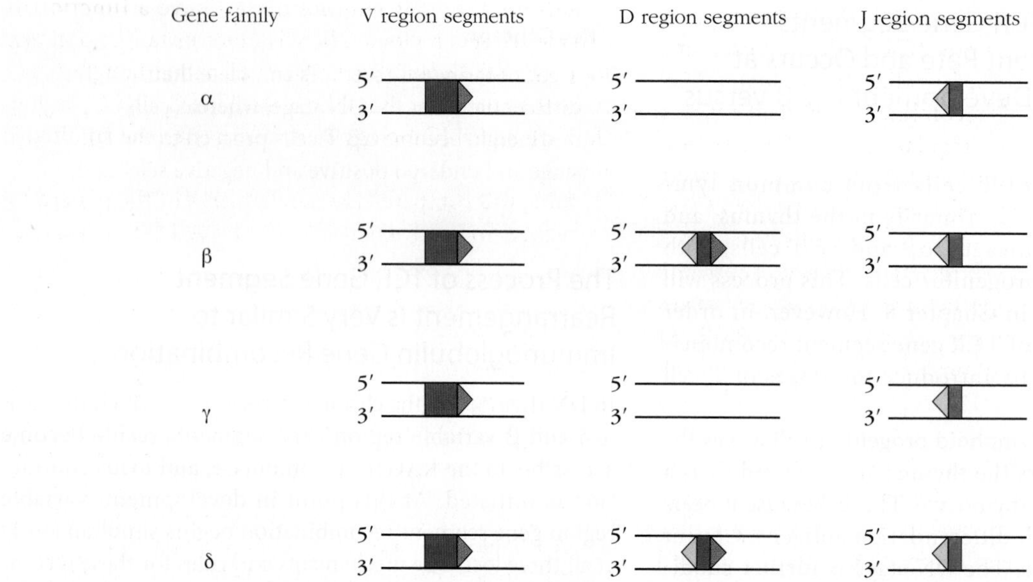

Gene family	V region segments	D region segments	J region segments
α	5' ▬▬ 3'	5' ——— 3'	5' ◁▬ 3'
β	5' ▬▬ 3'	5' ◁▬▷ 3'	5' ◁▬ 3'
γ	5' ▬▬ 3'	5' ——— 3'	5' ◁▬ 3'
δ	5' ▬▬ 3'	5' ◁▬ 3'	5' ◁▬ 3'

FIGURE 6-20 Locations of the 12-bp and 23-bp RSS spacers in TCR genes. Note that the spacers on either side of the D_β and D_δ gene segments are different, allowing for the occurrence of recombined genes bearing zero, one, or two copies of the D_δ gene segments. Duplication of D_β segments has not been observed in vivo.

12/23 rule at this locus, are not yet fully understood. However, investigators have suggested that the precise sequence of the relevant RSSs, the occurrence of germ-line transcription only at limited D_β loci, and potentially the binding of particular transcription factors to the D_β RSSs with resultant increase in their use in D_β-J_β recombination may be factors in the selection of single D_β regions for recombination.

In contrast, the situation is more complex for TCR δ-chain genes. Scientists noted early on that many TCR V_δ genes were considerably longer than TCR V_β genes; furthermore, the increased sequence length seemed to occur primarily in those parts of the gene encoding the CDR3 residues that make antigen contact. Subsequent analysis showed that many TCR V_δ genes have incorporated not one, but two, D region segments, and demonstrated that this additional segment is primarily responsible for the observed increase in gene length—an increase that places the length of V_δ regions close to that of immunoglobulin heavy-chain variable regions. In some V_δ genes, further length is added to the TCR δ-chain gene by N-nucleotide addition at the D-D joint.

An additional important question concerns the pathway of differentiation of the nascent T cells: If the β, γ, and δ variable region genes are all undergoing rearrangement at the same time in these early thymocytes, how does the cell determine whether it is going to differentiate into one of the majority (90%) αβ-bearing, T cells or instead become a γδ T cell?

It appears that a developing T cell allows the RAG recombinase a single cycle of attempted rearrangement involving the V_δ gene segments. If this is successful, and if the cell also generates a recombined γ chain, the functionality of the resulting γδ receptor is tested in the thymus. If the receptor is found to be acceptable (i.e., does not give rise to an autoimmune T-cell specificity) the cell is fixed as a double-negative, γδ T cell and leaves the thymus at this point.

However, if β-gene rearrangement is successful, a different fate ensues. In this case, the developing thymocyte transcribes and translates the V_β gene, expressing the protein on the thymocyte surface in combination with the pre-Tα protein to form the pre-T-cell receptor, or pre-TCR (**Figure 6-21**). This is directly analogous to the

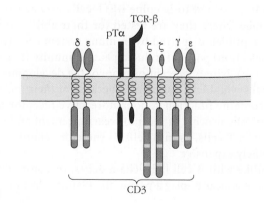

FIGURE 6-21 The pre-TCR: the TCR β chain is expressed on the T-cell surface in combination with the pre-Tα chain. Signaling through the pre-TCR halts further β-chain rearrangement and signals for several rounds of proliferation, followed by α-chain rearrangement.

expression of the completed immunoglobulin heavy chain on the B-cell surface as the pre-BCR, with the VpreB and λ5 proteins (see Figure 6-7). Signaling from the pre-TCR complex via CD3 turns off recombination at the β, γ, and δ loci by inducing the down-regulation of RAG1/2 expression and the degradation of RAG2. Pre-TCR signaling also results in the initiation of several rounds of proliferation that maximize the use of the successfully rearranged β-chain gene. This is again functionally parallel to the proliferative activity that marks successful Ig heavy-chain gene recombination and expression of the pre-BCR. At the end of several rounds of proliferation, the nascent cell makes the transition to the DP stage and initiates α-chain rearrangement.

α-Chain gene rearrangement is notably well ordered. Rearrangement is initiated at the most proximal V_α segments and, if these rearrangements are unsuccessful, they are followed by rearrangements involving distally located V_α segments. Computer modeling estimates suggest that an individual TCR α allele undergoes an average of three to five cycles of V_α-to-J_α rearrangement. If the resultant αβ TCR is successfully tested against intrathymic antigens during the process of negative selection (see Chapter 8), rearrangement is terminated, *RAG* gene expression is ended, and thymocyte development is completed. At this point, the DP thymocyte loses either CD4 or CD8 expression and becomes a mature, single-positive ($CD4^+CD8^-$ or $CD4^-CD8^+$) thymocyte. If the T cell either fails positive selection, or expresses an autoreactive phenotype and therefore fails negative selection, it will undergo programmed cell death.

Testing of the TCRs takes time, as the rearranged receptor genes must be transcribed and translated and the resultant receptors expressed on the cell surface before their reactivity can be assessed. This suggests that there is a mechanism that pauses V_α-J_α recombination after each cycle to allow this to occur. The protein *a*taxia *t*elangiectasia *m*utated (ATM) has been implicated in temporarily suppressing V(D)J *RAG* gene expression in B cells, and may be operating in the same way in T cells.

Two additional points of comparison between the rearrangement events in B and T cells should be considered. Recall that in the mouse, Ig light chains incorporate few N nucleotides because of the developmental down-regulation of expression of TdT in B cells by the time of light-chain rearrangement. In contrast, N nucleotides are seen at similar frequencies in all four of the TCR chains. It is tempting to speculate that this additional measure of diversity in the TCR makes up for the absence of somatic hypermutation following antigen contact in T cells. Note that the presence of multiple D_δ gene segments in some γδ TCRs provides more junctions at which N nucleotides can be added, dramatically increasing the diversity of both sequence and length of the antigen-contacting CDR3 regions of the γδ TCR.

Second, the TCR families do not have a functionally diverse set of C regions. There is one constant region gene for each of the α and δ gene families, and although there is more than one gene encoding each of the C_β and C_γ regions, they are highly homologous and do not appear to differ in function.

Table 6-3 compares the mechanisms of the generation and expression of diversity in BCR versus TCR molecules.

Key Concepts:

- TCR genes use the same RSSs as BCR genes, but the arrangement of RSSs allows D-D rearrangement in TCR genes (only in δ chains), enhancing the diversity in δ CDR3 regions.

- The TCR V_α and V_δ gene segments are interspersed in a single chromosomal region. Rearrangement of a functional α chain causes the deletion of δ D, J, and C gene segments, thus precluding the rearrangement of the δ chain.

- Rearrangement of the β, δ, and γ genes of the TCR begins at the same time at the DN stage of thymocyte development. If successful γ and δ V gene rearrangements occur, the T cell leaves the thymus as a γδ T cell. If δ gene rearrangement is unsuccessful but the cell generates a functional β-chain gene, the cell undergoes several rounds of proliferation and then attempts α gene rearrangement. If this step is successful and the receptor passes through thymic selection, it leaves the thymus as an αβ T cell.

TCR Expression Is Controlled by Allelic Exclusion

Allelic exclusion at the TCR β- and γ-chain loci occurs almost as efficiently as among Ig heavy- and light-chain genes. In contrast, approximately 30% of αβ T cells bear more than one successfully recombined TCR α chain (see Chapter 8). However, it would be unlikely, although not impossible, that a T cell bearing two α chains would be able to recognize more than one antigen, given the complexity of T-cell antigen recognition. And it should be noted that the presence of even one αβ TCR capable of binding self-antigens at high affinity would result in the elimination of the cell, irrespective of the specificity of the receptor formed using an alternative α chain.

Key Concept:

- Allelic exclusion is not as complete in T cells as in B cells, in that some T cells bear more than one α chain.

Conclusion

From early in the twentieth century, when it was first appreciated that antibody molecules were capable of specifically recognizing and binding to an enormously diverse array of antigens, immunologists have sought to comprehend how a finite amount of genetic information can possibly encode such a large number of specific receptor molecules in the lymphocytes of the adaptive immune response. We now understand that B- and T-cell receptor molecules are encoded in families of short gene segments, the members of which are recombined uniquely in different lymphocytes to encode the adaptive immune system receptor repertoire.

Receptors in T and B cells are made up of two different chains, which can be recombined in different ways. Furthermore, when two receptor gene segments come together, additional diversity is generated by the nontemplated addition of varying numbers of nucleotides at the junctions between the segments. These extremely variable sequences at the junctions of gene segments correspond to the regions on the antigen receptors that contact the antigen, the complementarity-determining regions. Because of the random addition and deletion of nucleotides at gene segment junctions, many recombined receptor genes do not encode viable proteins, and when this happens, the nascent B and T cells are destroyed. The extraordinary receptor variability that is the hallmark of the adaptive immune system therefore requires considerable sacrifice in terms of cellular energy. The timing of these recombination events is exquisitely regulated during T- and B-cell development. Although the overall outline of the process of generating the receptor genes is the same in B and T cells, subtle differences in the details tailor the receptors to the functions of the specific cells.

REFERENCES

Carmona, L. M., S. D. Fugmann, and D. G. Schatz. 2016. Collaboration of RAG2 with RAG1-like proteins during the evolution of V(D)J recombination. *Genes and Development.* **30**:909.

Chien, Y., D. M. Becker, T. Lindsten, M. Okamura, D. I. Cohen, and M. M. Davis. 1984. A third type of murine T-cell receptor gene. *Nature.* **312**:31.

Deng, Z., J. Liu, and X. Liu. 2015. RAG1-mediated ubiquitylation of histone H3 is required for chromosomal V(D)J recombination. *Cell Research.* **25**:181.

Desiderio, S. V., et al. 1984. Insertion of N regions into heavy-chain genes is correlated with expression of terminal deoxytransferase in B cells. *Nature.* **311**:752.

Dreyer, W. J., and J. C. Bennett. 1965. The molecular basis of antibody formation: a paradox. *Proceedings of the National Academy of Sciences USA.* **54**:864.

Early, P., H. Huang, M. Davis, K. Calame, and L. Hood. 1980. An immunoglobulin heavy chain variable region gene is generated from three segments of DNA: V_H, D and JH. *Cell.* **19**:981.

Early, P., et al. 1980. Two mRNAs can be produced from a single immunoglobulin μ gene by alternative RNA processing pathways. *Cell.* **20**:313.

Feeney, A. J. 2011. Epigenetic regulation of antigen receptor gene rearrangement. *Current Opinion in Immunology.* **23**:171.

Fugmann, S. D. 2010. The origins of the RAG genes: from transposition to V(D)J recombination. *Seminars in Immunology.* **22**:10.

Hedrick, S. M., D. I. Cohen, E. A. Nielsen, and M. M. Davis. 1984. Isolation of cDNA clones encoding T cell–specific membrane-associated proteins. *Nature.* **308**:149.

Hozumi, N., and S. Tonegawa. 1976. Evidence for somatic rearrangement of immunoglobulin genes coding for variable and constant regions. *Proceedings of the National Academy of Sciences USA.* **73**:3628.

Jhunjhunwala, S., et al. 2008. The 3D structure of the immunoglobulin heavy-chain locus: implications for long-range genomic interactions. *Cell.* **133**:265.

Kavaler, J., M. M. Davis, and Y. Chien. 1984. Localization of a T-cell receptor diversity-region element. *Nature.* **310**:421.

Kim, M.-S., M. Lapkouski, W. Yang, and M. Gellert. 2015. Crystal structure of the V(D)J recombinase, RAG1-RAG2. *Nature.* **518**:507.

Kumari, G., and R. Sen. 2015. Chromatin interactions in the control of immunoglobulin heavy chain gene assembly. *Advances in Immunology.* **128**:41.

Ma, Y., U. Pannicke, K. Schwarz, and M. R. Lieber. 2002. Hairpin opening and overhang processing by an Artemis/DNA-dependent protein kinase complex in nonhomologous end joining and V(D)J recombination. *Cell.* **108**:781.

Oettinger, M. A., D. G. Schatz, C. Gorka, and D. Baltimore. 1990a. RAG-1 and RAG-2, adjacent genes that synergistically activate V(D)J recombination. *Science.* **248**:1517.

Rother, M. B., et al. 2015. Nuclear positioning rather than contraction controls ordered rearrangements of immunoglobulin loci. *Nucleic Acids Research.* **44**:175.

Saito, H., D. M. Kranz, Y. Takagaki, A. C. Hayday, H. N. Eisen, and S. Tonegawa. 1984. A third rearranged and expressed gene in a clone of cytotoxic T lymphocytes. *Nature.* **312**:36

Schatz, D. G., and Y. Ji. 2011. Recombination centers and the orchestration of V(D)J recombination. *Nature Reviews Immunology.* **11**:251.

Shetty, K., and D. G. Schatz. 2015. Recruitment of RAG1 and RAG2 to chromatinized DNA during V(D)J recombination. *Molecular and Cellular Biology.* **35**:3701.

Shimazaki, N., A. G. Tsai, and M. R. Lieber. 2009. H3K4me3 stimulates the V(D)J RAG complex for both nicking and hairpinning in *trans* in addition to tethering in *cis*: implications for translocations. *Molecular Cell.* **34**:535.

Teng, G., and D. G. Schatz. 2015. Regulation and evolution of the RAG recombinase. *Advances in Immunology.* **128**:1.

Yanagi, Y., Y. Yoshikai, K. Leggett, S. P. Clark, I. Aleksander, and T. W. Mak. 1984. A human T cell–specific cDNA clone encodes a protein having extensive homology to immunoglobulin chains. *Nature.* **308**:145.

Useful Websites

http://imgt.org "IMGT" stands for *international ImMuno-GeneTics information system.* The creators of this website monitor the literature for information on the numbers of genes encoding Ig and TCRs in a variety of species, and regularly update information regarding the numbers of gene segments that have been sequenced. Also includes information on other gene families related to the immune system, such as the MHC. Up-to-date maps of immune-related genes, including Ig and TCR genes, are provided for several species.

www.imgt.org/IMGTeducation/ This is a subset of the above site that provides a wealth of teaching resources about immunogenetics in multiple languages.

www.cdc.gov/bam/diseases/immune/ The Centers for Disease Control and Prevention have a website entirely devoted to the immune system. Good for up-to-date information on how the immune system responds to infectious disease.

www.cellular-immunity.blogspot.com/2007/12/vdj-recombination.html A clear and brief explanation of V(D)J recombination, with clickable links.

www.hhmi.org/biointeractive The Howard Hughes Medical Institute has a wealth of educational materials available at its website, including some superb lectures on basic immunology.

www.ncbi.nlm.nih.gov The National Center for Biotechnology Information (NCBI) site offers library tools as well as numerous sequence analysis and protein structure resources (under the "Resources" tab) pertinent to immunology. In particular, detailed information can be located about the mouse immunoglobulin heavy-chain, κ-chain, and λ-chain genetic loci at https://www.ncbi.nlm.nih.gov/gene/111507, -243469, and -111519, respectively. Similar up-to-date information can be derived about the TCR α/δ, β, and γ loci at the same URL, https://www.ncbi.nlm.nih.gov/gene/21473, -21577, and -110067, respectively.

http://what-when-how.com/molecular-biology/gene-rearrangement-molecular-biology/ A useful, and very clear explanation, with diagrams, of V(D)J recombination.

STUDY QUESTIONS

1. Indicate whether each of the following statements is true or false. If you think a statement is false, explain why.

a. In generating a B-cell receptor gene, V_κ segments sometimes join to C_λ segments.

b. In generating a T-cell receptor gene, V_α segments sometimes join to C_δ.

c. The switch in constant region use from IgM to IgD is mediated by DNA rearrangement.

d. Although each B cell carries two alleles encoding the immunoglobulin heavy and light chains, only one allele is expressed.

e. Like the variable regions of the heavy chain of the B-cell Ig receptor, TCR variable genes are all encoded in three segments.

2. a. For the immunoglobulin κ light chain, sketch the arrangement of its genetic elements (V, J, and C segments) and show how that arrangement is altered in the mature B cell.

b. Do the same for the immunoglobulin heavy chain.

c. Now repeat for the TCR β chain, being sure to indicate the positions of both C_β gene segments relative to their upstream D_β units.

d. For the TCR α chain, repeat the sketch, being sure to demonstrate the effect of the rearrangement of TCR α-chain variable region gene segments on the potential for TCR δ-chain expression.

3. Explain why a V_H segment cannot join directly with a J_H segment in heavy-chain gene rearrangement.

4. For each incomplete statement below, select the phrase(s) that correctly completes the statement. More than one choice may be correct.

a. Recombination of Ig gene segments serves to:

  (1) Promote Ig diversification

  (2) Assemble a complete Ig coding sequence

  (3) Allow changes in coding information during B-cell maturation

  (4) Increase the affinity of Ig for antibody

  (5) All of the above

b. κ and λ light-chain genes:

  (1) Are located on the same chromosome

  (2) Associate with only one type of heavy chain

  (3) Can be expressed by the same B cell

  (4) All of the above

  (5) None of the above

   c. Generation of combinatorial diversity within the variable regions of Ig involves

     (1) mRNA splicing
     (2) DNA rearrangement
     (3) Recombination signal sequences
     (4) 12/23 joining rule
     (5) Switch sites

   d. The mechanism that permits Ig to be synthesized in either a membrane-bound or secreted form is:

     (1) Allelic exclusion
     (2) Codominant expression
     (3) Class switch recombination
     (4) The 12/23 joining rule
     (5) Differential RNA processing

   e. During Ig V_H recombination, the processes that contribute to additional diversity at the third complementarity-determining region of Ig variable regions include:

     (1) Introduction of the D gene segments into the heavy-chain V gene
     (2) mRNA splicing of the membrane form of the C region of the constant chain
     (3) Exonuclease cleavage of the ends of the gene segments
     (4) P-nucleotide addition
     (5) N-nucleotide addition

5. Compare and contrast the roles of the following enzymes/enzyme complexes in V(D)J recombination:

RAG1/2

TdT

Artemis

DNA PKcs

DNA ligase IV

6. You have been given a cloned mouse myeloma cell line that secretes IgG with the molecular formula $\gamma_2\lambda_2$. Both the heavy and light chains in this cell line are encoded by genes derived from allele 1 (i.e., the first of the two homologous alleles encoding each type of chain). Indicate the form(s) in which each of the genes listed below would occur in this cell line, using the following symbols: G = germ-line form; P = productively rearranged form; NP = nonproductively rearranged form. State the reason for your choice in each case.

   a. Heavy-chain allele 1
   b. Heavy-chain allele 2
   c. κ-chain allele 1
   d. κ-chain allele 2
   e. λ-chain allele 1
   f. λ-chain allele 2

7. You have identified a B-cell lymphoma that has made nonproductive rearrangements for both heavy-chain alleles. What is the arrangement of its light-chain DNA? Why?

8. The random addition of nucleotides by TdT is a wasteful and potentially risky evolutionary strategy. State why it may be disadvantageous to the organism and why, therefore, you think it is sufficiently useful to have been retained during vertebrate evolution.

9. Describe three ways in which the structure and organization of the immunoglobulin and TCR variable region genes in native chromatin aid in regulating the order in which those genes are rearranged.

10. Are there any differences in the genetic strategies used to generate complete V genes in TCRs compared with BCRs? If so, what are they?

11. In V(D)J rearrangement, we describe how segments of DNA are moved within the chromosome to create an active immunoglobulin or TCR gene. Can you think of an instance in which the expression of an immunoglobulin gene is controlled at the level of RNA splicing? Describe one such instance. What are the alternative outcomes in terms of immunoglobulin protein expression and how are they generated by RNA splicing?

12. What known features of the TCR did Hedrick, Davis, and colleagues use in their quest to isolate the TCR genes?

13. The following figure describes the end of a V region sequence and the beginning of the D region sequence to which it is about to be joined. Arrows mark the cleavage points where the RAG1/RAG2 complex will make the cut and recombination will be targeted.

   a. Is this a heavy-chain or a light-chain sequence? How do you know?
   b. What DNA sequence structure would you find just downstream of the AGCATC sequence immediately adjacent to the 3′ end of the V segment?
   c. Here is one possible V-D joint structure formed after recombination between these two gene segments:

     5′ —V— AGCATCGACGCCGTATCGA —D— 3′
     3′ — TCGTAGCTGCGGCATAGCT —

     (1) Which residue(s) *may* be P nucleotide(s)?
     (2) Can we know for certain that this residue is a P nucleotide?
     (3) Which residues *must* have been added by TdT, and therefore must be N nucleotides?
     (4) Can we know for certain whether a residue is an N nucleotide?

14. Which of the following two statements is true? Please explain the biological rationale that underlies your answer:

   a. Following recombination of the Ig heavy-chain genes, the B cell divides several times before commencing light-chain gene rearrangement.
   b. Following recombination of the Ig light-chain genes, the B cell divides several times before commencing heavy-chain gene rearrangement.

15. Only 55% of Ig heavy-chain gene rearrangements are successful. Explain.

16. In the following table, put a "+" in any space where the descriptor at the top of the table accurately describes the process or molecules shown on the left-hand side, and a "–" where the description does not apply.

	V(D)J rearrangement	V-J rearrangement	D-D joining	12/23 rule obeyed	N-nucleotide addition	More than one C region	Allelic exclusion perfect
Ig heavy chain							
Ig light chain							
TCR α							
TCR β							
TCR γ							
TCR δ							

ANALYZE THE DATA

In Classic Experiment Box 6-1, Figure 1, we see the pattern of bands that Hozumi and Tonegawa's experiment would have revealed, using the particular cell line that they used. Below, see a pair of gels that represent the results of a hypothetical experiment performed using the same general protocol. In this hypothetical experiment, our probes correspond to either the V region or the C region. Furthermore, the investigators used a different tumor cell line and a different restriction endonuclease.

a. Why are there two C region bands in the germ-line blot?
b. Why are these two bands still present in the myeloma blot, and why are there two bands recognized by the V region probe on the myeloma blot?
c. From this plot, would you hypothesize that the cell had achieved a successful arrangement at the first κ allele? Why or why not?
d. How would you prove your answer to part c?

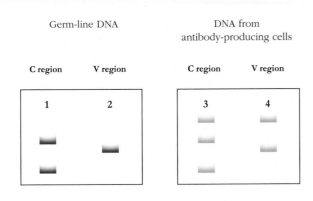

Germ-line DNA   DNA from antibody-producing cells

C region   V region   C region   V region

1   2   3   4

The Major Histocompatibility Complex and Antigen Presentation

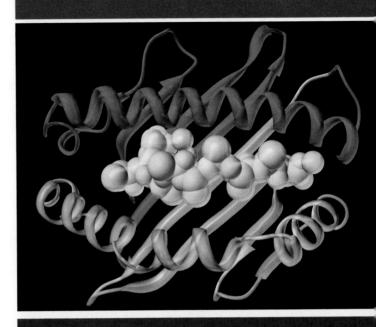

Learning Objectives

After reading this chapter, you should be able to:

1. Illustrate and categorize the structures, expression patterns, and source of antigen for MHC class I and II molecules, identifying connections between these molecules and receptors on the T cells that recognize them.

2. Describe how MHC genes are inherited in groups (haplotypes) that encode a set of alleles unique to each individual, and use this information to predict the outcome of tissue transplantation between individuals who are not completely MHC matched.

3. Recognize the ways in which diversity is built into the MHC locus, how this diversity complements the diversity of MHC binding partners (T-cell receptors; Chapter 6), and how it will influence the selection of T cells in the thymus (T-cell development; Chapter 8).

4. Predict how variations in the MHC binding groove influence the array of antigen fragments that will be presented to the immune system and therefore how MHC diversity contributes to survival of the individual and the species.

Although both T and B cells use surface molecules to recognize antigen, they accomplish this in very different ways. In contrast to antibodies or B-cell receptors,

Ribbon diagram of a human MHC class I molecule (light and dark blue) with a space-filled peptide (orange) held in the binding groove.

which can recognize free antigen, T-cell receptors only recognize pieces of antigen that are first positioned on the surface of other cells. These antigen pieces, or antigenic peptides, are held within the binding groove of a cell-surface protein called the **major histocompatibility complex (MHC) molecule**. MHC molecules are encoded by a cluster of closely associated genes collectively called the **MHC locus**, or simply the **MHC**. The peptide fragments that bind to MHC molecules are generated inside the cell following antigen digestion, and the complex of the antigenic peptide plus MHC molecule is transported to the cell surface. MHC molecules thus act as cell-surface vessels for holding and

Key Terms

Major histocompatibility complex (MHC) molecule

MHC locus

MHC class I molecules

MHC class II molecules

Antigen-presenting cells (APCs)

Antigen processing

Antigen presentation

β_2-Microglobulin

Anchor residues

Histocompatibility antigens

Human leukocyte antigen (HLA) complex

H2 complex

Polymorphic

Haplotype

Codominant

Polygenic

Professional antigen-presenting cells (pAPCs)

Endogenous pathway (of antigen processing/presentation)

Exogenous pathway (of antigen processing/presentation)

Immunoproteasome

TAP (transporter associated with antigen processing)

Invariant chain (Ii)

Cross-presentation

displaying fragments of antigen so that approaching T cells can engage with these molecular complexes via their T-cell receptors. This may be surprising but as we will see, these MHC molecule–peptide complexes can be found all over the body and T cells must make constant judgment calls about whether and how to react when they encounter these complexes.

The MHC got its name from the fact that the genes in this region encode proteins that determine whether a tissue transplanted between two individuals will be accepted or rejected. The pioneering work of Benacerraf, Dausset, and Snell helped to characterize the functions controlled by the MHC locus, specifically, organ transplant fate and the immune responses to antigen. This work resulted in the 1980 Nobel Prize in Physiology or Medicine for the trio (see Table 1-2). Follow-up studies by Rolf Zinkernagel and Peter Doherty illustrated that the proteins encoded by these genes play a seminal role in adaptive immunity by showing that T cells need to recognize MHC proteins, as well as antigen. Structural studies done by Don Wiley and others showed that different MHC proteins bind and present different antigen fragments. Most MHC genes in the population exist in multiple alternative forms called alleles, and the specific alleles one inherits can play a significant role in response to infection, susceptibility to disease, and the development of autoimmunity. This family of molecules exerts a strong influence on the development of immunity to nearly all types of antigens. In fact, study of the MHC locus has become a major theme in immunology, far beyond its origins in the field of transplantation biology.

There are three classes of MHC molecules: class I, class II, and class III. Members of the first two classes have a similar shape and both are responsible for displaying antigen to T cells, although they differ in their roles and in the way in which their quaternary or final three-dimensional structures are generated. Other molecules classified as MHC class III also play roles in the immune response, although these roles are more varied and generally do not involve direct presentation of antigen fragments to T cells.

This chapter will begin by discussing the structure of MHC class I and II molecules, followed by inheritance patterns and the genetic organization of the DNA region that encodes these proteins. We then turn to the function of the MHC in regulating immunity, and regulation of MHC gene expression, while exploring several seminal studies in these areas. This is followed by a detailed discussion of the cellular pathways that lead to antigen degradation and association of peptide with each type of MHC molecule (**antigen processing**) and the appearance of these MHC molecule–peptide complexes on the cell surface for recognition by T cells (**antigen presentation**). In some of these pathways, as well as in later chapters, the very diverse MHC class III proteins will also make an appearance. Of course, the role of particular **antigen-presenting cells (APCs)** in these processes will also be presented (no pun intended!). The chapter concludes with a discussion of some unique processing and presentation pathways, such as cross-presentation and the handling of nonpeptide antigens by the immune system.

The Structure and Function of MHC Class I and II Molecules

MHC class I molecules and **MHC class II molecules** are membrane-bound glycoproteins that are closely related in both structure and function (**Overview Figure 7-1**). Both classes of MHC molecule have been isolated and purified, and the three-dimensional structures of their extracellular domains have been resolved by x-ray crystallography. These membrane glycoproteins function as highly specialized antigen-presenting molecules with grooves that form unusually stable complexes with peptide ligands, displaying them on the cell surface for recognition via T-cell receptor (TCR) engagement. In contrast, MHC class III molecules are a group of unrelated proteins that do not share structural similarity or function with class I and II molecules, although many of them do participate in other aspects of the immune response.

Class I Molecules Consist of One Large Glycoprotein Heavy Chain Plus a Small Protein Light Chain

Two polypeptides assemble to form a single MHC class I molecule: a 45-kilodalton (kDa) α chain and a much smaller 12-kDa β chain, called **β₂-microglobulin** (see Overview Figure 7-1a, top and middle). The α chain is organized into three external domains (α1, α2, and α3), each approximately

---

(Figure on next page)

Top: The peptide-binding groove is formed by the membrane-distal domains (light and dark blue) in both class I and class II molecules. The membrane-proximal domains (teal and salmon) possess the immunoglobulin domain structure; thus, both MHC class I and II molecules, as well as β₂-microglobulin, are classified as members of the immunoglobulin superfamily. *Middle*: Side-view ribbon diagrams based on x-ray crystallography of the MHC class I molecule (*left*) and MHC class II molecule (*right*), in which the β strands are depicted as thick arrows and the α helices as spiral ribbons. Each disulfide bond is shown as two interconnected yellow spheres. The α1 and α2 domains of class I, and the α1 and β1 domains of class II, interact to form the peptide-binding groove. *Bottom*: The peptide-binding grooves of MHC class I (*left*) and MHC class II (*right*) as viewed from the top, consisting of a base of antiparallel β strands and sides of α helices for each molecule. This groove can accommodate peptides of 8 to 10 residues for class I and 13 to 18 residues for class II.

Schematic Diagrams of MHC Class I (a) and MHC Class II (b) Molecules, Showing the External Domains, Transmembrane Segments, Cytoplasmic Tails, and Peptide-Binding Groove

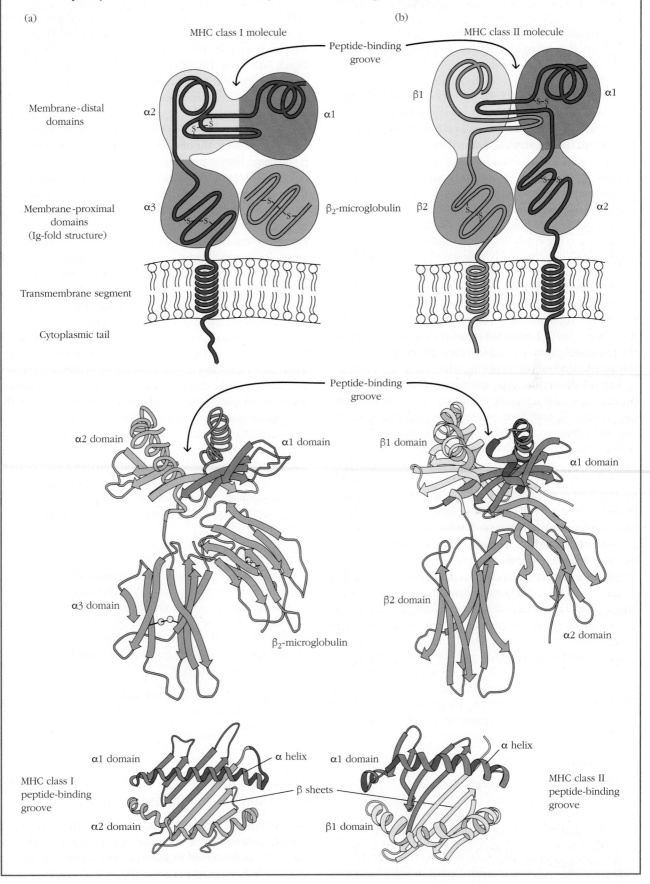

(a)

MHC class I molecule

Peptide-binding groove

Membrane-distal domains

α2   α1

Membrane-proximal domains (Ig-fold structure)

α3   β₂-microglobulin

Transmembrane segment

Cytoplasmic tail

(b)

MHC class II molecule

Peptide-binding groove

β1   α1

β2   α2

Peptide-binding groove

α2 domain   α1 domain

β1 domain   α1 domain

α3 domain

β₂-microglobulin

β2 domain

α2 domain

MHC class I peptide-binding groove

α1 domain   α helix

α2 domain   β sheets

MHC class II peptide-binding groove

α1 domain   α helix

β1 domain   β sheets

90 amino acids long; a transmembrane domain of about 25 hydrophobic amino acids followed by a short stretch of charged (hydrophilic) amino acids; and a cytoplasmic anchor segment of 30 amino acids. Its companion, β_2-microglobulin, is similar in size and organization to the $\alpha3$ domain. β_2-Microglobulin does not contain a transmembrane region and is noncovalently bound to the MHC class I α chain. Sequence data reveal strong homology between the $\alpha3$ domain of MHC class I, β_2-microglobulin, and the constant region domains found in immunoglobulins.

The $\alpha1$ and $\alpha2$ domains interact to form a floor of eight antiparallel β strands rimmed by two long α helices (see Overview Figure 7-1a, bottom). The structure forms a groove, or cleft, with the long α helices on the sides and the β strands as the bottom. This *peptide-binding groove* is located on the top surface of the MHC class I molecule, and it is large enough to bind a peptide of 8 to 10 amino acids. During the x-ray crystallographic analysis of class I molecules, small noncovalently associated peptides that had cocrystallized with the protein were found in the groove. The big surprise came when these peptides were later discovered to be fragments of processed *self-proteins* and not the foreign antigens that were expected. We'll see later in this chapter, as well as in subsequent chapters, that MHC molecule/self-peptide complexes are ubiquitous and also play an essential role in regulating tolerance to self-proteins.

The $\alpha3$ domain and β_2-microglobulin share a similar structure characteristic of immunoglobulin domains (see Chapter 3): both form two antiparallel β pleated sheets (teal and salmon in Overview Figure 7-1a, middle). Because of this structural similarity (not surprising, given the considerable sequence similarity with the immunoglobulin constant region) MHC class I α chain and β_2-microglobulin are classified as members of the immunoglobulin superfamily (see Figure 3-7). The $\alpha3$ domain appears to be highly conserved among MHC class I molecules and contains a sequence that interacts strongly with the CD8 cell-surface molecule found on T_C cells.

These three molecules (class I α chain, β_2-microglobulin, and peptide), associated solely through noncovalent interactions, are all essential to the proper folding and expression of the MHC-peptide complex on the cell surface. This was demonstrated in vitro using Daudi tumor cells, which are unable to synthesize β_2-microglobulin. These tumor cells produce MHC class I α chains but do not express them on the cell membrane. However, if Daudi cells are transfected with a functional gene encoding β_2-microglobulin, class I molecules with peptide will appear on the cell surface.

Key Concept:

- MHC class I molecules consist of one large transmembrane glycoprotein α chain with an inherent binding pocket that accommodates peptides of approximately 8 to 10 amino acids, plus a much smaller, noncovalently associated β chain, called β_2-microglobulin.

Class II Molecules Consist of Two Nonidentical Membrane-Bound Glycoprotein Chains

MHC class II molecules contain two different glycoprotein chains of similar size, a 33-kDa α chain and a 28-kDa β chain, which associate by noncovalent interactions (see Overview Figure 7-1b, top and middle). Like the class I α chain, both MHC class II chains are membrane-bound glycoproteins that contain external domains, a transmembrane segment, and a cytoplasmic anchor segment. Each chain in a class II molecule contains two external domains: $\alpha1$ and $\alpha2$ domains in one chain and $\beta1$ and $\beta2$ domains in the other. The membrane-proximal $\alpha2$ and $\beta2$ domains, like the membrane-proximal $\alpha3/\beta_2$-microglobulin domains of MHC class I molecules, bear sequence similarity to the immunoglobulin fold structure (see Overview Figure 7-1b, middle). For this reason, MHC class II molecules are also classified as belonging to the immunoglobulin superfamily. The membrane-distal $\alpha1$ and $\beta1$ domains form the peptide-binding groove for processed antigen (see Overview Figure 7-1b, middle and bottom). Although similar to the peptide-binding groove of MHC class I, the groove in MHC class II molecules is formed by the association of two separate chains; an important distinction between the class I and class II structures.

X-ray crystallographic analysis reveals the similarity between these two classes of molecule, strikingly apparent when class I and class II molecules are superimposed (**Figure 7-2**). Interestingly, despite the fact that these two structures are encoded quite differentially (one chain versus two), the final quaternary structure is similar and retains the same overall function: the ability to bind antigen and present it to T cells. The peptide-binding groove of class II molecules, like that found in class I molecules, is composed of a floor of eight antiparallel β strands and two sides of antiparallel α helices. Class II–associated peptides are frequently slightly larger than class I peptides, typically ranging from 13 to 18 amino acids.

Key Concepts:

- MHC class II molecules are composed of two noncovalently associated transmembrane glycoproteins (an α chain and a β chain) with similar structures, where their extracellular domains together form a binding pocket accommodating peptides of approximately 13 to 18 residues in length.

- MHC class I and II molecules are structurally and functionally similar, despite being encoded differently.

Class I and II Molecules Exhibit Polymorphism in the Region That Binds to Peptides

Several hundred different allelic variants of MHC class I and II molecules have been identified in the human population. However, as described in the next section of this chapter, any one individual expresses only a small subset of these

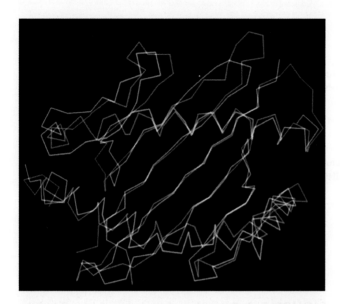

FIGURE 7-2 The peptide-binding groove of a human MHC class II molecule (blue), superimposed over the corresponding regions of a human MHC class I molecule (red). Overlapping regions are shown in white. These two molecules create very similar binding grooves such that most of the structural differences (long stretches, kinks, bends, etc.) lie outside the peptide-binding regions. *[Permission from Macmillan Publishers Ltd: from Brown, J. H., et al. 1993. Three-dimensional structure of the human class II histocompatibility antigen HLA-DR1. Nature 364:33, Fig 2B. Permission conveyed through Copyright Clearance Center, Inc.]*

specificity characteristic of the antigen-specific receptors discussed in Chapter 6 (T- and B-cell receptors, as well as antibodies). Instead, *any given MHC molecule can bind numerous different peptides, and some peptides can bind to several different MHC molecules.* Because of this broad specificity, the binding between a peptide and an MHC molecule is often referred to as "promiscuous." This promiscuity of peptide binding allows many different peptides to match up with the MHC binding groove and also allows exchange of peptides to happen on occasion, unlike the relatively stable, high-affinity cognate interactions of antibodies, B-cell receptors, and T-cell receptors (antigen-specific receptors) with their ligands.

Given the similarities in the structures of the peptide-binding grooves in MHC class I and II molecules, it is not surprising that they exhibit some common peptide-binding features (**Table 7-1**). In both types of MHC molecule, peptide ligands are held in a largely extended conformation that runs the length of the groove. The peptide-binding groove in class I molecules is blocked at both ends, whereas the ends of the groove are open in class II molecules (**Figure 7-3**). As a result of this difference, class I molecules bind shorter peptides than do class II molecules.

MHC Class I–Peptide Interaction

MHC class I molecules bind peptides from intracellular sources and present these to CD8$^+$ T cells. The bound peptides isolated from various class I molecules have two distinguishing features: they are 8 to 10 amino acids in length (most are 9, and are called *nonamers*), and they contain specific amino acid residues at key locations in the sequence. The ability of an individual MHC class I molecule to bind to a diverse spectrum of peptides is due to the presence of the same or similar amino acid residues at these key positions along the peptides (**Figure 7-4**). Because these amino acid residues *anchor* the peptide into the groove of the MHC molecule, they are called

molecules—up to six different class I molecules and about 12 different class II molecules. Yet we know that this limited variety of MHC molecules within an individual is able to present an enormous array of different antigenic peptides to T cells, permitting the immune system to respond specifically to a wide variety of antigenic challenges. Thus, peptide binding by class I and II molecules does not exhibit the fine

TABLE 7-1	Peptide binding by MHC class I and II molecules	
	Class I molecules	**Class II molecules**
Peptide-binding domain	α1/α2	α1/β1
Nature of peptide-binding groove	Closed at both ends	Open at both ends
General size of bound peptides	8–10 amino acids	13–18 amino acids
Peptide motifs involved in binding to MHC molecule	Anchor residues at both ends of peptide; generally hydrophobic at carboxy-terminus	Conserved residues distributed along the length of the peptide anchor
Nature of bound peptide	Extended structure in which both ends interact with MHC groove but middle arches up and away from MHC molecule	Extended structure that is held at a constant elevation above the floor of the MHC groove

(a)

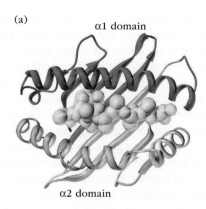

α1 domain

α2 domain

(b)

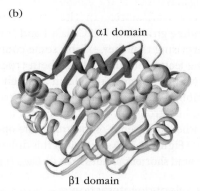

α1 domain

β1 domain

FIGURE 7-3 Peptide-binding groove of MHC class I and class II molecules, with bound peptides. (a) Ribbon model of human MHC class I molecule (α1 domain in dark blue, α2 domain in light blue) with a peptide from the HIV-1 envelope protein GP120 (space-filling, orange) in the binding groove. (b) Ribbon model of human MHC class II molecule, with the α1 domain shown in dark blue and the β1 domain in light blue. The peptide (space-filling, orange) in the binding groove is from influenza hemagglutinin (amino acid residues 306–318). [(a) PDB ID 1HHG; (b) PDB ID 1DLH.]

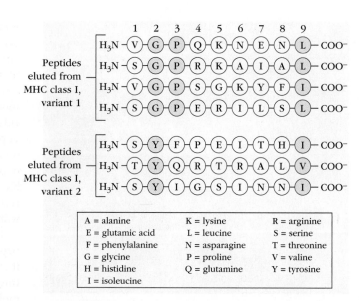

FIGURE 7-4 Examples of anchor residues (blue) in nonameric peptides eluted from two different MHC class I molecules. Anchor residues, at positions 2 (and/or 3) and 9, that interact with the MHC class I molecule tend to be hydrophobic amino acids. The two MHC proteins shown bind peptides with different anchor residues. *[Data from Engelhard, V. H. 1994. Structure of peptides associated with MHC class I molecules. Current Opinion in Immunology 6:13.]*

anchor residues. The side chains of the anchor residues in the peptide are complementary with surface features of the binding groove of the MHC class I molecule. The amino acid residues lining the binding sites vary among different class I allelic variants, and this is what determines the chemical identity of the anchor residues that can interact with a given class I molecule.

Each human or mouse cell can express several unique MHC class I molecules, each with slightly different rules for peptide binding at these anchor residues. Because a single nucleated cell expresses about 10^5 copies of each of these unique class I molecules each with its own peptide promiscuity rules, a wide range of different peptides can be expressed simultaneously on the surface of a nucleated cell. This means that although many of the peptide fragments of a given foreign antigen will be presented to CD8$^+$ T cells, *the particular set of MHC class I alleles inherited by each individual will determine which specific set of peptide fragments from a larger protein will be presented.*

In a critical study of peptide binding by MHC molecules, peptides bound by two different MHC class I proteins (two allelic variants) were released chemically and analyzed by high-performance liquid chromatography (HPLC) mass spectrometry. More than 2000 distinct peptides were found among the peptide ligands released from these two MHC class I molecules. Since there are approximately 10^5 copies of each class I protein per cell, it is estimated that each of the 2000 distinct peptides is presented with a frequency of between 100 and 4000 copies per cell. Evidence suggests that even a single MHC-peptide complex may be sufficient to target a cell for recognition and lysis by a cytotoxic T lymphocyte with a receptor specific for that target structure.

All peptides examined to date that bind to class I molecules contain a carboxyl-terminal anchor (see position 9 in Figure 7-4). These anchors are generally hydrophobic residues (e.g., leucine, isoleucine), although a few charged amino acids have been reported. Besides the anchor residue found at the carboxyl terminus, another anchor is often found at the second, or second and third, positions at the amino-terminal end of the peptide. In general, any peptide of the correct length that contains the same or chemically similar anchor residues will bind to the same MHC class I molecule. Knowledge of these key positions and the chemical restrictions for amino acids at these positions may allow for more nuanced clinical design studies, such as for future vaccines targeted at eliciting protective immunity to particular pathogens.

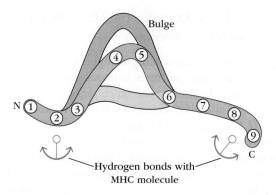

FIGURE 7-5 **Conformation of peptides bound to MHC class I molecules.** Schematic diagram depicting, in a side view, the conformational variance in MHC class I–bound peptides of different lengths. Longer peptides bulge in the middle, presumably interacting more with T-cell receptors, whereas shorter peptides lay flat in the groove. Contact with the MHC molecule is by hydrogen bonds to anchor residues 2 (and/or 3) and 9.

X-ray crystallographic analyses of MHC class I–peptide complexes have revealed how the peptide-binding groove in a given MHC molecule can interact stably with a broad spectrum of different peptides. The anchor residues at both ends of the peptide are buried within the binding groove, thereby holding the peptide firmly in place (see Figure 7-3a). Between the anchors, the peptide can arch away from the floor of the groove in the middle (**Figure 7-5**), allowing peptides that are slightly longer or shorter to be accommodated. Amino acids in the center of the peptide that arch away from the MHC molecule are more exposed and presumably can interact more directly with the T-cell receptor.

MHC Class II–Peptide Interaction

MHC class II molecules also bind a variety of peptides, presenting these to CD4$^+$ T cells. Peptides recovered from MHC class II–peptide complexes generally contain 13 to 18 amino acid residues, longer than the nonameric peptides that most commonly bind to class I molecules. The peptide-binding groove in class II molecules is open at both ends (see Figure 7-3b), allowing longer peptides to extend beyond the ends, like an extra-long hot dog in a bun. Peptides bound to MHC class II molecules maintain a roughly constant elevation on the floor of the binding groove, another feature that distinguishes peptide binding to class I and class II molecules.

The open ends and less conserved nature of the class II binding groove, compared with class I, allow for greater variability in the sequence and length of class II–bound peptides. Peptide-binding studies and structural data for class II molecules indicate that a central core of 13 amino acids determines the ability of a peptide to bind class II. Longer peptides may be accommodated within the class II groove,

but the binding characteristics are determined by the central 13 residues. The peptides that bind to a particular class II molecule often have internal conserved sequence motifs, but unlike class I–binding peptides they appear to lack conserved anchor residues (see Table 7-1). Instead, hydrogen bonds between the backbone of the peptide and the class II molecule are distributed throughout the binding site rather than being clustered predominantly at the ends of the site, as is seen in class I–bound peptides. Peptides that bind to MHC class II molecules contain an internal sequence of 7 to 10 amino acids that provide the major contact points. In general, this sequence has an aromatic (ring-shaped) or hydrophobic residue at the amino terminus and three additional hydrophobic residues in the middle portion and carboxyl-terminal end of the peptide. In addition, over 30% of the peptides eluted from class II molecules contain a proline residue at position 2 and another cluster of prolines at the carboxyl-terminal end. This relative flexibility contributes to peptide binding promiscuity.

Key Concepts:

- MHC class I and II molecules come in many allelic forms and exhibit semipromiscuous peptide binding, allowing them to form stable, noncovalent interactions with a range of different peptide ligands.

- The MHC class I binding pocket is smaller, more closed at the ends, and shows some amino acid anchor residue preferences in bound peptides, while the class II pocket can accommodate longer peptides and shows less restricted amino acid sequence preferences.

The Organization and Inheritance of MHC Genes

Every vertebrate species studied to date possesses the tightly linked cluster of genes that constitute the MHC. As we have just discussed, MHC molecules have the important job of deciding which fragments of an antigen, both foreign and self, will be "seen" by the host T cells. In general terms, MHC molecules face a ligand-binding challenge similar to that faced, collectively, by B-cell and T-cell receptors: they must be able to bind a wide variety of antigens and they must do so with relatively strong affinity. However, these immunologically relevant molecules meet this challenge using very different strategies. Although B- and T-cell receptor diversity is generated through genomic rearrangement and gene editing (described in Chapter 6), MHC molecules use a combination of peptide-binding promiscuity (discussed previously) and the expression of several different MHC molecule variants on every cell (described in this section). Using this clever combined strategy, the immune system has evolved a way of maximizing the chances that many different regions, or *epitopes*, of an antigen will be recognized.

Studies of the MHC gene cluster originated from two observations related to the study of transplantation that coalesced in the early twentieth century. The first was that human red blood cells (RBCs) carried biochemical surface markers that differed between individuals, and these molecules could be detected by antibodies in the serum of others. The second was that cancer cells transplanted between animals of different genetic backgrounds were quickly rejected. The biochemical markers on RBCs were identified by Karl Landsteiner as the ABO blood group antigens, a discovery that paved the way for a medical breakthrough in blood transfusions and resulted in a Nobel Prize for Landsteiner in 1930 (see Table 1-2).

Blood group antigens were also being studied by scientists interested in why some tumor transfers between animals were rejected and others were not. As luck and experimental rigor would have it, British scientist Peter Gorer took up this question and happened to have three inbred strains of mice at his disposal. He identified four groups of genes that encode blood-cell antigens, designated I through IV. Work carried out in the 1940s and 1950s by Gorer and George Snell established that antigens encoded by the genes in group II were responsible for the rejection of transplanted tumors and other tissue, as these antigens were present throughout the body (with another Nobel Prize, this time for Snell in 1980; see Table 1-2). Snell called these *histocompatibility genes*; leading to their current designation as histocompatibility-2 (H2) genes. Thus Gorer's original group II blood-cell antigens are today synonymously referred to **histocompatibility antigens**, MHC antigens, or MHC molecules, regardless of species.

The MHC Locus Encodes the Three Major Classes of MHC Molecules

The major histocompatibility complex is a collection of genes arrayed within a long continuous stretch of DNA on chromosome 6 in humans and on chromosome 17 in mice. The MHC is also referred to as the **human leukocyte antigen (HLA) complex** in humans and as the **H2 complex** (previously H-2) in mice, the two species in which these regions have been most studied. Although the arrangement of genes is somewhat different in the two species, in both cases the MHC genes are organized into regions encoding three classes of molecules (**Figure 7-6**). Within these regions are genes that are considered to be prototypic examples of each class, which are called **classical MHC genes**, as well as some that are considered **nonclassical MHC genes**. We describe each in turn in the following sections.

Classical MHC Genes

The first of the identified and characterized groups of MHC genes, and those that will receive the most attention in any immunology textbook, are the classical MHC genes. These are divided into three groups or classes:

- **MHC class I genes** encode glycoproteins expressed on the surface of nearly all nucleated cells; the major function of the class I gene products is presentation of endogenous (cytosolic) peptide antigens to CD8$^+$ T cells.
- **MHC class II genes** encode glycoproteins expressed predominantly on APCs (macrophages, dendritic cells, and B cells), where they primarily present exogenous (extracellular) peptide antigens to CD4$^+$ T cells.
- **MHC class III genes** encode a diverse set of proteins, some of which have immune functions, but that do not play a direct role in presenting antigen to T cells.

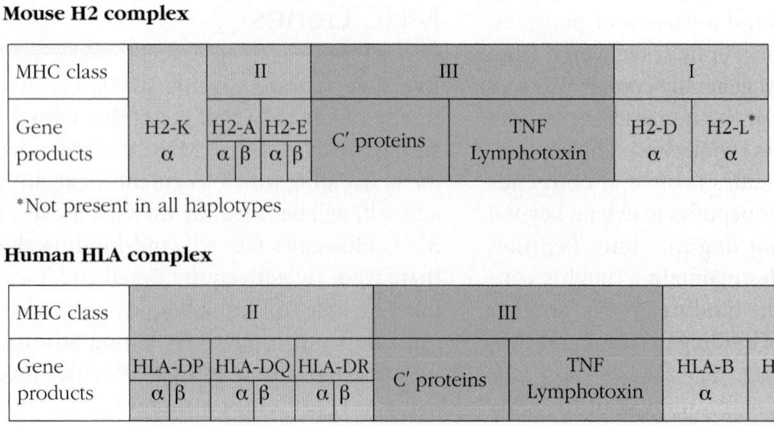

Mouse H2 complex

MHC class	II		III		I	
Gene products	H2-K α	H2-A α β | H2-E α β	C' proteins	TNF Lymphotoxin	H2-D α	H2-L* α

*Not present in all haplotypes

Human HLA complex

MHC class	II			III		I		
Gene products	HLA-DP α | β	HLA-DQ α | β	HLA-DR α | β	C' proteins	TNF Lymphotoxin	HLA-B α	HLA-C α	HLA-A α

FIGURE 7-6 Comparison of the organization of the major histocompatibility complex (MHC) in mice and humans. The MHC is referred to as the *H2 (formerly H-2) complex* in mice and as the *HLA complex* in humans. In both species, the MHC is organized into a number of regions encoding class I (pink), class II (blue), and class III (green) gene products. The class I and II gene products shown in this figure are considered to be the classical MHC molecules. The class III gene products include other immune function–related compounds such as the complement proteins (C') and tumor necrosis factors (TNF and lymphotoxin).

MHC class I molecules were the first to be discovered and are expressed in the widest range of cell types. The protein products from this region of the MHC can be *found on virtually every nucleated cell in humans and mice.* Unlike in humans, this region in mice is noncontinuous, interrupted by the class II and III regions (**Figure** 7-7; also see Figure 7-6). Recall that there are two chains to the MHC class I molecule: the more variable and antigen-binding α chain, and the common β$_2$-microglobulin chain. The α-chain molecules are encoded by the *H2-K* and *H2-D* genes in mice, with an additional *H2-L* locus found in some strains. Likewise, the *HLA-A, HLA-B,* and *HLA-C* loci in humans encode MHC class I genes. β$_2$-Microglobulin is encoded outside the MHC locus, on a separate chromosome altogether (chromosome 15 in humans and chromosome 2 in mice). Collectively, these α chains are referred to as *classical class I molecules*; all possess the functional capability of presenting protein fragments of antigen to CD8$^+$ T cells.

MHC class II proteins show a more limited tissue distribution and more varied expression levels (as we will see shortly). In general, class II molecules are primarily found on the surface of professional APCs, cells with the unique capacity to alert T cells to the presence of a specific antigen.

Recall that MHC class II molecules are composed of two polymorphic chains, α and β, both of which contribute to creating a peptide-binding site. The class II region of the MHC contains three loci in humans, called *HLA-DP, HLA-DQ,* and *HLA-DR.* Within each of these three loci is the coding sequence for both an α chain and a β chain. In mice there are two loci that encode MHC class II α and β chains, called *H2-A* and *H2-E.* In some cases multiple genes are present for either or both chains within a locus. For example, individuals can possess more than one DR β-chain gene on one or both chromosomes, and all of these are expressed simultaneously in the cell. Furthermore, any DR α-chain gene product can pair with any DR β-chain product. Since the antigen-binding groove of class II is formed by a combination of the α and β chains, this can create several unique antigen-presenting MHC class II DR molecules on the cell.

MHC class III molecules bear no structural or functional similarity to classical class I and II molecules, all of which (so far) seem to have direct roles in antigen processing and presentation. Proteins encoded in the class III region are involved in other critical immune functions. For instance, the complement components C4, C2, and factor B

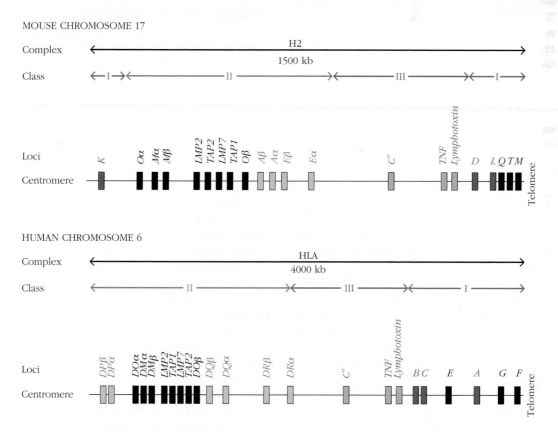

FIGURE 7-7 Simplified map of the mouse and human MHC loci. The classical MHC class I genes are colored red, classical MHC class II genes are colored blue, and genes in MHC class III are colored green. Nonclassical MHC genes are labeled in black. *Note:* The concept of classical and nonclassical does not apply to class III. Only some of the nonclassical class I genes are shown; the functions of only some of their proteins are known.

(see Chapter 5), as well as several inflammatory cytokines, such as the tumor necrosis factor proteins (TNF and lymphotoxin), lie within the class III region. While the genes in the class III region are much less polymorphic than those in the class I and II regions, allelic variants of some class III genes have been linked to certain diseases. For example, polymorphisms within the *TNF* gene, which encodes a cytokine involved in many immune processes (see Chapter 3), have been linked to susceptibility to certain infectious diseases and some forms of autoimmunity, including Crohn's disease and rheumatic arthritis. Despite its differences from class I and class II genes, the class III gene cluster is conserved in all species with an MHC locus, suggesting similar evolutionary pressures have come to bear on this cluster of genes.

Finally, it is worth noting that the nomenclature of the MHC locus is an ever-changing and complicated business. For example, the mouse MHC class II regions called *H2-A* and *H2-E* were previously designated *H2-IA* and *H2-IE*. The human MHC *class II* regions are called *DP*, *DQ*, and *DR*, but the *D* locus in mice encodes MHC *class I* molecules. Likewise, the mouse *class II* region has an *A* locus (formerly called *IA*), easily confused with the *A* locus in the *class I* region of humans (see Figure 7-6). Clearly, like biological evolution, the evolution of nomenclature is messy business!

Nonclassical MHC Genes

Additional genes within the MHC locus encode nonclassical molecules. These have a more restricted tissue expression, less diversity, and more varied roles in immunity. Unlike classical molecules, *nonclassical class I* proteins are expressed only on the surface of specific cell types with more specialized functions. In some cases these molecules play a role in self/nonself discrimination. One example is the HLA-G class I molecule in humans (see Figure 7-7). These are present on the surface of fetal cells at the maternal-fetal interface and are credited with inhibiting rejection by maternal CD8$^+$ T cells by protecting the fetus from identification as foreign, which may occur when paternally derived antigens begin to appear on the developing fetus.

As with the class I region, additional *nonclassical class II* molecules exhibiting little polymorphism and specialized immune functions are encoded within the class II region. For instance, the human *DM* genes are expressed in all antigen-presenting cells and encode a class II–like molecule (HLA-DM) that facilitates the intracellular loading of antigenic peptides into MHC class II molecules. Likewise, class II DO molecules are also involved in antigen processing and presentation. These molecules have a more limited expression profile, found only in specific regions of the thymus, in B cells during certain stages of development, and (discovered more recently) in subsets of dendritic cells. In all cases, DO molecules are believed to serve as an intracellular regulator of class II antigen

processing and presentation. In fact, DM and DO are known to associate with one another and seem to appear only at intracellular locations. It is now believed that DO may be involved in inhibiting or otherwise modifying the function of DM in cells that express both of these proteins. We will encounter these nonclassical MHC molecules again shortly when we turn to antigen processing and presentation.

Key Concepts:

- Classical MHC class I (α-chain) and MHC class II (α and β-chain) molecules are encoded within the MHC locus, function as presenters of antigen, and have similar protein structures but different tissue distribution; class I can be found on all nucleated cells while class II is restricted to APCs.

- The MHC class III region encodes a diverse set of proteins with little polymorphism and no structural or functional similarity to class I and II, although these proteins do play roles in several important immune pathways.

- Within the MHC locus are genes that encode nonclassical MHC proteins with structures similar to class I and class II but that show much less polymorphism, are more limited in tissue distribution (both in time and space), and display more nuanced roles in antigen processing and presentation.

Allelic Forms of MHC Genes Are Inherited in Linked Groups Called Haplotypes

As we have already mentioned, genes that reside within the MHC locus are highly **polymorphic**; that is, many alternative forms of each gene, or **alleles**, exist within the population. This polymorphism clusters in the DNA regions that encode amino acids likely to lie in the antigen-binding groove and therefore likely to come into contact with peptide. Genetic variations outside this region, for instance, in the membrane-proximal or transmembrane domains, are very rare.

The individual genes of the MHC locus (class I, II, and III) lie so close together that their inheritance is linked. Crossover, or recombination between genes, is more likely when genes are far apart. The recombination frequency within the mouse H2 complex (i.e., the frequency of chromosome crossover events during meiosis, indicative of the distance between given genes) is only 0.5%. Thus, crossover occurs only about once in every 200 meiotic cycles. For this reason, most individuals inherit all the alleles encoded by these genes as a set (known as *linkage disequilibrium*). This set of linked alleles is referred to as a **haplotype**. An individual inherits one haplotype from the mother and one haplotype from the father, or two sets of alleles.

In outbred populations, such as humans, the offspring are generally heterozygous at the MHC locus, with different alleles contributed by each of the parents. If, however, a population is highly inbred the offspring can become homozygous, expressing identical MHC molecules because the maternal and paternal haplotypes are identical. Certain mouse strains have been intentionally inbred in this manner (to be homozygous at the MHC) and are employed as prototype strains; that is, sets or strains of mice that are homozygous at the MHC locus for particular alleles. The unique MHC haplotype (in this case, also the genotype) expressed by each of these strains is designated by an arbitrary italic superscript after the H2 designation for the MHC region of mice (e.g., $H2^a$, $H2^b$, etc.). These haplotype designations are shorthand for the entire set of inherited H2 alleles within a strain without having to individually list the specific allele at each locus (**Table 7-2**). Different inbred mouse strains may share the same set of alleles, or MHC haplotype, with another strain (e.g., CBA, AKR, and C3H) but will differ in genes outside the H2 complex.

Mice in a traditional inbred mouse strain are said to be **syngeneic**, or identical at all genetic loci. Two strains are **congenic** if they are bred to be genetically identical everywhere *except* at a single genetic region. For example, MHC congenic strains have the same genotype everywhere *except* at the MHC locus, where they differ. These mice are produced by a series of crosses and backcrosses between two inbred strains that differ at the MHC locus, with selection for particular MHC alleles in the progeny. For example, a frequently used congenic strain, designated B10.A, is derived from B10 mice (normally $H2^b$) genetically manipulated to possess the $H2^a$ haplotype at the MHC locus. Other than at the MHC locus, B10 and B10.A mice are genetically identical. Therefore, any phenotypic differences between these congenic strains must come from the protein products of the MHC locus. Examples of common H2 congenic mouse strains are shown in the list in Table 7-2.

> **Key Concept:**
> • MHC genes are both highly polymorphic and tightly linked, so that the set of alleles within the entire MHC locus is generally passed down as one unit; one such linked set of MHC alleles inherited from a parent is called a *haplotype*.

MHC Molecules Are Codominantly Expressed

The genes within the MHC locus exhibit a **codominant** form of expression. This means that both maternal and paternal gene products (both haplotypes) are expressed at the same time and in the same cells. Therefore, if two mice from inbred strains possessing different MHC haplotypes are mated, the F_1 generation inherits both parental haplotypes and will express all these MHC alleles; twice as many as either parent. For example, if an $H2^b$ strain is crossed with an $H2^k$ strain, then the F_1 generation inherits both parental sets of alleles and is said to be $H2^{b/k}$ (**Figure 7-8a**).

TABLE 7-2	H2 haplotypes of some mouse strains						
				H2 ALLELES			
Prototype strain	Other strains with the same haplotype	Haplotype	K	A	E	D
CBA	AKR, C3H, B10.BR, C57BR	*k*	*k*	*k*	*k*	*k*
DBA/2	BALB/c, NZB, SEA, YBR	*d*	*d*	*d*	*d*	*d*
C57BL/10 (B10)	C57BL/6, C57L, C3H.SW, LP, 129	*b*	*b*	*b*	*b*	*b*
A	A/He, A/Sn, A/Wy, B10.A	*a*	*k*	*k*	*k*	*d*
B10.A (2R)*		*h2*	*k*	*k*	*k*	*b*
B10.A (3R)		*i3*	*b*	*b*	*k*	*d*
B10.A (4R)		*h4*	*k*	*k*	*b*	*b*
A.SW	B10.S, SJL	*s*	*s*	*s*	*s*	*s*
A.TL		*t1*	*s*	*k*	*k*	*d*
DBA/1	STOLI, B10.Q, BDP	*q*	*q*	*q*	*q*	*q*

*The "R" designates a recombinant haplotype, in this case between the $H2^a$ and $H2^b$ types. Gene contribution from the *a* strain is shown in yellow, and that from the *b* strain in red.

(a) Mating of inbred mouse strains with different MHC haplotypes

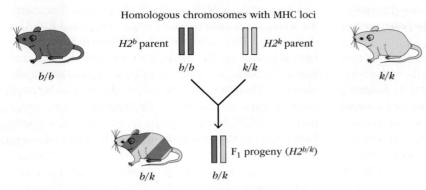

(b) Skin transplantation between inbred mouse strains with same or different MHC haplotypes

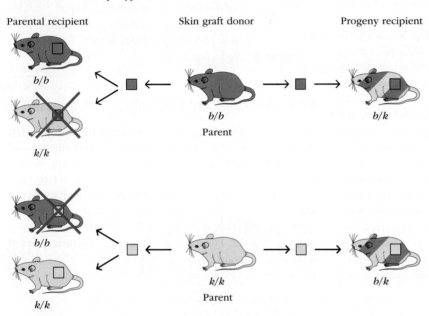

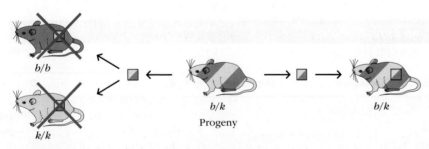

FIGURE 7-8 Illustration of inheritance of MHC haplotypes in inbred mouse strains and in humans. (a) The letters *b/b* designate a mouse homozygous for the *H2^b* MHC haplotype, *k/k* a mouse homozygous for the *H2^k* haplotype, and *b/k* a heterozygote. Because the MHC genes are closely linked and inherited as a set, the MHC haplotype of F$_1$ progeny from the mating of two different inbred strains can be predicted easily. (b) Acceptance or rejection (indicated by the large blue X) of skin grafts is controlled by the MHC type of the inbred mice. The progeny of the cross between two inbred strains with different MHC haplotypes (*H2^b* and *H2^k*) will express both haplotypes (*H2^b/k*) and will accept grafts from either parent and from one another. However, neither parent strain will accept grafts from the offspring. (c) Inheritance of HLA haplotypes in a hypothetical human family. For ease, the human paternal HLA haplotypes are arbitrarily designated A and B, maternal as C and D. Note that a new haplotype, R (recombinant), can arise from rare recombination of a parental haplotype (maternal shown here).

(c) Inheritance of HLA haplotypes in a typical human family

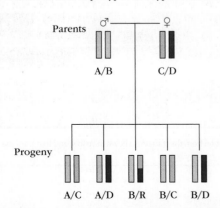

Because such an F_1 generation expresses the MHC proteins of both parental strains on its cells, it is said to be histocompatible, or MHC matched, with both parental strains. This means offspring are able to accept grafts from either parental source, each of which expresses MHC alleles viewed as "self" (Figure 7-8b). However, neither of the inbred parental strains can accept a graft from its F_1 offspring because half of the MHC molecules (those coming from the *other* parent) will be viewed as "nonself," or foreign, and thus subject to recognition and rejection by the immune system.

In an outbred population such as humans, each individual is generally heterozygous at each locus, and all alleles are expressed simultaneously. As in mice, the human HLA complex is highly polymorphic, MHC genes are closely linked, and alleles are inherited as a haplotype. When the father and mother have different haplotypes, as in the example shown in Figure 7-8c, there is a one-in-four chance that any two siblings will inherit the same paternal *and* maternal haplotypes, making them histocompatible with one another (i.e., genetically identical at the MHC loci, or MHC matched).

Although the rate of recombination by crossover is low within the HLA complex, it still contributes significantly to the diversity of the loci in human populations. Genetic recombination can generate new allelic combinations, and thus new haplotypes (see haplotype R in Figure 7-8c), and the high number of intervening generations since the appearance of humans as a species has allowed extensive recombination. As a result of recombination and other mechanisms for generating mutations, it is rare for any two unrelated individuals to have identical sets of HLA alleles. This makes transplantation between individuals who are not identical twins quite challenging! To address this, clinicians begin by looking for family members who will be at least partially histocompatible with the patient, or they rely on donor databases. Even with partial matches, physicians still need to administer heavy doses of immunosuppressive drugs to inhibit the strong rejection responses that typically follow tissue transplantation (see Chapter 16).

Key Concept:

- MHC genes show codominant expression patterns, meaning any cell expressing that class of MHC molecules transcribes DNA from that genetic locus on both chromosomes in the pair (maternal and paternal) simultaneously.

Class I and Class II Molecules Exhibit Diversity at Both the Individual and Species Levels

As noted earlier, any particular MHC molecule can bind many different peptides (called *promiscuity*), which gives the host an advantage in responding to pathogens. Rather than relying on just one gene for this task, the MHC region has evolved to include multiple genetic loci encoding proteins with the same function. In humans, class I HLA-A, -B, or -C molecules can all present peptides to CD8$^+$ T cells and class II HLA-DP, -DQ, or -DR molecules present to CD4$^+$ T cells. The MHC region is thus said to be **polygenic** because it contains multiple genes with the same function but with slightly different structures. Since the MHC alleles are also codominantly expressed, heterozygous individuals will express the gene products encoded by both alleles at each MHC gene locus. *In a fully heterozygous individual this amounts to six unique classical class I molecules on each nucleated cell.* An F_1 mouse, for example, expresses the H2-K, -D, and -L class I molecules from each parent (six different MHC class I molecules) on the surface of each of its nucleated cells (**Figure 7-9**). *The expression of so many individual MHC class I molecules, each with its own promiscuity of binding, allows a cell to display or present a large number of different peptides.*

MHC class II molecules have even greater potential for diversity. Each of the classical MHC class II molecules is composed of two different polypeptide chains encoded by different loci, which come together to form one class II binding pocket. Therefore, a heterozygous individual can

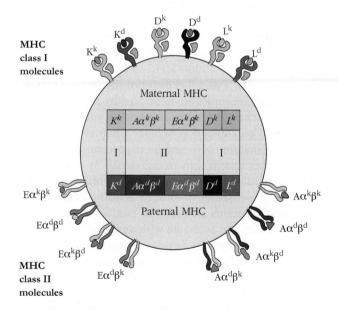

FIGURE 7-9 Diagram illustrating the various MHC molecules expressed on antigen-presenting cells of a heterozygous *H2$^{k/d}$* mouse. Both the maternal and paternal MHC genes are expressed (codominant expression). Because the class II molecules are heterodimers, new molecules containing one maternal-derived and one paternal-derived chain are also produced, increasing the diversity of MHC class II molecules on the cell surface. The β$_2$-microglobulin component of class I molecules (salmon) is encoded by a nonpolymorphic gene on a separate chromosome and may be derived from either parent.

express α-β combinations that originate from the same chromosome (maternal only or paternal only) as well as class II molecules arising from unique chain pairing derived from separate chromosomes (new maternal-paternal α-β combinations). For example, an $H2^k$ mouse expresses A^k and E^k class II molecules, whereas an $H2^d$ mouse expresses A^d and E^d molecules. The F₁ progeny resulting from crosses of these two strains express four parental class II molecules (identical to their parents) and also four new molecules that are mixtures from their parents, containing one parent's α chain and the other parent's β chain (as shown in Figure 7-9). Since the human MHC contains three classical class II loci (*DP*, *DQ*, and *DR*), a heterozygous individual expresses six class II molecules identical to the parents and six molecules containing new α and β chain combinations, allowing for the construction of a total of 12 different class II molecules. In some instances the number of different class II molecules expressed by an individual can be further increased by the occasional presence of an additional α- or β-chain gene within a given locus. The diversity generated by these new MHC molecules likely increases the number of different antigenic peptides that can be presented and is therefore advantageous to the organism in fighting infection.

The variety of peptides displayed by MHC molecules echoes the diversity of antigens bound by antibodies and T-cell receptors. This evolutionary pressure to diversify comes from the fact that both need to be able to interact with antigen fragments they have never before seen, or that may not yet have evolved. However, the strategy for generating diversity within MHC molecules and the antigen receptors on T and B cells is not the same. Antibodies and T-cell receptors are generated by several somatic processes, including gene rearrangement and the somatic mutation of rearranged genes (see Chapters 6 and 11). Thus, the generation of T- and B-cell receptors is dynamic, changing over time within an individual. By contrast, the MHC molecules expressed by an individual are fixed. However, promiscuity of antigen binding ensures that even "new" proteins are likely to contain at least some fragments that can associate with any given MHC molecule. Collectively, this builds in enormous flexibility within the host for responding to unexpected environmental changes that might arise in the future—an elegant evolutionary strategy.

Countering this limitation on the range of peptides that can be presented by any one individual is the vast array of peptides that can be presented at the species level, thanks to the diversity of the MHC in any outbred population. The MHC possesses an extraordinarily large number of different alleles at each locus and is one of the most polymorphic genetic complexes known in higher vertebrates. These alleles differ in their DNA sequences by 5% to 10%. This means

TABLE 7-3	Genetic diversity of MHC loci in the human population	
MHC region	**HLA locus**	**Number of allotypes (proteins)**
Class I	*A*	2480
	B	3221
	C	2196
	E	8
	F	4
	G	18
Class II	*DMα*	4
	DMβ	7
	DOα	3
	DOβ	5
	DPα1	22
	DPβ1	591
	DQα1	34
	DQβ1	678
	DRα	2
	DRβ1	1440
	DRβ3	106
	DRβ4	42
	DRβ5	39

Data from http://hla.alleles.org, a website maintained by the HLA Informatics Group based at the Anthony Nolan Trust in the United Kingdom, with up-to-date information on the numbers of HLA alleles and proteins. Data as of January 2017.

that the number of amino acid differences between MHC alleles can be quite significant, with up to 20 amino acid residues contributing to the unique structural nature of each allele. As of January 2017, analysis of the human HLA locus has identified over 15,000 alleles (**Table 7-3** shows the number of protein products for each gene; not all alleles encode expressed proteins). In mice, the polymorphism is similarly enormous. There has been a significant increase in the number of recognized HLA alleles in the past 5 years. This may be due to recent efforts to collect data from a broader range of worldwide populations. Earlier HLA allele estimates were primarily based on individuals of European descent. There is an effort underway to collect more representative data sets, specifically from India and Africa, regions where the currently available genome data are not representative of the population size. Thus, while contemporary calculations of diversity in the human HLA may be a better approximation of diversity, they may still significantly underestimate the true degree of polymorphism in the human MHC locus.

This disparity in available data can have significant impacts in the clinical realm. For instance, black Americans wait longer for and have lower success rates following kidney transplantation than their Caucasian counterparts. In addition to racial bias and socioeconomic factors, differences in HLA polymorphism rates among the African American population as well as the clinical guidelines routinely used by physicians for HLA matching may contribute to these discrepancies. Important changes in policy and practice can come from a greater appreciation for and implementation of more inclusive guidelines.

This enormous polymorphism results in a tremendous diversity of MHC molecules within a species. Even just considering the most polymorphic of the HLA class I genes (*A*, *B*, and *C*), the theoretical number of potential class I haplotypes in the human population is over *40 billion* (see Table 7-3). If the most polymorphic class II loci are considered, the numbers are even more staggering, with over 10^{13} different possible class II haplotypes. Because each HLA haplotype contains *both* class I and class II genes, theoretically, there could be as many as 10^{23} possible ways to combine HLA class I and II alleles within the human population. Combined with the promiscuity of peptide binding for each of these MHC molecules, this represents an enormous number of different vantage points for TCRs to engage with antigen.

Some evidence suggests that a reduction in MHC polymorphism within a species may predispose that species to disease (see **Evolution Box 7-1**). In one example, captive cheetahs and certain other wild cats, such as Florida panthers, show very limited genome diversity, theoretically due to episodes of past genetic bottlenecks. An apparent increased susceptibility of captive cheetahs to various viruses, to which close cousin species of cats show very low mortality, may result from a reduction in the number of different MHC molecules and a corresponding limitation in the range of processed antigens with which these MHC molecules can interact. Interestingly, this was first discovered when veterinarians noticed that captive cheetahs could receive skin grafts from unrelated members of their species without rejection! Wild cheetah populations do not appear to exhibit this same degree of MHC homogeneity and are also less susceptible to infectious disease. This suggests that this conservation of high levels of MHC polymorphism in most outbred species, including humans, may provide a survival advantage. Although some individuals within a species may not be able to develop a robust immune response to a given pathogen and therefore will be susceptible to infection, polymorphism in the population ensures that at least some members of a species will be resistant to that disease. In this way, MHC diversity at the population level may help protect the species as a whole from extinction due to infectious diseases.

Key Concepts:

• MHC genes are polymorphic (many alleles exist for each gene in the population), polygenic (several different MHC genes for MHC class I and II molecules exist in an individual), and codominantly expressed (both maternal and paternal copies).

• Species diversity at the MHC locus imparts an evolutionary survival advantage against mortality from infectious disease.

MHC Polymorphism Is Primarily Limited to the Antigen-Binding Groove

Although the sequence divergence among alleles of the MHC within a species is very high, this variation is not randomly distributed along the entire polypeptide chain. Instead, polymorphism in the MHC is clustered in short stretches, largely within the membrane-distal α1 and α2 domains of class I molecules (**Figure 7-10a**). Similar patterns of diversity are observed in the α1 and β1 domains of class II molecules.

Structural comparisons have located the polymorphic residues within the three-dimensional structure of the membrane-distal domains in MHC class I and II molecules and have related allelic differences to functional differences (Figure 7-10b). For example, of 17 amino acids previously shown to display significant polymorphism among HLA-A molecules, 15 were shown by x-ray crystallographic analysis to be in the peptide-binding groove of this molecule. The location of so many polymorphic amino acids within the binding site for processed antigen strongly suggests that allelic differences contribute to the observed differences in the ability of MHC molecules to interact with a given peptide ligand. Polymorphisms that lie outside these regions and might affect basic domain folding are rare. This clustering of polymorphisms around regions that make contact with antigen also suggests possible reasons why certain MHC genes or haplotypes can become associated with certain diseases (see **Clinical Focus Box 7-2**).

The preceding discussion points to additional parallels between MHC molecules and lymphocyte antigen receptors. The somatic hypermutations seen in B-cell receptor genes are also not randomly arrayed within the molecule, but instead are clustered in the regions most likely to interact directly with peptide (see Chapter 11), providing yet another example of how the immune system has solved a similar functional dilemma using a very different strategy.

BOX 7-1

The Sweet Smell of Diversity

As early as the mid-1970s, mate choice in mice was shown to be influenced by genes at the MHC (H2) locus. Research in a range of vertebrate species, including birds, reptiles, amphibians, and some fish, has revealed that in addition to mate selection, sensing of MHC type may influence other social behaviors, such as kinship cooperation, parent-progeny detection, and pregnancy blockade. Thanks to results from multiple studies conducted in the last 20 years, it looks like humans might be added to this list of vertebrates.

In terms of evolutionary pressure, local pathogens play a significant role in maintaining MHC diversity in the population and in selecting for specific alleles. As we now appreciate, this is because the MHC influences immune responsiveness. Because of its role in selecting the peptide fragments that will be presented, the inheritance of specific alleles at particular loci can predispose individuals to either enhanced susceptibility or resistance to specific infectious agents and immune disorders (see Clinical Focus Box 7-2). Over long time periods, endemic local pathogens exert evolutionary pressures that drive higher- or lower-than-expected rates of certain MHC alleles in a population, as well as overrepresentation of other resistance-associated genes.

The degree of diversity at the MHC locus clearly influences susceptibility to disease in populations; witness the enhanced viral susceptibility seen in cheetahs (see p. 263) and some devastating human stories. For instance, the European introduction of the smallpox virus to New World populations is credited with wiping out large Native American groups. This may be due to a lack of past evolutionary pressure for conservation of resistance-associated MHC alleles, which would therefore be rare or nonexistent in this population, as well as to a lack of any individuals with immunity from prior infection.

But how does an individual evaluate how well a potential mate will contribute "new" MHC alleles to one's offspring,

leading to greater diversity and the potential for enhanced fitness? The primary candidate is odor: the MHC is known to influence odor in many vertebrate species. For example, the urine of mice from distinct MHC congenic lines can be distinguished by both humans and rodents. Mice show a distinct preference for mating with animals that carry MHC alleles that are dissimilar to their own. In terms of maximizing the range of peptides that can be presented, this makes clear evolutionary sense, as increased diversity at the MHC should increase the number of different pathogenic peptides that can be "seen" by the immune system, increasing the likelihood of effective anti-pathogen responses. To back this up, the advantage of overall MHC diversity has been shown experimentally in mice, where most simulated epidemic experiments have found a survival advantage for H2 heterozygous animals over their homozygous counterparts. In humans, research in HIV-infected individuals has shown that extended survival and a slower progression to AIDS are correlated with full heterozygosity at the HLA class I locus, as well as absence of certain AIDS-associated *HLA-B* and *HLA-C* alleles. This specific link to MHC class I is not surprising in light of the key role of CD8[+] T cells in combating viral infections.

Human studies of attraction and mating also point to preferences for individuals with dissimilar MHC alleles. In one key study involving what is commonly known as the "sweaty T-shirt test," college-age volunteers were asked to rate their preference or sexual attraction to the odor of T-shirts worn by individuals of the opposite sex. In general, both males and females preferred the odor of T-shirts worn by individuals with dissimilar HLA types. The one key exception was seen in women who were concurrently using an estrogen-based birth control pill; they instead showed a preference for the odors of MHC-similar individuals, suggesting that this hormone not only interferes with this response

but that it potentially shifts the outcome. (To date, we are aware of no such studies conducted in same sex couples or with transgender individuals, leaving the MHC of attraction an open question in these instances.)

But how does this work? Soluble forms of MHC molecules have been found in many bodily fluids, including urine, saliva, sweat, and plasma. However, these molecules are unlikely to be small or volatile enough to account for direct olfactory detection. The current hypotheses for how MHC influences odor is via olfactory recognition of natural ligands carried by MHC molecules. In essence, the shape of the MHC binding groove is detected on the basis of the shape of the mirror-image antigenic peptides released from this pocket. Studies in mice have shown that the sensory neurons in their vomeronasal organ, which work like an auxiliary olfactory system and detect chemical stimuli, are differentially sensitive to self versus nonself peptides. In fact, the patterns of recognition mimic binding specificity in the immune response; amino acid substitutions at anchor residue sites resulted in a modified response, but not substitutions made to nonanchor residues.

As one might imagine, social and cultural factors also influence the outcome. Nonetheless, it appears that we humans may also use smell as an evolutionary strategy for promoting a diverse and robust immune response in our offspring.

REFERENCES

Ruff, J. S., A. C. Nelson, J. L. Kubinak, W. K. Potts. 2012. MHC signaling during social communication. *Advances in Experimental Medicine and Biology* **738:**290.

Leinders-Zufall, T., et al. 2004. MHC class I peptides as chemosensory signals in the vomeronasal organ. *Science* **306:**1033.

Wedekind, C., and S. Füri. 1997. Body odour preferences in men and women: do they aim for specific MHC combinations or simply heterozygosity? *Proceedings Biological Sciences* **264:**1471.

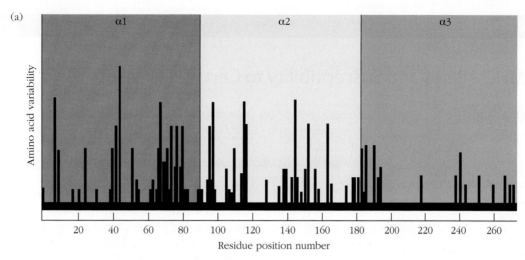

(a)

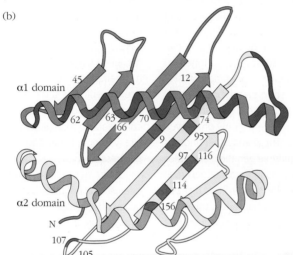

(b)

FIGURE 7-10 Variability in the amino acid sequences of allelic HLA class I molecules. (a) Most of the variability between MHC class I molecules lies in the membrane-distal α1 and α2 domains. (b) Polymorphic amino acid residues (red) in the α1/α2 domain of a human MHC class I molecule map to the peptide binding pocket. [Data from Sodoyer, R., et al. 1984. Complete nucleotide sequence of a gene encoding a functional human class I histocompatibility antigen (HLA-CW3). EMBO Journal 3:879.]

Key Concept:

• Polymorphic residues in MHC alleles cluster in the peptide-binding pocket, influencing the fragments of antigen that are presented to the immune system, and thereby influencing susceptibility to a number of diseases.

The Role and Expression Pattern of MHC Molecules

As we have just discussed, several genetic features help ensure a diversity of MHC molecules in outbred populations, including polygeny, polymorphism, and codominant expression. All this attention paid to maximizing the number of different binding grooves suggests that variety within

the MHC plays an important role in survival (see Clinical Focus Box 7-2). In fact, in addition to fighting infection, MHC expression throughout the body plays a key role in maintaining homeostasis and health even when no foreign antigen is present.

Although the presentation of foreign antigen to T cells by MHC molecules garners much attention (and space in this book!), most MHC molecules spend their lives presenting self-peptides. There are several reasons why an MHC-peptide molecule on the surface of a cell is important. In general, these include the following:

• To display self-MHC class I and self-peptide to demonstrate that the cell is healthy
• To display a foreign peptide in class I to show that the cell is infected and to engage with T_C cells
• To display a foreign peptide in class II to show the body is infected and activate T_H cells

MHC Alleles and Susceptibility to Certain Diseases

Some HLA alleles occur at a much higher frequency in people suffering from certain diseases than in the general population. The diseases associated with particular MHC alleles include autoimmune disorders, certain viral diseases, disorders of the complement system, some neurologic disorders, and several different allergies. In humans, the association between an HLA allele and a given disease may be quantified by determining the frequency of that allele expressed by individuals afflicted with the disease, and then comparing these data with the frequency of the same allele in the general population. Such a comparison allows calculation of an individual's **relative risk (RR)**:

$$RR = \frac{\textit{frequency of disease in the allele}^{+}\textit{ group}}{\textit{frequency of disease in the allele}^{-}\textit{ group}}$$

An RR value of 1 means that the HLA allele is expressed with the same frequency in disease-afflicted and general populations, indicating that this allele confers no increased risk for the disease. An RR value substantially above 1 indicates an association between the HLA allele and the disease. For example, individuals with the *HLA-B27* allele are 90 times more likely (RR = 90) to develop the autoimmune disease ankylosing spondylitis, an inflammatory disease of vertebral joints characterized by destruction of cartilage, than are individuals who lack this *HLA-B* allele. Other disease associations with a significantly high RR include *HLA-DQB1* and narcolepsy (RR = 130) and *HLA-DQ2* with celiac disease (RR = 50), the latter involving an allergy to gluten. A few HLA alleles have also been linked to relative protection from disease or clinical progression. This is seen in the case of individuals inheriting *HLA-B57*, which is associated with greater viral control and a slower progression to AIDS in HIV-infected individuals.

When the associations between MHC alleles and disease are weak, reflected by low RR values, it is possible that multiple genes influence susceptibility, of which only one lies within the MHC locus. The genetic origins of several autoimmune diseases, such as multiple sclerosis (associated with *DR2*; RR = 5) and rheumatoid arthritis (associated with *DR4*; RR = 10) have been studied in depth. The observation that these diseases are not inherited by simple Mendelian segregation of MHC alleles can be seen clearly in identical twins, where both inherit the same MHC risk factor, but frequently only one develops the disease. These and other findings suggest that multiple genetic factors plus one or more environmental factors are at play in the development of this disease. As we highlight in Chapter 16, this combined role for genes and the environment in the development of autoimmunity is not uncommon.

The existence of an association between an MHC allele and a disease should not be interpreted to imply that the expression of the allele has caused the disease. The relationship between MHC alleles and the development of disease is more complex, thanks in part to the genetic phenomenon of linkage disequilibrium. The fact that some of the MHC class I alleles are in linkage disequilibrium with the class II and class III alleles can make their contribution to disease susceptibility appear more pronounced than it actually is. If, for example, *DR4* contributes to the risk for a disease, and it also occurs frequently in combination with *HLA-A3* because of linkage disequilibrium, then *A3* might incorrectly appear to be associated with the disease. Improved genomic mapping techniques make it possible to analyze the fine linkage between genes within the MHC and various diseases more fully and to assess the contributions from other loci. In the case of ankylosing spondylitis, for example, it has been suggested that alleles of *TNF* and *lymphotoxin* may produce protein variants that are involved in the destruction of cartilage, and these alleles happen to be linked to certain *HLA-B* alleles. In the case of *HLA-B57* and AIDS progression, the allele has been linked more directly. It is believed that this class I allele is particularly efficient at presenting important components of the virus to circulating T_C cells, leading to increased destruction of virally infected cells and delayed disease progression.

Other hypotheses have also been offered to account for a direct role of particular MHC alleles in disease susceptibility. In some rare cases, certain allelic forms of MHC genes can encode molecules that are recognized as receptors by viruses or bacterial toxins, leading to increased susceptibility in the individuals who inherit these alleles. As will be explored further in Chapters 15 through 19, in many cases complex interactions between multiple genes, frequently including the MHC, and particular environmental factors are required to create a bias toward the development of certain diseases.

REFERENCES

Miyadera, H., and K. Tokunaga. 2015. Associations of human leukocyte antigens with autoimmune diseases: challenges in identifying the mechanism. *Journal of Human Genetics* **60:**697.

Trowsdale, J., and J. C. Knight. 2013. Major histocompatibility complex genomics and human disease. *Annual Review of Genomics and Human Genetics* **14:**301.

- To display a self-peptide in class I and II to test developing T cells for autoreactivity (in primary lymphoid organs)
- To display a self-peptide in class I and II to maintain tolerance to self-proteins (in secondary lymphoid organs)

It is worth noting that although some of these instances lead to immune activation, many do not. It is also important to note that the cell type, tissue location, and timing of expression vary for each of these situations. For instance, developing T cells first encounter MHC class I and II molecules presenting *self-peptides* in the thymus, where these signals are designed to *inhibit* the ability of T cells to attack self-structures later (see Chapter 8). Others occur only after certain types of exposure, such as during an immune response to extracellular pathogens or when cells in the body are infected with viral invaders, and are designed to *activate* T cells to act against the pathogen (see Chapter 10).

To help set the stage for this discussion, we turn now to the where, when, and why of MHC expression. This is followed by a description of the how, detailing the pathways that lead to peptide placement in the binding groove of each class of MHC molecule. As we will see in greater detail in the following sections, the type of MHC molecule(s) expressed by a given cell is linked to the role of the cell and the function of each class of molecule.

MHC Molecules Present Both Intracellular and Extracellular Antigens

In very general terms, *the job of MHC class I molecules is to collect and present antigens that come from intracellular locations.* This is a form of ongoing surveillance of the internal happenings of the cell—in essence, a window for displaying on the surface of the cell snippets of what is occurring inside. Nearly all cells in the body need this form of check and balance, and this is reflected in the fairly ubiquitous nature of MHC class I expression in the body (there are a few notable exceptions, which we will touch on later). Often, nothing other than normal cellular processes are occurring in the cytosol, and in these instances our cells present self-peptides in the grooves of MHC class I molecules. The expression of self-MHC class I with self-peptides typically signals that a cell is healthy, while absence of self-MHC class I (as can occur in virus-infected and tumor cells) can target that cell for destruction by natural killer (NK) cells (see Chapter 12). When foreign proteins are present in the cytosol and begin to appear in the grooves of MHC class I molecules on the cell surface, this alerts CD8$^+$ T cells to the presence of an unwelcome visitor, targeting that cell for destruction. In this case, the cell is called a **target cell** because it becomes a target for lysis by cytotoxic T cells.

Conversely, *MHC class II molecules primarily display peptides that have come from the extracellular spaces of the body.* Since this sampling of extracellular contents is not a form of policing that all cells need to perform, only specialized leukocytes possess this ability. These cells are collectively referred to as antigen-presenting cells (APCs), because their job is to present extracellular antigen to CD4$^+$ T cells, charging them with the ultimate job of coordinating the elimination of extracellular invaders.

> **Key Concept:**
> - In most cases, class I molecules present processed endogenous (intracellular) antigen to CD8$^+$ T$_C$ cells and class II molecules present processed exogenous (extracellular) antigen to CD4$^+$ T$_H$ cells.

MHC Class I Expression Is Found Throughout the Body

Classical MHC class I molecules are expressed constitutively on almost all nucleated cells of the body. However, the level of expression differs among different cell types, with the highest levels of class I molecules found on the surface of lymphocytes. On these cells, class I molecules may constitute approximately 1% of the total plasma membrane proteins, or some 5×10^5 MHC class I molecules per cell. In contrast, cells such as fibroblasts, muscle cells, liver hepatocytes, and some neural cells express very low to undetectable levels of MHC class I molecules. This low-level expression on liver cells may contribute to the relative success of liver transplants, reducing the likelihood of graft rejection when T$_C$ cells of the recipient recognize the foreign tissue of the donor. A few cell types (e.g., subsets of neurons and sperm cells at certain stages of differentiation) appear to lack MHC class I molecules altogether. However, nucleated cells without MHC class I expression are very rare. Nonnucleated cells, such as red blood cells in mammals, do not generally express any MHC molecules, making them poor targets for T$_C$ cells.

In normal, healthy cells, MHC class I molecules on the surface of the cell will display self-peptides resulting from normal turnover of self-proteins inside the cell. In cells infected with a virus, viral peptides as well as self-peptides will be displayed. Therefore, a single virus-infected cell can be envisioned as having various class I molecules on its membrane, some displaying a subset of viral peptides derived from the viral proteins being manufactured within. Because of individual allelic differences in the peptide-binding grooves of the MHC class I molecules, different individuals within a species will have the ability to bind and present different sets of viral peptides (like displaying different pieces of a jigsaw puzzle that represents the whole of that pathogen). In addition to virally infected cells, altered self-cells such as cancer cells, aging body cells, or cells from an allogeneic graft (tissue from a genetically different individual), also can serve as target cells because of their expression of defective or foreign MHC proteins, and can be lysed by T_C cells. The importance of constitutive expression of class I is highlighted by the response of NK cells to somatic cells that lack MHC class I, as can occur during some viral infections. NK cells can kill a cell that has stopped expressing MHC class I on its surface, presumably because this suggests that the cell is no longer healthy or has been altered by the presence of an intracellular invader.

Key Concept:

- MHC class I molecules are expressed on most nucleated cells, although expression levels can vary by cell type, and present pieces of proteins found in the cytosol of that cell; these proteins can be self or foreign in origin.

Expression of MHC Class II Molecules Is Primarily Restricted to Antigen-Presenting Cells

MHC class II molecules are found on a much more restricted set of cells than class I, and sometimes only after an inducing event. As mentioned previously, antigen-presenting cells (APCs) display peptides associated with MHC class II molecules to CD4$^+$ T_H cells, and these cells are primarily specific types of leukocytes. APCs are specialized for their ability to alert the immune system to the presence of an invader and can induce the activation of T-cell responses. APCs thus play a policing role in the body, with unique authorization to activate an immune response to extracellular infection, as needed.

A variety of cells can function as bona fide APCs. Their distinguishing feature is their ability to express MHC class II molecules and to deliver a costimulatory, or second activating signal, to T cells. Three cell types are known to have these characteristics and are thus often referred to as **professional antigen-presenting cells (pAPCs)**: dendritic cells, macrophages, and B lymphocytes. These cells differ from one another in their mechanisms of antigen uptake, in whether they constitutively express MHC class II molecules, and in their costimulatory activity, as follows:

- Dendritic cells are considered the most powerful and most efficient of the pAPCs. These cells constitutively express high levels of MHC class II molecules and have inherent costimulatory activity, allowing them to quickly activate naïve T_H cells.

- Macrophages must be activated before they express MHC class II molecules or costimulatory membrane molecules such as CD80/86 (see Chapter 10).
- B cells constitutively express MHC class II molecules, although at low levels, and possess antigen-specific surface receptors. This makes them particularly efficient at capturing and presenting their cognate antigen, or the specific epitope recognized by their BCR.

Among the various pAPCs, marked differences in the level of MHC class II expression are observed. In some cases, class II expression depends on the cell's differentiation stage or level of activation. APC activation usually occurs following interaction with a pathogen (e.g, via BCR or PRRs) and/or through cytokine signaling, which then induces gene expression changes, including significant increases in MHC class II expression.

Several other cell types, classified as nonprofessional APCs, can be induced to express MHC class II molecules and costimulatory signals under certain conditions (**Table 7-4**). These cells can be deputized for professional antigen presentation for short periods and in particular situations, such as during a sustained inflammatory response.

Key Concept:

- Expression of class II molecules and the ability to activate T_H cells is primarily restricted to B cells, macrophages, and dendritic cells, collectively referred to as professional antigen-presenting cells (pAPCs).

MHC Expression Can Change with Changing Conditions

As we've noted, MHC class I is constitutively expressed by most cells in the body, whereas class II is expressed only under certain conditions and in a very limited number of cell types. The different roles of the two molecules help explain this observation, and suggest that in certain

TABLE 7-4	Antigen-presenting cell types	
Professional	**Nonprofessional**	
Dendritic cells	Fibroblasts (skin)	Thymic epithelial cells
Macrophages	Glial cells (brain)	Intraepithelial lymphocytes
B cells	Pancreatic beta cells	Vascular endothelial cells

instances specific changes in MHC expression may prove advantageous. The MHC locus can respond to both positive and negative regulatory pressures. For instance, MHC class I production can be disrupted or depressed by some pathogens. MHC class II expression on APCs is already quite variable, and the microenvironment surrounding an APC can further modulate expression, usually serving to enhance the expression of these molecules. APCs in particular are conditioned to respond to local cues leading to their activation and heightened MHC expression, enabling them to arm other cells in the body for battle. As one might imagine, this activation and arming of APCs must be carefully regulated, lest these cells orchestrate unwanted aggressive maneuvers against self or benign foreign compounds, as occurs in clinical conditions such as autoimmunity or allergy, respectively. The mechanisms driving changes in expression of MHC molecules are described here.

Genetic Regulatory Components

The presence of internal or external triggers, such as intracellular invaders or cytokines, can induce a signal transduction cascade that leads to changes in MHC gene expression. Research aimed at understanding the mechanism of control of MHC expression has been advanced by the now complete sequence of the mouse genome. Both MHC class I and II genes are flanked by 5′ promoter sequences that bind sequence-specific transcription factors. The promoter motifs and the transcription factors that bind to these motifs have been identified for a number of MHC genes, with examples of regulation mediated by both positive and negative elements. For both MHC class I and II, members of the NOD-like receptor family (NLRs; see Chapter 4) function as core components of the MHC *transcriptional activators*, called CITA and CIITA, respectively. CIITA and another transcription factor called RFX have both been shown to activate the promoter of MHC class II genes. Defects in these transcription factors cause one form of *bare lymphocyte syndrome*. Patients with this disorder lack MHC class II molecules on their cells and suffer from severe immunodeficiency, highlighting the central role of class II molecules in T-cell maturation and activation.

Viral Interference

Certain viruses can interfere with MHC class I expression in the cells they infect and thus avoid easy detection by CD8$^+$ T cells. These viruses include human cytomegalovirus, hepatitis B virus, and adenovirus 12. In the case of cytomegalovirus infection, a viral protein binds to β_2-microglobulin, preventing assembly of MHC class I molecules and their transport to the plasma membrane. Adenovirus 12 infection causes a pronounced decrease in transcription of the transporter genes *TAP1* and *TAP2*.

As described in the following section on antigen processing, the *TAP* gene products play an important role in peptide transport from the cytoplasm into the rough endoplasmic reticulum (RER). Blocking of *TAP* gene expression inhibits peptide transport; as a result, MHC class I molecules cannot be assembled or transported to the cell membrane. These observations are especially important because decreased expression of MHC class I molecules, by whatever mechanism, is likely to block the immune system from detecting changes within a cell. For instance, MHC class I expression is also reduced or absent in many cancerous cells. One interesting example is the recent outbreak of contagious cancer in the Tasmanian devil. This disease is caused by a block in MHC class I expression and has decimated the population of this marsupial, found only in Tasmania (see **Clinical Focus Box 7-3**).

Cytokine-Mediated Signaling

The expression of MHC molecules is externally regulated by various cytokines. Leading among these are the interferons (α, β, and γ) and tumor necrosis factor (TNF), each of which has been shown to increase expression of MHC class I molecules on cells. Typically, phagocytic cells that are involved in innate responses, or locally infected cells, are the first to produce these MHC-regulating cytokines. In particular, IFN-α (produced by a cell following viral or bacterial infection) and TNF (secreted by APCs after activation) are frequently the first cytokines to kick off an MHC class I up-regulation event. In the later stages of infection, IFN-γ, secreted by activated T_H cells as well as other cell types, also contributes to increased MHC expression.

Binding of these cytokines to their respective receptors induces intracellular signaling cascades that activate transcription factors and alter expression patterns. These factors bind to their target promoter sequences and coordinate increased transcription of the genes encoding the class I α chain, β_2-microglobulin, and other proteins involved in antigen processing and presentation. IFN-γ has also been shown to induce expression of the class II transcriptional activator (CIITA), thereby increasing expression of MHC class II molecules on a variety of cells, including non-APCs (e.g., some fibroblasts, intestinal epithelial cells, vascular endothelium, and pancreatic beta cells), further enhancing the adaptive immune response.

Other cytokines influence MHC expression only in certain cell types. For example, interleukin (IL)-4 increases expression of class II molecules in resting B cells, turning them into more efficient APCs. Conversely, expression of class II molecules by B cells is down-regulated by IFN-γ. Corticosteroids and prostaglandins can also decrease expression of MHC class II molecules. These naturally

CLINICAL FOCUS

BOX 7-3

Without MHC, Cancer Can Be Devilishly Hard to See

A recent and fascinating example of the power of MHC molecules to control disease comes from an experiment of nature in the small Australian island state of Tasmania. In some regions the population of Tasmanian devils, a marsupial native only to this corner of the world, has fallen by more than 95% in the past 20 years (**Figure 1**, left). In 2008 this species was elevated to endangered status. The cause: a tumor of the face and mouth that kills 100% of its victims, most within 6 months of appearance (Figure 1, right). Even more puzzling is the observation that this disease, called devil facial tumor disease (DFTD), is a contagious form of cancer with no pathogen to blame.

Tasmanian devils are carnivorous animals, about the size of a small dog, with powerful jaws and sharp teeth. These teeth are used for more than feeding; most devils have scars from the bites of others of their species, whether during feeding or mating defense. It appears that this cancer first arose for unknown reasons in a female devil over 20 years ago. Although she is long dead her cancer cells live on, being passed from devil to devil during this biting behavior. Genotyping has established that all DFTD cancers arose from this same original devil. As we know from the early studies that helped characterize MHC molecules (then called *transplantation antigens*), cancer cells passed between members of a species should be rejected based on the presence of different MHC proteins (allorecognition). If that is the case, what has allowed these cancer cells to be passed from one devil to another for over two decades?

Initial hypotheses focused on the possibility that the Tasmanian devil

FIGURE 1 Tasmanian devils. *Left:* Uninfected. *Right:* Infected with DFTD. *[Left: Bruce Miller/Alamy. Right: Dave Watts/Alamy.]*

population might have such low MHC diversity that they were in essence clones of one another. However, unlike the example of cheetahs mentioned earlier, while MHC diversity amongst devils is low, this has been true for thousands of years. And also unlike cheetahs, transplant experiments showed that skin grafts between animals were quickly rejected, as would be expected in most outbred populations. Finally, there was no evidence of immune compromise in diseased devils.

In the end, it turned out that the hosts were not missing anything—but the cancer cells were. The cells from multiple independently isolated tumors were all found to lack surface MHC class I expression, allowing them to avoid destruction by CD8[+] T cells. Transcripts for key proteins involved in class I synthesis and assembly (TAP1, TAP2, and β_2-microglobulin) were all down-regulated in these cancers. The mechanism for this suppression was clarified when cancer cells began to express surface MHC class I molecules after treatment with trichostatin A, a histone deacetylase inhibitor. The cancer cells that cause DFTD exhibited high levels of histone acetylation at these MHC class I–associated loci, enough to silence gene expression. We now believe that without

surface MHC molecules these cancer cells were accepted much like an isograft, passed from one animal to another during biting incidents. Their transformed character and rapid growth then allowed them to thrive in, and eventually kill, their new host. No one yet understands why NK cells, which should notice these missing MHC molecules, don't destroy these tumors.

A recent vaccine aimed at protecting Tasmanian devils from DFTD has shown initial promise. That said, this experiment of nature may be resolving itself naturally. Unvaccinated wild Tasmanian devil populations are on the rise. This could be due to a few lucky devils with immunity against this contagious cancer, or it could be thanks to an evolution in behavior. It looks like fighting and biting are losing favor, and the less devilish devils may be taking over.

REFERENCES

Siddle, H. V. 2013. Reversible epigenetic down-regulation of MHC molecules by devil facial tumour disease illustrates immune escape by a contagious cancer. *Proceedings of the National Academy of Sciences USA* **110**:5103.

Woods, G. M., et al. 2015. Immunology of a transmissible cancer spreading among Tasmanian devils. *Journal of Immunology* **195**:23.

occurring, membrane-permeable compounds bind to intracellular receptors and are some of the most potent suppressors of adaptive immunity, primarily based on their ability to inhibit MHC expression. This property is exploited in many clinical settings, where these compounds are used as treatments to suppress overly zealous immune events, such as in allergic responses or during transplant rejection (see Chapters 15 and 16).

MHC Alleles Play a Critical Role in Immune Responsiveness

MHC haplotype plays a strong role in the outcome of an immune response, as these alleles determine which fragments of protein will be presented. Recall that MHC class II molecules present foreign antigen to CD4$^+$ T$_H$ cells, which go on to activate B cells to produce antibodies. Early studies by Benacerraf in which guinea pigs were immunized with simple synthetic antigens were the first to show that the ability of an animal to mount an immune response, as measured by the production of antibodies, is determined by its MHC haplotype. Later experiments by H. McDevitt, M. Sela, and colleagues used congenic mouse strains to specifically map the control of **immune responsiveness** to MHC class II genes. In early reports, the genes responsible for this phenotype were designated *Ir* or *immune response* genes, which led to mouse class II products being called *IA* and *IE* (the *I* was later dropped). We now know that the dependence of immune responsiveness on the genes within the MHC class II locus reflects the central role of these molecules in determining which specific peptide fragments of a foreign protein will be presented as antigen to T$_H$ cells.

Two explanations have been proposed to account for this variability in immune responsiveness observed among different haplotypes. According to the **determinant selection model**, different MHC class II molecules differ in their ability to bind particular processed antigens. In the end, some peptides may be more crucial to eliminating the pathogen than others. *The ability of an organism to present these crucial fragments would give them an advantage.* A separate hypothesis, termed the **holes-in-the-repertoire model**, postulates that T cells bearing receptors that recognize certain foreign antigens, which happen to closely resemble self-antigens, may be eliminated during T-cell development, leaving the organism without these cells/receptors that may be important for future responses to particular foreign molecules. These models are not mutually exclusive, and in fact both appear to be correct. That is, the absence of an MHC molecule that can bind to and present a particular peptide (on the pAPC), and the absence of T-cell receptors that can recognize a given MHC-peptide molecule complex (on the responding T cell), both result in decreased immune responsiveness to a given foreign substance. This interaction between MHC molecule, antigenic peptide, and T cell is like a trimolecular sandwich; lack of the right groove to hold important peptide fragments or absence of the TCR to recognize the MHC-peptide pair means that crucial antigenic fragments can get missed by the immune system, accounting for the observed relationship between MHC haplotype and the ability to respond to particular exogenous antigens.

Seminal Studies Demonstrate That T Cells Recognize Peptide Presented in the Context of Self-MHC Alleles

In the 1970s a series of experiments were carried out to further explore the relationship between the MHC and the immune response. These investigations contributed two crucial discoveries: (1) that both CD4$^+$ and CD8$^+$ T cells can recognize antigen *only* when it is presented in the groove of an MHC molecule, and (2) that the MHC haplotype of the APC and the T cell must *match*. This happens naturally in the host, where host T cells develop alongside host APCs, both expressing only that individual's MHC molecules (see Chapter 8). This implies a dual specificity of T cells for antigen plus the self-MHC molecule presenting that antigen.

In 1973, A. Rosenthal and E. Shevach showed that antigen-specific proliferation of T$_H$ cells occurs only in response to antigen presented by macrophages of the same MHC class II haplotype as the T cells recognizing the antigen. Less than 2 years later, R. Zinkernagel and P. Doherty demonstrated that CD8$^+$ T cells are similarly self-MHC class I restricted (see **Classic Experiment Box 7-4** for a description of both). Zinkernagel and Doherty were awarded the Nobel Prize in Physiology or Medicine for their seminal studies more than a decade later (see Table 1-2 for a list of Nobel Prizes related to immunology). This dependence of T-cell receptors on self-MHC is called **MHC restriction** and occurs as T cells are developing in the thymus, where their survival is dependent on engagement with the self-MHC molecules present there (Chapter 8).

This work was followed by studies, published in the 1980s by K. Ziegler and E. Unanue, showing that an intracellular processing step by APCs was required to activate T cells. These researchers observed that T$_H$-cell activation by large bacterial protein antigens was prevented by treating the APCs with paraformaldehyde prior to antigen exposure (in essence,

Demonstration of Self-MHC Restriction

By around the middle of the twentieth century it was clear that B lymphocytes could recognize specific antigen in soluble form and that this encounter could lead to proliferation and secretion of antibodies. No other cells were required to accomplish this feat. However, the mechanism by which T cells engaged with antigen, and whether other cells or structures were required, was highly controversial. T-cell receptors had not yet been discovered and characterization of MHC molecules was still in its early stages. A relationship between MHC molecules and T-cell responses was not established. This view changed dramatically in the early to mid-1970s with the discovery of self-MHC restriction. Two different groups, one working on CD4$^+$ and the other on CD8$^+$ T cells, almost simultaneously helped to bring this process at the center of adaptive immunity into clearer focus.

The seminal work of A. Rosenthal and E. Shevach was published in the *Journal of Experimental Medicine* in 1973, in a pair of articles showing that CD4$^+$ T cells recognize foreign peptides only when presented by self-MHC molecules. This was a major discovery, as the mechanism by which the MHC directed the immune response was still poorly understood. In their experimental system, guinea pig macrophages from one strain (called "strain 2") were incubated with an antigen (**Figure 1**). After these "antigen-pulsed" macrophages had processed the antigen and presented it on their surface, they were mixed with T cells from the same strain, a different strain (called "strain 13"), or F$_1$ progeny of these two strains ("2 × 13"). Next, the magnitude of T-cell proliferation in response to the antigen-pulsed macrophages was measured. The results of these experiments showed that antigen-pulsed macrophages from strain 2 could activate T cells from both strain 2 and F$_1$ mice, but not T cells from strain 13 animals. Similarly, strain 13 antigen-pulsed macrophages activated strain 13 and F$_1$ T cells but not strain 2 T

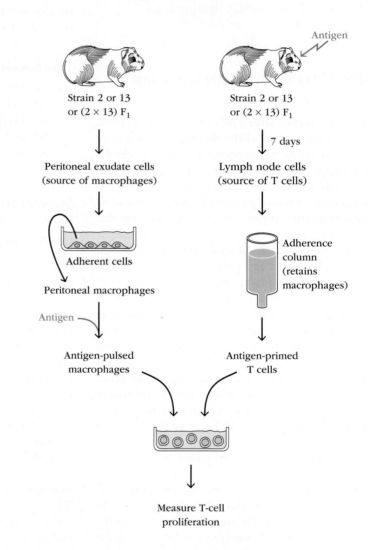

Antigen-primed T cell	Antigen-pulsed macrophages		
	Strain 2	Strain 13	(2 × 13) F$_1$
Strain 2	+	−	+
Strain 13	−	+	+
(2 × 13) F$_1$	+	+	+

FIGURE 1 Experimental demonstration of self-MHC restriction in cells. Peritoneal exudate cells from strain 2, strain 13, or (2 × 13) F$_1$ guinea pigs were incubated in plastic Petri dishes, allowing enrichment of macrophages, which are adherent cells. The peritoneal macrophages were then incubated with antigen. These "antigen-pulsed" macrophages were incubated in vitro with T cells from strain 2, strain 13, or (2 × 13) F$_1$ guinea pigs, and the degree of T-cell proliferation was assessed (positive [+] versus negative [−]). The results indicated that T cells could proliferate only in response to antigen presented by macrophages that shared MHC alleles/histocompatibility antigens. Thus, antigen recognition by CD4$^+$ T$_H$ cells is MHC class II restricted.

(continued)

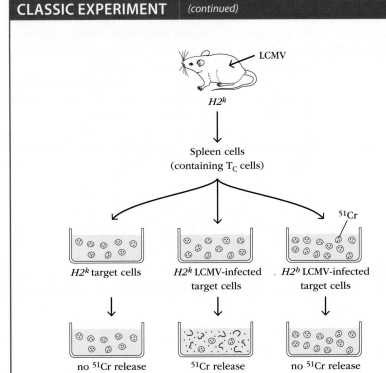

FIGURE 2 **Experimental demonstration that antigen recognition by T_C cells exhibits MHC restriction.** An *H2^k* mouse was primed with the lymphocytic choriomeningitis virus (LCMV) to induce cytotoxic T lymphocytes (CTLs) specific for the virus. Spleen cells from this LCMV-primed mouse were then added to target cells of various H2 haplotypes that were intracellularly labeled with ^{51}Cr (black dots) and either infected or not with LCMV. CTL-mediated killing of the target cells, as measured by the release of ^{51}Cr into the culture supernatant, occurred only if the target cells were infected with LCMV and had the same MHC haplotype as the CTLs.

cells. Subsequently, congenic and recombinant congenic strains of mice, which differed from each other only in selected regions of the MHC, were used as the source of macrophages and T cells. These experiments confirmed that the CD4$^+$ T cell is activated and proliferates only in the presence of macrophages that have been pre-treated with extracellular antigen and that share the T_H cell's MHC class II alleles. Thus, the idea of T cells as **MHC restricted** was born. The clarification that these proliferating lymphocytes are CD4$^+$ T_H cells and that the antigen-loaded histocompatibility antigens are in fact MHC class II molecules would come a bit later.

In 1975, R. Zinkernagel and P. Doherty also published an article in the *Journal of Experimental Medicine*, demonstrating that CD8$^+$ T cells are likewise restricted to recognizing antigen in the context of self-MHC molecules. They took advantage of recently derived inbred strains of mice, a virus that causes widespread neurological damage in infected animals, and new assays that allowed quantification of cytotoxic T-cell responses.

In their experiments, mice were first immunized with lymphocytic choriomeningitis virus (LCMV), a pathogen that results in central nervous system inflammation and neurological damage. Several days later, the animals' spleen cells, which included T_C cells specific for the virus, were isolated and incubated with LCMV-infected target cells of the same or different haplotype (**Figure 2**). The assay relied on measuring the release of a radioisotope (chromium-51, or ^{51}Cr) from labeled target cells (called a chromium release assay; see Chapter 20). They found that the T_C cells killed syngeneic virus-infected target cells (or cells with matched MHC alleles) but not uninfected cells or infected cells from a donor that shared no MHC alleles with these cytotoxic cells.

Later studies with congenic and recombinant congenic strains showed that the T_C cell and the virus-infected target cell must specifically share class I molecules encoded by the *K* or *D* region of the MHC. Thus, antigen recognition by CD8$^+$ cytotoxic T cells is **MHC class I restricted**. In 1996, Doherty and Zinkernagel were awarded the Nobel Prize in Physiology or Medicine for their major contribution to understanding the role of the MHC in cell-mediated immunity. Zinkernagel and Doherty's discovery is all the more revolutionary when you consider that little was known about cellular immunity and no T-cell receptor had yet been discovered!

These milestones in immunologic understanding set the stage for the development of two key models of how T cells respond to foreign antigen: altered self and dual recognition. The altered self hypothesis posited that histocompatibility molecules that associate with foreign particles, such as viruses, may appear to be altered forms of self-proteins. The dual recognition model proposed that T cells must be capable of simultaneous recognition of both the foreign substance and these self-histocompatibility molecules. We now appreciate the validity of both of these models, and the role of MHC restriction in the immune response to both infectious agents and allogeneic transplants.

REFERENCES

Doherty, P. C., and R. M. Zinkernagel. 1975. H-2 compatibility is required for T-cell mediated lysis of target cells infected with lymphocytic choriomeningitis virus. *Journal of Experimental Medicine* **141**:502.

Rosenthal, A. S., and E. M. Shevach. 1973. Function of macrophages in antigen recognition by guinea pig T lymphocytes. I. Requirement for histocompatible macrophages and lymphocytes. *Journal of Experimental Medicine* **138**:1194.

Shevach, E. M., and A. S. Rosenthal. 1973. Function of macrophages in antigen recognition by guinea pig T lymphocytes. II. Role of the macrophage in the regulation of genetic control of the immune response. *Journal of Experimental Medicine* **138**:1213.

killing the cells and immobilizing the current set of proteins in the membrane; **Figure 7-11a**). However, if the APCs were first given time to ingest the antigen and were fixed with paraformaldehyde 1 to 3 hours later, T_H-cell activation still occurred (Figure 7-11b). During that crucial interval of 1 to 3 hours, the APCs had taken up the antigen, digested it into peptide fragments, and displayed these fragments on the cell membrane in a form capable of activating T cells.

Subsequent experiments by J. Kappler and P. Marrack showed that internalization and processing could be bypassed if APCs were exposed to antigen in the form of already digested peptide fragments instead of the native antigen (Figure 7-11c). In these experiments, APCs were treated with glutaraldehyde (this chemical, like paraformaldehyde, fixes the cell, rendering it metabolically inactive) and then incubated with native ovalbumin or with ovalbumin that had been subjected to partial enzymatic digestion. The digested ovalbumin was able to interact with the MHC molecules on the surface of glutaraldehyde-fixed APCs,

thereby activating ovalbumin-specific T_H cells. However, the native ovalbumin failed to do so. These results suggest that digestion of the protein antigens into smaller peptides is required for the presentation of antigen in MHC class I and recognition by ovalbumin-specific T_H cells.

At about the same time, several investigators, including W. Gerhard, A. Townsend, and their colleagues, began to identify the proteins of influenza virus that were recognized by T_C cells. Contrary to their expectations, they found that internal proteins of the virus, such as polymerase and nucleocapsid proteins, were often recognized by T_C cells better than the more exposed envelope proteins found on the surface of the virus. Moreover, Townsend's work revealed that T_C cells recognized short linear peptide sequences of influenza proteins. In fact, when noninfected target cells were incubated in vitro with synthetic peptides corresponding to only short sequences of internal influenza proteins, these cells could be recognized by T_C cells and subsequently lysed just as well as target cells that had been infected with whole,

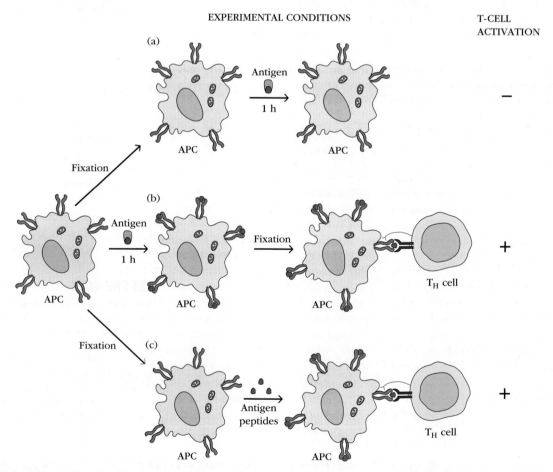

EXPERIMENTAL CONDITIONS

T-CELL ACTIVATION

FIGURE 7-11 Experimental demonstration that antigen processing is necessary for T_H-cell activation. (a) When antigen-presenting cells (APCs) are fixed before exposure to antigen, they are unable to activate T_H cells. (b) In contrast, APCs fixed at least 1 hour after antigen exposure can activate T_H cells. (This simplified figure does not show costimulatory molecules needed for T-cell activation.) (c) When APCs are fixed before antigen exposure and incubated with peptide digests of the antigen (rather than native antigen), they also can activate T_H cells. T_H-cell activation is determined by measuring a specific T_H-cell response (e.g., cytokine secretion).

live influenza virus. These findings, along with those presented shortly, suggest that *antigen processing is a metabolic process that digests proteins into small peptides, which can then be displayed on the cell surface together with an MHC class I or II molecule.*

Key Concept:

- Protein antigens must be processed into peptide fragments and presented by self-MHC molecules on the surface of host cells in order to be recognized by antigen-specific receptors (TCRs) on responding T cells.

Evidence Suggests Distinct Antigen Processing and Presentation Pathways

We now know that the immune system employs different pathways to eliminate intracellular and extracellular antigens, with some overlap as we shall see. As a general rule, endogenous antigens (those generated within the cell) are processed via the **cytosolic** or **endogenous pathway** and presented on the membrane with *MHC class I molecules* (**Figure 7-12**, left). Exogenous antigens (those taken up from the extracellular environment by endocytosis) are typically processed via the **exogenous pathway** and presented on the membrane with *MHC class II molecules* (Figure 7-12, right).

Experiments carried out by L. A. Morrison and T. J. Braciale provided an excellent demonstration that the antigenic peptides presented by MHC class I and II molecules follow different routes during antigen processing. To do this, they used a set of T-cell clones specific for an influenza virus antigen; some recognized the antigen presented by MHC class I and others recognized the *same antigen* presented by MHC class II. Examining the T-cell responses, they derived the following general principles about the two pathways:

- Class I presentation requires internal (cytosolic) synthesis of viral proteins; the target cell must be infected with live virus and class I presentation on the cell surface is impaired when protein synthesis is blocked by the inhibitor emetine.
- Class II presentation can occur with either live or replication-incompetent virus; protein synthesis inhibitors had no effect, indicating that new protein synthesis is not a necessary condition for class II presentation.
- Class II, but not class I, presentation is inhibited by treatment of the cells with an agent that blocks endocytic processing within the cell (e.g., chloroquine).

These studies support the distinction between the processing of exogenous and endogenous antigens. They also suggest a preferential, but not absolute, *association of exogenous/extracellular antigens with MHC class II molecules and of endogenous/intracellular antigens with class I molecules.* What these experiments do not show is that the type

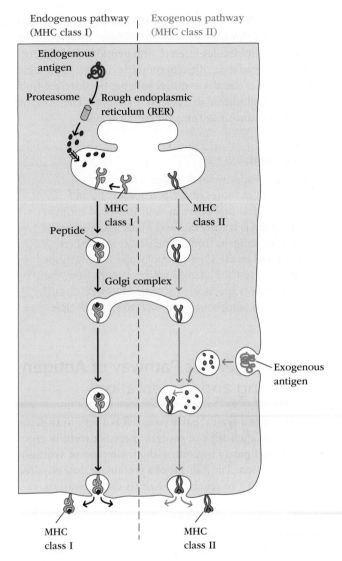

FIGURE 7-12 Overview of endogenous and exogenous pathways for processing antigen. In the endogenous pathway (*left*), antigens in the cytosol are degraded by the proteasome, converting proteins into smaller peptides. In the exogenous pathway (*right*), extracellular antigens are engulfed into endocytic compartments, where they are degraded by acidic pH–dependent endosomal and lysosomal enzymes. The antigenic peptides from proteasome cleavage and those from endocytic compartments associate with MHC class I or II molecules, respectively, and the MHC-peptide complexes are then transported to the cell membrane. It should be noted that the ultimate fate of most peptides in the cell is neither of these pathways; rather, most are degraded completely into amino acids.

of APC, nature of the antigen, and cues in the microenvironment can alter this outcome, as will be seen later when these rules are subverted during cross-presentation. In the experiments described here, association of viral antigen with MHC class I molecules required replication of the influenza virus and viral protein synthesis within the target cells, but association with class II did not. These findings suggested that the peptides presented by MHC class I and II molecules are trafficked through separate intracellular compartments; MHC class I molecules interact with peptides derived from cytosolic degradation of endogenously synthesized proteins, while class II molecules associate with peptides derived from endocytic degradation of exogenous antigens. The next two sections examine these two pathways in detail.

Key Concept:

- There is experimental evidence of at least two distinct routes of antigen processing and presentation that differ in their source of antigen, intracellular trafficking, and MHC association. In general, antigens from intracellular sources are presented in MHC class I molecules (endogenous pathway) to CD8+ T cells while antigens from the extracellular spaces are presented in MHC class II molecules (exogenous pathway) to CD4+ T cells.

The Endogenous Pathway of Antigen Processing and Presentation

The cytosol of a typical eukaryotic cell is a very busy place. The level of each type of protein there is carefully regulated, at least partly by controlling their rates of synthesis and degradation. The half-life of a protein, or time required for that protein to reach one-half its concentration, varies widely but can range from minutes to days (or longer, in a few cases). Some proteins, like those involved in formation of the nucleus, tend to have long half-lives, while others, such as transcription factors, cyclins, and key metabolic enzymes, typically have very short half-lives. Denatured, misfolded, or otherwise abnormal proteins also are degraded rapidly. In particular, defective ribosomal products (called DRiPs), which can come from premature translation termination, protein misfolding, or defects in multisubunit protein assembly, constitute a rather large subset of protein products in the cytosol. The consequence of this steady turnover of both normal and defective proteins is a constant pool of proteins and their fragments that are no longer needed by the cell. While many will be reduced to their constituent amino acids and recycled, some persist in the cytosol as peptides. The cell constantly samples from this pool of peptides and presents some fragments on the plasma membrane in association with MHC class I molecules, where cells of the immune system can likewise sample these peptides to keep a lookout for foreign proteins lurking inside of host cells. The pathway by which these endogenous peptides are generated for presentation with MHC class I molecules utilizes mechanisms similar to those involved in the normal turnover of intracellular proteins.

Peptides Are Generated by Protease Complexes Called Proteasomes

Intracellular proteins are degraded into short peptides by the proteasome, a cytosolic proteolytic system present in all cells (**Figure 7-13a**). The large (20S) proteasome is composed of multiple α and β subunits arranged in concentric rings; the α subunits make up the top and bottom rings while the β subunits construct the middle two rings. There are a total of 14 β subunits arrayed in this barrel-like structure of symmetrical rings.

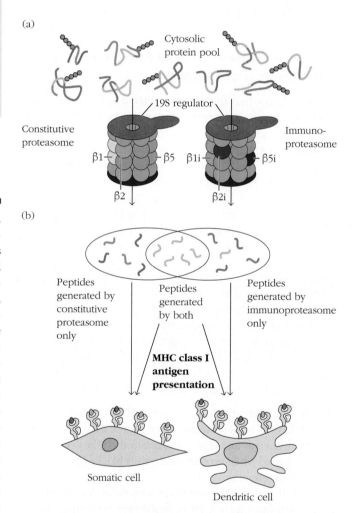

FIGURE 7-13 Proteolytic system for degradation of intracellular proteins. (a) Endogenous proteins in all cells are targeted for degradation by ubiquitin conjugation. These proteins are degraded by the 26S proteasome complex, which includes the 20S constitutive proteasome and a 19S regulator, generating a pool of peptides for MHC class I loading. In activated APCs, several proteins in the constitutive proteasome (β1, β2, and β5) are replaced by proteins encoded by the *LMP* genes (β1i, β2i, and β5i). (b) This generates an immunoproteasome with increased efficiency for creating unique peptides with a proclivity for assembly with MHC class I molecules.

We know that many proteins are targeted for proteolysis when a small protein called **ubiquitin** is attached to them. These ubiquitin-protein conjugates enter the proteasome complex, consisting of the 20S base and an attached 19S regulatory component, through a narrow channel at the 19S end. The proteasome complex cleaves peptide bonds in an ATP-dependent process. Degradation of ubiquitin-protein complexes is thought to occur within the central hollow core of the proteasome.

The immune system adds a twist to the general pathway of protein degradation to specifically produce small peptides optimized for binding to MHC class I molecules. In addition to the standard 20S proteasomes resident in all cells, a distinct proteasome of the same size, the **immunoproteasome**, can be found in pAPCs and some infected cells. It has the same basic structure as the traditional proteasome with some unique subunit substitutions. In most cells, these new subunits are not constitutively expressed like the other components of the proteasome but are induced by exposure to certain cytokines, such as IFN-γ or TNF. *LMP2* and *LMP7*, genes that are located within the class II region (see Figure 7-7) and are responsive to these cytokines, encode *replacement catalytic protein subunits* that convert standard proteasomes into immunoproteasomes, increasing the efficiency with which cytosolic proteins are cleaved into peptide fragments that specifically bind to MHC class I molecules (Figure 7-13b). In fact, a subset of the peptides created in the presence of immunoproteosomes is not found in cells lacking these structures. The half life of an immunoproteasome is shorter than that of a standard proteasome, possibly because the increased level of protein degradation in its presence may have negative consequences beyond the targeting of infected cells. It is possible that in some cases autoimmunity results from increased processing of self-proteins in cells with high levels of immunoproteasomes.

Key Concept:

- While all cells have constitutive proteasomes involved in the processing and presentation of cytosolic proteins in MHC class I molecules, infected cells or activated APCs can temporarily express immunoproteasomes, which generate peptide fragments that are optimized for MHC class I binding.

Peptides Are Transported from the Cytosol to the Rough Endoplasmic Reticulum

Insight into the cytosolic or endogenous processing pathway came from studies of cell lines with defects in peptide presentation by MHC class I molecules. One such mutant cell line, called RMA-S, expresses about 5% of the normal levels of MHC class I molecules on its membrane.

Although RMA-S cells synthesize normal levels of class I α chain and β₂-microglobulin, few MHC class I complexes appear on the membrane. A clue to the mutation in the RMA-S cell line was the discovery, by Townsend and colleagues, that "feeding" peptides to these cells restored the level of membrane-associated MHC class I molecules to normal. These investigators suggested that peptides might be required to stabilize the interaction between the class I α chain and β₂-microglobulin. The ability to restore expression of MHC class I molecules on the membrane by supplying the cells with predigested peptides suggested that the RMA-S cell line might have a defect in peptide processing or transport.

Subsequent experiments showed that the defect in the RMA-S cell line occurs in the protein that transports peptides from the cytoplasm into the rough endoplasmic reticulum (RER), where the transmembrane MHC class I molecules are synthesized. When RMA-S cells were transfected with a functional gene encoding this transporter protein, the cells began to express class I molecules on the membrane at near normal levels. The transporter protein, designated **TAP** (*t*ransporter *a*ssociated with *a*ntigen *p*rocessing), is an ER-membrane-spanning heterodimer consisting of two proteins: TAP1 and TAP2 (**Figure 7-14a**). In addition to their transmembrane segments, the TAP1 and TAP2 proteins each have a domain projecting into the lumen of the RER and an ATP-binding domain that projects into the cytosol. Both TAP1 and TAP2 belong to the family of ATP-binding cassette proteins found in the membranes of many cells, including bacteria. This family of proteins mediates ATP-dependent transport of amino acids, sugars, ions, and peptides across membranes.

Peptides generated in the cytosol by the proteasome are translocated by TAP into the RER by a process that requires the hydrolysis of ATP (Figure 7-14b). TAP has affinity for peptides containing 8 to 16 amino acids. The optimal peptide length for final association with MHC class I is nine amino acids. It was later discovered that longer peptides can also bind but are trimmed by enzymes present in the lumen of the ER, such as **ERAP** (*e*ndoplasmic *r*eticulum *a*mino*p*eptidase). In addition, TAP appears to favor peptides with hydrophobic or basic carboxyl-terminal amino acids, the preferred anchor residues for MHC class I molecules (see Figure 7-4). Thus, TAP is pre-optimized to transport peptides that are likely to interact with MHC class I molecules. The *TAP* and *LMP* genes (encoding immunoproteosome components) map within the MHC class II region. A few different allelic forms of these genes exist within the population and can account for modest differences in antigen processing and MHC class I presentation, some of which have been linked to disorders such as psoriasis, a skin disease with autoimmune features. TAP deficiencies can lead to more severe disease syndromes, many of which share aspects of both

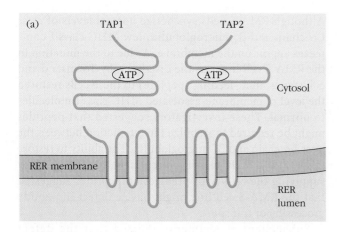

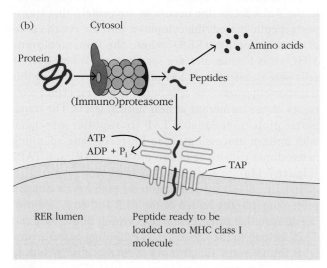

FIGURE 7-14 TAP (transporter associated with antigen processing). (a) Schematic diagram of TAP, a heterodimer anchored in the membrane of the rough endoplasmic reticulum (RER). The two chains are encoded by *TAP1* and *TAP2*. The cytosolic domain in each TAP subunit contains an ATP-binding site, and peptide transport depends on the hydrolysis of ATP. (b) In the cytosol, ubiquitinated proteins are directed to a constitutive or immunoproteasome, where they are digested into peptide fragments. These peptides are translocated by TAP into the RER lumen, where, in a process mediated by several other proteins, they will associate with MHC class I molecules.

immune deficiency (poor immune responses) and auto-immunity (overactive immune responses to self).

Key Concept:

- A transmembrane protein in the RER called TAP (*transporter* associated with *antigen* *processing*) is required to transport peptides of approximately the right size (that have been cleaved by cytosolic proteasomes/immunoproteasomes) into the lumen of the RER, where they can associate with newly formed MHC class I molecules.

Chaperones Aid Peptide Assembly with MHC Class I Molecules

Like other proteins destined for the plasma membrane, the α chain and β₂-microglobulin components of the MHC class I molecule are synthesized on ribosomes on the RER. Assembly of these components into a stable MHC class I molecular complex that can exit the RER requires the presence of a peptide in the binding groove of the class I molecule. The assembly process involves several steps and includes the participation of *molecular chaperones* that facilitate the folding of polypeptides.

The first molecular chaperone involved in MHC class I assembly is **calnexin**, a resident membrane protein of the ER. ERp57, a protein with enzymatic activity, and calnexin associate with the class I α chain and promote its folding (**Figure 7-15**). When β₂-microglobulin binds to the α chain, calnexin is released and the class I-ERp57 complex associates with the chaperones **calreticulin** and **tapasin**. Tapasin (*TAP-as*sociated prote*in*) brings the TAP transporter into proximity with the class I molecule and allows it to acquire an antigenic peptide. The TAP protein then promotes peptide capture by the class I molecule before the peptides are exposed to the luminal environment of the RER. In 2000 the tapasin-related protein, TAPBPR (for *TAP-b*inding *p*rotein *r*elated), was found to bind MHC class I molecules much like tapasin does. However, TAPBPR appears to be more than just an analogue of tapasin. This conclusion comes from the observation that while tapasin overexpression in cells leads to *increased* class I surface expression, TAPBPR overexpression has the opposite effect, resulting in *decreased* MHC class I surface expression. The final role for TAPBR in the endogenous pathway is yet to be resolved.

Some ER peptides are too long to bind efficiently to class I molecules, and exoproteases can act on these proteins. One ER aminopeptidase, ERAP1, removes the amino-terminal residue from peptides to achieve optimum class I binding size (see Figure 7-15). In fact, ERAP1 has little affinity for peptides shorter than eight amino acids in length. As a consequence of productive peptide binding, the class I molecule displays increased stability and can dissociate from the complex with calreticulin, tapasin, and ERp57 (also known as the *peptide loading complex*). The class I molecule can then exit from the RER, proceeding to the Golgi complex and exocytic vesicles, before ultimately reaching the cell surface.

Key Concept:

- In the RER, chaperone proteins and proteases assist with the loading and processing of peptide fragments as they associate with MHC class I molecules, stabilizing this protein complex and allowing peptide-loaded MHC class I molecules to move out of the RER and toward the cell surface.

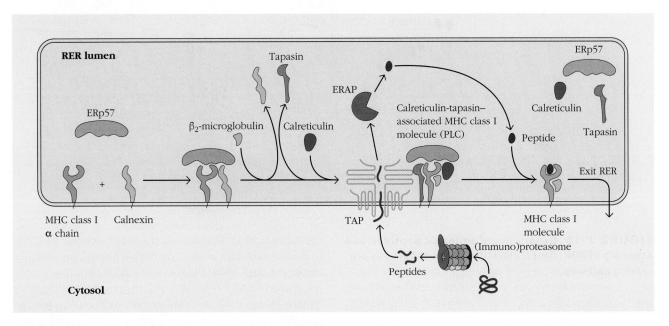

FIGURE 7-15 Assembly and stabilization of MHC class I molecules. Within the rough endoplasmic reticulum (RER) membrane, a newly synthesized class I α chain associates with calnexin (a molecular chaperone) and ERp57 until β₂-microglobulin binds to the α chain. The binding of β₂-microglobulin releases calnexin and allows binding to calreticulin and to tapasin, which is associated with the peptide transporter TAP. This association creates a protein loading complex (PLC) that promotes binding of an antigenic peptide. Antigens in the RER can be further processed via aminopeptidases such as ERAP, producing fragments ideally suited for binding to class I. Peptide association stabilizes the class I molecule–peptide complex, allowing it to be transported from the RER to the plasma membrane.

The Exogenous Pathway of Antigen Processing and Presentation

Extracellular material can gain access to vesicular compartments inside an APC through several means. Professional APCs can internalize particulate material by simple phagocytosis (also called "cell eating"), where material is engulfed by pseudopods of the cell membrane; or by receptor-mediated endocytosis, where the material first binds to specific surface receptors, followed by clathrin-mediated internalization. Macrophages and dendritic cells internalize antigen by both processes. Most other APCs, whether professional or not, demonstrate little or no phagocytic activity and therefore typically internalize exogenous antigen only by endocytosis (either receptor-mediated endocytosis or by pinocytosis, i.e., nonspecific "cell drinking"). B cells, for example, internalize antigen very effectively by receptor-mediated endocytosis, using their antigen-specific membrane immunoglobulin as the receptor. *The one thing that all these pathways have in common is that the internalized components gain access to the cell but remain bound by a phospholipid bilayer (membrane) structure.*

Peptides Are Generated from Internalized Antigens in Endocytic Vesicles

Once an antigen is internalized in this fashion it is typically degraded into peptides within compartments that make up the endocytic processing pathway. As the experiment shown in Figure 7-11 demonstrated, internalized antigen takes a mere 1 to 3 hours to travel through the endocytic pathway and ultimately appear at the cell surface in the form of MHC class II–peptide complexes. This endocytic antigen-processing and presentation pathway appears to involve several increasingly acidic compartments, including early endosomes (pH 6.0–6.5); late endosomes, or endolysosomes (pH 4.5–5.0); and lysosomes (pH 4.5). Internalized antigen progresses through these membrane-enclosed compartments, encountering hydrolytic enzymes and a lower pH in each successive compartment (**Figure 7-16**). Antigen-presenting cells have a unique form of late endosome, the MHC class II–containing compartment (MIIC), in which final protein degradation and peptide loading into MHC class II proteins occurs. Within the compartments of the endocytic pathway, antigen is degraded into short peptides of about 13 to 18 residues that ultimately meet up with and bind to preformed MHC class II molecules in late endosomes. Because the hydrolytic enzymes are optimally active under acidic conditions (low pH), antigen processing can be inhibited by chemical agents that increase the pH of the compartments (e.g., chloroquine) as well as by protease inhibitors (e.g., leupeptin).

The mechanism by which internalized antigen moves from one endocytic compartment to the next has not been conclusively demonstrated. However, it has been suggested that early endosomes from the periphery move inward to become late endosomes and eventually lysosomes. Alternatively, small transport vesicles may carry antigens from one compartment to the next. Eventually the

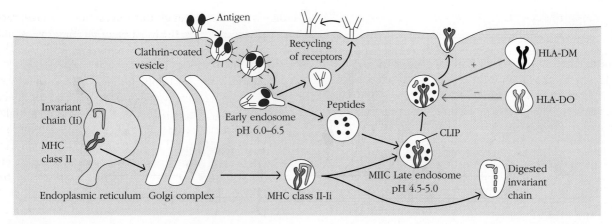

FIGURE 7-16 Generation of antigenic peptides and assembly of MHC class II molecules in the exogenous processing pathway. In the cell shown here, a B cell, exogenous antigen is internalized by receptor-mediated endocytosis (*top left*), with the membrane-bound antibody functioning as an antigen-specific receptor. Internalized antigen moves through several acidic compartments ending in specialized MIIC late endosomes, where it is degraded into peptide fragments. Within the rough endoplasmic reticulum (*bottom left*), a newly synthesized MHC class II molecule binds an invariant chain. The bound invariant chain prevents premature binding of peptides to the class II molecule and helps to direct the complex to endocytic compartments containing processed peptides derived from exogenous antigens. Digestion of the invariant chain leaves CLIP, a small fragment remaining in the binding groove of the MHC class II molecule. HLA-DM, a nonclassical MHC class II molecule present within the MIIC compartment, mediates exchange of antigenic peptides for CLIP. The nonclassical class II molecule HLA-DO may act as a negative regulator of class II antigen processing by binding to HLA-DM and inhibiting its role in the dissociation of CLIP from class II molecules.

endocytic compartments, or portions of them, return to the cell periphery, where they fuse with the plasma membrane. In this way, the surface receptors are recycled.

Key Concept:

• Phagocytosis of extracellular antigens by pAPCs results in transport of these antigens through a series of protease-containing intracellular vesicles, each with increasingly lower pH, where proteins are partially degraded into peptide fragments and eventually meet up with MHC class II molecules in vesicles coming from the RER.

The Invariant Chain Guides Transport of MHC Class II Molecules to Endocytic Vesicles

Since APCs express both MHC class I and II molecules, newly formed MHC molecules of both classes reside simultaneously within the membrane of the RER. Therefore, some mechanism must exist to prevent MHC class II molecules from binding to the antigenic peptides destined for MHC class I molecules. When MHC class II molecules are synthesized within the RER they associate with a protein called the **invariant chain** (**Ii**, or CD74). This conserved, non–MHC-encoded protein interacts with the peptide-binding groove formed by the combination of the α and β chains of MHC class II, in essence blocking any endogenously derived peptides from binding while the class II molecule is still in the RER (see Figure 7-16). The invariant chain also appears to be involved in the folding of the class II α and β chains, their exit from the RER, and the subsequent routing of class II molecules to the endocytic processing pathway from the trans-Golgi network.

The role of the invariant chain in the routing of class II molecules has been demonstrated in transfection experiments with cells that lack genes encoding both the MHC class II molecules and the invariant chain. Immunofluorescence labeling of these cells transfected only with MHC class II genes revealed that, in the absence of invariant chain, class II molecules remain primarily in the ER and do not transit past the cis-Golgi, or their final compartment before breaking off into vesicles. However, in cells transfected with both the MHC class II genes and the Ii-encoding gene, the class II molecules were localized in the cytoplasmic vesicular structures of the endocytic pathway. The invariant chain contains sorting signals in its cytoplasmic tail that direct the transport of the MHC class II complex from the trans-Golgi network to the endocytic compartments.

Key Concept:

• The invariant chain (Ii) serves as a chaperone, assisting in the assembly and transport of peptide-empty MHC class II molecules from the RER to late endosomes, where they encounter exogenous peptide fragments derived from phagocytosed antigens.

Peptides Assemble with MHC Class II Molecules by Displacing CLIP

Experimental observations indicate that most MHC class II–invariant chain complexes are transported from the RER, where they are formed, through the Golgi complex and trans-Golgi network. From there they proceed through the endocytic pathway, moving from early endosomes to the MIIC late endosomal compartment in some cases. As the proteolytic activity increases in each successive compartment, the invariant chain itself is gradually degraded. However, a short fragment of the invariant chain, termed **CLIP** (for *class* II–*associated invariant chain peptide*), remains bound to the class II molecule after the majority of the invariant chain has been cleaved within the endosomal compartment. Like antigenic peptide, CLIP physically occupies the peptide-binding groove of the MHC class II molecule, preventing any premature binding of antigen-derived peptide.

A nonclassical MHC class II molecule called HLA-DM is required to catalyze the exchange of CLIP with antigenic peptides (see Figure 7-16). The *DMα* and *DMβ* genes are located near the *TAP* and *LMP* genes in the MHC complex of humans, with similar genes in mice (see Figure 7-7). Like other MHC class II molecules, HLA-DM is a heterodimer of α and β chains. However, unlike other class II molecules it is relatively nonpolymorphic and is not normally expressed at the cell membrane but is found predominantly within the endosomal compartment. HLA-DM has been found to associate with the MHC class II β chain and to function in removing or "editing" peptides, including CLIP, that associate transiently with the binding groove of classical class II molecules. Peptides that make especially strong molecular interactions with MHC class II, creating long-lived complexes, are harder for HLA-DM to displace, and thus become the repertoire of peptides that ultimately make it to the cell surface as MHC-peptide complexes.

As with MHC class I molecules, peptide binding is required to maintain the structure and stability of class II molecules. Once a peptide has bound, the MHC class II–peptide complex is transported to the plasma membrane, where the neutral pH appears to enable the complex to assume a compact, stable form. Peptide is bound so strongly in this compact form that it is difficult to replace a class II–bound peptide on the membrane with another peptide under physiologic conditions.

One additional nonclassical member of the MHC class II family, HLA-DO, also is relatively nonpolymorphic and associates with classical class II molecules. However, it appears to act as a negative regulator of antigen binding, modulating the function of HLA-DM and changing the repertoire of peptides that preferentially bind to classical class II molecules. In cells that express both HLA-DO and HLA-DM, these two molecules strongly associate in the ER and maintain this interaction all the way to the endosomal compartments. Although this interaction has been recognized for many years, the function of this negative regulator and the impact of this changed peptide repertoire has been only more recently resolved.

Originally observed only in B cells and in the thymus, the cellular expression profile of HLA-DO has recently been expanded to dendritic cells (DCs), where it is thought to play a role in the maintenance of self-tolerance (discussed further in Chapter 16). This phenomenon was studied by L. K. Denzin and colleagues, using diabetes-prone mice engineered to express human HLA-DO. DCs are known to be essential in the establishment of self-tolerance, as well as in the presentation of autoantigens to self-reactive T cells. Self-reactive T cells develop in these nonobese diabetic animals, and these cells ultimately destroy pancreatic beta cells, causing type 1 diabetes. In this study, the development of diabetes was blocked by the presence of the *HLA-DO* transgene in DCs. In addition, using specific monoclonal antibodies, the authors showed that the repertoire of peptides being presented to the autoreactive T cells was significantly altered, resulting in reduced efficiency in presenting key self-peptides. This and other work suggests that normal HLA-DO expression may play an important role in modulating HLA-DM behavior to ensure the presentation of a self-peptide repertoire that encourages tolerance to self-antigens. Interestingly, in wild-type mice and humans, HLA-DO expression is down-regulated following DC activation by antigen, releasing HLA-DM to carry out its normal function of encouraging the presentation of a diverse array of peptides, many of which will presumably be derived from the foreign proteins that stimulated these APCs.

Overview Figure 7-17 illustrates the endogenous pathway (left side) and compares it with the separate exogenous pathway (right side). Whether an antigenic peptide associates with class I or with class II molecules is partially dictated by the mode of entry into the cell, either exogenous or endogenous; the site of processing; the cell type; and the microenvironment surrounding that cell. However, in the next section, we will see that these assignments are not absolute.

Key Concepts:

- The invariant chain is gradually processed into a smaller fragment, called CLIP, that blocks the peptide-binding groove of classical class II molecules in late endosomes, where HLA-DM (a nonclassical class II molecule) mediates exchange of CLIP for exogenous antigen fragments in the endosome.

- The presence of HLA-DO, another nonclassical class II molecule, in some pAPCs (specifically DCs) may regulate the activity of HLA-DM and encourage class II presentation of self-peptides in a manner that encourages self-tolerance.

Separate Antigen-Presenting Pathways Are Utilized for Endogenous (Green) and Exogenous (Red) Antigens

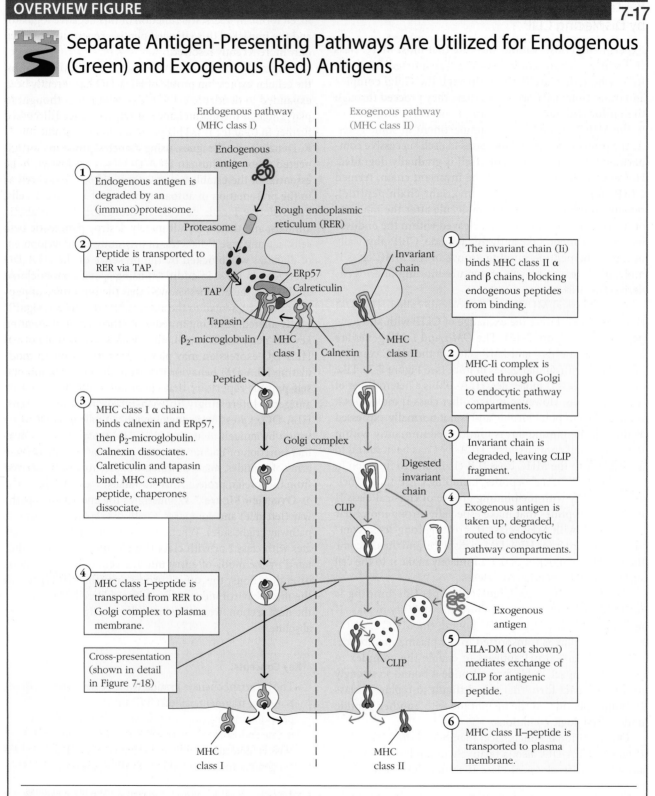

Endogenous pathway (MHC class I)

Exogenous pathway (MHC class II)

Endogenous antigen

1 Endogenous antigen is degraded by an (immuno)proteasome.

Proteasome

Rough endoplasmic reticulum (RER)

2 Peptide is transported to RER via TAP.

TAP

ERp57 Calreticulin

Invariant chain

Tapasin

β₂-microglobulin

MHC class I

Calnexin

MHC class II

1 The invariant chain (Ii) binds MHC class II α and β chains, blocking endogenous peptides from binding.

Peptide

3 MHC class I α chain binds calnexin and ERp57, then β₂-microglobulin. Calnexin dissociates. Calreticulin and tapasin bind. MHC captures peptide, chaperones dissociate.

Golgi complex

CLIP

Digested invariant chain

2 MHC-Ii complex is routed through Golgi to endocytic pathway compartments.

3 Invariant chain is degraded, leaving CLIP fragment.

4 Exogenous antigen is taken up, degraded, routed to endocytic pathway compartments.

4 MHC class I–peptide is transported from RER to Golgi complex to plasma membrane.

Exogenous antigen

Cross-presentation (shown in detail in Figure 7-18)

CLIP

5 HLA-DM (not shown) mediates exchange of CLIP for antigenic peptide.

MHC class I

MHC class II

6 MHC class II–peptide is transported to plasma membrane.

The mode of antigen entry into cells and the site of antigen processing determine whether antigenic peptides associate with MHC class I molecules in the rough endoplasmic reticulum or with class II molecules in endocytic compartments. Under certain circumstances, exogenous antigens may be assembled with MHC class I molecules via cross-presentation.

Unconventional Antigen Processing and Presentation

Now that you understand the general rules for processing and presentation of protein antigens (endogenous peptides are destined for MHC class I and exogenous peptides for MHC class II) it is time to mention the exceptions. It is now clear, based on both experimental and clinical observations, that peptide fragments from exogenous antigens can be presented by MHC class I, and cytosolic protein fragments have been found associated with MHC class II molecules. In an example of the latter, recent studies in mice have shown that protection from influenza is dependent on a yet to be defined crossover pathway of antigen processing and presentation, where viral proteins synthesized in the cytosol of infected pAPCs associate with MHC class II molecules and activate naïve CD4$^+$ T cells. Likewise, autophagy, the homeostatic process whereby cytosolic components become enclosed in vesicles that traffic through lysosomal compartments, can result in cytosolic peptides being presented by MHC class II molecules. While some of the molecular details for these unconventional MHC class II processing and presentation pathways are still unclear, crossover pathways that lead to MHC class I presentation of exogenous antigen are beginning to come into focus.

We have seen that after pAPCs internalize extracellular antigen they can process and present these antigens via the conventional exogenous pathway, leading to MHC class II association. Because pAPCs also express costimulatory molecules, engagement of CD4$^+$ T cells with these MHC class II–peptide complexes can lead to activation of T helper responses. On the other hand, infected cells will generally process and present cytosolic peptides via the endogenous pathway, leading to MHC class I–peptide complexes on their surface. However, unless these infected cells express the costimulatory molecules, they cannot activate naïve CD8$^+$ T cells. This leaves us with a dilemma: how does the immune system activate CD8$^+$ T cells to eliminate *intracellular* microbes unless a professional antigen-presenting cell happens to become infected? By the conventional route, a pAPC that phagocytoses a virus from *extracellular* sources should send these viral proteins through endocytic vesicles to associate with MHC class II molecules, activating CD4$^+$ T cells rather than the CD8$^+$ cytotoxic T lymphocytes (CTLs) that are required to combat this type of infection.

The answer to this dilemma is a process called **cross-presentation**, blending the exogenous and endogenous pathways in a process that is still being fully resolved. We do know that in some instances certain APCs will divert antigen obtained by endocytosis (exogenous antigen) to a pathway that leads to MHC class I loading and peptide presentation to CD8$^+$ T cells (like in the endogenous pathway)—in other words, crossing the two pathways. First reported by Michael Bevan and later described in detail by Peter Cresswell and colleagues, *the phenomenon of cross-presentation requires that antigens acquired from extracellular sources, normally handled by the exogenous pathway leading to MHC class II presentation, are redirected to a class I peptide loading pathway.* When this form of antigen presentation leads to the activation of a naïve CD8$^+$ T cell it is referred to as **cross-priming**; when it leads to the induction of tolerance in a CD8$^+$ T cell, such as when the APCs are not activated, it is called **cross-tolerance**.

Dendritic Cells Can Cross-Present Exogenous Antigen via MHC Class I Molecules

The concept of cross-presentation was first recognized as early as 1976, but the relevant cell types and mechanisms of action remained a mystery until later. Light was shed on this pathway following studies focused on subsets of DCs, now recognized to be the most efficient of the cross-presenters. For example, in vivo depletion or inactivation of DCs compromises cross-presentation. The most potent of the cross-presenting DCs appear to reside in secondary lymphoid organs, where they are believed to receive antigen from extracellular sources by handoff from migrating APCs or from dying infected cells. A few other cell types, such as B cells, macrophages, neutrophils, and mast cells, have been found capable of cross-presentation in vitro, although evidence that they cross-present in vivo or that they have the ability to prime CTL responses (i.e., activate naïve CD8$^+$ T cells) is still lacking.

Key Concept:

- In some cases, exogenous antigens internalized by dendritic cells can gain access to the endogenous presentation pathway in a process called *cross-presentation*, leading to peptide association with MHC class I and engagement with CD8$^+$ T cells.

Cross-Presentation by APCs Is Essential for the Activation of Naïve CD8$^+$ T Cells

Two possible models have been proposed for how DCs accomplish cross-presentation. The first hypothesizes that cross-presenting cells possess special antigen-processing machinery that allows loading of exogenously derived peptides onto MHC class I molecules. The second theory postulates specialized endocytosis machinery that can send internalized antigen directly to an organelle (such as a phagosome or early endosome), where peptides from those antigens are then loaded onto MHC class I molecules using conventional machinery. These two proposed mechanisms are not mutually exclusive, and evidence for both exists.

How the antigen achieves the crossover from its exogenous or extracellular origins to the endogenous pathway has not been conclusively resolved. For example, in some instances cross-presented antigen from external sources has been found to enter the cytoplasm. In the early 1990s, it was shown that bead-conjugated antigens captured by cross-presenting DCs could reach the cytosol of these cells. Later, M. L. Lin and colleagues showed that exogenously added cytochrome *c*, which causes programmed cell death when present in the cytosol, could induce cross-presenting DCs to undergo apoptosis. It is now proposed that retro-translocation, or movement of endocytosed proteins *out of* endocytic compartments and into the cytosol, can occur via TAP molecules present in these endocytic membranes—a sort of backflow out of these vesicles.

On the other hand, TAP-independent cross-presentation of some antigens has also been observed, suggesting multiple mechanisms may exist for targeting externally derived peptides to class I molecules. Recent data support recruitment of the cross-presenting machinery to the phagosomal compartment from multiple intracellular sources; with MHC class I molecules coming from endosomal recycling vesicles and the peptide loading complex, including TAP, coming from the ER or intermediate Golgi compartments. Ligand binding to pattern recognition receptors, especially Toll-like receptors (TLRs; see Chapter 3), can induce the maturation of DCs as well as trigger the transport MHC class I molecules from intracellular stores to these antigen-laden vesicles arising from extracellular sources.

Regardless of mechanism, the ability of DCs to cross-present antigens from extracellular sources has great advantage for the host. It allows these APCs to capture antigen, such as viral proteins, from the extracellular environment or from dying cells, process these antigens, and activate CTLs that can then seek out and attack virus-infected cells, inhibiting further spread of the infection. One outstanding question remains: if this pathway is so important, why don't all APCs cross-present? The answer may lie in the fact that cross-presenting cells could quickly become targets of lysis themselves. Another dilemma concerns how cross-presenting DCs handle the multitude of self-peptides constantly presented in MHC class I molecules. Shouldn't these cells break tolerance by presenting extracellularly derived *self-peptides* to CD8⁺ T cells, activating anti-self CTL responses?

To help explain how cross-presentation is regulated and tolerance maintained, scientists have proposed that DCs might first need to be "licensed" before they can cross-present. The cell type postulated to supply this licensing role is activated CD4⁺ T cells. Here is how we believe this works:

(a) DC licensing by T_H cell

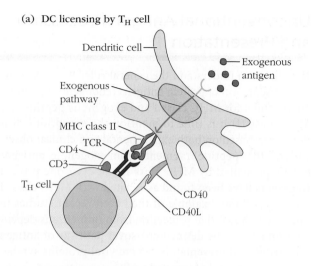

(b) DC cross-presentation and activation of CTL

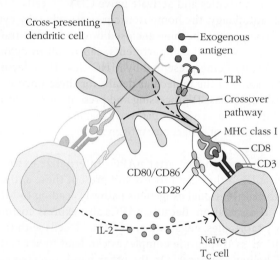

FIGURE 7-18 Activation of naïve T_C cells by exogenous antigen requires DC licensing and cross-presentation.
(a) Dendritic cells (DCs) first internalize and process antigen through the exogenous pathway, presenting antigen to CD4⁺ T_H cells via MHC class II molecules and activating these cells through, among other things, CD40-CD40L engagement. (b) These activated T_H cells can then serve as a bridge to help activate CTL responses; they provide local IL-2 to help activate the CTL and they also license the DCs to cross-present internalized antigen in MHC class I, up-regulate costimulatory molecules, and down-regulate their inhibitory counterparts. DC licensing creates an ideal situation for the stimulation of antigen-specific CD8⁺ T-cell responses. When the TLRs on these DCs are engaged, this further activates these cells, providing added encouragement for cross-presentation. The dashed arrow indicates antigen directed for cross-presentation.

first, the classical exogenous pathway of antigen processing in DCs leads to presentation of antigen to CD4⁺ T cells via class II, leading to activation of these cells (**Figure 7-18a**). These activated helper cells might then return the favor by inducing costimulatory molecule expression in the DC and by secreting stimulatory cytokines (e.g., IL-2). In principle, this would supply a "second opinion" that, respectively, licenses the DC to cross pathways and present internalized antigens via MHC class I, activating the more recalcitrant naïve $CD8^+$ T cells (Figure 7-18b). This requirement for licensing by a T_H cell could help avoid accidental induction of CTLs to nonpathogenic antigens or self-proteins. If this is the case, any *unactivated and unlicensed* DCs that happen to be cross-presenting antigens (e.g., self-peptides) may serve the opposite and equally important purpose: tolerance. Since they have not received the go-ahead from the T_H cell to activate CTL responses, they may lack costimulatory molecules and induce tolerance in the $CD8^+$ T cells they encounter. This would help to dampen reactivity to self-antigens and maintain self-tolerance.

Key Concept:

- Cross-presentation is the process by which pAPCs, primarily DCs that have been activated and licensed by antigen-specific T_H cells, divert exogenously acquired antigen into MHC class I molecules (crossing endogenous and exogenous pathways), allowing activation of $CD8^+$ T_C cells; the responses of these $CD8^+$ T_C cells are responsible for destroying other virally infected target cells that express this same peptide-MHC combination.

Presentation of Nonpeptide Antigens

To this point, the discussion of the presentation of antigens has been limited to protein antigens and their presentation by classical MHC class I and II molecules. However, it is well known that some nonprotein antigens are also recognized by T cells. As early as the 1980s, T-cell proliferation was detected in the presence of nonprotein antigens derived from infectious agents. Mycolic acid derived from pathogens such as *Mycobacterium tuberculosis* is one classic example. These and other small lipid-containing antigens are presented by a small group of structurally similar proteins encoded outside the MHC locus: a group of nonclassical class I molecules, including the CD1 family of proteins and the MHC class I–related protein (MR1).

CD1 and MR1 molecules share structural similarity with classical MHC class I but have more of a functional overlap with MHC class II. Five human *CD1* genes and one *MR1* gene have been identified. Most of the proteins encoded by these genes form a transmembrane heavy chain composed of three extracellular α domains, and associate noncovalently with β_2-microglobulin, much like classical MHC class I. However, unlike classical MHC molecules, CD1 and MR1

display very limited polymorphism. Since, during evolution, mice lost most of the *CD1* genes found in humans, research in this area has been slower to evolve. In terms of trafficking and expression profile, CD1 and MR1 molecules resemble MHC class II proteins, moving intracellularly to endosomal compartments, where they associate with exogenous antigen. Again, like MHC class II molecules, these nonclassical class I proteins are expressed by many immune cell types, including thymocytes, B cells, and DCs, although some members of the family have also been found on hepatocytes and epithelial cells.

The ligands for CD1 and MR1 include several different lipids or lipid-linked molecules, small molecules, and metabolites of vitamin B_2. Unlike their peptide counterparts, these moieties fit into deep pockets within the CD1/MR1 binding groove. Crystal structures have demonstrated that CD1 contains a binding groove that is both deeper and narrower than that of classical MHC molecules, and is lined with nonpolar amino acids that can easily accommodate hydrophobic structures. This means that antigen binds to CD1 and MR1 via a deep groove through a narrow opening—like a foot sliding into a shoe. **Figure 7-19** illustrates a comparison of the binding groove of CD1b holding a lipid antigen (left) with a classical class I molecule complexed with a peptide antigen (right).

It is believed that short-chain self-lipids with relatively low affinity are loaded onto CD1 molecules in the ER, shortly after translation, and allow proper CD1 protein folding. These self-antigen–loaded molecules then travel to the cell surface, where in some cases exogenous lipids and small molecules may be exchanged for these low-affinity self-antigens. In some cases, like with MHC class II molecules, after endocytosis and movement to lower pH environments, CD1 proteins may exchange low-affinity binding partners for higher-affinity, exogenously derived antigens resident in phagolysosomes or late endosomes. These newly loaded nonclassical MHC molecules then return to the cell surface for recognition by CD1- or MR1-restricted T cells.

A variety of T cells, both αβ and γδ, are known to bind these nonclassical MHC molecules, all of which share the property of invariant or semi-invariant TCRs. Tetramer technology (see Advances Box 12-2) has helped researchers to better characterize these T cells and their role in the immune response. We now understand that natural killer (NK) T cells, many skin and mucosal γδ T cells, and T cells responsible for recognizing *Mycobacterium tuberculosis* recognize CD1 molecules presenting lipid antigens. These and other "invariant" T cells are abundant in the body, especially in mucosal tissues, where they play a long-standing evolutionary role. Recently, the ligand for MR1 was identified as a microbially produced metabolite of the vitamin B_2. These small-molecule derivatives of vitamin B_2 are then thought to associate with host MR1 molecules, which present them to mucosal-associated invariant T cells (MAIT

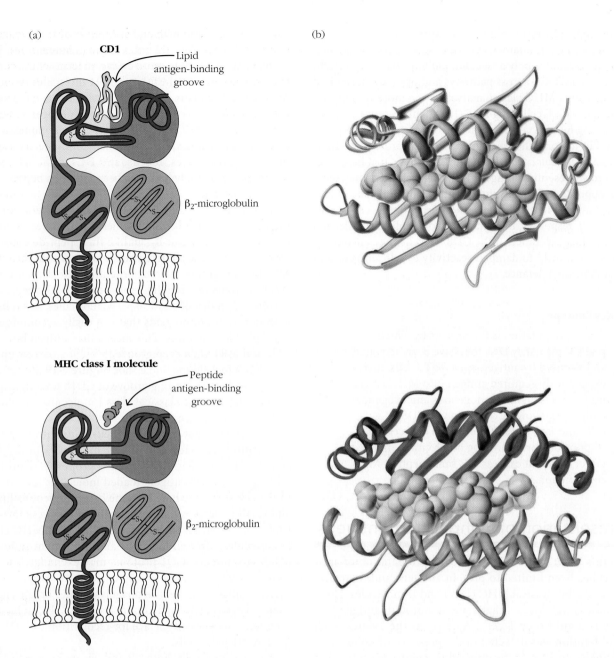

FIGURE 7-19 Lipid antigen binding to the CD1 molecule. (a) Schematic of the deep binding pockets that accommodate lipid antigens in a foot-in-shoe fashion in the members of the CD1 family of nonclassical class I molecules (*top*), compared with the shallower peptide-binding groove of classical MHC class I (*bottom*). (b) Ribbon diagram of the binding groove or pocket of human CD1b complexed with lipid (*top*) compared with a classical class I molecule binding peptide antigen (*bottom*). [(b) PDB IDs 1GZQ (left) and 1HHG (right).]

cells; see Chapter 13). The responding T cells play a role in both immune homoeostasis and mucosal control against infectious diseases, especially certain classes of bacterial and fungal pathogens (Chapter 17). In fact, these and other pathways may explain how gut microbes and commensals "tune" the host immune system, a process that when faulty, such as when the gut flora is abnormal or disrupted by medication like antibiotics (dysbiosis), can promote immune-mediated diseases such as allergy and autoimmunity (Chapter 13).

Key Concepts:

- Nonclassical MHC molecules like CD1 and MR1 share structural similarity with class I but function more like class II, presenting nonprotein (lipid and small-molecule) antigens to αβ and γδ T cells.

- CD1 and MR1 have limited variability, and serve to regulate immune homoeostasis as well as control some infectious agents at mucosal surfaces.

Conclusion

If antigen-presenting cells are the link between innate and adaptive immunity, then MHC molecules are the tool these cells use to create that link. MHC molecules hold antigenic fragments and present these to T-cell receptors, allowing the associated T cell to become activated and initiating the first steps in the adaptive response. These transmembrane molecules are present on the cell surface and are ubiquitously expressed in the body. MHC molecules display vast diversity, both at the individual and population levels, due to evolutionary pressure from pathogens that has favored gene duplication, polymorphism, and codominant expression patterns. To associate with MHC molecules, antigens must first be cleaved into smaller fragments (processed) and transported to locations in the cell where they can bind with and stabilize the MHC structure before they appear on the cell surface (presentation). The shape of the MHC antigen-binding groove, which comes from the alleles we inherit at this locus, thus determines the shape and chemical structure of the antigenic objects that can be held and presented to T cells. This, in essence, determines which parts of infectious agents will be available for recognition and therefore controls which naïve T cells will become activated. Since most B cells, which do not require MHC involvement to recognize antigen, rely on T-cell help to achieve their full function, this places MHC presentation of antigen at the fulcrum of adaptive immunity. For this reason, diversity at this locus is beneficial for individual hosts, and survival at the species level is favored when a population maintains diversity within its MHC gene pool.

REFERENCES

Amigorena, S., and A. Savina. 2010. Intracellular mechanisms of antigen cross presentation in dendritic cells. *Current Opinion in Immunology* **22:**109.

Blander, J. M. 2016. The comings and goings of MHC class I molecules herald a new dawn in cross-presentation. *Immunological Reviews* **272:**65.

Blum, J. S., P. A. Wearsch, and P. Cresswell. 2013. Pathways of antigen processing. *Annual Review of Immunology* **31:**443.

Brown, J. H., et al. 1993. Three-dimensional structure of the human class II histocompatibility antigen HLA-DR1. *Nature* **364:**33.

den Haan, J. M. M., S. M. Lehar, and M. J. Bevan. 2000. CD8$^+$ but not CD8$^-$ dendritic cells cross-prime cytotoxic T cells in vivo. *Journal of Experimental Medicine* **192:**1685.

Herman, C., Trowsdale J., Boyle LH. 2015. TAPBPR: a new player in the MHC class I presentation pathway. *Tissue Antigens* **85:**155.

Horton, R., et al. 2004. Gene map of the extended human MHC. *Nature Reviews Genetics* **5:**1038.

Kelley, J., et al. 2005. Comparative genomics of major histocompatibility complexes. *Immunogenetics* **56:**683.

Klein, J. 1986. Seeds of time: fifty years ago Peter A. Gorer discovered the H-2 complex. *Immunogenetics* **24:**331.

Kurts, C., B. W. S. Robinson, and P. A. Knolle. 2010. Cross-priming in health and disease. *Nature Reviews in Immunology* **10:**403.

Li, X. C., and M. Raghavan. 2010. Structure and function of major histocompatibility complex class I antigens. *Current Opinions in Organ Transplantation* **15:**499.

Madden, D. R. 1995. The three-dimensional structure of peptide-MHC complexes. *Annual Review of Immunology* **13:**587.

Mellins, E., and L. Stern. 2014. HLA-DM and HLA-DO, key regulators of MHC-II processing and presentation. *Current Opinion in Immunology* **26:**115.

Miller, M. A., et al. 2015. Endogenous antigen processing drives the primary CD4$^+$ T cell response to influenza. *Nature Medicine* **21:**1216.

Moody, D. B., D. M. Zajonc, and I. A. Wilson. 2005. Anatomy of CD1–lipid antigen complexes. *Nature Reviews in Immunology* **5:**387.

Nair-Gupta, P., et al. 2014. TLR signals induce phagosomal MHC-I delivery from the endosomal recycling compartment to allow cross-presentation. *Cell* **158:**506.

Roche, P., and K. Furuta. 2015. The ins and outs of MHC class II–mediated antigen processing and presentation. *Nature Reviews Immunology* **15:**203.

Rock, K. L., et al. 2004. Post-proteasomal antigen processing for major histocompatibility complex class I presentation. *Nature Immunology* **5:**670.

Salio, M., and V. Cerundolo. 2015. Regulation of lipid specific and vitamin specific non-MHC restricted T cells by antigen presenting cells and their therapeutic potentials. *Frontiers in Immunology* **6:**388.

Shimonkevitz, R., J. Kappler, P. Marrack, and H. Grey. 1983. Antigen recognition by H-2–restricted T cells. I. Cell-free antigen processing. *Journal of Experimental Medicine* **158:**303.

Strominger, J. L. 2010. An alternative path for antigen presentation: group 1 CD1 proteins. *Journal of Immunology* **184:**3303.

Trowsdale, J., and J. C. Knight. 2013. Major histocompatibility complex genomics and human disease. *Annual Review of Genomics and Human Genetics* **14**:301.

Van Rhijn, I., D. Godfrey, J. Rossjohn, and D. Moody. 2015. Lipid and small-molecule display by CD1 and MR1. *Nature Reviews Immunology* **15**:643.

Vartabedian, V. F., P. B. Savage, and L. Teyton. 2016. The processing and presentation of lipids and glycolipids to the immune system. *Immunological Reviews* **272**:109.

Yaneva, R., et al. 2010. Peptide binding to MHC class I and II proteins: new avenues from new methods. *Molecular Immunology* **47**:649.

Yi, W., et al. 2010. Targeted regulation of self-peptide presentation prevents type I diabetes in mice without disrupting general immunocompetence. *Journal of Clinical Investigation* **120**:1324.

Ziegler, K., and E. R. Unanue. 1981. Identification of a macrophage antigen-processing event required for I-region-restricted antigen presentation to T lymphocytes. *Journal of Immunology* **127**:1869.

Useful Websites

www.jax.org The Jackson Laboratory maintains hundreds of inbred laboratory mouse strains and a database that lists H2 haplotypes and disease model systems.

www.bshi.org.uk The British Society for Histocompatibility and Immunogenetics home page contains information on tissue typing and transplantation and links to worldwide sites related to the MHC.

www.ebi.ac.uk/imgt/hla The international ImMunoGeneTics (IMGT) database section contains links to sites with information about HLA gene structure and genetics and up-to-date listings and sequences for all HLA alleles officially recognized by the World Health Organization HLA nomenclature committee.

www.hla.alleles.org A website maintained by the HLA Informatics Group based at the Anthony Nolan Trust in the United Kingdom, with up-to-date information on the numbers of HLA alleles and proteins.

STUDY QUESTIONS

1. Indicate whether each of the following statements is true or false. If you think a statement is false, explain why.

a. A monoclonal antibody specific for β_2-microglobulin can be used to detect both MHC class I K and D molecules on the surface of cells.

b. Antigen-presenting cells express both MHC class I and II molecules on their membranes.

c. MHC class III genes encode membrane-bound proteins.

d. In outbred populations, an individual is more likely to be histocompatible with one of its parents than with its siblings.

e. MHC class II molecules typically bind to longer peptides than do class I molecules.

f. All nucleated cells express MHC class I molecules.

g. The majority of the peptides displayed by MHC class I and II molecules on cells are derived from self-proteins.

2. You cross a BALB/c ($H2^d$) mouse with a CBA ($H2^k$) mouse. What MHC molecules will the F_1 progeny express on (a) its liver cells and (b) its macrophages?

3. The SJL mouse strain, which has the $H2^s$ haplotype, has a deletion of the $E\alpha$ locus.

a. List the classical MHC molecules that are expressed on the membrane of macrophages from SJL mice.

b. If the class II $E\alpha$ and $E\beta$ genes from an $H2^k$ strain are transfected into SJL macrophages, what additional classical MHC molecules would be expressed on the transfected macrophages?

4. Draw diagrams illustrating the general structure, including the domains, of MHC class I molecules, MHC class II molecules, and membrane-bound antibody on B cells. Label each chain and the domains within it, the antigen-binding regions, and regions that have the immunoglobulin-fold structure.

5. One of the characteristic features of the MHC is the large number of different alleles at each locus.

a. Where are most of the polymorphic amino acid residues located in MHC molecules? What is the significance of this location?

b. How is MHC polymorphism thought to be generated and maintained?

6. As a student in an immunology laboratory class, you have been given spleen cells from a mouse immunized with the lymphocytic choriomeningitis virus (LCMV). You determine the antigen-specific functional activity of these cells with two different assays. In assay 1, the spleen cells are incubated with macrophages that have been briefly exposed to LCMV; the production of interleukin (IL)-2 is a positive response. In assay 2, the spleen cells are incubated with LCMV-infected target cells; lysis of the target cells represents a positive response in this assay. The results of the assays using macrophages and target cells of different haplotypes are presented in the table below. Note that the experiment has been set up in a way to exclude alloreactive responses (reactions against nonself MHC molecules).

a. The activity of which cell population is detected in each of the two assays?

b. The functional activity of which MHC molecules is detected in each of the two assays?

c. From the results of this experiment, which MHC molecules are required, in addition to LCMV, for specific reactivity of the spleen cells in each of the two assays?

d. What additional experiments could you perform to unambiguously confirm the MHC molecules required for antigen-specific reactivity of the spleen cells?

e. Which of the mouse strains listed in the table could have been the source of the immunized spleen cells tested in the functional assays? Give your reasons.

Mouse strain used as source of macrophages and target cells	MHC haplotype of macrophages and virus-infected target cells				Response of spleen cells	
	K	A	E	D	IL-2 production in response to LCMV-pulsed macrophages (assay 1)	Lysis of LCMV-infected cells (assay 2)
C3H	k	k	k	k	+	−
BALB/c	d	d	d	d	−	+
(BALB/c × B10.A) F_1	d/k	d/k	d/k	d/k	+	+
A.TL	s	k	k	d	+	+
B10.A (3R)	b	b	b	d	−	+
B10.A (4R)	k	k	—	b	+/−	+

7. A T_C-cell clone recognizes a particular measles virus peptide when it is presented by H2-D^b. Another MHC molecule has a peptide-binding groove identical to the one in H2-D^b but differs from H2-D^b at several other amino acids in the α1 domain. Predict whether the second MHC molecule could present this measles virus peptide to the T_C-cell clone. Briefly explain your answer.

8. Human red blood cells are not nucleated and do not express any MHC molecules. Why is this property fortuitous for blood transfusions?

9. The hypothetical allelic combination *HLA-A99* and *HLA-B276* carries a relative risk of 200 for a rare, and yet unnamed, disease that is fatal to preadolescent children.

a. Will every individual with *A99/B276* contract the disease?

b. Will everyone with the disease have the *A99/B276* combination?

c. How frequently will the *A99/B276* allelic combination be observed in the general population? Do you think that this combination will be more or less common

than predicted by the frequency of the two individual alleles? Why?

10. Explain the difference between the terms *antigen-presenting cell* and *target cell*, as they are commonly used in immunology.

11. Define the following terms:

a. Self-MHC restriction
b. Antigen processing
c. Endogenous antigen
d. Exogenous antigen
e. Anchor residues
f. Immunoproteasome

12. Ignoring cross-presentation, indicate whether each of the cell components or processes listed is involved in the processing and presentation of exogenous antigens (EX), endogenous antigens (EN), or both (B). Briefly explain the function of each item.

a. _____ MHC class I molecules
b. _____ MHC class II molecules
c. _____ Invariant (Ii) chains
d. _____ Lysosomal hydrolases
e. _____ TAP1 and TAP2 proteins
f. _____ Vesicular transport from the RER to the Golgi
g. _____ Proteasomes
h. _____ Phagocytosis or endocytosis
i. _____ Calnexin
j. _____ CLIP
k. _____ Tapasin

13. Antigen-presenting cells have been shown to present lysozyme peptide 46–61 together with the class II A^k molecule. When $CD4^+$ T_H cells are incubated with APCs and native lysozyme or the synthetic lysozyme peptide 46–61, T_H-cell activation occurs.

a. If chloroquine is added to the incubation mixture, presentation of the native protein is inhibited, but the peptide continues to induce T_H-cell activation. Explain why this occurs.

b. If chloroquine addition is delayed for 3 hours, presentation of the native protein is not inhibited. Explain why this occurs.

14. Cells that can present antigen to T_H cells have been classified into two groups: professional and nonprofessional APCs.

a. Name the three types of professional APCs. For each type, indicate whether it expresses MHC class II molecules and a costimulatory signal constitutively or must be activated before doing so.

b. Give three examples of nonprofessional APCs. When are these cells most likely to function in antigen presentation?

15. Predict whether T_H-cell proliferation or CTL-mediated cytolysis of target cells will occur with the following mixtures of cells. The $CD4^+$ T_H cells are from lysozyme-primed mice, and the CD8$^+$ CTLs are from influenza-infected mice. Use *R* to indicate a response and *NR* to indicate no response.

a. _____ $H2^k$ T$_H$ cells + lysozyme-pulsed $H2^k$ macrophages
b. _____ $H2^k$ T$_H$ cells + lysozyme-pulsed $H2^{b/k}$ macrophages
c. _____ $H2^k$ T$_H$ cells + lysozyme-primed $H2^d$ macrophages
d. _____ $H2^k$ CTLs + influenza-infected $H2^k$ macrophages
e. _____ $H2^k$ CTLs + influenza-infected $H2^d$ macrophages
f. _____ $H2^d$ CTLs + influenza-infected $H2^{d/k}$ macrophages

16. Molecules of the CD1 family were recently shown to present nonpeptide antigens.

    a. What is a major source of nonpeptide antigens?
    b. Why are CD1 molecules not classified as members of the MHC family even though they associate with β_2-microglobulin?
    c. What evidence suggests that the CD1 pathway is different from that utilized by classical MHC class I molecules?

17. A slide of macrophages was stained by immunofluorescence, using a monoclonal antibody for the TAP1/TAP2 complex. Which of the following intracellular compartments would exhibit positive staining with this antibody?

    a. Cell surface
    b. Endoplasmic reticulum
    c. Golgi apparatus
    d. The cells would not be positive because the TAP1/TAP2 complex is not expressed in macrophages.

18. HLA determinants are used not only for tissue typing of transplant organs, but also as one set of markers in paternity testing. Given the following phenotypes, which of the potential fathers is most likely the actual biological father? Indicate why each could or could not be the biological father.

	HLA-A	HLA-B	HLA-C
Offspring	A3, 43	B54, 59	C5, 8
Biological mother	A3, 11	B59, 78	C8, 8
Potential father 1	A3, 33	B54, 27	C5, 5
Potential father 2	A11, 43	B54, 27	C5, 6
Potential father 3	A11, 33	B59, 26	C6, 8

19. Assuming the greatest degree of heterozygosity, what is the maximum number of different classical HLA class I molecules that one individual can employ to present antigens to CD8$^+$ T cells? Assuming no extra β-chain genes, what is the maximum number of different class II molecules (isoforms) that one individual can use to present antigen to CD4$^+$ T cells?

20. Define the following terms and give examples using the human HLA locus: polygeny, polymorphism, and codominant expression. How exactly does each contribute to ensuring that a diversity of antigens can be presented by each individual?

21. What is the cellular phenotype that results from inactivation or mutation in both copies of the gene for the invariant chain?

22. In general terms, describe the process of cross-presentation. Which cell types are most likely to be involved in cross-presentation, and what unique role does this process play in the activation of naïve CD8$^+$ T cells?

CLINICAL FOCUS QUESTION

Patients with deficiency in the transporter for antigen processing (TAP) have partial immunodeficiency as well as autoimmune manifestations. First, which cells would be affected by a deficiency in TAP expression? Second, why do you think you see both immune deficiency and autoimmunity in people with this disorder? Third, would it be possible to design a gene therapy treatment for this disease, given the fact that a single gene defect is implicated? Finally, connecting this example of TAP deficiency with Clinical Focus Box 7-3 and infectious cancer in Tasmanian devils, why do you think we don't see any immune deficiency or autoimmunity in Tasmanian devils with DFTD?

ANALYZE THE DATA

Smith and colleagues (2002, *Journal of Immunology* **169**:3105) examined the ability of two peptides to bind two different allotypes—L^d and L^q—of the mouse MHC class I molecule. They looked at (1) the binding of a murine cytomegalovirus peptide (MCMV; amino acid sequence YPHFMPTNL) and (2) a synthesized peptide, tum$^-$ P91A$_{14-22}$ (TQNHRALDL). Before and after pulsing the cells with the target peptides, the investigators measured the amount of peptide-free MHC molecules on the surface of cells expressing either L^d, L^q, or a mutant L^d in which tryptophan had been mutated to arginine at amino acid 97 (L^d W97R). The ability of the peptide to decrease the relative number of open forms on the cell surface reflects increased peptide binding to the MHC class I molecules. Answer the following questions, based on the data and what you have learned from reading this book.

a. Are there allotypic differences in the binding of peptide by native MHC class I molecules on the cell surface?
b. Is there a difference in the binding of MCMV peptide to L^d after a tryptophan (W)-to-arginine (R) mutation in the L^d molecule at position 97? Explain your answer.
c. Is there a difference in the binding of the tum peptide to L^d after a tryptophan (W)-to-arginine (R) mutation in the L^d molecule at position 97? Explain your answer.

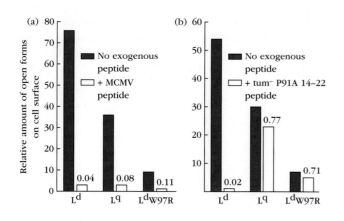

d. If you wanted to successfully induce a CD8$^+$ T-cell response against the tum peptide, would you inject a mouse that expresses L^q or L^d? Explain your answer.

e. The T cells generated against the MCMV peptide have a different specificity than the T cells generated against the tum peptide when L^d is the restricting MHC molecule. Explain how different peptides can bind the same L^d molecule yet restrict/present peptide to T cells with different antigen specificities.

T-Cell Development

Learning Objectives
After reading this chapter, you should be able to:

1. Outline the major events that transform a hematopoietic stem cell into a mature, naïve T cell.

2. Describe the microenvironments of the thymus where each stage of T-cell development takes place.

3. Describe the changes in expression of CD4, CD8, and TCR that occur during T-cell development.

4. Define and describe the importance of, and the basis for, positive and negative selection. Articulate the relationship between positive selection and MHC restriction, as well as the relationship between negative selection and central tolerance.

5. Describe the affinity model of selection and recognize that there are other perspectives on how positive and negative selection are achieved.

6. Recognize the variety of T cells that are generated in the thymus and articulate a basic understanding of the events that lead to the development of each lineage, particularly the CD4$^+$, CD8$^+$, and T$_{REG}$ lineages.

Thymocytes in the cortex of the thymus. *[CNRI/Science Source.]*

O rganisms are exposed to a remarkably diverse universe of pathogens and generate a remarkably diverse set of strategies to engage them. Each one of the several million T and B cells circulating in the body expresses a novel and unique antigen receptor. T cells generate these receptors via DNA rearrangements as they mature in the thymus, an organ in our chest cavity that is both required for and dedicated to the development of T lymphocytes from precursors that arrive from the bone marrow. Cells entering the thymus are not yet committed to becoming a lymphocyte and express no antigen-specific receptors. However, cells leaving the thymus are functional, mature T cells expressing unique and specific T-cell receptors (TCRs) that are both tolerant to self and restricted to self-MHC.

How do we generate such a diverse group of T cells that are also both self-restricted and self-tolerant? This question has intrigued immunologists for decades and inspired and required experimental ingenuity. Innovations have paid off and, although our understanding is still not comprehensive, we now have a fundamental appreciation for the remarkable strategies employed by the thymus, the nursery of immature T cells or **thymocytes**, to produce a functional, safe, and useful T-cell repertoire.

In this chapter we describe these strategies, dividing our discussion of T-cell development into two phases: *early thymocyte development*, during which a dizzyingly diverse TCR$^+$ population of immature T cells is generated, and *selection events* that depend on TCR interactions to shape this population so that only those cells that are self-restricted and self-tolerant will leave the thymus to populate the periphery (**Overview Figure 8-1**). Early thymocyte development

Key Terms

Thymocytes

Thymic cortex

Thymic medulla

Double-negative (DN) thymocytes

Pre-T-cell receptor (pre-TCR)

Double-positive (DP) thymocytes

MHC restriction

Self-tolerance

Positive selection

Negative selection

Apoptosis

Affinity model of selection

Single-positive (SP) thymocytes

Cortical thymic epithelial cells (cTECs)

Medullary thymic epithelial cells (mTECs)

Lineage commitment

Regulatory T cells (T$_{REG}$ cells)

Development of T Cells in the Thymus

(a)

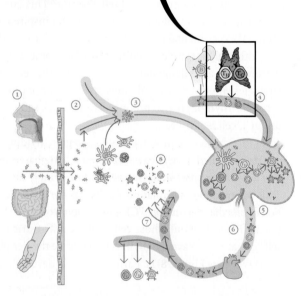

(c)

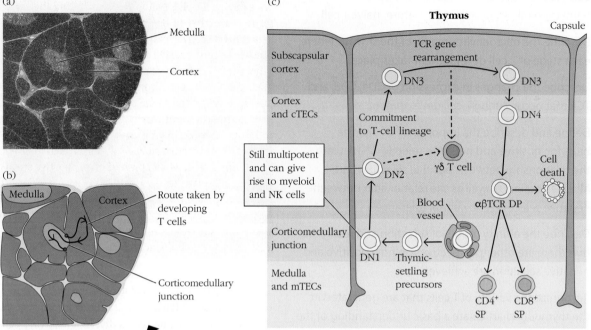

the outer cortex and inner medulla. (b) Thymocytes enter the thymus via blood vessels at the corticomedullary junction and then travel into the cortex, proliferating first in the region just below the capsule (the subcapsular cortex). As they mature they migrate from the cortex into the medulla and ultimately exit via vessels in the corticomedullary junction. (c) This schematic summarizes the developmental progression in more detail. The most immature, CD4⁻CD8⁻ (double negative, DN) thymocytes pass through several stages (DN1-DN4), during which they commit to the T-cell lineage and begin to rearrange their T-cell receptor (TCR) gene loci. Those that successfully rearrange their TCR β chain proliferate, initiate rearrangement of their TCR α chains, and become CD4⁺CD8⁺ (double positive, DP) thymocytes, which dominate the thymus. DP thymocytes undergo negative and positive selection in the thymic cortex. Positively selected thymocytes continue to mature and migrate to the medulla, where they are subject to another round of negative selection to self antigens that include tissue-specific proteins. Mature T cells express either CD4 or CD8 (single positive, SP) and leave the thymus with the potential to initiate an immune response. Although most thymocytes develop into conventional αβ TCR CD4⁺ or CD8⁺ T cells, some thymocytes develop into other cell lineages, including lymphoid dendritic cells, γδ TCR T cells, natural killer T (NKT) cells, regulatory T (T$_{REG}$) cells, and intraepithelial lymphocytes (IELs), each of which has a distinct function (see text and Overview Figures 8-3 and 8-5 for additional details). *[Part (a) Jose Luis Calvo/Shutterstock]*

T-cell precursors from the bone marrow travel to the thymus via the bloodstream, undergo development to mature T cells, and are exported to the periphery, where they can undergo antigen-induced activation and differentiation into effector cells and memory cells. Each stage of development occurs in a specific microenvironment of the thymus and is characterized by specific intracellular events and distinctive cell-surface markers. (a) The thymus is an organ surrounded by a fibrous capsule that encloses multiple lobes, each of which is separated into two major regions:

is largely T-cell receptor–independent and brings cells through several uncommitted CD4⁻CD8⁻ (double negative, DN) stages to the T lymphocyte–committed, T-cell receptor–expressing, CD4⁺CD8⁺ (double-positive, DP) stage. The specific events in this early phase include the following:

1. Commitment of hematopoietic precursors to the T-cell lineage,

2. Initiation of antigen receptor gene rearrangements, and

3. Expansion of cells that have successfully rearranged one of their T-cell receptor genes (a process called β-*selection*).

The second phase of T-cell development is largely dependent on T-cell receptor interactions and brings cells to maturity from the CD4⁺CD8⁺ stage to the CD4⁺ or CD8⁺ (single-positive, SP) stage. The events include the following:

1. *Positive selection*: Selection *for* those cells whose T-cell receptors engage self-MHC,

2. *Negative selection*: Selection *against* those cells whose T-cell receptors bind too strongly to self-peptide/MHC combinations, and

3. *Lineage commitment*: Commitment of thymocytes to effector cell lineages, including CD4⁺ helper or CD8⁺ cytotoxic populations, as well as CD4⁺ regulatory T cells.

Understanding T-cell development has been a challenge for students and immunologists alike. It helps to keep in mind the central purpose of the process: to generate a large population of T cells that express a diverse array of receptor specificities that *interact* with self-MHC/peptide complexes, but *do not react* against them.

Early Thymocyte Development

T-cell development occurs in the thymus, the primary lymphoid organ in our chest cavity (see Chapter 2). Thymocytes are produced throughout our lifetime, although after puberty the thymus shrinks (involutes) and produces fewer and fewer T cells over time.

Development begins with the arrival of small numbers of immature lymphocyte precursors migrating from the bone marrow, through the blood, and into the thymus. Here they proliferate, differentiate, and undergo selection processes that reduce their numbers by 95%, yet generate a diverse population of naïve CD4⁺ and CD8⁺ T cells.

The precise identity of the bone marrow cell precursors that give rise to T cells is still debated. However, it is clear that these precursors are directed to the thymus via chemokine receptors and that they are multipotent. Even after they arrive, thymocyte precursors, also known as *thymus-settling progenitors* (TSPs), retain the potential to give rise to B cells,

natural killer (NK) cells, dendritic cells (DCs), and some other myeloid cells. These cells become fully committed to the T-cell lineage only in the late DN2 stage of T-cell development (see Overview Figure 8-1).

The commitment of stem cells to the T-cell lineage is dependent on **Notch**, a receptor. Notch regulates several decisions during T-cell development, including the early choice between whether to become a T or B lymphocyte. When a constitutively active version of Notch1 (one of our versions of Notch) is overexpressed in hematopoietic stem cells, T cells rather than B cells develop in the bone marrow. Reciprocally, when the Notch1 gene is knocked out in hematopoietic precursors, B cells rather than T cells develop in the thymus.

The importance of Notch in T-cell commitment was also underscored by an in vitro system for studying T-cell development. Bone marrow stem cells were known to develop into B cells, not T cells, when cultured on supportive stromal cells growing in a plate. In the early 2000s, J. C. Zúñiga-Pflücker and colleagues modified the stromal cells by transfecting them with Notch ligand. This was precisely what was needed to support the development of hematopoietic stem cells (HSCs) to the T-cell lineage (**Figure 8-2**). This assay system has been invaluable in defining interactions that control early T-cell development and has helped investigators reveal, for instance, that the transcription factor GATA-3 is a key participant in Notch-mediated T-cell commitment (see Chapter 2 for more discussion of transcriptional regulators of blood cell development).

Thymocytes Progress through Four Double-Negative Stages

T-cell development is elegantly organized, spatially and temporally. The various stages of development take place in distinct microenvironments within the thymus; these microenvironments provide both membrane-bound and soluble signals that regulate maturation. After arriving in the thymus from the bone marrow via blood vessels at the corticomedullary boundary, T-cell precursors encounter Notch ligands, which are abundantly expressed by the thymic epithelium. T-cell precursors then travel to the outer **thymic cortex**, where they proliferate and begin to express their T-cell receptors. T cells that survive the stringent selection processes mature and migrate into the **thymic medulla** before exiting where they arrived, at the corticomedullary junction (see Overview Figure 8-1).

During the time it takes cells to develop to maturity in the thymus (1 to 3 weeks), thymocytes pass through a series of stages defined by changes in their surface phenotype (see Overview Figure 8-1c). The earliest T cells lack detectable CD4 and CD8 and are therefore referred to as **double-negative (DN) cells**. DN T cells can be further divided into four subsets, DN1-DN4, based on the presence or absence of other cell surface molecules, including **c-Kit (CD117)**,

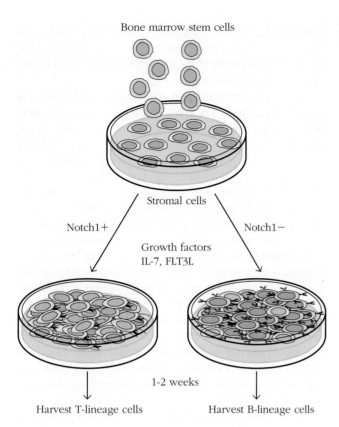

Bone marrow stem cells

Stromal cells

Notch1+   Notch1−

Growth factors
IL-7, FLT3L

1-2 weeks

Harvest T-lineage cells   Harvest B-lineage cells

FIGURE 8-2 Development of T cells from hematopoietic stem cells on bone marrow stromal cells expressing the Notch ligand. Investigators can induce lymphoid development from hematopoietic stem cells in vitro, using a combination of stromal cell lines and soluble cytokines and growth factors, as indicated. Investigators discovered that Notch signaling was the key to inducing development to the T- rather than B-lymphocyte lineage. After transfecting the stromal cell line with a gene encoding the Notch ligand (Notch1), lymphoid precursors adopted the T-cell lineage. Otherwise, they would develop into B cells.

the receptor for stem cell growth factor; **CD44**, an adhesion molecule; and **CD25**, the α chain of the interleukin (IL)-2 receptor (**Table 8-1** and **Overview Figure 8-3**).

Thymus-settling progenitors are at the **DN1** stage of development (c-Kit^{++}CD44^{+}CD25^{-}). Although still

multipotent, they receive Notch signals immediately on entering the thymic environment and begin to restrict themselves to the T-cell lineage. They travel to the cortex, where they proliferate and gain expression of CD25, becoming **DN2** thymocytes (c-Kit^{++}CD44^{+}CD25^{+}). During this critical stage of development, the genes for the TCR γ, δ, and β chains begin to rearrange (see Chapter 6). The TCR α-chain locus, however, remains inaccessible to the recombinase machinery and does not yet rearrange. At the late DN2 stage, T-cell precursors fully commit to the T-cell lineage and reduce expression of both c-Kit and CD44.

Cells in transition from the DN2 to **DN3** (c-Kit^{+}CD44^{-}CD25^{+}) stages continue to rearrange their TCR γ, TCR δ, and TCR β chains and make the first major decision in T-cell development: whether to join the TCR-$\gamma\delta$ or TCR-$\alpha\beta$ T-cell lineage. Those DN3 T cells that successfully rearrange their β chain typically commit to the TCR-$\alpha\beta$ T-cell lineage. They lose expression of CD25, halt proliferation, and enter the **DN4** stage of development (c-Kit$^{low/-}$CD44^{-}CD25^{-}), which matures directly into CD4^{+}CD8^{+} DP thymocytes.

Key Concepts:

- Uncommitted white blood cell progenitors enter the thymus from the bone marrow. Notch/Notch ligand interactions are required for T-cell commitment.

- Immature T cells are called *thymocytes*, and progress through several stages of development defined broadly by the expression of the coreceptors CD4 and CD8.

- The earliest thymocytes express neither CD4 nor CD8 and enter the thymus at the corticomedullary junction. They mature further in the thymic cortex, and then finalize their maturation in the thymic medulla; they exit as mature cells where they entered, via the corticomedullary junction.

- CD4^{-}CD8^{-} (DN) thymocytes progress through four stages of development (DN1-DN4) defined by CD44, CD25, and c-Kit expression. During these stages they proliferate and rearrange the T-cell receptor (TCR) β, δ, and γ genes.

TABLE 8-1	Double-negative thymocyte development		
	Phenotype	Location	Description
DN1	c-Kit (CD117)$^{++}$, CD44^{+}, CD25^{-}	Bone marrow to thymus	Migration to thymus
DN2	c-Kit (CD117)$^{++}$, CD44^{+}, CD25^{+}	Subcapsular cortex	TCR γ-, δ-, and β-chain rearrangement; T-cell lineage commitment
DN3	c-Kit (CD117)$^{+}$, CD44^{-}, CD25^{+}	Subcapsular cortex	Expression of pre-TCR; β-selection
DN4	c-Kit (CD117)$^{low/-}$, CD44^{-}, CD25^{-}	Subcapsular cortex to cortex	Proliferation, allelic exclusion of β-chain locus; α-chain locus rearrangement begins; becomes DP thymocyte

Changes in T-cell Receptor Expression and Function through T-cell Development

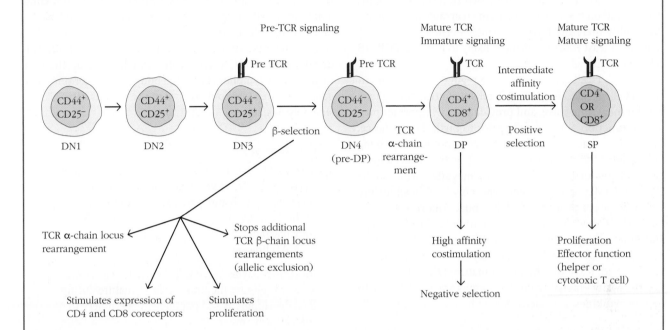

DN1 and DN2 thymocytes do not express any T-cell receptor proteins on their surface. The pre-TCR is assembled during the transition from the DN2 to the DN3 stage of development, when a successfully rearranged TCR β chain dimerizes with the nonvariant pre-Tα chain. Like the mature αβ TCR dimer, the pre-TCR is noncovalently associated with the CD3 complex. Successful assembly of this complex results in intracellular signals at the DN3 stage that induce a variety of processes, including the maturation to the DP stage and rearrangement of the TCR α chain. Once a thymocyte has successfully rearranged a TCR α chain (in transition between the DN4 and DP stages), this chain dimerizes with the TCR β chain, replacing the pre-Tα chain and generating a mature αβ TCR. The αβ TCR

expressed by DP thymocytes also complexes with CD3 and can generate signals that lead to either positive or negative selection (differentiation or death, respectively), depending on the affinity of its interaction with the MHC/peptide complexes it encounters in the thymic cortex and medulla. Although the αβ TCR/CD3 complex expressed by mature SP T cells is structurally the same as that expressed by DP thymocytes, the signals it generates are distinct. It responds to high-affinity engagement not by dying, but by initiating cell proliferation, activation, and the expression of effector functions. Low-affinity signals generate survival signals. The basis for the differences in consequence of signals generated by DP and SP TCR complexes is still unknown.

Thymocytes Express Either αβ or γδ T Cell Receptors

Vertebrates generate two broad categories of T cells: those that express TCR α and β chains and those that express TCR γ and δ chains. TCR-αβ cells are the dominant participants in the adaptive immune response in secondary lymphoid organs. However, TCR-γδ cells also play an

important role, particularly in protecting our barrier tissues from outside infection. Both types of cells are generated in the thymus, but how does a cell make the decision to become one or the other? To a large extent, the choice to become a γδ or αβ T cell is dictated by how quickly the genes that encode each of the four receptor chains successfully rearrange.

Recall from Chapter 6 that TCR genes are generated by shuffling (rearranging) V and J (and sometimes D) segments. Rearrangement of the β, γ, and δ loci begins during the DN2 stage. To become an αβ T cell, a cell must generate a TCR β protein chain—an event that depends on a single in-frame V(D)J rearrangement. To become a γδ cell, however, a thymocyte must generate two functional proteins that depend on two separate in-frame rearrangement events. Probability therefore favors the single event and, in fact, T cells are at least three times more likely to become TCR-αβ cells than TCR-γδ cells.

TCR-γδ T-cell generation is also regulated developmentally. TCR-γδ cells are the first T cells that arise during fetal development, in a wave, and provide a very important protective function perhaps even prior to birth. For instance, studies show that γδ T cells are required to protect very young mice against the protozoal pathogen that causes coccidiosis. However, production of γδ T cells declines after birth, and the TCR-γδ T-cell population represents only 0.5% of all mature T cells in the periphery of an adult animal (**Figure 8-4**).

Most TCR-γδ T cells are quite distinct in phenotype and function from conventional TCR-αβ T cells. Most do not go through the DP stage of thymocyte development and, instead, leave the thymus as mature DN (CD4⁻CD8⁻) T cells. Many emerge from the thymus with the ability to secrete cytokines, a capacity gained by most TCR-αβ cells

only after they encounter antigen in secondary lymphoid tissues. In addition, the TCR-γδ T-cell population, as a whole, expresses receptors that are not as diverse as TCR-αβ T cells. They appear to recognize unconventional antigens, including lipids associated with unconventional MHC molecules. Many take up long-term residence in mucosal tissues and skin and join innate immune cells in providing a first line of attack against invading microbes, as well as the response to cellular stress.

For the rest of this chapter, we will focus on the development of TCR-αβ T cells, which make up the vast majority of T lymphocytes in the body and are continuously generated, even after puberty, when the thymus shrinks considerably in size.

Key Concepts:

- Thymocytes mature into T cells that express either αβ or γδ TCRs. Thymocytes that successfully rearrange both δ and γ receptor genes mature to the TCR-γδ lineage, and those that successfully rearrange the β receptor gene commit to the TCR-αβ lineage.

- Most thymocytes become TCR-αβ T cells. However, TCR-γδ cells are the first T cells to mature during fetal development and populate the periphery in several early waves.

- TCR-γδ and TCR-αβ cells are functionally distinct. Most TCR-γδ cells have limited receptor diversity and antigen specificity. They are very important first responders to pathogens at mucosal surfaces and the skin. TCR-αβ T cells have more diverse specificities, anatomical ranges, and functions.

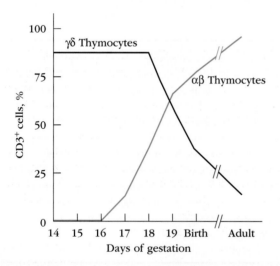

FIGURE 8-4 Time course of appearance of γδ thymocytes and αβ thymocytes during mouse fetal development. The graph shows the percentage of cells in the thymus that are double negative (CD4⁻CD8⁻) and bear the γδ T-cell receptor (black line) or are double positive (CD4⁺CD8⁺) and bear the αβ T-cell receptor (blue line). Fetal animals generate more γδ T cells than αβ T cells, but the proportion of γδ T cells generated drops off dramatically after birth. This early dominance of γδ TCR cells may have adaptive value: a large portion of these cells express nondiverse TCR specificities for common pathogen proteins and can mount a quick defense before the more traditional adaptive immune system has fully developed.

DN Thymocytes Undergo β-Selection, Which Results in Proliferation and Differentiation

Double-negative (DN) thymocytes that have successfully rearranged their TCR β chains are valuable, and are identified and expanded via a process known as **β-selection** (see Overview Figures 8-1, 8-3). This process involves a protein that is uniquely expressed at this stage of development: a 33-kDa invariant glycoprotein known as the **pre-Tα chain**. Pre-Tα acts as a surrogate for the real TCR α chain, which has yet to rearrange, and assembles with a successfully rearranged and translated β chain, as well as CD3 complex proteins. This immature TCR/CD3 complex is known as the **pre-T-cell receptor (pre-TCR)** (see Overview Figure 8-3) and acts as a sensor by initiating a signal transduction pathway. The signaling that the pre-TCR complex initiates is dependent on many of the same T cell–specific kinases used by a mature TCR (discussed in Chapter 10). Whether or not it binds ligands is still not clear. In fact, little if any of the complex is expressed on the cell surface, and some studies suggest that successful assembly of the complex may be sufficient to activate the signaling events. Others raise the possibility that the pre-TCR actually does interact with MHC/peptide complexes, although with lower affinity than mature TCR-αβ combinations. As you will learn in Chapter 9, B cells also express an immature receptor complex (the pre-BCR) that is coded by distinct genes, but is fully analogous to the pre-TCR.

Pre-TCR signaling at the DN3 stage triggers the following sequence of events:

1. Maturation to the DN4 stage (c-Kit$^{low/-}$CD44$^-$CD25$^-$)

2. Rapid proliferation in the subcapsular cortex of the thymus

3. Suppression of further rearrangement of TCR β-chain genes, resulting in allelic exclusion of the β-chain locus

4. Development to the CD4$^+$CD8$^+$ **double-positive (DP) stage**

5. Cessation of proliferation

6. Initiation of TCR α-chain rearrangement

It is important to note that the proliferative phase prior to α-chain rearrangement enhances receptor diversity considerably by generating clones of cells with the same TCR β-chain rearrangement. Each of the cells within a clone can then rearrange a different α-chain gene, thereby generating an even more diverse population than if the original cell had undergone rearrangement at both the β- and α-chain loci prior to proliferation. TCR α-chain gene rearrangement does not begin until double-positive thymocytes stop proliferating.

Most T cells successfully rearrange and express a TCR β chain from only one of their two TCR β alleles. Once a TCR β protein is generated, rearrangement at the other allele is shut down, a phenomenon known as **allelic exclusion**. The details of the mechanisms responsible for allelic exclusion are still being investigated. However, negative feedback signals from a successfully assembled TCR-β/pre-Tα complex clearly play a role. Signals appear to alter the accessibility of the gene locus to RAG activity and may also downregulate RAG expression [recall from Chapter 6 that RAG1 and RAG2 are enzymes that catalyze V(D)J recombination]. Other events, including the proliferative burst that follows β-selection, which dilutes RAG protein levels, may also play a role.

RAG levels continue to change after β-selection. They are restored after the proliferative burst and allow TCR-α rearrangement to occur. They decrease once again after expression of a successfully assembled TCR-αβ dimer. Interestingly, allelic exclusion at the TCR α-chain locus is not as efficient as it is at the TCR β-chain locus. A significant percentage of T cells express two different TCR α protein chains.

Once a young double-positive (DP) thymocyte successfully rearranges and expresses a TCR α chain, this chain has the opportunity to associate with the already produced TCR β chain, taking the place of the surrogate pre-TCR α chain, which is no longer actively expressed (see Overview Figure 8-3). At this point, several "goals" have been accomplished by the thymus: hematopoietic cell precursors have expanded in the subcapsular cortex, committed to the T-cell lineage, and rearranged a set of TCR genes. Each cell has also "chosen" to become either a TCR-αβ or a TCR-γδ T cell. The TCR-αβ population now expresses both CD4 and CD8, and is ready for the second stage of T-cell development: selection.

> **Key Concepts:**
>
> • Those thymocytes that successfully rearrange the TCR β chain are detected by a process called β-*selection*. β-Selection is initiated by assembly of the TCR β protein with a surrogate, invariant TCR α chain and CD3 complex proteins, forming the pre-TCR.
>
> • Pre-TCR signaling results in commitment to the TCR-αβ lineage, another burst of proliferation, maturation to the DP stage of development, and initiation of TCR-α rearrangement.

Positive and Negative Selection

CD4$^+$CD8$^+$ (DP) thymocytes are small, nonproliferating, and dominate the thymic cortex, comprising more than 80% of cells in the thymus as a whole. Significantly, they are the first subpopulation of thymocytes that express a fully mature surface TCR-αβ/CD3 complex and are therefore the primary targets of thymic selection (see Overview Figure 8-1). Thymic selection profoundly shapes the TCR repertoire of DP thymocytes, based on the affinity of their T-cell receptors for the MHC/peptides they encounter as they browse the surfaces of thymic epithelial cells.

Why is thymic selection necessary? The most distinctive property of a mature T cell is that its TCR recognizes peptide antigens only when combined with self-MHC molecules. This is a phenomenon known as **MHC restriction**, and was one of the most important immunological discoveries of the twentieth century. The other distinctive property of the population of mature T cells is that it is largely tolerant to self-MHC/self-peptide complexes. Even though T cells must recognize self-MHC with some affinity, they do not initiate a response to self-MHC/self-peptides, a phenomenon referred to as **self-tolerance**.

Given that TCR gene rearrangement is a random process and that the TCRs generated are unique for each cell, the probability that any given TCR will exhibit both self-restriction and self-tolerance is low. The selection events that occur in the thymus effectively screen the large population of newly generated DP thymocytes for those few cells whose TCRs have both these properties.

Two distinct selection processes are required:

- **Positive selection**, which selects for those thymocytes bearing receptors capable of binding self-MHC molecules with low affinity, resulting in MHC restriction.
- **Negative selection**, which selects against thymocytes bearing receptors with high affinity for self-MHC/self-peptide complexes, resulting in self-tolerance.

The vast majority of DP thymocytes (~98%) never meet the selection criteria and die by **apoptosis** within the thymus (**Overview Figure 8-5**). The bulk of DP thymocyte death (~95%) occurs among thymocytes that fail positive selection because their receptors do not specifically recognize self-MHC molecules, a process known as *death by neglect*. A small percentage of cells (2% to 5%) are eliminated by negative selection. Only 2% to 5% of DP thymocytes actually exit the thymus as mature T cells.

Positive and negative selection ensure that these mature T cells express TCRs with low affinity for self-peptide/self-MHC complexes, as long as they were presented in the thymus. On the other hand, there is no guarantee that these mature T cells express receptors that will be useful during any given infection. In fact, the ability of the adaptive immune system to respond to an infection depends entirely on the gamble that at least one of the millions of T cells that survive thymic selection with a randomly generated TCR will bind and react to at least one of the MHC/foreign-peptide combinations generated by an antigen-presenting cell (APC). This is a gamble that has paid off handsomely in our evolutionary history.

In this section, we briefly describe the experimental evidence for thymic involvement in MHC restriction. We also describe what is currently known about the positive and negative selection processes, as well as the models that have been advanced to explain what is known as the "thymic paradox."

Thymocytes "Learn" MHC Restriction in the Thymus

Early evidence for the role of the thymus in selection of the T-cell repertoire came from chimeric mouse experiments performed by R. M. Zinkernagel and colleagues. Recall that Zinkernagel won the Nobel Prize for showing that mature T cells are MHC restricted (i.e., they need to recognize antigen in the context of the MHC of their host to respond; see Classic Experiment Box 7-4). However, he and his colleagues were curious to know *how* T cells became "restricted" as such. They considered two possibilities: (1) either the T cell

and APC simply had to have matching MHC types (i.e., an "A" strain T cell had to see an "A" strain antigen-presenting cell); or (2) T cells, regardless of their own MHC type, "learned" the MHC type of their host sometime during development. Zinkernagel and colleagues thought that such learning could take place in the thymus, the T-cell nursery. To determine whether T cells could be "taught" to recognize the host MHC, they removed the thymus (thymectomized) and irradiated (A × B) F_1 mice so they had no functional immune system (**Figure 8-6**). They then reconstituted the hematopoietic cells with an intravenous infusion of (A × B) F_1 bone marrow cells, but replaced the thymus with one from a B-type mouse. (To be certain that the thymus graft did not contain any mature T cells, they irradiated it before transplantation.)

In this experimental system, T-cell progenitors from the (A × B) F_1 bone marrow matured within a thymus that expresses only B-haplotype MHC molecules on its stromal cells. Would these (A × B) F_1 T cells now be MHC restricted to the B haplotype of the thymus in which they developed? Or, because they expressed both A and B MHC, would they be able to recognize both A and B MHC haplotypes? To answer this question, the investigators infected the chimeric mice with lymphocytic choriomeningitis virus (LCMV, the antigen) and removed the immunized, mature T cells to see which LCMV-infected target cells (APCs) they could kill. They tested these T cells against infected APCs from strain A and strain B mice. As shown in Figure 8-6, T cells from the chimeric mice could only lyse LCMV-infected target cells from strain B mice. Thus, the MHC haplotype of the thymus in which T cells develop determines their MHC restriction. T cells "learned" which MHC haplotype they are restricted to during their early days in the thymus. Although once we referred to this process as "T-cell education," we now know that it is the consequence of a brutal selection process.

Key Concepts:

- DP thymocytes are the most abundant subpopulation in the thymus and are the first cells to express a mature αβ TCR. They browse the thymic cortex and are targets of both positive and negative selection, which are responsible for self-MHC restriction and self-tolerance, respectively.

- Zinkernagel and colleagues showed that thymic positive selection is responsible for self-MHC restriction by transplanting mice expressing one set of MHC proteins with a thymus expressing a distinct set of MHC proteins. The T cells that developed in these mice were only restricted to the MHC expressed by the thymus.

OVERVIEW FIGURE 8-5

Positive and Negative Selection of Thymocytes in the Mouse

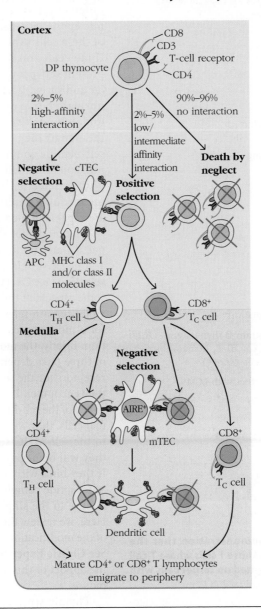

Thymic selection involves multiple interactions of DP and SP thymocytes with both cortical and medullary thymic stromal cells, as well as dendritic cells and macrophages. Selection results in a mature T-cell population that is both self-MHC restricted and self-tolerant. DP thymocytes that express new TCR αβ dimers browse the MHC/peptide complexes expressed by the cortical thymic epithelial cells (cTECs) as well as cortical dendritic cells (DCs) and macrophages. The large majority of DP thymocytes die in the cortex by neglect because of their failure to bind MHC/peptide combinations with sufficient affinity. The small percentage whose TCRs bind MHC/peptide with high affinity, particularly on the surface of DCs, die by clonal deletion (negative selection). Those DP thymocytes whose receptors bind to MHC/peptide with low to intermediate affinity are positively selected and mature to single-positive (CD4+ or CD8+) T lymphocytes. These migrate to the medulla, where they are exposed to AIRE+ medullary thymic epithelial cells (mTECs), which express tissue-specific antigens and can also mediate negative selection. Medullary dendritic cells can acquire antigens from mTECs and also mediate negative selection.

CONTROL

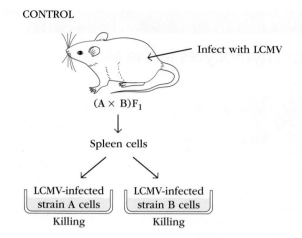

Infect with LCMV

(A × B)F₁

↓

Spleen cells

↙   ↘

LCMV-infected
strain A cells
Killing

LCMV-infected
strain B cells
Killing

EXPERIMENT

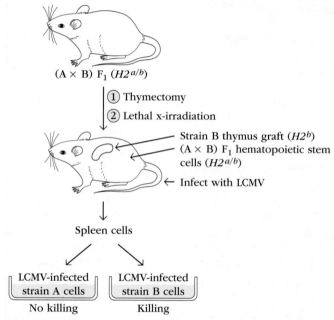

(A × B) F₁ (*H2ᵃ/ᵇ*)

① Thymectomy
② Lethal x-irradiation

Strain B thymus graft (*H2ᵇ*)
(A × B) F₁ hematopoietic stem
cells (*H2ᵃ/ᵇ*)
← Infect with LCMV

↓

Spleen cells

↙   ↘

LCMV-infected
strain A cells
No killing

LCMV-infected
strain B cells
Killing

FIGURE 8-6 Experimental demonstration that the thymus selects for maturation only those T cells whose T-cell receptors recognize antigen presented on target cells with the haplotype of the thymus. Control (A × B) F₁ mice infected with LCMV produce mature T cells that are able to kill both infected-strain A cells and infected-strain B cells, demonstrating that these T cells are restricted to MHC molecules expressed by both strain A (*H2ᵃ*) and strain B (*H2ᵇ*). In order to determine the involvement of the thymus in the restriction specificity of T cells, investigators grafted thymectomized and lethally irradiated (A × B) F₁ (*H2ᵃ/ᵇ*) mice with a strain-B (*H2ᵇ*) thymus and reconstituted it with (A × B) F₁ bone marrow cells. After infection with LCMV, the CD8⁺ cytotoxic (CTL) cells from this reconstituted mouse were assayed for their ability to kill ⁵¹Cr-labeled strain A or strain B target cells infected with LCMV. Only strain B target cells were lysed, suggesting that the *H2ᵇ* grafted thymus had selected for maturation only those T cells that could recognize antigen combined with *H2ᵇ* MHC molecules.

T Cells Undergo Positive and Negative Selection

In the thymus, thymocytes probe the surfaces of multiple different cell types, including cortical and medullary thymic epithelial cells, dendritic cells, and even B cells. Most of these cells express high levels of MHC class I and class II molecules on their surface. These self-MHC molecules present self-peptides, which are derived from intracellular and/or extracellular proteins that are degraded in the normal course of cellular metabolism. DP thymocytes undergo positive and negative selection, depending on the signals they receive when they encounter self-MHC/self-peptide combinations with their TCRs. Cortical thymic epithelial cells provide unique signals that mediate positive selection and several different cell types, including medullary epithelial cells, dendritic cells, and B cells, are capable of mediating negative selection.

One model for thymocyte development that provides a particularly useful framework for understanding positive and negative selection is known as the **affinity model of selection** (see Overview Figure 8-5). Although we know that some aspects of this model are too simplistic, its core principles remain relevant and are an excellent starting point for a more sophisticated understanding of thymic selection. Briefly, the model states that DP thymocytes have one of three fates depending on the affinity of their new T-cell receptors for the self-MHC/self-peptide combinations that they encounter. If their newly generated TCRs do not bind to any of the self-MHC/self-peptides they encounter on stromal cells, they will die by neglect. If they bind too strongly to the self-MHC/self-peptide complexes they encounter, they will be negatively selected, and ultimately destroyed. If they bind with a low, "just right" affinity to self-MHC/self-peptide complexes, they will be positively selected and mature to the **single-positive (SP) stage** of development. Here, we review the origins of the model and discuss some of the modifications that have been suggested by recent data. See **Classic Experiment Box 8-1** for experimental evidence in support of this model from the earliest TCR transgenic mouse.

Thymic stromal cells include cortical and medullary epithelial cells, multiple types of dendritic cells, and thymic B cells. They play essential roles in positive and negative selection and should be part of our visualization of thymic selection events. Young DP thymocytes are in intimate contact with these stromal cells and as they migrate through the thymus, they "browse" the many MHC/self-peptide complexes displayed on the stromal cell surfaces. In addition to expressing high levels of MHC class I and class II proteins, some stromal cells have the ability to generate and present unique peptides—a feature we will discuss shortly. Some extend long processes that contact many developing thymocytes. Some also express costimulatory ligands, including CD80 (B7-1) and CD86 (B7-2) (see Chapter 10).

Key Concepts:

- DP thymocytes have one of three fates depending on the affinity of their TCR for self-MHC/self-peptide complexes expressed by thymic stromal cells.

- Most thymocytes die because they fail to bind MHC/peptide with sufficient affinity (a failure of positive selection or death by neglect). Some (2% to 5%) die because they bind MHC/peptides with high affinity (negative selection). A similarly small percentage of thymocytes mature successfully by binding MHC/peptides with just the right intermediate affinity (positive selection).

- Although the affinity model of selection does not explain all aspects of negative and positive selection, it remains a useful starting point for understanding these processes.

Positive Selection Ensures MHC Restriction

In the thymic cortex, newly generated CD4$^+$CD8$^+$ thymocytes first browse the surfaces of **cortical thymic epithelial cells (cTECs)**, which are unique in their ability to mediate positive selection. If a DP thymocyte expresses a TCR that recognizes a self-MHC/self-peptide complex on the cTEC surface, it will undergo positive selection, a process that induces both survival and differentiation of DP thymocytes. Remarkably, over 90% of CD4$^+$CD8$^+$ thymocytes never engage the MHC/peptides they encounter with their TCRs. Either they have not generated a functional TCR-αβ combination, or the combination does not bind MHC/peptide complexes with sufficient affinity. These cells have "failed" the positive selection test, do not receive survival signals through their TCRs and, within 3 to 4 days, undergo a form of apoptosis referred to as **death by neglect**.

TCR/MHC/self-peptide interactions that initiate positive selection are thought to be at least three times lower in intensity than interactions that initiate negative selecting signals (**Figure 8-7**). We still do not fully understand how low-affinity positive-selecting TCR signals initiate a maturation program. The signaling protein, Themis, is expressed specifically by thymocytes and appears to be required for positive selection. Understanding its role is likely to lead to a more complete understanding of the signals that induce maturation of DP thymocytes.

Although the importance of cortical thymic epithelial cells in this process is established, the molecular basis for their unique capacity to mediate positive selection also remains an interesting mystery. Recent work shows that their antigen

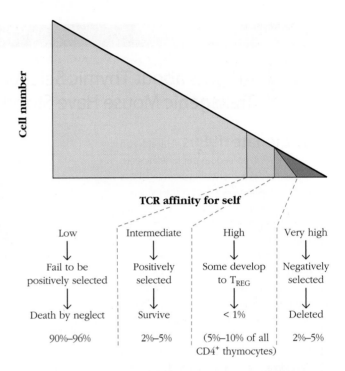

FIGURE 8-7 Relationship between TCR affinity and selection. This figure schematically illustrates the association between thymocytes' fate and the affinity of their TCR for self-MHC/peptide complexes that they encounter in the thymus. Fewer than 5% of thymocytes produce TCRs that bind to MHC/peptide complexes with high affinity. Most of these will be deleted by negative selection. Some (with high but not too-high affinity) will become regulatory T cells and other specialized cell types. More than 90% of thymocytes generate TCRs that either do not bind to MHC/peptide complexes or bind them with very low affinity. These die by neglect. Fewer than 5% generate TCRs that bind with just the right intermediate affinity to self-MHC/peptide complexes. These will survive and mature into CD4$^+$ and CD8$^+$ T cells.

presentation machinery is unique and that they generate a distinct array of peptides. These peptides may have features that enhance low-affinity engagements with TCR.

What is the adaptive value of positive selection? Why didn't we evolve a system that negatively selects but leaves the T-cell receptor repertoire otherwise intact? Some have suggested that the paring down of the repertoire is important to increasing the efficiency of negative selection, as well as to increasing the efficiency of peripheral T-cell responses. Without positive selection, the system would be cluttered with a great many cells that are unlikely to recognize anything, greatly reducing the probability that a T cell will find "its" antigen in a reasonable period of time. This is a reasonable speculation; however, positive selection may also offer more subtle advantages that have often inspired discussion in the field.

CLASSIC EXPERIMENT   BOX 8-1

Insights about Thymic Selection from the First TCR Transgenic Mouse Have Stood the Test of Time

In the late 1980s, Harald von Boehmer and colleagues published a series of seminal experiments that provided direct and compelling evidence for the influence of T-cell receptor interactions on positive and negative selection. They recognized that in order to understand how positive and negative selection worked, they would need to be able to trace and compare the fate of thymocytes with defined T-cell receptor specificities. But how could one do that if every one of the hundreds of millions of normal thymocytes expresses a different receptor? The researchers decided to develop a system in which all thymocytes expressed a T-cell receptor of known specificity, and thereby generated one of the first TCR transgenic mice.

Transgenic animals are made by injecting a gene under the control of a defined promoter into a fertilized egg (a *zygote;* see Chapter 20). The gene—in fact, often many copies of the gene—will be incorporated randomly into the genome of the zygote. Therefore, the gene will be present in all cells of the mouse that develop from that zygote; however, only those cells that can activate the promoter will express the gene.

von Boehmer's team developed the first TCR transgenic animals by isolating fully rearranged TCR α and TCR β genes of a mature CD8$^+$ T-cell line that was known to be specific for an antigen expressed only in male mice (the "H-Y" antigen) in the context of H2-D^b MHC class I molecules (**Figure 1**). They generated a genetic construct that included both rearranged genes under the control of a T-cell specific promoter and injected this into a mouse zygote. Since the receptor transgenes were already rearranged, other TCR gene rearrangements were suppressed in the transgenic mice (due to the phenomenon of *allelic exclusion;* see text and Chapter 6); therefore, a very high percentage of the thymocytes in the transgenic mice expressed the αβ TCR combination

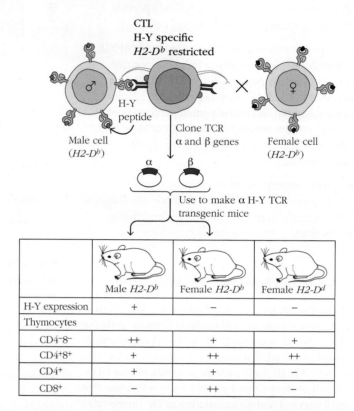

	Male *H2-D^b*	Female *H2-D^b*	Female *H2-D^d*
H-Y expression	+	−	−
Thymocytes			
CD4$^-$8$^-$	++	+	+
CD4$^+$8$^+$	+	++	++
CD4$^+$	+	+	−
CD8$^+$	−	++	−

FIGURE 1 Experimental demonstration that negative selection of thymocytes requires both self antigen and self-MHC, and positive selection requires self-MHC. In this experiment, *H2-D^b* male and female transgenics were generated by injecting TCR transgenes specific for H-Y antigen plus the D^b MHC molecule into zygotes. The H-Y antigen is expressed only in males. *H2-D^d* mice were also generated by backcrossing these transgenics onto an *H2^d* strain (e.g., BALB/c). FACS analysis of thymocytes from the transgenics showed that mature CD8$^+$ T cells expressing the transgene were absent in the male *H2-D^b* mice but present in the female *H2-D^b* mice, suggesting that thymocytes that bind with high affinity to a self antigen (in this case, H-Y antigen in the male mice) are deleted during thymic development. Results also show that DP but not CD8$^+$ thymocytes develop in female *H2-D^d* mice, which do not express the proper self-MHC. These findings indicate that positive selection and maturation require self-MHC interactions. [Data from von Boehmer, H., and P. Kisielow. 1990. Self–nonself discrimination by T cells. *Science* **248:**1369.]

specific for H-Y/H2-D^b. (It turns out that not all thymocytes expressed this TCR combination. Why not? Recall that allelic exclusion does not operate well for the TCR α-chain locus in particular.)

The team's choice of specificities was very clever. Because all thymocytes expressed an anti-male antigen TCR specificity, the researchers could directly compare the phenotype of autoreactive thymocytes and nonautoreactive

thymocytes simply by comparing male and female mice in the same litter. Because the MHC restriction of the TCR was known, they could also observe the influence of MHC haplotype on thymocyte development simply by breeding the mice to other inbred strains.

A comparison of CD4$^+$ versus CD8$^+$ expression among thymocytes in *H2-D^b* male versus female mice is shown in Figure 1. The primary data from one of the

(continued)

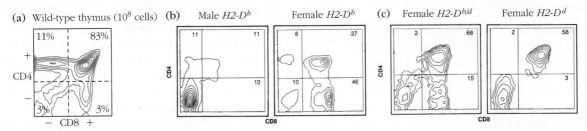

(a) Wild-type thymus (10^8 cells)   **(b)** Male *H2-D^b*   Female *H2-D^b*   **(c)** Female *H2-D$^{b/d}$*   Female *H2-D^d*

FIGURE 2 **Primary data from experiments summarized in Figure 1.** The relative staining by an anti-CD8 antibody is shown on the x axis, and the relative staining by an anti-CD4 antibody appears on the y axis. DP, DN, CD4 SP, and CD8 SP phenotypes are divided into quadrants, and percentages are given by quadrant. (a) CD4 versus CD8 phenotype of a wild-type thymus. (b) CD4 versus CD8 phenotype of thymocytes from H-Y TCR transgenic *H2^b* male and female. (c) CD4 versus CD8 phenotype of thymocytes from H-Y TCR transgenic *H2$^{b/d}$* and *H2$^{d/d}$* females. *[Reprinted by permission from American Association for the Advancement of Science, from Harald von Boehmer and Pawel Kisielow, Science, "Self-nonself discrimination by T cells." Science, 1990, June 15; **248**(4961):1369–1373, Figure 1. Permission conveyed through Copyright Clearance Center, Inc.]*

first publications are shown in **Figure 2** and depict flow cytometric data of CD4 and CD8 expression of thymocytes. The flow cytometric profile (see Chapter 20) of a typical wild-type thymus is shown in Figure 2a and can be used for comparison. In this profile, the CD8 expression status of a cell is shown on the x axis, the CD4 expression status on the y axis. The data are quantified in the upper right corner of the profile. As you can see, more than 80% of normal thymocytes are DP, 10% to 15% are CD4 single positive, 3% to 5% are CD8 single-positive cells, and only a small percentage of cells are DN. How do the CD4 versus CD8 phenotypes of male and female H-Y TCR transgenic mice (Figure 2b) differ, and what does this tell you?

Thymocytes in female mice are in all stages of development: DN, DP, and SP. In fact, they seem to have an abundance of mature CD8$^+$ SP thymocytes. We'll come back to this shortly. The male mice, however, have few if any anti–H-Y TCR transgenic DP thymocytes. These data show that in an environment where DP thymocytes are exposed to the antigen for which they are specific (in this case the male H-Y antigen), they are eliminated. This result was fully consistent with the concept of negative selection and showed that self-reactive cells were removed from the developing repertoire, perhaps at the DP stage. However, the mice offered insights into many more aspects of thymic selection.

Evidence for the role of TCR interactions in positive selection, for instance, came

from a different experiment in which the investigators asked what would happen if the "correct" restricting element, H2-D^b, was not present in the mouse. To do this they performed back crosses of their TCR transgenic mice to mice that expressed a different H2 haplotype, an *H2-D^d* mouse. (For those who want to brush up on their genetics, determine what crosses you would make to do this, and include a plan for assessing the genotype of your mice.) Figure 2c shows the CD4 versus CD8 phenotype of H-Y TCR transgenic thymocytes from female *H2-D$^{b/d}$* mice and *H2-D^d* mice. What did they find?

H2-D^d females had no mature T cells, indicating that in the absence of the MHC haplotype to which the TCR was restricted, thymocytes could not mature beyond the DP stage. These data provided direct evidence for the necessity of a TCR/self-MHC interaction for thymocyte maturation (positive selection) to occur.

These "simple" experiments also revealed a third feature of thymic development and initiated a controversy that still reverberates today. Return to Figure 2b and take a look at the female mouse data. Unlike wild-type thymocytes, which typically include 3% to 5% CD8$^+$ T cells, 46% of developing thymocytes in the anti–H-Y TCR transgenic were mature CD8$^+$ cells. What did this mean? Recall that the T-cell line that the TCR genes came from originally was a CD8$^+$ T-cell line, restricted to MHC class I (*H2-D^b*). These data showed that the MHC restriction specificity of the

TCR (in this case, a restriction for class I) influenced its CD4 versus CD8 lineage choice. In other words, the results suggested that thymocytes with new TCRs that preferentially bound class II would become CD4$^+$ SP mature T cells, and thymocytes with TCRs that preferentially bound class I would develop into CD8$^+$ SP mature T cells.

The fundamental conclusions of this now classic set of experiments have stood the test of time and have been supported by many different, equally clever experiments: (1) Negative selection of autoreactive cells results in their elimination at the DP stage and beyond. (It later became clear that the promoter driving these first H-Y TCR transgenes permitted earlier-than-normal expression of the TCR, which resulted in elimination of autoreactive cells at an earlier-than-usual point in development. Experiments now indicate that negative selection can occur at multiple stages, not just the DP but also the early SP stages of development.) (2) Positive selection involves a low-affinity interaction between thymocyte TCR and self-MHC/peptide interactions. (3) Commitment to the CD4 versus CD8 lineage is determined by the preference of the TCR for MHC class I versus class II. Precisely how this last event happens generated intense controversy (see text).

REFERENCE

von Boehmer, H., H. Teh, and P. Kisielow. 1989. The thymus selects the useful, neglects the useless and destroys the harmful. *Immunology Today* **10**:57.

Key Concepts:

- Cortical thymic epithelial cells (cTECs) mediate positive selection.

- Low/intermediate-affinity signals between DP thymocytes and MHC/self-peptides on cTECs result in positive selection, which triggers a maturation program that sends cells to the medulla and initiates their commitment to CD4$^+$ or CD8$^+$ single-positive lineages. Positive selection requires the thymocyte specific signaling molecule, Themis.

- The large majority of thymocytes (over 90%) do not interact with any MHC/self-peptides expressed by the thymic epithelium and die by neglect.

Negative Selection (Central Tolerance) Ensures Self-Tolerance

Autoreactive CD4$^+$CD8$^+$ thymocytes with high-affinity receptors for self-MHC/self-peptide combinations are potentially dangerous to an organism, and many are killed by negative selection in the thymus. Negative selection is defined broadly as any process that rids a repertoire of autoreactive clones and is responsible for central tolerance. Errors in the negative selection process are responsible for a host of autoimmune disorders, including type 1 diabetes (see **Clinical Focus Box 8-2**).

Most negative selection occurs via a process known as **clonal deletion**, where high-affinity TCR interactions directly induce apoptosis. Clonal deletion of DP thymocytes is mediated by a

---

CLINICAL FOCUS   **BOX 8-2**

How Do T Cells That Cause Type 1 Diabetes Escape Negative Selection?

We have an extensive appreciation of the pain caused by autoimmune disease and its clinical progression. We have also gained a deeper understanding of the mechanisms behind immune tolerance, but still know little about the precise origins of autoimmune disorders. Indisputably, the primary cause of autoimmune disease is the escape of self-reactive lymphocytes—B cells, T cells, or both—from negative selection. Surprisingly little is known about the reasons for this escape, or the specificity and phenotype of the escapees that cause even the best-characterized disorders, including type 1 diabetes (T1D), multiple sclerosis (MS), rheumatoid arthritis (RA), and systemic lupus erythematosus (SLE).

Recent work on a mouse strain, the nonobese diabetic (NOD) mouse, which is markedly susceptible to type 1 diabetes, shed light on the features of self-reactive T cells that cause the disease, and revealed some interesting reasons for their escape from central tolerance. T1D is a T cell–mediated autoimmune disease caused by T cells that kill insulin-producing beta islet cells in the pancreas. Many of the autoreactive T cells actually recognize a specific peptide from the insulin protein itself.

When Emil Unanue and colleagues examined the fine specificity of T cells that had infiltrated the islet cells in diabetes-prone mice, they found two

types of cells: T cells that could respond to the insulin peptide generated via intracellular MHC class II processing pathways by dendritic cells ("type A" T cells), and an unexpected population of T cells that responded to the insulin peptide only if it were added exogenously to the MHC class II molecules on dendritic cells ("type B" T cells). The investigators speculated that the endogenously processed MHC/insulin peptide combination and the exogenously formed MHC/insulin peptide complex differ in conformation (shape) and may therefore activate different T-cell receptor specificities.

These observations also suggested an elegant and precise reason for the escape of autoreactive T cells from the thymus. Dendritic cells in the pancreas, but not other tissues, appear uniquely capable of forming MHC class II/insulin peptides from exogenous sources, presumably because they are in an environment replete with extracellular insulin secreted by the beta islet cells. Thymic medullary epithelial cells or dendritic cells would not have this capacity, and therefore "type B" T cells would sneak through the negative selection process in the thymus.

Why "type A" T cells also escape is less clear. They are autoreactive to more conventionally processed insulin/MHC class II complexes, which are clearly present

on thymic medullary cells. Perhaps type A cells expressed too low an affinity for insulin/MHC class II peptides in the thymus, but were inspired by the high levels of insulin and inflammatory signals in a diabetic islet? Perhaps they escaped negative selection because the level of insulin peptide expression on medullary epithelium was too low in the NOD mouse strain? In fact, recent data from both mice and humans suggest that the level of insulin expression significantly influences the progression of disease and the efficiency of negative selection in the thymus.

Another recent study indicates that biological mimicry could play a role in activating autoreactive escapees. Peptides processed from common dairy and poultry microbes were shown to be highly homologous to an insulin peptide and were as potent at activating insulin-specific T cells.

Understandably, most current therapies for autoimmune disease focus on inhibiting the secondary but most proximal cause of autoimmune disease: the peripheral activation of autoreactive T and B cells that escaped negative selection. However, by also defining the molecular reasons for self-reactive lymphocyte escape, we may find even better ways to treat failures of immune tolerance.

variety of cell types present in the thymus. The same interactions that activate mature T cells (high-affinity TCR engagement coupled with costimulatory signals; see Chapter 10) are known to induce apoptosis in immature DP T cells. Thymic dendritic cells and macrophages, which are found in multiple areas of the thymus, clearly have the ideal features to mediate negative selection, but interestingly, so do **medullary thymic epithelial cells** (**mTECs**), which express high levels of the costimulatory ligands CD80 and CD86. Therefore, cells in both the cortex and the medulla have the potential to induce negative selection, and negative selection has been shown to occur in both microenvironments (see Overview Figure 8-5). Why strong TCR signals result in death of immature T cells, but proliferation and differentiation of mature T cells (see Overview Figure 8-3), remains an active area of investigation.

How Do We Delete Thymocytes Reactive to Tissue-Specific Antigens?

As you know from Chapter 7, almost all cells are capable of presenting endogenous, self-peptides with MHC on their surfaces. Yet only a fraction of cell types—thymocytes, cortical and medullary thymic epithelial cells, dendritic cells, B cells, and other antigen-presenting cells—reside in the thymus. How do we establish tolerance to *tissue-specific antigens* (*TSAs*), such as insulin or myelin basic protein?

This question bothered immunologists for a long time. For a while, we assumed that other mechanisms of tolerance in the periphery took care of autoreactivity to tissue-specific self-proteins, but investigations in the late 1990s surprised us all and revealed that *the thymus had an extraordinary capacity to express and present proteins from all over the body.* Bruno Kyewski and colleagues showed that this capacity was a unique feature of medullary thymic epithelial cells (mTECs). Diane Mathis, Christophe Benoist, and their colleagues went on to show that mTECs express a unique protein, **AIRE**, that allows mTECs to express, process, and present proteins that are ordinarily found only in specific organs.

This group discovered AIRE while studying the molecular basis for autoimmunity by examining human disease. They traced a mutation that caused autoimmune polyendocrinopathy syndrome 1 (APS1, otherwise known as APECED; see Chapter 18) to this gene, and called it *AIRE* for "*auto*immune *re*gulator." The title of the article that announced their discovery, "Projection of an Immunological Self Shadow within the Thymus by the AIRE Protein," aptly described AIRE's powerful function.

AIRE's mechanism of action is now better understood. It does not bind DNA directly; rather, it binds to epigenetic marks on histones associated with closed chromatin. AIRE then recruits transcription factors to these silenced gene promoters, allowing RNA polymerase to gain access. Recent studies indicate that AIRE is not the only transcriptional regulator in mTECs that controls expression of tissue-specific antigens (TSAs). Members of the FEZ family of zinc finger proteins may play a similar role.

Thus, mTECs play a particularly important and unique role in screening our developing T-cell repertoire and ridding it of T cells that could initiate a variety of tissue-specific autoimmune diseases. Other thymic stromal cells, including dendritic cells and B cells, are also capable of triggering clonal deletion (see Overview Figure 8-5). By presenting self-proteins involved in antigen presentation and B-cell function, they clear the repertoire of T cells that could react against immune cells themselves. Finally, migrating dendritic cells also have the potential to expose developing T cells to self-proteins they have processed outside the thymus and reduce the probability that autoreactive T cells escape clonal deletion.

Although cTECs may also be able to mediate negative selection, they may not be as efficient as dendritic cells or mTECs at inducing clonal deletion. However, cTECs may play a role in preventing the further development of self-reactive immature thymocytes, as discussed below.

Other Mechanisms of Central Tolerance

Other mechanisms of thymic negative selection that do not necessarily involve cell death have been proposed. They include clonal arrest, where thymocytes that express autoreactive T-cell receptors are prevented from maturing; clonal anergy, where autoreactive cells are inactivated, rather than deleted; and clonal editing, where autoreactive cells are given a second or third chance to rearrange a TCR α gene. Each of these mechanisms has some experimental support, but clonal deletion is probably the most common mechanism responsible for thymic negative selection. The generation of regulatory T cells from autoreactive thymocytes can certainly be considered of importance among central tolerance mechanisms and will be discussed shortly.

Superantigens

Superantigens, viral or bacterial proteins that bind simultaneously to the V_β domain of a T-cell receptor and to the α chain of an MHC class II molecule (outside the peptide-binding groove), induce activation of all mature T cells that express that particular family of V_β chains (see Figure 10-7). Superantigens are also expressed in the thymus of mice and humans and influence thymocyte maturation. Because they mimic high-affinity TCR interactions, superantigens induce clonal deletion of all DP thymocytes whose receptors possess V_β domains targeted by the superantigen. However, because we continuously generate T cells with a wide range of T-cell receptor specificities, this loss does not appear to have major clinical consequences.

Key Concepts:

- High-affinity TCR/MHC/peptide interactions result in negative selection, typically by initiating apoptosis of autoreactive thymocyte cells (clonal deletion).

- Medullary thymic epithelial cells (mTECs) express the transcription factor AIRE, which is in large part responsible for their unique capacity to express tissue-specific antigens.

- Thymic antigen-presenting cells, including dendritic cells and B cells, are also capable of mediating negative selection in both the cortex and medulla of the thymus.

The Selection Paradox: Why Don't We Delete All Cells We Positively Select?

A complete understanding of the process of positive and negative selection requires an appreciation of the following paradox: if positive selection allows only thymocytes reactive with self-MHC molecules to survive, and negative selection eliminates self-MHC–reactive thymocytes, then how do any T cells escape selection and mature? Other factors must operate to prevent these two MHC-dependent processes from eliminating the entire repertoire of MHC-restricted T cells.

The most straightforward model advanced to explain the paradox of MHC-dependent positive and negative selection is the affinity model of selection, which was introduced earlier.* This model asserts that differences in the *strength* of TCR-mediated signals received by thymocytes undergoing positive and negative selection determine the outcome of the interaction. Double-positive thymocytes that receive low-affinity signals are positively selected, those that receive high-affinity (autoreactive) signals are negatively selected, and those that receive no signal at all die by neglect (**Figure 8-8**; see Overview Figure 8-5).

How has this hypothesis been tested? One of the best experimental models for examining the connection between TCR affinity and thymic selection was devised by Kristin Hogquist and colleagues. They generated a transgenic mouse expressing the OT-I TCR, whose peptide/MHC specificity was precisely known: it bound a specific chicken ovalbumin (OVA) peptide in the context of the H2-K^b MHC class I molecule. The OT-I TCR's affinity for a variety of peptides that differed in sequence was also known and ranged from low to high.

To determine the influence of affinity on T-cell development, investigators took advantage of two additional immunological tools: (1) the fetal thymic organ culture (FTOC) system, where thymic development can be followed in vitro (see Figure 8-8a), and (2) mice defective in MHC class I antigen processing. In this particular experiment, the investigators used *TAP1* knockout mutant mice (TAP1$^-$), which cannot load their MHC class I molecules with peptide (see Chapter 7). Cells in these mutant mice express MHC class I on their surfaces, but these MHC molecules are "empty" and do not hold their shape. However, they can be loaded with exogenous peptides, which stabilize their conformation. Therefore, by incubating fetal thymic organs from these mutant mice with peptides of choice, investigators were able to precisely control the type of peptide/MHC complex that developing T cells would encounter.

What did investigators find? First, as expected, thymocytes in control OT-I$^+$/TAP1$^+$ fetal thymic organ cultures were positively selected and matured, predominantly into CD8$^+$ cells. Why CD8$^+$ T cells? Recall that OT-I TCR transgenic thymocytes express a receptor restricted to MHC class I. In normal (TAP1$^+$) fetal thymic organ cultures, the MHC class I molecules are occupied by a variety of self-peptides. OT-I$^+$ thymocytes that engage MHC class I/peptide combinations with an appropriate (intermediate) affinity mature. Because they recognize MHC class I, they mature specifically into CD8$^+$ T cells (see Classic Experiment Box 8-1).

Second, and also as expected, CD8$^+$ mature T cells failed to develop in OT-I$^+$/TAP1$^-$ thymic organ cultures. Without TAP1 there is no normal MHC class I and, therefore, OT-I$^+$ thymocytes have nothing to engage and will not receive a TCR signal. They die by neglect (see Figure 8-8b, top row).

Investigators now could perform the key experiments and add different peptides to the system and observe what happened to the thymocytes. When they added a low-affinity peptide to the OT-I$^+$/TAP1$^-$ culture, the MHC class I was able to load itself with peptide and assume a normal conformation (see Figure 8-8b, middle row). Positive selection occurred, and CD8$^+$ T cells developed successfully. Addition of peptides that were known to generate high-affinity OT-I TCR/MHC interactions resulted not in positive selection, but in deletion of DP cells in OT-I$^+$/TAP1$^-$ thymic organ cultures (see Figure 8-8b, bottom row). Interestingly, low concentrations of high-affinity peptides also induced positive selection, although the cells that developed were not fully functional. These results and many others provided clear support for the affinity model, showing that *the strength of TCR/MHC/peptide interaction does, indeed, influence cell fate.*

The OT-I system continues to be useful to address other questions and assumptions about positive and negative selection. For instance, it has allowed investigators to estimate the range of TCR affinities that define positive versus negative selection outcomes. It appears as if negative selection occurs at affinities that are threefold higher than those that induce positive selection.

The OT-I system has also been valuable in efforts to track the behavior of live thymocytes undergoing positive and negative selection (**Figure 8-9** and **Video 8-9v**). The movements of thymocytes were recorded, and their tracks are shown in red in the figure. In the absence of the OVA peptide, thymocytes expressing the OT-I TCR migrate among dendritic cells (yellow), make brief contacts, and move on (see Figure 8-9, left). In the presence of the OVA peptide (for which they have high affinity), OT-I thymocytes slow down

---

*Note that a similar version of this hypothesis is also sometimes referred to as the *avidity hypothesis*. Avidity and affinity do have distinct meanings, which can differ for different investigators. For the sake of simplicity, we consider *affinity* to mean the strength of interaction between the TCR and its MHC/peptide ligand. For a more nuanced discussion of the differences between affinity and avidity and their influence on thymic development, see the review by Klein et al. listed in the reference section of this chapter.

(a) Experimental procedure—fetal thymic organ culture (FTOC)

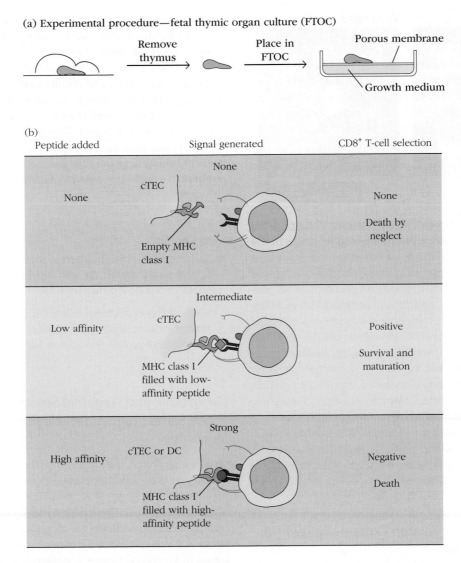

(b)

FIGURE 8-8 **Experimental support for the role of TCR affinity in thymic selection.** Fetal thymocyte populations have not yet undergone positive and negative selection and can be easily manipulated to study the development and selection of single-positive (CD4+CD8− and CD4−CD8+) T cells. (a) Outline of the experimental procedure for in vitro fetal thymic organ culture (FTOC). Fetal thymi are cultured on a filter disc at the interface between medium and air. Reagents added to the medium are absorbed by the thymi. In this experiment, peptide is added to the medium of FTOC from *TAP1* knockout (TAP1−) mice, which are unable to form peptide/MHC class I complexes unless peptide is added exogenously to the culture medium. (b) The development and selection of CD8+CD4− class

I–restricted T cells depends on TCR peptide/MHC class I interactions. The mice used in this study were transgenic for OT-I TCR α and β chains, which recognize an ovalbumin peptide when associated with MHC class I; the proportion of CD8+ T cells that develop in these thymi is higher than normal because all thymocytes in the mouse express the MHC class I–restricted specificity. Different peptide/MHC complexes interact with the TCR with different affinities. Varying the peptide added to FTOC from OT-I+/TAP1− mice revealed that peptides that interact with low affinities (middle row) resulted in positive selection, and levels of CD4−CD8+ cells that approached that seen in OT-I+/TAP1+ mice. Peptides that interact with high affinities (bottom row) cause negative selection, and fewer CD4−CD8+ T cells appeared.

and maintain contact with the dendritic cell (see Figure 8-9, right). These observations satisfyingly confirm assumptions that negative selection is the result of sustained APC/thymocyte interaction. These and other studies of thymocyte interactions with cTECs also reveal that positive selection is accomplished by surprisingly brief, and perhaps cumulative TCR contacts.

Key Concepts:

- The affinity model for thymocyte selection provides one elegant explanation for the thymic selection paradox.

- T-cell receptor transgenic mice, including the OT-I system, provide support for the affinity model of thymocyte selection.

Wild type

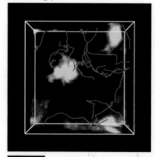

Wild type + OVA

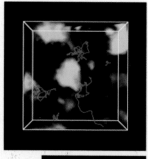

20 μm

DCs cell tracks

**FIGURE 8-9 AND ACCOMPANYING VIDEO 8-9V
Imaging live DP thymocytes undergoing selection in the
thymus.** This figure is taken from videos of live thymocytes (red)
as they probe dendritic cells (yellow) in the cortex of thymic tissue.
These thymocytes are specific for the OVA peptide, which is present in the thymi filmed on the right, but not the left. Their routes are
tracked and shown in red. OVA-specific thymocytes slow down and
commit to long-term interactions with dendritic cells (right), whereas
thymocytes that don't find their high-affinity peptide make brief contacts and then move on (left). The long-term interactions ultimately
lead to thymocyte death (negative selection) and the short contacts
with dendritic cells and cortical thymic epithelial cells (not labeled)
lead to maturation and positive selection. *[Reprinted by permission from
American Association for the Advancement of Science, from Melichar, H. J. et al.,
2013. Distinct temporal pattern of T cell receptor signals during positive versus
negative selection in situ. Science Signaling* **6:** *ra92, Figure 4. Permission conveyed
through Copyright Clearance Center, Inc.]*

An Alternative Model Can Explain the Thymic Selection Paradox

Philippa Marrack, John Kappler, and their colleagues
thoughtfully offered an alternative possibility to explain the
"thymic paradox"—that is, why we don't negatively select all
cells that we positively select. These investigators advanced
the **altered peptide model**, a suggestion that cortical thymic
epithelial cells, which induce positive selection, make peptides that are unique and distinct from peptides made by
thymic cells that induce negative selection. Thus, those thymocytes selected on these unique peptide/MHC complexes
would not automatically be negatively selected when they
browse the medulla and other negatively selecting cells.

Although early experiments did not support this model,
advances in our ability to analyze peptides presented on
cell surfaces in detail suggest the model has merit. Cortical
thymic epithelial cells do, indeed, process peptides differently, and present a different array of peptides to developing T cells. Recall that peptides are processed intracellularly
by proteasomes (see Figure 7-13). Interestingly, the proteasome expressed by cortical thymic epithelial cells (the

thymoproteasome) has a unique component. cTECs also
express a unique version of the cathepsin protease. These
features, together, allow cTECs to produce a population of
peptides that no other cells can generate.

The altered peptide and affinity models are by no means
mutually exclusive. Investigations have firmly established
that affinity plays an important role in distinguishing positive from negative selection. That the cortex also produces
novel peptides suggests that evolution favored multiple ways
to ensure the generation of a diverse T-cell pool—the best
response to a continuously evolving universe of antigens and
pathogens.

Key Concepts:

• Two models for thymic selection, the affinity model and
altered peptide model, are both supported by experimental evidence and may work together.

• Cortical thymic epithelial cells (cTECs) express a distinct proteasome system that produces and presents a
unique set of peptides for positive selection.

Do Positive and Negative Selection Occur at the Same Stage of Development, or in Sequence?

Strictly interpreted, the affinity model assumes that positive
and negative selection operate on exactly the same target
population (DP thymocytes) and in the same microenvironment of the thymus (the thymic cortex). However, although
positive selection occurs in the cortex, negative selection can
occur at a variety of developmental stages and sites (illustrated in Overview Figure 8-5). Dendritic cells are particularly good at mediating clonal deletion and are found in
all microenvironments of the thymus, including the cortex.
As we discussed above, cortical thymic epithelial cells may
arrest or anergize self-reactive DP thymocytes. DP thymocytes that are positively selected travel to the medulla and
develop into semimature SP thymocytes. These can be clonally deleted by multiple cells in the medulla, including B
cells, dendritic cells, and medullary epithelial cells, the only
cell type that expresses tissue-specific self antigens.

Migration of positively selected thymocytes to the
medulla is now known to be dependent on expression of
the CCR7 chemokine receptor, which is induced by TCR
signaling. Interestingly, although thymocytes from CCR7-
deficient mice fail to migrate from cortex to medulla, they
still mature and find their way outside the thymus. However,
they have a tendency to cause autoimmunity—an observation that, again, underscores the central importance of the
medulla in removing autoreactive, tissue-specific cells from
the T-cell repertoire.

Lineage Commitment

While thymocytes are being screened on the basis of TCR affinity for self antigens, they are also being guided in their lineage decisions. Specifically, a positively selected double-positive thymocyte must decide whether to join the CD8$^+$ cytotoxic T-cell lineage or the CD4$^+$ helper T-cell lineage. **Lineage commitment** requires changes in genomic organization and gene expression that result in (1) silencing of one coreceptor gene (CD4 or CD8) as well as (2) expression of genes associated with a specific lineage function. Immunologists continue to debate the precise cellular and molecular mechanisms responsible for lineage commitment. We will review the most current perspectives.

Several Models Have Been Proposed to Explain Lineage Commitment

When early studies with TCR transgenic mice revealed that affinity for MHC class I versus class II dictated the CD8$^+$ versus CD4$^+$ fate of developing thymocytes (see Classic Experiment Box 8-1), investigators advanced two simple testable models to explain how developing T cells "matched" their TCR preference for MHC class I or MHC class II with their coreceptor expression.

In the **instructive model**, TCR/CD4 and TCR/CD8 co-engagement generate unique signals that directly initiate distinct developmental programs (**Figure 8-10a**). For example, if a thymocyte randomly generated a TCR with an affinity for MHC class I molecules, the TCR and CD8 would bind a MHC class I molecule together, and generate a signal that specifically initiated a program that silenced CD4 expression and induced expression of genes specific for cytotoxic T-cell lineage function. Likewise, TCR/CD4 co-engagement would generate a unique signal that initiated CD8 silencing and the helper T-cell developmental program.

An alternative model, known as the **stochastic model**, proposed that a positively selected thymocyte randomly down-regulates CD4 or CD8 (Figure 8-10b). Only those cells that express the "correct" coreceptor—the ones that can co-engage MHC with the TCR—generate a TCR signal strong enough to survive to mature. In this model, TCR/CD4 and TCR/CD8 co-engagement does not necessarily generate distinct signals. Unfortunately, studies that followed the consequences of such mismatches confounded researchers by providing evidence in support of both models! Clearly they were too simplistic.

Further experiments challenged some of the assumptions on which these early efforts were based, and indicated that the strength and duration of the T-cell receptor/coreceptor signal experienced by a thymocyte play a more important role in dictating cell fate than its specificity for MHC class I or class II. By inhibiting TCR signaling, investigators could coax cells that normally commit to the CD4 lineage to become CD8$^+$ cells. By enhancing TCR signaling, investigators could coax MHC class I–restricted cells to become CD4$^+$. This **strength of signal model** represented an advance in our understanding, but it, too, did not incorporate all data.

Alfred Singer and colleagues have proposed a model that has the strongest experimental backing to date. The **kinetic signaling model** incorporates historical data, our improved understanding of changes in CD8 expression during positive selection, as well as recent advances in the understanding of the complexity of coreceptor gene silencing (Figure 8-10c). Briefly, it proposes that thymocytes commit to the CD4$^+$ T-cell lineage if they receive a continuous signal in response to TCR/coreceptor engagement, but commit to the CD8 lineage if the TCR signal is interrupted. We now know that all CD4$^+$CD8$^+$ thymocytes down-regulate surface levels of CD8 in response to positive selection. Given this response, only MHC class II–restricted T cells will maintain continuous TCR/CD4/MHC class II interaction, and therefore develop to the CD4 lineage. However, with the loss of CD8 expression, MHC class I–restricted T cells will lose the ability to maintain TCR/CD8/MHC class I interactions. Singer and colleagues provide evidence that these thymocytes, interrupted from their TCR/coreceptor engagement, are subsequently rescued by IL-7, a key cytokine that promotes differentiation into the CD8$^+$ lineage.

- Thymocytes whose TCR preferentially interacts with (is restricted to) MHC class II generate a continuous signal that initiates a CD4$^+$ helper T-cell developmental program.

- Thymocytes whose TCR preferentially interacts with (is restricted to) MHC class I cannot generate a continuous signal because surface levels of the coreceptor CD8 are down-regulated on all cells in response to positive selection. This reduction or interruption in signaling initiates a CD8 developmental program that is enforced by further stimulation in the presence of IL-7.

Transcription Factors Th-POK and Runx3 Regulate Lineage Commitment

Th-POK and Runx3 have taken center stage as transcription factors required for CD4 and CD8 commitment, respectively. Both Th-POK and Runx3 act, at least in part, by suppressing genes involved in differentiation to the other lineage. Th-POK activates expression of CD4 and inhibits expression of genes that regulate CD8 differentiation, including genes encoding CD8 itself and Runx3. Reciprocally, Runx3 activates expression of CD8 and inhibits expression of genes encoding CD4 and Th-POK. The sequence of

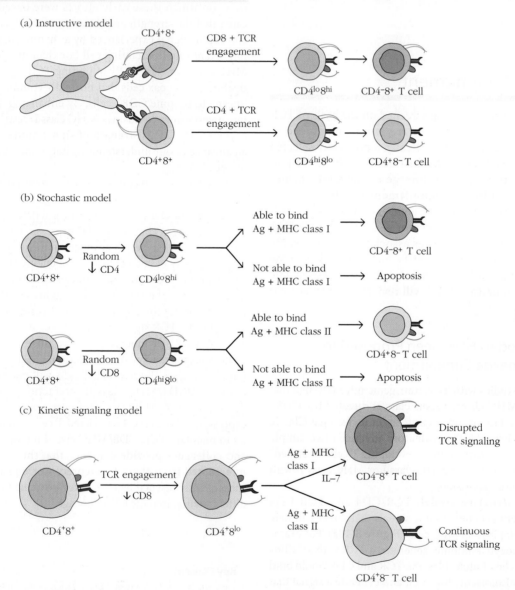

FIGURE 8-10 Proposed models of lineage commitment, the decision of double-positive thymocytes to become helper CD4$^+$ or cytotoxic CD8$^+$ T cells. (a) According to the instructive model, interaction of a coreceptor with the MHC molecule for which it is specific results in down-regulation of the other coreceptor. (b) According to the stochastic model, down-regulation of CD4 or CD8 is a random process. (c) According to the kinetic signaling model, the decision to commit to the CD4 or CD8 lineage is based on the continuity of the TCR signal that a thymocyte receives. Positive selection results in down-regulation of CD8 on all thymocytes. This will not alter the intensity of a TCR/CD4/MHC class II signal, and cells receiving this signal will continue development to the CD4 SP lineage. However, down-regulation of CD8 diminishes (interrupts) a TCR/CD8/MHC class I signal, an experience that sends a cell toward the CD8 lineage. IL-7 signals are required to "seal" CD8 lineage commitment.

events and full cast of players that regulate lineage commitment at the genomic level are still under investigation; there is clearly more to come.

Key Concept:

- Transcription factors Th-POK and Runx3 play key roles in CD4$^+$ helper cell and CD8$^+$ cytotoxic cell commitment, respectively. They work in part by reciprocally regulating each other.

Double-Positive Thymocytes May Commit to Other Types of Lymphocytes

Small populations of DP thymocytes also give rise to other T-cell types, including the NKT cell, regulatory T cell, and intraepithelial lymphocyte (IEL) lineages. **NKT cells** play a role in innate immunity (see Chapter 4) and express a TCR that includes an invariant TCR α chain (V$_\alpha$14). The TCR of NKT cells (sometimes referred to as *invariant* or *i*NKT cells) interacts not with classical MHC, but with the related molecule CD1. CD1 presents glycolipids, not peptides (see Chapter 7), and is expressed not by the epithelium, but by DP thymocytes themselves. **Intraepithelial lymphocytes** (IELs), most of which are CD8$^+$, also have features of innate immune cells and patrol barrier tissues. Regulatory T cells (T$_{REG}$ cells), a CD4$^+$ subset discussed in more detail later in this chapter, quench adaptive immune reactions. All three of these cell types can develop from DP thymocytes in response to autoreactive, high-affinity TCR interactions—the same interactions, in fact, that mediate negative selection. What determines whether a thymocyte undergoes negative selection or an alternative developmental pathway remains a topic of much interest.

Key Concept:

- For a small percentage of thymocytes, high-affinity TCR interactions do not lead to deletion, but instead drive development to specialized cell lineages, including NKT cells, IELs, and regulatory CD4$^+$ T cells.

Exit from the Thymus and Final Maturation

It takes less than 3 days for a newly formed DP thymocyte to mature to the SP stage, if positively selected. SP cells spend a longer period (from 4 to 12 days) in the medulla, browsing the surface of epithelial and dendritic cells for self antigen before being given permission to leave the thymus.

The cellular and molecular basis for thymocyte egress was unknown until investigators gained a better understanding of the molecules that regulate cell migration. As mentioned earlier, one set of investigators discovered that thymocytes failed to enter the medulla if they were deficient for the CCR7 chemokine receptor. However, these cells still were

able to leave the thymus directly from the cortex, indicating that at least one other receptor controls thymic exit. The identity of this receptor was discovered when investigators found that few if any T cells made it out of the thymus in a *sphingosine 1-phosphate receptor 1* (*S1P receptor 1* or *S1P1*) knockout mouse. We now know that mature thymocytes up-regulate S1P receptor 1 as part of the developmental program initiated by positive selection. The S1P receptor 1 interacts with its ligand, S1P, at the corticomedullary junction, facilitating thymocyte egress into the perivascular space and bloodstream.

Current observations indicate that these final stages of intrathymic maturation are controlled by the following cascade of events: positive-selecting TCR signals up-regulate Foxo1, a transcription factor that controls expression of several genes related to T-cell function. Foxo1 regulates expression of Klf2, which, in turn, up-regulates the S1P receptor 1. Foxo1 also up-regulates both the IL-7 receptor (IL-7R), which helps to maintain mature T-cell survival, and CCR7, the chemokine receptor that directs mature T-cell traffic to the lymph nodes.

Newly mature cells that exit the thymus are referred to as **recent thymic emigrants (RTEs)**. RTEs can be distinguished from the majority of peripheral naïve T cells because their levels of expression of several surface proteins (e.g., the maturation markers HSA and Qa-2, as well as IL-7R) are more like those of their immature thymocyte ancestors than their fully mature T-cell descendants. Interestingly, recent thymic emigrants are still not functionally mature: they do not proliferate or secrete cytokines as vigorously as most naïve T cells in response to T-cell receptor stimulation. Although research is ongoing, studies suggest that maturation of RTEs is completed by both MHC- and non−MHC-dependent interactions in secondary lymphoid tissue.

Investigators are particularly interested in understanding the behavior of RTEs because they are an important source of mature T cells in individuals who are *lymphopenic* (i.e., have a reduced pool of functional lymphocytes), including people who have undergone chemotherapy, newborns, and the aged.

Key Concepts:

- Maturing thymocytes initiate a cellular program that enhances their survival and up-regulates receptors that facilitate their migration from the thymus and through the blood.

- The exit of SP thymocytes from the thymus depends on expression of the sphingosine 1-phosphate receptor 1 (S1P receptor 1) and its interaction with S1P at the corticomedullary junction.

- Mature thymocytes that have just left the thymus are referred to as recent thymic emigrants. They are not yet optimally functional and undergo a post-thymic phase of maturation in secondary lymphoid tissue that fully licenses them as naïve T cells.

Other Mechanisms That Maintain Self-Tolerance

As we have seen, negative selection of thymocytes can rid a developing T-cell repertoire of cells that express a high affinity for both ubiquitous self antigens and, thanks to the activity of AIRE, tissue-specific antigens. However, negative selection in the thymus (central tolerance) is not perfect. Autoreactive T cells do escape, either because they have too low an affinity for self to induce clonal deletion, or because they happen not to have browsed the "right" tissue-specific antigen/MHC combination. The body has evolved several other mechanisms to avoid autoimmunity, including what has become a major focus of interest for immunologists: the development in both the thymus and the periphery of a fascinating group of cells known as *regulatory T cells*.

T_{REG} Cells Negatively Regulate Immune Responses

Regulatory T cells (T_{REG} cells) inhibit the proliferation of other T-cell populations both directly and indirectly, effectively suppressing autoreactive immune responses. They express surface CD4 as well as CD25, the α chain of the IL-2 receptor. However, T_{REG} cells (T_{REG}s) are more definitively identified by their expression of a master transcriptional regulator, FoxP3, the expression of which is necessary and sufficient to induce differentiation to the T_{REG} lineage.

Many T_{REG}s develop in the thymus and appear to represent an alternative fate for autoreactive T cells. As we have seen, most thymocytes that express receptors with high affinity for self antigen die via negative selection. However, a small fraction commits to the regulatory T-cell lineage and leaves the thymus to patrol the body and thwart autoimmune reactions.

What determines whether a self-reactive thymocyte dies or differentiates into a T_{REG} cell is still not fully understood. Thymocytes that experience strong, but not too strong, TCR signals appear more likely to commit to the T_{REG} lineage than undergo apoptosis (see Figure 8-7). However, other factors, such as subtle differences in maturation state or stochastic differences in chromatin status, may also have an influence. Although all types of antigen-presenting cells, including mTECs and dendritic cells, can mediate T_{REG} cell development, some studies suggest that T_{REG}s encounter these cells in a unique microenvironmental niche. Such a niche may provide T_{REG} precursors with the cytokines they need (IL-2 and IL-15) to survive and complete maturation.

T_{REG} cells that develop in the thymus are referred to as *thymic* or tT_{REG} cells. *Peripheral* or pT_{REG} cells develop from conventional mature T cells exposed to antigen in the context of TGF-β and IL-10 cytokines (see Chapters 10 and 13). Although tT_{REG} and pT_{REG} cells both quell immune responses, they may also adopt distinct inhibitory functions in various tissues.

Animal studies show that members of the $FoxP3^+$ T_{REG} population inhibit development of autoimmune diseases such as experimentally induced inflammatory bowel disease, experimental allergic encephalitis, and autoimmune diabetes. Suppression by these regulatory cells is antigen specific because it depends on activation through the T-cell receptor. Exactly how T_{REG}s quell responses is still debated, although they probably do so via a variety of means: directly inhibiting an antigen-presenting cell's ability to activate T cells, directly killing T cells, indirectly inhibiting T-cell activity by secreting inhibitory cytokines IL-10 and TGF-β, and/or depleting the local environment of stimulatory cytokines such as IL-2 (**Figure 8-11**). See Chapter 14 for a description of one experiment that directly demonstrated that T_{REG}s use more than one approach to inhibit autoreactive responses.

The existence of regulatory T cells that specifically suppress immune responses has clinical implications. Depletion or inhibition of T_{REG} cells before immunization may enhance immune responses to conventional vaccines. Elimination of T cells that suppress responses to tumor antigens may also facilitate the development of antitumor immunity. Conversely, increasing the suppressive activity of regulatory T-cell populations could be useful in the treatment of allergic or autoimmune diseases. The ability to increase the activity of regulatory T-cell populations might also be useful in suppressing organ and tissue rejection. Regulatory T-cell immunotherapy has recently become a reality in clinical trials. Investigators are currently testing T_{REG} effectiveness in preventing graft-versus-host disease (GvHD) after bone marrow transplantation, and clinicians hope to use them soon to quell the immune response that causes type 1 diabetes.

Key Concepts:

- A fraction of thymocytes that experience high-affinity TCR interactions do not die by negative selection but develop into $FoxP3^+$ regulatory T cells (T_{REG} cells) that inhibit T-cell responses outside the thymus. These cells require help from cytokines, including IL-2 and IL-15, to complete their maturation.

- Regulatory T cells that develop in the thymus are called *thymic* T_{REG} cells. Regulatory T cells that differentiate in the periphery are called *peripheral* T_{REG} cells. The function and location of both T_{REG} populations overlap, although they may have distinct responsibilities in different tissues.

- T_{REG}s inhibit immune responses in several ways. They generate cytokines (IL-10 and TGF-β) that inhibit immune cells and can also kill activated T cells directly.

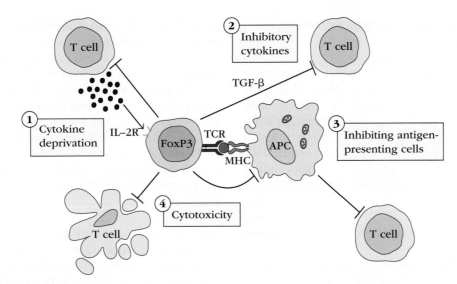

FIGURE 8-11 How regulatory T cells (T$_{REG}$s) inactivate traditional T cells. Some possible mechanisms of T$_{REG}$ activity are illustrated in this schematic. These may all contribute to quelling immune responses in vivo. (1) Cytokine deprivation: T$_{REG}$s express relatively high levels of high-affinity IL-2 receptors and can compete for the cytokines that activated T cells need to survive and proliferate. (2) Cytokine inhibition: T$_{REG}$s secrete several cytokines, including IL-10 and TGF-β, which bind receptors on activated T cells and reduce signaling activity. (3) Inhibition of antigen-presenting cells: T$_{REG}$s can interact directly with MHC class II–expressing antigen-presenting cells and inhibit their maturation, leaving them less able to activate T cells. (4) Cytotoxicity: T$_{REG}$s can also display cytotoxic function and kill cells by secreting perforin and granzyme.

Peripheral Mechanisms of Tolerance Also Protect against Autoreactive Thymocytes

Although the thymus screens the developing T-cell repertoire remarkably efficiently, it is not perfect. Autoreactive T cells do escape. The body has evolved several other mechanisms to manage the autoreactive escapee in the periphery. Briefly, many antigens are "hidden" from autoreactive T cells because only a subset of cells (professional APCs) express the right costimulatory molecules needed to initiate the immune response. Autoreactive naïve T cells that can see an MHC/self-peptide combination on a nonprofessional antigen-presenting cell will not receive the correct costimulatory signals and, therefore, will not divide or differentiate. For example, if a thymocyte specific for a peptide made by a kidney cell escaped from the thymus, it would not be activated unless that peptide were first presented on a professional antigen-presenting cell. Kidney cells do not express the costimulatory ligands required for activating a CD4$^+$ or a CD8$^+$ T cell. High-affinity interactions with MHC/peptide combinations on cells that do not express costimulatory ligands can also lead to T-cell anergy—another peripheral tolerance mechanism that is described in more detail in Chapter 10. The clinical consequences of failures of central and peripheral tolerance are discussed in Chapter 16.

Key Concepts:

- Mechanisms of central tolerance do not remove all autoreactive T cells from the developing T-cell repertoire. Mechanisms that enforce T-cell tolerance in the periphery—peripheral tolerance—provide an important backup (see Chapter 10).

- Regulatory T cells contribute to peripheral T-cell tolerance, as does the strict requirement for costimulatory interactions in order to activate a T cell. Costimulation can be provided only by professional antigen-presenting cells, whose activity is highly regulated.

Conclusion

Mature T lymphocytes have a diverse TCR repertoire that is tolerant to self antigens, yet restricted to self-MHC. They achieve this balance by passing a series of stringent tests in the thymus, reminiscent of natural selection processes in evolution. Developing T cells (thymocytes) arise from multipotent CD4$^-$CD8$^-$ precursors that migrate from the bone marrow to the thymus, where Notch signals commit them to the T-cell lineage. Immature thymocytes proliferate, up-regulate CD4 and CD8, and undergo random T-cell receptor gene rearrangements, thus generating a large and diverse pool of DP thymocytes, each expressing a distinct TCR.

The fate of a DP thymocyte depends on the affinity of this TCR for self-peptide/MHC complexes it encounters while browsing stromal cells in the two major microenvironments of the thymus: the cortex and medulla. If a DP thymocyte fails to bind peptide/MHC complexes with enough affinity, it dies by neglect—the fate of the large majority (>90%) of DP thymocytes. If a DP thymocyte binds peptide/MHC complexes with intermediate affinity, it undergoes positive selection and is given permission to travel from the cortex to the medulla, to complete maturation to a single-positive (SP) CD4$^+$ or CD8$^+$ lineage. If a DP thymocyte binds peptide/MHC complexes with very high affinity, it undergoes negative selection.

Positive selection is mediated exclusively by interactions between thymocytes and cortical thymic epithelial cells (cTECs). Negative selection, however, can be mediated by multiple cells in both the cortex and medulla, and can target thymocytes at both the DP and SP stages. Importantly, only medullary thymic epithelial cells (mTECs) have the capacity to present antigens expressed by other tissues and are responsible for removing tissue-specific autoreactive T cells from the repertoire. Mechanisms that remove autoreactive T cells during development—central tolerance—are not infallible, and are reinforced in the periphery by a variety of mechanisms, including the activity of regulatory T cells. Some regulatory T cells arise in the thymus in response to high-affinity TCR interactions—an exception to the "rule" that high-affinity interactions drive negative selection.

The development of positively selected thymocytes to the CD4$^+$ or CD8$^+$ lineage is also determined by TCR signaling, and is best explained by the kinetic signaling model of lineage commitment. CD4$^+$ and CD8$^+$ thymocytes that survive positive and negative selection are allowed to migrate from the thymus into the bloodstream and complete their maturation in the periphery.

REFERENCES

Anderson, M., et al. 2002. Projection of an immunological self shadow within the thymus by the Aire protein. *Science* **298**:1395.

Anderson, M. S., and M. A. Su. 2016. AIRE expands: new roles in immune tolerance and beyond. *Nature Reviews Immunology* **16**:247.

Baldwin, T., K. Hogquist, and S. Jameson. 2004. The fourth way? Harnessing aggressive tendencies in the thymus. *Journal of Immunology* **173**:6515.

Carpenter, A., and R. Bosselut. 2010. Decision checkpoints in the thymus. *Nature Immunology* **11**:666.

Caton, A., et al. 2004. CD4$^+$ CD25$^+$ regulatory T cell selection. *Annals of the New York Academy of Sciences* **1029**:101.

De Obaldia, M. E., and A. Bhandoola. 2015. Transcriptional regulation of innate and adaptive lymphocyte lineages. *Annual Review of Immunology* **33**:607.

Drennan, M., D. Elewaut, and K. Hogquist. 2009. Thymic emigration: sphingosine-1-phosphate receptor-1–dependent models and beyond. *European Journal of Immunology* **39**:925.

Gascoigne, N. 2010. CD8$^+$ thymocyte differentiation: T cell two-step. *Nature Immunology* **11**:189.

Germain, R. 2008. Special regulatory T-cell review: a rose by any other name: from suppressor T cells to T$_{REGS}$, approbation to unbridled enthusiasm. *Immunology* **123**:20.

Hogquist, K., M. Gavin, and M. Bevan. 1993. Positive selection of CD8$^+$ T cells induced by major histocompatibility complex binding peptides in fetal thymic organ culture. *Journal of Experimental Medicine* **177**:1469.

Hogquist, K., et al. 1994. T cell receptor antagonist peptides induce positive selection. *Cell* **76**:17.

Hogquist, K., S. Jameson, and M. Bevan. 1995. Strong agonist ligands for the T cell receptor do not mediate positive selection of functional CD8$^+$ T cells. *Immunity* **3**:79.

Hogquist, K., T. Baldwin, and S. Jameson. 2005. Central tolerance: learning self-control in the thymus. *Nature Reviews Immunology* **5**:772.

Kisielow, P., H. Blüthmann, U. Staerz, M. Steinmetz, and H. von Boehmer. 1988. Tolerance in T-cell-receptor transgenic mice involves deletion of nonmature CD4$^+$8$^+$ thymocytes. *Nature* **333**:742.

Klein, L., M. Hinterberger, G. Wirnsberger, and B. Kyewski. 2009. Antigen presentation in the thymus for positive selection and central tolerance induction. *Nature Reviews Immunology* **9**:833.

Klein, L., B. Kyewski, P. M. Allen, and K. A. Hogquist. 2014. Positive and negative selection of the T cell repertoire: what thymocytes see (and don't see). *Nature Reviews Immunology* **14**:377.

Kyewski, B., and P. Peterson. 2010. Aire: master of many trades. *Cell* **140**:24.

Li, R., and D. Page. 2001. Requirement for a complex array of costimulators in the negative selection of autoreactive thymocytes in vivo. *Journal of Immunology* **166**:6050.

Littman, D. R. 2016. How thymocytes achieve their fate. *Journal of Immunology* **196**:1983.

Marrack, P., J. McCormack, and J. Kappler. 1989. Presentation of antigen, foreign major histocompatibility complex proteins and self by thymus cortical epithelium. *Nature* **338**:503.

Marrack, P., L. Ignatowicz, J. Kappler, J. Boymel, and J. Freed. 1993. Comparison of peptides bound to spleen and thymus class II. *Journal of Experimental Medicine* **178**:2173.

Mathis, D., and C. Benoist. 2009. Aire. *Annual Review of Immunology* **27**:287.

Melichar, H. J., J. O. Ross, P. Herzmark, K. A. Hogquist, and E. A. Robey. 2013. Distinct temporal patterns of T cell receptor signaling during positive versus negative selection in situ. *Science Signaling* **6**:ra92.

Mohan, J., et al. 2010. Unique autoreactive T cells recognize insulin peptides generated within the islets of Langerhans in autoimmune diabetes. *Nature Immunology* **11**:350.

Page, D., L. Kane, J. Allison, and S. Hedrick. 1993. Two signals are required for negative selection of CD4$^+$CD8$^+$ thymocytes. *Journal of Immunology* **151**:1868.

Park, J., et al. 2010. Signaling by intrathymic cytokines, not T cell antigen receptors, specifies CD8 lineage choice and promotes the differentiation of cytotoxic-lineage T cells. *Nature Immunology* **11**:257.

Punt, J., B. Osborne, Y. Takahama, S. Sharrow, and A. Singer. 1994. Negative selection of CD4$^+$CD8$^+$ thymocytes by T cell receptor–induced apoptosis requires a costimulatory signal that can be provided by CD28. *Journal of Experimental Medicine* **179**:709.

Sambandam, A., et al. 2005. Notch signaling controls the generation and differentiation of early T lineage progenitors. *Nature Immunology* **6**:663.

Singer, A. 2010. Molecular and cellular basis of T cell lineage commitment: an overview. *Seminars in Immunology* **22**:253.

Stadinski, B., et al. 2010. Diabetogenic T cells recognize insulin bound to IAg7 in an unexpected, weakly binding register. *Proceedings of the National Academy of Sciences USA* **107**:10978.

Teh, H., et al. 1988. Thymic major histocompatibility complex antigens and the αβ T-cell receptor determine the CD4/CD8 phenotype of T cells. *Nature* **335**:229.

Uematsu, Y., et al. 1988. In transgenic mice the introduced functional T cell receptor β gene prevents expression of endogenous β genes. *Cell* **52**:831.

Vacchio, M. S. and R. Bosselut. 2016. What happens in the thymus does not stay in the thymus: how T cells recycle the CD4$^+$–CD8$^+$ lineage commitment transcriptional circuitry to control their function. *Journal of Immunology* **196**:4848.

Venanzi, E., C. Benoist, and D. Mathis. 2004. Good riddance: thymocyte clonal deletion prevents autoimmunity. *Current Opinion in Immunology* **16**:197.

von Boehmer, H., H. Teh, and P. Kisielow. 1989. The thymus selects the useful, neglects the useless and destroys the harmful. *Immunology Today* **10**:57.

Wang, L., and R. Bosselut. 2009. CD4–CD8 lineage differentiation: Thpok-ing into the nucleus. *Journal of Immunology* **183**:2903.

Wang, L., et al. 2008. Distinct functions for the transcription factors GATA-3 and ThPOK during intrathymic differentiation of CD4$^+$ T cells. *Nature Immunology* **9**:1122.

Yui, M. A., and E. V. Rothenberg. 2014. Developmental gene networks: a triathlon on the course to T cell identity. *Nature Reviews Immunology* **14**:529.

Useful Websites

https://www.youtube.com/watch?v=08H5CmDaRjU This is one of the wonderful Handwritten Tutorials. It includes very basic information about cell progression and doesn't go into detail about the selection processes, but is a good video for beginners.

https://youtu.be/9E_UxnC_L2o A lovely 3-D animation of T-cell development generated as part of a master's project by Janice Yau (also found at http://janiceyau.com/research.html). This is from the University of Toronto at Mississauga's Biomedical Communications Graduate Program, where other videos of other biological processes can be found.

www.bio.davidson.edu/courses/movies.html A full list of animations assembled, and in many cases generated, by individuals associated with Davidson College. See T Cell Development and Selection (http://www.bio.davidson.edu/courses/Immunology/Flash/Main.html). (Requires FLASH.)

STUDY QUESTIONS

1. Each of the following statements is false in one or more ways. Correct them (and explain your correction[s]).

a. Knockout mice lacking MHC class I molecules fail to produce CD4$^+$ mature thymocytes.
b. β-Selection initiates negative selection.
c. Negative selection to tissue-specific antigens occurs only in the cortex of the thymus.
d. Most thymocytes successfully mature to the CD4$^+$ or CD8$^+$ T-cell lineage.
e. Precursors of T cells express both CD4 and CD8 and first enter the medulla of a thymus.
f. Thymocytes that bind to peptide/MHC complexes with high affinity are positively selected.
g. DN thymocytes progress through several stages distinguished by changes in expression of immunoglobulin and CD25.
h. All thymocytes with autoreactive T-cell receptors undergo negative selection.

i. Regulatory T cells help to maintain central tolerance.

j. Commitment to the CD4⁺ T-cell lineage is regulated by Runx3.

2. Fill in the blanks: Precursors of thymocytes enter the thymus at the _____ junction. Interactions with _____ ligands are required to commit them to the T-cell lineage. If positively selected, DP thymocytes travel from the thymic cortex to the _____. Up-regulation of _____ allows them to leave the thymus and enter circulation.

3. Whereas the majority of T cells in our bodies express an αβ TCR, up to 5% of T cells express the γδ TCR instead. Explain the difference between these two cell types, in terms of development and antigen recognition.

4. You have fluorescein-labeled anti-CD4 and phycoerythrin-labeled anti-CD8. You use these antibodies to stain thymocytes and lymph node cells from normal mice and from *RAG-1* knockout mice. In the forms below, draw the fluorescence-activated cell sorting (FACS) plots that you would expect.

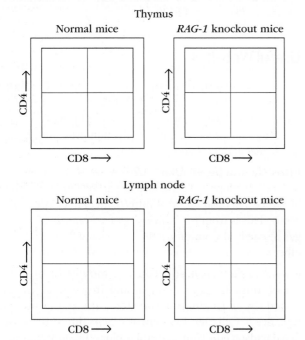

5. What stages of T-cell development (DN1, DN2, DN3, DN4, DP, CD4 SP, or CD8 SP) would be affected in mice with the following genetic modifications? Justify your answers.

a. Mice that do not express MHC class II.

b. Mice that do not express AIRE.

c. Mice that do not express the TCR α chain.

6. You stain thymocytes with phycoerythrin (PE; red)-conjugated anti-CD3 antibodies and fluorescein isothiocyanate (FITC; green)-conjugated anti-TCR β chain antibodies. Most cells stain with both. However, you find a proportion of cells that stain with neither antibody. You also find a small population that stain with anti-CD3, but not with anti-TCR β antibodies. What thymocyte populations might each of these populations represent?

7. You immunize an *H2ᵏ* mouse with chicken ovalbumin (a protein against which the mouse will generate an immune response) and isolate a CD4⁺ mature T cell specific for an ovalbumin peptide. You clone the αβ TCR genes from this cell line and use them to prepare transgenic mice with the *H2ᵏ* or *H2ᵈ* haplotype.

a. What approach can you use to distinguish immature thymocytes from mature CD4⁺ thymocytes in the transgenic mice?

b. Would thymocytes from a TCR transgenic mouse of the *H2ᵏ* background have a proportion of CD4⁺ thymocytes that is higher, lower, or the same as in a wild-type mouse?

c. Would thymocytes from a TCR transgenic mouse of the *H2ᵈ* background have a proportion of CD4⁺ thymocytes that is higher, lower, or the same as in a wild-type mouse? Speculate and explain your answer.

d. You find a way to "make" the medullary epithelium of an *H2ᵏ* TCR transgenic mouse express and present the ovalbumin peptide for which your T cell is specific. What would the CD4-versus-CD8 profile of a TCR transgenic thymus look like in these mice?

e. You also find a way to "make" the cortical epithelium express this ovalbumin peptide in its MHC class II dimer. What might the CD4-versus-CD8 profile of this TCR transgenic thymus look like?

8. In his classic chimeric mouse experiments, Zinkernagel took bone marrow from a mouse of one MHC haplotype (mouse 1) and the thymus from a mouse of another MHC haplotype (mouse 2) and transplanted them into a third mouse, which was thymectomized and lethally irradiated. He then immunized this reconstituted mouse with the lymphocytic choriomeningitis virus (LCMV) and examined the activity of the mature T cells isolated from the spleen and lymph nodes of the mouse.

He was specifically interested to see if the mature CD8⁺ T cells in these mice could kill target cells infected with LCMV with the MHC haplotype of mouse 1, 2, or 3. The results of two such experiments using *H2ᵇ* strain C57BL/6 mice and *H2ᵈ* strain BALB/c mice as bone marrow and thymus donors, respectively, are shown in the following table.

Lysis of LCMV-infected target cells						
Experiment	Bone marrow donor	Thymus donor	Thymectomized x-irradiated recipient	H2ᵈ	H2ᵏ	H2ᵇ
A	C57BL/6 (*H2ᵇ*)	BALB/c (*H2ᵈ*)	C57BL/6 × BALB/c	+	–	–
B	BALB/c (*H2ᵈ*)	C57BL/6 (*H2ᵇ*)	C57BL/6 × BALB/c	–	–	+

a. Why were the $H2^b$ target cells not lysed in experiment A but lysed in experiment B?

b. Why were the $H2^k$ target cells not lysed in either experiment?

9. You have a CD8$^+$ CTL clone (from an $H2^k$ mouse) that has a T-cell receptor specific for the H-Y antigen. You clone the αβ TCR genes from this cloned cell line and use them to prepare transgenic mice with the $H2^k$ or $H2^d$ haplotype.

a. How can you distinguish the immature thymocytes from the mature CD8$^+$ thymocytes in the transgenic mice?

b. For each of the following transgenic mice, indicate with (+) or (−) whether the mouse would have immature double-positive and mature CD8$^+$ thymocytes bearing the transgenic T-cell receptor: $H2^k$ female, $H2^k$ male, $H2^d$ female, $H2^d$ male.

c. Explain your answers for the $H2^k$ transgenics.

d. Explain your answers for the $H2^d$ transgenics.

10. To demonstrate positive thymic selection experimentally, researchers analyzed the thymocytes from normal $H2^b$ mice, which have a deletion of the class II *H2-E* gene, and from $H2^b$ mice in which the class II *H2-A* gene had been knocked out.

a. What MHC molecules would you find on antigen-presenting cells from the normal $H2^b$ mice?

b. What MHC molecules would you find on antigen-presenting cells from the *H2-A* knockout $H2^b$ mice?

c. Would you expect to find CD4$^+$ T cells, CD8$^+$ T cells, or both in each type of mouse? Why?

11. You wish to determine the percentage of various types of thymocytes in a sample of cells from mouse thymus using the indirect immunofluorescence method.

a. You first stain the sample with goat anti-CD3 (primary antibody) and then with rabbit FITC-labeled anti-goat immunoglobulin (secondary antibody), which emits a green color. Analysis of the stained sample by flow cytometry indicates that 70% of the cells are stained. Based on this result, how many of the thymus cells in your sample are expressing antigen-binding receptors on their surface? Would all be expressing the same type of receptor? Explain your answer. What are the remaining unstained cells likely to be?

b. You then separate the CD3$^+$ cells with the fluorescence-activated cell sorter (FACS) and restain them. In this case, the primary antibody is hamster anti-CD4, and the secondary antibody is rabbit PE-labeled anti-hamster immunoglobulin, which emits a red color. Analysis of the stained CD3$^+$ cells shows that 80% of them are stained. From this result, can you determine how many T$_C$ cells are present in this sample? If yes, then how many T$_C$ cells are there? If no, what additional experiment would you perform to determine the number of T$_C$ cells that are present?

CLINICAL FOCUS QUESTIONS

1. The susceptibility to autoimmune diseases often has a genetic basis and has been linked to many different gene loci. Identify two genes that could be involved in an increased susceptibility to autoimmune disease. Explain your reasoning.

2. Susceptibility to many autoimmune diseases has been linked to MHC gene variants. One of the best examples of such a linkage is provided by multiple sclerosis (MS), a human autoimmune disease caused by autoreactive T cells whose activity ultimately damages the myelin sheaths around neurons. Susceptibility to MS has been consistently associated with variants in the *HLA-DR2* gene. Although this link was first recognized in 1972, we still don't fully understand the basis for this susceptibility. One perspective on the reasons for the link between MHC variations and autoimmune disease was offered in a recent review article. The authors state, "The mechanisms underlying MHC association in autoimmune disease are not clearly understood. One long-held view suggests a breakdown in immunological tolerance to self antigens through aberrant class II presentation of self- or foreign peptides to autoreactive T lymphocytes. Thus, it seems likely that specific MHC class II alleles determine the targeting of particular autoantigens resulting in disease-specific associations." (Fernando, M. M., et al. 2008. Defining the role of the MHC in autoimmunity: a review and pooled analysis. *PLoS Genetics* 4:e1000024.)

a. Paraphrase this perspective, using your own words. What, specifically, might the authors mean by "aberrant class II presentation . . . to autoreactive T lymphocytes"?

b. (**Advanced question.**) Although this speculation has some merit, it does not resolve all questions. Why? Pose one question that this explanation inspires or does not answer.

c. (**Very advanced question.**) Offer one addition to the explanation (or an alternative) that helps resolve the question you posed above.

B-Cell Development

Learning Objectives

After reading this chapter, you should be able to:

1. Understand the experimental approaches used to identify and order the various cellular stages of B-cell development.

2. Explain the mechanisms that drive the progression of cells through the various stages of B-cell development in the bone marrow and spleen.

3. Describe how heavy- and light-chain gene rearrangement events are programmed into defined stages of B-cell development.

4. Compare and contrast the first and second checkpoints in B-cell development in terms of their stages, signaling, consequences, and importance.

5. Identify *four* processes ensuring self-tolerance that operate at the immature B and transitional B stages.

6. Examine the basic similarities and differences between B-2, B-1a, B-1b, and marginal zone B cells.

CXCL12⁺ B220⁺flt-3⁺

IL-7⁺ B220⁺c-kit⁺

B cells at various stages of development seek contact with stromal cells expressing CXCL12 (pre-pro-B cells, *left*) or IL-7 (pro-B cells, *right*). [Republished with permission of Elsevier, from Tokoyoda, K., et al., "Cellular niches controlling B lymphocyte behavior within bone marrow during development." Immunity 2004, June; 20(6) 707–718, Figure 2. Permission conveyed through Copyright Clearance Center, Inc.]

Millions of B lymphocytes are generated in the adult bone marrow every day and exported to the periphery. The rapid and unceasing generation of new B cells occurs in a carefully regulated sequence of events. Cell transfer experiments (similar to those described in Chapter 2) that identified hematopoietic stem cells (HSCs), in which genetically marked donor HSCs are injected into unmarked recipients, have indicated that B-cell development from HSC to mature B cell takes from 1 to 2 weeks.

B-cell development begins in the bone marrow with the asymmetric division of an HSC and continues through a series of progressively more differentiated progenitor stages to the production of common lymphoid progenitors (CLPs),

which can give rise to B cells, T cells, or innate lymphoid cells. These early stages in hematopoiesis and lymphocyte development were described in Chapter 2 (see Figures 2-1 and 2-3). Progenitor cells destined to become T cells migrate to the thymus, where they complete their maturation (see Chapter 8). The majority of CLPs that remain in the bone marrow enter the B-cell development pathway (**Overview Figure 9-1**). As differentiation proceeds, developing B cells express a precisely controlled sequence of cell-surface receptors and adhesion molecules. Some of the signals received from these receptors induce the differentiation of the developing B cell; others trigger its proliferation at particular stages of development; and yet others direct its movements within the bone marrow environment. These signals collectively allow differentiation of the CLP through the early B-cell stages to form the **immature B cell** that leaves the marrow to complete its differentiation in the

Key Terms

Immature B cell	E2A	VpreB	Immature B-cell (second) checkpoint
Pre-B-cell receptor	Foxo1	λ5	
B-2 B cells	Early B-cell factor 1 (EBF1)	Surrogate light chain (SLC)	Anergic
B-1 B cells	Pre-pro-B cell		Transitional B cells (T1, T2, T3)
Marginal zone (MZ) B cells	Runx1	Pre-B cell (precursor B cell)	
Epigenetic	SWI/SNF	Large pre-B cells	BAFF receptor (BAFF-R)
Ikaros	Pro-B cell (progenitor B cell)	Pre-B-cell (first) checkpoint	BAFF
Purine box factor 1 (PU.1)	PAX5	Small pre-B cells	

B-Cell Development Begins in the Bone Marrow and Is Completed in the Periphery

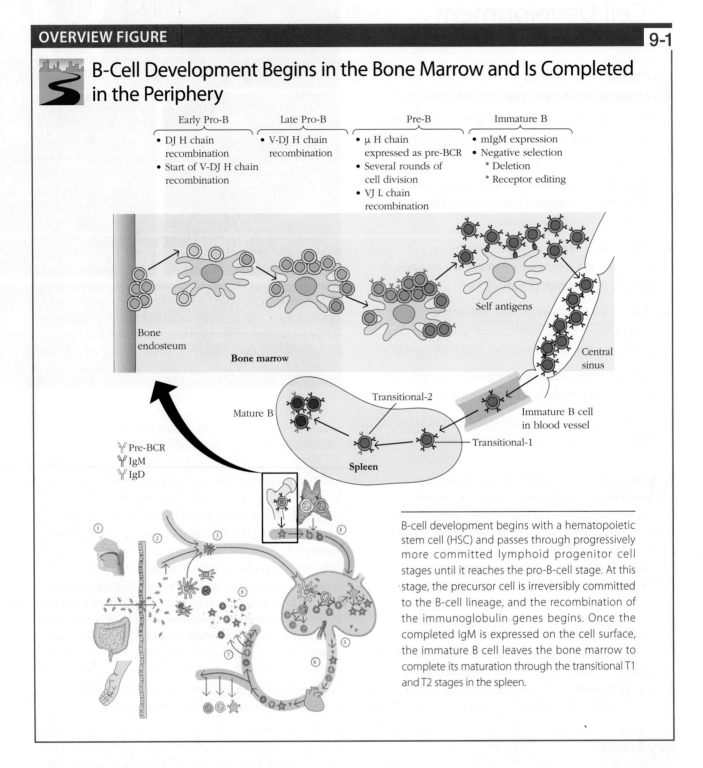

Early Pro-B	Late Pro-B	Pre-B	Immature B
• DJ H chain recombination • Start of V-DJ H chain recombination	• V-DJ H chain recombination	• μ H chain expressed as pre-BCR • Several rounds of cell division • VJ L chain recombination	• mIgM expression • Negative selection * Deletion * Receptor editing

B-cell development begins with a hematopoietic stem cell (HSC) and passes through progressively more committed lymphoid progenitor cell stages until it reaches the pro-B-cell stage. At this stage, the precursor cell is irreversibly committed to the B-cell lineage, and the recombination of the immunoglobulin genes begins. Once the completed IgM is expressed on the cell surface, the immature B cell leaves the bone marrow to complete its maturation through the transitional T1 and T2 stages in the spleen.

spleen. For the investigator, the expression of different cell-surface molecules at each stage of B-cell maturation provides an invaluable experimental tool with which to recognize and isolate B cells poised at discrete points in their development.

The primary function of mature B cells is to detect pathogens and other potentially harmful foreign antigens and to differentiate into plasma cells secreting antibodies that protect the host against the invaders. Therefore, the most critically important events occurring during B-cell

development are the rearrangements of immunoglobulin receptor heavy- and light-chain gene segments to form the B-cell receptor for antigen, determining its specificity. Recall from Chapter 6 that immunoglobulin gene rearrangements begin with heavy-chain D-to-J_H gene segment rearrangement, followed by the stitching together of the V_H and DJ_H segments to allow synthesis of an intact μ heavy chain. This chain initially pairs with a surrogate light chain, allowing the cell-surface expression of the *pre-B-cell receptor*.

This initial form of membrane immunoglobulin (Ig) activates several rounds of cell division, followed by rearrangement of the light-chain V and J gene segments, allowing the immature B cell to express membrane IgM (mIgM).

As is true for T cells (see Chapter 8), developing B cells must solve the problem of creating a diverse repertoire of receptors capable of recognizing an extensive array of antigens, while ensuring that self-reactive B cells are either eliminated or inactivated. Processes responsible for self-tolerance occur at several stages of B-cell development. Development of B cells is somewhat simpler than that of T cells, in that B-cell receptors recognize intact antigens, not antigen fragments presented by MHC proteins, and hence do not need to be selected for those that recognize self-MHC molecules. Also unlike T-cell development, B-cell development is almost complete by the time the B cell leaves the bone marrow; in mammals there is no thymus equivalent in which B-cell development is accomplished. Instead, immature B cells are released to the periphery, where they complete their developmental program in the spleen.

In this chapter, we will follow B-cell development from its earliest stages in the primary lymphoid organs to the generation of fully mature B cells in the secondary lymphoid tissues. Most of this chapter will focus on the predominant (or conventional) B-cell population, known as **B-2 B cells** (or follicular B cells), derived through HSC-initiated hematopoiesis. As for T cells, however, several B-cell subsets exist, and later in this chapter we will briefly address how the processes of differentiation of the minority subsets, that is, **B-1 B cells** and *marginal zone* **(MZ) B cells**, differ from the developmental program followed by the conventional B-2 B cells. We will conclude with a brief comparison of the maturational processes of T and B lymphocytes.

B-Cell Development in the Bone Marrow

While during embryonic development hematopoiesis and B-cell formation occur in several structures, including the fetal liver and spleen, beginning around the time of birth and continuing through adulthood these key developmental pathways occur in the bone marrow. The structure of bone and bone marrow was presented in Chapter 2 (Figure 2-10). As described there, the bone marrow contains microenvironments, or *niches*, populated by hematopoietic cells and various bone marrow stromal cells. The stromal cells express proteins (including cell-surface ligands, cytokines, and chemokines) that support the long-term survival and division of HSCs and the developmental pathways leading to the formation of mature blood cells.

At various points in their development, the precursors of B cells must interact with stromal cells expressing particular proteins that induce the developing cells to differentiate and move in an orderly progression from location to location within the bone marrow. For example, as illustrated in **Figure 9-2a**, HSCs begin their life in close contact with osteoblasts in an area near the lining of the endosteal (bone marrow) cavity. This *endosteal niche*, as it is known, provides an environment supportive of long-term maintenance of HSCs (see Figure 2-10). HSCs and early progenitor cells express the receptor c-kit (CD117), which binds the ligand stem cell factor (SCF, expressed both as a membrane and secreted protein), maintaining the cells in this niche and influencing their differentiation into progenitor cells. Once it differentiates to the pre-pro-B-cell stage, a developing B cell requires signals from the chemokine CXCL12, secreted by certain stromal cells, in order to progress to the pro-B-cell stage. Pro-B cells then require signaling from the cytokine interleukin (IL)-7, which is secreted by yet another stromal cell subset, to mature to the pre-B stage (Figure 9-2b). Many of these stromal cell factors serve to induce the expression of specialized transcription factors important in B-cell development.

Changes in Cell-Surface Markers, Gene Expression, and Immunoglobulin Gene Rearrangements Define the Stages of B-Cell Development

Full characterization of a developmental pathway requires that scientists understand the phenotypic and functional characteristics of each cell type in that pathway. Cells at particular stages of differentiation can be characterized by their surface molecules, which include adhesion molecules and receptors for chemokines and cytokines. They are also defined by the array of active transcription factors that determine which genes are expressed at each step in the developmental process. Finally, in the case of B cells, the developmental stages are also defined by the status of the rearranging heavy- and light-chain immunoglobulin genes. B-cell development is not yet completely understood; however, most of the important cellular intermediates have been defined, and developmental immunologists are steadily filling in the gaps in our knowledge.

Investigators have employed several general experimental approaches to characterize B-cell development. First, they generated antibodies against molecules (antigens or markers) present on the surface of bone marrow cells. They then determined which of these molecules were present at the same time as other antigens, and which combinations of antigens appeared to define unique cell types. Second, by sorting cells bearing particular combinations of cell-surface markers, and analyzing those cell populations for what daughter cells they gave rise to, as well as for the occurrence of immunoglobulin gene rearrangements, scientists were able to confirm the order in which cell populations and their gene rearrangements occur. Third, investigators used the power of knockout genetics to determine the effects of eliminating the expression of a particular

(a)

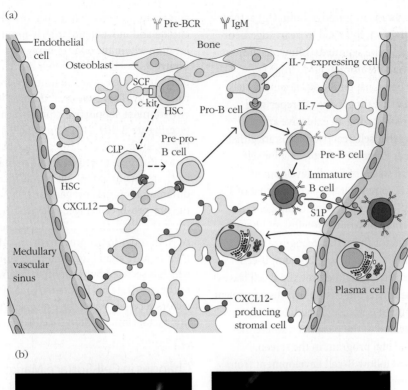

(b)

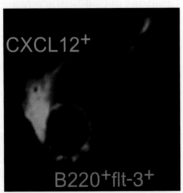
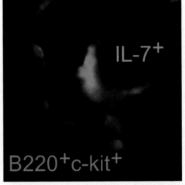

FIGURE 9-2 HSCs and B-cell progenitors make contact with various sets of bone marrow cells as they progress through their developmental program. (a) HSCs begin their developmental program close to the osteoblasts (*top*). An HSC is also shown entering from the blood (*left-hand side*), illustrating the fact that HSCs are capable of recirculation in the adult animal. Progenitor cells then move to gain contact with CXCL12-expressing stromal cells, where they mature into pre-pro-B cells. By the time differentiation has progressed to the pro-B-cell stage, the developing cell has moved to receive signals from IL-7–producing stromal cells. After leaving the IL-7–expressing stromal cell, the pre-B cell completes its differentiation and leaves the bone marrow as an immature B cell. Immature B cells express the S1P receptor, which recognizes the lipid chemoattractant sphingosine 1-phosphate (S1P) in the blood. CXCL12 is shown in purple; IL-7 in blue; S1P in red. (b) B cells at various stages of development seek contact with stromal cells expressing CXCL12 (pre-pro-B cells, *left*) or IL-7 (pro-B cells, *right*). [*Republished with permission of Elsevier, from Tokoyoda, K., et al., "Cellular niches controlling B lymphocyte behavior within bone marrow during development." Immunity 2004, June; 20(6) 707-718, Figure 2. Permission conveyed through Copyright Clearance Center, Inc.*]

gene, such as for a transcription factor, on B-cell development. One drawback of the knockout approach, however, is that it defines only the first stage in differentiation at which the transcription factor is required. More recent variations include *conditional knockouts,* where a gene is deleted only in a specific cell type or developmental stage, and *knock-in* experiments, where a gene with an easily detectable product (such as green fluorescent protein, GFP) is inserted into the genome under the same regulatory control as the gene of interest (e.g., a transcription factor). In such a knock-in animal, every cell in which the transcription factor is expressed can be detected by fluorescence.

Recent research on the control of development has also uncovered critical roles played by **epigenetic** changes (changes affecting a gene's expression that don't affect the DNA sequence of the gene itself). These findings have emerged from molecular analyses of the DNA and chromatin associated with specific genes and their modifiers. Epigenetic changes include DNA modification by methylation, chromatin alterations such as histone modification and chromatin structure remodeling, and the effects of microRNAs, which can inhibit protein expression by decreasing the stability of mRNAs and their translation into proteins. One approach used to study epigenetic changes is *chromatin immunoprecipitation* (*ChIP*). After fragmenting chromatin into small pieces, an antibody to a transcription factor or chromatin-modifying protein is used to immunoprecipitate fragments of chromatin containing that protein. The isolated fragments can then be analyzed for the genes to which that protein has bound and for associated DNA methylations, histone modifications, and chromatin-modifying complexes. **Advances Box 9-1** describes some of the epigenetic changes that have recently been shown to control key steps in lymphoid development, in particular, the regulation of progression through the stages of B-cell development. This is currently an extremely active research area.

> **Key Concepts:**
>
> - Beginning around the time of birth and extending through adult life, hematopoiesis, including B-cell development, occurs in the bone marrow and is influenced by various niches established by stromal cells.
>
> - Stages of B-cell development are defined by the presence of sets of cell-surface markers (which include cytokine and chemokine receptors and adhesion molecules), expression of specific transcription regulators, and the rearrangement status of immunoglobulin genes.
>
> - The stages of B-cell development are controlled by networks of transcription factors and by epigenetic changes that influence the expression of key genes. The cells become increasingly committed to becoming B lymphocytes.

The Earliest Steps in Lymphocyte Differentiation Culminate in the Generation of a Common Lymphoid Progenitor

In this section, we will briefly review the process by which a hematopoietic stem cell (HSC) in the bone marrow develops into a common lymphoid progenitor (CLP), described more fully in Chapter 2.

As noted in Chapter 2, HSCs express a unique set of surface proteins, some of which play key roles in initiating hematopoiesis. One of these, c-kit, is the receptor for stem cell factor (SCF); their interaction triggers key signals that help to induce differentiation into multipotent progenitor cells (MPPs). HSCs also express the stem cell–associated

antigen-1 (Sca-1, also known as Ly6D). Both c-kit and Sca-1 are expressed in parallel on early progenitor cells, but the levels of their expression drop as the cells commit to the lymphoid cell lineage (see Figure 2-3).

Multipotent progenitors (MPPs) generated following c-kit signaling lose the capacity for extensive self-renewal that characterizes HSCs, but retain the potential to differentiate into several different hematopoietic lineages. In progenitor cells bound for a lymphoid fate, the transcription factors **Ikaros**, ***Purine box factor 1*** (**PU.1**), and **E2A** participate in the earliest stages of lymphocyte development. Ikaros recruits chromatin-remodeling complexes to particular regions of the DNA and ensures the accessibility of genes necessary for B-cell development. The levels of PU.1 determine lymphoid versus myeloid differentiation: low levels favor lymphoid differentiation, whereas higher levels favor myeloid differentiation. The level of PU.1 protein expressed is in turn regulated by the transcriptional repressor Gfi1, which down-regulates the expression of PU.1 to the levels necessary for progression down the lymphoid pathway. PU.1 acts at least in part by initiating nucleosome remodeling, which is followed by specific histone modification of genomic regions associated with gene expression (see Advances Box 9-1). Ikaros and PU.1 combine to induce expression of E2A, which plays critical roles in subsequent stages of B-cell development (**Figure 9-3**).

These progenitors also begin to express the Fms-related tyrosine kinase 3 receptor (FLT-3). FLT-3 binds to the membrane-bound FLT-3 ligand on bone marrow stromal cells and signals the progenitor cell to begin synthesizing the **IL-7 receptor** (**IL-7R**) α chain, which pairs with the common γ chain found in receptors for various class 1 cytokines. As cells become increasingly committed to the lymphoid lineage and start preparing to rearrange their antigen receptor genes, they begin to express RAG1/2 and *t*erminal *d*eoxynucleotidyl*t*ransferase (TdT), which define the cell as an early lymphoid progenitor cell (ELP). A subset of ELPs migrates out of the bone marrow to seed the thymus and serve as T-cell progenitors (discussed in Chapter 8). Other ELPs remain in the bone marrow as B-cell progenitors. As the levels of the IL-7R increase, expression of c-kit and Sca-1 proteins decreases, and the ELP develops into a common lymphoid progenitor (CLP).

At the CLP stage, the progenitor on its way to B-cell commitment has lost myeloid potential but still retains the potential to mature along pathways leading to T cells (in the thymus), natural killer (NK) cells, and conventional dendritic cells (see Figure 2-3). Signals received through the IL-7R, together with transcription factors E2A and **Foxo1** (induced by E2A), activate expression of the key transcription factor **early B-cell factor 1** (**EBF1**), which is required for later steps in the B-cell differentiation pathway (see Figure 9-3). Signaling through the IL-7R occurs via a JAK/STAT pathway (introduced in Chapter 4 in the context of interferon signaling). In this case, IL-7 binding to the

Roles of Epigenetic Changes in the Control of B-Cell Development

In the immune system, like any developing system, the control of expression of certain genes and their protein products is of key importance to the step-wise differentiation of stem and progenitor cells into mature differentiated cells. How each step in the progression of cells from early hematopoiesis through B-cell development is controlled is a major question occupying immunologists and is of interest to clinicians as well, as defects in this process can lead to immunodeficiencies or malignancies (leukemias and lymphomas). Several transcription factors have been identified in the last few decades that control gene expression during B-cell development, as described in the text. But it has become clear through research in many developing systems that other regulatory mechanisms control gene transcription and translation. Collectively these are considered epigenetic changes; they include the effects of noncoding RNAs, such as microRNAs (also abbreviated as miRs), on the stability and translation of mRNAs, and the modification of DNA and histones through chromatin modifiers. Adding to the complexity of B (and T)-cell development is the necessity to rearrange immunoglobulin (or TCR) gene segments, processes in which chromatin structure, transcription, and DNA-binding proteins also play key roles.

Geneticists have long known that only a small fraction of chromosomal DNA specifies protein sequences, and early papers relegated the non-protein-coding DNA segments to the somewhat ignominiously described status of "junk DNA." In 1993, however, scientists studying the genome of the nematode *Caenorhabditis elegans* described groundbreaking investigations of some of the non-protein-coding sequences that they had

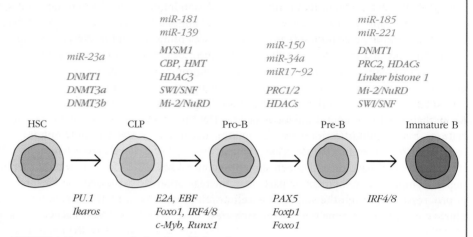

FIGURE 1 Factors regulating B-cell development. The main stages of B-cell development in the bone marrow are shown. Below the arrows (in black) are key transcription factors. Above the arrows are microRNAs (in red) and DNA methylases, histone modifiers, and chromatin remodelers (in green) that positively or negatively regulate that developmental progression. Specific examples are described in this box (for the HSC → CLP transition) and in the text for stages of B-cell development.

identified as having been transcribed but not translated. They showed that these primary transcripts were processed into small pieces of RNA, 18 to 30 nucleotides in length, that were capable of exerting control over mRNA expression levels.

The biosynthesis of these microRNAs follows a similar path in eukaryotes as diverse as *C. elegans* and humans. A series of ribonucleases, including Drosha in the nucleus and Dicer in the cytoplasm, along with RNA-binding proteins, convert a noncoding RNA transcript into an 18- to 30-nucleotide microRNA duplex, consisting of the mature microRNA and its antisense strand. In a final step, the mature microRNA, now single stranded, associates with a protein complex called the *RNA-induced silencing complex*, or RISC. The mature microRNA operates by complementary binding of a so-called "seed" region of 6 to 8 nucleotides at its 5′ end to a region on its target mRNA. Once the microRNA has bound, three things can happen: the target mRNA can be slated for cleavage; the mRNA can be destabilized, shortening its lifetime; or translation from the mRNA can be repressed. A single microRNA can target the synthesis of

many proteins, and each mRNA can be the target of more than one microRNA, thus adding to both the flexibility and complexity of this mode of control over gene expression.

From a theoretical standpoint, it is clear that the developmental changes that occur as B cells mature require rapid changes in the concentrations of such important proteins as transcription factors and pro- and anti-apoptotic molecules, among other regulatory proteins. The need for such rapid alterations in protein concentrations can be met efficiently by the type of post-transcriptional control mechanisms mediated by microRNAs. **Figure 1** shows examples of transcription factors, microRNAs, and chromatin modifiers that regulate stages of hematopoiesis and B-cell development. One example involves PU.1, which at high levels triggers HSCs and progenitor cells to give rise to myeloid cells, and at low levels induces commitment to the lymphoid lineage. PU.1 at high concentrations works at least in part by activating production of the miR-23a cluster of microRNAs; they antagonize lymphoid development and instead allow the generation of myeloid

(continued)

BOX 9-1

cells. Other microRNAs enhance or inhibit certain developmental transitions.

Figure 1 also lists some epigenetic enzymes and chromatin-remodeling proteins that play critical roles in regulating hematopoiesis and B-cell development. DNA methylation of cytosine, generating 5-methylcytosines, in CpG-rich genomic regions usually inhibits gene transcription. Histone modifiers act on the N terminus of histone tails through acetylation, methylation, phosphorylation, and other post-translational modifications. Some of these modifications, including acetylation and methylation of lysine-4 of histone H3 ("H3K4," followed by 1, 2, or 3 to indicate the number of methyl groups added to a lysine amino group), promote

active transcription, whereas addition of methyl groups on other lysines represses transcription. Other proteins shown in Figure 1 are chromatin remodelers, which change chromatin conformation and composition, thus regulating transcription (as well as DNA replication and repair). As one example, DNA methyltransferase-1 (DNMT1) plays essential roles in maintaining the capacity of HSCs to self-renew through cell division and in promoting lymphoid versus myeloerythroid development. The latter may reflect the activity of PU.1: it induces chromatin remodeling followed by H3K4me1 methylation, affecting numerous genes. PU.1 is then able to bind to those genomic regions, helping to induce gene expression. Roles of some

of the other methyltransferases and chromatin remodelers shown in Figure 1 in later stages of B-cell development are discussed in the text.

REFERENCES

Baltimore, D., M. P. Boldin, R. M. O'Connell, D. S. Rao, and K. D. Taganov. 2008. MicroRNAs: new regulators of immune cell development and function. *Nature Immunology* **9:**839.

Bao, Y., and X. Cao. 2016. Epigenetic control of B cell development and B-cell-related immune disorders. *Clinical Reviews in Allergy and Immunology* **50:**301.

Vasilatou, D., S. Papageorgiou, V. Pappa, E. Papageorgiou, and J. Dervenoulas. 2010. The role of microRNAs in normal and malignant hematopoiesis. *European Journal of Haematology* **84:**1.

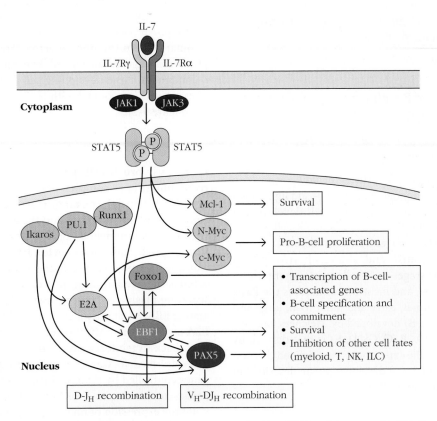

FIGURE 9-3 The interplay of transcription factors during early B-cell development. In early hematopoietic progenitors the transcription factors Ikaros and PU.1 induce expression of the transcription factor E2A. IL-7 binding to its receptor, which is expressed by the common lymphoid progenitor (CLP) (see Figure 2-3), activates JAK1 and JAK3 kinases, which activate STAT5. STAT5 stimulates B progenitor survival, by activating the anti-apoptotic protein Mcl-1, and proliferation, by activating the proliferative control proteins N-Myc

and c-Myc. STAT5 collaborates with E2A, Runx1 and Foxo1 proteins to promote the expression of early B-cell factor 1 (EBF1). EBF1 in turn feeds back to enhance expression of E2A and Foxo1. E2A, EBF1, PU.1 and Ikaros all promote the expression of PAX5. Together E2A, EBF1, PAX5, and Foxo1 activate many genes leading to B-cell lineage specification and commitment. E2A, EBF1, and PAX5 participate in positive feedback loops that enhance the levels of all three transcription factors. See text for details.

IL-7R activates JAK1 and JAK3 protein kinases that then phosphorylate and activate STAT5, which dimerizes and enters the nucleus, where it acts as a transcription factor. In addition to activating expression of EBF1, STAT5 signaling up-regulates expression of the anti-apoptotic molecule **Mcl-1**, thus supporting cell survival. Activated STAT5 and E2A also increase expression of the *c-Myc* and *N-Myc* genes, whose protein products will activate the cell proliferation that will occur later, at the pro-B-cell stage (see Figure 9-3).

As a CLP destined to differentiate along the B-cell pathway matures, the chromatin containing the immunoglobulin heavy-chain locus becomes increasingly accessible, and the developing lymphocyte approaches the point at which it is irrevocably committed to the B-cell lineage.

Key Concepts:

- Bone marrow stromal cell–derived proteins, including stem cell factor (SCF) and IL-7, induce early hematopoietic cells to become increasingly committed to the lymphoid lineage.

- A network of sequentially activated transcription regulators drives hematopoietic differentiation, generating common lymphoid progenitors that give rise to the B-lymphocyte lineage.

The Later Stages of B-Cell Development Result in Commitment to the B-Cell Phenotype and the Stepwise Rearrangement of Immunoglobulin Genes

Figure 9-4 lists important properties—including expression of key proteins and immunoglobulin heavy- and light-chain gene rearrangements—for stages of B-cell development beginning with the first cell committed to the B-cell lineage: the pre-pro-B cell. These stages have been defined by more than one group of scientists and, as a result, two systems of nomenclature for stages of B-cell development are in common use. The first, and most widely used, is the Basel nomenclature (pre-pro-B, early and late pro-B, large and small pre-B, immature B) developed by Melchers and colleagues. The second (A, B, C, C′, D, E) is that defined by Hardy and colleagues; the experiments that defined this system of classification of B-cell development are described in detail in **Classic Experiment Box 9-2**.

Pre-Pro-B Cells

With the acquisition of the B-cell lineage–specific marker B220 (CD45R), and the expression of increasing levels of the transcription factor EBF1, the developing common lymphoid progenitor enters the **pre-pro-B-cell** stage. EBF1 is an important transcription factor in lymphoid development, and therefore transcription of the *Ebf1* gene is itself under the control of multiple transcription factors. In addition to STAT5, E2A, and Foxo1, all mentioned in the previous section, **Runx1** also regulates *Ebf1* transcription (see Figure 9-3). At the pre-pro-B-cell stage, EBF1, along with E2A, binds to the immunoglobulin heavy-chain locus, promoting accessibility of the D-J_H gene segments and preparing the cells for the first step of Ig gene recombination. Epigenetic chromatin modifications also control Ig gene rearrangements in another way, as RAG2 must recognize methylated histone H3 in order for the RAG1/2 complex to bind to the recombination signal sequence (RSS) and mediate gene rearrangements (see Chapter 6).

EBF1 facilitates the activation of B-lineage genes that were previously silenced epigenetically through functional interactions with the **SWI/SNF** chromatin-remodeling complex (see Advances Box 9-1). EBF1 also contributes to the commitment of cells to the B-cell lineage by inhibiting the expression of Notch1 and GATA-3, which support T-cell development, and of the transcription factor ID2, which promotes the development of NK and other innate lymphoid cells (ILCs) and antagonizes B- and T-cell development.

Pre-pro-B cells remain in contact with CXCL12-secreting stromal cells in the bone marrow. However, at the early pro-B-cell stage, characterized by the onset of D-to-J_H gene recombination, the developing cell moves within the bone marrow, seeking contact with IL-7–secreting stromal cells (see Figure 9-2).

Pro-B Cells

In the early **pro-B-cell (progenitor B cell)** stage, D-to-J_H recombination is completed and the cell begins to prepare for V_H-to-DJ_H joining. However, this final heavy-chain recombination event and the establishment of stable B-lineage commitment awaits the expression of the quintessential B-cell transcription factor, **PAX5**, which will then be expressed throughout B-cell development until the mature B cell is activated by antigen to differentiate into antibody-secreting plasma cells (see Chapter 11). The *Pax5* gene is among EBF1's transcriptional targets, and Ikaros, PU.1, and E2A also support PAX5 expression (see Figure 9-3). PAX5 feeds back to reinforce EBF1 expression, thus generating a powerful feed-forward regulatory loop. Transcription of genes controlled by the PAX5 transcription factor denotes passage to the late pro-B-cell stage of development, at which point the expression of non–B-lineage genes is permanently blocked. Like EBF1, PAX5 can also act as a transcriptional repressor, blocking *Notch1* gene expression and thus any residual potential of the pro-B cell to develop along the T-cell lineage.

	Cell stage	Status of Ig genes	Membrane Ig receptor expression	RAG1,2	TdT	IL-7R	B220 (CD45R)	CD43	Igα,β	CD19
⬭	Pre-pro-B (Fr. A)[1]	GL[2]	None	+	+	+	+	+	−	−
⬤	Early Pro-B (Fr. B)	DJ_H	None	+	+	+	+	+	+	−
⬤	Late Pro-B (Fr. C)	Some V_HDJ_H	None	+	+	+	+	+	+	+
1st checkpoint										
⬤	Large Pre-B (Fr. C')	V_HDJ_H	Pre-BCR	−	+	+	+	+	+	+
⬤	Small Pre-B (Fr. D)	V_HDJ_H V_LJ_L rearrangement begins	Decreasing levels of pre-BCR	+	−	−	+	−	+	+
2nd checkpoint										
⬤	Immature B (Fr. E)	V_HDJ_H V_LJ_L	IgM	− / +	−	−	+	−	+	+

Ψ Pre-BCR
Ψ IgM

[1]Fractions (Fr.) A to E refer to the "Hardy nomenclature," described in Classic Experiment Box 9-2.
[2]GL = germ line arrangement of heavy-chain and/or light-chain V region segments.

FIGURE 9-4 Immunoglobulin gene rearrangements and expression of marker proteins during B-cell development. Shown are the expression of selected marker proteins and the timing of immunoglobulin gene rearrangements during stages of B-cell development, from the pre-pro-B cell stage to the immature B-cell stage. See text for details.

Many important B-cell genes are turned on at the pro-B-cell stage, under the control of PAX5 and other transcription factors. Among these are the genes encoding Igα and Igβ (CD79α,β) and CD19, which we first encountered in Chapter 3 (see Figure 3-14). Igα and Igβ, turned on in early pro-B cells, are the signaling chains of the membrane immunoglobulin B-cell receptors, and CD19, expressed in late pro-B cells, is one of the components of the B-cell coreceptor. Detailed studies of the control of expression of the *mb-1* gene, which encodes Igα, have revealed complex epigenetic changes underlying its induction. EBF1 and E2A contribute to the demethylation of the *mb-1* promoter, allowing PAX5 to bind, and changes mediated by several chromatin-remodeling complexes enable those transcription factors to activate gene expression. Induction of CD19 expression is also controlled epigenetically. Chromatin remodeling of the *Cd19* locus enhancer facilitates recruitment of E2A, followed by binding of EBF1 and PAX5, which finally activates *Cd19* gene transcription. Expression of Igα,β and CD19 will be essential for signaling through membrane Ig receptors and progression through later stages of B-cell development.

In addition to inducing transcription of key B-cell genes, PAX5 promotes V_H-to-DJ_H recombination by "contracting" the Ig_H locus, bringing the distant V_H gene segments closer to the rearranged DJ_H gene segments (see Chapter 6). B cells deficient in PAX5 are able to undergo D-to-J_H Ig gene rearrangement, but are not capable of rearranging a V_H to the already rearranged DJ_H Ig gene segment, indicating that PAX5 is essential to the second step of Ig gene rearrangement. By the late pro-B-cell stage, most cells have initiated V_H-to-DJ_H Ig gene segment recombination, which is completed by the onset of the early **pre-B cell (precursor B cell)** stage, allowing them to express μ heavy-chain protein.

The Stages of B-Cell Development: Characterization of the Hardy Fractions

Richard Hardy's laboratory was one of the first to combine flow cytometry and molecular biology in experiments designed to analyze lymphocyte maturation. In this feature, we describe what those researchers did and how they generated a model of the sequencing of the stages of B-cell development from their data. It provides an excellent example of how this approach can be employed to identify distinct developmental stages from a mixed population.

When Hardy and colleagues began their characterization of B-cell lineage development in the early 1980s, prior work using molecular analysis of long-term bone marrow cell lines had already established the sequential rearrangement of heavy-chain and light-chain immunoglobulin genes. In addition, the expression of a number of cell-surface markers on bone marrow cells had been measured, and several of these antigens had been shown to be co-expressed with B220 (CD45R), which had already been established as a marker on all B-lineage cells. Hardy's approach was to characterize the sequence of expression of cell-surface markers found on the same cells as B220. The hypothesis was that some of these markers might be expressed on early B-cell progenitors and might therefore help to generate a scheme of B-cell development. In order to place cells expressing different combinations of markers into a developmental lineage, Hardy and colleagues sorted cells bearing each combination of their selected markers, and placed them into cocultures with a bone marrow stromal cell line. After defined times in culture, they harvested the subpopulations and recharacterized their surface marker expression. They also characterized heavy- and light-chain genes to determine when the various gene rearrangements occur.

The markers used in these experiments included B220 (CD45R) and CD43 (leukosialin), which had previously been shown to be expressed on granulocytes and all T cells but was not present on mature B cells, with the exception of plasma cells. In addition, their experiments employed antibodies directed against heat-stable antigen, or HSA (CD24) and BP-1 (aminopeptidase A), an antigen on bone marrow cells. Both HSA and BP-1 had been previously shown to be differentially expressed at various stages of lymphoid differentiation.

The first experiments analyzed cells for both B220 and CD43 (**Figures 1 and 2a**). While most B220$^+$ bone marrow cells do not express CD43, a small population is CD43$^+$. The expression of HSA and BP-1 on these CD43$^+$ cells was then examined. The flow cytometry plot demonstrated that the B220$^+$CD43$^+$ cells neatly resolved into three discrete subpopulations, or fractions. The first, labeled A in Figures 1 and 2b, expressed neither HSA nor BP-1. The second, labeled B, expressed HSA but not BP-1, and the third expressed both of these antigens. Analysis of Ig gene rearrangements in these populations (isolated by cell sorting) revealed that no gene rearrangements occurred in fraction A (now known as pre-pro-B cells) but that D-to-J$_H$ gene segment rearrangements had begun in fraction B (now known as early pro-B cells). Subsequent work has shown that V$_H$-to-DJ$_H$ rearrangements occur in fraction C (late pro-B cells, although at the time, the method of analysis that Hardy and colleagues used failed to reveal this rearrangement).

Culture of fraction C cells yielded cells that expressed membrane (m) μ heavy chains; similarly, culture of fraction B cells also yielded daughter cells expressing μ heavy chains, but at a lower frequency than fraction C cells, suggesting that cells in fraction C were further along the

differentiation pathway to B cells. Furthermore, the three different fractions displayed differential dependence on the need to adhere to the stromal cell layer. Cells from fraction A required stromal cell contact for survival. Fraction B cells survived best in contact with the stromal cells, but were able to survive in a culture in which they were separated from the stromal cells by a semipermeable membrane. Under these conditions, they could still receive soluble factors generated by the stromal cells, but were prevented from generating adhesive interactions with stromal cell-surface-bound growth factors. Fraction C cells survived and proliferated in the absence of stromal cell contact. Analysis of the factors secreted by the stromal cells that were necessary for survival and proliferation of the fraction B and C cells revealed one of them to be interleukin 7.

Hence, using the criteria of Ig gene rearrangements and phenotypic analysis of cultured cell populations, Hardy and colleagues were able to place the three fractions in sequence; cells of fraction A gave rise to cells of fraction B, which in turn mature into cells of fraction C. Careful analysis of the contour graph of fraction C reveals that it, in turn, can be subdivided on the basis of the levels of expression of HSA. That population of cells bearing higher levels of HSA, as well as BP-1, is now defined as fraction C', which corresponds to early or large pre-B cells.

Hardy and colleagues next turned their attention to those cells that expressed B220 but had lost CD43, and measured their cell-surface expression of mIgM (Figure 2c). Three populations of cells were again evident, which they labeled D, E, and F. Cells belonging to fraction D expressed zero to low levels of mIgM, showed complete heavy-chain rearrangement and some light-chain rearrangement, and correspond to small

(continued)

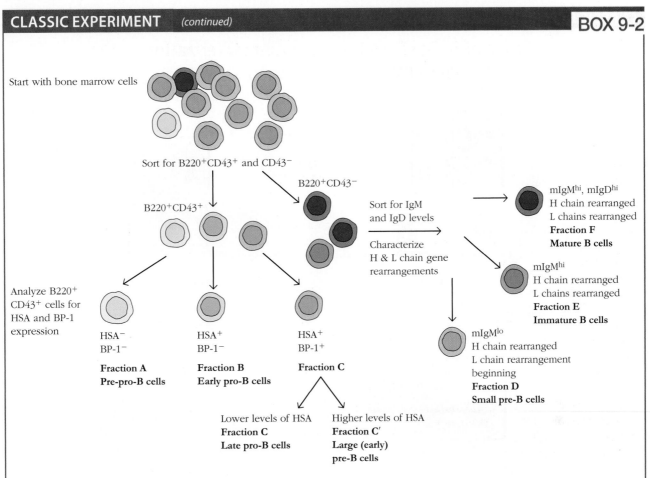

FIGURE 1 **Experimental approach for the isolation of Hardy's fractions from bone marrow.** Bone marrow cells were stained with antibodies to B220 and CD43 tagged with different fluoro-chromes and sorted for cells bearing B220 and CD43. The CD43$^+$ cells were then analyzed for their expression of the cell-surface markers HSA and BP-1, revealing populations A, B, C, and C'. The CD43$^-$ cells were analyzed for different levels of expression of mIgM and IgD (not shown), revealing populations D, E, and F.

pre-B cells. Cells belonging to fraction E displayed high levels of mIgM as well as of B220, complete heavy-chain rearrangement, and most of the cells in that fraction also displayed light-chain gene rearrangement. Fraction E cells are thus immature B cells ready to leave the bone marrow. Subsequent further characterization of fraction F cells showed that, in addition to surface IgM, some of these cells also expressed surface IgD and therefore represented fully mature B cells, presumably recirculating through the bone marrow.

Thus, Hardy's experiments revealed that the pool of progenitor and precursor B cells in the bone marrow represents a complex mixture of cells at different stages of development, with varying requirements for stromal cell contact and interleukin support.

These elegant experiments still had one more story to tell that did not appear in the original paper, but that emerged in later publications. Single-cell PCR analysis of fraction C cells showed that many of them had nonproductive rearrangements on both heavy-chain chromosomes (see Figure 2). In contrast, all the cells from the C' fraction demonstrated productive rearrangements on one of the heavy-chain chromosomes. Fraction C cells therefore represent B cells that have been unsuccessful in productively rearranging either of their heavy-chain genes and that will therefore eventually die by apoptosis. The C' fraction also included the highest proportion of cells in the cell cycle (i.e., dividing and getting ready to divide), proportionally more cells than in all of the B220$^+$ B-cell stages in the marrow. This is consistent with the notion that the new heavy chain has associated with the surrogate light chain at the C' stage, forming the pre-B-cell receptor complex that is expressed on the cell surface and triggers the period of clonal expansion of B cells described in this chapter.

(continued)

CLASSIC EXPERIMENT   *(continued)*   **BOX 9-2**

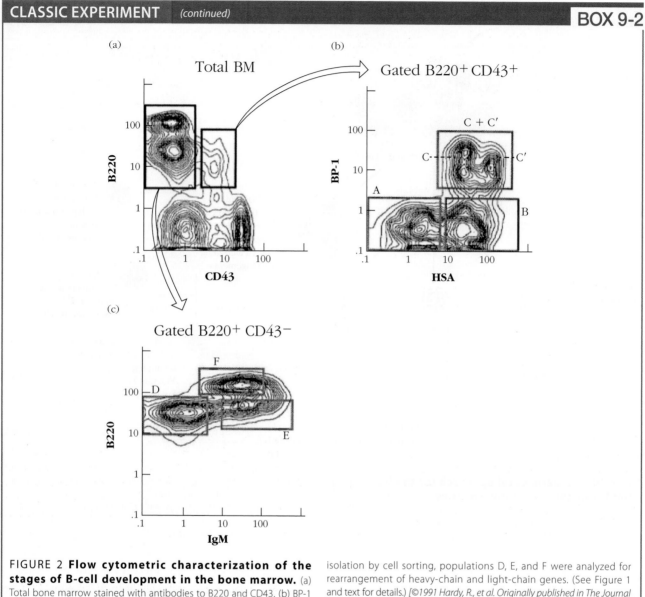

FIGURE 2 **Flow cytometric characterization of the stages of B-cell development in the bone marrow.** (a) Total bone marrow stained with antibodies to B220 and CD43. (b) BP-1 and HSA expression on the B220⁺CD43⁺ population from panel (a). (c) B220 and IgM expression on the B220⁺CD43⁻ population. After isolation by cell sorting, populations D, E, and F were analyzed for rearrangement of heavy-chain and light-chain genes. (See Figure 1 and text for details.) *[©1991 Hardy, R., et al. Originally published in The Journal of Experimental Medicine. 173:1213–1225. DOI:10.1084/jem.173.5.1213, Figures 1 and 2.]*

REFERENCES

Hardy, R. R., C. E. Carmack, S. A. Shinton, J. D. Kemp, and K. Hayakawa. 1991. Resolution and characterization of pro-B and pre-pro-B cell stages in normal mouse bone marrow. *Journal of Experimental Medicine* **173**:1213.

Hardy, R. R., P. W. Kincade, and K. Dorshkind. 2007. The protean nature of cells in the B lymphocyte lineage. *Immunity* **26**:703.

Pre-B Cells

Among the genes turned on in pro-B cells by EBF1 and PAX5 are those encoding **VpreB** and **λ5**, which together comprise the **surrogate light chain (SLC)**. Two SLCs pair with two μ heavy chains, forming the **pre-B-cell receptor (pre-BCR)**. The genes for VpreB and λ5 are not included in the light-chain gene families. VpreB is homologous to a light-chain V-region domain, and λ5 is homologous to

λ light-chain J+C sequences. Both, however, have additional sequences: VpreB has an extra 25 amino acids (including multiple acidic [negatively charged] amino acid residues) at its C terminus, whereas λ5 has an extra 50 amino acids (enriched in positively charged arginine residues) at its N terminus (**Figure 9-5a**). These unusual polypeptide regions contribute to the β sheets in the Ig-like domains of the surrogate light chain and also extend over what would

(a)

(b)

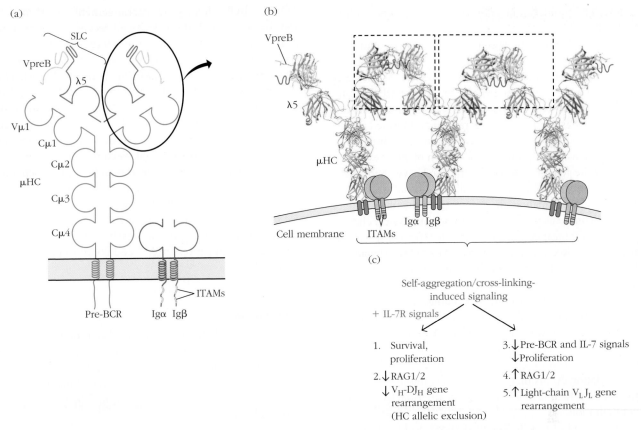

(c)

Self-aggregation/cross-linking-
induced signaling

+ IL-7R signals

1. Survival,
   proliferation
2. ↓RAG1/2
   ↓V_H-DJ_H gene
   rearrangement
   (HC allelic exclusion)

3. ↓Pre-BCR and IL-7 signals
   ↓Proliferation
4. ↑RAG1/2
5. ↑Light-chain $V_L J_L$ gene
   rearrangement

FIGURE 9-5 The pre-B-cell receptor. (a) Diagram showing the composition of the pre-BCR: two μ heavy chains (blue) associate with two surrogate light chains, each made up of λ5 (a light-chain J+C-like domain; red) and VpreB (a light-chain V-like domain; yellow), with extra peptide tails at the C terminus of VpreB and at the N terminus of λ5. (b) Models of oligomers of pre-BCR, suggested by X-ray crystallography and electron microscopy, involving the charged VpreB and λ5 tails, creating dimers at two different angles (shown in dashed boxes). This cross-linking induces signaling through the associated Igα/Igβ chains. (c) Sequence of events in pre-B cells triggered by pre-BCR signaling activated by aggregation/cross-linking; events 1 and 2 also require signals through the IL-7 receptor (in mice but not in humans). See text for details. *[Part (b) data from Bankovich, A. J., et al. 2007. Structural insight into pre-B cell receptor function. Science **316**:291.]*

be the antigen-binding site, covering the heavy-chain complementarity-determining region 3 (CDR3) and preventing it from recognizing any antigen. Studies revealed that multiple pre-BCRs self-aggregate, probably initially in the endoplasmic reticulum, and they are aggregated on the pre-B-cell surface (Figure 9-5b). The aggregation appears to be caused by charge interactions between the oppositely charged tails of VpreB and **λ5.** This ligand-independent self-aggregation initiates signaling through the Igα/Igβ signaling chains, with contributions from the CD19 coreceptor, similar to that activated by the BCR of mature B cells (see Chapter 3). Pre-BCR signals initiate a sequence of events essential for B-cell development. In mice, signals through the IL-7R are also required; animals deficient in the expression of either the pre-B-cell receptor or the signaling components Igα/Igβ fail to progress to the pre-B-cell stage. In humans, signals from the IL-7R are not required for B-cell development, but defects in SLC and the resulting absence of the pre-BCR block B-cell development and result in severe immunodeficiency.

Events triggered by self-aggregation of the pre-BCR are shown in Figure 9-5c. Signaling through the pre-BCR combined (in mice) with signals following binding of IL-7 to the IL-7R induce several (two to seven) rounds of proliferation. The large size of proliferating cells has led these early pre-B cells to be called **large pre-B cells.** Cells that have not productively rearranged and expressed a μ heavy chain, one that can associate with both another μ heavy chain and surrogate light chains to form a pre-BCR that transmits signals, undergo apoptosis. Thus, expression of a functional pre-BCR constitutes the **pre-B-cell (first) checkpoint** in B-cell development (see Figure 9-4).

The proliferation of pre-B cells that have achieved productive heavy-chain expression generates a pool of daughter cells, all with the same heavy chain, but each will undergo distinct light-chain rearrangement events. Thus, many different receptor specificities will be generated from each successful heavy-chain rearrangement. Recall from Chapter 8 that an analogous process of β-chain rearrangement, followed by proliferation prior to α rearrangement, occurs in T cells.

Pre-B-cell receptor signaling induces the transient down-regulation of RAG1/2 (both by inhibiting further transcription of *RAG* genes and by proliferation-associated degradation of RAG protein) and the loss of TdT activity. Together these events ensure that, as soon as one heavy-chain gene has productively rearranged and formed a functional pre-BCR, no further heavy-chain recombination is possible. This results in the phenomenon of *allelic exclusion*, whereby the genes of only one of the two heavy-chain alleles can be expressed in a single B cell. As a result of this pre-B-cell receptor signaling, the chromatin at the unrearranged heavy-chain locus undergoes a number of physical changes that render it incapable of participating in further rearrangement events. Recall that IL-7 provided one of the signals that brought the V_H, D, and J_H loci into close apposition with one another at the beginning of $V_H DJ_H$ recombination. A reduction in IL-7R expression due to changes in transcription factors at the pre-B-cell stage now reverses that initial locus contraction, resulting in the physical separation of the V_H, D, and J_H gene segments in the unrearranged heavy-chain locus. This decontraction is then followed by chromatin deacetylation events that deactivate the unused heavy-chain locus and return it to a heterochromatic (inactive, closed) configuration.

At the end of pre-BCR-activated cell proliferation, several changes lead to the loss of pre-BCR expression (see Figure 9-5c). Surrogate light-chain gene transcription is terminated by a negative feedback round of signaling through the pre-B-cell receptor, which causes the displacement of EBF1 from the λ5 and VpreB promoters. In addition, pre-existing pre-BCR proteins on the cells' surfaces are diluted out by cell divisions. IL-7R signaling also declines. Once the cells cease proliferating, they enter the late or **small pre-B-cell** stage.

Cells then commence light-chain gene recombination. In the mouse, κ light-chain V and J gene segments rearrange first, before λ. These gene segments had undergone pre-BCR-induced epigenetic changes typical of active, expressed chromatin, such as histone modifications. Newly expressed transcription factors IRF4 and IRF8 and FOXO1 stimulate the initiation of light-chain rearrangement, made possible by the re-expression of the RAG1/2 proteins, a late consequence of pre-BCR signaling. If the first κ-chain gene segment rearrangement is nonproductive, rearrangement on the second chromosome commences. If neither κ-chain rearrangement is successful, rearrangement is then successively attempted on each of the λ-chain chromosomes (see Figure 6-15). In humans, rearrangement is initiated randomly at either the κ or the λ loci. Very little TdT activity remains at the small pre-B-cell stage (see Figure 6-15), and therefore N region addition occurs much less frequently in light chains than in heavy chains.

Once a light-chain gene rearrangement has been successfully completed, the IgM B-cell receptor is expressed on the cell surface and signals the cell (apparently spontaneously, without ligand binding or self-aggregation) to terminate any further light-chain gene rearrangements. The cell is now an **immature B cell**, defined by the expression of membrane IgM. If the attempts at light-chain immunoglobulin gene rearrangement (at both κ and both λ loci) are not successful, the nascent cell dies by apoptosis; this constitutes the **immature B-cell (second) checkpoint** (see Figure 9-4). However, given the availability of four separate chromosomes on which to attempt light-chain rearrangement, and the opportunity for light-chain editing in the case of unproductive rearrangement (discussed shortly), most pre-B cells that have successfully rearranged their heavy chains will express mIgM and go on to form immature B cells.

> **Key Concepts:**
>
> - Stages of B-cell development can also be defined by the status of immunoglobulin gene rearrangements. The heavy-chain V genes rearrange first in pre-pro- and pro-B cells, with D-to-J_H recombination occurring initially, followed by V_H-to-DJ_H recombination.
>
> - The heavy chain is then expressed on the cell surface in combination with the surrogate light chain, which is made up of VpreB and λ5. Together they form the pre-B-cell receptor, which is expressed on the cell surface along with the Igα/Igβ signaling complex.
>
> - Signals from the pre-B-cell receptor and, in mice, the IL-7 receptor stop V_H gene rearrangement (ensuring heavy-chain allelic exclusion), confer a survival signal, and activate several rounds of cell division, followed by light-chain gene rearrangement. The cell divisions allow multiple B cells to use the same successfully rearranged heavy chain in combination with many different light chains. Expression of the pre-BCR and initiation of these events constitutes the pre-B-cell (first) checkpoint in B-cell development.
>
> - After light-chain rearrangement at the small pre-B-cell stage, expression of the completed mIgM B-cell receptor on the cell surface of immature B cells shuts down further light-chain gene rearrangement and confers a survival signal, constituting the immature B-cell (second) checkpoint in the formation of mature B cells.

Immature B Cells in the Bone Marrow Are Exquisitely Sensitive to Tolerance Induction through the Elimination of Self-Reactive Cells

Immature B cells bear a functional receptor in the form of membrane IgM but have not yet begun to express membrane IgD (present along with membrane IgM on mature naïve B cells) or any other class of immunoglobulin. They continue to express B220 and CD19 (see Figure 9-4).

Once the functional BCR is assembled on the B-cell membrane the receptor must be tested for whether it binds self antigens, in order to ensure that as few as possible autoreactive B cells emerge from the bone marrow. Those immature B cells that bear autoreactive receptors undergo one of three fates. Some are lost from the repertoire prior to leaving the bone marrow by the BCR-mediated induction of apoptosis, resulting in clonal deletion. Other autoreactive B cells reactivate their *RAG* genes to initiate the process of light-chain receptor editing (see Chapter 6). The loss of B cells bearing self-reactive receptors *within the bone marrow* by either of those mechanisms is referred to as **central tolerance**. As we will see later, some autoreactive B cells that recognize soluble self antigens within the bone marrow may survive to escape the bone marrow environment, but become **anergic**, or unresponsive, to any further antigenic stimuli.

Our understanding of how the immune system eliminates or neutralizes autoreactivity has been facilitated by the development of transgenic animals that express both deliberately introduced auto-antigens and the receptors that recognize them. Immature B cells are very susceptible to the induction of apoptosis, at least in part because they express low levels of the anti-apoptotic proteins Bcl-2 and Bcl-x$_L$. It has long been known that cross-linking the IgM receptors of immature B cells in vitro (performed experimentally by treating the cells with antibodies against the receptor μ chain) results in death by apoptosis. In contrast, performing the same experiments with mature B cells results in B-cell activation. David Nemazee and colleagues set out to test whether the apoptotic response of immature B cells in vitro reflected what happens in the bone marrow in vivo when an immature B cell meets a self antigen.

The approach taken by Nemazee and colleagues was conceptually simple, although experimentally complex, particularly for the time period in which the work was done (1989). They generated mice transgenic for both a heavy chain and a light chain specific for the MHC molecule H2-K^k. All the B cells in this mouse therefore had BCRs specific only for H2-K^k and could make only anti-H2-K^k-specific antibodies. If immature B cells undergo selection to prevent autoimmunity, these cells would be selected against in a mouse that expresses the H2-K^k MHC class I protein. By appropriate breeding, they introduced the H2-K^k-specific heavy- and light-chain transgenes into mice of two different MHC haplotypes (genotypes).

In mice in the first group (**Figure 9-6a**), whose cells expressed H2-K^d but not H2-K^k proteins, the investigators were able to detect the transgenic BCR at high frequency on the surface of B cells and as serum antibodies at high concentrations (**Table 9-1**). This makes sense, as the transgenic BCR would not recognize H2-K^d molecules, and so the B cells producing it would not be negatively selected. However, when these animals were bred to mice of the *H2^k* haplotype (Figure 9-6b), only low levels of membrane-bound BCR and secreted anti-H2-K^k antibodies could be detected, suggesting that all immature B cells bearing the potentially autoimmune receptor antibodies had been deleted by apoptosis in the bone marrow, as would be expected for negative selection.

Interestingly, in *H2$^{k/d}$* heterozygous mice, not all B cells were deleted, even though all B cells in these mice should bear the transgenic μ and light chains encoding the anti-H2-K^k receptor (Figure 9-6c). Closer examination revealed that some of the residual B cells in the bone marrow had undergone *light-chain receptor editing* (see Chapter 6), rearranging V and J gene segments upstream (5′) and downstream (3′), respectively, of the originally rearranged V and J, which were deleted. The new light chains changed the antigen specificity of the BCRs so that they no longer bound the H2-K^k self protein in the bone marrow. Recent experiments suggest that receptor editing (of either light chains or heavy chains; see Chapter 6) is more prevalent in vivo than clonal deletion as the mechanism by which self-reactive B cells in the bone marrow are eliminated prior to the release of immature B cells into the periphery. These studies suggest that those cells that fail to replace their heavy or light chain are the ones that undergo apoptosis, which thus represents a back-up mechanism.

In normal animals, not all potentially autoimmune B cells are lost to clonal deletion or altered via receptor editing within the bone marrow. Those mechanisms of central tolerance occur when the immature B cells receive strong signals through their mIgM BCR, such as when the receptors are cross-linked by binding to self antigens on the surface of stromal cells (this would be the case with MHC class I proteins in the Nemazee experiments). As we will see in the next section, some self-reactive immature B cells (those that recognize soluble self antigens and those that recognize self antigens not found in the bone marrow) are released to the periphery and are subject to additional processes that ensure self-tolerance.

(a)

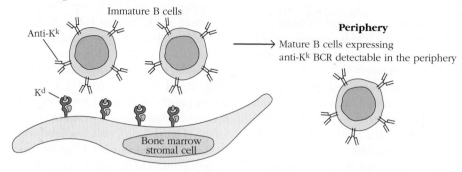

Bone Marrow

$H2^{d/d}$ mice transgenic for anti-K^k-specific BCR:
no recognition of self K^d

Immature B cells

Anti-K^k

K^d

Bone marrow
stromal cell

Periphery

Mature B cells expressing
anti-K^k BCR detectable in the periphery

(b)

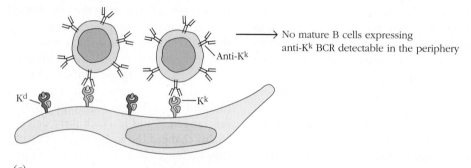

$H2^{k/d}$ mice transgenic for anti-K^k-specific BCR:
recognition of self K^k

Anti-K^k

K^d K^k

No mature B cells expressing
anti-K^k BCR detectable in the periphery

(c)

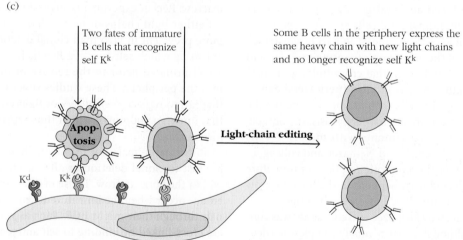

Two fates of immature
B cells that recognize
self K^k

Apop-
tosis

K^d K^k

Light-chain editing

Some B cells in the periphery express the
same heavy chain with new light chains
and no longer recognize self K^k

FIGURE 9-6 Experimental evidence for negative selection (clonal deletion) and light-chain editing of self-reactive immature B cells in the bone marrow. The presence or absence of mature peripheral B cells expressing a transgene-encoded IgM BCR that reacts with the H2 MHC class I molecule K^k was determined in $H2^{d/d}$ mice (a) and in $H2^{d/k}$ mice (b and c). (a) In the $H2^{d/d}$ mice transgenic for the K^k-specific BCR, there was no self antigen for the immature B cells to bind to and consequently they went on to mature, so that splenic B cells expressing the transgene-encoded anti-K^k as membrane Ig were found in the periphery. (b) In the $H2^{d/k}$ heterozygous mice transgenic for the K^k-specific BCR, there were no mature B cells in the periphery expressing the BCR specific for the self antigen K^k protein. (c) More detailed analysis revealed two fates of the transgenic B cells in the $H2^{d/k}$ mice. Recognition of self-K^k induced many to undergo apoptosis. There were also some peripheral B cells that expressed the transgene-encoded μ chain but a different light chain, resulting from light-chain editing in some of the immature B cells. With the new light chains these B cells no longer bound the K^k molecule and consequently escaped negative selection.

TABLE 9-1	Expression of transgene-encoded IgM antibody specific for H2-K^k MHC class I protein		
		EXPRESSION OF ANTI-H2K^K IgM TRANSGENE	
Experimental animal	**Number of animals tested**	**As membrane Ig**	**As secreted Ab (μg/ml)**
Nontransgenics	13	(−)	<0.3
$H2^d$ transgenics	7	(+)	93.0
$H2^{d/k}$ transgenics	6	(−)	<0.3

[Data from Nemazee, D. A., and K. Bürki. 1989. Clonal deletion of B lymphocytes in a transgenic mouse bearing anti-MHC class I antibody genes. Nature **337**:562.]

Key Concepts:

- Immature B cells are very sensitive to the induction of self-tolerance through the elimination of self-reactive cells.

- Self-tolerance mechanisms in B cells were investigated using mice transgenic for different self antigens and for BCRs recognizing self antigens.

- Immature B cells whose BCRs are strongly activated by self antigens present in the bone marrow are induced to undergo receptor editing to change the specificity of their BCRs.

- Self-reactive B cells that have not undergone receptor editing are deleted by apoptosis.

- These processes constitute central tolerance mechanisms for B cells.

Completion of B-Cell Development in the Spleen

Immature B cells are recruited to leave the bone marrow by their expression of the S1P receptor, which recognizes the lipid chemoattractant *sphingosine 1-p*hosphate (S1P) in the blood (see Figure 9-2). These cells then migrate to the spleen, where they complete their development into mature B cells.

The study of B-cell development in the periphery, like that in the bone marrow, has benefited significantly from the use of flow cytometry to define specific cell populations based on their expression of distinct marker proteins. This enabled the separation of cells derived from immature B cells into two subpopulations of **transitional B cells** (T1 and T2). These transitional B cells differentiate in a series of steps into fully mature B cells (note: these are conventional, or B-2 cells; the formation of B-1 cells will be described in the next section). A third population of peripheral B cells, called T3 transitional B cells, seems to be a group of self-reactive cells that are unresponsive (anergic) to peripheral self antigens.

T1 and T2 Transitional B Cells Form in the Spleen and Undergo Selection for Survival and against Self-Reactivity

T1 and T2 transitional B cells in the spleen were initially characterized on the basis of their cell-surface expression of immunoglobulin receptors and other membrane markers (**Table 9-2**). T2 B cells differ from T1 B cells in having higher levels of membrane IgD and in expressing CD21 (the complement receptor and B-cell coreceptor; see Figure 3-14) and CD23. T2 B cells also express the **BAFF-receptor (BAFF-R)**, a receptor for B-cell survival factor **BAFF** (*B*-cell *a*ctivating *f*actor belonging to the tumor necrosis factor *f*amily); BAFF-R expression is dependent on signals received through the BCR. As B cells differentiate from the T2 transitional state to full maturity, their levels of mIgD increase still further, while the expression of mIgM decreases. Mature B cells also cease to express CD24 and CD93.

TABLE 9-2	Surface marker expression on transitional T1 and T2 B cells and on mature B-2 B cells		
Marker	**T1**	**T2**	**Mature B-2 cells**
mIgM	High	High	Intermediate
mIgD	−/low	Intermediate	High
CD24	+	+	−
CD93	+	+	−
CD21	−	+	+
CD23	−	+	+
BAFF receptor	+/−	+	+

Note: CD93 is also found on monocytes, granulocytes, and endothelial cells. CD24 is otherwise known as the *heat-stable a*ntigen (HSA). CD23 is a low-affinity receptor for IgE. CD21 is a receptor for complement and part of the B-cell coreceptor.

[Data from Allman, D., and S. Pillai. 2008. Peripheral B cell subsets. Current Opinion in Immunology **20**:149; and others.]

T1 B cells that have been labeled and transferred into recipient mice develop into T2 B cells, and both T1 and T2 B cells were shown to be capable of differentiating into mature B cells. These experiments therefore showed that the order of the developmental sequence progresses from T1 to T2 to mature B cell. The time in transit of a T1 cell to a mature B cell has been measured to be approximately 3 to 4 days. Most T1 B cells differentiate to T2 B cells within the spleen, but a minority (about 25%) of transitional B cells emerge from the bone marrow already in the T2 state. The increased level of maturity of T2 B cells correlates with changes in the expression of chemokine and cytokine receptors, such that T2 B cells, but not T1 B cells, are capable of recirculating among the blood, lymph nodes, and spleen. T2 B cells, but not T1 B cells, can enter B-cell follicles in the lymph nodes and spleen.

Figure 9-7 shows the path of the developing mouse B cell as it leaves the bone marrow, enters the spleen through the central arteriole, and is deposited in the marginal sinuses, just inside the outer marginal zone. (In humans, the anatomy of the spleen is slightly different, and the cells arrive in the spleen in a perifollicular zone.) From there, the T1 B cells percolate through the spleen to the T-cell zone, where

some fraction of the T1 cells will mature into the T2 stage. T2 B cells are then able to enter the follicles, where they complete their development into fully mature, recirculating conventional (B-2) B lymphocytes. Some T2 B cells instead enter the marginal zone and become marginal zone B cells (see Figure 9-7), whose properties will be discussed shortly.

In Chapter 6, we learned that mature B cells bear on their surfaces two classes of membrane-bound immunoglobulins—IgM and IgD—and that the expression of mIgD along with mIgM requires carefully regulated alternative mRNA splicing events. It is at the point of transition between the T1 and T2 stages of development that we observe the onset of splicing that generates the mRNA for the δ heavy chain. Mature B cells bear almost 10 times more mIgD than mIgM, and so mIgD expression results in significant up-regulation in the number of B-cell immunoglobulin receptors.

As mentioned previously, immature B cells whose receptors recognize peripheral self antigens can emerge from the bone marrow and potentially be autoreactive, so mechanisms that eliminate or inactivate them are needed to maintain self-tolerance. The effect of strong BCR engagement with a multivalent, or membrane-bound, self antigen depends on the maturational status of the transitional B cell

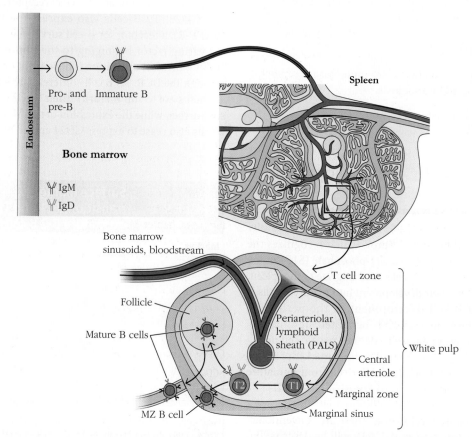

FIGURE 9-7 T2, but not T1, transitional B cells can enter splenic B-cell follicles and recirculate. Immature B cells leave the bone marrow via the blood. As T1 transitional immature B cells they enter the splenic marginal sinuses, percolating into the T-cell zones. There they differentiate into T2 transitional B cells, which gain the ability to enter the B-cell follicles, where they complete their differentiation into mature follicular B cells and recirculate in the blood. Marginal zone cells have also been shown to derive from T2 B cells.

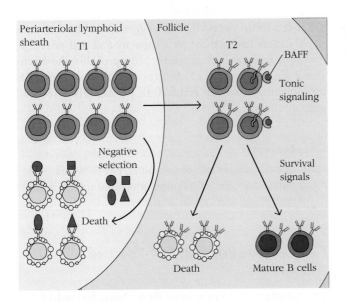

FIGURE 9-8 Transitional B cells undergo positive and negative selection in the spleen. T1 transitional B cells that recognize antigen, including self antigen, with high affinity in the spleen are eliminated by negative selection and never reach the splenic follicles. Those T1 B cells that escape negative selection enter the follicles and differentiate into T2 B cells. In the follicles their BCRs tonically deliver a stimulatory survival signal, either without binding a ligand or from interactions with an unknown molecule(s). T2 B cells that have received this survival signal up-regulate their BAFF receptors and bind BAFF, also contributing to survival. Survival due to these two sets of signals constitutes positive selection. Those T2 B cells that fail to receive these stimulatory signals die in the spleen. Selecting antigens are shown as violet shapes; T1 and T2 B cells are green. Light green cells represent dead cells that were either negatively selected or failed positive selection.

TABLE 9-3	Responses to strong BCR signaling in T1, T2, and mature B-2 B cells		
Nature of response	**T1**	**T2**	**Mature B-2 B cells**
Formation of lipid rafts	+/−	+	+
Increase in cytoplasmic Ca^{2+} ion concentrations	+	+	+
Increase in diacylglycerol concentrations	−	+	+
Induction of $Bcl\text{-}x_L$	−	+	+
Induction of apoptosis	+	−	−

(**Figure 9-8** and **Table 9-3**). Self-reactive T1 B cells are eliminated by apoptosis in response to a strong BCR signal, in a process reminiscent of the negative selection of thymocytes and of bone marrow immature B cells recognizing membrane self antigens. Thus this constitutes one major form of peripheral tolerance for B cells. Recent experiments have suggested that in healthy adults, fully 55% to 75% of T1 transitional B cells are lost by this process. In contrast, once the B cell has matured into a T2 transitional B cell, it becomes resistant to antigen-induced apoptosis, reminiscent of thymocytes that have reached the single-positive stage of development. This resistance to receptor-induced cell death results in part from the fact that T2 B cells have increased their expression of the anti-apoptotic molecule Bcl-x_L.

Using conditional knockout genetic techniques (see Chapter 20), animals can be generated that lack mIg receptor expression at various stages of B-cell development. Animals that fail to express a BCR during the immature B-cell stage lose the capacity to make any mature B cells at all. This indicates that some low level of *tonic* (i.e., antigen-independent)

signaling through the BCR is required for continued generation and survival of immature B cells. This signal appears to be constitutive and spontaneous, not dependent on the binding of any antigen to the cells' BCRs. B cells unable to receive this signal die at the T2 stage (see Figure 9-8). Those T2 B cells that receive the follicular signal up-regulate the expression of the receptor for the B-cell survival factor BAFF.

Given the different outcomes of signaling through the BCR for T1 versus T2 B cells, it is clear that there must be *differences in the signaling pathways* downstream of the BCR in the two transitional B-cell types, and such differences have been observed. Specifically, BCR-mediated signaling in T1 B cells results in calcium release without significant production of diacylglycerol, and provides an apoptotic signal. In contrast, receipt of BCR signals by T2 B cells induces both an increase in the concentration of intracytoplasmic calcium and in diacylglycerol production. This combination of intracellular second messengers delivers both maturational and survival signals to the cells, and suggests the involvement of a diacylglycerol-activated protein kinase in survival signaling (see Chapter 3).

But what causes this difference in the signal transduction pathways between T1 and T2 B cells? A partial answer to this question may lie in differences in the composition of the lipid membranes of the two types of cells. Immature T1 B cells contain approximately half as much cholesterol as their more mature counterparts, and this reduction in cholesterol levels appears to prevent efficient clustering of the B-cell receptor into lipid rafts during BCR stimulation. This may cause a reduction in the strength of BCR signaling in T1 versus T2 immature B cells.

The development of B cells through the transitional phase is absolutely dependent on signaling through the BAFF receptor. BAFF-R expression is first detected in T1 B cells and increases steadily thereafter. BAFF is then required constitutively throughout the life of mature B cells. Signaling through the BAFF/BAFF-R axis promotes survival of transitional B cells by inducing the synthesis of anti-apoptotic factors such as Bcl-2, Bcl-x$_L$, and Mcl-1, as well as by interfering with the function of the pro-apoptotic molecule Bim.

The discovery of a B-cell survival signal mediated by BAFF/BAFF-R interactions, distinct from survival signals transmitted from the BCR, extends our thinking about how B cells are selected for survival in the periphery. In the presence of high levels of BAFF, B cells that may not otherwise receive sufficient quantities of survival signals via the BCR may survive a selection process that would otherwise eliminate them. In this way, BAFF can provide plasticity and flexibility in the process of B-cell deletion. BAFF is produced by macrophages and dendritic cells in response to certain cytokines, and hence BAFF levels may increase during infections. That would be a situation where sustaining the numbers of B cells that potentially could respond to the pathogens would be helpful. However, this may be accomplished at the cost of maintaining potentially autoreactive cells.

Key Concepts:

- Immature B cells that migrate from the bone marrow to the spleen are called transitional 1 (T1) B cells. Interaction with self antigens in the spleen can induce these cells to undergo apoptosis, thus representing negative selection.

- T1 B cells then enter the follicles where the level of IgD expression increases, and they become T2 B cells.

- Signals from the cytokine BAFF binding to the BAFF receptor are necessary for the survival of transitional and mature B cells.

T2 B Cells Give Rise to Mature Follicular B-2 B Cells

Fully mature conventional B-2 cells express high levels of IgD and intermediate levels of IgM on their cell surfaces (see Table 9-2). Mature B cells recirculate between the blood and the lymphoid organs, entering the B-cell follicles in the lymph nodes and spleen; hence B-2 B cells are often called *follicular B cells*. In the follicles they respond to antigen encounter, in the presence of T-cell help, by generating antibody responses (Chapter 11). Approximately 10 to 20 million B cells are produced in the bone marrow of a mouse each day, but only about 10% of this number ever take up residence in the periphery and only 1% to 3% will ever enter the recirculating follicular B-2 B-cell pool. Some of the peripheral B cells are lost to the process of clonal deletion, but others are perfectly harmless B cells that nonetheless fail to thrive. Experimental depletion of the mature B-cell population, either chemically or by irradiation, followed by in vivo reconstitution, results in rapid replenishment of

the B-cell follicular pool. This suggests that the follicular B-cell niches have a maximum capacity and that once full, they turn away additional B cells. Most probably, the mechanism for this homeostatic control of B-cell numbers relies on competition for survival factors, particularly BAFF and related proteins.

Key Concept:

- T2 B cells mature into conventional B-2 cells that populate the follicles of lymph nodes and spleen, and hence are called *follicular B cells*.

T3 B Cells Are Primarily Self-Reactive and Anergic

T3 transitional B cells were first characterized in the blood and lymphoid organs by flow cytometry, and were described as being CD93$^+$mIgD^{high}mIgM^{low}CD23$^+$. Recent experiments have suggested that the T3 population may represent transitional B cells that have been rendered anergic by contact with soluble self antigen in the spleen (and possibly elsewhere) but have not yet been eliminated from the B-cell repertoire.

A transgenic system developed by Goodnow and colleagues first placed the concept of B-cell anergy, or unresponsiveness, onto a firm experimental footing. Anergic lymphocytes clearly recognize their antigens, as shown by the identification of low levels of molecular signals generated within the cells after binding to antigen. However, rather than being activated by antigen contact, anergic B cells fail to divide, differentiate, or secrete antibody after stimulation, and many die a short time after receipt of the antigenic signal.

Goodnow and colleagues developed the two groups of transgenic mice illustrated in **Figure 9-9a**. One group of mice carried a *hen egg-white lysozyme* (HEL) transgene linked to a metallothionein promoter, which placed transcription of the *HEL* gene under the control of zinc levels in the animals' diet. This allowed the investigators to alter the levels of soluble HEL expressed in the experimental animals by changing the concentration of zinc in their food. Under these experimental conditions, HEL was expressed in the periphery of the animal, but not in the bone marrow. The other group of transgenic mice carried rearranged immunoglobulin heavy-chain and light-chain transgenes encoding an anti-HEL antibody. In these transgenic mice, the rearranged anti-HEL transgene is expressed by 60% to 90% of the mature peripheral B cells. Goodnow then mated the two groups of transgenics to produce "double-transgenic" offspring carrying both the HEL and anti-HEL transgenes (Figure 9-9b) and asked what effect peripheral HEL expression would have on B cells expressing the anti-HEL BCR. They found that the double-transgenic mice continued to

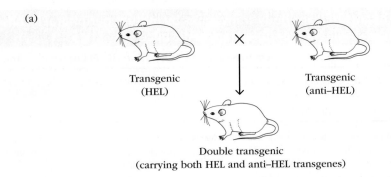

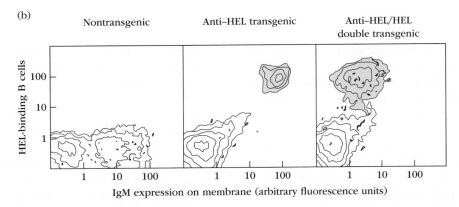

FIGURE 9-9 Goodnow's experimental system for demonstrating clonal anergy in mature peripheral B cells. (a) Production of double-transgenic mice carrying transgenes encoding HEL (hen egg-white lysozyme) and anti-HEL antibody. (b) Flow cytometric analysis of peripheral B cells for levels of membrane IgM and HEL binding. The number of B cells binding HEL was measured by determining how many cells bound fluorescently labeled HEL. Levels of membrane IgM were determined by staining the cells with anti-mouse IgM antibody tagged with a fluorescent label different from that used to label HEL. Normal, nontransgenic mice (*left*) had many B cells that expressed high levels of surface IgM but almost no B cells that bound HEL above the background level (none visible in this plot). Both anti-HEL transgenics (*middle*) and anti-HEL/HEL double transgenics (*right*) had large numbers of B cells that bound HEL (blue); however, the level of membrane IgM was about 20-fold lower in the double transgenics. The data in Table 9-4 suggest that the B cells expressing anti-HEL in the double transgenics cannot mount a humoral response to HEL.

generate mature, peripheral B cells bearing anti-HEL membrane immunoglobulin of both the IgM and IgD classes, indicating that the B cells had fully matured. However, these B cells were functionally nonresponsive, or anergic.

Flow cytometric analysis of B cells from the double-transgenic mice showed that although large numbers of anergic anti-HEL cells were present, they expressed membrane IgM at levels about 20-fold lower than anti-HEL single transgenics (see Figure 9-9b). When these mice were immunized with HEL, few anti-HEL plasma cells were induced and the serum anti-HEL titer was very low (**Table 9-4**). Furthermore, when antigen was presented to these anergic B cells in the presence of T-cell help, many of the anergic B cells responded by undergoing apoptosis. Additional analysis of the anergic B cells demonstrated that they had a shorter half-life than normal B cells and appeared to be excluded from the B-cell follicles in the lymph nodes and spleen. These properties were dependent on the continuing presence of antigen, as the B-cell half-lives were restored to normal lengths on adoptive transfer of the transgene-bearing B cells to an animal that was not expressing HEL. The induction of anergy appears to occur in vivo when immature B cells meet a soluble self antigen, with the potential of sustained exposure to significant levels of the antigen.

More recent experiments have focused on defining the differences between the signal transduction events leading to anergy versus activation. Anergic B cells show much less antigen-induced tyrosine phosphorylation of signaling molecules, when compared with their nonanergic counterparts, and antigen-stimulated calcium release from storage vesicles into the cytoplasm of the anergic B cells was also dramatically reduced. Anergic B cells also require higher levels of the cytokine BAFF for continued survival, and it is likely that their reduced lifetimes result from unsuccessful competition with normal B cells for limiting amounts of this survival molecule. One of the outcomes of BAFF signaling is a reduction in the cytoplasmic levels of the pro-apoptotic molecule Bim; as might be expected, anergic B cells show

TABLE 9-4	Expression of anti-HEL antibody transgenes by mature peripheral B cells and plasma cells in single- and double-transgenic mice				
Experimental group		HEL level	Membrane anti-HEL BCR	Anti-HEL PFC/spleen*	Anti-HEL serum titer
Anti-HEL single transgenics		None	+	High	High
Anti-HEL/HEL double transgenics		10^{-9} M	+	Low	Low

* Experimental animals were immunized with hen egg-white lysozyme (HEL). Several days later hemolytic plaque assays for the number of plasma cells secreting anti-HEL antibody were performed and the serum anti-HEL titers were determined. PFC = plaque-forming cells (an assay for plasma cells secreting specific antibodies).
[Data from Goodnow, C. C. 1992. Transgenic mice and analysis of B-cell tolerance. Annual Review of Immunology 10:489.]

higher-than-normal levels of Bim and a correspondingly increased susceptibility to apoptosis.

The conclusion from these experiments is that even after B cells have exited the bone marrow and entered the periphery, mechanisms exist that minimize the risk that B cells will make antibodies to soluble self proteins expressed outside the bone marrow. B cells reactive to such proteins respond to receptor stimulation in the absence of appropriate T-cell help by anergy and eventual apoptosis, a second important mechanism of peripheral tolerance.

What might be the function, if any, of these anergic cells? One possibility is that they serve to absorb excess self antigens that might otherwise be able to deliver activating signals to high-affinity B cells and thus lead to autoimmune reactions. Another is that they represent cells destined for apoptosis that do not yet display the characteristic microanatomy of apoptotic cells. Yet another is that these cells will eventually develop into B-regulatory cells (see Chapter 11). As is so often the case in the immune system, it is more than possible that all of these functions are subsumed within this intriguing cell population, which remains the subject of intensive current investigation.

Key Concepts:

- Another population of transitional cells, called T3, appears to consist of B cells anergized by exposure to weakly stimulating peripheral self antigens, such as soluble proteins.

- Anergic B cells have a shorter half-life than normal B cells and appear to be excluded from the B-cell follicles in the lymph nodes and spleen.

The Properties and Development of B-1 and Marginal Zone B Cells

So far, this chapter has focused on the development of those B cells that belong to the largest and best characterized B-cell subpopulation—the B-2, or follicular, B cells.

Mature B-2 B cells recirculate between the blood and the lymphoid organs and can be found in large numbers in the B-cell follicles of the lymph nodes and spleen. However, other subsets of B cells have been recognized that have distinct functions, occupy distinct anatomical locations, and arise through different developmental programs. This section of the chapter will therefore address the properties and development of B-1 B cells (so named because their progenitor cells arise earlier during embryonic development than the HSCs that generate B-2 cells) and of splenic marginal zone (MZ) B cells. **Figure 9-10** shows a comparison of these three cell types.

B-1a, B-1b, and MZ B Cells Differ Phenotypically and Functionally from B-2 B Cells

Cells with the B-1 phenotype can be divided into B-1a and B-1b cells, based on the expression of the CD5 (Ly-1) surface marker by the former, but not the latter. Although the two types of B-1 cells have similar properties, the CD5-expressing B-1a cells are the prototype B-1 cell and will be the focus of most of our discussion.

B-1a B cells are phenotypically and functionally distinct from B-2 B cells in a number of important ways (see Figure 9-10). They occupy different anatomical niches from B-2 B cells, constituting 30% to 50% of the B cells in the pleural and peritoneal cavities of mice, and representing about 1 million cells in each space. A similar number of B-1a B cells can also be found in the spleen, but there they represent a much smaller fraction (around 2%) of the splenic B-cell population. B-1a B cells have only a relatively limited receptor repertoire. Some heavy- and light-chain gene segments are predominantly rearranged and expressed by B-1a cells. In addition, as TdT is minimally expressed in the precursors of B-1a cells, nontemplated (N)-nucleotide diversity is more limited than in B-2 cells. Thus the BCRs of B-1a B cells express considerably less receptor diversity, especially in their heavy-chain CDR3 regions, than do B-2 cells.

The antibody repertoire of B-1a cells seems to have been selected during evolution to recognize conserved self

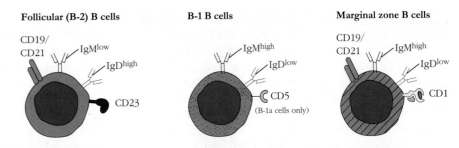

FIGURE 9-10 **The three major populations of mature B cells in the periphery.** The cell-surface properties and functions of B-2, B-1, and marginal zone (MZ) cells are shown. Conventional (follicular) B-2 cells were so named because they develop after B-1 B cells.

Attribute	Follicular (B-2) B cells	B-1 B cells	Marginal zone B cells
Major sites	Secondary lymphoid organs	Peritoneal and pleural cavities	Marginal zones of spleen
Progenitors first appear in mice	HSC: on day E10.5	Progenitor: on day E9.5	From HSCs; also earlier progenitor?
Source of new B cells in adults	From HSC in bone marrow	Self-renewing (division of existing B-1 cells)	From HSCs in bone marrow, long-lived
Dependence on IL-7 and BAFF	Yes	No	Yes
V-region diversity	Highly diverse	Restricted diversity: limited V_H and V_L usage and N nucleotide addition	Somewhat restricted
Somatic hypermutation	Extensive	Some	Unclear
Requirements for T-cell help	Yes	No	Variable
Isotypes produced	High levels of IgG	Primarily IgM; some IgG	Primarily IgM; some IgG
Response to carbohydrate antigens	Possibly	Yes	Yes
Response to protein antigens	Yes	Possibly	Yes
Memory	Yes	Some	Unknown

carbohydrate and lipid antigens, such as phosphatidylcholine (PtC) exposed on aged erythrocytes and apoptotic cells. PtC is recognized by numerous B-1a cells with BCR utilizing the V_H11 gene segment (rare in B-2 cells) and a light-chain $V_\kappa 9$ gene segment. In these antibodies there are few if any N nucleotides in the heavy-chain CDR3 region. Other B-1a cells recognize other lipid and polysaccharide antigens, generating antibodies that help clear dead cells and debris, and also have been shown to provide valuable protection against *Streptococcus pneumoniae* and *Francisella tularensis* bacteria and influenza virus. In expressing an evolved oligoclonal (few, as opposed to many, clones) repertoire of B-cell receptors that bind to antigens shared among many microbes, B-1a B cells occupy a functional niche that bridges the innate and adaptive immune systems.

Even in the absence of antigenic stimulation these cells secrete antibodies to these self and microbial antigens, the so-called **natural antibodies**. These serum antibodies (mostly IgM, but also some IgG and IgA) provide a first line of protection against invasion by many types of microorganisms. B-1a B cells can generate antibodies in the absence of T-cell help, although the addition of helper T cells enhances antibody secretion and allows for some degree of heavy-chain class switching and possibly also somatic hypermutation.

Work in a number of different transgenic mouse systems has suggested that developing B-1a B cells are selected for those that recognize self antigens; relatively strong BCR engagement by self antigens provides *positive selection* rather than the *negative selection* that we have described for immature B-2 cells and transitional T1 cells.

Marginal zone (MZ) B cells take their name from their location in the marginal zone, the outer regions of the white pulp of the spleen (see Figure 9-7). MZ cells are characterized by relatively high levels of membrane IgM and the complement receptor/B-cell coreceptor CD21 (see Chapters 3 and 5), but low levels of membrane IgD and the Fc receptor CD23. Some also express CD1 proteins that can present lipid antigens to T cells (see Figure 9-10). They also display phospholipid receptors and adhesion molecules that "glue" them to other cells within the marginal zone, holding them in place. MZ cells are long lived and may self-renew in the periphery.

In the spleen the fast-moving arterial flow spreads out into the marginal sinuses, decreasing the rate of blood flow and allowing blood-borne antigens to interact with cells residing within the marginal zone. Indeed, MZ B cells appear to be specialized for recognizing blood-borne antigens. They are capable of responding to both protein and carbohydrate antigens and produce natural antibodies. Like B-1a B cells, some (but maybe not all) MZ B cells can produce antibodies without the need for T-cell help.

Key Concepts:

- In addition to the predominant population of conventional B cells that arises in the bone marrow, called *B-2* or follicular B cells, other subsets of B cells (B-1 and marginal zone [MZ] B cells) arise through other developmental programs, occupy distinct anatomical locations, and have distinct functions.

- B-1 B cells (B-1a and B-1b) and MZ B cells differ from B-2 cells in surface markers, and in having a more limited repertoire of antibody specificities, which include cross-reactions between self and microbial antigens. They spontaneously generate natural antibodies and rapidly generate antibody responses to microbial infection without T-cell help.

B-1a B Cells Are Derived from a Distinct Developmental Lineage

We learned earlier in this chapter that conventional B-2 follicular B cells are generated throughout life from the developmental pathways that start with long-term self-renewing hematopoietic stem cells. The B-1a cell population in the adult has different origins, as it is self-renewing in the periphery from progenitor cells dispersed during embryonic development. This means that new daughter B-1a B cells are generated continually from pre-existing B-1a B cells in the peritoneal and pleural cavities and in other parts of the body in which B-1a B cells reside.

For many years, the preeminent issue debated by those interested in the development of B-1a B cells was whether they constituted a separate developmental lineage, or whether they are derived from the same progenitors as B-2 B cells. Elegant but challenging studies in recent years have involved isolating the different populations from embryonic, fetal, neonatal, and adult tissues and testing their differentiative potential in cell culture or after cell transfer. These experiments have shown that B-1a and B-2 B cells derive from distinct lineages of progenitor cells. Several lines of evidence support this conjecture:

- Progenitors of B-1a B cells appear before the HSCs that generate B-2 B cells during fetal development. Whereas the first long-term HSCs appear in the yolk sac of the mouse embryo on day 10.5 of gestation, progenitors capable of giving rise to B-1a B cells (but not B-2 cells) appear on embryonic day 9.5 (E9.5) in the para-aortic splanchnopleura (P-Sp) and yolk sac.

- Mice with several mutations that block the formation of HSCs and hence lack B-2 cells do have B-1 progenitor cells detectable in the fetal liver.

- B-cell progenitors of the $CD19^+CD45R^{low/-}$ phenotype transferred into an immunodeficient mouse were able to repopulate the B-1a, but not the B-2, B-cell compartments. Conversely, $CD19^+CD45R^{hi}$ B-cell progenitors gave rise to B-2, B-1b, and MZ B cells but not to B-1a daughter cells in an immunodeficient recipient mouse, supporting the notion that B-1a cells derive from a different lineage of progenitor cells than other B cells. Most convincingly, when *single HSCs* were transferred into recipients, they repopulated all mature blood cell types except B-1a B cells. Similar results were obtained with HSCs from fetal liver. In contrast, isolated B-1a progenitors repopulated B-1a cells but not B-2 or MZ cells.

- Whereas B-2 B cells must be constantly replenished by the emergence of newly generated cells from the bone marrow, B-1a B cells are constantly regenerated in the periphery of the animal. Bone marrow ablation therefore leaves the mouse with a depleted B-2 pool, but with a fully functional B-1a population. Other experiments using conditional *RAG2* knockout animals, in which shutting off RAG expression in otherwise healthy adult animals blocks all new B-cell development, indicate that follicular B-2 B cells have a half-life of approximately 4.5 months. In contrast, since B-1a B cells can self-renew in the periphery, their numbers are unaffected in this experimental knockout animal.

- In contrast with early stages of lymphoid development and B-2 cell differentiation from HSCs in mice, the development of B-1a B cells is not IL-7–dependent, nor do B-1a cells require interaction with BAFF during the transitional stage of development for survival.

The development of B-1a and B-2 cells from their progenitors is controlled by different transcriptional regulators and microRNAs. The story is different for the B-1b ($CD5^-$) B cells. Although they have some similar properties and functions to B-1a cells, including generating natural antibodies to self and microbial antigens, most of them seem to be derived from HSCs through conventional B-cell development pathways. However, some B-1b cells do develop from the early E9 yolk sac and from P-Sp progenitors that generate B-1a cells, so they may arise from multiple developmental lineages.

Marginal zone B cells also are derived from HSCs and, like follicular B-2 B cells, arise from the T2 transitional population (see Figure 9-7). MZ cells, like B-2 B cells, are also reliant on the B-cell survival factor BAFF. However, like developing B-1 B cells, developing MZ cells appear to require relatively strong signaling through the BCR in order to survive. Surprisingly, unlike any other B-cell subset so far described, the differentiation of MZ B cells also requires signaling through ligands of the Notch pathway, as is true of developing T cells in the thymus (see Chapter 8). Loss of the Notch2 receptor or deletion of the Notch2 ligand, Delta-like 1 (DLL1), results in the selective deletion of MZ B cells. Both MZ and B-1 B-cell populations are enriched in cells that express self antigen–specific receptors; relatively strong signaling of T2 B cells by binding of self antigens through the BCR is also necessary for MZ B-cell differentiation.

Key Concepts:

- Recent studies have provided strong evidence that B-1a B cells are derived from early embryonic precursors, distinct from and preceding HSCs, that seed out into the pleural and peritoneal cavities and are capable of self-renewal, generating B-1a cells throughout life.

- In contrast, B-1b and MZ B cells seem to be derived mostly from HSCs and B-2 lymphoid progenitors. MZ B cells inhabiting the splenic marginal zone arise from T2 B cells.

Comparison of B- and T-Cell Development

In closing this chapter, it is instructive to consider the many points of comparison in the development of the two arms of the adaptive immune system: B cells and T cells (**Table 9-5**). Both types of lymphocytes include two distinct cell lineages, of which one (γδ T cells and B-1 B cells) disperses to its own particular peripheral niches during fetal development, there to function as a self-renewing population until the death of the host. Beginning around the time of birth and continuing through adult life, the development of conventional B cells (B-2, follicular) and T cells (αβ) commences in the bone marrow, starting with hematopoietic stem cells. B and T cells share the early phases of their developmental programs, as they pass through progressively more differentiated stages as MPPs, ELPs, and CLPs. At the ELP and CLP stages, T-cell progenitors leave the bone marrow and migrate to the thymus to complete their development, leaving B-cell progenitors behind. In mammals, B cells do not have an organ analogous to the thymus in which to develop into mature, functioning cells, although birds do possess such an organ—the bursa of Fabricius.

Both B and T cells initiate V(D)J rearrangement for one of their chains, which are expressed with a surrogate second chain. The resulting pre-BCR or pre-TCR signals cells that the first chain rearranged productively, and to go ahead and rearrange the second receptor gene family.

B cells expressing mIgM BCR and T cells expressing αβ TCR both must pass through stages of positive selection, in which those cells capable of receiving survival signals are retained at the expense of those that cannot. For T cells positive selection occurs only if the TCR recognizes peptide bound to self MHC, needed to generate a TCR repertoire that will be useful in responding to foreign peptide–MHC complexes. The process of positive selection of B-2 cells, while not fully understood, may involve spontaneous tonic signaling, independent of any ligand binding. Conventional B and T cells must also survive the process of negative selection, in which lymphocytes with high affinity for self antigens are deleted, induced to undergo receptor editing (B-2 B cells only), or inactivated, before they become mature functional cells in the periphery. In contrast, B-1 and marginal zone B cells are positively selected by low-affinity recognition of self antigens.

With the expression of high levels of IgD on the cell surface and the necessary adhesion molecules to direct their recirculation, development of the mature, follicular B-2 cell is complete and, for a few weeks to months, it will recirculate, ready for antigen contact in the context of T-cell help and subsequent differentiation to antibody production. For the final, antigen-stimulated stages of B-cell differentiation, the reader is directed to Chapter 11.

As with many physiological systems in the body, the functions of the immune system gradually decline during aging. Changes in B-cell development and responses, including diminished production of certain natural antibodies that may be beneficial, are described in **Clinical Focus Box 9-3**.

Key Concepts:

- There are many similarities between B-cell and T-cell development, including the gradual commitment to the cell lineage and the sequential rearrangement of antigen receptor genes, with checkpoints that test for productive rearrangement and expression of receptor polypeptides and select for cells with appropriate specificities.

- The results of these developmental processes usually are mature B and T cells with appropriate antigen receptor repertoires that will protect us from infections and avoid undesirable autoreactivity.

TABLE 9-5   Comparison between T-cell and B-cell development

Structure or process	B cells	T cells
Lineage seeds the periphery during fetal development	+ (B-1a)	+ (γδ)
Develop from HSCs in the bone marrow	+ (B-2, MZ)	+ (αβ)
Development continues in the thymus	−	+
Ig heavy-chain or TCR β-chain gene rearrangement begins with D-J and continues with V-DJ recombination	+	+
The H chain (BCR) or β chain (TCR) is expressed with a surrogate form of the second chain on the cell surface. Signaling from this pre-BCR or pre-TCR is necessary for development to continue	+	+
Signaling through the pre-B-cell receptor or pre-TCR results in proliferation and initiation of the rearrangement of the second chain κ/λ (BCR) or α (TCR)	+	+
The κ/λ (BCR) or α/γ (TCR) chain bears only V and J segments	+	+
Signaling from the completed receptor is necessary for survival (positive selection)	+	+
Positive selection requires recognition of self components	+ (true for the minority B-1a and MZ subsets)	+ (low-affinity binding of self MHC and self peptide in the thymus is necessary for positive selection)
Receptor editing of κ/λ (BCR) or α/γ (TCR) chain	+ (to eliminate self-reactive BCR)	+ (rearrangements of remaining TCR α V and J segments may continue until a TCR is formed that can recognize self peptide/MHC to allow positive selection)
Immature cells bearing high-affinity autoreactive receptors are eliminated by apoptosis (negative selection)	+ (immature and T1 B cells)	+
Negative selection involves expression of peripheral tissue self antigens in the primary lymphoid organs	−	+ (peripheral tissue self antigens are expressed in the thymus under the influence of the AIRE transcription regulator)
Negative selection involves recognition of MHC-presented peptides	−	+
Antigen receptor allelic exclusion	+ (heavy and light chains)	+ (TCR β chain; many T cells display more than one TCR α chain if the first that was expressed does not generate a TCR recognizing self MHC)
>90% of lymphocytes are lost prior to export to the periphery	+	+
Processes in the periphery allow establishment of tolerance to antigens not expressed in the primary lymphoid organs	+	+

Note: Unless noted otherwise, the comparisons in this table refer to the predominant αβ T cell and follicular B-2 cell subsets.

BOX 9-3

B-Cell Development and Function in the Aging Individual

People of retirement age and older represent a growing segment of the population, and these older individuals expect to remain active and productive members of society. However, physicians and immunologists have long known that the elderly are more susceptible to infection than are young men and women, that vaccinations are less effective in older individuals, and that the elderly suffer from conditions that increase with age, including atherosclerosis and dementia, both of which have immune/inflammatory components. In this feature, we explore the differences in B-cell development between younger and older vertebrates, which may account for some of these immunological disparities between adult and older individuals.

Aging individuals display deficiencies in many aspects of B-cell function, including poor antibody responses to vaccination, inefficient generation of memory B cells, and an increase in the expression of autoimmune disorders. Current research demonstrates that aging individuals display a range of shortcomings in developing B cells that affect their immune status.

Experiments employing reciprocal bone marrow chimeras—in which aging HSCs were transplanted into young recipients or HSCs from young mice were injected into aging recipients—have shown that suboptimal processes of B-cell development in aging individuals result from deficiencies in both the aging stem cells and in the supporting stromal cells. For example, bone marrow stromal cells from aging mice secrete lower levels of IL-7 than do stromal cells from younger animals. However, a study of isolated, aging B-cell progenitors reveals that they also respond less efficiently to IL-7 than do B cells from younger mice, and so the IL-7 response in aging individuals is affected at both the producing and recipient-cell levels.

Indeed, the problems encountered by developing B cells from aging individuals start at the very beginning of their developmental program. The epigenetic regulation of HSC genes in aging mice is compromised, resulting in diminished levels of HSC self-renewal. Furthermore, the balance between the production of myeloid versus lymphoid progenitors is shifted in older individuals, with down-regulation of genes associated with lymphoid specification and a correspondingly enhanced expression of genes specifying myeloid development. The net effect of these changes in the HSC population is a reduction in the numbers of early B-cell progenitors, which is reflected in a decrease in the numbers of pro- and pre-B-cell precursors at all stages of development.

Detailed studies of the expression of particular genes important in B-cell development demonstrate that the expression of important transcription factors, such as E2A, is reduced in older animals. Furthermore, the genes for the RAG and λ5 surrogate light-chain proteins are down-regulated in older animals, contributing to the reduction in bone marrow output of immature B cells.

Thus, multiple mechanisms help to explain why the numbers of B cells released from the bone marrow are smaller in aging than in younger individuals. But is the antigen recognition repertoire—the quality—as well as the quantity of B cells different between the two populations? The answer to this question has come from the development of techniques that enable a global assessment of repertoire diversity. Study of the sizes and sequences of CDR3s from large numbers of human B cells suggests that in aging individuals the size of the repertoire (the number of different B-cell receptors an individual expresses) is drastically diminished and that this decrease in repertoire diversity correlates with a reduction in the health of the aging patient. The mechanisms for this age-related repertoire truncation are the subject of ongoing studies.

Intriguing recent results also suggest that reductions in human B-1 cells and their natural antibody products may contribute to aging-associated conditions including atherosclerosis and neurodegenerative conditions such as Alzheimer's.

Healthy individuals produce natural IgM antibodies that react with oxidized low-density lipoproteins (oxLDL) that contribute to plaque formation in blood vessels. These IgM antibodies appear to be protective: the levels of these IgM antibodies in people in their 60s were found to be inversely correlated with development of cardiovascular problems. In the second example, neurodegeneration, healthy people often have natural antibodies that react with amyloid and tau proteins, which can form abnormal plaques and tangles, respectively, that have been implicated in contributing to Alzheimer's. Such natural antibodies, which may help to clear the aggregated proteins and improve cell survival, appear to be diminished in individuals with Alzheimer's. Passive transfer of such natural antibodies into animals has reduced the abnormal amyloid and tau proteins and improved brain pathology and behavior.

These and other findings suggest that reduction in natural antibodies may lead to increased susceptibility to certain age-related conditions; indeed, age-associated declines in the number of B-1 cells have been observed in mice and humans. These results also raise the exciting possibility that passive transfer of natural antibodies reactive with appropriate targets may help restore a natural mechanism by which our immune system might protect us from tissue damage.

REFERENCES

Cancro, M. P., et al. 2009. B cells and aging: molecules and mechanisms. *Trends in Immunology* **30**:313.

Dorshkind, K., E. Montecino-Rodriguez, and R. A. Signer. 2009. The ageing immune system: is it ever too old to become young again? *Nature Reviews Immunology* **9**:57.

Labrie, J. E. III, et al. 2004. Bone marrow microenvironmental changes underlie reduced RAG-mediated recombination and B cell generation in aged mice. *Journal of Experimental Medicine* **200**:411.

Van der Put, E., et al. 2003. Aged mice exhibit distinct B cell precursor phenotypes differing in activation, proliferation and apoptosis. *Experimental Gerontology* **38**:1137.

Conclusion

The first essential challenge that must be accomplished in B-cell development is the generation of B cells with a vast repertoire—many billions—of B-cell receptor specificities that are sufficient to ensure responses to virtually anything foreign that enters the body. The antibody diversity generated by gene rearrangements, junctional diversification, and different combinations of heavy and light chains (discussed in Chapter 6) is amplified by the fact that millions of new B cells are generated every day. B cells whose antibody specificities aren't needed turn over, replaced by new B cells generated in the bone marrow by the processes of hematopoiesis and B-cell development.

Progression through the sequence of stages of hematopoiesis, commitment to the lymphoid lineage, and early B-cell development in the bone marrow resulting in the formation of immature B cells is driven by networks of transcription factors. Critically important is the key E2A → EBF1 → PAX5 transcription factor chain that constitutes a feed-forward regulatory network, with the end factor, PAX5, turning on the genes that determine the B lymphocyte phenotype that the cells retain until activated by antigen and other signals to differentiate into antibody-secreting plasma cells. The complex web of transcription factors modulates and is modulated by a variety of epigenetic changes that control the gene transcription and protein expression that characterize each stage.

The ordered and successful recombination of heavy and light-chain genes is programmed into and in some cases drives progression through the stages of B-cell development, with checkpoints along the way to ensure that the rearrangements are good ones that will eventually yield functional BCRs. So after the V_H-DJ recombination, the μ heavy chain is tested for its ability to pair up and to associate with the surrogate light-chain polypeptide; if all goes well the resulting pre-BCR signals the pre-B cell to stop rearranging heavy-chain genes. The resulting heavy-chain allelic exclusion is important to ensuring that the B cell will not have a variety of mixed-chain BCRs that would have different specificities. Signals from the pre-BCR drive the pre-B cell to divide, generating a pool of cells expressing that good heavy chain, and then each of those cells will begin rearranging light-chain genes, starting (in the mouse) with κ before trying λ. All it takes is one productive rearrangement out of the four light-chain rearrangements possible, and the cell now expresses mIgM, marking it as an immature B cell. Expression of the receptor—without ligand binding—is sufficient to signal the cell to terminate light-chain rearrangements, ensuring that the cell expresses only one heavy chain and one light chain. This constitutes the second checkpoint in B-cell development.

Immature B cells now have to deal with the second challenge of B-cell development, making sure that they are not autoreactive. This is accomplished by the induction of apoptosis in response to strong signals coming from mIgM receptors that have been cross-linked by binding to multiple self antigens in their bone marrow surroundings. If they receive such a signal, before they die they are given the opportunity to rearrange any remaining light-chain genes (either on a different chromosome or using V and J segments flanking the rearranged and expressed light-chain gene). If their attempts at receptor editing fail to provide a light chain that does not form a self-reactive BCR, the cell undergoes apoptosis.

The surviving immature B cells now leave their bone marrow birthplace for the spleen, where they become T1 transitional cells and are tested for recognition of peripheral self antigens, which can also induce apoptosis. Passing that test, the cells become T2 B cells; if they then receive survival signals from BAFF they go on to traffic to the follicles of the spleen and lymph nodes where they constitute an army of mature B cells with diverse non-autoreactive BCR specificities, ready to respond to foreign invaders along with their helper T-cell partners. Other T2 B cells may be induced to become anergic by weakly stimulating self antigens such as soluble self proteins and be characterized as T3 cells.

While the follicular B cells (also called B-2 cells) described above are the major (or conventional) population of B cells in mice and probably humans as well, several other B-cell populations exist as variations on that theme. B-1 B cells and marginal zone (MZ) B cells seem to have been positively selected for low affinity recognition of some self antigens; they have non-random BCR specificities (in some cases using specific conserved V region genes) which seem to have evolved to recognize certain polysaccharide and lipid antigens that cross-react with microbial antigens. MZ B cells are derived from T2 B cells; instead of trafficking to lymphoid follicles they take up residence in the marginal zone of the spleen, where they are exposed to blood borne antigens entering through the marginal sinus. As they often do not require T-cell help, they can respond more quickly to antigen than can follicular B-2 cells.

B-1a (CD5$^+$) B cells are the most unique population of B cells, as they are not derived from HSCs. Instead, most appear to have their origins in early embryonic progenitor cells that form before HSCs appear; the B-1a progenitors seed the pleural and peritoneal cavities where they self-renew throughout life. In addition to sharing specificities for polysaccharide and lipid antigens on self-components and bacteria, B-1a and MZ B cells both spontaneously produce natural antibodies that have some protective benefits. Thus, while B cells don't have quite the functional diversity exhibited by T cells, there is now strong evidence for the existence of B-cell subsets with different developmental origins and functions.

REFERENCES

Bankovich, A. J., et al. 2007. Structural insight into pre-B cell receptor function. *Science* **316**:291.

Bao, Y., and X. Cao. 2016. Epigenetic control of B cell development and B-cell–related immune disorders. *Clinical Reviews in Allergy and Immunology* **50**:301.

Cambier, J. C., S. B. Gauld, K. T. Merrell, and B. J. Vilen. 2007. B-cell anergy: from transgenic models to naturally occurring anergic B cells? *Nature Reviews Immunology* **7**:633.

Casola, S. 2007. Control of peripheral B-cell development. *Current Opinions in Immunology* **19**:143.

Dorshkind, K., and E. Montecino-Rodriguez. 2007. Fetal B-cell lymphopoiesis and the emergence of B-1-cell potential. *Nature Reviews Immunology* **7**:213.

Ghosn, E. E., and Y. Yang. 2015. Hematopoietic stem cell-independent B-1a lineage. *Annals of the New York Academy of Sciences* **1362**:23.

Goodnow, C. C. 1992. Transgenic mice and analysis of B-cell tolerance. *Annual Review of Immunology* **10**:489.

Hoek, K. L., et al. 2006. Transitional B cell fate is associated with developmental stage-specific regulation of diacylglycerol and calcium signaling upon B cell receptor engagement. *Journal of Immunology* **177**:5405.

Kee, B. L. 2009. E and ID proteins branch out. *Nature Reviews Immunology* **9**:175.

Koralov, S. B., et al. 2008. Dicer ablation affects antibody diversity and cell survival in the B lymphocyte lineage. *Cell* **132**:860.

Kurosaki, T., H. Shinohara, and Y. Baba. 2010. B cell signaling and fate decision. *Annual Review of Immunology* **28**:21.

Liao, D. 2009. Emerging roles of the EBF family of transcription factors in tumor suppression. *Molecular Cancer Research* **7**:1893.

Mackay, F., and P. Schneider. 2009. Cracking the BAFF code. *Nature Reviews Immunology* **9**:491.

Malin, S., et al. 2010. Role of STAT5 in controlling cell survival and immunoglobulin gene recombination during pro-B cell development. *Nature Immunology* **11**:171.

Melchers, F. 2015. Checkpoints that control B cell development. *Journal of Clinical Investigation* **125**:2203.

Monroe, J. G., and K. Dorshkind. 2007. Fate decisions regulating bone marrow and peripheral B lymphocyte development. *Advances in Immunology* **95**:1.

Nagasawa, T. 2006. Microenvironmental niches in the bone marrow required for B-cell development. *Nature Reviews Immunology* **6**:107.

Nemazee, D. 2006. Receptor editing in lymphocyte development and central tolerance. *Nature Reviews Immunology* **6**:728.

Nemazee, D. A., and K. Bürki. 1989. Clonal deletion of B lymphocytes in a transgenic mouse bearing anti-MHC class I antibody genes. *Nature* **337**:562.

Nutt, S. L., and B. L. Kee. 2007. The transcriptional regulation of B cell lineage commitment. *Immunity* **26**:715.

Pillai, S., and A. Cariappa. 2009. The follicular versus marginal zone B lymphocyte cell fate decision. *Nature Reviews Immunology* **9**:767.

Rothenberg, E. V. 2014. Transcriptional control of early T and B cell developmental choices. *Annual Review of Immunology* **32**:283.

Somasundaram, R., M. A. J. Prasad, J. Ungerback, and M. Sigvardsson. 2016. Transcription factor networks in B-cell differentiation link development to acute lymphoid leukemia. *Blood* **126**:144.

Srivastava, B., W. J. Quinn III, K. Hazard, J. Erikson, and D. Allman. 2005. Characterization of marginal zone B cell precursors. *Journal of Experimental Medicine* **202**:1225.

Tokoyoda, K., T. Egawa, T. Sugiyama, B. I. Choi, and T. Nagasawa. 2004. Cellular niches controlling B lymphocyte behavior within bone marrow during development. *Immunity* **20**:707.

Übelhart, R., M. Werner, and H. Jumaa. 2016. Assembly and function of the precursor B-cell receptor. *Current Topics in Microbiology and Immunology* **393**:3.

von Boehmer, H., and F. Melchers. 2010. Checkpoints in lymphocyte development and autoimmune disease. *Nature Immunology* **11**:14.

Whitlock, C. A., and O. Witte. 1982. Long-term culture of B lymphocytes and their precursors from murine bone marrow. *Proceedings of the National Academy of Sciences USA* **79**:3608.

Xiao, C., and K. Rajewsky. 2009. MicroRNA control in the immune system: basic principles. *Cell* **136**:26.

Yin, T., and L. Li. 2006. The stem cell niches in bone. *Journal of Clinical Investigation* **116**:1195.

Useful Website

www.bio.davidson.edu/courses/immunology/Flash/Bcellmat.html An unusual animation of B-cell development.

STUDY QUESTIONS

1. You wish to study the development of B-1 B cells in the absence of the other two major B-cell subsets. You have a recipient $Rag1^{-/-}$ mouse that you have already repopulated with T cells. What would you choose to be your source of B-1 progenitors and why? From which anatomical sites would you expect to harvest the B-1 B cells?

2. Describe the phenotypic and functional differences between T1 and T2 immature B cells.

3. Following expression of the pre-B-cell receptor on the progenitor B-cell surface, the B cell undergoes a few rounds of cell division. What purpose do these rounds of cell division serve in the development of the B-cell repertoire?

4. Immature B cells bearing potentially autoimmune receptors can be managed in three ways to minimize the probability of disease. Describe these three strategies, noting whether they are shared by T-cell progenitors.

5. You suspect that a new transcription factor is expressed at the pre-pro-B-cell stage of development. How would you test your hypothesis? What is the status of heavy- and light-chain rearrangement at this stage of development and how would you test it?

6. How would you determine whether a particular stage of B-cell development occurs in association with a stromal cell that expresses CXCL12?

7. Describe the order in which B-cell receptor genes undergo rearrangement, indicating at what steps you might expect to see the B cell express one or both chains on the cell surface. In what sense(s) does this gene rearrangement process mimic the analogous progression in $\alpha\beta$ T cells, and in what ways do the two processes differ?

8. In addition to the experiment shown in Figure 9-9 and Table 9-4, Goodnow and colleagues set up another experimental group. In this case a transgene was created that allowed HEL to be expressed as a *membrane-associated protein* throughout the body, including in the bone marrow. These membrane HEL-transgenic mice were crossed to mice transgenic for the anti-HEL-specific antibody. What do you think happens to the B cells in the double-transgenic mice? Do you think B cells expressing an HEL-specific BCR would be found in the periphery? Do you think these mice would generate an anti-HEL antibody response after immunization with HEL?

ANALYZE THE DATA

1. The two columns of data in the following figure are flow cytometric plots that describe the levels of antigens denoted on the x and y axes. Left: Antigens present on spleen (a) and bone marrow (b) from wild-type (genetically normal) animals. The plots represent all lymphocytes in the spleen (a) or B-cell progenitor and precursor cells in the bone marrow (b). Right: The same plots for animals in which the Dicer gene has been knocked out. As you will recall, Dicer is required for the maturation of microRNAs.

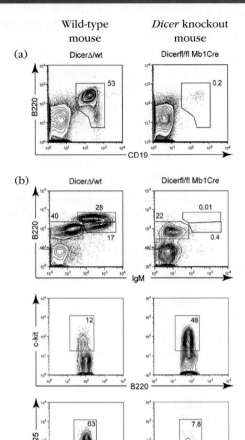

Republished with permission of Elsevier, from Koralov, S.B., et al., from "Dicer ablation affects antibody diversity and cell survival in the B lymphocyte lineage." Cell, 2008 March; 132(5) 860-874, Figure 1A and 1B. Permission conveyed through Copyright Clearance Center, Inc.

a. For each pair of plots, describe the differences in the cell populations, indicating whether the differences reflect losses or gains in particular developing B-cell populations.

b. At what point(s) in B-cell development do you think microRNAs are functioning?

2. The following figure is derived from the same article as the figure in the preceding question. In this case the data are expressed as histograms, in which the y axis represents the number of cells binding the molecule shown on the x axis, annexin A5. Annexin A5 binds to phosphatidylserine on the outer leaflet of cell membranes. Phosphatidylserine is found on the outer leaflet only in cells about to undergo apoptosis. The top two panels represent cells from a wild-type animal, and the bottom two panels represent cells from animals in which the *Dicer* gene has been knocked out.

a. Does the presence of Dicer have an effect on the fraction of pro-B cells undergoing apoptosis? Explain your reasoning.

b. Does the presence of Dicer have an effect on the fraction of pre-B cells undergoing apoptosis? Again explain your reasoning.

c. Describe one function that you now think microRNAs fulfill in B-cell development.

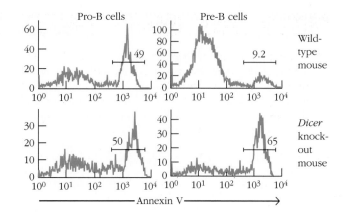

10

T-Cell Activation, Helper Subset Differentiation, and Memory

Learning Objectives

After reading this chapter, you should be able to:

1. Describe the two main signals needed to activate a naïve T cell and the difference between costimulatory and coinhibitory signals.

2. Describe the extracellular influences (polarizing cytokines) and intracellular influences (master gene regulators) that drive differentiation of naïve CD4⁺ T cells into helper T-cell lineages, as well as what distinguishes the various lineages functionally (effector cytokines).

3. Outline the major differences between type 1 and type 2 immune responses and understand how distinct helper T-cell subsets contribute to these responses.

4. Identify the four main memory T-cell subsets, understand their functions, and speculate about their origin.

5. Recognize some of the questions that still remain as we work to understand T-cell activation, differentiation, and memory.

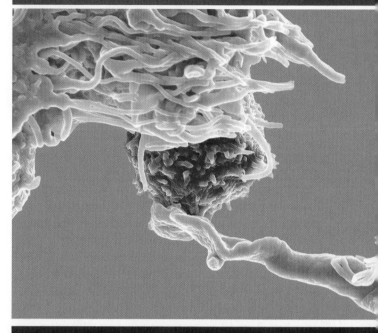

Dendritic cells (blue-green) interacting with a T cell (pink). [Dr. Olivier Schwartz, Institut Pasteur/Science Source.]

The interaction between a naïve T cell and an antigen-presenting cell (APC) is *the* initiating event of the adaptive immune response. Prior to this, the innate immune system was alerted to infection or tissue damage, and antigen-presenting cells were activated via pattern recognition receptors. These APCs may have engulfed extracellular (or opsonized intracellular) pathogens, or they may have been infected with an intracellular pathogen. In either case, they have processed and presented peptides from these pathogens in complex with surface MHC class I and class II molecules, and migrated to secondary lymphoid tissue, including local (draining) lymph nodes, Peyer's patches (see Chapter 13), and/or the spleen. Within these secondary lymphoid tissues APCs settled in the T-cell zones, joining networks of resident APCs, to be continually scanned by roving naïve CD8⁺ and CD4⁺ T cells, which recognize MHC class I–peptide and MHC class II–peptide complexes, respectively. (Chapter 14 provides an overview of the initiation of an immune response in secondary lymphoid tissue.)

We have seen that each mature T cell expresses a unique antigen receptor assembled via random gene rearrangement (Chapter 6) during T-cell development

Key Terms

Naïve
Two-signal hypothesis
Avidity
Anergy
Costimulatory receptors
Coinhibitory receptors
CD28

Programmed cell death protein-1 (PD-1 or CD279)
Polarizing cytokines
Type 1 responses
Type 2 responses
Adjuvants
Master gene regulator

Effector cytokines
T helper type 1 (T_H1) cells
T helper type 2 (T_H2) cells
T helper type 17 (T_H17) cells
Peripheral T_REG (pT_REG) cells
T follicular helper (T_FH) cells
T helper type 9 (T_H9) cells

Central memory T cells (T_CM)
Effector memory T cells (T_EM)
Resident memory T cells (T_RM)
Stem cell memory T cells (T_SCM)

in the thymus. Positive and negative selection in the thymus ensures that the mature, naïve T cells that enter the circulation are largely tolerant to self antigens, and restricted to self MHC (Chapter 8). Some naïve T cells commit to the CD8$^+$ cytotoxic T-cell lineage, others to the CD4$^+$ helper T-cell lineage. If, during their migration through secondary lymphoid organs, naïve CD8$^+$ or CD4$^+$ T cells bind tightly to an MHC-peptide complex expressed by an activated dendritic cell, they become activated by signals generated through their T-cell receptors (TCRs). As we will learn in this chapter, TCR signals in concert with costimulatory and cytokine signals stimulate naïve T cells to proliferate and differentiate into effector T cells.

In this chapter, we specifically review the cellular and molecular events that activate naïve T cells and investigate the variety of costimulatory interactions that play an important role in determining the outcome of T cell–APC interactions. We then discuss the end result of naïve T-cell activation—the development of distinct effector and memory T-cell subsets—focusing primarily on the various fates and functions of the CD4$^+$ helper T-cell (T$_H$) lineages that drive both B- and T-cell adaptive responses (**Overview Figure 10-1**). As you know, naïve CD8$^+$ T cells differentiate into cytotoxic cells in response to engagement of MHC class I–peptide combinations. Although we describe the initial activation and differentiation of CD8$^+$ T cells in this chapter, we spend more time on their cytotoxic effector functions in Chapter 12.

Here, we explore the surprising number of helper lineages that a naïve CD4$^+$ T cell can adopt (T$_H$1, T$_H$2, T$_H$17, T$_{FH}$, pT$_{REG}$, T$_H$9, and more), and broadly discuss their functions. We also examine what regulates a CD4$^+$ T cell's lineage decision, and how it is influenced by the response of the innate immune system to specific pathogens and antigens. We close the chapter with a discussion of the generation of T-cell memory, introducing the multiple memory subsets and current thinking about their origin and function.

A Classic Experiment box describes the basic research behind the discovery of the costimulatory molecule CD28, an essential participant in naïve T-cell activation. The first Clinical Focus box describes the development of an exciting new immunotherapy for cancer that took advantage of the discovery of a relative of CD28 that inhibited rather than enhanced T-cell activation. Together, these boxes illustrate the powerful connection between basic research and clinical advances. This partnership is the basis for *translational research*, an effort to bring the bench to the "bedside" that has captured the imagination of many biomedical investigators and clinicians.

The Advances box describes recent insights into the origin of regulatory T cells that play a role in maintaining maternal-fetal tolerance during pregnancy. The second Clinical Focus box discusses how a disease, an "experiment of nature," has helped us to better understand the basic biology and physiological function of one of the helper cell subsets (T$_H$17) introduced in this chapter.

T-Cell Activation and the Two-Signal Hypothesis

CD4$^+$ and CD8$^+$ T cells leave the thymus and enter the circulation as **naïve** T cells. Although relatively mature (see Chapter 8), they have not yet encountered antigen. Their chromatin is condensed, they have very little cytoplasm, and they exhibit little transcriptional activity. However, they are mobile cells and recirculate continually among the blood, lymph, and secondary lymphoid tissues, including the lymph nodes, browsing for antigen. It is estimated that each naïve T cell recirculates from blood through lymph nodes and back again every 12 to 24 hours. Because only about 1 in 10^5 naïve T cells is likely to be specific for any given antigen, this large-scale recirculation increases the chances that a particular T cell will find "its" antigen.

If a naïve T cell does not bind any of the MHC-peptide complexes it encounters as it browses the surfaces of antigen-presenting cells in the secondary lymphoid tissue, it exits and rejoins the circulation to try again in another tissue. If it has been browsing through a lymph node or Peyer's patch, it exits via the efferent lymphatics, ultimately draining into the thoracic duct and reentering the blood via the vena cava (see Figures 2-12 and 2-13). If it has been browsing the T-cell zones of the spleen, it exits directly into the blood (see Figure 2-15). However, if a naïve T cell does encounter an APC expressing an MHC-peptide complex to which it binds with high affinity, it stops migrating and initiates an activation program that produces a diverse array of cells that orchestrate the short-term and long-term responses to infection.

Early investigations revealed the need for not just one, but two signals to activate naïve cells (the **two-signal hypothesis**). Signal 1 is triggered by TCR engagement and Signal 2 by engagement of costimulatory molecules, such as CD28. Even though it is often still referred to as the two-signal hypothesis, full T-cell activation actually requires a third set of signals, provided by local cytokines (Signal 3), that directs the differentiation of T cells into distinct effector cell types (**Overview Figure 10-2**).

A successful T cell–APC interaction results in the stable organization of signaling molecules into an *immune synapse* (**Figure 10-3**). The TCR/MHC-peptide complexes, which

T-Cell Activation and Differentiation

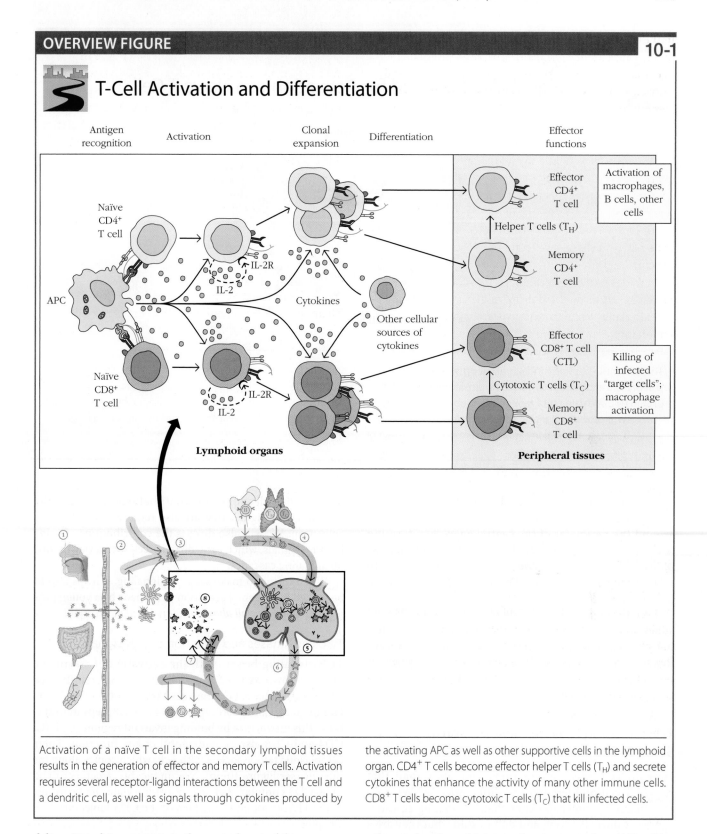

Activation of a naïve T cell in the secondary lymphoid tissues results in the generation of effector and memory T cells. Activation requires several receptor-ligand interactions between the T cell and a dendritic cell, as well as signals through cytokines produced by the activating APC as well as other supportive cells in the lymphoid organ. CD4⁺ T cells become effector helper T cells (T$_H$) and secrete cytokines that enhance the activity of many other immune cells. CD8⁺ T cells become cytotoxic T cells (T$_C$) that kill infected cells.

deliver Signal 1, aggregate in the central part of this synapse (the *central supramolecular activating complex*, or cSMAC). The intrinsic affinity between the TCR and MHC-peptide surfaces is actually quite low (K_d ranges from 10^{-4} M to 10^{-7} M). Signal 1, in fact, is stabilized by the activity of several other molecules, which together increase the **avidity** (the combined affinity of all cell-cell interactions) of the cellular interaction. The coreceptors CD4 and CD8, which are also found in the cSMAC, are key participants and bind MHC class II and MHC class I molecules, respectively.

Three Signals Are Required for Activation of a Naïve T Cell

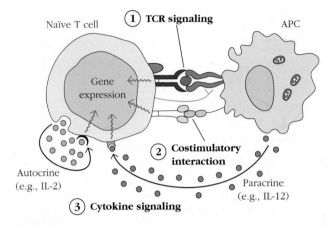

The TCR/MHC-peptide interaction, along with CD4 and CD8 coreceptors and adhesion molecules, provide Signal 1. Costimulation by a separate set of molecules, including CD28 (or ICOS, not shown) provides Signal 2. Together, Signal 1 and Signal 2 initiate a signal transduction cascade that results in activation of transcription factors and cytokines (Signal 3) that direct T-cell proliferation (IL-2) and differentiation (polarizing cytokines). Cytokines can act in an *autocrine* manner, by stimulating the same cells that produce them, or in a *paracrine* manner, by stimulating neighboring cells.

In contrast, adhesion molecules and their ligands (e.g., LFA-1/ICAM-1 and LFA-3/CD2) organize themselves around the perimeter of the central aggregate, forming the *peripheral supramolecular activating complex*, or pSMAC. Interactions between these molecules help to sustain the signals generated by allowing long-term cell interactions.

Even the increased avidity offered by CD4 (or CD8) and adhesion molecules is still not sufficient to fully activate a naïve T cell. Interactions between costimulatory receptors on T cells, including CD28, and costimulatory ligands on antigen-presenting cells, including CD80/86, provide a required second signal (Signal 2).

TCR Signaling Provides Signal 1 and Sets the Stage for T-Cell Activation

How do TCR-MHC interactions generate signals that trigger T-cell proliferation and differentiation? What do costimulatory molecules contribute? Signaling cascades can be at once daunting and dull to study, but they are a vital part of the immune response, which is shaped by the information it receives through these pathways. Here, we describe a few general features of TCR signaling (Signal 1). This outline also provides a useful framework for understanding many other receptor signaling pathways. The next section will address the contributions of Signal 2, the costimulatory signals that are also required for full T-cell activation. We will also discuss the contributions of coinhibitory signals, which provide the T cell with a way to turn off signaling cascades.

When trying to make sense of signaling, it helps also to be guided by a keen observation once made by a young professor: *Signaling is all about location, location, location!*

Tyrosine Kinases and Initiation of TCR Signaling

TCR signaling begins with the activation of a tyrosine kinase known as Lck, a member of the Src family (see Chapter 3). Once a TCR engages MHC-peptide on the surface of an APC, the coreceptor CD4 or CD8 steps in to stabilize this interaction by binding invariant regions of MHC. CD4 and CD8 also provide another key function. Their cytoplasmic tails associate with Lck, which comes along for the ride (see Figure 10-3b). Now in the direct vicinity of the TCR-CD3 complex, Lck can phosphorylate the tyrosines in the ITAMs found in the tails of the CD3 molecules (see Chapter 3).

(An important side note is that Lck, itself, needs to be activated before it is able to act. The details behind this activation are still debated, but, essentially, clustering of molecules triggered by TCR engagement also brings Lck closer to the membrane-associated tyrosine phosphatase CD45.

(a)

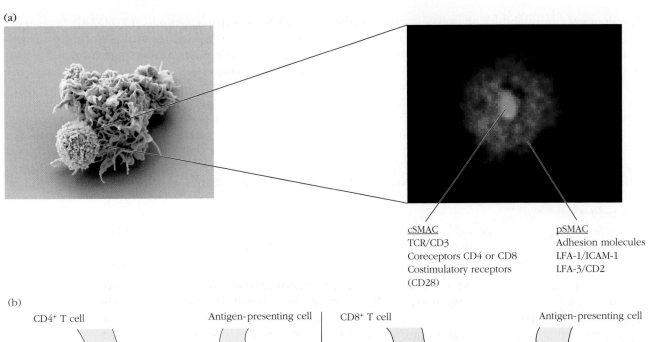

cSMAC
TCR/CD3
Coreceptors CD4 or CD8
Costimulatory receptors
(CD28)

pSMAC
Adhesion molecules
LFA-1/ICAM-1
LFA-3/CD2

(b)

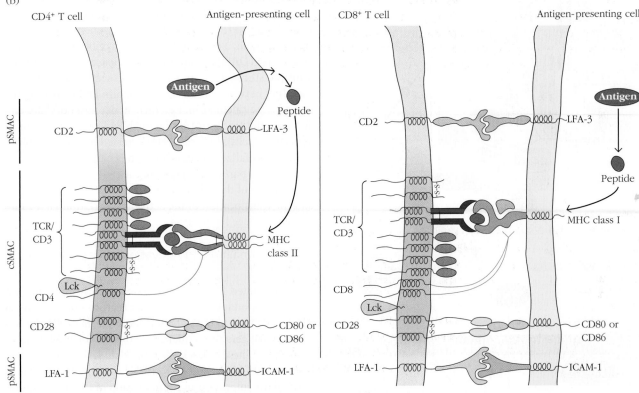

FIGURE 10-3   Surface interactions responsible for T-cell activation. (a) A successful T-cell/dendritic-cell interaction results in the organization of signaling molecules into an immune synapse. A scanning electron micrograph *(left)* shows the binding of a T cell (artificially colored yellow) and dendritic cell (artificially colored blue). A fluorescence micrograph *(right)* shows a cross-section of the immune synapse, where the TCR is stained with fluorescein (green) and adhesion molecules (specifically LFA-1) are stained with phycoerythrin (red). Other molecules that can be found in the central part of the synapse (central supramolecular activation complex [cSMAC]) and the peripheral part of the synapse (pSMAC) are listed. (b) A schematic of the interactions between a CD4$^+$ T cell *(left)* or CD8$^+$ T cell *(right)* and its activating dendritic cell. Dendritic cells (on the *right* in each diagram) process antigens and present peptides associated with MHC class II to CD4$^+$ T cells and present peptides associated with MHC class I to CD8$^+$ T cells. Binding of TCR to MHC-peptide is enhanced by the binding of coreceptors CD4 and CD8 to MHC class II and class I, respectively. CD4 and CD8 also associate with Lck, which helps initiate signaling. CD28 interactions with CD80/86 provide the required costimulatory signals. Adhesion molecule interactions, two of which (LFA-1/ICAM-1, CD2/LFA-3) are depicted, markedly strengthen the connection between the T cell and APC or target cell so that signals can be sustained. [(a) Left: *Dr. Olivier Schwartz, Institut Pasteur/Science Source.* Right: *Republished with permission of American Society of Clinical Investigation, from Michael L. Dustin, "Membrane domains and the immunological synapse: keeping T cells resting and ready," Journal of Clinical Investigation, 2002, January 15; 109(2): 155-160, Figure 1. Permission conveyed through Copyright Clearance Center, Inc.]*

CD45 removes an inhibitory phosphate group on Lck, which is then phosphorylated at an activating tyrosine site by neighboring Lck proteins. Only then can it phosphorylate the CD3 tails.)

Tyrosine phosphorylation of ITAMs generates new docking sites for proteins with SH2 domains, including ZAP-70, another T cell–specific tyrosine kinase. ZAP-70 joins the signaling complex and is phosphorylated and activated by Lck, its new neighbor. ZAP-70 is now in a position to phosphorylate multiple neighboring proteins. The adapter molecule LAT (*linker protein of activated T cells*) and its associate, SLP-76, are among its most important substrates (**Figure 10-4**).

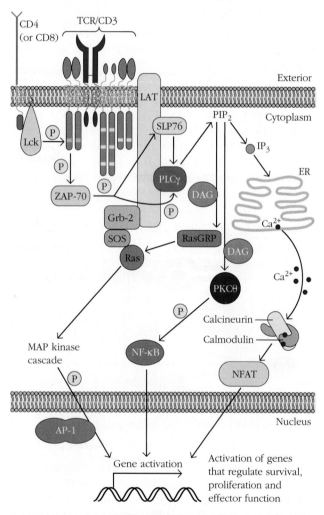

FIGURE 10-4   Schematic of T-cell receptor signaling.
The tyrosine kinases Lck and ZAP-70 initiate TCR signaling and phosphorylate the CD3 tail as well as adapter molecules LAT and SLP-76. Phosphorylation generates docking sites for the assembly and organization of signaling molecules, which lead to the activation of transcription factors. Signaling routes initiated by PLCγ and Ras are highlighted here (see text), but many others are triggered, too.

By phosphorylating LAT and SLP-76, ZAP-70 generates yet another set of new docking sites, permitting the assembly of new complexes of proteins. Phosphorylated adapter molecules provide a critical physical framework for the initiation of the next set of downstream signaling events. The network of signals initiated by the adapter molecule assemblies are varied and complex. Here, we will describe two of the major signaling cascades that contribute to activation of transcription factors responsible for many of the changes associated with T-cell activation. Many of the molecules in these TCR signaling pathways are common to all cells, and may be familiar.

PLCγ Signaling and Activation of Transcription Factors NF-κB and NFAT

One of the proteins that binds phosphorylated LAT via its SH2 domain is the enzyme phospholipase C gamma (PLCγ), which is now in proximity to its substrates, phospholipids in the plasma membrane. PLCγ splits the "head" of the phospholipid PIP_2 from its lipid tails, generating two new signaling molecules: soluble IP_3 and membrane-bound diacylglycerol (DAG). IP_3 induces the release of calcium from multiple stores. Ca^{2+} is a common second messenger in signaling networks, binding calmodulin and activating the phosphatase calcineurin. Calcineurin dephosphorylates and activates the transcription factor NFAT (*nuclear factor of activated T cells*). In T cells, DAG created by PIP_2 hydrolysis binds a specialized form of protein kinase C called PKC-θ (theta). This part of the signaling cascade culminates in the degradation of the inhibitors of transcription factor NF-κB and the translocation of the active transcription factor into the nucleus (see Figure 10-4).

Ras-ERK Signaling and Activation of Transcription Factor AP-1

Phosphorylated LAT also associates with the SH2 domain of Grb2, the adapter molecule that recruits components of the Ras pathway to the signaling complex. In T cells, the Ras pathway triggers the sequential activation of MAP kinases or MAPKs. These kinases phosphorylate serine residues and include RAF, MEK, and ERK. ERK has many targets, but is particularly important in the activation of the transcription factor AP-1 (see Figure 10-4).

As we have mentioned, successful activation of naïve T cells requires simultaneous engagement of the TCR and costimulatory molecules like CD28. As long as costimulatory molecules are engaged, just a few TCR-MHC interactions are needed to trigger a signaling cascade that culminates in (1) *cell survival*, (2) *proliferation*, and (3) *cell differentiation*.

Just as the adapter proteins provide a scaffold for the immune system to organize its signaling proteins, we hope that this section provides a similar scaffold for the organization of the reader's thoughts on the complexities of the first signal required for T-cell activation. We now turn our attention to Signal 2, which coordinates with TCR-generated signals in both time and space.

Key Concepts:

- Three distinct signals are required to induce naïve T-cell activation, proliferation, and differentiation. Signal 1 is generated by the interaction of the TCR-CD3 complex with an MHC-peptide complex on an antigen-presenting cell.

- TCR signaling is initiated by two tyrosine kinases called Lck and ZAP-70, which phosphorylate multiple proteins, generating new docking sites for signaling proteins. Adapter proteins like LAT and SLP-76 use these docking sites to organize new signaling cascades, which activate transcription factors.

- Transcription factors downstream of TCR signaling regulate expression of genes that induce T-cell survival, proliferation, and differentiation into effector cells.

Costimulatory Signals Are Required for Optimal T-Cell Activation Whereas Coinhibitory Signals Prevent T-Cell Activation

What evidence pointed to a requirement for a second signal, the costimulatory signal? In 1987, Helen Quill and Ron Schwartz recognized that high-affinity TCR-MHC interactions in the absence of functional APCs led to T-cell nonresponsiveness rather than activation—a phenomenon they called T-cell **anergy**. They advanced the simple, but powerful two-signal hypothesis. We now know that Signal 2 is generated by interactions between specific **costimulatory receptors** on T cells and costimulatory ligands that are expressed only by professional antigen-presenting cells (**Table 10-1**). When a T cell receives both Signal 1 and Signal 2, it produces cytokines that enhance entry into the cell cycle and proliferation (see Overview Figure 10-2).

Recall from Chapter 4 that antigen-presenting cells, including dendritic cells (DCs), are activated by antigen binding to pattern recognition receptors (PRRs) to express costimulatory ligands (e.g., CD80 and CD86) and produce cytokines that enhance their ability to activate T cells. CD28 is the most commonly cited example of a costimulatory receptor, but other, related molecules that provide costimulatory signals during T-cell activation have since been identified and are also described here.

Coinhibitory receptors, once referred to as *negative costimulatory receptors*, counter the activity of costimulatory receptors and inhibit T-cell activation. They play important roles in (1) maintaining peripheral T-cell tolerance and (2) reducing inflammation both after the natural course of an infection and during responses to chronic infection. In this section we introduce several coinhibitors that have a major impact on T-cell biology.

Costimulatory Receptors: CD28

CD28, a 44-kDa glycoprotein expressed as a homodimer, was the first costimulatory molecule to be discovered (**Classic Experiment Box 10-1**). Expressed by all naïve

TABLE 10-1	T-cell costimulatory and coinhibitory receptors and their ligands	
Receptor on T cell	**Ligand**	**Activity**
Costimulatory receptors		
CD28	CD80 (B7-1) or CD86 (B7-2) *Expressed by professional APCs (and medullary thymic epithelium)*	Activation of naïve T cells
ICOS	ICOS-L *Expressed by B cells, some APCs, and T cells*	Maintenance of activity of differentiated T cells; a feature of T-/B-cell interactions
Coinhibitory receptors		
CTLA-4	CD80 (B7-1) or CD86 (B7-2) *Expressed by professional APCs (and medullary thymic epithelium)*	Negative regulation of the immune response (e.g., maintaining peripheral T-cell tolerance; reducing inflammation; contracting T-cell pool after infection is cleared)
PD-1	PD-L1 or PD-L2 *Expressed by professional APCs, some T and B cells, and tumor cells*	Negative regulation of the immune response, regulation of T_{REG} differentiation
BTLA	HVEM *Expressed by some APCs, and T and B cells*	Negative regulation of the immune response, regulation of T_{REG} differentiation (?)

CLASSIC EXPERIMENT

BOX 10-1

Discovery of the First Costimulatory Receptor: CD28

In 1989, Navy immunologist Carl June took the first step toward filling (and revealing) the considerable gaps in our understanding of T-cell proliferation and activation by introducing a new actor: CD28. CD28 had been recently identified as a dimeric glycoprotein expressed on all human CD4$^+$ T cells and half of human CD8$^+$ T cells, and preliminary data suggested that it enhanced T-cell activation. June and his colleagues specifically wondered if CD28 might be related to the Signal 2 that was known to be provided by APCs (sometimes referred to as "accessory cells" in older literature).

June and his colleagues isolated T cells from human blood by density gradient centrifugation and by depleting a T-cell–enriched population of cells that did not express CD28 (negative selection). They then measured the response of these CD28$^+$ T cells to TCR stimulation in the presence or absence of CD28 engagement (**Figure 1a**).

To mimic the TCR-MHC interaction, June and colleagues used either monoclonal antibodies to the CD3 complex or the mitogen phorbol myristate acetate

(PMA), a protein kinase C (PKC) activator. To engage the CD28 molecule, they used an anti-CD28 monoclonal antibody. They included two negative controls: one population that was cultured in growth medium with no additives and another population that was exposed to phytohemagglutinin (PHA), which was known to activate T cells only in the presence of APCs. This last control was clever and would be used to demonstrate that the researchers' isolated populations were not contaminated with APCs, which would express their own sets of ligands and confound interpretation.

June and colleagues measured the proliferation of each of these populations by measuring incorporation of a radioactive (tritiated) nucleotide, [^{3}H] uridine, which is incorporated by cells that are synthesizing new RNA (and, hence, are showing signs of activation). Their results were striking, particularly when responses to PMA were examined (Figure 1b). As expected, T cells grown without stimulation or with incomplete stimulation (PHA)

remained quiescent, exhibiting no nucleotide uptake. Cell groups treated with stimuli that were known to cause T-cell proliferation—anti-CD3 and PMA—showed evidence of activation, with CD3 engagement producing relatively more of a response than PMA at the time points examined.

The cells treated with anti-CD28 only were just as quiescent as the negative control samples, indicating that engagement of CD28, alone, could not induce activation. However, when CD28 was engaged at the same time cells were exposed to PMA, incorporation of [^{3}H]uridine increased markedly. T cells cotreated with anti-CD28 and anti-CD3 also took up more [^{3}H]uridine than those treated with anti-CD3 alone.

What new RNA were these cells producing? Using Northern blot analysis and functional assays to characterize the cytokines in culture supernatant of the stimulated cells, June and colleagues went on to show that CD28 stimulation induced anti-CD3–stimulated T cells to produce higher levels of cytokines involved in antiviral, antitumor, and proliferative activity,

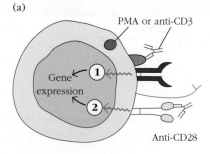

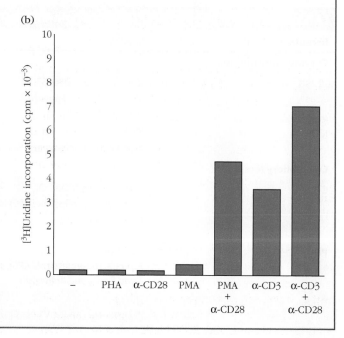

FIGURE 1 Evidence that CD28 is costimulatory ligand for T cell proliferation. (a) June's experimental setup used anti-CD3 monoclonal antibodies or a mitogen, PMA, to provide Signal 1, and anti-CD28 antibodies to provide Signal 2. (b) June measured [^{3}H]uridine incorporation, an indicator of RNA synthesis, in response to various treatments. Addition of stimulating anti-CD28 antibody increased RNA synthesis in response to activation by PMA or anti-CD3. *[Part (b) data from Thompson, C. B., et al. 1989. CD28 activation pathway regulates the production of multiple T-cell-derived lymphokines/cytokines. Proceedings of the National Academy of Sciences USA **86**:1333.]*

(continued)

generating an increase in T-cell immune response.

When this article was published, June did not know the identity of the natural ligand for CD28, or even if the homodimer could be activated in a natural immune context. We now know that CD28 binds to CD80/86 (B7), providing the critical Signal 2 during naïve T-cell activation, a signal required for optimal up-regulation of IL-2 and the IL-2 receptor. Finding this second switch capable of modulating T-cell activation was only the beginning of a landslide of discoveries of additional costimulatory signals—positive and negative—involved in T-cell activation, and the recognition that T cells are even more subtly perceptive than we once appreciated.

Based on a contribution by undergraduate Harper Hubbeling, Haverford College, 2011.

REFERENCE

Thompson, C. B., et al. 1989. CD28 activation pathway regulates the production of multiple T-cell-derived lymphokines/cytokines. *Proceedings of the National Academy of Sciences USA* **86:**1333.

---

and activated human and murine CD4$^+$ T cells, all murine CD8$^+$ T cells, and, interestingly, only 50% of human CD8$^+$ T cells, it markedly enhances TCR-induced proliferation and survival by cooperating with T-cell receptor signals to induce expression of the pro-proliferative cytokine IL-2 and the prosurvival Bcl-2 family member, Bcl-x$_L$.

CD28 binds to two distinct ligands of the B7 family of proteins: CD80 (B7-1) and CD86 (B7-2). These are members of the immunoglobulin superfamily, which have similar extracellular domains. Interestingly, their intracellular regions differ, suggesting that they might not simply act as passive ligands; rather, they may have the ability to generate signals that influence the APC, a view that has some experimental support.

Although most T cells express CD28, most cells in the body do not express its ligands. In fact, only professional APCs have the capacity to express CD80/86. Mature dendritic cells, the best activator of naïve T cells, appear to constitutively express CD80/86, and macrophages and B cells have the capacity to up-regulate CD80/86 after they are activated by an encounter with pathogen (see Chapter 4).

How does CD28 signaling add to the effects of TCR signaling described previously? It appears to make both quantitative and qualitative contributions—enhancing the strength of signal, but also assembling a unique group of signaling molecules. PI3 kinase (PI3K) is one of the most important molecules recruited by CD28. Once brought in proximity to the membrane, it activates other downstream kinases that, in turn, regulate many aspects of cell metabolism, survival and division. These pathways work in concert with those generated by the TCR to prepare a T cell for its role as an effector cell.

Costimulatory Receptors: ICOS

Since the discovery of CD28, several other structurally related receptors have been identified. Like CD28, the closely related *inducible costimulator* (ICOS) provides positive costimulation for T-cell activation. However, rather than binding CD80 and CD86, ICOS binds to another member of the growing B7 family, ICOS-ligand (ICOS-L), which is also expressed on a subset of activated APCs.

Differences in expression patterns of CD28 and ICOS indicate that these positive costimulatory molecules play distinct roles in T-cell activation. Unlike CD28, ICOS is not expressed on naïve T cells; rather, it is expressed on memory and effector T cells. Investigations suggest that CD28 plays a key costimulatory role during the *initiation* of activation and ICOS plays a key role in *maintaining* the activity of already differentiated effector and memory T cells.

Coinhibitory Receptors: CTLA-4 and PD-1

The discovery of **CTLA-4** (**CD152**), the second member of the CD28 family to be identified, caused a stir. Although closely related in structure to CD28 and also capable of binding both CD80 and CD86, CTLA-4 did not act as a positive costimulator. Instead, it antagonized T-cell–activating signals and was the first coinhibitory receptor to be identified.

CTLA-4 is not expressed constitutively on resting T cells. Rather, it is induced within 24 hours after activation of a naïve T cell and peaks in expression within 2 to 3 days post-stimulation. Peak surface levels of CTLA-4 are lower than peak CD28 levels, but because it binds CD80 and CD86 with markedly higher affinity, CTLA-4 competes very favorably with CD28.

Interestingly, CTLA-4 expression levels increase in proportion to the amount of CD28 costimulation, suggesting that CTLA-4 acts to "put the brakes on" the pro-proliferative influence of TCR-CD28 engagement. The importance of this inhibitory function is underscored by the phenotype of CTLA-4 knockout mice, whose T cells proliferate without control, leading to lymphadenopathy (greatly enlarged lymph nodes), splenomegaly (enlarged spleen), autoimmunity, and death within 3 to 4 weeks after birth.

Programmed cell *death-1* (PD-1 or **CD279)** is a coinhibitory receptor expressed by both B and T cells. It binds to two ligands, PD-L1 (B7-H1) and PD-L2 (B7-DC), which are also members of the CD80/86 family. PD-L2 is expressed

Checkpoint Inhibitors: Breakthrough in Cancer Therapy

One of the most exciting and promising breakthroughs in our centuries-long effort to combat cancer has been the development of checkpoint inhibitors, a clinical advance that took full advantage of the remarkable fruits of basic research.

Because tumors are variations of our own cells, the immune response to them is typically not as robust as it is to infection by pathogens. As you have seen in Chapters 8 and 9, our immune system has evolved multiple mechanisms to enforce tolerance to self. All of these, including central tolerance, regulatory T cells, and immune ignorance, impede the response to tumors.

Promisingly, over the last few decades some investigators have successfully identified tumor-specific antigens (TSAs) that theoretically could provoke an immune response. Several experiments in mice revealed the potential for the immune system to react to tumors. However, it was difficult to translate these observations into clinically useful results; most tumors

remained stubbornly resistant to immune cell activity and investigators began to wonder whether our immune system could ever be a useful partner in efforts to treat cancer. And frankly, many wondered whether a cure for cancer, which includes a very heterogeneous set of diseases, was even a realistic goal.

The discovery of coinhibitor molecules in the 1990s not only inspired breakthroughs in T-cell research but also inspired a new excitement, even among the most skeptical, about the possibility of treating cancer. Dr. Jim Allison, one of the discoverers of CTLA-4, had witnessed the ability of the immune system to attack tumors in some of his earlier experiments. He wondered if tumors may be using T-cell coinhibitors to their advantage. If tumor cells or cells associated with tumors expressed ligands for T-cell coinhibitory molecules, they might be inactivating existing tumor-specific T cells. If you removed this "brake," you might be able to unleash T-cell activity against the tumor.

Allison and his colleagues first tested the possibility in mice. They injected antibodies that blocked CTLA-4 interactions into mice with tumors. The results, reported first in *Science* in 1996, were dramatic. Although it took time to work, the antibody enhanced rejection of both developing and established tumors. Tumors shrunk and, in some cases, even disappeared, whether they expressed CTLA-4 ligands or not.

The remarkable success of CTLA-4 blockade in the treatment of many different tumors in mice led Allison and colleagues to shift their attention to human patients. They persuaded one brave company, Medarex, to take a chance on the approach and developed a humanized antibody against CTLA-4 (**Figure 1**). This antibody, now called ipilimumab (or *ippy*), was entered into a clinical trial and at first did not look promising. But patience was rewarded when it became clear in subsequent trials that, just as in mice, it took time for the antibody to re-activate T cells.

(continued)

---

predominantly on APCs; however, PD-L1 is expressed more broadly and may help to mediate T-cell tolerance in nonlymphoid tissues. Recent data suggest that interactions between PD-1 and PD-L1/2 may also regulate the differentiation of regulatory T cells.

The discovery of CTLA-4 and PD-1 inspired the development of checkpoint inhibitors, a very promising new immunotherapy for cancer (**Clinical Focus Box 10-2**). Molecules that block the interaction between these coinhibitory receptors and their ligands have been shown to enhance latent T-cell activity against tumors and are now being used in the clinic.

A Coinhibitory Receptor Expressed by Many Immune Cell Types: BTLA

B and *T* *l*ymphocyte *a*ttenuator (BTLA or CD272) is a coinhibitory receptor that is more broadly expressed: not only has it been found on conventional T_H cells as well as $\gamma\delta$ T cells and regulatory T cells, but it is also expressed on NK cells, some macrophages and dendritic cells, and most highly

on B cells. Interestingly, BTLA's primary ligand appears not to be a B7 family member, but a TNF receptor family member known as *h*erpes *v*irus *e*ntry *m*ediator (HVEM), which is also expressed on many cell types. Studies on the role of this interesting coinhibitory receptor-ligand pair are ongoing, but there are indications that BTLA-HVEM interactions also play a role in down-regulating inflammatory and autoimmune responses.

As you can imagine, the expression and activity of costimulatory and coinhibitory molecules must be carefully regulated, both temporally and spatially. Naïve T cells, for example, do not express coinhibitory receptors, allowing them to be activated in secondary lymphoid tissue during the initiation of an immune response. On the other hand, effector T cells up-regulate coinhibitory receptors at the end of an immune response, when proliferation is no longer advantageous. However, these generalizations belie the complexity of regulation of this highly important costimulatory network, and investigators are still working to understand the details.

CLINICAL FOCUS (continued)   **BOX 10-2**

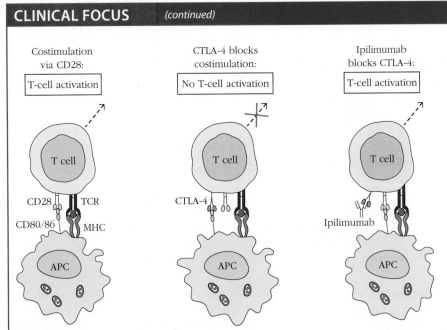

Costimulation via CD28:
T-cell activation

CTLA-4 blocks costimulation:
No T-cell activation

Ipilimumab blocks CTLA-4:
T-cell activation

FIGURE 1 **How the checkpoint inhibitor ipilimumab works.** Naïve T cells are activated when they engage both TCR and CD28 costimulatory receptors on antigen-presenting cells. Activated T cells up-regulate the coinhibitory molecule CTLA-4, which binds CD28 ligands even more strongly. This interaction inactivates the T cell. However, when the CTLA-4 engagement is blocked with the antibody ipilimumab, the T cell is reactivated.

The antibody was first tested in patients with malignant melanoma, a fatal cancer of pigment-producing skin cells. Although not all responded to the treatment, those who did experienced long-lasting and often dramatic remission.

It became the first therapy for the deadly cancer that showed a survival advantage. The U.S. Food and Drug Administration approved the antibody for the treatment of melanoma in 2011. It now joins an arsenal of other antibodies that block the

engagement of coinhibitory molecules, including antibodies to PD-1, one of the other potent coinhibitors. Together these therapeutic antibodies are referred to as *checkpoint inhibitors*, to describe their ability to inhibit the brakes (or "checkpoints") on the immune system. They represent the best of bench-to-bedside (translational) research, as well as the triumph of imagination, innovation, and sheer doggedness by individuals in the research community. In 2015, Jim Allison was the recipient of one of the most prestigious scientific honors, the Lasker Award.

The effectiveness of checkpoint inhibition continues to impress the medical community. However, not every patient and not every tumor responds and investigators are working furiously to understand why. The identification of other coinhibitory molecules expressed by T cells, including Lag-3, Tim-3, and TIGIT, may offer new hope and new targets for immunotherapy.

REFERENCE

Leach, D. R., M. F. Krummel, and J. P. Allison. 1996. Enhancement of antitumor immunity by CTLA-4 blockade. *Science* **271:**1734.

As the genome continues to be explored, additional costimulatory and coinhibitory molecules—both positive and negative in influence—have been identified. Understanding their regulation and function will continue to occupy the attention of the immunological community and has already provided the clinical community with new tools for manipulating the immune response during transplantation and disease (see Clinical Focus Box 10-2).

Key Concepts:

• Signal 2 is produced by costimulatory ligands expressed by APCs interacting with costimulatory receptors (CD28 or ICOS) expressed by T cells.

• Coinhibitory receptors, such as PD-1 and CTLA-4, are expressed by activated T cells and bind to ligands on antigen-presenting cells. When coengaged with the TCR, they send signals that inhibit T-cell activation.

Clonal Anergy Results If a Costimulatory Signal Is Absent

Experiments with cultured cells show that if a naïve T cell's TCR is engaged (Signal 1) in the absence of a suitable costimulatory signal (Signal 2), that specific T cell clone becomes unresponsive to subsequent stimulation, a state referred to as clonal *anergy* (**Figure 10-5**). There is good evidence that both CD4[+] and CD8[+] T cells can be anergized, but most studies of anergy have been conducted with CD4[+] T_H cells.

Anergy can be demonstrated experimentally with systems designed to engage the TCR in the absence of costimulatory molecule engagement. For instance, T cells specific for an MHC-peptide complex can be induced to proliferate in vitro by incubation with activated APCs that express both the appropriate MHC-peptide combination and CD80/86. However, exposing T cells to fixed APCs that cannot up-regulate CD80/86, or physically blocking the CD28-CD80/86 interaction with antibodies, renders

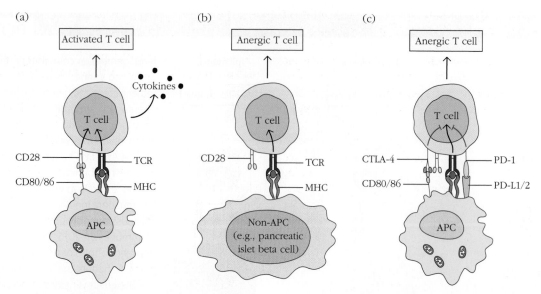

FIGURE 10-5 **Signals that lead to clonal anergy versus clonal expansion.** (a) Resting T cells that engage antigen and costimulatory ligands CD80/86 are activated and expand. (b) T cells that engage antigen on the surface of a non-antigen-presenting cell (such as a pancreatic islet beta cell) will not receive costimulatory signals and, instead, are inactivated or anergized. (c) Anergy can also be induced when a T cell engages antigen and receives inhibitory rather than costimulatory signals through coinhibitory receptors such as CTLA-4 and PD-1.

T cells unresponsive. Anergic T cells are no longer able to secrete cytokines or proliferate in response to subsequent stimulation.

The requirement for costimulatory ligands to activate a T cell decreases the probability that autoreactive T cells that have escaped the thymus will be activated and become dangerous. For instance, a naïve T cell expressing a T-cell receptor specific for an MHC class I–insulin peptide complex would be rendered nonresponsive if it encountered a pancreatic islet beta cell expressing this MHC class I–peptide complex. Why? Islet beta cells cannot be induced to express costimulatory ligands, and the encounter would result in T-cell anergy, preventing an immune attack by these T cells (see Figure 10-5b).

Interactions between coinhibitory receptors and ligands can also induce anergy (see Figure 10-5c). This phenomenon, which applies only to activated T cells that have up-regulated coinhibitory receptors, could help curb T-cell proliferation when antigen is cleared. Coinhibition may also contribute to the T-cell "exhaustion" during chronic infection, such as that caused by mycobacteria, HIV, hepatitis virus, and even exposure to tumor antigens. T cells specific for these pathogens and antigens express high levels of PD-1 and BTLA, and become functionally anergic. Recent therapies designed to block coinhibitory interactions have successfully reactivated T cells and are the basis for checkpoint inhibitor drugs that are considered one of the most exciting recent breakthroughs in cancer therapy (see Clinical Focus Box 10-2).

The biochemical pathways that lead to anergy are still under investigation. Chronic TCR stimulation may lead to intracellular changes in signaling molecules, and regulatory T cells may also contribute. Although the in vivo importance of anergy has been contested, recent studies clearly show that anergic $CD4^+$ T cells help maintain maternal tolerance to a fetus during pregnancy. Interestingly, anergic $CD4^+$ cells may also differentiate into regulatory T cells, which play a key role in maintaining immune tolerance in all vertebrates (see Chapters 8 and 16).

> **Key Concept:**
>
> • In the absence of costimulatory signals (Signal 2) or in the presence of coinhibitory signals, T-cell receptor engagement results in T-cell inactivity or clonal anergy.

Cytokines Provide Signal 3

We have now seen that naïve T cells are activated when they simultaneously engage both MHC-peptide complexes and costimulatory ligands on antigen-presenting cells, particularly dendritic cells. However, the outcome of T-cell activation is critically shaped by the activity of soluble cytokines produced by both APCs and T cells. These assisting cytokines are referred to, by some, as *Signal 3* (see Overview Figure 10-2).

Cytokines bind surface cytokine receptors, stimulating a cascade of intracellular signals that enhance both proliferation and/or survival. **IL-2** is one of the best-known cytokines involved in T-cell activation and plays a key role in inducing optimal T-cell proliferation, particularly when antigen and/or costimulatory ligands are limiting. TCR and costimulatory signals induce transcription of genes for both IL-2 and the α chain (CD25) of the high-affinity IL-2 receptor. These signals together also enhance the stability of IL-2 mRNA. The combined increase in IL-2 transcription and improved IL-2 mRNA stability results in a 100-fold increase in IL-2 production by the activated T cell. Secretion of IL-2 and its subsequent binding to the high-affinity IL-2 receptor induces activated naïve T cells to proliferate vigorously.

As we will see shortly, Signal 3 also includes another important set of cytokines, known as **polarizing cytokines**. These are produced by a variety of cell types, including APCs, T cells, and innate lymphoid cells (ILCs), and play central roles in determining what types of effector cells naïve T cells will become.

> **Key Concept:**
> - Signal 3 is provided by soluble cytokines and includes IL-2, which enhances T-cell proliferation, as well as polarizing cytokines, which play a key role in determining the type of effector cell that a T cell becomes.

Antigen-Presenting Cells Provide Costimulatory Ligands and Cytokines to Naïve T Cells

Which cells are capable of providing both Signal 1 and Signal 2 to a naïve T cell? Although almost all cells in the body express MHC class I, only professional APCs—dendritic cells, activated macrophages, and activated B cells—express the high levels of MHC class II molecules that are required for naïve CD4$^+$ T-cell activation (**Figure 10-6**). Professional APCs are also capable of expressing costimulatory ligands. (Only one other cell type, the medullary thymic epithelial cell, is known to share this capacity; see Chapter 8.) In order to initiate a T-cell response, all professional APCs must first be activated by microbial components via their pattern recognition receptors (PRRs). This encounter enhances antigen presentation activity, and up-regulates expression of MHC and costimulatory ligands.

Professional APCs are more diverse in function and origin than originally imagined, and each subpopulation differs both in the ability to display antigen and in the expression of costimulatory ligands (see Figure 10-6). DCs appear to be the most potent activators of naïve T cells and come in many different varieties. Conventional dendritic cells, which

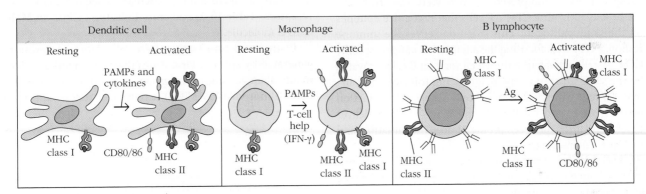

DCs and macrophages

Key features
- Phagocytic
- Express receptors for apoptotic cells, DAMPs, and PAMPs
- Localize to tissues
- Localize to T-cell zone of lymph nodes following activation (DCs)
- Constitutively express high levels of MHC class II molecules and antigen-processing machinery
- Express costimulatory molecules following activation
- Variety of subtypes, some migrating, some resident, some better at cross-presenting
- Specific subtypes may specialize in activating different T cells

B cells

Key features
- Internalize antigens via BCRs
- Constitutively express MHC class II molecules and antigen-processing machinery
- Express costimulatory molecules following activation

FIGURE 10-6 Comparison of professional antigen-presenting cells that induce T-cell activation. This figure describes general features of three major classes of professional APCs. Dendritic cells are the best activators of naïve T cells. This may be due, in part, to their relatively high levels of expression of MHC and costimulatory molecules when they are mature and activated. Activated B cells interact most efficiently with differentiated T$_{FH}$ cells that are specific for the same antigen that activated them. Macrophages play several different roles, processing and distributing antigen in secondary lymphoid tissues as well as interacting with effector cells in the periphery. It is important to recognize that the distinctions shown are rules of thumb only. Functions among the APC classes overlap, and investigators now recognize many different subsets within each major group of APCs, each of which may act independently on different T-cell subsets. This diversity may be a consequence of activation by different innate immune receptors or may reflect the existence of independent cell lineages. Note that activation of effector and memory T cells is not as dependent on costimulatory interactions.

arise from bone marrow stem cells, include migratory cells that are activated by antigen in peripheral tissues and travel to regional (draining) lymph nodes to expose themselves to naïve T cells. Conventional dendritic cells also include cells that do not circulate as vigorously. These resident DCs process antigen that comes into the lymph node from the blood. They can also acquire antigen from other antigen-presenting cells. Langerhans cells are another type of migratory DC found exclusively in the skin and discovered in the nineteenth century by an observant young medical student, Paul Langerhans. They originally arise from embryonic hematopoietic stem cells and replenish themselves not from bone marrow stem cells, but simply by dividing. Once activated by interactions with antigen, Langerhans cells migrate to both local and distal lymph nodes to meet and activate T cells (see Chapter 13).

Resting B cells residing in follicles also gain the capacity to activate T cells, although at the late stages of an immune response. Once they bind antigen through their B-cell receptor (BCR), they up-regulate MHC class II and CD80/86, and present antigen to activated CD4$^+$ T cells they encounter at the border between the follicle and T-cell zone (see Chapter 14). Because of their unique ability to internalize pathogen via specific BCRs and present them in MHC class II, B cells are best at activating CD4$^+$ T cells that recognize epitopes on the same pathogen. This situation serves the immune response very well, focusing the attention of antigen-specific CD4$^+$ T cells activated in the T-cell zone on B cells activated by the same antigen in the neighboring follicle. The pairing of B cells with their helper T cells occurs at the junction between the B- and T-cell zones and allows T cells to deliver the help required for B-cell proliferation, differentiation, and memory generation (see Chapter 11).

Several other antigen-presenting cells play a role in the primary immune response, including macrophages that line the sinuses of lymph node and spleen. These filter antigen and transfer it to other antigen-presenting cells in the T-cell zone. Some of the major antigen-presenting cell types are shown in Table 7-4 (and see Chapter 13). This list is not complete and is likely to be joined by even more subtypes as research continues.

Key Concepts:

- Several different professional antigen-presenting cells can provide Signals 1, 2, and 3 to a naïve T cell, although dendritic cell subsets appear particularly potent.

- Professional APCs differ by location and migratory behavior. Some circulate actively and others reside for long periods of time in specific tissues and organs.

- B cells use their capacity as antigen-presenting cells to solicit help from CD4$^+$ T cells during an immune response in secondary lymphoid tissue.

Superantigens Are a Special Class of T-Cell Activators

Superantigens are viral or bacterial proteins that bind simultaneously to specific V_β regions of T-cell receptors and to the α chain of MHC class II molecules. V_β regions are encoded by over 20 different V_β genes in mice and 65 different genes in humans. Each superantigen displays a "specificity" for one of these V_β versions, which can be expressed by up to 5% of T cells, regardless of their antigen specificity. This clamp-like connection mimics a strong TCR-MHC interaction and induces activation, bypassing the need for TCR antigen specificity (**Figure 10-7**). Superantigen binding, however, does not bypass the need for costimulation; professional APCs are still required for full T-cell activation by these microbial proteins.

Both endogenous superantigens and exogenous superantigens have been identified. *Exogenous* superantigens are soluble proteins secreted by bacteria. Among them are a variety of **exotoxins** secreted by gram-positive bacteria, such as staphylococcal enterotoxins, toxic shock syndrome toxin, and exfoliative dermatitis toxin. Each of these exogenous superantigens binds particular V_β sequences in T-cell receptors (**Table 10-2**) and cross-links the TCR to an MHC class II molecule.

Endogenous superantigens are cell-membrane proteins generated by specific viral genes that have integrated into mammalian genomes. One group, encoded by mouse mammary tumor virus (MMTV), a retrovirus that is integrated

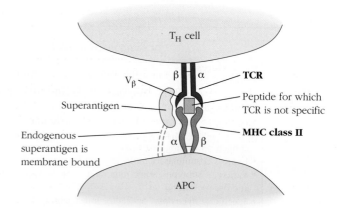

FIGURE 10-7   Superantigen-mediated cross-linkage of T-cell receptor and MHC class II molecules. A superantigen binds to all TCRs bearing a particular V_β sequence regardless of their antigen specificity. Exogenous superantigens are soluble secreted bacterial proteins, including various exotoxins. Endogenous superantigens are membrane-embedded proteins produced by certain viruses; they include Mls antigens encoded by mouse mammary tumor virus.

TABLE 10-2   Exogenous superantigens and their V$_\beta$ specificity

Superantigen	Disease*	V$_\beta$ SPECIFICITY	
		Mouse	**Human**
Staphylococcal enterotoxins			
SEA	Food poisoning	1, 3, 10, 11, 12, 17	ND
SEB	Food poisoning	3, 8.1, 8.2, 8.3	3, 12, 14, 15, 17, 20
SEC1	Food poisoning	7, 8.2, 8.3, 11	12
SEC2	Food poisoning	8.2, 10	12, 13, 14, 15, 17, 20
SEC3	Food poisoning	7, 8.2	5, 12
SED	Food poisoning	3, 7, 8.3, 11, 17	5, 12
SEE	Food poisoning	11, 15, 17	5.1, 6.1–6.3, 8, 18
Toxic shock syndrome toxin (TSST1)	Toxic shock syndrome	15, 16	2
Exfoliative dermatitis toxin (ExFT)	Scalded skin syndrome	10, 11, 15	2
Mycoplasma arthritidis supernatant (MAS)	Arthritis, shock	6, 8.1–8.3	ND
Streptococcal pyrogenic exotoxins (SPE-A, B, C, D)	Rheumatic fever, shock	ND	ND

*Disease results from infection with bacteria that produce the indicated superantigens.

ND = not determined.

into the DNA of certain inbred mouse strains, produces proteins called **minor lymphocyte-stimulating (Mls) determinants**, which bind particular V$_\beta$ sequences in T-cell receptors and cross-link the TCR to MHC class II molecules. Four Mls superantigens, originating from distinct MMTV strains, have been identified.

Because superantigens bind outside the TCR antigen-binding cleft, any T cell expressing that particular V$_\beta$ sequence will be activated by a corresponding superantigen. Hence, the activation is polyclonal and can result in massive T-cell activation, resulting in overproduction of T$_H$-cell cytokines and systemic toxicity. Food poisoning induced by staphylococcal enterotoxins and toxic shock induced by toxic shock syndrome toxin are two examples of disorders caused by superantigen-induced cytokine overproduction.

Given that superantigens directly activate the host T-cell response, it is difficult to imagine what value they have for the pathogens that make them. There is some evidence that such antigen-nonspecific T-cell activation and inflammation hampers the development of a coordinated antigen-specific response. Some speculate that the large-scale proliferation and cytokine production that results from superantigen exposure harms the cells and microenvironments that are required to start a normal response; others argue that these events induce T-cell tolerance to the pathogen.

Key Concepts:

- Superantigens are products of bacteria and viruses that activate T cells in an antigen-nonspecific manner.

- Superantigens bind directly to select TCR β chains. This interaction activates all T cells expressing these chains, regardless of the TCR α chain they express. This large-scale T-cell activation leads to inflammation that can cause disease, such as toxic shock syndrome.

Helper CD4$^+$ T-Cell Differentiation

One to two days after successful engagement with a dendritic cell in the T-cell zone of a secondary lymphoid organ, a naïve T cell enlarges into a blast cell and undergoes repeated rounds of cell division. Signals 1 plus 2 induce up-regulation of expression and activity of prosurvival genes (e.g., *Bcl-2*), as well as the transcription of genes for both IL-2 and the α chain (CD25) of the high-affinity IL-2 receptor (**Figure 10-8**). The combined effect on a naïve T cell results in robust proliferation. Activated T cells divide two or three times per day for 4 to 5 days, generating a clone of progeny cells.

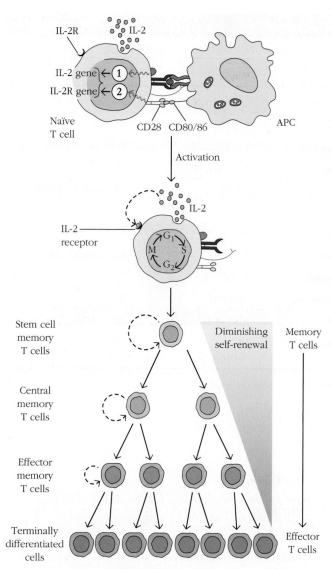

FIGURE 10-8   Activation and differentiation of naïve T cells into effector and memory T cells. Signal 1 and Signal 2 cooperate to enhance the transcription and stability of mRNA for IL-2 and the high-affinity IL-2R chain (also referred to as CD25). Secreted IL-2 binds the IL-2R, which generates signals that enhance the entry of the T cell into the cell cycle. T cells undergo many rounds of proliferation, during which they differentiate into memory cell subsets and effector helper or cytotoxic cells.

Activated T cells and their progeny gain unique functional abilities, becoming *memory* and *effector* helper or cytotoxic T cells (see Figure 10-8). These coordinate efforts with other immune cells to both indirectly and directly clear infections. CD8$^+$ cytotoxic T cells leave the secondary lymphoid tissues and circulate to sites of infection, where they bind and kill infected cells. CD4$^+$ helper T cells secrete cytokines that orchestrate the activity of several other cell types, including B cells, macrophages, and other T cells. Some CD4$^+$ T cells, particularly those that help B cells and those

that become central memory T cells, stay within the secondary lymphoid tissue. Others return to the sites of infection and enhance the activity of macrophages and cytotoxic cells. Still others circulate to other tissues to join the first lines of attack against re-infection.

Whereas naïve T cells can live for months, effector cells tend to be short-lived and have life spans that range from a few days to a few weeks. Memory cells typically live longer, and some extend their lives by dividing over the many months or even years that they are present in an organism.

Effector T cells come in many more varieties than originally anticipated, and each subset plays a specific and important role in the immune response. The first effector cell distinction to be recognized, of course, was between CD8$^+$ T cells and CD4$^+$ T cells. Activated CD8$^+$ T cells acquire the ability to induce the death of target cells, becoming "killer" or "cytotoxic" T lymphocytes (CTLs; also called T$_C$ cells). Because cytotoxic CD8$^+$ T cells recognize peptide bound to MHC class I, which is expressed by almost all cells in the body, they are perfectly poised to clear the body of cells that have been internally infected with the pathogen that resulted in their activation. On the other hand, activated CD4$^+$ T cells, or **T helper (T$_H$) cells**, acquire the ability to secrete factors that enhance the activation and proliferation of other cells. Specifically, they regulate the activation and antibody production of B cells; enhance the phagocytic, antimicrobial, cytolytic, and antigen-presenting capacity of macrophages; and are required for the development of B-cell and CD8$^+$ T-cell memory.

As immunologists developed and adopted more tools to distinguish proteins expressed and secreted by T cells, it became clear that CD4$^+$ helper T cells are particularly diverse, differentiating into several different subtypes, each of which secretes a signature set of cytokines. The cytokines secreted by CD4$^+$ T$_H$ cells act either directly on the same cell that produced them (i.e., in an *autocrine* fashion) or indirectly by binding to receptors on cells in the vicinity (i.e., in a *paracrine* fashion). Below we describe the major features and functions of the best-characterized CD4$^+$ helper T-cell subsets. Although CD8$^+$ cytotoxic T cells also secrete cytokines, and there are indications that CD8$^+$ T cells may also differentiate into more than one killer subtype, the diversity of CD8$^+$ T-cell effector functions is clearly more restricted than that of CD4$^+$ T$_H$ cells. The generation and activity of CTLs are described in more detail in Chapters 12 and 13.

Helper T Cells Can Be Divided into Distinct Subsets and Coordinate Type 1 and Type 2 Responses

Tim Mosmann, Robert Coffman, and colleagues can be credited with one of the earliest and most elegant experiments that demonstrated the diversity of helper CD4$^+$

T cells. Earlier investigations revealed that helper T cells produced a diverse array of cytokines. Although this could mean that each cell made many different cytokines, Mosmann and Coffman tested a different possibility: that the $CD4^+$ cell population was, itself, heterogeneous and made of distinct helper subtypes that made distinct sets of cytokines.

Specifically, they isolated individual T-cell clones from the polyclonal mixture of $CD4^+$ T cells generated in the spleen of an immunized mouse. At the time, the immunology community did not have the tools to identify most cytokines directly. However, Mosmann and Coffman developed clever (and elaborate) strategies to define which cytokines each clone generated. To the surprise of many, they showed that the individual clones did not make all of the cytokines; instead, they fell into two distinct categories, which they named T_H1 and T_H2.

Because T_H1 and T_H2 subsets were originally identified from in vitro studies of cloned T-cell lines, some researchers doubted that they represented true in vivo subpopulations. They suggested instead that these subsets might represent different maturational stages of a single lineage. In addition, the initial failure to locate either subset in humans led some to believe that T_H1, T_H2, and other subsets of $CD4^+$ T helper cells were not present in our species. Further research corrected these views. In many in vivo systems, the full commitment of populations of T cells to either a T_H1 or T_H2 phenotype occurs late in an immune response. Hence, it was difficult to find clear T_H1 and T_H2 subsets in studies on healthy human subjects, where these cells would not have developed.

These experiments set the stage for the discovery that naïve $CD4^+$ T cells can adopt not just two but six or more distinct helper fates, depending on the signals they receive during activation. T_H1 and T_H2 subpopulations were quickly joined by the T_H17, T_{FH}, and T_{REG} subpopulations, each of which produces a distinct cytokine profile and regulates distinct activities within the body (**Overview Figure 10-9**). The genotype, phenotype, and function of these five populations are well described and will be the focus of this section. However, other helper subpopulations continue to reveal themselves. Whether these merit their own category or are variants within the major subgroups is not always clear. Nonetheless, two additional $CD4^+$ T helper subpopulations, T_H9 and T_H22, have recently gained prominence and will be discussed briefly.

In retrospect, we probably should have anticipated the heterogeneity of helper responses, which allows an organism to "tailor" a response to a particular type of pathogen. The type of effector T_H cell that a naïve T cell (also called a T_H0 cell) becomes depends largely on the kind of infection that occurs. For example, worm (helminth) infections stimulate the differentiation of activated $CD4^+$ T cells into T_H2 cells, which help B cells to produce IgE antibodies that bind the parasite and trigger the release of damaging agents from granulocytes.

On the other hand, infection with an intracellular virus or bacterium induces differentiation of $CD4^+$ T cells into T_H1 helpers that enhance the cytolytic activity of macrophages and $CD8^+$ T cells, which then kill infected cells. Responses to fungi stimulate the differentiation of other helper responses than the responses to extracellular bacteria, and so on.

The reality is, of course, even more complex. Infections evoke the differentiation of more than one helper subtype and recruit other T-cell subsets and ILCs (see Chapters 4 and 13) that both overlap in function and add unique features to the immune response. Investigators are now finding it useful to divide the complex responses into two major types (type 1 and type 2). **Type 1 responses** are triggered by viral and many bacterial infections and polarize $CD4^+$ T cells to the T_H1 and T_H17 helper subsets. These work with other immune cells (including ILC1s and ILC3s) to generate protective cytotoxic responses. **Type 2 responses** are triggered by larger parasites, including worms, protozoa, and allergens. These polarize $CD4^+$ T cells to T_H2 and T_H9 helper subsets, which work with other populations of immune cells (including ILC2s) to generate an IgE response.

What regulates the differentiation of helper T cells to each effector subset and their participation in type 1 or type 2 responses is still being actively investigated. We describe the fundamentals of our current understanding in this section.

Key Concepts:

- Activation of naïve T cells leads to the differentiation of effector cells, which regulate the response to pathogen, and of long-lived memory T cells, which are responsible for the stronger and quicker response to future infections by the same pathogen.

- Although each T-cell effector subset is unique, they share some functions with other cells, including ILCs, and work together to mount responses that are divided into two major categories.

- Type 1 responses to viruses and many bacteria involve the activation of effector subsets that coordinate cytotoxic responses.

- Type 2 responses to worms, protozoa, and allergens involve the activation of effector subsets that coordinate IgE and eosinophilic responses.

The Differentiation of Helper T-Cell Subsets Is Regulated by Polarizing Cytokines

As we have seen, T-cell activation requires TCR and costimulatory receptor engagement, both of which are supplied by an activated APC. It is now clear that the functional fate of activated T cells is determined by signals they receive from additional cytokines generated during the response.

OVERVIEW FIGURE   10-9

T Helper Subset Differentiation

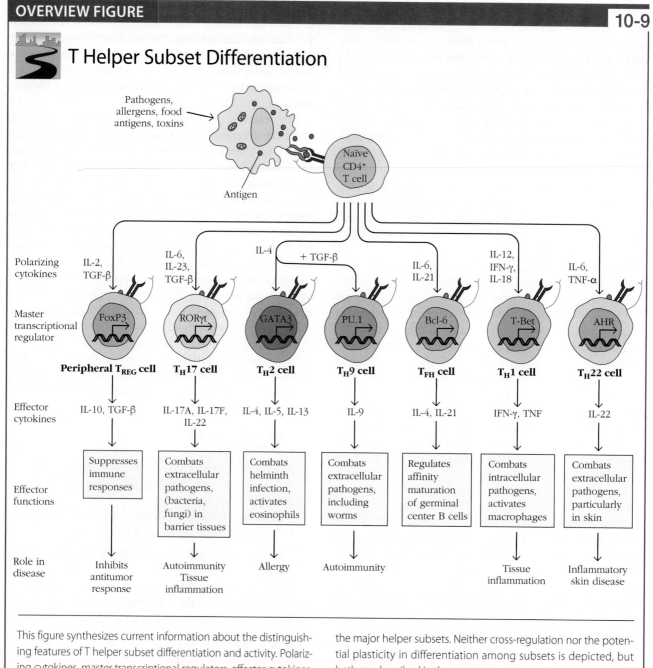

This figure synthesizes current information about the distinguishing features of T helper subset differentiation and activity. Polarizing cytokines, master transcriptional regulators, effector cytokines, and broad functions in health and disease are depicted for each of the major helper subsets. Neither cross-regulation nor the potential plasticity in differentiation among subsets is depicted, but both are described in the text.

These cytokines (Signal 3) are referred to as **polarizing cytokines** because they are responsible for guiding a helper T cell toward one of several different effector fates (**Figure 10-10**).

For example, T cells that are activated in the presence of IL-12 and IFN-γ tend to differentiate, or polarize, to the T_H1 lineage, whereas T cells that are activated in the presence of IL-4 and IL-6 polarize to the T_H2 lineage (**Figure 10-11**). Overview Figure 10-9 and **Table 10-3** show the polarizing cytokines responsible for the development of each CD4+ helper T-cell subset. These will be discussed in greater detail shortly.

Polarizing cytokines can be generated by the stimulating APC itself, or by neighboring innate and adaptive immune cells that have also been activated by antigen. Which polarizing cytokines are produced during an immune response depends on which cell has been alerted (DC, macrophage, B cell, NK cell, etc.), what pathogen has alerted it, and what pattern recognition receptors have been stimulated, and its tissue context. Innate interactions therefore play a critical role in determining the outcome of adaptive responses, by shaping the information that an antigen-presenting cell conveys to naïve T cells (see Figure 4-23).

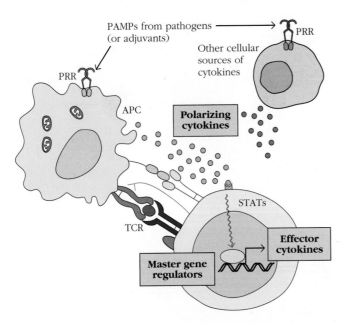

FIGURE 10-10 General events and factors that drive T$_H$ subset polarization. Interaction of pathogen with pattern recognition receptors (PRRs) on dendritic cells and other neighboring immune cells determines which polarizing cytokines are produced and, hence, into which T helper subset a naïve T cell will differentiate. In general, polarizing cytokines that arise from dendritic cells or other neighboring cells interact with cytokine receptors and generate signals that induce transcription of unique master gene regulators. These master regulators, in turn, regulate expression of various genes, including effector cytokines, which define the function of each subset.

Recall from Chapter 4 that APCs and other innate immune cells are activated by interaction with pathogens bearing pathogen-associated molecular patterns (PAMPs). These PAMPs bind PRRs, including (but not limited to) Toll-like receptors (TLRs). PRR interactions activate dendritic cells by stimulating the up-regulation of MHC and costimulatory proteins. They also determine the type of cytokine(s) that dendritic cells and other immune cells secrete. These polarizing cytokines, in turn, dictate the fate a T cell will adopt following activation (see Overview Figure 10-9).

For example, double-stranded RNA, a product of many viruses, binds TLR3 receptors on dendritic cells, initiating a signaling cascade that results in the production of IL-12, which directly promotes T$_H$1 differentiation and a type 1 response. On the other hand, worms stimulate PRRs on innate immune cells, including mast cells, which generate IL-4. IL-4 directly promotes polarization of activated T cells to the T$_H$2 subset, which coordinates a type 2 response and IgE-mediated killing of worms (see Figure 10-11). In this case, the main polarizing cytokine is not made by the activating dendritic cell, but is generated by a neighboring immune cell. The pathogen interactions that give rise to the polarizing cytokines that drive T helper cell differentiation are often complex and continue to be actively investigated.

Adjuvants, which have been used for decades to enhance the efficacy of vaccines, are now understood to exert their influence on the innate immune system by regulating the expression of costimulatory ligands and polarizing cytokines by APCs, events that ultimately shape the consequences of T-cell activation. PAMPs and cytokines such as IL-12, produced by APCs themselves, are considered natural adjuvants. Dead mycobacteria, which clearly activate many PRRs, have long been used as a very potent adjuvant for immune responses in mice. Very few adjuvants are approved for human vaccination, but given our new and evolving understanding of the molecules that stimulate PRRs and the consequences of that stimulation, investigators expect that we will one day be able to shape the effector response to vaccine antigens by varying the adjuvants—natural and/or synthetic—included in vaccine preparations (see Chapter 17).

Key Concepts:

- The helper lineage adopted by a naïve CD4$^+$ T cell is determined by the set of cytokines it encounters during activation. These are referred to as *polarizing cytokines*.

- The cytokines that an activating APC delivers are tailored to the type of antigen it encounters and determined by the pattern recognition receptors (PRRs) the antigen activates.

- Adjuvants enhance vaccine effectiveness in large part by inducing APCs to produce costimulatory ligands and polarizing cytokines. In theory, they can be adjusted to generate the effector population of choice.

Each Effector Helper T-Cell Subset Has Unique Properties

Each helper T-cell subset is defined by an array of features, the details of which can rapidly overwhelm a new (or old) student of immunology. Understanding these specifics, however, helps clarify the role each subset plays in resolving infection and causing disease. And having a ready reference to them makes it easier to decipher the primary literature describing advances. However, some generalizations provide a useful conceptual framework for organizing some of these details. Consider the following:

- Each of the major T helper cell subsets is characterized by (1) a distinct set of *polarizing cytokines* that induce the expression of (2) a **master gene regulator** that regulates expression of (3) a signature set of **effector cytokines** the T-cell population produces once it is fully differentiated.

- Which effector subset an activated helper cell becomes a member of depends on the quality and quantity of signals its naïve cell precursor receives from APCs in a secondary lymphoid organ; that activity, in turn, depends

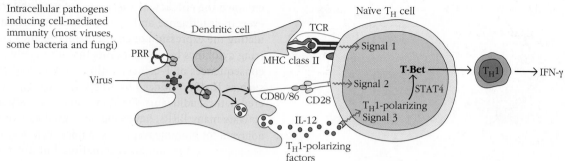

(a) **Type 1 response**

Intracellular pathogens inducing cell-mediated immunity (most viruses, some bacteria and fungi)

(b) **Type 2 response**

Pathogens inducing humoral immunity, particularly extracellular parasites (e.g., worms)

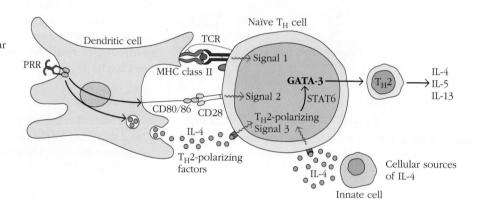

FIGURE 10-11   Initiation of T_H1 and T_H2 responses by pathogens. (a) Intracellular pathogens activate a cascade of signals that polarize cells to the T_H1 lineage and initiate type 1 responses (which can also activate T_H17 cells). For example, viruses interact with PRRs (e.g., TLR3) that induce dendritic cells to generate IL-12. This binds to receptors on naïve T cells (which have also engaged MHC-peptide and costimulatory ligands on the DC), activating a cytokine signal transduction pathway mediated by STAT4 that induces expression of the master regulator T-Bet. T-Bet, in turn, activates expression of effector cytokines, including IFN-γ, which define the T_H1 subset's functional capacities as a type 1 responder. (b) Extracellular pathogens activate signal cascades that polarize naïve T cells to the T_H2 lineage and initiate a type 2 response. Parasitic worms interact with PRRs on neighboring immune cells (such as mast cells, basophils, or germinal center B cells), triggering the release of the signature T_H2 polarizing cytokine IL-4. This interacts with receptors on T cells that activate STAT6, up-regulating expression of the transcriptional regulator GATA-3. GATA-3, in turn, induces expression of the T_H2 and type 2 effector cytokines, including IL-4, IL-5, and IL-13.

on the nature of the pathogen the APC encountered at the site of infection.

• Broadly speaking, T_H1 cells regulate immunity to intracellular bacteria and viruses, T_H17 cells regulate immunity to extracellular bacteria and fungi, T_H2 (and perhaps T_H9) cells regulate the response to worms, and T_{FH} cells regulate humoral immunity (B cells). It is important to recognize that multiple CD4$^+$ effector T-cell subsets may have the potential to provide help to B cells. T_H1 and T_H17 subsets generally encourage B cells to produce antibodies that contribute to cell-mediated immunity (e.g., isotypes like IgG2a that can "arm" NK cells for cytotoxicity; see Chapter 12). T_H2 cells encourage B cells to produce antibodies that mediate the clearance of extracellular pathogens (e.g., isotypes like IgE that induce the release of molecules that harm

extracellular parasites). T_{REG} cells are inhibitory and play a role in terminating healthy immune responses and inhibiting autoimmunity.

• Helper T-cell subsets often "cross-regulate" each other. The cytokines they secrete typically enhance their own differentiation and expansion, while inhibiting commitment to other helper T-cell lineages. This is particularly true of the T_H1 and T_H2 subset pair, as well as the T_H17 and T_{REG} subset pair.

• Helper T cell lineages may not be fixed; some subsets can assume the cytokine secretion profile of other subsets if exposed to a different set of cytokines, particularly early in the differentiation process.

• The precise biological function and sites of differentiation and activity of each subset continue to be actively investigated. Much remains unknown.

	Polarizing cytokines	**Master gene regulators**	**Effector cytokines**	**Functions**
T$_H$1	IL-12 IFN-γ IL-18	T-Bet	IFN-γ TNF	Enhances APC activity Enhances T$_C$ activation Protects against intracellular pathogens Involved in delayed type hypersensitivity, autoimmunity
T$_H$2	IL-4	GATA-3	IL-4 IL-5 IL-13	Protects against extracellular pathogens (particularly IgE responses) Involved in allergy
T$_H$9	IL-4 TGF-β	PU.1	IL-9	Protects against extracellular pathogens Involved in mucosal autoimmunity
T$_H$17	TGF-β IL-6 (IL-23)	RORγt	IL-17A IL-17F IL-22	Protects against fungal and extracellular bacterial infections Contributes to inflammation, autoimmunity
T$_H$22	IL-6 TNF-α	AHR	IL-22	Protects against extracellular pathogens Involved in inflammatory skin disease
T$_{REG}$	TGF-β IL-2	FoxP3	IL-10 TGF-β	Inhibits inflammation Inhibit antitumor responses
T$_{FH}$	IL-6 IL-21	Bcl-6	IL-4 IL-21	B-cell help in follicles and germinal centers

TABLE 10-3   Regulation and function of T helper subtypes

We start our deeper discussion of helper cell subset characteristics by focusing on the first two subsets to be identified: T$_H$1 and T$_H$2 cells. They provide an illustrative example of the features that distinguish T helper cells as well as the relationship between subsets. We follow this section with summaries of what is currently understood about the more recently characterized helper subsets: T$_H$17, T$_{REG}$, T$_{FH}$, T$_H$9, and T$_H$22. Overview Figure 10-9 and Table 10-3 provide useful summaries of the information presented here, which probes some of the complexities of origin and function that are being continually updated in this rapidly moving field.

The Differentiation and Function of T$_H$1 and T$_H$2 Cells

The key polarizing cytokines that induce differentiation of naïve T cells into **T helper type 1 (T$_H$1) cells** are IL-12, IL-18, and IFN-γ (see Overview Figure 10-9 and Figure 10-11). IL-12 is produced by dendritic cells after an encounter with pathogens via PRRs, including TLR4 and TLR3. IL-12 expression is also up-regulated in response to IFN-γ, which is generated by activated T cells and activated NK cells. IL-18, which is also produced by dendritic cells, promotes proliferation of developing T$_H$1 cells and enhances their own production of IFN-γ. These polarizing cytokines trigger signaling pathways in naïve T cells

that up-regulate expression of the master gene regulator T-Bet. This master transcription factor induces expression of signature type 1 effector cytokines, including IFN-γ and TNF, and differentiation of CD4$^+$ T cells to the T$_H$1 lineage.

IFN-γ is a particularly potent type 1 effector cytokine. It activates macrophages, stimulating these cells to increase microbicidal activity, up-regulate the level of MHC class II molecules, and, as mentioned above, secrete cytokines such as IL-12, which further enhance T$_H$1 differentiation. IFN-γ secretion also induces antibody class switching in B cells to IgG classes (such as IgG2a in mice) that support phagocytosis and fixation of complement. Finally, IFN-γ secretion promotes the differentiation of fully cytotoxic T$_C$ cells from CD8$^+$ precursors by activating the dendritic cells that engage naïve T$_C$ cells. These combined effects make the T$_H$1 subset particularly well suited to respond to viral infections and other intracellular pathogens. They also contribute to the pathological effects of T$_H$1 cells, which are also involved in the delayed-type hypersensitivity response to poison ivy (see Chapter 15).

Just as differentiation to the T$_H$1 subset is promoted by IL-12 and IFN-γ, differentiation of naïve T cells into **T helper type 2 (T$_H$2) cells** is promoted by a defining and potent polarizing cytokine, IL-4 (see Overview Figure 10-9 and Figure 10-11). Exposing naïve helper T cells to IL-4 at the beginning of an immune response causes them to differentiate into T_H2 cells and, interestingly, T_H2 development is greatly favored over T_H1 development. Even in the presence of IFN-γ and IL-12, naïve T cells will differentiate into T_H2 effectors if IL-4 is present. IL-4 triggers a signaling pathway within the T cell that up-regulates the master gene regulator GATA-3, which, in turn, regulates expression of type 2–specific cytokines, including IL-4, IL-5, and IL-13.

Investigators are still working to understand which cells produce the polarizing cytokine IL-4. Several may be responsible, depending on the origin and type of infection. Recent data suggest that type 2 innate lymphoid cells (ILC2) are a major source of IL-4 in lymphoid tissue responding to an intestinal worm infection. However, other studies suggest that ILC2s may not be required. Many innate cells, such as mast cells, basophils, eosinophils, $\gamma\delta$ T cells, and natural killer T (NKT) cells, can be induced to make IL-4 after exposure to pathogen and could influence helper T-cell development if they make their way to secondary lymphoid tissue. Cells that are more routinely found in secondary lymphoid tissue, including germinal-center B cells, T_{FH} cells, and a specialized DC subset, can also produce IL-4 and may also play a role in polarizing naïve T cells. Finally, T_H2 cells themselves are an excellent source of additional IL-4 and may enhance polarization events of neighboring naïve T cells. Clarity will require more research.

The type 2 effector cytokines produced by T_H2 cells help clear extracellular parasitic infections, including those caused by worms. IL-4, the defining T_H2 effector cytokine, acts on both B cells and eosinophils. It induces eosinophil differentiation, activation, and migration and promotes B-cell activation and class switching to IgE. These effects act synergistically because eosinophils express IgE receptors (FcϵR) that, when cross-linked, release inflammatory mediators that are particularly good at attacking roundworms. Thus, IgE antibody can form a bridge between the worm and the eosinophil, binding to worm antigens via its variable regions and binding to FcϵR via its constant region. IgE antibodies also mediate allergic reactions, and the role of T_H2 activity in these pathological responses is described in Chapter 15.

IL-5 can also induce B-cell class switching to IgG subclasses that do not activate the complement pathway (e.g., IgG1 in mice), and IL-13 functions largely overlap with IL-4. Finally, IL-4 itself can suppress the expansion of T_H1-cell populations.

T_H1 and T_H2 Cross-Regulation

The major effector cytokines produced by T_H1 and T_H2 subsets (IFN-γ and IL-4, respectively) not only influence the overall immune response, but also influence the fate and function of helper T cells, themselves. First,

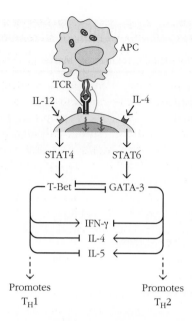

FIGURE 10-12   Cross-regulation of T helper cell subsets by transcriptional regulators. GATA-3 and T-Bet reciprocally regulate differentiation of T_H1 and T_H2 lineages. IL-12 promotes the expression of the T_H1-defining transcription factor, T-Bet, which induces expression of T_H1 effector cytokines, including IFN-γ. At the same time, T-Bet represses expression of the T_H2-defining master transcriptional regulator, GATA-3, as well as expression of the effector cytokines IL-4 and IL-5. Reciprocally, IL-4 promotes expression of GATA-3, which up-regulates the synthesis of IL-4 and IL-5, and at the same time represses expression of T-Bet and the T_H1 effector cytokine IFN-γ.

they promote the growth and enhance the polarization of the subset that produces them; second, they inhibit the development and activity of the opposite subset, an effect known as *cross-regulation* (**Figure 10-12**). For instance, IFN-γ (secreted by the T_H1 subset) inhibits proliferation of the T_H2 subset, and IL-4 (secreted by the T_H2 subset) down-regulates the secretion of IL-12 by APCs, thereby inhibiting T_H1 differentiation. Furthermore, IL-4 enhances T_H2-cell development by making T_H cells less susceptible to the T_H1-promoting cytokine signals (and vice versa).

Similarly, these cytokines have opposing effects on target cells other than T_H subsets. In mice, where the T_H1 and T_H2 subsets have been studied most extensively, the cytokines have distinct effects on the type of antibody made by B cells. For example, antibody isotype IgG2a enhances cell-mediated immunity by arming NKT cells, whereas the isotypes IgG1 and IgE enhance humoral immunity by their activities on extracellular pathogens. IFN-γ secreted by the T_H1 subset promotes IgG2a production by B cells but inhibits IgG1 and IgE production. On the other hand, IL-4 secreted by the T_H2 subset promotes production of

IgG1 and IgE and suppresses production of IgG2a. Thus, these effects on antibody production are consistent with the T_H1 and T_H2 subsets' overall tendencies to promote cell-mediated versus humoral immunity, respectively.

IL-10 was once considered a signature effector cytokine for T_H2 cells. However, we now know that many other cell types and multiple helper subsets also produce IL-10. IL-10 inhibits T_H1-cell differentiation, although not directly. Instead, IL-10 acts on monocytes and macrophages, interfering with their ability to activate the T_H1 subset by abrogating their activation, specifically by (1) inhibiting expression of MHC class II molecules, (2) suppressing production of bactericidal metabolites (e.g., nitric oxide) and various inflammatory mediators (e.g., IL-1, IL-6, IL-8, GM-CSF, G-CSF, and TNF-β), and even by (3) inducing apoptosis.

The master regulators T-Bet and GATA-3 also play an important role in cross-regulation (see Figure 10-12). Specifically, the expression of T-Bet drives expression of genes required for T-cell differentiation into the T_H1 lineage and represses genes involved in differentiation along the T_H2 pathway. Expression of GATA-3 does the opposite, promoting T_H2-specific gene expression and repressing genes associated with T_H1 differentiation. Consequently, cytokine signals that induce one of these transcription factors set in motion a chain of events that represses the other.

T_H17 Cells

The discovery of the **T helper type 17 (T_H17) cell** subset was inspired in part by the recognition that IL-12, one of the polarizing cytokines that induces T_H1 development, was a member of a larger family of cytokines that shared the same subunit (p40), including IL-23. At first investigators assumed that the defects in helper T-cell activity in p40 knockout mice were due entirely to a deficiency of T_H1 cells. However, once it was understood that p40 was required not only for IL-12, but also for IL-23 production, investigators quickly realized that p40 knockout mice also failed to produce a T-cell subset that required IL-23. Some of the functions originally attributed to IL-12–induced T_H1 cells were actually carried out by an IL-23–induced T helper-cell subpopulation now referred to as the T_H17 lineage because of their ability to secrete IL-17.

Several cytokines are involved in polarizing naïve T cells to the T_H17 lineage, including TGF-β, IL-6, and IL-23 (see Overview Figure 10-9). T_H17 cells have dual roles and can adopt both a protective and pathogenic function; the cytokine mixtures that lead to both types are still being clarified. Activation of naïve T cells in the presence of TGF-β and IL-6 appears to induce development to the more protective form of T_H17, which can produce IL-10 and patrol our barrier tissues (**Clinical Focus Box 10-3**; and see Chapter 13). Activation in the presence of IL-23,

however, drives development of a more inflammatory version of T_H17 cells. IL-23 is produced by dendritic cells in response to antigens produced by some bacteria and fungi. Alone, IL-23 cannot polarize naïve T cells to the T_H17 lineage; rather, it enhances their ability to produce inflammatory cytokines. Inflammatory T_H17 is an important participant in the protective response against fungi and some bacteria at mucosal barriers (see Chapter 13). Not only do T_H17 cells produce IL-17 (also called IL-17A), but they also produce IL-17F and IL-22. Like all of our effector T-cell subsets, they can also go off-script and are one of several culprits behind chronic inflammatory and autoimmune disorders, including inflammatory bowel disease, arthritis, and multiple sclerosis.

T_H17-cell differentiation is controlled by the master transcriptional regulator, RORγt, an orphan steroid receptor. RORγt knockout mice are less susceptible to autoimmune disease, including experimental autoimmune encephalitis (EAE, a mouse model of multiple sclerosis), largely because of the reduction in pathogenic T_H17 cells.

Because of the pathogenic potential of inflammatory T_H17 cells, many researchers are interested in identifying approaches to inactivate them. IL-23 inhibitors are already in clinical trials. Dectin-1, a PRR that binds fungal antigens and triggers IL-23 production by T_H17 cells, may also be a promising new drug target.

Peripheral T_{REG} Cells

Another major CD4$^+$ T-cell subset negatively regulates T-cell responses and plays a critically important role in peripheral tolerance by limiting autoimmune T-cell activity. This subset of T cells, called **peripheral T_{REG} (pT_{REG}) cells** (previously known as *induced T_{REG} [iT_{REG}] cells*), is similar in function to the thymic T_{REG} cells (tT_{REG}s) (previously known as *natural T_{REG} [nT_{REG}] cells*) that differentiate to this lineage in the thymus (see Chapter 8). Peripheral T_{REG} cells, however, arise from naïve T cells that are activated in secondary lymphoid tissue in the presence of TGF-β (see Overview Figure 10-9). The vitamin A metabolite retinoic acid (RA) also contributes to T_{REG} polarization.

TGF-β triggers a signaling cascade that up-regulates expression of FoxP3, the master transcriptional regulator responsible for pT_{REG} commitment. pT_{REG} cells secrete the effector cytokines IL-10 and TGF-β, which down-regulate inflammation via their inhibitory effects on APCs, and can also exert their suppressive function by interacting directly with T cells. The depletion of pT_{REG} cells in otherwise healthy animals leads to multiple autoimmune outbreaks, revealing that even healthy organisms are continually warding off autoimmune responses. A reduction in pT_{REG} cells also increases women's susceptibility to pre-eclampsia, a pregnancy complication thought to be due to an autoimmune response that impairs blood flow at the placenta.

What a Disease Revealed about the Physiological Role of T$_H$17 Cells

Experiments that reveal the inner workings of normal healthy immune cells offer invaluable insights into what can and does go wrong. However, at times we need "experiments of nature"—the diseases themselves—to clarify how the immune system works in healthy individuals. Job syndrome, a rare disease in which patients suffer from defects in bone, teeth, and immune function, helped us solve the mystery of the physiological function of T$_H$17 cells.

People with Job syndrome suffer from recurrent infections, typically of the lungs and skin. The painful abscesses that are often a feature of the disease and the trials endured by its patients explain its name, which comes from a biblical story

of a man (Job) who is subject to horrific hardship as a test of his faith. Another cardinal feature of the disease, elevated IgE levels in serum, is the basis for its more formal name, *hyper-IgE* syndrome (HIES). HIES comes in two major forms. Job syndrome, the most common, is also referred to as type 1 or autosomal dominant (AD)-HIES. Patients with type 2 HIES, which we will not discuss here, do not have trouble with bone or dental development.

The abundance of IgE originally suggested to investigators who were savvy about the roles of helper T-cell subsets that type 1 HIES symptoms may be caused by an imbalance among helper T-cell responses. Initial work did not reveal a difference in the activities of T$_H$1 and

T$_H$2, but as the diversity of T helper subsets was revealed, investigators looked to other subsets. A number of groups from Japan and America independently discovered that lymphocytes from patients with HIES were unable to respond to select cytokines, including IL-6, IL-10, and IL-23. An analysis of genes that could explain this signaling failure revealed a dominant negative mutation in *STAT3*, a transcription factor activated by multiple cytokine receptors.

The absence of signaling through these cytokines suggested a specific problem with polarization to the T$_H$17 helper subset. Indeed, circulating T$_H$17 cells were absent in patients with HIES with *STAT3* mutations. The investigators

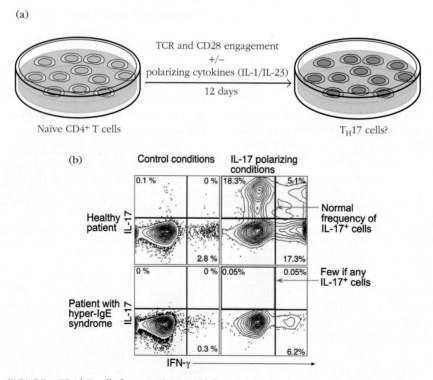

(a)

Naïve CD4$^+$ T cells → TCR and CD28 engagement +/− polarizing cytokines (IL-1/IL-23) 12 days → T$_H$17 cells?

(b)

Control conditions | IL-17 polarizing conditions

Healthy patient — IL-17: 0.1 %, 0 %, 18.3%, 5.1% — Normal frequency of IL-17$^+$ cells; 2.8 %, 17.3%

Patient with hyper-IgE syndrome — IL-17: 0 %, 0 %, 0.05%, 0.05% — Few if any IL-17$^+$ cells; 0.3 %, 6.2%

IFN-γ →

FIGURE 1 CD4$^+$ T cells from patients with hyper-IgE syndrome do not differentiate into T$_H$17 cells. (a) Conditions used to polarize CD4$^+$ T cells to the T$_H$17 lineage in vitro. The ability of the cells to make IL-17 (a feature of T$_H$17 cells) or IFN-γ (a property of T$_H$1, not T$_H$17 cells) after they were polarized was assessed by staining for intracellular cytokines. (b) Flow cytometric analysis of intracellular staining. The boxes outlined in red indicate the quadrant where IL-17$^+$ (T$_H$17) cells would appear. *[Part (b) reprinted with permission from Macmillan Publishers Ltd, from Milner J. D. et al., "Impaired T(H)17 cell differentiation in subjects with autosomal dominant hyper-IgE syndrome." Nature, 2008, April 10; 452(7188): 773–6, Fig. 2. Permission conveyed through Copyright Clearance Center, Inc.]*

(continued)

directly tested their hypothesis that this absence was due to a failure of CD4$^+$ T cells to polarize normally. Specifically, they removed T cells from blood, and exposed them to conditions that would ordinarily polarize them to the T_H17 lineage: T-cell receptor stimulation (Signal 1) in the presence of costimulatory signals (CD28; Signal 2) and cytokines known to drive human T_H17 differentiation (IL-6, TGF-β, and IL-23; Signal 3) (**Figure 1a**). They stained the cells for intracellular expression of T_H17's signature cytokine, IL-17, and performed flow cytometry (Figure 1b; also see Chapter 20). Although CD4$^+$ T cells from patients with HIES were able to develop into other helper lineages (and, as you can see from the flow cytometry contour plots, were also able to make IFN-γ, indicating they could become T_H1 cells), they could not be induced to secrete IL-17. Specifically,

whereas 18.3% of T cells from healthy patients that were subject to these conditions stained with antibodies against IL-17, 0.05% (essentially none) of the T cells from patients with HIES stained with the same antibodies.

These striking observations suggest that the recurrent infections that patients with HIES experience are caused at least in part by the absence of T_H17 cells. Reciprocally, they indicate that T_H17 cells play an important role in controlling the type of infection that afflicts these patients, including *Staphylococcus aureus* skin infections and pneumonia. The critical role of T_H17 in controlling bacterial and fungal infections at epithelial surfaces has now been confirmed.

We now know that STAT3 is involved in the generation of other lymphocyte subsets, including T_{FH} cells, as well as memory B and CD8$^+$ T cells. Alterations in these

cell functions help to explain some of the other features of Job disease, such as the increased susceptibility to viral infection. Ironically, the precise cause of aberrant production of IgE is still being debated. Recent work suggests that it is related to deficits in IL-21 signaling, which also depends on STAT3. Research specifically points to the possibility that STAT3 inhibits B-cell class switching to IgE; however, the scientific jury about this signature feature of Job's disease is still out.

REFERENCES

Milner, J. D., et al. 2008. Impaired T_H17 cell differentiation in subjects with autosomal dominant hyper-IgE syndrome. *Nature* **452**:773.

Kane, A., A. Lau, R. Brink, S. G. Tangye, and E. K. Deenick. 2016. B-cell-specific STAT3 deficiency: insight into the molecular basis of autosomal-dominant hyper-IgE syndrome. *Journal of Allergy and Clinical Immunology* **138**:1455.

Advances Box 10-4 describes discoveries that revealed the importance of this subset of T_{REG} cells in maintaining a mother's tolerance to her developing fetus.

T_H17 and T_{REG} Cross-Regulation

Just as T_H1 and T_H2 cells reciprocally regulate each other, T_{REG} and T_H17 cells also cross-regulate each other. TGF-β plays a role in the differentiation of both T_{REG} and T_H17 cells. Alone, it induces a program that leads cells to the T_{REG} fate. When accompanied by IL-6, however, TGF-β induces T_H17 differentiation. How is this possible? TGF-β appears to up-regulate expression of both master regulators FoxP3 and RORγt, which control T_{REG} and T_H17 differentiation, respectively. In combination with signals generated by IL-6, TGF-β actually inhibits FoxP3 expression, letting RORγt dominate and induce T_H17 development. Other variables play a role, too, including the presence of IL-21 and IL-23, which favor the differentiation of T_H17 cells. High levels of TGF-β and the availability of retinoic acid favor the differentiation of T_{REG} cells.

The cross-regulation of anti-inflammatory and inflammatory cell populations is elegantly adaptive, particularly at barrier tissues. Anti-inflammatory T_H17 and pT_{REG} populations dominate the barrier tissues of healthy animals, where innate cells generate immunosuppressive cytokines and metabolites (like TGF-β and RA). Activation of antigen-presenting cells by invading pathogens induces the release of a variety of inflammatory cytokines, which shift the development of cells away from pT_{REG}s and anti-inflammatory

T_H17s toward the pro-inflammatory T_H17 cells, so a proper defense can be mounted.

T_{FH} Cells

T follicular helper (T_{FH}) cells are now firmly established as a distinct helper T-cell subset. Whereas T_H2 cells focus their help on the response to worms and enhance IgE production, T_{FH} cells provide cognate help to a wide range of B cells and enhance switching to a variety of antibody classes. T_{FH} cells are required for the formation of germinal centers and provide the signals that drive affinity maturation of B cells.

Cytokines that polarize activated T cells toward the T_{FH} lineage include IL-6 and IL-21, both of which are produced by a variety of activated antigen-presenting cells. Together with signals produced by the TCR and costimulatory molecules, these cytokines induce the expression of Bcl-6, a repressor that is the master transcriptional regulator of T_{FH} cells (see Overview Figure 10-9). Cross-regulation is also a feature of T_{FH} function: Bcl-6 expression inhibits T-Bet, GATA-3, and RORγt expression, thus inhibiting T_H1, T_H2, and T_H17 differentiation, respectively, while inducing T_{FH} polarization. On the other hand, IL-2 production inhibits differentiation to the T_{FH} lineage.

Although both T_{FH} and T_H2 cells secrete IL-4, T_{FH} cells are best characterized by their secretion of IL-21, which contributes to B-cell differentiation. They also express high levels of CD40L, which is required for cognate B-cell help (**Figure 10-13**). T_{FH} cells are also uniquely characterized by the expression of CXCR5, the chemokine receptor that

ADVANCES   **BOX 10-4**

Tolerance for Two: Jumping Genes, T_{REG}s, and the Evolution of Immune Tolerance during Pregnancy

We, like many mammals, face a serious immunological problem. We carry our embryos internally and nourish them via a placenta, which provides babies with access to maternal blood and also exposes baby and mom to each other's antigens (**Figure 1**). Embryos express paternal antigens, many of which, from the mom's perspective, are foreign and can activate an immune response that can reject the developing fetus the way we reject unmatched transplanted organs. Fetuses, in fact, can be considered an allograft—or, more accurately, a "semi-allograft." But they are not rejected.

How do mothers avoid rejecting their babies? As early as the 1950s, one of the founders of immunology, Peter Medawar, recognized the problem and

posed several possible explanations. Perhaps the uterus was an immune-privileged site and did not allow cellular contact between mother and fetus? Perhaps the fetus was too immature to produce antigen capable of inducing immune responses?

Most of these early speculations turned out to be wrong, and it took decades of research into the cells and molecules of the immune system for the mechanism behind maternal tolerance of the fetus to reveal itself. It is not surprising that we have evolved several strategies to protect mother and fetus. Some strategies were predicted. For instance, maternal and fetal expression of indoleamine 2,3-dioxygenase (IDO) suppresses T-cell responses. Some, however, were wholly unanticipated. For instance, placental cells express a unique

form of MHC molecule, HLA-G, that has an unexpectedly broad inhibitory effect on maternal NK and CD8$^+$ cytotoxic cells. (For a recent review on the role of HLA-G in maternal/fetal immune tolerance, see Ferreira et al., *Trends in Immunology* 2017;**38:**272.)

When CD4$^+$ regulatory T cells were discovered in the 1990s, they seemed ideal candidates for a role in fetal-maternal tolerance. T_{REG} cells have the potential to be antigen specific, not just generally immunosuppressive, and they could travel to sites of potential inflammation. They also have the potential to become memory cells and enhance the safety of future pregnancies.

Studies in the last decade confirmed this suspicion. In fact, maternal T_{REG}s are activated as soon as the new embryo is implanted—an event that establishes the placenta and exposes the mother to

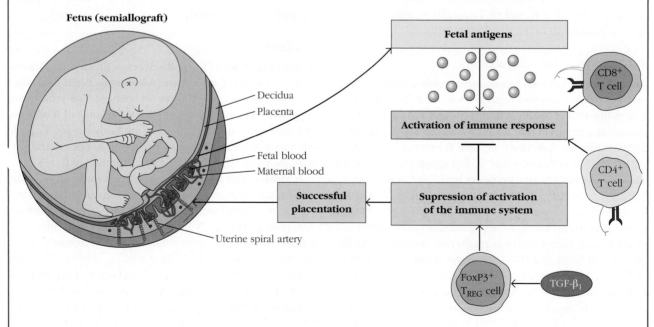

FIGURE 1 The anatomy and cell biology of the human placenta. Figure 1 shows the general structure of the human placenta, where maternal blood supply (uterine spiral arteries) and maternal uterine lining (decidua) are in intimate contact with fetal blood supply and fetal antigens. Fetal antigens on the surface

of placental cells (trophoblasts) and/or generated from cell debris are processed and presented by maternal APCs, which activate CD8$^+$ and CD4$^+$ T cells. TGF-β, which can be produced by both maternal and fetal cells in the placental environment, polarizes CD4$^+$ T cells to the T_{REG} lineage.

(continued)

ADVANCES *(continued)* **BOX 10-4**

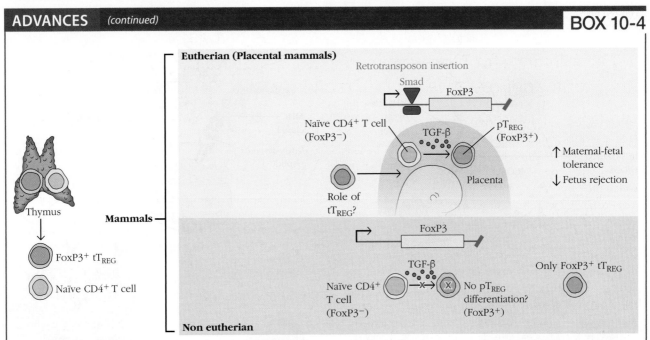

FIGURE 2 **Genetic differences between the *FoxP3* enhancer in placental (eutherian) and nonplacental animals.** The enhancer in placental animals *(left)* includes a sequence that comes from a retrotransposon. This sequence binds the TGF-β–associated transcription factor Smad, which regulates expression of FoxP3 in naïve T cells. While both placental and nonplacental animals can generate FoxP3$^+$ T$_{REG}$ cells in the thymus, only placental animals can generate peripheral, FoxP3$^+$ T$_{REG}$ cells in response to TGF-β.

paternal alloantigens (see Figure 1). They are also recruited to the uterus by chemokines and pregnancy hormones themselves. Their importance is underscored by observations in mice and women that reductions in the numbers of T$_{REG}$s increase the risk of pre-eclampsia, a proinflammatory and possibly autoimmune complication of pregnancy that can harm both fetus and mother.

Recall that regulatory T cells are generated in two ways—thymic T$_{REG}$s, or *tT$_{REG}$s*, are made as T cells develop within the thymus from immature thymocytes in response to very high-affinity interactions with self antigen (**Figure 2**). Peripheral T$_{REG}$s, or *pT$_{REG}$s*, are generated from naïve cells that have left the thymus and are activated in the presence of polarizing cytokines, including TGF-β.

Which regulatory T-cell population helps to maintain maternal tolerance—and does it matter? Fascinating work from the Rudensky lab suggests that it does. His team was interested in understanding what controlled expression of FoxP3, the master gene regulator of T$_{REG}$s.

When they compared the sequences of enhancers across species they made an unexpected discovery. The *FoxP3* enhancer of placental (eutherian) mammals included a "piece" that was not present in the *FoxP3* enhancer of marsupials, mammals whose embryos develop externally, in pouches (like kangaroos and opossums). This extra sequence seemed to have appeared suddenly, in evolutionary terms.

When they took a closer look at the function of this piece, they showed that it was uniquely responsible for the development of peripheral T$_{REG}$s. How did this relatively simple change in sequence have such a dramatic effect? The answer, quite satisfyingly, makes studying this chapter worthwhile! The sequence addition made the *FoxP3* gene in naïve T cells sensitive to the polarizing cytokine TGF-β. Specifically, it provided a spot for the transcription factor, Smad, just downstream of TGF-β receptor signaling, to bind and induce FoxP3 expression (Figure 2). Marsupial *FoxP3* genes don't respond to TGF and they may not even generate peripheral T$_{REG}$s.

Where did this piece come from? Like the *RAG* genes (see Evolution Box 6-3), this piece may have "jumped" into the enhancer region as a transposable element. Although it was not necessarily a huge leap to suggest that T$_{REG}$s played a role in maternal-fetal tolerance, it is fair to say that few anticipated the small jump that would permit them to have an impact on the evolution of placental animals.

REFERENCES

Ferreira, L. M. R., T. B. Meissner, T. Tilburgs, and J. L. Strominger. 2017. HLA-G: at the interface of maternal-fetal tolerance. *Trends in Immunology* **38**:272.

Samstein, R. M., S. Z. Josefowicz, A. Arvey, P. M. Treuting, and A. Y. Rudensky. 2012. Extrathymic generation of regulatory T cells in placental mammals mitigates maternal-fetal conflict. *Cell* **150**:29.

Tower, C., I. Crocker, D. Chirico, P. Baker, and I. Bruce. 2011. SLE and pregnancy: the potential role for regulatory T cells. *Nature Reviews Rheumatology* **7**:124.

(a) B-cell help

(b) T-cell help

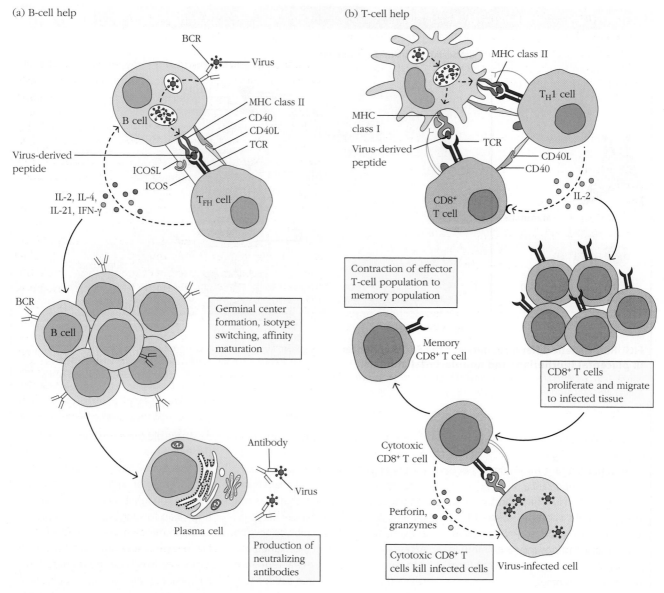

FIGURE 10-13   Examples of how T$_{FH}$ and T$_H$1 T cells provide help in the immune response. (a) T$_{FH}$ cells interact directly with B cells and generate effector cytokines, such as IL-21 and IL-4, which induce B-cell proliferation and differentiation into antibody-producing plasma cells. (b) T$_H$1 cells provide indirect help to CD8$^+$ T cells by interacting with antigen-presenting cells and producing effector cytokines, such as IFN-γ, that license the antigen-presenting cell to finalize CD8$^+$ T-cell differentiation into cytotoxic T cells. T$_H$1 cells also produce IL-2, which enhances CD8$^+$ T-cell proliferation.

attracts them to the B-cell follicle, where they can help establish germinal centers.

Interestingly, CD28 isn't the only costimulatory molecule involved in the activation of naïve T cells destined to differentiate into the T$_{FH}$ lineage. ICOS (discussed earlier) plays an additional and unique role. CD28 appears to play the key role in early activation events that result in up-regulation of Bcl-6. ICOS cooperates with the TCR later in the developmental program and stabilizes the cell's commitment to the T$_{FH}$ lineage. Specifically, ICOS appears to contribute signals that relieve repression of T$_{FH}$-specific genes, including CXCR5. In its absence, T$_{FH}$

cells leave the germinal center and lose their signature T$_{FH}$ functions.

T$_H$9 Cells

Although T$_H$2 cells were once thought to be the source of both IL-4 and IL-9, investigators have discovered that these cytokines originate from distinct helper T-cell subsets. The helper cells that produce IL-9 are now known as **T helper type 9 (T$_H$9) cells.** Cytokines that polarize naïve T cells to the T$_H$9 lineage include IL-2, and the combination of TGF-β and IL-4. (IL-1 also plays a role in enhancing polarization.) These cytokines induce expression of the transcriptional regulators IRF4 and PU.1, which are both responsible for driving T_H9 differentiation. IL-9 plays a role in expelling worms and contributes to some antitumor responses. It is also involved in a variety of allergic, asthmatic, and autoimmune responses and may be a promising target for therapy.

Other Helper T-Cell Subsets

Another helper T-cell subset that may become a definitive member of the helper T-cell family is the **T helper type 22 (T_H22) cell** subset. As its name suggests, this subset's signature effector cytokine is IL-22. IL-22 can also be made by other T-cell subsets including T_H17. However, T_H22 cells appear quite distinct from T_H17 cells and do not express RORγt or make IL-17. In addition to IL-22, they secrete IL-13 and express homing receptors for the skin, where they join the cells that protect the epithelium. What polarizes naïve T cells to the T_H22 subset? TNF, IL-6, and IL-23 appear to act in concert to induce up-regulation of the *aryl hydrocarbon receptor* (AHR), which may be the master transcriptional regulator for T_H22 cells. The function of T_H22 cells in healthy animals is still not clear; however, they contribute to the inflammation associated with multiple autoimmune disorders.

Ongoing investigations may identify even more helper T-cell subsets. Distinguishing bona fide new lineages from functional or developmental variants of currently defined subsets remains a challenge. As you have seen above, some effector cytokines are secreted by more than one subset and some cytokines contribute to the polarization of more than one lineage. Adding to the complication is the growing awareness that the relationship among subsets is plastic and the identity of a cell may not be stable (as we'll see shortly). Our understanding of the developmental relationship among effector subtypes will continue to evolve as new technologies allow us access to information about complex cell populations in vivo.

Key Concepts:

- $CD4^+$ T cells differentiate into at least six main subpopulations of effector cells: T_H1, T_H2, T_H9, T_H17, pT_{REG}, and T_{FH} cells. Each subpopulation is characterized by (1) a unique set of polarizing cytokines that initiate differentiation, (2) a unique master transcriptional regulator that regulates the production of helper-cell-specific genes, and (3) a distinct set of effector cytokines that they secrete to regulate the immune response.

- $CD4^+$ T_H1 cells regulate the response to intracellular pathogens, including viruses. T_H17 cells regulate responses to fungi and extracellular bacteria, particularly at our barrier organs. T_{FH} cells are responsible for helping B cells during affinity maturation in the germinal centers. T_H2 and T_H9 cells regulate our response to worms (helminths). Peripheral T_{REG} cells inhibit T-cell responses and help quell autoimmunity.

Helper T Cells May Not Be Irrevocably Committed to a Lineage

As indicated earlier, investigations now suggest that the relationship among T_H-cell subpopulations may be more plastic than previously suspected. At early stages in differentiation, at least, helper cells may be able to shift their commitment and produce another subset's signature cytokine(s). For example, when exposed to IL-12, young T_H2 cells can be induced to express the signature T_H1-cell cytokine, IFN-γ. Young T_H1 cells can also be induced to express the signature T_H2 cytokine, IL-4, under T_H2 polarizing conditions.

Interestingly, once they commit to their lineage T_H1 and T_H2 cells seem rather inflexible and are unable to adopt T_H17 or pT_{REG} characteristics. On the other hand, T_H17 and pT_{REG} cells are more versatile and appear to be able to adopt the cytokine expression profiles of other subsets, including T_H1 and T_H2. The T_H17 subset may be the most unstable or "plastic" lineage and can be induced to express type 1 (IFN-γ) and type 2 (IL-4) cytokines, depending on environmental input. Some pT_{REG} cells can also be induced to express IFN-γ, and some can be redirected toward a T_H17 phenotype if exposed to IL-6 and TGF-β. T_{FH} cells appear to lose their subset identity after a B-cell response has concluded. These are just some examples of the plasticity that has been observed, which adds to the complexity of determining the independence of specific helper cell lineages.

Key Concepts:

- The adoption of a helper T-cell lineage is not always a lifelong commitment.

- Multiple helper T-cell subsets, including T_H17 and pT_{REG} lineages, have the ability to differentiate into other helper subsets. T_H1 and T_H2 lineages may be more stable.

Helper T-Cell Subsets Play Critical Roles in Immune Health and Disease

Helper T-cell differentiation is typically initiated in secondary lymphoid tissue, but can also be completed at the site of infection, where these effector cells are most needed. How do helper T cells actually provide help? T cells that provide B-cell help interact directly with the B cell, providing what is known as *cognate* help (see Figure 10-13 and Chapter 11). This interaction is similar to the interaction between a naïve T cell and its activating APC, and also involves the formation of an immune synapse.

A helper T cell, often a T_{FH} cell, first recognizes peptide-MHC complexes that have been processed by the B cell and at the same time engages costimulatory ligands on the B-cell surface. TCR/costimulatory receptor co-engagement induces the helper T cell to release its effector cytokines. Helper T cells also express CD40L that binds CD40 on the

B-cell surface. Without this critical engagement, B cells do not differentiate into high-affinity, long-lived plasma cells. Figure 10-13a shows the interaction between a virus-specific B cell and helper T_{FH} cell, which produces several effector cytokines, including IL-21 and IL-4.

T_H1 cells deliver a different and more indirect kind of help to CD8$^+$ cytotoxic T cells, as shown in Figure 10-13b. They interact with and license macrophages, which in turn provide cytokines and signals that finalize CD8$^+$ T-cell activation. T_H1 cells can also provide CD40L directly to a neighboring CD8$^+$ T cell.

Studies in both mice and humans also show that the balance of activity among T-cell subsets can significantly influence the outcome of the immune response. A now classic illustration of the influence of T-cell subset balance on disease outcome is provided by leprosy, which is caused by *Mycobacterium leprae*, an intracellular pathogen that can survive within the phagosomes of macrophages. Leprosy is not a single clinical entity; rather, the disease presents as a spectrum of clinical responses, with two major forms of disease, tuberculoid and lepromatous, at each end of the spectrum. In **tuberculoid leprosy**, cell-mediated immune responses destroy most of the mycobacteria. Although skin and peripheral nerves are damaged in tuberculoid leprosy, it progresses slowly and patients usually survive. In contrast, **lepromatous leprosy** is characterized by a humoral response; cell-mediated immunity is depressed. The humoral response sometimes results in markedly high levels of immunoglobulin (hypergammaglobulinemia). This response is not as effective in inhibiting disease, and mycobacteria are widely disseminated in macrophages, often reaching numbers as high as 10^{10} per gram of tissue. Lepromatous leprosy progresses into disseminated infection of the bone and cartilage with extensive nerve and tissue damage.

The development of lepromatous or tuberculoid leprosy depends, in part, on the balance between T_H1 and T_H2 responses (**Figure 10-14**). In tuberculoid leprosy, the immune response is characterized by a T_H1-type response and high circulating levels of IL-2, IFN-γ, and lymphotoxin-α (LT-α). In lepromatous leprosy, there is a T_H2-type immune response, with high circulating levels of IL-4 and IL-5. IL-10 is also produced and may come from other helper subsets, including T_{REG} cells. This cytokine profile explains the diminished cell-mediated immunity and increased production of serum antibody in lepromatous leprosy.

Presumably each of these patients was exposed to the identical bacterial pathogen. Why do some patients develop an effective T_H1 response while others do not? Studies suggest that genetic differences among human hosts play a role. For example, differences in susceptibility may correlate with individual differences in the expression of PRRs (specifically, TLR1, TLR2, and NOD2) by innate cells. This makes sense given that interactions between pathogen and innate immune cells determine the cytokine environment that influences the outcome of T-cell polarization. Changes in

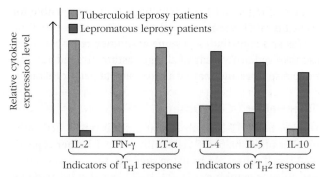

FIGURE 10-14   Correlation between type of leprosy and relative T_H1 or T_H2 activity. Messenger RNA isolated from lesions from patients with tuberculoid and lepromatous leprosy was analyzed by Southern blotting, using the cytokine probes indicated. The results are represented in this graph. Cytokines characteristic of T_H1 cells and type 1 responses (e.g., IFN-γ and LT-α [also known as TNF-β]) predominate in the patients with tuberculoid leprosy, whereas cytokines characteristic of T_H2 cells and type 2 responses (e.g., IL-4) predominate in the patients with lepromatous leprosy. [Data from Sieling, P.A., and Modlin, R.L., "Cytokine patterns at the site of mycobacterial infection," Immunobiology, 1994, October, 191(4–5):378–87.]

PRR expression or activity could alter either the quality or quantity of cytokines produced.

Progression of HIV infection to AIDS is also characterized by upsets in the balance among T-cell subsets. All CD4$^+$ T cells are susceptible to HIV infection. In the early stages of infection, our bodies are still able to generate new CD4$^+$ T cells to replace those that have died. However, inflammation induced by HIV infection generates cytokines and other factors that favor polarization of these new T cells to the T_{REG} lineage, reducing the body's ability to fight the infection. Even as total T-cell numbers decline, the proportion of T_{REG} cells remains high in infected patients.

Some pathogens have evolved to directly influence the activity of the T_H subsets. The Epstein-Barr virus, for example, produces a homolog (mimic) of human IL-10 called *viral IL-10* (vIL-10). Like cellular IL-10, vIL-10 suppresses helper T-cell and macrophage activity in a variety of ways, reducing the cell-mediated response to the Epstein-Barr virus and giving it a survival advantage.

T_H17 cells first received attention because of their association with chronic autoimmune disease. Mice that were unable to make IL-23, a cytokine that contributes to T_H17 polarization, were found to be resistant to autoimmunity. T_H17 cells and their defining effector cytokine, IL-17, are often found in inflamed tissue associated with rheumatoid arthritis, inflammatory bowel disease, multiple sclerosis, psoriasis, and asthma. The role T_H17 cells played in protecting organisms from infection was not immediately obvious until studies of individuals with an autosomal dominant form of a disease known as *hyper-IgE syndrome* or *Job syndrome* confirmed that T_H17 cells were important in controlling infections by extracellular bacteria and fungi (see Clinical Focus Box 10-3).

These are just some examples of the profound influence helper T-cell subsets have on disease progression. It is important to recognize that our current perspectives on the roles of helper subsets in disease and health remain simplistic. Our appreciation of the complex interplay among subsets will continue to improve and add more subtlety to our explanations in the future.

Key Concepts:

- Helper T cells can provide direct, cognate help to B cells and influence their differentiation via expression of CD40L and the secretion of effector cytokines.

- Helper T cells can provide indirect help to CTLs and other neighboring immune cells by interacting with APCs and producing cytokines that influence cytotoxic and inflammatory activity of multiple cell types.

- Different helper T-cell subsets deliver effector cytokines that are tailored to the pathogen that initiated the immune response.

- Helper T-cell subsets can also exacerbate inflammatory diseases and can participate in autoimmunity and allergy.

T-Cell Memory

T-cell activation results in a proliferative burst, effector cell generation, and then a dramatic contraction of cell number. At least 90% of effector cells die by apoptosis after pathogen is cleared, leaving behind an all-important population of antigen-specific **memory T cells**. Memory T cells are generally longer-lived than effector cells, and are quiescent. They represent about 35% of circulating T cells in a healthy young adult, rising to 60% in individuals over 70 years old.

Memory cells respond with heightened reactivity to a subsequent challenge with the same antigen. This **secondary immune response** is both faster and more robust, and hence more effective than a primary response.

Like naïve T cells, most memory T cells are metabolically quiet. They generate their energy from lipids via oxidative phosphorylation and are typically in the G_0 stage of the cell cycle. However, memory cells appear to have less stringent requirements for activation than naïve T cells. For example, naïve T cells are activated almost exclusively by dendritic cells, whereas memory T cells can be activated by macrophages, dendritic cells, and B cells. Memory cells express different patterns of surface adhesion molecules and costimulatory receptors that allow them to interact effectively with

a broader spectrum of APCs. They also appear to be more sensitive to stimulation and respond more quickly. This may, in part, be due to epigenetic changes that enhanced access to genes required for activation. Finally, memory cells display recirculation patterns that differ from those of naïve or effector T cells. Some stay for long periods of time in the lymph nodes and other secondary lymphoid organs, some circulate among tissues, and some travel to and remain in peripheral organs, anticipating the possibility of another infection with the antigen to which they are specific.

Our ability to distinguish memory cells from naïve and effector cells, using surface markers, has improved dramatically and we now distinguish four broad subsets of memory T cells: **central memory T cells (T_{CM})**, **effector memory T cells (T_{EM})**, **resident memory T cells (T_{RM})**, and **stem cell memory T cells (T_{SCM})**. These differ in location, phenotype, and function. Recent work has also revealed a great deal of diversity within these subsets, whose relationships are still being clarified (**Figure 10-15**). We describe some useful generalizations below, and close with the many questions that remain.

Naïve, Effector, and Memory T Cells Can Be Distinguished by Differences in Surface Protein Expression

Four surface markers have been used to broadly distinguish naïve, effector, and various memory T-cell subsets in mice (**Table 10-4**):

- CD62L, an adhesion protein that regulates homing to secondary lymphoid organs
- CCR7, a chemokine receptor that also regulates homing to secondary lymphoid organs
- CD44, which increases in response to TCR-mediated activation signals
- CD69, a C-type lectin whose interactions prevent immune cells from leaving a tissue

The patterns of expression of these four molecules on T-cell subsets have a certain logic. Naïve T cells express low levels of CD44, reflecting their unactivated state, and high levels of the adhesion molecule CD62L, which directs them to the lymph nodes or spleen. In contrast, effector helper and cytotoxic T cells have the reciprocal phenotype. They express high levels of CD44, indicating that they have received TCR signals, and low levels of CD62L, which prevents them from recirculating to secondary lymphoid tissue, allowing them to thoroughly probe sites of infection in the periphery.

What about memory cells? Like effector T cells, all memory T cells express CD44, indicating that they are antigen experienced (i.e., have received signals through their TCR). Like naïve T cells, central memory cells (T_{CM}) express CD62L and the chemokine receptor CCR7, consistent with their residence in secondary lymphoid organs. Effector memory cells (T_{EM}), which are found in a variety of tissues,

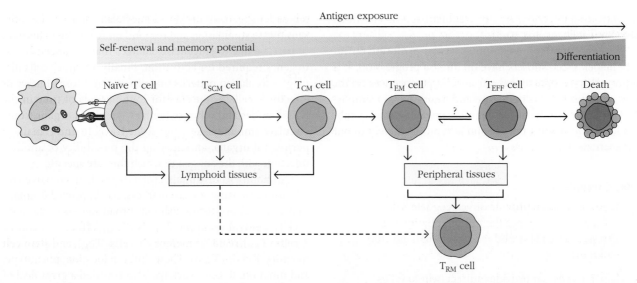

FIGURE 10-15   One possible model for the development of memory T-cell subsets. This model, which is supported by multiple lines of evidence, suggests that activated T cells proliferate and differentiate progressively into stem cell memory T cells (T_{SCM}), central memory T cells (T_{CM}), and effector memory T cells (T_{EM}), which ultimately terminally differentiate into effector cells (T_{EFF}). As cells differentiate along this pathway they lose stem cell potential and become more differentiated. Resident memory T cells (T_{RM}) arise from several precursors and likely complete their differentiation in their tissue of residence. It is possible that some effector memory T cells also arise from fully differentiated effector T cells.

express varying levels of CD62L depending on their locale; however, *they do not express CCR7*, reflecting their travels through and residence in nonlymphoid tissues. Resident memory cells (T_{RM}) are distinguished by their expression of CD69, which prevents them from leaving a tissue. Those that reside in lymphoid tissue express CCR7; those that reside in the periphery do not.

The existence of memory T cells with stem cell potential (stem cell memory T cells or T_{SCM}) had been predicted for many years, but was confirmed only recently. These are rare cells that appear to be the least differentiated of all memory subsets. They share some features with naïve T cells, including low expression of CD44 and high expression CCR7 and CD62L. However, they also express markers that have been associated with memory cells, including Fas (CD95) and IL-2Rβ (CD122).

These surface markers represent only a starting point for understanding memory T-cell subsets (see Figure 10-15).

TABLE 10-4	Surface proteins that are used to distinguish naïve, effector, and memory T cells					
	Naïve	**T_{EFF}**	**T_{SCM}**	**T_{CM}**	**T_{EM}**	**T_{RM}**
CCR7	+	–	+	+	–	–
CD62L	+	–	+	+	Variable	–
CD44	–	+	–	+	+	+
CD69	–	–	–	+	Variable	+
CD45RO (human)	–	–	–	+	+	+
CD45RA (human)	+	+	+	–	–	–
Fas (human)	–	+	+	+	+	+

Many others have also proven useful. For instance, human memory T cells are distinguished from naïve T cells by the expression of the RO, but not RA, variant of CD45, as well as by the expression of CD95 (see Table 10-4). $CD8^+$ resident memory T cells often express CD103, an integrin that interacts with the epithelial membranes of barrier tissues. Other markers continue to reveal heterogeneity within the memory cell subsets, which promise to be as complex and interesting as the effector T-cell population.

Key Concepts:

- Memory T cells, which are more easily activated than naïve cells, are responsible for secondary responses. They can be distinguished from naïve and effector cells by differences in surface protein expression and function.

- Four types of memory T cells have been described: stem cell memory T cells (T_{SCM}), central memory T cells (T_{CM}), effector memory T cells (T_{EM}), and resident memory T cells (T_{RM}).

Memory Cell Subpopulations Are Distinguished by Their Locale and Effector Activity

Where are memory cells found? In general, T_{CM} cells reside in and travel between secondary lymphoid tissues. They live longer and have the capacity to undergo more divisions than their T_{EM} counterparts. They generate a high amount of IL-2, but much lower amounts of other effector cytokines. When they re-encounter their cognate antigen in secondary lymphoid tissue, they are rapidly activated and have the capacity to differentiate into a variety of effector T-cell subtypes, depending on the cytokine environment.

On the other hand, T_{EM} cells circulate among peripheral tissues (including skin, lungs, liver, and intestine). They secrete low levels of IL-2, but generate high levels of effector cytokines. They exhibit their effector functions rapidly when they engage their cognate antigen and are important early responders to re-infection.

T_{RM} cells are even better situated to respond immediately to re-infection. Permanent residents of tissues that have experienced infection, they patrol very actively, ready to respond as soon as there is a sign of re-infection (see Chapter 14). $CD8^+$ T_{RM} cells have been best characterized, and can be found in multiple tissues, including skin, mucosa, and brain (see Chapter 13). $CD4^+$ T_{RM} cells proved more difficult to locate and may circulate more readily. However, they have been found in multiple tissues, including the lungs and bone marrow.

Stem cell memory T cells are found in secondary lymphoid tissue and can develop into all other memory T-cell subsets. Importantly, as their name indicates, they are also self-renewing and provide a long-lasting source of T cells with proven utility.

Key Concepts:

- Memory T-cell subsets differ in location, circulation, and function.

- Central memory cells and memory stem cells are found in secondary lymphoid organs and have not committed to any particular effector lineage. They respond quickly and flexibly to re-infection.

- Effector and tissue-resident memory cells together provide a pool of rapid responders to peripheral tissues and quickly display their original effector functions after re-infection.

Many Questions Remain Surrounding Memory T-Cell Origins and Functions

Antigen-specific memory is the unique product of the adaptive immune system. It extends lives and is the biological basis for the success of vaccination. The better we understand memory, the better we might be able to harness it, and the better we may be able to design vaccines for diseases that still endanger humans and other animals. Yet immune memory has been slow to reveal its secrets and multiple questions remain.

How and When Do Memory Cells Arise?

Naïve $CD4^+$ and $CD8^+$ T cells proliferate robustly during the first few days of the immune response in secondary lymphoid tissue. Only a small proportion of the progeny of a naïve T cell commit to the memory cell lineages. Although multiple explanations have been advanced to explain how memory T-cell subsets arise, several lines of evidence favor a linear, progressive model of development (see Figure 10-15). In this scheme, naïve T cells give rise to a small number of memory stem cells, most of which continue to proliferate and give rise to central memory cells, most of which continue to proliferate and differentiate into effector memory cells, most of which terminally differentiate into effector cells. This model is consistent with the observations that the stem cell capacity and life span of memory subsets decrease progressively—from T_{SCM}, which are bona fide stem cells, through T_{CM}, which still have some ability to self-renew, to T_{EM}, the shortest lived memory population with little or no self-renewal capacity.

Some data raise other possibilities. Naïve cells may give rise directly to a heterogeneous population of effector and memory stem cells. Effector T cells may differentiate directly into effector memory T cells, which may also be precursors of central memory cells. Only time will tell which model(s), which are not necessarily mutually exclusive, holds up to scientific scrutiny.

The source of resident memory T cells is also not yet clear. They could arise from either or both central and effector memory T cells and are likely to complete their commitment in the peripheral tissues they end up in.

What Signals Induce Memory Cell Commitment?

Most investigators agree that T-cell help is critical to generating long-lasting memory of $CD8^+$ T cells. Although $CD8^+$ T cells can be activated in the absence of $CD4^+$ T-cell help, this "helpless" activation event does not yield long-lived memory $CD8^+$ T cells.

The relative importance of other influences in driving memory development is still under investigation. IL-7, IL-15, and the Wnt signaling pathway may all play a role in generating and maintaining T_{SCM} cells. IL-2 is important in generating effector memory T cells. Although strength and duration of antigen stimulation play an important role in memory cell commitment, recent data also suggest that even low-affinity interactions can generate memory T cells. All studies agree that the more proliferation a response inspires, the better the memory pool.

Do Memory Cells Reflect the Heterogeneity of Effector Cells Generated during a Primary Response?

We have seen that naïve T cells differentiate into a wide variety of effector T-cell subpopulations, largely determined by the cytokine signals they receive during activation. Studies indicate that the memory cell response is also very diverse, in terms of both the T-cell receptor specificities and the array of cytokines produced. However, the cellular origin of this diversity is still under investigation. Specifically, does this diverse memory response strictly reflect the functional effector diversity generated during the primary response? Or does it develop anew from central memory T cells responding to different environmental cues during rechallenge? The answer is likely to be "both," but investigations continue.

How Do CD4$^+$ and CD8$^+$ Memory T Cells Differ?

Memory $CD8^+$ T cells are more abundant than memory $CD4^+$ T cells. This is partly because $CD8^+$ T cells proliferate more robustly and therefore generate proportionately more memory T cells. It may also be due to differences in the life span of memory T cells: $CD4^+$ memory T cells are not as long-lived as $CD8^+$ memory T cells. $CD4^+$ and $CD8^+$ memory cells are likely to express different homing receptors and populate different layers of our barrier tissues. $CD8^+$ memory T cells tend to be found in the epithelial layers (e.g., the epidermis of the skin and epithelium of the gut), whereas $CD4^+$ memory T cells seem to prefer the deeper layers (e.g., the dermis of the skin and lamina propria of the gut). $CD4^+$ memory T cells may also be more mobile. How these subpopulations coordinate their efforts during a secondary immune response is a topic of much interest.

How Are Memory Cells Maintained over Many Years?

Whether memory cells can persist for years in the absence of antigen remains controversial, although evidence seems to favor the possibility that they do. Regardless, it does seem that memory persistence depends on the input of cytokines that induce occasional divisions, a process known as *homeostatic proliferation*, which maintains the pool size by balancing apoptotic events with cell division. Both IL-7 and IL-15 appear important in enhancing homeostatic proliferation, but $CD4^+$ and $CD8^+$ memory T-cell requirements may differ.

Key Concepts:

- The generation of $CD4^+$ and $CD8^+$ memory T cells requires T-cell help. The more a naïve lymphocyte proliferates after activation, the better the memory response.

- Activated naïve T cells are thought to give rise directly to memory T-cell subpopulations that proliferate and differentiate in a linear fashion, culminating in the production of effector T cells. T_{SCM} give rise to T_{CM}, which give rise to T_{EM}, which terminally differentiate into effector T cells. Resident memory cells may arise from both T_{CM} and T_{EM}.

- The self-renewing potential and life span of memory T cells decrease as they differentiate from T_{SCM} to T_{EM}. T_{SCM} and T_{CM} circulate among secondary lymphoid tissue. T_{EM} circulate among peripheral tissues and T_{RM} take up permanent residence in peripheral tissues, ready to respond immediately if exposed to their cognate antigen.

- The regulation of memory T-cell development and response remains a very active area of investigation.

Conclusion

The fate of a mature, naïve T cell varies depending on the signals it receives. Most naïve T cells die within days or a few weeks after leaving the thymus, because they fail to bind to MHC-peptide complexes as they browse the surface of APCs during their circulation through lymphoid tissues. In order to survive and differentiate into effector cells, T cells must receive two signals from activated dendritic cells: one through the TCR and another through a costimulatory receptor, such as CD28.

The effector fate of an activated T cell depends on a third set of signals: polarizing cytokines that are produced by APCs. Polarizing cytokines induce expression of master transcriptional regulators that program the cell to adopt specific functions, including secretion of effector cytokines.

CD4$^+$ T cells differentiate into one of several helper T-cell subsets including T$_H$1, T$_H$2, T$_H$9, T$_H$17, and T$_{FH}$. These work together with many other cells to mount type 1 and type 2 responses, each of which is characterized by the activities of a distinct network of helper T-cell subsets, effector cytokines, as well as other immune cell types, including ILCs. CD4$^+$ T cells can also differentiate into regulatory T cells that help suppress autoimmune reactions.

Activated T cells not only differentiate into effector cells, but also into distinct memory cell subsets, which differ by their locale, their circulation patterns, and their effector function. These are responsible for the quick effector responses that characterize secondary responses. Many questions remain about their origin, relationships, and the molecular and cellular basis for their development.

REFERENCES

Ahmed, R., M. Bevan, S. Reiner, and D. Fearon. 2009. The precursors of memory: models and controversies. *Nature Reviews Immunology* **9**:662.

Belz, G. T., and S. L. Nutt. 2012. Transcriptional programming of the dendritic cell network. *Nature Reviews Immunology* **12**:101.

Chen, L., and D. B. Flies. 2013. Molecular mechanisms of T cell co-stimulation and co-inhibition. *Nature Reviews Immunology* **13**:227.

Craft, J. E. 2012. Follicular helper T cells in immunity and systemic autoimmunity. *Nature Reviews Rheumatology* **8**:337.

DuPage, M., and J. A. Bluestone. 2016. Harnessing the plasticity of CD4$^+$ T cells to treat immune-mediated disease. *Nature Reviews Immunology* **16**:149.

Farber, D. L, N. A. Yudanin, and N. P. Restifo. 2014. Human memory T cells: generation, compartmentalization and homeostasis. *Nature Reviews Immunology* **14**:24.

Gagliani, N., et al. 2015. T$_H$17 cells transdifferentiate into regulatory T cells during resolution of inflammation. *Nature* **523**:221.

Gattinoni, L., et al. 2011. A human memory T cell subset with stem cell–like properties. *Nature Medicine* **17**:1290.

Gattinoni, L., C. A. Klebanoff, and N. P. Restifo. 2012. Paths to stemness: building the ultimate antitumour T cell. *Nature Reviews Cancer* **12**:671.

Hoyer, K. K., W. F. Kuswanto, E. Gallo, and A. K. Abbas. 2009. Distinct roles of helper T-cell subsets in systemic autoimmune disease. *Blood* **113**:389.

Jameson, S., and D. Masopust. 2009. Diversity in T cell memory: an embarrassment of riches. *Immunity* **31**:859.

Jelley-Gibbs, D., T. Strutt, K. McKinstry, and S. Swain. 2008. Influencing the fates of CD4$^+$ T cells on the path to memory: lessons from influenza. *Immunology and Cell Biology* **86**:343.

Kaiko, G. E., J. C. Horvat, K. W. Beagley, and P. M. Hansbro. 2007. Immunological decision making: how does the immune system decide to mount a helper T-cell response? *Immunology* **123**:326.

Kapsenberg, M. L. 2003. Dendritic cell control of pathogen-driven T-cell polarization. *Nature Reviews Immunology* **3**:984.

Kassiotis, G., and A. O'Garra. 2009. Establishing the follicular helper identity. *Immunity* **31**:450.

Khoury, S., and M. Sayegh. 2004. The roles of the new negative T cell costimulatory pathways in regulating autoimmunity. *Immunity* **20**:529.

King, C. 2009. New insights into the differentiation and function of T follicular helper cells. *Nature Reviews Immunology* **9**:757.

Korn, T., E. Bettelli, M. Oukka, and V. Kuchroo. 2009. IL-17 and T$_H$17 cells. *Annual Review of Immunology* **27**:485.

Lefrancois, L., and D. Masopust. 2009. The road not taken: memory T cell fate "decisions." *Nature Immunology* **10**:369.

Linsley, P., and S. Nadler. 2009. The clinical utility of inhibiting CD28-mediated costimulation. *Immunological Reviews* **229**:307.

Malissen, B. 2009. Revisiting the follicular helper T cell paradigm. *Nature Immunology* **10**:371.

Mazzoni, A., and D. Segal. 2004. Controlling the Toll road to dendritic cell polarization. *Journal of Leukocyte Biology* **75**:721.

McGhee, J. 2005. The world of T$_H$1/T$_H$2 subsets: first proof. *Journal of Immunology* **175**:3.

Miossec, P., T. Korn, and V. K. Kuchroo. 2009. Interleukin 17 and type 17 helper T cells. *New England Journal of Medicine* **361**:888.

Mosmann, T., H. Cherwinski, M. Bond, M. Giedlin, and R. Coffman. 1986. Two types of murine helper T cell clone. I. Definition according to profiles of lymphokine activities and secreted proteins. *Journal of Immunology* **136**:2348.

Mueller, S. N., T. Gebhardt, F. R. Carbone, and W. R. Heath. 2013. Memory T cell subsets, migration patterns, and tissue residence. *Annual Review of Immunology* **31**:137.

Murphy, K., C. Nelson, and J. Sed. 2006. Balancing co-stimulation and inhibition with BTLA and HVEM. *Nature Reviews Immunology* **6**:671.

O'Shea, J. J., and W. E. Paul. 2010. Mechanisms underlying lineage commitment and plasticity of helper CD4$^+$ T cells. *Science* **327**:1098.

Palmer, M. T., and C. T. Weaver. 2010. Autoimmunity: Increasing suspects in the CD4$^+$ T cell lineup. *Nature Immunology* **11**:36.

Pennock, J. D., et al. 2013. T cell responses: naïve to memory and everything in between. *Advances in Physiology Education* **37**:273.

Pepper, M., and M. K. Jenkins. 2011. Origins of CD4$^+$ effector and memory T cells. *Nature Immunology* **12**:467.

Quill, H., and R. H. Schwartz. 1987. Stimulation of normal inducer T cell clones with antigen presented by purified Ia molecules in planar lipid membranes: specific induction of a long-lived state of proliferative nonresponsiveness. *Journal of Immunology* **138**:3704.

Readinger, J., K. Mueller, A. Venegas, R. Horai, and P. Schwartzberg. 2009. Tec kinases regulate T-lymphocyte development and function: new insights into the roles of Itk and Rlk/Txk. *Immunological Reviews* **228**:93.

Reiner, S. 2008. Inducing the T cell fates required for immunity. *Immunological Research* **42**:160.

Riley, J. 2009. PD-1 signaling in primary T cells. *Immunology Reviews* **229**:12.

Rudd, C., A. Taylor, and H. Schneider. 2009. CD28 and CTLA-4 coreceptor expression and signal transduction. *Immunological Reviews* **229**:114.

Sallusto, F., and A. Lanzavecchia. 2009. Heterogeneity of CD4$^+$ memory T cells: functional modules for tailored immunity. *European Journal of Immunology* **39**:2076.

Sharpe, A. 2009. Mechanisms of costimulation. *Immunological Reviews* **229**:5.

Smith-Garvin, J., G. Koretzky, and M. Jordan. 2009. T cell activation. *Annual Review of Immunology* **27**:591.

Thomas, R. 2004. Signal 3 and its role in autoimmunity. *Arthritis Research & Therapy* **6**:26.

Thompson, C., et al. 1989. CD28 activation pathway regulates the production of multiple T-cell-derived lymphokines/cytokines. *Proceedings of the National Academy of Sciences of the United States of America* **86**:1333.

Wang, S., and L. Chen. 2004. T lymphocyte co-signaling pathways of the B7-CD28 family. *Cellular & Molecular Immunology* **1**:37.

Weaver, C., and R. Hatton. 2009. Interplay between the T$_H$17 and T$_{REG}$ cell lineages: a (co)evolutionary perspective. *Nature Reviews. Immunology* **9**:883.

Weber, J. P., et al. 2015. ICOS maintains the T follicular helper cell phenotype by down-regulating Kruppel-like factor 2. *Journal of Experimental Medicine* **212**:217.

Yu, D., et al. 2009. The transcriptional repressor Bcl-6 directs T follicular helper cell lineage commitment. *Immunity* **31**:457.

Zhou, L., M. Chong, and D. Littman. 2009. Plasticity of CD4$^+$ T cell lineage differentiation. *Immunity* **30**:646.

Zhu, J., and W. Paul. 2010. Heterogeneity and plasticity of T helper cells. *Cell Research* **20**:4.

Useful Websites

The following are examples of well-organized websites developed by students, teachers, and even writers for reputable companies that sell immunological reagents, all of whom have worked hard to simplify a complex topic: helper T-cell subset differentiation. The websites may not be continually updated, so, as always with Internet sources, double-check the date and the accuracy of the information.

Searching Google Images for "T-cell subsets," "memory T-cell subsets," and "helper T-cell subsets" also can be very helpful, as long as you keep an eye on the date of publication and its source. Pictures are powerful teachers.

http://docs.abcam.com/pdf/immunology/t_cells_the_usual_subsets.pdf A poster made in collaboration between *Nature Reviews Immunology* and the company Abcam, which sells high-quality antibodies for flow cytometry and more. Other posters can be found on this site, for example, **www.abcam.com/pathways/scientific-pathway-poster-library**

https://www.cellsignal.com/common/content/content.jsp?id=pathways-tcell&pathway=T-Cell%20Receptor%20Signaling Cell Signaling Technology page on T-cell receptor signaling. Explore their other pages, which are updated responsibly and are good representations of the current thinking on signaling. One can also order their posters for display.

www.rndsystems.com/resources/posters/t-cell-subsets R&D Systems' brief description of multiple T-cell subsets.

www.lonza.com/products-services/bio-research/primary-cells/hematopoietic-cells/hematopoietic-knowledge-center/cd4-t-cell-subsets.aspx Lonza's version of T-cell subset development.

www.biology-pages.info/T/Th1_Th2.html A relatively up-to-date and well-written summary of T-cell subsets from this main site, an online open access biology textbook by Dr. John Kimball: **http://users.rcn.com/jkimball.ma.ultranet/BiologyPages**

http://microbewiki.kenyon.edu/index.php/Host_Dependency_of_Mycobacterium_leprae You will find here a posting from MicrobeWiki, "the student-edited microbiology resource" originating from Kenyon College.

The following websites link you to entertaining and informative interviews with and mini-biographies of Dr. Jim Allison, one of the individuals responsible for developing checkpoint inhibitors (see Clinical Focus Box 10-2), which are revolutionizing cancer therapy.

https://www.jci.org/articles/view/84236#SEC1

www.houstonchronicle.com/news/health/article/The-scientist-who-just-might-cure-cancer-5376864.php

www.cancerresearch.org/our-strategy-impact/people-behind-the-progress/scientists/james-allison

STUDY QUESTIONS

1. Which of the following conditions would lead to T-cell anergy?

  a. A naïve T-cell interaction with a dendritic cell in the presence of CTLA-4 Ig

  b. A naïve T cell stimulated with antibodies that bind both the TCR and CD28

  c. A naïve T cell stimulated with antibodies that bind only the TCR

  d. A naïve T cell stimulated with antibodies that bind only CD28

2. A virus enters a cut in the skin of a mouse and infects dendritic cells, stimulating a variety of PRRs both on and within the dendritic cells that induce them to produce IL-12. The mouse subsequently mounts an immune response that successfully clears the infection. Which of the following statements is (are) likely to be true about the immune response that occurred? Correct any that are false.

  a. The infected dendritic cells up-regulated CD80/CD86 and MHC class II.

  b. The dendritic cells encountered and activated naïve T cells in the skin of the mouse.

  c. Naïve T cells activated by these dendritic cells generated signals that released internal Ca^{2+} stores.

  d. Naïve T cells activated by these dendritic cells were polarized to the T_H2 lineage.

  e. Only effector memory T cells were made in this mouse.

  f. This response is likely to be a type 2 response.

3. Your lab acquires mice that do not have the *GATA-3* gene (*GATA-3* knockout mice). You discover that this mouse has a difficult time clearing helminth (worm) infections. Why might this be?

4. You isolate naïve T cells from your own blood and want to polarize them to the T_H1 lineage in vitro. You can use any of the following reagents to do this. Which would you choose?

Anti-TCR antibody
CTLA-4 Ig
IL-12
IL-4
Anti-CD80 antibody
IL-17
IFN-γ
Anti-CD28 antibody

5. The following sentences are all false. Identify the error(s) and correct.

  a. Macrophages activate naïve T cells better than dendritic cells do.

  b. ICOS enhances T-cell activation and is called a *coinhibitor*.

  c. Virtually all cells in the body express costimulatory ligands.

  d. CD28 is the only costimulatory receptor that binds to B7 family members.

  e. Signal 3 is provided by negative costimulatory receptors.

  f. Toxic shock syndrome is an example of an autoimmune disease.

  g. Superantigens mimic TCR–MHC class I interactions.

  h. $CD4^+$ T cells interact with MHC class I on $CD8^+$ T cells.

  i. Naïve T cells produce IFN-γ.

  j. T-Bet and GATA-3 are effector cytokines.

  k. Polarizing cytokines are produced only by APCs.

  l. Bcl-6 is involved in the delivery of costimulatory signals.

  m. T_H17- and T_{FH}-cell subsets are the major sources of B-cell help.

  n. pT_{REG} cells enhance inflammatory disease.

  o. Effector cytokines act exclusively on T cells.

  p. Central memory T cells tend to reside at the site of infection.

  q. Like naïve T cells, effector memory cells express CCR7.

6. Like T_H1 and T_H2 cells, T_H17 and T_{REG} cells cross-regulate each other. Which of these two statements about this cross-regulation is (are) true? Correct either, if false.

  a. TGF-β is a polarizing cytokine that stimulates up-regulation of each of the master transcriptional regulators that polarize T cells to the T_H17 and T_{REG} lineages.

  b. IL-6 inhibits polarization to the T_{REG} lineage by inhibiting expression of RORγt.

7. What is the role of tyrosine kinases in the early stages of TCR signaling?

8. List the following intracellular signaling molecules in order of involvement/activity after TCR engagement: LAT, Lck, Ca^{2+}, ZAP-70, NFAT.

9. The helper T-cell subset, T_H22, has been recently identified. It secretes IL-22 and appears to play a role in protecting skin from infection. What other information about this subset would give you confidence that it should be considered an independent helper T-cell lineage?

10. Stem cell memory T cells (T_{SCM}) have caught the attention of many interested in transplanting tumor-specific T cells into patients to fight cancer. Why might these cells be potentially more attractive than effector memory T cells (T_{EM}) for this purpose? Why might they also be riskier to use?

CLINICAL AND EXPERIMENTAL FOCUS QUESTION

Multiple sclerosis is an autoimmune disease in which T_H cells participate in the destruction of the protective myelin sheath around neurons in the central nervous system. Each person with this disease has different symptoms, depending on which neurons are affected, but the disease can be very disabling. Recent work in a mouse model of this disease suggests that transplantation of cell precursors of neurons may be a good therapy. Although these immature cells may work because they can develop into neuronal cells that replace the lost myelin sheath, some investigators realized that they play another, perhaps even more important role. These scientists showed that the neuronal cell precursors secrete a cytokine called *leukemia inhibiting factor* (LIF). In fact, the administration of this factor, alone, ameliorated symptoms.

These investigators were curious to know if this cytokine had an effect on T-cell activity. They added LIF to cultures of (normal) T cells that were being stimulated under various polarizing conditions (i.e., TCR and CD28 engagement in the presence of cytokines that drive differentiation to distinct T helper subsets). They stained T cells for cytokine production and analyzed the results by flow cytometry. The data below show their results from normal mouse T cells polarized to the lineage indicated in the absence (top, control) or presence (bottom) of LIF.

Which T helper lineage(s) is(are) most affected by the addition of LIF? Explain your answer. Why might these results explain the beneficial effect of LIF on the disease? Knowing what you know now about the molecular events that influence T-cell differentiation, speculate on the molecular basis for the activity of LIF.

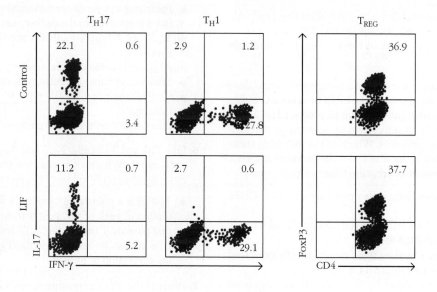

B-Cell Activation, Differentiation, and Memory Generation

Learning Objectives

After reading this chapter, you should be able to:

1. Describe the essential tenets of the clonal selection hypothesis.

2. Draw a lymph node, label the B-cell follicles and the T-cell areas. Show the path of migration of a B cell as it participates in a T-dependent response, indicating the sites of contact between antigen and T cells and the chemokine receptor–ligand interactions that drive these migrations.

3. Compare and contrast the processes of class switch recombination and somatic hypermutation with respect to the time after antigen stimulation at which they occur, their requirements for activation-induced cytidine deaminase, the need for an organized germinal center structure, and their outcomes.

4. Explicate the differences between T-independent and T-dependent B-cell responses, describing the B-cell subsets that participate in each type of response and the biochemical nature of the antigens that evoke them.

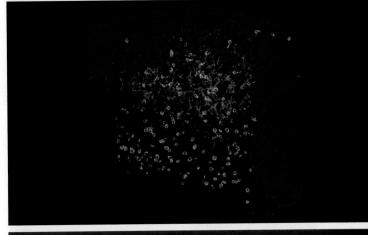

The germinal center of the lymph node contains B cells (green) in the dark and light zones and follicular dendritic cells (red) in the light zone. Naïve B cells (blue) define the follicular mantle zone. [*Republished with permission of Elsevier, from Victora, G. D., et al, "Germinal center dynamics revealed by multiphoton microscopy with a photoactivatable fluorescent reporter," from Cell, 2010, November 12, 143(4):592–605, Figure 6. Permission conveyed through Copyright Clearance Center, Inc.*]

The function of a B cell is to give rise to plasma cells that secrete antibodies capable of binding to an organism or molecule that poses a threat to the host. The secreted antibodies have antigen-binding sites identical to those of the receptor molecules on the B-cell surface. Antibodies belong to the class of proteins known as *immunoglobulins*, and once secreted they can protect the host against the pathogenic effects of invading viruses, bacteria, and parasites in a variety of ways, as described in Chapters 1 and 12.

Our current understanding of B-cell clonal selection, activation, proliferation, and deletion finds its beginnings in a theoretical paper in the *Australian Journal of Science*, written by Sir Frank Macfarlane Burnet (**Figure 11-1**) over the course of a single weekend in 1957. In this paper, building on prior work by Niels Jerne, David Talmage, Peter Medawar, and others, Burnet laid out the **clonal selection hypothesis**, which provides the conceptual underpinnings for the entire field of adaptive immunity (see Figure 1-6 and **Table 11-1**). This hypothesis suggested for the first time that the receptor molecule on the lymphocyte surface

Key Terms

Clonal selection hypothesis
B-2 B cells
T-dependent (TD) antigens
T-independent (TI) antigens
CD40 and CD40L
Primary foci

Plasmablasts
Sphingosine 1-phosphate (S1P)
Follicular helper T (T_{FH}) cells
Dark zone
Light zone

Centroblasts
Centrocytes
Activation-induced cytidine deaminase (AID)
Switch (S) regions
B-1 B cells

Marginal zone (MZ) B cells
Immunoreceptor tyrosine-based activation motif (ITAM)
Memory B cells

FIGURE 11-1 Sir Frank Macfarlane Burnet, 1899–1985. Sir Frank Macfarlane Burnet, an Australian and the author of the Clonal Selection Hypothesis, shared the 1960 Nobel Prize for Physiology or Medicine with Sir Peter Medawar of Britain for "the discovery of acquired immunological tolerance." *[Sir Frank Macfarlane Burnet 1960–61, by Sir William Dargie. Oil on composition board. Collection: National Portrait Gallery, Canberra. Purchased 1999.]*

TABLE 11-1	Components of the clonal selection hypothesis

- Immature B lymphocytes bear immunoglobulin (Ig) receptors on their cell surfaces. All receptors on a single B cell have identical specificity for antigen.

- On antigen stimulation, the B cell will mature and migrate to the lymphoid organs, where it will replicate. Its clonal descendants will bear the same receptor as the parental B cell and secrete antibodies with an identical specificity for antigen.

- At the close of the immune response, more B cells bearing receptors for the stimulating antigen will remain in the host than were present before the antigenic challenge. These memory B cells will then be capable of mounting an enhanced secondary response.

- B cells with receptors for self antigens are deleted during embryonic development.

that "The theory requires at some stage in early embryonic development *a genetic process for which there is no available precedent* [italics added]. In some way we have to picture a 'randomization' of the coding responsible for part of the specification of gamma globulin molecules, so that after several cell generations in early mesenchymal cells there are specifications in the genomes for virtually every variant that can exist as a gamma globulin molecule."

Just 4 years after the elucidation of the double-helical structure of DNA, Burnet was postulating that novel genetic mechanisms were required to create the antigen receptor repertoire!

The paper ends with a final flourish, in which Burnet predicted the need for clonal deletion in the developing B-cell population, in order to eliminate those cells bearing receptors with specificity for self antigens. His formulation of the clonal selection hypothesis, along with the brilliant experimental work he performed with others on the generation of immunological tolerance, resulted in his being awarded the Nobel Prize for Physiology or Medicine in 1960, along with Peter Medawar. The essential tenets of the clonal selection hypothesis are summarized in Table 11-1, and a visual representation of the events he predicted are shown in **Figure 11-2**.

As discussed in Chapter 9, there are two major types of B-cell responses, which are elicited by structurally distinct types of antigens (**Figure 11-3**). The first type of response is generated following recognition of protein antigens and requires the participation of CD4$^+$ helper T cells (see Figure 11-3a); this class of B-cell response is therefore known as a **T-dependent (TD) response**. It is mediated by **B-2 B cells** binding to **TD antigens**. Because B-2 B cells represent the majority of B cells, we will routinely refer to the B-2 B-cell

and the antibody products secreted by that cell had identical antigen-binding specificities. Furthermore, it posited that stimulation of a single lymphocyte would result in the generation of a clone of cells having the identical receptor specificity as the original cell. The daughter cells within each clone would be able to secrete large amounts of specific antibodies, and some progeny cells would also remain viable and available to neutralize a secondary infection by the same pathogen.

In other words, in one brilliant paper Professor Burnet gave generations of immunologists the complete conceptual framework for thinking about B-cell receptor diversity, lymphocyte proliferation, and immune memory—all this at a time when practitioners in the field were still determining the differences between T and B cells.

Burnet's paper went on to predict the generation of the vast array of antibody specificities known to exist today. It is difficult for us now, in the twenty-first century, to even begin to appreciate the prescience of Burnet's suggestion

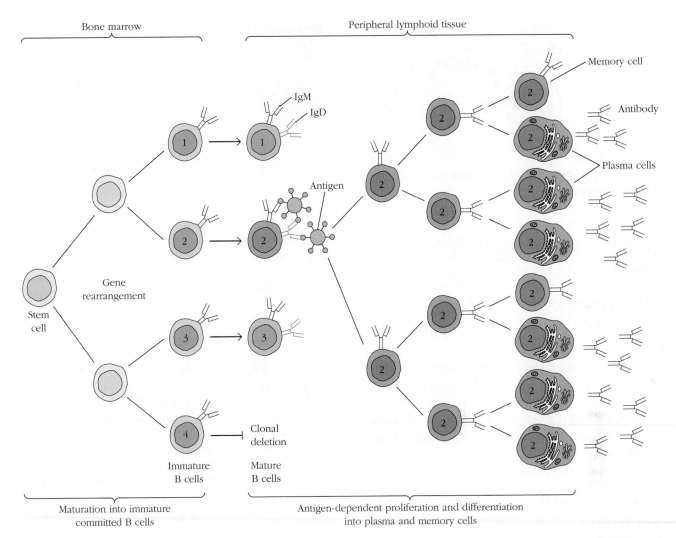

Bone marrow

Peripheral lymphoid tissue

Stem cell

Gene rearrangement

IgM
IgD

Antigen

Memory cell

Antibody

Plasma cells

Clonal deletion

Immature B cells

Mature B cells

Maturation into immature committed B cells

Antigen-dependent proliferation and differentiation into plasma and memory cells

FIGURE 11-2 Maturation and clonal selection of B lymphocytes. B-cell development in the bone marrow produces immature B cells bearing IgM receptors. Each B cell bears receptors of only one specificity. Any B cell with receptors specific for antigens expressed in the bone marrow is deleted at the immature B-cell stage (indicated by the immature B cell labeled "4"). Those B cells that do not express self-reactive receptors are released into the periphery, where they mature to express both IgM and IgD receptors. There, they recirculate among the blood, lymph, and lymphoid organs. Clonal selection occurs when an antigen binds to a B cell with a receptor specific for that antigen. Proliferation of an antigen-activated B cell (mature B cell labeled "2" in this diagram) leads to a clone of effector B cells and memory B cells; all cells in the expanded clone are specific for the original antigen. The plasma cell daughters of the original B cell secrete antibodies reactive with the activating antigen.

subset simply as "B cells," and distinguish the other B-cell subclasses by their particular names, such as B-1 B cells, marginal zone B cells, and so on. It is the B-2 B cells that are responsible for the high-affinity, **memory B cells** of secondary and later antibody responses and that protect the organism following an initial encounter with antigen including a protein component. And it is these B-2 B cells that undergo the complex processes of **somatic hypermutation (SHM)** and **class switch recombination (CSR)**, which result in the production of high-affinity antibodies of isotypes other than IgM.

The second type of response, which we will describe later in this chapter, is directed toward multivalent or highly polymerized antigens and does not require T-cell help. This type of response is referred to as a **T-independent response**, and the antigens that elicit such responses are ***T-independent* (TI) antigens**. TI-1 antigens (see Figure 11-3b) bind to innate receptors on B cells, are mitogenic, and (at high antigen concentrations) elicit a polyclonal, antibody-secreting response. In contrast, TI-2 antigens (see Figure 11-3c) are highly multivalent and bind only to Ig receptors. Their ability to cross-link a

(a) TD antigen

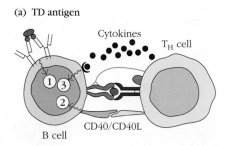

(b) TI-1 antigen   (c) TI-2 antigen

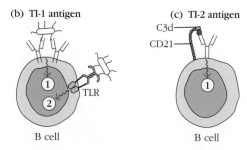

FIGURE 11-3 Different types of antigens signal through different receptor units. (a) T-dependent (TD) antigens bind to the Ig receptor of B cells. Some of the antigen is processed and presented to helper T cells. T cells bind to the MHC-peptide antigen, and deliver further activating signals to the B cell via interaction between CD40L (on the T cells) and CD40 (on the B cells). In addition, T cells secrete activating cytokines, such as IL-2 and IL-4, which are recognized by receptors on the B-cell surface. Cytokines deliver differentiation, proliferation, and survival signals to the B cells.

(b) *T-independent-type 1* (TI-1) antigens bind to B cells through both Ig and innate immune receptors. For example, LPS from gram-negative organisms binds to B cells via both membrane-bound immunoglobulin (mIg) and TLR4, resulting in signaling from both receptors. (c) *T-independent-type 2* (TI-2) antigens are frequently bound by C3d complement components and cross-link both mIg and CD21 receptors on B cells. Cross-linking of between 12 and 16 Ig receptors by TI-2 antigens has been shown to be sufficient to deliver an activating signal.

large fraction of the Ig receptors on the surface of a B cell allows them to deliver an activation signal in the absence of T-cell help. Most T-independent responses in the mouse are mediated by **B-1 B cells** and by a third subset of cells found in the marginal zone of the spleen, **marginal zone B-cells**.

T-Dependent B-Cell Responses: Activation

Following the completion of their maturation program (see Chapter 9), B cells migrate to the lymphoid follicles, where one of two things can occur. The B cell can interact with antigen and become activated. Or, in the absence of immediate antigen stimulation, the B cell recirculates through the blood and lymphatic systems and back to the lymphoid follicles. This latter process can recur many times. B-cell survival is dependent on access to the TNF family member cytokine *B*-cell *a*ctivating *f*actor (BAFF), which is secreted by lymphoid stromal cells, including *f*ollicular *d*endritic *c*ells (FDCs), as well as by several types of innate immune cells, such as neutrophils, macrophages, monocytes, and dendritic cells. Mature B cells unable to secure a sufficient supply of BAFF die by apoptosis. Recirculating mature B cells have a half-life of approximately 4.5 months.

At the start of a T-dependent B-cell response, the B cell binds antigen via its Ig receptors (signal 1 in Figure 11-3a). Some of the bound antigen is internalized into specialized vesicles, where it is processed and re-expressed in the form of peptides in the antigen-binding groove of MHC class II molecules (see Figure 7-16). Signal 2 is provided by an

activated T cell, which binds to the B cell both through its antigen receptor and via a separate interaction between **CD40** on the B cell and **CD40L** (CD154) on the activated T_H cell. On binding to the B cell, the T cell releases its activating cytokines (signal 3) directly into the T-cell/B-cell interface as described in Chapter 3. The nature of the response is also affected by cytokines released by other cells in the vicinity of the antigen encounter, as described later in this section.

The experiments proving that B cells required "help" from T cells in order to complete their differentiation were performed by Miller, Mitchell, Mitchison, and others in the 1960s, using the technique of **adoptive transfer**. Specifically, mice were lethally irradiated in order to eliminate their own lymphocytes; the investigators then attempted to reconstitute the irradiated mouse's ability to generate an immune response by adding back purified cells of various types from genetically identical mouse donors.

The results of these experiments showed that a mouse must receive cells derived from both the bone marrow and the thymus of a healthy donor animal in order to generate an antibody response. Neither thymus-derived nor bone marrow–derived cells were capable of reconstituting the responding animal's immune system on their own (**Figure 11-4**). We now know, of course, that the thymus-derived cells active in this response were helper T cells, whereas the bone marrow–derived, antibody-producing cells were mature B cells, recirculating through the bone marrow. In this way it was demonstrated that the antibody response to protein antigens required both B and T cells.

Contemporary work has shown us that signals delivered from the T cell to the antigen-activated B cell stimulate the

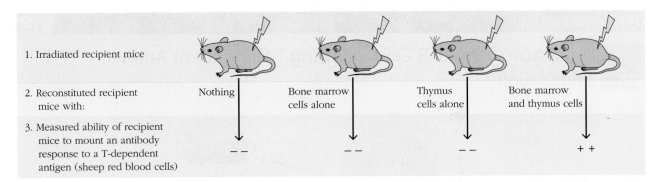

1. Irradiated recipient mice				
2. Reconstituted recipient mice with:	Nothing	Bone marrow cells alone	Thymus cells alone	Bone marrow and thymus cells
3. Measured ability of recipient mice to mount an antibody response to a T-dependent antigen (sheep red blood cells)	– –	– –	– –	+ +

FIGURE 11-4 Adoptive transfer experiments demonstrated the need for two cell populations during the generation of antibodies to T-dependent antigens. Early adoptive transfer experiments reconstituted irradiated mice with syngeneic bone marrow cells, thymus-derived cells, or a mixture of bone marrow- and thymus-derived cells. These mice were then challenged with a T-dependent antigen. Only recipient mice reconstituted with both bone marrow- and thymus-derived cells were able to mount an antibody response.

B cell to differentiate along one of three pathways (**Overview Figure 11-5**): it can proliferate to form a "primary focus" of antibody-secreting plasma cells that provide the initial IgM antibodies of the primary response; it can develop directly into an IgM-bearing memory cell, or it can enter the germi-nal *c*enter (GC) and undertake one of the most extraordi-nary differentiation programs in all of biology.

Having introduced the major players and briefly described the landscape in which the cells are operating, we will now step through a T-dependent B-cell response.

Naïve B Cells Encounter Antigen in the Lymph Nodes and Spleen

When antigen is introduced into the body, it becomes con-centrated in various peripheral lymphoid organs. Blood-borne antigen is filtered by the spleen, whereas antigen from tissue spaces drained by the lymphatic system is filtered by regional lymph nodes. Here, we will focus on antigen presentation to B cells in the lymph nodes. The specialized process of antigen sampling by B cells in the gut lymphoid system is described in Chapter 13.

Antigen enters the lymph nodes either alone or associated with antigen-transporting cells. Recall that, unlike T cells, B cells are capable of recognizing antigenic determinants on native, unprocessed antigens. In addition, as described in Chapter 5, antigen is often covalently modified with com-plement fragments, and CR2 (CD21) on B cells plays an important role in binding complement-coupled antigen.

The mechanism of B-cell antigen acquisition varies according to the size of the antigen. Soluble antigens picked up by the afferent lymphatic vessels flow into the subcap-sular sinus cavity of the lymph node. From there, antigens with a molecular weight less than 70 kDa (corresponding to a hydrodynamic radius of less than 4 nm) enter a system of

conduits that originate in the base of the subcapsular sinus (SCS) (**Figure 11-6a**). These conduits are produced by fibro-blasts and consist of highly organized bundles of collagen fibers, ensheathed by a basement membrane and surrounded by fibroblast reticular cells in the T-cell zone. (These reticu-lar cells may be replaced by follicular dendritic cells in the B-cell follicles during lymph node development.) Since the cellular sheaths are somewhat leaky, dendritic cells, macro-phages, and B cells can gain access to the antigens carried in these conduits by extending processes through the basement membrane.

In addition to carrying antigen into the lymph node, these conduits have also been demonstrated to serve as transport routes for chemokines, which attract cells to the conduit contents. Many of these conduits terminate in or near the high endothelial venules, which act as ports of entry for lym-phocytes arriving at the lymph node, and so some interac-tions between the B cell and low-molecular-weight antigens may occur immediately as the B cell enters the node.

Larger, more complex antigens take a different route into the lymph node. The subcapsular sinus macrophages (SCSMs), which lie within the layer of endothelial cells lin-ing the subcapsular sinus (see Figure 11-6a), are a distinc-tive subpopulation of macrophages with limited phagocytic ability. They express high levels of cell-surface molecules able to bind and retain unprocessed antigen. Microscopic studies have shown that SCSMs protrude into the subcapsu-lar sinus cavity and are therefore well positioned to capture native antigen or antigen in the form of immune complexes for presentation to B cells. For example, bacteria, viruses, particulates, and other complex antigens that have been covalently linked to complement components are held by complement receptors on the surfaces of these macrophages. B cells in the lymph node follicles migrate to regions directly underlying the capsule and acquire antigen directly from the

Alternative Fates of B Cells following T-Dependent Antigen Stimulation

Following T-dependent stimulation with antigen, B cells may differentiate into antibody-producing plasma cells in the primary focus or into early, low-affinity memory cells. Alternatively, they may enter the follicles to participate in the germinal center reaction.

SCSMs, subsequently returning to the follicle. Antigens also bind to the surfaces of dendritic cells and follicular dendritic cells via complement and other receptors and can be passed from these cell types to B cells.

Immunization or infection with antigens that the host has encountered previously, and to which antibodies already exist, results in the formation of immune complexes of antigens and antibodies. These are picked up by Fc receptors (receptors that bind to the non–antigen-binding regions of antibodies) on a variety of cells, including noncognate B cells (B cells whose native receptors are not specific for the antigens) and macrophages and enter

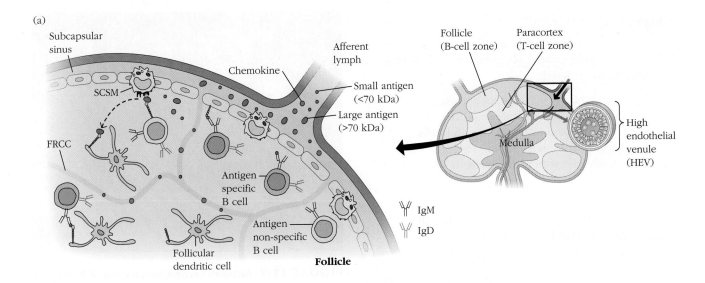

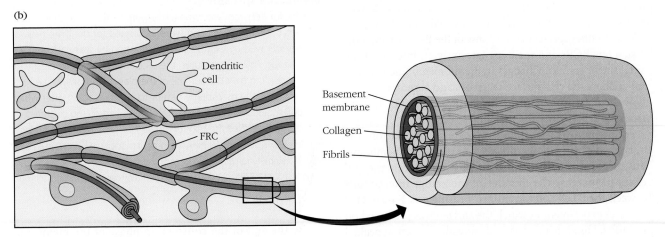

FIGURE 11-6 Antigen presentation to follicular B cells in the lymph node. (a) Lymphatic fluid containing antigens (red) and cytokines and chemokines (blue) reaches the lymph node through the afferent lymph vessel and enters the subcapsular sinus (SCS) region. The SCS region is lined with a porous border of SCS macrophages (SCSMs) that lie on or just under the lymphatic endothelial border and that prevent the free diffusion of the lymph fluid into the lymph node. Larger antigens are bound by surface receptors, such as complement and Fc receptors, on the SCSMs and then presented directly to B cells. Smaller antigens, and chemokines less than approximately 70 kDa in molecular weight, access B cells in the follicles either by diffusion or by passage through conduits emanating from the sinus. (b) These conduits are formed by fibroblastic reticular cells (FRCs) wrapped around collagen fibers, and B cells are able to access their contents through pores in the sides of the conduits.

the lymph nodes passively on these cells. Once in the lymph node, these antigen-transporting cells traffic to the follicles under the influence of chemokines, where their antigen can be recognized by B cells bearing the appropriate *B-cell receptor* (BCR).

Any discussion of antigen presentation to B cells would be incomplete without consideration of the role of the *follicular dendritic cells* (FDCs). Early electron microscope investigations of lymph node–derived cells revealed the dendrites of FDCs to be studded with antigen-antibody complexes, retained on the surface of the FDC through interaction either with Fc or with complement receptors. Because of the high surface density of antigen on FDCs,

scientists have long postulated that FDCs play an important role in antigen presentation. Current evidence suggests that their main function is to provide a reservoir of antigen for B cells to bind as they undergo mutation, selection, and differentiation during germinal center differentiation. The half-life of antigen on the surface of FDCs is relatively long; FDCs can retain antigen for weeks, whereas SCSMs have only been shown to display antigen over periods of hours. In addition, as previously mentioned, FDCs secrete factors that ensure the survival of B cells within the lymph node. Finally, it is worth noting that marginal zone B cells (see Chapter 9) have also been implicated in antigen transport and presentation in the spleen.

B-Cell Recognition of Cell-Bound Antigen Culminates in the Formation of an Immunological Synapse

Prior to antigen contact, the majority of B-cell receptors (BCRs) are expressed on the B-cell surface in tiny nanoclusters. Interaction of BCRs with multivalent, cell-bound antigens induces a rather spectacular response of the B-cell membrane. First, a few BCRs and their cognate antigens interact at the initial site of contact. Changes in the submembrane network that anchors the cell-surface receptors and coreceptors then allow the formation of microclusters of 50 to 100 BCRs with associated coreceptors and signaling molecules.

Following this successful microcluster formation, the B-cell membrane rapidly spreads over the target membrane. This membrane-spreading response is quite dramatic and serves to increase the number of molecular interactions between the B cell and the antigen-bearing cell. **Figure 11-7** shows an experiment in which the antigen was presented on an artificial lipid membrane. This spreading reaction peaked around 2 minutes after antigen contact. After maximal spreading, the area of contact between the cell and the artificial lipid membrane began to contract, and by approximately 10 minutes after antigen contact, the antigen-receptor complex was gathered into a central, defined cluster with an area of approximately 16 μm².

The biological relevance of this membrane-spreading and contraction response is supported by the finding that B cells in which the spreading response is deficient accumulate lower concentrations of antigen from their presenting cells. These B cells are then disadvantaged relative to B cells that accumulate higher antigen loads when seeking T-cell help (see later).

During this antigen-induced oligomerization of receptor molecules, the BCR complex moves transiently into parts of the membrane characterized by highly ordered, detergent-insoluble, sphingolipid- and cholesterol-rich regions, designated as **lipid rafts**. Association of the BCR with lipid rafts brings the *immunoreceptor tyrosine-based activation motifs* (ITAMs) of the Igα and Igβ components of the BCR into close apposition with the raft-tethered, Src-family member tyrosine kinase, Lyn, and allows for initiation of the BCR signaling cascade.

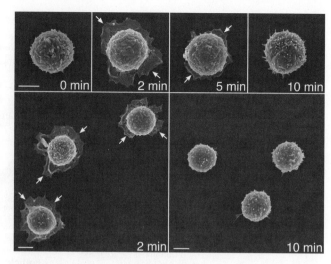

FIGURE 11-7 Antigen recognition by the BCR triggers membrane spreading. Transgenic B cells expressing a BCR specific for *hen egg lysozyme* (HEL) were settled onto planar lipid bilayers containing HEL and were incubated for the time periods shown, followed by fixation and visualization by scanning electron microscopy. At 2 to 4 minutes, the B-cell membrane can clearly be seen spreading over the surface of the planar lipid bilayer (arrows). By 5 minutes, the membrane begins to contract again, after which the BCR molecules (not shown in this figure) are found clustered on the cell surface. (See the text for details.) *[Republished with permission of the American Association for the Advancement of Science, Fleire, S.J., et al.,"B cell ligand discrimination through a spreading and contraction response," Science 2006 May 5; **312**(5774): 738–741, Figure 1. Permission conveyed through Copyright Clearance Center, Inc.]*

By the end of the contraction phase of the membrane response, the BCR microclusters have collapsed into a single central cluster of receptors. This cluster of BCRs (the *central supramolecular activation cluster*, or cSMAC) is surrounded by a ring of adhesion molecules, including the integrin LFA-1, which is referred to as the *peripheral supramolecular activation cluster*, or pSMAC. The pSMAC is in turn encircled by an actin ring forming the *distal*, or dSMAC (**Figure 11-8**). The integrins promote adhesion of the B cells to the antigen-presenting cells, lowering the threshold of antigen-binding affinity required for B-cell activation. This arrangement corresponds to that formed on T cells following recognition of antigen-presenting cells and is known as an "immunological synapse."

(a)

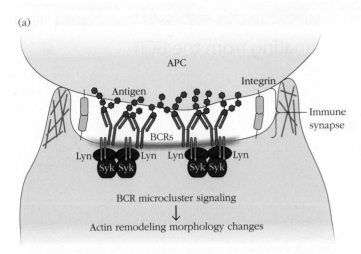

FIGURE 11-8 The B-cell immunological synapse includes a central core of receptor, surrounded by adhesion molecules, and is corraled by an actin ring. (a) Antigen binding results in clustering of receptors, receptor-associated signaling molecules, and adhesion molecules into the B-cell immunological synapse. The B-cell synapse is made up of three

(b)

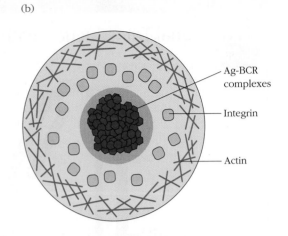

concentric rings. Antigen-BCR complexes are clustered at the center, forming the cSMAC (central supramolecular activation cluster), and are surrounded by a peripheral supramolecular activation cluster (pSMAC) made up of adhesion molecules such as LFA-1. Polymerized actin forms the distal or dSMAC. (b) Signaling through the clustered BCR triggers remodeling of the actin skeleton.

- Receptor clustering leads to the formation of an immunological synapse between the B cell and its antigen. On the B-cell surface, an inner ring of BCRs forms the central part of the supramolecular activating cluster, or cSMAC, which is surrounded by a ring of adhesion molecules, referred to as the pSMAC, encircled by a distal ring of polymerized actin, the dSMAC.

Antigen Binding to the BCR Leads to Activation of a Signal Transduction Cascade within the B Cell

Overview Figure 11-9 shows a general outline of some of the B-cell signaling pathways that are triggered on antigen binding. Antigen stimulation leads eventually to the activation of several transcription factors including NF-κB, NFAT, Egr-1, and Elk-1, which act together and with other factors to alter the cell's transcriptional program. Other molecular pathways trigger changes in membrane motility, in the expression of adhesion molecules and chemokine receptors, and in the production of anti-apoptotic molecules.

Members of three different protein tyrosine kinase families, the Src, Syk, and Tec families, are implicated in the early steps of B-cell activation. Immediately following antigen-induced clustering of the receptors into lipid rafts, raft-associated phosphatases begin the activation of the BCR-associated Src family kinases, Lyn and Fyn. Lyn then autophosphorylates, completing its own activation, and goes on to phosphorylate the ITAM-motif tyrosine residues

on the Igα/β receptor–associated molecules. This provides attachment sites for the SH2 motifs of the adapter protein BLNK and for the kinase Syk. After binding to the Igα/β ITAMs, Syk is also activated by autophosphorylation and in turn phosphorylates the adapter protein BLNK, creating docking sites for multiple downstream components of the signaling pathway. These downstream molecules include the Tec family kinase, Bruton's tyrosine kinase (Btk), and phospholipase Cγ2 (PLCγ2). Btk and PLCγ2 are then themselves phosphorylated and activated.

The term "signalsome" is frequently used to refer to the signal-transducing molecular complex that is formed in activated B cells. The B-cell signalsome contains the BCR; the Src family kinases Lyn, Fyn, and Blk; Syk; Btk; CD19; the adapter proteins BLNK and BCAP; and the signaling enzymes PLCγ2 and PI3 kinase. From this complex, further signals activate multiple cascades, as illustrated in Overview Figure 11-9. These cascades together signal changes in gene expression, cytoskeletal organization, and metabolism.

Signal transduction pathways initiated in the BCR signalsome interact closely with other signaling pathways activated through cytokine receptors, such as those recognizing IL-4 and IL-21, through coreceptors such as CD40 and via survival factor receptors such as BAFF-R. These pathways also interface with negative signals from cell-surface molecules such as CD22 and CD32 (see later in this chapter). Such interactions enable the same BCR to participate in directing many different cellular outcomes. Furthermore, the maturation status of the cell and the magnitude and duration of antigen stimulation also affect the quality of the transmitted signals.

OVERVIEW FIGURE   11-9

Signal Transduction Pathways Emanating from the BCR

Antigen-mediated receptor clustering into the lipid raft regions of the membrane leads to Lyn-mediated phosphorylation of the signal transduction mediators Igα/β and the coreceptor CD19, and to the recruitment of the Syk and Tec family kinases, Syk and Btk. Syk and Btk are activated by both trans- and autophosphorylation steps, the latter indicated by circular arrows. Phosphorylation of the adapter proteins BCAP (and BLNK) by Syk allows recruitment of phosphatidylinositol 3-kinase (PI$_3$ kinase) to the membrane with generation of phosphatidylinositol trisphosphate (PIP$_3$), which allows membrane localization of PDK1 and Akt and activation of Akt. Akt phosphorylates and inactivates the proapoptotic molecules Bax and Bad.

Activation of the B-cell isoform of PLC, PLCγ2, occurs on binding to the phosphorylated adapter molecule BLNK and phosphorylation of PLCγ2 by Syk. PLCγ2 breaks down phosphatidylinositol bisphosphate (PIP$_2$) to form diacylglycerol (DAG) and inositol trisphosphate (IP$_3$). IP$_3$ releases Ca^{2+} from intracellular stores with resultant activation of the NFAT pathway as shown. DAG interacts with PKC as well as with the GTP exchange factor RasGRP, leading to activation of MAP kinase pathways. These result in increases in Elk-1– and Egr-1–mediated transcription as well as CREB- and Jun-directed transcription. Cytoskeletal reorganization is mediated via Vav, activated on binding to the adapter protein BLNK.

How can we determine which molecules are implicated in the signal relay for which cellular processes? The generation of cell lines and animals lacking a particular enzyme or adapter molecule often provides important information. For example, examination of mutant forms of various B-cell lines, followed by confirmatory experiments in primary mouse B-cell lines, has demonstrated that the kinases Lyn and Syk are required for initiating the membrane-spreading response.

In conclusion, antigen binding at the BCR leads to multiple changes in transcriptional activity, as well as in the localization and motility of B cells, which together result in their enhanced survival, proliferation, differentiation, and eventual antibody secretion.

Key Concepts:

- Antigen engagement by the BCR induces phosphorylation of tyrosine residues in the ITAMs of Igα and Igβ by Src family kinases. This phosphorylation initiates a response leading to the formation of a cytoplasmic signal-transducing complex called a *signalsome*.

- Multiple outcomes are possible following BCR antigen binding, which depend on the strength and duration of antigen binding and are further regulated by interactions between the BCR and other cell receptors, including CD21, CD40, IL-4R, IL-21R, and BAFF-R.

B Cells Also Receive and Propagate Signals through Coreceptors

Following antigen binding, immunoprecipitation experiments show that the immunoglobulin receptor on the B-cell membrane is noncovalently associated with three transmembrane molecules: CD19, CD21, and CD81 (the latter also called *TAPA-1*) (see Figure 3-14). Antigens are sometimes presented to the BCR already covalently bound to complement proteins, in particular to the complement component C3d. (The complement cascade is discussed in Chapter 5.) The B-cell coreceptor CD21, otherwise known as CR2, specifically binds to C3d on C3d-coated antigens. This coengagement of the BCR and CD21 brings the coreceptor and the BCR into close apposition with one another.

When this happens, tyrosine residues on the cytoplasmic face of the CD19 coreceptor become phosphorylated by the activated Src family kinases, providing sites of attachment for PI3 kinase. Localization of PI3 kinase to the coreceptor enhances cell survival (see Overview Figure 11-9) and also results in alterations in the transcriptional program. CD19-mediated signaling is also vital in the membrane-spreading response.

Key Concept:

- The coreceptor CD21 can bind C3d complement fragments that are covalently associated with antigen, bringing the BCR into close contact with its coreceptor and enhancing antigen signaling.

B Cells Use More Than One Mechanism to Acquire Antigen from Antigen-Presenting Cells

How does the antigen presented on the surface of an antigen-presenting cell end up being internalized by the responding B cell and subsequently presented on the B-cell surface to a cognate T cell? We now know that there are two nonexclusive mechanisms by which this occurs.

When a B cell forms a synapse with an antigen-presenting cell, the B cell polarizes the microtubule-organizing center toward the antigen contact site and lysosomes move toward the immunological synapse. On reaching the plasma membrane, the lysosomes spill their contents into the synaptic junction, acidifying the junction and allowing their proteolytic enzymes to cleave the bonds between the antigen and the antigen-presenting cell. In some cases, direct loading of protease-generated peptides onto B-cell surface MHC class II proteins has been observed. More usually, the B cell internalizes the proteolytically generated antigen in complex with the BCR into the endosomal pathway and antigen presentation occurs via the traditional exogenous route (Chapter 7). The BCR continues to associate with downstream signaling molecules during the process of endocytosis and, indeed, the efficient regulation of B-cell signaling is dependent on internalization of the receptor molecule.

The second method of antigen extraction involves a more "muscular" response by the B cell (**Figure 11-10**). Actomyosin fibers within the B cell exert force on the BCR that in turn pulls on the antigen located on the antigen-presenting cell. If the affinity of the interaction between the BCR and the antigen is sufficiently high, the antigen is actually pulled out of the antigen-presenting cell membrane and into membrane invaginations of the B cell. From there, BCR-antigen complexes enter the endosomal system and are processed for antigen presentation.

Key Concept:

- Antigen to be endocytosed by B cells may either be cleaved from the surface of the antigen-presenting cell by B cell–derived lysosomal proteases or "tugged" from the surface by binding to the BCR.

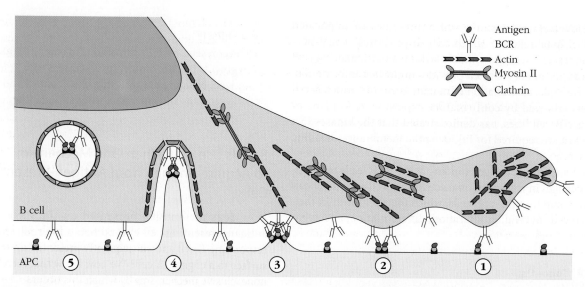

FIGURE 11-10 B cells extract antigen from the antigen-presenting cell membrane, using active contractions of the actomyosin skeleton. (1) During the spreading of the B-cell membrane over the antigen surface, actin fibers in the lamellipodia polymerize and facilitate close contact between the B-cell membrane and the antigen. (2) Microcluster formation is followed by actomyosin-mediated movements that pull the clusters toward the center of the contact. (3) Activity of myosin II initiates invagination of the B-cell membrane. (4) Clathrin-coated pits form, leading to endocytosis of the antigen still associated with fragments of the antigen-presenting cell membrane. (5) Endocytosis is complete.

Antigen Receptor Binding Induces Internalization and Antigen Presentation

Following antigen binding, most of the antigen-occupied BCR molecules are internalized, leaving just a few copies of the receptor on the cell surface.

The endocytic vesicles carrying the BCR-antigen complex fuse with a specialized vesicle in which antigen-derived peptides, generated by endosomal proteases, are loaded onto the MHC class II molecules (Chapter 7). The peptide-loaded MHC class II molecules are then transported back to the B-cell surface for presentation to T cells. The more antigen that is acquired by the B cell, the more effective the B cell will be at eliciting T-cell help, a concept we will return to shortly.

B-cell antigen processing and presentation results in the expression of peptide-loaded MHC class II molecules on the B-cell surface. In addition, BCR-mediated signaling causes increased expression of the costimulatory molecules CD80, CD86, and CD40 on the B-cell surface. Recall that CD80 and CD86 both bind to the T-cell molecule CD28 and that CD40 binds to the T-cell molecule CD40L (CD154). Changes in the expression of these three costimulatory molecules prepare the B cells for their subsequent interactions with T cells.

A B cell that has taken up its specific antigen via BCR-mediated endocytosis is about 10,000-fold more efficient in presenting antigen to cognate T cells than is a non–antigen-specific B cell that can only acquire the antigen by nonspecific pinocytosis. Effectively, this means that *the only*

B cells that present high levels of antigen to T cells are those B cells that can make antibody to that antigen.

> **Key Concepts:**
>
> - BCR-mediated endocytosis transports antigen into vesicles, where it is broken down into peptides. These peptides are subsequently loaded onto MHC class II molecules and returned to the B-cell surface.
>
> - Antigen engagement results in up-regulation of the expression of CD40, CD80, and CD86 on the B-cell surface. These molecules bind with coreceptors on T cells, further facilitating productive B-cell interactions with cognate T cells.

The Early Phases of the T-Dependent Response Are Characterized by Chemokine-Directed B-Cell Migration

Now that the B cell is ready to present antigen to T cells, we must ask how a B cell, with a receptor that is normally expressed at extremely low frequency within the receptor repertoire, can possibly find and bind to a T cell specific for the same antigen, which is also present at low frequency among all the available T cells.

We owe much of our understanding of the movement of cells and antigens through the lymph nodes to recent advances in imaging cells within their biological context.

Specifically, lymph nodes can be brought outside the bodies of living, anesthetized animals, and the circulation of fluorescently tagged T, B, and antigen-presenting cells through these nodes can be visualized in real time. This technique is known as **intravital fluorescence microscopy** (see Chapters 14 and 20). In **Advances Box 11-1** we describe additional technical developments that have furthered our understanding of the migration and differentiation of activated B cells. What follows is a distillation of the information gained from a large number of such experiments using various antigen and transgenic model systems.

B cells are highly motile cells, programmed to respond to chemoattractant signals. **Table 11-2** describes some key B-cell chemokine receptors and their ligands. **Figure 11-11** illustrates how chemokine receptor expression is temporally modulated during B-cell activation and shows the effects of this differential receptor expression on B-cell migration.

The chemokine CXCL13 is made by follicular stromal cells and binds to heparan sulfate on stromal cells and collagen fibers in the follicle. When a B cell enters the lymph node, the B-cell chemokine receptor, CXCR5, responds to CXCL13 signaling and migrates to the follicle. Within the follicle, B cells move around at approximately 6 μm per minute, contacting follicular dendritic cells (FDCs) and other stromal cells. Contact with FDCs allows B cells to receive survival signals. If B cells are merely transporting noncognate antigen attached to non-Ig receptors, such as complement receptors, the higher levels of complement receptors on follicular dendritic cells will usually strip the antigen from the B cells for future presentation to antigen-specific B cells. As B cells move through the follicle, they may also come into contact with antigen presented by subcapsular sinus macrophages (SCSMs) and dendritic cells or even with soluble or cell-bound antigen that enters the follicle through the high endothelial venules.

When the B cell meets an antigen capable of binding to its BCR (its "cognate" antigen), its previously random walk through the follicles begins to acquire a recognizable pattern, organized by successive modulations in chemokine receptors.

The initial set of chemokine-directed migrations occurs in the first 1 to 3 hours after antigen contact (see Figure 11-11, first two columns), and takes the B cell to the outer regions of the follicle under the influence of the chemokine receptor EBI2, whose ligand, 7α,25-dihydroxycholesterol, accumulates in the outer and interfollicular regions. The EBI2 receptor is up-regulated within the first hour following B-cell activation, in response to the transcription factor NF-κB (see Table 11-2). The biological rationale for this migration is uncertain, but it may have evolved to allow the B cell access to any additional antigens that have accumulated in association with the subcapsular sinus macrophages. EBI2 expression remains elevated for the first 2 to 3 days following B-cell activation.

A few hours after antigen encounter, B cells up-regulate CCR7 whose ligands, CCL19 and CCL21, are secreted by stromal cells in the T-cell zones (see Table 11-2). In addition, CCL21 expression also extends in decreasing concentrations from the T-cell zone into the follicle. Since the B cell continues to express CXCR5 and EBI2 as well as CCR7, antigen-engaged B cells move under the instructions of these three chemokine receptors to the boundary of the B- and T-cell zones. This migration occurs at around 6 hours post-stimulation (see Figure 11-11, third column). By this time, the B cell has internalized and processed its antigen, and therefore displays antigenic peptides on its cell surface for presentation to T cells.

At around 2 to 3 days after antigen recognition, CCR7 expression is reduced, and the EBI2 receptor once more redirects B-cell traffic to the outer and interfollicular regions (see Figure 11-11, fourth column). Interactions between antigen-activated T and B cells continue to occur throughout this stage, with conjugate pairs of T cells readily observable during intravital microscopy experiments.

Once contact with an antigen-specific, activated T cell is made, antigen-stimulated B cells engage with their conjugate T-cell partners over extended periods of time, ranging from a few moments to several hours (see Chapter 14). During this period of interaction, the T-cell receptor and its microtubule-organizing center reorient toward the synapse (point of contact) with the B cell, and the T cell begins to secrete interleukins, such as IL-4 and IL-21, that enable the B cell's differentiation program to progress. Activated B cells up-regulate the expression of receptors for those cytokines. The receptor for IL-4, an important B cell–specific cytokine,

TABLE 11-2	Chemokine receptors and ligands controlling B-cell migration in secondary lymphoid organs	
Receptor	**Ligand**	**Ligand production**
CXCR5	CXCL13	B-cell follicles
CCR7	CCL19, CCL21	T-cell zone
CXCR4	CXCL12	GC dark zone, splenic red pulp, lymph node medullary cords
EBI2	7α,25-OHC	Interfollicular and outer follicular areas, marginal zone–bridging channels

Data from Gatto, D., and R. Brink. 2013. B cell localization: regulation by EBI2 and its oxysterol ligand. *Trends in Immunology* **34**:336, Table 1.

How Did Scientists Track the Movements of B Cells between the Dark and Light Zones of the Germinal Center?

Activated GC B cells divide within the dark zone (DZ) of the GC, and then enter the light zone (LZ) to test their affinity for antigen. But how can the migrations between the two zones be quantified so that we can know what fraction of cells move between the two zones over a defined time period? This was the problem that Victora and colleagues sought to solve in the experiments described in this box.

Victora and colleagues used a photoactivatable label incorporated into all hematopoietic cells of an animal that had been genetically manipulated to express only a single, "B1-8" antibody heavy chain. Why did they select this particular heavy chain? C57BL/6 mice respond to the hapten 4-hydroxy-3-nitrophenylacetic acid (NP), conjugated to a protein such as ovalbumin (NP-OVA), using antibodies bearing the λ1 light chain; the B1-8 heavy chain is known to combine with the intrinsic λ1 light chain to create an anti-NP B-cell receptor. In their knock-in mouse, the B1-8 heavy chain is expressed by every B cell, since the presence of a previously rearranged heavy chain prevents further DNA rearrangement (as described in Chapter 6). This results in an anti-NP response that is essentially monoclonal, and consists of B cells secreting a B1-8 heavy chain and a λ1 light chain.

These mice were then crossed with a second transgenic mouse, in which all hematopoietic cells express

photo activatable green fluorescent protein (PA-GFP). PA-GFP is a chemical variant of the commonly used GFP, in which the peak excitation wavelength shifts from 415 to 495 nm following two-photon illumination at 720 to 840 nm. Subsequent, one-photon excitation at 495 nm (or two-photon excitation at 940 nm) will induce fluorescence *only in those cells that have already been irradiated at one of the activating wavelengths.* Victora and colleagues were able to show that PA-GFP–expressing cells could be photoactivated within intact lymph nodes. Furthermore, they proved that the resolution of their technique was such that one lymphocyte could be activated, while a second just 10 μm (or approximately one cell diameter) distant would be untouched by the laser beam. Therefore, in the intact lymph node, a cell that had previously been irradiated by long-wavelength light (and therefore was photoactivated) could be identified by fluorescence following two-photon excitation at 940 nm.

Recipient C57BL/6 mice were primed with the protein ovalbumin, in order to ensure plenty of T-cell help for the subsequent B-cell response. The scientists then transferred PA-GFP/B1-8 B cells into these mice and challenged them with a subcutaneous injection of NP-OVA. GCs were generated in skin-draining lymph nodes and allowed to mature.

The light zones of the GCs were distinguished from the dark zones by another clever trick; the mice were injected with NP conjugated to the red fluorescent protein, tdTomato. NP-tdTomato formed immune complexes with the anti-NP antibodies generated during the primary immune response, and these complexes bound to follicular dendritic cells (FDCs) in the LZ, turning them red under illumination.

Now tracking the GC cells between the two zones was merely a matter of photoactivating NP-specific LZ or DZ cells in the popliteal lymph nodes, and then tracking their migration by illuminating the slides at 940 nm. Migration could be tracked within a living lymph node for 6 hours or more without loss of viability. **Figure 1** shows antigen-specific B cells, labeled with PA-GFP, in a germinal center at a series of different time points post-photoactivation. Figure 1 clearly shows that 6 hours after labeling a group of GC DZ cells engaged in an antigen-specific immune response, many of those cells had moved into the LZ.

This experiment demonstrated that DZ B cells migrated rapidly into the LZ, with up to 50% of cells reaching the LZ by 4 hours after photoactivation. In contrast, over a longer time period, only about 15% of the activated LZ cells made it back into the DZ. Statistical analysis of data drawn from the movement of many cells indicated that 15% of DZ cells per

| 5 min | 1 hour | 3 hours | 6 hours |

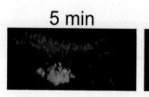  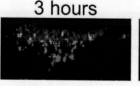 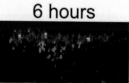

FIGURE 1 **Visualization of antigen-specific B cell movements in the germinal center.** The LZ is labeled with NP-tdTomato, which forms immune complexes that bind to the FDCs in the LZ. DZ GC B cells with specificity for NP are transgenic for PA-GFP and have been photoactivated at time 0. The four images illustrate the locations of the labeled dark zones at the indicated times after photoactivation. *[Republished with permission of Elsevier, from Victora, G. D., et al., "Germinal center dynamics revealed by multiphoton microscopy with a photoactivatable fluorescent reporter," from Cell, 2010, November 12, **143**(4):592–605, Figure 5B. Permission conveyed through Copyright Clearance Center, Inc.]*

(continued)

ADVANCES *(continued)* BOX 11-1

hour move to the LZ, and only 3% of LZ cells move in the opposite direction. Taking into account the proliferative events in the dark zone, this suggests that on the order of 30% of the cells in

the LZs of these animals are selected to re-enter the DZ. This allowed scientists for the first time to model the cellular dynamics of an ongoing response in a living lymph node.

REFERENCE

Victora, G. D., et al. 2010. Germinal center dynamics revealed by multiphoton microscopy with a photoactivatable fluorescent reporter. *Cell* **143:**592.

---

can be detected as early as 6 hours after antigen contact and reaches maximal levels of cell-surface expression at 72 hours. On naïve B cells, the receptor for IL-21 is undetectable until the second day after antigen stimulation, and continues to increase in expression through the beginning of the germinal center response on day 4. Of interest, IL-21 receptor expression occurs much more rapidly on memory B cells, and is already detectable on these cells by 12 hours after antigen encounter.

Interaction between CD40L (CD154) on T cells and CD40 on B cells is key to continued B-cell proliferation and differentiation. Many other important cell-surface changes characteristic of B lymphocyte activation can be observed at 2 to 3 days post-stimulation, such as increased levels of MHC class II antigens and of the two costimulatory molecules, CD80 and CD86.

The experiment depicted in **Figure 11-12** clearly illustrates these cellular migrations. Figure 11-12a is included as a reminder of the relative locations of the B-cell follicles and the T-cell zone in a lymph node. In this experiment, a transgenic mouse has been immunized with the T-dependent antigen *nitrophenacetyl-ovalbumin* (NP-OVA). Antigen-specific B cells express GFP, and are therefore labeled green;

antigen-specific T cells are labeled red; and the B-cell follicles are stained blue. In the naïve animal, T cells can be seen clearly localized in the T-cell zones, and occasional green B cells can be seen scattered throughout the follicles. With time postimmunization, B and T cells migrate into the boundaries of the T- and B-cell zones and into the interfollicular regions (days 1 and 2 postimmunization). By day 3, B and T cells are clustered in the outer as well as the interfollicular regions, and by 4 days postimmunization most, but not all, of the B cells have entered the follicles and the beginning of germinal center development can be seen.

After this period of intense communication with antigen-responsive T cells, at around 4 days postimmunization, some activated B cells down-regulate CCR7 and EBI2, enter the interior regions of the B-cell follicle, and begin the establishment of germinal centers (see Figure 11-11, fifth column). Meanwhile, other activated B cells within the same node retain EBI2 expression while decreasing CXCR5 expression and up-regulating the levels of cell-surface CXCR4 (see Table 11-2 and see Figure 11-11, fifth column). These cells move toward the splenic red pulp or the lymph node medullary cords, forming **primary foci** of proliferating B cells that will rapidly differentiate into **plasmablasts**.

Time since antigen exposure		0 hours	1–3 hours	6 hours–1 day	2–3 days	4–7 days
Events		Antigen surveillance	Antigen capture	T-cell encounter	Clonal expansion	Plasma cell and germinal center development
Expression level	EBI2	++	+++	+++	+++	−/+
	CXCR5	++	++	++	++	++
	CCR7	+	+	+++	+	−

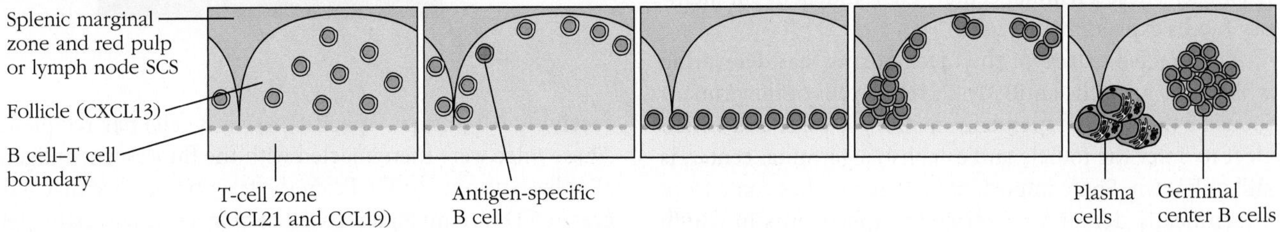

Splenic marginal zone and red pulp or lymph node SCS

Follicle (CXCL13)

B cell–T cell boundary

T-cell zone (CCL21 and CCL19)

Antigen-specific B cell

Plasma cells

Germinal center B cells

FIGURE 11-11 Differential chemokine receptor expression controls B-cell migration during the T-dependent immune response. The relative levels of expression of the chemokine receptors EBI2 (binds 7α,25-dihydroxycholesterol), CXCR5 (binds CXCL13), and CCR7 (binds CCL19 and CCL21) are shown on the B-cell surface at different times after antigen encounter. See text for details. [*Data from Cyster, J. G., E. V. Dang, A. Reboldi, and T. Yi. 2014. 25-Hydroxycholesterols in innate and adaptive immunity.* Nature Reviews Immunology *14:731, Figure 6.*]

(a)

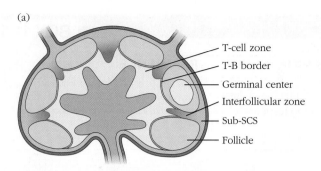

- T-cell zone
- T-B border
- Germinal center
- Interfollicular zone
- Sub-SCS
- Follicle

FIGURE 11-12 Movement of antigen-specific T and B cells within the lymph node after antigen encounter. (a) Section of the lymph node in diagrammatic form, showing the various regions of the lymph node. (b) The locations of antigen-specific B cells (green) and antigen-specific T cells (red) within the lymph node were visualized at specified times after antigen stimulation. The antigen used in the experiment was nitrophenacetyl (NP) conjugated to ovalbumin (Ova). Fluorescently labeled NP-specific B cells and Ova-specific T cells were transferred into a recipient animal, which was immunized with NP-Ova in the footpads 2 days after cell transfer. Draining popliteal lymph nodes were excised at the stated times. Anti-B220 (blue) was used to identify the B-cell follicles. The times refer to the days postimmunization. Day 1 and Day 2: Antigen-specific B-T cell pairs could be found both at the border of the T- and B-cell zones and in the interfollicular regions. Day 3: T cells have begun to enter the follicle, and many B cells can be seen just underneath the subcapsular sinus of the lymph node. Day 4: B cells have taken up residence in the follicle, and the formation of the germinal center can be seen. *[(b) Republished with permission of Elsevier Science and Technology Journals, from Kerfoot, S. M., et al., "Germinal center B cell and T follicular helper cell development initiates in the interfollicular zone," Immunity, 2011, June 24; 34(6):947–60, Figure 1a. Permission conveyed through Copyright Clearance Center, Inc.]*

These are differentiated B cells that have begun to secrete antibodies, have not yet lost the capacity to proliferate, and still bear cell-surface BCRs. These cells will eventually become the nonproliferating plasma cells that secrete the unmutated, predominantly IgM antibodies of the early immune response. In addition, it is now clear that some memory cells are generated very early in a primary immune response. These germinal center (GC)–independent memory B cells express mainly IgM.

The precise nature of the mechanisms that determine which cells form the antibody-secreting cells of the primary focus or the GC-independent memory B cells, and which elect to enter the follicle and establish a germinal center, is still unknown. Some important information has come from a technically demanding series of experiments in which extremely small numbers of antigen-responsive B cells were injected into recipient mice (**Figure 11-13**). The allotypic marker, CD45.1, was used to distinguish the injected B cells (CD45.1) from the recipient B cells (CD45.2). Donor cell numbers were calculated to generate recipient mice receiving 0.3 antigen-specific, naïve B cells per recipient. These mice were then injected with the fluorescent antigen, allophycocyanin (APC). From the fraction of mice that generated CD45.1-mediated B-cell responses, it was estimated that 91% of the observed responses derived from only a single CD45.1[+] antigen-specific B cell per mouse.

Seven days after antigen injection, the nature of the CD45.1[+], APC-specific B cells was analyzed in each of the recipients. Seventy-four recipients were analyzed and of

(b)

Naïve

RFP (OTII T cells)
GFP (B18 B cells)
B220

Day 1

Day 2

Day 3

Day 4

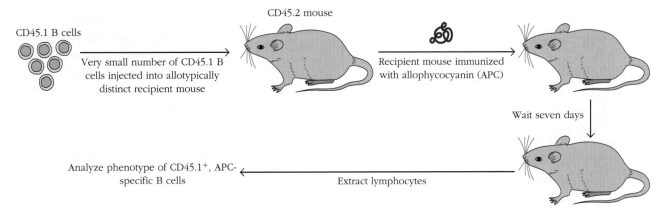

CD45.1 B cells

Very small number of CD45.1 B cells injected into allotypically distinct recipient mouse

CD45.2 mouse

Recipient mouse immunized with allophycocyanin (APC)

Wait seven days

Analyze phenotype of CD45.1⁺, APC-specific B cells

Extract lymphocytes

FIGURE 11-13 Experiment showing that a single B cell can give rise to plasmablasts, germinal center B cells, or memory B cells. Limiting numbers of CD45.1 allotype B cells are injected into CD45.2 allotype recipient mice, such that each mouse responds with a single clone of antigen-specific CD45.1 B cells to a subsequent injection of antigen (allophycocyanin, APC). Responding B cells were analyzed to ascertain the phenotypes of the responding daughter cells.

these, 35 contained *only* plasma cells, *only* germinal center cells, or *only* memory cells; four clonal populations contained all three subsets as well as undifferentiated antigen-specific B cells; and the rest contained two or three of these subsets. These experiments therefore suggest that an individual B cell is capable of generating daughter cells that have differentiated into antibody-producing plasma cells, GC-independent memory cells, and germinal center B cells. However, they also indicated a somewhat biased differentiation of B cells bearing receptors of varying affinities into the different cell types; higher affinity B-cell clones showed a higher frequency of differentiated plasma cells, and lower affinity B-cell clones tended to contain more memory B cells and more cells that entered the germinal center. This distinction may result from the fact that higher affinity B cells express higher levels of antigen on their surface, which may attract better and longer interactions with helper T cells.

Specification of the Stimulated B-Cell Fate Depends on Transcription Factor Expression

The transcription factors that control whether antigen-stimulated B cells differentiate along the plasma cell or germinal center route are linked in a mutually regulatory network (**Figure 11-14**). Pax-5 and Bcl-6, along with *low* levels of IRF-4, favor the generation of proliferating, germinal center cells. Conversely, the expression of BLIMP-1 and of *high* levels of IRF-4 support the generation of antibody-secreting cells. Other transcription factors, including IRF-8 and Bach2, also play modulating roles that will not be discussed here.

Key Concepts:

- Chemokine interactions with their receptors on B cells direct B-cell migration through the lymph node in the absence as well as in the presence of antigen.

- T cells interact with their cognate B cells by binding to the processed antigen with their T-cell receptor (TCR), as well as by interactions between T-cell CD28 with B-cell CD80 and CD86, and between T-cell CD40L and B-cell CD40. These interactions facilitate the directional secretion of T-cell cytokines that are necessary for full B-cell activation, such as IL-4 and IL-21.

- Individual B cells may differentiate into plasma cells, early memory cells, or cells that enter the germinal center.

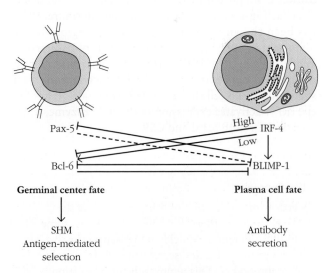

Pax-5

Bcl-6

High
IRF-4
Low

BLIMP-1

Germinal center fate

SHM
Antigen-mediated selection

Plasma cell fate

Antibody secretion

FIGURE 11-14 A regulatory network of transcription factors controls the germinal center B cell/plasma cell decision point. The transcription factors that control germinal center B-cell versus plasma-cell states of differentiation are related to one another through a mutually regulatory network. (See the text for details.)

The importance of the Bcl-6 protein to the generation of the germinal center phenotype is evidenced by the finding that *bcl-6* gene knockout animals are incapable of forming germinal centers. Bcl-6 expression is important in both B and T cells in the germinal center, and in B cells, the Bcl-6 protein inhibits DNA repair, thus facilitating the process of somatic hypermutation (see below). Expression of the *bcl-6* gene is stimulated following interaction of cytokines IL-21 and IL-6 with their receptors on the surface of B cells.

Conversely, BLIMP-1 expression controls many of the important functions conducted by antibody-secreting B cells. For example, BLIMP-1 supports the alternative splicing of Ig mRNA that generates the secreted form of Ig. BLIMP-1 also represses MHC class II expression on the B-cell surface.

Given the opposing roles played by BLIMP-1 and Bcl-6 proteins, it makes sense that Bcl-6 protein actively represses BLIMP-1 expression, which is also indirectly suppressed by Pax-5. (Specifically, Pax-5 induces the expression of a transcriptional suppressor, which in turn represses BLIMP-1.)

The role of the transcription factor **interferon regulatory factor 4 (IRF-4)** during B-cell differentiation is particularly interesting because its activity depends on its concentration (see Figure 11-14). The *irf-4* gene is expressed before *blimp-1* expression is turned on, and high concentrations of IRF-4 protein bind to elements upstream of the *blimp-1* gene, up-regulating *blimp-1* transcription as well as repressing Bcl-6 expression. Thus, it was concluded that high levels of IRF-4 aid in driving B-cell differentiation to the plasma cell stage. This proposition is supported by the finding that IRF-4–deficient mice lack Ig-secreting plasma cells. Furthermore, IRF-4 has been shown to bind to Ig enhancer regions and is important in the generation of high levels of Ig secretion. While *high* levels of IRF-4 promote the *plasma cell* phenotype, *low* concentrations of IRF-4 have recently been shown to drive B cells to a *germinal center* fate. At low levels, IRF-4 activates expression of the AID enzyme, which is critical for both SHM and CSR.

Key Concepts:

- Following stimulation of primary B cells at the T-cell/B-cell border within the lymph node, some B cells differentiate quickly into plasma cells that form primary foci and secrete an initial wave of IgM antibodies. This requires the up-regulation of the plasma cell transcription factors IRF-4 and BLIMP-1.

- Other B cells from the antigen-stimulated clones migrate to the primary follicles and form germinal centers, an action that requires the transcription factor Bcl-6 and low levels of IRF-4.

T-Dependent B-Cell Responses: Differentiation and Memory Generation

Following activation by antigen in the presence of T cells, B cells can differentiate into plasma cells, memory cells, or activated germinal center B cells that secrete the high-affinity antibodies of the late primary response. In this section, we first describe the processes that lead to the generation of the plasma cells (antibody-forming cells, or AFCs) in primary foci and then go on to discuss the extraordinary fate that awaits those activated B cells that enter the follicles, and eventually develop into germinal center B cells. We will then discuss the formation of memory B cells, both outside and within the germinal center.

Some Activated B Cells Differentiate into Plasma Cells That Form the Primary Focus

B cells mediate many immune functions, including antigen presentation and the secretion of regulatory cytokines, but by far their most important role is the production of antibodies. As described earlier, following antigen encounter, the first B cells to produce antibodies are the antibody-forming cells (AFCs) in the extrafollicular primary foci of lymph nodes and spleen (**Figure 11-15**). Plasmablasts formed on B-cell stimulation move to the medullary cords in the lymph nodes or to the border between the white and red pulp in the spleen, where they complete their differentiation to IgM-producing plasma cells (see Figure 11-11, final column).

The differentiation process that starts with a naïve B cell and culminates in the formation of a mature plasma

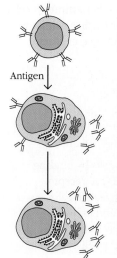

Naïve B cells. Bear cell surface IgM. Do not secrete antibody.

Plasmablasts. Differentiated B cells that have begun to secrete antibodies, have not yet lost the capacity to proliferate, and still bear cell surface BCRs.

Plasma cells. Differentiated B cells that can no longer divide, bear little to no cell surface immunoglobulin, and rapidly secrete large numbers of antibody molecules.

Also referred to as antibody forming cells or AFCs

FIGURE 11-15 Terminology describing antibody-secreting cells. Cells that have begun to secrete antibodies, but have not yet terminally differentiated, are called *plasmablasts*. Plasmablasts are actively dividing, BCR-bearing cells. Plasma cells are terminally differentiated cells that bear very little cell membrane BCR. Plasmablasts and plasma cells are both also called *antibody-forming* or *antibody-secreting cells* (AFCs or ASCs).

cell requires the coordinated expression and repression of hundreds of genes. Plasma cells depend for their growth and survival on the cytokines IL-6 and APRIL (*a proliferation-inducing, anti-apoptotic ligand*), which are produced by dendritic cells and monocytes within the lymph node. These cytokines induce the formation of anti-apoptotic molecules including Bcl-2 and Bcl-X_L, which are important in the generation of the extrafollicular antibody-forming foci, and Mcl-1 and/or A1, which are necessary for plasma cell maintenance.

The kinetics of AFC focus generation varies according to the antigen and its mode of delivery, but in general, foci of AFCs can be seen in spleen and/or lymph nodes by around 3 days postimmunization. The AFCs initially produce only IgM antibodies, but IgG antibodies can be detected by 5 to 6 days after immunization, clearly illustrating that entry into the follicle is *not* necessary for the process of class switch recombination (CSR). However, neither IgM, nor IgG antibodies from the AFC, display evidence of somatic hypermutation (SHM), demonstrating conversely that the process of SHM occurs *only* in the germinal center. SHM and CSR will be described in detail later in this chapter.

Plasma cells achieve remarkably high rates of Ig secretion. These cells therefore provide high concentrations of specific antibodies that can neutralize or opsonize antigen by the first 5 to 6 days of an immune response and thus are the major source of protective humoral immunity early after antigen contact. Since antibodies can diffuse freely throughout the lymph node, these newly formed antibodies can also bind free antigen within the node, helping to drive affinity maturation (see later). In addition, the immune complexes (antigen-antibody complexes) formed by this binding can be bound by follicular dendritic cells within the node and serve as an antigen reservoir.

The size of extrafollicular AFC foci peaks around days 7 to 8 after antigen encounter, after which the foci decline in size; they are barely detectable by day 14. Most of the AFCs of the primary foci die by apoptosis. However, as we will see later, plasma cells generated within the follicles during the germinal center reaction may live much longer than this.

Since the antigen-binding sites of the primary focus antibodies have not yet been optimized by somatic hypermutation and antigen selection, their affinity for antigen is relatively low compared with that achieved in the later stages of the response. However, the multivalent nature of IgM antibodies enables them to lower microbial numbers until the high-affinity antibodies generated by the extraordinary events of the germinal center reaction are released.

Key Concept:

- Plasma cells of the primary focus secrete large quantities of nonmutated IgM and IgG antibodies that provide early, protective humoral immunity and help to drive affinity maturation.

Other Activated B Cells Enter the Follicles and Initiate a Germinal Center Response

Those stimulated B cells that entered the follicles following an encounter with antigen begin to divide rapidly and undergo further differentiation, resulting in the formation of specialized structures called **germinal centers (GCs) (Figure 11-16)**. Although GCs consist primarily of rapidly dividing B cells, they also contain follicular dendritic cells (FDCs), *T follicular helper*

(a)

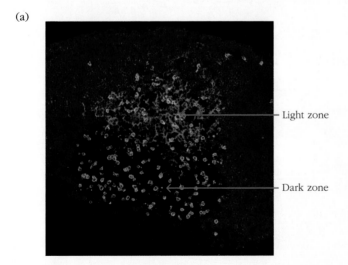

— Light zone

— Dark zone

Follicular mantle zone

(b)

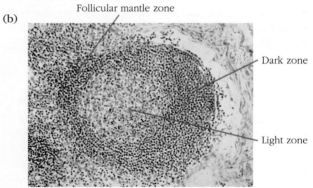

— Dark zone

— Light zone

FIGURE 11-16 The germinal center. (a) A mouse lymph node germinal center 7 days after secondary immunization, in which 10% of the germinal center B cells express green fluorescent proteins (and are therefore labeled green). These cells are seen most intensely in the dark zone of the germinal center, but are also present less frequently in the light zone. Follicular dendritic cells, labeled red with an antibody to the FDC marker CD35, mark the light zone of the germinal center. The blue staining marks IgD-bearing B cells not specific for the immunizing antigen. (b) Histochemical staining of a germinal center, illustrating the remarkably high centroblast cell density within the dark zone. [*(a) Republished with permission of Elsevier, from Victora, G. D., et al., "Germinal center dynamics revealed by multiphoton microscopy with a photoactivatable fluorescent reporter," from Cell, 2010, November 12, **143**(4):592–605, Figure 6. Permission conveyed through Copyright Clearance Center, Inc. (b) Courtesy Dr. Roger C. Wagner, Professor Emeritus of Biological Sciences, University of Delaware.*]

(T_{FH}) cells, and macrophages. Depending on the nature of the antigen, the size of the GC peaks around 7 to 12 days after antigen stimulation, and GCs normally resolve within 3 to 4 weeks.

In the germinal centers, B cells undergo a period of intense proliferation, and their Ig genes are subjected to some of the most extraordinary processes in biology. First, the Ig variable region genes undergo extremely high rates of mutation, on the order of one mutation per thousand base pairs per generation. (Contrast this with the background cosmic ray–induced mutation rate of one mutation per hundred million base pairs per generation.) If the mutated genes can still encode functional Ig receptor molecules, these BCRs are then expressed on B cells within the GC, and the receptors are tested to see if their affinities for antigen have been altered for the better. Those B cells bearing receptors with higher affinity than those on the parental cells are selected for further rounds of mutation and selection, while B cells bearing receptors with no, or decreased, affinity for antigen are allowed to die by apoptosis.

Given the active mutational and selection events that occur within the germinal center, it should come as no surprise that it has been referred to as a "Darwinian microcosm."[1]

Second, sometimes within, but also outside of, the germinal center (as described earlier) the constant regions of the Ig genes undergo class switch recombination: the replacement of the μ constant region gene segments with segments encoding other classes of constant regions.

We will begin here by describing the formation and biology of the germinal center. In the next section, we will address the molecular mechanisms that underlie the unique genetic processes of somatic hypermutation and the Ig class switching. In reading this section, students may find it useful to refer to **Overview Figure 11-17**.

---

[1]Kelsoe, G. 1998. V(D)J hypermutation and DNA mismatch repair: vexed by fixation. *Proceedings of the National Academy of Sciences USA* **95**:6576.

OVERVIEW FIGURE   **11-17**

B-Cell Differentiation Events Occur in Different Anatomical Locations

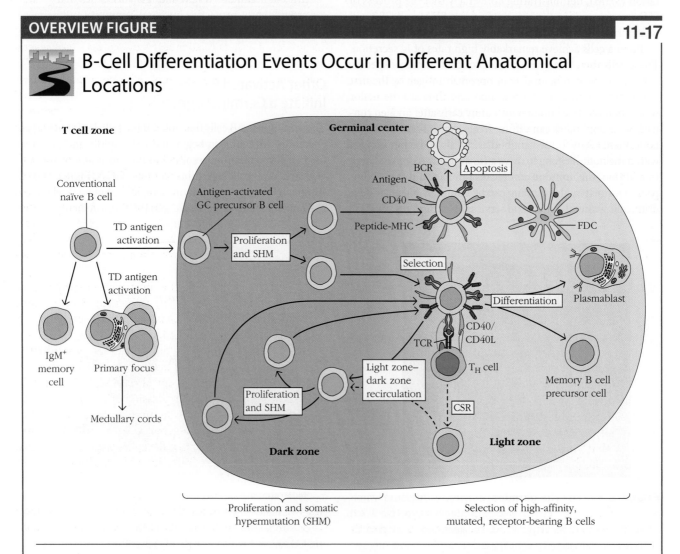

Following T cell–dependent antigen stimulation, a B cell may undergo differentiation to a plasma cell, forming a primary focus and secreting the low-affinity antibodies of the early primary response; it may immediately differentiate to form a memory B cell; or it may migrate to the B-cell follicle, forming a germinal center and undergoing somatic hypermutation and antigenic selection.

The Cell Biology of Germinal Center Formation and Maintenance

As previously described, the exact signals that determine whether an activated B cell will differentiate into a primary focus AFC, or enter a follicle and begin GC development, are still unclear. However, in recent years, immunologists have made enormous progress in their understanding of the creation and maintenance of the germinal center and the cells from which it is formed.

An increase in the level of the master transcription factor Bcl-6 is key to the differentiation of both B and T cells to a GC phenotype; mice with deficiencies in Bcl-6 completely fail to form GCs. Bcl-6 up-regulation causes a decrease in the expression of the chemoattractant receptor EBI2, which allows GC B cells to move away from the stromal cells around the edge of the follicle and into its center. Bcl-6 also induces expression of *activation-induced cytidine deaminase* (AID), which is responsible for both somatic hypermutation (SHM) and class switch recombination (CSR) and also turns down the expression of DNA damage response genes that might interfere with the genetic changes generated during SHM.

An additional chemoattractant signal contributes to the clustering of B cells in the center of the B-cell follicles by *inhibiting* their movement toward the *outer* regions of the GC. The metabolic intermediate, **sphingosine 1-phosphate (S1P)**, is rapidly degraded by B cells, so the S1P concentration inside the follicles is low, while that immediately outside the follicle is high. Germinal center B cells bear an *inhibitory* **S1P₂-type receptor** for S1P that induces cells to move *away* from high concentrations of the ligand. Hence, S1P₂-mediated signaling will favor the formation of tightly clustered GCs.

Once the antigen-activated B cells enter the follicles, they colocalize with FDCs and begin to proliferate. As the numbers of activated, proliferating B cells within the GC increase, the naïve B cells that originally filled the follicles are displaced to the periphery, where they form a **mantle zone**. Since naïve B cells bear both IgM and IgD, whereas stimulated B cells lose IgD on antigen activation, the GC mantle zone structure can be visualized by staining for IgD-bearing cells (see Figure 11-16a).

Beginning in the center of B-cell follicles as just a few rapidly dividing B cells, a GC can grow to include as many as 10,000 cells in a matter of a few days. GC B cells are highly susceptible to apoptosis, and some of the anti-apoptotic signals that they require to survive are provided by **follicular helper T (T_FH) cells** (see Chapter 10). Survival signals are delivered to the GC B cells by T cells through interactions between CD40L (CD154) on the T-cell membrane and CD40 on the B-cell surface when the T cell interacts with the antigen presented on the surface of the B cell.

B cells also interact closely with FDCs and must do so in order to survive. As described earlier in this chapter, FDCs hold antigen on their surfaces for long periods of time. This antigen can be found in the form of antigen-antibody complexes, or as antigen covalently bound to complement fragments secured to the FDC surface by FDC complement receptors. B cells interacting with FDC-bound antigen receive survival signals from the cells that help to maintain the integrity of the GC structure.

Characteristics of Germinal Center Dark and Light Zones

The developing germinal center quickly resolves into two, histologically distinct zones, termed the **dark zone (DZ)** and the **light zone (LZ)** on the basis of their histological appearance. Much of the light zone is occupied by processes extending from the FDCs, whereas few FDC processes extend into the dark zone, in which lymphocytes are closely packed (see Figure 11-16). The DZ is closer to the T-cell area, and its rapidly dividing B cells are called **centroblasts**. In contrast, B cells in the LZ proliferate less rapidly and are referred to as **centrocytes**. The fully differentiated GC can be observed by approximately 7 days after antigen contact.

The distribution of B cells between the two zones is dependent on their relative levels of expression of several chemokine receptors, which are identified in **Table 11-3**. The dark zone stromal cells consist of reticular cells that express CXCL12, the ligand for CXCR4, which is highly expressed on centroblasts (dark zone B cells). The interaction between CXCR4 and CXCL12 holds the dark zone B cells away from the FDCs of the light zone.

Gene expression analysis has shown that centroblasts display high expression levels of AID as well as of error-prone DNA polymerases such as DNA polymerase eta (η) (which introduces point mutations into Ig variable regions while repairing AID-induced DNA lesions). The presence of these two proteins strongly suggests that the dark zone is the site of somatic hypermutation of Ig genes (see below).

The light zone consists of a more diverse array of cells than the dark zone, including light zone B cells, some infiltrating naïve B cells, FDCs, and a small but critically important set of T_FH cells. FDCs secrete the chemokine CXCL13. This is detected by the CXCR5 receptor, which is present on all GC B cells, but that is expressed at higher levels on centrocytes than on centroblasts. Light zone B cells also show higher levels of expression of many B-cell activation markers, such as CD86, than do centroblasts. The presence of T_FH cells and antigen, and the activation status of centrocytes, all suggest that the light zone of the germinal center may be the site of selection of high-affinity variants resulting from SHM.

Now that we have a clearer sense of the cell biology of the two specialized zones of the germinal center, let's take a closer look at the biochemistry of the remarkable functions that take place within the germinal center.

Somatic Hypermutation and Affinity Selection

The term *somatic hypermutation* describes one of the most extraordinary processes in biology. The word *somatic* tells

TABLE 11-3	Responsiveness of B-cell populations to EBI2-, CCR7-, CXCR5-, and CXCR4-mediated chemotaxis			
	Receptor			
	EBI2	CCR7	CXCR5	CXCR4
Naïve B cells	++	+	+++	+
Activated B cells (2–6 h)	+++	+++	++	+
Activated B cells (2–3 d)	+++	+	++	+
Dark zone GC B cells (centroblasts)	–	–	++	+++
Light zone GC B cells (centrocytes)	+	–	++	+
Plasmablasts and plasma cells	+	–	–	+++

Data from Gatto, D., and R. Brink. 2013. B cell localization: regulation by EBI2 and its oxysterol ligand. *Trends in Immunology* **34**:336, Table 2.

us that the mutational processes are occurring outside of the germ line (egg and sperm) cells. *Hypermutation* alludes to the fact that the mutational processes occur extremely rapidly. Somatic hypermutation in mice and humans occurs only following antigen contact, affects only the variable regions of the antibody heavy and light chains, and requires the engagement of T cells via CD40-CD40L binding. However, it should be noted that hypermutation in some species occurs even in advance of antigen contact to diversify the primary repertoire of BCRs.

The possibility that SHM of Ig genes might play a role in antibody diversification was first implied by amino acid sequencing studies conducted in the 1970s by Cesari, Weigert, and Cohn. Their investigations focused on antibodies produced by mouse myeloma tumors expressing λ light chains. As described in Chapter 6, the mouse λ locus has been severely truncated, and as a result, mice have very few different λ-chain variable regions. These investigators were therefore able to compare each of their myeloma light-chain sequences with a known germ-line λ chain sequence and definitively show that the myeloma tumors expressed point mutations that were restricted to the variable regions of the λ chains and clustered in the complementarity-determining regions. With the advent of nucleic acid sequencing technology, these data were confirmed and extended by others who demonstrated that SHM affects the variable regions, but not the constant regions, of both the heavy and light chains of Igs; that the frequency of somatic hypermutations increases with time postimmunization; and that somatic hypermutation, followed by antigen selection, results in an increase in the affinity of secreted antibodies for the immunizing antigen.

These experiments defined the *process* of SHM, but left open the question of *where* it occurred. That question

was answered by a series of papers published in the early 1990s. Ig genes from individual B cells were isolated from stained germinal centers as well as from surrounding areas of the lymph node at various time points following antigen encounter, and their variable regions were amplified and sequenced. The results showed that somatic hypermutation occurs only within the germinal centers of an active lymph node.

For somatic hypermutation and affinity selection to work, we must posit the existence of a mechanism by which mutations are introduced into the antibody variable region genes, followed by an opportunity for the B cell to meet up again with the same antigen and test the affinity of its newly mutated receptor. Any efficient mechanism for generating B cells with improved affinity should also include a strategy whereby higher affinity B cells would be better able to access proliferative and survival signals. We now know that the germinal center provides this optimal environment for Ig mutation and affinity selection.

Having discussed earlier the pathway of the migration of B cells prior to their entry into the follicle, we describe here the path taken by activated B cells within the different zones of the GC.

The Pathway of Activated B Cells within the Germinal Center

As described above, germinal centers are established by proliferating B cells. B cells therefore begin their time in the GC with a centroblast phenotype in what then becomes the DZ of the GC. Somatic hypermutation, as well as B-cell proliferation, occur within the DZ, as suggested by the DZ expression of the AID and DNA polymerase η enzymes. Once the GC is fully established, up to 50% of GC B cells transition from the DZ to the LZ every 4 to 6 hours, which suggests that very little time elapses between the occurrence of a mutation and when it is tested for increased antigen binding.

How does affinity testing occur? On entering the GC DZ, B cells lose expression of MHC class II molecules, and hence much of the antigen they had acquired originally and processed for presentation to T_{FH} cells is lost. However, on cycling to the LZ of the GC, they re-express MHC class II and therefore they must now seek out fresh antigen within the light zone for presentation to T_{FH} cells. Those B cells with higher affinity receptors will have a competitive advantage over their sibling B cells and be able to gather and express higher levels of antigen on their cell surfaces.

T_{FH} cells in the light zone constantly browse GC B cells, forming the most extensive contacts with those B cells that display the highest antigen density. As the T_{FH} and GC B cells interact, molecular contacts between CD40 on B cells and CD154 (CD40L) on T_{FH} cells have been shown to be important to the delivery of helper signals.

Clearly, an LZ B cell with a receptor that can no longer bind antigen will be unable to attract T-cell help and will die by apoptosis. But what is the mechanism by which two B cells, with differing affinities for antigen, leave their interactions with T_{FH} cells with different instructions? Recent experiments in which B-cell divisions have been counted following interactions with T_{FH} cells have provided a potential answer to this question. B cells that enjoyed a longer interaction period with T_{FH} cells were programmed to divide more times on re-entry into the dark zone. Specifically, investigators demonstrated that T-cell help actually increases the speed of the cell cycle in GC B cells by shortening the S phase, thus allowing more proliferative cycles per unit time. Therefore, high-affinity B cells both spend longer in the dark zone and divide faster, out-competing their lower affinity neighbors.

In a feed-forward loop, the highest affinity GC B cells then receive repeated instructions to re-enter the DZ for further rounds of mutation and proliferation, resulting eventually in those B cells with the highest affinity dominating the population. As antibodies from the primary focus, and from newly generated plasmablasts, continue to accumulate within the GC and compete with GC B cells for antigen binding, the selective pressure mounts and so the affinity of the antibodies produced increases.

Observation of labeled cells in living lymph nodes, followed by sophisticated mathematical modeling, tells us that approximately 10% to 30% of the B cells that arrive in the LZ from the DZ are selected to reenter the DZ for further rounds of mutation and selection. The remaining B cells either die by apoptosis or exit the GC. Eventually, these exiting B cells will either differentiate into plasmablasts, creating the later, higher-affinity antibodies of the mature immune response, or form memory B cells that are ready for a second encounter with the same antigen. In some GCs, the iterative rounds of mutation and selection result in a clear "winner": a single clone of B cells with high affinity that crowds out all its competitors by binding all the available antigen and excluding its competitors from gaining any T-cell help. This type of outcome occurs most frequently in responses to simple haptens. In contrast, in responses to complex antigens, the clonality of B cells within the GC can remain diverse throughout the response.

What happens when a mutation in a BCR gene results in the formation of an autoimmune B cell? Early experiments showed that GC B cells undergo apoptosis within 4 to 8 hours of encountering a soluble antigen, thus protecting against recognition of self, soluble, circulating proteins. If a self antigen is expressed on cells within the GC, some activation and B-cell differentiation may occur, but the B cells do not accumulate. Most probably, this is because suitable T_{FH} help is not available to enable B-cell survival, because of the prior elimination of self-reactive T cells. However, neither of these mechanisms protects against the emergence of new BCRs capable of recognizing self antigens that are rare or tissue-specific, and for which T-cell help may be available in the form of receptors that cross-react with foreign antigens. Such autoimmune GC B cells may account for the common finding of low-affinity self-reactive antibodies in the aftermath of severe infectious disease.

Key Concepts:

- Following T-dependent stimulation by antigen, B cells enter the follicles and divide rapidly. The B cells, together with associated T_{FH} cells and follicular dendritic cells, form germinal centers.

- Germinal centers are made up of dark zones, in which B cells divide rapidly and undergo somatic hypermutation, and light zones, where the B cells interact with T_{FH} and follicular dendritic cells, and B cells bearing high-affinity, mutated receptors are selected.

- B cells in the dark zone are referred to as *centroblasts*; those in the light zone are referred to as *centrocytes*. Movement between the two zones is orchestrated by modifications in cell-surface chemokine receptors.

- GC B cells are prone to apoptosis. Survival requires productive interactions with T_{FH} cells as well as with follicular dendritic cells. High-affinity B cells take up and present higher levels of antigen to T_{FH} cells and thus receive a greater quantity of survival signals.

The Mechanisms of Somatic Hypermutation and Class Switch Recombination

Once investigators knew *where* somatic hypermutation was occurring, their next question, inevitably, was *how*? This question presented huge experimental challenges, and for almost 30 years investigators made only incremental progress in defining the mutational mechanism. As so often happens, the breakthrough came from an unexpected direction and simultaneously offered insights into the mechanisms of class switch recombination (CSR) and SHM.

When attempting to identify proteins that mediate a specific process, scientists often begin by isolating cell lines, or mutant animals, that fail to carry it out. By comparing cells that do, and cells that fail to perform the process under investigation, the genes or proteins that are missing or mutated in the defective cells can be identified. This was the approach taken by Tasuku Honjo and colleagues, who elected to determine which enzymes mediated the process of CSR. They first isolated a B-cell line in which the antibody genes failed to undergo CSR. Next, they compared cells that could and could not perform CSR in order to identify those molecules responsible for the process.

However, the investigators quickly noticed something completely unexpected; the B-cell lines that had been isolated on the basis of their inability to allow CSR *were also unable to create somatic mutations in the variable regions of their Ig genes!* Could it be that the same molecular machinery that performs CSR also mediates SHM?

In 1854, in a lecture at the University of Lille, Louis Pasteur famously said: "In the fields of observation, chance favors only the prepared mind." It would have been easy for Honjo's laboratory to continue to study only CSR with a laser-like focus —after all, they had found their mutant and the unraveling of the mechanism of CSR was a big story! However, the minds of the scientists in Honjo's lab were "well prepared"—they knew exactly how important a finding they had made and they pursued their studies of the mechanisms of both SHM and CSR with equal vigor.

They quickly identified the mutated gene in this cell line as encoding the enzyme **activation-induced cytidine deaminase (AID)**. But now the scientists were faced with a host of new questions: How does AID induce hypermutation? How is its mutating action restricted to antibody variable regions? And how does this same enzyme bring about the quite different process of CSR? As we will see below, the unifying feature of these two processes is the creation of a break in the DNA strands encoding the Ig genes, by first deaminating cytidine residues at key locations in the genes and then repairing the resulting damage in a way that gives rise to mutations and/or double-strand breaks. The differences between the two outcomes of SHM and CSR rely on the positioning and frequency of DNA sequences that act as targets for AID activity and the mechanisms by which the DNA damage is repaired.

AID is a small (24-kDa) protein that deaminates deoxycytidine residues found in stretches of single-stranded DNA, yielding deoxyuridine (**Figure 11-18**). The targeted deoxycytidine residues are always found in the context of particular, short DNA sequences. For decades, immunologists had known that some of the regions of active Ig genes

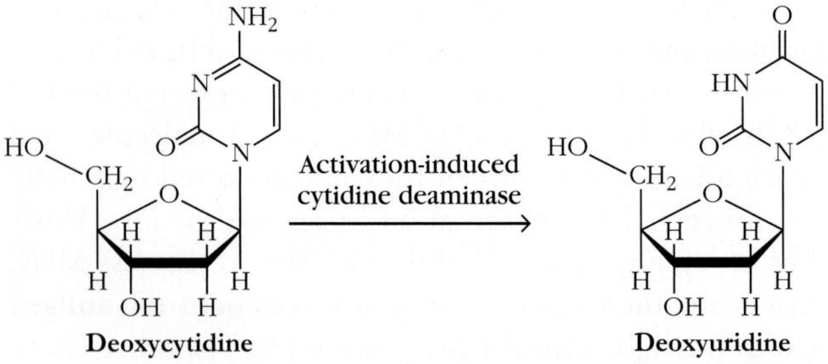

FIGURE 11-18 Activation-induced cytidine deaminase (AID) mediates the deamination of deoxycytidine and the formation of deoxyuridine.

are transcribed into sterile (i.e., untranslated) transcripts. However, the reason for this transcription now became clear; transcription demands the transient separation of the coding and noncoding strands and thus yields a single-stranded sequence target for AID to bind. The switch regions where CSR occurs had also been shown to be regions where sterile transcription occurred, and SHM occurs in antigen-activated B cells that are actively transcribing their genes. Here, we will first address the mechanism of SHM and then show how many of the same rules apply to the description of CSR. Note that CSR can occur outside as well as within the GCs, whereas GCs are the only locations that support SHM.

AID-Mediated Somatic Hypermutation

Although the details of the SHM mechanism are not yet completely understood, many aspects of the process are now well established. AID-induced formation of deoxyuridine, as shown in Figure 11-18, creates a U-G mismatch in the double-stranded DNA. Several alternative mechanisms then come into play that participate in the resolution of the original mismatch, and lead to the creation of shorter or longer stretches of mutated DNA (**Figure 11-19**).

The simplest manner in which the situation can be handled is by high-fidelity repair (pathway 1 in Figure 11-19), using the enzymes of either the base excision repair pathway or the mismatch repair pathway. This simply replaces the newly generated uridine residue with a fresh cytidine residue, resulting in no net change in the DNA product. In pathway 2, we visualize the outcome of a simple replication step over the mismatch. This results in a transition mutation at the original C-G base pair, in which the original C is eventually replaced by a T. In this case, one of the daughter cells would have an A-T pair instead of the original G-C pair found in the parent cell. The other modes of resolution are more complex and involve mutagenic DNA repair.

The mismatched uridine can also be excised by uracil-DNA glycosylase (UNG; pathway 3), which is a component of the base excision repair pathway. This creates an abasic site on one of the DNA strands. Error-prone polymerases from the short-patch base excision repair mechanism then fill the gap ("N" refers here to any nucleotide; see

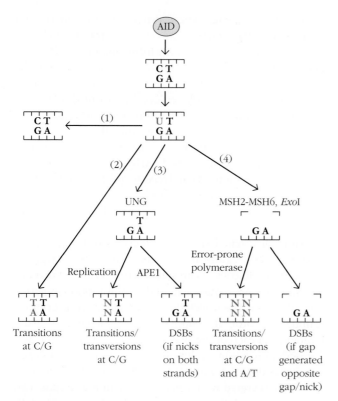

FIGURE 11-19 The generation of somatic cell mutations in Ig genes by AID. AID deaminates a deoxycytidine residue, creating a uridine-guanosine (U-G) mismatch. Resolution of this mismatch may be mediated by any one of several pathways: (1) The DNA mismatch may be repaired with high fidelity by proteins of the *base excision repair (BER)* or *mismatch repair (MMR)* pathway. (2) The deoxyuridine residue may be interpreted by the DNA replication machinery as if it were a deoxythymidine, resulting in the creation of an A-T pair in place of the original G-C pair in one of the daughter cells. (3) The mismatched uridine may be excised by uridine DNA glycosylase, leaving an abasic site that is filled in by any of the four bases, in a reaction known as *short-patch BER*, catalyzed by one of a number of error-prone polymerases. (4) Enzymes of the MMR pathway, such as MSH2-MSH6, may recognize the gap and *Exo*I excises a long stretch of the DNA surrounding the U-G couple. Error-prone polymerases are then recruited to the hypermutable site, and these polymerases can introduce a number of mutations around the original mismatch. Thus, depending on the repair mechanism, a mutation may occur only in the originally altered base or in one or more bases surrounding it.

Figure 11-19). Alternatively, the AP endonuclease 1 enzyme (APE1) may create a nick at the abasic site, which is then processed to create a single-nucleotide gap. This gap is filled by the error-prone DNA polymerase β and the DNA strand is then sealed by DNA ligase 1. If a nick also occurs nearby on the opposite strand, this may result in a double-strand break, a situation that we will encounter again as we discuss the process of CSR. Sometimes, rather than using short-patch base excision repair, the cell will instead turn to a

long-patch base excision repair pathway. Under these circumstances, after the DNA is nicked by APE1, one of the error-prone DNA polymerases (β, δ, or ε) will displace the cut strand and polymerize a tract of DNA approximately 2 to 10 bp in length. The displaced strand is then removed by a specialized endonuclease and the remaining nick is sealed by DNA ligase 1.

Sometimes the enzymes of the mismatch repair pathway, rather than those of the base excision repair pathway, operate on the deaminated cytidine (pathway 4 in Figure 11-19). Under these circumstances, the MSH2-MSH6 heterodimer detects the mismatch and recruits a protein complex that includes exonuclease 1 to excise a region of DNA surrounding the mismatch. The excised strand is then repaired by error-prone DNA polymerases, such as DNA polymerase η, leading to a lengthier series of mutations in the region of the original mismatch. Again, if the exonuclease is targeted to both strands of DNA close to the same location, a double-strand break can result.

Targeting of the Mutational Apparatus to Antibody Variable Regions

Since the rate of hypermutation is orders of magnitude higher in Ig variable region DNA than in other genes in germinal center B cells, some mechanism must exist to direct the mutational machinery to the correct chromosomal location. We have already mentioned that AID targets genes that are actively undergoing transcription, but since many genes are transcribed in B cells, and mutation is restricted to the variable regions of Ig genes, additional targeting mechanisms must exist.

Careful analysis of germline Ig variable region sequences revealed that some sequence motifs were more likely than others to be targeted by the AID mutational apparatus. These sequences are referred to as **mutational hot spots**. Early experiments suggested that mutations preferentially accumulated at the G and C residues embedded within the sequence motif AGCT. Further analysis expanded this motif to the sequence

$$DGYW$$
$$HCRW$$

where:

$$D = A/G/T$$
$$Y = C/T$$
$$R = A/G$$
$$W = A/T$$
$$H = T/C/A$$

Comparison of the sequences in the antigen-binding, complementarity-determining regions (CDRs) with the rest of the amino-terminal variable regions of antibodies indeed showed a higher concentration of mutational target sequences in CDRs than in the framework sequences. Other factors that may also influence the highly selective frequency of CDR mutations include the sequence context of the hot-spot regions, the high rate of transcription, the frequency with which RNA polymerase stalls during transcription, and the potential for differential repair of particular sequences.

AID-Mediated Class Switch Recombination

In Chapter 6, we noted that naïve B cells could simultaneously express both membrane-bound IgM (mIgM) and mIgD: both proteins are encoded on the same long transcript, and the decision to translate μ (IgM) versus δ (IgD) heavy chains is made at the level of RNA splicing. In contrast, the molecular machinery that allows the cell to switch from expressing μ to expressing any heavy-chain class other than μ or δ operates at the level of DNA recombination, and the process by which it occurs is referred to as *class switch recombination* (CSR). The switch from μ to the expression of any heavy-chain class other than δ results in the irreversible loss of the intervening DNA (**Figure 11-20**).

The Ig heavy-chain locus is approximately 200 kb in length. In a B cell producing IgM antibodies, the rearranged V_H gene lies approximately 8 kb 5′ of the Cμ gene. The formation of γ, ε, and α heavy-chain genes requires cutting and rejoining of the heavy-chain DNA in such a way that the V_H region is removed to a similar location 5′ of Cα, Cγ, or Cε. Class switching occurs by the induction of recombination between **donor** and **acceptor switch (S) regions** located 2 to 3 kb upstream from each C_H region, with the exception of Cδ (see Figure 11-20). The donor S region is defined as that closest to the V_H region. Switch regions are 2 to 10 kb in length and consist of simple, short sequences repeated in tandem. For example, mouse Sμ, Sε, and Sα sequences are composed of variations of pentamers such as GGGGT, GAGCT, and GGGCT. The Sγ sequences are a little more complex and are made up of repeats of a 49- or 52-base pair sequence. These switch regions contain high densities of AID-targeting sites on both strands, such that there is a high probability of the formation of AID-induced double-strand breaks. CSR can occur anywhere within the S regions.

During CSR, AID-induced cytidine-to-uridine lesions are created on both strands within a switch region, and these lesions are then converted into double-stranded breaks by members of the base excision repair or mismatch repair pathway, as described earlier. Ligation between two broken S regions is catalyzed by proteins of the nonhomologous end-joining pathway. But how does the cell make the choice as to which S regions will be broken and rejoined?

We have learned that AID operates only on single-stranded DNA regions that are undergoing transcription, and so it should come as no surprise that, prior to the onset of CSR, cytokine signals lead to the transcription of germline DNA over the targeted switch regions. Transcription is initiated at promoters upstream of a noncoding "I" exon, located 5′ of each of the switch regions (see Figure 11-20). Transcription continues through the associated S region and terminates downstream of the relevant C_H exons. Different cytokines secreted by T cells or other immune cells, can stimulate transcription from different I region promoters. In this way, T cell (or other immune cell) signals direct CSR to generate the class of Ig most suitable for eliminating the current pathogen (**Table 11-4**).

B cells must also receive costimulatory signals from CD40 or B-cell Toll-like receptors in order to engage in CSR. The importance of CD40-CD40L interactions in the mediation of CSR is illustrated in patients suffering from **X-linked hyper-IgM syndrome**, an immunodeficiency disorder in which T_H cells fail to express CD40L. Patients suffering from

(a)

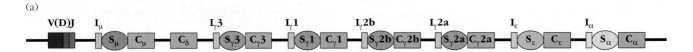

(b)

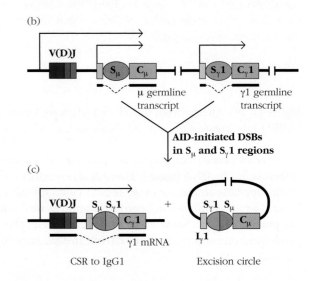

(c)

FIGURE 11-20 Class switch recombination from a Cμ to a Cγ1 heavy-chain constant region gene. (a) The Ig locus is drawn to show regions important in class switch recombination (CSR). The transcription required for generating the single-stranded DNA sequences for AID binding is initiated by promoters located upstream of each I exon. The activation-induced cytidine deaminase (AID) enzyme initiates CSR by deaminating cytidine residues within the designated switch (S) regions upstream of the donor and acceptor switch sites. (b) Active transcription upstream of Cμ and Cγ1 readies the region for CSR from μ to γ1 heavy-chain expression. (c) Because of the frequency of AID-targeting sequences in these switch regions, this leads to the formation of double-strand breaks within both S regions, which are then resolved by DNA repair mechanisms, with the loss of the intervening DNA sequence, as described in text.

TABLE 11-4	Specific cytokine signals causing B cells to undergo CSR to various heavy-chain classes
Cytokine signal	Isotype synthesized by target B cell
IL-4	IgG1, IgE
TGF-β	IgA, IgG2b
IL-5	IgA
IFN-γ	IgG3, IgG2a

this disorder express predominantly IgM, with only very low levels of IgG and IgA. Such patients also fail to form germinal centers, or to generate memory B-cell populations, and their antibodies do not show evidence of SHM.

CSR can occur more than once during the lifetime of the cell. For example, an initial CSR event can switch a cell from making IgM to synthesizing IgG1, and a second CSR can switch it to making IgE or IgA.

It is worth emphasizing that the process of CSR begins rapidly after antigen stimulation and occurs prior to the formation of germinal centers. Therefore, although CSR can occur in GCs (CSR to IgA production frequently occurs in the germinal centers in Peyer's patches, for example), its occurrence is not restricted to GCs. CSR can also occur in T-independent responses (see later), in which SHM is extremely rare. In this context, it is worth noting that signals for CSR need not emanate from T cells (as is the case for SHM), but can also be mediated by eosinophils and antigen-presenting cells, including monocytes and dendritic cells. For example, in the presence of IL-10 or TGF-β, the B-cell proliferation and survival factors BLyS (B lymphoid-stimulating factor) and APRIL are capable of inducing CSR from Cμ to Cα.

Key Concepts:

- Somatic hypermutation is initiated by a cytidine deamination reaction catalyzed by the enzyme AID. DNA repair mechanisms then cause alterations of the sequence of the variable region of immunoglobulin genes.

- AID activity is targeted to transcribed sequences that have high levels of specific targeting sequences. These targeting sequences are found at significant frequencies in areas of the Ig genes involved in SHM and CSR.

- Class switch recombination results in the replacement of an antibody heavy-chain constant region, with a different constant region, originally located 3′ to the first. Regulation of which heavy chain constant regions will be targeted is mediated by cytokines secreted either by T cells or by other immune system cells.

- SHM occurs only in the germinal centers, and in mice and humans only after T-dependent antigen stimulation. CSR may occur in the GC, but it can also occur elsewhere and may occur in the presence or absence of help from T cells.

Memory B Cells Recognizing T-Dependent Antigens Are Generated Both within and outside the Germinal Center

B-cell memory is the immunological phenomenon we have known the longest and have used the most extensively in the clinic. Referred to by Thucydides in his *History of the Peloponnesian War* (see Chapter 1), the ability of an animal to respond more rapidly and effectively with antibody production on a second (or later) exposure to an antigen than on first encounter is the principle that underlies the process of vaccination. Once thought to be a uniform population of class-switched B cells that rapidly secreted high-affinity, mutated antibodies on secondary antigen challenge, it is now clear that the memory B cells of the recall response include a variety of cell types with different and specialized functions in the long-term maintenance of immunity. Understanding when and where these cell populations are generated during the course of a primary immune response, as well as how they are recalled on subsequent antigen encounter, has important consequences for vaccine design. Some general properties of memory cells as compared with naïve B cells are described in **Table 11-5**.

Phenotypically Distinct Memory B-Cell Subsets

As classically defined, memory B cells have participated in prior, T-dependent contact with antigen, after which they underwent differentiation to form long-lived, recirculating, secondary B cells. These cells respond more rapidly and more strongly than primary B cells on subsequent antigenic activation. Memory B cells can be subdivided into two subsets.

The first subset is generated early in the immune response, before the germinal center has formed, and comprises primarily (but not exclusively) IgM-bearing B cells with unmutated receptors. The generation of these early memory B cells may continue for several weeks after antigen contact. These early memory B cells form a reservoir of antigen-binding cells that retain the potential to enter the GC if early mutational events fail to clear the stimulating pathogen. Because of their relatively low affinity for the immunizing antigen, activation of this population is readily inhibited by the presence of soluble antibodies.

With the onset of germinal center formation, at around 4 days after antigen encounter, the bulk of memory B-cell generation switches location to the GCs, where the second class of mutated, high-affinity memory B cells is generated. In the GC, memory B cells gradually shift from expressing chiefly IgM to predominantly other classes of antibodies, and these antibodies

TABLE 11-5	Functional differences between primary and secondary B cells	
	Naïve B cell	**Memory B cell**
Lag period after antigen administration	4–7 days (depends on antigen)	1–3 days (depends on antigen)
Time of peak response	7–10 days	3–5 days
Magnitude of peak antibody response	Varies, depending on antigen	Generally 10 to 1000 times higher than primary response
Antibody isotype produced	IgM predominates in early primary response	IgG predominates (IgA in the mucosal tissues)
Antigens	Thymus independent and thymus dependent	Primarily thymus dependent
Antibody affinity	Low	High
Life span of cells	Short-lived (days to weeks)	Long-lived, up to life span of animal host
Recirculation	Yes	Yes

begin to accumulate mutations in their variable regions. This is a time-dependent process, and consequently, those memory B cells generated early in the GC response (days 6 to 8 after antigen contact) have a higher proportion of IgM-bearing than IgG-bearing cells and have lower numbers of mutations than those generated later. Some of these memory B cells remain in and near the GCs, although many migrate to sites of antigen drainage, such as the marginal zone of the spleen and the mucosal epithelium of the gut, lungs, tonsils.

The Formation of Long-Lived Plasma Cells

Although not technically defined as memory cells, a second type of B-cell progeny also provides long-term humoral immunity. Once established in a long-term niche, often the bone marrow, *long-lived plasma cells* (LLPCs) produce antigen-specific antibodies for a very long period after antigen stimulation, apparently without the need for further antigenic stimulation. Indeed, measurements of the half-life of bone marrow plasma cells show that they can be very long-lived indeed. For example, smallpox-specific serum antibodies have been identified in human subjects 75 or more years after immunization with smallpox vaccine, suggesting that the plasma cells secreting these antibodies may persist for the lifetime of the host!

Starting at around 10 days after encountering antigen, a significant fraction of germinal center B cells up-regulate the expression of transcription factors that drive plasma cell fate. Thus, LLPC differentiation in the GC begins after the bulk of conventional memory cells have already been generated. Interactions between the T_{FH} PD-1 receptor and its ligands, PD-L1 and PD-L2, on GC B cells are important in the formation of LLPCs.

As the GC B cell differentiates into a fully mature plasma cell, the level of expression of the chemokine receptor CXCR5 decreases and that of CXCR4 increases, enabling the cell to leave the follicles and enter the peripheral circulation. Many LLPCs, as many as 10% to 20%, take up residence in

the bone marrow, whereas others enter the gut and lung mucosal tissues. Germinal center–derived LLPCs differ from those generated in the primary focus in that their Ig genes have undergone SHM and affinity maturation. Membrane BCR expression is significantly reduced on LLPCs versus on conventional memory B cells. However, some plasma cells continue to express low levels of cell-surface IgM and therefore may be capable of being further stimulated by antigen, even after taking up residence in distant locations.

The niches occupied by fully differentiated plasma cells differ from those inhabited by developing B cells (**Figure 11-21**). Within the bone marrow, LLPCs receive cytokine signals provided by mesenchymal stromal cells (CXCL12) and by eosinophils and megakaryocytes (APRIL). These cytokines induce the up-regulation of anti-apoptotic molecules such as Mcl-1, that support long-term survival of the LLPCs. Plasma cell longevity also appears to depend on specialized cellular metabolism; for example, plasma cells engage in autophagy, thus enhancing their supply of nutrients.

In addition to the differences in life span, plasma cells can also be distinguished from naïve B cells on the basis of their histology (they are larger and more complex than their B-cell parents), their proliferative potential, and their expression of different cell-surface markers and transcription factors (**Table 11-6**). Additional information on the specialized nature of plasma cells in the mucosal tissues can be found in Chapter 13.

The temporal separation in the formation of LLPCs compared with conventional memory cells also suggests that earlier models for the generation of memory cells in the GC are unlikely to be correct. These simpler models relied on the notion that a B cell divided asymmetrically on initial antigen encounter to create one LLPC and one memory cell. Clearly, the bulk of cell divisions that occur soon after antigen stimulation give rise to conventional memory B cells, whereas the majority of later divisions throw off plasmablasts, which give rise to LLPCs.

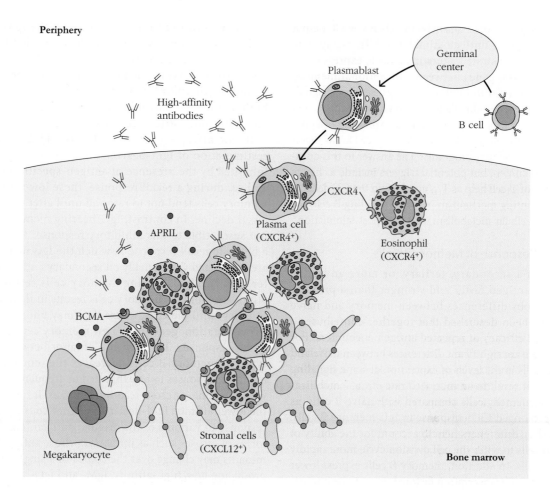

FIGURE 11-21 **The bone marrow niche occupied by plasma cells is supported by eosinophils and megakaryocytes, as well as by mesenchymal stromal cells.** Plasmablasts and plasma cells that have passed through the germinal center reaction and entered the circulation take up residence in the bone marrow, where they seek niches adjacent to eosinophils and megakaryocytes, as well as to the traditional, mesenchymally derived stromal cells. The eosinophils and megakaryocytes provide the long-lived plasma cell with the survival factor APRIL, a TNF family member recognized by the receptor BCMA, whereas the stromal cells release CXCL12, recognized by the receptor CXCR4 on plasma cells.

TABLE 11-6	Phenotypic modifications in the transformation of naïve B cells into plasma cells		
	Naïve B cell	**Plasmablast**	**Long-lived plasma cell**
BCR membrane expression	+++	++	+/−?
Life span	++	+	++++
Proliferation	−	++	−
CD138 and CXCR4 expression	−	+	+++
CD19 and MHC class II expression	+++	++	+/−
Location	Lymphoid organs	Lymphoid organs and blood	Bone marrow
BLIMP-1 expression	−	+	++

Data from Nutt, S. L., P. D. Hodgkin, D. M. Tarlinton, and L. M. Corcoran. 2015. The generation of antibody-secreting plasma cells. *Nature Reviews Immunology* **15**:160, Table 1.

What might be the mechanism for this change in GC output with time following antigen stimulation? Transcriptome analysis has demonstrated alterations in the expression of approximately 100 genes between conventional memory B cells as compared with LLPCs. These genes encode signaling molecules, transcription factors, chromatin modifiers, and other molecules that are important for the expression of a phenotypic change in cell type. But how does the cell know that it is time to switch cellular programs? The answer to this question is still unknown, but potential triggers include a change in the quality of T-cell help as T_{FH} numbers in the GC decline, a division counting mechanism in the responding B cells, or alterations in cellular metabolism with time post-stimulation.

The Recall Response of Memory B Cells

What makes a secondary, tertiary, or more advanced response to antigen faster and stronger than a primary one? Numerous differences between memory and naïve B cells have been described that together contribute to the increased efficacy of repeated antigen encounters. For example, there are significant differences between naïve and memory B cells in the levels of expression of some signaling molecules that result in an increased rate of Ca^{2+} mobilization in IgM memory cells compared with naïve B cells, as well as a prolonged Ca^{2+} response in IgG memory B cells. Such signaling differences help to account for the ability of memory B cells to enter the cell division cycle more rapidly than naïve cells. In addition, memory B cells express lower concentrations of transcription factors known to be important in the maintenance of the cellular resting state (quiescence); they are therefore more effectively "poised" to enter the cell cycle. Memory B cells also constitutively express higher levels of molecules that mediate signaling interactions with T cells, such as CD40, CD80, and CD86; and their expression of receptors for the survival- and proliferation-inducing ligands BAFF and TACI is also enhanced, as are their intracellular levels of anti-apoptotic molecules such as Bcl-2, Bcl-X_L, A1, and Mcl-1. Finally, investigators have shown that some memory B cells undergo proliferation in response to innate stimuli such as CpG, alone, whereas naïve B cells require concomitant stimulation through the BCR and TLR. This suggests that innate stimuli may play a role in the maintenance of memory B-cell populations.

Memory and naïve B-cell populations also differ in their requirements for help from cell types other than T cells. Some virus-specific memory B cells have been shown to entirely lose the requirement for T-cell help on restimulation, although most memory B cells still need some help from memory T_{FH} cells for efficient recall responses. Memory T_{FH} cells retain low, but detectable levels of expression of the CXCR5 receptor and as a result, memory T_{FH} cells in the lymph nodes and spleen accumulate in the follicles and at the T-cell/B-cell border, and are immediately available to interact with B cells. Follicular dendritic cells also help in the maintenance of memory B cells, supplying survival factors and/or serving as antigen reservoirs.

Analysis of the recall responses of the IgM- versus IgG-bearing memory cell pools has revealed that they have different sensitivities to inhibition by serum antibody, which results in part from differences in the affinities of their respective receptor molecules—high-affinity, somatically hypermutated receptors will compete more effectively for antigen binding than will low-affinity receptors (**Figure 11-22**). Antigen restimulation of IgM-bearing memory B cells is readily inhibited by the presence of antigen-specific antibodies and so, during a recall response, these low-affinity IgM memory cells tend not to respond until after the antibody levels decline. In contrast, IgG-bearing memory cells are not susceptible to this inhibitory response. This suggests a biphasic recall response in which the IgG memory cells are stimulated very quickly on secondary antigen encounter, and the IgM-bearing memory cells respond later. Stimulation of IgM memory cells results in their entry (or re-entry) into the GCs, where they may undergo somatic hypermutation, generating new memory cells for subsequent responses. This division of labor between the two types of memory B cells decreases the tendency of the memory responses to be exhausted by frequent encounters with the same, or cross-reacting antigens. It further allows the lower affinity, IgM-bearing memory cells to mutate in response to variants of the original antigen (e.g., mutated forms of viruses such as influenza, whose antigenic determinants may change over the course of a single infection), producing new, high-affinity, IgM- and IgG-bearing cells.

Key Concepts:

- After initial antigen exposure to a T-dependent antigen, secondary and later encounters with the same antigen will give rise to faster and stronger recall responses, referred to as "memory" responses.

- IgM-bearing memory B cells are generated early in the primary response, prior to the onset of somatic hypermutation, and secrete nonmutated antibodies of lower affinity. Their response to antigen is easily inhibited by pre-existing serum antibodies.

- IgG-bearing memory cells generated later in the primary response, have mutated and selected receptors and yield recall responses that deliver high-affinity antibodies.

- Memory cells differ from naïve cells in their dependence on other cells for stimulation, their expression of cell-surface antigens and chemokine receptors, their metabolism, their concentrations of anti-apoptotic molecules, and the speed with which they can enter the cell cycle on antigen restimulation.

- Long-lived plasma cells are generated late in a primary response, and survive for very long periods in the bone marrow, the mucosa, and other locations.

(a) Early memory response

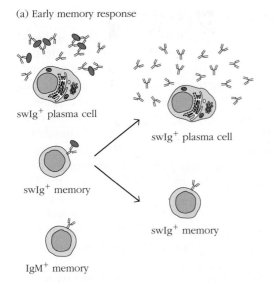

(b) Late memory response

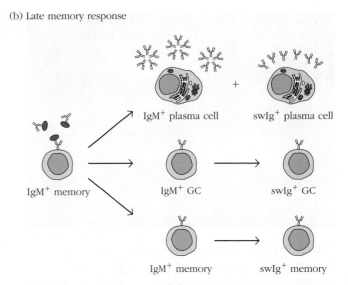

FIGURE 11-22 Temporal separation of recall responses from IgG$^+$ and IgM$^+$ memory responses. (a) In the presence of high concentrations of serum antibody, high affinity memory cells bearing class switched Ig receptors (swIg$^+$ memory) can respond, but lower affinity IgM$^+$ memory cells are unable to compete for antigen. High affinity class-switched memory cells form plasma cells and new memory cells, but do not re-enter the germinal centers. (b) As serum antibody levels decline, low affinity IgM$^+$ memory cells are able to bind antigen and are stimulated to form IgM$^+$ and swIg$^+$ plasma cells, as well as new IgM$^+$ and swIg$^+$ memory cells. In addition, these IgM$^+$ memory cells are able to enter the germinal center, where they can undergo somatic hypermutation, creating high affinity B cells.

Most Newly Generated B Cells Are Lost at the End of the Primary Immune Response

Between 14 and 18 days after its initiation, the primary immune response winds down, and the immune system is faced with a problem of excess. Rapid proliferation of antigen-specific B cells over the course of the immune response leads to the generation of expanded clones of cells, and if all the newly generated cells were allowed to survive at the close of every immune response, there would soon be no room in the follicles for new B cells emerging from the bone marrow.

Although we know that most B cells are lost by apoptosis at the end of the immune response, the exact mechanism by which this occurs has not yet been fully characterized. As antigen levels wane, the balance between survival signals and death signals in lymph node and splenic B cells may be tipped in favor of apoptosis as they cease to receive survival signals via BCR-mediated antigen binding. Animals deficient in the genes encoding Fas have excess numbers of B cells, implicating the Fas-FasL interaction in the control of B-cell numbers. But we do not yet know the whole story.

Key Concept:

- At the end of the immune response, most newly generated B cells die by apoptosis.

T-Independent B-Cell Responses

Not all antibody-producing responses require the participation of T cells, and certain subsets of B cells have evolved mechanisms to secrete antibodies to particular classes of antigens, without T-cell help (see Figures 11-3b and c). Antigens capable of eliciting T-independent antibody responses tend to have polyvalent, repeating determinants that are shared among many microbial species, thus echoing characteristics of many PAMPs (Chapter 4). Many of these antigens are recognized by the B-1a, B-1b, and marginal zone B cell subsets (discussed below). B-1a and B-1b cells secrete mainly IgM antibodies that are not subject to SHM. Because of the shared nature of their antigens and their oligoclonal (meaning "few clones") antibody responses, B-1 B cells are classified as being an "innate-like" category of lymphocytes. Our understanding of the physiology of marginal zone B cells is still developing, but study of these cells suggests that they may receive help from cell types other than T cells, as is discussed below.

T-Independent Antigens Stimulate Antibody Production in the Absence of T-Cell Help

The nude (*nu/nu*) mouse is one of the most bizarre accidents of nature. Devoid of body hair, its ears seem oversized, and it appears absurdly vulnerable (**Figure 11-23**). These mice

FIGURE 11-23 The nude mouse. Nude mice have a mutation in the *Foxn1* gene that results in hair loss (alopecia) and interferes with thymic development, such that they possess very few T cells and only a thymic rudiment. Nude mice provided useful early models for the exploration of a T cell–depleted immune response. *[David A. Northcott/Corbis.]*

have a mutation in the gene for the transcription factor Foxn1 that, in addition to affecting hair growth, also results in athymia: mice and humans with this mutation have only thymic rudiments and possess few mature recirculating T cells.

Using nude mice as well as mice whose thymuses were surgically removed early in life, immunologists showed that most *protein* antigens fail to elicit an antibody response in such animals. In contrast, the response to many *carbohydrate* antigens was unaffected. Those antigens capable of generating antibody responses in athymic mice were referred to as T-independent antigens. Since T-cell interactions are required for the induction of AID and hence for SHM, TI antigens elicit predominantly low-affinity antibody responses from B cells expressing mainly the IgM isotype. We now know that T-independent antigens fall into two subclasses (**Table 11-7**).

TI-1 Antigens

The first class of *T-i*ndependent antigens (TI-1 antigens) is exemplified by *lipopolysaccharide* (LPS), expressed by gram-negative bacteria. TI-1 antigens bind to innate immune receptors on the surface of all B cells (including the majority, B-2 B-cell population), and are capable, at high antigen doses, of being mitogenic for all B cells bearing the relevant innate receptors. Since B-cell stimulation in this instance is occurring through the innate receptor (TLR4/MD-2/CD14), only a small minority of the antibodies produced will be able to bind directly to the TI-1 antigen. The dramatic in vivo polyclonal TI-1 responses generated in response to high levels of gram-negative organisms can be catastrophic for an individual and are associated with the phenomenon we know as *septic shock* (see Chapter 4).

At lower doses, the innate immune receptors are unable to bind sufficient antigen to be stimulatory for the B cell. However, in those B cells that bind to the TI-1 antigen through their Ig receptors, the TI-1 antigen cross-links the Ig and innate receptors, thereby eliciting an oligoclonal B-cell response that remains independent of T cells. Under these low-dose circumstances, all the secreted antibodies will be specific for the TI-1 antigen.

TI-2 Antigens

Unlike TI-1 antigens, type *2 T-i*ndependent antigens (TI-2 antigens), such as capsular bacterial polysaccharides or polymeric flagellin, do not activate B cells via their innate immune receptors and therefore do not elicit a polyclonal response at high concentrations. Rather, their capacity to activate B cells in the absence of T-cell help results from their ability to present antigenic determinants in a remarkably multivalent array, causing extensive cross-linking of the BCR. In addition, most naturally occurring TI-2 antigens are characterized by the ability to bind the complement fragments C3d and C3dg (see Chapter 5). As a result, TI-2 antigens activate B cells by cross-linking high numbers

TABLE 11-7	Properties of thymus-dependent and thymus-independent antigens		
		TI antigens	
Property	**TD antigens**	**Type 1**	**Type 2**
Chemical nature	Soluble protein	Bacterial cell-wall components (e.g., LPS)	Polymeric protein antigens; capsular polysaccharides
Humoral response			
Isotype switching	Yes	No	Limited
Affinity maturation	Yes	No	No
Immunologic memory	Yes	No	Limited
Polyclonal activation	No	Yes (high doses)	No

of BCR and CD21 (CR2) receptors on the B-cell surface (see Figure 11-3c). Marginal zone B cells are highly represented among B cells that bind TI-2 antigens through BCR and CD21 binding.

TI-2 antigens can only partially stimulate B cells in the complete absence of help from other cells. Monocytes, macrophages, and dendritic cells have been shown to facilitate B-cell responses to TI-2 antigens by secreting the B-cell survival molecule **BAFF**, which is bound by mature B cells through the BAFF and TACI (*t*ransmembrane *a*ctivator and *CAML i*nteractor) receptors. This interaction activates important transcription factors, including NF-κB, that promote B-cell survival, maturation, and antibody secretion. Although not necessary for the response, T cells may also enhance the activation of B cells by TI-2 antigens, by producing cytokines that help to push TI-2 antigen–activated B cells from activation to antibody production. Recognition by B cell–bound receptors of cytokines secreted by a number of cell types may also stimulate the B cells to secrete antibody classes other than IgM in response to TI-2 antigen stimulation. Unlike TI-1 antigens, TI-2 antigens cannot stimulate immature B cells and do not act as polyclonal activators.

Key Concepts:
- T-independent responses generated by B-1 and marginal zone (MZ) B cells give rise to relatively low-affinity, primarily IgM antibodies.

- Most TI antigens are carbohydrate in nature. TI-1 antigens interact with B cells via both the BCR and innate immune receptors, whereas TI-2 receptors are highly polymerized antigens that do not bind to innate immune receptors.

- Both types of T-independent responses are enhanced by interactions with other cell types, including T cells, macrophages, and monocytes, and neutrophils.

Two Novel Subclasses of B Cells Mediate the Response to T-Independent Antigens

All B cells bear Ig receptors and secrete antibodies, but recent research has demonstrated that there are multiple subpopulations of B cells that vary in location, phenotype, and function. Some of these subpopulations (the *transitional* B-cell populations, T1 and T2) represent temporal stages of B-cell development that occur after the B cell leaves the bone marrow. An additional transitional subset, T3, appears to represent an anergic subpopulation of B cells (see Chapter 9). The others (B-1a, B-1b, B-2, and marginal zone B cells) represent different subpopulations of mature B cells, each characterized by a preferential location and range of functional capacities (**Table 11-8**). The T-dependent B-cell response described above is conducted by members of the B-2 cell population, also known as follicular B cells.

TABLE 11-8	Functional differences among mature B-cell subsets		
Attribute	**Conventional B-2 B cells**	**B-1a and B-1b B cells**	**Marginal zone B cells**
Major sites	Secondary lymphoid organs	Pleural and peritoneal cavity; also spleen	Marginal zones of spleen in mice; primates also have MZ cells in other locations
Variable region diversity	Highly diverse	More restricted diversity	Moderate diversity
Rapidity of antibody response	Slow; plasma cells appear 7–10 days post-stimulation	Rapid; plasma cells appear as early as 3 days after stimulation	Rapid; plasma cells appear as early as 3 days after stimulation
Surface IgD?	High levels of IgD	Low levels of IgD	Low levels of IgD
Somatic hypermutation	Yes	No	Yes in primates; possibly in rodents
Requirements for help	Provided by T cells	T-cell help not required; T and other cells can enhance response	T-cell help not required; T cells, dendritic cells, and neutrophils can enhance response
Participate in germinal center reaction?	Yes	No	Possibly, although with slower kinetics than follicular B cells
Isotypes produced	All isotypes	Predominantly IgM	Predominantly IgM
Immunological memory	Yes	Very little	Source of IgM-producing memory cells in humans

The presence of B cells with properties distinct from those of the majority B-2 subset was first suggested when scientists noticed that B cells responding to T-independent antigens differed in several important ways from B cells recognizing T-dependent antigens. Those B cells with specificity for T-independent antigens can be divided into two major subtypes that differ in aspects of their development, their anatomical location, and their cell-surface markers. We will briefly describe each of them in turn.

B-1 B Cells

B-1 B cells occupy a functional niche midway between cells of the innate and adaptive immune systems; many B-1 B cells are located outside the classical secondary lymphoid tissues, and the repertoire of B-1 B-cell antigen receptors is less diverse than that of conventional B-2 B cells. B-1 B cells are self-renewing in the periphery, in that the population of B-1 B cells is maintained without the need to be continuously reseeded from bone marrow–derived precursors. Finally, B-1 B cells respond rapidly to antigen challenge with relatively low-affinity responses that are primarily of the IgM class (see Table 11-8).

The existence of the B-1 B-cell subpopulation was first described in 1983, when the lab of Leonard and Leonore Herzenberg discovered a set of B cells bearing the **CD5 antigen**, whose expression had previously been thought to be restricted to T cells. These CD5$^+$ B cells were termed *B-1 B cells* to distinguish them from the conventional B-2 B-cell population. B-1 B cells appear before B-2 B cells during embryonic development (see Chapter 9).

In humans and mice, B-1 B cells make up only about 5% of B cells, although in some species such as rabbits and cattle, B-1–like cells represent the major B-cell subset. However, even in humans and mice, B-1 cells are the most abundant B cells in the peritoneal cavity, and it is probable that their major function is to protect body cavities from bacterial infection. They are also found at a low, but detectable frequency in the spleen. B-1 B cells appear to have evolved to generate rapid responses to TI antigens, and most antibodies secreted by B-1 B cells recognize common, repeated antigenic determinants expressed by gut and respiratory system bacteria, such as phosphorylcholine (a component of pneumococcal cell walls) and lipopolysaccharide. B-1 B cells do not undergo somatic hypermutation, since T-cell help is required for AID activation. Therefore, B-1 B cells primarily secrete IgM antibodies of relatively low affinity, and the antibodies they secrete are also significantly less diverse than the antibodies secreted by B-2 B cells.

Because of their priority in the B-cell developmental sequence, B-1 B cells are found at relatively high frequency among B cells present during fetal and neonatal life. Cells having the functional characteristics of B-1 cells but lacking expression of the CD5 molecule were identified at a later stage and termed **B-1b B cells**.

Very early in the history of immunology, investigators noted that the serum of unimmunized mice and humans contains so-called *natural IgM antibodies* that bind a broad spectrum of antigens with relatively low affinity and express an unusually high level of cross-reactivity between different antigens. These antibodies derive mainly from B-1 B cells, and their presence in unimmunized animals suggests that B-1 B cells may exist in a partially activated state and constitutively secrete low levels of natural antibodies. In addition, a relatively high proportion of these IgM antibodies display autoimmune reactivities, although their affinities for self antigens are sufficiently low that they rarely induce disease. Investigators have suggested that low-level interactions with self antigens during development may in fact be important in the development of B-1 B-cell function and in the maintenance of their partially activated phenotype.

Although antibody secretion by B-1 B cells is not dependent on T-cell help, it can be enhanced in the presence of T cell–derived cytokines. Furthermore, recent work has shown that the functional distinctions between B-1 and B-2 B cells may not be as absolute as was once thought. Indeed, in the presence of T-cell help, some B-1b cells may express certain attributes of B-2 cells, such as Ig class switching (with the resultant production of IgA antibodies), low levels of SHM, and the production of both long-lived memory cells resident in the peritoneal cavity and LLPCs in the bone marrow. Given the relevance of carbohydrate antigens in vaccine generation, the study of TI responses by B-1 B cells is a clinically important area of current investigation.

Marginal Zone B Cells

Marginal zone (MZ) B cells, which reside in the marginal zone of the spleen (**Figure 11-24**), also fall into the "innate-like" category of B cells capable of responding to TI antigens. The unique location of this subset gives it rapid exposure to antigens that are carried to the spleen via the blood. Typically, antigens that are presented to MZ B cells are captured by metallophilic or marginal zone macrophages in the marginal sinus, or alternatively are picked up by neutrophils or dendritic cells in the circulation and hence brought into the MZ.

Investigators have shown that differences in the microvasculature of the spleen exist between humans and mice, which impact the structure and function of the MZ. In the mouse, the splenic blood supply is connected to the general circulation through the splenic artery. This artery branches into central arterioles, shown in Figure 11-24, that are surrounded by the periarteriolar lymphoid sheath (PALS, or T-cell area of the spleen). These central arterioles further branch into smaller follicular arteries that cross the PALS and the follicles to end as capillaries in the marginal sinus and red pulp. The marginal sinus separates the MZ from the PALS and the follicles of the white pulp (see Figure 11-24). The walls of the sinus are fenestrated (leaky), and so the blood that is contained in the marginal sinus flows slowly

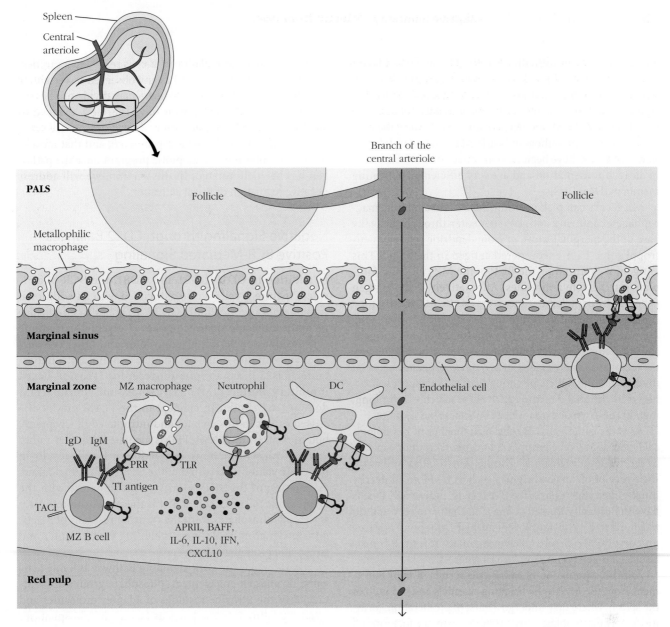

FIGURE 11-24 The marginal zone of the mouse spleen. This figure shows the relative locations within the spleen of the B cell follicles, the PALS, the marginal sinus, marginal zone, and the splenic red pulp. In the mouse, antigen leaves the blood in the marginal sinus and directly accesses the marginal zone via fenestrations in the marginal sinus. In the marginal zone, antigen is picked up by macrophages and dendritic cells and presented to marginal zone B cells, which receive help in the form of cytokines released by myeloid cells (and occasionally, T cells). Marginal zone B cells are effective at antigen processing and presentation. They have also been shown to bind complement-bound antigen via complement receptors and to transport that antigen into the B-cell follicles of the spleen.

through the MZ and subsequently drains back into the red pulp to join the general circulation. Because of its place in the circulatory pattern, the MZ is essentially "open" to the splenic blood supply.

In contrast, the human spleen lacks a histologically defined marginal sinus. Instead, the central arterioles branch into smaller arterioles that drain directly into the capillaries of the red pulp and the perifollicular zone. Both the perifollicular zone and the red pulp consist of an open circulatory system of blood-filled spaces known as *splenic cords*. These are in close contact with the venous sinusoidal vessels of the red pulp.

Given the histological differences between the MZ regions of mouse and human spleens, it should not be surprising that differences exist between the functions of mouse and human MZ B cells. Specifically, human MZs consist largely of IgM$^+$ memory cells. In contrast, mouse MZ cells are thought to represent a B-cell lineage that is distinct from follicular B cells or their differentiated daughters. We will focus this discussion on mouse MZ B cells.

Mouse MZ B cells are larger than follicular B cells and express higher levels of IgM, CD21, CD80, and CD86, but lower levels of IgD. MZ B cells differentiate into plasmablasts more rapidly than follicular B cells. The increased levels of CD21 enable the MZ B cells to bind very efficiently to antigens covalently conjugated to C3d or C3dg. Since one characteristic of TI-2 antigens is an enhanced tendency to bind C3d, MZ B cells are particularly important in the host protection against pathogens bearing TI-2 antigens. In addition, MZ B cells have been demonstrated to be effective cells in antigen presentation and are very efficient at activating memory $CD4^+$ T cells.

MZ B cells can pick up antigen from metallophilic macrophages, capturing antigens that enter through the leaky base of the marginal sinus, or from dendritic cells and neutrophils that have encountered antigen in the blood. This antigen, if captured by the BCR of the MZ cell, may stimulate it to secrete antibodies. Alternatively, if the antigen is bound by complement receptors on the MZ B cells, they can also shuttle antigens into the B-cell follicles, where they are transferred to FDCs. The importance of this antigen-shuttling function to the immune system is highlighted by the fact that as many as 20% of MZ B cells exchange between the MZ and the follicle per hour. Antigen transport is facilitated by the shape of the MZ B cells, which exhibit long, membrane extensions, as well as by their rapid motility.

As is the case for B-1 B cells, maintenance of physiological levels of MZ B cells appears to depend on their capacity to receive low-level signals through the BCR as well as on their receipt of survival signals such as BAFF from nearby innate-like cells. Again, like B-1 B cells, mouse MZ B cells have the capacity for self renewal in the periphery and do not need to be constantly replenished from bone marrow precursors. MZ B cells derive originally from the transitional T2 B-cell population (see Chapter 9).

Antigen stimulation of MZ B cells results in their movement from the MZ to the bridging channels and red pulp of the spleen, where they undergo a burst of proliferation, forming foci of plasmablasts not unlike the primary foci formed in the lymph nodes by B-2 B cells. These cells produce high levels of antigen-specific IgM within 3 to 4 days after antigenic stimulation. Of interest is the recent finding that MZ B cells may be helped in antibody secretion, SHM, and CSR by neutrophils, as well as by other cells, including T cells.

> **Key Concept:**
> • Mouse marginal zone B cells are exposed directly to antigen entering the spleen through the blood supply and are an important first line of defense. Neutrophils participate in regulating their differentiation.

Negative Regulation of B Cells

Up until this point, we have been discussing how B cells are activated, and the functional correlates of that activation.

However, antigen stimulation of B cells results in a proliferative response that is as rapid as any observed in vertebrate organisms; an activated lymphocyte may divide once every 6 hours. Control mechanisms have therefore evolved to ensure that B-cell proliferation is slowed down once sufficient specific B cells have been generated, and that most of the B cells enter into an apoptotic program once the pathogen has been eliminated. In this section, we will address negative regulation of B-cell activation.

Negative Signaling through CD22 Balances Positive BCR-Mediated Signaling

In addition to CD19/CD21 and CD81, the BCR of activated, mature B cells is also associated with an additional transmembrane molecule, CD22. CD22 is a cell-surface receptor molecule that recognizes *N*-glycolylneuraminic acid residues on serum glycoproteins and other cell surfaces and can thus double as an adhesion molecule. Importantly, CD22 bears an **immunoreceptor *t*yrosine-based *i*nhibitory *m*otif (ITIM)** on its cytoplasmic domain. ITIM domains are similar in concept to the ITAM motifs discussed earlier in this chapter and introduced in Chapter 3, but, in contrast to ITAMs, ITIMs mediate inhibitory rather than activating functions. Activation of B cells results in phosphorylation of tyrosine in the CD22 ITIM, that allows association of the SHP-1 tyrosine phosphatase with the cytoplasmic tail of CD22. SHP-1 is then in position to strip activating phosphates from the tyrosine residues of neighboring signaling molecules, including from adapter and enzymatic proteins.

For as long as the BCR signaling pathway is being activated by antigen engagement, phosphate groups are reattached to the tyrosine residues of adapter molecules and other signaling intermediates as fast as the phosphatases can strip them off. However, once antigen levels begin to decrease, receptor-associated tyrosine kinase activity slows down, and signaling through CD22 can then induce the removal of any residual activating phosphates. CD22 thus functions as a negative regulator of B-cell activation, and its presence and activity ensure that signaling from the BCR is shut down when antigen is no longer bound to the BCR. Consistent with this negative feedback role, levels of B-cell activation are elevated in CD22 knockout mice, and aging CD22 knockout animals have increased levels of autoimmunity.

It is clear, however, that CD22 functions as more than merely a simple negative regulator. $CD22^{-/-}$ mice have lower numbers of MZ B cells than wild-type mice, and are deficient in their responses to TI antigens. Further, recent work indicates that $CD22^{-/-}$ animals are deficient in the generation of GC-dependent memory B cells. Additional experiments are clearly needed to allow us to fully understand all of the roles filled by this interesting molecule.

Negative Signaling through the Receptor FcγRIIb Inhibits B-Cell Activation

It has long been known that binding circulating, antigen-IgG complexes to the B-cell surface inhibits further B-cell activation, and this phenomenon has now been explained at the molecular level by the characterization of the receptor **FcγRIIb** (also known as CD32). FcγRIIb recognizes immune complexes containing IgG and, like CD22, bears a cytoplasmic ITIM domain. Coligation of the B cell's FcγRIIb molecules with the BCR by a specific antigen-antibody immune complex results in activation of the FcγRIIb signaling cascade, and phosphorylation of FcγRIIb's ITIM, interfering with pathways downstream from the BCR and resulting in the inhibition of B-cell signaling.

Negative signaling through FcγRIIb makes intuitive sense, as the presence of immune complexes containing the antigen for which a B cell is specific is indicative of high levels of antigen-specific antibodies, and hence a reduced need for further B-cell differentiation to antibody secretion.

CD5 Acts as a Negative Regulator of B-Cell Signaling

First defined as a pan T-cell marker, and later as a marker for the mouse B-1a B-cell subset, CD5 has also been demonstrated to function as a negative regulator in both BCR and TCR signaling pathways. CD5 can be induced on conventional B-2 B cells following simultaneous engagement of both the BCR and CD40, and some polyclonal activators also up-regulate its expression on human B-2 B cells.

Experiments that knocked out CD5 expression on B-1a B cells established CD5 as a negative regulator of BCR-mediated signaling; $CD5^{-/-}$ B cells become hyperresponsive to stimulation through the B-cell receptor, although the magnitude of signaling through other receptors is not affected. The mechanism for this negative regulation remains unclear.

Many $CD5^+$ B cells also secrete IL-10, which acts as a prosurvival cytokine, but also serves to down-regulate immune responsiveness.

B-10 B Cells Act as Negative Regulators by Secreting IL-10

An unusual population of B cells has been discovered that appears to be capable of negatively regulating inflammatory immune responses by secreting the cytokine IL-10.

In two mouse autoimmune models, investigators demonstrated that B cells capable of secreting the immunoregulatory cytokine IL-10 could alleviate the symptoms of mice suffering from autoimmune disease. Recall from Chapter 10 that IL-10 is a cytokine normally associated with regulatory T cells. It has pleiotropic effects on other immune system cells, which include the suppression of T-cell production of the cytokines IL-2, IL-5, and TNF-α. Furthermore, IL-10 interacts with antigen-presenting cells in such a manner as to reduce the cell-surface expression of MHC antigens.

The finding that B cells could be capable of secreting this immunoregulatory cytokine represents the first indication that B cells, in addition to T cells, might have the capacity to down-regulate the function of other immune system cells. A small population of splenic B cells appears to account for almost all of the B cell–derived IL-10. However, at this point, we do not know whether this IL-10–secreting B-cell population represents a single developmental B-cell lineage. For example, B cells producing IL-10 have been identified among both B-1 and B-2 B-cell populations. In addition, some, but not all B cells producing IL-10 bear markers typical of the transitional T2 B-cell subset.

Importantly, all B-10 B cells appear to demonstrate the capacity to secrete a diverse repertoire of antibodies, with specificities for both foreign and auto-antigens. It is thought that the function of these cells may be to limit and control inflammation during the course of an ongoing immune response. A great deal of work is still required to tease out the lineage relationships among these various IL-10–secreting and other B-cell subpopulations.

Conclusion

B cells are characterized by the presence of a membrane-bound immunoglobulin receptor that binds antigen. Antigen binding, along with other auxiliary signals, instructs the B cell to secrete soluble antibody molecules. There are at least four subsets of B cells—B-1a, B-1b, B-2 (follicular), and marginal zone (MZ) B cells—that differ from one another according to their anatomical locations, the types of antigens that they recognize, and their requirements for T-cell help. B-1 B cells primarily protect the body cavities, most particularly the peritoneal cavity. They are capable of generating antibodies on antigen stimulation without the need for T-cell help, although T cell–derived signals can enhance their responses. B-1 B cells are self-renewing in the periphery and secrete mainly IgM antibodies, many of which bind carbohydrate antigens. B-1a and B-1b cells can be distinguished from one another by the expression on B-1a cells of CD5 molecules. Marginal zone B cells are located in the marginal zones of the spleen and are particularly effective in responding to TI-2 antigens. The locations of the B-1 and MZ B-cell subpopulations at antigen entry points, as well as their oligoclonality and relative cross-reactivities among different microbes, places them functionally at the interface of the innate and adaptive immune systems.

B-2 (follicular) B cells are the most common subset of B cells and can respond to antigens only if they are provided with help by CD4$^+$ T cells. Early in an immune response B-2 B cells can differentiate into IgM-secreting plasma cells and IgM-bearing memory cells. B-2 B cells also undergo class switch recombination, a reaction that also requires help from CD4-bearing T cells. Some B-2 B cells enter the follicles and develop into germinal center B cells that collaborate with T$_{FH}$ cells to undergo somatic hypermutation followed by antigen-mediated selection. These cells generate the high-affinity antibodies of the late primary and subsequent responses. Some of these cells differentiate into long-lived plasma cells and high-affinity memory cells. Both somatic hypermutation and class switch recombination require the action of the enzyme AID (activation-induced cytidine deaminase), which deaminates cytidine residues in the context of particular sequences in the Ig DNA. Repair of the deaminated cytidine residues by components of the base excision repair and mismatch repair pathways results in either mutation or class switch recombination, depending on the locations and frequencies of the targeted cytidines within the Ig locus.

In addition to receiving positive signals that instruct B cells to divide and differentiate, B cells can also receive negative signals via cell-surface receptors such as CD22 and CD32, that inform the B cell that the immune response is completed. Recently, B cells have been implicated as mediators, as well as recipients of negative signals, in that a subset of B cells has been shown to secrete the regulatory cytokine IL-10.

Although in the last few decades we have enjoyed an explosion of new discoveries about B-cell biology, there is still much to be learned. Questions remain regarding antigen uptake and processing, and cell interactions. For years, immunologists were convinced that T cells were the only relevant helper cell populations for B cells, but recent data have shown that this is not necessarily the case, particularly for MZ cells, and we must ask questions about the effects of other innate cell signals on other B-cell subsets. Study of B-1 B cells has taught us that B cells are simultaneously signaled via innate and Ig receptors, but we are still learning how these signals interact with one another and with those mediated by negative regulators. Furthermore, we are only just beginning to understand the molecular decisions that underlie differentiation of B cells into early and late memory cells and antibody-producing cells. And there is much we do not know regarding the shutting down of a vigorous B-cell response once sufficient antibody and memory B-cell levels have been reached.

As more T-cell subpopulations are identified, generation of precise knowledge about which T cells aid in the activation of which B cells, and which classes of antibodies or other B cell–derived molecules are then secreted, remains an active area of research. In Chapter 13, you will read about the B cells of mucosal immunity and how they interact with one another and other immune cells in the barrier tissues. The study of mucosal immunity has taught us a great deal about how the immune system balances itself between tolerance of commensal bacteria and antigens that we take in with food, but there is still much we do not know.

REFERENCES

Bannard, O., and J. G. Cyster. 2017. Germinal centers: programmed for affinity maturation and antibody diversification. *Current Opinion in Immunology* **45**:21.

Berek, C. 2016. Eosinophils: important players in humoral immunity. *Clinical and Experimental Immunology* **183**:57.

Burnet, F. M. 1957. A modification of Jerne's theory of antibody production using the concept of clonal selection. *Australian Journal of Science* **20**:67.

Cerutti, A., M. Cols, and I. Puga. 2013. Marginal zone B cells: virtues of innate-like antibody-producing lymphocytes. *Nature Reviews Immunology* **13**:118.

Crotty, S., R. J. Johnston, and S. P. Schoenberger. 2010. Effectors and memories: Bcl-6 and Blimp-1 in T and B lymphocyte differentiation. *Nature Immunology* **11**:114.

Cyster, J. G. 2010. B cell follicles and antigen encounters of the third kind. *Nature Immunology* **11**:989.

Cyster, J. G., and S. R. Schwab. 2012. Sphingosine-1-phosphate and lymphocyte egress from lymphoid organs. *Annual Review of Immunology* **30**:69.

De Silva, N. S., and U. Klein. 2015. Dynamics of B cells in germinal centres. *Nature Reviews Immunology* **15**:137.

Germain, R. N., et al. 2008. Making friends in out-of-the-way places: how cells of the immune system get together and how they conduct their business as revealed by intravital imaging. *Immunological Reviews* **221**:163.

Gitlin, A. D., et al. 2015. Humoral immunity: T cell help controls the speed of the cell cycle in germinal center B cells. *Science* **349**:643.

Harwood, N. E., and F. D. Batista. 2010. Early events in B cell activation. *Annual Review of Immunology* **28**:185.

Hwang, J. K., F. W. Alt, and L. S. Yeap. 2015. Related mechanisms of antibody somatic hypermutation and class switch recombination. *Microbiology Spectrum* **3**:MDNA3-0037-2014.

Jones, D. D., J. R. Wilmore, and D. Allman. 2015. Cellular dynamics of memory B cell populations: IgM^+ and IgG^+ memory B cells persist indefinitely as quiescent cells. *Journal of Immunology* **195**:4753.

Kerfoot, S. M., et al. 2011. Germinal center B cell and T follicular helper cell development initiates in the interfollicular zone. *Immunity* **34**:947.

Kuraoka, M., et al. 2016. Complex antigens drive permissive clonal selection in germinal centers. *Immunity* **44**:542.

Kurosaki, T., K. Kometani, and W. Ise. 2015. Memory B cells. *Nature Reviews Immunology* **15**:149.

Lian, J., and A. D. Luster. 2015. Chemokine-guided cell positioning in the lymph node orchestrates the generation of adaptive immune responses. *Current Opinion in Cell Biology* **36**:1.

Matthews, A. J., S. Zheng, L. J. DiMenna, and J. Chaudhuri. 2014. Regulation of immunoglobulin class-switch recombination: choreography of noncoding transcription, targeted DNA deamination, and long-range DNA repair. *Advances in Immunology* **122**:1.

Mesin, L., J. Ersching, and G. D. Victora. 2016. Germinal center B cell dynamics. *Immunity* **45**:471.

Moens, L., A. Kane, and S. G. Tangye. 2016. Naïve and memory B cells exhibit distinct biochemical responses following BCR engagement. *Immunology and Cell Biology* **94**:774.

Muramatsu, M., H. Nagaoka, R. Shinkura, N.A. Begum, and T. Honjo. 2007. Discovery of activation-induced cytidine deaminase, the engraver of antibody memory. *Advances in Immunology* **94**:1.

Natkanski, E., et al. 2013. B cells use mechanical energy to discriminate antigen affinities. *Science* **340**:1587.

Nera, K. P., M. K. Kyläniemi, and O. Lassila. 2015. Regulation of B cell to plasma cell transition within the follicular B cell response. *Scandinavian Journal of Immunology* **82**:225.

Shulman, Z., et al. 2014. Dynamic signaling by T follicular helper cells during germinal center B cell selection. *Science* **345**:1058.

Tas, J. M., et al. 2016. Visualizing antibody affinity maturation in germinal centers. *Science* **351**:1048.

Taylor, J. J., M. K. Jenkins, and K. A. Pape. 2012. Heterogeneity in the differentiation and function of memory B cells. *Trends in Immunology* **33**:590.

Taylor, J. J., K. A. Pape, H. R. Steach, and M. K. Jenkins. 2015. Humoral immunity: apoptosis and antigen affinity limit effector cell differentiation of a single naïve B cell. *Science* **347**:784.

Victora, G. D., and M. C. Nussenzweig. 2012. Germinal centers. *Annual Review of Immunology* **30**:429.

Yeap, L. S., et al. 2015. Sequence-intrinsic mechanisms that target AID mutational outcomes on antibody genes. *Cell* **163**:1124.

Useful Websites

www.bio-alive.com/seminars/immunology.htm The Bio Alive website, an excellent source of lectures by accomplished immunologists, can be downloaded for personal use. The series contains lectures pertinent to this chapter, including seminars entitled *Immunological Memory, Regulating B Cell Immunity, Moviemaking and Modeling* (this one by Ronald Germain, one of the pioneers of the application of sophisticated imaging technology to the study of the immune system), *Molecular Mechanisms of Leukocyte Migration*, and *Somatic Hypermutation*.

www.sciencedirect.com/science/article/pii/ S1074761304002389 Two movies from a paper (Catron, D. M., et al. 2004. Visualizing the first 50 hr of the primary immune response to a soluble antigen. *Immunity* **21**:341) that is discussed in Chapter 14 can be found at this website. The first portrays the movements of T cells and dendritic cells prior to antigen entry into the observed lymph node. The second shows the movements of T cells, B cells, and dendritic cells through a nearby lymph node, following injection of antigen into the footpad of a mouse.

https://www.khanacademy.org/science/biology/ human-biology/immunology/v/b-lymphocytes- b-cells A useful Khan Academy introduction to B cells.

www.hhmi.org/scientists/browse?kw=&&field_sci- entist_classification%5B0%5D=17367&field_ scientist_research_areas%5B1%5D=17708 Many young scientists engaged in their first course in a subject are faced with the requirement to write a research paper. Or perhaps, they've finished that first and even subsequent courses in the field, are fascinated by the material, and want to know where to look to learn more about the work being done by leaders in the field. One useful website to browse is that of the Howard Hughes Medical Institute, which funds the research of many outstanding immunologists, and brief descriptions of the work in the labs of each of the HHMI Investigators can be found at this site.

www.cellsignal.com An immensely helpful company website that provides a useful and up-to-date pathway description for B-cell signaling along with accompanying references. On the main page of the site you will see a pull-down option to choose "Content" rather than "Products." Choosing "Content" and then searching on the term "B-cell signaling" will open a menu of interactive signaling pathways.

STUDY QUESTIONS

1. Name one distinguishing feature of TI-1 and TI-2 antigens.

2. In the following flow cytometric dot plots, draw a circle where you expect to see the designated cell populations, labeled (a), (b), (c), and (d) below.

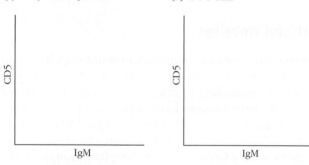

(a) B2 (follicular) B cells   (b) B-1 B cells

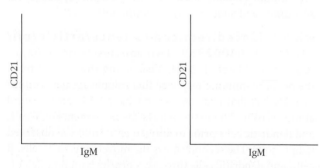

(c) B2 (follicular) B cells   (d) MZ B cells

3. You have generated a knockout mouse that does not express the CD40 molecule. On stimulation with a T-dependent antigen, do you expect this mouse to be able to:

a. Secrete IgG antibodies specific for the antigen?
b. Secrete antigen-specific antibodies that have undergone somatic hypermutation?

Explain your answer.

4. You have immunized two mice with a T-dependent antigen. One of them is a wild-type mouse, and the other belongs to the same strain as the first but is a knockout mouse that does not have the gene encoding activation-induced cytidine deaminase (AID). On primary immunization, both mice express similar titers (concentrations) of IgM antibodies. You allow the mice to rest for 6 weeks and then reimmunize. Do you expect the secondary antibodies to be similar or different between the two animals? Explain your answer.

5. Draw the biochemical reaction catalyzed by AID, and describe how subsequent DNA repair mechanisms can lead to somatic hypermutation (SHM) and to class switch recombination (CSR).

6. Describe one mechanism by which higher-affinity B cells may gain a selective survival advantage within the germinal center.

7. The presence of circulating antigen-antibody complexes has been shown to result in the down-regulation of B-cell activation. State why this would be advantageous to the organism, and offer one mechanism by which the down-regulation may occur.

8. Recent evidence suggests that not all regulatory cytokines secreted by lymphocytes derive from T cells. Explain.

9. Describe three mechanisms by which antigen can enter the lymph node and make contact with the B-cell receptor.

ANALYZE THE DATA

Biochemical analysis of the cytoplasmic domains of the BCR μ and γ1 heavy chains revealed that the γ1 heavy chains have an extra segment that interacts with intracellular signaling molecules. Investigators postulated that this extra segment might be responsible for the enhanced memory responses of IgG1-bearing compared with that of IgM-bearing memory B cells. To address this question, a group of scientists transferred the nuclei from IgG1-expressing, NP-specific B cells into embryonic stem cells. When the mice grew to adulthood, they were found to contain both IgM- and IgG1-bearing antigen-specific cells that had never encountered antigen. The presence of an antigen-naïve IgG1-bearing cell would never normally be encountered in nature, and allowed the investigators to ask whether the longer IgG1 heavy chain is responsible for conferring memory characteristics on the B cell, or whether the B cell actually needs to encounter antigen before it can act like a memory cell. The heavy chain encoded by this transgene combines with λ1 light chains to create an antibody that binds to the hapten, NP.

In the next step of the experiment, they took cells from this mouse and adoptively transferred them into a group of wild-type mice that had been primed with the protein chicken γ globulin, or CGG. Another set of wild-type mice was injected with naïve,

IgM-bearing cells carrying the same, transgenic heavy-chain variable region as the first. A third set of mice was injected with transgenic, IgG1-bearing cells, but these IgG1-bearing cells were derived from a mouse that had previously encountered antigen. So now, the investigators were able to judge whether the "memory" properties of IgG1-bearing B cells resulted from the biochemistry of the γ1 chain, or actually required prior antigen contact.

After adoptive transfer of the three cell types, the mice were immunized with antigen and then their spleens were analyzed several days later for the presence of plasma cells, which have the staining characteristics B220 (CD45) low, CD138 (syndecan) high. This is what they found:

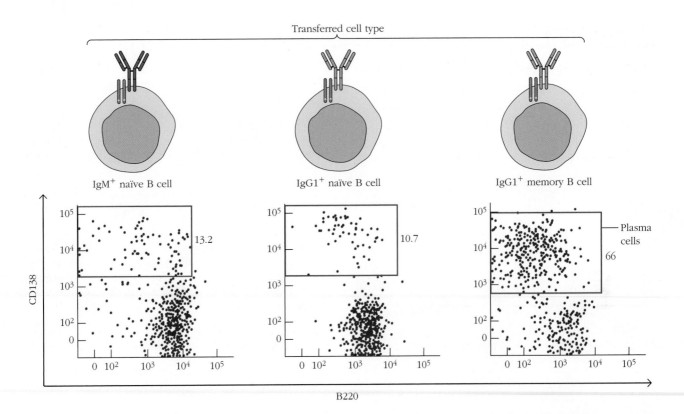

Note that the numbers on the plots refer to the fraction of cells that fall within the marked boxes, labeled "plasma cells."

a. Explain what you see in each of the three panels. (Really looking at data and describing it carefully is an important skill that is too often glossed over by the early, enthusiastic student.)

b. Is the response mounted by the mouse that received the naïve IgG1 cells more like that of the mouse that received the naïve IgM-bearing cells, or does it more closely resemble that of the mouse that received the antigen-experienced IgG-bearing cells?

c. What is your conclusion regarding whether the cytoplasmic region of the receptor alone is sufficient to confer memory status on a B cell?

d. Do you see any flaws in the way in which this experiment was conducted? If so, what were they?

e. What is the next experiment you would do? (There are many "right" answers here.)

Effector Responses: Antibody- and Cell-Mediated Immunity

Learning Objectives

After reading this chapter, you should be able to:

1. Explain how humoral immune responses protect us against pathogens, toxins, or malignancy.

2. For antibody responses, recognize how the multiple immunoglobulin classes and subclasses can have common, but also distinct, mechanisms for antigen inactivation, elimination, tissue distribution, and other biological activities.

3. Describe the multistep process required for inducing the differentiation of T_C precursors into effector cytotoxic T lymphocytes (CTLs).

4. Compare and contrast the mechanisms by which CTLs and natural killer (NK) cells recognize and kill infected or tumor target cells.

5. Explain how NK cells and NKT cells each have some properties of innate immune cells and adaptive immune cells.

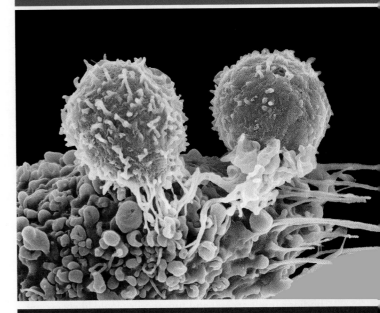

Two cytotoxic T lymphocytes bind to a tumor-specific antigen on the surface of a cancer cell, deliver the "kiss of death," and induce apoptosis. *[Steve Gschmeissner/Science Source.]*

The previous chapters focused on how the immune response is initiated. Here, we will finally describe how the targets of the immune system—pathogens, infected cells, and even tumor cells—are actually cleared from the body.

We have already seen how the innate immune system initiates the response to a pathogen and alerts the adaptive immune system to the presence and nature of that pathogen (Chapter 4). In Chapters 8 through 11, we described the development and activation of the antigen-specific cells of the adaptive immune system, the B and T lymphocytes. You were also introduced to the differentiation and activity of helper T lymphocytes, a type of effector cell that regulates the activity and function of cytotoxic T cells, B cells, and other antigen-presenting cells. This chapter focuses on the effector molecules and cells of both the antibody-mediated (humoral) and cell-mediated immune responses that directly rid the organism of pathogens and abnormal cells. These effector responses are arguably the most important manifestations of immune responses: they rid the host both of foreign pathogens that have breached its innate defenses, as well as any of the host's own cells that have become dangerous through infection or malignant transformation.

Key Terms

Antibody-dependent cell-mediated cytotoxicity (ADCC)

Degranulation

Neutralization

Agglutination

Opsonization

Antibody-dependent cellular phagocytosis (ADCP)

Poly-Ig receptor (polyIgR)

Fc receptors (FcRs)

Neonatal Fc receptor (FcRn)

Cytotoxic T lymphocytes (T_C cell or CTLs)

CTL precursors (CTL-Ps)

MHC tetramers

Perforin

Granzymes

Natural killer (NK) cell

Missing self model

Balanced signals model

NK receptors

KIR

Ly49

NKT cell

The humoral and cell-mediated branches of the immune system assume different, although overlapping, roles in clearing infection from a host. The effector molecules of the humoral branch are antibodies, the secreted versions of the highly specific membrane immunoglobulin receptors on the surface of B cells. Antibodies secreted into extracellular spaces and carried in the body secretions are exquisitely antigen specific and have several methods at their disposal to rid a body of pathogens. How an antibody contributes to clearing infection depends on its class (its heavy-chain isotype), which controls some of its effector functions, including whether it can recruit complement (recall Chapter 5). The class of an antibody also determines which receptors can bind it. Antibody-binding receptors, which bind the constant regions of antibodies and are therefore called Fc receptors or FcRs, determine which cells an antibody can activate to aid in its protective mission, as well as whether it can gain access to certain locations in the body.

If antibodies were the only agents of immunity, then intracellular pathogens (which occupy spaces that antibodies cannot access) and tumor cells would likely escape the immune system. Fortunately there is another effector branch of our immune system, cell-mediated immunity, which detects and kills cells that harbor intracellular pathogens or have otherwise become altered and potentially harmful. Cell-mediated immunity is carried out by helper T ($CD4^+$ T_H) cells and several types of cytotoxic cells. As you saw in Chapter 10, T_H cells exert their effector functions indirectly, by contributing to the activation of antigen-presenting cells, B cells, and cytotoxic T cells via receptor-ligand interactions and soluble cytokines and chemokines. The functions of T_H cells in activating macrophages to kill their intracellular pathogens and to contribute to delayed-type hypersensitivity will be discussed in Chapter 15. On the other hand, cytotoxic cells exert their effector functions directly, by attacking infected cells and, in some cases, the pathogens themselves.

Effector cytotoxic cells arise from both the adaptive and innate immune systems and therefore include both antigen-specific and -nonspecific cells. Antigen-nonspecific (innate immune) cells that contribute to the clearance of infected cells include natural killer (NK) cells and myeloid cell types such as macrophages, neutrophils, and eosinophils. Antigen-specific cytotoxic cells include $CD8^+$ cytotoxic T lymphocytes (CTLs or T_C cells), as well as the $CD4^+$ NKT cell subpopulation, which, although derived from the T-cell lineage, displays some useful features of innate immune cell types, too. Even some populations of $CD4^+$ T_H cells can be cytotoxic, producing cytokines that cause cell death.

The humoral and cell-mediated immune systems also cooperate effectively. Cells such as macrophages, neutrophils, eosinophils, and NK cells all express Fc receptors, which can induce phagocytosis of antibody-antigen complexes and/or the direct killing of target cells via a process known as antibody-dependent cell-mediated cytotoxicity.

In this chapter we will first focus on antibody-mediated effector functions, describing how they protect us from infections and how they get to certain sites in the body where they are needed. We will then describe the cytotoxic effector mechanisms mediated by cells of the innate and adaptive immune systems.

Last, the chapter also includes Clinical Focus, Advances, and Classic Experiment features. The Clinical Focus box describes how knowledge of antibody structure and function along with techniques for generating large amounts of homogeneous antibodies of desired specificities has led to a growing arsenal of therapeutic antibodies for a variety of conditions. The Advances box describes approaches that allow the detection, characterization, and isolation of antigen-specific T cells. The Classic Experiment box discusses recent findings showing that B and T lymphocytes are not the only cell types that can develop immunological memory; natural killer cells, members of the innate immune system, also have this capacity.

Antibody-Mediated Effector Functions

Antibodies generated by activated B cells (see Chapter 11) protect the body against pathogens through six antibody-mediated effector functions. They can **neutralize** (inactivate) pathogens or toxins; they can **agglutinate** pathogens, facilitating their clearance; they can **opsonize** pathogens or altered cells by binding and recruiting phagocytic cells; and they can activate **complement** by binding to pathogen and initiating the complement cascade, which leads to several protective processes (see Chapter 5). In addition, they can cooperate with the cellular branch of the immune system by binding and recruiting the activities of cytotoxic cells, most often natural killer (NK) cells, in a process called *antibody-dependent cell-mediated cytotoxicity (ADCC)*. Last, anti-pathogen antibodies can trigger mediator release from granulocytes through a process called **degranulation**. These antibody effector mechanisms are all illustrated in **Figure 12-1**.

The ability of antibodies to mediate these responses is dependent not only on their antigen specificity, but also on their isotype (heavy-chain class or subclass) (see Chapter 3 and **Table 12-1**). For example, an antibody's isotype determines

The Six Broad Categories of Antibody Effector Functions

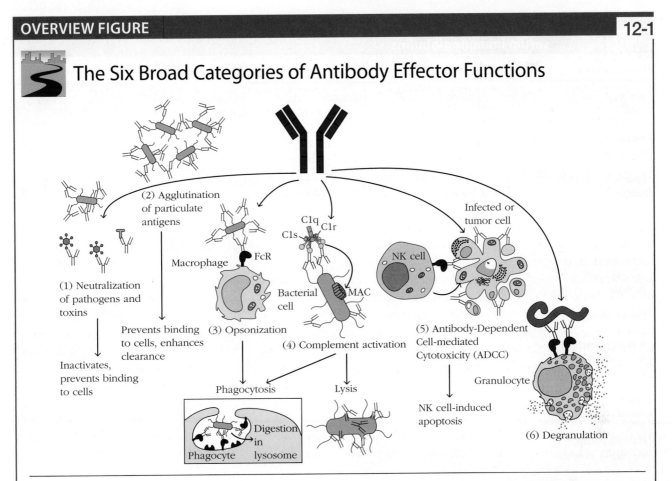

Antibody binding to pathogens or protein antigens such as toxins can enhance their inactivation or clearance by (1) neutralization—binding to pathogens or toxins and preventing them from binding to cells and leading to infection or damage; (2) agglutination—cross-linking particulate antigens such as bacteria, blocking them from binding to cells or tissues and enhancing their clearance; (3) opsonization—binding to Fc receptors on the surface of phagocytic cells and inducing phagocytosis and killing; (4) complement activation—resulting in the generation of complement proteins that can enhance phagocytosis via complement receptors or can directly kill pathogens via the membrane attack complex; (5) antibody-dependent cell-mediated cytotoxicity (ADCC)—cell-bound antibody binding to Fc receptors on cytotoxic cells (e.g., NK cells; also macrophages, neutrophils, eosinophils), activating secretion of cytotoxic proteins that induce apoptosis of or kill infected cells, tumor cells, and/or pathogens; and (6) degranulation—antibody-antigen complexes binding to Fc receptors triggers fusion of prepackaged secretory granules with the plasma membrane, allowing release of mediators that have protective effects, such as damaging helminth parasites.

whether it will activate complement, to which Fc receptor(s) it will bind, and in some cases, whether it can be transported across certain tissue layers. Collectively, these varied mechanisms result in the inhibition or destruction of pathogens, toxins, and cells in our body that have become harmful, and hence are critically important contributors to adaptive immunity.

Antibodies Provide Protection against Pathogens, Toxins, and Harmful Cells in a Variety of Ways

Antibodies are versatile effector molecules that, through a variety of mechanisms, play direct roles in protecting us from pathogens, pathogen-derived toxins, and cells that have become dangerous through infection or malignant transformation.

Neutralization

Most viruses and some bacteria gain entry into a cell or tissue by binding specifically to one or more cell-surface proteins, stimulating endocytosis. For example, the human immunodeficiency virus (HIV) expresses a coat protein, gp120, that specifically binds the CD4 receptor on helper T cells. The influenza virus expresses a protein, hemagglutinin, that binds to sugar-modified proteins on epithelial cells. Antibodies that bind such pathogen proteins and block them from binding to cells or tissues are particularly potent effector

TABLE 12-1	Properties and biological activities* of classes and subclasses of human serum immunoglobulins								
	IgG1	**IgG2**	**IgG3**	**IgG4**	**IgA1**	**IgA2**	**IgM§**	**IgE**	**IgD**
Molecular weight (Da)†	150,000	150,000	150,000	150,000	150,000–600,000	150,000–600,000	900,000	190,000	150,000
Heavy-chain component	γ1	γ2	γ3	γ4	α1	α2	μ	ε	δ
Normal serum level (mg/ml)	9	3	1	0.5	3.0	0.5	1.5	0.0003	0.03
In vivo serum half-life (days)	23	23	8	23	6	6	5	2.5	3
Activates classical complement pathway	+	+/−	++	−	−	−	++	−	−
Crosses placenta	+	+/−	+/−	+	−	−	−	−	−
Present on membrane of mature naïve B cells	−	−	−	−	−	−	+	−	+
Binds to Fc receptors of phagocytes	++	+/−	++	+	+	+	+	−	−
Mucosal transport via poly-Ig receptor	−	−	−	−	++	++	+	−	−
Induces mast cell and/or basophil degranulation	−	−	−	−	−	−	−	+	+

*Activity levels are indicated as follows: ++, high; +, moderate; +/−, minimal; −, none.

§IgM is the first isotype produced by the neonate and during a primary immune response.

†IgG, IgE, and IgD always exist as monomers; IgA can exist as a monomer, dimer, trimer, or tetramer. Membrane-bound IgM is a monomer, but secreted IgM in serum is a pentamer.

molecules because they can prevent a pathogen from ever initiating an infection. They are referred to as **neutralizing antibodies (Figure 12-2)**.

Neutralizing antibodies can also block the entry of toxins into cells. For example, tetanus toxin, a product of the *Clostridium tetani* bacteria, is a neurotoxin that can result in uncontrolled muscle contraction and death. The tetanus vaccine contains an inactive version of this toxin that stimulates B-cell production of anti-tetanus toxin antibodies. These antibodies bind and potently inhibit the entry of the toxin into nerve cells. Neutralizing antibodies have also been raised against snake venom toxins and are effective therapies for some snakebites.

Physiologically, **neutralization** is the most effective mode of protection against infection or the damaging effects of toxins, and antibodies of all immunoglobulin isotypes can mediate neutralization. However, because microbes proliferate rapidly, they can generate genetic variants able to evade neutralizing antibodies. For example, influenza viral variants expressing modified hemagglutinins that do not bind circulating antibodies have a distinct reproductive advantage and can rapidly take over an infection. HIV is one of the most adept viruses at evading antibody responses. In theory, gp120 is an ideal target for neutralizing antibodies; however, few individuals generate antibodies that broadly neutralize HIV's many distinct and rapidly evolving strains (see Chapter 18).

Agglutination

As antibodies have multiple antigen-binding sites per molecule (two per IgG, four per IgA dimer, and 10 per IgM pentamer; see Figure 3-12), each antibody molecule can bind and cross-link multiple pathogens, resulting in

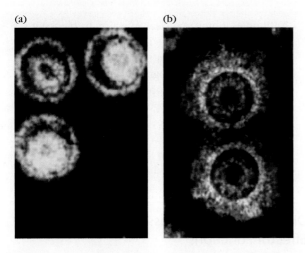

FIGURE 12-2 Neutralization of virus particles. Electron micrographs of negatively stained preparations of Epstein-Barr virus. (a) Virus particles without antibody. (b) Antibody-coated particles. *[From Cooper, N. R., and G. R. Nemerow. 1986. In: Immunobiology of the Complement System, G. D. Ross, ed. Academic Press, Orlando, FL.]*

agglutination (**Figure 12-3**). This can result in enhanced neutralization (it's even harder for a clump of pathogens to bind to and infect cells or tissues than it is for a single pathogen) and more efficient clearance of the pathogens from the body. For example, agglutinated bacteria in the intestine get trapped in mucus and are more efficiently cleared out by peristalsis. Similarly, agglutinated bacteria in the respiratory tract are more susceptible to being transported by ciliary movement. Recent studies have shown that people immunized with a vaccine against *Streptococcus pneumoniae* (a major cause of pneumonia) generate anti–*S. pneumoniae* polysaccharide antibodies able to agglutinate the bacteria in their respiratory tract, preventing bacterial colonization, which is the first step leading to disease.

Opsonization and Phagocytosis

From the Greek word for "to make tasty," **opsonization** refers to the ability of antibodies to promote and/or enhance the engulfment of antigens by phagocytes. Opsonizing antibodies bind antigen via their antigen-binding sites and are then bound by Fc receptors (FcRs) expressed on phagocytic cells, triggering phagocytosis.

Thus, phagocytic cells of the innate immune system (e.g., macrophages and neutrophils) have several sets of receptors capable of binding pathogens and initiating phagocytosis. Phagocytes can bind pathogens directly, using innate pattern recognition immune receptors that trigger phagocytosis, or they can bind pathogens via receptors that recognize opsonins bound to the pathogens (see Table 4-5). Fc receptors on phagocytes that recognize Fc regions of IgG antibodies and trigger phagocytosis of antibody-coated pathogens are an example of opsonin receptors. If the antigen is a cellular antigen like a virus-infected or tumor cell, the process is often referred to as **antibody-dependent cellular phagocytosis (ADCP)**.

Complement Activation

As we saw in Chapter 5, antigen-antibody complexes can also activate the classical **complement** pathway. Specifically, when antibodies that associate with complement bind to the surface of bacteria and some (enveloped) viruses, they can initiate a cascade of reactions (see Figure 5-5) that results in the generation of the *membrane attack complex* (MAC), creating pores in pathogen membranes and killing the microbe. This capacity to stimulate complement-mediated membrane damage is a property of specific antibody classes, including IgM and some IgG subclasses.

Activation of the early part of the complement cascade can also result in binding of complement components that protect the host by opsonizing pathogens. This complement function does not require the later components (C5–C9) of

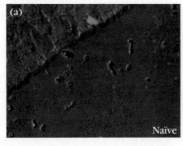

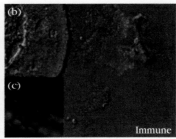

FIGURE 12-3 Agglutination of *Streptococcus pneumoniae* by antibodies in nasal secretions. Normal mice, and mutant mice unable to produce antibodies, were either left un-immunized or immunized three times with *S. pneumoniae*. The mice were infected nasally with the bacteria. After 30 minutes nasal tissue samples were obtained and sections were stained with fluorescent antibodies to the bacteria (red) and with a general stain for nasal cells (blue); tissue autofluorescence shows up as green. (a) In nonimmunized (naïve) normal mice, which had extremely low levels of antibodies that reacted with *S. pneumoniae*, the bacteria are mostly individual (nonagglutinated). (b) In immunized normal mice, which were shown to have high levels of antibodies in their nasal fluids and serum, the bacteria are highly agglutinated. (c) In mutant mice unable to produce antibodies after immunization, the bacteria are mostly individual (nonagglutinated). *[Reprinted with permission from Nature Publishing Group: from Roche, A. M., A. L. Richard, J. T. Rahkola, E. N. Janoff, and J. N. Weiser, "Antibody blocks acquisition of bacterial colonization through agglutination," 2015 Jan, Mucosal Immunology 8(1): 176–185, Figure 6. Permission conveyed through Copyright Clearance Center, Inc.]*

the complement cascade. Instead, complement components C1q, C3b (and various fragments of it), and C4b, when bound to pathogens or other antigens, can be recognized by various complement receptors on phagocytes and trigger enhanced phagocytosis (see Figure 5-13 and Table 5-5). Interestingly, binding of C3b-bound pathogen by a red blood cell results in delivery of the pathogen to the liver or spleen, where the pathogen is removed from the red blood cell, without affecting the viability of the red blood cell, followed by phagocytosis of the pathogen by a macrophage (see Figure 5-15).

Antibody-Dependent Cell-Mediated Cytotoxicity

Antibodies bound to foreign antigens on the surface of the body's cells, such as tumor antigens or virus proteins on an infected cell, can activate the killing activity of several types of cytotoxic cells. Some examples are NK cells (a major class of innate lymphoid cells; see Chapter 4 and later in this chapter) and granulocytes. Unlike cytotoxic T lymphocytes, NK cells do not express antigen-specific T-cell receptors. However, NK cells do express receptors for IgG Fc (specifically, FcγRIII; discussed shortly) and can use these to bind the antibodies already bound to antigens on infected cells, triggering the release of cytotoxic proteins (see Figure 12-1). The mechanism by which these proteins kill the target cells will be discussed later in this chapter. This process, referred to as **antibody-dependent cell-mediated cytotoxicity** (**ADCC**), is under intense investigation because of its therapeutic potential. Monoclonal antibodies (mAbs) to receptors and other markers on cancer cells that have been developed as cancer treatments are thought to kill tumor cells in part by triggering ADCC by NK cells. As with complement fixation, the ability to mediate ADCC is antibody class dependent, and murine IgG2a and human IgG1 are most potent in this activity.

Antibody-Activated Degranulation and Mediator Release

Granulocytes such as neutrophils, mast cells, basophils, and eosinophils express Fc receptors for various immunoglobulin classes. When antibodies are bound by granulocyte Fc receptors and then are cross-linked by some pathogens, such as parasitic worms (helminths), fusion of the secretory granules with the plasma membrane can be triggered, releasing a variety of soluble mediators, some of which can be harmful to the pathogen. Degranulation will be discussed in more detail in Chapter 15 in the context of degranulation activated by IgE antibodies.

Key Concepts:

- The humoral arm of the immune system refers to the activities of antibodies secreted by B lymphocyte–derived plasma cells. Antigen specificity, class or subclass, and FcR binding are all important features of antibodies that mediate their effector functions.

- Antibodies inhibit and clear infection by blocking the ability of pathogens to infect (neutralization), by cross-linking them (agglutination), by coating pathogens so that they are recognized and phagocytosed by innate immune cells (opsonization), by recruiting complement to the pathogen (complement activation), by directing other cells of the innate immune system to kill infected cells (antibody-dependent cell-mediated cytotoxicity), and by inducing degranulation and release of mediators from granulocytes.

Different Antibody Classes Mediate Different Effector Functions

As you learned in Chapters 3 and 11, activated B cells develop into plasma cells, which are remarkably productive antibody-producing cellular factories. Antibody specificity is determined by the heavy- and light-chain variable regions in the Fab fragments, and the isotype or class is defined by the heavy-chain constant regions (see Figure 3-10). Depending on the type of T-cell help they received during activation, as well as the cytokines to which they were exposed, B cells and plasma cells will synthesize one of five classes of antibody: IgM, IgG, IgA, IgD, or IgE. The structure of each of these antibodies is summarized in Chapter 3 (see Figure 3-12). Each has distinct properties and plays distinct roles in thwarting and clearing infection. The properties of the human immunoglobulins (which include four subclasses of IgG and two subclasses of IgA) are listed in Table 12-1 and are described in this section. Some of these roles are carried out by the binding of the antibody's Fc region by an Fc receptor (FcR) on cells. Each receptor is specific for the Fc of one or several heavy-chain classes. Fc receptor structures are shown in **Figure 12-4**, and their functions and expression on various cell types are described in **Figure 12-5** and **Table 12-2**. The varied functions of Fc receptors are discussed further in the next section.

Effector Functions of IgM Antibodies

IgM antibodies are the first class of antibodies to be produced during a primary immune response. They tend to be low-affinity antibodies because the B cells that produce them have not gone through affinity maturation. However, because they are decavalent (i.e., have 10 antigen-binding sites) they bind pathogens with high avidity. Interestingly, many of the circulating IgM antibodies come not from conventional B cells but from B-1 cells, which constitutively produce IgM antibodies without being exposed to pathogen (see Chapter 9). These constitutively produced IgM antibodies appear to protect us from common pathogens, particularly those that can come from the gut and other mucosal areas; they are frequently called "natural antibodies."

IgM antibodies are very good at activating complement and efficiently induce the lysis of pathogens to which they bind.

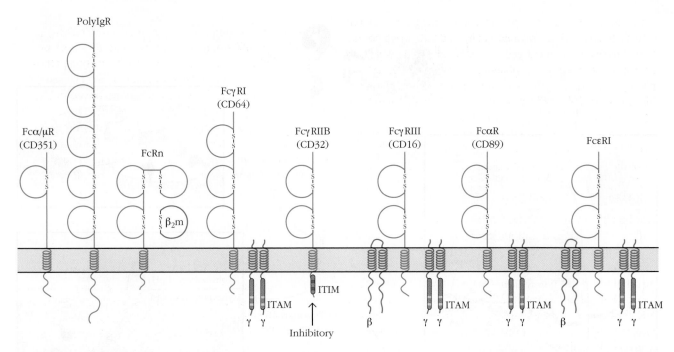

FIGURE 12-4 Structure of human Fc receptors. The antibody-binding polypeptides are shown in blue and, where known and present, accessory signal-transducing polypeptides are shown in green. Most FcRs are activating receptors and are associated with signaling proteins that contain the *immunoreceptor tyrosine activation motif* (ITAM) or include ITAM(s) in their own intracellular regions. FcγRIIB is an inhibitory FcR and includes an *immunoreceptor tyrosine inhibition motif* (ITIM) in its intracellular region. All FcRs shown in this figure are members of the immunoglobulin supergene family; their extracellular regions include two or more immunoglobulin folds. These molecules are expressed on the surface of various cell types as cell-surface antigens and, as indicated in the figure, some have been assigned cluster of differentiation (CD) designations (for clusters of differentiation, see Appendix I). Fcα/μR binds both IgM and IgA. Not shown is the low-affinity receptor for IgE, FcεRII (CD23) (see Figure 15-3), or the IgM receptor FcμR.

They are also good at agglutinating pathogens, which then are efficiently phagocytosed by macrophages.

Effector Functions of IgG Antibodies

IgG antibodies are the most common antibody class in the serum. They are also the most diverse, including several subclasses (IgG1, IgG2, IgG3, and IgG4 in humans; IgG1, IgG2a, IgG2b, and IgG3 in mice), each of which has distinct effector capabilities. All of the IgG subclasses bind to Fc receptors and can enhance phagocytosis by macrophages (opsonization). Human IgG1 and IgG3 are particularly good at activating complement. Mouse IgG2a and human IgG1 are particularly good at mediating ADCC by NK cells.

Effector Functions of IgA Antibodies

Although IgA antibodies are found in the circulation, they are the major antibody class present in secretions, including mucus in the gut, respiratory, and reproductive tracts, and in tears, saliva, and milk from mammary glands. Of the two subclasses of IgA found in humans, IgA1 and IgA2, IgA1 is more prevalent in the serum while IgA2 is more prevalent in secretions. In the blood both subclasses can stimulate phagocytosis, mediate ADCC by binding FcRs on NK cells, and trigger degranulation of granulocytes.

IgA does exist in monomeric form (particularly in the blood) but forms dimers and polymers in mucosal tissues, with the monomers connected by the J chain (see Figure 3-12). In secretions IgA can neutralize toxins and pathogens and enhance their clearance, continuously binding to the resident (commensal) bacteria that colonize our mucosal surfaces and preventing them from entering the bloodstream. Because IgA cannot fix complement, these interactions do not induce inflammation, which is advantageous given that IgA acts continuously on antigens and pathogens that typically pose no threat. IgA antibodies are highly glycosylated, enabling them—with bound pathogens—to associate with the mucus layer, facilitating their clearance. In addition, the multimeric IgA antibodies are particularly effective in agglutinating pathogens, enhancing their clearance from the body through trapping in mucus and peristalsis.

One key early question intrigued scientists about IgA: how is it able to pass through the layer of tightly connected epithelial cells and into the mucosal and glandular secretions? The answer came from the observation that IgA in secretions has an extra polypeptide chain, called the secretory component (SC), that is not part of IgA in the blood. Surprisingly, this extra chain was subsequently found not to be synthesized by IgA-producing plasma cells in the secretory tissues, but by the epithelial cells as a membrane-bound

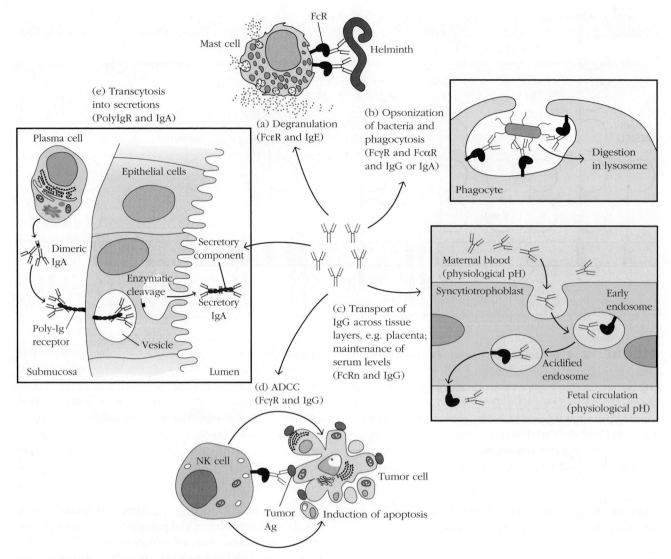

FIGURE 12-5 Functions of Fc receptors. Fc receptors (FcRs) come in a variety of types and are expressed by many different cell types. Some of their major functions are illustrated here. (a) When bound by IgE-pathogen (e.g., helminth) complexes, FcεRs expressed by granulocytes are activated, leading to release of the contents of granules (degranulation), which include histamine and proteases. (b) When bound by antibody-pathogen complexes, FcαR and FcγR can induce macrophage activation and phagocytosis. (c) The neonatal FcR (FcRn) transports IgG across cell layers, such as across the placenta from the maternal to fetal circulation; it also mediates uptake by vascular endothelial cells. (d) When bound by IgG antibodies coating infected or tumor cells, FcγRs activate the cytolytic activity of natural killer (NK) cells (antibody-dependent cell-mediated cytotoxicity, ADCC). (e) Polymeric Ig receptors (PolyIgR) expressed by the inner (basal) surface of epithelial cells (facing the inside of the body) of mucosal and secretory tissues bind dimers and multimers of IgA and IgM antibodies, initiating transcytosis: endocytosis and transport through a cell to its apical (outer) surface, where the vesicles fuse with the plasma membrane, the polyIgRs are cleaved, and the antibodies with associated secretory component are released into the lumen of the organ (e.g., the GI tract). This process is responsible for the accumulation of IgA antibodies in bodily secretions.

receptor at their basolateral (inner) surfaces. This Fc receptor, called the **poly-Ig receptor (polyIgR)**, binds IgA or IgM, triggering endocytosis and transport (**transcytosis**) of these antibodies across the epithelial cell to the apical (outer) side of the epithelial cell. At this point the polyIgR is cleaved near the membrane, releasing a large fragment, which remains bound (as secretory component) to the IgA, now in the lumen (see Figure 12-5). While IgM is also polymeric and

can bind to the polyIgR, there are many more IgA plasma cells in the tissue underlying mucosal and glandular epithelia (such as the lamina propria of the intestine), hence the predominance of IgA in the secretions. Between 3 and 5 grams of IgA are secreted into the intestinal lumen every day! IgA's half-life in secretions is relatively long because the amino acid sequence of the Fc region is resistant to many of the proteases that are present, and the secretory component

TABLE 12-2	Expression and function of FcRs		
FcR	**Isotype(s) they bind**	**Cells that express them**	**Function**
FcγRI (CD64)	IgG2a in mice, IgG1 and IgG3 in humans High-affinity receptor	Dendritic cells, monocytes, macrophages, granulocytes, B lymphocytes	Phagocytosis Cell activation Activation of respiratory burst ADCC
FcγRII (three forms: A, B1, B2) (CD32)	IgG	Dendritic cells, monocytes, macrophages, granulocytes, B lymphocytes, some immature lymphocytes	A: Phagocytosis and degranulation B: Inhibitory receptor; blocks B-cell activation Traps antigen-antibody complexes in germinal center
FcγRIII (two forms: A and B) (CD16)	IgG1, IgG2a, and IgG2b in mice; IgG1 in humans Low-affinity receptor Only FcR (along with FcRn) that binds mouse IgG1	Dendritic cells, monocytes, macrophages, granulocytes, B lymphocytes Only FcR expressed by NK cells	ADCC Cell activation for microbial killing and cytokine production
FcγRIV (in mice, with some similarity to human FcγRIIIA)	IgG2a and IgG2b in mice	Monocytes, macrophages, neutrophils Not found on lymphocytes	ADCC Cell activation
FcεRI	IgE (high affinity)	Eosinophils, basophils, mast cells, monocytes, Langerhans cells	Degranulation of granulocytes, including eosinophils, basophils, mast cells Phagocytosis
FcεRII (CD23)	IgE (low affinity)	B lymphocytes, eosinophils, Langerhans cells	Regulation of B-cell production of IgE IgE transport across intestinal epithelium, and of IgE immune complexes into B-cell follicles
FcαR (CD89)	IgA	Monocytes, macrophages, granulocytes, dendritic cells	Phagocytosis Cell activation ADCC
FcμR	IgM	B, T, and NK cells; low expression on granulocytes	Regulation of cell activation
Fcα/μR (CD351)	IgM (high affinity) IgA (intermediate affinity)	B cells, macrophages, follicular dendritic cells, kidney mesangial cells	Endocytosis, phagocytosis
PolyIgR	IgA and IgM	Multiple epithelial cells	Transport of antibody from blood to the lumens of GI, respiratory, and reproductive tracts (transcytosis) and secretory glands
FcRn (neonatal FcR)	IgG	Syncytiotrophoblasts of the placenta Epithelial cells (including intestinal epithelium) Endothelial cells of mature animals Monocytes, macrophages, dendritic cells	Transport of IgG across the placenta into fetus Transport of IgG from milk in neonatal intestine to blood Transport of IgG into mucosal secretions Transport of antibody-pathogen complexes from gut lumen to mucosal immune tissue Maintenance of levels of serum IgG and albumin Phagocytosis

also helps protect the IgA from degradation. Interestingly, several microbes, including those that cause gonorrhea and strep infections, produce proteases that do degrade IgA, allowing these pathogens to evade this form of immune protection.

Effector Functions of IgE Antibodies

IgE antibodies are best known for their role in allergy and asthma. However, research suggests that they also play an important role in protection against parasitic worms (helminths) and protozoa, and in immunity to a variety of venoms. IgE is made in very small quantities but has very potent effects, inducing degranulation of eosinophils and basophils and release of molecules such as histamine and proteases that damage pathogens. How IgE accomplishes these roles will be discussed in Chapter 15.

Effector Functions of IgD Antibodies

IgD is a minor immunoglobulin in the blood, constituting only 0.2% of circulating antibodies. However, it is present at higher levels in secretions in the upper respiratory tract, where it reacts with respiratory bacterial and viral pathogens. IgD was recently found to bind to basophils and mast cells and, after antigen binds to the IgD antibodies, these cells are activated to release antimicrobial peptides, which attack the pathogens. The activated cells also release cytokines (including interleukin [IL]-4, tumor necrosis factor [TNF], and IL-1), the B-cell survival factor BAFF, and chemokines. The receptor that binds IgD has not been well characterized.

> **Key Concept:**
>
> - Each antibody class has different effector functions: IgM and some IgG antibodies fix complement; IgG antibodies mediate ADCC; several classes and subclasses induce phagocytosis; IgE antibodies induce release of inflammatory molecules from granulocytes to kill parasites; IgA antibodies are a major class in bodily secretions that block entry of bacteria and toxins to the tissues; and IgD antibodies stimulate release of protective molecules from basophils in the respiratory tract.

Fc Receptors Mediate Many Effector Functions of Antibodies

Although several antibody classes have a direct effect on the viability of a pathogen by initiating complement-mediated lysis, many of the effector functions of the antibodies described in this chapter depend on their ability to interact with receptors on cells of the innate and adaptive immune systems as well as cells in various epithelial and endothelial tissues. These antibody receptors were originally discovered over 40 years ago by early antibody biochemists, who referred to them as **Fc receptors** (**FcRs**) because they bound specifically to the Fc portion of the antibody.

FcRs allow nonspecific immune cells to take advantage of the exquisite specificity of antibodies to focus their cellular functions on specific antigens and pathogens. They also provide a physical bridge between the humoral and cell-mediated immune systems. When bound by antibody alone, FcRs usually do not trigger signals; however, when cross-linked by multiple antibody-antigen complexes, FcRs initiate effector responses (see Figure 12-5).

The diversity of FcRs, most of which are members of the immunoglobulin superfamily (see Figure 12-4), is now fully appreciated. They differ by (1) the antibody class(es) that they bind, (2) the cells that express them (e.g., macrophages, granulocytes, NK cells, B cells), and (3) the responses that they activate. Together, these properties define distinct effector functions that antibody classes can induce through their binding to Fc receptors (see Table 12-2).

Interestingly, some FcRs do not stimulate effector responses but, rather, *inhibit* cell activity. Indeed, a cell can express varying levels of both activating and inhibiting FcRs. The cell's response is determined by the balance between the positive and negative signals it receives when antibody-antigen complexes bind these two types of receptors. We will discuss these events in more detail here in this section, where we describe the function of individual classes of FcRs.

FcR Signaling

A single antibody will not activate an Fc receptor; instead, multiple FcRs need to be cross-linked or oligomerized by binding to the multiple antibodies coating a single pathogen. This cross-linking results in the generation of either a positive or negative signal that enhances or suppresses the effector function of that cell.

Whether engagement of an FcR generates positive or negative signals depends on whether that FcR is associated with an *immunoreceptor tyrosine activation motif* (ITAM) or an *immunoreceptor tyrosine inhibition motif* (ITIM), respectively (see Figure 12-4). Although some activating FcRs include their own ITAM motif in their cytoplasmic region, many do not and, instead, associate at the membrane with a common gamma (γ) chain, which provides the services of an ITAM to the stimulating receptor complex.

The ITAM amino acid sequence is a common activating motif found in many signaling proteins in the immune system. Recall from Chapter 3 that ITAM motifs are found in the cytoplasmic domains of molecules associated with T-cell receptors (TCRs) and B-cell receptors (BCRs): the CD3 complex and the Igα,Igβ heterodimer, respectively. They are also features of NK cell receptors. ITAMs act as substrates for tyrosine kinases, which are recruited after receptor engagement. The signaling pathways initiated by FcR ITAMs are similar to those initiated by the BCR and TCR, including the activation of additional kinases that phosphorylate downstream signaling intermediates, leading to the generation of potent second messengers such as diacylglycerol (DAG) and calcium ion (Ca^{2+}). These events and others result in activation of the FcR-expressing cell.

Although the general signaling pathways are similar, the precise consequences of FcR signaling differ depending on the type of receptor that is activated, the versions of the downstream signaling enzymes that are present, and the genes that are accessible in the specific cell type that is stimulated. For example, if the cell is a macrophage, the signaling pathways stimulate and enhance its phagocytic capacity and can induce expression and secretion of cytokines (e.g., the inflammatory cytokines TNF-α, IL-1, and IL-6). Granulocytes such as eosinophils, mast cells, and basophils are induced to degranulate, releasing potent mediators that have a variety of effects including allergic responses, as we will see in Chapter 15. Binding of NK Fcγ receptors to IgG antibodies that have recognized cellular antigens induces the release of mediators that induce apoptosis of the target cell (ADCC).

Instead of an ITAM, an ITIM is found in the intracellular regions of the FcγRIIB, the main inhibitory receptor among the FcR family. As you may remember from Chapter 11, ITIMs are also found in the B-cell transmembrane molecule CD22, which plays a role in shutting off BCR signaling once sufficient B cells have been generated and pathogen has been eliminated. ITIM motifs associate not with kinases, but with phosphatases, such as SHIP (SH2-containing inositol phosphatase), which help to assemble complexes that inhibit signaling cascades, in part by dephosphorylating what was phosphorylated by activating signals.

But what is the role of an inhibitory FcR? It turns out that they are important in tuning the threshold of activation of a cell. FcγRIIB, the main inhibitory FcR, binds several different IgG subclasses and is expressed by many cell types, including B cells. When bound by antibody-antigen complexes, FcγRIIB sends negative signals through the BCR that inhibit further B-cell activation, alerting the B cell that sufficient antibodies are present and no further antibody production is necessary. FcγRIIB is also expressed by dendritic cells; binding of IgG-coated antigens inhibits activation and maturation of a dendritic cell. This appears to prevent the dendritic cell from becoming activated "spontaneously" by low levels of antibody-antigen complexes that may always be circulating in the serum. Therefore, in order to be activated, a dendritic cell will need to receive signals from other receptors (e.g., Toll-like receptors [TLRs] and other pattern recognition receptors [PRRs]) that are strong enough to overcome the inhibitory signals—a clever way to determine how authentic an infection may be. Experimental evidence supports this perspective. Dendritic cells in FcγRIIB-deficient mice activate T-cell responses at lower antigen thresholds. FcγRIIB-deficient mice also develop autoimmunity, suggesting that negative signals play an important role in maintaining immune tolerance.

Interestingly, cells can co-express both inhibitory and activating FcRs and tune their response according to their integration of positive and negative signals. Which FcRs a cell expresses sometimes depends on the signals that the cell receives during an immune response. For instance, transforming growth factor (TGF)-β and IL-4 up-regulate the expression of inhibitory FcRs, whereas inflammatory cytokines (TNF and interferon [IFN]-γ) up-regulate the expression of some activating FcRs.

FcγR

FcγRs are the most diverse group of FcRs and are important mediators of antibody functions in the body. Expressed by a wide range of cells, they bind antibodies of the IgG class, often several different IgG subclasses. Most are activating receptors and can induce phagocytosis if expressed by monocytes, macrophages, and dendritic cells. Phagocytosis of IgG antibody–antigen complexes can lead to antigen degradation and presentation of antigen-derived peptides on MHC class II proteins (see Chapter 7). FcγRs on granulocytes can also induce degranulation and release of mediators. NK cell FcγRs can recognize antibodies bound to cellular antigens and activate ADCC.

The four families of FcγR (FcγRI [CD64], FcγRII [CD32], FcγRIII [CD16], and FcγRIV) are conserved across mammals. Three of these, FcγRI, FcγRIII, and FcγRIV, are activating receptors and vary in their binding affinities and cellular expression. FcγRI binds antibodies with high affinity, and prefers to bind IgG2a in mouse and IgG1 and IgG3 in humans. FcγRIII binds with lower affinity to several IgG isotypes. Most immune cells, except mature T cells, express both of these receptors. NK cells express only FcγRIII.

FcγRIV was discovered only recently. Investigators wishing to study the role of FcγRs in immune responses developed mice in which the two major FcγRs (FcγRI and FcγRIII) were absent. They were puzzled when IgG antibodies still exhibited effector functions, but recognized that this probably meant mice expressed other, as yet unidentified, FcγRs. Indeed, they found a new protein, FcγRIV, expressed exclusively on innate immune cells (including monocytes, macrophages, and neutrophils), which bound IgG antibodies with high affinity. This receptor also shows an isotype preference, binding murine IgG2a and IgG2b best. Its human equivalent may be FcγRIIIA, although recent data suggest that it also binds mouse IgE and functionally resembles one of the human FcεRs (FcεRI).

While FcγRIIA is an activating receptor, as mentioned previously FcγRIIB (CD32) is the major inhibitory FcγR and is expressed on all immune cells except mature T and NK cells. When antibody-antigen complexes are bound, it generates inhibitory signals that increase the activation threshold of a cell, helping to prevent B cells, dendritic cells, and macrophages from responding excessively. In its absence, mice are more susceptible to autoantibody production and can develop a lupus-like disease.

FcεR

Two types of FcεR are recognized: the high-affinity FcεRI that is similar in structure to most activating FcRs and is expressed by mast cells, eosinophils, and basophils; and the lower-affinity FcεRII (CD23) that is a member of the C-type lectin family, not the immunoglobulin superfamily. This low-affinity form is expressed by B cells and eosinophils, and its function is still under investigation (see Chapter 15). The high-affinity FcεRI binds IgE antibodies from the circulation, and after the antibodies bind and are cross-linked by an antigen—often parasites or allergens—degranulation is triggered. The mediators released can kill parasites but also can initiate symptoms that we recognize as allergic responses, which will be discussed in detail in Chapter 15.

FcαR

The Fcα receptor (CD89) is expressed by myeloid cells, including monocytes, macrophages, granulocytes, and dendritic cells. Like many activating FcRs, when cross-linked by IgA-containing immune complexes, FcαR contributes to pathogen destruction by triggering phagocytosis, ADCC, and activation of cells, resulting in degranulation and release of cytokines and inflammatory mediators, and generation of superoxide free radicals, which help kill internalized pathogens. Interestingly, monomeric IgA present in the serum conveys an inhibitory signal through the FcαR, which dampens excessive inflammatory responses resulting from FcR cross-linking induced by immune complexes.

FcμR and Fcα/μR

FcμR and Fcα/μR have been recently described. The FcμR is expressed by B, T, and NK cells, where it regulates cell activation. Another FcR that binds IgM is the Fcα/μR, which binds IgM with high affinity and IgA with intermediate affinity. It is expressed on B cells, macrophages, and follicular dendritic cells; binding of IgM antibody–coated antigens induces phagocytosis.

PolyIgR

The *poly*meric *i*mmunoglobulin *r*eceptor (polyIgR) is expressed by epithelial cells and has a unique function. As discussed previously, it initiates the transcytosis (endocytosis and transport) of IgA and IgM across the epithelial layer, from the blood and underlying tissue to the lumen (the inside of passageways or ducts) of mucosal and glandular tissues (see Figure 12-5e). These include the gastrointestinal (GI), respiratory, and reproductive tracts, and the salivary, lacrimal, and mammary glands. IgA is the main class of antibody transported into the secretions of these tissues, providing protection at these key locations that are connected to the body's openings.

FcRn

Although the **neonatal Fc receptor (FcRn)** can be considered another FcγR, it is unique in structure and function. Related structurally to MHC class I proteins, it, too, associates with β$_2$-microglobulin. Expressed on the surface of many different cell types, including epithelial and vascular endothelial cells, it binds not only IgG, but also albumin, inhibiting the degradation of both molecules and thus contributing importantly to the long half-life—up to several weeks—of some IgG subclasses (see Table 12-1). Vascular endothelial cells can take proteins up nonspecifically by pinocytosis. Under ordinary circumstances these proteins would be degraded in acidic lysosomes. However, endothelial cells also express FcRns, which bind tightly to antibodies and albumin within the acid environment of the lysosomes. They sort these important proteins into vesicles that are carried back to the bloodstream, where the higher pH of blood permits their release. Thus an important role of FcRn in adult animals may be to maintain levels of circulating IgG and albumin.

FcRn also plays important roles in protecting human fetuses and newborns from infection. It is expressed in the placenta and can carry IgG antibodies from the maternal circulation into that of the fetus, giving the fetus a dose of whatever antibodies the mother may have been making against pathogens in her environment. Given that IgG half-lives can be up to 3 weeks, these neonatally acquired antibodies can protect the newborn while its immune system gets established (see Figure 12-5c). FcRn is also responsible for carrying IgG antibodies ingested from mother's milk across epithelial cells of an infant's intestine to the bloodstream—the opposite direction of the antibodies carried by the polyIgR. Even though its expression levels drop after an animal is weaned, FcRn still plays an important role in managing intestinal infections and can transport antibody-pathogen complexes from the intestinal lumen into mucosal immune tissue, where antigen can be processed and presented to T cells. FcRn can also augment the role of polyIgR (which carries IgA into organ secretions) and carry IgG antibodies into some secretions, including milk and GI, respiratory, and vaginal secretions. When expressed on phagocytes, it can also play a more conventional role: stimulating internalization and destruction of pathogens.

> **Key Concepts:**
>
> - Fc receptors, which bind specific immunoglobulin classes or subclasses, are responsible for many of the effector functions of antibodies. These include phagocytosis of antibody-antigen complexes by macrophages; transcytosis of antibodies through epithelial cell layers; lysis of antibody-bound infected cells by NK cells (ADCC); protection of serum antibodies from degradation; induction of degranulation (release of mediators from granulocytes); and regulation of the activities of innate immune cells.
>
> - FcRs are expressed by many cell types in the body and generate signals when bound to antibody-antigen complexes. FcR-generated signals can either activate or inhibit the activity of cells. Whether an FcR is activating or inhibiting depends on whether it includes or associates with an ITAM (activating) or an ITIM (inhibiting) protein motif.

Protective Effector Functions Vary among Antibody Classes

Now that we have an understanding of the mechanisms and molecular players involved in humoral immunity, we can begin to conceptualize antibody effector functions in the context of an immune response. After infection, B cells encounter a pathogen in the secondary lymphoid organs. B cells become activated by antigen binding and additional signals, such as those from helper T cells (or the pathogen itself, if it contains T-independent antigen), and differentiate into antibody-producing plasma cells. The plasma cells late in primary responses, and those in secondary responses elicited from memory cells, may have undergone somatic mutation and heavy-chain class switching in the germinal center (as discussed in Chapter 11). The antibody classes produced by plasma cells depend not only on the timing of the response—early (usually IgM) versus late (other classes, following class switching)—but also the nature of the pathogen, which affects T_H cytokines that influence class switching to different heavy chains. A single infection can induce the production of antibodies with the same specificity but multiple different heavy-chain classes.

All classes of antibodies can neutralize pathogens and some, particularly the polymeric IgA and IgM antibodies, agglutinate pathogens, enhancing their clearance. IgM antibodies, produced early in the response, can initiate the complement cascade, lysing pathogen directly. IgG antibodies, produced later in the response, may not only activate complement and opsonize pathogens, but also enter tissues to interact with innate immune cells that express FcRs (e.g., neutrophils), enhancing their inflammatory activity and inducing mediator release from granulocytes. Whereas extracellular pathogens bias heavy-chain class switching to IgG subclasses that can fix complement and opsonize pathogen, intracellular pathogens bias class switching to antibodies that can activate cytotoxic cells, including NK cells, which release their granule contents to kill an infected cell (ADCC). By binding to specific receptors that mediate antibody transcytosis across tissues layers, IgA and IgM (via the polyIgR) and IgG (via the FcRn) can provide protection in the secretions of mucosal and glandular tissues exposed to infections through the body's openings. IgE and IgD antibodies promote degranulation of mast cells and basophils, releasing mediators that are protective against certain pathogens.

Antibodies Have Many Therapeutic Uses in Treating Diseases

When the method for generating mouse monoclonal antibodies was developed by Köhler and Milstein in 1975, immunologists quickly recognized their value as highly specific reagents with desired specificities that could be produced in large amounts from immortal B-cell hybridoma cell lines (see Chapter 20). Clinicians soon recognized the potential of monoclonal antibodies as treatment modalities. The first to be approved for clinical use, an anti-CD3 monoclonal antibody that depletes T cells (muromonab-CD3, marketed as Orthoclone OKT3), became available in 1986 for inhibiting kidney transplant rejection.

However, mouse monoclonal antibodies produced by the Köhler and Milstein approach had many disadvantages as therapeutics, most importantly because they are immunogenic in humans. So over the ensuing years the biopharmaceutical industry has developed an array of sophisticated approaches for generating partly or fully human monoclonal antibodies. Recombinant DNA techniques can be used to generate *chimeric antibodies*, with the mouse monoclonal antibody's variable regions joined to human constant regions. Even more sophisticated techniques can replace the mouse variable region framework regions with human, leaving only the hypervariable (or complementarity-determining) regions that determine antibody specificity from the original mouse monoclonal antibodies; these are called *humanized antibodies*. Today, *fully human* therapeutic antibodies can be made by several approaches. One utilizes two very special mouse strains, the XenoMouse (from Abgenix/Amgen) and the HuMAb-Mouse (from Medarex/Bristol-Myers Squibb), in which the mouse heavy- and light-chain gene families have been replaced by the human genes. The mouse B cells in these strains generate intact human antibodies that can undergo somatic hypermutation and affinity maturation. A very different and totally in vitro set of approaches utilizes screening of phage or yeast displays of libraries of human antibody sequences for recognition of the desired antigen. Both approaches have yielded totally human antibodies approved for clinical use.

With the ability to generate large amounts of monoclonal antibodies that specifically target particular proteins and are not recognized as foreign, the use of antibodies for treating a variety of conditions has dramatically increased in recent years. Investigators are making excellent use of our growing knowledge of IgG effector functions and are tailoring the subclasses of monoclonal antibodies used in therapies to enhance their effects. Reflecting the long lifetimes and diverse effector functions of IgG antibodies, some mediated by FcγRs, all therapeutic antibodies approved for human use thus far have been IgGs. For example, IgG1 antibodies are most commonly used in tumor therapy because they can both activate complement and mediate ADCC by NK cells. Recent data suggest that IgG2 is a potent mediator of ADCC by myeloid cells (including neutrophils) and may offer some therapeutic advantages. The diverse clinical applications of therapeutic antibodies and how they are being optimized for greatest efficacy are described in **Clinical Focus Box 12-1.**

Therapeutic Antibodies for the Treatment of Diseases

Therapeutic antibodies are being used to treat a growing array of conditions, in some cases revolutionizing the treatment of serious diseases. **Table 1** shows examples of therapeutic antibodies in use in three clinical contexts: cancer, autoimmune and inflammatory conditions, and infectious diseases. The first five therapeutic antibodies being used in cancer treatment target specific molecules expressed by tumor cells. For example, rituximab (Rituxan) targets CD20, a B-cell protein expressed on B lymphomas. Considerable research has shown that the antibodies kill the tumor cells by two mechanisms. When bound to tumor cells, these IgG1 antibodies are recognized by the Fcγ receptors on macrophages, activating phagocytosis (in the cancer literature this is called antibody-dependent cellular phagocytosis, ADCP). In addition, FcγRIII on NK cells binds the tumor cell–bound antibodies, triggering apoptosis of the tumor cell by antibody-dependent cell-mediated cytotoxicity (ADCC). Another therapeutic antibody with a specific target, trastuzumab (Herceptin), reacts with the epidermal growth factor receptor (EGFR, also known as HER-2) found on some advanced breast cancers. EGF is required for tumor growth, and blocking the receptor deprives the cells of this essential activator. Herceptin also has been shown to provide protection by inducing both ADCP and ADCC.

Other cancer therapeutic antibodies work through very different mechanisms. For example, bevacizumab (Avastin), which is beneficial in colorectal cancer, neutralizes vascular endothelial growth factor (VEGF), a growth factor required for vascularization of solid tumors. More recently developed, the final three cancer therapeutic antibodies in Table 1 work with the patients' immune systems to eliminate the tumors. They react with two coinhibitory receptors (CTLA-4 and PD-1) or the coinhibitory ligand PD-L1 on tumor cells. Antibody blockade of these inhibitory proteins

allows any existing tumor-specific T cells to destroy what can otherwise be very hard-to-treat cancers (see Chapter 19).

A second general use of therapeutic antibodies is to block cytokines that are causing harmful inflammatory conditions by neutralizing the cytokines or blocking their receptors. A prime example is adalimumab (Humira), a fully human anti–TNF-α receptor antibody that was isolated through a phage display library approach. Initially approved in 2002 to reduce the inflammatory joint damage of rheumatoid arthritis, over subsequent years it has been approved for a growing list of autoimmune diseases. From 2012 to 2016 it was the #1 top-selling pharmaceutical product, with global sales of $16 billion in 2016 alone. A second anti-cytokine antibody, secukinumab (Cosentyx), neutralizes IL-17, a cytokine that contributes to the skin inflammation in the autoimmune inflammatory condition psoriasis. Mepolizumab (Nucala) neutralizes IL-5, a cytokine that contributes to the recruitment and activation of eosinophils, whose mediators contribute to asthma. The final example in this group is daclizumab (Zinbryta, Zenapax), which recognizes the IL-2Rα chain (CD25). This therapeutic antibody is very interesting in that it was derived from the original mouse monoclonal antibody to the mouse IL-2R generated in 1981. It was humanized in 1991 and then approved some years later for use in preventing T cell–mediated organ transplant rejection. Not finding much demand for that use, it subsequently was shown to be a beneficial treatment for relapsing multiple sclerosis.

The third group of therapeutic antibodies in Table 1 consists of recently approved antibodies that are protective against bacterial toxins and should reduce the adverse effects of infection. Obiltoxaximab (Anthim) neutralizes the toxin of *Bacillus anthracis* and is indicated in adult and pediatric patients for the treatment of inhalation anthrax. Bezlotoxumab (Zinplava) neutralizes toxin B of *Clostridium*

difficile, which causes recurring serious intestinal infections in some individuals.

These are just a subset of the several dozen therapeutic antibodies that have been approved for use in patients thus far. Several hundred more are in various stages of clinical development. The biopharmaceutical industry is also working to improve the efficacy of therapeutic antibodies. These next-generation or "biobetter" antibodies will be improved in their delivery (preferably allowing subcutaneous rather than intravenous injection; note that oral administration isn't possible, as antibody proteins would be degraded in the digestive system). Improvements are also being sought in the antibodies' lifetimes after injection and in their efficacy. Antibodies intended to inhibit or kill cells are being coupled to inhibitory or toxic drugs, which then would be delivered to the target cells by the antibodies; however, making sure the drug goes only to the desired cells is an ongoing challenge.

A safer approach to improving therapeutic antibodies takes advantage of the functional differences among the various IgG subclasses, the result of variations in γ-chain constant region sequences (see Table 12-1). Amino acids in the various hinge and Fc regions have been identified that are responsible for differences in (1) hinge flexibility; (2) proteolytic cleavage of the hinge region; (3) complement activation; (4) binding to Fc receptors that activate phagocytosis and ADCC; (5) binding to the FcγRIIB, which inhibits B-cell activation (potentially useful in inhibiting autoantibody production); and (6) binding to the FcRn, which controls IgG lifetime in the circulation. As is seen from Table 1, IgG1 is the most common subclass used in therapeutic antibodies; it binds to most Fcγ receptors, elicits ADCC and ADCP, and activates complement (though less well than IgG3). IgG4 does not kill cells via ADCC or complement, but does activate phagocytosis.

Alteration of hinge and Fc sequences by mutation is being carried out to

(continued)

CLINICAL FOCUS *(continued)*   **BOX 12-1**

enhance the desired functional properties of these antibodies. For example, Fc regions have been engineered to produce antibodies with a 50-fold higher binding affinity for the C1q complement component, enabling a single IgG antibody to activate complement-mediated killing of tumor cells. Other amino acid substitutions in the Fc region recognized by the FcRn on vascular endothelial cells can increase the half-life of IgG in the circulation. For many therapeutic antibodies, prolonging their lifetimes in the body

will greatly benefit patients in terms of antibody effectiveness, less frequent injections, and lower treatment costs.

This Clinical Focus has barely touched on the exciting new approaches being pursued in this field. The partnerships between immunologists, biochemists, physicians, and the biopharmaceutical industry that have led to this explosion of therapeutic antibodies will surely continue, producing a growing arsenal of new treatment approaches for an expanding array of conditions.

REFERENCES

Brezski, R. J., and G. Georgiou. 2016. Immunoglobulin isotype knowledge and application to Fc engineering. *Current Topics in Immunology* **40**:62.

Elgundi, Z., M. Reslan, E. Cruz, V. Sifniotis, and V. Kayser. 2016. The state-of-play and future of antibody therapeutics. *Advanced Drug Delivery Reviews* (DOI: 10.1016/j.addr.2016.11.004).

Sondermann, P., and D. E. Szymkowski. 2016. Harnessing Fc receptor biology in the design of therapeutic antibodies. *Current Topics in Immunology* **40**:78.

TABLE 1   Examples of therapeutic antibodies in clinical use*

Therapeutic antibody: generic name (trade name)	Nature of antibody	Target (antibody specificity)	Condition(s) used for
Cancer therapeutics			
Rituximab (Rituxan)	Chimeric IgG1	CD20 (mouse B-cell antigen)	Relapsed or refractory non-Hodgkin's lymphoma
Trastuzumab (Herceptin)	Humanized IgG1	Human EGFR2 (HER-2)	HER-2 receptor–positive advanced breast cancers
Alemtuzumab (Campath)	Humanized IgG1	CD52	Chronic lymphocytic leukemia, cutaneous T-cell lymphoma
Dinutuximab (Unituxin)	Chimeric IgG1	CD2	Neuroblastoma (pediatric)
Daratumumab (Darzalex)	Fully human IgG1	CD38	Multiple myeloma
Bevacizumab (Avastin)	Humanized IgG1	Vascular endothelial growth factor (VEGF)	Colorectal cancer
Ipilimumab (Yervoy)	Humanized IgG1	CTLA-4	Melanoma
Pembrolizumab (Keytruda)	Humanized IgG4	PD-1	Advanced melanoma, non–small cell lung cancer, head and neck squamous cell carcinoma
Atezolizumab (Tecentriq)	Humanized IgG1	PD-L1	Non–small cell lung cancer, bladder cancer
Cytokine modulators			
Adalimumab (Humira)	Fully human IgG1	TNF-α receptor	Rheumatoid arthritis, psoriatic arthritis, ankylosing spondylitis, Crohn's disease, ulcerative colitis, chronic psoriasis, hidradenitis suppurativa, and juvenile idiopathic arthritis
Daclizumab (Zinbryta, Zenapax)	Humanized IgG1	IL-2Rα (CD25)	Relapsing multiple sclerosis
Secukinumab (Cosentyx)	Fully human IgG1	IL-17	Psoriasis
Mepolizumab (Nucala)	Humanized IgG1	IL-5	Asthma
Infectious diseases			
Obiltoxaximab (Anthim)	Humanized IgG3	*Bacillus anthracis* exotoxin	Anthrax
Bezlotoxumab (Zinplava)	Fully human IgG1	*Clostridium difficile* toxin B	*C. difficile* infection

*In some cases other antibodies have been developed that are specific for the same target antigen; only one is included in the table, which focuses on some of the early monoclonal antibodies available for clinical use and others approved recently that target a range of antigens and conditions. See text for definitions of chimeric, humanized, and fully human antibodies.

Cell-Mediated Effector Responses

Cytotoxic effector cells include the CD8$^+$ cytotoxic T lymphocytes (CTLs) of the adaptive immune system; NKT lymphocytes, which bridge the innate and adaptive immune systems; and NK cells, which were once associated strictly with the innate immune system but are now known to share intriguing functional features with their adaptive immune lymphocyte relatives. CTL, NKT, and NK effectors all induce cell death by triggering apoptosis in their target cells. Not only do these cytotoxic cells eliminate targets infected with intracellular pathogens (viruses and bacteria), but they also play a critical role in eliminating tumor cells and cells that have been stressed by extreme temperatures or trauma (**Table 12-3**). CTL and NK cells also play a less desirable role in rejecting cells from allogeneic organ transplants. Essentially, the cell-mediated immune response is prepared to recognize and attack any cell that exhibits "nonself" or "altered-self" characteristics.

Cytotoxic T Lymphocytes Recognize and Kill Infected or Tumor Cells Via T-Cell Receptor Activation

The importance of cytotoxic T lymphocytes in the cell-mediated immune response to pathogens is underscored by the pathologies that afflict those defective in T-cell

generation. Children with DiGeorge syndrome are born without a thymus and therefore lack the T-cell component of the cell-mediated immune system. They are generally able to cope with extracellular bacterial infections because their innate and T cell–independent antibody responses are intact, but they cannot effectively eliminate intracellular pathogens such as viruses, intracellular bacteria, and fungi. The severity of the cell-mediated immunodeficiency in these children is such that even the attenuated virus present in a vaccine, which is capable of only limited growth in normal individuals, can produce life-threatening infections.

T cell–mediated immune responses can be divided into two major categories according to the different effector populations that are mobilized: cytotoxic T cells and helper T cells. The differentiation and activity of helper T-cell subsets was discussed in detail in Chapter 10, and their roles in macrophage activation and delayed-type hypersensitivity will be described in Chapter 15. Here, we focus on the immune response mediated by **cytotoxic T lymphocytes (T$_C$ cells or CTLs)**, which can be divided into two phases. In the first phase, naïve T$_C$ cells undergo activation and differentiation into functional effector CTLs within secondary lymphoid tissue. As you learned from Chapter 10, differentiation of a naïve CD8$^+$ T cell into an effector CTL involves recognition of MHC class I–peptide complexes on an activated dendritic cell as well as help from T$_H$1 effector CD4$^+$ T cells. This help is provided both indirectly, via the ability of CD4$^+$ T cells to enhance dendritic cell function, and directly, via the release of cytokines by helper CD4$^+$ T cells. These events usually occur in secondary lymphoid tissues, such as lymph nodes and the spleen.

In the second phase, effector CTLs recognize MHC class I–peptide antigen complexes on specific target cells in the periphery, such as virus-infected or tumor cells, an event that ultimately induces the apoptosis of the target cells. Since virtually all nucleated cells in the body express MHC class I molecules, a CTL can recognize and eliminate almost any cell in the body that displays the specific foreign antigen–derived peptide recognized by that CTL in association with an MHC class I molecule.

Effector CTL Generation from CTL Precursors

Naïve T$_C$ cells are incapable of killing target cells and are therefore also referred to as **CTL precursors (CTL-Ps)**

Cell type	Effector molecules produced	Mechanism of killing
TABLE 12-3	**How cytotoxic lymphocytes kill**	
CTL (typically CD8$^+$ T cell)	Cytotoxins (perforins and granzymes), IFN-γ, TNF, Fas ligand (FasL)	Cytotoxic granule release; and FasL-Fas interactions
NKT cell	IFN-γ, IL-4, GM-CSF, IL-2, TNF, FasL	FasL interactions predominantly; can activate NK cells indirectly via cytokines
NK cell	Cytotoxins (perforins and granzymes), IFN-γ, TNF, FasL	Cytotoxic granule release; and FasL-Fas interactions

to denote their functionally immature state. Only after a CTL-P has been activated does the cell differentiate into a functional CTL with cytotoxic activity. As is true for naïve CD4$^+$ T cells (see Chapter 10), three signals are required for CTL-P activation (**Figure 12-6**):

- *Signal 1:* An antigen-specific signal transmitted by the TCR complex on recognition of an MHC class I–peptide

complex on a "licensed" APC, usually a dendritic cell (*licensing* is explained below)

- *Signal 2:* A costimulatory signal transmitted when CD28 on the CTL-P surface binds CD80/86 (B7) on the dendritic cell
- *Signal 3:* A signal provided by the CTL-P's high-affinity IL-2 receptor binding to IL-2, generated by helper T cells as well as by the CD8$^+$ T cell itself.

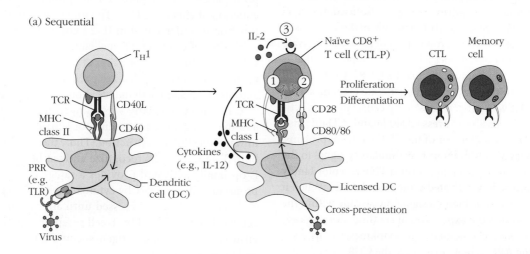

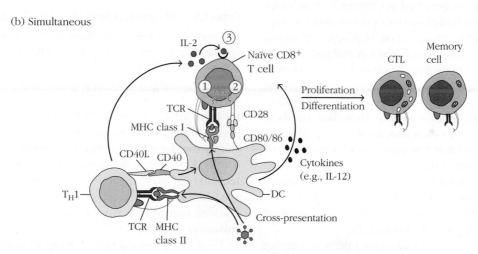

FIGURE 12-6 Generation of effector CTLs. The differentiation of a naïve CD8$^+$ T cell (a CTL precursor, or CTL-P) into a functional CTL requires several events. The CTL precursor, in this case specific for a viral antigen, must engage MHC class I–peptide complexes and costimulatory ligands on a "licensed" antigen-presenting cell (dendritic cell). Licensing of dendritic cells occurs either through engagement with an activated, CD40L$^+$ helper T cell (e.g., a T$_H$1 cell) or through signals from pattern recognition receptors (e.g., via Toll-like receptors). (a) Sequential activation. *Left:* CD4$^+$ T-cell activation of an unlicensed dendritic cell can occur prior to naïve CD8$^+$ T-cell engagement by the dendritic cell. *Right:* The licensed dendritic cell then activates a naïve CD8$^+$ T cell. (b) Simultaneous activation. Alternatively, activation can occur at the same time as the CTL-P engages the dendritic cell. Dendritic cells play a key role in the activation of

naïve CTL-Ps because licensing enables dendritic cells to cross-present peptides from internalized viral particles on MHC class I proteins that then are recognized by the CTL-P's T-cell receptors, providing signal 1 for T-cell activation. In addition, licensing induces dendritic cell expression both of membrane CD80/86, which provides signal 2 to the CTL-P, and of secreted cytokines (IL-12), which contribute to T-cell activation. Simultaneous engagement of a dendritic cell by both helper CD4$^+$ and precursor CD8$^+$ T cells would allow delivery to the pre-CTL of IL-2 generated by the helper CD4$^+$ T cells. In response to TCR stimulation, precursor CTLs begin to up-regulate expression of the IL-2 receptor and begin to make more of their own IL-2. IL-2, which provides signal 3 to the T cells, is critical for the induction of proliferation and successful differentiation into functional CTL and memory cells.

These three signals induce the proliferation and differentiation of the antigen-activated CTL-P into effector CTLs and memory cells.

Differentiation of a CTL-P is initiated by its recognition of antigen on an APC presented in the context of an MHC class I molecule, but other factors are also required. For CTL-Ps to mature into cytotoxic cells they need to recognize peptide–MHC class I complexes presented by a **licensed APC**, usually a dendritic cell. A dendritic cell can be licensed in several ways—by a CD4$^+$ helper T cell (T$_H$) usually of the T$_H$1 subset, or by engagement with microbial products, which activate Toll-like receptors. What does a CD4$^+$ T cell provide that is so important for optimal activation of a CD8$^+$ T cell? Recall from Chapter 10 that helper T cells generate cytokines (e.g., IFN-γ) that activate antigen-presenting cells. However, activated helper T cells also express CD40 ligand (CD40L, also known as CD154), a member of the TNF family of proteins, which provides an all-important costimulatory signal to the APC. CD40L is recognized by CD40, a TNF receptor family member expressed by activated professional APCs. When it binds CD40L, CD40 initiates a signaling cascade within the APC that increases the expression of costimulatory ligands (CD80 and CD86), chemokines, and cytokines, significantly enhancing the APC's ability to activate the CD8$^+$ T cell.

In addition, licensing enables the dendritic cell to carry out cross-presentation. As described in Chapter 7, in order for naïve CTL-Ps to be activated by viral or tumor antigens, peptide fragments must be presented on MHC class I proteins. If the dendritic cell is not itself infected or malignant, which will be the case most of the time, it will be taking up virus or antigens released from tumor cells from the environment. But exogenous antigens are typically processed and presented in the context of MHC class II molecules. How, then, can those antigens be processed in a way that allows their peptides to be presented by MHC class I proteins? Recall from Chapter 7 that cross-presentation is the process by which endocytosed exogenous proteins are degraded and their peptides loaded onto MHC class I proteins, rather than on MHC class II. Only licensed dendritic cells can carry out the cross-presentation necessary for the activation of naïve CTL-Ps.

Investigators envision a simultaneous interaction among three cells that results in CD8$^+$ T-cell activation: the T$_H$ cell, which interacts with and licenses the APC, which, in turn, interacts with and activates the CTL-P. In fact, fluorescence imaging studies have provided direct evidence of the formation of three-cell complexes of a dendritic cell, a CD4$^+$ T cell, and a CD8$^+$ T cell during the T-cell response to viral antigen (see Figure 12-6b and Figure 14-17). It is important to realize, however, that the interactions between the dendritic cell and the two different T cells may not have to be simultaneous; rather, a dendritic cell "licensed" by a T$_H$ cell could retain its capacity to activate CTL-Ps for some period after the T$_H$ cell disengages (see Figure 12-6a). In this case, specific T$_H$ cells could move on to sequentially activate other dendritic cells, amplifying the activation response.

Consequences of CTL Activation

CTL differentiation is accompanied by several changes. CTL precursors do not express the high-affinity IL-2 receptor α chain (CD25), nor do they produce much IL-2. They do not proliferate, and do not display cytotoxic activity. Signals 1 and 2 induce expression of both the IL-2Rα chain (generating the high-affinity IL-2 receptor) and IL-2 itself (the principal cytokine required for full proliferation and differentiation of effector CTLs). The importance of IL-2 in CTL function is underscored in IL-2 knockout mice, which are markedly deficient in CTL-mediated cytotoxicity. Fully activated CTLs turn on expression of proteins, including granzyme B and perforin, that are packaged into lytic granules and, when released, will induce apoptosis of the target cell.

CTLs are potentially very dangerous to an organism, and the stringent requirements for activation help prevent unwanted cellular destruction. Thus, the requirement that both T$_H$ and T$_C$ cells recognize antigen before the T$_C$ cell is optimally activated provides a safeguard against inappropriate self-reactivity by cytotoxic cells. The fact that the IL-2 receptor is not expressed until after a naïve CD8$^+$ T cell has been activated by T-cell receptor engagement also ensures that only the antigen-specific CTL-Ps proliferate and become cytotoxic.

Expansion of Antigen-Specific CD8$^+$ T Cells

For many years the study of the CTL response to viruses or tumors was hampered by the inability to identify CD8$^+$ T cells specific for viral or tumor antigens. As these T cells recognize antigen only as peptide fragments bound to MHC proteins, and as TCRs have much lower affinity for their MHC-peptide ligands compared with the much higher affinity of BCRs for intact antigens, one could not simply use fluorescent antigens to detect antigen-specific T cells. Fortunately, a technique was developed that uses tetramers of MHC-peptide complexes to identify T cells specific for an MHC-peptide complex. This technique allowed identification of antigen-specific T cells (see **Advances Box 12-2**).

With this powerful tool, one can directly measure the increase in antigen-specific CD8$^+$ T cells in response to exposure to pathogens such as viruses or cancer-associated antigens, and trace their tissue distribution. For instance, researchers infected mice with vesicular stomatitis virus (VSV) and, using tetramer technology, examined all organs for the presence of CD8$^+$ cells specific for a VSV-derived peptide–MHC complex. This study demonstrated that during acute infection with VSV, VSV-specific CD8$^+$ cells migrate away from the lymphoid system and are distributed widely, with large numbers found in the liver and kidneys, but antigen-specific cells present virtually everywhere (**Figure 12-7a**). Follow-up studies showed that, regardless of the original site of infection, effector CD8$^+$ T cells distribute themselves throughout the body. Similar analyses over the course of an

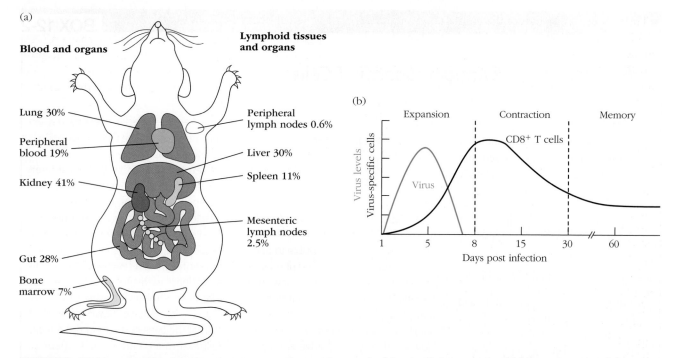

FIGURE 12-7 Localizing antigen-specific CD8[+] T-cell populations in vivo. (a) Mice were infected with *vesicular stomatitis virus* (VSV), and during the course of the acute stage of the infection cell populations were isolated from the tissues indicated. These cells were then incubated with fluorescent tetramers containing VSV peptide–MHC complexes. Flow cytometric analysis allowed determination of the percentages of CD8[+] T cells that were VSV specific in each of the populations examined. (b) General pattern of changes in levels of virus and numbers of virus peptide–MHC tetramer–positive CD8[+] T cells in mice following acute viral infections. *[(a) Data from Klenerman, P., V. Cerundolo, and P. R. Dunbar. 2002. Tracking T cells with tetramers: new tales from new tools.* Nature Reviews Immunology **2:***263. (b) Data from Kaech, S. M., and W. Cui. 2012. Transcriptional control of effector and memory CD8[+] T cell differentiation.* Nature Reviews Immunology **12:***749.]

infection have shown that the dramatically increased numbers of virus-specific CD8[+] T cells that arise soon after infection, many of which are CTL effector cells, decline as the virus is eliminated. Most effector CTLs have short half-lives, with only 5% to 10% remaining after the virus is cleared; the bulk of the surviving virus-specific cells are an expanded population of memory cells (Figure 12-7b).

How CTLs Kill Cells

A CTL can kill a target in two major ways: either via the directional release of granule contents, or by a Fas-FasL membrane signaling interaction. Rather than inducing cell lysis, both of these processes induce the target cell to undergo apoptosis, typically within an hour of contact with the cytotoxic cell.

Regardless of which method is employed, CTL killing involves a carefully orchestrated sequence of events that begins when the attacking cell binds to the target cell (**Figure 12-8**) and forms a cell-cell conjugate, an event so intimate that it is sometimes referred to as the "kiss of death." Formation of a CTL–target cell conjugate is followed within several minutes by a Ca^{2+}-dependent, energy-requiring step in which, ultimately, the CTL induces death of the target cell. The CTL then dissociates from the target cell and may go on to bind another target cell.

The specific signaling events involved in establishing the CTL–target cell interaction are very similar to those associated with an activating T cell–APC interaction. The TCR-CD3 membrane complexes on a CTL recognize peptide antigen in association with MHC class I molecules on the target cell. This recognition event triggers the development of a highly organized immunological synapse (see Chapter 10) characterized by a central ring of TCR molecules, surrounded by a peripheral ring of adhesion molecules, formed primarily by interactions between the integrin receptor LFA-1 on the CTL membrane and the *intracellular adhesion molecules* (ICAMs) on the target cell membrane (see Figure 10-3).

TCR signals directly enhance the adhesion between killer and target by converting LFA-1 from a folded, low-affinity state to an extended, high-affinity state (**Figure 12-9a**). This change is a consequence of "inside-out signaling," in which an intracellular signal cascade (generated by the TCR in this example, but can also be activated by some cytokine and chemokine receptors) acts on the intracellular portion of LFA-1 and induces a conformational change that straightens out the extracellular region of LFA so that its high-affinity face is accessible to ICAM. LFA-1 persists in its high-affinity state for only 5 to 10 minutes after antigen-mediated activation, and then it returns to the low-affinity state. This downshift in LFA-1 affinity may facilitate dissociation of the CTL from the target cell.

Detection of Antigen-Specific T Cells

To fully understand the basis for a successful (or unsuccessful) immune response, immunologists realized that they would need to be able to identify and track the behavior of antigen-specific T lymphocytes in an organism. While possible for B cells, which bind fluorescent probe–tagged antigens with high affinity and can be enumerated by flow cytometry even at low frequencies, this was a challenge for T cells, whose TCRs recognize antigen as peptide fragments bound to MHC proteins. This is a low-affinity recognition (K_D ~1–200 µM), 1,000- to 200,000-fold weaker than a typical antibody–antigen interaction, so the binding of single fluorescent peptide–MHC complexes to antigen-specific T cells could not be detected. The answer to this problem was fluorescent MHC-peptide tetramers, or **MHC tetramers** for short: a novel and clever technology developed in the 1990s. MHC tetramers made it possible to identify, isolate, and follow the behavior of T cells specific for an antigen of choice. Now used routinely, MHC tetramer staining has been an invaluable tool in our efforts to understand the complexities of lymphocyte behavior in space and time.

MHC tetramers are laboratory-generated complexes of four identical MHC molecules (either class I or class II), each containing a bound peptide and linked to a fluorescent protein (**Figure 1**). The higher avidity of interaction that comes from multiple TCRs binding the multiple MHC-peptide complexes on a tetramer confers stability to the binding. The MHC-peptide combination used to generate these complexes depends on the antigen of interest. For example, one laboratory has developed tetramers that include four HLA-A2 molecules (human MHC class I) complexed with a peptide from HIV that is known to stimulate a CTL response. A given MHC class I tetramer–peptide complex will be bound only by

those CD8+ T cells with TCRs specific for that particular peptide-MHC complex. Thus, when a particular MHC-peptide tetramer is added to a diverse T-cell population, cells that bear TCRs specific for the tetramer will bind it and become fluorescently labeled. These cells can then be detected by flow cytometry. With these technologies, it is now possible to determine the number, locale, and phenotype of cells in a population that have TCRs specific for that particular antigen (i.e., that particular MHC-peptide combination) before, during, and after an infection. Flow cytometry is sensitive enough to

detect antigen-specific T cells even when their frequency in the CD8+ population is as low as 0.1%.

Tetramer-based studies have been exceptionally useful in tracing the phenotype and fate not just of pathogen-specific T cells, but also of autoreactive T cells. For example, various MHC–insulin peptide tetramers have been used to identify insulin-reactive CD8+ T cells in patients with type 1 (autoimmune) diabetes. Although these studies are ongoing, some have shown that CD8+ T cells at the earliest stages of the disease respond to a different insulin peptide

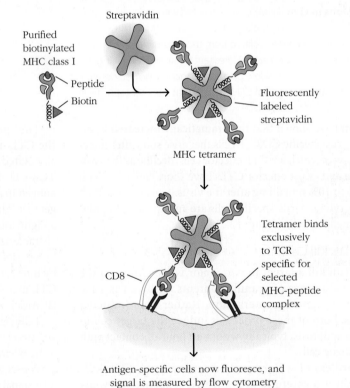

FIGURE 1 MHC-peptide tetramers. A homogeneous population of MHC class I molecules with a bound peptide (e.g., an HIV-derived peptide bound to HLA-A2) is conjugated to biotin and mixed with fluorescently labeled streptavidin, which has four high-affinity biotin-binding sites, forming a tetramer. Addition of the fluorescent tetramer to a population of T cells results in binding of the fluorescent tetramer only to those CD8+ T cells with TCRs that are specific for the MHC-peptide complexes of the tetramer. The subpopulation of T cells that are specific for the target antigen is now fluorescently tagged, making the cells readily detectable by flow cytometry.

(continued)

BOX 12-2

than CD8$^+$ T cells at later stages of the disease. The investigators speculate that insulin therapy itself may influence which autoreactive CD8$^+$ T cells become dominant in the disease.

The tools for detecting antigen-specific T cells have continued to advance. Tetramers of MHC class II proteins with bound peptides have been developed that enable investigators to identify antigen-specific CD4$^+$ T cells, including T$_H$ and T$_{REG}$ cells specific for

pathogen- or autoantigen-derived peptides. In addition, reagents with higher numbers of MHC-peptide complexes, such as dodecamers (12-mers), have made it possible to detect antigen-specific T cells with lower levels of TCR, including immature T cells in the thymus and naïve peripheral T cells. Utilizing these MHC-peptide reagents together with antibodies to other T-cell markers has enabled immunologists to describe in detail the numbers of antigen-specific

cells in various T-cell subsets and their changes over time.

REFERENCES

Altman, J. D., et al. 1996. Phenotypic analysis of antigen-specific T lymphocytes. *Science* **274**:94.

Huang, J., et al. 2016. Detection, phenotyping, and quantification of antigen-specific T cells using a peptide-MHC dodecamer. *Proceedings of the National Academy of Sciences USA* **113**:E1890.

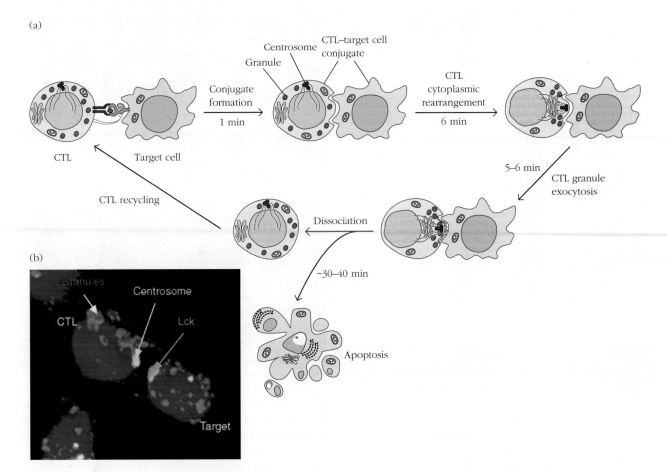

FIGURE 12-8 Stages in CTL-mediated killing of target cells. (a) T-cell receptors on a CTL interact with processed antigen–MHC class I complexes on an appropriate target cell, leading to formation of a CTL–target cell conjugate. The centrosome (also called the *micro*tubule *o*rganizing *c*enter, or MTOC) polarizes to the site of interaction, repositioning the Golgi stacks and granules toward the point of contact with the target cell, where the granules' contents are released by exocytosis onto the surface of the target cell, which is induced to apoptose. After dissociation of the conjugate, the CTL is recycled and the target cell dies by apoptosis. (b) Fluorescence image of a CTL–target cell conjugate, showing the enrichment of the protein kinase Lck (green) under the CTL membrane at the synapse between the two cells, where it is contributing to activation of the CTL; the centrosome (yellow), which is located under the synapse; and secretory granules (red), which are migrating on microtubule tracks up to the junction between the cells. (Note: the red granules in the target cell are probably lysosomes. Nuclei are blue.) *[(b) Republished with permission of Elsevier, from Jenkins M. R. and G. M. Griffiths, "The synapse and cytolytic machinery of cytotoxic T cells," 2010 June, Current Opinion in Immunology **22**(3):308–13, Figure 2. Permission conveyed through Copyright Clearance Center, Inc.]*

The importance of TCR signals in promoting CTL adhesion was demonstrated by an experiment in which purified ICAM protein was coated onto plastic and the ability of resting CTLs versus TCR-stimulated CTLs to adhere was measured. As you can see from Figure 12-9b, more than 10 times the number of TCR-stimulated CTLs (versus unstimulated CTLs) bound to the ICAM-coated plastic. Antibodies that could block the interaction between LFA-1 and ICAM

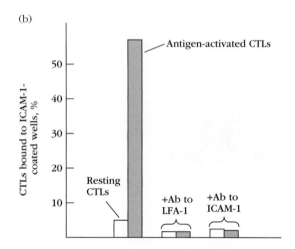

abrogated the effect of TCR stimulation, showing that the adhesion was, in fact, LFA-1/ICAM specific.

Granzyme- and perforin-mediated cytolysis

Many CTLs initiate killing of their targets via the delivery of proapoptotic molecules. These molecules are packaged within granules that can be visualized by microscopy (**Figure 12-10**). Investigators originally isolated CTL granules by subcellular fractionation and showed that they could induce target cell damage directly. Analysis of their contents revealed 65-kDa monomers of a pore-forming protein called

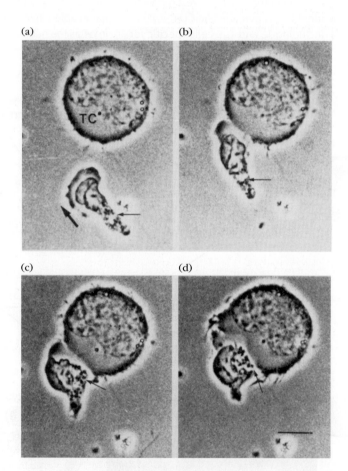

FIGURE 12-9 Effect of antigen activation on the ability of CTLs to bind to the intercellular cell adhesion molecule ICAM-1. (a) TCR signals induce a conformational change in LFA-1 molecules from a folded state to an extended state that allows them to bind to ICAM with high affinity. (b) The importance of TCR signaling in inducing LFA-1–mediated adhesion is illustrated by an experiment in which resting mouse CTLs were first incubated with anti-CD3 antibodies. Cross-linkage of CD3 molecules on the CTL membrane by anti-CD3 has the same activating effect as interaction with MHC class I–peptide complexes on a target cell. Adhesion was assayed by binding radiolabeled CTLs to microwells coated with ICAM-1. Antigen activation increased CTL binding to ICAM-1 more than 10-fold. The presence of excess monoclonal antibody to LFA-1 or ICAM-1 in the microwell abolished binding, demonstrating that both molecules are necessary for adhesion. *[Part (b) data from Dustin, M. L., and T. A. Springer. 1989. T-cell receptor cross-linking transiently stimulates adhesiveness through LFA-1.* Nature **341**:619.]

FIGURE 12-10 Formation of a conjugate between a CTL and a target cell and reorientation of CTL cytoplasmic granules as recorded by time-lapse photography. (a) A motile mouse CTL (thin arrow) approaches an appropriate target cell (TC). The thick arrow indicates the direction of movement. (b) Initial contact of the CTL and target cell has occurred. (c) Within 2 minutes of initial contact, the membrane-contact region has broadened and the rearrangement of dark cytoplasmic granules within the CTL (thin arrow) is underway. (d) Further movement of dark granules toward the target cell is evident 10 minutes after initial contact. *[From J. R. Yannelli, et al., "Reorientation and fusion of cytotoxic T lymphocyte granules after interaction with target cells as determined by high resolution cinemicrography," 1986, Jan 1, The Journal of Immunology,* **136**(2), 377–382. Copyright © 1986 by American Association of Immunologists.]

perforin and several serine proteases called **granzymes.** CTL-Ps lack cytoplasmic granules; however, once activated, CTLs begin to form cytoplasmic granules that include perforin monomers and granzyme molecules.

Almost immediately after conjugate formation, CTL granules containing granzyme and perforin are brought to the site of interaction between a killer and target (see Figure 12-8a) via the activity of the centrosome (also called the microtubule organizing center), which polarizes to this synapse in response to TCR stimulation. The secretory granules migrate on the microtubules up to the site of contact with the target cell and the vesicles fuse with the outer membrane, releasing perforin monomers and granzyme proteases into the space at the junction between the two cells.

As the perforin monomers contact the target cell membrane, they undergo a conformational change, exposing an amphipathic domain that inserts into the target cell membrane; the monomers then polymerize (in the presence of Ca^{2+}) to form cylindrical pores with an internal diameter of 5 to 20 nm (**Figure 12-11**). Perhaps not surprisingly, perforin exhibits some sequence homology with the terminal C9 component of the complement system, which forms the membrane attack complex that causes complement-mediated lysis. The importance of perforin to CTL-mediated killing is demonstrated in perforin-deficient knockout mice, which are unable to eliminate *lymphocytic choriomeningitis virus* (LCMV) from the body even though they mount a significant $CD8^+$ cell immune response to virally infected cells.

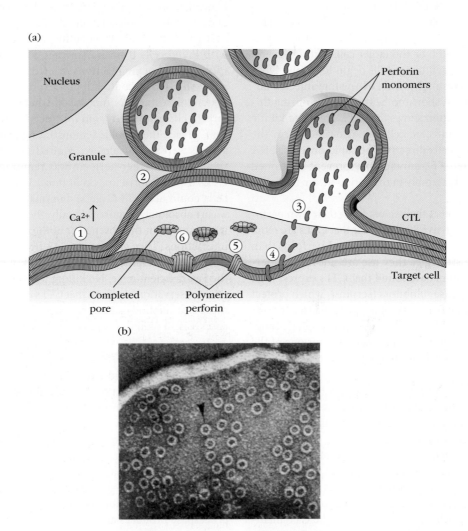

(a)

(b)

FIGURE 12-11 CTL-mediated pore formation in target cell membrane. (a) In this model, a rise in intracellular Ca^{2+} triggered by CTL–target cell interaction (1) induces exocytosis, in which the granules fuse with the CTL cell membrane (2) and release monomeric perforin into the small space between the two cells (3). The released perforin monomers undergo a Ca^{2+}-induced conformational change that allows them to insert into the target cell membrane (4). In the presence of Ca^{2+}, the monomers polymerize within the membrane (5), forming cylindrical pores (6). (b) Electron micrograph of perforin pores on the surface of a rabbit erythrocyte target cell. The arrow indicates a single pore. *[(b) Reprinted by permission from Nature Publishing Group, from Podack, E. R., and G. Dennert, "Assembly of two types of tubules with putative cytolytic function by cloned natural killer cells," 1983 March, Nature **302**:442–445. Permission conveyed through Copyright Clearance Center, Inc.]*

Although granzyme B was once thought to gain entry into the target cell via surface perforin pores, it is now thought that it enters mostly via endocytic processes. Many target cells express the mannose 6-phosphate receptor on their surface, which binds granzyme B. Complexes of granzyme B bound to mannose 6-phosphate receptor are internalized and appear inside endosomal vesicles. Perforin internalized at the same time then forms pores that release granzyme B from the endosomal vesicle into the cytoplasm of the target cell.

Regardless of the mechanism of entry, once in the cytoplasm, granzyme B and other granzymes initiate a cascade of reactions that result in the fragmentation of target cell DNA into oligomers of 200 base pairs (bp); this type of DNA fragmentation is typical of apoptosis. Granzyme proteases do not directly mediate DNA fragmentation. Rather, they activate an apoptotic pathway within the target cell. This apoptotic process does not require mRNA or protein synthesis in either the CTL or the target cell. Within several minutes of CTL contact, target cells begin to exhibit DNA fragmentation. Interestingly, viral DNA within infected target cells has also been shown to be fragmented during this process. This observation shows that CTL-mediated killing not only kills virus-infected cells but also can destroy the viral DNA in those cells directly. The rapid onset of DNA fragmentation after CTL contact may prevent continued viral replication and assembly in the period before the target cell is destroyed.

How do CTLs protect themselves from their own perforin and granzyme activity? Studies show that CTLs are more resistant to the activities of granzyme and perforin than their targets. The strategies they use are not fully understood, but investigators have found that CTLs express high levels of *serine protease inhibitors* (serpins), which protect them from granzyme B activity. In support of a role for serpins in protecting CTLs from their own proapoptotic functions, CTLs do not survive in mice deficient for one of the most common serpins expressed by CD8$^+$ T cells, Spi-1.

Fas/FasL-mediated cytolysis

Some potent CTL lines have been shown to lack perforin and granzymes. In these cases, cytotoxicity is mediated by Fas (CD95). This transmembrane protein, expressed by many cell types, is a member of the TNF receptor family and can deliver a death signal when cross-linked by its natural ligand, a member of the TNF family called Fas ligand (FasL). FasL is synthesized by activated CTLs and localizes in the granule membrane, which becomes part of the CTL membrane after granule fusion. The interaction of FasL with Fas on a target cell triggers apoptosis.

Fas mutations lead to multiple disorders in both mice and humans. Mice that are homozygous for the *lpr* (*lympho*proliferation) mutation express little or no Fas on their cell membranes, and have remarkably large lymph nodes. In more rigorous terms, they are afflicted with a

lymphoproliferative disease characterized by the accumulation of mature, activated T and B lymphocytes in their lymph nodes. CTLs in *lpr* mutant mice can be induced to express FasL; however, these CTLs cannot kill targets, as they express no Fas. These mice also develop autoimmune diseases because peripheral autoreactive lymphocytes are not eliminated. A very similar disorder, now known as *auto*immune *lympho*proliferative *s*yndrome (ALPS or Canale-Smith syndrome), has been identified in human patients who have genetic defects in Fas, Fas ligand, or Fas signaling pathways.

The critical importance of both the perforin and the Fas-FasL systems in CTL-mediated cytolysis was revealed by experiments with two types of mutant mice: perforin knockout mice and the Fas-deficient *lpr* strain (**Figure 12-12**). Investigators wanted to determine the relative importance of each of these molecules and developed a system in which they generated CTLs by coculturing T cells from *H2^b* mice with killed cells from *H2^k* mice. CTLs generated in such mixed-lymphocyte reactions reacted strongly against allogeneic MHC and could kill H2^k-expressing cells. With this system, investigators first asked whether CTLs could kill target cells generated from *lpr* mice (which expressed no Fas). In fact, they could, demonstrating that Fas-FasL interactions were not absolutely required for CTL activity. Next they asked whether perforin was required and generated CTLs from perforin knockout mice. They found that these cells could also kill targets, demonstrating that perforin wasn't absolutely required. Finally, they asked whether killing occurred in the absence of both perforin *and* Fas-FasL interactions. They generated CTLs from perforin knockout mice and determined their ability to kill target cells from the Fas-deficient mice. No killing was observed. These and other observations indicate that CTLs employ these and only these two methods—perforin-mediated and Fas-mediated apoptosis—to kill their targets.

Induction of apoptosis

Both the perforin and Fas-FasL killing strategies activate signaling pathways in the target cell that induce apoptosis (**Figure 12-13**). A central feature of cell death by apoptosis is the involvement of the caspase family of cysteine proteases, which cleave after an aspartic acid residue. The term *caspase* incorporates each of these elements (*c*ysteine, *asp*artate, prote*ase*). Normally, caspases are present in the cell as inactive proenzymes—procaspases—that require proteolytic cleavage for conversion to the active forms. Cleavage of a procaspase produces an active caspase, which cleaves other procaspases, thereby activating their proteolytic activity, which results in a cascade of events that systematically disassemble the cell—the hallmark of apoptosis. More than a dozen different caspases have been found, each with its own specificity. They are typically divided into two categories: those that initiate the caspase cascade (initiator caspases) and those that directly initiate apoptosis (effector caspases).

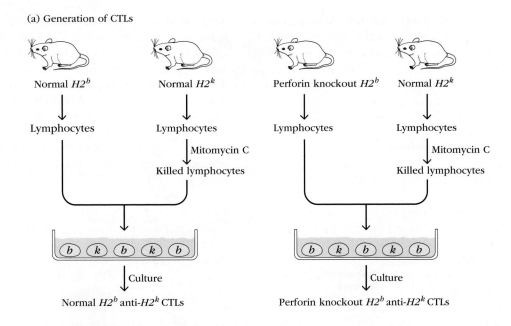

(a) Generation of CTLs

Normal $H2^b$ Normal $H2^k$ Perforin knockout $H2^b$ Normal $H2^k$

Lymphocytes   Lymphocytes   Lymphocytes   Lymphocytes

Mitomycin C   Mitomycin C

Killed lymphocytes   Killed lymphocytes

Culture   Culture

Normal $H2^b$ anti-$H2^k$ CTLs   Perforin knockout $H2^b$ anti-$H2^k$ CTLs

(b) Interaction of CTLs with Fas⁺ and Fas⁻ targets

CTLs	Target cells	
	Normal $H2^k$	*lpr* mutant $H2^k$ (no Fas)
Normal $H2^b$ anti-$H2^k$	Killed	Killed
Perforin knockout $H2^b$ anti-$H2^k$	Killed	Survive

FIGURE 12-12 Experimental demonstration that CTLs use Fas and perforin pathways. (a) Lymphocytes were harvested from mice of $H2^b$ and $H2^k$ MHC haplotypes. $H2^k$ haplotype cells were killed by treatment with mitomycin C and cocultured with $H2^b$ haplotype cells to stimulate the generation of anti-H2^k CTLs. If the $H2^b$ lymphocytes were derived from normal mice, they gave rise to CTLs that had both perforin and Fas ligand. If the CTLs were raised by stimulation of lymphocytes from perforin knockout (KO) mice, they expressed Fas ligand but not perforin. (b) Interaction of CTLs with Fas⁺ and Fas⁻ targets. Normal $H2^b$ anti-$H2^k$ CTLs that express both Fas ligand and perforin kill normal $H2^k$ target cells and $H2^k$ *lpr* mutant cells, which do not express Fas. In contrast, $H2^b$ anti-$H2^k$ CTLs from perforin KO mice kill Fas⁺ normal cells by engagement of Fas with Fas ligand but are unable to kill the *lpr* cells, which lack Fas.

What strategies do CTLs use to initiate caspase activation in target cells? The engagement of Fas on a target cell by Fas ligand on a CTL first induces the activation of an initiator caspase in the target cell. Fas is associated with a protein known as *Fas-associated protein with death domain* (FADD), which in turn associates with procaspase-8 (see Figure 12-13a). On Fas cross-linking, procaspase-8 is converted to caspase-8 and initiates an apoptotic caspase cascade.

The CTL-derived granzymes, which enter target cells through perforin pores in endosomal vesicles as mentioned earlier, are proteolytic and have several targets (see Figure 12-13b). They can directly cleave procaspase-3, an event that appears to only partially activate this effector caspase. They can also cleave the protein Bid, which induces mitochondrial release of cytochrome *c*. The latter assembles with Apaf-1 (*a*poptosis *p*rotease *a*ctivating *f*actor-*1*), ATP, and

procaspase-9 to form an apoptosome, activating caspase-9. Caspase-9 cleaves procaspase 3, activating its apoptosis-inducing functions. Both granzyme activities—activating caspase-3 directly and indirectly via the apoptosome—appear to be required for optimal proapoptotic activity.

The end result of both perforin-granzyme and Fas-mediated pathways is, therefore, the activation of apoptotic programmed cell death pathways that are present in the target cell. As one immunologist has so aptly put it, CTLs don't so much kill target cells as persuade them to commit suicide.

T$_C$1 and T$_C$2: Two Subsets of Effector CTLs

As you know from Chapter 10, CD4⁺ effector T helper cells can differentiate into several subsets, each of which secretes a distinct panel of cytokines. Effector CD8⁺ cytotoxic cells are not as diverse, but can develop into two distinct subsets: **T$_C$1 cells** and **T$_C$2 cells**. These subtypes loosely resemble T$_H$1 and

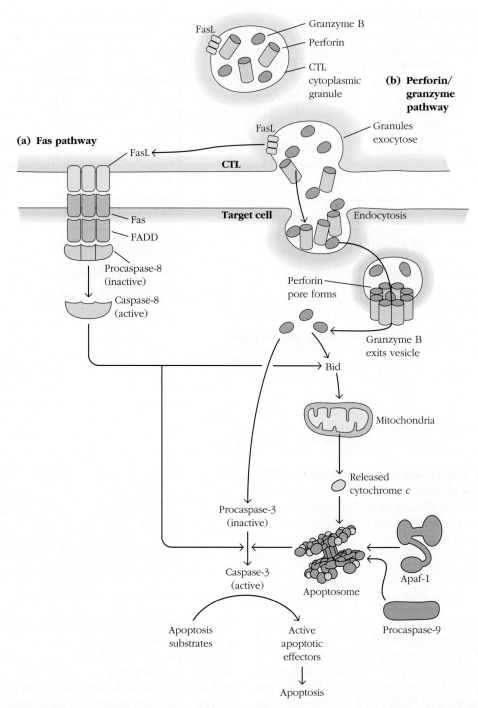

FIGURE 12-13 Two pathways of CTL-activated target cell apoptosis. (a) The Fas pathway. Ligation of the trimeric Fas protein by the CTL's Fas ligand (FasL) delivered by the granule membrane leads to the association of Fas with the adapter molecule FADD, which in turn results in a series of reactions that activate a caspase cascade, leading to apoptosis of the target cell. (b) The perforin-granzyme pathway. Granule exocytosis releases granzymes and perforin from the CTL into the space between the CTL and the target cell. Granzyme B enters the target cell by endocytosis and then passes into the cytoplasm through perforin pores. Granzyme B can cleave and activate the proapoptotic Bcl-2 family member Bid, which stimulates mitochondria to release cytochrome *c*. Cytochrome *c*, a molecule called Apaf, and procaspase-9 assemble into an apoptosome, leading to caspase-9 activation and cleavage of procaspase-3, which activates death pathways. Granzyme B can also cleave and partially activate caspase-3. Release of cytochrome *c* and activation of caspase-3 are both required to initiate the caspase cascade that leads to target cell apoptosis.

T_H2 cells in terms of the cytokines they generate as well as the cytokines that promote their development (see Table 10-3). CTLs are biased toward becoming T_C1 cells, which secrete IFN-γ but not IL-4. In the presence of IL-4, CTLs develop into T_C2 cells, which secrete much more IL-4 and IL-5 than IFN-γ. Both subsets are potent killers, although T_C1 cells can use both perforin/granzyme- and FasL-mediated strategies, whereas T_C2 cells appear to use only perforin and granzymes. Studies suggest that these two subsets play different roles in regulating disease, but results are not yet sufficient to generate confidence in one particular model.

Key Concepts:

- Compared with naïve T_H and T_C cells, CTL effector cells are more easily activated, express higher levels of cell-adhesion molecules, exhibit different trafficking patterns, and produce both soluble and membrane effector molecules.

- In order to become functional CTLs, naïve T_C cells (CTL precursors, or CTL-Ps) must engage with APCs (usually dendritic cells) that have been previously activated (licensed). Licensing often is carried out by CD4$^+$ helper T cells that recognize MHC class II with bound peptide presented by the dendritic cell; DCs can also be activated through pattern recognition receptors.

- T-cell help is not absolutely necessary for this first activation step, but is required for optimal proliferation and generation of CTL and memory cells.

- Antigen-specific T-cell populations, including CTLs, can be identified and tracked by labeling with MHC tetramers containing bound antigen-derived peptides.

- CTLs induce cell death via two mechanisms: the perforin-granzyme pathway and the Fas-FasL pathway. Both trigger apoptosis in target cells.

- The cytotoxic effector function of CTLs involves several steps: TCR/MHC-mediated recognition of target cells; formation of CTL–target cell conjugates and immune synapse formation; repositioning of CTL cytoplasmic granules toward the target cell; granule fusion with the plasma membrane, release of mediators, and insertion of FasL into the CTL plasma membrane where it can bind Fas on the target cell; initiation of apoptosis; dissociation of the CTL from the target; and death of the target cell.

Natural Killer Cell Activity Depends on the Balance of Activating and Inhibitory Signals

Another cell type involved in cell-mediated immunity, the **natural killer (NK) cell**, initiates apoptotic pathways in target cells using very similar mechanisms as CTLs, but via very different receptors. NK cells were discovered essentially by accident when immunologists were measuring the cytolytic ability of lymphocytes isolated from mice with tumors. They originally predicted that these lymphocytes would exhibit a specific ability to kill the tumor cells to which they had been exposed. To do their experiments they included multiple controls, comparing the activity of these lymphocytes of interest with those taken from mice that had no tumors, as well as mice that had unrelated tumors. Much to their surprise, the investigators discovered that even the control lymphocytes, which either had not been exposed to any tumor or had been exposed to a very different type of tumor, were able to kill the tumor cells. In fact, the killing they were seeing did not seem to be following the rules of specificity that are the hallmark of a lymphocyte response to conventional antigens and pathogens. Further study of this nonspecific tumor-cell killing revealed that, in fact, neither T nor B lymphocytes were involved at all. Instead, a population of larger, more granular lymphocytes was responsible. Similar nonspecific and rapid tumor-cell killing was observed with human peripheral blood lymphocytes, even from people without cancer.

The cells, called "natural killer" (NK) cells for their ability to kill tumor cells without prior exposure, make up 5% to 10% of the circulating lymphocyte population. Despite the absence of antigen-specific receptors (i.e., membrane antibodies, T-cell receptors), they play a major role in immune defenses against infected cells, stressed cells, and tumor cells. They can also contribute to autoimmunity when dysregulated. More versatile than originally anticipated, NK cells also play a regulatory role in both innate and adaptive immune responses to conventional antigens by secreting cytokines that regulate immune responses. They have also recently been shown to be critically important for the development of a normal placenta, not via their cytotoxic ability, but via their ability to recognize the presence of a different (paternal) MHC and initiate the remodeling of blood vessels. As introduced in Chapter 4, NK cells are innate lymphoid cells (ILCs) that play a variety of roles in early protection against infection, in stimulating and regulating adaptive immune responses, and in tissue development and remodeling (see Table 4-6).

The importance of NK cells in our defense against infections is compellingly illustrated by the case of a young woman with a disorder that resulted in a complete absence of these cells. Even though this patient had normal T- and B-cell counts, she suffered severe *Varicella* virus (chickenpox) infections and a life-threatening cytomegalovirus infection. We now know that NK cells are a critical first line of defense against infection with intracellular pathogens (viruses and some bacteria) by killing infected cells early and thereby controlling pathogen replication during the 7 days it takes CTL-Ps to develop into functional CTLs. NK cell cytotoxic activity is stimulated by the innate immune cytokines IFN-α, IFN-β, and IL-12, which all rise rapidly during the early stages of a viral infection. The wave of NK cell activity peaks subsequent to this rise, about 3 days after infection (**Figure 12-14**).

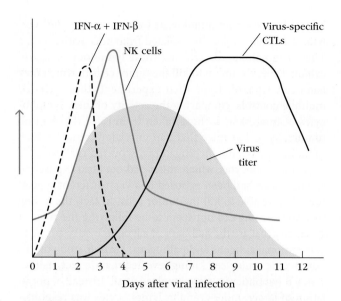

FIGURE 12-14 Time course of responses to viral infection.
IFN-α and IFN-β (dashed curve) are released from virus-infected cells soon after infection. These cytokines stimulate the NK cells, quickly leading to a rise in the NK-cell population (blue curve) from the basal level. NK cells help contain the infection during the period required for the generation of CTLs (black curve). Once the CTL population reaches a peak, the virus titer (blue area) rapidly decreases.

As we mentioned, NK cells also produce an abundance of immunologically important cytokines that can indirectly but potently influence both innate and adaptive immune responses. IFN-γ production by NK cells enhances the phagocytic, microbicidal, and antigen presentation activities of macrophages. IFN-γ derived from NK cells also influences the differentiation of CD4⁺ T helper subsets, stimulating T_H1 development via induction of IL-12 production by macrophages and dendritic cells and inhibiting T_H2 proliferation (see Chapter 10). NK cells also secrete TNF-α, GM-CSF, and chemokines that attract and activate macrophages, contributing to the local innate immune response.

NK cells are potent enough that, in conjunction with other protective mechanisms provided by the innate immune system, they can protect animals totally lacking in adaptive immunity. This is nicely illustrated by *RAG1* knockout mice, which have no antigen-specific B or T lymphocytes yet are healthy, active, and able to fend off many infections. These animals do not fare nearly as well when NK-cell development is also impaired. Interestingly, humans appear to be more dependent on their adaptive immune systems and suffer more without B and T lymphocytes than do their murine relatives.

Phenotype of NK Cells

Where do NK cells come from, and what do they look like? Like B cells, T cells, and other ILCs, NK cells are derived from the *common lymphoid progenitor* (CLP) in the bone marrow. Although some NK cells develop in the thymus, this organ is not required for NK cell maturation. Nude mice, which lack a thymus and have few or no T cells, have functional NK-cell populations. Unlike T cells and B cells, NK cells do not undergo rearrangement of receptor genes; NK cells still develop in mice in which the recombinase genes *RAG1* and *RAG2* have been knocked out, preventing the rearrangement and expression of antibody and TCR genes.

NK cells are quite heterogeneous. Various subpopulations can be distinguished on the basis of differences in expression and secretion of specific immunologically relevant molecules. Whether this heterogeneity reflects different stages in their activation or maturation or truly distinct subpopulations remains unclear.

Most murine NK cells express CD122 (the 75-kDa β subunit of the IL-2 receptor), NK1.1 (a member of the NKR-P1 family), and CD49b (an integrin). NK cells also typically express CD2 and FcγRIII (CD16). In fact, cell depletion with monoclonal anti-FcγRIII antibody removes almost all NK cells from the circulation. Human NK cells also express IL-2 receptors and FcγRIII, but do not express NK1.1. Rather, they are distinguished from other lymphocytes by expression of the adhesion molecule CD56, which varies in expression depending on the maturation and activity state of an NK cell (CD56 high expressers tend to produce cytokines and may differentiate into CD56 low expressers, which exhibit more cytolytic activity).

Perhaps the most distinctive phenotypic characteristic of NK cells is their expression of a set of unique activating and inhibiting NK receptors (NKRs). These receptors are responsible for determining which targets NK cells will kill. Interestingly, NK cells from mice and humans use mostly distinct sets of receptors to accomplish the same thing; however, the principles driving NK activation remain the same. In addition, the number and type of activating and inhibitory receptors expressed by NK cells vary widely even within an individual, and we know now that it is the balance of signals received through these receptors that determines whether or not an NK cell will kill a target cell.

How NK Cells Recognize Targets: The Missing Self and Balanced Signals Models

As NK cells do not express antigen-specific receptors, the mechanism by which these cells recognize tumor or infected cells and distinguish them from normal body cells baffled immunologists for years. Klas Kärre advanced an interesting hypothesis that formed the foundations for our understanding of how NK cells distinguish self from nonself (or *altered self*). He proposed that NK cells kill when they do not perceive the presence of normal self-MHC proteins on a cell; this was known as the **missing self model**. The implied corollary to the proposal is that recognition of self inhibits the ability to kill.

Clues to the origins of such inhibitory signals came from early studies looking at which tumor targets NK cells killed best. Investigators examined multiple variables associated with NK-cell preferences and discovered that killing was inversely correlated with levels of MHC class I molecules expressed on tumor cells. In one study, they examined the ability of CTLs and NK cells to kill B cells that were transformed into tumor cells by infection with *Epstein-Barr virus* (EBV). CTLs were unable to recognize and lyse these B cells. However, NK cells were very effective killers of these tumor cells. Ultimately, investigators realized that EBV infection down-regulated MHC class I, allowing the cells to evade CD8$^+$ T-cell recognition. However, this absence of class I made them perfect targets for NK cells, which responded to the "missing self." The investigators "clinched" a role for MHC class I by transfecting the B-cell tumors with human MHC class I genes. NK cells were no longer able to kill the cells. The subsequent discovery of receptors on NK cells that produce inhibitory signals when they recognize MHC molecules on potential target cells provided direct support for this model. These inhibitory receptors were shown to prevent NK-cell killing, proliferation, and cytokine release.

As many virus-infected and tumor cells have decreased MHC expression, the missing self model (i.e., basing the decision to kill on whether a target expresses sufficient levels of MHC class I, a ubiquitous self protein) makes good physiological sense. Although the fundamental paradigm has stood the test of time, not surprisingly the real situation is more complicated (**Figure 12-15**). It turns out that NK cells express two different categories of receptors: one that delivers signals that inhibit the cytotoxic activity of NK cells (receptors that recognize MHC class I proteins) and another that delivers signals that stimulate cytotoxic activity (receptors that recognize ligands up-regulated on infected, stressed, and tumor cells). NK cells distinguish healthy cells from infected or cancerous ones by monitoring and integrating both sets of signals; whether or not a target is killed depends on the balance between activating and inhibitory ligands that it expresses. Inhibitory signals, in general, trump activating signals; thus, NK cells are tolerant of cells that express normal levels of unaltered self MHC class I molecules (see Figure 12-15a). However, consistent with the original missing self model, if a cell has lost MHC class I expression, such as through infection with a virus that down-regulates MHC class I as part of its immune evasion strategies, then expression on the target cell of a ligand for an activating receptor can trigger activation and killing or cytokine production (see Figure 12-15b). More recent findings that most NK cells express multiple inhibitory and activating receptors have led to a modified model for the control of NK cell activation, called the **balanced signals model**. In this model it is the balance of inhibitory-versus-activating signals coming in from receptors recognizing a variety of ligands on the target cell that determines whether the NK cell is activated (see Figure 12-15c).

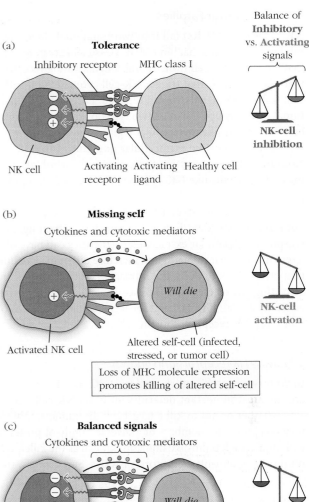

FIGURE 12-15   How NK cytotoxicity is restricted to altered self cells: missing self model and balanced signals model. The balance of signals from activating and inhibitory receptors determines whether an NK cell will kill a target cell. (a) Tolerance: no killing. An activating receptor on NK cells interacts with its ligand on normal cells, inducing an activation signal. However, engagement of inhibitory NK-cell receptor(s) by self MHC class I molecules delivers inhibitory signals that counteract the activation signal, and the cell is not killed. (b) Missing self model. Because MHC class I expression is often decreased on altered self cells, such as virus-infected and tumor cells, the activating signal predominates, leading to target cell destruction. (c) As some cells that are killed by NK cells are not always MHC class I-negative and express multiple activating and inhibitory receptors, the balanced signals model was developed. Infected, stressed, or tumor cells up-regulate expression of certain proteins that are recognized by activating receptors on the NK cell. If the activating signals are more extensive than inhibitory receptors, the NK cell becomes activated.

NK Cell Receptor Families

NK receptors (**NKRs**) fall into two functional categories: inhibitory receptors that bind MHC class I and create signals that block the NK cell from killing, and activating receptors that induce signals that trigger NK-cell cytotoxicity if sufficient inhibitory signals are not received. Each of these functional NKR groups includes two types of receptors: members with lectin-like extracellular regions and members with immunoglobulin-like extracellular domains (**Table 12-4**). Note that although lectins typically bind carbohydrates, most of the lectin-like NK-cell receptors actually recognize proteins.

The extracellular structure of the NK-cell receptor does not immediately identify it as inhibitory or activating. NK receptors with similar extracellular structures can have different intracellular domains and therefore different signaling properties. As with FcRs, the intracellular sequences of the activating NK receptors or their associated signaling chains generally contain ITAMs, and the intracellular sequences of inhibitory NK receptors contain ITIMs.

Inhibitory NK receptors and ligands

Inhibitory receptors, which bind to MHC class I molecules, are the most important determinant of an NK cell's decision whether or not to kill a target cell. In humans, inhibitory receptors are members of a diverse family of proteins with immunoglobulin-like domains, known as the *killer-cell immunoglobulin-like receptors*, or **KIRs**. Surprisingly, the KIR family appears to have evolved extremely rapidly. Functional KIRs, found in primates, do not exist in rodents. Mice use a distinct family of receptors, the lectin-like **Ly49** family, to achieve the same function as KIRs, namely inhibition of NK-cell cytolytic activity through binding to MHC class I molecules on healthy cells. Functional Ly49 receptors do not exist in humans.

Both KIR and Ly49 family members are highly diverse and polymorphic. These inhibitory receptors are generally specific for polymorphic regions of MHC class I molecules (H2-K or H2-D in mice; particular alleles of HLA-A, HLA-B, or HLA-C in humans). **Figure 12-16a** shows the human KIR3DL1 inhibitory receptor bound to the HLA-B*5701 class I allelic protein (allotype). Note that while the HLA-B*5701 class I protein has a bound peptide, as do all MHC proteins expressed on the cell surface, recognition by the KIR3DL1 receptor is only partially specific for the peptide. Only a fraction of MHC class I and class I–like ligands for these diverse receptors have been identified. Given the genetic polymorphism of both KIRs and their MHC class I ligands, it is not surprising that individuals can inherit KIR–MHC class I combinations that are not ideal and could increase their susceptibilities to disease.

An exception to the rule of thumb that human NK inhibitory receptors are Ig-like molecules is the lectin-like inhibitory receptor CD94-NKG2A, a disulfide-bonded heterodimer made up of two glycoproteins: CD94 and NKG2A. The inhibitory receptor CD94-NKG2A is a member of the NKG2 family; the other members of this family

TABLE 12-4	Features of NK cell receptors				
Family	**Species**	**Structure**	**Ligands**	**Activating or inhibitory**	**Examples**
NKG2A-D	Mouse and human	Lectin-like	MHC class I–like	Most (but not all) activating	NKG2D (activating) binds MHC class I–like MICA, MICB, and ULPB family members (human) or H60, Mult1, Rae-1 family members (mouse) CD94-NKG2A (inhibiting, human) binds HLA-E with bound signal peptide
Natural cytotoxicity receptors (NCRs)	Human	Immunoglobulin family	Viral Ags and proteins up-regulated on infected and tumor cells	Activating	NKp30 recognition of BAT3/BAG6 on tumor cells NKp44 and NKp46 recognition of influenza virus HA and Sendai virus HN
KIR (many)	Human	Immunoglobulin family	MHC class I (HLA-B and HLA-C)	Both activating and inhibitory	KIR2DL1 (inhibitory) interacts with HLA-C KIR2DS4 (activating) interacts with HLA-Cw4 and other non-HLA proteins in tumors
Ly49A-P	Mouse	Lectin-like	MHC class I and homologs	Both activating and inhibitory	Ly49A,C (inhibitory) recognize various MHC class I Ly49H (activating) recognizes MCMV protein m157, a viral homolog of MHC class I

(a) NK Inhibitory receptor

(b) NK Activating receptor

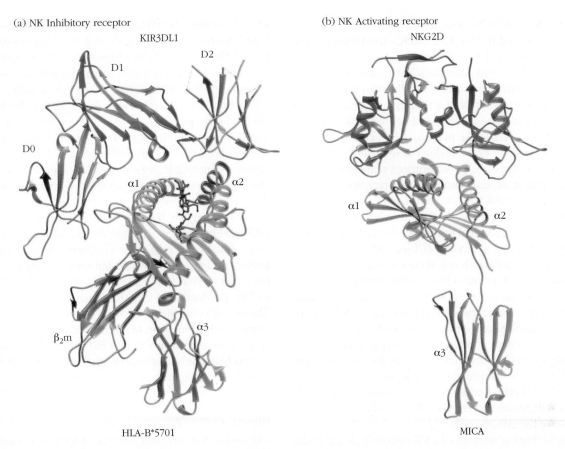

FIGURE 12-16 **Structures of NK inhibitory and activating receptors bound to their ligands.** (a) Human KIR3DL1 inhibitory receptor bound to a normal MHC class I protein, HLA-B*5701 with a bound peptide. The three Ig-like domains (D0, D1, D2) bind to various portions of the MHC class I molecule; recognition is not peptide-specific. (b) Human NKG2D activating receptor bound to MICA, a nonclassical MHC class I protein that does not have bound peptides. *[(a) PDB ID 3VH8. (b) PDB ID 1HYR.]*

are all activating receptors (see Table 12-4). Whereas KIRs typically recognize polymorphisms of HLA-B or HLA-C, CD94-NKG2A receptor recognizes the MHC class I–related protein HLA-E on potential target cells. HLA-E is not transported to the surface of a cell unless it has bound a peptide derived from the HLA-A, HLA-B, or HLA-C proteins: the leader (or signal) peptide cleaved from the nascent MHC leader peptide as it enters the rough endoplasmic reticulum. Thus the amount of HLA-E on the surface serves as an indicator of the overall level of MHC class I biosynthesis in the cells. The inhibitory CD94-NKG2A receptor recognizes surface HLA-E and sends inhibitory signals to the NK cell, with the net result that killing of potential target cells is inhibited if they are expressing adequate levels of MHC class I.

Unlike the antigen receptors expressed by B cells and T cells, NK receptors are not subject to allelic exclusion, and NK cells can express several different KIRs or Ly49 receptors, each specific for a different MHC molecule or for a set of closely related MHC molecules. Individual human NK cells expressing the CD94-NKG2A receptor and as many as six different KIRs have been found. The ability of

each NK cell to express multiple KIRs or NKG2A receptors improves its chances of recognizing the polymorphic MHC class I variants expressed by an individual's cells, therefore preventing NK cells from killing healthy cells.

Activating NK receptors and ligands

Most activating receptors expressed by murine and human NK cells are structurally similar and many are C-lectin-like—so named because they have domains related to calcium-dependent carbohydrate recognition domains, although the NK receptors recognize protein determinants. One major family in both humans and mice is the NKG2 family. **NKG2D** has emerged as one of the most important activating receptors in humans and mice; it acts through a signaling cascade similar to that initiated by CD28 in T cells. The ligands for NKG2D are nonpolymorphic MHC class I–like molecules that do not associate with β_2-microglobulin. These ligands are often induced on cells undergoing stress, such as that caused by DNA damage or infection. Their binding to NKG2D activates NK-cell functions—cytotoxicity and cytokine production at sites of infection, malignancy, and tissue damage. The structure of the NKG2D activating receptor bound to MICA, an MHC class I–like protein encoded in the class I region of the *HLA* gene complex, is shown in Figure 12-16b. One particularly interesting example of an activating lectin-like NK receptor is Ly49H, a member of a second family of activating receptors, the Ly49 receptors, that are found in mice but not in humans. Ly49H binds an MHC class I–like protein, m157, encoded by the mouse virus *murine cytomegalovirus* (MCMV). Thus this activating receptor directly recognizes a pathogen-derived activating ligand. In the absence of this receptor, mice are poorly protected from this common virus.

Some receptors that are not limited to NK cells also can serve as activating receptors to enhance NK activity. Perhaps the most important is FcγRIII (CD16), which binds IgG antibodies associated with cell-surface antigens (including viral and tumor antigens). This binding provides key activating signals that trigger NK cytotoxicity, an important example of antibody-dependent cell-mediated cytotoxicity (ADCC) by the NK cell. As described earlier in this chapter, ADCC is a potent immune effector response to infection and malignancy. Cells infected with virus, for instance, often express viral envelope proteins on their surface. Antibodies produced by B cells that responded to these proteins will bind the cell surface and recruit NK cells, which will induce apoptosis of the infected cell.

Other activating receptors on NK cells include CD2 (the receptor for the adhesion molecule LFA-3) and receptors for inflammatory cytokines. The involvement of these proteins as activating receptors again makes biological sense, enabling NK cells to participate in clearance of infected or tumor cells.

How NK Cells Induce Apoptosis of Their Targets

Regardless of which receptors are involved in regulating NK-cell lytic activity, NK cells kill targets by processes similar to those employed by CTLs (see Table 12-3). Like CTLs, the cytoplasm of NK cells also has numerous granules containing perforin and granzymes. NK cells develop an organized immunological synapse at the site of contact with a target cell, after which degranulation occurs, with release of perforin and granzymes at the junction between the interacting cells. Perforin and granzymes are thought to play the same roles in NK cell–mediated induction of target cell apoptosis as they do in the CTL-mediated killing process. In addition, also similar to CTLs, NK cells express FasL and readily induce death of Fas-bearing target cells.

NK Cell Licensing and Regulation

Even newly formed NK cells have large granules in their cytoplasm, and it was traditionally thought that NK cells were capable of killing from the moment they matured. It is now thought that most NK cells need to be licensed before they can use their cytotoxic machinery on any target. **NK cell licensing** is thought to occur via a first engagement of their inhibitory, MHC class I–binding receptors. This event can be considered a way for the immune system to test an NK cell's ability to restrain its killing activity when interacting with normal cells and an important strategy for maintaining NK-cell tolerance to self. Specifically, licensing allows only those cells that have the capacity to be disarmed via inhibitory interactions, usually from the recognition of self MHC class I proteins, to become armed and ready to kill.

Once licensed, NK cells are thought to continuously browse tissues and potential target cells via their multiple inhibitory and activating receptors. Engagement of activating ligands on the surface of a tumor cell, virus-infected cell, or otherwise stressed cell signals NK cells to kill the target cell. If the NK cells' inhibitory receptors detect normal levels of MHC class I on potential target cells, these inhibitory signals override the activation signals. This would not only prevent the death of the target cell but would also abrogate NK-cell proliferation and the production of cytokines such as IFN-γ and TNF-α. The overall consequence of this strategy is to spare cells that express critical indicators of normal self, the MHC class I molecules, and to kill cells that lack indicators of self and/or also express high levels of activating ligands that indicate that they are infected, malignant, or dangerous in other ways.

NK-Cell Memory

The distinctions between NK cells and B and T lymphocytes continue to blur. Clearly, B and T lymphocytes are unique in their generation of clonally restricted antigen-specific receptors through V(D)J gene rearrangements. But B and T cells were also once considered the only cells in the immune system that could generate memory responses. However, recent data indicate that NK cells can also generate a memory response to antigen. The evidence for this important and unexpected property came from experiments showing that NK cells expressing a receptor that binds a viral protein could transfer memory of this antigen exposure to naïve animals that had not been previously exposed or infected (see **Classic Experiment Box 12-3**).

These observations show that at least some NK cells possess the developmental machinery to become memory cells, in other words to increase their number and longevity (although perhaps not as long as B and T lymphocytes) and thereby improve their responses over time. Recent studies have also documented the induction of memory NK cells in humans. For example, acute infection with human cytomegalovirus induces expansion and persistence of NKG2C$^+$ NK cells that are activated not only by that virus but also by hantavirus. Evidence for NK memory has also been gathered for influenza, herpes simplex, and vaccinia viruses. This raises the exciting idea that it may be possible to immunize people to enhance their NK-cell memory against viruses and possibly even against certain tumors, potentially providing an expanded army of NK cells for early innate protection against infection or malignancy.

Key Concepts:

- NK cells induce apoptosis of tumor cells and virus-infected cells by mechanisms similar to those of CTLs (involving perforin/granzymes and FasL-Fas interactions), but are regulated by distinct receptors.

- In general, NK-cell killing is regulated by a balance between positive signals generated by the engagement of activating NK receptors and negative signals from inhibitory NK receptors. Only if the level of activating signals exceeds the level of inhibitory signals is an NK cell activated to kill the target cell.

- NK-cell receptors fall into two major structural groups based on their extracellular regions: the lectin-like receptors and the Ig-like receptors. Whether receptors are activating or inhibitory depends on their intracellular regions: ITAM-containing receptors are activating, ITIM-containing receptors are inhibitory.

- Many NK inhibitory receptors bind MHC class I proteins. The expression of relatively high levels of MHC class I molecules on normal cells protects them against NK cell–mediated killing by engaging inhibitory receptors. NK cells kill those cells that have lost or reduced their levels of MHC class I and express ligands for NK activating receptors that are up-regulated on infected or stressed cells.

- NK cells can also kill target cells via antibody-dependent cell-mediated cytotoxicity (ADCC), which results from the binding of cell-associated IgG antibodies to FcγRIII (CD16), which functions as an activating receptor.

- The capacity of NK cells to distinguish between normal and altered and potentially dangerous cells develops through the process of NK-cell licensing, in which only those cells that express inhibitory receptors for self MHC class I proteins develop the capacity to kill targets.

- NK cells have recently been shown to have memory responses, the expansion of NK cells capable of responding to cells expressing particular ligands for activating receptors.

NKT Cells Bridge the Innate and Adaptive Immune Systems

Thus far, this chapter's discussions of cell-mediated immunity covered the CTL, an important component of adaptive immunity that expresses an antigen-specific TCR; and the NK cell, a cell with both innate and adaptive properties that bears receptors that recognize inhibitory self ligands and activating ligands on altered cells. A third type of cytolytic lymphocyte has been identified with characteristics shared by both the CTL and the NK cell. This cell type, designated the **NKT cell** to reflect its hybrid quality, develops in the thymus, and, strictly speaking, is a member of the adaptive immune system. It undergoes antigen-receptor gene rearrangements and expresses an αβ TCR complex on its surface. However, it also exhibits characteristics of cells in the innate immune system:

- The T-cell receptor on many human NKT cells is invariant, with the TCR α and TCR β chains encoded by specific gene segments ($V_\alpha 24$-$J_\alpha 18$ and $V_\beta 11$, respectively) within the germ-line DNA; the cells expressing this αβ TCR combination are therefore sometimes referred to as **invariant NKT (iNKT) cells**. Similar cells occur in mice.

- The TCR on NKT cells does not recognize MHC-bound peptides but rather glycolipid antigens presented by the nonpolymorphic MHC class I–related CD1 molecule, as described in Chapter 7 (Figure 7-19).

- NKT cells can act both as helper cells (secreting cytokines that shape responses) and as cytotoxic cells (killing target cells).

- NKT cells include both $CD4^+$ and $CD4^-$ subpopulations, which may also differ in their cytokine production.

- NKT-cell killing appears to depend predominantly on FasL-Fas interactions.

- NKT cells do not form memory cells.

- NKT cells do not express a number of markers characteristic of T lymphocytes but do express multiple proteins characteristic of NK cells. For example, mouse NKT cells express the NK1.1 protein.

The exact role of NKT cells in immunity remains to be defined. However, experiments show that mice lacking NKT cells are deficient in their response to certain low-dose infections of bacteria that express glycolipids that can be recognized by NKT-cell receptors (e.g., *Sphingomonas* and *Ehrlichia*). Interestingly, high-dose *Sphingomonas* infection leads to sepsis and death in wild-type mice, but those lacking NKT cells survive this challenge, suggesting that the NKT cells may also contribute to pathology if they secrete excessive levels of inflammatory cytokines. (See Chapter 15 for a description of the role of proinflammatory cytokines in the onset of sepsis and the exacerbation of disease.)

Other data implicate NKT cells in immunity to tumors and suggest that NKT cells recognize lipid antigens specific to tumor cells. Finally, NKT cells also appear to contribute to viral immunity, despite the fact that viruses do not typically express glycolipids. NKT cells may play an indirect role in shaping the viral immune responses via the production of cytokines, including IFN-γ, IL-2, TNF-α, and IL-4.

Rethinking Immunological Memory: NK Cells Join B and T Lymphocytes as Memory-Capable Cells

Memory has a mystique in immunological circles, partly because it is the fundamental basis for the success of vaccination, which empowered societies by radically changing our relationship to disease, and partly because its molecular basis still remains elusive. As one of the hallmarks of the adaptive immune system, memory has always been considered an exclusive feature of B cells and T cells.

NK cells, the underappreciated lymphocyte cousins, have no antigen-specific receptors and until recently were assumed to have no capacity for memory. As recognition of their importance in responding to infection, killing tumor cells, and sculpting adaptive immune systems has grown, assumptions about their activities have been reexamined. To our surprise, the results of several clever experiments have revealed that NK cells do, indeed, share a capacity for antigen-specific memory with B and T cells. O'Leary and colleagues

showed in 2006 that memory of an antigen that causes hypersensitivity can be transferred by liver NK cells into a mouse that had never seen this antigen. Sun and colleagues showed in 2009 that memory of infection with *murine cytomegalovirus* (MCMV) could be transferred by NK cells. Here, we discuss aspects of this latter study.

To see whether NK cells had any memory capacity, the investigators wisely chose to examine the NK response to MCMV. NK cells are known to be an important component of protection against infection with MCMV, and a large fraction of NK cells (up to 50%) express an activating receptor specific for MCMV (Ly49H). A schematic of their experimental approach and results is shown in **Figure 1**, and two key pieces of their original data are shown in **Figure 2**.

The investigators devised a rigorously controlled system by which they could track MCMV-specific NK cells. Specifically, they

adoptively transferred (introduced a population of cells from one mouse to another) purified NK cells from wild-type mice into mice whose NK cells were not capable of expressing or signaling through Ly49H. (These mice are deficient in an ITAM-containing signaling molecule, DAP12, which associates with Ly49 receptors and initiates downstream signaling.) The donor NK cells also differed from the host NK cells by expression of CD45.1, a CD45 allelic variant that can be easily identified by flow cytometry (see Chapter 20). This way, the investigators could control the number of Ly49H$^+$ NK cells in the mice and could specifically trace those they introduced (Figure 2a).

They first found that the population of NK cells that were specific for MCMV (Ly49H$^+$CD45.1$^+$) proliferated during the 7 days following infection and then declined over the weeks after viral infection (Figure 2b). This expansion and contraction precisely paralleled the behavior of

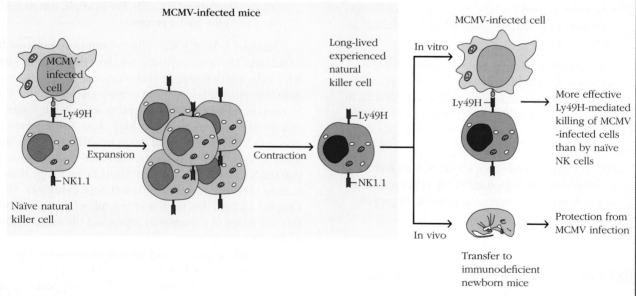

FIGURE 1 **The investigators' experimental approach.** To examine the possibility that NK cells exhibited hallmark characteristics of memory cells, investigators first looked to see whether NK cells with the Ly49H activating receptor, which binds to the MCMV virus, would expand and persist. They next looked to see whether the NK cells that persisted after infection could protect another animal from infection with MCMV.

They adoptively transferred the NK cells that had been generated in the infected mice into uninfected mouse pups. Then they infected the mouse pups with MCMV and looked for signs of disease. Finally, they examined the behavior of these persistent NK cells in vitro, to see whether they responded faster and better to signals through Ly49H. The results of these experiments are described in the text and some are shown in Figure 2.

(continued)

conventional antigen-specific T cells (specifically, CD4$^+$ T cells) in a primary response.

They next asked a critical question: did this antigen-specific NK-cell response result in memory? Specifically, did MCMV-specific NK cells persist over time? And if they did, would they exhibit a more robust response if rechallenged with MCMV? The investigators showed that even 70 days after infection, a population of Ly49H$^+$ (MCMV-specific), CD45.1$^+$ (donor-derived) cells could be detected. When the activity of these cells was compared with that of cells from mice that hadn't been infected, the investigators found that they were, indeed, more active: twice as many NK cells from infected mice ("memory NK cells") could make IFN-γ compared with NK cells from uninfected mice.

However, to show definitively that this memory was physiologically relevant, the investigators asked whether memory NK cells could rescue mice from infection. They isolated and transferred naïve and "memory" NK cells in varying numbers into neonatal mice, which are naturally immunodeficient and die if infected with MCMV (Figure 2c). Sun and colleagues asked whether the memory NK cells would enhance survival of infected neonatal mice. Figure 2d shows their results as a survival curve, in which the percentage of infected mice that remained alive is plotted versus time (in days) after infection. In the absence of antigen-specific NK cells (PBS control), 100% of mice died within 2 weeks of a primary infection with MCMV. However, when Ly49H$^+$ NK cells were transferred into the neonatal mice, their survival prospects increased. In fact, in the presence of Ly49H$^+$ NK cells isolated from previously infected mice (memory NK cells), 75% of the mice lived. Introduction of Ly49H$^+$ cells from mice that had not been previously infected (naïve NK cells) also improved survival, but it took at least 10-fold more of these naïve NK cells to provide equal protection.

Memory NK cells have more recently been induced in responses to haptens (small organic molecules) applied to the skin, and evidence of them in people has also been obtained in virus-infected individuals. These and other studies show that B and T lymphocytes must share the immunological stage with at least some NK cells as actors with excellent memories.

REFERENCES

O'Leary, J. G., M. Goodarzi, D. L. Drayton, and U. H. von Andrian. 2006. T cell- and B cell-independent adaptive immunity mediated by natural killer cells. *Nature Immunology* **7**:507.

Sun, J. C., J. N. Beilke, and L. L. Lanier. 2009. Adaptive immune features of natural killer cells. *Nature* **457**:557.

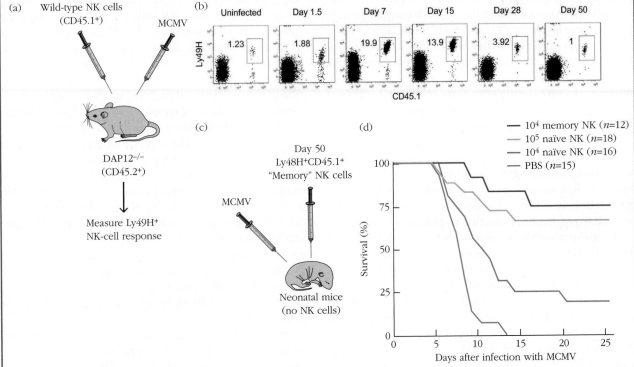

FIGURE 2 **Experimental results.** (a) This experimental design allowed the investigators to ask how NK cells reactive to MCMV responded to infection. The investigators took advantage of strain differences in CD45 allotypes to trace responsive NK cells. Naïve NK cells from mice that expressed the CD45.1 allotype were introduced into mice that expressed the CD45.2 allotype. (CD45 is expressed by all blood cells.) These hosts were then infected with MCMV. Ly49H$^+$ NK cells that expressed the CD45.1 marker were then identified by flow cytometry. (b) The flow cytometric profile of spleen cells isolated at various times after infection and stained with fluorescent antibodies to Ly49H and CD45.1; the percentage of cells staining for both markers (boxed) is indicated in each panel. The meaning of these results is discussed in the text. (c) The NK cells from part (b) that persisted through day 50 after MCMV exposure were then isolated. Varying numbers of these "memory" NK cells and varying numbers of naïve, freshly isolated NK cells were transferred into neonatal animals (which have no NK cells) that had not been exposed to virus. These animals were then infected with MCMV. Part (d) shows the percent survival of the mice in each group. The meaning of these results is discussed in the text. *[Data from Sun, J. C., J. N. Beilke, and L. L. Lanier. 2009. Adaptive immune features of natural killer cells.* Nature ***457**:557]*

Key Concepts:

- NKT cells have characteristics common to both T lymphocytes and NK cells. Many express an invariant TCR that recognizes lipid antigens bound to CD1 proteins, as well as markers common to NK cells.

- NKT cells exhibit both helper and cytotoxic activity and kill cells predominantly via FasL-Fas interactions.

Conclusion

The adaptive immune system is well known for its tremendous diversity of antibody and T-cell receptor specificities. The uniqueness of the mechanisms generating antibody and T-cell receptor diversity—V(D)J rearrangements and somatic hypermutation—has been recognized and appreciated by scientists in other areas of biology. But few who have not been trained in immunology understand another critically important kind of diversity: the wide array of immune effector mechanisms, both antibody- and cell-mediated, that actually provide the protection.

For humoral responses, this diversity in biological properties of antibodies—including differences in structure and mechanisms of eliminating pathogens, in passage across tissue layers into various body fluids, and in resistance to degradation and lifetimes in the circulation—all reflect sequence variation in the constant regions of the classes and subclasses of heavy chains. These biological differences arose during vertebrate evolution because of the adaptive benefit of being able to produce antibodies that can inactivate pathogens and infected and tumor cells by multiple distinct mechanisms. These mechanisms include neutralizing and agglutinating antigens, enhancing phagocytosis through opsonization, activating complement and its various protective mechanisms including cell lysis, inducing antibody-dependent cell-mediated cytolysis, and triggering degranulation and mediator release.

To make possible the production of antibodies with such diverse properties, the immune system's third unique gene-altering molecular process—heavy-chain class switch recombination, or CSR—evolved. CSR makes it possible for naïve B cells with IgM and IgD BCRs to generate antibody-secreting plasma cells making antibodies of other classes that are effective against an invading pathogen. As described in earlier chapters, the regulation of which heavy chain will ultimately be used in an activated B cell reflects earlier events, including pathogen recognition by innate immune cells, which produce signals that influence naïve CD4$^+$ T-cell commitment to a T$_H$ subset producing cytokines that control CSR. Thus the classes of antibodies produced in a particular immune response represent the culmination of a series of recognition and differentiation events that have evolved to generate antibody responses that will be effective against the particular invader.

Among T cells, most of the functional diversity resides in the T$_H$ subsets, largely mediated by the cytokines they produce. Some of these cytokines are instrumental in pathogen elimination. CD8$^+$ T cells have one main function after activation by virus-infected or tumor cells: to generate CTLs that kill those altered self cells by inducing apoptosis. Given the potential damage that rogue killer T cells could cause if directed at normal cells, their formation is tightly controlled by a process that has two checkpoints: naïve CD8$^+$ T cells must be activated by antigen presented by a licensed dendritic antigen-presenting cell, which itself previously had to be activated through mechanisms that depend on the presence of pathogen. CD8$^+$ T$_C$1 and T$_C$2 cells can also make T$_H$1- or T$_H$2-like cytokines and so contribute to the regulation of nearby immune responses.

NK cells provide parallel protection against infected and tumor cells, also inducing their apoptosis through mechanisms similar to those of CTLs. However, as part of the early innate response, most NK cells are not antigen-specific. Instead, NK cells recognize that a cell has become infected or malignant through the up-regulation on these stressed cells of proteins not expressed by normal cells. NK cells have evolved the expression of activating receptors that recognize these stress-induced proteins; another activating receptor on NK cells is an Fc receptor that binds IgG antibodies, so a cell coated with antibodies specific for surface viral or tumor antigens will be killed by NK-mediated ADCC. Again, controls have evolved to protect normal cells from NK cell–mediated killing. NK cells express inhibitory receptors recognizing proteins common to all cells in the body, usually MHC class I proteins. During NK cell development, only NK cells with inhibitory receptors recognizing MHC class I on our own cells become functional killers, providing necessary checks and balances. An additional cell type, the NKT cell, has properties of both T cells and NK cells. The TCR on many NKT cells is invariant, recognizing lipid associated with MHC class I–like CD1 proteins, and activating both cytokine production and cytotoxicity.

With this varied arsenal of innate and adaptive immune effector mechanisms, enhanced by memory B, T, and NK cells, our immune systems are able to protect us against the vast majority of pathogens to which we are exposed.

REFERENCES

Bevan, M. J. 2004. Helping the CD8$^+$ T-cell response. *Nature Reviews Immunology* **4**:595.

Bourgeois, C., and C. Tanchot. 2003. Mini-review: CD4 T cells are required for CD8 T cell memory generation. *European Journal of Immunology* **33**:3225.

Bournazos, S., and J. V. Ravetch. 2015. Fcγ receptor pathways during active and passive immunization. *Immunological Reviews* **268**:88.

Bournazos, S., T. T. Wang, R. Dahan, J. Maamary, and J. V. Ravetch. 2017. Signaling by antibodies: recent progress. *Annual Review of Immunology* **35**:285.

Brooks, C. G. 2008. Ly49 receptors: not always a class I act? *Blood* **112**:4789.

Cerwenka, A., and L. L. Lanier. 2016. Natural killer cell memory in infection, inflammation, and cancer. *Nature Reviews Immunology* **16**:112.

Che, K., and A. Cerutti. 2011. The function and regulation of immunoglobulin D. *Current Opinion in Immunology* **23**:345.

Cruz-Muñoz, M. E., and A. Veillette. 2010. Do NK cells always need a license to kill? *Nature Immunology* **11**:279.

de Saint Basile, G., G. Ménasché, and A. Fischer. 2010. Molecular mechanisms of biogenesis and exocytosis of cytotoxic granules. *Nature Reviews Immunology* **10**:568.

Diana, J., and A. Lehuen. 2009. NKT cells: friend or foe during viral infections? *European Journal of Immunology* **39**:3283.

Gould, H. J., and B. J. Sutton. 2008. IgE in allergy and asthma today. *Nature Reviews Immunology* **8**:205.

Jenkins, M. R., and G. M. Griffiths. 2010. The synapse and cytolytic machinery of cytotoxic T cells. *Current Opinion in Immunology* **22**:308.

Kärre, K., H. G. Ljunggren, G. Piontek, and R. Kiessling. 1986. Selective rejection of H-2-deficient lymphoma variants suggests alternative immune defence strategy. *Nature* **319**:675.

Mescher, M. F., et al. 2007. Molecular basis for checkpoints in the CD8 T cell response: tolerance versus activation. *Seminars in Immunology* **19**:153.

Mestas, J., and C. C. W. Hughes. 2004. Of mice and not men: differences between mouse and human immunology. *Journal of Immunology* **172**:2731.

Mitsi, E., et al. 2017. Agglutination by anti-capsular polysaccharide antibody is associated with protection against experimental pneumococcal carriage. *Mucosal Immunology* **10**:385.

Moretta, L. 2010. Dissecting CD56^{dim} human NK cells. *Blood* **116**:3689.

Orr, M. T., and L. L. Lanier. 2010. Natural killer cell education and tolerance. *Cell* **142**:847.

Parham, P. 2005. MHC class I molecules and KIRs in human history, health and survival. *Nature Reviews Immunology* **5**:201.

Petricevic, B., et al. 2013. Trastuzumab mediates antibody-dependent cell-mediated cytotoxicity and phagocytosis to the same extent in both adjuvant and metastatic HER2/neu breast cancer patients. *Journal of Translational Medicine* **11**:307.

Rajagopalan, S., and E. O. Long. 2005. Understanding how combinations of HLA and KIR genes influence disease. *Journal of Experimental Medicine* **201**:1025.

Salfield, J. G. 2007. Isotype selection in antibody engineering. *Nature Biotechnology* **25**:1369.

Stapleton, N. M., H. K. Einarsdóttir, A. M. Stemerding, and G. Vidarsson. 2015. The multiple facets of FcRn in immunity. *Immunological Reviews* **268**:253.

Trambas, C. M., and G. M. Griffiths. 2003. Delivering the kiss of death. *Nature Immunology* **4**:399.

Vivier, E., S. Ugolini, D. Blaise, C. Chabannon, and L. Brossay. 2012. Targeting natural killer cells and natural killer T cells in cancer. *Nature Reviews Immunology* **12**:239.

Useful Websites

YouTube offers many animations of CTL and NK killing. Use your knowledge from this book to critique their accuracy. The following link provides one example: www.youtube.com/watch?v=yRnuwTDR1og

www.signaling-gateway.org The "molecule pages" are accessible and up-to-date descriptions of the characteristics of many signaling receptors (including the FcRs and NK receptors described in this chapter).

www.nature.com/nri/posters/nkcells/nri1012_nkcells_poster.pdf This is a poster of our current understanding of the diversity of NK receptors, offered for free by *Nature Reviews Immunology*.

www.en.wikipedia.org/wiki/Fc_receptor This Wikipedia site offers a credible and particularly well-organized description of the Fc receptors. Wikipedia is a very useful first source for information relevant to this and other chapters. However, it is important to be aware that not all information on the site is validated. It is always important to go to the primary sources to verify what you read.

STUDY QUESTIONS

1. Investigators have developed antibodies that bind to FcγRIII and CD32 and potently block the interactions of these receptors with their ligands. Which of the following humoral immune response effector mechanisms would these antibodies block? Explain.

   a. Neutralization
   b. Opsonization
   c. Complement fixation
   d. ADCC

2. Indicate whether each of the following statements is true or false. If you believe a statement is false, explain why.

   a. Fcγ receptor engagement always results in internalization and destruction of antibody-antigen complexes.
   b. IgE mediates ADCC.
   c. IgM and certain subclasses of IgG are the only antibodies that activate complement.
   d. IgA is the only immunoglobulin class that can be transported across epithelial layers into the body's secretions.
   e. Both CTLs and NK cells release perforin after interacting with target cells.
   f. Antigen activation of naïve CTL-Ps requires a costimulatory signal delivered by interaction of CD28 and CD80/86.
   g. CTLs use a single mechanism to kill target cells.
   h. CD4$^+$ T cells are absolutely required for the activation of naïve CD8$^+$ T cells.
   i. The ability of an NK cell to kill a target is determined by signals received via both activating and inhibiting receptors.
   j. NK cells must express inhibitory receptors for self MHC class I proteins in order to develop the capacity to kill altered self cells.

3. You have a monoclonal antibody specific for LFA-1. You test the ability of a CTL clone to kill target cells for which the clone is specific, in the presence and absence of this antibody. Predict the relative amounts of target cell killing in the presence or absence of the anti–LFA-1 antibodies. Explain your answer.

4. A new virus, named Zobola, is causing an epidemic in South and Central America, and considerable resources are being invested into developing a vaccine. Your lab is testing a Zobola vaccine in mice, and finds that mice immunized with the vaccine are protected against subsequent infection a month later with the Zobola virus. Briefly, how would you determine whether the protection was due to one or more of the following immune responses?

   a. The presence of antibodies to Zobola virus induced by the immunization
   b. The presence of CTLs (or CTL memory cells) specific for Zobola virus activated by the immunization
   c. The presence of NK cells (or memory NK cells) activated by the immunization that can kill Zobola virus–infected cells

5. Indicate whether each of the properties listed below is exhibited by NK cells, CTLs, NKT cells, several, all, or none.

   a. _____ can make IFN-γ
   b. _____ can make IL-2
   c. _____ is MHC class I restricted
   d. _____ expresses CD8
   e. _____ is required for B-cell activation
   f. _____ is cytotoxic for target cells
   g. _____ expresses NK1.1
   h. _____ expresses CD4
   i. _____ expresses CD3
   j. _____ recognizes lipid antigens presented by an MHC-like protein
   k. _____ can express the IL-2 receptor
   l. _____ expresses the αβ T-cell receptor
   m. _____ expresses NKGD2 receptors
   n. _____ responds to soluble antigens alone
   o. _____ produces perforin
   p. _____ expresses FasL

6. Mice from several different inbred strains were infected with LCMV, and several days later their spleen cells were isolated. The ability of the primed spleen cells to kill LCMV-infected target cells from four different strains was determined, for example by using the ^{51}Cr-release assay. In the following table, indicate with a (+) or (−) whether the spleen cells listed in the left-hand column would kill the target cells listed in the headings across the top of the table.

Killing of LCMV-infected target cells				
Source of primed spleen cells	B10.D2 (*H2^d*)	B10 (*H2^b*)	B10.BR (*H2^k*)	(BALB/c × B10) F$_1$ (*H2$^{b/d}$*)
B10.D2 (*H2^d*)				
B10 (*H2^b*)				
B10.BR (*H2^k*)				
(BALB/c × B10) F$_1$ (*H2$^{b/d}$*)				

7. A mouse is infected with influenza virus. How could you assess whether the mouse has T$_H$ and T$_C$ cells specific for influenza?

8. NK cells do not express TCR molecules, yet they bind to MHC class I molecules on potential target cells. Explain how NK cells lacking TCRs can specifically recognize infected cells.

9. Consider the following three mouse strains:

- $H2^d$ mice in which perforin has been knocked out
- $H2^d$ mice in which Fas ligand has been knocked out
- $H2^d$ mice in which both perforin and Fas ligand have been knocked out

Each strain is immunized with LCMV. One week after immunization, T cells from these mice are harvested and tested for cytotoxicity. Which of the following targets would T cells from each strain be able to lyse?

a. Target cells from normal LCMV-infected $H2^b$ mice
b. Target cells from normal $H2^d$ mice
c. Target cells from $H2^d$ mice in which both perforin and Fas have been knocked out
d. Target cells from LCMV-infected normal $H2^d$ mice
e. Target cells from $H2^d$ mice in which both perforin and Fas have been knocked out

10. You wish to determine the frequency of MHC class I–restricted T cells in an HIV-infected individual that are specific for a peptide generated from gp120, a component of the virus. Assume that you know the HLA type of the subject. What method would you use, and how would you perform the analysis? Be as specific as you can.

11. Indicate whether each of the following statements regarding Fas-mediated or perforin-mediated programmed cell death is true or false. If you believe a statement is false, explain why.

a. Both mechanisms induce apoptosis of the target cell.
b. Target cells must express Fas ligand to be killed via the Fas-mediated pathway.
c. Only the perforin-mediated pathway stimulates a caspase cascade.
d. Both pathways require granzyme to induce apoptosis.
e. Both pathways ultimately activate caspase-3.
f. Granzyme is responsible for the assembly of membrane pores.

CLINICAL FOCUS QUESTIONS

1. One inherited combination of *KIR* and *MHC* genes leads to increased susceptibility to a form of arthritis. Would you expect this to be caused by increased or decreased activation of NK cells? Could an increase in susceptibility to diabetes be explained using the same logic? Explain.

2. Griscelli syndrome (type 2) is a rare and fatal disease caused by a loss of function of the *Rab27A* gene. It is also characterized by both loss of pigment (partial albinism) and immunodeficiency, which has been associated with a failure in cytotoxic T-cell activity. Look up the function of Rab27A online. Why might a deficiency in Rab27A lead to both of these symptoms? What other cell types might you expect to be affected?

ANALYZE THE DATA

Jäger and colleagues isolated an antigen from a tumor in a patient with cancer who expressed HLA-A2. To characterize the cell-mediated immune response of the patient to this tumor antigen, the investigators generated a series of peptide fragments from the tumor antigen and pulsed antigen-presenting cells with

	Peptide fragments derived from antigen isolated from tumor		
Number	**Peptide sequence**	**Position**	**Specific lysis***
1	SLAQDAPPLPV	108–118	0
2	SLLMWITQCFL	157–167	55
3	QLSISSCLQQL	146–156	1
4	QLQLSISSCL	144–153	1
5	LLMWITQCFL	158–167	15
6	RLTAADHRQL	136–145	5
7	FTVSGNILTI	126–135	1
8	ITQCFLPVFL	162–171	1
9	SLAQDAPPL	108–116	3
10	PLPVPGVLL	115–123	1
11	WITQCFLPV	161–169	5
12	SLLMWITQC	157–165	78
13	RLLEFYLAM	86–94	7
14	SISSCLQQL	148–156	6
15	LMWITQCFL	159–167	4
16	QLQLSISSC	144–152	1
17	CLQQLSLLM	152–160	1
18	QLSLLMWIT	155–163	84
19	NILTIRLTA	131–139	2
20	GVLLKEFTV	120–128	3
21	ILTIRLTAA	132–140	12
22	TVSGNILTI	127–135	4
23	GTGGSTGDA	7–15	9
24	ATPMEAELA	97–105	1
25	FTVSGNILT	126–134	1
26	LTAADHRQL	137–145	9

*Specific lysis is a relative measure of lytic activity and can range from 0 to 100%.

these peptides. They then measured the patient's CTL response against each of these targets. Answer the following questions based on the data in the table and what you have learned from multiple chapters in this book. ("Pulsing" refers to a technique in which antigen-presenting cells are exposed to high concentrations of peptides in solution. They exchange these peptides for those that were in the grooves of their surface MHC molecules. This is a convenient way to generate antigen-presenting cells that express specific MHC-peptide complexes of interest.)

a. Which epitope(s) of the tumor antigen was/were recognized by T cells? Explain your answer.

b. Immunologists have identified anchor residues on HLA-A2 molecules that are the most important for antigen presentation. What amino acids are most likely to bind HLA-A2 anchor residues if these amino acids must be conserved? Explain your answer.

c. Some of the peptides inspected by these investigators are poorly immunogenic. One explanation is that less immunogenic peptides may bind anchor residues ineffectively. Propose a different but reasonable hypothesis to explain why some of the peptides are poor immunogens.

d. The investigators pulsed antigen-presenting cells with peptides and used CTLs from the patient's peripheral blood to perform CTL assays. What HLA-A molecule is expressed by these antigen-presenting cells? Explain your answer.

e. In what way is peptide 2 an unusual epitope?

Barrier Immunity: The Immunology of Mucosa and Skin

Learning Objectives

After reading this chapter, you should be able to:

1. Identify the barrier organs in the human body and describe the general strategies they use to maintain a homeostatic relationship with the commensal microbiome.

2. Describe two examples that demonstrate how the microbiome and the immune system influence each other in biological dialog.

3. Using the intestinal immune system as an example, describe the diversity of cell types in secondary lymphoid tissue that help barrier organs to maintain homeostasis.

4. Distinguish between homeostatic and inflammatory responses in barrier tissues and outline the general events that initiate type 1 and type 2 responses at these surfaces.

5. Outline distinctions and similarities in immune systems that monitor our intestinal and respiratory tracts as well as our skin.

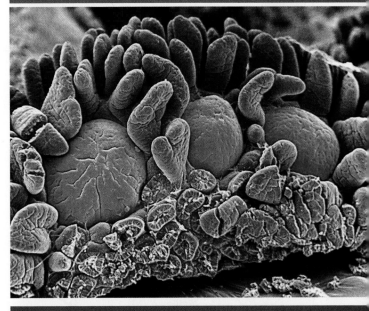

Digitally colorized scanning electron micrograph (SEM) of the mouse small intestine (ileum), with villi in blue and Peyer's patches in green. [SPL/Science Source]

The power of poop: "Eating poop pills can make you thinner!" "Probiotics: A diabetes cure?" "Parasitic worms may prevent Crohn's disease." Are any of these headlines based on fact? Can poop really benefit our health? Although healthy skepticism is critical, some of these claims, indeed, are supported by experimental evidence. And all are based on a relatively new understanding of our relationship with the vast community of microbes that live with us, on us, and in us. More specifically, they are based on our understanding of the dialog between our immune system and the microbes that live on the epithelial surfaces of our **barrier organs**—our intestinal, respiratory, and reproductive tracts, as well as our skin, all of which provide a boundary between the external world and our internal environment (**Overview Figure 13-1**). This chapter provides readers with a foundation not only to interpret the literature in this relatively new field of **barrier immunity**, but also to evaluate claims and coverage in the news.

Although introductory immunology courses have traditionally focused on cellular development and interactions in primary and secondary immune organs, a vast number

Key Terms

Barrier organs
Barrier immunity
Mucosal tissues
Commensal microbiome
Mucosal-associated lymphoid tissue (MALT)
Antimicrobial peptides (AMPs)

Peyer's patches
Isolated lymphoid follicles (ILFs)
Lamina propria
Transcytosis
Goblet cells
Microfold (M) cells

Paneth cells
Tuft cells
Intraepithelial lymphocytes (IELs)
Alarmins
Inflammatory bowel disease (IBD)

Mucociliary boundary
Alveolar macrophages (dust cells)
Epidermis
Dermis
Keratinocytes

Barrier Immune Tissues

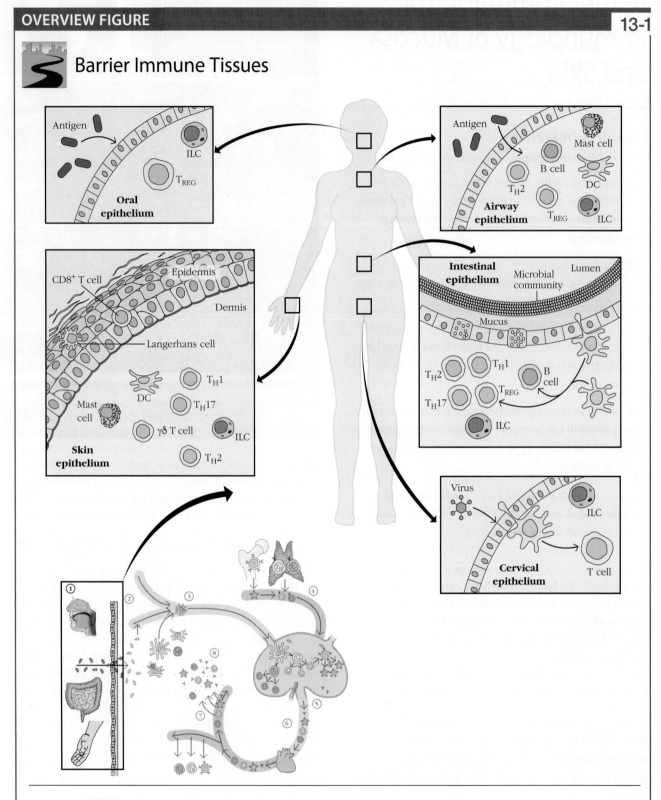

The surfaces of the human body—gastrointestinal tract (mouth and intestine), respiratory tract (airway), reproductive tract (cervix), and skin—are populated by microorganisms and monitored by barrier immune systems. Each is lined by one or more layers of epithelial cells, which interact with a variety of immune cells at and below these layers. The interaction between microbiome and innate and adaptive immune cells regulates the balance between tolerance and inflammation, and, ultimately, between health and disease.

of immune cells not only populate but also initiate immune responses at our body surfaces. In fact, the intestine alone harbors more immune cells than any other tissue, including 50 billion lymphocytes. We now recognize that our relationship with the microbes that inhabit our surfaces fundamentally influences our susceptibility and resistance to asthma, autoimmunity, cancer, and even mood and neurological disorders. The barrier immune systems mediate our relationship with microbes and are charged with both cultivating harmonious interactions with benign microbes and responding aggressively to harmful microbes.

The biologists Scott Gilbert and Jan Sapp and the philosopher Alfred Tauber recently reminded us that "we have never been individuals."[1] Instead, we are better understood as superorganisms or "holobionts." For example, approximately 40 trillion bacteria share our human bodies, which consist of about 30 trillion cells. Viruses, fungi, and worms are also important members of the community of organisms that co-exist in a healthy human body.

In this chapter, we begin by introducing common characteristics of barrier immune systems. Focusing primarily on the intestinal immune system, where our knowledge is most advanced, we introduce the cells, molecules, and microenvironments that help barrier immune communities maintain healthy relationships with our microbiome as well as respond to invasion.

We investigate the reciprocal relationship between the microbiome and the barrier immune system, which are in an authentic dialog, responding to and tuning each other during development and throughout our life span. We also touch on the surprising influence barrier immune responses have on distal organ function and systemic immunity.

The focus on intestinal immunity establishes a framework for our exploration of immune systems in other barrier organs, including the skin and respiratory tract. These barrier immune systems share many features, yet also have unique strategies that reveal the breadth and flexibility of our immune system.

Finally, we describe advances in our understanding of the relationship between barrier immunity and inflammatory disease associated with the intestine and skin. We also include three boxes, the first of which provides an overview of the many innate and adaptive immune cells that participate in barrier immunity. Next, a Clinical Focus box discusses the intriguing relationship between the intestinal response to the microbiome and brain activity—a two-way communication that may influence mood disorders, as well as susceptibilities to other neurological disorders, including schizophrenia. Last, an Advances box describes one of the first

observations that differences in animals' immune responses can be attributed to differences in their microbiomes.

Common Themes in Barrier Immune Systems

The typical human being is only 1.5 m tall and less than 0.1 m "thick." However, the combined surface area of all tissues exposed to the external environment is over 400 m^2. The epithelial surfaces of our **mucosal tissues**—intestinal, respiratory, reproductive, and urinary tracts—account for most of this total surface area, while the skin accounts for only 2 m^2. Each of these surfaces harbors communities of microorganisms, which include bacteria, viruses, protozoa, fungi, and even worms. Collectively, these microorganisms are referred to as our **commensal microbiome**. Most of our current understanding of the relationship between microbiomes and health comes from research on bacterial species; however, we are rapidly gaining an appreciation for the diversity and importance of viruses that cohabit our surfaces, as well as the influence of fungi, protozoa, and worms. Some very respectable studies are even examining the beneficial effect of worm antigens, which may help quell autoimmune and inflammatory reactions.

Although the skin and mucosal immune systems differ in some important details, they share common themes and strategies that provide a useful introductory framework for the more specific descriptions and discussions in this chapter. We begin by reviewing vocabulary terms commonly applied to discussions of these immune systems, as well as common strategies the immune systems use to respond to the microbiome.

Barrier immunity refers generally to immune systems associated with the skin and mucosal tissues. The immune systems associated specifically with mucosal tissues are often referred to as **mucosal-associated lymphoid tissue**, or **MALT**. Their surfaces are moist, mucus-rich, and are often covered by a single epithelial cell layer. Because the skin is topped by multiple layers of dead, dying, and dry keratinocytes, its immune network is not considered a MALT. However, its immune system has much in common with other barrier tissues, which, in all cases, are protected by a combination of physical, chemical, and cellular activities.

Barrier immune systems are continually exposed to microbes and are constantly stimulated, even in the healthy individual (see Overview Figure 13-1). Microbes that live in relative harmony at our barrier surfaces are collectively called *commensal organisms*. The healthy equilibrium achieved between microbes and the vertebrate immune systems at barriers is referred to as **homeostasis** and is maintained by a combination of mechanisms that inhibit inflammation and promote tolerance. Because homeostasis involves immune cell activity (it is not a quiescent state) it is also referred to as "physiological inflammation."

---
[1] Gilbert, S. F., J. Sapp, and A. I. Tauber. 2012. A symbiotic view of life: we have never been individuals. *The Quarterly Review of Biology* **87**:325.

All Barrier Surfaces Are Lined by One or More Layers of Epithelial Cells

Each of our barrier organs is separated from the external environment by at least one layer of epithelial cells, which not only provide a physical barrier but are active participants in our response to the microbiome. The epidermis of our skin, mouth, and reproductive and urinary tracts is made up of multiple layers of epithelial cells, while our intestinal and respiratory tracts are each separated from the outside world by a single layer of epithelial cells.

Connected to each other by tight junctions, epithelial cells are diverse in phenotype and function and have multiple strategies to prevent external microbes from contacting and invading barrier tissues (**Advances Box 13-1** and **Table 13-1** describe the various cell types of barrier immune systems and their functions; see also Chapter 4). Some

---

ADVANCES   **BOX 13-1**

Cells Involved in Barrier Immunity

Here, we briefly describe the main innate and adaptive immune cells that monitor our barrier organs (**Figure 1**). **Figure 2** illustrates some fundamental interactions among these cells that generate type 1 and type 2 responses (see Chapter 10) to pathogens. This schematic is fleshed out in the main text, where we describe how these cells cooperate to maintain tolerance to or mount a protective immune response against microorganisms at specific barrier tissues. The network of connections among these cells and their association with our microbiome is remarkably and appropriately complex; our understanding is evolving rapidly.

Innate Immune Cells

Epithelial cells

Epithelial cells are considered part of the innate immune system, although they are not derived from hematopoietic stem cells. They vary in phenotype and function and play active roles in defense. Epithelial cells in the gut (see Figure 1, *right*) include absorptive cells known as *enterocytes*, the

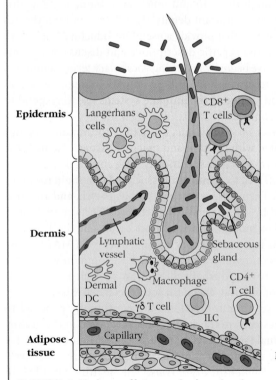

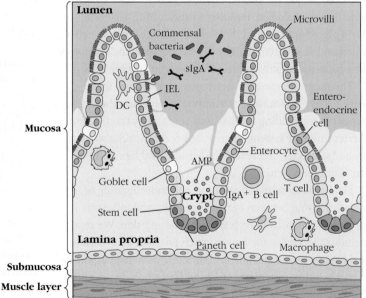

FIGURE 1 Major cell types in barrier immune systems. Figure 1 compares the microanatomy of two barrier tissues: the skin (*left*) and the intestine (*right*). Both are separated from the external environment by one or more epithelial layers. The skin epidermis consists of multiple layers of epithelial cells called *keratinocytes*, which interact with Langerhans cells (a specialized subset of DCs) and multiple lymphocytes, including tissue-resident CD8⁺ memory T cells. The intestine has only a single layer of epithelial cells, which are diverse in phenotype and function (see text). These cells also interact with hematopoietic cells including macrophages, dendritic cells, and lymphocytes. The dermis of the skin and the lamina propria of the intestine lie just below the epithelial surface and contain most of the immune cells involved in maintaining tolerance and/or mounting an inflammatory response. These include resident and migratory APCs, T cells, B cells, and ILCs. During an inflammatory immune response, other types, including granulocytes, are recruited to the barrier tissue.

(continued)

BOX 13-1

main cell type responsible for absorbing digested food. They also include a diverse population of secretory cells: *enteroendocrine cells*, which secrete a variety of hormones that regulate digestion; *goblet cells* that secrete mucus; *Paneth cells*, which secrete antimicrobial peptides and other cytokines; and rare chemosensory *tuft cells* that secrete cytokines, particularly IL-25. The epithelial *cells* in the gut also include specialized *microfold (M) cells* that carry antigen via transcytosis from the lumen to the lamina propria, where they interact directly with lymphoid tissue, including Peyer's patches. Finally, gut epithelial cells that populate the crypts include stem cells that continually divide and replace all epithelial subtypes when needed.

Although intestinal and respiratory mucosal tissues are lined by single layers of epithelium, other barrier tissues are protected by multiple layers. The skin includes several layers of epithelial cells or *keratinocytes*, which make up the epidermis (see Figure 1, *left*). Skin stem cells are found at the bottom layer and continually replace epithelial cell layers that mature as they move toward the surface. Several layers of dead keratinocytes protect our skin surface and contribute to its water-proof nature. Several other mucosal tissues, including the mouth, reproductive tracts, and urinary tracts, are also covered by several layers of epithelium. However, these surfaces are not "keratinized" by dead cell layers; rather, they generate mucus and are constantly moist.

Hematopoietic innate immune cells

Antigen-presenting cells associate closely with epithelial cells in barrier tissues. These include multiple subsets of *dendritic cells* (*DCs*) and *macrophages,* which express pattern recognition receptors (PRRs) that interact with microorganisms at barrier surfaces. They also act as professional antigen-presenting cells and, in response to PRR engagement, secrete cytokines that polarize helper T-cell subsets and active ILCs.

Dendritic cells and macrophages are unexpectedly diverse in phenotype and function, and may be interrelated. Efforts to standardize nomenclature and classification are ongoing; we will touch briefly on categories that seem to be currently useful in describing these APCs.

Some DCs and macrophages appear to be permanent residents in barrier tissue and extend processes between epithelial cells to sample external microbes (see Figure 1, *right*). Some, however, are migratory and convey antigen to local lymph nodes to initiate an immune response. Most resident DCs and macrophages help to maintain tolerance to the microbiome. Migratory APCs can both reinforce tolerizing conditions or initiate inflammatory responses.

Macrophages can be distinguished from DCs by their expression of CD64. In the intestine, macrophages that express the chemokine receptor CX_3CR1 are resident and extend processes (transepithelial dendrites) between epithelial cells to sample antigen in the lumen, particularly in the small intestine. These antigen-presenting cells have been implicated in

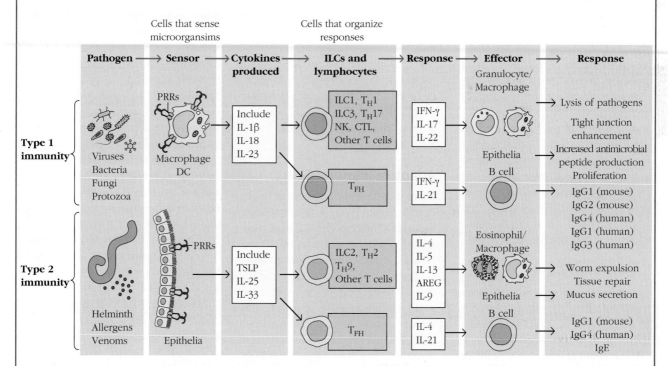

FIGURE 2 **Schematic showing how some of the major barrier tissue immune cells interact to produce type 1 and type 2 responses.** Cells that sense pathogen via PRRs include epithelial cells, themselves, and APCs that extend processes across the epithelial barriers. APCs migrate to local lymphoid tissue and both APCs and epithelial cells produce cytokines and other molecules that induce activity of ILCs and helper T cells, which help coordinate an effector response (refer to the main text for more details). B cells, cytotoxic lymphocytes, NK cells, and granulocytes participate in the effector response to clear the pathogen.

(continued)

the development of regulatory, tolerizing CD8$^+$ T cells.

DC populations vary by expression of CD103, CD11b, and CD8α. In the intestine, dendritic cells that express the integrin CD103 (or CD141 in humans) tend to be more migratory. CD103 expression allows these dendritic cells to interact with integrins on the basolateral epithelial cell surface and positions them to receive antigen that is transcytosed. Once activated, however, these CD103$^+$ DCs travel to local lymph nodes to promote the differentiation of naïve T cells into one of a number of different helper T-cell subsets. Some produce cytokines (e.g., IL-10) that polarize naïve CD4$^+$ T cells to the regulatory FoxP3$^+$ T$_{REG}$ lineage and are critical in maintaining tolerance in the barrier organs. Others produce cytokines that polarize naïve CD4$^+$ cells to the distinct helper cell lineages (T$_H$17, T$_H$1, T$_H$2, and T$_H$9), which contribute to type 1 and type 2 responses to pathogens. DCs coming from barrier tissues also have the capacity to induce T cells to express specific homing receptors that help them find their way back to the site of infection.

Plasmacytoid dendritic cells (pDCs) can also be found in barrier tissues. Best known for their ability to produce the antiviral type I IFNs, they also influence B-cell differentiation and produce cytokines (e.g., BAFF and APRIL) that enhance IgA production and support B-cell survival and proliferation. They also stimulate T cells to produce the anti-inflammatory cytokine IL-10 and promote tolerance at barrier tissues.

Although variations of these populations are found at all barrier tissues, *Langerhans cells* are a specialized DC subset found only in the skin (see Figure 1, *left*).

Innate lymphoid cells

Our appreciation for the role that innate lymphoid cells (ILCs) play in all aspects of immunity has increased exponentially in the last few years—and their role in the barrier immune system is especially important. The nomenclature describing the different subsets finally seems to be settling. As discussed in Chapter 4, we now recognize three main groups: ILC1, ILC2, and ILC3 (see Table 4-6). These subtypes

are distinguished largely by the cytokines they secrete, which mirror those secreted by T$_H$1, T$_H$2, and T$_H$17 T-cell subsets, respectively. The ILC1 group also includes NK cells, which (like other ILC1 and T$_H$1 cells) produce IFN-γ, but are also cytotoxic.

ILCs are particularly abundant in healthy barrier tissues, where they participate in all aspects of barrier immunity (see Figure 2). These hematopoietic cells do not express antigen-specific receptors, but respond to cytokines produced by epithelial cells, dendritic cells, and macrophages. Most ILC subsets mirror helper T-cell subsets in the combinations of cytokines they secrete. Like T$_H$1 cells, ILC1 cells generate type 1 cytokines, including IFN-γ, and contribute to inflammatory immune responses to intracellular pathogens, including viruses and bacteria. They can also play a role in chronic inflammatory diseases. Like T$_H$2 cells, ILC2 cells coordinate a type 2 response designed to expel worms and repair the damage they do in the gut. Like T$_H$17 cells, ILC3 cells play dual roles. They work together with dendritic cells to maintain tolerance to the commensal microbiome (see main text), but also are an important contributor to the inflammatory response to pathogens. Our awareness of the importance of ILCs in initiating and customizing the barrier immune response continues to grow, and new roles are likely to be identified.

Adaptive Immune Cells

Conventional B and T lymphocytes reside in and migrate through barrier tissues. Regulatory T lymphocytes (T$_{REGS}$) are overrepresented in the mucosa and skin. They help quell immune responses to antigens and microbes that are not posing a threat. T helper type 17 (T$_H$17) T cells are also overrepresented in some mucosal immune systems. Although they are often associated with aggressive responses to invading pathogen, they also contribute to barrier homeostasis. T$_{FH}$ cells are found in follicles associated with barrier tissue, particularly in the intestine. These cells help induce and maintain IgA production by B cells. Other CD4$^+$ T helper cells, including

T$_H$1, T$_H$2, and T$_H$9 cells, are recruited to barrier tissue as part of coordinated responses to pathogens (see Figure 2).

A large pool of predominantly CD8$^+$ T cells are found intercalated between epithelial cells of barrier tissues in all organisms. These are collectively referred to as *intraepithelial lymphocytes* (*IELs*) and include a mixture of CD8$^+$ T$_{RM}$ cells, other conventional T cells, and invariant T cells. These cells often express the CD8$\alpha\alpha$ rather than the CD8$\alpha\beta$ dimer. Although IELs are thought to play important roles in maintaining the health of the epithelial barrier, their specific functions continue to be debated and probably vary by cell type. Tissue-resident memory T cells (T$_{RM}$ cells), particularly CD8$^+$ T$_{RM}$ cells, are found in relatively high numbers in skin epithelial layers and are poised to respond rapidly to invading pathogens. CD4$^+$ T$_{RM}$ cells tend to be found below the epithelial layers and circulate more readily.

Specialized T lymphocytes that express antigen-specific receptors of limited diversity (invariant lymphocytes), including invariant $\gamma\delta$ T cells, CD8$^+$ **mucosal-associated invariant T cells (MAIT cells)**, and NK1.1$^+$ T cells, are also found in relatively high frequencies in barrier tissue. These unconventional lymphocytes join resident innate lymphoid cells in efforts to shape the response to common microbes and distinguish between helpful commensals and harmful pathogens.

Finally, IgA-secreting B cells are critically important residents in barrier tissues and help maintain the healthy separation between commensal microbes and epithelial cells. Secreted IgA is an antibody class that limits inflammatory responses. It is typically generated as a dimer and has the unique ability to be transported across epithelial cell barriers. Some IgA antibodies have a broad specificity for common microbes, while others are exquisitely antigen specific.

REFERENCE

Iwasaki, A., and R. Medzhitov. 2015. Control of adaptive immunity by the innate immune system. *Nature Immunology* **16**:343.

TABLE 13-1	Soluble factors that regulate the barrier immune response: their source and function		
Secreted molecule(s)	**Cells producing them**	**Effect**	**Tolerogenic (homeostatic) or inflammatory responses?**
IL-10, RA, TGF-β	Multiple cell types in homeostatic environment, including resident APCs, epithelial cells, T cells, and B cells	Enhance T_{REG} and IgA B-cell generation, inhibits inflammatory responses	Tolerogenic
APRIL, BAFF	Epithelial cells, resident APCs	Enhance IgA production	Tolerogenic
TSLP	Epithelial cells	Stimulates APCs to produce BAFF, APRIL Enhances IgA production	Tolerogenic
AMPs, including REG3	Epithelial cells	Kill bacteria	Both
IL-23	APCs	Induces T_H17 cells and ILC3s to make IL-17, IL-22	Both
IL-17, IL-22	T_H17 and ILC3s (and other T-cell subsets)	Induce epithelial cells to make AMPs and chemokines that attract granulocytes	Both
IL-1β, IL-18	APCs (activated by caspase-1 following pathogen binding to NLRs)	Induce epithelial cells to generate chemokines; enhance cytotoxic responses	Type 1 inflammatory responses
IFN-γ	T_H1 cells and ILC1s	Many effects on many immune cells (APCs, T cells, ILCs) that help coordinate a cytotoxic response	Type 1 inflammatory responses
TSLP, IL-33, IL-25	Epithelial cells and tuft cells (sole source of IL-25)	Alarmins that respond to worms and protozoal parasites	Initiation of type 2 inflammatory responses
IL-4, IL-5, IL-13	T_H2 cells and ILC2s as well as some granulocytes	Enhance IgE production and eosinophil activity to attack worms and protozoal parasites	Type 2 inflammatory responses

produce a protective mucus layer, some secrete **antimicrobial peptides** (**AMPs**) that kill or inactivate bacteria, and some have beating cilia that sweep pathogens away. In general, epithelial cells play a far more active role in communicating information about the external environment to the immune system than was previously appreciated, and are considered to be very important innate immune cell participants. Indeed, many epithelial cell types express innate immune receptors and interact directly with microbes, communicating the outcome of these interactions to innate and adaptive immune cells in tissue layers below. When damaged, epithelial cells send specific signals to these immune cells that initiate our inflammatory immune response to invasion. In turn, immune cells enhance the health of epithelial barriers, and assist in their repair. We will go into further detail about this fascinating cross-talk between epithelial and hematopoietic cells when we focus on specific barrier tissues in the upcoming sections.

Key Concepts:

- All barrier organs are separated from the external environment by one or more layers of epithelial cells, which are considered part of the innate immune system.

- Epithelial cells are diverse in phenotype and function and play active roles in barrier immunity.

Barrier Organs Are Populated by Innate and Adaptive Immune Cells That Interact with Epithelium and Secondary Lymphoid Tissue

Cells of both the innate and adaptive immune systems reside in and migrate through barrier tissues. These include epithelial cells, dendritic cells (DCs), macrophages, innate lymphoid cells, invariant and conventional T lymphocytes, and B cells (see Advances Box 13-1).

Like all organs and tissues, barrier tissues are served by local lymph nodes (**Figure 13-2**; see also Chapter 2). Once activated, migratory dendritic cells from barrier organs travel via lymphatics to draining lymph nodes, where they establish systemic responses. **Mesenteric lymph nodes** specifically drain the intestine. Antigen-presenting cells (APCs) that come from barrier tissues often induce lymphocytes to express adhesion molecules (**addressins**; see Chapter 14) and chemokine receptors that direct them to the barrier tissue at the initial site of infection. However, some activated lymphocytes travel throughout the body and take up residence at other barrier tissues.

Mucosal tissue is also associated with unique secondary lymphoid tissues that offer T and B lymphocytes even more direct access to microbial antigens. Specific names are occasionally applied to some of the specialized lymphoid tissues that serve each barrier site (see Figure 13-2). For example,

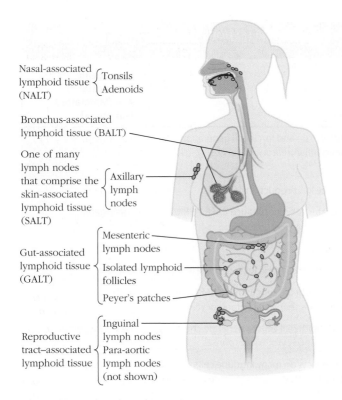

Nasal-associated lymphoid tissue (NALT) { Tonsils, Adenoids

Bronchus-associated lymphoid tissue (BALT)

One of many lymph nodes that comprise the skin-associated lymphoid tissue (SALT) { Axillary lymph nodes

Gut-associated lymphoid tissue (GALT) { Mesenteric lymph nodes, Isolated lymphoid follicles, Peyer's patches

Reproductive tract–associated lymphoid tissue { Inguinal lymph nodes, Para-aortic lymph nodes (not shown)

FIGURE 13-2 Lymphoid tissues associated with barrier organs. Shown are examples of lymph nodes and other lymphoid tissues associated with barrier organs, including the gut, lung, skin, and reproductive tract. See text for details. Note that the bronchus-associated lymph node tissue is most developed in nonprimate animals, including cat and rabbit.

the lung is served by bronchus-associated lymphoid tissue (BALT), which is most prominent in nonprimates. The upper airways are served by nasal-associated lymphoid tissue (NALT), which includes the tonsils and adenoids, and the skin is served by less organized skin-associated lymphoid tissue (SALT).

Gut-associated lymphoid tissue, or GALT, is best understood and provides a good example of the types of lymphoid tissue that can serve barrier organs (**Figure 13-3**). These range from well-organized **Peyer's patches** of the small intestine, which share many anatomical and functional features with traditional lymph nodes, to the more loosely organized **isolated lymphoid follicles** (**ILFs**), which are associated with epithelial surfaces along the entire intestinal tract. These unique entities constitute single B-cell follicles that develop from cell clusters known as **cryptopatches**. The locale and size of ILFs change relatively quickly in response to changes in the intestinal microbiome and cytokine milieu. In the absence of microbes, in fact, ILFs do not develop fully. Most importantly, they are a primary source of IgA-secreting B cells. Both Peyer's patches and ILFs are positioned strategically just beneath a layer of specialized epithelial cells that deliver antigen from the lumen to underlying antigen-presenting cells that initiate immune responses (described shortly).

Key Concepts:

• Healthy barrier tissues are populated by a diverse array of innate and adaptive immune cells that sample antigen and, together with epithelial cells, regulate the response to the microbiome.

• Barrier tissues are associated with draining lymph nodes, and mucosal tissues are associated with specialized secondary lymphoid tissue, including Peyer's patches (small intestine), isolated lymphoid follicles, and cryptopatches.

Barrier Immune Systems Initiate Both Tolerogenic and Inflammatory Responses to Microorganisms

The epithelial and immune cells in barrier tissues are constantly sampling and assessing the external environment and the microbiome. When a barrier is healthy and in homeostatic balance, the immune system is in a *tolerogenic* mode (**Overview Figure 13-4**, left). Cell-cell interactions and cell-microbe interactions trigger production of anti-inflammatory molecules and cytokines, including TGF-β and IL-10, which enhance the production of regulatory T cells and IgA-secreting B cells. IgA interacts with commensal microbes, preventing them from penetrating the epithelial barriers and initiating an inflammatory immune response. Epithelial cells help maintain this barrier by secreting antimicrobial peptides (AMPs), which kill bacteria that get too close.

When the epithelial barrier is damaged and microorganisms find a way into the subepithelial space, innate and adaptive immune cells change their tune and generate type 1 or type 2 immune responses (introduced in Chapter 10), tailored to the type of invading pathogen (Overview Figure 13-4, right; see also Advances Box 13-1, Figure 2). Intracellular pathogens and some bacteria stimulate APCs to release IL-1, IL-18, and IL-23, which, in turn, activate T_H1 and T_H17 cells, as well as their innate lymphoid cell type 1 (ILC1) and ILC3 counterparts. Worms and other extracellular antigens induce type 2 responses and the release of cytokines that activate T_H2, T_H9, and their ILC2 counterparts (see Table 13-1). In these cases, B cells that produce IgG, which are more likely to promote inflammatory cell activity, and proinflammatory cells, including neutrophils and specific macrophage subsets, are called to the site to help clear infection. In all cases, hematopoietic immune cells also produce molecules that help repair and strengthen epithelial cell barriers.

Because each barrier is confronted with distinct environmental challenges and microbes, each also has specific ways of managing its relationship to commensal and pathogenic bacteria. We will focus most of this chapter on describing specific strategies and responses of the intestinal immune system because this is where our understanding of the dialog between the microbiome and our immune system is most developed. We will then summarize current knowledge of strategies taken by other barrier organs, including the respiratory tract and skin.

(a)

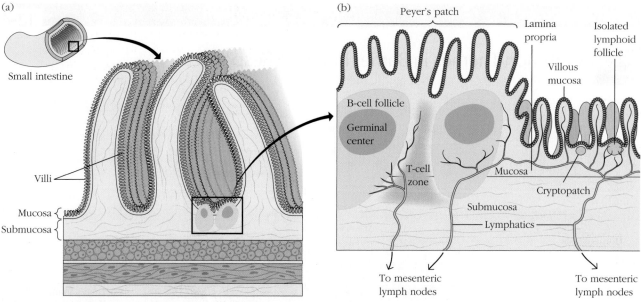

(c)

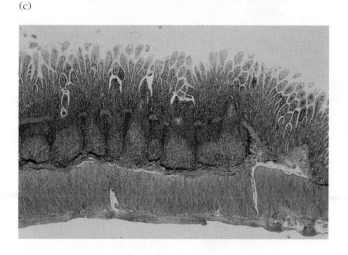

FIGURE 13-3 Secondary lymphoid tissue associated with the small intestine. The small intestinal surface is deeply folded, as shown in (a). Several types of secondary lymphoid tissue are associated with the mucosa, the intestinal layer that includes the epithelium and the tissue just beneath, the lamina propria (see text). (b) Loosely organized B-cell follicles associate closely with the epithelial layer, including small cryptopatches, which give rise to isolated lymphoid follicles (ILFs). The most organized and stable lymphoid tissues are called *Peyer's patches* and also lie just beneath the epithelium along some parts of the small intestine. These embedded, tiny lymph nodes are also shown in (c), an H&E (hematoxylin and eosin)–stained section of the ileum. Lymphatic vessels convey cells to draining, peripheral lymph nodes, including mesenteric lymph nodes. *[(c) Biophoto Associates/Science Source]*

Key Concepts:

- Under homeostatic conditions, innate and adaptive immune cells in barrier tissues generate a tolerizing response dominated by TGF-β and IL-10 cytokines, regulatory T cells, and IgA-producing B cells.

- When damaged or alerted to an invading pathogen, innate and adaptive immune cells cooperate to generate both type 1 and type 2 immune responses.

- Precisely what criteria barrier immune systems use to determine whether to mount tolerogenic or proinflammatory responses is still an area of active investigation.

Intestinal Immunity

The immune response of the intestine and most barrier organs was largely ignored in the early days of immunology. In part, this was because the field was justifiably preoccupied with gaining a basic understanding of immune cells and organs that were experimentally accessible: blood, lymph nodes, spleen, and bone marrow. In part it is due to the difficulty of identifying and isolating immune cells from mucosal tissues, once a daunting task only a few stalwart investigators tackled before the 1980s. However, it was also due to our collective obliviousness to what is now obvious—that those areas that are most exposed to microbes must have a well-developed immune system. The Society for Mucosal Immunology was finally formed in 1985, out of a recognition that the subfield deserved unique attention and its members needed a community to share and advance ideas.

As fetuses, we are sterile organisms. However, from the moment we are born, we begin establishing our microbiome. Maternal microbes acquired during birth and from breast-feeding are important pioneers, but other microbes from contact with skin, milk, and ultimately solid food rapidly join in the colonization effort. For instance, although Peyer's patches begin to form at the end of gestation, they don't fully develop until animals switch from milk to a mixed food diet, indicating

OVERVIEW FIGURE   **13-4**

Common Themes in Barrier Immune Responses

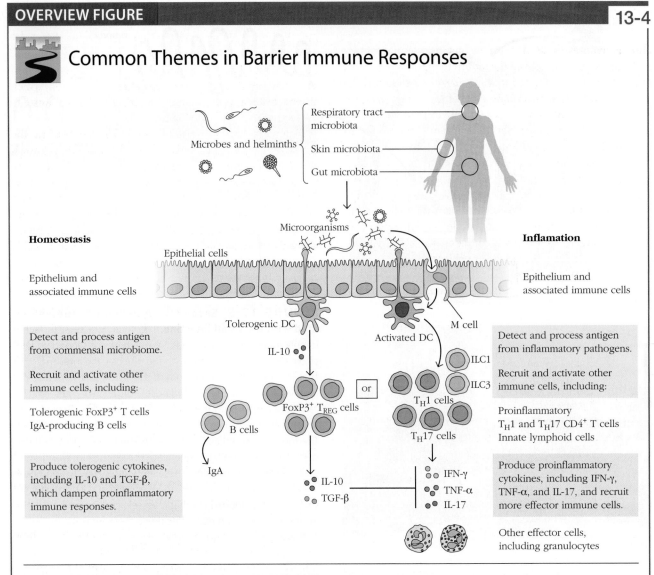

Homeostasis

Inflamation

Epithelium and associated immune cells

Epithelium and associated immune cells

Detect and process antigen from commensal microbiome.

Recruit and activate other immune cells, including:

Tolerogenic FoxP3⁺ T cells
IgA-producing B cells

Produce tolerogenic cytokines, including IL-10 and TGF-β, which dampen proinflammatory immune responses.

Detect and process antigen from inflammatory pathogens.

Recruit and activate other immune cells, including:

Proinflammatory T_H1 and T_H17 CD4⁺ T cells
Innate lymphoid cells

Produce proinflammatory cytokines, including IFN-γ, TNF-α, and IL-17, and recruit more effector immune cells.

Other effector cells, including granulocytes

This figure shows features of the two major responses of barrier immune systems to the microbiome: tolerogenic responses to commensal bacteria (homeostasis) and inflammatory responses to pathogenic microorganisms. Tolerogenic responses are induced by interactions between microbes and antigen-presenting cells that generate cytokines (including IL-10) that enhance the generation and activity of regulatory T cells and IgA-producing B cells. Inflammatory responses are initiated by invasion of microorganisms into the subepithelial spaces that, depending on the type of microorganism, induces a type 1 or type 2 response. The figure illustrates the main features of a typical type 1 inflammatory response to intracellular pathogens, which activates specific T-cell and ILC subsets, including T_H1, T_H17, ILC1 cells, and ILC3 cells. These produce cytokines that recruit other effector cells (granulocytes, killer T cells, NK cells, and more) to clear infection. A type 2 response to worms (not shown) activates T_H2, T_H9, and ILC2 cells that release cytokines that recruit eosinophils and other granulocytes.

---

that exposure to antigens—from food and from microbes—helps drive the maturation of the mucosal immune system.

Although some of these microbes cause disease, many provide benefits we cannot get elsewhere. Benign colonizers provide us with competitive protection from virulent microbes. Certain bacterial species provide us with key nutrients, like vitamins B_{12} and K. Others digest food that our own cells cannot break down. Some generate metabolites that directly enhance the health of our epithelium. And most importantly to this chapter, our microbial communities tune our own immune system, helping it to mature and establish an all-important balance between tolerance and responsiveness. And perhaps even more intriguingly, the relationship cultivated between our microbiome and our barrier immune systems influences tissues well beyond our barrier organs, and may help regulate autoimmune and inflammatory diseases, as well as neurological health and disease (**Clinical Focus Box 13-2**).

The Gut Is Organized into Different Anatomical Sections and Tissue Layers

Our **gastrointestinal (GI) tract**, or "gut," is essentially a tube that runs continuously from the mouth to the anus; the inside of this long tube is referred to as the *lumen*. Well known for its role in digestion and absorption of food and water, the GI tract is now also known for its role in maintaining the well-being of our commensal microbiome and regulating both local and systemic immune responses (**Figure 13-5**).

The GI tract is teeming with billions of organisms, including viruses, bacteria, and in most parts of the world, parasitic worms. The composition of the total intestinal microbiome is determined by a complex interplay of developmental, environmental, genetic, and immunological influences. An understanding of its gross and cellular anatomy will help our discussion of its immune system considerably.

The small intestine, which receives partially digested food from the stomach, consists of three segments—the *duodenum*, the *jejunum*, and the *ileum*. Digestive enzymes from the pancreas and bile from the gallbladder are excreted into the lumen of the short, C-shaped duodenum. The jejunum and ileum are long and looping—and the major sites of digestion and absorption of food. What we cannot digest in the small intestine enters the large intestine, or colon, which absorbs water as well as some vitamins, and delivers waste to the rectum and anus (see Figure 13-5).

The surfaces of the jejunum and, to a lesser extent, the ileum, are deeply folded and covered with projections known as *villi* (see Figure 13-3). The folds and villi greatly increase the surface area available for absorption and digestion, as well as the opportunities for our own cells to interact with the microbiome. The large intestine is shorter in length, has fewer folds, much shorter villi, and produces more mucus, which provides an additional protective layer from possible pathogens and lubricates the exit of waste (**Figure 13-6**).

The walls of the intestinal tract consist of two major layers: the *mucosa*, the layer in contact with the inside of the intestinal tube, which is known as the lumen, and the *submucosa*, the layer between the mucosa and the outer, muscular wall of the intestine (see Figures 13-3 and 13-6). The mucosa itself includes two layers: a single layer of diverse epithelial cells, and the **lamina propria**, a layer of connective tissue populated by resident and migrating immune cells as well as a network of capillaries and lymph vessels. Most immune cells of interest in the intestine are found in the lamina propria and epithelial layers, although some immune tissue also extends into the submucosa. A large number of plasma cells, most of which produce IgA against intestinal antigens, are present in the lamina propria. Most of these IgA-producing B cells are generated in isolated lymphoid follicles (ILFs) (**Figure 13-7a**).

The folds that form the intestinal villi also form the intestinal *crypts*, the valleys between the villi "mountains." These crypts contain distinct epithelial cell subsets, including the epithelial stem cells that continually replace the short-lived epithelial cell population (see Figure 13-6).

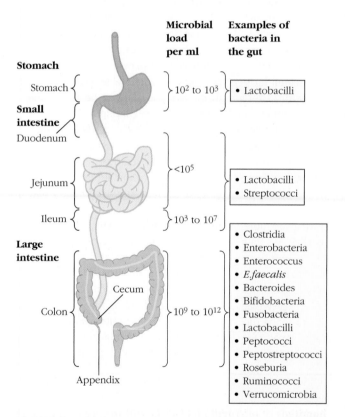

	Microbial load per ml	Examples of bacteria in the gut
Stomach		
Stomach	10^2 to 10^3	• Lactobacilli
Small intestine		
Duodenum		
Jejunum	$<10^5$	• Lactobacilli • Streptococci
Ileum	10^3 to 10^7	
Large intestine		• Clostridia • Enterobacteria • Enterococcus • *E.faecalis* • Bacteroides • Bifidobacteria • Fusobacteria • Lactobacilli • Peptococci • Peptostreptococci • Roseburia • Ruminococci • Verrucomicrobia
Colon	10^9 to 10^{12}	
Cecum		
Appendix		

FIGURE 13-5 Gross anatomy of the gastrointestinal (GI) tract. Shown are the major parts of our GI tract, which is a single tube starting at the esophagus (not shown), extending through the stomach, the small intestine, and the large intestine. The small intestine is divided into three main sections: the short duodenum, which receives digestive enzymes from the pancreas, and the longer jejunum and ileum. The ileum connects to the large intestine, also known as the colon. Also shown are examples of commensal bacteria that inhabit different parts of the GI tract, as well as their quantities, which vary between sections and are most abundant in the large intestine.

Key Concepts:

- The gastrointestinal tract contains trillions of microorganisms that influence our health in many different ways.

- The small intestine is the site of most digestion and absorption and has three sections (duodenum, jejunum, and ileum). The large intestine, or colon, is responsible for absorbing water and expelling waste and includes the highest number and diversity of bacteria.

- The small intestine is lined by microscopic folds called *villi* and *crypts*.

- The lamina propria is the tissue layer just under the gut epithelial layer and is the site of most immune cell activity.

BOX 13-2

The Gut-Brain Axis

Psychobiotics—this is the new term for microbes that some claim can influence our mood and behavior. The *Lactobacillus* in yogurt may be one of these and, although skepticism about the sweeping benefits of psychobiotic therapy is warranted, there is clear experimental evidence for a biological and potentially powerful dialog between the brain, the gut, and the microbes. This relationship is referred to as the *microbiota-gut-brain axis*, and as you can imagine, is a topic of intense study.

Our microbiome and our neurons have multiple opportunities to communicate (**Figure 1**). First, our gut is directly connected to our central nervous system via a network of neurons known as the *enteric nervous system*. The *vagus nerve* is the longest autonomic nerve in the body and one that includes both motor and sensory neurons. It conveys signals about energy status, pain, and pressure to the brain. The brain, in turn, sends signals back to the intestine via the vagus and other neurons, which directly and indirectly influence muscle cells, epithelial cells, and the microbiome itself. Perhaps even more interestingly, sensory neurons also express pattern recognition receptors, such as TLRs, and sense infection. LPS, a common proinflammatory molecule produced by gram-negative bacteria that activates TLR4, has direct and indirect effects on neuron activity. When epithelial barriers are compromised during infection, LPS can find its way to

the bloodstream and the brain. Increased levels of blood LPS have been associated with depression.

Second, microbes in our gut produce metabolites that have both direct and indirect influences on neurons. These include tryptophan and tyrosine, which are precursors of neurotransmitters. They also include neurotransmitters themselves, such as acetylcholine and dopamine. In fact, *Lactobacillus* produces GABA, a potent inhibitory neurotransmitter and one of the reasons *Lactobacillus* has been referred to as a psychobiotic.

Third, some secretory gut epithelial cells generate hormones and "neuroactive" peptides that directly influence brain activity and regulate sensations of fullness and hunger. Short-chain fatty acids (SCFAs) made by microbes fermenting dietary fiber induce mucosal epithelial cells to produce peptides like 5-HT (serotonin) that can have a direct impact on mood. These observations have prompted claims that eating fiber may improve memory and mental health. Experiments, however, have generated conflicting results—and we await more clarity about the effects of SCFAs and dietary fiber on mood.

Fourth, and most relevant to this chapter, the immune system plays a role in the dialog between gut and brain. Given that the immune system helps to shape the composition of our microbes, it naturally follows that it would have an indirect

impact on the microbe-gut-brain axis. However, the mucosal immune system can play an even more direct role. Cytokines released by innate immune cells in the gut, particularly monocytes, influence the development of the enteric nervous system as well as the mature nervous system. Neurons have receptors for inflammatory cytokines, including the classic IL-1, IL-6, and TNF-α trio. Inflammation is associated with depression, and some studies even suggest a causal relationship. Reciprocally, stress hormones like cortisol, which are produced in response to sympathetic nervous system activity activated, in part, by stimuli received from our senses and processed in our limbic system, have a dampening effect on immune cells. Neuropeptides released by neurons, themselves, also influence innate immune cells (e.g., LTi cells, a population of ILC3 cells) that are essential for the generation of secondary lymphoid tissue.

Is there any relationship between barrier immunity and autoimmune diseases of the nervous system? There may be. Autoimmune diseases, including multiple sclerosis (in which the myelin sheath of neurons is attacked), may be triggered by infections in peripheral tissues, including mucosal surfaces. Autoreactive T cells may be activated by antigens from pathogens that mimic self antigen, including epitopes on myelin basic protein. Inflammatory subsets of T cells, including CD4$^+$ T$_H$1 and T$_H$17 cells,

(continued)

Gut Epithelial Cells Vary in Phenotype and Function

The epithelial cells in the intestinal mucosa are more diverse in function and form than originally appreciated. Not only are they part of the digestive system, but they are now recognized as a key part of our innate immune system. The various gut epithelial cell types described here are shown in Figure 13-6.

Most epithelial cells are polarized—in other words, they have distinct tops and bottoms. A cell's *apical surface* is in contact with the lumen of the gut and is invaginated with

hundreds of *microvilli*. Its *basolateral surface* is in contact with the lamina propria. Apical and basolateral membranes include very distinct surface proteins, segregated from each other by the tight junctions that connect the epithelial cells to each other. These distinctions help the epithelial cell to convey information and molecules in precise directions. For instance, polymeric immunoglobulin receptors (polyIgRs) on the basolateral surface of epithelial cells capture and internalize IgA antibodies made in the lamina propria (Figure 13-7b). These antibodies are then conveyed via vesicles to the apical surface and released into the intestinal lumen, where they

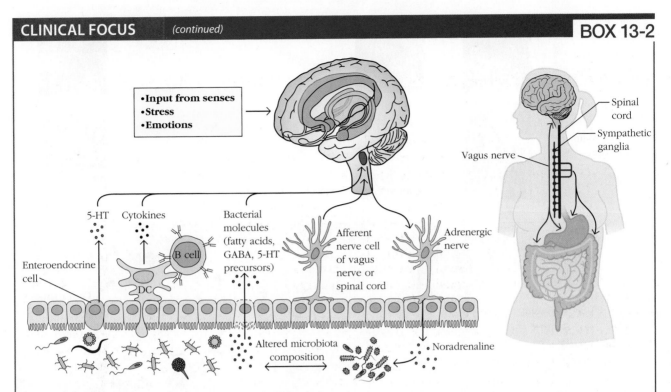

FIGURE 1 Ways in which intestinal cells and commensal microbes influence our brain. First, microorganisms, gut epithelial cells, and other intestinal immune cells produce molecules including cytokines, fatty acids, and neurotransmitters (5-HT and GABA) that travel to the brain via the bloodstream. These have the potential to influence not just appetite, but also mood. Second, the gut is directly connected to the brain by the vagus nerve and indirectly by the enteric nervous system, both of which receive and send signals from the gut to the brain. Finally, the brain, itself, receives input from sensory organs that are processed and communicated to the gut via the autonomic (parasympathetic and sympathetic) systems. Stress, for instance, sets off a cascade of signals that result in the release of hormones (e.g., noradrenaline) that can have direct effects on intestinal cells as well as microorganisms.

as well as CD8$^+$ MAIT (mucosal-associated invariant T) cells, accumulate in the barrier organs of individuals with multiple sclerosis. Intriguingly, tolerogenic responses to gut microbiota may also ameliorate experimental autoimmune encephalitis, a mouse model for multiple sclerosis. However, there is as yet no evidence for such an effect in humans.

What about an association between mucosal immunity and complex psychological disorders like schizophrenia, bipolar disease, and autism? These disorders are often accompanied by disorders of the gut, and it is intriguing to speculate about a potential relationship among microbiome communities, mucosal immunity, and these diseases. Alterations in the gut microbiome have influenced the severity and presentation of autistic disorders in mice. However, correlation is not causation, and mice are not humans. While there seems to be a relationship between what we eat, what microbes colonize our gut, and our mental health, there is not, as yet, definitive evidence that changing our microbiome will cure or change our susceptibility to mental health disorders.

REFERENCES

Collins, S. M., M. Surette, and P. Bercik. 2012. The interplay between the intestinal microbiota and the brain. *Nature Reviews Microbiology* **10:**735.

Smith, P. A. 2015. The tantalizing links between gut microbes and the brain. *Nature* **526:**312.

are largely resistant to breakdown by digestive enzymes and help keep bacteria at a healthy distance from the epithelial membranes. This process of shuttling molecules through epithelial cells is known as **transcytosis**. As we will see, transcytosis of antibodies and antigens across the intestinal epithelium is mediated by several epithelial cell types, and is critical to healthy immune function.

The most common cell in the epithelium of the small intestine, and the one traditionally associated with its digestive function, is the **enterocyte**. These cells are responsible for transporting nutrients from digested food from the apical layer to the capillaries and lymph vessels present in each villus. We now know that they also have the capacity to respond to microbes via pattern recognition receptors (PRRs) and send signals via cytokines to immune cells in the lamina propria. For example, although enterocytes do not seem to express a lot of surface Toll-like receptor 4 (TLR4), a receptor that binds to products of gram-negative bacteria

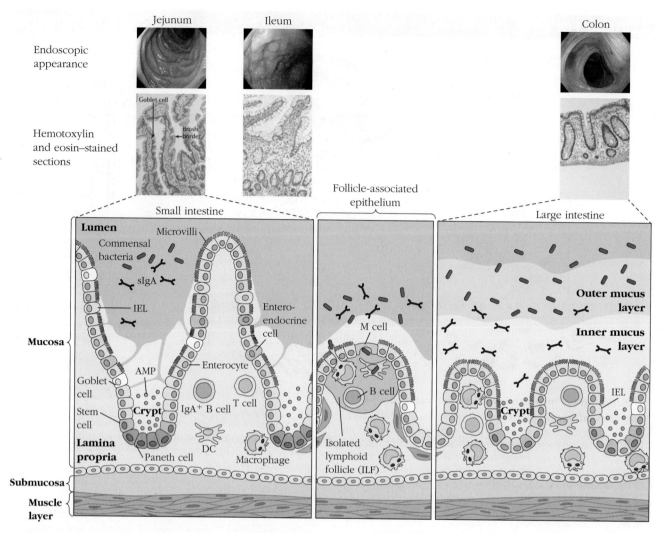

FIGURE 13-6 Cellular anatomy of the small and large intestines. Illustrated here are some of the differences between the small and large intestines, as well as the cells and molecules that populate the immune system of these barrier tissues. See text for details. Briefly, the small intestine has longer villi and deeper crypts. The large intestinal epithelial cells produce more mucus. Epithelial layers in both tissues include a variety of cell types and associate with a similar variety of innate and adaptive immune cells, most of which inhabit the lamina propria and some of which extend processes between the epithelial cells to sample the microorganisms at the surface (see text). Both sections include lymphoid follicles, which are often lined with M cells that deliver antigen to APCs via transcytosis. Peyer's patches are present only in the small intestine. *[Photos reprinted by permission from Macmillan Publishers Ltd., from Allan M. Mowat, William W. Agace, "Regional specialization within the intestinal immune system," Nature Reviews Immunology October 2014;* **14(10):** *667–685, Figure 1. Permission conveyed through Copyright Clearance Center, Inc.]*

(see Chapter 4), they do express TLR4 internally. This allows them to ignore some of the commensal bacteria in the lumen, but to alert the immune system if the bacteria have invaded and penetrated their membranes.

Goblet cells are distributed throughout the intestine, but found at highest concentration in the large intestine. Classically characterized by their ability to produce mucus, they also have other important immune functions. They secrete antimicrobial peptides (AMPs) that inhibit the activity of luminal microbes that come too close to their membranes. Indeed, the combination of mucus and AMPs provides a potent barrier between microbes and epithelial surfaces. Goblet cells also sense and transport antigen and live microbes from the lumen to antigen-presenting cells in the lamina propria. Finally, they have the ability to secrete regulatory cytokines.

Microfold (M) cells are highly specialized for the transcytosis of antigen across the epithelium. Their unique structure allows them more intimate contact than most epithelial cells with both the gut lumen and the immune cells in the lamina propria. Their apical surfaces are smoother than those of most epithelial cells because they have blunter microvilli and are covered by only a thin layer of mucus. Their basolateral surfaces are distinct and include a large pocket, or invagination, which is often occupied by immune cells. Most M cells form part of the epithelial layer directly above Peyer's patches and isolated

(a)

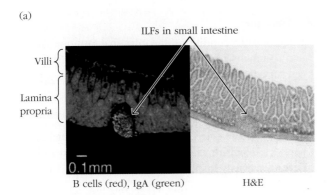

(b)

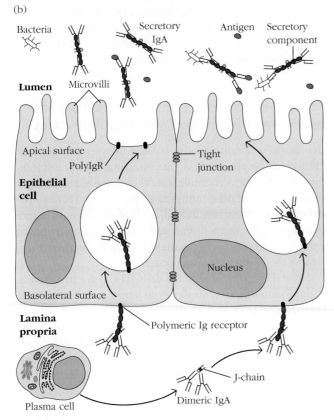

FIGURE 13-7 IgA shuttled from the lamina propria to the lumen of the intestine. (a) Sections of the small intestine stained with green fluorescent anti-IgA antibodies (*left*) and hematoxylin and eosin (H&E; *right*). The ILF that helps generate IgA B cells is indicated in both panels. (b) IgA antibody made by plasma cells binds to polymeric Ig receptors (polyIgRs) and is transcytosed by epithelial cells as a dimer from the lamina propria to the lumen of the intestine. [*Photos in (a) from Takashi Shikina et al., "IgA class switch occurs in the organized nasopharynx- and gut-associated lymphoid tissue, but not in the diffuse lamina propria of airways and gut," The Journal of Immunology May 15, 2004;* **172(10):** *6259–6264, Figure 1.*]

lymphoid follicles, and transfer antigens from the lumen of the intestine directly to dendritic cells within the follicle.

Paneth cells are secretory cells that inhabit the intestinal crypts. They play at least two critical roles. First, they act as supportive companions of stem cells and secrete factors that

sustain them. Second, they secrete a variety of AMPs that protect the gut epithelium. Defects in the ability of Paneth cells to produce AMPs have been associated with inflammatory bowel disease (IBD).

Tuft cells are also highly specialized and bear a resemblance to the sensory taste cells found in our tongues. Typically very rare in the epithelial layer, tuft cells expand in number in response to worm infection. Recent work, discussed later, shows that tuft cells are a critical source of a cytokine that initiates the successful immune response to worms.

Finally, stem cells are found at the bottom of the crypts and give rise to all epithelial cell types, which turn over rapidly in the harsh environment of the gut. With the assistance of Paneth cells, stem cells proliferate and differentiate. Their progeny move smoothly up the walls of the crypt to the top of the villi, continually replacing damaged and dying cells that slough into the intestinal lumen.

Multiple hematopoietic cells are also intimately associated with gut epithelial cells (see Advances Box 13-1). Some antigen-presenting cells (dendritic cells and macrophages) extend processes between epithelial cells and directly sample antigen in the intestinal lumen. **Intraepithelial lymphocytes (IELs)** are also present in large numbers, particularly in the upper small intestine (the jejunum). Many of these express CD8, and some have been identified as tissue-resident memory cells that can respond rapidly, in an antigen-specific manner.

Key Concepts:

- Both small and large intestines are separated from the gut lumen by a single layer of epithelial cells that are diverse in phenotype and function.

- Many epithelial cells respond directly to microorganisms via PRRs. They cooperate with other innate and adaptive immune cells to sense, sample, and respond flexibly to the microbiome.

- Some epithelial cells secrete mucus and antimicrobial peptides to maintain an additional boundary between the microbiome and cell surfaces.

- Some hematopoietic cells, such as intraepithelial lymphocytes and antigen-presenting cells, are also intimately associated with the epithelial layers and join epithelial cells in communicating with immune cells in the lamina propria.

Setting the Stage: Maintaining Immune Homeostasis in the Intestine

In a healthy individual, the intestinal or gut-associated immune system is working constantly. It is responsible for maintaining a healthy distance between epithelial membranes and luminal microbes. At the same time, it is responsible for generating and maintaining tolerance to beneficial commensal microbes. To do this, it relies on several cell types and networks that establish communication between the gut lumen and the cells in the lamina propria.

The Gut Immune System Maintains a Barrier between the Microbiome and the Epithelium

The intestinal immune system employs three major strategies to prevent microbes from penetrating the gut epithelium without permission, and we touched on these earlier when describing the epithelial cell types. First, specialized epithelial cells produce mucus that inhibits bacterial mobility. In the large intestine, goblet cells are particularly abundant and produce a very thick layer of protective mucus that microbes have difficulty penetrating. In the small intestine, both goblet cells and Paneth cells are responsible for mucus production, but create a much thinner layer (see Figure 13-6).

Second, epithelial cells also generate molecules that have direct effects on the microbial communities. Paneth cells in the small intestine produce a wide range of antimicrobial peptides (AMPs) and other proteins, including **defensins**, which generate pores in bacterial membranes; lysozyme, which digests bacterial cell walls; and members of the regenerating islet-derived protein (REG3) family, which are especially lethal for gram-positive bacteria. Enterocytes throughout the intestine also have the ability to produce AMPs, which are concentrated in the mucus layer just above the epithelial cells.

Finally, plasma cells in the healthy lamina propria secrete large amounts of IgA, which binds to commensal organisms and dietary antigens (see Figure 13-7b). Secreted IgA (sIgA) antibodies are typically dimers joined by a small J-chain protein. IgA dimers are transcytosed across the epithelial barrier by binding polyIgRs expressed on the basolateral surfaces of multiple cell types in the intestinal epithelium. After transcytosis, the IgA dimer leaves with a portion of the polyIgR called the *secretory component* (SC), which helps to stabilize the dimer in the intestinal lumen. The protective role of antibody transcytosis is underscored by studies showing that mice lacking polyIgR are more susceptible to *Salmonella typhimurium* and *Helicobacter pylori* infections.

Some IgA antibodies are broadly specific for microbe-associated molecular patterns (MAMPs) and others bind very specifically to antigens encountered by the adaptive immune system at an earlier time. Regardless of specificity, IgA binding has several effects. It inhibits microbes from penetrating the epithelial cell, and at the same time discourages inflammatory responses to any of the antigens it binds to. Secretory IgA also plays a role in transporting antigen from the lumen to the subepithelial space and delivering it to antigen-presenting cells in a manner that promotes tolerance. We will discuss the source of the IgA antibodies shortly, when we describe the activities of the adaptive immune response.

Key Concepts:

- Under healthy conditions, the intestinal immune system maintains a healthy distance between epithelial cell surfaces and the commensal microbiome.

- Goblet cells and Paneth cells create mucus, which provides a physical barrier, particularly in the large intestine.

- Paneth cells and enterocytes generate antimicrobial peptides, defensins, and REG3 family members, which have the capacity to kill microbes that come too close.

- B cells in the intestinal mucosa secrete IgA, which is transcytosed from the lamina propria into the lumen where it interacts with commensal microbes. IgA binding inhibits contact with the epithelium and also has anti-inflammatory properties.

Antigen Is Delivered from the Intestinal Lumen to Antigen-Presenting Cells in Multiple Ways

In order to maintain a healthy balance between immune tolerance and responsiveness, our intestinal immune system must find a way to communicate the presence and identity of microbial and other antigens in the lumen. Our intestinal immune system has, indeed, evolved an elaborate system of sampling the microbiome and transferring microbes to antigen-presenting cells, so that they can be further evaluated by the adaptive immune system.

Several intestinal epithelial cells (IECs) are involved in antigen sampling and transfer from lumen to lamina propria (**Figure 13-8**). As we have discussed previously, M cells are highly specialized for this purpose. Although they are found throughout the intestine, they are most abundant in the small intestine and are typically associated with underlying secondary lymphoid tissue, such as isolated lymphoid follicles (ILFs) and Peyer's patches. Although best known for their ability to transcytose antigens and microbes nonspecifically, M cells also express receptors that allow them to convey specific classes of microbes to the lamina propria. They and other intestinal epithelial cells express receptors for antibody that allow them to carry IgA-antigen complexes back from the lumen into the lamina propria. Goblet cells are able to convey small, soluble antigens from lumen to lamina propria. And finally, some resident antigen-presenting cells extend processes between epithelial cells and sample antigen directly from the lumen.

Key Concept:

- The intestine has evolved several strategies to convey microbial antigens from the intestinal lumen to antigen-presenting cells: M cells and goblet cells carry antigens across the barrier by transcytosis; cells expressing receptors for IgA carry IgA-antigen complexes across the barrier; and APCs that extend processes into the lumen bind and process antigen directly.

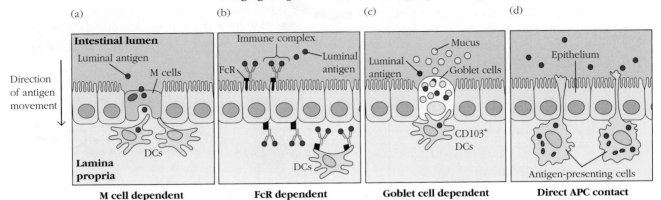

FIGURE 13-8 How antigen is delivered from the lumen to antigen-presenting cells. (a) Antigen and whole microorganisms can be delivered to antigen-presenting cells (macrophages and dendritic cells) by M cells, which transcytose antigen and sometimes microbes from the lumen. (b) Antigen-antibody complexes can be carried across epithelial cells by Fc receptors (FcR). (c) Small antigens can also be transcytosed by goblet cells and delivered to underlying dendritic cells. Finally (d) antigen can interact directly with antigen-presenting cells that have extended their processes into the intestinal lumen.

Immune Homeostasis in the Intestine Is Promoted by Several Innate and Adaptive Cell Types

Epithelial cells and antigen-presenting cells both produce molecules that influence the outcome of antigen presentation in the intestine (**Figure 13-9**; see also Table 13-1). Under healthy conditions, epithelial cells are stimulated by commensal microbes that interact with PRRs, often Toll-like receptors (TLRs). These interactions result in the production of TGF-β, the vitamin A metabolite retinoic acid (RA), and thymic stromal lymphopoietin (TSLP). This trio of molecules maintains a tolerogenic environment in part by programming antigen-presenting cells to polarize T-cell differentiation toward the regulatory T cell (T_{REG}) lineage, an event that also depends on IL-10.

Some epithelial cells and antigen-presenting cells also produce BAFF (B-cell activating factor) and APRIL (a proliferation-inducing ligand), which promote IgA production by lamina propria B cells in both T-independent and T-dependent manners. IL-10 is also involved in IgA production by B cells.

IL-10, a potent anti-inflammatory cytokine, is prevalent in the healthy mucosa. Once thought to be made only by T cells, it is now known to be produced by many different cell types, including macrophages, dendritic cells, and B cells. Its importance is underscored by the observation that individuals with lower levels of IL-10 are susceptible to inflammatory bowel disease (IBD).

Intestinal antigen-presenting cells that produce or stimulate the production of IL-10 are particularly important in maintaining immune tolerance in the gut. Which innate immune cells play this role during homeostasis remains a topic of controversy and active study. It is fair to say that the extent of phenotypic diversity among gut APCs has surprised investigators (see Advances Box 13-1). Figure 13-9 shows examples of APCs that play important roles in maintaining gut tolerance: CX_3CR1^+ resident macrophages, which extend processes between epithelial cells and sample microbial antigen directly; and $CD103^+$ migratory dendritic cells, which receive signals from epithelial cells and travel to local secondary lymphoid tissue to interact with naïve T and B cells. Both of these cell types are likely sources of IL-10, as well as other molecules that influence homeostasis.

Epithelial cells and antigen-presenting cells communicate what they learn from sampling the microbiome to both **innate lymphoid cells** (**ILCs**) as well as conventional T and B lymphocytes. The cell subsets that play the most prominent role in maintaining gut homeostasis are T_{REG} cells, IgA-secreting B cells, T follicular helper (T_{FH}) cells, T_H17 cells, ILC3s, and intraepithelial lymphocytes (IELs) (see Advances Box 13-1). Conventional B and T lymphocytes are activated in the secondary lymphoid tissues associated with the gut, while ILCs are activated in the lamina propria.

Generating T_{REG} Cells

Activated dendritic cells (DCs) travel via the lymphatics to mesenteric lymph nodes or Peyer's patches, where they meet circulating naïve T cells. If DCs first meet their antigen in the tolerizing cytokine environment described earlier, they skew T-cell differentiation toward the FoxP3$^+$ T_{REG} lineage. Whereas only 10% of circulating T cells are of the T_{REG} subtype, T_{REG} cells represent 30% of all T cells in the healthy gut lamina propria. Individuals with conditions that reduce T_{REG} populations are, in fact, susceptible to a variety of immune-mediated diseases, including colitis.

Regulatory T cells have two distinct sources. Thymic T_{REG}s commit to this lineage during development in the thymus. Peripheral T_{REG}s are generated from naïve T cells activated by tolerogenic antigen-presenting cells, including gut $CD103^+$ dendritic cells. Peripheral T_{REG}s activated by gut APCs are often induced to express gut homing receptors

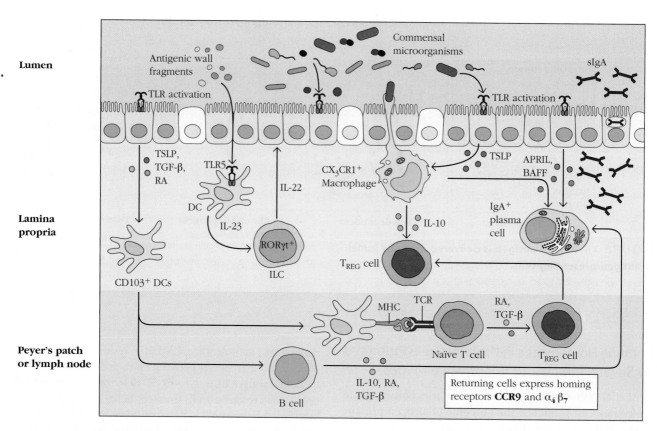

FIGURE 13-9 Maintaining homeostasis and tolerance to the microbiome at the intestinal surface. Innate and adaptive immune cells coordinate a response to gut commensal microorganisms that results in the generation of tolerogenic regulatory T cells and protective IgA-secreting B cells. See text for details.

and travel back to the intestinal mucosa where they release cytokines, such as IL-10, that reinforce tolerance. Whether the origin of T_{REG}s makes a difference in their function in the gut remains a question.

Generating IgA-Secreting B Cells

IgA-producing B cells are uniquely abundant in intestinal tissues and produce a hefty 3 grams of IgA every day. As we have discussed, IgA antibody is central to the intestine's strategy to maintain a healthy distance from commensal luminal microbes. It is a first line of defense against pathogenic microbes and toxins and also is involved in transporting antigen from lumen to lamina propria for further evaluation. Where do they come from?

Some of our IgA comes from conventional B cells activated in B-cell follicles in the Peyer's patches of the small intestine. Much also comes from B cells in isolated lymphoid follicles as well as B cells activated in draining mesenteric lymph nodes. What favors class switching to IgA? Interestingly, IgA class switching occurs in both T-dependent and T-independent manners and depends on the unique cytokine milieu generated by intestinal immune cells (**Figure 13-10**).

T-dependent class switching to IgA occurs in the traditional manner described in Chapter 11. Briefly, B cells are activated by antigen in follicles of mesenteric lymph nodes or

Peyer's patches. They generate germinal centers, where they undergo somatic hypermutation and class switching. T_{FH} cells are particularly important at this stage and provide the B cells with the combination of signals that favor IgA switching, including CD40L and the key cytokine TGF-β. Interestingly, intestinal T_H17 and T_{REG} cells both have the capacity to convert into T_{FH} cells that induce IgA class switching. B cells that successfully mature in lymphoid tissue generate plasma cells that express the adhesion molecule $\alpha_4\beta_7$ and chemokine receptor CCR9, which cause them to home to the mucosa (see Figure 13-10). It takes about 7 days to produce IgA antibodies in this manner. They also undergo somatic hypermutation and tend to have high affinities for their antigens.

IgA antibodies can also be produced in a less traditional, but quicker, T-independent manner. This route is dependent on the cytokines BAFF and APRIL, which are produced by stimulated epithelial cells, as well as mucosal dendritic cells. It occurs not only in ILFs, but even in the lamina propria itself. IgA antibodies produced in this way tend to be of lower affinity.

Role of ILC3 and T_H17 Cells in Intestinal Immune Homeostasis

The ILC3 subset plays a particularly important role in maintaining tolerance in the gut mucosa. In response to a

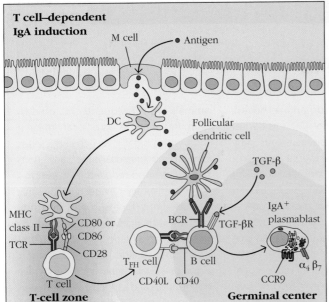

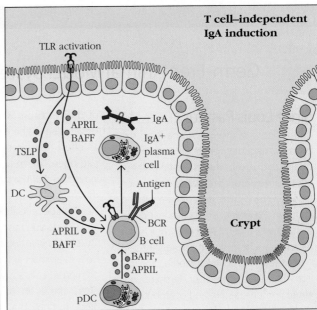

FIGURE 13-10 Generating IgA-producing B cells. IgA-secreting B cells can be generated in a T cell–dependent (a) and T cell–independent (b) manner (see text). (a) Briefly, antigen delivered to migratory dendritic cells travels to lymphoid tissue and stimulates naïve T cells to become T_{FH} cells, which provide cognate help to B cells that have also received signals from cytokines, such as TGF-β, that encourage their development into IgA-secreting plasma cells. These express integrins ($\alpha_4\beta_7$) and chemokine receptors (CCR9) that direct them back to the mucosa. (b) Intestinal epithelial cells and several different APCs in the mucosa can also be induced by interaction with the microbiome to generate cytokines, including BAFF and APRIL, which stimulate lamina propria B cells to class switch to IgA.

variety of signals produced by innate immune cells, including IL-23 and RA, ILC3 cells produce IL-17 and IL-22 (see Figure 13-9). In fact, ILC3s are now thought to be the major source of intestinal IL-22, which stimulates epithelial cells to secrete antimicrobial peptides, principally REG3. ILC3s also directly enhance B-cell production of IgA, by producing molecules that stimulate B-cell differentiation, and indirectly by inducing the development of isolated lymphoid follicles. In fact, a subset of ILC3 cells (called lymphoid tissue inducer or *LTi cells*) are required for the development of isolated lymphoid follicles and their cryptopatch precursors.

ILC1s and ILC2s are also found in small numbers in the healthy gut, but their role in intestinal homeostasis is unclear. They appear to play more important roles in the gut response to pathogens.

T_H17 cells were initially recognized for the detrimental role they play in inflammatory and autoimmune disorders (see Chapter 10). However, like ILC3 cells, they are also abundant in the healthy gut lamina propria. They, too, help maintain barrier immunity by secreting IL-22 and enhancing the health and activity of epithelial cells. An unexpected series of observations, described in **Advances Box 13-3**, led to the recognition that their development is encouraged by specific commensal microbes, including segmented filamentous bacteria (SFB). The beneficial effects of gut T_H17 cells are underscored by the observation that, in their absence, mice are more susceptible to invasion by bacteria, including *Citrobacter*.

The cytokines produced by ILC3s and T_H17 cells can also enhance inflammation and cause disease. Chronic production of IL-17A and IL-17F, for instance, contribute to colitis and psoriasis, respectively. How ILC3s and T_H17 cells balance their health-promoting and inflammation-promoting functions is still not fully understood. The cytokine milieu probably plays a key role. Whereas the tolerogenic IL-10 and TGF-β cytokines promote the development of anti-inflammatory T_H17 cells, the presence of the proinflammatory cytokine IL-6, which is produced by multiple cell types after infection, can induce T_H17 cells to take on more inflammatory roles. It is, indeed, possible that these two subsets offer the gut a uniquely flexible defense mechanism with the ability to quickly convert from quiet inhabitants to proinflammatory defenders.

Role of Intraepithelial Lymphocytes (IELs) in Intestinal Immune Homeostasis

TCR γδ and TCR αβ intraepithelial lymphocytes (IELs) also play important roles in intestinal immunity and typically insert themselves between epithelial cells, below their tight junctions (see Advances Box 13-1). They can be recruited in a conventional immune response to an invading pathogen; however, they also participate in homeostasis by maintaining the integrity of the epithelial layer, producing antimicrobial peptides and immunosuppressive cytokines. Some, the natural IELs, come directly from the thymus and take up residence in the intestine; these tend to be CD4⁻CD8⁻.

Germ-Free Animal Model Systems

Since Louis Pasteur advanced the germ theory of disease, humankind has been working to obliterate germs from our bodies and environments. Some of these efforts are appropriate, particularly when trying to prevent and treat disease-causing infection. However, as we have learned in this chapter, a germ-free existence is not necessarily a healthy one. The relationship we cultivate with our microbiome, probably from the earliest time in our life, shapes our entire immune system and contributes to both our physical and mental health. Even antibiotics, which have had a profoundly positive influence on animal health, also have a darker side in their counter-effects on our commensal microbiome.

Germ-free animal models, also called *axenic* models, have been an invaluable resource for studying the influence of the microbiome on animal health. Developed in the early 1900s as part of an effort to address diseases caused by pathogens, germ-free rodents have provided some of the strongest evidence for the association of specific microbial species with beneficial biological outcomes.

How do you raise a germ-free mouse? This is not trivial and requires a facility that isolates animals and prevents them from coming in contact with all microorganisms in the air, on surfaces, in food, on handlers, and on other animals (**Figure 1a**). A combination of barriers, radiation, high-temperature treatment, air filters, and very careful handling has made this possible. In addition, founder animals must be delivered sterilely, via cesarean section, to avoid contact with maternal microbes in the vagina/birth canal. Ideally, these founder animals are subsequently fostered by germ-free mothers. As one can imagine, food, which is naturally rich in microbes, presents a unique challenge. Early efforts to sterilize food reduced its nutritional value dramatically. New technologies that stabilize important nutrients and supplement with vitamins that are made by microbes have helped. Although germ-free mouse facilities are now common, vigilance is always important. Microbes are opportunists and can too easily find ways through barriers. Differences in experimental results from laboratory to laboratory always raise the possibility that some facilities are more germ-free than others.

What is the difference between a germ-free and a normally raised mouse? First, they simply look different. Littermates raised apart in germ-free versus

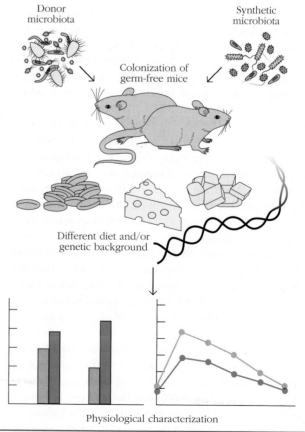

FIGURE 1 **Maintaining germ-free mice.** (a) Example of an isolation chamber that maintains a sterile barrier between the external environment, the handler, and germ-free mice. (b) Multiple approaches used to assess the impact of microorganisms on health. Single microorganisms or communities of microorganisms from the laboratory, from other mice, or even from other donors, including humans, can be introduced. Diet can also be controlled and manipulated. *[(a) Image courtesy of Chriss J. Vowles and J. Wayne Jones, Germ Free Laboratory, University of Michigan.]*

(continued)

(a)

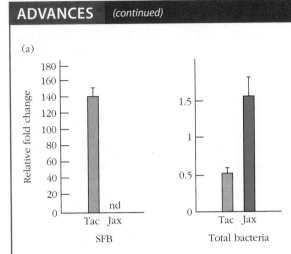

(b)

FIGURE 2 Mice from different laboratories harbor different microorganisms. (a) A section of the small intestine was evaluated for the presence of segmented filamentous bacteria (SFB) by performing quantitative PCR of 16S rRNA sequences. The difference in amount of SFB-specific 16S rRNA sequences between mice from Taconic Biosciences (Tac) versus the Jackson Laboratory (Jax) are shown on the left (nd, not detectable). Right: The amount of total bacteria in the intestine of both sets of mice (a control). (b) Scanning electron microscopy of the small intestine shows that the Taconic mice have a much higher concentration of filaments on the epithelial surfaces. *[(b) Republished with permission of Elsevier, from Ivanov II, Atarashi K. et al., "Induction of intestinal Th17 cells by segmented filamentous bacteria." Cell October 30, 2009; 139 (3): 485–98. Figure 2. Permission conveyed through Copyright Clearance Center, Inc.]*

normal conditions differ in weight gain; those in germ-free environments tend to be skinnier and, in fact, need 30% more calories to maintain their weight. This is likely due to the contributions the microbiome makes to an organism's nutritional status, by digesting food that the host organism cannot and generating vitamins and other predigested nutrients.

The differences in the mucosal immune systems of the two types of animal are also striking. Mucosal immune cells of germ-free mice are dramatically reduced in number and in secondary lymphoid tissues, such as Peyer's patches, which are dramatically reduced in size. IgA secretion is diminished, as are the numbers of T_H17 cells. Although one might expect germ-free mice to have fewer regulatory T cells, too, results have been conflicting and may reflect variations in diet, as antigens provided in the food may also induce regulatory T-cell development. Interestingly, although germ-free mice appear less anxious, they also do not seem to have as high a cognitive or memory capacity.

Germ-free mice have been very useful in understanding the effects of the microbiome on multiple body systems including the immune system. By repopulating these mice with selected and characterized populations of microorganisms, investigators gain understanding of the influence of specific microbial species and diet on many aspects of animal health (Figure 1b).

One of the first experiments to definitively reveal a specific effect of a microbial species on mucosal immune functions came out of an intriguing observation

that laboratory mice purchased from two different vendors had different mucosal immune responses. Specifically, the investigators noticed that inbred C57BL/6 mice from the Jackson Laboratory had fewer gut T_H17 cells than the same mouse strain from Taconic Biosciences. Quantification of 16S RNA sequences specific for bacterial species revealed a difference in the gut microbiome (**Figure 2**). Strikingly, the Jackson Laboratory mice had little or no segmented filamentous bacteria (SFB). These investigators hypothesized that this microbe played a specific role in maintaining gut T_H17 cells and tested their idea using germ-free animals. Before reading on, see if you can devise a well-controlled experiment that would test their hypothesis.

The results of one of their experiments are shown in **Figure 3**. In this experiment, they introduced segmented filamentous bacteria (SFB) into germ-free (GF) mice and measured the production of IL-17 by small intestinal CD4+ T cells. They also assessed IL-17 production by CD4+ T cells from the intestines of specific pathogen–free (SPF) mice as a control. (These mice are raised so that they are not exposed to a subset of potential pathogens; however, they have a fairly healthy commensal microbiome.) What do you conclude, and what other controls may be helpful in drawing a definitive conclusion?

These results were the first to establish the importance of SFB as commensal

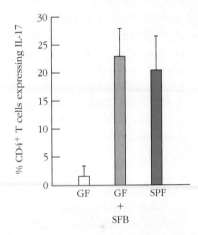

FIGURE 3 SFB colonization affects IL-17 production by intestinal helper T cells. Briefly, CD4+ T cells were isolated from the intestine of germ-free (GF) mice, GF mice colonized specifically with SFB (GF + SFB), and specific pathogen–free (SPF) mice that have a healthy commensal microbiome. Each set of cells was stimulated under conditions that promote IL-17 production, and the percentage of T cells making IL-17 was measured by flow cytometry. See text for full explanation.

bacteria and their influence on the health of one of our gut T-cell subsets. See Clinical Focus Box 16-1 for another example of the use of germ-free systems in understanding the immune response.

REFERENCE

Ivanov, I. I., et al. 2009. Induction of intestinal T_H17 cells by segmented filamentous bacteria. *Cell* **139:**485.

Induced IELs, on the other hand, develop from naïve T cells that have been activated in gut-associated lymphoid tissue. These are typically CD4$^+$ TCR $\alpha\beta^+$ cells that express the unusual CD8$\alpha\alpha$ homodimer, which interacts with MHC class I. CD4$^+$ IELs appear to repress inflammatory T-cell activity and may develop from mucosal FoxP3$^+$ T$_{REG}$ cells.

Key Concepts:

- Tolerance to the intestinal commensal microbiome is achieved through interactions between the microbiome, epithelial cells, and immune cells in the lamina propria.

- The homeostatic environment of the gut is dominated by anti-inflammatory cytokines including IL-10, TGF-β, and RA, as well as cells that maintain barrier health, including T$_{REG}$ cells, IgA-producing B cells, ILC3 and T$_H$17 cells, and intraepithelial lymphocytes.

- Antigen-presenting cells in the healthy intestine sample antigen, travel to secondary lymphoid tissue, and polarize naïve T cells to the T$_{REG}$ lineage. Many of these cells take up residence in the lamina propria and help maintain tolerance to the microbiome.

- IgA-producing B cells are generated in both T cell–dependent and T cell–independent manners. The latter depends on the presence of several molecules, including BAFF and APRIL, which are generated by epithelial cells and DCs in response to direct interaction with the microbiome.

- ILC3 and T$_H$17 cells are also abundant in the intestinal lamina propria and play dual roles. They help maintain homeostasis by producing IL-22, which stimulates epithelial cells to produce protective AMPs. However, they can also contribute to inflammatory responses.

The Immune Systems in the Small and Large Intestines Differ

Although the small and large intestines share general strategies for managing their relationship with the microbiome, it is important to emphasize that the physiology of each segment differs, as do the communities of microbes and responding immune cells that reside in them (see Figures 13-5 and 13-6).

The small intestine harbors a smaller and less diverse commensal community than the large intestine, which is the site of the most abundant and diverse microbial community in the body. Whereas the lumen of the small intestine has fewer than a million commensal bacteria per milliliter of fluid, the lumen of the large intestine can contain between a billion and a trillion bacteria per milliliter.

The small and large intestines are also vulnerable to different pathogens. For example, *Clostridium difficile*, a bacterium associated with diarrhea outbreaks in hospitals and care centers, is found in the large intestine. Norovirus, which is associated with outbreaks of diarrhea on cruise ships,

infects the small intestine. The large intestine is the site of infection for whipworm (*Trichuris trichiura*), whereas the common roundworm (*Ascaris lumbricoides*) does its damage in the small intestine. Similarly, the small and large intestines are susceptible to different inflammatory disorders and cancers. Celiac disease results from inflammatory reactions in the small intestine, and ulcerative colitis is a damaging inflammation of the large intestine. Finally, tumors are also more common in the large versus the small intestine. These are only a few of the biological distinctions between the two locations, which suggest parallel distinctions in the immune strategies adopted by each intestinal site.

As we discussed earlier, the relative concentrations of particular epithelial cell types also vary from small to large intestine. The small intestine has many Paneth cells, the large intestine has very few. The large intestine has many goblet cells that generate a much thicker and more effectively protective mucus layer. Peyer's patches are present only in the small intestine, and their associated M cells are more prevalent there, too. Isolated lymphoid follicles are more common in the large intestine, which has no villi and, unlike the small intestine, does not have to occupy itself with the absorption and digestion of food. More distinctions continue to be identified and understanding the differences and their relationship to disease remains a topic of great interest.

Key Concepts:

- The small and large intestines contain different communities of microorganisms and distinct immune cell populations.

- The large intestine has shorter villi, more microorganisms, and more goblet cells, which produce more protective mucus. It is also more susceptible to tumors.

- The small intestine, but not the large, has Peyer's patches. The large intestine has a higher concentration of ILFs.

Commensal Microbes Help Maintain Tolerogenic Tone in the Intestine

The research community is still working to understand all the criteria the gut immune system uses to decide whether to generate a tolerogenic or an immune and/or inflammatory response to the microbes and antigens it encounters. However, it is clear that both evolutionary and developmental events are at play. Microbes that co-evolved with us influenced the evolution of our immune receptors, including invariant T-cell receptors that have fixed specificities for common microbial antigens. Microbes that colonize our guts in early life ("old friends") tune and tolerize our immune system by inducing the development of regulatory T cells and the production of IgA specific for commensal bacteria.

The availability of germ-free animal models (see Advances Box 13-3) has allowed investigators to more precisely

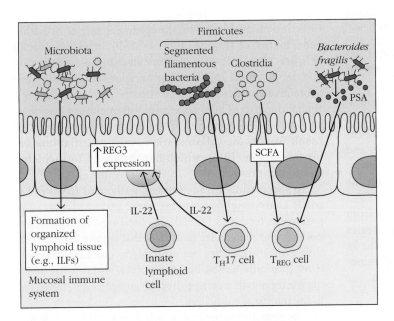

FIGURE 13-11 **Effect of commensal bacteria on intestinal immune responses.** Shown are several bacterial species that are common members of our commensal microbiome. These express and/or produce molecules (e.g., PSA, SCFA) that interact with epithelial cells (and APCs) and stimulate the development and activity of ILCs and helper T-cell subsets, each of which contributes differently to intestinal immune homeostasis (see text).

determine the influence of specific microbial species. Investigators have found that several species of commensal bacteria are particularly good at stimulating T_{REG} accumulation and tolerance (**Figure 13-11**). These include bacteria in the phylum Firmicutes, such as *Clostridium*; and bacteria in the phyla Actinobacteria and Bacteroidetes, all of which are prominent members of the healthy human microbiome. Firmicutes bacteria and others produce short-chain fatty acids (SCFAs) by fermenting dietary fiber. SCFAs directly influence dendritic cells in the gut and enhance the development of regulatory T cells. Segmented filamentous bacteria (SFB), another member of the Firmicutes phylum, colonize the small intestine and alone enhance IgA production and T_H17 development. Remarkably, recent data show that even a single molecule of polysaccharide A (PSA) produced by a single intestinal bacterial species (*Bacteroides fragilis*) helps to maintain the systemic regulatory T-cell pool.

Although we still have much to learn about the viral communities that inhabit our gut, recent work shows that virus exposure also has beneficial effects on intestinal development and systemic immune function. Remarkably, when germ-free mice, whose intestinal epithelial barrier is more porous and whose mucosa is depleted of immune cells, are exposed to a single viral species (e.g., norovirus), immune cell populations are restored, epithelial cell connections are strengthened, and immune homeostasis in the gut is re-established. This is, in part, due to the interaction of the virus with pattern recognition receptors, which triggers the cascade of events responsible for maintaining healthy, tolerogenic immune activity in the intestine.

Finally, and importantly, it turns out that what happens in the gut does not simply stay in the gut. Not only does our commensal gut microbiome help tolerize our gut immune system, but it influences the entire immune system. When tolerogenic responses are impaired in the gut, autoimmune responses are more common elsewhere. For instance, patients with systemic lupus erythematosus (SLE) appear to

have a microbiome that promotes inflammatory responses, and may be helped by supplementation with a bacterium, *Bifidobacterium*, from the phylum Actinobacteria. *Bifidobacterium* is now a relatively common probiotic and may act in part by promoting regulatory T-cell development.

Fecal transplantation, otherwise known as bacteriotherapy, was inspired by studies that show the beneficial influence of the microbiome. It has been performed by large-animal veterinarians for over 100 years and was first used in humans over 50 years ago. Most often used today in patients with *C. difficile* infections and the diarrhea it produces, fecal transplantation involves the introduction of a healthy donor's colonic microbiome to an infected individual. The re-introduced commensal bacteria can outcompete disease-causing bacteria and restore epithelial integrity, although more work needs to be done to standardize therapy and to understand which specific microbes are beneficial.

Key Concepts:

- Several bacterial species help tune the intestinal immune system and enhance its ability to protect from pathogenic organisms by stimulating the production of anti-inflammatory intestinal immune cells (e.g., T_{REG} cells, IgA$^+$ B cells, and T_H17 cells).

- In the absence of microorganisms, the intestine does not develop normally and the immune system is underdeveloped both locally and systemically.

- Single microorganisms and even single microbial antigens can restore intestinal and immune health in germ-free mice.

- Introducing specific microbes through eating (probiotics) or fecal transplantation could enhance intestinal health, as well as the intestinal and systemic immune systems.

Springing into Action: Intestinal Immune System Response to Invasion

Most of our commensal organisms are not pathogenic and contribute to the tolerogenic environment of the intestine at steady state. Nevertheless, the gut immune system maintains a healthy distance between epithelial surfaces and even beneficial commensal bacteria, by secreting IgA antibodies and generating mucus. These strategies help us fend off the few commensal bacteria, termed **pathobionts**, that do have the capacity to cause disease in compromised hosts.

Most dangerous microorganisms, however, come from the outside. Our gut has evolved several different strategies to recognize, respond to, and expel them. We describe in this section some of the current thinking behind inflammatory immune responses of the intestine, recognizing that our understanding is still evolving rapidly.

The Gut Immune System Recognizes and Responds to Harmful Pathogens

While we still do not fully understand all the variables that influence the intestinal decision to initiate an inflammatory rather than a tolerogenic response,

infection of, or damage to, the epithelial barrier by pathogens, toxins, or trauma often is the distinguishing factor (**Figure 13-12**). Inflammation can cause discomfort, of course, but most inflammatory responses are ultimately protective and designed to clear pathogen and repair epithelial integrity. These successful inflammatory responses should be distinguished from immune responses that result in chronic inflammation, such as inflammatory bowel disease (IBD), which has increased in incidence and media presence.

To successfully invade the gut, pathogens first need to battle commensal bacteria for space and nutrients. Some pathogens feed off of metabolites released by commensal bacteria and metabolize molecules that commensals cannot, thus exploiting niches not available to commensal organisms. Ironically, antibiotics, themselves, also provide invading bacteria with a competitive advantage and better access to the intestinal epithelium.

The response to pathogens that invade the gut is divided into two phases—the *inductive phase*, when the response is initiated and shaped, and the *effector phase*, when mature immune cells actively work to clear the pathogen and repair damage (**Figure 13-13**).

The inductive phase shares many features with other immune responses. It is initiated by epithelial cells and

(a)

(b)

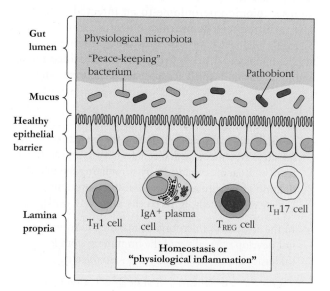

FIGURE 13-12 Conditions that cause switch from homeostatic (a) to inflammatory (b) immune responses. Environmental, physiological, and genetic factors can compromise the integrity of the epithelium and/or alter the balance between beneficial and virulent microorganisms (see text). Antibiotics, diet, and ingestion of infectious organisms can alter the commensal

microbiome directly. Pollutants and invading microorganisms can trigger inflammatory responses by the epithelium and antigen-presenting cells. Hormones produced by stress can have direct and indirect effects on the microbiome and epithelial integrity. Inherited mutations in immune molecules (e.g., IL-10 and IL-23 receptors) can also compromise our ability to maintain homeostasis.

(a) Inductive response to *Salmonella*

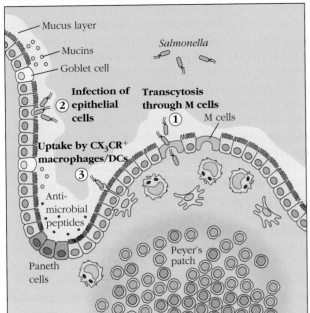

(b) Effector response to *Salmonella*

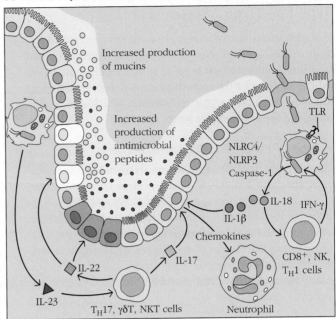

FIGURE 13-13 **Intestinal immune system response to *Salmonella* bacterial infection: an example of a type 1 response.** (a) The inductive phase; (b) the effector phase. See text for details.

antigen-presenting cells that have been alerted to the presence of microbes via pattern recognition receptors that generate proinflammatory rather than tolerogenic signals. For instance, interactions with TLR5, a receptor for bacterial flagellin, tend to induce inflammation, whereas interactions with TLR2 can be tolerizing and important for resolving infection and epithelial integrity. Pathogens that activate NOD-like receptors (NLRs; see Chapter 4) and those that trigger inflammasome pathways in macrophages and epithelial cells also shift the immune system away from tolerogenic responses and trigger protective inflammatory responses (the inflammasome is described in Advances Box 4-2).

The intracellular location of PRRs also provides immune cells with information that allows them to distinguish between healthy and unhealthy interactions with microbes. For instance, TLR5 is located only on the basolateral and not the apical surface of epithelial cells in the large intestine. Therefore, bacteria will only trigger this PRR if the mucosal layer is damaged or the bacteria is invasive.

Activated APCs alert naïve T cells to the presence and general identity of the antigen. Rather than polarize naïve T cells to the regulatory lineage, APCs activated by pathogens polarize naïve T cells to the T_H1 and inflammatory T_H17 lineages or T_H2 and T_H9 lineages, depending on the microorganism encountered (see Advances Box 13-1, Figure 2, and specific examples discussed here). Rather than induce IgA class switching, the T cells and APCs generated after an encounter with pathogens encourage class switching of

B cells to more proinflammatory IgG classes, which protect the body from the spread of the infection.

The effector phase involves the recruitment of cells and strategies that clear or expel the invading organism, and varies depending on the identity of the invading organism. For example, some bacterial infections are thwarted by the activation of a type 1 response, where inflammatory ILC3 and T_H17 cells recruit phagocytic neutrophils. Bacterial infections can also induce ILCs to secrete IL-22, which stimulates epithelial cells to produce antimicrobial peptides that kill bacteria. Worm invasion stimulates type 2 responses that include ILC2 and T_H2 activation, which lead, in turn, to the recruitment of eosinophils and enhanced mucus production and peristalsis, the muscle activity that ripples through intestines during digestion and helps expel worms. Resolution of the infection also requires the repair of epithelial damage.

Key Concepts:

- The intestinal immune response to infection includes an inductive phase, when cells are alerted to the presence of a pathogen and assemble their cytokine and cell arsenal, and an effector phase when these cells work to clear the infection.

- In order to successfully invade the intestine, pathogens must compete successfully with commensal organisms and evade or withstand the inflammatory response induced by pattern recognition receptors.

The Intestinal Immune System Can Mount Both Type 1 and Type 2 Responses

The intestinal immune responses to specific pathogens follow these general themes. However, they also reveal additional complexities, some of which we illustrate below, where we describe the response to two classes of pathogens: the single-cell prokaryote *Salmonella*, which causes diarrhea, and the multicellular worm *Ascaris*.

A Type 1 Response to *Salmonella* Bacteria

Bacterial species that cause severe diarrhea include *Salmonella* species, *Clostridium difficile*, *Citrobacter*, and enterohemorrhagic *Escherichia coli* (EHEC). We will focus on the immune response to *Salmonella typhimurium*, which is a common and highly contagious bacterium that causes fever, cramps, and diarrhea in all vertebrates. The infection is usually cleared after a week or so, but some individuals, particularly those who have compromised immune systems, may suffer for much longer.

Salmonella is transferred by food, water, or feces from infected individuals. It spreads easily and rapidly and, once swallowed, successfully finds its way to the small and large intestines, where it can invade the epithelium. As you know, the gut has established several barriers that must be breached for infection to occur. *Salmonella* must first compete with the healthy commensal bacteria community and find a way around the defensins, the antimicrobial proteins made by the gut epithelium. Some, but not all, antimicrobial peptides (e.g., defensins) are particularly good at killing *Salmonella*, and these can sometimes eliminate an intestinal infection all by themselves. Pre-existing IgA with broad specificity also limits the access of *Salmonella* to the epithelial barrier.

However, *Salmonella* has multiple other means of entry into intestinal tissue (see Figure 13-13a). It is often transcytosed by M cells, it can use its own secretory machinery to enter epithelial cells, and it can be engulfed by resident macrophages. During infection, *Salmonella* meets the next line of defense—our pattern recognition receptors (PRRs). Surface Toll-like receptors (e.g., TLR4 and TLR5) and internal TLR9 engage multiple antigens on this gram-negative pathogen, including lipopolysaccharide (LPS), flagellin, and CpG. Internal NOD-like receptors (NLRs) bind *Salmonella* products, too, and trigger inflammasome activity.

Several innate immune cells in the intestine express TLRs and NLRs and contribute to the inductive response to *Salmonella*. These include epithelial cells, antigen-presenting cells, and ILCs, which cooperate in developing a cytokine milieu that encourages a type 1 response (see Figure 13-13b; see also Figure 4-16).

Pattern recognition receptor engagement results in the expression of IL-23, a particularly potent cytokine that enhances the production of IL-17 and IL-22 by T_H17 and ILC3 cells. IL-22 is protective and up-regulates the production of antimicrobial proteins, including but not limited to

REG3. IL-17 also enhances the production of chemokines that attract neutrophils. Neutrophils are particularly effective at clearing *Salmonella* and also produce additional IL-22 and IL-17 cytokines. This series of reactions, known as the **IL-23–T_H17 cell axis**, works within 2 to 3 days of a *Salmonella* infection to restore epithelial integrity.

Pathogen interactions with NLRs induce inflammasome activity in macrophages and activate caspase-1. Caspase-1, in turn, activates the cytokines IL-1β and IL-18, which contribute to the type 1 response by inducing T-cell and ILC subsets to produce IFN-γ (see Figure 13-12b). Finally, antibodies are also critical in controlling *Salmonella* infection and dissemination. Although IgA antibodies may play a role in the initial defense against *Salmonella*, antibodies of the IgG class are more potent in managing an active infection. Antigen sampled by M cells is relayed to the B-cell follicles where it activates antigen-specific B cells. These are induced to class switch by appropriate T helper cells. The IgG2a antibody class seems particularly effective against *Salmonella* infection. Unlike IgA, IgGs are not secreted into the lumen of the intestine unless the epithelial layer is physically damaged.

Salmonella has also evolved very clever defenses against these protective responses. It actually thrives on some of the carbohydrates that decorate proteins on epithelial surfaces, including sialic acid. If internalized by a macrophage, it takes advantage of the acidic pH of endosomal vesicles to activate its own virulence genes. It seems more resistant to some antimicrobial proteins than some commensal bacteria, and may even enjoy a growth advantage in the presence of these proteins. Hence, a successful immune response is a battle of time and evolutionary wits between host and microbe.

A Type 2 Response to Worm Infection

Billions of people are infected with parasitic worms. Intestinal worms include hookworms, roundworms, whipworms, tapeworms, and *Trichinella*, and can cause disorders and diseases that range from the very mild to the completely devastating. Ironically, a growing group of individuals in industrialized nations are gaining interest in therapeutic worm infection as a way to avoid autoimmune diseases and asthma. Although skepticism is warranted, there is also experimental evidence for the health benefits of some worm infections.

Just as our immune system has co-evolved with our commensal bacteria, it has also adapted over millions of years to a co-existence with intestinal worms. Some worms are treated as commensal organisms and generate tolerogenic responses that enhance the production of regulatory T cells and the generation of IL-10 and TGF-β cytokines (**Figure 13-14a**). Worms that invade our gut, however, meet a well-organized defense that reflects our long evolutionary history with these organisms.

Infectious worms trigger a type 2 immune response, which is characterized by the activities of ILC2 and $CD4^+$ T_H2 cells and the generation of the classic type 2 cytokine

(a)

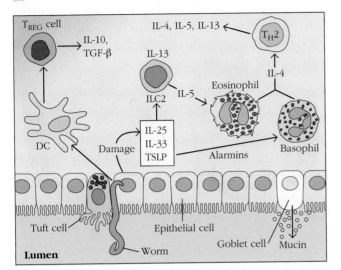

(b)

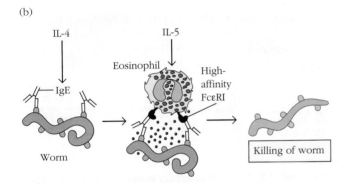

FIGURE 13-14 Intestinal immune system response to worm infection: an example of a type 2 response. (a) The major inductive events that occur in the intestine after contact of epithelial tuft cells with a worm. (b) The effects of cytokines IL-4 and IL-5 on the effector response to the worm. See text for details. See also Videos **13-14v1** and **13-14v2**. These two videos, both taken by Michael Patnode, show human eosinophils coating a *Caenorhabditis elegans* larval nematode. *(Access the YouTube videos at https://youtu.be/fw_I21RnBWg and https://youtu.be/uvPQY3arzro.)*

trio IL-4, IL-5, and IL-13. In fact, investigators speculate that our type 2 immune response evolved specifically as a strategy to manage helminth infections (see Figure 13-14a).

Type 2 responses in the gut are initiated by a specific subset of cytokines called **alarmins**. These include TSLP, IL-25, and IL-33, which are released by gut epithelial cells. These act on a variety of type 2 immune cells, including ILC2s, mast cells, basophils, and ultimately CD4$^+$ T$_H$2 cells, to induce the release of IL-4, IL-5, and IL-13.

How are intestinal cells alerted to the presence of worms? What stimulates them to secrete alarmins? Until recently, we did not know. However, investigators have now shown that tuft cells play a key role (see Advances Box 13-1). Tuft cells, also called *brush cells*, were identified in the 1960s but their

function was not well understood. They are rare cells and share intriguing features with the sensory cells in the tongue that respond to umami and bitter tastes.

In fact, using sensory receptors, tuft cells seem to "taste" products of multiple parasites, including worms and some protozoa. This induces the production of the alarmin IL-25. IL-25 stimulates lamina propria ILC2 cells to secrete the key type 2 cytokine IL-13. IL-13 has many effects, including a positive feedback influence on tuft cell development. It acts on stem cells in the intestinal epithelium, inducing the development and proliferation of more tuft cells. Mice without tuft cells fail to produce enough IL-25 and fail to generate an optimal type 2 response to worms.

Type 2 cytokines have multiple additional effects on the intestinal immune system. IL-13 enhances goblet cell production and secretion of mucus, which helps to clear worms physically from the gut. IL-4 produced by eosinophils and basophils helps polarize naïve T cells to the T$_H$2 subtype. T$_H$2 cells enhance B-cell class switching to IgE. Worm-specific IgE antibodies bind via their antigen-binding sites to worms and via their constant regions to granulocytes such as eosinophils, and induce the release of granule mediators; these granule mediators have potent toxic effects that can be fatal to worms (Figure 13-14b and associated videos).

Some of these mediators, including histamines, are well known to those suffering from allergy. The allergic response is thought to be a consequence of inappropriate activation of this ancient system, which evolved to manage worms, protozoa, and other complex parasites.

The recognition that the incidence of allergy and autoimmunity was increased in industrialized areas, where people were less exposed to the full array of microbes available to our ancestors, led David Strachan and others to advance the hygiene hypothesis (see Chapter 1, Clinical Focus Box 1-3). Briefly, this is a proposal that exposure to microbes, including worms, tunes the immune system toward a more tolerogenic, less inflammatory state. The hypothesis now has experimental support, although perspectives on the biological and cellular bases for the phenomenon are evolving.

Intriguingly and somewhat paradoxically, worm exposure also has a powerful dampening effect on the immune system. Not only do parasitic worms inspire type 2 immune responses, but they also inspire tolerogenic responses characterized by the development of T$_{REG}$ cells and the activities of IL-10 and TGF-β (see Figure 13-14a). How these responses are regulated and coregulated is an active area of investigation. However, it is clear that the tolerogenic state has systemic consequences. Anecdotal evidence that worm exposure helps to quell autoimmune and allergic states has inspired some individuals to lobby for worm therapy to ameliorate autoimmune symptoms. While some research supports these observations, some highly publicized results remain anecdotal and much more needs to be done to validate such a therapeutic approach. Even if the notion gains more support, it is also important to remember that regulatory T cells can interfere with our immune response, particularly type 1 responses to intracellular pathogens.

It is also important to recognize that this description of a worm infection is an oversimplification. Worms are very diverse metazoans with complex life cycles and the ability to traverse different parts of our bodies. Different worms affect our immune system in unique ways. They inspire multilayered immune responses that promise to yield deeper insights into our intimate immune relationship with microbes.

Key Concepts:

- Pattern recognition receptors and NLRs expressed by epithelial cells, APCs, and ILCs trigger a type 1 response to the gram-negative bacterium *Salmonella*, which is coordinated by several cytokines, including IL-23 and IL-17, and several cell types, including T_H1 cells and ILCs. It culminates in the recruitment of phagocytic neutrophils.

- Worms are sensed by tuft cells in the epithelium, which generate alarmins and cytokines, including IL-25, that trigger a type 2 effector response coordinated by ILC2 and T_H2 cells, which produce the trio of type 2 cytokines, IL-4, IL-5, and IL-13. The response culminates in the production of IgE and the degranulation of eosinophils, which kills the worms directly.

Dysbiosis, Inflammatory Bowel Disease, and Celiac Disease

The immune activity represented in responses to commensal and pathogenic microbes comes at a cost. Even successful immune responses and protective inflammation may cause damage to the epithelium that must be repaired and replaced by ever-active stem cells, after pathogens are successfully cleared. If a pathogen cannot be cleared, inflammation can become chronic, with a lasting effect on the structure and cellular components of the intestine and its immune system.

Inflammatory bowel disease, or **IBD**, is one of the most common chronic disorders of the intestine (**Figure 13-15a**). It is really two diseases—**Crohn's disease** and **ulcerative colitis**—characterized by chronic inflammation of the gut mucosa that results in pain, diarrhea, and dramatically reduces quality of life.

Crohn's disease is a form of IBD that can affect any part of our digestive tract. It is broadly associated with an inappropriate type 1 response and is characterized by patchy aggregates of inflammatory macrophages in the mucosa. These can form granulomas that are difficult to resolve and distort the microenvironment of the small intestine. Ulcerative colitis is a form of IBD that afflicts only the descending part of the large intestine. It is broadly associated with an inappropriate type 2 response that results in active inflammation of

(a) Inflammatory bowel disease (IBD)

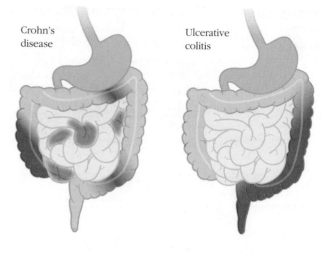

Crohn's disease

Ulcerative colitis

(b) Prevalence of IBD in 2015

■ Highest
■ Intermediate
■ Lowest
□ Uncharted

FIGURE 13-15 Inflammatory bowel disease (IBD) distribution in the gut and around the world. (a) The intestinal distribution of the two different variations of IBD. Crohn's disease can affect any and all areas of the GI tract, whereas ulcerative colitis is found only in the large intestine. (b) The distribution of IBD in world populations. Its prevalence, where measured, is highest in North America, England, Scandinavia, and Australia and lowest in Asia and Brazil. This pattern suggests that diet may have a major influence.

the mucosa with infiltration of neutrophils, reduced mucus boundaries, and direct damage to the epithelium (ulcers).

The causes (etiology) of IBD are not well understood and likely vary from individual to individual. Genetic, environmental and infectious influences all contribute to the development and trajectory of the disorder. Reductions in mucus production, antimicrobial peptide production, tight junction maintenance, regulatory T-cell production, and levels of IL-10 have all been associated with IBD. Alterations in any of these activities can contribute to the loss of tolerance and compromised epithelial integrity featured in these diseases. Interestingly, IBD is also associated with increased levels of IL-23, which activates T_H17 and ILC3 cells. As we discussed above, these cell subtypes can contribute to both homeostasis and inflammation, depending on other, yet unclear, environmental influences.

The composition of the commensal intestinal microbiome—and the interplay between microbiome and mucosal immunity—clearly plays a key role in initiating, sustaining and treating IBD. **Dysbiosis**, the disruption of a healthy microbiome, is now a well-accepted contributing factor to IBD. The rising incidence of IBD in industrialized nations (Figure 13-15b) further suggests that the influence of diet on the microbiome is a root cause.

However, it is difficult to distinguish initiating causes from consequences of chronic intestinal inflammation. Genome-wide association studies (GWAS) have revealed several gene variations associated with increased susceptibility to IBD that may provide better clues. Some of the genetic variations point to defects in pattern recognition receptors, and hence dysfunctional relationships between the gut microbiome and the innate immune system.

For instance, as described in Clinical Focus Box 4-3, individuals who inherit specific variants of the *NOD2* gene can be forty times more susceptible to developing IBD. NOD2 is a member of the NLR family of cytosolic PRRs, expressed by multiple innate immune cells in the intestine. It recognizes intracellular bacterial MAMPs and activates inflammatory signaling pathways that generate antimicrobial peptides and enhance the immune activity of antigen-presenting cells (see discussion of our immune response to *Salmonella* in the previous section).

Mice with defective *CARD9* and humans with a variation in the *CARD9* gene sequence also are more susceptible to IBD. CARD9 is part of the signaling cascade generated by certain pattern recognition receptors. Interestingly, a defect in *CARD9* has an indirect effect on the composition of the gut microbiome, reducing the overall production of intestinal tryptophan. Why is tryptophan important? Its metabolites interact with receptors on gut epithelium, stimulating production of the protective cytokine IL-22. In fact, tryptophan levels—provided only by food and the microbiome because our cells cannot synthesize this amino acid—can partially ameliorate the symptoms caused by *CARD9* mutation. This phenomenon is only one of many that illustrate the importance of the dialog between microbiome and host immunity in gut inflammatory disease. There is much more to come, and clarity will certainly inspire new therapies.

Individuals with IBD often suffer from inflammatory disorders of other tissues, including the skin. Canker sores and oral ulcers are common and painful accompaniments. These extra-intestinal manifestations of IBD are thought to arise from a systemic defect in immune tolerance. As we have discussed above, the commensal intestinal microbiome helps tune the immune system and encourages the generation of anti-inflammatory cells, including regulatory T cells. Not all of these cells remain in the intestinal mucosa; many populate other organs and tissues and are thought to contribute to a systemic tolerizing affect. Dysbiosis may not only impair the development of tolerizing influences but can result in the activation of proinflammatory T_H17 cell subsets that travel to other organs and wreak havoc.

Celiac disease (CD) is an autoimmune disorder that also affects our intestinal mucosa. People with CD share some symptoms (diarrhea, gastrointestinal discomfort, bloating) with those affected by IBD. However, the specific cause of CD is known and is distinct from the complex causes of IBD. CD is triggered by an immune response to dietary glutens (and other molecules) associated with several grains, including wheat. The immune response to gluten results in the production of IL-15, which activates IELs resulting in epithelial cell death and physical damage to the barrier. Gluten peptides gain access to the lamina propria and trigger predominantly T_H1 (versus T_H17) responses, as well as NK-cell and B-cell activity.

Gastrointestinal viral infection often precedes CD symptoms, and genetic variations in MHC molecules (as well as many other gene loci) may enhance susceptibility to CD. Finally, microbiomes of people with CD differ from those of healthy individuals, but it is unclear whether this is a cause or consequence of the associated inflammation.

Treatments for IBD and CD differ. A life-long gluten-free dietary regimen helps most, but not all, patients with CD. Immunosuppression and immunotherapy with antibodies that block inflammatory molecules (such as anti–TNF-α) help many, but not all, patients with IBD. The complexity of the disease and the individuality of immune responses contribute to therapeutic challenges that face physicians and patients alike, and there is room for much more research.

Key Concepts:

- Several diseases of the immune system are associated with dysregulation of our intestinal immune response. Dysbiosis—alteration in the gut microbiome—may be both cause and consequence.

- Inflammatory bowel disease includes two diseases: Crohn's disease and ulcerative colitis. They are characterized by both a decrease in tolerogenic and increase in inflammatory (type 1 and type 2) immune responses in the intestine. Causes are not fully understood but may be associated with diet, the microbiome, and a failure to induce homeostasis conditions, which also may have a genetic component. It is typically treated with immunosuppressive drugs and anti-inflammatory therapies.

- Celiac disease is an autoimmune response to dietary glutens (and other wheat-associated molecules) that have penetrated the epithelial barrier. It is characterized by a damaging T_H1 response and is typically treated with a life-long gluten-free diet.

Other Barrier Immune Systems

For many good reasons, the intestinal immune system has been the most intensively studied mucosal immune tissue. In sheer size and cell density, it is the largest immune organ in our bodies. It is also relatively easy to access for experimental work. Although we still have much more to learn about the complex networks of cells and cytokines that regulate intestinal immunity, what we do know provides an excellent framework for exploring other barrier immune systems.

In this section, we introduce some of the fundamental features of two other barrier immune systems—the respiratory system and the skin. Many of the cellular and molecular players are the same as those found in the intestine and should be very familiar at this point. Therefore, we will focus most on features and questions that are unique to each.

The Respiratory Immune System Shares Many Features with the Intestinal Immune System

Like the GI tract, the respiratory tract is exposed directly to the external environment. Its total surface area is second in size only to the intestine (**Figure 13-16a**). The *upper respiratory tract* includes the nose and mouth. The *lower respiratory tract* starts with the *trachea*, which is separated from the oral cavity by a flap called the *glottis*. The trachea

branches into *bronchi*, and then into smaller and smaller *bronchioles*, which ultimately dead end in clusters of microscopic sacs called *alveoli*. These are in intimate contact with capillary beds and are the site of gas exchange.

As in the intestinal tract, the walls of the lower respiratory tract are organized into multiple layers, including a single epithelial cell layer and the underlying lamina propria, which together constitute the respiratory mucosa. As one progresses down the respiratory tract, the walls thin. The height of the epithelial cells, the density of epithelial cell cilia, and the thickness of the lamina propria all diminish (Figure 13-16b). The alveolar walls are lined by very thin epithelium, allowing free passage of gases into surrounding capillaries.

The epithelial layer of the respiratory tract shares many of the same inhabitants as the intestinal tract (Figure 13-16c). These include goblet cells, M cells, and transepithelial processes of antigen-presenting cells. Like the intestinal epithelium, the respiratory epithelium also includes multipotent stem cells that replace damaged and dead epithelial cells. Respiratory epithelium also has some unique features. The apical surfaces of most respiratory epithelial cells have cilia.

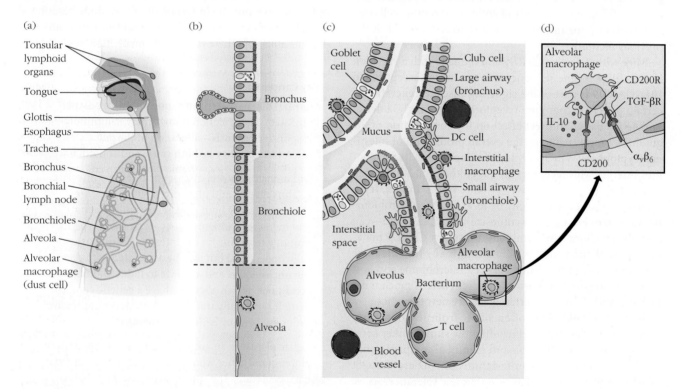

FIGURE 13-16 Gross and cellular anatomy of the respiratory tract. (a) The main anatomical structures of the respiratory tract, which starts in the nose and mouth, extends through the trachea, and branches into bronchi that serve both lungs. The bronchi divide into smaller areas called *bronchioles* and end in grapelike bunches of sacs called *alveoli*, where gas exchange with blood capillaries occurs. Many lymph nodes serve the respiratory tract, including the tonsils and bronchial lymph nodes. The airways

are lined by a single epithelial layer (b), which gradually reduces in height and thickness as it reaches the alveoli, where it provides only a very thin layer of protection. The epithelial cells (c) are diverse in phenotype and function (see text) and in the alveoli are joined by a key immune participant, the alveolar macrophage or dust cell (a, c, and d). The dust cell interacts with epithelial cells via a variety of receptors (d) and helps to maintain respiratory homeostasis in part by secreting IL-10.

Together with the mucus produced by goblet cells, these form a *mucociliary boundary* that actively helps sweep away and expel microbes and particulates that have descended into the airways. The respiratory epithelial layer also includes a unique secretory cell called the **club cell**, formerly known as a *Clara cell*,[2] which has a variety of protective functions. It is most prevalent in the lower airways of humans and throughout the airways of mice. Club cells also appear to have the ability to act as stem cells.

The alveoli are also the home of specialized **alveolar macrophages** (**dust cells**) that monitor the lower airways and alveoli for infection and work with respiratory epithelial cells to regulate the balance between tolerogenic and inflammatory responses in this very delicate site (Figure 13-16d).

The integrity of the respiratory epithelial barrier function is maintained by various antimicrobial proteins and peptides secreted by respiratory mucosal cells and by secretory IgA. Antigen-specific IgA is generated as described for the intestine, and its production depends, as expected, on the previous activation of both innate and T lymphoid cells.

The upper respiratory tract is home to commensal communities of microbes that contribute to immune protection and inspire tolerogenic activities, such as the development of regulatory T cells. The lamina propria of this portion of the respiratory system is home to many of the same subsets of innate and adaptive immune cells found in the intestine. In contrast, a healthy lower respiratory tract (from the trachea down) does not support a community of commensal bacteria and, instead, focuses on ridding itself of microbe visitors.

Secondary lymphoid tissue and isolated follicles can be found in the walls of the entire respiratory tract and is most developed in the nasal tissue (see Figure 13-16a). The **nasal-associated lymphoid tissue**, or **NALT**, is part of the ring of tissue that we often refer to as *tonsils* and *adenoids*. Comparable in organization to the Peyer's patches of the intestine, it supports the activation of T cells and B cells triggered by innate cells in the respiratory mucosa. In some animals, such as rabbits and cats, well-organized lymphoid tissue is present in the deeper part of the lungs and is referred to as **bronchus-associated lymphoid tissue**, or **BALT**. In other animals, such as humans and mice, this tissue is very loosely organized and requires antigenic stimulation to develop fully. It is therefore referred to as inducible BALT (iBALT).

The lamina propria of the respiratory mucosa is home to populations of innate and lymphoid cells that are also found in the intestine, including $CD103^+$ dendritic cells, CX_3CR1^+ macrophages, ILCs, regulatory T cells, IgA-secreting plasma cells, and more (**Figure 13-17**). Natural killer (NK) cells, also known as cytotoxic ILC1 cells, may even be more abundant in the lung than in the intestinal lamina propria. These cells are very effective at identifying and killing virally

infected cells. As in the gut, ILC2s also play an important role in maintaining the integrity of the respiratory epithelium. They express the growth factor **amphiregulin**, which interacts with epidermal growth factor (EGF) receptors on the epithelium, enhancing their health and growth.

As in the intestine, interactions of upper airway epithelial and immune cells with commensal microbes stimulate tolerogenic responses. They enhance the production and activity of regulatory T cells and class switching of B cells to the anti-inflammatory IgA phenotype. Under healthy conditions, alveolar macrophages in the lower airways receive anti-inflammatory signals from the epithelium that expresses ligands for TGF-β (e.g. integrins $\alpha_v\beta_6$) that interact with TGF-β receptors on the macrophage (see Figure 13-16d).

Interactions with invading microbes trigger a sequence of events roughly similar to those that occur in the intestine (see Figure 13-17). Bacterial and viral microbes engage pattern recognition receptors, including NLRs expressed by epithelium and antigen-presenting cells. This initiates a cascade of proinflammatory signals and cytokines, including IL-17, IL-23, and IL-1β, that program the antigen-presenting cells to induce type 1 responses and activate ILC1, T_H1, and T_H17 cells, as well as other subsets. As in the intestine, worms and allergens stimulate epithelial cells to produce alarmins, including IL-25, IL-33, and TSLP. These trigger a type 2 response, activating ILC2 and T_H2 cells to produce IL-4, IL-5, and IL-15 and recruit effector eosinophils, basophils, and mast cells.

As in the intestine (see Figure 13-9), antigen-specific lymphocytes generated in lymph nodes draining the lung are induced to express homing receptors that send them back to respiratory tissues. CCR4 is one chemokine receptor that directs immune cells back to the lung, which expresses the CCR4 ligands CCL17 and CCL22. Interestingly, other barrier tissues, including the skin, can also attract $CCR4^+$ cells. This behavior offers one mechanistic explanation for how antigen exposure in the lung (or intestine or skin) can have both local and systemic effects. It is also the basis for our hope that mucosal vaccination may help enhance general human health.

Example: Respiratory Immune Response to a Virus

Influenza virus is among the most common invaders of the respiratory epithelium. It gains access when its surface hemagglutinin protein binds to the terminal sugar, sialic acid, of surface carbohydrate side chains on epithelial and other innate immune cells in the epithelium. This interaction induces endocytosis of the virus. The virus fuses its membranes with those of the endocytic vesicle, releasing its single-stranded RNA genome into the cytoplasm of the infected cell.

As expected, cells are alerted to the virus's presence via pattern recognition receptors (see Chapter 4). For instance, TLR7 and RIG-I bind flu's RNA genome and NLRs bind to flu-specific surface proteins. These interactions set off

---

[2] The change in name was prompted by the recognition by respiratory investigators that *Clara* referred to an individual who was a Nazi sympathizer responsible for immoral acts during World War II.

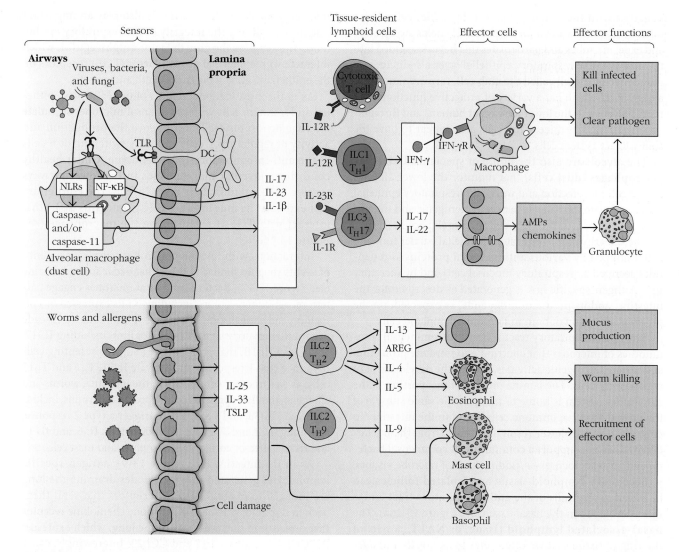

FIGURE 13-17   Type 1 and type 2 immune responses in the respiratory tract. Events are very similar to those described for the intestine. Briefly, viruses, and some bacteria and fungi, interact with PRRs (TLRs and NLRs) on antigen-presenting cells to trigger a type 1 response that activates ILC1 and T_H1 cells, ILC3 and T_H17 cells. These produce type 1 cytokines that activate effector cells, like cytotoxic cells and macrophages, which kill infected cells and engulf pathogens. Type 2 responses are triggered by worms and some allergens. As in the intestine, they induce the production of alarmins by epithelial cells, which activate ILC2, T_H2, and T_H9 cells, which generate type 2 cytokines and amphiregulin (AREG). See text for details.

signaling cascades that result in the production of inflammatory cytokines, including IL-1β, type I IFNs, and the alarmin IL-33. NK cells (cytotoxic ILC1 cells) specifically recognize and kill infected cells that display flu's hemagglutinin proteins on their surface, an event that happens just before the budding of new viral particles. This, together with early type I IFN production, limits the initial spread of infection.

In addition, antigen-presenting cells that have been activated by exposure to the flu migrate to the NALT and iBALT, where they activate T and B cells that induce type 1 cytotoxic and humoral responses, locally and systemically. Memory T and B cells are also generated and are fundamental to the success of flu vaccines, which are designed to activate APCs

either locally (in the case of nasal delivery) or systemically (in the case of intramuscular injection).

Allergy and Asthma in the Respiratory Tract

Asthma is an inappropriate respiratory immune response to nonpathogenic antigens and conditions. It is one example of a respiratory allergic response. Our understanding of its causes is still incomplete, even as its incidence continues to increase in large parts of the world.

Asthma can be initiated by microbes, pollens, pollution, obesity, and even cold temperatures. Antigen-inspired asthma triggers a type 2 immune response that, interestingly, involves many of the same players responsible for the immune response to worms in the intestine. T_H2 cells, ILC2s, eosinophils, basophils, and IgE-producing B cells all contribute to the response, which is mediated by the classic type 2 cytokines, including IL-4, IL-5, and IL-13 (see Figure 13-17). Activation of tissue mast cells by IgE bound to allergens or by other signals triggers degranulation, with release of mediators such as histamine that were prepackaged in granules. The effects of these mediators on bronchial smooth muscle cells, mucus secretion, and blood vessels cause the classic symptoms of allergic responses (see Chapter 15). Chronic exposure to these stimuli induces more permanent changes in the bronchial tissue, resulting in asthma.

Intriguingly, exposure to intestinal worms may ameliorate allergic responses in the airways, including asthma. If you recall from our previous discussion, parasitic worm infection also has the paradoxical benefit of inducing a tolerogenic response in the gut. This response seems to have systemic effects that quell immune responses in distal sites. Attempts to replicate this influence in humans with worm antigens or worms themselves have been less successful than they have been in mice and remain a topic of active research.

Intranasal Vaccines

Mucosal vaccines represent a promising effort to translate basic knowledge of mucosal immunity to the clinics. Intranasal vaccination may be particularly useful because of the relative ease of delivery, as well as its ability to induce both local and systemic immune responses. As shown in **Figure 13-18**, exposure of respiratory tract APCs to antigen (e.g., either live attenuated virus or lipid-associated protein antigen particles) can result in both local and systemic adaptive immune memory. The APCs travel to draining lymph nodes, stimulating the generation of mucosal tissue–homing T cells and B cells.

An intranasal, live attenuated vaccination for flu virus, known as "FluMist," has been marketed for several years. Although considered effective for earlier flu outbreaks, particularly in young children, it was not recommended for the 2016–2017 flu season. Some speculate that its effectiveness varies by flu strain and that H1N1 flu preparations, in particular, do not stimulate protective immunity.

Intranasal vaccines are common in veterinary medicine. Those who have dogs might be familiar with the intranasal vaccine for kennel cough, which is caused by the gram-negative *Bordetella* bacterium. This intranasal vaccine often includes nonvirulent versions of three different organisms—*Bordetella* and two viruses (a parainfluenza and adenovirus)—and protects dogs from kennel cough and other respiratory infections. Immunity is not necessarily long-lasting, and efforts to improve immune memory are ongoing. Intranasal sprays are also used successfully to vaccinate chicken flocks against avian flu and other poultry diseases.

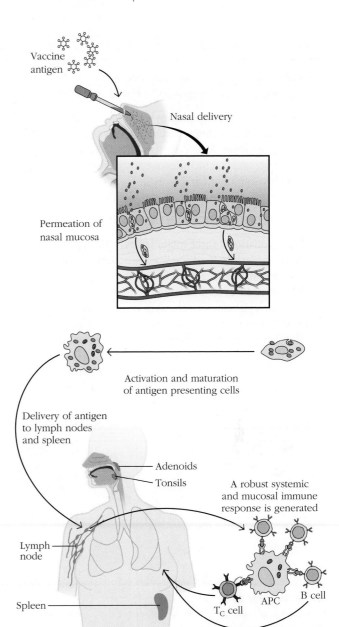

FIGURE 13-18 Nasal vaccination: exploiting basic knowledge for therapy. Introducing pathogen antigens nasally can trigger immune responses that theoretically can lead to both systemic and local (mucosal) immune protection and memory. Nasal vaccines for viral infections (flu) and bacterial infections (canine *Bordetella*) have had mixed success. See text for details.

Key Concepts:

- The respiratory immune system is very similar to the intestinal immune system, although some cell types found in their respective epithelia are distinct.

- Communities of microbes populate the upper but not lower airways and stimulate immune tolerance via similar mechanisms (IL-10 production, regulatory T cells, and IgA-producing B cells).

- Macrophages in the upper and lower airways cooperate with the epithelium and underlying immune cells to mount type 1 responses to bacteria and viruses and type 2 responses to worms and allergens.

- Some activated lymphocytes home specifically to the respiratory epithelium; others travel to other barrier immune tissues. Respiratory immune responses influence and can be influenced by immune responses at other barrier organs.

The Skin Is a Unique Barrier Immune System

The skin consists of three major layers—the upper epithelial layers or *epidermis*, the area directly underneath the epithelium or *dermis*, and the *hypodermis*, the deepest layer of our skin (**Figure 13-19a**). Unlike the epithelium of the mucosal organs, the epithelium of the skin is multilayered and does not produce mucus. The cells and lymphoid tissue that make up the skin's immune system are therefore not considered part of the mucosal-associated lymphoid tissue (MALT). Not surprisingly, however, the skin's immune system shares many features with that of the intestinal and respiratory tracts.

Skin epithelial cells, or **keratinocytes**, are densely packed in stratified layers in the epidermis. New keratinocytes are generated at the border between epidermis and dermis. These new cells displace the older keratinocyte layers, which shift upward. The surface of our skin, in fact, is a 10- to 15-μm layer of dead keratinocytes and lipids, known as the *stratum corneum*, which forms a very protective, waterproof, first barrier to infection. Immune cells are present in the epidermis, as well as in the dermis, which also includes hair follicles, sebaceous glands, capillaries, and nerve endings. The hypodermis is a fatty layer occupied by arteries, veins, and sweat glands.

Like epithelial cells of the intestinal and respiratory systems, keratinocytes do not simply provide a physical barrier. Rather, they are active members of the innate immune system; they constitutively produce antimicrobial substances (such as **psoriasin**) and also recognize and respond to pathogens via pattern recognition receptors (see Chapter 4). **Langerhans cells**, a type of dendritic cell, are specialized antigen-presenting cells that are found in epidermis. They and other antigen-presenting cells extend processes between keratinocytes and act as sentinels that carry information about microbes from the epidermis to immune cells in the dermis. In addition, they carry skin antigens to the draining

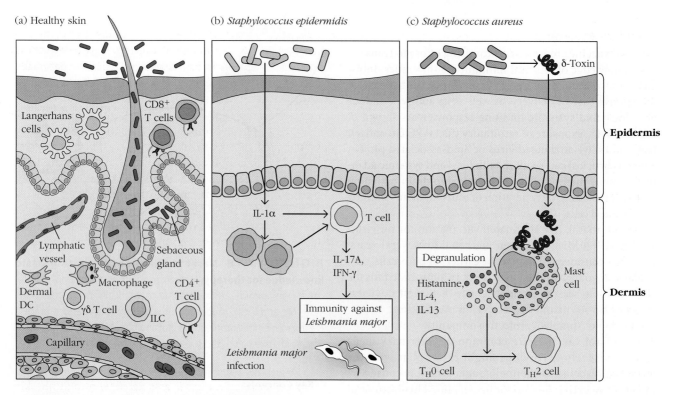

(a) Healthy skin   (b) *Staphylococcus epidermidis*   (c) *Staphylococcus aureus*

FIGURE 13-19 Immune responses in the skin. (a) The skin is covered by multiple layers of specialized epithelial cells known as *keratinocytes*, which compose the epidermis. A variety of innate and adaptive immune cells, including Langerhans cells and CD8⁺ T_RM cells, reside in the epidermis, but most of the immune cells are in the dermis, the layer just beneath the epithelium. Like the lamina propria of the intestine and respiratory tracts, the dermis includes a now familiar host of innate and adaptive immune cells that coordinate homeostatic and inflammatory responses. Unlike mucosal immune systems, the dermis does not include organized secondary lymphoid tissue (although some follicles can form in inflammatory states). (b and c) Two examples of the relationship between microorganisms on the skin surface and the immune system below: (b) *Staphylococcus epidermidis* interacts directly and indirectly with dermal myeloid cells to stimulate a type 1 T-cell response that protects us from *Leishmania* infection. (c) *Staphylococcus aureus*, on the other hand, produces a toxin that stimulates a type 2 response that enhances allergic dermatitis. See text.

lymph nodes, where they activate naïve T and B cells to generate immune responses. Investigators have marveled at the ability of Langerhans cells to migrate surprisingly great distances. Indeed, Langerhans cells activated by skin antigens have shown up in the mesenteric lymph nodes and consequently have the potential to exert an influence on immune responses at other mucosal tissues.

Lymphocytes also reside in the epidermis and human epidermal tissues, including a large population of resident memory CD8$^+$ T cells (CD8$^+$ T$_{RM}$ cells). Although $\gamma\delta$ T cells are abundant in the mouse epidermis and extend processes that resemble those of dendritic cells, $\gamma\delta$ T cells are only thinly scattered throughout the human epidermis. Interestingly, there are no organized lymphoid follicles in the skin as there are in the intestine.

Most other immune cells are found in the dermal layer (see Figure 13-19a), which includes the full complement of innate and adaptive immune cells present in the lamina propria of mucosal immune tissue. Whereas CD8$^+$ memory cells are present in the epidermis and appear to be long-term residents, CD4$^+$ memory cells are found in the dermis and appear to circulate. Dermal $\gamma\delta$ T cells respond to IL-23 produced by dermal APCs and appear to be a major source of IL-17 in the skin, contributing to both protective immunity as well as inflammatory skin diseases, like psoriasis. They play a similar role in skin immune homeostasis as ILC3 and T$_H$17 cells in the gut and respiratory immune systems.

Regulatory T cells abound in mouse and human dermis. They tend to be long-term residents and accumulate over the life span of an individual. As in other barrier sites, they regulate tolerance to commensal bacteria and ameliorate inflammatory skin reactions. However, recent studies suggest that their origin and the timing of their development may differ considerably from those of intestinal and respiratory T$_{REG}$s.

Rather than develop continually in response to constant sampling of luminal antigens by tolerogenic antigen-presenting cells, as happens in the gut, skin T$_{REG}$ cells may arise in one wave early in development and then take up long-term residence in the skin (**Figure 13-20**). Recent studies show that tolerance to commensal bacteria arises in a narrow window of time during neonatal development. In fact, mice were not able to generate tolerizing T$_{REG}$s if they were colonized with commensal bacteria when mature. This observation underscores the possibility that exposure to microbes in infants has a profound influence on our immune health as adults.

Although this time-dependent effect of T$_{REG}$ development seems most dramatic in the skin, it is likely that time of exposure to commensal microbes and antigens has an effect on tolerance in all barrier immune systems. For instance, exposure to antigens in the lung during the neonatal period may help individuals avoid IgE hypersensitivities, an observation that provides scientific support for the hygiene hypothesis (see Chapter 1, Clinical Focus Box 1-3).

Interestingly, recent studies also show that skin T$_{REG}$ cells concentrate in hair follicles, which are also host to commensal microbes. As we have learned from the intestinal immune system, interactions between commensal microbes, epithelial

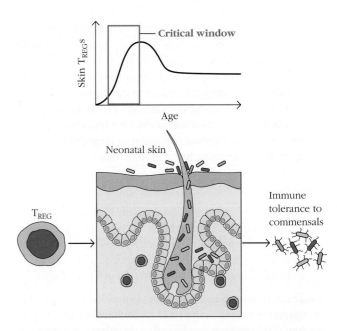

FIGURE 13-20 Developmental regulation of T$_{REG}$ cells in the skin. The generation of skin-specific T$_{REG}$ cells is age dependent and occurs when newborns (neonates) are first colonized by microorganisms (see text).

cells, and immune cells often favor T$_{REG}$ development. It is tempting to speculate that similar mechanisms are at work in the skin. Perhaps some T$_{REG}$ development occurs continually, too, supplementing the more long-term T$_{REG}$ population that arose during neonatal periods. There is still much to learn.

The surface of human skin is replete with communities of microorganisms including bacteria, which interact with the immune system in many ways (Figure 13-19b and c). Recent work shows that colonization of skin with specific bacterial species can protect animals from infection with *Leishmania*—a protozoal parasite that is spread by the bite of a sandfly.

Staphylococcus epidermidis is one commensal bacterial species that provides protection and appears to do so by interacting with pattern recognition receptors that induce IL-1 production by dermal (not epidermal!) dendritic cells (see Figure 13-19b). IL-1, in turn, induces IL-17 production by dermal CD8$^+$ T cells, which contribute to a successful immune response against *Leishmania* parasites. *S. epidermidis* also protects against other bacterial infections by inducing antimicrobial peptide production by keratinocytes. It even produces its own antimicrobial peptides that can fend off more pathogenic bacteria at the skin surface.

Staphylococcus aureus is associated with multiple skin and systemic disorders, including allergic dermatitis. It can generate inflammation by producing a molecule, δ-toxin, that penetrates the epidermal layer. δ-Toxin interacts with dermal mast cells, causing them to release cytokines that stimulate a type 2 inflammatory reaction (see Figure 13-19c).

Finally, Edward Jenner, an eighteenth century physician who is now considered an early founder of immunological thinking, was among the first to exploit the ability of the skin's immune system to generate not just a local but a systemic immune response. He successfully generated protective immunity to smallpox, a strictly human disease and scourge, by scratching live cowpox virus into the skin of his patients (and his son). This approach, later dubbed *vaccination* by Louis Pasteur, was responsible for the worldwide eradication of this fatal and deforming disease. How a local skin response induces systemic effects is still under investigation. However, the ability of the Langerhans cell to travel large distances, carrying antigen to draining lymph nodes, offers one clue.

Key Concepts:

- The skin immune system shares many similarities with mucosal immune systems but is distinguished by its multilayered epithelium (epidermis), which is associated with a unique population of dendritic cells, as well as a high concentration of tissue-resident $CD8^+$ T cells.

- The dermis includes most of the immune cell types found in the lamina propria of mucosal tissues, circulating $CD4^+$ memory T cells, and an abundance of regulatory T cells that help to maintain immune homeostasis.

- Commensal bacteria interact with both epidermal and dermal cells to generate immune protection to other organisms, including *Leishmania* parasites.

- Immune responses in the skin can have systemic protective effects.

Conclusion

The barrier immune systems, which include tissues and cells in the intestinal, respiratory, reproductive, and urinary tracts (mucosal-associated lymphoid tissue or MALT), as well as in the skin, monitor and protect areas of our body that are exposed to the outside world. Epithelial cells represent the first layer of innate immunity, and each barrier tissue is covered with one or more epithelial layers that cooperate with other innate and adaptive immune cells to maintain a harmonious relationship with the rich community of microorganisms that cohabit our bodies. The interaction between the microbiome and our immune system enhances the integrity of our epithelial barriers and establishes optimal conditions for warding off dangerous pathogens. Each barrier tissue has unique attributes, but they share general strategies that promote tolerance to commensal microorganisms via the maintenance of regulatory T cells and IgA-producing B-cell activity as well as strategies that generate type 1 and type 2 inflammatory responses to organisms that damage barrier tissues. Striking the right balance between maintaining tolerance and mounting an inflammatory response to microbes is a central challenge, which is met by a variety of molecular and cellular immune strategies that we are just beginning to understand.

REFERENCES

Behnsen, J., A. Perez-Lopez, S. P. Nuccio, and M. Raffatellu. 2015. Exploiting host immunity: the *Salmonella* paradigm. *Trends in Immunology* **36**:112.

Belkaid, Y., and S. Tamoutounour. 2016. The influence of skin microorganisms on cutaneous immunity. *Nature Reviews Immunology* **16**:353.

Broz, P., M. B. Ohlson, and D. M. Monack. 2012. Innate immune response to *Salmonella typhimurium*, a model enteric pathogen. *Gut Microbes* **3**:62.

Cadwell, K. 2015. Expanding the role of the virome: commensalism in the gut. *Journal of Virology* **89**:1951.

Cerf-Bensussan, N., and V. Gaboriau-Routhiau. 2010. The immune system and the gut microbiota: friends or foes? *Nature Reviews Immunology* **10**:735.

Cerovic, V., C. C. Bain, A. M. Mowat, and S. W. Milling. 2014. Intestinal macrophages and dendritic cells: what's the difference? *Trends in Immunology* **35**:270.

Cheroutre, H., F. Lambolez, and D. Mucida. 2011. The light and dark sides of intestinal intraepithelial lymphocytes. *Nature Reviews Immunology* **11**:445.

Couper, K. N., D. G. Blount, and E. M. Riley. 2008. IL-10: the master regulator of immunity to infection. *Journal of Immunology* **180**:5771.

Ellis, J. A., et al. 2016. Comparative efficacy of intranasal and oral vaccines against *Bordetella bronchiseptica* in dogs. *Veterinary Journal* **212**:71.

Gallo, R. L. 2015. *S. epidermidis* influence on host immunity: more than skin deep. *Cell Host & Microbe* **17**:143.

Garrett, W. S., J. I. Gordon, and L. H. Glimcher. 2010. Homeostasis and inflammation in the intestine. *Cell* **140**:859.

Gause, W. C., T. A. Wynn, and J. E. Allen. 2013. Type 2 immunity and wound healing: evolutionary refinement of adaptive immunity by helminths. *Nature Reviews Immunology* **13**:607.

Heath, W. R., and F. R. Carbone. 2013. The skin-resident and migratory immune system in steady state and memory: innate lymphocytes, dendritic cells and T cells. *Nature Immunology* **14**:978.

Hooper, L. V., D. R. Littman, and A. J. Macpherson. 2012. Interactions between the microbiota and the immune system. *Science* **336**:1268.

Huang, B. L., S. Chandra, and D. Q. Shih. 2012. Skin manifestations of inflammatory bowel disease. *Frontiers in Physiology* **3**:13.

Ivanov, I. I., et al. 2009. Induction of intestinal T_H17 cells by segmented filamentous bacteria. *Cell* **139**:485.

Iwasaki, A., and R. Medzhitov. 2015. Control of adaptive immunity by the innate immune system. *Nature Immunology* **16**:343.

Iwasaki, A., E. F. Foxman, and R. D. Molony. 2017. Early local immune defences in the respiratory tract. *Nature Reviews Immunology* **17**:7.

Kernbauer, E., Y. Ding, and K. Cadwell. 2014. An enteric virus can replace the beneficial function of commensal bacteria. *Nature* **516**:94.

Littman, D. R., and A. Y. Rudensky. 2010. T_H17 and regulatory T cells in mediating and restraining inflammation. *Cell* **140**:845.

Manichanh, C., N. Borruel, F. Casellas, and F. Guarner. 2012. The gut microbiota in IBD. *Nature Reviews Gastroenterology & Hepatology* **9**:599.

Mantis, N. J., N. Rol, and B. Corthesy. 2011. Secretory IgA's complex roles in immunity and mucosal homeostasis in the gut. *Mucosal Immunology* **4**:603.

Merad, M., P. Sathe, J. Helft, J. Miller, and A. Mortha. 2013. The dendritic cell lineage: ontogeny and function of dendritic cells and their subsets in the steady state and the inflamed setting. *Annual Review of Immunology* **31**:563.

Mirzaei, M. K., and C. F. Maurice. 2017. Ménage à trois in the human gut: interactions between host, bacteria and phages. *Nature Reviews Microbiology* **15**:397.

Mowat, A. M., and W. W. Agace. 2014. Regional specialization within the intestinal immune system. *Nature Reviews Immunology* **14**:667.

Mueller, S. N., and L. K. Mackay. 2016. Tissue-resident memory T cells: local specialists in immune defence. *Nature Reviews Immunology* **16**:79.

Nestle, F. O., P. Di Meglio, J. Z. Qin, and B. J. Nickoloff. 2009. Skin immune sentinels in health and disease. *Nature Reviews Immunology* **9**:679.

Parigi, S. M., M. Eldh, P. Larssen, S. Gabrielsson, and E. J. Villablanca. 2015. Breast milk and solid food shaping intestinal immunity. *Frontiers in Immunology* **6**:415.

Pascual, V., et al. 2014. Inflammatory bowel disease and celiac disease: overlaps and differences. *World Journal of Gastroenterology* **20**:4846.

Pasparakis, M., I. Haase, and F. O. Nestle. 2014. Mechanisms regulating skin immunity and inflammation. *Nature Reviews Immunology* **14**:289.

Perez-Lopez, A., J. Behnsen, S. P. Nuccio, and M. Raffatellu. 2016. Mucosal immunity to pathogenic intestinal bacteria. *Nature Reviews Immunology* **16**:135.

Peterson, L. W., and D. Artis. 2014. Intestinal epithelial cells: regulators of barrier function and immune homeostasis. *Nature Reviews Immunology* **14**:141.

Powell, N., M. M. Walker, and N. J. Talley. 2017. The mucosal immune system: master regulator of bidirectional gut-brain communications. *Nature Reviews Gastroenterology & Hepatology* **14**:143.

Ramirez-Valle, F., E. E. Gray, and J. G. Cyster. 2015. Inflammation induces dermal $V\gamma4^+$ $\gamma\delta T17$ memory-like cells that travel to distant skin and accelerate secondary IL-17–driven responses. *Proceedings of the National Academy of Sciences USA* **112**:8046.

Rivera, A., M. C. Siracusa, G. S. Yap, and W. C. Gause. 2016. Innate cell communication kick-starts pathogen-specific immunity. *Nature Immunology* **17**:356.

Rooks, M. G., and W. S. Garrett. 2016. Gut microbiota, metabolites and host immunity. *Nature Reviews Immunology* **16**:341.

Shikina, T., et al. 2004. IgA class switch occurs in the organized nasopharynx- and gut-associated lymphoid tissue, but not in the diffuse lamina propria of airways and gut. *Journal of Immunology* **172**:6259.

Steele, S. P., S. J. Melchor, and W. A. Petri, Jr. 2016. Tuft cells: new players in colitis. *Trends in Molecular Medicine* **22**:921.

Steinbach, E. C., and S. E. Plevy. 2014. The role of macrophages and dendritic cells in the initiation of inflammation in IBD. *Inflammatory Bowel Disease* **20**:166.

von Hertzen, L., I. Hanski, and T. Haahtela. 2011. Natural immunity: biodiversity loss and inflammatory diseases are two global megatrends that might be related. *EMBO Reports* **12**:1089.

Useful Websites

https://www.nature.com/ni/multimedia/index.html Multiple, lovely animations of barrier immune systems in gut, skin, and lung.

www.anatomybox.com AnatomyBox—an educational website/blog that provides useful images and enlightening commentaries relevant to human anatomy (including the anatomy of barrier organs).

https://www.taconic.com/taconic-insights/microbiome-and-germ-free/lab-mice-gut-microbiota-source-matters.html Article posted by Taconic Biosciences showing how different the commensal microbiome is depending on the origin of mice used for research.

www.theibdimmunologist.com This site is run by Dr. Mary E. Morgan, an immunologist who has studied the role of immune cells and probiotics in IBD. It describes intestinal immunology clearly, accurately, and accessibly and includes information of interest to those suffering from IBD, too. (She also runs a blog about the microbiome called *The Beneficial Bacteria Site,* at **www.beneficialbacteria.net.**)

https://www.bio-rad-antibodies.com/mucosal-Immunology-minireview.html A very clear review of mucosal immune systems by the company Bio-Rad, which sells experimental reagents and equipment to investigators.

STUDY QUESTIONS

1. As you know from Advances Box 13-3, the intestinal epithelium and immune system are defective in germ-free mice. Colonizing these mice with bacteria from other mice (or humans) can restore the health of the immune system and strengthen epithelial health and function. Even a single bacterial protein or single bacterial species can do this. Investigators Kernbauer, Ding, and Cadwell (*Nature* 2014; **516**:94) were interested to see whether viruses, which are also part of our commensal microbiome, played a similar role. They colonized germ-free (GF) mice with a strain of norovirus. (Norovirus is associated with diarrhea that spreads on cruise ships, but not all strains are harmful.) Here is a summary of some of their experimental results. Explaining briefly, what conclusions can be drawn from these data? What additional questions could you ask?

Feature	GF mice	GF mice + norovirus	Conventional mice
Villus width	–	++	++
T-cell numbers in lamina propria and mesenteric lymph nodes	–	++	++
IFN-γ expression by T cells	–	+	++
Antibody levels	–	++	++
Intestinal ILC2s	++	–	–

2. You are a pathologist (a doctor who analyzes data from patients to better understand what is causing their symptoms) asked to review some stained tissue sections (biopsies) taken from the skin of a patient with atopic dermatitis. You put the slides under the microscope and this is what you see:

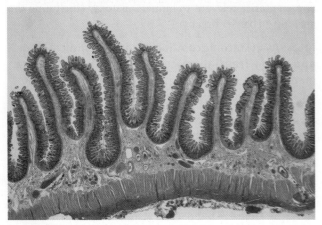

Innerspace Imaging/Science Source

You immediately call the intern in charge of the patient and let her know that you have the wrong sample. Why do you know this is not from the patient you have been asked to help? Identify at least two reasons.

3. Briefly explain to a high school student who has just finished their first year of biology why antibiotics can have harmful as well as have helpful effects.

4. The following is a list of cell types found in the epithelial layers of skin, intestine, and respiratory tissues:

Paneth cell
M cell
Enterocyte
Tuft cell
Intraepithelial lymphocyte (IEL)
Goblet cell
Eosinophil
Langerhans cell
Club cell
Innate lymphoid cell (ILC)
Keratinocyte
Dendritic cell

a. Which cell types produce mucus?
b. Which cell types are made from hematopoietic stem cells?
c. Which cell types make IL-25 in response to worms?
d. Which cell types can present antigen to CD4$^+$ T cells?
e. Which cell types express T-cell receptors?
f. Which cell types express PRRs?
g. Which cell types are found in the intestine? The skin? The respiratory tract?

5. The following is a list of cytokines that regulate the immune response in barrier tissues:

IL-4
IL-10
IFN-γ
IL-13
TSLP
IL-33

a. Which is/are made by T cells?
b. Which is/are part of a type 2 response?
c. Which is/are alarmins?
d. Which is/are involved in maintaining tolerance?

6. Which of the following activities do epithelial cells engage in? (More than one answer is possible.)

a. Secreting cytokines
b. Migrating to lymph nodes
c. Interacting via PRRs with microbes
d. Producing mucus

7. Which of the following does *not* contribute to the tolerogenic environment of a healthy barrier immune system?

a. IgA-secreting B cells
b. Regulatory T cells
c. Eosinophils
d. IL-10

8. Which of the following molecules, cells, or anatomical features are not involved in the production of IgA antibodies?

a. Goblet cells
b. T$_{FH}$ cells
c. Peyer's patches
d. Macrophages
e. Tuft cells
f. B cells

14

The Adaptive Immune Response in Space and Time

Learning Objectives

After reading this chapter, you should be able to:

1. Visualize the behavior of innate and adaptive immune cells before, during, and after a response to antigen in living tissue; recognize the role dynamic imaging approaches have played experimentally.

2. Outline the events that lead to activation and differentiation of APCs, CD4$^+$ T cells, CD8$^+$ T cells, and B cells.

3. Provide examples of the behavior of effector lymphocytes in peripheral tissues.

4. Identify the multiple variables that influence immune cell movement, positioning, and localization.

Imagine trying to understand American or European football simply from snapshots taken of players. You could certainly learn something about how the games are played, particularly if you carefully noted differences in the color of jerseys, the directions players are facing in snapshots, implied motion of feet relative to the position of a ball, and contact between players. Imagine then having the capacity to pull one or two players off the field and examine them, with and without a ball. This would add to your understanding. However, you may still miss the relevance of each position on the team, the significance of substitutions, and the rationale behind a stop in play. You are even more likely to misinterpret the antics of players who have just sacked a quarterback or flopped in dramatic agony in front of the goal. Now imagine that you can watch and listen not just to one entire game, but to many games, in real time. With that luxury, you have a real chance of understanding the rules behind, and meanings of, each activity.

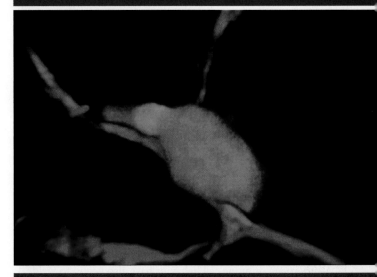

T cell interacting with fibroblastic reticular cell network. A T cell (cyan) interacts closely with the FRC network (stained red and green) as it scans the surface of dendritic cells whose processes are also associated with the network (not shown). [*Republished with permission of Elsevier, from Bajénoff, M., et al., "Stromal cell networks regulate lymphocyte entry, migration, and territoriality in lymph nodes," Immunity, 2006, December; **25**(6):989–1001, Figure 1C. Permission conveyed through Copyright Clearance Center, Inc.*]

Understanding the rules governing cell activities during an immune response is even more daunting. Immune cells are the most mobile, dynamic cells in the body. They browse every tissue, and organize responses in nearly every niche. They are also among the most diverse collection of cells that form a tissue, and their individual phenotypes and functions change over the course of the immune response. Immune cell choreography, indeed, can seem impenetrable. However, some of the most elegant cellular, molecular genetic, and biochemical experiments in biology have successfully revealed many of the rules behind the movements of immune cells in space and time. New technologies that allow us to trace the dynamic behavior of individual cells in living organisms have

Key Terms

Dynamic imaging
Selectins
Adhesion molecules
Addressin
Extravasation
Diapedesis

Sphingosine
  1-phosphate (S1P)
Homing receptors
Chemokines
Chemokine receptors
Subcapsular sinus (SCS)

B-cell follicle
Germinal center
T-cell zone
Paracortex
Periarteriolar lymphoid
  sheath (PALS)

Marginal zone
Fibroblastic reticular cell
  (FRC)
Follicular dendritic cell
  (FDC)

ushered in a new experimental era that is filling gaps in our understanding of the immune response.

Dynamic imaging techniques, including two-photon (2P) intravital fluorescence microscopy (see Chapter 20), allow investigators to directly visualize immune cells responding to antigen in living organisms. Single-cell transcriptome sequencing allows investigators to follow changes in gene expression of thousands of individual cells over the course of an infection. Teams of immunologists, geneticists, biophysicists, biochemists, and computer scientists continue to develop other innovative approaches that bring us ever closer to the ability to visualize and interpret the complex choreography of an immune response in its physiologic context.

This chapter takes advantage of these recent advances to cap our exploration into the basic immunology of an immune response by offering glimpses of immune cells in the tissues where they reside. Because foundational investigations first mapped the behavior of lymphocytes, we focus largely on the adaptive immune response. Investigations into innate immune cell behavior have also begun to yield intriguing results; stay tuned for future updates and insights.

We start with a general summary of the dynamic behavior of innate and adaptive immune cells in healthy tissue (in **homeostasis**). We then summarize what we know about the behavior of the innate and adaptive immune cells when they first encounter antigen. First, we examine the behavior of innate immune cells in tissues that have been breached by antigen, and then we discuss the choreography of innate and adaptive cells during the induction of a primary response in secondary lymphoid tissues. We focus primarily on the behavior of cells during the adaptive immune response in lymph nodes, the secondary lymphoid organ that has been most accessible to imaging techniques. We then address the activity of mature effector and memory lymphocytes that emerge from the primary immune response, describing current examples of the in vivo behavior of cells as they actively respond to infection or transplantation.

Because the regulation of cell movement is so important in studies of immune cell dynamics in living tissue, we also include an Advances box that describes the molecular basis for cell trafficking in more detail. It can be used as a springboard to explore and understand more advanced literature.

Immune Cells in Healthy Tissue: Homeostasis

Hematopoietic stem cells, which are found in the bone marrow of adults, give rise to all cells involved in an immune response throughout the lifetime of an organism (see Chapter 2). Myeloid cells, part of the innate immune system, include antigen-presenting cells (dendritic cells and macrophages) and granulocytes (neutrophils, basophils, and eosinophils). Once mature, these cells exit the bone marrow. Some continuously circulate through the body, while others take up long-term residence in our many tissues and organs. Lymphoid cells, part of the adaptive immune system, also begin their development in the bone marrow, but complete it in distinct niches. B cells mature among the osteoblasts of the bone marrow, and some complete maturation in the spleen. T-cell precursors leave the bone marrow quite early and mature in a distinct organ, the thymus. During maturation, B and T cells rearrange genes that encode the B-cell receptor and T-cell receptor, and each generates a unique BCR or TCR (see Chapter 6). Immature T cells (thymocytes) and B cells undergo negative selection to rid their antigen receptor repertoires of autoreactive cells. Thymocytes also undergo positive selection for T-cell receptor specificities that recognize self-MHC (*major histocompatibility complex*)–peptide complexes with some affinity. The few immature lymphocytes that survive these selection events exit from the thymus and bone marrow as mature, naïve T lymphocytes and naïve B lymphocytes (see Chapters 8 and 9).

Naïve Lymphocytes Circulate between Secondary and Tertiary Lymphoid Tissues

About 30 minutes after entering the bloodstream, nearly half of all mature, naïve lymphocytes produced by the thymus and bone marrow travel directly to the spleen, where they browse for approximately 5 hours (**Overview Figure 14-1**). Most of the remaining lymphocytes enter various peripheral lymph nodes, where they spend 12 to 18 hours scanning cellular networks for antigen. A small number of lymphocytes (about 10%) migrate to barrier immune tissues, including the skin and gastrointestinal, pulmonary, and genitourinary mucosa, where they are in close contact with the external environment (see Chapter 13).

How do the millions of newly generated naïve lymphocytes find their antigenic match, if there is one? The odds that the tiny percentage of lymphocytes capable of interacting with an antigen (one in 10^5) actually makes contact with that particular antigen are improved considerably by a variety of strategies. First, lymphocytes circulate through secondary lymphoid tissues continuously. An individual lymphocyte may make a complete circuit from blood to tissues to lymph and back again once or twice per day, probing earnestly for antigen in secondary lymphoid tissues. Second, as we will see shortly, the organization of lymphocytes within secondary lymphoid tissues profoundly increases the probability that a lymphocyte will make contact with "its" antigen (**Overview Figure 14-2**).

Naïve lymphocytes destined for lymph nodes exit the blood at the **high-endothelial venules** (**HEVs**) in the cortex

Lymphocyte Recirculation Routes

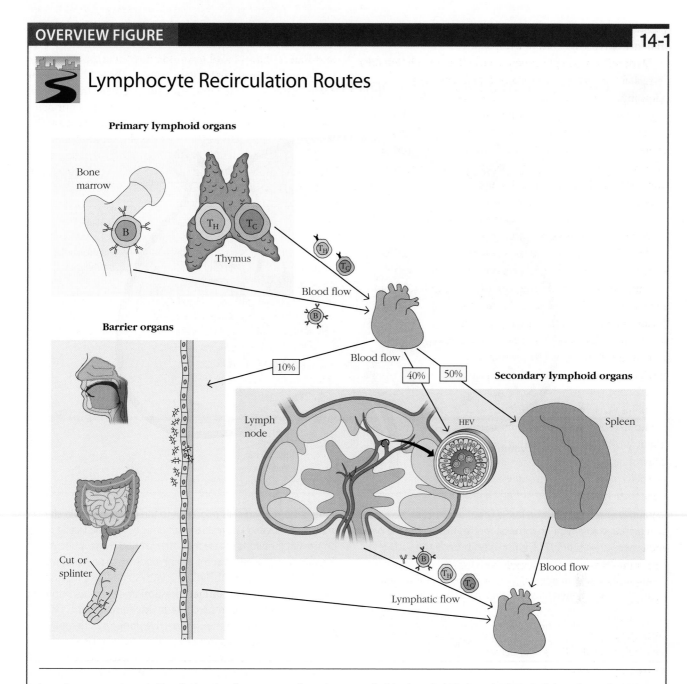

Lymphocytes and myeloid cells develop from stem cells in the bone marrow. T lymphocytes complete their maturation in the thymus. Once mature, naïve B and T cells enter the blood and distribute themselves throughout the body, traveling predominantly to the spleen (50%), lymph nodes (40%), and lymphoid tissue in barrier organs (10%). Lymphocytes enter lymph nodes and Peyer's patches via specialized regions of capillaries called *high-endothelial venules* (HEVs). If they do not bind antigen, naïve lymphocytes exit the lymph nodes via efferent lymphatic vessels. They return to the blood via the thoracic duct to begin their search for antigen again. HEVs are not present in the spleen, where cells exit from the end of open arterioles. Some naïve lymphocytes may leave the spleen via efferent lymphatics, but most return directly into the blood.

of the lymph nodes (**Figure 14-3a**). These specialized post-capillary venules are lined with distinct endothelial cells that have a plump, cuboidal ("high") shape (see Figure 14-3b). This contrasts sharply in appearance with the flattened endothelial cells that line the rest of the blood vessel. As many as 3×10^4 lymphocytes move through a single lymph node's HEVs every second.

Interestingly, HEVs fail to develop in animals raised in a germ-free environment. Investigators tested the possibility that HEV formation was dependent on antigen by surgically

OVERVIEW FIGURE   14-2

Cell Traffic in a Resting Lymph Node

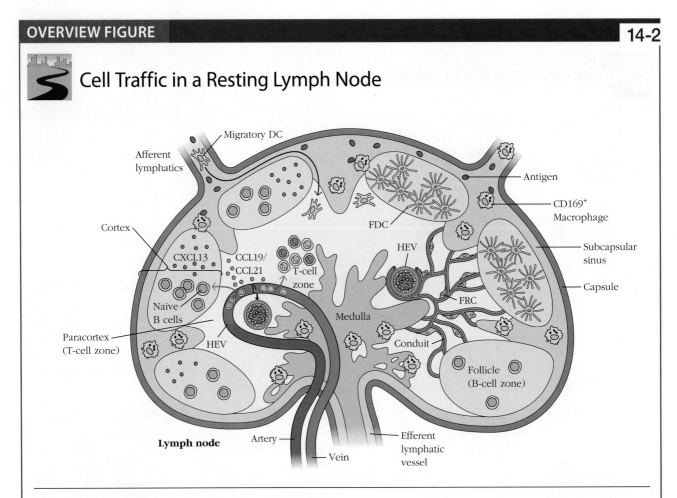

T and B lymphocytes travel into a lymph node from the blood, migrating across high-endothelial venules (HEVs) into the paracortex (extravasation). T lymphocytes browse the surfaces of cells on the *fibroblastic reticular cell* (FRC) network, and if they do not bind to antigen within 16 to 20 hours, they leave the lymph node via efferent lymphatics in the medulla. B lymphocytes travel into the follicles and browse the surfaces of follicular dendritic cells. If they do not bind antigen, they, too, leave the node via the efferent lymphatics. Dendritic cells and other antigen-presenting cells migrate into the lymph node via afferent lymphatics and join resident dendritic cells in the T-cell zone on the FRC network, where they are scanned by lymphocytes. Antigen typically arrives at the lymph node via the afferent lymphatics, entering the subcapsular sinuses either as processed peptide presented by antigen-presenting cells or as unprocessed protein that can be coated with complement. T and B cells are directed to their respective zones by chemokine receptor–chemokine interactions that are discussed in the text.

blocking the afferent lymphatic vessels, where antigen typically enters. Within a surprisingly short period, HEVs stopped functioning as access points for lymphocytes and in days the endothelial cells reverted to a more flattened morphology. These results show that endothelial cells receive and respond to signals from the lymph node.

How do naïve cells know that HEVs are the right site for entry into lymph nodes? The answer illustrates a common theme in immune cell migration: where cells go is determined by the array of surface protein receptors they express and the set of ligands expressed by cells they browse. Several different sets of interactions regulate lymphocyte homing to specific tissues before, during, and after

an immune response. These include chemokines and their corresponding chemokine receptors, **selectins**, **integrins**, and **adhesion molecules**. We will consider some of these below and describe them more thoroughly in **Advances Box 14-1**.

HEVs express a specific combination of ligands, collectively known as an **addressin**, that is recognized by CD62L, a selectin that is expressed on the surface of newly generated naïve T and B cells (see Advances Box 14-1, Figure 1). Endothelial cells associated with other tissues, such as the intestinal mucosa, skin, or brain, express distinct addressins that allow entry of different subsets of white blood cells, including activated lymphocytes.

(a)

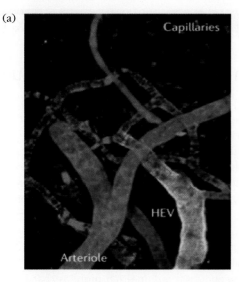

(b)

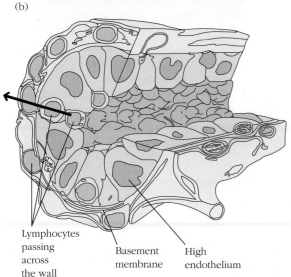

Lymphocytes passing across the wall

Basement membrane

High endothelium

(c)

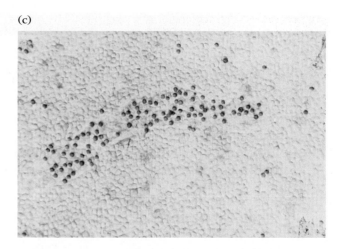

FIGURE 14-3 Lymphocyte migration through HEVs.
(a) An HEV in a lymph node as visualized by two-photon microscopy. Blood vessels are stained red and the HEV appears orange. (b) Schematic cross-sectional diagram of a lymph-node high endothelial venule (HEV). Lymphocytes (blue) are shown in various stages of attachment to the HEV and in migration through the endothelial cells (tan) into the cortex of the lymph node in the direction of the arrow. (c) Micrograph of frozen sections of lymphoid tissue. Some 85% of the lymphocytes (darkly stained) are bound to HEVs (in cross-section), and constitute only 1% to 2% of the total area of the tissue section. *[Part (a) reprinted with permission from Macmillan Publishers Ltd: from Girard, J-P., et al., "HEVs, lymphatics and homeostatic immune cell trafficking in lymph nodes,"* Nature Reviews Immunology, *2012, November;* **12***(11): 762-773, Figure 1b. Permission conveyed through Copyright Clearance Center, Inc.; part (c) republished with permission of Elsevier, from Rosen, S. D. "Lymphocyte homing: progress and prospects,"* Current Opinion in Cell Biology, *1989, October;* **1***(5):913-919, Figure C. Permission conveyed through Copyright Clearance Center, Inc.]*

When naïve lymphocytes engage HEV addressins, they slow down and begin to roll along the blood vessel wall. This initiates a sequence of events that ultimately induce cells to leave the blood vessel by squeezing between endothelial cells, a process called **extravasation**.

> **Key Concept:**
>
> • Naïve B and T lymphocytes leave the blood and enter the lymph nodes at high-endothelial venules (HEVs) via a process called *extravasation.*

Extravasation Is Driven by Sequential Activation of Surface Molecules

Extravasation is divided into four steps, each of which is regulated by distinct families of molecules: (1) *rolling*, mediated by selectins; (2) *activation* by chemokines; (3) *arrest*

and adhesion, mediated by integrin interaction with Ig family members; and, finally, (4) **diapedesis**, transendothelial migration (**Figure 14-4**).

As naïve T and B cells approach the HEV in a lymph node, their surface CD62L latches on specifically to the adhesion molecule GlyCAM, which is expressed by endothelial cells. Because this interaction is relatively weak, the cell doesn't adhere tightly to the blood vessel wall, but rather tumbles and rolls as the blood flows by. This slows the cell down long enough to allow new interactions to form. Chemokines decorating the surface of the endothelial cells interact with the chemokine receptor CCR7, which is expressed by T and B cells. Chemokine receptor signals induce a change in conformation of the lymphocytes' *integrins* (e.g., LFA-1) that enhance their ability to bind tightly to ICAMs on the endothelial cell. Even the shear forces generated by blood flow contribute to enhanced binding and lymphocyte arrest. The naïve lymphocyte is now prepared

ADVANCES   BOX 14-1

Molecular Regulation of Cell Migration between and within Tissues

The molecules that control movement of the cellular players within and between immune tissues include a number of important receptor/ligand families: selectins and integrins (collectively referred to as *cell-adhesion molecules [CAMs]*), and chemokines and chemokine receptors. These molecules regulate extravasation, the transit of cells from blood to tissue, as well as trafficking within tissues.

Cell-Adhesion Molecules: Selectins, Mucins, Integrins, and Immunoglobulin Superfamily Proteins

Cell-adhesion molecules (CAMs) are versatile molecules that play a role in all cell-cell interactions. In the immune system, CAMs play a role in helping leukocytes adhere to the vascular endothelium prior to extravasation. They help secure the location of resident memory cells among innate cells in barrier tissues. They also increase the strength of the functional interactions between cells of the immune system, including T$_H$ cells and APCs, T$_H$ cells and B cells, and CTLs and their target cells.

Most CAMs belong to one of four protein families: the selectin family, the mucin-like family, the integrin family, and the immunoglobulin (Ig) superfamily (**Figure 1**).

Selectins and Mucin-Like Proteins

The *selectin* family of membrane glycoproteins has an extracellular lectin-like domain that enables these molecules to bind to specific carbohydrate groups that decorate glycosylated *mucin-like proteins*.

The selectin family includes *L-, E-,* and *P-selectin*. L-selectin, also called *CD62L*, is expressed by naïve lymphocytes and helps initiate extravasation at the HEVs of lymph nodes. *Cutaneous leukocyte antigen (CLA)* is another selectin expressed by memory T cells. It binds E-selectin and regulates homing to the skin.

Mucin-like proteins are a group of heavily glycosylated serine- and threonine-rich proteins that bind to selectins. Those relevant to the immune system include CD34 and GlyCAM-1, found on HEVs in peripheral lymph nodes, and MAdCAM-1, found on endothelial cells in the intestine.

Selectins can bind to themselves as well as to sulfated carbohydrate moieties, including the sialyl-Lewis^x carbohydrate "cap" that decorate selectins.

(a) General structure of CAM families

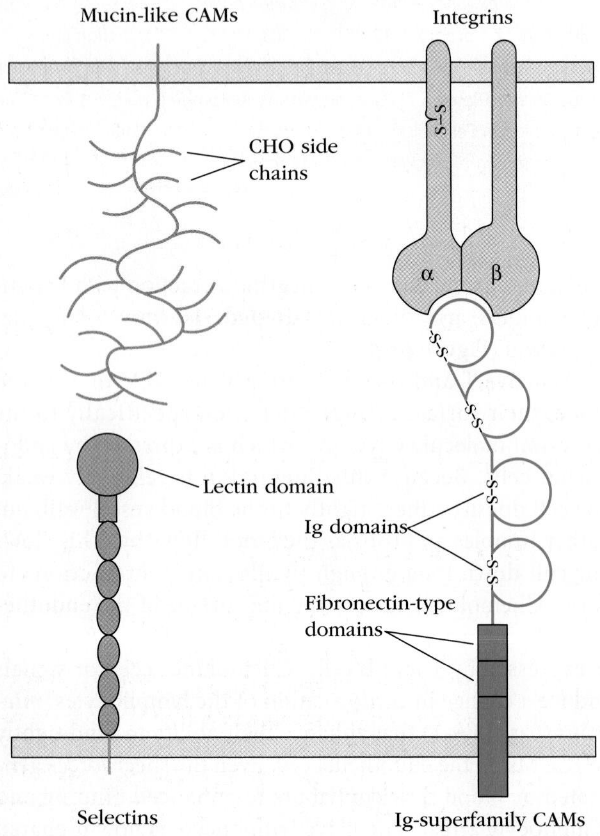

Mucin-like CAMs   Integrins

CHO side chains

α   β

Lectin domain

Ig domains

Fibronectin-type domains

Selectins   Ig-superfamily CAMs

(b) Selected CAMs belonging to each family

Mucin-like CAMs:	Selectins:
GlyCAM-1	L-selectin (CD62L)
CD34	P-selectin
PSGL-1	E-selectin
MAdCAM-1	Cutaneous leukocyte antigen (CLA)

Ig-superfamily CAMs:	Integrins:
ICAM-1, -2, -3	$\alpha_4\beta_1$ (VLA-4, LPAM-2)
VCAM-1	$\alpha_4\beta_7$ (LPAM-1)
LFA-2 (CD2)	$\alpha_6\beta_1$ (VLA-6)
LFA-3 (CD58)	$\alpha_L\beta_2$ (LFA-1)
MAdCAM-1	$\alpha_M\beta_2$ (Mac-1)
	$\alpha_X\beta_2$ (CR4, p150/95)
	$\alpha_E\beta_7$ (CD103)

FIGURE 1 **The four families of cell-adhesion molecules.**
(a) Schematic diagrams depict the general structures of members of the four families of cell-adhesion molecules. The lectin domain in selectins interacts primarily with carbohydrate (CHO) moieties on mucin-like molecules (collectively referred to as *PNAd* and *PSGL-1*). In integrin molecules, the α and β subunits combine to form the binding site, which interacts with an Ig domain in CAMs belonging to the Ig superfamily. These CAMs also include protein domains similar to those found in fibronectin. MAdCAM-1 (not shown), which is on the surface of endothelial cells that allow entry into the gut mucosa, contains both mucin-like and Ig-like domains and can therefore bind to both selectins and integrins. (b) A list of representative CAMs from each family.

(continued)

ADVANCES *(continued)*   **BOX 14-1**

The L-selectin that is expressed on naïve lymphocytes (CD62L) specifically interacts with carbohydrate residues referred to generally as *peripheral node addressin* (*PNAd*), which is found associated with multiple molecules (e.g., CD34, GlyCAM-1, and MAdCAM-1) on endothelial cells.

Integrins

Integrins are heterodimeric proteins consisting of noncovalently associated α and β chains. Integrins bind extracellular matrix molecules as well as cell-surface adhesion molecules. Leukocytes express several important integrins. The β_2 integrins (or CD18 integrins) combine with CD11 α chains to bind to members of the *Ig* superfamily *cellular-adhesion molecules* (ICAMs). *Lymphocyte function–associated antigen-1* (LFA-1 or CD11a/CD18) is one of the best-characterized integrins and regulates many immune cell interactions, including naïve T-cell/APC encounters. It initially binds weakly to ICAM-1, but a signal from the T-cell receptor will alter its conformation and result in stronger binding (see Figure 1). Such *inside-out signaling* is an important feature of multiple integrins and allows cells to probe other cell surfaces before making a full commitment to an interaction. The β_7 integrins (e.g., CD103 [$\alpha_E\beta_7$] and $\alpha_4\beta_7$) are a particularly important subset of integrins that allows

leukocytes to mingle with epithelial cells in our barrier tissues. CD103 interacts with *E-cadherin (CD324)* on epithelial cells and is expressed by gut and skin T_{REG} cells, T_{RM} cells, and APCs.

Immunoglobulin Superfamily CAMs

The Ig superfamily CAMs (ICAMs) are a diverse group of proteins, which act as ligands for β_2 integrins. ICAMs interact with LFA-1 (see above), but also exhibit *homotypic binding*—meaning they interact with themselves. MAdCAM-1 is expressed on the endothelium of blood vessels in the gut and other mucosa and has both Ig-like domains and mucin-like domains. It helps direct cell entry into the mucosa, binding to integrins and selectins expressed by lymphocytes induced to home to the gut, respiratory, and urogenital tracts.

Chemokines and Chemokine Receptors

Chemokines are small polypeptides that play a major role in regulating immune cell trafficking. Most consist of 90 to 130 amino acid residues (see Appendix III for the complete list of chemokines and a table showing the differences in chemokine receptor expression that govern leukocyte migration). Not only do they act as chemoattractants, but they also play a role in adhesion and activation of leukocytes. "Housekeeping" chemokines

are constitutively produced in lymphoid organs and tissues or in nonlymphoid sites such as the skin, where they direct normal trafficking of lymphocytes between primary and secondary lymphoid organs (e.g., from bone marrow to spleen, and from thymus to lymph node). These include CCL19 and CCL21, which decorate the FRC network and direct lymphocytes to T-cell zones, and CXCL13, which decorates the FDC network and directs cells to B-cell zones.

In contrast, *inflammatory chemokines* are typically induced in response to infection and proinflammatory cytokines, such as TNF-α. IL-8, also known as CXCL8, is one of the best examples of inflammatory chemokines and is made by multiple cells in response to antigen invasion. It attracts neutrophils, which express the receptors for IL-8: CXCR1, and CXCR2.

Extravasation is an excellent example of how adhesion molecules and chemokine receptors organize the migration of leukocytes. We discussed the extravasation of naïve lymphocytes across high-endothelial venules in the text of this chapter. **Table 1** highlights the molecules and events involved in extravasation of these and other leukocytes, neutrophils, and monocytes, which typically cross blood vessels to join the immune response against pathogens.

TABLE 1	**Molecules involved in extravasation of leukocytes**			
Leukocyte	**Molecules involved in rolling**	**Chemokines involved in activation**	**Molecules involved in adhesion**	**Comments**
Neutrophils	L-selectin and PSGL-1	IL-8 and macrophage inflammatory protein 1β (MIP-1β [also called CCL4])	LFA-1 and MAC-1	First to the site of inflammation: responds to C5a, bacterial peptides containing *N*-formyl peptides, and leukotrienes within inflamed tissues
(Inflammatory) monocytes	L-selectin	Monocyte-chemoattractant protein-1 (MCP-1 [also called CCL2])	VLA-4	Home to tissues after neutrophils; respond to bacterial peptide fragments and complement fragments within inflamed tissues
Naïve lymphocytes	L-selectin, LFA-1, VLA-4 (in low-affinity forms)	CCL21, CCL19, CXCL12 (T cells), and CXCL13 (B cells)	LFA-1 and VLA-4	Travel across high-endothelial venules to enter the lymph node

(a)

Rolling and extravasation

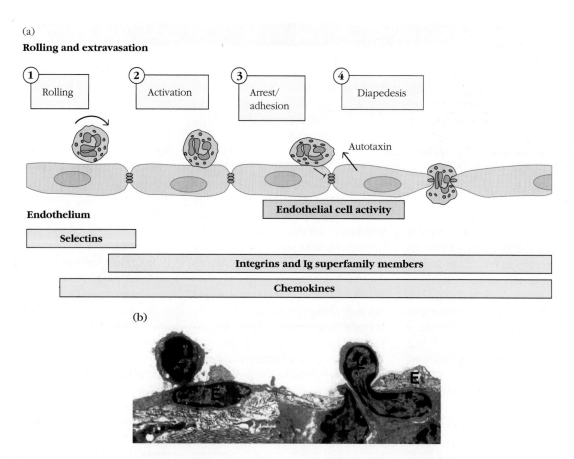

(b)

FIGURE 14-4   The steps of leukocyte extravasation.
(a) Schematic of the major events regulating extravasation. A neutrophil is depicted, but the events depicted are applicable to all leukocytes. Tethering and rolling are mediated by binding of selectin molecules to sialylated carbohydrate moieties on mucin-like CAMs. Chemokines then bind to G protein–linked receptors on the leukocyte, triggering an activating signal. This signal, combined with the shear force produced by blood flow, induces a conformational change in the integrin molecules, enabling them to adhere firmly to Ig superfamily molecules on the endothelium. The arrest of a leukocyte also generates signals within endothelial cells that generate factors (autotaxin) that enhance leukocyte motility, loosen adhesive connections between cells, and induce extension of processes that help draw the leukocyte across. Ultimately, leukocytes crawl between endothelial cells into the underlying tissue (transmigration). (b) Transmission electron micrograph capturing lymphocytes (L) migrating through the endothelial cell (E) layer. *[Part (b) reprinted by permission from Macmillan Publishers Ltd, from Imhof, B. A., et al., "Adhesion mechanisms regulating the migration of monocytes," Nature Reviews Immunology, 2004 June; **4(6):** 432–444, Figure 3a. Permission conveyed through Copyright Clearance Center, Inc.]*

to crawl between endothelial cells, driven in addition by attraction to higher concentrations of chemokines inside the lymph node.

Lymphocyte motility is enhanced by the activity of an enzyme, autotaxin, released by HEV cells themselves. In fact, endothelial cells are very active participants. They recognize the binding of an arrested lymphocyte and send signals that loosen adhesions with neighboring endothelial cells. They also reorganize their cytoskeleton and extend processes that help draw the lymphocyte between. All in all, it takes about 2 to 3 minutes for lymphocytes to pass between endothelial cells.

Naïve T and B cells are not the only cells that extravasate, of course. Any circulating leukocyte that recognize addressins expressed by particular endothelial cells exit the blood and enter the neighboring tissues. Only cells that express the right set of receptors for the addressins are allowed to cross. Effector cells, for instance, recognize addressins induced by inflammatory responses at the site of infection. However, they will pass by healthy tissues, ignoring endothelial cells that do not express appropriate addressins.

Interestingly, the spleen appears to have no HEVs. Arterioles release immune cells directly into the marginal sinus. (See **Figure 14-5** for a comparison of approaches taken by spleen and lymph node.) Furthermore, the cues lymphocytes use to home to the white pulp of the spleen differ from those they use to enter the cortex of the lymph node and are not yet fully understood. Entry appears to require the coordinated activity of chemokines, integrins, and autotaxin, but is independent of selectins such as CD62L, which are important for lymph-node entry.

(a)

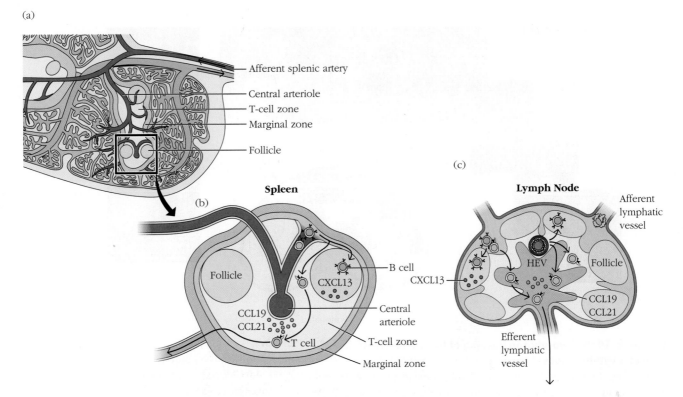

FIGURE 14-5 Lymphocyte migration in the spleen.
(a) Schematic of spleen microanatomy showing the central arteriole, which releases circulating lymphocytes into the marginal zone, a microenvironment that separates blood from the white pulp. (b) T cells then migrate into the T-cell zone (PALS) and B cells migrate into follicles. Cells that do not meet their antigen usually exit via the splenic vein. In the lymph node (c), these migrations are regulated by chemokine receptor–chemokine interactions and guided by FRC and FDC networks, which are described in more detail in the text.

Key Concepts:

- Extravasation involves four steps: rolling, activation, arrest/adhesion, and, finally, migration between endothelial cells lining the blood vessel. It is the result of an organized sequence of interactions between molecules expressed by leukocytes and their ligands on endothelial cells, including selectins and addressins, chemokine receptors and chemokines, and integrins.

- Addressins are collections of diverse adhesion molecules that vary depending on the tissue a blood vessel serves. They specify which cells can enter that tissue.

- Lymphocytes gain entry into the splenic white pulp via distinct homing mechanisms. The spleen has no HEVs, and selectins do not appear to play a role.

Naïve Lymphocytes Browse for Antigen along the Reticular Network of Secondary Lymphoid Organs

After squeezing between HEV cells, naïve B and T lymphocytes enter the cortex of the lymph node (see Chapter 2), where they are guided by different chemokine interactions to distinct microenvironments. The movements of naïve lymphocytes that enter and scan the lymph node are remarkable to watch. When these were first visualized, the idea that lymphocytes were rather dull and inert (an impression based on cells fixed on microscope slides) was immediately reversed (see **Figure 14-6**; see also **Video 14-6v** for an example; direct links to original sources for all videos can be found at the end of this chapter, on p. 545). These initial videos also inspired the discovery of the fibroblastic reticular cell network. Investigators recognized that fluorescently tagged lymphocytes did not move freely but were influenced by "invisible" structures. These were ultimately identified as the fibroblastic reticular cell (FRC) network of conduits and cells that provide roadways for naïve lymphocytes and antigen-presenting cells (see Figure 2-14).

Naïve B and T cells spend many hours probing for antigen: B cells in B-cell follicles and T cells in T-cell zones (also called the *paracortex*) (see Figure 14-5). Naïve T lymphocytes, which express the chemokine receptor CCR7, dive in and out of the lymph-node parenchyma, using FRC tracts (see chapter opening image–associated **Video 14-Ov**). They are specifically attracted to this network because it is decorated with the CCR7 ligands CCL21 and CCL19. Naïve B cells that transit across HEVs into the lymph-node cortex

(a)

Follicles

Paracortex

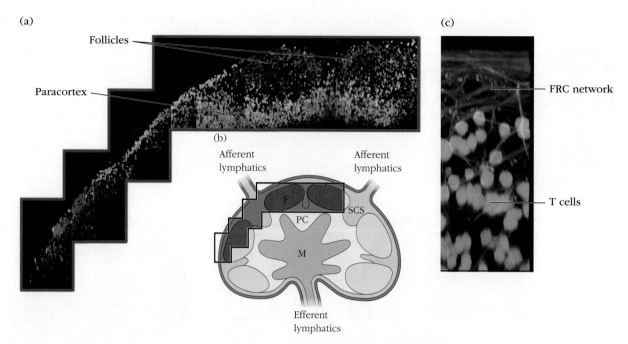

(b)

Afferent lymphatics   Afferent lymphatics

F

PC   SCS

M

Efferent lymphatics

(c)

FRC network

T cells

FIGURE 14-6 Two-photon imaging of live T and B lymphocytes within a mouse lymph node. Fluorescently labeled T lymphocytes (green) and B lymphocytes (red) were injected into a mouse and visualized by two-photon microscopy after they homed to the inguinal lymph node. (a and b) T and B cells localize to distinct regions of the lymph node: T cells in the paracortex (PC) and B cells in the follicles (F). Antigen and APCs enter the subcapsular sinus (SCS). Cells leave via efferent lymphatics from the medulla (M). (c) A magnified image of T cells (fluorescing green) interacting with the fibroblastic reticular cell (FRC) network (stained red). *[Parts (a) and (c) republished with permission of the American Association for the Advancement of Science, from Miller, M. J., et al., "Two-photon imaging of lymphocyte motility and antigen response in intact lymph node," Science 2002, June 7;* ***296(5574):*** *1869–73, Figures 1B and 1G. Permission conveyed through Copyright Clearance Center, Inc.]*

also begin their travels along FRC fibers. However, because they express different chemokine receptors, including CXCR5, they soon change allegiance to the fibers established by the follicular DCs, which are decorated with the corresponding chemokine CXCL13.

The movements of naïve T and B cells in the spleen are guided by the same chemotactic cues and conduit network (see Figure 14-5). Once they find their way into the white pulp, naïve B cells are attracted to CXCL13 in the follicle and T cells are attracted to CCL19 and CCL21 in the periarteriolar lymphoid sheath (PALS).

Antigen-presenting cells are found in every secondary lymphoid tissue and every microenvironment (**Figure 14-7**). Some are long-term residents and others actively migrate within and between tissues. Figure 14-7 shows dendritic cells in several different areas of a lymph node. **Video 14-7v** shows

(a) Subcapsular cortex

Dendritic cells

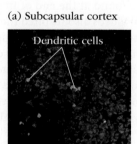

(b) Paracortex

Dendritic cells

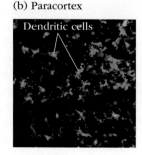

(c) B-cell follicle

Dendritic cells

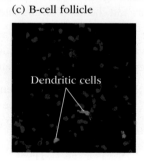

FIGURE 14-7 Antigen-presenting cells are present in all lymph-node microenvironments. Intravital microscopy of inguinal lymph nodes of anesthetized mice, all of whose dendritic cells fluoresce green. Several areas of the lymph node are visualized (a-c). In the subcapsular sinus (a), green dendritic cells are surrounded by macrophages, which have taken up a red fluorescent dye. In the T-cell zone or paracortex (b), dendritic cells mingle with T cells (red). In the follicle (c), dendritic cells are not as numerous but can be found among the B cells (blue). *[Reprinted with permission from Macmillan Publishers Ltd, from Lindquist, Randall, et al., "Visualizing dendritic cell networks in vivo." 2004 November; Nature Immunology 5, 1243–1250, Fig. 4a, b, f. doi:10.1038/ni1139. Permission conveyed through Copyright Clearance Center, Inc.]*

the activity of DCs (green) traveling among a bed of more sessile, long-term resident macrophages (red) in the **subcapsular sinus** of a lymph node. A population of migratory DCs also enters the deeper T-cell zone of the lymph node, where they crawl less vigorously and become part of the FRC network (see Figure 14-7b). They extend their long processes along the conduits to allow naïve T cells to scan their surfaces during their travels. A few DCs are even found in B-cell follicles (see Figure 14-7c), where they also present antigen.

Early estimates using static imaging techniques suggested that up to 500 T cells probe the surface of a single dendritic cell (DC) per hour. However, more accurate dynamic imaging data reveal that one DC can be surveyed by more than 5,000 T cells per hour! Suddenly it is much easier to imagine that one in 100,000 antigen-specific T cells can find the MHC-peptide complex to which it can bind.

Naïve B cells spend a similar amount of time scanning for antigen. They travel on distinct cellular pathways, specifically the follicular DC networks in the follicles (**Figure 14-8**). They probe for antigen on the surface of the follicular DCs themselves, and can also bind soluble antigen that has entered follicles from draining afferent lymphatic vessels.

(a)

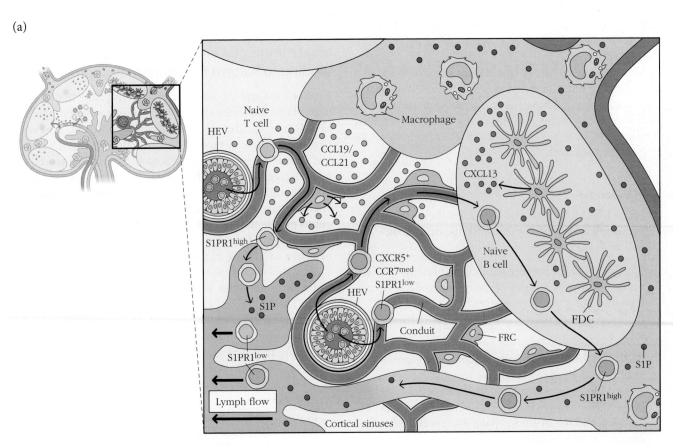

(b)

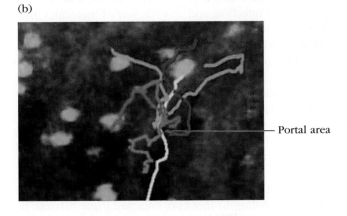

Portal area

FIGURE 14-8 Lymphocytes exit the lymph node through portals in the cortical and medullary sinuses. (a) Schematic of lymphocyte traffic in and egress from a lymph node. T cells and B cells follow chemokine cues as they probe the FRC and FDC networks, respectively. If naïve lymphocytes do not find antigen within a certain time period they up-regulate S1PR1 and enter efferent lymphatic vessels from the cortex (B cells) or the medulla (T cells). (b) The movements of fluorescently labeled T cells (green) were visualized by two-photon microscopy in a lymph node where the medullary sinus stains red. The movements of several T cells are traced with colored lines and indicate that they leave at discrete sites called *portals*. See Video 14-8v. *[Part (b) reprinted by permission from Macmillan Publishers Ltd, from Wei, S.H., et al., "Sphingosine 1-phosphate type 1 receptor agonism inhibits transendothelial migration of medullary T cells to lymphatic sinuses," Nature Immunology 2005 December; **6(12):**1228–35, Figure 5c. Permission conveyed through Copyright Clearance Center, Inc.]*

If they do not find and bind antigen on their travels, naïve B and T cells up-regulate the **sphingosine 1-phosphate** (S1P) receptor, S1PR1, which allows them to leave (egress) secondary lymphoid tissues (see Figure 14-8a). Naïve lymphocytes circulating through lymph nodes leave via efferent lymphatics in both the cortical and medullary sinuses (see Overview Figure 14-2). Cells lining these sinuses express S1P, the ligand for S1PR1. S1PR1-S1P interactions induce naïve lymphocyte migration via specific portals (see Figure 14-8b and **Video 14-8v**). Once they have left the node, they return to the blood via the thoracic duct and resume their search for antigen in other lymph nodes. Interestingly, S1PR1 mediates lymphocyte egress from a wide range of other tissues, including the bone marrow and thymus.

Although the details of egress from the spleen are still being worked out, the red pulp is rich in S1P and clearly plays a role in regulating the exit of naïve B and T cells. However, most lymphocytes circulating through the spleen appear to exit into the bloodstream directly from the white pulp or marginal zone.

Key Concepts:

- Naïve T and B lymphocytes both enter the cortex of the lymph nodes and are then attracted by chemokines to distinct microenvironments.

- T cells browse the surfaces of antigen-presenting cells in the paracortex of the lymph node, traveling along the FRC network. B cells browse for antigen along follicular DC networks in the follicle.

- Naïve lymphocytes that do not encounter antigen leave the lymph node via efferent lymphatics after about 12 to 18 hours and re-enter the circulation to probe another lymph node. Naïve lymphocytes scan for antigen in the spleen for about 5 hours and exit directly into the bloodstream.

- S1PR1, the receptor for the S1P ligand, regulates lymphocyte exit from many different tissues including the lymph nodes, spleen, thymus, and bone marrow.

Immune Cell Response to Antigen: The Innate Immune Response

Infectious agents range in size and complexity from viruses, bacteria, eukaryotic protozoa, and fungi to large multicellular pathogens, such as worms. Antigens also come from other sources, including tumor cells, environmental toxins, foods that we eat, and pollens that we breathe. The immune response is tailored to each type

of antigen—a phenomenon that is possible only because of the coordinated activities of innate immune cells that make first contact with antigen.

As you know from Chapters 4 and 13, the innate immune system plays a critical role in our first response to invading pathogens. Epithelial cells and myeloid cells at barrier tissues work together and respond quickly to clear infection, recruit adaptive immune cells, and shape the adaptive immune response. In this section, we briefly review their activities in the context of host tissues, focusing particularly on the role of the host in delivering antigen to secondary lymphoid tissue.

Innate Immune Cells Are Activated by Antigen Binding to Pattern Recognition Receptors

Extracellular and intracellular pathogens gain entry to the body via breaks in the skin or mucosal surfaces. Pathogens, as well as molecules released by damaged tissues, are recognized by *pattern recognition receptors* (PRRs), which are expressed by many different innate immune cells. PRR engagement sets off signaling cascades that contribute in multiple ways to the defense against pathogens. For example, TLR engagement induces enterocytes in the gut to proliferate, release antimicrobial peptides (AMPs), and transport IgA into the intestinal lumen. The gram-negative, diarrhea-causing bacteria *Salmonella* binds TLR4, expressed by enterocytes of the intestinal epithelium, while herpes simplex virus (HSV) binds to both external (TLR2) and internal (TLR9) Toll-like receptors expressed by microglial cells, which are myeloid-lineage APCs of the nervous system. *Staphylococcus aureus*, an extracellular bacterium that typically gains entry via the skin, binds to TLR2 on dendritic cells in dermal (skin) tissue. The specific PRR or set of PRRs engaged by a pathogen shapes the type of response.

Many bacterial infections trigger release of chemoattractants like IL-8 that call neutrophils to the site of infection. Neutrophils, in fact, rapidly swarm to sites of damage and infection, and, in turn, secrete molecules that attract additional innate immune cells. **Figure 14-9** and **Video 14-9v** show a vivid example of live neutrophils (red) swarming to a site of damage (dotted blue circle) in the ear of a mouse. Monocytes (green) also migrate to the site of damage, following the swarm, although much more slowly. Traces of the movements of neutrophils (blue) and monocytes (yellow) are shown in the bottom panels of Figure 14-9. Chapter 13 provides a more thorough description of the activities of the innate immune system at the initial site of infection.

Although the innate immune system can be very effective at clearing pathogens that have breached tissue barriers, it plays another critical role by communicating the presence of infection to the adaptive immune system and sending emissaries to local lymphoid tissue.

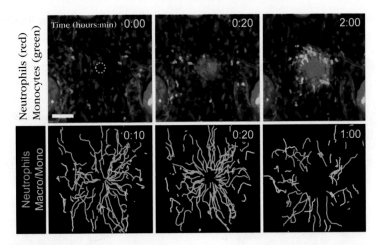

FIGURE 14-9 Neutrophils and monocytes migrate to a site of inflammation. The ear of a mouse was scratched with a laser at the site indicated by the blue circle at time zero, and the trafficking of neutrophils (red) and monocytes (green) was imaged in the dermis by intravital time-lapse microscopy. Within minutes, neutrophils swarm to the site of damage; monocytes follow more slowly. *[Reprinted with permission from Macmillan Publishers Ltd, from Lämmermann, T., et al., "Neutrophil swarms require LTB4 and integrins at sites of cell death in vivo," Nature, 2013, June 20;* ***498:****371–375, Figure 1. Permission conveyed through Copyright Clearance Center, Inc.]*

Key Concepts:

- Innate immune cells, including granulocytes and antigen-presenting cells (APCs), are the first to respond to pathogen. Innate immune cells bind pathogen via pattern recognition receptors (PRRs), which generate signals that result in the release of inflammatory signals, including chemokines and cytokines.

- Chemokines attract other innate immune cells to the site of infection. Neutrophils are generally the first cell type to move from the bloodstream into inflammatory sites; they can swarm around pathogens as they attack. They also produce more chemokines that attract APCs, including dendritic cells (DCs), which are the most efficient activators of naïve T cells.

Antigen Travels in Two Different Forms to Secondary Lymphoid Tissue via Afferent Lymphatics

In order to initiate an adaptive immune response and generate immune memory, the pathogens and antigens that penetrate our barrier tissue must make contact with the naïve T and B lymphocytes, which are not in barrier tissues, but are circulating among lymph nodes and spleen (**Figure 14-10**). B cells will be looking for unprocessed antigen in the follicle, while T cells will be scanning antigen-MHC complexes on the surfaces of antigen-presenting cells in T-cell zones. Therefore, antigen must travel to lymph nodes or spleen and arrive in two different forms—as whole antigen and in the context of MHC complexes. How are multiple forms of antigen delivered from the site of infection to secondary lymphoid tissues?

First, antigen-presenting cells that have been activated by PRR signals are capable of processing antigen at the site of infection. After activation, APCs also gain the ability to migrate to local (draining) lymph nodes, traveling via afferent lymphatic vessels. Second, some pathogens and particulate antigen travel via lymphatic vessels directly to the lymph nodes. Some of this antigen remains unprocessed and is relayed to follicles, where it can be scanned by B cells. Some is further processed by lymph node–resident APCs and presented with MHC.

Recent imaging studies that tracked immature malarial parasites (sporozoites) confirm that whole pathogen can travel directly to lymph nodes from the site of infection. Minutes after injection into the skin by a mosquito bite, sporozoites travel rapidly and directly to the draining lymph nodes. Lymph node–resident APCs, rather than skin dendritic cells, take up the task of processing and presenting parasite antigen (**Figure 14-11** and **Video 14-11v**).

Key Concepts:

- APCs at the site of infection engulf antigen, presenting processed peptide in MHC class II and cross-presenting it in MHC class I. They up-regulate chemokine receptors that allow them to travel to draining lymph nodes via the afferent lymphatics to alert the adaptive immune system of the presence of pathogen.

- Some antigen travels directly to lymph nodes. Unprocessed antigen is relayed to follicles, and some is processed by APCs in the lymph node.

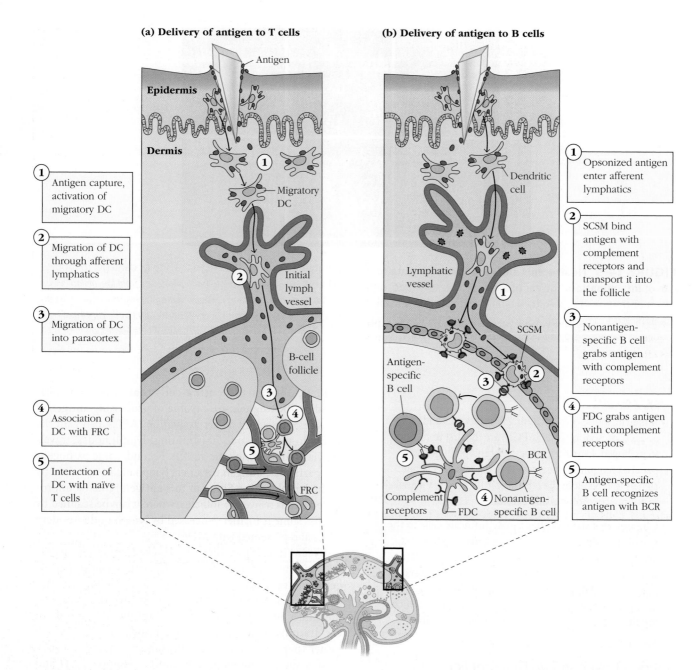

(a) Delivery of antigen to T cells

① Antigen capture, activation of migratory DC

② Migration of DC through afferent lymphatics

③ Migration of DC into paracortex

④ Association of DC with FRC

⑤ Interaction of DC with naïve T cells

(b) Delivery of antigen to B cells

① Opsonized antigen enter afferent lymphatics

② SCSM bind antigen with complement receptors and transport it into the follicle

③ Nonantigen-specific B cell grabs antigen with complement receptors

④ FDC grabs antigen with complement receptors

⑤ Antigen-specific B cell recognizes antigen with BCR

FIGURE 14-10   How antigen travels into a lymph node. Antigen is delivered from tissues (e.g., a wound in the skin) to the lymph node via the afferent lymphatics. (a) *Delivery of antigen to T cells.* Antigen-presenting cells, including DCs, that have processed antigen at the site of infection (1) travel via afferent lymphatic vessels (2) to the paracortex (3) where they position themselves on the fibroblastic reticular cell (FRC) network (4) and are scanned by naïve CD4$^+$ and CD8$^+$ T cells (5). Soluble antigen can also be processed by resident antigen-presenting cells in the cortex and paracortex. (b) *Delivery of antigen to B cells.* Soluble antigen (red) that is opsonized by circulating antibodies or by complement (shown) travels via afferent lymphatics (1) where it binds to complement receptors (blue) on CD169$^+$ subcapsular sinus macrophages (SCSMs). SCSMs carry antigen to the follicles of the lymph node (2) either by directly migrating into the follicle or by transferring opsonized antigen to other cells with complement receptors, including nonantigen-specific (noncognate) B cells, shown here (3). Nonantigen-specific B cells travel through the follicle and hand off the antigen to follicular dendritic cells (4), which are continuously scanned by naïve B cells and recognized by antigen-specific naïve B cells (5).

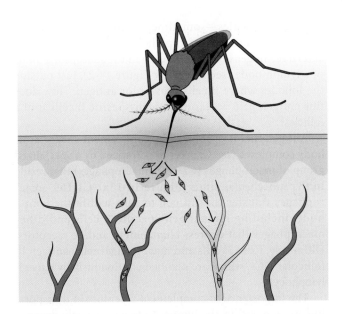

FIGURE 14-11 Mosquitos carry malarial parasites (sporozoites) and transmit them into tissue, blood, and lymphatics after a bite.

Antigen-Presenting Cells Presenting Processed Antigen Travel to the T-Cell Zones of Secondary Lymphoid Tissue

Whether antigen is processed by APCs before or after it enters a lymph node, it still needs to find its way to sites where T cells circulate. In the lymph node, T cells circulate on the FRC network in the paracortex. In the spleen, they travel along FRC conduits in the PALS. How do APCs find their way to T-cell zones? Activation of APCs by PRR signaling not only enhances their antigen-presenting abilities but also up-regulates the expression of distinct chemokine receptors, allowing them access to new microenvironments.

Specifically, activated APCs in barrier tissues process antigen and up-regulate the chemokine receptor CCR7, which allows them to follow trails of chemokines. CCL21 seems particularly important and decorates the endothelial cells that line lymphatic vessels. Indeed, if you block CCR7-CCL21 interactions, APCs stop crawling toward the lymph node (see Figure 14-4a). Lymph flow assists them at the very end of their transit to a lymph node, flushing APCs into the subcapsular sinus (**Figure 14-12** and **Video 14-12v**).

Once in the lymph node, APCs also use CCR7 to find their way to the paracortex, where they settle themselves along FRC conduits to be scanned by up to 5,000 naïve T cells per hour (see Figure 14-10). Because they are the most efficient activators of naïve T lymphocytes, DCs play an especially important role in the primary response.

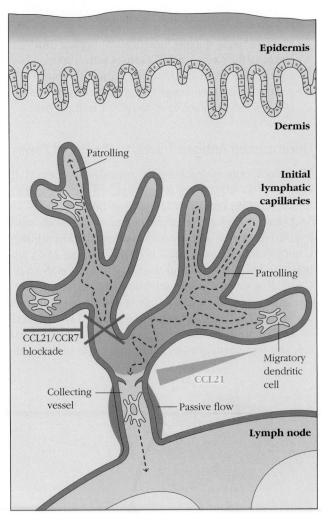

FIGURE 14-12 Migration of antigen-presenting cells from tissue to lymph node through efferent lymphatics. Schematic of the trajectory of migratory dendritic cells in efferent lymphatics. The lymphatic endothelium generates a CCL21 chemokine gradient (orange) in response to tissue inflammation. Patrolling dendritic cells, which crawl along the endothelium, sense the gradient with their CCR7 receptors and migrate toward the higher concentrations of CCL21 (darker orange). Toward the end of their journey, they are flushed into the subcapsular sinus of the lymph node by the flow of lymph. APC migration to the lymph node can be inhibited by molecules that block the binding of CCR7 with CCL21 (CCR7/CCL21 blockade)—a strategy that can be helpful in reducing immune response to transplanted tissue.

Naïve CD4$^+$ and CD8$^+$ T cells appear to be activated by distinct subsets of APCs in distinct subregions of the paracortex. Naïve CD8$^+$ T cells interact best with APCs that are able to cross-present the antigen they capture in MHC class I molecules (see Chapter 7). These include migratory CD103$^+$ and CD8$^+$ DCs. Naïve CD4$^+$ T cells tend to be activated by CD11b$^+$ DCs, including Langerhans cells, which can travel from the skin to draining lymph nodes.

Key Concept:

- Processed antigen in the form of MHC-peptide complexes on the surface of APCs arrives within hours and travels to the T-cell zone (paracortex). APCs position themselves along the fibroblastic reticular cell network to be scanned by up to 5,000 T cells per hour.

Unprocessed Antigen Travels to the B-Cell Zones

Whereas T cells are probing for processed antigen-MHC complexes on APCs in the lymph nodes, B cells are looking for unprocessed antigen. Where do they find this?

As mentioned above, unprocessed antigen can reach lymph-node tissue directly, and can do so within minutes, if not seconds, following infection. (In contrast, APCs take hours and even days to find their way to the lymph node.) Depending on its size and solubility, unprocessed antigen can gain access to a lymph node in several ways. If small and soluble enough, whole antigen can travel directly to the lymph node via the blood. Larger antigen particles and pathogens are carried to the lymph node via the afferent lymphatics—sometimes on the "backs" of APCs that have bound opsonized pathogen. Unprocessed small and large antigens typically end up in the subcapsular sinus of a lymph node (see Overview Figure 14-2), just outside the follicles where B cells browse for antigen. How does this antigen gain access to antigen-specific B cells in the follicles?

Antigen that is opsonized by complement and/or antibody complexes (see Chapter 5) is trapped by a specialized group of macrophages (CD169$^+$ macrophages) that reside in the subcapsular sinus (**Figure 14-13a**). CD169$^+$ macrophages transfer the antigen to cells within the lymph node, including nonantigen-specific B cells and other macrophages that express complement and Fc receptors. Ultimately, these B cells and macrophages relay antigen to follicular DCs, which are scanned continuously by naïve B lymphocytes.

The importance of complement receptors in transporting antigen within the lymph node was directly demonstrated by an elegant study that compared the abilities of normal versus complement receptor–deficient B cells to capture antigens from subcapsular sinus macrophages

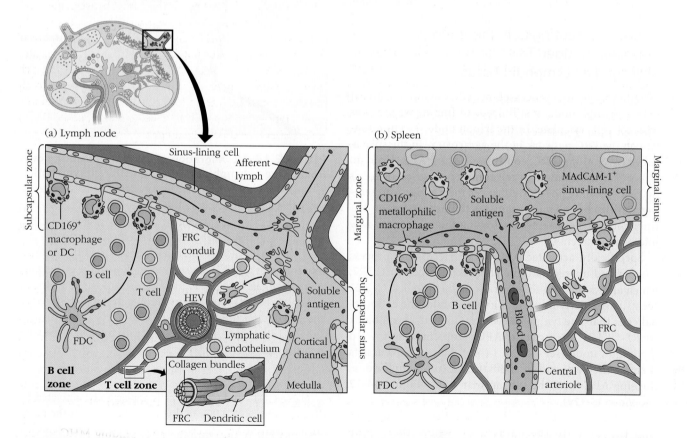

FIGURE 14-13 Antigen entry to lymph node and spleen.
(a) Processed and unprocessed antigen enters draining lymph nodes via afferent lymphatics. Unprocessed antigen is relayed to the follicles and recognized by B lymphocytes. Antigen-presenting cells make their way to the FRC network and present antigen to scanning T cells.

(b) In contrast, processed and unprocessed antigen enters the spleen directly from the blood, which empties into the sinus bordering the marginal zone. From here, unprocessed antigen is relayed to follicles and antigen-presenting cells migrate to T-cell zones via processes similar to those that occur in lymph nodes.

(**Video 14-13v**). Normal B cells can be seen sampling antigen present on the surface of macrophages (right panel). They grab and run away with chunks of the red protein. Complement receptor–deficient B cells also probe the macrophages, but come away completely empty (left panel). The B cells that grab antigen via their complement receptors ultimately relay the antigen to follicular dendritic cells in the follicle, which have complement receptors of their own and are continuously scanned by naïve B cells.

> **Key Concept:**
> - Small soluble antigens enter the follicle quickly and directly. Unprocessed antigen opsonized with complement is trapped by macrophages lining the sinuses and relayed to follicular DCs through the activity of nonantigen-specific B cells (and others).

Blood-Borne Antigen Is Captured by Specialized APCs at the Marginal Zone of the Spleen

How does antigen gain access to naïve T and B cells in the spleen? As you know from Chapter 2, the spleen is composed of the red pulp and white pulp, which are separated by the marginal zone. Much like lymph nodes, the white pulp is organized into T-cell zones (PALS) and B-cell follicles. The spleen primarily captures antigen coming from the bloodstream, which empties directly into the marginal zone (see Figure 14-13b). Innate immune cells in the marginal zone of the spleen respond to and process antigens. Activated APCs migrate to the T-cell zone of the white pulp, where they position themselves to be scanned by circulating naïve T cells. Whole or particulate antigen also enters the white pulp either directly or via a relay system, involving CD169$^+$ macrophages, that operates in a manner similar to what we described for the lymph node (see Figure 14-13a).

> **Key Concept:**
> - Antigen enters the spleen directly from the blood. Unprocessed antigen is relayed to follicles by macrophages lining the marginal zone sinus. Processed antigen is carried by APCs across the marginal zone to the T-cell zone of the white pulp.

First Contact between Antigen and Lymphocytes

In 2004, the Jenkins laboratory published two remarkable simulations of the activities of T cell, B cell, and dendritic cell interactions in a lymph node before and after the introduction of antigen. Although not live images, their animations were based directly on published data about leukocyte behavior and trafficking. Even as more studies fill in details (**Figure 14-14**), these simulations remain among the best

introductions to the behavior of three main cell types (T cell, B cell, and APC) during the first 50 hours of a primary immune response in the lymph node.

As described above, in the absence of infection (**Video 14-14v1**) naïve T cells (green) and B cells (red) browse for antigen in their own niches: B cells in the follicles and T cells in the T-cell zones (paracortex). Dendritic cells (purple) also crawl slowly through the lymph node—predominantly in the paracortex but also between the follicles.

After antigen is introduced (**Video 14-14v2**), the behavior of each cell type changes markedly. An activated, migrating antigen-presenting DC (depicted in brighter purple) enters the lymph node from the site of infection and travels to the T-cell zones. Within a short period of time, the DC is discovered by a probing antigen-specific T cell. This naïve T cell engages MHC-peptide on the surface of the DC and becomes active (turning more yellow). Within hours it begins to divide in the cortex.

Antigen-specific B cells in the follicle encounter soluble antigen and are activated by BCR-antigen signaling, becoming a brighter white. B cells activated in the follicle change their patterns of movement markedly and crawl toward the T-cell zone. Here they cavort with antigen-specific T cells, which have also been drawn to the border between the T-cell zone and the follicle. Once an activated B cell is engaged by an activated T cell, this B cell has a variety of options, one of which is to return to the follicle to generate a germinal center.

Below, we describe the behavior of leukocytes in a primary immune response in more detail, focusing most closely on the travels and interactions of the three major lymphocyte subpopulations—CD4$^+$ T cells, B cells, and CD8$^+$ T cells—after they encounter antigen.

Naïve CD4$^+$ T Cells Arrest Their Movements after Engaging Antigens

As you have learned from Chapter 10, CD4$^+$ T cells are required for the initiation of optimal primary and secondary humoral and cellular adaptive immune responses. They are activated when they engage MHC-peptide on the surface of DCs that have been exposed to pathogen. Depending on the specific combination of signals they receive from DCs, activated CD4$^+$ T cells differentiate into distinct effector helper cells. Some helper T-cell subsets assist in type 1 responses, licensing APCs to induce the differentiation of killer CD8$^+$ T cells. Other effector T-cell subsets assist in type 2 responses, helping activated B cells to class switch and differentiate.

Studies of live, naïve CD4$^+$ T cells interacting with antigen in a lymph node show that they travel relatively quickly through the paracortex (10 μm/min or more) prior to engaging antigen. Within minutes after binding MHC-peptide complexes, they "arrest," slowing down to 4 to 6 μm/min. Arrested T cells form both transient and stable engagements with DCs. The kinetics of these early encounters differs depending on the quality, quantity, and availability of antigen as well as the activation state of a DC. In order to proliferate

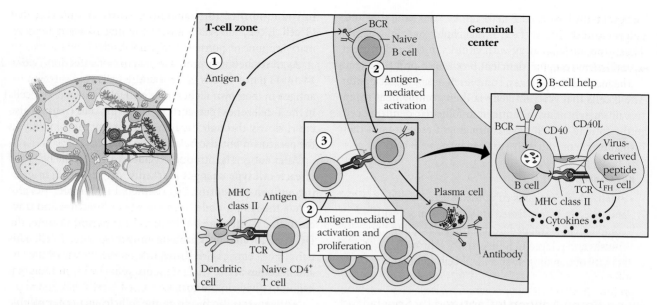

FIGURE 14-14 Activation of CD4⁺ T cells and B cells in a lymph node during a primary immune response. (1) T and B cells make contact with antigen in the paracortex and follicle of the lymph node. (2) Antigen-specific CD4⁺ T cell interacts with the APC in the paracortex, proliferating and differentiating into helper T-cell subsets. Some activated helper T cells, such as T_FH cells, migrate to the border of follicles. B cells that have recognized antigen in the follicle migrate to the follicle border, where they interact with and receive help from antigen-specific T cells. (3) Helper T cells bind B cells via TCR/MHC engagement, generating an immunologic synapse. T cells deliver help through CD40/CD40L interactions and by secreting cytokines.

and differentiate optimally, it appears as if naïve CD4⁺ T cells become involved in committed, long-term relationships, lasting 8 hours or more, with an antigen-presenting DC. Cells can experience up to 10 divisions over the following 4 to 5 days.

A compelling example of the effect of antigen on T-cell movements is shown in **Video 14-14v3**. Within minutes after antigen-presenting DCs (green) are introduced into a lymph node, T cells specific for that antigen (red) crawl more slowly, "arrest," and form stable cell-cell interactions with the DCs.

Key Concepts:

- Naïve CD4⁺ and CD8⁺ T lymphocytes that bind antigen in T-cell zones of secondary lymphoid tissue slow down considerably and "arrest." Ultimately, they form stable contacts with APCs.

- T cell–APC interactions last between 1 and 24 hours, and, if productive, result in proliferation and differentiation into helper T-cell subsets.

B Cells Seek Help from CD4⁺ T Cells at the Border between the Follicle and Paracortex of the Lymph Node

Naïve B cells need two signals to become fully activated (see Figure 14-14). They are partially activated after they find and bind antigen during their travels through the B-cell follicle. They then need to engage an activated helper T cell,

which provides ligands (CD40L) that stimulate CD40 signaling and releases cytokines that promote differentiation and memory formation. Several different effector helper cells—T_H1, T_H2, and T_FH subpopulations, for example—can provide the help B cells need to proliferate and differentiate. Each helper subset encourages B cells to produce different antibody isotypes, tailoring the response to the type of pathogen that initiated the activity. For instance, T_H2 cells secrete IL-4, which encourages class switching to IgE.

Activated B cells seek T-cell help at the border between the follicle and the paracortex of the lymph node. Hours after binding antigen via their BCR, B cells in the follicle move purposefully toward the T-cell zone (**Video 14-14v4**). What is responsible for this remarkable change in trajectory?

Naïve B cells express chemokine receptors such as CCR4 and CCR5, which attract them to chemokines made in the follicle. Activated B cells, however, up-regulate CCR7, which allows B-cell chemotaxis to the chemokines generated in the paracortex. Reciprocally, activated helper T cells up-regulate CCR5, which attracts them to chemokines made in the follicle. The recruitment of both B and T cells to the border between follicle and paracortex dramatically increases the probability that they will encounter each other and deliver and receive appropriate help.

Encounters between an activated B cell and activated helper CD4⁺ T cell have been captured on film (**Video 14-14v5**). Within a day after antigen is introduced to an organism, antigen-specific T and B cells form stable antigen-specific B-cell/T-cell pairs at the border between the follicle and paracortex.

Unexpectedly, the B cell appears in full control of the antigen-specific T cell, which trails behind "helplessly." Only antigen-specific lymphocytes exhibit this behavior; B cells that express an irrelevant BCR specificity remain in the follicle.

The intimate, long-term interaction between B and T cell at the follicular border reflects the nature of the help T cells must deliver. They establish an immunologic synapse where clustered TCR, CD28, and CD40L molecules engage a B cell's MHC class II–peptide, CD80/86, and CD40 molecules, respectively. Into this synaptic space, the T cell also delivers cytokines (e.g., IL-4) that stimulate B-cell differentiation and proliferation.

After receiving T-cell help, activated B cells follow other chemotactic clues and travel to the outer edges of the follicle. Some of these activated B cells differentiate directly into short-lived, IgM-producing plasma cells and leave the lymph node; others return to the interior of the follicle, seeding a germinal center, where they undergo more rounds of proliferation, somatic hypermutation, and class switching in the presence of additional T-cell help.

> **Key Concepts:**
>
> - Naïve B cells that bind antigen in follicles up-regulate CCR7 and travel by chemotaxis from the center of the follicle toward its border with the T-cell zone, where they wait for T-cell help. Helper T cells also up-regulate chemokines that attract them to the follicle border. Antigen-specific T-cell/B-cell interactions are stable for several hours and the T-cell/B-cell pair moves actively at the follicular/T-cell border.
>
> - Some activated B cells differentiate directly into short-lived plasma cells. Others return to the follicle and form germinal centers, where they undergo somatic mutation and class switching.

Dynamic Imaging Adds New Perspectives on B- and T-Cell Behavior in Germinal Centers

Recall from Chapter 11 that the germinal center is divided into two major microenvironments, based on their appearance in stained sections under a light microscope (see Figure 11-16). The dark zone contains proliferating B cells and is thought to be the main site of somatic hypermutation. The light zone has fewer B cells, but is replete with follicular DCs, antigen, and T_{FH} cells. Here, B cells test the affinity of their mutated receptors for antigen. Those that bind antigen most effectively are thought to compete best for T-cell help and generate stable interactions with T_{FH} cells. B cells that receive T-cell help survive and return to the dark zone to undergo somatic mutation with the hopes of generating even higher-affinity specificities. The highest-affinity clones arise after several iterations of somatic hypermutation in the dark zone, and antigen sampling with T-cell help in the light zone.

Dynamic imaging of germinal center activity has largely supported this model of positive selection (**Figure 14-15**),

but has also added some interesting twists. Germinal center B and T cells are unusually motile, traveling between 9 and 12 μm/min. Germinal center B cells are also unusual in appearance and extend long processes, resembling dendritic cells (**Video 14-15v1**). Germinal center B cells also appear to sample antigen more actively and effectively than do other B cells.

In contrast to initial predictions, only a small percentage of B cells traffic between the light and dark zones each hour, and even the highest-affinity B cells spend less time in contact with helper T cells than originally predicted. The brevity of the encounter may give high-affinity B cells, which express higher densities of MHC-peptide, a competitive advantage. T-cell help not only increases B-cell survival, but also increases the speed at which B cells divide. Those that receive help therefore begin to dominate the dark zone population.

Immunologists once thought that each germinal center was composed of a single B-cell clone. However, imaging and sequencing experiments show that B cells that first populate a germinal center are very diverse. Over the multiple days it takes for a germinal center to mature, specific clones begin to dominate, although many germinal centers retain a more diverse population of B-cell clones than expected.

Finally, imaging studies showed that, while germinal center B cells stay within the center they populated, antigen-specific T_{FH} cells are more adventurous. They travel between germinal centers and generously distribute their help (**Video 14-15v2**).

> **Key Concepts:**
>
> - Germinal center B cells are unusually motile and distinct in morphology (they look more like DCs). A small percentage traffic from the dark zone, where they proliferate and mutate their receptors, to light zones, where they sample antigen and receive T-cell help. Their contact with helper T cells is surprisingly brief, but potent. It enhances B-cell survival and allows them to divide faster.
>
> - Germinal center helper T cells (T_{FH} cells) are also mobile and travel between germinal centers, distributing their help.
>
> - One germinal center is initiated by many different B-cell clones. Those with the highest affinity for antigen ultimately dominate the GC.

CD8$^+$ T Cells Are Activated in the Lymph Node via a Multicellular Interaction

Naïve CD8$^+$ T cells are the precursors of cytotoxic T lymphocytes (CTLs). As major participants in the cellular immune response, fully mature CTLs rove the body for infected cells, which they can kill very efficiently by inducing apoptosis. Like naïve CD4$^+$ T cells, naïve CD8$^+$ T cells are activated by interacting with MHC-peptide complexes on the surface of DCs. Like B cells, they also require CD4$^+$

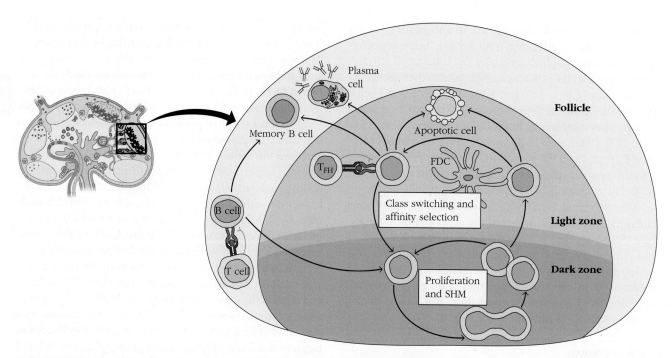

FIGURE 14-15 B-cell activity in the germinal center. B cells that receive T-cell help have one of two fates. Some become plasma cells, which exit the lymph node and typically produce IgM. Some re-enter the follicle and join other activated B cells to form a germinal center, where cells undergo affinity maturation and class switching. B cells proliferate and undergo somatic hypermutation (SHM) in the dark zone. They migrate to sample antigen and seek T-cell help in the light zone. Those that successfully compete for T-cell help return to the dark zone, proliferate more rapidly, and repeat the process. Those that do not receive help undergo apoptosis. High-affinity, class-switched B-cell clones ultimately differentiate into plasma cells or memory cells and exit the lymph node.

T-cell help to be optimally activated and to generate memory cells (**Figure 14-16**).

The movements of naïve CD8$^+$ T cells after infection resemble those of CD4$^+$ T cells: CD8$^+$ T cells arrest when they contact MHC-antigen complexes on APCs and engage in both transient and stable associations. CD8$^+$ T cells actually may form stable contacts with APCs more quickly than CD4$^+$ T cells, even at low antigen concentrations. Recent studies suggest that naïve CD8$^+$ and CD4$^+$ T cells are activated in distinct microenvironments by different types of APCs. APCs specialized at cross-presenting extracellular antigen in MHC class I, including XCR1$^+$ dendritic cells, are especially good at activating CD8$^+$ T cells.

Like CD4$^+$ T cells, CD8$^+$ T cells divide up to 10 times over a 4- to 5-day period after antigen engagement. CD8$^+$ T cells gain the attributes of cytotoxic effector cells very rapidly, in some cases even prior to the first cell division. In order to divide and differentiate optimally, and to develop robust memory, CD8$^+$ T cells need additional help from CD4$^+$ T$_H$ cells.

Elegant dynamic imaging techniques have shed light on a question that plagued early students of immunology and investigators alike. How do CD4$^+$ T cells provide help, given that they cannot directly engage CD8$^+$ T cells? CD8$^+$ T cells, of course, interact with MHC class I–peptide complexes. CD4$^+$ T cells interact with MHC class II–peptide complexes,

which are not expressed by CD8$^+$ cells. The simplest model for the delivery of CD4$^+$ T-cell help envisions an interaction among three cells: a single DC that simultaneously presents peptides in both MHC class I and class II to a CD8$^+$ T cell and a CD4$^+$ T cell, respectively.

The probability that three cells with the appropriate antigen specificities and complexes could find each other in a physiologic context seems extraordinarily low. However, to the surprise and delight of investigators, imaging experiments confirmed that just such an interaction occurs in a living lymph node. Cellular trios consisting of an antigen-expressing DC and antigen-specific CD4$^+$ T and CD8$^+$ T cells were observed by investigators 20 hours after T cells were exposed to antigen (**Figure 14-17a** and **Video 14-17v**). Further data suggest these interactions may not be as improbable as supposed. In fact, they are facilitated by chemokine interactions. Specialized DCs are first licensed by an interaction with antigen-specific CD4$^+$ T$_H$ cells. This licensing results in the production of chemokines (CCL3, CCL4, CCL5) that specifically attract pre-activated CD8$^+$ T cells that express chemokine receptors CCR4 and CCR5 (see Figure 14-17b). Questions remain about the precise timing and sequence of events, and there may be more than one way to activate a CD8$^+$ T cell. Nevertheless, the discovery of the *tricellular complexes* helped clarify confusion.

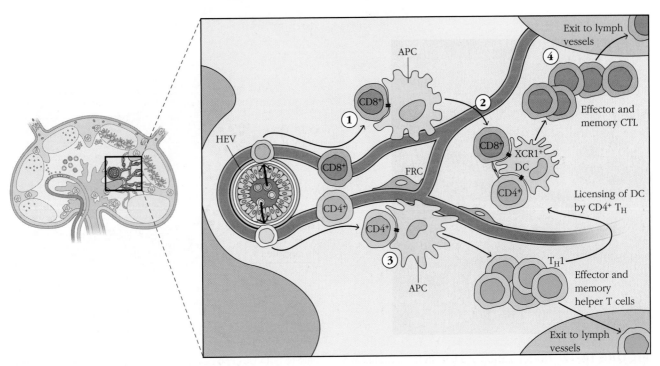

FIGURE 14-16 CD8⁺ T-cell activation in the lymph node.
The inset shows the interactions that optimally activate a naïve T cell during a primary response. The naïve CD8⁺ T cell, like the B cell, appears to require two stimulating interactions. The first encounter (1) with MHC class I–antigen on the surface of an antigen-presenting cell induces partial activation of the CD8⁺ T cell. This includes up-regulation of chemokine receptors that allow the CD8⁺ T cell to migrate (2) to an APC that has been licensed to help by an interaction with a previously (3) activated helper T cell (e.g., T_H1 or T_H17). These tricellular complexes have been observed in vivo. Both the APC and the helper T cell provide help in the form of cytokines and CD40 engagement. Fully activated CD8⁺ T cells proliferate and differentiate into effector CTLs and memory cells (4). XCR1⁺ DCs appear to be especially effective at activating CD8⁺ T cells.

Key Concepts:

- CD8⁺ T cells exhibit signs of differentiation into cytotoxic cells early after activation. Like B cells, they also need T-cell help for optimal differentiation and to develop into memory cells.

- To receive help, activated CD8⁺ T cells up-regulate chemokine receptors that attract them to dendritic cells that have been licensed (activated) by a CD4⁺ helper T cell. Stable tricellular complexes that include a CD8⁺ T cell, a CD4⁺ helper T cell, and a dendritic cell have been directly observed in dynamic images.

A Summary of the Timing of a Primary Response

A summary of the timing and organization of the fundamental T, B, and DC behaviors during a primary adaptive immune response in a lymph node is shown schematically in **Overview Figure 14-18**. Because investigations into the dynamics of the immune response are still ongoing, not every detail is likely to be correct—and not every cellular player is represented. However, this graphic overview should provide a useful reference and reminder of the fundamental events that govern the development of the adaptive immune response in space and time.

Briefly, naïve lymphocytes—B cells, CD4⁺ T cells, and CD8⁺ T cells—continuously circulate through secondary lymphoid tissue, browsing for antigens that they can bind. After entering a lymph node via HEVs, lymphocytes are guided by fibrous networks as they randomly explore their microenvironments. B cells scan the surface of follicular DCs in the follicle (B-cell zone) for unprocessed antigen that they might bind. T cells scan the processed peptide-MHC complexes on the surface of DCs in the paracortex.

Antigen arrives in a lymph node in waves. Soluble, unprocessed, and opsonized antigen can arrive within minutes and is passed from cell to cell to the follicular dendritic cells that are scanned by naïve B cells. DCs that have processed antigen at the sites of infection arrive in the paracortex hours and sometimes even days after infection.

When a probing lymphocyte engages an antigen, its movements slow down as it begins to commit to the interactions that induce both differentiation into an effector cell

(a)

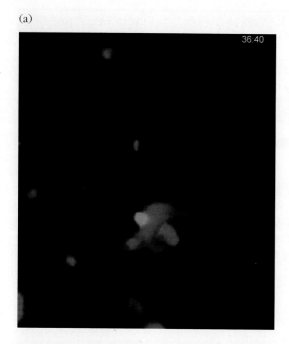

(b)

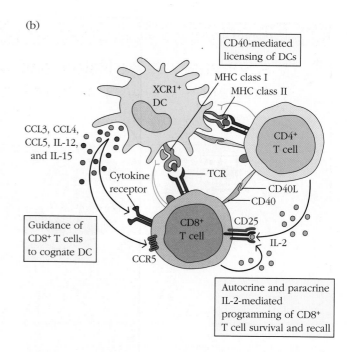

FIGURE 14-17 The formation of a tricellular complex in a lymph node during CD8⁺ T-cell activation. (a) The lymph node of a live, anesthetized mouse had first been injected subcutaneously with antigen-expressing dendritic cells (blue), and subsequently injected intravenously with antigen-specific CD4⁺ T cells (red) and CD8⁺ T cells (green). This static image shows one example of a stable complex that includes all three cells. (b) Some of the key molecular interactions regulating the tricellular complex. CD4⁺ T cells provide

help to both DCs and CD8⁺ T cells by engaging CD40. Both T cells engage MHC-peptide complexes expressed by the DC. The activated DC produces cytokines and chemokines that attract the CD8⁺ T cell, and the CD4⁺ T cell delivers IL-2 to the recruited CD8⁺ T cell. *[Part (a) from Castellino, F., et al. "Chemokines enhance immunity by guiding naïve CD8⁺ T cells to sites of CD4⁺ T cell–dendritic cell interaction," Nature, 2006, April 13; 44:0890-895. Supplemental Movie 5.]*

and proliferation, processes that start early but continue over multiple days, peaking typically on day 7. A CD4⁺ T cell that engages MHC class II–peptide complexes on the surfaces of DCs may differentiate into one of several types of effector helper cells. Which effector cell type it will become depends on both quantitative variables (affinity) and qualitative variables (costimulatory molecule engagements and cytokine interactions). Some CD4⁺ T cells gain the ability to help B cells to differentiate into antibody-producing cells. Others gain the ability to enhance cytotoxic cell differentiation and activity.

Within hours after successfully engaging antigen in follicles with their BCRs, B cells process and present antigen peptides in their own MHC molecules and migrate to the edges of a follicle, where they seek contact with activated CD4⁺ T cells that have differentiated into effectors that help B cells. Some differentiate directly into plasma cells; others re-enter the follicle and establish a germinal center. In the germinal center, B cells undergo affinity maturation and class switch with additional CD4⁺ T-cell help. Within roughly the same period of time, CD8⁺ T cells engage in multicellular complexes with DCs and helper CD4⁺ T cells and start to differentiate into bona fide cytotoxic lymphocytes.

The differentiation of antigen-specific lymphocytes into effector and memory helper CD4⁺ T cells, cytotoxic CD8⁺ T

cells, and antibody-producing B cells takes place during the proliferative burst, which is most robust by 24 to 48 hours after antigen encounter but continues for 4 days and more.

Key Concept:

- The differentiation of naïve lymphocytes into effector lymphocytes (helper T cells, cytotoxic T cells, plasma cells) takes place as cells expand over the first 4 to 7 days of the response.

Differentiation into Central Memory T Cells Begins Early in the Primary Response

How and when lymphocytes decide to become effector versus memory cells is still an area of active investigation. Do cells choose these fates early or late in the response? In the first division or after many? Are they instructed to become memory cells by extrinsic signals or is the fate randomly assigned? Advanced imaging techniques have provided some clues, most recently about central memory cell development during the response to flu virus.

It takes about 7 days for memory and effector CD8⁺ T cells to develop after a flu infection. Flu-specific naïve CD8⁺ T cells begin to divide a few days after they make contact

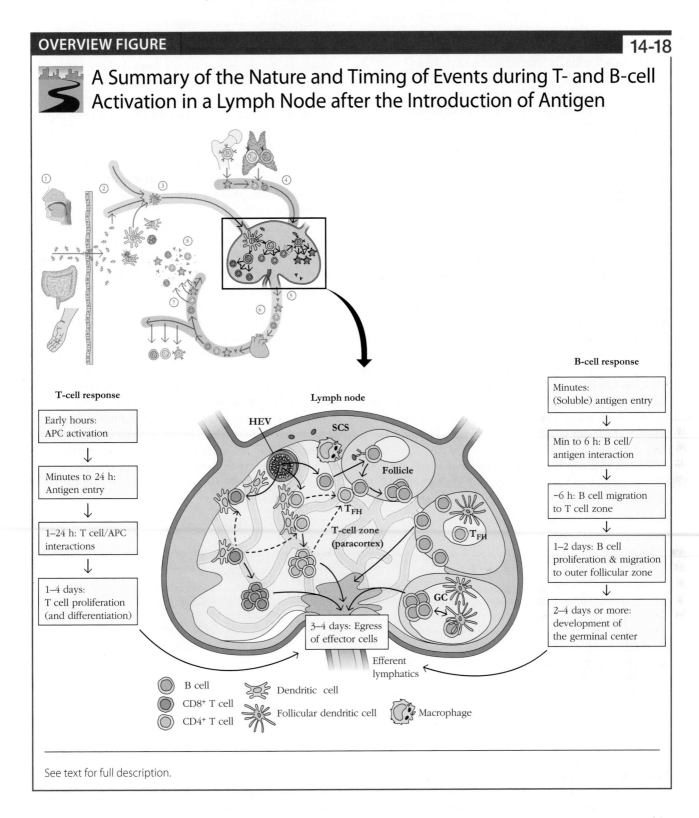

OVERVIEW FIGURE 14-18

A Summary of the Nature and Timing of Events during T- and B-cell Activation in a Lymph Node after the Introduction of Antigen

T-cell response

Early hours: APC activation

↓

Minutes to 24 h: Antigen entry

↓

1–24 h: T cell/APC interactions

↓

1–4 days: T cell proliferation (and differentiation)

Lymph node

HEV

SCS

Follicle

T_{FH}

T-cell zone (paracortex)

T_{FH}

GC

3–4 days: Egress of effector cells

Efferent lymphatics

B-cell response

Minutes: (Soluble) antigen entry

↓

Min to 6 h: B cell/ antigen interaction

↓

~6 h: B cell migration to T cell zone

↓

1–2 days: B cell proliferation & migration to outer follicular zone

↓

2–4 days or more: development of the germinal center

B cell
CD8+ T cell
CD4+ T cell
Dendritic cell
Follicular dendritic cell
Macrophage

See text for full description.

with APCs in the draining lymph nodes. Studies show that after about nine or 10 divisions, a subpopulation of flu-specific CD8+ T cells stops proliferating as quickly as the main population. Cells within this slowly dividing population differentiate into central memory cells and ultimately separate themselves from the main pool of rapidly dividing CTLs. These central memory cells up-regulate chemokine receptors, including CXCR3, that attract them to the subcapsular sinus of the lymph node, where they position themselves to be among the first responders to re-infection. Central memory cells in the spleen also relocate to sites where antigen is likely to enter first.

Key Concept:

- The basis for the decision to become a memory versus an effector cell is still under investigation, but data suggest that central memory cells separate themselves in behavior and position early in the response, after nine or 10 divisions.

The Immune Response Contracts within 10 to 14 Days

Lymphocytes have a remarkable ability to expand during an immune response and can increase in number 1000-fold. This expansion is followed by as dramatic a drop in lymphocyte number during the period referred to as **contraction** (see **Figure 14-19a**). At the end of this period only 5% to 10% of the lymphocyte pool remains. Most of these are central and effector memory cells.

What regulates lymphocyte contraction? Although the underlying basis is still being debated, both cell-intrinsic and cell-extrinsic factors are involved (see Figure 14-19b). Most immune cells have limited life spans and depend on external stimulation to maintain their survival. Half-lives vary widely—from hours for neutrophils to decades for memory cells—and effector CD4$^+$ and CD8$^+$ T cells live only 2 to 3 days in the absence of stimulation.

Stimuli that contribute to the survival of effector cells include antigen and cytokines like IL-2, IL-7, and IL-15. As an infection is cleared, these stimuli become scarce and cells compete for access. Those that can't engage antigen or cytokine receptors die as a consequence of the withdrawal of these stimuli. In the absence of signals from these receptors, the balance between prosurvival and prodeath Bcl-2 members shifts, triggering cell-intrinsic apoptosis (see Figure 14-19b).

Studies tracing T cells undergoing apoptosis during an immune response confirmed the importance of antigen accessibility in maintaining T-cell survival. As antigen started to disappear, about 5 to 8 days into the response, T-cell numbers also diminished. Importantly, T cells during the contraction phase could be rescued from death by restimulation with antigen, indicating that competition for limiting antigen plays an important role in the reduction of T-cell numbers. Other studies show that CD4$^+$ T cells appear to contract faster than CD8$^+$ T cells, an interesting observation that is not yet fully understood.

Paradoxically, under some circumstances T-cell receptor engagement up-regulates expression of death receptor ligands such as FasL. Activated T cells expressing FasL can engage Fas on neighboring cells and trigger cell-extrinsic apoptosis, a form of death that is referred to as *activation-induced cell death* (AICD). Given that all T cells and some antigen-presenting cells express Fas, this would seem to be a potent way to limit T-cell expansion. However, its in vivo importance is still unclear. Some evidence suggests that AICD operates to contract responses during chronic infections where there is a particularly high level of antigen. Interestingly, and for reasons still unknown, T$_H$1 cells appear more vulnerable to AICD than T$_H$2 cells.

(a)

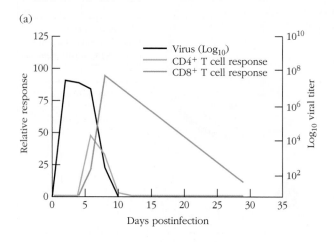

(b)

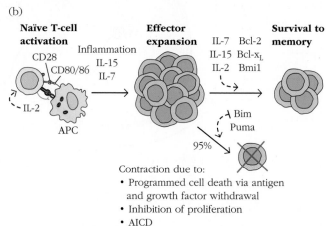

FIGURE 14-19 The contraction of an immune response. (a) The CD4$^+$ and CD8$^+$ T-cell response expands during a primary response to flu virus in the lung, and then contracts. Changes in flu virus titer are also depicted. Virus clearance is coincident with increases in CD4$^+$ and CD8$^+$ cell response and occurs between days 6 and 10 after infection. In this particular infection, the CD4$^+$ T-cell response contracts faster than the CD8$^+$ T-cell response, which remains evident in the lung weeks after peak infection. (b) Expansion and contraction of the T-lymphocyte population are influenced by several growth factors and cytokines. Withdrawal of antigen and growth factors may play the most profound role in reducing cell numbers and results in programmed cell death. The activity of regulatory T cells, as well as expression of negative costimulatory molecules such as CTLA-4, also contributes to contraction by inhibiting T cell proliferation. Finally, Fas-FasL interactions that stimulate activation-induced cell death (AICD) may also play a role in reducing lymphocyte number, particularly in chronic infections.

Withdrawal of stimuli and AICD reduce lymphocyte numbers by triggering apoptosis. Contraction of immune effector cells is also aided by interactions that inhibit proliferation. Negative costimulatory molecules or coinhibitors, like CTLA-4 and PD-1 (see Chapter 10), are up-regulated by activated T cells late in the response. CTLA-4 outcompetes CD28 for binding to CD80/86 and sends inhibitory rather than activating signals to the T cell. PD-1 binding to PD-L2 on APCs also sends signals that block T-cell proliferation and cytokine secretion. Finally, T_{REG} cells also help quell T-cell activity, in part by releasing inhibitory cytokines such as IL-10 and TGF-β (see Chapter 10).

Thus, a combination of pro-apoptotic and antiproliferative stimuli reduce the effector cell numbers by 95% so that approximately 10 to 14 days after initiation of the adaptive immune response the microenvironments of the lymph node have returned to their antigen-naïve state and structure. All that is left is a memory, in the form of potent effector and central memory T and B cells.

Key Concepts:

- Lymphocyte activation and expansion are followed by dramatic contraction, during which 95% of the lymphocytes die. Contraction is due to a combination of pro-apoptotic and antiproliferative influences. Withdrawal of signals generated by antigen and cytokines is a principal cause and triggers cell-intrinsic apoptosis.

- Effector cells exit the lymph node via efferent lymphatics to fight infection in other tissues. Some memory cells stay in the lymph node (central memory cells) and others leave to circulate or take up residence in other tissue (effector memory and resident memory cells).

The Effector and Memory Cell Response

Most effector and memory lymphocytes leave secondary lymphoid tissues and recirculate in the periphery. What tissues they visit and how long they stay there depend both on the signals they received during activation and on the cells and molecules they encounter during circulation. In this section, we will review the molecular signals that guide the migration of lymphocytes and describe some insights that we have gained from imaging techniques used to monitor the activities of living cells in tissues. Much remains to be learned about what variables regulate the paths lymphocytes follow. Imaging techniques have only begun to penetrate their dynamic journey through our bodies.

Activated Lymphocytes Exit the Lymph Node and Recirculate through Various Tissues

After leaving a lymph node or spleen, effector and memory B, CD8$^+$ T, and CD4$^+$ T cells distribute themselves throughout the body, following cues from the chemokines and addressins they encounter (**Figure 14-20**).

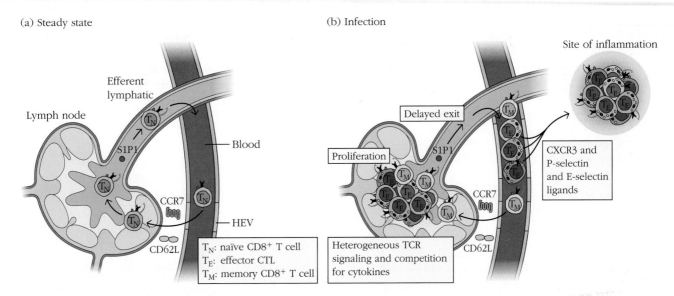

(a) Steady state (b) Infection

FIGURE 14-20 Effector and memory lymphocytes leave the lymph node via efferent lymphatics and circulate to sites of infection. This schematic compares the trafficking of naïve and effector or memory T cells. (a) Naïve lymphocytes express CD62L and CCR7, which allow them to enter and recirculate among secondary lymphoid tissues. (b) Effector and effector memory lymphocytes lose expression of CCR7 and CD62L, but gain expression of other selectins and chemokine receptors that allow them access to peripheral tissues and sites of infection that express the appropriate addressins and chemokines.

Antibody-producing B cells (plasma cells) travel to several sites, depending on the isotype of the antibody that they are producing. Many IgM producers take up residence in the medulla of the lymph nodes, where they release antibodies into the efferent lymphatics. IgG producers tend to go to the bone marrow, and IgA producers return to barrier tissues, including gut-associated and lung-associated lymphoid tissues. Pathogen-specific cytotoxic CD8$^+$ T cells follow chemokine cues to the original sites of infection, where they kill infected cells expressing MHC class I–peptide complexes.

The trafficking of mature helper CD4$^+$ T cells varies according to effector subtypes, and the molecules that regulate their travels are still being investigated. Some CD4$^+$ T cells that provide signals promoting B-cell and CD8$^+$ T-cell differentiation stay in the lymph nodes and continue to stimulate lymphocyte differentiation. Others travel to sites of infection to enhance the ability of tissue phagocytes to clear opsonized pathogen. Some effector CD4$^+$ T cells may even complete their differentiation or change their functions at the sites of infection, responding with plasticity to the collection of signals and needs of the local response.

Central memory cells (T$_{CM}$) stay in secondary lymphoid tissue, and effector memory cells (T$_{EM}$) circulate among peripheral tissues throughout the body. Tissue-resident memory cells (T$_{RM}$) take up long-term residence in tissues where they share sentinel functions with residing innate immune cells and are among the first responders to reinfection (see Chapter 10 for a review of memory cell subsets). Resident memory cells are found in especially high concentrations in barrier tissues and compose the intraepithelial lymphocytes (IELs) of the gut as well as many of the lymphocytes in the outer layers of the skin. In fact, the skin contains at least twice as many T cells than blood and we now know that most of these are resident memory T cells (see Chapter 13). Resident memory cells are also present in many internal tissues such as brain, kidney, and joints (**Figure 14-21**).

Key Concepts:

- Once generated in a primary immune response, effector and memory T cells home to distinct tissues and tissue microenvironments.

- Effector B cells (plasma cells) distribute themselves in several areas, including the medulla of lymph nodes and the bone marrow.

- Three types of memory cells populate different regions of the body. Effector memory cells circulate between tissues, and resident memory cells take up long-term residence in particular tissues and are particularly abundant in barrier tissues. Central memory cells stay in secondary lymphoid tissue and position themselves at the periphery, where antigen first arrives (e.g., the subcapsular sinus of lymph nodes).

Chemokine Receptors and Adhesion Molecules Regulate Homing of Memory and Effector Lymphocytes to Peripheral Tissues

Not surprisingly, the trafficking of effector and memory lymphocytes within and between tissues is regulated by changes in expression of the combination of *cell-a*dhesion *m*olecules (CAMs) and chemokine receptors. In turn, tissue microenvironments express unique sets of adhesion molecules and chemokines that attract specific effector subsets.

In general, effector cells and effector memory cells avoid recirculation back to secondary lymphoid tissues by decreasing expression of CCR7 and L-selectin (CD62L), which prevents them from entering via HEVs. Instead, effector cells up-regulate expression of adhesion molecules and chemokine receptors that coordinate their homing to relevant tissues.

Where effector and effector memory cells travel is partly determined by where they came from. For instance, if generated in a lymph node that drains the skin, effector cells up-regulate **homing receptors** that send them back to the skin. Specifically, they up-regulate expression of *cut*aneous *l*eukocyte *a*ntigen (CLA) and LFA-1, which bind to E-selectin and ICAMs on dermal venules of the skin (**Figure 14-22**). They also home to chemokines CCL17, CCL27, and CCL1 produced by keratinocytes. Resident memory cells also up-regulate an additional protein, CD69, a lectin that helps retain them in tissues.

Alternatively, if generated in gut-associated lymph nodes, lymphocytes express high levels of the integrins $\alpha_4\beta_7$ (LPAM-1) and CD11a/CD18 (LFA-1, $\alpha_L\beta_2$), which bind to MAdCAM and various ICAMs on intestinal lamina propria venules (see Figure 14-22). Cells homing to the gut mucosa also express the chemokine receptor CCR9, which binds to CCL25 in the small intestine. IgA-secreting B cells are among those recruited into gut tissue via the chemokines CCL25 and CCL28.

Key Concepts:

- Homing of effector and memory lymphocytes is regulated by interactions between cell-adhesion molecules (CAMs) and chemokine/chemokine receptors.

- Effector T lymphocytes tend to home to the tissue where their stimulating antigen-presenting cells originated. Thus, T lymphocytes activated in skin draining lymph nodes tend to return to the skin. T lymphocytes activated in lymph nodes that serve the intestine return to the intestine.

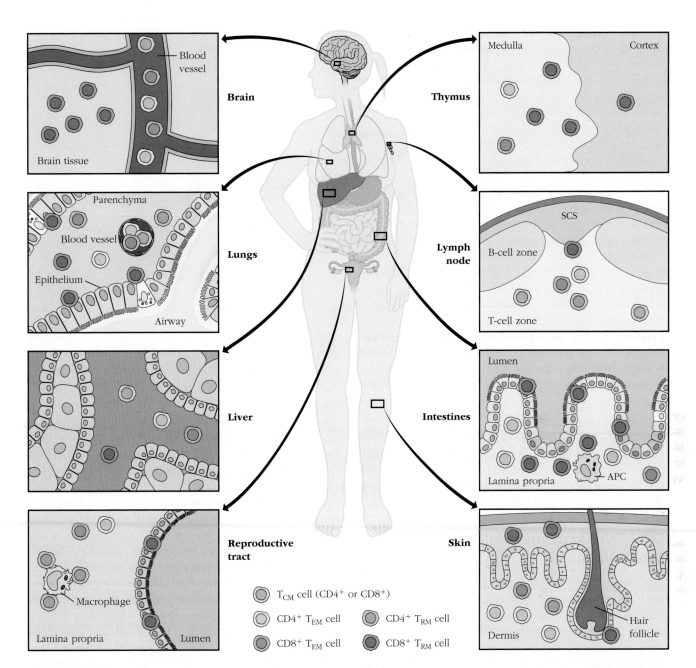

FIGURE 14-21 Where memory T cells can be found. Whereas central memory T cells (gray, T_{CM}) are found predominantly in secondary lymphoid tissue, effector and resident memory CD4$^+$ (blue) and CD8$^+$ (red) T cells are found in a wide variety of tissues. CD8$^+$ resident memory cells (dark red) are abundant in the epithelial layers of barrier organs, including skin, intestines, lungs, and reproductive tract. Our understanding of the fate of CD4$^+$ memory cells is still murky and this figure will undoubtedly be modified in the future.

The Immune Response: Case Studies

Dynamic imaging techniques have been especially useful in helping us visualize the immune response to real pathogens and insults in vivo. Below, we describe studies that monitor the behavior of cells responding to physiologic insults in living tissue. We explore three in some depth. One follows the behavior of T cells in response to infection with the protozoa that causes toxoplasmosis (*Toxoplasma gondii*), a pathogen that is transmitted from cats to humans (and other animals) via feces and produces an infection that is often asymptomatic. However, "Toxo" can harm the fetus of a pregnant woman. The second study examines the role of resident memory T cells in the rapid response to herpes simplex virus (HSV) infection in the skin. The third traces the behavior of T cells responding to—and rejecting—an allogeneic skin graft. The observations offer therapeutic insights that may help to inhibit graft rejection.

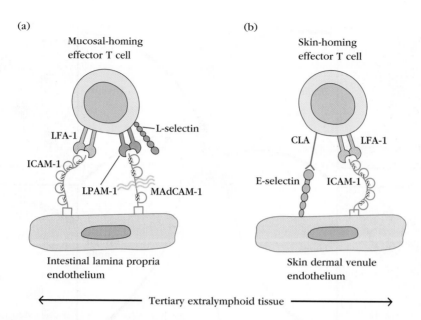

FIGURE 14-22 Examples of homing receptors and vascular addressins involved in selective trafficking of naïve and effector T cells. Various subsets of effector T cells express high levels of particular homing receptors that allow them to home to endothelium in particular tertiary extralymphoid tissues. The initial interactions in homing of effector T cells to (a) mucosal and (b) skin sites are illustrated.

We close with a collection of shorter, tantalizing descriptions of interactions between immune cells and physiologic antigens revealed by dynamic imaging studies. These observations are a powerful reminder of the influence of cell context on immune and inflammatory responses, something that is difficult to mimic in vitro.

CD8$^+$ T-Cell Response to Infection with *Toxoplasma gondii*

Investigators have used dynamic imaging methods to trace the behavior of the T cell specific for *Toxoplasma gondii* (Toxo), a protozoal parasite with three infectious stages (cysts, oocysts, and rapidly dividing tachyzoites), which infects many different tissues, including the brain. More than half of the world's population has been infected with Toxo, which is acquired by ingesting contaminated raw meat, soil, or litter exposed to feces from infected cats (**Figure 14-23a**). Most of us never know that we have been invaded, and some of us harbor the parasite in cysts for long periods of time. In those with weakened immune systems, however, Toxo tachyzoites can infect the brain and eyes. The pathogen also can cross the placenta and cause disease in fetuses, whose immune systems are underdeveloped. Understanding our response to this pathogen is an important step in controlling it.

In order to follow the immune response to Toxo, investigators cleverly modified the parasite so that it expresses an antigen (an *ova*lbumin [OVA] peptide) that can be recognized by TCR-transgenic (OT-1) CD8$^+$ T cells, which can be more easily traced and manipulated (see Chapter 20). The Hunter laboratory imaged CD8$^+$ T cells responding to this infection in both the lymph nodes and the brain (see John et al. 2009. *PLoS Pathogens* 5:e1000505). Their work shows that antigen-specific CD8$^+$ T cells in the lymph node respond very rapidly (within 36 hours) after intravenous injection of *T. gondii*, forming stable contacts with DCs that peak between 36 and 48 hours. As predicted by work described in this chapter, crawling naïve antigen-specific T cells slow down considerably (from 6–8 to 4 μm/min) after encounter with DCs that have been exposed to antigen. Their movements pick up again after 48 hours, when they are actively proliferating and seem to rely on transient contacts, again as predicted by investigations with noninfectious systems.

Unexpectedly, responding CD8$^+$ T cells and DCs appear to collect in the lymph node's subcapsular sinus (see Figure 14-23b), and investigators now speculate that this may be an active site of inflammation and effector cell activity in the lymph nodes.

CD8$^+$ T cells reactive to *T. gondii* antigens began to enter the brain by day 3. Their influx peaked on day 22, coincident with peak proliferation of CD8$^+$ T cells in lymph nodes. These killer T cells were surprisingly motile when they arrived. Some formed stable clusters near infected cells, and appeared to be interacting with brain-resident APCs. Unfortunately, the ability of infiltrating *T. gondii*–reactive T cells to control the infection waned over time (>40 days), a loss of function associated with the up-regulation of expression of PD-1, a negative costimulatory molecule (see Chapter 10). This observation is consistent with other data showing that chronic infection "exhausts" lymphocytes, reducing their ability to clear pathogens.

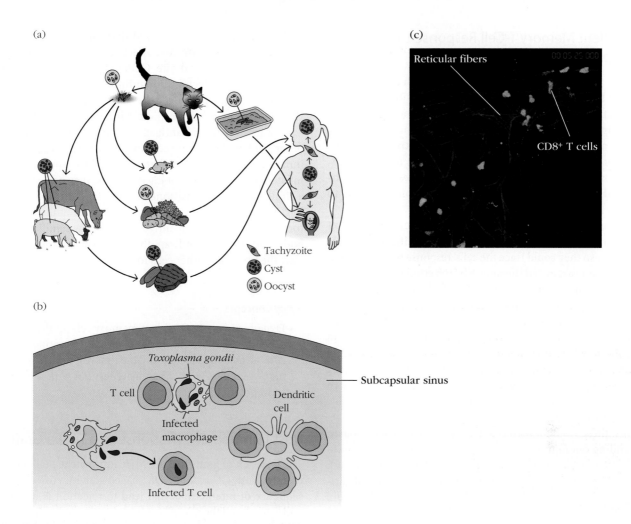

(a)

(c)

Reticular fibers

CD8⁺ T cells

(b)

Toxoplasma gondii

T cell

Infected
macrophage

Dendritic
cell

Infected T cell

Subcapsular sinus

Tachyzoite
Cyst
Oocyst

FIGURE 14-23 The immune response to *Toxoplasma gondii*. (a) *Toxoplasma gondii* is a very common parasite that can infect all warm-blooded animals. It reproduces in cats that have eaten prey containing cysts. Infectious oocysts are then spread via cat feces. Humans can be infected by eating contaminated food or handling feces. Although infection typically doesn't cause serious disease, the actively dividing form of the parasite, tachyzoites, can travel to the brain or to developing fetuses, causing damage. (b) Tachyzoites have been shown to gain direct access to the subcapsular sinus of lymph nodes, where it attracts and infects multiple cell types including activated T cells and antigen-presenting cells. (c) CD8⁺ T cells (green) respond to *Toxoplasma gondii* infection in the brain, which, over time, establishes a reticular network (blue) reminiscent of the FRC network of a lymph node. *[Part (c) republished with permission of Elsevier, Wilson, E. H., et al., "Behavior of parasite-specific effector CD8 T cells in the brain and visualization of a kinesis-associated system of reticular fibers," Immunity, 2009 Feb 20; **30:**(2):300–311, Figure 7. Permission conveyed through Copyright Clearance Center, Inc.]*

The infection altered the microenvironment of both the lymph nodes and the brain in several unanticipated ways. In the lymph nodes, inflammation (antigen specific and nonspecific) disrupted the reticular network and decreased the levels of the chemokine CCL21—taking away the cues needed for naïve T-cell migration and limiting responses to new infections. In the brain, however, infection induced the appearance of a new reticular network decorated with CCL21. Although unexpected, this made biologic sense. A new reticular network can help organize T-cell entry, trafficking, and responses to antigen in the brain (see Figure 14-23c and **Video 14-23v**). Other studies suggest that reticular networks are routinely established in other tertiary tissues after infection.

Key Concept:

• Dynamic imaging of cells responding to a protozoal parasite that infects the brain reveals that inflammation can temporarily alter the microenvironment (reticular structure) of a lymph node and can remodel the tissue that is the site of infection by establishing reticular systems for lymphocyte trafficking.

Resident Memory T-Cell Response to Herpes Simplex Virus Infection

Most people in the world are familiar with cold sores, or fever blisters, which are caused by herpes simplex virus 1. Investigators have been impressed by how rapidly our skin and mucosal surfaces resolve HSV infections; sometimes they are cleared within hours. It is now known that the majority of T cells in the skin are resident memory cells, and one group of investigators set out to test the hypothesis that they play a major role in the rapid response to herpesvirus.

They introduced, into a mouse, green fluorescent protein-positive (GFP$^+$) CD8$^+$ T cells specific for an HSV protein so they could trace the cells' response by microscopy. They then vaccinated the mouse by injecting DNA encoding this protein into the skin. As expected, the naïve HSV CD8$^+$ T cells expanded and approximately 2 weeks later, a whopping 5% of circulating T cells were GFP$^+$ and HSV specific. Some of these found their way back to the site of infection in the skin, where they joined Langerhans cells and epidermal γδ T cells. There they took up residence, gained a memory phenotype and up-regulated CD103, which interacts with integrins on epithelial cells and helps retain cells in the skin.

Although they confined their movements to the epidermis, these resident memory T cells migrated continuously, sending out long searching processes that allowed them to explore a wide surface area (**Figure 14-24** and **Video 14-24v**). This movement was not dependent on the presence of antigen and continued for many weeks after the initial infection.

When the investigators reintroduced HSV antigen into the skin weeks later, the HSV-specific resident memory T cells responded remarkably quickly. Within hours, they established contact with antigen-expressing epithelial cells, slowed down, and rounded up.

The investigators discovered that this antigen-specific interaction was broadly and effectively protective. Antigen engagement induced the CD8$^+$ T$_{RM}$ cells to release inflammatory molecules, including IFN-γ that enhanced the ability of neighboring cells to resist infection. In fact, IFN-γ production enhanced the skin's ability to protect itself against many different antigens, not simply HSV! This and other studies have inspired some to describe T$_{RM}$ cells as innate-like. Not only have they adopted the appearance of innate cells with their probing dendrites; they have acquired a similar function. Positioned on the front lines of attack, T$_{RM}$ cell antigen receptors act like pattern recognition receptors. They recognize an antigen that is now familiar to the body, and they respond rapidly by releasing cytokines that alert the tissue and other cells to attack.

Key Concept:

- Dynamic images of resident memory cells in the skin show that they are continuously probing keratinocyte surfaces and are among the first responders to herpes simplex virus. They share features with the most potent innate immune cells.

Host Immune Cell Response to a Tissue Graft

The rejection of grafts that are MHC mismatched (allografts) requires both CD4$^+$ and CD8$^+$ T-cell activity. However, the sequence of interactions that lead to rejection is not fully understood, and the role of host (recipient) versus graft (donor) APCs in mediating the activation is still disputed. Specifically, in order to be activated, naïve CD4$^+$ and CD8$^+$ T cells must interact with APCs that express the graft's alloantigens. Are the activating APCs coming from the graft or from the host?

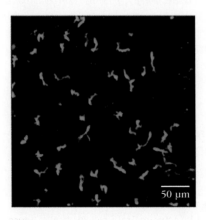 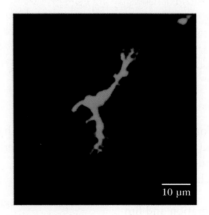 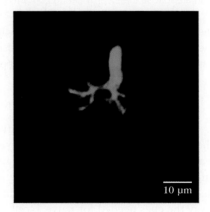

FIGURE 14-24 The remarkable activities of resident memory T cells in the response to herpes simplex virus. These images show examples of the unique morphology and motility of resident memory T cells (green) in the dermis of a mouse. They send out processes and are continuously mobile, probing for antigen, in this case a peptide from herpes simplex virus (gB). They patrol for many weeks, if not months, after infection; the pictures were taken 1 month following infection with HSV. Their rapid response to infection has led some to compare them with innate cells in activity and potency. See Video 14-24v. [Ariotti, S., et al., "Tissue-resident memory CD8+ T cells continuously patrol skin epithelia to quickly recognize local antigen," Proceedings of the National Academy of Sciences USA 2012, Nov 27; **109**(48):19739–44, Figure 1c.]

The Bousso laboratory performed an elegant set of dynamic imaging experiments to address this and other questions. MHC-mismatched skin was grafted onto a mouse ear and imaged by two-photon intravital microscopy. The mismatched skin was from a mouse whose DCs fluoresced yellow (a CD11c-YFP reporter mouse), so their motions could be traced. These fluorescent DCs exited the graft and were gone within 6 days. The investigators looked for these donor (graft)-derived DCs in the draining lymph nodes of the recipient mouse and found them as early as 3 days after engraftment. However, they had the look of dying cells and never reappeared.

The investigators then asked, which cells came into the skin graft from the host mouse? To answer this, they put the graft onto mice whose cells all express green fluorescent protein (GFP). They found that CD11b$^+$ (myeloid) green fluorescent cells infiltrated the graft within 3 days. These cells were dominated by neutrophils at first, but over time they included more and more monocytes and DCs. Interestingly, such myeloid cells were also found in grafts from control, MHC-matched mice (isografts), indicating that they were recruited in an antigen-nonspecific manner, probably in response to inflammation caused by surgery. The investigators took advantage of their system to look more closely at the locations of these cells and found that GFP$^+$ myeloid cells infiltrated the dermis of both allografts and isografts. However, only the allografts showed evidence of GFP$^+$ cells infiltrating all layers of the donor skin graft, including the outer layer (the epidermis), indicating that extensive infiltration was antigen dependent.

To see if the infiltrating cells carried alloantigen back to draining lymph nodes, the investigators introduced a clever twist to their system (**Figure 14-25a**). First, they put their

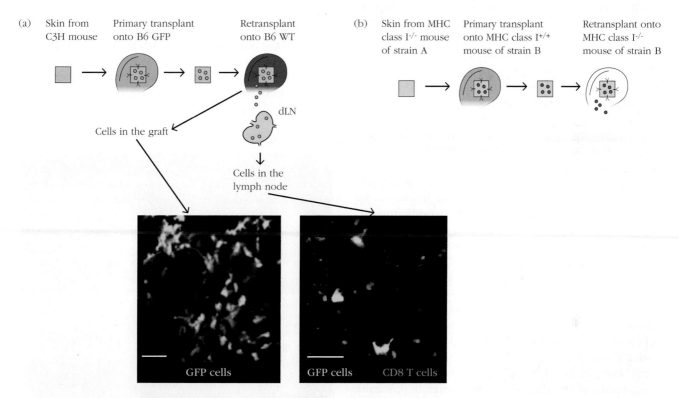

FIGURE 14-25 Imaging the immune response to an antigen-mismatched skin graft. This figure shows the approaches taken by investigators to see which antigen-presenting cells—host or graft (donor) APCs—activate the CD8$^+$ T cells that will reject a skin graft. (a) Skin from a mouse of one strain (C3H) was grafted onto the ear of a mouse of an MHC-mismatched strain (B6) in which all cells expressed GFP. The cells that infiltrated the skin from this B6-GFP mouse (colored yellow) were predominantly myeloid (neutrophils, monocytes, macrophages). After time was allowed for infiltration, the investigators removed the graft and placed it on a wild-type B6 mouse whose CD8$^+$ T cells fluoresced green. This strategy allowed investigators to see where the infiltrating cells (yellow) traveled. They found the cells in the *draining lymph node* (dLN), where they were interacting with host CD8$^+$ T cells. These host CD8$^+$ T cells ultimately migrated to the graft, killing the MHC-mismatched cells. (b) Skin from a strain A mouse that expressed no MHC class I was transplanted on the ear of an antigen-mismatched (strain B) mouse that expressed MHC class I. In this situation, *none of the donor graft cells and only the infiltrating recipient cells could present antigen to CD8$^+$ T cells.* The investigators let this graft accumulate recipient cells, then retransplanted it onto an antigen-matched class I$^{-/-}$ mouse. The T cells from this mouse responded to the strain B cells that had infiltrated the primary graft and picked up the strain A antigen. *[Photos reprinted by permission from Macmillan Publishers Ltd, from Celli, S., et al., "Visualizing the innate and adaptive immune responses underlying allograft rejection by two-photon microscopy," Nature Medicine 2011 June; **17**(6):744–749. Figure 3a. Permission conveyed through Copyright Clearance Center, Inc.]*

grafts on MHC-mismatched hosts with GFP⁺ cells and let the green, myeloid cells (neutrophils and monocytes) infiltrate the graft over a 6-day period. Then they removed each graft, with its new green host cell infiltrates, and put it on another MHC-mismatched mouse, but one that did not have any GFP⁺ cells. They could then trace the fate of the green infiltrates. Three days later, they found some of these GFP⁺ cells in the T-cell zones of the draining lymph nodes, suggesting that infiltrating cells from the host could play a role in initiating the T-cell response to the graft. This is consistent with suggestions that host APCs *cross-present* the graft's alloantigens to CD8⁺ T cells. (Recall that cross-presentation is the process by which APCs pick up extracellular antigens and present them as MHC class I–peptide complexes to CD8⁺ T cells.)

They strengthened their case for the role of host APCs in cross-presenting alloantigen via another clever, albeit complex scheme. Essentially, they grafted skin from a "strain A" MHC class I–deficient mouse onto an antigen-mismatched ("strain B") mouse that expressed normal MHC class I (see Figure 14-25b). In this system, the skin graft's own APCs were completely unable to present antigen to MHC class I–restricted CD8⁺ T cells; only the infiltrating cells from the recipient would be able to do so. They allowed strain B cells to infiltrate the graft, and then removed it, transplanting it onto another strain B mouse, but one that was MHC class I deficient. If and only if the infiltrating (strain B) cells from the first host successfully cross-presented foreign antigen from the strain A graft would the CD8⁺ T cells in this second recipient respond (and initiate rejection). Indeed, in vivo imaging revealed that antigen-specific CD8⁺ T cells from this second recipient responded to the graft—confirming that host DCs that infiltrated the graft pick up and present foreign antigen to CTLs, which are responsible for graft rejection.

Finally, the investigators visualized the activity of the graft-rejecting cytotoxic CD8⁺ T cells. Dividing T cells could be found in the draining lymph node as early as 2 days after graft transplantation. However, such cells were not found in the graft itself until day 8, and seemed to enter from the graft edges. CD8⁺ T cells were found at the border between the dermis (deep) and epidermis (more superficial) border of the allograft and were often in proximity to dying cells, which were likely their targets. CD8⁺ T cells were slower moving in the allografts (versus isografts), which again is consistent with indications that antigen-specific cells arrest when they meet their antigen. By day 10, when the tissue was undergoing active rejection, CD8⁺ T cells were found throughout the graft.

Key Concept:

- Dynamic images of cells responding to an MHC-mismatched skin graft show that the APCs of the recipient are capable of sampling and presenting alloantigen to host CD8⁺ T cells, which ultimately distribute themselves throughout the graft and destroy it.

Dendritic Cell Contribution to *Listeria* Infection

As you know, DCs are important for initiating an adaptive immune response against infection; however, in some cases they are responsible for maintaining infection! *Listeria monocytogenes* is an intracellular bacterium that resides in soil and can cause food-associated illness and fatalities (**Figure 14-26**; also see www.dnatube.com/video/2506/Intracellular-Listeria-Infection for an illustrative video). In 2011, one of the worst food-borne disease outbreaks in the United States was traced to *Listeria* associated with cantaloupe melons.

Recall that the spleen is the site of response to blood-borne pathogens: all white blood cells circulate through sinuses in the red pulp, which also is a site where DCs and monocytes can sample antigen (see Figure 2-15). White blood cells enter and scan the spleen's version of the T-cell zone, the PALS. B cells enter and scan follicles, which are very similar to those found in lymph nodes. The area that separates the red pulp from the white pulp, the *m*arginal zone (MZ), is a relatively unique microenvironment that is the first line of defense against blood-borne pathogens. It is also the site of residence of DCs and monocytes as well as a unique group of B cells (MZ B cells) that do not circulate.

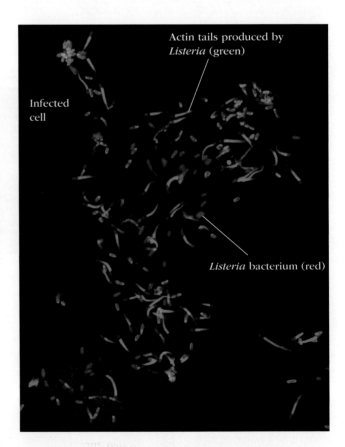

FIGURE 14-26 Intracellular *Listeria* infection. *Listeria* bacteria (red) use the host cell actin (green) to propel themselves through the cytoplasm of an infected cell (cometlike tails of green actin can be seen trailing the red bacteria). *[Courtesy Alain Charbit.]*

Dynamic imaging of fluorescently tagged *Listeria* bacteria showed that they arrive in the red pulp of the spleen within seconds of intravenous injection. (During a physiologic infection, when bacteria are ingested, they probably take longer to arrive in the red pulp—either within white blood cells that they infected in the mucosa of the gut or directly from blood that they penetrated during infection.)

Within minutes, *Listeria* associates with and is presumably phagocytosed by the DCs in the red pulp sinuses, thereby infecting them. *Listeria*-specific CD8$^+$ T cells accumulate in these sinuses, forming stable contacts with the DCs that activate them. Investigators found that DCs in these sinuses recruit other innate immune cells, including monocytes, and unwittingly pass their *Listeria* infection to these cells (via their extensive processes, down which *Listeria* travel). These circulating cells, which are critical for controlling the pathogen, also become vehicles of infection that spread the bacteria throughout the body.

Key Concept:

• Imaging experiments show that *Listeria* bacteria travel to the red pulp of the spleen and infect dendritic cells that engulf them. These cells unwittingly carry the infection to other immune cells, including T cells, which spread *Listeria* through the body.

T-Cell Response to Tumors

The immune response to tumors, especially solid tumors, is notoriously poor. In principle, cytotoxic cells, including CD8$^+$ T cells, should be able to kill tumor cells, which often express novel, nonself antigens. However, they rarely seem effective enough to shrink a tumor. Is this because they don't penetrate the tumor? Even if they do penetrate, do they have the capacity to kill cells effectively? Dynamic imaging experiments show that a few antigen-specific, activated T cells actually do gain access to tumors (**Figure 14-27**). There they migrate vigorously and even exhibit effective cytolytic behavior, associating for long periods of time (6 hours or more) with tumor cells and inducing apoptosis. However, the tumors failed to regress. In this particular experiment, the T-cell response was limited by a number of factors including (1) poor access to the tissue, and (2) poor antigen presentation by tumor-associated cells, rather than by the availability of activated CTLs or their ability to mediate cytolysis. These studies suggest that strategies to improve antigen presentation and tumor accessibility may be more effective than strategies that simply increase antigen-specific cytotoxic T-cell number.

Key Concept:

• Imaging experiments show that activated, tumor-specific CD8$^+$ T cells have difficulty penetrating solid tumors and suggest that therapies that enhance accessibility may improve tumor response.

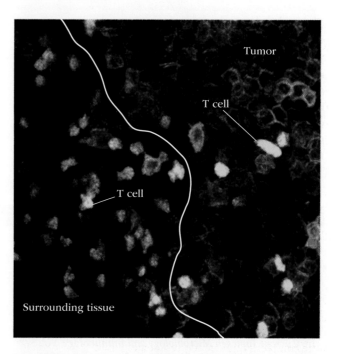

FIGURE 14-27 T cells infiltrating a tumor. The image shows interactions between antigen-specific T cells (yellow) and antigen-expressing tumor cells (red). Some T cells infiltrate the solid tumor, but many gather at the borders of the tumor. *[Republished with permission of Elsevier, from Deguine, J., et al., "Intravital imaging reveals distinct dynamics for natural killer and CD8$^+$ T cells during tumor regression," Immunity 2010, October 29; **33**(4):632-644. Permission conveyed through Copyright Clearance Center, Inc.]*

Regulatory T Cells Inhibit the Immune Response in Multiple Ways

Dynamic imaging has also allowed us to see when and where regulatory T cells exert their suppressive activity, and the results satisfyingly confirm predictions made from "static" experiments, suggesting that regulatory T cells can suppress immune responses in more than one way (see Chapter 10). In a model of type 1 diabetes (the result of T cell–mediated destruction of beta cells of the pancreatic islet), investigators traced the activities of fluorescently labeled antigen-specific regulatory T cells introduced prior to disease onset. They exhibited two distinct behaviors. Some prevented diabetogenic effector cells from making productive clustering contacts with cells presenting auto-antigen in the pancreatic islets, consistent with proposals that regulatory T cells can inhibit effector T-cell function by quelling the activation potential of resident APCs. Other regulatory T cells formed stable connections with effector CD8$^+$ T cells and inhibited their ability to kill pancreatic target cells, an effect associated with TGF-β secretion by regulatory T cells.

Key Concept:

• Regulatory T cells can suppress an immune response by influencing both APCs and CD8$^+$ T cells.

Conclusion

We end our many-chaptered exploration of the cellular and molecular biology of the immune response by highlighting studies revealing the choreography of immune cells in living tissue. Not only have these images and videos confirmed predictions made by immunologists who did not have the luxury to visualize the cells they were studying, but they have revealed unanticipated features of immune cells: the dramatic swarming behavior of neutrophils, the extensive dendritic processes of germinal center B cells, the dance of B- and T-cell pairs at the follicular boundary, the unusual appearance and activity of resident memory T cells in barrier tissues, to name just a few. We have concluded the chapter by introducing studies that are exploring the behavior of immune cells in pathologic contexts, the featured topic of the next set of chapters, which focus on our growing understanding of the role of immune cells in combating and even causing disease.

REFERENCES

Ariotti, S., et al. 2012. Tissue-resident memory CD8$^+$ T cells continuously patrol skin epithelia to quickly recognize local antigen. *Proceedings of the National Academy of Sciences USA* **109**:19739.

Arnon, T. I., and J. G. Cyster. 2014. Blood, sphingosine-1-phosphate and lymphocyte migration dynamics in the spleen. In: *Current Topics in Microbiology and Immunology*, Vol. 378: *Sphingosine-1-Phosphate Signaling in Immunology and Infectious Diseases*, M. B. H. Oldstone and H. Rosen, eds. pp. 107–128. Springer International, Cham, Switzerland.

Bastista, F. D., and N. E. Harwood. 2009. The who, how and where of antigen presentation to B cells. *Nature Reviews Immunology* **9**:15.

Beuneu, H., et al. 2010. Visualizing the functional diversification of CD8$^+$ T-cell responses in lymph nodes. *Immunity* **33**:412.

Bousso, P. 2008. T-cell activation by dendritic cells in the lymph node: lessons from the movies. *Nature Reviews Immunology* **8**:675.

Cahalan, M. D. 2011. Imaging transplant rejection: a new view. *Nature Medicine* **17**:662.

Cahalan, M. D., and I. Parker. 2008. Choreography of cell motility and interaction dynamics imaged by two-photon microscopy in lymphoid organs. *Annual Review of Immunology* **26**:585.

Catron, D. M., A. A. Itano, K. A. Pape, D. L. Mueller, and M. K. Jenkins. 2004. Visualizing the first 50 hr of the primary immune response to a soluble antigen. *Immunity* **21**:341.

Celli, S., M. L. Albert, and P. Bousso. 2011. Visualizing the innate and adaptive immune responses underlying allograft rejection by two-photon microscopy. *Nature Medicine* **17**:744.

Coombes, J. L., and E. A. Robey. 2010. Dynamic imaging of host-pathogen interactions in vivo. *Nature Reviews Immunology* **10**:353.

Cyster, J. G. 2010. Shining a light on germinal center B cells. *Cell* **143**:503.

Finlay, D., and D. Cantrell. 2011. Metabolism, migration and memory in cytotoxic T cells. *Nature Reviews Immunology* **11**:109.

Girard, J.-P., C. Moussion, and R. Forster. 2012. HEVs, lymphatics and homeostatic immune cell trafficking in lymph nodes. *Nature Reviews Immunology* **12**:762.

Jenkins, M. K. 2008. Imaging the immune system. *Immunological Review* **221**:5.

John, B., et al. 2009. Dynamic imaging of CD8$^+$ T cells and dendritic cells during infection with *Toxoplasma gondii*. *PLoS Pathogens* **5**:e1000505.

Kinjyo, I., et al. 2015. Real-time tracking of cell cycle progression during CD8$^+$ effector and memory T-cell differentiation. *Nature Communications* **6**:6301.

Kitano, M., et al. 2016. Imaging of the cross-presenting dendritic cell subsets in the skin-draining lymph node. *Proceedings of the National Academy of Sciences USA* **113**:1044.

Kurosaki, T., K. Kometani, and W. Ise. 2015. Memory B cells. *Nature Reviews Immunology* **15**:149.

Laidlaw, B. J., J. E. Craft, and S. M. Kaech. 2016. The multifaceted role of CD4$^+$ T cells in CD8$^+$ T cell memory. *Nature Reviews Immunology* **16**:102.

Lämmermann, T., et al. 2013. Neutrophil swarms require LTB4 and integrins at sites of cell death in vivo. *Nature* **498**:371.

MacHeyzer-Williams, M., S. Okitsu, N. Wang, and L. McHeyzer-Williams. 2012. Molecular programming of B cell memory. *Nature Reviews Immunology* **12**:24.

Mueller, S. N., and H. D. Hickman. 2010. In vivo imaging of the T cell response to infection. *Current Opinion in Immunology* **22**:293.

Mueller, S. N., and L. K. McKay. 2016. Tissue-resident memory T cells: local specialists in immune defence. *Nature Reviews Immunology* **16**:79.

Pentcheva-Hoang, T., T. R. Simpson, W. Montalvo-Ortiz, and J. P. Allison. 2014. CTLA-4 blockade enhances antitumor immunity by stimulating melanoma-specific T-cell motility. *Cancer Immunology Research* **2**:970.

Pereira, J. P., L. M. Kelly, and J. G. Cyster. 2010. Finding the right niche: B-cell migration in the early phases of T-dependent antibody responses. *International Immunology* **22**:413.

Qi, H., W. Kastenmuller, and R. N. Germain. 2014. Spatiotemporal basis of innate and adaptive immunity in secondary lymphoid tissue. *Annual Review of Cell and Developmental Biology* **30**:141.

Radtke, A. J., et al. 2015. Lymph-node resident CD8a[+] dendritic cells capture antigens from malaria, sporozoites and induce CD8[+] T-cell responses. *PLoS Pathogens* **11**: e1004637.

Russo, E., et al. 2016. Intralymphatic CCL21 promotes tissue egress of dendritic cells through afferent lymphatic vessels. *Cell Reports* **14**:1723.

Schwartzberg, P. L., K. L. Mueller, H. Qi, and J. L. Cannons. 2009. SLAM receptors and SAP influence lymphocyte interactions, development and function. *Nature Reviews Immunology* **9**:39.

Shulman, Z., et al. 2013. T follicular helper cell dynamics in germinal centers. *Science.* **341**:673.

Wilson, E. H., et al. 2009. Behavior of parasite-specific effector CD8[+] T cells in the brain and visualization of a kinesis-associated system of reticular fibers. *Immunity* **30**:300.

Useful Websites

www.artnscience.us/index.html Immunologist Art Anderson has developed a lovely website that appreciates the importance of integrating excellent immunologic information with images.

https://www.youtube.com/watch?v=FzcTgrxMzZk "The Inner Life of the Cell" is an artistic and informative animation of the intercellular and intracellular molecular events associated with leukocyte extravasation. A shorter version is also available on Youtube (https://www.youtube.com/watch?v=wJyUtbn0O5Y).

www.nature.com/ni/multimedia/index.html This is an archive of images and videos of live immune cells from articles published in *Nature* and other Nature Publishing Group journals.

www.atlasofscience.org/the-thymus-a-small-organ-with-a-mighty-big-function/ An accessible and visually appealing description of the migratory patterns of T cells.

www.youtube.com/watch?v=XOeRJPlMpSs YouTube hosts many videos of immune cell behavior—in vitro and in vivo. This YouTube video, for instance, shows fluorescently labeled dendritic cells migrating through skin.

Video Links

Hyperlinks to each video are provided in our e-Book. Alternatively, they can be found by accessing the associated reference online. Most of the videos associated with figures in this chapter can also be found in the online LaunchPad associated with *Kuby Immunology* (Chapter 14). Note that some links work better with different browsers.

Video 14-0v T lymphocytes migrate along the fibroblastic reticular cell network. Supplementary Movie 5 from Bajénoff, M., et al. 2006. Stromal cell networks regulate lymphocyte entry, migration, and territoriality in lymph nodes. *Immunity* **25**:989.

https://dx.doi.org/10.1016/j.immuni.2006.10.011

Video 14-6v Two-photon imaging of live T and B lymphocytes within a mouse lymph node. Movie S3 from Miller, M. J., S. H. Wei, I. Parker, and M. D. Cahalan. 2002. Two-photon imaging of lymphocyte motility and antigen response in intact lymph node. *Science* **296**:1869.

https://doi.org/10.1126/science.1070051

Video 14-7v Dendritic cells (DCs) are present in all lymph-node microenvironments. Supplementary Video 5 from Lindquist, R. L., et al. 2004. Visualizing dendritic cell networks in vivo. *Nature Immunology* **5**:1243.

https://doi.org/10.1038/ni1139

Video 14-8v Lymphocytes exit the lymph node through portals. Supplementary Video 8 from Wei, S. H., et al. 2005. Sphingosine 1-phosphate type 1 receptor agonism inhibits transendothelial migration of medullary T cells to lymphatic sinuses. *Nature Immunology* **6**:1228.

https://doi.org/10.1038/ni1269

Video 14-9v Neutrophils (red) swarming and monocytes (green) following. Video 1 from Lämmermann, T, et al. 2013. Neutrophil swarms require LTB4 and integrins at sites of cell death in vivo. *Nature* **498**:371.

https://doi.org/10.1038/nature12175

Video 14-11v Malarial sporozoites travel to a lymph node and are engulfed by an APC. Movie S2 from Radtke, A. J., et al. 2015. Lymph-node resident CD8[+] dendritic cells capture antigens from malaria, sporozoites and induce CD8[+] T-cell responses. *PLoS Pathogens* **11**:e1004637.

https://doi.org/10.1371/journal.ppat.1004637

Video 14-12v Dendritic cells crawl toward lymph nodes along lymphatic vessels. Movie S3 from Russo, E., et al. 2016. Intralymphatic CCL21 promotes tissue egress of dendritic cells through afferent lymphatic vessels. *Cell Reports* **14**:1723.

https://doi.org/10.1016/j.celrep.2015.12.067

Video 14-13v B cells capture antigen from macrophages in the subcapsular sinus of the lymph node. Supplementary Movie 5 from Phan, T. G., I. Grigorova, T. Okada, and J. G. Cyster. 2007. Subcapsular encounter and complement-dependent transport of immune complexes by lymph-node B cells. *Nature Immunology* **8**:992.

https://doi.org/10.1038/ni1494

Videos 14-14v1 and 14v2 Simulation of movements of DC and naïve lymphocytes in the absence (Movie 1) and presence (Movie 2) of antigen. From Catron, D. M., A. A. Itano,

K. A. Pape, D. L. Mueller, and M. K. Jenkins. 2004. Visualizing the first 50 hr of the primary immune response to a soluble antigen. *Immunity* **21**:341.

http://dx.doi.org/10.1016/j.immuni.2004.08.007

Video 14-14v3 T cells arrest after antigen encounter. Movie S10 from Celli, S., F. Lemaître, and P. Bousso. 2007. Real-time manipulation of T cell–dendritic cell interactions in vivo reveals the importance of prolonged contacts for CD4 T cell activation. *Immunity* **27**:625.

http://dx.doi.org/10.1016/j.immuni.2007.08.018

Video 14-14v4 B cells travel to the border between the follicle and paracortex after being activated by antigen. Video S1 from Okada, T., et al. 2005. Antigen-engaged B cells undergo chemotaxis toward the T zone and form motile conjugates with helper T cells. *PLoS Biology* **3**:e150.

https://doi.org/10.1371/journal.pbio.0030150

Video 14-14v5 Antigen-specific B and T cells interact at the border between the follicle and paracortex. Video S6 from Okada, T., et al. 2005. Antigen-engaged B cells undergo chemotaxis toward the T zone and form motile conjugates with helper T cells. *PLoS Biology* **3**:e150.

https://doi.org/10.1371/journal.pbio.0030150

Video 14-15v1 Germinal center B cells differ in their behavior from naïve B cells. Movie S1 from Allen, C. D., T. Okada, H. L. Tang, and J. G. Cyster. 2007. Imaging of germinal center selection events during affinity maturation. *Science* **315**:528.

https://doi.org/10.1126/science.1136736

Video 14-15v2 T follicular helper cells exiting a germinal center. Movie S1 from Shulman, Z., et al. 2013. T follicular helper cell dynamics in germinal centers. *Science* **341**:673.

https://doi.org/10.1126/science.1241680

Video 14-17v The formation of a tricellular complex in a lymph node during $CD8^+$ T-cell activation. Supplementary Movie 5 from Castellino, F., et al. 2006. Chemokines enhance immunity by guiding naïve $CD8^+$ T cells to sites of $CD4^+$ T cell–dendritic cell interaction. *Nature* **440**:890.

https://doi.org/10.1038/nature04651

Video 14-23v A reticular network is established at the site of infection by *Toxoplasma gondii*. Movie S11 from Wilson, E. H., et al. 2009. Behavior of parasite-specific effector $CD8^+$ T cells in the brain and visualization of a kinesis-associated system of reticular fibers. *Immunity* **30**:300.

http://dx.doi.org/10.1016/j.immuni.2008.12.013

Video 14-24v Resident memory cells patrolling the dermis of mouse. Movie S1 from Ariotti, S., et al. 2012. Tissue-resident memory $CD8^+$ T cells continuously patrol skin epithelia to quickly recognize local antigen. *Proceedings of the National Academy of Sciences USA* **109**:19739.

https://doi.org/10.1073/pnas.1208927109

STUDY QUESTIONS

1. You want to track the behavior of T cells specific for the influenza virus in a mouse lymph node. You have a mouse whose cells all express *yellow fluorescent protein* (YFP). You decide to isolate T cells from this mouse and introduce them into a mouse that has been immunized with influenza, as well as into a control mouse that was given no antigen. You look at the lymph nodes of both mice, expecting to see a difference in the behavior of the cells. However, you do not see much of a difference. At first you wonder if all you read is true—perhaps T cells do not arrest when they encounter antigen! But then you realize that your experimental design was flawed. What was the problem?

2. Indicate whether each of the following statements is true or false. If you think a statement is false, explain why.
   a. Chemokines are chemoattractants for lymphocytes but not other leukocytes.
   b. T cells, but not B cells, express the chemokine receptor CCR7.
   c. Antigen can come into the lymph node only if it is associated with an antigen-presenting cell.
   d. Lymphocytes increase their motility after they engage dendritic cells (DCs) expressing an antigen to which they bind.
   e. T cells crawl along the follicular DC network as they scan DCs for antigen in the lymph node.
   f. Lymphocytes make use of reticular networks only in secondary lymphoid organs.
   g. Leukocyte extravasation follows this sequence: adhesion, chemokine activation, rolling, transmigration.
   h. All secondary lymphoid organs contain high-endothelial venules (HEVs).

3. Provide an example of lymphocyte chemotaxis during an immune response.

4. Describe where $CD8^+$ T cells and B cells receive T-cell help within a secondary lymphoid organ.

5. How might the behavior of an antigen-specific $CD8^+$ T cell differ in the lymph node of a CCL3-deficient animal versus a wild-type animal?

6. Extravasation of neutrophils and of lymphocytes occurs by generally similar mechanisms, although some differences distinguish the two processes.

   a. List in order the basic steps in leukocyte extravasation.
   b. Which step requires chemokine activation and why?

c. Naïve lymphocytes generally do not enter tissues other than the secondary lymphoid organs. What confines them to this system?

7. Naïve T and naïve B-cell subpopulations migrate preferentially into different parts of the lymph node. What is the basis for this compartmentalization? Identify both structural and molecular influences.

8. True or false? Germinal center B cells differ in morphology and motility from other B cells in the follicle.

9. Place a check mark next to the molecules that interact with each other.

   a. _____ Chemokine and L-selectin
   b. _____ E-selectin and L-selectin
   c. _____ CCL19 and CCR7
   d. _____ ICAM and chemokine
   e. _____ Chemokine and G protein–coupled receptor
   f. _____ BCR and MHC
   g. _____ TCR and MHC

10. Predict how a deficiency in each of the following would affect T-cell and B-cell trafficking in a lymph node during a response to antigen. How might they affect an animal's health?

    a. CD62L
    b. CCR7
    c. CCR5
    d. S1P1

11. Match the following type of cells with their location in the body. Note that (1) more than one cell type can be associated with a location and (2) a cell type might be found in more than one location.

    Cell types: Resident memory T cell, central memory T cell, effector T cell, plasma cell, naïve lymphocyte, dendritic cell, CD169$^+$ macrophage, T$_{FH}$ cell
    Locations: Secondary lymphoid organs, barrier organs, T-cell zones of secondary lymphoid organs, sinuses of secondary lymphoid organs, bone marrow, follicle, blood, lymphatic vessels, brain

ANALYZE THE DATA A paper (Gebhardt, T., et al. 2011. Different patterns of peripheral migration by memory CD4$^+$ and CD8$^+$ T cells. *Nature* **477**:216) presented some of the first dynamic imaging data directly comparing the behavior of memory CD4$^+$ and CD8$^+$ T cells in response to infection. Their results were surprising.

Gebhardt et al.'s experimental strategy: The investigators infected the skin of mice with herpes simplex virus and adoptively transferred CD4$^+$ T cells specific for the virus as well as CD8$^+$ T cells, both of which expressed T-cell

receptors specific for the virus. They observed the movements of these cells in the infected skin during the effector phase of the immune response (8 days after infection) as well as 5 or more weeks later when the only cells in the tissue would be memory cells. Remember that skin has two layers: an outer layer (epidermis) and an inner layer (dermis).

Their results: During the effector phase, both CD4$^+$ and CD8$^+$ T cells initially moved similarly through the dermis. However, gradually, these two cell populations distributed themselves differently: CD4$^+$ T cells stayed in the dermis, CD8$^+$ T cells moved to the epidermis (!).

The investigators then looked at the memory cell populations in infected skin weeks later. What they saw is depicted in the supplementary movie 4 (memory CD8$^+$ T cells [red], memory CD4$^+$ T cells [green]): www.nature.com/nature/journal/v477/n7363/full/nature10339.html#supplementary-information

Your assignment: Take a look at this time-lapse video and read its legend provided in the original paper.

a. Describe what you see, identifying at least two specific differences between CD4$^+$ and CD8$^+$ T-cell behavior.
b. Propose a molecular difference that could explain one of these distinctions.
c. Advance one hypothesis about the adaptive value of the difference(s) you observed. That is, what advantage (if any) may such a difference in CD4$^+$ and CD8$^+$ T-cell behavior provide an animal responding to a skin infection?

EXPERIMENTAL DESIGN QUESTION You want to directly test claims that CCR5 is important for the localization of naïve B cells to the follicles of a lymph node. You have all the reagents that you need to perform dynamic imaging in live tissue, including (1) an anti-CCR5 antibody that you know will block the interactions between CCR5 and its ligand (and can be injected), and (2) a CCR5-deficient (CCR5$^{-/-}$) mouse. Design an experiment that will definitively test these claims. Define what you will measure, and sketch one figure predicting your results.

CLINICAL FOCUS QUESTION *Leukocyte adhesion deficiency I* (LAD I) is a rare genetic disease that results from a defect or deficiency in CD18. Patients with this condition usually do not live past childhood because they cannot fight off bacterial infections. Why? Given what you have learned in this chapter, advance a proposal explaining how this disease impairs the immune system's ability to fight bacterial infections. Be specific and concise. What approaches might be taken to treat this disease?

Allergy, Hypersensitivities, and Chronic Inflammation

15

Learning Objectives
After reading this chapter, you should be able to:

1. Distinguish between the four different types of hypersensitivities, and understand the immunological mechanisms behind each of them.

2. For each of the four types of hypersensitivities, recognize the *beneficial functions* of the underlying immune responses in the elimination of pathogens as well as the *harmful effects* of these immune responses when they become hypersensitivity reactions.

3. Discuss the roles of environmental factors and genetics in the predisposition to allergies, particularly in the context of the model discussed for the induction of allergic responses.

4. Describe how beneficial local inflammatory responses may become harmful chronic inflammatory responses, and provide examples induced by infectious and noninfectious causes.

The same immune reactions that protect us from infection can also inflict a great deal of damage to our own cells and tissues. As you have learned, the immune response uses multiple strategies to reduce damage to self by turning off responses once pathogen is cleared, and by avoiding reactions to self antigens. However, these checks and balances can break down, leading to immune-mediated reactions that are more detrimental than protective. Some immune-mediated disorders are caused by a failure of immune tolerance. These autoimmune disorders will be discussed in Chapter 16. Others are caused by an inappropriately vigorous innate and/or adaptive response to antigens that pose little or no threat. These

Young girl sneezing in response to flowers. *[Flashpop/Getty Images]*

disorders, called **hypersensitivities**, are one of the main focuses of this chapter.

Two French scientists, Paul Portier and Charles Richet, were the first to recognize and describe hypersensitivities. In the early twentieth century, as part of their studies of the responses of bathers in the Mediterranean to the stings of Portuguese man-o'-war jellyfish (*Physalia physalis*), they demonstrated that the toxic agent in the sting was a small protein. They reasoned that eliciting an antibody response that could neutralize the toxin may serve to protect the host. Therefore, they injected low doses of the toxin into dogs to elicit an immune response, and followed with a booster injection a few weeks later. However, instead of generating a protective antibody response, the unfortunate dogs responded immediately to the second injection with vomiting, diarrhea, asphyxia, and death. Richet coined the term *anaphylaxis*, derived from the Greek and translated loosely as "against protection" to describe this overreaction of the immune system,

Key Terms

Immediate hypersensitivity
Delayed-type hypersensitivity (DTH)
Types I, II, III, and IV hypersensitivity

Allergy
Chronic inflammation
Atopy
FcεRI receptor
FcεRII receptor

Degranulation
Histamine
Leukotrienes and prostaglandins
Atopic (allergic) march

Anaphylaxis
Hygiene hypothesis
Desensitization

the first description of a hypersensitivity reaction. Richet was subsequently awarded the Nobel Prize in Physiology or Medicine in 1913.

Since that time, immunologists have learned that there are multiple types of hypersensitivity reactions. **Immediate hypersensitivity** reactions result in symptoms that manifest themselves within very short time periods after the immune stimulus, like those described above. Other types of hypersensitivity reactions take 1 to 3 days to manifest themselves, and are referred to as *delayed-type hypersensitivity* **(DTH)** reactions. In general, immediate hypersensitivity reactions result from antibody-antigen reactions, whereas DTH is caused by T-cell reactions.

As it became clear that different immune mechanisms give rise to distinct hypersensitivity reactions, two immunologists, P. G. H. Gell and R. R. A. Coombs, proposed a classification scheme to discriminate among the various types of hypersensitivity. Hypersensitivities are classically divided into four categories (types I–IV) that differ by the immune molecules and cells that cause them, and the way they induce damage (**Figure 15-1**). **Type I hypersensitivity** reactions are mediated by IgE antibodies that bind to mast cells or basophils and induce mediator release; these reactions include the most common responses to respiratory allergens, such as pollen and dust mites, and to food allergens, such as peanuts and shellfish. **Type II hypersensitivity** reactions result from the binding of IgG or IgM to the surface of host cells, which are then destroyed by complement- or cell-mediated mechanisms. For example, this is the fate for transfused red blood cells in transfusions between people differing in ABO blood types. In **type III hypersensitivity** reactions, antigen-antibody complexes

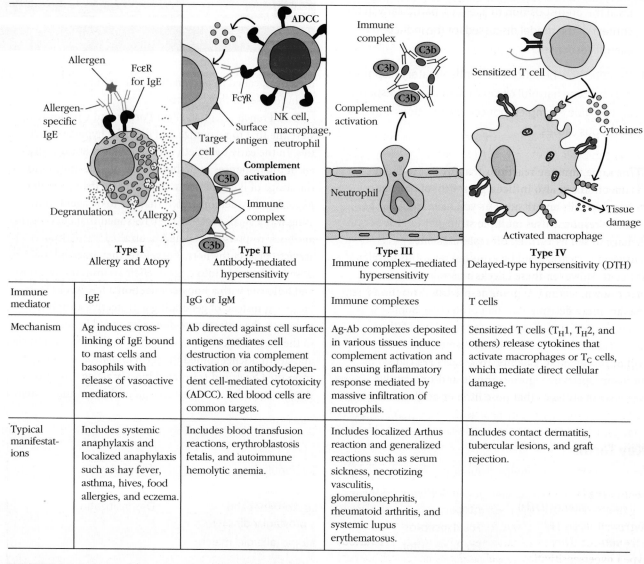

	Type I Allergy and Atopy	**Type II** Antibody-mediated hypersensitivity	**Type III** Immune complex–mediated hypersensitivity	**Type IV** Delayed-type hypersensitivity (DTH)
Immune mediator	IgE	IgG or IgM	Immune complexes	T cells
Mechanism	Ag induces cross-linking of IgE bound to mast cells and basophils with release of vasoactive mediators.	Ab directed against cell surface antigens mediates cell destruction via complement activation or antibody-dependent cell-mediated cytotoxicity (ADCC). Red blood cells are common targets.	Ag-Ab complexes deposited in various tissues induce complement activation and an ensuing inflammatory response mediated by massive infiltration of neutrophils.	Sensitized T cells (T_H1, T_H2, and others) release cytokines that activate macrophages or T_C cells, which mediate direct cellular damage.
Typical manifestations	Includes systemic anaphylaxis and localized anaphylaxis such as hay fever, asthma, hives, food allergies, and eczema.	Includes blood transfusion reactions, erythroblastosis fetalis, and autoimmune hemolytic anemia.	Includes localized Arthus reaction and generalized reactions such as serum sickness, necrotizing vasculitis, glomerulonephritis, rheumatoid arthritis, and systemic lupus erythematosus.	Includes contact dermatitis, tubercular lesions, and graft rejection.

FIGURE 15-1 The four types of hypersensitivity reactions.

(such as those generated by the injection of foreign serum proteins) deposited on host cells or tissues activate complement or the release of mediators from granulocytes, often causing inflammatory responses. **Type IV hypersensitivity** reactions result from excessive and sometimes inappropriate T-cell activation. Common examples are the skin reactions caused by poison oak or poison ivy. It should be noted that although this classification method has proven to be a useful analytical and descriptive tool, many clinical hypersensitivity disorders include molecular and cellular contributions from components belonging to more than one of these categories. The subdivisions are not as frequently evoked in real-world clinical settings as they once were.

The term **allergy** first appeared in the medical literature in 1906, when the pediatrician Clemens von Pirquet noted that the response to some antigens resulted in damage to the host, rather than in a protective response. Although most familiar respiratory and food allergies result from the generation of IgE antibodies toward the eliciting agent, and are therefore type I hypersensitivity reactions, other common reactions that are sometimes thought of as allergies, such as the response to poison ivy, result from T cell–mediated, type IV responses.

Hypersensitivities are not the only examples of pathological conditions that are caused by normally beneficial immune responses. Some disorders are caused by a failure to turn off innate or adaptive responses, resulting in a chronic inflammatory state; we will close this chapter with a discussion of the causes and consequences of **chronic inflammation**, which, for example, can result from persistent infections, from autoimmunity, and also from metabolic disorders. Chronic inflammation is of interest to many because of its intriguing association with the current obesity epidemic.

Allergies: Type I Hypersensitivity

More than 30% of adults and 40% of children in the United States suffer from type I hypersensitivity reactions, or allergies. Type I hypersensitivity reactions are mediated by IgE antibodies, and include the most common allergic reactions, such as allergic rhinitis (hay fever), asthma, atopic dermatitis (eczema), and food allergies. The incidence of allergy continues to rise in the human population, and understanding immune mechanisms behind the response has already led to new therapies. In this section, we describe the molecular and cellular participants and processes in the various type I hypersensitivities, as well as the rationale behind current treatments.

IgE Antibodies Are Responsible for Type I Hypersensitivity

Type I hypersensitivity reactions (allergies) are initiated by the interaction between IgE antibodies and the multivalent antigens with which they react. **Classic Experiment Box 15-1** describes the brilliant series of experiments by K. Ishizaka and T. Ishizaka in the 1960s and 1970s that led to the identification of IgE as the class of antibody responsible for allergies. In normal individuals, the level of IgE in serum is the lowest of any of the immunoglobulin classes, making physicochemical studies of this molecule particularly difficult. It was not until the discovery of an IgE-producing myeloma by Johansson and Bennich in 1967 that extensive analyses of IgE could be undertaken.

Key Concept:
- IgE was discovered in the 1960s and shown to cause allergic reactions.

Many Allergens Can Elicit a Type I Response

Individuals without allergies generate IgE antibodies only in response to parasitic infections. However, individuals who genetically are highly susceptible to allergies, a condition known as **atopy**, are predisposed to generate IgE antibodies against common environmental antigens, such as those listed in **Table 15-1**. Chemical analyses revealed that most, if not all, allergens are highly soluble proteins or glycoproteins, usually with multiple *epitopes*, or antigenic determinants.

TABLE 15-1	Common allergens associated with type I hypersensitivity
Plant pollens	**Drugs**
Rye grass	Penicillin
Ragweed	Sulfonamides
Timothy grass	Local anesthetics
Birch tree	Salicylates
Foods	**Insect products**
Nuts	Bee venom
Seafood	Wasp venom
Eggs	Ant venom
Peas, beans	Cockroach calyx
Milk	Dust mites
Other allergens	
Animal hair and dander	
Latex	
Mold spores	

BOX 15-1

The Discovery and Identification of IgE as the Carrier of Allergic Hypersensitivity

In a stunning series of papers published between 1966 and the mid-1970s, Kimishige Ishizaka and Teruko Ishizaka, working with a number of collaborators, identified a new class of immunoglobulins, which they called IgE, as being the major effector molecules in type I antibody–mediated hypersensitivity reactions (allergic reactions).

The Ishizakas built on work performed in 1921 by C. Prausnitz and H. Küstner, who injected serum from an allergic person into the skin of a nonallergic individual. When the appropriate antigen was later injected into the same site, a wheal-and-flare reaction (swelling and reddening) was detected. This reaction, called the *P-K reaction* after its originators, was the basis for the first biological assay for IgE activity.

In their now classic experiments, published just over 50 years ago, the Ishizakas assayed for the presence of allergen-specific antibody, using the wheal-and-flare reaction. As an additional assay, they

also employed *radioimmunodiffusion*, which tested the ability of radioactive "allergen E" derived from ragweed pollen to bind to and precipitate pollen-specific antibodies; the antibodies formed a radioactive precipitate on binding to the ragweed allergen. (Note that both the antigen and the immunoglobulin class are designated "E" in this series of experiments.)

The Ishizakas reasoned that the best starting material for purifying the protein responsible for the P-K reaction would be the serum of an allergic individual who displayed hypersensitivity to ragweed pollen E. To purify the serum protein responsible for the allergic reaction, they took whole human serum and subjected it to *ammonium sulfate precipitation* (in which different proteins precipitate out at varying concentrations of ammonium sulfate), and *ion-exchange chromatography*, which separates proteins on the basis of their intrinsic charge.

Different fractions from the chromatography column were tested by radioimmunodiffusion for their ability to bind to radioactive antigen E. Portions of the different fractions were also injected at varying dilutions into volunteers, along with allergen E, to test for a P-K reaction. Finally, each fraction was also tested semiquantitatively for the presence of IgG and IgA antibodies. The results of these experiments are shown for the serum from one of the three individuals they tested (see **Table 1**).

From the results in this table, it is clear that the ability of proteins in the various fractions to induce a P-K reaction did not correlate with the amounts of either IgG or IgA antibodies, but it did correlate with the amounts of antibodies that could be detected in an immunodiffusion reaction with radioactive antigen E. This suggested that another, previously unidentified antibody class was responsible for the line of immunoprecipitation on the immunodiffusion gel (see Chapter 20).

TABLE 1	Data from original paper identifying the immunoglobulins responsible for skin sensitization				
			RELATIVE AMOUNT OF:		
Serum donor	Fraction	Minimum dose for P-K reaction	IgE	IgG	IgA
U	A	0.04	+	+	−
	B	**0.008**	++	+	−
	C	0.26	+	−	±
	D	>0.9	−	−	−
A	A	0.002	++	++	−
	B	**0.0006**	+++	+	±
	C	0.0014	++	+	±
	D	0.005	++	+	+
	E	0.017	++	−	+
	F	0.13	+	−	+

Modified from original table entitled "Distribution of skin-sensitizing activity and of γG (IgG) and γA (IgA) globulin following diethylaminoethyl (DEAE) Sephadex column fractionation," in Ishizaka, K., and T. Ishizaka. 1967. Identification of γE-antibodies as a carrier of reaginic activity. *Journal of Immunology* **99**:1187.

(continued)

The fractions containing the highest concentration of protein able to bind to allergen E were further purified by gel chromatography, which separates proteins on the basis of size and molecular shape. Again, the presence of the protein causing antigen-specific allergic reactions was detected on the basis of its ability to bind to radioactive antigen E and to induce a P-K reaction in the skin of a test subject who had been injected with antigen E.

The resulting protein still contained small amounts of IgG and IgA antibodies, which were eliminated by mixing the fractions with antibodies directed toward those human antibody subclasses, and then removing the resultant immunoprecipitate. The Ishizakas' final protein product was 500 to 1000 times more potent than the original serum in its ability to generate a P-K reaction; the most active preparation generated a positive skin reaction at a dilution of 1:8000. None of its activities correlated with the presence of any of the other known classes of antibody, and so a new class of antibody was named, IgE, based on its ability to bind to allergens and bring about a P-K reaction.

In Table 1, which is a modified version of the original data in this classic 1967 paper, serum protein fractions from two separate donors were evaluated for the relative amounts of IgA or IgG (referred to as γA and γG, respectively, as immunoglobulins were originally called g globulins), using rabbit antisera against the two immunoglobulin subclasses, and for the presence of the putative gE (later changed to IgE), using radioimmunodiffusion. IgG is the most common class of immunoglobulin in serum, and IgA was included because early experiments had suggested that IgA may be the antibody responsible for the wheal-and-flare reaction. The "Minimum dose for P-K reaction" column refers to the quantity of the fraction required to yield a measurable inducing antibody in the fraction (i.e., fraction B had the highest amount of the allergenic antibody).

It can readily be seen that the B fractions, which showed the strongest P-K responses, also had the highest amounts of IgE as measured by radioimmunodiffusion. P-K reactions did not correlate with either IgG or IgA levels in the serum from this donor, or from two other donors.

The level of IgE in the serum is the lowest of all the antibody classes, falling within the range of 0.1 to 0.4 µg/ml, and was thus a challenge to purify and characterize. Although allergic (atopic) individuals can have as much as 10 times this concentration of IgE in their circulation, that's still a small amount and difficult to work with. However, in 1967, Johansson and Bennich discovered an IgE-producing myeloma, which enabled a full biochemical analysis of the molecule. The structure of IgE is described in Chapter 3.

REFERENCES

Finkelman, F. D. 2017. Pillars of immunology: identification of IgE as the allergy-associated Ig isotype. *Journal of Immunology* **198**:3.

Ishizaka, K., and T. Ishizaka. 1967. Identification of γE-antibodies as a carrier of reaginic activity. *Journal of Immunology* **99**:1187.

Johansson, S. G., and H. Bennich. 1967. Immunological studies of an atypical (myeloma) immunoglobulin. *Immunology* **13**:381.

For many years, scientists tried unsuccessfully to find any structural commonalities among molecules that induced allergic reactions in susceptible individuals. Recently, several features shared by many allergens have begun to provide clues to the biological basis of their activity.

First, many allergens have intrinsic enzymatic activity that contributes to the allergic response. For example, allergens from dust mites (the allergenic component of house dust), cockroaches, pollen, fungi, and bacteria have protease activity. Some of these proteases have been shown to be capable of disrupting the integrity of epithelial cell junctions, allowing allergens to access the underlying cells and molecules of the innate and adaptive immune systems. Others, including a protease (Der p 1) produced by the dust mite (*Dermatophagoides pteronyssinus*), cleave and activate complement components at the mucosal surface. Still others cleave and stimulate protease-activated receptors on the surfaces of immune cells, enhancing inflammation. Several cleave the inactive form of the IL-33 cytokine produced by epithelial cells, generating active IL-33 that contributes to allergic responses, as will be discussed shortly. Thus, one factor that distinguishes allergenic from nonallergenic antigens may be the presence of protease activity that affects the cells and molecules of the immune system.

Second, many allergens contain potential *pathogen-associated molecular patterns*, or PAMPS (see Chapter 4), capable of interacting with receptors of the innate immune system and initiating a cascade of responses that contribute to an allergic response. Third, many allergens enter the host via mucosal tissues at very low concentrations, which tend to induce T_H2 responses. The IL-4 and IL-13 produced by T_H2 cells induce heavy-chain class switching to IgE during the generation both of plasma cells secreting allergen-specific antibodies and of allergen-specific memory B cells (see Chapter 11).

Key Concepts:

- Allergens are highly soluble proteins or glycoproteins, usually with multiple epitopes.

- Many allergens are proteases and/or contain PAMPs, which result in stimulation of the immune system.

- Some allergens activate T_H2 cells, which induce heavy-chain class switching to IgE.

IgE Antibodies Act by Binding Antigen, Resulting in the Cross-Linking of Fcε Receptors

IgE antibodies alone do not cause harmful responses. Instead, they cause hypersensitivity by binding to Fc receptors, called **FcεRIs**, specific for their constant regions. FcεRIs are expressed by a variety of innate immune cells, including mast cells and basophils (Chapter 2). The binding of IgE antibodies to FcεRI activates these granulocytes, inducing a signaling cascade that causes cells to release the contents of intracellular granules into the blood or tissue, a process called **degranulation** (**Figure 15-2**). The contents of granules vary from cell to cell, but typically include histamine, heparin, and proteases. Together with other molecules (leukotrienes, prostaglandins, chemokines, and cytokines) that are synthesized by these activated granulocytes, these mediators act on surrounding tissues and other immune cells, causing allergy symptoms. A second Fcε receptor with lower affinity for IgE, **FcεRII**, regulates the production of IgE by B cells.

The High-Affinity IgE Receptor, FcεRI

Mast cells and basophils constitutively express high levels of the IgE receptor FcεRI, which binds IgE with an exceptionally high affinity constant of $10^{10}\,\mathrm{M}^{-1}$. This high affinity helps overcome the difficulties associated with binding IgE antibodies, which are present at an extremely low concentration in the serum ($1.3 \times 10^{-7}\,\mathrm{M}$). Eosinophils, Langerhans cells, monocytes, and platelets also express FcεRI, although at lower levels. The serum half-life of IgE is only 2 to 3 days, the shortest of the various immunoglobulin classes (see Table 12-1). The low levels and short lifetime of IgE in the serum reflect the low numbers of IgE-producing plasma cells in most individuals, as well as IgE being bound up by the high-affinity Fcε receptors on cells. However, when bound to its receptor on a granulocyte, IgE is stable for weeks. That FcεRI is the receptor that induces type I hypersensitivity reactions was confirmed by experiments conducted in mice genetically engineered to lack one of the chains of FcεRI. These mice do not develop localized or systemic allergic responses, despite having normal numbers of mast cells.

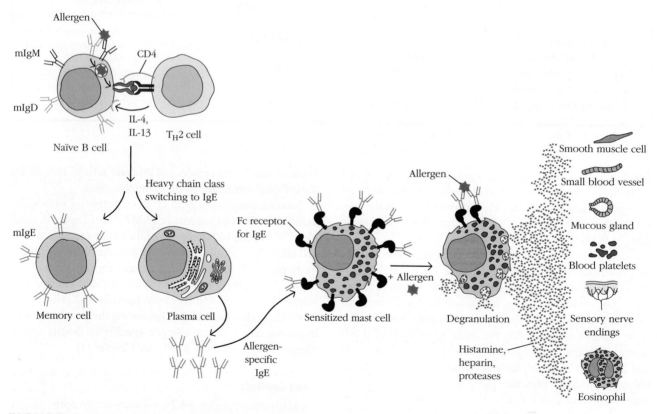

FIGURE 15-2 General mechanism underlying an immediate type I hypersensitivity reaction. Exposure to an allergen activates T$_H$2 cells that stimulate B cells to proliferate, undergo heavy-chain class switching to IgE, and differentiate into IgE-secreting plasma cells and memory B cells expressing membrane IgE B-cell receptors (mIgE). The secreted IgE molecules bind to IgE-specific Fc receptors (FcεRI) on mast cells and blood basophils. (Many molecules of IgE with various specificities for this and other allergens can bind to FcεRI.) A second exposure to the allergen leads to cross-linking of the bound IgE, triggering the release of pharmacologically active mediators from mast cells and basophils. The mediators cause numerous effects, including smooth muscle contraction, increased vascular permeability, and vasodilation.

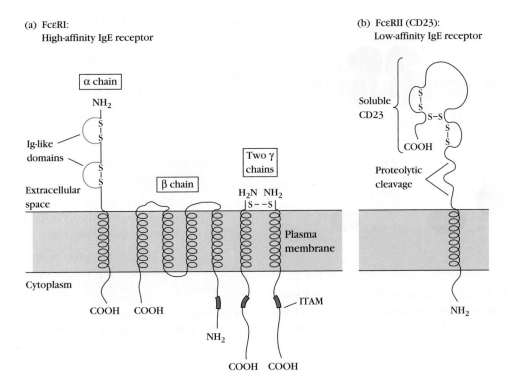

FIGURE 15-3 Schematic diagrams of the high-affinity FcεRI and low-affinity FcεRII receptors that bind the Fc region of IgE. (a) FcεRI consists of an α chain that binds IgE, a β chain that participates in signaling, and two disulfide-linked γ chains that are the most important component in signal transduction. The β and γ chains contain cytoplasmic ITAMs, a motif also present in the Igα/Igβ (CD79α/β) heterodimer of the B-cell receptor and in the CD3 chains of the T-cell receptor complex. (b) The single-chain FcεRII is unusual because it is a type II transmembrane protein, oriented in the membrane with its NH$_2$ terminus directed toward the cell interior and its COOH terminus directed toward the extracellular space.

The standard form of FcεRI is a tetramer that includes an α chain and β chain and two identical disulfide-linked γ chains (**Figure 15-3a**). Monocytes and platelets express an alternative form lacking the β chain. The α chain of the FcεRI, a member of the immunoglobulin superfamily, directly binds the IgE Fc region (via a single FcεRI binding site on the paired Cε3 domains of IgE heavy chains), whereas the β and γ chains are responsible for signal transduction. The β and γ chains contain *i*mmunorecep*t*or *t*yrosine-based *a*ctivation *m*otifs (ITAMs) (see Chapter 3) that are phosphorylated in response to IgE and FcεRI cross-linking by bound antigens.

IgE-mediated signaling begins when an allergen or other antigen is bound by and cross-links IgE antibodies that previously had been captured by the surface FcεRI receptors (see Figure 15-2). Although the specific biochemical events that follow cross-linking of the FcεRI receptors vary among cells and modes of stimulation, the FcεRI-activated signaling cascade generally resembles that initiated by antigen receptors and growth factor receptors (Chapter 3). Briefly, IgE cross-linking induces the aggregation and migration of receptors into membrane lipid rafts, followed by phosphorylation of ITAM motifs by associated tyrosine kinases. Adapter molecules then latch onto the phosphorylated tyrosine residues and initiate signaling cascades culminating in

enzyme and/or transcription factor activation. In addition, a brief increase in cyclic AMP (cAMP) levels is induced, followed by a decline; that spike in cAMP levels and increased levels of intracellular Ca^{2+} are important for inducing degranulation. Also, phospholipase A initiates metabolism of the lipid arachidonic acid, producing inflammatory lipid mediators. **Figure 15-4** identifies just a few of the signaling events specifically associated with the activation of mast cells and basophils.

FcεRI signaling leads to several mast cell and basophil responses: (1) degranulation—the fusion of vesicles containing multiple inflammatory mediators with the plasma membrane and release of their contents (**Figure 15-5a**), (2) synthesis of inflammatory cytokines, and (3) conversion of arachidonic acid into leukotrienes and prostaglandins, two important lipid mediators of inflammation. These mediators have multiple short-term and long-term effects on tissues that will be described in more detail shortly (Figure 15-5b).

The Low-Affinity IgE Receptor, FcεRII

The other IgE receptor, designated FcεRII and also known as CD23, has much lower affinity for IgE (1 × 10^6 M^{-1}) (Figure 15-3b). CD23 is structurally distinct from FcεRI (it is a lectin and a type II membrane protein) and exists in two

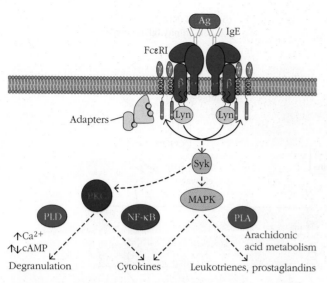

FIGURE 15-4 Signaling pathways initiated by IgE allergen cross-linking of FcεRI receptors. By cross-linking FcεRI receptors, IgE initiates signals that lead to mast cell degranulation and release of prepackaged mediators, cytokine production, and leukotriene and prostaglandin generation. The signaling cascades initiated by the FcεRI are generally similar to those initiated by antigen receptors. This figure illustrates only a few of the players in the complex signaling network. Briefly, cross-linking of FcεRI activates the tyrosine kinase Lyn, which phosphorylates the receptor ITAMs and activates the tyrosine kinase Syk, which phosphorylates adapter molecules that organize signaling responses. Multiple kinases are activated, including *protein kinase C* (PKC) and various *mitogen-activated protein kinases* (MAPKs). These, in turn, activate transcription factors (e.g., NF-κB) that regulate cytokine production. They also activate lipases, including *phospholipase D* (PLD), and stimulate an increase in intracellular free calcium ions and a transient increase in cAMP, all of which induce degranulation. *Phospholipase A* (PLA) is activated, initiating the production of leukotrienes and prostaglandins from the metabolism of arachidonic acid.

isoforms that differ only slightly in the N-terminal cytoplasmic domain. CD23a is found on activated B cells, whereas CD23b is induced on various cell types by the cytokine IL-4. Both isoforms also exist as membrane-bound and soluble forms, the latter being generated by proteolysis of the surface molecule. Interestingly, CD23 binds not only to IgE, but also to the complement receptor CD21.

The outcome of CD23 ligation depends on which ligand it binds (IgE or CD21) and whether it does so as a soluble or membrane-bound molecule. For example, when soluble CD23 (sCD23) binds to CD21 on the surface of IgE-synthesizing B cells, IgE synthesis is increased. However, when membrane-bound CD23 binds soluble IgE, further IgE synthesis is suppressed. This is a negative feedback pathway whose function seems to be to limit the amount of IgE that is synthesized. CD23 has another role in food allergies: it triggers transport of IgE-allergen complexes across the

intestinal epithelium. CD23 is also expressed by macrophages and some dendritic cells. Atopic individuals express relatively high levels of surface and soluble CD23. It also triggers macrophages to release inflammatory cytokines TNF, IL-1, IL-6, and GM-CSF.

> **Key Concepts:**
>
> - IgE antibodies bind to antigen via their variable regions and to one of two types of Fc receptor via their constant regions.
>
> - Mast cells, basophils, and to a lesser extent, eosinophils express the high-affinity FcεRI, and are the main mediators of allergy symptoms.
>
> - Cross-linking of FcεRI receptors by antigen-IgE complexes initiates multiple signaling cascades that resemble those initiated by antigen receptors.
>
> - Mast cells, basophils, and eosinophils that are stimulated by FcεRI cross-linking release their granular contents (including histamine and proteases) in a process called *degranulation*. They also generate and secrete inflammatory cytokines and lipid inflammatory molecules (leukotrienes and prostaglandins).
>
> - The second IgE receptor, the low-affinity FcεRII (also known as CD23), is found on IgE-expressing B cells and on other cells; it helps regulate IgE responses, transports IgE across the intestinal epithelium, and induces inflammatory cytokine production by macrophages.

IgE Receptor Signaling Is Tightly Regulated

Given the powerful effects of the molecular mediators released by mast cells, basophils, and eosinophils following FcεRI signaling, it should come as no surprise that the responses are subject to complex systems of regulation. Here, we offer just a few examples of ways in which signaling through the FcεRI receptor can be inhibited.

Recall from earlier chapters that the intracellular regions of some lymphocyte receptors, including FcγRIIB, bear *immunoreceptor tyrosine inhibitory motifs* (ITIMs, distinct from ITAMs; see Chapter 11 and Figure 12-4). Cross-linking of these receptors leads to *inhibition* of cellular responses, rather than to their *activation*. Interestingly, mast cells express both the activating FcεRI and the inhibitory FcγRIIB. Therefore, if an allergen binds both IgG and IgE molecules, it will trigger signals through both Fc receptors. The inhibitory signal dominates. This phenomenon is part of the reason that eliciting IgG antibodies against common allergens through desensitization therapies can help atopic patients, as we will see later in the chapter. The more allergen-specific IgG antibodies they have, the higher the probability that allergens will cocluster FcεRI receptors with inhibitory FcγRIIB receptors.

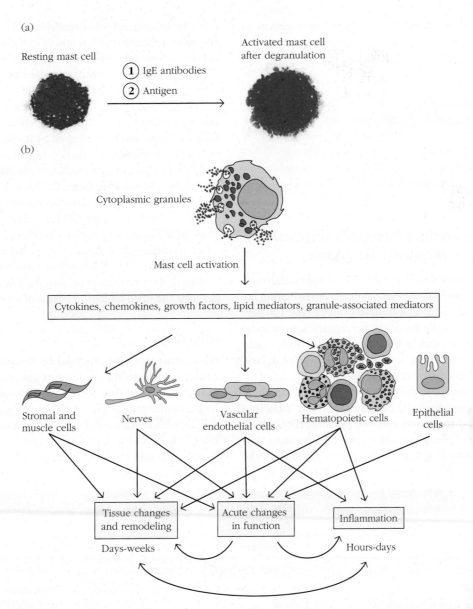

FIGURE 15-5 Effects of mast cell activation. (a) Mast cells before (left) and after degranulation (right) induced by IgE antibodies and antigen that is bound by the antibodies. Resting mast cells have numerous granules (secretory vesicles) stored in their cytoplasm. After the mast cell is activated by addition of IgE antibodies (which bind to FcεRI) and an antigen that cross-links the IgE antibodies, the granules fuse with the plasma membrane, releasing their contents. Extra membrane from the granules is seen in the cell's plasma membrane after degranulation. (b) Mast cell mediators and their effects. Various stimuli activate mast cells to secrete different types and/or amounts of products. Activated mast cells immediately release preformed, granule-associated inflammatory mediators (including histamine, proteases, and heparin) and are induced to generate lipid mediators (such as leukotrienes and prostaglandins), chemokines, cytokines, and growth factors (some of which can also be packaged in granules). These mediators act on different cell types, and have both acute and chronic effects. When produced over long periods of time, mast cell mediators have a significant influence on tissue structure by enhancing proliferation of fibroblasts and epithelial cells, increasing production and deposition of collagen and other connective tissue proteins, stimulating the generation of blood vessels, and more. *[(a) Lorentz A., Baumann A., Vitte J., Blank U., "The SNARE machinery in mast cell secretion,"* Frontiers in Immunology, *2012 Jun 5;* **3**:143. doi: 10.3389/fimmu.2012.00143. eCollection 2012. Figure 1.]

Because many of the reactions in the activation pathway downstream from the FcεRI pathways are phosphorylations, multiple phosphatases, such as SHPs, SHIPs, and PTEN, can play an important role in dampening receptor signaling. For example, FcγRIIB activates the SHIP phosphatase, which removes key activating phosphate groups from signaling intermediates. In addition, the tyrosine kinase Lyn can play a negative role as well as a positive one in FcεRI signaling by phosphorylating and activating the inhibitory FcγRIIB. Finally, FcεRI signaling through Lyn and Syk kinases also activates E3 ubiquitin ligases, including c-Cbl. c-Cbl ubiquitinylates Lyn and Syk, as well as FcεRI itself, triggering their degradation. Thus, FcεRI activation contributes to its own demise, thereby limiting the duration of the response.

> **Key Concept:**
> • Mast cell and basophil activation by FcεRI signaling can be down-regulated by inhibitory signals, including inhibitory FcγRIIB signaling, phosphatases that remove key phosphate residues from signaling intermediates, and ubiquitinylation and degradation of signaling molecules.

Granulocytes Produce Molecules Responsible for Type I Hypersensitivity Symptoms

The molecules released by degranulation of mast cells, basophils, and eosinophils in response to FcεRI cross-linking are responsible for the clinical manifestations of type I hypersensitivity. These inflammatory mediators act on local tissues as well as on populations of secondary effector cells, including other eosinophils, neutrophils, T lymphocytes, monocytes, and platelets (see Figure 15-2).

When generated in response to parasitic infection, these mediators initiate beneficial defense processes, including vasodilation and increased vascular permeability, which brings an influx of plasma mediators and inflammatory cells to attack the pathogen. They inflict direct damage on the parasite. In contrast, mediator release induced by allergens results in unnecessary increases in vascular permeability and inflammation that lead to tissue damage with little benefit.

The molecular mediators can be classified as either primary or secondary (**Table 15-2**). Primary mediators are preformed and stored in granules prior to cell activation, whereas secondary mediators are either synthesized after target-cell activation or released by the breakdown of membrane phospholipids during the degranulation process. The most significant primary mediators are histamine, proteases, eosinophil chemotactic factor (ECF), neutrophil chemotactic factor (NCF), and heparin. Secondary mediators include platelet-activating factor (PAF), leukotrienes, prostaglandins, bradykinin, and various cytokines and chemokines. The varying manifestations of type I hypersensitivity in different tissues and species reflect variations in the primary and secondary mediators present. In this section, we briefly describe the main biological effects of several key mediators.

Histamine

Histamine, which is formed by decarboxylation of the amino acid histidine, is a major component of mast cell granules, accounting for about 10% of granule weight. Its biological effects are observed within minutes of mast cell activation. Once released from mast cells, histamine binds one of four different histamine receptors, designated H_1, H_2, H_3, and H_4. These receptors have different tissue

TABLE 15-2	Principal mediators involved in type I hypersensitivity
Mediator	**Effects**
	Primary
Histamine, heparin	Increased vascular permeability; smooth muscle contraction
Serotonin (rodents)	Increased vascular permeability; smooth muscle contraction
Eosinophil chemotactic factor (ECF-A)	Eosinophil chemotaxis
Neutrophil chemotactic factor (NCF-A)	Neutrophil chemotaxis
Proteases (tryptase, chymase)	Bronchial mucus secretion; degradation of blood vessel basement membrane; generation of complement split products
	Secondary
Platelet-activating factor	Platelet aggregation and degranulation; contraction of pulmonary smooth muscles
Leukotrienes (slow reactive substance of anaphylaxis, SRS-A)	Increased vascular permeability; contraction of pulmonary smooth muscles
Prostaglandins	Vasodilation; contraction of pulmonary smooth muscles; platelet aggregation
Bradykinin	Increased vascular permeability; smooth muscle contraction
Cytokines IL-1 and TNF-α	Systemic anaphylaxis; increased expression of adhesion molecules on venous endothelial cells
IL-4 and IL-13	Induction of T_H2 cells, increased IgE production
IL-3, IL-5, IL-6, IL-8, IL-9, IL-10, TGF-β, and GM-CSF	Various effects (see text)

distributions and mediate different effects on histamine binding. Serotonin is also present in the mast cells of rodents and has effects similar to histamine.

Most of the biological effects of histamine in allergic reactions are mediated by the binding of histamine to H_1 receptors. This binding induces contraction of intestinal and bronchial smooth muscles, increased permeability of venules (small veins), and increased mucus secretion. Interaction of histamine with H_2 receptors increases vascular permeability (due to contraction of endothelial cells) and vasodilation (by relaxing the smooth muscle of blood vessels), stimulates exocrine (secretory) glands, and increases the release of acid in the stomach. Binding of histamine to H_2 receptors on mast cells and basophils suppresses degranulation; thus, histamine ultimately exerts negative feedback on the further release of mediators. The H_4 receptor mediates chemotaxis of mast cells, while H_3 is less involved in type I hypersensitivity responses; it modulates neurotransmitter activity in the central nervous system.

Leukotrienes and Prostaglandins

As secondary mediators, the **leukotrienes** and **prostaglandins** are not formed until the mast cell undergoes degranulation and phospholipase signaling initiates the enzymatic breakdown of phospholipids in the plasma membrane. An ensuing enzymatic cascade generates the prostaglandins and the leukotrienes.

In a type I hypersensitivity–mediated asthmatic response, the initial contraction of human bronchial and tracheal smooth muscles is at first mediated by histamine; however, within 30 to 60 seconds, further contraction is signaled by leukotrienes and prostaglandins. Active at nanomolar levels, the leukotrienes are approximately 1000 times more effective at mediating bronchoconstriction than is histamine, and they are also more potent stimulators of vascular permeability and mucus secretion. In humans, the leukotrienes are thought to contribute significantly to the prolonged bronchospasm and buildup of mucus seen in people with asthma.

Cytokines and Chemokines

Adding to the complexity of type I hypersensitivity reactions is the variety of cytokines released from mast cells and basophils. Mast cells, basophils, and eosinophils secrete several cytokines, including IL-4, IL-5, IL-8, IL-9, IL-13, GM-CSF, and TNF-α. These cytokines alter the local microenvironment and lead to the recruitment and activation of inflammatory cells such as neutrophils and eosinophils. In addition, IL-4 and IL-13 stimulate a T_H2 response and thus increase IgE production by B cells. IL-5 is especially important in the recruitment and activation of eosinophils. IL-8, the chemokine CXCL8, attracts additional neutrophils, monocytes, mast cells, basophils, and various subsets of T cells to the site of the hypersensitivity response. IL-9 increases the number and activity of mast cells. The high concentrations of the potent inflammatory cytokine TNF-α

secreted by mast cells may contribute to shock in systemic anaphylaxis. GM-CSF stimulates the production and activation of myeloid cells, including granulocytes and macrophages. Chemokines secreted at the same time recruit other cells to the site of the reaction.

Key Concepts:

- Degranulation of activated mast cells, basophils, and eosinophils releases preformed primary mediators: histamine, proteases, eosinophil chemotactic factor, neutrophil chemotactic factor, and heparin.

- Activation also induces production of leukotrienes and prostaglandins, from membrane lipid metabolism, as well as cytokines and chemokines.

- These mediators produce the symptoms of allergic responses. Among its many effects, histamine induces contraction of intestinal and bronchial smooth muscles and increased vasodilation, vascular permeability, and secretion of mucus and other body fluids. Chemoattractants recruit additional granulocytes to the site of the response. Leukotrienes and prostaglandins are potent activators of bronchial and tracheal smooth muscle contraction and increased vasodilation and vascular permeability.

Type I Hypersensitivities Are Characterized by Both Early and Late Responses

Type I hypersensitivity responses are divided into an immediate early response and one or more late-phase responses, as shown for asthma in **Figure 15-6**. The early response occurs within minutes of allergen exposure and, as described above, results from the release of histamine, leukotrienes, and prostaglandins from local mast cells.

However, hours after the immediate phase of a type I hypersensitivity reaction begins to subside, mediators released during the course of the reaction induce localized inflammation, called the *late-phase reaction*. Cytokines released from mast cells, particularly TNF-α and IL-1, increase the expression of chemokines and cell adhesion molecules on venous endothelial cells, thus facilitating the influx of neutrophils, eosinophils, and T_H2 cells that characterizes this phase of the response.

Eosinophils play a principal role in the late-phase reaction. Eosinophil chemotactic factor (ECF), released by mast cells during the initial reaction, attracts large numbers of eosinophils to the affected site. Cytokines released at the site, including IL-3, IL-5, and GM-CSF, contribute to the growth and differentiation of these cells, which are then activated by binding of antigen-antibody complexes. This leads to degranulation and further release of inflammatory mediators that contribute to the extensive tissue damage typical of the

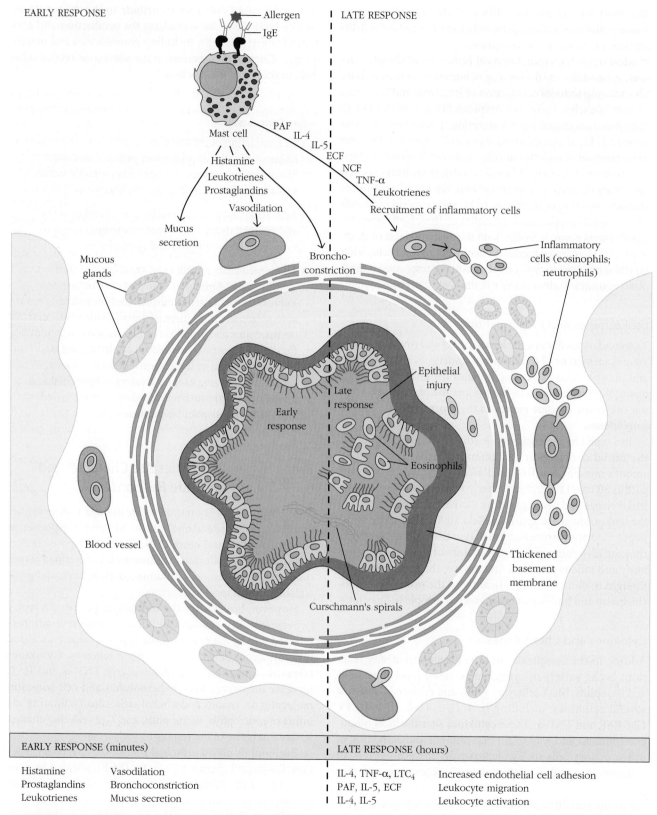

FIGURE 15-6 The early and late inflammatory responses in asthma. Early responses mediated by IgE binding to mast cells, triggering degranulation and mediator release, occur within about 30 minutes of allergen inhalation, while late responses may take 6 to 12 hours. The effects of various mediators on an airway, represented in cross-section, are illustrated in the center and also described in the text. Chronic asthma can result in more serious effects, including disruption of the epithelial barrier, thickening of the basement membrane and smooth muscle layer, and increase in mucus-secreting cells.

late-phase reaction. Neutrophils, another major participant in late-phase reactions, are attracted to the site of an ongoing type I reaction by neutrophil chemotactic factor (NCF) released from degranulating mast cells. Once activated, the neutrophils release their granule contents, including lytic enzymes, platelet-activating factor, and leukotrienes.

Recently, a third phase of type I hypersensitivity has been described in models of type I hypersensitivity reactions in the skin. This third phase starts around 3 days after antigen challenge and peaks on day 4. It is also characterized by massive eosinophil infiltration but, in contrast to the second phase, requires the presence of basophils that have been recruited from the blood into the site by mediators produced in the first and second phases of the response. Cytokines and proteases released from basophils act on tissue-resident cells such as fibroblasts. These fibroblasts then secrete chemokines that are responsible for the recruitment of larger numbers of eosinophils and neutrophils to the skin lesion. Subsequent degranulation of the eosinophils and neutrophils adds to the considerable tissue damage at the site of the initial allergen contact. These experiments illustrate how multiple granulocyte subsets can cooperate in the induction of chronic allergic inflammation.

> **Key Concepts:**
>
> - Type I hypersensitivity responses include an early phase within minutes of allergen exposure, including the release of histamine, leukotrienes, and prostaglandins, and a late-phase reaction mediated by inflammatory cytokines and chemokines.
>
> - Eosinophils play a major role in the late-phase response, including recruitment of neutrophils. Degranulation by both cell types induces further inflammation and tissue damage.
>
> - A third phase of response has been identified at some sites, including the skin, involving basophils and fibroblasts, which then recruit other cells that promote continued inflammation.

There Are Several Categories of Type I Hypersensitivity Reactions

The clinical manifestations of type I reactions can range from localized reactions, such as hay fever and atopic dermatitis, to life-threatening conditions, such as systemic anaphylaxis and severe asthma. The nature of the clinical symptoms depends on the route by which the allergen enters the body, as well as on its concentration and on the prior allergen exposure of the host. Genetics also plays a role, as we will see later. The particular symptoms may evolve in an allergic individual over time. What may start out as atopic dermatitis (which affects 15% to 30% of children early in life) may be followed by the development of asthma and/or food allergies, a process called the **atopic (allergic) march**. In this section, we describe the pathology of the various type I hypersensitivity reactions.

Systemic Anaphylaxis

The most severe type of allergic response, **anaphylaxis**, is a systemic, often fatal state that occurs within minutes of exposure to an allergen. It is usually initiated by an allergen introduced directly into the bloodstream or absorbed into the circulation from the gut or skin. Symptoms include a precipitous drop in blood pressure leading to *anaphylactic shock*, followed by contraction of smooth muscles leading to defecation, urination, and bronchiolar constriction causing labored respiration. This can lead to asphyxiation, which can cause death within 2 to 4 minutes of exposure to the allergen. These symptoms are all due to rapid and widespread IgE antibody-mediated degranulation of mast cells and basophils and the systemic effects of their contents.

A wide range of allergens has been shown to trigger this reaction in susceptible humans, including the venom from bee, wasp, hornet, and ant stings; drugs such as penicillin, insulin, and antitoxins; foods such as seafood and nuts; and latex. If not treated quickly, these reactions can be fatal. Epinephrine, the drug of choice for treating systemic anaphylactic reactions, counteracts the effects of mediators such as histamine and the leukotrienes, relaxing the smooth muscles of the airways and reducing vascular permeability. Epinephrine also improves cardiac output, which is necessary to prevent vascular collapse during an anaphylactic reaction. Many individuals with severe allergic responses carry syringes of injectable epinephrine at all times, enabling them to quickly arrest this severe reaction.

Localized Hypersensitivity Reactions

In localized hypersensitivity reactions, the effects are limited to a specific target site in a tissue or organ, often occurring at the epithelial surfaces first exposed to allergens. These localized allergic reactions include a wide range of IgE-mediated reactions: allergic rhinitis (hay fever), asthma, atopic dermatitis (eczema), urticaria (hives), angioedema (deep tissue swelling), and food allergies.

The most common localized hypersensitivity reaction is *allergic rhinitis* or hay fever. Symptoms result from the inhalation of common airborne allergens (pollens, dust, animal dander, mold spores), which are recognized by IgE antibodies bound to sensitized mast cells in the conjunctivae and nasal mucosa. Allergen cross-linking of the receptor-bound IgE induces the release of histamine and other mediators from tissue mast cells, which then cause vasodilation, increased capillary permeability, and production of secretions in the eyes, nasal passages, and respiratory tract. Tearing, runny nose, sneezing, and coughing are the main symptoms. The term *asthma* was introduced by Hippocrates to describe attacks of breathlessness and wheezing. Like hay fever, allergic asthma is triggered by activation and degranulation of mast cells, with subsequent release of inflammatory mediators, but instead of occurring in the nasal mucosa, the reaction develops deeper in the lower respiratory tract. Contraction of the bronchial smooth muscles, mucus secretion, and swelling of the tissues surrounding the airway all contribute to bronchial constriction and airway obstruction. With chronic asthma, over time more serious changes can occur in the airway passages, including damage to the epithelial layers, thickened basement membrane, increases in mucus-producing cells, and accumulation of inflammatory cells (neutrophils, eosinophils, mast cells, and lymphocytes) (see Figure 15-6). In addition to allergic (atopic) asthma, which may reflect a genetic predisposition to allergic responses (extrinsic asthma), in other individuals an asthma attack can be induced by exercise or cold, apparently independently of allergen stimulation (intrinsic asthma).

Allergic conjunctivitis is an inflammation of the eye surface initiated by IgE-activated mast cell mediator release caused by airborne allergens such as pollen. Early-phase symptoms include itching, tearing, edema, and redness, which can be followed by eosinophilia (high local levels of eosinophils) and inflammation.

Atopic dermatitis (*allergic eczema*) is an allergic inflammatory disease of skin. It is observed most frequently in young children, often developing during infancy. Serum IgE levels are usually elevated. The affected individual develops a rash, erythematous (red) skin eruptions that can fill with pus if there is an accompanying bacterial infection. Unlike skin delayed-type hypersensitivity reactions (discussed later), which often involve T_H1 or T_H17 cells, the skin lesions in atopic dermatitis contain T_H2 cells and an increased number of eosinophils.

Atopic urticaria, or hives, results when there is histamine release in the skin; this induces leakage of small blood vessels, producing localized areas of redness and swelling. A more serious skin allergic reaction sometimes seen together with hives is *angioedema*, swelling in deep layers of skin. Angioedema can also occur in the eyelids, mouth (especially with food allergies), and genitals.

Food Allergies

Food allergies, whose incidence is on the rise, are another common type of atopy. In children, food allergies account for more anaphylactic responses than any other type of allergy because of the transport of food allergens across the gut wall and into the circulation. They are highest in frequency among infants and toddlers (6% to 8%) and decrease slightly with maturity. Approximately 4% of adults display reproducible allergic reactions to food.

The most common food allergens for children are found in cow's milk, eggs, peanuts, tree nuts, soy, wheat, fish, and shellfish. Among adults, nuts, fish, and shellfish are the predominant culprits. Most major food allergens are water-soluble glycoproteins that are relatively stable to heat,

acid, and proteases and, therefore, are digested slowly. Some food allergens (e.g., the major glycoprotein in peanuts, Ara h 1 [from the Latin name for peanuts, *Arachis hypogaea*]) are also capable of acting directly as adjuvants and promoting a T_H2 response and IgE production in susceptible individuals.

Allergen cross-linking of IgE on mast cells along the upper or lower gastrointestinal tract can induce localized smooth muscle contraction and vasodilation, resulting in such symptoms as nausea, abdominal pain, vomiting, and/or diarrhea. Symptoms can develop within minutes to 2 hours of allergen ingestion. Some individuals also have oral hypersensitivity, leading to tingling and angioedema of the lips, palate, and throat. Mast-cell degranulation along the gastrointestinal tract can increase the permeability of the mucous epithelial layer so that the allergen enters the bloodstream, where it may lead to anaphylaxis. Basophils in the blood may also play a significant role in acute food allergy symptoms. Some individuals may develop hives when a food allergen is carried to sensitized mast cells in the skin. In some cases food allergy reactions may be caused directly by cells as well as or instead of IgE-induced release of mediators from mast cells and basophils. Granulocytes may be directly activated by allergens to degranulate, and products of T cells activated by allergens may also contribute to local responses. These and other reactions that are caused by food allergy are listed in **Table 15-3**.

> **Key Concepts:**
> - Allergy symptoms vary depending on where the IgE response occurs and whether it is local or systemic. Asthma, atopic dermatitis, allergic conjunctivitis, and urticaria are examples of local allergic responses.
> - Anaphylaxis refers to a serious systemic IgE response and can be caused by systemic (e.g., through the blood) introduction of the same allergen that induces local responses. Effects on multiple organs may be fatal.
> - Food allergies may cause local (e.g., mouth and throat, GI tract) responses as well as anaphylaxis.

Susceptibility to Type I Hypersensitivity Reactions Is Influenced by Both Environmental Factors and Genetics

While it has long been recognized that there is a genetic component to allergies and asthma, as they tend to run in families, it has become clear in recent years that exposure to certain environmental factors, especially early in life, can also lead individuals to become more susceptible to these conditions (**Figure 15-7**). Together, both genetics and environmental factors can affect the epithelial barriers that control infections and entry of allergens into the body, as well as the resulting local immune responses.

TABLE 15-3	Immune basis for food allergy reactions		
Disorder	**Symptoms**	**Common triggers**	**Notes about mechanism**
IgE mediated (acute)			
Skin effects: hives (urticaria), itching, angioedema	Redness, swelling triggered by ingestion or skin contact	Multiple foods (e.g., milk, egg, wheat, soy, peanut, tree nuts, shellfish, fish)	Allergen, IgE antibody, and mast cell mediated
Oral allergy	Itchiness, swelling of mouth	Multiple foods	Allergen, IgE antibody, and mast cell mediated. Inhaled pollens may induce IgE that cross-reacts with food proteins
Gastrointestinal effects	Nausea, vomiting, intestinal pain	Multiple foods	Allergen, IgE antibody, and mast cell mediated
Wheezing, asthma, rhinitis	Bronchial constriction, mucus production	Inhalation of aerosolized food proteins	Allergen, IgE antibody, and mast cell mediated
Anaphylaxis	Rapid, multiorgan inflammation that can result in cardiovascular failure	Peanuts, tree nuts, fish, shellfish, milk	Response to systemic distribution of allergen and IgE antibodies
Exercise-induced anaphylaxis	As above, but occurs when one exercises after eating trigger foods	Wheat, shellfish, celery	May be due to changes in gut absorption associated with exercise
IgE and cell mediated (chronic)			
Atopic dermatitis	Rash (often in children)	Egg, milk, wheat, soy	Skin T_H2 cells may contribute
Gastrointestinal inflammation	Pain, weight loss, edema, and/ or obstruction	Multiple foods	Eosinophil mediated
Cell mediated (chronic)			
Intestinal inflammation brought about by dietary protein (e.g., enterocolitis, proctitis)	Most often seen in infants: diarrhea, poor growth, and/or bloody stools	Cow's milk (directly or via breast milk), soy, grains, egg	T cells and TNF-α have been implicated

Information from Sicherer, S. H., and H. A. Sampson. 2009. Food allergy. *Annual Review of Medicine* **60**:261; and Yu, W., D. M. H. Freeland, and K. C. Nadeau. 2016. Food allergy: immune mechanisms, diagnosis and immunotherapy. *Nature Reviews Immunology* **16**:751.

Environmental Factors Affecting Predisposition to Allergies

Various features of our environment help to determine whether individuals will develop allergies. They include substances in the environment around us, such as air pollution, as well as things we take into our body, such as food. Microbes to which we are exposed also play a number of roles.

Air pollution

Air pollutants that have been associated with inducing allergies include smoke from factories and diesel truck exhaust. These pollutants are associated with early atopic dermatitis, which then can lead to asthma. Recent studies have revealed a mechanism by which polluted air can induce atopic dermatitis. Polycyclic aromatic hydrocarbons, the main organic compounds in particulates of polluted air, penetrate through the outer layer of skin and bind to and activate the aryl hydrocarbon receptor (AhR) of keratinocytes. AhR-activated cells then produce substances that promote hypersensitivity and itching. These responses, plus the scratching

they incite, damage the skin epithelial barrier, allowing other allergens to enter the body and inducing atopic dermatitis. The AhR also turns on genes for two cytokines, thymic stromal lymphopoietin (TSLP) and IL-33, which promote local allergic responses, as we will see.

Second-hand cigarette smoke is another form of air pollution. Exposure early in life to second-hand smoke has been known for some time to be a risk factor for asthma and possibly for eczema. A recent study of several thousand Swedish children up to age 16 showed that children whose parents who smoked when the children were 2 months old were more likely to develop symptoms of food allergies, especially to eggs and peanuts, than were children whose parents did not smoke.

Farm animals and bacteria

On the other side of the equation, exposure to farm animals and their bacteria can protect against allergies, including hay fever, atopic dermatitis, and asthma. Comparative studies in Europe, Australia, New Zealand, and the United States of populations that differ in exposure to farm animals have

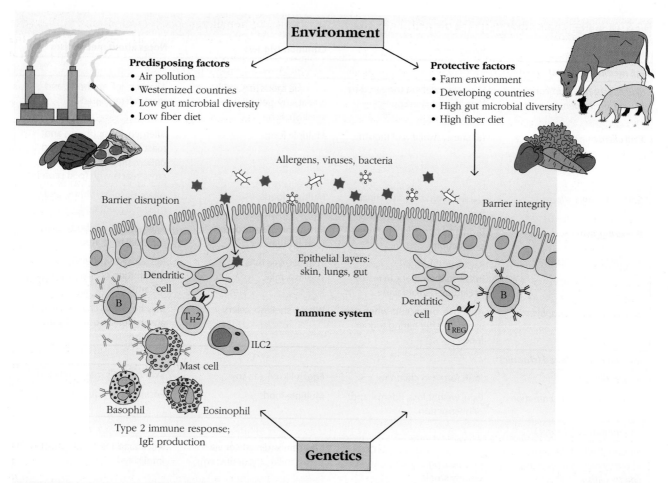

FIGURE 15-7 Environmental factors and genetics influence predisposition to allergies. Exposure to environments with significant microbial content, such as those including farm animals, and diets high in fiber are protective. In contrast, environmental factors that enhance development of type I hypersensitivities include air pollution (such as smoke from trucks and factories, second-hand cigarette smoke, and ozone typical of Western industrialized countries), low fiber diet, and low microbial exposure (e.g., in areas with good sanitation and use of vaccines and antibiotics). These factors contribute to the disruption of the integrity of epithelial barrier layers. Genetic differences also influence the mucosal immune system. Development of type 2 immunity, involving dendritic cells, ILC2 cells, and T$_H$2 cells—which all promote IgE antibody production—leads to allergic responses.

shown that living with farm animals protects against the development of childhood asthma. Dust from such farms, which contains bacteria from the animals, was able to protect mice from induced airway hypersensitivity in an experimental model of asthma. Other studies showed that chronic exposure to farm dust or to low doses of bacterial endotoxin (LPS) protects mice from developing house dust mite–induced asthma. Having pets may also protect children from developing allergies.

In general, greater diversity of intestinal microflora seems to be protective against the development of type I hypersensitivities. Healthy commensal bacteria are important for normal intestinal immune function, including the development of tolerance to food substances (see Chapter 13). Thus, extensive use of antibiotics that reduce or alter commensal bacterial populations, more common in developed countries, appears to leave children more susceptible to allergies. A recent study showed that taking antibiotics (especially multiple times) during the first year of life is associated with increases in food allergies in young children. Conversely, probiotic supplements may be beneficial in reducing the likelihood of allergies; there is some evidence that giving children probiotics protects against the development of eczema. This is a very active area of current research.

Diet

The diet to which a child is exposed before and after birth constitutes an important set of environmental influences, though the effects of dietary components on the control of susceptibility to allergies are complex. The diet of a pregnant woman may affect the allergic predisposition of the newborn, and some studies have indicated that nursing is beneficial to the newborn in preventing allergies. Diets high in plant fiber from fruits and vegetables, which are more common in developing than developed countries, protect against childhood asthma and allergies. Fiber supports microbial diversity, including commensal bacteria that digest the fiber, generating short-chain fatty acids (acetate, butyrate, and propionate). These compounds induce intestinal macrophages and dendritic cells to produce proteins, such as IL-10, that inhibit the type 2 immune responses generated by T_H2 cells and ILC2 cells (discussed shortly; also see Chapter 13). Cytokines secreted by these cells promote IgE production and allergic responses. The complex effects of diet on allergic predisposition are shown by the findings that various types of fats appear to differentially affect allergies, and that certain vitamins that typically promote immune responses (e.g., vitamins A and D) have given mixed results. This continues to be a fertile and important area of research.

The hygiene hypothesis

The early findings that susceptibility to type I hypersensitivity conditions is high in Western industrialized societies and reduced in environments where there is early exposure to microbes, as is found in many developing countries, led to the **hygiene hypothesis**. This hypothesis proposes that exposure to some pathogens during infancy and childhood benefits individuals by stimulating immune responses other than the type 2 responses that induce IgE responses and allergies. There is some evidence that while the immune system of newborn babies may be biased in the T_H2 direction by the uterine environment, that bias diminishes during the first few years of life in nonallergic individuals, but becomes stronger in allergic children. The skewing of responses away from T_H2 responses appears to occur as children become exposed to childhood infections, reflecting either the induction of T_H1 responses or the suppression of T_H2 responses by T_{REG} cells. In Westernized countries, where microbial infections are reduced as a result of sanitation, vaccinations, antibiotics, and less exposure to farm animals, a child's immune system may be less likely to undergo the exposure to infections that would otherwise reorient it away from T_H2 responses.

Studies testing various predictions of the hypothesis are ongoing, and it continues to be modified in its particulars. As we will see shortly, immunologists are learning more about the pathogenesis of allergies and the roles that microbes and other environmental factors play in this process; this work will help clarify the validity of the hygiene hypothesis.

Genetic Effects on Susceptibility to Allergies

The first evidence that genetics influences susceptibility to allergies came in 1916, when a study showed that 48% of allergic individuals had a family history of allergies, whereas only 14% of nonallergic people were from families with allergies. A role for genetics has also been indicated by the higher concordance (co-occurrence) of allergies in identical twins than in nonidentical twins.

Of great interest has been the identification of the actual genes that control atopic responses, as that should provide clues about the mechanisms that lead to allergic responses. Several approaches have led to the identification of a number of possible predisposing loci. Many of these loci encode proteins involved in maintaining the integrity of the epithelial barrier, in the generation and regulation of immune responses (e.g., innate immune receptors, cytokines and chemokines and their receptors, MHC proteins, and transcription factors), and in the activity of molecules involved in triggering allergic responses (e.g., FcεRI, growth factors, and proteolytic enzymes). As described in **Advances Box 15-2**, the involvement of these genes in allergic responses supports the emerging model for the pathogenesis of allergies.

Induction of Allergic Responses

The growing understanding of genetic and environmental influences on susceptibility to allergies and the results of research on allergic responses in humans and mice have led to a model for the induction of allergic responses. A critical initiating factor is likely to be disruption of the integrity of epithelial layers, from genetic deficiencies or environmental damage (such as that from pathogens or air pollution), allowing the initial entry of allergens. This is illustrated in **Figure 15-8** for food allergies, where initial sensitization to food allergens often appears to take place through the skin. This model is supported by a study that showed that infants who developed eczema in their first 3 months were 6 and 11 times more likely to develop egg or peanut allergies, respectively, by 12 months of age than were infants without eczema. The heightened susceptibility to food allergies of children with inherited deficiencies of filaggrin, a protein that helps maintain the integrity of the skin epithelial barrier, supports the role of allergen entrance through the skin in the pathogenesis of food allergies (see Advances Box 15-2).

In response to damage, pathogens, or allergens, skin epithelial cells produce the innate cytokines TSLP (*t*hymic *s*tromal *l*ympho*p*oietin), IL-33, and IL-25; these cytokines, through effects on antigen-presenting dendritic cells, induce the differentiation of allergen-specific T_H2 cells (see Figure 15-8). T_H2 cytokines IL-4 and IL-13 induce B cells to switch to IgE, generating circulating IgE antibodies that are carried in the blood to the intestinal tissue. In the intestine the epithelial cells, ILC2 cells, and T_H2 and T_H9 cells produce cytokines that recruit, support, and activate mast cells and basophils. These cells are activated by binding IgE and allergen to degranulate, releasing mediators that cause the symptoms of food allergies—abdominal pain, vomiting, diarrhea, and occasionally anaphylaxis.

BOX 15-2

The Genetics of Asthma and Allergy

Control of susceptibility to allergies and asthma is clearly complex, with contributions from environmental factors, genetics, and epigenetics. With this degree of complexity, it is not surprising that, while of great interest, identification of the genes involved in controlling an individual's vulnerability to hypersensitivity responses has proven to be a difficult task. However, since the late 1980s, the geneticist's toolkit has markedly expanded with the availability of genome-wide sequence information in addition to libraries of single-nucleotide polymorphisms (SNPs). These tools, along with more classical genetic approaches, have been used to identify hypersensitivity susceptibility genes.

One approach to determining which genes are associated with a particular pathological state is to use knowledge of the disease to develop and then genetically test a hypothesis (i.e., an "educated guess") regarding potential candidate genes. For example, we know that allergies and asthma are associated with high numbers of T$_H$2 cells, which among other activities induce expression of IgE antibodies. High levels of IL-4 expression induce the differentiation of activated CD4$^+$ T cells toward the T$_H$2 cell fate. It was therefore hypothesized

that asthma sufferers may exhibit polymorphisms in structural or regulatory regions of the *IL4* gene, leading to unusually high levels of IL-4 production.

Using this theoretical framework, geneticists focused on a region of human chromosome 5, 5q31-33, for detailed analysis. This region contains a cluster of cytokine genes, among which are the genes for IL-3, IL-4, IL-5, IL-9, IL-13, and GM-CSF. Careful study of this region revealed a polymorphism associated with the predisposition to asthma that maps to the promoter region of the IL-4 gene—a confirmation of the hypothesis advanced above. In addition, two alleles of the IL-9 gene that are associated with a predisposition to atopy were also identified.

A second approach starts with a statistically based search for genes associated with particular disease states; this type of search is referred to as a *genome-wide association survey* (GWAS). The genomes of individuals who do have the disease in question, and of those who do not, are analyzed for the presence of SNPs. Statistically significant association of disease with a particular polymorphism then leads to detailed sequence analysis in the region of the SNP and a search for likely candidate genes. Cloning of genes begins

in the region identified by the candidate SNP and then proceeds by a sequential search of contiguous sequences until a gene of interest is identified. This technique is referred to as *positional cloning*, and several genes important in asthma and atopy have been identified in this way.

A third approach is to look at effects of single-gene mutations causing immune system defects that lead to allergic conditions and/or excessive IgE levels. An example is an autosomal dominant hyper-IgE syndrome that involves high levels of IgE and high numbers of eosinophils (eosinophilia). It is caused by a mutation in STAT3, which is involved in signaling activated by several cytokine receptors.

Table 1 lists some of the candidate genes that have been shown to be associated with one or several allergy-associated conditions, including high IgE levels, asthma, allergic rhinitis, atopic dermatitis/eczema, and eosinophilia. Some of the genes on this list may be tentative for several reasons. For some of the conditions, there may occasionally be non-allergic causes (such as cold-induced asthma). Second, if a gene is closely linked to a second gene, it may be the second gene that is primarily associated with the

TABLE 1	Susceptibility genes for allergic diseases
Likely site of activity	**Genes**
Epithelial barrier integrity	*FLG, DSG1, CDSN, KIF3A, SPINK5, CAPN14, LPP, OVOL1, CLLorf30*
Pathogen sensing, innate responses	*TLR1/6/10, TLR7, TLR9, PYHIN1, CDHR3, DEFB1, CD14*
Chemokines	*CCL5, CCL6, CCL11, CCL24, CCL26*
Cytokines/receptors activating ILC and CD4$^+$ T cells; other T-cell functions	*TSLP, IL25, IL33, IL13, IL4, IL5, IL9, IL2, TNFA, IL1RL1, IL2RB, IL4R, IL6R, IL18R, DENND1B*
Transcription factors in immune cells	*STAT3, STAT6, BCL6, NFATC2, GATA2, GATA3, RORA, IKZF4, MYC, KLF13*
Immune tolerance (T$_{REG}$s)	*TGFBR1, TGFBR2, LRRC32, SMAD3, IL10, FOXP3, FOXA1*
Cell activation and signaling (including T cells, mast cells, and eosinophils)	Multiple *HLA* genes, *TCR, FcεR1A/B, PLCG2, PLCL1, PTGER4, PTEN, ORMDL3, ADAM33*
Other (cytoskeleton, glycosylation, protease, cell death)	*DOCK8, WAS, COTL1, PGM3, GSDMA/B*

References: Bonnelykke, K., R. Sparks, J. Waage, and J. D. Milner. 2015. Genetics of allergy and allergic sensitization: common variants, rare mutations. *Current Opinion in Immunology* **36:**115; and Portelli, M. A., E. Hodge, and I. Sayers. 2015. Genetic risk factors for the development of allergic disease identified by genome-wide association. *Clinical and Experimental Allergy* **45:**21.

(continued)

condition. However, with the assumption that these genes indeed do affect atopy, here we describe how genetic variants of some of the gene products may lead to increased susceptibility to atopic conditions. The proposed functions of other genes in Table 1 can be found through web searches.

As shown in the table, the products of these genes act at several levels to promote IgE production and allergic responses. Some are important for the differentiation and integrity of epithelial barrier layers; defects in these proteins allow the entry of allergens into the body. An important example is filaggrin (encoded by the *FLG* gene), a protein made by keratinocytes that interacts with keratin to contribute to the flattening and adhesion of skin cells. Filaggrin also generates breakdown products that maintain skin homeostasis. Mutations and polymorphisms interfering with filaggrin function are the strongest risk factors for atopic dermatitis in European and Asian populations. Carriers of certain SNPs in filaggrin can have about a 3-fold increased risk of developing atopic dermatitis. As a second example, loss-of-function mutations in the *DSG1* gene, which encodes the desmoglein 1 protein critical for the formation of desmosomes (which cement skin keratinocytes together), cause a severe atopy syndrome that includes dermatitis and food allergies. Mutations in the other genes in this set also interfere with the skin's barrier function and increase susceptibility for one or several allergy-associated conditions.

Polymorphisms in proteins involved in innate responses at epithelial surfaces are also associated with atopic conditions. Genetic variants of the closely linked genes *TLR1*, *TLR6*, and *TLR10* (encoding TLRs 1, 6, and 10) predispose the individual to IgE responses to a variety of allergens as well as to asthma. The *CDHR3* gene encodes a protein that is the receptor for rhinovirus C, one of the forms of the cold virus. Infections with rhinovirus early in life can lead to wheezing and the development of allergic asthma.

Genetic variants of many genes encoding cytokines and cytokine receptors are associated with a wide array of allergic conditions. Prominent in this group are genes for three innate cytokines made by epithelial cells in response to microbial stimulation or damage— TSLP, IL-25, and IL-33—and the *IL1RL1* gene, which encodes a chain of the IL-33 receptor. These cytokines, which can be produced by epithelial cells in the skin, lungs, and intestine, are pivotal in the induction of type 2 immune responses, leading to a variety of allergic conditions (see Figure 15-8).

The ILC2 and T_H2 cells that arise in allergic individuals produce IL-4, IL-5, and IL-13, which drive production of more T_H2 cells, heavy-chain class switching in B cells to IgE, and recruitment and activation of eosinophils. Polymorphisms in all three of the genes for those cytokines and in the *IL4R* gene are all associated with IgE production and numerous allergic conditions. These cytokines and receptors activate specific transcription factors that drive type 2 responses, and polymorphisms in some of their genes (e.g., *STAT3, STAT6, BCL6, GATA3*) also predispose to allergies. STAT6 mediates responses to IL-4 and IL-13, which not only induce naïve T cells to become T_H2 cells but also induce B cells to switch to IgE. Genes controlling other cytokine, cytokine receptor, and transcription factors that modulate other immune system cells also are associated with atopy.

Not surprisingly, variants of genes whose protein products suppress the generation or function of type 2 responses and allergies also are associated with allergic conditions. These include FOXP3, the master transcriptional regulator promoting the formation of regulatory T cells. Among the consequences of defective FOXP3 and loss of T_{REG}s, which causes the IPEX (*immune dysregulation, polyendocrinopathy, enteropathy, X*-linked) autoimmune syndrome, are elevated IgE levels and eosinophilia. Possibly also related to the role of T_{REG}s in suppressing allergic responses, polymorphisms in the genes for IL-2 and the IL-2R, which are essential

for the generation and maintenance of T_{REG}s, are associated with allergic conditions. The cytokine TGF-β is important both for the generation and inhibitory effects of T_{REG}s on immune responses. Variants of genes for TGF-β receptors, including *LRRC32*, which is expressed by T_{REG}s, and for the transcription factor they activate, SMAD3, also predispose to atopic conditions. Polymorphisms in the gene encoding IL-10, another inhibitory cytokine made by regulatory T cells, also are associated with allergies.

As would be expected, polymorphisms of genes controlling T-cell activation are among those that predispose to allergies. Just as for other immune-related diseases, such as autoimmune diseases, the strongest associations with essentially all allergic conditions are with genes encoding HLA MHC proteins, both class I and class II, which control the presentation of specific peptides from pathogens and allergens. Mutations affecting TCR genes are also associated with allergies, as are polymorphisms in genes encoding transcription factors that induce T-cell responses, such as *NFATC2*.

Given the centrality of IgE-mediated release of mediators from mast cells and basophils in causing allergic reactions, it is not surprising that polymorphisms in the genes for the two chains of FcεRI, which triggers degranulation after IgE antibody and allergen binding, are associated with allergic conditions. Signaling molecules downstream of that and other immune receptors, such as the gene for phospholipase Cγ2, also show associations with allergies.

As genetic studies are now underway of populations of different origins that may vary in their array of genetic polymorphisms, additional genes associated with allergic conditions will probably be discovered. Knowledge of the genes that regulate allergic responses may contribute both to our understanding of the mechanisms that control the generation of allergic responses as well as to the development of new therapeutics for preventing or treating allergic conditions.

(a) Allergen sensitization through skin

(b) Allergen challenge in intestine

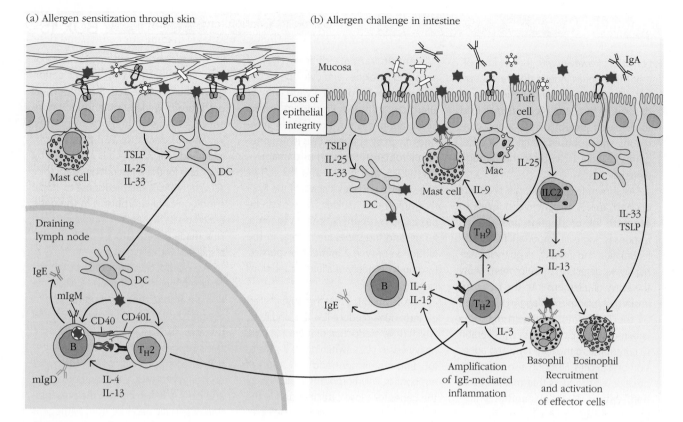

FIGURE 15-8 **Induction of IgE-mediated food allergy response.** (a) Allergen sensitization. In the skin (or intestine), the integrity of the epithelial layer is disrupted by pathogens, food allergens, or genetic deficiencies. Epithelial cells respond to these stimuli by producing innate cytokines thymic stromal lymphopoietin (TSLP), IL-25, and IL-33. These innate type 2 cytokines influence dendritic cells that have picked up allergen, so that when they present allergen peptides to naïve T cells in the draining lymph node, those cells are induced to become T_H2 cells. IL-4 and IL-13 produced by the T_H2 cells induce B cells to produce IgE antibodies.

(b) Allergen challenge in the intestine. Intestinal epithelial cells respond to damage, pathogens, and/or allergen by producing TSLP, IL-33, and IL-25. These activate local ILC2s (see Chapter 13) to produce type 2 cytokines IL-13 and IL-5, which recruit and activate basophils and eosinophils. IL-25 also stimulates T_H9 cells to produce IL-9, which supports mast cells. Allergen-specific memory T_H2 cells continue to activate allergen-specific B cells, increasing production of allergen-specific IgE antibodies, which continue to stimulate mediator release by mast cells, basophils, and eosinophils, leading to the symptoms of food allergies.

Key Concepts:

- Environmental factors (including air pollution, exposure to farm animals and their bacteria, and diet) and genetics both influence susceptibility to allergies.

- The hygiene hypothesis, which has been advanced to explain increases in asthma and allergy incidence in the developed world, proposes that early exposure to microorganisms inhibits the development of allergy by preventing T_H2 cell–mediated responses that induce IgE antibodies and allergic responses.

- Among the genes that have variants associated with predisposition to allergies and asthma are genes that affect the integrity of the epithelial barrier, cytokines and chemokines, proteins controlling regulatory T cells, transcription factors, and receptors and signaling proteins.

- A model for the induction of allergic responses consistent with environmental and genetic influences on allergic susceptibility starts with the entry of allergen through a disrupted epithelial barrier, leading to the activation of T_H2 cells and IgE-producing B cells, which then promote allergic responses at subsequent sites of exposure to the allergen.

Diagnostic Tests and Treatments Are Available for Allergic Reactions

IgE-mediated immediate hypersensitivity is commonly assessed by skin testing, an inexpensive and relatively safe diagnostic approach that allows screening of a wide range of antigens at once. Small amounts of potential allergens are introduced at specific skin sites (e.g., the forearm or back), either by intradermal injection or by dropping onto a site of

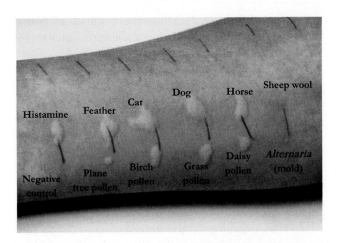

FIGURE 15-9 Skin testing for hypersensitivity. This photograph shows an example of a skin test for a variety of antigens. These were introduced by superficial injection and read after 30 minutes. The positive control for a response is histamine; the negative control is typically just saline. This individual is clearly atopic; the skin test reveals responses to multiple animal and plant allergens. *[Photo: Southern Illinois University/Getty Images]*

a superficial scratch. Thirty minutes later, the sites are reexamined. Redness and swelling (the result of local mast cell degranulation) indicate an allergic response (**Figure 15-9**). Less commonly, physicians may elect to measure the serum levels of either total or allergen-specific IgE, using ELISA or Western blot technologies (see Chapter 20).

Treatment for allergies always begins with measures to avoid exposure to the allergens. However, no one can avoid contact with aeroallergens such as pollen, and a number of immunological and pharmaceutical interventions are now available to either alleviate the symptoms of allergic responses or prevent them from occurring in the first place.

Drugs That Reduce the Symptoms of Allergic Responses

Decongestants that reduce the runny noses of allergic rhinitis by shrinking swollen nasal blood vessels and tissues come in pills, liquids, nose drops, and nasal sprays. They include oxymetazoline, phenylephrine, and pseudoephedrine. These drugs do not alleviate other symptoms.

For many years antihistamines have been the most useful drugs for the treatment of allergic rhinitis. These drugs inhibit histamine activity by binding and blocking histamine receptors on target cells. The H_1 receptors are blocked by first-generation antihistamines such as diphenhydramine and chlorpheniramine, which are quite effective in controlling the symptoms of allergic rhinitis. Unfortunately, as they are capable of crossing the blood-brain barrier, they also act on H_1 receptors in the nervous system and can have multiple side effects. Because these first-generation drugs bind to muscarinic acetylcholine receptors, they can also induce dry mouth, urinary retention, constipation, slow

heartbeat, sedation, and drowsiness. Second-generation antihistamines such as fexofenadine, loratadine, and desloratadine, developed in the early 1980s, exhibit significantly less cross-reactivity with muscarinic receptors and hence have fewer side effects. Some medications, such as Claritin and Zyrtec, combine a decongestant and an antihistamine.

Leukotriene antagonists, specifically montelukast, have also been used to treat type I hypersensitivities and are comparable in effectiveness to antihistamines.

Finally, corticosteroids (often just referred to as *steroids*) can reduce inflammation associated with many forms of allergic reactions. Inhalation therapy with low-dose corticosteroids such as Flonase and Nasacort (now available without a prescription) reduces inflammation by inhibiting innate immune cell activity and has been used successfully to reduce the frequency and severity of asthma attacks. Also available as pills or liquids, corticosteroids can help with other serious allergic conditions. As they must be taken regularly and with long-term use can cause side effects, ingested corticosteroids usually require prescriptions. However, for skin allergic reactions (such as for insect bites), topical creams (usually hydrocortisone) are available over-the-counter.

Medications Used to Limit Allergic Asthma and Anaphylaxis

Asthma attacks and anaphylaxis are two of the more severe allergic reactions, and drugs can be prescribed that alleviate their symptoms. In particular, drugs that enhance production of the second messenger cAMP help to prevent the degranulation of mast cells and to counter the bronchoconstriction occurring in asthma attacks. Epinephrine (adrenalin), or epinephrine agonists like albuterol, do this by binding to G protein–coupled receptors, which initiate signals that generate cAMP. Theophylline, another commonly used drug in the treatment of asthma, does this by antagonizing *phosphodiesterase* (PDE), an enzyme that normally breaks down cAMP. Recall that a brief transient elevation of cAMP levels is required for mast cell degranulation, so treatments that either prolong the elevation of cAMP levels or break down cAMP and prevent its transient increase can block degranulation. For asthma these are available as inhalants by prescription.

Finally, for anaphylaxis—the systemic allergic responses that can occur in some individuals exposed to food, drug, and insect sting allergens—a shot of epinephrine can interrupt the response before the consequences (including bronchoconstriction and a drop in blood pressure) can cause shock and potentially death. Therefore people susceptible to these severe systemic allergic reactions are advised to carry a syringe of epinephrine for use in an emergency.

Immunotherapeutics

Therapeutic anti-IgE antibodies have been developed; one such antibody, omalizumab, has been approved by the U.S. Food and Drug Administration and is available as a pharmacological agent. Omalizumab binds the Fc region of IgE and prevents the IgE from binding to FcεRI and triggering mast cell degranulation. This reagent has been used to treat both allergic rhinitis and allergic asthma. However, for the treatment of allergic rhinitis, omalizumab is no more effective than second-generation antihistamines and is rarely prescribed because of its higher cost and the necessity of administering it by injection. Other monoclonal antibody reagents are also being evaluated for their clinical value.

Desensitization

For many years, physicians have been treating allergic patients with repeated exposure (via injection, application on the skin or under the tongue, or ingestion) to increasing doses of allergens, in a regimen termed **desensitization** or immunotherapy. This mode of treatment attacks the disease mechanism of the allergic individual at the source and, when it works, is an effective way to manage allergies. Initial approaches used were "allergy shots," injections for reducing allergic rhinitis caused by environmental allergens such as pollen, dust mites, insect stings, mold, and pet dander. These treatments always start in a doctor's office in case anaphylaxis occurs. Administration of the allergen under the tongue with drops or tablets now is common, allowing patients to treat themselves at home, an easier and safer approach as oral administration runs less of a risk of anaphylaxis than injections. After a maintenance dose is reached, which may take 3 years, allergic rhinitis responses may be eliminated for several years—even up to 12 years in one study. Immunotherapy may also prevent the development of asthma.

More recently, considerable effort has been invested in developing desensitization protocols for food allergies, as these allergies to common foods make life difficult for affected individuals and their families and responses can be severe and even fatal. *Oral immunotherapy* (OIT) consists of feeding children gradually increasing doses (beginning with extremely small amounts) of the food allergens with the goal of establishing desensitization, reduced reactivity to the allergen that is maintained through regular ingestion of the food. The initial and increasing doses of the food allergen are always given in a doctor's office because of the possible risk of adverse reactions, which could include anaphylaxis. The long-term goal of OIT is unresponsiveness that would be sustained even in the absence of eating the food substance, with the hope of achieving long-term tolerance to the food.

Clinical trials of OIT are being conducted for peanut, egg, cow's milk, and wheat allergens. Tolerance to some amounts of food allergens has been obtained in most of these trials, but thus far only a subset of the patients has achieved sustained unresponsiveness and even fewer long-term tolerance. Not surprisingly, these latter patients seem to be individuals who started with lower levels of allergen-specific IgE antibodies. A recent modification of OIT has been to combine it with injections of omalizumab (the anti-IgE antibody

mentioned earlier) to reduce the amount of mast cell–bound IgE antibodies available to bind food allergens. Initial results suggest that the addition of omalizumab to the desensitization protocol can make OIT more quickly effective, even for combinations of food allergens, and also safer for patients at high risk of severe reactions.

How does desensitization work? Based on extensive research, several mechanisms have been proposed both for immunotherapy for allergic rhinitis to airborne allergens and for food allergies (**Figure 15-10**). Exposure to gradually increasing doses of the allergen provides protection by skewing the T-cell response away from T_H2 cells, instead inducing T_H1 and T_{REG} cells. T_H1 cytokines induce heavy-chain class switching to IgG4 instead of IgE. The IgG antibodies block IgE binding to the allergens but they also bind the inhibitory FcγRIIB receptor; both effects reduce mast cell and basophil degranulation. T_{REG} cytokines IL-10 and TGF-β both down-regulate T_H2 responses and inhibit the recruitment of basophils and eosinophils, reducing local inflammatory responses. At higher doses immunotherapy could induce the apoptosis or anergy of T_H2 cells.

Key Concepts:

- Skin tests are an effective way to diagnose allergies.

- Allergic reactions can be treated with pharmacological inhibitors of cellular and tissue responses and inflammation, including antihistamines, leukotriene inhibitors, and corticosteroids. An anti-IgE antibody also can be effective, though expensive and difficult to administer.

- Immunotherapeutic approaches include attempts to desensitize allergic individuals by exposing them to increasing levels of their allergen. Immunotherapy by injection or sublingual administration of airborne allergens such as pollen, dust, insect venom, and animal dander proteins has been successful in preventing allergic rhinitis. Clinical trials are underway to desensitize children with food allergies by feeding increasing doses of allergen, which might work by inducing regulatory T cells and T_H1 cells instead of T_H2 cells, and the production of IgG4 instead of IgE antibodies.

Why Did Allergic Responses Evolve?

Given how dangerous allergic responses can be, why IgE, the FcεRI receptor, and the degranulation process evolved to cause type I hypersensitivity reactions has been a puzzle. While it is impossible to know for sure why evolution has led to where we are now for any biological response, examples of beneficial roles of this response provide some clues. In the response to helminth worm parasite infections, degranulation of eosinophils by IgE antibodies cross-linked

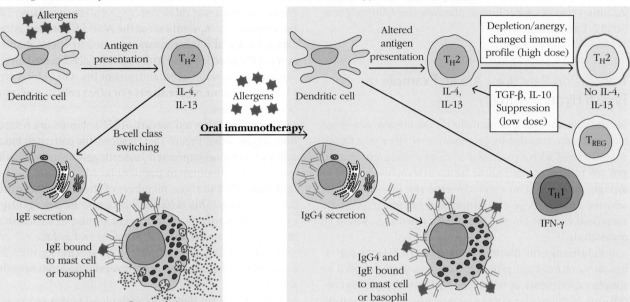

(a) Allergic immune response

(b) Immunotherapy-induced desensitization

FIGURE 15-10 Mechanisms underlying immunotherapy-induced desensitization. Depending on how exposure occurs, allergens can induce either allergic responses or desensitization. (a) As described earlier in this chapter, exposure to an allergen early in life can lead to the formation of T_H2 cells and production of allergen-specific IgE that induces degranulation and allergic symptoms (see Figure 15-8 for sensitization to food allergens). (b) Repeated injection or ingestion of low or increasing doses of antigen may lead to immune tolerance via the induction of regulatory T cells that block the formation or activity of T_H2 cells or employ other mechanisms for inhibiting T_H2 cells. Alternatively or in addition, immunotherapy may induce the generation of IgG4 antibodies that either compete with IgE for binding to antigen or induce coclustering of FcεRI with inhibitory FcγRIIB receptors (see chapter text), thus preventing mast cell and basophil degranulation.

by antigens on the surface of the worm releases enzymes that damage the worms. Recent research also suggests that IgE antibodies may provide protection against venoms from reptiles (including snakes and Gila monsters), insects such as bees, and jellyfish. Degranulation of mast cells and basophils triggered by IgE anti-venom antibodies releases proteases that degrade the venoms. In one study the immunization of mice with bee venom induced IgE antibodies that protected the mice against challenge with a lethal dose of the venom, but mice genetically deficient in IgE or the FcεRI receptor were not protected. It may be that there is some chemical or structural relationship between helminth and venom antigens and more benign antigens like pollen proteins that has caused the latter to induce IgE antibody responses. Interestingly, some degranulation is activated by helminth and venom proteins on their own, suggesting the adaptive benefit of degranulation and release of these enzymes, and the role of IgE may have evolved to enhance that response.

Key Concept:

- IgE-mediated type I hypersensitivity reactions may have evolved because of their protective roles against helminth worm parasites and insect and animal venoms.

Antibody-Mediated (Type II) Hypersensitivity

Type II hypersensitivity reactions involve antibody-mediated destruction of cells by IgG and IgM immunoglobulins. Antibody bound to a cell-surface antigen can induce death of the antibody-bound cell by three distinct mechanisms (see Chapters 5 and 12). First, certain immunoglobulin subclasses can activate the complement system, creating pores in the membrane of a foreign cell. Second, antibodies can mediate cell destruction by *antibody-dependent cell-mediated cytotoxicity* (ADCC), in which cytotoxic cells bearing Fc receptors bind to the Fc region of antibodies on target cells and promote killing of the cells. Finally, antibody bound to a foreign cell also can serve as an opsonin, enabling phagocytic cells with Fc receptors or (after complement has been activated by the bound antibodies) receptors for complement fragments such as C3b to bind and phagocytose the antibody-coated cell. However, when excessive or misdirected, these responses can be damaging, and this is the focus of this section. Here, we examine three examples of type II hypersensitivity reactions. Certain auto-immune diseases involve autoantibody–mediated cellular destruction by type II mechanisms and will be described in Chapter 16. These antibody-mediated killing mechanisms are also important in the use of antibodies to tumor antigens for some types of cancer immunotherapy (see Clinical Focus Box 12-1 and Chapter 19).

Transfusion Reactions Are an Example of Type II Hypersensitivity

Several proteins and glycoproteins on the membranes of red blood cells are encoded by genes with multiple allelic forms. An individual with a particular allele of a blood-group antigen can recognize other allelic forms in transfused blood as foreign, and mount an antibody response. Blood types are referred to as A, B, or O, and the surface antigens that are associated with the blood types are identified as A, B, and H, respectively.

Interestingly, the blood type antigens (ABH) are carbohydrates, rather than proteins. This was demonstrated by simple experiments in which the addition of high concentrations of particular simple sugars was shown to inhibit antibody binding to red blood cells bearing particular types of red blood cell antigens. These inhibition reactions revealed that antibodies directed to group A antigens predominantly bound to *N*-acetylglucosamine residues, those to group B antigens bound to galactose residues, and those directed toward the so-called H antigens bound to fucose residues (**Figure 15-11a**). Note that the H antigen is present in all blood types.

The A, B, and H antigens are synthesized by a series of enzymatic reactions catalyzed by glycosyltransferases. The final step of the biosynthesis of the A and B antigens is catalyzed by A and B transferases, encoded by alleles *A* and *B* at the *ABO* genetic locus. Although initially detected on the surface of red blood cells, antigens of the ABO blood type system also occur on the surface of other cells as well as in bodily secretions.

Antibodies directed toward ABH antigens are termed *isohemagglutinins*. Figure 15-11b shows the pattern of blood cell antigens and expressed isohemagglutinins normally found within the human population. Most adults possess IgM antibodies to those members of the ABH family they do not express. This is because common microorganisms express carbohydrate antigens very similar in structure to the carbohydrates of the ABH system and induce a B-cell response. B cells generating antibodies specific for the ABH antigens expressed by the host, however, undergo negative selection.

For example, an individual with blood type A recognizes B-like epitopes on microorganisms and produces isohemagglutinins to the B-like epitopes. This same individual does not respond to A-like epitopes on the same microorganisms because they have been tolerized to self-A epitopes. If a type A individual is transfused with blood containing type B cells, a *transfusion reaction* occurs in which the preexisting anti-B isohemagglutinins bind to the B blood cells and initiate their destruction via complement-mediated lysis.

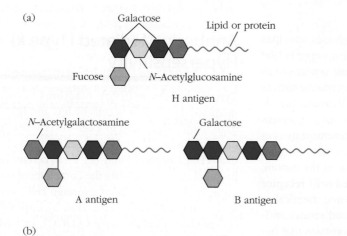

(a)

Galactose

Lipid or protein

Fucose   *N*–Acetylglucosamine

H antigen

N–Acetylgalactosamine   Galactose

A antigen   B antigen

(b)

Genotype	Blood–group phenotype	Antigens on erythrocytes *(agglutinins)*	Serum antibodies *(isohemagglutinins)*
AA or AO	A	A	Anti–B
BB or BO	B	B	Anti–A
AB	AB	A and B	None
OO	O	H	Anti–A and anti–B

FIGURE 15-11 ABO (ABH) blood groups. (a) Structure of terminal sugars, which constitute the distinguishing epitopes of the A, B, and H blood antigens. All individuals express the H antigen, but not all individuals express the A or B antigens. The blood group of those who express neither A or B antigens (but, like all people express the H antigen) is referred to as O. (b) ABO genotypes, corresponding phenotypes, agglutinins (antigens), and isohemagglutinins (antibodies that react to nonhost antigens).

Individuals with blood type O express only the H antigen. Although they can donate blood to anyone, they have antibodies that will react to both A-type or B-type blood. Their anti-A– and anti-B–producing B cells were never exposed to A or B antigens and therefore were never deleted.

The clinical manifestations of transfusion reactions result from massive intravascular hemolysis (destruction) of the transfused red blood cells by antibody plus complement. These manifestations may be either immediate or delayed. Reactions that begin immediately are most commonly associated with ABO blood-group incompatibilities, which lead to complement-mediated lysis triggered by the IgM isohemagglutinins. Within hours, free hemoglobin can be detected in the plasma; it is filtered through the kidneys, resulting in hemoglobinuria, that is, excessive free hemoglobin in the blood. As the hemoglobin is degraded, the porphyrin component is metabolized to bilirubin, which at high levels is toxic to the organism. Typical symptoms of bilirubinemia include fever, chills, nausea, clotting within blood vessels, pain in the lower back, and hemoglobin in the urine. Treatment involves prompt termination of the transfusion and maintenance of urine flow with a diuretic, because the accumulation of hemoglobin in the kidney can cause acute damage to the kidney tubules (tubular necrosis).

Antibodies to other blood-group antigens such as Rh factor (see the next section) may result from repeated blood transfusions because minor allelic differences in these antigens can stimulate antibody production. These antibodies are usually of the IgG class. These incompatibilities typically result in delayed hemolytic transfusion reactions that develop between 2 and 6 days after transfusion. Because IgG is less effective than IgM in activating complement, complement-mediated lysis of the transfused red blood cells is incomplete. Free hemoglobin is usually not detected in the plasma or urine in these reactions. Rather, many of the transfused cells are destroyed at extravascular sites by agglutination, opsonization, and subsequent phagocytosis by macrophages. Symptoms include fever, increased bilirubin, mild jaundice, and anemia.

Key Concepts:

- Transfusion reactions are caused by antibodies that bind to A, B, or H carbohydrate antigens, which are expressed on the surface of red blood cells. Individuals with different blood types (A, B, or O) express different carbohydrate antigens. They are tolerant to their own antigens, but generate antibodies against the antigen (A or B) that they do not express. All individuals express antigen H, so no antibodies are generated to this carbohydrate.

- Transfusion across differences in other blood-group antigens stimulate production of IgG antibodies, which cause delayed and less severe reactions.

Hemolytic Disease of the Newborn Is Caused by Type II Reactions

Hemolytic disease of the newborn develops when maternal IgG antibodies specific for fetal blood-group antigens cross the placenta and destroy fetal red cells. The consequences of such transfer can be minor, serious, or lethal. Severe hemolytic disease of the newborn, called *erythroblastosis fetalis,* most commonly develops when the mother and fetus express different alleles of the *Rh*esus (Rh) antigen. Although there are actually five alleles of the Rh antigen, expression of the *D* allele elicits the strongest immune response. Individuals bearing the *D* allele of the Rh antigen are therefore designated as Rh$^+$.

An Rh$^-$ mother pregnant by an Rh$^+$ father is in danger of developing a response to the Rh antigen that the fetus may have inherited from the father and rejecting the Rh$^+$ fetus. During pregnancy, fetal red blood cells are separated from the mother's circulation by a layer of cells in the placenta called the *trophoblast.* During her first pregnancy with an Rh$^+$ fetus, an Rh$^-$ woman is usually not exposed to enough fetal red blood cells to activate her Rh-specific B cells. However, at the time of delivery, separation of the placenta from the uterine wall allows larger amounts of fetal umbilical cord blood to enter the mother's circulation. These fetal red blood cells stimulate Rh-specific B cells to mount an immune response, resulting in the production of Rh-specific plasma cells and memory B cells in the mother. The secreted IgM antibody clears the Rh$^+$ fetal red cells from the mother's circulation, but memory cells remain, a threat to any subsequent pregnancy with an Rh$^+$ fetus. Importantly, since IgM antibodies do not pass through the placenta, IgM anti-Rh antigens are no threat to the fetus.

However, activation of IgG-expressing memory cells in a subsequent pregnancy results in the formation of IgG anti-Rh antibodies, which can cross the placenta and damage the fetal red blood cells (**Figure 15-12**). Mild to severe anemia can develop in the fetus, sometimes with fatal consequences. In addition, conversion of hemoglobin to bilirubin can present an additional threat to the newborn because the lipid-soluble bilirubin may accumulate in the brain and cause brain damage. Because the blood-brain barrier is not complete until after birth, very young babies can suffer fatal brain damage from bilirubin. Fortunately, bilirubin is rapidly broken down on exposure of the skin to *ultraviolet* (UV) light, and babies who display the telltale jaundiced appearance that signifies high levels of blood bilirubin are treated by exposure to UV light in their cribs (**Figure 15-13**).

Hemolytic disease of the newborn caused by Rh incompatibility in a second or later pregnancy can be almost entirely prevented by administering antibodies against the Rh antigen to the mother at around 28 weeks of her first pregnancy and within 24 to 48 hours after the first delivery. Anti-Rh antibodies are also administered to pregnant women after amniocentesis. These antibodies, marketed as

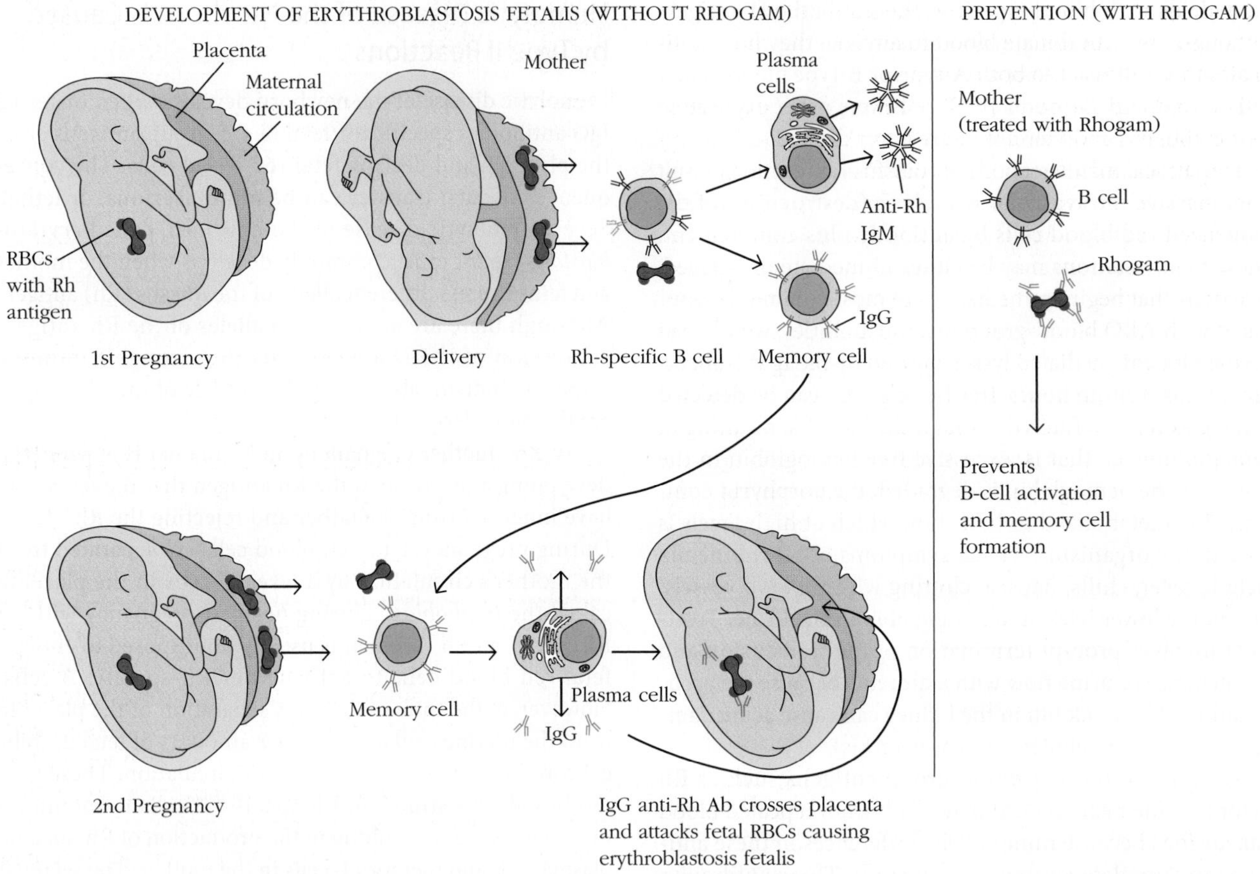

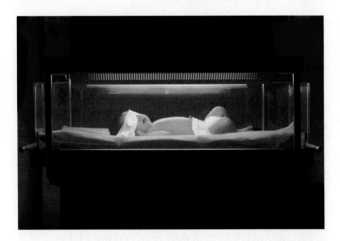

DEVELOPMENT OF ERYTHROBLASTOSIS FETALIS (WITHOUT RHOGAM)

PREVENTION (WITH RHOGAM)

1st Pregnancy

Delivery

Rh-specific B cell

Memory cell

2nd Pregnancy

IgG anti-Rh Ab crosses placenta and attacks fetal RBCs causing erythroblastosis fetalis

FIGURE 15-12 Destruction of Rh-positive red blood cells during erythroblastosis fetalis. Development of erythroblastosis fetalis (hemolytic disease of the newborn) is caused when an Rh⁻ mother carries an Rh⁺ fetus (*left*). The effect of treatment with anti-Rh antibody (RhoGAM), usually after the first pregnancy, is shown on the *right*.

FIGURE 15-13 Ultraviolet light is used to treat bilirubin-emia of the newborn. [*Stephanie Barbary/Shutterstock*]

RhoGAM, bind to any fetal red blood cells that may have entered the mother's circulation and facilitate their clearance before B-cell activation and ensuing memory-cell production can take place. In a subsequent pregnancy with an Rh⁺ fetus, a mother who has been treated with RhoGAM is unlikely to produce IgG anti-Rh antibodies; thus, the fetus is protected from the damage that would occur when these antibodies cross the placenta.

The development of hemolytic disease of the newborn caused by Rh incompatibility can be detected by testing maternal serum at intervals during pregnancy for antibodies to the Rh antigen. A rise in the titer of these antibodies as pregnancy progresses indicates that the mother has been exposed to Rh antigens and is producing increasing amounts of antibody. Treatment depends on the severity of the reaction. For a severe reaction, the fetus can be given an intrauterine blood-exchange transfusion to replace fetal Rh⁺ red blood cells with Rh⁻ cells. These transfusions are given every 10 to 21 days until delivery. In less severe cases, a blood-exchange transfusion is not given until after birth, primarily to remove bilirubin; the infant is also exposed to low levels of UV light to break down the bilirubin and prevent cerebral damage. The mother can also be treated during the pregnancy by *plasmapheresis*. In this procedure, a cell separation machine is used to separate the mother's blood into two fractions: cells and plasma. The plasma containing the anti-Rh antibody is discarded, and the cells are reinfused into the mother in an albumin or fresh plasma solution.

The majority of cases (65%) of hemolytic disease of the newborn, however, are caused by ABO blood-group incompatibility between the mother and fetus and are not severe. Type A or B fetuses carried by type O mothers most commonly develop these reactions. A type O mother can develop IgG antibodies to the A or B blood-group antigens through exposure to fetal blood-group A or B antigens in successive pregnancies. Usually, the fetal anemia resulting from this incompatibility is mild; the major clinical manifestation is a slight elevation of bilirubin, with jaundice. Exposure of the infant to low levels of UV light is often enough to break down the bilirubin and avoid cerebral damage. In severe cases, transfusion may be required.

Key Concepts:

- Hemolytic disease of the newborn is caused by maternal antibody reaction to the Rh antigen, which can happen if the mother is Rh⁻ and the father is Rh⁺. As red blood cells from a fetus enter the maternal circulation during pregnancy and birth, the mother will develop Rh antibodies that can cause hemolytic disease in subsequent pregnancies. This can be prevented by several approaches to eliminate fetal red blood cells or the maternal antibodies.

- Similar immunization of the mother against A or B blood-group antigens of the fetus may also occur; blood-group antigen antibodies cause less severe hemolytic disease of the newborn.

Hemolytic Anemia Can Be Drug Induced

Certain antibiotics (e.g., penicillin, cephalosporins, and streptomycin), as well as other well-known drugs (including ibuprofen and naproxen), can adsorb nonspecifically to proteins on red blood cell membranes, forming a drug-protein complex. In some patients, such drug-protein complexes induce the formation of antibodies. These antibodies then bind to the adsorbed drug on red blood cells, inducing complement-mediated lysis and thus progressive anemia. When the drug is withdrawn, the hemolytic anemia disappears. Penicillin is notable in that it can induce all four types of hypersensitivity with various clinical manifestations (**Table 15-4**).

Key Concept:

- Drug-induced hemolytic anemia is caused by antibody responses to red blood cells that have bound drug molecules or metabolites.

TABLE 15-4	Penicillin-induced hypersensitivity reactions	
Type of reaction	**Antibody or lymphocyte induced**	**Clinical manifestations**
I	IgE	Urticaria, systemic anaphylaxis
II	IgM, IgG	Hemolytic anemia
III	IgG	Serum sickness, glomerulonephritis
IV	T cells	Contact dermatitis

Immune Complex–Mediated (Type III) Hypersensitivity

The reaction of antibody with antigen generates immune complexes. In general, these antigen-antibody complexes facilitate the clearance of antigen by phagocytic cells and red blood cells (see Chapter 12). In some cases, however, the presence of large numbers and networks of immune complexes can lead to tissue-damaging type III hypersensitivity reactions. The magnitude of the reaction depends on the levels and size of immune complexes, their distribution within the body, and the ability of the phagocyte system to clear the complexes and thus minimize the tissue damage. Failure to clear immune complexes may also result from peculiarities of the antigen itself, or disorders in phagocytic machinery. The deposition of immune complexes in the blood vessels or tissues initiates reactions that result in the recruitment of complement components and neutrophils to the site, with resultant tissue injury.

Immune Complexes Can Damage Various Tissues

The formation of antigen-antibody complexes occurs as a normal part of an adaptive immune response. It is usually followed by Fc receptor–mediated recognition of the complexes by phagocytes, which engulf and destroy them; by binding to red blood cells for clearance in the spleen or kidney; and/or by complement activation that results in the lysis of the cells on which the immune complexes are found. However, under certain conditions, immune complexes are inefficiently cleared and may be deposited in the blood vessels or tissues, setting the stage for a type III hypersensitivity response. These conditions include (1) the presence of antigens capable of generating particularly extensive antigen-antibody lattices, (2) a high intrinsic affinity of antigens for particular tissues, (3) the presence of highly charged antigens (which can affect immune complex engulfment), and (4) a compromised phagocytic system. All have been associated with the initiation of type III responses.

Uncleared immune complexes bind to mast cells, neutrophils, and macrophages via Fc receptors, triggering the release of vasoactive mediators and inflammatory cytokines, which interact with the capillary epithelium and increase the permeability of the blood vessel walls. Immune complexes then move through the capillary walls and into the tissues, where they are deposited and set up a localized inflammatory response. Complement activation results in the production of the anaphylatoxin chemokines C3a and C5a, which attract more neutrophils and macrophages (see Chapter 5). These in turn are further activated by immune complexes binding to their Fc receptors to secrete proinflammatory chemokines and cytokines, prostaglandins, and proteases. Proteases attack the basement membrane proteins collagen and elastin, as well as cartilage. Tissue damage is further mediated by oxygen free radicals released by the activated neutrophils. In addition, immune complexes interact with platelets and induce the formation of tiny clots. Complex deposition in the tissues can give rise to symptoms such as fever, urticaria (rashes), joint pain, lymph node enlargement, and protein in the urine. The resulting inflammatory lesion is referred to as *vasculitis* if it occurs in a blood vessel, *glomerulonephritis* if it occurs in the kidneys, or *arthritis* if it occurs in the joints.

Key Concept:

- Uncleared immune complexes can induce degranulation of mast cells and inflammation, and can be deposited in tissues and capillary beds where they induce more innate immune activity, blood vessel inflammation (vasculitis), and tissue damage, such as glomerulonephritis in the kidneys or arthritis in the joints.

Immune Complex–Mediated Hypersensitivity Can Resolve Spontaneously

If immune complex–mediated disease is induced by a single large bolus of antigen that is then gradually cleared, it can resolve spontaneously. Spontaneous recovery is seen, for example, when glomerulonephritis is initiated following a streptococcal infection. Streptococcal antigen–antibody complexes bind to the basement membrane of the kidney and set up a type III response, which resolves as the bacterial load is eliminated. Similarly, patients being treated for various conditions by injections of antibody can develop immune responses to the foreign antibody and generate large immune complexes. This was seen initially during the use of horse anti-diphtheria toxin antibodies in the treatment of diphtheria in the early 1900s. On repeated injections with the horse antibodies, patients developed a syndrome known as *serum sickness*, which resolved as soon as the antibodies were withdrawn. Serum sickness is an example of a systemic form of immune complex disease, which resulted in arthritis, skin rash, and fever.

A more recent manifestation of the same problem occurred in patients who received therapeutic mouse-derived monoclonal antibodies designed to treat cancers. After several such treatments, some patients generated their own antibodies against the foreign monoclonals and developed serum sickness–like symptoms. We know now that injection of the mouse antibodies caused a generalized type III reaction, and in many cases the therapeutic antibodies were actually cleared before they could reach their pathogenic target. To avoid this response, current therapeutic antibodies are genetically engineered to replace the mouse-specific regions of antibody proteins with the corresponding human sequences (they are then called *humanized* antibodies; see Chapter 12).

Key Concept:

- If antibodies are present, a single bolus of the antigen may produce immune complexes that may be cleared without problems, but repeated exposure (e.g., injection with antibodies from a different species) can cause serum sickness.

Auto-Antigens Can Be Involved in Immune Complex–Mediated Reactions

If the antigen in the immune complex is an auto-antigen (self antigen), it cannot be permanently eliminated; hence type III hypersensitivity reactions cannot be easily resolved. In such situations, chronic type III responses develop. For example, in systemic lupus erythematosus, persistent antibody responses to auto-antigens such as DNA and various nuclear proteins are an identifying feature of the disease, and complexes are deposited in the joints, kidneys, and skin of patients. Examples of conditions resulting from type III hypersensitivity reactions are found in **Table 15-5**.

TABLE 15-5	Examples of conditions involving type III hypersensitivity reactions
Autoimmune diseases	
Systemic lupus erythematosus	
Rheumatoid arthritis	
Drug reactions	
Allergies to penicillin and sulfonamides	
Infectious diseases	
Poststreptococcal glomerulonephritis	
Meningitis	
Hepatitis	
Mononucleosis	
Malaria	
Trypanosomiasis	

Key Concept:

- Chronic exposure to immune complexes against auto-antigens can lead to chronic type III hypersensitivity reactions and tissue damage.

Arthus Reactions Are Localized Type III Hypersensitivity Reactions

One example of a localized type III hypersensitivity reaction has been used extensively as an experimental tool. If an animal or human subject is injected intradermally with an antigen to which large amounts of circulating antibodies exist (or have been recently introduced by intravenous injections), antigen will diffuse into the walls of local blood vessels and large immune complexes will precipitate close to the injection site. This initiates an inflammatory reaction that peaks approximately 4 to 10 hours postinjection and is known as an Arthus reaction. Inflammation at the site of an Arthus reaction is characterized by swelling and localized bleeding, followed by fibrin deposition (**Figure 15-14**). Though not commonly used now, this was used as an in vivo assay to detect the presence of antigens and/or antibodies,

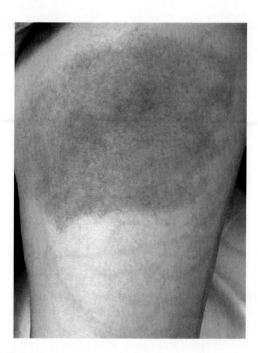

FIGURE 15-14 An Arthus reaction. This photograph shows an Arthus reaction on the thigh of a 72-year-old woman. This occurred at the site of injection of a chemotherapeutic drug, 3 to 4 hours after the patient received a second injection (15 days after the first). This response was accompanied by fever and significant discomfort. [Reproduced with permission from BMJ Publishing Group, from Boura, P., et al.,"Eosinophilic cellulitis (Wells' Syndrome) as a cutaneous reaction to the administration of adalimumab," Annals of the Rheumatic Diseases, 2006, June; **65(6)**: 839-840. doi:10.1136/ard.2005.044685. Figure 1. Permission conveyed through Copyright Clearance Center, Inc.]

especially in situations in which the antibodies or antigen had not been purified.

A sensitive individual may react to an insect bite with a rapid, localized type I allergic reaction, which can be followed, some 4 to 10 hours later, by the development of a typical Arthus reaction, characterized by pronounced erythema and edema. Intrapulmonary Arthus-type reactions in the lung induced by bacterial spores, fungi, or dried fecal proteins in people with antibodies to these antigens can also cause pneumonitis or alveolitis. These reactions are known by a variety of common names reflecting the source of the antigen. For example, farmer's lung develops after inhalation of actinomycetes from moldy hay, and pigeon fancier's disease results from inhalation of a serum protein in dust derived from dried pigeon feces.

Key Concept:

- Arthus reactions are examples of immune complex (type III) hypersensitivity reactions and can be induced by insect bites, as well as by inhalation of fungal or animal protein in individuals with antibodies to those antigens. Deposition of immune complexes in blood vessels can cause local and sometimes severe inflammation of blood vessels in the skin and other tissues.

Delayed-Type (Type IV) Hypersensitivity

Type IV hypersensitivity, commonly referred to as delayed-type hypersensitivity (DTH), is the only hypersensitivity category that is purely cell mediated rather than antibody mediated. In 1890, Robert Koch observed that individuals infected with Mycobacterium tuberculosis developed a localized inflammatory response when injected intradermally (i.e., via the skin) with a filtrate derived from a mycobacterial culture. He therefore named this localized skin reaction a tuberculin reaction. Later, as it became apparent that a variety of other antigens could induce this cellular response (**Table 15-6**), its name was changed to delayed-type, or type IV, hypersensitivity. The hallmarks of a type IV reaction are its initiation by T cells (as distinct from antibodies), the delay required for the reaction to develop (usually 1 to 2 days), and the recruitment of macrophages (as opposed to neutrophils or eosinophils) as the primary cellular component of the infiltrate that surrounds the site of inflammation.

The most common type IV hypersensitivity is the contact dermatitis that occurs after exposure to Toxicodendron species, which include poison ivy, poison oak, and poison sumac. This is a significant public health problem. Approximately 50% to 70% of the U.S. adult population is clinically sensitive to exposure to Toxicodendron; only 10% to 15% of

TABLE 15-6	Intracellular pathogens and contact antigens that induce delayed-type (type IV) hypersensitivity	
Intracellular bacteria	**Intracellular viruses**	
Mycobacterium tuberculosis	Herpes simplex virus	
Mycobacterium leprae	Variola (smallpox)	
Brucella abortus	Measles virus	
Listeria monocytogenes		
Intracellular fungi	**Contact antigens**	
Pneumocystis carinii	Picryl chloride	
Candida albicans	Hair dyes	
Histoplasma capsulatum	Nickel salts	
Cryptococcus neoformans	Poison ivy	
	Poison oak	
Intracellular parasites		
Leishmania sp.		

the population is tolerant. Some responses can be severe and require hospitalization.

The Initiation of a Type IV DTH Response Involves Sensitization by Antigen

A DTH response begins with an initial sensitization by antigen, followed by a period of at least 1 to 2 weeks during which antigen-specific T cells are activated, clonally expanded, and mature into effector T cells (**Figure 15-15a**). Various *antigen-presenting cells* (APCs) are involved in the induction of a DTH response, including macrophages, dendritic cells, and Langerhans cells (dendritic cells found in the epidermis) if it is a skin reaction. These cells pick up antigen and transport it to regional lymph nodes, where T cells are activated. In some species, including humans, the vascular endothelial cells express MHC class II molecules and can also function as APCs in the development of the DTH response. In general, the T cells activated during the sensitization phase of a traditional DTH response are CD4$^+$, primarily of the T$_H$1 subtypes. However, recent studies indicate that T$_H$17 and CD8$^+$ cells can also play a role.

Key Concept:

• In the sensitization phase, T cells are activated by antigen-presenting cells. The T cells are primarily of the T$_H$1 subtypes, but can also be T$_H$17, T$_H$2, and CD8$^+$ cells.

(a) Sensitization phase

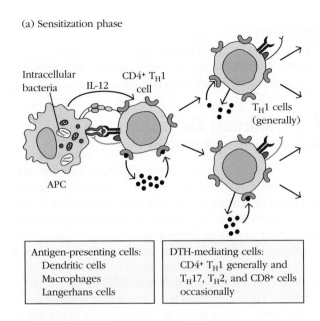

Antigen-presenting cells: Dendritic cells Macrophages Langerhans cells	DTH-mediating cells: CD4$^+$ T$_H$1 generally and T$_H$17, T$_H$2, and CD8$^+$ cells occasionally

(b) Effector phase

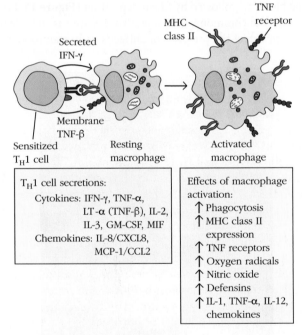

T$_H$1 cell secretions: Cytokines: IFN-γ, TNF-α, LT-α (TNF-β), IL-2, IL-3, GM-CSF, MIF Chemokines: IL-8/CXCL8, MCP-1/CCL2	Effects of macrophage activation: ↑ Phagocytosis ↑ MHC class II expression ↑ TNF receptors ↑ Oxygen radicals ↑ Nitric oxide ↑ Defensins ↑ IL-1, TNF-α, IL-12, chemokines

FIGURE 15-15 The DTH response. (a) In the sensitization phase after initial contact with antigen (e.g., peptides derived from intracellular bacteria), naïve CD4$^+$ cells proliferate and differentiate into T$_H$1 cells. Cytokines secreted by these T cells are indicated by the black dots. (b) In the effector phase after subsequent exposure of sensitized effector T$_H$ cells to antigen, T$_H$1 cells secrete a variety of cytokines and chemokines, including IFN-γ. These factors attract and activate macrophages and other nonspecific inflammatory cells. Activated macrophages are more effective in presenting antigen, thus perpetuating the DTH response, and function as the primary effector cells in this reaction. The T$_H$17 helper T-cell subset and CD8$^+$ T cells also contribute to DTH responses.

The Effector Phase of a Classical DTH Response Is Induced by Second Exposure to a Sensitizing Antigen

A second exposure to the sensitizing antigen induces the effector phase of the DTH response (Figure 15-15b). In the effector phase, previously activated T_H1 cells are stimulated to secrete a variety of cytokines, including interferon-γ (IFN-γ), TNF-α, and lymphotoxin-α (TNF-β), which recruit and activate macrophages and other inflammatory cells. A DTH response normally does not become apparent until an average of 24 hours after the second contact with the antigen and generally peaks 48 to 72 hours after this stimulus. The delayed onset of this response reflects the time required for the cytokines to induce localized influx and activation of macrophages. Once a DTH response begins, a complex interplay of nonspecific cells and mediators is set in motion that can result in extensive amplification of the response. By the time the DTH response is fully developed, only about 5% of the participating cells are antigen-specific T cells; the remainder are macrophages and other innate immune cells.

T_H1 cells often are important initiators of DTH, but the principal effector cells of the DTH response are activated macrophages. Cytokines elaborated by helper T cells, including IFN-γ and TNF-α and -β, induce blood monocytes to adhere to vascular endothelial cells, migrate from the blood into the surrounding tissues, and differentiate into activated macrophages. As described in Chapter 2, activated macrophages exhibit enhanced phagocytosis and an increased ability to kill microorganisms. They produce cytokines, including TNF-α and IL-1β, and chemokines that lead to the recruitment of more monocytes and neutrophils, and enhance the activity of T_H1 cells, amplifying the response.

The heightened phagocytic activity and the buildup of lytic enzymes from macrophages in the area of infection lead to nonspecific destruction of cells and thus of any intracellular pathogens, such as the mycobacteria discussed earlier. In addition, while *Mycobacterium tuberculosis* can survive inside macrophage endosomes by blocking their fusion with lysosomes, in macrophages activated by T cell–derived cytokines (in particular IFN-γ) this block can be alleviated so that the bacteria can be killed. Usually, any pathogens targeted by immune responses are cleared rapidly with little tissue damage. However, in some cases, and especially if the antigen is not easily cleared, a prolonged DTH response can develop that becomes destructive to the host, causing a visible granulomatous reaction. Granulomas develop when continuous activation of macrophages induces them to adhere closely to one another. Under these conditions, macrophages assume an epithelioid shape and sometimes fuse to form multinucleated giant cells (**Figure 15-16a**). These giant cells displace the normal tissue cells, forming palpable nodules, and releasing high concentrations of lytic enzymes, which destroy surrounding tissue. The granulomatous response can damage blood vessels and lead to extensive tissue necrosis.

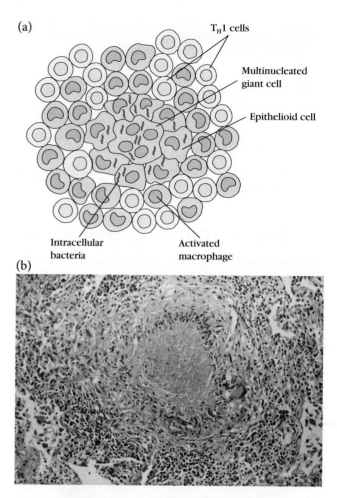

(a)

T_H1 cells

Multinucleated giant cell

Epithelioid cell

Intracellular bacteria

Activated macrophage

(b)

FIGURE 15-16 A prolonged DTH response can lead to formation of a granuloma, a nodule-like mass. (a) Lytic enzymes released from activated macrophages in a granuloma can cause extensive tissue damage. (b) Stained section of a granuloma associated with tuberculosis, showing a central region devoid of normal healthy cells. *[(b) Biophoto Associates/Getty Images]*

The response to *M. tuberculosis* illustrates the double-edged nature of the DTH response. The macrophage activation that occurs is essential to eliminate the infection, but if the response does not clear the infection, the DTH response persists and activated macrophages wall off the organism in the lung, attempting to contain it within a granuloma-type lesion called a *tubercle* (Figure 15-16b). Often, however, the release of concentrated lytic enzymes from the activated macrophages within the tubercle damages the very lung tissue that the immune response aims to preserve.

Key Concepts:

- In the effector phase, sensitized T cells are re-activated by an antigen-presenting cell, which produces cytokines (e.g., IFN-γ) that activate macrophages.

- Macrophages produce inflammatory cytokines and other mediators that produce local hypersensitivity responses.

- During infections with *Mycobacterium tuberculosis*, activation of the macrophages by T cells in the DTH reaction can be beneficial, as it can lead to macrophages killing the intracellular *M. tuberculosis*. However, if the infection is not cleared, continuing macrophage activation can lead to cell death and tissue damage in the lungs, forming granulomas.

The DTH Reaction Can Be Detected by a Skin Test

The presence of a DTH reaction can be measured experimentally by injecting antigen intradermally into an animal and observing whether a characteristic skin lesion develops days later at the injection site. A positive skin-test reaction indicates that the individual has a population of sensitized T_H1 cells specific for the test antigen. For example, to determine whether an individual has been exposed to *M. tuberculosis*, purified protein derivative (PPD) derived from the cell wall of this mycobacterium is injected intradermally. Development of a red, slightly swollen, firm lesion at the site between 48 and 72 hours later indicates previous exposure (**Figure 15-17**).

Note, however, that a positive test does not allow one to conclude whether the exposure was due to a pathogenic form of *M. tuberculosis* or to vaccination with a related *Mycobacterium*, which is used in some parts of the world.

Key Concept:

- Prior sensitization to *M. tuberculosis* can be detected by a skin test, in which a small amount of *M. tuberculosis* protein is injected into the skin. If sensitized T cells are present, a localized hypersensitivity response occurs, assessed after 48 hours.

Contact Dermatitis Is a Type IV Hypersensitivity Response

Contact dermatitis is one common manifestation of a type IV hypersensitivity response. The simplest form of contact dermatitis occurs when a reactive chemical compound contacts the skin and binds chemically to skin proteins. Peptides with the modified amino acid residues are presented to T cells in the context of the appropriate MHC antigens. The reactive chemical may be a pharmaceutical, a component of a cosmetic or a hair dye, an industrial chemical such as

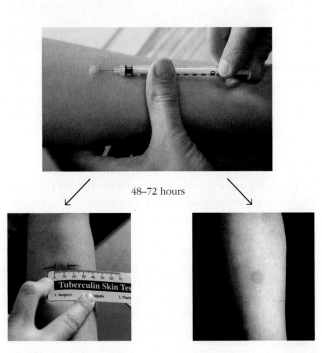

FIGURE 15-17 Tuberculin skin test. This test, also known as the Mantoux test, is carried out by injecting a small amount of purified protein derivative from *M. tuberculosis* into the skin. The skin is examined 2 to 3 days later for redness and swelling, indicative of a type IV hypersensitivity (DTH) reaction due to the activation of memory *M. tuberculosis*–specific T cells. [Photos: (top and bottom left) Andrew Aitchison/Getty Images; (bottom right) Medical Images RM/Bob Tapper]

(a)

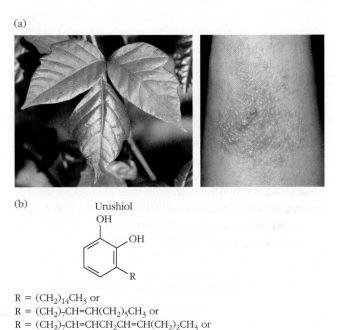

(b)   Urushiol

R = $(CH_2)_{14}CH_3$ or
R = $(CH_2)_7$CH=CH$(CH_2)_5$CH$_3$ or
R = $(CH_2)_7$CH=CHCH$_2$CH=CH$(CH_2)_2$CH$_3$ or
R = $(CH_2)_7$CH=CHCH$_2$CH=CHCH=CHCH$_3$ or
R = $(CH_2)_7$CH=CHCH$_2$CH=CHCH$_2$CH=CH$_2$
and others

FIGURE 15-18 Poison ivy causes contact dermatitis due to its toxin, urushiol. (a) Poison ivy and the contact dermatitis form of DTH that it causes in many people. (b) The structures of the alkylcatechols with varying R-groups, alkyl chains of 15–17 carbon atoms, that comprise the urushiol toxin family. [(a):(left) Stuart Monk/Shutterstock; (right) John Kaprelian/Science Source]

formaldehyde or turpentine, an artificial hapten such as flu-orodinitrobenzene, a metal ion such as nickel, or the active compound from poison ivy, poison oak, and related plants. For example, nickel ions, a common cause of contact dermatitis that affects 15% of the population (it is often present in cheap jewelry), can bind to histidine amino acid residues, generating modified peptides (neoantigens) to which T cells are not self tolerant. T cells are activated by the modified peptides and generate a DTH response in the skin. Nickel also binds directly to histidine residues in Toll-like receptor 4 (TLR4), inducing innate and inflammatory responses independent of T cell–mediated hypersensitivity.

The classical DTH response to *Mycobacterium* antigens described earlier is mediated by CD4$^+$ T$_H$1 T cells. However, we now know that DTH reactions can also be mediated by other T-cell types, including T$_H$17 and CD8$^+$ T cells. A good example is the contact dermatitis induced by the toxins found in plants in the genus *Toxicodendron*, including poison oak and poison ivy (**Figure 15-18a**). The toxins, a family of related alkylcatechols, are known collectively as urushiol (Figure 15-18b). Urushiol has been shown to activate DTH-inducing T$_H$1 cells, CD8$^+$ cells, and T$_H$17 cells (**Figure 15-19**). After oxidation in the body, urushiol binds covalently to skin proteins, which can be taken up by skin

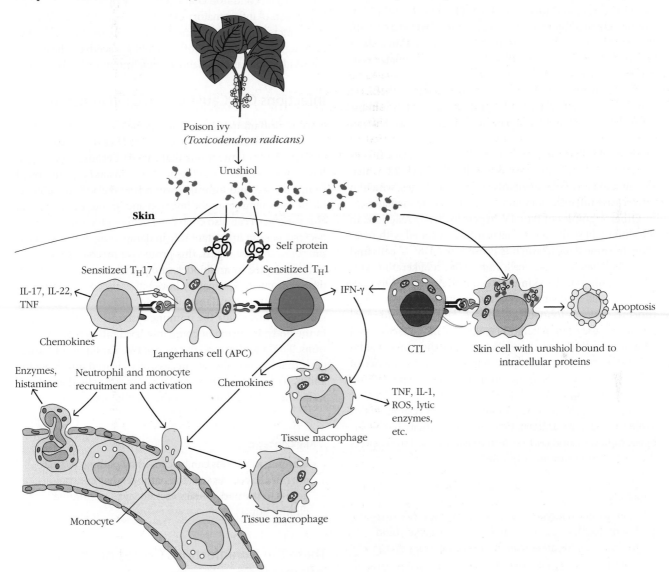

FIGURE 15-19 Induction of contact dermatitis by urushiol can be mediated by T$_H$1, T$_H$17, and CTL effector T cells. Skin DTH reactions are caused by three different effector T cells: effector T$_H$1 cells that recognize urushiol-modified peptides bound to MHC class II proteins on Langerhans cells; effector T$_H$17 cells that recognize urushiol bound to CD1a on Langerhans cells; and CD8$^+$ CTL effector cells that recognize urushiol-modified peptides presented by MHC class I proteins on skin cells. After activation, the T cells produce chemokines and cytokines that recruit and activate macrophages and neutrophils to release inflammatory cytokines, enzymes, and ROS that cause local tissue damage. CTLs may also kill skin cells expressing urushiol-modified peptides bound to MHC class I proteins. See text for more on the generation of these effector cells from naïve T cells.

dendritic cells and carried to the draining lymph node, where they can be degraded into peptides, presented bound to MHC class II proteins, and induce the formation of T_H1 cells. These sensitized effector cells can go back to the skin and release chemokines that recruit leukocytes to the site and cytokines, such as IFN-γ and TNF-α, that activate macrophages to release inflammatory cytokines, lytic enzymes, and reactive oxygen species (ROS) that cause tissue damage. Urushiol can also enter cells, where it can bind to cytoplasmic proteins that may be degraded into peptides that enter the endoplasmic reticulum and bind to MHC class I. $CD8^+$ T cells can be activated by the modified peptides bound to MHC class I and form effector CTLs, which in the skin can be activated by skin cells expressing MHC class I with the urushiol-bound peptides to either kill those skin cells or release cytokines including IFN-γ, a major macrophage activator. Recently, T_H17 cells have also been shown to generate DTH responses to urushiol. Recall from Chapter 7 that CD1 proteins are MHC class I–like proteins with hydrophobic pockets that bind lipid antigens (see Figure 7-19a). Human CD1a, expressed by skin Langerhans dendritic cells, binds urushiol, and that complex activates T_H17 cells. These T cells secrete proinflammatory cytokines IL-17 and IL-22, which recruit and activate neutrophils and macrophages, which in turn release inflammatory and tissue-damaging mediators.

Other examples of type IV hypersensitivity reactions in the skin include the severe dermatitis associated with reactions to some drugs, which is caused by $CD8^+$ T cells and NK cells. These cytotoxic cells induce death of keratinocytes and sloughing of the skin or mucous membrane. Diseases in which this mechanism is active include erythema multiforme, Stevens-Johnson syndrome, and toxic epidermal necrolysis; they can be fatal. The allergen in these cases can be associated with drugs as common as the nonsteroidal anti-inflammatory medication ibuprofen. The incidence of such complications is higher in males than in females and usually occurs in young adults.

At present, the best way to avoid a DTH response is to avoid the causative antigen. Once hypersensitivity has developed, topical or oral corticosteroids can be used to suppress the destructive immune response.

Key Concepts:

- Contact dermatitis is a skin DTH response. Sensitizing molecules, including metals such as nickel, bind to skin proteins and create modified peptides that can be recognized by T cells.

- The plant lipid toxin urushiol induces contact dermatitis by activating three types of effector T cells. It binds to extracellular and intracellular proteins in the skin and activates T_H1 and $CD8^+$ cells specific for urushiol-modified peptides. It also binds to the CD1a protein on skin Langerhans cells and activates urushiol-specific T_H17 cells.

Chronic Inflammation

Regardless of whether an immune response is activated appropriately or inappropriately, a more widespread immune response makes one feel ill, due to such symptoms such as fever and muscle aches, largely because of the activity of inflammatory mediators released by innate immune cells. In most cases, the misery subsides when the insult (antigen, allergen, or toxin) is cleared. However, in some circumstances, an inflammatory stimulus persists, generating a chronic inflammatory response that has systemic effects. Chronic inflammation is a pathological condition characterized by persistent, increased expression of inflammatory cytokines. One example, unfortunately increasing in incidence, is type 2 diabetes. Recent studies also suggest that chronic inflammation exacerbates heart disease, kidney disease, Alzheimer's, autoimmunity, and cancer.

Infections Can Cause Chronic Inflammation

Chronic inflammatory conditions have a variety of causes, some of which are still being identified (**Figure 15-20**). Some are the result of infections that persist because a pathogen has continuous access to the body. For instance, periodontal (gum) disease and unhealed wounds make a body vulnerable to continuous microbe invasion and immune stimulation. Gut pathogens can also contribute. Although our commensal bacteria play an important role in dampening our reaction to microbes that we ingest, this protective mechanism can fail or be disrupted by antibiotics, and gut microbes can contribute to chronic inflammatory bowel diseases (see Chapter 13).

Some chronic inflammatory conditions are caused by pathogens that evade the immune system and remain active in the body, inspiring ongoing low-level inflammatory reactions. Fungi and mycobacteria are two examples of pathogens that are not always successfully cleared and have the ability to continuously stimulate immune cells that release inflammatory cytokines and other mediators.

Key Concept:

- Chronic infections can be caused by pathogens that are not cleared because the pathogen has continuous access to the body or successfully evades immune elimination.

There Are Noninfectious Causes of Chronic Inflammation

Interestingly, pathogens are not the only causes of chronic inflammation. Physical damage to tissue also releases molecules (*damage-associated molecular patterns*, or DAMPs; see Chapter 4) that induce the secretion of inflammatory cytokines and other mediators typical of innate immune responses. If tissue damage is not resolved the inflammatory stimulus persists. Tumors, autoimmune disorders,

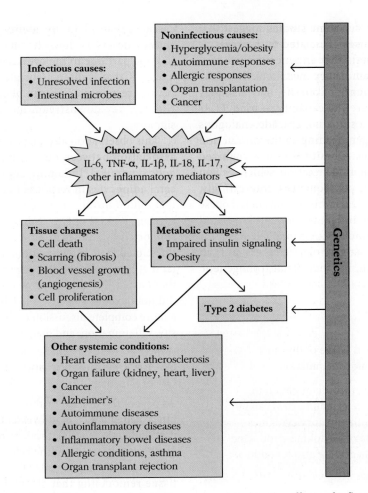

FIGURE 15-20 Causes and consequences of chronic inflammation. Chronic inflammation has infectious and non-infectious causes, including obesity. Chronic inflammatory conditions, regardless of cause, have common systemic consequences, some of which are related to the effects of inflammatory mediators on metabolism (type 2 diabetes) and some of which are related to the effects of inflammatory mediators on tissue organization and cell proliferation (e.g., cancer). Other disorders have also been associated with chronic inflammation, although the mechanisms behind the association may be indirect and are still being studied. Genetic variation affects the induction and effects of chronic inflammation.

atherosclerosis, heart disease, and Alzheimer's, each of which results in tissue damage that stimulates innate and sometimes adaptive immune responses, are examples of noninfectious causes of chronic inflammation. The biomedical community was startled, however, to discover that one of the most common noninfectious causes of chronic inflammation today is obesity, a condition that did not, at first, suggest a relationship to inflammation.

Key Concepts:

- Chronic inflammation can be caused by a variety of noninfectious conditions that lead to tissue damage, including tumors and autoimmunity.

- DAMPs released by damaged tissue induce the secretion of inflammatory cytokines and other mediators, and can lead to chronic inflammation if tissue damage persists.

Obesity Is Associated with Chronic Inflammation

Obesity has long been associated with a constellation of metabolic and systemic disorders, including type 2 diabetes. The biological mechanisms responsible for these associations are still being investigated. However, recent work suggests that many of the systemic effects of obesity are mediated by inflammation.

What does fat have to do with inflammation? It turns out that the immune system is not the only source of inflammatory cytokines. Visceral adipocytes, that is, the fat cells that surround organs (as opposed to subcutaneous adipocytes, which are located under the skin), are very active, responsive cells. Not only do they generate hormones such as leptin that regulate metabolism, but they also secrete a variety of proinflammatory mediators, including TNF-α and IL-6, and can recruit macrophages to adipose tissue.

What triggers this release? Some studies suggest that intracellular stress responses associated with excessive lipid buildup induce signals that enhance the production of cytokines and inflammatory mediators. Free fatty acids, a common consequence of obesity, may also play a role. They appear to have the capacity to bind TLRs on adipocytes, initiating a signaling cascade analogous to that induced by pathogen, leading to the production of inflammatory cytokines. Obesity is now recognized as a major cause of chronic inflammation, which, as you will see shortly, has severe consequences. Interestingly, approximately 6% of individuals who are considered obese by weight do not generate inflammatory cytokines and show few signs of metabolic dysfunction. The basis for the ability of these individuals to tolerate excess fat is an area of active investigation, but may result from genetic differences.

Key Concepts:

- Obesity is now recognized as one of the most common causes of chronic inflammation.

- Obesity can result in chronic inflammation in part because visceral fat cells (adipocytes) can be stimulated to produce inflammatory cytokines directly. Other inflammatory cytokine–producing cells, including macrophages, are also found in adipose tissue.

Chronic Inflammation Can Cause Systemic Disease

The specific consequences of chronic inflammation vary with the tissue of origin as well as the sex, age, and health status of the individual. However, given that those who suffer from chronic inflammation all exhibit increased circulation of inflammatory mediators, including the classical proinflammatory cytokine trio (IL-1, IL-6, and TNF-α), it is not surprising that many also suffer from similar systemic disorders (see Figure 15-20).

Chronic Inflammation and Insulin Resistance

Type 2 diabetes is one of the most common consequences of chronic inflammation. Diabetes results from a failure in insulin signaling, a failure that leads to general metabolic dysfunction. Type 1 diabetes, discussed in Chapter 16, is caused by the autoimmune-mediated destruction of pancreatic islet cells that make insulin. Type 2 diabetes, however, is caused by a failure of cells to respond to insulin, a state known as *insulin resistance*, which interferes with proper regulation of glucose levels. What does inflammation have to do with insulin resistance? Inflammatory cytokines, particularly TNF-α and IL-6, induce signaling

cascades that inhibit the ability of the insulin receptor to activate necessary downstream events. This interference is in large part due to the cytokines' ability to activate JNK, a MAPK that is often associated with stress and inflammatory responses. JNK can phosphorylate and inactivate IRS-1, a key downstream mediator of insulin receptor signaling.

This observation also helps explain the long-recognized association between obesity and type 2 diabetes (*metabolic syndrome*). Inflammatory cytokines released by visceral adipocytes in response to excess lipid induce signals that inhibit insulin signaling, leading to insulin resistance, a primary cause of type 2 diabetes. This model is directly supported by studies in mouse models where obesity was uncoupled from inflammation. Wild-type mice fed a high-fat diet became obese and developed type 2 diabetes. Mice that lacked JNK also became obese on the same diet, but did not develop diabetes. See **Clinical Focus Box 15-3** for a more complete discussion of the association between obesity, inflammation, and diabetes.

Chronic Inflammation and Susceptibility to Other Diseases

As part of the normal healing process, inflammatory cytokines also enhance blood vessel flow and blood vessel formation (angiogenesis), induce proliferation and activation of fibroblasts and immune cells, and regulate the death of infected or damaged cells. Together, these events induce tissue remodeling that gives immune cells better access to pathogens, and scarring that heals wounds. However, continuous stimulation of this healing process has deleterious consequences. Overstimulation of fibroblasts leads to excessive tissue scarring (fibrosis), which can severely impair organ function. Continuous stimulation of cell proliferation enhances the probability of mutations and may contribute to tumor formation or growth. Enhancement of blood vessel formation can also enhance survival of cells in solid tumors.

Although focusing on similarities between acute and chronic inflammation is helpful in understanding some of the consequences described, it is also important not to oversimplify the relationship. For instance, although neutrophil infiltration is a cardinal feature of acute inflammation, monocytes, macrophages, and lymphocytes accumulate during chronic inflammation. Fibroblasts associated with chronic inflammation secrete distinct cytokines and may represent distinct lineages. Genetic variability also impacts inflammatory conditions. Variants of genes that encode molecules involved in innate and inflammatory responses are associated with some inflammatory conditions, such as inflammatory bowel diseases and autoinflammatory diseases (see Chapter 4). These factors and others underscore the importance of thoughtfully considering and customizing approaches to ameliorating inflammatory conditions in different situations.

BOX 15-3

Type 2 Diabetes, Obesity, and Inflammation

As late as the 1960s, scientists and physicians searching for information on type 2 diabetes, obesity, or inflammation would have looked in separate chapters of physiology or pathology books. Obesity was viewed as a problem of poor nutritional management or resources, psychology, or (in the case of rare hormonal disorders) endocrinology. Type 2 diabetes was known to be a result of the inability to effectively use insulin, resulting in high blood glucose levels (hyperglycemia) and was, therefore, seen solely as the province of endocrinologists. However, knowledge about how inflammatory mediators work and the cells from which they are released has helped us to understand the linkages between obesity and type 2 diabetes. Here we will explore the intersecting biological pathways that connect obesity, inflammation, and type 2 diabetes.

Obesity and type 2 diabetes currently represent a public health problem of stunning proportions in the United States. As of 2016, 36.5% of adults in the United States are obese (defined as having a body mass index [BMI] of 30 or above), and childhood obesity rates are rising rapidly, with 17% of children and adolescents aged from 2 to 19 years falling into this category. Nor is the problem restricted to the United States. The World Health Organization reports that 300 million adults are obese and as many as 1 billion are reported to be overweight worldwide. But why should this be a problem, and what does it have to do with immunology?

In type 2 diabetes, patients experience a state of insulin resistance, in which the body still makes insulin, but the responses to it are dulled and the amount of insulin in the circulation is unable to do its job of driving dietary sugar out of the bloodstream and into the waiting cells. The first indication

that type 2 diabetes may result from, or at least be exacerbated by, inflammatory signals came almost a hundred years ago, when it was discovered that patients receiving salicylate (aspirin) for pain or inflammatory conditions showed an increase in insulin sensitivity. Studies published in the early years of the twenty-first century further demonstrated that patients suffering from a variety of infectious diseases, including hepatitis C and HIV, as well as those with autoimmune diseases such as rheumatoid arthritis, displayed insulin resistance. These diseases share the common feature of inducing an active inflammatory response. In each case, the insulin resistance was improved on treatment with anti-inflammatory drugs.

What do we know about the mechanism by which inflammation leads to insulin resistance? Insulin signals a cell to import glucose by binding to a cell surface receptor that is a member of the *receptor tyrosine kinase* (RTK) family. When insulin binds to the receptor on the external surface of the cell, a signal is transmitted to the intracellular part of the receptor, activating the intrinsic tyrosine kinase activity of the receptor (see Chapter 3). The two halves of the insulin-bound dimeric receptor phosphorylate one another on tyrosine residues, and these residues then act as docking sites for other proteins in the signaling cascade. A set of six proteins termed the *insulin receptor substrate* (IRS) proteins are among the early, pivotally important substrates of the insulin receptor, and they bind to it. The IRS proteins are then phosphorylated by the RTKs and subsequently act as adapter molecules for transmitting the insulin signal to downstream molecules such as the kinases PI3 kinase and Fyn, and the docking proteins Grb2 and SHP2. However, if the IRS proteins are phosphorylated on serine residues by IRS serine/

threonine kinases, their signaling capacity is inhibited (**Figure 1**).

Inflammatory cytokines, such as IL-6, IL-1β and TNF-α, bind to cell-surface receptors and signal the activation of kinases, including JNK, which phosphorylate IRS-1 on serine residues, inhibiting its activity. Thus, inflammatory cytokines act to inhibit the insulin signal, leading to insulin resistance. However, what is the source of the excess inflammatory cytokines released in individuals with type 2 diabetes?

Patients with type 2 diabetes are frequently (although not always) obese, and so investigators began to explore the relationships between obesity and the generation of inflammatory cytokines. Mice fed high-fat diets, or those with a genetic predisposition to obesity, were found to develop chronically elevated levels of inflammatory mediators such as TNF-α, IL-1β, and IL-6, and increased local concentrations of chemokines such as CCL2, which draw immune cells, particularly macrophages, into adipose tissue. Investigators therefore advanced the hypothesis that the nutrients themselves might be activating signaling pathways leading to the release of these mediators. But with what receptors are the nutrients interacting?

One clue has come from genetically modified mice. Mice in which the gene encoding the TLR4 receptor has been eliminated are protected from the insulin resistance engendered by eating a high-fat diet. This suggests that TLRs may be recognizing the excess nutrients and initiating the inflammatory response. Indeed, TLR4 and TLR2 have both been shown to be responsive to high levels of free fatty acids. A second observation that implicates TLR4 in the sensing of a nutrient-rich environment is that in mice, serum levels of the TLR4 ligand *lipopolysaccharide* (LPS), a bacterial cell wall component, are increased

(continued)

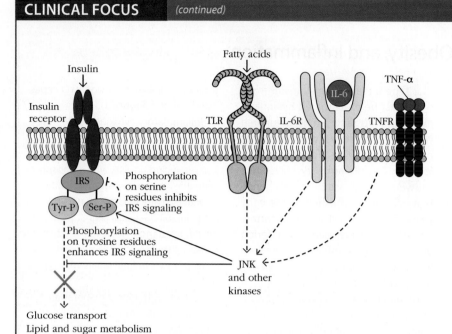

FIGURE 1 Signaling events that link obesity and inflammation to insulin resistance. Several factors trigger signaling cascades that interfere with insulin receptor signaling cascades. Inflammatory cytokines, including TNF-α, IL-6, and IL-1β, are produced by immune cells, as well as by adipocytes themselves. These trigger signaling events in multiple cells that activate kinases, including JNK, which inactivate insulin receptor substrates (IRSs) by phosphorylating serine residues. Thus, downstream insulin signaling and normal glucose regulation are impaired. JNK can also be activated (and thus insulin signaling inactivated) by the interaction between TLRs and free fatty acids, which are increased in obesity.

after feeding. It is possible that in animals living in the nutrient-rich environment characteristic of obesity, intestinal permeability remains relatively high long after a meal is completed, allowing bacteria or their LPS to enter the system, and thus these animals are chronically exposed to low levels of inflammatory signals.

If TLRs are signaling the release of inflammatory cytokines, what TLR-expressing cells are responding? A significant number of studies have demonstrated the presence of macrophages and mast cells in adipose (fat) tissue, and depletion of cells bearing the CD11c marker (macrophages, dendritic cells, and neutrophils) increases insulin sensitivity. In addition, macrophages newly recruited into expanding adipose tissue may differentiate into a more potent proinflammatory phenotype than macrophages normally

resident in lean adipose tissue. Finally, recent work shows that some adipocytes (fat cells) themselves are capable of generating inflammatory cytokines, including TNF-α and IL-6. Adipocytes also express TLRs that trigger cytokine release. Excess internal lipid also appears to stimulate internal stress responses by adipocytes that enhance cytokine production.

The adaptive immune system may also play a role in controlling the inflammatory milieu in adipose tissue. Lean visceral adipose tissue supports an anti-inflammatory state, with production of IL-4 and IL-13 cytokines characteristic of T_H2 and ILC2 cells; there are also ample T_{REG}s, which produce IL-10 that inhibits inflammatory responses. In contrast, in obese visceral adipose tissue the hormone leptin induces the formation of IFN-γ–producing T_H1 cells; IFN-γ activates macrophage inflammatory

cytokine production and also induces MHC class II expression and antigen presentation functions in macrophages, dendritic cells, and even adipose cells themselves, which further activates the T cells to produce more proinflammatory IFN-γ. IFN-γ–producing CD8[+] T cells are also found. T_H17 and γδ T cells producing the proinflammatory IL-17 cytokine may also contribute to the inflammatory environment in obese adipose tissue. The nature of the antigens recognized by these T cells remains to be determined, but bacterial antigens or superantigens from gut flora are likely candidates.

Once an animal starts down the road to obesity, its problems can become self-perpetuating. Adipocytes that are full of fat will tend to leak free fatty acids into the circulation, and these in turn induce further inflammation. Indeed, free fatty acids and cellular stress have also been shown to be additional triggers for IRS protein kinases. In addition, high levels of proinflammatory cytokines block the formation of new adipocytes and reduce the secretion of adiponectin, an important regulator of adipocyte production. As the obese adipocyte expands, it approaches its mechanical limit. Cellular responses to mechanical stress, or death, lead to the release of cytokines and additional fatty acids into the circulation. However, the problems of the type 2 diabetic are not confined just to the adipocytes. In the liver, normally the site of glucose homeostasis, increased levels of inflammatory cytokines also help to induce insulin resistance. Gluconeogenesis (the formation of new glucose) is normally inhibited by insulin, but under inflammatory conditions gluconeogenesis is no longer suppressed and the high blood glucose levels characteristic of the type 2 diabetic are further increased. In the pancreas, the site of insulin production, the high blood glucose levels initially

(continued)

CLINICAL FOCUS   *(continued)*   BOX 15-3

induce hyperproliferation of the pancreatic beta cells, but eventually apoptosis of the insulin-producing cells occurs, further exacerbating the state of high blood glucose. In the brain of animals fed a high-fat diet, inflammatory pathways are also activated in the hypothalamus, leading to resistance to the effects of both insulin and leptin, a hormone that normally signals satiety, thus setting up a positive feedback loop: the fatter the animal, the more it needs to eat to achieve satiety.

In summary, current research suggests that obese animals exhibit a state of chronic inflammation resulting from the release of nutrient-stimulated inflammatory mediators by adipocytes themselves, as well as by macrophages and mast cells. These inflammatory mediators in turn act on adipocytes and other cells to reduce their sensitivity to insulin, leading to the syndrome we know as type 2 diabetes.

REFERENCES

Majdoubi, A., O. A. Kishta, and J. Thibodeau. 2016. Role of antigen presentation in the production of pro-inflammatory cytokines in obese adipose tissue. *Cytokine* **82:**112.

Wensveen, F. M., et al. 2015. The "Big Bang" in obese fat: events initiating obesity-induced adipose tissue inflammation. *European Journal of Immunology* **45:**2446.

Key Concepts:

- Inflammatory cytokines associated with chronic inflammation contribute to insulin resistance (type 2 diabetes) by interfering with the activity of enzymes downstream of the insulin receptor.

- Whereas they have beneficial activities as part of local inflammatory responses, cytokines produced during chronic inflammation can induce tissue scarring, which can lead to organ dysfunction, as well as cell proliferation and angiogenesis, which may contribute to tumor development.

- Genetics can contribute to inflammatory responses, as indicated by the association of some genes encoding proteins involved in innate and inflammatory responses with inflammatory diseases.

Conclusion

Immune responses can be a double-edged sword. Their roles in protecting us from infections are essential to life, as evidenced by the fatal consequences of immunodeficiency diseases left untreated (to be discussed in Chapter 18). To provide effective protection, a wide variety of innate and adaptive immune mechanisms has evolved that usually enable us to generate a response that will be effective against the type of pathogen that has entered our body. But immune responses are by their very nature destructive, and if those responses are excessive, persistent, or reactive with the wrong targets, they can do damage to the body. Several of these conditions—the four classes of hypersensitivity reactions and chronic inflammation—have been the focus of this chapter.

Type I hypersensitivity reactions, what we recognize as common allergies, are caused by IgE antibodies that are bound by FcεRI receptors on mast cells, basophils, and eosinophils and then are cross-linked by antigens they recognize. This induces degranulation, and the mediators that are released cause the common symptoms of allergies, which include local responses in the respiratory tract for airborne allergens and in the GI tract for food allergens, but can also be systemic if the allergen gets into the bloodstream (as can occur with insect stings, drugs such as penicillin, and foods). While IgE and the granulocyte degranulation responses it triggers probably evolved to combat parasitic worms and animal and insect venoms, and while many allergic symptoms, such as hay fever, are usually just inconveniences, anaphylaxis and asthma are clear examples of maladaptive responses.

Type II and type III hypersensitivity responses are caused by normal IgM and IgG antibody-antigen interactions that can be harmful if misdirected or excessive. Type II reactions result from extensive cell death; examples are transfusion reactions in which antibodies and complement attack transfused blood cells or fetal cells differing in blood-group antigens; this can cause excessive red blood cell death from complement-mediated lysis and toxic levels of bilirubin from the released hemoglobin. Fortunately, blood-group testing can prevent mismatched blood transfusions, and treatments are available to prevent hemolytic disease of newborns. Binding of penicillin and other drugs to red blood cells can cause similar problems if antibodies to the drugs are present. Type III hypersensitivity results from excessive levels of immune complexes; deposition in tissues and complement activation can trigger local inflammatory responses such as vasculitis, glomerulonephritis, and arthritis.

Type IV hypersensitivity reactions are mediated by T cells, generally T_H1, T_H17, and $CD8^+$ cells, which activate inflammatory responses. These responses can be triggered by intracellular bacteria and cause tissue damage if not resolved, as in tuberculosis. Another example is contact dermatitis induced in the skin by the lipid toxins of poison oak and poison ivy, which induce sensitized T cells to produce chemokines and proinflammatory cytokines and may also involve CD8$^+$ T-cell killing of cells modified by the toxin.

Chronic inflammatory responses constitute another class of beneficial immune responses gone bad. A wide array of persistent infectious and noninfectious causes can lead to ongoing innate and adaptive responses that result in chronic local inflammation, such as that which causes the lung damage in tuberculosis or the joint damage in arthritis, or chronic systemic inflammation, such as the inflammatory link between obesity and type 2 diabetes.

Considerable progress has been made in recent years in understanding the causes of hypersensitivity reactions and chronic inflammation. That information is leading to approaches for preventing, diagnosing, and treating these undesirable immune system responses.

REFERENCES

Abramson, J., and I. Pecht. 2007. Regulation of the mast cell response to the type I Fcε receptor. *Immunological Reviews* **217**:231.

Acharya, M., et al. 2010. CD23/FcεRII: molecular multitasking. *Clinical and Experimental Immunology* **162**:12.

Adam, J., W. J. Pichler, and D. Yerly. 2011. Delayed drug hypersensitivity: models of T-cell stimulation. *British Journal of Clinical Pharmacology* **71**:701.

Bonnelykke, K., R. Sparks, J. Waage, and J. D. Milner. 2015. Genetics of allergy and allergic sensitization: common variants, rare mutations. *Current Opinion in Immunology* **36**:115.

Boura-Halfon, S., and Y. Zick. 2009. Phosphorylation of IRS proteins, insulin action, and insulin resistance. *American Journal of Physiology Endocrinology and Metabolism* **296**:E581.

Cabanillas, B., and N. Novak. 2016. Atopic dermatitis and filaggrin. *Current Opinion in Immunology* **42**:1.

Cavani, A., and A. De Luca. 2010. Allergic contact dermatitis: novel mechanisms and therapeutic perspectives. *Current Drug Metabolism* **11**:228.

Donath, M. Y., and S. E. Shoelson. 2011. Type 2 diabetes as an inflammatory disease. *Nature Reviews Immunology* **11**:98.

Forsberg, A., et al. 2016. Pre- and probiotics for allergy prevention: time to revisit recommendations? *Clinical and Experimental Allergy* **46**:1506.

Gladman, A. C. 2006. *Toxicodendron* dermatitis: poison ivy, oak, and sumac. *Wilderness and Environmental Medicine* **17**:120.

Graham, M. T., S. Andorf, J. M. Spergel, T. A. Chatila, and K. C. Nadeau. 2016. Temporal regulation by innate type 2 cytokines in food allergies. *Current Allergy and Asthma Reports* **16**:75.

Gregor, M. F., and G. S. Hotamisligil. 2011. Inflammatory mechanisms in obesity. *Annual Review of Immunology* **29**:415.

Hardman, C., and G. Ogg. 2016. IL-33, friend and foe in type-2 immune responses. *Current Opinion in Immunology* **42**:16.

Hidaka, T., et al. 2017. The aryl hydrocarbon receptor AhR links atopic dermatitis and air pollution via induction of the neurotrophic factor artemin. *Nature Immunology* **18**:64.

Julia, V., L. Macia, and D. Dombrowicz. 2015. The impact of diet on asthma and allergic diseases. *Nature Reviews Immunology* **15**:308.

Kim, H. H., et al. 2016. CD1a on Langerhans cells controls inflammatory skin disease. *Nature Immunology* **17**:1159.

Love, B. L., et al. 2016. Antibiotic prescription and food allergy in young children. *Allergy, Asthma and Clinical Immunology* **12**:41.

McKee, A. S., and A. P. Fontenot. 2016. Interplay of innate and adaptive immunity in metal-induced hypersensitivity. *Current Opinion in Immunology* **42**:225.

Mehta, P., A. M. Nuotio-Antar, and C. W. Smith. 2015. γδ T cells promote inflammation and insulin resistance during high fat diet–induced obesity in mice. *Journal of Leukocyte Biology* **97**:121.

Minai-Fleminger, Y., and F. Levi-Schaffer. 2009. Mast cells and eosinophils: the two key effector cells in allergic inflammation. *Inflammation Research* **58**:631.

Mortuaire, G., et al. 2017. Specific immunotherapy in allergic rhinitis. *European Annals of Otorhinolaryngology, Head and Neck Diseases* **134**:253.

Mukai, K., M. Tsai, P. Starkl, T. Marichal, and S. J. Galli. 2016. IgE and mast cells in host defense against parasites and venoms. *Seminars in Immunopathology* **38**:581.

Mullane, K. 2011. Asthma translational medicine: report card. *Biochemical Pharmacology* **82**:567.

Portelli, M. A., E. Hodge, and I. Sayers. 2014. Genetic risk factors for the development of allergic disease identified by genome-wide association. *Clinical and Experimental Allergy* **45**:21.

Rosenwasser, L. J. 2011. Mechanisms of IgE inflammation. *Current Allergy and Asthma Reports* **11**:178.

Schuijs, M. J., et al. 2015. Farm dust and endotoxin protect against allergy through A20 induction in lung epithelial cells. *Science* **349**:1106.

Smits, H. H., L. E. van der Vlugt, E. von Mutius, and P. S. Hiemstra. 2016. Childhood allergies and asthma: new insights on environmental exposures and local immunity at the lung barrier. *Current Opinion in Immunology* **42**:41.

Yamamoto, F. 2004. Review: ABO blood group system—ABH oligosaccharide antigens, anti-A and anti-B, A and B glycosyltransferases, and *ABO* genes. *Immunohematology* **20**:3.

Yu, W., D. M. Hussey Freeland, and K. C. Nadeau. 2016. Food allergy: immune mechanisms, diagnosis, and immunotherapy. *Nature Reviews Immunology* **16**:751.

Useful Websites

www.aaaai.org This is the website of the American Academy of Allergy, Asthma, and Immunology. It has descriptions of various types of allergic response, current treatment recommendations, and a variety of resources.

www.webmd.com/allergies Contains information for the lay public on types of allergic reactions and their management, including treatments.

https://chriskresser.com/how-inflammation-makes-you-fat-and-diabetic-and-vice-versa This is an interesting and credible series of commentaries by Chris Kresser, who did not go to medical school, but graduated from an alternative medicine program and is open about his interest in examining the assumptions that underlie medical practices. His online articles on obesity and inflammation are informed and clearly written.

Four excellent animations about the four types of hypersensitivity:

https://www.youtube.com/watch?v=2tmw9x2Ot_Q— Type I

https://www.youtube.com/watch?v=kLaUz58CBMc— Type II

https://www.youtube.com/watch?v=SyxzU2Sl_Yw— Type III

https://www.youtube.com/watch?v=C3E5COZ1XC8— Type IV

STUDY QUESTIONS

1. You have in your possession five mouse strains. The mice of each strain lack a specific gene (i.e., they are gene knockout animals). How might the type I hypersensitivity response of each knockout strain (a–e) differ from that wild-type mice? How might the type II hypersensitivity response differ? Explain your answers.

   Strain a: Mice are unable to generate an ε heavy chain.
   Strain b: Mice are unable to generate a high-affinity FcεRI receptor.
   Strain c: Mice are unable to generate a low-affinity FcεRII receptor.
   Strain d: Mice are deficient in the ability to generate the complement attack complex.
   Strain e: Mice are unable to express CD21.

2. What is the difference between primary and secondary pharmacological mediators in the type I hypersensitivity response? Name two of each.

3. How does histamine suppress its own release?

4. Describe two mechanisms by which desensitization through allergy shots or oral immunotherapy are thought to reduce IgE responses to allergens.

5. A mother has an Rh^+ blood type and the father has an Rh^- blood type. Under these circumstances, the family pediatrician is not worried about the possibility of a type II hypersensitivity reaction. However, if the converse is true, and the mother is Rh^- and the father Rh^+, the pediatrician does worry and asks the obstetrician to inject the mother with antibodies toward the end of her first pregnancy. Explain her reasoning in both cases.

6. A mother has an Rh^- blood type and the father has an Rh^+ blood type. The first baby born to the parents was Rh^+. However, the parents elect for the mother not to receive RhoGAM. Are all future babies of this couple at risk for type II hypersensitivity reactions? Why or why not?

7. Describe type III hypersensitivity, describing both the initiating cells and molecules and the cells and molecules that bring about the pathological effects, and indicate two triggers for this type of response.

8. Indicate which type(s) of hypersensitivity reaction (I–IV) apply to the following characteristics. Each characteristic can apply to one, or more than one, type.

   a. Is an important defense against intracellular pathogens.
   b. Can be induced by penicillin.
   c. Involves histamine as an important mediator.
   d. Can be induced by poison oak in sensitive individuals.
   e. Can lead to asthma.
   f. Occurs as a result of mismatched blood transfusion.
   g. Systemic form of reaction is treated with epinephrine.
   h. Can be induced by pollens and certain foods in sensitive individuals.
   i. May involve cell destruction by antibody-dependent cell-mediated cytotoxicity.
   j. One form of clinical manifestation is prevented by RhoGAM.
   k. Localized form characterized by wheal-and-flare reaction.

9. Describe how infections can lead to chronic inflammation. Give two examples. (*Hint:* One was extensively discussed in an earlier section of this chapter!)

10. As described in the text, a small number of obese individuals (6% or so) do not suffer from the chronic inflammatory state characteristic of most forms of obesity.

   a. These individuals do not develop type 2 diabetes. Why not? Provide a specific, molecular-based answer.
   b. Some have suggested that these individuals express genetic polymorphisms that make them less susceptible to obesity-generated inflammation. Describe one possible genetic polymorphism that could uncouple obesity from inflammation.

ANALYZE THE DATA

A group in Finland has been investigating the association between exposure to various bacterial species and the development of allergies. They showed previously that children in homes surrounded with forests and agriculture are less likely to have allergies than those in other environments, such as along the sea. The researchers examined the children for bacterial species that might be correlated with reduced allergies and found that children from forest and farm areas had higher levels of skin bacteria of the genus *Acinetobacter* than did children from other regions. To attempt to establish a causal relationship between exposure to *Acinetobacter* and protection from allergies, they used a mouse model in which intranasal exposure to the allergen ovalbumin can cause respiratory allergic responses. (Fyhrquist, N., et al. 2014. *Acinetobacter* species in the skin microbiota protect against allergic sensitization

and inflammation. *Journal of Allergy and Clinical Immunology* **134:**1301.)

Mice were injected intradermally several times over 3 weeks with the diluent phosphate-buffered saline (PBS) alone, or with the allergen ovalbumin (OVA) without or with *Acinetobacter lwoffii* (Al) or, as controls, two other skin bacteria, *Staphylococcus aureus* (Sa) or *Staphylococcus epidermidis* (Se), which had been shown not to be associated with reduced allergies in children. A week after this initial sensitization to OVA the mice were given three daily intranasal exposures to ovalbumin, after which bronchoalveolar fluid, lung tissue samples, and sera were obtained. The results of various assays are shown in the figure: (a) numbers of eosinophils in lung fluid, and levels of IL-5 and IL-13 mRNAs in lung tissue; (b) levels of IgE and IgG2a antibodies specific for OVA in the serum; and (c) levels of IL-10 and IFN-γ in the skin.

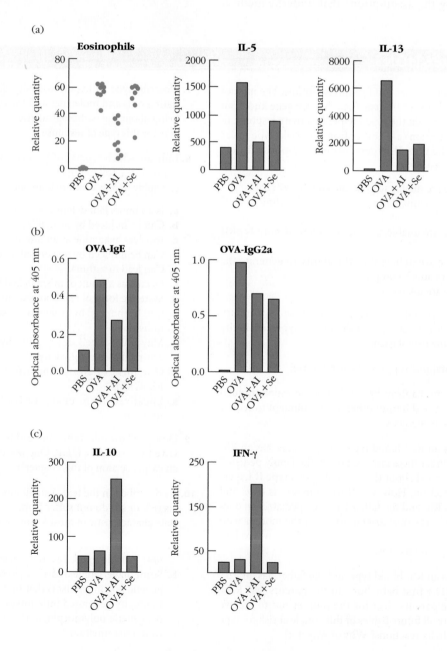

1. What do the three panels in part (a) indicate about the effect of co-injection of *A. lwoffii* on the response to OVA?

2. What do the two panels in part (b) indicate about the effect of co-injection of *A. lwoffii* on the nature of the antibody response to OVA?

3. What do the two panels in part (c) indicate about the effect of *A. lwoffii* co-injection on the skin cytokine environment?

4. Based on the data (and what you have learned about the regulation of immune and allergic responses), provide an overall mechanism by which co-injection with *A. lwoffii* may reduce the allergic response to OVA in this study.

5. Do these findings support the hygiene hypothesis; if so, how?

16

Tolerance, Autoimmunity, and Transplantation

Learning Objectives

After reading this chapter, you should be able to:

1. Incorporate the principles from earlier chapters, specifically those related to MHC structure and diversity, lymphocyte development, and immune regulation, with the concepts of tolerance induction and maintenance, the development of autoimmunity, and the immune events leading to rejection of an allograft.

2. Distinguish between the events and immune players involved in central versus peripheral tolerance pathways, and predict the impact that selected mutations will have on each pathway.

3. Given details of specific autoimmune syndromes, categorize or group like diseases by their effector cell/molecule types as well as their targets (organ-specific versus systemic), and explain your rationale.

4. Design, defend, and assess the effectiveness of a given therapy for the treatment of an autoimmune syndrome by applying basic knowledge of immunologic principles presented here and in earlier chapters.

5. Use your understanding of primary and secondary immune responses to create a sequence for the immune events that occur during the sensitization and effector phases of allograft rejection, and explain how specific therapeutic interventions can alter steps in this process.

6. Explain the relationship between the three topics in this chapter (tolerance, autoimmunity, and transplantation) and why they form a natural grouping.

Late in the nineteenth century, many European scientists and clinicians began to realize that in some instances, the immune system might work against us.

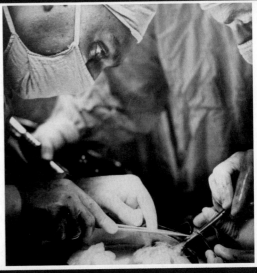

Dr. Samuel Kountz at Stanford. Kountz performed the first allogeneic kidney transplant in 1961. *[University of California, San Francisco, Archives & Special Collections.]*

Instead of limiting its attack to foreign antigens, it occasionally targeted the host. There was much controversy over the validity of this claim at the time. In fact, Paul Ehrlich himself was so disturbed by and reluctant to accept this notion that he coined the term *horror autotoxicus* to describe the repugnant idea of the body attacking itself. For decades afterward, many scientists and scholars argued against the concept. In fact, publication of results that supported this concept were delayed, and several followers of Ehrlich persisted, long after his death, in refuting the existence of compounds in the host that could react against self structures. Interestingly, Ehrlich's initial arguments were less focused on the claim that compounds specific for self components *could not exist*, but rather on his belief that the immune system would *exert control over them*. Incredibly, this latter interpretation is very close to our present day understanding of immune processes!

Key Terms

Autoimmunity
Tolerance
Negative selection
Regulatory T (T$_{REG}$) cells
Tolerogens
Immunogens

Anergy
Immunomodulatory
Molecular mimicry
Alloreactivity
Autograft
Isograft

Allograft
Xenograft
Histocompatible
Graft-versus-host disease (GvHD)
Tissue typing

Cross-matching
Hyperacute rejection
Acute rejection
Chronic rejection

Today, the clinical syndrome referred to as **autoimmunity**, in which the immune system attacks self tissues, is all but incontrovertible. This response, directing humoral and/or T-cell–mediated immune activity against self components, is the cause of a host of **autoimmune diseases** such as rheumatoid arthritis (RA), multiple sclerosis (MS), systemic lupus erythematosus (SLE, or lupus), inflammatory bowel disease (IBD), and certain types of diabetes. Simply stated, autoimmunity results from a disruption in, or a failure of, the host's immune system to protect self structures.

Although on the rise, autoimmunity is still rare, serving as a reminder that mechanisms to protect us from anti-self immune attack must exist, and they do. This process and the mechanisms that control it are collectively referred to as **tolerance**, or **self tolerance**. The mechanisms that maintain self tolerance do so partly by establishing what belongs to "us," making the "them" clearer. However, many benign and even beneficial companions in our evolutionary history are also tolerated by the immune system, such as food and gut commensals. Instead of either ignoring or attacking these appendages, homeostatic immunologic mechanisms are constantly at work to maintain a delicate balance; safely recognizing and protecting self components while also orchestrating inflammatory attacks against pathogenic invaders. Our understanding of the mechanisms that control tolerance have really blossomed in the last decade or two, giving rise to new ways of understanding the immune system in health and yielding novel treatments for autoimmune disease.

When self-tolerance processes are working correctly host tissues and commensals should remain undisturbed by the immune system, and only antagonistic foreign invaders should be attacked. The mechanisms that maintain self tolerance also therefore, quite naturally, interpret the introduction of foreign organs or cells that carry new proteins as potentially harmful, leading to an immune assault. **Transplantation** refers to the act of transferring cells, tissues, or organs from one site to another, or from donor to recipient. The development of new clinical practices and surgical techniques has removed many of the previously intractable barriers to successful transplantation, and many life-threatening diseases can now be treated or cured with this approach. Unfortunately, a continual worldwide shortage of organs for transplantation leaves tens of thousands of individuals waiting for a transplant, sometimes for many years. And yet the most formidable barrier to greater application of tissue and cell transplantations to treat organ failure is still the immune system, and its inherent drive to maintain self tolerance.

In this chapter, we first describe our current understanding of the general mechanisms that establish and maintain immune tolerance. When these mechanisms fail or are disrupted, autoimmunity becomes likely, the second major topic of this chapter. Several common human autoimmune diseases resulting from failures of these mechanisms are described, divided into organ-specific and multiorgan (systemic) categories. A few experimental animal models of autoimmunity that have helped us to understand these disorders and to design therapies are also included, as are the most common forms of treatment. In the final part of the chapter we turn to the topic of transplantation, or situations in which self tolerance works against us. Here we discuss some of the characteristics of the most commonly transplanted tissues, the immunologic processes governing graft rejection, and therapeutic modalities for suppressing these responses as a means to improve graft acceptance.

Establishment and Maintenance of Tolerance

The term *tolerance* applies to the many layers of protection imposed by the immune system to prevent the reaction of its cells and antibodies against host components. In other words, individuals should tolerate—or not respond aggressively against—their own antigens, although their immune systems will attack pathogens or even cells from another individual. Until fairly recently, tolerance was thought to be mediated primarily by the elimination of cells that were autoreactive—those with antigen-specific receptors that recognize self structures. Contemporary studies of tolerance provide evidence for a much more nuanced and active role of immune cells in the selective recognition of self antigens and commensal microbes. *In other words, rather than simply ignoring self, the immune system recognizes and protects self compounds and beneficial commensals.* The characterization of regulatory T cells, lymphocytes that recognize self proteins with high affinity and inhibit the immune response, has opened up new lines of investigation in the fields of tolerance, autoimmunity, and transplantation science. It has also led to a new focus of clinical attention on the manipulation and use of these cells as immunotherapy.

We begin with a discussion of how location and sequestration can play a role in protecting some sites and the tissue-specific antigens found there from exposure to the immune system (*evasion*). This is followed by a description of mechanisms that remove many self-reactive lymphocytes before they can do damage (*elimination*), while cultivating certain self-reactive cells for the protection of self structures (*engagement*). Collectively, these processes help create an environment that maintains a delicate balance of self tolerance alongside vigilance against pathogens. While many exciting advances have been made in this field in recent decades, plenty of unanswered questions remain.

Antigen Sequestration, or Evasion, Is One Means to Protect Self Antigens from Attack

One effective means to avoid self reactivity is sequestration, or partitioning, of antigens away from immune cells. For example, the anterior chamber and lens of the eye are considered sequestered sites, with little or no lymphatic drainage. The tissue-specific antigens that are expressed in these privileged sites are at least partially isolated from interaction with many elements of the immune system. Sequestration allows these antigens to *evade* encounter with reactive lymphocytes under normal circumstances; if the antigen is not exposed to immune cells, there is little possibility of reactivity. However, one possible consequence of sequestration is that the antigen may rarely if ever be involved in life-long peripheral tolerance pathways, as we discuss shortly. In this case, when the barriers between immune cells and the sequestered antigens are breached (by trauma, for example), the newly exposed antigen may be seen as foreign and aggressively attacked. Trauma to an eye is a good example, where the sudden entry of immune cells can lead to locally damaging inflammation, tissue destruction, and impaired vision. Interestingly, in these cases *the other eye* may also become inflamed. This is likely due to the sudden entry of clones of recently activated immune cells recognizing some newly discovered tissue-specific antigens.

The above example of sudden inflammation in the contralateral eye, along with recent data concerning the central nervous system (CNS), suggests that our notion of sequestered or "immune-privileged" locales may be oversimplified. For example, research has shown that the CNS does get regular visits from some circulating lymphocytes, and that lymphatic drainage is present. Likewise, there are antigen-presenting cell types that are residents of the CNS, and many self-reactive lymphocytes, even those with receptors that recognize components of the nervous system, can be found in healthy individuals. Collectively, this suggests that active suppression of anti-self responses must be fairly ubiquitous, including in sites previously thought to have limited access to immune circulation. These "protected sites" may have only partial barriers to the influx of immune cells or partitions that can be opened and closed as needed, as we see with the blood-brain barrier. It is also possible, even likely, that microenvironments where inflammation can be highly destructive, such as the CNS, are structured in ways that, under normal conditions, bias immune engagement toward tolerance rather than assault.

Key Concept:

- In some cases, tolerance may be favored by the partial partitioning of tissue-specific antigens in sensitive or immune-privileged sites, away from most immune circulation and potentially harmful inflammatory mediators (*evasion*).

Central Tolerance Processes Occur in Primary Lymphoid Organs

Beyond a partitioning of self antigen away from the immune system, several processes work coordinately to allow self structures to live intimately and in harmony with elements of the immune system. In the earliest developmental steps of this process, a phenomenon termed **central tolerance** occurs in primary lymphoid organs (PLOs): the thymus for T cells (Chapter 8) and the bone marrow for B cells (Chapter 9). Worth noting, central tolerance is mediated by mechanisms that both foster the destruction (*elimination*) and cultivation (*engagement*) of selected self-reactive lymphocytes in primary lymphoid organs. In general, the outcomes for the selected cells include apoptosis, anergy, or the capacity to later inhibit selected immune responses in the periphery.

Elimination is the first developmental step in central tolerance. As you may recall from Chapter 6, the mechanisms that generate diversity in T- and B-cell receptors include genetic rearrangement of DNA at the variable region, plus the addition of random nucleotides at the junctions between gene segments. This means that variable regions that can react with self antigens are inevitable. If all these T and B cells were allowed to develop into mature, naïve lymphocytes, autoimmune disease might be fairly common. Instead, most developing lymphocytes with receptors that recognize self antigens are eliminated in the thymus and bone marrow before they are allowed to mature (**Figure 16-1a**). This process, described in detail in Chapter 8, is called **negative selection** and results in the induction of apoptosis in many developing lymphocytes with high-affinity TCRs or BCRs that recognize antigen expressed in PLOs. Thanks to the transcription factor AIRE, many tissue-specific self antigens found only in particular organs are also expressed in medullary epithelial cells of the thymus. This expression mediates deletion of the potentially harmful self-reactive T cells recognizing these antigens. In fact, mutations in the *AIRE* gene, along with other components that control tolerance, can lead to a range of autoimmune syndromes with systemic, or whole-body, consequences.

An illustration of central tolerance in action comes from the classic experiment performed by Christopher Goodnow and colleagues, described in Chapter 9 (see Figure 9-9). They mated transgenic mice expressing *hen egg lysozyme* (HEL) with mice expressing a transgenic immunoglobulin specific for HEL, to demonstrate that self-reactive lymphocytes are removed or inactivated after encounter with self antigen, rendering these mice incapable of attacking HEL antigens. David Nemazee and colleagues showed that some developing B cells can undergo **receptor editing** as a salvage pathway during central tolerance (see Figure 9-6). In this process, the antigen-specific V region is "edited" or switched for a different V-region gene segment via additional V-J recombination at the light chain loci, sometimes

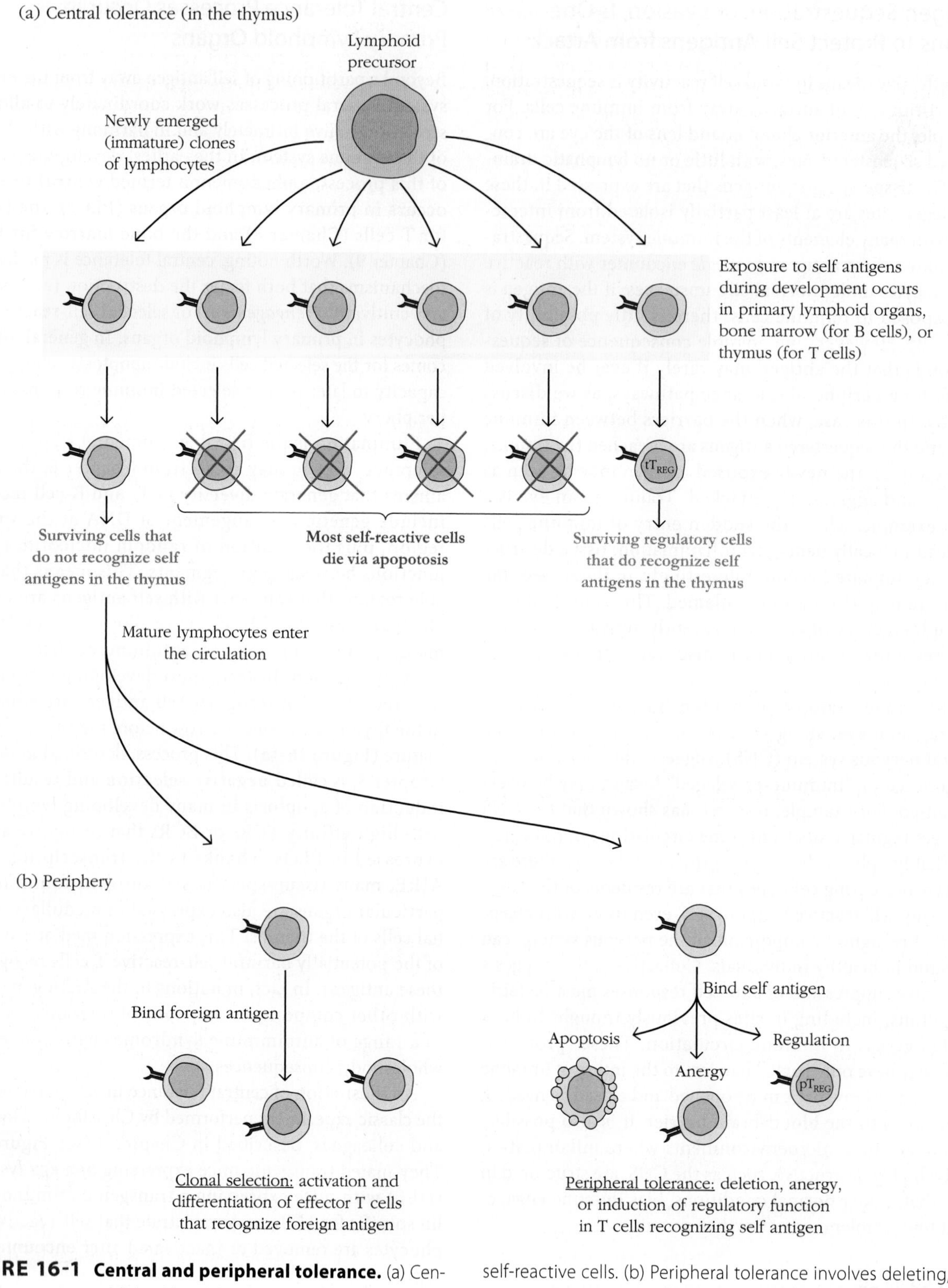

(a) Central tolerance (in the thymus)

Lymphoid precursor

Newly emerged (immature) clones of lymphocytes

Exposure to self antigens during development occurs in primary lymphoid organs, bone marrow (for B cells), or thymus (for T cells)

tT_REG

Surviving cells that do not recognize self antigens in the thymus

Most self-reactive cells die via apopotosis

Surviving regulatory cells that do recognize self antigens in the thymus

Mature lymphocytes enter the circulation

(b) Periphery

Bind self antigen

Bind foreign antigen

Apoptosis

Anergy

Regulation

pT_REG

Clonal selection: activation and differentiation of effector T cells that recognize foreign antigen

Peripheral tolerance: deletion, anergy, or induction of regulatory function in T cells recognizing self antigen

FIGURE 16-1 Central and peripheral tolerance. (a) Central tolerance is established by deletion of lymphocytes in primary lymphoid organs (thymus for T cells and bone marrow for B cells) if they possess receptors that can react with self antigens, or by the emergence of regulatory T cells (tT_REG) that can inhibit self-reactive cells. (b) Peripheral tolerance involves deleting, rendering anergic, or actively suppressing (via induction of pT_REG cells) escaped lymphocytes that possess receptors that react with self antigens. This process occurs primarily in secondary lymphoid organs.

producing a less autoreactive receptor and allowing the cell in question to avoid elimination. These central tolerance processes of negative selection and receptor editing work to eliminate many autoreactive lymphocytes in the thymus and bone marrow prior to their maturation. In addition, some self-reactive lymphocytes may be released from the primary lymphoid organs in an anergic state, and later deleted via apoptosis in the periphery.

On the more proactive side of the equation, some lymphocytes with high affinity for self antigens are instead *selected for survival* (*engaged*) during development. This aspect of central tolerance is most well studied for T cells in the thymus. These self-reactive cells selected for survival in the thymus express the FoxP3 transcription factor, a hallmark of this cell type. Their role, once they depart the thymus, is to suppress or regulate autoimmune responses to self antigens in the periphery; thus their name, **regulatory T cells (T$_{REG}$ cells)**. In the upcoming section on regulatory cells we discuss the mechanism of action of these and other immunosuppressive cells. Current nomenclature refers to these *t*hymic emigrants as **tT$_{REG}$ cells** although previous convention listed these as *n*atural or nT$_{REG}$ cells.

Interestingly, the TCRs of these regulatory cells display high affinity for self antigen, not unlike their eliminated counterparts mentioned above. What accounts for this difference in fate? Based on an accumulation of studies, mostly in mice, a combination of both positive and negative signaling events is likely involved. Intercellular interactions such as CD28 with CD80/86 or CD40 with CD40L, as well as the presence of certain cytokines, may favor one outcome over the other. In terms of TCR engagement, a "hit and run" model has been proposed, suggesting that short, high-affinity but transient engagement of the TCR with MHC-antigen in the thymic medulla favors the generation of regulatory cells (engagement). On the other hand, more sustained, high-affinity TCR engagement in the thymus seems to favor deletion of the self-reactive lymphocytes (via elimination).

Surviving tT$_{REG}$ cells migrate out of the thymus and are capable of suppressing reactions to self antigens in the periphery, which we discuss shortly. CD8$^+$ tT$_{REG}$ cells are likely a very small population, compared with the number of CD4$^+$ tT$_{REG}$ cells generated in the thymus. In studies using TCR ovalbumin-specific transgenic mice, in which all the T cells are specific for this nonself antigen, no FoxP3-expressing CD8$^+$ tT$_{REG}$ cells were identified. In other experimental systems, only rare CD8$^+$ tT$_{REG}$ cells emigrating from the thymus with this natural regulatory phenotype could be detected. That said, in AIRE-deficient mice, CD8$^+$ tT$_{REG}$ cells were found to be crucial for inhibiting autoimmune colitis. While much less is known about this population than about CD4$^+$ tT$_{REG}$ cells, CD8$^+$ regulatory T cells are clearly also involved in suppressing anti-self and/or anti-commensal immune engagement.

Cells That Mediate Peripheral Tolerance Are Generated Outside Primary Lymphoid Organs

Despite this elaborate system of central tolerance, some potentially self-destructive lymphocytes can find their way out of PLOs or evolve during development of adaptive immunity. In fact, it is now clear that mature, naïve lymphocytes with specificity for self antigens are not uncommon in the periphery. Two factors contribute to this: (1) not all self antigens are expressed in the central lymphoid organs where negative selection occurs (even with wild-type *AIRE* expression), and (2) there is a threshold requirement for affinity to self antigens before clonal deletion is triggered, allowing some weakly self-reactive clones to survive the weeding-out process. Multiple additional safeguards limit or redirect the activity of these anti-self cells outside primary lymphoid organs. These processes are collectively referred to as **peripheral tolerance** and are believed to occur mainly in secondary lymphoid organs or at the tissue site where the relevant self antigen is expressed (Figure 16-1b).

Like central tolerance, the mechanisms mediating peripheral tolerance involve specific engagement with self antigen leading to immunosuppression, rather than activation. Typically, encounters between mature, naïve lymphocytes and antigen lead to stimulation of the immune response. However, presenting the antigen in certain contexts or in specific locations/microenvironments can *instead* lead to tolerance. Antigens that induce tolerance are called **tolerogens** rather than **immunogens**. Here, context is important; the same chemical compound can be both an immunogen and a tolerogen.

A T cell engaged by a tolerogen or in a tolerogenic setting has at least two possible fates besides apoptosis: **anergy** (unresponsiveness) or regulation (engagement leading to suppression). Naïve T cells that enter the regulatory pathway in the *p*eriphery can be induced to express FoxP3 and become **pT$_{REG}$ cells** (previously known as *i*nduced or iT$_{REG}$s), acting as antigen-specific inhibitors of activation. This could occur thanks to lack of costimulation, the presence of inhibitory cytokines or surface molecules, or in connection with the time and place of exposure. For instance, during fetal and early neonatal stages, when the immune system is still immature, exposure to antigen can lead to inhibitory responses. Likewise, when some antigens are introduced orally, tolerance can be the result, whereas the same antigen given as an intradermal or subcutaneous injection can be immunogenic. In other instances, mucosally administered antigens provide protective immunity, such as in the case of Sabin's oral polio vaccine. It's complicated! There is one universal principle for tolerogens: they are antigen specific. The inactivation of an immune response does not result in general immune suppression, but rather inhibition specific for the tolerogenic antigen.

Other than fetal exposure, factors that promote tolerance rather than stimulation of the immune system by a given antigen include the following:

- High doses of antigen
- Long-term persistence of antigen in the host
- Intravenous or oral introduction
- Absence of adjuvants (compounds that enhance the immune response to antigen)
- Low levels of costimulation
- Presentation of antigen by immature or unactivated antigen-presenting cells (APCs)

Finally, the possibility of damage from self-reactive lymphocytes is further limited by the need for coordination between multiple cell types. During a proinflammatory response, an activated APC must present antigen to a naïve $CD4^+$ T_H cell, which can coordinate with B cells and $CD8^+$ T cells specific for the same antigen, propagating the adaptive response. When one or more of these cells are missing or present in an inactivated state, tolerance may be the outcome. The level of interaction between immune cells means that it can be hard to distinguish the inhibitory chicken from the egg! It also means that one inhibitory link in the chain of immune cell interactions can help regulate or control the other members.

Key Concept:

- Peripheral tolerance processes occur after lymphocyte development, when immune cells are induced to act as inhibitors of self reactivity for antigen that is presented in a nonimmunogenic context.

Multiple Immune Cell Types Work in the Periphery to Inhibit Anti-Self Responses

Regulatory immune cells act in secondary lymphoid tissues and at sites of inflammation. A range of immune cell types can serve as regulatory cells. Some bear antigen-specific receptors (e.g., TCRs or BCRs) for self structures while some function in more of an accessory role (e.g., inhibitory pAPCs). What they all have in common is that they *down-regulate* immune processes when they engage in immune activity, and they are antigen specific. In fact, most of the circulating B and T cells with specificity for self antigens likely possess regulatory or immune-inhibitory function.

While regulatory T cells are the most well characterized, subsets of B cells, macrophages, and dendritic cells, among others, can also participate in regulation. In any event, full maintenance of peripheral tolerance is likely a matter of balance and teamwork, with more members of the team still yet to be identified. Here, we describe some of the best characterized of the immune regulatory cell types.

Regulatory T Cells

Regulatory T cells are currently the focus of much research attention, where subsets of both $CD4^+$ and $CD8^+$ lymphocytes have been found to possess immune-dampening capabilities. As we saw in Chapter 10, in order for T cells to become activated, the TCR must bind antigen presented by self *major histocompatibility complex* (MHC) molecules (signal 1), while at the same time the T cell must undergo costimulatory engagement (signal 2). Signal 1 without signal 2, or with different cytokines in the microenvironment, can lead to different outcomes. Early experiments by Marc Jenkins and colleagues showed that when $CD4^+$ T-cell clones are stimulated in vitro through the TCR alone, without costimulation, they become anergic. Subsequent data showed that the interaction between CD28 on the T cell and CD80/86 (B7) on the APC provided the costimulatory signal required for T-cell activation. This led to a careful examination of costimulation, revealing the existence of other molecules that could bind to CD80/86 and the discovery of a related molecule, called CTLA-4. This molecule *inhibits* rather than stimulates T-cell activation after binding CD80/86. We now appreciate that many such molecules deliver supplementary signals during T-cell engagement, and the group of molecules that regulate T-cell behavior are now often referred to as **immunomodulatory**, to cover both costimulatory and inhibitory behavior. Mice lacking CTLA-4 display massive proliferation of lymphocytes and widespread autoimmune disease, suggesting an essential role for this molecule in maintaining peripheral tolerance.

Phenotypically, regulatory T cells are diverse. Nonetheless, a few regulatory cell hallmarks have emerged from the study of $CD4^+$ T_{REG} cells (see Chapter 8). These include expression of the FoxP3 transcription factor and CTLA-4, plus high levels of the IL-2R α chain (CD25). As mentioned earlier, these cells can come from the thymus (tT_{REG} cells) or they can be generated in the periphery from naïve, mature $FoxP3^-$ precursors (pT_{REG} cells). The importance of FoxP3 expression, which appears to be both essential and sufficient for the induction of immunosuppressive function, can be seen in humans inheriting a mutated form of this X-linked gene, which causes a multiorgan autoimmune disease (see the section Autoimmunity, below, and also Chapter 18).

The regulatory $CD8^+$ T-cell population is less well characterized than the $CD4^+$ subset, fewer in number, and phenotypically more diverse. In addition to FoxP3 and CTLA-4, they often express $CD8^+\alpha\alpha$ (as opposed to the more common α and β chains), plus both the high- and low-affinity receptors for IL-2 (CD25 and CD122, respectively), dendritic cell markers (CD11c), as well as a number of other surface molecules. Some mouse CD8$^+$ T$_{REG}$ cells, like their CD4$^+$ counterparts, recognize antigen using conventional MHC molecules. However, the best-characterized CD8$^+$ T$_{REG}$ cells are restricted to the nonclassical MHC class I molecules: Qa-1 in mice and HLA-E in humans. These nonclassical MHC molecules play key roles in presenting lipid antigens (see Chapter 7). Studies using mice engineered to lack Qa-1-restricted CD8$^+$ T$_{REG}$ cells showed that these animals develop aggressive autoimmune reactions against self antigens, suggesting that this population is involved in regulating CD4$^+$ T-cell responses to self antigens to maintain peripheral tolerance. Likewise, adoptive transfer of a subset of CD8$^+$ T$_{REG}$ cells has been shown to induce tolerance to allogeneic heart transplants in recipient rats and protect mice from the autoimmune disease *experimental autoimmune encephalomyelitis* (EAE), a model of multiple sclerosis.

In studying the mechanisms by which T$_{REG}$ cells inhibit immune responses, both contact-dependent and contact-independent processes have been observed. T$_{REG}$ cells have been shown to kill APCs or effector T cells directly, by means of granzyme and perforin, as well as modulate the function of other cells responding to antigen via surface receptor engagement. One prime example of this is the inhibitory molecule CTLA-4, expressed at high levels in T$_{REG}$ cells. As shown in **Figure 16-2**, interaction of CTLA-4 on T$_{REG}$ cells with CD80/86 on an APC can lead to inhibition of APC function, including reduced expression of costimulatory molecules (e.g., CD28) and proinflammatory cytokines such as IL-6 and *tumor necrosis factor-α* (TNF-α). T$_{REG}$ cells themselves also secrete immune inhibitory cytokines, especially IL-10, but also TGF-β and IL-35, suppressing the activity of other nearby T cells and APCs. Finally, because T$_{REG}$ cells express high levels of CD25, the high-affinity IL-2 receptor, they can act as a sponge, absorbing this growth- and survival-promoting cytokine and further discouraging expansion of local immunostimulatory effector T cells.

Evidence that CD4$^+$ T$_{REG}$ cells can control the immune response to self antigens has now been demonstrated in many

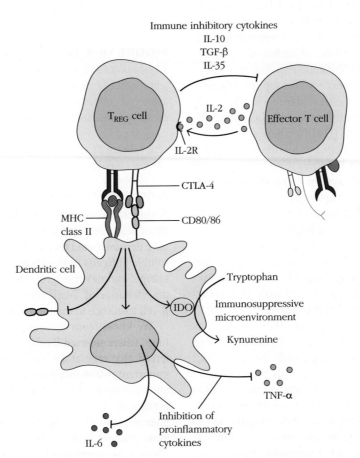

FIGURE 16-2 CTLA-4–mediated inhibition of APCs by T$_{REG}$ cells. One of the proposed mechanisms used by T$_{REG}$ cells to inhibit APCs involves signaling through the CD80/CD86 (B7) receptor. In the APC, this engagement results in decreased expression of CD80/86, activation of indoleamine 2,3-dioxygenase (IDO; an enzyme that converts tryptophan to kynurenine, creating an immunoinhibitory microenvironment), and changes in transcription leading to decreased expression of IL-6 and TNF-α. At the same time, the T$_{REG}$ cell absorbs local IL-2 via the CD25 molecule and makes its own inhibitory cytokines that act on nearby T cells.

experimental settings. In experiments in *nonobese diabetic* (NOD) mice and *BioBreeding* (BB) rats, two strains prone to the development of autoimmune-based diabetes, the onset of diabetes was delayed when these animals were injected with normal CD4$^+$ T cells from histocompatible donors. Further characterization of the CD4$^+$ T cells from non–disease-prone donor mice revealed that a subset expressing high levels of CD25 was responsible for the suppression of diabetes. This population was further characterized using transgenic reporter mice expressing *green fluorescent protein* (GFP) fused with the transcription factor FoxP3 (FoxP3-GFP mice). The GFP$^+$ T cells but not the GFP$^-$ T cells from these mice could be used to transfer the immune-suppressive activity, identifying this transcription factor as a major regulator controlling the development of these cells.

CD4$^+$ T$_{REG}$ cells have also been found to suppress responses to some *nonself* antigens. For example, these cells may control allergic responses against innocuous environmental substances and/or responses to the commensal microbes that make up the normal gut flora. In mice experimentally manipulated to lack CD4$^+$ T$_{REG}$ cells (approximately 5% to 10% of their peripheral CD4$^+$ T-cell population), inflammatory bowel disease is common. In strains of mice that are genetically resistant to the induction of the autoimmune disease EAE (a murine model of MS), depletion of this CD4$^+$ T-cell subset renders the mice susceptible to disease, suggesting that these regulatory cells play a role in suppressing autoimmunity. There is now significant evidence that pT$_{REG}$ cells and other regulators of immunity are present throughout the *gut-associated lymphoid tissue* (GALT). These cells are continuously exposed to gut microbes and food-borne antigens, which themselves may play a significant role in fine-tuning the immune response, with systemic repercussions (see **Clinical Focus Box 16-1** and Chapter 13).

It is important to note that the pathway of inhibition by regulatory T cells is believed to be highly antigen specific. T$_{REG}$ cells inhibit APCs presenting *their cognate antigen* or effector T cells that *share their same antigen specificity* and not T cells with a different specificity. However, again taking advantage of FoxP3-GFP mice bred with TCR transgenic animals, it was shown that it is possible for FoxP3$^+$ T cells to inhibit T cells recognizing *other* antigens, as occurs when both the T$_{REG}$ cell and the "bystander" T cell recognizing another antigen interact with the same APC. The result is inhibition of the APC, via both contact-dependent and -independent pathways, as well as inhibition of the bystander T cell through soluble inhibitory factors and decommissioning of the APC (**Figure 16-3**). This simultaneous processing and presentation of different antigens might happen naturally in vivo when the antigens in question are parts of the same pathogen, although based on the findings in these experimental systems this was not required. This phenomenon, termed *linked suppression*, has now been seen in multiple experimental systems and may represent another way that T$_{REG}$ cells support local

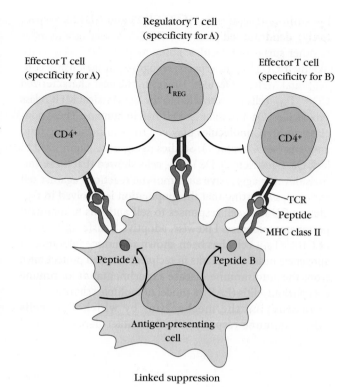

FIGURE 16-3 Linked suppression mediated by T$_{REG}$ cells. When a single antigen-presenting cell (APC) engages simultaneously with T cells of different specificity, inhibitory signals meant for one can be transmitted to both and lead to a "spreading" of immune suppression to include other antigens.

self tolerance in tissues lacking any pathogen-induced danger signals.

Regulatory B Cells and B-Cell Tolerance

Peripheral tolerance in B cells appears to follow a similar set of rules. For instance, experiments with transgenic mice have demonstrated that when mature B cells encounter most soluble antigens in the absence of T-cell help, they become anergic and never migrate to germinal centers. In this way, maintenance of T-cell tolerance to self antigens enforces B-cell tolerance to the same antigens.

We know from various experimental and clinical settings, where selected B cells are depleted or adoptively transferred, that regulatory B cells do exist and that they play an important role in maintaining tolerance. The surface markers observed on B cells with suppressing or regulatory function present no clear pattern. At present, there are no consensus phenotypic markers used to characterize these cells, although most produce high levels of the inhibitory cytokine IL-10, which is frequently used to define them (B cells that produce only this cytokine are called *B10 cells*). Because of this diversity of surface markers and likely phenotypes, the mechanisms of action of regulatory B cells are difficult to nail down.

Nonetheless, a number of experimental systems have highlighted an in vivo role for regulatory B cells, sometimes

BOX 16-1

It Takes Guts to Be Tolerant

In the past decade, a significant appreciation has developed for the role played by microflora, especially in the gastrointestinal tract, in regulating the immune system. Specifically, the commensal organisms that comprise the gut microbiota appear to work actively at inducing tolerance to themselves. Maybe even more important, the gut and the organisms that reside there seem to play key roles in maintaining a level of systemic homeostasis (immunologic balance) that allows for the development of tolerance to self and that sets the stage for effective identification and elimination of pathogens (see Chapter 13 and Clinical Focus Box 1-3).

Formerly thought to be "ignored" by immune cells, the host microbiota is now known to participate in a two-way communication with the immune system. This cross-talk results in advantages for both the host and the microbes. For the microbes, this interaction drives immune tolerance to the bugs, allowing them to continue to thrive in their home. For the host, there appear to be multiple advantages to immune health, depending somewhat on the composition of the microbe(s) in question. For instance, germ-free mice have been found to harbor defects in both humoral and adaptive immunity, and in some strains there is an increased susceptibility to the development of autoimmunity. In one recent study looking at the early stages of human rheumatoid arthritis, a paucity or absence of specific bacterial species were found in the intestinal flora of afflicted individuals. Collectively, these observations in humans and in animal models suggest that communication between commensal microbes and the host immune system may influence the induction or severity of some autoimmune diseases.

The antibody responses of germ-free mice to exogenous antigens are depressed, as are responses to conconavalin A (ConA), a strong T-cell mitogen. Germ-free animals were found to exhibit a T_H2 cytokine bias in response to antigenic challenge, which could be restored to balance by colonization with just *a single microbe,* specifically *Bacteroides fragilis.* Further, this restored immune balance was most strongly associated with microbial expression of one particular molecule: the polysaccharide A. This association clearly suggests that single microorganisms, and in some cases maybe even single molecules expressed by these microbes, can have systemic effects on the balance of immunity in the host.

This increased awareness of gut-associated immune regulation has spawned several interesting hypotheses related to tolerance. One, called the *microflora* or *altered microbiota hypothesis,* posits that changes in gut microflora due to dietary modifications and/or increased antibiotic use have disrupted normal microbially mediated pathways important for regulating immune tolerance. There is already a clear link between intestinal microbiota and health of the gut, including a role for certain microbes in regulating inflammatory syndromes of the bowel. Further evidence for this comes from transplantation studies, where individuals treated with immune ablation therapy and high-dose antibiotics show overcolonization with certain, sometimes pathogenic, microbes. These individuals frequently manifest correlating immune defects, which can be reversed by directly manipulating the gut microflora.

But how do our commensal microbes influence the immune balance? Some posit that intestinal epithelial cells (IECs) and mucosal dendritic cells (DCs), which express several key innate receptors, including Toll-like receptors and NOD-like receptors (TLRs and NLRs), may be involved. Mucosal DCs in the gut are known to sample the contents of the intestinal lumen. These cells could be another connection between commensal microbes and the maintenance of tolerance. In a study by Ruslan Medzhitov and colleagues, mice engineered to lack the *MyD88* gene, which encodes an intracellular signaling molecule, were more susceptible to intestinal injury and autoimmunity. This suggests that signaling through these innate receptors can in some instances induce tolerance rather than an inflammatory response. This could be explained by the delivery of tolerogenic and homeostatic signals by as yet undefined antigens carried by the gut microbiota. Experiments to identify the specific ligands and receptors important for this cross-talk between microflora and the host immune system are ongoing. In some of these studies, gut microflora is transferred between animals as a means to adjust immune status. Likewise, in humans, fecal transplants have become one of many new tools for clinicians who hope to rebalance immune activity in their patients.

In case you were thinking of taking this role of the microflora in immune balance with a grain of salt, you might want to think again. Sodium chloride may be a new addition to the list of dietary contributors to immune pathway development. In vitro studies have shown that the addition of sodium chloride to cultures of $CD4^+$ T cells can drive the development of T_H17 cells. Expansion of this cell type has been linked to the development of certain autoimmune diseases.

It now appears that the saying "We are what we eat" extends to the immune system too, which has a discriminating palate of its own.

REFERENCES

Rakoff-Nahoum, S., J. Paglino, F. Eslami-Varzaneh, S. Edberg, and R. Medzhitov. 2004. Recognition of commensal microflora by Toll-like receptors is required for intestinal homeostasis. *Cell* **118:**229.

Round, J., R. O'Connell, and S. Masmanian. 2010. Coordination of tolerogenic immune responses by the commensal microbiota. *Journal of Autoimmunity* **34:**220.

referred to as B_{REG} *cells*. For example, B_{REG} cells have been shown to suppress inflammatory cascades associated with IL-1. In mouse models of autoimmune diseases such as multiple sclerosis and rheumatoid arthritis, animals engineered with B cells *incapable* of secreting IL-10 showed chronic T_H1-cell activation and a worsening of the disease. While B cells are certainly not the only cells that can make IL-10, it is clear that B_{REG} cells are important inhibitors of adaptive immunity.

Accessory Cells with Regulatory Function

The hypothesis that regulatory T cells work primarily by "decommissioning" pAPCs has received much attention of late. For example, T_{REG} cells can signal APCs, through surface molecules and cytokine secretion, to reduce their costimulatory potential. A reduction in CD80/86 expression on the APC may be even more efficient in favoring immunosuppression than destruction of APCs; it not only bars that APC from stimulating T cells but also encourages further production of regulatory cells when naïve, mature T cells interact with the APC (see Figure 16-2). In fact, a reduction in the expression of various costimulatory molecules on APCs, but not MHC class I or class II, has been seen in several experimental systems designed to study T_{REG} cells.

Another regulatory immune cell worth mention is the *myeloid-derived suppressor cell*, or MDSC. These cells were first identified in vivo as immune suppressor cells found in pro-tumor microenvironments associated with poor prognosis (see Chapter 19). More recently, MDSCs have gained recognition in situations of chronic inflammation, including some autoimmune diseases. MDSCs are a heterogeneous group of immature myeloid cells that can accumulate at sites of infection or immune activity, suppressing local antigen-specific T-cell responses. They have been reported to secrete inhibitory compounds, such as IL-10, indoleamine 2,3-dioxygenase (IDO), arginase-1, and inducible nitric oxide synthase (iNOS), and also to express immunosuppressive surface markers that negatively regulate T-cell proliferation, such as PD-L1.

Key Concepts:

- T_{REG} cells, which inhibit immune reactivity against their cognate antigen in the periphery, can be CD4$^+$ or CD8$^+$, and typically express CD25 (IL-2R α chain), CTLA-4 (coinhibitory receptor), plus the master regulator transcription factor FoxP3.

- Regulatory T cells dampen immune responses by inhibiting, decommissioning, or killing other immune cells that respond to their cognate antigen, including T cells, B cells, and pAPCs. They do this through immune-suppressing cytokines (IL-10, TGF-β, IL-35), expression of inhibitory surface molecules (like CTLA-4), absorbing local IL-2 (with CD25), and via cellular cytotoxicity.

- Regulatory populations of B cells (B_{REG} cells) and macrophages (MDSCs) can also suppress inflammation, often by secreting compounds like IL-10 and by acting as immune-inhibiting APCs.

Autoimmunity

Simply stated, autoimmune disease is caused by failure of the tolerance processes described in the previous section. According to the NIH, up to 8% of the population is affected by one of more than 80 different autoimmune diseases. In certain cases the damage to self cells or organs is caused by antibodies; in other cases, T cells, or both T cells and antibodies, are the culprit. Often chronic and debilitating, these diseases can lead to morbidity and mortality from complications, including prolonged medical intervention.

Autoimmune diseases result from the destruction of self proteins, cells, and organs by auto-antibodies or self-reactive T cells. In 1957, Deborah Doniach (**Figure 16-4**) and

FIGURE 16-4 Deborah Doniach. Doniach and colleagues were the first to show that antibodies against normal components of the thyroid were to blame for Hashimoto's thyroiditis. *[Photo courtesy of Tabitha Doniach.]*

TABLE 16-1   Some autoimmune diseases in humans

Disease	Self antigen/Target gene	Immune effector
ORGAN-SPECIFIC AUTOIMMUNE DISEASES		
Addison's disease	Adrenal cells	Auto-antibodies
Autoimmune hemolytic anemia	RBC membrane proteins	Auto-antibodies
Goodpasture's syndrome	Renal and lung basement membranes	Auto-antibodies
Graves' disease	Thyroid-stimulating hormone receptor	Auto-antibodies (stimulating)
Hashimoto's thyroiditis	Thyroid proteins and cells	T_H1 cells, auto-antibodies
Idiopathic thrombocytopenic purpura	Platelet membrane proteins	Auto-antibodies
Type 1 diabetes mellitus	Pancreatic beta cells	T_H1 cells, auto-antibodies
Myasthenia gravis	Acetylcholine receptors	Auto-antibodies (blocking)
Myocardial infarction	Heart	Auto-antibodies
Pernicious anemia	Gastric parietal cells; intrinsic factor	Auto-antibodies
Poststreptococcal glomerulonephritis	Kidneys	Immune complexes
Spontaneous infertility	Sperm	Auto-antibodies
SYSTEMIC AUTOIMMUNE DISEASES		
Ankylosing spondylitis	Vertebrae	Immune complexes
Multiple sclerosis	Brain or white matter	T_H1 cells and T_C cells, auto-antibodies
Rheumatoid arthritis	Connective tissue, IgG	Auto-antibodies, immune complexes
Scleroderma	Nuclei, heart, lungs, gastrointestinal tract, kidneys	Auto-antibodies
Sjögren's syndrome	Salivary glands, liver, kidneys, thyroid	Auto-antibodies
Systemic lupus erythematosus (SLE)	DNA, nuclear protein, RBC and platelet membranes	Auto-antibodies, immune complexes
Immune dysregulation, polyendocrinopathy, enteropathy, X-linked (IPEX) syndrome	Multiorgan/loss of FoxP3 gene	Missing regulatory T cells
Autoimmune polyendocrine syndrome type 1 (APS-1)	Multiorgan/loss of AIRE gene	Defective central tolerance

colleagues working in London were the first to theorize, and later show, that serum antibodies from patients with Hashimoto's thyroiditis reacted against normal thyroid components. This was the first organ-specific autoimmune disease to be characterized, breaking the spell surrounding the controversy over whether self reactivity was even possible. Doniach and co-workers were also responsible for identifying an autoimmune factor that was involved in juvenile diabetes, also known as type 1 diabetes (T1D), another organ-specific autoimmune disease. Rheumatoid arthritis (RA), multiple sclerosis (MS), and systemic lupus erythematosus (SLE, or lupus) are other examples of all too common autoimmune diseases. **Table 16-1** lists several of the more prevalent autoimmune disorders, as well as their primary immune mediators.

Autoimmune diseases are often categorized as either organ-specific or systemic, depending on whether they affect a single organ or multiple systems in the body. Another method of grouping involves the immune component that does the bulk of the damage: T cells versus antibodies. In this section, we describe several examples of both organ-specific and systemic autoimmune disease. In each case, we discuss the antigenic target (when known), the causative process (either cellular or humoral), and the resulting symptoms. When available, examples of animal models used to study these disorders are also considered (**Table 16-2**). Finally, we touch on the factors believed to be involved in induction or control of autoimmunity, and treatments for these conditions.

TABLE 16-2	Experimental animal models of autoimmune diseases		
Animal model	Possible human disease counterpart	Inducing antigen	Disease transferred by T cells
SPONTANEOUS AUTOIMMUNE DISEASES			
Nonobese diabetic (NOD) mouse	*Type 1 diabetes* (T1D)	Unknown	Yes
(NZB × NZW) F₁ mouse	*Systemic lupus erythematosus* (SLE)	Unknown	Yes
Obese-strain chicken	Hashimoto's thyroiditis	Thyroglobulin	Yes
EXPERIMENTALLY INDUCED AUTOIMMUNE DISEASES*			
Experimental autoimmune myasthenia gravis (EAMG)	Myasthenia gravis	Acetylcholine receptor	Yes
Experimental autoimmune encephalomyelitis (EAE)	*Multiple sclerosis* (MS)	*Myelin basic protein* (MBP); *proteolipid protein* (PLP)	Yes
Autoimmune arthritis (AA)	*Rheumatoid arthritis* (RA)	*Mycobacterium tuberculosis* (proteoglycans)	Yes
Experimental autoimmune thyroiditis (EAT)	Hashimoto's thyroiditis	Thyroglobulin	Yes

*These diseases can be induced by injecting appropriate animals with the indicated antigen in complete Freund's adjuvant. Except for autoimmune arthritis, the antigens used correspond to the self antigens associated with the human disease counterpart. Rheumatoid arthritis involves reaction to proteoglycans, which are self antigens associated with connective tissue.

Some Autoimmune Diseases Target Specific Organs

Autoimmune diseases are caused by immune-stimulatory lymphocytes or antibodies that recognize self components, resulting in cellular lysis and/or an inflammatory response in the affected organ. Gradually, the damaged cellular structure is replaced by connective tissue (fibrosis), and the function of the organ declines. In an organ-specific autoimmune disease, the immune response is usually directed to a target antigen unique to a single organ or gland, so the manifestations are largely limited to that organ. The cells of the target organs may be damaged directly by humoral or cell-mediated effector mechanisms. Alternatively, anti-self antibodies may overstimulate or block the normal function of the target organ.

Hashimoto's Thyroiditis

In Hashimoto's thyroiditis, an individual produces autoantibodies and sensitized T_H1 cells that are specific for thyroid antigens. This disease is much more common in women, often striking in middle age (see **Clinical Focus Box 16-2** for a discussion of sex differences in autoimmune disease). Antibodies are formed against a number of thyroid proteins, including thyroglobulin and thyroid peroxidase. Binding of the auto-antibodies to these proteins interferes with iodine uptake, leading to decreased thyroid function and hypothyroidism (decreased production of thyroid hormones). The

resulting *delayed-type hypersensitivity* (DTH) response is characterized by an intense infiltration of the thyroid gland by lymphocytes, macrophages, and plasma cells, which form lymphocytic follicles and germinal centers (see Chapter 15). These collections of leukocytes can sometimes coalesce into spontaneous lymph node–like assemblies, called *tertiary lymphoid organs*. The ensuing inflammatory response causes a goiter, or visible enlargement of the thyroid gland, a physiological response to local inflammation caused by antibodies against thyroid-specific proteins. This immune attack leads to decreased function of the gland and symptoms such as fatigue, lethargy, and unexplained weight gain. Replacement therapy involving daily administration of thyroxine, the hormone secreted by the thyroid gland, usually yields good results and allows people to live a normal life. Again, this autoimmune disease is more common in women and the cause is largely unexplained.

Type 1 Diabetes

Type 1 diabetes (T1D), also known as **insulin-dependent diabetes mellitus (IDDM)**, affects almost 2 in 1000 children in the United States; roughly double the incidence observed just 20 years ago. It is seen mostly in youth under the age of 14 and is less common than type 2, or non–insulin-dependent, diabetes mellitus. T1D is caused by an autoimmune attack against insulin-producing cells (beta cells) scattered throughout the pancreas, which results in decreased production of insulin and consequently increased

(a)

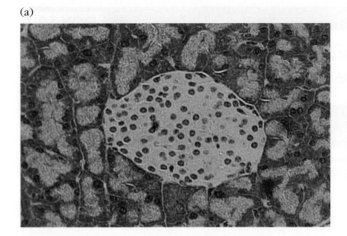

(b)

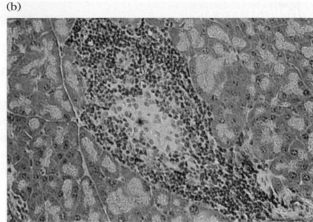

FIGURE 16-5 Insulitis in Type 1 diabetes. Photomicrographs of an islet of Langerhans in (a) the pancreas from a normal mouse and (b) the pancreas from a mouse with a disease resembling insulin-dependent diabetes mellitus. Note the lymphocyte infiltration into the islet (insulitis) in (b). *[Courtesy Noel Maclaren MD, Weill Cornell Medicine, and William Riley, University of Texas, Medical Branch at Galveston.]*

levels of blood glucose. The disease begins with *cytotoxic T lymphocyte* (CTL) infiltration and activation of macrophages, frequently referred to as *insulitis* (**Figure 16-5**), which leads to a cell-mediated DTH response, with resulting cytokine release and the production of auto-antibodies. The subsequent beta-cell destruction is thought to be mediated by cytokines released during the DTH response and by lytic enzymes released from the activated macrophages. Auto-antibodies specific for beta cells may contribute to cell destruction by facilitating either antibody-mediated complement lysis or *antibody-dependent cell-mediated cytotoxicity* (ADCC) (see Chapter 12).

The abnormalities in glucose metabolism associated with T1D result in serious metabolic problems that include ketoacidosis (accumulation of ketone, a breakdown product from fat) and increased urine production. The late stages of the disease are often characterized by atherosclerotic vascular lesions (which cause gangrene of the extremities due to impeded vascular flow), renal failure, and blindness. If untreated, death can result. The most common therapy for T1D is daily administration of insulin. Although this is helpful, sporadic doses are not the same as metabolically regulated, continuous, and controlled release of the hormone. Unfortunately, T1D can remain undetected for many years, allowing irreparable loss of pancreatic tissue to occur before treatment begins. Recently, novel transplantation approaches studied in mice have yielded exciting advances in the control of glycemia. If this can be translated to the clinic it could hold promise as a cure for this disease.

One of the best-studied animal models of this disease is the *nonobese diabetic* (NOD) mouse, which spontaneously develops a form of diabetes that resembles human T1D. This disorder also involves lymphocytic infiltration of the pancreas and destruction of beta cells, and is strongly associated with certain MHC alleles. Disease is mediated by bone marrow–derived cells; normal mice reconstituted with an injection of bone marrow cells from NOD mice will develop diabetes, and healthy NOD mice that have not yet developed disease can be spared by reconstitution with bone marrow cells from MHC-matched normal mice. NOD mice housed in germ-free environments show a higher incidence of diabetes compared with those in regular housing, suggesting that a diverse flora (most likely intestinal) may help block development of autoimmune disease. There is also evidence in humans that perturbations of gut flora are associated with T1D. In genome-wide scans, over 20 *insulin-dependent diabetes* (*Idd*) loci associated with disease susceptibility have been identified, including at least one member of the TNF receptor family.

Myasthenia Gravis

Myasthenia gravis is the classic example of an autoimmune disease mediated by blocking antibodies. A patient with this disease produces auto-antibodies that bind the *acetylcholine receptors* (AChRs) on the motor end plates of muscles, blocking the normal binding of acetylcholine and inducing complement-mediated lysis of the cells. The result is a progressive weakening of the skeletal muscles (**Figure 16-6**). Ultimately, the antibodies cause the destruction of the cells bearing ACh receptors. The early signs of this disease include drooping eyelids and inability to retract the corners of the mouth. Without treatment, progressive weakening of the muscles can lead to severe impairment in eating as well as problems with movement. However, with appropriate treatment,

BOX 16-2

Why Are Women More Susceptible Than Men to Autoimmunity? Sex Differences in Autoimmune Disease

Of the nearly 50 million individuals in the United States believed to be living with autoimmune disease, approximately 80% are women. As shown in **Table 1**, female-biased predisposition to autoimmunity is more apparent in some diseases than others. For example, the female-to-male ratio of individuals who suffer from diseases such as *multiple sclerosis* (MS) or *rheumatoid arthritis* (RA) is approximately two or three to one, while there are up to 19 women for every one man afflicted with Hashimoto's thyroiditis. That women are more susceptible to autoimmune disease has been recognized for many years. The reasons are not entirely understood, although recent advances are helping scientists to develop testable hypotheses concerning these sex differences.

Although it may seem unlikely, considerable evidence suggests significant widespread sex-based differences in immune responses. In general, females mount more vigorous innate and adaptive responses, both humoral and cell-mediated, clearing infectious disease faster than their male counterparts. Immune cell activation, cytokine secretion after infection, numbers of circulating CD4$^+$ T cells, and mitogenic responses are all higher in women than men. Immunization studies conducted in mice and humans show that females produce a higher titer of antibodies than do males; this is true during both primary and secondary responses. In transplantation, women also suffer from a higher rate of graft rejection. As one might guess, this enhanced immunity in females means that males, in general, are slightly more prone to infections and more likely to die of cancer or infectious disease. It also means that they are less likely to be diagnosed with autoimmune disorders.

The prevailing view is that sex hormones also account for at least part of this observed difference in the rates of autoimmunity between men and women. The evidence for this comes from observations made in SLE, where young women of child-bearing age are at greatest risk for the disease. Lupus flares during pregnancy (a high estrogen state) and increased rates of remission following menopause (a low estrogen state) also point to sex hormones as potential regulators of this autoimmune disease. The general consensus is that estrogens, the more female-specific hormones, are associated with enhanced immunity whereas androgens, or male-based hormones, are associated with its suppression.

In mice, where sex differences are easier to study, a large body of literature documents sexually dimorphic immune responses. Female mice are much more likely than male mice to develop T_H1 responses and, in infections for which proinflammatory T_H1 responses are beneficial, are more likely to be resistant to the infection. An excellent example is infection with viruses such as *vesicular stomatitis virus* (VSV), *herpes simplex virus* (HSV), and *Theiler's murine encephalomyelitis virus* (TMEV). Clearance of these viruses is enhanced by T_H1 responses. In other cases, however, a proinflammatory response can be deleterious. For example, a T_H1 response to *lymphocytic choriomeningitis virus* (LCMV) correlates with more severe disease and significant pathology. Thus, female mice are more likely to succumb to infection with LCMV. The importance of sex in LCMV infection is underscored by experiments demonstrating that castrated male mice behave immunologically like females and are more likely to experience autoimmune disease than are uncastrated males. Also in mice, recent work has shown that sex hormones may affect gut microflora, and that these microbes have a strong influence on systemic immunity. Nonobese diabetic (NOD) mice show a significant female bias toward the development of spontaneous T1D. However, germ-free NOD mice show no such sexual dimorphism in the development of diabetes, and transfer of male-specific microbiota to the guts of female NOD mice protected them from diabetes.

What maintains this dichotomy between the sexes? One hypothesis posits that this increased risk of autoimmunity in women is a by-product of the evolutionary role of women as bearers of children. Pregnancy may give us a clue as to how sex plays a role in regulating immune response. During pregnancy, it is

TABLE 1	Sex prevalence ratios for selected autoimmune diseases	
Autoimmune disease		**Ratio (female/male)**
Hashimoto's thyroiditis/hypothyroidism		19:1
Sjögren's syndrome		16:1
Systemic sclerosis		12:1
Systemic lupus erythematosus		9:1
Graves' disease/hyperthyroidism		7:1
Multiple sclerosis		3:1
Rheumatoid arthritis		3:1
Myasthenia gravis		2:1
Type 1 diabetes mellitus		1:1.2
Ankylosing spondylitis		1:3

Data from Brunelleschi (2016).

(continued)

critical that the mother tolerate the fetus, which is a type of foreign semi-allograft. This makes it very likely, maybe even crucial for successful implantation and fetal development, that the female immune system undergo important modifications during pregnancy. Recall that females normally tend to mount more T_H1-like responses than T_H2 responses. During pregnancy, however, women mount more T_H2-like responses. It is thought that pregnancy-associated levels of sex steroids may promote an anti-inflammatory environment. In this regard, it is notable that diseases enhanced by T_H2-like responses, such as SLE, which has a strong antibody-mediated component, can be exacerbated during pregnancy, whereas diseases that involve inflammatory responses, such as RA and MS, sometimes are ameliorated in pregnant women.

Another effect of pregnancy is the exchange of cells between mother and fetus, allowing for a state of bidirectional *microchimerism* that can last for decades or more. There is interesting but still limited evidence that long-lived allogeneic fetal cells in the mother and maternal cells in offspring may play a role in immune regulatory balance and thus the development of autoimmune disease. For instance, autoimmune disease is more common in women after child-bearing years. In fact, post-partum, women with MS, Hashimoto's thyroiditis, and Graves' disease all show higher rates of fetal microchimerism. These cells can be found in peripheral blood as well as multiple organs, where they have even been shown to contribute to wound healing, tissue regeneration, and inflammation. While a conclusive role and

mechanism for fetal cell participation in maternal autoimmune disease remains uncertain, these data do suggest that in addition to the many gifts of pregnancy, risks to womens' health may linger for many years. Of course, this has to be balanced with an advantage against infectious disease; maybe "man flu" is for real?

REFERENCES

Brunelleschi, S. 2016. Immune response and auto-immune diseases: gender does matter and makes the difference. *The Italian Journal of Gender-Specific Medicine* **2(1):**5.

Rubtsova, K., P. Marrack, and A. Rubtsov. 2015. Sexual dimorphism in autoimmunity. *Journal of Clinical Investigation* **125:**2187.

Jeanty, C., S. C. Derderian, and T. C. MacKenzie. 2014. Maternal-fetal cellular trafficking: clinical implications and consequences. *Current Opinion in Pediatrics* **26(3):**377.

this disease can be managed quite well and afflicted individuals can lead a normal life. Treatments are aimed at increasing acetylcholine levels (e.g., using cholinesterase inhibitors), decreasing antibody production (using corticosteroids or other immunosuppressants), and/or removing antibodies (via plasmapheresis: the removal and exchange of blood plasma).

One of the first autoimmune disease animal models was discovered serendipitously in 1973, when rabbits immunized with AChRs purified from electric eels suddenly became paralyzed. (The original aim was to generate monoclonal antibodies specific for eel AChRs that could be used for research.) These rabbits developed antibodies against the foreign AChR that cross-reacted with their

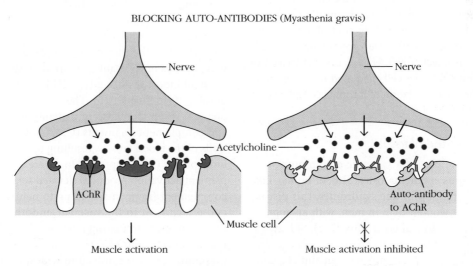

BLOCKING AUTO-ANTIBODIES (Myasthenia gravis)

FIGURE 16-6 Mechanism of myasthenia gravis induction. In myasthenia gravis, binding of auto-antibodies to the acetylcholine receptor (AChR) *(right)* blocks the normal binding of acetylcholine (burgundy dots) and subsequent muscle activation. In addition, the anti-AChR auto-antibody activates complement, which damages the motor end plate of the muscle; the number of acetylcholine receptors declines as the disease progresses.

own AChRs. These auto-antibodies then blocked muscle stimulation by acetylcholine at the synapse and led to progressive muscle weakness. Within a year, study of this animal model, called *e*xperimental *a*utoimmune *m*yasthenia *g*ravis (EAMG), led to the discovery that auto-antibodies to the AChR were also the cause of myasthenia gravis in humans.

Key Concepts:

- Hashimoto's thyroiditis is an example of an organ-specific autoimmune disease where auto-antibodies and T cells recognize thyroid-specific proteins, leading to organ destruction, hypothyroidism, and inflammatory accumulation.

- Type 1 diabetes is an organ-specific autoimmune disease that usually begins early in life and is caused by T cells recognizing components of the beta cells of the pancreas, leading to beta cell destruction, loss of insulin production, and the life-long need for treatment with insulin.

- Myasthenia gravis is a tissue-specific autoimmune disease caused by antibodies that recognize the acetylcholine receptors at the motor end plates of muscles, blocking the binding of this neurotransmitter and resulting in progressive muscle weakness.

Some Autoimmune Diseases Are Systemic

With systemic autoimmune diseases, autoreactive cells recognize a target antigen or antigens found in multiple tissues or organs. This leads to inflammation and physiologic disruptions at multiple, sometimes unconnected, locations in the body. Alternatively, systemic autoimmune disease can be caused by a disruption in regulation or control of tolerance. These diseases arise because the ever-present anti-self cells are no longer held in check. In either case, tissue damage can be widespread and driven by cell-mediated immune activity, auto-antibodies, accumulation of immune complexes, or combinations of these.

Systemic Lupus Erythematosus

One of the best examples of a systemic autoimmune disease is **systemic *l*upus *e*rythematosus (SLE,** or **lupus)**. Like several of the other autoimmune syndromes, this disease is more common in women than in men, with an approximately 9:1 ratio (see Clinical Focus Box 16-2). SLE symptoms typically appear between 20 and 40 years of age, and occur more frequently in African American and Hispanic women than in Whites, for unknown reasons. In identical twins where one suffers from SLE, the other has up to a 60% chance of developing SLE, suggesting a genetic component.

FIGURE 16-7 Systemic lupus erythematosus (SLE or lupus). Representational drawing of a man with characteristic lupus "butterfly" rash over the cheeks and nose. *[Wellcome Images/ Science Source]*

However, although close relatives of a patient with SLE are 25 times more likely to contract the disease, still only 2% of these individuals ever develop SLE.

Affected individuals may produce auto-antibodies to a vast array of cells or common cellular components, such as DNA and histones, as well as clotting factors, RBCs, platelets, and even leukocytes. Signs and symptoms include fever, weakness, arthritis, kidney dysfunction, and frequently skin rashes, especially the characteristic butterfly rash across the nose and cheeks (**Figure 16-7**). Auto-antibodies specific for RBCs and platelets can lead to complement-mediated lysis, resulting in hemolytic anemia and thrombocytopenia, respectively. When immune complexes of auto-antibodies with various nuclear antigens are deposited along the walls of small blood vessels, a type III hypersensitivity reaction develops (see Chapter 15). The complexes activate the complement system and generate membrane-attack complexes and complement fragments (C3a and C5a) that damage the walls of the blood vessels, resulting in vasculitis and glomerulonephritis (see Chapter 5). In severe cases, excessive complement activation produces elevated serum levels of certain complement fragments, leading to neutrophil aggregation and attachment to the vascular endothelium. Over time, the number of circulating neutrophils declines (neutropenia) and occlusions of various small blood vessels develop (vasculitis), which can lead to widespread tissue damage. Laboratory diagnosis of SLE involves detection of antinuclear antibodies (a common feature), which can be directed against DNA, nucleoprotein, histones, or nucleolar RNA.

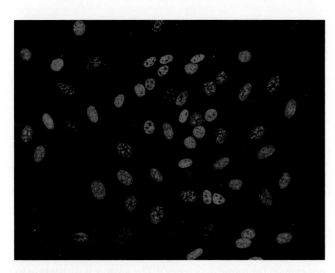

FIGURE 16-8 Diagnostic test for SLE. Serum dilutions from a patient are mixed with cells attached to a glass slide. Fluorescently labeled secondary antibodies directed against human antibodies are then added and reveal staining of the nucleus, and thus presence of anti-nuclear antibodies, under a fluorescence microscope. *[Courtesy ORGENTEC Diagnostika GmbH.]*

Indirect immunofluorescence staining, using serum from patients with SLE, produces a characteristic nuclear-staining pattern (**Figure 16-8**) and yields a presumptive positive diagnosis.

The study of animal models of SLE has yielded much insight into this disease, including the role of specific genes in creating a balance of regulatory-to-effector lymphocytes. While there are several excellent models, the New Zealand mouse remains the oldest and most well-characterized spontaneous model system. F_1 hybrids of *New Zealand Black* (NZB) and *New Zealand White* (NZW) mice spontaneously develop a severe autoimmune syndrome that closely resembles human SLE, although each of the parent strains displays only mild or variable autoimmune symptoms. NZB/W F_1 mice develop autoimmune hemolytic anemia between 2 and 4 months of age, at which time various auto-antibodies can be detected, including antibodies to erythrocytes, nuclear proteins, DNA, and T lymphocytes. As in human SLE, the incidence of autoimmunity in F_1 mice is greater in females than in males.

Multiple Sclerosis

Multiple sclerosis (MS) is the most common cause of neurologic disability associated with disease in Western countries. MS occurs in women approximately 3 times more frequently than in men (see Clinical Focus Box 16-2). This difference between the sexes represents an increase from two decades ago, when the ratio was more like 2:1, suggesting that recent increases in the incidence of MS have been primarily in women. Like SLE, MS frequently develops in young to middle adulthood, from 20 to 40 years of age, although the incidence in younger women is also on the rise.

Individuals with this disease produce autoreactive CD4$^+$ T cells, with T_H17 cells and the IL-17 they secrete as a hallmark. These cells recruit other cells to the site, encouraging inflammatory foci along the myelin sheath of nerve fibers in the brain and spinal cord. Since myelin functions to insulate the nerve fibers, a breakdown in the myelin sheath leads to numerous progressive neurologic dysfunctions, ranging from numbness in the limbs to paralysis or loss of vision. The most frequent form of the disease manifests as "relapsing and remitting"—flare-ups, interspersed with periods of partial recovery—although a chronic, progressive form of the disease is also seen.

MS has both genetic and environmental associations, although the cause of the disease is not well understood. In addition to alleles at the *DRB1* locus of MHC class II, which have long been recognized for their association, genome-wide studies in MS-affected families have identified many other potential loci connected to this autoimmune disease, some with immune function. Genetic influence is strong but not absolute; while the average person in the United States has about a 1 in 1000 chance of developing MS, this increases to 1 in 20 for siblings and to 1 in 4 for an identical twin. Epidemiological studies indicate that MS is most common in the Northern Hemisphere, especially in the United States. Relocation from low- to high-incidence regions during early years imparts higher risk of disease. These data suggest that, in addition to strong genetic affects, an environmental component early in life has profound impact on the risk of contracting MS. Infection with certain viruses, especially *Epstein-Barr virus* (EBV), has long been cited as a possible explanation for this geographic gradient, although recent data suggest more region-specific or lifestyle factors, some related to the hygiene hypothesis (see Chapters 1 and 15). The environmental factors that may impact MS susceptibility include diet, smoking, obesity or BMI, and exposure to sunlight. The latter is likely associated with vitamin D levels and could explain the increasing south-to-north gradient of susceptibility. Interestingly, vitamin D is a known immune modulator that can promote anti-inflammatory responses. Many leukocytes, including T cells, have vitamin D receptors.

Experimental autoimmune encephalomyelitis (EAE) is one of the best-studied animal models of autoimmune disease, and a rodent model for MS. This disease is mediated solely by T cells and can be induced in a variety of species by immunization with *myelin basic protein* (MBP) or *proteolipid protein* (PLP)—both components of the myelin sheaths surrounding neurons in the CNS. Within 2 to 3 weeks, the animals develop cellular infiltration of the CNS, resulting in demyelination and paralysis. Most of the animals will become paralyzed and can die. Others have milder symptoms, some with a chronic form of the disease that resembles relapsing and remitting MS in humans. Recent work in this model has highlighted the increasingly appreciated role of diet and the microbiome in disease susceptibility. Germ-free mice are less susceptible to induction of EAE, and addition of particular gut commensals has been shown to either exacerbate or attenuate disease. Studies in patients with MS have likewise noted disruption in the gut flora associated with flare-ups. This small-animal model of the disease has provided an important system for testing causal hypotheses as well as treatments for MS.

Rheumatoid Arthritis

Rheumatoid arthritis (RA) is a fairly common autoimmune disorder, most often diagnosed between the ages of 40 and 60 and, again, more frequently seen in women. This autoimmune disease also has a strong genetic susceptibility component and, as with SLE, the HLA-DRB1 locus is implicated, along with many non-MHC genes. The major symptom is chronic inflammation of the joints (**Figure 16-9**), although the hematologic, cardiovascular, and respiratory systems are also frequently affected. Most individuals with RA produce antibodies that react with citrullinated protein antigens (where an arginine residue is converted to the nonstandard amino acid citrulline), as well as a group of auto-antibodies called **rheumatoid factors (RFs)**. The latter are specific for the Fc region of IgG—in other words, antibodies against antibodies! The classic rheumatoid factor is an IgM antibody, although any isotype can participate. When RFs bind to normal circulating IgG, immune complexes form and are deposited in the joints. These can activate the complement cascade, resulting in a type III hypersensitivity reaction and chronic inflammation of the joints. Like SLE and MS, this systemic autoimmune disease has recently been connected with body flora. In particular, RA is associated with gum disease and the bacteria that cause gingivitis, as well as with smoking. Some believe that these environmental triggers may influence the level of citrullination of proteins in the mucosa, triggering or exacerbating the production of these anti-self antibodies in individuals who are already susceptible to RA.

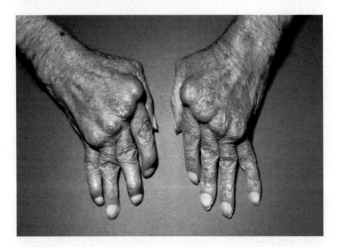

FIGURE 16-9 Rheumatoid arthritis. Swollen and painful joints are a common symptom of rheumatoid arthritis. [James Stevenson/Science Source.]

Systemic Autoimmunity Due to Disruptions in Immune Regulation

In the past two decades, a whole new category of disease with autoimmune association has been characterized, linked to genetic disruptions in the regulation of immune function. The two classic examples are **autoimmune polyendocrine syndrome type-1 (APS-1)** (previously called APECED) and **immune dysregulation, polyendocrinopathy, enteropathy, X-linked (IPEX) syndrome.** These two share many features, despite disruption to separate genes: both are monogenic disorders (caused by a single gene), impact multiple organs, and display the same range of immunopathologies (endocrine dysfunction, autoimmunity, and primary immune deficiency). How can disruption to a single gene result in both *too much* immune response (directed against self; autoimmunity) and *not enough* (immune deficiency)? The answer is disruption to homeostatic immune regulation, or a break in self tolerance. APS-1 is caused by mutations in the *AIRE* gene, critical for central tolerance in the thymus (and possibly elsewhere). The AIRE transcription factor ensures that tissue-specific antigens are expressed in the thymus during T-cell development. This ensures both the elimination of T cells with receptors that recognize tissue-specific antigens and engagement or selection of regulatory T (T_{REG}) cells. IPEX is caused by mutations in the *FoxP3* gene, the master transcriptional regulator associated with regulatory T cells, which are generated both during central (tT_{REG} cells) and peripheral tolerance (pT_{REG} cells). In other words, absence of regulatory T-cell function and a disruption to self tolerance are the common denominator.

Key Concepts:

- Systemic lupus erythematosus (lupus) is a whole-body autoimmune disease that results from the production of antibodies against common cellular components (especially nuclear proteins), leading to type III hypersensitivity with immune complexes blocking blood vessels, and ultimately progressive organ failure.

- Multiple sclerosis is a neurologic autoimmune disease caused by autoreactive T cells that recognize components of the CNS, leading to destruction of the myelin sheath surrounding nerves and progressive neurological dysfunction.

- Rheumatoid arthritis is a systemic autoimmune disease that usually develops later in life, caused by antibodies against citrullinated protein antigens and the Fc region of IgG (called *rheumatoid factor*), leading to immune complex deposition in joints and progressive joint destruction via complement.

- Defects in the *AIRE* and *FoxP3* genes can lead to systemic syndromes (APS-1 and IPEX, respectively) affecting a myriad of antigens, manifesting both autoimmune and immune deficiency–like symptoms, all due to disruption to T_{REG} cells.

Both Intrinsic and Extrinsic Factors Can Favor Susceptibility to Autoimmune Disease

Overzealous immune activation can lead to autoimmunity, while suboptimal immune stimulation results in insufficiency that can allow microbes to take control. But what tips the balance toward a break in tolerance and the development of autoimmunity? Experiments with germ-free mouse models, discordance data in identical twins, and epidemiologic studies of geographic associations all suggest roles for both the environment and genes in susceptibility to the development of autoimmunity. In the gene category, alleles in the HLA locus take top prize, as well as several repeat offenders in the immune response category. In terms of environment, lifestyle influences have increasingly entered the picture. Of late, the association of a range of autoimmune diseases with metabolic disruptions (linked to diet, exercise, obesity, and even stress) and an imbalance of microbial flora, especially early in life, are taking center stage.

The Role of Genes in Autoimmunity

Genetically, we are all very, very similar. However, a few small nucleotide differences at key locations can sometimes make a big difference. In the field of autoimmunity, that key location is the MHC locus, thanks to the primary role of its products in determining what fragments of antigens will be presented to T cells. For this reason, particular class I and II alleles or mutations are commonly associated with particular autoimmune diseases. However, we also know that these MHC variants are not the whole story, because two individuals can have exactly the same set of MHC alleles (monozygotic, or identical, twins) and still be discordant for the development of autoimmune disease. This means that while genetics may predispose us to autoimmune susceptibility, one or more factors in the environment must pull the trigger.

The strongest association between an HLA allele and autoimmunity is seen in ankylosing spondylitis (AS). AS is an inflammatory disease of vertebral joints, which is associated with expression of the allele *HLA-B27*. Expression of this MHC class I allele increases the risk of developing AS by 90-fold. This does not imply causation; most individuals who express *HLA-B27* will not develop AS, suggesting additional susceptibility factors and/or triggers. Research conducted in AS-affected families has unearthed additional associations with genes that encode three cytokines connected to T_H17 function (IL-12, IL-17, and IL-23), suggesting a potential role for this cell subset in the development of AS. Interestingly, unlike many other autoimmune diseases, AS is more common in men.

It is no surprise that other non-MHC immune genes are also associated with autoimmune disease. As noted previously, inactivating mutations in two genes involved in establishing and maintaining tolerance, *AIRE* and *FoxP3*, have strong associations with systemic autoimmune disease (see also Chapters 8 and 18). Many other genes with more subtle or even cumulative effects on susceptibility to autoimmunity have also been discovered. Genes for cytokines and their receptors, antigen processing and presentation, C-type lectin receptors, signaling pathways, adhesion molecules, and costimulatory or inhibitory receptors have all been linked to specific autoimmune diseases (**Table 16-3**). In many instances, multiple genes plus compounding environmental factors collaborate in autoreactive responses. In some cases, a single gene can heighten susceptibility to multiple different disorders. For instance, a mutant form of the tyrosine phosphatase PTPN22, which dampens TCR signaling capacity, has been linked to T1D, RA, and SLE. It is believed that attenuation of TCR signaling during positive and negative selection in the thymus may be what predisposes carriers of this allele to autoimmune disorders.

Environmental Factors Favoring the Development of Autoimmune Disease

Genetic background, especially immune-related, and sex clearly play a role in susceptibility to autoimmune disease.

TABLE 16-3	Examples of genetic associations with autoimmune disease				
Disease	C-type lectin	Cytokines, their receptors and regulators	Innate immune response	Adhesion and costimulation	Antigen processing and presentation
Type 1 diabetes (T1D)	CLEC16A	IL-2R		CTLA4	VNTR-Ins, PTPN22
Rheumatoid arthritis (RA)	DCIR	STAT4	REL, C5-TRAFI	CD40	PTPN22, MHC2TA
Ankylosing spondylitis (AS)		IL-1A, IL-23R	KIR complex		ERAP1
Multiple sclerosis (MS)	CLEC16A	IL-2RA, IL-7R		CD40	
Systemic lupus erythematosus (SLE)		STAT4, IRF5	TNFAIP3	TNFSF4	PTPN22
Crohn's disease	CLEC16A	IL-23R	NOD2, NCF4	TNFSF15	PTPN2

Data from Thomas, R. 2010. The balancing act of autoimmunity: central and peripheral tolerance versus infection control. *International Reviews of Immunology* **29**:211, Table 1.

Likewise, some role of the environment in tipping this genetic predisposition toward disease is fairly universal. In the last decade, in addition to the other environmental suspects (obesity, smoking, infection, etc.), diet and the mucosal flora have jumped onto the stage.

Autoimmune syndromes are more common in certain geographic locations or in particular climates. This suggests a link between environmental and/or lifestyle factors, and the development of autoimmune disease. For example, we know that certain gut microbes (sometimes called our "old friends") or their secreted products make contact with immune cells in the intestinal mucosa and elsewhere on body surfaces. This communication and negotiation between gut microflora and host cells does not stay local, it "goes viral," so to speak, helping to regulate peripheral tolerance and suppress autoimmune disease in distant locales. For example, animals maintained in germ-free environments, where even healthy, benign commensals are absent, show heightened susceptibility to autoimmune disease compared with their "microbe-laden" counterparts. How commensal microbes might influence tolerance and autoimmunity is an exciting and fruitful new area of research, and the topic of Clinical Focus Box 16-1.

Infections may also influence the induction of autoimmunity, which could explain some of the geographic disparities. For example, tissue pathology following infection may result in the release of sequestered self antigens that are presented in a way that fosters immune activation rather than tolerance induction. Likewise, the molecular structures of certain microbes may share chemical features with self components, resulting in the activation of immune cells with cross-reactive potential. Since local climate, flora, and fauna help determine which infectious agents can thrive in that environment, geography could be a surrogate for certain infections that serve as autoimmune triggers in the right genetic background.

The Role of Certain T Helper Cell Types in Autoimmunity

In both organ-specific and systemic autoimmunity, CD4$^+$ more than CD8$^+$ T cells have been linked to instigation of disease. As we know from Chapter 10, characteristics of the antigen, the status of the APC to make first encounter, and the surface receptors used during this engagement set the stage for which pathway will be chosen in the transition from innate to adaptive immunity. In this transition, the cytokine milieu will help determine which subsets of a T$_H$ cell predominate. The induction of autoimmunity is likewise a complex process, where even experimental models of the same human disease can be induced by different means, making outcomes in each case difficult to correlate. Nevertheless, a few themes have emerged from both human and animal studies of autoimmune disease.

Much of the initial data collected from various studies of autoimmune disease supported a role for autoreactive

T$_H$1 cells and IFN-γ. For example, IFN-γ levels in the CNS of mice with EAE correlate with the severity of disease, and treatment with this cytokine exacerbates MS in humans. Likewise, adoptive transfer of IFN-γ–producing CD4$^+$ T$_H$ cells from mice with EAE can induce disease in naïve hosts. On the flip side, elimination of IFN-γ, by means of neutralizing antibodies or by removal of the gene, does not protect animals from EAE; in fact, it worsens the symptoms.

These conflicting results led to the study of other cytokines or T-cell types that may be involved in the induction of autoimmunity, especially those connected to IFN-γ. Recall from Chapter 10 that IL-12 and IL-23, which can be produced by APCs during activation, encourage the production of other cytokines, such as either IFN-γ or IL-17, favoring T-cell development along the T$_H$1 or T$_H$17 pathway, respectively. Studies have shown that mice engineered to lack the gene for the p40 subunit of IL-12, which happens to be shared with IL-23, are protected from EAE. This protection is due to inhibition of IL-23, a cytokine required to sustain T$_H$17 cells. Mice lacking IL-17A are *less susceptible* to both EAE and collagen-induced arthritis, a model for human RA. In subsequent studies of patients with MS, RA, and psoriasis (another autoimmune disorder), elevated IL-17 expression has been found at the site of inflammation, and increased serum levels of IL-17 and IL-23 have been observed in patients with SLE. Collectively, these findings support the notion that T$_H$17 cells may be an important driver of multiple autoimmune diseases. Not surprisingly, this has led to new trials of immunotherapy designed to manipulate the T$_H$17 pathway in individuals affected by these autoimmune syndromes.

Key Concepts:

- Genetic variants of one or more immune-related genes are associated with predisposition to most autoimmune diseases, most notably certain MHC alleles and immune-regulatory genes (e.g., *AIRE* and *FoxP3*).

- Environmental factors, such as diet, obesity, smoking, infection, and mucosal microflora, have all been associated with risk of autoimmune disease, typically in an already susceptible genetic background.

- Of late, much attention has focused on T$_H$17 cells and the cytokines they produce, namely IL-17 and IL-23, linked to several autoimmune syndromes in both animal models and in humans.

What *Causes* Autoimmunity?

As we have just seen, specific genetic and environmental predisposing factors have strong and predictable impacts on susceptibility to autoimmunity. Sex differences in autoimmune susceptibility, preferentially affecting women, must also be considered (see Clinical Focus Box 16-2). Factors that may account for this include hormonal differences between the sexes, plus the impact of conception and pregnancy. Of late, differences between the sexes in terms of microflora, especially after puberty, are also under study. A heightened appreciation for the evolutionarily conserved role of our microbiome in tuning systemic immunity should not be underestimated (see Chapter 13), and advances in this area are likely to have profound impacts on our understanding of and treatments for many autoimmune diseases. Finally, random (stochastic) events also take their toll. Although not very satisfying, in most cases a complex and nuanced combination of these is likely at fault for autoimmune induction.

Recall that V(D)J recombination is *random*. Even in identical twins these processes will play out in unique ways. This means we all live with a very different set of antigen-specific T-cell and B-cell receptors, even monozygotic twins. For T cells, activation requires pAPCs presenting self MHC in association with unique antigen fragments, which will depend on the haplotype of the individual. Over half of all antigen-specific receptors recognize self proteins, yet not all of the T cells bearing them are deleted during negative selection. Many will be selected as tT$_{REG}$ cells, all following a nuanced set of interconnected pathways in the thymus. At the end of this, surviving self-reactive T and B cells in the periphery *should* be held in check by anergic or regulatory mechanisms, but of course they are also susceptible to daily environmental influences. Exposure to carcinogens or infectious agents that favor DNA damage or polyclonal activation can interfere with this regulation and/or lead to the expansion and survival of rare T- or B-cell clones with autoimmune potential (**Table 16-4**). Acquired mutations in genes that could favor expansion include those encoding antigen receptors, signaling molecules, costimulatory or inhibitory molecules, apoptosis regulators, or growth factors (see Table 16-3). Gram-negative bacteria, cytomegalovirus, and Epstein-Barr virus are all known polyclonal activators, inducing the proliferation of numerous clones of B cells, different in every person, that express IgM in the absence of T-cell help. If B cells reactive to self antigens are activated by this mechanism, auto-antibodies can appear. Finally, since body flora, especially at mucosal surfaces, is a known regulator of systemic immune cells, from birth to death, the fluctuations in our microbiome can be creating fluctuations in our immune status. Collectively, this regulated chaos suggests that the control over autoreactivity may be a daily, uphill battle!

An association between specific microbial agents and autoimmunity has been postulated for several reasons beyond the potential for DNA damage or polyclonal activation. As discussed earlier, some autoimmune syndromes are associated with certain geographic regions, and immigrants to an area can acquire enhanced susceptibility to the

TABLE 16-4	Common proinflammatory environmental factors in autoimmune diseases

Group	Examples	Disease association examples
Infection	Viral	Type 1 diabetes
	Bacterial	Reiter's syndrome
	Fungal	Autoimmune polyendocrine syndrome type 1 (APS-1)
Toxins	Smoking	Rheumatoid arthritis
	Fabric dyes	Scleroderma
	Iodine	Thyroiditis
Stress	Psychological	Multiple sclerosis, systemic lupus erythematosus (SLE)
	Oxidative, metabolic	Rheumatoid arthritis
	Ultraviolet light	SLE
	Endoplasmic reticulum (ER) stress	Ulcerative colitis
Food	Gluten	Celiac disease
	Breastfeeding cessation	Type 1 diabetes
	Gastric bypass	Spondyloarthropathy

Data from Thomas, R. 2010. The balancing act of autoimmunity: central and peripheral tolerance versus infection control. *International Reviews of Immunology* 29:211, Table 2.

disorder associated with that region. This, coupled with the fact that a number of viruses and bacteria possess antigenic determinants that are similar in structure to host-cell components, led to a hypothesis known as **molecular mimicry**. This hypothesis posits that some pathogens express protein epitopes resembling self components, either in conformation or primary sequence, and when these enter the body they inadvertently activate self-reactive cells in a proinflammatory microenvironment, bypassing immune regulation.

For instance, rheumatic fever, an autoimmune disease that can lead to destruction of heart muscle cells and joints, can develop a few weeks *after* infection with group A *Streptococcus*. In this case, antibodies to streptococcal antigens have been shown to cross-react with the heart muscle proteins, resulting in a type II hypersensitivity reaction with immune complex deposition and complement activation (see Chapter 15). In a separate study, 600 different monoclonal antibodies specific for 11 different viruses were evaluated for their reactivity with normal tissue antigens. More than 3% of the virus-specific antibodies tested also bound to normal host tissue, suggesting that molecular similarity between foreign and host antigens may be fairly common. In these cases, susceptibility may also be influenced by the MHC haplotype of the individual, since certain MHC class I and MHC class II molecules may be more effective than others in presenting cross-reactive peptides for T-cell activation.

Release of sequestered antigens is another proposed path to autoimmune initiation, one that may also be connected with environmental exposures. The induction of self tolerance in T cells results from exposure of immature thymocytes to self antigens in the thymus, followed by clonal deletion or inactivation of most self-reactive cells (see Chapter 8). Tissue antigens that are not expressed in the thymus will not engage with developing T cells and will thus not induce self tolerance. In thymic medullary epithelial cells expressing the AIRE transcription factor, tissue-specific antigens are expressed randomly and typically at very low levels, and some may not be expressed at all. Peripheral tolerance should cover these as well, but trauma to tissues following an accident, surgery, or an infection can release these tissue-specific antigens. For instance, the release of heart muscle antigens following myocardial infarction (heart attack) can lead to the formation of auto-antibodies that target healthy heart muscle cells. Studies involving injection of normally sequestered antigens directly into the thymus of susceptible animals support this proposed mechanism: injection of CNS myelin proteins or pancreatic beta cells can inhibit the development of EAE or diabetes, respectively. In these experiments, exposure of immature T cells to self antigens normally not present in the thymus, or present at low levels, presumably boosted central tolerance to these antigens.

In summary, it is worth reiterating that, although certain events may be associated with the development of autoimmunity, a complex combination of genotype, environmental exposures, and random events likely influences the balance of self tolerance versus autoimmune disease induction.

Key Concept:

• In a background of environmental and/or genetic predisposition, induction of autoimmunity is further influenced by sex/hormones, microflora, infection (via polyclonal activation and molecular mimicry), trauma, carcinogens, and plain old randomness.

Treatments for Autoimmune Disease Range from General Immune Suppression to Targeted Immunotherapy

Ideally, treatment for autoimmune diseases should reduce or suppress only the autoreactive cells and molecules, leaving the rest of the immune system unaffected. Implementing this precision approach has proven difficult. Current therapies to treat autoimmune disease fall into three categories: broad-spectrum immunosuppressive treatments, immunosuppression directed at specific cells or pathways, and targeted immunotherapy aimed at guiding the host immune cells toward a new and more beneficial pathway (**Table 16-5**).

In this section, these are discussed in relative order, from the most general to the most targeted. As one might expect, the magnitude and range of side effects follow suit.

Broad-Spectrum Therapies

Most first-generation therapies for autoimmune diseases are broad-spectrum immune suppressants. These are not cures but do reduce symptoms and provide the patient with an acceptable quality of life. Corticosteroids, azathioprine, cyclophosphamide, and methotrexate are all strong anti-inflammatory drugs. They suppress lymphocytes fairly indiscriminately, by inhibiting their survival and proliferation, or by killing rapidly dividing leukocytes. While this can work to inhibit overall inflammation throughout the body, side effects are significant and sometimes long-lasting. These include general cytotoxicity, often to all rapidly dividing cells (e.g., hair follicles, intestinal lining, blood cells), an increased risk of uncontrolled infection, and even the development of cancer.

In some autoimmune diseases, removal of a specific organ or set of toxic compounds can alleviate symptoms. Patients with myasthenia gravis often have thymic abnormalities (e.g., thymic hyperplasia or thymomas), in which case thymectomy can increase the likelihood of remission. Plasmapheresis may also provide significant if short-term benefit for diseases involving antigen-antibody complexes (e.g., myasthenia gravis, SLE, and RA), where removal of a patient's plasma antibodies temporarily eliminates autoreactive antibodies and the resulting immune complexes.

Strategies That Target Specific Cell Types

When antibodies and/or immune complexes are heavily involved in autoimmune pathology, strategies aimed at killing or blocking B cells can improve clinical symptoms. For example, a monoclonal antibody against the B cell–specific antigen CD20 (rituximab) depletes a subset of B cells and provides short-term benefit for patients with RA. However, most cell type–specific agents used to treat autoimmune disorders also need to target T cells or their products, because these cells are either directly pathogenic or provide help to autoreactive B cells.

The first anti–T-cell antibodies used to treat autoimmune disease targeted the CD3 molecule and were designed to deplete T cells without signaling through this receptor. Although somewhat effective in the treatment of T1D, this method still induced broad-spectrum immune suppression. The moderately more specific anti-CD4 antibodies successfully reversed MS and arthritis in animal models, although human trials of this treatment have shown no efficacy. A possible reason for this failure is that anti-CD4 also likely interferes with the activity of $CD4^+CD25^+$ regulatory T cells, a cell type we know is key to regulating tolerance and inhibiting autoimmunity.

With this in mind, and with the discovery of the T_H17 subset, scientists are beginning to target specific helper T-cell

TABLE 16-5	Drugs currently approved by the FDA or undergoing clinical trials to treat autoimmune disease or suppress the immune response, arranged according to mechanism of action		
Name	**Brand name**	**Mechanism of action**	**Target disease/Syndrome**
T- OR B-CELL–DEPLETING AGENTS			
Lymphocyte immune globulin (horse), anti-thymocyte globulin (rabbit)	ATGAM (horse), Thymoglobulin (rabbit)	Depleting horse/rabbit poly-clonal anti-thymocyte antibody	Renal transplant rejection; aplastic anemia
Muromonab (OKT3)	Orthoclone OKT3	Mouse anti-human CD3 mAb; depleting	Acute transplant rejection; graft-versus-host disease (GvHD)
Zanolimumab	HuMax-CD4	Human anti-CD4 mAb, partially depleting	Rheumatoid arthritis (RA)
Rituximab (IDEC-C2B8)	Rituxan	Chimeric anti-CD20 mAb; depleting	RA
TARGETING TRAFFICKING/ADHESION			
Fingolimod (FTY720)	Gilenya	S1P receptor agonist; stimulating	Relapsing/remitting multiple sclerosis (MS); renal transplant rejection
TARGETING TCR SIGNALING			
Cyclosporin A	Gengraf, Neoral, Sandimmune	Calcineurin inhibitor	Transplant rejection; active RA; severe plaque psoriasis
Tacrolimus (FK506)	Prograf (systemic), Protopic (topical)	Calcineurin inhibitor	Transplant rejection; atopic der-matitis; ulcerative colitis (UC); RA; myasthenia gravis; GvHD
TARGETING COSTIMULATORY AND ACCESSORY MOLECULES			
Abatacept (BMS-188667)	Orencia	Fc fusion protein with extracel-lular domain of CTLA-4; blocks CD28-CD80/86 interaction	RA; lupus nephritis; inflammatory bowel disease (IBD); juvenile idiopathic arthritis (JIA)
Belatacept (BMS-224818, LEA29Y)	Nulojix	Same as Abatacept, higher affinity	Transplant rejection
TARGETING CYTOKINES/CYTOKINE SIGNALING			
Sirolimus	Rapamune	mTOR inhibitor	Renal transplant rejection; GvHD

Data from Steward-Tharp, S. M., Y.-J. Song, R. M. Siegel, and J. J. O'Shea. 2010. New insights into T cell biology and T cell–directed therapy for autoimmunity, inflammation, and immunosuppression. *Annals of the New York Academy of Sciences* **1183:**123, Table 1.

pathways. In several mouse models of autoimmunity, including MS, T1D, SLE, and IPEX, the transfer of T_{REG} cells can clearly inhibit disease pathogenesis. The greatest difficulty with translating this from mouse to human is in selecting a population of T_{REG} cells to transfer, as FoxP3 in humans does not correlate well with immunosuppressive activity. Therefore, most of the emphasis in the clinical applications of this approach is currently directed toward mimicking T_{REG}-like mechanisms of suppression (e.g., using IL-10) or inhibiting the T_H17 cells because of their known role in autoimmune disease (e.g., by using IL-17 or IL-23 blocking antibodies).

Therapies That Block Specific Steps in the Inflammatory Process

Since chronic inflammation is a hallmark of debilitating autoimmune disease, individual steps in the inflammatory process are potential targets for intervention. These would be more targeted then broad-spectrum anti-inflammatories and might therefore spare some arms of the immune response to still work to protect us from foreign invaders. Drugs that block TNF-α, one of the early mediators in most autoimmune inflammatory processes, are widely used to treat RA, psoriasis, and Crohn's disease. An IL-1 receptor antagonist is also approved for treatment of RA, as are antibodies directed against the IL-6 receptor and IL-15. Other anti-inflammatory, cytokine-based experimental treatments for autoimmunity include targeting the IL-2 receptor (CD25 and CD122), IL-1, and various IFNs. All of these have some side effects that overlap with broad-spectrum anti-inflammatory drugs, plus they are much more expensive.

More broadly, the class of drugs designated as *statins*, and used by millions to reduce cholesterol levels, has been found to lower serum levels of C-reactive protein (CRP). This acute-phase protein is an indicator of inflammation. Reduced levels of CRP can inhibit proinflammatory cytokines and decrease expression of adhesion molecules on endothelial cells, reducing many of the symptoms of most autoimmune diseases. Clinical trials of statins for the treatment of RA and MS have shown encouraging results. The addition of such drugs with prior FDA approval and extensive safety testing is a tremendous advantage, considering that 95% of agents fail human trials because of safety concerns.

Compounds that block the chemokine or adhesion molecule signals controlling lymphocyte movement into sites of inflammation can also thwart autoimmune processes. The most well-characterized inhibitor of cell trafficking is fingolimod (or FTY720). This compound is an analog of sphingosine and induces internalization of sphingosine 1-phosphate (S1P) receptors in the body, which are involved in the migration of lymphocytes into the blood and lymph nodes (see Chapters 8 and 9). This inhibits egress of all subsets of T cells, resulting in a reduction of up to 85% in circulating blood lymphocytes. So far it has been effective in treating MS, where it has also been reported to inhibit T_H1 and T_H17 cells, and to enhance T_{REG}-cell activity. A compound with similar outcomes, natalizumab, is a monoclonal antibody specific for the adhesion molecule α_4 integrin and has also been approved for the treatment of MS. This molecule is involved in lymphocyte homing to the brain. However, natalizumab is not without its problems; at least 500 patients, most of whom were treated for 2 years or more, have developed a life-threatening CNS infection. This side effect is believed to stem from enhanced mobilization and infection of B cells harboring the JC virus, which causes this brain infection.

Therapies That Interfere with Costimulation

As we saw in Chapter 10, T cells require both antigenic stimulation via the TCR (signal 1) and costimulation (signal 2) to become fully activated. Without costimulation, T cells undergo apoptosis, become anergic, or are selected as regulatory cells to become specific immune inhibitors. Therefore, one way to control the activation of specific T cells would be to regulate or block costimulation. CTLA-4 is a potent inhibitor of T-cell activity, binding to CD80/86 with an affinity that is approximately 20 times greater than that of CD28, its costimulatory counterpart. To this end, a fusion protein was generated consisting of the extracellular domain of CTLA-4 combined with the human IgG1 constant region; it is called CTLA-4Ig. This therapeutic fusion protein drug, Abatacept (marketed as Orencia), was approved for the treatment of RA and is designed to block CD80/86 on APCs from engaging with CD28 on T cells, inhibiting costimulation. This drug has also been studied with limited success in patients with MS and inflammatory bowel disease, but results in patients with SLE have been more disappointing.

Antigen-Specific Immunotherapy

The holy grail of immunotherapy to treat autoimmune disease is a strategy that specifically targets just the autoreactive cells, sparing all other leukocytes and their actions. In this ideal world, clinical therapies that could specifically *induce tolerance* to an auto-antigen might reverse the course of autoimmune disease. In order to manipulate the antigen-specific tolerance process one would need to introduce the specific antigen in question, ideally in a tolerogenic form or context. However, even when the auto-antigen is well characterized, as with T1D and MS, introduction of a self antigen "reworked" to be viewed as a tolerogen carries the risk of *exacerbating* the disease. In fact, this was seen in some of the first trials using this approach. Nonetheless, glatiramer acetate (GA, or Copaxone), a polymer of four basic amino acids found commonly in myelin basic protein (MBP), has been approved for treating MS. Although it seems to shift the T_H population toward an increase in the number of T_{REG} cells and to modulate the function of APCs, it has shown only modest improvement over standard therapies for this disease.

Several other compounds that act as immunotherapy treatments for autoimmune disorders are in earlier stages of development, and some look quite promising. For example, Francisco Quintana and co-workers in Boston are exploring a novel immunotherapy approach for the treatment of T1D, using nanoparticles coated with both proinsulin, one of the target antigens in T1D, and a molecule that can deliver a tolerogenic signal to APCs. In vitro, treatment of murine DCs with these nanoparticles induced a tolerogenic phenotype, suppressed proinflammatory cytokine production, and led to differentiation of T cells into a regulatory phenotype. More importantly, they found that administration of these coated nanoparticles to NOD mice blocked spontaneous development of diabetes. If strategies such as this can be successfully applied to humans, there may be new hope for antigen-specific immunotherapy in the treatment of a host of autoimmune diseases.

Key Concepts:

- The most common broad-spectrum therapies involve drugs that suppress systemic inflammation and/or preferentially kill certain leukocytes and their precursors, providing some relief from symptoms but with significant side effects.

- Depending on the relevant cell type or cytokine, therapies that block B cells, T cells, specific pathways, and subsets of cells or their products have shown some promise in the treatment of specific autoimmune diseases.

- Drugs aimed at manipulating the immune response by blocking costimulation, such as a CTLA-4Ig fusion protein, are being used in the treatment of RA and are currently being explored in other autoimmune disorders.

- Treatment for autoimmune disease that induces tolerance to the auto-antigen(s) in question and/or redirects the damaging effector response away from its target without impacting other immune processes would be ideal, but is still in the investigational stages.

Transplantation Immunology

The first accounting of tissue transplantation comes from India, in the fifth century B.C. The legendary surgeon Sushruta (or Susruta) describes nose reconstruction that required collection and transfer of skin from one site to another. Although there is no documentation of success rates, today we know that this type of transfer, called an *autograft*, is the most likely to succeed. It was not until 1908 that Alexis Carrel produced a systematic study of kidney transplantation in cats, some of which maintained urinary output for 25 days, establishing that a transplanted organ could carry out its normal function in a new recipient. Then, in 1954, Joseph Murray and colleagues in Boston successfully transplanted a kidney between identical twins. However, it was not until 1961 that success across the histocompatibility barrier was achieved, tackling the obstacle of MHC molecules, or **alloreactivity**. A team of surgeons led by Dr. Samuel Kountz (**Figure 16-10**), an African American transplant surgeon at Stanford, completed the first nontwin, living human transplant: a kidney, from mother to daughter. Their pioneering work with immune suppressants and a new kidney perfusion technique heralded a leap forward in the ability of physicians to imagine transplantation as a cure for disease. Today, the transfer of various organs and tissues between nonidentical individuals is performed with ever-increasing frequency and success. Presently, besides a need to generate more specific immune suppressants, the greatest obstacle is organ availability.

In 2016, over 30,000 solid organ transplants were performed in the United States, in addition to more than 46,000 corneal tissue grafts. Although the clinical outcomes have improved considerably in the past decade, major obstacles still exist. Immunosuppressive drugs greatly increase the

FIGURE 16-10 Dr. Samuel Kountz, Stanford University. Along with colleagues, Kountz performed the first transplant between individuals expressing different MHC molecules, a kidney from mother to daughter. *[University of California, San Francisco, Archives & Special Collections.]*

short-term survival of the transplant, but medical problems arise from their use. Exciting new treatments that promise more specific tolerance to the transferred tissue without crippling immune function are under development. Even with greater short-term success, later chronic rejection remains a lingering problem. This creates a need for second or third transplants, exacerbating the shortage of organs and making research into artificial organs or the use of other species increasingly attractive (**Clinical Focus Box 16-3**). This section describes the clinical terrain of transplantation and the specific immune mechanisms that underlay graft rejection, followed by novel strategies to manipulate the immune response in order to enhance graft survival and maintain quality of life.

Demand for Transplants Is High, but Organ Supplies Remain Low

For a number of medical conditions, a transplant may be the only means of therapy. Ailments that can lead to this conclusion are varied, and occur in both young and old. For instance, transplantation of healthy tissue is the only long-term relief for a multitude of genetic anomalies or birth defects, severe burns, certain types of trauma (of the eye, for

CLINICAL FOCUS

BOX 16-3

Xenotransplantation: Science Fiction Turned Fact

Unless organ donations increase drastically, most of the over 118,000 U.S. patients still on the waiting list for a transplant at the end of 2016 will not receive one. In fact, fewer than 30,000 donor organs will become available in time to save these waiting patients. While donation rates have been on the rise for the last 5 years, the disparity between the number of individuals on the waiting list for a transplant and the number of available organs grows every year, making it increasingly unlikely that human organs will ever fill this need. One solution to this shortfall is to utilize animal organs, a process called *xenotransplantation*. Clinical attempts at using nonhuman primates as donors led to some short-term success, including successful functioning of a baboon liver for 70 days and of a chimpanzee kidney for 9 months in human recipients. Although there are advantages based on the phylogenetic similarity between primate species, the use of nonhuman primates as organ donors has several important disadvantages. This similarity carries with it an elevated risk for the transfer of pathogenic viruses, not to mention the impracticalities and ethical concerns that arise in the use of these close cousins.

The use of pigs to supply organs for humans has been under serious consideration for many years. Pigs breed rapidly, have large litters, can be housed in pathogen-free environments, and share considerable anatomic and physiologic similarity with humans. In fact, cardiac valves from pigs have been used in humans for years. However, balancing the advantages of pigs as organ donors are several serious difficulties. For example, if a pig kidney were implanted into a human by techniques standard for human transplants, it would likely fail in a rapid and dramatic fashion due to hyperacute rejection. This antibody-mediated response is due to the presence on the pig cells of a disaccharide antigen called galactose-α-1,3-galactose (α-Gal, for short). The presence of this antigen on many microorganisms means that nearly all of us have been exposed to and formed antibodies against this antigen. The pre-existing antibodies in our bodies cross-react with pig cells, which are then lysed rapidly by complement. The absence of human regulators of complement activity on the pig cells, including human *decay accelerating factor* (DAF) and human *membrane cofactor protein* (MCP), intensifies the complement lysis cycle (see Chapter 5 for descriptions of DAF and MCP).

How can this major obstacle be circumvented? Strategies for blocking these antibody-mediated reactions have been largely unsuccessful. A more elegant solution involves genetically engineered pigs. In one case, pigs were engineered with a gene knockout for the enzyme responsible for the addition of α-Gal to proteins. These *gal*actosyl*transferase gene *knock*out (GalT-KO) pigs have been used as heart or kidney donors for baboons in experimental systems. Kenji Kuwaki and colleagues transplanted GalT-KO pig hearts into baboons immunosuppressed with anti-thymocyte globulin and an anti-CD154 monoclonal antibody (the CD40 ligand found mostly on T cells) and then maintained with commonly used immunosuppressive drugs. The mean survival time was 92 days, and one GalT-KO pig heart transplant survived in a baboon for 179 days. Kazuhiko Yamada and co-workers demonstrated kidney function in recipients of GalT-KO pig kidneys for up to 83 days, using a regimen of simultaneous thymus transplantation in an attempt to establish tolerance in the baboon recipients. Although these studies were not conclusive, some promising results have encouraged further exploration of the use of pigs for xenotransplantation in a clinical setting.

Even if all issues of antigenic difference were resolved, additional concerns remain for those considering pigs as a source of transplanted tissue. Pig endogenous retroviruses introduced into humans as a result of xenotransplantation could cause significant disease. Opponents of xenotransplantation raise the specter of another HIV-type epidemic resulting from human infection with a new animal retrovirus. Continuing work on the development of pigs free of endogenous pig retroviruses could reduce the possibility of this bleak outcome.

Another breakthrough in the area of cross-species transplantation came in the past year. In 2017 two different teams of researchers successfully created chimeric (interspecies) animals as proof of concept: human cells growing in a pig embryo and a rat embryo with viable mouse cells (**Figure 1**). Both feats are a first. These chimeric organisms were created by collecting adult stem cells from the donor animal, reprogramming them to behave like embryonic cells, and then injecting these embryonic cells into developing fetuses. So far, these chimeric fetuses

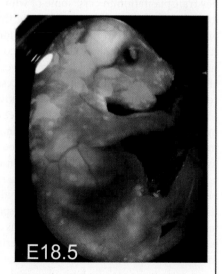

E18.5

FIGURE 1 Embryonic rat at 18.5 days of gestation, expressing mouse cells (red). [Republished with permission of Elsevier, from Jun Wu, et al., "Interspecies chimerism with mammalian pluripotent stem cells," Cell, 2017, January 26; 168(3): 473-486.e15. 10.1016/j.cell.2016.12.036. Figure 1A. Permission conveyed through Copyright Clearance Center, Inc.]

(continued)

CLINICAL FOCUS   *(continued)*   **BOX 16-3**

have been allowed to survive for only a few weeks, skirting regulatory restrictions. However, with the current demand for donor tissues, this approach is being heralded by some as a major new advance.

Although kidneys are the most sought-after organ at present, other organs and cells from specially bred and engineered animals could find use if they are proven to be safe and effective. A statement issued in 2000 by the American Society of Transplantation and the American Society of Transplant Surgeons endorses

the use of xenotransplants if certain conditions are met, including the demonstration of feasibility in a nonhuman primate model, proven benefit to the patient, and lack of infectious disease risk. Although certain barriers still remain in the field of human xenotransplantation, science fiction is slowly evolving into science fact.

REFERENCES

Ekser, B., and D. Cooper. 2010. Overcoming barriers to xenotransplantation: prospects for the future. *Expert Reviews in Clinical Immunology* **6**:219.

Kuwaki, K., et al. 2005. Heart transplantation in baboons using α1,3-galactosyltransferase gene-knockout pigs as donors: initial experience. *Nature Medicine.* **11**:29.

Yamada, K., et al. 2005. Marked prolongation of porcine renal xenograft survival in baboons through the use of α1,3-galactosyltransferase gene-knockout donors and the cotransplantation of vascularized thymic tissue. *Nature Medicine* **11**:32.

Yamaguchi, T., et al. 2017. Interspecies organogenesis generates autologous functional islets. *Nature* **542**:191.

Wu, J., et al. 2017. Interspecies chimerism with mammalian pluripotent stem cells. *Cell* **168**:473.

---

example), organ failure from infection or chronic disease, and some types of cancer. Alas, many of these ailments are on the rise, especially those related to cancer and chronic disease.

Demand for donor tissue increases every year, continually outstripping the supply by almost 4 to 1. The American Transplant Foundation estimates that 20 people per day die waiting for a transplant. As of March 2018, almost 115,000 individuals still needed an organ. (see http://optn.transplant.hrsa.gov for real-time data). The majority of those individuals require a kidney, for which the median waiting period ranges from 3 to 5 years: a very long time when you need hours-long blood dialysis three times a week, or more.

> **Key Concept:**
>
> • Birth defects, infection, burns, trauma, or chronic disease can lead to organ failure requiring correction through tissue or organ transplantation, but many candidates do not survive the years-long wait due to the short supply of available organs.

Antigenic Similarity between Donor and Recipient Improves Transplant Success

The degree and type of immune response to a transplant varies with the type and source of the grafted tissue. The following terms denote different types of transplants:

- **Autograft:** Self tissue transferred from one body site to another in the same individual.
- **Isograft:** Tissue transferred between genetically identical individuals, as with identical twins or an inbred strain of mice, when the donor and recipient are syngeneic.

- **Allograft:** Tissue transferred between genetically different members of the same species. This is the most common type of tissue graft, occurring between nonidentical humans or different strains of mice.
- **Xenograft:** Tissue transferred between different species.

Autografts and isografts are usually accepted, owing to the genetic identity between donor and recipient (**Figure 16-11a**). Tissues that share sufficient antigenic similarity, allowing transfer without immunologic rejection, are said to be **histocompatible**. This is the case when the transfer occurs between identical twins or between members of the same mouse strain. Tissues that display significant antigenic differences are *histoINcompatible* and typically induce an immune response leading to tissue rejection, as in the case for most allografts in the absence of immune suppression. Of course, there are many shades of gray in the degree of histocompatibility between a donor and recipient. The antigens involved are encoded by more than 40 different loci, but the loci responsible for the most vigorous allograft rejection reactions are located within the MHC. The organization of the MHC—called the *H2 complex* in mice and the *HLA complex* in humans—was described in detail in Chapter 7 (see Figures 7-6 and 7-7). Because the genes in the MHC locus are closely linked, they are usually inherited as a complete set from each parent, called a **haplotype**. Obviously, xenografts between different species exhibit the greatest genetic and antigenic disparity, engendering a rapid and vigorous graft rejection response.

In inbred strains of mice, offspring inherit the same haplotype from each parent, meaning they are homozygous at the MHC locus. When mice from two different inbred strains are mated, the F_1 progeny each inherit one maternal and one paternal haplotype (see Figure 7-8, where parental strains are *b/b* and *k/k*). These heterozygous F_1 offspring express the MHC type from both parents (*b/k* in Figure 7-8), which means they are tolerant to the alleles from both

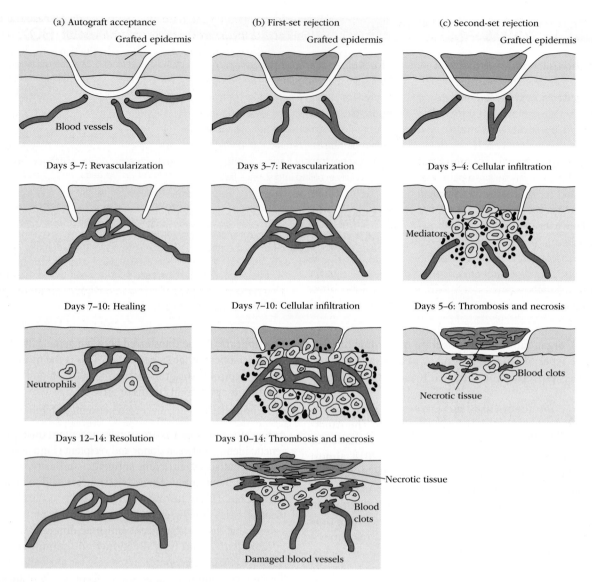

(a) Autograft acceptance

(b) First-set rejection

(c) Second-set rejection

Grafted epidermis

Grafted epidermis

Grafted epidermis

Blood vessels

Days 3–7: Revascularization

Days 3–7: Revascularization

Days 3–4: Cellular infiltration

Mediators

Days 7–10: Healing

Days 7–10: Cellular infiltration

Days 5–6: Thrombosis and necrosis

Neutrophils

Blood clots

Necrotic tissue

Days 12–14: Resolution

Days 10–14: Thrombosis and necrosis

Necrotic tissue

Blood clots

Damaged blood vessels

FIGURE 16-11 Schematic diagrams of the process of graft acceptance and rejection. (a) Acceptance of an autograft is completed within 12 to 14 days. (b) First-set rejection (based on a primary response) of an allograft begins 7 to 10 days after grafting, with full rejection occurring by 10 to 14 days. (c) Second-set rejection (based on a secondary response) of an allograft begins within 3 to 4 days, with full rejection by 5 to 6 days. The cellular infiltrate that invades an allograft (b and c) contains lymphocytes, phagocytes, and other inflammatory cells.

haplotypes and can accept grafts from either parent. However, neither of the parental strains can accept grafts from the F_1 offspring because each parent lacks one of the F_1 haplotypes and will therefore reject these MHC antigens.

In outbred populations, there is a high degree of heterozygosity at most loci, including the MHC. In matings between members of an outbred species, there is only a 25% chance that any two offspring will inherit identical MHC haplotypes unless the parents share a haplotype. Therefore, for purposes of organ or bone marrow grafts, it can be assumed that there is a 25% chance of MHC identity between any two siblings. With parent-to-child grafts, the donor and recipient will always have at least half their HLA alleles in common (50% match), which is why these grafts are so common. However,

in this case, the donor and recipient are still nearly always mismatched for the MHC alleles inherited from the other parent, providing a target for the immune system.

Key Concept:

- Grafts that are syngeneic (isografts and autografts) are usually accepted because of their antigenic compatibility, but grafts from other species (xenografts) are quickly rejected and in-species allografts with partial MHC matches show varying degrees of success, correlated to the amount of antigen overlap and the immune suppression protocol.

Some Organs Are More Amenable to Transplantation Than Others

Since the first kidney transplant was performed in the 1950s, it is estimated that over 500,000 kidneys have been transplanted worldwide. The next most frequently transplanted solid organ is the liver, followed by the heart, the lung, and the pancreas. **Figure 16-12** displays the number of completed transfers in 2016 for the most commonly transplanted organs. Certain combinations of organs, such as heart and lung or kidney and pancreas, are being transferred simultaneously with increasing frequency and, interestingly, better odds of success than either alone. The reason for this is unclear, but the hypothesis is that introduction of more donor cells (blood cells, for example) resident in the organ can boost tolerance mechanisms, favoring acceptance. In fact, the decrease in acute and chronic rejection seen in the recipients of two organs from the same donor is lost if these organs come from two different donors, lending support to this theory.

We do know that in addition to the degree of antigenic similarity between donor and recipient, tissue damage from lack of oxygen is a major limitation in transplant situations. Cell death causes the release of damage- or danger-associated molecular patterns (DAMPS; see Chapter 4), resulting in a proinflammatory microenvironment even before the organ is transplanted. If damage and inflammation can be reduced by shortening the time an organ spends outside the body and/or by improving preservation techniques during that time, we could see lower rates of rejection and improved graft survival.

The frequency with which a given organ or tissue is transplanted depends on a number of factors, including alternative treatment options, organ availability, and the level of difficulty of the procedure. Several factors contribute to the kidney being the most commonly transplanted organ. Many increasingly common diseases (e.g., diabetes) result in kidney failure, and a patient requires dialysis until a suitable organ can be identified. Because kidneys come in pairs and we can survive with only one, this organ is available from living as well as deceased donors. Mechanically, surgical procedures are simpler and time that the organ can survive outside the body is longer for kidneys than for liver or heart. Also, the long track record with this organ means that patient-care procedures and effective immunosuppressive regimens are well established.

Bone marrow (BM) or hematopoietic stem cell (HSC) transfer involves transplanting cells rather than a solid organ from donor to recipient. This procedure is most often used to treat hematologic diseases, including leukemia, lymphoma, and inherited immune deficiencies, especially *severe combined immunodeficiency* (SCID; see Chapter 18). Although the supply of BM/HSCs, which is a renewing resource, is less of a problem to procure than for

Numbers of Transplants Performed in 2016 for Selected Organs

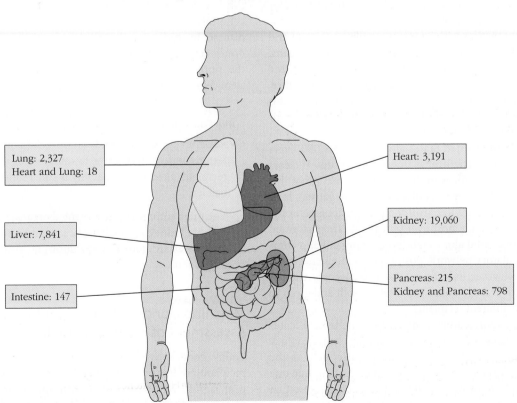

Lung: 2,327
Heart and Lung: 18

Heart: 3,191

Liver: 7,841

Kidney: 19,060

Intestine: 147

Pancreas: 215
Kidney and Pancreas: 798

FIGURE 16-12   Solid organ transplant numbers for 2016. *[Data from https://optn.transplant.hrsa.gov/data/.]*

most solid organs, finding a matched donor is a more formidable obstacle. Here the donor and recipient must share MHC alleles, with some loci more critical than others. Current tissue-typing techniques can quickly identify donors with at least a partial HLA match, but too few individuals are included in the bone marrow registry and some alleles, especially among ethnic minorities, are very difficult to find. Since bone marrow transplant recipients are typically highly immune suppressed before transfer, once a match is found graft rejection is rare. However, the opposite becomes a real problem: **graft-versus-host disease (GvHD)**. The introduction of immunocompetent cells from a foreign donor means that the transplant can attack *host* alloantigens, or foreign MHC molecules present on recipient tissues. Present-day procedures that deplete T cells before transfer have decreased the incidence of GvHD, but this syndrome remains a major obstacle to the success of bone marrow transfers.

Perhaps the most dramatic forms of transplantation involve the transfer of heart, lung, or both: situations where the recipient must be kept alive via artificial means during surgery. The human heart can remain viable for only a few hours in ice-cold buffer solutions, which delay tissue damage. Recent technological advances that keep the heart warm and oxygenated for transport (called "heart in a box") may expand this time frame significantly. However, eventually the lack of oxygen (ischemia) and the resulting deprivation of ATP lead to irreversible organ damage. The surgical methods for implanting a heart have been available since the first heart transplant was carried out in 1964 in South Africa by Dr. Christian Barnard. Today, the 1-year survival rate for heart transplants has climbed to greater than 80%. Brain-dead accident victims with an intact circulatory system and a functioning heart are the typical source of these organs. HLA matching is often not possible because of the limited supply of these organs and the urgency of the procedure. However, MHC matching could improve outcomes, and this may be possible if the heart can survive outside the body for longer periods.

In 2016 in the United States alone, almost 8,000 livers were transplanted. The liver is important because it clears and detoxifies substances in the body. Malfunction of this organ can be caused by viral diseases (e.g., hepatitis), liver cancers, and exposure to harmful chemicals (e.g., chronic alcoholism), although most liver transplants are performed to correct congenital abnormalities. This organ has a complicated circulatory network, posing some unique technical challenges. However, its large size also presents opportunity; the liver from a single donor can often be split and given to two or more different recipients.

One of the more common diseases in the United States is type 1 diabetes. This disease is caused by immune attack on or malfunction of insulin-producing islet cells in the pancreas, making the receipt of new pancreatic tissue one possible solution. Newer protocols that avoid whole-organ transfer involve harvesting donor islet cells and transferring

them into the recipient's liver, where they become permanently established in the liver sinusoids. Results indicate that more than half the recipients are insulin independent after such a transplant, some for up to 2 years. Several factors favor the survival of functioning pancreatic cells, the most important being the condition of the islet cells used for implantation.

Most skin transplants are autografts, with healthy tissue moved from one site to another in the same individual. However, after severe burns, foreign skin may be required. In this case, fresh tissue is not always available and frozen grafts may be used. For instance, Sir Peter Medawar, who won a Nobel Prize in 1960 for his work on immune tolerance (see Chapter 1 for a list of Nobel Prizes related to immunology), made use of frozen skin samples to treat burn victims during the Second World War. These thawed grafts generally act as biologic dressings because the cells are no longer able to proliferate. Nonetheless, they can foster healing and protect delicate tissue underneath. In fact, in some countries with less access to preserved human skin, alternatives are being explored. Brazilian scientists are currently pioneering the use of sterilized tilapia fish skin, which has been shown to protect underlying tissue and reduce wound-healing times in individuals with severe burns.

This list of commonly transplanted organs and tissue is in no way comprehensive, and will surely expand with time. As we describe in the following sections, success with improved procedures for inducing tolerance, controlling rejection, and expanding sources for transplanted tissue would only add to this list. For instance, the recent use of intracerebral neural cell grafts has restored function in victims of Parkinson's disease. In studies conducted thus far, the source of neural donor cells has primarily been human embryos. However, the more ethically straightforward possibility of using adult stem cells for this is under investigation.

Key Concepts:

- Allograft survival depends on the health of the transferred tissue, as well as the immune status of the recipient, although availability is still a substantial barrier.

- The most commonly transplanted organs or tissues are kidney (75%), liver, heart, lung, and bone marrow, based both on procedural and availability considerations.

Matching Donor and Recipient Involves Prior Assessment of Histocompatibility

Today, geography and compatibility between donor and recipient are less of a hurdle to transplantation than in the past, before we had a host of immunosuppressive drugs or major advances in organ collection and preservation.

Nonetheless, three important pretransplantation tests must be completed prior to transfer: (1) blood group matching (also called blood typing), (2) MHC matching (also known as tissue typing), and (3) cross-matching. The results of these tests are used to determine compatibility, and can predict subsequent graft success rates.

ABO Blood Group Matching

The first human kidney transplant, attempted in 1933 by a Russian surgeon, failed because a mismatch in blood type between donor and recipient caused almost immediate rejection. In fact, the most intense graft rejection reactions are frequently due to ABO blood group differences between the donor and recipient. Blood group antigens are expressed on red blood cells (RBCs), epithelial cells, and endothelial cells, requiring the donor and recipient to first be screened for ABO compatibility. If the recipient carries antibodies to any of the donor's blood-group antigens, the transplanted tissue will undergo rapid antibody-mediated lysis in a process known as *hyperacute rejection* (discussed below). For this reason, most transplantations are conducted between individuals with a matching ABO blood type. This is the "type" part of a "type and cross-match" procedure we receive in hospitals prior to, for example, a blood transfusion. The "cross-match" refers to a complementary procedure that we describe shortly.

MHC Matching

Next, depending somewhat on the type of tissue and difficulty of preservation, MHC compatibility between potential donors and a recipient will be evaluated. In situations where MHC matching is particularly important and a live donation is possible, the first choice for donors is usually parents or first-order siblings, followed by other family members. Given our current success with immunosuppression and immune tolerance induction protocols, solid organ transplantations between individuals with significant or even total HLA mismatch can be successful, and this is still the course of action with tissues like the heart or lungs that do not survive for more than a few hours outside the body. This window of time is too short to transport MHC-matched organs any significant distance, requiring semilocal transplants of heart and lung. The most rigorous MHC testing is conducted in bone marrow transplants, where at least partial HLA matching is crucial to ensure that the grafted immune cells such as APCs and T cells will be able to recognize the recipient's MHC alleles in order to coordinate an immune response.

Either serologic or molecular tests can be used to determine the HLA compatibility, collectively called **tissue typing**. The choice is somewhat dependent on the organ or tissue in question, and the time available. Molecular assays using sequence-specific primers to establish which HLA alleles are expressed by the recipient and potential donors have become more common in recent years, especially in bone marrow transplantation. Molecular assays provide greater specificity and higher resolution than assays that characterize MHC molecules serologically, using antigen-antibody interactions alone, a previously common practice that can sometimes miss subtle differences.

The MHC makeup of donor and recipient is not the sole factor determining tissue acceptance. Even when MHC antigens are identical, the transplanted tissue can be rejected because of differences at various other loci, including the **minor histocompatibility locus**. As described in Chapter 7, nonself MHC molecules can be recognized directly by TCRs on T_H and T_C cells, a phenomenon termed **alloreactivity**. In contrast, minor histocompatibility antigens are recognized only when peptide fragments of them are presented in the context of some self MHC molecule. Rejection based only on minor histocompatibility differences is usually less vigorous, but can still lead to graft rejection. Therefore, even in cases of HLA-identical matches, some degree of immune suppression is usually still required.

Cross-Matching

The presence of any preformed antibodies against potential donor alloantigens must also be evaluated in the recipient. We generate antibodies against nonself HLA proteins for a number of reasons, but transplant recipients who have received prior blood transfusions or allografts are especially likely to possess them. This type of testing is called **cross-matching** and is the most important level of compatibility testing that occurs prior to solid organ transfer; a positive cross-match means that the recipient has antibodies against HLA proteins expressed by the donor and that these are likely to lead to rapid (hyperacute) rejection. Unlike expression of ABO or MHC antigens, this status can change in an individual. For example, someone previously lacking antibodies against certain MHC molecules can begin to make these after a recent exposure (e.g., from a blood transfusion or previous transplant). For this reason cross-matching tests are always performed shortly before a potential donor tissue is transferred.

The most common technique used today for cross-matching is the Luminex assay. This employs fluorochrome-labeled microbeads impregnated with specific HLA proteins, where each HLA protein is associated with a fluorochrome of a different intensity. These HLA-impregnated beads are mixed with recipient serum, allowing clinicians to determine more precisely which donor-specific anti-HLA antibodies are present in the recipient prior to transplantation. The importance of careful cross-matching was shown in a seminal 1969 study, where up to 80% of kidney transplant patients with a positive cross-match experienced almost immediate transplant rejection, while only 5% of patients with a negative cross-match had this outcome.

Allograft Rejection Follows the Rules of Immune Specificity and Memory

The rate of allograft rejection varies according to the tissue involved; skin grafts are generally rejected faster than other tissues, such as kidney or heart. Despite these time differences, the immune response culminating in graft rejection always displays the attributes of specificity and memory. For example, if an inbred mouse of strain A is grafted with skin from strain B, primary graft rejection, known as *first-set rejection*, occurs (see Figure 16-11b). The skin first becomes revascularized between days 3 and 7. As the reaction develops, the vascularized transplant becomes infiltrated with inflammatory cells. There is decreased vascularization of the transplanted tissue by 7 to 10 days, visible necrosis by 10 days, and complete rejection by 12 to 14 days. This is an example of a primary response to nonself antigens.

Immunologic memory is demonstrated when a second strain-B graft is transferred to a previously engrafted strain-A mouse, inducing a secondary response. In this case, the anti-graft reaction develops more quickly, with complete rejection occurring within 5 to 6 days, mainly due to the recall of graft-specific T and B cells. This is called *second-set rejection* (see Figure 16-11c). Specificity can be demonstrated by grafting skin from an unrelated mouse of strain C at the same time as the second strain-B graft. Rejection of the strain-C graft proceeds according to the slower, first-set rejection/primary response kinetics, whereas the strain-B graft is rejected in an accelerated second-set fashion.

Graft Rejection Takes a Predictable Clinical Course

The exact timing of immune rejection reactions varies depending on the nature and type of tissue, as well as on the health of the recipient. Nonetheless, these reactions follow a fairly predictable course. Hyperacute rejection reactions typically occur within the first 24 hours after transplantation, acute rejection happens in the first few weeks after transplantation, while chronic rejection reactions can occur from months to years after transplantation. Careful blood group matching, tissue typing, and cross-matching can avoid most cases of early, hyperacute rejection, and current immunosuppressive agents have greatly advanced our ability to reduce instances of acute rejection in the first months or even years after transplantation. However, later, chronic rejection episodes still remain a major problem.

Hyperacute Rejection

In rare instances, a transplant is rejected so quickly that the grafted tissue never becomes vascularized. These **hyperacute rejection** episodes are caused by pre-existing host serum antibodies specific for unique antigens found on the graft, sometimes also called *antibody-mediated rejection* (AMR). The most common targets are ABO blood group antigens or MHC alloantigens. Pre-existing recipient antibodies bind to foreign antigens on the endothelial cells lining graft capillaries, leading to accumulation of neutrophils, deposition of complement, and a severe inflammatory reaction. This results in endothelial damage that obstructs capillaries, preventing vascularization of the graft (**Figure 16-13**). AMR, although most associated with hyperacute rejection, can occur at any time during the clinical course of allograft rejection.

Why do we have pre-existing antibodies against these nonself components? Antibodies for A or B antigens on blood cells are believed to arise from exposure to cross-reactive oligosaccharides in nature, including their presence on many gut commensals. Several mechanisms can account for the presence of antibodies specific for MHC alloantigens, including past blood transfusions that induced antibodies to MHC antigens expressed on allogeneic white blood cells in the blood; past pregnancies, in which women develop antibodies against paternal alloantigens of the fetus; exposure to infectious agents, which can elicit MHC cross-reactive antibodies; or a previous transplant, which results in high levels of antibodies to the allogeneic MHC antigens present in that graft. However, careful pretransplantation testing can all but eliminate hyperacute rejection.

Acute Rejection

As we have just seen, hyperacute rejection is customarily caused by pre-existing antibodies against blood-group or MHC antigens. When graft rejection occurs in the absence of this pre-existing immunity it is called **acute rejection**, and occurs in two stages, similar to hypersensitivity responses: a *sensitization* phase and an *effector* phase. During the *sensitization phase*, $CD4^+$ and $CD8^+$ T cells recognize alloantigens expressed on cells of the foreign graft and proliferate in response. Both major and minor histocompatibility alloantigens can be recognized. In general, the response to minor histocompatibility antigens is weak, although the combined response to several minor

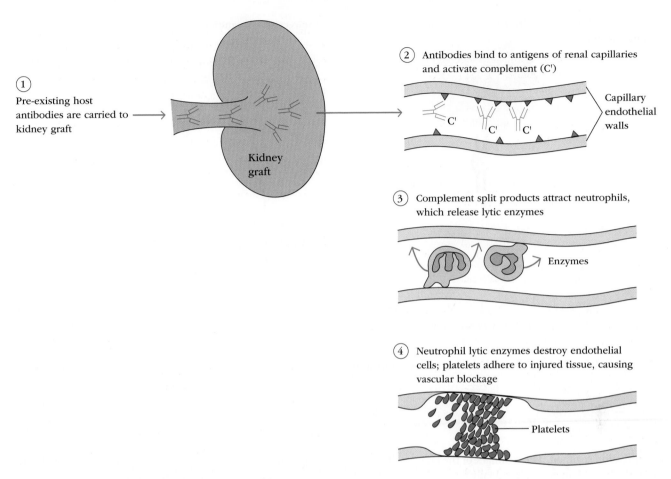

① Pre-existing host antibodies are carried to kidney graft

Kidney graft

② Antibodies bind to antigens of renal capillaries and activate complement (C')

Capillary endothelial walls

C' C' C'

③ Complement split products attract neutrophils, which release lytic enzymes

Enzymes

④ Neutrophil lytic enzymes destroy endothelial cells; platelets adhere to injured tissue, causing vascular blockage

Platelets

FIGURE 16-13 Steps in the hyperacute rejection of a kidney graft.

differences can be quite vigorous. The response to MHC antigens can involve recipient T-cell recognition of *donor MHC molecules* expressed on the surface of cells in the transplant (*direct presentation*) or recognition of processed peptides from donor HLA proteins presented in the cleft of the recipient's own APCs via *self MHC molecules* (*indirect presentation*), as shown in **Figure 16-14**.

As we know from Chapter 10, activation of naïve T cells requires presentation by an APC that expresses the appropriate antigenic ligand/MHC molecule and provides the requisite costimulatory signal. Because DCs are found in most tissues and constitutively express high levels of MHC class II molecules, activated allogeneic donor DCs carried in the transplanted tissue can mediate direct presentation in grafts or draining lymph nodes, to which they often migrate. Likewise, APCs of host origin can also migrate into a graft and endocytose the foreign alloantigens (both major and minor histocompatibility molecules), where they become activated and present these antigens indirectly as processed peptides bound to self MHC molecules. This process is further encouraged by the damage that can occur during the transfer of the organ or tissue between the recipient and donor, resulting in the release of danger and stress signals that can act as adjuvants. The cross-presentation ability of

DCs (see Chapter 7) also allows them to present endocytic antigens in the context of MHC class I molecules to CD8$^+$ T cells, which can then participate in allograft rejection. In addition to DCs, other cell types have been implicated in alloantigen presentation and immune activation during sensitization, including Langerhans cells and endothelial cells lining the blood vessels. Both of these cell types can express MHC class I and class II antigens.

Interestingly, as we know from Chapter 10, T cells that respond to antigen via the TCR in the *absence* of costimulation or danger signals can become tolerant. This may help to explain a long-standing clinical observation: transfusion of donor blood into a graft recipient *prior* to transplantation can facilitate acceptance of a subsequent graft from that donor. This suggests that exposure to donor cells in this non-inflammatory context encourages tolerance to donor alloantigens. Newer immunomodulation protocols based on this observation, as well as related experimental studies, are currently underway to design techniques to effectively induce donor-specific tolerance prior to engraftment.

This sensitization phase takes some time, which is why the effector phase of acute rejection typically manifests 7 to 10 (or more) days later, depending on the immune

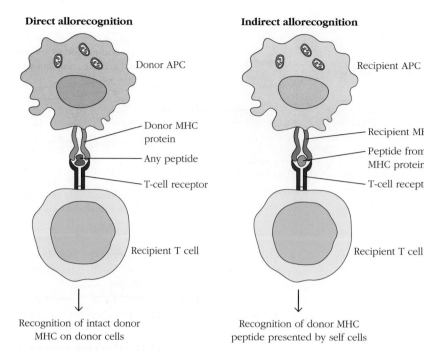

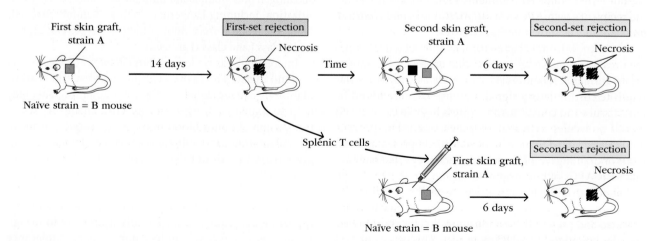

Direct allorecognition

Donor APC

Donor MHC protein

Any peptide

T-cell receptor

Recipient T cell

Recognition of intact donor MHC on donor cells

Indirect allorecognition

Recipient APC

Recipient MHC protein

Peptide from donor MHC protein

T-cell receptor

Recipient T cell

Recognition of donor MHC peptide presented by self cells

FIGURE 16-14 Direct versus indirect presentation of allogeneic MHC. During foreign tissue recognition, donor APCs transferred during the operation and bearing foreign MHC molecules can engage directly with host T cells (direct allorecognition, *left*), or donor cells and cellular debris can be taken up by host APCs and processed, allowing fragments of foreign MHC peptides to be presented by recipient APCs bearing self MHC (indirect allorecognition, *right*).

suppression regimen. The hallmark of the effector phase is a large influx of leukocytes, especially CD4$^+$ T cells and macrophages. We know that T cells are crucial for acute rejection from transplant studies in model animals. For instance, nude mice, which lack a thymus and consequently lack functional T cells, were found to be *incapable* of allograft rejection; these mice even accept xenografts. In other studies, T cells derived from an allograft-primed mouse were shown to transfer second-set allograft rejection to an unprimed

syngeneic recipient as long as that recipient was grafted with the same allogeneic tissue (**Figure 16-15**).

Histologically, the infiltrate seen during an acute rejection resembles that observed during a DTH response, in which cytokines produced by T cells promote inflammatory cell infiltration (see Figure 15-15). Although probably less important, direct recognition by host CD8$^+$ T cells of either foreign MHC class I alloantigens on the graft or alloantigenic peptides cross-presented in the context of self MHC class I

First skin graft, strain A

14 days → First-set rejection — Necrosis

Time →

Second skin graft, strain A

6 days → Second-set rejection — Necrosis

Naïve strain = B mouse

Splenic T cells

First skin graft, strain A

6 days → Second-set rejection — Necrosis

Naïve strain = B mouse

FIGURE 16-15 Experimental demonstration that T cells can transfer allograft rejection. When T cells derived from an allograft-primed mouse are transferred to an unprimed syngeneic mouse, the recipient mounts a second-set rejection to an initial allograft from the original allogeneic strain.

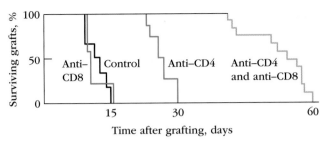

FIGURE 16-16 The role of CD4$^+$ and CD8$^+$ T cells in allograft rejection is demonstrated by the curves showing survival times of skin grafts between mice mismatched at the MHC. Animals in which the CD8$^+$ T cells were removed by treatment with an anti-CD8 monoclonal antibody (red) showed little difference from untreated control mice (black). Treatment with monoclonal anti-CD4 antibody (blue) improved graft survival significantly, and treatment with both anti-CD4 and anti-CD8 antibody prolonged graft survival most dramatically (green). [*Data from Cobbold, S. P., G. Martin, S. Qin, and H. Waldmann. 1986. Monoclonal antibodies to promote marrow engraftment and tissue graft tolerance. Nature **323**:164.*]

by DCs (indirect presentation; see Figure 16-14) can lead to CTL-mediated killing. In one study, mice were injected with monoclonal antibodies to deplete one or both types of T cells, and then the rate of graft rejection was measured. As shown in **Figure 16-16**, removal of the CD8$^+$ T-cell population alone had no effect on graft survival, and the graft was rejected at the same rate as in control mice (15 days). Removal of the CD4$^+$ T-cell population alone prolonged graft survival from 15 days to 30 days. However, removal of both CD4$^+$ and CD8$^+$ T cells resulted in long-term survival (up to 60 days) of the allografts. This study indicated that both CD4$^+$ and CD8$^+$ T cells participated in rejection and that the collaboration of the two resulted in more pronounced graft rejection.

Cytokines secreted by T$_H$ cells play a central role in acute rejection. For example, IL-2 and IFN-γ produced by T$_H$1 cells have been shown to be important mediators of graft rejection. These two cytokines promote T-cell proliferation (including CTLs), DTH responses, and the synthesis of IgG by B cells, with resulting complement activation. A number of cytokines that encourage the expression of MHC class I and class II molecules (e.g., the interferons and TNFs) increase during graft rejection episodes, inducing a variety of cell types within the graft to increase surface expression of these proteins. Many of the cytokines most closely associated with T$_H$2 and T$_H$17 cells have also been implicated in graft rejection. Elevations in IL-4, -5, and -6, responsible for B-cell activation and eosinophil accumulation in allografts, along with increases in IL-17, have all been linked to transplant rejection. Recent studies showing that neutralization of IL-17 could extend the survival of cardiac allografts in mice have generated much interest in this cytokine and in the role of T$_H$17 cells in graft rejection. Finally, AMR, although less frequent, is still a major issue in the acute phase of rejection

and is sustained by T-cell-dependent maintenance of alloreactive B cells. See **Figure 16-17** for a graphic summary of the pathways involved in acute rejection.

Chronic Rejection

Chronic rejection can develop months or years after acute rejection reactions have subsided. The mechanisms include both humoral and cell-mediated responses. Although immunosuppressive drugs and advanced tissue-typing techniques have dramatically increased survival of allografts during the first years, less progress has been made in episodes of chronic rejection, or late stages of transplant failure. In data collected in the United States for kidney transplant recipients, 1-year patient survival rates approached 97%. However, even in cases of a living donor—the most ideal scenario—10-year survival rates drop as low as 60% (based on procedures performed in 2000). Immunosuppressive drugs usually do little to manage chronic rejection, which not infrequently necessitates another transplant. Plus, these drugs take a toll on the recipient, as described in the next section. In up to 60% of cases in which there is some form of chronic allograft dysfunction, anti-donor antibodies can be found in the recipient, suggesting that in addition to cell-mediated responses, AMR may also be involved in chronic rejection, including via *antibody-dependent cell-mediated cytotoxicity* (ADCC) (see Chapter 12). Alas, the inducers of chronic rejection episodes are not well understood.

> **Key Concepts:**
> - The earliest type of rejection, occurring within hours, is called *hyperacute rejection* and is caused by damage to transplanted capillaries by preformed antibodies recognizing foreign antigens, including those from the ABO blood group and HLA.
>
> - Acute rejection is induced by the action of T cells (especially CD4$^+$ T cells), APCs, and their cytokines after an initial sensitization to donor alloantigens via either direct or indirect presentation to host T cells, leading to inflammation and cell death, usually weeks or months after transplantation.
>
> - Chronic rejection, which can occur months or years after resolution of earlier rejection responses, follows the same course as acute rejection, has unknown inducers, and is more resistant to reversal by standard immunosuppression.

Immunosuppressive Therapy Can Be Either General or Target-Specific

Allogeneic transplantation always requires some degree of immune suppression in the recipient if the transplant is to survive. Most immunosuppressive treatments are nonspecific,

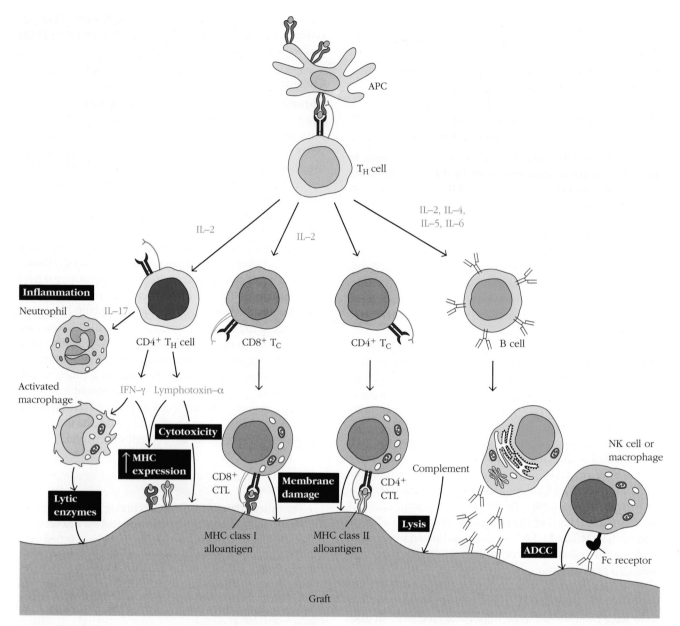

FIGURE 16-17 Effector mechanisms involved in allograft rejection. The generation or activity of various effector cells depends directly or indirectly on cytokines (blue) secreted by activated T_H cells. ADCC = *a*ntibody-*d*ependent *c*ell-mediated *c*ytotoxicity.

resulting in generalized suppression of all immune responses to all antigens, not just those of the allograft. This places the recipient at increased risk of infection and cancer for the duration of the time they are receiving treatment, which can be decades or more. In fact, infection is the most common cause of transplant-related death. Many immunosuppressive measures slow the proliferation of activated lymphocytes, thus affecting all rapidly dividing cells (e.g., gut epithelial cells or bone marrow hematopoietic stem cells), leading to serious or even life-threatening complications. Patients receiving long-term immunosuppressive therapy are also at increased risk of hypertension and metabolic bone disease. Fine-tuning of their immunosuppressive cocktail, and eventual weaning

off these drugs is a nuanced and ongoing process for most transplant recipients.

Total Lymphoid Irradiation to Eliminate Lymphocytes

Because lymphocytes are extremely sensitive to x-rays, irradiation can be used to eliminate them in the transplant recipient just before grafting. Although not a part of most immunosuppressive regimens, this is often used prior to bone marrow or HSC transplantation. In total lymphoid irradiation, the recipient receives multiple x-ray exposures to the thymus, spleen, and lymph nodes before the transplant, and the recipient is engrafted in this immunosuppressed state. Because the bone marrow is not x-irradiated, lymphoid stem cells proliferate and renew the population of recirculating lymphocytes. These newly formed lymphocytes appear to be more likely to become tolerant to the antigens of the graft. It can also be used to treat graft-versus-host disease (GvHD), in which the graft rejects the host.

Generalized Immunosuppressive Therapy

In 1959, Robert Schwartz and William Dameshek reported that treatment with 6-mercaptopurine suppressed immune responses in animal models. Joseph Murray and colleagues then screened a number of its chemical analogs for use in human transplantation. One, azathioprine, when used in combination with corticosteroids, dramatically increased survival of allografts. Murray received a Nobel Prize in 1990 for this clinical advance, and the developers of the drug, Gertrude Elion and George Hitchings, received the Nobel Prize in 1988.

Azathioprine (Imuran) is a potent mitotic inhibitor often given just before and after transplantation to diminish both B- and T-cell proliferation. Other mitotic inhibitors that are sometimes used in conjunction with immunosuppressive agents are cyclophosphamide and methotrexate. Cyclophosphamide is an alkylating agent that inserts into the DNA helix and becomes cross-linked, leading to disruption of the DNA chain. It is especially effective against rapidly dividing cells and is therefore sometimes given at the time of grafting to block T-cell proliferation. Methotrexate acts as a folic acid antagonist to block purine biosynthesis. Because mitotic inhibitors act on all rapidly dividing cells, they cause significant side effects, especially affecting the gut and liver, in addition to their target, bone marrow–derived cells. Most often, these mitotic inhibitors are combined with immunosuppressive drugs such as corticosteroids (e.g., prednisone and dexamethasone). These potent anti-inflammatory agents exert their effects at many levels of the immune response but especially in repressing the activity of NFκB, a transcription factor essential for the induction of a multitude of immune response genes, such as cytokines, involved in graft rejection.

More specific immune suppression became possible with the development of several fungal metabolites, including *cyclosporin A* (CsA), FK506 (tacrolimus), and rapamycin (also known as sirolimus). Although chemically unrelated, these exert similar effects, blocking the activation and proliferation of resting T cells. Some of these also prevent transcription of several genes encoding important T-cell activation molecules, such as IL-2 and the high-affinity IL-2 receptor (IL-2Rα). By inhibiting T_H-cell proliferation and cytokine expression, these drugs reduce the subsequent activation of various effector populations involved in graft rejection, making them a mainstay in heart, liver, kidney, and bone marrow transplantation. In one study of 209 kidney transplants from deceased donors, the 1-year survival rate was 80% among recipients receiving CsA and 64% among those receiving other immunosuppressive treatments.

Similar results have been obtained with liver transplants. Despite these impressive results, CsA does have some side effects, most notably toxicity to the kidneys. FK506 and rapamycin are 10 to 100 times more potent immunosuppressants than CsA and therefore can be administered at lower doses and frequently with fewer side effects.

Specific Immunosuppressive Therapy

The ideal immunosuppressant would be antigen-specific, inhibiting the immune response to the alloantigens present in the graft while preserving the recipient's ability to respond to other foreign antigens. Although this goal has not yet been achieved, several more targeted immunosuppressive agents have been developed. Most involve the use of *monoclonal antibodies* (mAbs) or soluble ligands that bind specific cell-surface molecules. One limitation of most first-generation mAbs came from their origin in animals. Recipients of these frequently developed an immune response to the nonhuman epitopes, rapidly clearing the mAbs from the body. This limitation has been overcome by the construction of humanized mAbs and mouse-human chimeric antibodies.

Many different mAbs have been tested in transplantation settings, and the majority work by either depleting the recipient of a particular cell population or by blocking a key step in immune signaling. *Anti-thymocyte globulin* (ATG), prepared from animals exposed to human lymphocytes, can be used to deplete lymphocytes in recipients prior to transplantation, but it has significant side effects. A more subset-specific strategy uses a mAb to the CD3 molecule of the TCR, called OKT3, and rapidly depletes mature T cells from the circulation. This depletion appears to be caused by binding of antibody-coated T cells to Fc receptors on phagocytic cells, which then phagocytose and clear the T cells from the circulation. In a further refinement of this strategy, a cytotoxic agent such as diphtheria toxin is coupled with the mAb. Antibody-bound cells then internalize the toxin and die. Another technique uses a mAb (basiliximab) specific for the high-affinity IL-2 receptor CD25. Since this receptor is expressed only on activated T cells, this treatment specifically blocks proliferation of T cells activated in response to the alloantigens of the graft. However, since T_{REG} cells also express CD25 and may aid in alloantigen tolerance, this strategy may have drawbacks. More recently, a mAb against CD20 (rituximab) has been used to deplete mature B cells and is aimed at suppressing AMR responses. Finally, in cases of bone marrow transplantation, mAbs against T cell–specific markers have been used to pretreat the donor's bone marrow to destroy immunocompetent T cells that may react with the recipient tissues, causing GvHD.

Because cytokines appear to play an important role in allograft rejection, these compounds can also be specifically targeted. Animal studies have explored the use of mAbs specific for the cytokines implicated in transplant rejection, particularly TNF-α, IFN-γ, and IL-2. In mice, anti-TNF-α mAbs prolong bone marrow transplants and reduce the

(a)

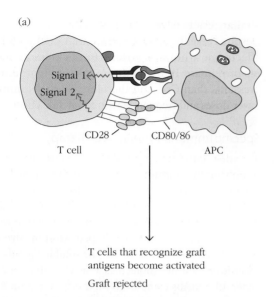

CD28   CD80/86
T cell   APC

T cells that recognize graft
antigens become activated

Graft rejected

(b)

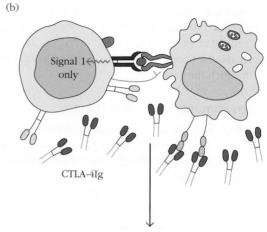

CTLA-4Ig

T cells that recognize graft antigens
lack costimulation and become anergic

Graft survives

FIGURE 16-18 Blocking costimulatory signals at the time of transplantation can cause anergy instead of activation of T cells reactive against a graft. T-cell activation requires both the interaction of the TCR with its ligand and the reaction of costimulatory receptors with their ligands (a). In (b), contact between one of the costimulatory receptors (CD28) on the T cell, and its ligand (CD80/86) on the APC, is blocked by reaction of CD80/86 with the soluble ligand CTLA-4Ig. The CTLA-4 is coupled to an immunoglobulin heavy chain, which slows its clearance from the circulation. This process specifically suppresses the induction of T cell responses against graft-specific antigens and improves graft survival.

incidence of GvHD. Antibodies to IFN-γ and to IL-2 have each been reported in some cases to prolong cardiac transplants in rats.

As described in Chapter 10, T_H-cell activation requires a costimulatory signal in addition to the signal mediated by the TCR. The interaction between CD80/86 on the membrane of APCs and the CD28 or CTLA-4 molecule on T cells provides one such signal (see Figure 10-3). Without this costimulatory signal, antigen-activated T cells become anergic (see Figure 10-5). CD28 is expressed on both resting and activated T cells, while CTLA-4 is expressed only on activated T cells and binds CD80/86 with a 20-fold-higher affinity. Scientists have demonstrated prolonged graft survival by blocking CD80/86 signaling in mice using a soluble fusion protein consisting of the extracellular domain of CTLA-4 fused to human IgG1 heavy chain (called CTLA-4Ig). This new drug, belatacept (Nulojix), was shown to induce anergy in T cells directed against the grafted tissue and has been approved by the FDA for the prevention of organ rejection in adult kidney transplant patients (**Figure 16-18**). The drug belatacept differs by only two amino acids from its close cousin, abatacept, another CTLA-4Ig molecule used to treat RA. However, this modification enhances the binding affinity for CD80/86, its costimulatory partner, making it superior for inhibiting transplant rejection. Some of the many treatments used to suppress transplant rejection in clinical settings are summarized in **Figure 16-19**, along with their sites of action.

Key Concepts:

- First-generation treatments to thwart transplant rejection, many of which are still used today, include drugs that block cell division or signaling as well as total lymphoid irradiation, all of which have a general suppressing effect on immunity.

- Recent additions to the antirejection inventory include several monoclonal antibodies that are more targeted and therefore have fewer side effects, such as those that block costimulation or specific cytokines, or that target certain immune cell types for destruction.

Immune Tolerance to Allografts Is Favored in Certain Instances

Sometimes, an allograft may be accepted with little or no use of immunosuppressive drugs. For instance, with tissues that lack alloantigens (e.g., cartilage or heart valves), no immunologic barrier to transplantation exists. However, in most instances an allograft is susceptible to immune attack unless preventive measures are taken. In these instances, allograft acceptance can be favored in one of two situations: when cells or tissue are grafted to a so-called privileged site that is more protected from immune system surveillance, or when a state of tolerance to donor alloantigens has

Transplant tissue

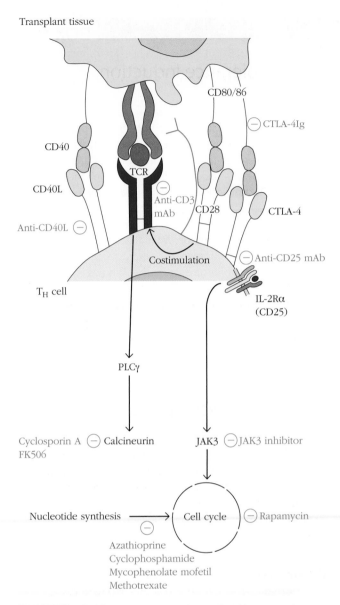

FIGURE 16-19 Site of action for various immunotherapy agents used in clinical transplantation.

been induced biologically in the recipient prior to transfer. One unique example of the latter can happen naturally when nonidentical twins share a placenta. In this case, each fetus is exposed to the alloantigens of the other in utero during formative stages of immune development, creating a state of histocompatibility between the two (**Classic Experiment Box 16-4**). Studies aimed at understanding how tolerance is induced and maintained throughout life, as well as methods for the clinical induction of tolerance in transplant settings, are collectively the subject of intense research and some recent advances in the field of transplantation science.

Immunologically Privileged Sites

An allograft placed in an immunologically privileged site, or an area without significant immune cell access (e.g., the anterior chamber of the eye, cornea, uterus, testes, and brain), is less likely to experience rejection. Each of these sites is characterized by a paucity of lymphatic vessels, and sometimes also blood vessels. Consequently, the recipient's T cells are less likely to become sensitized to the alloantigens of the graft, and the graft has an increased likelihood of acceptance even when HLA antigens are not matched. The privileged status of the cornea has allowed corneal transplants to be highly successful. Ironically, the successful transplantation of allogeneic pancreatic islet cells into the thymus in a rat model of diabetes suggests that the thymus is either immunologically privileged or can foster tolerance of antigens found there.

Presumably, improved graft survival in these so-called immunologically privileged sites occurs because they are effectively sequestered from the cells of the immune system. This suggests that the physical distancing of grafted cells may be a way to avoid attack. In one study, pancreatic islet cells were encapsulated in semipermeable, nonimmunogenic membranes and then transplanted into diabetic mice. The islet cells survived and produced insulin. The transplanted cells were not rejected, because the recipient's immune cells could not penetrate the membrane. This novel transplant method may have application for the treatment of diabetes.

Cells and Cytokines Associated with Graft Tolerance

There is now significant evidence that FoxP3-expressing T_{REG} cells play a role in transplantation tolerance. In *clinical operational tolerance*, where the graft survives despite the removal of all immunosuppressive therapy, there is an increase in the number of T_{REG} cells in the circulation as well as in the graft. These cells are believed to inhibit alloreactive cells by a combination of direct contact and expression of immunosuppressive cytokines, such as TGF-β, IL-10, and IL-35. Recent advances have been made that aim to induce alloantigen-specific T_{REG} cells prior to the transfer of a new organ. Likewise, several groups have reported improved graft survival in clinical trials that involve collecting regulatory T cells from transplant recipients, expanding these ex vivo, and then infusing patients with a boost of their own T_{REG} cells shortly before they receive their transplant!

Inducing Transplantation Tolerance

Methods for inducing tolerance to allow acceptance of allografts have been studied extensively in animal models, with some discoveries now entering human trials. The current favorite involves the induction of a state of mixed hematopoietic chimerism, where donor and recipient

Early Life Exposure to Antigens Favors Tolerance Induction

In 1945, Ray Owen, an immunologist working at University of Wisconsin (and later, CalTech), reported a novel observation in cattle. His discovery would advance our understanding of immune tolerance and provide grist for many transplant immunologists who followed in his footsteps. He noticed that nonidentical, or *dizygotic*, cattle twins retained the ability to accept cells or tissue from their genetically distinct sibling throughout their lives. This was not true for the nonidentical twins of other mammalian species that did not share a placenta in utero. In cattle, the shared placenta allowed free blood circulation from one twin to the other throughout the embryonic and fetal period. Although the twins may have inherited distinct paternal and maternal antigens, they did not recognize those of their placental partner as foreign and could therefore later accept grafts from them. He hypothesized that exposure to the alloantigens of their placental sibling during this early stage of life somehow

induced a lifelong state of immune tolerance to these antigens; in other words, these alloantigens were treated as self.

In 1953, Owen's observations were extended in a seminal paper by Rupert Billingham, Leslie Brent, and Peter Medawar. They showed that inoculation of fetal mice with cells from a genetically distinct donor mouse strain led to subsequent acceptance of skin grafts from donor mice of the same strain. This and other work led to the hypothesis that fetal development is an immunologically privileged period, during which exposure to an antigen induces tolerance to that antigen later in life. Peter Medawar and Sir Frank Macfarlane Burnet won the 1960 Nobel Prize in Physiology or Medicine, for their shared work in the discoveries that led to our understanding of acquired immunological tolerance.

Although no experimental data are available to demonstrate such specific tolerance in humans, there is anecdotal evidence. For example, transplants in very

young children show a slightly higher success rate than those in older individuals, suggesting that early life exposure to antigens may bias toward tolerance induction in humans as well. There are also clinical examples in adults where allografts mismatched at a single HLA locus are accepted with little or no immune suppression. When this mismatched antigen happens to be expressed by the transplant recipient's mother, it is possible that this occurred via perinatal exposure. We now appreciate that bidirectional movement of cells during pregnancy is not uncommon. The resulting state of microchimerism, which can last for decades in both parties, may thus influence immune tolerance to alloantigens.

REFERENCES

Billingham, R. E., L. Brent, and P. B. Medawar. 1953. Actively acquired tolerance of foreign cells. *Nature* **172:**603.

Owen, R. D. 1945. Immunogenetic consequences of vascular anastomoses between bovine twins. *Science* **102:**400.

---

hematopoietic cells coexist in the host prior to transplantation. The seed for this strategy originated from animal studies and observations in humans. For instance, transplant recipients who underwent total myeloablative therapy followed by donor bone marrow transfer *prior* to receiving a solid organ from the same donor displayed enhanced later tolerance when the solid organ was engrafted. A modified protocol involving a less intense nonmyeloablative procedure followed by bone marrow transfer resulted in mixed chimerism that, even when quite transient, was still associated with improved graft outcomes. The mechanism for this induction of tolerance is still unclear; both central deletion of alloreactive T cells and an enhancement of immune suppression by T_{REG} cells are both hypothesized.

Key Concept:

- Short of exposure to alloantigens in utero, tolerance can be favored when engrafted cells are placed outside the reach of the immune system or when T_{REG} cells against alloantigens are intentionally induced or expanded prior to transplantation.

Conclusion

Our understanding of the principles underlying immune tolerance has made huge advances in the past decade. Tolerance was once viewed primarily as the elimination of all autoreactive cells—the "ignorance is bliss" model. Our understanding of tolerance is now much more nuanced. Scientists now recognize that while some structures are clearly kept off the radar of the immune system (evasion) and many of the most aggressive anti-self lymphocytes are deleted (elimination), certain self-recognizing regulatory lymphocytes are essential for dampening anti-self immune responses (engagement). Without this regulatory component, the delicate balance falters. The mechanisms underlying both central and peripheral tolerance have been demonstrated in animal models and are now being applied to manipulate immune tolerance in humans. Various forms of immunotherapy are used to treat autoimmune disease and to block immune rejection of allografts, illustrating some of the most promising new bench-to-bedside applications of the principles of immune tolerance.

REFERENCES

Abdelnoor, A. M., et al. 2009. Influence of HLA disparity, immunosuppressive regimen used, and type of kidney allograft on production of anti–HLA class-I antibodies after transplant and occurrence of rejection. *Immunopharmacology and Immunotoxicology* 31:83.

Anderson, M. S., and M. A. Su. 2016. AIRE expands: new roles in immune tolerance and beyond. *Nature Reviews Immunology* 16(4):247.

Chinen, J., and R. H. Buckley. 2010. Transplantation immunology: solid organ and bone marrow. *Journal of Allergy and Clinical Immunology* 125(2 Suppl 2):S324.

Costa, V. S., T. C. Mattana, and M. E. da Silva. 2010. Unregulated IL-23/IL-17 immune response in autoimmune diseases. *Diabetes Research and Clinical Practice* 88:222.

Damsker, J. M., A. M. Hansen, and R. R. Caspi. 2010. Th1 and Th17 cells: Adversaries and collaborators. *Annals of the New York Academy of Sciences* 1183:211.

Feng, Y., et al., 2015. A mechanism for expansion of regulatory T-cell repertoire and its role in self-tolerance. *Nature* 528:132.

Giles, J. R., et al., 2017. Autoreactive helper T cells alleviate the need for intrinsic TLR signaling in autoreactive B cell activation. *JCI Insight.* 2(4):e90870.

Gorantla, V. S., et al. 2010. T regulatory cells and transplantation tolerance. *Transplantation Reviews (Orlando)* 24:147.

Groth, C. G. 1972. Landmarks in clinical renal transplantation. *Surgery, Gynecology and Obstetrics* 134(2):327.

Issa, F., A. Schiopu, and K. J. Wood. 2010. Role of T cells in graft rejection and transplantation tolerance. *Expert Review of Clinical Immunology* 6:155.

Kansupada, K. B., and J. W. Sassani. 1997. Sushruta: the father of Indian surgery and ophthalmology. *Documenta Ophthalmologica* 93:159.

Kapp, J. A., and R. P. Bucy. 2008. CD8$^+$ suppressor T cells resurrected. *Human Immunology* 69:715.

Kunz, M., and S. M. Ibrahim. 2009. Cytokines and cytokine profiles in human autoimmune diseases and animal models of autoimmunity. *Mediators of Inflammation* 2009:979258.

Li, M. O., and A. Y. Rudensky. 2016. T cell receptor signalling in the control of regulatory T cell differentiation and function. *Nature Reviews Immunology* 16:220.

Lu, L., and H. Cantor. 2008. Generation and regulation of CD8$^+$ regulatory T cells. *Cellular and Molecular Immunology* 5:401.

Panduro, M., et al., 2016. Tissue Tregs. *Annual Review of Immunology* 34:609.

Round, J. L., R. M. O'Connell, and S. K. Mazmanian. 2010. Coordination of tolerogenic immune responses by the commensal microbiota. *Journal of Autoimmunity* 34:J220.

Sakaguchi, S. 2004. Naturally arising CD4$^+$ regulatory T cells for immunologic self-tolerance and negative control of immune responses. *Annual Review of Immunology* 22:531.

Sethi, S., et al., 2017 Desensitization: overcoming the immunologic barriers to transplantation. *Journal of Immunology Research* 2017:6804678.

Steward-Tharp, S. M., Y. J. Song, R. M. Siegel, and J. J. O'Shea. 2010. New insights into T cell biology and T cell–directed therapy for autoimmunity, inflammation, and immunosuppression. *Annals of the New York Academy of Sciences* 1183:123.

Tao, J. H., et al., 2017. FoxP3, Regulatory T cell, and autoimmune diseases. *Inflammation* 40(1):328.

Thomas, R. 2010. The balancing act of autoimmunity: central and peripheral tolerance versus infection control. *International Reviews of Immunology* 29:211.

Turka, L. A., K. Wood, and J. A. Bluestone. 2010. Bringing transplantation tolerance into the clinic: lessons from the ITN and RISET for the Establishment of Tolerance consortia. *Current Opinion in Organ Transplantation* 15:441.

Ulges, A., E. Schmitt, C. Becker, T. Bopp. 2016. Context- and tissue-specific regulation of immunity and tolerance by regulatory T cells. *Advances in Immunology* 132:1.

von Boehmer, H., and F. Melchers. 2010. Checkpoints in lymphocyte development and autoimmune disease. *Nature Immunology* 11:14.

Wing, K., and S. Sakaguchi. 2010. Regulatory T cells exert checks and balances on self tolerance and autoimmunity. *Nature Immunology* 11:7.

Yamaguchi, T. et al.,. 2017. Interspecies organogenesis generates autologous functional islets. *Nature* 542:191.

Useful Websites

https://www.immunetolerance.org This website, run by the United States–based Immune Tolerance Network, is aimed at translating basic research findings in tolerance induction into therapy for autoimmunity, allergy, and transplantation.

https://www.niaid.nih.gov/diseases-conditions/autoimmune-diseases A website run by a branch of the National Institutes of Health related to immunology and autoimmune diseases.

https://www.nature.com/subjects/autoimmune-diseases A site maintained by the journal *Nature* and dedicated to the latest scientific research published on autoimmune diseases and their treatment.

https://www.unos.org The United Network for Organ Sharing site has information concerning solid organ transplantation for patients, families, doctors, and teachers, as well as up-to-date numbers on waiting patients.

https://bethematch.org/ The National Marrow Donor Program website contains information about all aspects of bone marrow transplantation.

https://optn.transplant.hrsa.gov/data The Organ Procurement and Transplantation Network site is run by the U.S. Department of Health and Human Services. It maintains real-time numbers on waiting patients, as well as data on organ transplants in the United States.

1. Explain why all self-reactive lymphocytes are not eliminated in the thymus or bone marrow. How are the surviving self-reactive lymphocytes prevented from harming the host?

2. Why is tolerance critical to the normal functioning of the immune system? Can you name and describe examples of human disorders in which there has been a breakdown in self tolerance?

3. What is the mechanism and importance of receptor editing in B-cell tolerance? What role, if any, does T-cell tolerance play in immune tolerance by B cells?

4. For each of the following autoimmune diseases (a–j), select the most appropriate characteristic (1–10) from the list. Also identify the immune cell/s or molecules most associated with the disease (e.g., Antibodies, T_H cells, CTLs and T_{REG} cells). It is possible that more than one effector cell/molecule is involved.

DISEASE

    a. _____ Experimental autoimmune encephalitis (EAE)
    b. _____ Immune dysregulation, polyendocrinopathy, enteropathy, X-linked (IPEX) syndrome
    c. _____ Systemic lupus erythematosus (SLE)
    d. _____ Type 1 diabetes (T1D)
    e. _____ Rheumatoid arthritis (RA)
    f. _____ Hashimoto's thyroiditis
    g. _____ Autoimmune polyendocrine syndrome type 1 (APS-1)
    h. _____ Myasthenia gravis
    i. _____ Multiple sclerosis (MS)
    j. _____ Autoimmune hemolytic anemia

CHARACTERISTIC

(1) Disease caused by immune recognition of the acetylcholine receptor

(2) Results from cell-mediated reaction to thyroid antigens

(3) Systemic autoimmune disorder caused by defects in the *AIRE* gene

(4) Caused by recognition and destruction of antigens on red blood cells

(5) Autoimmune disease that results in immune response to myelin proteins

(6) Induced by injection of *myelin basic protein* (MBP) plus complete Freund's adjuvant

(7) Caused by a B-cell response to self IgG

(8) Symptoms included anti-DNA and DNA-associated protein immune reaction

(9) Individuals with this disorder demonstrate little or no FoxP3 expression

(10) A disease caused by immune reactions to pancreatic beta cells

5. Experimental autoimmune encephalitis (EAE) has proved to be a useful animal model of autoimmune disorders.

    a. Describe how this animal model is made.
    b. What is unusual about the animals that recover from EAE?
    c. How has this animal model indicated a role for T cells in the development of autoimmunity?

6. Molecular mimicry is one mechanism proposed to account for the development of autoimmunity. Use rheumatic fever and group A streptoccocus to explain how the principle of molecular mimicry has been used to explain the induction of some forms of autoimmune disease.

7. Describe at least three different mechanisms by which a localized viral infection might contribute to the development of an organ-specific autoimmune disease.

8. Monoclonal antibodies have been administered for therapy in various autoimmune animal models. Which specific monoclonal antibodies have been used, and what is the rationale for these approaches?

9. Indicate whether each of the following statements is true or false. If you think a statement is false, explain why.

    a. T_H1 cells have been associated with development of autoimmunity.
    b. Immunization of mice with IL-12 prevents induction of EAE after injection of MBP plus adjuvant.
    c. The presence of the *HLA B27* allele is diagnostic for ankylosing spondylitis, an autoimmune disease affecting the vertebrae.
    d. A defect in the gene encoding Fas can reduce programmed cell death by apoptosis and can be associated with autoimmune disease.
    e. In the absence of FoxP3 expression, central tolerance pathways will not occur.

10. For each of the following autoimmune disorders (a–f), indicate which of the following treatments (1–7) may be appropriate.

DISEASE

    a. Hashimoto's thyroiditis
    b. Systemic lupus erythematosus (SLE)
    c. Immune dysregulation, polyendocrinopathy, enteropathy, X-linked (IPEX) syndrome
    d. Myasthenia gravis
    e. Type 1 diabetes (T1D)
    f. Rheumatoid arthritis (RA)

TREATMENT

    (1) Cyclosporin A
    (2) Thymectomy
    (3) Plasmapheresis

(4) Anti-CD20 mAb
(5) Kidney transplant
(6) Pancreatic islet transfer
(7) Thyroid hormones

11. Which of the following are examples of mechanisms for the development of autoimmunity? For each possibility, give an example.

    a. Polyclonal B-cell activation
    b. Tissue damage
    c. Viral infection
    d. Increased expression of TCR molecules
    e. Increased expression of MHC class II molecules

12. Indicate whether each of the following statements is true or false. If you think a statement is false, explain why.

    a. Acute rejection is mediated by pre-existing host antibodies specific for antigens on the grafted tissue.
    b. Second-set rejection is a manifestation of immunologic memory.
    c. Host dendritic cells can migrate into grafted tissue and act as APCs.
    d. All allografts between individuals with identical HLA haplotypes will be accepted.
    e. Cytokines produced by host T_H cells activated in response to alloantigens play a major role in graft rejection.

13. Indicate whether a skin graft from each donor to each recipient listed in the following table would result in rejection (R) or acceptance (A). If you believe a rejection reaction would occur, indicate whether it would be a first-set rejection (FSR), occurring in 12 to 14 days, or a second-set rejection (SSR), occurring in 5 to 6 days. All the mouse strains listed have different H2 haplotypes.

Donor	Recipient
BALB/c	C3H
BALB/c	Rat
BALB/c	Nude mouse
BALB/c	C3H, had previous BALB/c graft
BALB/c	C3H, had previous C57BL/6 graft
BALB/c	BALB/c
BALB/c	(BALB/c × C3H)F$_1$
BALB/c	(C3H × C57BL/6)F$_1$
(BALB/c × C3H)F$_1$	BALB/c
(BALB/c × C3H)F$_1$	BALB/c, had previous F$_1$ graft

14. Graft-*versus*-*h*ost *d*isease (GvHD) frequently develops after certain types of transplantations.

    a. Briefly outline the mechanisms involved in GvHD.
    b. Under what conditions is GvHD likely to occur?
    c. Some researchers have found that GvHD can be diminished by prior treatment of the graft with monoclonal antibody plus complement or with monoclonal antibody conjugated with toxins. List at least two cell-surface antigens to which monoclonal antibodies could be prepared and used for this purpose, and give the rationale for your choices.

15. What is the biologic basis for attempting to use soluble CTLA-4Ig or anti-CD40L to block allograft rejection? Why might this be better than treating a graft recipient with cyclosporin A or FK506?

16. Immediately after transplantation, a patient is often given strong doses of immunosuppressive or anti-rejection drugs, which are then tapered off as time passes. Describe the specific mechanisms of action of the commonly used antirejection drugs azathioprine, cyclosporin A, FK506, and rapamycin. Why is it possible to decrease the use of some of these drugs at some point after transplantation?

17. Like allergy, autoimmunity can run in families. Using each of the following genes/proteins as an example, explain how inheritance of a specific *variant* of this gene could predispose members of a family to autoimmune disease. In your answer, be specific about the phenotype you expect from this variant form (e.g., enhanced expression) and how it alters the immune response in a way that could lead to autoimmunity.

    a. IL-2 receptor
    b. CTLA-4
    c. CD40
    d. ERAP1
    e. FoxP3

18. Beginning with sensitization, describe the immune events that you expect to take place following the transplantation of a kidney between nonidentical siblings who share half their MHC molecules. You can assume that they are matched for ABO blood group antigens and have successfully passed the cross-match (i.e., they are antibody negative in a cross-match test).

CLINICAL FOCUS QUESTIONS

1. What are some of the possible sex differences that may make females more susceptible to autoimmune diseases than males? Do these same features provide females with any immune advantages or disadvantages in nature?

2. Some immunotherapy regimens aim to treat either autoimmune disease or transplant rejection by inducing self tolerance. Name one specific example of a drug or regimen for each type of clinical situation (autoimmune disease and transplantation). In these instances, is the goal of the physician to induce central or peripheral tolerance, or both? Please explain.

3. What features would be desirable in an ideal animal donor for xenotransplantation? How would you test your model prior to moving into clinical trials in humans? Are there any ethical hurdles related to your ideal model, and how could you begin to address these?

Infectious Diseases and Vaccines

17

Learning Objectives

After reading this chapter, you should be able to:

1. Apply previous immunology knowledge when presented with a new pathogen, including location of infection and specific pathogen characteristics, to identify specific innate immune response elements (e.g., NLRs versus CLRs) that would be most effective in the initial recognition of the infection, and adaptive response elements (e.g., antibodies versus T_C cells) most appropriate for detection and elimination.

2. Describe and categorize the various methods used by different infectious organisms to evade host immune defense, and specifically explain how each subverts particular elements of host immunity.

3. Use backward design to create a hypothetical vaccine, first by identifying correlates of immune protection, and then by selecting the technique(s), administration route(s), and adjuvant(s) most appropriate for inducing the desired immune response, all while applying rational immunologic justification for your choices.

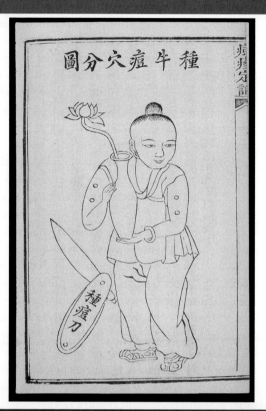

Woodcut showing sites for cowpox inoculation, from an 1888 edition of *Definitive Treatise on Pox Diseases* by Zhu Chunxia (Qing period, 1644–1911). *[Chinese C19 woodcut: "Cowpox inoculation." Credit: Wellcome Collection. CC BY]*

Infectious disease has probably been around longer than humans have. Some scientists even posit that infectious agents contributed to the demise of the dinosaur. We do know that bubonic plague, smallpox, and measles are just a few of the communicable diseases credited with killing millions in centuries past, as well as the downfall of nations. There are likely other infectious diseases that have been lost to history, as the evolutionary arms race between microbe and man is an ongoing process. So maybe it is not surprising that, rather than letting fate take its course, ancient civilizations employed practices that increased their chance of survival. Take smallpox, for example. Over two millennia ago, humans realized

that those who survived smallpox would not contract the disease again. As early as 430 B.C.E., Thucydides in ancient Greece tells of enlisting only smallpox survivors to nurse the sick, applying the principle of immunologic memory long before the science was understood. At least as early as the tenth century, intentional manipulation of the immune response was practiced, again as a tool to survive death from smallpox. In ancient China, material from smallpox pustules of individuals who experienced

Key Terms

Immunization	Immunogens	Passive immunity	Toxoids
Vaccination	Correlates of immune protection	Active immunity	Adjuvant
Vector-borne infections		Herd immunity	Antigenic drift
Original antigenic sin	Vaccine	Sterilizing immunity	Antigenic shift

mild disease was dried and inhaled by nonimmune individuals to induce protection, with some risk but generally positive results. During roughly the same period, reports from India, Africa, and Persia recount medical practices or religious ceremonies likewise involving intentional exposure to smallpox as a means to survive later outbreaks. These practices, sometimes referred to as scarification or *inoculation* (Latin for "grafting"), are the early predecessors of present-day **immunization** through the practice of **vaccination**. While Jenner gets much worthy credit for developing and disseminating this practice in modern history (see Chapter 1), the stage was set independently in multiple societies many centuries earlier.

Since those initial vaccination trials of Edward Jenner and Louis Pasteur, vaccines have been developed against many infectious agents that were once major afflictions of humankind. For example, the incidence of diphtheria, measles, mumps, pertussis (whooping cough), rubella (German measles), poliomyelitis, and tetanus, which collectively once claimed the lives of approximately one of every four children, has declined dramatically as vaccination has become more common. We may know intuitively that vaccination is a life-saving weapon, but recent reports have highlighted the cost-effectiveness of these measures in dollars. In terms of a monetary investment that can have long-term impacts on health and productivity, immunization outranks many other possible investments (**Figure 17-1**).

While vaccines against all of the early childhood killers mentioned above are available, their use globally is not uniform. This necessitates continual campaigns to increase distribution. And, of course, the evolution of new infectious agents or the increased spread of existing microbes sustains the need for new vaccines. These and other public health concerns led to the development of agencies, such as the World Health Organization (WHO) and the United States–based Centers for Disease Control and Prevention (CDC), to help organize the accumulating data concerning infectious disease. These organizations monitor public health and disease, guide health care policy discussions, respond to sudden infectious disease outbreaks, and report regularly on their findings. Although the local and international expenditures on these practices are questioned at times, there is no doubt that these and present-day biomedical advances have led us to an age in which rapid and often effective response to sudden infectious disease outbreaks is commonplace. These health-focused agencies also facilitate a better appreciation for the social conditions and policies, both locally and globally, that can either sustain or limit such outbreaks of infectious disease.

Although vaccination can provide critical defense against many pathogens, infectious diseases still kill millions each

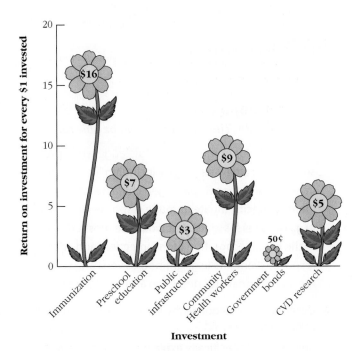

FIGURE 17-1 The economics of vaccination in 94 low- and middle-income countries. Dollar for dollar, investment in immunization returns more in terms of savings in health care costs, lost salaries, and low productivity due to illness than these other education- and health-related investments, all of which also pay back more than is spent on the programs. CVD = cardiovascular disease. *[Data from* Journal of Health Affairs, *2016, and GAVI.]*

year. While the number varies greatly by region, about 25% of deaths worldwide are associated with communicable diseases, which kill up to 12 million people each year. Communicable diseases and their sometimes chronic aftereffects are the underlying issue for most of the top 10 causes of death in children under age 5 worldwide (**Figure 17-2**). The fact that malnutrition weakens the immune response only adds to this toll, especially in developing nations. In fact, there is no question that access to sanitation, antibiotics, vaccines, and nutrient-rich foods has hugely reduced the impact of infectious disease. Even so, lower respiratory infections are still the leading cause of death in low-income economies. And for countries like the United States, where vaccine use is prevalent, communicable diseases both new and old still affect us. The recent spread of Zika virus has prompted anxiety but also much progress in the development of a vaccine (described in **Clinical Focus Box 17-1**). And there are still many "old" infectious killers out there that we need to focus on. In 2015, our longtime companion tuberculosis surpassed HIV as the leading cause of death from infectious disease.

In this chapter, the concepts of immunity described throughout the text are applied to selected infectious

Top 10 causes of death worldwide among children under 5

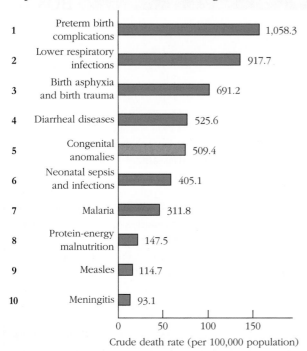

Cause group
- ■ Communicable, maternal, perinatal and nutritional conditions
- ■ Noncommunicable diseases

FIGURE 17-2 Illness related to infectious disease contributes to most of the top 10 causes of death worldwide in children under 5. Communicable diseases (along with nutrition and maternal/fetal conditions; blue) are associated with nine of 10 of the major causes of death in young children across the globe. Numbers 2, 4, 6, 7, 9, and 10 are directly associated with communicable diseases. *[Data from 2017 WHO Global Health Observatory Data website: www.who.int/gho.]*

diseases caused by the four main types of pathogens (viruses, bacteria, fungi, and parasites). We focus on particular infectious diseases that affect large numbers of people, that illustrate specific immune concepts, or that use novel strategies to subvert the immune response, as well as some diseases that have warranted recent headlines. The chapter concludes with a section on vaccines, organized by the type of vaccine design being applied and including examples of specific pathogens that have been successfully targeted using these strategies.

The Importance of Barriers and Vectors in Infectious Disease

First and foremost, in order for a pathogen to establish an infection in a susceptible host, it must breach physical and chemical barriers. One of the first and most important of these barriers consists of the epithelial surfaces of the skin and the lining of the gut. The difficulty of penetrating these surfaces ensures that most pathogens never gain productive entry into the host. In addition, the epithelia produce chemicals that are useful in preventing infection. The secretion of gastric enzymes by specialized epithelial cells lowers the pH of the stomach and upper gastrointestinal tract, and other specialized cells in the gut produce antibacterial peptides. In addition, normal commensal flora present at mucosal surfaces (the gastrointestinal, urogenital, and respiratory tracts) can competitively inhibit the binding of pathogens to host cells. When the host is otherwise healthy, and pathogen dose and virulence are minimal, these barriers can often block productive infection altogether.

Sometimes infectious agents get help from other organisms to circumvent host barriers. In these cases a third party, called a *vector*, helps to carry the infection from one organism to another. These vectors, or *intermediate hosts*, can transmit a pathogen between one infected human and another, or from an infected animal to a human. Infectious disease vectors are most often blood-sucking arthropods (e.g., ticks, fleas, flies, or mosquitoes), which breach natural barriers like the skin with their bite and introduce the pathogens they carry directly into a susceptible host. These **vector-borne infections** account for approximately one out of every six instances of human infectious disease and are typically restricted to areas in which the intermediate host is found. Examples include malaria and Zika fever, both discussed later in this chapter.

Interventions that introduce barriers to infection in these *intermediate* hosts can be used as an indirect strategy to disrupt the cycle of infectious disease in humans. Recent studies in dengue virus, transmitted by the bite of an infected mosquito and the cause of an often fatal hemorrhagic fever in humans, suggest that it may be possible to engineer mosquitoes that are resistant to infection with the virus. When these engineered mosquitoes were released into the wild, they began to supplant the wild-type, virus-susceptible mosquito population, potentially breaking the cycle of transmission. This and other exciting new avenues of research that target animal disease vectors could advance infectious disease eradication without the requirement to intervene with the human immune response. Of course, this strategy is not a possibility with most infectious diseases, for which there is no arthropod vector.

Of course, even infectious agents that penetrate these barriers will need to face the first responders of innate immunity, which in many instances take care of them without the need for a full-scale adaptive response. These early responses are often tailored to the type of pathogen, using molecular pattern recognition receptors (see Chapter 4). Some bacteria produce endotoxins such as *lipopolysaccharide* (LPS), which stimulate macrophages or endothelial cells to produce cytokines, including IL-1, IL-6, and *tumor necrosis factor-α* (TNF-α). These cytokines can activate nearby innate cells, encouraging

CLINICAL FOCUS
BOX 17-1

The Emerging Story of Zika Virus

Pregnancy is typically a period of both excitement and anxiety for expectant parents. However, in certain regions of the world, fears around pregnancy and health of the developing fetus have increased significantly thanks to the Zika virus. Despite these new fears, the virus itself is not new. First identified and named in 1947, Zika was originally isolated from a macaque in captivity in the Zika Forest of Uganda. Since then there have been many reports of outbreaks of human infection, likely spurred by mosquito transfer from a wild animal reservoir, but with only mild symptoms. In fact, in areas where the mosquito vector for this virus is endemic, transmission rates can be quite high, occurring through both vector-borne transfer and sexual contact. In a study conducted in New York in 2016, 5% of individuals who had recently traveled out of the country to regions at risk of Zika exposure possessed virus-specific antibodies, despite little or no memory of illness. As far as we know, Zika virus itself only arrived inside U.S. borders in 2016, with reports so far of local virus infection in Puerto Rico, Florida, Texas, and parts of California. **Figure 1** shows a map of regions affected by Zika, as of February 2017. No one knows for certain why the geographic region for this virus seems to be spreading, although expansion of the animal reservoir and/or

insect vector habitat, possibly associated with changes in climate, is one hypothesis.

Based on recent findings, it appears that Zika virus takes its biggest toll on those who first encounter it in utero. The first sign of this was the 2015 observation of a 20-fold increase over previous years in the number of infants born in Brazil with microcephaly. Some of these were later linked to Zika infection of the mothers, especially during the first or second trimester. Later, other cases of neurological defects were associated with similar Zika infection timing. A study done in Brazil to track the impact of virus exposure on developing fetuses found that up to 55% of the infants exposed in utero during the first 6 months of gestation displayed neurological delays. A similar United States–based study found a much lower level of 6%. Differences in the way the studies were conducted and among the women who were recruited (symptomatic versus nonsymptomatic) may at least partially explain these variations, although many questions remain. Nonetheless, from this work we have learned that congenital Zika syndrome, as it is now called, occurs mainly when susceptible (nonimmune) women are infected shortly before conception or during the first 6 months of pregnancy. Individuals who have recovered from previous infection (2 months for women and a

little longer for men) do not carry live virus and are therefore not a source of transfer to others, including any child they conceive.

For several reasons, this virus and associated disease are ripe for vaccine development. First, scientists have experience with developing effective vaccines against flaviviruses, like yellow fever and dengue, the group to which Zika belongs. Second, we know that natural Zika infection produces rapid protective immunity in humans, and usually without significant associated disease. This means that our immune systems know how to fight this virus and researchers know what correlates of immune protection to look for. Finally, bluntly, it hit the right places. Both Brazil and the United States, where Zika is currently impacting health and well-being, are relatively rich countries with the infrastructure to move forward quickly on vaccine design. National health agencies and government money have already fueled the early stages of vaccine development, insulating interested drug companies from financial loss. And financial gains from a successful vaccine are likely to be large—very large. Collectively, this means that there is no lack of interest from pharmaceutical companies, many of which are already racing for the finish line.

Early anti-Zika vaccine initiatives have produced promising results, exploring several different approaches including

(continued)

phagocytosis of the bacteria. The cell walls of many gram-positive bacteria contain a peptidoglycan that activates the alternative complement pathway, leading to opsonization and phagocytosis or lysis (see Chapter 5). Viruses commonly induce the production of interferons, which can inhibit viral replication by inducing an antiviral response in neighboring cells. Viruses are also controlled by *natural killer* (NK) cells, which frequently form the first line of defense in these infections (see Chapter 4). In many cases, these innate responses can lead to the resolution of infection. If these innate measures are not sufficient to eradicate the pathogen, the more specific adaptive immune response will come into play.

During this later, very pathogen-specific stage of the immune response, final eradication of the foreign invader

often occurs, typically leaving a memory response capable of halting secondary infections. However, just as adaptive immunity in vertebrates has evolved over many millennia, pathogens have evolved a variety of strategies to escape destruction by the immune system. Some pathogens reduce their own antigenicity either by growing within host cells, where they are sequestered from immune attack, or by shedding their membrane antigens. Other pathogen strategies include camouflage (expressing molecules with amino acid sequences similar to those of host cell membrane molecules or acquiring a covering of host membrane molecules); suppressing the immune response selectively or directing it toward a pathway that is ineffective at fighting the infection; and continual variation in microbial surface

Zika travel recommendation:
- High risk
- Lower risk
- No known Zika
- Reported cases (Texas and Florida)

FIGURE 1 **Zika virus–affected regions as of 2017.** This global map shows the major risk zones for Zika virus infection (purple) and new regions where initial reports show limited activity (purple outline), as of February 2017. *Source: https://wwwnc.cdc.gov/travel/page/world-map-areas-with-zika*

live attenuated, killed, and DNA-based vaccines. To date, five clinical trials using a purified inactivated virus (ZPIV) are already in process. The ZPIV vaccine is prepared by inactivating whole virus, using techniques that leave important surface structures intact and available for immune recognition of conformational epitopes. Earlier animal studies with this form of the vaccine produced promising neutralizing antibody responses that delivered immune protection. Ongoing human trials are aimed at evaluating the optimal dose, timing, and number of required exposures, as well as any cross-reactivity with other flaviviruses (dengue is a strong candidate). One advantage of this technique is that it has been tested previously and

found to be safe. One disadvantage is that production and quality control testing can take significant time. For this reason, DNA vaccines against Zika, which have advantages in terms of speed, efficiency, and ease of delivery, are also being explored. In fact, some of the ZPIV clinical trials will also evaluate a Zika virus DNA vaccine prime followed by a ZPIV protein boost. Because this virus seems to produce such mild symptoms when encountered outside the womb, some groups are also exploring live attenuated versions of the virus as a vaccine. The good news is that most of the scientists leading these efforts are confident that a protective vaccine is not far off.

Zika virus also has some real advantages as an immunology teaching tool. What

better way to study much of the material in this chapter and the field, all in one place? Zika virus research includes emerging infectious diseases, viruses, arthropod vectors, animal reservoirs, expanding disease geography, sexual transmission, antiviral immune response, neutralizing antibodies, correlates of immune protection, multiple vaccine designs (live, killed, and DNA), and of course, a quick and robust ongoing global response.

REFERENCES

Centers for Disease Control and Prevention (CDC). 2017. Zika virus. https://www.cdc.gov/zika/

Pierson, T. C., and B. S. Graham. 2016. Zika virus: immunity and vaccine development. *Cell* **167**:625.

antigens. Examples of these evasion strategies will be highlighted throughout the chapter as we discuss the four different classes of pathogen—viral, bacterial, fungal, and parasitic—and the various adaptive responses that are most effective against them.

Key Concepts:
- Barriers such as the skin and mucous-lined surfaces serve as a buffer between host and infectious agents.

- Some infectious agents cross these barriers with the help of the bite from arthropod vectors, such as the mosquito, that may transmit vector-borne disease from one infected individual to another.

- The first responders of the innate immune system are an additional barrier to infection, sometimes clearing pathogen without the need for an adaptive response.

The Link between Location and Immune Effector Mechanism

There are many different places that infectious agents can "be" in or on the body. Infections can occur on body surfaces (skin or mucosa) or they can penetrate epithelial layers, breaching skin, urogenital, or gut linings. Once they have crossed these barriers, infectious agents may remain local or spread from the site of entry. Some infectious agents can be found in the interstitial fluid that bathes our tissues, where they should then be carried to draining lymph nodes. Infections that actually enter the bloodstream (called *sepsis*) are rare. When this happens, the infection can spread quickly throughout the body, and the resulting immune response can do more damage than the pathogen itself (e.g., septic shock). Finally, some pathogens spend all or part of their lives inside host cells, occupying either membrane-enclosed (vesicular) or cytosolic and nuclear spaces. *Importantly, the entry site and ultimate location of an infectious agent in or on the body will determine which immune tools are available and best suited for pathogen detection and elimination.* **Overview Figure 17-3** shows common infection entry sites and some mechanisms for breaching epithelial barriers, as well as the types of spaces infectious agents may occupy in the body and the various immune tools required for their detection. The following discussion focuses on the key immune response effectors based on the space occupied by the pathogen, rather than on the pathogen type itself.

Mucosal or Barrier Infections Are Typically Controlled by T_H2-Type Responses

The majority of, but not all, infectious agents enter through mucosal routes: the mouth, nose, eyes, or urogenital tract (M in Overview Figure 17-3). This site is somewhat immunologically unique, in that regular encounter with foreign substances like food and commensal microorganisms is expected and essential for survival. Therefore, dangerous infectious agents need to be limited or eradicated before they can induce damage or penetrate the epithelial cell lining of the body, while still maintaining tolerance and even collaboration with commensals (see Chapter 13 for a more complete discussion of this balance).

When a would-be pathogen enters the digestive system, it must survive stomach acid and successfully make its way through the intestinal tract. As we know from Chapter 13, there is an entire system designed to deal with pathogenic invaders in these regions. The immune effectors available here include soluble antimicrobial proteins secreted by epithelial cells, innate lymphoid cells, and a dispersed set of lymph node–like structures, collectively referred to as *m*ucosal-*a*ssociated *l*ymphoid *t*issue (MALT).

With a few exceptions, pathogens that reside in or near the surface of the body are most effectively controlled by T_H2-type responses. These include the activation of ILC2s, T_H2-specific cytokines (e.g., IL-4 and IL-13), and IgE capable of recognizing surface epitopes of the pathogen. In fact, immunity against many of the most common metazoan parasites (helminths, or worms) correlates with high antigen-specific IgE:IgG ratios. For this reason, cells expressing the high-affinity IgE receptor (e.g., mast cells, basophils, and eosinophils) are also important players in this type-2 response. In some cases, activation of these effector responses leads to expulsion of helminths from body cavities or surfaces.

Another important player in the health of barrier surfaces is dimeric IgA (dIgA). This particular isotype is found most abundantly at mucosal surfaces, after being carried across epithelial cells via transcytosis by the poly-Ig receptor (see Chapter 12). Surface IgA serves an important role in neutralizing potential pathogens and maintaining barrier integrity. Binding of IgA to infectious agents can block attachment to epithelial cells (neutralization) and thus aid in elimination of the infectious organism passively, without induction of inflammation. In fact, we know that chronic inflammation of the gut correlates with dysbiosis, disruption of barrier integrity, systemic immune imbalance, and inflammatory disease. Thus, these mechanisms that foster a "quiet exit" of surface pathogens via noninflammatory means can actually prevent further disease.

> **Key Concept:**
> • T_H2-type responses are important for the control of infections that arise at the body surfaces, especially anti-helminth immunity.

Extracellular Pathogens Must Be Recognized and Attacked Using Extracellular Tools

When infectious organisms do breach epithelial barriers, they can enter the relatively sterile environment of the body. At this point they may either hitch a ride inside host cells (intracellular infections, discussed shortly) or enter the fluid-filled extracellular spaces of the body, namely interstitial fluid or the bloodstream (E in Overview Figure 17-3). Extracellular infections can remain local or spread through the body, via either the circulatory or lymphatic system. The immune mediators located in these extracellular spaces vary somewhat depending on the specific microenvironment, but can include pattern recognition receptors (PRRs; see Chapter 4) on phagocytic cells, complement, antimicrobial compounds, cytokines (which activate immune cells), and antibody (especially IgG, mIgA, and IgM). Often, T_{FH}, T_H17, and/or T_H2 cells are the T-cell drivers behind the adaptive elements of the response against extracellular infections. It is worth noting that all classes of pathogen that breach epithelial barriers will be found in the extracellular spaces (even viruses) for at least some of their lifetime in the host. Therefore, these extracellular immune effectors can and do play a role in recognition, neutralization, and eradication whenever the pathogen is present in these spaces.

The Entry Point and In Vivo Microenvironment of Infectious Agents

(a)

Entry points:

After breaching epithelial barriers:

Contact
Conjunctiva

Inhalation
Respiratory tract

Ingestion
Gastrointestinal tract

Wound or Arthropod bite
Skin

Sexual exposure
Urogenital tract

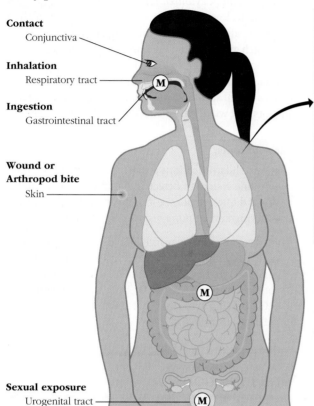

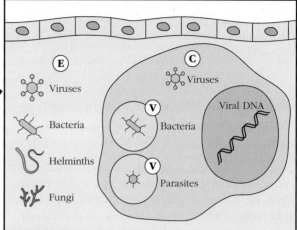

Viruses
Bacteria
Helminths
Fungi
Viruses
Bacteria
Parasites
Viral DNA

(M) Mucosal
(E) Extracellular
(V) Vesicles (intracellular)
(C) Cytosol (intracellular)

(b)

(M) **Mucosal** T_H2	(E) **Extracellular** T_H2 and T_H17	(V) **Vesicular** (intracellular) T_H1	(C) **Cytosolic** (intracellular) CTL and T_H1
Viruses	Bacteria	Bacteria	Viruses
Bacteria	Helminths	Parasites	Some bacteria
Helminths	Fungi		
Fungi			
Innate lymphoid cells	T_H2 or T_H17 cells	T_H1 cells	T_H1 cells
MALT B and T cells	B cells	Cytokines	CTLs
Soluble dIgA	mIgA, IgM, IgG	Activated macrophages	Cytokines
Mast cells with IgE	Phagocytic cells	NK cells	NK cells and ADCC
Antimicrobial compounds	Antimicrobial compounds		

(a) Most infectious organisms enter through mucosal surfaces (M), like the respiratory, gastrointestinal, and urogenital tracts. Alternatively, they may breach skin barriers through a wound or via an insect bite. Once they have crossed this barrier they exist in the sterile microenvironment of the body, where they can remain local, or spread through the blood and lymphatic circulation. Examples of all four types of pathogen can be found in the extracellular fluid (E), while intracellular parasites and bacteria may be engulfed by or enter host cells and remain vesicular (V). Viruses fuse with the plasma membrane and enter the cytosol (C), from where they may insert their genetic material into host chromosomes in the nucleus. (b) Each of these locations allows for access to different detection molecules and may necessitate different effectors for elimination.

Key Concept:

- Extracellular infections within the body are controlled by a combination of innate and adaptive effector mechanisms present in extracellular fluid: phagocytosis and pAPC activation (via PRR binding and cytokines), complement activation, and antibody binding.

Mechanisms That Recognize Infected Host Cells Are Required to Combat Intracellular Infections

Infectious organisms that reside inside host cells are the most difficult for the immune response to find and to destroy. Examples include intracellular bacteria and intracellular parasites (both of which are often smaller than their extracellular counterparts) and all viruses, which are obligate, intracellular pathogens. As one might imagine, identifying and eradicating infectious agents in these locations requires a different set of tools. For instance, antibodies do not recognize a pathogen hidden inside a cell, a space to which they do not have access, although they can effectively neutralize this infectious agent before it binds to host cells. Most bacteria and parasites that are intracellular spend at least some time within intracellular vesicles (Overview Figure 17-3, location V), where they might trigger endosomal PRRs, such as certain TLRs (Toll-like receptors). Often, *eradication of infectious organisms that occupy endosomal spaces inside host cells requires a strong T_H1 pathway response.* This classic *delayed-type hypersensitivity* (DTH) response relies on cytokines secreted by T_H1 cells (e.g., IFN-γ) to activate macrophages that can digest these vesicle-bound agents and overcome common pathogen-induced immune evasion of phagosome-lysosome fusion. *Mycobacterium* species, including the agent that causes tuberculosis, are an example (**Clinical Focus Box 17-2**).

All viruses and some parasitic agents eventually enter the cytosol of their host cell, where they will spend much of their time (Overview Figure 17-3, location C). Once there, these agents can be detected by cytosolic PRRs such as the NOD-like receptors (NLRs) and RIG-I-like receptors (RLRs). This leads to cytokine secretion, sometimes associated with inflammasome activation, and ultimately the induction of cytotoxic cells capable of killing infected host cells; namely, NK cells and CTLs (see Chapter 4). NK cells can kill infected host cells via antibody-dependent cell-mediated cytotoxicity (ADCC; see Chapter 12) or by detecting cell-surface changes that are warning signs of infection (e.g., decreased MHC class I expression or changes in the expression of killer inhibitory receptors). CTLs, activated by cross-presentation of antigen by T_H1-licensed DCs, are the primary adaptive mediators of infected target cell killing. In some instances, cytotoxic T_H1 cells may also kill infected target cells via Fas–Fas ligand binding. *The eradication of cytosolic infections*

therefore requires strong and comprehensive cell-mediated immunity; activated pAPCs (frequently, dendritic cells), CD4$^+$ T cells that can license DCs for cross-presentation (often T_H1 type), and finally, cytotoxic CD8$^+$ T cells.

In the following sections, we discuss each of the four classes of pathogen in turn, focusing on specific characteristics of each group and on the immune mechanism(s) required for detection and elimination. In this and the following section on vaccines, Overview Figure 17-3 should help connect the location of specific pathogens with key immune response effectors that will be most effective for the job at hand.

Key Concepts:

- Intracellular infections are the most difficult for the immune system to detect and eradicate; these can be divided into membrane-bound and cytosolic, with differing key immune response mediators.

- Intracellular vesicular infections are most effectively eradicated via macrophages activated by the cytokines secreted by T_H1 cells, while cytosolic infection requires host cell lysis by CTLs (generated with DC-licensing help from T_H1 cells), cytotoxic T_H1 cells, or NK cells.

Viral Infections

Viruses constitute small segments of nucleic acid with a protein or lipoprotein coat, and require host resources for their replication. Passage across the mucosa of the respiratory, urogenital, or gastrointestinal tract accounts for most instances of viral transmission. Viruses may also gain entrance through broken skin, such as during an insect bite or puncture wound. These obligate intracellular pathogens normally enter their host cell via one or more specific cell-surface receptor/s for which they have affinity. For example, influenza virus binds to sialic acid residues in cell membrane glycoproteins and glycolipids, rhinovirus binds to *inter*cellular *a*dhesion *mole*cules (ICAMs), and *E*pstein-*B*arr *v*irus (EBV) binds to type 2 complement receptors on B cells. Once inside a cell, the virus diverts cell biosynthetic machinery to replicate itself. The influenza virus genome replication step is often error prone, generating numerous mutations (interestingly, this is not true for many viruses). Because large numbers of new influenza viral particles (virions) are produced in a replication cycle, many different mutants, some with survival and immune evasion advantages, can arise.

The Antiviral Innate Response Provides Key Instructions for the Later Adaptive Response

A number of specific innate immune effector mechanisms, together with nonspecific defense mechanisms, can prevent or eliminate many viral infections even before adaptive

BOX 17-2

Lessons Learned from Tuberculosis: Importance of T$_H$1-type Responses in Fighting Intracellular Bacterial Infections

Roughly one-third of the world's population is infected with *Mycobacterium tuberculosis*, the bacteria that cause tuberculosis. Although tuberculosis was once believed to be eliminated as a public health problem in the United States, the disease re-emerged in the early 1990s, particularly in areas where HIV infection levels are high. This disease is still the leading killer of individuals with AIDS.

M. tuberculosis spreads easily, and pulmonary infection usually results from inhalation of small droplets of respiratory secretions containing a few bacilli. The inhaled bacilli are engulfed by alveolar macrophages in the lung, where they are able to survive and multiply intracellularly by inhibiting the formation of phagolysosomes. When the infected macrophages ultimately burst, large numbers of bacilli are released.

The most common clinical pattern of infection with *M. tuberculosis*, seen in 90% of infected individuals, is pulmonary tuberculosis. In this pattern, CD4$^+$ T cells are activated within 2 to 6 weeks of infection and secrete cytokines that induce the infiltration of large numbers of activated macrophages. These cells wall off the organism inside a **granuloma,** called a *tubercle* (**Figure 1**), a cluster of small lymphocytes surrounding infected macrophages. The localized concentrations of lysosomal enzymes in these granulomas can cause extensive tissue necrosis. The massive activation of macrophages that occurs within tubercles often results in the concentrated release of lytic enzymes. These enzymes destroy nearby healthy cells, resulting in circular regions of necrotic tissue, which eventually form a lesion with a caseous (cheeselike) consistency. As these lesions heal, they become calcified and are readily visible on x-rays of the lungs as a defined shadow. Much of the tissue damage seen with *M. tuberculosis* is thus actually due to pathology associated with the cell-mediated immune response.

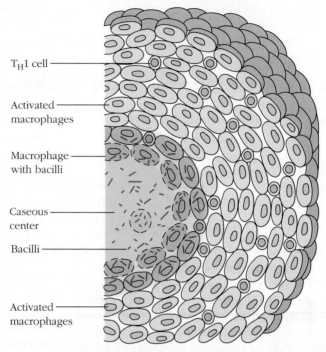

FIGURE 1   A tubercle formed in pulmonary tuberculosis. During the immune response to pulmonary tuberculosis, T$_H$1 cells will recruit and activate macrophages, which surround dead and dying infected cells (mostly macrophages) that harbor the bacilli, walling off the bacteria from spreading further and, in some cases, clearing the debris.

Because the activated macrophages suppress proliferation of the phagocytosed bacilli, infection is contained. *In other words, the intracellular infection is contained even if it is not completely eliminated.* This is just like the DTH response that occurs as part of a type IV hypersensitivity reaction, and usually does not involve CD8$^+$ CTLs killing bacterially infected cells. Cytokines produced by CD4$^+$ T cells (T$_H$1 subset) play an important role in the response by activating macrophages so that they are able to kill the bacilli or inhibit their growth.

Tuberculosis has traditionally been treated for long periods of time with several different antibiotics, sometimes in combination. The tubercle and intracellular growth of *M. tuberculosis* makes it difficult for drugs to reach the bacilli, necessitating up to 9 months of daily treatment. At present, the only vaccine for *M. tuberculosis* is an attenuated strain

of *Mycobacterium bovis* called *bacillus Calmette-Guérin* (BCG). This vaccine is fairly effective against extrapulmonary tuberculosis but less so against the more common pulmonary tuberculosis; in some cases, BCG vaccination has even increased the risk of infection. Moreover, after BCG vaccination the tuberculin skin test cannot be used as an effective monitor of exposure to wild-type *M. tuberculosis*. Because of these and other drawbacks this vaccine is not used in the United States, but it is used in several other countries. However, the alarming increase in multidrug-resistant strains has stimulated renewed efforts to develop a more effective tuberculosis vaccine.

REFERENCE

Cardona, P. J. 2017. What we have learned and what we have missed in tuberculosis pathophysiology for a new vaccine design: searching for the "pink swan." *Frontiers in Immunology* **8:**556.

The following labels appear in Figure 1: T$_H$1 cell; Activated macrophages; Macrophage with bacilli; Caseous center; Bacilli; Activated macrophages

TABLE 17-1   Mechanisms of humoral and cell-mediated immune responses to viruses

Response type	Effector molecule or cell	Activity
Humoral	Antibody (especially secretory IgA)	Blocks binding of virus to host cells, thus preventing infection or reinfection
	IgG, IgM, and IgA antibody	Block fusion of viral envelope with host cell's plasma membrane
	IgG and IgM antibody	Enhance phagocytosis of viral particles (opsonization)
	IgM antibody	Agglutinates viral particles
	Complement activated by IgG or IgM antibody	Mediates opsonization by C3b and lysis of enveloped viral particles by membrane-attack complex
Cell mediated	IFN-γ secreted by T_H or T_C cells	Has direct antiviral activity
	Cytotoxic T lymphocytes (CTLs)	Kill virus-infected self cells
	NK cells and macrophages	Kill virus-infected cells by antibody-dependent cell-mediated cytotoxicity (ADCC)

immunity is engaged (**Table 17-1**). These include complement activation via innate pathways and antimicrobial peptides, as well as the recognition of *pathogen-associated molecular patterns* (PAMPs) by PRRs expressed by phagocytic cells. For example, *double-stranded RNA* (dsRNA) molecules and other virus-specific structures are detected by one of several intracellular or endosomal membrane PRRs; specific NRLs and TLRs, respectively. Signaling through these receptors induces the expression of type I interferons (IFN-α and IFN-β), the assembly of intracellular inflammasome complexes, and the activation of NK cells. When type I interferons bind to the IFN-α/β receptors, antiviral activity and resistance to viral replication is the outcome, acting through the JAK-STAT pathway that leads to production of new transcripts, one of which encodes an enzyme that leads to viral RNA degradation (see Figure 4-16). IFN-α/β binding also induces ds*RNA-dependent protein kinase* (PKR), which leads to inactivation of protein synthesis, thus blocking viral replication in neighboring infected cells. The binding of type I interferon to NK cells induces lytic activity, making them very effective in killing virally infected cells. This activity is enhanced by IL-12, a cytokine that is produced by dendritic cells very early in the response to viral infection. The specific cytokines and other effector mechanisms triggered during this innate response provide the early directives that will guide the adaptive response that follows, when additional measures are required to complete pathogen eradication.

Key Concept:

- Innate response elements commonly engaged by encounter with viral PAMPs, such as secretion of type I interferons, inflammasome and NK-cell activation, as well as IL-12 production, can help eliminate the virus but also provide crucial instructions for the adaptive response that will follow.

Many Viruses Are Neutralized by Antibodies

While cell-mediated responses, like those involving *cytotoxic T lymphocytes* (CTLs), are most commonly associated with virus resolution, antibodies specific for viral surface antigens can be crucial in blocking spread and secondary infection. Ideally, natural exposure or vaccination will induce the production of neutralizing antibodies, especially isotypes that localize to areas where the virus is found; surface or mucosal IgA and circulating IgG are often most protective. These antibodies are generated during a primary response when whole virions or individual viral components are recognized in extracellular spaces. This occurs as virus spreads from cell to cell or when infected cells burst. During a secondary response, antibodies are most effective if they are already localized to the site of viral entry, and if they bind to key viral surface structures in a way that interferes with their ability to attach to host cells (called *neutralizing antibodies*). For instance, the advantage of the attenuated oral polio vaccine, discussed later in this chapter, is that it induces production of secretory IgA, which effectively blocks attachment of poliovirus to epithelial cells lining the gastrointestinal tract.

Viral neutralization by antibody can also involve mechanisms that operate after viral attachment to host cells. For example, antibodies may block viral penetration by binding to epitopes that are necessary to mediate fusion of the viral envelope with the plasma membrane. If the antibody is of a complement-activating isotype, lysis of enveloped virions can ensue. Antibody or complement can also agglutinate viral particles and function as an opsonizing agent to facilitate Fc or C3b receptor–mediated phagocytosis of the free virions (see Chapter 5). Finally, some isotypes of antibody bound to infected target cells can trigger ADCC by NK cells.

Key Concept:

- Neutralizing antibodies, especially those at the sites of infection, as well as circulating antibodies that foster opsonization, complement activation, and phagocytosis, protect the host by blocking or eliminating virus in the extracellular spaces, although they cannot eliminate virally infected cells.

Cell-Mediated Immunity Is Important for Viral Control and Clearance

Although antibodies have an important role in controlling virus during the acute phases of infection, they cannot typically eliminate established infection once the viral genome is integrated into host chromosomal DNA. Once such an infection is established, cell-mediated immune mechanisms are required to complete the job. *In general, both CD8$^+$ T$_C$ cells and CD4$^+$ T$_H$1 cells are required components of this cell-mediated antiviral defense.* Activated T$_H$1 cells produce a number of cytokines, including IL-2, IFN-γ, and TNF-α, which defend against viruses either directly or indirectly. IFN-γ acts directly by inducing an antiviral state in nearby cells. IL-2 acts indirectly by helping to activate CTL precursors, generating an effector population of cytotoxic cells. Both IL-2 and IFN-γ activate NK cells, which play an important role in host defense and lysis of infected cells, especially prior to specific CTL responses. And finally, T$_H$1 cells directed against the same pathogen (but not always the same epitopes) are required in order to license pAPCs for cross-presentation, allowing activation of naïve CD8$^+$ T cells in the first place (see Chapter 7).

During the immune response to a viral infection, specific CTL activity usually arises within 3 to 4 days of infection, peaks by 7 to 10 days, and then declines gradually over the following weeks or months. In many cases, virions are eliminated in those first 7 to 10 days, paralleling the development of CTLs. CTLs specific for the virus eliminate virus-infected self cells and thus eliminate potential sources of new virus. Virus-specific memory CD8$^+$ T cells then confer protection against that virus in the future, and have been shown to protect nonimmune recipients following adoptive transfer. This memory response is highly pathogen specific, as transfer of a CTL clone specific for influenza virus strain X protects mice against strain X but usually not influenza virus strain Y.

Key Concept:

- In order to eliminate an established infection, where host cells harbor intracellular virus, virus-specific CD8$^+$ T cells must be activated to kill infected cells, which requires the assistance of helper T cells (often T$_H$1 type) that recognize the same pathogen, providing cytokines and pAPC licensing for cross-presentation.

Viruses Employ Several Strategies to Evade Host Defense Mechanisms

Despite their restricted genome size, most viruses encode several genes that interfere with innate and/or adaptive levels of host defense. Importantly, they have had the time (millennia), resources, and numbers to hone their evasion strategies; they take over the host cell machinery, and a host cell can produce thousands of virions in short order. As described earlier, the induction of type I interferon is a major innate defense against viral infection. Not surprisingly, some viruses have developed strategies to evade the action of IFN-α/β. For instance, hepatitis C virus overcomes the antiviral effect of the interferons by blocking or inhibiting the action of PKR, a protein kinase essential to signal transduction (see Figure 4-16).

Another mechanism for evading host responses is inhibition of antigen presentation by infected host cells. *Herpes simplex virus* (HSV) produces a protein that very effectively inhibits the human transporter molecule needed for antigen processing (TAP; see Figure 7-14). Inhibition of TAP blocks antigen delivery to MHC class I molecules in HSV-infected cells, trapping empty MHC class I molecules in the endoplasmic reticulum and effectively shutting down presentation of HSV antigens to CD8$^+$ T cells and CTL recognition of infected cells. Likewise, adenovirus and *cytomegalovirus* (CMV) use distinct molecular mechanisms to reduce the surface expression of MHC class I molecules, again inhibiting antigen presentation to CD8$^+$ T cells.

A number of viruses escape immune attack by constantly changing their surface antigens. The influenza virus is a prime example (**Clinical Focus Box 17-3**). For this reason, a new vaccine is created yearly, prepared for each new season of the flu. Nowhere is antigenic variation greater than in HIV, the causative agent of AIDS, estimated to accumulate mutations 65 times faster than the influenza virus. We discuss vaccine progress against HIV later in this chapter, plus an entire section of Chapter 18 is dedicated to HIV and AIDS.

Some viruses, such as Epstein-Barr virus (EBV) and HIV, can cause generalized or specific immunosuppression, which also works as a means of evasion. In the case of HIV, viral infection of lymphocytes or macrophages can either destroy the immune cells or alter their function. In other cases, immunosuppression is the result of a cytokine imbalance or pathogen-induced diversion toward less effective immune response pathways. For instance, EBV, the cause of mononucleosis, produces a protein that is homologous to IL-10; like IL-10, this protein suppresses cytokine production by the T$_H$1 subset, resulting in inhibition of the antiviral inflammatory response.

Key Concept:

- Viruses have evolved several strategies to evade or subvert the host immune response, including expression of immune-blocking or -inhibiting compounds, suppression of MHC class I expression, regularly changing surface antigens, and the delivery of instructions that misdirect the host immune response.

Influenza Has Been Responsible for Some of the Worst Pandemics in History

The influenza virus infects the upper respiratory tract and major central airways in humans, horses, birds, pigs, and even seals. Between 1918 and 1919, the largest influenza pandemic (worldwide epidemic) in recent history occurred, killing between 20 million and 50 million people. Two other, less major pandemics occurred in the twentieth century, caused by influenza strains that were new or had not circulated in the recent past, catching most people with little protective immunity. To understand where these come from we must first discuss some of the general characteristics of the flu virus.

Influenza virus is an enveloped virus, meaning that virions are surrounded by a lipid bilayer or envelope derived from the plasma membrane of the infected cell. Embedded in this envelope are two key viral glycoproteins: **hemagglutinin (HA)** and **neuraminidase (NA)**. HA trimers are responsible for attachment of the virus to host cells, binding to the sialic acid groups on host-cell glycoproteins and glycolipids. NA is an enzyme that cleaves *N*-acetylneuraminic (sialic) acid from nascent viral glycoproteins and host-cell membrane glycoproteins, facilitating viral budding from the infected host cell. Thus these two structures are essential for viral attachment and for exit of new virus from infected cells—so important, in fact, that we track and name new strains of influenza on the basis of their antigenic subtypes of HA and NA (e.g., H1N1 versus H5N1 virus). Within the envelope, an inner layer of matrix protein surrounds the nucleocapsid, which contains the eight, single-stranded *RNA* (ssRNA) molecules that make up the virus genome. Each RNA strand encodes one or more different influenza proteins.

To date, there are 18 different antigenic subtypes for HA and 11 for NA. Antigenic variation in these structures is generated by two different mechanisms: antigenic drift and antigenic shift. **Antigenic drift** involves a series of spontaneous point mutations that occur gradually, resulting in minor changes in HA and NA over time. **Antigenic shift** describes the sudden emergence of a new subtype of influenza, where the structures of HA and/or NA are considerably different from that of the virus present in a preceding year (**Figure 1**).

The immune response contributes to the emergence of these antigenically distinct influenza strains. In a typical year, the predominant virus strain undergoes antigenic drift, generating minor antigenic variants (see Figure 1a). As individuals infected with influenza mount an effective immune response, they will eliminate that strain. However, the accumulation of point mutations sufficiently alters the antigenicity of some variants so that they are able to escape immune elimination and become a new variant of influenza that is transmitted to others, causing another local epidemic cycle. The role of antibody in such immunologic selection can be demonstrated in the laboratory by mixing an influenza strain with a monoclonal antibody specific for that strain and then culturing the virus in cells. The antibody neutralizes all unaltered viral particles, and only those viral particles with mutations resulting in altered antigenicity escape neutralization and are able to continue the infection. Within a short time in culture, a new influenza strain emerges, just as it does in nature. In this way, influenza evolves during a typical flu season, such that the dominant strains at the start and end of the season are antigenically distinct. This is why we are offered a new flu vaccine each year. The vaccine formulation is based on carefully constructed models tracking the dominant variant(s) from the end of the previous season.

Episodes of antigenic shift are thought to occur through a different mechanism. The primary mechanism is genetic reassortment between influenza virions from humans and those from various animals (see Figure 1b). The fact that the influenza genome contains eight separate strands of ssRNA makes possible the mixing of individual RNA strands (like mini-chromosomes) of human and animal virions within a secondary (nonhuman) host cell infected with both viruses—in other words, shuffling of the DNA segments derived from the animal and human strains. For example, pigs and birds can harbor human influenza A viruses, as well as their own and sometimes those of other species, making them perfect conduits for in vivo genetic reassortment between influenza A viruses. This is where the terms "swine flu" or "avian flu" come from. As one might imagine, all the proteins encoded on one of these mini-chromosome ssRNA fragments derived from animal influenza are frequently new to humans who will have little or no immunity, sparking pandemics.

The most virulent and devastating of the pandemic strains of influenza virus in recent history was seen in 1918 and 1919. Worldwide deaths from that so-called "Spanish flu" strain may have reached 50 million in less than 1 year, compared with the roughly 10,000 to 15,000 who die yearly from nonpandemic strains. Approximately 675,000 of the victims of Spanish flu were located the United States, with certain areas, such as Alaska and the Pacific Islands, losing more than half of their population during the outbreak. Mortality rates for the 1918 pandemic flu were surprisingly high, especially among young and healthy individuals, reaching 2.5% in infected individuals compared with less than 0.1% during other flu epidemics. Most of these deaths were the result of a virulent pneumonia, which felled some patients in as little as 5 days.

A research team led by Jeffery Taubenberger determined the entire genetic sequence of the deadly 1918 flu virus—a controversial move. Their results were

(continued)

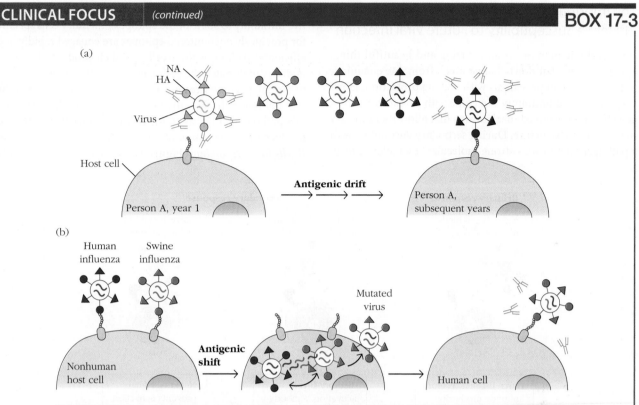

FIGURE 1 **Two mechanisms generate variations in influenza surface antigens.** (a) In antigenic drift, the accumulation of point mutations eventually yields a variant protein that is no longer recognized by antibody to the original antigen. (b) Antigenic shift may occur by reassortment of an entire ssRNA between human and animal virions infecting the same cell. The surface proteins on the new influenza subtype are so different that human antibodies no longer recognize them and humans thus have no immunity. Only two of the eight RNA strands are depicted.

made possible following isolation of viral RNA from Spanish flu victims, using formalin-fixed lung autopsy samples and tissue collected from an Inuit woman who was buried in permafrost in Alaska. Analysis of the sequence revealed an H1N1 virus significantly different from that of any contemporary H1N1 strains derived from an avian virus, making it the most "bird-like" of the influenza strains ever isolated from humans. Word of the imminent publication of the 1918 influenza sequence caused a scientific and public controversy. On the one hand, many researchers were eager to glimpse this sequence for clues to what determinants might play a role in virulence and what measures could be taken to avoid this in the future. On the other hand, some feared that the sequence information could be used for evil rather than for good, resulting in the reconstruction of a weaponized version of the influenza virus.

In the end, the sequences were published and follow-up studies using reconstructed virus in animals shed much light

on the history of influenza. In mouse studies, researchers found that the reconstructed virus spread rapidly in the respiratory tract and produced high numbers of progeny, causing pervasive damage in the lungs. Using recombinant virus strains, they found that three polymerase genes and the HA gene appeared to account for the high lethality of the virus.

Subsequent studies in a nonhuman primate model showed similar results in terms of pathogenesis and spread. These scientists also examined the immune response to the reconstructed 1918 flu and found profound innate immune response differences between "Spanish flu" and a control influenza virus. Most notably, serum levels of IL-6 were elevated from 5- to 25-fold over control virus by day 8 of infection, and closely correlated with virus replication and symptoms like fever. Interestingly, other key innate response elements were suppressed or missing. Compared with controls, the animals infected with the reconstructed 1918 flu showed marked reduction in

type I interferon responses, a typical early and positive immune indicator of virus eradication and disease resolution. As we know from reports of people infected with this virus, the 1918 infection triggered debilitating inflammatory responses that resulted in rapid respiratory distress. It appears that this robust innate response was, however, selective in detrimental ways, missing IFN response elements that we now know are effective against influenza.

One piece of really good news from this recreation of that pandemic event is that scientists believe present-day antiviral drugs and vaccine preparations would be effective against the 1918 version of the flu. Phew!

REFERENCES

Tumpey, T. M., et al. 2005. Characterization of the reconstructed 1918 Spanish influenza pandemic virus. *Science* **310**:77.

Kobasa, D., et al. 2007. Aberrant innate immune response in lethal infection of macaques with the 1918 influenza virus. *Nature* **445**:319.

The Imprinting of a Memory Response Can Influence Susceptibility to Future Viral Infection

Immunologic memory is an amazing and beautiful thing. For pathogens that don't change much from one encounter to the next, it can provide us with lifelong protection. However, preformed immunity can come with caveats for pathogens that have evolved mechanisms that allow them to vary their antigenic structure. During secondary encounters with a pathogen that bears a strong molecular resemblance to an

agent seen in the past (i.e., we have already developed adaptive immunity to some of the epitopes), memory cells specific for previously encountered epitopes are engaged rapidly and efficiently. As long as these cells and their products, like antibodies, can dispatch the pathogen efficiently, there is no need to mount a *primary* response to any *new epitopes* carried by that pathogen. In fact, the presence of antibodies attached to a pathogen, either as residuals from a recent infection or produced by reactivation of memory B cells, *will divert naïve B cells from responding* (**Figure 17-4a**). This occurs when the

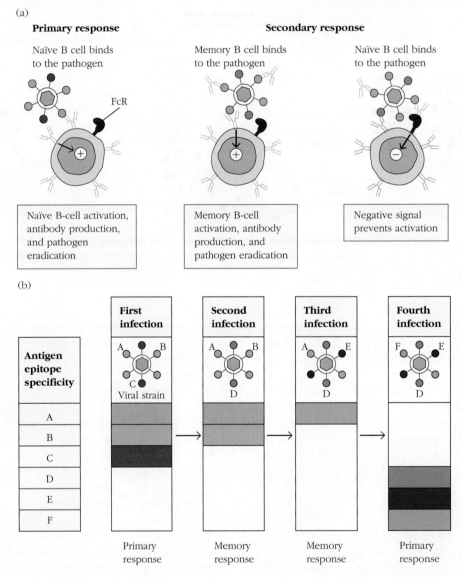

FIGURE 17-4 The presence of preformed antibody inhibits primary responses to a pathogen. (a) During a primary response, naïve B cells are activated and produce antibodies specific to epitopes on the pathogen. During a secondary response to a variant of that pathogen, memory B cells specific to epitopes encountered in the past will be reactivated and help to eradicate the pathogen. The Fc regions of antibodies bound to the surface of the pathogen will bind to the FcRs on naïve B cells and inhibit them from responding to new epitopes on the pathogen. (b) This inhibition of primary responses against unique epitopes on pathogens

that elicit memory cell responses is called *original antigenic sin*. No immune response is mounted to each new epitope during subsequent exposures to the pathogen until the pathogen expresses a significant number of unique epitopes and memory cells can no longer eradicate the organism. In this case, a new primary response is mounted, leading to adaptive effectors associated with both symptoms and resolution of infection. Once cleared, this encounter and the memory response then resets the immune focus to these new antigenic structures.

Fc region of the pathogen-associated antibody binds with Fc receptors on naïve B cells, inducing anergy.

This means that if there is a way to take care of an infection using immune memory, this will be the default pathway. This concept is referred to as **original antigenic sin**, or the tendency to focus an immune attack on those antigens that were present during the original, or primary, encounter with a pathogen and for which we have established memory. Most commonly studied in relation to antiviral memory, our immune systems effectively ignore the subtle changes occurring each year in pathogens that drift antigenically, like influenza virus (Figure 17-4b). Once the organism has drifted sufficiently that there are only "new" epitopes, or insufficient numbers of key epitopes to effectively dispatch with existing memory cells, a new primary response is mounted. In such a year, we might experience more significant symptoms from virus infection. Since we all begin our journeys of original antigenic sin at different times and in response to different antigenic variants, we don't necessarily all experience this in the same year; the exceptions are pandemic influenza years when the virus has evolved new virulence characteristics (see Clinical Focus Box 17-3).

Key Concept:

- *Original antigenic sin* is a term used to describe the observation that we rely first on memory responses before activating naïve lymphocytes; in other words, when we encounter an infectious agent with some epitopes for which we have memory (original antigens), we enlist these memory cells or effectors rather than activate naïve cells that target new, unique epitopes present on this infectious agent.

Bacterial Infections

Bacteria can enter the body through a number of natural routes (e.g., the respiratory, gastrointestinal, and urogenital tracts) or through normally inaccessible routes opened up by breaks in mucous membranes or skin. Depending on the number of organisms entering and their virulence, different levels of host defense are enlisted. If the inoculum size and the virulence are both low, then localized tissue phagocytes may be able to eliminate the bacteria via nonspecific innate defenses. Larger inocula, organisms with greater virulence, and intracellular bacteria typically require antigen-specific adaptive immune responses.

It is worth noting that in some instances, disease symptoms are caused not by the pathogen itself but by the immune response. In the case of some bacteria, pathogen-stimulated overproduction of cytokines, or non-discriminate and systemic expression, can lead to the symptoms associated with bacterial septic shock, food poisoning, and toxic shock syndrome.

Immune Responses to Extracellular and Intracellular Bacteria Differ

Immunity to bacterial infections is usually achieved by a combination of humoral and cell-mediated immunity, depending somewhat on the type of pathogen. That said, *the humoral immune response is the main protective response against extracellular bacteria*. Exposure to extracellular bacteria induces production of antibodies, which are ordinarily secreted by plasma cells in regional lymph nodes or the submucosa of the respiratory and gastrointestinal tracts. These antibodies act in several ways to protect the host from the invading organisms (**Figure 17-5**). Extracellular bacteria typically induce a local inflammatory response. In some cases this is triggered by the presence of immunogenic toxins. These toxins can be integral components of the bacterial cell wall (**endotoxins**), such as lipopolysaccharide (LPS), or secreted proteins that are toxic (**exotoxins**). Both tetanus and diphtheria are disorders caused by exotoxins made by bacteria: *Clostridium tetani* and *Corynebacterium diphtheria*, respectively.

Antibody that binds to antigens on the surface of a bacterium can, together with the C3b component of complement, act as an opsonin to increase phagocytosis and clearance of the bacterium. In the case of some bacteria—notably, the gram-negative organisms—complement activation can lead directly to lysis of the organism. Antibody-mediated activation of the complement system can also induce localized production of immune effector molecules that help to develop an amplified and more effective inflammatory response. For example, the complement fragments C3a and C5a act as anaphylatoxins, inducing local mast-cell degranulation and thus vasodilation and the extravasation of lymphocytes and neutrophils from the blood into tissue spaces (see Chapter 5 and Figure 17-5). Other complement components serve as chemotactic factors for neutrophils and macrophages, thereby contributing to the buildup of phagocytic cells at the site of infection. Antibody to a bacterial toxin may bind to the toxin and neutralize it; the antibody-toxin complexes are then cleared by phagocytic cells in the same manner as any other antigen-antibody complex.

Intracellular (vesicular) bacteria present a different challenge due to their residency inside host cells. In addition to activation of TLRs in membranes, intracellular bacteria can also activate NK cell–mediated killing, which in turn provides an early defense against these organisms. *Ultimately, to be effective, the cellular response to intracellular bacteria requires T_H1 cell–mediated immune responses*, such as DTH (see Chapter 15). In this response, cytokines secreted by $CD4^+$ T_H cells are crucial—most notably IFN-γ—for activating macrophages to phagocytose and kill these bacteria more effectively. The most notable example of this is in the *Mycobacterium* family of intracellular pathogens, where strong T_H1 responses have been shown to protect against or clear a typically recalcitrant intracellular *M. tuberculosis* infection (see Clinical Focus Box 17-2).

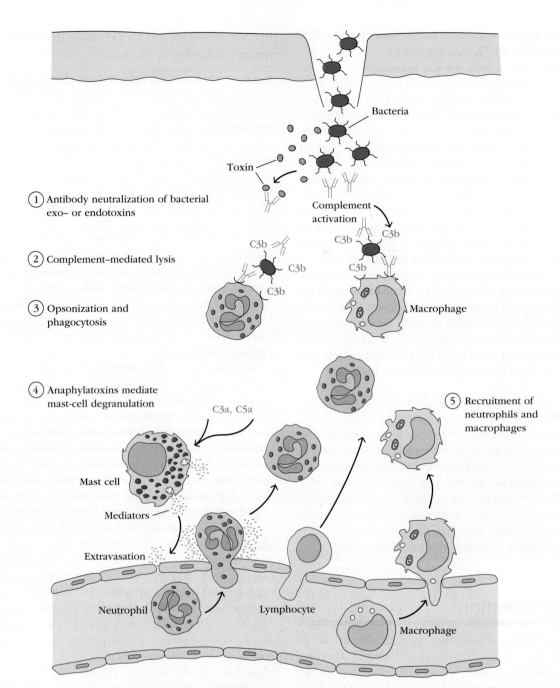

1. Antibody neutralization of bacterial exo– or endotoxins

2. Complement–mediated lysis

3. Opsonization and phagocytosis

4. Anaphylatoxins mediate mast-cell degranulation

5. Recruitment of neutrophils and macrophages

Bacteria

Toxin

Complement activation

C3b

C3b

C3b

C3b

C3b

Macrophage

C3a, C5a

Mast cell

Mediators

Extravasation

Neutrophil

Lymphocyte

Macrophage

FIGURE 17-5 Antibody-mediated mechanisms for combating infection by extracellular bacteria. (1) Antibody neutralizes bacterial toxins. (2) Complement activation on bacterial surfaces leads to complement-mediated lysis of bacteria. (3) Antibody and the complement split product C3b bind to bacteria, serving as opsonins to increase phagocytosis. (4) C3a and C5a, generated by antibody-initiated complement activation, induce local mast-cell degranulation, releasing substances that mediate vasodilation and extravasation of lymphocytes and neutrophils. (5) Other complement products are chemotactic for neutrophils and macrophages.

Key Concept:

- Both humoral and cell-mediated immunity are usually required to fight bacterial infections, with antibody the primary effector for extracellular bacteria and CD4$^+$ T_H1 cytokine responses and activation of macrophages, the main effectors for intracellular (vesicular) bacteria.

Bacteria Can Evade Host Defense Mechanisms at Several Different Stages

There are four primary steps in most bacterial infections: attachment to host cells, proliferation of the bacterium, invasion of host tissue, and (in some cases) toxin-induced damage to host cells. Host-defense mechanisms can act at

TABLE 17-2	Host immune responses to bacterial infection and bacterial evasion mechanisms	
Infection process	**Host defense**	**Bacterial evasion mechanisms**
Attachment to host cells	Blockage of attachment by secretory IgA antibodies	Secretion of proteases that cleave secretory IgA dimers (*Neisseria meningitidis, N. gonorrhoeae, Haemophilus influenzae*)
		Antigenic variation in attachment structures (pili of *N. gonorrhoeae*)
Proliferation	Phagocytosis (Ab- and C3b-mediated opsonization)	Production of surface structures (polysaccharide capsule, M protein, fibrin coat) that inhibit phagocytic cells
		Mechanisms for surviving within phagocytic cells
		Induction of apoptosis in macrophages (*Shigella flexneri*)
	Complement-mediated lysis and localized inflammatory response	Generalized resistance of gram-positive bacteria to complement-mediated lysis
		Insertion of membrane-attack complex prevented by long side chain in cell-wall LPS (some gram-negative bacteria)
Invasion of host tissues	Ab-mediated agglutination	Secretion of elastase that inactivates C3a and C5a (*Pseudomonas*)
Toxin-induced damage to host cells	Neutralization of toxin by antibody	Secretion of hyaluronidase, which enhances bacterial invasiveness

each of these steps, and many bacteria have evolved ways to circumvent most of them (**Table 17-2**).

Some bacteria express molecules that enhance their ability to attach to host cells. A number of gram-negative bacteria, for example, have pili (long hairlike projections), which enable them to attach to the membrane of the intestinal or urogenital tract (**Figure 17-6**). Other bacteria, such as *Bordetella pertussis*, the cause of whooping cough, secrete adhesion molecules that help the bacterium adhere to ciliated epithelial cells of the upper respiratory tract. Secretory IgA specific for such bacterial structures can block attachment to epithelial cells and are the main host defense against many of these bacterial strains. However, some bacteria, such as the species of *Neisseria* that cause gonorrhea and meningitis, evade the IgA response by secreting proteases that cleave secretory IgA at the hinge

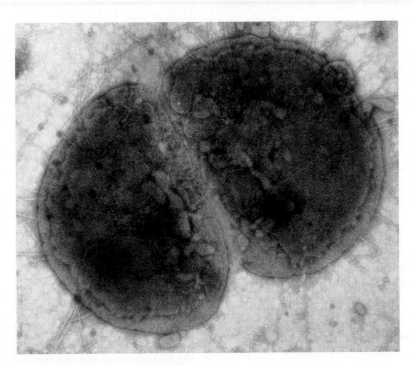

FIGURE 17-6 *Neisseria gonorrhoeae.* Pili extend from the gonococcal surface and mediate the attachment to target cells. *[Kwangshin Kim/Getty Images.]*

region; the resulting Fab and Fc fragments have a shortened half-life in mucous secretions and are not able to agglutinate microorganisms.

These antibody responses made by the host can be evaded by some bacteria that undergo frequent changes to their surface antigens. In *Neisseria gonorrhoeae*, for example, pilin (the protein component of the pili) has a highly variable structure, generated by gene rearrangements in the coding sequence. This process generates enormous antigenic variation, which may contribute to the pathogenicity of *N. gonorrhoeae* by increasing the likelihood that expressed pili will go undetected by antibody, allowing them to bind firmly to epithelial cells and avoid neutralization by IgA.

Bacteria may also possess surface structures that inhibit phagocytosis. A classic example is *Streptococcus pneumoniae*, whose polysaccharide capsule prevents phagocytosis very effectively. Likewise, because there are 84 serotypes of *S. pneumoniae* which differ from one another by distinct capsular polysaccharides, the host only produces antibody against the infecting serotype, not the others. This antibody protects against reinfection with the same serotype but will not protect against infection by most of the other 83 serotypes. In this way, genetic variants of *S. pneumoniae* can cause disease many times in the same individual. Finally, some pathogenic staphylococci are able to assemble a protective coat from host blood proteins. These bacteria secrete a coagulase enzyme that precipitates host-generated fibrin, making a coat around them and shielding them from phagocytic cells.

Mechanisms for interfering with the complement system help other bacteria survive. In some gram-negative bacteria, long side chains on the lipid A moiety of the cell wall core polysaccharide help to resist complement-mediated lysis. *Pseudomonas* secretes an enzyme, elastase, that inactivates both the C3a and C5a anaphylatoxins, thereby diminishing the localized inflammatory reaction.

A number of bacteria escape host-defense mechanisms through their ability to survive within phagocytic cells. Bacteria such as *Listeria monocytogenes* escape from the phagolysosome to the cytoplasm, a favorable environment for their growth. Other bacteria, such as members of the *Mycobacterium* genus, block lysosomal fusion with the phagolysosome or resist the oxidative attack, allowing them to remain and replicate in endosomal vesicles.

Key Concept:
- Bacteria have evolved a number of evasion strategies that work at all stages of the infection cycle, including attachment to host cells, blocking IgA, inhibiting complement, changing antigenic structures, and inhibiting phagocytosis or phagosome-lysosome fusion and thus avoiding intracellular degradation.

Parasitic Infections

Infections caused by parasites account for an enormous disease burden worldwide, especially in developing countries within tropical or subtropical regions. In these locales sanitation and living conditions are not always ideal, increasing the spread of all types of infectious disease. Thanks to climate, tropical regions are also common breeding grounds for the arthropod vectors that carry parasitic infection, such as mosquitoes, flies, and ticks. Further complicating this system, many of these parasites can infect nonhuman primates and other mammals, allowing both human-to-human and animal-to-human (*zoonotic*) spread.

The term *parasite* encompasses a vast array of infectious protozoans (unicellular) and metazoans (helminths or worms). The diversity of the parasitic universe makes it difficult to generalize about this group. For instances, most protozoan parasites, although eukaryotic, inhabit intracellular spaces in their human host for at least one of their life cycle stages. Conversely, helminths are multicellular eukaryotes that can be quite large in their adult stages; up to 1 m in length! These organisms typically live and reproduce exclusively *outside* host cells (region E in Overview Figure 17-3), sometimes occupying host body cavities, like the gut.

One of the biggest challenges posed to the immune response by most parasites is their complicated life cycle, leading to changes in antigenic structure and location over time. Therefore, the most effective immune response will depend on the type of organism, the location of the infection, and the life cycle stage of the parasite.

Protozoan Parasites Are a Diverse Set of Unicellular Eukaryotes

Many of the most burdensome and least treatable tropical diseases are caused by protozoan parasites, itself a broad category of all parasitic infections. The only common features of this group are that all are unicellular eukaryotes, and many are motile. Some, but not all, are pathogenic. Many can be free-living and found in contaminated water (e.g., *Giardia* or *Toxoplasma*). Other protozoan parasites move from their arthropod vector hosts, such as mosquitoes and flies (e.g., the parasites that cause malaria and African sleeping sickness, respectively), to their mammalian hosts when the infected insects draw blood during feeding. These often complicated gymnastics of movement between hosts or environmental sites, combined with multiple life stages within any one host, make immune detection and eradiation extremely challenging.

There is no one common protozoan parasitic infection cycle. However, there are some protozoan parasites with particular significance to human health and disease that have been well characterized. For instance, many protozoan parasites progress through multiple antigenic forms and/or locations during their life cycle in the human host, leaving the immune response one step behind. When parasites are in the bloodstream, gut, or interstitial fluid of their human host, humoral immunity is the most effective response. However, these stages can be very transient or include evasion strategies, presenting little opportunity for clonal selection of lymphocytes or antibody attachment. Those parasites that undergo intracellular life cycle stages require cell-mediated immune reactions as a defense. However, these can be but short stops in a series of life cycle "jumps" to another site, and each presents the host with new antigenic structures to attack and new immune pathways to initiate. This challenges not only the immune response but also our ability to design effective treatments and vaccines.

Some important immunologic lessons have been learned from the study of protozoan parasites. For example, the trypanosomes that cause African sleeping sickness use a novel evasive strategy that employs up to 1000 possible variants of protein coat to outrun the immune response (**Clinical Focus Box 17-4**). The individual immune response to another trypanosome, leishmania, can head in one of two polarized directions, depending on host and pathogen characteristics; a T_H1-driven response that effectively limits pathology or a T_H2-mediated pathway leading to rampant dissemination and progressive disease. Finally, our struggle against malaria, arguably the protozoan parasite that has taken the greatest toll in recent memory, is confounded by a complicated life cycle involving multiple extracellular and intracellular stages of infection, like a red blood cell stage that is particularly refractory to immune detection. This pathogen illustrates several unique challenges presented to the immune system and highlights many of the obstacles to vaccine design common to protozoan parasites.

Key Concept:

- The term *parasite* is a very broad category, including unicellular protozoan eukaryotes that live inside host cells to macroscopic worms (helminths), and thus the mode of immune detection and elimination will depend on the parasite stage and the location of the infection.

Parasitic Worms (Helminths) Typically Generate Weak Immune Responses

Metazoan parasites, or helminths (worms), are responsible for a range of diseases in humans and animals. Adult forms of helminths are large, multicellular organisms that can often be seen with the naked eye. The three main types of parasitic worms are nematodes (roundworms), cestodes (tapeworms), and trematodes (flukes). Most enter their animal hosts through the intestinal tract; helminth eggs can contaminate food, water, feces, and soil. Some, like schistosomes, are transmitted directly through the skin (**Clinical Focus Box 17-5**).

Although helminths are exclusively extracellular and therefore more accessible to the immune system than protozoans (see Overview Figure 17-3, region E versus V), most infected individuals carry relatively few individual parasites at any one time. Also, unlike protozoan parasites, helminths do not multiply within their human hosts. This results in fewer foreign epitopes that can be recognized by the immune system and weak engagement to each, generating relatively poor immune reactivity. Adult helminths are also too big for phagocytic cells to engulf. This means that the best approach may be expulsion rather than the typical humoral opsonization and digestion response. In that case, IgE-mediated responses that result in mast cell degranulation can help, ejecting the worm from the body via the release of histamines and leukotrienes that induce muscle contractions and mucus production (e.g., coughing, vomiting, or explosive diarrhea). One common feature of effective immune responses against metazoan parasites is a reliance on T_H2-type responses, including ILC2s, IL-4 production, T_H2-cell activation, and the production of IgE over IgG. Interestingly, a paucity of early life exposures to helminthic parasites, as occurs in highly developed and urban settings, is credited with a lower threshold for T_H2-type response induction, resulting in overproduction of IgE: heightened type I, IgE-mediated, allergic responses to random, benign environmental antigens (see Chapter 15).

Key Concept:

- Natural immunity to helminth infections is generally weak, although T_H2-type responses, including ILC2s, and IL-4 and IgE production, are associated with the most protective immunity against this type of pathogen.

Fungal Infections

Fungi are a diverse and ubiquitous group of organisms, neither plant nor animal but with characteristics seen in both; they possess a cell wall but gain nutrients from external sources (are heterotrophic). In a kingdom of their own, fungi occupy many environmental niches and perform many beneficial services for humans, including the fermentation of bread, cheese, wine, and beer, as well as the production of penicillin. As many as a million species of fungi are known to exist, but only about 400 are potential agents of human disease. Infections may result from introduction of exogenous organisms due to injury or inhalation, or during host disruptions that allow endogenous organisms such as the commensals to induce disease. Since fungi are ubiquitous in our environment, widespread fungal infections are often a sign of reduced immune competence in the host. In these cases, fungal agents may penetrate mucosal barriers and gain access to extracellular spaces deeper in the body (see Overview Figure 17-3, regions M and E).

BOX 17-4

African Sleeping Sickness: Novel Immune Evasion Strategies Employed by Trypanosomes

Two species of African trypanosomes, a protozoan parasite, cause African sleeping sickness, a chronic, debilitating disease transmitted to humans and cattle by the bite of the tsetse fly. In the bloodstream, this flagellated protozoan differentiates into a long, slender form that continues to divide every 4 to 6 hours. The disease progresses from an early, systemic stage in which trypanosomes multiply in the blood to a neurologic stage in which the parasite infects cells of the central nervous system, leading to meningoencephalitis and eventual loss of consciousness—thus the name.

The surface of the *Trypanosoma* parasite is covered with a *variable surface glycoprotein* (VSG). Several unusual genetic processes generate extensive variation in these surface structures, enabling the organism to escape immunologic clearance. An individual trypanosome carries a large repertoire of VSG genes, each encoding a different VSG primary sequence, but the trypanosome expresses only a single VSG gene at one time. *Trypanosoma brucei*, for example, carries more than 1000 VSG genes in its genome. Activation of a VSG gene results in duplication of the gene and its transposition to a transcriptionally active *expression site* (ES) at the telomeric end of a specific chromosome (**Figure 1a**). Activation of a new VSG gene displaces the previous gene from the telomeric ES. Trypanosomes have multiple transcriptionally active ES sites, so that a number of VSG genes can potentially be expressed; unknown control mechanisms limit expression to a single VSG expression site at any time.

As parasite numbers increase after infection, an effective humoral response develops to the VSG covering the surface of the parasites. These antibodies eliminate most of the parasites from the bloodstream, both by complement-mediated lysis and by opsonization and subsequent phagocytosis. However,

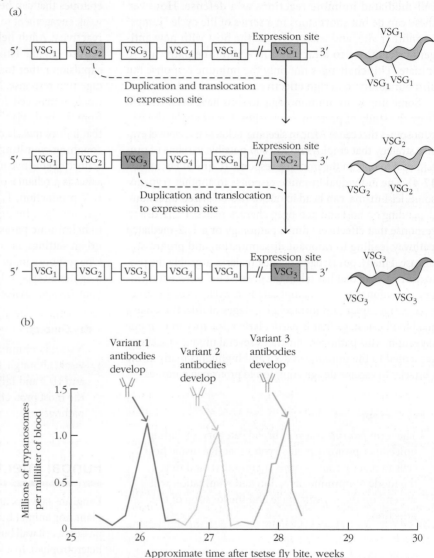

FIGURE 1 **Successive waves of parasitemia after infection with *Trypanosoma* result from antigenic shifts in the parasite's variant surface glycoprotein (VSG).** (a) Antigenic shifts in trypanosomes occur by the duplication of gene segments encoding variant VSG molecules and their translocation to an expression site located close to the telomere. (b) Antibodies develop against each variant as the numbers of these parasites rise, but each new variant that arises is unaffected by the humoral antibodies induced by the previous variant. *[Part (b) data from Donelson, J. E. 1988. Unsolved mysteries of trypanosome antigenic variation. In: The Biology of Parasitism: A Molecular and Immunological Approach (MBL Lectures in Biology series), P. T. Englund and A. Sher, eds. pp. 371–400. Alan R. Liss, New York.]*

about 1% of the organisms bear an antigenically different VSG because of transposition of a different VSG gene into the ES. These parasites escape the initial antibody response, begin to proliferate in the bloodstream, and go on to populate the next wave of parasitemia in the host. The successive waves of

(continued)

parasitemia reflect a unique mechanism of antigenic shift by which the trypanosomes sequentially evade the immune response to their surface antigens. Each new variant that arises in the course of a single infection escapes the humoral antibodies generated in response to the preceding variant, and so waves of parasitemia recur (Figure 1b). The new variants arise not by clonal outgrowth from a single escape variant cell, but from the expansion of multiple cells that have activated the same VSG gene in the current wave of parasitic growth. It is not known how this process is coordinated. This continual shifting of surface epitopes has made vaccine development extremely difficult.

REFERENCE

Geiger, A., et al. 2016. Escaping deleterious immune response in their hosts: lessons from trypanosomatids. *Frontiers in Immunology* **7**:212.

Fungal diseases, or **mycoses**, are classified on the basis of three criteria: the site of infection, the route of acquisition, and level of virulence. These criteria and their subcategories are described in **Table 17-3**. Cutaneous infections include attacks on skin, hair, and nails; examples are ringworm, athlete's foot, and jock itch. Subcutaneous infections are normally introduced by trauma and accompanied by inflammation; when inflammation is chronic, extensive tissue damage may ensue. Deep mycoses involve the lungs, the central nervous system, bones, and the abdominal viscera. These infections can occur through ingestion, inhalation, or inoculation into the bloodstream. A very rare and deadly outbreak of fungal meningitis in 2012 was linked to *Exserohilum rostratum*, a fungal contaminant in a preparation of corticosteroids used in epidural steroid injections, most often used to treat chronic back and joint pain.

Virulence types can be divided into primary, indicating the rare agents with high pathogenicity, and opportunistic, denoting weakly virulent agents that primarily infect individuals with compromised immunity. Most fungal infections of healthy individuals are resolved rapidly, with few clinical signs. The most commonly encountered and best-studied human fungal pathogens are *Cryptococcus neoformans*, *Aspergillus fumigatus*, *Coccidioides immitis*, *Histoplasma capsulatum*, and *Blastomyces dermatitidis*. Diseases caused by these fungi are named for the agent; for example, *C. neoformans* causes cryptococcosis and *B. dermatitidis* causes blastomycosis. In each case, infection with these environmental agents is aided by predisposing conditions that include AIDS, immunosuppressive drug treatment, and malnutrition.

Innate Immunity Controls Most Fungal Infections

Physical barriers and agents involved in innate immunity control infection by most fungi. The presence of commensal organisms also helps control the growth of potential pathogens. This has been demonstrated by long-term treatment with broad-spectrum antibiotics, which destroy normal mucosal bacterial flora and often lead to oral or vulvovaginal infection with *Candida albicans*, an opportunistic fungal agent. Phagocytosis by neutrophils is a strong defense against most fungi, and therefore people with neutropenia (low neutrophil count) are generally more susceptible to fungal disease.

Resolution of infection in normal, healthy individuals is often rapid and initiated by recognition of common fungal cell wall PAMPs by PRRs, especially those in the C-type lectin receptor (CLR) family. The three most immunologically relevant cell wall components include β-glucans (polymers of glucose), mannans (long chains of mannose), and chitin (a polymer of *N*-acetylglucosamine). The importance of certain PRRs for resolving fungal infection has been

TABLE 17-3	Classification of fungal diseases	
Site of infection:	Superficial	Epidermis, no inflammation
	Cutaneous	Skin, hair, nails
	Subcutaneous	Wounds, usually inflammatory
	Deep or systemic	Lungs, abdominal viscera, bones, CNS
Route of acquisition:	Exogenous	Environmental, airborne, cutaneous, or percutaneous
	Endogenous	Latent reactivation, commensal organism
Virulence:	Primary	Inherently virulent, infects healthy host
	Opportunistic	Low virulence, infects immunocompromised host

Schistosomiasis: Low Antigenicity and Large Size Pose Unique Challenges to Immune Detection and Elimination of Helminths

More than 300 million people are infected with the helminthic parasite *Schistosoma species*, which causes the chronic, debilitating, and sometimes fatal disease **schistosomiasis**. Infection occurs through contact with free-swimming infectious larvae that are released from an infected snail and bore into the skin, frequently while individuals wade through contaminated water. As they mature, these parasites migrate in the body, with the final site of infection varying by species. The females produce eggs, some of which are excreted and infect more snails. Most symptoms of schistosomiasis are initiated by the eggs, which invade tissues and cause hemorrhage. A chronic state can develop in which the unexcreted eggs induce cell-mediated DTH reactions, resulting in large granulomas that can obstruct the venous blood flow to the liver or bladder.

An immune response does develop to the schistosomes, but it is usually not sufficient to eliminate the adult worms. Instead, the worms survive for up to 20 years, causing prolonged morbidity. Adult schistosome worms have several unique mechanisms that protect them from immune defenses. These include decreasing the expression of antigens on their outer membrane and enclosing themselves in a glycolipid-and-glycoprotein coat derived from the host, masking the presence of their own antigens. Among the antigens observed on the adult worm are the host's own ABO blood-group and histocompatibility antigens. The immune response is, of course, diminished by this covering made of the host's self antigens, which contributes to the lifelong persistence of these organisms.

The major contributors to protective immunity against schistosomiasis are controversial. The immune response to infection with *S. mansoni*, the most common cause of the disease, is dominated by T_H2-like mediators, with high titers of anti-schistosome IgE antibodies, localized increases in degranulating mast cells, and an influx of eosinophils (**Figure 1**, *top*). These cells can then bind the antibody-coated parasite, using their Fc receptors for IgE or IgG, inducing degranulation and death of the parasite via *antibody-dependent cell-mediated cytotoxicity* (ADCC; see Chapter 12). One eosinophil mediator, called *basic protein*, has been found to be particularly toxic to helminths. However, immunization studies in mice suggest that a T_H1 response, characterized by IFN-γ and macrophage accumulation, may actually be *more* effective for inducing protective immunity (Figure 1, *bottom*). In fact, inbred strains of mice with deficiencies in mast cells or IgE can still develop protective immunity to *S. mansoni* following vaccination. Based on these observations, it has been suggested that the ability to induce an *ineffective* T_H2-like response may have evolved in schistosomes as a clever defense mechanism to ensure that IL-10 and other T_H1 inhibitors are produced in response to infection, therefore blocking initiation of a more *effective* T_H1-dominated pathway.

REFERENCE

Wilson, R. A., and Coulson, P. S. 2009 Immune effector mechanisms against schistosomiasis: looking for a chink in the parasite's armour. *Trends in Parasitology* **25**:423.

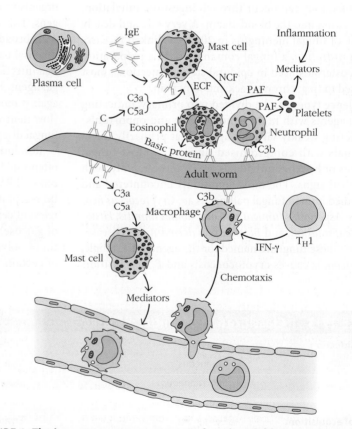

FIGURE 1 **The immune response generated against *Schistosoma mansoni*.** The response includes an IgE humoral component (top) and a cell-mediated component involving CD4⁺ T cells (bottom). C = complement; ECF = eosinophil chemotactic factor; NCF = neutrophil chemotactic factor; PAF = platelet-activating factor.

demonstrated by the increased susceptibility to mycoses seen in individuals with particular alleles at the relevant genetic loci. For instance, certain molecular variants of dectin-1, a C-type lectin receptor (see Chapter 4), are associated with chronic mucocutaneous candidiasis. Toll-like receptors 2, 4, and 9, as well as *complement receptor 3* (CR3), are also involved in the innate response to fungi. In sum, recognition of these fungal cell wall components leads to the activation of complement (via both alternative and lectin pathways) along with the induction of phagocytosis and destruction of fungal cells. The key role of CR3, which recognizes complement deposited on the β-glucans of fungal cells, was confirmed by the fact that mortality from experimental infections of mice with *Cryptococcus* increased after an antibody to CR3 was administered.

Like other microbes, fungi have evolved mechanisms to evade the innate immune response. These include production of a capsule, as in the case of *C. neoformans,* which blocks PRR binding. Another evasion strategy employed by this organism involves fungi-induced expulsion from macrophages after phagocytosis. Because this does not kill host cells it avoids induction of inflammation and further activation of immunologic attention.

Key Concept:

- Most fungal infections are controlled by innate responses, especially PRR recognition of common surface structures, neutrophils, and complement, as evidenced by increased susceptibility to fungal infection in individuals with defects in one or more of these components, including during systemic immunosuppression.

Immunity against Fungal Pathogens Can Be Acquired

The most convincing demonstration of acquired immunity against any infectious agent is the presence of memory, or protection against subsequent attacks following an infection. This protection is not always obvious for fungal disease because primary infection often goes unnoticed. However, positive skin reactivity (secondary recall responses) against fungal antigens are one indicator of prior infection and the presence of memory. For instance, a granulomatous inflammation response, like that seen against *M. tuberculosis*, also controls the spread of *C. neoformans* and *H. capsulatum* in most individuals, indicating the presence of acquired cell-mediated immunity. However, also like tuberculosis, the infectious organism may remain in a latent state within the granuloma, reactivating if the host becomes immunosuppressed.

The presence of specific antibodies is another sign of prior exposure and lasting immunity, and antibodies against *C. neoformans* are commonly found in healthy subjects. However, probably the most convincing argument for pre-existing immunity against fungal pathogens comes from the frequency of normally rare fungal diseases in patients with compromised immunity. Patients with AIDS suffer increased incidence of mucosal candidiasis, histoplasmosis, coccidioidomycosis, and cryptococcosis. These observations in T cell–compromised patients with AIDS, and data showing that B cell–deficient mice have *no* increased susceptibility to fungal disease, are strong indications that cell-mediated rather than humoral mechanisms of adaptive immunity likely control most fungal pathogens.

Strong T_H1 responses and the production of IFN-γ, important for optimal macrophage activation, are most commonly associated with protection against fungi. Conversely, T_H2-cell and T_{REG}-cell responses, or their products, are associated with susceptibility to mycoses. This is apparent in patients displaying distinct T helper responses to coccidioidomycosis, where T_H1 immune activity is associated with a mild, asymptomatic infection and T_H2 responses result in a severe and often relapsing form of the disease. Although the role for other cell types is less certain, recently a regulatory role for T_H17 cells in controlling adaptive immunity against fungi has been postulated, where these cells are hypothesized to help support T_H1-cell and discourage T_H2-cell activation.

Key Concept:

- Both humoral and cellular pathways are engaged against fungal infections, as evidenced by memory responses and increased susceptibility in immune-compromised individuals, where the primary modes of clearance of mycoses appear to be mediated by T_H1, and possibly T_H17, cells.

Emerging and Re-emerging Infectious Diseases

At least yearly, it seems, an outbreak of a new or old infectious disease makes the news, accompanied by reports of severe illness and even death. While no one consensus definition exists, the CDC defines *emerging infectious diseases* as those that have arisen or increased in the human population in the past 2 decades. Examples include news-makers like SARS and Zika virus, as well as some that fly under the radar, such as the recently identified *Candida auris* yeast (discussed shortly). Emerging infectious disease (EID) outbreaks may seem to come from nowhere (e.g., Zika virus) and in some cases this represents significant geographic spread of known human pathogens.

On the other hand, *re-emerging pathogens* are those that were formerly rare, viewed as largely under control or exhibiting reduced infection rates, but that have recently

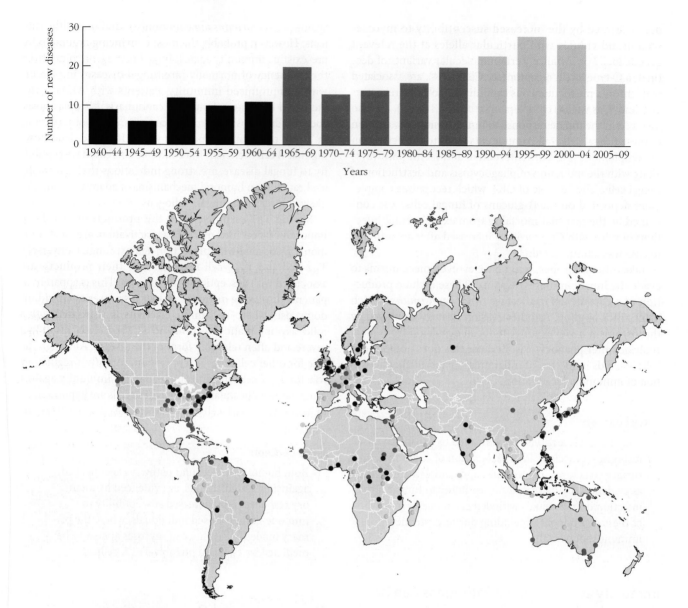

FIGURE 17-7 Emerging and re-emerging infectious disease origin points over the past 75 years. Global map with a blue-to-yellow transition showing the point of origin of newly emerged infectious diseases in 5-year increments from 1940 to 2010. *[Data from EcoHealth alliance data and NPR; https://eidr. ecohealthalliance.org/event-map.]*

begun to resurge. This can be due to the development of drug resistance (e.g., tuberculosis), acquisition of new virulence factors (e.g., MRSA), or environmental changes that enhance transmission rates or geographic range (e.g., Ebola and dengue virus). **Figure 17-7** illustrates the global regions where infectious disease emerged or re-emerged in the past 75 years, highlighting the expanding geography of this issue.

Some Noteworthy New Infectious Diseases Have Appeared Recently

New infectious diseases come in all categories, infecting the young and old, the well and unwell. For example, Zika virus is spread via mosquitoes and sexual contact, but has

minimal impact in adult hosts. However, significantly compromised neurological development has been seen in the fetuses and newborns of some women who become infected while pregnant. In Brazil alone, Zika is now estimated to have generated billions in lost income, added significantly to entrenched systems of economic inequality, and increased the burden on an already stressed health care system (see Clinical Focus Box 17-1).

Contaminated surfaces are likely to blame for a recent outbreak of a new and highly virulent strain of yeast, *Candida auris*. This fungal pathogen has so far remained isolated to health care facilities, taking advantage of those who are elderly, ill, and often immune compromised. This newly emerging infectious disease is particularly troubling because it is difficult to identify and has acquired resistance to many of the common antifungal drugs in use.

In November 2002, an unexplained atypical pneumonia was seen in the Guangdong Province of China, proving resistant to any treatment. A physician who had cared for some of these patients traveled to Hong Kong and infected guests in his hotel, who then seeded a multinational outbreak that lasted until May 2003. By the time the disease, called *severe acute respiratory syndrome* (SARS), was contained, 8096 cases had been reported, with 774 deaths. A rapid response by the biomedical community identified the etiologic agent as a coronavirus, so named because the spike proteins emanating from these viruses give them a crownlike appearance (**Figure 17-8**). This virus was soon traced to several marketplace animals (especially exotic cats) and ultimately to its likely animal reservoir in nature—bats. SARS is a member of the coronaviruses family, known for many years, primarily as the cause of a mild form of the common cold. This newly emerged variant, likely due to a mutation that allowed spread from marketplace animals to vendors, had not been seen previously. Animal models for SARS showed that antibodies to the viral spike protein could thwart replication of the virus, leading to the rapid development of an intranasal vaccine that could induce protective immunity. This was an example of rapid and efficient global detection, containment, and characterization of a new infectious disease, with a very short vaccine pipeline.

West Nile virus (WNV), first identified in Uganda in 1937, was not seen outside Africa or western Asia until 1999, when it suddenly appeared in the New York City area. By 2016 it had been reported in all 50 states with the exception of Alaska. WNV is a flavivirus that replicates very well in certain species of birds and is carried by mosquitoes from infected birds to so-called dead-end hosts such as horses and humans.

Transmission between humans via mosquitoes is inefficient because the titer of virus in human blood is low and the amount of blood transferred by the insect bite is small. WNV may, however, be transferred from human to human by blood transfusion and may be passed from infected pregnant mothers to their newborns. Its major impact appears to be in individuals with compromised immune function, in whom it can cross the blood-brain barrier and cause life-threatening encephalitis and meningitis. The primary public health control measure to combat WNV remains education of the public concerning mosquito control.

Are emerging infectious diseases actually occurring with increased frequency? Heightened awareness and increased spread of EIDs do appear to be the new normal and may not be surprising if we consider three important recent factors: increased access to international travel, climate change, and greater human contact with wild animals (due mostly to deforestation and encroachment). For instance, international travel is what helped Ebola virus spread from one African nation to another and finally across U.S. borders. Tropical diseases are no longer isolated to the tropics. Thanks to rapid travel between countries and spreading habitats for vectors and intermediate hosts, much of the southern United States can now also lay claim to some tropical fever. We know from both current changes taking place and historical data, including recorded El Niño/La Niña patterns, that microbes responsible for malaria, dengue fever, and Lyme disease are likely to spread to new regions if we remain on the current climate trajectory. Finally, the plague of our time, HIV, likely made its leap to humans thanks to the increased association between man and nonhuman primates around the middle of the twentieth century. Alas, none of these factors are likely to recede any time soon, leading to an anticipated increase in EIDs in the future.

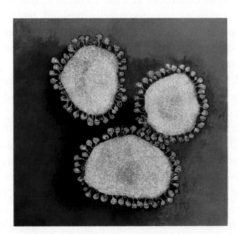

FIGURE 17-8 The coronavirus that caused the outbreak of severe acute respiratory syndrome, or SARS. The virus is studded with spikes that in cross-section give it the appearance of a crown, hence the name *coronavirus*. [*Dr. Linda Stannard, University of Cape Town/Science Source.*]

Key Concepts:

- Examples of current emerging infectious agents include the fungi *C. auris* and several diseases caused by viruses, including SARS, WNV, and Zika.

- The emergence of these infectious agents is linked to opportunistic situations: expanding animal reservoirs and/or arthropod vector habitats (associated with climate change), increased travel by human hosts, and enhanced contact between wild animals that harbor these infections and humans, linked to factors such as population growth, encroachment, and deforestation.

Diseases May Re-emerge for Various Reasons

Tuberculosis is a re-emerging disease now receiving considerable attention. Twenty years ago, public health officials were convinced that tuberculosis would soon disappear as a major health consideration in the United States. A series of events conspired to interrupt that trend, including the AIDS epidemic and other immunosuppressive conditions, allowing *Mycobacterium* strains to regain a foothold and even evolve resistance to the conventional battery of antibiotics. Infected individuals then passed on newly emerged, antibiotic-resistant strains of *M. tuberculosis* to others. Despite the disappointment of not eradicating this disease in the United States, rates of tuberculosis have been declining annually by about 1.5% per year since 2000. Still, worldwide, tuberculosis remains one of the top 10 leading causes of death and is the culprit for more than one-third of all fatalities associated with AIDS.

The first recorded case of Ebola, one of the most deadly infectious agents, occurred after an outbreak in Africa in 1976, although it likely predates this documented incident. By 1977, the causative virus had been isolated and classified as a filovirus, a type of RNA virus that includes the similarly deadly Marburg virus, a close relative of Ebola. The most pathogenic strain, Ebola-Zaire, causes a particularly severe hemorrhagic fever, killing between 50% and 90% of those infected, often within days of the onset of symptoms. Ironically, the short incubation, debilitating illness, and high death rate normally blunt human-to-human spread of this virus—a factor that may have contained the early episodes of its outbreak. Ebola is an example of a zoonotic pathogen with the fruit bat as its likely primary host. Infected bats appear to be largely unharmed, but they do create a local "reservoir" for spread of the virus. Encroachment of humans into bat habitats and contact with infected monkeys is credited with almost all of the initial outbreaks of Ebola. One such naturally acquired infection that began in Guinea in late 2013 was transmitted through human contact to other African countries and ultimately to the United States. The 2014–2016 cross-continent West African outbreak was the largest in history, and the associated international focus brought much-needed attention and money to the study of this disease. Thanks in part to this, previously stalled vaccine efforts have now yielded two promising vaccine candidates, generating memory responses lasting at least 1 year (see Recombinant Vector Vaccines, in the next section). There is speculation that the first Ebola vaccine may be on the near horizon.

Laxity in adherence to established vaccination programs can also lead to re-emergence of diseases that were nearly eradicated. For example, diphtheria began to re-emerge in parts of the former Soviet Union in 1994, where it had almost vanished thanks to European vaccination programs. By 1995, over 50,000 cases were reported and thousands died. The social upheaval and instability that came with the breakup of the Soviet Union, leading to lapses in vaccination and public health programs, was almost certainly a major factor in the re-emergence of this disease. Likewise, poliomyelitis has been on the verge of worldwide eradication for decades. Social unrest and war have delayed progress, although today we are closer than ever to meeting this goal. The only remaining countries still reporting wild polio (cases not associated with vaccination; see below) are Afghanistan and Pakistan, with only 22 cases of naturally acquired polio reported for 2017. However, this kind of progress can sometimes lead to less urgency in vaccination routines. Even in the United States, an increasing trend in some regions to delay or opt out of childhood vaccination has led to sporadic local outbreaks in previously rare childhood diseases such as measles and whooping cough. In some states there has been a backlash to this laxity, resulting in greater enforcement and even legislative mandates, leading to interesting ethical debates.

> **Key Concept:**
>
> - The reasons why previously controlled infectious agents might re-emerge include increasing populations with immune compromise, the acquisition of antibiotic resistance by infectious agents, and a decrease in vaccination rates in some areas.

Vaccines

Preventive vaccines have led to the control or elimination of many infectious diseases that once claimed millions of lives. Since October 1977, not a single naturally acquired smallpox case has been reported anywhere. On the heels of the global victory over smallpox, the program to eradicate polio went into overdrive. That campaign started in 1998 and, led largely by the WHO and several large philanthropic donors, has reduced the number of polio cases worldwide by over 99%. Worldwide vaccination campaigns can also be credited with the control of at least 10 other major infectious diseases (measles, mumps, rubella, typhoid, tetanus, diphtheria, pertussis, influenza, yellow fever, and rabies), many of which previously affected and killed many, mostly babies and young children.

Still, a need remains for vaccines against many other diseases, including malaria, tuberculosis, and AIDS, among others. More work is needed for existing vaccines as well: to improve the safety and efficacy of some, or to lower the cost and ensure delivery of existing vaccines to those most in need, especially in developing countries. Even today, based on WHO data, millions of infants still die of diseases that could be prevented by existing vaccines. The good news is that decades spent on basic research aimed at characterizing the mammalian immune system and the recent "omics" boom (genomics, transcriptomics, proteomics, and even immunomics) are bearing clinical fruit. We have entered a new age of "rational design" for drugs and vaccines, aimed at maximizing the impact on immune function.

In this section, we describe the most common vaccine strategies, some vaccines presently in use, and new avenues of research under development. Keep in mind: no one strategy, additive, or administration route is likely to work for all infectious agents, or even for all members of one type of pathogen, and many of these approaches can be applied in an à la carte fashion, depending on the situation.

Basic Research and Rational Design Advance Vaccine Development

Development of effective new vaccines is a long, complicated, and costly process, rarely even reaching the final stage of years-long clinical trials. Many vaccine candidates that were successful in laboratory and animal studies fail to prevent disease in humans, have unacceptable side effects, or worsen the disease they were meant to prevent. Stringent testing is an absolute necessity, because approved vaccines will be given to large numbers of healthy people. Clear information for consumers about adverse side effects (even those that occur at very low frequency), contraindications (in what situations the vaccine is ill-advised), and potential interactions with other drugs must be made available and carefully balanced with the potential benefit of protection by the vaccine.

Vaccine development begins with years of basic research. For example, characterization of the SARS virus would never have moved with such speed if not for decades of previous work understanding other, less pathogenic, coronaviruses. Along with rhinovirus, coronavirus is one of several causes for the coldlike symptoms we all experience from time to time. Intensive research such as this has led to an appreciation of the features of **immunogens**, epitopes on a pathogen that can be recognized by T and B cells. This has enabled immunologists to design vaccine candidates that maximize activation of key cellular and humoral elements that recognize these immunogens. In animal models, new adjuvants, or additives, are being tested as a means to maximize antigen presentation, overcome original antigenic sin, or to activate the most productive immune pathways. Newer targeting strategies to elicit protection at mucosal surfaces, the most common site of infection, are also underway.

No matter the approach, the first crucial step in the path to a new vaccine is to define specific immunologic targets. These targets are called **correlates of immune protection**, and they represent the specific immunologic goals or markers that scientists believe will result in protection (immunity) from infection or disease upon natural encounter with a pathogen. For instance, sometimes high circulating levels of IgG against a specific surface protein may be required to protect the host from infection, while other times mucosal IgA is more protective. To be immune, we may need activated macrophages to assist in the destruction of a vesicular infection, and not cytotoxic T cells looking to kill infected cells. In other words, we must first ask the following question: *what specific memory response do we need to have on hand before we encounter the real pathogen in order to be protected*? The aim of the latter phases of clinical trials is to determine this empirically; have we in fact immunized individuals against this infectious agent? Of course, we are more likely to hit our targets (often, decades later) if we actually aim at them in the first place!

Key Concept:
- The development of a new vaccine requires much basic science research, huge investments of time and money in development (with more failures than successes), and a clear prior delineation of the specific immune targets of a vaccination program, called *correlates of immune protection*.

Protective Immunity Can Be Achieved by Active or Passive Immunization

Immunization is the process of eliciting a state of protective immunity against a disease-causing pathogen. Exposure to the live pathogen followed by recovery is one route to immunization. However, while highly effective, this can also be dangerous. **Vaccination**, or *intentional exposure to modified forms or parts of a pathogen that do not cause disease* (a **vaccine**), is another. In an ideal world, both will engage antigen-specific lymphocytes and result in the generation of memory cells, providing long-lived protection. However, vaccination does not always ensure immunity. Thus, vaccination is an event, whereas immunization (the development of a protective memory response) is a potential outcome of that event.

A state of at least temporary immune protection can also be achieved by means other than infection or vaccination: for example, the transfer of antibodies from mother to fetus or the injection of antiserum against a pathogen or a toxin to provide immune protection (passive immunization). Without the development of memory B or T cells specific to the organism, however, this state of immunity is only temporary. In this section, we describe current use of immunization techniques, both passive and active.

Passive Immunization by Delivery of Preformed Antibody

Edward Jenner and Louis Pasteur are recognized as the pioneers of vaccination for their documented attempts to induce active immunity, although prior civilizations had employed similar protective strategies (see the chapter introduction). Recognition also is due to Emil von Behring and Kitasato Shibasaburō for their contributions to **passive immunity**. These latter two investigators were the first to show that immunity elicited in one animal can be transferred to another by injecting serum taken from the first.

Passive immunization, in which preformed antibodies are transferred to a recipient, occurs naturally when maternal IgG crosses the placenta to the developing fetus. Maternal antibodies to diphtheria, tetanus, streptococci, measles, mumps, and poliovirus all afford passively acquired protection to the developing fetus and for months in the newborn. Maternal antibodies present in breast milk can also provide passive immunity to the infant in the form of maternally produced IgA. The latter, however, enters the baby's digestive tract and therefore has a different and complementary effect to maternal IgG circulating in the blood.

Passive immunization can also be achieved by injecting a recipient with preformed antibodies, called *antiserum*, from other immune individuals. Before vaccines and antibiotics became available, passive immunization was the only effective therapy for some otherwise fatal diseases, such as diphtheria, providing much needed humoral defense (see Chapter 1, Clinical Focus Box 1-2). Currently, several conditions still warrant the use of passive immunization, including the following:

- Immune deficiency, especially congenital or acquired B-cell defects
- Toxin or venom exposure with immediate threat to life
- Exposure to pathogens that can cause death faster than an effective immune response can develop

Babies born with congenital immune deficiencies are frequently treated by passive immunization, as are children experiencing acute respiratory failure caused by *respiratory syncytial virus* (RSV). Passive immunity is used in unvaccinated individuals exposed to the organisms that cause botulism, tetanus, diphtheria, hepatitis, measles, and rabies (**Table 17-4**), or to protect travelers and health care workers who anticipate or experience exposure to pathogens for which they lack protective immunity. Antiserum also provides an antidote against the poisonous venom in some snake and insect bites. In all these instances, it is important to remember that *passive immunization does not activate the host's natural immune response.* It serves as a buffer between the pathogen, or a toxin, and the host but generates no memory response, so protection is transient.

Although passive immunization may be effective, it should be used with caution because certain risks are associated with the injection of preformed antibody. If the antibody was produced in another species, such as a horse (one of the most common animal sources), the recipient can mount a strong response to the isotypic determinants of the foreign antibody, or the parts of the antibody that are unique to the horse species (typically constant-region domains). This anti-isotype response can cause serious complications. Some individuals will produce IgE antibody against horse-specific determinants. High levels of these IgE–horse antibody immune complexes can induce pervasive mast-cell degranulation, leading to systemic anaphylaxis (see Chapter 15). Other individuals produce IgG or

TABLE 17-4	Common agents used for passive immunization
Disease	**Agent**
Black widow spider bite	Horse antivenin
Botulism	Horse antitoxin
Cytomegalovirus	Human polyclonal Ab
Diphtheria	Horse antitoxin
Hepatitis A and B	Pooled human immunoglobulin
Measles	Pooled human immunoglobulin
Rabies	Human or horse polyclonal Ab
Respiratory disease	Monoclonal anti-RSV[*]
Snake bite	Horse antivenin
Tetanus	Pooled human immunoglobulin or horse antitoxin
Varicella zoster virus	Human polyclonal Ab

[*]Respiratory syncytial virus.

Adapted from Casadevall, A. 1999. Passive antibody therapies: progress and continuing challenges. *Clinical Immunology* **93**:5.

IgM antibodies specific for the foreign antibody, resulting in complement-activating immune complexes. The deposition of these complexes in the tissues can lead to type III hypersensitivity reactions. Even when purified human antiserum or human gammaglobulin is used (a mixture of IgG from many different human B cells), the recipient can generate an anti-allotype response. This recognition of foreign human immunoglobulin (within-species antigenic differences) can cause some of the same symptoms, although its intensity is usually much less than that of an anti-isotype response.

Active Immunization to Induce Immunity and Memory

The goal of active immunization is to trigger the adaptive immune response in a way that will elicit protective immunity and long-lived immunologic memory. When active immunization is successful, a subsequent exposure to the infectious agent elicits a *secondary immune response* that successfully eliminates the pathogen or prevents disease mediated by its products. Active immunization can be achieved by natural exposure to the infectious agent or a similar agent (e.g., cowpox exposure can protect against smallpox) or it can be acquired artificially by administration of a vaccine. An example of the former might be the "chickenpox parties" of the past, where parents invited unprotected children to come play with their pox-ridden youngster as a means to generate a natural immune response early in life. Chickenpox in adults can be more serious with more complications, so immunization while young is advisable. In **active immunity**, as the name implies, the immune system plays an active role—proliferation of antigen-reactive T and

B cells is induced and results in the formation of protective memory cells. This is the primary goal of vaccination.

Vaccination programs have played an important role in the reduction of deaths from infectious diseases, especially among children. In the United States, vaccination of children begins at birth. The American Academy of Pediatrics sets nationwide recommendations (updated in 2017) for childhood immunizations in this country, as outlined in **Table 17-5**. The program recommends or requires 10 vaccines for children from birth to age 6.

In adolescents between the age of 11 and 12, vaccination against the sexually transmitted *human papillomavirus* (HPV), the primary cause of cervical cancer in women, is also recommended (see Clinical Focus Box 19-1). Vaccines against meningitis, as well as boosters for tetanus and influenza, are recommended for all on a regular schedule throughout adulthood (for the latest, see https://www.cdc.gov/vaccines/schedules/).

As illustrated in Table 17-5, children typically require boosters (repeated vaccinations or inoculations) at appropriately timed intervals to achieve protective immunity against many of the common pathogens. In the first months of life, the reason for this may be persistence of circulating maternal antibodies in the young infant. For example, passively acquired maternal antibodies can bind to epitopes on the DTaP vaccine and block adequate activation of the immune system; therefore, in order to achieve protective immunity this vaccine must be given more than once *after* we expect all maternal antibody has been cleared from an infant's circulation (6–12 months). Passively acquired maternal antibody is also known to interfere with the effectiveness of the measles vaccine; for this reason, the combined measles/mumps/rubella (MMR) vaccine is not given before 12 months of age. However, in some developing countries, the measles vaccine is administered at 9 months, a little early to ensure clearance of all maternal antibodies but crucial because 30% to 50% of young children in these countries contract the disease *before* 15 months of age. Multiple immunizations with the polio vaccine are required to ensure that an adequate immune response is generated to each of the strains of poliovirus that make up the vaccine.

The widespread use of vaccines for common, life-threatening diseases in the United States has led to a dramatic decrease in the incidence of these diseases. As long as these immunization programs are maintained, especially in young children, the incidence of these diseases typically remains very low. However, the occurrence or even the suggestion of possible side effects to a vaccine, as well as general trends toward reduced vaccination in children, can cause a drop in vaccination rates that leads to re-emergence of the disease. For example, rare but significant side effects from the original pertussis attenuated bacterial vaccine included seizures, encephalitis, brain damage, and even death. Decreased usage of the vaccine led to an increase in the incidence of whooping cough. In 1991 and on the heels of this, an *a*cellular *p*ertussis vaccine (the *aP* in DTaP) was introduced; it was equally effective but with fewer side effects.

TABLE 17-5	Recommended childhood immunization schedule in the United States, 2017										
							AGE					
Vaccine	Birth	1 mo	2 mo	4 mo	6 mo	9 mo	12 mo	15 mo	18 mo	19–23 mo	2–3 yr	4–6 yr
Hepatitis B	Hep B	← Hep B →			←		Hep B		→			
Rotavirus			RV	RV	RV							
Diphtheria, tetanus, pertussis			DTaP	DTaP	DTaP			← DTaP →				DTaP
Haemophilus influenzae type b			Hib	Hib	Hib		← Hib →					
Pneumococcal			PCV	PCV	PCV		← PCV →					
Inactivated poliovirus			IPV	IPV	←		IPV		→			IPV
Influenza					←		Influenza (yearly)					→
Measles, mumps, rubella							← MMR →					MMR
Varicella							← Varicella →					Varicella
Hepatitis A							←	Dose 1	→		← HepA series →	

Recommendations in effect as of January 2017. Any dose not administered at the recommended age should be administered at a subsequent visit, when indicated and feasible. A combination vaccine is generally preferred over separate injections of equivalent component vaccines.

Data from 2017 American Academy of Pediatrics recommendations, found on CDC website: www.cdc.gov/vaccines/schedules/index.html.

It is important to remember that vaccination is not 100% effective. With any vaccine, a small percentage of recipients may not respond and therefore remain (unknowingly) susceptible. In other cases, individuals may have to forgo or delay vaccination thanks to other health issues, including immune deficiency. This is not usually a serious problem if the majority of the population is immune to an infectious agent, reducing the pathogen reservoir and therefore chances of spread. In this case, the chance of a susceptible individual contacting an infected individual is very low. This phenomenon is known as **herd immunity**. The appearance of some recent measles epidemics, such as among college students or preschool-age children, is a testament to the power of herd immunity and likely occurred partly thanks to decreased vaccination rates in these populations. One recent example is a measles outbreak centered around an amusement park in California during the 2014-15 holiday season, highlighting the potential fallout from rising numbers of unvaccinated individuals and falling levels of herd immunity. Crowding, international travel, and reduced vaccination rates contributed to over 50 reported cases of measles amongst those who visited the park or came in contact with these individuals during this time, and may have contributed to the first U.S. death from measles in 12 years. The year 2014 saw more than three times the number of measles cases than in previous years. And with almost 50,000 cases in 2012, you would need to go all the way back to 1955 to

find a worse year for pertussis (whooping cough) infections (**Figure 17-9**). In addition to overall sickness and hospitalization, this led to 20 deaths in U.S. children. Of note, fatality from whooping cough occurs predominantly in infants under 3 months of age, when they are either too young to receive the pertussis vaccine or before protective immunity kicks in, making them highly susceptible to transfer of an otherwise mild case from unvaccinated individuals. Despite the safety record of this vaccine and the frightening rise in this potentially deadly disease, some parents still take their chances and elect not to vaccinate their children (see Chapter 1, Clinical Focus Box 1-1).

Recommendations for vaccination of adults depend on the risk group. Vaccines for influenza are recommended yearly, while those that protect against meningitis and pneumonia are offered to groups living in close quarters (e.g., military recruits and incoming college students) or those with reduced immunity (e.g., the elderly). Depending on their destination, international travelers are also routinely immunized against diseases endemic to their destination, such as cholera, yellow fever, rabies and typhoid. Immunization against the deadly disease anthrax had been reserved for workers coming into close contact with infected animals or animal products. Concerns about the potential use of anthrax spores by terrorists or in biological warfare has widened use of the vaccine to military personnel and civilians in areas believed to be at risk.

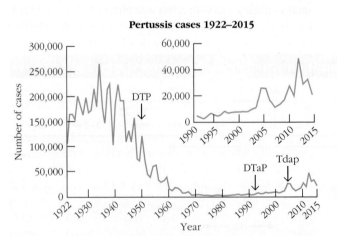

FIGURE 17-9 Introduction of the pertussis vaccine in 1950 led to a dramatic decrease in the annual incidence of this disease in the United States: pertussis cases from 1922 through 2015. Despite huge reductions in outbreaks of whooping cough (caused by pertussis) since introduction of the vaccine, the past decade (inset) has seen several outbreaks. Most are linked to travelers from other countries, where disease spreads to unvaccinated individuals, mostly young children. While many immune-competent individuals above age 1 can resolve this infection, the spread to immune-compromised individuals and babies under 6 months of age can be life-threatening and has resulted in several deaths. [Data from Centers for Disease Control and Prevention.]

Key Concepts:

- Passive immunity, which is only temporary and does not engage the host's immune response or generate memory, can be acquired naturally (e.g., in utero transplacental IgG) or delivered artificially to protect individuals from subsequent infectious disease or recent venom exposures, and in those who lack humoral responses.

- Active immunity can be triggered by either natural infection or artificial exposure to some form of a pathogen, such as a vaccine, with a goal of inducing a memory response that will be protective in the future.

- The introduction of vaccine campaigns, especially in children, has vastly reduced the risk of death from infectious disease worldwide.

There Are Several Vaccine Strategies, Each with Unique Advantages and Challenges

Three key factors must be kept in mind in the development of a successful vaccine: the vaccine must be safe, it must be effective in preventing infection, and the strategy should be reasonably achievable given the population in question.

Population considerations can include geographical locale, access to the target group (which may require several vaccinations), complicating co-infections or nutritional states, and of course, cost. Critical for success is the branch or branches of the immune system that are activated, and therefore vaccine designers must target specific humoral and/or cell-mediated pathways. Protection must also reach the relevant site of infection. In some cases, mucosal surfaces are the prime candidates. An additional factor is the development of long-term immunologic memory. For example, a vaccine that induces a protective primary response may fail to induce the formation of memory cells, leaving the host unprotected after the primary response subsides. Some vaccines generate long-term memory and others, like tetanus, require regular reminders in the form of booster shots or memory responses will wane.

Before vaccines can progress from the laboratory bench to the bedside, they must go through rigorous testing in animals and humans. Most vaccines in development never progress beyond animal testing. The type of testing depends on what animal model systems are available, but frequently involves rodents and/or nonhuman primates. When these animal studies prove fruitful, follow-up clinical trials in humans can be initiated. Phase I clinical trials assess human safety; a small number of volunteers are monitored closely for adverse side effects. Only once this hurdle has been successfully passed can a trial move on to phase II, where effectiveness against the pathogen in question is evaluated. In phase II trials, volunteers are tested for measurable immune response to the immunogen. However, even when volunteers test positive for one or more of the targeted correlates of immune protection, it does not necessarily mean that a state of protective immunity has been achieved or that long-term memory is established. Again, many vaccines fail at this stage. Phase III clinical trials are run in expanded volunteer populations, where natural evidence of protection against "the real thing" is the desired outcome, and where safety, the evaluation of several measurable immune markers, and the incidence of wild-type infections with the relevant pathogen are all monitored carefully over time. Finally, phase IV trials occur after marketing and distribution, and are used to monitor safety, effectiveness, and any long-term impacts.

In some cases circulating effector cells/molecules, in addition to memory cells, are required in order to establish protection. This can depend on the incubation period of the pathogen. For influenza virus, which has a very short incubation period (1 or 2 days), disease symptoms are normally already underway by the time memory cells would be reactivated. Effective protection against disease from influenza therefore depends on maintaining high levels of neutralizing antibody via regular immunizations; those at highest risk are immunized each year, usually at the start of the flu season. For pathogens with a longer incubation period the presence of detectable neutralizing antibody at the time of infection is not always necessary. The poliovirus, for example, requires more than 3 days to begin to infect the central nervous system. An incubation period of this length gives the memory B cells time to respond by producing high levels of serum antibody. Thus, the vaccine for polio is designed to induce high levels of protective immunologic memory that can be recalled and reactivated once the virus is encountered. After immunization with the Salk vaccine (an inactivated form of polio), serum antibody levels peak within 2 weeks and then decline. However, the memory cell response continues to climb, reaching maximal levels 6 months postvaccination and persisting for many years. If an immunized individual is later exposed to the poliovirus, these memory cells will respond by differentiating into plasma cells that produce high levels of serum antibody, which defend the individual from the effects of the virus, even if they don't block the virus from entering the body. In other words, **sterilizing immunity**, or the presence of immune effectors that can *block infection*, is not always required in order to *thwart disease*: poliomyelitis in this case.

In the remainder of this section, various approaches to the design of vaccines—both currently used vaccines and experimental ones—are described, with an examination of the ability of the vaccines to induce humoral and cell-mediated immunity and memory cells. As **Table 17-6** indicates, the common vaccines currently in use consist of live but attenuated organisms, inactivated (killed) bacterial cells or viral particles, as well as protein or carbohydrate fragments (subunits) of the target organism. In addition, the recent Ebola vaccine relies on recombinant vector technology. The primary characteristics of each type, as well as some advantages and disadvantages, are also included.

Live Attenuated Vaccines

In terms of a matched exposure to the real thing, live attenuated vaccines produce the most robust response, although not without risk. For these vaccines, microorganisms are attenuated (disabled) so that they lose their ability to cause significant pathogenicity (disease) but retain their capacity for slow and transient growth within an inoculated host. This allows the immune system a taste of the real thing, but also the upper hand against a pathogen-like organism with only temporary residency. Some agents are naturally attenuated by virtue of their inability to cause disease in a given host, even while having the ability to immunize. The first vaccine used by Jenner is of this type: vaccinia virus (cowpox) inoculation of humans confers immunity to smallpox but does not cause smallpox.

Attenuation can often be achieved in the laboratory by growing a pathogenic bacterium or virus for prolonged periods under abnormal culture conditions. This selects mutants that are better suited for growth in the abnormal culture conditions than in the natural host. For example, an attenuated strain of *Mycobacterium bovis*, called **bacillus Calmette-Guérin (BCG),** was developed by growing *M. bovis* on a medium containing increasing concentrations

TABLE 17-6	Classification of common vaccines for humans		
Vaccine type	**Diseases**	**Advantages**	**Disadvantages**
WHOLE ORGANISMS			
Live attenuated	Measles Mumps Polio (Sabin vaccine) Rotavirus Rubella Tuberculosis Varicella Yellow fever	Strong immune response; often lifelong immunity with few doses	Requires refrigerated storage; may mutate to virulent form
Inactivated or killed	Cholera Influenza Hepatitis A Plague Polio (Salk vaccine) Rabies Zika	Stable; safer than live vaccines; refrigerated storage not required	Weaker immune response than live vaccines; booster shots usually required
PURIFIED MACROMOLECULES			
Toxoid (inactivated exotoxin)	Diphtheria Tetanus	Immune system becomes primed to recognize bacterial toxins	May require booster shots
Subunit	Hepatitis B Pertussis Streptococcal pneumonia	Specific antigens lower the chance of adverse reactions	Difficult to develop
Conjugate	*Haemophilus influenzae* type b Streptococcal pneumonia	Primes infant immune systems to recognize certain bacteria	
OTHER			
Recombinant vector	Ebola (in clinical testing)	Mimics natural infection, resulting in strong immune response	Too early to tell
DNA	HPV (in clinical testing) Zika virus (in clinical testing)	Strong humoral and cellular immune response; relatively inexpensive to manufacture	Not yet available

of bile. After 13 years, this strain had adapted to growth in strong bile and had become sufficiently attenuated that it was suitable as a vaccine for tuberculosis. Because of variable effectiveness, relatively low prevalence, and difficulties in follow-up monitoring, BCG is not used in the United States, although it is commonly used in countries where tuberculosis is pervasive. Likewise, the Sabin form of the polio vaccine and the measles vaccine both consist of attenuated viral strains. Finally, a new weakened form of *Plasmodium falciparum* is being used as a vaccine for malaria (called PfSPZ). In clinical trials conducted in Mali, Africa aimed at testing this vaccine during high transmission periods, only 66% of vaccinees contracted malaria after five doses of the vaccine, compared with 93% of control subjects: a modest but significant improvement.

Attenuated vaccines have obvious advantages. Because of their capacity for growth, even transient growth, such vaccines provide prolonged immune system exposure to the epitopes (immunogens) on the attenuated organism and more closely mimic the growth patterns of the "real" pathogen. This often results in increased immunogenicity and more efficient production of highly effective memory cells. Thus, these vaccines often require only a single immunization, a major advantage in developing countries, where studies show that a significant number of individuals fail to return for boosters. This ability of attenuated vaccines to replicate within host cells thus makes them particularly suitable for inducing cell-mediated responses.

One good example of a live attenuated vaccine that has been in use for decades worldwide is the *oral polio vaccine* (OPV) designed by Albert Sabin. In its original form, OPV consists of three attenuated strains of poliovirus and is administered orally to children in regions where risk of polio is still relatively high. The attenuated viruses colonize the intestine and induce production of secretory IgA, an important defense against naturally acquired poliovirus. The vaccine also induces IgM and IgG classes of antibody and ultimately protective immunity to all three strains of virulent poliovirus. Unlike most other attenuated vaccines, OPV requires boosters, because the three strains of attenuated poliovirus can interfere with each other's replication in the intestine (although there are now also mono- and divalent forms of OPV without this drawback). With the first immunization, one strain will predominate, inducing immunity to that strain. With the second immunization, the immunity generated by the previous immunization will limit the growth of the previously predominant strain in the vaccine, enabling one of the two remaining strains to colonize the intestines and induce immunity. Finally, with the third immunization, immunity to all three strains is typically achieved.

Despite these advantages, the major disadvantage of attenuated vaccines is that these live forms can sometimes mutate and revert to a more virulent form in the host—a major drawback. In the case of polio, this can therefore risk paralytic disease in the vaccinated individual, or in unprotected individuals who come in contact with these more virulent forms shed in feces. The rate of reversion of the OPV to a virulent form is extremely low: about 1 case in 2.4 million doses of vaccine. This is arguably considered an acceptable risk in areas where the danger from wild-type polio is high, so risk of paralysis is high, but maybe not in regions where the threat is minimal. This reversion can also allow pathogenic forms of the virus to find their way into the water supply and spread through a community, especially in areas where sanitation is not rigorous or wastewater is recycled. Despite less ideal immune protection, this possibility has led to the exclusive use of the safer but less effective inactivated polio vaccine in the United States since 2000 (see Table 17-5).

Attenuated vaccines also may be associated with complications similar to those seen in the natural disease. A small percentage of recipients of the measles vaccine, for example, develop postvaccination encephalitis or other complications, although the risk of vaccine-related complications is still significantly lower than risks from infection. An independent study showed that 75 million doses of measles vaccine were given between 1970 and 1993, with 48 cases of vaccine-related encephalopathy (approximately 1 per 1.5 million). This low incidence compared with the rate of infection-induced encephalopathy argues for the efficacy of the vaccine. An even more convincing argument for vaccination is the high death rate from measles—10% or more in regions where nutrition and healthcare are inadequate.

In addition to culturing methods, genetic engineering provides a means to attenuate a virus irreversibly, by selectively removing genes that are necessary for virulence or for growth in the host. This has been done with a herpesvirus vaccine for pigs, in which the thymidine kinase gene was removed. Because thymidine kinase is required for the virus to grow in certain types of cells (e.g., neurons), removal of this gene rendered the virus incapable of causing disease. A live attenuated vaccine against influenza has been developed under the name FluMist. For this, the virus was grown at lower-than-normal temperatures until a cold-adapted strain, unable to grow at human body temperature of 37°C, arose. This live attenuated virus can be administered intranasally and causes a transient infection in the upper respiratory tract, sufficient to induce a strong immune response. The virus cannot spread beyond the upper respiratory tract because of its inability to grow at the elevated temperatures of the inner body. Immunologically, this approach should provide better protection. However, changes in the formulation (going from a trivalent to a quadrivalent formula) and some resulting reports of weaker than expected responses in the United States have led to a pullback on the use of this approach, despite its continued use in the UK and Canada.

Inactivated or "Killed" Vaccines

Another common means to make a pathogen safe for use in a vaccine is by treatment with heat or chemicals. This kills the pathogen, making it incapable of replication, but still allows it to induce an immune response to at least some of the immunogens (antigens) contained within the organism. It is critically important to maintain the structure of key epitopes on surface antigens during inactivation. Heat inactivation is often unsatisfactory because it causes extensive denaturation of proteins; thus, any epitopes that depend on higher orders of protein structure are likely to be altered significantly. Chemical inactivation with formaldehyde or various alkylating agents has been more successful. The Salk inactivated polio vaccine (IPV) is produced by formaldehyde treatment of the poliovirus.

Although live attenuated vaccines generally require only one dose to induce long-lasting immunity, killed vaccines often require repeated boosters to achieve a protective immune status. Because they do not replicate in the host, killed vaccines typically induce a predominantly humoral/antibody response and are less effective than attenuated vaccines in inducing cell-mediated immunity or in eliciting a secretory IgA response, key components of many ideal protective and mucosally based responses.

Even though the pathogens they contain are killed, inactivated whole-organism vaccines still carry certain risks. A serious complication with the early Salk vaccines arose when proper inactivation procedures were not followed by some manufacturers and some of the virus in two vaccine lots were not killed, leading to paralytic polio in a significant percentage of the recipients. This led to temporary suspension of the vaccine, plus changes in all subsequent inactivation and quality control procedures for inactivated vaccines. Risk is also encountered by those who grow and inactivate the virus during vaccine manufacturing. Large quantities of the infectious agent must be handled prior to inactivation, and those exposed to the process are at risk of infection. However, in general, the safety of inactivated vaccines is greater than that of live attenuated vaccines. Inactivated vaccines are commonly used against both viral and bacterial diseases, including the classic yearly flu vaccine and vaccines for hepatitis A and cholera. In addition to their relative safety, their advantages also include stability, and ease of storage and transport.

Subunit Vaccines

Many of the risks associated with attenuated or killed whole-organism vaccines can be avoided with a strategy that uses only specific, purified macromolecules derived from the pathogen, otherwise known as a *subunit approach*. The three most common applications of this strategy are inactivated pathogen exotoxins (called **toxoids**), isolated capsular polysaccharides or surface glycoproteins, and purified key recombinant protein antigens (see Table 17-6). One limitation of some subunit vaccines, especially polysaccharide vaccines, is their inability to activate T_H cells. Instead, they typically activate B cells in a *thymus-independent type 2* (TI-2) manner, resulting in IgM production but little class switching, no affinity maturation, and little, if any, development of memory cells. This can be avoided in vaccines that conjugate a polysaccharide antigen to a protein carrier, which induces T_H-cell responses against both the protein and polysaccharide, discussed below.

Some bacterial pathogens, including those that cause diphtheria and tetanus, produce exotoxins that account for all or most of the disease symptoms resulting from infection. Diphtheria and tetanus vaccines have been made by purifying the bacterial exotoxin and then inactivating it with formaldehyde to form a toxoid. Vaccination with the toxoid induces anti-toxoid antibodies, which are capable of binding to the toxin and neutralizing its effects. Conditions for the production of toxoid vaccines must be closely controlled and balanced to avoid excessive modification of the epitope structure while also accomplishing complete detoxification. As discussed previously, passive immunity can also be used to provide temporary protection in unvaccinated individuals exposed to organisms that produce these exotoxins, although no long-term protection is achieved in this case.

The virulence of some pathogenic bacteria depends primarily on the antiphagocytic properties of their hydrophilic polysaccharide capsule. Coating the capsule with antibodies and/or complement greatly increases the ability of macrophages and neutrophils to phagocytose such pathogens. These findings provide the rationale for vaccines consisting of purified capsular polysaccharides. The current vaccine for *Streptococcus pneumoniae* (the organism that causes pneumococcal pneumonia) consists of 13 antigenically distinct capsular polysaccharides (PCV13). The vaccine induces formation of opsonizing antibodies and is now on the list of vaccines recommended for all infants (see Table 17-5). The vaccine for *Neisseria meningitidis*, a common cause of bacterial meningitis, also consists of purified capsular polysaccharides. Surface glycoproteins on pathogens can be more difficult to detect (e.g., HIV envelope glycoprotein), and some have failed as vaccine candidates. However, a vaccine based on glycoprotein-D from HSV-2 has recently been shown to prevent genital herpes, suggesting that this may be a viable approach for future antiviral vaccines.

Theoretically, the gene encoding any immunogenic protein can be cloned and expressed in cultured cells using recombinant DNA technology, and this technique has been applied widely in the design of many types of subunit vaccines. For example, the safest way to produce sufficient quantities of the purified toxins that go into the generation of toxoid vaccines involves cloning the exotoxin genes from pathogenic organisms into easily cultured cells. A number of genes encoding surface antigens from viral, bacterial, and protozoan pathogens have also been successfully cloned into cellular expression systems for use in vaccine development. The first such recombinant antigen vaccine approved for human use is the hepatitis B vaccine, developed by cloning the gene for the major *hepatitis B surface antigen* (HBsAg) and expressing it in yeast cells. The recombinant yeast cells are grown in large fermenters, allowing HBsAg to accumulate in the cells. The yeast cells are harvested and disrupted, releasing the recombinant HBsAg, which is then purified by conventional biochemical techniques. Recombinant hepatitis B vaccine induces the production of protective antibodies and holds much worldwide promise for protecting against this human pathogen.

Recombinant Vector Vaccines

Recall that live attenuated vaccines prolong immunogen delivery and encourage cell-mediated responses, but have the disadvantage that they can sometimes revert to pathogenic forms. Recombinant vectors maintain the advantages of the live attenuated vaccine approach while avoiding this major disadvantage of reversion. Individual genes that encode key antigens of especially virulent pathogens can be introduced into safe attenuated viruses or bacteria that are used as live carriers. The attenuated organism serves as a vector, replicating within the vaccinated host and expressing the individual gene product/s it carries from the pathogen. Since most of the genome of the pathogen is missing, reversion potential is virtually eliminated. Recombinant vector vaccines have been prepared utilizing existing licensed live attenuated vaccines and adding to them genes encoding antigens present on newly emerging pathogens. These part old–part new chimeric virus vaccines often move more quickly through testing and approval than an entirely new product. A recent example of this is the yellow fever vaccine that was engineered to express antigens from West Nile virus. A number of organisms are under investigation as vectors in such preparations, including vaccinia virus, the canarypox virus, attenuated poliovirus, adenoviruses, attenuated strains of *Salmonella*, the BCG strain of *Mycobacterium bovis*, and certain strains of *Streptococcus* that normally exist in the oral cavity.

Vaccinia virus, the attenuated vaccine used to eradicate smallpox, has been widely employed as a vector for the design of new vaccines. This large, complex virus, with a genome of about 200 genes, can be engineered to carry several dozen foreign genes without impairing its capacity to infect host cells and replicate. The procedure for producing a vaccinia vector that carries a foreign gene from another pathogen is outlined in **Figure 17-10**. The genetically engineered vaccinia expresses high levels of the inserted gene product, which can then serve as a potent immunogen in an inoculated host. Like the smallpox vaccine, genetically engineered vaccinia vector vaccines can be administered simply by scratching the skin, causing a localized infection in host cells. If the foreign gene product expressed by the vaccinia vector is a viral envelope protein, it is inserted into the membrane of the infected host cell, inducing development of both cell-mediated and antibody-mediated immunity.

This strategy, using the vesicular stomatitis virus (VSV) as a vaccine vector, was given a test run in 2015 during the West African Ebola outbreak. A gene for the Ebola virus surface protein was inserted into VSV, a harmless virus that infects humans and cattle. This vaccine, called rVSV-ZEBOV, was still in the early stages of development. Initial tests showed it was protective in animals and elicited an immune response in humans. The in-development vaccine was quickly marshaled for use at the tail end of the 2013-2016 West African outbreak, using a ring vaccination approach like those used successfully during the smallpox eradication campaign. To do this, individuals who came in contact with a patient with Ebola, plus their contacts, were identified and designated as one cluster, or a ring. Some ring groups were offered the experimental Ebola vaccine immediately, and others, 3 weeks after identification. It quickly became apparent that individuals in the immediate vaccination rings were protected, leading to vaccine being offered to all contacts. In an illustration of the principle of herd immunity, even contacts within a vaccine cluster *that did not receive the vaccine* appeared to be more protected from infection by Ebola. Final estimates from this ad hoc trial of rVSV-ZEBOV are that the vaccine is 75–100% effective. Because of the circumstances, no immune response studies were conducted in those who received the vaccine, although follow-up immunologic studies are under way in vaccinees.

Other attenuated vectors may also prove useful in vaccine preparations. A relative of vaccinia, the canarypox virus, is also large and easily engineered to carry multiple genes. Unlike vaccinia, it does not appear to be virulent, even in individuals with severe immune suppression. Another possible vector is an attenuated strain of the bacterium *Salmonella typhimurium*, which has been engineered with genes from the bacterium that causes cholera. The advantage of this vector is that *Salmonella* infects cells of the mucosal lining of the gut and therefore will induce secretory IgA production. Similar strategies are underway for organisms that enter via oral or respiratory routes, targeting bacteria that are normal flora at these sites as vectors for the addition of

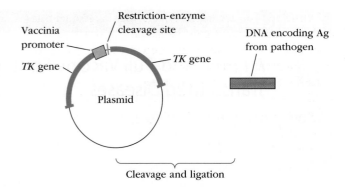
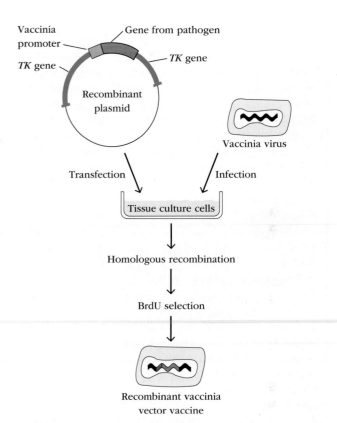

FIGURE 17-10 Production of vaccine, using a recombinant vaccinia vector. *Top*: The gene that encodes the desired antigen (orange) is inserted into a plasmid vector adjacent to a vaccinia promoter (pink) and flanked on either side by the vaccinia thymidine kinase (*TK*) gene (green). *Bottom*: When tissue culture cells are incubated simultaneously with vaccinia virus and the recombinant plasmid, DNA encoding the antigen is inserted into the vaccinia virus genome by homologous recombination at the site of the nonessential TK gene, resulting in a TK⁻ recombinant virus. Cells containing the recombinant vaccinia virus are selected by addition of *bromodeoxyuridine* (BrdU), which kills TK⁺ cells.

pathogen-specific genes. Eliciting immunity at the mucosal surface could provide excellent protection at the portal of entry for many common infectious agents, such as cholera and gonorrhea (see **Advances Box 17-6**).

A Prime-and-Pull Vaccine Strategy for Preventing Sexually Transmitted Diseases

Most pathogens enter our body at one of several mucosal surfaces. The most common sites are the gastrointestinal, respiratory, and genital tracts. We know that these mucosal sites have an immunologic infrastructure of their own (see Chapter 13), and that responses mounted at these locations help protect us from future infection with the same pathogen at the same or similar sites. Most vaccines are designed to elicit strong adaptive responses that will provide protective immunity. But most vaccines are administered intramuscularly (IM) or subcutaneously (called SC, sub-cu, or sub-Q), neither of which mimics mucosal exposures.

Are we missing something here? Some scientists think so. The intentional positioning of protective immunity at mucosal sites of entry is used, for example, with the live polio vaccine, and is being studied in the design of vaccines to protect against *sexually transmitted infections* (STIs), a major public health problem. According to the WHO, there are more than 30 different sexually acquired pathogens, including viral, bacterial, and parasitic organisms. It is estimated that up to 1 million people become infected with an STI *every day*! STIs are one of the top five reasons for individuals to seek health care in developing countries, and 30% to 40% of female infertility is linked to post-infection damage of the fallopian tubes. The most common STIs are chlamydia, gonorrhea, syphilis, hepatitis B, and HIV. Worldwide, these STIs take their greatest toll on young women of childbearing age. Although infection can be transferred in either direction, based on anatomy, women or receptive partners are more likely to become infected during sexual exposure.

While bacterial STIs like chlamydia, gonorrhea, and syphilis are dangerous bacterial agents and can lead to increased susceptibility to other STIs, they are largely treatable with antibiotics. This is not the case for most viral infections. To be effective, a vaccine that protects the vaginal or cervical mucosa needs to ensure the relevant effector cells and and their products arrive at the battlefield. Antigen-specific splenic B cells, serum IgG, and circulating central memory cells are the primary effectors elicited during conventional vaccine administration; none of these will meet that goal of mucosal coverage.

Establishing and maintaining memory cells that reside in or home to the genital tract and protect against STIs is a challenging task. In experimental systems, intranasal or intravaginal immunogen delivery has been shown to elicit antigen-specific CD4+ T-cell and B-cell responses that traffic to the female genital tract, where protective IgA can also be found. However, as in the case of HIV, the ideal vaccine might need to recruit antigen-specific CD8+ T cells and/or IgA-secreting B cells to the genital tract, but not activated CD4+ T cells. The latter are, alas, potential targets for new infections and could therefore inadvertently enhance transmission rates in this particular situation. This means that we might want to be able to pick and choose: we want this type A cell or type B cell, but not type C.

Akiko Iwasaki's group at Yale University has recently applied a new vaccine

(continued)

DNA Vaccines

DNA vaccines are based on plasmid DNA encoding antigenic proteins; the plasmid DNA is injected directly into the muscle of the recipient. This strategy relies on the host cells to take up the DNA and produce the immunogenic protein in vivo, thus directing the antigen through endogenous MHC class I presentation pathways, theoretically helping to activate better CTL responses. The DNA appears either to integrate into host chromosomal DNA or to be maintained for transient periods of time in an episomal form. The hope is to deliver the DNA plasmid to antigen-presenting cells (APCs) such as dendritic cells in or near the injection area. Since muscle cells express low levels of MHC class I molecules and do not express costimulatory molecules, direct or indirect delivery to local APCs is crucial to the development of antigenic responses to these vaccines. Tests in animal models have shown that DNA vaccines are able to induce protective immunity against a number of pathogens, including influenza and rabies viruses. The addition of a follow-up booster shot with protein antigen (in combination, called a *DNA prime and protein boost strategy*), or inclusion of supplementary DNA sequences in the vector (sometimes called *genetic adjuvants*; see the Adjuvants section, below), may enhance the immune response to DNA vaccines. One sequence that has been added to some vaccines is the CpG DNA motif commonly found in many pathogens; recall that this sequence is the ligand for TLR9 (see Chapter 4).

DNA vaccines offer some potential advantages over many of the existing vaccine approaches. Since the encoded protein is expressed in the host in its natural form—there is no denaturation or modification—the immune response is directed to the antigen in a three-dimensional structure similar to that seen in the pathogen, inducing both humoral and cell-mediated immunity. Strong stimulation of both arms of

strategy called "prime and pull" in a mouse model of genital herpes that may be able to achieve this goal. They used a conventional subcutaneous injection of attenuated Herpes Simplex Virus (HSV)-2 to activate a systemic T cell response in mice—the prime. This was followed by a topical application of the chemokine CXCL9 to the vaginal canal of the mice—the pull. This chemokine resulted in the specific recruitment of effector T cells with a memory phenotype to the mucosal tissues of the vagina. All mice treated using the prime-and-pull strategy survived a lethal challenge with live herpesvirus, compared with 36% of control mice that received only the prime part of the vaccine (**Figure 1**). The team also found that local, tissue-resident memory CD8[+] T cells were crucial, but that circulating CD8[+] T cells were dispensable. These resident T cells produced IFN-γ, which was required for this protection, and the resident CD8[+] cells associated closely with vaginal CD301[+] DCs, the cell type that likely delivers local antigen presentation via MHC class I.

If this scheme proves safe and effective, the pull arm of this approach could theoretically be used as an "add on" to conventional vaccines. This would

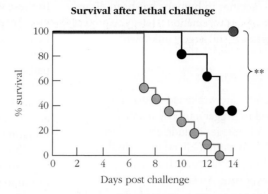

Survival after lethal challenge

- ○ Unimmunized
- ● Subcutaneous immunization (prime)
- ◐ Subcutaneous immunization with topical CXCL9 (prime + pull)

FIGURE 1 A prime-and-pull vaccine strategy protects mice against lethal challenge with HSV. Compared with unimmunized mice, animals treated with a subcutaneous vaccine against HSV, followed by topical treatment of the vagina with the CXCL9 chemokine, showed full protection from lethal challenge with live virus. ** = statistically significant difference.

shorten the lengthy pipeline for most new vaccines. Likewise, the chemotactic agent could be modified to recruit different target cells, or applied to other mucosal tissues, "reeling in" the previously primed cells of choice. Regardless, this technique of training the army first and then selectively bringing them to the battlefield makes good immunologic sense, and it is an exciting new addition to rational vaccine design.

REFERENCES

Shin, H., and A. Iwasaki. 2012. A vaccine strategy that protects against genital herpes by establishing local memory T cells. *Nature* **491**:463.

Shin, H., Y. Kumamoto, S. Gopinath, and A. Iwasaki. 2016. CD301b[+] dendritic cells stimulate tissue memory CD8[+] T cells to protect against HSV-2. *Nature Communications* **7**:13346.

Iwasaki, A. 2016. Exploiting mucosal immunity for antiviral vaccines. *Annual Review of Immunology* **34**:575.

the adaptive immune response typically requires immunization with a live attenuated or recombinant vector preparation, which incurs additional risk. DNA vaccines should also theoretically induce prolonged expression of the antigen, enhancing the induction of immunological memory. While in practice this has not proven true in most settings, newer technology has begun to chip away at this hurdle, producing expression in animals that can now last for months.

DNA vaccines present some practical advantages. No refrigeration of the plasmid DNA is required, eliminating long-term storage challenges. In addition, the same plasmid vector can be custom tailored to insert DNA encoding a variety of proteins, which allows the simultaneous manufacture of a variety of DNA vaccines for different pathogens, saving time and money. Methods for administering DNA vaccines include coating microscopic gold beads with the plasmid DNA and delivery via gene gun, electroporation, mucosal administration via DNA-containing liposomes, or bacterial delivery of DNA. To date, all have advantages and disadvantages, with no prevailing method that appears applicable to humans.

It has now been almost 20 years since the principle behind the use of DNA as a vaccine was proven viable. Those intervening two decades have generated many advances and multiple aborted trials but few overall successes at the end of the day. Initially, safety concerns were the primary hurdle to greater implementation, although more recently variable delivery doses, transient gene expression, and poor immunogenicity have hampered application of this strategy.

As of March 2018, there are still no DNA vaccines licensed for use in humans, although one such vaccine has been approved for veterinary use: a West Nile virus vaccine in horses. This vaccine has also been tested in humans, and after three doses, most volunteers demonstrated titers of neutralizing antibody similar to those seen in horses, as well as CD8[+] and CD4[+] T-cell responses against the virus. Currently, the most promising and farthest along of the human trials of a DNA vaccine are actually for the treatment of cancer (see

Chapter 19); this is likely the first place we will see the licensing of this technology for human use. Since the widespread development of DNA vaccines in humans has been disappointing, the usefulness of this vaccination strategy to protect us against infectious disease is still largely unknown.

Key Concepts:

- Live attenuated vaccines are weakened forms of the infectious agent used to trigger a robust adaptive immune response. Advantages include effector cell types well matched with natural infection (including CTLs) and strong, long-lived, protective memory after even a single exposure. An important disadvantage is the potential for reversion to more virulent forms that can cause disease.

- An infectious organism can be killed or chemically inactivated, and then administered as a means to trigger protective adaptive immune responses. Advantages include the low risk of reversion and disease, plus rapid production times. Disadvantages include lower immunogenicity, poor cell-mediated immune responses (especially CTLs), and a weaker protection that can necessitate booster shots.

- Subunit vaccines use just a part of an inactivated or killed infectious agent; thus they share many of the same advantages and disadvantages, with the added advantage of simpler, safer, and more rapid production time.

- Viruses and other microbes that replicate inside cells but cause no disease can be used as live delivery vehicles for fragments of pathogens. These recombinant vector vaccines retain many of the same advantages of live attenuated vaccines (cell-mediated responses, more robust immunity) and fewer of the disadvantages (low or nonexistent reversion potential and moderate production times).

- DNA vaccines aim to deliver selected genetic material from an infectious agent into host cells as a means for host-based gene expression and immune activation. Currently, there is one DNA vaccine licensed to prevent WNV in horses, plus several in the pipeline as a part of human vaccine regimens that include a DNA prime followed by a protein boost, and for the treatment of some cancers.

Adding a Conjugate or Multivalent Component Can Improve Vaccine Immunogenicity

One of the major drawbacks to the vaccine techniques that do not utilize a live component is that they typically induce weak immune responses due to poor immunogenicity.

To address this, schemes have been developed that employ the fusing of a highly immunogenic protein (a conjugate) to these weak vaccine immunogens. Alternatively, extraneous proteins associated with strong immune activation can be added to the vaccine (multivalent) to enhance or supplement immune reactivity against a weakly immunogenic pathogen-associated antigen.

One example is the vaccine against *Haemophilus influenzae* type *b* (Hib), a major cause of bacterial meningitis and infection-induced deafness in children. A conjugate formulation of a Hib vaccine is included in the recommended childhood regimen (see Table 17-5). This consists of type b capsular polysaccharide covalently linked to a strongly immunogenic protein carrier, tetanus toxoid (**Figure 17-11**). Introduction of the conjugate Hib vaccines has resulted in a rapid decline in Hib cases in the United States and other countries that have introduced this vaccine. The polysaccharide-protein conjugate is considerably more immunogenic than the polysaccharide alone; and because it activates T_H cells, it enables class switching from IgM to IgG. Although this type of vaccine can induce memory B cells, it cannot induce memory T cells specific for the pathogen. In the case of the Hib vaccine, it appears that the memory B cells can be activated to some degree in the absence of a population of memory T_H cells, thus accounting for its efficacy.

A separate study used a similar technique to protect against fungal infection. Immunization with β-glucan isolated from brown alga was conjugated to diphtheria toxoid, and in animal models this vaccine induced antibodies in mice and rats that protected these animals against challenge with both *Aspergillus fumigatus* and *Candida albicans*. The protection was transferred by serum or vaginal fluid from

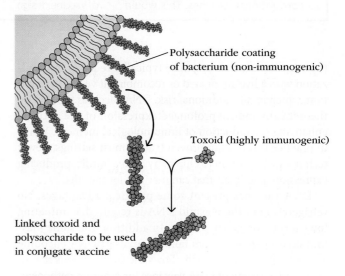

Polysaccharide coating of bacterium (non-immunogenic)

Toxoid (highly immunogenic)

Linked toxoid and polysaccharide to be used in conjugate vaccine

FIGURE 17-11 A conjugate vaccine protects against *Haemophilus influenzae* type b (Hib). The vaccine is prepared by conjugating the surface polysaccharide of Hib (non-immunogenic) to a protein molecule such as tetanus toxoid (highly immunogenic), making the vaccine more immunogenic than either alone.

the immunized animals, indicating that the immunity is antibody based. Infections with fungal pathogens are a serious problem for immunocompromised individuals. The availability of immunization or antibody treatment could circumvent problems with toxicity of antifungal drugs and the emergence of resistant strains, an issue that is especially important in hospital settings.

Since subunit polysaccharide or protein vaccines tend to induce humoral but not cell-mediated responses, a method is needed for constructing vaccines that contain both immunodominant B-cell and T-cell epitopes. Furthermore, if a CTL response is desired, the vaccine must be delivered intracellularly so that the peptides can be processed and presented via MHC class I molecules. One innovative means of producing a multivalent vaccine that can deliver many copies of the antigen into cells is to incorporate antigens (or DNA) into lipid vesicles called *liposomes*, or immunostimulating complexes (**Figure 17-12a**). Likewise, virus envelopes,

called *virosomes*, can serve as a similar delivery vehicle. Protein-containing liposomes are prepared by mixing the proteins with a suspension of phospholipids under conditions that form lipid bilayer vesicles; the proteins are incorporated into the bilayer with the hydrophilic residues exposed. *Immunostimulating complexes* (ISCOMs) are phospholipid monolayers which likewise carry antigenic proteins. Membrane proteins from various pathogens, including influenza virus, measles virus, hepatitis B virus, and HIV, have been incorporated into liposomes and ISCOMs, and are being assessed as potential vaccines. In addition to their increased immunogenicity, liposomes and ISCOMs appear to fuse with the plasma membrane to deliver their antigens intracellularly, where it can be processed by the endogenous pathway, leading to CTL responses (Figure 17-12b).

Key Concept:

- When subunit antigens are poor immune stimulators they can be combined with strong immune activators, either as a fused protein (conjugate) or in a mixture of proteins (multivalent), allowing the strong immunogen to act as a carrier for the weaker antigen, inducing a more robust immune response against both.

(a)

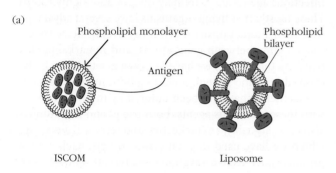

(b) ISCOM delivery of antigen into cell

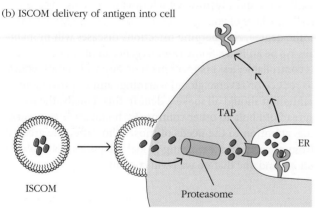

FIGURE 17-12 Multivalent subunit vaccines. (a) Liposomes and *immunostimulating complexes* (ISCOMs) can all be prepared from detergent-extracted antigens or antigenic peptides. ISCOMs are composed of phospholipid monolayers and liposomes are made of phospholipid bilayers. In either case, the immunogen of interest can be inserted as a transmembrane protein or packaged inside these vesicles. (b) ISCOMs and liposomes can deliver agents inside cells, so they mimic endogenous antigens. Subsequent processing by the endogenous pathway and presentation with MHC class I molecules induces a cell-mediated response. ER = endoplasmic reticulum; TAP = transporter associated with antigen processing.

Adjuvants Are Included to Enhance the Immune Response to a Vaccine

In an ideal case, vaccines mimic most of the key immunologic events that occur during a natural infection, eliciting strong and comprehensive immune responses, but without the risks associated with "the real thing." A discussion of vaccines would be incomplete without mentioning the importance of **adjuvants**. Derived from the Latin term "to help or aid," adjuvants are substances that are added to vaccine preparations to enhance the immune response to the antigens with which they are mixed: in other words, to enhance immunogenicity. This is not completely dissimilar to the discussion above about conjugate and multivalent vaccines and can be employed for almost any vaccine strategy being employed. This is especially important to consider when the vaccine preparation is a pathogen subunit or other nonliving form of the organism, where immunogenicity is typically quite low. Nonlive vaccine preparations often also lack triggers for innate immunity, which are inherent in most live and some killed vaccines. *Adjuvants primarily target these innate response elements.* When mixed with the pathogen-associated antigens, these additives can also help with delivery of the vaccine to the immune system and enhance general immune responsiveness.

For almost 80 years, the only adjuvant used in human vaccines has been aluminum salts (called *alum*). It turns out that alum is a fairly good enhancer of T_H2 responses but a weaker stimulator of T_H1 pathways. Despite its long-term use and inclusion in many different vaccine preparations, the mechanism of action of alum is still poorly understood. As an adjuvant, alum is mixed in an emulsion with the immunogen and is thought to work primarily by creating slow-release delivery of the antigen at the injection site, which helps in sustained stimulation of the immune response. It may also help to recruit APCs and encourage the formation of large antigen complexes that are more likely to be phagocytosed by these cells.

In recent years, two new types of adjuvants have been licensed for use in human vaccines. One, called a *virosome*, is a reconstituted virus envelope containing phospholipids and virus glycoproteins but without any genetic information. This type of adjuvant is used in the Inflexal V influenza vaccine and the hepatitis A vaccine. The other, AS04, is an alum salt containing a derivative of lipopolysaccharide (LPS), found naturally on the surface of gram-negative bacteria, and a strong TLR4 agonist. Use of this adjuvant triggers PRR signaling that helps to encourage T_H1 pathway responses. AS04 is currently used in vaccines against HPV and hepatitis B. All of these newer adjuvants have been found to enhance the production of antibodies as compared with unadjuvanted vaccine preparations, and may have the advantage of eliciting much greater cell-mediated immune responses than alum alone.

Some creative uses of existing adjuvants and next-generation adjuvants or immune modulators are in more basic stages of development. These include compounds designed to stimulate certain PRRs—specifically, several TLRs and at least one NOD-like receptor. Although adjuvants have improved humoral immunity, few if any actually enhance cellular immunity, especially $CD8^+$ T-cell responses. Recently, a combination of two already licensed adjuvant components was used in mice treated with an influenza virus peptide and was found to generate protective $CD8^+$ T-cell memory responses. This strategy is attractive, as it utilizes adjuvants with an already established track record of safety and efficacy in humans. Finally, Akiko Iwasaki and colleagues at Yale have created a novel approach aimed at protection against sexually transmitted disease and treatment of some forms of cancer caused by these infectious agents. Her strategy uses a subunit vaccine to activate the immune response, followed by local chemokine administration to recruit memory cells; this approach is called *prime and pull* (see Advances Box 17-6).

> **Key Concept:**
>
> - Adjuvants are additives that can be mixed in with a vaccine to stimulate innate signals that help to instruct and enhance the adaptive response, and are modeled on natural PAMPs known to help direct specific adaptive pathways.

Conclusion

Infectious agents are extremely diverse and highly robust. These mostly free-living organisms have several advantages over their human hosts: much more evolutionary time, a shorter time between generations, and remarkable plasticity. As their hosts, we have our own assets, like a highly evolved immune system, including both innate and adaptive elements that have been tailored by their interactions with these infectious agents, both friend and foe. Humans also have, arguably, an intellectual and technical advantage, which we have used to great effect to fight back: first in primitive but effective ways and more recently in the form of antibiotics and vaccines, which now save many lives, especially the lives of young children. However, emerging and, to some degree, re-emerging infectious diseases will probably always be an issue. Some re-emergence of infectious disease is controllable; less encroachment on animal habitats, efforts to reduce or reverse global warming, and improvement of sanitation should all make a dent in this. Finally, the recent spread of Ebola to other continents should serve as a warning; attention to the needs of those who suffer the greatest from poverty and increasing global inequities is a problem for all, and one that no wall will contain.

REFERENCES

Bachmann, M. F., and G. T. Jennings. 2010. Vaccine delivery: a matter of size, geometry, kinetics and molecular patterns. *Nature Reviews Immunology* **11**:787.

Bloom, B. R., ed. 1994. *Tuberculosis: Pathogenesis, Protection and Control.* ASM Press, Washington, DC.

Borst, P., et al. 1998. Control of VSG gene expression sites in *Trypanosoma brucei. Molecular and Biochemical Parasitology* **91**:67.

Coffman, R. L., A. Sher, and R. A. Seder. 2010. Vaccine adjuvants: putting innate immunity to work. *Immunity* **33**:492.

Grencis, R. 2015. Immunity to helminths: resistance regulation, and susceptibility to gastrointestinal nematodes. *Annual Review of Immunology* **33**:201.

Gross, C. P., and K. A. Sepkowitz. 1998. The myth of the medical breakthrough: smallpox, vaccination, and Jenner reconsidered. *International Journal of Infectious Diseases* **3**:54.

Hussaarts, L., M. Yazdanbakhsh, and B. Guigas. 2014. Priming dendritic cells for Th2 polarization: lessons learned from helminths and implications for metabolic disorders. *Frontiers in Immunology* **5**:499.

Iwasaki, A. 2016. Exploiting mucosal immunity for antiviral vaccines. *Annual Review of Immunology.* **34**:575.

Kaufmann, S. H., A. Sher, and R. Ahmed, eds. 2002. *Immunology of Infectious Diseases.* ASM Press, Washington, DC.

Knodler, L. A., J. Celli, and B. B. Finlay. 2001. Pathogenic trickery: deception of host cell processes. *Nature Reviews Molecular Cell Biology* **2**:578.

Li, F., et al. 2005. Structure of SARS coronavirus spike receptor-binding domain complexed with receptor. *Science* **309**:1864.

Liu, M. A. 2011. DNA vaccines: an historical perspective and view to the future. *Immunology Reviews* **239**:62.

Lorenzo, M. E., H. L. Ploegh, and R. S. Tirabassi. 2001. Viral immune evasion strategies and the underlying cell biology. *Seminars in Immunology* **13**:1.

Macleod, M. K., et al. 2011. Vaccine adjuvants aluminum and monophosphoryl lipid A provide distinct signals to generate protective cytotoxic memory CD8 T cells. *Proceedings of the National Academy of Sciences USA* **108**:7914.

Merrell, D. S., and S. Falkow. 2004. Frontal and stealth attack strategies in microbial pathogenesis. *Nature* **430**:250.

Morens, D. M., G. K. Folkers, and A. S. Fauci. 2008. Emerging infections: a perpetual challenge. *Lancet Infectious Diseases* **8**:710.

National Center for Emerging and Zoonotic Infectious Diseases. 2016. Top 10 Accomplishments 2016. Atlanta, GA: Centers for Disease Control and Prevention. Available from: https://www.cdc.gov/ncezid/pdf/accomplishments-2016-top10.pdf (accessed December 14, 2017).

Romani, L. 2011. Immunity to fungal infections. *Nature Reviews Immunology* **11**:275.

Saade, F., and N. Petrovsky. 2012. Technologies for enhanced efficiency of DNA vaccines. *Expert Review of Vaccines* **11**:189.

Schofield, L., and G. E. Grau. 2005. Immunological processes in malaria pathogenesis. *Nature Reviews Immunology* **5**:722.

Skowronski, D. M., et al. 2005. Severe acute respiratory syndrome. *Annual Review of Medicine* **56**:357.

Torosantucci, A., et al. 2005. A novel glyco-conjugate vaccine against fungal pathogens. *Journal of Experimental Medicine* **202**:597.

Yang, Y.-Z., et al. 2004. A DNA vaccine induces SARS coronavirus neutralization and protective immunity in mice. *Nature* **248**:561.

Useful Websites

https://www.niaid.nih.gov The National Institute of Allergy and Infectious Diseases is the institute within the National Institutes of Health that sponsors research in infectious diseases, and its website provides a number of links to other relevant sites.

www.who.int This is the home page of the World Health Organization, the international organization that monitors infectious diseases worldwide.

https://www.cdc.gov The Centers for Disease Control and Prevention (CDC) is a U.S. government agency that tracks infectious disease outbreaks and vaccine research in the United States.

www.upmchealthsecurity.org The Johns Hopkins Bloomberg School of Public Health Center for Health Security website providing information about select agents and emerging diseases that may pose a biosafety or health security threat.

www.gavi.org The Global Alliance for Vaccines and Immunization (GAVI) is a source of information about vaccines in developing countries and worldwide efforts at disease eradication. This site contains links to major international vaccine information sites.

www.ecbt.org Every Child by Two offers useful information on childhood vaccination, including recommended immunization schedules.

STUDY QUESTIONS

1. Describe the nonspecific defenses that operate when a disease-producing microorganism first enters the body.

2. Describe specific immune defense mechanisms, both innate and adaptive, that the immune response focuses on when combating each of the major types of pathogens (viruses, bacteria, fungi, and parasites).

3. Explain the phenomenon of herd immunity. How does it relate to the appearance of certain epidemics?

4. What is the role of the humoral response in protection from influenza?

5. Describe the unique mechanisms each of the following pathogens has for escaping the immune response: (a) African trypanosomes and (b) influenza virus.

6. Michael F. Good and co-workers analyzed the effect of MHC haplotype on the antibody response to a key malarial surface protein, the *circumsporozoite* (CS) peptide antigen, in several recombinant congenic mouse strains. Their results are shown in the following table:

Strain	H2 alleles					Antibody response to CS peptide
	K	A	E	S	D	
B10.BR	k	k	k	k	k	<1
B10.A (4R)	k	k	b	b	b	<1
B10.HTT	s	s	k	k	d	<1
B10.A (5R)	b	b	k	d	d	67
B10	b	b	b	b	b	73
B10.MBR	b	k	k	k	q	<1

Source: Data from Good, M. F., et al. 1988. The T cell response to the malaria circumsporozoite protein: an immunological approach to vaccine development. *Annual Review of Immunology* **6:**663.

a. Based on the results of this study, which MHC molecule(s) are best at presenting this peptide antigen (i.e., which demonstrate MHC restriction)?

b. Since antigen recognition by B cells is not MHC restricted, why is the humoral antibody response influenced by the MHC haplotype?

7. Fill in the blanks in the following statements.

a. The current vaccine for tuberculosis consists of an attenuated strain of *M. bovis* called _____.

b. Variation in influenza surface proteins is generated by _____ and _____.

c. Variation in pilin, which is expressed by many gram-negative bacteria, is generated by the process of _____.

d. The mycobacteria causing tuberculosis are walled off in granulomatous lesions called _____, which contain a small number of _____ and many _____.

e. The diphtheria vaccine is a formaldehyde-treated preparation of the exotoxin, called a _____.

f. A major contribution to nonspecific host defense against viruses is provided by _____ and _____.

g. The primary host defense against viral and bacterial attachment to epithelial surfaces is _____.

h. Two cytokines of particular importance in the response to infection with *M. tuberculosis* are _____, which stimulates development of T_H1 cells, and _____, which promotes activation of macrophages.

8. Despite the fact that there are no licensed vaccines for them, life-threatening fungal infections are not a problem for the general population. Why? Who may be at greatest risk for these types of infection?

9. Discuss the factors that contribute to the emergence of new pathogens or the re-emergence of pathogens previously thought to be controlled in human populations.

10. Which of the following are strategies used by pathogens to evade the immune system? Give a specific example when possible.

a. Changing the antigens expressed on their surfaces
b. Going dormant in host cells
c. Secreting proteases to inactivate antibodies
d. Having low virulence
e. Developing resistance to complement-mediated lysis
f. Allowing point mutations in surface epitopes, resulting in antigenic drift
g. Increasing phagocytic activity of macrophages

11. Which of the following is a characteristic of the inflammatory response against extracellular bacterial infections?

a. Increased levels of IgE
b. Activation of self-reactive $CD8^+$ T cells
c. Activation of complement
d. Swelling caused by release of vasodilators
e. Degranulation of tissue mast cells
f. Phagocytosis by macrophages

12. Your mother may have scolded you for running around outside without shoes. This is sound advice because of the mode of transmission of the helminth *Schistosoma mansoni*, the causative agent of schistosomiasis.

a. If you disobeyed your mother and contracted this parasite, what cells of your immune system would fight the infection?

b. If your doctor administered a cytokine to drive the immune response, which would be a good choice, and how would this supplement alter maturation of plasma cells to produce a more helpful class of antibody?

13. Usually, the influenza virus changes its structure very slightly from one year to the next. However, although we are being exposed to these "modified" influenza strains every year, we do not always come down with the flu, even when the virus successfully breaches physical barriers. However, some years we do get a really bad case of the flu, despite the fact that we presumably have memory cells left from an earlier primary response to influenza (via vaccine or natural acquisition). Aside from higher doses of virus and the possibility of a particularly pathogenic strain, why is it that some years we get very sick and other years we do not? For example, I might get a bad case of the flu while you experience no disease, and yet we are both being exposed to the exact same virus strain that year. What is happening here? You can assume that you are not receiving the yearly influenza vaccine.

14. Indicate whether each of the following statements is true or false. If you think a statement is false, explain why.

a. Transplacental transfer of maternal IgG antibodies against measles confers short-term immunity on the fetus.

b. Attenuated vaccines are more likely to induce cell-mediated immunity than killed vaccines are.

c. One disadvantage of DNA vaccines is that they don't generate significant immunologic memory.

d. Macromolecules generally contain a large number of potential epitopes.

e. A DNA vaccine only induces a response to a single epitope.

15. What are the advantages and disadvantages of using live attenuated organisms as vaccines?

16. A young girl who had never been immunized to tetanus stepped on a rusty nail and got a deep puncture wound. The doctor cleaned out the wound and gave the child an injection of tetanus antitoxin.

a. Why was antitoxin given instead the tetanus vaccine?

b. If the girl receives no further treatment and steps on a rusty nail again 3 years later, will she be immune to tetanus?

17. What are the advantages of the Sabin polio vaccine compared with the Salk vaccine? Why is the Sabin vaccine no longer recommended for use in the United States?

18. Why doesn't the live attenuated influenza vaccine (FluMist) cause respiratory infection and disease?

19. In an attempt to develop a synthetic peptide vaccine, you have analyzed the amino acid sequence of a protein antigen for (a) hydrophobic peptides and (b) strongly hydrophilic peptides. How might peptides of each type be used as a vaccine to induce different immune responses?

20. You have identified a bacterial protein antigen that confers protective immunity to a pathogenic bacterium and have cloned the gene that encodes it. The choices are either to express the protein in yeast and use this recombinant protein as a vaccine or to use the gene for the protein to prepare a DNA vaccine. Which approach would you take and why?

21. Explain the relationship between the incubation period of a pathogen and the vaccine approach and correlates to immune protection that are required.

22. List the three types of purified macromolecules that are currently used as vaccines.

23. Some parents choose not to vaccinate their infants. Reasons include religion, allergic reactions, fear that the infant will develop the disease the vaccine is raised against, and, recently, a fear, unsupported by research, that vaccines can cause autism. What would be the consequence if a significant proportion of the population was not vaccinated against childhood diseases such as measles or pertussis?

24. For each of the following diseases or conditions, indicate what type of vaccination is used:

a. Polio
b. Chickenpox
c. Tetanus
d. Hepatitis B
e. Cholera
f. Measles
g. Mumps

1. Inactivated
2. Attenuated
3. Inactivated exotoxin
4. Purified macromolecule

25. While on a backpacking trip you are bitten by a poisonous snake. The medevac comes to airlift you to the nearest hospital, where you receive human immunoglobulin treatment (gammaglobulin or antiserum) against the poisonous snake venom. You recover from your snakebite and return home for some TLC. One year later, during an environmental studies field trip, you are bitten once again by the same type of snake. Please answer the following questions:

a. Since you fully recovered from the first snakebite, are you protected from the effects of the poison this second time (i.e., did you develop adaptive immunity)?

b. Immunologically, what occurred the first time you were bitten and treated for the bite?

c. Compared with the first snakebite, are you more sensitive, less sensitive, or equally sensitive to the venom from the second bite?

CLINICAL FOCUS QUESTION

1. The effect of the MHC on the immune response to peptides of the influenza virus nucleoprotein was studied in $H2^b$ mice that had been previously immunized with live influenza virions. The CTL activity of primed lymphocytes was determined by in vitro CML assays using $H2^k$ fibroblasts as target cells. The target cells had been transfected with different $H2^b$ MHC class I genes and were either infected with live influenza or incubated with nucleoprotein synthetic peptides. The results of these assays are shown in the following table.

Target cell ($H2^k$ fibroblast)	Test antigen	CTL activity of influenza-primed $H2^b$ lymphocytes (% lysis)
(A) Untransfected	Live influenza	0
(B) Transfected with class I D^b	Live influenza	60
(C) Transfected with class I D^b	Nucleoprotein peptide 365–380	50
(D) Transfected with class I D^b	Nucleoprotein peptide 50–63	2
(E) Transfected with class I K^b	Nucleoprotein peptide 365–380	0.5
(F) Transfected with class I K^b	Nucleoprotein peptide 50–63	1

a. Why was there no killing of the target cells in system A even though the target cells were infected with live influenza?

b. Why was a CTL response generated to the nucleoprotein in system C, even though it is an internal viral protein?

c. Why was there a good CTL response in system C to peptide 365–380, whereas there was no response in system D to peptide 50–63?

d. If you were going to develop a synthetic peptide vaccine for influenza in humans, how would these results obtained in mice influence your design of a vaccine?

2. A connection between the new pneumococcus vaccine and a relatively rare form of arthritis has been reported. What data would you need to validate this report? How would you proceed to evaluate this possible connection?

ANALYZE THE DATA

Kim and co-workers (Kim, T. W., et al. 2003. Enhancing DNA vaccine potency by combining a strategy to prolong dendritic cell life with intracellular targeting strategies. *Journal of Immunology* **171**:2970) investigated methods to enhance the immune response against *human papillomavirus* (HPV)-16 E7 antigen. Groups of mice were vaccinated with the following antigens incorporated in DNA vaccine constructs:

- + HPV E7 antigen
- + E7 + heat-shock protein 70 (HSP70)
- + E7 + calreticulin
- + E7 + sorting *signal* of *lysosome-associated membrane protein 1* (Sig/LAMP-1)

A second array of mice received the same DNA vaccines and was coadministered an additional DNA construct incorporating the anti-apoptosis gene *Bcl-xL*. To test the efficacy of these DNA vaccine constructs in inducing a host response, spleen cells from vaccinated mice were harvested 7 days after injection and the cells were incubated overnight in vitro with MHC class I–restricted E7 peptide (amino acids 49–57), and then the cells were stained for both CD8 and IFN-γ [part (a) of the figure, left]. In another experiment Kim and colleagues determined how effective their vaccines were if mice lacked CD4[+] T cells [part (b) of the figure, below.].

a. Which DNA vaccine(s) is (are) the most effective in inducing an immune response against papillomavirus E7 antigen? Explain your answer.

b. Propose a hypothesis to explain why expressing calreticulin in the vaccine construct was effective in inducing CD8[+] T cells.

c. Propose a mechanism to explain the data in part (a) of the figure.

d. If you were told that the +E7 +Sig/LAMP-l construct is the only one that targets antigen to the MHC class II processing pathway, propose a hypothesis to explain why antigen that would target MHC class II molecules enhances a CD8[+] T-cell response. Why do you think a special signal was necessary to target antigen to MHC class II?

e. What four variables contribute to the E7-specific CD8[+] T lymphocyte response in vitro as measured in part (b) of the figure?

(a)

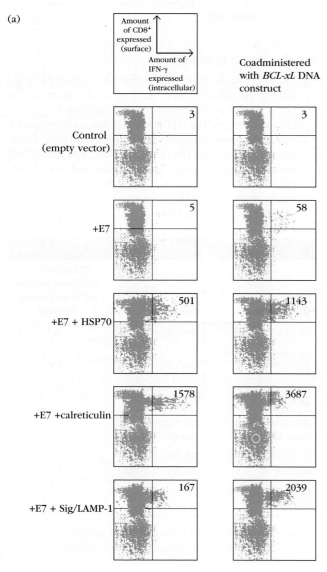

(b)

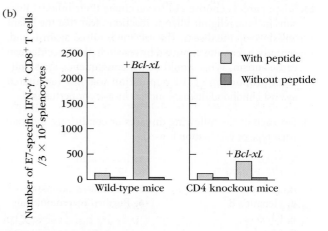

(a) Intracellular cytokine staining followed by flow cytometric analysis to determine the E7-specific CD8[+] T-cell response in mice vaccinated with DNA vaccines using intracellular targeting strategies. A DNA construct including the anti-apoptosis gene *Bcl-x$_L$* was coadministered to the group on the right. The number at top right in each graph is the number of cells represented in the top right quadrant.

(b) E7-specific CD8[+] T lymphocyte response in CD4 knockout mice vaccinated with the +E7 +Sig/LAMP-1 DNA construct, with or without the Bcl-x$_L$ DNA construct.

Immunodeficiency Diseases

18

Learning Objectives

After reading this chapter, you should be able to:

1. Compare and contrast primary and secondary immunodeficiency diseases; provide examples of each.

2. Explain why some primary immunodeficiencies are called severe combined immunodeficiencies, including what is common about their effects on the immune system, while others are not; provide examples of both types.

3. Describe three treatments currently employed to treat primary immunodeficiency diseases and the types of immunodeficiencies for which they are used.

4. List four aspects of the biology of HIV that make it so difficult for both the immune system and antiretroviral therapy to eliminate the virus in an infected individual.

5. Identify two reasons why it is so difficult to generate an effective vaccine for HIV.

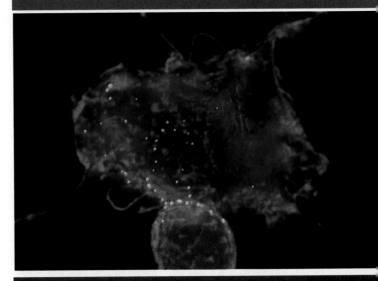

Interaction between dendritic cell and T cell, indicating passage of HIV-1 (green dots) between the cells. *[Courtesy of Thomas J. Hope, Northwestern University.]*

As with any complex multicomponent system, the immune system is subject to failures of some or all of its parts. These failures can have dire consequences. When the system loses its sense of self and begins to attack the host's own cells, the result is autoimmunity, described in Chapter 16. When the system errs by failing to protect the host from disease-causing agents, the result is **immunodeficiency**, the subject of this chapter.

Immunodeficiency resulting from an inherited genetic defect in the immune system is called a **primary immunodeficiency**. In such a condition, the defect is present at birth and often creates health problems early in life. These diseases can be caused by defects in virtually any gene involved in immune development or function—innate or adaptive, humoral or cell-mediated. As one can imagine,

the roles of the component that is missing or defective determine the degree and type of the immune defect; some immunodeficiency disorders are relatively minor, requiring little or no treatment, but others can be life-threatening and necessitate major interventions.

Secondary immunodeficiency, also known as *acquired immunodeficiency*, is the loss of immune function that results from exposure to an external agent, including an infection. These agents include various diseases and infections, medical treatments such as with immunosuppressive drugs following organ transplantation, and social conditions that cause malnutrition. But by far the best known secondary immunodeficiency is acquired immunodeficiency syndrome (AIDS), which results from infection with the human immunodeficiency virus (HIV).

The first part of this chapter describes the most common primary immunodeficiency diseases, examines progress in identifying new defects that can lead to these types of

Key Terms

Immunodeficiency
Primary immunodeficiency
Secondary immunodeficiency
Combined immunodeficiencies (CIDs)
Severe combined immunodeficiency (SCID)
Agammaglobulinemia
Human immunodeficiency virus (HIV)
Acquired immunodeficiency syndrome (AIDS)
Retrovirus
Reverse transcriptase (RT)
Seroconversion
Antiretroviral therapy (ART)

disorders, and considers approaches to their study and treatment. The rest of the chapter describes acquired immunodeficiency, with a focus on HIV infection and AIDS, including the current status of therapeutic and prevention strategies for combating this disorder, which is fatal if untreated.

Primary Immunodeficiencies

To date, more than 300 different genetic defects have been identified that cause primary (inherited) immunodeficiency. Theoretically, any component important to immune function that is defective can lead to some form of immunodeficiency. Collectively, *primary immunodeficiency diseases* (PIDs) have helped immunologists to appreciate the importance of specific proteins and cellular processes that are required for proper immune system function. While most of these disorders are monogenic (caused by defects in a single gene) and are rare, overall the prevalence of immunodeficiency has been estimated as one per ~500 individuals in the United States. Primary immunodeficiency diseases vary in severity from mild to fatal if the defect is not corrected or if infections are not prevented. They can be loosely categorized as affecting either innate or adaptive immunity and are often grouped by the specific components of the immune system most affected: B cells (humoral immunity), T cells (cell-mediated immunity), combined B and T cells, phagocytes and innate immunity, and complement. However, because of the complex interconnections of immune responses, defects in one component or pathway can also manifest in other arms of the immune response, and different gene defects can produce the same phenotype, making strict categorization complicated.

The consequences of a particular gene disruption depend on the role of the affected protein in the immune system, the nature of the mutation (i.e., whether it prevents protein expression or just interferes with the protein's functions), and the impact of the mutation on immune responses (**Overview Figure 18-1**). Defects that interrupt early hematopoietic cell development affect everything downstream of this step; other genetic defects block production of some or all lymphocyte populations. These defects lead to forms of severe combined immunodeficiency affecting both humoral and cell-mediated immunity. Mutations in proteins required for antigen receptor V(D)J rearrangements, such as the RAG proteins, can totally or partially block T- and B-cell development, depending on whether the mutant protein retains any function. In general, defects in T cells tend to have a greater overall impact on the immune response than do genetic mutations that affect only B cells or innate responses. This is due to the pivotal role of T cells in directing immune responses; defects inhibiting T-cell development or function usually affect both humoral and cell-mediated immunity.

Decreased production of phagocytes, such as neutrophils and macrophages, or reduced killing of phagocytosed microbes, typically manifest as increased susceptibility to bacterial or fungal infections.

Mutations affecting downstream components of the immune system, such as deficiencies in production of interferons, complement components, or certain immunoglobulin classes, tend to have more specific and less severe consequences. More recently, a new category of immunodeficiency syndrome has come to light, illustrating the importance of immune regulation, "the brakes" of the immune system. APECED (autoimmune polyendocrinopathy with candidiasis and ectodermal dysplasia) and IPEX (immune dysregulation, polyendocrinopathy, enteropathy, X-linked syndrome) are both immunodeficiency disorders caused by single-gene defects that result in the failure of normal self-tolerance mechanisms: thymic negative selection and the formation of regulatory T cells, respectively. The presence of self-reactive T cells in these conditions results in autoimmunity, as we learned in Chapter 16. Some of the best-characterized primary immunodeficiency disorders with known genetic causes are listed in **Table 18-1**, along with their specific gene defects and resulting immune impairment.

The nature of the immune defect will determine which groups of pathogens are most challenging to individuals who inherit these immunodeficiency disorders (**Table 18-2**). Inherited defects that impair B cells, resulting in depressed expression of one or more of the antibody classes or isotypes, are typically characterized by recurring bacterial infections. These symptoms may be similar to those exhibited by some of the individuals who inherit mutations in genes that encode complement components, which work with antibodies to clear extracellular pathogens. Phagocytes are so important for the removal of fungi and bacteria that individuals with disruptions of phagocytic function suffer from more of these types of infections. Finally, the pivotal role of T cells in orchestrating immune responses means that disruptions to the numbers or performance of this cell type can have wide-ranging effects, including impaired cell-mediated cytotoxicity, depressed antibody production, and dysregulation of cytokine production. In some instances, such as when T- and B-cell responses to self are not properly regulated, autoimmunity can become a major symptom; this is particularly true for defects in regulatory T cells.

The first part of this section discusses how primary immunodeficiencies are detected. We then present diseases caused by defects in adaptive immunity, starting with the most severe cases, which result from mutations affecting T cells, B cells, or both. This is followed by a discussion of disruptions to innate responses, including cells of the myeloid lineage, receptors important for innate immunity, and complement defects. The autoimmune consequences stemming from dysregulation of the immune system are

OVERVIEW FIGURE 18-1

Primary Immunodeficiencies Resulting from Inherited Defects Affect Specific Cell Types

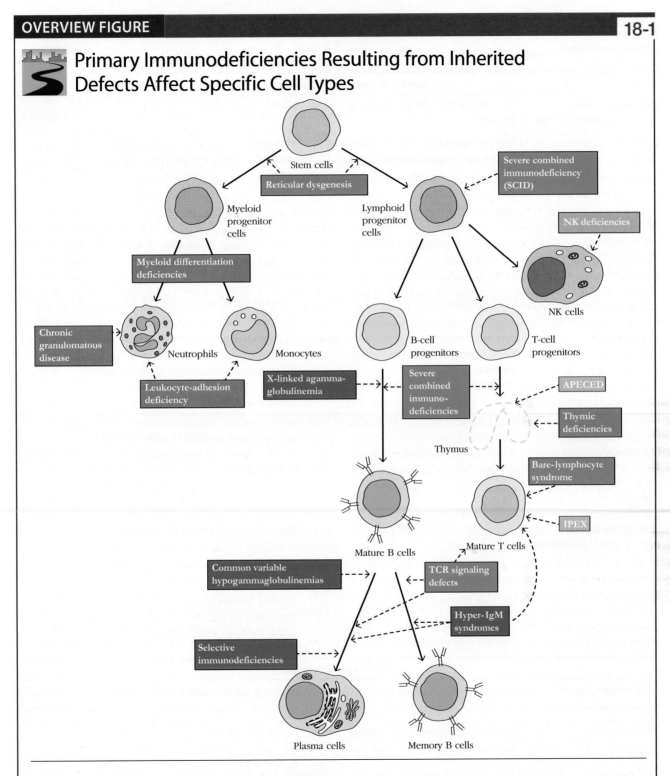

Box colors: Teal = combined immunodeficiencies, or defects that affect more than one cell lineage; orange = phagocyte deficiencies; green = humoral deficiencies; pink = deficiencies in processes of T-cell tolerance; and purple = NK deficiencies.

APECED = autoimmune polyendocrinopathy with candidiasis and ectodermal dystrophy; IPEX = immune dysregulation, polyendocrinopathy, enteropathy, X-linked syndrome.

TABLE 18-1	Some primary human immunodeficiency diseases and underlying genetic defects		
Immunodeficiency disease	Specific defect	Impaired immune function	Inheritance mode*
Combined immunodeficiencies (CIDs)			
Severe combined immunodeficiency (SCID)	Deficiency in RAG1, RAG2, Artemis, or DNA-PK	No TCR or Ig gene rearrangement	AR
	Reticular dysgenesis: Defect in mitochondrial enzyme AK2 required for hematopoiesis	Block in differentiation of hematopoietic stem cells into myeloid and lymphoid lineages	AR AR
	ADA deficiency PNP deficiency	Toxic purine metabolites in lymphoid progenitor cells, leading to loss of B, T, and NK cells	AR XL
	Common γ-chain deficiency JAK3 deficiency IL-7Rα deficiency	Defective signals from IL-2, -4, -7, -9, -15, -21 lead to blocks in development of T and NK cells	AR AR AR
	CD3δ, ε, or ζ deficiency	Defective signals from TCR block T-cell development	AR
Leaky SCID	Hypomorphic mutations in proteins required for V(D)J gene rearrangements, AK2, common γ chain	T cells present but with reduced repertoire; B cells normal or decreased; Igs reduced but may have increased IgE. May include oligoclonal T cells and autoimmunity (Omenn syndrome)	AR, AD, or XL
Bare-lymphocyte syndrome (BLS): MHC deficiencies	Class I: β2-Microglobulin, TAP peptide transporter, tapasin Class II: Defect in transcription factors activating *MHCII* gene expression	No/reduced MHC class I expression, reducing number of CD8+ T cells No MHC class II molecules, reducing numbers of CD4+ T cells	AR AR
Wiskott-Aldrich syndrome (WAS)	Cytoskeletal protein (WASP)	Defective T cells and platelets	XL
Mendelian susceptibility to mycobacterial diseases (MSMD)	IFN-γR, IL-12, IL-12R, STAT1	Impaired immunity to mycobacteria	AR or AD
DiGeorge syndrome	Thymic aplasia	Decreased numbers of T cells	AD
Hyper-IgM syndrome	Defective CD40 ligand or CD40	Elevated IgM due to loss of isotypes other than IgM; defective APC activation leading to reduced T-cell responses to intracellular pathogens	XL AR
Agammaglobulinemias			
X-linked (Bruton's) agammaglobulinemia	Defective Bruton's tyrosine kinase (Btk)	Block in B-cell development; no mature B cells	XL
Reduction in one or more immunoglobulin isotypes	Defective activation-induced cytidine deaminase (AID) or B cell surface proteins	No/low IgG, IgA, IgE—various effects on IgM, IgG, or IgA; elevated or normal IgM; no APC or T-cell deficiency	AR, AR
Common variable immunodeficiency	Complex/unknown	Low IgG, IgA; variable IgM	Complex
Selective IgA deficiency	Complex/unknown	Low or no IgA	Complex

(continued)

TABLE 18-1 *(continued)*

Immunodeficiency disease	Specific defect	Impaired immune function	Inheritance mode*
Innate immune deficiencies			
Chronic granulomatous disease	Mutations in phagosome NADPH oxidase subunits	No ROS or RNS for killing of phagocytosed pathogens	XL, AR
Chédiak-Higashi syndrome	Defective lysosomal trafficking regulator protein (LYST)	Decreased CTL- and NK-mediated cytolytic activity	AR
Leukocyte adhesion deficiencies	Defective integrins	Leukocyte extravasation and chemotaxis	AR
Selective NK-cell deficiencies	Mutations in *STAT5b*, *GATA2*, *MCM4*, and *RTEL1* genes	Absent or decreased total NK cells or NK subsets; increased susceptibility to viral infections	AR
Immune regulation deficiencies			
Autoimmune polyendocrinopathy with candidiasis and ectodermal dystrophy (APECED)	AIRE defect	Reduced thymic negative selection and T_{REG} formation; autoimmunity	AR
Immune dysregulation, polyendocrinopathy, enteropathy, X-linked (IPEX) syndrome	FoxP3 defect	Absence of T_{REG} cells; autoimmunity	XL

*AD = autosomal dominant; AR = autosomal recessive; XL = X linked; "complex" inheritance modes include conditions for which precise genetic data are not available and that may involve several interacting loci.

TABLE 18-2 Patterns of infection and autoimmunity associated with primary immunodeficiency diseases

	Disease	
Immunodeficiency	Opportunistic infections	Other symptoms
Antibody	Ear, sinus, lung infections (staphylococcal, streptococcal, pneumococcal, *Haemophilus influenzae* bacteria); gastrointestinal (enterovirus, *Giardia*)	Autoimmune diseases (autoantibodies, inflammatory bowel disease)
Cell-mediated immunity	Pneumonia (pyogenic bacteria, *Pneumocystis jirovecii*, viruses); tuberculosis and other mycobacterial infections, gastrointestinal (viruses), mycoses of skin and mucous membranes (fungi)	Variable; autoimmune diseases
Complement	Sepsis and other blood-borne infections (streptococci, pneumococci, *Neisseria*)	Autoimmune diseases (systemic lupus erythematosus, glomerulonephritis)
Phagocytes	Skin abscesses, reticuloendothelial infections (staphylococci, enteric bacteria, fungi, mycobacteria)	Variable
Regulatory T cells	N/A	Autoimmune diseases

Data from Lederman, H. M. 2000. The clinical presentation of primary immunodeficiency diseases. *Clinical Focus on Primary Immune Deficiencies* **2**:1 [available from https://primaryimmune.org/publication/healthcare-professionals/idf-clinical-focus-clinical-presentation-primary]; and Immune Deficiency Foundation. 2013. *Patient & Family Handbook for Primary Immunodeficiency Diseases*, 5th ed. Immune Deficiency Foundation, Towson, MD.

also described. Finally, we look at the current treatment options available to affected individuals and the use of animal models of primary immunodeficiency in basic immunology research.

Primary Immunodeficiency Diseases Are Often Detected Early in Life

Traditionally, primary immunodeficiencies have been detected early in life when recurring illnesses and related health problems are brought to the attention of primary care physicians, often by parents worried about their baby's frequent infections. Certain warning signs suggest immunodeficiencies to physicians (**Figure 18-2**). Evidence of early autoimmunity, which accompanies many conditions in which immune regulation is affected, may suggest an

immunodeficiency. Diagnosing primary immunodeficiencies is complex, in part because they do not always manifest themselves in the first few months of life: infants are born with circulating maternal antibodies that initially help to protect them from many infections. The possibility of immunodeficiency can be evaluated by determining levels of antibodies and of various immune cell types, by monitoring an infant's response to vaccines, and by testing for genetic variants associated with particular conditions (especially if there is a family history of immunodeficiency). Confirming the diagnosis of a specific immunodeficiency condition may take some time, however. This may delay the diagnosis several months, thus increasing the risk to newborns during a time when they may become increasingly susceptible to infections and when treatments to address the defect may have the best chance of success.

In recognition of the importance of early detection of the most serious primary immunodeficiencies, which block the development of lymphocytes, a screening test has been developed for newborns. It utilizes the standard blood samples collected from newborns via heel or finger pricks and employs a rapid *polymerase chain reaction* (PCR)–based assay to look for evidence of *V-J* gene rearrangements in T-cell receptor α-chain genes. During the rearrangement of TCR genes, the deleted DNA between the *V* and *J* gene segments forms circles (see Chapter 6). These DNA circles, called *T-cell receptor excision circles* (TRECs), are often retained in T cells circulating in the newborn's blood. Recommendations to screen every newborn for defective TCR gene rearrangements along with the other defects were approved in 2010 and adopted by many states. Today more than 75% of babies born in the United States receive this standard newborn screening, before live viral vaccines are administered that could cause infections in immunodeficient babies and at a time when the implementation of aggressive therapy is most beneficial. Recent data from newborn screening in California indicate that immunodeficiencies detected by this method affect one in 66,250 live births.

FIGURE 18-2 Primary immunodeficiency warning signs. This poster for physicians and other health care workers was created by the Jeffrey Modell Foundation, which funds immunodeficiency disease research and patient information resources. Patients with two or more of these warning signs (and autoimmunity, not in this list) should be evaluated for underlying immunodeficiency. *[These warning signs were developed by the Jeffrey Modell Foundation Medical Advisory Board. Consultation with Primary Immunodeficiency experts is strongly suggested. ©2016 Jeffrey Modell Foundation]*

> **Key Concepts:**
> - Immunodeficiency diseases often are detected early in life, as they may cause recurring infections.
> - Today most newborns in the United States are tested at birth for severe immunodeficiencies by screening blood samples for evidence of T-cell receptor gene rearrangements.

Combined Immunodeficiencies Disrupt Adaptive Immunity

Among the most severe forms of inherited immunodeficiency are a group of disorders termed **combined immunodeficiencies** (**CIDs**): diseases resulting from an absence of

T cells or significantly impaired T-cell function, combined with some disruption of antibody responses. Defects within the T-cell compartment generally also affect the humoral system because T_H cells are typically required for complete B-cell activation, antibody production, heavy-chain class switching, and affinity maturation. Therefore, some depression in the level of one or more antibody isotypes and an associated increase in susceptibility to bacterial infection are common with CIDs. T-cell impairment can also lead to a reduction in both delayed-type hypersensitivity responses and cell-mediated cytotoxicity, resulting in increased susceptibility to almost all types of infectious agents, but especially viruses, protozoa, and fungi. For instance, infections with species of *Mycobacterium* are common in patients with CID, reflecting the importance of T cells in eliminating intracellular pathogens. Likewise, viruses that are otherwise rarely pathogenic (such as cytomegalovirus or even live, attenuated measles and chicken pox vaccines) may be life-threatening for individuals with CIDs. The following section first discusses the most severe CIDs, such as when there is an absence of both T and B cells, followed by less severe forms of the disease, in which more minor disruptions to particular components of the T- and B-cell compartments are observed.

Severe Combined Immunodeficiency

The most extreme forms of CID make up a family of disorders termed **severe combined immunodeficiency** (**SCID**). These stem from genetic defects that lead to a complete or almost complete lack of functional T cells in the periphery. As a general rule, these defects target steps that occur early in hematopoiesis, in the lymphoid lineage, or in early stages of T-cell (and sometimes B-cell) development. The five general categories of defects that have been found to result in SCID include the following (**Figure 18-3**):

1. In reticular dysgenesis, defective differentiation of myeloid and lymphoid lineages from hematopoietic stem cells due to a defect in a mitochondrial adenylate kinase.

2. Premature death of the lymphoid lineage due to accumulation of toxic metabolites, caused by defects in the purine metabolism pathways, such as in the enzyme adenosine deaminase (ADA).

3. Defective V(D)J rearrangement in developing T and B lymphocytes, caused by mutations in the genes for RAG1, RAG2, or other proteins involved in the rearrangement process.

4. Defective cytokine signaling in T-cell progenitors, caused by mutations in certain cytokines, cytokine receptors, or regulatory molecules that control their expression.

5. Disruptions in pre-TCR or TCR signaling during development, caused by mutations in CD3 chains.

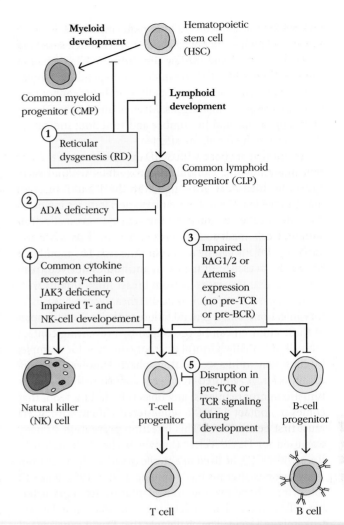

FIGURE 18-3 Defects in lymphocyte development and signaling can lead to severe combined immunodeficiency (SCID) in humans. SCID may result from (1) defects preventing differentiation of myeloid and lymphoid cells (reticular dysgenesis) or of cells of the lymphoid lineage [(2) defects in purine metabolism such as ADA deficiency]; (3) SCID also results from defects in the V(D)J rearrangement process producing the functional pre-BCR and pre-TCR, required for development of T and B lymphocytes; examples include mutations in recombination-activating genes (*RAG1* and *RAG2*) and the DNA excision repair pathway (e.g., *Artemis*); (4) defects in the common γ chain of receptors for IL-2, -4, -7, -9, -15, and -21, required for the development of T lymphocytes, or JAK3, which transduces signals from those receptors, can also lead to SCID as well as NK-cell deficiency; and (5) defects in the CD3 signaling chains of the pre-TCR and TCR can prevent T-cell development.

Depending on the underlying genetic defect, an individual with SCID may have a loss of only T cells (T^-B^+ SCID) or both T and B cells (T^-B^- SCID). In either case, both cellular and humoral immunity are either severely depressed or absent. Clinically, SCID is characterized by a very low number of circulating lymphocytes and a failure to mount immune responses mediated by or requiring T cells. In many cases, the thymus will not fully develop without a sufficient number of T cells, and the few circulating T cells present in some patients with SCID often do not respond to stimulation by mitogens, indicating that they probably cannot proliferate in response to antigens. Myeloid and erythroid cells often appear normal in number and function, indicating that only lymphoid cells are affected.

Infants that are born with SCID experience severe recurrent infections that can quickly prove fatal without early aggressive treatment. Although both the T and B lineages may be affected, the initial manifestation in these infants is typically infection by fungi or viruses that are normally dealt with by T cell–mediated immune responses. This is because antibody deficits can be masked in the first few months of life by the presence of maternal antibodies derived from transplacental transport or from breast milk. Infants with SCID often suffer from chronic diarrhea, recurrent respiratory infections, and a general failure to thrive. The life span of these children can be prolonged by preventing contact with all potentially harmful microorganisms—for example, by confinement in a sterile atmosphere. However, extraordinary effort is required to prevent contact with all opportunistic microorganisms; any object, including food, that comes in contact with the sequestered patient with SCID must first be sterilized. This infection prevention strategy was used for David Vetter (known as the "Bubble Boy"). Born with SCID, he lived in a sterile isolation chamber from a few seconds after his birth until he died in 1984 at age 12 following a bone marrow transplantation from his sister; her cells contained an undetected virus that caused a rare leukemia in David. The challenges that David experienced raised many ethical issues, but his case also led to important research findings about SCID as well as better public knowledge about immunodeficiency diseases. Today such isolation for babies with SCID is used mostly as a temporary measure, pending replacement therapy treatments and/or bone marrow transplantation or gene therapy (more on these below).

In a fairly rare but very impactful form of SCID, genetic defects lead to perturbations in hematopoiesis. In *reticular dysgenesis (RD)*, the initial stages of hematopoietic cell development are blocked by defects in the *adenylate kinase 2 gene (AK2)*, which interfere with energy production in mitochondria. This results in apoptosis of myeloid and lymphoid precursors, leading to severe reductions in circulating leukocytes (see Figure 18-3). The resulting general failure leads to impairment of both innate and adaptive immunity, resulting in susceptibility to infection with all types of microorganisms. Without aggressive treatment, babies with this very rare form of SCID usually die in early infancy from uncontrolled infection.

A more common defect affecting early hematopoietic development and resulting in SCID is *adenosine deaminase (ADA)* deficiency. Adenosine deaminase catalyzes conversion of adenosine or deoxyadenosine to inosine or deoxyinosine, respectively. Its deficiency results in the intracellular accumulation of toxic adenosine metabolites, which interfere with purine metabolism and DNA synthesis. This housekeeping enzyme is found in all cells, so these toxic compounds also produce neurologic and metabolic symptoms, including deafness, behavioral problems, and liver damage. However, early lymphoid progenitors are particularly affected by toxic metabolites, undergoing apoptosis and leading to deficiencies in all lymphoid populations—T, B, and NK cells (see Figure 18-3). Deficiency in another purine salvage pathway enzyme, *purine nucleoside phosphorylase* (PNP), produces a similar phenotype via much the same mechanism.

Deficiency in cytokine signaling is at the root of the most common forms of SCID, and defects in the gene encoding the common gamma (γ) chain of the IL-2 receptor (*IL2RG*; see Figure 3-5) are the most frequent culprits. This particular form of immunodeficiency is often referred to as *X-linked SCID* (or SCIDX1) because the affected gene is located on the X chromosome; thus the disorder is more common in males. This is the form of SCID that Bubble Boy David Vetter had. Defects in this chain impede signaling not only through the IL-2R but also through receptors for IL-4, -7, -9, -15, and -21, all of which use this chain in their structures. This leads to a form of T^-B^+ SCID. Although the common γ chain was first identified as a part of the IL-2 receptor, impaired IL-7 signaling is likely the source of T-cell developmental defects. Unlike mice, where IL-7 signaling is required for the development of both T cells and B cells, in humans signaling through the IL-7R is required for the development of T cells but not B cells. However, T cell–dependent antibody responses will be affected. As signals from IL-21 derived from T_{FH} cells are required for induction of normal B-cell responses in germinal centers, the lack of IL-21R signaling interferes with the generation of antibody responses. Reduced numbers of NK cells also are common in SCIDX1, as a result of the absence of signaling through the IL-15R, necessary for key steps in NK-cell development (see Figure 18-3). Defects in the kinase JAK3, which associates with the cytoplasmic region of the common γ chain and initiates signaling from these receptors, can produce a phenotype similar to the common γ-chain deficiency (though not X-linked) (see Appendix II, Table 3).

Defects in the pathways involved in the V(D)J recombination events that produce membrane immunoglobulin B-cell receptors and T-cell receptors highlight the importance of early signaling through these receptors for lymphocyte development and survival. Mutations in the *recombinase activating genes* (*RAG1* and *RAG2*) and genes encoding proteins involved in the DNA excision-repair pathways employed during gene rearrangement (e.g., *Artemis*) can lead to SCID (see Chapter 6 and Figure 18-3). In these cases, rearrangement and expression of TCR and immunoglobulin genes, necessary for forming the pre-TCR and pre-BCR, are blocked at the early stages of T- and B-cell development, leading to a virtual absence of functioning T and B cells. Only the numbers and function of NK cells remain largely intact (see Clinical Focus Box 6-2). Defects in the CD3 signaling chains associated with the pre-TCR and TCR also can block early T-cell development and cause SCID.

Leaky SCID

Some mutations in typical SCID genes (such as genes for the RAG and common γ-chain proteins) are leaky, that is, they only partially affect the expression or function of the proteins. Known as *hypomorphic mutations*, these mutations can lead to partially reduced numbers of T cells, impaired T-cell and antibody responses, and in some cases to some generalized effects on DNA repair processes. These conditions are now referred to as *leaky SCID*. In some individuals the leaky mutations result in oligoclonal (expanded) populations of autoreactive T cells that cause chronic inflammation of the skin, swollen lymphoid tissues and liver, and high levels of eosinophils and IgE. This version of leaky SCID is called *Omenn syndrome*.

MHC Defects That Can Resemble SCID

A failure to express MHC class I or class II molecules can lead to general failures of immunity that generally are not as severe as SCID. For example, without MHC class II molecules, positive selection of CD4$^+$ T cells in the thymus is impaired, and peripheral T helper cell responses are limited. Deficiencies in MHC class I or class II are called *bare-lymphocyte syndrome*, reflecting the absence of these proteins from cell surfaces. The important and ubiquitous role of MHC class I molecules is highlighted in patients with defective expression of all class I proteins. This rare immunodeficiency disorder can be caused by mutations in β$_2$-microglobulin, a subunit of MHC class I proteins, or in the TAP proteins, which are vital to antigen presentation by MHC class I molecules (see Figure 7-14). These defects affecting MHC class I proteins result in impaired positive selection of CD8$^+$ T cells, depressed cell-mediated immunity, and heightened susceptibility to viral infection.

Developmental Defects of the Thymus

Some immunodeficiency syndromes affecting T cells are grounded in failure of the thymus to undergo normal development. These thymic malfunctions can have subtle to profound outcomes on T-cell function, depending on the nature of the defect. *DiGeorge syndrome (DGS)*, also called *velocardiofacial syndrome*, is one example. This disorder typically results from various deletions in a region on chromosome 22 containing up to 50 genes, with the *T-box* transcription factor (*TBX1*) thought to be most influential. This transcription factor is highly expressed during particular stages of embryonic development, when facial structures, heart, thyroid, parathyroid, and thymus tissues are forming. For

this reason, the syndrome is sometimes also called the third and fourth pharyngeal pouch syndrome. Not surprisingly, patients with DGS present with symptoms of immunodeficiency, abnormal facial features, hypoparathyroidism, and congenital heart anomalies, with the latter typically being the most critical. Although most DGS sufferers show some degree of immunodeficiency, it varies widely. In very rare cases of complete DGS, where no thymic tissue develops, severe depression of T-cell numbers and poor antibody responses due to lack of T-cell help leave patients susceptible to all types of opportunistic pathogens. Thymic transplantation and passive antibody treatment can be of value to these individuals, although severe heart disease can limit long-term survival even when the immune defects are corrected. In the majority of patients with DGS, in whom some residual thymic tissue develops and some functional T cells are found in the periphery, treatments to avoid bacterial infection, such as antibiotics, are often sufficient to compensate for the immune defects.

Wiskott-Aldrich Syndrome and Other Deficiencies in T-Cell Signaling

Although SCID is caused by genetic defects that result in the loss or major impairment of T cells, a number of other CIDs can result from less severe disruptions to T-cell function. The defect in patients suffering from *Wiskott-Aldrich syndrome (WAS)* occurs in an X-linked gene named for this disease (*WASP*), which encodes a cytoskeletal protein highly expressed in hematopoietic cells (see Table 18-1). The WAS protein (WASP) is required for assembly and reorganization of actin filaments in cells of the hematopoietic lineage, events critical to proper immune synapse formation and intracellular signaling. Clinical manifestations, which usually appear early in the first year of life, vary widely. Severity depends on the specific mutation, but eczema and thrombocytopenia (low platelet counts and smaller than normal platelets, which can result in near fatal bleeding) are both common. Impaired cell-mediated immunity and humoral immune defects, including lower than normal levels of IgM, are common features. Patients with WAS often experience recurrent bacterial infections, especially by encapsulated strains such as *Streptococcus pneumoniae*, *Haemophilus influenzae* type b (Hib), and *Staphylococcus aureus*. As the disease develops, autoimmunity and B-cell malignancy are not uncommon, suggesting that regulatory T-cell functions are also impaired. Mild forms of the disease can be treated by targeting the symptoms—transfusions for bleeding and passive antibodies and/or antibiotics for bacterial infections—but severe cases require hematopoietic stem cell transfer.

Defects in several other proteins involved in signaling pathways downstream of TCRs also adversely affect T-cell development, peripheral T-cell activation, and effector functions such as cell-mediated cytotoxicity. These include mutations in the protein kinases Lck, ZAP-70, and ITK in other proteins required for cytoskeletal reorganization, and in several proteins in the NF-κB activation pathway. Antibody responses may also be affected, and autoimmunity may occur because of immune dysregulation.

Hyper-IgM Syndrome

An inherited deficiency in either CD40 ligand (CD40L), which is expressed by T cells, or in CD40, which is expressed by B cells and antigen-presenting cells such as dendritic cells, monocytes, and macrophages, leads to impaired interactions between T cells and cells that are presenting antigen. In the case of CD40L deficiency (an X-linked disorder), or the less common CD40 deficiency (controlled by an autosomal gene), the normal interactions between CD40L on T_H cells and CD40 on B cells or antigen-presenting cells cannot occur. This costimulatory engagement is required for B cells to be activated by T_H cells (see Chapter 11). In the absence of this costimulatory signal B cells cannot mount normal responses to T-dependent antigens, including the formation of germinal centers, somatic hypermutation, the heavy-chain class switching that is necessary for production of IgG, IgA, and IgE antibodies, and the production of memory cells (**Figure 18-4**). B-cell responses to T-independent antigens, however, are unaffected, accounting for the presence of IgM antibodies in these patients, which range from normal to abnormally high levels and give the disorder its common name, *hyper-IgM (HIM) syndrome.*

As CD40L-CD40 interactions are also required for T_H activation of DC maturation and IL-12 secretion, defects in this pathway result in diminished cell-mediated immunity and increased susceptibility to intracellular pathogens. For example, licensing of dendritic cells to be able to activate naïve $CD8^+$ CTL precursors requires interactions between the CD40L of T_H and the CD40 of DC (see Figure 12-6). Therefore children with CD40L or CD40 defects suffer from a range of recurrent viral and fungal infections, especially in the respiratory tract. As this form of immunodeficiency affects the responses of both T cells and B cells it is classified as a CID.

Hyper-IgE Syndrome (Job Syndrome)

Another primary immunodeficiency is characterized by skin abscesses, recurrent pneumonia, eczema, and elevated levels of IgE, accompanied by facial abnormalities and bone fragility. This multisystem disorder, known as *hyper-IgE (HIE) syndrome*, also known as *Job syndrome*, is most frequently caused by an autosomal dominant mutation in the *STAT3* gene. STAT3 is involved in the intracellular signaling cascade activated by IL-6, IL-10, and IL-21 ligation, and is important for T_H17 and T_{FH} cell differentiation (see Figure 10-9 and Appendix II, Table 3). The absence of STAT3 signaling prevents induction of IL-17, IL-10, IL-22, and TGF-β in response to antigenic stimulation, and patients with Job syndrome have lower-than-normal levels of circulating T_H17 cells. Depressed T_H17 responses, which are important for clearance of fungal and extracellular bacterial infections, explain the susceptibility of these patients to fungal and bacterial infections, including those caused by *Candida albicans* and *S. aureus*. Reduced T_{FH} cell activity may be responsible for the lower numbers of antigen-specific memory B cells in these individuals. STAT3 defects also inhibit IL-10 signaling and the development of regulatory T cells, reducing induced T_{REG} cells in these patients. Although STAT3 is involved in the signal transduction of many cytokines and therefore could play a role in the elevation of IgE in these patients, no clear mechanism for this connection has been defined. Surprisingly, individuals with hyper-IgE syndrome do not show abnormal levels of allergic responses, perhaps because STAT3 also normally is required for the activation of IgE-mediated mast cell and basophil degranulation.

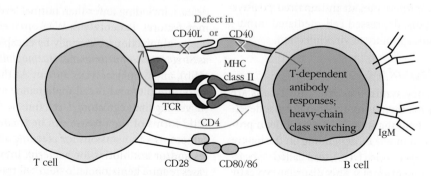

FIGURE 18-4 Defects in CD40L on T cells or CD40 on B cells and other APCs can give rise to the primary immunodeficiency known as hyper-IgM syndrome. An important component of T-cell help to B cells is the binding of the CD40L protein, which is induced when T cells are activated through their TCR, to the CD40 protein on B cells. This results in antibody responses to T-dependent antigens and heavy-chain class switching. A defect in either CD40 or CD40L (indicated by red X's) results in a deficiency in T cell–dependent antibody responses and heavy-chain class switching, leading to elevated levels of IgM and the absence of other isotypes.

B-Cell Immunodeficiencies Exhibit Depressed Production of One or More Antibody Isotypes

Immunodeficiency disorders caused by B-cell defects make up a diverse spectrum of diseases ranging from the complete absence of mature recirculating B cells, plasma cells, and immunoglobulin, to the selective absence of only certain classes of immunoglobulins. Patients with inherited B-cell defects are usually subject to recurrent bacterial infections but display normal immunity to most viral and fungal infections because the T-cell branch of the immune system is largely unaffected. In patients with these types of immunodeficiencies, the most common infections are caused by encapsulated bacteria, such as staphylococci, streptococci, and pneumococci, because antibody is critical for the opsonization and clearance of these organisms. Although the underlying defects have been identified for some of these conditions, several of the more common deficiencies, such as common variable immunodeficiency and selective IgA deficiency, appear to involve multiple genes and exhibit a continuum of phenotypes.

X-Linked Agammaglobulinemia

X-linked agammaglobulinemia (X-LA), or Bruton's hypogammaglobulinemia, is characterized by extremely low immunoglobulin levels in most patients. Babies born with this particular form of **agammaglobulinemia** (absence of immunoglobulins in the blood) have virtually no peripheral B cells (<1% of normal) and suffer from recurrent bacterial infections. X-LA is caused by a defect in Bruton's tyrosine kinase (Btk), which is required for signal transduction through the pre-BCR and BCR (see Figure 11-9 and Clinical Focus Box 3-4). Without functional Btk, B-cell development in the bone marrow is arrested when they express the pre-BCR, and the B cells in these patients remain in the pre-B stage, with heavy chains rearranged but light chains in their germ-line configuration. Present-day use of antibiotics and replacement therapy in the form of passively administered antibodies can make this disease quite manageable.

Other Disorders with Profoundly Reduced B Cells and Immunoglobulins

Defects in pre-BCR and BCR polypeptide chains also block the formation of mature B cells, preventing antibody responses. Mutations with these profound consequences have been detected in the genes for the μ heavy chain, the $\lambda 5$ surrogate light-chain, and the Igα and Igβ signaling chains of pre-BCR and BCR. As with X-LA, individuals with these mutations are plagued by severe bacterial infections.

Common Variable Immunodeficiency Disorders

The defects underlying the complex group of diseases belonging to this category are more different than they are similar. However, sufferers of *common variable immunodeficiency disorders* (CVIDs) do share recurrent infection resulting from immunodeficiency, marked by reduction in the levels of one or more antibody isotypes and impaired B-cell responses to antigen, all with no other known cause. This condition can manifest in childhood or later in life, when it is sometimes called late-onset hypogammaglobulinemia. Respiratory tract infection with common bacterial strains is the most common symptom; it can be controlled by administration of immunoglobulin. Most cases of CVID have undefined genetic causes, and most patients have normal numbers of B cells, suggesting that B-cell development usually is not the underlying defect. Reflecting the diversity of this set of diseases, inheritance can follow autosomal recessive or autosomal dominant patterns. Several different proteins involving various steps of the B-cell activation cascade have been implicated in recent years.

Selective IgA Deficiency

A number of immunodeficiency states are characterized by significantly lowered amounts of specific immunoglobulin isotypes. Of these, IgA deficiency is by far the most common, affecting approximately 1 in 700 individuals. Individuals with selective IgA deficiency typically exhibit normal levels of other antibody isotypes and may enjoy a full life span, troubled only by a greater-than-normal susceptibility to infections of the respiratory, intestinal, and genitourinary tracts, the primary sites of IgA secretion. Family association studies have shown that IgA deficiency sometimes occurs in the same families as CVID, suggesting some overlap in causation. The spectrum of clinical symptoms of IgA deficiency is broad; most of those affected are asymptomatic (up to 70%), whereas others may suffer from an assortment of serious complications. Problems such as intestinal malabsorption, allergic disease, and autoimmune disorders can be associated with low IgA levels. The reasons for this variability in the clinical profile are not clear but may be related to the ability of some, but not all, patients to substitute IgM for IgA as a mucosal antibody. The defect in IgA deficiency reflects the inability of IgA-expressing B cells to undergo normal differentiation to the plasma-cell stage. A gene or genes outside of the immunoglobulin gene complex is suspected of being responsible for this fairly common syndrome.

Disorders Involving Severe Reductions in at Least Two Immunoglobulin Isotypes

Defects in genes coding for enzymes required for heavy-chain class switch recombination, such as activation-induced cytidine deaminase (AID; see Chapter 11) can block the production of IgG, IgA, and IgE, accompanied by normal or high levels of IgM. While this resembles hyper-IgM syndrome (discussed earlier), there is no effect on T-cell responses, so this condition is not included with the combined immunodeficiencies. Mutations in several B-cell surface proteins prevent normal B-cell responses and can lead to low levels of IgG, IgA, and occasionally IgM as well. Defects in proteins that provide costimulatory signals during B-cell activation (CD19, CD20, CD21, and CD81) can lead to these consequences, as can defects in the receptors for B-cell survival factors BAFF and APRIL.

> **Key Concepts:**
> - B-cell immunodeficiencies, which are associated with susceptibility to bacterial infection, can range from total disruptions of immunoglobulin production to defects in the production of individual isotypes.
> - Selective immunoglobulin isotype deficiencies are milder forms of B-cell deficiency.

Disruptions to Innate Immune Components May Also Impact Adaptive Responses

Some innate immune defects are caused by problems in myeloid cells or in components of the complement activation pathways (see Figure 18-1). Many of these defects result in depressed numbers of phagocytic cells or defects in the phagocytic process that are manifested by recurrent microbial infection of varying severity. The phagocytic processes may be faulty at several stages, including cell motility, adherence to and phagocytosis of organisms, and intracellular killing by macrophages. Examples of other mutations interfering with innate responses, described in Chapter 4, include those in signaling components activated by pattern recognition receptors such as TLRs or by the receptors for type I interferons (see Clinical Focus Box 4-3).

Leukocyte Adhesion Deficiency

As described in Chapter 14, cell-surface molecules belonging to the integrin family of proteins function as adhesion

molecules and are required to facilitate cellular interaction. Three of these, LFA-1, Mac-1, and gp150/95 (CD11a, b, and c, respectively), have a common β chain (CD18) and are variably present on different myeloid cells; CD11a is also expressed on B cells. An immunodeficiency related to dysfunction of these adhesion molecules is rooted in a defect localized to the common β chain and affects expression of all three of the molecules that use this chain. This defect, called *leukocyte adhesion deficiency* (LAD), causes susceptibility to infection with both gram-positive and gram-negative bacteria as well as various fungi. Impairment of adhesion of leukocytes to vascular endothelium limits recruitment of cells to sites of inflammation. Viral immunity is somewhat impaired, as would be predicted from the defective T–B-cell cooperation arising from the adhesion defect and the important role of LFA-1 in the binding of cytotoxic T and NK cells to their infected target cells (see Chapter 12). LAD varies in its severity; some affected individuals die within a few years, whereas others survive into their forties. The reasons for the variable disease phenotypes in this disorder are not known.

Chronic Granulomatous Disease

Chronic granulomatous disease (CGD), the prototype of immunodeficiency that impacts phagocytic function, occurs in at least two distinct forms: an X-linked form in about 70% of patients and an autosomal recessive form found in the rest. This group of disorders is rooted in a defect in the *nicotinamide adenine dinucleotide phosphate* (NADPH) oxidase enzyme by which phagocytes generate superoxide radicals and the reactive oxygen and nitrogen species (ROS and RNS) that kill phagocytosed pathogens (see Figure 4-20). For this reason, patients with CGD suffer from infections with bacterial and fungal pathogens, as well as excessive inflammatory responses that lead to the formation of granulomas (small masses of inflamed tissue). Genetic causes have been mapped to several missing or defective NADPH oxidase subunits that participate in this pathway. Standard treatment includes the use of antibiotics and antifungal compounds to control infections. In recent years the addition of IFN-γ to this regimen has been shown to improve CGD symptoms in both humans and animal models. In vitro studies have shown that treatment with IFN-γ, a macrophage activator, induces TNF-α and the production of nitric oxide, and enhances the phagocytosis of inflammation-inducing dead cells, which could play a role in inhibiting the formation of granulomas during inflammation in these patients.

Chédiak-Higashi Syndrome

This rare autosomal recessive disease is an example of a lysosomal storage and transport disorder. *Chédiak-Higashi syndrome* (CHS) is characterized by recurrent bacterial infections as well as defects in blood clotting, pigmentation, and neurologic function. Immunodeficiency hallmarks include neutropenia (depressed numbers of neutrophils) as well as impairments in T cells, NK cells, and granulocytes.

CHS is associated with oculocutaneous albinism, or light-colored skin, hair, and eyes, accompanied by photosensitivity. The underlying cause has been mapped to mutations in the *lysosomal trafficking regulator* (*LYST*) gene that cause defects in the LYST protein, which is important for transport of proteins into lysosomes as well as for controlling lysosome size, movement, and function. Disruption of this and related organelles, such as the melanosomes of skin cells (melanocytes), results in enlarged organelles and defective transport functions. Affected phagocytes produce giant granules, a diagnostic hallmark, but are unable to kill engulfed pathogens, and melanocytes fail to transport melanin (responsible for pigmentation). Similar enlarged lysosome-like structures in platelets and nerve cells are also thought to interfere with blood clotting and neurologic function, respectively. Defects are also seen in CTL and NK cell–mediated cytotoxicity; their cytotoxic granules share some properties with lysosomes, including the LYST protein's role in delivery of granule proteins such as perforin and granzymes. Without early antimicrobial therapy followed by bone marrow transplantation, patients often die of opportunistic infections before reaching 10 years of age. However, no therapies are currently available to treat the defects in non–bone marrow–derived cells, so even when immune function is restored, neurologic and other complications continue to progress.

Mendelian Susceptibility to Mycobacterial Diseases

Recently, a set of immunodeficiency disorders has been grouped into a mixed-cell category based on the shared characteristic of single-gene (*Mendelian*) inheritance of susceptibility to *mycobacterial diseases* (MSMDs). Discovery of the underlying defects in MSMDs highlights the connections between innate and adaptive immunity, as well as the key role played by IFN-γ in fighting infection by mycobacteria, intracellular organisms that can cause tuberculosis and leprosy. During natural mycobacterial infection, macrophages in the lungs or DCs in the draining lymph nodes recognize these bacteria through pattern recognition receptors (PRRs), such as TLR2 and TLR4, triggering cell migration to lymph nodes followed by APC activation and differentiation. In the presence of strong costimulation, such as engagement of CD40 on an APC with CD40L on a T cell, these activated APCs produce significant amounts of IL-12 and IL-23, which can bind to their receptors on T$_H$ cells and NK cells, respectively. This leads to the production of cytokines such as IFN-γ, IL-17, and TNF-α. In a positive feedback loop, T$_H$ cells in this environment differentiate into T$_H$1-type cells, which produce additional IFN-γ. On binding to the IFN-γ receptor on APCs, this cytokine induces a signaling cascade, involving Janus kinases and STAT1, which results in enhanced phagocytosis and optimal phagosome-lysosome fusion, effectively killing engulfed bacteria. This T$_H$1 cell–mediated delayed-type hypersensitivity response is essential for protection against *Mycobacterium tuberculosis* and *M. leprae* (see Figure 15-15).

In the context of this story of mycobacterial infection, the defective genes or proteins implicated in MSMDs should come as no surprise (**Figure 18-5**). To date, six genes within

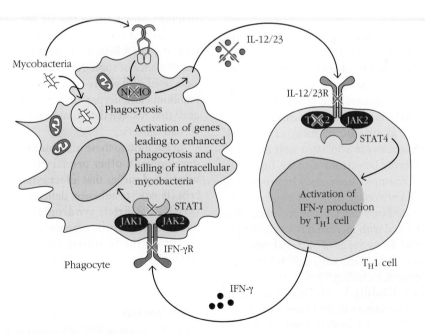

FIGURE 18-5 Genetic defects resulting in Mendelian susceptibility to mycobacterial diseases (MSMDs). Many primary immunodeficiencies associated with increased susceptibility to mycobacterial infection are caused by defects in genes encoding proteins (marked by a red X) in either the IL-12 activation and signaling pathway (NEMO [downstream of TLR], IL-12, IL-12Rβ chain, and the associated TYK2 signaling molecule) or the IFN-γ pathway (e.g., IFN-γR or the related STAT1 signaling molecule). The production of IFN-γ by T$_H$1 cells, important for clearing intravesicular infections, is activated by APC production of IL-12.

the IFN-γ/IL-12/IL-23 pathways have been linked to MSMDs, including those encoding IL-12, IL-12R, TYK2 (the JAK kinase activated by IL-12 binding to its receptor; see Appendix II, Table 3), IFN-γR (both chains), and STAT1. Another gene linked to MSMDs encodes the NEMO protein, which is part of the IKK complex in the NF-κB activation pathway (see Chapter 4); activation of NF-κB is involved in TLR- and CD40-activated production of IL-12. However, most mutations in the *NEMO* gene lead to more widespread immune defects and susceptibility patterns than are seen in typical patients with MSMD, as NF-κB is an essential transcription factor in activating innate immune responses (see Chapter 4) as well as in T- and B-cell activation. The specific gene that is defective and the type of mutation determine which other pathogens, if any, also pose a risk to patients with MSMD and influence the prognosis as well as treatment options.

Key Concept:

- Many immunodeficiencies affecting innate immunity are due to defects affecting myeloid cells, especially impaired functions of phagocytes. These include impaired phagocyte migration, ROS and RNS production, lysosomal function, and macrophage activation by microbes. Affected individuals suffer from increased susceptibility to infections, especially from bacteria.

Complement Deficiencies Are Relatively Common

Immunodeficiency diseases resulting from defects in the complement system, which is activated by innate as well as adaptive triggers, are described in Chapter 5. Depending on the specific component that is defective, these immunodeficiencies can manifest as a generalized failure to activate complement (e.g., C4 defects) or failures of discrete pathways or functions (e.g., alternative pathway activation). Most complement deficiencies are associated with increased susceptibility to bacterial infections and/or immune-complex diseases. For example, deficiency in properdin, which stabilizes the C3 convertase in the alternative complement pathway, is caused by a defect in an X-linked gene and is specifically associated with increased risk of infection with species of *Neisseria*. These types of bacterial infection are also more common in individuals with defects in the late components of complement, including C5–C9.

Defects in *mannose-binding lectin* (MBL) result in increased susceptibility to a variety of infections with bacterial or fungal agents. Recall from Chapters 4 and 5 that MBL is an important component of the innate immune response to many organisms, as it functions as an opsonin and also as a key initiator of the lectin pathway of complement activation, leading to complement attack on many pathogens.

Key Concept:

- Complement deficiencies are relatively common and vary in their clinical impact, but are generally associated with increased susceptibility to bacterial infections.

NK-Cell Deficiencies Increase Susceptibility to Viral Infections and Cancer

Reflecting the important roles of NK cells in innate responses, including production of type I interferons and cytolytic activity against virus-infected and tumor cells, mutations that interfere with NK-cell development and function can lead to immunodeficiency and susceptibility to viral infections and some cancers. Several of the mutations causing devastating SCID, including reticular dysgenesis, purine metabolism defects, common cytokine γ-chain deficiency, and JAK3 deficiency, lead to blocks in the development of NK as well as T and B cells. Deficiencies in the $γ_c$ chain, JAK3, and STAT5b proteins prevent signaling downstream of the IL-15R, essential for NK-cell development from early lymphocyte progenitors. Mutations selectively affecting NK cells are rare. Defects in the GATA2 transcription factor, necessary for the final differentiation step leading to mature NK cells, cause immunodeficiency. This defect was first identified in a teenage girl with recurring infections, including varicella (chicken pox) and cytomegalovirus pneumonia, associated with very low NK-cell numbers. Individuals with some *GATA2* mutations may have detectable NK cells, though with reduced cytotoxic function. Mutations in the *MCM4* gene also uniquely affect NK cells and not T and B cells. This gene encodes a subunit of a DNA helicase important for the proliferation and maintenance of some NK subsets. These NK deficiencies are associated with severe viral infections, especially with human papilloma and herpes viruses, and increasing evidence suggests that NK-cell defects are associated with the incidence of certain cancers.

In addition to these immunodeficiencies affecting NK-cell development, other primary immunodeficiency diseases mentioned earlier that affect synapse formation with target cells (leukocyte adhesion deficiency), lymphocyte activation (Wiskott-Aldrich syndrome), and the formation and function of cytotoxic granules (Chédiak-Higashi syndrome) also affect the activation and functions of both NK cells and CTLs.

Key Concepts:

- Deficiencies in NK cells occur in forms of SCID interfering with the development of lymphoid lineages or may affect NK-cell development more specifically.

- Cell-mediated cytotoxic functions of NK cells as well as CTLs are blocked by genetic defects affecting synapse formation with target cells, lymphocyte activation, and the formation and function of cytotoxic granules.

- NK-cell deficiencies lead to increased susceptibility to viral infections and some cancers.

Immunodeficiency Disorders That Disrupt Immune Regulation Can Manifest as Autoimmunity

In addition to recognizing and eliminating foreign antigens, the adaptive immune system must learn to recognize self MHC proteins and to be proactive in suppressing reactions to self antigens in the host. These key processes are carried out by the induction of tolerance in the thymus and by the surveillance activities of regulatory T cells (T$_{REG}$ cells; see Chapters 8 and 16). While two defects produce clear-cut deficiencies in T-cell tolerance mechanisms, in many other immunodeficiency disorders the defects lead to immune dysregulation that results in some autoimmunity.

Autoimmune Polyendocrinopathy and Ectodermal Dystrophy

Individuals with a defect in the autoimmune regulatory gene *AIRE*, discussed in detail in Chapters 8 and 16, suffer from a disease called *autoimmune polyendocrinopathy with candidiasis and ectodermal dystrophy* (APECED), also called *autoimmune polyendocrine syndrome-1* (APS-1). The AIRE protein is expressed in medullary epithelial cells of the thymus, where it acts as a transcriptional regulator to control expression of a wide array of tissue-restricted antigens. Proper expression of these peripheral tissue proteins in the thymus facilitates the negative selection of autoreactive T cells before they can exit into the circulation and the generation of thymic T$_{REG}$ cells. It appears that depressed expression of *AIRE* in these individuals results in reduced levels of tissue-specific antigens in medullary thymic epithelial cells, allowing the escape of autoreactive T cells into the periphery, where they precipitate organ-specific autoimmunity. Patients with APECED experience disruption of endocrine function, including hypoadrenalism, hypoparathyroidism, and hypothyroidism, along with chronic candidiasis. Autoimmune responses against antigens present in these endocrine organs, as well as the adrenal cortex, gonads, and pancreatic beta cells, are observed in these individuals. Although autoantibodies to these tissues are also observed, these may result from the tissue destruction mediated by pathogenic T cells.

Immune Dysregulation, Polyendocrinopathy, Enteropathy, X-linked Syndrome

Although many T cells with the ability to recognize self antigens are destroyed in the thymus during negative selection,

one class of self-reactive CD4$^+$ T cells with regulatory capabilities survives and actively inhibits reactions to these self antigens in the periphery. The development and function of these T$_{REG}$ cells are controlled by a master regulator and transcription factor, called FoxP3 (see Chapters 8 and 16). FoxP3 also is central to the differentiation of naïve CD4$^+$ T cells into T$_{REG}$ cells in the periphery. Patients with *immune dysregulation, polyendocrinopathy, enteropathy, X-linked (IPEX) syndrome* have inherited a mutated *FoxP3* gene and lack expression of this protein, leading to a near absence of T$_{REG}$ cells. Without these regulatory cells in the periphery, autoreactive T cells that have escaped central tolerance in the thymus go unchecked, leading to systemic autoimmune disease. Affected infants exhibit immune destruction of the bowel, pancreas, thyroid, and skin, and they often die in the first 2 years of life, due to sepsis and failure to thrive.

Similar symptoms occur in patients with autosomal recessive mutations in CD25, the α chain of the IL-2 receptor, which is required for high-affinity IL-2 binding. IL-2 enhances the functions of T$_{REG}$ cells and absence of the high-affinity IL-2R interferes with peripheral self tolerance. Mutations in the coinhibitory protein CTLA4, which contributes to T$_{REG}$ inhibitory functions, also may be associated with autoimmunity.

Autoimmune Consequences of Other Immunodeficiency Disorders

Autoimmunity accompanies many other immunodeficiency disorders, reflecting a range of defects in immune responses and their regulation. Mutations that result in low numbers of lymphocytes (*lymphopenia*) often lead to autoimmunity as well as to immunodeficiency. Examples are hypomorphic (reduced function) mutations in RAG and Artemis proteins. As mentioned earlier, these mutant proteins generate some immunoglobulin and TCR gene rearrangements, allowing the formation of reduced numbers and diversity of B and T cells. The low numbers of B cells stimulate up-regulation of the B-cell survival factor BAFF, which allows survival of autoreactive immature and peripheral B cells. In the thymus the disruption in T-cell development caused by limited TCR rearrangements results in reduced AIRE expression, leading to the survival of autoreactive cells, which enter the circulation. Survival of autoreactive T cells also occurs in some individuals with partial DiGeorge syndrome, in which enough thymus structure is present to generate low numbers of T cells.

Immunodeficiencies associated with reduced clearance of immune complexes and apoptotic cells may also lead to autoimmunity. Complete deficiencies in C1q, C1r/s, C2, or C4, which are quite rare, show an associated risk for systemic lupus erythematosus (SLE) of 90%, 50%, 30%, and 70%, respectively. These early complement components are strong opsonins for immune complexes and apoptotic cells, as well as for certain bacteria (see Chapter 4). The reduced clearance of immune complexes and apoptotic cell components leads to an increased chance that nuclear self antigens induce IFN-α production and break self tolerance of autoreactive B and T cells (see Chapter 16).

Key Concepts:

- Immunodeficiencies that disrupt immune regulation can lead to overactive immune responses that manifest as autoimmune syndromes. This is especially true of defects in self tolerance in the thymus or in the generation of regulatory T cells.

- Other immunodeficiencies that lead to low numbers of lymphocytes or to reduced clearance of immune complexes can result in autoimmunity.

Immunodeficiency Disorders Are Treated by Replacement Therapy

Although there are no surefire cures for immunodeficiency disorders, there are several treatment possibilities. In addition to the use of antimicrobial agents and the drastic option of total isolation from exposure to any opportunistic pathogens, immunodeficiencies can be treated by replacement therapy targeting missing proteins, cells, or genes.

Protein Replacement Therapy

For disorders that impair antibody production, the classic course of treatment is administration of the missing immunoglobulins. The injection of pooled human gammaglobulin, known as *intravenous immunoglobulin (IVIG)*, protects against recurrent infection in many types of immunodeficiency. Maintenance of reasonably high levels of serum immunoglobulin (5 mg/ml serum) will prevent most common infections in the agammaglobulinemic patient. Advances in the preparation of human monoclonal antibodies and in the ability to genetically engineer chimeric antibodies with mouse V regions and human-derived C regions make it possible to prepare injectable antibodies specific for important pathogens (see Chapters 12 and 20).

To generate large amounts of purified proteins for reconstituting other protein deficiencies, such as enzymes, cytokines, and complement components, the genes can be expressed in vitro, using bacterial or eukaryotic expression systems. The availability of such proteins allows new modes of therapy in which immunologically important proteins may be replaced or their concentrations increased in the patient. For example, recombinant adenosine deaminase has been successfully administered to patients with ADA-deficient SCID, and recombinant IFN-γ has proven effective for patients with chronic granulomatous disease.

Cell Replacement Therapy

Cell replacement as therapy for some immunodeficiencies has been made possible by progress in bone marrow

or hematopoietic stem cell (HSC) transplantation (see Chapter 16). This currently is the primary potential long-term cure for patients with SCID. Transfer of cell populations containing HSCs from an immunocompetent donor allows development of a functional immune system (see Clinical Focus Box 2-2). Success rates are highest when transplantations are done in the first 3.5 months of life and before infections occur. The degree of HLA matching is also critical; survival rates of greater than 90% have been reported for those who are fortunate enough to have an HLA-identical donor, usually a sibling. These procedures can also be relatively successful in infants with SCID when haploidentical (complete match of one HLA haplotype) donor bone marrow is used. In this case, T cells are depleted to avoid graft-versus-host reactions, where donor-derived T cells attack the recipient; this is the major adverse complication of HSC transplantation. Another strategy is to enrich for CD34$^+$ stem cells from the donor bone marrow prior to transfer. A variation of bone marrow transplantation is the injection of parental CD34$^+$ cells in utero when the birth of an infant with SCID is expected.

HSC transplantation is still evolving as a general approach for treating immunodeficiencies, given the complexity of hematopoiesis, the many and varied consequences of immunodeficiencies, the importance of reconstituting all of the blood cell lineages, and graft-versus-host disease. Another issue that affects the ability of the donor HSCs to repopulate the recipient is whether there is room in the bone marrow and thymus for the donor-derived cells to expand and differentiate. Prior to the transfer, many recipients receive *myeloablative conditioning* treatments that deplete HSCs and bone marrow cells to make room for the normal donor-derived HSCs and progenitor and precursor cells. The drugs that are used have side effects, since they are, by their very nature, toxic to cells. And not all cells are reconstituted with equal efficiency after transplantation. For example, haploidentical HSC transplantation of infants with X-linked SCID achieves more than 70% long-term survival. But in two-thirds of these patients, although T cells engraft and are functional after several months, B cells or NK cells rarely reconstitute, so these patients must continue on IVIG.

Gene Replacement Therapy

A recent and growing alternative for treating some immunodeficiency diseases is gene therapy. If a single-gene defect has been identified, as in adenosine deaminase (ADA) or common γ-chain (SCIDX1) defects, replacement of the defective gene may be a treatment option. During the last several decades, clinical tests of gene therapy for these two types of SCID have been undertaken, with mixed results. In these trials, CD34$^+$ HSCs were first isolated from the bone marrow or umbilical cord blood of HLA-identical or haploidentical donors, or from the immunodeficient patient him/herself (an *autologous* transplant). These cells are transduced with the corrected gene, using a retroviral vector, which leads to the insertion of a copy of the gene into the cells' chromosomes, and then are introduced into the patient, in some cases after myeloablative conditioning.

We are now more than 25 years out from these initial trials. In general, gene therapy for the ADA defect has been very successful, with most of the more than two dozen patients with ADA-SCID receiving this treatment attaining significant long-lasting reconstitution of their T cells and B cells and a dramatic improvement in their immune responses. Importantly, no episodes of serious adverse reactions have occurred. For those with SCIDX1, of the first 20 babies with SCIDX1 around the world who were treated by gene therapy, 18 are alive, and in 17 of these 18 children, gene therapy alone was sufficient to restore the development of T lymphocytes and immune function, and no other treatment was needed. Unfortunately, while the SCID was cured, five of these patients developed leukemia because the transferred gene inserted in the vicinity of oncogenes. In four of the children the leukemia was cured, but one child died. Better results have recently been obtained with a vector from a lentivirus, a different type of retrovirus. Five patients with SCIDX1 who received this vector have had effective reconstitution of B, T, and NK cells with no evidence of toxicity, and so this approach seems very promising. Gene therapy trials for correction of the defects in chronic granulomatous disease and Wiskott-Aldrich syndrome are underway.

A new and exciting approach to gene therapy is the correction of the defective gene in the patient's own HSCs either in vitro or even in vivo in the patient. This approach takes advantage of the recently developed CRISPR/Cas9 technique, which can replace a mutant DNA sequence with the normal sequence (see Chapter 20). The CRISPR/Cas9 approach has been used recently to correct the defective gene in HSCs from a patient with one form of chronic granulomatous disease. The HSCs were then transferred into immunodeficient mice and shown to generate normally functioning myeloid and lymphoid cells. The next important step will be to see whether corrected HSCs can repopulate immunodeficient individuals and correct their defects.

Key Concepts:

- Immunodeficiency disorders can be treated by replacement of defective or missing proteins (by injection), cells (through bone marrow or HSC transplantation), or genes (through gene therapy).

- Administration of human immunoglobulin is a common treatment, especially for those disorders that disrupt antibody responses.

Animal Models of Immunodeficiency Have Been Used to Study Basic Immune Function

There is a good reason PIDs are sometimes called nature's experiments. Many of the molecular details of how the immune system works have come from discovering and studying the effects of mutations in humans and animals. Experimental animals with spontaneous or engineered primary immunodeficiencies have provided fertile ground for manipulating and studying basic immune processes. By comparing the phenotypes of animals with and without these defects in certain components of the immune system, scientists have been able to tease out many details of normal immune processes. Two commonly used animal models of primary immunodeficiency are the athymic, or nude, mouse and the SCID mouse. However, the development of other genetically altered animals in which a single target immune gene is knocked out or mutated has also yielded valuable information about the roles of these genes in the development of immune system cells and in immune responses, and has highlighted some unexpected connections between the immune system and other systems in the body.

Nude (Athymic) Mice

A genetic mutation, designated *nu* (now called *Foxn1^{nu}*), in a recessive gene on chromosome 11, was discovered in 1962 by Norman Roy Grist. Mice homozygous for this trait (*nu/nu*, or nude mice) are hairless and have a vestigial thymus (**Figure 18-6**). Heterozygotic *nu/wt* littermates have hair

FIGURE 18-6 A nude mouse (*Foxn1^{nu}*/*Foxn1^{nu}*). This defect, which interferes with the development of thymus and skin epithelia, leads to the absence of a thymus or a vestigial thymus, preventing T-cell development, and to deficiency in cell-mediated immunity. [*Courtesy of the Jackson Laboratory, Bar Harbor, Maine.*]

TABLE 18-3   Immunodeficient mutant mouse strains

Mouse name	Mutant gene(s)	Affected protein(s)	Immunodeficiency
Nude	*Foxn1*	Foxn1 transcription factor expressed in skin and thymic epithelial cells	No thymus; no T cells
SCID	*PRKDC*	Catalytic subunit, DNA-dependent protein kinase	Defective Ig and TCR gene rearrangement; no/few B and T cells (leaky)
RAG knockout	*RAG1* and/or *RAG2*	RAG1 and/or RAG2	No B or T cells
RAG/γ_c chain knockout	*RAG1* and/or *RAG2* and *IL2RG*	RAG1 and/or RAG2 and the cytokine receptor common γ chain	No B, T, or NK cells

and a normal thymus. We now know that the gene defective in these mice, *FOXN1*, encodes a transcription factor, expressed mainly in the thymic epithelium and skin epithelial cells, that plays a role in cell differentiation and survival (Table 18-3). Thus the hair loss and immunodeficiency are caused by the same defect. Like humans born with severe immunodeficiency, these mice do not survive for long without intervention, and 50% or more die within the first 2 weeks of birth from opportunistic infection if housed under standard conditions. When these animals are to be used for experimental purposes, precautions include the use of sterilized food, water, cages, and bedding. The cages are protected from dust by placing them in a laminar flow rack or by using cage-fitted air filters.

Nude mice have now been studied for many years and have been developed into a tool for biomedical research. For example, because these mice can permanently tolerate both allografts and xenografts (tissues from another species), they have a number of practical experimental uses in the study of transplantation and cancer. Hybridomas (immortalized B or T cells) or solid tumors of any origin can be grown in nude mice, allowing their propagation and the evaluation of new tumor imaging techniques or pharmacological treatments for cancer in these animals.

SCID Mouse

In 1983, Melvin and Gayle Bosma and their colleagues described an autosomal recessive mutation in mice from their animal colony that caused severe deficiency in mature lymphocytes. They designated the trait SCID because of its similarity to human severe combined immunodeficiency. The SCID mouse was shown to have early B- and T-lineage cells but a virtual absence of lymphoid cells in the spleen, lymph nodes, and gut tissue, the usual locations of functional T and B cells. Precursor cells in the SCID mouse appeared to be unable to differentiate into mature functional B and T lymphocytes. Inbred mouse lines carrying this defect, which have now been propagated and studied in great detail, do not produce antibodies or carry out T cell–mediated delayed-type hypersensitivity or graft rejection reactions. Lacking much of their adaptive response, the mice succumb to infection early in life if not kept in extremely pathogen-free

environments. Hematopoietic cells other than lymphocytes develop normally in SCID mice; red blood cells, monocytes, and granulocytes are present and functional. Like humans, SCID mice may be rendered immunologically competent by transplantation of stem cells from a matched donor.

The mouse SCID mutation was found to be in a gene called *protein kinase, DNA activated, catalytic polypeptide (PRKDC)*, which was later shown to participate in the double-stranded DNA break–repair pathway important for immunoglobulin and TCR gene rearrangements in developing B and T cells (see Table 18-3). This defect is a leaky mutation: a certain number of SCID mice do produce immunoglobulin, and about half of these mice can also reject skin allografts, suggesting that some components of both humoral and cell-mediated immunity are present. Mutations in this same gene cause SCID in humans (see Table 18-1).

RAG Knockout Mice

The potential utility of a mouse model that lacks adaptive immunity, or certain components of adaptive responses, led to the engineering of mice with specific targeted mutations. Arguably, the most widely used have been mice with deletions of one of the recombination-activating enzymes, RAG1 and RAG2, responsible for the rearrangement of immunoglobulin and T-cell receptor genes. Unlike nude or SCID mice, RAG knockout mice exhibit "tight" defects in both B-cell and T-cell compartments; precursor cells cannot rearrange the genes for antigen-specific receptors or proceed along a normal developmental path, and thus both B and T cells are absent (see Table 18-3). With a SCID phenotype, RAG knockout mice can be used as an alternative to nude or conventional SCID mice. Their applications include experimental cancer and infectious disease research, as well as more targeted investigations of immune gene function. RAG knockout mice can be the background strain for the production of transgenic mice carrying specific rearranged T-cell or B-cell receptor genes of desired antigen specificity. For example, since these transgenes have already rearranged, T cells expressing them will not require the RAG enzymes, and can develop "normally" in the thymus, allowing immunologists to study the events that occur during positive and negative selection while observing the behavior of a million or more T cells with the same TCR. Although the degree to which this represents truly typical in vivo development of a T cell is questionable, this model has been widely used to ask and to answer many important questions related to T-cell selection and tolerance.

Recently, mice have been bred that have mutant *RAG* genes along with a defective *IL2RG* gene that produces no cytokine receptor common γ chain (see Table 18-3). These doubly mutant mice are totally deficient in B, T, and NK cells and are the choice for many studies, including the development of chimeric mice reconstituted with a humanized immune system (called SCID-hu). Human hematopoietic stem cells can differentiate in a normal fashion and, as a result, SCID-hu mice contain B, T, and NK cells and immunoglobulins of human origin. In one important application, these mice can be infected with HIV, a pathogen that does not infect mouse cells. This provides an animal model in which to test therapeutic or prophylactic strategies against HIV infection.

Key Concepts:

- Animal models for immunodeficiency include nude and SCID mice possessing one or several defects and that have varying levels of immunodeficiency.

- Certain SCID mouse strains have been generated with multiple mutations so that they are deficient in B, T, and NK cells. These are usually the mice of choice for generating mice with humanized immune systems following transfer of human hematopoietic stem cells.

Secondary Immunodeficiencies

As described above, a variety of defects in the immune system can give rise to immunodeficiency. In addition to the inherited primary immunodeficiencies, there are also acquired (secondary) immunodeficiencies. Although AIDS resulting from HIV infection is the best known of these, other factors, such as immunosuppressive drug treatment, metabolic disease, or malnutrition, can also impact immune function and lead to secondary deficiencies. As in primary immunodeficiencies, symptoms include heightened susceptibility to common infectious agents, opportunistic infections, and certain cancers, conditions that don't occur in people with healthy immune systems. The effect depends on the degree of immune suppression and inherent host susceptibility factors, but can range from no clinical symptoms to almost complete collapse of the immune system, as in HIV-induced AIDS. In most cases, withdrawal of the external condition causing the deficiency can result in restoration of immune function (not yet possible with HIV/AIDS).

The first part of this section will cover secondary immunodeficiency due to some non-HIV causes, and the remainder will deal with HIV/AIDS.

Secondary Immunodeficiencies May Be Caused by a Variety of Factors

One secondary immunodeficiency that has been recognized for some time but has an unknown cause is *acquired hypogammaglobulinemia*. This condition is sometimes confused with CVID, a condition that shows genetic predisposition (see above). Symptoms include recurrent infections, and the condition typically manifests in young adults who have very low but detectable levels of total immunoglobulin with normal T-cell numbers and function. However, some cases do involve T-cell defects, which may grow more severe as the disease progresses. The disease is generally treated by immunoglobulin therapy, allowing patients to live a relatively normal life. Unlike the similar primary deficiencies described above, there is no evidence for genetic transmission of this disease; mothers with acquired hypogammaglobulinemia deliver normal infants. However, at birth these infants will be deficient in circulating immunoglobulin due to the lack of IgG in the maternal circulation that can be passively transferred to the infant.

Another class of secondary immunodeficiency, called *agent-induced immunodeficiency*, results from exposure to any of a number of environmental agents that induce an immunosuppressed state. These include immunosuppressive drugs, including corticosteroids, used following organ transplantation to prevent rejection or to combat autoimmune diseases such as rheumatoid arthritis. The mechanisms of action of these immunosuppressive agents vary, as do the resulting defects in immune function, although T cells or B cells are common targets. As described in Chapter 16, recent efforts have been made to use more specific means of inducing tolerance to allogeneic tissue transplants to circumvent the unwanted side effects of general immunosuppression. Other agents of acquired immunodeficiency include cytotoxic drugs or radiation treatments given to treat various forms of cancer. These cancer treatments frequently damage rapidly dividing cells in the body, including those of the immune system, inducing a state of temporary immunodeficiency as an unwanted consequence. Patients undergoing such therapy must be monitored closely and treated with antibiotics and/or immunoglobulin if infection appears.

Extremes of age are also natural factors affecting immune function. The very young and elderly suffer from impairments to immune function not typically seen during the remainder of the life span. Neonates, and especially premature babies, can be very susceptible to infection, with the degree of prematurity linked to the extent of immune dysfunction. Although all the basic immune components are in place in full-term, healthy newborns, the complete range of innate and adaptive immune functions take some time to mature. Along with the presence of passive maternal antibody for about the first 6 months of life, this is part of the reason for a gradual vaccination schedule against the common childhood infectious diseases, with most immunizations recommended for ages 2 to 15 months (see Chapter 17). In later life, individuals again experience an increasing risk of infection, especially with bacteria and viruses, as well as more malignancies. Cell-mediated immunity is generally depressed, with reduced diversity of T cells, and although there are increases in memory B cells and circulating IgG, the diversity of the B-cell repertoire is diminished.

The single most common cause of acquired immunodeficiency, even dwarfing the number of individuals worldwide affected by AIDS, is severe malnutrition, which affects both innate and adaptive immunity. Sustained periods with very low protein-calorie diets (hypoproteinemia) are associated with reduced T-cell numbers and functions, while deleterious effects on B cells may take longer to appear. The reason for this is unclear, although some evidence suggests a bias toward anti-inflammatory immune pathways (e.g., involving IL-10 and T_{REG} cells) when protein is scarce. In addition to protein, insufficient micronutrients, such as zinc and ascorbic acid (vitamin C), likely contribute to the general immunodeficiency and increased susceptibility to opportunistic infection that occurs with malnutrition. This can be further complicated by stress and infection, both of which may contribute to diarrhea, further reducing nutrient absorption in the gut. Deficiency in vitamin D, required for calcium uptake and bone health, has also been linked to an inhibition in the ability of macrophages to act against intracellular pathogens, such as *M. tuberculosis*, endemic in many regions of the world where people are at greatest risk of malnutrition. Severe malnourishment thus ranks as one of the most preventable causes of poor immune function in people who otherwise should be healthy, and when combined with chronic infection (as with HIV/AIDS, tuberculosis, cholera, malaria, or other parasitic diseases) can be all the more deadly.

Key Concept:

- Secondary or acquired immunodeficiency can result from immunosuppressive drugs or other agents, age, and malnutrition.

HIV/AIDS Has Claimed Millions of Lives Worldwide

In recent years, all other forms of immunodeficiency have been overshadowed by a pandemic (global epidemic) of severe immunodeficiency caused by the infectious agent **human immunodeficiency virus** (**HIV**). HIV causes **acquired immunodeficiency syndrome** (**AIDS**), which was first recognized because of opportunistic infections in a cluster of individuals on both coasts of the United States in June 1981. This group of patients displayed unusual infections, including with the opportunistic fungal pathogen *Pneumocystis jirovecii* (formerly called *Pneumocystis carinii*), which causes pneumocystis pneumonia (PCP) in people with immunodeficiency. Previously, these infections had been limited primarily to individuals taking immunosuppressive drugs. In addition to PCP, some of those early patients had Kaposi's sarcoma, an extremely rare skin cancer caused by a virus, as well as other rarely encountered opportunistic infections. More complete evaluation showed that all patients had a marked deficiency in cell-mediated immune responses and a significant decrease in the subpopulation of T cells that carry the CD4 marker (T helper cells).

The majority of patients with this new syndrome were homosexual males. In those early days before the cause or transmission route was known, and as the number of AIDS cases climbed throughout the world, people thought to be at highest risk for AIDS were homosexual males, promiscuous heterosexual individuals of either sex and their partners, intravenous drug users, people who received blood or blood products prior to 1985, and infants born to HIV-infected mothers. We now know that all these initial patients had intimate contact with an HIV-infected individual or exposure to HIV-tainted blood.

Since its discovery in the early 1980s, AIDS has increased to pandemic proportions throughout the world. As of 2016, approximately 36.7 million people were living with HIV infection, 1.1 million in the United States. Although reporting of AIDS cases is mandatory in the United States, many states do not require reporting of cases of HIV infection that have not yet progressed to AIDS, and it is estimated that one in seven infected individuals does not know he or she is infected. These uncertainties make the count of HIV-infected individuals an estimate.

The toll of HIV/AIDS in the United States is dwarfed by figures for other parts of the world. The global distribution of those afflicted with HIV is shown in **Figure 18-7**. In sub-Saharan Africa, the region most affected, an estimated 25.5 million people were living with HIV in 2016, and another 5 million infected individuals were in South and Southeast Asia. The country with the biggest AIDS burden is South Africa, with 7.1 million infected individuals, a devastating 13% of its population. Epidemiologic statistics estimate that since the beginning of the pandemic more than 80 million people have been infected with HIV and 36 million people worldwide have died of AIDS. Despite a better understanding of how HIV is transmitted and how people can protect themselves from infection, estimates indicate that there were 1.8 million new HIV infections in 2016.

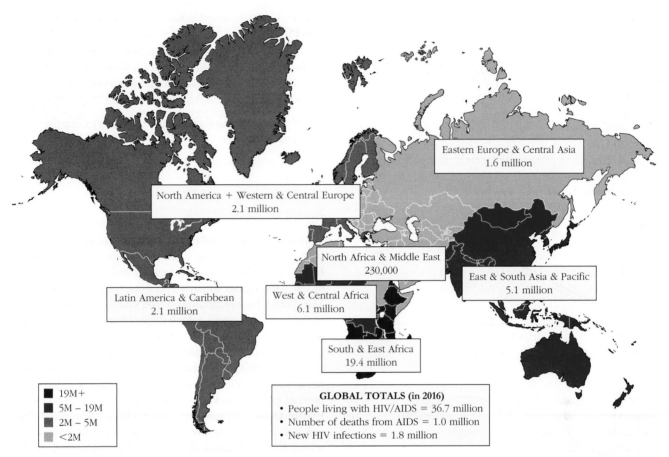

FIGURE 18-7 Global AIDS epidemic. The status of the AIDS epidemic globally and by region is presented, as of the end of 2016. [Data from *http://www.unaids.org/en/resources/documents/2017/2017_data_book/*]

Fortunately, the numbers have improved somewhat in recent years. The rate of increase in numbers of people living with HIV infection has slowed since 2000 (**Figure 18-8a**), and the number of infected children has actually decreased, reflecting declines in children contracting the infection from their mothers (see Figure 18-8b). The number of AIDS-related deaths in 2015 was about half of the peak number of annual deaths from AIDS, in 2005 (see Figure 18-8c), and the number of new infections annually has slowed since its peak in the late 1990s (see Figure 18-8d). With the decline in AIDS-related deaths, the number of children orphaned annually by AIDS—one of the most devastating consequences of this disease—has plateaued and may be declining (see Figure 18-8e).

These and other gains are attributable partly to the United Nations Declaration of Commitment on HIV/AIDS, signed in 2001, and subsequent international agreements that have paved the way for stepped-up prevention, treatment, and education programs around the world, including expanded worldwide access to antiretroviral drugs. In 2016 it was estimated that 18.2 million people,

almost half of all infected individuals, had access to these lifesaving drugs. That antiretroviral drugs enable many infected individuals to live longer lives has led to increasing numbers of individuals living with AIDS, despite the decline in new cases. Despite this progress there is still no indication that an end to the pandemic is in sight, and the rate of decline in new cases annually has slowed down (see Figure 18-8d). About 5000 people are still being infected per day. Expanded efforts of the international community will be needed to fully meet its stated goal of ending the pandemic by 2030.

In the remainder of this chapter we will see why HIV is such a dangerous pathogen. Most importantly, it is because HIV infects and kills CD4$^+$ T cells, eventually leading to profound immunodeficiency. We will also see that other properties of the virus (including its high mutation rate and its ability to latently infect cells) contribute to the challenges this virus presents. The chapter will close with a discussion of advances in treatments that slow progression to AIDS and reduce the infection of others, as well as the possibility of preventing or curing this disease.

(a) Number of people living with HIV

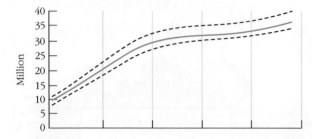

(b) Number of children living with HIV

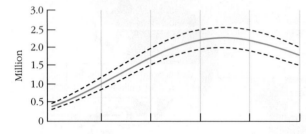

(c) Number of AIDS-related deaths

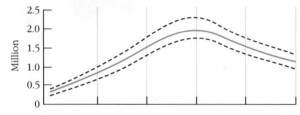

(d) Number of new HIV infections

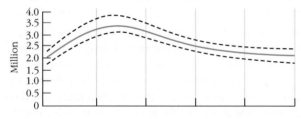

(e) Number of orphans due to AIDS

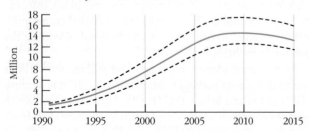

FIGURE 18-8 Trends in the HIV/AIDS epidemic. Annual numbers from 1990 to 2015 are shown for (a) people living with HIV infection, (b) children living with HIV, (c) AIDS-related deaths, (d) new HIV infections, and (e) children orphaned because of AIDS. These numbers are estimates, within the ranges indicated by the dashed lines. [*Data from AIDS Data, UNAIDS Reference 2016. http://www.unaids.org/sites/default/files/media_asset/2016-AIDS-data_en.pdf*]

Key Concepts:

• The HIV/AIDS pandemic has claimed millions of lives. While the number of people dying each year has gone down, thousands of people globally are still being infected each day.

• HIV presents many challenges to efforts to end the pandemic.

The Retrovirus HIV-1 Is the Causative Agent of AIDS

Within a few years after the recognition of AIDS as an infectious disease, the causative agent, now known as HIV-1, was discovered and characterized in the laboratories of Luc Montagnier in Paris and Robert Gallo in Bethesda, Maryland. The infectious agent was found to be a **retrovirus** of the lentivirus genus, which display long incubation periods (*lente* is Latin for "slow"). Retroviruses carry their genetic information in the form of RNA, and when the virus enters a cell this RNA is reverse-transcribed (RNA to DNA, rather than the other way around) by a virally encoded polymerase enzyme, **reverse *t*ranscriptase** (**RT**) (**Overview Figure 18-9**). This complementary cDNA copy of the viral genome is then integrated into the cell's chromosomes, where it is replicated along with the cell's DNA. If the provirus is actively transcribed to form large numbers of new virions (virus particles), the cell lyses. Alternatively, the provirus may remain latent in the cell until some activation signal starts the expression process.

About 5 years after the discovery of HIV-1, a close retroviral cousin, HIV-2, was isolated from some AIDS sufferers in Africa. Unlike HIV-1, its prevalence is limited mostly to areas of western Africa, and disease progresses much more slowly, if at all. In Guinea-Bissau, where HIV-2 is most common, up to 8% of the population may be persistently infected and yet most of these individuals experience a nearly normal life span. There is some hope that scientists can gain a better understanding of HIV-1 from the study of the more benign cohabitation of HIV-2 and its human host.

Viruses related to HIV-1 have been found in nonhuman primates, and some of these are believed to be the original sources of HIV-1 and HIV-2 in humans. These viruses, variants of *s*imian *i*mmunodeficiency *v*irus (SIV), can cause immunodeficiency disease in certain infected monkeys. Typically, SIV strains cause no disease in their natural hosts but may produce immunodeficiency similar to AIDS when injected into another primate species. There is good evidence that HIV-1 evolved from strains of SIV that jumped the species barriers from African chimpanzees and gorillas to humans in the West African country of Cameroon, while HIV-2 is thought to have arisen from a similar transfer of SIV from sooty mangabey monkeys. These cross-species transfer events are believed to have occurred some time

OVERVIEW FIGURE   **18-9**

Structure of HIV

(a)

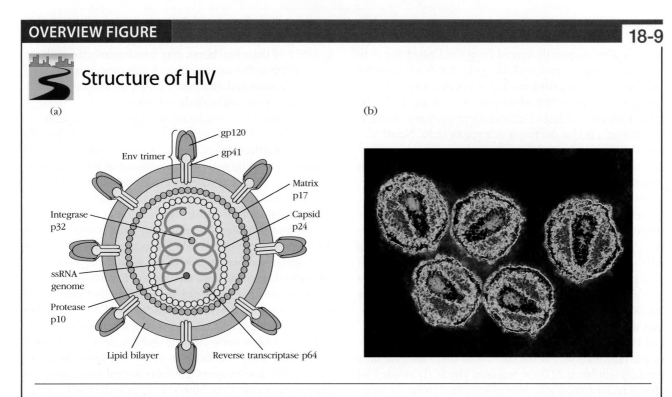

(b)

(a) Cross-sectional schematic diagram of HIV. Each virion carries a lipid bilayer derived from the host cell, containing about 14 glycoprotein projections composed of trimers of gp120 and gp41: gp41 is a transmembrane molecule that crosses the lipid bilayer of the viral envelope, and gp120 is associated with gp41. These proteins, encoded by the *env* gene, are cleaved from the large Env precursor, and the trimers are referred to as *Env spikes*. Their function is to bind to the target receptor (CD4) and coreceptor (CXCR4 or CCR5) on host cells, initiating infection. Within the envelope are several proteins, encoded by the *gag* gene, that form the matrix, capsid, and core around the RNA genomes. Inside the capsid are two copies of the HIV genome made of single-stranded RNA (ssRNA), along with multiple molecules of each of three enzymes: reverse transcriptase (p64), which also has ribonuclease H activity, a protease (p10), and an integrase (p32); these are products of the *pol* gene. Not all of the proteins present inside the virus are shown. (b) Electron micrograph of HIV virions magnified 200,000 times. The glycoprotein projections are faintly visible as "knobs" extending from the periphery of each virion. *[(b) BSIP/Getty Images]*

during the early twentieth century, making HIV a relatively new pathogen for the human population.

Molecular analyses of the relationships of primate SIV to HIV-1, based on blood samples taken from people in West Central Africa from the 1950s to the present, have revealed a fascinating story of the origin and spread of the virus that has caused the AIDS pandemic. SIV appears to have crossed from chimpanzees to humans in Cameroon at least twice around 1920, giving rise to the current HIV-1 groups M (for the *m*ain virus group) and N, while HIV-1 groups O and P appear to have arisen from transfers into humans of SIV from western lowland gorillas in the same region. Group M viruses are responsible for the global pandemic, having caused 99% of HIV-1 infections worldwide, while N, O, and P viruses are still largely limited to Cameroon and neighboring countries in West Africa. Transmission into humans probably occurred during the hunting and butchering of these animals for food; this "bushmeat trade" is thought to have grown in the early 1900s. Recent studies suggest that individuals infected with the group M HIV-1 virus traveled from Cameroon via the Congo River system to where the initial major expansion started, in Leopoldville in the Belgian Congo (what is now Kinshasa in the Democratic Republic of the Congo [DRC]). Group M virus subsequently spread through the DRC, apparently facilitated by colonial era changes that included expansion of the train routes of the area, which allowed easy travel, as well as the growing sex trade in developing population centers and the possible re-use of needles in health clinics.

Beginning in about 1960 group M viruses spread dramatically through Africa and diverged into different subtypes, of which 10 are known today. Subtype B was carried to Haiti around 1967 by Haitians who had worked in Kinshasa; shortly thereafter this strain spread from Haiti to the United States.

The first known death in the United States from an AIDS-related condition, Kaposi's sarcoma, occurred in 1969. Subtype B subsequently spread from the United States to Western Europe, Japan, and Australia, but it accounts for only about 12% of global cases. A separate subtype, C, has been responsible for 50% of the pandemic, in particular the large numbers of infected individuals in southern and eastern Africa; it is also the major subtype in India, Nepal, and parts of China. Most of the other subtypes are found primarily in parts of Africa, with some appearing also in South America and parts of Asia.

As HIV does not replicate in standard laboratory animals, model systems for studying it are few. Only the chimpanzee supports infection with HIV-1 at a level sufficient to be useful in vaccine trials, but infected chimpanzees rarely develop AIDS, which limits the value of this model in the study of viral pathogenesis. In addition, the number of chimpanzees available for such studies is low, and both the cost and the ethical issues involved preclude widespread use of chimpanzees in research. Studies of SIV or of SHIV (a hybrid of SIV and HIV) in rhesus macaque monkeys have provided some useful information, but the conditions they cause are only partially similar to AIDS in humans. The SCID mouse (see above) reconstituted with human lymphoid tissue for infection with HIV-1 has been useful for certain studies of HIV-1 infection, especially for the development of drugs to combat viral replication.

Key Concepts:

- AIDS was shown to be caused by the retrovirus HIV-1, which is more infectious and causes more severe disease than its cousin HIV-2.

- Both HIV-1 and HIV-2 evolved from SIV and entered the human population from chimpanzees and gorillas (HIV-1) and monkeys (HIV-2) in the early 1900s.

- HIV-1 group M virus is responsible for the HIV/AIDS pandemic; its several subtypes are found in various regions of the world.

HIV-1 Is Spread by Intimate Contact with Infected Body Fluids

Epidemiological data indicate that the most common means of transmission of HIV-1 (which we will subsequently refer to as HIV) include vaginal and anal intercourse, receipt of infected blood or blood products, and passage from HIV-infected mothers to their infants in utero, during birth, or from mother's milk. Before routine tests for HIV were in place, patients who received blood transfusions and hemophiliacs who received blood products were at risk for HIV infection. Exposure to infected blood accounts for the high incidence of AIDS among intravenous drug users, who often share hypodermic needles. Infants born to mothers who are infected with HIV are at high risk of infection; without prophylaxis, between 15% and 45% of these newborns may become infected with the virus. However, antiretroviral treatment programs for HIV$^+$ pregnant women and their newborns are making a real dent in these numbers, as will be discussed below.

In the worldwide pandemic, it is estimated that approximately 75% of the cases of HIV transmission are attributable to sexual practices. Infection with other *s*exually *t*ransmitted *d*iseases (STDs) increases the likelihood of HIV transmission, and in situations where STDs flourish, such as unregulated prostitution, these infections probably represent a powerful cofactor for sexual transmission of HIV. The open lesions and activated inflammatory cells (some of which may express receptors for HIV) associated with STDs favor the transfer and penetration of the virus through mucosal epithelial layers during intercourse. Estimates of transmission rates per exposure vary widely and depend on many factors, such as the presence of STDs and number of virions being transferred. However, when one partner is infected, transfer between male and female during vaginal intercourse is approximately twice as risky to the female as to the male, and receptive partners in anal intercourse are even more at risk. Data from studies in India and in Africa indicate that men who are circumcised are at significantly lower risk of acquiring HIV-1 via sexual contact, possibly because foreskin provides a source of cells that can become infected or harbor the virus. However, this does not work in reverse: circumcised males were found to be equally likely to transmit HIV-1 to their sexual partners. Education about the benefits of male circumcision is part of the international efforts to reduce new infections and end the AIDS epidemic.

Identifying the initial events that take place during HIV transmission from an infected to an uninfected individual is very challenging. Nonetheless, hypotheses concerning the most likely sequence of events have been pieced together, based on observations in humans and animals, including in vitro studies using explanted human tissue and in vivo studies in macaques. On the basis of these observations, both free virus and virus-infected cells in vaginal secretions and semen are thought to contribute to infection through the reproductive and gastrointestinal tracts. Direct infection of the many resting memory CD4$^+$ T cells present within the vaginal mucosal epithelial layer is likely the initial site of infection in the female genital tract (the most studied location). In macaques, initial association of the virus with cells can be established in as little as 30 to 60 minutes, and high numbers of infected CD4$^+$ cells are seen within 1 day of exposure. Replication of HIV in the vaginal mucosa was also shown to help activate local CD4$^+$ T cells, providing yet more targets for the virus and creating a vicious cycle. In addition, the local inflammation associated with STDs is thought to enhance the number of T$_H$ cells and their susceptibility to infection.

How does the virus get to the CD4$^+$ T cells in the mucosal epithelium? Lesions, abrasions, or tears in the epithelial

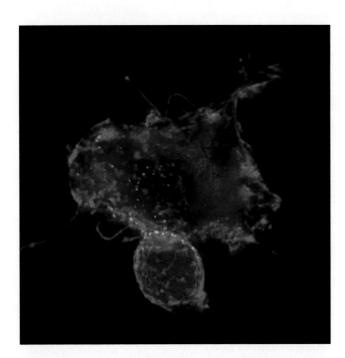

FIGURE 18-10 Interaction between a dendritic cell and T cell, indicating passage of HIV (green dots) between the cells. Note that virus particles cluster at the interface between the large dendritic cell and the smaller T cell; this interface is sometimes referred to as an *infectious synapse*. [Courtesy of Thomas J. Hope, Northwestern University.]

layer occurring during intercourse provide an easy route of entry. Free virus may squeeze between epithelial cells and penetrate through them by transcytosis (endocytic transport from the luminal to the basal surface of a cell). *Lang-erhans cells* (LCs), a type of intraepithelial DC with long processes that reach close to the surface of the epithelial layer, have also been shown to capture and take up virus, although they may not become infected. These and other DCs may transport intact infectious virus within their endocytic compartments and are able to transfer this virus to CD4$^+$ T cells via an *infectious synapse* (**Figure 18-10**). The role of macrophages in these early events, as transporters or targets of infection, is somewhat controversial, but macrophages are not suspected to be a major contributor. Whether free or cell-associated, the virus then migrates through the submucosa into the lymphatic vessels and is filtered out in the draining lymph nodes, where the adaptive immune response can be initiated. However, once in the lymph nodes further viral spread is facilitated, some through cell-to-cell hand-off via infectious synapses, as many more cells with the proper surface receptors are encountered. Emerging evidence based on the analysis of the sequences of viruses from individual patients suggests that a single HIV virion may be responsible for all or most of the systemic infection in many male-to-female transfers.

Because transmission of HIV infection requires direct contact with infected blood, milk, semen, or vaginal fluid,

preventive measures can be taken to block these events. Scientific researchers and medical professionals who take reasonable precautions, which include avoiding exposure of broken skin or mucosal membranes to fluids from their patients, significantly decrease their chances of becoming infected. When exposure does occur, rapid administration of antiretroviral drugs can often prevent systemic infection. The use of condoms when having sex with individuals of unknown infection status also significantly reduces chances of infection. One factor contributing to the spread of HIV is the long period after infection during which no clinical signs may appear but during which the infected individual may infect others. Thus, universal use of precautionary measures is important whenever infection status is uncertain.

Key Concepts:

- HIV-1 infection is spread by contact with infected body fluids such as occurs during sex (the most common mode of transmission), direct transfers to the blood such as from using contaminated needles, and from mother to infant during pregnancy, childbirth, or breast-feeding.

- During sexual practices, free HIV and infected cells in semen and vaginal fluids cross the mucosal epithelium through lesions, tears, or abrasions, by transcytosis of free virus, or by virus binding to extensions of dendritic cells. The virus is then transported to draining lymph nodes, where other cells can be infected.

In Vitro Studies Have Revealed the Structure and Life Cycle of HIV

The structure of HIV has been well characterized (see Overview Figure 18-9). It has a lipid bilayer envelope with embedded viral proteins that mediate infection of host cells. Inside the virus are two copies of a single-stranded RNA genome and a series of proteins that are essential for virus replication. The HIV genome and encoded proteins have been fairly well characterized, and the functions of most of these proteins are known (**Figure 18-11**). HIV carries three structural genes (*gag*, *pol*, and *env*) and six regulatory or accessory genes (*tat*, *rev*, *nef*, *vif*, *vpr*, and *vpu*). The *gag* gene encodes several proteins, including the capsid and matrix, which enclose the viral genome and associated proteins. The *pol* gene codes for the three main enzymes that are required for the viral life cycle: reverse transcriptase, integrase, and protease. In fact, the protease enzyme is required to process the large Gag and Pol precursor proteins into smaller proteins. As we will see shortly, these uniquely viral enzymes are some of the main targets for therapeutic intervention. The final structural gene, *env*, is the source of the surface proteins gp120 and gp41, responsible for attachment of the

(a)

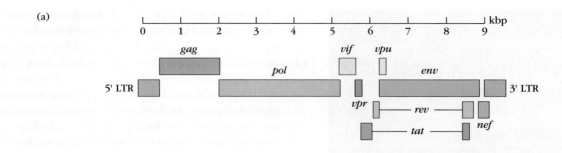

(b)

Gene	Protein product	Function of encoded proteins
gag	53-kDa precursor ↓	*Nucleocapsid proteins*
	p17	Forms outer core-protein layer (matrix)
	p24	Forms inner core-protein layer (capsid)
	p9	Is component of nucleoid core
	p7	Binds directly to genomic RNA
env	160-kDa precursor ↓	*Envelope glycoproteins*
	gp41	Is transmembrane protein associated with gp120 and required for fusion
	gp120	Associated with gp41 on virion surface; forms trimers that bind CD4 and coreceptor
pol	Precursor ↓	*Enzymes*
	p64	Has reverse transcriptase and RNase H activity
	p51	Has reverse transcriptase activity
	p10	Is protease that cleaves *gag* and *pol* precursors
	p32	Is integrase
		Regulatory proteins
tat	p14	Strongly activates transcription of proviral DNA
rev	p19	Allows export of unspliced and singly spliced mRNAs from nucleus
		Auxiliary proteins
nef	p27	Supports T-cell activation; reduces CTLA4, CD4, and MHC expression
vpu	p16	Is required for efficient viral assembly and budding Promotes extracellular release of viral particles, degrades CD4 in ER
vif	p23	Promotes maturation and infectivity of viral particle
vpr	p15	Promotes nuclear localization of preintegration complex; inhibits cell division

FIGURE 18-11 Genetic organization of HIV-1 (a) and functions of encoded proteins (b). The three major genes— *gag, pol,* and *env*—encode polypeptide precursors that are cleaved to yield the nucleocapsid proteins, enzymes required for replication, and envelop proteins, respectively. The remaining six genes (*tat, rev, nef, vif, vpu,* and *vpr*) play various roles in the regulation of HIV transcription, RNA transport, virus assembly and release from the cell, and immune evasion. The long terminal repeats (LTRs) are important for integration of the DNA copy of the viral genome into the host chromosome. In addition, the 5′ LTR contains sequences to which various regulatory proteins bind, activating transcription. The HIV-2 and SIV genomes are very similar except that the *vpu* gene is replaced by *vpx* in both of them.

OVERVIEW FIGURE

HIV Infection of Target Cells and Virus Replication

(a) Infection of target cell

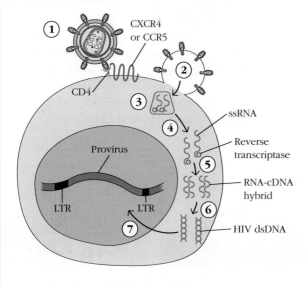

(b) Activation of provirus

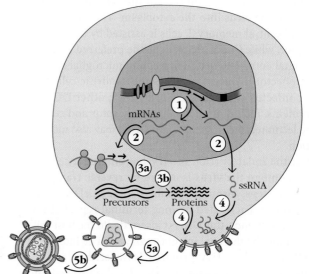

(1) HIV gp120 binds to CD4 and then CXCR4 or CCR5 on target cell.

(2) HIV gp41 causes fusion with the target cell membrane.

(3) Nucleocapsid containing viral genome and enzymes enters the cell.

(4) Viral genome and enzymes are released following removal of core proteins.

(5) Viral reverse transcriptase catalyzes reverse transcription of ssRNA, forming RNA-cDNA hybrids.

(6) Original RNA template is partially degraded by ribonuclease H, followed by synthesis of second DNA strand to yield HIV dsDNA.

(7) The viral dsDNA is then translocated to the nucleus and integrated into the host chromosomal DNA by the viral integrase enzyme.

(1) Transcription factors stimulate transcription of proviral DNA into genomic ssRNA and, after processing, several mRNAs.

(2) Viral RNA is exported to cytoplasm.

(3a) Host-cell ribosomes catalyze synthesis of viral precursor proteins.

(3b) Viral protease cleaves precursors into viral proteins.

(4) HIV ssRNA and proteins assemble beneath the host-cell membrane, into which gp41 and gp120 are inserted.

(5a) The membrane buds out, forming the viral envelope.

(5b) Released viral particles complete maturation; remaining precursor proteins are cleaved by viral protease present in viral particles.

(a) Following binding of HIV to the receptor and coreceptor, entry of the HIV capsid into cells, and reverse transcription of the viral RNA genome into dsDNA, integration of the viral DNA into the host-cell genome creates the provirus. (b) The provirus remains latent until the infected cell is activated, leading to the transcription and generation of new viral ssRNA genomes and mRNAs, translation and processing of viral proteins, assembly of components under the membrane, and formation and release of virus particles.

virus to the CD4$^+$ viral receptor and its coreceptor, either CXCR4 or CCR5. The regulatory genes expressed by HIV have functions such as modulating CD4 and MHC class I expression, inactivating host proteins that interfere with viral transcription, and facilitating intracellular viral transport.

Much has been learned about the life cycle of HIV from in vitro studies, where cultured human T cells have been used to map out virus attachment and postattachment intracellular events (**Overview Figure 18-12a**). HIV infects cells that express the CD4 protein on their surface; in addition to T$_H$ cells, these can include macrophages, dendritic cells, and brain microglial cells. This preference for CD4$^+$ cells is due to the high-affinity binding of gp120 to the CD4 molecule on the host cell. However, this interaction alone is not sufficient for viral entry and productive infection. Expression of another cell-surface molecule, called a coreceptor, is required for HIV infection of the cell. Both of the known coreceptors for HIV, CCR5 and CXCR4, are chemokine receptors. The normal role of chemokine receptors on leukocytes is to bind chemokines, chemoattractants that guide the migration of cells (see Chapter 14 and Appendix III). When the virus's gp120 protein binds CD4, the gp120 undergoes a conformational change that exposes a binding site for either CXCR4 or CCR5. Binding to the coreceptor brings the fusion-promoting region of the gp41 protein into contact with the membrane, triggering fusion of the virus's lipid bilayer with the membrane of the cell, releasing the contents of the virus into the cytoplasm. The infection of naïve and central memory T cells is assisted by the CXCR4 coreceptor, while CCR5 seems to be the preferred coreceptor for viral entry into macrophages and microglial cells, as well as effector memory T cells. Some dendritic cells may also be infected through CD4 and CCR5; other DC surface proteins, including the mannose receptor and certain C-type lectin receptors, such as DC-SIGN, may also mediate infection.

After the inner portion of HIV has entered a cell, the RNA genome of the virus is copied by reverse transcriptase in two steps, forming a double-stranded cDNA copy of its genome. Facilitated by *long terminal repeat* (LTR) sequences at its ends, this DNA is inserted into the host cell's chromosomes by the viral integrase enzyme. The integrated viral DNA, called the HIV **provirus**, is transcribed, and the various viral RNA messages are spliced and translated into proteins that, along with new full-length RNA copies of the HIV genome, assemble into new viral particles (see Overview Figure 18-12b). Some large viral protein precursors are cleaved by the viral protease into the smaller proteins that make up the nuclear capsid and the three enzymes in mature viruses. The gp120 and gp41 envelope proteins are cleaved from a large precursor in the ER and are transported to the plasma membrane. The other viral proteins and two copies of the RNA genome gather under the membrane, and newly formed virions bud from the surface of the infected cell (**Figure 8-13a**).

When the viral genome is actively transcribed, a large number of virus particles bud from the plasma membrane (see Figure 18-13b). If too much of the host cell's biosynthetic machinery and plasma membrane are involved in virus production, this can cause cell lysis. However, HIV can also become latent, or remain unexpressed, for long periods of time in an infected cell. This period of dormancy, during which the cells are not expressing any viral proteins, makes the immune system's task of finding and eliminating these latently infected cells especially difficult. Latent infection results in the establishment of HIV reservoirs, safe havens for the virus where both antiviral immunity and antiretroviral drug therapy can have little impact.

- HIV infects CD4$^+$ cells by binding to the CD4 proteins as well as a chemokine receptor (CCR5 or CXCR4). Once inside the cell, the RNA genome is reverse transcribed into cDNA by the viral reverse transcriptase; the cDNA copy is integrated by the viral integrase into the host cell's chromosomes, where it is called a provirus.

- When the cell is activated, the provirus is transcribed into new RNA genomes and mRNAs; the latter are translated into viral proteins, some of which are processed by the viral protease. Viral components collect under the cell's plasma membrane and new virus particles bud from the plasma membrane.

(a)

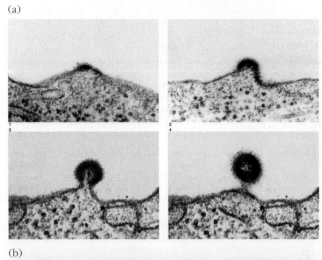

(b)

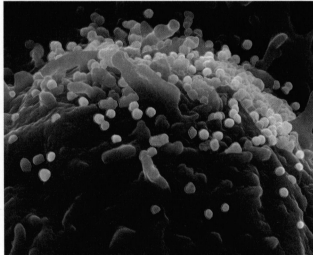

FIGURE 18-13 Budding of new virus particles from the surface of an infected T cell. (a) The stages of formation of a new virus particle and its separation from the plasma membrane have been captured by transmission electron microscopy. (b) The budding of large numbers of virions from an infected T cell, shown by scanning electron microscopy. *[(a) Eye of Science/Science Source. (b) Dr. Olivier Schwartz, Institut Pasteur/Science Source.]*

Key Concepts:

- HIV has a lipid bilayer and two copies of an RNA genome, which contains three structural and six regulatory genes that encode the proteins necessary for infection and propagation in host cells.

HIV Variants with Preference for CCR5 or CXCR4 Coreceptors Play Different Roles in Infection

During the course of an infection the *tropism* of the virus particles (i.e., which cells they will infect) often changes. The chemokine receptors CCR5 and CXCR4 serve as the major coreceptors for HIV. CCR5 is expressed by effector memory CD4$^+$ T cells (which give rise to effector T cells after activation), macrophages, brain microglial cells, and dendritic cells, while CXCR4 is expressed on naïve CD4$^+$ T cells and central memory T cells (which proliferate extensively after activation before generating effector T cells). This explains the distinct cell tropism of different variants of HIV: those that bind the CCR5 coreceptor, called R5 variants (formerly M-tropic), are able to infect macrophages, dendritic cells, and effector memory T cells, while those that bind the CXCR4 coreceptor, called X4 variants (formerly T-tropic), preferentially infect naïve and central memory T cells (**Figure 18-14**).

The preferences of virus variants for CCR5 or CXCR4 also help explain some roles of chemokines in regulating virus infection and replication. It is known from in vitro studies that certain chemokines, such as RANTES (CCL5; see Appendix III), have a negative effect on virus replication. CCR5 and CXCR4 cannot bind simultaneously to HIV and to their natural chemokine ligands. Competition for the receptor between the virus and the natural chemokine ligand can thus block viral entry into the host cell. Early enthusiasm for the use of these chemokines as antiviral agents was dampened when significant RANTES expression was observed in some HIV-infected individuals who progress to disease, with no obvious antiviral effect. Despite this, an antagonist of CCR5 is now used as a therapeutic inhibitor of HIV infection of cells (see below).

The two types of HIV variants, R5 viruses and X4 viruses, play different roles in HIV infection. As most of the CD4$^+$ cells in the epithelial layers exposed to virus-containing body fluids during intercourse (the vagina, cervix, foreskin, rectum) or infection of infants (the upper GI tract) express the CCR5 coreceptor, it is the R5 viruses that usually are responsible for the initial infection. These epithelial tissues have numerous macrophages, dendritic cells, and T cells because of constant microbial exposure. CCR5 also may play several roles in transporting the virus through the epithelial layers. It is expressed by some epithelial cells and may facilitate the transcytosis of virus through the epithelial layer. CCR5 expressed on the long extensions of dendritic cells between epithelial cells may contribute to pulling viruses from the surface into the epithelial layer. R5 variants may then be delivered to CD4$^+$ T cells in the epithelium or be carried to the draining lymph nodes, where they can replicate. Thus the R5 viruses are likely the initial infecting virus and are responsible for dissemination to other lymphoid tissues throughout the body. As the virus mutates over time, X4 variants arise late in the course of HIV infection and are responsible for infection and elimination of larger numbers of T cells, leading to full-blown AIDS. Studies of the viral envelope protein gp120 identified a region that determines whether it will bind the CCR5 or CXCR4 coreceptor; a single amino acid difference in gp120 may be sufficient to determine whether CCR5 or CXCR4 is used.

Interestingly, there is a deletion mutation in the *CCR5* gene that imparts nearly total resistance to infection with the R5 strains of HIV that are most commonly transmitted during sexual encounters. Individuals who are homozygous for this mutation express no CCR5 on the surface of their cells, making them resistant to viral variants that require this coreceptor but, remarkably, they are otherwise apparently unperturbed by the loss of this chemokine receptor. This has led to some new ideas and hopes for HIV elimination. One HIV-infected patient (Timothy Ray Brown, known as the Berlin patient) who received a bone marrow transplant in 2007 (to treat his leukemia) from a donor who had the mutation preventing CCR5 expression has been virus-free for at least 10 years without taking antiretroviral drugs. Thus far he is the only known HIV-infected individual to have experienced what seems to be a complete cure of his infection. Confirmation and long-term follow-up of this finding, as well as the development of other techniques to exploit coreceptor blockade as a method of virus elimination, are ongoing.

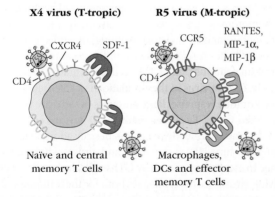

FIGURE 18-14 CXCR4 and CCR5 serve as coreceptors for HIV infection of different cell types. After gp120 binds CD4, it must bind a coreceptor for virus entry and infection. R5 (T cell–tropic) variants of HIV use the coreceptor CXCR4 expressed on naïve and central memory T cells, whereas the X4 (macrophage-tropic) strains use CCR5 to infect macrophages, dendritic cells, and effector memory T cells. Both are receptors for chemokines, and their normal ligands (SDF-1 [also known as CXCL12; see Appendix III], RANTES [CCL5], and MIP-1α/β [CCL3/4]) can block HIV infection of the cell.

Key Concepts:

- Viruses that cause infections, called R5 viruses, usually use the CCR5 coreceptor, found on effector memory T cells, macrophages, and dendritic cells common in mucosal epithelia. These R5 viruses are the major virus type through much of the early infection period.

- As the infection progresses, R5 viruses may mutate to preferring the CXCR4 coreceptor, enabling them to infect naïve as well as central memory T cells. These X4 viruses contribute to the later significant decline in numbers of CD4$^+$ T cells.

Infection with HIV Leads to Gradual Impairment of Immune Function

Although the precise course of HIV-1 infection and disease onset varies considerably in different individuals, a general scheme for the progression to AIDS in the absence of antiretroviral treatment can be outlined (**Figure 18-15**). First, there is the acute, or primary, stage of infection. This is the period immediately after infection, when there usually are no detectable anti-HIV antibodies. Estimates vary, but some reports find that more than half of the individuals undergoing

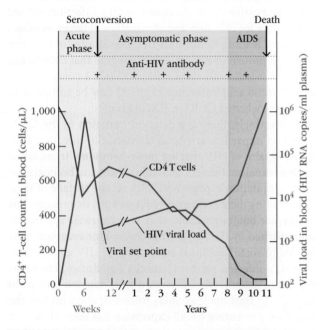

FIGURE 18-15 Typical course of HIV infection in an untreated patient. Soon after infection, viral RNA is detectable in the serum. However, HIV infection is most commonly detected by the presence of anti-HIV antibodies after seroconversion, which normally occurs within 2 months of infection. The *viral set point* refers to the level of virus in the blood at the time of rebound, when the immune response begins to control virus levels. Clinical symptoms indicative of AIDS generally do not appear for 1 to 20 years after infection, but this interval is variable and is extended by antiretroviral therapy. The onset of clinical AIDS is usually signaled by a decrease in CD4$^+$ T-cell numbers to below 200/ml and a sharp increase in viral load. Patients become very susceptible to opportunistic infections and other health problems. *[Data from Fauci, A. S., G. Pantaleo, S. Stanley, D. Weissman. 1996. Immunopathogenic mechanisms of HIV infection. Annals of Internal Medicine **124**:654.]*

primary infection experience flu-like symptoms, including fever, lymphadenopathy (swollen lymph nodes), and malaise for several weeks after exposure. During this acute phase, HIV infection is spreading rapidly among CCR5-expressing T cells, and the *viral load* (number of virions) in the blood as well as in other body fluids can be quite high, elevating the risk of infecting others. The initial appearance of antibody against HIV antigens (**seroconversion**) usually occurs 1 to 3 months after infection. The most commonly used test for HIV-specific antibodies is an ELISA (see Chapter 20) to detect the presence of antibodies directed against HIV proteins. Because of the delay in seroconversion, some HIV ELISA tests also look for HIV antigens in the blood, which may be detectable 2 to 6 weeks after infection. Positive ELISA results are confirmed using either the more specific Western blot technique, which detects the presence of antibodies against several HIV proteins, or PCR assays for HIV RNA.

With the high levels of circulating virus (viremia) during the acute phase, the number of CCR5$^+$ CD4$^+$ cells declines dramatically (see Figure 18-15). This is followed by a rebound that reflects the development, around the time of seroconversion, of HIV-specific cytotoxic T cells that begin to control the infection. The level of virus in the blood at the time of the rebound is called the *viral set point*; higher set points usually lead to more rapid progression of the disease.

A long asymptomatic period (see Figure 18-15) follows during which there is a gradual decline in CD4$^+$ T cells but usually no outward symptoms of disease. This slow progression reflects both antibody and CTL responses that keep viral replication in check. Antibodies provide protection through neutralization, opsonization of virus, and binding to gp120 on the surface of infected cells, leading to their elimination by ADCC and phagocytosis. But each generation of antibodies loses effectiveness as the virus mutates due to unrepaired replication errors made by reverse transcriptase. The duration of the asymptomatic phase varies greatly, likely due to a combination of host and viral factors. Although the infected individual normally has no clinical signs of disease at this stage, viral replication continues, and CD4$^+$ cell levels gradually fall. Even when the level of virus in the circulation is stable, kept in check by the immune response, large amounts of virus are produced by infected CD4$^+$ T cells; as many as 10^9 virions are released every day and continually infect and destroy additional T cells. During this time HIV is mutating, and viruses whose antigens change to escape recognition by CTLs will survive. The rapidly evolving virus presents challenges both for the immune system to keep up with these virus escape variants and for the development of drugs and vaccines to treat or prevent progression of the disease, as will be discussed below.

Without treatment, most HIV-infected patients eventually progress to AIDS (see Figure 18-15), where the hallmark is opportunistic infection. Three stages in disease progression have been defined by the Centers for Disease Control and Prevention: the first two are defined by diminished numbers or proportions of CD4$^+$ T cells (**Table 18-4**, top), and

TABLE 18-4   Stage definition for HIV infection among adults and adolescents

Stage*	CD4+ T-cell count		CD4+ T-cell percentage		Clinical evidence
1	≥500/μl	or	≥26%	and	No AIDS-defining condition
2	200–499/μl	or	14–25%	and	No AIDS-defining condition
3 (AIDS)	<200/μl	or	<14%	or	Presence of an AIDS-defining condition

AIDS-Defining Conditions

- Bacterial infections, multiple or recurrent

- Candidiasis of bronchi, trachea, or lungs

- Candidiasis of esophagus

- Cervical cancer, invasive

- Coccidioidomycosis, disseminated or extrapulmonary

- Cryptococcosis, extrapulmonary

- Cryptosporidiosis, chronic intestinal (>1 month duration)

- Cytomegalovirus disease (other than liver, spleen, or nodes)

- Cytomegalovirus retinitis (with loss of vision)

- Encephalopathy, HIV related

- Herpes simplex: chronic ulcers (>1 month in duration) or bronchitis, pneumonitis, or esophagitis

- Histoplasmosis, disseminated or extrapulmonary

- Isosporiasis, chronic intestinal (>1 month in duration)

- Kaposi's sarcoma

- Lymphoma, Burkitt (or equivalent term)

- Lymphoma, immunoblastic (or equivalent term)

- Lymphoma, primary, of brain

- *Mycobacterium avium* complex or *Mycobacterium kansasii*, disseminated or extrapulmonary

- *Mycobacterium tuberculosis* of any site, pulmonary, disseminated, or extrapulmonary

- *Mycobacterium*, other species or unidentified species, disseminated or extrapulmonary

- *Pneumocystis jirovecii* pneumonia

- Pneumonia, recurrent

- Progressive multifocal leukoencephalopathy

- *Salmonella* septicemia, recurrent

- Toxoplasmosis of brain

- Wasting syndrome attributed to HIV

* All require laboratory confirmation of HIV infection.

Data from Selik, R.M., et al; Centers for Disease Control and Prevention. Revised surveillance case definition for HIV infection - United States, 2014. MMWR April 11, 2014/63(RR03):1-10. (https://www.cdc.gov/mmwr/preview/mmwrhtml/rr6303a1.htm)

the third by a reduction in CD4$^+$ T cells to <200 cells/ μl of blood <20% of the normal number) or one or more AIDS-defining conditions, which include opportunistic infections (see Table 18-4, bottom). A rise in the level of circulating HIV in the plasma (viremia) and a concomitant drop in the number of CD4$^+$ T cells generally precede this first appearance of symptoms. The first overt indication of AIDS is often opportunistic infection with the yeast *Candida albicans*, which causes the appearance of sores in the mouth (thrush) and, in women, a vulvovaginal yeast infection that does not respond to treatment. A persistent hacking cough caused by *P. jirovecii* infection of the lungs is another early indicator of AIDS. Patients with late-stage AIDS generally succumb to tuberculosis and other mycobacterial infections, pneumonia, severe wasting diarrhea, viral diseases, or various malignancies. Without treatment, the time between acquisition of the virus and death from the immunodeficiency averages 8 to 10 years.

Key Concepts:

- Infection with HIV causes gradual and severe impairment of immune function, marked by depletion of CD4$^+$ T cells, and if untreated, usually results in death from opportunistic infections or cancers.

- Early HIV-specific antibody and cytotoxic T-cell responses partially control virus levels, but the virus mutates due to errors made by reverse transcriptase, and viral escape variants arise that are resistant to existing antibodies and CTLs.

- An individual is said to have AIDS when the number of CD4$^+$ T cells in the blood drops below 200/μl; the individual then becomes susceptible to opportunistic diseases.

Changes over Time Lead to Progression to AIDS

Immunologists are especially interested in the events that take place between the initial encounter with HIV and the takeover and collapse of the host immune system. Understanding how the immune system holds HIV in check during the asymptomatic phase could aid in the design of effective therapeutic and preventive strategies. For this reason, the handful of HIV-infected individuals who remain asymptomatic for very long periods without treatment (called *long-term nonprogressors*), estimated to be <2% of HIV$^+$ individuals in the United States, are the subject of intense study. Another well-studied group of individuals consists of those in high-risk groups, such as prostitutes, some of whom do not contract HIV infections. In addition to the discovery of the CCR5 deletion (see above), several interesting findings have emerged

from studying high-risk populations, including the presence of strong CD8$^+$ T-cell and NK-cell responses against HIV-infected cells in many of these individuals, as well as association of particular HLA and NK receptor alleles with delayed progression to AIDS.

Although the viral load in blood plasma remains fairly stable throughout the period of chronic HIV infection, examination of the lymph nodes and gastrointestinal (GI) tract tissue reveals a different picture. Fragments of lymph nodes obtained by biopsy from infected subjects show high levels of infected cells at all stages of infection; in many cases, the structure of the lymph node is completely destroyed by virus long before the plasma viral load increases above the steady-state level. Studies have shown that the gut may be the main site of HIV replication and CD4$^+$ T-cell depletion, the latter starting as early as the acute stage of infection (**Figure 18-16**). Numbers of CCR5$^+$ CD4$^+$ T cells in the intestine may drop by 60% to 80% within the first 3 weeks after infection. Subsequent investigations of the association between the GI tract and HIV have suggested that intestinal T$_H$17 cells, which express both the CCR5 and CXCR4 coreceptors, are a primary target for infection and destruction. Normally these T$_H$17 cells are thought to play an important role in homeostatic regulation of the innate and adaptive responses to microbial flora in the gut. Destruction of these cells and disruption of the integrity of the mucosal barrier in the GI tract allow the passage of microbes and their products across the epithelial layer into the body, explaining some of the rampant immune stimulation and inflammation that is characteristic of HIV infection. In a deadly feedback loop, this immune stimulation generates yet more activated CD4$^+$ cells, the favored targets for HIV infection and replication.

The severe decrease in CD4$^+$ T cells is a clinical hallmark of AIDS, and several explanations have been advanced for the death of uninfected as well as infected CD4$^+$ T cells. In addition to the lysis of cells actively replicating HIV, abortive HIV infection of resting T cells (i.e., production of viral cDNA without release of viral particles) may result in cell death, as accumulation of cDNA in the cytoplasm may activate innate responses, including inflammasome activation and cell death by pyroptosis (see Chapter 4). Other processes that may contribute to CD4$^+$ cell depletion include the killing of virus-infected cells by virus-specific CTLs; killing of anti-gp120 antibody-coated cells by phagocytosis, complement-mediated lysis, or NK cell-mediated ADCC; cell fusion mediated by binding of an infected cell's gp120 to an uninfected cell's CD4 protein; apoptosis due to induction of FasL; and reduced generation of T cells by the thymus.

This depletion of CD4$^+$ T cells is the primary cause of immunodeficiency in HIV-infected individuals. Memory T-cell responses, such as to influenza virus, decline early in disease progression. Loss of T$_H$1 cells results in a decrease

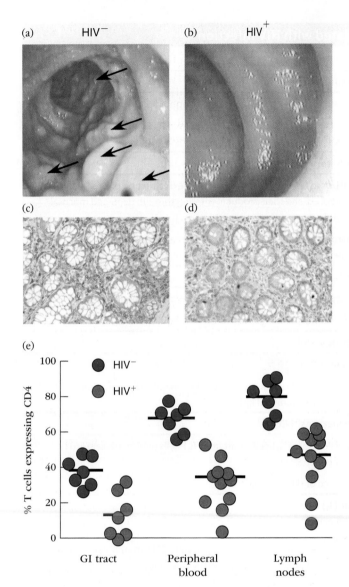

FIGURE 18-16 Endoscopic and histologic evidence for depletion of CD4⁺ T cells in the GI tract of patients with AIDS. (a and c) The intestinal tract of a normal uninfected individual and a stained section from a biopsy of the same area (terminal ileum) with obvious large lymphoid aggregates (arrows, a) and CD4⁺ T cells stained with antibody (brown color, c). (b and d) Similar analyses of samples from an HIV⁺ patient in the acute stage of infection indicate the absence of normal lymphoid tissue and sparse staining for CD4⁺ T cells. (e) Comparison of CD4⁺ T-cell numbers in samples from the GI tract, peripheral blood, and lymph nodes of HIV-positive and -negative individuals. *[Data from Brenchley, J. M., et al. 2004. CD4⁺ T cell depletion during all stages of HIV disease occurs predominantly in the gastrointestinal tract.* Journal of Experimental Medicine **200**:749. Photos ©2004 Brenchley et al. Originally published in The Journal of Experimental Medicine **200**:749–759. DOI: https://doi.org/10.1084/jem.20040874, Figs 2a-d.]*

or absence of delayed-type hypersensitivity to intracellular pathogens that a normal immune system defeats, contributing, for example, to increased susceptibility to tuberculosis and other mycobacterial infections. Other effects on both adaptive and innate immune functions can be observed during the progression to AIDS. Chronic exposure to HIV and to intestinal pathogens entering through the damaged mucosal epithelium produces ongoing systemic inflammation. Inflammatory mediators can cause cell death and damage to lymphoid organs. Invading gut microbes and their products initially induce polyclonal B-cell activation; however, with the decline over time in T$_H$ cells, this is followed by reduced antibody responses to T-dependent antigens, especially of the IgG and IgA classes. Innate responses are also impacted, including dendritic cell functions. **Table 18-5** lists some immune abnormalities common to HIV/AIDS.

Individuals infected with HIV often display dysfunction of the central and peripheral nervous systems, especially in the later stages of infection. Macrophages and microglial cells in the brain can be infected and support viral replication; they also produce molecules that are toxic to brain neurons and astrocytes. Quantitative comparison of specimens from brain, lymph node, spleen, and lung of AIDS patients with progressive encephalopathy indicated that the brain was heavily infected. A frequent complication in later stages of HIV infection, HIV-associated neurocognitive disorder (also known as *AIDS dementia complex* and *HIV/AIDS encephalopathy*) is a neurologic syndrome characterized by abnormalities in cognition, motor performance, and behavior. It is estimated that 10% to 24% of untreated patients with AIDS develop this condition, which often contributes to AIDS-associated mortality.

Key Concepts:

- Decreases in CD4⁺ T cells are earliest and most significant in the intestine, leading to disruptions in tissue structure that result in infections with gut pathogens and resulting systemic inflammation.

- As HIV infection continues, many aspects of innate and adaptive immunity are affected.

Antiretroviral Therapy Inhibits HIV Replication, Disease Progression, and Infection of Others

The development of drugs that block the ability of HIV to infect or replicate in cells has had a huge impact on the outcome of HIV infection. **Antiretroviral therapy (ART)** has made HIV/AIDS a manageable disease, now allowing infected individuals to live essentially normal life spans.

TABLE 18-5	Immunologic abnormalities associated with HIV infection
Stage of infection	**Typical abnormalities observed**
	Lymphoid tissue structure
Early	Some structural disruption, especially to gastrointestinal tract–associated lymphoid tissues
Late	Extensive damage and tissue necrosis; loss of follicular dendritic cells and germinal centers; inability to trap antigens or support activation of T and B cells
	Innate and inflammatory responses
Early	Infection of dendritic cells and transport of HIV to draining lymph nodes; some destruction of DCs and ILCs; inflammatory responses induced by HIV, microbes, and their products entering through damaged mucosal barriers, dead cells, and proinflammatory cytokines
Late	Ongoing systemic inflammation; inflammatory cytokines TNF-α and IL-1β can cause cell death; chronic cell activation contributes to tissue damage; infection of microglia in brain can result in neurological disorders
	T helper (T_H) cells
Early	Depletion of CD4$^+$ T cells, especially memory T cells in the gut, where T_H17 cells are targeted
Late	Further decrease in CD4$^+$ T-cell numbers and corresponding T_H activities; shift from T_H1 to T_H2 responses
	Antibody production
Early	Enhanced polyclonal IgG and IgA production
Late	Reduced memory and marginal zone B cells. Reduced responses to antigens. Few broadly neutralizing anti-HIV antibodies. Decreased class switching, and therefore reduced IgG and IgA
	Delayed-type hypersensitivity
Early	Highly significant reduction in proliferative capacity of T_H1 cells; shift from T_H1 cells (which mediate DTH responses) to T_H2 cells and reduction in skin-test reactivity
Late	Elimination of DTH response; complete absence of skin-test reactivity
	T cytotoxic (T_c) cells
Early	Normal reactivity
Late	Reduction but not elimination of CTL activity due to impaired ability to generate CTLs from T_c cells resulting from reduced numbers of T_H1 cells and increased T_{REG} cells, and reduced thymus function

There are several targets for effective antiviral drugs that take advantage of the life cycle of HIV (**Figure 18-17**). The keys to success for such therapies are that they must be specific for HIV and interfere minimally with normal cell processes, be effective in blocking HIV replication, and have few side effects. Thus far, antiviral agents targeting five separate steps in the viral life cycle have proven effective.

The first success came with drugs that interfere with the reverse transcriptase enzyme that copies viral RNA into cDNA (step ③ in Figure 18-17). There are two possible strategies for developing pharmaceutical agents that can interfere with reverse transcription. The first uses *nucleoside reverse transcriptase inhibitors (NRTIs)*, which basically are competitive inhibitors of normal nucleoside binding, and the second uses *nonnucleoside reverse transcriptase inhibitors (NNRTIs)*, which are inhibitors of reverse transcriptase that do not bind in the active site. The prototype of the drugs that interfere with reverse transcription was zidovudine, or

*azido*thymidine (AZT), which was approved in 1987. The incorporation of AZT, a nucleoside analogue and competitive inhibitor of the enzyme (and hence an NRTI), into the growing cDNA chain of the retrovirus causes termination of the chain. AZT is effective in some but not all patients, and its efficacy is further limited because long-term use has adverse side effects and because resistant viral mutants develop in treated patients. The administered AZT is used not only by the HIV reverse transcriptase but also by human DNA polymerase. The incorporation of AZT into the DNA of host cells kills them. Precursors of red blood cells and other rapidly dividing cells are especially sensitive to AZT, resulting in anemia and other side effects. Other, less toxic NRTIs have been developed more recently.

The second class of reverse transcriptase inhibitors, NNRTI drugs, inhibits the action of reverse transcriptase by binding to the enzyme away from the active site. These noncompetitive inhibitors of reverse transcriptase have less of

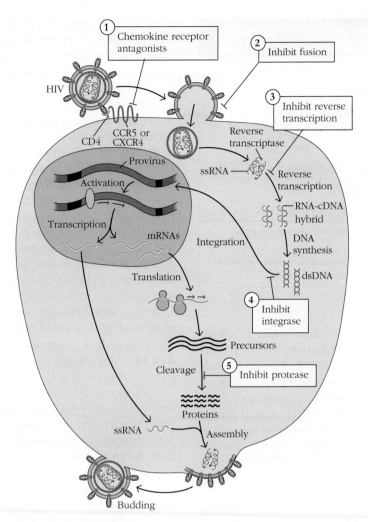

FIGURE 18-17 Stages in viral replication cycle that provide targets for therapeutic antiretroviral drugs. In order of the steps in the viral life cycle that they inhibit, antiretroviral drugs that inhibit HIV replication interfere with (1) binding of the coreceptor (CCR5 or CXCR4) to the Env spike, (2) viral fusion with the plasma membrane, (3) copying of the viral RNA chromosome into cDNA by the viral reverse transcriptase, (4) insertion of the viral double-stranded cDNA into the host cell's chromosomes by the viral integrase, and (5) processing of large viral precursor proteins into mature viral proteins by the viral protease.

an adverse effect on host proteins, and therefore fewer side effects. However, they are still susceptible to the development of resistance as the virus mutates, and for this reason are typically used only in combination with other anti-HIV drugs that have different targets. Nevirapine, licensed in 1996, was the first NNRTI designed to treat HIV, but since then more NNRTI drugs have come on the market.

The second generation of drugs inhibits the viral protease (step ⑤ in Figure 18-17) required to cleave precursor proteins into the mature proteins needed for assembly of new mature virions. The first protease inhibitor, saquinavir, came to market in the mid-1990s. This was followed by the development of a viral gp41 protein inhibitor that blocks fusion of the virus with the host cell membrane (step ② in Figure 18-17), and thus inhibits infection of host cells. The next type of antiviral agent to become available also keeps viruses from infecting cells by blocking access to the CCR5 chemokine coreceptor used by the virus (step ① in Figure 18-17). The final class of antiretroviral agents to be developed interferes with the HIV integrase (step ④ in Figure 18-17) required for insertion of the cDNA copy of the viral genome into host cell DNA. **Table 18-6** lists the currently available categories of anti-HIV therapies, along with the year in which they were first approved for use. As of 2017 more than 25 antiretroviral drugs in the six categories shown in Table 18-6 have been approved for use.

A major concern for those undergoing antiretroviral therapy is the emergence of viruses resistant to the anti-HIV drugs. Viral mutants arise continually, due to reverse transcriptase replication errors. Some of the mutations generate variants of enzymes and other viral proteins that render the virus resistant to the drugs. To counteract

TABLE 18-6	Categories of HIV-1 drugs in clinical use	
Category		**FDA approval date***
Nucleoside/nucleotide analogues		1987
Nonnucleoside reverse transcriptase inhibitors		1996
Protease inhibitors		1995
Fusion/attachment inhibitors		2003
Chemokine coreceptor antagonists		2007
Integrase inhibitors		2007

*Year of first FDA approval for a drug to treat HIV-1 infection in that drug category.

Data from HIV InSite. 2012. Antiretroviral drug profiles. University of California, San Francisco. http://hivinsite.ucsf.edu/InSite?page=ar-drugs

this issue, HIV-infected individuals now take combinations of drugs that inhibit multiple steps in the HIV life cycle. The recommendation for ART for someone newly diagnosed with HIV infection includes two NRTIs plus a third drug that is either an NNRTI, an integrase inhibitor, or a protease inhibitor. The specific cocktail chosen is influenced by a variety of factors, including side effects and the ease of following the treatment regimen. It is currently recommended that treatment should begin immediately following the HIV diagnosis. Other combinations of medications may be recommended for individuals who have been infected for longer periods of time, depending on the levels of virus in the blood, what drugs their virus may be resistant to, their CD4$^+$ T-cell count, and their health status.

The use of combination drug therapy regimens usually prevents the virus from rapidly producing resistant mutants, because it is unlikely that individual viruses become resistant to all of the drugs. In most cases, these combination therapies are effective in lowering the plasma viral load to levels that are not detectable by current methods, slowing down the destructive effects on the immune system, and improving the health of patients with HIV/AIDS. The decrease in the numbers of AIDS deaths in the United States in recent years is largely attributed to this advance in therapy. Despite the optimism engendered by success with combination therapy, drawbacks include side effects and the need for consistent adherence to these regimens, lest drug-resistant viral mutants be favored in the patient. The recent development of single daily pills containing multiple drugs has made combination therapy easier for patients to manage.

The success of combination antiretroviral therapy has led researchers to wonder whether it might be possible to eradicate all virus from an infected individual and thus actually cure someone with AIDS. This goal faces a significant

obstacle: the persistence of latently infected CD4$^+$ T cells and macrophages, which can serve as reservoirs of infectious virus if and when the provirus becomes reactivated. Even with a viral load beneath the level of detection by PCR assays, the immune system may not recover sufficiently to clear virus should these cells begin to produce virus in response to some activation signal. In addition, virus may persist in sites, such as the brain, that are not readily penetrated by the antiretroviral drugs or immune responses. There is considerable interest in finding a way to re-activate virus replication in the latently infected cells so that those cells could be attacked by the immune response or drug treatment. As mentioned earlier, transcription of the provirus normally occurs only in activated T cells. However, a treatment that causes systemic T-cell activation would also activate previously uninfected T cells, which would make them targets to be infected and thus could do more harm than good.

Antiretroviral therapy not only slows the progression of HIV/AIDS in infected individuals, but it also can significantly reduce transmission to others. **Clinical Focus Box 18-1** documents the dramatic reduction globally in transmission of HIV from infected women to their infants, due to treatment of both pregnant women and their newborns by ART. And there is considerable evidence that antiretroviral drugs may prevent HIV transmission between adults participating in high-risk activities. Studies have shown that ART results in a reduction in transmission of HIV to an uninfected partner by up to 96% in HIV-discordant couples (where one of the two is infected). In addition, antiretroviral therapy taken prophylactically can be utilized to prevent HIV infection. This *pre-exposure prophylaxis* (*PrEP*) treatment was approved by the FDA in 2012 as a preventive measure for healthy, HIV-negative individuals at high risk of infection. The current recommendation is to take a daily single pill, Truvada, that contains two reverse transcriptase inhibitors. For people engaging in high-risk behaviors, PrEP has been shown to reduce the risk of HIV infection by 92%.

Key Concepts:

- Treatment of HIV infection with antiretroviral drugs that target specific steps in the viral life cycle, especially in combination, can lower the viral load, provide relief from some symptoms of infection, and prevent infection of others.

- Antiretroviral therapy of pregnant women and newborns has greatly reduced mother-to-child transmission of HIV worldwide.

- Antiretroviral drugs can also be taken prophylactically to prevent infection of individuals engaging in risky behaviors.

A Vaccine May Be the Only Way to Stop the HIV/AIDS Pandemic

Despite the progress that has been made in slowing the HIV/AIDS pandemic, it appears that the best option to stop the spread of HIV/AIDS is a safe, effective vaccine that prevents infection and/or progression to disease. The cause of AIDS was discovered over 35 years ago. Why don't we have an AIDS vaccine by now?

The best way to approach an answer to this question is to examine the specific challenges presented by HIV-1. There are many HIV variants, and even within one infected individual HIV mutates rapidly, creating a moving target for both the immune response and any vaccine design. We know from HIV-seropositive individuals that the development of humoral immunity during a natural infection, or even the presence of neutralizing antibodies, does not necessarily inhibit viral spread, in part because of viral escape mutants. The same is true for cytotoxic T-cell responses. Still, valuable lessons have been learned from nearly two decades of HIV vaccine initiatives, both from studies in nonhuman primates and from clinical trials in humans (**Table 18-7**). The earliest of these trials in humans was aimed at eliciting humoral immunity to neutralize incoming virus. This approach used purified gp120 envelope protein from HIV as an immunogen. These studies, completed in the United States and Thailand in 2003, showed weak neutralizing antibody responses and no protection from infection. This was followed by a wave of new vaccine designs aimed at eliciting cellular immunity to the virus, using recombinant viral vectors that infect cells and would better mimic natural infection. The initial vector chosen was *ade*novirus serotype 5 (Ad5), carrying recombinant DNA derived from the *gag, pol,* and *nef* genes of HIV. Ad5 is derived from a naturally occurring human virus to which between 30% and 80% of individuals (depending on geographic location) have previously been exposed, appearing to make this a safe vector choice. These trials included two stages of vaccination: an initial priming dose followed by a vaccine boost using the same vector. These human trials began in 2003 but were halted prematurely in 2007, midway through the trial period, because the rate of infection actually appeared to be higher in the vaccinated trial group than in the placebo control population, especially among individuals who had pretrial immune responses to adenovirus. Memory T cells specific for adenovirus antigens that would have been activated by the vaccine may have provided a population of easily infectable cells. This was a resounding blow to the AIDS research community and sent many vaccine design teams back to the drawing board.

These disappointing results have led to new thinking in terms of targets for the next wave of HIV vaccine design: vaccines that elicit both humoral and cellular immunity. One trial using this strategy, called RV144, that was completed in 2009 showed some promise. It used a prime-boost combination with a recombinant DNA vector containing HIV genes (the prime) followed by protein derived from the virus (the boost). The vector chosen was canarypox, a virus that infects cells but does not replicate in humans and that was engineered to carry HIV DNA from *env, gag,* and the protease portion of *pol.* The protocol also included a boost with a booster vaccine consisting of an engineered version of HIV gp120 protein in an adjuvant. Results from this study, involving 16,000 volunteers from communities in Thailand, were modest but promising, demonstrating a 31% reduction in infection rates among the vaccine-treated group compared with the matched placebo control population 42 months after immunization. Some follow-up analyses indicated that the RV144 vaccine induced some antibodies that correlated with reduced risk of infection. Although statistically

TABLE 18-7	Major HIV-1 vaccine trials			
Vaccine design		**Study name**	**Status**	**Results**
Purified protein (gp120)		VAX003, VAX004	Completed in 2003	No protection
Recombinant adenovirus vector containing *gag/pol/nef* genes		HVTN 502 STEP HVTN 503 Phambili	Terminated in 2007	No protection
Recombinant canarypox vector containing *env/gag/protease* genes + Env gp120 protein boost		RV144	Completed in 2009	30% reduction
Recombinant adenovirus 5 vector containing *env/gag/pol/nef* genes + Gag/Pol and Env protein boost		HVTN 505	Completed in 2013	No protection
Recombinant canarypox vector (*env/gag/protease*) + 3 canarypox vector and Env gp120 protein boosts; contains HIV components from HIV subtype C		HVTN 702	Began in 2016	Expected 2020

HVTN = HIV vaccine trials network.

Data from Barouch, D. H., and B. Korber. 2010. HIV-1 vaccine development after STEP. *Annual Review of Medicine* **61:153**; https://www.nih.gov/news-events/news-releases/first-new-hiv-vaccine-efficacy-study-seven-years-has-begun

Prevention of Mother-to-Child Transmission of HIV

One of the greatest tragedies of the HIV/AIDS pandemic has been the large number of children infected by their HIV+ mothers. The overall rate of infection of infants born to untreated HIV+ mothers is estimated to be about 37%, and it is estimated that globally almost 500,000 infants became infected with HIV through mother-to-child transmission in 2002 alone. These infections result from transmission of virus from HIV-infected mothers *in utero* (across the placenta), during childbirth, or by transfer of virus from milk during breastfeeding. Fortunately, one of the triumphs of antiretroviral drug therapy has been the dramatic reduction in the infection of infants. As shown in **Figure 1**, by 2015 the number of such infections worldwide had dropped by about two-thirds; cumulatively it is estimated that 1.6 million infections of children have been averted by treating pregnant and breast-feeding women and their infants with antiretroviral drugs.

In the United States mother-to-child transmission declined by more than 90% after the prescription of antiretroviral therapy for pregnant women and their babies began in the early 1990s. Today the Centers for Disease Control and Prevention (CDC) recommends a set of standard practices, starting with combination therapy of HIV-infected women throughout pregnancy, using two nucleoside reverse transcriptase inhibitors (NRTIs) plus an integrase inhibitor or a protease inhibitor; these drugs cross the placenta to protect the fetus *in utero*. As the baby is exposed to the mother's blood and other body fluids during a normal birth, delivery by C-section may be considered. An extra dose of antiretroviral drugs may be given to the mother at the time of birth. Immediately or as early after birth as possible the infant should begin a 4- to 6-week course of treatment with zidovudine or combinations of multiple drugs. The CDC also recommends that HIV+ mothers not breastfeed their infants and instead use

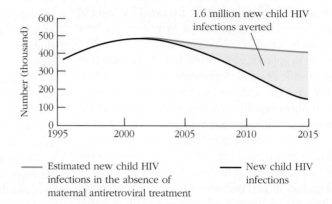

FIGURE 1 New HIV infections worldwide among children with and without the provision of antiretroviral medicines to prevent mother-to-child transmission, from 1995 to 2015. Numbers of new infections are shown for children, 0 to 14 years of age, compared with estimates of infections that would have occurred in the absence of antiretroviral treatment of pregnant and breast-feeding women and their infants. *[Data from: UNAIDS 2016 Reference: AIDS Data. http://www.unaids.org/en/resources/documents/2017/AIDSdata2016]*

readily available infant formulas. Following these guidelines has reduced the incidence of mother-to-child transmission in the United States to 1% or less, a remarkable achievement.

However, most HIV infections of infants worldwide occur in sub-Saharan Africa and other less developed areas, where many obstacles exist to preventing mother-to-child transmission. These include a higher proportion of women who are infected (as heterosexual transmission of HIV is the major route of infection in Africa), pregnant women not knowing their HIV status, poor access to health care for HIV testing and drug administration, cost and availability of drugs (which must be safe and stable in rural conditions), the importance of breastfeeding for the health of babies, and cultural barriers.

Fortunately, several clinical trials have demonstrated the practicality and effectiveness of antiretroviral therapy in sub-Saharan Africa. A 1999 clinical trial of the antiretroviral drug nevirapine (an NNRTI) brought hope for a practical way to combat HIV transmission at birth under less-than-ideal conditions of clinical care. Some mothers were given a single dose of nevirapine at the onset of labor, while

others received a short course; infants were given a single dose 1 day after birth. The highly encouraging results of this study revealed infection in only 13.5% of the babies in the single nevirapine dose group when tested at 16 weeks of age. Of those given a short course of zidovudine, 22.1% were infected at this age compared with 40.2% in a small placebo group.

These promising results led to World Health Organization (WHO) recommendations that this nevirapine regimen be used in all instances where mother-to-child transmission of HIV was a danger. But what about the risks associated with breastfeeding? In 2007, it was estimated that up to 200,000 infants become infected with HIV annually through breast milk. In resource-limited countries, poor early nutrition and susceptibility to disease are key factors in infant death rates; breastfeeding significantly reduces these risks. Therefore, a follow-up investigation called the Breast-feeding, Antiretrovirals, and Nutrition (BAN) Study was conducted, from 2006 to 2008 in Malawi. This investigation employed a dose of nevirapine for the mother and child at birth, followed by 7 days of treatment of the pair with two NRTIs. The 2369 mother-infant pairs were then randomized to one of three

(continued)

postdelivery prophylaxis groups. Mothers in the maternal regimen group received a triple-drug cocktail for 28 weeks postdelivery, including two NRTIs and either an NNRTI or a protease inhibitor. The babies in the infant treatment group received nevirapine daily for the same period of time, while initial participants in the control population received no additional treatment. All mothers were encouraged to breastfeed for 6 months and wean before 7 months. The study revealed a 53% protective effect for the maternal regimen and a 74% protective effect in the infant treatment group.

Collectively, the conclusions from these and other trials led to worldwide recommendations for protecting newborns from maternally transmitted HIV. Current WHO guidelines, published in 2015, have several key components. First, all pregnant women should know their HIV status; expansion of healthcare resources that provide testing has made this possible in many locations. And all HIV⁺ pregnant and breastfeeding women, independent of the stage of their HIV infection, should receive antiretroviral drug therapy *for their entire lifetime*. These guidelines have three interconnected benefits: they improve the health of mothers, they prevent mother-to-child transmission, and they prevent transmission of HIV to any sexual partners. As shown in **Figure 2**, in addition to high-income countries like the United States that already provide antiretroviral therapy to infected mothers and their infants, most low- and middle-income countries of the world are implementing these guidelines. With this buy-in, 77% of all pregnant women living with HIV were receiving antiretroviral therapy as of 2015, and that number should continue to improve. As of 2017 four countries—Armenia, Belarus, Cuba, and Thailand—have officially eliminated mother-to-child transmission and other countries are nearing that goal.

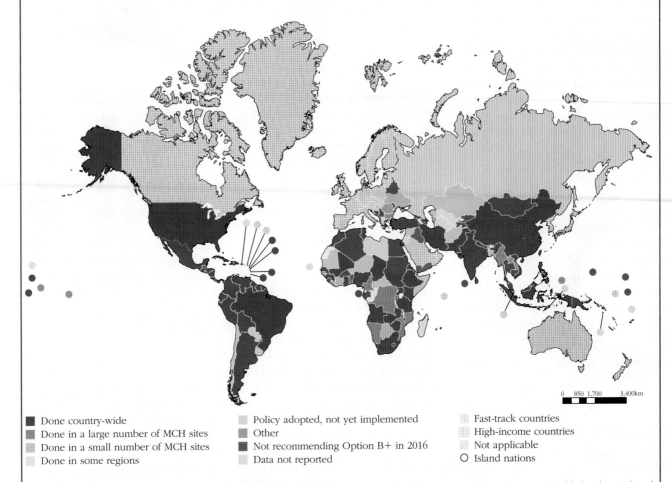

- ■ Done country-wide
- ■ Done in a large number of MCH sites
- ■ Done in a small number of MCH sites
- ■ Done in some regions
- ■ Policy adopted, not yet implemented
- ■ Other
- ■ Not recommending Option B+ in 2016
- ■ Data not reported
- ▥ Fast-track countries
- ▦ High-income countries
- □ Not applicable
- ○ Island nations

FIGURE 2 **Most countries are providing lifelong antiretroviral therapy to pregnant and breastfeeding women living with HIV.** Compliance with the 2015 WHO guidelines (see text) is shown for low- and middle-income countries and Fast-Track countries (Fast-Track countries have committed to providing data and working toward a set of 2014 UNAIDS goals that seek to end the HIV pandemic by 2030). Compliance by some higher income countries is not shown; some may follow other guidelines. Countries highlighted in pink and red have not implemented the full guidelines including lifelong therapy (option B⁺). MCH countries are those benefiting from USAID Maternal and Child Health programs, which focus on 24 countries that have more than 70% of maternal and child deaths. [*Data from: World Health Organization Progress Report 2016: Prevent HIV, Test and Treat All. http://www.who.int/hiv/pub/progressreports/2016-progress-report/en/*]

BOX 18-2

Broadly Neutralizing Antibodies to HIV

The generation of HIV escape variants resistant to neutralizing antibodies allows the virus to persist and spread within an infected individual, and therefore in most people antibodies do not generate long-lasting protection. In recent years it has become apparent that some infected individuals (10% to 50%) develop broadly neutralizing antibodies (bNAbs) capable of neutralizing many HIV-1 variants, but only a small percentage develop highly potent bNAbs capable of neutralizing most forms of HIV. Given the exciting possibility of developing a vaccine that would induce such antibodies, extensive research has been done to understand the specificities of these antibodies, why they are

rare, and how a vaccine strategy might be designed to induce bNAbs.

Several dozen bNAbs have been identified by screening serum antibodies or memory B cells from infected individuals for reactivity with many HIV strains, cloning their heavy- and light-chain genes, and expressing the antibody proteins in large amounts. They have been found to recognize six "vulnerable" sites on the envelope (Env) spike, which is a trimer of three gp120/gp41 heterodimers. These six sites are vulnerable because each is critical for maintaining the Env trimer structure or the ability to infect cells. The sites are as follows: the apex of the Env trimer (made up of V1V2 loops of all three gp120 molecules); the base of the V3 loop and adjacent high-mannose glycan (also called the high-mannose patch; it overlaps with the CCR5-binding site); the CD4-binding site; the interface between gp120 and gp41; a region containing the gp41 fusion peptide; and the membrane-proximal external region of gp41 (**Figure 1**).

These bNAbs must also negotiate another major challenge posed by HIV: the high degree of glycosylation of the Env trimer spike. About 50% of the mass of the Env trimer is made up of the glycans, which are attached to all or some of the approximately 90 potential asparagine-linked glycosylation sites on the trimer (**Figure 2a**). The Env trimer is one of the most highly glycosylated proteins known, undoubtedly another adaptation of the virus for avoiding immune elimination. These glycan chains are synthesized by the cells' normal glycosylation pathways and hence are self in nature, and therefore the only epitopes that antibodies can recognize are the patches of Env that show in between the glycans. The binding of two bNAbs to the CD4-binding site and the V3 loop with the nearby high-mannose glycan is shown in Figure 2a. In many cases only small areas of the protein surface are available to be recognized, protected by what have been called "glycan shields." Thus only rare B-cell receptors

V1V2-glycan (apex)
V3-glycan (high-mannose patch)
gp120
CD4 binding site
gp120-gp41 interface
Fusion peptide
gp41
Membrane-proximal external region
Transmembrane domain
Cytoplasmic domain

FIGURE 1 Env spike trimer, made up of three heterodimers of gp120 (gray) and gp41 (white), showing sites of binding of broadly neutralizing antibodies (bNAbs) that recognize most groups and variants of HIV-1. The six sites are as follows: the V1V2 region of the protein with nearby glycan (also called the apex site, pink); the base of the V3 loop with nearby high-mannose glycan (also called the high-mannose patch, green); the CD4-binding site (gold); the interface between gp120 and gp41 (magenta); the fusion peptide (red); and the membrane-proximal external region of gp41 (aqua).

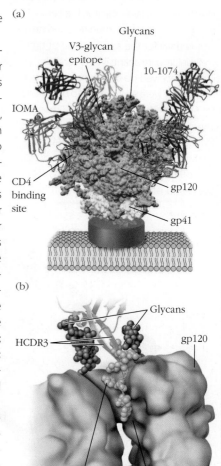

(a)
Glycans
V3-glycan epitope
10-1074
IOMA
CD4 binding site
gp120
gp41

(b)
Glycans
HCDR3
gp120
Electronegative HCDR3 side chains
Electropositive epitope

FIGURE 2 (a) Env trimer showing extensive coverage by glycans (orange), with Fabs of bNAbs IOMA (aqua), which reacts with CD4-binding site, and 10-1074 (blue), which reacts with V3-glycan (high-mannose patch) region, showing how antibody-binding sites must fit between glycans to recognize determinants on the Env protein. Three of each Fab are bound to the epitopes on the three gp120 proteins in the trimer. (b) The long heavy-chain CDR3 (pink ribbon) of bNAb PGT145, which recognizes the V1V2 epitope at the apex of the Env trimer, extends between the glycans at the top of the molecule (small gray spheres) to contact residues of all three gp120 subunits (gray surfaces). Acidic residues of CDR3 interact with basic residues (purple) of the three gp120s facing the hole between the proteins.

(continued)

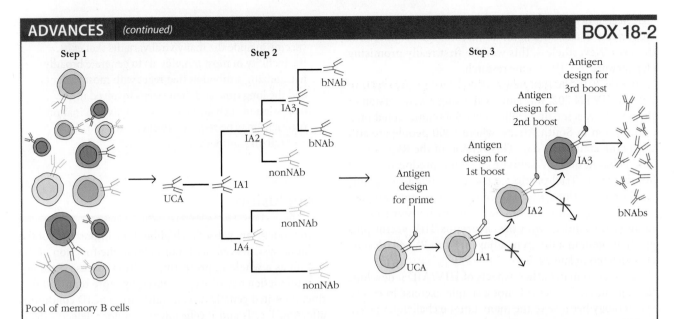

Step 1 **Step 2** **Step 3**

Pool of memory B cells

FIGURE 3 An immunization approach for stimulating production of broadly neutralizing antibodies to the HIV-1 Env spike. Step 1: Broadly neutralizing antibodies from HIV-infected individuals are isolated from memory B cells and sequenced. Step 2: The phylogenetic relationships of the antibodies are established, allowing identification of the probable unmutated (germline) common ancestor (UCA) and intermediate antibodies (IA) leading to the bNAbs; some mutated antibodies are non-neutralizing (nonNAbs). Step 3: The UCA and IA antibodies are synthesized and used as templates to develop a series of immunogen Env spike proteins that, through sequential immunizations, will selectively activate B cells with BCRs evolving through somatic hypermutation toward broadly neutralizing activity.

would have binding sites that could recognize these shielded sites and be activated to produce antibody.

To penetrate through the glycan barriers to bind the epitopes on the protein, about two-thirds of bNAbs have extra-long heavy-chain (H) CDR3 regions (see Chapter 3). For example, some of the antibodies that recognize the gp120 V1V2 regions at the apex of the trimer have long HCDR3 regions of 30 to 39 amino acid residues (compared with more normal lengths of 10 to 15 residues), enabling them to penetrate in between the glycans to the protein surface. Figure 2b shows one such antibody, PGT145; its HCDR3 region is 33 residues, at the end of which are several acidic amino acid residues that interact with basic residues on the surfaces of the three gp120s. Other residues in the PGT145 antigen-binding site interact with the glycans, but the long HCDR3 is key to the antibody's ability to bind to this site.

In addition to the long HCDR3s (generated by unusual VDJ rearrangements) that help some of the antibodies reach through the glycan shield to bind the protein, most bNAbs have undergone extensive somatic hypermutation (SHM). Up to 30% of the nucleotides in the antibodies' CDRs have been somatically mutated during the HIV-driven expansion and selection of activated

B cells in germinal centers (see Chapter 11). Thus, as the virus evolves to escape certain antibodies, rare B cells undergo many cycles of mutation in their BCRs, involving serial activation and selection by evolving viruses, and more mutation. In this way the resulting bNAbs acquire the ability to recognize many virus variants. This long process explains why bNAbs do not usually appear until 2 to 4 years after infection.

These characteristics present great challenges for the development of immunization protocols that will stimulate the production of bNAbs. How does one immunize if HIV-specific antibodies are not broadly neutralizing unless they have acquired many somatic mutations over considerable time? Considerable research is focusing on this issue. One general scheme, illustrated in **Figure 3**, is to isolate bNAbs from memory B cells of an individual, sequence the heavy and light chains, and develop a lineage phylogeny that will point to an unmutated common ancestor antibody [i.e., the germline V(D)J sequences] and intermediate antibodies in the evolution of the bNAbs. These ancestral antibodies would be synthesized and used to design a series of Env protein vaccine immunogens that can be used sequentially to "encourage" the immune system to develop bNAbs.

In addition to providing information about possible immunization protocols for inducing bNAbs, the bNAbs that have been isolated may have a more immediate and practical use: passive immunization by transfer of bNAb protein to prevent or treat HIV infection. Experiments with an SHIV hybrid virus model in macaques have provided promising results, and recently clinical trials have begun with two bNAbs. While antibody protein therapeutics are expensive, passive transfer of bNAbs may complement antiretroviral drugs as part of the arsenal against HIV/AIDS.

REFERENCES

Bonsignori, M., et al. 2017. Antibody-virus co-evolution in HIV infection: paths for HIV vaccine development. *Immunological Reviews* **275**:145.

Burton, D. R., and L. Hangartner. 2016. Broadly neutralizing antibodies to HIV and their role in vaccine design. *Annual Review of Immunology* **34**:635.

Gristick, H. B., et al. 2016. Natively glycosylated HIV-1 Env structure reveals new mode for antibody recognition of the CD4-binding site. *Nature Structural Biology* **23**:906.

Lee, J. H., et al. 2017. A broadly neutralizing antibody targets the dynamic HIV envelope apex via a long, rigidified, and anionic β-hairpin structure. *Immunity* **47**:690.

significant, these results are still far from the near 100% protection rate that is the aim for all vaccines against infectious diseases. Nevertheless, this was the first really promising step forward in HIV vaccine research.

In late 2016 the first new HIV vaccine efficacy trial in 7 years, HVTN 702, was launched, using a new version of the RV144 vaccine. It enrolled over 5000 uninfected men and women in South Africa, where 1000 people are still newly infected every day. This version of the RV144 vaccine is based on HIV subtype C, which predominates in South Africa. Trial participants are receiving two priming immunizations with the canarypox vector carrying HIV *env*, *gag*, and *pol* genes, followed by booster shots at months 3, 6, and 12 with the canarypox/HIV vector plus gp120 protein in a different adjuvant from RV144. Results are expected in late 2020.

As with so many other aspects of HIV/AIDS, development of an HIV vaccine is not a simple exercise in classic vaccinology because of the many unique challenges posed by HIV. As mentioned earlier, reverse transcriptase generates mutations in the viral genome, leading to virus escape mutants not recognized by antibodies or cytotoxic T cells, or by the memory B and T cells that would be induced by a vaccine. Also, very few of the antibodies to gp120, the major protein on the outside of the virus, are able to recognize and neutralize multiple subtypes that exist around the world or the variants that often arise within an individual. For a vaccine to be effective in inducing long-term immunological memory that protects against infection over time, antibodies (and memory B cells) have to be induced that are broadly neutralizing, that is, react with most or all viral subtypes and variants. As is discussed in **Advances Box 18-2**, excellent progress has been made recently in understanding the challenges the immune system must surmount to be able to produce broadly neutralizing antibodies (bNAbs) and how those obstacles are overcome in individuals who generate such antibodies. Translating that knowledge into the development of a vaccine or immunization protocol remains a major challenge.

While the world waits for an HIV vaccine, progress must continue to be made in the United States and globally in reducing the spread of HIV. Essential for this effort are the development and availability of new and improved antiretroviral drugs and other treatments and education about how individuals can protect themselves.

Key Concepts:

- The best prospect for halting the HIV pandemic is a vaccine.

- While most HIV vaccine trials to date have shown no efficacy, one trial in Thailand showed about 30% protection and is being followed up in subsequent trials.

- Challenges to the development of an effective vaccine include the many viral variants that exist, the inability of most individuals to generate broadly neutralizing antibodies that react with most variants, and the long time and many steps involved in the formation of such antibodies. Complex vaccination regimens may be required to elicit such broadly neutralizing antibodies.

Conclusion

Immunodeficiency diseases highlight the importance of the immune system's cells and molecules to the function of the system as a whole in protecting us from disease. Primary immunodeficiency diseases, caused by more than 300 distinct inherited genetic defects, range from SCID conditions affecting T cells and B cells down to milder defects affecting the production of individual immunoglobulin classes or complement components. The most severe SCID conditions block the development of all hematopoietic lineages, of T and B cells, or T cells alone—which also impacts antibody production because of the important roles of helper T cells in many antibody responses. Early screening methods now allow most forms of SCID to be detected at birth, allowing the newborn to be protected from infection and for therapies to be initiated. The correction of some of these defects is now possible through bone marrow or HSC transplantation, with gene therapy an emerging approach.

Other primary immunodeficiencies affecting narrower segments of the immune system, such as antibodies or complement components, may be more easily managed by restoring the missing immune protein, such as through intravenous administration of immunoglobulins or complement components. But reduced numbers of B or T cells can lead to immune dysregulation, explaining the apparent paradox of immunodeficiency being accompanied by autoimmunity. More clear-cut are the processes underlying the severe autoimmune conditions of APECED and IPEX caused, respectively, by defects in self tolerance in the thymus or in the generation of regulatory T cells.

Secondary, or acquired, immunodeficiencies result from conditions adversely affecting immune responses that arise during life, including malnutrition, treatment with immunosuppressive drugs, and infection with HIV. Several aspects of HIV epidemiology and biology have led to its status as a huge global problem. First, HIV can be transmitted through natural human activities, including sexual intercourse, pregnancy, birth, and breastfeeding. Second, the fact that the virus infects and leads to the death of CD4$^+$ cells means that the virus is destroying many of the cells needed for its elimination; if untreated this eventually leads to the opportunistic diseases that cause the death of patients with AIDS. Third, the virus is continually mutating; some of the virions that are generated will by chance be resistant to HIV-specific antibodies and cytotoxic T cells, as well as to the antiretroviral drugs used to block virus replication. The use of combinations of three or four drugs targeting different events in the viral life cycle makes it difficult for viruses to become resistant to all of them and survive. These combination antiretroviral therapies have led to the long-term survival of many infected individuals and have prevented many new infections—including reduced mother-to-child transmission. However, antiretroviral therapy cannot cure HIV infections because of the fourth problematic property of HIV: its ability to latently infect cells without virus expression, forming a reservoir of virus that is not affected either by the immune system or by antiretroviral drugs.

Hope for ending the HIV/AIDS pandemic largely rests with the development of an effective HIV vaccine. The tremendous heterogeneity of HIV strains, including the continual formation of new variants through mutation, presents a high hurdle for the development of a vaccine whose goal would be to induce memory T and B cells that would respond to future infections. The discovery and characterization of broadly neutralizing antibodies produced by some individuals have generated ideas for immunization regimens that, over time, might lead to the induction of memory cells that would be capable of protective responses to most viral variants.

REFERENCES

Azizi, G., et al. 2016. Autoimmunity in primary antibody deficiencies. *International Archives of Allergy and Immunology* **171**:180.

Bansal, K., H. Yoshida, C. Benoist, and D. Mathis. 2017. The transcriptional activator Aire binds to and activates super enhancers. *Nature Immunology* **18**:263.

Becerra, J. C., L. S. Bildstein, and J. S. Gach. 2016. Recent insights into the HIV/AIDS pandemic. *Microbial Cell* **3**:451.

Belizário, J. E. 2009. Immunodeficient mouse models: an overview. *Open Immunology Journal* **2**:79.

Burgener, A., I. McGowan, and N. R. Klatt. 2015. HIV and mucosal barrier interactions: consequences for transmission and pathogenesis. *Current Opinion in Immunology* **36**:22.

Chasela, C. S., et al. 2010. Maternal or infant antiretroviral drugs to reduce HIV-1 transmission. *New England Journal of Medicine* **362**:2271.

Faria, N. R., et al. 2014. HIV epidemiology: the early spread and epidemic ignition of HIV-1 in human populations. *Science* **346**:56.

Fischer, A., L. D. Notarangelo, B. Neven, M. Cavazzana, and J. M. Puck. 2015. Severe combined immunodeficiencies and related disorders. *Nature Reviews Disease Primers* **1**:15061.

Griffith, L. M., et al. 2016. Primary Immune Deficiency Treatment Consortium (PIDTC) update. *Journal of Allergy and Clinical Immunology* **138**:375.

Grimbacher, B., et al. 2016. The crossroads of autoimmunity and immunodeficiency: lessons from polygenic traits and monogenic defects. *Journal of Allergy and Clinical Immunology* **137**:3.

Haynes, B. F., and D. R. Burton. 2017. Developing an HIV vaccine. *Science* **355**:1129.

Hui-Qi, Q., S. P. Fisher-Hoch, and J. B. McCormick. 2011. Molecular immunity to mycobacteria: knowledge from the mutation and phenotype spectrum analysis of Mendelian susceptibility to mycobacterial diseases. *International Journal of Infectious Diseases* **15**:e305.

Kienzler, A.-K., C. E. Hargreaves, and S. Y. Patel. 2017. The role of genomics in common variable immunodeficiency disorders. *Clinical and Experimental Immunology* **188**:326.

O'Keefe, A. W., et al. 2016. Primary immunodeficiency for the primary care provider. *Paediatric Child Health* **21**:e10.

Piacentini, L., et al. 2008. Genetic correlates of protection against HIV infection: the ally within. *Journal of Internal Medicine* **265**:110.

Picard, C., et al. 2015. Primary immunodeficiency diseases: an update on the classification from the International Union of Immunological Societies Expert Committee for Primary Immunodeficiency 2015. *Journal of Clinical Immunology* **35**:696.

Stephenson, K. E., H. T. D'Couto, and D. H. Barouch. 2016. New concepts in HIV-1 vaccine development. *Current Opinion in Immunology* **41**:39.

Thrasher, A. J., and D. A. Williams. 2017. Evolving gene therapy in primary immunodeficiency. *Molecular Therapy* **25**:1132.

Volberding, P. A. 2017. HIV treatment and prevention: an overview of recommendations from the IAS-USA Antiretroviral Guidelines Panel. *Topics in Antiviral Medicine* **25**:17.

Voss, M., and Y. T. Bryceson. 2017. Natural killer cell biology illuminated by primary immunodeficiency syndromes in humans. *Clinical Immunology* **177**:29.

Woodham, A. W., et al. 2016. Human immunodeficiency virus immune cell receptors, coreceptors, and cofactors: implications for prevention and treatment. *AIDS Patient Care and STDs* **30**:291.

World Health Organization. 2016. *Consolidated Guidelines on the Use of Antiretroviral Drugs for Treating and Preventing HIV Infection: Recommendations for a Public Health Approach*, 2nd ed. Available from http://apps.who.int/iris/bitstream/10665/208825/1/9789241549684_eng.pdf?ua=1

Useful Websites

https://www.niaid.nih.gov/diseases-conditions/primary-immune-deficiency-diseases-pidds The National Institute of Allergy and Infectious Diseases (NIAID) maintains a website for learning more about primary immunodeficiency diseases, their treatment, and current research.

www.primaryimmune.org The website of the Immune Deficiency Foundation; this national patient foundation is dedicated to improving the diagnosis, treatment, and quality of life of people with primary immunodeficiency diseases through advocacy, education, and research.

www.usidnet.org The U.S. Immunodeficiency Network (USIDNET) is an NIH-funded research program of the Immune Deficiency Foundation. It was established to advance research in primary immune deficiency diseases and to maintain repositories of patient samples and data.

https://www.niaid.nih.gov/diseases-conditions/hiv-vaccine-development Another NIAID website that includes information about AIDS vaccines and links to documents about vaccines in general.

www.scid.net This website contains links to periodicals and databases with information about SCID.

http://hivinsite.ucsf.edu This HIV/AIDS information site is maintained by the University of California, San Francisco, one of many centers of research in this field.

www.unaids.org/en The most extensive and current information about the global AIDS pandemic, including updated data and international goals and programs, can be accessed from this site.

www.hiv.lanl.gov This website, maintained by the Los Alamos National Laboratory, contains all available sequence data on HIV and SIV along with up-to-date reviews on topics of current interest to AIDS research.

www.aidsinfo.nih.gov This website, maintained by the National Institutes of Health, contains information and national guidelines on the treatment and prevention of AIDS, including current guidelines on antiretroviral therapy.

https://www.avert.org/ AVERT is an organization focused on empowering people to protect themselves and others from HIV and AIDS. This website provides global information on HIV and AIDS, including about transmission and prevention, testing, living with HIV, sexual practices, and sexually transmitted infections.

STUDY QUESTIONS

1. Indicate whether each of the following statements is true or false. If you think a statement is false, explain why.

a. Complete DiGeorge syndrome is a congenital birth defect resulting in absence of the thymus.

b. X-linked agammaglobulinemia (XLA) is a combined B-cell and T-cell immunodeficiency disease.

c. The hallmark of a phagocytic deficiency is increased susceptibility to viral infections.

d. In chronic granulomatous disease, the underlying defect is in phagosome oxidase or an associated protein.

e. Injections of immunoglobulins are given to treat individuals with X-linked agammaglobulinemia.

f. Multiple defects have been identified as causing human SCID.

g. Mice with the SCID defect lack functional B and T lymphocytes.

h. Mice with a SCID-like phenotype can be produced by knockout of *RAG* genes.

i. Children born with SCID often manifest increased infection by encapsulated bacteria in the first months of life.

j. Failure to express MHC class II molecules in bare-lymphocyte syndrome affects cell-mediated immunity only.

2. For each of the following immunodeficiency disorders listed on the left, indicate which treatment(s) listed on the right would be appropriate.

Immunodeficiency disorder	Treatment
a. Chronic granulomatous disease _____	1. Bone marrow or HSC transplantation
b. ADA-deficient SCID _____	2. Intravenous immunoglobulin
c. X-linked agammaglobulinemia _____	3. Recombinant IFN-γ
d. Common cytokine receptor γ chain–deficient SCID _____	4. Recombinant adenosine deaminase
e. Common variable immunodeficiency _____	5. Thymus transplant in an infant

3. Patients with X-linked hyper IgM syndrome express normal genes for other antibody subtypes but fail to produce IgG, IgA, or IgE. Explain how the defect in this syndrome accounts for the lack of other antibody isotypes.

4. Patients with DiGeorge syndrome are born with either no thymus or a severely defective thymus. In the severe form, the patient cannot develop mature helper, cytotoxic, or regulatory T cells. In contrast, if an adult loses his or her thymus through accident or injury, little or no T-cell deficiency is seen. Explain this discrepancy.

5. Infants born with SCID experience severe recurrent infections. The initial manifestation in these infants is typically fungal or viral infections, and only rarely bacterial ones. Why are bacterial infections less of an issue in these newborns?

6. Granulocytes from patients with leukocyte adhesion deficiency (LAD) express greatly reduced amounts of CD11a, b, and c adhesion molecules.

  a. What is the nature of the defect that results in decreased expression of these adhesion molecules in patients with LAD?

  b. What is the normal function of the integrin molecule LFA-1? Give specific examples.

7. How can an inherited defect in the IL-2 receptor cause the demise of developing B cells, as well as T cells, if B cells do not possess receptors for IL-2 signaling?

8. Bare-lymphocyte syndrome mutations that greatly reduce levels of MHC class I are often in genes for proteins involved in loading peptides into the MHC class I peptide–binding groove. Why are these mutations not found in the genes encoding MHC class I proteins?

9. Primary immunodeficiency results from a defective component of the immune response. Typically, this manifests as failure to fight off one or more types of pathogen. Very occasionally, immunodeficiency results in autoimmune syndromes, where the immune response attacks self tissues. Explain how this might occur.

10. Gene therapy—correcting a genetic defect in a patient's own HSCs—is being developed as a treatment for some primary immunodeficiency diseases. What is one advantage of this approach compared with transplantation of donor bone marrow?

11. Indicate whether each of the following statements is true or false. If you think a statement is false, explain why.

  a. HIV-1 and HIV-2 are both believed to have evolved following the species jump of SIV from chimpanzees to humans.

  b. HIV-1 causes immune suppression in both humans and chimpanzees.

  c. SIV is endemic in the African green monkey.

  d. All antiretroviral drugs either prevent viral binding or entry into the cell or the reverse transcription of the viral genome.

  e. T-cell activation increases transcription of the HIV proviral genome.

  f. HIV-infected patients meet the criteria for AIDS diagnosis as soon as their CD4$^+$ T-cell count drops below 500 cells/μl.

  g. The polymerase chain reaction is a sensitive test used to detect antibodies to HIV.

  h. If antiretroviral therapy is successful, the viral load will typically decrease.

12. Various mechanisms have been proposed to account for the decrease in the numbers of CD4$^+$ T cells in HIV-infected individuals. What are several reasons for depletion of CD4$^+$ T cells?

13. Would you expect the viral load in the blood of HIV-infected individuals in the early years of the asymptomatic phase of HIV-1 infection to vary significantly (assuming no drug treatment)? What about CD4$^+$ T-cell counts? Why?

14. If the viral load begins to increase in the blood of an HIV-infected individual and the level of CD4$^+$ T cells decrease, what would this indicate about the infection?

15. Clinicians often monitor the level of skin-test reactivity, or delayed-type hypersensitivity, to commonly encountered infectious agents in HIV-infected individuals. Why do you think these immune reactions are monitored, and what change might you expect to see in skin-test reactivity with progression into AIDS?

16. Certain chemokines have been shown to suppress infection of cells by HIV, and proinflammatory cytokines enhance cell infection. What is the explanation for this?

17. Treatments with combinations of anti-HIV drugs have reduced virus levels significantly in some treated patients and can delay the onset of AIDS. If a patient with AIDS becomes free of opportunistic infection and has no detectable virus in the circulation, can that person be considered cured?

18. Suppose you are a physician who has two HIV-infected patients. Patient B. W. has a bacterial infection of the skin (*S. aureus*), and patient L. S. has a mycobacterial infection. The CD4$^+$ T-cell counts of both patients are about 250 cells/μl. Would you diagnose either patient or both of them as having AIDS?

19. What are two significant challenges to the development of an effective vaccine for HIV?

ANALYZE THE DATA

Common variable immunodeficiency (CVID) causes low concentrations of serum Igs and leads to frequent bacterial infections in the respiratory and gastrointestinal tracts. People with CVID also have an increased prevalence of autoimmune disorders and cancers. Isgrò and colleagues (*Journal of Immunology* 2005;**174**:5074) examined the bone marrow of several individuals with CVID. They looked at the T-cell phenotypes of patients with CVID (as shown in the following table) as well as some of the cytokines made by these individuals (see accompanying graphs).

(a)

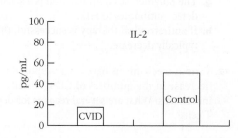

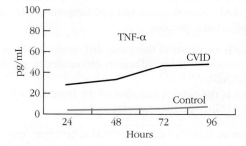

(b)

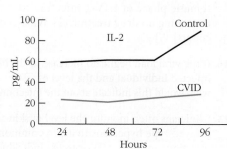

a. What is the impact of CVID on T helper cells?

b. Naïve CTLs require IL-2 production from T helper cells in order to become activated (see Chapter 12). How might CVID impact the generation of CTLs?

c. True or false? CVID inhibits cytokine production. Explain your answer. Speculate on the physiological impact of the cytokine pattern of patients with CVID.

d. Would you predict an effect of CVID on the humoral immune response?

		Patients											**Control**
		1	**2**	**5**	**6**	**7**	**8**	**9**	**10**	**11**	**CVID**	**($n = 10$)**
CD4+	%	47	36	28	28	27	19	19	32	57	34	47.5
	cells/µl	296	234	278	1652	257	361	289	248	982	351	1024
Naïve CD4+	%	4	6.8	25.8	4	11.6	7.9	14	12	2	10	52
	cells/µl	12	16	72	66	30	29	40	30	20	31	519
Activated CD4+	%	2	5	22	3	9	8	12	9	1.5	8	37
	cells/µl	6	12	61	50	23	29	35	22	15	25	385
CD8+	%	31	38	30	57	44	56	47	45	21	39	20
	cells/µl	195	247	298	3363	420	1065	714	348	362	414	404
Naïve CD8+	%	25	30	43.9	7	16.9	12.9	20	13	15	22	58
	cells/µl	49	74	131	235	71	138	143	45	54	88	233

CLINICAL FOCUS QUESTION

The spread of HIV/AIDS from infected mothers to infants can be reduced by antiretroviral therapy for both the mother and the infant. Why are combination therapies recommended for the mother, and why is it recommended that newborns begin antiretroviral therapy as soon as possible after birth and proceed for several weeks or longer?

Cancer and the Immune System

Learning Objectives

After reading this chapter, you should be able to:

1. Demonstrate how specific innate and adaptive immune response pathways and molecules are involved in the recognition and eradication of cancer cells, and compare this with other pathways and molecules that might foster a more protumor microenvironment.

2. Explain the concept of central and peripheral immune tolerance as it relates to cancer cells, comparing the impact of this on the immune recognition of tumor-associated antigens versus tumor-specific antigens.

3. Explain the various phases of cancer immunoediting, and be able to predict the stage or prognosis when given a new example, designing a test to confirm your prediction.

4. Critically evaluate and estimate the effectiveness of a new immune-based anticancer therapy in terms of recognition and destruction by the immune system.

5. Formulate a novel immunotherapy that might be effective against a specific type of cancer, using one or more of the strategies and approaches described.

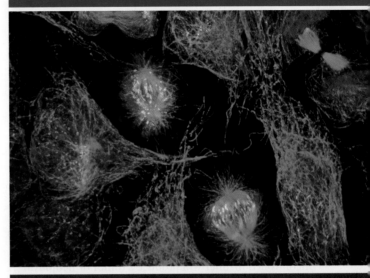

A cluster of proliferating cervical carcinoma cells, stained to visualize the DNA (blue), centromeres (red), and cytoskeleton (green). *[Jennifer C. Waters/Science Source.]*

As the death toll from infectious disease has declined in the Western world, cancer has become the second leading cause of death, topped only by heart disease. Current estimates project that almost half of all men and one in three women in the United States will develop cancer at some point in their lifetimes, and that 20% to 25% of us will die of some form of this disease. From an immunologic perspective, cancer cells can be viewed as altered self cells that have escaped normal growth-regulating mechanisms. This makes recognition and response by the immune system challenging; while we want to be rid of these cells, we also do not want to mistakenly stimulate an autoimmune attack on healthy tissues. The good news is that surveillance for cancer cells is one of the regular maintenance roles of the immune system. The bad news is that cancers, as a rule, find ways to avoid this immune recognition.

In most organs and tissues of a mature animal, a constant balance is struck between cell renewal and cell death. Most mature cells in the body have a given life span (which varies enormously). As cells die, they are replaced by new cells formed from the proliferation and differentiation of local adult stem cells. This cell growth and proliferation are essential for wound healing and homeostasis. Under normal circumstances in the adult, the production of new cells is regulated so that

Key Terms

Neoplasm
Malignant
Metastasis
Transformation
Proto-oncogene
Oncogenes

Tumor-suppressor genes
Tumor-specific antigens (TSAs)
Tumor-associated antigens (TAAs)

Immunoediting
Immunotherapy
Adjuvant cancer therapy
Neoadjuvant cancer therapies

Neoantigens
Checkpoint blockade
CARs (CAR T cells)

the number of any particular type of cell remains fairly constant. Occasionally, however, a cell arises that no longer responds to normal growth control mechanisms; it proliferates in an unregulated manner and avoids apoptotic signals, eventually giving rise to cancer.

This chapter opens with some background on cancer and introduces the most common types of the disease. This is followed by extensive discussions of how the immune response deals with cells that become cancerous, and how these cells avoid engagement with the immune system. Finally, we touch on the vast range of therapies available to treat cancer, focusing on a host of new immunotherapies.

Terminology and the Formation of Cancer

Cells that give rise to progeny with the ability to expand in an uncontrolled manner will produce a tumor or **neoplasm**. A tumor that is not capable of indefinite growth and does not invade the healthy surrounding tissue is said to be **benign**. A tumor that continues to grow and becomes progressively more invasive is called **malignant**. The term *cancer* refers specifically to a *malignant tumor*. In addition to uncontrolled growth, most malignant tumors eventually exhibit **metastasis** or spread, whereby small clusters of cancerous cells dislodge from the original tumor, invade the blood or lymphatic vessels, and are carried to distant sites where they

take up residence and continue to proliferate. In this way, a primary tumor at one site can give rise to a secondary tumor at another site (**Overview Figure 19-1**).

Cancers are classified according to the embryonic origin of the tissue from which they arise. Most (80% to 90%) are **carcinomas**, tumors that develop from epithelial origins such as skin, gut, or the epithelial lining of internal organs and glands. Skin cancers and the majority of cancers of the colon, breast, prostate, and lung are carcinomas. **Sarcomas** arise less frequently and are derived from mesodermal connective tissues, such as bone, fat, and cartilage. These represent a small minority of cancers (about 1%) and are grouped into soft tissue sarcomas and osteosarcomas, or bone cancers. Last, there are cancers derived from blood cells, which make up approximately 9% of all malignancies. These can arise at multiple differentiated or undifferentiated stages of development.

Lymphomas, **myelomas**, and **leukemias** are all considered blood cell cancers, or malignancies arising in one of the many cell types derived from hematopoietic stem cells. What distinguishes them is the stage of differentiation, or where along the pathway from HSC to mature blood cell they arise. This can get complicated because, as we know from earlier chapters, leukocytes spend time in the bone marrow, secondary lymphoid tissues, the blood circulation, and even as mature residents in tissues. For this reason there is some overlap in the cell types and symptoms for these three types of blood cell cancer. In short, leukemias arise from cells that are still in their early stages of development in the bone marrow. These are classified as either myelogenous or lymphocytic, depending on which branch of the common

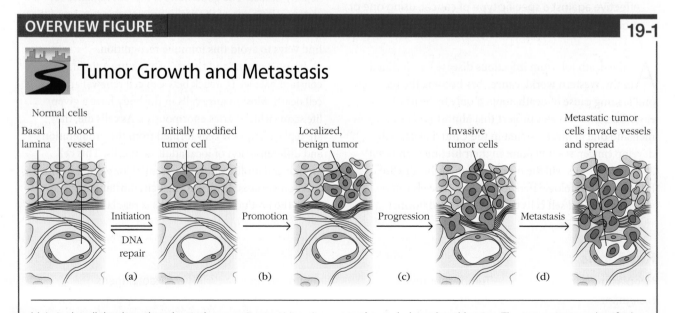

OVERVIEW FIGURE 19-1

Tumor Growth and Metastasis

(a) A single cell develops altered growth properties at a tissue site, which may be corrected via DNA repair. (b) The altered cell proliferates, forming a mass of localized tumor cells, or a benign tumor. (c) The tumor cells become progressively more invasive, spreading to the underlying basal lamina. The tumor is now classified as malignant. (d) The malignant tumor metastasizes by generating small clusters of cancer cells that dislodge from the tumor and are carried by the blood or lymph to other sites in the body.

progenitor they derive from. They are also classified as acute or chronic according to the clinical progression of the disease. Acute leukemia appears suddenly and progresses rapidly, whereas chronic leukemia is much less aggressive and develops slowly as a mild, barely symptomatic disease. These clinical distinctions apply to untreated leukemias; with current treatments, the acute leukemias often have a good prognosis, and permanent remission is possible.

Lymphomas and myelomas arise at later stages of blood cell development, typically after progenitors have migrated out of the bone marrow (although some of these cancerous cells will return to the bone marrow later). Lymphomas tend to spread through the lymphatic system and proliferate in secondary lymphoid organs; a common early symptom is one or more swollen lymph nodes. Myelomas arise from uncontrolled proliferation of fully differentiated B cells (plasma cells), producing a mutated form of antibody that can be found in the blood, called *M protein*. These cancers migrate to the bone marrow where they take up residence, often leading to bone pain and anemia. The term *multiple myeloma* comes from the fact that most affected individuals will present with multiple bone lesions, or more than one location where these cancerous plasma cells have taken up residence in bones.

Much has been learned about cancer from in vitro studies of primary cells, which are cells that have been isolated directly from the host and have a limited life span. Treatment of normal cultured cells with specific chemical or physical agents, irradiation, and certain viruses can alter their morphology and growth properties. In some cases, this process leads to unregulated growth and produces cells capable of growing as tumors when they are injected into animals. Such cells are said to have undergone **transformation**, or malignant transformation, and the agents that lead to cellular transformation are commonly referred to as **carcinogens**. Transformed cells often exhibit properties in vitro similar to those of cancer cells that form in vivo. For example, they have decreased requirements for the survival factors needed by most cells (such as growth factors and serum), are no longer anchorage or cell-contact dependent, and grow in a density-independent fashion with the ability to avoid apoptotic signals. Moreover, both cancer cells and transformed cells can be subcultured in vitro indefinitely; that is, for all practical purposes, they are immortal. Because of the similar properties of cancer cells and in vitro–transformed cells, the process of malignant transformation has been studied extensively as a model of cancer induction.

Accumulated DNA Alterations or Translocation Can Induce Cancer

Transformation can be induced by various chemical substances (such as formaldehyde, benzene, and some pesticides), physical agents (e.g., asbestos), and ionizing radiation; all are linked to DNA mutations. Infection with certain viruses, most of which share the property of

TABLE 19-1	Common human infectious carcinogens		
Viral agent		**Type**	**Cancer**
HTLV-1 (human T-cell leukemia virus-1)		RNA	Adult T-cell leukemia or lymphoma
HHV-8 (human herpesvirus-8)		DNA	Kaposi's sarcoma (especially in HIV⁺ patients)
HPV (human papillomavirus)		DNA	Cervical carcinoma
HBV and HCV (hepatitis B and C viruses)		DNA	Liver carcinoma
EBV (Epstein-Barr virus)		DNA	Burkitt's lymphoma and nasopharyngeal carcinoma

Data sources: The National Institute for Occupational Safety and Health (NIOSH) and the Centers for Disease Control and Prevention (CDC) (see http://www.cdc.gov/niosh/topics/cancer/).

integrating into the host cell genome and disrupting chromosomal DNA, can also lead to transformation. **Table 19-1** lists the most common viruses associated with cancer. Although the activity of each of these agents is associated with cancer initiation, thanks to the variety of DNA repair mechanisms present in our cells, exposure to carcinogens does not always lead to cancer. Instead, a confluence of factors must occur before enough changes to normal cellular genes arise to induce malignant transformation, as we discuss further in the following sections.

Some transformations arise when chromosomal translocations alter the expression and regulation of certain genes. For instance, a number of B- and T-cell leukemias and lymphomas arise from translocations involving immunoglobulin or T-cell receptor loci, very transcriptionally active regions in these cells, to other areas of the DNA. One of the best characterized is the translocation of *myc* from its position on chromosome 8 to the immunoglobulin heavy-chain enhancer region on chromosome 14, accounting for 75% of Burkitt's lymphoma cases (**Figure 19-2**). In the remaining patients with Burkitt's lymphoma, *myc* remains on chromosome 8 but the κ or λ light-chain genes are translocated to the tip of chromosome 8, near the *myc* gene region and inducing a similar change. As a result of these translocations, synthesis of the Myc protein, which functions as a transcription factor controlling the behavior of many genes involved in cell growth and proliferation, increases and allows unregulated cellular growth.

Key Concept:

• Cancer-causing agents (carcinogens) and random events that damage DNA, including some viral infections and chromosomal translocations, can accumulate into a condition that allows unregulated cell proliferation and cellular transformation.

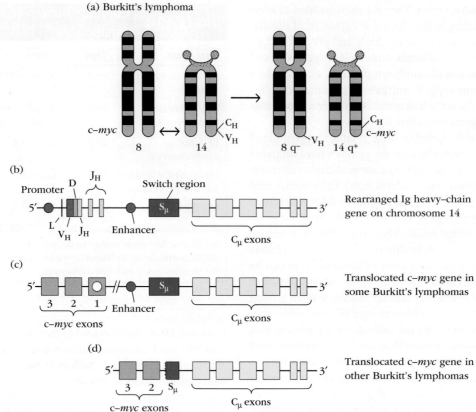

FIGURE 19-2 Chromosomal translocations resulting in Burkitt's lymphoma. (a) Most cases of Burkitt's lymphoma arise following a chromosomal translocation that moves part of chromosome 8, containing the c-myc gene, to the Ig heavy-chain locus on chromosome 14, shown in more detail in (b). (c) In some cases, the entire c-myc gene is inserted near the Ig heavy-chain enhancer. (d) In other cases, only the coding exons (2 and 3) of c-myc are inserted at the μ switch site. The proto-oncogene myc encodes a transcription factor involved in the regulation of many genes with roles in cell growth, proliferation, and inhibition of apoptosis.

Genes Associated with Cancer Control Cell Proliferation and Survival

Normal tissues maintain homeostasis through a tightly controlled process of cell proliferation balanced by regulated cell death, or apoptosis. An imbalance at either end of the scale can encourage development of a tumor. The genes involved in these homeostatic processes work by producing proteins that either *encourage* or *discourage* cellular proliferation and survival. Not surprisingly, it is the disruption of these same growth-regulating genes that accounts for most, if not all, forms of cancer. The activities of these proteins can occur anywhere in the cell cycle pathway, from signaling events at the surface of the cell, to intracellular signal transduction processes and nuclear events.

Normal cellular genes that are associated with the formation of cancer fall into three major categories based on their activities (**Table 19-2**):

- **Oncogenes**—sequences that encourage growth and proliferation

- **Tumor-suppressor genes**—sequences that discourage or inhibit cell proliferation
- **Apoptosis genes**—sequences that control programmed cell death

Oncogenes are involved in cell growth–promoting processes, while tumor-suppressor genes play the opposite role in homeostasis: dampening cellular growth and proliferation. Unlike oncogenes, which become the villain when their activity is enhanced, tumor-suppressor genes, also known as anti-oncogenes, become involved in cancer induction when they *fail*. Finally, many of the genes involved in apoptosis are also associated with cancer, as these genes either enforce or inhibit the cell death signals.

Oncogenes

A **proto-oncogene** is a normal cellular gene involved in some aspect of cell growth and proliferation. When proto-oncogenes are mutated or subject to dysregulation, the change in expression can lead to uncontrolled cell proliferation,

TABLE 19-2	Functional classification of cancer-associated genes
Type/name	**Nature of gene product**
CATEGORY I: PROTO-ONCOGENES THAT INDUCE CELLULAR PROLIFERATION	
Growth factors	
sis	A form of platelet-derived growth factor (PDGF)
Growth factor receptors	
fms	Receptor for colony-stimulating factor 1 (CSF-1)
erbB	Receptor for epidermal growth factor (EGF)
neu	Protein (HER2) related to EGF receptor
erbA	Receptor for thyroid hormone
Signal transducers	
src	Tyrosine kinase
abl	Tyrosine kinase
Ha-*ras*	GTP-binding protein with GTPase activity
N-*ras*	GTP-binding protein with GTPase activity
K-*ras*	GTP-binding protein with GTPase activity
Transcription factors	
jun	Component of transcription factor AP1
fos	Component of transcription factor AP1
myc	DNA-binding protein
CATEGORY II: TUMOR SUPRESSOR GENES, INHIBITORS OF CELLULAR PROLIFERATION*	
Rb	Suppressor of retinoblastoma
TP53	Nuclear phosphoprotein that inhibits formation of small-cell lung cancer and colon cancers
DCC	Suppressor of colon carcinoma
APC	Suppressor of adenomatous polyposis
NF1	Suppressor of neurofibromatosis
WT1	Suppressor of Wilms' tumor
CATEGORY III: GENES THAT REGULATE PROGRAMMED CELL DEATH OR APOPTOSIS	
bcl-2	Suppressor of apoptosis
Bcl-x$_L$	Suppressor of apoptosis
Bax	Inducer of apoptosis
Bim	Inducer of apoptosis
Puma	Inducer of apoptosis

*The activity of the normal products of the category II genes inhibits progression of the cell cycle. In most cases, loss or inactivation of both copies of a tumor suppressor gene is required for development of cancer.

or cancer. The mutated form of a proto-oncogene that can induce cancer is called an **oncogene**, from the Greek word *ónkos* (which means "mass" or "tumor").

One category of proto-oncogenes encodes growth factors and growth factor receptors. In normal cells, the expression of growth factors and their receptors is carefully regulated. Inappropriate expression of either a growth factor or its receptor can result in uncontrolled proliferation. For instance, in some breast cancers, increased synthesis of the growth factor receptor encoded by c-*neu* has been linked with cancer development and disease progression. Included within this category are the genes *fms, erbA,* and *erbB,* all of which encode growth factor receptors (see Table 19-2). Mutations in cancer-associated genes may be a major mechanism by which chemical carcinogens or x-irradiation convert a proto-oncogene into a cancer-inducing oncogene. For instance, single-point mutations in c-*ras*, which encodes a GTPase, have been detected in carcinomas of the bladder, colon, and lung. Alterations that result in overactivity of the *ras* oncogene, part of the epidermal growth factor (EGF) receptor signaling pathway, are seen in up to 30% of all human cancers and nearly 90% of all pancreatic cancers.

Likewise, signal transduction processes are another potential target for disruptions that can lead to cancer. The *src* and *abl* oncogenes encode tyrosine kinases. The products of these genes act as signal transducers, upstream of transcription factors acting on their DNA targets. Not surprisingly, *myc*, *jun*, and *fos* are all oncogenes that encode transcription factors involved in cell proliferation and cell cycle progression. Overactivity of any of the genes in these common pathways, from ligand binding to signal transduction and transcription factor translocation to the nucleus, can thus result in unregulated cell proliferation.

Finally, some infectious agents can themselves act as initiators of uncontrolled cell proliferation. In particular, viral integration into the host-cell genome may serve to transform a proto-oncogene into an oncogene. For example, *avian leukosis virus* (ALV) is a retrovirus (RNA virus) that does not carry any viral oncogenes, yet it is able to transform B cells into lymphomas. This particular retrovirus has been shown to integrate within the c-*myc* proto-oncogene, resulting in increased synthesis of the Myc protein. Some DNA viruses have also been associated with malignant transformation, such as certain serotypes of the *human papillomavirus* (HPV), which are linked with cervical cancer (see Table 19-1).

Tumor-Suppressor Genes

Tumor-suppressor genes, or anti-oncogenes, encode proteins that *inhibit cell proliferation*. In their normal state, tumor-suppressor genes prevent cells from progressing through the cell cycle inappropriately, functioning like brakes on a car. A release of this inhibition is what can lead to cancer induction. The prototype of this category of oncogenes is *Rb*, the retinoblastoma gene (see Table 19-2). Hereditary retinoblastoma is a rare childhood cancer in which tumors develop from neural precursor cells in the immature retina. The affected child inherits a mutated *Rb* allele; later, somatic inactivation of the remaining *Rb* allele is what leads to tumor growth. Unlike oncogenes where a single allele alteration can lead to unregulated growth, tumor-suppressor genes require a "two-hit" disabling sequence, as one functional allele is typically sufficient to suppress the development of cancer.

Probably the single most frequent genetic abnormality in human cancer, found in 60% of all tumors, is a mutation in the *TP53* gene. This tumor-suppressor gene encodes p53, a nuclear phosphoprotein with multiple cellular roles, including growth arrest, DNA repair, and apoptosis. Over 90% of small-cell lung cancers and over 50% of breast and colon cancers have been shown to express mutations in *TP53*.

Apoptosis Genes

A third category of cancer-associated genes is sequences involved in programmed cell death, or apoptosis. Genes associated with apoptosis can act as either inhibitors or promoters of this pathway. Pro-apoptotic genes act like tumor suppressors, normally inhibiting cell survival, whereas anti-apoptotic genes behave more like oncogenes, promoting cell survival. Thus, a failure of the former or overactivity of the latter can encourage neoplastic transformation of cells.

Included in this category are genes such as *bcl-2*, an anti-apoptosis gene (see Table 19-2). This oncogene was originally discovered because of its association with B-cell follicular lymphoma. Since its discovery, *bcl-2* has been shown to play an important role in regulating cell survival during hematopoiesis and in the survival of selected B cells and T cells during maturation. Interestingly, the Epstein-Barr virus, which can cause infectious mononucleosis, contains a gene that has sequence homology to *bcl-2* and may act in a similar manner to suppress apoptosis. Like tumor suppressors, in many cases the products of these genes function as "checkpoints" in the cell cycle.

Key Concepts:

- Oncogenes are typically DNA sequences that encode proteins involved in promoting cell growth and proliferation, such as transcription factors, growth factors (and their receptors), or intracellular signaling molecules.

- Tumor suppressors act naturally as anti-oncogenes or inhibitors of cell growth and proliferation, becoming agents of transformation when they fail to function, such as during DNA repair or in blocking progression through cell cycle checkpoints.

- When the DNA sequences involved in apoptosis, or programmed cell death, are not functioning properly they also can aid in cellular transformation, encouraging the survival and proliferation of neoplastic cells.

Malignant Transformation Involves Multiple Steps

Promotion of a cancerous state doesn't just happen overnight. Instead, it arises from a series of small changes, each taking the cell one step closer to uncontrolled replication. This multistep process of clonal evolution is driven by a series of somatic mutations. In fact, the early genetic changes that begin this cascade are sometimes called "driver" mutations, which are later followed by "passenger" mutations that come along for the ride but also foster the development of cancer. Clinical investigations in humans have suggested that as few as two and as many as eight key genes may serve as the original drivers for the development of cancer, with a long list of possible passenger mutations that, while less critical to initiation, also contribute to malignancy. Collectively, these genetic changes progressively convert the cell from normal growth to a malignant state where each of the regular checkpoints that control proliferation has been surmounted.

Sequential Genetic Disruptions

Induction of malignant transformation appears to evolve in distinct phases, referred to as initiation, promotion, progression, and metastasis. *Initiation* involves changes or mutations in the genome that alter cell proliferation potential but do not, in themselves, lead to malignant transformation. The next stage, *promotion*, occurs when preneoplastic cells gradually begin to accumulate. At this stage tumor size is generally small and the cells are still amenable to repair mechanisms. Importantly, these first two stages in the transformation process can last for long periods and the cells are still susceptible to immune-mediated detection and chemopreventive agents, with the potential to reverse the course of disease.

Progression tends to move more quickly than the previous two phases. Genetic alterations occurring here allow for rampant cell proliferation and the acquisition of new mutations to potential cancer-promoting genes, exacerbating the cycle. As expected, tumor sizes can grow rapidly in this stage. When one or more of these rapidly dividing cells acquire mutations that allow invasion of nearby tissue, the situation has progressed to the final stage, *metastasis*. By definition, metastatic cancers that come from solid tissues have lost adhesion with neighboring cells and no longer exhibit contact inhibition. This allows them to move outside the original site and even enter blood or lymphatic vessels, where they can then spread through the body. One example that clearly illustrates the type of natural progression that can lead to cellular transformation comes from human colon cancer, which typically develops in a series of well-defined morphologic stages (**Figure 19-3**). Colon cancer begins as small, benign tumors called *adenomas* in the colorectal epithelium. These precancerous

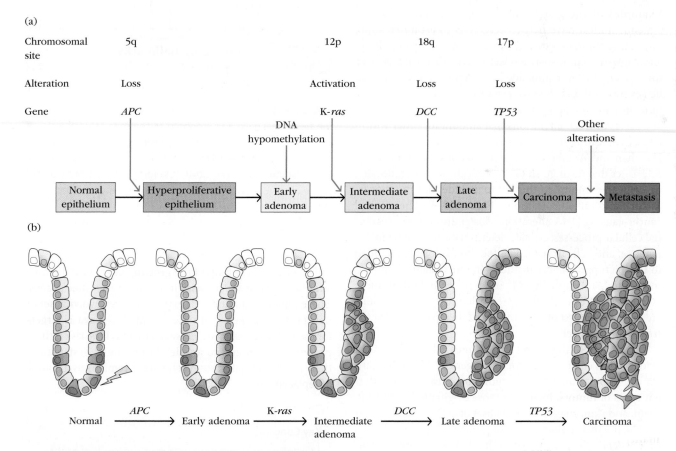

FIGURE 19-3 Model of sequential genetic alterations leading to metastatic colon cancer. Each of the stages indicated in (a) is morphologically distinct, as illustrated in (b), allowing researchers to determine the sequence of genetic alterations. In this sequence, benign colorectal polyps can progress to carcinoma following mutations resulting in the inactivation or loss of three tumor-suppressor genes (*APC*, *DCC*, and *TP53*) and the activation of one oncogene linked to cellular proliferation (*K-ras*).

tumors grow, gradually displaying increasing levels of intracellular disorganization until they acquire the malignant phenotype, or carcinoma (see Figure 19-3b). The morphologic stages of colon cancer have been correlated with a sequence of gene changes (see Figure 19-3a) involving inactivation or loss of three tumor-suppressor genes (*APC*, *DCC*, and *TP53*) and activation of one cellular proliferation oncogene (K-*ras*).

Studies with transgenic mice also support the role of multiple steps in the induction of cancer. Transgenic mice expressing high levels of Bcl-2, a protein encoded by the anti-apoptotic gene *bcl-2*, develop a population of small resting B cells (derived from secondary lymphoid follicles) that have greatly extended life spans. Gradually, these transgenic mice develop lymphomas. Analysis of lymphomas from these animals has shown that approximately half have a c-*myc* translocation (a proto-oncogene) to the immunoglobulin H-chain locus. The synergism of Myc and Bcl-2 is highlighted in double-transgenic mice produced by mating the *bcl-2*⁺ transgenic mice with *myc*⁺ transgenic mice. These mice develop leukemia very rapidly.

Hallmarks of Cancer

Examples of the genetic mutations typical of cellular transformation have helped scientists to establish some interesting common denominators for cancer. By definition, all neoplastic cells display a selective growth advantage over their peers. However, imbedded within this simple statement lie several unifying themes. Data from genomic studies in humans and animals have coalesced into a picture of cancer as a series of DNA alterations that amount to "order within chaos," all leading to the same ultimate endpoint. In 2000, Hanahan and Weinberg suggested a set of six defining characteristics that distinguish the cancer cell, and a decade later, in 2011, they modified this list to include four more overarching and linked conditions (**Figure 19-4**). The original cancer-causing DNA alterations cluster around three essential cellular processes: cell fate determination, genome maintenance, and cell survival. The four conditions later added to this picture of cancer include genome instability, altered metabolic pathways, chronic inflammation, and immune avoidance patterns. As discussed below, these observations are linked to a burst of immune-based therapies aimed at the treatment of cancer.

In the face of these unifying themes related to cancer, evidence has emerged also for heterogeneity within the cancer cell population. Clinical studies of at least three different types of tumors, including those originating in the gut, brain, and skin, suggest that a subset of cells within a tumor may be the real engines of tumor growth. This subset, called *cancer stem cells*, displays true unlimited regenerative potential and is the major producer of new cells to feed the tumor. Nonstem cells each undergo unique mutations and differentiate in novel ways. This gives cancer an advantage in terms of renewal potential and immune evasion. Nonstem

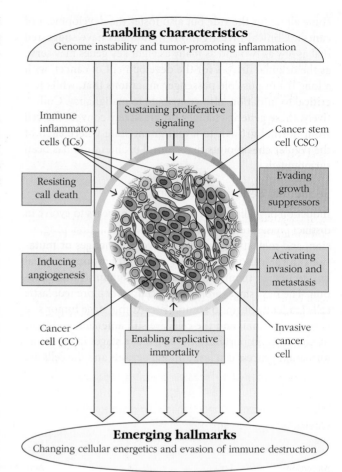

FIGURE 19-4 Hallmarks of cancer. Shown in the center are the six original hallmarks proposed as common characteristics of cancer. Later, two enabling characteristics were added as an overlay or microenvironmental factor: genome instability and protumor inflammation. Finally, two newly emerging hallmarks were added to the observed cancer profile: changes in cellular energetics and immune evasion.

cells constitute the bulk of the growing edges of the tumor and thus serve as the primary immune targets—like decoys. These rapidly mutating cells express an ever-evolving set of new protein markers, with the potential to serve as targets for the immune response. However, there is little risk if these proteins are recognized and lead to immune destruction; their undifferentiated stem cell parent remains as a source of replacement.

> **Key Concepts:**
> • Cellular transformation occurs as the result of multiple gene mutations that accumulate in several genes over time, and gradually subvert the normal checks on cell growth and survival.

- Malignant cells display alterations in key cellular processes and microenvironmental conditions: cell fate decisions, genome maintenance, cell survival, genetic instability, metabolic changes, and immune response patterns.

Tumor Antigens

As menacing as they may be for the host, neoplastic cells are still self cells. As such, all or most of the antigens associated with these cells are subject to the same tolerance-inducing processes that maintain homeostasis and inhibit the development of autoimmunity elsewhere in the body. However, in some instances, cancer cells may produce unique or inappropriately expressed antigens that can be detected by the immune system. Collectively, these are called *tumor antigens*. Most tumor antigens give rise to peptides that are recognized by the immune system via presentation by self MHC class I molecules (endogenous pathway; see Chapter 7). In fact, many of these antigens have been identified by their ability to induce the proliferation of antigen-specific CTLs. Tumor antigens recognized by human T cells fall into one of four groups based on their source:

- Antigens encoded by genes exclusively expressed by tumors
- Antigens encoded by variant forms of normal genes that are altered by mutation
- Antigens normally expressed only at certain stages of development
- Antigens that are overexpressed in particular tumors

These four sources of tumor antigens can be categorized by their level of uniqueness into two groups that have relevance for later forms of therapy: **tumor-specific antigens (TSAs;** the first two above) and **tumor-associated antigens (TAAs;** the second two above). **Table 19-3** lists several categories of common tumor antigens. As one can imagine, many clinical research studies aim to utilize these antigens as diagnostic or prognostic indicators, as well as therapeutic targets for tumor elimination. As one might expect, TSAs can help us target cancer for destruction with little collateral damage. Alas, these are in the minority in terms of actionable tumor antigen targets.

Tumor-Specific Antigens Contain Unique Sequences

Tumor-specific antigens are unique proteins that may result from DNA mutations in tumor cells that generate altered proteins and, therefore, *new nonself antigens or epitopes*. Cytosolic processing of these proteins then gives rise to novel peptides that are presented with MHC class I molecules

TABLE 19-3	Examples of common tumor antigens	
Category	**Antigen(s)**	**Associated cancer type(s)**
Tumor-specific antigens (TSAs)		
Viral	HPV: L1, E6, E7	Cervical carcinoma
	HBV: HBsAg	Hepatocellular carcinoma
	SV40: TAg	Malignant pleural mesothelioma (cancer of the lung lining)
Tumor-associated antigens (TAAs)		
Over-expression	MUC1	Breast, ovarian
	MUC13/ CA-125	Ovarian
	HER-2/neu	Breast, melanoma, ovarian, gastric, pancreatic
	MAGE	Melanoma
	PSMA	Prostate
	TPD52	Prostate, breast, ovarian
	PSA	Prostate
	PAP	Prostate
Differentiation Stage	CEA	Colon
	gp100	Melanoma
	AFP	Hepatocellular carcinoma
	Tyrosinase	Melanoma

Abbreviations: SV40, simian virus 40; L, late gene; E, early gene; HBsAg, hepatitis B surface antigen; TAg, large tumor antigen; MUC, mucin; MAGE, melanoma-associated antigen; HER2/neu, human epidermal receptor/neurological; PSMA, prostate-specific membrane antigen; TP, tumor protein; PSA, prostate-specific antigen; PAP, prostatic acid phosphatase.

Data source: Table 1 in Aldrich, J. F., et al. 2010. Vaccines and immunotherapeutics for the treatment of malignant disease. *Clinical and Developmental Immunology* **2010**:697158.

(**Figure 19-5**, cell on the left), which can lead to induction of a cell-mediated response by tumor-specific CD8$^+$ T cells. In this way, TSAs contain some "foreign" or new peptides, and thus become natural targets for immune recognition.

TSAs can be found in some virally induced tumors, where sequences from the infecting virus are recognized by the immune system. These unique sequences will be shared by all tumors induced by the same virus, making their characterization simpler. For example, when mice are injected with killed polyoma virus–induced tumor cells from a syngeneic mouse (see Chapter 7) the recipients are protected against subsequent challenge with live tumor cells from any polyoma-induced tumors. This suggests that the mice mounted an immune response against virus-specific antigens present on these tumor cells. Likewise, when lymphocytes are transferred from mice with a virus-induced tumor into normal syngeneic recipients, the recipients reject subsequent transplants of all syngeneic

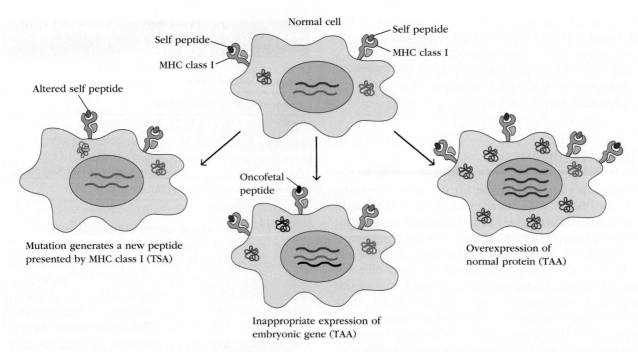

FIGURE 19-5 Different mechanisms generate tumor-specific antigens (TSAs) and tumor-associated antigens (TAAs). TSAs are unique to tumor cells, and can result from DNA mutations that lead to expression of altered self proteins (called *neoantigens*) or, as is more common, from expression of viral antigens (not shown) in transformed cells. TAAs are more common and represent normal cellular proteins displaying unique expression patterns, such as an embryonic protein expressed in the adult or overexpression of self proteins. Both types of tumor antigens can be detected by the immune system following presentation in MHC class I.

tumors induced by the same virus. This adoptive transfer experiment shows that lymphocytes taken from the tumor-bearing mouse recognize and kill cells expressing a virally derived TSA.

In some cases, the presence of virus-specific tumor antigens is an indicator of neoplastic transformation. In humans, Burkitt's lymphoma cells have been shown to express a nuclear antigen of the Epstein-Barr virus that may indeed be a tumor-specific antigen for this type of tumor. HPV E6 and E7 proteins are found in more than 80% of invasive cervical cancers, and they provide the clearest example of a virally encoded TSA. In fact, the first clinically approved cancer vaccine is one used to prevent infection with HPV and block the emergence of cervical cancer (see **Clinical Focus Box 19-1**). This is arguably the closest we have come to a "cancer cure."

Demonstrating the presence of TSAs on spontaneously occurring or chemically induced tumors has been more difficult. Naturally, these antigens can be quite diverse and are identified only by their ability to induce T cell–mediated rejection. The immune response to such tumors typically eliminates all of the tumor cells bearing sufficient numbers of these unique antigens, and thus selects for cells bearing few or no antigens (i.e., TSA-negative cells). Nonetheless, experimental methods have been developed to facilitate the characterization of TSAs that have been shown to differ

from normal cellular proteins by as little as a single amino acid. Characterization of numerous TSAs has demonstrated that many of these antigens are not cell membrane proteins; rather, they are short peptides derived from cytosolic proteins, processed and presented together with MHC class I molecules on the cell surface.

Key Concept:

• Tumor-specific antigens are nonself proteins that arise from DNA mutation or cancer-causing viruses, yielding peptides recognized by the immune system as foreign.

Tumor-Associated Antigens Are Normal Cellular Proteins with Unique Expression Patterns

In contrast to TSAs, tumor-associated antigens, or TAAs, are not unique to neoplastic cells. Instead, these represent normal cellular proteins and thus are prone to the usual self-tolerance mechanisms. Most TAAs are proteins expressed only during specific developmental stages, such as in the fetus, or at extremely low levels, but which become induced or up-regulated in tumor cells (see the middle cell in Figure 19-5). Those derived from reactivation of certain fetal or embryonic genes, called **oncofetal tumor antigens**, normally only appear early in embryonic development, before the immune system acquires immunocompetence. When transformation of cells causes these fetal proteins to appear at later stages of development on the neoplastic cells of the adult, they can be recognized as aberrant and induce an immunologic response.

Two well-studied oncofetal antigens are α-**fetoprotein (AFP)** and **carcinoembryonic antigen (CEA)**. AFP, the most abundant fetal protein, ranges from 100 μg/ml in fetal serum to an average of less than 6 ng/ml in adults. Elevated levels of this glycoprotein can also be found in women, especially during the early stages of pregnancy. Significantly elevated AFP levels in nonpregnant adults is seen in certain types of cancer, especially liver, ovarian, and testicular cancers, where serum levels above 500 ng/ml can be indicative of small lesions even in asymptomatic individuals. Monitoring AFP levels can help clinicians make prognoses and evaluate treatment efficacy, especially in liver cancer.

CEA is another oncofetal membrane glycoprotein found on gastrointestinal and liver cells of 2- to 6-month-old fetuses. Approximately 90% of patients with advanced colorectal cancer and 50% of patients with early colorectal cancer have increased levels of CEA in their serum; some patients with other types of cancer also exhibit increased CEA levels. However, because AFP and CEA can be found in trace amounts in some normal adults and in some noncancerous disease states, the presence of these oncofetal antigens is not necessarily diagnostic of tumors but can still be used to monitor tumor growth. If, for example, a patient has had surgery to remove a colorectal carcinoma, CEA levels are monitored after surgery; an increase in the CEA level is an indication that tumor growth may have resumed.

In addition to embryonic antigens, the category of TAAs also includes the products of some oncogenes, such as growth factors and growth factor receptors. These proteins, although transcribed in the healthy adult, are normally tightly regulated and expressed only at low levels. For instance, a variety of tumor cells express the *epidermal growth factor* (EGF) receptor at levels 100 times greater than in normal cells. Another, melanotransferrin, designated p97, has fibroblast growth factor–like activities. Whereas normal cells express fewer than 8000 molecules of p97 per cell, melanoma cells express 50,000 to 500,000 molecules per cell. The gene that encodes p97 has been cloned, and a recombinant vaccinia virus vaccine has been prepared that carries the cloned gene. When this vaccine was injected into mice, it induced both humoral and cell-mediated immune responses, which protected the mice against live melanoma cells expressing the p97 antigen. Targeting such oncogene products has yielded some significant clinical success, such

as in the case of breast cancers overexpressing HER2 (discussed further in the final section of this chapter).

Key Concept:

- Cancer-associated antigens are normal cellular proteins that display abnormal expression patterns and thus are not foreign. In some instances these can be recognized by the immune system, or used clinically to monitor or therapeutically target the cancer.

The Immune Response to Cancer

For over a century controversy raged over whether and how the immune system participates in cancer recognition and destruction. Some said that the immune response played no role, since these cancer cells arise from self and are therefore protected from immune recognition by ongoing tolerance processes. Others insisted that one of the essential roles of the immune response was to protect us from cancer. Data collected in the past few decades from both animal models and clinical studies have clearly defined a role for the immune response in tumor cell identification and eradication. The evolving data concerning the relationship between the immune system and cancer will be covered in this section.

However, before we discuss the role of the immune system it is worth mentioning that there are several intrinsic and extrinsic mechanisms designed to *prevent* cancer. One example of an intrinsic system is the nucleotide excision repair (NER) pathway, which encourages cell senescence (permanent cell cycle arrest) and DNA repair, or even apoptosis, at the first signs of unregulated growth. Should this or other intrinsic systems fail there are still extrinsic or cell external control mechanisms at work to inhibit transformation. At their most basic, these extrinsic mechanisms involve environmental signals that instruct a cell to activate internal pathways leading to growth arrest and/or apoptosis in order to prevent neoplastic cell spread. For instance, disruption of epithelial cell associations with the extracellular matrix due to malignant transformation triggers death signals that block proliferation and spread of these contact-dependent cells. Thus, these extracellular attachments normally serve as inhibitors of cell death, which when broken set off a safety mechanism promoting apoptosis. However, if unregulated growth continues despite this safety net, identification and rejection of tumor cells by components of the immune system may help salvage homeostasis. Although several key immune cell types and effector molecules that participate in this response have been identified in recent years, much still remains to be learned about natural mechanisms of antitumor immunity and how best to harness or induce these in clinical settings.

CLINICAL FOCUS   BOX 19-1

A Vaccine to Prevent Cervical Cancer, and More

Globally, cervical cancer is the third leading cause of death among women, and second only to breast cancer in terms of cancer deaths in women. Over 500,000 women each year develop cervical cancer (80% of them are in developing countries), and approximately 275,000 women die of the disease annually. Periodic cervical examination (using the Papanicolaou test, or Pap smear) to detect abnormal cervical cells significantly reduces the risk for women. However, a health care program that includes regular Pap smears is commonly beyond the means of the less advantaged and is largely unavailable in many developing countries.

Human papillomavirus (HPV), the most common sexually transmitted infection, is implicated in over 99% of cervical cancers. HPV is also associated with most cases of vaginal, vulval, anal, penile, and oropharyngeal (head and neck) cancers, as well as genital warts. Among the hundred-plus genotypes of HPV, approximately 40 are associated with genital or oral infections. Two of these, types 16

and 18, account for more than 70% of all instances of cervical cancer, whereas types 6 and 11 are most often involved in HPV-associated genital warts.

It is estimated that the majority of sexually active men and women become infected with HPV at some point in their lives. A study of female college students at the University of Washington, published in 2003, showed that after 5 years more than 60% of study participants (all of whom were HPV negative when enrolled in the study) became infected. Most infections are resolved without disease; it is persistent infection leading to cervical or anal intraepithelial neoplasia that is associated with high cancer risk.

Preventing cervical cancer, therefore, appears to be a matter of preventing HPV infection. Gardasil (manufactured by Merck), the first vaccine ever approved for the prevention of cancer, was licensed in 2006 for the prevention of infection with HPV and potential development of cervical cancer or genital warts. This quadrivalent formulation targets HPV

types 6, 11, 16, and 18. Three years later, GlaxoSmithKline received a license for Cervarix, a vaccine to prevent cervical cancer that targets only HPV types 16 and 18. These vaccines are between 95% and 99% effective in preventing infection with HPV. Conclusive evidence that this will translate into significantly reduced rates of cervical cancer in women, which can take many years to develop, will not be available until long-term follow-up studies have been completed.

In June 2006, the federal *Advisory Committee on Immunization Practices* (ACIP) recommended routine HPV vaccination for girls ages 11 to 12, and catch-up immunizations for females ages 13 to 26 who have not already received the vaccine. Although the committee did not recommend routine immunization for boys at that time, it did suggest that Gardasil be *offered* to males ages 9 to 26. As of 2007, 25% of 13- to 17-year-old girls in the United States reported receiving at least one dose of this vaccine. In 2011, this number rose to 53% in girls, still far

(continued)

Immunoediting Can Both Protect Against and Promote Tumor Growth

To date, there are three proposed mechanisms by which the immune system is thought to control or inhibit cancer:

- By destroying viruses that are known to transform cells
- By rapidly eliminating pathogens and regulating inflammation
- By actively identifying and eliminating transformed cells

The first two constitute the typical purview of immunity outside of a role in cancer: find and destroy foreign infectious agents. The third mechanism, involving tumor cell identification and eradication, is termed **immunosurveillance**. It posits that the immune system continually monitors for and destroys neoplastic cells. Not a small feat, as the immune system should not attack self cells. However, significant evidence from animal models, immune deficiency disorders, and induced immune suppression regimens in humans

supports the notion that immunosurveillance is an important inhibitor of cancer. For instance, immune-suppressed individuals, such as patients with AIDS or transplant recipients receiving immunosuppressive drugs, have a much higher incidence of several types of cancer than do individuals with fully competent immune systems. Likewise, in animal models, almost any form of immunosuppression leads to increases in the incidence of both spontaneous and induced cancers. These observations and experiments clearly demonstrate the power of the immune system to hold cancer at bay.

Exactly how the immune system recognizes and targets neoplastic cells is the topic of the following section. However, recent data also point to potential *protumor influences* of the immune response on cancer. For instance, chronic inflammation and immune-mediated selection for malignant cells may actually contribute to cancer cell spread and survival. Contemporary studies of immunity to cancer have now generated a more nuanced hypothesis of immune involvement

short of targeted numbers (about 80%) and significantly lower than the rates of compliance for most other routine childhood vaccines (somewhere around 90%, depending on the age of the child).

In 2011, boys ages 11 to 13 were added to the ACIP list of recommendations for routine HPV vaccination. The hope is that this will curb the rising tide of anal and oropharyngeal cancers among men, but also cut back the infection cycle and impact rates of cervical cancer in women. However, HPV vaccination rates among young men remained at only 8% at the end of 2011. The idea was that, with Gardasil in particular, the ability to reduce the incidence of unsightly genital warts might provide added incentive for male vaccination.

With a safe and effective vaccine against a common and deadly cancer available for several years, why are the rates of immunization in young people still so low? The answer depends somewhat on the country in question, as well as social and economic factors. Especially in developing countries, cost and ease of use are major barriers. Development is currently underway for a second generation of HPV vaccines, which are more cost

effective, easier to administer, and produce longer-lived immunity to a broader range of HPV genotypes.

Controversy based on social and ethical issues, as well as misinformation, are also high on the list of reasons why HPV vaccination rates are believed to remain so low. In United States–based studies of factors that influence decisions to vaccinate adolescents against HPV, mother's attitudes, physician recommendations, and misunderstandings in all groups were highlighted. For instance, since HPV is a sexually transmitted infection, most parents prefer to consider this an issue for "the future," assuming that their children are not sexually active and that there is plenty of time before a vaccine for a sexually transmitted disease should be considered. In fact, based on the Centers for Disease Control and Prevention (CDC) surveillance data from 2011, 47% of high school students in the United States have had sexual intercourse; greater than 6% beginning before the age of 13. The HPV vaccine regimen, which involves three intramuscular injections administered over a 6-month period, is most effective when completed *prior* to exposure,

and produces the most robust immune response in 11- to 12-year-olds, the target population.

Physician recommendations are also key to making a dent in the rates of HPV vaccination. In a 2011 study, women 19 to 26 years old were asked whether they had received an HPV vaccine. In the group that had received a provider recommendation, 85% were immunized, compared with only 5% among women who did not receive a physician recommendation. More public and professional information concerning the advantages of this vaccine before the onset of sexual activity, as well as the lifetime risk of disease caused by HPV, may help drive down the cycle of infection and worldwide deaths due to this sexually transmitted killer.

REFERENCES

Eaton, D. K., et al.; Centers for Disease Control and Prevention (CDC). 2012. Youth risk behavior surveillance—United States, 2011. *Morbidity and Mortality Weekly Report Surveillance Summaries* **61**:1.

Winer, R. L., et al. 2003. Genital human papillomavirus infection: incidence and risk factors in a cohort of female university students. *American Journal of Epidemiology* **157**:218.

in neoplastic regulation, including both tumor-inhibiting and tumor-enhancing processes.

In the mid-1990s, research using animal models of cancer suggested that natural immunity could eliminate tumors. Armed with this understanding, researchers identified some of the key cell types and effector molecules involved. Experimental studies showed that mice lacking intact T-cell compartments or IFN-γ signaling pathways were more susceptible to chemically induced or transplanted tumors, respectively. Likewise, RAG2 knockout mice, which lack the enzyme necessary to undergo somatic recombination and thus fail to generate T, B, or NKT cells (no adaptive immunity), were found to be more likely to spontaneously develop cancer as they age and were more susceptible to chemical carcinogens.

However, the real surprise came when scientists used a chemical carcinogen to induce tumors in RAG2 knockout and wild-type mice of the same strain. Catherine Koebel and colleagues adoptively transferred induced tumors from wild-type and $RAG2^{-/-}$ mice into syngeneic wild-type

recipients. (See Chapters 11 and 20 for further discussion of adoptive transfer experiments.) The tumors coming from wild-type animals grew aggressively in their new hosts as expected. However, up to 40% of the tumors taken from immunodeficient $RAG2^{-/-}$ mice were rejected by healthy syngeneic recipients. This suggested that *tumors growing in immune-deficient environments are more immunogenic,* that is, they are easier for the immune system to recognize than those arising in an immunocompetent environment. These observations led to the idea that the immune system exerts a dynamic influence on cancer, inhibiting some tumor cells but also *sculpting or editing* them in a Darwinian process of selection: those that survive immune winnowing are better able to outwit the immune response and thus have a survival advantage. Thus was born the term **immunoediting**, to describe how the immune system engages in both positive (antitumor) and negative (protumor) actions that help to sculpt the tumor, determining which cells will be eliminated and which will remain.

The Three Stages of Cancer Immunoediting

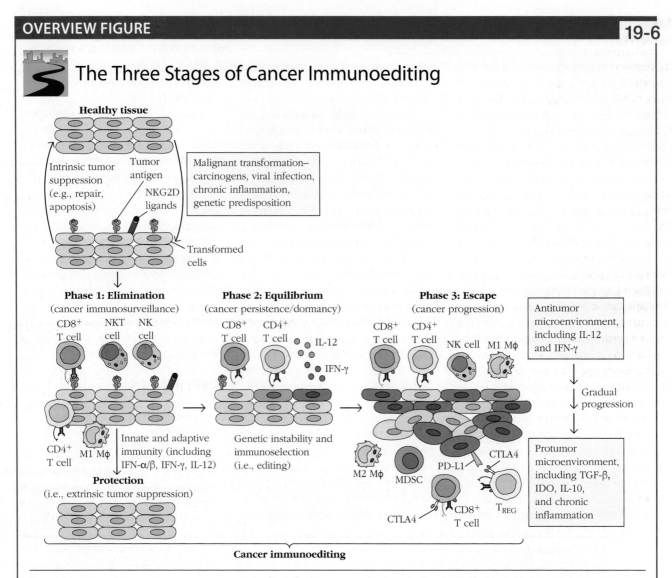

Healthy tissue (top) is constantly exposed to potential DNA damage; such damage can either be repaired or lead to apoptosis, followed by cellular replacement. When this is in balance, unhealthy cells are eliminated and new cells take their place. However, once this balance is tipped in favor or more cellular expansion than loss, tumors can arise. Recognition and targeting of tumor cells by the immune system is believed to occur in three phases. Phase I (*elimination*): Cancer cells are recognized by the immune system (e.g., via tumor antigens or NKG2D ligands), and targeted for destruction. In the process, some cells acquire mutations that allow them to resist immune destruction. Phase II (*equilibrium*): Low levels of abnormal cells persist, but their proliferation and spread are held in check by the adaptive immune response. Phase III (*escape*): Further mutation in the surviving tumor cells leads to the capacity for immortal growth and metastasis. Over time, inhibitory immune responses begin to dominate and immune activity shifts from anti- to protumor. Tan cells are normal; pink to red cells represent progressive development of decreased immunogenicity in tumor cells. Abbreviations: Mϕ, macrophage; MDSC, myeloid-derived suppressor cells; PD-L1, programmed death ligand-1.

Cancer immunoediting can be divided into three sequential phases: elimination, equilibrium, and escape (**Overview Figure 19-6**). The first phase, *elimination*, is the traditional view of the role of the immune system in cancer, roughly analogous to immunosurveillance—identification and destruction of newly formed cancer cells. *Equilibrium* is the next phase, characterized by a state of balance between moderate destruction of neoplastic cells with survival of a small number of cancer cells. Ample clinical evidence now suggests that the equilibrium phase can continue for decades after the emergence of a tumor, whether it is treated or not. In fact, equilibrium states of cancer may be much more common than previously appreciated. During this process of survival of the fittest, environmental conditions that influence immune reactivity can have a significant role in lengthening or shortening this time period. This is likely the period in which lifestyle choices come most to bear. Clinical identification of transformed cells at this stage can be tricky, despite the fact that this likely represents a fruitful window of time for intervention.

Immune *escape* is the final phase of immunoediting, when the most aggressive and least immunogenic of the residual tumor cells begin to thrive and spread, often thanks to help from immune pathways. Most basic research studies have focused on the role of the immune system in the elimination phase, where both innate and adaptive processes identify and target transformed cells for destruction, sometimes setting the stage for what will occur later during the equilibrium and escape phases. The following sections discuss our current understanding of the roles that the immune system can play in cancer eradication and survival of neoplastic cells during cancer immunoediting.

Key Concept:

- Immunoediting is a model that describes how the developing tumor-specific immune response identifies and kills cells (elimination, or immune surveillance), maintains a balance between death and survival of cancer cells (equilibrium), and also fosters the selection and survival of cells that develop immunoevasive mutations (escape).

Innate and Adaptive Pathways Participate in Cancer Detection and Eradication

As long ago as the 1860s, Rudolf Virchow observed that leukocytes infiltrate the site of a solid tumor and he proposed a possible link between inflammation and cancer *induction*. However, as infiltrating leukocytes can be found in both cancers that progress and those that resolve, the role of inflammation in cancer eradication was obscure. More recent scientific advances have allowed a detailed analysis of the location and type of cells involved, leading to the identification of key indicators of cancer regression. This newfound knowledge has led to many exciting new advances in cancer diagnostic tools, prognostic indicators, and therapies to treat the disease.

Much of the research aimed at probing the relationship between the immune system and cancer has employed mouse model systems. Most mice rendered immune deficient via specific targeted gene knockouts or neutralizing antibodies display some increased incidence of cancer. Likewise, spontaneous or induced tumors in immune-competent mice have allowed scientists to model typical human cancer

induction and test hypotheses concerning specific cell types and immune pathways in cancer eradication. Although many of the elements of both innate and adaptive immunity can be linked in some way to tumor cell recognition and destruction, certain components appear to play key roles in immune-mediated cancer control.

Studies in mice and humans have led to the awareness that there are both "good" and "bad" forms of inflammation in the response to cancer (e.g., see Overview Figure 19-6, bottom right). On the good or antitumor side are innate responses dominated by immune-activating macrophages (called *M1*), cross-presenting *d*endritic *c*ells (DCs; see Chapter 7), and NK cells. These cells and the cytokines they produce help elicit strong T_H1 and CTL responses, which are associated with a good prognosis and tumor regression. Conversely, in tumors that are more likely to progress and metastasize, the immune cell infiltrates include anti-inflammatory macrophages (M2) and *m*yeloid-*d*erived *s*uppressor *c*ells (MDSCs). Concomitantly, adaptive responses to cancer dominated by the T_H2 pathway (and in some cases, also T_H17 or T_{REG} cells) are associated with poorer clinical outcomes and reduced survival times. In the following sections, we go into greater detail concerning our developing awareness of the tumor-specific immune response pathways.

Innate Inhibitors of Cancer

NK cells were among the first cell types to be recognized for their inherent ability to destroy tumor cells, from which their name derives. Mice rendered deficient in NK cells, either by gene knockout or neutralizing antibodies, show an increased incidence of lymphomas and sarcomas. The importance of NK cells in tumor immunity is highlighted by the mutant mouse strain called beige and by Chédiak-Higashi syndrome in humans. In both, a genetic defect causes marked impairment of NK cells, and in each case an associated increase in certain types of cancer.

NK cell recognition mechanisms use a series of surface receptors that respond to a combination of activating and inhibiting signals delivered by self cells (see Chapter 12). Since many transforming viruses can induce the downregulation of MHC expression, detecting "missing self" is likely at least one of the ways in which NK cells participate in tumor cell identification and eradication. Likewise, inducing signals on tumor cells in the form of changes in protein expression, like danger- or damage-associated molecular patterns (DAMPs; see Chapter 4), can also engage NK cell–activating receptors (e.g., NKG2D) delivering what is referred to as an "altered" or "induced-self" signal (see Overview Figure 19-6, top left). Various forms of cellular stress, including viral infection, heat shock, UV radiation, and other agents that induce DNA damage, can trigger expression of the ligands for these activating NK-cell receptors. When activating signals are induced by DNA damage pathways, NK cells may be able to distinguish cancerous or precancerous cells from healthy neighboring cells. Once engaged, these cells use cytolytic granules, which include compounds like perforin, to kill target tumor cells. In fact, perforin deficiency in NK cells and CTLs is linked to increased cancer susceptibility. Indirectly, NK cells may also participate in cancer eradication by secreting IFN-γ, a potent anticancer cytokine that encourages MHC expression in DCs, which in turn stimulates strong CTL responses in vitro (see Adaptive Cellular Processes, below).

Numerous observations indicate that activated macrophages also play a significant role in the immune response to tumors. For example, macrophages are often observed to cluster around tumors, and the presence of proinflammatory macrophages, such as the M1 type, is correlated with tumor regression. Like NK cells, macrophages express Fc receptors, enabling them to recognize antibodies bound to tumor antigens and to mediate *antibody-dependent cell-mediated cytotoxicity* (ADCC; discussed further below). The antitumor activity of activated macrophages is likely mediated by lytic enzymes, as well as reactive oxygen and nitrogen intermediates. In addition, activated macrophages secrete *tumor necrosis factor alpha* (TNF-α), a cytokine with potent antitumor activity.

Recently, a previously unsuspected role for the eosinophil in cancer immunity has come to light. Mice engineered to lack eotaxin-1 (CCL11), a chemoattractant for eosinophils, and mice lacking IL-5, a stimulatory cytokine for this same cell type, were found to be more susceptible to carcinogen-induced cancers than were wild-type mice. In addition, IL-5 transgenic animals, which display higher levels of circulating eosinophils, are more resistant to chemically induced sarcomas.

Adaptive Cellular Processes and the Cells Involved in Cancer Eradication

In experimental animals, tumor antigens induce humoral and cell-mediated immune responses that lead to the destruction of transformed cells expressing these proteins. Animals that lack either $\alpha\beta$ or $\gamma\delta$ T cells are more susceptible to a number of induced and spontaneous tumors. Several tumors have been shown to induce CTLs that recognize tumor antigens presented by MHC class I on these neoplastic cells. In fact, strong antitumor CTL activity correlates significantly with tumor remission and is primarily credited with maintaining the equilibrium phase of cancer immunoediting, where neoplastic cell survival and death are fairly balanced (see Overview Figure 19-6). This observation has led to the development of some cancer vaccines aimed at inducing strong antitumor CTL responses against tumor antigens: a form of immunotherapy. Here, the vaccine is being used to generate an immune response that is therapeutic rather than prophylactic, the latter being most often the case with vaccine usage. As we know from Chapter 10, CD8$^+$ T cells rarely work alone, suggesting that activated DCs and T$_H$1-type cells must also be involved in fostering this CTL response.

Evidence for the protective role of adaptive immune responses against cancer cells comes from studies in mice treated with low-dose carcinogens. In the fraction of animals that do not develop cancer, rendering the animals immunodeficient can result in the sudden onset of cancer. Importantly, blocking NK-cell responses does not result in the eruption of occult cancer, but blocking CD8$^+$ T-cell responses or IFN-γ (produced by NK cells, T$_H$1 cells, and CTLs) does, highlighting the importance of adaptive immunity during this stage of cancer. The pressures of adaptive immunity on cancerous cells during this relatively long stage are believed to continuously sculpt tumors (thus the *immunoediting* name), driving the selective survival of neoplastic cells that are less immunogenic and which have accumulated mutations that are favorable for immune evasion. One example of this is increased expression of PD-L1 on tumor cells, which binds to PD-1 on CTLs and inhibits engagement (see Overview Figure 19-6, bottom right).

In clinical studies of cancer, the frequency of *tumor-infiltrating lymphocytes* (TILs)—a combination of T cells, NKT cells, and NK cells—at the site of the tumor correlates with a prognosis of cancer regression. For instance, in one seminal study of ovarian cancer, 38% of women with high numbers of TILs compared with 4.5% of women with low numbers of TILs survived more than 5 years past diagnosis. However, beyond numbers, the type of infiltrating cells may be even more crucial and in some cases can have more prognostic power than clinical cancer staging. In general, a high frequency of tumor-specific CD8$^+$ T cells, or an elevated ratio of CTL to T$_{REG}$ cells, is associated with enhanced survival. In several mouse models, the specific depletion or disruption of T$_{REG}$ cells resulted in cancer regression, and in some cases this was a sustained and durable response. This has led to ongoing studies aimed at the disruption of regulatory T cells in clinical settings.

B cells respond to tumor-specific antigens by generating anti-tumor antibodies that can foster tumor cell recognition and lysis. Using their Fc receptors, NK cells and macrophages participate in this response, mediating ADCC (see Chapter 12). However, some anti-tumor antibodies, called *enhancing antibodies*, serve a more detrimental role, blocking CTL access to tumor-specific antigens and enhancing survival of the cancerous cells. For this reason, a clear positive or negative role for B cells in cancer immunity is less obvious.

The Role of Cytokines in Cancer Immunity

Animal models in which cytokines or cytokine response pathways are eliminated have helped us to identify the role of specific cytokines in tumor cell eradication. In general, the cytokines most associated with the T$_H$1 pathway and CTL responses are also associated with cancer elimination. IFN-γ and the regulatory components of this pathway are clearly important in cancer elimination, as mice lacking these are more susceptible to a number of different tumors. IFN-γ can exert direct antitumor effects on transformed cells, including enhanced MHC class I expression, making neoplastic cells better targets for CD8$^+$ T-cell recognition and destruction (see Overview Figure 19-6). Both type I (α/β) and type II (γ) interferons have immune cell–enhancing activities that can foster tumor cell removal.

The cytokine IL-12 has received much attention for its ability to enhance antitumor immunity. For example, administration of exogenous IL-12 protected mice from chemically induced tumors. Similarly, mice that are genetically deficient for IL-12 develop more papillomas (a type of epithelial cell cancer) than do wild-type animals. This may be due in part to IL-12 driving the development of T-cell pathways: this cytokine encourages DCs to activate strong T$_H$1 and CTL responses, which help to create an antitumor microenvironment. Nonetheless, this highlights the strong antitumor potential of IL-12, which is currently a component of some cancer trials.

The cytokine TNF-α was named for its anticancer activity. When injected into tumor-bearing animals, TNF-α induces hemorrhage and necrosis of the tumor. However, TNF-α was later shown to have both tumor-inhibiting and tumor-promoting effects. Various carcinogen-treated TNF-$\alpha^{-/-}$ mice have been found to display either more sarcomas or fewer skin carcinomas than do wild-type mice, depending on the mouse strain and the means of cancer induction, suggesting that TNF-α has a complex role in tumor immunity.

Key Concepts:

- Several nonadaptive cells and processes are involved in cancer recognition and removal, including NK cells, M1-type macrophages, and possibly eosinophils, along with the cytokines produced by these cells, such as IFN-γ, TNF-α, and IL-5, respectively.

- Antitumor CTLs are the adaptive cell type most associated with cancer eradication, although activated DCs and T$_H$1-type cells likely also assist in this pathway.

- Anti-tumor antibodies can bind to the surface of cancer cells, allowing cells with Fc receptors, such as NK cells and macrophages, to induce ADCC in tumor targets.

- The cytokines most connected with antitumor responses include both type I and type II IFNs, TNF-α, and IL-12; all are associated with strong T$_H$1-cell and CTL responses.

Some Immune Response Elements Can Promote Cancer Survival

As we know, the immune response involves a balance of activating and inhibiting pathways—both some gas and some brakes. Without the brakes, uncontrolled immune

activity can lead to pathologic inflammation and even autoimmunity. In the immune response to cancer, inflammatory responses can serve a positive role, as we have seen above, but they also have the potential to promote cancer and create protumor microenvironments, such as occurs during the escape phase of cancer immunoediting (see Overview Figure 19-6, bottom right). In this section we discuss some specific immune response components that either support tumor growth or that direct immunity toward pathways of natural immunosuppression.

Chronic Inflammation

Chronic or ongoing inflammation is believed to create a protumor microenvironment via several mechanisms. First, sustained inflammatory responses increase cellular stress signals and can lead to genotoxic stress (e.g., reactive oxygen species, or ROS; see Chapter 4), increasing mutation rates in cells and thus fostering tumorigenesis. Second, the growth factors and cytokines secreted by leukocytes often induce cellular proliferation, and during mutation events, nonimmune tumor cells can acquire the ability to respond to these growth stimulators. In this way, some immune cells and the factors they produce can help sustain and advance cancer growth. Finally, inflammation is proangiogenic and prolymphangiogenic, increasing the growth of local vessels. This directs oxygen-supplying blood vessels to the site of a solid tumor and aids tumor cell invasion into surrounding tissues via transport through newly constructed lymphatic vessels. These are both hallmarks of cancer.

Tumor-Enhancing Antibodies

Antibodies can be produced against Tumor-Specific Antigens, and these may be important flags for tumor cell eradication. Based on this discovery, attempts were made to protect animals against tumor growth by active immunization with tumor antigens or by passive immunization with anti-tumor antibodies. Much to the surprise of the researchers involved, these immunizations usually did not protect against the tumor; in some cases, they actually enhanced tumor growth.

The tumor-enhancing ability of immune sera was studied via *cell-mediated lympholysis* (CML) reactions, which measure in vitro lysis of target cells by CTLs. CML assays performed in the presence of serum taken from animals with progressive tumor growth *blocked* the CTL lysis of targets, whereas serum taken from animals with shrinking or regressive tumors did not. In 1969, Ingegerd and Karl Erik Hellström extended these findings by showing that children with progressive neuroblastoma had high levels of some kind of blocking factor in their serum, whereas children with regressive neuroblastoma did not have these serum factors. Since these first reports, serum blocking factors have been found to be associated with a number of human tumors.

In some cases, anti-tumor antibody itself acts as a serum blocking factor. Presumably the antibody binds to tumor-specific antigens and masks the antigens from cytotoxic T cells. However, in many cases the blocking factors are not antibodies alone, but rather antibodies complexed with free tumor antigens. Although these immune complexes have been shown to block CTL responses, the mechanism of this inhibition is not clear. These complexes also may inhibit ADCC by binding to Fc receptors on NK cells or macrophages and blocking their activity. Therefore, although some clinical studies today utilize antibodies to treat cancer (see Clinical Focus Box 12-1), for this reason most of these antibodies are not directed against tumor-specific antigens, especially in the case of solid tumors.

Immunosuppression in the Tumor Microenvironment

Soluble factors secreted by tumor cells or the immune cells that infiltrate a tumor can sometimes encourage the development of a local immunosuppressive microenvironment. For instance, increased expression of TGF-β and indoleamine-2,3-dioxygenase (IDO), which inhibit T_H1 responses, has been found in various human cancers (see Overview Figure 19-6, bottom right). Another immunosuppressive cytokine, IL-10, may play a duplicitous role in cancer immunity. IL-10 has tumor-promoting, tolerance-inducing properties in some situations but can also encourage anticancer innate immune responses in others. Likewise, TGF-β expressed during the latter stages of cancer encourages progression, possibly by blocking local DC activation and inhibiting T-cell function, although the presence of this cytokine during early tumor growth can be tumor inhibiting. These microenvironment effects appear to be quite localized to the primary tumor site. Studies in mice with a primary tumor have shown that additional cancer cells of the same type introduced to a new site can be rejected by the immune system, even while the primary tumor remains intact, highlighting the importance of the tumor microenvironment.

In both animal models and human cancer studies, much recent attention has focused on the role of various naturally immunosuppressive cell types in tumor responses. The general consensus from most animal studies is that an abundance of T_{REG} cells, myeloid-derived suppressor cells (MDSCs), M2 macrophages, and T_H2 cells confers a local state of immunosuppression or immunity directed away from the more effective T_H1 pathway, allowing tumor cell evasion of the immune response (see Overview Figure 19-6, bottom right).

Key Concepts:

- Leukocyte infiltration is important for tumor eradication; however, prolonged chronic inflammation and tumor-enhancing antibodies correlate negatively with survival.

- The presence of immunosuppressive cells (e.g., MDSCs, M2 macrophages, and T_{REG} cells) and the cytokines they produce (e.g., IL-10 and TGF-β) in the tumor microenvironment fosters tumor survival and immune evasion.

Tumor Cells Evolve to Evade Immune Recognition and Apoptosis

Although the immune system clearly can respond to tumor cells, many of which express tumor antigens, the fact that so many individuals die each year from cancer suggests that the immune response to tumors is often ineffective. The selective pressure applied by the antitumor immune response can select for escape mutants, or cells that evade the immune response. This is the essence of immunoediting. In the following sections, we describe several pathways that are common in tumor cell evasion of the immune response.

Reduced MHC Expression in Tumor Cells

Defects in antigen processing and presentation are common among the escape mutants arising in many tumors. These could include mutations that lead to reduced MHC expression, secretion of TSAs (rather than surface expression), defective transporter associated with antigen processing (TAP) or β2-microglobulin, and IFN-γ insensitivity. Each of these types of mutations results in decreased MHC class I presentation of tumor antigens and profound inhibition of $CD8^+$ T-cell recognition (**Figure 19-7**). NK cells should recognize these cells lacking MHC class I. However, decreased expression of ligands that bind activating receptors on NK cells, also common among tumors, allows these cells to avoid NK cell–mediated killing. In fact, the absence of MHC molecules on tumor cells is generally an indication of cancer progression and carries a poor prognosis.

Tumor Cell Resistance to Apoptotic Signals

As a part of the evolution of cancer, neoplastic cells begin to develop resistance to the induction of programmed cell death, or apoptosis. This, of course, leads to an imbalance of proliferation and cell death, with expansion of the tumor cell population. This can arise due to changes in the extrinsic response system, namely cell-surface death receptors, as well as intrinsic mechanisms based on mitochondrial pathways of apoptosis. For instance, the Fas cell death receptor (CD95) and TRAILR, as well as the NKG2D engaged by NK cells, all receive external signals for the induction of programmed cell death. Defects in any of these components will therefore inhibit the induction of this pathway. Likewise, similar effects are seen during changes in the mitochondria-associated intracellular components that regulate cell death signals. Here, decreases in proapoptotic signaling components (typified by members of the Bcl-2 superfamily), or increases in anti-apoptotic players (inhibitors of apoptosis, such as survivin), both encourage enhanced cell survival and cancer progression. As mentioned earlier, faulty DNA repair mechanisms in transformed cells combined with immune-mediated pressures (e.g., selective destruction of cells expressing MHC class I) encourage the accumulation of tumor cells with these types of survival-enhancing mutations.

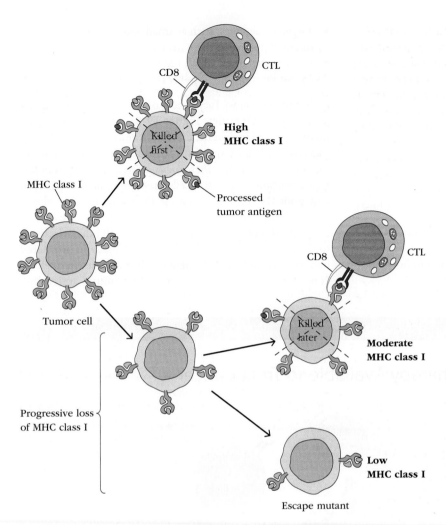

FIGURE 19-7 Down-regulation of MHC class I expression on tumor cells may allow for tumor escape mutants. The immune response itself may play a role in selection for tumor cells that express lower levels of MHC class I molecules, by preferentially eliminating those cells expressing high levels of class I molecules first. Malignant tumor cells that express fewer MHC molecules may thus escape CTL-mediated destruction.

Poor Costimulatory Signals and Immunosuppressive Microenvironments

As we know from Chapter 10, complete T-cell activation requires two signals: an activating signal (signal 1), triggered by recognition of a peptide-MHC molecule complex by the T-cell receptor, and a costimulatory signal (signal 2), supplied by the interaction of CD80/86 (B7) on *antigen-presenting cells* (APCs) with molecules such as CD28 on T cells. Both signals are needed to induce IL-2 production and proliferation of T cells. By virtue of their status as self somatic cells, tumors have fairly poor immunogenicity and tend to lack costimulatory molecules. Without sufficient numbers of APCs in the immediate vicinity of a tumor and with few stimulators to drive the activation of these cells, responding T cells may receive only a partial activating signal. This can lead to clonal anergy and immune tolerance. In the absence of activation, T cells interacting with these cells can be induced to act as immunosuppressors (e.g., T_{REG} cells), expressing immune-inhibiting cytokines and acting via comodulatory molecules, such as CTLA-4, as impediments to immune activation against the cancer. Some of the most cutting-edge cancer therapies specifically aim to enhance the costimulation provided to antitumor T cells.

Key Concept:
- Transformed cells employ several strategies to evade the immune response, including reduced MHC class I expression, anti-apoptotic responses, and poor or blocked costimulation of T-cell responses, generating a microenvironment of immune suppression.

Anticancer Immunotherapies

Present-day cancer treatments take many forms. In the case of discrete, solid tumors, surgical removal and local radiation are often employed. Frequently added to this are various drugs or other small-molecule inhibitors designed to target residual or metastatic tumor cells, called **adjuvant cancer therapy**. The drug or therapy used will depend on the type of cancer and specific characteristics of the tumor cells. While surgery was once considered the starting point, newer **neoadjuvant cancer therapies** (drugs first, surgery later) have taken hold in certain cases. The latter allows physicians to monitor the effectiveness of the adjuvant treatment in reducing the size of the primary tumor prior to its removal, and therefore is prognostic for the effectiveness of this adjuvant at destroying distant, metastatic cells that may not be detectable via imaging techniques. Chemical or drug therapy for cancer falls broadly into four categories, presented below in increasing order of specificity for tumor cells (and, by default, decreasing order of collateral damage to the host).

- Chemotherapies, aimed at blocking DNA synthesis and cell division
- Hormonal therapies, which can interfere with hormone receptor–positive tumor cell growth

- Targeted therapies, such as small-molecule inhibitors of tumors sensitive to these drugs
- Immunotherapies, which induce or enhance specific antitumor immune responses

In this section we focus most of our attention on **immunotherapies** (**Overview Figure 19-8**). These are any treatment designed to specifically *revive, initiate, or supplement* in vivo antitumor immune responses. *Unlike the other therapies listed above, these focus on activating an immune response against the cancer cells*, creating signals or cells that will guide the immune system in the right direction, allowing the body to take care of the rest. This can take the form of humanized monoclonal antibodies that will bind selectively to cancer cells, peptides from tumor antigens administered in a way that induces CTL activity, or molecules that release cancer-generated blocks to natural immune activation.

OVERVIEW FIGURE **19-8**

Types of Immunotherapy Available to Treat Cancer

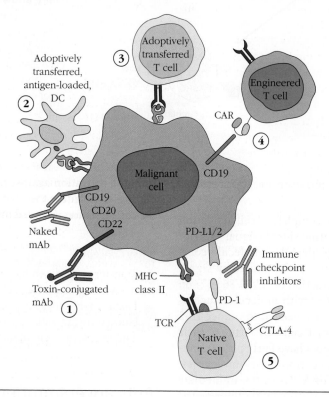

Immunotherapy to treat cancer can take many forms, including (1) monoclonal antibodies that target specific surface molecules (either naked or with toxins conjugated); (2) adoptively transferred autologous DCs that have been loaded with TAAs and expanded in vitro, followed by reinfusion into the patient; (3) adoptively transferred T cells that have been collected from the patient and expanded or modified in vitro, and then reintroduced; (4) CAR T cells generated by adding a chimeric receptor recognizing a tumor antigen to autologous T cells that are expanded and later re-infused (see Clinical Focus Box 19-2); and (5) checkpoint blockade, involving the use of mAbs specific for one or more of the surface molecules (CTLA-4, PD-1, or PD-L1) involved in regulating or dampening immune activity.

While no one immunotherapy will fit all situations, some forms of immune-based treatment can be applied to groups of cancers that share similar immunoevasive characteristics. Frequently, immunotherapy is used as an immune-boosting component, added to either experimental or standard regimens of surgery, radiation, and/or chemotherapy. The following sections describe several immune-based molecules or immunotherapeutic agents under investigation or licensed to fight cancer.

Before we begin this discussion, however, some historical perspective is warranted. Harnessing the immune system to fight cancer is not a new idea. In 1891, the cancer surgeon William B. Coley first experimented with this approach, injecting bacteria directly into the inoperable tumor of one of his patients with bone cancer, leading to surprising success. This was an era prior to the development of chemotherapy or radiation treatment for cancer, and clinicians then had few choices. Coley and a few other physicians continued this novel form of immunotherapy for many years, using bacteria or bacterial products that became known as "Coley's toxins" to treat some highly aggressive cancers. However, published reports of remission and even elimination of tumors induced by this technique met with much skepticism and variable results. This treatment was later replaced with chemo- and radiotherapy. Nonetheless, today we believe that Coley's observations and successes were likely based on sound immunologic reasoning; the immune-stimulatory characteristics of his toxins may have boosted natural anticancer immune activity and led to tumor regression in some of his patients.

In the following sections we describe four types of immunotherapy, each based on a different element of the immune response. These include the administration of specific antibodies or T cells, thereby supplying the body with a boost in humoral or cellular immunity against the cancer. Next, we discuss the use of antigenic peptides as therapeutic vaccines, designed to induce tumor-specific T-cell responses (i.e., signal 1; see Chapter 10). We end with a discussion of the most recent and perhaps most exciting new area of immunotherapy: manipulation of the comodulatory signals responsible for stimulation or inhibition of T-cell activation (i.e., signal 2; see Chapter 10), or immune activation checkpoints.

Monoclonal Antibodies Can Be Used to Direct the Immune Response to Tumor Cells

Monoclonal antibodies (mAbs) (see Clinical Focus Box 12-1 and Chapter 20) have long been used as immunotherapeutic agents for treating cancer. Once imagined as "magic bullets," the idea was that mAbs recognizing tumor-specific surface markers would selectively attach to these malignant cells, mark them for destruction by leukocytes, and deliver a payload of antibody-conjugated toxins. At present, there are more than a dozen different mAbs licensed for the treatment of cancer, primarily directed against cell-type specific surface molecules. **Table 19-4** lists many of these, as well as the cancers for which they are approved.

In one early success of mAb treatment, Ron Levy and his colleagues at Stanford treated a 64-year-old man with terminal B-cell lymphoma that had metastasized to the liver, spleen, and bone marrow. Because this was a B-cell cancer, the membrane-bound immunoglobulin on all the cancerous cells had the exact same idiotype (antigenic specificity). By the procedure outlined in **Figure 19-9**, these researchers produced mouse mAb specific for the idiotype (the antigen-binding region) of this B-lymphoma. When this anti-idiotype antibody was injected into the patient, it bound specifically only to the cancerous B-lymphoma cells that expressed that particular immunoglobulin. Since these cells are susceptible to complement- or ADCC-mediated lysis, the mAb activated the destruction of these cells without harming other cells. After four injections with this anti-idiotype mAb, the tumors began to shrink and the patient entered an unusually long period of remission.

A custom approach such as this, targeting idiotypes in B-cell lymphomas specific to each patient with cancer, was very costly and time-consuming. (Of note, this idea was resurrected in 2015, when the cost and time involved became more reasonable thanks to new technology! For more information, see the podcast link at the end of this chapter.) The same research group initiated a like-minded strategy, using a more general therapy for B-cell lymphoma based on the fact that most B cells, whether cancerous or not, express many copies of lineage-distinctive antigens on their surface. For example, therapeutic mAbs that target the B-cell marker CD20, such as rituximab, are widely used to treat non-Hodgkin's lymphoma. While rituximab treatment does lead to the destruction of noncancerous B cells, we know the immune system is capable of regenerating new B cells from hematopoietic stem cells, somewhat blunting this side effect.

Likewise, antibodies can be coupled with toxic substances, creating "guided missiles" that are called *antibody-drug conjugates* (ADCs). Examples of conjugates include radioactive isotopes, chemotherapy drugs, or potent toxins, which can be delivered at high local concentrations directly to cancer cells, while sparing most nonmalignant cells. When the conjugate is a toxin (e.g., diphtheria toxin), these are called **immunotoxins**. The first FDA-approved ADC was Mylotarg, an anti-CD33 molecule conjugated to a cytotoxin; it was used to treat acute myeloid leukemia. However, Mylotarg was pulled from the market due to poor results and increased mortality among patients. Two other cytotoxin-conjugated ADCs are now FDA approved and in clinical use. One targets CD30, found on the surface of several blood cell cancers, and is commonly used to treat relapsed Hodgkin's lymphoma. The other is used for some forms of metastatic breast cancer and is described next.

A variety of tumors express significantly increased levels of growth factors or their receptors, which are promising targets for anti-tumor mAbs. For example, in 25% to 30%

TABLE 19-4	Monoclonal antibodies approved by the FDA and licensed for cancer treatment				
mAb name	Trade name	Target	Used to treat:		Approved in:
Rituximab	Rituxan	CD20	Non-Hodgkin's lymphoma		1997
			Chronic lymphocytic leukemia (CLL)		2010
Trastuzumab	Herceptin	HER2	Breast cancer		1998
			Stomach cancer		2010
Gemtuzumab ozogamicin*	Mylotarg	CD33	Acute myelogenous leukemia (AML)		2000[†]
Alemtuzumab	Campath	CD52	CLL		2001
Ibritumomab tiuxetan*	Zevalin	CD20	Non-Hodgkin's lymphoma		2002
[131]I-Tositumomab*	Bexxar	CD20	Non-Hodgkin's lymphoma		2003
Cetuximab	Erbitux	EGFR	Colorectal cancer		2004
			Head and neck cancers		2006
Bevacizumab	Avastin	VEGF	Colorectal cancer		2004
			Non–small cell lung cancer		2006
			Breast cancer		2008
			Glioblastoma and kidney cancer		2009
Panitumumab	Vectibix	EGFR	Colorectal cancer		2006
Ofatumumab	Arzerra	CD20	CLL		2009
Denosumab	Xgeva	RANK ligand	Cancer spread to bone		2010
Ipilimumab	Yervoy	CTLA-4	Melanoma		2011
Brentuximab vedotin*	Adcetris	CD30	Hodgkin's lymphoma and one type of non–Hodgkin's lymphoma		2011
Pembrolizumab	Keytruda	PD-1	Melanoma, non–small cell lung cancer, kidney cancer, and Hodgkin's lymphoma[‡]		2014
Atezolizumab	Tecentriq	PD-L1	Bladder cancer, non–small cell lung cancer, and Merkel cell carcinoma		2016

*Conjugated monoclonal antibodies.

[†]General approval withdrawn in 2010 and now used only as a part of ongoing clinical trials.

[‡]Approved in 2017 for the treatment of all solid tumors with microsatellite instability–high and mismatch repair–defective genotypes.

Data sources: American Cancer Society, www.cancer.org; and Table 2 from Aldrich, J. F., et al. 2010. Vaccines and immunotherapeutics for the treatment of malignant disease. *Clinical and Developmental Immunology* **2010:**697158.

of women with metastatic breast cancer, a genetic alteration of the tumor cells results in the increased expression of *h*uman *e*pidermal growth factor–like *r*eceptor 2 (HER2), which is encoded by the *neu* gene. HER2 is expressed in only trace amounts in normal adults. Because of this difference in HER2 protein levels, a humanized mAb against HER2, called *Herceptin*, has been successfully used to treat breast cancers that overexpress *neu*. An immunotoxin form of Herceptin, called *Kadcyla* or *T-DM1*, was recently approved for the selective treatment of certain types of metastatic breast cancer. Like Herceptin, this drug binds to HER2$^+$ cancer cells and introduces a toxin called *DM1*, which binds to tubulin in these cells, blocking the cell division machinery. Several other mAbs that target specific growth factors or their receptors have also been approved for clinical use and are included in Table 19-4.

Key Concept:

- Monoclonal antibodies (mAbs), with or without cytotoxin conjugates, that bind to cell type-specific or overexpressed surface markers can be used to target these cancer cells for recognition and elimination by the toxin or phagocytic cells.

Tumor-Specific T Cells Can Be Expanded, or Even Created

Early observations of lymphocyte infiltration into solid tumors suggested that these cells might be a possible source of tumor-specific immunity. Subsequent research has shown that in many cases, tumor-reactive T cells (called *TILs,* or

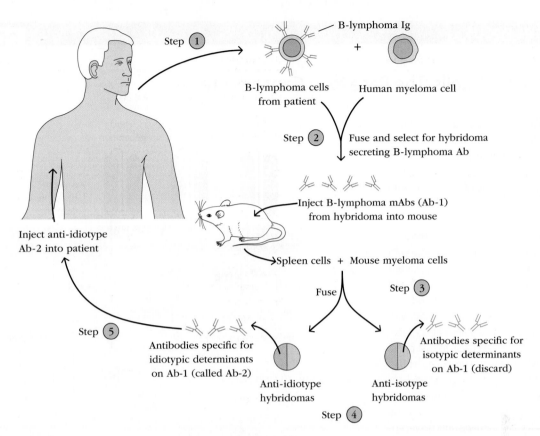

FIGURE 19-9 Development of a monoclonal antibody specific for idiotypic determinants on B-lymphoma cells. Because all the B-lymphoma cells in a patient are derived from a single transformed B cell, they all express the same membrane-bound antibody (Ab-1) with the same idiotype (i.e., the same antigenic specificity). In the procedure illustrated, a monoclonal anti-idiotypic antibody (Ab-2) against the B-lymphoma membrane-bound antibody is produced ex vivo (steps 1–4). This anti-idiotype antibody is then injected into the patient (step 5), where it binds selectively to the idiotypic determinants on the immunoglobulin of B-lymphoma cells, making these cells susceptible to complement-mediated lysis.

tumor-infiltrating lymphocytes) can be isolated from the peripheral blood, lymph nodes, and solid tumors of patients with cancer. Despite the presence of these cells in most solid tumors, however, cancer persists. This suggests that these cells are not able to perform their antitumor effector functions in vivo or they are not *antitumor* cells at all.

In one multicenter study, the TILs from patients with metastatic melanoma were collected and expanded in the presence of IL-2 in vitro, ostensibly to overcome their anergic state. Following lymphodepleting treatments designed to create a niche for new cells to take hold, patients were infused with large numbers of these activated, autologous TILs. Almost half of the patients saw significant tumor regression, with approximately 10% of the patients experiencing long-lived or complete remission. Despite these promising results in some, at least half the patients showed little or no response. This type of disparate outcome is not uncommon in cancer clinical trials, where differences in both the cancer and the immune response between individuals can mean the difference between complete remission and no response. This divergent outcome could be influenced by an abundance of

T_{REG} cells in nonresponding patients. These immune-suppressing cells express the high-affinity IL-2 receptor and thus have a competitive advantage against immune-activating T cells in the presence of IL-2. Most such cell transfer therapies are now conducted with peripheral cells rather than TILs, called *adoptive cell transfers*, partly to avoid expanding suppressor cells. One big advantage of this form of treatment over mAbs (described above) is that autologous T cells can be fairly long-lived in the patient.

The newest addition to adoptive T-cell therapy for cancer is CARs—a fast-moving field (sorry, couldn't resist!). Chimeric *a*ntigen-*r*eceptor (CAR) T cells begin with autologous T cells isolated from a patient with cancer. In vitro, scientists add the gene for a chimeric tumor antigen–specific receptor, followed by infusion of these cells back into the patient (**Clinical Focus Box 19-2**). Early studies showed mixed results and were limited primarily to the treatment of nonsolid cancers, such as leukemia. The concept, first pioneered by Israeli immunologist Zelig Eshhar, has now moved into newer second- and third-generation high-potency CARs, designed with a range of immunomodulatory and costimulating features.

CAR T Cells: The Race for a Cancer Cure

In 1989, Gideon Gross and colleagues at the Weizmann Institute of Science in Israel described a new, "artificial" T-cell receptor. This group published the first description of the use of a chimeric antigen-specific receptor—part immunoglobulin and part T-cell receptor. The idea behind this was to harness the antigen-binding advantages that come with a B-cell receptor (non–MHC-restricted engagement with native antigen) and combine it with the power behind activated T cells (cytokines associated with cell-mediated immunity and cells capable of inducing target cell lysis). To that point, no one had successfully produced a functioning receptor of this type, capable of actually activating a T cell. Ultimately, this synthetic receptor would come to be known as a CAR (chimeric *a*ntigen *r*eceptor) and the cells expressing these receptors became **CAR T cells**.

First-generation CARs had a simple design: an extracellular antigen-binding region connected via a transmembrane domain to an intracellular signaling region, usually CD3-ζ (see **Figure 1**). The variable region was designed to recognize a known cell type-specific antigen, such as the CD19 molecule found on the surface of B-cell leukemias (as well as healthy B cells). Next, T cells were collected from the patient and transfected with the CAR construct. After in vitro expansion, these CAR-expressing self T cells were re-infused into the patient, where they acted as cancer cell search-and-destroy missiles. Unfortunately, early trials with these first-generation CAR T cells yielded less than ideal results. The engineered cells did not persist in patients and the antitumor effects were similarly short-lived.

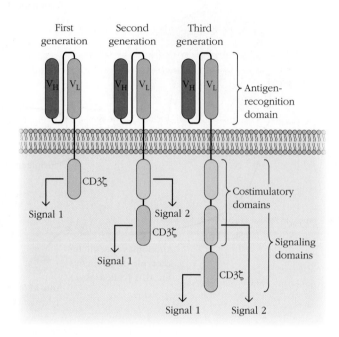

FIGURE 1 Evolution of the chimeric antigen receptor used in CAR T cells, from signal 1-only versions to those with one or more costimulatory signals.

This led to second-generation CARs with more accessories. These receptors included the basic design principles of their predecessors, with the addition of one or more costimulatory signals (signal 1 plus signal 2). This single-chain model, containing antigen-binding, intracellular signaling, and costimulation domains, all in one transmembrane receptor, yielded more promising results. Second-generation CAR T cells persisted long enough to demonstrate significant destruction of malignant B cells—90% of patients previously diagnosed with relapsed or drug-resistant disease responded to treatment—a highly significant improvement over standard therapies.

Unfortunately, severe side effects and drug toxicity dampened the initial enthusiasm. Side effects from the first two generations of CAR T cells included neurological symptoms and a life-threatening systemic cytokine storm, coined *cytokine release syndrome* (*CRS*). This systemic response to the activation, proliferation, and cytokine secretion of CAR T cells resulted in high fever and flu-like symptoms, with potential neurologic impacts, including several patient deaths. Subsequent trials have shown that CRS can be reversed in most patients with the use of anti–IL-6 monoclonal antibodies, which allowed trials of this new drug to resume.

Nonetheless, miraculous improvements for many patients with a previously poor prognosis and limited chances for survival added fuel for a third (and fourth, and fifth. . .) generation of CARs. These updated iterations include one or more additions to the model 2 version. Many of the exciting

(continued)

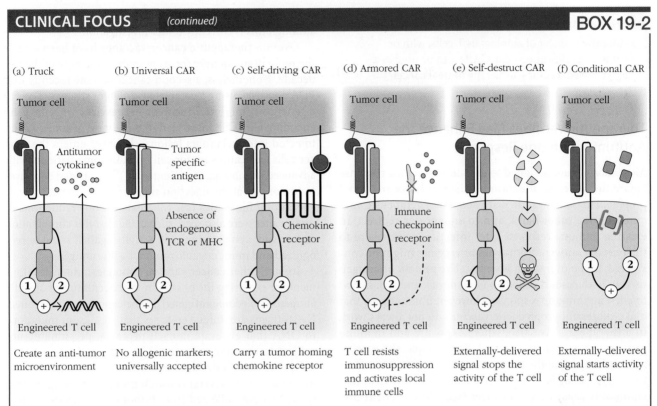

(a) Truck

(b) Universal CAR

(c) Self-driving CAR

(d) Armored CAR

(e) Self-destruct CAR

(f) Conditional CAR

Create an anti-tumor microenvironment

No allogenic markers; universally accepted

Carry a tumor homing chemokine receptor

T cell resists immunosuppression and activates local immune cells

Externally-delivered signal stops the activity of the T cell

Externally-delivered signal starts activity of the T cell

FIGURE 2 Examples of specialized accessories included in some third generation and beyond CAR T cells under investigation as cancer immunotherapies.

new accessories available, and the flashy names to match, are illustrated in **Figure 2**. For instance: (1) TRUCKs (*T cells redirected for universal cytokine killing*) express antitumor cytokines along with their CAR; (2) universal cells with CARs no longer express endogenous TCR or HLA molecules, making them the universal donor CAR T cells; (3) self-driving CARs express a chemokine receptor along with their CAR, to find their way around; (4) armored CARs are engineered to resist immunosuppression because they lack coinhibitory molecules (such as CTLA-4 or PD-1) and express immune stimulatory molecules (such as CD40L and IL-12); (5) self-destructing CARs are designed with suicide triggers so that they can be turned off as needed during severe side effects; and (6), the opposite, conditional CARs, are designed to be in the "off" position until triggered for activation by addition of a drug.

Of course, all of these accessories come at a price. The cost of up to $800,000 per patient caused some initial sticker shock, despite its similarity to the expense of other adoptive cell transfer therapies, like bone marrow transplantation. Both are examples of potentially curative therapies and both are considered "living drugs," thanks to their potential for self-renewal and thus life-long impact. Since many of the most successful CAR T-cell trials have required only a single injection and have resulted in high cure rates, these costs are now placed in better perspective.

The first CAR T-cell therapy was approved by the FDA in August 2017. Tisagenlecleucel-T (Kymriah) was approved for use in children and young adults (3 to 25 years) with relapsed or refractory acute lymphoblastic leukemia. In a multicenter preliminary clinical trial, 83% of patients had complete remission of disease within

the first 3 months of treatment. This, in a patient population with previously poor clinical prognosis and few other treatment options. Importantly, no deaths were reported in these trials.

Clearly, CARs are having a big impact on cancer treatment. What's next in CAR T-cell therapy? With all the excitement and everyone wanting a ride, don't be surprised if CAR pools are the next model.

REFERENCES

1. Jackson, H. J., S. Rafiq, and R. J. Brentjens. 2016. Driving CAR T-cells forward. *Nature Reviews Clinical Oncology* **13:**370.

2. Fesnak, A. D., C. H. June, and B. L. Levine. 2016. Engineered T cells: the promise and challenges of cancer immunotherapy. *Nature Reviews Cancer* **16:**566.

Therapeutic Vaccines May Enhance the Antitumor Immune Response

Most vaccines are designed to initiate an immune response *before* the onset of infection or disease; these are called *prophylactic* vaccines, or just vaccines. Therapeutic vaccines, on the other hand, aim to enhance or redirect an existing immune response after infection or exposure to the relevant antigen. Both types of vaccines might be considered a type of immunotherapy. The protein or antigen used in both cases is called an *immunogen*, and is designed to elicit an immune response. However, immunogens that make successful prophylactic vaccines do not always work in a therapeutic scenario, where exposure and immune response have already occurred. For example, the vaccine against human papillomavirus (HPV) is up to 99% effective at *preventing infection* with the strains of this virus that commonly cause cervical cancer (see Clinical Focus Box 19-1). However, this same vaccine is ineffective in women who are already infected with HPV. Thus, vaccines for use in patients with existing cancer-causing infections or resident neoplastic cells must be designed specifically to *reroute* an already engaged immune response; a unique challenge.

With this idea in mind, initial cancer vaccine studies in mice investigated methods for altering or enhancing tumor-specific antigen presentation. In one study, mouse DCs were cultured in myeloid-stimulating growth factor GM-CSF and incubated with tumor fragments, and then infused into the mice. The DCs activated both T_H cells and CTLs specific for the tumor antigens found on the tumor fragments. When the mice were subsequently challenged with live tumor cells, they displayed tumor immunity. Employing a similar strategy, sipuleucel-T (Provenge) became the first approved therapeutic cancer vaccine, specific for prostate cancer. This therapy used autologous patient DCs treated in vitro with a fusion protein consisting of the tumor-associated *prostatic acid phosphatase* (PAP) antigen linked to GM-CSF (**Figure 19-10**). These stimulated autologous APCs were then re-infused into the patient. However, the modest survival increases seen in treated patients combined with the steep cost ($93,000 for the recommended three infusions) significantly dampened interest in this drug. However, recent findings that sipuleucel-T showed significantly better outcomes in African American men with prostate cancer compared with matched white male counterparts is an exciting new development, especially given that African Americans have traditionally fared more poorly in terms of prognosis and treatment response. As yet, there

is no indication of a possible mechanism for this difference, although this is currently under investigation.

Overall, therapeutic cancer vaccines have been one of the more disappointing forms of immunotherapy of the last decade. Nonetheless, a second cancer vaccine recently met FDA approval, using a completely different strategy. T-vec is a genetically modified oncolytic herpesvirus with the gene encoding GM-CSF integrated into the viral genome. When injected into melanoma lesions, the virus enters the cancer cells and causes tumor cell lysis. In some patients with advanced melanoma, treatment with T-vec yielded decreases in tumor size at the injection site, and in combination with checkpoint blockades (see the next section) reductions in distant sites were also seen, suggesting that this combination may trigger systemic immune activity against some unrecognized and immunogenic tumor-specific antigens.

In fact, other cancer vaccine strategies under development are exploring the possibility that occult tumor-specific antigens (called **neoantigens**) can be exposed and exploited. Neoantigens are unique epitopes that arise from mutations to DNA that generate nonself proteins, presumably not subject to immune tolerance. These neoantigens are thus unique to each individual and also highly specific to the malignant cells. Several research groups are using protein lysate or RNA collected from tumor cells, introduced in a context designed to elicit strong immune activation. These individualized strategies are showing some promising results although none are yet available in the clinic.

Manipulation of Comodulatory Signals, Using Checkpoint Blockade

As discussed in the section on the escape phase of immunoediting, the microenvironments surrounding aggressive cancers can be inhospitable to immune activation. Cancer cells lack the costimulatory signals required for full T-cell activation. Recall from Chapter 10 that anergy is the outcome for T cells activated in the presence of signal 1 (TCR engagement) without signal 2 (CD28 costimulation), with the latter typically supplied by CD80/86 on pAPCs. This led to the hypothesis that manipulation of costimulation might encourage host T-cell attack on cancer cells. Early support for this hypothesis was presented in 1992, when Peter Linsley and colleagues demonstrated complete tumor regression in 40% of tumor-bearing mice injected with melanoma cells that had been transfected with CD80/86. The next year,

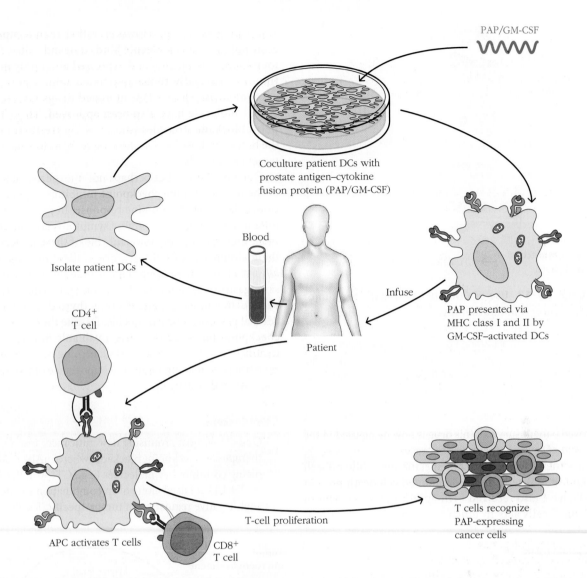

PAP/GM-CSF

Coculture patient DCs with
prostate antigen–cytokine
fusion protein (PAP/GM-CSF)

Blood

Isolate patient DCs

Infuse

PAP presented via
MHC class I and II by
GM-CSF–activated DCs

Patient

CD4+
T cell

APC activates T cells

CD8+
T cell

T-cell proliferation

T cells recognize
PAP-expressing
cancer cells

FIGURE 19-10 Mechanism of action of sipuleucel-T, a prostate cancer vaccine. Autologous DCs are isolated from a patient's blood and cultured with a fusion protein consisting of the prostate cancer–specific antigen PAP and the APC-activating cytokine GM-CSF (PAP/GM-CSF). DCs take up and process these antigens, after which they are infused into the patient in order to stimulate a T-cell response against PAP expressed on tumor cells in the prostate. Abbreviations: APC, antigen-presenting cell; DCs, dendritic cells; GM-CSF, granulocyte-macrophage colony-stimulating factor; PAP, prostatic acid phosphatase.

Sarah Townsend and James Allison used a similar approach to prophylactically vaccinate mice against malignant melanoma. Normal mice were first immunized with irradiated, CD80-transfected melanoma cells, and then later challenged with malignant melanoma cells lacking this costimulatory molecule (**Figure 19-11**). This "vaccine" was found to protect almost 90% of the mice when they were challenged with the wild-type, malignant cancer cells.

In order to move this idea to the clinic, a new approach was needed—one that did not depend on the transfer of neoplastic cells into patients. As we saw in Chapter 10, co-inhibitory molecules are the other side of the costimulation coin. These T-cell surface molecules serve as a checkpoint, dampening or regulating the immune response. Therefore, instead of engaging T-cell activation "gas" (CD28-B7 costimulation), could we get a similar effect if we release the "brakes" (blocking CTLA-4–B7 co-inhibitory interactions)? Groundbreaking studies, also led by James Allison and colleagues, showed proof of this concept in mice. They found that a mAb against CTLA-4 could induce cancer rejection in animals with existing tumor burdens. In 2011, after years of clinical trials, the FDA approved ipilimumab, the first in a new class of anticancer drugs called **checkpoint inhibitors**, or immune checkpoint blockers. This particular mAb is capable of blocking human CTLA-4. Originally licensed for use against malignant melanoma, where it significantly

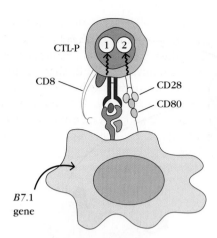

Tumor cell transfected → CTL activation → Tumor destruction
with *B7.1* gene

FIGURE 19-11 Use of CD80 (B7.1)-transfected tumor cells for cancer immunotherapy. Tumor cells transfected with the *B7.1* gene express the costimulatory CD80 molecule, enabling them to provide both activating signal (1) and costimulatory signal (2) to CTL-Ps (CTL precursors). As a result of the combined signals, the CTL-Ps differentiate into effector CTLs, which can mediate tumor destruction. In effect, the transfected tumor cells act as APCs.

improved patient survival, this drug is now being used in the treatment of several other types of cancer.

Since then, two other mAbs specific for another T-cell checkpoint molecule, the programmed cell death protein (PD-1), have also received FDA approval. Engagement of PD-1 on T cells acts in much the same way as CTLA-4,

dampening activation. However, rather than compete for costimulation, this molecule binds a ligand called *PD-L1*, found naturally on DCs and expressed aberrantly in many tumors, especially those associated with a poor clinical prognosis. Three FDA-approved drugs targeting the PD-L1 molecule have also been approved. These **checkpoint blockade** strategies (shown in **Figure 19-12**) release the brakes on host T cells, allowing recognition and lysis of target cancer cells.

One unfortunate but perhaps not unexpected side effect of the use of immune checkpoint inhibitors is generalized dysregulation of the immune response. In particular, acute inflammatory and autoimmune symptoms are not uncommon, highlighting the powerful role of these molecules in the maintenance of self tolerance. Called *immune-related adverse events*, these symptoms include immune attacks to skin, gastrointestinal tract, liver, and endocrine glands. In particular, endocrinopathies of the thyroid, pituitary, and adrenal glands can be irreversible. Despite these side effects, checkpoint blockade is clearly a valuable new tool in the treatment of cancer. It also marks a turning point in the cancer battle; one that highlights the importance of the immune response in this struggle.

Key Concept:

• One of the most promising new anticancer immunotherapies is checkpoint blockade, where mAbs recognizing co-inhibitory molecules (e.g., CTLA-4, PD-1, or PD-L1) are used alone or in combination to interrupt negative regulation of tumor-specific T cells.

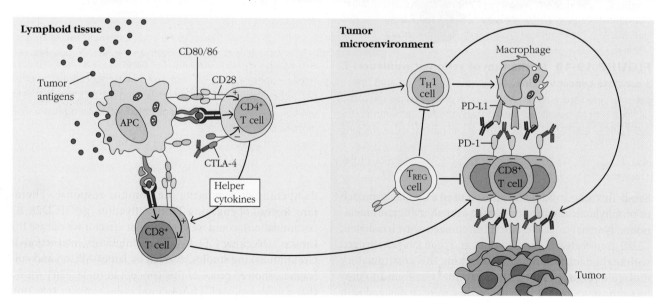

FIGURE 19-12 Using checkpoint blockade therapy to treat cancer. The most recent cancer immunotherapy involves using mAbs to block surface molecules involved in dampening the immune response of T cells. Specifically, antibodies against CTLA-4 or PD-1, both inhibitory receptors on T cells, as well as antibodies against the PD-L1 ligand (frequently expressed on cancer cells) can block these checkpoint molecules. These antibodies work in lymphoid tissues (to enhance immune stimulatory signals) and at the site of the tumor (blocking inhibitory pathways), allowing activated CD8$^+$ T cells free rein to migrate to tumor sites and carry out their destructive programming, with less interference from regulatory cells and other immunosuppressive cues.

Conclusion

Our understanding of the causes and genetic mediators of cancer has come a long way in the past 50 years. This has been followed, with a slight lag, by our appreciation for the role of the immune system in cancer, seen now for both its positive and negative impacts on the neoplastic microenvironment. Basic science research related to antigen presentation, T-cell activation, and immunomodulatory signaling, combined with our understanding of cancer genetics, has led to clinical breakthroughs in the fight against cancer that now focus more on inducing or boosting natural immune pathways of cancer cell eradication. While there is still no magic bullet or cure on the immediate horizon, thanks to our ability to observe and manipulate the tumor-specific immune response, clinicians now have an expanded toolbox and many cancer patients have a more promising prognosis than just a decade ago.

REFERENCES

Aisenberg, A. C. 1993. Utility of gene rearrangements in lymphoid malignancies. *Annual Review of Medicine* **44**:75.

Aldrich, J. F., et al. 2010. Vaccines and immunotherapeutics for the treatment of malignant disease. *Clinical and Developmental Immunology* **2010**:697158.

Allison, J. P., A. A. Hurwitz, and D. R. Leach. 1995. Manipulation of costimulatory signals to enhance antitumor T-cell responses. *Current Opinion in Immunology* **7**:682.

Boon, T., P. G. Coulie, and B. Van den Eynde. 1997. Tumor antigens recognized by T cells. *Immunology Today* **18**:267.

Chen, D. S., and I. Mellman. 2013. Oncology meets immunology: the cancer-immunity cycle. *Immunity* **39**:1.

Cho, H.-J., Y.-K. Oh, and Y. B. Kim. 2011. Advances in human papilloma virus vaccines: a patent review. *Expert Opinion on Therapeutic Patents* **21**:295.

Coulie, P. G., et al. 1994. A new gene coding for a differentiation antigen recognized by autologous cytolytic T lymphocytes on HLA-A2 melanomas. *Journal of Experimental Medicine* **180**:35.

Hanahan, D., and R. A. Weinberg. 2011. Hallmarks of cancer: the next generation. *Cell* **144**:646.

Houghton, A. N., J. S. Gold, and N. E. Blachere. 2001. Immunity against cancer: lessons learned from melanoma. *Current Opinion in Immunology* **13**:134.

Hsu, F. J., et al. 1997. Tumor-specific idiotype vaccines in the treatment of patients with B-cell lymphoma. *Blood* **89**:3129.

Jacquelot, N., et al. 2017. Predictors of responses to immune checkpoint blockade in advanced melanoma. *Nature Communications* **8**:592.

Kamran, N., M. Chandran, P. R. Lowenstein, and M. G. Castro. 2016. Immature myeloid cells in the tumor microenvironment: implications for immunotherapy. *Clinical Immunology* doi: 10.1016/j.clim.2016.10.008 (in press).

Lesterhuis, W. J., J. B. Haanen, and C. J. Punt. 2011. Cancer immunotherapy—revisited. *Nature Reviews Drug Discovery* **10**:591.

Makkouk, A., and G. Weiner. 2015. Cancer immunotherapy and breaking immune tolerance—new approaches to an old challenge. *Cancer Research* **75**:5.

Li, M. O., and A. Y. Rudensky. 2016. T cell receptor signalling in the control of regulatory T cell differentiation and function. *Nature Reviews Immunology* **16**: 220.

Moynihan, K. D., et al. 2016. Eradication of large established tumors in mice by combination immunotherapy that engages innate and adaptive immune responses. *Nature Medicine* **22**:1402.

Sahin, U., O. Tureci, and M. Pfreundschuh. 1997. Serological identification of human tumor antigens. *Current Opinion in Immunology* **9**:709.

Sharma, P., K. Wagner, J. D. Wolchok, and J. P. Allison. 2011. Novel cancer immunotherapy agents with survival benefit: recent successes and next steps. *Nature Reviews Cancer* **11**:805.

Townsend, S. E., and J. P. Allison. 1993. Tumor rejection after direct costimulation of CD8+ T cells by B7-transfected melanoma cells. *Science* **259**:368.

van der Burg, S. H., R. Arens, F. Ossendorp, T. van Hall, and C. J. Melief. 2016. Vaccines for established cancer: overcoming the challenges posed by immune evasion. *Nature Reviews Cancer* **16**:219.

Vesely, M. D., M. H. Kershaw, R. D. Schreiber, and M. J. Smyth. 2011. Natural innate and adaptive immunity to cancer. *Annual Review of Immunology* **29**:235.

Vogelstein, B., et al. 2013. Cancer genome landscapes. *Science* **339**:1546.

Useful Websites

https://seer.cancer.gov/statfacts/html/all.html Surveillance, Epidemiology, and End Results (SEER), which provides United States–based cancer statistics, is a site maintained by the National Cancer Institute.

https://www.oncolink.org OncoLink offers comprehensive information about many types of cancer and is a good source of information about cancer research and advances in cancer therapy. The site is regularly updated and includes many useful links to other resources.

https://www.cancer.org The website of the American Cancer Society contains a great deal of information on the incidence, treatment, and prevention of cancer. The site also highlights significant achievements in cancer research.

www.iarc.fr The International Agency for Research on Cancer (IARC) is an extension of the World Health Organization that aims to identify the causes of cancer and promote international collaborations surrounding cancer research.

cdn.sandiegouniontrib.com//audio/2016/04/24/levy-lymphomapodcastfinal.mp3 A podcast and interview by the *San Diego Union Tribune* with Ron Levy in 2016 about his work using monoclonal antibodies to treat B-cell lymphoma.

STUDY QUESTIONS

1. Indicate whether each of the following statements is true or false. If you think a statement is false, explain why.

   a. Hereditary retinoblastoma results from overexpression of a cellular oncogene.

   b. Translocation of the *c-myc* gene is found in many patients with Burkitt's lymphoma.

   c. Multiple copies of cellular oncogenes are sometimes observed in cancer cells.

   d. Viral integration into the cellular genome may convert a proto-oncogene into a transforming oncogene.

   e. All oncogenic retroviruses carry viral oncogenes.

   f. The immune response against a virus-induced tumor protects against another tumor induced by the same virus.

2. You are a clinical immunologist studying *acute lymphoblastic leukemia* (ALL). Leukemic cells from most patients with ALL have the morphology of lymphocytes but do not express cell-surface markers characteristic of mature B or T cells. You have isolated cells from patients with ALL that do not express membrane Ig but do react with mAb against a normal pre-B-cell marker (B-200). You therefore suspect that these leukemic cells are pre-B cells. How would you use genetic analysis to confirm that the leukemic cells are committed to the B-cell lineage?

3. In a recent experiment, melanoma cells were isolated from patients with early or advanced stages of malignant melanoma. At the same time, T cells specific for tetanus toxoid antigen were isolated and cloned from each patient.

   a. When early-stage melanoma cells were cultured together with tetanus toxoid antigen and the tetanus toxoid–specific T-cell clones, the T-cell clones were observed to proliferate. This proliferation was blocked by addition of chloroquine, a drug that accumulates in lysosomes, or by addition of mAb to HLA-DR. Proliferation was not blocked by addition of mAb to HLA-A, -B, -DQ, or -DP. What might these findings indicate about the early-stage melanoma cells in this experimental system?

   b. When the same experiment was repeated with advanced-stage melanoma cells, the tetanus toxoid T-cell clones failed to proliferate in response to the tetanus toxoid antigen. What might this indicate about advanced-stage melanoma cells?

   c. When early and advanced malignant melanoma cells were fixed with paraformaldehyde and incubated with processed tetanus toxoid, only the early-stage melanoma cells could induce proliferation of the tetanus toxoid

   T-cell clones. What might this indicate about early-stage melanoma cells?

   d. How might you confirm the hypotheses in (a), (b), and (c) experimentally?

4. Describe three likely sources of tumor antigens.

5. Various cytokines have been evaluated for use in tumor immunotherapy. Describe four mechanisms by which cytokines mediate antitumor effects and the cytokines that induce each type of effect.

6. Infusion of transfected melanoma cells into patients with cancer is a promising immunotherapy.

   a. Name two genes that have been transfected into melanoma cells for this purpose. What is the rationale behind the use of each of these genes?

   b. Why might use of such transfected melanoma cells also be effective in treating other types of cancers?

7. For each of the following descriptions, choose the most appropriate term:

Description	Term
a. A benign or malignant tumor	1. Sarcoma
b. A tumor that has arisen from endodermal tissue	2. Carcinoma
c. A tumor that has arisen from mesodermal connective tissue	3. Metastasis
d. A tumor that is invasive and continues to grow	4. Neoplasm
e. Tumor cells that have separated from the original tumor and grow in a different part of the body	5. Malignant
f. A tumor that is noninvasive	6. Leukemia
g. A tumor that has arisen from lymphoid cells	7. Transformation
h. A permanent change in the genome of a cell that results in abnormal growth	8. Lymphoma
i. Cancer cells that have arisen from hematopoietic cells that do not grow as a solid tumor	9. Benign

8. Which of the following parts of the immune response are *not* believed to be involved in cancer cell eradication

(i.e., have not been observed as a part of an effective antitumor response)?

   a. CTLs
   b. IFN-γ
   c. NK cells
   d. M2 macrophages
   e. IL-12

9. In the case of cancer, inflammation at the site of a solid tumor is a good sign, suggestive of a more positive prognostic outcome. Explain why you think this statement is true or false.

10. Describe a specific anticancer therapy that uses each of the following principles, and explain its proposed mechanism of action:

   a. Adoptive transfer of modified T cells
   b. Monoclonal antibodies
   c. Checkpoint blockade
   d. Therapeutic vaccine

CLINICAL FOCUS QUESTIONS

1. In the late nineteenth century, before radio- or chemotherapy was available, a physician named William Coley was among the first to record experiments of immune-based cancer treatment. He treated patients by injecting bacteria directly into aggressive and inoperable cancers. His technique, though controversial, met with some success and was coined "Coley's toxins." Using your understanding of immunoediting in cancer, explain specifically what you think was happening in instances where Dr. Coley was successful at inducing cancer remission or elimination.

2. Why is cervical cancer a likely target for a vaccine that can prevent cancer? Can the approach being investigated for cervical cancer be applied to all types of cancer? Why might a prophylactic vaccine against HPV not work as a therapeutic vaccine in HPV$^+$ women with cervical cancer?

Experimental Systems and Methods

Learning Objectives
After reading this chapter, you should be able to:

1. Understand the principles behind the methods used by practicing immunologists in order to be able to read more knowledgeably in the immunological literature.

2. Evaluate the appropriateness of the experimental choices made by investigators.

3. Grasp the range of methods that can be used to analyze particular immunological questions so as to be able to design appropriate experiments.

The days when an experimental methods chapter in an immunology textbook could neatly describe all the techniques used by practitioners of the subject are long gone. Modern immunologists use tools derived from the experimental arsenals of structural biologists, physicists and biophysicists, biochemists and chemists, cell biologists, anatomists, microbiologists, computer scientists, mathematicians, and physiologists. In return, the field of immunology has donated an extensive toolbox of antibody-based techniques to the sciences and to biomedicine.

In this chapter, we attempt to provide students with the capacity to understand the methodological choices made by immunological researchers. We hope that students will learn some of the advantages and limitations of many of the techniques they encounter, and we have tried to provide students with the tools to understand the context in which particular methods or techniques are applicable. These tools should also prove useful as students design their own experiments.

Fearlessness in following important scientific questions wherever they lead, and in being willing to learn whatever methods are required to answer those questions, is an

Mouse inflamed lung tissue stained for activated airway epithelial cells (green), infiltrating macrophages (red), and cell nuclei (blue). *[Courtesy Meera Nair, University of California, Riverside, Laurel Monticelli and David Artis, Perelman School of Medicine, University of Pennsylvania.]*

attribute shared by most great scientists. In this chapter, therefore, we offer students some insight into the broad array of technical possibilities available. We have included some classical techniques, but we have also worked to maintain currency so that this chapter can provide insight into the technical aspects of specific experiments described in previous chapters and in related reading in the current literature.

Space does not permit a detailed description of every technique; this chapter is designed to provide a conceptual sense of the applicability of various methods, rather than specific protocols for pursuing them. Students who wish to delve further into the details of any particular method can then locate specific protocols in various sources, many of which are noted in the Useful Websites section at the end of the chapter. For purposes of concision, we have chosen

Techniques

Antibody generation

Immunoprecipitation assays

Agglutination assays

Solid-phase support assays

Affinity quantification

Microscopy

Immunofluorescence-based imaging

Chromatin analysis

Flow cytometry

Magnetic-activated cell sorting

Cell cycle analysis

Cell death assays

CRISPR-Cas9

Whole-animal experimental systems

not to describe common methods of molecular biology or biochemistry (cloning, PCR, Southern and Northern blots, etc.) as they are well described elsewhere.

We begin by enumerating those methodologies that are used to generate antibodies, and then describe some of the many ways in which the interactions between antibodies and antigens can be analyzed. We then briefly describe some of the methods used to visualize cellular and subcellular structures in the immune system, before moving on to a discussion of various magnetic- and fluorescence-based techniques used for cell sorting and cellular analysis at the population level. Assays that analyze the cell cycle and measure cell death are then addressed, and we complete the chapter with a brief description of a number of commonly used whole-animal experimental systems.

Antibody Generation

From the early days of immunology, investigators and clinicians have made use of the ability of animals to respond to immunization by the production of antibodies directed toward injected antigens, such as viruses, bacteria, fungi, peptides, or even simple chemicals from the laboratory shelf. Antibodies harvested from the serum of immunized animals are the secreted products of many clones of B cells and are thus referred to as *polyclonal* antibodies. With subsequent immunizations using the same antigen, the average affinity of this polyclonal antibody mixture for the antigen increases, as a result of the process of affinity maturation, described in Chapter 11.

Polyclonal Antibodies Are Secreted by Multiple Clones of Antigen-Specific B Cells

Polyclonal antibodies are generated by immunizing an experimental animal or a human subject with antigen one or more times, drawing a blood sample, and purifying the antibodies from the subject's **serum**. (Serum is what remains when both the cellular components and the clotting factors have been removed from the subject's blood.) Because only a small fraction of the subject's blood is withdrawn each time, the same animal or person may be asked to provide multiple serum samples over an extended time period.

The addition of **adjuvants** to the immunizing preparation elicits a stronger immune response by deliberately engaging the innate immune system to help in the activation of antigen-specific B and T cells. Traditionally, Freund's or alum adjuvants were used to maximize mouse antibody responses to antigens and were typically mixed with the antigen prior to injection. Incomplete Freund's adjuvant is a water-and-oil emulsion, into which the antigen was mixed. Presentation in the emulsion allowed for a longer-lasting, slower release of antigen. Complete Freund's adjuvant (CFA) also includes fragments of dead mycobacteria that effectively stimulate the innate immune system. The use of CFA is now discouraged on ethical grounds because the localized inflammation and ulceration around the site of injection cause unnecessary suffering to the animal. Aluminum (alum) adjuvants are currently used in human vaccines against papillomavirus and hepatitis B virus. Antigens presented with these adjuvants for vaccination are normally adsorbed onto preformed aluminum hydroxide or aluminum phosphate complexes, which are sometimes chemically altered to form gels. The mechanism of action of alum adjuvants is still being debated, but most probably their immuno-enhancing effects result from the formation of antigen depots, analogous to those created following the use of Freund's adjuvant. Alternatively, it is possible that the adjuvant effect results from the enhanced stimulation of antigen uptake by phagocytic cells and the subsequent generation of an inflammatory response.

More current adjuvant approaches employ Toll-like receptor (TLR) agonists such as polyinosinic-polycytidylic acid [poly(I:C)] or CpG nucleotides. Poly(I:C) is a synthetic, double-stranded RNA that activates TLRs in a manner that mimics an antiviral response. Synthetic oligonucleotides that express unmethylated CpG motifs similarly trigger innate responses from cells that express TLR9, including plasmacytoid dendritic cells and B cells. CpG adjuvants promote T_H1 responses and elicit the secretion of proinflammatory cytokines. They are currently being used in a variety of clinical trials to promote immune responses against tumors and infectious agents.

Because polyclonal antibodies are a mixture of antibodies directed toward a variety of different epitopes of the immunizing antigen, they are particularly useful for techniques such as agglutination or immunoprecipitation, which rely on the ability of the antibody to form a large antigen-antibody complex. The disadvantage of using a polyclonal preparation is that some of the antibodies in the mixture may have ill-defined cross-reactivities with related antigens. Furthermore, since the antibody response matures with time postimmunization (Chapter 11), the range of cross-reactivities of different preparations of polyclonal antibodies for related antigens may vary among different blood drawings (or "bleeds"), even when the sera are derived from the same donor animal.

Key Concepts:
- Polyclonal antibodies are generated by immunizing an animal with an antigen (usually complexed with an adjuvant) one or more times, and then collecting the antiserum.
- Polyclonal antibodies are particularly useful for applications that require cross-linking multiple antigen molecules, such as in immunoprecipitation or Western blotting.
- Polyclonal antibodies suffer from the disadvantage that their cross-reactivities are not reproducible from one bleed to the next.

A Monoclonal Antibody Is the Product of a Single Stimulated B Cell

The disadvantages of unpredictable cross-reactivities of polyclonal antibodies with related (or sometimes apparently unrelated) antigens are eliminated when using *monoclonal antibodies* (mAbs), because mAbs are the product of a single, stimulated B cell and therefore do not alter with time. In 1975, Georges Köhler and César Milstein discovered how to generate large quantities of antibodies derived from a single B-cell clone (**Figure 20-1**). By fusing a normal, activated, antibody-producing B cell with a myeloma cell (a cancerous plasma cell), they were able to generate a **hybridoma** that possessed the immortal growth properties of the myeloma-cell parent and secreted the unique antibody produced by the B-cell parent. Myeloma-cell partners were quickly generated that had lost the ability to synthesize their own immunoglobulins, thus ensuring that the only antibodies secreted into the culture medium were those from the B-cell fusion partner.

The original fusions used Sendai virus to disrupt the plasma membranes of the cells; nowadays, chemical fusogens such as polyethylene glycol are used instead. In general, fusions between three or more cells are unstable, and the vast majority of fused cells growing out of these cultures are the products of the hybridization of two parent cells.

Hybrids formed by the fusion of two antibody-producing B cells will not grow out of these cultures because B cells have a relatively short half-life in vitro and die within a few days. However, hybrids formed by the fusion of two or more cancer cells would have the potential to grow out from the initial fusions, and compete successfully for nutrients with the B cell–myeloma hybrids. A method therefore had to be devised to eliminate these tumor-tumor hybrids from the cultures of fused cells. Köhler and Milstein solved this problem by using myeloma cells lacking the enzyme *hypoxanthine guanine phosphoribosyltransferase* (HGPRT).

To understand why this mutation is important in fusion partners, it is important to know that there are two potential pathways of DNA synthesis: the *de novo* pathway

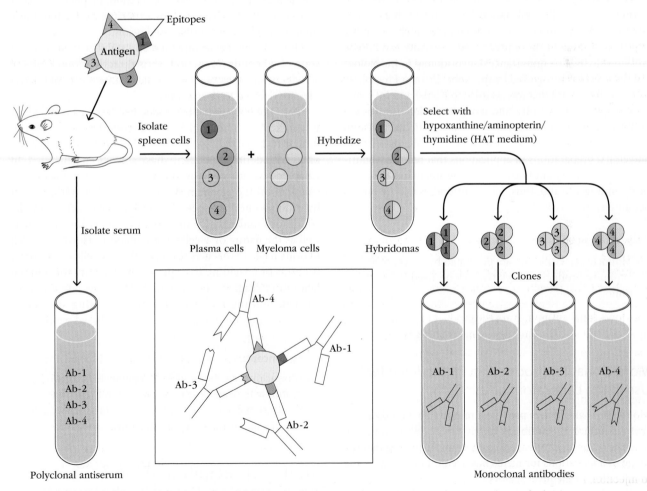

FIGURE 20-1 The generation of polyclonal and monoclonal antibodies. Conventional polyclonal antisera produced in response to a complex antigen contain a mixture of monoclonal antibodies, each specific for one of the four epitopes shown on the antigen (inset). In contrast, a monoclonal antibody, which is derived from a single plasma cell, is specific for one epitope on a complex antigen. The outline of the basic method for obtaining a monoclonal antibody is illustrated here.

(from new, or from scratch) and the *salvage* pathway. De novo DNA synthesis from its basic building blocks can be inhibited using the antibiotic aminopterin. Therefore, any cells treated with aminopterin must use the salvage pathway of DNA synthesis. The enzyme HGPRT catalyzes a required step in the salvage pathway, and therefore denying the cells access to this enzyme means that they must use the de novo pathway. By making the hybrids with myeloma cells lacking HGPRT, and then growing the hybrid cultures in the presence of aminopterin, Köhler and Milstein ensured that the mutant tumor cells and mutant tumor-tumor hybrids would be unable to synthesize new DNA by either the salvage or the de novo pathway and would eventually die.

In contrast, in the hybridomas formed by fusion between B cells and tumor cells, the B-cell parent provides the HGPRT, and so these hybrids would survive in the selection medium. Because the medium containing aminopterin is normally supplemented with hypoxanthine and thymidine to support nucleotide synthesis, the selective medium is known as "HAT medium."

The resulting clones of hybridoma cells randomly lose chromosomes over the first few days following fusion, but eventually they stabilize and can be cultured indefinitely, secreting large quantities of mAbs of predefined specificity and known cross-reactivity. The importance of hybridomas to the biological sciences was recognized by the Nobel Prize in Physiology or Medicine in 1984 that was awarded to Köhler and Milstein.

The specificity, affinity, and cross-reactivity of mAbs are entirely stable with time, and mAbs are particularly useful for diagnostic purposes. However, mAbs are less useful than polyclonal preparations in applications that require agglutination, since every antibody in the experiment has the identical binding site and, thus, fewer antibodies will be bound per antigen molecule.

> **Key Concepts:**
>
> - Monoclonal antibodies are generated by fusion of an antibody-producing B cell with a long-lived B-cell tumor.
>
> - Monoclonal antibodies are the product of a single B cell and so their binding site never changes.

Monoclonal Antibodies Can Be Modified for Use in the Laboratory or the Clinic

Monoclonal antibodies provide a reproducible binding site that will attach the antibody to any target cell or molecule that can act as an antigen, and so it is not surprising that the range of applications for mAbs has been limited only by the imagination of the investigators.

A mAb can be genetically altered so that only the binding site is retained, and the rest of the molecule is replaced. For example, because injection of large amounts of xenogeneic antibodies (antibodies from a species that is different from the recipient) can induce inflammation, mouse hybridomas

secreting mAbs to be used in immunotherapy are often subject to genetic manipulations, in which the binding sites of the original mouse mAb are cut and pasted onto the constant regions of human antibodies. In addition, some antibodies are modified by conjugation to toxins designed to kill any cells bound by the antibody. Several such chimeric and toxin-conjugated antibodies are now in regular clinical use. For example, the toxin-conjugated antibody trastuzumab emtansine (Kadcyla) is used as a therapy for HER-2–positive metastatic breast cancer and inotuzumab ozogamicin (Besponsa) is employed to treat CD22-positive B-cell precursor acute lymphoblastic leukemia, or ALL.

Many antibodies are now available to which other molecules have been covalently conjugated in ways that do not interfere with antigen binding. For example, some antibodies are modified by the attachment of either biotin or enzymes, for use in ELISAs (see below). Others are modified by conjugation with fluorescent dyes (or again, biotin) for use in immunofluorescence applications such as microscopy or flow cytometry. Yet others are attached by their constant regions to various types of synthetic beads or particles that enable their use in immunoprecipitation, magnetic cell separation, or electron microscopy experiments. Each of these different applications is discussed below.

Other mAbs specifically bind and stabilize the transition state of a chemical reaction, thus directly mimicking the activity of enzymes. Such antibodies with enzyme-like activities are referred to as **abzymes**.

Once the hybridoma technique had been established for B lymphocytes, immunologists recognized its potential usefulness to T-cell biology, and many long-term T-cell hybridomas with defined specificity have since been generated. In Chapter 3, we learned how one such hybridoma was used to characterize the biochemistry of the αβ TCR. Other T-cell hybridoma lines have been invaluable in characterizing the conditions under which different cytokines are secreted as well as the nature of T-cell subpopulations. However, T-cell hybrids have not been as useful as their B-cell counterparts in terms of providing therapeutic, diagnostic, and general laboratory reagents, because they do not secrete a soluble antigen-binding molecule and their binding specificity requires recognition of a complex, MHC-based peptide.

> **Key Concept:**
>
> - Monoclonal antibodies can be conjugated with biotin, with beads, with enzymes, with toxins, with fluorescent dyes, or with other reagents to allow their use in a broad variety of applications.

Immunoprecipitation- and Agglutination-Based Techniques

The multivalency of antibodies has allowed the development of techniques in which antibody-bound molecules can be precipitated from solution, or otherwise separated from nonbound molecules for further analysis. Some of these techniques are quite venerable; other applications are brand new. Nonetheless, all rely on the ability of antibodies to bind to more than one antigenic determinant on a single antigen, thus forming a large complex that will fall out of solution.

Immunoprecipitation Can Be Performed in Solution

When antibodies and soluble antigen are mixed in solution, the bi- or multivalent nature of immunoglobulins allows for a single antibody molecule to bind to more than one antigen (**Figure 20-2**). If the antigen is polyvalent (has more than one antibody binding site per antigen molecule, represented in Figure 20-2 by the red sites on the brown antigens), it may in its turn bind multiple different antibodies. Eventually, the resulting cross-linked complex becomes so large that it falls out of solution as a precipitate. This precipitate can be spun out of the solution and the antigen separated from the precipitating antibodies by biochemical means.

Solution immunoprecipitation can be used to purify antigenic molecules from a heterogeneous mixture of soluble molecules, or to remove particular antigens from a solution. The most efficient immunoprecipitation occurs only when the antibody and antigen concentrations are essentially equivalent (see Figure 20-2, middle). When either the antigen (Figure 20-2, left) or antibody (Figure 20-2, right) is present in excess, monovalent binding is favored that does not result in the formation of a precipitate. Recall from Chapter 3 that Kabat used immunoprecipitation with the ovalbumin antigen to remove anti-ovalbumin antibodies from solution, followed by electrophoresis; this experiment characterized antibodies as belonging to the γ-globulin class of serum proteins.

Key Concepts:
- When bi- or multivalent antibodies are mixed with antigen, they can form a cross-linked matrix, resulting in the formation of an immunoprecipitate.
- Immunoprecipitation can be used to purify soluble proteins from tissue fluids.

Immunoprecipitation of Soluble Antigens Can Be Performed in Gel Matrices

Immune precipitates can form not only in solution but also in an agar matrix. When antigen and antibody diffuse toward one another in a gel matrix, a visible line of precipitation will form. As in a precipitation reaction in solution, visible precipitation occurs when the concentrations of antibody and antigen are equivalent to one another.

Immunodiffusion in gels is rapid, easy to perform, and surprisingly accurate. In the Ouchterlony method, the most frequently employed variation of gel immunoprecipitation, both antigen and antibody diffuse radially from wells toward each other, thereby establishing a concentration gradient. At the relative antibody-antigen concentrations at which lattice formation is maximized, termed "equivalence," a visible line of precipitation, or "precipitin line," forms in the gel. More sophisticated analyses of Ouchterlony gels can offer information regarding the extent of cross-reactivity of antibody preparations with related antigens (see **Figure 20-3**).

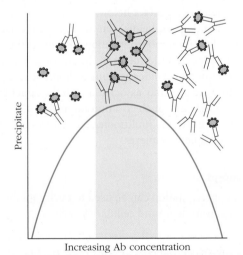

FIGURE 20-2 Immunoprecipitation in solution. When bi- or multivalent antibodies are mixed in solution with antigen, the antibodies can form cross-linkages with two or more antigen molecules, leading to the formation of a cross-linked precipitate (middle portion of graph). Precipitate formation requires that neither antigen (left-hand portion of graph) nor antibody (right-hand portion of graph) molecules are in excess. In either of these two cases, primarily monovalent binding takes place, as shown.

FIGURE 20-3 Immunodiffusion in agar gels can be used to assay for the presence of antibodies and determine cross-reactivity patterns between complex antigens and antibody samples. A polyclonal antiviral antiserum has been placed in the lower well and viral antigens in the upper two wells. On the left the two antigens are the same, as shown by the smooth precipitin curve. In the middle, the two antigens are partially identical, but the sample on the top left contains antigens not shared by the well on the top right. On the right, two different viral antigens are presented in the two top wells, that are both recognized by the polyclonal antiserum sample. *[Obtained from Amrita School of Biotechnology, Amrita Vishwa Vidyapeetham. www.amrita.edu.]*

TABLE 20-1	Sensitivity of various immunoassays	
Assay	**Sensitivity* (μg antibody/ml)**
Precipitation reaction in fluids	20–200
Precipitation reaction in gels	
Ouchterlony double immunodiffusion	20–200
Agglutination reactions	
Direct	0.006–0.06
Agglutination inhibition	0.3
Radioimmunoassay (RIA)	0.0006–0.006
Enzyme-linked immunosorbent assay (ELISA)	~0.0001–0.01
ELISA using chemiluminescence	~0.00001–0.01[†]
Immunofluorescence	1.0
Flow cytometry	0.006–0.06

*The sensitivity depends on the affinity of the antibody used for the assay as well as the epitope density and distribution on the antigen.

[†]Note that the sensitivity of chemiluminescence-based ELISAs can be made to match that of RIA.

Source: Updated and adapted from Rose, N. R., et al., eds. 1997. *Manual of Clinical Laboratory Immunology,* 5th ed. Washington, DC: American Society for Microbiology.

Although various modifications of precipitation reactions were, at one time, the major types of assay used in immunology, other, more sensitive methods are now available for antigen and antibody measurement and are described below. However, Ouchterlony assays are still used in both the research laboratory and the clinic, because of their technical ease and reproducibility. **Table 20-1** presents a comparison of the sensitivity, or minimum amount of antibody detectable, of a number of immunoassays.

Key Concept:

• Immunoprecipitation reactions in gel matrices are used to analyze the presence and quantity of antibodies or antigens.

Immunoprecipitation Enables Isolation of Specific Molecules from Cell and Tissue Extracts

Immunoprecipitation is frequently used to isolate protein antigens from cell and tissue samples. A detergent extract of cells or tissues is mixed with antibodies to the protein of interest to form an antigen-antibody complex. The detergent is carefully selected to minimize disruption of the antigen-antibody bond. To facilitate efficient retrieval of the antigen-antibody complexes, a secondary antibody or other protein that binds to the primary antibody (such as the bacterial protein A or G) may be added to the mixture. These secondary reagents all bind specifically to the Fc region of the first antibody and are usually pre-attached

to a solid-phase support, such as a synthetic bead. In some assays, the primary antibody may be attached directly to the solid-phase support. Since the beads can easily be spun down, the antibody-antigen-bead complex can be collected by centrifugation. Following centrifugation, the protein of interest can be separated from the precipitating antibodies by SDS gel electrophoresis. In a variant of this technique, the beads attached to the secondary reagents may be magnetic, in which case the protein of interest is purified by passage over a magnetic column (see below).

Western blotting (see below) can then be used to ascertain the efficacy of the immunoprecipitation, to estimate the relative abundance of the bound protein in the cell or tissue sample, and to determine which other proteins co-immunoprecipitated and therefore are probably associated with the target protein in its cellular location. Such co-immunoprecipitation studies were the first clue to the multimolecular natures of the TCR and BCR coreceptor complexes.

Key Concept:

• Immunoprecipitation can be used to purify soluble proteins from tissue and cellular extracts.

Hemagglutination Reactions Can Be Used to Detect Any Antigen Conjugated to the Surface of Red Blood Cells

The cross-linking that occurs between di- or multivalent antibodies and multivalent, cellular antigens can result in visible clumping of the complexes formed between the

FIGURE 20-4 Demonstration of hemagglutination, using antibodies against sheep red blood cells (SRBCs). The control well (well 10) contains only SRBCs, which settle into a solid "button." Experimental wells 1 to 9 contain a constant number of SRBCs plus serial twofold dilutions of anti-SRBC serum. The spread pattern in the experimental series indicates positive hemagglutination through well 3.

antigens and their cognate (binding) antibodies. This clumping reaction is called **agglutination**. Agglutination reactions are identical in principle to precipitation reactions; the only difference is that the antigen being bound is associated with a cell surface and the cross-linked product is therefore visible to the naked eye because of the larger size of the particle that contains the antigen.

When antibodies bind antigens on the surface of *red blood cells* (RBCs), the resultant clumping reaction is referred to as **hemagglutination**. In the example shown in **Figure 20-4**, control buffer was added to well 10 of the microtiter tray. Antibodies to *sheep red blood cells* (SRBCs) were added to well 1 of this tray, and then this antiserum was serially diluted into wells 2 through 9, such that the concentration of antibodies in well 2 was half that in well 1, and so on. The same number of SRBCs was then added to each well.

In well 10, in the absence of any agglutinating antibody, the SRBCs settle into a tight "button" at the bottom of the well. This tight button represents a *negative* result in a hemagglutination assay. In well 1, the high concentration of anti-SRBC antibodies induced cross-linking of the SRBCs, so that they form a cross-linked clump of cells that is too misshapen to fall down to the bottom of the well. The diffuse shading of RBCs seen in well 1 represents a *positive* interaction between the antibodies and the SRBC surface antigen. The concentration of anti-SRBC antibodies in wells 2 and 3 remains high enough to allow hemagglutination, but once the antibodies have been diluted eightfold (well 4), there are too few antibodies to generate cross-links and the SRBCs can again settle into the bottom of the well. The responses in wells 1, 2, and 3 therefore represent a positive hemagglutination reaction.

Hemagglutination reactions are routinely performed to type RBCs. With tens of millions of blood-typing determinations run each year, this is one of the world's most frequently used immunoassays. In typing for human ABO antigens, human RBCs are mixed with antisera to the A or B blood-group antigens. If the antigen is present on the cells, they agglutinate, forming a visible clump on the slide.

The ease and sensitivity of hemagglutination reactions and the fact that they do not require sensitive instrumentation for data analysis mean that hemagglutination assays can

be adapted to measure antibodies directed against any antigen that can be attached to the RBC surface.

> **Key Concept:**
> - Hemagglutination reactions measure the presence of antibodies to antigens located on red blood cells.

Hemagglutination Inhibition Reactions Are Used to Detect the Presence of Viruses and of Antiviral Antibodies

Hemagglutination inhibition reactions are also useful tools in the clinic and in the laboratory for the detection of viruses and of antiviral antibodies. Some viruses (most notably influenza) bear multivalent proteins or glycoproteins on their surfaces that interact with macromolecules on the RBC surface, and induce agglutination. Specifically, the influenza virus envelope bears a trimeric glycoprotein, hemagglutinin (HA). This HA molecule is subjected to mutation and selection, such that different strains of influenza bear different HA types, which in turn are bound by different antibodies. However, all HA molecules bind in a multivalent manner to the sialic acid residues on RBCs and agglutinate them.

To determine whether a patient has antibodies to a particular strain of influenza virus, a technician would perform a serial dilution of the patient's antiserum in a microtiter plate. The technician would then add the relevant virus and RBCs to each well, at concentrations known to allow hemagglutination. If the patient's antiserum has anti-HA antibodies that bind the particular influenza strain being tested, the antibodies will attach to the HA molecules on the surface of the virus and prevent those molecules from inducing hemagglutination. The more antibodies in the patient's serum, the more the serum can be diluted without loss of hemagglutination inhibition.

Other viruses that can cause hemagglutination, and that therefore can be tested by the hemagglutination inhibition assay, include adenoviruses, parvoviruses, togaviruses, some coronaviruses, picornaviruses, other orthomyxoviruses, and paramyxoviruses.

> **Key Concepts:**
> - Hemagglutination reactions measure the presence of antibodies to antigens located on red blood cells.
> - Hemagglutination inhibition reactions measure the presence of antibodies to those viruses that induce hemagglutination, notably influenza.

Bacterial Agglutination Can Be Used to Detect Antibodies to Bacteria

A bacterial infection often elicits the production of antibacterial antibodies, and such antibodies can be detected by bacterial agglutination reactions. The principle of bacterial agglutination is identical to that for hemagglutination, but in this case the visible pellet is made up of bacteria, cross-linked by antibacterial antibodies directed against antigens on the surfaces of the bacterial cells.

Agglutination reactions can also provide quantitative information about the concentration of antibacterial antibodies in a patient's serum. The patient's serum is serially diluted (titrated), as described above. The last well in which agglutination is visible tells us the **agglutinin titer** of the patient, defined as *the reciprocal of the greatest serum dilution that elicits a positive agglutination reaction*. The agglutinin titer of an antiserum can be used to diagnose a bacterial infection. Patients with typhoid fever, for example, show a significant rise in the agglutination titer to *Salmonella typhi*. Agglutination reactions also provide a way to type bacteria. For instance, different species of the bacterium *Salmonella* can be distinguished by agglutination reactions with a panel of typing antisera.

> **Key Concept:**
> • Bacterial agglutination reactions measure the presence of antibodies that bind specific bacterial strains.

Antibody Assays Based on Molecules Bound to Solid-Phase Supports

Many antibody-based assays now rely on antibodies or antigens that are bound to solid-phase supports, such as microtiter plates, microscope slides, or beads of various kinds. These technical modifications allow for increased automation and faster delivery of test results.

Radioimmunoassays Are Used to Measure the Concentrations of Biologically Relevant Proteins and Hormones in Body Fluids

Although many *radioimmunoassays* (RIAs) have now been replaced by enzyme-based immunoassays, some measurements of hormones are still made using this technology. Furthermore, the historical significance of RIAs necessitates a brief introduction here.

Prior to the development of the RIA, the concentrations of many biologically relevant proteins and hormones in body fluids were too low to detect by *any* known methods. Therefore, the development of the radioimmunoassay made it possible to measure substances that had previously been completely undetectable. The first description of an RIA was published in 1960 by two endocrinologists, Solomon A. Berson and Rosalyn S. Yalow, who designed an exquisitely sensitive technique to determine levels of insulin/anti-insulin complexes in diabetic patients. Their technique was rapidly adopted for measuring a variety of hormones, serum proteins, drugs, and vitamins at levels that were orders

FIGURE 20-5 Rosalyn Sussman Yalow. Creator, along with Solomon Berson, of the radioimmunoassay technique and Nobel Laureate, 1977. *[Courtesy of the Department of Chemistry, Michigan State University.]*

of magnitude lower than had previously been detectable. Their accomplishment was recognized in 1977 (several years after Berson's death), by the award of a Nobel Prize to Yalow (**Figure 20-5**).

Since the initial description of Yalow's assay, many technical variations have been developed to make the method more rapid and reliable. However, all variations depend on the availability of radioactively labeled antibody or antigen, and a method by which to separate antigen-antibody complexes from unbound reagents. We will describe one such assay here, but there are many ways to use this methodology to accomplish the desired experimental goal. Most RIAs still in use are based on the binding of antibody or antigen to a solid-phase support, such as the polystyrene or polyvinylchloride wells of a microtiter plate. The radioactive label that is most commonly used is ^{125}I, which binds to exposed tyrosine residues on proteins, with little effect on their overall structure.

Let's assume that we want to determine the concentration of a particular cytokine in the blood of a patient. First, the wells of a microtiter plate are coated with a constant amount of antibody specific for the cytokine. The surface of the plastic binds tightly, and nonspecifically, to proteins, and so the antibodies stick irreversibly to the plastic surface. An irrelevant protein is then added to block any unused protein binding sites and excess protein is washed away.

For purposes of quantitation, a standard curve based on unlabeled cytokines is generated by adding increasing, known concentrations of cytokine to the wells of one row of the antibody-coated plate. Then, a known, constant amount of radiolabeled cytokine is added to each well of that row. As more unlabeled antigen competes with the labeled antigen, less and less radiolabeled cytokine will bind. After a predetermined incubation period, the amount of plate-bound radiolabeled material is assessed by washing off the unbound material and measuring the remaining, antibody-bound radioactivity in individual wells. An example of a standard curve generated in this way is shown in **Figure 20-6**.

Measurement of the amount of cytokine in experimental samples is accomplished by treating unknown samples in exactly the same way as the standard curve. The investigator then compares the amount of radioactivity bound to the plate in the experimental wells with the radioactive signal obtained in the standard curve wells containing known amounts of unlabeled cytokine.

The total number of counts bound will depend on the fraction of antibodies labeled with the iodine, the number of iodine atoms per antibody, the age of the label, the timing of the various incubations, and the efficiency of the washing steps used to remove excess, unbound label and cytokines. Since these will vary from experiment to experiment, each experiment must therefore have its own standard curve, and the addition of the standard samples to the wells must be performed simultaneously with the addition of the experimental samples.

One can measure either antigen or antibody concentrations by using variations on this basic technique, which is extremely powerful and capable of sensitivities in the picogram range. However, with the proliferation of RIAs came increasing levels of concern about the amount of waste radioactivity generated by research and clinical laboratories and the associated risks to the scientists and to the environment. The next development in antibody-antigen solid-phase support assays was designed to use the principles of the RIA, but to eliminate the risks associated with the use of radioactivity. These new assays were called *enzyme-linked immunosorbent assays*, or ELISAs.

> **Key Concept:**
> • Radioimmunoassays provide sensitive means to measure antibody or antigen concentrations, using radioactivity.

ELISAs Use Antibodies or Antigens Covalently Bound to Enzymes

ELISAs are similar in principle to RIAs but, instead of using antibodies conjugated to radioisotopes, they use antibodies covalently bound to enzymes. The conjugated enzymes are selected on the basis of their ability to catalyze the conversion of a substrate into a colored, fluorescent, or chemiluminescent product. These assays match the sensitivity of RIAs and have the advantage of being safer and, often, less costly.

A number of variations of the basic ELISA have been developed (**Figure 20-7**). Each type of ELISA can be used qualitatively to detect the presence of antibody or antigen. Alternatively, a standard curve based on known concentrations of antibody or antigen can be prepared and used to determine the concentration of a sample.

Indirect ELISA

Antibody can be detected, or its concentration determined, with an *indirect ELISA* (see Figure 20-7a). Serum or some other sample containing primary antibody (Ab_1) is added to an antigen-coated microtiter well and allowed to react with the antigen attached to the well. After any free Ab_1 is washed away, the antibody bound to the antigen is detected by adding an enzyme-conjugated secondary antibody (Ab_2) that binds to Ab_1. Any free Ab_2 is again washed away, and a substrate for the enzyme is added. The amount of colored, fluorescent, or luminescent reaction product that forms after incubation of the enzyme with its substrate is measured using a specialized plate reader and compared with the amount of product generated when the same set of reactions is performed using a standard curve of known Ab_1 concentrations. This assay works on the principle that, the more primary antibody that is present, the more secondary antibody that will be present and hence the more enzyme activity. (A *direct* ELISA would detect the amount of antigen on the plate using enzyme-coupled antibodies, and is rarely used.)

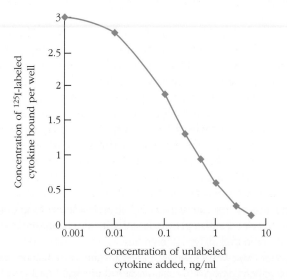

FIGURE 20-6 Competitive, solid-phase radioimmunoassay (RIA) to measure cytokine concentrations in serum. Anti-cytokine antibody is used to coat an RIA plate. A standard curve is obtained using known quantities of cytokine. As the unlabeled cytokine outcompetes the labeled form for plate binding, the amount of radioactivity per well drops. The amount of cytokine in the experimental samples is determined by interpolation from the standard curve. See text for details.

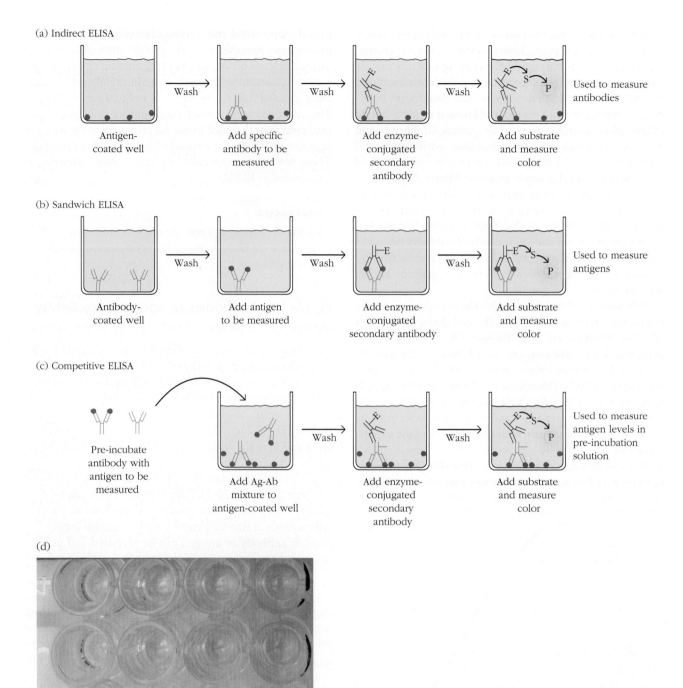

FIGURE 20-7 Variations in enzyme-linked immuno-sorbent assay (ELISA) technique allow determination of antibody or antigen. Each assay can be used qualitatively or quantitatively by comparison with standard curves prepared with known concentrations of antibody or antigen. Antibody can be determined with an indirect ELISA (a), whereas antigen can be determined with a sandwich ELISA (b) or competitive ELISA (c). In the competitive ELISA, which is an inhibition-type assay that is identical in principle to the competition RIA described in Figure 20-6, the concentration of antigen is inversely proportional to the color produced. (d) Typical control results from a sandwich ELISA: the bottom row, with known, increasing antigen concentrations from left to right, provides the data yielding the standard curve, and the top row contains the negative controls. E = enzyme; S = substrate; P = product. *[Part (d) Judith Owen.]*

This version of ELISA is the method of choice to detect the presence of serum antibodies against *human immunodeficiency virus* (HIV), the causative agent of AIDS. In this assay, recombinant envelope and core proteins of HIV are adsorbed as solid-phase antigens to microtiter wells. Individuals infected with HIV will produce serum antibodies to epitopes on these viral proteins. In general, serum antibodies to HIV can be detected by ELISA as soon as 6 weeks after infection.

Sandwich ELISA

Antigen can be detected or measured by a sandwich ELISA (see Figure 20-7b). In this technique, an antigen-specific monoclonal antibody (rather than the antigen) is immobilized on a microtiter well. A sample containing unknown amounts of antigen is allowed to react with the immobilized antibody. After the well is washed, a second enzyme-linked antibody specific for a different epitope on the antigen is added and allowed to react with the bound antigen. After any free secondary antibody is removed by washing, substrate is added, and the colored reaction product is measured. An example of what negative controls and a standard curve might look like in such an assay is shown in Figure 20-7d.

A common variant on this assay uses a biotin-linked secondary antibody and then adds enzyme-linked avidin in an additional step (see below). Sandwich ELISAs have proven particularly useful for the measurement of soluble cytokine concentrations in tissue culture supernatants, as well as in serum and body fluids. Note that, for this assay to work, the two antibodies used for the antigen immobilization (capture) and detection phases, respectively, must bind to different determinants (epitopes) on the antigen. Sandwich ELISAs therefore routinely use a pair of monoclonal antibodies termed "capture" and "detection" antibodies that bind to different regions on the antigen.

Competitive ELISA

The competitive ELISA provides another extremely sensitive variation for measuring amounts of antigen (see Figure 20-7c). In this technique, constant amounts of antibodies are first incubated in solution with samples containing variable amounts of antigen. The antigen-antibody mixture is then added to an antigen-coated microtiter well. The more antigen that is present in the initial solution-phase sample, the less free antibody will be available to bind to the antigen-coated well. After washing off the unbound antibody, an enzyme-conjugated Ab_2 specific for the isotype of Ab_1 can be added to determine the amount of Ab_1 bound to the well. In the competitive assay, the higher the concentration of antigen in the original sample, the lower the final signal, just as in the cytokine-specific RIA described above.

Modifications of ELISAs Using Biotin-Streptavidin Bonding Interactions

The original design of ELISAs involved the addition of a primary antibody to the antigen, followed by a secondary, enzyme-conjugated antibody specific for the Fc region of the primary antibody. This required that each laboratory purchase a separate set of enzyme-conjugated antibodies specific for each class of primary antibody. Investigators therefore quickly realized the advantages that would accrue if the enzyme could be bound to the primary antibody, using a more standardized method.

Biotin is a water-soluble B complex vitamin, which may have remained in chemical obscurity but for one significant property: it binds to the bacterial protein streptavidin with an affinity that is almost unparalleled in biology. Indeed, the K_d of the biotin-streptavidin interaction is on the order of 10^{-14} M, making it one of the strongest, naturally occurring, noncovalent interactions in nature. Furthermore, this interaction is stable under a wide variety of conditions, including in the presence of both organic and nonorganic solvents, denaturants, and detergents, and in extremes of temperature. Streptavidin is a tetrameric protein capable of binding four molecules of biotin per molecule of streptavidin.

Many chemical derivatives of biotin have been synthesized that enable covalent conjugation to antibodies, with minimal effect on the structure of the antibody or of biotin. Thus, one can use a variety of primary, biotin-conjugated antibodies with just one enzyme-conjugated stock of streptavidin. Biotin/streptavidin–based steps are now the norm in many immunological assays. The use of biotin-streptavidin binding in an ELISA also adds to the sensitivity of the assay, since each molecule of primary antibody is capable of being conjugated by more than one molecule of biotin and hence can bind more than one streptavidin molecule, thereby amplifying the signal from antibody binding. Similarly, each enzyme molecule will process multiple substrates into product.

In-Cell ELISAs

In-cell ELISAs are a relatively new addition to the immunological toolbox and are capable of measuring intracellular proteins without the need for cell disruption. Cells are grown and treated experimentally in 96-well microplates and then fixed and permeabilized. Following treatment with one or more primary and secondary antibody mixtures, the absorbance in each well is read by a microplate reader. The cells can then be stained with whole-cell stain and the absorbance again measured. This latter step allows the enzyme-based assay to be normalized to a per-cell basis.

Because the in-cell ELISA is so new, the advantages and disadvantages of in-cell ELISAs have yet to be fully appreciated on an experimental level. However, it is easy to see that multiple replicates of each sample will be more easily generated if cells do not have to be lysed and proteins extracted for individual sample measurements. In addition, proteins that tend to precipitate on extraction with some detergents will be retained in the in-cell procedure. Furthermore, quantitative analysis of the in-cell ELISA is automated. However, some antigens may be masked in whole-cell preparations, and the range and relative sensitivity of the in-cell ELISA methods will therefore need to be established separately for each protein-antibody pair that is analyzed.

Key Concepts:

- Enzyme-linked immunosorbent assays (ELISAs) use enzyme-conjugated antibodies to measure antigen and antibody concentrations without the need for radioisotopes.

- Different enzyme and substrate systems allow adaptation of ELISA technology to increase sensitivity. Substrates can be chromogenic, fluorogenic, or chemiluminogenic.

- Biotin-conjugated antibodies can be used with streptavidin-conjugated enzymes to increase the flexibility of ELISAs and other assays.

- In-cell ELISA methods allow measurement of proteins by antibody binding in permeabilized cells, prior to cell disruption.

ELISPOT Assays Measure Molecules Secreted by Individual Cells

A modification of the ELISA, called the ELISPOT assay, allows the quantitative determination of the number of cells in a population that are producing a particular type of molecule, such as a cytokine. The end result of an ELISPOT assay is a spot of colored precipitate at the location originally occupied by a cell secreting the cytokine that is being measured (**Figure 20-8**).

For example, in an ELISPOT assay for interferon (IFN)-γ, assay plates are coated with a monoclonal anti–IFN-γ antibody. This antibody is referred to as the "capture antibody" because its role is to capture IFN-γ as it is secreted by individual cells, before it has had time to diffuse. A known number of cells is then added to each well of the coated plates and incubated with stimulating agents. The cells settle onto the surface of the plate, and IFN-γ that is secreted by the stimulated cells is bound by the capture antibodies on the plate, creating a ring of immobilized IFN-γ–antibody complexes around each IFN-γ-producing cell. After incubation, the plate is washed to remove the cells and any excess reagents. An enzyme-linked antibody, the "detection antibody," specific for a different antigenic determinant on IFN-γ from that recognized by the capture antibody, is added. After another washing step, an ELISPOT substrate is added. Substrates for ELISPOT assays are normally colorless but, when acted on by their cognate enzyme, they produce a solid, colored product that precipitates out of solution, leaving a clearly demarcated, colored "spot" wherever the enzyme-conjugated antibody bound. The investigator then

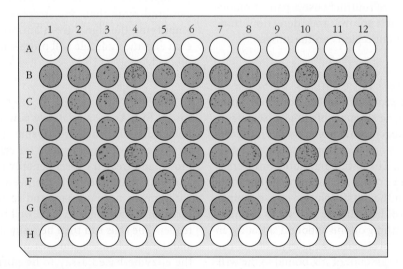

FIGURE 20-8 ELISPOT measurements of interferon (IFN)-γ secretion by NKT cells. A capture antibody to IFN-γ was used to coat the wells of an ELISA plate, and purified mouse splenocytes were added. After an 18-hour stimulation by peptide-pulsed antigen-presenting cells, the cells were washed off and a biotin-conjugated, IFN-γ–specific detection antibody was added and allowed to bind. Following removal of excess detecting antibody a color-changing substrate was added and allowed to develop. After a wash, the spots were counted via an automated plate reader and analyzed with its software.

counts the number of spots per well, either by hand or using specialized instrumentation, and calculates the fraction of cells in the original population that secreted the cytokine of interest.

Key Concept:
- ELISPOT assays measure the number of cytokine-secreting cells in a population.

Western Blotting Is an Assay That Can Identify a Specific Protein in a Complex Protein Mixture

Western blotting identifies and provides preliminary quantitation of a specific protein in a complex mixture of proteins. In Western blotting, a protein mixture is first subjected to SDS-polyacrylamide gel electrophoresis to separate the different protein species on the basis of their molecular weights (**Figure 20-9**). In order to prevent diffusion of the proteins that have been tightly focused by the electric field, the bands are then electrophoretically transferred onto a nitrocellulose or *polyvinylidene fluoride* (PVDF) membrane. The individual protein bands are subsequently identified by flooding the membrane with specific, enzyme-linked antibodies. In an alternative version of the protocol that should now be familiar to the reader, the membrane may first be incubated with a biotin-conjugated antibody, followed by washing and addition of a streptavidin-conjugated enzyme. The protein-antibody complexes that form on the membrane are visualized, just as for the ELISPOT assay, by the addition of a chromogenic substrate that produces a highly colored and insoluble product at the site of the target protein. Even greater sensitivity can be achieved if a precipitable chemiluminescent, fluorescent, or phosphorescent compound with suitable enhancing agents is used to produce light at the antigen site, which is detected by the appropriate instrumentation.

FIGURE 20-9 Western blotting uses antibodies to identify protein bands following gel electrophoresis. In Western blotting, a protein mixture is (a) treated with SDS, a strong denaturing detergent, and then (b) separated by electrophoresis in an SDS-polyacrylamide gel, which separates the components according to their molecular weight; lower-molecular-weight components migrate farther than higher-molecular-weight components. (c) The gel is removed from the apparatus and applied to a protein-binding sheet of nitrocellulose or *polyvinylidene fluoride* (PVDF), and the proteins in the gel are transferred to the sheet by the passage of an electric current. (d) Enzyme-linked antibodies are added, which detect the antigen of interest, and (e) the position of the antibodies is visualized by means of an ELISA reaction that generates a highly colored insoluble product that is deposited at the site of the reaction. Alternatively, a chemiluminescent ELISA can be used to generate light that is readily detected by exposure of the blot to a piece of photographic film.

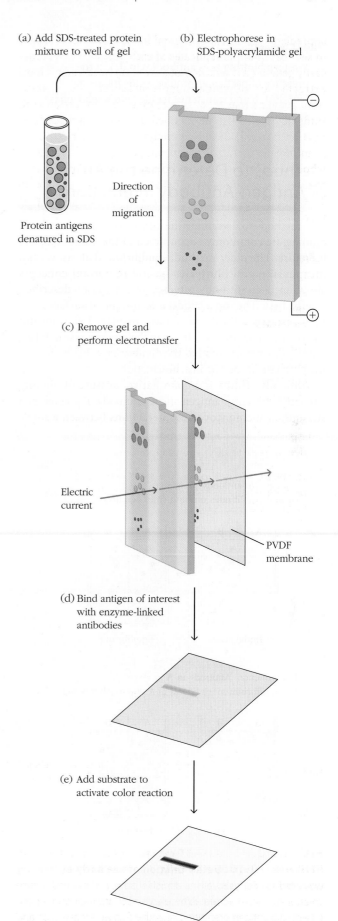

(a) Add SDS-treated protein mixture to well of gel

(b) Electrophorese in SDS-polyacrylamide gel

Protein antigens denatured in SDS

Direction of migration

(c) Remove gel and perform electrotransfer

Electric current

PVDF membrane

(d) Bind antigen of interest with enzyme-linked antibodies

(e) Add substrate to activate color reaction

Methods to Determine the Affinity of Antigen-Antibody Interactions

Two methods to determine the affinity of antigen-antibody binding are commonly encountered in the immunological literature. The older method, equilibrium dialysis, is easy, inexpensive, and illustrates several important concepts about antigen-antibody interactions. It will be described briefly, first. The more modern technique of *surface plasmon resonance* (SPR) has almost displaced equilibrium dialysis as the method of choice in the modern research laboratory (see below), but has the disadvantage that it requires the purchase of specific instrumentation.

Antibody affinity is a quantitative measure of binding strength between an antigen and an antibody. The combined strength of the noncovalent interactions between a *single*

antigen-binding site on an antibody and a *single* epitope is the **affinity** of the antibody for that epitope and can be described by the dissociation constant of the interaction, in units of molarity (see Chapter 3). (Note that the association constant is simply the reciprocal of the dissociation constant.)

$$K_{\mathrm{d}} = \frac{[S][L]}{[SL]} \qquad (1)$$

where

[S] = the concentration of antibody-binding sites
[L] = the concentration of free ligand
[SL] = the concentration of bound complexes

Equilibrium Dialysis Can Be Used to Measure Antibody Affinity for Antigen

Equilibrium dialysis uses a chamber containing two compartments separated by a semipermeable membrane. Antibody is placed in one compartment (A), and a radioactively (or otherwise) labeled ligand, small enough to pass through the semipermeable membrane, is placed in the other compartment (B) (**Figure 20-10**). In the absence of antibody, ligand added to compartment B will equilibrate on both sides of the membrane (see Figure 20-10a). In the presence of antibody, however, some of the labeled ligand molecules will be

(a)

Control: No antibody present
(ligand equilibrates on both sides equally)

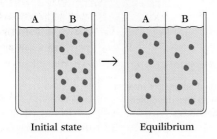

Initial state   Equilibrium

Experimental: Antibody in A
(at equilibrium more ligand in A due to Ab binding)

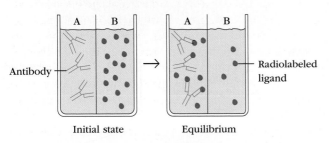

Antibody   Radiolabeled
                            ligand

Initial state   Equilibrium

(b)

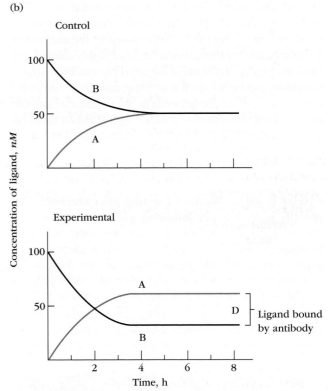

FIGURE 20-10 Determination of antibody affinity by equilibrium dialysis. (a) The dialysis chamber contains two compartments (A and B) separated by a semipermeable membrane. In the experimental chamber, antibody is added to one compartment and a radiolabeled ligand to another. At equilibrium, the concentration of

radioactivity in both compartments is measured. (b) Plot of concentration of ligand in compartments A and B with time. At equilibrium, the difference in the concentration of radioactive ligand in the two compartments represents the concentration of ligand bound to antibody: [SL] = the concentration of bound complexes.

bound to the antibody at equilibrium, trapping the ligand on the antibody side of the vessel, whereas unbound ligand will be equally distributed in both compartments. Thus, the total concentration of ligand will be greater in the compartment containing antibody (see Figure 20-10b, red line) than in the compartment with no antibody (Figure 20-10b, black line). The difference in ligand concentration in the two compartments represents the concentration of ligand bound to the antibody (i.e., the concentration of Ag-Ab complex). The higher the affinity of the antibody, the more ligand is bound.

Since the concentration of antibody placed into compartment A is known, and the concentration of bound antigen (and therefore bound antibody) and free antigen can be deduced from the amounts of radioactivity in the antibody and nonantibody compartments, respectively, the dissociation constant can be calculated.

Key Concept:

- Equilibrium dialysis is an inexpensive and relatively easy way to measure the affinity of antibody for antigens.

Surface Plasmon Resonance Is Now Commonly Used for Measurements of Antibody Affinity

Since the mid-1990s, equilibrium dialysis has been superseded as a method for affinity determination by **surface _plasmon_ resonance (SPR)**, which is more rapid and sensitive and can also provide information about antigen-antibody reaction rates (**Figure 20-11**). SPR works by detecting changes in the reflectance properties

(a)

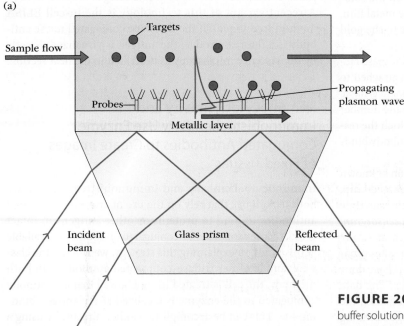

(b)

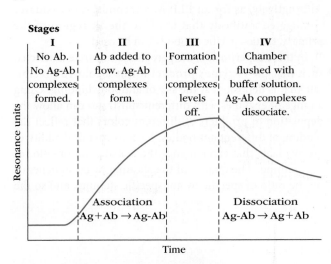

FIGURE 20-11 Surface plasmon resonance (SPR). (a) A buffer solution containing antibody is passed through a flow chamber, one wall of which contains a layer of immobilized antigen. As explained in text, the formation of antigen-antibody complexes on this layer causes a change in the resonance angle of a beam of polarized light against the back face of the layer. A sensitive detector records changes in the resonance angle as antigen-antibody complexes form. (b) Interpretation of a sensorgram. There are four stages in the plot of the detector response (expressed as resonance units, which represent a change of 0.0001 degree in the resonance angle) versus time. Stage I: Buffer is passed through the flow chamber. No Ag-Ab complexes are present, establishing a baseline. Stage II: Antibody is introduced into the flow and Ag-Ab complexes form. The ascending slope of this curve is proportional to the forward rate of the reaction. Stage III: The curve plateaus when all sites that can be bound at the prevailing antibody concentration are filled. The height of the plateau is directly proportional to the antibody concentration. Stage IV: The flow cell is flushed with buffer containing no antibody and the Ag-Ab complexes dissociate. The rate of dissociation is proportional to the slope of the dissociation curve. The ratio of the slopes, ascending over descending, equals $k_1/k_2 = k_a$.

at the surface of an antigen-coated metal sensor when it binds antibody and relies on the occurrence of electromagnetic waves, called *surface plasmon waves*, that propagate at the interface of a metal and a solvent. The nature of the wave is sensitive to any alteration in this boundary, such as the adsorption of molecules to the metal surface.

Although the physics underlying SPR is rather sophisticated, the actual experimental measurements are quite straightforward (see Figure 20-11a). A beam of polarized light is directed through a prism onto a chip coated with a thin gold film on one side and with antigen on the opposite side. The light is then reflected off the gold film toward the sensor. At a unique angle, some incident light is absorbed by the gold layer, and its energy is transformed into surface plasmon waves. A sharp dip in the reflected light intensity can be measured at that angle, which is called the **resonance angle**. The size of the resonance angle depends on the color of the light, the thickness and conductivity of the metal film, and the optical properties of the material close to the gold layer's surfaces.

The SPR method takes advantage of the last of these factors, as the binding of antibodies to the antigen attached to the film produces a detectable change in the resonance angle, and the amount of the change is proportional to the number of bound antibodies. By measuring the rate at which the resonance angle changes, the rate of the antigen-antibody binding reaction can be determined.

Operationally, this is done by passing a solution of known concentration of antibody over the antigen-coated chip. A plot of the changes in the resonance angle versus time measured during an SPR experiment is called a *sensorgram* (see Figure 20-11b). In the course of an antigen-antibody reaction, the sensorgram plot rises until all of the sites capable of binding antibody (at a given concentration) have done so. Beyond that point, the sensorgram plateaus. The data from these measurements can be used to calculate k_1, the *association rate constant*, for the antibody-antigen binding reaction.

Once the plateau has been reached on the sensorgram plot, solution containing no antibody can be passed through the chamber. Under these conditions, the antigen-antibody complexes dissociate, allowing calculation of k_2, the *dissociation rate constant*. Measurement of k_1 and k_2 allows determination of the association constant, K_a, since $K_a = k_1/k_2$. (Recall that the dissociation constant, K_d, is simply the reciprocal of the association constant.)

Key Concept:

- Surface plasmon resonance measures forward and reverse rate constants of antibody binding, allowing calculation of association constants.

Antibody-Mediated Microscopic Visualization of Cells and Subcellular Structures

Imaging of cells and tissues can be accomplished using antibodies specific for antigens present in tissues. In these experiments, the antibodies are conjugated to other molecules that can be made visible in a variety of ways. For example, if the molecule bound to the antibody is an enzyme, the presence of the antigens in the tissue sample can be visualized by the use of substrates that are acted on by the antibody-conjugated enzymes to create insoluble colored precipitates at the precise sites of antibody binding. Techniques that rely on this basic approach include immunocytochemistry and immunohistochemistry, and these methods use simple compound microscopes, coupled with computer-controlled systems to create images of fixed tissues. A recent variant of this technology is the in-cell ELISA method (see above). If the molecule conjugated to the antibody is a fluorescent dye, the antigen is visualized directly by fluorescence microscopy as described in the next section.

Immunocytochemistry and Immunohistochemistry Use Enzyme-Conjugated Antibodies to Create Images of Fixed Tissues

Immunocytochemistry and immunohistochemistry are both techniques that rely on the use of enzyme-conjugated antibodies to bind to proteins or other antigens in intact cells. A variety of enzyme-conjugated antibodies is available at this point; in explaining this strategy, we will use the classic example of peroxidase-conjugated antibodies. In both methods, the cell is treated in such a way that an antibody conjugated to the enzyme is localized at the antigen binding site. This can be accomplished either directly, by using a peroxidase-conjugated antibody to bind to the cellular antigen, or more usually, indirectly, by using a biotin-conjugated primary antibody and streptavidin-conjugated peroxidase. Alternatively, as for an ELISA, a secondary, peroxidase-conjugated antibody that binds to the Fc region of the primary, tissue-specific antibody can be used.

In these experiments, the peroxidase substrate is selected such that the enzyme's product forms a colored precipitate that is deposited at the site of antibody binding. Using a range of enzymes with different substrates can allow the deposition of products of different colors that reflect the binding of different antibodies. Various chemical additives are available that can enhance the density and/or color of the staining. The quality of the staining reaction depends on the ratio of specific to nonspecific staining, and so the

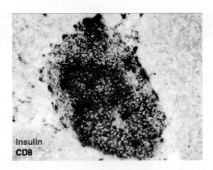

FIGURE 20-12 Enzyme-conjugated antibodies can be used to create immunohistochemical images. The pancreas from a female mouse of the nonobese diabetic strain is stained for insulin (blue) and for CD8$^+$ T cells (brown-red). This slide shows a pancreas close to the onset of clinical disease, when the infiltration of CD8$^+$ T cells is significant. *[Reprinted with permission of American Physiological Society, from van Belle, T. L., Coppieters, K. T., and von Herrath, M. G., 2011. Type I diabetes: etiology, immunology and therapeutic strategies. Physiological Reviews 2011, January; **91**(1): 79–118, Figure 3B.]*

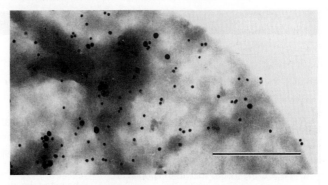

FIGURE 20-13 Immunoelectron microscopy. Immuno-electron micrograph of the surface of a B-cell lymphoma stained with two antibodies: one against MHC class II molecules labeled with 30-nm gold particles and another against MHC class I molecules labeled with 15-nm gold particles. The density of class I molecules exceeds that of class II on this cell. Magnification bar: 500 nm. *[A. Jenei et al., PNAS (1997) 94:7269-7274. Copyright 1997 National Academy of Sciences, U.S.A. Courtesy of Attila Jenei and Sandor Damjanovich, University Medical School of Debrecen, Hungary. Copyright (1997) National Academy of Sciences, U.S.A. Fig.2A.]*

investigator usually performs the staining reaction in the presence of relatively high concentrations of nonspecific proteins, such as nonfat dry milk (really!) in order to minimize nonspecific antibody binding. In expert hands, the images obtained from these techniques can be extraordinarily detailed and also aesthetically pleasing (**Figure 20-12**).

Immunocytochemistry and immunohistochemistry differ from one another in the nature of the sample being analyzed. In immunohistochemistry, the samples are prepared by sectioning intact tissue and the stained cells are therefore localized in their biological context. In contrast, immunocytochemistry is performed on isolated cells, often those grown in tissue-culture suspension. If staining is designed to detect intracellular targets, the cells must first be permeabilized by organic fixatives, or by detergent.

Key Concept:

- Immunocytochemistry and immunohistochemistry use antibodies covalently conjugated to enzymes to visualize cells and tissues. Once the antibodies are bound, substrates are added that are converted to products, which form colored precipitates that are deposited at the site of antibody binding.

Immunoelectron Microscopy Uses Gold Beads to Visualize Antibody-Bound Antigens

In order to resolve fine structural details at high levels of magnification, scientists may need to use electron, rather than light, microscopy. In such experiments, antibodies are coupled to electron-dense particles such as colloidal gold particles. Wherever antibodies bind, gold deposits will be detected by the electron microscope. By using differently sized gold particles conjugated to particular antibodies, the subcellular relationship between two or more different antigenic epitopes can be analyzed (**Figure 20-13**). Although electron microscopy provides the highest degree of resolution currently available to the average laboratory, the elaborate fixation and mounting procedures required to create samples limit its usefulness.

Key Concept:

- Immunoelectron microscopy uses antibodies conjugated to gold beads in order to visualize antibody-bound structures at high resolution.

Immunofluorescence-Based Imaging Techniques

The last decade has seen a veritable explosion in the manner in which fluorescence microscopy has been applied to studies of the immune system. Since students will be exposed to many of these techniques in the current primary literature, we present here a summary of the common variations of immunofluorescence. See also **Advances Box 20-1**, which introduces students to dynamic imaging techniques that visualize cells in real time during an ongoing immune response and offers hints on how to view the resulting movies with an educated and critical eye.

BOX 20-1

Dynamic Imaging Techniques, or How to Watch a Movie

As described in this chapter, dynamic imaging, otherwise known as intravital microscopy, allows investigators to image living tissues such as the popliteal or inguinal lymph node, or a section of the intestine of a living, anesthetized mouse. Videos of cell trafficking in intact tissue are now relatively common additions to published papers. However, videos, no less than data that appear in figures and tabular form, must be viewed with a critical eye. We suggest that students will benefit considerably if they ask the following questions, when viewing movies provided as part of immunological papers:

- **Which cells are labeled?**

  For technical reasons, investigators can typically label only two or three cell types at a time, and this means you are seeing only a fraction of the cellular interactions that are occurring in the imaged tissues. Apparently empty areas in a video may, in reality, be dense with unlabeled cells and extracellular material. For example, it was first thought that T and B cells moved through lymph nodes simply by following soluble gradients of chemokines. However, both static and dynamic imaging data have now revealed networks of reticular cells and fibers on which the cells travel in ways that are assisted by chemokine gradients. Therefore, always ask yourself what isn't there; not just what is.

- **What fraction of the population of interest carries the label?**

  In some experiments, labeling all the cells of interest would lead simply to a blur of label, so investigators sometimes elect to label only a small fraction of their input population. Although the activity of this fraction is most probably representative of all similar cells, the picture we see in our image should be modified in our heads to reflect reality. In addition, any

subsequent calculations regarding the ratios of the relative numbers of cooperating cells (e.g., T cells, or dendritic cells, reacting with B cells) will have to take into account the fraction of cells that carry the original label.

- **How are cells labeled and how stable is the label?**

  If cells are labeled with a fluorescent protein such as GFP or YFP, make sure you understand the properties of the promoter/enhancer system that drives its activity. In **Figure 1a**, the promoter that is driving YFP drives expression of the CD11c antigen, so any cell that bears CD11c will also glow yellow, even if the figure legend tells you that only DCs are labeled. Alternatively, if the cells are labeled with fluorescent antibodies before infusion, information should be obtained about the specificity and cross-reactivity of the antibody that is used in the labeling procedure, so you can understand

(a) CD11c-YFP dendritic cells imaged in mouse ear

exactly what cells and tissues will be labeled. Another variable to consider is the stability of the label. The apparent loss of fluorescent cells could mean the fading of a stain, particularly if illumination is prolonged and the dye is subject to quenching. Alternatively, the loss of fluorescence could result from the reduction in expression of a fluorescent protein after cell proliferation or proteasomal destruction, and not necessarily the loss of a whole cell population. Figure 1b shows an experiment in which T cells were labeled with CFSE prior to injection, allowing investigators to track not only the location of the injected T cells, but also the frequency of their cell division.

- **When and how are cells introduced?**

  Most current dynamic imaging systems involve the introduction of fluorescently tagged cell populations into a recipient mouse. However, manipulations of cells performed in vitro, prior to introduction, can alter cell function and subsequent cell migration.

(b) CFSE staining

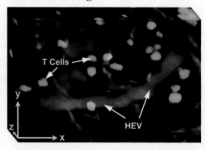

FIGURE 1 Fluorescence labeling of immune cells. (a) In this transgenic mouse, the gene for YFP is downstream of the CD11c promoter. Two-photon intravital microscopy was used to visualize the movements of dendritic cells (yellow) in the skin after infection with *Leishmania major*. The extracellular matrix can be seen because it produces what are known as "second harmonic" fluorescence signals under the illumination of the incident light. The migration of these cells can be visualized in this YouTube video: www.youtube.com/watch?v5XOeRJPIMpSs. (b) Cells (e.g., T cells) can be labeled with membrane-permeable dyes that fluoresce green (CFSE) or red (CMTMR) and injected intravenously (adoptively transferred). Miller and colleagues were among the first to visualize CFSE-labeled T-cell behavior in a lymph node, a static example of which is shown here with HEV labeled red. *[Part (a) from Ng LG, et al., Migratory dermal dendritic cells act as rapid sensors of protozoan parasites. PLoS Pathogens 2008, November; 4(11):e1000222. p. 4, Figure 2a. Part (b) from Mark J. Miller, et al., Autonomous T cell trafficking examined in vivo with intravital two-photon microscopy. PNAS, Vol. 100, No. 5:2604–2609. ©2003 National Academy of Sciences, USA.]*

(continued)

Antigen-presenting cells, for instance, can be nonspecifically activated in culture, which can change their trafficking patterns and cell interactions. Awareness of the potential artifacts introduced by in vitro manipulations will improve your ability to understand and critique the videos. In addition, you should always look for, and understand the rationale behind, the timing and sequence with which cells are introduced into an animal.

- ### When and how is antigen introduced?

  The ability to track cells during an immune response requires that antigen be introduced. Antigen can be noninfectious—for example, a foreign protein such as ovalbumin, for which transgenic T and B cells are specific—or infectious (e.g., *Toxoplasma gondii; lymphocytic choriomeningitis virus,* LCMV). Antigen introduction by infection with living pathogens will inspire a natural, full-fledged innate immune response that can in turn alter cell migration patterns. Antigens can also be introduced by a variety of routes, and the mode of introduction can dramatically influence the quality and outcome of the immune response. For example, respiratory system immunity is best generated by introducing antigen intranasally. Antigen may also be given in the presence or absence of adjuvant or cytokines. Be particularly careful to take the nature and mode of introduction of antigen into consideration when these are **not** the primary topics of interest of the investigators, as it is then that potential artifacts are most likely to occur.

- ### How are investigators visualizing antigen-specific immune cells?

  In order to understand the behavior of antigen-specific lymphocytes, one has to be able to distinguish them from the millions of other lymphocytes that are not specific for the particular antigen of interest. Investigators will often introduce fluorescent lymphocytes whose antigen specificity is known (e.g., cells isolated from mice that express transgenic T-cell receptors or B-cell receptors). This approach allows one to specifically trace the behavior of antigen-specific cells. However, transgenic systems are not available for every antigen of interest and can introduce their own artifacts (e.g., receptor levels tend to be high, and high cell frequencies do not mimic physiological responses). Investigators have found some clever ways around this by introducing into the genome of the pathogen an antigen against which receptor transgenics react (e.g., ovalbumin or ovalbumin peptide.) They have also found ways to infer antigen specificity from behavior after introducing T cells that are generally fluorescent.

- ### What type of tissue is being examined, and under what conditions?

  Lymph nodes are easier to image by two-photon intravital microscopy than is the spleen, which has more structures that autofluoresce (i.e., generate a fluorescence signal in the absence of addition of a label). The temperature, humidity, and culture conditions for the lymph node will have a significant influence on cell trafficking, and therefore isolated organs must be maintained in perfusion chambers that stabilize these conditions. However, even with these variables accounted for, cellular traffic patterns in explanted organs may not reflect all that is physiological since the surgery required to image the node can itself inspire inflammation and stress that in turn influence trafficking.

Questions Dynamic Imaging Allows Us to Ask and Answer

Dynamic imaging is now a vital component of an immunologist's experimental arsenal. It can offer visually stunning confirmation or refutation of predictions suggested by more indirect and static approaches, and it can also provide information about the sequencing of immune cell interactions within a microanatomical context that could not otherwise be gained. Dynamic imaging has already revealed unexpectedly important sites of immune-cell interaction (e.g., the subcapsular sinus of a lymph node), unexpected participants (neutrophils) in lymph node activity, and unique shape changes and motility patterns (e.g., within germinal centers) that challenge or enhance conventional immunological wisdom. It has allowed investigators to directly address centrally important questions: Which cells do naïve T cells encounter first during an immune response? What types of interactions are associated with activation, death, or tolerance? And ever more specific questions can be asked. For instance, do somatically mutating dark-zone germinal-center B cells travel often to the light zone to test their antigen specificity? Furthermore, dynamic imaging is also moving beyond the lymph node into studies of barrier immunity and intestinal vasculature in a way that opens up new opportunities to investigate intestinal pathogens.

However, even these striking imaging experiments suffer from technical limitations. Observations can be made only in one or two organs of a mouse in a single experiment, and only for 20 to 60 minutes at a time. Dynamic imaging is dependent on experimental systems that have been extensively manipulated, and the antigen-specific cell frequencies observed in such experiments are usually significantly higher than they are under physiological situations.

Nonetheless, dynamic imaging cannot be overestimated as an educational tool that inspires as effectively as it teaches. The immune system is a marvelous tangle of multidimensional complexity and organization, and a simple video can clarify in minutes what our rich yet imperfect language cannot easily articulate. Richard Avedon, a twentieth-century American photographer, perhaps said it best: "All photos are accurate. None of them is the truth." However, moving images of cells in multidimensional time and space may inspire us to move ever closer to the truths inherent in our discipline.

Fluorescence Can Be Used to Visualize Cells and Molecules

The phenomenon of fluorescence results from the property of some molecules (fluorescent dyes) to absorb light at one wavelength and emit it at a longer wavelength. If the emitted light has a wavelength in the visible region of the spectrum, the fluorescent dye can be used to detect any molecules bound by that dye.

Some fluorescence imaging experiments take advantage of the fact that particular dyes specifically bind to particular macromolecules. For example, the blue dye 4′,6-diamidi-no-2-phenylindole (DAPI) specifically stains DNA. Other protocols utilize the affinities of easily obtained proteins (which can be readily conjugated with fluorescent dyes) to bind to biologically important molecules. For example, the protein phalloidin, which specifically binds filamentous actin, can easily be conjugated to fluorescent probes. Similarly, the soluble protein annexin A5 binds to phosphatidylserine, which is exposed on the outer surface of cells undergoing apoptosis; fluorescently labeled annexin A5 is easy to obtain and use as a measure of apoptosis. In immunofluorescence measurements, antibodies or streptavidin can be artificially conjugated to a host of dyes. These various approaches to fluorescence labeling can be combined with one another to provide spectacular images of cellular and subcellular structures (**Figure 20-14a**).

Such images can be visualized by fluorescence microscopy, which uses short-wavelength light to excite the fluorescent dyes. A series of filters and mirrors that can be adjusted by the investigator can then be employed to select which wavelengths of light will reach the eyepiece and therefore which molecules will be visualized (see Figure 20-14b). Modern instruments use combinations of several filters and mirrors that allow the investigator to detect light emitted at multiple different fluorescence wavelengths. The different colored images generated from the various dyes can then be overlaid by the instrument's software to provide a single representation in which the locations of the antibody-bound molecules can be compared.

In addition to artificially synthesized dyes, some naturally occurring proteins such as GFP and RFP (green and red fluorescent protein, respectively) contain fluorescent chromophores. **Figure 20-15** shows a striking image of three neonatal mice engineered to express GFP under the control of an actin promoter, along with their non-GFP littermates. In some experiments, investigators will place the fluorescent protein genes under the control of particular promoters or make fusion proteins that contain both the native protein and GFP sequences. These adaptations allow researchers to use fluorescence imaging to determine where and when proteins under the control of those same promoters are expressed.

Key Concept:

- Antibodies covalently conjugated to fluorescent probes provide vivid images of structures under the fluorescence microscope.

(b)

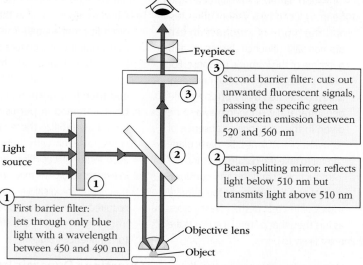

Second barrier filter: cuts out unwanted fluorescent signals, passing the specific green fluorescein emission between 520 and 560 nm

Beam-splitting mirror: reflects light below 510 nm but transmits light above 510 nm

First barrier filter: lets through only blue light with a wavelength between 450 and 490 nm

(a)

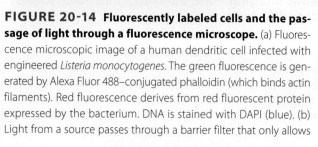

FIGURE 20-14 Fluorescently labeled cells and the passage of light through a fluorescence microscope. (a) Fluorescence microscopic image of a human dendritic cell infected with engineered *Listeria monocytogenes*. The green fluorescence is generated by Alexa Fluor 488–conjugated phalloidin (which binds actin filaments). Red fluorescence derives from red fluorescent protein expressed by the bacterium. DNA is stained with DAPI (blue). (b) Light from a source passes through a barrier filter that only allows passage of blue light of particular wavelengths. The light is then directed onto the sample by a dichroic mirror that reflects light of short wavelengths (below approximately 510 nm) but allows passage of higher wavelengths. When the blue light interacts with the sample, any fluorescent molecules excited by it emit fluorescence that then passes through the dichroic mirror, through a second barrier filter, and then is transmitted to the eyepiece. *[(a) Image courtesy of Dr. Keith Bahjat, Earle A. Chiles Research Institute.]*

FIGURE 20-15 Fluorescence labeling of whole animals with green fluorescent protein. Three neonatal transgenic mice express GFP under the control of an actin promoter. *[Hiroshi Kubota, et al., "Growth factors essential for self-renewal and expansion of mouse spermatogonial stem cells." PNAS Nov. 23, 2004, Vol.101, No.47, 16489-16494. ©2004 National Academy of Sciences, USA. Courtesy James Hayden, RBP, Hiroshi Kubota and Ralph Brinster: School of Veterinary Medicine, University of Pennsylvania. Copyright (2004) National Academy of Sciences, U.S.A. Figure 4c.]*

Confocal Fluorescence Microscopy Provides Three-Dimensional Images of Extraordinary Clarity

One of the limiting factors in obtaining clear images by fluorescence microscopy is that fluorescent molecules lying above and below the focal plane can contribute to the light that reaches the objective, leading to a blurred image. In confocal microscopy, that artifact is eliminated by using an objective lens that focuses the light from the desired focal plane directly onto a pinhole aperture in front of the detector (**Figure 20-16**). Light emitted from molecules located at other levels within the sample is stopped at the perimeter of the pinhole, and remarkably clear images of a single plane within the sample can thus be generated. In **laser scanning confocal microscopy**, investigators use lasers to provide the exciting light and computing power to move the focal plane in all three dimensions, thus enabling them to scan an *x, y* plane at different depths of focus, and reconstitute powerful three-dimensional images.

> **Key Concept:**
> - By using a pinhole to allow images only from a particular depth of field, confocal microscopy enables the visualization of tissues at different focal planes. Software can then be used to re-create a three-dimensional picture of the tissue.

Multiphoton Fluorescence Microscopy Is a Variation of Confocal Microscopy

Two-photon and multiphoton microscopy are variations on confocal microscopy that offer even greater resolution in the development of three-dimensional images. In standard confocal microscopy, excitation of the fluorescent probes (dyes) occurs along the whole path of the laser beam through the tissue (**Figure 20-17a**). This means that, although the emission beams are derived only from a single level within the sample, fluorescent probes are being excited throughout many levels of the tissue. Since fluorescent probes will eventually photobleach (i.e., cease to emit light) after extensive

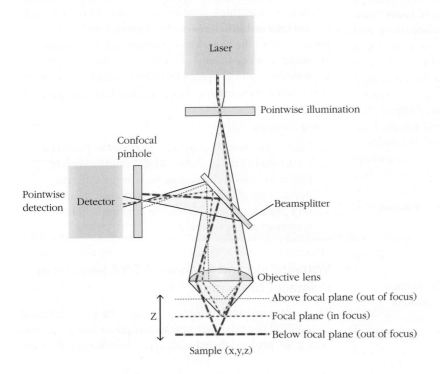

FIGURE 20-16 The principle of confocal microscopy. The sample is illuminated by a laser beam that excites fluorescence from dyes in several different focal planes, represented here by the green, red, and purple lines. However, passage of light through the confocal pinhole (shown on the left-hand side of the image) filters out light emitted from all but a single focal plane, resulting in an extraordinarily clear image. Computer control of the exact plane from which light can be received through the pinhole allows the development of images from a number of focal planes and the creation of a composite three-dimensional representation.

(a) 1-photon

(b) 2-photon

Fluorescence from out-of-focus planes →

← Fluorescence from focus spot only

FIGURE 20-17   Fluorescence excitation by one-photon versus two-photon laser excitation. (a) On the left-hand side, we visualize excitation by a short-wavelength, high-energy laser that excites fluorescence from all molecules in its path. (b) On the right-hand side, we see the fluorescence that results when molecules are excited with near simultaneity by two beams of longer wavelength, lower energy laser light. Since excitation can occur only in the region where the two beams intersect, it is limited to a single focal plane. *[Photos by Steven Ruzin, College of Natural Resources Biological Imaging Facility, University of California, Berkeley.]*

excitation, this limits the useful life of the sample. It also means that some additional light is emitted from the sample that must be filtered out of the final image.

In multiphoton fluorescence microscopy, long-wavelength lasers are used that emit in the infrared region of the spectrum. The laser beams are relatively low energy, and so more than one photon must impinge on a fluorescent molecule in order to provide sufficient energy to excite the electrons. The low energy of these infrared lasers minimizes the extent of photobleaching and enhances the useful lifetime of the sample, since those parts of the sample that interact with only a single laser beam are not typically damaged. Furthermore, the fact that at least two beams of light are required to bring about fluorescence excitation ensures that *excitation occurs only within the plane of intersection of the laser beams* (see Figure 20-17b). By moving the focal point of excitation within the x and y planes, information about a full optical section can be generated, and that whole process can then be repeated on additional z levels, thus giving rise to a three-dimensional image. In Chapter 14, you have seen some of the powerful images developed with this technique.

Key Concept:

- Two- or multiphoton microscopy provides further resolution by requiring that two or more photons simultaneously impinge on a fluorescent probe before emission is possible.

Intravital Imaging Allows Observation of Immune Responses in Vivo

Intravital imaging takes advantage of the ability to maintain both the blood and the lymphatic circulations within lymph nodes, sections of intestine, or other organs, after they have been gently lifted out of an anesthetized donor and onto a warmed perfused microscope stage. Using a multiphoton microscope, three-dimensional images of fluorescently labeled cells and structures can be generated and information gleaned about the behavior of immune cells and molecules essentially in vivo. Advances Box 20-1 describes this approach in more detail and Chapter 14 describes significant advances that have been made using this, and related technologies.

Key Concept:

- Whole organs such as lymph nodes can be placed on a warmed microscope stage while maintaining lymphatic and blood circulation. By labeling different populations of cells in vivo with particular fluorescent probes, intravital images of ongoing immune responses can be captured.

Visualization and Analysis of DNA Sequences in Intact Chromatin

Three-dimensional fluorescence in situ hybridization (3-D FISH) is a relatively new technique in which fluorescent RNA probes labeled with several different colors are generated that hybridize with known DNA sequences. For example, probes have been developed that hybridize with V_H sequences known to be situated farthest from the D_H regions (distal V_H sequences), or with V_H sequences closest to the D_H region (proximal V_H sequences). Investigators have used combinations of such probes to study the relative arrangements of BCR and TCR genes in cell nuclei at defined stages of lymphoid development. Sophisticated analytical tools then use the distances measured between the individual probes to develop topological maps of the chromosomes. **Figure 20-18** is a striking image that allows us to see with our own eyes how the distance between D-J_H distal and D-J_H proximal V_H genes is altered in pro-B cells as the Ig-encoding chromosome undergoes contraction during V-D recombination.

Key Concept:

- Using polynucleotides labeled with a fluorescent probe enables investigators to visualize the three-dimensional arrangement of particular nucleotide sequences in situ.

Flow Cytometry and Cell Sorting

Flow cytometry is the *sine qua non* (without which, nothing) of the modern immunologist's toolbox. It was developed in the early 1980s by Leonard and Leonore Herzenberg and colleagues and found some of its earliest applications in analyzing lymphocyte subpopulations derived from patients

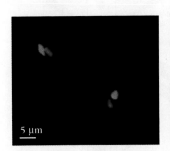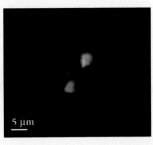

FIGURE 20-18 Three-dimensional fluorescence in situ hybridization (3-D FISH). 3-D FISH is used to show the distances between proximal (V_H7183—red) and distal (V_HJ558—green) genes in EL4 T cells (left) relative to those in pro-B cells (right). On the left-hand side, we see that, in T cells, the proximal and distal gene families (red and green, respectively) are at a distance where they can be readily discriminated from one another. In contrast, in pro-B cells, where the Ig chromosome is undergoing contraction, the two gene families are closer together and the two colors can barely be resolved, if at all. [*Republished with permission of American Association for the Advancement of Science, from Kosak, S. T, et al. Subnuclear compartmentalization of immunoglobulin loci during lymphocyte development. Science, 2002, April 5; 296(5565); pp. 158–162.*]

suffering from HIV infection. Flow cytometry is an *analytical* technique that quantifies the frequencies of cells that are made fluorescent by binding to fluorescent antibodies or to other probes.

Flow cytometers also simultaneously measure the light that is scattered by each cell. The amount of light scattered in a "forward" direction, that is, the same direction of the laser beam, provides a rough measure of the size of the cell that caused the light scattering. The amount of light scattered perpendicular to the laser light beam provides information about the presence or absence of intracellular membranous structures, such as endoplasmic reticulum, intracellular vesicles, and so on. Thus, light-scattering measurements allow the investigator to learn about the size and intracellular complexity of each cell within a population, and this information can be integrated with that provided from fluorescence measurements.

When a **flow cytometer** is adapted to *sort* cell subpopulations on the basis of their fluorescence and light-scattering properties, it is referred to as a ***fluorescence-activated cell sorter*** (FACS). In a FACS instrument, cells that bear the desired combination of light-scattering properties and fluorescent markers are tagged with an electrical charge, and can then be deflected into a separate tube for further analysis. Monoclonal antibody and FACS technologies were developed at around the same time, and the two technological breakthroughs proved synergistic: the more antibodies that were available for cell typing and sorting, the more informative flow cytometric experiments became.

In fluorescence microscopy, the sample is stationary, and the investigator seeks to learn a great deal about relatively few cells or structures. In contrast, in flow cytometry, the goal is to analyze hundreds of thousands, or even millions of cells, that flow past the laser light beam, in order to ascertain the frequency at which particular subpopulations of cells occur within the sample. The earliest experiments simply used two different antibodies, labeled with red and green fluorescent probes, to measure $CD4^+$ versus $CD8^+$ T cells in the blood of patients suspected of being infected with HIV. (Since the virus attaches to CD4, a reduced frequency of $CD4^+$ cells is indicative of active, or advanced disease.) Nowadays, sophisticated instruments measure 20 or more parameters per cell, using instruments equipped with multiple laser beams and cells tagged with numerous fluorescent labels.

Clearly, the engineering required to enable an investigator to integrate all the of the fluorescence and light-scattering information generated from an individual cell streaming past a laser beam is extremely complex, and the software and hardware requirements for running modern instruments are formidable. Nonetheless, most modern instruments have user interfaces that readily allow well-trained immunologists to learn a great deal about the biology of the populations being studied.

The Flow Cytometer Measures Scattered and Fluorescent Light from Cells Flowing Past a Laser Beam

The flow cytometer uses laser beams to provide fluorescence excitation and a series of detectors to identify both scattered light and fluorescence signals from single cells flowing in a focused stream past the laser. The simplest cytometers use a single argon laser that emits a high-intensity beam at 488 nm, although high-end cytometers now use lasers that can deliver eight or more distinct excitation wavelengths. The path taken by the scattered and emitted fluorescent light in a typical argon laser cytometer is shown in **Figure 20-19a**. The nature of the filters and mirrors shown in this light path is explained in detail in **Advances Box 20-2**. Cells in a well-mixed suspension are introduced to the instrument at the sample injection port (SIP) (see Figure 20-19b). Within the SIP tube, a rapidly flowing saline sheath encircles the narrow stream of cells, focusing the sample and ensuring that cells pass in front of the laser beam one at a time. Using a fast-flowing annular (donut-shaped) stream of liquid to focus the cell suspension in this way is referred to as "hydrodynamic focusing."

Every time a cell passes in front of the laser beam, some of the light is scattered, as described above. The wavelength of the scattered light remains unchanged from that of the incident beam. However, in addition to being scattered, some of the incident light interacts with the fluorescent dyes on the cell surfaces or within their cytoplasm and is re-emitted at a wavelength characteristic of each fluorescent dye. The raw data collected from each cell are referred to as *list mode data*, and must be further processed in order to generate the plots found in the literature. All of the attributes of every cell are kept in the original list mode file, and so the data can be analyzed in a great variety of ways, long after the cells have been discarded.

When a cell passes in front of a laser beam, most of the light is scattered forward, that is, within 5 to 10 degrees of the direction of the light path of the laser beam. This light is detected by a photodiode, which is not particularly sensitive and so is well-suited to this task. As alluded to earlier, larger cells scatter more light in the forward direction than do smaller ones; the more *f*orward light *s*catter (FSC), the larger the cell. In contrast to the low sensitivity of the photodiode, most light detectors in flow cytometers are more sensitive detectors called *photomultiplier tubes* (PMTs), which

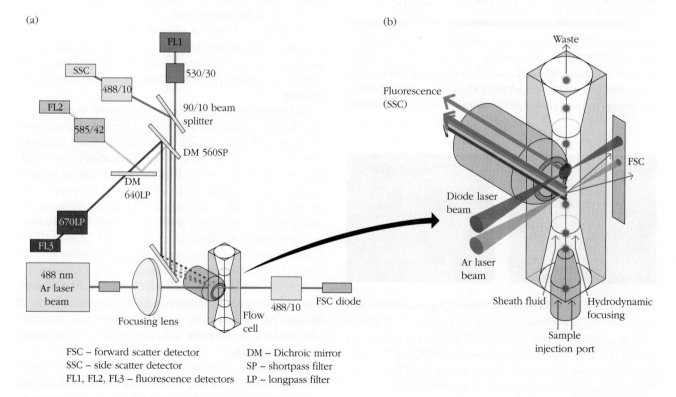

(a)

FL1
SSC
488/10
530/30
FL2
585/42
90/10 beam splitter
DM 560SP
DM 640LP
670LP
FL3
488 nm Ar laser beam
Focusing lens
Flow cell
488/10
FSC diode

(b)

Waste
Fluorescence (SSC)
FSC
Diode laser beam
Ar laser beam
Sheath fluid
Hydrodynamic focusing
Sample injection port

FSC – forward scatter detector
SSC – side scatter detector
FL1, FL2, FL3 – fluorescence detectors

DM – Dichroic mirror
SP – shortpass filter
LP – longpass filter

FIGURE 20-19 A simple flow cytometry setup. (a) Cells passing through the flow cell are interrogated by the laser. Scattered and fluorescent light is directed through the series of mirrors and filters to the appropriate photo-diode or PMT. There, the induced voltages are digitized and represented by the software in graphical form. Since each parameter of light scatter or fluorescence is recorded for each cell detected, results can be displayed that include any combination of parameters for the cell population being studied. A variety of display styles are available, depending on the graphing software used by the investigator. See text for further details. For clarity, the red diode laser shown in part (b) is omitted in part (a). (b) Cells entering the flow cell are focused by the encircling sheath fluid and exposed to the laser beam one cell at a time. Scattered and fluorescent light beams are visualized leaving the flow chamber.

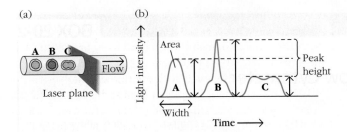

FIGURE 20-20 Nature of the voltage pulse is determined by the shape of the emitting structure. The size and shape of the voltage pulse are determined by the parameters of the cells and particles that are emitting the light signal. (See text for further details.)

record a pulse of electrons each time they receive a light signal. Each pulse has a *height* (H), *width* (W), and *area* (A), which can be measured, digitized, and recorded by the instrument (**Figure 20-20**). A second light-scattering detector is placed at approximately 90 degrees to the path of the laser beam. The amount of *side-scattered* (SSC) light offers an indication of the extent of intracellular complexity of the scattering cells. For example, the more intracellular membranous structures, such as *endoplasmic reticulum* (ER) and mitochondria, are present in the scattering cell the higher the level of side scatter.

Key Concept:

• Flow cytometry allows quantitative measurements of the frequencies of cells binding to particular fluorescent antibodies or substrates and also provides information about their forward- and side-scattering properties.

Sophisticated Software Allows the Investigator to Identify Individual Cell Populations within a Sample

Figure 20-21a shows a typical FSC-versus-SSC plot from human peripheral blood. Neutrophils are distinguished by high forward and side scatter and can therefore be found high and to the right on an FSC-to-SSC scatter plot. In contrast, unactivated lymphocytes, which are small and do not have very many internal membranes, are found at the low, left-hand side of the scatter plot. Flow cytometric software can draw a "gate" around any population the investigator wishes to analyze. Many experiments that immunology students will analyze will begin with a gated population of lymphocytes, an example of which is shown in Figure 20-21a as the area labeled R1.

The dyes of fluorescently tagged antibodies bound to surface antigens of particular cells can be excited by the laser. After excitation, the fluorescent dyes emit light that is recorded by a series of photomultiplier tubes located at a right angle to the laser beam (in order to minimize

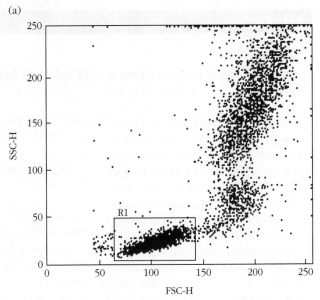

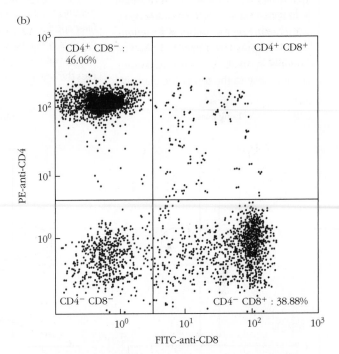

FIGURE 20-21 Typical dot plots of cytometric data. (a) A typical FSC-versus-SSC plot of peripheral white blood cells. (b) A fluorescence dot plot showing the distribution of CD4+ versus CD8+ cells in peripheral blood. CD4+ cells were stained with phycoerythrin (PE, red) dye and CD8+ cells with fluorescein (FL, green) dye. See text for details.

interference from the bright, forward-scattered light). Each photomultiplier tube is placed behind a series of filters and mirrors, as shown in Figure 20-19a and further described in Advances Box 20-2, so that it only receives and detects light within a particular range of wavelengths. Flow cytometers count each cell as it passes in front of the laser beam and record the level of emitted fluorescence at each

Optical Components of the Flow Cytometer

As is described in the accompanying section, use of the flow cytometer requires that the instrument can separate and resolve light of multiple different wavelengths that emanates from each cell. This is accomplished with a complex array of beam splitters, mirrors, and filters. Examples of the application of each of these are shown in Figure 20-19a.

Beam splitters perform precisely the function specified by their name—they split a beam of light into two or three beams that travel in different directions, without altering the nature of the wavelengths represented in that light. For example, in Figure 20-19a we see a beam splitter, close to the top of the diagram,

that is sending different fractions of the beam to the SSC photodiode and the FL1 photodiode.

The *filters* used to sort and guide the light coming from a cell fall into several categories, which are illustrated in **Figure 1**.

Band pass (BP) filters only allow light within a certain range of wavelengths to pass. For example, a BP filter might only allow passage of wavelengths from 520 to 550 nm. BP filters are routinely described by a pair of numbers, the first of which denotes the central wavelength of the BP filter and the second of which shows the range of wavelengths on either side of that central wavelength that will be allowed to pass through the filter (e.g., 530/30).

Long pass (LP) filters allow passage of any light of wavelengths longer than the filter specification but reflect (or stop) light of shorter wavelengths. A 640 LP filter, for example, would allow any wavelengths longer than 640 nm to pass through, but would reflect light of wavelengths shorter than 640 nm.

Short pass (SP) filters operate in a converse manner, letting wavelengths lower than the filter specification pass, but reflecting higher wavelengths.

Dichroic mirrors are very useful; they reflect light of specific wavelengths, allowing all other light to pass, and can also be referred to as SP or LP, depending on their characteristics. In the example shown in

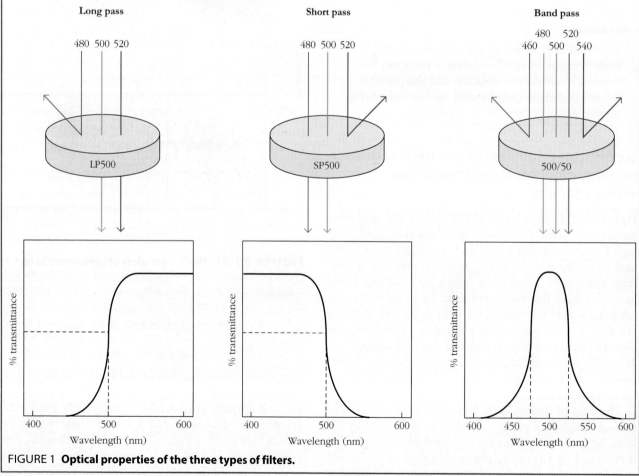

FIGURE 1 **Optical properties of the three types of filters.**

(continued)

Figure 20-19a, the DM 560nm SP dichroic mirror immediately above the collection lens allows the passage of light of wavelengths shorter than 560 nm through to the FL1 and SSC detectors, but reflects light of longer wavelengths. Similarly, the DM 640LP collects the reflected light from the first mirror and allows light of wavelengths longer than 640 nm through to the FL3 detector, while reflecting light between wavelengths 560 and 640 nm to the FL2 detector. Dichroic mirrors are often placed at a 45 degree angle and split the light coming to them in this way, so that it can be directed to different photon detectors.

Let us consider the path of the light, through the optical system of a flow cytometer, that is generated by a single CD8$^+$ lymphocyte labeled with an antibody conjugated to the green dye fluorescein. The lymphocyte has entered through the sample injection port tube and passed in front of the argon laser. This light path will be described with reference to Figure 20-19a.

Some nonfluorescent 488-nm light from the laser will be bounced to the side, after interacting with intracellular membranes and granules. This side-scattered (SSC) light will pass through the first dichroic mirror, which allows passage of light of wavelengths shorter than 560 nm. This includes both the 488-nm light

scattered from the laser and the green fluorescence, which peaks at around 525 to 530 nm. A beam splitter then sends some of this light to the SSC photomultiplier tube (PMT), via a *band pass* (BP) filter that is set to allow through light of the same wavelength as the original illuminating laser. This BP filter is labeled 488/10 in Figure 20-19a. A voltage pulse will be generated in the SSC PMT, as the cell passes by. The magnitude of the voltage pulse will be recorded and stored.

Similarly, light scattered in a forward direction will be detected by the forward scatter (FSC) photodiode, and the magnitude of the forward scatter signal will also be recorded and stored. We note that a second 488/10 BP filter is placed immediately before the FSC diode. Not shown in Figure 20-19a is a filter block that is placed in the path of the laser beam and that prevents any direct light from the laser from impinging on the FSC photodiode. Most flow cytometric applications will require a graph or "dot plot" of forward scatter versus side scatter of the cell population under study, such as we see in Figure 20-21a. Each dot on these plots represents a single "event," which usually represents a single cell passing in front of the laser beam.

Recall that the cell that has scattered this light is a CD8-bearing cell. Because

this CD8$^+$ cell is bound by antibodies labeled with fluorescein that will be activated by the 488 nm laser, the cell will also emit green light with a peak wavelength of approximately 525 to 530 nm. We will follow the path of this green light (which is emitted in all directions) as it travels vertically upward through our diagram of the optics of the instrument until it impinges on the relevant PMT, which is seen at the top of Figure 20-19a, labeled FL1.

First, it will be allowed to pass through the dichroic mirror (labeled DM 560SP—*dichroic mirror, 560-nm short pass*), since its wavelength is lower than the 560-nm cutoff. Then, after having some of the light split off to the SSC photodiode by a beam splitter, it will pass through the BP filter (530/30) in front of the green PMT, finally generating another voltage pulse in the green PMT. The shape of this pulse will be determined by the shape of the structure that is emitting the fluorescence. For instance, stains that render only the nuclei fluorescent give rise to a narrower voltage pulse than do dyes that stain the entire cell surface (compare B with A in Figure 20-20b). Signals recorded from this green PMT are referred to as emanating from the FL1 channel. Similar combinations of beam splitters, mirrors, and filters separate out light of other colors and send them to the correct photodiode.

specified wavelength, as well as the amounts of forward- and side-scattered light for each cell, storing the data in a comprehensive, list mode file. This technology is advancing very quickly, and flow cytometers capable of detecting up to 20 fluorescence and light-scattering parameters are used in sophisticated clinical and research laboratories.

Figure 20-21b is a representative plot of the fluorescence intensity of cells derived from the gated lymphocyte region, R1, shown in Figure 20-21a. These cells have been stained with a fluorescein-conjugated antibody to CD8 (FITC anti-CD8) so that CD8$^+$ T cells will fluoresce in the green part of the spectrum. (The abbreviation FITC derives from the fact that most antibodies are treated with a form of the dye called *fluorescein isothiocyanate* in order to form the antibody-dye conjugate.) Green fluorescence is detected in the FL1, or green channel of the detector, and the amount of fluorescence for each cell is plotted on the abscissa of the graph in Figure 20-21b. Similarly, CD4$^+$ T cells have been

stained with an antibody conjugated with the fluorochrome, *phycoerythrin* (PE), to form PE anti-CD4. PE-labeled antibodies are detected in the FL2, or red channel and FL2 levels of the cells in the sample are plotted on the ordinate of the graph. This plot shows us that, of the cells falling into the lymphocyte gate, R1, 46.06% were red-staining CD4-bearing cells and 38.88% were green-staining CD8-bearing cells. At the bottom left-hand side of the plot we see that some of the cells failed to stain for CD4 or CD8. Since they fell within the lymphocyte gate and did not stain for T-cell markers, this suggests that they are most probably B lymphocytes, a hypothesis that could be tested by staining with fluorescent antibodies for a B-cell marker such as CD19. At the top right-hand side of the plot are a few cells that apparently stained for both CD4 and CD8. This could reflect the escape of immature double-positive T cells into the blood, or more likely, indicates that nonspecifically bound antibodies were insufficiently washed during sample preparation.

As the technology to detect more, different parameters per cell has become more sophisticated, computer scientists have had to develop increasingly sophisticated ways to represent complex data.

For students wishing to acquire a deeper understanding of the optics of the flow cytometer, Advances Box 20-2 describes more clearly how these data are generated, and breaks down the pathway of the light from the sample to each of the relevant detectors.

Key Concept:

- Flow cytometers use systems of mirrors and filters to analyze the scattered and fluorescent light emanated from individual cells flowing past a laser beam.

Flow Cytometers and Fluorescence-Activated Cell Sorters Have Important Clinical Applications

In many medical centers, the flow cytometer is used to detect and classify leukemias in order to make educated decisions regarding future treatment. Likewise, as described above, the rapid measurement of T-cell subpopulations, an important prognostic indicator in AIDS, is routinely done by flow cytometric analysis. When the number of $CD4^+$ T cells in the blood of a patient with AIDS falls below a certain level, the patient is at high risk for opportunistic infections.

Sophisticated flow cytometers are also used clinically in the analysis of purified stem cell populations during bone marrow transplantation and in the diagnosis of most disorders involving blood cell populations.

Key Concept:

- Flow cytometers and fluorescence-activated cell sorters are important clinical tools.

The Analysis of Multicolor Fluorescence Data Has Required the Development of Increasingly Sophisticated Software

The first few generations of flow cytometers contained analysis software that was directly associated with the programs used to run the instruments. As use of the instruments became more widespread, it became inconvenient to delay running new samples while data from prior experiments were being analyzed. Furthermore, as investigators began to use more fluorescent colors, computer scientists began to develop increasingly powerful and creative solutions for visualizing the huge amounts of data that were generated in each experiment. At this point, many different programs are available for data analysis and presentation.

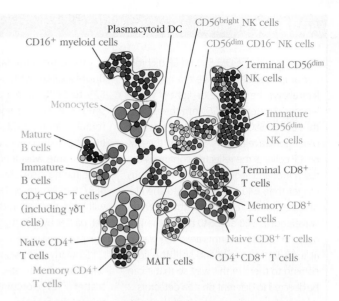

FIGURE 20-22 Analysis of multicolor fluorescence data. Multicolor analysis has required the development of new algorithms with which to accurately display complex data. Shown is an example of high-dimensional, multiparameter flow cytometric visualization. Data from http://ki.se/en/medh/niklas-bjorkstrom-group.

The industry standard for many years has been the FlowJo program, which analyzes and prepares for display data acquired on most, if not all flow cytometers and FACS machines. However, now that investigators use as many as 12 to 14 stains per experiment, analysis of the complex patterns of data that are generated has required the development of algorithms of similar complexity. One example of a program that is frequently used when analyzing complex, multicolor experiments is SPADE, which uses a hierarchical clustering methodology to generate representations such as that shown in **Figure 20-22.**

As increasing numbers of monoclonal antibodies and matching numbers of dyes are used per experiment, cytometer designs have also become extremely complex. The essential problem is that no dye absorbs or emits light at a single wavelength; instead, each dye absorbs and emits light over a well-defined spectrum of wavelengths. If the emission wavelengths of two dyes overlap, then it can become impossible to tell which antibody is being represented by its fluorescent surrogate. In addition, if one dye absorbs light in the region of the spectrum in which another dye on the same cells emits light, the amount of emitted light that is detected by the PMT will be correspondingly reduced, leading the investigator to underestimate the amount of the corresponding protein on the cell surface. (In fact, the use of dyes with overlapping emission and absorption wavelengths can be used to determine the distance between two molecules, each of which is labeled with one of the dyes, using a technique known as *fluorescence resonance energy transfer.*)

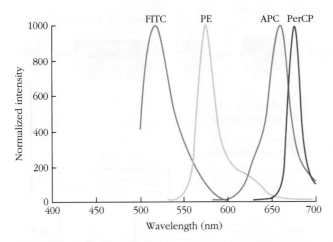

FIGURE 20-23 Emission spectra of commonly used dyes have considerable overlap. The spectral overlap requires the use of a complex array of dichroic mirrors and filters to isolate the fluorescence output from individual dyes.

For the purposes of measuring the concentrations of particular proteins on the cell surface, however, overlapping spectra provide a serious problem. **Figure 20-23** illustrates the nature of the problem by showing the amount of emission spectral overlap among four commonly used dyes. To a certain extent, one can compensate electronically for overlapping spectral emissions, and current flow cytometric analysis allows one to resolve approximately 18 proteins per cell. However, as the numbers of potential monoclonal antibodies applied to the surface of a single cell have increased, and demand for the technology has expanded, investigators have searched for new methodologies that allow them to increase the number of parameters that can be resolved.

Key Concept:

• The development of instruments capable of simultaneous measurement of many fluorescent dyes has mandated the development of increasingly complex computer programs to discriminate among the multiple fluorescent spectra and to display the resultant data in meaningful and cogent ways.

CyTOF Uses Antibodies to Harness the Power of Mass Spectrometry

Time of flight mass cytometry (CyTOF) combines the antibody-based approach of flow cytometry with the resolution power of mass spectrometry to allow the analysis of at least 45 simultaneous parameters. In CyTOF, instead of being coupled to fluorescent dyes, antibodies are conjugated with stable isotopes of rare earth metals. In addition, rhodium- or iridium-conjugated DNA intercalators can also be used

with this technology and can provide additional information about cell number, DNA content, and cell viability.

In CyTOF (**Figure 20-24**), cells first pass individually through a nebulizer, which ensures the cells enter the instrument in uniform droplet form. They then enter a high-temperature plasma, created by an inductive coil system (labeled ICP in Figure 20-24). In the plasma, which reaches temperatures of approximately 7000 K, each cell vaporizes, all molecular bonds are broken, and one electron is stripped from each atom, creating a cloud of elemental ions representing each cell. From this cloud, the relative abundance of the contributing isotopes (and hence of the presence of the markers on each cell) can be computed, leading to the creation of the familiar scatter plot.

Although CyTOF can increase the number of measurable parameters per cell over that of flow cytometry, this increase comes with a practical cost. The number of cells that can be analyzed with a state-of-the-art flow cytometer approaches 10,000 per second. In contrast, because of the lifetime of the ionized cloud, and the number of times each cloud must be scanned for accurate results, analysis of more than 1000 cells per second by CyTOF is precluded. Furthermore, the light-scattering measurements that can be used so successfully to gate cell populations of different sizes and intracellular complexity in flow cytometric experiments are not currently available with CyTOF.

Key Concept:

• Time of flight mass cytometry (CyTOF) uses heavy metal isotopes instead of fluorescence tags to label monoclonal antibody, and allows simultaneous analysis of up to 45 different parameters.

Magnets Can Be Used in a Gentle, Sterile Method for Sorting Cells

Fluorescence is not the only type of marker that immunologists can use to separate cells. As early as 1981, a methodology to separate cells on the basis of their magnetic properties was described in the immunological literature. From these humble beginnings has sprung an array of tools and techniques by which cells can be separated from one another after binding to antibodies covalently attached to a variety of magnetic molecules and particles.

In the most common variation of magnetic-activated cell sorting (MACS), a fine wool mesh made of ferromagnetic metal is localized in a short column. Application of a magnetic field across the column will ensure that any magnetized material will stick to the mesh. By conjugating antibodies to magnetic beads or molecules, and then passing the antibody-bound cells through the column, those cells that bound to the magnetic beads can be held onto the mesh in the column. In contrast, any cells that did not bind the magnetic antibodies will simply flow through. After washing off

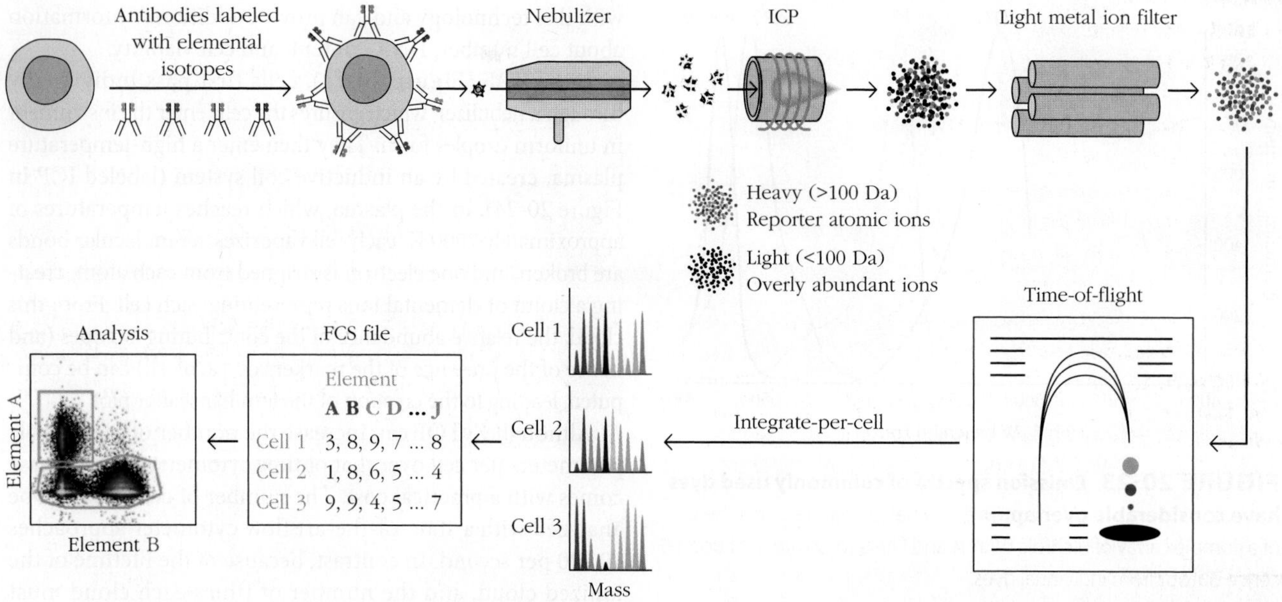

FIGURE 20-24 CyTOF enables the measurement of up to 45 different parameters. Antibodies are conjugated with stable, heavy metal isotopes and used to label the cell population under study. Cells are passed through a nebulizer, to ensure efficient droplet formation, and then introduced into a plasma at approximately 7000 K; this vaporizes each cell one at a time, breaking all molecular bonds and stripping one electron per metal atom. Light metal ions that may confound measurements are filtered out. The time taken for each heavy metal ion to fly to the detection screen under the influence of a magnetic field is measured, and the time-of-flight data are converted into information that records the presence or absence of each metal ion per cell.

any nonspecifically bound cells with a stream of buffer, the magnetic bead–conjugated cells can then be released simply by removing the magnet from the outside of the column and passing buffer through.

Magnetic cell separation is extremely gentle, and has proven to be particularly useful for batch separation of large numbers of cells. In contrast, fluorescence-activated cell sorting, by virtue of sorting one cell at a time, makes fewer mistakes but is significantly slower (although newer flow cytometers sort much more rapidly than their predecessors). In the laboratory or the clinic, immunologists will often perform a batch sort using magnetic-activated cell sorting, and then follow it up with a fluorescence-activated cell sort, in order to maximize the accuracy of cell separation. During bone marrow stem cell transplantation, technicians will typically use MACS as a sterile method to separate out the stem cell population from donor bone marrow and then check the purity of their cell populations and/or further purify the stem cell population by FACS.

Key Concept:

• Magnetic-based cell sorting can be used to sort cells coated with antibodies coupled to magnetic beads or molecules.

Cell Cycle Analysis

One of the first responses of most lymphocytes to an immunological stimulus is to divide. Immunologists have therefore been at the forefront of developing methodologies for cell cycle analysis. We will describe several methods (classic as well as more modern) that are commonly used by immunologists to analyze the cell cycle status of populations of immune cells.

Tritiated Thymidine Uptake Was One of the First Methods Used to Assess Cell Division

Tritiated thymidine ($[^{3}H]$thymidine) uptake was the first method to be used routinely to measure cell division in lymphocyte cultures. This technique relies on the fact that dividing cells synthesize DNA at a rapid pace, and radioactive thymidine in the culture fluid will therefore be quickly incorporated into high-molecular-weight DNA. In a $[^{3}H]$thymidine uptake assay, cells cultured in the presence of tritiated thymidine are subjected to proliferative signals and lysed at defined periods post-stimulation. Their DNA is then extracted and precipitated onto filters that bind to the high-molecular-weight nucleic acids, but allow unincorporated thymidine to wash straight through. The amount of radioactivity retained on the filters after washing provides

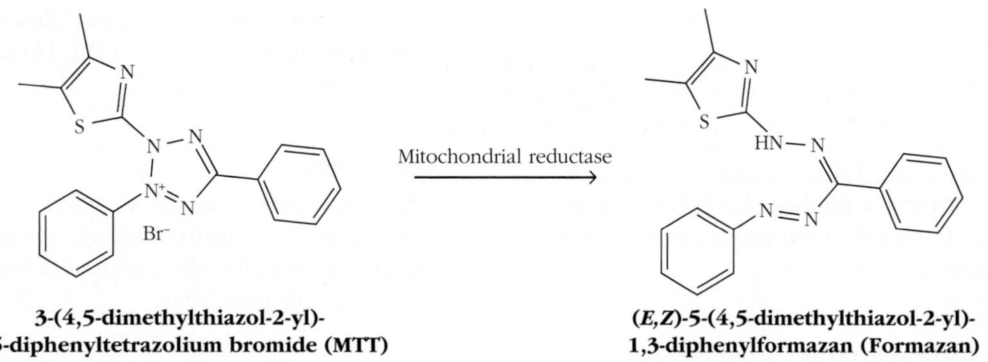

**3-(4,5-dimethylthiazol-2-yl)-
2,5-diphenyltetrazolium bromide (MTT)**

**(E,Z)-5-(4,5-dimethylthiazol-2-yl)-
1,3-diphenylformazan (Formazan)**

FIGURE 20-25 The MTT assay is used to measure the number of viable cells in a suspension. Mitochondrial oxidoreductases in metabolically active cells convert the yellow MTT substrate to a purple formazan product that can be solubilized and read in a plate reader.

a measure of the amount of newly synthesized DNA and hence the number of cells undergoing division in the culture.

> **Key Concept:**
> • Dividing cells take up tritiated thymidine and incorporate it into high-molecular-weight DNA; the radioactivity of the DNA can be measured to provide an indicator of cell division.

Colorimetric Assays for Cell Division Are Rapid and Eliminate the Use of Radioactive Isotopes

Motivated by reasons of safety and environmental responsibility to move away from the use of radioactivity-based measurements, scientists have developed a number of different assays in which metabolically active cells cleave colorless substrates into colored, often insoluble products that can then be measured spectrophotometrically. In one such assay, the tetrazolium compound MTT [3-(4,5-dimethylthiazol-2-yl)-2,5-diphenyltetrazolium bromide], a yellow tetrazole, is reduced by mitochondrial NAD(P)H-dependent oxidoreductases to form insoluble, purple formazan dye crystals. The purple crystals can be solubilized and the absorbance of each cell sample can then be read directly in the culture wells at 570 nm, the absorbance peak of formazan (**Figure 20-25**). Under a wide range of conditions, the amount of enzyme activity is directly proportional to the number of metabolically active cells present and so this assay provides a readout of the number of live cells as a function of time. Since it measures the number of live cells at the end of an experiment, the MTT assay can measure cell proliferation or cell death.

> **Key Concept:**
> • The number of actively metabolizing cells in a culture can be measured by the MTT assay.

Bromodeoxyuridine-Based Assays for Cell Division Use Antibodies to Detect Newly Synthesized DNA

When introduced into cells, *bromodeoxyuridine* (BrdU), an analogue for deoxythymidine (dT) (**Figure 20-26**), is rapidly phosphorylated to bromodeoxyuridine triphosphate, an analogue for deoxythymidine triphosphate, and is incorporated in its place into newly synthesized DNA. Cells that divide following BrdU incorporation can then be identified using antibodies to BrdU. In addition to serving as a label for newly divided cells, BrdU can also mark cells for light-induced cell death. If cells that have incorporated a high level of BrdU are exposed to light, they will photolyse; this property has been used to selectively kill newly dividing cells. In recent years, other chemical analogues that mimic BrdU's functions as a marker of dividing cells have been generated, including 5-ethynyl-2′-deoxyuridine (EdU), which can be detected directly using specific fluorescent reagents that react in situ with a reactive terminal alkyne group in the EdU molecule.

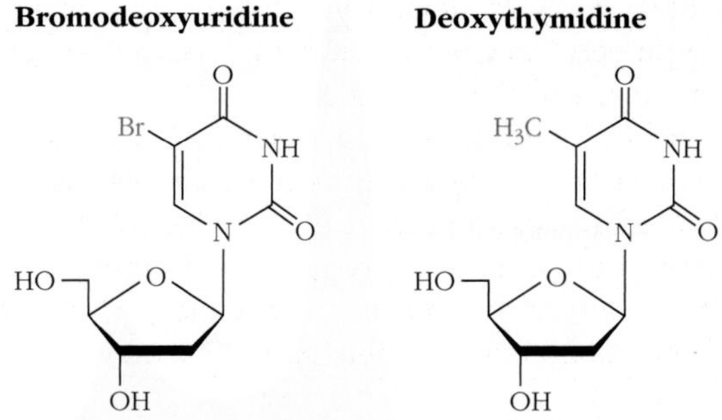

Bromodeoxyuridine   **Deoxythymidine**

FIGURE 20-26 Bromodeoxyuridine incorporates into DNA in place of deoxythymidine during DNA synthesis. Bromodeoxyuridine (BrdU) is a thymidine analog, as the large bromine group serves to mimic the size and shape of the methyl group of thymidine. It is incorporated into DNA instead of thymidine and can be detected by anti-BrdU antibodies.

Key Concept:

• Bromodeoxyuridine (BrdU) is a thymidine analogue that is incorporated into actively replicating DNA. Antibodies to BrdU can be used to ascertain whether a cell divided after the addition of BrdU to the culture. BrdU also sensitizes a cell to photolysis. Recently, additional chemical analogs of BrdU have been developed.

Propidium Iodide Enables Analysis of the Cell Cycle Status of Cell Populations

Propidium iodide (PI) is a fluorescent dye with a flat, planar structure that slides between the rungs of (intercalates into) the DNA ladder in a quantitative manner. Its fluorescence can be detected in the red (FL2) channel of a flow cytometer. When using PI to make cell cycle measurements, the investigator typically analyzes the area of the voltage pulse, rather than its height, which is used in most other flow cytometric measurements. Typically, G_2/M cells are considerably larger than G_1 cells, and G_1 cells will have half the DNA of G_2 cells about to enter mitosis; cells that are currently replicating DNA and are therefore in S phase will have an intermediate value. Apoptotic cells and fragments that have begun to break down their DNA will appear as events with less than G_1 amounts of DNA (**Figure 20-27**). Cells that have stuck together in pairs (doublets) or larger clumps must first be excluded from analysis because they have the same amount of DNA per doublet as a cell in the G_2 or M phase of the cell cycle; failure to exclude doublets will therefore result in an overestimate of the fraction of dividing cells.

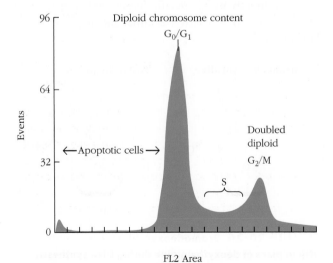

FIGURE 20-27 Propidium iodide intercalates into DNA and acts as a cell cycle and apoptosis indicator. A histogram of fluorescence measured in the FL2 channel shows cells bearing amounts of DNA characteristic of apoptotic cells, and cells in the G_0/G_1, S, and G_2/M phases of the cell cycle.

Other dyes that bind to DNA and allow for similar types of cell cycle analysis include DAPI, Hoechst 33342, and 7-*amino*actinomycin *D* (AAD).

Key Concept:

• Propidium iodide intercalates into the DNA helix in a quantitative manner. Flow cytometric profiles of propidium iodide allow analysis of the cell cycle status of a cell population.

Carboxyfluorescein Succinimidyl Ester Can Be Used to Follow Cell Division

Carboxyfluorescein succinimidyl ester (CFSE) is more correctly named *carboxyfluorescein diacetate succinimidyl ester* (CFDASE), but we will use the more common CFSE abbreviation in this section. The uncharged acetyl groups, seen at the top right and left of the molecule shown in **Figure 20-28a**, enable CFSE to enter a cell, and are then cleaved by intracellular esterases, so that the CFSE remains trapped within the cytoplasm. In the cytoplasm, molecules of CFSE are efficiently and covalently attached to intracytoplasmic proteins, with the succinimidyl ester (seen at bottom left of the molecule in Figure 20-28a) acting as a leaving group. Somewhat surprisingly, attachment of CFSE to intracytoplasmic proteins occurs without deleterious effects on cellular metabolism or cell division.

The power of CFSE labeling lies in the fact that the amount of fluorescence emitted is cut in half each time a labeled cell divides. This is illustrated in Figures 20-28b and c. The left side of Figure 20-28b shows in cartoon histogram form how each time a cell divides, the fluorescence per cell is cut in half. On the right side of Figure 20-28b we see how those data look in an actual experiment, noting that, in this experiment, significant numbers of cells have divided up to five times, with smaller numbers of outliers who have divided six and seven times. Figure 20-28c shows the same data viewed as a dot plot. In each case, the right-hand peak represents those cells that did not divide after CFSE incorporation. The peak to its immediate left represents cells that have divided once, the next one to the left represents cells that have divided twice, and so on. One can use the sorting capacity of the flow cytometer to physically separate those cells that have not divided, or have divided once, twice, or more times, and then analyze the separate cell populations for the expression of particular genes.

Key Concept:

• Carboxyfluorescein diacetate succinimidyl ester (CFSE) binds to the cytoplasmic proteins of cells in a quantitative manner; its intracellular concentration is cut in half with each cell division. CFSE fluorescence can therefore be used to measure the number of times a cell has divided since addition of the CFSE.

(a)

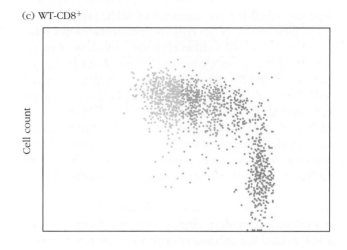

(b)

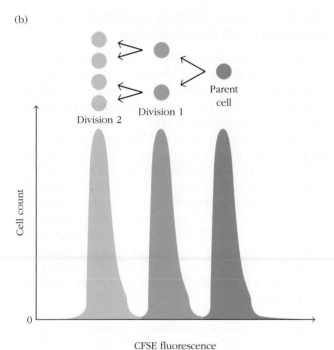

CFSE fluorescence

(c) WT-CD8+

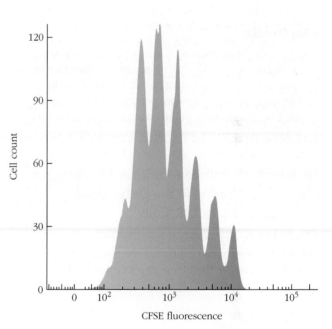

CFSE fluorescence

FIGURE 20-28 CFSE labeling can be used to determine the frequency of cells that have divided a defined number of times. (a) Structure of carboxyfluorescein diacetyl succinimidyl ester (CFSE). The acetyl groups that enable the molecule to pass through the cell membranes are highlighted in red, and the succinimidyl ester that is cleaved is shown in green. The chemistry of the molecule is such that the succinimidyl ester derivatizes fluorescein at either the 5 or the 6 position of the ring in approximately equal amounts. Hence the ester is shown midway between the 5 and the 6 position on the ring. (b) The left-hand theoretical plot shows a histogram illustrating how CFSE fluorescence is cut in half with each cell division. The peak on the right of this plot represents those cells that have not divided since addition of the CFSE. The next peak to the left represents cells that have divided once, and the peak farthest left, cells that have divided twice. The right-hand plot shows real data from an experiment in which significant numbers of cells have divided one, two, three, four, or five times, with a few outlying cells that have divided six or seven times. (c) A similar experiment to that shown on the right-hand side of part b is represented as a scatter plot. Again, peaks of cell density represent cells that have divided different numbers of times.

Assays of Cell Death

At the close of an immune response, most of the activated immune cells die. Furthermore, the outcome of many immune responses is the death of infected or affected cells, and so immunologists utilize a battery of methodologies to test for cell death.

The ^{51}Cr Release Assay Was the First Assay Used to Measure Cell Death

The ^{51}Cr release assay was, for decades, the method of choice for measuring cytotoxic T cell– and natural killer cell–mediated killing and was used in the Nobel Prize–winning experiments performed by Doherty and Zinkernagel that first described the phenomenon of MHC restriction in T-cell recognition. Target cells are first incubated in a solution of ^{51}Cr-labeled sodium chromate, which is taken up into the cells. Excess chromium is washed out of the cell suspension and the radioactively labeled targets are mixed with the killer-cell population at defined effector-to-target cell ratios. Death of the target cells is indicated by the release of ^{51}Cr into the supernatant of the mixed cell culture, and is quantified by comparing the ^{51}Cr release from the test cells with that from detergent-treated (maximum release control) and untreated (minimal release control) cells. ^{51}Cr is a γ-emitting radioactive isotope, and the radioactivity in the assay supernatants can therefore be readily measured with a γ counter. A modern alternative to this technique uses CFSE to label the cells and quantifies the release of fluorescent material into the supernatant with a fluorescence plate reader. These two assays measure all forms of cell death.

Key Concept:

• In the ^{51}Cr release assay, ^{51}Cr is released from prelabeled cells on cell death. ^{51}Cr release can therefore be used to measure the extent of cell death within the labeled population. Modern versions of this assay use fluorescent, rather than radioactive, reagents that are released into the supernatant on cell death.

Fluorescently Labeled Annexin A5 Measures Phosphatidylserine in the Outer Lipid Envelope of Apoptotic Cells

In cells undergoing apoptosis (programmed cell death), but not other modes of cell death, the membrane phospholipid *phosphatidylserine* flips from the interior to the exterior side of the plasma membrane phospholipid bilayer. Annexin A5 is a protein that binds to phosphatidylserine in a calcium-dependent manner; fluorescently labeled annexin A5 can therefore be used to tag apoptotic cells for detection using immunofluorescence microscopy, flow cytometry, or a fluorescence plate reader.

Key Concept:

• Fluorescently labeled annexin A5 binds to phosphatidylserine exposed on the surface of cells undergoing apoptosis and can therefore be used as a measure of apoptosis in immunofluorescence or flow cytometric experiments.

The TUNEL Assay Measures Apoptotically Generated DNA Fragmentation

Another common method for the detection of apoptotic cell death uses the fact that apoptotic cells undergo a process of progressive DNA degradation, which results in the generation of short DNA fragments within the nucleus. The TUNEL assay relies on the use of the enzyme *terminal deoxynucleotidyl transferase* (TdT) (see Chapter 6 for a detailed explanation of this enzyme's activity) to add bases onto the broken ends of DNA sequences in a nontemplated manner. The classic variation of the TUNEL method uses TdT to add BrdU (see **Figure 20-29**) to fixed and permeabilized cells. BrdU is incorporated into the newly synthesized DNA, and is then detected with fluorescently labeled anti-BrdU antibodies. More recent iterations of this method use the incorporation of short DNA segments prelabeled with a molecule that then binds to a fluorescent tag under very gentle conditions.

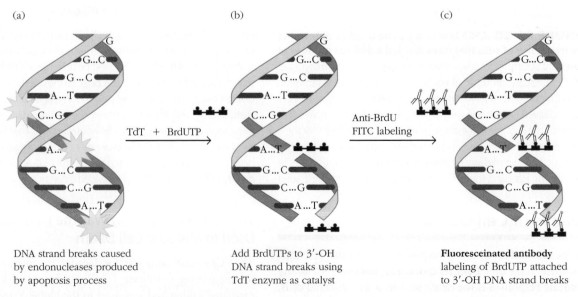

(a)   (b)   (c)

TdT + BrdUTP   Anti-BrdU
                              FITC labeling

DNA strand breaks caused   Add BrdUTPs to 3′-OH   **Fluorescinated antibody**
by endonucleases produced   DNA strand breaks using   labeling of BrdUTP attached
by apoptosis process   TdT enzyme as catalyst   to 3′-OH DNA strand breaks

FIGURE 20-29 Assessment of apoptosis, using a TUNEL assay. (a) Apoptosis results in DNA fragmentation by intracellular nucleases. (b) BrdU-labeled nucleotide triphosphates are added onto the broken ends of fragmented DNA in fixed and permeabilized apoptotic cells, using the enzyme TdT. (c) Fluoresceinated antibodies specific for BrdU can then be used to detect apoptotic cells.

Caspase Assays Measure the Activity of Enzymes Involved in Apoptosis

Caspases are a family of *c*ysteine proteases that cleave proteins after *asp*artic acid residues. Different members of the caspase family are activated during apoptotic cascades, depending on how the cascades are initiated. For example, caspase-8 is activated on engagement of the Fas receptor by Fas ligand. Several different types of caspase assays are now commercially available, including caspase detection kits that yield fluorescent products on caspase-mediated cleavage, as well as kits that detect the cleaved and active forms of caspases using Western blot methodology.

Analysis of Chromatin Structure

Three major approaches to the question of how DNA is folded within its native chromatin structure and which proteins interact with it at various developmental stages have been extensively used by immunologists in the last decade or so, and their application to the study of TCR and BCR

gene structure was described in Chapter 6. The first, 3-D FISH, was described above in the section on immunofluorescence-based imaging techniques. Here, we briefly explain two other methods that have allowed investigators to understand how DNA and chromatin structure alters during recombination events in intact cells.

Chromatin Immunoprecipitation Experiments Characterize Protein-DNA Interactions

In a chromatin immunoprecipitation (ChIP) experiment, chromatin is treated with formaldehyde, which covalently cross-links DNA to any bound proteins. The chromatin is then isolated and the bound DNA is sheared into small pieces. Antibodies to a protein that is hypothesized to be interacting with DNA are used to immunoprecipitate that protein along with any bound DNA. After protein digestion, the released DNA fragment is amplified by PCR and sequenced, leading to knowledge of which proteins specifically bind to which DNA sequences.

Chromosome Conformation Capture Technologies Analyze Long-Range Chromosomal DNA Interactions

The technique of *c*hromosome *c*onformation *c*apture (3C) was first described in 2002. The object of this process is to analyze long-range interactions between DNA sequences within a chromosomal context. One of the most recent variants of the 3C technique is called Hi-C (see **Figure 20-30**).

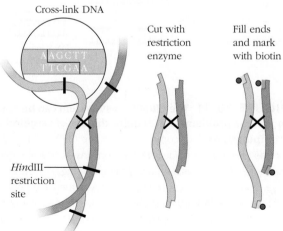
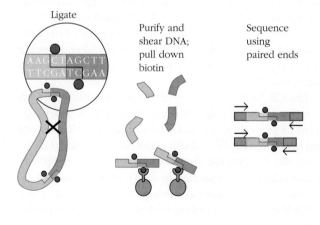
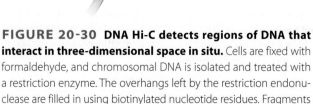

FIGURE 20-30 DNA Hi-C detects regions of DNA that interact in three-dimensional space in situ. Cells are fixed with formaldehyde, and chromosomal DNA is isolated and treated with a restriction enzyme. The overhangs left by the restriction endonuclease are filled in using biotinylated nucleotide residues. Fragments that were initially close in the chromosome are now close enough to be ligated, creating a genome-wide library of ligation products. The DNA is sheared, and the residual biotin-containing sequences are purified and sequenced in order to create a map of sequences that were associated in the original chromatin.

In this form of the assay, cells are fixed with formaldehyde to cross-link macromolecules that are in geographical proximity to one another. Chromosomal DNA is then isolated and treated with a restriction enzyme. The overhangs left by the restriction endonuclease are filled in with nucleotide residues labeled with biotin, which marks the end of each fragment. Blunt-end ligation of fragments cross-linked by the same proteins then creates a genome-wide library of ligation products, with the demarcation between individual products indicated by the biotin residues. The DNA is then sheared, the biotin-containing sequences are purified, and those DNA sequences that are close to one another in the native chromatin structure can be identified by high-throughput sequencing and analysis. The eventual goal is to generate an interaction matrix of DNA sequences. Such matrices have been invaluable in revealing features of 3-D chromosomal organization.

Key Concepts:

- Significant advances have been made in the last decade in our abilities to analyze the structural relationships between particular DNA sequences in native chromatin.

- Hi-C technologies allow the creation of an interaction map of DNA sequences that are in geographical proximity in chromosomes.

CRISPR-Cas9

It is appropriate that our discussion of the CRISPR-Cas9 system falls at the interface of in vitro and in vivo methodologies, since CRISPR-Cas9 has been used effectively in both in vitro and in vivo systems.

Few technological innovations have had a greater, more rapid, and more wide-ranging impact on the progress of biology as the development of CRISPR-Cas9 methodology. In this chapter, we will address the role of one of the subtypes of CRISPR, CRISPR-Cas9, as a powerful and flexible tool for the modulation of genetic information, both in vitro and in vivo.

The CRISPR system evolved to provide bacteria with adaptive immunity against bacteriophages. A short segment of bacteriophage DNA (the spacer) is clipped out and inserted into the CRISPR region of the bacterial chromosome. This region consists of a set of bacteriophage spacer sequences, separated by short, repeated sequences (repeats). New spacers are integrated at one end of the CRISPR locus. Transcription of the entire locus is typically followed by enzymatic digestion of the RNA pieces into mature, guiding CRISPR RNAs (crRNAs). These crRNAs are bound by CRISPR-associated (Cas) proteins to form targeting complexes. Each targeting complex contains a single spacer, or guiding RNA sequence, which will bind to the

corresponding DNA sequence in the target genome. Unfortunately for the target bacteriophage, the Cas proteins that are noncovalently associated with the binding crRNA are nucleases that then create double-strand breaks in the foreign DNA. In the laboratory, this means that any sequence that can be incorporated into a CRISPR complex can be targeted for genetic manipulation.

There are several types of CRISPR complexes. The type selected for technological development is a type II CRISPR complex that uses a single protein, Cas9, to provide its nuclease activity. In addition to the Cas9 protein and the guiding crRNA sequence, CRISPR Cas9 also uses a second RNA molecule, *trans-activating crRNA* (tracrRNA), that forms a double-stranded stem with the 3' end of the crRNA and facilitates recruitment of the Cas9 protein to the crRNA (see **Figure 20-31**).

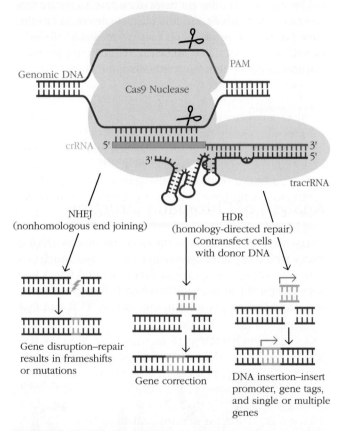

FIGURE 20-31 The CRISPR-Cas9 system can be applied to various problems that require the use of targeted DNA manipulations. The components of the CRISPR-Cas9 system include the crRNA, or guide RNA, that contains the spacer sequence; the tracr-RNA that facilitates Cas9 binding; and the nuclease enzyme, Cas9. In addition to recognizing the DNA sequence homologous to the spacer RNA, Cas9 must also recognize the protospacer adjacent motif (PAM). Various applications of CRISPR-Cas9 require use of the nonhomologous end-joining DNA repair pathway, leading to gene disruption (left), or the homologous DNA repair pathway to repair the cleavage using a corrected (middle) or new (right) DNA sequence. For details, see text.

One might ask how an organism circumvents the knotty problem of committing suicide by CRISPR-Cas9 recognition of the CRISPR complex in its own genome. It turns out that, in order to target foreign DNA, the CRISPR-Cas9 system must recognize not only the DNA sequence corresponding to the spacer; it must also see a short sequence motif, called the *protospacer adjacent motif* (PAM), located close by the target sequence. The PAM required by Cas9, for example, is the 5′-NGG-3′ PAM. Since this sequence is never incorporated by the bacterium during the process of spacer acquisition, host organism DNA is protected from CRISPR attack.

How is this targeting system used in the laboratory? CRISPR-Cas9 allows investigators to edit genomic DNA inside cells in a number of ways depending on the experimental protocol that is employed.

The simplest application is designed to inactivate the gene of interest (see Figure 20-31, left). Most investigators will elect to achieve heterologous expression of the Cas9, together with the two RNA sequences, using lentiviral transfection, transient plasmid transfection, or direct microinjection. Cas9 targets the locus of interest and induces a double-strand break that is repaired by the cell's endogenous machinery via a nonhomologous end-joining (NHEJ) pathway (see Chapter 6). Because the NHEJ pathway is error-prone, this process results in the presence of small insertions or deletions in the target gene, known as *indels*. The open reading frame is disrupted and the gene is inactivated. Such inactivated genes have been employed in genome-wide screens that investigated the effects of knocking out one or more genes.

Alternatively, if the investigator wishes to produce a particular change in the target gene, including correcting a pre-existing mutation, they may do so by biasing the system toward using homology-directed repair. In this case, along with the genes encoding the two RNA species and Cas9, the scientist will also introduce a piece of donor DNA that bears homology to sequences flanking the target site. The donor DNA then will be used by homology-directed double-strand break repair systems as a template. This mode of DNA repair has been used both to correct existing mutations (Figure 20-31, middle) and to introduce epitope tags, or disease-causing mutations (Figure 20-31, right).

Prior systems designed to alter host DNA had been limited to introducing one change at a time. CRISPR-Cas9 also provides the capacity for multiplexing, in which several different crRNAs are introduced during the transfection process.

Recently, investigators have exploited the ability of the CRISPR-Cas9 system to target various molecules to precise DNA sequences, even in the absence of subsequent gene modification. Cas9 contains two conserved nuclease domains, and inactivating both of these domains results in a system (dead Cas9, or dCas9) that can still use the ability of the crRNA to target binding to a particular DNA sequence, without inducing subsequent DNA cleavage. Multiple uses have been described for these dCas9 systems. For example, when dCas9 is expressed in concert with its guide RNA, it can block the binding of RNA polymerase and prevent transcription of particular sequences. A dCas9 fusion with GFP has been used to image precise sequences in chromatin. Finally, dCas9 fusions with enzymes that introduce epigenetic marks can perturb epigenetic regulation and allow its investigation in new and precise ways.

The power of CRISPR-Cas9 technology has been harnessed in vivo to modulate DNA in both somatic and germ-line cells. In adult mice, a G-to-A mutation causing a hereditary tyrosinemia was corrected by hydrodynamic (high-pressure) injection of a donor DNA template and plasmid DNA that encoded both Cas9 and the required RNA species. The CRISPR system injected into the tail was delivered to the liver and corrected the mutation (which is fatal in humans). Liver cells in which the mutation was corrected were found to have a selective advantage; they expanded and repopulated the liver. Correction of the mutation in mice that causes Duchenne muscular dystrophy in humans was recently achieved following direct injection of the requisite CRISPR-Cas9 components into a mouse germ line. In addition, a clinical trial that used CRISPR-Cas9–modified T cells in the treatment of lung cancer has emerged from China.

It is clear that the power, flexibility, and technical accessibility of this system have much to offer and that its applications will continue to increase rapidly over the coming years. Combination of CRISPR-Cas9 technology with several of the technologies described below is likely to reap considerable rewards.

Key Concepts:

- CRISPR-Cas9 technologies allow investigators to make targeted changes in nucleotide sequences in whole cells or even in vivo.

- CRISPR-Cas9 can created mutated sequences, and hence knockout genes, correct mutations, insert sequences, and even tag particular sequences for imaging.

- CRISPR-Cas9 is already in use in animal model systems; and clinical trials using CRISPR-Cas9–modified cells that will be re-injected into human patients are ongoing.

Whole-Animal Experimental Systems

Many animal species have been used in the study of immunology, with the choice of animal species selected for any one study being dictated by the needs of the particular investigation. To test the effectiveness of particular vaccines against viruses or bacteria that affect only primates, primate animal models must be used. Studies of horses, goats, sheep, dogs, and rabbits have yielded much information about immune responses and have provided us with numerous reagents for the study of human and mouse immunology. However, the species that has been used most frequently and with the greatest effectiveness in modeling the human immune system is the mouse. Mice are easy to handle, are genetically well characterized, and relatively easy to manipulate genetically, and have a rapid breeding cycle. In this section, we will outline various types of murine animal models and clarify some of the more confusing nomenclature that students are likely to encounter as they read in the immunological literature. We will first briefly address the ethical questions that must and should arise in the minds of those who work with whole animals and the regulations that have been developed to protect nonhuman research subjects.

Animal Research Is Subject to Federal Guidelines That Protect Nonhuman Research Species

Our understanding of the immune response owes a great deal to work performed in animal models, and huge advances have been made using a variety of different species. Concerns about animal welfare, however, have accompanied these advances and have led many countries to adopt laws that use ethical guidelines to regulate animal use. In the United States, institutions and investigators that perform research on animals must comply with the Animal Welfare Act of 1966. Institutions receiving federal funds must establish an Institutional Animal Care and Use Committee (IACUC) that regulates and oversees all experiments that utilize animals, and all investigators at such institutions must receive approval of their experimental protocols from this committee. The constitution and activity of all institutional IACUCs are carefully regulated, and institutional compliance is enforced by the U.S. Department of Agriculture (USDA). Failure to adhere to the guidelines described in the publication *Guide for the Care and Use of Laboratory Animals*, published by the National Research Council of the National Academies, is punishable by loss of the privilege to pursue any animal research by scientists in an entire institution.

Standards for the ethical treatment of animals are continually evaluated and updated as our awareness of animal biology, alternative technologies, and ethical concerns develops. Since it was signed into law, the Animal Welfare Act has been amended multiple times in order to reflect these advances. In 2010, the European Union passed a law that updated and strengthened standards outlined in its 1986 animal welfare laws. In response to concerns that animal research was poorly described in the literature, the influential journal *Nature* established new policies in 2012 that require authors to include more detailed descriptions of approaches and standards associated with animal experimentation. Finally, a growing number of investigators have formally embraced the "Three R's" principles originally articulated in 1954 by Hume and described fully by Russell and Burch. These describe a commitment to *refine*, *reduce*, and *replace* approaches that use animals in research. These principles were communicated to scientists first by a committee that included the Nobelist and immunologist Peter Medawar, and have been the focus of several prominent conferences in recent years.

Many individuals and groups feel that regulations still do not fully address or respect the needs of animals, and creative tension continues to exist between ethical concerns and the desire to advance knowledge; although it inspires conflict at times, it also inspires continuing efforts to refine and improve policies.

Key Concept:

- Animal research must be undertaken in a manner consistent with high ethical standards and is subject to federal guidelines.

Inbred Strains Reduce Experimental Variation

To control experimental variation caused by differences in the genetic backgrounds of experimental animals, immunologists often work with inbred strains of mice. The rapid breeding cycle of mice makes them particularly well suited for the production of inbred strains, in which the heterozygosity of alleles that is normally found in randomly outbred mice is replaced by homozygosity at all loci. Repeated inbreeding for 20 generations yields an inbred strain whose progeny are homozygous and identical (**syngeneic**) at more than 99% of all loci. Approximately 500 different inbred strains of mice are available, each designated by a series of letters and/or numbers (**Table 20-2**), and most of these strains are commercially available. Inbred strains have also been produced in rats, guinea pigs, hamsters, rabbits, and domestic fowl.

Recombinant inbred strains of mice are those in which two inbred strains (e.g., strains A and B) have been mated, resulting in a recombination within an interesting locus (e.g., within the MHC locus). Subsequent inbreeding of the animals with this recombination results in an inbred strain of mice bearing part of its MHC from strain A and the other MHC part from strain B. Animals of this recombinant strain

TABLE 20-2	Some common inbred mouse strains used by immunologists	
Strain	**Common substrains**	**Characteristics**
A	A/He A/J A/WySn	High incidence of mammary tumors in some substrains
AKR	AKR/J AKR/N AKR/Cum	High incidence of leukemia *Thy-1.2* allele in AKR/Cum, and *Thy-1.1* allele in other substrains (*Thy* gene encodes a T-cell surface protein)
BALB/c	BALB/cJ BALB/c AnN BALB/cBy	Sensitivity to radiation Used in hybridoma technology Many myeloma cell lines were generated in these mice
CBA	CBA/J CBA/H CBA/N	Gene (*rd*) causing retinal degeneration in CBA/J Gene (*xid*) causing X-linked immunodeficiency in CBA/N
C3H	C3H/He C3H/HeJ C3H/HeN	Gene (*rd*) causing retinal degeneration High incidence of mammary tumors in many substrains (these carry a mammary-tumor virus that is passed via maternal milk to offspring)
C57BL/6	C57BL/6J C57BL/6By C57BL/6N	High incidence of hepatomas after irradiation High complement activity
C57BL/10	C57BL/10J C57BL/10ScSn C57BL/10N	Very close relationship to C57BL/6 but differences in at least two loci Frequent partner in preparation of congenic mice
C57BR	C57BR/cdJ	High frequency of pituitary and liver tumors Very resistant to x-irradiation

Adapted from Altman, P. L., and D. D. Katz, eds. 1979. *Biological Handbooks*, Vol. III: *Inbred and Genetically Defined Strains of Laboratory Animals, Part 1: Mouse and Rat.* Federation of American Societies for Experimental Biology, Bethesda, MD.

can then be used to determine which subregions of a locus contribute to which properties in the immune system of the animal. Such strains were used to delineate the functions of class 1 versus class 2 proteins encoded by the MHC locus.

Key Concept:
- Different types of mouse strains have contributed immeasurably to immunological research. Inbred and recombinant inbred strains of mice have each provided investigators with invaluable information.

Congenic Strains Are Used to Study the Effects of Particular Gene Loci on Immune Responses

Two strains of mice are **congenic** if they are genetically identical at all but a single genetic locus or region. Any phenotypic differences that can be detected between congenic

strains must therefore be encoded in the genetic region that differs between the two strains. Congenic strains that are identical with each other except at the MHC were critical to the development of our understanding of transplantation rejection and were pivotal to the description of the role of the MHC in antigen presentation. Sometimes the reader will encounter the nomenclature "congenic resistant" strains. The term "resistant" here refers to the fact that some congenic strains were originally selected on the basis of their ability to "reject" tumors of a particular genetic origin.

Key Concept:
- Congenic strains of mice have provided investigators with invaluable information about the contribution of particular genes to specific traits.

Adoptive Transfer Experiments Allow in Vivo Examination of Isolated Cell Populations

Lymphocyte subpopulations isolated from one animal can be injected into another animal of the same inbred strain without eliciting a rejection reaction. This type of experimental system is referred to as "adoptive transfer" and permitted immunologists to demonstrate for the first time that lymphocytes from an antigen-primed animal could transfer immunity to an unprimed syngeneic recipient. Adoptive transfer experiments using sorted lymphocyte subpopulations also proved the need for both T and B cells in the generation of an antibody response.

Adoptive transfer systems permit the in vivo examination of the functions of isolated cell populations. More sophisticated adoptive transfer models involve the transfer of cells between animals that differ in their expression of an allotypic marker. Such markers are genetic variants of proteins which do not elicit a rejection reaction, but which enable the investigator to follow the fate of the injected cells using anti-marker antibodies.

Another commonly used adoptive transfer protocol involves the transfer of cells that have been fluorescently labeled into a recipient animal so that their fate can be followed using in vivo imaging technologies.

In some adoptive transfer protocols, it is important to eliminate the immune responsiveness of the host by exposing it to x-rays that kill host lymphocytes, prior to donor cell injection. If the host's hematopoietic cells might influence an adoptive transfer experiment, then high x-ray levels (900 to 1000 rads) are used to eliminate the entire hematopoietic system. Mice irradiated with such doses will die unless reconstituted with bone marrow from a syngeneic donor.

Key Concept:

- Adoptive transfer techniques introduce cells taken from one animal into a second animal so that they can be studied in a defined context.

Transgenic Animals Carry Genes That Have Been Artificially Introduced

Development of techniques to introduce cloned foreign genes (**transgenes**) into mouse embryos has permitted immunologists to study the effects of many isolated genes on the immune response in vivo. If the introduced gene integrates stably into the germ-line cells, it will be transmitted to the offspring.

The first step in producing transgenic mice is the injection of foreign cloned DNA into a fertilized egg. In this technically demanding process, fertilized mouse eggs are held under suction at the end of a pipette and the transgene is microinjected into one of the pronuclei with a fine needle. In some fraction of the injected cells, the transgene integrates into the chromosomal DNA of the pronucleus and is passed on to the daughter cells of eggs that survive the process. The eggs, or early embryos, are then implanted into the oviduct of "pseudopregnant" females, and transgenic pups are born after 19 or 20 days of gestation (**Figure 20-32**).

With transgenic mice, immunologists have been able to study the expression patterns and functions of a large number of transgenes within the context of living animals. By constructing a transgene with an inducible promoter, researchers can also artificially control the expression of the transgene. For example, the expression of genes controlled by the metallothionein promoter is activated by zinc, and transgenic mice carrying a transgene linked to a metallothionein promoter will express the transgene only when zinc is added to their water supply. Other promoters are functional only in certain tissues; the CD19 promoter, for instance, promotes transcription only in B lymphocytes.

If a transgene is integrated into the chromosomal DNA within the one-cell mouse embryo, it will be integrated into both somatic cells and germ-line cells. The resulting transgenic mice can then transmit the transgene to their offspring as a Mendelian trait. In this way, it has been possible to produce lines of transgenic mice in which every member of a line contains the same transgene. A variety of such transgenic lines are currently available commercially, or via collaborations with the producing labs.

Key Concept:

- Transgenic animals carry genes that have been artificially introduced.

Knock-in and Knockout Technologies Replace an Endogenous with a Nonfunctional or Engineered Gene Copy

One limitation of transgenic mice that are generated as described above is that the transgene is integrated randomly within the genome. This means that some transgenes insert in regions of DNA that are not transcriptionally active, and hence the genes are not expressed. Alternatively, insertions at other sites may disrupt vital genes or induce tumor formation. To circumvent this limitation, researchers have developed technology to target the desired gene to specific sites within the germ line of an animal, using homologous DNA recombination. This technique can be used to replace the endogenous gene with a truncated, mutated, or otherwise altered form of that gene, or alternatively to completely replace the endogenous gene with a DNA sequence of choice. For example, a nonfunctional form of a gene may be used to replace the normal allele, in order to determine the effects of losing the expression of the gene in the intact animal. Alternatively, knock-in technology may be used to determine when and where the promoter for a particular gene is activated. In this latter case, a gene for a fluorescent

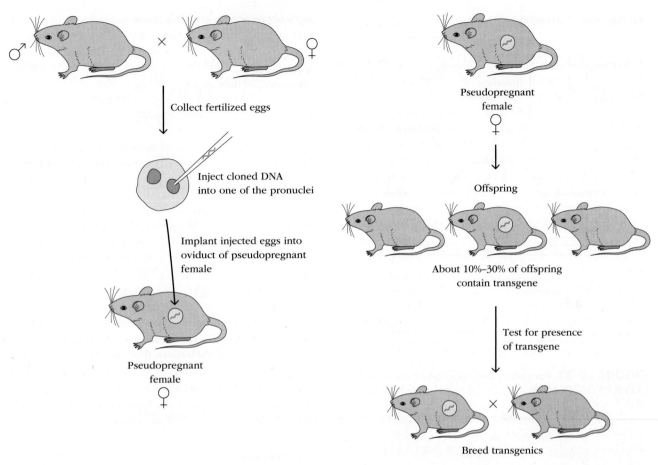

Collect fertilized eggs

Inject cloned DNA into one of the pronuclei

Implant injected eggs into oviduct of pseudopregnant female

Pseudopregnant female ♀

Pseudopregnant female ♀

Offspring

About 10%–30% of offspring contain transgene

Test for presence of transgene

Breed transgenics

FIGURE 20-32 General procedure for producing transgenic mice. Fertilized eggs are collected from a pregnant female mouse. Cloned DNA (referred to as the *transgene*) is microinjected into one of the pronuclei of a fertilized egg. The eggs are then implanted into the oviduct of pseudopregnant foster mothers (obtained by mating normal females with a sterile male). The transgene will be incorporated into the chromosomal DNA of about 10% to 30% of the offspring and will be expressed in all of their somatic cells. If a tissue-specific promoter is linked to a transgene, then tissue-specific expression of the transgene will result.

protein, such as green fluorescent protein, can be specifically engineered into a site downstream from the promoter of the gene of interest. Every time that promoter is activated, the cells in which the promoter is turned on will glow green. How might this be accomplished?

We will describe one method for the generation of knock-out mice using homologous DNA recombination. The same principles apply to the generation of knock-in mice, which would simply use different gene segments bounded by the homologous stretches of DNA. Production of gene-targeted knockout mice involves the following steps:

- Isolation and culture of **embryonic stem (ES) cells** from the inner cell mass of a mouse blastocyst

- Generation of the desired, altered form of the gene, bounded by sufficient DNA sequence from the native gene to facilitate homologous recombination

- Introduction of the desired gene into the cultured ES cells and selection of homologous recombinant cells in which the gene of interest has been incorporated.

Sensitive *p*olymerase *c*hain *r*eaction (PCR) techniques can be used to determine which ES cell colonies have incorporated the desired gene into the correct location

- Injection of homologous recombinant ES cells into a recipient mouse blastocyst and surgical implantation of the blastocyst into a pseudopregnant mouse

- Mating of chimeric offspring heterozygous for the disrupted gene to produce homozygous knockout mice

The ES cells used in this procedure are obtained by culturing the inner cell mass of a mouse blastocyst in the presence of specific growth factors and on a feeder layer of fibroblasts. Under these conditions, the stem cells grow but remain pluripotent. One of the advantages of ES cells is the ease with which they can be genetically manipulated. Cloned DNA containing a desired gene can be introduced into ES cells in culture by various transfection techniques; the introduced DNA will be inserted by recombination into the chromosomal DNA of a small fraction of these.

(a) Formation of recombinant ES cells

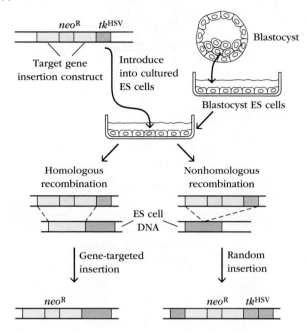

(b) Selection of ES cell carrying knockout gene

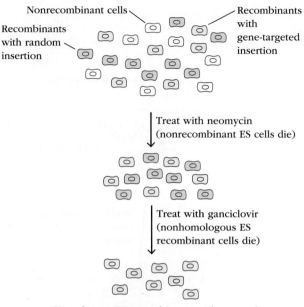

Homologous ES recombinants with targeted disruption in gene *X* survive

FIGURE 20-33 Formation and selection of mouse recombinant ES cells in which a particular target gene is disrupted. (a) In the engineered insertion construct, the target gene is disrupted with the *neo*^R gene, and the thymidine kinase *tk*^{HSV} gene is located outside the target gene. The construct is transfected into cultured ES cells. Recombination occurs in only about 1% of the cells, with nonhomologous recombination much more frequent than homologous recombination. (b) Selection with the neomycin-like drug G418 will kill any nonrecombinant ES cells because they lack the *neo*^R gene. Selection with ganciclovir will kill the nonhomologous recombinants carrying the *tk*^{HSV} gene, which confers sensitivity to ganciclovir. Only the homologous ES recombinants will survive this selection scheme.

In one model of generating knockout mice, the insertion constructs introduced into ES cells contain three genes: the target gene of interest and two selection genes, such as *neo*^R, which confers neomycin resistance, and the thymidine kinase gene from herpes simplex virus (*tk*^{HSV}), which confers sensitivity to ganciclovir, a cytotoxic nucleotide analogue (**Figure 20-33a**). The construct in this example is engineered with the target gene sequence disrupted by the *neo*^R gene and with the *tk*^{HSV} gene at one end, beyond the sequence of the target gene. Most constructs will insert at random by nonhomologous recombination rather than by gene-targeted insertion through homologous recombination. As shown in Figure 20-33a, those cells in which the construct inserts at random retain expression of the *tk*^{HSV} gene, whereas those cells in which the construct inserts by homologous recombination lose the *tk*^{HSV} gene and hence the sensitivity to ganciclovir. As illustrated in Figure 20-33b, a two-step selection scheme is used to obtain those ES cells that have undergone homologous recombination, whereby the disrupted gene replaces the target gene. The desired cells are resistant to both ganciclovir and neomycin.

Other selection schemes exist, and individual scientists choose the one that best suits their needs, but this example protocol illustrates some of the general principles involved in generating and selecting those cells with the desired genetic alteration.

ES cells obtained by this procedure will be heterozygous for the knockout mutation in the target gene. These cells are clonally expanded in cell culture and injected into a mouse blastocyst, which is then implanted into a pseudopregnant female. The transgenic offspring that develop are chimeric, composed of cells derived from the genetically altered ES cells and cells derived from normal cells of the host blastocyst. When the germ-line cells are derived from the genetically altered ES cells, the genetic alteration can be passed on to the offspring. If the recombinant ES cells are homozygous for black coat color (or another visible marker) and they are injected into a blastocyst homozygous for white coat color, then the chimeric progeny that carry the heterozygous knockout mutation in their germ line can be easily identified (**Figure 20-34**). When these are mated with each other, some of the offspring will be homozygous for the knockout mutation.

Key Concept:

• Knock-in and knockout gene technologies enable the introduction of altered or inactive forms of genes into specific locations in the genome.

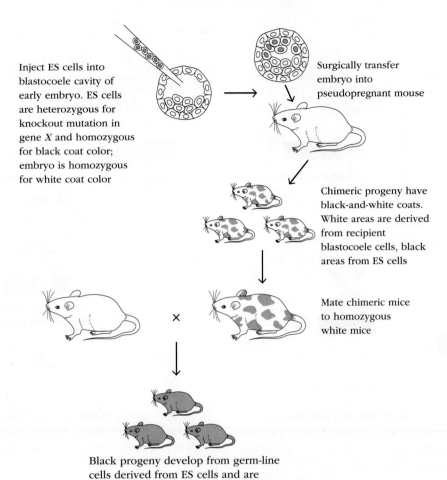

Inject ES cells into blastocoele cavity of early embryo. ES cells are heterozygous for knockout mutation in gene *X* and homozygous for black coat color; embryo is homozygous for white coat color

Surgically transfer embryo into pseudopregnant mouse

Chimeric progeny have black-and-white coats. White areas are derived from recipient blastocoele cells, black areas from ES cells

Mate chimeric mice to homozygous white mice

×

Black progeny develop from germ-line cells derived from ES cells and are heterozygous for disrupted gene *X*

FIGURE 20-34 General procedure for producing homozygous knockout mice. ES cells homozygous for a marker gene (e.g., black coat color) and heterozygous for a disrupted target gene are injected into an early embryo homozygous for an alternate marker (e.g., white coat color). The chimeric transgenic offspring, which have black-and-white coats, then are mated with homozygous white mice. The all-black progeny from this mating have ES-derived cells in their germ line, which are heterozygous for the disrupted target gene. Mating of these mice with each other produces animals homozygous for the disrupted target gene—that is, knockout mice.

The Cre/*lox* System Enables Inducible Gene Deletion in Selected Tissues

In addition to the deletion of genes by gene targeting, experimental strategies have been developed that allow the specific deletion of a gene of interest only in selected tissues. This enables investigators to determine the effects of losing gene activity only in, for example, the tissues of the immune system, even if expression of those genes in other tissues is necessary for viability of the organism. These technologies rely on the use of site-specific recombinases from bacteria or yeast.

The most commonly used recombinase is Cre, isolated from bacteriophage P1. Cre recognizes a specific 34-bp site in DNA known as *loxP* and catalyzes a recombination event between two *loxP* sites such that the DNA between the two sites is deleted. Animals that ubiquitously express Cre recombinase will therefore delete all *loxP*-flanked sequences, whereas animals that express Cre only in certain tissues will delete *loxP*-flanked sequences only in those tissues. If the expression of the Cre recombinase gene is placed under the control of a tissue-specific promoter, then tissue-specific deletion of any DNA that is flanked by *loxP* sites will occur. For example, one could express Cre in B cells by using the immunoglobulin promoter, and this would result in the targeted deletion of *loxP*-flanked DNA sequences only in B cells.

This technology is particularly useful when the targeted deletion of a particular gene in the whole animal would have lethal consequences. For example, the DNA polymerase β gene is required for embryonic development, and deletion of this gene in the whole animal would therefore result in embryonic lethality. In experiments designed to delete the DNA polymerase β gene only in thymic tissues, scientists flanked the mouse DNA polymerase β gene with *loxP* and

(a)

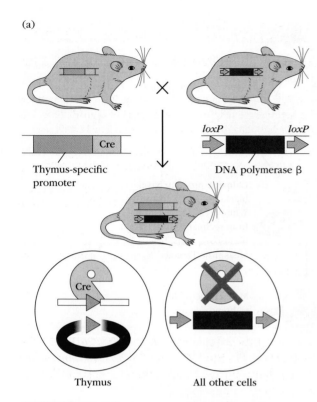

Thymus-specific promoter

DNA polymerase β

loxP   *loxP*

Cre

Thymus   All other cells

(b)

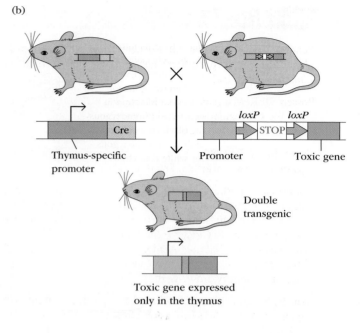

Thymus-specific promoter

Promoter   Toxic gene

loxP   *loxP*

STOP

Double transgenic

Toxic gene expressed only in the thymus

FIGURE 20-35 Gene targeting with Cre/*lox*. (a) Conditional deletion by Cre recombinase. The targeted DNA polymerase β gene is modified by flanking the gene with *loxP* sites (for simplicity, only one allele is shown). Mice are generated from ES cells by standard procedures. Mating of the *loxP*-modified mice with a Cre transgenic will generate double-transgenic mice in which the *loxP*-flanked DNA polymerase β gene will be deleted in the tissue where Cre is expressed. In this example, Cre is expressed in thymus tissue, so that deletion of the *loxP*-flanked gene occurs only in the thymus of the double transgenic. Other tissues and organs still express the *loxP*-flanked gene. (b) Activation of gene expression, using Cre/*lox*. A *loxP*-flanked translational STOP cassette is inserted between the promoter and the potentially toxic gene, and mice are generated from ES cells by standard procedures. These mice are mated to a transgenic line carrying the Cre gene driven by a tissue-specific promoter. In this example, Cre is expressed in the thymus, so that mating results in expression of the toxic gene (blue) solely in the thymus. Using this strategy, one can determine the effects of expression of the potentially toxic gene in a tissue-specific fashion.

mated these mice with mice carrying a Cre transgene under the control of a T-cell promoter (**Figure 20-35a**). The results of this mating are offspring that express the Cre recombinase specifically in T cells, allowing scientists to examine the effects of specifically deleting the enzyme DNA polymerase β in T cells.

The Cre/*lox* system can also be used to turn on gene expression in a particular tissue. Just as the lack of a particular gene may be lethal during embryonic development, the expression of a gene can be toxic. To examine tissue-specific expression of such a gene, it is possible to insert a translational stop sequence flanked by *loxP* into an intron at the beginning of the gene (Figure 20-35b). Using a tissue-specific promoter driving Cre expression, the stop sequence may be deleted in the tissue of choice and the expression of the potentially toxic gene examined in this tissue.

Some investigators have combined this technology with the use of an artificial inducer of Cre activity in order to control precisely when the gene is lost. Transgenic mice have been developed that express fusion proteins in which the Cre recombinase is linked to a second protein—for example, an altered estrogen receptor designed to respond to the drug tamoxifen (Nolvadex). This Cre fusion protein is designed such that Cre is not active unless tamoxifen is present. Thus, one can place the Cre fusion protein expression under the control of a tissue-specific promoter, and precisely time the knockout of the gene in that tissue by the administration of tamoxifen. Other, similar fusion proteins have been developed that allow Cre expression to come under the control of various antibiotics. These modifications of gene-targeting technology have been pivotal in the determination of the effects of particular genes in cells and tissues of the immune system.

Key Concept:

• The Cre/*lox* recombinase system allows investigators to target altered genes into specific locations in the genome in an inducible manner.

REFERENCES

Bendall, S. C., G. P. Nolan, M. Roederer, and P. K. Chattopadhyay. 2012. A deep profiler's guide to cytometry. *Trends in Immunology* **33**:323.

Bossen, C., R. Mansson, and C. Murre. 2012. Chromatin topology and the regulation of antigen receptor assembly. *Annual Review of Immunology* **30**:337.

Capecchi, M. R. 1989. The new mouse genetics: altering the genome by gene targeting. *Trends in Genetics* **5**:70.

Committee for the Update of the Guide for the Care and Use of Laboratory Animals, Institute for Laboratory Animal Research, Division on Earth and Life Studies, National Research Council. *Guide for the Care and Use of Laboratory Animals*, 8th ed. National Academies Press, Washington, DC. https://grants.nih.gov/grants/olaw/Guide-for-the-Care-and -use-of-laboratory-animals.pdf.

Hsu, P., E. Lander, and H. Zhang. 2014. Development and applications for CRISPR-Cas9 for genome engineering. *Cell* **157**:1262.

Jhunjhunwala, S., et al. 2008. The 3D structure of the immunoglobulin heavy-chain locus: implications for long-range genomic interactions. *Cell* **133**:265.

Köhler, G., and C. Milstein. 1975. Continuous cultures of fused cells secreting antibody of predefined specificity. *Nature* **256**:495.

Kosak, S. T., et al. 2002. Subnuclear compartmentalization of immunoglobulin loci during lymphocyte development. *Science* **296**:158.

Malmqvist, M. 1993. Surface plasmon resonance for detection and measurement of antibody-antigen affinity and kinetics. *Current Opinion in Immunology* **5**:282.

Miltenyi, S., W. Muller, W. Weichel, and A. Radbruch. 1990. High gradient magnetic cell separation with MACS. *Cytometry* **11**:231.

Nair, N., et al. 2015. Mass cytometry as a platform for the discovery of cellular biomarkers to guide effective rheumatic disease therapy. *Arthritis Research and Therapy* **17**:127.

Owen, C. S., and N. L. Sykes. 1984. Magnetic labeling and cell sorting. *Journal of Immunological Methods* **73**:41.

Parks, D. R., and L. A. Herzenberg. 1984. Fluorescence-activated cell sorting: theory, experimental optimization, and applications in lymphoid cell biology. *Methods in Enzymology* **108**:197.

Rajewsky, K., et al. 1996. Conditional gene targeting. *Journal of Clinical Investigation* **98**:600.

Rich, R. L., and D. G. Myszka. 2003. Spying on HIV with SPR. *Trends in Microbiology* **11**:124.

Russell, W. M. S., and R. L. Burch. 1959. *The Principles of Humane Experimental Technique*. Methuen, London; reprinted by Universities Federation for Animal Welfare, Potters Bar, UK, 1992.

Saeys, Y., S. V. Gassen, and B. N. Lambrecht. 2016. Computational flow cytometry: helping to make sense of high-dimensional immunology data. *Nature Reviews Immunology* **16**:449.

Sternberg, S. H., and J. A. Doudna. 2015. Expanding the biologist's toolkit with CRISPR-Cas9. *Molecular Cell* **58**:568.

Weiss, A. J. 2012. Overview of membranes and membrane plates used in research and diagnostic ELISPOT assays. *Methods in Molecular Biology* **792**:243.

Wright, A. V., J. K. Nuñez, and J. A. Doudna. 2016. Biology and applications of CRISPR systems: harnessing nature's toolbox for genome engineering. *Cell* **164**:29.

Useful Websites

www.currentprotocols.com/WileyCDA/CurPro3Title/isbn-0471142735.html *Current Protocols in Immunology* is a frequently updated compendium of most of the techniques used by immunologists.

Many of the most useful sources of protocols and product information are those found on websites and product inserts from the manufacturers of relevant reagents. The following is a selection of some of the most useful websites, but students are encouraged to surf the Web and compare and contrast protocols from different manufacturers before finalizing their experimental designs:

www.miltenyibiotec.com/en/NN_628_Protocols .aspx Descriptions of protocols for use with Miltenyi magnetic beads are found here.

www.bdbiosciences.com/home.jsp The home page of BD Biosciences provides a wealth of information about flow cytometry.

www.bu.edu/flow-cytometry/files/2010/10/BD-Flow-Cytom-Learning-Guide.pdf This website gives a terrific introduction to flow cytometry for the beginner.

https://www.jax.org The home page of Jackson Laboratories provides information about many inbred, transgenic, and other useful mouse strains.

https://cre.jax.org/introduction.html An introduction to Cre/*lox* technology provided by Jackson Laboratories.

https://www.addgene.org/crispr/guide/ A useful introduction to the practical application of CRISPR-Cas9 technology.

https://www.wikimedia.org Wikipedia and Wikimedia often provide useful and updated information and links to protocols.

https://bitesizebio.com ("Brain Food for Biologists") This website provides protocols and useful reviews of modern and competing technologies.

www.virology.ws/2009/05/27/influenza-hemagglutination-inhibition-assay Refer to this site for information on viral hemagglutination or inhibition of hemagglutination.

https://www.rockefeller.edu/bioimaging/ Rockefeller University's Bio-Imaging Resource Center site provides useful information and some beautiful microscopy images.

The following websites point to multiple applications including molecular biology, biochemistry, and flow cytometry:

https://www.expertcytometry.com

http://www.thermofisher.com/us/en/home.html This website provides access to information formerly found on the InVitrogen website.

https://www.promega.com

https://www.sigmaaldrich.com/united-states.html

https://www.jove.com/ "JoVE" stands for "Journal of Visualized Experiments." It provides scientifically reviewed online videos of how to perform a range of scientific experiments. Although YouTube also provides some useful videos, in contrast to those found on the JoVE website, they are not subject to prior review.

STUDY QUESTIONS

1. When might you elect to use a polyclonal rather than a monoclonal antibody preparation, and why?

2. You successfully used one batch of polyclonal serum to immunoprecipitate a protein of interest. Analysis of the precipitated protein by polyacrylamide gel electrophoresis showed a single band, indicating a pure protein. However, when you repeated the experiment using a second batch of the serum derived from the same animal, but at a later date, your protein precipitate was contaminated with other proteins that did not coprecipitate the first time around. Why?

3. For the following applications, would you elect to use a polyclonal antibody preparation, a monoclonal antibody, or more than one monoclonal antibody to detect your antigen? Explain your answer.

   a. Bacterial agglutination
   b. Immunoprecipitation
   c. Western blotting
   d. Detection of a cytokine, using a solid-phase ELISA
   e. Diagnostic tissue typing

4. The following figure illustrates a hemagglutination inhibition assay. It shows the change with time in antibody titer of a newborn baby who survived the H1N1 influenza epidemic in 2009 in Thailand. The numbers along the bottom of the plate represent the dilution of the sera used in the experiment. The most concentrated serum is a 10-fold dilution from the patient's serum. In this experiment, sera obtained from different times and different sources have been serially diluted as indicated. Influenza virus and red blood cells have then been added to each well of the plate, and the ability of the various antisera to inhibit hemagglutination assessed.

The top row of this plate shows the hemagglutination inhibition that occurs when a positive-control anti-influenza antibody sample is added to the combination of virus and RBCs. It shows that the serum can be diluted out to 1 in 320 and still bind to the virus sufficiently to inhibit hemagglutination. The second row shows the hemagglutination that occurs when the virus is added in the absence of any neutralizing antibodies; it represents the negative control in this experiment.

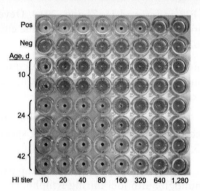

[Dulyachai, W., et al. Perinatal pandemic (H1N1) 2009 infection, Thailand [letter]. Emerging Infectious Diseases [serial on the Internet]. 2010 Feb. http://wwwnc.cdc.gov/eid/article/16/2/09-1733.htm]

The bottom six rows represent the hemagglutination inhibition capacity of the serum of the baby who is the subject of the experiment. The serum diluted in rows 3 and 4 (duplicate) was drawn from the baby when she was 10 days of age. The serum diluted in rows 5 and 6 was taken when the baby was 24 days old, and that in rows 7 and 8 when the baby was 42 days old.

   a. Did the baby's serum contain any influenza-specific antibodies when she was 10 days old?
   b. How old was the baby before her serum displayed the same hemagglutination capacity as the positive-control sample?
   c. Can we conclude from this experiment that the baby is making these antibodies herself?

5. Why might you elect to use an RIA rather than an ELISA with conventional chromogenic substrates? Would an ELISA with a chemiluminogenic substrate solve the problem?

6. The substrates used in an ELISPOT assay differ from those employed in a conventional ELISA, but may be similar to, or identical to, those used in Western blots. Why?

7. a. You have just started up your laboratory and are working on a shoestring budget until you hear about your first grant. You need to measure the affinity of a monoclonal antibody you're working on, and you have it available in radioactively labeled form. What method would you use, and why?

b. Your experiment indicates that you now need to know the association rate of the antibody-antigen interaction. You receive good news about your grant. How do you proceed?

8. What advantages are offered by two- or multiphoton microscopy over more traditional confocal microscopy?

9. When might you elect to purify cell populations with magnetic-activated cell sorting, rather than fluorescence-activated cell sorting, and when might you choose fluorescence-based separation over magnetic-based methods?

10. Experimental Design Question: You posit that under conditions of infection with a particularly virulent virus, B-1 B cells are entering inflamed regions of the lung. How might you test this hypothesis?

11. Experimental Design Question: You wish to knockout the expression of a particular gene, only in B cells, and you want to do the knockout only after you have exposed your B cells to antigen. What genetic constructs do you need to generate for this experiment?

Appendix I: CD Antigens

The following table presents information about the nature, cellular distribution, and function of the cluster of differentiation (CD) antigens that you will come across on a regular basis. A full listing of all CD antigens can be found online at http://www.macmillanhighered.com/launchpad/immunology8e. For the most part, CD antigens are plasma membrane–associated molecules, although some are also expressed on internal membranes and others may also occur in secreted forms. These molecules serve a huge variety of functions within the immune system. Because they are often expressed selectively on particular cell types, and sometimes alter their expression according to the developmental and/or activation state of those cells, they can serve as markers for particular cell types at specific stages of their differentiation.

Many CD antigens are known by a variety of names, and therefore some synonyms have been indicated in addition to the official CD designations in the table that follows. We have tried to place the most commonly used terms early in each list. The molecular weights that are shown are those of the human versions of the proteins, as listed on the UniProt and BioLegend websites (see below). Note that the molecular weight is calculated on the basis of the protein sequence and does not include the molecular weight of any associated carbohydrate. This means that the molecular weight shown in this table may differ substantially from that observed on SDS-PAGE. A useful note: many of the websites developed by the companies that sell antibodies (see below) recognizing these CD antigens will display examples of SDS-PAGE gels that demonstrate approximately where the relevant proteins may be expected to run.

Many of the proteins on this list are expressed in more than one isoform. Where there are two isoforms, both molecular weights are shown; where there are more than two, only the first molecular weight is shown in the table, and the reader is alerted to the number of known isoforms. No molecular weight information is listed for the CD antigens that are carbohydrates or lipids.

Those CD antigens that have an enzymatic function also have a number designated by the Enzyme Commission that describes the type of reaction that they catalyze. The explanation for these numbers can be found at http://www.sbcs.qmul.ac.uk/iubmb/enzyme/. Enzyme Commission numbers are listed after the letters "EC" for the relevant CD antigens. Recall that type 1 membrane proteins present the amino terminus to the external face of the membrane, and type 2 membrane proteins conversely present the carboxyl terminus to the external membrane face. Both type 1 and type 2 proteins cross the membrane only a single time. Multipass proteins may cross the membrane many times, but most cross the membrane seven times (e.g., chemokine receptors) or four times (tetraspanins).

In summarizing the expression patterns for CD antigens, we have concentrated on cells of the immune system (leukocytes). However, many of these antigens are also expressed on other cell types, and we also give some (but not all) examples of these. The description of the functions of each CD antigen also focuses on cells of the immune system, with an occasional example of functions in other cell types. The function is recorded as "not known" if there is no immunologically relevant function known for the CD marker. CD antigens are similarly named and share many properties in humans and mice, although there are some differences. The table presents information for human CD antigens. Recall that these numbers are based on antigenic determinants detected by monoclonal antibodies. Most determinants have been confirmed by the binding of more than one monoclonal antibody. However, determinants defined by the binding of a single antibody are awarded provisional status and designated by a CDw number, for example, CDw293.

Responsibility for naming and describing CD antigens rests with the *Human Cell Differentiation Molecules (HCDM)* organization (www.hcdm.org), which runs the *Human Leukocyte Differentiation Antigens (HLDA) Workshops*. The mission of HCDM is to characterize the structure, function, and distribution of leukocyte surface molecules and other molecules of the immune system. The website and international workshops are the product of a collaborative effort by numerous researchers and biomedical supply companies around the world and are updated regularly. The list of CD antigens was most recently updated at the HLDA10 workshop, held in Australia in 2014; the new CD antigens are described on the HLDA website at http://www.hcdm.org/index.php/hlda10-workshop/new-cd-molecules.

A list of most of the CD protein molecules is available through the UniProt knowledge base at https://www.uniprot.org/docs/cdlist. This searchable database, which is frequently updated, is sponsored and maintained by the Swiss Institute of Bioinformatics and provides invaluable information about the structure, sequence, cellular distribution, and functions of the proteins, as well as genetics, with links to relevant references.

The websites of several bioscience supply companies also include lists of CD antigens in both humans and mice, with links to antibody reagents the companies sell for detecting these markers. You can find the CD antigen lists of three of these companies at the following websites:

BD Biosciences: www.bdbiosciences.com/documents/cd_marker_handbook.pdf

Abcam: http://docs.abcam.com/pdf/immunology/Guide-to-human-CD-antigens.pdf

Miltenyi Biotec: www.miltenyibiotec.com/~/media/Files/Navigation/Cell%20analysis/fluorochrome%20tech%20specs/Human%20CD%20Marker%20Poster.ashx

An additional source of up-to-date information is PubMed, a website of the National Library of Medicine found at https://www.ncbi.nlm.nih.gov/pubmed.

PubMed allows searches of published articles by keywords as well as other variables, such as author name or year of publication. A keyword search using the name of a particular CD antigen will provide a list of the most recent journal articles that address it. For a more comprehensive search on a particular antigen, the synonyms listed in this table can be included as alternative keywords.

This table was updated on the basis of information in the websites mentioned above, along with some recent literature references.

CD Antigen; MW; Synonyms; Properties	Expression	Function
CD2; 39.4 kDa; LFA-2, T11, Leu-5, Tp50; CD58-binding adhesion molecule, sheep red blood cell (SRBC) receptor	T cells, cortical thymocytes, NK cells, some B cells, some monocytes	Interacts with LFA-3 and CD48/BCM1 to mediate adhesion between T cells and other cell types; contributes to T-cell activation
CD3; composed of three polypeptide chains: γ (20.5 kDa), δ (18.9 kDa), ε (23.1 kDa); T3; type 1 membrane protein	Thymocytes, T cells	Signaling chains of the TCR. Essential roles in cell surface expression of the TCR and TCR signal transduction
CD4; 51.1 kDa; T4, Leu-3, L3T4, Ly-4 (mouse), Ox38; type 1 membrane protein	MHC class II–restricted T cells, some thymocytes, monocytes/macrophages	Coreceptor for MHC class II–restricted T-cell activation; thymic differentiation marker for T cells; receptor for HIV
CD5; 54.6 kDa; Leu-1, T1, Ly-1 (mouse), Ox19; type 1 membrane protein	Mature T cells, cortical thymocytes, mature B-cell subset (B-1a B cells)	Positive or negative modulation of TCR and BCR signaling, depending on the type and developmental stage of cell displaying it
CD8; membrane-bound dimer of two chains, an $\alpha\beta$ heterodimer or an $\alpha\alpha$ homodimer: α, 25.7 kDa; β, 23.7 kDa; multiple isoforms of CD8β, some of which can be secreted. T8, Leu-2, Lyt-2 (mouse); type 1 membrane proteins	MHC class I–restricted T cells, some thymocytes, subset of dendritic cells, NK cells	Coreceptor for MHC class I–restricted T-cell activation, thymic differentiation marker for T cells
CD11a; 128.8 kDa (three isoforms); integrin α_L chain, associates with integrin β_2 chain (CD18) to form leukocyte function–associated molecule-1 (LFA-1); type 1 membrane protein	All leukocytes	Subunit of LFA-1, a membrane glycoprotein that provides cell-cell adhesion by interaction with ICAMs 1–4 (intercellular adhesion molecules 1–4, CD54); functions in leukocyte–endothelial cell interaction, cytotoxic T cell–mediated killing, and antibody-dependent killing by granulocytes and monocytes
CD11b; 127.2, 127.3 kDa (two isoforms); integrin α_M chain, α chain of CR3 complement receptor, formerly known as MAC-1. Associates with β_2-integrin (CD18) to form CR3. Ly-40 (mouse), Ox42; type 1 membrane protein	Granulocytes, monocytes, macrophages, NK cells, subsets of T and B cells, myeloid dendritic cells	Implicated in various adhesive interactions of monocytes, macrophages, and granulocytes as well as in mediating the uptake of complement-coated particles. Identical to CR3, the receptor for the iC3b fragment of the third complement component. It probably recognizes the R-G-D peptide in C3b. Integrin α_M/β_2 is also a receptor for fibrinogen, factor X, and ICAM-1

CD Antigen; MW; Synonyms; Properties	Expression	Function
CD14; 40.1 kDa; LPS receptor (LPS-R); GPI-anchored membrane receptor	Monocytes, macrophages, granulocytes (weak expression), Langerhans cells	Receptor for endotoxin (lipopolysaccharide [LPS]) bound to LPS-binding protein (LBP), activating innate immune responses; transfers the complex to TLR4. May also be involved in binding of peptidoglycans and lipoproteins to TLR2
CD19; 61.1 kDa, 61.2 kDa (two isoforms). B4, Leu-12; type 1 membrane protein	B cells from earliest recognizable B-lineage cells during development to B-cell blasts but lost on maturation to plasma cells, follicular dendritic cells	Part of B-cell coreceptor with CD21 and CD81; a critical signal transduction molecule that assembles with the BCR and regulates B-cell development, activation, and differentiation. Commonly used marker for mature B cells and target of immunotherapeutic antibodies in B-cell lymphoma treatment
CD20; 33.1 kDa, 14.6 kDa (two isoforms); B1, Bp35, Leu-16; multipass membrane protein with lipid anchor	B cells, T-cell subsets	Ligation activates signaling pathways; may have a role in regulating B-cell activation, proliferation, and differentiation
CD21; 112.9 kDa (four isoforms); complement receptor 2 (CR2), C3d receptor, Epstein-Barr virus (EBV) receptor; with CD19 and CD81 forms the B-cell coreceptor; type 1 membrane protein, but a soluble form of CD21 can be shed from the cell surface	Mature B cells, subset of T cells, follicular dendritic cells, astrocytes	Receptor for C3d, C3dg, and iC3b; with CD19 and CD81, part of the B-cell coreceptor complex that contributes to B-cell activation. Also serves as the receptor for EBV
CD25; 30.8 kDa (this protein is further processed into the mature form); IL-2Rα (IL-2 receptor α chain), Tac antigen, p55; type 1 membrane protein	Activated B cells and T cells, regulatory T cells, immature thymocytes, activated monocytes, macrophages, NK cells, dendritic cell subset	Low-affinity IL-2 receptor, associates with β and γ chains to form high-affinity IL-2R; activation marker; induces activation and proliferation of T cells, NK cells, B cells, and macrophages; thymocyte differentiation marker; tT$_{REG}$ marker
CD28; 25.1 kDa (this is further processed into a mature form; there are seven isoforms of this molecule, which is often expressed as a homodimer); Tp44; type 1 membrane protein	Mature thymocytes, most peripheral T cells, plasma cells, NK cells	Costimulation of T-cell proliferation and cytokine production on binding CD80 or CD86
CD32; 34 kDa (five isoforms); low-affinity IgG Fc region receptor IIb, CDw32, FcγRIIb, FcRIIb; type 1 membrane protein	The most broadly distributed Fcγ receptor. Expressed in monocytes, neutrophils, macrophages, basophils, eosinophils, Langerhans cells, B cells, platelets, and placenta (endothelial cells)	Binds Fc region of IgG. Involved in phagocytosis of immune complexes and modulation of antibody production by B cells. Binding to this receptor results in down-modulation of previous state of cell activation triggered via antigen receptors on B cells (BCR), T cells (TCR), or via another Fc receptor. Different isoforms variously trigger phagocytosis
CD34; 40.7, 35.0 kDa (two isoforms); hematopoietic progenitor cell antigen; type 1 membrane protein	Hematopoietic stem and progenitor cells, small-vessel endothelial cells	Bound by L-selectin. Adhesion molecule mediating hematopoietic stem cell adhesion to bone marrow stromal cells or extracellular matrix
CD35; 223.7 kDa (this sequence is further processed into the mature form); complement receptor type 1 (CR1), C3b/C4b receptor (C3b/4bR); type 1 membrane protein	B cells, some T-cell subsets, neutrophils, monocytes, eosinophils, follicular dendritic cells, and erythrocytes	Receptor for C3b/C4b-coated particles, mediating their adherence and phagocytosis; facilitator of C3b and C4b cleavage, thus limiting complement activation

(continued)

CD Antigen; MW; Synonyms; Properties	Expression	Function
CD40; 30.6, 22.3 kDa (two isoforms); tumor necrosis family receptor superfamily member 5 (TNFRSF5), B-cell surface antigen CD40, CD40L receptor, Bp50, CDw40; type 1 membrane protein and secreted forms	Mature B cells but not plasma cells, activated monocytes, macrophages, dendritic cells, epithelial and endothelial cells, fibroblasts, keratinocytes, CD34$^+$ hematopoietic cell progenitors	Binds to CD40 ligand (CD154). Provides essential costimulatory signals for B-cell activation, proliferation, differentiation, and isotype switching; apoptosis rescue signal for germinal center B cells. Stimulates cytokine production by macrophages and dendritic cells and up-regulates adhesion molecules on dendritic cells. Regulates cell-mediated, as well as antibody-mediated, immunity
CD45; 147 kDa (many different isoforms with different molecular weights are generated by alternative splicing of three exons that can be inserted immediately after an amino-terminal sequence of 8 amino acids found on all isoforms); leukocyte common antigen (LCA), T200 on T cells and B220 on B cells, receptor-type protein tyrosine phosphatase C (PTPRC), EC 3.1.3.48; type 1 membrane protein. *See the online full listing of CD antigens for isoforms CD45RA, CD45RB, CD45RC, and CD45RO.*	All hematopoietic cells except erythrocytes and platelets; especially high on lymphocytes (10% of their surface area comprising CD45); different isoforms characteristic of differentiated subsets of various hematopoietic cells	Regulates activation of a variety of cellular processes, including cell growth, differentiation, mitotic cycle, and oncogenic transformation. Essential role in T- and B-cell antigen receptor–mediated activation; possible role in receptor-mediated activation in other leukocytes
CD50; 59.5 kDa; intercellular adhesion molecule-3 (ICAM-3); type 1 membrane protein	Most leukocytes, including immature thymocytes and Langerhans cells; endothelial cells	Ligand for LFA-1 (CD11a/CD18) and is involved in integrin-dependent adhesion and cell migration. Recognized by CD209 (DC-SIGN) on dendritic cells. Acts as a costimulatory molecule for T-cell activation. Regulates leukocyte morphology
CD54; 57.8 kDa; intercellular adhesion molecule-1 (ICAM-1); type 1 membrane protein	Activated T and B cells, monocytes, activated endothelial cells	Ligand for CD11a/CD18 or CD11b/CD18, shown to bind to fibrinogen and hyaluronan; receptor for rhinoviruses and for RBCs infected with malarial parasites. Adhesion molecule; contributes to antigen-specific T-cell activation by antigen-presenting cells; contributes to the extravasation of leukocytes from blood vessels, particularly in areas of inflammation. Also, soluble form may inhibit the activation of CTL or NK cells by malignant cells
CD55; 41.4 kDa (seven isoforms); decay-accelerating factor (DAF); current information indicates that Isoform 1 is a type 1 membrane protein; isoforms 2, 6, and 7 are GPI-anchored membrane proteins; isoforms 3, 4, and 5 are secreted	Most cell types; a soluble form exists in plasma and body fluids	Member of the regulator of complement activation (RCA) family of proteins. Protective barrier against inappropriate complement activation and deposition on plasma membranes. Also binds CD97. May contribute to lymphocyte activation. Also serves as a receptor for echovirus and coxsackie B virus
CD59; 14.2 kDa; protectin, membrane attack complex inhibition factor (MACIF), homologous restriction factor-20 (HRF20), membrane inhibitor of reactive lysis (MIRL); exists as GPI-anchored and secreted forms	Most hematopoietic and nonhematopoietic cell types	Potent inhibitor of the complement membrane attack complex (MAC). Binds to C8 and/or C9, thereby preventing incorporation of C9 into the structure of the osmolytic pore. Interacts with CD2 and Src kinases and involved in T-cell signal transduction and activation

CD Antigen; MW; Synonyms; Properties	Expression	Function
CD62E; 66.7 kDa; E-selectin, endothelial leukocyte adhesion molecule-1 (ELAM-1), leukocyte–endothelial cell adhesion molecule-2 (LECAM-2); type 1 membrane protein	Acutely activated vascular endothelium, chronic inflammatory lesions of skin and synovium	C-type lectin endothelial adhesion molecule that mediates leukocyte (e.g., neutrophil) rolling on activated endothelium at inflammatory sites through interaction with PSGL1 through the sialyl Lewis X (CD15s) carbohydrate, CD43, and other leukocyte antigens; may also participate in angiogenesis and tumor-cell adhesion during metastasis via the blood
CD62L; 42.2, 43.6 kDa (two isoforms); L-selectin, leukocyte adhesion molecule-1 (LAM-1), leukocyte–endothelial cell adhesion molecule-1 (LECAM-1); type 1 membrane protein	Most peripheral blood B cells, T-cell subsets, NK-cell subsets, cortical thymocytes, monocytes, granulocytes	C-type lectin adhesion molecule that mediates adherence (initial tethering and rolling, through interaction with PSGL1 through the sialyl Lewis X [CD15s] carbohydrate) of lymphocytes for homing to high endothelial venules of peripheral lymphoid tissue, and also leukocyte adhesion and rolling on activated endothelium at inflammatory sites
CD62P; 90.8 kDa; P-selectin, granule membrane protein-140 (GMP-140), platelet activation–dependent granule-external membrane (PADGEM) protein, C-type lectin; single-chain type 1 glycoprotein	Activated platelets and endothelial cells	C-type lectin adhesion molecule that binds to PSGL1 through the sialyl Lewis X (CD15s) carbohydrate on neutrophils and monocytes and mediates tethering and rolling of leukocytes on the surface of activated endothelial cells, the first step in leukocyte extravasation and migration toward sites of inflammation; mediates adherence to platelets; may contribute to inflammation-associated tissue destruction, atherogenesis, and thrombosis
CD69; 22.6 kDa; early activation antigen 1 (EA 1), activation-inducer molecule (AIM), very early activation antigen (VEA), MLR3, Leu-23; type 2 membrane protein	Activated T, B, and NK lymphocytes; thymocyte subsets, monocytes, macrophages, granulocytes, Langerhans cells, platelets	Involvement in early events of lymphocyte, monocyte, and platelet signaling and activation; may provide costimulatory signals in lymphocytes
CD79a; 25.0, 20.8 kDa (two isoforms); Ig-α, MB1, B-cell antigen receptor complex–associated protein α chain; type 1 membrane glycopeptide, disulfide-linked heterodimer with CD79b	B cells	Component of B-cell antigen receptor analogous to CD3; required for cell surface expression and signal transduction
CD79b; 26.0 kDa (three isoforms); Ig-β, B29, B-cell antigen receptor complex–associated protein β chain; type 1 glycopeptide, disulfide-linked heterodimer with CD79a	B cells	With CD79a, part of the B-cell antigen receptor
CD80; 33.0 kDa (three isoforms); B7.1, B7, BB1, T-lymphocyte activation antigen CD80, Ly-53; type 1 glycoprotein	Activated B and T cells, macrophages, low levels on resting peripheral blood monocytes and dendritic cells	Binds CD28 and CD152/CTLA-4; costimulation of T-cell activation with CD86 when bound to CD28; inhibits T-cell activation when bound with CD152
CD81; 26 kDa; target for antiproliferative antibody-1 (TAPA-1), tetraspanin 28 (T-span28), single-chain type 3, 4-span protein; member of the four-transmembrane-spanning protein superfamily (TM4SF)	Broadly expressed on hematopoietic cells; expressed by endothelial and epithelial cells; absent from erythrocytes, platelets, and neutrophils	Member of CD19/CD21 signal transduction complex; mediates signal transduction events involved in the regulation of cell development, growth, and motility; participates in early T-cell development. Binds the E2 glycoprotein of hepatitis C virus

(continued)

CD Antigen; MW; Synonyms; Properties	Expression	Function
CD86; 37.7 kDa (six isoforms); B7.2, B70, T-lymphocyte activation antigen CD86; type 1 membrane glycoprotein	Dendritic cells, memory B cells, germinal center B cells, monocytes, activated T cells, and endothelial cells	Major T-cell costimulatory molecule, interacting with CD28 (stimulatory) and CD152/CTLA4 (inhibitory)
CD90; 17.9 kDa; Thy-1 membrane glycoprotein (Thy-1); single-chain, GPI-anchored glycoprotein	Hematopoietic stem cells, neurons, connective tissue, thymocytes, peripheral T cells, human lymph node HEV endothelium	Possible involvement in lymphocyte costimulation; possible inhibition of proliferation and differentiation of hematopoietic stem cells
CD95; 37.7 kDa (seven isoforms); APO-1, apoptosis-mediating surface antigen (Fas), TNF receptor superfamily member 6 (TNFRSF6), FasL receptor; type 1 glycoprotein	Activated T and B cells, monocytes, fibroblasts, neutrophils, NK cells	Binds Fas ligand (FasL, CD178) and initiates apoptotic pathway
CD117; 110 kDa (three isoforms); proto-oncogene tyrosine-protein kinase c-KIT, mast/stem cell growth factor receptor (SCFR); single-chain type 1 glycoprotein	Hematopoietic stem cells and progenitor cells, mast cells	Stem-cell factor receptor, tyrosine kinase activity; early-acting hematopoietic growth factor receptor, necessary for the development of hematopoietic progenitors. Capable of inducing proliferation of mast cells and is a survival factor for primordial germ cells
CD135; 113, 108 kDa (two isoforms); receptor-type tyrosine-protein kinase FLT3, FMS-like tyrosine kinase 3 (Flt3), Flk-2 (in mice), stem cell tyrosine kinase-1 (STK-1), EC 2.7.10.1, tyrosine kinase receptor, type 3; single-chain type 1 glycoprotein	Multipotential, myelomonocytic, and primitive B-cell progenitors	Growth factor receptor for early hematopoietic progenitors, tyrosine kinase
CD138; 32.5 kDa; syndecan-1 (SYND-1), heparin sulfate proteoglycan, type 1 glycoprotein; type 1 membrane protein (also found in secreted form)	Pre-B cells, immature B cells and plasma cells, but not mature circulating B lymphocytes; basolateral surfaces of epithelial cells, embryonic mesenchymal cells, vascular smooth muscle cells, endothelium, neural cells, breast cancer cells	Binds to many extracellular matrix proteins and mediates cell adhesion and growth
CD152; 24.7 kDa (five isomeric forms); cytotoxic T lymphocyte– associated protein-4 (CTLA-4); disulfide-linked homodimeric type 1 membrane glycoprotein	Activated T cells and some activated B cells	Negative regulator of T-cell activation; binds CD80 and CD86
CD154; 29.3 kDa; CD40 ligand (CD40L), T-cell antigen gp39, T-BAM; tumor necrosis factor ligand superfamily member 5, TNF-related activation protein (TRAP); TNF superfamily, homotrimeric type 2 glycoprotein	Activated CD4$^+$ T cells, small subset of CD8$^+$ T cells and $\gamma\delta$ T cells; also activated basophils, platelets, monocytes, mast cells	Ligand for CD40, inducer of B-cell proliferation and activation, antibody class switching, and germinal center formation; costimulatory molecule and a regulator of T$_H$1 generation and function; role in negative selection and peripheral tolerance
CD178; 31.5, 14.0 kDa (two isoforms); CD95 ligand (CD95-L), Fas antigen ligand (Fas ligand, FasL), apoptosis antigen ligand (APTL), tumor necrosis factor ligand superfamily member 6; type 2 membrane protein, located on plasma and on internal membranes	Constitutive or induced expression on many cell types, including T cells, NK cells, microglial cells, neutrophils, nonhematopoietic cells such as retinal and corneal parenchymal cells	Ligand for the apoptosis-inducing receptor CD95 (Fas/APO-1); key effector of cytotoxicity and involved in Fas/FasL interaction, apoptosis, and regulation of immune responses. Proposed to transduce a costimulatory signal for CD8$^+$ and naïve CD4$^+$ T-cell activation
CD179a; 16.6 kDa; protein VPreB1, V(pre)B protein (VpreB), immunoglobulin ι (iota) chain, external side of plasma membrane	Selectively expressed in pro-B and early pre-B cells	Associates with CD179b to form surrogate light chain of the pre–B-cell receptor; involved in early B-cell differentiation

CD Antigen; MW; Synonyms; Properties	Expression	Function
CD179b; 22.9, 8.6 kDa (two isoforms); λ5, immunoglobulin λ-like polypeptide-1, Ig λ5; external side of plasma membrane	Selectively expressed in pro-B and early pre-B cells	Associates with CD179a to form surrogate light chain of the pre–B-cell receptor; involved in early B-cell differentiation
CD227; 122.1 kDa (17 isomers); mucin 1 (MUC1), polymorphic epithelial mucin (PEM), episialin, tumor-associated mucin, carcinoma-associated mucin; heavily glycosylated glycoprotein A type 1 transmembrane protein, although some isoforms are secreted. Cleaved into two subunits, α and β. The β subunit can be found in the nucleus and cytoplasm	Epithelial cells, follicular dendritic cells, monocytes, subsets of lymphocytes, B and stem cells, some myelomas; some hematopoietic cell lineages	Involved in adhesion and signaling
CD247; 18.7 kDa (three isoforms); T-cell surface glycoprotein CD3 ζ (zeta) chain, zeta chain of CD3, CD3ζ, T3ζ, TCRζ; single-pass type 1 membrane protein	NK cells during thymopoiesis and mature T cells in the periphery	T-cell activation
CD252; 21.1, 15.4 kDa (two isoforms); OX-40 ligand (Ox40L), tumor necrosis factor ligand superfamily member 4, glycoprotein 34 (gp34), transcriptionally activated glycoprotein 1 (TAX), TNF superfamily; type 2 membrane protein	Activated B cells, cardiac myocytes	CD40 ligand; T-cell costimulation
CD256; 27.4 kDa (six isoforms); a proliferation-inducing ligand (APRIL), TNF- and APOL-related leukocyte-expressed ligand 2 (TALL-2), TNF-related death ligand-1 (TRDL1), tumor necrosis factor ligand superfamily member 13; TNF superfamily; secreted and membrane-bound forms exist	Monocytes and macrophages	Binds TACI and BCMA; involved in B-cell proliferation
CD257; 31.2 kDa (three isoforms); B-cell activator factor (BAFF), B-lymphocyte stimulator (BLyS), TNF- and APOL-related leukocyte-expressed ligand 1 (TALL-1), dendritic cell–derived TNF-like molecule, tumor necrosis factor ligand superfamily member 13B, TNF superfamily; type 2 membrane protein and secreted forms occur	Activated monocytes; exists as a soluble form	B-cell growth and differentiation factor; costimulation of Ig production
CD267; 31.8 kDa (three isoforms); transmembrane activator and CAML interactor (TACI), tumor necrosis factor receptor superfamily member 13B (TNFRSF13B); type 1 membrane protein	B cells, activated T cells	Binds BAFF and APRIL; regulates B cells
CD268; 118.9, 18.9 kDa (two isoforms); B-cell activating factor receptor (BAFF-R), tumor necrosis factor receptor superfamily member 13C (TNFRSF13C); type 1 membrane protein	B cells	Binds BAFF; involved in B-cell survival
CD269; 20.2, 14.8 kDa (two isoforms); B-cell maturation protein (BCMA), tumor necrosis factor receptor superfamily member 17 (TNFRSF17); type 1 membrane protein	Mature B cells	Binds BAFF; involved in B-cell survival and proliferation

(continued)

CD Antigen; MW; Synonyms; Properties	Expression	Function
CD273; 31 kDa (three isoforms); programmed cell death 1 ligand 2 (PDL2, PD-1 ligand 2, PDCD1 ligand 2), butyrophilin B7-DC (B7-DC); a glycoprotein expressed on plasma and endomembrane systems (type 1 protein) and also secreted	Dendritic cells, macrophages, monocytes, activated T cells	Second ligand for PD-1 (CD279); this interaction, like that of PD-L1/PD-1, inhibits TCR-mediated proliferation and cytokine production
CD274; 33.2 kDa (three isoforms); programmed cell death 1 ligand 1 (PD-L1, PDCD1 ligand-1), B7 homolog 1 (B7H1); single-pass type 1 membrane glycoprotein. Isoform 2 is located on endosomal membranes	Macrophages, epithelial and dendritic cells, NK cells, activated T cells, and monocytes	Ligand for PD-1 (CD279); involved in the costimulatory signal, essential for T-cell proliferation and production of IL-10 and IFN-γ. Interaction with PDCD1 inhibits T-cell proliferation and cytokine production
CD275; 33.3 kDa (three isoforms); inducible T-cell costimulator ligand (ICOSL), B7 homolog 2 (B7H2); type 1 membrane protein	Macrophages, dendritic cells, weak T and B cells, activated monocytes	Costimulation of T cells; reported to promote proliferation and cytokine production
CD278; 22.6, 19.1 kDa; inducible T-cell costimulator (ICOS); isoform 1 is a single-pass membrane protein and isoform 2 is secreted	T_H2 cells, thymocyte subsets, activated T cells	Plays a critical role in costimulating T-cell activation, development, proliferation, and cytokine production
CD279; 31.6 kDa; programmed cell death protein 1 (PD1, PDCD1, hPD-1); single-chain type 1 glycoprotein	Subset of thymocytes, activated T and B cells	Inhibits TCR-mediated proliferation and cytokine production
CD280; 166.7 kDa; C-type mannose receptor 2, macrophage mannose receptor 2 (MRC2), urokinase receptor–associated protein (uPARAP), endocytic receptor 180 (ENDO180); type 1 membrane protein	Myeloid progenitor cells, fibroblasts, chondrocytes, osteoclasts, osteocytes, subsets of endothelial and macrophage cells	May affect cell motility and remodeling of the extracellular matrix
CD339; 133.8, 116.6 kDa (two isoforms); protein jagged-1 (Jagged-1), JAG1, JAGL1, hJ1; type 1 membrane protein	Stromal cells, epithelial cells, myeloma cells	Notch ligand; involved in hematopoiesis
CD340; 137.9 kDa (six isoforms); receptor tyrosine-protein kinase erbB-2, HER2, ERBB2, C-erbB-2, NEU proto-oncogene (NEU); EC 2.7.10.1. Depending on isoform, may be located at the plasma membrane, in the perionuclear region of the cytoplasm, throughout the cytoplasm, or in the nucleus. The membrane protein is single-pass, type 1	Tumor tissues, epithelial cells	Cell proliferation and differentiation, tumor cell metastasis
CD363; 42.8 kDa (sequence is further processed); sphingosine 1-phosphate receptor (S1P receptor 1, S1P1), endothelial differentiation G protein–coupled receptor 1, sphingosine 1-phosphate receptor Edg-1 (S1P receptor Edg-1); multipass membrane protein. Found associated with lipid rafts. Also expressed on endosomal membranes	T cells, B cells, NK cells; induced on endothelial cells by stimulation with phorbol myristoyl acetate	Chemotactic receptor recognizing sphingosine 1-phosphate

Appendix II: Cytokines and Associated JAK-STAT Signaling Molecules

Table 1 displays the biological activities and cellular sources of cytokines of immunological interest. In most cases, protein mass is given, usually for human cytokines. In some instances, a given cytokine may have biological activities in addition to those listed here or may be produced by other sources as well as the ones cited here. (Note that cytokines that are not closely identified with the immune system—e.g., growth hormone—are not listed. Also, with the exception of IL-8, chemokines are not included in this compilation.) **Table 2** displays select groups of cytokines that are functionally related and **Table 3** displays the Janus kinase (JAK)–signal transducer and activator of transcription (STAT) molecules that are activated by major cytokine/cytokine receptor interactions.

Major references used for the information in this appendix:

GeneCards Human Gene Database: www.genecards.org

PeproTech: https://www.peprotech.com

PubMed: https://www.ncbi.nlm.nih.gov/pubmed

R&D Systems: https://www.rndsystems.com/research-area/cytokines-and-growth-factors

UniProt: https://www.uniprot.org

Wikipedia: https://www.wikipedia.org/

TABLE 1	Cytokine Sources and Biological Activities	
Cytokine; MW; Synonyms	**Sources**	**Activity**
Interleukin 1 (IL-1) IL-1α: 17.5 kDa IL-1β: 17.3 kDa Lymphocyte-activating factor (LAF); mononuclear cell factor (MCF); endogenous pyrogen (EP)	Many cell types, including monocytes, macrophages, dendritic cells, NK cells, and non–immune system cells such as epithelial and endothelial cells, fibroblasts, adipocytes, astrocytes, and some smooth muscle cells	Displays a wide variety of biological activities on many different cell types, including T cells, B cells, monocytes, eosinophils, and dendritic cells, as well as fibroblasts, liver cells, vascular endothelial cells, and some cells of the nervous system. The in vivo effects of IL-1 include induction of local inflammation and systemic effects such as fever, the acute-phase response, and stimulation of neutrophil production
Interleukin 2 (IL-2) 15–20 kDa T-cell growth factor (TCGF)	Activated T cells	Stimulates proliferation and differentiation of T and B cells; activates NK cells
Interleukin 3 (IL-3) 15.1 kDa (monomer) 30 kDa (dimer) Multipotential colony-stimulating factor; hematopoietic cell growth factor (HCGF); mast-cell growth factor (MCGF)	Activated T cells, mast cells, basophils, and eosinophils	Growth factor for hematopoietic cells; stimulates colony formation in neutrophil, eosinophil, basophil, mast cell, erythroid, megakaryocyte, and monocytic lineages but not of lymphoid cells
Interleukin 4 (IL-4) 15–19 kDa B cell–stimulatory factor-1 (BSF-1)	T cells (particularly those of the T_H2 subset), ILC2, mast cells, basophils, and bone marrow stromal cells	Involved in type 2 responses. Promotes naïve T-cell differentiation to T_H2 cells. Stimulates the growth and differentiation of B cells. Induces class switching to IgE. Involved in defense against worms and protozoa, as well as pathology of asthma and allergy (with IL-5, -9, and -13)

(continued)

Cytokine; MW; Synonyms	Sources	Activity
Interleukin 5 (IL-5) 15 kDa Eosinophil differentiation factor (EDF); eosinophil colony-stimulating factor (E-CSF)	T cells (particularly those of the T_H2 subset), ILC2, mast cells, eosinophils	Involved in type 2 responses. Induces eosinophil formation and differentiation. Stimulates B-cell growth and differentiation. Involved in defense against worms and protozoa, as well as pathology of asthma and allergy (with IL-4, -9, and -13)
Interleukin 6 (IL-6) 26 kDa B-cell stimulatory factor 2 (BSF-2); hybridoma/plasmacytoma growth factor (HPGF); hepatocyte-stimulating factor (HSF)	Some T cells and B cells, several nonlymphoid cells, including macrophages, bone marrow stromal cells, fibroblasts, endothelial and muscle cells, adipocytes, and astrocytes	Regulates B- and T-cell functions; in vivo effects on hematopoiesis. Induces inflammation and the acute-phase response
Interleukin 7 (IL-7) 20–28 kDa Pre-B-cell growth factor; lymphopoietin-1 (LP-1)	Bone marrow and thymic stromal cells, intestinal epithelial cells	Growth factor for T-cell, B-cell, and ILC progenitors
Interleukin 8 (IL-8) 6–8 kDa Neutrophil-attractant/activating protein (NAP-1); neutrophil-activating factor (NAF); granulocyte chemotactic protein 1(GCP-1); CXCL8 chemokine	Many cell types, including monocytes, macrophages, lymphocytes, granulocytes, and non–immune system cells such as fibroblasts, endothelial and epithelial cells, and hepatocytes	*Chemokine* that functions primarily as a chemoattractant and activator of neutrophils; also attracts basophils and some subpopulations of lymphocytes; has angiogenic activity
Interleukin 9 (IL-9) 32–40 kDa P40; T-cell growth factor III	Some activated T-helper-cell subsets (T_H9), ILC2	Involved in type 2 responses. Stimulates proliferation of T lymphocytes and hematopoietic precursors, involved in defense against worms and protozoa, and pathology of asthma and allergy, with IL-4, -5, and -13
Interleukin 10 (IL-10) 35–40 kDa Cytokine synthesis inhibitory factor (CSIF)	Activated subsets of CD4$^+$ and CD8$^+$ T cells, macrophages, and dendritic cells	Anti-inflammatory; antagonizes generation of the T_H1 subset of helper T cells; stimulates IgA synthesis and secretion by human B cells. Enhances proliferation of B cells, thymocytes, and mast cells. Cooperates with TGF-β
Interleukin 11 (IL-11) 23 kDa	Bone marrow stromal cells and IL-1–stimulated fibroblasts	Growth factor for plasmacytomas, megakaryocytes, and macrophage progenitor cells
Interleukin 12 (IL-12) Heterodimer containing a p35 subunit of 30–35 kDa and a p40 subunit of 35–44 kDa NK cell stimulatory factor (NKSF); cytotoxic lymphocyte maturation factor (CLMF)	Macrophages, B cells, and dendritic cells	Induces differentiation of the T_H1 subset of helper T cells; induces IFN-γ production by T cells and NK cells and enhances NK and cytotoxic T-cell activity
Interleukin 13 (IL-13) 10 kDa	Activated T cells (particularly those of the T_H2 subset), ILC2, mast cells, and NK cells	Involved in type 2 responses. Role in T_H2 responses; up-regulates synthesis of IgE and suppresses inflammatory responses. Involved in defense against worms and protozoa, as well as pathology of asthma and some allergies

Cytokine; MW; Synonyms	Sources	Activity
Interleukin 14 (IL-14) 60 kDa High-molecular-weight B-cell growth factor (HMW-BCGF)	Activated T cells	Enhances B-cell proliferation; inhibits antibody synthesis
Interleukin 15 (IL-15) 14–15 kDa	Many cell types but primarily dendritic cells and cells of the monocytic lineage	Stimulates NK-cell and T-cell proliferation and development; helps to activate NK cells
Interleukin 16 (IL-16) Homotetramer: 60 kDa Monomer: $\approx$17 kDa Lymphocyte chemoattractant factor (LCF)	Activated T cells and some other cell types	Chemoattractant for CD4$^+$ T cells, monocytes, and eosinophils. Binding of IL-16 by CD4 inhibits HIV infection of CD4$^+$ cells
Interleukin 17 (IL-17) 28–31 kDa CTLA-8 (cytotoxic T lymphocyte–associated antigen 8) Family members IL-17A–F	CD4$^+$ T cells (particularly those of the T$_H$17 subset), ILC3, CD8$^+$, $\gamma\delta$ T cells, NK cells, intraepithelial lymphocytes, and some other cells	Involved in immunity to extracellular bacteria. Promotes inflammation by increasing production of proinflammatory cytokines (IL-1, IL-6, TNF-α, G-CSF, GM-CSF) by epithelial, endothelial, and fibroblast cells, as well as chemokines that attract monocytes and neutrophils. Can also have anti-inflammatory effects in barrier tissues
Interleukin 18 (IL-18) 18.2 kDa Interferon γ–inducing factor (IGIF)	Cells of the monocytic lineage and dendritic cells	IL-1 family member. Promotes differentiation of T$_H$1 subset of helper T cells. Induces IFN-γ production by T cells and enhances NK-cell cytotoxicity
Interleukin 19 (IL-19) Homotetramer: 35–40 kDa	LPS-stimulated monocytes and B cells	A member of the IL-10 family of cytokines, induces reactive oxygen species and proinflammatory cytokines, which may promote apoptosis. Promotes T$_H$2 differentiation by inhibiting IFN-γ and augmenting IL-4 and IL-13 production
Interleukin 20 (IL-20) 18 kDa	Monocytes and keratinocytes	A member of the IL-10 family of cytokines; has effects in epidermal tissues and psoriasis. Like IL-19, shown to promote T$_H$2 differentiation
Interleukin 21 (IL-21) 15 kDa	Activated CD4$^+$ T cells	Enhances cytotoxic activity and IFN-γ production by activated NK and CD8+ T cells. Contributes to B-cell and follicular helper T cell activation in germinal centers
Interleukin 22 (IL-22) Homodimer: 25 kDa IL-10–related T cell–derived inducible factor (IL-TIF)	CD4$^+$ T cells (particularly those of the T$_H$17 subset), ILC3 cells	A member of the IL-10 family with roles in barrier immunity and pathogenesis. Induces epithelial cell release of antimicrobial peptides and enhances tight junction integrity. Has both pro- and anti-inflammatory effects; involved in immunity to extracellular bacteria
Interleukin 23 (IL-23) Heterodimer consisting of p40 subunit of IL-12 (35–40 kDa) and p19 (18.7 kDa)	Activated dendritic cells and macrophages, epithelial cells of barrier tissues	Induces T$_H$17 and ILC3 differentiation; involved in several pathologies of barrier tissues, including psoriasis and Crohn's disease

(continued)

Cytokine; MW; Synonyms	Sources	Activity
Interleukin 24 (IL-24) 35–40 kDa IL-10B; melanoma differentiation–associated protein 7 (MDA7)	Melanocytes, NK cells, B cells, subsets of T cells, fibroblasts, and melanoma cells	Member of the IL-10 family. Induces TNF-α and IFN-γ and low levels of IL-1β, IL-12, and GM-CSF in human PBMCs. Induces selective anticancer properties in melanoma cells by inhibiting proliferation and in breast carcinoma cells by promoting apoptosis
Interleukin 25 (IL-25) 20 kDa IL-17E; stroma-derived growth factor (SF20)	T_H2 subset of helper T cells, mast cells, basophils, eosinophils, intraepithelial lymphocytes, epithelial cells of barrier tissues	Proinflammatory member of the IL-17 family and alarmin. Produced by epithelial cells when barriers are breached. Together with IL-33 and TSLP induces production of type 2 cytokines by T_H2 and ILC2 cells; suppresses T_H17 and ILC3 cytokines and eotaxin
Interleukin 26 (IL-26) Homodimer: 36 kDa AK155	Subset of T and NK cells	Member of the IL-10 family. May have similar functions to IL-20
Interleukin 27 (IL-27) Heterodimer (27.5 kDa) composed of EBI3 (IL-27β) and p28 (IL-27α)	Produced by dendritic cells, macrophages, endothelial cells, and plasma cells	Shown to induce clonal expansion of naïve CD4$^+$ T cells, to synergize with IL-12 to promote IFN-γ production from CD4$^+$ T cells, and to induce CD8$^+$ T cell–mediated antitumor activity
Interleukin 28A/B (IL-28A/B) 22.3/21.7 kDa Interferon-λ2/3 (IFN-L2/3)	Monocyte-derived dendritic cells	Type III interferon (with IL-29), coexpressed with IFN-β, participates in the antiviral immune response, and shown to induce increased level of both MHC class I and II; induces regulatory T-cell proliferation
Interleukin 29 (IL-29) 22 kDa Interferon-λ1 (IFN-L1)	Monocyte-derived dendritic cells, T_H17 cells	Type III interferon (with IL-28) that functions similarly to IL-28A/B. Gene not found in mice
Interleukin 30 (IL-30) 28 kDa IL-27p28	Antigen-presenting cells	Subunit of IL-27 heterodimer; functions same as IL-27
Interleukin 31 (IL-31) 18 kDa	Mainly activated T_H2 T cells; can be induced in activated monocytes	Proinflammatory and may be involved in recruitment of polymorphonuclear cells, monocytes, and T cells to a site of skin inflammation; also regulates hematopoietic progenitors
Interleukin 32 (IL-32) 27 kDa NK4	Activated NK cells and PBMCs	IL-1 family member. Proinflammatory cytokine, has mitogenic properties, induces TNF-α and IL-8 expression
Interleukin 33 (IL-33) 31 kDa Nuclear factor in high endothelial venules (NF-HEV)	Epithelial cells of barrier tissues; high endothelial venule and smooth muscle cells	IL-1 family member and alarmin produced by epithelial cells when barriers are breached. Together with IL-25 and TSLP induces T_H2 cytokine production by T cells, mast cells, eosinophils, and basophils
Interleukin 34 (IL-34) 27.5 kDa	Many cell types	Promotes growth and development of myeloid cells

Cytokine; MW; Synonyms	Sources	Activity
Interleukin 35 (IL-35)	Regulatory T cells	IL-12 family member. Induces and activates regulatory T cells. Suppresses inflammatory responses
Interleukin 36α, β, γ (IL-36α, β, γ) ?18 kDa	Dendritic cells, monocytes, T cells, epithelial cells of barrier tissues	IL-1 family members. Induce dendritic cells to produce proinflammatory cytokines and to express MHC class II, CD80, and CD86. Induce T cells to produce IFN-γ, IL-4, and IL-17; involved in skin inflammation
Interleukin 37 (IL-37) 24 kDa	Monocytes, macrophages, dendritic cells, epithelial cells	IL-1 family member. Inhibits innate immunity and inflammatory responses
APRIL (a proliferation-inducing cytokine) 27 kDa Tumor necrosis factor ligand superfamily member 13 (TNFSF13)	T cells, monocytes, macrophages, and dendritic cells	Member of the TNF family. Promotes B- and T-cell proliferation. Induces class switch recombination to IgA
BAFF (human B-cell–activating factor) 18 kDa BLyS (B-lymphocyte stimulator); tumor necrosis factor ligand superfamily member 13B (TNFSF13B); TALL-1 (TNF and apoptosis ligand-related leukocyte-expressed ligand 1)	T cells, monocytes, macrophages, and dendritic cells	Member of the TNF family, occurs in membrane-bound and soluble form. Promotes B-cell differentiation and survival factor for immature B cells. Promotes activation and proliferation of mature B cells
Cardiotrophin-1 (CT-1) 21.5 kDa	Many cell types, including heart and skeletal muscle	A member of the IL-6 family shown to stimulate hepatic expression of the acute-phase proteins; induces cardiac myocyte hypertrophy; increases monocyte adhesion; involved in metabolic syndrome
Ciliary neurotrophic factor (CNTF) 24 kDa Membrane-associated neurotransmitter stimulating factor (MANS)	Schwann cells and astrocytes	A member of the IL-6 family that induces the expression of acute-phase proteins in the liver and has been shown to function as an endogenous pyrogen. Also shown to function in ontogenesis and promote survival and regeneration of nerves
Granulocyte colony-stimulating factor (G-CSF) 22 kDa	Bone marrow stromal cells and macrophages	Essential for growth and differentiation of neutrophils
Granulocyte-macrophage colony-stimulating factor (GM-CSF) 22 kDa	T cells, macrophages, fibroblasts, and endothelial cells	Growth factor for hematopoietic progenitor cells and differentiation factor for granulocytic and monocytic cell lineages
Macrophage colony-stimulating factor (M-CSF) Disulfide-linked homodimer of 45–90 kDa Colony-stimulating factor 1 (CSF-1)	Many cell types, including lymphocytes, monocytes, fibroblasts, epithelial cells, and others	Growth, differentiation, and survival factor for macrophage progenitors, macrophages, and granulocytes
Interferon alpha (IFN-α) 16–27 kDa Type 1 interferon; leukocyte interferon; lymphoblast interferon	Cells activated by viral and other microbial components: macrophages, dendritic cells, and lymphocytes, virus-infected cells	Induces resistance to virus infection. Inhibits cell proliferation. Increases expression of MHC class I molecules on nucleated cells

(continued)

Cytokine; MW; Synonyms	Sources	Activity
Interferon beta (IFN-β) 22 kDa Type 1 interferon; fibroblast interferon	Cells activated by viral and other microbial components: fibroblasts, dendritic cells, and some epithelial cells, virus-infected cells	Induces resistance to virus infection. Inhibits cell proliferation. Increases expression of MHC class I molecules
Interferon gamma (IFN-γ) Monomer: 17.1 kDa Dimer: 40 kDa Type 2 interferon; immune interferon; macrophage-activating factor (MAF); T-cell interferon	T_H1 cells, ILC1 cells including NK cells, and some CD8$^+$ T cells	The key type 1 cytokine. Supports T_H1 differentiation, induces class switching to IgG subclasses. Activates macrophages and induces MHC class II expression. Weak antiviral and antiproliferative activities
Interferon lambda (IFN-λ) Same as IL-28 and IL-29 (type 3 interferon)		
Leukemia inhibitory factor (LIF) 45 kDa Differentiation-inhibiting activity (DIA); differentiation-retarding factor (DRF)	Many cell types, including T cells, cells of the monocytic lineage, fibroblasts, liver, and heart	A member of the IL-6 family. Major experimental application: keeps cultures of ES cells in undifferentiated state to maintain their proliferation. In vivo, in combination with other cytokines, promotes hematopoiesis, stimulates acute-phase response of liver cells, affects bone resorption, enhances glucose transport and insulin resistance, alters airway contractility, and causes loss of body fat
Lymphotoxin alpha (LT-α) 25 kDa Tumor necrosis factor beta (TNF-β); cytotoxin (CTX); differentiation-inducing factor (DIF); TNF ligand superfamily member 1 (TNFSF1)	Activated T cells, B cells, ILC1 cells, including NK cells, macrophages, virus-infected hepatocytes	Cytotoxic for some tumor and other cells. Required for development of lymph nodes and Peyer's patches and for formation of splenic B- and T-cell zones and germinal centers. Induces inflammation. Activates vascular endothelial cells and induces lymphangiogenesis. Required for NK-cell differentiation
Macrophage migration inhibitory factor (MIF) Monomer: 12 kDa; forms biologically active multimers	Small amounts by many cell types; major producers are activated T cells, hepatocytes, monocytes, macrophages, and epithelial cells	Activates macrophages and inhibits their migration
Oncostatin M (OSM) 28–32 kDa Onco M; ONC	Activated T cells, monocytes, and adherent macrophages	Many functions, including inhibition of growth of tumor cell lines; regulation of the growth and differentiation of cells during hematopoiesis, neurogenesis, and osteogenesis. Shown to enhance LDL uptake and also stimulates synthesis of acute-phase proteins in the liver
Stem-cell factor (SCF) 36 kDa Kit ligand (kitL); steel factor (SLF)	Bone marrow stromal cells, cells of other organs such as brain, kidney, lung, and placenta	Roles in development of hematopoietic, gonadal, and pigmental lineages; active in both membrane-bound and secreted forms
Thrombopoietin (THPO) 60-70 kDa Megakaryocyte colony-stimulating factor; thrombopoiesis-stimulating factor (TSF)	Liver, kidney, and skeletal muscle	Megakaryocyte lineage–specific growth and differentiation factor that regulates platelet production

Cytokine; MW; Synonyms	Sources	Activity
Thymic stromal lymphopoietin (TSLP) 140 Da	Epithelial cells of barrier tissues, basophils	With IL-25 and IL-33 acts as an alarmin produced when epithelial barriers are breached. Acts on dendritic cells and CD4$^+$ T cells to induce type 2 cytokine production by T$_H$2 and ILC2 cells; supports B-cell proliferation and differentiation
Transforming growth factor beta (TGF-β) ≈25 kDa Differentiation-inhibiting factor	Some T cells (especially T$_{REG}$s), macrophages, platelets, and many other cell types	Inhibits growth, differentiation, and function of a number of cell types, including T and B cells, ILCs, and monocytes/macrophages. Inhibits inflammation and enhances wound healing. Induces class switching to IgA
Tumor necrosis factor alpha (TNF-α) 52 kDa Cachectin; TNF ligand superfamily member 2 (TNFSF2)	Monocytes, macrophages, and other cell types, including activated T cells, NK cells, neutrophils, and fibroblasts	Strong mediator of inflammatory and immune functions. Regulates growth and differentiation of a wide variety of cell types. Cytotoxic for many types of transformed and some normal cells. Promotes angiogenesis, bone resorption, and thrombotic processes. Suppresses lipogenic metabolism
Tumor necrosis factor beta (TNF-β) Same as lymphotoxin-α		

TABLE 2	Functional Groups of Cytokines	
Group	**Cytokine Members**	**Function**
Type 1 cytokines	IL-2, IFN-γ, IL-12, TNF-β (LT-α)	Regulate response to invasion by intracellular organisms (viruses, intracellular bacteria)
Type 2 cytokines	IL-4, IL-5, IL-9, IL-13, and sometimes IL-10	Regulate response to worms and other extracellular pathogens
Type 17 cytokines	IL-22, IL-17, GM-CSF	Promote immune-mediated diseases
Proinflammatory cytokines	IL-1, IL-6, TNF-α, as well as some type 1 and type 2 cytokines	Mediate inflammatory response
Anti-inflammatory cytokines	IL-10, TGF-β	Secreted by regulatory T cells and typically inhibit inflammation
Cytokines that act as alarmins	IL-25, IL-33, TSLP	Alert the immune system to cell damage and/or stress
Antiviral cytokines	IFN-α, IFN-β	Enhance antiviral responses
Hematopoietic cytokines	IL-3, IL-7, IL-9, IL-11, G-CSF, M-CSF, GM-CSF	Regulate maturation of red and white blood cells

Some cytokines work in "teams" when responding to pathogens or damage. Some cytokine groups and their associated functions are shown here (see also Overview Figure 10-9). Note that this table is not comprehensive; investigators are not in full agreement about groupings. Cytokines can also fall into more than one category.

TABLE 3	STATs and JAKs Associated with Cytokine/Cytokine Receptor Interactions	
Cytokine Receptor	**Janus Kinase**	**STAT**
IFN-α/-β	JAK1, Tyk 2*	STATs 1 and 2
IFN-γ	JAK1, JAK2	STAT1
IL-2	JAK1, JAK3	Mainly STATs 3 and 5; also STAT1
IL-4	JAK1, JAK3	Mainly STAT6; also STAT5
IL-6	JAK1, JAK2	STAT3
IL-7	JAK1, JAK3	STATs 5 and 3
IL-12	JAK2, Tyk2	STATs 2, 3, 4, and 5
IL-15	JAK1, JAK3	STAT5
IL-21	JAK1, JAK3	Mainly STATs 1 and 3; also STAT5

Each cytokine receptor signals through a distinct JAK-STAT pathway. Specific JAKs and STATs associated with common cytokines are shown here. JAKs may operate as either homo- or heterodimers and phosphorylate STATs, which then dimerize, enter the nucleus, and regulate gene expression (see Chapter 3).

*Despite its name, Tyk2 is also a Janus kinase.

Appendix III: Chemokines and Chemokine Receptors

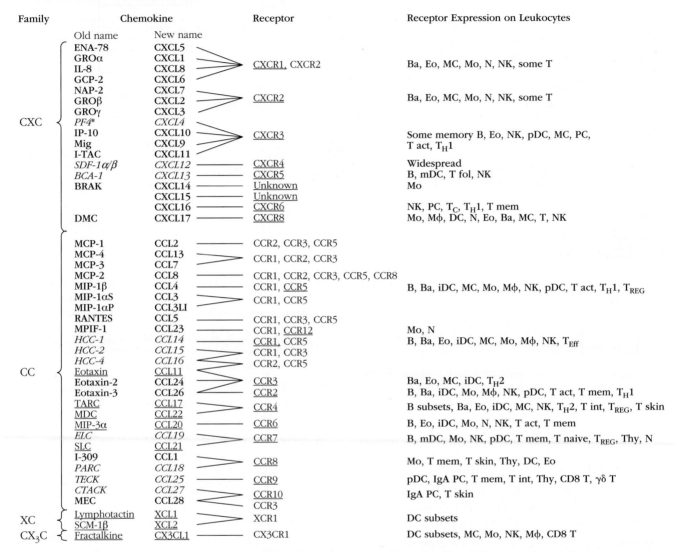

Family	Chemokine		Receptor	Receptor Expression on Leukocytes
	Old name	New name		

FIGURE 1 The chemokine system: an overview. Chemokines (family, old and new nomenclature), their receptors, and predominant receptor repertoires in various leukocyte populations are listed. In general, the information refers to human chemokines and their receptors. Chemokine names in **bold** identify inflammatory chemokines, names in *italics* identify homeostatic chemokines, and underlined names refer to molecules belonging to both realms. Chemokine acronyms are as follows: BCA, B cell–activating chemokine; BRAK, breast and kidney chemokine; CTACK, cutaneous T cell–attracting chemokine; DMC, dendritic cell– and monocyte-attracting chemokine-like protein; ELC, Epstein-Barr virus–induced receptor ligand chemokine; ENA-78, epithelial cell–derived neutrophil-activating factor (78 amino acids); GCP, granulocyte chemoattractant protein; GRO, growth-related oncogene; HCC, hemofiltrate CC chemokine; I (I-309), intercrine; IL, interleukin; IP, IFN-inducible protein; I-TAC, IFN-inducible T-cell α chemoattractant; MCP, monocyte chemoattractant protein; MDC, macrophage-derived chemokine; MEC, mucosa-associated epithelial chemokine; Mig, monokine induced by γ interferon; MIP, macrophage inflammatory protein; MPIF, myeloid progenitor inhibitory factor; NAP, neutrophil-activating protein; PARC, pulmonary and activation-regulated chemokine; PF, platelet factor*; RANTES, regulated upon activation normal T cell-expressed and secreted; SCM, single C motif; SDF, stromal cell–derived factor; SLC, secondary lymphoid tissue chemokine; TARC, thymus and activation-related chemokine; TECK, thymus-expressed chemokine. Chemokine receptors to which the leukocyte expression lists pertain are underlined. Leukocyte acronyms are as follows: B, B cells; Ba, basophils; DC, dendritic cells; Eo, eosinophils; iDCs, immature dendritic cells; MC, mast cells; mDCs, mature DCs; Mo, monocytes; Mϕ, macrophages; N, neutrophils; NK, natural killer cells; PC, plasma cells; T act, activated T cells; TC, cytotoxic T cells; T Eff, T effector cells; T fol, T cells in follicles; T$_H$1, type 1 helper T cells; T$_H$2, type 2 helper T cells; T int, intestine-homing T cells; T mem, memory T cells; T naïve, naïve T cells; T$_{REG}$, regulatory T cells; T skin, skin-homing T cells; Thy, thymocytes.

* Not a chemoattractant; signaling affects proliferation and various other functions.

TABLE 1	Chemokine Receptors Expressed by Leukocyte Subpopulations
Cell Type	**Chemokine Receptor Expression**
Neutrophils	CXCR1, CXCR2, CXCR4
Eosinophils	CCR1, CCR3
Resting B cells	CCR5
Activated B cells	CCR7
Resting T cells	CCR7
Activated T cells	CCR1, CCR4, CCR5, CCR8, CXCR3
T_H1 cells	CCR5, CXCR3
T_H2, T_{REG} cells	CCR4, CCR8
T_H17 cells	CCR6

REFERENCES

1. Bonecchi, R., et al. 2009. Chemokines and chemokine receptors: an overview. *Frontiers in Bioscience* **14**:540.

2. Kasper, B., and F. Petersen. 2011. Molecular pathways of platelet factor 4/CXCL4 signaling. *European Journal of Immunology* **90**:521.

3. Miao, Z., et al. 2007. Proinflammatory proteases liberate a discrete high-affinity functional FPRL1 (CCR12) ligand from CCL23. *Journal of Immunology* **178**:7395.

4. Bachelerie, F., et al. 2015. Chemokine receptors: introduction. In: IUPHAR/BPS Guide to Pharmacology database [last modified on October 8, 2015; accessed on September 7, 2017]. IUPHAR database (IUPHAR-DB).

5. Schall, T. J., and A. E. I. Proudfoot. 2011. Overcoming hurdles in developing successful drugs targeting chemokine receptors. *Nature Reviews Immunology* **11**:355.

6. Seth, S., et al. 2011. CCR7 essentially contributes to the homing of plasmacytoid dendritic cells to lymph nodes under steady-state as well as inflammatory conditions. *Journal of Immunology* **186**:3364.

7. Viola, A., and A. D. Luster. 2008. Chemokines and their receptors: drug targets in immunity and inflammation. *Annual Review of Pharmacology and Toxicology* **48**:171.

8. Lacalle, R. A., et al. 2017. Chemokine receptor signaling and the hallmarks of cancer. *International Review of Cell and Molecular Biology* **331**:181.

9. Schulz, O., et al. 2016. Chemokines and chemokine receptors in lymphoid tissue dynamics. *Annual Review of Immunology* **34**:203.

10. Bachelerie, F., et al. 2015. Chemokine receptors. In: IUPHAR/BPS Guide to Pharmacology database [last modified on October 8, 2015; accessed on September 7, 2017]. http://www.guidetopharmacology.org/GRAC/FamilyDisplayForward?familyId=14

11. Shore, D. M., and P. H. Reggio. 2015. The therapeutic potential of orphan GPCRs, GPR35 and GPR55. *Frontiers in Pharmacology* **6**:69.

ABO blood-group antigen Antigenic determinants of the blood-group system defined by the agglutination of red blood cells exposed to anti-A and anti-B antibodies.

Abzyme A monoclonal antibody that has catalytic activity.

Acquired immunity See **adaptive immunity**.

Acquired immunodeficiency syndrome (AIDS) A disease caused by **human immunodeficiency virus (HIV)** that is marked by significant depletion of CD4$^+$ T cells and that results in increased susceptibility to a variety of opportunistic infections and cancers.

Activation-induced cytidine deaminase (AID) An enzyme that removes an amino group from deoxycytidine, forming deoxyuridine. This is the first step in the processes of both **somatic hypermutation** and **class switch recombination**.

Active immunity Adaptive immunity that is induced by natural exposure to a pathogen or by **vaccination**.

Acute lymphocytic leukemia (ALL) A form of cancer in which there is uncontrolled proliferation of a cell of the lymphoid lineage. The proliferating cells usually are present in the blood.

Acute myelogenous leukemia (AML) A form of cancer in which there is uncontrolled proliferation of a cell of the myeloid lineage. The proliferating cells usually are present in the blood.

Acute phase protein One of a group of serum proteins that increase in concentration in response to inflammation. Some **complement** components and **interferons** are acute phase proteins.

Acute phase response (APR) The production of certain proteins that appear in the blood shortly after many infections, often induced by proinflammatory cytokines generated at the site of infection. It is part of the host's early innate response to infection.

Acute phase response proteins Proteins synthesized in the liver in response to inflammation; serum concentrations of these proteins increase in inflammation.

Acute rejection Refers to the process of allo- or xenograft recognition and rejection that occurs after hyperacute rejection and that involves the action of both activated T and B cells. This stage of rejection can begin as early as 7 days after engraftment (following sensitization and effector cell engagement) and can continue for a year or more. This stage is usually amenable to control via immunosuppressive therapy.

Adapter Proteins Proteins that connect to other effector proteins in a signaling pathway and create a signaling scaffold.

Adaptive immunity Host defenses that are mediated by B cells and T cells following exposure to antigen and that exhibit specificity, diversity, memory, and self-nonself discrimination. See also **innate immunity**.

Addressin A cell-surface protein or set of cell-surface proteins that are ligands for specific homing receptors on immune cells; they help guide immune cell trafficking.

Adenosine deaminase w(ADA) deficiency An immune deficiency disorder that is characterized by defects in adaptive immunity and is caused by the intracellular accumulation of toxic adenosine metabolites, especially in hematopoietic cells, which interferes with purine metabolism and DNA synthesis.

Adhesion molecules Families of surface proteins that regulate cell interactions with tissues, vessels, and with each other; they are important players in immune cell trafficking.

Adjuvant cancer therapy A supplement or secondary treatment for cancer applied after the primary treatment (typically, surgical removal), which can include radiation and/or chemical/drug therapy meant to target residual tumor cells.

Adjuvants Factors that are added to a vaccine mixture to enhance the immune response to antigen by activating innate immune cells. Dead mycobacterium were among the original adjuvants, but more refined preparation include alum, cytokines, and/or lipids.

Adoptive transfer The transfer of the ability to make or participate in an immune response by the transplantation of cells of the immune system.

Affinity The strength with which a monovalent ligand interacts with a binding site. It is represented quantitatively by the affinity constant K_a.

Affinity constant The ratio of the forward (k_1) to the reverse (k_{-1}) rate constant in an antibody-antigen reaction. Equivalent to the association constant in biochemical terms ($K_a = k_1/k_{-1}$).

Affinity model of selection A proposal stating that the fate of a developing T cell depends on the affinity of the interaction between its T cell receptor (TCR) and MHC-peptide ligand(s) it encounters in the thymus. High affinity interactions result in death by negative selection, lower affinity interactions in positive selection and maturation, and very low or no affinity interactions result in death by neglect.

Affinity maturation The increase in average antibody affinity for an antigen that occurs during the course of an immune response or in subsequent exposures to the same antigen.

Agammaglobulinemia Lack of immunoglobulin in the blood, causing immunodeficiency.

Agent-induced immunodeficiency A state of immune deficiency induced by exposure to an environmental agent/s.

Agglutination inhibition The reduction of antibody-mediated clumping of particles by the addition of the soluble forms of the epitope recognized by the agglutinating antibody.

Agglutination The aggregation or clumping of particles (e.g., latex beads) or cells (e.g., red blood cells).

Agglutinin A substance capable of mediating the clumping of cells or particles; in particular, a hemagglutinin (HA) causes clumping of red blood cells.

Agglutinin titer The reciprocal of the greatest serum dilution that elicits a positive agglutination reaction.

AIM2-like receptors (ALRs) Cytosolic receptors that bind DNA from bacteria and viruses. They are characterized by the presence of one or two HIN (hematopoietic expression, interferon inducibility, nuclear localization) domains at their carboxyl terminus.

AIRE A protein that regulates expression of tissue specific antigens in the thymus. It is expressed by a subset of medullary epithelial cells and regulates transcription.

Alarmins A diverse group of molecules, released in response to cellular stress, that summon protective inflammatory responses. Members of the IL-1 family of cytokines, including IL-33, are alarmins that are particularly important in regulating barrier immunity; they are released by damaged epithelial cells and help organize an immune response to the damaging pathogen.

Alleles Two or more alternative forms of a gene at a particular **locus** that confer alternative characters. The presence of multiple alleles results in **polymorphism**.

Allelic exclusion   A process that permits expression of only one of the allelic forms of a gene. For example, a B cell expresses only one allele for an antibody heavy chain and one allele for a light chain.

Allergy   A **hypersensitivity** reaction that can include hay fever, asthma, **serum sickness**, systemic **anaphylaxis**, or contact dermatitis.

Allogeneic   Denoting members of the same species that differ genetically.

Allograft   A tissue transplant between **allogeneic** individuals.

Alloreactivity   Immunologic reactions directed against the nonself MHC molecules of another member of the same species. Up to 10% of circulating T cells are thought to have the ability to recognize foreign MHC within a species.

Allotypes   A set of **allotypic determinants** characteristic of some but not all members of a species.

Allotypic determinant   An antigenic determinant that varies among members of a species or between different inbred strains of animals. The constant regions of antibodies possess allotypic determinants.

Allotypic marker   A genetic marker that defines the presence of an allele on one strain of mouse that is not shared by other strains. Normally refers to allelic variants of immunoglobulin heavy chains.

Alpha-feto protein (AFP)   See **oncofetal tumor antigen**.

Altered peptide model   A proposal stating that developing T cells encounter different sets of peptides in the cortical region versus the medullary region of the thymus. Advanced to help explain differences in the subsets of cells that undergo positive versus negative selection.

Alternative pathway of complement activation   A pathway of complement activation that is initiated by spontaneous hydrolysis of the C3 component of complement, resulting in the formation of a fluid-phase C3 convertase enzyme. This spontaneous initiation distinguishes the alternative pathway from the classical and lectin-mediated pathways that are both initiated by specific antigen binding by either antibodies or lectins respectively. However, one recently-discovered branch of the alternative pathway may begin with Properdin binding to the surface of bacteria from the *Neisseria* genus.

Alternative tickover pathway   The alternative pathway of complement activation that is initiated by spontaneous hydrolysis of C3 molecule in the serum.

Alveolar macrophages (dust cells)   Macrophages found in the alveoli of the lung.

Alveoli   The clusters of sacs at the end of the bronchiolar branches where gas exchange takes place; alveoli are lined by a single layer of epithelial cells and are in contact with capillaries.

Amphiregulin   A protein factor, produced by epithelial cells as well as several types of immune cells, which contributes to tolerance and promotes healthy epithelial barriers; in the skin, amphiregulin helps promote keratinocyte proliferation.

Anaphylactic shock   An acute, life threatening (Type I) whole-body allergic response to an antigen (e.g. drugs, insect venom). See also **Anaphylaxis**.

Anaphylatoxins   The **complement** split products C3a and C5a, which mediate **degranulation** of mast cells and basophils, resulting in release of mediators that induce contraction of smooth muscle and increased vascular permeability.

Anaphylaxis   An immediate type I hypersensitivity reaction, which is triggered by IgE-mediated mast cell. Systemic anaphylaxis leads to shock and is often fatal. Localized anaphylaxis involves various types of **atopic** reactions.

Anchor residues   The amino acid residues at key locations in a peptide sequence that fit into pockets of an make close molecular associations with

complementary amino acids in the groove of an MHC molecule and which help to determine the peptide-binding specificity of particular MHC molecules.

Anergic, anergy   Unresponsive to antigenic stimulus.

Antagonist, antagonize   A molecule that inhibits the effect of another molecule.

Anti-allotype antibodies   Antibodies directed towards **allotypic determinants**.

Anti-Fab antibodies   Antibodies directed towards the **Fab regions** of other antibodies.

Anti-Fc antibodies   Antibodies specific for the **Fc regions** of other antibodies.

Anti-idiotypic antibodies   Antibodies directed towards antigenic determinants located in the antigen binding site of other antibodies.

Anti-isotype antibodies   Antibodies directed towards antigenic determinants located in the constant regions of antibodies, that are shared among all members of a species.

Anti-oncogenes   Another name for tumor suppressor genes.

Antibodies   Immunoglobulin proteins consisting of two identical heavy chains and two identical light chains, that recognize a particular **epitope** on an antigen and facilitates clearance of that antigen. Membrane-bound antibody is expressed by B cells that have not encountered antigen; secreted antibody is produced by **plasma cells**. Some antibodies are multiples of the basic four-chain structure.

Antibody molecule   See **antibodies**.

Antibody-dependent cell-mediated cytotoxicity (ADCC)   A cell-mediated reaction in which nonspecific cytotoxic cells that express **Fc receptors** (e.g., NK cells, neutrophils, macrophages) recognize bound antibody on a target cell and subsequently cause lysis of the target cell.

Antibody-dependent cellular phagocytosis (ADCP)   Phagocytosis of antigens opsonized by antibodies.

Antigen   Any substance (usually foreign) that binds specifically to an antibody or a T-cell receptor; often is used as a synonym for **immunogen**.

Antigen presentation   See **antigen processing**.

Antigen processing   Degradation of antigens by one of two pathways yielding antigenic peptides that are displayed bound to MHC molecules on the surface of antigen-presenting cells or altered self cells.

Antigen-presenting cell (APC)   Any cell that can process and present antigenic peptides in association with **MHC class II molecules** and deliver a **costimulatory signal** necessary for T-cell activation. Macrophages, dendritic cells, and B cells constitute the professional APCs. Nonprofessional APCs, which function in antigen presentation only for short periods include thymic epithelial cells and vascular endothelial cells.

Antigenic determinant   The site on an antigen that is recognized and bound by a particular antibody, TCR/MHC-peptide complex, or TCR-ligand-CD1 complex; also called **epitope**.

Antigenic drift   A series of spontaneous point mutations that generate minor antigenic variations in pathogens and lead to strain differences. See also **Antigenic shift**.

Antigenic peptide   In general, a peptide capable of raising an immune response, for example, in a peptide that forms a complex with MHC that can be recognized by a T-cell receptor.

Antigenic shift   Sudden emergence of a new pathogen subtype, frequently arising due to genetic reassortment that has led to substantial antigenic differences. See also **Antigenic drift**.

Antigenic specificity See **specificity, antigenic**.

Antigenically committed The state of a mature B cell displaying surface antibody specific for a single immunogen.

Antigenicity The capacity to combine specifically with antibodies or T-cell receptor/MHC.

Antimicrobial peptides Peptides/small proteins, such as defensins, less than 100-amino acids long that are produced constitutively or after activation by pathogens.

Antimicrobial proteins Enzymes and other proteins that directly damage pathogens, induce phagocytosis, or inhibit pathogen infectivity or replication.

Antiretroviral therapy (ART) Treatment with drugs that inhibit the replication of HIV.

Antiserum Serum from animals immunized with antigen that contains antibodies to that antigen.

Apical surface The portion of the membrane of an epithelial cell that faces the lumen of a tissue.

Apoptosis A process, often referred to as **programmed cell death**, where cells initiate a signaling pathway that results in their own demise. Apoptosis requires ATP and is typically dependent on the activation of internal caspases. In contrast to **necrosis**, it does not result in damage to surrounding cells.

Apoptosome A wheel-like assemblage of molecules that regulate cell death initiated via the mitochondrial (intrinsic) pathway. Includes cytochrome-c, ATP, Apaf-1, and caspase-9.

APRIL A member of the Tumor Necrosis Factor family of cytokines, important in B cell development and homeostasis.

Artemis An enzyme that is a member of the Non-homologous End Joining (NHEJ) DNA repair pathway. During V(D)J recombination, Artemis opens the hairpin loops formed after RAG1/2-mediate cleavage of the immunoglobulin genes.

Association constant (Ka) See **affinity constant**.

Atopic Pertaining to clinical manifestations of type I (IgE-mediated) hypersensitivity, including allergic rhinitis (hay fever), eczema, asthma, and various food allergies.

Atopic (allergic) march The natural history or typical progression of allergic diseases that often begin early in life, starting with atopic dermatitis (eczema) and progressing to food allergy, allergic rhinitis (hay fever), and possibly asthma.

Atopy The genetic tendency to develop allergic diseases such as allergic rhinitis, asthma, and atopic dermatitis (eczema); typically associated with heightened immune responses to common allergens, especially inhaled and food allergens.

Attenuate To decrease the virulence of a pathogen and render it incapable of causing disease. Many vaccines are composed of attenuated bacteria or viruses that induce protective immunity without causing harmful infection.

Autocrine A type of cell signaling in which the cell acted on by a cytokine is the source of the cytokine.

Autograft Tissue grafted from one part of the body to another in the same individual.

Autoimmune diseases A group of disorders caused by the action of ones own antibodies or T cells reactive against self proteins.

Autoimmune polyendocrine syndrome type 1 (APS-1) An immune deficiency disorder in which depressed expression of *AIRE* results in reduced levels of tissue-specific antigens in thymic epithelial cells, allowing the escape of autoreactive T cells into the periphery, where they precipitate organ-specific autoimmunity.

Autoimmune polyendrocrinopathy and ectodermal dystrophy (APECD) An immune deficiency disorder in which depressed expression of *Aire* results in reduced levels of tissue-specific antigens in thymic epithelial cells, allowing the escape of autoreactive T cells into the periphery, where they precipitate organ-specific autoimmunity.

Autoimmunity An abnormal immune response against self antigens.

Autologous Denoting transplanted cells, tissues, or organs derived from the same individual.

Autophagosome Membrane-bound vesicle formed by autophagy.

Autophagy Elimination of intracellular pathogens and organelles by envelopment by intracellular membranes and fusion of resulting autophagosomes with lysosomes.

Avidity The strength of antigen-antibody binding when multiple epitopes on an antigen interact with multiple binding sites of an antibody. See also **affinity**.

B cell See **B lymphocytes**.

B lymphocytes (B cells) Lymphocytes that mature in the bone marrow and express membrane-bound antibodies. After interacting with antigen, they differentiate into antibody-secreting plasma cells and memory cells.

B-1 B cells A subclass of B cells that predominates in the peritoneal and pleural cavity. B-1 B cells in general secrete low affinity IgM antibodies and do not undergo class switch recombination or somatic hypermutation. They thus occupy a niche between the innate and adaptive immune responses. Most, but not all, B-1 B cells express CD5 on their surface.

B-1b B cells A subclass of B-1 cells that does not express the antigen CD5 on its cell surface, like most B-1 B cells.

B-2 B cells The predominant class of B cells that are stimulated by antigens with T cell help the generate antibodies of multiple heavy chain classes whose genes undergo somatic hypermutation.

B-cell coreceptor A complex of three proteins (CR2 (CD21), **CD19**, and TAPA-1) associated with the B-cell receptor. It is thought to amplify the activating signal induced by cross-linkage of the receptor.

B-cell receptor (BCR) Complex comprising a membrane-bound immunoglobulin molecule and two associated signal-transducing Igα/Igβ molecules.

B-cell-specific activator protein (BSAP) A transcription factor encoded by the gene Pax-5 that plays an essential role in early and later stages of B-cell development.

B-lymphocyte-induced maturation protein 1 (BLIMP-1) Transcription factor vital to differentiation of B cells into plasma cells.

Bacillus Calmette-Guérin (BCG) An attenuated form of *Mycobacterium bovis* used as a vaccine against another member of the genus, *M. tuberculosis*, the cause of tuberculosis. BCG can also be found as an adjuvant component in other vaccines.

Bacteremia An infection in which viable bacteria are found in the blood.

BAFF B-cell survival factor; a membrane-bound homolog of tumor necrosis factor, to which mature B cells bind though the **TACI** receptor. This interaction activates important transcription factors that promote B-cell survival, maturation, and antibody secretion.

BAFF receptor (BAFF-R) Receptor for **BAFF**, a cytokine belonging to the tumor necrosis factor family that is important in B cell development and homeostasis.

Balanced signals model   Model for how NK cell killing is targeted at virus-infected, stressed, or tumor cells. The balance of signals from NK inhibitory and activating receptors determines whether the NK cell will induce apoptosis in the target cells.

Bare-lymphocyte syndrome (BLS)   An immunodeficiency syndrome in which, without MHC class II molecules, positive selection of CD4$^+$ T cells in the thymus is impaired and, with it, peripheral T helper cell responses.

Barrier immunity   The system of immune cells, tissues, and responses that protects barrier tissues from damage and infection.

Barrier organs   Tissues lined by epithelial cells, which are directly exposed to the external environment; includes the gastrointestinal, respiratory, reproductive, and urinary tracts, as well as the skin.

Barrier tissues   Tissues and organs that form protective boundaries between the external and internal environments of a body; include skin and the mucosal tissues (gastrointestinal, respiratory, reproductive, urogenital tracts) as well as distinct immune cells and systems.

Basolateral surface   The portion of the membrane of an epithelial cell that faces the mucosal layer of a tissue (and is oriented away from the lumen).

Basophil   A nonphagocytic granulocyte that expresses **Fc receptors** for IgE. Antigen-mediated cross-linkage of bound IgE induces **degranulation** of basophils.

BCG   See **Bacillus Calmette-Guérin**.

Bence-Jones proteins   Monoclonal light chains secreted by plasmacytoma tumors. Found in high concentrations in the urine of patients with multiple myeloma.

Benign   Pertaining to a nonmalignant form of a neoplasm or a mild form of an illness.

β-selection   The process during the **DN3** stage of T cell development where the functionality of thymocytes' rearranged TCRβ chains is tested. Only those thymocytes that have successfully rearranged a TCRβ chain and expressed it as a protein that can interact with pre-TCRα will deliver signals that ensure its survival, maturation to the CD4$^+$CD8$^+$ (DP) stage, and induce its proliferation.

β2-microglobulin   Invariant subunit that associates with the polymorphic α chain to form **MHC class I molecules**; it is not encoded by MHC genes.

Bispecific antibody   Hybrid antibody made either by chemically cross-linking two different antibodies or by fusing hybridomas that produce different monoclonal antibodies.

Bone marrow   The living tissue found within the hard exterior of bone.

Booster   Inoculation given to stimulate and strengthen an immunologic memory response.

Bradykinin   An endogenously produced peptide that produces an **inflammatory response**.

Bronchus-associated lymphoid tissue (BALT)   Secondary lymphoid microenvironments in the lung mucosa system that support the development of the T and B lymphocyte response to antigens that enter the lower respiratory tract. Part of the mucosa associated lymphoid tissue system (MALT).

C (constant) gene segment   The 3′ coding of a rearranged immunoglobulin or T-cell receptor gene. There are multiple C gene segments in germ-line DNA, but as a result of gene rearrangement and, in some cases, RNA processing, only one segment is expressed in a given protein.

c-Kit (CD117)   Receptor for **stem cell factor (SCF)**.

C-reactive protein (CRP)   An acute phase protein that binds to phosphocholine in bacterial membranes and functions in opsonization; an increased level of serum CRP is an indicator of inflammation.

C-type lectin receptor (CLR)   A family of pattern-recognition receptors that contains C-type lectin carbohydrate-binding domains.

C3 convertase   Enzyme that breaks down the C3 component of complement into C3a and C3b.

C3aR   A cell-surface receptor for the complement component and anaphylatoxin C3a.

C5 convertase   Enzyme that breaks down the C5 component of complement into C5a and C5b.

C5aR and C5L2   Cell-surface receptors for the complement component and anaphylatoxin C5a.

Calnexin   A protein resident of the ER that serves, along with **calreticulin**, as a molecular chaperone to assist in MHC class I molecule assembly.

Calreticulin   A protein resident of the ER that serves, along with **calnexin**, as a molecular chaperone to assist in MHC class 1 molecule assembly.

Cancer stem cells   A subset of cells within a tumor that has the stem-cell-like ability to give rise to all cells within that tumor and the ability to self-renew indefinitely. They are thought to be responsible for tumor growth.

CAR T cells   See **chimeric antigen receptor T cells.**

Carcinoembryonic antigen (CEA)   An oncofetal antigen (found not only on cancerous cells but also on normal cells) that can be a tumor-associated antigen.

Carcinogen   Any chemical substances, physical agents or types of radiation that can induce DNA mutations and lead to the development of cancer.

Carcinoma   Tumor arising from endodermal or ectodermal tissues (e.g., skin or epithelium). Most cancers (>80%) are carcinomas.

Carrier   An immunogenic molecule containing antigenic determinants recognized by T cells. Conjugation of a carrier to a nonimmunogenic **hapten** renders the hapten immunogenic.

Carrier effect   A **secondary immune response** to a hapten depends on use of both the **hapten** and the **carrier** used in the initial immunization.

Cascade induction   The property of cytokines that pertains to their ability to induce one cell to release cytokines that then act upon another to induce the release of other cytokines and growth factors.

Caspase   A family of cysteine proteases that cleave after an aspartate residue. The term *caspase* incorporates these elements (*c*ysteine, *asp*artate, *prote*ase*), which play important roles in the chain of reactions that leads to **apoptosis**.

Caspase recruitment domains (CARD)   Protein domain that binds caspase proteases.

Cathelicidin   A type of antimicrobial peptide secreted by epithelial cells and found in lysosomes of phagocytic cells; it disrupts pathogen membranes and has other toxic effects.

CC subgroup   A subgroup of chemokines in which a disulfide bond links adjacent cysteines.

CD19   A quintessential B-cell marker, often used as such in flow cytometry experiments.

CD21   The B cell co-receptor molecule that also serves as a co-receptor for the complement components C3d and C3dg. Also known as CR2.

CD25   The high affinity IL-2 receptor chain (IL-2α) expressed on the surface of multiple immune cells, including some developing T cells, activated T cells, and many FoxP3$^+$ T cells.

CD28   A costimulatory receptor that cooperates with the TCR to activate naïve T cells; binding to CD80 and CD86 ligands results in an increase in IL-2 expression.

CD3   A polypeptide complex containing three dimers: a γε heterodimer, a εδ heterodimer, and either a ξξ homodimer or a ξη heterodimer. It is associated with the T-cell receptor and functions in signal transduction.

CD4   A glycoprotein that serves as a co-receptor on MHC class II–restricted T cells. Most helper T cells are CD4$^+$.

CD40   Member of the tumor necrosis factor receptor family. Signaling through **CD40L** on T cells to CD40 on B cells is necessary for **germinal center** formation, **somatic hypermutation** and **class switch recombination**.

CD40L   Ligand for CD40. CD40L is a member of the Tumor Necrosis Factor family of molecules and CD40 a member of the TNF receptor family. CD40:CD40L interactions are indispensable during T cell mediated B cell differentiation. B cells bear CD40 and T cell, CD40L.

CD44   Surface protein involved in cell-cell adhesion that is expressed by multiple immune cells, including some developing T cells, and some activated T cells. Differences in CD44 and CD25 expression distinguish very early stages of T cell development. CD44 is also associated with immune cell activation.

CD5 antigen   An antigen found on most B-1 B cells, (B-1a B cells), as well as on many T cells.

CD59   Regulatory protein that binds to and inhibits the membrane attack complex of complement.

CD8   A dimeric protein that serves as a co-receptor on MHC class I–restricted T cells. Most cytotoxic T cells are CD8$^+$.

CD80/86   Molecules whose expression on the cell surfaces of antigen-presenting cells and B cells is up-regulated when the cells take up and process antigen. They serve as ligands for the T-cell-surface molecule CD28.

CDR3   The third complementarity-determining region, (or hypervariable region) of the immunoglobulin or TCR molecules.

Cell adhesion molecules (CAMs)   A group of cell surface molecules that mediate intercellular adhesion. Most belong to one of four protein families: the **integrins**, **selectins**, mucin-like proteins, and **immunoglobulin superfamily**.

Cell line   A population of cultured tumor cells or normal cells that have been subjected to chemical or viral **transformation**. Cell lines can be propagated indefinitely in culture.

Cell-mediated immune response   Host defenses that are mediated by antigen-specific T cells. It protects against intracellular bacteria, viruses, and cancer and is responsible for graft rejection. Transfer of primed T cells confers this type of immunity on the recipient. See also **humoral immune response**.

Cell-mediated immunity   See **cell-mediated immune response**.

Cell-mediated lympholysis (CML)   In vitro lysis of allogeneic cells or virus-infected syngeneic cells by T cells; can be used as an assay for CTL activity or MHC class I activity.

Cellular innate immune responses   Cell responses activated by binding of conserved pathogen components to cell-surface or intracellular receptors.

Cellular oncogene   See **proto-oncogene**.

Central memory T cells (T$_{CM}$)   A memory T cell subset that localizes to and resides in secondary lymphoid tissue. It participates in the secondary response to antigen and can give rise to new effector T cells. T$_{CM}$ may arise from effector T cells and/or from T cells that have been stimulated towards the end of an immune response.

Central tolerance   Elimination of self-reactive lymphocytes in primary generative organs such as the bone marrow and the thymus (see also **peripheral tolerance**).

Centroblasts   Rapidly dividing B cells that have recognized antigen, migrated into the follicles, and formed a germinal center. Centroblasts reside in the dark zone of the germinal center.

Centrocytes   B cells that have recognized antigen and migrated into the follicles, where they undergo somatic hypermutation followed by antigen-induced selection. Centrocytes are located in the light zone of the germinal center.

Checkpoint blockade   In immunotherapy, any of a number of cancer treatments that employ immune checkpoint inhibitors to prevent inhibitory signaling in T cells, theoretically releasing these T cells to mount anti-tumor cell responses.

Checkpoint inhibitors   Anticancer drugs usually in the form of monoclonal antibodies that interfere with co-inhibitory signaling (checkpoints) in the activation of T cell responses, such as via CTLA-4 or PD-1.

Chediak-Higashi syndrome   An autosomal recessive immune deficiency disorder caused by a defect in lysosomal granules that impairs killing by NK cells.

Chemical barriers   Tissue layer that provides innate immune protection against infection by chemical means, such as low pH and presence of degradative enzymes.

Chemoattractant   A substance that attracts cells. Some chemoattractants also cause significant changes in the physiology of cells that bear receptors for them.

Chemokine receptors   Surface proteins expressed by immune cells that guide their migration among tissues and localization within tissues. They generate signals that regulate motility and adhesion when bound to chemokines secreted by a variety of immune and stromal cells.

Chemokines   Any of several secreted low-molecular-weight cytokines that mediate **chemotaxis** in particular leukocytes via receptor engagement and that can regulate the expression and/or adhesiveness of leukocyte **integrins** (see Appendix III).

Chemotactic factor   An agent that can cause leukocytes to move up its concentration gradient.

Chemotaxis   The induction of cell movement by the secretion of factors that either attract or repel the cell through the mediation of receptors for those factors.

Chimera   An animal or tissue composed of elements derived from genetically distinct individuals. The **SCID-human mouse** is a chimera. Also, a chimeric antibody that contains the amino acid sequence of one species in one region and the sequence of a different species in another (for example, an antibody with a human constant region and a mouse variable region).

Chimeric antibody   See **chimera**.

Chimeric antigen receptor T (CAR T) cells   T cells that have been modified to encode a gene for a composite antigen-specific receptor that is a fusion of BCR sequences recognizing a target antigen (H and L chain variable regions) with TCR-specific sequences (CD3 and sometimes costimulatory domains), allowing the T cell to recognize and kill tumor cells expressing the antigen without the requirement for MHC presentation.

Chromogenic substrate   A colorless substance that is transformed into colored products by an enzymatic reaction.

Chronic granulomatous disease   Immunodeficiency caused by a defect in the enzyme NADPH (phagosome) oxidase resulting in failure to generate reactive oxygen species in neutrophils.

Chronic inflammation   Inflammation that may have a rapid or slow onset but is characterized primarily by its persistence and lack of clear resolution; it occurs when the tissues are unable to overcome the effects of the injuring agent; it involves a progressive change in the types of cells present at the site of inflammation.

Chronic lymphocytic leukemia (CLL)   A type of leukemia in which cancerous lymphocytes are continually produced.

Chronic myelogenous leukemia (CML)   A type of leukemia in which cancerous lymphocytes of the myeloid lineage are continually produced.

Chronic rejection   Transplant/graft rejection reactions that begin months or years after engraftment and can sometimes continue or recur for the lifetime of the patient. The effector cells and molecules are the same as those involved in acute rejection; however, this stage is more difficult to treat and accounts for most of the incidents of graft failure after the initial weeks and months post transfer.

Cilia   Hairlike projections on cells, including epithelial cells in the respiratory and gastrointestinal tracts; cilia function to propel mucus with trapped microbes out of the tract.

Class (isotype) switching   The change in the antibody class that a B cell produces.

Class 1 cytokines   The largest of the cytokine families, typified by Interleukin 2 (IL-2).

Class I MHC genes   See **MHC class I genes**.

Class I MHC molecules   See **MHC class I molecules**.

Class II MHC genes   See **MHC class II genes**.

Class II MHC molecules   See **MHC class II molecules**.

Class III MHC genes   See **MHC class III genes**.

Class III MHC molecules   See **MHC class III molecules**.

Class switch recombination (CSR)   The generation of antibody genes for heavy chain isotypes other than μ or δ by DNA recombination.

Class   The property of an antibody that is defined by the nature of its heavy chain (μ, δ, γ, α, or ε).

Classical MHC molecules   A set of highly polymorphic surface proteins encoded in the major histocompatibility locus that are characterized by an alpha and beta chain which binds peptide for presentation to classical a/b T cells (CD4$^+$ or CD8$^+$) and essential for activation of adaptive responses. Examples of classical Class I molecules include HLA-A, –B and –C (human) or H2-D, –L and –K (mouse). Class II examples include HLA-DP, –DQ and –DR (human) or H2-A and –E (mouse).

Classical pathway of complement activation   That pathway of **complement** activation that is initiated by antibody binding to antigen.

CLIP   A protein that binds to the groove of MHC class II as it is assembled and carried to the cell surface. It prevents other peptides from associating with MHC class II until it encounters endocytosed proteins, when CLIP is digested and removed from the groove.

Clonal anergy   A physiological state in which cells are unable to be activated by antigen.

Clonal deletion   The induced death of members of a clone of lymphocytes with inappropriate receptors (e.g., those that strongly react with self during development).

Clonal selection hypothesis   This hypothesis states that antigen interacting with a receptor on a lymphocyte induces division and differentiation of that lymphocyte to form a clone of identical daughter cells. All daughter cells will bear the same receptor as the stimulated cell, and antibodies produced by B cells stimulated in this way will share the antigen-binding site with the membrane receptor of the stimulated cell. Following antigen elimination, representatives of the stimulated clone remain in the host as a source of immunological memory. Those clones of B cells that meet antigen at an immature stage of development will be eliminated from the repertoire.

Clonal selection   The antigen-mediated activation and proliferation of members of a clone of B cells that have receptors for the antigen (or for complexes of MHC and peptides derived from the antigen, in the case of T cells).

Clone   Cells arising from a single progenitor cell.

Clot   Coagulated mass; usually refers to coagulated blood, in which conversion of fibrinogen in the plasma to fibrin has produced a jelly-like substance containing entrapped blood cells.

Cluster of differentiation (CD)   A collection of monoclonal antibodies that all recognize an antigen found on a particular differentiated cell type or types. Each of the antigens recognized by such a collection of antibodies is called a CD marker and is assigned a unique identifying number.

Coding joints   The nucleotide sequences at the point of union of coding sequences during V(D)J rearrangement to form rearranged antibody or T-cell receptor genes.

Codominant   The expression of both the maternal and the paternal copy of a gene in a heterozygote.

Coinhibitory receptors   Receptors expressed on the surface of some T cells that send signals that inhibit T cell activation. CTLA-4 is a common negative costimulatory molecule that is expressed on some activated T cells and helps to downregulate immune responses when antigen is cleared.

Collectins   Family of calcium-dependent carbohydrate-binding proteins containing collagen-like domains.

Combined immunodeficiencies (CID)   Any of a number of immune deficiency disorders resulting from an absence of T cells or significantly impaired T-cell function, combined with some disruption of antibody responses.

Commensal microbiome   The diverse community of microbes (bacteria, viruses, fungi, and worms) that coexist with our barrier tissues without causing damage; commensal microbes provide diverse benefits to barrier tissues.

Common lymphoid progenitor (CLP)   An immature blood cell that develops from the hematopoietic stem cell and gives rise to lymphocytes, including B and T cells and NK cells.

Common myeloid-erythroid progenitor (CMP)   An immature blood cell that develops from the hematopoietic stem cell and gives rise to all red blood cells and myeloid cells, including monocytes, macrophages, and granulocytes.

Complement   A group of serum and cell membrane proteins that interact with one another and with other molecules of innate and adaptive immunity to carry out key effector functions leading to pathogen recognition and elimination.

Complement system   See **complement**.

Complementarity-determining region (CDR)   Portions of the variable regions of antibody molecules that contain the antigen-binding residues.

Confocal microscopy   A type of fluorescence microscopy that, like two-photon microscopy, allows one to image fluorescent signals within one focal plane within a relatively thick tissue sample.

Conformational determinants   Epitopes of a protein that are composed of amino acids that are close together in the three-dimensional structure of the protein but may not be near each other in the amino acid sequence.

Congenic   Denoting individuals that differ genetically at a single genetic locus or region; also called coisogenic.

Constant (C) region   The nearly invariant portion of the immunoglobulin molecule that does not contain antigen-binding domains. The sequence of amino acids in the constant region determines the isotype (α, γ, δ, ε, and μ) of heavy chains and the type (κ and λ) of light chains.

Constant (C$_L$) That part of the light chain that is not variable in sequence.

Contraction The final phase in the immune response after infection has been cleared and the number of responsive lymphocytes decreases via apoptosis.

Coreceptor A cell-surface molecule that is noncovalently associated with an antigen receptor, and that binds to molecules on antigen-presenting cells that are associated either covalently or noncovalently with the antigen or antigen-associated molecules.

Correlates of immune protection Specific immunologic effector memory responses, such as neutralizing antibodies or epitope-specific CTLs, that recognize particular structures present in/on an infectious agent and that when present will protect the individual against subsequent infection or disease from infection with that same organism.

Cortex The outer or peripheral layer of an organ.

Cortical thymic epithelial cells (cTECs) Stromal cells of epithelial origin that populate the cortex of the thymus and mediate positive selection of thymocytes.

Costimulatory receptors Receptors expressed on the surface of T cells that deliver one of two signals required for T cell activation (Signal 2). They are activated when engaged by ligands, which are typically expressed by professional APC. The most common costimulator receptor is CD28.

Costimulatory signal Additional signal that is required to induce proliferation of antigen-primed T cells and is generated by interaction of CD28 on T cells with CD80/86 on antigen-presenting cells. In B-cell activation, an analogous signal is provided by interaction of CD40 on B cells with CD40L on activated T$_H$ cells.

CR1 Complement Receptor 1. Expressed on both erythrocytes and leukocytes and binds to C3b, C4b and their breakdown products. CR1 expression on erythrocytes is important in the clearance of immune complexes in the liver.

CR3 A cell-surface receptor for complement component iC3b and factor H.

CR4 A cell-surface receptor for complement component iC3b.

CRIg A cell-surface receptor for complement components C3b, iC3b, and C3c.

Crohn's disease A form of inflammatory bowel disease (IBD) that can afflict any part of the intestine and is characterized by accumulation of inflammatory cells and granuloma formation, which interferes with digestion.

Cross-matching Refers to pretransplantation testing of the serum of the recipient for any preexisting antibodies capable of recognizing the allogeneic MHC molecules of a potential donor. Even when the recipient has never before received a transplant, most will have some antibodies capable of recognizing allogenic MHC molecules, most likely due to our regular exposure to food and microbes expressing cross-reactive epitopes.

Cross-presentation A protein processing and presentation pathway that occurs in some pAPCs where antigen acquired by endocytosis is redirected from the exogenous to the endogenous pathway, such that peptides associate with MHC class I molecules for presentation to CD8$^+$ T cells.

Cross-priming The activation of CTL responses to antigens processed and presented via cross-presentation.

Cross-reactivity Ability of a particular antibody or T-cell receptor to react with two or more antigens that possess a common epitope.

Cross-tolerance The induction of CD8$^+$ T cell tolerance to an antigen processed and presented via cross-presentation.

Cryptopatches Small aggregates of lymphoid cells found in the intestinal wall that are precursors of isolated lymphoid follicles (ILFs).

Crypts The epithelial cell–lined valleys between the villi of the gastrointestinal tract; these are lined by specialized epithelial cells and include the stem cells that continually replenish the epithelial lining of the intestine.

CTLA-4 (CD152) A coinhibitory receptor that blocks T-cell activation, acting as a checkpoint that down-regulates immune responses; up-regulated by T cells after activation.

CTL precursors (CTL-Ps) Naïve CD8$^+$ T cells that have not yet been activated by antigen recognition. They do not yet express the cytotoxic machinery associated with fully mature killer T cells.

CXC subgroup A family of chemokines that contain a disulfide bridge between cysteines separated by a different amino acid residue (X).

Cyclooxygenase 2 (COX2) Enzyme responsible for the formation from arachidonic acid of prostaglandins and other pro-inflammatory mediators; target of non-steroidal anti-inflammatory drugs.

Cyclosporin A A fungal product used as a drug to suppress allograft rejection. The compound blocks T-cell activation by interfering with transcription factors and preventing gene activation.

Cytokine storms The pathological secretion of extremely high levels of cytokines induced by massive infection with particular pathogens. Typical symptoms include increased capillary permeability with resultant loss of blood pressure and shock, sometimes leading to death.

Cytokine-binding Homology Region (CHR) A protein motif common to the cytokine binding receptors of several families.

Cytokines Any of numerous secreted, low-molecular-weight proteins that regulate the intensity and duration of the immune response by exerting a variety of effects on lymphocytes and other immune cells that express the appropriate receptor (see Appendix II).

Cytosolic pathway See **Endogenous pathway**.

Cytotoxic natural killer (NK) cells A subset of ILCs within the ILC1 category that have cytotoxic rather than helper potential.

Cytotoxic T lymphocytes (CTLs, or T$_c$ cells) An effector T cell (usually CD8$^+$) that can mediate the lysis of target cells bearing antigenic peptides complexed with a MHC class I molecule.

Damage-associated molecular patterns (DAMPs) Components of dead/dying cells and damaged tissues that are recognized by pattern-recognition receptors.

Danger or damage model A model of immune response theory which posits that the immune response distinguishes between dangerous and non-dangerous structures rather than the classical view of differentiation between self and nonself structures.

Dark zone A portion of the **germinal center** that is the site of rapid cell division by forms of B cells called centroblasts.

Death by neglect Apoptosis of developing T cells (typically CD4$^+$CD8$^+$ thymocytes) that results when they do not receive TCR signals of adequate affinity. Most (90% or more) developing T cells undergo death by neglect.

Death domains Protein motifs found in the cytoplasmic region of **Fas** and other proapoptotic signaling molecules. They engage the domains on other signaling molecules and initiate the formation of the **Death-Inducing Signaling Complex (DISC)**.

Death-Inducing Signaling Complex (DISC) An intracellular signaling aggregate formed in response to engagement of death receptors, including Fas. It includes the cytoplasmic tail of **Fas**, FADD, and procaspase-8, and initiates **apoptosis**.

Defensins   Antimicrobial peptides (AMPs) generated by many cell types, including some T cells, epithelial cells of barrier tissue, and other innate immune cells; defensins can compromise bacteria, fungi, and certain viruses.

Degranulation   Discharge of the contents of cytoplasmic granules by **basophils** and **mast cells** following cross-linkage (usually by antigen) of bound IgE. It is characteristic of **type I hypersensitivity**.

Delayed-type hypersensitivity (DTH)   A type IV hypersensitive response mediated by sensitized T_H cells, which release various **cytokines** and **chemokines**. The response generally occurs 2 to 3 days after T_H cells interact with antigen. It is an important part of host defense against intracellular parasites and bacteria.

Dendritic cells (DCs)   Bone-marrow-derived cells that descend through the myeloid and lymphoid lineages and are specialized for antigen presentation to helper T cells.

Dermis   Layer of skin under the epidermis that contains blood and lymph vessels, hair follicles, nerves, and nerve endings.

Desensitization   Reduction in allergic responses achieved by repeated low or increasing doses of allergens administered through injections of allergens (allergy shots), sublingual application (under the tongue), and oral immunotherapy (feeding the allergen).

Determinant-selection model   A hypothesis proposed to explain the variability in immune responsiveness to different MHC haplotypes. This model states that each MHC molecule binds a unique array of antigenic peptides, and some peptides are more successful in eliciting an effective immune response than others. See also **Holes-in-the-repertoire model**.

Diacylglyerol (DAG)   A lipid molecule generated upon cleavage of **phosphatidyl inositol bisphosphate** that is important in cell signaling.

Diapedesis   The process by which a cell crosses from the lumen of a vessel, between endothelial cells, and into the surrounding tissue.

Differentiation antigen   A cell surface marker that is expressed only during a particular developmental stage or by a particular cell lineage.

DiGeorge syndrome (DGS)   Congenital thymic aplasia (partial or total absence of the thymus) caused by deletion of a sequence on chromosome 22 during embryonic life. Consequences include immunodeficiency, facial abnormalities, and congenital heart disease.

Direct staining   A variation of fluorescent antibody staining in which the primary antibody is directly conjugated to the fluorescent label.

Dissociation constant K_d, the reciprocal of the **association constant** $(1/K_a)$.

Diversity (D) segment   One of the gene segments encoding the immunoglobulin heavy chain or the TCR β or δ chains or its protein product.

DN1   The first in the four stages in the development of the most immature (CD4⁻CD8⁻ or double negative) thymocytes. DN1 cells express CD44 but not CD25 and are the progenitors that come from the bone marrow and have the potential to give rise to multiple lymphoid and myeloid cell lineages.

DN2   The second in the four stages in the development of the most immature (CD4⁻CD8⁻ or double negative) thymocytes. Commitment to the T cell lineage and rearrangement of the first TCR receptor genes occur among DN2 cells, which express both CD44 and CD25.

DN3   The third in the four stages in the development of the most immature (CD4⁻CD8⁻ or double negative) thymocytes. Only those DN3 cells that express a functional TCRβ chain continue to mature to the CD4⁺CD8⁺ stage and proliferate (β-**selection**). DN3 cells express CD25, but not CD44.

DN4   The last of the four stages in the development of the most immature (CD4⁻CD8⁻ or double negative) thymocytes. DN4 cells express neither CD44 nor CD25 and are in transition to the CD4⁺CD8⁺ (double positive or DP) stage of development.

Double immunodiffusion   A type of precipitation in gel analysis in which both antigen and antibody diffuse radially from wells toward each other, thereby establishing a concentration gradient. As equivalence is reached, a visible line of precipitation, a precipitin line, forms.

Double-negative (DN) cells   A subset of developing T cells (thymocytes) that do not express CD4 or CD8. At this early stage of T-cell development, DN cells do not express the TCR.

Double-positive (DP) stage   A subset of developing T cells (thymocytes) that express both CD4 and CD8. DP cells are an intermediate stage of developing thymocytes that express TCRs.

Downstream   (1) Towards the 3′ end of a gene; (2) Further away from the receptor in a signaling cascade.

Duodenum   A section of the small intestine; the duodenum is short and the site of secretion of digestive enzymes.

Dynamic imaging   An approach and set of techniques that allows the visualization of living cells in living tissue.

Dysbiosis   An alteration in the microbiome that negatively affects the balance of microbial communities typically found in healthy, commensal communities of microbes.

E2A   A transcription factor that is required for the expression of the **recombination-activating genes (RAG)** as well as the expression of the λ5 (lambda 5) component of the **pre-B-cell receptor** during B-cell development. It is essential for B-cell development.

Early B-cell factor 1 (EBF1)   Key transcription factor at the common lymphoid progenitor stage of B-cell development. It is induced by E2A, Foxo1, Runx1, and signals through the IL-7R.

Early lymphoid progenitor cell (ELP)   A progenitor cell capable of dividing to give rise to either T or B lymphocyte progenitors.

Edema   Abnormal accumulation of fluid in intercellular spaces, often resulting from a failure of the lymphatic system to drain off normal leakage from the capillaries.

Effector caspases   The subset of caspase enzymes directly responsible for the cell apoptosis. Their cleavage activity results both in the breakdown of structural molecules (e.g. actin) or the activation of destructive molecules (e.g. endonucleases). Caspase-3 and caspase-7 are two well-characterized effector caspases.

Effector cell   Any cell capable of mediating an immune function (e.g., activated T_H cells, CTLs, and plasma cells).

Effector cytokines   Cytokines released by effector lymphocytes, including helper T-cell subsets and cytotoxic T cells, that regulate the immune response to pathogens.

Effector memory T cells (T_{EM})   A memory T cell subset that circulates among or resides in peripheral, non-lymphoid tissue. It is generated during the primary response and participates in the secondary response to antigen, exhibiting effector functions and proliferating more quickly than immune cells.

Effector phase   The stage of immune activity when effector cells work to resolve pathogen infection at barrier tissues; preceded by the inductive phase.

Effector response   Immune cell action that contributes to the clearance of infection. It includes responses mediated by helper T cells, which secrete cytokines that enhance the activity of several other immune cell subsets, by cytotoxic cells, including CD8⁺ T cells and NK cells, and by antibody, which recruits soluble proteins (complement) and cells that can kill and clear pathogen. Also called effector function.

ELISA   See **enzyme-linked immunosorbent assay**.

Embryonic stem (ES) cell   Stem cell isolated from early embryo and grown in culture. Mouse ES cells give rise to a variety of cell types and are used to develop transgenic or knockout mouse strains.

Endocrine   Referring to regulatory secretions such as hormones or cytokines that pass from producer cell to target cell by the bloodstream.

Endocytosis   Process by which cells ingest extracellular macromolecules by enclosing them in a small portion of the plasma membrane, which invaginates and is pinched off to form an intracellular vesicle containing the ingested material.

Endogenous pathway   Intracellular route taken by antigen that is processed for presentation by MHC class I, typically associated with proteins generated in the cytosol.

Endosteal niche   Microenvironment in the bone marrow that fosters the development of hematopoietic stem cells and is postulated to associate specifically with self-renewing, long-term hematopoietic stem cells.

Endotoxins   Certain **lipopolysaccharide (LPS)** components of the cell wall of gram-negative bacteria that are responsible for many of the pathogenic effects associated with these organisms. Some function as **superantigens**.

Enterocyte   The absorptive epithelial cell that lines the small intestine; these cells are polarized with an apical surface folded into microvilli that faces the lumen of the intestine and a basolateral surface that faces the intestinal wall. They are responsible for absorbing food, but also play an active role in protecting the epithelial layer from infection.

Enzyme-linked immunosorbent assay (ELISA)   An assay for quantitating either antibody or antigen by use of an enzyme-linked antibody and a substrate that forms a colored reaction product.

Eosinophils   Motile, somewhat phagocytic granulocytes that can migrate from blood to tissue spaces. They have large numbers of IgE receptors and are highly granular. They are thought to play a role in the defense against parasitic organisms such as roundworms.

Epidermis   The outermost layer of the skin.

Epigenetic   Describes modifications to genes other than changes in the DNA sequence that affect gene expression; examples include chromatin modifications such as histone methylation.

Epitope mapping   Localization of sites (epitopes) on an antigen molecule that are reactive with different antibodies or T-cell receptors.

Epitope   The portion of an antigen that is recognized and bound by an antibody or TCR-MHC combination; also called **antigenic determinant**.

Equilibrium dialysis   An experimental technique that can be used to determine the affinity of an antibody for antigen and its **valency**.

ERAP   Endoplasmic reticulum aminopeptidase. An enzyme responsible for trimming amino acids from peptides in the ER in order to reach an optimal length for binding to MHC class I molecules.

Erythroblastosis fetalis   A type II hypersensitivity reaction in which maternal antibodies against fetal **Rh antigens** cause hemolysis of the erythrocytes of a newborn; also called *hemolytic disease of the newborn*.

Erythrocytes   Red blood cells.

Exocytosis   Process by which cells release molecules (e.g., cytokines, lytic enzymes, degradation products) contained within a membrane-bound vesicle by fusion of the vesicle with the plasma membrane.

Exogenous pathway   Intracellular route taken by antigen that is processed for presentation by MHC class II, typically associate with proteins that are endocytosed.

Exotoxins   Toxic proteins secreted by gram-positive and gram-negative bacteria; some function as **superantigens**. They cause food poisoning, toxic shock syndrome, and other disease states. See also **immunotoxin**.

Extravasation   Movement of blood cells through an unruptured blood vessel wall into the surrounding tissue, particularly at sites of inflammation.

F (ab′)₂ fragment   Two Fab units linked by disulfide bridges between fragments of the heavy chain. They are obtained by digestion of antibody with pepsin.

Fab (fragment antigen binding) region   Region at the N-terminus of the antibody molecule that interact with antigen. This antibody fragment, consisting of one light chain and part of one heavy chain, linked by an interchain disulfide bond, is obtained by brief papain digestion.

Factor B   A protein component of the alternative pathway of complement activation that is cleaved by factor D.

Factor D   An enzyme of the alternative pathway of complement activation that cleaves factor B into Ba and Bb only when it is bound to either $C3(H_2O)$ or to C3b.

Fas (CD95)   A member of the Tumor Necrosis Factor Receptor family. On binding to its ligand, **FasL**, the Fas-bearing cell will often be induced to commit to an apoptotic program. Occasionally, however, Fas ligation leads to cell proliferation.

Fas ligand (FasL)   FasL is a member of the Tumor Necrosis Factor family of molecules and interacts with the Fas receptor, which is a member of the Tumor Necrosis Factor Receptor family. Signals delivered from FasL to Fas usually result in the death by **apoptosis** of the Fas-bearing cell.

Fc (fragment crystallizable) region   Region at the C terminus of the antibody molecule that interacts with Fc receptors on other cells and with components of the complement system. This crystallizable antibody fragment consists of the carboxyl-terminal portions of both heavy chains and is obtained by brief papain digestion.

Fc receptor (FcR)   Cell-surface receptor specific for the Fc portion of certain classes of immunoglobulin. It is present on lymphocytes, mast cells, macrophages, and other accessory cells.

FcεRI receptor   The high-affinity receptor for the Fc of IgE that induces degranulation of mast cells, basophils, and eosinophils.

FcεRII (CD23) receptor   The low-affinity receptor for IgE. Regulates B-cell activation, growth, and IgE synthesis; triggers transport of IgE and IgE/food allergen complexes across the intestinal epithelium; activates macrophages to release TNF, IL-1, IL-6, and GM-CSF.

FcγRIIb   A receptor that binds to the **Fc region** of antibodies engaged in antigen:antibody complexes. Signals through this receptor down-regulate B cell division and differentiation.

Fibrin   A filamentous protein produced by the action of thrombin on fibrinogen; fibrin is the main element in blood clotting.

Fibrinopeptide   One of two small peptides of about 20 amino acids released from fibrinogen by thrombin cleavage in the conversion to **fibrin**.

Fibroblast reticular cells (FRCs)   Stromal cells in secondary lymphoid tissue (and at some immune response sites in the periphery) that extend processes which provide the surface networks on which dendritic cells position themselves and T and B lymphocytes migrate as they probe for antigen. Associated with chemokines and cytokines that help guide cell movements.

Fibrosis   A process responsible for the development of a type of scar tissue at the site of chronic inflammation.

Ficolin   Member of a family of carbohydrate-binding proteins that contain a fibrinogen-like domain and a collagen-like domain.

Flow cytometer   An instrument that users lasers along with sophisticated optics to measure multiple fluorescent and light scattering parameters from thousands of cells as flow rapidly, one-by-one in front of the laser beam.

Fluorescence Activated Cell Sorter (FACS)   A flow cytometer equipped with the ability to sort cells sharing particular fluorescence and light scattering properties into different containers.

Fluorescence microscopy   A microscopic technique that allows the visualization of fluorescent signals generated from cells tagged with fluorescent antibodies or proteins.

Fluorescent antibody   An antibody with a **fluorochrome** conjugated to its Fc region that is used to stain cell surface molecules or tissues; the technique is called **immunofluorescence**.

Fluorochrome   A molecule that fluoresces when excited with appropriate wavelengths of light. See **immunofluorescence**.

fms-related tyrosine kinase 3 receptor (flt-3)   Binds to the membrane-bound flt-3 ligand on bone marrow stromal cells and signals the progenitor cell to begin synthesizing the **IL-7 receptor**.

Follicles   Microenvironments that specifically support the development of the B lymphocyte response in lymph nodes, spleen, and other secondary lymphoid tissue. They also become the site of development of the germinal center when a B cell is successfully activated.

Follicular dendritic cell (FDC)   A cell with extensive dendritic extensions that is found in the follicles of lymph nodes. Although they do not express MHC class II molecules, they are richly endowed with receptors for complement and Fc receptors for antibody. They are of a lineage that is distinct from MHC class II–bearing dendritic cells.

Follicular helper T (T$_F$H) cells   Helper CD4$^+$ T cell subset that supports the development of B lymphocytes in the follicle and germinal center and expresses the master transcriptional regulator Bcl-6.

Follicular mantle zone   Zone of naïve, IgD-bearing B cells that surrounds the central region of a follicle engaged in a germinal center reaction. The non-antigen-specific, IgD-bearing cells are slowly pushed to the outside of the follicle as they are displaced by dividing cells in the germinal center.

Foxo1   Transcription factor important in B-cell development. It is induced by E2A and contributes to the activation of EBF1.

Fragmentin   Enzymes present in the granules of cytotoxic lymphocytes that induce DNA fragmentation.

Framework region (FR)   A relatively conserved sequence of amino acids located on either side of the hypervariable regions in the variable domains of immunoglobulin heavy and light chains.

Freund's complete adjuvant (CFA)   A water-in-oil emulsion to which heat-killed mycobacteria have been added; antigens are administered in CFA to enhance their immunogenicity.

Freund's incomplete adjuvant   Freund's adjuvant lacking heat-killed mycobacteria.

G-Protein–Coupled Receptors (GPCRs)   Ligand receptors that interact with G proteins on the cytoplasmic side of the plasma membrane. G proteins are signal-transducing molecules that are activated when the receptor binds to its ligand. Receptor:ligand binding induces a conformational change in the G protein that induces it to exchange the GDP (which is in its binding site in the resting state), for GTP. The activated G protein:GTP complex then transduces the signal. GPCRs have a shared structure in which the proteins passes through the member a total of seven times.

γ (gamma)-globulin fraction   The electrophoretic fraction of serum that contains most of the immunoglobulin classes.

Gastrointestinal (GI) tract   A mucosal tissue that is part of the barrier organ system; the GI tract, or gut, is responsible for digesting and distributing food, cultivating beneficial relationships with the commensal microbiome, and protecting the body from infection. It includes the mouth, esophagus, stomach, small intestine (duodenum, jejunum, ileum), and large intestine (colon). All of the intestinal tract except the mouth is lined by a single layer of epithelial cells.

***GATA-2* gene**   A gene encoding a transcription factor that is essential for the development of several hematopoietic cell lineages, including the lymphoid, erythroid, and myeloid lineages.

Gene conversion   Process in which portions of one gene (the recipient) are changed to those of another gene (the donor). Homologous gene conversion is a diversification mechanism used for immunoglobulin V≈genes in some species.

Gene segments   Germ-line gene sequences that are combined with others to make a complete coding sequence; Ig and TCR genes are products of V, D, J gene segments.

Gene therapy   General term for any measure aimed at correction of a genetic defect by introduction of a normal gene or genes.

Generation of diversity   The generation of a diverse repertoire of antigen-binding receptors on B or T lymphocytes that occurs in the bone marrow or thymus, respectively.

Genome wide sequence   The sequence of all DNA (the entire genome) present in a cell.

Genotype   The combined genetic material inherited from both parents; also, the **alleles** present at one or more specific loci.

Germ-line theories   Classical theories that attempted to explain antibody diversity by postulating that all antibodies are encoded in the host chromosomes.

Germinal centers (GCs)   A region within lymph nodes and the spleen where T-dependent B-cell activation, proliferation, and differentiation occur. Germinal centers are sites of intense B-cell somatic mutation and selection.

Goblet cells   Specialized cells in the epithelium that lines mucosal tissues; they are responsible for secreting mucus and are particularly abundant in the large intestine but are present in all mucosal tissues.

Graft-versus-host (GVH) reaction   A pathologic response to tissue transplantation in which immune cells in the transplanted tissue (graft) react against and damage host cells.

Graft-versus-host disease (GVHD)   A reaction that develops when a graft contains immunocompetent T cells that recognize and attack the recipient's cells.

Granulocytes   Any **leukocyte** that contains cytoplasmic granules, particularly the basophil, eosinophil, and neutrophil.

Granuloma   A tumor-like mass or nodule that arises because of a chronic **inflammatory response** and contains many activated macrophages, T$_H$ cells, and multinucleated giant cells formed by the fusion of macrophages.

Granzyme (fragmentin)   One of a set of enzymes found in the granules of T$_C$ cells that can help to initiate apoptosis in target cells.

Grave's disease   An autoimmune disease in which the individual produces auto-antibodies to the receptor for thyroid-stimulating hormone TSH.

GTP-binding proteins   Proteins that bind Guanosine Tri-phosphate.

GTPase Activating Proteins (GAPs)   G proteins have an intrinsic GTPase activity that serves to limit the time during which G proteins can actively transduce a signal. GAPs enhance this GTPase activity, and thus further limit the signal through a GPCR.

Guanine-nucleotide Exchange Factors (GEFs) Small proteins that catalyze the exchange of GTP for GDP in the guanine nucleotide binding sites of small and trimeric G proteins.

Gut-associated lymphoid tissue (GALT) Secondary lymphoid microenvironments in the intestinal (gut) system that support the development of the T and B lymphocyte response to antigens that enter gut mucosa. Part of the **mucosa associated lymphoid tissue system** (MALT).

H2 complex Term for the **MHC** in the mouse.

HAART See **Highly active antiretroviral therapy**.

Haplotype The set of **alleles** of linked genes present on one parental chromosome; commonly used in reference to the **MHC** genes.

Hapten A low-molecular-weight molecule that can be made immunogenic by conjugation to a suitable carrier.

Hapten-carrier conjugate A covalent combination of a small molecule (**hapten**) with a large carrier molecule or structure.

Heavy (H) chain The larger polypeptide of an antibody molecule; it is composed of one variable domain V_H and three or four constant domains (C_H1, C_H2, etc.) There are five major classes of heavy chains in humans (α, γ, δ, ε, and μ), which determine the **isotype** of an antibody.

Heavy-chain Joining segment (J_H) One of the gene segments encoding the immunoglobulin heavy chain or its protein product.

Heavy-chain Variable region That part of the immunoglobulin heavy chain protein that varies from antibody to antibody and is encoded by the V, D, and J gene segments.

Heavy-chain Variable segment (V_H) (1) One of the gene segments encoding the immunoglobulin heavy chain gene, or its protein product.

Helper T (T_H) cells See **T helper (T_H) cells**.

Hemagglutination The process of sticking together red blood cells using multivalent cells, viruses or molecules that bind to molecules on the red blood cell surface. Viruses such as influenza or antibodies are routinely measured by hemagglutination assays.

Hemagglutinin (HA) Any substance that causes red blood cells to clump, or agglutinate. Most commonly the virally-derived glycoprotein found on the surface of influenza virus that binds to sialic acid residues on host cells causing them to agglutinate. See also **agglutinin**.

Hematopoiesis The formation and differentiation of blood cells.

Hematopoietic stem cell (HSC) The cell type from which all lineages of blood cells arise.

Hemolysis Alteration or destruction of red blood cells, which liberates hemoglobin.

Heptamer A conserved set of 7 nucleotides contiguous to each of the V, D, and J gene segments of all immunoglobulin and TCR gene segments. It serves as the recognition signal and binding site of the RAG1/2 protein complex.

Herd immunity When the majority of the population is immune to an infectious agent, thus significantly reducing the pathogen reservoir due to the low chance of a susceptible individual contacting an infected individual.

Heteroconjugates Hybrids of two different antibody molecules.

Heterotypic An interaction between two molecules where the interacting domains have different structures from one another.

High-endothelial venule (HEV) An area of a capillary venule composed of specialized cells with a plump, cuboidal ("high") shape through which lymphocytes migrate to enter various lymphoid organs.

Highly active antiretroviral therapy (HAART) A form of drug therapy used to treat infection with HIV that utilizes a combination of three or more anti-HIV drugs from different classes to inhibit viral replication and avoid selection of drug-resistant mutants.

Hinge The flexible region of an immunoglobulin heavy chain between the C_H1 and C_H2 domains that allows the two binding sites to move independently of one another.

Histamine A small mediator released during degranulation of mast cells, basophils, and eosinophils that causes symptoms of allergic reactions, including vasodilation, increased vascular permeability, mucus secretion, and smooth muscle contraction.

Histiocyte An immobilized (sometimes called "tissue fixed") macrophage found in loose connective tissue.

Histocompatibility antigens Family of proteins that determines the ability of one individual to accept tissue or cell grafts from another. The major histocompatibility antigens, which are encoded by the **MHC**, function in antigen presentation.

Histocompatible Denoting individuals whose major histocompatibility antigens are identical. Grafts between such individuals are generally accepted.

HLA (human leukocyte antigen) complex Term for the **MHC** in humans.

Holes-in-the-repertoire model The concept that immune tolerance results from the absence of receptors specific for self antigens.

Homeostasis, homeostatic Pertaining to processes that contribute to the maintenance and stability of a system, in this case, the immune system, under normal conditions.

Homing receptor A receptor that directs various populations of lymphocytes to particular lymphoid and inflammatory tissues.

Homing The differential migration of lymphocytes or other leukocytes to particular tissues or organs.

Homotypic An interaction between two molecules where the interacting domains have identical or very similar structures to one another.

Human immunodeficiency virus (HIV) The retrovirus that causes acquired immune deficiency syndrome (AIDS).

Human leukocyte antigen (HLA) complex See **HLA complex**.

Humanized antibody An antibody that contains the antigen-binding amino acid sequences of another species within the framework of a human immunoglobulin sequence.

Humoral immune response Host defenses that are mediated by antibody present in the plasma, lymph, and tissue fluids. It protects against extracellular bacteria and foreign macromolecules. Transfer of antibodies confers this type of immunity on the recipient. See also **cell-mediated immune response**.

Humoral immunity See **humoral immune response**.

Humoral Pertaining to extracellular fluid, including the plasma, lymph, and tissue fluids.

Hybridoma A clone of hybrid cells formed by fusion of normal lymphocytes with **myeloma cells**; it retains the properties of the normal cell to produce antibodies or T-cell receptors but exhibits the immortal growth characteristic of myeloma cells. Hybridomas are used to produce **monoclonal antibody**.

Hygiene hypothesis Hypothesis that a lack of early childhood exposure to infectious agents, certain symbiotic microorganisms (such as the gut flora or probiotics), and parasites increases susceptibility to allergic diseases by suppressing the natural development of some components of the immune system.

Hyper IgE syndrome (HIE)   An immune deficiency syndrome characterized by over expression of IgE and most frequently caused by mutations in the gene encoding STAT3. Also known as **Job syndrome**.

Hyper IgM syndrome (HIM)   An immune deficiency disorder that arises from inherited deficiencies in **CD40L**, resulting in impaired T cell-APC communication and a lack of isotype switching, manifesting as elevated levels of IgM but an absence of other antibody isotypes.

Hyperacute rejection   The earliest of the stages of graft rejection, occurring within minutes or hours of transfer; it involves the action of preformed antibodies that can recognize antigens (usually MHC molecules) on the engrafted organ/tissue/cells.

Hypersensitivity   Exaggerated immune response that causes damage to the individual. **Immediate hypersensitivity** (types I, II, and III) is mediated by antibody or immune complexes, and delayed-type hypersensitivity (type IV) is mediated by T_H cells.

Hypervariable   Those parts of the variable regions of the BCR and TCR that exhibit the most sequence variability and interact with the antigen. Otherwise known as the **complementarity determining regions**.

Hypogammaglobulinemia   Any immune deficiency disorder, either inherited or acquired, characterized by low levels of gammaglobulin (IgG).

Iccosomes   Immune-complex-coated cell fragments often found coating the spines of **follicular dendritic cells**.

Idiotope   A single **antigenic determinant** in the variable domains of an antibody or T-cell receptor; also called idiotypic determinant. Idiotopes are generated by the unique amino acid sequence specific for each antigen.

Idiotype   The set of antigenic determinants (**idiotopes**) characterizing a unique antibody or T-cell receptor.

IgD   Immunoglobulin D. An antibody class that serves importantly as a receptor on naïve B cells.

IgM   Immunoglobulin M. An antibody class that serves as a receptor on naïve B cells. IgM is also the first class of antibody to be secreted during the course of an immune response. Secreted IgM exists primarily in pentameric form.

Ikaros   A transcription factor required for the development of all lymphoid cell lineages.

IKK   *See* IκB kinase.

IKKα   IκB kinase α subunit.

IKKβ   IκB kinase β subunit.

IL-1 Receptor Activated Kinase (IRAK)   A family of kinases that participates in the signaling pathway from IL-1. IRAKs are also important in **TLR** signaling.

IL-2   A cytokine produced by activated T cells that promotes T-cell division in both autocrine and paracrine fashions.

IL-10   A member of the interferon family of cytokines, that usually mediates an immuno-suppressive effect.

IL-17 family   A family of cytokines implicated in the early stages of the immune response. Most members of this family are pro-inflammatory in action.

IL-23–T_H17 cell axis   A cytokine/T-cell system that regulates inflammation and is responsible for several immune-mediated and inflammatory diseases; IL-23 induces differentiation of T_H17 and ILC3 subsets, both of which can produce several potent proinflammatory cytokines, including IL-17, IL-22, and GM-CSF.

IL-7 receptor   Receptor for the cytokine Interleukin 7, which is important for lymphocyte development.

Ileum   A section of the small intestine; the jejunum and ileum are responsible for final digestion and absorption of food.

Immature B cell   Immature B cells express a fully-formed IgM receptor on their cell surface. Contact with antigen at this stage of B cell development results in tolerance induction rather than activation. Immature B cells express lower levels of IgD and higher levels of IgM than do mature B cells. They also have lower levels of anti-apoptotic molecules and higher levels of Fas than mature B cells, reflective of their short-half lives.

Immediate hypersensitivity   An exaggerated immune response mediated by antibody (type I and II) or antigen-antibody complexes (type III) that manifests within minutes to hours after exposure of a sensitized individual to antigen.

Immune complex   A complex of antibody with bound antigen.

Immune deficiency   See **immunodeficiency**.

Immune dysregulation, polyendocrinopathy, enteropathy, X-linked (IPEX) syndrome   An inherited immune deficiency disorder that manifests as an autoimmune syndrome caused by a lack of FoxP3 expression and near absence of regulatory T cells (T_{REG} cells).

Immune imbalance   A bias of the immune response toward overly robust immune activity (or occasionally, underactivity) that is nontypical and nonspecific, most often manifesting as a generalized decrease in the immuno-inhibitory elements of immune recognition of self and/or benign foreign antigens.

Immune responsiveness   The ability of an organism to respond to a particular antigen in a manner that leads to an adaptive immune response. Immune responsiveness is determined by the genes within the MHC locus, especially class II, and is based on the affinity of MHC molecules for the antigen in question.

Immunity   A state of protection from a particular infectious disease.

Immunization   The process of producing a state of immunity in a subject. See also **active immunity** and **passive immunity**.

Immunocompetent   Denoting a mature lymphocyte that is capable of recognizing a specific antigen and mediating an immune response; also an individual without any immune deficiency.

Immunodeficiency   Any deficiency in the immune response, whether inherited or acquired. It can result from defects in phagocytosis, humoral immunity, cell-mediated responses, or some combination of these.

Immunodominant   Referring to epitopes that produce a more pronounced immune response than others under the same conditions.

Immunoediting   A recently formulated theory concerning the role of the immune system in responding to cancer. It includes three phases (elimination, equilibrium, and escape) and incorporates both positive (anti-tumor) and negative (pro-tumor) processes mediated by the immune system in responding to malignancy.

Immunoelectron microscopy   A technique in which antibodies used to stain a cell or tissue are labeled with an electron-dense material and visualized with an electron microscope.

Immunoelectrophoresis   A technique in which an antigen mixture is first separated into its component parts by electrophoresis and then tested by **double immunodiffusion**.

Immunofluorescence   Technique of staining cells or tissues with **fluorescent antibody** and visualizing them under a fluorescent microscope.

Immunogen   A substance capable of eliciting an immune response. All immunogens are **antigens**, but some antigens (e.g., haptens) are not immunogens.

Immunogenicity The capacity of a substance to induce an immune response under a given set of conditions.

Immunoglobulin (Ig) Protein consisting of two identical heavy chains and two identical light chains, that recognize a particular **epitope** on an antigen and facilitates clearance of that antigen. There are 5 types: IgA, IgD, IgE, IgG, and IgM. Also called **antibody**.

Immunoglobulin domains Three dimensional structures characteristic of immunoglobulin and related proteins including T cell receptors, MHC proteins and adhesion molecules. Consists of a domain of 100 – 110 amino acids folded into two β -pleated sheets, each containing three of four antiparallel β strands and stabilized by an intrachain disulfide bond.

Immunoglobulin fold Characteristic structure in immunoglobulins that consists of a domain of 100 to 110 amino acids folded into two β-pleated sheets, each containing three or four antiparallel β strands and stabilized by an intrachain disulfide bond.

Immunoglobulin superfamily Group of proteins that contain **immunoglobulin-fold** domains, or structurally related domains; it includes immunoglobulins, T-cell receptors, MHC molecules, and numerous other membrane molecules.

Immunologic memory The ability of the immune system to respond much more swiftly and with greater efficiency during a second, or later exposure to the same pathogen.

Immunomodulatory Any compound or inducer that changes the behavior of immune cells. This is most often used to refer to the activity of costimulatory (e.g., CD28) or coinhibitory (e.g., CTLA-4) molecules but encompasses both categories more broadly.

Immunoproteasome A variant of the standard 20S proteasome, found in pAPCs and infected cells, that has unique catalytic subunits specialized to produce peptides that bind efficiently to MHC class I proteins.

Immunoreceptor tyrosine-based inhibitory motif (ITIM) An amino acid sequence containing tyrosine residues in conserved sequence relationships with one another that serves as a docking site for downstream signaling molecules that will send an inhibitory signal to the cell. More formally known as immuno-receptor tyrosine-based inhibitory motif.

Immunoreceptor tyrosine-based activation motif (ITAM) Amino acid sequence in the intracellular portion of signal-transducing cell surface molecules that interacts with and activates intracellular kinases after ligand binding by the surface molecule.

Immunosurveillance A theory concerning anti-cancer responses that is now part of the immunoediting hypothesis (elimination phase) which posits that cells of the immune system continually survey the body in order to recognize and eliminate tumor cells.

Immunotherapy Any of a variety of medical therapies or drug treatments designed to revive, initiate, or supplement the ability of the existing immune response to target cancer cells for elimination.

Immunotoxin Highly cytotoxic agents sometimes used in cancer treatment that are produced by conjugating an antibody (for instance, specific for tumor cells) with a highly toxic agent, such as a bacterial toxin.

Incomplete antibody Antibody that binds antigen but does not induce agglutination.

Indirect staining A method of immunofluorescent staining in which the primary antibody is unlabeled and is detected with an additional fluorochrome-labeled reagent.

Induced T_{REG} (iT_{REG}) cells $CD4^+$ T cell subset that negatively regulates immune responses and is induced to develop by specific cytokine interactions in secondary lymphoid tissue that upregulate the master regulator FoxP3.

Inducible costimulator (ICOS or CD278) A costimulatory receptor that cooperates with the TCR to activate some T cells; it binds the ICOS ligand, which is known by several names including B7-H2.

Inducible nitric oxide synthetase (iNOS) An inducible form of NOS that generates the antimicrobial compound nitric oxide from arginine.

Inflamed Manifesting redness, pain, heat, and swelling. See also **Inflammation**.

Inflammasomes Multiprotein complex that promotes inflammation by processing inactive precursor forms of pro-inflammatory cytokines such as IL-1 and IL-18.

Inflammation Tissue response to infection or damage which serves to eliminate or wall-off the infection or damage; classic signs of acute inflammation are heat (calor), pain (dolor), redness (rubor), swelling (tumor), and loss of function.

Inflammatory bowel disease (IBD) Chronic intestinal inflammation that can be subdivided into two types: Crohn's disease and ulcerative colitis.

Inflammatory response A localized tissue response to injury or other trauma characterized by pain, heat, redness, and swelling. The response includes both localized and systemic effects, consisting of altered patterns of blood flow, an influx of phagocytic and other immune cells, removal of foreign antigens, and healing of the damaged tissue.

Inductive phase The stage of immune activity when the response to pathogens encountered in barrier tissues is initiated; followed by the effector phase.

Inhibitor of NF-κB (IκB) A small protein that binds to the transcription factor **NF-κB** that inhibits its action, in part by retaining it in the cytoplasm.

Initiator caspases The subset of caspase enzymes that initiate the process leading to cell apoptosis. Initiator caspases typically cleave and activate effector caspases, although they can also cleave other molecules that indirectly activate effector caspases (e.g. Bid). Caspase-8 is a well-characterized initiator caspase that is associated with the death receptor, **Fas**, and is cleaved and activated when Fas is engaged.

Innate immunity Non-antigen specific host defenses that exist prior to exposure to an antigen and involve anatomic, physiologic, endocytic and phagocytic, anti-microbial, and inflammatory mechanisms, and which exhibit no adaptation or memory characteristics. See also **adaptive immunity**.

Innate immunity effect or mechanisms Chemical and cellular mechanisms by which the innate immune system eliminates pathogens.

Innate lymphoid cells (ILCs) Lymphocytes that lack antigen-specific B- and T-cell receptors and contribute to both innate and adaptive immune responses. Three subsets of ILCs (ILC1, ILC2, ILC3) are recognized, most of which share features with helper T-cell subsets (T_H1, T_H2, T_H17, respectively).

Inositol trisphosphate (IP₃) A phosphorylated six-carbon sugar that binds to receptors in the membranes of the endoplasmic reticulum, leading to the release of Ca^{2+} ions into the cytoplasm.

Instructive model A model advanced to explain the molecular basis for **lineage commitment**, the choice of a $CD4^+CD8^+$ thymocyte to become a $CD4^+$ versus a $CD8^+$ T cell. This model proposes that DP thymocytes that interact with MHC class II receive a distinct signal from DP thymoctyes that interact with MHC class I. These distinct signals induce differentiation into the helper $CD4^+$ or the cytotoxic $CD8^+$ lineage, respectively. This model is no longer accepted. See also **Kinectic signaling model**.

Insulin-dependent diabetes mellitus (IDDM)   An autoimmune disease caused by T cell attack on the insulin-producing beta cells of the pancreas, necessitating daily insulin injections.

Integrins   A group of heterodimeric cell adhesion molecules (e.g., LFA-1, VLA-4, and Mac-1) present on various leukocytes that bind to **Ig-superfamily CAMs** (e.g., ICAMs, VCAM-1) on endothelium.

Intercellular adhesion molecules (ICAMs)   Cellular adhesion molecules that bind to integrins. ICAMs are members of the immunoglobulin superfamily.

Interferon regulatory factor 4 (IRF-4)   Transcription factor important to the initiation of plasma cell differentiation.

Interferon regulatory factors (IRFs)   Transcription factors induced by signaling downstream of pattern-recognition, interferon, and other receptors that activate interferon genes.

Interferons (IFNs)   Several glycoprotein **cytokines** produced and secreted by certain cells that induce an antiviral state in other cells and also help to regulate the immune response.

Interleukin-1 (IL-1) family   Interleukin-1 was the first cytokine to be discovered. Members of this family interact with dimeric receptors to induce responses that are typically pro-inflammatory.

Interleukins (ILs)   A group of **cytokines** secreted by leukocytes that primarily affect the growth and differentiation of various hematopoietic and immune system cells (see Appendix II).

Interstitial fluid   Fluid found in the spaces between cells of an organ or tissue.

Intraepidermal lymphocytes   T cells found in epidermal layers.

Intraepithelial lymphocytes (IELs)   T cells found in the epithelial layer of organs and the gastrointestinal tract.

Intravital fluorescence microscopy   A type of microscopy that allows one to image cell activity within living tissue and live organisms.

Invariant (Ii) chain   Component of the MHC class II protein that shows no genetic polymorphism. The Ii chain stabilizes the class II molecule before it has acquired an antigenic peptide.

Invariant NKT (iNKT) cells   A cytotoxic T cell subset that develops in the thymus and expresses very limited TCRαβ receptor diversity (one specific TCRα paired with only a few TCRβ chains) and recognize lipids associated with CD1, an MHC-like molecule.

IRAK1   See **IL-1 receptor–activated kinase**.

Isograft   Graft between genetically identical individuals.

Isolated lymphoid follicles (ILFs)   Dynamic secondary lymphoid tissue aggregates found in the wall of the small intestine that assemble and disassemble in response to antigen exposure; follicles in ILFs are key reservoirs of IgA-producing B cells.

Isotype   (1) An antibody class that is determined by the constant-region sequence of the heavy chain. The five human isotypes, designated IgA, IgD, IgE, IgG, and IgM, exhibit structural and functional differences. Also refers to the set of **isotypic determinants** that is carried by all members of a species. (2) One of the five major kinds of heavy chains in antibody molecules (α, γ, δ, ε, and μ).

Isotype switching   Conversion of one antibody class (**isotype**) to another resulting from the genetic rearrangement of heavy-chain constant-region genes in B cells; also called class switching.

Isotypic determinant   An **antigenic determinant** within the immunoglobulin constant regions that is characteristic of a species.

iT$_{REG}$ cells   A type of T cell that, following antigen exposure in the periphery, is induced to express FoxP3 and acquire regulatory functions, suppressing immune activity against specific antigen. See also **nT$_{REG}$ cells**.

IκB kinase (IKK)   The enzyme that phosphorylates the inhibitory subunit of the transcription factor **NF-κB**. Phosphorylation of **IκB** results in

its release from the transcription factor and movement of the transcription factor into the nucleus.

J (joining) chain   A polypeptide that links the heavy chains of monomeric units of polymeric IgM and di- or trimeric IgA. The linkage is by disulfide bonds between the J chain and the carboxyl-terminal cysteines of IgM or IgA heavy chains.

J (joining) gene segment   The part of a rearranged immunoglobulin or T-cell receptor gene that joins the variable region to the constant region and encodes part of the **hypervariable region**. There are multiple J gene segments in germ-line DNA, but gene rearrangement leaves only one in each functional rearranged gene.

JAK   See **Janus Activated Kinase**.

Janus Activated Kinase (JAK)   Kinase that typically transduce a signal from a Type 1 or Type 2 cytokine receptor to a cytoplasmically-located transcription factor belonging to the STAT (**Signal Transducer and Activator of Transcription**) family. On cytokine binding to the receptor, the JAK kinases are activated and phosphorylate the receptor molecule. This provides docking sites for a pair of STAT molecules which are phosphorylated, dimerize, and translocate to the nucleus to effect their transcriptional programs.

Jejunum   A section of the small intestine; the jejunum and ileum are responsible for final digestion and absorption of food.

Job syndrome   See **Hyper IgE syndrome**.

Joining (J) segment   One of the gene segments encoding the immunoglobulin heavy or light chain or any of the four TCR chains or its protein product

Junctional flexibility   The diversity in antibody and T-cell receptor genes created by the imprecise joining of coding sequences during the assembly of the rearranged genes.

K_d   See **dissociation constant**

Kappa (κ) light chain   One of the two types of immunoglobulin **light chains** that join with heavy chains to form the B cell receptor and antibody heterodimer. Lambda (λ) is the other type.

Keratinocytes   Specialized epithelial cells of the skin that contribute to a many-layered epithelial barrier; they are part of the epidermis—the top layer of the skin.

Kinetic signaling model   A model advanced to explain the molecular basis for **lineage commitment**, the choice of a CD4$^+$CD8$^+$ thymocyte to become a CD4$^+$ versus a CD8$^+$ T cell. This model proposes that all DP thymocytes receiving T cell receptor signals decrease expression of CD8. Those thymocytes whose TCR binds MHC class II will continue to receive a signal stabilized by CD4-MHC class II interactions and will progress to the CD4$^+$ lineage. However those thymocytes whose TCR binds MHC class I will have this signal disrupted by the reduction in stabilizing CD8-MHC class I interactions. These cells require rescuing by cytokines (IL-7 or IL-15) which promote their development to the CD8$^+$ lineage.

KIR   Immunoglobulin-like receptors expressed by human natural killer cells that bind MHC class I molecules and inhibit cytotoxicity.

Knock-in genetics   A genetic manipulation that results in the insertion of a desired mutant form of a gene or a marker gene in a pre-selected site in the genome.

Knockout genetics   A genetic manipulation that results in the elimination of a selected gene from the genome.

Kupffer cell   A type of tissue-fixed macrophage found in liver.

λ5   A polypeptide that associates with **Vpre-B** to form the **surrogate light chain** of the **pre-B-cell receptor**.

Lambda (λ) chain   One of the two types of immunoglobulin **light chains** that join with heavy chains to form the B cell receptor and antibody heterodimer. Kappa (κ) is the other type.

Lamina propia Layer of loose connective tissue under the intestinal epithelium where immune cells are organized. The site of the GALT and part of the mucosal immune system.

Langerhans cells Specialized, highly mobile antigen-presenting cells found specifically in the skin; Langerhans cells process antigen and travel from skin to lymph nodes, where they present antigen to naïve T cells.

Large pre-B cells The stage in B-cell development at which the BCR heavy chain first appears on the cell surface in combination with the surrogate light chain, made up of VpreB and λ5, forming the pre-BCR. Signals through the pre-BCR activate several rounds of cell division.

Laser scanning confocal microscopy Microscopy that uses lasers to focus on a single plane within the sample.

Lck A tyrosine kinase that operates early in the TCR signaling cascade. Associates non-covalently with the T cell co-receptor.

Leader (L) peptide A short hydrophobic sequence of amino acids at the N-terminus of newly synthesized immunoglobulins; it inserts into the lipid bilayer of the vesicles that transport Ig to the cell surface. The leader is removed from the ends of mature antibody molecules by proteolysis.

Lectin pathway Pathway of complement activation initiated by binding of serum protein MBL to the mannose-containing component of microbial cell walls.

Lectins Proteins that bind carbohydrates.

Leishmaniasis The disease caused by the protozoan parasite *Leishmania major*.

Lepromatous leprosy A disease caused by the intracellular bacteria *Mycobacterium leprae* and *Mycobacterium lepromatosis*, where skin, nerves, and upper respiratory tract are severely and chronically damaged by infection that is not successfully regulated by the immune response. This form of disease is associated with the production of a T_H2 rather than T_H1 response.

Leucine-rich repeats (LRRs) Protein structural domains that contain many repeats of the leucine-containing 25-amino acid repeat sequence xLxxLxLxx.; the repeats stack up on each other. Found in TLR pattern-recognition receptors and in VLRs of jawless fish.

Leukemia Cancer originating in any class of hematopoietic cell that tends to proliferate as single cells within the lymph or blood.

Leukocyte A white blood cell. The category includes lymphocytes, granulocytes, platelets, monocytes, and macrophages.

Leukocyte adhesion deficiency (LAD) An inherited immune deficiency disease in which the leukocytes are unable to undergo adhesion-dependent migration into sites of inflammation. Recurrent bacterial infections and impaired healing of wounds are characteristic of this disease.

Leukocytosis An abnormally large number of leukocytes, usually associated with acute infection. Counts greater than $10,000/mm^3$ may be considered leukocytosis.

Leukotrienes Several lipid mediators of inflammation and type I hypersensitivity, also called slow reactive substance of anaphylaxis (SRS-A). They are metabolic products of arachidonic acid.

Licensed APC An antigen-presenting cell, usually a dendritic cell, that has been activated by interactions with helper T cells or signals from pattern-recognition receptors; it expresses costimulatory receptors (e.g., CD80/86) and can carry out cross-presentation.

Ligand A molecule that binds to a receptor.

Light (L) chains Immunoglobulin polypeptides of the lambda or kappa type that join with heavy-chain polypeptides to form the antibody heterodimer.

Light zone A region of the germinal center that contains numerous **follicular dendritic cells**.

Lineage commitment The development of a cell that can give rise to multiple cell types (multipotent) into one of those cell types. (1) In T cell development, the choice a $CD4^+CD8^+$ thymocyte makes to become a helper $CD4^+$ versus cytotoxic $CD8^+$ T cell. (2) In hematopoiesis, the choice a pluripotent stem cell makes to become either myeloid or lymphoid, as well as the subsequent choices to become specific immune cell subtypes.

Lipid rafts Parts of the membrane characterized by highly ordered, detergent-insoluble, sphingolipid- and cholesterol-rich regions.

Lipopolysaccharide (LPS) An oligomer of lipid and carbohydrate that constitutes the endotoxin of gram-negative bacteria. LPS acts as a polyclonal activator of murine B cells, inducing their division and differentiation into antibody-producing plasma cells.

Locus The specific chromosomal location of a gene.

Lower respiratory tract That part of the respiratory system that includes the smaller branches of the bronchi and the alveoli.

LPS tolerance State of reduced responsiveness to LPS following an initial exposure to low/sublethal dose of LPS.

LRRs See **leucine-rich repeats**.

Lumen The inside of a tube or sac, such as the inside of the gastrointestinal system.

Ly49 Receptors in the C-lectin protein family expressed by murine natural killer cells that bind MHC class I and typically inhibit cytotoxicity.

Lymph Interstitial fluid derived from blood plasma that contains a variety of small and large molecules, lymphocytes, and some other cells. It circulates through the lymphatic vessels.

Lymph node A small **secondary lymphoid organ** that contains lymphocytes, macrophages, and dendritic cells and serves as a site for filtration of foreign antigen and for activation and proliferation of lymphocytes. See also **germinal center**.

Lymphatic system A network of vessels and nodes that conveys lymph. It returns plasma-derived interstitial fluids to the bloodstream and plays an important role in the integration of the immune system.

Lymphatic vessels Thinly walled vessels through which the fluid and cells of the lymphatic system move through the lymph nodes and ultimately into the thoracic duct, where it joins the bloodstream.

Lymphoblast A proliferating lymphocyte.

Lymphocyte A mononuclear leukocyte that mediates humoral or cell-mediated immunity. See also **B cell** and **T cell**.

Lymphoid progenitor cell A cell committed to the lymphoid lineage from which all lymphocytes arise. Also known as **common lymphoid progenitor** (CLP).

Lymphoid-primed, multipotent progenitors (LMPPs) Progenitor hematopoietic cells with the capacity to differentiate along either the lymphoid or the myeloid pathways.

Lymphoma A cancer of lymphoid cells that tends to proliferate as a solid tumor.

Lymphotoxin-α (LT-α) Also known as TNF-β, this cytokine is a member of the Tumor Necrosis Family. It is produced by activated lymphocytes and delivers a variety of signals to its target cells, including the induction of increased levels of MHC class II expression.

Lyn A tyrosine kinase important in lymphocyte signaling.

Lysosome A small cytoplasmic vesicle found in many types of cells that contains hydrolytic enzymes, which play an important role in the degradation of material ingested by **phagocytosis** and **endocytosis**.

Lysozyme An enzyme present in tears, saliva, and mucous secretions that digests mucopeptides in bacterial cell walls and thus functions as a nonspecific antibacterial agent. Lysozyme from hen egg white (HEL) has frequently been used as an experimental antigen in immunological studies.

M cells   Specialized cells of the intestinal mucosa and other sites, such as the urogenital tract, that deliver antigen from the apical face of the cell to lymphocytes clustered in the pocket of its basolateral face.

Macrophages   Mononuclear phagocytic leukocytes that play roles in adaptive and innate immunity. There are many types of macrophages; some are migratory, whereas others are fixed in tissues.

Major histocompatibility complex (MHC) proteins   Proteins encoded by the major histocompatibility complex and classified as class I, class II, and MHC class III molecules. See also **MHC**.

Malignancy, malignant   Refers to cancerous cells capable of uncontrolled growth.

Mannose-binding lectin (MBL)   A serum protein that binds to mannose in microbial cell walls and initiates the lectin pathway of complement activation.

Mantle zone   See **Follicular mantle zone.**

MAP kinase cascade   Mitogen activated protein kinase cascade. A series of reactions initiated by a cellular signal that results in successive phosphorylations of intra-cellular kinases and culminates in activation of transcription factors within the nucleus and often the initiation of cellular locomotion.

Marginal zone   A diffuse region of the spleen, situated on the periphery of the **periarteriolar lymphoid sheath (PALS)** between the **red pulp** and **white pulp**, that is rich in B cells.

Marginal zone (MZ) B cells   Noncirculating B cells expressing high IgM and low IgD found in the splenic marginal zone; they have somewhat restricted receptor diversity. They generate T-independent antibody responses similar to B-1 B cells, and respond readily to blood-borne antigens entering the spleen.

Mast cell   A bone-marrow-derived cell present in a variety of tissues that resembles peripheral blood basophils, bears **Fc receptors** for IgE, and undergoes IgE-mediated **degranulation**.

Master gene regulator   A gene at the top of a hierarchy that regulates cell development; in immunology it often refers to the genes that initiate programs responsible for the differentiation of CD4$^+$ T cells to distinct helper T-cell subsets; their induction is dependent on the activity of specific polarizing cytokines.

Mcl-1   Myeloid leukemia cell differentiation protein-1; anti-apoptotic protein that contributes to cell survival.

Medulla   The innermost or central region of an organ.

Medullary thymic epithelial cells (mTECs)   Stromal cells of epithelial origin that populate the medulla of the thymus and mediate clonal deletion of autoreactive thymocytes, particularly those expressing TCRs that recognize tissue-specific antigens.

Megakaryocytes   Hematopoietic cells in the myeloid system that give rise to platelets.

Membrane-attack complex (MAC)   The complex of complement components C5–C9, which is formed in the terminal steps of the classical, lectin, and alternative complement pathways and mediates cell lysis by creating a membrane pore in the target cell.

Membrane-bound immunoglobulin (mIg)   A form of antibody that is bound to a cell as a transmembrane protein. It acts as the antigen-specific receptor of B cells.

Memory B cell   An antigen-committed, persistent B cell. B-cell differentiation results in formation of **plasma cells**, which secrete antibody, and **memory cells**, which are involved in the **secondary responses**.

Memory cells   Lymphocytes generated following encounters with antigen that are characteristically long lived; they are more readily stimulated than naïve lymphocytes and mediate a **secondary response** to subsequent encounters with the antigen.

Memory response   See **memory, immunologic**.

Memory T cells   T cells generated during a primary immune response that become long lived and more easily stimulated by the antigen to which they are specific. They include two main subsets, central and effector memory T cells, and are among the first participants in the faster more robust secondary immune response to antigen.

Memory, immunologic   The attribute of the immune system mediated by **memory cells** whereby a second encounter with an antigen induces a heightened state of immune reactivity.

Mesenteric lymph nodes   The set of lymph nodes that drain the intestine; they can number from 100 to 200 and are situated in the membrane that connects the intestine to the wall of the abdomen (the mesentery).

Metastasis   The movement and colonization by tumor cells to sites distant from the primary site.

MHC (major histocompatibility complex) or MHC locus   A group of genes encoding cell-surface molecules that are required for antigen presentation to T cells and for rapid graft rejection. It is called the H-2 complex in the mouse and the HLA complex in humans.

MHC class I genes   The set of genes that encode MHC class I molecules, which are glycoproteins found on nearly all nucleated cells.

MHC class I molecules   Heterodimeric membrane proteins that consist of an α chain encoded in the MHC, associated noncovalently with **β$_2$-microglobulin**. They are expressed by nearly all nucleated cells and function to present antigen to CD8$^+$ T cells. The classical class I molecules are H-2 K, D, and L in mice and HLA-A, -B, and -C in humans.

MHC class II genes   The set of genes that encode MHC class II molecules, which are glycoproteins expressed by only professional antigen presenting cells.

MHC class II molecules   Heterodimeric membrane proteins that consist of a noncovalently associated α and β chain, both encoded in the MHC. They are expressed by **antigen-presenting cells** and function to present antigen to CD4$^+$ T cells. The classical class II molecules are H-2 IA and IE in mice and HLA-DP, -DQ, and -DR in humans.

MHC class III genes   The set of genes that encode several different proteins, some with immune function, including components of the complement system and several inflammatory molecules.

MHC class III molecules   Various proteins encoded in the **MHC** but distinct from MHC class I and class II molecules. Among others, they include some complement components and TNF-α and Lymphotoxin-α.

MHC restriction   The characteristic of T cells that permits them to recognize antigen only after it is processed and the resulting antigenic peptide is displayed in association with either a **MHC class I** or a **MHC class II molecule**.

MHC tetramers   A soluble cluster of four MHC-peptide complexes used as a research tool to identify and trace antigen specific T cells in vitro and in vivo.

Microfold (M) cells   Specialized epithelial cells that cover Peyer's patches and sample antigen from the intestinal lumen, delivering it to associated antigen-presenting cells in the intestinal mucosa. They are morphologically distinct, with a smooth apical surface and a pocket on the basolateral surface that allows intimate association with immune cells in the mucosa.

Microglial cell   A type of macrophage found in the central nervous system.

Microvilli   Folds in the membrane of epithelial cells that line mucosal tissues.

Minor histocompatibility loci   Genes outside of the MHC that encode antigens contributing to graft rejection.

Minor lymphocyte-stimulating (Mls) determinants   Antigenic determinants encoded by endogenous retroviruses of the murine mammary tumor virus family that are displayed on the surface of certain cells.

Missing self model   A model proposing that NK cell cytotoxicity is inhibited as long as they engage MHC class I with their receptors. However, when tumor cells and some virally infected cells reduce MHC class I expression, in other words, when they are "missing self" they are no longer protected from NK cytotoxicity.

Mitogen Activated Protein Kinase (MAPK)   The first kinase in a MAP kinase cascade.

Mitogens   Any substance that nonspecifically induces DNA synthesis and cell division. Common lymphocyte mitogens are concanavalin A, phytohemagglutinin, **lipopolysaccharide (LPS)**, pokeweed mitogen, and various **superantigens**.

Mixed-lymphocyte reaction (MLR)   In vitro T-cell proliferation in response to cells expressing allogeneic MHC molecules; can be used as an assay for MHC class II activity.

Molecular mimicry   One hypothesis used explain the induction of some autoimmune diseases, positing that some pathogens express antigenic determinants resembling host self components which can induce anti-self reactivity.

Monoclonal antibody   Homogeneous preparation of antibody molecules, produced by a single clone of B lineage cells, often a hybridoma, all of which have the same antigenic specificity.

Monoclonal   Deriving from a single clone of dividing cells.

Monocytes   A mononuclear phagocytic leukocyte that circulates briefly in the bloodstream before migrating into the tissues where it becomes a **macrophage**.

Mucin   A group of serine- and threonine-rich proteins that are heavily glycosylated. They are ligands for **selectins**.

Mucociliary boundary   A protective layer of mucus produced by goblet cells in the epithelial lining of the respiratory tract; it traps foreign material, which is swept toward the mouth by ciliated epithelial cells.

Mucosa   The layer of the intestinal wall that includes the epithelial boundary and the lamina propria; where most immune cells in the gastrointestinal tract are found.

Mucosal-associated lymphoid tissue (MALT)   Lymphoid tissue situated along the mucous membranes that line the digestive, respiratory, and urogenital tracts.

Mucosal tissues   The subset of barrier organs whose epithelial tissues are covered with a mucus layer that provides additional protection as well as moisture; includes all barrier organs except the skin. All mucosal tissues have a mucosal layer and a submucosal layer. The mucosal layer includes the epithelial cells and the lamina propria.

Multiple myeloma   A plasma-cell cancer.

Multiple sclerosis (MS)   An autoimmune disease caused by auto-reactive T cells specific for components of the myelin sheaths which surround and insulate nerve fibers in the central nervous system. In Western countries, it is the most common cause of neurologic disability caused by disease.

Multipotential   Can divide to form daughter cells that are more differentiated than the parent cell and that can develop along distinct blood-cell lineages.

Multipotent progenitors (MPPs)   The stage of lymphoid differentiate that immediately precedes the **LMPP** stage. MPPs have lost the capacity for self-renewal that characterizes the true stem cell, but they retain the capacity to differentiate along many different hematopoietic lineages, including lymphoid, myeloid, erythroid, and megakaryocytic.

Multivalent   Having more than one ligand binding site.

Mutational hot spots   DNA sequences that are particularly susceptible to somatic hypermutation. Found in the variable regions of the immunoglobulin heavy and light chains.

Myasthenia gravis   An autoimmune disease mediated by antibodies that block the acetylcholine receptors on the motor end plates of muscles, resulting in progressive weakening of the skeletal muscles.

Mycoses   Any disease caused by fungal infection.

MyD88   Myeloid differentiation factor 88; adaptor protein that binds to all the IL-1 receptors and all TLRs except TLR3 and activates downstream signaling.

Myeloid progenitor cell   A cell that gives rise to cells of the myeloid lineage. Also known as a **common myeloid progenitor** (CMP).

Myeloma   A malignant tumor arising from cells of the bone marrow, specifically B cells.

Myeloma cell   A cancerous plasma cell.

N-nucleotides, N-region nucleotides   See **Non-templated (N) nucleotides**.

NADPH oxidase enzyme complex   Phagosome oxidase, activated by pathogen binding to pattern-recognition receptors on phagocytic cells, that generates reactive oxygen species from oxygen.

Naïve   Denoting mature B and T cells that have not encountered antigen; synonymous with *unprimed* and *virgin*.

Naïve BCR repertoire   The set of B-cell receptors that exists in an individual organism prior to its interaction with antigen.

Nasal-associated lymphoid tissue (NALT)   Secondary lymphoid microenvironments in the nose that support the development of the T and B lymphocyte response to antigens that enter nasal passages. Part of the **mucosa associated lymphoid tissue system** (MALT).

Natural antibodies   Antibodies spontaneously produced by B-1 and marginal zone B cells.

Natural killer (NK) cells   A class of large, granular, cytotoxic lymphocytes that do not have T- or B-cell receptors. They are antibody-independent killers of tumor cells and also can participate in **antibody-dependent cell-mediated cytotoxicity**.

Necrosis   Morphologic changes that accompany death of individual cells or groups of cells and that release large amounts of intracellular components to the environment, leading to disruption and atrophy of tissue. See also **apoptosis**.

Negative selection   The induction of death in lymphocytes bearing receptors that react too strongly with self antigens.

Neoadjuvant cancer therapies   Often compared to adjuvant cancer therapy, this treatment strategy involves administration of drugs or small molecule inhibitors prior to surgical removal of the tumor, with the goal of targeting metastatic cells and assessing drug efficacy by monitoring changes to the primary tumor evident at surgery.

Neoantigens   Unique epitopes that arise from mutations to DNA that are associated with cancer and that generate new, non-self proteins, which may be capable of eliciting an immune response.

Neonatal Fc receptor (FcR$_N$)   An MHC class I–like molecule that controls IgG and albumin half-life and transports IgG across the placenta.

Neoplasm   Any new and abnormal growth; a benign or malignant tumor.

NETosis   Neutrophil cell death following release of neutrophil extracellular traps (NETs).

Neuraminidase (NA)   An enzyme that cleaves N-acetylneuraminic acid (sialic acid) from glycoproteins. Most commonly a virally-expressed enzyme found on the surface of influenza virus that facilitates viral attachment to and budding from host cells.

Neutralize   The ability of an antibody to prevent pathogens from infecting cells.

Neutralizing antibodies   Antibodies that bind to pathogen and prevent it from infecting cells. Our most powerful vaccines induce neutralizing antibodies.

Neutrophil   A circulating, phagocytic granulocyte involved early in the **inflammatory response**. It expresses **Fc receptors** and can participate in **antibody-dependent cell-mediated cytotoxicity**. Neutrophils are the most numerous white blood cells in the circulation.

Neutrophil extracellular traps (NETs)   Filaments of chromatin and proteins released by activated neutrophils that entrap and kill pathogens.

NF-κB   An important transcription factor, most often associated with pro-inflammatory responses.

NK cell licensing   The activation of an NK cell's competence to kill appropriate target cells; it occurs following binding of the NK cell's inhibitory receptors to MHC class I proteins, ensuring that functional NK cells have inhibitory receptors that recognize self MHC.

NKG2D   An activating receptor on NK cells that sends signals that enhance NK cytotoxic activity.

NK receptors (NKRs)   Activating and inhibitory receptors expressed by NK cells.

NKT cell   A subset of cytotoxic T cells that have features of both lymphocytes and innate immune cells. Like NK cells, they express NK1.1 and CD16, and like T cells they express TCRs, but these bind to glycolipids associated with an MHC class-I like molecule called CD1. iNKT cells are a subset of NKT cells that express very limited TCRαβ diversity.

Nod-like receptors (NLR)   Family of cytosolic pattern-recognition receptors with nuclear-binding/oligomerization and leucine-rich repeat domains; they have a nuclear-binding domain.

Non-nucleoside reverse transcriptase inhibitors (NNRTIs)   An antiretroviral drug inhibits the viral reverse transcriptase enzyme in a non-competitive fashion, by binding outside the substrate binding site and inducing a conformation change that inhibits DNA polymerization.

Non-templated (N) nucleotides   Nucleotides added to the V-D and D-J junctions of immunoglobulin and TCR genes by the enzyme **TdT**.

Nonamer   A sequence of nine nucleotides that is partially conserved and serves, with the absolutely conserved **heptamer**, as a binding site for the RAG1/2 recombination enzymes. The heptamer:nonamer sequences are found down- and up-stream of each of the V, D, and J segments encoding the immunoglobulin and T cell receptors.

Nonclassical MHC molecules   A set of mostly nonpolymorphic proteins with some structural similarity to classical MHC molecules but with more restricted tissue expression and more limited roles in the immune response, some associated with the innate response. Examples of nonclassical Class I molecules include HLA-E, -F, and -G (human) or H2-Q, -M and -T (mouse). Class II examples include HLA-DM & -DO (human) or H2-M and -O (mouse).

Nonproductive rearrangement   Rearrangement in which gene segments are joined out of phase so that the triplet-reading frame for translation is not preserved.

Northern blotting   Common technique for detecting specific mRNAs, in which denatured mRNAs are separated electrophoretically and then transferred to a polymer sheet, which is incubated with a radiolabeled DNA probe specific for the mRNA of interest.

NOS2   see **Inducible nitric oxide synthetase**.

Notch   A surface receptor that when bound is cleaved to release a transcriptional regulator that regulates cell fate decisions. Notch activation is required for T cell development and determines whether a lymphocyte precursor becomes a B versus T cell.

nT$_{REG}$ cells   A type of T cell that is induced to express FoxP3 and acquire regulatory function during development in the thymus, and is responsible for suppressing immune activity against specific antigen. See also **iT$_{REG}$ cells**.

Nuclear Factor of Activated T cells (NFAT)   A transcription factor activated by TCR ligation. NFAT is held in the cytoplasm by the presence of a bound phosphate group. Activation through the TCR results in the activation of a phosphatase that releases the bound phosphate and facilitates entry of NFAT into the nucleus.

Nucleoside reverse transcriptase inhibitors (NRTIs)   An antiretroviral drug composed of nucleoside analogues which compete with native cellular nucleosides for binding to the viral reverse transcriptase enzyme but which lead to DNA chain termination and therefore halt viral DNA synthesis.

Nucleotide oligomerization domain/leucine-rich repeat-containing receptors (NLR)   see **Nod-like receptors (NLR)**.

Nude mouse   Homozygous genetic defect (*nu/nu*) carried by an inbred mouse strain that results in the absence of the thymus and consequently a marked deficiency of T cells and cell-mediated immunity. The mice are hairless (hence the name) and can accept grafts from other species.

Oncofetal tumor antigen   An antigen that is present during fetal development but generally is not expressed in tissues except following transformation. Alpha-feto protein (AFP) and carcinoembryonic antigen (CEA) are two examples that have been associated with various cancers.

Oncogene   A gene that encodes a protein capable of inducing cellular **transformation**. Oncogenes derived from viruses are written v-onc; their counterparts (**proto-oncogenes**) in normal cells are written c-onc.

One-turn recombination signal sequences   Immunoglobulin gene-recombination signal sequences separated by an intervening sequence of 12 base pairs.

Opportunistic infections   Infections caused by ubiquitous microorganisms that cause no harm to immune competent individuals but that pose a problem in cases of **immunodeficiency**.

Opsonin   A substance (e.g., an antibody or C3b) that promotes the phagocytosis of antigens by binding to them.

Opsonization, opsonize   Deposition of opsonins on an antigen, thereby promoting a stable adhesive contact with an appropriate phagocytic cell.

Original antigenic sin   The concept that a secondary response relies on the activity of memory cells rather than the activation of naïve cells, focusing on those structures that were present during the original, or primary, encounter with a pathogen and ignoring any new antigenic determinants.

Osteoclast   A bone macrophage.

Oxidative burst   Metabolic reactions that use large amounts of oxygen to generate toxic oxygen metabolites such as hydrogen peroxide, oxygen free radicals, hypochlorous acid, and various oxides of nitrogen. These metabolites are generated within specialized vesicles in neutrophils and macrophages and are used to kill invading pathogens.

P-addition   See **P-nucleotide addition**.

P-K reaction   Prausnitz-Kustner reaction, a local skin reaction to an allergen by a normal subject at the site of injected IgE from an allergic individual. (No longer used because of risk of transmitting hepatitis or AIDS.)

P-nucleotide addition   Addition of nucleotides from cleaved hairpin loops formed by the junction of V-D or D-J gene segments during Ig or TCR gene rearrangements.

P-region nucleotides   See **Palindromic (P) nucleotides**.

Palindromic (P) nucleotides   Nucleotides formed at the V-D and D-J junctions by asymmetric clipping of the hairpin junction formed by RAG1/2-mediated DNA cleavage.

PALS   See **Periarteriolar lymphoid sheath**.

Paneth cells   Specialized cells in the epithelium of the small intestine that secrete molecules that regulate the interaction with microbes, including antimicrobial peptides (AMPs) such as defensins; Paneth cells also help maintain stem cells in the intestinal crypt.

Papain   A proteolytic enzyme derived from papayas and pineapples. Papain cleavage of immunoglobulins releases two **Fab** and one **Fc** fragment per IgG.

Paracortex   An area of the lymph node beneath the cortex that is populated mostly by T cells and interdigitating dendritic cells.

Paracrine   A type of regulatory secretion, such as a cytokine, that arrives by diffusion from a nearby cellular source.

Passive immunity   Temporary adaptive immunity conferred by the transfer of immune products, such as antibody (antiserum), from an immune individual to a nonimmune one. See also **active immunity**.

Passive immunotherapy   Treatment of an infectious disease by administration of previously generated antibodies specific for the infectious pathogen.

Pathobionts   Members of the commensal microbiome in healthy animals that are typical harmless, but have the potential to cause disease under certain conditions (e.g., if the immune system is perturbed).

Pathogen   A disease-causing infectious agent.

Pathogen-associated molecular patterns (PAMPs)   Molecular patterns common to pathogens but not occurring in mammals. PAMPs are recognized by various **pattern-recognition receptors** of the innate immune system.

Pathogenesis   The means by which disease-causing organisms attack a host.

Pattern recognition receptors (PRRs)   Receptors of the innate immune system that recognize molecular patterns or motifs present on pathogens but absent in the host.

Pattern recognition   The ability of a receptor or ligand to interact with a class of similar molecules, such as mannose-containing oligosaccharides.

PAX5 transcription factor   A quintessential B cell transcription factor that controls the expression of many B cell specific genes.

Pentraxins   A family of serum proteins consisting of five identical globular subunits; CRP is a pentraxin.

Perforin   Cytolytic product of CTLs that, in the presence of Ca^{2+}, polymerizes to form transmembrane pores in target cells.

Periarteriolar lymphoid sheath (PALS)   A collar of lymphocytes encasing small arterioles of the spleen.

Peripheral T_{REG} (pT_{REG}) cells   A class of T cell that, following antigen exposure in the periphery, is induced to express FoxP3 and acquire regulatory functions, suppressing immune activity against specific antigen. See also **Thymic T_{REG} (tT_{REG}) cells**.

Peripheral tolerance   Process by which self-reactive lymphocytes in the circulation are eliminated, rendered **anergic**, or otherwise inhibited from inducing an immune response.

Peyer's patches   Lymphoid follicles situated along the wall of the small intestine that trap antigens from the gastrointestinal tract and provide sites where B and T cells can interact with antigen.

PH domain   See **Pleckstrin homology (PH) domain**.

Phage display library   Collection of bacteriophages engineered to express specific V_H and V_L domains on their surface.

Phagocytes   Cells with the capacity to internalize and degrade microbes or particulate antigens; neutrophils and monocytes are the main phagocytes.

Phagocytosis   The cellular uptake of particulate materials by engulfment; a form of endocytosis.

Phagolysosome   An intracellular body formed by the fusion of a phagosome with a lysosome.

Phagosome   Intracellular vacuole containing ingested particulate materials; formed by the fusion of pseudopodia around a particle undergoing **phagocytosis**.

Phagosome oxidase (phox)   See **NADPH oxidase enzyme complex**.

Phosphatidyl Inositol bis-Phosphate (PIP$_2$)   A phospholipid found on the inner leaflet of cell membranes that is cleaved on cell signaling into the sugar **inositol trisphosphate** and diacylglycerol.

Phosphatidyl Inositol tris-Phosphate (PIP$_3$)   The product of **PIP$_2$** phosphorylation. PIP$_3$ is bound by signaling proteins bearing **PH domains**.

Phosphatidyl Inositol-3-kinase (PI3 kinase)   A family of enzymes that phosphorylate the inositol ring of phosphatidyl inositols at the 3 position. On immune signaling, the usual substrate is **PIP$_2$**.

Phosphatidyl serine   Membrane phospholipid. Normally located on the inner membrane leaflet, but flips to the outside leaflet in apoptotic cells.

Phospholipases Cγ (PLCγ)   A family of enzymes that cleave **phosphatidyl inositol bis-phosphate** into the sugar **inositol tris-phosphate** and the lipid **diacylglycerol**.

Physical barriers   Tissue layers, especially epithelia, whose physical integrity blocks infection.

PKCθ   A serine/threonine protein kinase activated by **diacylglycerol** and important in TCR signaling.

Plasma cell   The antibody-secreting effector cell of the B lineage.

Plasma   The cell-free, fluid portion of blood, which contains all the clotting factors.

Plasmablasts   Cells that have begun the pathway of differentiation to plasma cells, but have not yet reached the stage of terminal differentiation and are therefore still capable of cell division.

Plasmacytomas   A plasma-cell cancer.

Plasmapheresis   A procedure that involves the separation of blood into two components, plasma and cells. The plasma is removed and cells are returned to the individual. This procedure is done during pregnancy when the mother makes anti-Rh antibodies that react with the blood cells of the fetus.

Plasmin   A serine protease formed by cleavage of plasminogen. Its major function is the hydrolysis of fibrin.

Platelets   Cells in the myeloid system that arise from megakaryocytes and regulate blood clotting.

Pleckstrin homology (PH) domain   A protein domain that binds specifically to **phosphatidyl inositol tris-phosphate** or **phosphatidyl inositol bis-phosphate**. These domain interactions are important in a wide variety of signaling cascades, including those initiated by ligand binding at the TCR and BCR.

Pleiotropic   Having more than one effect. For example, a cytokine that induces both proliferation and differentiation.

Poly-Ig receptor   A receptor for polymeric Ig molecules (IgA or IgM) that is expressed on the basolateral surface of most mucosal epithelial cells. It transports polymeric Ig across epithelia.

Polarizing cytokines   Cytokines that regulate which helper subset naïve CD4$^+$ T cells will differentiate into; they are produced by immune cells as well as a variety of other cell types.

Polyclonal antibody   A mixture of antibodies produced by a variety of B-cell clones that have recognized the same antigen. Although all of the antibodies react with the immunizing antigen, they differ from each other in amino acid sequence.

Polygenic   The presence of multiple genes within the genome that encode proteins with the same function but slightly different structures.

Polymorphic, polymorphism   The presence of multiple allelic forms of a gene (**alleles**) within a population, as occurs within the major histocompatibility complex.

Positional cloning   A technique to identify a gene of interest that involves using known genetic markers to identifying a specific region in the genome associated with a characteristic of interest (e.g. disease susceptibility). This region is then cloned and sequenced to discover specific gene or genes that may be directly involved.

Positive selection   A process that permits the survival of only those T cells whose T-cell receptors recognize self MHC.

Potential candidate genes   List of genes that could be involved in the condition of interest, identified as a result research and educated guesses.

Pre-B-cell (first) checkpoint   Developing B cells are tested at the pre-B cell stage to determine whether they can express a functional BCR heavy chain protein, in combination with the **VPreB** and **λ5** proteins, to form the pre-B-cell receptor. Those B cells that fail to form a functional pre-B-cell receptor are eliminated by apoptosis and are referred to as having failed to pass through the pre-B-cell checkpoint.

Pre-B-cell receptor   A complex of the Igα,Igβ heterodimer with membrane-bound Ig consisting of the μ heavy chain bound to the **surrogate light chain** Vpre-B/λ5.

Pre-pro-B cell   Earliest stage in B-lineage development; initial expression of the B-cell marker B220.

Pre-T-cell receptor (pre-TCR)   A complex of the CD3 group with a structure consisting of the T-cell receptor β chain complexed with a 33-kDa glycoprotein called the pre-Tα chain.

Pre-B cell (precursor B cell)   The stage of B-cell development that follows the pro-B-cell stage. Pre-B cells produce cytoplasmic μ heavy chains and most display the **pre-B-cell receptor**.

pre-Tα chain   An invariant protein homolog of the TCRα chain that is expressed in early T cell development and pairs with newly rearranged TCRβ to form the pre-T cell receptor. This receptor delivers a signal that induces proliferation and differentiation to the CD4$^+$CD8$^+$ stage, when the TCRα chain is rearranged and if successfully translated into a protein takes the place of the pre-Tα.

Precipitin   An antibody that aggregates a soluble antigen, forming a macromolecular complex that yields a visible precipitate.

Primary foci   Clusters of antigen-stimulated dividing B cells found at the borders of the T and B cell areas of the lymph nodes and spleen that secrete IgM quickly after antigen stimulation.

Primary follicle   A lymphoid follicle, prior to stimulation with antigen, that contains a network of **follicular dendritic cells** and small resting B cells.

Primary immunodeficiency   An inherited genetic or developmental defect in some component/s of the immune system.

Primary lymphoid organs   Organs in which lymphocyte precursors mature into antigenically committed, immunocompetent cells. In mammals, the bone marrow and thymus are the primary lymphoid organs in which B-cell and T-cell maturation occur, respectively.

Primary response   Immune response following initial exposure to antigen; this response is characterized by short duration and low magnitude compared to the response following subsequent exposures to the same antigen (**secondary response**).

Pro–B cell (progenitor B cell)   The earliest distinct cell of the B-cell lineage.

Productive rearrangement   The joining of V(D)J gene segments in phase to produce a VJ or V(D)J unit that can be translated in its entirety.

Professional antigen-presenting cell (pAPC)   A myeloid cell with the capacity to activate T lymphocytes. Dendritic cells, macrophages, and B cells are all considered professional APCs because they can present antigenic peptide in both MHC class I and class II and express costimulatory ligands necessary for T cell stimulation.

Progenitor cell   A cell that has lost the capacity for self renewal and is committed to the generation of a particular cell lineage.

Programmed death-1 (PD-1 or CD279)   A coinhibitory receptor that blocks T-cell activation, acting as a checkpoint that down-regulates immune responses; up-regulated by T cells after activation.

Proinflammatory   Tending to cause inflammation; TNF-α, IL-6 and IL-1 are examples of proinflammatory cytokines.

Properdin   Component of the alternative pathway of complement activation that stabilizes the alternative pathway **C3 convertase** C3bBb.

Prostaglandins   A group of biologically active lipid derivatives of arachidonic acid. They mediate the **inflammatory response** and type I hypersensitivity reaction by inhibiting platelet aggregation, increasing vascular permeability, and inducing smooth-muscle contraction.

Protease inhibitors   The common name for a class of antiviral medications that inhibit virus-specific proteases and therefore interfere with viral replication. Most often associated with anti-HIV drug therapy.

Proteasome   A large multifunctional protease complex responsible for degradation of intracellular proteins.

Protectin   See **CD59.**

Protein A   An F$_C$-binding protein present on the membrane of *Staphylococcus aureus* bacteria. It is used in immunology for the detection of antigen-antibody reactions and for the purification of antibodies.

Protein A/G   A genetically engineered F$_C$-binding protein that is a hybrid of protein A and protein G. It is used in immunology for the detection of antigen-antibody reactions and for the purification of antibodies.

Protein G   An F$_C$-binding protein present on the membrane of Streptococcus bacteria. It is used in immunology for the detection of antigen-antibody reactions and for the purification of antibodies.

Protein kinase C (PKC)   A family of serine/threonine protein kinases activated by **diacylglycerol.**

Protein scaffold   A set of proteins that binds together in a defined way to bring together molecules that would otherwise not come into contact. Protein scaffolds usually include a number of **adapter proteins** that then serve to bring enzyme and substrate proteins into apposition with the resultant activation or inhibition of the substrate protein.

Proto-oncogene   A cancer-associated gene that encodes a factor that normally regulates cell proliferation, survival, or death; these genes are required for normal cellular functions. When mutated or produced in inappropriate amounts, a proto-oncogene becomes an oncogene, which can cause transformation of the cell.

Provirus   Viral DNA that is integrated into a host-cell genome in a latent state and must undergo activation before it is transcribed, leading to the formation of viral particles.

Pseudogene   Nucleotide sequence that is a stable component of the genome but is incapable of being expressed. Pseudogenes are thought to have been derived by mutation of ancestral active genes.

Pseudopodia   Membrane protrusions that extend from motile and phagocytosing cells.

Psoriasin   A protein produced by keratinocytes that has potent antimicrobial properties, particularly against *Escherichia coli*.

Purine box factor 1 (PU.1)   A transcription factor important in B cell development and survival.

Radioimmunoassay (RIA)   A highly sensitive technique for measuring antigen or antibody that involves competitive binding of radiolabeled antigen or antibody.

RAG1/2 (recombination-activating genes 1 and 2)   The protein complex of RAG1 and RAG2 that catalyzes V(D)J recombination of B- and T-cell receptor genes. These proteins operate in association with a number of other enzymes to bring about the process of recombination, but RAG1/2, along with **TdT**, represent the lymphoid-specific components of the overall enzyme complex.

RAG1 (Recombination Activating Gene 1)   See **RAG1/2**.

RAG2 (Recombination Activating Gene 2)   See **RAG1/2**.

Ras   Small, monomeric G protein important in signaling cascades.

Reactive nitrogen species   Highly cytotoxic antimicrobial compounds formed by the combination of nitric oxide and superoxide anion within phagocytes such as neutrophils and macrophages.

Reactive oxygen species (ROS)   Highly reactive compounds such as superoxide anion $\cdot O_2^-$, hydroxyl radicals $(OH\cdot)(OH^-)$, hydrogen peroxide (H_2O_2), and hypochlorous acid $(HClO)$ that are formed from oxygen under many conditions in cells and tissues, including microbe-activated innate responses of phagocytic cells; have anti-microbial activity.

Recent thymic emigrants (RTEs)   Newly developed single positive ($CD4^+$ or $CD8^+$) thymocytes that been allowed to leave the thymus. These new T cells still bear features of immature T cells and undergo further maturation in the periphery before joining the circulating fully mature, naïve T cell pool.

Receptor   A molecule that specifically binds a **ligand**.

Receptor editing   Process by which the T- or B-cell receptor sequence is altered after the initial recombination event, in order to reduce affinity for self antigens.

Recombinant inbred strains   Mouse strains created by the mating of two inbred strains, with the formation of a recombination in an interesting locus such as the MHC. The mice bearing the recombination are then inbred to create a new inbred strain.

Recombination signal sequences (RSSs)   Highly conserved heptamer and nonamer nucleotide sequences that serve as signals for the gene rearrangement process and flank each germ-line V, D, and J segment.

Recombination-activating genes   See **RAG1/2**.

Red pulp   Portion of the spleen consisting of a network of sinusoids populated by macrophages and erythrocytes. It is the site where old and defective red blood cells are destroyed.

Redundant   Having the same effect as another signal.

Regulated cell death   Activated or induced cell death, including apoptosis, necroptosis, pyroptosis, and NETosis.

Regulatory T cell (T_{REG})   A type of $CD4^+$ T cell that that negatively regulates immune responses. It is defined by expression of the master regulator FoxP3 and comes in two versions, the induced T_{REG}, which develops from mature T cells in the periphery, and the natural T_{REG}, which develops from immature T cells in the thymus.

Relative risk   Probability that an individual with a given trait (usually, but not exclusively, a genetic trait) will acquire a disease compared with those in the same population group who lack that trait.

Resident memory T cells (T_{RM})   A class of memory T cell that homes to and spends long periods of time in mucosal tissues and skin, where they participate with innate cells in the first line of defense against reinfection.

Resonant angle   A property measured when using surface plasmon resonance to assess the affinity of the interaction between two molecules.

Respiratory burst   A metabolic process in activated phagocytes in which the rapid uptake of oxygen is used to produce reactive oxygen species that are toxic to ingested microorganisms.

Reticular dysgenesis (RD)   A type of **severe combined immunodeficiency** (SCID) in which the initial stages of hematopoietic cell development are blocked by defects in the adenylate kinase 2 gene (*AK2*), favoring apoptosis of myeloid and lymphoid precursors and resulting in severe reductions in circulating leukocytes.

Retinoic acid-inducible gene-I-like receptors (RLRs)   A family of cytosolic pattern-recognition receptors named for one member, RIG-I; known members are RNA helicases with CARD domains.

Retrovirus   A type of RNA virus that uses a reverse transcriptase to produce a DNA copy of its RNA genome. HIV, which causes AIDS, and HTLV, which causes adult T-cell leukemia, are both retroviruses.

Reverse transcriptase (RT)   The polymerase of retroviruses, such as HIV, which copies the virus's RNA genome into a complementary DNA copy.

Rh antigen   Any of a large number of antigens present on the surface of blood cells that constitute the Rh blood group. See also **erythroblastosis fetalis**.

Rheumatoid arthritis   A common autoimmune disorder, primarily diagnosed in women 40 to 60 years old, caused by self-reactive antibodies called **rheumatoid factors**, which mediate chronic inflammation of the joints.

Rheumatoid factors   Auto-antibodies found in the serum of individuals with rheumatoid arthritis and other connective-tissue diseases.

Rhogam   Antibody against **Rh antigen** that is used to prevent **erythroblastosis fetalis**.

Rous sarcoma virus (RSV)   A retrovirus that induces tumors in avian species.

Runx1   Transcription factor in hematopoiesis that contributes to the activation of the EBF1 transcription factor during B-cell development.

Sarcoma   Tumor of supporting or connective tissue.

Schistosomiasis   A disease caused by the parasitic worm *Schistosoma*.

SCID   See **Severe combined immunodeficiency**.

SCID-human mouse   Immunodeficient mouse into which elements of a human immune system, such as bone marrow and thymic fragments, have been grafted. Such mice support the differentiation of pluripotent human hematopoietic stem cells into mature immunocytes and so are valuable for studies on lymphocyte development. See also **Severe combined immunodeficiency**.

SDS-polyacrylamide gel electrophoresis (SDS-PAGE)   An electrophoretic method for the separation of proteins. It employs SDS to denature proteins and give them negative charges; when SDS-denatured proteins are electrophoresed through polymerized-acrylamide gels, they separate according to their molecular weights.

Secondary follicle   A **primary follicle** after antigenic stimulation; it develops into a ring of concentrically packed B cells surrounding a **germinal center**.

Secondary immune response   The immune response to an antigen that has been previously introduced and recognized by adaptive immune cells. It is mediated primarily by memory T and B lymphocytes that have differentiated to respond more quickly and robustly to antigenic stimulation than the **primary response**.

Secondary immunodeficiency   Loss of immune function that results from exposure to an external agent, often an infection.

Secondary lymphoid organs (SLOs)   Organs and tissues in which mature, immunocompetent lymphocytes encounter trapped antigens and are activated into **effector cells**. In mammals, the **lymph nodes**, **spleen**, and **mucosal-associated lymphoid tissue (MALT)** constitute the secondary lymphoid organs.

Secondary response   See **Secondary immune response**.

Secreted immunoglobulin (sIg)   The form of antibody that is secreted by cells of the B lineage, especially plasma cells. This form of Ig lacks a transmembrane domain. See also **Membrane immunoglobulin (mIg)**.

Secretory component   A fragment of the **poly-Ig receptor** that remains bound to Ig after transcytosis across an epithelium and cleavage.

Secretory IgA   J chain–linked dimers or higher polymers of IgA that have transited epithelia and retain a bound remnant of the **poly-Ig receptor**.

Selectin   One of a group of monomeric **cell adhesion molecules** present on leukocytes (L-selectin) and endothelium (E- and P-selectin) that bind to mucin-like CAMs (e.g., GlyCAM, PSGL-1).

Self-MHC restriction   The property of recognizing antigenic peptides only in the context of self MHC molecules.

Self-renewing   Can divide to create identical copies of the parent cell.

Self-tolerance   Unresponsiveness to self antigens.

Sepsis   Infection of the bloodstream, frequently fatal.

Septic shock   Shock induced by septicemia.

Septicemia   Blood poisoning due to the presence of bacteria and/or their toxins in the blood.

Seroconversion   The stage of HIV infection at which anti-HIV antibodies first are detected in the serum.

Serum   Fluid portion of the blood, which is free of cells and clotting factors.

Serum sickness   A type III hypersensitivity reaction that develops when antigen is administered intravenously, resulting in the formation of large amounts of antigen-antibody complexes and their deposition in tissues. It often develops when individuals are immunized with antiserum derived from other species.

Severe combined immunodeficiency (SCID)   A genetic defect in which adaptive immune responses do not occur, due to a lack of T cells and possibly B and NK cells.

SH2 domain   A protein domain that binds phosphorylated tyrosine residues.

SH3 domain   A protein domain that binds proline-rich peptides.

SIGN-R1   A cell-surface receptor for the complement component C1q.

Signal   A molecule that elicits a respond in a cell bearing a **receptor** for that signal. The response may be cell movement, cell division, activation of cell metabolism, or even cell death.

Signal joints   In V(D)J gene rearrangement, the nucleotide sequences formed by the union of recombination signal sequences.

Signal peptide   A small sequence of amino acids, also called the leader sequence, that guides the heavy or light chain through the endoplasmic reticulum and is cleaved from the nascent chains before assembly of the finished immunoglobulin molecule.

Signal Transducer and Activator of Transcription (STAT)   Transcription factors that normal reside in the cytoplasm. Phosphorylation of cytokine receptors by **Janus Activated Kinases** results in the generation of binding sites on those receptors for the STATs, which relocate to the cytoplasmic regions of the receptor and are themselves phosphorylated by JAKs. Phosphorylated STATs them dimerize. Phosphorylation and dimerization expose nuclear localization signals and the STATs move to the nucleus where they act as transcription factors.

Signal transduction   The process by which a molecular signal is passed from one part of a cell to another, usually by the binding of a signaling molecule to a receptor at the cell surface, followed by a conformational change in the receptor that causes a cascade of intracellular events, culminating in an alteration in transcription of one or more genes.

Signal transduction pathway   A sequence of intra-cellular events brought about by receipt of a **signal** by a specific receptor.

Signaling   Intracellular communication initiated by receptor-ligand interaction.

Single nucleotide polymorphisms (SNPs)   A single base-pair variation in a DNA sequence among individuals in a population (i.e. an allelic variation). SNPs are often found in non-coding regions of the genome and can be used as genetic markers to locate genes that may underlie differences in individual responses to disease.

Single-positive (SP) stage   Either $CD4^+CD8^-$ or $CD8^+CD4^-$ mature T lymphocytes.

Skin-associated lymphoid tissue (SALT)   The immune cells and immune system associated specifically with the skin; includes Langerhans cells, a skin-specific antigen-presenting cell.

Slow-reacting substance of anaphylaxis (SRS-A)   The collective term applied to leukotrienes that mediate inflammation.

Small pre-B-cell   The stage in B-cell development, following the large pre-B-cell stage, where the pre-B-cell receptor is lost from the B-cell surface and light-chain recombination begins in the genome.

Somatic hypermutation (SHM)   The induced increase of mutation, 10^3- to 10^6-fold over the background rate, in the regions in and around rearranged immunoglobulin genes. In animals such as humans and mice, somatic hypermutation occurs in **germinal centers**.

$S1P_2$-type receptor   A receptor for sphingosine 1-phosphate that acts as a chemokine in the lymph nodes.

Specificity, antigenic   Capacity of antibody and T-cell receptor to recognize and interact with a single, unique **antigenic determinant** or **epitope**.

Sphingosine 1-phosphate (S1P)   A sphingolipid that plays a role in immune cell trafficking. Interactions between S1P and its membrane receptors, including S1P1, are required for immune cell egress from lymphoid tissue, including lymph nodes and thymus.

Spleen   Secondary lymphoid organ where old erythrocytes are destroyed and blood-borne antigens are trapped and presented to lymphocytes in the **PALS** and **marginal zone**.

Splenectomy   Surgical removal of the spleen.

Splenic artery   The blood vessel that provides the spleen with nutrients and oxygen; it is the source of immune cells that populate the splenic environment.

Splenic vein   Vein that drains blood from the spleen and the site of egress of many splenic white blood cells.

Spliceosome   The complex of enzymes that mediates splicing of RNA.

Src-family kinases   A family of tyrosine kinase critically important in the early stages of signaling pathways in many cell types, including lymphocytes.

STAT5   See **Signal Transducer and Activator of Transcription**.

Stem cell   A cell from which differentiated cells derive. Stem cells are classified as totipotent, pluripotent, multipotent, or unipotent depending on the range of cell types that they can generate.

Stem cell associated antigen-1 (Sca-1)   An antigen present on hematopoietic stem cells.

Stem cell factor (SCF)   A cytokine that exists in both membrane-bound and soluble forms. The SCF/c-Kit interaction is critical for the development, in adult animals, of **multipotential progenitor cells (MPPs)**.

Stem-cell memory T cells (T_{SCM})   A class of memory T cell that has self-renewing properties and, when stimulated, can give rise to all effector cell populations, as well as more stem cells.

Stem cell niches   Cellular microenvironments in the bone marrow (and some other tissues) that support the development of hematopoietic stem cells. Two are recognized: the **endosteal niche**, in proximity to bone cells (osteoblasts) and the **vascular niche**, in proximity to cells that line the blood vessels (endothelial cells).

Sterilizing immunity   A memory response or immune status that blocks the ability of a potential pathogen to infect an individual during exposure to the infectious agent in question.

Stochastic model   A model advanced to explain the molecular basis for **lineage commitment**, the choice of a CD4$^+$CD8$^+$ thymocyte to become a CD4$^+$ versus a CD8$^+$ T cell. This model proposes that all DP thymocytes receiving T cell receptor signals randomly decrease expression of either co-receptor CD4 or CD8. Only those thymocytes expressing the co-receptor that stabilizes the TCR-MHC interaction they experience will mature successfully. For example, a DP thymocyte that expresses a TCR with an affinity for MHC class I but randomly down-regulates CD8 will not receive adequate stimulation and will not mature. However, if that thymocyte randomly down-regulates CD4, it will maintain a TCR signal and mature to the CD8 lineage.

Strength of signal model   A model for thymocyte lineage commitment proposing that strong TCR signals favor development to the CD4 lineage and weak TCR signals favor development to the CD8 lineage.

Streptavidin   A bacterial protein that binds to biotin with very high affinity. It is used in immunological assays to detect antibodies that have been labeled with biotin.

Stromal cell   A nonhematopoietic cell that supports the growth and differentiation of hematopoietic cells.

Sub-isotypes   A particular antibody subclass, e.g., IgG1 or IgA2.

Subcapsular sinus (SCS) macrophages   Macrophages that line the subcapsular sinus of the lymph nodes.

Subclasses   Variant sequences of the constant regions of antibodies of the IgG and IgA classes. There are four common variants of IgG and two of IgA in both mice and humans.

Submucosa   The layer of connective tissue in the intestinal wall just beneath the mucosa; the site of a network of blood vessels, lymphatics, and nerves.

Superantigens   Any substance that binds to the V_β domain of the T-cell receptor and residues in the chain of MHC class II molecules. It induces activation of all T cells that express T-cell receptors with a particular V_β domain. It functions as a potent T-cell **mitogen** and may cause food poisoning and other disorders.

Surface plasmon resonance (SPR)   An instrumental technique for measurement of the affinity of molecular interactions based on changes of reflectance properties of a sensor coated with an interactive molecule.

Surfactants   In innate immunity, collectins that associate with microbe surfaces and promote phagocytosis.

Surrogate light chain   The polypeptides **Vpre-B** and **λ5** that associate with μ heavy chains during the pre-B-cell stage of B-cell development to form the **pre-B-cell receptor**.

SWI/SNF   A chromatin-modifying complex; it remodels nucleosomes by destabilizing the interactions of histones with DNA, providing access to transcription factors and facilitating gene expression.

Switch (S) regions   In class switching, DNA sequences located upstream of each C_H segment (except C_δ).

Synergy   The property of two separate signals having an effect that is greater than the sum of the two signals.

Syngeneic   Denoting genetically identical individuals.

Systemic lupus erythematosus (SLE)   A multi-system autoimmune disease characterized by auto-antibodies to a vast array of tissue antigens such as DNA, histones, RBCs, platelets, leukocytes, and clotting factors.

T cell   See **T lymphocyte**.

T cytotoxic (T_C) cells   See **cytotoxic T lymphocytes**.

T helper (T_H) cells   T cells that are stimulated by antigen to provide signals that promote immune responses.

T helper type 1 (T_H1) cells   Helper CD4$^+$ T cell subset that enhances the cytotoxic immune response against intracellular pathogens and expresses the master transcriptional regulator T-Bet.

T helper type 17 (T_H17) cells   Helper CD4$^+$ T cell subset that enhances an inflammatory immune response against some fungi and bacteria and expresses the master transcriptional regulator RORγ.

T helper type 2 (T_H2) cells   Helper CD4$^+$ T cell subset that enhances B cell production of IgE and the immune response to pathogenic worms. It expresses the master transcriptional regulator GATA-3.

T lymphocyte   A lymphocyte that matures in the thymus and expresses a T-cell receptor, CD3, and CD4 or CD8.

T-cell hybridoma   An artificially generated T cell tumor formed by fusing a single T cell with a naturally-occurring thymoma cell.

T-cell receptor (TCR)   Antigen-binding molecule expressed on the surface of T cells and associated with the **CD3** molecule. TCRs are heterodimeric, consisting of either an α and β chain or a γ and δ chain.

T-dependent (TD) response   An antibody response elicited by a T-dependent antigen.

T-helper type 9 (T$_H$9) cells   A helper T-cell subset that generates the effector cytokine IL-9 and regulates the response to worms; they also contribute to allergy and asthma symptoms.

T-helper type 22 (T$_H$22) cells   A helper T-cell subset that generates the effector cytokines IL-22 and TNF-a; they play a major role in skin immune responses and may regulate wound healing as well as contribute to chronic skin inflammatory conditions.

T-independent (TI) response   An antibody response elicited by a T-dependent antigen.

TAB1   TAK1-binding protein 1.

TAB2   TAK1-binding protein 2.

TACI receptor   Transmembrane activator and CAML interactor.

TAK1   See **Transforming growth factor-beta-kinase-1**.

TAP (transporter associated with antigen processing)   Heterodimeric protein present in the membrane of the rough endoplasmic reticulum (RER) that transports peptides into the lumen of the RER, where they bind to MHC class I molecules.

Target cell   Any self cell expressing MHC-peptide that is recognizable by a cytotoxic T cell (whether due to viral infection, cancer or defective), that becomes a target for recognition and lysis by Tc cells.

Tapasin   TAP-associated protein. Found in the ER that brings the TAP transporter into proximity with the MHC class I molecule, allowing it to associate with antigenic peptide.

TATA box   A Thymidine and Adenine-rich sequence upstream of many genes that serves as part of the RNA polymerase recognition and binding sequence.

T$_C$1 cells   Cytotoxic T cells that produce IFN-γ and induce target cell apoptosis by perforin and FasL-mediated pathways.

T$_C$2 cells   Cytotoxic T cells that produce IL-4 and IL-5, and induce target cell apoptosis by the perforin pathway.

TD antigen   See **Thymus-dependent (TD) antigen**.

Terminal deoxynucleotidyl transferase (TdT)   Enzyme that adds untemplated nucleotides at the V-D and D-J junctions of B and T cell receptor genes.

Tertiary lymphoid tissue   Aggregates of organized immune cells in organs that were the original site of infection.

T$_H$1 subset   See **T helper type 1 (T$_H$1) cells**.

T$_H$2 subset   See **T helper type 2 (T$_H$2) cells**.

Thoracic duct   The largest of the lymphatic vessels. It returns lymph to the circulation by emptying into the left subclavian vein near the heart.

Thromboxane   Lipid inflammatory mediator derived from arachidonic acid.

Thymic cortex   The outer region of the thymic organ that is just under the capsule. This is populated by cortical thymic epithelial cells and immature double-positive (CD4$^+$CD8$^+$) thymocytes. It is the major site of positive selection.

Thymic medulla   The inner region of the thymic organ that is populated by medullary thymic epithelial cells and mature, single-positive (CD4$^+$ or CD8$^+$) thymocytes. It is the major site of negative selection against tissue-specific antigens.

Thymic T$_{REG}$ (tT$_{REG}$) cells   A T-cell subset that is induced to express FoxP3 and acquires regulatory function during development in the thymus, and is responsible for suppressing immune activity against specific antigen. See also **Peripheral T$_{REG}$ (pT$_{REG}$) cells**.

Thymocytes   Developing T cells present in the thymus.

Thymus   A primary lymphoid organ, in the thoracic cavity, where T-cell maturation takes place.

Thymus-dependent (TD) antigen   Antigen that is unable to stimulate B cells to antibody production in the absence of help from T cells; response to such antigens involves isotype switching, **affinity maturation**, and memory-cell production.

T-independent (TI) antigens   Antigens capable of eliciting a response from B cells in the absence of help from T cells.

TI-1 antigens   T-Independent antigens, Type 1. Antigens that are capable of eliciting an antibody response in the absence of T cell help. TI-1 antigens are capable of being mitogenic.

TI-2 antigens   T-Independent antigens, Type 2. Antigens that are capable of eliciting an antibody response in the absence of T cell help. TI-2 antigens are not capable of being mitogenic.

TIR domain   Toll-IL-1R domain. Cytoplasmic domain of the IL-1 receptor and TLRs. Also present in the adaptors **MyD88** and **TRIF**, which interact with the receptors via TIR/TIR domain interactions.

Tissue typing   Refers to the determination of which MHC alleles are expressed or carried by an individual. The test for this (the "type" part of blood "type and matching") can be either serologic, where the surface MHC allotypes are determined, or molecular, where probes are used to determine which MHC alleles are present in an individual. This assay is used to establish histocompatibility prior to transplantation of foreign tissues.

Titer   A measure of the relative strength of an antiserum. The titer is the reciprocal of the last dilution of an antiserum capable of mediating some measurable effect such as precipitation or agglutination.

TNF-α   Tumor Necrosis Factor α. A pro-inflammatory cytokine.

Tolerance   A state of immunologic unresponsiveness to particular antigens or sets of antigens. Typically, an organism is unresponsive or tolerant to self antigens.

Tolerogenic   A response that down-regulates immune activity.

Tolerogens   Antigens that induce tolerance rather than immune reactivity. See also **immunogen**.

Toll-like receptors (TLRs)   A family of cell-surface receptors found in invertebrates and vertebrates that recognize conserved molecules from many pathogens.

Toxoid   A toxin that has been altered to eliminate its ability to cause disease but that still can function as an **immunogen** in a vaccine preparation.

Trabeculae   Extensions of connective tissue (in this case, cartilage and bone) that provide a structural site and surface for the development of blood cells. Found at the ends of long bones (femur, tibia, humerus) as well as the edges of other bones.

TRAF6   TNF receptor-associated factor 6. Key signaling intermediate in IL-1R and TLR pathways. Activated by IRAK kinases and essential for activating downstream components.

Trafficking   The differential migration of lymphoid cells to and from different tissues.

Transcytosis   The movement of antibody molecules (polymeric IgA or IgM) across epithelial layers mediated by the poly-Ig receptor.

Transduce   To pass on a signal from one part of the cell or one signaling molecule to the next in a signaling cascade.

Transformation   Change that a normal cell undergoes as it becomes malignant, normally mediated by DNA alternations; also a permanent, heritable alteration in a cell resulting from the uptake and incorporation of foreign DNA into the genome.

Transforming growth factor-beta-kinase-1 (TAK1)   Protein kinase downstream of IL-1R, TLRs, and other pattern-recognition receptors; part of complex with TAB1, TAB2 that is activated by TRAF6 and phosphorylates the IKK complex.

Transfusion reaction   Type II hypersensitivity reaction to proteins or glycoproteins on the membrane of transfused red blood cells.

Transgene   A cloned foreign gene present in an animal or plant.

Transitional B cells (T1, T2)   Immature B cells that express the BCR but are not yet fully immunologically competent.

Transplantation   The process of moving an organ, tissue, or group of cells from one individual (donor) to another (recipient). The resulting tissue is sometimes called a "graft."

TRIF   TIR-domain-containing adaptor protein. Adaptor recruited by TLR3 and TLR4 that mediates signaling pathway activating IRFs and interferon production.

T3 transitional B cells   B-cell stages in the spleen between bone marrow immature B cells and mature peripheral B cells. T1 cells express IgM and no/low levels of IgD and give rise to T2 cells, which express intermediate levels of IgD. Some enter follicles, where they give rise to mature B-2 cells; others enter the marginal zone and become MZ B cells. T3 cells express reduced levels of IgM and are anergic to self antigens.

Tuberculoid leprosy   A disease caused by the intracellular bacteria, *Mycobacterium leprae* and *Mycobacterium lepromatosis*, where the immune response successfully encapsulates the bacteria and prevents widespread tissue damage. This more controlled form of disease is associated with the production of a T_H1 rather than T_H2 response.

Tuft cells   Specialized sensory cells in the epithelium of mucosal tissues that play a role in the defense against worms; tuft cells are the sole source of IL-25.

Tumor necrosis factor (TNF) family   A family of cytokines whose members interact with trimeric cell-surface receptors and deliver a variety of signals to receptor-bearing cells, ranging from the induction of apoptosis to the elicitation of differentiation and cell division. They may be either soluble or membrane-bound.

Tumor-associated antigens (TSAs)   Antigens that are expressed on particular tumors or types of tumors that are not unique to tumor cells. They are generally absent from or expressed at low levels on most normal adult cells. Formerly referred to as tumor-associated transplantation antigens (TATAs).

Tumor-specific antigens (TSAs)   Antigens that are unique to tumor cells. Originally referred to as tumor-specific transplantation antigens (TSTAs).

Tumor-suppressor genes   Genes that encode products that inhibit excessive cell proliferation or survival. Mutations in these genes are associated with the induction of malignancy.

Two-photon microscopy   A type of microscopy that, like confocal microscopy, allows one to visualize fluorescent signals in one focal plane of a relatively thick tissue sample. It causes less damage than confocal microscopy and can be coupled with intravital imaging techniques.

Two-signal hypothesis   The proposal, now well established experimentally, that in order to be activated a naïve T cell must receive signals through its TCR (signal 1) as well as costimulatory receptors such as CD28 (signal 2).

Two-turn recombination signal sequences   Immunoglobulin gene recombination signal sequences separated by an intervening sequence of 23 base pairs.

Tyk   A kinase that belongs to the **JAK** family of kinases.

Type 1 diabetes (T1D)   An autoimmune disease caused by T-cell attack on the insulin-producing beta cells of the pancreas, necessitating daily insulin injections.

Type 1 responses   Immune responses to intracellular pathogens; they are typically dominated by T_H1 and ILC1 lymphocytes, activate cytotoxic cells, and heighten phagocytic activity.

Type 2 responses   Immune responses to worms and some other extracellular pathogens; they are typically dominated by T_H2 and ILC2 lymphocytes, as well as IgE and the cytokines IL-4, IL-5, and IL-13.

Type I hypersensitivity   A pathologic immune reaction to non-infectious antigens mediated by IgE. It is the basis for allergy and atopy.

Type I interferons   A group of cytokines belonging to the Interferon family of cytokines that mediates anti-viral effects. Type I interferons are released by many different cell types and are considered part of the innate immune system.

Type II hypersensitivity   A pathologic immune reaction to non-infectious antigens mediated by IgG and IgM, which recruit complement or cytotoxic cells. It underlies blood transfusion reactions, Rh factor responses, and some hemolytic anemias.

Type II interferon   A cytokine belonging to the interferon family that is normally secreted by activated T cells. Also known as Interferon γ.

Type III hypersensitivity   A pathologic immune reaction to non-infectious antigens mediated by antibody-antigen immune complexes. It underlies damage associated with several disorders, including rheumatoid arthritis and systemic lupus erythematosus.

Type III interferons   A subset of the Interferon family of cytokines, also known as lambda interferons (IFN-λ).

Type IV hypersensitivity   A pathologic immune reaction to non-infectious antigens mediated by T cells. It underlies the response to poison ivy.

Ubiquitin   A small signaling peptide that can either tag a protein for destruction by the proteasome, or, under some circumstances, activate that protein.

Ulcerative colitis   A form of inflammatory bowel disease (IBD) limited to the large intestine and characterized by an inappropriate type 2 response, accumulation of inflammatory cells, and damage to the epithelial layer.

Unproductive   An unproductive BCR or TCR gene rearrangement is one in which DNA recombination leads to a sequence containing a stop codon, that cannot therefore be transcribed and translated to form a functional protein.

Upper respiratory tract   That part of the respiratory system that includes the mouth, trachea, and major bronchi.

Upstream   (1) Towards the 5′ end of a gene; (2) Closer to the receptor in a signaling cascade.

V (variable) gene segment   The 5′ coding portion of rearranged immunoglobulin and T-cell receptor genes. There are multiple V gene segments in germ-line DNA, but gene rearrangement leaves only one segment in each functional gene.

V (D)J recombinase   The set of enzymatic activities that collectively bring about the joining of gene segments into a rearranged V(D)J unit.

Vaccination   Intentional administration of a harmless or less harmful form of a pathogen in order to induce a specific adaptive immune response that protects the individual against later exposure to the pathogen.

Vaccine   A preparation of immunogenic material used to induce immunity against pathogenic organisms.

Valence, valency Numerical measure of combining capacity, generally equal to the number of binding sites. Antibody molecules are bivalent or multivalent, whereas T-cell receptors are univalent.

Variability (Antibody) variability is defined by the number of different amino acids at a given position divided by the frequency of the most common amino acid at that position.

Variable (V) region Amino-terminal portions of immunoglobulin and T-cell receptor chains that are highly variable and responsible for the **antigenic specificity** of these molecules.

Variable (V_L) The variable region of an antibody **light chain**.

Vascular addressins Tissue-specific adhesion molecules that direct the extravasation of different populations of circulating lymphocytes into particular lymphoid organs.

Vascular niche Microenvironment in the bone marrow that fosters the development of hematopoietic stem cells and is postulated to associate specifically with hematopoietic stem cells that have begun to differentiate into mature blood cells.

Vector-borne infections Infections that are transmitted to their hosts via an intermediate vector species, such as a mosquito, tick, or fly.

Villi Folds in the wall of the intestinal tract that enhance the surface area of the intestinal lining.

Viral load Concentration of virus in blood plasma; usually reported as copies of viral genome per unit volume of plasma.

Viral oncogene Any cancer-promoting sequence carried by a virus that can induce transformation in infected host cells.

VpreB A polypeptide chain that together with λ5 forms the **surrogate light chain** of the **pre-B-cell receptor**.

Western blotting A common technique for detecting a protein in a mixture; the proteins are separated electrophoretically and then transferred to a polymer sheet, which is flooded with radiolabeled or enzyme-conjugated antibody specific for the protein of interest.

Wheal and flare reaction A skin reaction to an injection of antigen that indicates an allergic response.

White pulp Portion of the spleen that surrounds the arteries, forming a periarteriolar lymphoid sheath (PALS) populated mainly by T cells.

Wiskott-Aldrich syndrome An X-linked immune deficiency disorder caused by inheritance of a mutated *WASP* gene, which encodes a cytoskeletal protein highly expressed in hematopoietic cells and essential for proper immune synapse formation and intracellular signaling.

X-linked Hyper-IgM syndrome An immunodeficiency disorder in which T_H cells fail to express CD40L. Patients with this disorder produce IgM but not other isotypes, they do not develop **germinal centers** or display somatic hypermutation, and they fail to generate **memory B cells**.

X-linked severe combined immunodeficiency (X-SCID) Immunodeficiency resulting from inherited mutations in the common γ chain of the receptor for IL-2, -4, -7, -9, and -15 that impair its ability to transmit signals from the receptor to intracellular proteins.

Xenograft A graft or tissue transplanted from one species to another.

Yolk sac Sac attached to the early embryo that provides nutrition for the embryo

Zymogen A protein that is activated on proteolytic cleavage by a protease.

Chapter 1

1. Jenner's method of using cowpox infection to confer immunity to smallpox was superior to earlier methods because it carried a significantly reduced risk of serious disease. The earlier method of using material from the lesions of smallpox victims conferred immunity but at the risk of acquiring the potentially lethal disease.

2. The Pasteur method for treating rabies consists of a series of inoculations with attenuated rabies virus. This process actively immunizes the recipient, who then mounts an immune response against the virus to stop the progress of infection. A simple test for active immunity would be to look for antibodies specific to the rabies virus in the recipient's blood at a time after completion of treatment, when all antibodies from a passive treatment would have cleared from the circulation. Alternatively, one could challenge the recipient with attenuated rabies to see whether a secondary response occurred (this test may be precluded by ethical ramifications).

3. The immunized mothers would confer passive immunity on their offspring because the anti-streptococcal antibodies, but not the B cells, cross the placental barrier and are present in the babies at birth. In addition, colostrum and milk from the mother would contain antibodies to protect the nursing infant from infection.

4. (a) H; (b) CM; (c) B; (d) H; (e) CM.

5. The four immunologic attributes are specificity, diversity, self/nonself recognition, and memory. *Specificity* refers to the ability of certain membrane-bound molecules on a mature lymphocyte to recognize only a single antigen (or a small number of closely related antigens). Rearrangement of the immunoglobulin genes during lymphocyte maturation gives rise to antigenic specificity; it also generates a vast array of different specificities, or *diversity*, among mature lymphocytes. The ability of the immune system to respond to nonself molecules, but (generally) not to self molecules (*self/nonself recognition*), results from the elimination during lymphocyte maturation of immature cells that recognize self antigens. After exposure to a particular antigen, mature lymphocytes reactive with that antigen proliferate, differentiate, and adapt, generating a larger and more effective population of memory cells with the same specificity; this expanded population can respond more rapidly and intensely after a subsequent exposure to the same antigen, thus displaying immunologic *memory.*

6. The secondary immune response is faster (because it starts with an expanded population of antigen-specific cells), more effective (because the memory cells have learned and adapted during the primary response), and reaches higher levels of magnitude than the primary response (again, because we begin with many more cells that have already honed their strategy).

7. Consequences of mild forms of immune dysfunction include sneezing, hives, and skin rashes caused by allergies. Asthma and anaphylactic reactions are more severe consequences of allergy and can result in death. Consequences of severe immune dysfunction include susceptibility to infection by a variety of microbial pathogens if the dysfunction involves immunodeficiency, or chronic debilitating diseases, such as rheumatoid arthritis, if the dysfunction involves autoimmunity. The most common cause of immunodeficiency is infection with the retrovirus HIV-1, which leads to AIDS.

8. (a) True. (b) True. (c) False. Most pathogens enter the body through mucous membranes, such as the gut or respiratory tract. (d) True. (e) False. Both are involved in each case. Innate immunity is deployed *first* during the primary response, and adaptive immunity begins later during that first encounter. During the secondary response, innate and adaptive immunity are again both involved. While innate responses are equally efficient, the second time around adaptive immunity uses memory cells to pick up where it left off at the end of the primary response and is therefore quicker and more effective in pathogen eradication during a secondary response. (f) False. These are two different types of disorder: autoimmunity occurs when the immune system attacks self, and immunodeficiency occurs when the immune system fails to attack nonself. The one caveat occurs in cases of immune deficiency involving immune-regulatory components. Just like broken brakes, these can result in an overzealous immune attack on self structures that thus presents as autoimmunity. (g) False. The intravenous immunoglobulin provides protection for as long as it remains in the body (up to a few weeks), but this individual has not mounted his or her own immune response to the antigen and will therefore not possess any memory cells. (h) True. (i) False. The genes encoding a T-cell receptor are rearranged and edited during T-cell development in the thymus so that each mature T cell carries a different T-cell receptor gene sequence. (j) False. The innate immune response does not generate memory cells (as the adaptive immune response does), so it is equally efficient during each infection. (k) False. Fragments of foreign antigen are not retained. Memory cells will continue to express the same T-cell receptor that originally bound antigen during the primary response, leading to clonal selection and expansion. Likewise, memory cells retain their differentiation profile and effector functions going into the secondary response (e.g., T_H1-specific cytokines and effector capabilities).

9. Pasteur had inadvertently immunized his chickens during the first inoculation, using an old, attenuated bacterial strain. The old strain was no longer virulent enough to be fatal, but it was still able to elicit an adaptive immune response that protected the chickens from subsequent infections with fresh, virulent bacteria of the same type.

10. Viruses live inside host cells and require the host cell's machinery to replicate. Fungi are extracellular and are often kept in check by the immune system, but they can be a problem for people with immune deficiencies. Fungi are the most homogeneous in form. Parasites are the most varied in form and can range in size from single-celled, intracellular microorganisms to large macroscopic intestinal worms. Some parasites also go through several life-cycle stages in their human host, altering their antigenic structures and location so significantly between stages that they require completely different immune eradication strategies. Bacteria can cause intracellular or extracellular infections, which require different immune targeting and elimination methods. Many bacteria express cell surface molecular markers (PAMPs) that are recognized by receptors that are part of our innate immune system (PRRs).

11. Herd immunity is what occurs when enough members of a population have protective immunity to a pathogen, either through vaccination or prior infection, that they act as a buffer against spread and help protect those without immunity. The efficacy of herd immunity depends on pathogen characteristics such as how the pathogen is transmitted (airborne, fecal/oral, etc.) and whether it can survive for long outside the host. Herd characteristics include how we interact with one another as a "herd" (e.g., crowding indoors versus sparse populations and outdoor encounters) and the primary target population (e.g., young children versus adults). (*Other answers are also acceptable.*)

12. Ehrlich's original theory had each cell expressing many different receptors and this was refined to many different cells, each one making many copies of a unique receptor. With Ehrlich's original conception, "selection" of this receptor would somehow need to trigger the cell to secrete only this receptor and not the others. In the refined and current version of the theory, once a receptor is selected that cell will make many copies to secrete. Since this is the only version of the antigen-specific receptor made by this cell, there is no confusion over which receptor received the signal and needs to be secreted.

13. (a) I; (b) A; (c) A; (d) A; (e) I; (f) A; (g) A; (h) A; (i) I; (j) A; (k) A.

14. Tolerance means that our immune system can discern between ourselves and foreign antigens, and does not attack self antigens. Lymphocytes learn tolerance by being exposed to self antigens during development, when most potentially self-reactive cells are either destroyed or inhibited from responding.

15. An antigen is anything that elicits an adaptive immune response—most commonly a part of some foreign protein or pathogen. An antibody is a soluble antigen-specific receptor molecule released by B cells that binds to an antigen and labels it for destruction. One interesting twist: antibodies can be antigens if they come from another species and are recognized as foreign by the host. This occurs in some instances of passive immunization, when antibodies from an animal such as a horse are given to humans, who then mount an adaptive immune response against horse-specific chemical patterns in the antibodies.

16. Pattern recognition receptors (PRRs) are germ line-encoded receptors expressed in a variety of immune cells. They are designed by evolution to recognize molecules found on common pathogens, and they initiate the innate immune response when they bind these molecules. In contrast, B- and T-cell receptors are expressed in lymphocytes, and the genes that encode these are produced by DNA rearrangement and editing, so that the receptor locus in each B or T cell's genome has a different sequence and encodes a different receptor. B- and T-cell receptors are diverse and have the ability to recognize a much greater assortment of antigens than PRRs, including those never before encountered by the immune system. B- and T-cell receptors are part of the adaptive immune response.

17. Cytokines are soluble molecular messengers that allow immune cells to communicate with one another. Chemokines are a subset of cytokines that are chemotactic, or have the ability to recruit cells to the site of infection.

18. (a) Yes. The problem is that T and B cells rearrange their DNA and each cell makes a unique receptor, which means they must have a unique DNA sequence for creating that molecule. (b) No. Since DNA rearrangement does not happen in germ cells, only in somatic cells (B and T cells, specifically), this is not passed on to progeny.

19. Only like making replicas of the weapons. There is no "saving" pieces or all of the pathogen to remember for later. The immune response saves the solution and not the problem.

20. No. Since memory occurs only in B and T cells and these cells are not passed on to the progeny, there is no inheritance of memory. Babies receive maternal antibodies in the womb and through breast milk, but not the cells that produce them. The antibodies themselves do not last more than a few months. This is why babies need to be vaccinated, to make their own memory B and T cells.

21. Extracellular bacteria and fungi are more alike. They are both extracellular and of similar size. Helminths are large extracellular worms, and viruses are small obligate intracellular pathogens. The immune system is likely to sort the latter two into very different bins based on where they are found, inside versus outside of cells, and their size or structure.

22. Clonal selection occurs in the lymphoid organs, not the site of infection (unless the site of infection is a lymphoid organ). The lymphoid organs are where innate and adaptive immunity meet, and where clonal selection occurs. There are sites in the body where the immune response is all but off limits. These include places like the central nervous system and the eye. Inflammation in these areas is likely to do more damage than good, and could result in permanent loss of function in the host if left unchecked.

23. Inflammation-mediated destruction of healthy tissue or possibly a blood cell cancer. In fact, uncontrolled replication of B or T cells leads to leukemia or lymphoma, cancers of the white blood cells and lymphatic system, respectively.

24. Immune imbalance disorders, especially allergy and autoimmunity. Both of these are overreactions of the immune

system (to benign foreign structures or self structures, respectively). What they have in common is a lack of or break in tolerance. This can be caused by a decrease in the immune-suppressing or inhibiting side of the immune balance equation (less brake).

25. The self/nonself theory of immune reactivity would suggest that we should recognize and attack microbes in our gut, including commensal (i.e., beneficial) bacteria. Typically this does not occur, although we will see later in the book that we do in fact recognize many of our "old friend" microbes without attacking them. The damage or danger theory suggests that we will not attack commensals unless (1) they cause damage to host cells, which then triggers an immune response, and/or (2) they induce host cells to express warning signs that set off an immune reaction. In most instances, this is true. We do not attack benign microbes in our body. Damage to the epithelial cells lining the gut can lead to reactions against otherwise benign microbes, even if the microbes themselves did not cause this damage, also consistent with the danger model.

Clinical Focus Answers

1. Considerations in developing nations vary as much as the nations themselves, but might include cost to create or purchase the vaccine, spaces to preserve the vaccine (usually cold), tools and expertise for administering the vaccine (e.g., needles and trained personnel), safety or religious concerns, mistrust of the industry and medical intervention, and so on. The concerns in developing nations, such as the United States, are very different and usually revolve around issues of personal choice or safety. Myth can be a barrier in both situations.

2. The target population is really the developing embryo or fetus, but that is a difficult stage to target. Administration of antibodies against Zika to women who are either trying to become pregnant or have just conceived might help to protect them from becoming infected, thereby saving the baby from complications. These antibodies might also cross the placenta and reach the child, which could also protect the baby in utero from the infection and consequences. Of course, neither the mother nor the child would be immune, so the risk of later infection and adverse outcomes is still present.

Chapter 2

1. (a) Although T cells complete their maturation in the thymus, not the bone marrow, mature $CD4^+$ and $CD8^+$ T cells will recirculate back to the bone marrow. (b) Pluripotent hematopoietic stem cells (HSCs) are rare, representing less about 0.05% of cells in the bone marrow. (c) HSCs can be mobilized from the bone marrow and circulate in the blood, which is now used as a source of stem cells for transplantation. (d) Macrophages will increase both MHC class I and class II expression after activation; however, T_H cells are $CD4^+$ and recognize antigenic peptide bound to MHC class II. (e) T cells develop in the thymus; B cells develop in the bone marrow and achieve full maturity in the spleen. (f) Lymphoid follicles are found in all secondary lymphoid tissues, including that associated with mucosal tissues (MALT).

(g) The follicular dendritic cell (FDC) network guides B cells within follicles. The follicular reticular cell (FRC) system guides T cells within the T-cell zone (although in some cases it may also help B cells to get to follicles, so aspects of this statement are true). (h) Infection and associated inflammation stimulates the release of cytokines and chemokines that enhance blood cell development (particularly to the myeloid lineage). (i) FDCs present soluble antigen on their surfaces to B cells, not T cells. (j) Dendritic cells can arise from both myeloid and lymphoid precursors. (k) B and T lymphocytes have antigen-specific receptors on their surface, but NK cells, which are also lymphocytes, do not. (l) B cells are generated outside the bone marrow in birds and ruminants. (m) All vertebrates have a thymus (even the jawless vertebrates have a rudimentary thymus). (n) Recent data suggest that at least one jawless vertebrate, the lamprey, has T- and B-like cells.

2. Myeloid: a, b, d, e. Lymphoid: c, f, g, h.

3. The primary lymphoid organs are the bone marrow (bursa of Fabricius in birds) and the thymus. These organs function as sites for B-cell and T-cell maturation, respectively. The secondary lymphoid organs are the spleen, lymph nodes, and barrier-associated lymphoid tissue (including skin and mucosal-associated lymphoid tissue, MALT). All these organs trap antigen and provide sites where lymphocytes can interact with antigen and subsequently undergo clonal expansion.

4. HSCs are stem cells, which, unlike mature fully differentiated cells, are (1) multipotent and (2) capable of self renewal.

5. The thymus helps us avoid autoimmune responses by negatively selecting thymocytes expressing T-cell receptors that bind to self peptide–MHC complexes with high affinity. Thymocytes with new TCRs scan the surfaces of epithelial cells in the cortex and medulla of the thymus; if they bind too tightly to surface MHC molecules presented by these cells they are eliminated (typically but not exclusively by clonal deletion).

6. In humans, the thymus reaches maximal size during puberty (b). During the adult years, the thymus gradually atrophies.

7. Immunodeficient mice are missing one or more immune cell type (either because of genetic mutations or chemotherapy). Injection of stem cells will restore these cell types—an easily measured outcome. As HSCs are successively enriched in a preparation, the total number of cells that must be injected to restore these cell populations decreases.

8. Monocytes are the blood-borne precursors of macrophages. Monocytes have a characteristic kidney bean–shaped nucleus and limited phagocytic and microbial killing capacity compared with macrophages. Macrophages are much larger than monocytes and undergo changes in phenotype to increase phagocytosis, antimicrobial mechanisms (oxygen dependent and oxygen independent), and secretion of cytokines and other immune system modulators. Tissue-specific functions are also found in tissue macrophages.

9. The bursa of Fabricius in birds is the primary site where B lymphocytes develop. Bursectomy would result in a lack of circulating B cells and humoral immunity, and it would probably be fatal.

10. (a) The spleen is classically considered an organ that "filters" antigens from blood. The lymph node receives antigens principally from the afferent lymphatics. (b) Both the lymph node paracortex and the splenic periarteriolar lymphoid sheath are rich in T cells. B cells are found primarily in the follicles. (c) Germinal centers are found wherever there are follicles, which are present in all secondary lymphoid tissue, including lymph node, spleen, and the wide variety of mucosal-associated lymphoid tissues. (d) True. (e) Afferent lymphatics are associated with lymph nodes, not the spleen. (Efferent lymphatics can be found in both.) (f) Ikaros is required for lymphoid development, which occurs primarily in the bone marrow (and thymus). However, lymphocytes populate the spleen and mount an immune response there, so the spleen would clearly be compromised in function in the absence of Ikaros and lymphocytes.

11. (1) d; (2) l; (3) c; (4) m, n; (5) g; (6) i; (7) n; (8) b; (9) o; (10) h; (11) c, f, g; (12) c, f; (13) c, e, f; (14) j, k; (15) e.

Research Focus Answer

This is an open-ended question that describes real observations; several answers could be considered correct. The model you advance "simply" has to be logical and consistent with all the data presented. Perhaps the most straightforward possibility is the one that turns out to be true: Notch regulates the decision of a progenitor (the common lymphoid progenitor [CLP], in fact) to become either a B cell or a T cell. If Notch is active, CLPs become T cells (even in the bone marrow). If Notch is turned off, CLPs become B cells (even in the thymus). Other possibilities? Notch could induce apoptosis of B cells. However, if this were the case, why would there be more T cells in the bone marrow? Alternatively, Notch could influence the microenvironments that the cells develop in and, when active, make the bone marrow niches behave like thymic niches. When off, it could allow the thymic niche to "revert" to a bone marrow–like environment. This would work (even if it is a more complex scenario) and deserves full credit as an answer. Experimental approach depends on the hypothesis proposed.

Clinical Focus Answer

(a) This outcome is unlikely; since the donor hematopoietic cells differentiate into T and B cells in an environment that contains antigens characteristic of both the host and donor, there is tolerance to cells and tissues of both. (b) This outcome is unlikely. T cells arising from the donor HSCs develop in the presence of the host's cells and tissues and are therefore immunologically tolerant of them. (c) This outcome is unlikely for the reasons cited in (a). (d) This outcome is a likely one because of the reasons cited in (a).

Chapter 3

1. (a) Cytoplasm. (b) Nucleus. (c) By dephosphorylation with the enzyme calcineurin phosphatase, activated by binding to the calcium-calmodulin complex. Activation of the cell results in an increase in intracytoplasmic calcium ion concentration. (d) There are many possible answers to this question. For example, there may be multiple forms of NFAT, and so these drugs may just interact with particular forms of the enzyme. Other cells may have alternative pathways that can bypass the need for calcineurin, whereas T cells may not. In fact, cyclosporin binds to the protein immunophilin, which is specifically expressed in activated T cells, and it is the complex of cyclosporin and immunophilin that inhibits the activity of calcineurin.

2. Immunoglobulin proteins and the T-cell receptor share a common motif: the immunoglobulin fold, which may be the binding target for these antibodies.

3. (a) False. Receptors and ligands bind through noncovalent interactions such as hydrogen bonds, ionic bonds, hydrophobic interactions, and van der Waals interactions. (b) True. Receptor-ligand binding can activate the receptor and result in phosphorylation and receptor cross-linking.

4. (a) Reduction enabled the breakage of disulfide bonds between the heavy and light chains and the pairs of heavy chains in the IgG molecule. Alkylation ensured that the bonds between the separated chains would not reform. Scientists could then separate the chains and figure out their molecular weight and how many chains belonged to each molecule. (b) Papain digested the molecule in the hinge region, releasing two antigen-binding fragments (Fabs) and one non–antigen-binding fragment, which spontaneously crystallized (Fc). This told investigators that there were two antigen-binding sites per molecule and suggested that part of the molecule was not variable in sequence (since it was regular enough in structure to crystallize). Pepsin isolated the divalent antigen-binding part of the molecule from the rest. (c) Antibodies were generated that specifically recognized the Fab and Fc fragments. Antibodies to Fab were found to bind to both heavy and light chains, thus indicating that the antigen-binding sites had components of both chains. Antibodies to Fc fragments bound only to the heavy chain.

5. An ITAM is an *i*mmunoreceptor *t*yrosine-based *a*ctivation *m*otif. It is a motif with a particular sequence containing phosphorylatable tyrosines in the cytoplasmic regions of several immunologically important proteins. Igα and Igβ are part of the signaling complex in B cells and are phosphorylated by the Src-family kinase Lyn, when the B-cell receptor (BCR) moves into the lipid raft regions of the membrane on activation.

6. An adapter protein bears more than one binding site for other proteins and serves to bring other proteins into contact with one another without, itself, having any enzymatic activity. Activation of the TCR results in activation of Lck, a Src family kinase. Lck phosphorylates the tyrosine residues on the ITAMs of the CD3 complex associated with the TCR. ZAP-70 then binds to the phosphorylated tyrosine (pY) residues via its SH2 regions, and is then itself phosphorylated and activated.

7. I would expect IgM to be able to bind more molecules of antigen than IgG, but perhaps not five times more. Steric hindrance/conformational constraints may prevent all 10 antigen-binding sites of IgM from being able to simultaneously bind to 10 antigenic sites.

8. Because you know that, in at least one case, activation of a cell can result in an alteration of the phenotype of the cytokine

receptor, resulting in a dramatic increase in its affinity for the cytokine. For example, the IL-2 receptor (IL-2R) exists in two forms: a moderate-affinity form consisting of a βγ dimer and a high-affinity form, synthesized only following cell activation that consists of an αβγ trimer.

9. Src family kinases have two tyrosine sites on which they can be phosphorylated: an inhibitory site and an activating site. In the resting state, the inhibitory site is phosphorylated by the kinase Csk, and the kinase folds up on itself, forming a bond between an internal SH2 group and the inhibitory phosphate, shielding the active site of the enzyme. Cleavage of the phosphate group from the inhibitory tyrosine allows the enzyme to open its structure and reveal the active site. Further activation of the enzyme then occurs when a second tyrosine is phosphorylated and stabilizes the activated state.

10. Since Lck lies at the beginning of the signaling pathway from the T-cell receptor, a T cell with an inactive form of Lck will not be able to undergo activation to secrete IL-2.

11. The signaling kinase Bruton's tyrosine kinase (Btk) is defective in 85% of cases of X-linked agammaglobulinemia. It is encoded on the X chromosome, and hence most cases of this disease occur in boys. Phosphorylation by Btk activates PLCγ, which cleaves phosphatidylinositol bisphosphate (PIP$_2$) into inositol trisphosphate (IP$_3$) and diacylglycerol (DAG) following activation of pre-B cells through the pre-B-cell receptor, or of B cells through the immunoglobulin receptor. This leads eventually to activation of the NFAT and MAP kinase transcription factor pathways, culminating in B-cell differentiation and proliferation, which cannot occur in the absence of a functioning Btk protein.

12. Both the BCR and the TCR are noncovalently associated on their respective cell membranes, with signal transduction complexes that function to transduce the signal initiated by antigen binding to the receptor and into the interior of the cell. In the case of the B-cell receptor, the signal transduction complex is composed of Igα/Igβ. ITAMs on Igα/Igβ are phosphorylated by tyrosine kinases that are brought into close proximity with the B-cell receptor complex on the oligomerization of the receptor and its movement into the lipid raft regions of the membrane, which occur on antigen binding. The phosphorylated tyrosines of Igα/Igβ then serve as docking points for downstream components of the signal transduction pathway. In the case of the TCR, the signal transduction complex is the CD3 set of proteins, which contains six chains and serves a similar function to Igα/Igβ. ITAMs on CD3 are phosphorylated by Src family kinases, particularly Lck, and then serve as docking points for downstream components of the TCR signaling pathway. Lck in turn is associated with the TCR coreceptors CD8 and CD4.

13. There are multiple possibilities for this answer, and we offer a few here: (a) The antibody is a Y-shaped molecule with the antigen-binding regions located at the two tips of the Y. At the junction of the three sections is a flexible hinge region that allows the two tips to move with respect to one another and hence to bind to antigenic determinants arranged at varying distances from one another on a multivalent antigen. (b) Within the variable region domains of the heavy and light chains, a common β-pleated sheet scaffold includes multiple antiparallel β strands in a conserved conformation. However, at the turns between the β strands, there are varying numbers and sequences of amino acids, corresponding to hypervariable regions in the immunoglobulin variable region sequences. These enable the creation of many different antigen-binding sites. (c) The constant regions of antibody molecules form the bridge between the antigen-binding region and receptors on phagocytic cells that will engulf antigen-antibody complex, or components of the complement system that will bind to the Fc regions of the antibody and aid in the disposal of the antigen. Different constant region structures bind to different Fc receptors and complement components.

14. Src family kinases such as Lck are located at the beginning of many signal transduction pathways, and their activities are subjected to rigorous control mechanisms. Lck is maintained in an inactive state by being phosphorylated on an inhibitory tyrosine residue. This phosphorylated tyrosine is then bound by an internal SH2 domain that holds Lck in a closed, inert conformation. If the DNA encoding this tyrosine residue has been mutated or eliminated, such that this phosphorylation cannot occur, then there will be a constitutive level of Lck activity. Since the enzyme that phosphorylates this inhibitory tyrosine is Csk, a reduction in Csk activity would have the same effect.

15. *Pleiotropy* is the capacity to bring about different end results in different cells.
Synergy is the ability of two or more cytokines affecting a cell to bring about a response that is greater than the sum of each of the cytokines.
Redundancy is the property that describes the fact that more than one cytokine can bring about the same effect.
Antagonism is the tendency for two cytokines binding to the same cell to bring about opposite effects, or to reduce/eliminate the response to the other.
Cascade induction is the ability of a cytokine to bind to one cell and to induce that cell to secrete additional cytokines.

16. A cytokine may induce the expression on the cell surface of new chemokine receptors and/or new adhesion molecules that would cause the cell to move to a new location and, once present, to be retained there.

17. When a type I interferon binds to its receptor on a virally infected cell, the interferon signal results in the activation of a ribonuclease that breaks down cytoplasmic RNA. It is particularly effective against double-stranded RNA.

18. (a) The IL-2R is made up of three components: α, β, and γ. Quiescent (unactivated) T cells bear only the βγ dimer, which binds IL-2, but at intermediate affinity insufficient to result in a signal under physiological conditions. On antigen stimulation, the T cell synthesizes the α subunit, which binds to the βγ dimer and converts it into a high-affinity receptor capable of mediating the IL-2 signal at physiological cytokine concentrations.

(b) See the following figure:

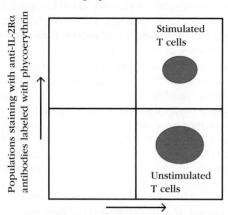

Populations staining with anti-IL-2Rβ and
γ antibodies labeled with fluorescein

19. The receptors for these cytokines share a common γ chain, which functions as a signal-transducing unit following recognition of each of these cytokines by cytokine-specific α-chain receptor components. The receptors for IL-2 and IL-15, but not the others, also contain a β chain, to form heterotrimeric receptors. The IL-4 and IL-7 receptors are heterodimeric.

Chapter 4

1. (a) Defensins, lysozyme, psoriasin; (b) phagocytosis, C-reactive protein, ficolins, MBL, surfactant proteins A and D, C-reactive protein; (c) phagosome NADPH oxidase, O_2, ROS, NO, iNOS, arginine, RNS; (d) interferons-α and -β, NK cells; (e) PRRs, PAMPs, antibodies, T-cell receptors; (f) TLR2, TLR4; (g) TLR3, TLR7, TLR9; (h) TLR4, MyD88, TRIF; (i) NLRs; (j) cGAS; (k) STING; (l) NF-κB, IRFs; (m) IL-1, PRRs, caspase-1, inflammasomes, NLRs; (n) IFIT proteins, Mx proteins, 2′, 5′-oligoadenylate A synthetase, protein kinase R; (o) PAMPs, PRRs, dendritic cells; (p) cytokines, dendritic cells; (q) PRRs, defensins.

2. The first line of defense is anatomical barriers. Examples: physical barriers of epithelial layers; mechanical mechanisms of pathogen elimination (peristalsis, sneezing, coughing); chemical barriers (acidic pH, enzymes, antimicrobial proteins and peptides). The second line of defense is induced cellular responses. Examples: induced production of antimicrobial proteins and peptides; phagocytosis and killing by enzymatic degradation and ROS and RNS; formation of NETs; regulated cell death.

3. CLRs (such as mannose receptor, dectin-1, and DC-SIGN) directly activate signaling pathways that induce phagocytosis. Scavenger receptors such as SR-A and SR-B also directly activate phagocytosis. Opsonin receptors bind opsonins that have bound to the bacteria. Examples are CD91/calreticulin that binds collectins, MBL, and so on; complement receptors that bind complement components; and Fc receptors that bind immunoglobulins.

4. B. Beutler showed that *lpr* mice were resistant to endotoxin (LPS) and that the genetic difference in these mice was lack of a functional TLR4 because of a single mutation in the *TLR4* gene. R. Medzhitov and C. Janeway demonstrated that a protein with homology to *Drosophila* Toll (which turned out to be TLR4) activated the expression of innate immunity genes when expressed in a human cell line.

5. Inflammation is characterized by redness, heat, swelling, pain, and sometimes loss of local function. Cytokines made by PRR-activated resident innate cells act on the vascular endothelium, causing vascular dilation (producing redness and heat) and increasing permeability, resulting in the influx of fluid and swelling (producing edema). Prostaglandins generated following the induced expression of COX2, together with mediators such as histamine, lead to the activation of local pain receptors. The swelling and local tissue damage can result in loss of function. The increased vascular permeability allows an influx of fluid containing protective substances, including opsonins and complement (as well as antibodies, if present). Local production of chemokines, together with induced expression of adhesion molecules on vascular endothelial cells, recruits to the site additional innate cells, such as neutrophils and macrophages, which contribute further to innate responses and pathogen clearance through phagocytosis and release of antimicrobial mediators. Proinflammatory cytokines made during this innate response may also act systemically, triggering the acute-phase response.

6. Regulated cell death is cell death induced by receptor-activated signaling pathways. One example in innate immunity is NETosis, the death of neutrophils due to the extrusion of neutrophil extracellular traps (NETs), which trap and kill bacteria. A second example is pyroptosis, the induced death of macrophages; it is beneficial in that it results in the killing of any intracellular bacteria and also allows the release of mature IL-1 and IL-18 following their processing by inflammasomes.

7. ILCs have a variety of functions. While NK cells kill (by inducing apoptosis) cells that have become altered due to malignancy, infection, or stress, most ILCs function through the production of cytokines that activate other cells to release mediators that are protective against pathogens. Some ILCs also produce factors that are important for the normal development of lymphoid tissues.

8. The innate immune system plays key roles in activating and regulating adaptive immune responses. Dendritic cells are key mediators of these roles. They bind pathogens at epithelial layers and deliver them to secondary lymphoid organs such as lymph nodes. After activation/maturation induced by TLR signaling, they present peptides from processed antigen to activate naïve CD4$^+$ and CD8$^+$ T cells. Depending on the pathogen and the PRRs to which it binds, dendritic cells are activated to produce certain cytokines that differentially induce naïve CD4$^+$ cells to differentiate into T_H subsets with different functions, usually appropriate for the particular pathogen. Also, PAMP binding to TLRs expressed on B and T lymphocytes can contribute to their activation by specific antigens to generate adaptive responses. Example of an adaptive response enhancing innate immune responses: The cytokine IFN-γ, produced by activated T_H1 cells, is a potent macrophage activator, including activating them to kill intracellular bacteria such as *Mycobacterium tuberculosis*.

9. A major potential disadvantage of the adaptive immune system, with the *de novo* generation of diverse antigen receptors in each individual's B and T cells, is the possibility of autoimmunity, which may result in disease. Adaptive responses are also slow. Conserved PRRs that have evolved to recognize PAMPs are less likely to generate destructive responses to self components, and innate responses are activated rapidly. The disadvantages of potential autoreactivity and slow response are overwhelmed by the advantages of having adaptive as well as innate immunity. While the innate response is rapid and helps initiate and regulate the adaptive response, innate immunity cannot respond to new pathogens that may have evolved to lack PAMPs recognized by PRRs. Also, in general (except for NK cells), there is no immunological memory, so the innate response cannot be primed by initial exposure to pathogens. Given their complementary advantages and disadvantages, both systems are essential to maintaining the health of vertebrate animals.

Clinical Focus Answer Children with genetic defects in MyD88 and IRAK4 are particularly susceptible to infections with *Streptococcus pneumoniae*, *Staphylococcus aureus*, and *Pseudomonas aeruginosa*, and children with genetic variants of TLR4 are susceptible to gram-negative urinary tract infections. Children with defects in the pathways activating the production or antiviral effects of IFN-α and IFN-β are susceptible to herpes simplex virus encephalitis. These individuals are thought not to be susceptible to a broader array of infectious diseases because other innate and adaptive immune responses provide adequate protection. The supporting evidence for the adaptive immune response protecting against infections is that this limited array of susceptibilities is seen primarily in children, who become less susceptible as they get older, presumably as they develop adaptive immunological memory to these pathogens.

Analyze the Data Answer

a. The innate responses are important for survival. From the high mortality in TLR7$^{-/-}$ and MyD88$^{-/-}$ mice, we can conclude that TLR7 and MyD88 are required to induce the innate immune response that plays a positive role in protecting mice from RV infection. It is very likely that viral single-stranded RNA (ssRNA) is the component that activates the immune response, as that is the PAMP that TLR7 recognizes, and TLR7 utilizes the MyD88 adaptor.

b. The virus is endocytosed by leukocytes in the inflammatory infiltrate of the infected bladder epithelium. TLR7 in endosomes following endocytosis and partial degradation binds the viral ssRNA and activates the MyD88 signaling pathway, leading to activation of IRF7, which induces IFN-α and IFN-β production. The IFNs bind to the IFN-α receptor (IFNAR) on virus-infected cells, activating expression of several proteins that inhibit viral replication.

c. Cytosolic dsRNA in the infected bladder epithelial cells should activate the RIG-I-like receptor (RLR) pathway, leading to the activation of IRF3 and IRF7 and type I interferon production. The fact that inflammatory infiltrate leukocytes are required for protection suggests that the RLR pathway is not being activated in the infected cells. RV probably has developed a mechanism to block this pathway in infected cells, such as the proteins expressed by other viruses that block dsRNA and prevent it from binding to RLR (see Table 4-7).

Chapter 5

1. (a) True. (b) True. (c) True. (d) True. (e) False. The outer membrane of a virus is derived from the outer membrane of the host cell and is therefore susceptible to complement-mediated lysis. However, some viruses have developed mechanisms that enable them to evade complement-mediated lysis. (f) True.

2. IgM undergoes a conformational change on binding to antigen, which enables it to be bound by the first component of complement C1q. In the absence of antigen binding, the C1q-binding site in the Fc region of IgM is inaccessible.

3. (a) The initiation of the classical pathway is mediated by the first component of complement, and C3 does not participate until after the formation of the active C3 convertase C4aC2b. The initiation of the alternate pathway begins with spontaneous cleavage of the C3 component, and therefore no part of the alternate pathway can operate in the absence of C3. (b) Clearance of immune complexes occurs only following opsonization of complexes by C3b binding, followed by phagocytosis or binding to the surface of erythrocytes via CR1 binding. Therefore, immune complex clearance is inhibited in the absence of C3. (c) Phagocytosis would be diminished in the absence of C3b-mediated opsonization. However, if antibodies specific for the bacteria are present, some phagocytosis would still occur.

4. Any combination of the following answers is acceptable. (1) Complement opsonizes pathogens, thus facilitating the binding of immune system cells via complement receptors. (2) The membrane attack complex (MAC) can induce lysis of pathogens. (3) C3a, C4a, and C5a act as chemoattractants to bring leukocytes to the site of infection, increasing inflammatory response. (4) Binding of C3 fragments by CD21 enhances B-cell activation by complement-coated antigen.

5. (a) The classical pathway is initiated by immune complexes involving IgM or IgG; the alternate pathway is generally initiated by binding of C3b to bacterial cell wall components, and the lectin pathway is initiated by binding of lectins (e.g., MBL to microbial cell wall carbohydrates). (b) The terminal reaction sequence after C5 convertase generation is the same for all three pathways. The differences in the first steps are described in part (a) above. For the classical and lectin pathways, the second step involves binding of the serine protease complexes C4b2a (classical) and MASP-1, MASP-2 (lectin). These each act as a C3 convertase in each respective pathway, and the two pathways are identical from that point on. The alternative pathway uses a different C3 convertase, C3bBb. The formation of Bb requires factor D, and the C3 convertase is stabilized on the cell surface by properdin. The alternative pathway C5 convertase is C3bBbC3b. (c) Healthy cells contain a variety of complement inhibitor proteins that prevent inadvertent

activation of the alternative pathway. Some of these proteins are expressed only on the surface of host (not on microbial) cells; others are in solution, but are specifically bound by receptors on host cells. Mechanisms of action are explained in the answer to question 7.

6. (a) Induces dissociation and inhibition of the C1 proteases C1r and C1s from C1q. Serine protease inhibitor. (b) Accelerates dissociation of the C4b2aC3, C3 convertase. Acts as a cofactor for factor I in C4b degradation. (c) Accelerates dissociation of C4b2a and C3bBb C3 convertases. (d) Acts as cofactor for factor I in degradation of C3b and C4b. (e) Binds C5b678 on host cells, blocking binding of C9 and formation of the MAC. (f) Cleaves and inactivates anaphylatoxins.

7. There are many options here and the readings describe several alternatives. The bacterium *Staphylococcus aureus* has evolved many ways to evade complement action. A small protein called *Staphylococcus* complement inhibitor (SCIN) binds and blocks the activity of the two C3 convertases, C4b2a and C3bBb. *Variola* and *Vaccinia* viruses express complement-inhibitory proteins that bind C3b and C4b and act as cofactors for factor I, thus preventing complement activation. Fungi have evolved mechanisms to destroy complement proteins. The opportunistic human pathogen *Aspergillus fumigatus*, a fungus that tends to attack patients who are already immunocompromised, secretes an alkaline protease, Alp1, that is capable of cleaving C3, C4, C5, C1q, and IgG.

8. Immune complexes are cleared from the body following opsonization by C3b. In the absence of the early components of complement, C3b convertases are not formed, and hence no C3b will be available.

9. a. False. Long before the emergence of the adaptive immune system, the early complement components were already functional, although the appearance of functional components of the MAC coincides approximately with the development of the adaptive immune system.

   b. False. The various gene families each have unique domain structures.

   c. True.

   d. True.

10. See the following table.

	Complement component knocked out						
	C1q	**C4**	**C3**	**C5**	**C9**	**Factor B**	**MASP-2**
Formation of classical pathway C3 convertase	A	A	NE	NE	NE	NE	NE
Formation of alternative pathway C3 convertase	NE	NE	A	NE	NE	A	NE
Formation of classical pathway C5 convertase	A	A	A	NE	NE	NE	NE
Formation of lectin pathway C3 convertase	NE	A	NE	NE	NE	NE	A
C3b-mediated opsonization	D	D	A	NE	NE	D	D
Neutrophil chemotaxis and inflammation	D	D	D	D	NE	D	D
Cell lysis	D	D	A	A	A	D	D

Clinical Focus Answer

(a) The two complement regulatory proteins DAF and protectin are both attached to plasma membrane surfaces by glycosylphosphatidylinositol linkages. In paroxysmal nocturnal hemoglobinuria (PNH), a defect in the enzyme PIG-A, which synthesizes these linkages, causes decreased surface expression of both of these proteins. (b) Defects in PIG-A tend to be expressed somatically in cells early in hematopoietic development. A given individual may express red blood cells that are wholly deficient, partially deficient, or wholly competent in *PIGA* expression. (c) Patients with PNH are essentially unable to express CD16 or CD66abce, indicating that those antigens are also most probably attached to the membranes using GPI linkages. Similarly, patients with PNH are unable to express CD14, so it is also probably a GPI-linked protein. In contrast, patients with PNH as well as normal control individuals are both able to express CD64, which is therefore most likely to be a transmembrane protein.

Analyze the Data Answers

1. The arrow shows the population of cells that is undergoing apoptosis. CD46 is rapidly lost from the surface of cells undergoing apoptosis. C1q binds to the DNA that appears on the surface of apoptotic cells, and in the absence of the regulatory component CD46, the cell is susceptible to C3b deposition and C3b-mediated opsonization.

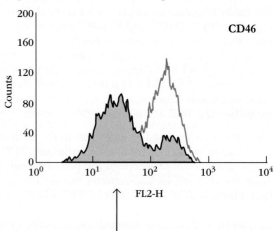

2. Each of the circles shown represents cell populations with the indicated surface expression of C1q and CD46. T cells undergoing apoptosis begin to express DNA on their cell surfaces, which is bound by the first component of the classical pathway, C1q. Healthy T cells, on the other hand, express relatively high levels of the regulatory complement component CD46, which is a cofactor for factor I.

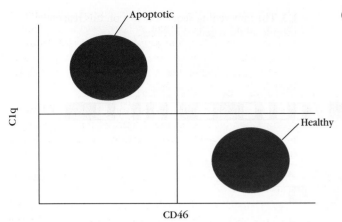

Chapter 6

1. (a) False. Vκ gene segments and Cλ are located on separate chromosomes and cannot be brought together during gene rearrangement. (b) True. (c) False. Naïve B cells produce a long primary transcript that carries the variable region and the mRNA for both the μ and δ constant regions. The switch in expression from the μ to the δ heavy chain occurs by mRNA splicing, not by DNA rearrangement. Switching to all other heavy-chain classes is mediated by DNA rearrangements. (d) True. (e) False. The variable regions of the β and δ TCR genes are encoded in three segments, analogous to the V, D, and J segments of the Ig heavy-chain variable region. The Vα and Vγ regions are each encoded in two segments.

2. (a) This is Figure 6-9a.

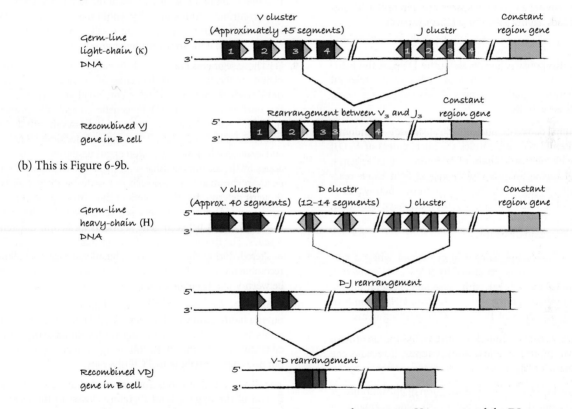

(b) This is Figure 6-9b.

(c) Start with Figure 6-19 (the mouse β locus). Recombination may occur between any D and J segments, and then between any V segment and the DJ segment.

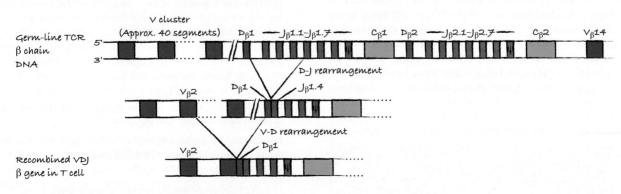

(d) Start with Figure 6-19, top line showing the α and δ loci. Draw a second line that shows $V_\alpha 2$ contiguous with $J_\alpha 3$. The intervening sequences lost on this rearrangement include all the δ gene sequences.

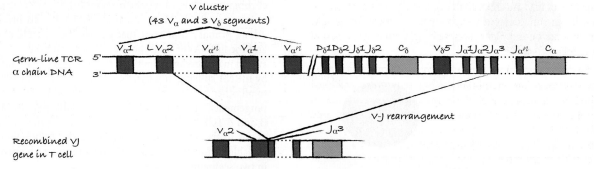

3. V_H and J_H gene segments cannot join because both are flanked by recombination signal sequences (RSSs) containing a 23-bp (two-turn) spacer (see Figure 6-8b). According to the one-turn/two-turn joining rule, signal sequences having a two-turn spacer can join only with signal sequences having a one-turn (12-bp) spacer.

4. (a) 1, 2, 3; (b) 5; (c) 2, 3, 4; (d) 5; (e) 1, 3, 4, 5.

5. RAG1/2—Responsible for breaking the DNA at the junction of the V, D, or J region sequences and the relevant recombination signal sequences. It also expresses some endonuclease activity.

   TdT (terminal deoxyribonucleotidyltransferase)—Randomly adds nontemplated nucleotides (N nucleotides) at V(D)J junctions in the heavy chain of Ig genes and TCR genes. Note that TdT activity is low during Ig light-chain rearrangement and absent in the fetal thymus, when waves of γδ T cells leave the thymus for the periphery. However, γδ T cells generated in the adult thymus do display pronounced N-region diversity.

   Artemis—Cleaves the hairpin loop formed when a single-strand DNA break generated by RAG1/2 is converted into a double-strand break with subsequent generation of a hairpin loop between the top and bottom DNA strands of the coding sequences of TCR and BCR gene segments.

   DNA PKcs—Forms a complex with Ku70/80. This protein kinase phosphorylates and actives Artemis. It recruits the ligation machinery.

   DNA ligase IV—Catalyzes the formation of phosphodiester bonds between the recombined V(D)J gene segments of BCRs and TCRs.

6. (a) P: Productive rearrangement of heavy-chain allele 1 must have occurred since the cell line expresses heavy chains encoded by this allele. (b) G: Allelic exclusion forbids the second heavy-chain allele from undergoing either productive or nonproductive rearrangement. (c) NP: In mice, the κ genes rearrange before the λ genes. Since the cell line expresses λ light chains, both κ alleles must have undergone nonproductive rearrangement, thus permitting λ-gene rearrangement to occur. (d) NP: Same reason as given in (c) above. (e) P: Productive rearrangement of the first λ-chain allele must have occurred since the cell line expresses λ light chains encoded by this allele. (f) G: Allelic exclusion forbids λ-chain allele 2 from undergoing either productive or nonproductive rearrangement (see Figure 6-15).

7. The κ-chain DNA must have the germ-line configuration because a productive heavy-chain rearrangement must occur before the light-chain (κ) DNA can begin to rearrange.

8. Random addition of N nucleotides at the D-J and V-DJ junctions contributes to the diversity within the CDR3 regions of heavy chains, but this process can result in a nonproductive rearrangement if the triplet reading frame is not preserved.

9. (1) Those segments not yet ready to engage in DNA recombination are located toward the center of the nucleus, whereas those ready to engage in DNA recombination move close to the nuclear membrane. (2) Epigenetic control: the histone mark H3K4me3 placed on chromatin increases the affinity of that chromatin for the RAG1/2 complex (specifically RAG2), and associated histone acetylation opens up the chromatin for access by the recombinational machinery. (3) The Ig and TCR genes in chromatin are arranged in rosettes, with specialized proteins determining the centers of those rosettes. Recombination can occur between gene segments located in the same rosettes, but not between gene segments located in different rosettes. The three-dimensional arrangement of these rosettes is altered during the DNA recombination process to allow recombination first between D and J regions, and then later, between V and D regions.

10. Whereas N-region addition occurs at the joints of Ig heavy- but not light-chain variable regions, all TCR variable region joints may include N-region nucleotides. Somatic mutation adds diversity to the BCR, following antigen stimulation, but does not contribute to TCR diversity.

11. The two heavy-chain constant regions that are closest to the 5′ end of the Ig gene [and therefore closest to the recombined V(D)J regions] are μ and δ. All constant regions can be expressed in both a membrane-bound form, which has a hydrophobic sequence at its 3′ end suitable for crossing the membrane in an α-helical conformation, and a soluble form, in which the membrane-crossing region has been replaced by a short segment of hydrophilic amino acids. IgM and IgD antibodies, which can be expressed as membrane-bound or soluble forms, are generated following differential RNA splicing as shown in Figure 6-17a. The decision to express membrane-bound IgM versus membrane-bound IgD is made by differentially splicing the RNA.

In contrast, the decision to express soluble IgM versus membrane-bound IgM is made by differentially splicing the RNA as shown in Figure 6-17b.

In order to generate antibodies of any heavy-chain class other than IgM or IgD, the DNA encoding the constant regions of upstream heavy chains is deleted and the recombined V(D)J segment is reassigned to a location in close proximity to the newly expressed heavy-chain constant region. However, in the case of IgM and IgD, all the DNA remains intact, and so a cell can simultaneously express all four forms of Ig: IgM membrane-bound, IgM secreted; IgD membrane-bound and IgD secreted.

12. They used the fact that the receptor is a *membrane-bound protein* to isolate membrane-bound polysomes and used the RNA associated with the polysomes to generate cDNA probes specific for membrane receptor genes. They used the fact that the TCR is expressed in T cells but not B cells to remove all cDNAs that were expressed in both B and T cells. They hypothesized that the gene for the TCR would be encoded in recombining segments and that the pattern of DNA fragments encoding the receptor genes would be differentially arranged in different T-cell clones.

13. (a) It must be a heavy chain because light chains do not have D regions. (b) The RSS. The heptamer of the heptamer-spacer-nonamer sequence directly abuts the end of the V region. (c) (1) P nucleotides are formed by asymmetric cleavage of the hairpin at the coding joint prior to DNA ligation. The italicized GA residues on the coding strand, and the CT residues on the noncoding strand, could have been generated by that mechanism. (c) (2) We cannot know for certain that they were formed by P-nucleotide addition, as they could just as easily have been randomly inserted by TdT. (c) (3) The residues shown in boldface have no place of origin in the original sequence, and so must have been added by N-nucleotide addition. (c) (4) Yes, we can, if no corresponding nucleotides can be found in the germ-line sequence, and they could not be accounted for by asymmetric hairpin joining.

14. Answer (a) is true. The heavy-chain gene segments are recombined prior to those of the light chain, enabling each recombined heavy chain to form many different antibodies with different light chains.

15. On average, two out of three attempts at rearrangements at the first heavy-chain chromosomal locus will result in an unproductively rearranged heavy chain. This is because the V- and J-chain gene segments must be read in the proper phase. The same is true for rearrangement at the second heavy-chain locus. Therefore, the probability of successfully generating a functional Ig heavy chain is only 1/3 (at the first allele) + (2/3 [probability the first rearrangement is not successful] × 1/3 [probability that the second rearrangement is successful]) = 0.55 or 55%.

16. See the completed table:

Analyze the Data Answer

(a) The restriction endonuclease must cut at a site within the constant region, as well as at sites upstream and downstream from it. (b) Recombination has occurred at only one of the two alleles. The germ-line bands complementary to both the constant and the variable region probes most probably derive from the other allele. (c) Given that the additional bands have remained in the same positions in the gel in both the germ-line and the myeloma DNA, it seems likely that there was a successful rearrangement at the first allele. (d) I would clone and sequence the DNA upstream from the Cκ region from both alleles and prove that one displayed a successful arrangement, although the other was still in the germ-line configuration.

Chapter 7

1. (a) True. (b) True. (c) False: MHC class III molecules are soluble proteins that do not function in antigen presentation. They include several complement components, TNF, and lymphotoxin. (d) False: The offspring of heterozygous parents inherit one MHC haplotype from each parent and thus will express some molecules that differ from those of each parent; for this reason, parents and offspring are histoincompatible. In contrast, siblings have a one in four chance of being histocompatible (see Figure 7-8c). (e) True. (f) False: Most nucleated cells express MHC class I molecules, but neurons, placental cells, and sperm cells at certain stages of differentiation appear to lack class I molecules. (g) True.

2. (a) Liver cells: Class I K^d, K^k, D^d, D^k, L^d, and L^k. (b) Macrophages: Class I K^d, K^k, D^d, D^k, L^d, and L^k. Class II $A\alpha^k\beta^k$, $A\alpha^d\beta^d$, $A\alpha^k\beta^d$, $A\alpha^d\beta^k$, $E\alpha^k\beta^k$, $E\alpha^d\beta^d$, $E\alpha^k\beta^d$, and $E\alpha^k\beta^d$.

3. (a) SJL macrophages express the following MHC molecules: K^s, D^s, L^s, and A^s. Because of the deletion of the Eα locus, E^s is not expressed by these cells. (b) The transfected cells would express one heterologous E molecule, $E\alpha^k\beta^s$, and one homologous E molecule, $E\alpha^k\beta^k$, in addition to the molecules listed in (a).

4. See Figure 7-1, Figure 3-7, and Figure 3-10.

5. (a) The polymorphic residues are clustered in short stretches primarily within the membrane-distal domains of the MHC class I and class II molecules (see Figure 7-10). These regions form the peptide-binding groove of MHC molecules, or the surfaces most in contact with antigen. (b) MHC polymorphism is thought to arise by gene conversion of short, nearly homologous DNA sequences within unexpressed pseudogenes in the MHC to functional class I or class II genes. MHC diversity in the population is likely

	V(D)J Rearrangement	V-J Rearrangement	D-D Joining	12/23 Rule Obeyed	N-Nucleotide Addition	More Than One C Region	Allelic Exclusion Perfect
Ig heavy chain	+	−	−	+	+	+	+
Ig light chain	−	+	−	+	(−)	+	+
TCR α	−	+	−	+	+	−	−
TCR β	+	−	−	+	+	+	+
TCR γ	−	+	−	+	+ (in adults)	+	+
TCR δ	+	−	+	+	+ (in adults)	−	+

maintained via selective pressure. If inheritance of more MHC molecules (heterogeneity) is favorable for survival, then this will lead to a bias toward diversity at the locus. Likewise, if mate selection is indeed influenced by the olfactory detection of MHC difference (see Evolution Box 7-1), then offspring with more diverse MHC alleles might be expected.

6. (a) The proliferation of T_H cells and IL-2 production by them are detected in assay 1, and the killing of LCMV-infected target cells by cytotoxic T lymphocytes (CTLs) is detected in assay 2. (b) Assay 1 is a functional assay for MHC class II molecules, and assay 2 is a functional assay for MHC class I molecules. (c) Class II A^k molecules are required in assay 1, and class I D^d molecules are required in assay 2. (d) You could transfect K cells with the A^k gene and determine the response of the transfected cells in assay 1. Similarly, you could transfect a separate sample of L cells with the D^d gene and determine the response of the transfected cells in assay 2. In each case, a positive response would confirm the identity of the MHC molecules required for LCMV-specific activity of the spleen cells. As a control in each case, L cells should be transfected with a different MHC class I or class II gene and assayed in the appropriate assay. (e) The immunized spleen cells express both A^k and D^d molecules. Of the listed strains, only the A.TL strain expresses both of these MHC molecules, and thus these are the only strains from which the spleen cells could have been isolated. A positive IL-2 and killing response for (BALB/c × B10.A) F_1 suggests these might be a possible source as well. However, if that were the case, they should be able to kill LCMV-infected targets that express K^k as in the infected C3H and B10.A (4R) cells. Since infected target cells from these strains are not killed in assay 2, these F_1 mice could not have been the original source.

7. It is not possible to predict. Since the peptide-binding cleft is identical, both MHC molecules should bind the same peptide. However, the amino acid differences outside the cleft might prevent recognition of the second MHC molecule by the T-cell receptor on the T_C cells.

8. If RBCs expressed MHC molecules, then extensive tissue typing would be required before a blood transfusion, and only a few individuals would be acceptable donors for a given individual.

9. (a) No. Although those with the *A99/B276* haplotype are at significantly increased relative risk, there is no absolute correlation between these alleles and the disease. (b) Nearly all of those with the disease will have the *A99/B276* haplotype, but depending on the exact gene or genes responsible, this may not be a requirement for development of the disease. If the gene responsible for the disease lies between the *A* and *B* loci, then weaker associations to *A99* and *B276* may be observed. If the gene is located outside of the *A* and *B* regions and is linked to the haplotype only by association in a founder, then associations with other MHC genes may occur. (c) It is not possible to know how frequently the combination will occur relative to the frequency of the two individual alleles; linkage disequilibrium is difficult to

predict. However, based on the data given, it may be speculated that the linkage to a disease that is fatal in individuals who have not reached reproductive years will have a negative effect on the frequency of the founder haplotype. An educated guess would be that the *A99/B276* combination would be rarer than predicted on the basis of the frequency of the *A99* and *B276* alleles.

10. By convention, antigen-presenting cells are defined as those cells that can display antigenic peptides associated with MHC class II molecules and can deliver a costimulatory signal to $CD4^+$ T_H cells. A target cell is any cell that displays peptides associated with MHC class I molecules to $CD8^+$ T_C cells.

11. (a) Self MHC restriction is the attribute of T cells that limits their response to antigen associated with self MHC molecules on the membrane of antigen-presenting cells or target cells. In general, $CD4^+$ T_H cells are MHC class II restricted, and $CD8^+$ T_C cells are MHC class I restricted, although a few exceptions to this pattern occur. (b) Antigen processing involves the intracellular degradation of protein antigens into peptides that associate with MHC class I or class II molecules. (c) Endogenous antigens are synthesized within altered self cells (e.g., virus-infected cells or tumor cells), are processed in the endogenous pathway, and are presented by MHC class I molecules to $CD8^+$ T_C cells. (d) Exogenous antigens are internalized by antigen-presenting cells, processed in the exogenous pathway, and presented by MHC class II molecules to $CD4^+$ T_H cells. (e) Anchor residues are the key locations (typically, positions 2/3 and 9) within an 8- to 10-amino-acid-long antigenic peptide that make direct contact with the antigen-binding cleft of MHC class I. The specific residues found at these locations distinguish the peptide fragments that can bind each allelic variant of class I. (f) An immunoproteasome is a variant of the classical proteasome, found in all cells, and is expressed in antigen-presenting cells and in infected target cells. The presence of this variant increases the production of antigenic fragments optimized for binding to MHC class I molecules.

12. (a) EN: Class I molecules associate with antigenic peptides and display them on the surface of target cells to $CD8^+$ T_C cells. (b) EX: Class II molecules associate with exogenous antigenic peptides and display them on the surface of APCs to $CD4^+$ T_H cells. (c) EX: The invariant chain interacts with the peptide-binding cleft of MHC class II molecules in the rough endoplasmic reticulum (RER), thereby preventing binding of peptides from endogenous sources. It also assists in folding of the class II α and β chains and in movement of class II molecules from the RER to endocytic compartments. (d) EX: Lysosomal hydrolases degrade exogenous antigens into peptides; these enzymes also degrade the invariant chain associated with class II molecules, so that the peptides and MHC molecules can associate. (e) EN: TAP, a transmembrane protein located in the RER membrane, mediates transport of antigenic peptides produced in the cytosolic pathway into the RER lumen, where they can associate with MHC class I molecules. (f) B: In the endogenous pathway, vesicles containing peptide–MHC class I complexes move from the RER to the Golgi complex and then on to the

cell surface. In the exogenous pathway, vesicles containing the invariant chain associated with MHC class II molecules move from the RER to the Golgi and on to endocytic compartments. (g) EN: Proteasomes are large protein complexes with peptidase activity that degrade intracellular proteins within the cytosol. When associated with LMP2 and LMP7, which are encoded in the MHC region, and LMP10, which is not MHC encoded, proteasomes preferentially generate peptides that associate with MHC class I molecules. (h) B: Antigen-presenting cells internalize exogenous (external) antigens by phagocytosis or endocytosis. (i) EN: Calnexin is a protein within the RER membrane that acts as a molecular chaperone, assisting in the folding and association of newly formed class I α chains and β_2-microglobulin into heterodimers. (j) EX: After degradation of the invariant chain associated with MHC class II molecules, a small fragment called CLIP remains bound to the peptide-binding cleft, presumably preventing premature peptide loading of the MHC molecule. Eventually, CLIP is displaced by an antigenic peptide. (k) EN: Tapasin (TAP-associated protein) brings the transporter TAP into proximity with the MHC class I molecule and allows the MHC molecule to acquire an antigenic peptide (see Figure 7-15).

13. (a) Chloroquine inhibits the exogenous processing pathway, so that the APCs cannot display peptides derived from native lysozyme. The synthetic lysozyme peptide will exchange with other peptides associated with class II molecules on the APC membrane, so that it will be displayed to the T_H cells and induce their activation. (b) Delay of chloroquine addition provides time for native lysozyme to be degraded in the endocytic pathway.

14. (a) Dendritic cells: constitutively express both MHC class II molecules and costimulatory signals. B cells: constitutively express class II molecules but must be activated before expressing the CD80/86 costimulatory signal. Macrophages: must be activated before expressing either class II molecules or the CD80/86 costimulatory signal. (b) See Table 7-4. Many nonprofessional APCs function only during sustained inflammatory responses.

15. (a) R; (b) R; (c) NR; (d) R; (e) NR; (f) R.

16. (a) Intracellular bacteria, such as members of the *Mycobacterium* family, are a major source of nonpeptide antigens; the antigens observed in combination with CD1 are lipid and glycolipid components of the bacterial cell wall. (b) Members of the CD1 family associate with β_2-microglobulin and have structural similarity to MHC class I molecules. They are not true MHC molecules because they are not encoded within the MHC, but rather on a different chromosome. (c) The pathway for antigen processing taken by the CD1 molecules differs from that taken by MHC class I molecules. A major difference is that CD1 antigen processing is not inhibited in cells that are deficient in TAP, whereas MHC class I molecules cannot present antigen in TAP-deficient cells.

17. (b) The TAP1-TAP2 complex is located in the endoplasmic reticulum.

18. The offspring must have inherited HLA-A3, HLA-B59, and HLA-C8 from the mother. Potential father 1 cannot be the biological father because although he shares HLA determinants with the offspring, the determinants are the same genotype inherited from the mother. Potential father 2 could be the biological father because he expressed the HLA genes expressed by the offspring that are not inherited from the mother (HLA-A43, HLA-B54, HLA-C5). Potential father 3 cannot be the biological father because although he shares HLA determinants with the offspring, the determinants are the same genotype inherited from the mother.

19. Considering HLA-A, HLA-B, and HLA-C only, a maximum of six different class I molecules are expressed in individuals who inherit unique maternal and paternal alleles at each locus. In the case of class II, considering only HLA-DP, -DQ, and -DR molecules where any α chain and β chain of each gene can pair to produce new maternal/paternal combinations, a maximum of 12 different class II molecules can be expressed (4 DP, 4 DQ, and 4 DR). Since humans can inherit as many as three functional DRβ genes, each of which is polymorphic, in practice fully heterozygous individuals have the ability to express more than four HLA-DR proteins.

20. Polygeny is defined as the presence of multiple genes in the genome with the same or similar function. In humans, MHC class I A, B, and C or class II DP, DQ, and DR are both examples of this (see Figure 7-6). Polymorphism is defined as the presence of multiple alleles for a given gene locus within the population. HLA-A1 versus HLA-A2 (e.g., see Table 7-3) are examples of polymorphic alleles at the class I locus. Codominant expression is defined as the ability of an individual to simultaneously express both the maternal and the paternal alleles of a gene in the same cell. This process is what allows a heterozygous individual to express, for instance, both HLA-Cw2 and HLA-Cw4 alleles (see Figure 7-9). Polygeny ensures that even MHC homozygous individuals express a minimum of three different class I and class II proteins, each with a slightly different antigen-binding profile, expanding their repertoire of antigens that can be presented. MHC polymorphism and codominant expression in outbred populations help facilitate the inheritance and expression of different alleles at each locus, further increasing the number of different antigens that one individual can present. Codominant expression at the class II loci carries an added bonus: since these proteins are generated from two separate genes/chains, new combinations of α and β chains can arise, further enhancing the diversity of class II protein isoforms, or the number of unique MHC class II antigen-binding clefts.

21. The invariant chain is involved in MHC class II folding and peptide binding. Cells without this protein primarily retain misfolded class II proteins in the RER and are therefore unable to express MHC class II molecules on the cell surface. Since APCs are the primary cell types that express class II, cells with this mutant phenotype would be incapable of presenting exogenously processed antigens to naïve CD4$^+$ T cells.

22. Cross-presentation is the process by which some APCs can divert antigens collected from extracellular sources (exogenous pathway) to processing and presentation via MHC class I proteins (typically the realm of the endogenous pathway). This process is important for activation of naïve $CD8^+$ T cells to generate CTLs capable of detecting and lysing virally infected target cells. Dendritic cells, or a subset of this cell type, are thought to be the major players in this process, although "licensing" by antigen-specific $CD4^+$ T_H cells may first be required in order for DCs to engage in cross-presentation.

Clinical Focus Answer

Human TAP deficiency results in a lack of class I molecules on the cell surface or a type I bare-lymphocyte syndrome. This leads to partial immunodeficiency in that antigen presentation is compromised, but there are NK cells and $\gamma\delta$ T cells to limit viral infection. Autoimmunity results from the lack of class I molecules that give negative signals through the killer-cell inhibitory receptor (KIR) molecules; interactions between KIR and class I molecules prevent the NK cells from lysing target cells. In their absence, self cells are targets of autoimmune attack on skin cells, resulting in the lesions seen in TAP-deficient patients. The use of gene therapy to cure those affected with TAP deficiency is complicated by the fact that class I genes are expressed in nearly all nucleated cells. Because the class I product is cell bound, each deficient cell must be repaired to offset the effects of this problem. Therefore, although the replacement of the defective gene may be theoretically possible, ascertaining which cells can be repaired by transfection of the functional gene and reinfused into the host remains an obstacle. In the case of Tasmanian devils and facial cancer, it was found that the cancer cells being transferred between devils lack surface MHC class I expression, allowing them to avoid destruction by $CD8^+$ T cells. Transcripts for several key proteins involved in class I synthesis and assembly (TAP1, TAP2, and β_2-microglobulin) were all found to be down-regulated in these cancer cells thanks to high levels of histone acetylation at these MHC class I–associated loci, enough to silence gene expression. Thus these cancer cells appear to be accepted much like an isograft, passed from one animal to another during biting incidents. No one is sure why NK cells in Tasmanian devils do not attack these cancer cells.

Analyze the Data Answer

(a) Yes. Comparing the relative amounts of L^d and L^q molecules without peptides, there are about half as many open L^q molecules as L^d molecules. The data suggest that L^q molecules form less stable peptide complexes than L^d molecules. (b) Part (a) in the figure shows that 4% of the L^d molecules don't bind MCMV peptide compared with 11% of the L^d after a W-to-R mutation. Thus, there appears to be a small decrease in peptide binding to L^d. It is interesting to note that nonspecific peptide binding increases severalfold after mutagenesis, based on the low amount of open-form L^d W97R (mutated L^d) versus native L^d. (c) Part (b) in the figure shows that 71% of the L^d molecules don't bind tum^∂ $P91A_{14-22}$ peptide after a W-to-R mutation, compared with 2% for native L^d molecules. Thus, there is very poor binding of tum^∂ $P91A_{14-22}$ peptide after a W-to-R mutation. (d) You would inject a mouse that expressed L^d because

only 2% of the L^d molecules were open forms after the addition of tum^- $P91A_{14-22}$ peptide compared with 77% free forms when L^q were pulsed with peptide. Therefore, L^d would present peptide better and probably activate T cells better than L^q. (e) Conserved anchor residues at the ends of the peptide bind to the MHC, allowing variability at other residues to influence which T-cell receptor engages the MHC class I–antigen complex.

Chapter 8

1. a. Knockout mice lacking MHC class II molecules fail to produce mature $CD4^+$ thymocytes, or those clacking MHC class I molecules fail to produce mature $CD8^+$ thymocytes, because at some level lineage commitment requires engagement between the MHC and the appropriate CD4/8 receptor.

   b. β-Selection initiates maturation to the DN4 stage, proliferation, allelic exclusion, maturation to the DP stage, and TCR α-chain locus rearrangement.

   c. Negative selection of tissue-specific antigens occurs only in the medulla of the thymus, by medullary thymic epithelial cells (mTECs) and some DCs that pick up antigens produced by mTECs.

   d. Most thymocytes (>90%) die of neglect in the thymus because they either did not produce viable TCR, or because they do not bind to self MHC.

   e. Thymocyte precursors express neither CD4 nor CD8 and enter the thymus from the bone marrow at the corticomedullary junction.

   f. Thymocytes that bind peptide-MHC complexes with high affinity are negatively selected.

   g. Double-negative (DN) thymocytes progress through several stages distinguished by expression primarily of CD44 and CD25.

   h. Some thymocytes with autoreactive T-cell receptors mature to become T_{REG} cells.

   i. Regulatory T cells help maintain peripheral tolerance.

   j. Commitment to the $CD4^+$ T-cell lineage is regulated by Th-Pok. Runx3 regulates commitment to the $CD8^+$ lineage.

2. Precursors of thymocytes enter the thymus at the **corticomedullary** junction. Interactions with **Notch** ligands are required to commit them to the T-cell lineage. If positively selected, double-positive (DP) thymocytes travel from the thymic cortex to the **medulla.** Upregulation of **S1P receptor** allows them to leave the thymus and enter circulation.

3. Many if not most $\gamma\delta$ T cells have receptors that have a more restricted (invariant) specificity and can recognize a variety of antigens, including lipids (they do not always require their antigens to be peptide presented by MHC). In this respect their response to antigen is more akin to that of innate immune cells, which use pattern recognition receptors to respond rapidly to antigen. $\gamma\delta$ T cells also develop differently—many are generated in waves during fetal development and populate mucosal tissues. The decision to become a $\gamma\delta$ T cell occurs early, during the DN3 stage,

and most γδ T cells do not go through the conventional positive and negative selection process in the thymus.

4. See the following figure:

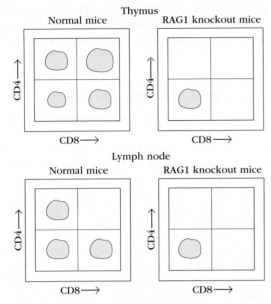

5. (a) There would be no CD4⁺ SP, but all other stages would be present. The absence of MHC class II would prevent positive selection and lineage commitment of CD4⁺ T cells. (b) All stages would be present, but some of the mature cells would be reactive to tissue-specific antigens. (This would be revealed only by functional experiments.) AIRE regulates the expression of self tissue–specific antigens by medullary epithelial cells. (c) All DN and DP cells would be present (β-selection would proceed unhindered). However, none of the DP cells would express normal TCR-αβ dimers and could not be positively selected. (For the advanced, TCR-γδ cell development would proceed normally—many of these are DN in phenotype, but a few are CD4⁺ and CD8⁺ SP.)

6. The first are CD3⁻ TCR β-chain⁻ thymocytes and could simply be immature DN thymocytes. The second group are CD3⁺ TCR β-chain⁻ and could be TCR-γδ T cells!

7. (a) Flow cytometry. (b) Higher. (c) Lower. (Positive selection would occur in the H2ᵏ but not H2ᵈ background [MHC class II].) Most cells in the TCR transgenic would have the receptor specific for this MHC haplotype, and therefore would get more than the normal number of CD4⁺ SP cells when positive selection occurs.) (d) No mature single-positive cells; may have a reduced number of DP cells (negative selection is going to occur).

Must speculate with this question—no absolutely clear answer. The cortical epithelium may not be able to mediate clonal deletion because it doesn't express the right costimulatory molecules. Investigators who have done experiments like this find evidence for negative selection of a sort, however. SP T cells develop, but they appear not to be easily activated.

8. (a) Thymocytes in experiment A developed in a thymus whose cortical epithelial cells expressed H2ᵈ MHC molecules and therefore became restricted to that MHC (via positive selection). They are not restricted to H2ᵈ, and thus

ignore targets that express this MHC. (b) Same reasoning as above.

9. (a) The immature thymocytes express both CD4 and CD8, whereas the mature CD8⁺ thymocytes do not express CD4. To distinguish these cells, the thymocytes are double-stained with fluorochrome-labeled anti-CD4 and anti-CD8 and analyzed by FACS. (b) See the following table.

H-Y TCR Transgenic	Immature Thymocytes	Mature CD8⁺ Thymocytes
H2ᵏ female	+	+
H2ᵏ male	+	−
H2ᵈ female	+	−
H2ᵈ male	+	−

(c) Because the gene encoding the H-Y antigen is on the Y chromosome, this antigen is not present in females. Thymocytes bearing the transgenic T-cell receptor, which is H2ᵏ restricted, would undergo positive selection in both male and female H2ᵏ transgenics. However, subsequent negative selection would eliminate thymocytes bearing the transgenic receptor, which is specific for H-Y antigen, in the male H2ᵏ transgenics. (d) Because the H2ᵈ transgenics would not express the appropriate MHC molecules, T cells bearing the transgenic T-cell receptor would not undergo positive selection.

10. (a) Class I K, D, and L molecules and class II IA molecules. (b) Class I molecules only. (c) The normal H2ᵇ mice should have both CD4⁺ and CD8⁺ T cells because both MHC class I and class II molecules would be present on thymic stromal cells during positive selection. H2ᵇ mice with knockout of the *IA* gene would express no class II molecules; thus, these mice would have only CD8⁺ cells.

11. (a) Because the pre-T-cell receptor, which does not bind antigen, is associated with CD3, cells expressing the pre-TCR as well as the antigen-binding T-cell receptor would stain with anti-CD3. It is impossible to determine from this result how many of the CD3-staining cells are expressing complete T-cell receptors. The remaining cells are even more immature thymocytes that do not express CD3. (b) No. Because some of the CD3-staining cells express the pre-TCR or TCR-αβ instead of the complete TCR-γδ, you cannot calculate the number of T_C cells by simple subtraction. To determine the number of T_C cells, you need fluorescent anti-CD8 antibody, which will stain only the CD8⁺ T_C cells.

Clinical Focus Answers

1. AIRE (regulates tissue-specific expression of antigens by mTECs), FoxP3 (involved in development of T_REG cells), any of the TCR signaling molecules (which regulate TCR signal strength and therefore the outcome of thymic selection), MHC (which presents self peptides), or anything that prevents T cells from getting to the medulla (e.g., CCR7).

2. (a) The authors appear to be saying that the MHC variant, which would be expressed by medullary epithelial cells and dendritic cells in the thymus, may not be able to present

(bind to) certain brain-specific self peptides (or does so inefficiently). Therefore, some autoreactive CD4$^+$ T cells in the thymus may not be deleted. (Note that HLA-DR is an MHC class II molecule.) (b) There is no one right question to pose. Here are two possibilities. (1) If this were the case, wouldn't this also mean that peripheral dendritic cells would not be able to present the peptide, and therefore wouldn't activate the autoreactive T-cell escapees? (2) If you let an autoreactive helper T-cell escape, don't you still need an autoreactive cytotoxic T cell to escape also? (This question depends on your knowing that autoimmune diseases like multiple sclerosis are caused in part by CD8$^+$ T-cell–mediated damage.) (c) Given that immunologists are still pondering the issue, this is a challenging question. However, here are a couple of possibilities. (1) One accepts the possibility that this is a problem with negative selection. However, to understand how this leads to autoimmune disease, one can propose, for instance, that there is a difference between antigen presentation in the thymus and antigen presentation in the periphery. Presentation of brain self peptides by this MHC variant may be inefficient and allow autoreactive CD4$^+$ T cells to escape. But in the periphery this inefficiency can be overcome by enhancements in costimulatory molecule expression, levels of MHC expression, and so on among antigen-presenting cells that have been stimulated by pathogen or cell damage. (2) In addition (or alternatively), one can propose that this MHC variant compromises the development of regulatory T cells, not just the deletion of autoreactive CD4$^+$ T cells. Specifically, if the MHC class II variant is unable to present the self peptides, then it will neither mediate deletion of autoreactive CD4$^+$ T cells *nor* select for suppressive, autoreactive regulatory T cells. Note that we need to assume, in all cases, the existence of autoreactive CD8$^+$ T-cell escapees. This is not a radical assumption. As you now know, negative selection in the thymus is never perfect, and our ability to maintain tolerance depends to a significant extent on peripheral mechanisms.

Chapter 9

1. Fetal liver cells; B-1 B-cell progenitors are highly enriched in the fetal liver and are the first B cells to populate the periphery. The B-1 B cells would be harvested from the peritoneal and pleural cavities, as well as the spleen.

2. T2 cells have intermediate levels of IgD, whereas T1 cells have no to low amounts of IgD. T2 cells bear both CD21 and CD23, whereas neither antigen is expressed on T1 cells. T2 cells have higher levels of the receptor for the B-cell survival factor BAFF than do T1 cells. Interaction of antigen with T1 cells results in apoptosis; interaction of antigen with T2 cells sends survival and maturation signals.

3. Cell division at this stage allows the repertoire to maximize its use of B cells in which a heavy chain has been productively rearranged. Each daughter cell can then rearrange a different set of light-chain gene segments, giving rise to multiple B-cell clones bearing the same heavy chain, but different light-chain genes.

4. (1) They can undergo apoptosis, in a process called *negative selection*. This occurs for B cells in the bone marrow and for T cells in the thymus. (2) They can become anergic—refractory to further stimulation—and eventually die. This occurs for both T and B cells. (3) Their receptors can undergo receptor editing. This occurs quite frequently in B cells. In T cells, the extent to which receptor editing occurs varies according to the nature of the animal (the nature of the transgene used to study editing), and therefore whether it is a meaningful mechanism in T-cell receptor development and selection is so far unclear.

5. *To test the hypothesis*: first, knock out the ability of the animal to express that transcription factor, and then analyze the bone marrow for the occurrence of B-cell progenitors at each stage of development, using flow cytometry. One would not expect to see any progenitors after the pre-pro-B-cell stage if the transcription factor is expressed then *and* is necessary for further B-cell development. Second, make a fusion protein in which the promoter of the transcription factor is fused to a fluorescent protein, such as green or yellow fluorescent protein. Correlate the expression of the fluorescent proteins with the cell-surface markers. One would expect to see the fluorescent protein show up first in cells bearing markers characteristic of the pre-pro-B-cell stage. *To test the status of heavy- and light-chain rearrangement*: Rearrangement has begun at this stage, with D-to-J$_H$ rearrangement occurring on the heavy chain. Test it by PCR, with primers complementary to sequences upstream of the D regions and downstream of the J regions, followed by sequencing, if necessary.

6. Create an animal in which the CXCL12 promoter is fused to a marker fluorescent protein, such as GFP. Make slides of bone marrow, taking care that the conditions did not break up cell attachments, and label the cells with markers characteristic of the target stage of development. Look for cell pairings between the CXCL12-labeled cells and progenitor cells labeled with the markers characteristic of the target stage of development.

7. Rearrangement starts first at the heavy-chain locus, beginning with D to J$_H$ and proceeding with V$_H$ to D. If the rearrangements at the first allele are not productive, then rearrangement starts again on the second heavy-chain locus and proceeds in the same order. Successful rearrangement at a heavy-chain locus results in the expression of a heavy chain at the surface of the B-cell progenitor in combination with the surrogate light chain, to form the pre-B-cell receptor. This occurs at the beginning of the large pre-B-cell stage. Expression of the heavy chain at the cell surface signals the cessation of further heavy-chain rearrangement.

At the light-chain locus in mice, rearrangement begins at one of the κ loci, and again, if it is not productive, it starts again on the other κ locus. If this is also not productive, the process repeats at the λ loci. In humans, the process is similar, but rearrangement may start at either the κ or the λ loci. Light-chain rearrangement is completed by the end of the small pre-B-cell stage, and the expression of the complete Ig receptor on the surface of the cell signals the beginning of the immature B-cell stage.

In T cells, rearrangement begins on successive β-chain loci. In possession of V, D, and J segments, the β-chain locus is analogous to the heavy-chain Ig locus. Successful rearrangement of the β chain results in expression of a pre-TCR on the cell surface, just as for the pre-BCR on B cells, coupled with the cessation of further β-chain gene segment recombination. Rearrangement at the α-chain locus follows. One major difference between the processes of rearrangement in T and B cells is that allelic exclusion at the T-cell α-chain locus is incomplete.

8. The membrane HEL can cross-link the HEL-specific BCR of immature B cells in the bone marrow, giving a strong negative selection signal. This should induce light-chain receptor editing to change the specificity of the BCR, and if not successful, the HEL-specific B cells probably would undergo apoptosis. No HEL-specific B cells should be found in the periphery (or were in the studies done), and hence these mice should not generate HEL-specific antibodies after immunization. (See the reference cited in Table 9-4.)

Analyze the Data Answer

1. (a) In the spleen, the *Dicer* knockout animal shows no mature B cells, indicated by the loss of the B220^{hi}CD19^{+} population. In the top bone marrow population, in the absence of *Dicer*, there are no sIgM-bearing B220^{hi} cells. The second bone marrow panel shows retention of the progenitor cell marker, c-Kit, in the *Dicer* knockout. The third panel shows that in the absence of *Dicer* there is a loss of CD25, a marker characteristic of the point in development at which the pre-BCR is expressed on the cell surface. This suggests that the cells cannot get past this checkpoint.

(b) Since no IgM is expressed on the cell surface of the Dicer knockout animals, miRNAs must be necessary for progression to the pre-B-cell stage, at which IgM is first expressed on the cell surface. The presence of c-Kit and of low amounts of CD25 suggests that miRNAs may be acting at the pre-BCR checkpoint.

2. (a) No. The fraction of cells labeled with annexin A5 (Annexin V in the figure) and therefore in the pre-apoptotic state is identical in the control and *Dicer* knockout populations. (b) Yes. The fraction of cells labeled with annexin A5 increases from 9.2% in the control to 65% in the *Dicer* knockout. (c) Pulling together the data from the preceding two questions, I would hypothesize that miRNAs aid in controlling expression of the pre-BCR on the cell surface, thereby allowing the cells to pass through the first checkpoint in development. Cells that cannot express pre-BCR die by apoptosis, and that is the process described in the second set of slides.

Chapter 10

1. (a) Anergy: Signal 1 (if TCR is engaged) without costimulatory signal 2 because CTLA-4 Ig will block the ability of CD28 to bind CD80/86. (b) No anergy: Signal 1 and signal 2 are both generated. (c) Anergy: Signal 1 without signal 2.

(d) No anergy, but no activation either: Neither signal 1 nor signal 2 is generated.

2. (a) Very likely: Any activated professional APC, like a dendritic cell, up-regulates MHC molecules and costimulatory ligands, making them ideal activators of T cells. (b) Very unlikely: Activated dendritic cells travel to the draining lymph nodes (or spleen) and encounter naïve T cells there, not in peripheral tissues. Naïve T cells travel among secondary lymphoid organs, not peripheral tissues. However, effector T cells, and some memory T cells, do travel to peripheral tissues and can be activated by dendritic cells there. (c) Very likely: TCR stimulation rapidly induces Ca^{2+} mobilization. (d) Very unlikely: The virus induced dendritic cells to make IL-12, one of the central polarizing cytokines for the T$_H$1 lineage. (e) Very unlikely: Central memory cells were certainly generated, too. (f) False: This response is likely to be a type 1 response.

3. A mouse without GATA-3, the master regulator for T$_H$2 lineage commitment, will be unable to generate T$_H$2 cells, which are instrumental in mounting the immune response to worm infections. T$_H$2 cells help B cells to produce IgE, which has potent antiparasite activity.

4. You will need to supply signal 1 (anti-TCR), signal 2 (anti-CD28), and signal 3 (IL-12). CTLA-4 Ig and anti-CD80 antibody both bind to the ligands for the costimulatory receptors and would not engage your T cells.

5. (a) Dendritic cells are best at activating naïve T cells—they express a high density of costimulatory ligands and MHC molecules. (b) ICOS is a positive costimulatory receptor (it is expressed on some effector T cells, including T$_{FH}$ cells). (c) Most cells do not express costimulatory ligands. Professional APCs (and thymic epithelial cells) are among the only cells that do. (d) ICOS and CTLA-4 also bind B7 family members (CD80 and CD86). PD1 also binds a B7-like molecule, PD-L1. (e) Signal 3 is provided by cytokines, which include the polarizing cytokines that induce helper T-cell lineage differentiation. (f) It is a disease caused by T-cell response to superantigens (bacterial and/or viral), not autoantigens. (g) They mimic some TCR–MHC class II interactions. (h) They do not have any receptor for MHC class I and do not interact directly with CD8^{+} T cells via their TCRs, which bind to MHC class II. (i) Naïve T cells do not produce any effector cytokines. (j) They are master transcriptional regulators of T helper cell lineage differentiation. (k) APCs can make some polarizing cytokines, but many of these cytokines originate from other cells, including other T cells, B cells, mast cells, and NK cells. (l) Bcl-6 is a master transcriptional regulator of T$_{FH}$ lineage differentiation. (m) T$_{FH}$ and T$_H$2 cells are classically the major sources of B-cell help, although all helper subsets can interact with B cells and influence Ig class switching. (n) They inhibit T-cell activation. (o) Effector cytokines have many different cellular targets, including B cells, endothelial cells, stromal cells in tissues, innate immune cells, and so on, as well as other T cells. (p) Central memory cells tend to reside in secondary lymphoid organs. (q) CCR7 attracts cells to secondary lymphoid tissue, and effector memory cells tend to rove the periphery. They typically down-regulate CCR7.

6. (a) True. It stimulates production of both FoxP3 and RORγt. (b) False. IL-6 in combination with TGF-β polarizes cells to the T_H17 lineage, an event that requires RORγt. IL-6 acts in part by inhibiting expression of FoxP3.

7. Tyrosine kinases initiate the TCR signaling cascade. Two of the first enzymes activated by TCR engagement are tyrosine kinases: Lck and ZAP-70. These phosphorylate several molecules, activating new kinases and providing sites for interaction with other signaling proteins.

8. Lck, ZAP-70, LAT, Ca^{2+}, NFAT

9. T_H22 differentiation should be regulated by a distinct transcriptional protein (**master regulator**), which is induced by distinct **polarizing cytokines** and should result in production of a unique panel of **effector cytokines**.

10. There are many different "right" answers. Benefits (in theory): T_{SCM} can provide protection to the individual indefinitely because they self-renew. They can also differentiate into several different effector subsets (they are not restricted to one type). Risks: Stem cells are often more vulnerable to transformation (they may become cancerous more easily); they may also differentiate into effector cells that might not be as useful.

Clinical and Experimental Focus Answer

The data show that LIF inhibits T_H17 polarization (in its presence, the frequency of IL-17$^+$ cells is reduced by 50% after cells are exposed to T_H17 polarization conditions). T_H1 differentiation appears unaffected (the same frequency of IFN-γ$^+$ cells are present after exposure to T_H1 polarization conditions). This suggests that LIF has a specific effect on the pathways that induce T_H17 differentiation or on those responsible for production of IL-17. It could interfere with any of the steps involved including (from outside to inside) (1) signaling induced by TGF-β or by IL-6 polarizing cytokines, (2) RORγt expression itself, or (3) IL-17 expression itself. A reduction in T_H17 cells could result in less inflammation and the amelioration in disease that is seen in this model.

(It turns out that LIF acts in opposition to IL-6 and blocks its downstream signaler, STAT3. This abrogates the inhibitory effect that IL-6 has on FoxP3 expression, shifting the balance to T_{REG} rather than T_H17 lineage commitment. So, disease amelioration is not just a consequence of fewer activated T cells, but a result of the increase in cells that quell T-cell responses.)

Chapter 11

1. TI-1 antigens are mitogenic and induce activation through both the BCR and innate immune receptors. TI-2 antigens bind tightly to the complement components C3d and C3dg, and so are bound by both the BCR and the complement receptor CD21 (CR2).

2. See the following figure:

(a) B-2 (follicular) B cells
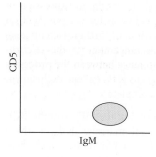

(b) B-1 B cells
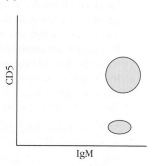

(c) B-2 (follicular) B cells
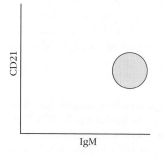

(d) MZ B cells
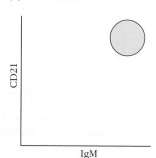

(a and b) B-2 (follicular) B cells bear relatively high levels of IgM but do not express CD5. The majority of B-1 B cells, known as the B-1a fraction, does express CD5. However, there is a minority fraction of B-1 B cells, the B-1b fraction, that does not express CD5. (c and d) B-2 B cells express normal levels of CD21, the complement receptor, whereas marginal zone (MZ) B cells express particularly high levels of CD21.

3. No to both. Both class switch recombination and somatic hypermutation require the ability of T cells and B cells to interact with each other through the binding of B-cell CD40 molecules by the CD40L molecule on T cells.

4. Since activation-induced cytidine deaminase (AID) is required for both class switch recombination and for somatic hypermutation, I would expect my knockout mouse to be unable to express any classes of antibody other than IgM. Furthermore, I would expect the average affinity of the antibodies produced by the knockout mouse to be unchanged between primary and secondary stimulation, since, in the absence of AID, the antibodies' genes will not be subject to somatic hypermutation.

5. See the following figure:

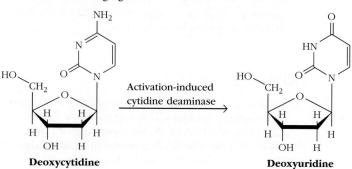

Deoxycytidine → Activation-induced cytidine deaminase → Deoxyuridine

During SHM, the deamination of cytidine on one strand of the DNA encoding antibody variable regions leads to the formation of a mismatched G-U pair. The mismatch can then be recognized by a number of DNA repair mechanisms in the cell and resolved in one of several different ways. The simplest mechanism is the interpretation of the deoxyuridine as a deoxythymidine by the DNA replication apparatus. In this case, one of the daughter cells would have an A-T pair instead of the original G-C pair found in the parent cell. Alternatively, the mismatched uridine could be excised by a uracil-DNA glycosylase enzyme. Error-prone polymerases would then fill the gap as part of the cell's short-patch base excision repair mechanism. Third, mismatch repair mechanisms could be induced that result in the excision of a longer stretch of DNA surrounding the mismatch. The excised strand could then be repaired by error-prone DNA polymerases, leading to a series of mutations in the region of the original mismatch.

In the case of class switch recombination (CSR), AID deaminates several cytidine residues in the switch (S) regions upstream of the two heavy-chain constant regions between which the class switch will occur (the donor and acceptor S regions). The resulting uridine residues are excised by uridine glycosylases, and the abasic sites are then nicked by endonucleases that create single-strand breaks at the abasic sites. These single-strand breaks are converted to double-strand breaks suitable for end joining by mismatch repair mechanisms. A constellation of enzymes then faithfully reconnects the two S regions, with the excision of the intervening DNA.

6. In order to survive, B cells need to receive signals from T cells. Since there are many more antigen-specific B cells than T cells within the germinal centers, B cells must compete with one another for T-cell binding. Since T cells are specific for peptide antigen displayed in the groove of MHC class II molecules, B cells that have internalized and displayed more antigen will have a selective advantage in attracting T-cell attention. B cells with higher-affinity receptors will bind, internalize, and display more antigen than B cells with lower-affinity receptors, and therefore compete successfully for T-cell help and survival signals. B cells with higher-affinity receptors have even been demonstrated to strip antigen from lower-affinity B cells.

7. The presence of circulating immune complexes serves as an indicator that the host organism has made a high concentration of antigen-specific antibodies and has succeeded in neutralizing the antigen. Therefore, no more antibody production is needed, and the host should not expend further energy in generating antibodies of this specificity. IgG-containing immune complexes are recognized by the Fc receptor FcγRIIb (CD32), and coligation of the immune complex by FcγRIIb and by the BCR results in phosphorylation of the ITIM domain on the cytoplasmic tail of FcγRIIb. Docking of the SHP phosphatase at this receptor molecule allows it to dephosphorylate PIP_3 to PIP_2. This interferes with transmission of antigen signals at the B-cell receptor, resulting in the down-regulation of B-cell activation.

8. B-10 B cells have recently been shown to secrete the immunosuppressive cytokine IL-10 on antigen stimulation.

9. (a) Small, soluble antigens can be directly acquired from the lymphatic circulation by follicular B cells, without the intervention of any other cells. These antigens enter the lymph node via the afferent lymph and pass into the subcapsular sinus (SCS) region. Some small antigens may diffuse between the SCS macrophages that line the sinus to reach the B cells in the follicles. (b) Other small antigens leave the sinus through a conduit network. Follicle B cells can access antigen through gaps in the layer of cells that form the walls of the conduits. (c) Larger antigens are bound by complement receptors on the surfaces of SCS macrophages. Antigen-specific B cells within the follicles can acquire the antigens directly from the macrophages and become activated.

Analyze the Data Answer

(a) The two left-hand plots show considerable numbers of cells that are high in B220 but low in CD138. This suggests that these mice have considerable numbers of B cells that have not formed plasma cells during the time allowed by the experiments. The plot on the right also shows some B220^{high} cells, but not as many as in the other two plots. Looking at the CD138^{high} populations, the fractions of cells that are CD138^{high} in the plots derived from the IgM$^+$ and IgG$^+$ naïve populations are very similar (13.2% and 10.7%, respectively) whereas the fraction of CD138^{high} cells that is derived from the IgG1$^+$ memory B-cell populations is much bigger (66%). I also note that the CD138^{high} population in the IgG1$^+$ memory B-cell fraction appears to be lower in B220 expression than the CD138^{high} cells in the other two populations.

(b) It more closely resembles that of the mouse that received the naïve, IgM-bearing cells.

(c) I would conclude that the cytoplasmic region of the IgG receptor alone is insufficient to confer memory status on a B cell.

(d) I would want to know what the cell populations looked like prior to antigen stimulation.

(e) There are many "right" answers here. For example, this experiment appears to confirm that the extra piece of the IgG receptor does not confer memory status on a cell, but I would like to engineer a B cell that carries a constant region heavy-chain γ1 gene that does not contain the extra segment and study its memory response.

Chapter 12

1. FcγRIII and CD23 are both Fc receptors (see Table 12-2). Neutralization and complement fixation are antibody functions that do not rely on Fc receptors (although FcR can help mediate the clearance of neutralized antibody-pathogen complexes). Opsonization and antibody-dependent cell-mediated cytotoxicity (ADCC) are mediated by cells that express activating FcRs, including FcγRIII. CD23, however, is an inhibitory FcR that regulates (inhibits) the activity of other activating FcRs. So, in short, antibodies to FcγRIII would block opsonization and ADCC, but not neutralization and complement fixation. Antibodies to CD23 would

not inhibit any process. (Because they block inhibitory signals they could theoretically enhance opsonization and ADCC; it is not clear, however, that this would occur in all contexts.)

2. (a) False: Some FcγRs, such as FcγRIIB, are inhibitory receptors. Others are expressed on cells that are not phagocytes and can function in transport across tissues (e.g., FcγRn transporting IgG across the placenta). (b) False: IgE mediates degranulation and activation of mast cells, basophils, and eosinophils. (c) True. (d) False: IgM is a polymeric immunoglobulin that can also be transported across epithelial layers. (e) True. (f) True. (g) False: There are two pathways by which cytotoxic T cells kill target cells. One pathway is perforin- and granzyme-dependent, and the other uses Fas ligand expressed by the CTL to induce death of Fas-expressing cells. (h) False: Dendritic cells can be licensed by TLR signals instead of by T_H cells. T-cell help is required for optimal proliferation and memory-cell generation. (i) True. (j) True.

3. The monoclonal antibody to LFA-1 should block formation of the CTL–target cell conjugate. This should inhibit killing of the target cell.

4. (a) If anti-Zobola antibodies contribute to protection against Zobola virus, then transferring antibodies to mice before infecting the mice with the virus should provide protection. As a control, transfer antibodies from a nonimmunized mouse.

(b) Isolate $CD8^+$ T cells from previously immunized and nonimmunized mice, using fluorescent anti-CD8 antibodies and a fluorescence-activated cell sorter, and transfer them into mice prior to infection with Zobola virus. If immunization had stimulated the production of CTLs (or memory CTLs) specific for Zobola, mice receiving $CD8^+$ cells from the immunized mice, but not from the nonimmunized mice, would be protected.

(c) Isolate NK cells from previously immunized and nonimmunized mice, using antibodies to the NK1.1 marker (or other markers) found on mouse NK cells and the fluorescence-activated cell sorter, and transfer them into mice prior to infection with Zobola virus. If immunization had stimulated increases in NK cells or NK memory cells effective against Zobola, then mice receiving NK cells from the immunized mice, but not from the nonimmunized mice, would be protected.

5. (a) All; (b) all; (c) CTLs; (d) CTLs; (e) none; (f) all; (g) some NK and NKT cells; (h) some NKT cells; (i) CTLs and NKT cells; (j) NKT cells; (k) all; (l) CTLs and NKT cells; (m) NK cells; (n) none; (o) all; (p) some CTLs.

6. See the following table:

Source of primed spleen cells	Killing of LCMV-infected target cells			
	B10.D2 ($H2^d$)	B10 ($H2^b$)	B10.BR ($H2^k$)	(BALB/c × B10) F_1 ($H2^{b/d}$)
B10.D2 ($H2^d$)	+	–	–	+
B10 ($H2^b$)	–	+	–	+
B10.BR ($H2^k$)	–	–	+	–
(BALB/c × B10) F_1 ($H2^{b/d}$)	+	+	–	+

T cells will lyse targets expressing peptides from the antigen to which they were primed (LCMV) and expressing the MHC to which they are restricted (syngeneic MHC). These requirements are met in all cases where there is a positive symbol. The very observant student might also recognize that T cells of one strain will also react to alloantigens (cells from another strain expressing a distinct MHC haplotype). In fact, this is true, and there will be some background death in all cases of MHC "mismatch" (in other words, you would also notice some "background" killing in conditions that are labeled with a "−"). However, because the T cells have been primed by immunization to LCMV, the LCMV-specific MHC-restricted response would be a secondary response and would dominate the primary alloresponse.

7. To determine T_C activity specific for influenza, perform a cell-mediated lympholysis (CML) reaction by incubating spleen cells from the infected mouse with influenza-infected syngeneic target cells. To determine T_H activity, incubate spleen cells from the infected mouse with syngeneic APCs presenting influenza peptides; measure IL-2 production.

8. The "missing self" model has been used to explain how NK cells detect infected or tumor cells. If a potential target cell expresses normal levels of MHC class I molecules, inhibitory receptors on the NK cell (KIR, CD94, NKG2A) induce a signal transduction cascade that abrogates NK lytic activity. These negative signals override prokilling signals generated via ligands binding to activating receptors on the NK cell (NKR-P1 and others). Some tumor cells and virally infected cells, however, reduce their expression of MHC class I and no longer stimulate NK inhibitory receptors. In this case, the activating (prokilling) receptor signal dominates.

9. There are two pathways by which cytotoxic T cells kill target cells: one that is perforin-dependent and one that uses Fas ligand (FasL) to induce death in Fas-expressing target cells. (And, as always, T cells will lyse only cells expressing the MHC-peptide combinations to which they are specific and restricted.) T cells from immunized perforin knockout $H2^d$ mice will be able to lyse (d). (These T cells will depend on FasL-Fas interactions to kill. Target cells don't need to express perforin to be susceptible, but they do need to express Fas.) T cells from immunized FasL knockout $H2^d$ mice will be able to lyse (d), but they will also be able to lyse (e). (These cells will depend on perforin-mediated pathways. Targets do not need to express perforin or Fas to be susceptible.) T cells from $H2^d$ mice in which both perforin and Fas ligand have been knocked out will not be able to lyse any of the cell types.

10. If the HLA (human MHC) type is known, MHC tetramers bound to the peptide generated from gp120 and labeled with a fluorescent tag can be used to specifically label all of the $CD8^+$ T cells in a sample that has T-cell receptors capable of recognizing this complex of HLA and peptide.

11. (a) True. (b) False: They need to express Fas, which transmits the proapoptotic signal. (c) False: Both mechanisms induce caspase activation. (d) False: Only the perforin-mediated pathway depends on granzyme activity. (e) True.

(f) False: Perforin is responsible for the development of surface membrane and endocytic membrane pores.

Clinical Focus Answers

1. Arthritis is characterized by inflammation of a joint leading to damaged tissue, swelling, and pain. Psoriatic arthritis is accompanied by skin lesions caused by immune attack (psoriasis). Association of *KIR-MHC* combinations with susceptibility to arthritic disease would likely stem from a deficiency of inhibitory signals (e.g., absence of *MHC* alleles that produce inhibitory ligands for specific KIR molecules), leading to damage of host cells and tissues. Diabetes is another autoimmune disease; exhibiting destruction of the host pancreatic islet cells that produce insulin, the same mechanisms predicted for arthritis could operate in diabetes. In both cases, the absence of inhibitory NK signals could lead to damage inflicted by NK cells, directly or through their recruitment of other effector cells.

2. Rab27A is a GTPase that regulates the transport of intracellular vesicles (granules) to the cell membrane. Transport is required for the release of vesicular contents into the extracellular space. Many cells depend on this ability to function, including cytotoxic T cells, which release perforin and granzyme from internal vesicles, and melanocytes, which release pigment from vesicles (melanosomes). Without this capacity, an individual will be unable to kill infected cells and will exhibit a form of albinism. Many other cells could be affected, including granulocytes (eosinophils, basophils, and mast cells), although it is important to recognize that some express other Rab variants that compensate for the loss of Rab27A function.

Analyze the Data Answer

(a) Epitopes 2, 12, and 18 generated high CTL activity, and epitopes 5 and 21 generated medium activity. (b) It is possible that different peptides use distinct anchor residues, which would make this prediction more difficult. However, if we make the assumption that the same amino acids would be bound by HLA-A2, it appears that a leucine (L) on the amino-terminal side separated by four amino acids from a threonine (T) is the only common motif for the five most immunogenic peptides (2, 12, 18, 5, and 21). All of the peptides that generate high CTL activity have two consecutive leucines on the amino-terminal side as well. The problem with the threonines serving as anchors is that peptide 2 has four amino acids at the carboxyl side of the T, which seems to result in the end of the peptide extending out of the binding pocket. This would be a very unusual configuration. Peptide 12 may have a less dramatic but similar problem. Therefore, a leucine in pocket 2 of HLA-A2 would be consistent with the data generated by M. Matsumura and colleagues in 1992 (Emerging principles for the recognition of peptide antigens by MCH class molecules. *Science* **257**:927). Thus, the main anchor may be a leucine residue at the amino end of the binding pocket, with possible contribution by threonine at the carboxyl end under some circumstances. (c) It is possible that there are no T cells specific for those peptides, even if they are presented in complex with MHC. Therefore, you would not see a CTL response. (d) CTLs recognize antigen only in the context of self MHC molecules. Therefore, in order to

assess CTL activity, the T2 cells also had to express HLA-A2. (e) MHC class I molecules typically bind peptides containing 8 to 10 residues (see Figure 7-5). Peptide 2 is 11 residues long, suggesting that it bulges in the middle when bound. Since it appears to be a major epitope for CTL killing, bulging does not seem to interfere with CTL interaction and may contribute.

Chapter 13

1. An open-ended question, but essentially one can conclude that the colonization of germ-free mice with norovirus almost fully restores intestinal epithelium health—and immune cell number and activity—to its normal state. Many questions can be asked to follow up on this observation. Does this virus restore the intestinal immune system as well as a bacterial species? What is the mechanism behind this? Is IFN-γ involved in restoring health? The original article (Kernbauer, E., et al. *Nature* 2014; **516**:94) is quite accessible and can be used to answer and explore questions.

2. This is an image of a stained section of the small intestine, not the skin. Several features give this away quickly, including the single, rather than multiple layers of epithelium, the villi, and the presence of goblet cells (white, mucus-filled cells in the epithelial layer).

3. Many possible answers.

4. (a) Goblet cells. (b) Langerhans cells, dendritic cells, ILCs, IELs, eosinophils. (c) Tuft cells. (d) Langerhans cells, dendritic cells. (e) IELs. (f) Probably all! (g) Paneth cells (intestine); M cells (intestine and respiratory tract); enterocytes (intestine); tuft cells (intestine and probably respiratory tract); IELs (intestine and probably respiratory tract; skin has resident lymphocytes, but they are not strictly referred to as IELs); goblet cells (intestine and respiratory tract); eosinophils (all); Langerhans cells (skin); club cells (respiratory tract); ILCs (all); keratinocytes (skin); dendritic cells (all).

5. (a) IL-4, IL-10, IFN-γ, IL-13. (b) TSLP and IL-33 initiate a type 2 response; IL-4 and IL-13 are effector cytokines. (c) TSLP and IL-33 are two alarmins (IL-25 is another). (d) IL-10.

6. (a), (c), and (d)

7. (c)

8. (a) and (e)

Chapter 14

1. You have not designed an experiment that allows you to focus on antigen-specific T cells—which will represent only a fraction of the whole population. You will need to find a better way to track these. (What are the possibilities? Use TCR transgenics specific for influenza [most, if not all, cells will be antigen specific], or modify the virus so that it expresses another antigen [e.g., OVA] that can be seen by TCR-transgenic cells. Or, try to isolate influenza-specific

T cells [by tetramer staining? Difficult, but possible in theory].)

2. (a) False: All leukocytes respond to chemokines—they are one of the central regulators of immune cell migration. (b) False: Naïve B cells do not express CCR7 (which helps to send cells to the paracortex), but when activated by antigen binding, they up-regulate it so that they travel to the paracortex to find T-cell help. (c) False: Small antigens (typically opsonized [by complement, for instance]) arrive on their own via the afferent lymphatics. (d) False: They "arrest" their migrating behavior. (e) False: They crawl along the fibroblastic reticular cell network. (f) False: It appears from recent data that these networks can be established at sites of infection. (g) False: Strong adhesion requires chemokine activation. Rolling is the first event (mediated by selectins). (h) True.

3. The movement of an antigen-activated B cell from the follicle to the border between the follicle and paracortex is a classic example. However, there are many others, including the response of innate cells to signals generated by inflammation at the site of infection (e.g., neutrophils are attracted to IL-8 produced by other innate cells, including other neutrophils).

4. Both receive help in the lymph node cortex. B cells travel to the interface between follicle and paracortex to receive T-cell help and remain loosely associated with the follicle during their interaction with the T cell. The CD8$^+$ T cells receive help in the paracortex, where they interact with APCs and CD4$^+$ T cells.

5. Recall that CCL3 is produced by antigen-presenting cells that have been activated by CD4$^+$ helper T cells in the lymph node. CCL3 attracts CD8$^+$ T cells to form a tricellular complex so that they can receive optimal help. Without this cytokine, CD8$^+$ T cells may not find their way to antigen-presenting cell/CD4$^+$ T-cell pairs and may not be optimally activated. On the other hand, other chemokines (e.g., CCL4) may be able to compensate.

6. (a) Rolling, chemokine interactions, adhesion, transmigration. (b) Adhesion: Adhesion molecules such as LFA-1 and VLA-4 are converted to their high-affinity forms by chemokine receptor signals (via an inside-out process). (c) The homing and chemokine receptors expressed by naïve lymphocytes attract them to secondary lymphoid tissues. For example, they express L-selectin (CD62L), which interacts with ligands on specialized endothelial structures (high-endothelial venules) located in the cortex of the lymph node.

7. Both the pattern of expression of chemokine receptors and chemokines regulate the compartmentalization of T and B cells in the lymph node. For example, naïve T cells express CCR7, which interacts with chemokines that decorate the fibroblastic reticular cell network in the paracortex. Naïve B cells express CCR5, which is expressed by cells in the follicle and by the follicular DC network. T-cell and B-cell movement is guided by the routes laid down by these networks.

8. True: Germinal center B cells are more motile and extend unexpectedly long processes within the germinal center.

9. (a) −; (b) ✓; (c) ✓; (d) −; (e) ✓; (f) −; (g) ✓.

10. (a) Naïve T and B cells would not home properly to the HEV. (The individual would be significantly immunocompromised, although possibly able to compensate with some innate immune activity and splenic T-cell and B-cell activity.) (b) Naïve T cells would not home properly to the paracortex. Activated B cells would not home properly to the follicle-paracortex border. Animal would not be able to develop optimal adaptive immune responses unless compensated by other chemokines. (The individual would be immunocompromised.) (c) Naïve B cells would not home properly to the follicle. Animal would not be able to develop T cell–dependent antibody responses unless compensated by other chemokines. (The individual would be partially, but still significantly, immunocompromised.) (d) Naïve T and B cells, effector T and B cells, and effector memory T and B cells would not be able to leave the lymph node (and other tissues). (The individual would be immunocompromised unless compensated by other egress regulators and extralymphoid immune cell activity.)

11. Note that all cells can transit via blood and lymphatics.

    Resident memory T cell: Barrier organs, possibly brain

    Central memory T cell: T-cell zones of secondary lymphoid organs, secondary lymphoid organs

    Effector T cell: Secondary lymphoid organs, barrier organs, brain (sometimes), T-cell zones and sinuses of secondary lymphoid organ (sometimes)

    Plasma cell: Bone marrow, sinuses of secondary lymphoid organ, follicles (for a short time)

    Naïve lymphocyte: Secondary lymphoid organs, bone marrow

    Dendritic cell: All sites

    CD169$^+$ macrophage: Sinuses of secondary lymphoid organs

    T$_{FH}$ cell: Follicle, secondary lymphoid organ, T-cell zone of secondary lymphoid organ (for a time)

Analyze the Data Answer

(a) This should be clear from the video. (b) Differences in expression of homing receptors and/or chemokine receptors would be reasonable hypotheses. Make sure to state specific possibilities (based on the information in Appendix III and examples in the text). (c) This requires creative (and rigorous) speculation on your part—we do not really know. Both these subpopulations are effector memory cells—consider what each will do if reactivated. Will they react with different kinetics? Will they stay in the same area of the tissue? Will they serve the same cell populations? Read the authors' discussion for their view of the possibilities.

Experimental Design Answer

The best experimental designs will include your question, your prediction, and your experimental design. The design must include controls (both positive and negative, if possible—and more than one at times). You must also identify what you will measure and how you will interpret that measurement.

Question: Do naïve B cells require CCR5 to localize to B-cell follicles? *Prediction:* Yes, they are absolutely dependent on this chemokine receptor; or no, they can use other chemokine receptors, although perhaps less efficiently.

Experimental Design: Fundamentally, you must compare the in vivo activities of B cells that can use CCR5 with those that cannot. *One possibility:* Compare the behavior of labeled, wild-type versus labeled, CCR5$^{-/-}$ B cells in two groups of mice. Alternatively, or in addition, track the behavior of labeled, wild-type B cells in the presence or absence of the CCR5-blocking antibody.

Once you decide on your design, you must develop a protocol using intravital two-photon microscopy (dynamic imaging). You must be able to trace naïve B-cell movements (and, ideally, identify the B-cell follicle, too). Therefore, you need to fluorescently label those B cells: in vitro CFSE staining (see Chapter 20) is probably the best approach in this case because you will be examining the behavior of B cells from different mice. It is not the only approach, however. (*Note:* To define the follicular area, you could also co-inject T cells that are labeled with a different color [few if any should be in the follicles] or come up with another more clever, original idea. Some investigators [as you may have noticed] do not directly label the follicle, but infer its location from the behavior of the cells.)

Isolate, label, and inject the cells. Wait a specified time (based on previous studies in the literature or on your own experiments), anesthetize the mice, and record cell behavior in an exposed lymph node over time.

What will you measure? Go back to your question. Identifying the number of B cells that end up in a follicle over a period of time would answer your question directly. The figure you sketch could be a bar graph comparing these numbers in each experimental condition. Direction and speed of the cells may be two other useful parameters that could be generated from an analysis of trajectories—and you can describe how they will contribute to your understanding of the question.

Clinical Focus Answer

There are many different possibilities, in theory. CD18 is part of the LFA-1 complex, which regulates extravasation of multiple subsets of leukocytes (see text and Advances Box 14-1). A CD18 deficiency could, therefore, inhibit the ability of innate immune cells to travel to the site of infection, naïve lymphocytes to enter secondary lymphoid organs, effector cells to recirculate effectively, and so on. All of these problems would severely decrease the ability of a child to fight off infection. Treatment could include genetic modification of bone marrow stem cells (reintroducing the *CD18* gene into hematopoietic stem cells), but should also include judicious antibiotic use. Check online resources for what is possible. (Wikipedia [https://www.wikipedia.org/] and any government- or university-based clinical site will likely provide good information.)

Chapter 15

1. (a) No type I (which is mediated by IgE/FcεRI interactions), normal type II (which is mediated by IgG or IgM). (b) Same as part (a). (c) May have some type I reaction, but given that this receptor is important in regulating (both enhancing and suppressing) B-cell production of IgE, the response may be abnormal. (d) Type II responses would be most impaired because IgG and IgM exert their effects, in part, by recruiting complement, as well as by inducing ADCC. (e) Type I responses are likely to be suppressed. When bound by soluble versions of FcεRII on B cells, CD21 enhances IgE production. In its absence, the animal may not be able to generate as much IgE antibody as a wild-type mouse.

(*Note:* All these answers assume that there are no other similar or redundant genes and proteins that could compensate for the absence of the gene in question.)

2. Primary mediators are found in mast cell and basophil granules and are released as soon as a mast cell is activated. These include histamine, proteases, serotonin, and so on (see Chapter 15 text). Secondary mediators are generated by mast cells and basophils in response to activation and are released later in the response. These include cytokines, leukotrienes, prostaglandins, and so on (see Chapter 15 text).

3. Histamine binds to at least four different histamine receptors. Binding to H$_2$ receptors inhibits mast cell degranulation and therefore inhibits its own release.

4. Through the induction of IgG antibodies instead of IgE antibodies and through the activation of regulatory T cells that inhibit the IgE response.

5. Rh-mismatched moms and dads can generate both Rh$^+$ and Rh$^-$ fetuses. An Rh$^+$ mother will be tolerant of the Rh antigen and will not produce antibodies that could harm either an Rh$^-$ or Rh$^+$ fetus. However, an Rh$^-$ mother has the potential to generate an antibody response against an Rh$^+$ fetus and could generate a harmful secondary response to a second, Rh$^+$ fetus. RhoGAM (anti-Rh antibodies) will clear B cells (and antibodies) generated during the first pregnancy, preventing such a secondary response.

6. See answer to question 5. Rh$^-$ babies are not at risk, but Rh$^+$ fetuses are.

7. Type III hypersensitivities are disorders brought about by immune complexes that cannot be cleared. They can activate innate immune cells that express Fc receptors and can activate complement, both of which induce inflammation. Such immune complex–mediated inflammation occurs in blood vessels, resulting in vasculitis, as well as in tissue where the complexes are deposited when they pass through inflamed, vasodilated capillaries. Multiple insults can cause these hypersensitivities, including insect bites and inhalation of fungal spores or animal protein (see Chapter 15 text).

8. (a) IV; (b) I, II, III, IV; (c) I, mainly, but III also results in mast cell activation and histamine release; (d) IV; (e) I; (f) II; (g) I, mainly (and others can benefit, too); (h) I; (i) II; (j) II; (k) I.

9. Chronic infections can cause damaging chronic inflammation. For example, viral hepatitis causes liver damage. The

text of Chapter 15 mentions gum disease due to periodontal bacteria, which can cause damage to gums and teeth. The example extensively discussed earlier is tuberculosis, where the unresolved *M. tuberculosis* infection results in recruitment of many inflammatory cells such as macrophages and tissue damage from their release of inflammatory mediators. TB can be a long-term chronic disease.

10. (a) In the absence of inflammation, insulin signaling will not be impaired by cytokine stimulation. Cytokine signals activate kinases, including JNK, that phosphorylate and inactivate *insulin receptor substrate* (IRS), a key downstream mediator of insulin receptor signaling. (Interestingly, this observation suggests that free fatty acids, alone, may not be enough to induce insulin resistance.) (b) This question requires speculation—there is no right answer. Some possibilities include (1) differences in IRS that make it less able to be phosphorylated at the serine residue, (2) differences in JNK that make it less likely to bind IRS, and (3) differences in gene regions that regulate cytokine production by adipocytes, and so on. See Chapter 15 text and use your imagination!

Analyze the Data Answer

1. Eosinophils are recruited to the sites of allergic responses (by eosinophil chemotactic factor, a primary mediator released by mast cell degranulation); the eosinophils release mediators that increase the local response. Co-injection of OVA with *Acinetobacter lwoffii* reduces the number of eosinophils relative to OVA alone or to OVA with *Staphylococcus epidermidis*. IL-5 and IL-13 are both T_H2 cytokines that play roles in allergic responses; the lower levels of mRNAs seen when OVA was co-administered with OVA indicate a reduced T_H2 response, which should lead to a less severe allergic response.

2. Mice that had been sensitized with OVA plus *A. lwoffii* had reduced IgE anti-OVA antibodies relative to those receiving OVA alone or OVA plus *S. epidermidis*. Injection of either bacterial species resulted in some reduction in IgG2a anti-OVA antibodies, but this is probably not relevant to the IgE response.

3. The induction of higher IFN-γ levels after co-injection with *A. lwoffii* indicates skewing of sensitized effector CD4$^+$ cells to the T_H1 subset (some of which will home back to the site of allergen exposure in the skin), rather than the T_H2 subset that induces IgE production and allergic responses. IL-10, which inhibits immune and inflammatory responses, can be made by monocytes, macrophages, dendritic cells, and T cells (especially T_{REG}s). In this case it may be suppressing the T_H2 response as well as the overall inflammatory response.

4. OVA alone induces differentiation of naïve T cells to the allergy-promoting subset T_H2. Co-injection of *A. lwoffii* with OVA probably induces dendritic antigen–presenting cell production of the T_H1-inducing cytokine IL-12. With lower numbers of T_H2 cells and hence lower levels of IL-4 and/or IL-13 cytokines, there is less heavy-chain class switching to IgE, necessary for activating the allergic response to OVA.

5. Yes. From the researchers' earlier studies it is known that children raised near farms and forests, who are less likely to develop allergies, have higher levels of *Acinetobacter* species on their skin. The results of this study in mice demonstrate that skin exposure to *Acinetobacter* during initial allergen sensitization to skewing immune responses away from the T_H2 responses that are necessary to drive IgE production and allergic responses.

Chapter 16

1. The process called *central tolerance* eliminates lymphocytes with receptors displaying affinity for self antigens in the thymus or in the bone marrow. A self-reactive lymphocyte may escape elimination in these primary lymphoid organs if the self antigen is not encountered there or if the affinity for the self antigen is less than what is needed to trigger the induction of apoptotic death. Self-reactive lymphocytes escaping central tolerance elimination are kept from harming the host by peripheral tolerance, which involves three major strategies: induction of cell death or apoptosis, induction of anergy (a state of nonresponsiveness), or induction of an antigen-specific population of regulatory T cells that keeps the self-reactive cells in check.

2. Tolerance is necessary to remove or regulate the many self-reactive B and T lymphocytes that we know escape negative selection in the thymus. We also need to control responses against some foreign substances that enter the body, which are either beneficial (e.g., gut commensals and components of food) or benign (e.g., tree pollen and animal dander). Without tolerance, which is defined as unresponsiveness to an antigen, unnecessary and pathogenic inflammation can arise against foreign or self antigens, such as during life-threatening hypersensitivity reactions (e.g., allergic asthma, anaphylaxis from food allergies, etc.) or any of a number of autoimmune diseases (e.g., SLE, lupus, RA, type 2 diabetes, etc.), respectively.

3. Receptor editing is a process by which B cells exchange their existing light-chain V region for another, via RAG-mediated recombination. This occurs late in B-cell development and leads to modified antigen specificity that, in some cases, will rescue a self-reactive B cell from negative selection or anergy. Because most B-cell responses are T-dependent, T-cell tolerance can override some B-cell self reactivity. If the relevant self antigen–specific T_H cells are absent or regulated, this will indirectly regulate or inhibit B-cell self reactivity.

4. (a) 6, T_H cells and CTLs; (b) 9, T_{REG} cells; (c) 8, antibodies; (d) 10, T_H cells and CTLs; (e) 7, antibodies; (f) 2, antibodies; (g) 3, T_{REG} cells; (h) 1, antibodies; (i) 5, T_H cells and CTLs; (j) 4, antibodies.

5. (a) EAE is induced by injecting mice or rats with myelin basic protein in complete Freund's adjuvant. (b) The animals that recover from EAE are now resistant to EAE. If they are given a second injection of MBP in complete Freund's adjuvant, they no longer develop EAE. (c) If T cells from mice with EAE are transferred to normal syngeneic mice, the mice will develop EAE.

6. This theory for the cause of autoimmune myocarditis posits that bacterial antigens found in group A streptococcus share similar molecular structures with host proteins found in cardiac muscle. This is believed to be the reason why some individuals develop a myocarditis shortly after infection with group A strep, presumably due to the activation of lymphocytes recognizing the streptococcus that cross react with host cell structures.

7. (1) A virus might express an antigenic determinant that cross-reacts with a self component. (2) A viral infection might induce localized expression of IFN-γ. IFN-γ might then induce inappropriate expression of MHC class II molecules on non–antigen-presenting cells, enabling self peptides presented together with the MHC class II molecules on these cells to activate T_H cells. (3) A virus might damage an organ, resulting in release of antigens that are normally sequestered from the immune system.

8. Anti-CD3 monoclonal antibodies have been used to block T-cell activity in type 1 diabetes mellitus (T1DM). Rituximab, a monoclonal antibody against the B cell–specific antigen CD20, depletes a subset of B cells and has been used to treat patients with rheumatoid arthritis (RA). Monoclonal antibodies against CD4, which deplete T_H cells, and one against IL-6, which blocks this pro-inflammatory cytokine, have also been used to treat RA. For psoriasis, a monoclonal antibody that recognizes the p40 subunit shared by IL-12 and IL-23 blocks this signaling pathway. Likewise, the fusion protein CTLA-4Ig blocks interactions between CD28 on T cells and CD80/86 on APCs, as treatment for RA, lupus, and inflammatory bowel disease. (See Table 16-5.)

9. (a) True. (b) False: IL-12, which promotes the development of T_H1 cells, increases the autoimmune response to MBP plus adjuvant. (c) False: The presence of *HLA-B27* is strongly associated with susceptibility to ankylosing spondylitis but is not the only factor required for development of the disease. Most HLA-B27$^+$ individuals *will not* develop the disease. (d) True. (e) False, or only partially true: The elimination of autoreactive T cells in the thymus, which occurs during negative selection, will still occur. However, the tT_{REG} cells generated in the thymus will be absent. Likewise, pT_{REG} cells, which are induced in the periphery and help to regulate or suppress immune reactivity, will also be absent. Therefore, *parts* of both central and peripheral tolerance will be missing.

10. (a) 7; (b) 3, 4, and 5; (c) 8; (d) 3 and 4; (e) 6; (f) 3 and 4.

11. (a) Polyclonal B-cell activation can occur as a result of infection with gram-negative bacteria, cytomegalovirus, or Epstein-Barr virus (EBV), which induce nonspecific proliferation of B cells; some self-reactive B cells can be stimulated in this process. (b) If normally sequestered antigens are exposed, self-reactive T cells may be stimulated. (c) The immune response against a virus may cross-react with normal cellular antigens, as in the case of molecular mimicry. (d) Increased expression of TCR molecules should not lead to autoimmunity; however, if the expression is not regulated in the thymus, self-reactive cells could be produced. (e) Increased expression of MHC class II molecules has been seen in T1DM and Graves' disease, suggesting that inappropriate antigen presentation may stimulate self-reactive T cells.

12. (a) False: Acute rejection is cell mediated and probably involves the first-set rejection mechanism (see Figures 16-11b and 16-13. (b) True. (c) False: Passenger leukocytes are donor dendritic cells that express MHC class I molecules and high levels of MHC class II molecules. They migrate from the grafted tissue to regional lymph nodes of the recipient, where host immune cells respond to alloantigens on them. (d) False: A graft that is matched for the major histocompatibility antigens, encoded in the HLA, may be rejected because of differences in the minor histocompatibility antigens encoded at other loci. (e) True.

13. See the following table:

Donor	Recipient	Response	Type of Rejection
BALB/c	C3H	R	FSR
BALB/c	Rat	R	FSR
BALB/c	Nude mouse	A	
BALB/c	C3H, had previous BALB/c graft	R	SSR
BALB/c	C3H, had previous C57BL/6 graft	R	FSR
BALB/c	BALB/c	A	
BALB/c	(BALB/c × C3H) F$_1$	A	
BALB/c	(C3H × C57BL/6) F$_1$	R	FSR
(BALB/c × C3H) F$_1$	BALB/c	R	FSR
(BALB/c × C3H) F$_1$	BALB/c, had previous F$_1$ graft	R	SSR

14. (a) Graft-versus-host disease (GvHD) develops as donor T cells recognize alloantigens on cells of an immune-suppressed host. The response develops as donor T_H cells are activated in response to recipient MHC-peptide complexes displayed on APCs. Cytokines elaborated by these T_H cells activate a variety of effector cells, including NK cells, CTLs, and macrophages, which damage the host tissue. In addition, cytokines such as TNF may mediate direct cytolytic damage to the host cells. (b) GvHD develops when the donated organ or tissue contains immunocompetent lymphocytes and when the host is immune suppressed. (c) The donated organ or tissue could be treated with monoclonal antibodies to CD3, CD4, or the high-affinity IL-2 receptor (IL-2R) to deplete donor T_H cells. The rationale behind this approach is to diminish T_H-cell activation in response to the alloantigens of the host. The use of anti-CD3 will deplete all T cells; the use of anti-CD4 will deplete all

T_H cells; the use of anti–IL-2R will deplete only the activated T_H cells.

15. The use of soluble CTLA-4 or anti–CD40 ligand to promote acceptance of allografts is based on the requirement of a T cell for a costimulatory signal when its receptor is bound. Even if the recipient T cell recognizes the graft as foreign, the presence of CTLA-4 or anti-CD40L will prevent the T cell from becoming activated because it does not receive a second signal through the CD40 or CD28 receptor (see Figure 16-18). Instead of becoming activated, T cells stimulated in the presence of these blocking molecules become anergic. The advantage of using soluble CTLA-4 or anti-CD40L is that these molecules affect only those T cells involved in the reaction against the allograft. These allograft-specific T cells will become anergic, but the general population of T cells will remain normal. More general immunosuppressive measures, such as the use of CsA or FK506, cause immunodeficiency and subsequent susceptibility to infection.

16. Azathioprine is a mitotic inhibitor used to block proliferation of graft-specific T cells. Cyclosporin A, FK506 (tacrolimus), and rapamycin (sirolimus) are fungal metabolites that block activation and proliferation of resting T cells. Ideally, if early rejection is inhibited by preventing a response by specific T cells, these cells may be rendered tolerant of the graft over time. Lowering the dosage of the drugs is desirable because of decreased side effects in the long term.

17. (a) IL-2 receptor—*underexpression* of IL-2 or the IL-2R would be associated with a reduction in the generation and maintenance of T_{REG} cells (a key polarizing cytokine for this cell type), favoring inflammation over regulation and the development of autoimmunity. (b) CTLA-4—this negative regulatory molecule is found on T cells and is involved in suppression of the immune response. Dysfunction in this molecule that biases toward *absence or underperformance* will lead to overactivation or prolonged activation of T cells, again resulting in a tendency toward sustained inflammation and autoimmune reactions. (c) CD40—this is a costimulatory molecule involved in activation of antigen-presenting cells on B cells. *Enhanced activity* of this molecule (a variant of which has been associated with RA and Graves' disease) would lower the threshold for costimulation in APCs and B cells, allowing increased APC activation and antibody production that can, at times, be directed against autoantigens. (d) ERAP1—this aminopeptidase is involved in antigenic trimming in the ER during MHC class I peptide loading. Therefore, variants of this gene can have a role in which peptides are presented to the immune system. In particular, when there is a strong association of a particular MHC allele and autoimmunity (e.g., ankylosing spondylitis), certain variants of ERAP could lead to more efficient trimming of self peptides into fragments that can be presented to the immune system and stimulate autoimmunity. Therefore, rather than more or less ERAP, it is the variant form of this enzyme combined with the MHC alleles of the host that can bias toward autoimmunity. (e) FoxP3—*suppression or inactivation* of this gene would result in a reduction or absence of both tT_{REG} and pT_{REG} cells. This tips the inflammation:regulation balance toward inflammation and could result in attacks on self tissues (autoimmunity) or benign foreigners (e.g., food or gut commensals).

18. Local innate inflammatory responses, including complement activation and phagocytosis, will lead to local inflammation, especially in the blood vessels of the newly engrafted organ. This can lead to occlusion of blood vessels and poor perfusion (low oxygen), with further cell death. Adaptive responses will likely occur both locally and in the draining lymph node, where direct (via donor APC) and indirect (via recipient APC) presentation of MHC alloantigens will lead to selection and activation of recipient T cells capable of recognizing foreign MHC (whole or processes, respectively). Without immune suppression, the damage and resulting danger signals that accompany surgery and cell death in the transplanted tissue will result in APC activation and enhanced MHC expression and costimulatory potential. These activated T cells will then produce cytokines (mostly T_H1) that recruit and further activate immune cells and that help encourage activation of B cells with immunoglobulins specific for allo- or tissue-specific antigens. Finally, T_H1 cells can produce cytokines (IL-2 and IFN-γ) and license DCs, allowing activation of alloantigen-specific CTLs. Finally, graft-specific CTLs, T_H1 cells, and antibody will collaborate to induce cell death and inflammation, which leads to the death of the engrafted tissue.

Clinical Focus Answers

1. The observations that women mount more robust immune responses and more T_H1 pathway–directed responses than men, as well as the effects of female sex hormones on the immune response, may in part explain sex-based differences in susceptibility to autoimmunity. Since the T_H1 type of response is proinflammatory, the development of autoimmunity may be enhanced. However, this also means that women should be, in general, less likely to mount T_H2-mediated reactions (recall from Chapter 10 that T_H1 and T_H2 pathways are antagonistic). This could result in a lower prevalence of allergy (hypersensitivity) in women, as well as a disadvantage during encounters with helminths and other parasites that require the action of IgE (see Chapter 17).

2. (Multiple answers are possible; here are a few examples.) In the case of most T cell–mediated autoimmune diseases, anything that boosts or induces T_{REG} cells (especially those specific for the autoantigen), or is aimed at building tolerance and suppressing inflammation against autoantigens, would be an example of induction of peripheral tolerance. Example drugs/procedures include addition of IL-10 (to encourage T_{REG} cell production or expansion, and suppression of the immune response), blocking T_H17 cytokines (such as IL-17 or IL-23) to enhanced action of T_{REG} cells, or interference with costimulation (e.g., using a CTLA-4 fusion protein, such as Abatacept and Belatacept). These same procedures could be used to treat transplant rejection, with the addition of specific tolerance-inducing procedures as the autoantigens in this case are known (alloantigens).

Donor leukocyte infusion is one example of a preparatory tolerizing procedure, again, likely working through peripheral tolerance, which can prime the immune system for allograft acceptance.

3. The ideal donor animal for xenotransplantation would have organ sizes roughly equivalent to those of humans, would grow quickly, and would either be similar in genetic make-up to humans (e.g., nonhuman primates) or highly amenable to genetic alteration (e.g., removal of antigens associated with rejection). It should be free of any disease that can be passed to humans, and this is favored by phylogenetic distance (e.g., nonhuman primates are not very distant but pigs are). The test of the organs must include transplantation into nonhuman primates first and observation periods that are sufficiently long to ascertain that the organ remains fully functional in the new host and that no disease is transmitted. You need to consider ethical controversies related to experimentation in animals, but especially nonhuman primates. Finally, even after jumping these hurdles, you must consider that some agricultural animals are forbidden as food or revered (e.g., pigs, cows, etc.), and their use might directly conflict with specific cultural and/or religious beliefs. By considering and addressing these issues at the outset, you are practicing greater cultural and religious inclusion and sensitivity; in other words, you are working toward a goal of application of this new technique or resource for everyone, not just a specific segment of society with which you are most familiar.

Chapter 17

1. Nonspecific host defenses include ciliated epithelial cells, bactericidal substances in mucous secretions, complement split products activated by the alternative pathway that serve both as opsonins and as chemotactic factors, and phagocytic cells.

2. Specific host defenses include humoral immunity, which targets the destruction of extracellular infections (bacterial, fungal, or parasitic) or neutralization of all types of pathogens during extracellular stages, CTLs that identify and eliminate virally infected host cells, and T helper cells that secrete cytokines to assist other leukocytes in the elimination of both intracellular and extracellular pathogens.

3. When the majority of a population is immune to a particular pathogen—that is, there is herd immunity—then the probability of the few susceptible members of the population contacting an infected individual is very low. Thus, susceptible individuals are not likely to become infected with the pathogen. If the number of immunized individuals decreases sufficiently, most commonly because of reduction in vaccination rates, then herd immunity no longer operates to protect susceptible individuals and infection may spread rapidly in a population, leading to an epidemic.

4. Humoral antibody peaks within a few days of infection and binds to the influenza HA glycoprotein, blocking viral infection of host epithelial cells. Because the antibody is strain specific, its major role is in protecting against re-infection with the same strain of influenza.

5. (a) African trypanosomes are capable of antigenic shifts in the *variant surface glycoprotein* (VSG). The antigenic shifts are accomplished as gene segments encoding parts of the VSG are duplicated and translocated to transcriptionally active expression sites. (b) Influenza is able to evade the immune response through frequent antigenic changes in its hemagglutinin and neuraminidase glycoproteins. The antigenic changes are accomplished by the accumulation of small point mutations (antigenic drift) or through genetic reassortment of RNA between influenza virions from humans and animals (antigenic shift).

6. (a) A^b. (b) Because antigen-specific, MHC-restricted T_H cells participate in B-cell activation.

7. (a) BCG (bacille Calmette-Guérin); (b) antigenic shift, antigenic drift; (c) gene conversion; (d) tubercles, T_H cells; activated macrophages; (e) toxoid; (f) IFN-α, IFN-γ; (g) secretory IgA; (h) IL-12, IFN-γ.

8. Most fungal infections prevalent in the general population do not lead to severe disease and are dealt with by innate immune mechanisms and lead to protective adaptive responses. Problematic fungal infections are more commonly seen in those with some form of immunodeficiency, such as patients with HIV/AIDS or those with immunosuppression caused by therapeutic measures.

9. One possible reason for the emergence of new pathogens is the crowding of the world's poorest populations into very small places within huge cities, because population density increases the spread of disease. Another factor is the increase in international travel. Other features of modern life that may contribute include mass distribution of food, which exposes large populations to potentially contaminated food, and unhygienic food preparation.

10. (a) Influenza virus changes surface expression of neuraminidase and hemagglutinin. (b) Herpesviruses remain dormant in nerve cells. Later emergence can cause outbreaks of cold sores or shingles (chickenpox virus). (c) *Neisseria* secretes proteases that cleave IgA. (d) False. (e) Several gram-positive bacteria resist complement-mediated lysis. (f) Influenza virus accumulates mutations from year to year. (g) False.

11. (a) No, IgE is raised against allergens and some parasites. (b) No, autoreactive T cells are activated only by intracellular infections. The statements in (c) through (f) are correct.

12. (a) Large, granular cells such as mast cells and eosinophils. Neutrophils and macrophages will also be involved. (b) Therapeutic cytokines such as IL-4 would help encourage IgE and the T_H2 response, which is already present. However, cytokines that drive a T_H1 response, such as IL-12 or IFN-γ, may be more beneficial to longer-term immunity.

13. The answer comes from the concept of original antigenic sin, which posits that we only mount a primary response once we have exhausted the potential to use memory cells to eradicate the infection. Since most of our first encounters

with influenza will vary, the years in which "all" of the key influenza epitopes are significantly "new" to each of us will also vary. It is only in these years that we experience a new primary response to influenza virus, and therefore symptoms of the flu are most severe.

14. (a) True. (b) True. (c) False: Because DNA vaccines allow prolonged exposure to antigen, they are likely to generate immunologic memory. (d) True. (e) False: A DNA vaccine contains the gene encoding an entire protein antigen, which most likely contains multiple epitopes.

15. Because attenuated organisms are capable of limited growth within host cells, they are processed by the cytosolic pathway and presented on the membrane of infected host cells together with MHC class I. These vaccines, therefore, usually can induce a cell-mediated immune response. The limited growth of attenuated organisms within the host often eliminates the need for booster doses of the vaccine. Also, if the attenuated organism is able to grow along mucous membranes, then the vaccine will be able to induce the production of secretory IgA. The major disadvantage of attenuated whole-organism vaccines is that they may revert to a virulent form. They also are more unstable than other types of vaccines, requiring refrigeration to maintain their activity.

16. (a) The antitoxin was given to inactivate any toxin that might be produced if *Clostridium tetani* infected the wound. The antitoxin was necessary because the girl had not been previously immunized and, therefore, did not have circulating antibody to tetanus toxin or memory B cells specific for tetanus toxin. (b) Because of the treatment with antitoxin, the girl would not develop immunity to tetanus as a result of the first injury. Therefore, after the second injury 3 years later, she will require another dose of antitoxin. To develop long-term immunity, she should be vaccinated with tetanus toxoid.

17. The Sabin polio vaccine is live and attenuated, whereas the Salk vaccine is heat killed and inactivated. The Sabin vaccine thus has the usual advantages of an attenuated vaccine compared with an inactivated one (see the answer to question 15). Moreover, since the Sabin vaccine is capable of limited growth along the gastrointestinal tract, it induces production of secretory IgA. The attenuated Sabin vaccine can cause life-threatening infection in individuals, such as children with AIDS, whose immune systems are severely suppressed. Now that polio is rarely if ever seen in the United States, continuing use of a vaccine with the potential to revert to a more virulent form introduces an unwarranted element of risk to both the vaccinee and others who might contract the disease from them.

18. The virus strains used for the nasally administered vaccines are temperature-sensitive mutants that cannot grow at human body temperature (37°C). The live attenuated virus can grow only in the upper respiratory tract, which is cooler, inducing protective immunity. These mutant viruses cannot grow in the warmer environment of the lower respiratory tract, where they could replicate and mutate into a disseminated influenza infection.

19. T-cell epitopes generally are internal peptides, which commonly contain a high proportion of hydrophobic residues. In contrast, B-cell epitopes are located on an antigen's surface, where they are accessible to antibody, and contain a high proportion of hydrophilic residues. Thus, synthetic hydrophobic peptides are most likely to represent T-cell epitopes and induce a cell-mediated response, whereas synthetic hydrophilic peptides are most likely to represent accessible B-cell epitopes and induce an antibody response.

20. In this hypothetical situation, the gene can be cloned into an expression system and the protein expressed and purified in order to test it as a recombinant protein vaccine. Alternatively, the gene can be cloned into a plasmid vector that can be injected directly and tested as a DNA vaccine. Use of the cloned gene as a DNA vaccine is more efficient, because it eliminates the steps required for preparation of the protein and its purification. However, the plasmid containing the gene for the protective antigen must be suitably purified for use in human trials. DNA vaccines have a greater ability to stimulate both the humoral and cellular arms of the immune system than protein vaccines do and thus may confer more complete immunity. The choice must also take into consideration the fact that recombinant protein vaccines are in widespread use whereas DNA vaccines for human use are still in early test phases.

21. Pathogens with a short incubation period (e.g., influenza virus) cause disease symptoms before a memory-cell response can be induced. Protection against such pathogens is achieved by repeated immunizations to maintain high levels of neutralizing antibody. For pathogens with a longer incubation period (e.g., polio virus), the memory-cell response is rapid enough to prevent development of symptoms, and high levels of neutralizing antibody at the time of infection are unnecessary.

22. Bacterial capsular polysaccharides, inactivated bacterial exotoxins (toxoids), and surface protein antigens. The latter two commonly are produced by recombinant DNA technology. In addition, the use of DNA molecules to direct synthesis of antigens on immunization is being evaluated.

23. A possible loss of herd immunity in the population. Even in a vaccinated population of children, a small percentage may have diminished immunity to the disease target due to differences among MHC molecule expression in a population, providing a reservoir for the disease. In addition, most vaccinated individuals, if exposed to the disease, will develop mild illness. Exposure of unvaccinated individuals to either source of disease would put them at risk for serious illness. Epidemics within adult populations would have more serious consequences, and infant mortality due to these diseases would increase.

24. (a) 1 or 2; (b) 2; (c) 3; (d) 4; (e) 1; (f) 2; (g) 2.

25. (a) No. The antiserum you received 1 year ago protected you temporarily, but those antibodies are now gone and you have no memory B cells to produce new antibodies during this second exposure. (b) Antibodies in the antiserum bound to the snake venom and neutralized its ability

to cause damage. Phagocytes then engulfed and destroyed this antibody-coated venom. Because the snake venom was coated with antibodies, naïve B cells were not activated during this first exposure and therefore no adaptive immune response was mounted. (c) Equally sensitive. There are no residual cells or antibodies that were involved in the original encounter with this snake venom and, therefore, no recall response.

Clinical Focus Answers

1. (a) Because the infected target cells expressed $H2^k$ MHC molecules, but the primed T cells were $H2^k$ restricted. (b) Because the influenza nucleoprotein is processed by the endogenous processing pathway and the resulting peptides are presented by MHC class I molecules. (c) Probably because the transfected MHC class I D^b molecule is only able to present peptide 365–380 and not peptide 50–63. Alternatively, peptide 50–63 may not be a T-cell epitope. (d) These results suggest that a cocktail of several immunogenic peptides would be more likely to be presented by different MHC haplotypes in humans and would provide the best vaccines for humans.

2. Any connection between vaccination and a subsequent adverse reaction must be evaluated by valid clinical trials involving sufficient numbers of subjects in the control group (those given a placebo) and experimental group (those receiving the vaccine). This is needed to give a statistically correct assessment of the effects of the vaccine versus other possible causes for the adverse event. Such clinical studies must be carried out in a double-blind manner; that is, neither the subject nor the caregiver should know who received the vaccine and who received the placebo until the end of the observation period. In the example cited, it is possible that the adverse event (increased incidence of arthritis) was caused by an infection occurring near the time when the new vaccine was administered. Determining the precise cause of this side effect may not be possible, but ascertaining whether it is likely to be caused by this vaccine is feasible by appropriate studies of the vaccinated and control populations.

Analyze the Data Answer

(a) pSG5DNA-Bcl-x_L (i.e., plasmid pSG5 encoding Bcl-x_L) targeting calreticulin and pSG5DNA-Bcl-x_L targeting LAMP-1 are the most effective vaccines at inducing $CD8^+$ T cells to make IFN-γ. pSG5DNA-Bcl-x_L targeting HSP70 also activated $CD8^+$ T cells. However, the pSG5 construct without the anti-apoptosis gene targeting calreticulin also induced a good $CD8^+$ T-cell response. (b) Calreticulin is a chaperone protein associated with partially folded MHC class I molecules in the endoplasmic reticulum. Associating E7 antigen with the chaperone may enhance loading of MHC class I molecules with E7, making the antigen more available to T cells once it is expressed on cells. (c) The DNA vaccines co-injected with pSG5DNA-Bcl-x_L were effective in inducing $CD8^+$ T cells, possibly because the expression of anti-apoptotic genes in dendritic cells allowed those cells to survive longer and present antigen to T cells for a longer time. The longer they presented

antigen, the longer the host would respond to produce antigen-specific T cells. (d) The data in part (b) of the figure (see Chapter 17 text) indicate that in the absence of $CD4^+$ helper T cells (in CD4 knockout mice), there is ineffective activation of $CD8^+$ T cells. Therefore, T-cell help is necessary to activate the CD8 response; by targeting antigen to MHC class II, you more efficiently activate helper T cells. The Sig/E7/LAMP-1 construct was necessary because most antigens presented by MHC class II molecules are processed by the endocytic pathway, and the Sig/E7/LAMP-1 construct targets antigen to the Golgi, where the E7 peptides can be exchanged with CLIP and inserted into MHC class II. (e) Helper T cells (poor response in the CD4 knockout mice), long-lived dendritic cells (immunization with pSG5DNA-Bcl-x_L improves the response), antigen (the absence of peptide failed to induce a response), and the targeting of antigen to MHC class II (immunizing with the Sig/E7/LAMP-1 construct is the only one that induces a $CD8^+$ T-cell response).

Chapter 18

1. (a) True. (b) False: X-linked agammaglobulinemia is characterized by a reduction in B cells and an absence of immunoglobulins. (c) False: Phagocytic defects result in recurrent bacterial and fungal infections. (d) True. (e) True. (f) True. (g) True. (h) True. (i) False: These children are usually able to eliminate common encapsulated bacteria with antibody plus complement but are susceptible to viral, protozoan, fungal, and intracellular bacterial pathogens, which are eliminated by the cell-mediated branch of the immune system. (j) False: Humoral immunity also is affected because class II–restricted T_H cells must be activated for an antibody response to occur.

2. (a) 1 and 3; (b) 1, 2, and 4; (c) 2; (d) 1; (e) 2.

3. The defect in X-linked hyper-IgM syndrome is in CD40L expressed on B cells. CD40L mediates binding of B cells to T cells and sends costimulatory signals to the B cell for class switching. Without CD40L, class switching does not occur, and the B cells do not express other antibody isotypes.

4. As discussed in Chapter 8, the thymus is the location for differentiation and maturation of helper and cytotoxic T cells. Positive and negative selection occur in this organ as well. Thus, the thymocytes produced in the bone marrow of patients with DiGeorge syndrome do not have the ability to mature into effector cell types. In Chapter 2, we noted that the thymus decreases in size and function with age. In the adult, effector-cell populations have already been produced (peak thymus size occurs during puberty); therefore, a defect after this stage would cause less severe T-cell deficiency.

5. As long as the mother is not immunocompromised, maternal antibodies in the mother's serum will be passively transferred to the baby in utero. After birth, these IgG molecules will supply the newborn with passive immune protection from many common bacterial infections, which can then be quickly dispatched by antibody-mediated mechanisms. In immune-competent babies, these maternal antibodies will eventually be replaced by the child's own immune

response to the infectious agents they encounter. In children with SCID, this does not occur, and they gradually become more susceptible to bacterial infections as their maternally derived antibodies disappear.

6. (a) Leukocyte adhesion deficiency results from biosynthesis of a defective β chain in LFA-1, CR3, and CR4, which all contain the same β chain. (b) LFA-1 plays a role in cell adhesion by binding to ICAM-1 expressed on various types of cells. Binding of LFA-1 to ICAM-1 is involved in the interactions between T_H cells and B cells, between CTLs and target cells, and between circulating leukocytes and vascular endothelial cells.

7. The high-affinity IL-2 receptor is composed of two chains: the α chain and the common γ chain. Later, the γ chain was discovered to be a component of the receptors for five other cytokines: IL-4, -7, -9, -15, and -21. During hematopoiesis, developing lymphocytes require IL-7 signaling, and therefore the complete IL-7R, in order to develop properly. Without the common γ chain, this does not occur and the development of lymphocytes is blocked.

8. As there are multiple MHC I genes (for HLA-A, -B, and -C), it is unlikely that a mutation would affect expression of all the genes on both chromosomes. Rare individuals lack MHC class I proteins due to deficiency in β_2-microglobulin.

9. Some components of the immune system are responsible for regulating or suppressing the activity of leukocytes (e.g., T_{REG} cells). When these pathways are defective, overactive immune responses can occur, leading to breaks in self tolerance that lead to attacks on self molecules, or autoimmune syndromes. One example is a disorder called immune dysregulation, polyendocrinopathy, enteropathy, X-linked (IPEX) syndrome, in which the gene encoding the transcription factor that controls development of T_{REG} cells, *Foxp3*, is defective. Another example is autoimmune polyendocrinopathy and ectodermal dystrophy (APECED), a disorder that arises from defects in the *AIRE* gene, encoding a transcription factor found in the thymus. The Aire protein is responsible for the expression of tissue-restricted antigens. Without this protein, the negative selection of T cells in the thymus that recognize self antigen is disrupted and autoreactive T cells emerge and instigate organ-specific autoimmune attacks.

10. As the individual's own HSCs are being used, matching HLA types is not necessary and there will be no graft-versus-host reaction.

11. (a) False: HIV-1 is believed to have evolved from a strain of SIV that jumped the species barrier from African chimpanzees to humans, although HIV-2 is thought to have arisen from a separate but similar transfer from SIV-infected sooty mangabeys. (b) False: HIV-1 infects chimpanzees but does not cause immune suppression. (c) True. (d) False: Zidovudine (or azidothymidine, AZT) acts at the level of reverse transcription of the viral genome, whereas saquinavir is an inhibitor of the viral protease. (e) True. (f) False: A diagnosis of AIDS is based on both a low $CD4^+$ T-cell count (below 200 cells/μl, or 14%) and the presence of certain AIDS-defining conditions (see Table 18-4).

(g) False: The PCR detects HIV proviral DNA in latently infected cells. (h) True.

12. Lysis of active virus-producing cells; killing by cytotoxic T cells; ADCC or antibody-mediated phagocytosis.

13. No. For most of the asymptomatic phase of HIV infection, viral replication and the $CD4^+$ T-cell number are in dynamic equilibrium; the level of virus and the percentage of $CD4^+$ T cells remains relatively constant.

14. An increase in the viral load and a decrease in $CD4^+$ T-cell levels indicate that HIV infection is progressing from the asymptomatic phase into AIDS. Often, infection with an opportunistic agent occurs when the viral load increases and the $CD4^+$ T-cell level drops. The disease AIDS in an HIV-1–infected individual is defined as a $CD4^+$ T-cell count less than 200 cells/μl and/or the presence of certain opportunistic infections (see Table 18-4).

15. Skin-test reactivity is monitored to indicate the functional activity of T_H cells. As AIDS progresses and $CD4^+$ T cells decline, there is a decline in skin-test reactivity to common antigens.

16. Receptors for certain chemokines, such as CXCR4 and CCR5, also function as coreceptors for HIV-1. The chemokine that is the normal ligand for the receptor competes with the virus for binding to the receptor and can thus inhibit cell infection by blocking attachment of the virus (see Figure 18-14). Cytokines that activate T cells stimulate infection because they increase expression of receptors used by the virus.

17. No. There are latently infected cells that reside in lymph nodes or in other sites. These cells can be activated and begin producing virus, thus causing a relapse of the disease. In a true cure for AIDS, the patient must be free of all cells containing HIV DNA.

18. Patient L. S. fits the definition of AIDS, whereas patient B. W. does not. The clinical diagnosis of AIDS among HIV-infected individuals depends on both the T-cell count, or percentage, and the presence of various indicator diseases. See Table 18-4.

19. One hurdle is the rapid generation of HIV variants through mutation, so that the variants that arise in an individual or to which s/he is exposed will not be recognized by memory B and T cells induced by the vaccine. A second challenge is presented by the extensive glycan coat that covers up the gp120/gp41 Env protein so that BCRs and antibodies have a hard time binding to the Env protein. A third challenge is that the generation of broadly neutralizing antibodies that will react with most viral strains requires many somatic hypermutations that need to occur and be selected for over time.

Analyze the Data Answer

(a) CVID lowers the number and percentage of T helper cells. (b) The table accompanying this question indicates that there are fewer naïve $CD8^+$ T cells in patients with

CVID. This could mean chronic activation in CVID has made these T cells less dependent on IL-2. Alternatively, the figure accompanying this question shows that bone marrow cells from patients with CVID make less IL-2. If this is also true of T_H cells, or there are fewer T_H cells (as seen in the table accompanying this question), then the generation of CTLs may be impaired. This example of the complexity of the immune system demonstrates that specific responses are not always easily predictable. (c) False: The data in the figure accompanying this question show that the kinetics and overall production of TNF-α is greater in CVID than in normal patients, but IL-2 production is lower. Thus, there is no consistent impact. Based on what we know about TNF-α activity, it might be responsible for increased pathology or cell death, perhaps explaining the loss of both CD4$^+$ and CD8$^+$ cells in patients with CVID. (d) According to information in Chapter 18, CVID lowers IgG, IgA, and IgM. However, based on the data presented in the table accompanying this question, one might predict that in the absence of T-cell help, there will be a significant impact on class switching.

Clinical Focus Answer

All individuals needing antiretroviral therapy are supposed to receive combination drugs to lessen the chance that resistant viral variants will survive. These drugs cross the placenta, which will prevent the fetus from becoming infected in utero. Newborns should begin antiretroviral therapy as soon as possible after birth to reduce the chance that infection will occur during birth from the mother's body fluids. Treatment should continue for several weeks, and longer if the infant is breastfeeding.

Chapter 19

1. (a) False: Hereditary retinoblastoma results from inactivation of both alleles of *Rb*, a tumor-suppressor gene. (b) True. (c) True. (d) True. (e) False: Some oncogenic retroviruses do not have viral oncogenes. (f) True.

2. Cells of the pre-B-cell lineage have rearranged the heavy-chain genes and express the μ heavy chain in their cytoplasm (see Figure 9-4). You could perform Southern blot analysis with a Cμ probe to see whether the heavy-chain genes have rearranged. You could also perform fluorescence staining with antibody specific for the cytoplasmic μ heavy chain.

3. (a) Early-stage melanoma cells appear to be functioning as APCs, processing the antigen by the exogenous pathway and presenting the tetanus toxoid antigen together with the MHC class II DR molecule. (b) Advanced-stage melanoma cells might have a reduction in the expression of MHC class II molecules, or they might not be able to internalize and process the antigen by the exogenous route. (c) Since the paraformaldehyde-fixed early melanoma cells could present processed tetanus toxoid, they must express MHC class II molecules on their surface. (d) Stain the early and advanced melanoma cells with fluorescent mAb specific for MHC class II molecules.

4. Tumor antigens may be encoded by genes expressed only by tumors, may be products of genes overexpressed by the tumor or of genes normally expressed only at certain stages of differentiation, or may be products of normal genes that are altered by mutation. In some cases, viral products may be tumor antigens.

5. IFN-α, IFN-β, and IFN-γ enhance the expression of MHC class I molecules on tumor cells, thereby increasing the CTL response to tumors. IFN-γ also increases the activity of CTLs, macrophages, and NK cells, each of which plays a role in the immune response to tumors. TNF-α and lymphotoxin-α have direct antitumor activity, inducing hemorrhagic necrosis and tumor regression. IL-2 activates *tumor-infiltrating lymphocytes* (TILs), which have antitumor activity.

6. (a) Melanoma cells transfected with the CD80/86 gene are able to deliver the costimulatory signal necessary for activation of CTL precursors to effector CTLs, which then can destroy tumor cells (Figure 19-11). Melanoma cells transfected with the GM-CSF gene secrete this cytokine, which stimulates the activation and differentiation of APCs, especially DCs, in the vicinity. The DCs then can present tumor antigens to T_H cells and CTL precursors, enhancing the generation of effector CTLs (see Figure 19-10). (b) Because some of the tumor antigens on human melanomas are expressed by other types of cancers, transfected melanoma cells might be effective against other tumors carrying identical antigens.

7. (a) 4; (b) 2; (c) 1; (d) 5; (e) 3; (f) 9; (g) 8; (h) 7; (i) 6.

8. Part D (M2 macrophages).

9. False: Inflammation can be a bad sign when T_H2, T_{REG} cells, or M2 macrophages dominate the response, creating a microenvironment that is protumor.

10. (a) Adoptive transfer of modified T cells—collection of tumor-infiltrating lymphocytes, which are then expanded ex vivo (e.g., with IL-2) and reinfused into the patient, is one example. While some patients with metastatic melanoma had a positive response to this therapy it did not help all, possibly due to the inadvertent expansion of regulatory T cells. More recently, chimeric antigen receptor (CAR) T cells have been developed for treating cancer. These are T cells collected from patients and transduced to express a chimeric receptor (part immunoglobulin and part T-cell receptor) that recognizes tumor antigens without the need for MHC presentation. This topic is covered in Clinical Focus Box 19-2. (b) Monoclonal antibodies—any mAb treatment listed in Table 19-4. For example, rituximab, an anti-CD20 mAb, can bind to B cells (which express this surface molecule) and induce cell death. This treatment can be used as a form of therapy for some blood cell cancers, such as non-Hodgkin's lymphoma. (c) Checkpoint blockade—mAbs that recognize CTLA-4 or PD-1/PD-L1 can be used to block the negative regulatory signals (PD-L1) or the receptors of these signals (PD-1 and CTLA-4); basically, blocking the brake pedal on T cells. Collectively, these checkpoint inhibitors allow endogenous T cells (T_H1 or CTLs) to perform their natural

anti–tumor cell activity, unhindered by regulation (see Figure 19-12). (d) Therapeutic vaccine—the prostate cancer therapy sipuleucel-T (Provenge) is one example (Figure 19-10). In this method, patient DCs are collected and treated with a fusion protein expressing a prostate-specific antigen and GM-CSF, a cell-stimulating cytokine/growth factor. These are expanded ex vivo and reinfused into the patient, as a means to activate T cells specific for this prostate-specific antigen. These T cells should then home to the site of the tumor and help to eradicate prostate cancer cells expressing these antigens at high levels. The anti-HPV vaccine would not be an example here, only because this is considered a *prophylactic* vaccine, or one that must be administered before infection with this cervical cancer–causing virus, rather than a *therapeutic* vaccine, or one that can be administered after the onset of disease.

Clinical Focus Answers

1. In some instances, Coley's toxins may have induced a local immune response that helped the immune system to recognize and fight the tumor cells. For instance, introduction of bacteria should elicit local innate responses when PAMPs on the bacteria are recognized by PRRs on immune cells. These danger or damage signals would help to recruit leukocytes to the site (inflammation) and activate local antigen-presenting cells. These activated APCs could then supply the necessary signal 2 components (e.g., up-regulation of MHC class II and CD80/86) to engage with T cells. The antigens presented by these cells might include both bacterial and local tumor antigens, acquired via phagocytosis of dead or damaged tumor cells. If antigens from the tumors were presented to naïve T cells, this could activate antitumor-specific adaptive responses. Even if there are no tumor-specific epitopes recognized by naïve T cells, if the introduced bacteria activate a local T_H1-like response the tumor microenvironment would begin to shift away from protumor (M2 macrophages, T_H2 cells, and immune-suppressing microenvironment) toward a more antitumor environment (M1 macrophages, IL-12, T_H1 cells, and CTLs). One might think of this as an early form of immunotherapy!

2. Because cervical cancer is linked to HPV infection, a preventive cancer vaccine may be possible; if HPV infection is prevented, then cervical cancer should also be prevented. Other cancers that may be targets for such prevention include adult T-cell leukemia/lymphoma and the liver cancer that is linked to hepatitis B infection. Most cancers have not been clearly linked to an agent of infection, and therefore a preventive vaccine is not an option.

Chapter 20

1. If I wished to precipitate my antigen, I might elect to use a polyclonal preparation, as it would contain antibodies toward multiple different determinants on the antigen and therefore many antibody molecules could bind per antigen molecule, maximizing the chances that at least some of the antibodies could bind more than one antigen and facilitate precipitation.

2. With time, the population of B-cell clones that respond to an antigen in an individual will change. Some B-cell clones will die, and different clones will predominate. Overall, the affinity of the antibodies in the serum will increase according to the methods described in Chapter 12. However, this means that the proteins with which individual antibodies will cross-react will change as the range of binding sites modulates, and this is what has happened in your experiment.

3. (a) Ideally, polyclonal. I will have the most effective agglutination if antibodies can cross-link multiple sites on the bacterial surface. However, a monoclonal antibody would also work, as most antigens are repeated many times on the bacterial surface. (b) Ideally, polyclonal. To form a precipitate, I will need to cross-link multiple proteins. If each only has a single site at which an antibody can bind, a bivalent antibody can cross-link only two proteins, and that would be insufficient to create a precipitate. If I am precipitating a protein with multiple copies of the same site, then monoclonal antibodies could still work. (c) Here, either would work. The antibody binds to the band, localizing an enzymatic reaction at the band and causing substrate conversion to product. A polyclonal antibody mixture would have the advantage that different antibodies could bind at different antigenic determinants on the target protein and therefore could give rise to a stronger signal. However, different bleeds of polyclonal sera would have different levels of cross-reactivity with other, structurally similar determinants on other proteins. Monoclonal antibodies will bind to predictable determinants, and, although they might still cross-react with structurally similar determinants on other proteins, those cross-reactivities are predictable and will not change from batch to batch. (d) Here, one would use two monoclonal antibodies directed toward different determinants on the cytokine. If the capture antibody were to be polyclonal, such that all the binding sites on the cytokine were bound, this would compete with a polyclonal detection antibody. (e) Monoclonal. Here, reproducibility of reactivity and cross-reactivity is demanded for clinical safety.

4. (a) Yes. There is some evidence of hemagglutination in the first well, which represents an antiserum dilution of 1 in 10. (b) By 42 days of age, the baby's serum has the same capacity for hemagglutination inhibition as the positive-control sample. (c) No. These could be maternal antibodies absorbed through breast milk or colostrum.

5. RIAs are orders of magnitude more sensitive than are ELISAs with chromogenic substrates. And yes, chemiluminogenic substrates allow more amplification and greater sensitivity, resulting in an assay that matches the RIA in its ability to measure low concentrations of antibodies or antigens.

6. ELISPOT assays measure the number of cells capable of secreting particular molecules, such as antibodies or cytokines. These assays therefore require that when substrate is converted to product, the product remains at the location where the substrate-to-product reaction occurred. Products in ELISPOT reactions are therefore insoluble. Similarly, Western blot assays use antibodies to determine

the location of particular bands on a gel. Again, one requires that the product of the enzymatic assay remains localized precisely where the enzyme is bound.

7. (a) Since cost is an issue, and I already have the antibody labeled, I would probably use equilibrium dialysis. (b) Surface plasmon resonance experiments will enable me to measure the association, as well as the dissociation rate constant, of the binding reaction between my antigen and antibody.

8. By using longer-wavelength, lower-energy light, two-photon microscopy induces less photobleaching of the tissue preparation. Further, since no fluorescence is emitted unless two exciting photons simultaneously impinge on the sample, the focal plane of the observed images can be more tightly defined.

9. Magnetic-activated cell sorting is most useful for batch separations of cell populations. It is faster than fluorescence-based methods, but not quite as precise. I would choose fluorescence-based sorting when I need to be sure there are no contaminating cells, because in FACS, cells are quite literally separated one at a time.

10. **Experimental Design Answer** There are many ways to test this idea, but we will offer just one here, which makes use of adoptive transfer and fluorescence labeling: Generate B-1 B cells in culture specific for the virus in question, and load them with CFSE. Inject the cells into the tail of a mouse infected with the virus, allow the cells to home for 12 to 24 hours, and then sacrifice the mouse and search for CFSE-labeled cells in the lung, using tissue sections and immunofluorescence.

11. **Experimental Design Answer** I will use a tamoxifen/Cre fusion protein whose expression is under the control of a B cell–specific promoter, such as that controlling the expression of CD19. Therefore, Cre will be active only in B cells and only if I add the tamoxifen and I can therefore control exactly when the gene targeting will occur in reference to antigen immunization.

I also need to generate an inactive, truncated form of the gene in question that is flanked by *loxP* sites. Ideally, the gene will have a selectable marker, such as neomycin resistance, and also will bear a sequence external to the *loxP* sites that will control for insertion of the gene into the correct location (see Figure 20-33a, left).

Page numbers followed by f indicate figures; those followed by t indicate tables.